MONOGRAPHS IN CONTACT ALLERGY
Anton C. de Groot

Volume 1: Non-Fragrance Allergens in Cosmetics
Volume 2: Fragrances and Essential Oils
Volume 3: Topical Drugs
Volume 4: Systemic Drugs

T0295403

MONOGRAPHS IN CONTACT ALLERGY
VOLUME 4

SYSTEMIC DRUGS

MONOGRAPHS IN CONTACT ALLERGY
VOLUME 4

SYSTEMIC DRUGS

Anton C. de Groot

CRC Press
Taylor & Francis Group
Boca Raton London New York

CRC Press is an imprint of the
Taylor & Francis Group, an **informa** business

First Edition published 2022
by CRC Press
6000 Broken Sound Parkway NW, Suite 300, Boca Raton, FL 33487-2742

and by CRC Press
4 Park Square, Milton Park, Abingdon, Oxon, OX14 4RN
CRC Press is an imprint of Taylor & Francis Group, LLC

Library of Congress CataloginginPublication Data
A catalog record has been requested for this book.

ISBN: 978-0-367-43649-0 (hbk)
ISBN: 978-0-367-74462-5 (pbk)
ISBN: 978-1-003-15800-4 (ebk)

DOI: 10.1201/9781003158004

Contents

PREFACE

One year after the release of Volume 3 of the *Monographs in contact allergy* series on 'Topical drugs', here is Volume 4, discussing 'Systemic drugs'. For those among you who wonder how a sizable book like this can be written by one author in one year's time: there is a simple explanation. In my original planning, Volume 3 would present both topical and systemic drugs. So I worked on it for a year and a half, *in which period I already wrote a considerable volume of copy on the systemic drugs.* At that point, I came to realize that the combined subjects would comprise so much information, that a book would be created too heavy to lift for most dermatologists and other scientists. Separating Topical drugs from Systemic drugs and presenting them in two volumes appeared to be a good option, so at that moment I progressed with Topical drugs only. And there was another reason why discussing the categories individually made sense: the manifestations of delayed-type hypersensitivity to systemic drugs are entirely different from those of topical drugs. While sensitization to the latter category leads to allergic contact dermatitis, delayed-type hypersensitivity to systemic drugs given orally or parenterally manifests as drug eruptions.

During the last 25 years, it has been found that a variety of cutaneous eruptions may result from drug hypersensitivity, including maculopapular eruptions, erythroderma, systemic allergic dermatitis (systemic contact dermatitis), symmetrical drug-related intertriginous and flexural exanthema (SDRIFE), baboon syndrome, fixed drug eruption, photosensitivity, delayed urticaria, acute generalized exanthematous pustulosis (AGEP), drug reaction with eosinophilia and systemic symptoms (DRESS), and – to a far lesser extent - Stevens-Johnson syndrome and toxic epidermal necrolysis (SJS/TEN). That these drug reactions, or at least a number of them, are mediated by delayed-type hypersensitivity is very likely, as evidenced by positive patch tests and positive delayed intradermal tests with the culprit drugs in patients suffering from these eruptions, the identification of drug-specific T cells and results of *in vitro* tests such as the lymphocyte transformation test. Major scientific progress on this subject has been made in recent decades, especially by investigators in France, Switzerland, Italy, Portugal and Germany.

Through extensive database searches and meticulous examination of the references in original articles and review articles (of which many have appeared lately, it is obviously a 'hot topic'), I have found 507 systemic drugs that have induced drug eruptions and/or occupational allergic contact dermatitis from delayed-type hypersensitivity as shown by positive patch tests. These are all presented in separate monographs in this book.

Although patients with drug eruptions such as AGEP, DRESS and SJS/TEN are frequently treated by non-dermatologists because these individuals are often ill and have internal organ malfunction from the hypersensitivity reaction, dermatologists should always be involved in diagnosing delayed-type hypersensitivity after complete recovery: they are the only experts in patch testing and know how to perform this very valuable diagnostic procedure properly. Importantly, in many cases, the culprit drugs can be identified by positive patch tests, and this is the only *in vivo* test to demonstrate sensitization that can safely be applied in all drug eruptions, even the most severe ones. A negative patch test does not exclude delayed-type hypersensitivity, and should therefore (generally) be followed by intradermal tests with delayed readings.

Unfortunately, many healthcare professionals do not investigate drug reactions with skin tests and rely on the history alone to make a diagnosis of drug allergy. This is obviously unreliable and may lead to unjustified use or avoidance of indicated drugs. As a consequence, in suspected cutaneous hypersensitivity drug reactions, diagnostic skin tests must in many cases be performed in order to institute proper preventive measures. This should, in the opinion of the author, always start with patch testing, which highlights the role for dermatologists in this exciting new field of medicine. The information in this book, which should also be very useful for allergists, internists and other physicians caring for patients with drug hypersensitivity, should enable each dermatologist to perform this role with confidence and optimal expertise.

Anton de Groot, MD, PhD
Wapserveen, The Netherlands, October 2021

ABOUT THE AUTHOR

Anton C. de Groot, MD, PhD (1951) received his medical and specialist training at the University of Groningen, The Netherlands. In 1980, he started his career as dermatologist in private practice in 's-Hertogenbosch. At that time, he had already become interested in contact allergy and in side effects of drugs by writing the chapter *Drugs used on the skin* with his mentor prof. Johan Nater, for the famous *Meyler's Side Effects of Drugs* series. Soon, the subject of this chapter in new Editions and the yearly *Side Effects of Drugs Annuals* would be expanded to include cosmetics and oral drugs used in dermatology (1980-2000). Contact allergy to cosmetics would become de Groot's main area of interest and expertise and in 1988, he received his PhD degree on his Thesis entitled *Adverse Reactions to Cosmetics*, supervised by prof. Nater.

Frustrated by the lack of easily accessible information on the ingredients of cosmetic products, and convinced that compulsory ingredient labeling of cosmetics (which at that time was already implemented in the USA) would benefit both consumers and allergic patients and would lead to only slight and temporary disadvantages to the cosmetics industry, De Groot approached the newly founded European Society of Contact Dermatitis and became Chairman of the Working Party European Community Affairs. The European Commission and its committees, elected legislators, national trade, health departments and the cosmetics industries were extensively lobbied. This resulted in new legislation by the Commission of the European Communities in 1991, making ingredient labeling mandatory for all cosmetic products sold in EC Member States by December 31, 1997.

Anton has been the chairman of the 'Contact Dermatitis Group' of the Dutch Society for Dermatology and Venereology from 1984 to 1998. In 1990, he was one of the founders of the *Nederlands Tijdschrift voor Dermatologie en Venereologie* (Dutch Journal of Dermatology and Venereology) and was Editor of this scientific journal for 20 years, of which he served 10 years as Editor-in-chief.

De Groot has authored nineteen book titles, eleven of which – all co-authored by Johan Toonstra MD, PhD – are general dermatology books in Dutch for medical students, general practitioners, 'skin therapists' (huidtherapeuten, a paramedical profession largely restricted to The Netherlands) and pedicures and podotherapists. He also co-authored a Dutch booklet for (parents of) patients with atopic dermatitis and published his scientific biography in 2020.

Anton has written seven (including the current one) international books, of which one has had three Editions: *Unwanted Effects of Cosmetics and Drugs used in Dermatology* (first Edition 1983, second 1985, third 1994). Of his best known book *Patch Testing*, 4 editions have been published (first Edition 1986, second 1994, third 2008, fourth 2018) (www.patchtesting.info).

After writing a book entitled *Essential Oils: Contact Allergy and Chemical Composition* with Erich Schmidt, which appeared in 2016, he started working on a series of *Monographs in Contact Allergy*. The first Volume discussing 'Non-fragrance allergens in cosmetics' was released in 2018, followed by 'Fragrances and essential oils' in 2019 and 'Topical drugs' in 2020. The current book presenting delayed-type hypersensitivity to systemic drugs is Volume 4.

In addition to these books, Anton has written over 70 book chapters (mostly in international books), 135 articles in international journals and some 235 articles in Dutch medical and paramedical journals. He served as board member of several journals including *Dermatosen* and is currently member of the Editorial Advisory Board of the Journal *Dermatitis*.

In 2019, the author received the American Contact Dermatitis Society Honorary Membership status for his 'vast contributions to contact dermatitis' and was awarded the Presidential Citation of this Society in 2021 for making the contents of his book *Patch Testing*, 4th edition, digitally available to all members of the ACDS for free.

Anton de Groot has retired from dermatology practice, but since 2008 regularly teaches general dermatology to junior medical doctors at the University of Groningen, which unfortunately is (still) temporarily interrupted by the covid-19 pandemic. He and his wife Janny have two daughters, both lawyers, and two grandchildren.

Chapter 1   INTRODUCTION

1.1 WHY A BOOK WITH MONOGRAPHS ON SYSTEMIC DRUGS?

After a first volume in the series *Monographs in Contact Allergy* on 'Non-fragrance allergens in cosmetics' (1), a second on 'Fragrances and essential oils' (2) and a third on 'Topical drugs' (pharmaceuticals) (3), the subject 'systemic drugs' may seem a logical choice for Volume 4. In fact, however, it is not, as there are major differences between the contents of this book and the previous volumes. The haptens/allergens presented previously are all applied to the skin or mucous membranes, where they can induce sensitization and, upon renewed or prolonged contact, elicit allergic contact dermatitis. Contact allergy to these substances is frequent. Indeed, in an estimated 10-15% of all patients with allergic contact dermatitis seen by dermatologists, topical drugs are the cause (iatrogenic allergic contact dermatitis) (3,4,5,6). Fragrance allergy is even more important: in the general adult population, up to 4.5% may be allergic to fragrance materials, and in consecutive patients patch tested for suspected contact dermatitis, the frequency may reach 20 to 25% (7).

The situation with systemic drugs, however, is entirely different. Systemic drugs are not applied to the skin or mucous membranes, but are given orally, as enema (infrequent) or parenterally by injections: intravenous, subcutaneous, intra-articular, intramuscular, or intrathecal. This is obviously a far less effective route for the induction of delayed-type hypersensitivity than dermal application, as the numbers of sensitizations to systemic drugs are far lower than those of the classic 'contact allergens' discussed in the previous 3 volumes of the *Monographs in Contact Allergy* series. Indeed, in relation to the widespread use of systemic drugs, proven delayed-type hypersensitivity reactions are rather infrequent, although it should be realized that only a minority of patients with presumed or possible hypersensitivity reactions are tested with patch or intradermal tests to detect sensitization, which will almost certainly result in underreporting.

Allergic contact dermatitis from accidental contact

The manifestations of delayed-type hypersensitivity to systemic drugs in the majority of patients are also quite different from those of topical drugs and other contact allergens, with one exception: delayed-type hypersensitivity resulting from accidental contact leading to allergic contact dermatitis. This occurs mainly in 2 situations, the first of which is contact in a professional setting. Occupational allergic contact dermatitis from systemic drugs is seen mainly in workers of pharmaceutical and chemical companies producing drugs and in health care professionals handling them, especially nurses and to a lesser degree pharmacists, dentists and physicians. In veterinary medicine, there is extensive use of drugs, which may sensitize veterinarians, animal caretakers, and farmers who prepare or apply medicaments for animals (Chapter 2.11). Sensitization to systemic drugs from accidental contact may also be observed in non-professional caregivers, e.g. a woman administering drugs to her disabled husband or a mother who prepares medicines for her child. This was formerly called 'connubial' or 'consort' allergic dermatitis and is currently termed allergic contact dermatitis/sensitization *by proxy*.

Sensitization in these two settings (occupational, *by proxy*) often manifests as contact dermatitis of the fingers, hands and forearms. When the drug is released in the air, e.g. by breaking tablets, air contamination with drug dust in the pharmaceutical industry, or preparing drugs for injection, this may lead to airborne occupational allergic contact dermatitis, manifesting as dermatitis of the face, primarily the eyelids. Occupational allergic contact dermatitis in the pharmaceutical industry was formerly frequent. When work hygiene measures in pharmaceutical companies were poor and there was heavy contamination of the factory area with dust of the drugs produced there, 25-40% of the workers exposed to beta-lactam antibiotics developed allergic reactions, mainly eczema. Stricter preventive measures later drastically lowered the risk of sensitization. In nurses, although occupational sensitization to systemic drugs is well known, they only account for a very small portion of occupational allergic contact dermatitis, in the order of 1% (Chapter 2.11).

Cutaneous drug hypersensitivity reactions

The oral route is the most frequent mode of systemic drug administration and this may – infrequently – lead to sensitization, delayed-type hypersensitivity. This may manifest in a broad range of skin eruptions, most frequently maculopapular eruptions. Other cutaneous adverse drug reactions, which have shown to be mediated by delayed-type hypersensitivity, include erythroderma and exfoliative dermatitis, systemic allergic dermatitis (systemic contact dermatitis), symmetrical drug-related intertriginous and flexural exanthema (SDRIFE)/baboon syndrome, fixed drug eruption, photosensitivity and delayed urticaria. Infrequent, but serious and potentially dangerous and even fatal manifestations of drug hypersensitivity, manifesting both in the skin and internal organs, are acute generalized exanthematous pustulosis (AGEP), drug reaction with eosinophilia and systemic symptoms (DRESS), Stevens-Johnson syndrome and toxic epidermal necrolysis (Chapters 2.2-2.9). However, a great many more or less well-defined other drug eruptions have been reported as result of hypersensitivity to systemic drugs, e.g. erythema multiforme-like, lichenoid and bullous eruptions (Chapter 2.10). Hypersensitivity to drugs given as subcutaneous, intramuscular, intra-articular or intralesional injection may manifest as localized allergic reactions with edema, erythema and sometimes vesicles or as more extensive drug eruptions (Chapter 2.10).

Risk factors

Which cutaneous adverse drug reactions develop – if any – in individual patients, depends *inter alia* on the nature of the drugs taken. Beta-lactams often induce maculopapular eruptions, whereas many cases of DRESS are caused by the aromatic anticonvulsants carbamazepine, phenobarbital and phenytoin (DRESS by these drugs was formerly called 'anticonvulsant hypersensitivity syndrome'). Allopurinol is also a frequent inducer of DRESS, but delayed-type hypersensitivity to this gout medication can never be established by patch testing.

Genetics also play an important role. All patients with abacavir (ABC) hypersensitivity, for example, carry the HLA-B*5701 allele, and nearly 50% of all HLA-B*5701-positive individuals may develop ABC hypersensitivity when treated with the drug, whereas HLA-B*5701-negative persons are never affected (Chapter 3.1 Abacavir). Another cofactor relevant in cutaneous adverse drug reactions is the presence of or previous infection with certain viruses. Epstein-Barr virus, causing mononucleosis infectiosa, for example, strongly favors the development of maculopapular eruptions from ampicillin and amoxicillin (Chapter 3.36 Amoxicillin). In many patients presenting with DRESS, there is re-activation of human herpes virus 6 (HHV-6) and of other herpesviruses including Epstein-Barr virus, HHS-7, and cytomegaly virus. Finally, other diseases in the patient must be mentioned as an important factor. Patients with hiv/aids, for example, have a strongly increased risk of developing drug hypersensitivity.

Pathophysiology

How sensitization exactly develops and which mechanisms are involved, is as yet largely unknown, although much progress has been made in this area in the past 25 years. Drug-specific T cells have been demonstrated in many investigations and positive patch and delayed intradermal tests are strong indicators of a delayed-type hypersensitivity mechanism being involved in many of these cutaneous adverse drug reactions. Most patients become sensitized from taking the drugs, which usually takes a minimum of 4-7 days to develop. Once sensitized, re-administration of the drug in these patients may result in a recurrence of the hypersensitivity reaction within a few hours to 1-2 days. Patients may also previously have become sensitized to the drug from topical application of this drug or a cross-reacting pharmaceutical to skin or mucosa. In these individuals, systemic administration may induce systemic allergic dermatitis (SAD, previously called systemic contact dermatitis) within a few hours to a few days. SAD may give rise to a variety of manifestations, mostly exacerbation of previous allergic contact dermatitis, re-emergence of previously positive patch test reactions to the drug(s) in question or maculopapular eruptions (Chapter 2.4 Systemic allergic dermatitis).

Why a book with monographs on systemic drugs?

Delayed-type hypersensitivity to systemic drugs is a rapidly evolving and important part of medicine, that appears to have been somewhat underexposed in dermatological literature thus far. Whereas the subject of contact allergy (as presented in the first 3 volumes of the *Monographs in Contact Allergy* series) is mostly published in specialized journals (*Contact Dermatitis, Dermatitis*) and other dermatological journals, drug hypersensitivity studies are often published in journals focusing on hypersensitivity (non-allergic), allergy (non-contact) and immunology such as *Journal of Allergy and Clinical Immunology, Current Opinion in Allergy and Clinical Immunology, Allergy, Expert Review of Clinical Immunology,* and *Annals of Allergy, Asthma and Immunology*, which are not widely read by dermatologists. They may also be published in journals of internal medicine (*AIDS*), or in non-English literature, especially Annales de Dermatologie et de Vénéréologie, in French (many publications on drug hypersensitivity come from France).

Because of the growing importance of drug hypersensitivity, underexposure of the topic in dermatological literature, as relevant information may often be difficult to trace in electronic databases, and no book focusing on the clinical aspects of delayed-type drug hypersensitivity reactions, the drugs responsible for them and their diagnosis by patch tests has as yet been published, 'Delayed-type hypersensitivity to systemic drugs' was – despite the differences with the previous volumes – considered to be a very suitable subject for this *Monograph in Contact Allergy*, Volume 4.

1.2 DATA PROVIDED IN THIS BOOK

1.2.1 SCOPE AND DATA COLLECTING

This book provides monographs of all systemic drugs that have caused delayed-type hypersensitivity reactions, *as shown by positive patch tests to the drugs*. For practical reasons, studies in which intradermal tests with delayed reading to drugs were positive, but patch tests were either not performed or negative, were not included. Systemic drugs were defined as pharmaceuticals (including vitamins) used in human and veterinary medicine that may be administered by the oral route, as enemas or as subcutaneous, intramuscular, intravenous, intra-articular, intrathecal or intralesional injections. Drugs given as vaginal tablets, as suppositories or by inhalation in the nose or lungs were (with few exceptions) excluded, as their systemic manifestations have been presented in the previous volume of this book series (3). Also excluded are vaccines and drugs used systemically but present in topical formulations. Veterinary drugs present in animal feed, however, are included; they may cause occupational allergic contact dermatitis in farmers and animal caretakers.

Any cutaneous adverse drug reaction caused by patch test proven delayed-type hypersensitivity to any systemic drug has been included, as are all cases of occupational allergic contact dermatitis and dermatitis from accidental contact to drugs in other situations that are used as systemic pharmaceuticals.

The main sources of information for this book are:
- the journals *Contact Dermatitis* and – to a lesser degree - *Dermatitis*, that were fully screened from their start in 1975 resp. 1990 through July 2021;
- selected chapters in the author's book *Unwanted Effects of Cosmetics and Drugs used in Dermatology*, 3rd edition (8) and in the most recent editions of the – digital - books *Contact Dermatitis*, 6th Edition, 2018 (9) and *Kanerva's Occupational Dermatology*, 3rd Edition, 2018 (10);
- relevant articles found in literature lists of journal publications used for the book; most of these could be accessed on-line through the Medical Library of the University of Groningen; a limited number of important articles that were not accessible on-line were requested from the library;
- targeted PubMed searches in all journals available on-line through the Medical Library of the University of Groningen.

After the author had made an inventory of the types of drug eruptions that can be caused by delayed-type hypersensitivity, he performed PubMed searches for 'DRESS', 'fixed drug eruption', 'baboon syndrome', 'symmetrical drug-related intertriginous and flexural exanthema', 'SDRIFE', 'systemic contact dermatitis', 'maculopapular eruption', 'acute generalized exanthematous pustulosis', 'Stevens-Johnson syndrome', 'toxic epidermal necrolysis', 'photoallergy' and 'occupational allergic contact dermatitis'. More targeted searches (including the search word 'patch test' and 'delayed-type hypersensitivity') proved to be less reliable, articles were missed that were found in literature lists of relevant articles. These lists proved to be very useful for identifying relevant literature.

It was decided that only cases with positive patch tests qualified for inclusion. However, in the great majority of drug eruptions from systemic drugs that may be caused by delayed-type hypersensitivity, no patch tests have been performed and in many where they were, patch tests proved to be negative. It was found that articles with positive patch tests were difficult to find, even with targeted search strategies. Therefore, and also because not all literature was available to the author, it is quite certain that the data provided in this book – although extensive and probably containing the great majority of information available – cannot be considered complete. Nevertheless, before finishing and closing a monograph (in the period June-July 2021), a final search was performed using the drug name AND 'patch test' in PubMed, and only very few additional relevant articles and data were identified. As studies with a positive delayed intradermal test, in which patch tests were either negative or not performed, were not included, more cases of drug eruptions from delayed-type hypersensitivity (as shown only by positive intradermal tests) are missing in this book. .

A number of articles could neither be accessed on-line nor were they requested. If some relevant information was available, e.g. from the Abstract or from being cited in other sources, it was included, in the latter situation mentioning that the information was cited in another article (often in non-English literature including Japanese, French or German). If data from articles are missing, this is indicated with 'data are unavailable to the author', 'no details available' or similar text in the Monographs.

1.2.2 DATA PROVIDED IN THE MONOGRAPHS
Searching the scientific literature as described above has resulted in the identification of 507 systemic drugs that have caused drug eruptions from delayed-type hypersensitivity, allergic contact dermatitis from accidental contact, or both, proven by patch tests. For each of these 507 pharmaceuticals, the data shown in table 1.1 are – where available – provided in each monograph. For most drugs, it has been attempted to provide a review as comprehensive as possible. However, for a number of drugs, especially 'historical allergens' (drugs that were formerly used, but have been [largely] discontinued, e.g. certain sulfonamides, chlorpromazine), drugs with a large number of articles in (very) early literature and for systemic drugs with a vast number of available data (e.g. amoxicillin, carbamazepine), a full review was not attempted.

Table 1.1 Information provided (when available and applicable) in the monographs

IDENTIFICATION
Description/definition   :
Pharmacological classes :
IUPAC name   :
Other names   :
CAS registry number   :
EC number   :
Merck Index monograph   :
Patch testing   :
Molecular formula   :
Structural formula

GENERAL

CONTACT ALLERGY FROM ACCIDENTAL CONTACT

CUTANEOUS ADVERSE DRUG REACTIONS FROM SYSTEMIC ADMINISTRATION CAUSED BY TYPE IV (DELAYED-TYPE)
HYPERSENSITIVITY (as demonstrated by positive patch tests)
Cutaneous adverse drug reactions from systemic administration of .... caused by type IV (delayed-type)
hypersensitivity have included ....

Case series with various or unknown types of drug reactions
Maculopapular eruption
Erythroderma, widespread erythematous eruption, exfoliative dermatitis
Acute generalized exanthematous pustulosis (AGEP)
Systemic allergic dermatitis (systemic contact dermatitis)
Symmetrical drug-related intertriginous and flexural exanthema (SDRIFE)/Baboon syndrome
Fixed drug eruption
Drug reaction with eosinophilia and systemic symptoms (DRESS)
Stevens-Johnson syndrome/toxic epidermal necrolysis (SJS/TEN)
Photosensitivity
Dermatitis/eczematous eruption
Other cutaneous adverse drug reactions

Cross-reactions, pseudo-cross-reactions and co-reactions

LITERATURE

IDENTIFICATION
For the monographs' titles, the names shown in the ChemIDplus database of the NIH U.S. National Library of
Medicine were used (https://chem.nlm.nih.gov/chemidplus/). In the section IDENTIFICATION the following data are
provided: Pharmacological classes, IUPAC (International Union of Pure and Applied Chemistry) name, other names
(synonyms), CAS (Chemical Abstract Service) registry number (www.cas.org; not validated by CAS), EC/EINECS
number (ECHA European Chemicals Agency, EC Inventory: https://echa.europa.eu/information-on-chemicals/ec-
inventory), Merck Index monographs number (The Merck Index online: https://www.rsc.org/merck-index) and their
molecular and structural formulas. Also, a general description/definition of the compounds is provided. Most of
these data were readily available from ChemIDPlus and from PubChem, the NIH National Library of Medicine,
National Center for Biotechnology Information online database (https://pubchem.ncbi.nlm.nih.gov/). Most structural
formulas were copied and adapted from ChemIDplus, some from other internet sources, notably chemicals supplying
companies.

Patch testing
Advice on how to patch test the individual drugs is provided under this heading based on the publications presented,
the author's book *Patch Testing*, 4th Edition (11), and commercial availability (Chemotechnique Diagnostics,
www.chemotechnique.se; SmartPractice Europe, www.smartpracticeeurope.com; and SmartPractice Canada,
www.smartpracticecanada.com).

It must be pointed out that in the great majority of cases, patients have been patch tested with the drugs (tablets or capsules) that they used, usually diluted to 10-30% in petrolatum, water, or both. The pure chemical was rarely tested, which means that some (but probably few) positive patch test reactions may in fact have been to one of the excipients rather than the active drug. Furthermore, where excellent studies have *inter alia* been provided by investigators from France, Switzerland, Italy, Portugal and Germany, quite a few other reports have been of low quality (i.e. as regards patch testing), where the investigators apparently lacked the skills to perform patch tests adequately (*lege artis*). Not infrequently, the only information provided was 'A patch test with the drug was positive'. Quite often, one or more (usually more) of the following relevant data were missing: nature of the test material (pulverized tablet or pure chemical, purity of the chemical), test concentration, test vehicle, concentration of the active material in the test material in case of using tablets provided by the patient, fixation time (24 hours, 48 hours), days of reading, strength of the reactions, number of controls and results in controls.

Guidelines advise reading the patch test results at D2, D3/4 and preferable a late reading at D7-D10. However, in many investigations, the patch test was read only once, usually at D2 (presumably after removing the patch test materials), which is most certainly unreliable, both when the reaction at that moment is positive and when it is negative. Many authors have not tested – or appear not to have tested – control patients, which implies that irritation from the test material (false-positive reaction) was not excluded. When test concentrations of 10% or higher were used, in most cases the drugs provided by the patient will have been used for patch testing. However, when lower concentrations were used, it was very often not mentioned whether the test materials contained tablets or pure chemicals. This means that some of the patch test concentrations mentioned in this book may be less reliable than they should be and the author would like them to be.

When no commercial drug patch test materials are available (Chapter 4 , table 4.3), extemporaneous test materials have to be prepared. Generally speaking, most systemic drugs can be tested at 10% pet. pure drug. If the pure chemical cannot be obtained, the test material should be prepared from intravenous powder, the content of capsules or – when also not available – from powdered tablets to achieve a final concentration of the active drug of 10% in petrolatum. The reliability of patch testing greatly benefits from using serial dilutions, e.g. 1%, 5% and 10% in pet. and water. At least 10 (preferably 20 or more) unexposed controls should also be tested to exclude irritancy of the test materials used, although the author realizes that this may be problematic in some countries. Always read the test twice (D2 and D3 or D4) and a late reading at D7-D10 is strongly recommended.

GENERAL
A general description of the index drug with indications, mechanism of action and sometimes additional relevant data are provided here. Important sources were ChemIDPlus, PubChem, DrugBank (https://go.drugbank.com/), Drug Central (https://drugcentral.org/), Drugs.com (https://www.drugs.com/), and Wikipedia (https://www.wikipedia.org/).

CONTACT ALLERGY FROM ACCIDENTAL CONTACT
Case series and case reports of contact allergy/allergic contact dermatitis from accidental contact with systemic drugs are presented here.

CUTANEOUS ADVERSE DRUG REACTIONS FROM SYSTEMIC ADMINISTRATION CAUSED BY TYPE IV (DELAYED-TYPE) HYPERSENSITIVITY (as demonstrated by positive patch tests)
Cutaneous adverse drug reactions from systemic administration of .... caused by type IV (delayed-type) hypersensitivity have included ....
All drug reactions caused by delayed-type hypersensitivity to the drug are summed up here (in the sequence as shown below) with references.

Case series with various or unknown types of drug reactions
Maculopapular eruption
Erythroderma, widespread erythematous eruption, exfoliative dermatitis
Acute generalized exanthematous pustulosis (AGEP)
Systemic allergic dermatitis (systemic contact dermatitis)
Symmetrical drug-related intertriginous and flexural exanthema (SDRIFE)/Baboon syndrome
Fixed drug eruption
Drug reaction with eosinophilia and systemic symptoms (DRESS)
Stevens-Johnson syndrome/toxic epidermal necrolysis (SJS/TEN)
Photosensitivity
Dermatitis/eczematous eruption
Other cutaneous adverse drug reactions

All hypersensitivity reactions found in literature for the drug in question are presented here under their relevant headings.

Cross-reactions, pseudo-cross-reactions and co-reactions
It may be difficult to determine whether positive patch test reactions to other pharmaceuticals besides the index drug are the result of cross-reactivity (contact allergic reaction to a structurally similar chemical to which the individual has not yet been exposed), of pseudo-cross-reactions (the index chemical and the other reacting chemical have the same allergenic constituent, contaminant, or metabolite) or co-reactions, which are independent of each other. In this monograph section, established cross-reactions are mentioned, as are co-reactions to drugs other than the index drug, that are *likely* or *possibly* the result of cross-reactivity. These include both reactions to systemic and to topical drugs.

LITERATURE

1   De Groot AC. Monographs in Contact Allergy Volume I. Non-Fragrance Allergens in Cosmetics (Part I and Part 2). Boca Raton, Fl, USA: CRC Press Taylor and Francis Group, 2018 (ISBN 978-1-138-57325-3 and 9781138573383)
2   De Groot AC. Monographs in Contact Allergy, Volume II. Fragrances and Essential Oils. Boca Raton, Fl, USA: CRC Press Taylor and Francis Group, 2019 (ISBN 9780367149802)
3   De Groot AC. Monographs in Contact Allergy, Volume III. Topical Drugs. Boca Raton, Fl, USA: CRC Press Taylor and Francis Group, 2021 (ISBN 978-0-367-23693-9)
4   Gilissen L, Goossens A. Frequency and trends of contact allergy to and iatrogenic contact dermatitis caused by topical drugs over a 25-year period. Contact Dermatitis 2016;75:290-302
5   Goossens A, Gonçalo M. Topical drugs. In: Johansen J, Mahler V, Lepoittevin JP, Frosch P, Eds. Contact Dermatitis, 6th Edition. Springer: Cham, 2020
6   De Groot AC. Allergic contact dermatitis from topical drugs: An overview. Dermatitis 2021;32(4):197-213
7   De Groot AC. Fragrances: Contact allergy and other adverse effects. Dermatitis 2020;31:13-35
8   De Groot AC, Nater JP, Weijland JW. Unwanted effects of cosmetics and drugs used in dermatology, 3rd Edition. Amsterdam: Elsevier Science, 1994 (ISBN 0444897755)
9   Johansen JD, Mahler V, Lepoittevin J-P, Frosch PJ, Eds. Contact Dermatitis, 6th Edition. Berlin Heidelberg: Springer-Verlag, 2018
10 John S, Johansen J, Rustemeyer T, Elsner P, Maibach H, Eds. Kanerva's Occupational Dermatology, 3rd Ed. Cham: Springer, 2018
11 De Groot AC. Patch Testing, 4th Edition. Wapserveen, The Netherlands: acdegroot publishing, 2018 (ISBN 978-90-813233-1-4) (www.patchtesting.info)

Chapter 2   THE SPECTRUM OF CUTANEOUS ADVERSE DRUG REACTIONS FROM SYSTEMIC DRUGS CAUSED BY DELAYED-TYPE HYPERSENSITIVITY

Contents

2.1 Introduction

Drug hypersensitivity reactions (DHRs) are adverse effects of drugs that clinically resemble allergic reactions. They are called drug *allergies* when a definite immunological mechanism (either drug-specific antibody or T cell) has been demonstrated (23). DHRs affect more than 7% of the population (209) and 5% of hospitalized patients, and are associated with significant morbidity and mortality (58). Women are generally more often affected. The most frequent reactions are maculopapular eruptions and urticaria/angioedema. Antibiotics are most frequently implicated, mostly aminopenicillins such as amoxicillin and ampicillin. Other important classes of drugs that cause cutaneous adverse drug reactions are non-steroidal anti-inflammatory drugs (NSAIDs), contrast media and – for the more severe reactions – antiepileptics and antivirals (226).

Based on the time between drug exposure and onset of symptoms/signs, reactions may be divided into immediate and nonimmediate (= delayed) hypersensitivity reactions. The mechanism underlying the former is thought to be IgE-mediated and the latter is primarily T cell-mediated, but there is some overlap (58). Immediate reactions tend to occur within minutes to one hour after drug administration, but may develop after 1-6 hour and in exceptional cases later. Symptoms are often confined to the skin and mucous membranes, for example, generalized urticaria (with or without angioedema) and may progress in some cases to more severe symptoms of bronchospasm, hypotension and anaphylactic shock (23). Nonimmediate hypersensitivity reactions (NIHR) occur mostly later than 6 hours, often 24 hours, after drug intake. NIHR show a diversity of clinical manifestations (table 2.1) with maculopapular eruptions being the most common presentation (20,50,58,209).

Nonimmediate drug hypersensitivity reactions are often divided into non-severe and severe cutaneous adverse drug reactions (CADRs) (table 2.1). To the *non-severe* category belong *inter alia* maculopapular eruption (Chapter 2.2), fixed drug eruption (FDE) (Chapter 2.6), symmetrical drug-related intertriginous and flexural exanthema (SDRIFE) (Chapter 2.5) and photoallergic reactions (Chapter 2.9). In some cases, patients have been presensitized to a drug from topical application and develop an exanthema after systemic administration of the drug or a cross-reacting chemical, which is called systemic allergic dermatitis or systemic allergic dermatitis (synonym/older name: systemic contact dermatitis) (Chapter 2.4). It may present with diverse manifestations, including reactivation of previous eczema or positive patch tests, maculopapular eruptions, extensive eczematous dermatitis, acrovesicular dermatitis, urticaria or symmetrical drug-related intertriginous and flexural exanthema, formerly (and also sometimes today by some authors) called the baboon syndrome (Chapter 2.5). The *severe* cutaneous adverse drug reactions (SCARs) are acute generalized exanthematous pustulosis (AGEP) (Chapter 2.3), drug reaction with eosinophilia and systemic symptoms (DRESS, formerly mostly presented as 'drug hypersensitivity syndrome') (Chapter 2.7), Stevens-Johnson syndrome/toxic epidermal necrolysis (SJS/TEN) (Chapter 2.8) and generalized bullous fixed drug eruption (Chapter 2.6.1). Excellent reviews of the classification of cutaneous manifestations of drug hypersensitivity and diagnosing them have recently been published (20,45,207,209).

All these delayed hypersensitivity reactions to drugs are the subject of this book. Another form of delayed-type hypersensitivity caused by systemic drugs (not from systemic use by the patient but from accidental contact) is occupational allergic contact dermatitis, which is observed mostly in health personnel and workers in the pharmaceutical industries. This topic is discussed in Chapter 2.11 and presented in detail in the monographs.

Table 2.1 Delayed (nonimmediate, non-IgE-mediated) cutaneous adverse drug reactions

Non-severe cutaneous adverse drug reactions
Eczematous eruption (Chapter 2.10)
Erythema multiforme-like eruption (Chapter 2.10)
Erythroderma, widespread erythematous eruption, exfoliative dermatitis (Chapter 2.2)
Fixed drug eruption (Chapter 2.6)
Lichenoid drug eruption (Chapter 2.10)
Localized hypersensitivity reactions to subcutaneous injections (e.g. heparins, local anesthetics) (Chapter 2.10)
Maculopapular eruption (Chapter 2.2)
Photosensitivity/photoallergic dermatitis (Chapter 2.9)
Symmetrical drug-related intertriginous and flexural exanthema (SDRIFE)/Baboon syndrome (Chapter 2.5)
Systemic allergic dermatitis (systemic contact dermatitis) (Chapter 2.4)
Urticaria/angioedema (delayed) (Chapter 2.10)
Miscellaneous drug eruptions (Chapter 2.10)

Severe cutaneous adverse drug reactions (SCARs)
Abacavir hypersensitivity syndrome (Chapter 3.1 Abacavir)
Acute generalized exanthematous pustulosis (AGEP) (Chapter 2.3)
Drug reaction with eosinophilia and systemic symptoms (DRESS) (Chapter 2.7)
Stevens-Johnson syndrome/toxic epidermal necrolysis (SJS/TEN) (Chapter 2.8)
Generalized bullous fixed drug eruption (Chapter 2.6.1)

Patch testing in delayed drug hypersensitivity eruptions is important, as overwhelming evidence suggests that they are caused by delayed-type, T cell-mediated hypersensitivity to the drugs. Patch testing may identify the culprit drug in a number of cases, the frequency of positive reactions depending *inter alia* on the nature of the drug (amoxicillin and carbamazepine frequently positive; allopurinol nearly always negative) and the nature of the CADR. Patch testing in cutaneous adverse drug reactions (and briefly prick tests, intradermal tests, *in vitro* tests and drug provocation tests) are discussed in Chapter 4.

Unfortunately, in many publications on cutaneous drug hypersensitivity reactions, no patch tests were performed or the results were negative. **Such articles are not discussed in this book, only investigations showing positive patch tests are presented.** Discussion of *immediate hypersensitivity reactions* to drugs falls outside the scope of this publication (and would probably create a similar-sized book) with the exception of immediate contact reactions (contact urticaria) (Chapter 5).

2.2 Maculopapular eruption, erythroderma and exfoliative dermatitis

Introduction
Maculopapular eruptions (MPEs) are the most frequent drug hypersensitivity reactions, which are often termed 'drug rashes' or 'drug eruptions'; the name 'morbilliform eruption' is also frequently used. MPEs are benign rashes which account for the large majority of all cutaneous adverse drug reactions (CADRs). Such exanthemas may also result from viral infections, especially in children.

Etiology and pathophysiology
The pathomechanism of maculopapular drug eruptions is not completely understood, but a type IV delayed cell-mediated immune mechanism involving drug-specific T lymphocytes is thought to be relevant. Viral infections (HIV and herpesviruses including Epstein-Barr virus [EBV], cytomegaly virus and HHV-6) may increase the incidence of MPEs, as seen in infectious mononucleosis caused by EBV under treatment with ampicillin or amoxicillin (93). MPEs can be caused by almost any drug, but the most frequent medications eliciting them are aminopenicillins, cephalosporins, antibacterial sulfonamides, antiepileptics, allopurinol and NSAIDs.

Clinical features
Maculopapular exanthemas include a heterogeneous variety of skin reactions that literally burst forth on the skin. The eruptions usually appear between four and 14 days after a new drug has been started. However, in a sensitized individual, initial symptoms may already appear within a few hours and develop into a typical exanthem after 1 or 2 days. MPEs can also arise a few days after the drug intake has been stopped. Erythematous macules and infiltrated papules are the primary lesions; more rarely, vesicles or pustules may be present and some lesions may develop into

bullae (183). The trunk and the proximal extremities are most often involved in a symmetric distribution. Widespread exanthems may generalize, become confluent and develop into erythroderma. Whereas in early phases typically no scaling occurs, desquamation is common in the later clearing phase (which may then sometimes be termed 'exfoliative dermatitis'). Mucous membranes are normally not involved. Pruritus is typical and fever and systemic involvement may occasionally occur but are very mild. Eosinophilia can be present (93,183).

Histology
Histological examination shows an interface dermatitis with vacuolar changes of keratinocytes at the basal cell layer and an upper dermal mononuclear cell infiltrate with some eosinophils (93).

Diagnosis
The diagnosis of drug-induced maculopapular exanthema is based on the clinical picture in combination with the chronology of new drug use. Patch tests should always be performed and may aid in identifying the culprit drug. It is important to realize that exanthems with macules and papules can also be the early presenting findings of severe cutaneous drug hypersensitivity reactions such as DRESS (Chapter 2.7) or SJS/TEN (Chapter 2.8), which usually become evident within 48 hours (209). Worrisome signs and symptoms of these life-threatening entities include high fever, mucous membrane involvement, lymphadenopathy, the development of bullae, and systemic involvement, as shown by elevated hepatic transaminases, hematologic changes, or renal manifestations (93).

Therapy
MPEs usually resolves in a few days to a week when the causative drug is stopped, but sometimes it may take longer. Progress despite discontinuation of the offending medication may sometimes be observed. Treatment includes prompt diagnosis with discontinuation of the offending medication and symptomatic care with topical corticosteroids, oral antihistamines (sedative, for the itching) and emollients. In severe cases, treatment with systemic corticosteroids over a short period of time is often initiated (93,183).

Results of patch testing in patients with maculopapular eruptions
Positive patch tests may be observed in 14-59% of drug-induced MPEs, depending on the medicament (Chapter 4, table 4.4). In cases of negative patch tests, prick or intradermal tests (read at D2-D3) may sometimes be positive and identify the offending drug. If these tests are also negative, oral provocation can be considered in low probability situations (50,57,83). See Chapter 4 for details on diagnostic tests (patch test, prick test, intradermal test, drug provocation test, *in vitro* tests) in suspected drug hypersensitivity reactions.

Over 120 systemic drugs that have caused maculopapular eruptions had a positive patch test (table 2.2). Medicaments that have been responsible for erythroderma, extensive/generalized erythema or exfoliative dermatitis and had a positive patch test are shown in table 2.3. Additional drugs that have caused maculopapular eruptions, as a manifestation of systemic allergic dermatitis (systemic contact dermatitis), may be found in table 2.7.

Table 2.2 Drugs that have caused maculopapular eruptions and showed a positive patch test [a,b,c]

Acetaminophen (paracetamol)	Cefazolin
Acetazolamide	Cefcapene pivoxil
Acexamic acid	Cefonicid
Acyclovir (from systemic allergic dermatitis)	Cefoxitin
Alendronate (limited eruption)	Ceftriaxone (202)
Ambroxol	Cefuroxime (202)
Aminocaproic acid	Celecoxib
Amitriptyline	Cetirizine
Amlexanox	Chlorambucil
Amoxicillin	Chloramphenicol
Amoxicillin-clavulanic acid	Cimetidine
Ampicillin (204)	Ciprofloxacin
Azathioprine	Clavulanic acid
Aztreonam	Clindamycin
Bacampicillin	Clobazam
Benznidazole	Clofazimine
Benzylpenicillin	Clonidine
Betamethasone	Clopidogrel
Captopril	Clorazepate
Carbamazepine (205)	Codeine
Cefalexin	Deflazacort

Table 2.2 Drugs that have caused maculopapular eruptions and showed a positive patch test (continued) [a,b,c]

Dexamethasone	Metronidazole
Dexamethasone sodium phosphate	Mexiletine
Diclofenac (202)	Miconazole
Dicloxacillin	Mirtazapine (202)
Dihydrocodeine	Morphine
Diltiazem	Nadroparin
Enoxaparin	Nevirapine
Ephedrine	Nimodipine
Eslicarbazepine	Norfloxacin (202)
Ethambutol	Nystatin
Fenofibrate	Omeprazole
Floxacillin (flucloxacillin) (202)	Oxcarbazepine
Fluconazole	Penicillin V
Flurbiprofen	Phenethicillin
Gabapentin (202)	Phenindione
Gadobutrol (202)	Phenobarbital
Garenoxacin	Phenylbutazone
Heparin, unfractionated	Phenytoin
Hydroxyzine (204)	Piperacillin
Ibandronic acid	Piperazine
Imipenem	Practolol
Iobitridol (202,217)	Prednisolone
Iodixanol	Prednisone
Iohexol (217)	Pregabalin (202)
Iomeprol (202)	Pristinamycin
Iopromide (202)	Proguanil
Ioxaglic acid (217)	Pseudoephedrine (204)
Ioxitalamic acid (217)	Rosuvastatin
Irbesartan	Sertraline
Isoniazid	Sildenafil
Isotretinoin	Spiramycin
Lamotrigine	Sulfamethoxazole-trimethoprim
Lansoprazole	Terbinafine
Levofloxacin	Tetrazepam
Lidocaine	Thioctic acid
Meprobamate	Triamcinolone
Meropenem	Valdecoxib
Metamizole	Valproic acid (202)
Methylprednisolone	Vancomycin
Methylprednisolone hemisuccinate	Zinc acexamate (acexamic acid)

[a] Details and additional references can be found in the monographs of the drugs mentioned in this table; [b] There may be some overlap with other chapters, notably Chapter 2.4 (Systemic allergic dermatitis) and Chapter 2.7 (Drug reaction with eosinophilia and systemic symptoms [DRESS]); [c] Included are 'morbilliform' eruptions/exanthemas

Table 2.3 Drugs that have caused erythroderma, extensive/generalized erythema or exfoliative dermatitis and showed a positive patch test [a]

Erythroderma or extensive/generalized erythema	
Acetaminophen (paracetamol)	Metamizole
Alendronate	Methylprednisolone acetate
Aminophylline	Methylprednisolone hemisuccinate
Amoxicillin	Mexiletine
Bacampicillin	Pantoprazole
Betamethasone	Paracetamol (acetaminophen)
Betamethasone sodium succinate	Paramethasone
Captopril	Phenylbutazone
Carbamazepine	Piperacillin
Ceftazidime	Piperazine
Cefuroxime	Piritramide
Chlorambucil	Prednisone
Chlorpheniramine	Pristinamycin
Clavulanic acid	Pseudoephedrine
Codeine	Talastine
Cyanamide	Tetrazepam
Dexamethasone sodium phosphate	Vancomycin
Diltiazem	
Doxycycline (from photoallergy)	**Exfoliative dermatitis**
Ephedrine	Carbamazepine
Flavoxate	Chlorambucil
Gentamicin	Cloxacillin
Gliclazide	Codeine
Hydrocortisone sodium phosphate	Dexamethasone
Indeloxazine	Diltiazem
Iodixanol	Lamotrigine
Iohexol	Phenobarbital
	Ranitidine

[a] Details and references can be found in the monographs of the drugs mentioned in this table

2.3 Acute generalized exanthematous pustulosis (AGEP)

Introduction

Acute generalized exanthematous pustulosis (AGEP) is a severe cutaneous reaction pattern characterized by the rapid development of non-follicular, sterile pustules on an erythematous base accompanied by fever. The disease is most frequently caused by drugs, notably antibiotics. AGEP is a rare adverse drug reaction with an incidence of one to five cases per million per year, but it might be underreported. It can occur at any age and seems to be more frequent in women (8). AGEP was originally considered to be a form of pustular psoriasis; in 1968, it was first suspected that it was actually a separate entity (1). The name 'acute generalized exanthematous pustulosis' was proposed to describe the disease in 1980 in a French publication (2).

In this chapter, the most important practical aspects of AGEP are presented, but it falls outside the scope of this book to provide a detailed discussion. Recent review articles, often combined with Drug reaction with eosinophilia and systemic symptoms (DRESS) and Stevens-Johnson syndrome/toxic epidermal necrolysis (SJS/TEN) (the SCARS, severe cutaneous adverse drug reactions) can be found in refs. 88,89,90,91,92,94,95,96,97,98,99,100,101,102,183 and 218.

Etiology and pathophysiology

AGEP is mostly (>90%) an adverse reaction to drugs, especially pristinamycin, aminopenicillins, quinolones, sulfonamides, hydroxychloroquine, terbinafine, diltiazem, ketoconazole, and fluconazole (7). In some cases, AGEP is induced by bacterial, viral or parasitic infections. AGEP is a sterile neutrophilic inflammatory response caused by a T cell dependent type IV hypersensitivity reaction (94). This concept is supported by positive patch tests and *in vitro* tests (1,8).

Clinical features

AGEP is clinically characterized by the rapid development of tens to hundreds small, sterile, non-follicular pustules on an erythematous edematous base. It starts in the main folds (axillary, inguinal and submammary areas) and spreads within a few hours on the trunk, arms and legs (1,8). During the early stage, pustule confluence can result in a positive Nikolsky sign, with superficial detachment (90). The time period from drug ingestion to reaction onset ranges from several hours to 11 days, but is usually within 48 hours, with antibiotics having a median of 24 hours. There is an itching or sometimes burning sensation. Mucous membrane involvement, if any, is usually confined to a single site, most often the lips or buccal mucosa. Sometimes patients have lymphadenopathy.

Systemic inflammation signs in the acute phase of the disease include fever (>38.0°C), leukocytosis (>10,000/ml), elevated levels of C-reactive protein (CRP) and mostly increased levels of neutrophils (>7000/ml) (1,8). In 15-20% of the patients, there may be internal organ involvement, notably hepatic, renal, and pulmonary dysfunction. Hepatic involvement may present as elevated liver enzymes, steatosis or hepatomegaly (3). Pulmonary involvement includes bilateral pleural effusion resulting in hypoxemia, requiring supplemental oxygen. Multiple organ dysfunction in AGEP may occasionally require treatment in an intensive care unit (3).

When the causative drug is discontinued, the cutaneous features typically disappear within a few days to two weeks with a very typical collarette-shaped desquamation over the affected areas. AGEP usually shows a mild course, but high fever, cutaneous superinfection or multiple organ dysfunction with disseminated intravascular coagulation can complicate the process and lead to severe illness and sometimes life-threatening situations, especially in (elderly) patients of poor general condition. The reported mortality is under 5%, probably 1-3% (1,8).

Histology

AGEP is characterized histologically by intracorneal, subcorneal, and/or intraepidermal pustules with papillary dermal edema containing neutrophilic and eosinophilic infiltrates. The majority of intraepidermal pustules are located in the upper epidermis, often contiguous with the subcorneal pustules. The pustules tend to be large and contain eosinophils. Spongiform changes may be observed in the intra- and subcorneal pustules. Epidermal changes also include spongiosis with exocytosis of neutrophils and necrotic keratinocytes (1,4).

Diagnosis

The diagnosis of AGEP is based on clinical and histologic criteria. An AGEP validation score has been developed by the EuroSCAR group. This is a standardized scheme based on morphology, clinical course, and histology that classifies patients with suspected AGEP as having definite, probable, possible, or no AGEP (5). Patch testing with all drugs is very useful, especially to identify the cause of AGEP when the responsible drug is unclear (many drugs used) and to confirm the suspected causality of a drug (6,9,11). See Chapter 4 for details on diagnostic tests (patch test, prick test, intradermal test, drug provocation test, *in vitro* tests) in suspected drug hypersensitivity reactions.

Therapy

The most important objective is to discontinue the use of the (suspected) causative agent promptly. Depending on the degree of fever, antipyretics may be advised. Topical steroids are often given and secondary bacterial infections should be treated. In the desquamative phase, rehydration measures may be appropriate. Systemic steroids are sometimes prescribed in very extensive eruptions, but there is no evidence that they reduce disease duration. Systemic manifestations should be identified and, whenever needed and possible, appropriately treated (8).

Acute localized exanthematous pustulosis (ALEP)

A rare variant of acute generalized exanthematous pustulosis (AGEP) is a *localized* reaction called acute localized exanthematous pustulosis (ALEP) (36,40,41). ALEP is usually located to the face, neck, or chest. The skin reaction arises quickly within a few hours, mostly 3-5 days after commencement of a culprit drug. It is usually accompanied by fever and neutrophilic leukocytosis. ALEP is described histologically as sterile subcorneal pustules rich in neutrophils. Analogous to AGEP, ALEP is thought to be due to drug-specific T cell-mediated immune processes. In 80% of cases, ALEP is caused by a systemic drug, mostly antibiotics, especially β-lactams and macrolides. Some cases have been linked with viral infections and insect bites (40,42). Patch tests have infrequently been performed. Positive patch test reactions have been observed to amoxicillin-clavulanate (44), bemiparin (36), iomeprol, metronidazole (46), and nimesulide. Clinical picture, histopathology, course and therapy are comparable with AGEP (40).

Results of patch testing in patients with AGEP

Percentages of positive reactions to culprit drugs in patients with AGEP have ranged from 0 to 65%, but mostly between 40 and 65% (Chapter 4, table 4.4). The positive patch test usually mimics the exanthema both clinically with erythema and pustules and histologically. Over 90 systemic drugs that have caused AGEP had a positive patch test (table 2.4).

Table 2.4 Drugs that have caused AGEP and showed a positive patch test [a]

Acetaminophen (paracetamol) (11)	Hydroxychloroquine
Acetazolamide	Hydroxyzine
Acyclovir	Ibuprofen
Amoxicillin (11)	Iobitridol
Amoxicillin-clavulanic acid (AGEP and ALEP) (11)	Iodixanol (11)
Ampicillin	Iohexol
Apronalide (allylisopropylacetylurea)	Iomeprol (AGEP and ALEP)
Bacampicillin	Iopamidol
Beclomethasone (11)	Iopromide
Bemiparin (ALEP) (36)	Ioversol (11)
Bendamustine	Isoniazid
Benznidazole	Labetalol
Benzylpenicillin	Lansoprazole
Betamethasone sodium phosphate	Levofloxacin
Bleomycin	Lincomycin
Buphenine (nylidrin)	Metamizole
Bupropion	Methoxsalen
Carbimazole	Methylprednisolone acetate or hemisuccinate
Cefixime	Metronidazole (AGEP and ALEP) (46)
Cefotaxime	Mexiletine
Cefpodoxime	Miconazole
Ceftriaxone (11)	Mifepristone
Cefuroxime (202)	Minocycline
Celecoxib	Morphine
Cetirizine	Nifuroxazide
Chloramphenicol	Nimesulide (ALEP)
Ciprofloxacin	Nystatin
Clavulanic acid	Oxacillin
Clindamycin (11,202)	Paracetamol (acetaminophen)
Cloxacillin	Phenobarbital
Dalteparin	Prednisolone
Dexamethasone sodium phosphate	Prednisolone sodium succinate
Dextropropoxyphene (11)	Prednisolone sodium tetrahydrophthalate
Dicloxacillin	Prednisone (11)
Diltiazem	Pristinamycin (11,205)
Enoxaparin (11)	Propacetamol
Eperisone	Propicillin (uncertain)
Eprazinone	Pseudoephedrine (11)
Ertapenem	Ranitidine
Erythromycin	Spiramycin (205)
Etoricoxib	Terbinafine
Floxacillin (flucloxacillin)	Tetrazepam (11)
Fluconazole	Ticlopidine
Fluindione (11)	Vancomycin
Gadobutrol	Varenicline (11)
Hydroquinidine	

[a] Details and additional references can be found in the monographs of the drugs mentioned in this table; ALEP: Acute *localized* exanthematous pustulosis

2.4 Systemic allergic dermatitis (systemic contact dermatitis)

Systemic allergic dermatitis (also termed systemic contact dermatitis, systemic allergic contact dermatitis) is a condition that occurs when an individual sensitized to a contact allergen is exposed to that same allergen or a cross-reacting molecule through a systemic route. Systemic exposure to allergens can include transcutaneous, transmucosal, oral, intravenous, intramuscular, intravesical, and inhalational routes, as well as implants (107,108, 109,110,111,112,113,114,329). Possible manifestations of systemic allergic dermatitis are shown in table 2.5 and include reactivation of previous eczema and positive patch tests, acrovesicular dermatitis, systemic symptoms and various drug exanthemas including maculopapular rashes, urticaria, erythema multiforme, photosensitivity and vasculitis.

Table 2.5 Symptoms and signs of systemic allergic dermatitis (103,104,111,112,323,327,328,329)

Reactivation of previous allergic contact dermatitis
Reactivation of previous positive patch test
Worsening of existing eczema
Vesicular dermatitis of the palms of the hands, sides of the fingers and soles of the feet, with or without erythema
Drug eruptions
-   generalized eczema
-   maculopapular eruption
-   erythroderma, widespread erythema
-   urticaria/angioedema
-   symmetrical drug-related intertriginous and flexural exanthema (SDRIFE, baboon syndrome)
-   erythema multiforme
-   purpura
-   photosensitivity
-   vasculitis
-   acute generalized exanthematous pustulosis (AGEP)
Systemic symptoms: fever, malaise, nausea, vomiting, diarrhea, headache, arthralgia, (rarely) syncope

Such reactions are most frequently caused by drugs given orally or parenterally. However, topical drugs absorbed through the skin or the mucosae may also induce systemic allergic dermatitis. Well-known causes are the vasodilator diltiazem used for anal fissure, the topical anesthetic dibucaine, which are absorbed through the anal mucosa and the inflamed perianal skin (115,118,119,121) and corticosteroids, mostly budesonide in inhalation preparations (107,113,114,117). Topical drugs (including in rectal and vaginal suppositories) that have caused systemic allergic dermatitis from resorption through the mucous membranes and skin are shown in table 2.6 (103,104). In this book, discussion of systemic allergic dermatitis is limited to cases caused by systemically administered drugs.

Table 2.6 Topical drugs that have caused systemic allergic dermatitis (103,104)

Acetarsone	Diltiazem	Nifuroxime
Amlexanox	Dimethindene	Nylidrin
Bacitracin	Dorzolamide	Nystatin
Benzydamine (systemic *photo*allergic dermatitis)	Ephedrine	Oxyphenbutazone
	Estradiol	Phenylbutazone
Budesonide	Eucaine	Phenylephrine
Bufexamac	Framycetin	Piperazine
Buprenorphine	Gentamicin	Prednisolone acetate
Carbarsone	Hydrocortisone aceponate	Promestriene
Chloral hydrate	Iodine	Pyrazinobutazone
Chloramphenicol	Iodoquinol	Sisomicin
Chlorquinaldol	Lidocaine	Stannous fluoride
Clioquinol	Methyl aminolevulinate	Testosterone
Clonidine	Methylphenidate	Tetracaine
Dibucaine (allergic and photoallergic dermatitis)	Neomycin	Triamcinolone acetonide
	Nicotine	Trimebutine

The most characteristic manifestation of systemic allergic dermatitis is the so-called baboon syndrome (116,118, 119). It presents as diffuse pink or dark violet erythema of the buttocks and inner thighs, like an inverted triangle or V-shaped, resembling the red bottom of a baboon, often accompanied by dermatitis in the axillae and sometimes other body folds (111). A similar eruption can be caused by systemic drugs, notably antibiotics, *without* previous sensitization (107,120). For this variant of what was previously also called the baboon syndrome, in 2004 the alternative term 'symmetric drug-related intertriginous and flexural exanthema' (SDRIFE) was coined (120). These drug exanthemas are discussed in Chapter 2.5.

Systemically administered drugs that have caused systemic allergic dermatitis (including reactions from oral provocation tests) are shown in table 2.7 with the previously sensitizing drug (previous sensitization must have been either established or have been very likely, to qualify for being accepted as 'systemic allergic dermatitis'), the clinical manifestations of the drug reaction and references.

Review articles on systemic allergic dermatitis, which can also be caused by materials other than medications (metals, botanicals, foods) can be found in refs. 109,110,111,112,323,327,328,329, and 330. The early literature on this subject has been discussed in 1986 (331) and 1966 (337).

Table 2.7 Systemically administered drugs that have caused systemic allergic dermatitis [a,b]

Drug	Previous sensitization to:	Clinical manifestations	Refs.
Acyclovir	Acyclovir	Urticaria	253,254
		Maculopapular rash, eczema	255
		Eczema	255
Albuterol (salbutamol)	Albuterol	Dermatitis	258
Alprenolol	Alprenolol	Dermatitis	259
Aminocaproic acid	Aminocaproic acid	Micropapular eruption	261
		Maculopapular rash	262
		Maculopapular eruption, possibly SDRIFE	262
Aminophylline [f]	Ethylenediamine	Various, including SDRIFE	165,166,305
		Lichenoid eruption	375
		Generalized dermatitis	376
		Erythroderma	376,378
		Maculopapular eruption	377
		For more cases see Chapter 3.29	
Amlexanox	Amlexanox	Erythema multiforme-like eruption	263
Amoxicillin	Topical penicillin (likely)	SDRIFE/Baboon syndrome	176
Bacampicillin	Bacampicillin [g]	Hand eczema, relapse of positive patch test, stomach pain, diarrhea	368
Betamethasone	Betamethasone/Other CS	SDRIFE/Baboon syndrome	172,173
	Hydrocortisone	Facial erythema	369
	Triamcinolone acetonide	Generalized papular dermatitis	12
Betamethasone acetate	Hydrocortisone	Facial erythema	369
Betamethasone dipropionate	Unknown (other) CS	Maculopapular exanthema + flare-up	137
		Generalized eczema + flare-up	137
		Unknown adverse drug reaction	137
Betamethasone sodium phosphate	Hydrocortisone	Facial erythema	369
	Unknown (other) CS	Unknown adverse drug reaction	137
		Maculopapular exanthema + flare-up	137
		Generalized eczema + flare-up	137
Carbutamide	Sulfanilamide	Relapse of dermatitis and patch test	333
Cefalexin	Cefalexin	Itching, erythema, scaling	264
Chloral hydrate	Chloral hydrate	'Eruption'	265
Chloramphenicol	Chloramphenicol	Facial edema, exudation	266
Chlorpheniramine	Dexchlorpheniramine	Generalized dermatitis	267
Chlorpromazine	Promethazine	Exacerbation of previous dermatitis	154
	Chlorpromazine	Edema of the hands, arms and face, vertigo, tendency to fainting	154
Chlorpropamide	Sulfanilamide	Relapse of dermatitis and patch test	333
Clioquinol	Clioquinol	Generalized dermatitis	268-270
		Flare-up of previous dermatitis	271
Clonidine	Clonidine	Flare-up of previous dermatitis	272
		Maculopapular exanthema	272
Cloprednol	Other unknown CS	SDRIFE/Baboon syndrome	173
Codeine		Various, see Chapter 3.152 Codeine	
Cyanocobalamin	Cobalt	Dermatitis	273

Table 2.7 Systemically administered drugs that have caused systemic allergic dermatitis (continued) [a,b]

Drug	Previous sensitization to:	Clinical manifestations	Refs.
Deflazacort	Three other CS	Maculopapular eruption	157
	Unknown other CS	Erythematous plaques with pustules	158
	Deflazacort [g]	SDRIFE/Baboon syndrome with fever, nausea, vomiting, and hypotension	159
	8 other CSs	Pruritic rash	307
	Betamethasone valerate	Generalized dermatitis	145
Dexamethasone	Unknown (other) CS	Maculopapular exanthema	325
		SDRIFE/Baboon syndrome	173
Dexamethasone disodium phosphate	Three other CS	Maculopapular eruption	157
Dexketoprofen	Ketoprofen	Systemic photoallergic dermatitis	390
Dimethindene	Dimethindene	Maculopapular and vesicular rash	274
Dimethyl sulfoxide	Dimethyl sulfoxide	Erythematous micropapular eruption	19
Diphenhydramine	Diphenhydramine	Generalized dermatitis	275
		Vesiculobullous eruption	276
		Photosensitivity (uncertain)	391
Disulfiram	Disulfiram	Rash, fever, vomiting	277
		Dermatitis	278,279,280, 392
		Edema of the feet + vesicular rash	281
Doxepin	Doxepin	Generalized dermatitis	282
Epirubicin	Mitomycin C (?)	Widespread dermatitis	283
Erythromycin	Erythromycin	Generalized dermatitis	301,302
Estradiol	Estradiol	Systemic pruritic rash	303
		Maculopapular rash	304
Estradiol (?)	Estradiol	Generalized eczema	304
Famciclovir	Acyclovir	Erythematous dermatitis	253
		Pruriginous rash	255
		Itchy erythematous patches	256
Fenofibrate	Ketoprofen [e]	Photodermatitis	127,284
Fluorouracil	Fluorouracil	Vesicular dermatitis	286
Fusidic acid	Fusidic acid	Micropapular exanthema	287
Gentamicin	Gentamicin	Eczematous eruption	289
		Generalized eczema	290
	Neomycin	Severe drug reaction	291
		Exfoliative erythroderma	288
Hydrocortisone	Hydrocortisone	Facial erythema	369
		Various manifestations	293,294
	Prednisolone (ester, salt?)	Maculopapular exanthema	111,137
	Other (unknown) CS	SDRIFE/Baboon syndrome	173
		Rash	292
		Generalized erythematous plaque-like lesions, flare-up patch tests	131
		See also Chapter 3.244 Hydrocortisone	
Hydrocortisone sodium phosphate	Hydrocortisone (ester?)	Erythematous rash	324
Hydroxychloroquine	Hydroxychloroquine [g]	Fever, generalized erythema	296
Hydroxyprogesterone	Unknown corticosteroid(s)	Itchy papular and vesicular eruption	17
Hydroxyzine	Piperazine [g]	Unknown	385
Ibuprofen	Ibuprofen	Eczematous eruption with nausea and fever	380
Iodoquinol	Iodoquinol or clioquinol	Exacerbation of contact dermatitis	334
Isoxsuprine	Nylidrin (buphenine)	Generalized dermatitis	297
Ketoconazole	Econazole, miconazole (?)	Generalized eczema	298
Ketoprofen	Ketoprofen	Photoallergic dermatitis	299
Mesna	Mesna [g]	Erythematous papulovesicular rash	300

Table 2.7 Systemically administered drugs that have caused systemic allergic dermatitis (continued) [a,b]

Drug	Previous sensitization to:	Clinical manifestations	Refs.
Methoxsalen	Methoxsalen	Photoallergic dermatitis	306,371
		Allergic dermatitis	306,371
Methylprednisolone	Unknown (other) CS	Maculopapular exanthema	325,137
		Unspecified exanthema, possibly SDRIFE	173
		Generalized eczema	137
		See also Chapter 3.313 Methylprednisolone	
	7 other CS	Pruritic rash	307
Methylprednisolone acetate	Hydrocortisone (ester?)	Erythema around the neck	324
Methylprednisolone sodium succinate	Hydrocortisone (ester?)	Generalized erythematous rash, probably SDRIFE	324
	Unknown (other) CS	Widespread erythema	369
Methylprednisolone, unspecified salt or ester	Unknown (other) CS	Maculopapular eruption	137
Metronidazole	Metronidazole	Maculopapular rash + facial eczema	307
Mitomycin C	Mitomycin C	Maculopapular rash from patch test	308
		Various manifestations including acrovesicular dermatitis and SDRIFE	Chapter 3.325
Mofebutazone	Mofebutazone	Generalized dermatitis	372
Morphine	Heroin (diacetylmorphine) [g]	Maculopapular rash with vesicles	341
	Morphine	Maculopapular rash	309
Neomycin	Neomycin	SDRIFE/Baboon syndrome	164
Nitrofurantoin	Nitrofurazone	Drug rash	340
Norfloxacin	Possibly clioquinol	Papulopustular, erythematous and edematous lesions	310
Nystatin	Nystatin	SDRIFE/Baboon syndrome	171
		Generalized eruption	311
		Maculopapular eruption with fever, arthralgia, malaise and diarrhea	147
Piperazine	Ethylenediamine	Morbilliform rash	153,386
		Erythroderma	386,388
		Angioedema	387
Prednisolone	Other corticosteroids	Various manifestations	Chapter 3.399
Prednisone	Probably hydrocortisone	Generalized dermatitis	312
	Prednisolone acetate	Generalized eczematous dermatitis	313
	Unknown (other) CS	Maculopapular exanthema	325
Pristinamycin	Virginiamycin	Various manifestations	Chapter 3.407
Procaine	Procaine/p-phenylene-diamine	Various manifestations	49
Promazine [e]	Promethazine	Photoallergic dermatitis	156
Promethazine	Promethazine	Reactivation of dermatitis, erythroderma, systemic manifestations	154,155
Propacetamol	Propacetamol [g]	Maculopapular exanthema	381
Pseudoephedrine	Phenylephrine	Erythroderma	382
Pyrazinobutazone	Phenylbutazone	Lesions on the hands and flare-up of previous patch test	314
Pyridoxine (vitamin B$_6$)	Pyridoxine [g]	Eczema in sun-exposed area	150
Ranitidine	Ranitidine [g]	Swelling and burning of the lips	348
Ribostamycin	Neomycin	Exfoliative erythroderma	374
Sevoflurane (inhalation)	Sevoflurane [g]	SDRIFE/Baboon syndrome	170
Succinylcholine [d]	Not mentioned	Eczematous rash	315
Sulfamethoxazole	Sulfanilamide	Clinical exacerbation	332
		Severe drug eruption	128
Sulfanilamide	Sulfanilamide	Various manifestations	49,326,335,336
Sulfapyridine	Sulfanilamide	Exacerbation of previous eczema	326

Table 2.7 Systemically administered drugs that have caused systemic allergic dermatitis (continued) [a,b]

Drug	Previous sensitization to:	Clinical manifestations	Refs.
Sulfathiazole	Sulfathiazole	Exacerbation of previous eczema and extensive erythematous macular eruption	148
Terbinafine [c]	Terbinafine	SDRIFE/Baboon syndrome	174
Thiamine (vitamin B$_1$)	Thiamine [g]	Exacerbation of) previous eczema	316,317
	Thiamine	Erythematous plaques, micropapular rash	318
Tiaprofenic acid	Ketoprofen	Photosensitive eruption	129
		Photosensitive vesiculobullous eruption	130
Tolbutamide	Sulfanilamide	Relapse of dermatitis and patch test	333
Tramadol	Buprenorphine	Exacerbation of patch test, fever	319
		Exacerbation of eczema, fever	320
Triamcinolone	Unknown (other) CS	Maculopapular exanthema	325
Triamcinolone acetonide	Budesonide or other CS	SDRIFE/Baboon syndrome	169
	7 other CSs	Generalized pruritic eruption	132
	Triamcinolone acetonide and 2 other CSs	SDRIFE with bullae resembling bullous pemphigoid	133
	Desoximetasone	Generalized papulovesicular eruption	134
	Clobetasol propionate, betamethasone valerate	Spreading eczema becoming vesicular	135
	Budesonide	Generalization of dermatitis	136
	Unknown (other) CS	Maculopapular eruption	137
Trimebutine	Trimebutine	Generalized urticaria	321
Valaciclovir	Acyclovir (probably)	Itchy exanthema on the face, trunk, arms and legs	256
Virginiamycin	Virginiamycin	Bullous eczema on the hands and elbows, facial edema, pruritus and generalized erythema	322

[a] Details and additional references can be found in the monographs of the drugs mentioned in this table; [b] There may be some overlap with other chapters, notably Chapter 2.2 (Maculopapular eruption, erythroderma and exfoliative dermatitis) and Chapter 2.5 (Symmetrical drug-related intertriginous and flexural exanthema [SDRIFE]/Baboon syndrome); [c] Atypical manifestations;
[d] As it is unknown whether the patient had been sensitized before, the authors' diagnosis 'Systemic allergic dermatitis' may be challenged; [e] Photocross-reaction; [f] Aminophylline = theophylline + ethylenediamine; [g] Occupational sensitization.
CS: Corticosteroid/Corticosteroids; EM: Erythema multiforme; MP: Maculopapular

2.5 Symmetrical drug-related intertriginous and flexural exanthema (SDRIFE)/Baboon syndrome

2.5.1 Introduction

The term 'baboon syndrome' was coined in 1984 to describe a highly characteristic skin eruption (116). It presents as diffuse pink or dark violet erythema of the buttocks and inner thighs, like an inverted triangle or V-shaped, resembling the red bottom of a baboon, often accompanied by dermatitis in the axillae and sometimes other body folds (116). The first case described was caused by topical ampicillin (116) and many, often pediatric, patients developed the baboon syndrome from sensitization to and inhalation or ingestion of mercury (106,116,122,177,178, 180, review in ref. 179). Most cases, however, have resulted from systemic administration of a drug. Many of these patients had been previously sensitized by the same drug from topical administration or by a similar, cross-reacting drug. These were in fact cases of 'systemic allergic dermatitis' (Chapter 2.4) manifesting as the baboon syndrome. However, it was found that an identical or very similar eruption could also be caused by systemic drugs without previous topical sensitization, notably antibiotics (107,120).

For this variant of what was previously also called the baboon syndrome, in 2004 the alternative term 'symmetrical drug-related intertriginous and flexural exanthema' (SDRIFE) was coined (120). One of the considerations of the authors was that the term baboon syndrome was ethically incorrect and possibly offensive (in my opinion baboons are more likely to be proud of than offended by referring to their distinctive red behind). Previous sensitization was excluded by the authors introducing this term (120). Since then, the term 'baboon syndrome' is used by some as a separate entity for cases with prior topical sensitization (107), others have replaced the term baboon syndrome with SDRIFE (125) and still other investigators now treat the baboon syndrome and SDRIFE as synonyms; the situation, therefore, is currently rather confusing (105,121).

The precise nosology of both conditions and clinical classification (162) is thus a matter of debate. As (i) not all patients with this eruption have the classic bright red buttocks of the baboon (for whom the description would not be quite accurate); (ii) the clinical manifestations are very similar and the conditions are diagnosed clinically; (iii) both are likely the result of type IV (delayed-type) hypersensitivity; (iv) both are mostly caused by systemic drugs; and (v) pre-sensitization in SDRIFE cannot always be excluded beforehand, the author of this book suggests that the term 'symmetrical drug-related intertriginous and flexural exanthema' (SDRIFE) also be used for cases of the baboon syndrome and irrespective of whether previous sensitization has occurred or is known. One might add 'with previous sensitization' or 'without previous sensitization' and, in cases with the characteristic red buttocks, 'presenting as the baboon syndrome'. Here, for practical reasons, SDRIFE *without* previous sensitization from a topical drug and SDRIFE *with* previous sensitization will be discussed separately, irrespective of whether the term baboon syndrome was used by the authors of the publications discussed. This is in fact the division into 2 distinct subgroups as suggested by the original 'SDRIFE' authors (120); unfortunately they forgot to give a proper name to the baboon syndrome *with* previous sensitization. However, meanwhile, one of the authors of the SDRIFE study (120), prof. Andreas Bircher, does *not* consider previous sensitization from cutaneous contact a criterium for excluding the diagnosis SDRIFE anymore (Email communication, March 2021).

The most important practical aspects of SDRIFE/baboon syndrome are presented here, but it falls outside the scope of this book to provide a detailed discussion. Review articles can be found in refs. 107,108,120,123,125,126, 162 and 179.

2.5.2 Symmetrical drug-related intertriginous and flexural exanthema (SDRIFE) without previous sensitization from a topical drug

Introduction
Symmetrical drug-related intertriginous and flexural exanthema (SDRIFE), formerly (27) and sometimes currently also called 'baboon syndrome', is a symmetrical erythematous rash on the gluteal and intertriginous areas observed after exposure to systemic drugs without previous sensitization to a topical drug. It is an uncommon condition. Up to 2018, 110 cases have been reported in literature, mostly as single case reports (126); the author found another 17 reported cases (of which 2 were dubious) up to January 2021. The largest series presented 18 patients with SDRIFE, but these were observed in 2 large hospitals in France in a period of 12 years (126).

Etiology and pathophysiology
The most common triggers for SDRIFE are antibiotics, especially the beta-lactam antibiotics, most frequently amoxicillin (120,126). A variety of other drugs have been implicated, including iodinated contrast media, terbinafine and chemotherapeutics/immunotherapeutics (126). The exact pathophysiology of SDRIFE is unknown, but it is generally assumed that the skin rash it is the result of a T cell-mediated, type IV hypersensitivity mechanism (123,126). The suggested reasons for the flexural predilection are occlusion, sweating, or the excretion of certain drugs or metabolites from the eccrine glands (124), but this is doubted by some (120).

Clinical features
SDRIFE may be observed in patients of any age (mean: 49 years) and occurs in both sexes, with a male:female ratio of 1.5 (120,126). The rash starts within a few hours to 5 days after the first or subsequent dose of a systemic drug and is characterized by well-demarcated erythema (maculopapular, plaques) symmetrically located in the gluteal, perianal, inguinal, or peri-genital area, sometimes with tiny papules, pustules and vesicles or bullae (120,176). Itching or burning may be present. One or more of the other large or smaller skin folds may be affected, most frequently the axillae and the submammary folds (126). SDRIFE typically does not present with mucosal involvement, and the face and palmoplantar surfaces are very uncommonly affected. The rash may sometimes spread to the trunk or limbs as a maculopapular eruption. Atypical manifestations with purpuric lesions have been described infrequently (123). There are no systemic symptoms such as malaise, fever, systemic signs of inflammation (leukocytosis, elevated erythrocyte sedimentation rate, elevated C-reactive protein) or internal involvement (lever, kidneys, lungs).

Histology
SDRIFE is characterized histologically by a vacuolar interface dermatitis induced by cytotoxic T lymphocytes and neutrophilic granulocytes. This pattern may be obscured by accompanying spongiotic, psoriasiform, or pustular features combined with a mixed superficial and sometimes deep dermal infiltrate.

Diagnosis

Criteria for the diagnosis of SDRIFE as suggested by the name givers of this condition are: (1) exposure to systematically administered drug either following the first or subsequent dose; (2) sharply demarcated erythema of the gluteal and/or inguinal area; (3) involvement of at least one other flexural localization; (4) symmetry of affected area; and (5) absence of systemic symptoms and signs (120). A sixth criterion was very recently proposed by French investigators: exclusion of another cause of flexural eruption, including AGEP, toxic erythema of chemotherapy (TEC; 108,144), psoriasis, fixed drug eruption, maculopapular exanthema with flexural reinforcement, and systemic allergic dermatitis presenting as the baboon syndrome), with skin biopsy in cases with an atypical clinical presentation (126).

Patch tests, lymphocyte transformation tests, and drug provocation tests can be useful for diagnosis, but the outcomes of these tests are highly variable. Patch testing is the preferred diagnostic method, and, according to some, the drugs are preferably tested on previously affected skin (123,124). Intradermal tests with delayed readings (D2,D3) may be helpful (18-80% positive; 126). When negative, a controlled drug provocation test can be performed, which is considered as the diagnostic gold standard and is positive in virtually all patients to suspected drugs and appears to be safe (120,126).

In patients with pustules, bullae, erosions, and atypical chronology, or in patients receiving chemotherapy, a skin biopsy should be performed to rule out another diagnosis and laboratory investigations (routine blood panel, liver and renal function) initiated.

See Chapter 4 for details on diagnostic tests (patch test, prick test, intradermal test, drug provocation test, *in vitro* tests) in suspected drug hypersensitivity reactions.

Therapy and prognosis

The most important aspect of therapy is to discontinue suspected drugs, after which SDRIFE is a self-limiting disease. Systemic or topical steroids are usually prescribed to speed up the healing process, and antihistamines are sometimes given for symptomatic management of itching (which is illogical, unless the drugs sedate). The prognosis is excellent. SDRIFE does rarely cause any complications and upon discontinuation of the offending drug all symptoms disappear within 1-18 days (126).

Results of patch testing

Patch tests have been performed in about half of the patients with SDRIFE reported in 47 publications between 2004 and 2018 (reviewed in ref. 126). The rates of positive patch test reactions have varied: 21% (126), 29% (literature review performed in ref. 126), 50% (120) and 40% (review of 5 cases between 2019 and 2021 by the author). This means that the majority of patch tests are negative. Drugs that have caused SDRIFE without previous sensitization and showed a *positive* patch test to culprit drugs are shown in table 2.8.

Table 2.8 Drugs that have caused SDRIFE/Baboon syndrome (without previous sensitization) and showed a positive patch test [a]

Aminocaproic acid	Iopromide
5-Aminosalicylic acid	Itraconazole
Amoxicilllin	Metronidazole
Amoxicillin-clavulanic acid	Penicillin V
Clarithromycin	Pivampicillin
Clindamycin (202)	Prednisolone
Deflazacort	Prednisone
Etonogestrel	Pristinamycin
Etoricoxib	Pseudoephedrine
Heparin, unfractionated	Secnidazole
Hydroxyzine	Sevoflurane
Iomeprol	Tacrolimus

[a] Details and additional references can be found in the monographs of the drugs mentioned in this table

2.5.3 Symmetrical drug-related intertriginous and flexural exanthema (SDRIFE) with previous sensitization from a topical drug

SFRIFE with previous sensitization (formerly and sometimes currently also called 'baboon syndrome') is one of the possible manifestations of systemic allergic dermatitis (Chapter 2.4). Patients have become sensitized to a contact allergen from topical administration to the skin or mucous membrane and later develop SDRIFE/baboon syndrome from systemic exposure to this or a related allergen, which can result from transcutaneous (116,119,121,163,167, 168), transmucosal, oral, intravenous, intramuscular, and inhalational routes. Oral administration is the usual route for inducing the skin rash, while culprit corticosteroids can also be administered by intramuscular or intravenous injections.

For general information the reader is referred to Chapter 2.5.2. There are insufficient data to establish distinct differences between SDRIFE *without* and SDRIFE *with* previous sensitization apart from their different routes of sensitization and diverging patch test results. Systemic drugs that have caused SDRIFE/baboon syndrome with previous sensitization are shown in table 2.9.Topical drugs can also induce SDRIFE, which has been fully reviewed in Volume 3 of the *Monographs in contact allergy* series (104).

Table 2.9 Drugs that have caused SDRIFE/Baboon syndrome with previous sensitization

Drug causing SDRIFE	Previous sensitization	Ref.
Aminophylline (intravenous) [a]	Ethylenediamine	165,166
Amoxicillin	Topical penicillin (no patch test performed but very likely)	176
Betamethasone	Betamethasone or other corticosteroid	172,173 [b]
Cloprednol	Other unknown corticosteroid	173 [b]
Deflazacort	Deflazacort (from occupational sensitization) (uncertain)	159
Dexamethasone	Other unknown corticosteroid	173 [b]
Hydrocortisone	Other unknown corticosteroid	173 [b]
Methylprednisolone sodium succinate	Hydrocortisone (ester?)	324 [d]
Mitomycin C	Mitomycin C (from intravesical instillation)	151
Neomycin	Neomycin	164
Nystatin	Nystatin	171
Sevoflurane (inhalation)	Sevoflurane	170
Terbinafine [c]	Terbinafine	174
Triamcinolone acetonide	Budesonide or other corticosteroids	169

[a] Aminophylline = theophylline + ethylenediamine; [b] Oral provocation test; [c] Atypical manifestations; [d] *Probably* SDRIFE/Baboon syndrome

2.6 Fixed drug eruption

Introduction

A fixed drug eruption (FDE) is an allergic drug reaction characterized by one or more erythematous or violaceous circular patches or plaques, sometimes with a vesicle or bulla, and with a dusky-grey center. Characteristically, when the culprit drug is reintroduced, the skin eruption recurs at the same site(s), often with more involved sites with each drug exposure (228,236). FDE is a delayed-type (type IV) hypersensitivity reaction. Based on clinical morphology, there are several types of FDE: localized pigmenting, localized bullous (about 1/3 of all patients [248]), mucosal, non-pigmenting (more frequent than previously thought [240]), generalized, and generalized bullous (rare). The localized pigmenting type is by far the most frequent and constitutes a benign self-limited drug hypersensitivity reaction. On the other end of the spectrum is the rare generalized bullous fixed drug eruption, which may be difficult to distinguish from SJS/TEN and may sometimes be lethal (228). Review articles on the subject of fixed drug eruptions have been published in 2020 (228), 2008 (241), 2006 (236) and 2000 (231); most of the text in the section below has been derived from these publications, notably from ref. 228.

Etiology and pathophysiology

Most cases of fixed drug eruptions are the result of a delayed-type hypersensitivity reaction to oral drugs; quinine in tonic water (and in gin and tonic) is also a well-known cause (251,252) and some cases have been induced by topical and intravaginal medications. Over 100 medications have provoked FDE; commonly responsible are analgesics/NSAIDs (acetaminophen [paracetamol], acetylsalicylic acid, piroxicam, mefenamic acid, etoricoxib, metamizole, oxyphenbutazone), antibiotics (trimethoprim-sulfamethoxazole [synonym: cotrimoxazole], metronidazole, tetracyclines, ciprofloxacin, levofloxacin, amoxicillin, rifampicin), laxatives (phenolphthalein; not used anymore), and barbiturates (228,236). Some cases occur after exposure to ultraviolet light in the UVA- or UVB-range (228). Oral mucosal lesions are often caused by naproxen (249) and trimethoprim-sulfamethoxazole (230); male genital FDE (mostly located on the glans penis) is likely caused by tetracyclines or trimethoprim-sulfamethoxazole, while the lesions on the vulva in females is often linked to NSAIDs. The subtype of non-pigmenting FDE is typically caused by pseudoephedrine (228,232).

The immunological and pathological reactions in FDE include damage of the basal layer of the epidermis by the offending drug, the release of cytotoxic mediators such as interferon-γ from resident epidermal memory CD8+ T cells, damage to melanocytes and keratinocytes by CD8+ and CD4+ T cells and neutrophils with leakage of melanin into the dermis. During regeneration, this melanin is taken up by macrophages which remain in the dermis and are responsible for the residual hyperpigmentation which is observed in most cases (228,250).

Clinical features

Fixed drug eruptions are the most common drug hypersensitivity reactions after maculopapular exanthemas, representing 14-22% of all such patients (228). FDE can develop at all ages, but most patients are in the age range of 20-40 years and women may be slightly more frequently affected (240). FDE begins as one or more erythematous to violaceous round or oval, sometimes edematous, plaques with a dusky-grey center, which may become vesicular or bullous. The appearance of FDE lesions is often preceded and/or accompanied by a sensation of itching or burning (183). Lesions of FDE usually measure between 0.5 and 5 cm (233). Characteristic for FDE is the reappearance of the lesions precisely over the previously affected site(s) when the offending agent is reused.

The eruption may be solitary, a localized cluster, or diffuse. Polycyclic, linear, zosteriform and band-liker patterns have been observed. Some individuals have a single plaque, others 2-5 lesions and – in nearly half of the affected individuals – more than five plaques are observed on presentation (233). Areas with thin skin, such as the lip mucosa (233), genitals (236), and perianal skin are frequently involved, but FDE can appear anywhere on the skin. Previous trauma such as insect bites, burns, and healed herpes simplex virus infection predisposes to development of FDE (228,241,246,247). Men more often have genital FDE and women lesions located on the hands and feet (240). Generalized fixed drug eruption is similar to localized pigmenting FDE but presents with multiple bilateral cutaneous macules on the trunk and limbs; the mucosae are usually not affected.

When the patient has taken the culprit drug for the first time, FDE may develop up to one week after the first drug exposure. On subsequent exposures, however, the lesions already appear between 30 minutes and 8 hours. Each recurrence of FDE is associated with an increased amount of inflammation and hyperpigmentation and may in rare cases lead to generalized bullous fixed drug eruption (GBFDE). After stopping the causative medication, FDE usually heals within 7-10 days. However, the post-inflammatory hyperpigmentation can last up to a few months or longer (228). FDE of the oral mucosa (mainly the hard palate and dorsum of the tongue) or genital mucosa often has bullous or erosive lesions without residual hyperpigmentation (230).

Histology

Histological changes include a vacuolar interface dermatitis, hydropic degeneration of the basal epidermal layer, spongiosis, a normal keratin layer, scattered individual keratinocyte necrosis, a perivascular lymphocytic and eosinophilic infiltrate, and pigmentary incontinence (228).

Diagnosis

On the basis of the history and the clinical picture, and notably the pathognomonic feature of recurrent lesions at the same sites, the diagnosis FDE can often be suspected. When patients use multiple drugs, it may be difficult to pinpoint the culprit. Patch testing may reveal the sensitizer, but generally this diagnostic method has a sensitivity of <60% and is more sensitive for NSAIDs than for antibiotics. This sensitivity depends on the reactivity of resident memory CD8+ T cells and drug permeability through the skin. Patch testing should be conducted on a (hyperpigmented) site in an area of previous FDE (postlesional skin), utilizing normal, previously unaffected skin as a control. To avoid a refractory period, patch tests should be performed not earlier than 2 weeks after resolution of the lesions (250). It may be advisable to patch test on more than one previously affected area, both hyperpigmented and non-hyperpigmented (141).

The use of DMSO as penetration-enhancing vehicle or ethanol instead of petrolatum may sometimes be useful in patch testing (232). Single or repeated open tests with the suspected drugs has also been successful (238,242,243, 244,245). In some cases, for example with sulfamethoxazole-trimethoprim, patch tests in petrolatum are (almost) always negative, whereas open testing with the constituents in DMSO yields positive reactions, for which high concentrations (10%, 20% and 50%) may be necessary (244). Open tests with the appearance of erythema with or without induration at an old FDE site that starts within 24 hours and lasts at least 6 hours is defined as a positive test reaction (242). Patch and open tests are ineffective on normal skin as it relies on the resident memory CD8+ T cells that lie primarily in the previously affected (postlesional) epidermis (228,250). In a few cases, patch testing may reactivate old lesions (248).

There appears to be no consensus on the safety of oral provocation tests when a patch test is negative. Some authors (50,57,83,233) propose a full dose provocation, whereas others consider this contraindicated due to the risk of widespread FDE or the development of generalized bullous fixed drug eruption (GBFDE) (229,230). Neverthe-less, in one large series of nearly 500 patients clinically diagnosed with FDE and provoked with half or full therapeutic oral dose, adverse effects associated with the oral challenge test were mild and infrequent (233). Oral provocation with a subtherapeutic dose has been recommended, with a cautious gradual increase (e.g. 1/8, 1/4, 1/2, or 1/10, 1/4, 1/2) until the full therapeutic dose is achieved, but only one drug should be tested at a time (230,234,236,239, 241).

A diagnostic biopsy is indicated in patients with an unclear diagnosis or associated systemic symptoms such as fever, malaise, or arthralgias, and in suspected cases of generalized FDE, generalized bullous fixed drug eruption, and mucosal FDE (228). See Chapter 4 for details on diagnostic tests (patch test, prick test, intradermal test, drug provocation test, *in vitro* tests) in suspected drug hypersensitivity reactions.

Therapy
The primary treatment of FDE is removal of the causative drug. Medical management is symptomatic to relieve pruritus or pain (228).

2.6.1 Generalized bullous fixed drug eruptions (GBFDE)

Introduction
Generalized bullous fixed drug eruption (GBFDE) is a particular form of fixed drug eruption (FDE), defined as the presence of typical FDE lesions with blisters involving at least 10% of the body surface area or at least three of six different anatomic sites (235).

Etiology and pathophysiology
GBFDE is caused by a delayed-type (type IV) hypersensitivity reaction to (mostly oral) drugs. Common causative medications include antibiotics (metronidazole, trimethoprim-sulfamethoxazole, rifampicin, erythromycin) and analgesics (e.g. ibuprofen, paracetamol) (228).

Clinical features
GBFDE is a rare drug hypersensitivity reaction which occurs mainly in patients of over 70 years of age; women are more frequently affected. GBFDE is characterized by multiple, sharply defined, deep-red or heavily pigmented oval or polycyclic macules or patches with non-tense blisters and erosions, displaying a bilateral, often symmetric, distribution. It develops abruptly and continues to increase in size and number several days after cessation of the offending drug (236). There are areas of intact skin between the blisters; thus, blister formation only affects a small percentage of the body surface. Patients with GBFDE do not usually have fever and are not ill or only have mild malaise. Most report a history of a similar, often local or more limited reaction. Recurrences from repeated intake of the offending drug can lead to more widespread skin detachment and therefore to a more severe disease course (95,228).

Diagnosis and prognosis
GBFDE is defined by the involvement of at least three of six separated body areas, including the head, neck, trunk, upper limbs, lower limbs, and genital area. GBFDE should be considered when a patient presents with a history of similar previous episodes, an onset of 30 minutes to 1 day after drug ingestion, and lack of or minimal mucosal involvement. The main differential diagnosis is Stevens-Johnson syndrome/toxic epidermal necrolysis (SJS/TEN). In GBFDE, the blistering usually affects only a small percentage of body surface area, and between the large blisters there are sizable areas of intact skin. In contrast with SJS/TEN, the mucosae are rarely affected and, when present, mild. An evaluation of a skin biopsy specimen is the preferred method to confirm the diagnosis. Patch testing may reveal the culprit drug or confirm clinical suspicion.

Despite the lack of mucosal or visceral involvement, GBFDE is still potentially life-threatening, and a surprisingly high 22% mortality rate among elderly patients (13 of 58) has been reported (235).

Therapy
The primary treatment for GBFDE is removal of the causative drug, along with supportive care. Moderate-dose oral corticosteroids are often given but have not been shown to be effective (228). Mild cases can be managed as outpatients, but in elderly subjects with severe GBFDE, intensive care and possible admission to a burns unit has been suggested (228).

Results of patch testing in patients with fixed drug eruption
Percentages of positive reactions to culprit drugs in patients with fixed drug eruptions have generally ranged from 20 to 79% (Chapter 4, table 4.4). Over 80 systemic drugs that have caused fixed drug eruptions (not specified for type) had a positive patch test (table 2.10).

Table 2.10 Drugs that have fixed drug eruptions and showed a positive patch test [a]

Aceclofenac	Ioversol
Acetaminophen (paracetamol)	Lamotrigine
Acyclovir	Levocetirizine
Adalimumab	Mefenamic acid (237)
Aminophylline	Meprobamate
Amlexanox	Mesalazine
Amoxicillin	Mesna
Amoxicillin-clavulanic acid	Metamizole
Antipyrine salicylate	Metronidazole
Apronalide (allylisopropylacetylurea)	Naproxen
Atenolol	Nimesulide
Atorvastatin	Norfloxacin
Bromhexine	Ofloxacin
Carbamazepine	Ornidazole
Carbocysteine	Oxcarbazepine
Celecoxib	Oxytetracycline
Cetirizine	Paracetamol (acetaminophen)
Chlormezanone	Pefloxacin
Ciprofloxacin	Pethidine
Citiolone	Phenazone
Clarithromycin	Phenobarbital
Codeine	Phenylephrine
Cotrimoxazole (sulfamethoxazole-trimethoprim)	Picosulfuric acid
Cyclophosphamide	Piroxicam
Dimenhydrinate	Pristinamycin
Dipyrone	Promethazine
Doxycycline (237)	Pseudoephedrine
Ephedrine	Rupatadine
Erythromycin	Scopolamine
Esomeprazole	Sulfadiazine
Ethenzamide	Sulfamethoxazole
Etoricoxib	Sulfamethoxazole-trimethoprim (cotrimoxazole)
Feprazone	Tenoxicam
Fluconazole	Tetracycline
Fulvestrant	Tetrazepam
Hydroxyzine	Ticlopidine
Ibuprofen	Tosufloxacin tosilate
Iohexol	Trazodone
Iomeprol	Trimethoprim
Iopamidol	Vinburnine
Iopromide	

[a] Details and references can be found in the monographs of the drugs mentioned in this table

2.7 Drug reaction with eosinophilia and systemic symptoms (DRESS)

Introduction

DRESS is an acronym for Drug Reaction with Eosinophilia and Systemic Symptoms. This term was first introduced in 1996, when the R (still) stood for Rash (182). Synonyms include Hypersensitivity syndrome (HSS) and Drug-induced hypersensitivity syndrome (DiHS, used especially in Japan) (184). Reactions to antiepileptic drugs are sometimes called the Anticonvulsant hypersensitivity syndrome (190). DRESS is an infrequently occurring disease with an estimated 10 cases per million per year (184,199). Its incidence in new users of antiepileptic drugs (e.g. carbamazepine or phenytoin) may be one per 1000 to one per 10.000 (90,190).

DRESS is a serious, sometimes fatal reaction to drugs with a non-specific rash, often of the maculopapular type, fever and organ involvement, notably of the liver and kidneys. Viral reactivation of herpesviruses characteristically follows onset of the disease. The mortality has been variably reported as ranging from 5 to 10% (88,90), but appears to be lower (2%) in strictly validated cases (209). A limited number of drugs cause DRESS, especially antiepileptics, anti-infective drugs and allopurinol. Prompt withdrawal of suspected drugs and administration of systemic prednisolone is crucial.

The most important practical aspects of DRESS are presented here (largely based on ref. 184), but it falls outside the scope of this book to provide a detailed discussion. Recent other review articles can be found in refs. 88 (focus on pathophysiology), 90,91,92,95,97 (focus on epidemiology),102,183,185,186,187,188 (focus on biomarkers of disease severity and HHV-6 reactivation), 189 (focus on epidemiology and risk factors), 190 (anticonvulsant hypersensitivity syndrome), 191 (management), and 199.

Etiology and pathophysiology

The pathogenesis of DRESS is unknown, but it is generally regarded as a T cell-mediated hypersensitivity reaction to drugs. Often implicated are antiepileptic drugs (carbamazepine, phenobarbital, phenytoin, lamotrigine, zonisamide), antibacterial and antiviral drugs (sulfonamides [notably sulfasalazine]), vancomycin, nevirapine, abacavir [clinical signs different from DRESS]), allopurinol, dapsone, minocycline and mexiletine (184). The development of the disease is independent of the dose given and may already occur during the first treatment cycle. Approximately half of the patients have had an episode of an infection within the previous month, particularly virus infections such as herpes zoster. Patients with DRESS have drug-specific T cells and it is assumed that viruses have a critical role in the generation and activation of these cells. In the vast majority of the affected individuals, there is re-activation of human herpes virus 6 (HHV-6) and of other herpesviruses including Epstein-Barr virus, HHS-7, and cytomegaly virus, in a sequential manner. How viral infections contribute to the pathogenesis of DRESS is, however, as yet unknown (184).

Clinical features

DRESS usually starts abruptly with a maculopapular morbilliform exanthema with fever of >38°C, 2-3 up to 12 weeks after the introduction of the causative drug. Sometimes, there may be an upper-airway infection-like prodrome. The cutaneous lesions usually begin as patchy erythematous macules, pustular, target-like or eczema-like lesions, which may be slightly purpuric and can become confluent. The lesions are symmetrically distributed on the trunk and extremities (184). Characteristic cutaneous lesions during the early phase are periorbital and facial edema with pinhead-sized pustules. Blisters are occasionally present, especially on the wrists. The mucosae, palms and soles are usually unaffected. Lymphadenopathy is often present, notably in the cervical, axillary or inguinal regions, as is bilateral swelling of the salivary glands. Some patients may also complain of dryness of the mouth due to severe xerostomia which makes swallowing or even taking food difficult (184).

DRESS syndrome-specific organ involvement results from specific eosinophil or lymphocyte tissue infiltration. Liver involvement is observed in more than 80% of patients: mainly hepatic cytolysis, sometimes cholestasis, and rarely hepatic failure. Kidney involvement is characterized by interstitial nephritis. The lungs are affected in up to 15% of cases, manifested by dyspnea, cough, eosinophilic pneumonitis, and rare respiratory failure. Heart involvement (myocarditis, pericarditis) with electrocardiogram, CT scan, or cardiac enzyme abnormalities, can be fatal (90). Worsening of clinical symptoms may occur 3-4 days after withdrawal of the causative drug and flare-ups can even be observed after weeks, either from reactivation of herpesviruses in various organs (184), rapid reduction of systemic steroids or administration of new drugs or from previously tolerated drugs after dose increase (201).

Unique laboratory features of the early phase of DRESS are leukocytosis with atypical lymphocytes and eosinophilia of various degree. Elevated liver enzymes are found in up to 80% of the patients in the acute phase. Human herpes virus 6 (HHV-6) reactivation is shown in the vast majority of the patients 2-3 weeks after onset by a significant increase in serum IgG titers to HHV-6 and the detection of HHV-6 DNA in leukocytes (184).

Histology

Various inflammatory patterns can be found in a single skin biopsy, namely interface dermatitis, lichenoid, eczematous, AGEP-like vascular damage, superficial perivascular infiltration, peri-appendage infiltration and erythema multiforme-like patterns. No single pathological finding is specific enough to confirm the diagnosis DRESS. However, the co-existence of three histopathological patterns in a skin specimen has a higher likelihood of being a definite case of DRESS and is correlated with clinical severity (88).

Diagnosis

The RegiSCAR validation score is most frequently used for diagnosing DRESS, both in daily practice and in publications. Based on multiple parameters (fever, enlarged lymph nodes, eosinophilia, atypical lymphocytes, skin involvement, organ involvement, time to resolution, and evaluation of other potential causes), each suspected DRESS reaction can be scored as no case, possible case, probable case, or definite case (22). Patch tests, which should be performed 6 months after complete healing (not earlier because of the risk of reactivation of viral infection [50]), may quite frequently identify the culprit drug, depending on the drug (carbamazepine and amoxicillin frequently positive, allopurinol never positive [142]). There is increasing evidence that intradermal tests may be safe

(50). Oral provocation tests are mostly contra-indicated because of the risk or recurrence of DRESS. See Chapter 4 for details on diagnostic tests (patch test, prick test, intradermal test, drug provocation test, *in vitro* tests) in suspected drug hypersensitivity reactions.

Table 2.11 Drugs that have caused DRESS and showed a positive patch test [a]

Abacavir (different from DRESS, see Chapter 3.1)	Iopromide
Acetaminophen (paracetamol) (196)	Ioversol
Acetylsalicylic acid	Ioxaglic acid (217)
Acyclovir (11)	Ioxitalamic acid
Amikacin (11,200,204)	Isoniazid (214)
Aminosalicylic acid	Lamotrigine (57,62)
Amoxicillin (11,201,202)	Lansoprazole (11)
Amoxicillin-clavulanic acid (11,216)	Levofloxacin
Atovaquone (205)	Meropenem (201,202)
Benznidazole	Metamizole
Benzylpenicillin	Mexiletine
Captopril	Miconazole
Carbamazepine (11,62,63,196,197,202,204)	Olanzapine (11)
Cefadroxil	Oxacillin
Cefotaxime	Oxcarbazepine (197)
Cefoxitin	Pantoprazole (11,202)
Ceftriaxone (11,201,202)	Paracetamol (acetaminophen)
Cefuroxime (201)	Penicillin V (205)
Celecoxib (11)	Phenindione (205, *possibly* DRESS)
Chloroquine	Phenobarbital (216)
Cilastatin, mixture with imipenem	Phenytoin (62)
Ciprofloxacin	Piperacillin (201,202)
Clarithromycin (201,202)	Piperacillin-tazobactam (206)
Clindamycin (216)	Potassium aminobenzoate
Clobazam (11)	Pristinamycin (11)
Cloxacillin (11)	Proguanil
Codeine	Propylthiouracil
Cyanamide	Pyrazinamide
Dabrafenib (201)	Pyrimethamine
Diclofenac (201)	Ranitidine
Dicloxacillin (11)	Rifampicin (214)
Diltiazem (11)	Spironolactone (11,62)
Enoxaparin (11,216)	Sulfamethoxazole (201,202)
Esomeprazole (11,193,204)	Sulfamethoxazole-trimethoprim (cotrimoxazole)
Ethambutol (214)	Sulfasalazine (positive photopatch test)
Ethosuximide	Teicoplanin
Floxacillin (flucloxacillin) (201,202)	Tenoxicam
Fluindione (11,204)	Tetrazepam (11)
Fluvoxamine (197)	Topiramate (62)
Fusidic acid (or TEN?)	Triazolam
Gadobutrol (201)	Tribenoside
Hydroxychloroquine	Valaciclovir
Iobitridol (201,202)	Valproic acid (201)
Iodixanol (11)	Vancomycin (11,201)
Iohexol	Zonisamide
Iomeprol (201)	

[a] Details and additional references can be found in the monographs of the drugs mentioned in this table

Therapy and prognosis

The prognosis of DRESS is highly variable and unpredictable. A composite score has been created using demographic data, medical history, and clinical variables, by which disease severity, and treatment efficacy can be assessed and disease progression to a more aggressive stage can be predicted (181). Complications leading to morbidity and mortality include myocarditis, *Pneumocystis jirovecii* pneumonia, sepsis, liver failure and gastrointestinal bleeding. Reactivation of cytomegaly virus may be the cause of some of these complications (184).

Systemic corticosteroids (prednisolone 40-50 mg/day) are the gold standard treatment in the acute phase. Rapid resolution of rashes and fever occur within several days after starting treatment. The steroids have to be tapered very slowly over 6-8 weeks or longer to prevent the relapse of various symptoms and the emergence of the so-called immune reconstitution inflammatory syndrome (IRIS). There is growing concern that chronic virus activation in DRESS may eventually contribute to the development of autoimmune diseases such as systemic sclerosis, lupus erythematosus, diabetes, or thyroiditis arising after DRESS remission (90,184).

Results of patch testing in patients with DRESS
Patch testing can be used to identify the causative drug in patients with DRESS and appears to be generally safe (68,190). As with fixed drug eruptions, the materials should not only be applied to the back, but also at the most highly affected cutaneous sites, as they may be positive on postlesional skin only (52,192). The sensitivity of patch testing in DRESS has ranged from 32 to 83%, mainly depending on the culprit drugs, e.g. 80-90% for carbamazepine, but only around 10-20% for phenobarbital (Chapter 4, table 4.4; 62,64,190). Allopurinol does (virtually) never give any positive patch test reactions (142), and neither does salazopyrine (11).

Multiple positive patch tests to drug from different classes are observed fairly frequently in patients with DRESS (11,202). Sensitization may occur to different drugs given simultaneously or after an episode of DRESS (196,197,201). This is called 'Multiple drug hypersensitivity (syndrome)' (142,197,203,) which has been reported in the literature with incidence rates between 10% and 18% when considering severe drug reactions proven by positive skin tests or *in vitro* tests (142). In the case of sensitization to a new drug after a period of DRESS, the new drug hypersensitivity reaction may be either (relapse of) DRESS or another type of cutaneous adverse drug reaction (197,202) such as maculopapular exanthema (142).

Over 90 systemic drugs that have caused DRESS had a positive patch test (table 2.11).

2.8 Stevens-Johnson syndrome and toxic epidermal necrolysis (SJS/TEN)
Stevens-Johnson syndrome and toxic epidermal necrolysis (SJS/TEN) are the most serious of the drug hypersensitivity reactions. They are considered as severity variants of the same disease entity, recently referred to as epidermal or epithelial necrolysis (209). The incidence of SJS/TEN is 1-2 persons per million populations per year (90,95). As SJS and TEN are so rare and patch testing infrequently yields positive results (with the possible exception of anticonvulsants, notably carbamazepine [63,93]), they are described only shortly here. Review articles can be found in refs. 59,209,210,211,212, and 213. This subject has also been discussed in several articles on severe cutaneous adverse drug reactions (SCARs) (88,89,90,91,92,93,95,97,102,183).

SJS/TEN in its classic form starts, sometimes after prodromal fever and upper respiratory tract symptoms, with small blisters arising on purple macules and atypical flat target lesions, which are widespread and usually predominant on the trunk. The skin may be initially painful. Bullous lesions develop rapidly, often within 12 hours, and both the skin and the mucous membranes (oral, nasal, conjunctival, genital, anal) are affected. The patients develop fever and are severely ill. When the area of confluent bullae leading to detachment of the skin is <10% (as calculated in burns) of the total body surface the disease is called SJS, with 10%-30% SJS/TEN overlap and when >30% is affected, TEN is diagnosed (32). Nikolsky sign is positive (lateral extension of a blister with light pressure from a finger). Histology shows subepidermal blisters and full thickness necrosis; immunofluorescence studies are negative (209).

The time latency between the first dose of drug and the onset of SJS/TEN generally ranges from 4 days to 4 weeks, but can be up to 8 weeks for drugs with a long half-life (e.g. allopurinol). The most commonly implicated drugs in SJS/TEN are allopurinol, antiepileptics (carbamazepine, lamotrigine, phenobarbital, phenytoin), antibacterial sulfonamides (e.g. sulfasalazine), the antiviral drug nevirapine and nonsteroidal anti-inflammatory drugs (NSAIDs) of the oxicam-type. The mortality is high: 9% in SJS, 29% in SJS/TEN overlap, and up to 48% in TEN. The prognosis mainly depends on the age of the patient, the extent of skin detachment and how soon the culprit drug is withdrawn (209). An algorithm for assessment of drug causality in Stevens-Johnson syndrome and toxic epidermal necrolysis is available (33).

See Chapter 4 for details on diagnostic tests (patch test, prick test, intradermal test, drug provocation test, *in vitro* tests) in suspected drug hypersensitivity reactions.

Results of patch testing in patients with SJS/TEN
In a few cases (n=49) of SJS, TEN or SJS/TEN overlap, patch tests have been positive to the culprit drugs. In one instance, a patch test (with sulfamethoxazole) was positive only on previously involved skin (225). Drugs that have caused SJS, SJS/TEN overlap or TEN and gave positive patch tests are shown in table 2.12.

Table 2.12 Drugs that have caused SJS, SJS/TEN or TEN and showed a positive patch test [a]

Aminophenazone (aminopyrine)	Lamivudine
Amoxicillin (11,202)	Lamotrigine (11)
Ampicillin	Lansoprazole
Benznidazole	Meropenem
Benzylpenicillin (penicillin G) (220,222)	Metamizole
Bortezomib	Metronidazole
Bromisoval	Pantoprazole
Bucillamine	Penicillin G (benzylpenicillin) (220,222)
Carbamazepine (63,93)	Phenobarbital (9,221)
Ceftriaxone	Phenylbutazone (222)
Cefuroxime	Phenytoin (222)
Chlorambucil	Pristinamycin (227)
Chlormezanone	Procaine benzylpenicillin
Clindamycin (202)	Propranolol
Cloxacillin	Pseudoephedrine
Diclofenac (doubtful case)	Pyrabital
2,3-Dimercapto-1-propanesulfonic acid	Pyrazinamide (214)
Emtricitabine	Ramipril (11)
Erythromycin	Stevens-Johnson syndrome
Esomeprazole (11)	Sulfamethoxazole (225)
Ethosuximide	Sulfonamide, unspecified (9)
Fexofenadine	Terbinafine
Ibuprofen	Tetrazepam (11)
Iohexol	Vancomycin (11)
Isoniazid (214)	

[a] Details and additional references can be found in the monographs of the drugs mentioned in this table

2.9 Photosensitivity

Introduction

Photosensitivity is a skin reaction to light, mostly to UV-light, due to the presence of endogenous or exogenous chromophores in the dermis or epidermis. Photoactive molecules in products applied to the skin or systemic drugs are frequently the cause of photosensitivity. Most reactions are phototoxic, involving non-specific cutaneous inflammation. Photoallergy is caused by hypersensitivity to the chemicals or their photoproducts, involving a T cell-mediated reaction (delayed-type hypersensitivity, type IV hypersensitivity) (357). Photoallergic eczematous reactions to systemic drugs appear to be rare, but photopatch testing facilities are not widely available, which may result in underdiagnosing (357).

Discussion of photosensitivity in this book is limited to photoallergy and photoallergic dermatitis caused by systemic drugs proven by positive photopatch tests. In early literature, when photopatch testing had not yet been standardized, many cases of 'photoallergy' may in fact have been phototoxic in nature, e.g. by using too high concentrations of the suspected drugs, or too high doses of UVA for irradiation. Such cases, and cases of photosensitivity diagnosed with phototesting or rechallenge but without a positive photopatch test, are excluded from presentation. Nevertheless, it should be realized that many systemic drugs can induce both phototoxicity and photoallergy, e.g. psoralens, phenothiazines and fluoroquinolones (357).

The *topical* use of certain systemic drugs discussed in this book has been responsible for many cases of photocontact allergy and photoallergic contact dermatitis, notably the use of ketoprofen, benzydamine, other NSAIDs and phenothiazines (especially promethazine). This subject has been discussed fully in the previous volume of the *Monographs in contact allergy* series (Volume 3, Topical drugs) and will not be repeated here (104). Useful recent reviews of drug-induced photosensitivity (363,365,366) and drug-induced phototoxicity (364) have been published.

Etiology and pathophysiology

When photoactive chemicals (chromophores) reach the skin (epidermis or dermis) after oral or parenteral administration of drugs, they may be transformed by UVA radiation into a stable photoproduct (photohapten) or enhanced to react with an endogenous peptide to form a photoallergen. Next, these new chemicals are captured by antigen-presenting cells, which become activated and present the newly formed hapten to T cells, resulting in sensitization of naïve T cells. When these sensitized T cells, which include memory and effector T cells, encounter the same chemical again, they will be activated and generate a specific T cell immune reaction, which is the same type IV (delayed-type) hypersensitivity as in allergic contact dermatitis (357).

Clinical features

The main presentation of photoallergy to systemic drugs is an acute or subacute eczema or, in severe cases, erythema multiforme-like lesions. The reaction usually involves, in a symmetrical distribution, the face, V-shaped area of the neck and upper chest, dorsum of the hands and forearms, and occasionally also the legs and dorsum of the feet. UV-shaded areas of the face and neck are spared, namely, the upper eyelids, upper lip, deep facial wrinkles, retroauricular areas, and a submandibular rhomboid area. Spreading of the dermatitis to non-UV-exposed sites is possible. Photoallergic reactions may become persistent (persistent light reactions) and eventually progress to chronic actinic dermatitis with extreme photosensitivity despite no further exposure to the culprit chemical (357).

Histology

The histology of photoallergic dermatitis shows spongiosis, vesicles, and keratinocytes transforming into sunburn cells (apoptotic cells) in the epidermis. There is migration of inflammatory cells, mainly lymphocytes and occasionally polymorphonuclear neutrophils and eosinophils, into the epidermis (exocytosis), accumulating in the spongiotic vesicles. Dermal edema is prominent and a dense mononuclear cell infiltrate is usually present around blood vessels throughout the dermis.

Diagnosis

Photopatch testing is the diagnostic method indicated for the etiologic diagnosis of systemic drug photoallergy, although less reliable (less often positive, less sensitive) than for photoallergic contact dermatitis (362). False-negative reactions may be caused by insufficient penetration of the photopatch tested drug into the epidermis or the fact that the culprit agent of the photosensitivity is not the native drug but one of its metabolites (367). Photopatch testing should be performed according to a standardized procedure (358) using the baseline photopatch tests series, the extended series, and additional substances according to patient exposure and suspected patients' own products and drugs (357,358,359,360,361).

All photopatch test agents are applied in two identical sets, one for irradiation and one as control, under occlusion to the skin of the back as one would apply patch tests. The length of application of photopatch test agents is either 24 or 48 hours. After this time, the agents are removed and the test site is visually inspected for any reactions. The next step is to cover the control set of photopatch test agents with a UV-opaque material. Then, the set of photopatches is irradiated with UVA at a distance of 20 cm and a dose of 5 J/cm^2, or less if the UVA MED (minimum erythema dose) is reduced. Readings of both sets of photopatch test agents are made 48 hours later. Further readings at 72 and 96 hours post-irradiation are desirable to enable detection of crescendo or decrescendo scoring patterns suggesting allergic and non-allergic mechanisms, respectively.

A positive reaction in the irradiated set only indicates photoallergy. A positive reaction seen in the irradiated set, and a reaction of equal grade at the same site in the control set indicates (plain) contact allergy. When a positive reaction is seen in the control set, and a positive reaction which is at least one ICDRG grade higher is seen at the same site in the irradiated set, photoaugmentation of contact allergy or contact allergy + photoallergy is diagnosed (357,358,360,361).

See Chapter 4 for details on diagnostic tests (patch test, prick test, intradermal test, drug provocation test, *in vitro* tests) in suspected drug hypersensitivity reactions.

Therapy

Therapy relies on the immediate withdrawal of any suspected causative drug, avoidance of sunlight and artificial UVA-sources such as sunbed and solarium (because some drugs can remain in the skin for some time), topical corticosteroids and, in cases of a serious and extensive photoallergic eczema, a short course of oral corticosteroids.

Results of patch testing in patients with photosensitivity

Over 70 systemic drugs have caused photoallergy (table 2.13), mostly only in a few cases or even a single patient. Several cases have been observed to for example piroxicam, lomefloxacin, fenofibrate, quinidine and flutamide. Whereas topical ketoprofen and phenothiazines (notably promethazine) are notorious photosensitizers (104), systemic administration hardly ever seems to induce photoallergy. Some cases of 'photoallergy' presented here may in fact have been phototoxic: the tests were virtually never repeated, more than one concentration of the drug hardly ever used and adequate control testing was in many cases not performed.

Additional drugs that have caused photosensitivity, as a manifestation of systemic photoallergic dermatitis (systemic photocontact dermatitis), may be found in table 2.7

Table 2.13 Systemic drugs that have caused photoallergic dermatitis and showed a positive photopatch test [a]

Acetylsalicylic acid	Hydroxychloroquine
Actarit	Hydroxyurea
Afloqualone	Ibuprofen (doubtful)
Althiazide (uncertain)	Isoniazid
Amantadine	Isotretinoin (photoaggravation)
Ambroxol	Ketoprofen
Amitriptyline	Levopromazine
Amoxicillin	Lomefloxacin
Ampiroxicam	Mequitazine
Benzydamine	Methoxsalen
Carbamazepine	Methyldopa
Carprofen	Naproxen
Chloroquine	Nicardipine
Chlorpromazine	Paroxetine
Ciprofloxacin	Pirfenidone
Clofibrate	Piroxicam
Clomipramine	Piroxicam betadex
Cyamemazine	Potassium aminobenzoate
Dapsone	Promazine
Dexketoprofen	Promethazine
Diltiazem	Pyridoxine
Diphenhydramine	Pyritinol
Doxycycline	Quinidine
Dronedarone	Quinine
Droxicam	Ramipril
Efavirenz	Simvastatin
Enoxacin	Spironolactone (uncertain)
Epirubicin	Sulfasalazine (photoallergy in DRESS)
Erlotinib (unconvincing case)	Tegafur
Esomeprazole (unconvincing case)	Tenofovir (uncertain)
Etretinate (uncertain)	Terbinafine
Fenofibrate	Tetrazepam
Flutamide	Thioridazine
Fluvoxamine	Tiaprofenic acid
Griseofulvin	Tilisolol
Hydrochlorothiazide	Triflusal
Hydroquinidine	Trimeprazine

[a] Details and references can be found in the monographs of the drugs mentioned in this table

2.10 Other cutaneous adverse drug reactions

In many publications, patients have presented with cutaneous adverse drug reactions from systemic drugs to which delayed-type hypersensitivity was demonstrated by a positive patch test, where the diagnosis or the description of the symptoms does not fit in any of the CADRs discussed in the previous sections of this chapter. These are the subject of this paragraph. Drugs that have caused eczematous, erythema multiforme-like or lichenoid drug eruptions or urticaria/urticarial exanthema/urticaria-like exanthema are shown in table 2.14 and bullous eruptions in table 2.15. Allergic reactions to injections (mostly subcutaneous and intra-articular, intravenous excluded) are shown in table 2.16. Drugs that have caused any other drug reaction can be found in table 2.17, where for each drug the diagnosis or description of the skin reactions as given in the relevant publications is provided. Details and references can be found in the monographs of these medicaments. Additional drugs that have caused eruptions as presented in this chapter (as a manifestation of systemic allergic dermatitis [systemic contact dermatitis]), may be found in table 2.7.

Table 2.14 Drugs that have caused eczematous, erythema multiforme-like or lichenoid drug eruptions or urticaria and showed a positive patch test [a]

Eczematous drug eruption

Amlexanox
Amoxicillin
Ascorbic acid
Benzylpenicillin (penicillin G)
Bupivacaine
Captopril
Carbamazepine
Carbimazole
Cefazolin
Cetirizine
Clomipramine (+ photoallergy)
Codeine
Cyanocobalamin
Danaparoid
Desloratadine
Dexamethasone
Dexamethasone phosphate
Dipyridamole
Disulfiram
Enoxaparin
Epirubicin
Ethambutol

Heparin, unfractionated
Hydromorphone
Hydroxyzine
Interferon (PEG-IFN-α2a)
Isoniazid
Lidocaine
Metamizole
Methoxsalen
Nadroparin
Nimodipine
Phenobarbital
Pravastatin
Prednisolone
Pristinamycin
Pseudoephedrine
Streptomycin
Succinylcholine
Tacrolimus
Tamsulosin
Tobramycin
Triamcinolone
Triflusal (UVB-aggravated)

Erythema multiforme-like drug eruption

Acetaminophen (paracetamol)
Amoxicillin (also in amoxicillin-clavulanic acid)
Carbamazepine
Ceftriaxone
Cefuroxime
Celecoxib
Chlorambucil
Clindamycin
Diltiazem
Hydroxyzine
Iohexol
Iopamidol
Mepivacaine
Methotrexate
Naproxen (photosensitive)

Pancreatin
Paracetamol (acetaminophen)
Paroxetine (photosensitive)
Phenazone
Phenytoin
Prasugrel
Prednisone
Risedronic acid
Sildenafil (uncertain)
Sorafenib
Sulfaguanidine
Tetrazepam
Tobramycin
Triamcinolone acetonide
Tribenoside

Lichenoid drug eruption

Acetazolamide
Aminophylline (from the component ethylenediamine)
Captopril

Carbamazepine
Cycloserine
Tiopronin

Urticaria/urticarial exanthema/urticaria-like exanthema

Acyclovir
Aminophylline (from subcutaneous injection)
Amoxicillin
Ampicillin
Bacampicillin
Benznidazole
Benzylpenicillin (penicillin G)
Ceftriaxone (diagnosed as maculopapular eruption)
Cefuroxime

Ciprofloxacin
Clavulanic acid
Clofazimine
Codeine
Deflazacort
Dexamethasone
Diclofenac
Hydrocortisone sodium phosphate
Interferon (PEG-IFN-α2a)

Table 2.14 Drugs that have caused eczematous, erythema multiforme-like or lichenoid drug eruptions or urticaria and showed a positive patch test (continued) [a]

Iohexol (217)	Oxcarbazepine
Iopamidol (uncertain) (217)	Omeprazole
Lidocaine	Piperazine
Methylprednisolone acetate	Pristinamycin
Methylprednisolone hemisuccinate	Tetrazepam
Nicomorphine	

[a] Details and references can be found in the monographs of the drugs mentioned in this table

Table 2.15 Bullous eruptions

Allopurinol	Papular exanthema with bullae
Amoxicillin	Bullous exanthema/erythema multiforme
Amoxicillin-clavulanic acid	Linear IgA disease; bullous exanthema
Benzylpenicillin (penicillin G)	Bullous exanthema/erythema multiforme; bullous pemphigoid
Ceftriaxone	Linear IgA bullous dermatosis; bullous exanthema/erythema multiforme
Metronidazole	Linear IgA bullous dermatosis; bullous exanthema/erythema multiforme; erythematous, partly urticarial exanthema with bullae
Penicillin V	Bullous pemphigoid
Sulfamethoxazole	Bullous drug eruption; bullous exanthema/erythema multiforme; maculopapular eruption with vesicles and bullae

[a] Details and references can be found in the monographs of the drugs mentioned in this table

Table 2.16 Allergic reactions to injections (except intravenous)

Articaine	Localized eczematous reaction from subcutaneous injection
Bemiparin	Localized eczematous plaques at the injection sites (Chapter 3.239 Heparins)
Bupivacaine	Localized allergic reaction to subcutaneous injection
Certoparin	Localized eczematous plaques at the injection sites (Chapter 3.239 Heparins)
Dalteparin	Localized eczematous plaques at the injection sites (Chapter 3.239 Heparins)
Danaparoid	Localized eczematous plaques at the injection sites (Chapter 3.239 Heparins)
Dutasteride	Angioedema-like contact dermatitis caused by mesotherapy
Enoxaparin	Localized eczematous plaques at the injection sites (Chapter 3.239 Heparins)
Gentamicin	Localized allergic dermatitis from intra-articular injection
Heparin, unfractionated	Generalized delayed-type skin reaction; erythema on both hands; local and generalized vesiculopustular eruption; erythematous and eczematous plaques at the injection sites
Hydromorphone	Localized allergic reaction from subcutaneous infusion spreading to generalized dermatitis
Interferon	Localized allergic reaction at the injection site
Lidocaine	Localized allergic reaction (from subcutaneous injection); micropapular eruption; generalized vesiculobullous exanthema; urticaria-like exanthema
Mepivacaine	Localized allergic reaction (from subcutaneous injection); localized maculopapular rash from sclerotherapy (intravenous injection)
Methylprednisolone acetate	Unspecified and localized skin eruption from intra-articular injection; chronic urticaria from combined immediate- and delayed-type hypersensitivity;
Nadroparin	Localized eczematous plaques at the injection sites (Chapter 3.239 Heparins)
Paramethasone	Localized allergic reaction from intralesional injections
Piperacillin	Localized allergic reaction from intramuscular injection
Prednisolone acetate	Local allergic reaction from intra-articular injection
Prilocaine	Localized allergic reactions from subcutaneous/mucosal injections
Procaine	Localized allergic reactions from subcutaneous/mucosal injections; exanthemas
Secukinumab	Localized and expanding eczematous reaction from subcutaneous injection
Tinzaparin	Localized eczematous plaques at the injection sites (Chapter 3.239 Heparins)
Triamcinolone acetonide	Localized allergic reaction from intralesional injection; reactions to intra-articular injections: erythema multiforme-like allergic dermatitis, morbilliform and partially persistent urticarial dermatitis, localized and generalized allergic reaction, and generalized erythema

[a] Details and references can be found in the monographs of the drugs mentioned in this table

Table 2.17 Drugs that have caused any other drug eruption and showed a positive patch test [a]

Drug	Clinical description/diagnosis
Acetaminophen	Micropapular eruption; generalized pruriginous rash
Allopurinol	Unspecified drug eruption
Aminophylline	Generalized rash consisting of erythematous papules
Amitriptyline	Generalized rash
Amlexanox	Pityriasis rosea-like exanthema
Amoxapine	Erythematous papular eruption on the trunk and limbs
Amoxicillin	Unspecified drug eruption
Ampicillin	Unspecified drug eruption
Apronalide (allylisopropylacetylurea)	Mucocutaneous ocular syndrome
Ascorbic acid	Papular exanthema
Azathioprine	Acneiform eruption
Bacampicillin	Unspecified drug eruption
Benzylpenicillin (penicillin G)	Generalized exanthema with fever; unspecified drug eruption
Betamethasone acetate	Erythema of the face and neck
Betamethasone sodium succinate	Erythema of the face and neck
Bucillamine	Red papules on the face, neck and arms; unspecified drug eruption
Captopril	Erythematous eruption on the cheeks
Carbamazepine	Erythema and edema of the face; 'allergic reactions'; unspecified drug eruption
Carbocromen	Erythematous, non-itchy eruption on the face, back and arms
Carbocysteine	Drug fever (fever without cutaneous eruption)
Carbromal	Purpuric rash
Cefaclor	Unspecified drug eruption
Cefadroxil	Unspecified drug eruption
Cefalexin	Unspecified drug eruption
Cefalotin	Unspecified drug eruption
Cefcapene pivoxil	Pruritic papules and erythema on the trunk and arms
Ceftriaxone	Exanthematous rash; undefined cutaneous eruption; erythematous, partly urticarial exanthema with blisters
Cefuroxime	Unspecified drug eruption
Chloramphenicol	Drug eruption
Chloroquine	Erythematous and mainly papular eruption
Clindamycin	Cutaneous vasculitis
Cloxacillin	Unspecified drug eruption
Clozapine	Erythematous, papular and pustular eruption with vasculitis
Codeine	Exanthema
Dexamethasone	Exanthema
Diazepam	Eczema of the hands and periorbital edema
Diltiazem	Psoriasiform eruption; generalized demarcated erythema with infiltration all over the body
Dimethindene	Widespread maculopapular and vesicular rash
Emtricitabine	Unspecified exanthema with palpebral edema
Ephedrine	Itchy dermatosis
Ethambutol	Desquamative, erythematous, papular rash, with painful crusts and excoriations, spreading from the face and hands to the rest of the body
Gabexate	Panniculitis with eosinophilic infiltration from intravenous administration
Gentamicin	Unspecified drug eruption
Indeloxazine	Eosinophilic pustular folliculitis (Ofuji's disease) (atypical)
Iodixanol	Erythema and deep edema
Iohexol	Facial edema and respiratory distress (217)
Isoniazid	Desquamative, erythematous, papular rash, with painful crusts and excoriations, spreading from the face and hands to the rest of the body; unspecified drug reaction
Lansoprazole	Maculopapular dermatitis and dyspnea
Levocetirizine	Multiple itchy, erythematous, and edematous plaques
Magnesium oxide	Erythematous eruption on the abdomen and back (uncertain)

Table 2.17 Drugs that have caused any other drug eruption and showed a positive patch test (continued) [a]

Drug	Clinical description/diagnosis
Meprobamate	Anaphylactoid reaction with cyanosis, dyspnea, circulatory failure and maculopapular eruption
Meropenem	Unspecified drug eruption
Metamizole	Diffuse edema of the face with breathing difficulty; widespread pruritic exanthema
Methyldopa	Generalized polymorphic eruption with excoriated papulovesicular lesions, plaques of nummular dermatitis and bullae on the palms and soles
Methylprednisolone hemisuccinate	Generalized skin rash; generalized erythema, urticaria and dyspnea
Methylprednisolone sodium succinate	Widespread macular exanthema
Metoprolol	Psoriasiform dermatitis
Mexiletine	Generalized pruritic eruption with papules, infiltrated erythematous patches and pustules
Nadroparin	Spreading of itchy erythematous plaques over the trunk
Nebivolol	Periorbital eczema
Nystatin	Generalized micropapular eruption with edema of the face; erythematous macules on the abdomen and thighs and eczema of a hand
Oxcarbazepine	Unspecified skin reaction
Oxycodone	Exanthema
Penicillin V	Unspecified drug eruption
Pheniramine	Urticarial and maculopapular eruption
Phenobarbital	Unspecified drug eruption
Phenytoin	EMPACT (erythema multiforme associated with phenytoin and cranial radiation therapy); unspecified drug eruption
Piroxicam	Acrovesicular (dyshidrosiform) dermatitis
Prednisolone	Pustular exanthema; generalized rash
Propylthiouracil	Leukocytoclastic vasculitis
Pseudoephedrine	Generalized papulovesicular eruption with mucosal involvement; recurrent erythema; generalized pruritic exanthematous eruption; pigmented purpuric dermatosis
Pyrazinamide	Pruriginous rash mainly on the thorax and abdomen
Streptomycin	Toxic erythema with generalized follicular pustules, later developing into spinous protrusions
Sulfamethoxazole	Unspecified exanthema
Sulfamethoxazole-trimethoprim	Unspecified eruption
Tenofovir	Unspecified exanthema with palpebral edema
Terfenadine	Unspecified drug exanthema
Tetrazepam	Papular and micropapular exanthema
Topiramate	Erythema of the neck, nausea, vomiting

[a] Details and references can be found in the monographs of the drugs mentioned in this table

2.11 Occupational allergic contact dermatitis

Occupational allergic contact dermatitis is a form of contact dermatitis that develops when individuals become sensitized (delayed-type hypersensitivity, contact allergy) to chemicals that they have contact with at and during their professional work and the contact with this hapten/allergen thereafter continues.

People at risk
Occupational allergic contact dermatitis to systemic drugs (i.e. drugs used systemically in humans or animals) is mainly observed in workers of pharmaceutical and chemical companies producing drugs and in health care professionals handling them (342,347). Inadvertent repeated contacts may result in sensitization and further exposure subsequently leads to allergic contact dermatitis. The 'pharmaceutical group' includes chemists and laboratory technicians involved in the development of new pharmaceuticals, workers exposed during the process of drug manufacturing and synthetization, technicians or cleaning staff employed in the maintenance of equipment, and individuals handling and packing the final products. This group may not only be exposed to final drugs, but also

to sensitizing precursor substances, intermediates, and other chemicals used in the synthesis. Human health care workers at risk of sensitization are especially nurses and to a lesser degree pharmacists, dentists and physicians. In veterinary medicine, there is extensive use of drugs, which may sensitize veterinarians, animal caretakers, and farmers who prepare or apply medicaments for animals (342).

Mode of contact
Contact of the skin with drugs can occur from direct contact, from airborne spread of the drugs in the form of powder or aerosols or from both exposures. Airborne contact may, for example, occur in nurses crushing tablets or preparing lyophilized drugs for injection, in pharmacists manufacturing hand-compounded medications and in farmers and caretakers preparing animal feed (342). Formerly, before work hygiene became stricter and more processes were automated, high levels of drug dust could often be demonstrated in the indoor environment in pharmaceutical factories (344).

Clinical picture
The clinical picture of occupational allergic contact dermatitis depends on the mode of contact. Direct contact usually results in allergic contact dermatitis of the hands and forearms. Airborne reactions manifest on exposed skin, notably the eyelids and the face. Airborne allergic contact dermatitis is symmetrical in most cases and can be acute or chronic, depending on the environmental conditions. Swelling and redness may then spread to the neck, hands, and forearms. There, dermatitis usually stops at the margins of the sleeves and collar, but entrapment of drug particles under clothing may lead to dermatitis at non-exposed sites. Because of the accumulation of dust and sweat, the elbow flexures and the skin under a tight collar are often accentuated or become, with prolonged contact, lichenified. Even generalized dermatitis or an exanthematous eruption occasionally occurs as a result of inhalation of the airborne drug (systemic allergic dermatitis) (342,343). Patients may find that the clinical symptoms improve during weekends, work interruption or holidays, and this is highly suggestive of an occupational cause for the dermatitis.
Sensitization may also occur in asymptomatic patients: one study reported a sensitization rate of 12% to penicillin by patch testing a group of nurses without dermatitis (353).

The responsible drugs
Nearly 200 drugs for systemic use have caused occupational allergic contact dermatitis, often only in a few cases or even a single patient. Most allergic reactions in health care professionals have been caused by antibiotics, notably penicillins. Streptomycin used to be a frequent occupational contact allergen in nurses. Formerly, when work hygiene measures in pharmaceutical companies were poor and there was heavy contamination of the factory area with dust of the drugs produced there, 25-40% of the workers exposed to beta-lactam antibiotics developed allergic reactions, mainly eczema. The risk increased with higher exposures (344,345,346). Other well-known causes of occupational allergic contact dermatitis from systemic drugs are propacetamol, ranitidine, cephalosporins (which may sometimes cross-react with beta-lactam penicillins), omeprazole, and tetrazepam (348,350). It should be appreciated, however, that, in healthcare workers, systemic drugs only account for a very small portion of occupational allergic contact dermatitis (349,350), in the order of 1% (351).
 Review articles on occupational allergy to pharmaceutical products have been published in 2020 (342,356), 2016 (352) and 2011 (355). The older literature on occupational drug dermatitis has been reviewed in 1992 (354).

Drugs that have caused occupational allergic contact dermatitis are shown in table 2.18. Details and references can be found in their monographs in this book. Included are cases of what formerly was called 'connubial' or 'consort' allergic contact dermatitis (13). This refers to cases where people using certain products (e.g. a fragrance or hair dye) do not become sensitized to these items *themselves*, but their 'consort', often their husband of wife. Later, the concept was broadened to include partners, children, parents, siblings or others in the close personal environment. In the case of systemic drugs, it is not the individual using the drugs who develops a hypersensitivity reaction, but the care-giver that handles the drugs becomes sensitized. Examples are a woman who breaks tablets and gives them to her disabled husband or a mother who administers inhalation drugs to her asthmatic child and develops airborne allergic contact dermatitis (i.e. the mother, not the child taking the drug). This concept of consort/connubial dermatitis is currently termed 'contact dermatitis *by proxy*' (257).

Table 2.18 Drugs for systemic use that have caused occupational allergic contact dermatitis [a,e]

Abacavir [b]	Chloral hydrate
Acemetacin [b]	Chlorambucil
Acetaminophen [b]	Chloramphenicol
Acetomenaphthone (vitamin K4) [b]	Chloroquine [b]
Acetylcysteine [b]	Chlorpromazine [b,d]
Albendazole	Chlorprothixene [b,d]
Alprazolam [b]	Chlortetracycline
Alprenolol [b]	Clavulanic acid
Althiazide	Clopidol
Amdinocillin pivoxil (pivmecillinam) [b]	Clorazepate [c]
Amikacin	Clotiazepam [b,c]
Aminophylline [b]	Cloxacillin
Amlodipine besylate [b]	Codeine [b]
Amoxicillin	Colistin
Amoxicillin in amoxicillin-clavulanic acid	Cotrimoxazole (sulfamethoxazole-trimethoprim) [d]
Ampicillin	Cyanocobalamin [b]
Amprolium	Dexlansoprazole [b]
Anileridine [c]	Diatrizoic acid
Apomorphine [b,c]	Diazepam [b]
Arecoline	Dihydrostreptomycin
Aripiprazole [b]	Dimetridazole
Atorvastatin	Diphenhydramine
Avoparcin [b]	Dipyridamole
Azaperone	Disulfiram [c]
Azathioprine [c]	Doxycycline
Azithromycin [b]	Enalapril [b]
Bacampicillin [b]	Ertapenem
Baclofen [b]	Esomeprazole [b]
Bendroflumethiazide	Ethambutol
Benzocaine	Famotidine
Benzylpenicillin	Floxacillin (flucloxacillin) [b]
Biperiden [b]	Flutamide
Bisoprolol [b]	Fluvastatin [b]
Bromazepam [b,c]	Furaltadone
Captopril [b,c]	Furazolidone
Carbenicillin	Fusidic acid
Carbimazole [d]	Gentamicin
Carbocromen	Griseofulvin
Carprofen [b,c]	Halquinol
Carvedilol [b]	Heroin [b]
Casanthranol	Hydralazine
Cefalotin [b]	Hydrocortisone
Cefamandole [b]	Hydroxychloroquine
Cefazolin [b]	Imatinib [b]
Cefmetazole	Imipenem (in the mixture with cilastatin)
Cefodizime [b]	Iomeprol
Ceforanide	Isoflurane [b]
Cefotaxime [b]	Isoniazid
Cefotetan	Ketoprofen (photoaggravated)
Cefotiam	Kitasamycin [b]
Cefoxitin [b]	Lamivudine
Cefradine [b]	Lansoprazole [b]
Ceftazidime [b]	Lidocaine
Ceftiofur [b]	Lincomycin
Ceftizoxime [b]	Lisinopril [b]
Ceftriaxone [b]	Lorazepam [b,c]
Cefuroxime	Lormetazepam [b]

Table 2.18 Drugs for systemic use that have caused occupational allergic contact dermatitis (continued) [a,e]

Meclofenoxate	Piroxicam (presenting as fixed drug eruption)
Meglumine diatrizoate (Chapter 3.171 Diatrizoic acid)	Pivampicillin [b]
Melphalan	Pivmecillinam (Chapter 3.23 Amdinocillin pivoxil) [b]
Menadione [b]	Prednisone
Meropenem	Pristinamycin [b]
Mesna [b]	Procaine
Metaproterenol [b]	Procaine benzylpenicillin [b]
Methacycline	Propacetamol [b]
Methotrexate (uncertain) [b]	Propanidid [b]
Methylprednisolone	Propiopromazine
Metoprolol [b]	Propranolol [b]
Mezlocillin	Pyrazinobutazone
Midecamycin [b]	Pyridoxine [b]
Mitomycin C (uncertain)	Pyritinol [b]
Monensin	Quinidine [b,d]
Morphine [b]	Quinine [b,d]
Nalmefene [b]	Rabeprazole [b]
Neomycin	Ranitidine [b]
Nicergoline [b]	Retinyl acetate [b]
Nifuroxazide	Rifampicin
Nitrofurazone [b]	Risedronic acid [b]
Nosiheptide	Risperidone [b]
Olanzapine [b]	Sevoflurane [b]
Olaquindox [b,d]	Simvastatin [b]
Omeprazole [b]	Sotalol [b]
Oxacillin	Spectinomycin
Oxolamine	Spiramycin [b]
Oxprenolol [b]	Spironolactone [b]
Oxybutynin [b]	Streptomycin [b]
Oxycodone [b]	Sulfamethoxazole-trimethoprim [d]
Oxytetracycline	Sulfathiazole
Pantoprazole [b]	Tetracaine
Penethamate	Tetracycline
Penicillin, unspecified [b]	Tetrazepam [b,c]
Penicillin G benzathine	Thiabendazole
Perazine	Thiamine [b]
Periciazine	Trazodone [b,c]
Perindropril [b]	Triamcinolone acetonide
Perphenazine [c,d]	Tylosin [b]
Phenoxybenzamine [b]	Vincamine [b]
Piperacillin [b]	Virginiamycin [b]
Piperazine	Ziprasidone
Pirmenol [b]	Zolpidem [b]

[a] Included are cases of sensitization *by proxy* ('connubial contact dermatitis', 'consort dermatitis'), e.g. a woman who becomes sensitized by tablets she is breaking to be used by her husband or child; [b] Airborne allergic contact dermatitis has been reported; [c] Sensitization *by proxy* has been reported (see [a]); [d] Photoallergic contact dermatitis has been reported; [e] Details and references can be found in the monographs of the drugs mentioned in this table

2.12 Diagnostic tests in suspected cutaneous adverse drug reactions from systemic drugs
Diagnostic tests in suspected cutaneous adverse drug reactions from systemic drugs (patch tests, prick tests, intradermal tests, drug provocation tests, *in vitro* tests) are discussed in Chapter 4.

2.13 References
References of this chapter are shown in Chapter 4 Diagnostic tests in suspected cutaneous adverse drug reactions from systemic drugs.

Chapter 3   MONOGRAPHS OF SYSTEMIC DRUGS THAT HAVE CAUSED CUTANEOUS ADVERSE DRUG REACTIONS FROM DELAYED-TYPE HYPERSENSITIVITY

3.0 INTRODUCTION

In this chapter, monographs of 507 systemic drugs that have caused cutaneous adverse drug reactions from delayed-type hypersensitivity *as shown by positive patch tests* are presented. The monographs, shown in alphabetical order, have a standardized format, which is explained and detailed in Chapter 1.2. The systemic drugs discussed here are shown in table 3.1. An overview of the clinical manifestations of delayed-type hypersensitivity to systemic drugs is given in Chapter 2. These manifestations can be either (a variety of) cutaneous drug reactions (eruptions, exanthemas), with or without internal organ involvement, or allergic contact dermatitis from accidental contact.

The cutaneous drug reactions most frequently seen are maculopapular eruptions. Other such drug reactions, which have shown to be mediated by delayed-type hypersensitivity, include erythroderma and exfoliative dermatitis, systemic allergic dermatitis (systemic contact dermatitis), symmetrical drug-related intertriginous and flexural exanthema (SDRIFE)/baboon syndrome, fixed drug eruption, photosensitivity and delayed urticaria. Infrequent, but serious and potentially dangerous and even fatal manifestations of drug hypersensitivity, manifesting both in the skin and internal organs, are acute generalized exanthematous pustulosis (AGEP), drug reaction with eosinophilia and systemic symptoms (DRESS), Stevens-Johnson syndrome and toxic epidermal necrolysis. All these and other possible drug reactions are presented in Chapter 2, each with an introduction and discussion of etiology and pathophysiology, clinical features, histology, diagnostics, and therapy. All systemic drugs which have caused these manifestations and showed positive patch tests are also presented there, in tabular format (Chapters 2.2-2.10). Details for all drugs can be found in the individual monographs in this chapter.

Sensitization from accidental contact may affect workers producing the drugs and healthcare professionals working with pharmaceuticals, and occasionally non-professional caregivers dispensing drugs to for example a disabled husband or a sick child (Chapter 2.11).

Table 3.1 Systemic drugs presented in monographs in this chapter

Chapter	Name of monograph	Chapter	Name of monograph
3.1	Abacavir	3.29	Aminophylline
3.2	Aceclofenac	3.30	Aminosalicylic acid
3.3	Acemetacin	3.31	Amitriptyline
3.4	Acetaminophen	3.32	Amlexanox
3.5	Acetazolamide	3.33	Amlodipine besylate
3.6	Acetomenaphthone	3.34	Amobarbital
3.7	Acetylcysteine	3.35	Amoxapine
3.8	Acetylsalicylic acid	3.36	Amoxicillin
3.9	Acexamic acid	3.37	Amoxicillin mixture with clavulanate potassium
3.10	Actarit		
3.11	Acyclovir	3.38	Ampicillin
3.12	Adalimumab	3.39	Ampiroxicam
3.13	Afloqualone	3.40	Amprolium
3.14	Albendazole	3.41	Anileridine
3.15	Albuterol	3.42	Antipyrine salicylate
3.16	Alendronic acid	3.43	Apomorphine
3.17	Allopurinol	3.44	Apronalide
3.18	Alprazolam	3.45	Arecoline
3.19	Alprenolol	3.46	Aripiprazole
3.20	Althiazide	3.47	Articaine
3.21	Amantadine	3.48	Ascorbic acid
3.22	Ambroxol	3.49	Atenolol
3.23	Amdinocillin pivoxil	3.50	Atorvastatin
3.24	Amikacin	3.51	Atovaquone
3.25	Amiloride	3.52	Avoparcin
3.26	Aminocaproic acid	3.53	Azaperone
3.27	Aminophenazone	3.54	Azathioprine
3.28	Aminophenazone hemibarbital	3.55	Azithromycin

Table 3.1 Systemic drugs presented in monographs in this chapter (continued)

Chapter	Name of monograph	Chapter	Name of monograph
3.56	Aztreonam	3.112	Ceftiofur
3.57	Bacampicillin	3.113	Ceftizoxime
3.58	Baclofen	3.114	Ceftriaxone
3.59	Bemiparin	3.115	Cefuroxime
3.60	Bendamustine	3.116	Celecoxib
3.61	Bendroflumethiazide	3.117	Certoparin
3.62	Benznidazole	3.118	Cetirizine
3.63	Benzydamine	3.119	Chloral hydrate
3.64	Benzylpenicillin	3.120	Chlorambucil
3.65	Betamethasone	3.121	Chloramphenicol
3.66	Betamethasone acetate	3.122	Chlormezanone
3.67	Betamethasone dipropionate	3.123	Chloroquine
3.68	Betamethasone sodium phosphate	3.124	8-Chlorotheophylline
3.69	Betamethasone sodium succinate	3.125	Chlorpheniramine
3.70	Biperiden	3.126	Chlorpromazine
3.71	Bismuth subcitrate	3.127	Chlorpropamide
3.72	Bisoprolol	3.128	Chlorprothixene
3.73	Bleomycin	3.129	Chlortetracycline
3.74	Bortezomib	3.130	Cilastatin mixture with imipenem
3.75	Brivudine	3.131	Cimetidine
3.76	Bromazepam	3.132	Ciprofloxacin
3.77	Bromhexine	3.133	Cisplatin
3.78	Bromisoval	3.134	Citalopram
3.79	Bucillamine	3.135	Citiolone
3.80	Bupivacaine	3.136	Clarithromycin
3.81	Bupropion	3.137	Clavulanic acid
3.82	Captopril	3.138	Clindamycin
3.83	Carbamazepine	3.139	Clioquinol
3.84	Carbenicillin	3.140	Clobazam
3.85	Carbimazole	3.141	Clofazimine
3.86	Carbocromen	3.142	Clofibrate
3.87	Carbocysteine	3.143	Clomipramine
3.88	Carbromal	3.144	Clonidine
3.89	Carbutamide	3.145	Clopidogrel
3.90	Carprofen	3.146	Clopidol
3.91	Carvedilol	3.147	Cloprednol
3.92	Casanthranol	3.148	Clorazepic acid
3.93	Cefaclor	3.149	Clotiazepam
3.94	Cefadroxil	3.150	Cloxacillin
3.95	Cefalexin	3.151	Clozapine
3.96	Cefalotin	3.152	Codeine
3.97	Cefamandole	3.153	Colistin
3.98	Cefazolin	3.154	Cyamemazine
3.99	Cefcapene pivoxil	3.155	Cyanamide
3.100	Cefixime	3.156	Cyanocobalamin
3.101	Cefmetazole	3.157	Cyclophosphamide
3.102	Cefodizime	3.158	Cycloserine
3.103	Cefonicid	3.159	Dabrafenib
3.104	Ceforanide	3.160	Dalteparin
3.105	Cefotaxime	3.161	Danaparoid
3.106	Cefotetan	3.162	Dapsone
3.107	Cefotiam	3.163	Deflazacort
3.108	Cefoxitin	3.164	Desloratadine
3.109	Cefpodoxime	3.165	Dexamethasone
3.110	Cefradine	3.166	Dexamethasone phosphate
3.111	Ceftazidime		

Table 3.1 Systemic drugs presented in monographs in this chapter (continued)

Chapter	Name of monograph	Chapter	Name of monograph
3.167	Dexamethasone sodium phosphate	3.222	Flutamide
3.168	Dexketoprofen	3.223	Fluvastatin
3.169	Dexlansoprazole	3.224	Fluvoxamine
3.170	Dextropropoxyphene	3.225	Fondaparinux
3.171	Diatrizoic acid	3.226	Foscarnet
3.172	Diazepam	3.227	Fulvestrant
3.173	Diclofenac	3.228	Furaltadone
3.174	Dicloxacillin	3.229	Furazolidone
3.175	Dihydrocodeine	3.230	Fusidic acid
3.176	Dihydrostreptomycin	3.231	Gabapentin
3.177	Diltiazem	3.232	Gabexate
3.178	Dimenhydrinate	3.233	Gadobutrol
3.179	2,3-Dimercapto-1-propanesulfonic acid	3.234	Garenoxacin
3.180	Dimethindene	3.235	Gentamicin
3.181	Dimethyl sulfoxide	3.236	Gliclazide
3.182	Dimetridazole	3.237	Griseofulvin
3.183	Diphenhydramine	3.238	Halquinol
3.184	Dipyridamole	3.239	Heparins
3.185	Disulfiram	3.240	Heroin
3.186	Doxepin	3.241	Hyaluronidase
3.187	Doxycycline	3.242	Hydralazine
3.188	Dronedarone	3.243	Hydrochlorothiazide
3.189	Droxicam	3.244	Hydrocortisone
3.190	Dutasteride	3.245	Hydrocortisone sodium phosphate
3.191	Efavirenz	3.246	Hydromorphone
3.192	Emtricitabine	3.247	Hydroquinidine
3.193	Enalapril	3.248	Hydroxychloroquine
3.194	Enoxacin	3.249	Hydroxyprogesterone
3.195	Enoxaparin	3.250	Hydroxyurea
3.196	Eperisone	3.251	Hydroxyzine
3.197	Ephedrine	3.252	Ibandronic acid
3.198	Epirubicin	3.253	Ibuprofen
3.199	Eprazinone	3.254	Imatinib
3.200	Erlotinib	3.255	Imipenem
3.201	Ertapenem	3.256	Indeloxazine
3.202	Erythromycin	3.257	Indomethacin
3.203	Eslicarbazepine	3.258	Interferons
3.204	Esomeprazole	3.259	Iobitridol
3.205	Estradiol	3.260	Iodixanol
3.206	Ethambutol	3.261	Iodoquinol
3.207	Ethenzamide	3.262	Iohexol
3.208	Ethosuximide	3.263	Iomeprol
3.209	Etonogestrel	3.264	Iopamidol
3.210	Etoricoxib	3.265	Iopentol
3.211	Etretinate	3.266	Iopromide
3.212	Famotidine	3.267	Ioversol
3.213	Fenofibrate	3.268	Ioxaglic acid
3.214	Feprazone	3.269	Ioxitalamic acid
3.215	Fexofenadine	3.270	Ipragliflozin
3.216	Flavoxate	3.271	Irbesartan
3.217	Floxacillin	3.272	Isepamicin
3.218	Fluconazole	3.273	Isoflurane
3.219	Fluindione	3.274	Isoniazid
3.220	Fluorouracil	3.275	Isotretinoin
3.221	Flurbiprofen	3.276	Isoxsuprine

Table 3.1 Systemic drugs presented in monographs in this chapter (continued)

Chapter	Name of monograph	Chapter	Name of monograph
3.277	Itraconazole	3.333	Naproxen
3.278	Kanamycin	3.334	Nebivolol
3.279	Ketoconazole	3.335	Neomycin
3.280	Ketoprofen	3.336	Nevirapine
3.281	Kitasamycin	3.337	Nicardipine
3.282	Labetalol	3.338	Nicergoline
3.283	Lamivudine	3.339	Nicomorphine
3.284	Lamotrigine	3.340	Nifuroxazide
3.285	Lansoprazole	3.341	Nimesulide
3.286	Levocetirizine	3.342	Nimodipine
3.287	Levofloxacin	3.343	Nitrofurantoin
3.288	Levomepromazine	3.344	Nitrofurazone
3.289	Lidocaine	3.345	Norfloxacin
3.290	Lincomycin	3.346	Nosiheptide
3.291	Lisinopril	3.347	Nylidrin
3.292	Lomefloxacin	3.348	Nystatin
3.293	Lorazepam	3.349	Ofloxacin
3.294	Lormetazepam	3.350	Olanzapine
3.295	Magnesium oxide	3.351	Olaquindox
3.296	Meclofenoxate	3.352	Omeprazole
3.297	Mefenamic acid	3.353	OnabotulinumtoxinA
3.298	Melphalan	3.354	Ornidazole
3.299	Menadione	3.355	Oxacillin
3.300	Meperidine	3.356	Oxcarbazepine
3.301	Mepivacaine	3.357	Oxolamine
3.302	Meprobamate	3.358	Oxprenolol
3.303	Mequitazine	3.359	Oxybutynin
3.304	Meropenem	3.360	Oxycodone
3.305	Mesalazine	3.361	Oxytetracycline
3.306	Mesna	3.362	Pancreatin
3.307	Metamizole	3.363	Pantoprazole
3.308	Metaproterenol	3.364	Paramethasone
3.309	Methacycline	3.365	Paroxetine
3.310	Methotrexate	3.366	Pefloxacin
3.311	Methoxsalen	3.367	Penethamate
3.312	Methyldopa	3.368	Penicillin G benzathine
3.313	Methylprednisolone	3.369	Penicillin V
3.314	Methylprednisolone acetate	3.370	Penicillins, unspecified
3.315	Methylprednisolone hemisuccinate	3.371	Perazine
3.316	Metoprolol	3.372	Periciazine
3.317	Metronidazole	3.373	Perindopril
3.318	Mexiletine	3.374	Perphenazine
3.319	Mezlocillin	3.375	Phenazone
3.320	Miconazole	3.376	Phenethicillin
3.321	Midecamycin	3.377	Phenindione
3.322	Minocycline	3.378	Pheniramine
3.323	Mirtazapine	3.379	Phenobarbital
3.324	Misoprostol	3.380	Phenoxybenzamine
3.325	Mitomycin C	3.381	Phenylbutazone
3.326	Mofebutazone	3.382	Phenylephrine
3.327	Monensin	3.383	Phenytoin
3.328	Morantel	3.384	Phytonadione
3.329	Morphine	3.385	Picosulfuric acid
3.330	Moxifloxacin	3.386	Piperacillin
3.331	Nadroparin	3.387	Piperacillin mixture with tazobactam
3.332	Nalmefene	3.388	Piperazine

Table 3.1 Systemic drugs presented in monographs in this chapter (continued)

Chapter	Name of monograph	Chapter	Name of monograph
3.389	Pirfenidone	3.444	Simvastatin
3.390	Piritramide	3.445	Sorafenib
3.391	Pirmenol	3.446	Sotalol
3.392	Piroxicam	3.447	Spectinomycin
3.393	Piroxicam betadex	3.448	Spiramycin
3.394	Pivampicillin	3.449	Spironolactone
3.395	Potassium aminobenzoate	3.450	Streptomycin
3.396	Practolol	3.451	Succinylcholine
3.397	Prasugrel	3.452	Sulfadiazine
3.398	Pravastatin	3.453	Sulfaguanidine
3.399	Prednisolone	3.454	Sulfamethoxazole
3.400	Prednisolone acetate	3.455	Sulfamethoxazole mixture with trimethoprim
3.401	Prednisolone hemisuccinate		
3.402	Prednisolone sodium succinate	3.456	Sulfanilamide
3.403	Prednisolone tetrahydrophthalate sodium salt	3.457	Sulfasalazine
		3.458	Sulfathiazole
3.404	Prednisone	3.459	Sulfonamide, unspecified
3.405	Pregabalin	3.460	Tacrolimus
3.406	Prilocaine	3.461	Talampicillin
3.407	Pristinamycin	3.462	Talastine
3.408	Procaine	3.463	Tamsulosin
3.409	Procaine benzylpenicillin	3.464	Tegafur
3.410	Proguanil	3.465	Teicoplanin
3.411	Promazine	3.466	Tenofovir
3.412	Promethazine	3.467	Tenoxicam
3.413	Propacetamol	3.468	Terbinafine
3.414	Propanidid	3.469	Terfenadine
3.415	Propicillin	3.470	Tetracycline
3.416	Propiopromazine	3.471	Tetrazepam
3.417	Propranolol	3.472	Thiabendazole
3.418	Propylthiouracil	3.473	Thiamine
3.419	Pseudoephedrine	3.474	Thioctic acid
3.420	Pyrazinamide	3.475	Thioridazine
3.421	Pyrazinobutazone	3.476	Tiaprofenic acid
3.422	Pyridoxine	3.477	Ticlopidine
3.423	Pyrimethamine	3.478	Tilisolol
3.424	Pyritinol	3.479	Tinzaparin
3.425	Quinapril	3.480	Tiopronin
3.426	Quinidine	3.481	Tobramycin
3.427	Quinine	3.482	Tolbutamide
3.428	Rabeprazole	3.483	Topiramate
3.429	Ramipril	3.484	Tosufloxacin tosilate
3.430	Ranitidine	3.485	Tramadol
3.431	Retinyl acetate	3.486	Tranexamic acid
3.432	Ribostamycin	3.487	Trazodone
3.433	Rifampicin	3.488	Triamcinolone
3.434	Risedronic acid	3.489	Triamcinolone acetonide
3.435	Risperidone	3.490	Triazolam
3.436	Ritodrine	3.491	Tribenoside
3.437	Rosuvastatin	3.492	Triflusal
3.438	Rupatadine	3.493	Trimebutine
3.439	Secnidazole	3.494	Trimeprazine
3.440	Secukinumab	3.495	Trimethoprim
3.441	Sertraline	3.496	Tylosin
3.442	Sevoflurane	3.497	Valaciclovir
3.443	Sildenafil	3.498	Valdecoxib

Table 3.1 Systemic drugs presented in monographs in this chapter (continued)

Chapter	Name of monograph	Chapter	Name of monograph
3.499	Valproic acid		
3.500	Vancomycin		
3.501	Varenicline		
3.502	Vinburnine		
3.503	Vincamine		
3.504	Virginiamycin		
3.505	Ziprasidone		
3.506	Zolpidem		
3.507	Zonisamide		

Chapter 3.1 ABACAVIR

IDENTIFICATION

Description/definition : Abacavir is the nucleoside reverse transcriptase inhibitor analog of guanosine that conforms to the structural formula shown below
Pharmacological classes : Anti-HIV agents; reverse transcriptase inhibitors
IUPAC name : [(1S,4R)-4-[2-Amino-6-(cyclopropylamino)purin-9-yl]cyclopent-2-en-1-yl]methanol
CAS registry number : 136470-78-5
EC number : Not available
Merck Index monograph : 1271
Patch testing : Abacavir sulfate 1% and 10% pet. (2); 10% pet. (16)
Molecular formula : $C_{14}H_{18}N_6O$

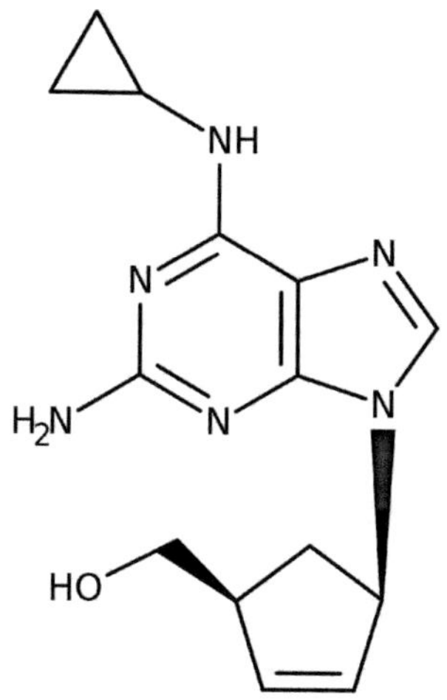

GENERAL

Abacavir is a powerful nucleoside reverse transcriptase inhibitor analog of guanosine with activity against Human Immunodeficiency Virus Type 1 (HIV-1). This drug is indicated, in combination with other antiretroviral agents, for treatment of HIV-1 infection. It decreases HIV viral loads, retards or prevents the damage to the immune system, and reduces the risk of developing AIDS. Abacavir is usually employed as abacavir sulfate (CAS number 188062-50-2, EC number not available, molecular formula $C_{28}H_{38}N_{12}O_6S$), sometimes as abacavir hydrochloride. In vivo, abacavir sulfate dissociates to its free base, abacavir (1).

CONTACT ALLERGY FROM ACCIDENTAL CONTACT

Two cases of occupational allergic contact dermatitis in technicians in a manufacturing plant caused by abacavir have been reported. A 49-year-old man developed eyelid dermatitis within a few months of starting his job. Patch testing showed positive reactions to abacavir glutarate (1% and 10% pet.) and abacavir sulfate (1% and 10% pet.). A 52-year-old man presented with a rash on his face, occipital neck, wrists and dorsa of the feet within a few months of being moved to the abacavir production site in the plant. He was patch tested and also showed positive reactions to abacavir glutarate 1% and 10% and abacavir sulfate 1% and 10% pet. The rashes of both patients cleared after they had been transferred to a different department (2). Both patients were positive for HLA-B*5701, which is strongly associated with abacavir hypersensitivity (3).

A few years earlier, two process operators, aged 48 and 51, working in a pharmaceutical plant, had developed occupational allergic contact dermatitis from abacavir. The primary sites of the dermatitis were the eyelids and the hands (8). It may be concluded that occupational allergic contact dermatitis from abacavir may both result from direct contact and from airborne exposure.

CUTANEOUS ADVERSE DRUG REACTIONS FROM SYSTEMIC ADMINISTRATION CAUSED BY TYPE IV (DELAYED-TYPE) HYPERSENSITIVITY (as demonstrated by positive patch tests)

Cutaneous adverse drug reactions from systemic administration of abacavir caused by type IV (delayed-type) hypersensitivity have included abacavir hypersensitivity syndrome. In contrast with hypersensitivity syndrome to other drugs (e.g. antibiotics, antiepileptics), this is not the same as drug reaction with eosinophilia and systemic symptoms (DRESS).

Abacavir hypersensitivity syndrome

General

Abacavir is used in combination with other antiretroviral drugs for the treatment of HIV. Approximately 4-8% of patients starting abacavir will experience a hypersensitivity syndrome (AHS) (13,15,17). This AHS is clinically characterized by at least 2 of the following manifestations: constitutional symptoms (fever, malaise, lethargy, headache, and myalgia), rash, gastrointestinal symptoms (nausea, vomiting, and diarrhea), and/or respiratory symptoms (dyspnea, cough). In about 30% of the patients, no skin eruption is present; when present, it is often a late feature (15). The reaction may occur from one week to several years after initiation of therapy, but typically (90%) develops within 6 weeks after drug introduction (6,11,12,17). The mortality rate in all treated patients is estimated at 0.03% (11). It should be realized that symptoms of hypersensitivity reactions to abacavir are not specific and are easily mistakable with concomitant infection, reaction to other drugs, or inflammatory disease, which may lead to clinical overdiagnosis of AHS (14,15). Results from randomized, placebo-controlled trials of abacavir, in which hypersensitivity has been diagnosed at rates of 2-7% in participants not receiving abacavir, illustrate this overdiagnosis (15).

AHS reactions are often mediated by delayed-type hypersensitivity, as shown by positive patch tests. The frequency of positive patch tests to abacavir in patients with *clinical signs* of the ABC hypersensitivity syndrome has ranged from 25-100%, depending on the number of patients (generally speaking, the larger the number, the lower the percentage of positive reactions) and – obviously – on the selection criteria (3,4,5,6,9). However, the diagnostic sensitivity of patch testing for 'real' abacavir hypersensitivity is estimated to be 87% (6). The presence of a skin exanthema does not seem to be a sensitive predictor for patch test positivity (3). Positive patch test reactions have been reproduced after 7-42 months (7) and patch test reactivity may persist for many years (10,18), supporting the durability of the abacavir patch test in correctly classifying AHS (7). Positive patch tests to abacavir may occur in the absence of skin rashes (3,5). Severe reactions such as shock and even death have been described with full dose rechallenge of abacavir after the occurrence of the hypersensitivity syndrome (5) and, therefore, rechallenge is considered to be contra-indicated.

All patients with a clinical syndrome compatible with hypersensitivity to abacavir who were identified as having a positive result on epicutaneous patch testing have carried the HLA-B*5701 allele (3,4,5,6,9). Nearly 50% of all HLA-B*5701-positive individuals may develop ABC hypersensitivity when treated with the drug (10,13). Screening for HLA-B*5701 in HIV patients and excluding them from abacavir treatment (which has become standard since 2008) has therefore significantly reduced the number of cases of hypersensitivity (6,11,19), Indeed, HLA-B*5701 has a negative predictive value of 100% for patch-test-confirmed AHS, which is specifically and exquisitely restricted to HLA-B*5701-positive patients and is mediated by CD8+ lymphocytes (13).

Case series

A group of 130 white and 69 black patients who had been treated with abacavir (ABC) were recruited for patch testing. They had been selected on the basis of experienced clinically suspected – self identified - hypersensitivity reactions to abacavir, consisting of at least 2 major symptoms (fever, rash, or gastrointestinal or constitutional symptoms) and with date within 6 weeks after initiating ABC therapy. The patients were patch tested with ABC 1% and 10% pet. in duplicate. They applied and removed the patches themselves and also read the reactions. Patch test kits were standardized, with detailed instruction regarding application and interpretation of the test. Test results were read and recorded at 24 and 48 hours after application, and photographic images were collected using standard cameras. Patch tests were recorded as positive in 32% of the white and 7% of the black patients, all of who were HLA-B*5701-positive. Patients with positive patch test reactions tended to have 3 or more signs of hypersensitivity more often than those with a negative patch test. However, rashes had occurred in about 60% of patch test positive patients, which was only slightly higher than in patch test negative subjects (3).

Seven patients were described who had positive patch tests to abacavir (ABC) within 4 months of a presumed ABC hypersensitivity reaction, defined as either having at least two intensifying symptoms of rash, fever, gastrointestinal complaints, headache and resolution within 24 hour of stopping the drug without an alternative explanation, or a positive rechallenge to abacavir. Three patients were patch tested with and reacted to ABC 1% and 10% pet. at D2 and D4. Four others were additionally tested with ABC 0.1%, 5%, 15% and 25% pet.; all 4 had positive patch tests at D2 and D4 to all concentrations. Two HIV-negative controls with no previous ABC exposure and five HIV-positive controls with previous ABC exposure matched for age, sex, race, CD4 cell count and viral load had negative patch tests at D2 and D4 to all concentrations. Biopsies of the patch tests showed lymphocytic vasculitis. Immunohisto-chemistry on skin biopsies from the skin rash in 3 patients matched those from positive patch test reactions, suggesting an identical pathophysiological process (5). These seven previously patch test-positive patients were again patch tested with abacavir 7-42 months later and all had a strong reaction after 24 hours. (7).

In a large international study, a study group of 802 HIV patients negative to HLA-B*5701 and a control group of 842 HIV patients not prescreened for HLA-B*5701 were both treated with abacavir in the same manner. Of the study group, 27 patients (3.4%) had clinical signs of a hypersensitivity reaction (control group: 7.8%), but none of them (0%) had positive patch tests, whereas in the control group 2.7% had positive patch tests. This indicates that pre-treatment HLA-B*5701 screening significantly reduces, if not abolishes, the risk of hypersensitivity reaction to abacavir (6).

In Italy, 20 patients suspected of having a drug hypersensitivity reaction to abacavir (ABC) (criteria: at least two intensifying symptoms [rash, fever, and gastrointestinal complaints, headache], followed by resolution within 24 hours of stopping the drug without an alternative explanation, or a positive challenge to ABC), of who only 7 had a skin reaction (all maculopapular exanthema) were patch tested with ABC 1% and 10% pet. and 10 (50%) had positive reactions. These were all HLA-B*5701-positive, whereas all 10 not reacting to the patch test were HLA-B*5701-negative (9).

In Portugal, 7 (4%) of 186 patients treated with ABC developed a (suspected) hypersensitivity reaction. Two had positive patch tests to ABC 1% and 10% pet. After the implementation of HLA-B*5701 screening and excluding positive patients, no suspected hypersensitivity reactions have been observed anymore (11).

LITERATURE

1   The data in the section 'General' may have been obtained from literature discussed in this chapter, but mostly also or exclusively from one or more of the following online sources: ChemIDPlus Advanced, PubChem, DrugBank, RxList, Drug Central, Drugs.com, and Wikipedia

2   Khalid A, Ghaffar S. Two cases of occupational allergic contact dermatitis caused by abacavir. Contact Dermatitis 2019;80:187-188

3   Saag M, Balu R, Phillips E, Brachman P, Martorell C, Burman W, et al. High sensitivity of human leukocyte antigen-B*5701 in immunologically-confirmed cases of abacavir hypersensitivity in white and black patients. Clin Infect Dis 2008;46:1111-1118

4   Milpied B. Drug eruptions to antiretroviral agents: use of patch testing. Dermatitis 2008;19:349

5   Phillips EJ, Sullivan JR, Knowles SR, Shear NH. Utility of patch testing in patients with hypersensitivity syndromes associated with abacavir. AIDS 2002;16:2223-2225

6   Mallal S, Phillips E, Carosi G, , Molina JM, Workman C, Tomazic J, et al. HLA-B*5701 screening for hypersensitivity to abacavir. N Engl J Med 2008;358:568-579

7   Phillips EJ, Wong GA, Kaul R, Shahabi K, Nolan DA, Knowles SR, et al. Clinical and immunogenetic correlates of abacavir hypersensitivity. AIDS 2005;19:979-981

8   Bennett MF, Lowney AC, Bourke JF. A study of occupational contact dermatitis in the pharmaceutical industry. Br J Dermatol 2016;174:654-656 (Abstract in Brit J Dermatol 2011;165:73)

9   Giorgini S, Martinelli C, Tognetti L, Carocci A, Giuntini R, Mastronardi V, et al. Use of patch testing for the diagnosis of abacavir-related hypersensitivity reaction in HIV patients. Dermatol Ther 2011;24:591-594

10   Schnyder B, Adam J, Rauch A et al. HLA-B*57:01(+) abacavir-naive individuals have specific T cells but no patch test reactivity. J Allergy Clin Immunol 2013;132:756-758

11   Carolino F, Santos N, Piñeiro C, Santos AS, Soares P, Sarmento A, et al. Prevalence of abacavir-associated hypersensitivity syndrome and HLA-B*5701 allele in a Portuguese HIV-positive population. Porto Biomed J 2017;2:59-62

12   Chaponda M, Pirmohamed M. Hypersensitivity reactions to HIV therapy. Br J Clin Pharmacol 2011;71:659-671

13   Phillips E, Mallal S. Successful translation of pharmacogenetics into the clinic: the abacavir example. Mol Diagn Ther 2009;13:1-9

14   Rauch A, Nolan D, Thurnheer C, Fux CA, Cavassini M, Chave JP, et al. Refining abacavir hypersensitivity diagnoses using a structured clinical assessment and genetic testing in the Swiss HIV Cohort Study. Antivir Ther 2008;13:1019-1028

15   Shear N, Milpied B, Bruynzeel DP, Phillips E. A review of drug patch testing and implications for HIV clinicians. AIDS 2008;22:999-1007

16   Brockow K, Garvey LH, Aberer W, Atanaskovic-Markovic M, Barbaud A, Bilo MB, et al.; ENDA/EAACI Drug Allergy Interest Group. Skin test concentrations for systemically administered drugs – an ENDA/EAACI Drug Allergy Interest Group position paper. Allergy 2013;68:702-712

17   Hughes CA, Foisy MM, Dewhurst N, Higgins N, Robinson L, Kelly DV, et al. Abacavir hypersensitivity reaction: an update. Ann Pharmacother 2008;42:387-396

18   Micozzi S, Rojas P, Rodriguez-Gamboa A, De Barrio M. Exuberant positive patch test to abacavir in a patient with the HLA-B*5701 Haplotype. J Allergy Clin Immunol Pract 2015;3:965-967

19   Young B, Squires K, Patel P, Dejesus E, Bellos N, Berger D, et al. First large, multicenter, open-label study utilizing HLA-B*5701 screening for abacavir hypersensitivity in North America. AIDS 2008;22:1673-1675

Chapter 3.2   ACECLOFENAC

IDENTIFICATION

Description/definition   : Aceclofenac is the monocarboxylic acid that conforms to the structural formula shown
                              below
Pharmacological classes   : Anti-inflammatory agents, non-steroidal
IUPAC name   : 2-[2-[2-(2,6-Dichloroanilino)phenyl]acetyl]oxyacetic acid
Other names   : 2-((2,6-Dichlorophenyl)amino)benzeneacetic acid carboxymethyl ester
CAS registry number   : 89796-99-6
EC number   : Not available
Merck Index monograph   : 1293
Patch testing   : 1% and 5% pet.
Molecular formula   : $C_{16}H_{13}Cl_2NO_4$

GENERAL

Aceclofenac, the carboxymethyl ester of diclofenac, is a nonsteroidal anti-inflammatory drug (NSAID) with marked
anti-inflammatory and analgesic properties. It is orally administered for the relief of pain and inflammation in
osteoarthritis, rheumatoid arthritis and ankylosing spondylitis. Aceclofenac is also reported to be effective in other
painful conditions such as dental and gynecological conditions (1). Aceclofenac is metabolized into diclofenac after
systemic administration (4). This NSAID is also widely used in topical form for acute soft trauma and inflammatory or
degenerative musculoskeletal disorders (3). In topical preparations, aceclofenac has caused contact allergy/allergic
contact dermatitis and photoallergic contact dermatitis, which has been fully reviewed in Volume 3 of the
Monographs in contact allergy series (2).

CUTANEOUS ADVERSE DRUG REACTIONS FROM SYSTEMIC ADMINISTRATION CAUSED BY TYPE IV
(DELAYED-TYPE) HYPERSENSITIVITY (as demonstrated by positive patch tests)

Cutaneous adverse drug reactions from systemic administration of aceclofenac caused by type IV (delayed-type)
hypersensitivity have included fixed drug eruption (4).

Fixed drug eruption

A 55-year-old woman had developed an erythematous and pruritic lesion on the nape 10 hours after one dose of
oral aceclofenac 100 mg as analgesic. It resolved in 6-7 days without any treatment, but a residual hyperpigmented
area persisted for some weeks. In the past 5 years, she had developed 5 identical episodes, always after aceclofenac
intake. Patch testing was carried out with aceclofenac 1.5% cream and diclofenac 1% gel. on both previously affected
and non-affected skin. Ten hours after having placed the patch with aceclofenac on the affected skin, the patient
took it off because of severe itching. At D2 and D4, the patch tests with the diclofenac gel and the aceclofenac cream
were positive on previously affected skin but negative on unaffected skin. The NSAIDs themselves were not tested,
but contact allergy to them is highly likely (4).

Cross-reactions, pseudo-cross-reactions and co-reactions

Cross-allergy to diclofenac in patients with aceclofenac allergy (4,5). Two patients with photocontact allergy to
diclofenac had photocross-reactions to aceclofenac (3).

LITERATURE

1   The data in the section 'General' may have been obtained from literature discussed in this chapter, but mostly also or exclusively from one or more of the following online sources: ChemIDPlus Advanced, PubChem, DrugBank, RxList, Drug Central, Drugs.com, and Wikipedia

2   De Groot AC. Monographs in contact allergy, volume 3: Topical Drugs. Boca Raton, Fl, USA: CRC Press Taylor and Francis Group, 2021 (ISBN 978-0-367-23693-9)

3   Fernández-Jorge B, Goday-Buján JJ, Murga M, Molina FP, Pérez-Varela L, Fonseca E. Photoallergic contact dermatitis due to diclofenac with cross-reaction to aceclofenac: two case reports. Contact Dermatitis 2009;61:236-237

4   Linares T, Marcos C, Gavilan MJ, Arenas L. Fixed drug eruption due to aceclofenac. Contact Dermatitis 2007;56:291-292

5   Pitarch Bort G, de la Cuadra Oyanguren J, Torrijos Aguilar A, García-Melgares Linares ML. Allergic contact dermatitis due to aceclofenac. Contact Dermatitis 2006;55:365-366

Chapter 3.3   ACEMETACIN

IDENTIFICATION

Description/definition : Acemetacin is the carboxymethyl ester of indomethacin that conforms to the structural formula shown below
Pharmacological classes : Anti-inflammatory agents, non-steroidal
IUPAC name : 2-[2-[1-(4-Chlorobenzoyl)-5-methoxy-2-methylindol-3-yl]acetyl]oxyacetic acid
Other names : Indomethacin carboxymethyl ester
CAS registry number : 53164-05-9
EC number : 258-403-4
Merck Index monograph : 1298
Patch testing : 0.1%, 1%, 10% and 30% pet. (2)
Molecular formula : $C_{21}H_{18}ClNO_6$

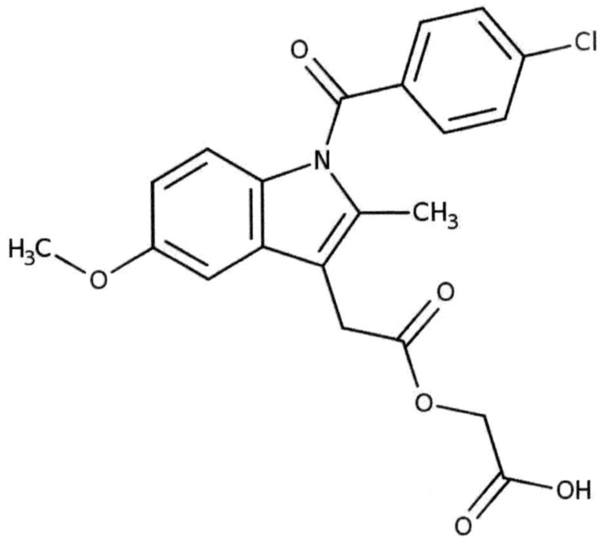

GENERAL

Acemetacin is a carboxymethyl ester of indomethacin. It is a potent non-steroidal anti-inflammatory drug, used in the treatment of rheumatoid arthritis, osteoarthritis, and low back pain, as well as for postoperative pain and inflammation. Acemetacin's activity is due to both acemetacin and its major metabolite, indomethacin (1).

CONTACT ALLERGY FROM ACCIDENTAL CONTACT

A 38-year-old woman had a 1-month history of recurrent pruritic erythematous patches on the eyelids, nasolabial folds, and chin, and papules on an erythematous base in the neck, frontal upper trunk, and flexor surface of the arms. The patient was a pharmaceutical employee who recently started working in the production process of acemetacin capsules. Patch tests with the contents of an acemetacin capsule in petrolatum was strongly positive (+++) on D3. Later, patch tests were positive to pure acemetacin 0.1%, 1%, 10% and 30% pet. and negative to the other ingredients of the capsule. Ten controls were negative. The patient was diagnosed with occupational airborne allergic contact dermatitis to acemetacin (2).

Cross-reactions, pseudo-cross-reactions and co-reactions

Not to indomethacin, of which acemetacin is the carboxymethyl ester (2).

LITERATURE

1   The data in the section 'General' may have been obtained from literature discussed in this chapter, but mostly also or exclusively from one or more of the following online sources: ChemIDPlus Advanced, PubChem, DrugBank, RxList, Drug Central, Drugs.com, and Wikipedia
2   Machado Á, Ferreira S, Lobo I, Sanches M, Selores M. Airborne allergic contact dermatitis due to acemetacin. Contact Dermatitis 2020;82:133-134

Chapter 3.4 ACETAMINOPHEN

IDENTIFICATION

Description/definition : Acetaminophen is the derivative of acetanilide that conforms to the structural formula shown below
Pharmacological classes : Analgesics, non-narcotic; antipyretics
IUPAC name : *N*-(4-Hydroxyphenyl)acetamide
Other names : Paracetamol; acetamide, *N*-(4-hydroxyphenyl)-
CAS registry number : 103-90-2
EC number : 203-157-5
Merck Index monograph : 1317
Patch testing : 10.0% pet. (Chemotechnique, SmartPracticeCanada)
Molecular formula : $C_8H_9NO_2$

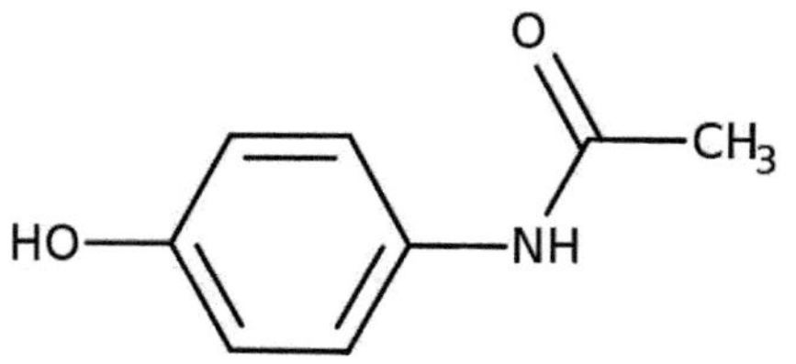

GENERAL

Acetaminophen, better known as paracetamol, is an acetanilide derivative with analgesic and antipyretic activities and weak anti-inflammatory properties. It is a widely used non-prescription drug for mild-to-moderate pain and fever. In fact, acetaminophen is the most commonly taken analgesic worldwide and is recommended as first-line therapy in pain conditions by the World Health Organization (1).

CONTACT ALLERGY FROM ACCIDENTAL CONTACT

A 44-year-old nurse presented with a 9-month-history of eczema on the hands and face, which had started 3 months after she was required to crush tablets daily during her work as a nurse in a clinic for disabled people. She was patch tested with all the drugs that she had contact with and reacted to 8 of these, including acetaminophen (paracetamol) 30% pet. (D2 +++, D3 +++). 126 Controls were negative (6). This was a case of occupational, partly airborne, allergic contact dermatitis to acetaminophen.

CUTANEOUS ADVERSE DRUG REACTIONS FROM SYSTEMIC ADMINISTRATION CAUSED BY TYPE IV (DELAYED-TYPE) HYPERSENSITIVITY (as demonstrated by positive patch tests)

Cutaneous adverse drug reactions from systemic administration of acetaminophen caused by type IV (delayed-type) hypersensitivity have included maculopapular eruption (17,18), generalized erythema (22), acute generalized exanthematous pustulosis (AGEP) (2,3,4,16), fixed drug eruption (7,8,9,10,11), drug reaction with eosinophilia and systemic symptoms (DRESS) (19), micropapular eruption (20), generalized pruriginous rash (20), and erythema multiforme (21).

Maculopapular eruption

A man, who had previously suffered from DRESS caused by carbamazepine and amoxicillin/clavulanic acid, five years later presented with a generalized maculopapular rash, partly confluent to plaques with neither fever nor lymph-node swelling. No lesions were found on the mucous membranes. Laboratory findings showed eosinophilia, no atypical lymphocytes and normal hepatic and renal function. One week earlier, he had taken one combination tablet containing acetaminophen, ascorbic acid and chlorpheniramine maleate for flu symptoms. Six weeks after complete resolution, a patch test was performed with the tablet 10% in water and in petrolatum, which was positive at D2. Later, patch tests were performed separately for acetaminophen and ascorbic acid, both prepared at 10% in water and in petrolatum. At D2, only the acetaminophen patch tests were positive; 2 controls were negative. Quite curiously, the 3rd component of the tablet was not patch tested (17). This was a case of multiple drug hypersensitivity syndrome.

An 18-month-old boy had suffered 3 episodes of a rash following administration of a acetaminophen-containing suppository. His parents described it as an itchy rash, starting at one particular site on the neck about 6 hours after use of the drug, spreading to the upper part of the chest, and later followed by desquamation of the skin during the following days, resolving after about one week. Physical examination showed facial edema and maculopapular erythema. Patch tests were positive to acetaminophen 10% pet. on the previously involved skin of the neck but negative on the back (where he had also had the skin eruption). An oral provocation test resulted in a widespread maculopapular exanthema, including on the particular site on the neck (18).

Erythroderma, widespread erythematous eruption, exfoliative dermatitis

A 62-year-old woman developed 4-5 hours after taking a tablet of acetaminophen a generalized exanthema. Three weeks later she took a tablet of the drug again, which resulted in generalized erythema with eyelid edema and associated with nausea, vomiting, and fever (40.4°C), regressing with exfoliation. Patch tests with acetaminophen in a dilution series from 20% down to 1% were ?+ on D3 and + on D3 after tape-stripping. An oral provocation test resulted in extensive confluent erythema and a lymphocyte transformation test was positive (22).

Acute generalized exanthematous pustulosis (AGEP)

In a multicenter investigation in France, of 45 patients patch tested for AGEP, 26 (58%) had positive patch tests to drugs, including 2 to paracetamol-dextropropoxyphene 30% pet.. The separate ingredients were not tested. It concerned a woman of 85 and a man of 31, but more clinical details were not provided (15).

A 48-year-old man was treated with acetaminophen and amoxicillin for otitis media. Within 3 days he developed an exanthema, which was progressive despite immediate cessation of both drugs and therapy with prednisolone. On day 7 he had a pustular nearly erythrodermic exanthema and oral erosions. Histology revealed subcorneal pustules filled with neutrophils and a mixed perivascular and interstitial infiltrate of neutrophils, eosinophils and lymphocytes in the papillary dermis. The skin lesions were accompanied by fever (>39°C), leukocytosis, neutrophilia (87%) and increased C-reactive protein level (250 mg/l). He had also elevated liver enzymes. Serology was negative for acute viral infections. The content of the pustules was sterile. A diagnosis of 'probable acute generalized exanthematous pustulosis (AGEP)' was made. Skin testing performed 4 months later revealed positive patch tests for both amoxicillin and acetaminophen (both 5% in petrolatum) with papulopustular infiltrates reminiscent of typical AGEP lesions (2).

An 83-year-old man was admitted with a disseminated erythematous rash, which had occurred 2 days after hip replacement. Within 48 hours the erythematous skin was covered by hundreds of small non-follicular pustules. A skin biopsy demonstrated subcorneal pustules; there was slight spongiosis, papillary edema and a perivascular infiltrate. Laboratory examination showed leukocytosis with neutrophilia, elevated erythrocyte sedimentation rate and C-reactive protein, abnormal liver tests and acute renal failure. For his hip operation, the patient had used many drugs including paracetamol-codeine, all of which were stopped. The eruption resolved within 5 days with desquamation and the abnormal laboratory findings also disappeared in 5 days. Patch tests with all drugs were positive only to acetaminophen medication and acetaminophen 5% and 20% in saline and petrolatum, producing a pustular eruption on an erythematous base. Ten controls were negative. The histology of a positive patch test showed subcorneal pustules. Involuntary rechallenge with intravenous propacetamol as single drug was responsible for recurrence of acute generalized exanthematous pustulosis one year later (3).

A pediatric case of AGEP in a 4-year-old girl with a positive pustular patch test to acetaminophen was reported from Taiwan in 2016 (4). In one patient, who had suffered AGEP from acetaminophen administration twice, patch tests to acetaminophen 1% and 10% pet. were negative on two occasions. However, on day 7 of the first test session and day 6 of the second, a symmetric versicular eruption appeared on the trunk, arms and legs (16).

Fixed drug eruption

Case series

In a university hospital in Tunisia, in the period 2004-2018, 41 patients with FDE were investigated. Seven cases were caused by acetaminophen. Six of these were patch tests, and in 4 (67%) there were positive reactions on postlesional skin. The bullous form of FDE was overrepresented in the acetaminophen group (7). In France, in the period 2005-2007, 59 cases of fixed drug eruptions were collected in 17 academic centers. There were 2 cases of FDE to acetaminophen with positive patch tests. Clinical details were not provided (8). In Finland, in the period 1989-2001, 826 patients with suspected cutaneous drug eruptions were patch tested and 89 had one or more positive reactions. Of these individuals, one with a fixed drug eruption reacted to acetaminophen (9).

Case reports

A 2-year-old girl was treated with a syrup containing acetaminophen for fever. Three hours after ingestion, the patient developed a bullous lesion, 0.6 cm in diameter, located on the third finger of the right hand. In a further few hours the lesion extended to cover almost all the dorsal surface of the finger reaching 2.4 cm. All laboratory investigations were normal. The bullous lesion was emptied and treated with a zinc oxide paste. No scarring or milia ensued. Patch tests with acetaminophen 5% pet. applied both on the back and on the site of past lesion (third finger of the right hand) were positive (++) on both unaffected and postlesional skin. This was a case of non-pigmenting bullous fixed drug eruption from acetaminophen (10).

A 46-year-old man presented with a bullous eruption of the lips, as well as a pruritic, discoid erythematous and hyperpigmented eruption of the left pretibial area together with several other areas of hyperpigmentation on both legs. The patient gave a history of multiple episodes occurring in different locations at a time when he took acetaminophen. Histopathology was consistent with fixed drug eruption. Patch testing of a quiescent lesion with 1% acetaminophen in ethyl alcohol revealed an eczematous reaction at D3, but was negative on normal skin. Two hyperpigmented lesions distal to the patch test site also flared, with bullae along one margin of each lesion (11).

There have been several other case reports of fixed and bullous fixed drug eruptions, but in these reports, patch tests were ether negative (5,11,12,13) or not performed (14) (only a few examples mentioned).

Drug reaction with eosinophilia and systemic symptoms (DRESS)

Three years after a 51-year-old man had suffered an episode of drug reaction with eosinophilia and systemic symptoms (DRESS, termed anticonvulsant hypersensitivity syndrome in the report), he developed an acute episode of fever, generalized pruritic rash, leukocytosis, and eosinophilia two days after he had taken an acetaminophen (acetaminophen) tablet. Patch tests were positive to acetaminophen and propacetamol (an ester prodrug of acetaminophen) 5% water. As it was not mentioned whether the patient also had systemic symptoms such as liver, kidney or lung involvement, it is uncertain whether this was also a case of DRESS, but in any case, probably a *forme fruste* (incompletely developed form) of it (19).

Other drug reactions

Three patients with hypersensitivity reactions to acetaminophen were reported from Spain in 1996. Patient 1: an 11-year-old girl developed a generalized micropapular eruption (one episode with edema of the face) after taking acetaminophen. Patient 2: at the age of 2 years, a 7-year-old girl developed generalized pruriginous exanthem several hours after administration of a suppository containing acetaminophen. Five years later, she experienced a similar reaction 12 hours after oral administration of this drug. Patient 3: a 4-year-old girl suffered two episodes of generalized micropapular eruption and facial edema while she was taking amoxicillin and acetaminophen for febrile pharyngitis. Later, she tolerated amoxicillin. In all three patients, patch tests were strongly positive to acetaminophen syrup, suppository and pure powder, all 30% pet. Six controls were negative (20).

A 19-year-old woman had previously been hospitalized for erythema multiforme after taking acetaminophen. She now presented with a history of recurrent rashes and vesicles for 2 years. Physical examination showed several erythematous macules 0.5–1 cm in length and patches with target-like bulla in the centre, mainly on her palms and arms. She had no mucosal lesions or other symptoms. The patient had taken acetaminophen and various other medications one day earlier. Patch tests with acetaminophen 0.1% pet. were negative, but one day after its application, erythema multiforme lesions started to develop on the palm-side of the patient's fingers and then spreading centripetally. The authors suggested that a metabolite of acetaminophen may have been the culprit (21). Alternatively, it may be argued, that the reaction was false-negative due to the very low concentration of acetaminophen used for patch testing (0.1% pet., advised test concentration: 10% pet.).

Cross-reactions, pseudo-cross-reactions and co-reactions

Acetaminophen (paracetamol) may cross-react to or from propacetamol, which is an ester prodrug of paracetamol (Chapter 3.413 Propacetamol).

LITERATURE

1   The data in the section 'General' may have been obtained from literature discussed in this chapter, but mostly also or exclusively from one or more of the following online sources: ChemIDPlus Advanced, PubChem, DrugBank, RxList, Drug Central, Drugs.com, and Wikipedia

2   Treudler R, Grunewald S, Gebhardt C, Simon J-C. Prolonged course of acute generalized exanthematous pustulosis with liver involvement due to sensitization to amoxicillin and paracetamol. Acta Derm Venereol 2009;89:314-315

3   Léger F, Machet L, Jan V, Machet C, Lorette G, Vaillant L. Acute generalized exanthematous pustulosis associated with paracetamol. Acta Derm Venereol 1998;78:222-223

4   Chen Y-C, Fang L-C, Wang J-Y. Paracetamol-induced acute generalized exanthematous pustulosis in a 4-year-old girl. Dermatologica Sinica 2016;34:49-51 (Available on-line)

5   Rojas-Pérez-Ezquerra P, Sánchez-Morillas L, Gómez-Traseira C, Gonzalez-Mendiola R, Alcorta Valle AR, Laguna-Martinez J. Selective hypersensitivity reactions to acetaminophen: a 13-case series. J Allergy Clin Immunol Pract 2014;2:343-345

6   Swinnen I, Ghys K, Kerre S, Constandt L, Goossens A. Occupational airborne contact dermatitis from benzodiazepines and other drugs. Contact Dermatitis 2014;70:227-232

7   Ben Fadhel N, Chaabane A, Ammar H, Ben Romdhane H, Soua Y, Chadli Z, et al. Clinical features, culprit drugs, and allergology workup in 41 cases of fixed drug eruption. Contact Dermatitis 2019;81:336-340

8 Brahimi N, Routier E, Raison-Peyron N, Tronquoy AF, Pouget-Jasson C, Amarger S, et al. A three-year-analysis of fixed drug eruptions in hospital settings in France. Eur J Dermatol 2010;20:461-464

9 Lammintausta K, Kortekangas-Savolainen O. The usefulness of skin tests to prove drug hypersensitivity. Br J Dermatol 2005;152:968-974

10 Nino M, Francia MG, Costa C, Scalvenzi M. Bullous fixed drug eruption induced by paracetamol: Report of a pediatric case. Case Rep Dermatol 2009;1:56-59

11 Guin JD, Haynie LS, Jackson D, Baker GF. Wandering fixed drug eruption: a mucocutaneous reaction to acetaminophen. J Am Acad Dermatol 1987;17:399-402 ook een negatieve plkaporoef (en een positieve)

12 Agarwala MK, Mukhopadhyay S, Sekhar MR, Peter CD. Bullous fixed drug eruption probably induced by paracetamol. Indian J Dermatol 2016;61:121

13 Gómez-Traseira C, Rojas-Pérez-Ezquerra P, Sánchez-Morillas L, González-Mendiola R, Rubio-Pérez M, Moral-Morales A, Juliolaguna-Martinez J. Paracetamol-induced fixed drug eruption at an unusual site. Recent Pat Inflamm Allergy Drug Discov 2013;7:268-270

14 Ayala F, Nino M, Ayala F, Balato N. Bullous fixed drug eruption induced by paracetamol: report of a case. Dermatitis 2006;17:160

15 Barbaud A, Collet E, Milpied B, Assier H, Staumont D, Avenel-Audran M, et al. A multicentre study to determine the value and safety of drug patch tests for the three main classes of severe cutaneous adverse drug reactions. Br J Dermatol 2013;168:555-562

16 Mashiah J, Brenner S. A systemic reaction to patch testing for the evaluation of acute generalized exanthematous pustulosis. Arch Dermatol 2003;139:1181-1183

17 Chadli Z, Ben Fredj N, Youssef M, Chaabane A, Boughattas NA, Zili JE, Aouam K. The rest of the story of the patient described in the letter to the editors: 'Hypersensitivity to amoxicillin after... (DRESS) to carbamazepine...: a possible co-sensitization'. Br J Clin Pharmacol 2016;81:784-785

18 Gilissen L, Mertens S, Goossens A, Bullens D, Morren MA. Delayed-type drug hypersensitivity caused by paracetamol in a 2-year-old boy, confirmed by a positive patch test reaction and oral provocation. Contact Dermatitis 2018;78:362-363

19 Gaig P, García-Ortega P, Baltasar M, Bartra J. Drug neosensitization during anticonvulsant hypersensitivity syndrome. J Investig Allergol Clin Immunol 2006;16:321-326

20 Ibanez MD, Alonso E, Munoz MC, Martinez E, Laso MT. Delayed hypersensitivity reaction to paracetamol (acetaminophen). Allergy 1996;51:121-123

21 Oh SW, Lew W. Erythema multiforme induced by acetaminophen: a recurrence at distant sites following patch testing. Contact Dermatitis 2005;53:56-57

22 Irion R, Gall H, Werfel T, Peter R U. Delayed-type hypersensitivity rash from paracetamol. Contact Dermatitis 2000;43:60-61

Chapter 3.5   ACETAZOLAMIDE

IDENTIFICATION

Description/definition : Acetazolamide is the thiadiazole sulfonamide that conforms to the structural formula shown below
Pharmacological classes : Carbonic anhydrase inhibitors; anticonvulsants; diuretics
IUPAC name : *N*-(5-Sulfamoyl-1,3,4-thiadiazol-2-yl)acetamide
Other names : 2-Acetylamino-1,3,4-thiadiazole-5-sulfonamide
CAS registry number : 59-66-5
EC number : 200-440-5
Merck Index monograph : 1322
Patch testing : Tablet 30% pet. (2,3); if the pure chemical is not available, prepare the test material from intravenous powder, the content of capsules or – when also not available – from powdered tablets to achieve a final concentration of the active drug of 10% pet.
Molecular formula : $C_4H_6N_4O_3S_2$

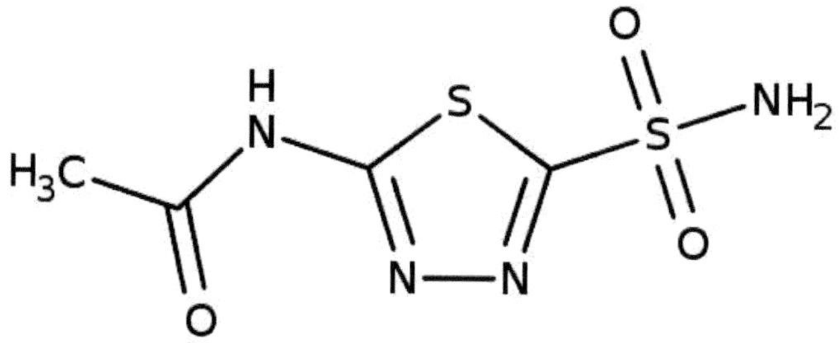

GENERAL

Acetazolamide is a sulfonamide derivative and non-competitive inhibitor of carbonic anhydrase with diuretic, antiglaucoma, and anticonvulsant properties. Acetazolamide is Indicated for the treatment of hypercapnia due to chronic obstructive pulmonary disease, idiopathic intracranial hypertension, prevention or treatment of postoperative intraocular pressure after cataract surgery, absence seizures and prophylaxis of acute mountain sickness (1,2). In pharmaceutical products, acetazolamide is employed as acetazolamide sodium (CAS number 1424-27-7, EC number not available, molecular formula $C_4H_5N_4NaO_3S_2$) (1). The drug is used both in topical and in systemic pharmaceutical applications. In topical preparations, acetazolamide has caused contact allergy/allergic contact dermatitis, which has been fully reviewed in Volume 3 of the *Monographs in contact allergy* series (4).

CUTANEOUS ADVERSE DRUG REACTIONS FROM SYSTEMIC ADMINISTRATION CAUSED BY TYPE IV (DELAYED-TYPE) HYPERSENSITIVITY (as demonstrated by positive patch tests)

Cutaneous adverse drug reactions from systemic administration of acetazolamide caused by type IV (delayed-type) hypersensitivity have included maculopapular exanthema (2,3) and acute generalized exanthematous pustulosis (AGEP (3).

Maculopapular eruption

Two days after receiving oral acetazolamide (2 tablets of 250 mg), a 76-year-old patient developed a pruriginous maculopapular rash of the large folds, initially localized in submammary, inguinal, and axillary areas; the skin lesions evolved into a fine "collar shape" desquamation. A skin biopsy revealed vacuolar changes, necrotic keratinocytes in the epidermis, and a bandlike cell infiltrate of lymphocytes in the upper dermis compatible with a lichenoid pattern. After two months, the patient developed hyperpigmented skin patches localized on previously affected areas of the large skin folds. Six months later, patch tests were positive to commercial acetazolamide 30% in petrolatum and in water; they later became hyperpigmented with the same histology as the first biopsy of the clinical lesions (2).

Case series with various or unknown types of drug reactions

In the period 2001-2011, seven patients with maculopapular exanthema (MPA) and three with acute generalized exanthematous pustulosis (AGEP) were investigated in four French dermatology and allergy departments. Most patients had received acetazolamide for cataract surgery as a prevention of peri- and postoperative ocular hypertension. In all cases except one, the time lag of appearance of the drug reaction was short, ranging from 6 to 48 hours. Patch tests with commercial acetazolamide 30% pet. were performed in all but one patient with AGEP 8 weeks to 3 years after the drug eruption. Eight of nine patients had positive patch tests at D2 and D4. One patient with MPA (who later developed AGEP from challenge with the drug) had negative patch tests, but showed positive prick and intradermal tests at D2 and D4. Patch tests with other sulfonamides, performed in 4 patients, were negative. Twelve controls were negative to acetazolamide 30% pet. (3).

LITERATURE

1   The data in the section 'General' may have been obtained from literature discussed in this chapter, but mostly also or exclusively from one or more of the following online sources: ChemIDPlus Advanced, PubChem, DrugBank, RxList, Drug Central, Drugs.com, and Wikipedia
2   Dequidt L, Milpied B, Chauvel A, Seneschal J, Taieb A, Darrigade AS. A case of lichenoid and pigmented drug eruption to acetazolamide confirmed by a lichenoid patch test. J Allergy Clin Immunol Pract 2018;6:283-285
3   Jachiet M, Bellon N, Assier H, Amsler E, Gaouar H, Pecquet C, et al. Cutaneous adverse drug reaction to oral acetazolamide and skin tests. Dermatology 2013;226:347-352
4   De Groot AC. Monographs in contact allergy, volume 3: Topical Drugs. Boca Raton, Fl, USA: CRC Press Taylor and Francis Group, 2021 (ISBN 978-0-367-23693-9)

Chapter 3.6 ACETOMENAPHTHONE

IDENTIFICATION

Description/definition : Acetomenaphthone is the naphthalene that conforms to the structural formula shown below

Pharmacological classes : Vitamins; anticoagulants

IUPAC name : (4-Acetyloxy-3-methylnaphthalen-1-yl) acetate

Other names : Menadiol diacetate; vitamin K_4; 2-methyl-1,4-naphthohydroquinone diacetate

CAS registry number : 573-20-6

EC number : 209-352-1

Merck Index monograph : 7168 (Menadiol)

Patch testing : 0.1% and 0.01% olive oil (2); 1% is irritant

Molecular formula : $C_{15}H_{14}O_4$

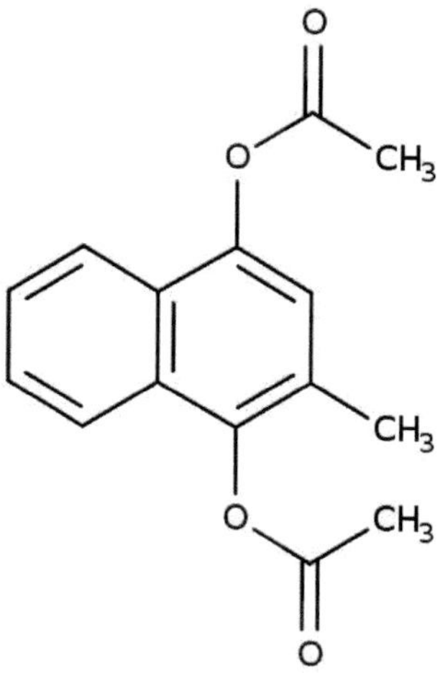

GENERAL

Acetomenaphthone is a vitamin K analog that is used to treat and prevent hypoprothrombinemia caused by vitamin K deficiency, which can result from long term intake of antimicrobials, malabsorption, obstructive jaundice, hepatitis, or liver cirrhosis (1).

CONTACT ALLERGY FROM ACCIDENTAL CONTACT

A 52-year-old woman, working in a pharmaceutical factory, developed acute dermatitis on the dorsal side of the fingers, followed by dermatitis of the forearms and erythema of the face. During her work, she would come in contact with many medicaments including vitamin K. There was an obvious causal relationship with her work. Patch tests were positive to 'vitamin K' (not further specified). Ten years later, the patient had a new work duty where she had again contact with vitamin K. Already on the first day, erythematous patches appeared on the face, followed by the neck and the hands, which soon evolved into acute oozing eczema. Patch tests were positive to vitamin K_4 commercial tablet and ampoule. In a second session, patch tests were positive to menadiol diacetate (acetomenaphthone, vitamin K_4) 0.1% in olive oil (negative to 0.01%), and to menadiol sodium diphosphate 0.1% water. Seven controls were negative. There was a cross-reaction to menadione (vitamin K_3) (2). Of course, it cannot be excluded that the patient previously already had contact with menadione.

Cross-reactions, pseudo-cross-reactions and co-reactions

Cross-reaction to menadione (vitamin K_3) in a patient sensitized to acetomenaphthone (vitamin K_4) (2).

LITERATURE

1   The data in the section 'General' may have been obtained from literature discussed in this chapter, but mostly also or exclusively from one or more of the following online sources: ChemIDPlus Advanced, PubChem, DrugBank, RxList, Drug Central, Drugs.com, and Wikipedia

2   Jirásek L, Schwank R. Berufskontaktekzem durch Vitamin K. Hautarzt 1965;16:351-353 (Article in German)

Chapter 3.7   ACETYLCYSTEINE

IDENTIFICATION

Description/definition   : Acetylcysteine is the synthetic *N*-acetyl derivative of the endogenous amino acid
L-cysteine; it conforms to the structural formula shown below
Pharmacological classes : Antiviral agents; expectorants; free radical scavengers
IUPAC name   : (2*R*)-2-Acetamido-3-sulfanylpropanoic acid
Other names   : L-α-Acetamido-β-mercaptopropionic acid; *N*-acetyl-L-cysteine; mercapturic acid
CAS registry number   : 616-91-1
EC number   : 210-498-3
Merck Index monograph : 1353
Patch testing   : 10% pet. (3)
Molecular formula   : $C_5H_9NO_3S$

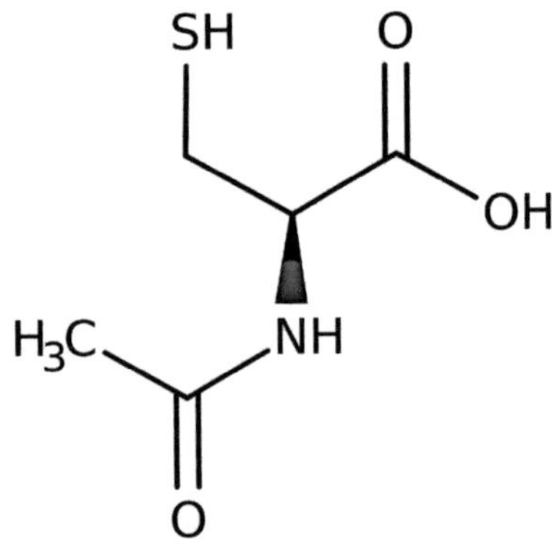

GENERAL

Acetylcysteine is a synthetic *N*-acetyl derivative of the endogenous amino acid L-cysteine, a precursor of the antioxidant enzyme glutathione. Acetylcysteine is used mainly as a mucolytic drug to reduce the viscosity of mucous secretions, in the management of paracetamol (acetaminophen) overdose, and as a protective agent for renal function in contrast medium-induced nephropathy. It has also been shown to have antiviral effects in patients with HIV due to inhibition of viral stimulation by reactive oxygen intermediates. Acetylcysteine is essentially a prodrug that is converted to cysteine in the intestine and absorbed there into the blood stream. In combination with hypromellose eyedrops, is commonly used to alleviate the chronic soreness associated with dry eyes (1,2).

In topical preparations, acetylcysteine has caused rare cases of contact allergy/allergic contact dermatitis, which subject has been reviewed in Volume 3 of the *Monographs in contact allergy* series (2).

CONTACT ALLERGY FROM ACCIDENTAL CONTACT

A 32-year-old nurse presented with itchy eczematous skin lesions on exposed areas of the hands, arms and face that had been present for several weeks. The symptoms had developed since she started working in a general ward, where she mixed *N*-acetylcysteine (NAC) with normal saline, and improved during the weekends. Patch tests were positive to NAC 10% pet., as were intradermal tests with NAC at 1 and 10 mg/ml in 0.9% saline, read at D2 (3). This was a case of occupational airborne allergic contact dermatitis.

LITERATURE

1   The data in the section 'General' may have been obtained from literature discussed in this chapter, but mostly also or exclusively from one or more of the following online sources: ChemIDPlus Advanced, PubChem, DrugBank, RxList, Drug Central, Drugs.com, and Wikipedia
2   De Groot AC. Monographs in contact allergy, volume 3: Topical Drugs. Boca Raton, Fl, USA: CRC Press Taylor and Francis Group, 2021 (ISBN 978-0-367-23693-9)
3   Kim JH, Kim SH, Yoon MG, Jung HM, Park HS, Shin YS. A case of occupational contact dermatitis caused by *N*-acetylcysteine. Contact Dermatitis 2016;74:373-374

Chapter 3.8   ACETYLSALICYLIC ACID

IDENTIFICATION

Description/definition   : Acetylsalicylic acid is the analgesic that conforms to the structural formula shown below
Pharmacological classes : Anti-inflammatory agents, non-steroidal; antipyretics; cyclooxygenase inhibitors; platelet aggregation inhibitors; fibrinolytic agents
IUPAC name   : 2-Acetyloxybenzoic acid
Other names   : Aspirin ®
CAS registry number   : 50-78-2
EC number   : 200-064-1
Merck Index monograph   : 2111
Patch testing   : 10.0% pet. (Chemotechnique)
Molecular formula   : $C_9H_8O_4$

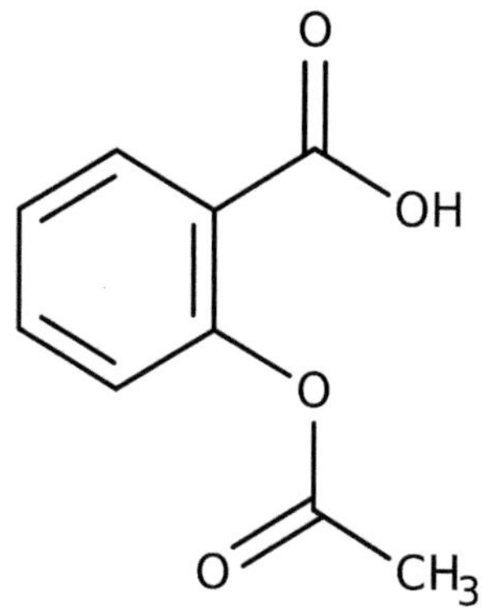

GENERAL

Acetylsalicylic acid is the prototypical nonsteroidal anti-inflammatory drug which has analgesic, antipyretic, anti-inflammatory and antirheumatic properties. It is used to relieve pain, fever, and inflammation associated with many conditions, including the flu, the common cold, neck and back pain, dysmenorrhea, headache, tooth pain, sprains, fractures, myositis, neuralgia, synovitis, arthritis, bursitis, burns, and various injuries. As acetylsalicylic acid inhibits platelet aggregation it is also indicated in the prevention of arterial and venous thrombosis (1).

CUTANEOUS ADVERSE DRUG REACTIONS FROM SYSTEMIC ADMINISTRATION CAUSED BY TYPE IV (DELAYED-TYPE) HYPERSENSITIVITY (as demonstrated by positive patch tests)

Cutaneous adverse drug reactions from systemic administration of acetylsalicylic acid caused by type IV (delayed-type) hypersensitivity have included drug reaction with eosinophilia and systemic symptoms (DRESS) (3) and photosensitivity (2,4).

Drug reaction with eosinophilia and systemic symptoms (DRESS)

A 2-year old Japanese boy was treated with acetylsalicylic acid for 4 weeks for Kawasaki disease, with good result. Soon thereafter, he developed a generalized maculopapular eruption, high fever, leukocytosis with eosinophilia, and an increased number of atypical lymphocytes, severe liver dysfunction, lymphadenopathy, and prominent increases in antihuman herpesvirus-6 immunoglobulin G titer. Hypersensitivity to acetylsalicylic acid was confirmed by positive patch tests to commercial Aspirin 10% and 20% pet. and by a positive lymphocyte stimulation test. The patient was diagnosed with drug reaction with eosinophilia and systemic symptoms (DRESS) caused by acetylsalicylic acid (3).

Photosensitivity

In Italy, before 1993, the members of the GIRDCA Multicentre Study Group diagnosed 102 patients (49 men, 53 women), aged 16 to 66 years (mean 37 years), with (photo)dermatitis induced by systemic or topical NSAIDs. Acetylsalicylic acid caused two contact allergic and zero photocontact allergic reactions (2). In the period 2003-2007, in Portugal, 30 individuals, a subgroup of 83 patients with suspected photoaggravated facial dermatitis or systemic photosensitivity, were photopatch tested with acetylsalicylic acid 1% pet. and there was one positive reaction, which was attributed to systemic photosensitivity (4).

LITERATURE

1   The data in the section 'General' may have been obtained from literature discussed in this chapter, but mostly also or exclusively from one or more of the following online sources: ChemIDPlus Advanced, PubChem, DrugBank, RxList, Drug Central, Drugs.com, and Wikipedia

2   Pigatto PD, Mozzanica N, Bigardi AS, Legori A, Valsecchi R, Cusano F, et al. Topical NSAID allergic contact dermatitis. Italian experience. Contact Dermatitis 1993;29:39-41

3   Kawakami T, Fujita A, Takeuchi S, Muto S, Soma Y. Drug-induced hypersensitivity syndrome: drug reaction with eosinophilia and systemic symptoms (DRESS) syndrome induced by aspirin treatment of Kawasaki disease. J Am Acad Dermatol 2009;60:146-149

4   Cardoso J, Canelas MM, Gonçalo M, Figueiredo A. Photopatch testing with an extended series of photoallergens: a 5-year study. Contact Dermatitis 2009;60:325-329

Chapter 3.9 ACEXAMIC ACID

IDENTIFICATION

Description/definition : Acexamic acid is the medium chain fatty acid that conforms to the structural formula shown below
Pharmacological classes : Anti-inflammatory agents, non-steroidal
IUPAC name : 6-Acetamidohexanoic acid
Other names : 6-Acetylaminocaproic acid; ε-acetamidocaproic acid; 6-(acetylamino)hexanoic acid
CAS registry number : 57-08-9
EC number : 200-310-8
Merck Index monograph : 1316
Patch testing : Acexamic acid 5% and 10% pet. (4); zinc acexamate 5% water (3)
Molecular formula : $C_8H_{15}NO_3$

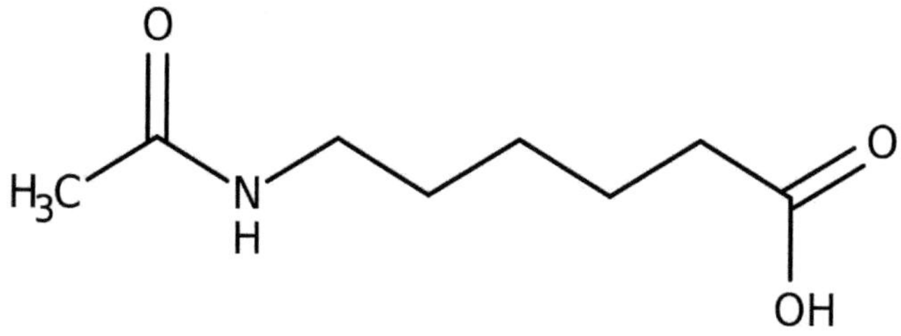

GENERAL

Acexamic acid is a drug used as a cicatrization helper. Its salt zinc acexamate has been employed in the treatment of gastric and duodenal ulcers and in the prevention of gastric ulcer induced by nonsteroidal anti-inflammatory drugs. In pharmaceutical products, acexamic acid may be used as sodium acexamate (CAS number 7234-48-2, EC number 230-635-0, molecular formula $C_8H_{14}NNaO_3$) or zinc acexamate (CAS number 70020-71-2, EC number not available, molecular formula $C_{16}H_{28}N_2O_6Zn$) (1). Acexamic acid is used both in topical and in systemic pharmaceutical applications. In topical preparations, it has caused contact allergy/allergic contact dermatitis, which has been fully reviewed in Volume 3 of the *Monographs in contact allergy* series (2).

CUTANEOUS ADVERSE DRUG REACTIONS FROM SYSTEMIC ADMINISTRATION CAUSED BY TYPE IV (DELAYED-TYPE) HYPERSENSITIVITY (as demonstrated by positive patch tests)

Cutaneous adverse drug reactions from systemic administration of acexamic acid caused by type IV (delayed-type) hypersensitivity have included maculopapular eruption (3).

Maculopapular eruption

Twelve hours after taking oral zinc acexamate (probably capsules, possibly oral suspension), a 60-year-old man had developed a generalized rash, which resolved with slight desquamation after several days of oral corticosteroids and antihistamines. Patch tests were positive to commercial zinc acexamate pure, zinc acexamate 5% water, peppermint oil (present in the commercial zinc acexamate) and negative to the other excipients. Ten controls were negative. Next, a double-blind placebo-controlled oral challenge test was carried out with pure zinc acexamate. After 12 hours, the patient developed a generalized eruption of pruriginous maculopapular lesions with urticarial characteristics, predominantly on the trunk. The positive patch test to zinc acexamate was reactivated by the oral challenge test. Biopsy of the flare-up patch test was compatible with eczema (3).

LITERATURE

1 The data in the section 'General' may have been obtained from literature discussed in this chapter, but mostly also or exclusively from one or more of the following online sources: ChemIDPlus Advanced, PubChem, DrugBank, RxList, Drug Central, Drugs.com, and Wikipedia
2 De Groot AC. Monographs in contact allergy, volume 3: Topical Drugs. Boca Raton, Fl, USA: CRC Press Taylor and Francis Group, 2021 (ISBN 978-0-367-23693-9)
3 Galindo PA, Garrido JA, Gómez E, Borja J, Feo F, Encinas C, et al. Zinc acexamate allergy. Contact Dermatitis 1998;38:301-302
4 De Groot AC. Patch testing, 4th edition. Wapserveen, The Netherlands: acdegroot publishing, 2018 (ISBN 9789081323345)

Chapter 3.10   ACTARIT

IDENTIFICATION
Description/definition   : Actarit is the anilide and member of acetamides that conforms to the structural formula shown below
Pharmacological classes   : Antirheumatic agents
IUPAC name   : 2-(4-Acetamidophenyl)acetic acid
CAS registry number   : 18699-02-0
EC number   : 242-511-3
Merck Index monograph   : 1394
Patch testing   : 1% and 10% pet. (1)
Molecular formula   : $C_{10}H_{11}NO_3$

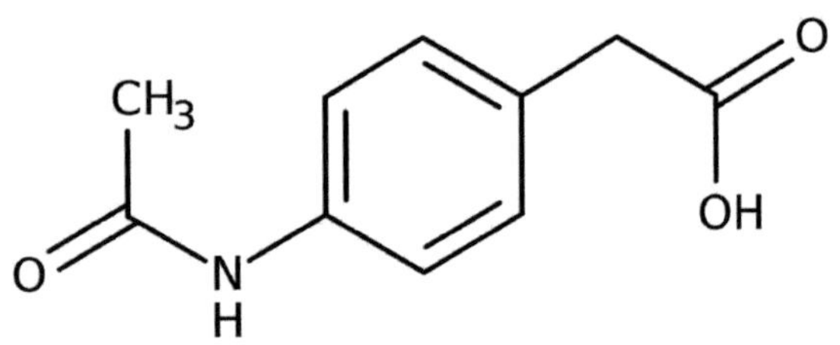

GENERAL
Actarit is a disease-modifying antirheumatic drug (DMARD) developed in Japan for use in rheumatoid arthritis. It is also available in China (drugs.com).

CUTANEOUS ADVERSE DRUG REACTIONS FROM SYSTEMIC ADMINISTRATION CAUSED BY TYPE IV (DELAYED-TYPE) HYPERSENSITIVITY (as demonstrated by positive patch tests)
Cutaneous adverse drug reactions from systemic administration of actarit caused by type IV (delayed-type) hypersensitivity have included photosensitivity (1).

Photosensitivity
A 52-year-old woman was seen with erythema and lichenification on sun-exposed areas that had started 2 months earlier. She had been taking actarit and doxycycline HCl orally because of rheumatoid arthritis and a chronic infectious ulcer of the right knee for 3 months. Patch and photopatch tests were performed with actarit and doxycycline (10%, 1%, and 0.1% pet.). Only the photopatch test with actarit was positive at 1% and 10%, but negative at 0.1% pet. Five controls were negative (1). The patient was diagnosed with photoallergy to actarit.

LITERATURE
1   Kawada A, Hiruma M, Miura Y, Noguchi H, Akiyama M, Ishibashi A. Photosensitivity due to actarit. Contact Dermatitis 1997;36:175-176

Chapter 3.11   ACYCLOVIR

IDENTIFICATION

Description/definition : Acyclovir is the synthetic analog of the purine nucleoside, guanosine, that conforms to the structural formula shown below
Pharmacological classes : Antiviral agents
IUPAC name : 2-Amino-9-(2-hydroxyethoxymethyl)-3H-purin-6-one
Other names : Acycloguanosine
CAS registry number : 59277-89-3
EC number : 261-685-1
Merck Index monograph : 1404
Patch testing : 10.0% pet. (Chemotechnique); scratch-patch tests may be performed in case of a ?+ reaction or a negative reaction to acyclovir with highly suspect history for allergy (3)
Molecular formula : $C_8H_{11}N_5O_3$

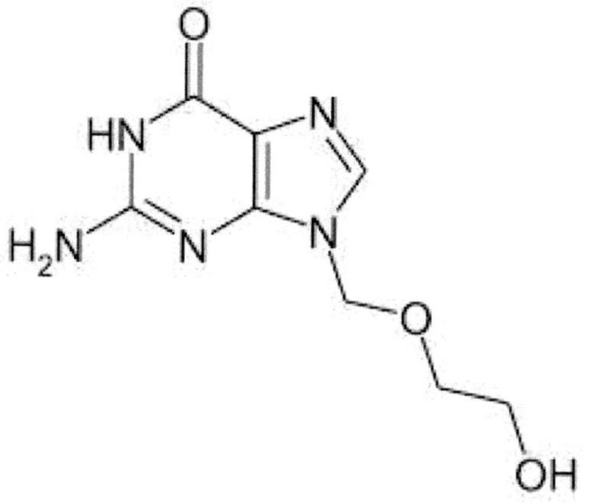

GENERAL

Acyclovir is a synthetic analog of the purine nucleoside, guanosine, with potent antiviral activity against *Herpes simplex* viruses type 1 and 2, *Varicella zoster* virus and other viruses of the herpesvirus family. After conversion *in vivo* to the active metabolite acyclovir triphosphate by viral thymidine kinase, acyclovir competitively inhibits viral DNA-polymerase by incorporating into the growing viral DNA chain and terminating further polymerization. Acyclovir is used for the treatment of herpes simplex virus infections, varicella and herpes zoster. In topical pharmaceutical products, acyclovir base is used; in powder for injection fluids, acyclovir sodium is employed (CAS number 69657-51-8, EC number 614-996-5, molecular formula $C_8H_{10}N_5NaO_3$) (1). In topical preparations, acyclovir has caused contact allergy/allergic contact dermatitis and possibly photocontact allergy, which has been fully reviewed in Volume 3 of the *Monographs in contact allergy* series (9).

CUTANEOUS ADVERSE DRUG REACTIONS FROM SYSTEMIC ADMINISTRATION CAUSED BY TYPE IV (DELAYED-TYPE) HYPERSENSITIVITY (as demonstrated by positive patch tests)

Cutaneous adverse drug reactions from systemic administration of acyclovir caused by type IV (delayed-type) hypersensitivity have included maculopapular eruption (11), acute generalized exanthematous pustulosis (AGEP) (2,10), systemic allergic dermatitis (4,5,6,7), fixed drug eruption (8), drug eruption with eosinophilia and systemic symptoms (DRESS) (12,13), and unspecified eruptions (14).

Maculopapular eruption

In a study from Nancy, France, 27 patients with maculopapular eruptions were patch tested and one had a positive reaction to acyclovir, which reproduced the eruption (11).

Acute generalized exanthematous pustulosis (AGEP)

Three weeks after starting acyclovir (intravenously during 2 weeks and orally for 7 days) for herpetic retinitis, a 53-year-old man developed a severe febrile eruption consisting of generalized erythema with numerous small pustules that initially appeared on major intertriginous areas. Laboratory evaluation disclosed leukocytosis (24,400/KL), neutrophilia, and an elevated C-reactive protein. A skin biopsy revealed multiple subcorneal spongiform pustules filled with neutrophils and a perivascular, lymphomononuclear inflammatory infiltrate in the upper dermis. A diagnosis of acute generalized exanthematous pustulosis (AGEP) was established. Patch tests were performed with acyclovir 10% pet., commercial creams containing acyclovir at 2% or 5%, and pure acyclovir in different vehicles (water, petrolatum, dimethyl sulfoxide, and propylene glycol) and concentrations (2%-10%). Positive reactions were obtained only with three commercial topical formulations of acyclovir with infiltrated erythema and micropustules from day 2 onward. A punch biopsy revealed histopathologic changes similar to those found during the acute episode. Patch tests with the excipients of the creams were negative. A lymphocyte stimulation test with pure acyclovir was negative (2). That AGEP in this case was caused by acyclovir, although not proven, seems highly likely.

A 44-year-old woman with a history of atopic dermatitis and herpes labialis was referred for eczema herpeticum. Topical therapy for 5 days with acyclovir cream resulted in worsening and therefore, oral acyclovir 200 mg 5 times a day for 5 days was initiated. After one day's treatment, she developed erythema with small pustules symmetrically on the trunk and proximal aspects of her arms and legs. The skin symptoms were accompanied by leukocytosis and low-grade fever. A punch biopsy revealed the histological features of AGEP. Patch tests were positive to acyclovir 1%, 5% and 10% pet. (10).

Systemic allergic dermatitis (systemic contact dermatitis)

A 23-year-old woman developed a severe itchy erythematous vesicular dermatitis of the upper lip and right cheek while treating recurrent herpes simplex of the lips with acyclovir 5% cream. A few months after the resolution of the dermatitis, 6 hours after beginning oral acyclovir, the patient developed urticaria with itchy, erythematous, edematous lesions on the trunk and extremities. Patch tests were positive to commercial acyclovir cream (50% pet.), acyclovir ophthalmic ointment ('as is'), and acyclovir tablet, probably 100% and crushed. The patient also reacted to commercial valaciclovir (tablet 30% pet., water, alcohol) and to propylene glycol, which is present in the cream. She had a negative patch test to famciclovir, but when oral administration of famciclovir, with progressively increased dosages, was carried out, an itchy erythematous dermatitis occurred on the trunk 15 hr after administration of the last dose of famciclovir 500 mg (4). This was a case of allergic contact dermatitis and systemic dermatitis from acyclovir with cross-reactivity to valaciclovir and famciclovir.

Several other case reports of systemic allergic dermatitis from oral or parenteral administration of acyclovir after previous sensitization from topical medicaments containing acyclovir have been reported (5,6,7). A 19-year-old woman had a lip dermatitis after application of 5% acyclovir cream to herpes labialis. Clinical suspicion that the herpetic infection initially diagnosed had worsened, therapy was given with 2 intravenous infusions of acyclovir 250 mg at an interval of 8 hours. These were both followed, some 15 minutes later, by diffuse urticaria. Patch tests were positive to acyclovir cream and ophthalmic ointment, to acyclovir 1%, 3% and 5% pet. and to propylene glycol 10% water (5).

A 29-year-old woman presented with labial eczema after application of 5% acyclovir cream, then with edematous eczema of the upper eyelid following the application of acyclovir ophthalmic ointment. A few months later, 24 hours after the intake of acyclovir tablets, she developed a pruriginous maculopapular rash, with secondary eczematization on the arms, trunk, and inner thigh. Patch tests were positive to propylene glycol 5% pet. and to acyclovir 10% water. There was a cross-reaction to valaciclovir but not to famciclovir (6). The authors also presented a 28-year-old woman who was sensitized to topical acyclovir and developed a widespread eczema that had started 6 hours after taking one acyclovir tablet. Patch tests showed the same pattern of sensitization as in the first patient: ++ to propylene glycol, acyclovir and valaciclovir and negative to famciclovir. A provocation test with famciclovir, however, elicited a pruriginous rash 12 hours after its first intake (6).

A 44-year-old woman used acyclovir cream for 2 weeks on her first attack of genital herpes without any improvement. Oral valaciclovir (500 mg 2x daily) was started. After the first 2 tablets, an itchy, symmetrical exanthem appeared on the face, trunk and extremities. One month later, the patient developed a labial herpes infection. She used acyclovir cream and vesicobullous lesions with erythema appeared in the labial and perioral skin, with an exanthem on the upper trunk and extremities. Patch tests were positive to cetearyl alcohol (present in the cream), acyclovir and valaciclovir as is, 20%, 10% and 1% water and pet., and ganciclovir 'as is' and 20% pet. (no lower concentrations tested). There was no reaction to famciclovir, but when orally challenged with famciclovir, itchy erythematous patches appeared on the upper lip and upper trunk after 325 mg famciclovir given in 3 doses over 2 days (7).

Fixed drug eruption

A 56-year-old woman was treated with topical acyclovir and later with oral acyclovir for herpes labialis. Two hours after she had taken the 3rd tablet, she felt intense pruritus on her left ankle. One hour later, an erythematous vesicular lesion developed at this site, which healed leaving a hyperpigmented geographic macule. One month later, herpes virus infection on the patient's lips recurred. Topical treatment resulted in an exacerbation and oral treatment was followed by a more extensive fixed eruption at the left ankle. Patch tests were positive to acyclovir cream and acyclovir 5% pet. on lesional but not on normal skin (8).

Drug reaction with eosinophilia and systemic symptoms (DRESS)

In a multicenter investigation in France, of 72 patients patch tested for DRESS, 46 (64%) had positive patch tests to drugs, including 2 to acyclovir (12). One case of DRESS with positive patch tests to acyclovir was reported from Portugal in 2017; after nearly 4 years, a repeat patch test was again positive (13).

Other cutaneous adverse drug reactions

In Finland, in the period 1989-2001, 826 patients with suspected cutaneous drug eruptions were patch tested and 89 (11%) had one or more positive reactions. Of these individuals, 2 reacted to acyclovir. The nature of the eruptions caused by acyclovir was not specified (14).

Cross-reactions, pseudo-cross-reactions and co-reactions

Cross-reactions between acyclovir and valaciclovir have been documented (4,6,7). One patient sensitized to acyclovir cross-reacted to ganciclovir (7). In acyclovir-allergic individuals who do *not* cross-react when patch tested with famciclovir (i.e., no positive patch test to famciclovir), oral provocation tests may yet be positive (4,6,7). Valaciclovir is the L-valine ester of acyclovir and is almost completely metabolized to acyclovir after oral administration. The chemical structure common to acyclovir, valaciclovir, ganciclovir and famciclovir is the 2-aminopurine nucleus, and this is probably the allergenic determinant of the molecules (4).

LITERATURE

1   The data in the section 'General' may have been obtained from literature discussed in this chapter, but mostly also or exclusively from one or more of the following online sources: ChemIDPlus Advanced, PubChem, DrugBank, RxList, Drug Central, Drugs.com, and Wikipedia

2   Serra D, Ramos L, Brinca A, Gonçalo M. Acute generalized exanthematous pustulosis associated with acyclovir, confirmed by patch testing. Dermatitis 2012;23:99-100

3   Nino M, Balato N, Di Costanzo L, Gaudiello F. Scratch-patch test for the diagnosis of allergic contact dermatitis to aciclovir. Contact Dermatitis 2009;60:56-57

4   Vernassiere C, Barbaud A, Trechot PH, Weber-Muller F, Schmutz JL. Systemic acyclovir reaction subsequent to acyclovir contact allergy: which systemic antiviral drug should then be used? Contact Dermatitis 2003;49:155-157

5   Gola M, Francalanci S, Brusi C, Lombardi P, Sertoli A. Contact sensitization to acyclovir. Contact Dermatitis 1989;20:394-395

6   Bayrou O, Gaouar H, Leynadier F. Famciclovir as a possible alternative treatment in some cases of allergy to acyclovir. Contact Dermatitis 2000;42:42

7   Lammintausta K, Mäkelä L, Kalimo K. Rapid systemic valaciclovir reaction subsequent to aciclovir contact allergy. Contact Dermatitis 2001;45:181

8   Montoro J, Basomba A. Fixed drug eruption due to acyclovir. Contact Dermatitis 1997;36:225

9   De Groot AC. Monographs in contact allergy, volume 3: Topical Drugs. Boca Raton, Fl, USA: CRC Press Taylor and Francis Group, 2021 (ISBN 978-0-367-23693-9)

10 Kubin ME, Jackson P, Riekki R. Acute generalized exanthematous pustulosis secondary to acyclovir confirmed by positive patch testing. Acta Derm Venereol 2016;96:860-861

11 Barbaud A, Reichert-Penetrat S, Tréchot P, Jacquin-Petit MA, Ehlinger A, Noirez V, et al. The use of skin testing in the investigation of cutaneous adverse drug reactions. Br J Dermatol 1998;139:49-58

12 Barbaud A, Collet E, Milpied B, Assier H, Staumont D, Avenel-Audran M, et al. A multicentre study to determine the value and safety of drug patch tests for the three main classes of severe cutaneous adverse drug reactions. Br J Dermatol 2013;168:555-562

13 Pinho A, Marta A, Coutinho I, Gonçalo M. Long-term reproducibility of positive patch test reactions in patients with non-immediate cutaneous adverse drug reactions to antibiotics. Contact Dermatitis 2017;76:204-209

14 Lammintausta K, Kortekangas-Savolainen O. The usefulness of skin tests to prove drug hypersensitivity. Br J Dermatol 2005;152:968-974

Chapter 3.12   ADALIMUMAB

IDENTIFICATION

Description/definition   : Adalimumab is a monoclonal human IgG1 antibody to tumor necrosis factor alpha (TNF-α)
Pharmacological classes : Anti-inflammatory agents; antirheumatic agents; monoclonal antibodies
IUPAC name   : Not available
Other names   : Immunoglobulin G1, anti-(human tumor necrosis factor) (human monoclonal D2E7 heavy
                           chain), disulfide with human monoclonal D2E7 light chain, dimer; Humira ®
CAS registry number   : 331731-18-1
EC number   : Not available
Merck Index monograph   : 1406
Patch testing   : Commercial preparation (50 mg/ml) undiluted (3)
Molecular formula   : Unspecified

GENERAL

Adalimumab is a humanized monoclonal IgG1 antibody that binds specifically to tumor necrosis factor alpha (TNF-α) and blocks its interaction with endogenous TNF-receptors to modulate inflammation. It is produced by recombinant DNA-technology using a mammalian cell expression system. Adalimumab is administered subcutaneously for the treatment of chronic debilitating diseases mediated by TNF-α including rheumatoid arthritis, psoriatic arthritis, Crohn's disease, ulcerative colitis, juvenile idiopathic arthritis, ankylosing spondylitis, plaque psoriasis, non-infectious intermediate, posterior and pan-uveitis, hidradenitis suppurativa and (off-label) pyoderma gangrenosum (1).

CUTANEOUS ADVERSE DRUG REACTIONS FROM SYSTEMIC ADMINISTRATION CAUSED BY TYPE IV (DELAYED-TYPE) HYPERSENSITIVITY (as demonstrated by positive patch tests)

Cutaneous adverse drug reactions from systemic administration of adalimumab caused by type IV (delayed-type) hypersensitivity have included fixed drug eruption (2).

Fixed drug eruption

A 40-year-old woman had been treated for ankylosing spondylitis with subcutaneous adalimumab for 5 years with good results. The drug was administered every 2 weeks to alternating thighs without history of local injection-site reactions. However, for the past 3 months, she had developed a solitary rash over her left anterior shin around 4 hours after every dose of adalimumab. The lesions were well demarcated, beginning with intense itching and burning, followed by a fixed erythematous oval plaque about 10 cm in diameter. This would last for a few days, healing with hyperpigmentation and recurring after each dose. The lesions were progressively getting worse with a localized vesicular reaction in the most recent episode. Two months later, patch tests with the commercial adalimumab from the autoinjector at 10% water were negative on previously uninvolved skin but strongly positive at D4 at the postlesional skin with reactivation of the lesion. The patient was diagnosed with fixed drug eruption from type IV hypersensitivity to adalimumab (2).

LITERATURE

1   The data in the section 'General' may have been obtained from literature discussed in this chapter, but mostly also or exclusively from one or more of the following online sources: ChemIDPlus Advanced, PubChem, DrugBank, RxList, Drug Central, Drugs.com, and Wikipedia
2   Li PH, Watts TJ, Chung HY, Lau CS. Fixed drug eruption to biologics and role of lesional patch testing. J Allergy Clin Immunol Pract 2019;7:2398-2399
3   Brockow K, Garvey LH, Aberer W, Atanaskovic-Markovic M, Barbaud A, Bilo MB, et al.; ENDA/EAACI Drug Allergy Interest Group. Skin test concentrations for systemically administered drugs – an ENDA/EAACI Drug Allergy Interest Group position paper. Allergy 2013;68:702-712

Chapter 3.13   AFLOQUALONE

IDENTIFICATION

Description/definition : Afloqualone is the quinazolinone that conforms to the structural formula shown below
Pharmacological classes : Muscle relaxants, central; photosensitizing agents
IUPAC name : 6-Amino-2-(fluoromethyl)-3-(2-methylphenyl)quinazolin-4-one
CAS registry number : 56287-74-2
EC number : Not available
Merck Index monograph : 1445
Patch testing : 1% pet. (2)
Molecular formula : $C_{16}H_{14}FN_3O$

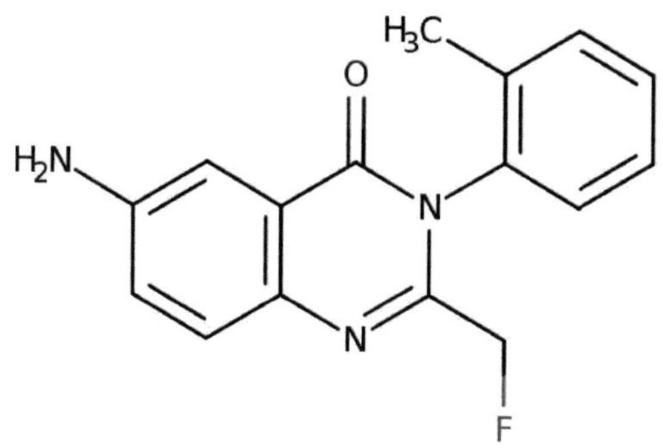

GENERAL

Afloqualone is a quinazolinone family GABAergic drug, which has sedative and centrally muscle-relaxant effects resulting from its agonist activity at the β subtype of the GABA-A receptor. It was never widely used due to photosensitivity (2) and skin irritation issues, although it has/had some popularity in Japan for conditions such as neck-shoulder-arm syndrome, lumbago, and spastic paralysis (1,2).

CUTANEOUS ADVERSE DRUG REACTIONS FROM SYSTEMIC ADMINISTRATION CAUSED BY TYPE IV (DELAYED-TYPE) HYPERSENSITIVITY (as demonstrated by positive patch tests)

Cutaneous adverse drug reactions from systemic administration of afloqualone caused by type IV (delayed-type) hypersensitivity have included photosensitivity (2).

Photosensitivity

A 71-year-old man had been treated with afloqualone and imipramine for cervical spondylosis for 2 months, when he noticed pruritic erythema on his face, neck, and the dorsa of his hands. Both drugs were ceased and with oral and topical corticosteroids, the cutaneous lesions almost disappeared. However, spotty pigmentations and depigmentations developed in sun-exposed areas shortly thereafter. Laboratory investigations were normal. Patch tests were negative to both drugs, but photopatch tests positive to afloqualone 1% pet. (negative to 0.1% pet.) and negative to imipramine. An oral challenge test with afloqualone and exposure in cloudy weather in the evening for 10 minutes evoked edematous erythema 24 hours later. The patient was diagnosed with photosensitivity to afloqualone presenting as photoleucomelanodermatitis. As the photopatch test to afloqualone 1% was positive but negative to 0.1%, it was uncertain whether this case represented phototoxicity or photoallergy (2).

The authors reviewed the Japanese literature on photosensitivity from afloqualone and found 24 such patients reported between 1986 and 1993, 10 of who had positive photopatch tests. Clinical presentations were erythematous/eczematous, erythema multiforme-like, lichenoid and one case of photoleucomelanodermatitis (2).

It has been suggested that afloqualone causes phototoxicity, but may in a number of cases also induce photoallergy (3).

LITERATURE
1   The data in the section 'General' may have been obtained from literature discussed in this chapter, but mostly also or exclusively from one or more of the following online sources: ChemIDPlus Advanced, PubChem, DrugBank, RxList, Drug Central, Drugs.com, and Wikipedia
2   Ishikawa T, Kamide R, Niimura M. Photoleukomelanodermatitis (Kobori) induced by afloqualone. J Dermatol 1994;21:430-433
3   Tokura Y, Ogai M, Yagi H, Takigawa M. Afloqualone photosensitivity: immunogenicity of afloqualone photomodified epidermal cells. Photochem Photobiol 1994;60:262-267

Chapter 3.14   ALBENDAZOLE

IDENTIFICATION

Description/definition   : Albendazole is the benzimidazole anthelmintic that conforms to the structural formula
                             shown below
Pharmacological classes : Tubulin modulators; anthelmintics; anticestodal agents; antiprotozoal agents
IUPAC name   : Methyl N-(6-propylsulfanyl-1H-benzimidazol-2-yl)carbamate
Other names   : Carbamic acid, [5-(propylthio)-1H-benzimidazol-2-yl]-, methyl ester; methyl 5-propylthio-
                             2-benzimidazolecarbamate
CAS registry number   : 54965-21-8
EC number   : 259-414-7
Merck Index monograph   : 1473
Patch testing   : 5% pet. (2)
Molecular formula   : $C_{12}H_{15}N_3O_2S$

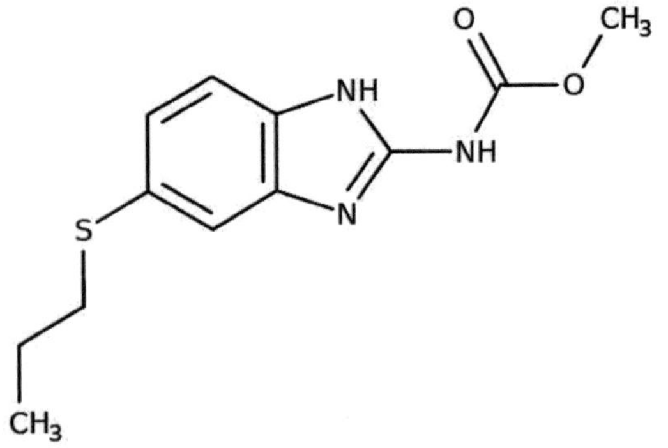

GENERAL

Albendazole is a broad-spectrum, synthetic benzimidazole-derivative anthelmintic. It is indicated for the treatment
of parenchymal neurocysticercosis due to active lesions caused by larval forms of the pork tapeworm, *Taenia solium*
and for the treatment of cystic hydatid disease of the liver, lung, and peritoneum, caused by the larval form of the
dog tapeworm, *Echinococcus granulosus* (1).

CONTACT ALLERGY FROM ACCIDENTAL CONTACT

Ten workers in a pharmaceutical plant manufacturing or – indirectly – exposed to albendazole, 9 with eyelid edema
and one with dermatitis, were patch tested on tape-stripped skin with albendazole suspension 1.9% and 3.8%, the
vehicle of the suspension, and albendazole 1% and 5% pet. Four patients were contact allergic to albendazole. They
all reacted to albendazole 5% pet. at D4, one of them also to 1% pet., two of them also to albendazole 3.8%
suspension, but none to the 1.9% suspension. One also had urticaria, 3 rhinitis, and 2 conjunctivitis. It was concluded
that 9 of the 10 patients had 'occupational disorders, with symptoms suggestive of contact urticaria and one with
dermatitis'. However, patch tests read at 1 hour had probably all been negative (possibly too late to detect contact
urticaria), no other tests to establish immediate-type reactions were performed and the 4 patients with contact
allergy to albendazole had the same symptoms as those who were suggested to have contact urticaria (2).

Cross-reactions, pseudo-cross-reactions and co-reactions

The question has been raised (but not answered) whether contact allergy to albendazole and other (nitro)imidazoles
may be overrepresented in patients allergic to methylchloroisothiazolinone/methylisothiazolinone (MCI/MI (3).

LITERATURE

1   The data in the section 'General' may have been obtained from literature discussed in this chapter, but mostly
     also or exclusively from one or more of the following online sources: ChemIDPlus Advanced, PubChem,
     DrugBank, RxList, Drug Central, Drugs.com, and Wikipedia
2   Macedo NA, Piñeyro MI, Carmona C. Contact urticaria and contact dermatitis from albendazole. Contact
     Dermatitis 1991;25:73-75
3   Stingeni L, Rigano L, Lionetti N, Bianchi L, Tramontana M, Foti C, et al. Sensitivity to imidazoles/nitroimidazoles in
     subjects sensitized to methylchloroisothiazolinone/methylisothiazolinone: A simple coincidence? Contact
     Dermatitis 2019;80:181-183

Chapter 3.15   ALBUTEROL

IDENTIFICATION

Description/definition   : Albuterol is the sympathomimetic agent that conforms to the structural formula shown below
Pharmacological classes : Bronchodilator agents; β_2-adrenergic receptor agonists; tocolytic agents
IUPAC name   : 4-[2-(*tert*-Butylamino)-1-hydroxyethyl]-2-(hydroxymethyl)phenol
Other names   : Salbutamol; 2-(*tert*-butylamino)-1-(4-hydroxy-3-hydroxymethylphenyl)ethanol
CAS registry number   : 18559-94-9
EC number   : 242-424-0
Merck Index monograph   : 1480
Patch testing   : 5% pet. (3)
Molecular formula   : $C_{13}H_{21}NO_3$

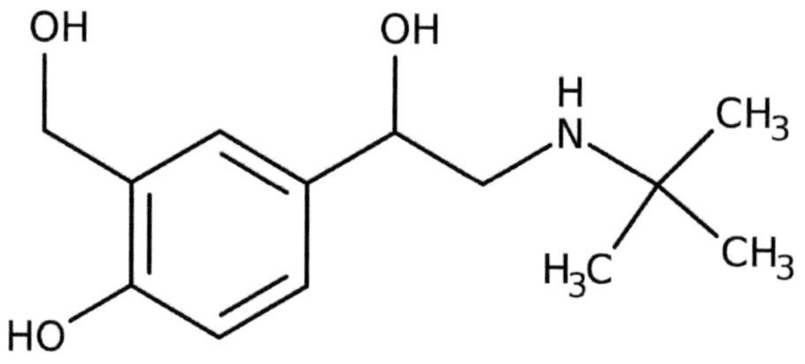

GENERAL

Albuterol (salbutamol) is a short-acting, selective β_2-adrenergic receptor agonists with bronchodilator activity. It is generally used for acute episodes of bronchospasm caused by bronchial asthma, chronic bronchitis and other chronic bronchopulmonary disorders such as chronic obstructive pulmonary disorder (COPD). It is also used prophylactically for exercise-induced asthma. In pharmaceutical products, albuterol is employed as albuterol sulfate (salbutamol hemisulfate) (CAS number 51022-70-9, EC number 256-916-8, molecular formula $C_{26}H_{44}N_2O_{10}S$).

CUTANEOUS ADVERSE DRUG REACTIONS FROM SYSTEMIC ADMINISTRATION CAUSED BY TYPE IV (DELAYED-TYPE) HYPERSENSITIVITY (as demonstrated by positive patch tests)

Cutaneous adverse drug reactions from systemic administration of albuterol caused by type IV (delayed-type) hypersensitivity have included systemic allergic dermatitis (2).

Systemic allergic dermatitis (systemic contact dermatitis)

A 74-year old man with chronic obstructive pulmonary disease presented with a 2½ year history of eczema of the central part of the face, corresponding with a mask through which he inhaled an aerosol of salbutamol, delivered from a nebulizer. He also had eczematous lesions on the arms and legs. Patch tests were positive to the inhalation fluid 0.5% and later to salbutamol 5% pet. (++/++) and 5% water (+/+) but negative to 0.5% water. Thirty controls were negative. Contact dermatitis and systemic allergic dermatitis from salbutamol was diagnosed (2).

LITERATURE

1   The data in the section 'General' may have been obtained from literature discussed in this chapter, but mostly also or exclusively from one or more of the following online sources: ChemIDPlus Advanced, PubChem, DrugBank, RxList, Drug Central, Drugs.com, and Wikipedia
2   Smeenk G, Burgers GJ, Teunissen PC. Contact dermatitis from salbutamol. Contact Dermatitis 1994;31:123

Chapter 3.16   ALENDRONIC ACID

IDENTIFICATION

Description/definition   : Alendronic acid is the second generation bisphosphonate and synthetic analog of
                             pyrophosphate that conforms to the structural formula shown below
Pharmacological classes   : Bone density conservation agents
IUPAC name   : (4-Amino-1-hydroxy-1-phosphonobutyl)phosphonic acid
Other names   : Alendronic acid is often termed 'alendronate'
CAS registry number   : 66376-36-1
EC number   : Not available
Merck Index monograph : 1493
Patch testing   : 10% and 20% pet. (3); perform adequate control testing!; lower concentrations (1% and
                             0.1% in water) may be suitable in some patients and are not irritant (4)
Molecular formula   : $C_4H_{13}NO_7P_2$

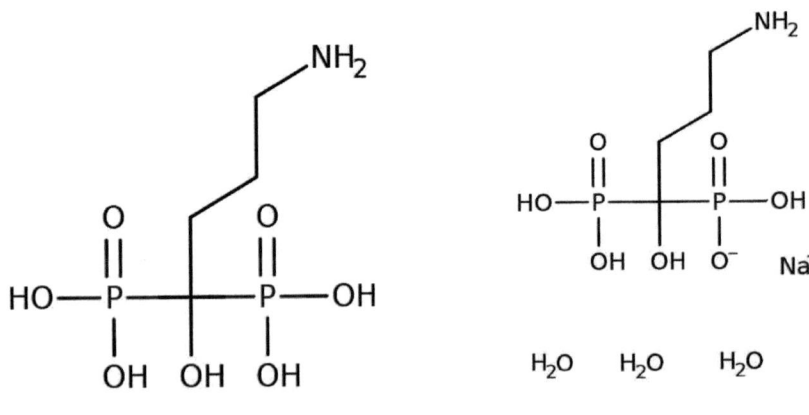

Alendronic acid   Alendronate sodium trihydrate

GENERAL

Alendronic acid is a second-generation bisphosphonate and synthetic analog of pyrophosphate with anti-bone-resorption activity. It is a nonhormonal medication that builds healthy bone, restoring some of the bone loss as a result of osteoporosis. Alendronic acid is indicated for the treatment and prevention of osteoporosis in men and postmenopausal women, treatment of glucocorticoid-induced osteoporosis, and Paget's disease of bone. In pharmaceutical products, alendronic acid is usually employed as alendronate sodium trihydrate (CAS number 121268-17-5, EC number 601-766-4, molecular formula $C_4H_{18}NNaO_{10}P_2$). Both alendronate sodium (anhydrous, CAS number 129318-43-0) and alendronate sodium trihydrate are called alendronate sodium (1).

CUTANEOUS ADVERSE DRUG REACTIONS FROM SYSTEMIC ADMINISTRATION CAUSED BY TYPE IV
(DELAYED-TYPE) HYPERSENSITIVITY (as demonstrated by positive patch tests)

Cutaneous adverse drug reactions from systemic administration of alendronic acid caused by type IV (delayed-type) hypersensitivity have included maculopapular eruption (2), erythematous rash (4) and papular petechial rash (3).

Maculopapular eruption

A 60-year-old woman with postmenopausal osteoporosis developed maculopapular skin lesions in the head and neck region 4 months after daily alendronate intake of 10 mg. Spontaneous regression occurred with some residual scaling within 6 weeks after cessation of the drug. Patch testing (scratch-chamber) with alendronate 50% in water and petrolatum showed crescendo reactions at D2 and D3. Control testing (closed patch tests with alendronate 50%, 10%, and 1% in water) on 10 individuals showed all concentrations to be - more than slightly - irritant. After 2 months of complete remission, oral challenge with alendronate in identical dosage for 3 months resulted in an identical maculopapular eruption as previously. In view of the late onset of cutaneous inflammation in springtime after a 3-month challenge of the drug, the authors could not exclude a pathogenetic influence of UV light (2). They suggest a cellular immune response to alendronate, but this appears far from proven.

Erythroderma, widespread erythematous eruption, exfoliative dermatitis

A 70-year-old man developed an erythematous rash while on alendronate. Patch tests were positive to a pulverized commercial alendronate tablet 0.1% water (+++), 1% water (+++) and 1% pet. (++) at D4. Of twenty controls, 19 were negative and one had a ++ reaction to alendronate 1% pet. but negative to 1% and 0.1% water (4).

Other cutaneous adverse drug reactions

A 72-year-old woman with osteoporosis was seen with numerous red papules and petechiae on the legs that had started 10 days before. She had been taking alendronate sodium hydrate 5 mg once daily orally for 3 weeks. Patch tests and scratch-patch tests with 20%, 10%,1%, and 0.1% pet. alendronate sodium hydrate and 3 excipients in the commercial alendronate tablet resulted in positive reactions to the 10% and 20% concentration. Six controls were negative. A lymphocyte stimulation test with alendronate sodium hydrate was positive (stimulation index 4.0, normal <1.8) (3).

LITERATURE

1   The data in the section 'General' may have been obtained from literature discussed in this chapter, but mostly also or exclusively from one or more of the following online sources: ChemIDPlus Advanced, PubChem, DrugBank, RxList, Drug Central, Drugs.com, and Wikipedia
2   Brinkmeier T, Kügler K, Lepoittevin JP, Frosch PJ. Adverse cutaneous drug reaction to alendronate. Contact Dermatitis 2007;57:123-125
3   Kimura M, Kawada A, Murayama Y, Murayama M. Drug eruption due to alendronate sodium hydrate. Contact Dermatitis 2003;48:116
4   Barrantes-González M, Espona-Quer M, Salas E, Giménez-Arnau AM. Bisphosphonate-induced cutaneous adverse events: the difficulty of assessing imputability through patch testing. Dermatology 2014;229:163-168

Chapter 3.17   ALLOPURINOL

IDENTIFICATION

Description/definition   : Allopurinol is the structural analog of the natural purine base hypoxanthine, that
                               conforms to the structural formula shown below
Pharmacological classes   : Free radical scavengers; gout suppressants; antimetabolites; enzyme inhibitors
IUPAC name   : 1,5-Dihydropyrazolo[3,4-d]pyrimidin-4-one
CAS registry number   : 315-30-0
EC number   : 206-250-9
Merck Index monograph : 1541
Patch testing   : Tablet 1%, 10% and 20% pet. (3); patch testing with allopurinol and its metabolite 8-
                               oxypurinol does not induce positive patch test reactions in patients with cutaneous
                               adverse drug reactions which are considered to be type-IV hypersensitivity reactions, such
                               as DRESS
Molecular formula   : $C_5H_4N_4O$

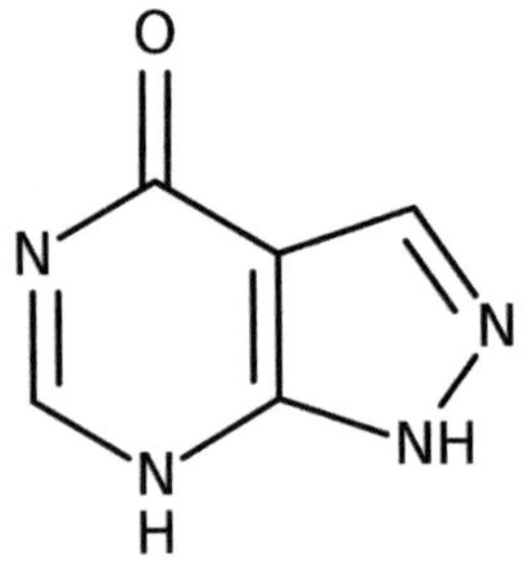

GENERAL

Allopurinol is an inhibitor of xanthine oxidase, an enzyme that converts oxypurines to uric acid. By blocking its production, this agent decreases serum and urine concentrations of uric acid. Allopurinol is frequently used in the treatment of chronic gout. Other indications include the management of patients with leukemia, lymphoma and malignancies who are receiving cancer therapy which causes elevations of serum and urinary uric acid levels and the management of patients with recurrent calcium oxalate calculi. In tablets, allopurinol base is used; for injection fluids, allopurinol sodium is employed (CAS number 17795-21-0, EC number 241-771-5, molecular formula $C_5H_4N_4NaO$) (1).

CUTANEOUS ADVERSE DRUG REACTIONS FROM SYSTEMIC ADMINISTRATION CAUSED BY TYPE IV (DELAYED-TYPE) HYPERSENSITIVITY (as demonstrated by positive patch tests)

Cutaneous adverse drug reactions from systemic administration of allopurinol caused by type IV (delayed-type) hypersensitivity have included papular exanthema with bullae (10) and unknown cutaneous adverse drug reactions (2).

General

Allopurinol is an important cause of drug reaction with eosinophilia and systemic symptoms (DRESS), together with anticonvulsant drugs, dapsone, minocycline and antiretroviral drugs like nevirapine and abacavir (3,4). The HLA haplotype HLA-B*5801 appears to predispose to DRESS from allopurinol. When administered orally, the drug is rapidly converted in the liver to its oxidative metabolite 8-oxypurinol, which is considered to be responsible for most of the actions of allopurinol. It may be the culprit in drug hypersensitivity to allopurinol, as these reactions appear to be primarily mediated by an oxypurinol-specific T cell response (7). However, patch tests are virtually always negative (3). In a study from a university hospital in Portugal, for example, 19 patients with DRESS very likely to have been caused by allopurinol were all patch test negative to allopurinol 1%, 10% and 20% pet. and this was also the case in 9 patients tested with oxypurinol 5% and 10% pet. The negativity of these patch tests could not exclude responsibility of allopurinol, as two of these patients, who were accidentally rechallenged orally, again developed a hypersensitivity reaction (3). In fact, in other studies, in which allopurinol was tested for DRESS or other drug reactions, the results were always negative (e.g. 5,9)

There is no definite explanation for the high number of false-negative results, but there may be several causes: (i) the final responsible agent is another drug metabolite that is not formed in the skin during patch testing; (ii) there is no immune mechanism involved; (iii) concomitant factors that are responsible in inducing transient oral drug intolerance, such as viral infection, are not present at the time of testing; and (iv) wrong choice of vehicle (limited skin penetration [6]), drug concentration, or exposure time (3). A recent hypothesis is that oxypurinol is rapidly bound in the peptide-binding groove of the HLAB*5801 molecule without requiring peptide processing. The net result is a novel drug peptide-HLA complex that drives the immunological reaction resulting in allopurinol-related severe cutaneous adverse reactions (8).

Other cutaneous adverse drug reactions

A patient treated with allopurinol developed a papular exanthema, in which later bullae appeared on the arms. A patch test was positive to allopurinol 200 mg/ml saline (10).

In Finland, in the period 1989-2001, 826 patients with suspected cutaneous drug eruptions were patch tested and 89 had one or more positive reactions. Of these individuals, one had a positive patch test to commercial allopurinol 20% or 30% pet. (of 10 tested) (2).

LITERATURE

1   The data in the section 'General' may have been obtained from literature discussed in this chapter, but mostly also or exclusively from one or more of the following online sources: ChemIDPlus Advanced, PubChem, DrugBank, RxList, Drug Central, Drugs.com, and Wikipedia

2   Lammintausta K, Kortekangas-Savolainen O. The usefulness of skin tests to prove drug hypersensitivity. Br J Dermatol 2005;152:968-974

3   Santiago F, Gonçalo M, Vieira R, Coelho S, Figueiredo A. Epicutaneous patch testing in drug hypersensitivity syndrome (DRESS). Contact Dermatitis 2010;62:47-53

4   Roujeau JC, Stern RS. Severe adverse cutaneous reactions to drugs. N Engl J Med 1994;331:1272-1285

5   Barbaud A, Collet E, Milpied B, Assier H, Staumont D, Avenel-Audran M, et al. A multicentre study to determine the value and safety of drug patch tests for the three main classes of severe cutaneous adverse drug reactions. Br J Dermatol 2013;168:555-562

6   Friedmann P S, Ardern-Jones M. Patch testing in drug allergy. Curr Opin Allergy Clin Immunol 2010;10:291-296

7   Yun J, Mattsson J, Schnyder K, Fontana S, LArgiader C, Pichler W, et al. Allopurinol hypersensitivity is primarily mediated by dose-dependent oxypurinol-specific T cell response. Clin Exp Allergy 2013;43:1246e55.

8   Stamp LK, Chapman PT. Allopurinol hypersensitivity: Pathogenesis and prevention. Best Pract Res Clin Rheumatol 2020;34:101501

9   Andrade P, Brinca A, Gonçalo M. Patch testing in fixed drug eruptions--a 20-year review. Contact Dermatitis 2011;65:195-201

10   Hari Y, Frutig-Schnyder K, Hurni M, Yawalkar N, Zanni MP, Schnyder B, et al. T cell involvement in cutaneous drug eruptions. Clin Exp Allergy 2001;31:1398-1408

Chapter 3.18   ALPRAZOLAM

IDENTIFICATION

Description/definition   : Alprazolam is the triazolobenzodiazepine that conforms to the structural formula shown below

Pharmacological classes : GABA modulators; hypnotics and sedatives; anti-anxiety agents

IUPAC name   : 8-Chloro-1-methyl-6-phenyl-4H-[1,2,4]triazolo[4,3-a][1,4]benzodiazepine

CAS registry number   : 28981-97-7

EC number   : 249-349-2

Merck Index monograph   : 1578

Patch testing   : Crushed tablet, 10% and 30% pet. (3); most systemic drugs can be tested at 10% pet.; if the pure chemical is not available, prepare the test material from intravenous powder, the content of capsules or – when also not available – from powdered tablets to achieve a final concentration of the active drug of 10% pet.

Molecular formula   : $C_{17}H_{13}ClN_4$

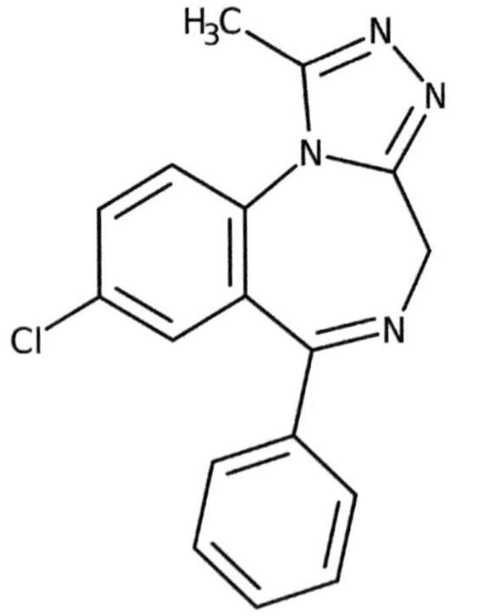

GENERAL

Alprazolam is a triazolobenzodiazepine agent with anxiolytic, sedative-hypnotic and anticonvulsant activities. It is indicated for the management of anxiety disorder, anxiety associated with depression, panic disorder, and panic disorder with agoraphobia. Alprazolam may also be prescribed off-label for insomnia, premenstrual syndrome, and depression (1).

CONTACT ALLERGY FROM ACCIDENTAL CONTACT

Case series

In Leuven, Belgium, in the period 2001-2019, 201 of 1248 health care workers/employees of the pharmaceutical industry had occupational allergic contact dermatitis. In 23 (11%) dermatitis was caused by skin contact with a systemic drug: 19 nurses, two chemists, one physician, and one veterinarian. The lesions were mostly localized on the hands, but often also on the face, as airborne dermatitis. In total, 42 positive patch test reactions to 18 different systemic drugs were found. In one patient, alprazolam was the drug/one of the drugs that caused occupational dermatitis (4, overlap with refs. 2 and 3).

In the same clinic in Leuven, Belgium, in the period 2007-2011, 81 patients have been diagnosed with occupational airborne allergic contact dermatitis. In 23 of them, drugs were the offending agents, including alprazolam in 2 cases (2, overlap with refs. 3 and 4).

Case reports

A 24-year-old woman presented with eyelid dermatitis, which had started with localized edema 4 months previously. Later, the area had become itchier, with redness and scaling. The patient suspected a relationship with her work as a pharmacy assistant, which involved breaking and crushing different types of tablets. She was patch tested with the crushed tablets that she had contact with at 10% pet. and showed positive reactions to alprazolam, 2 other benzodiazepines, 3 ACE-inhibitors and 4 beta-blockers. Twenty-one controls were negative to alprazolam 10% pet. Cosmetic allergy was excluded and a diagnosis of occupational airborne allergic contact was made (3).

A 30-year-old woman presented with an itchy, scaly skin reaction on the face, which was most pronounced on the eyelids. She reported a clear relationship with her work as a geriatric nurse, during which she was required to crush tablets for the elderly on a daily basis, mostly benzodiazepine drugs. She had positive patch tests to alprazolam (crushed tablet 30% pet.; 21 controls were negative), 4 other benzodiazepines (bromazepam, diazepam, lorazepam, tetrazepam) and 3 unrelated drugs. The patient was diagnosed with occupational airborne allergic contact dermatitis from drugs (3, overlap with refs. 2 and 4).

Cross-reactions, pseudo-cross-reactions and co-reactions

The possibility of cross-reactions between benzodiazepines, especially with primary sensitization to tetrazepam, is considered likely, but the pattern is unknown and may differ between sensitivity from (occupational) contact with the drugs and systemically induced reactions (3).

LITERATURE

1   The data in the section 'General' may have been obtained from literature discussed in this chapter, but mostly also or exclusively from one or more of the following online sources: ChemIDPlus Advanced, PubChem, DrugBank, RxList, Drug Central, Drugs.com, and Wikipedia

2   Swinnen I, Goossens A. An update on airborne contact dermatitis: 2007-2011. Contact Dermatitis 2013;68:232-238

3   Swinnen I, Ghys K, Kerre S, Constandt L, Goossens A. Occupational airborne contact dermatitis from benzodiazepines and other drugs. Contact Dermatitis 2014;70:227-232

4   Gilissen L, Boeckxstaens E, Geebelen J, Goossens A. Occupational allergic contact dermatitis from systemic drugs. Contact Dermatitis 2020;82:24-30

Chapter 3.19   ALPRENOLOL

IDENTIFICATION

Description/definition   : Alprenolol is the secondary alcohol that conforms to the structural formula shown below
Pharmacological classes : Anti-arrhythmia agents; sympatholytics; antihypertensive agents; β-adrenergic
                           antagonists
IUPAC name   : 1-(Propan-2-ylamino)-3-(2-prop-2-enylphenoxy)propan-2-ol
CAS registry number   : 13655-52-2
EC number   : 237-140-9
Merck Index monograph   : 1579
Patch testing   : Aqueous solution 10 mg/ml (2,3)
Molecular formula   : $C_{15}H_{23}NO_2$

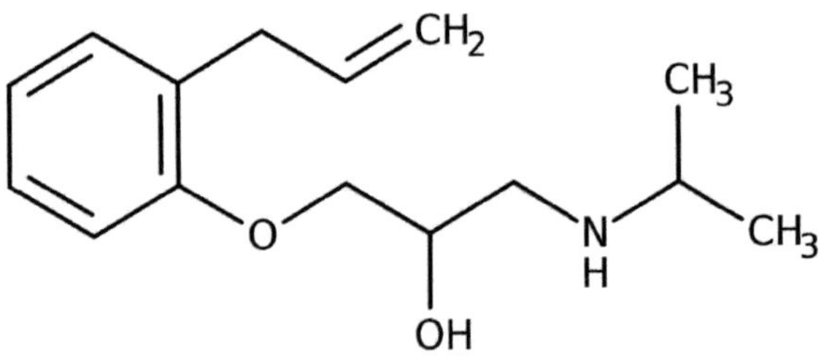

GENERAL

Alprenolol is a non-selective β-adrenergic antagonist and sympatholytic agent used as an antihypertensive, anti-anginal, and anti-arrhythmic agent. Its current use is probably limited. In pharmaceutical products, alprenolol is employed as alprenolol hydrochloride (CAS number 13707-88-5, EC number 237-244-4, molecular formula $C_{15}H_{24}ClNO_2$) (1).

CONTACT ALLERGY FROM ACCIDENTAL CONTACT

Case series

In a plant manufacturing alprenolol in Sweden, in the second half of the 1970s, about 200 employees working at the factory and 100 at the laboratories of this company were estimated to come in contact with alprenolol during their daily work. 32 of these people had, or previously had, skin changes possibly related to work and they were patch tested with an aqueous solution of alprenolol in the dilutions 10, 5, 2.5 and 1.25 mg per ml. Fourteen of them, 12 men and 2 women, had positive patch tests (not specified). Twenty controls were negative. They had shown signs of contact dermatitis after a varying time from 3 weeks to more than 5 years of contact with alprenolol. Two of the seven alprenolol-positive patients reacted when tested with metoprolol, neither of who had come into contact with metoprolol at work. One man received oral alprenolol 100 mg and demonstrated pruritus and a widespread dermatitis after 7 hours (systemic allergic dermatitis). A skin biopsy showed dermatitis. The majority of the sensitized individuals worked in departments with extensive alprenolol dust generation. After reducing dust exposure, only a few new cases of alprenolol sensitization have appeared (2).

Five workers from a pharmaceutical plant developed eczema localized to various parts of the body and two workers had eczema and urticaria. These individuals handled various pharmaceutical products, including alprenolol and quinidine. All patients were patch tested with alprenolol hydrochloride: 10, 5, 2.5, and 1.25 mg/ml water. Six (86%) had positive reactions. The lowest concentration inducing positivity was 1.25 mg/ml in 5 individuals and 2.5 in one patient. Thirty controls were negative to all concentrations (3).

CUTANEOUS ADVERSE DRUG REACTIONS FROM SYSTEMIC ADMINISTRATION CAUSED BY TYPE IV (DELAYED-TYPE) HYPERSENSITIVITY (as demonstrated by positive patch tests)

Cutaneous adverse drug reactions from systemic administration of alprenolol caused by type IV (delayed-type) hypersensitivity have included systemic allergic dermatitis (2).

Systemic allergic dermatitis (systemic contact dermatitis)

See the section 'Contact allergy from accidental contact' above, ref. 2

Cross-reactions, pseudo-cross-reactions and co-reactions

Of 7 patients sensitized to alprenolol and tested with metoprolol, 2 (28%) reacted to metoprolol (2). In a patient who had a psoriasiform eruption from metoprolol and a positive patch test to metoprolol tartrate, a cross-reaction to alprenolol (test concentration/vehicle not mentioned) was observed (4).

LITERATURE

1   The data in the section 'General' may have been obtained from literature discussed in this chapter, but mostly also or exclusively from one or more of the following online sources: ChemIDPlus Advanced, PubChem, DrugBank, RxList, Drug Central, Drugs.com, and Wikipedia

2   Ekenvall L, Forsbeck M. Contact eczema produced by a beta-adrenergic blocking agent (alprenolol). Contact Dermatitis 1978;4:190-194

3   Stejskal VD, Olin RG, Forsbeck M. The lymphocyte transformation test for diagnosis of drug-induced occupational allergy. J Allergy Clin Immunol 1986;77:411-426

4   Neumann HA, Van Joost T, Westerhof W. Dermatitis as side-effect of long-term metoprolol. Lancet 1979;2(8145):745

Chapter 3.20   ALTHIAZIDE

IDENTIFICATION

Description/definition   : Althiazide is the thiazide and sulfonamide derivative that conforms to the structural formula shown below

Pharmacological classes : Diuretics

IUPAC name   : 6-Chloro-1,1-dioxo-3-(prop-2-enylsulfanylmethyl)-3,4-dihydro-2H-1λ^6,2,4-benzothiadiazine-7-sulfonamide

Other names   : Altizide

CAS registry number   : 5588-16-9

EC number   : 226-994-8

Merck Index monograph : 1584

Patch testing   : Tablet 10% pet. (3); most systemic drugs can be tested at 10% pet.; if the pure chemical is not available, prepare the test material from intravenous powder, the content of capsules or – when also not available – from powdered tablets to achieve a final concentration of the active drug of 10% pet.

Molecular formula   : $C_{11}H_{14}ClN_3O_4S_3$

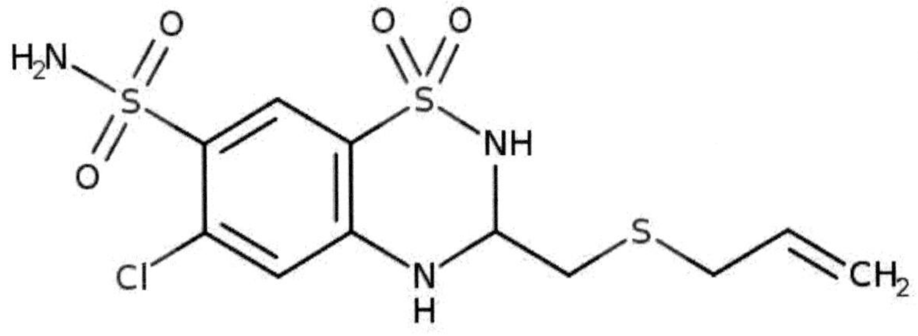

GENERAL

Althiazide, also known as altizide, is an agent belonging to the class of thiazide diuretics with antihypertensive activity. This aldosterone antagonist is often combined with spironolactone (1).

CONTACT ALLERGY FROM ACCIDENTAL CONTACT

In Leuven, Belgium, in the period 2001-2019, 201 of 1248 health care workers/employees of the pharmaceutical industry had occupational allergic contact dermatitis. In 23 (11%) dermatitis was caused by skin contact with a systemic drug: 19 nurses, two chemists, one physician, and one veterinarian. The lesions were mostly localized on the hands, but often also on the face, as airborne dermatitis. In total, 42 positive patch test reactions to 18 different systemic drugs were found. In 2 patients, Aldactazine ® (althiazide/spironolactone), tested 10% pet., was the drug/one of the drugs that caused occupational dermatitis. Althiazide and spironolactone were not tested separately (2).

CUTANEOUS ADVERSE DRUG REACTIONS FROM SYSTEMIC ADMINISTRATION CAUSED BY TYPE IV (DELAYED-TYPE) HYPERSENSITIVITY (as demonstrated by positive patch tests)

Cutaneous adverse drug reactions from systemic administration of altizide caused by type IV (delayed-type) hypersensitivity have included (possible) photosensitivity (3).

Photosensitivity

A 70-year-old woman had been treated for 15 years with a combination tablet of althiazide and spironolactone for hypertension, when she developed a photodistributed burning and itching erythematous, papulosquamous eruption involving the face (eyelids, cheeks), lateral neck, dorsum of both hands and proximal phalanges. On withdrawal of the medication, the lesions disappeared within 4 weeks. Patch and photopatch tests were performed in triplicate with the commercial tablet 10% aq. and 10% pet. Patch tests and photopatch tests irradiated with 5 J/cm² of UVA were negative. However, the site irradiated with a suberythemal UVB dose (0.75 UVB-MED) was positive (++) one and 2 days after irradiation. The patient was diagnosed with althiazide UVB photoallergy (3). This was, however, not substantiated by photopatch tests, as althiazide itself was not tested. In addition, no controls were tested to exclude phototoxicity.

LITERATURE

1 The data in the section 'General' may have been obtained from literature discussed in this chapter, but mostly also or exclusively from one or more of the following online sources: ChemIDPlus Advanced, PubChem, DrugBank, RxList, Drug Central, Drugs.com, and Wikipedia

2 Gilissen L, Boeckxstaens E, Geebelen J, Goossens A. Occupational allergic contact dermatitis from systemic drugs. Contact Dermatitis 2020;82:24-30

3 Schwarze HP, Albes B, Marguery MC, Loche F, Bazex J. Evaluation of drug-induced photosensitivity by UVB photopatch testing. Contact Dermatitis 1998;39:200

Chapter 3.21   AMANTADINE

IDENTIFICATION

Description/definition   : Amantadine is the synthetic tricyclic amine that conforms to the structural formula shown
                            below
Pharmacological classes   : Antiparkinson agents; dopamine agents; analgesics, non-narcotic; antiviral agents
IUPAC name   : Adamantan-1-amine
Other names   : Tricyclo(3.3.1.1(3,7))-decan-1-amine
CAS registry number   : 768-94-5
EC number   : 212-201-2
Merck Index monograph   : 1638
Patch testing   : No data available; most systemic drugs can be tested at 10% pet.; if the pure chemical is
                            not available, prepare the test material from intravenous powder, the content of capsules
                            or – when also not available – from powdered tablets to achieve a final concentration of
                            the active drug of 10% pet.
Molecular formula   : $C_{10}H_{17}N$

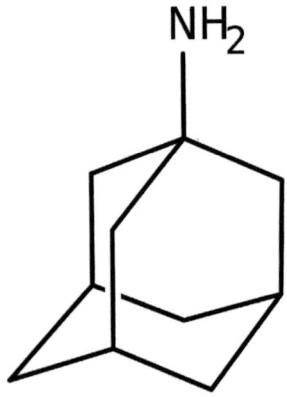

GENERAL

Amantadine is a synthetic tricyclic amine with antiviral, antiparkinsonian, and antihyperalgesic activities. The drug is
indicated for the prophylaxis and treatment of signs and symptoms of infection caused by various strains of influenza
A virus. Amantadine is also used for the treatment of parkinsonism (usually combined with L-DOPA), extrapyramidal
reactions and for postherpetic neuralgia. In pharmaceutical products, amantadine is employed as amantadine
hydrochloride (CAS number 665-66-7, EC number 211-560-2, molecular formula $C_{10}H_{18}ClN$) (1).

CUTANEOUS ADVERSE DRUG REACTIONS FROM SYSTEMIC ADMINISTRATION CAUSED BY TYPE IV (DELAYED-TYPE) HYPERSENSITIVITY (as demonstrated by positive patch tests)

Cutaneous adverse drug reactions from systemic administration of amantadine caused by type IV (delayed-type)
hypersensitivity have included systemic allergic dermatitis (5) and photosensitivity (2,3)

Systemic allergic dermatitis (systemic contact dermatitis)

Oral administration of amantadine may result in systemic allergic dermatitis in patients previously sensitized to
tromantadine (5).

Photosensitivity

A 74-year-old man, who had been treated for Parkinson's disease with amantadine 100 mg daily for several weeks,
developed dermatitis on the sun-exposed areas of the body. A photopatch test with amantadine was weakly positive
after 2 days and strongly positive at D3. Details on the test material, test concentration and vehicle were not
provided, and ordinary patch tests were probably not performed, nor were control tests to exclude irritancy /
phototoxicity done (2).
    A similar case of photoallergic dermatitis had been presented a decade earlier by the same investigators (3).

Cross-reactions, pseudo-cross-reactions and co-reactions
Most patients sensitized to the topical antiviral agent tromantadine cross-react to amantadine (4,6). Contact allergy to tromantadine been fully reviewed in Volume 3 of the *Monographs in contact allergy* series (7).

LITERATURE

1   The data in the section 'General' may have been obtained from literature discussed in this chapter, but mostly also or exclusively from one or more of the following online sources: ChemIDPlus Advanced, PubChem, DrugBank, RxList, Drug Central, Drugs.com, and Wikipedia

2   Van den Berg WH, van Ketel WG. Photosensitization by amantadine (Symmetrel®). Contact Dermatitis 1983;9:165

3   Van Ketel WG, Goedhart-van Dijk B. Fotosensitization by amantadine (Symmetrel). Dermatologica 1974;148:124-126

4   Przybilla B. Allergic contact dermatitis to tromantadine. J Am Acad Dermatol 1983;9:165

5   Van Ketel WG. Systemic contact-type dermatitis by derivatives of adamantane? Derm Beruf Umwelt 1988;36:23-24 (Article in German)

6   Przybilla B, Wagner-Grösser G, Balda BR. Kontaktallergische Kreuzreaktion von Tromantadin und Amantadin. Dtsch Med Wochenschr 1983;108:172-175 (Article in German)

7   De Groot AC. Monographs in contact allergy, volume 3: Topical Drugs. Boca Raton, Fl, USA: CRC Press Taylor and Francis Group, 2021 (ISBN 978-0-367-23693-9)

Chapter 3.22   AMBROXOL

IDENTIFICATION

Description/definition   : Ambroxol is the aromatic amine and metabolite of bromhexine that conforms to the
                            structural formula shown below
Pharmacological classes   : Expectorants
IUPAC name   : 4-[(2-Amino-3,5-dibromophenyl)methylamino]cyclohexan-1-ol
CAS registry number   : 18683-91-5
EC number   : 242-500-3
Merck Index monograph   : 1650
Patch testing   : 10% pet. (4)
Molecular formula   : $C_{13}H_{18}Br_2N_2O$

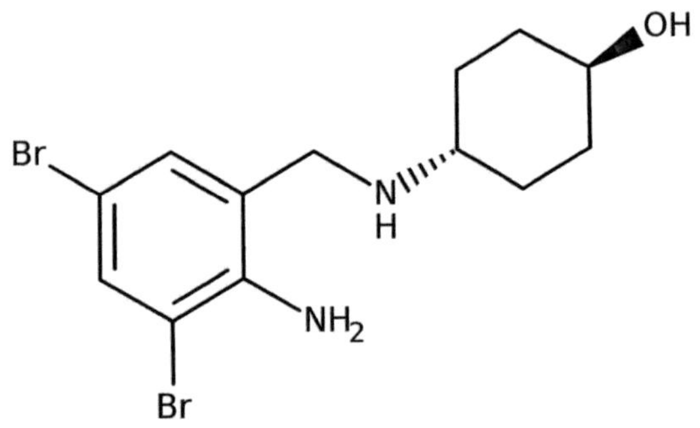

GENERAL

Ambroxol is an aromatic amine and a metabolite of bromhexine that stimulates mucociliary action and clears the air passages in the respiratory tract. As a secretolytic agent it is indicated for treatment of bronchopulmonary diseases with abnormal or excessive mucus secretion and transport. It allows the mucus to be more easily cleared and eases a patient's breathing. In pharmaceutical products, ambroxol is employed as ambroxol hydrochloride (CAS number 23828-92-4, EC number 245-899-2, molecular formula $C_{13}H_{19}Br_2ClN_2O$) (1). Ambroxol is used both in topical and in systemic pharmaceutical applications. In topical preparations, it has caused contact allergy/allergic contact dermatitis, which has been fully reviewed in Volume 3 of the *Monographs in contact allergy* series (3).

CUTANEOUS ADVERSE DRUG REACTIONS FROM SYSTEMIC ADMINISTRATION CAUSED BY TYPE IV (DELAYED-TYPE) HYPERSENSITIVITY (as demonstrated by positive patch tests)

Cutaneous adverse drug reactions from systemic administration of ambroxol caused by type IV (delayed-type) hypersensitivity have included maculopapular eruption (4) and photosensitivity (2).

Maculopapular eruption

A 79-year-old woman presented with a maculopapular rash with intense pruritus. She had taken ambroxol, acetaminophen and codeine for 4 days. Clinical examination showed a generalized maculopapular exanthematous eruption, with furfuraceous desquamation and intense erythema. The mucosa was spared. She was treated with antihistamines and oral corticosteroids and the skin lesions resolved within a week. After this episode the patient tolerated acetaminophen and acetylsalicylic acid. Patch tests with ambroxol (10% pet.) and codeine (10% pet.) gave a positive reaction to ambroxol (++/+++).Ten controls were negative (4).

Photosensitivity

An 82-year-old man, who used various drugs for hypertension and upper respiratory inflammation, presented with a 10-day history of a rash on sun-exposed areas. Examination showed a pruritic, edematous, scaly erythema on his face and 'V' area of his neck and a papulosquamous eruption on the dorsum of his hands. There was clear sparing of the finger webs. An UVB phototest at 10 mJ/cm² provoked an indurated, papular, severe erythema. After stopping all drugs, this test became within the normal range. Patch tests and photopatch tests with all medicines (10% pet.) using 5 J/cm² of UVA irradiation produced a positive patch test with ambroxol, not enhanced by UVA. Patch and photopatch tests with ambroxol in a dilution series ranging from 10% to 0.05% gave positive patch tests, with weaker reactions at the lower concentrations and slight erythema at 0.05%, enhanced by UVB radiation but not UVA. An oral photochallenge test with the normal dose of ambroxol for 3 days was negative to UVA, but UVB irradiation again caused an indurated erythema two days after the oral photo-challenge test. Histology of this test site showed histological changes similar to the patient's lesions at the first visit (2). The authors conclude that this is a case of photosensitivity to ambroxol and UVB, but the patient also had 'plain' delayed-type hypersensitivity to ambroxol.

LITERATURE

1 The data in the section 'General' may have been obtained from literature discussed in this chapter, but mostly also or exclusively from one or more of the following online sources: ChemIDPlus Advanced, PubChem, DrugBank, RxList, Drug Central, Drugs.com, and Wikipedia

2 Fujimoto N, Danno K, Wakabayashi M, Uenishi T, Tanaka T. Photosensitivity with eosinophilia due to ambroxol and UVB. Contact Dermatitis 2009;60:110-113

3 De Groot AC. Monographs in contact allergy, volume 3: Topical Drugs. Boca Raton, Fl, USA: CRC Press Taylor and Francis Group, 2021 (ISBN 978-0-367-23693-9)

4 Monzón S, Del Mar Garcés M, Lezaun A, Fraj J, Asunción Dominguez M, Colás C. Ambroxol-induced systemic contact dermatitis confirmed by positive patch test. Allergol Immunopathol (Madr) 2009;37:167-168

Chapter 3.23   AMDINOCILLIN PIVOXIL

IDENTIFICATION

Description/definition   : Amdinocillin pivoxil is the pivaloyloxymethyl ester of the beta-lactam penicillin
                            amdinocillin that conforms to the structural formula shown below
Pharmacological classes   : Anti-bacterial agents; anti-infective agents, urinary
IUPAC name   : 2,2-Dimethylpropanoyloxymethyl (2S,5R,6R)-6-(azepan-1-ylmethylideneamino)-3,3-
                            dimethyl-7-oxo-4-thia-1-azabicyclo[3.2.0]heptane-2-carboxylate
Other names   : Pivmecillinam
CAS registry number   : 32886-97-8
EC number   : 251-276-6
Merck Index monograph : 1654
Patch testing   : Amdinocillin pivoxil 10% pet. (4)
Molecular formula   : $C_{21}H_{33}N_3O_5S$

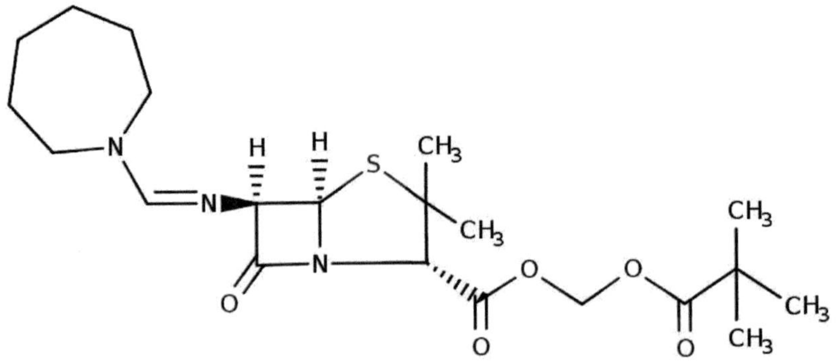

GENERAL

Amdinocillin pivoxil is an ester and a prodrug of the beta-lactam penicillin amdinocillin that is well absorbed orally, but broken down to amdinocillin in the intestinal mucosa. It is active against gram-negative organisms, e.g. in the treatment of urinary tract infections, salmonellosis and typhoid fever. In pharmaceutical products, amdinocillin pivoxil is employed as amdinocillin pivoxil hydrochloride (CAS number 32887-03-9, EC number 251-278-7, molecular formula $C_{21}H_{34}ClN_3O_5S$) (1). The classification and structures of beta-lactam antibiotics are discussed in Chapter 3.36 Amoxicillin.

CONTACT ALLERGY FROM ACCIDENTAL CONTACT

In Denmark, before 1986, 45 people working in a Danish factory producing the semisynthetic beta-lactam antibiotics pivampicillin and pivmecillinam (amdinocillin pivoxil) developed dermatitis, mainly on the hands, arms, calves and face. Nineteen of these patients also had hay fever (n=17) and/or asthma (n=5), probably provoked by contact with airborne penicillin at the place of employment. Nearly half of the workers developed symptoms of sensitization between 1 and 4 weeks after first exposure. The factory area was highly contaminated with penicillin dust. Patch tests were performed with pivampicillin base and HCl 1% and 5%, pivmecillinam base and HCl 1% and 10%, pivampicillin and pivmecillinam combined (both 0.5% and both 2.5%), ampicillin sodium 1% and 5%, and various other penicillins, all in petrolatum. There were 31 positive reactions to pivampicillin and 27 to pivmecillinam; 18 reacted to both antibiotics (2). These were cases of (partly airborne) occupational allergic contact dermatitis. The – rather complicated – results of retesting these patients were reported one year later (3).

   In another study by the same authors (but probably in the same patients), of 18 patients reacting to amdinocillin pivoxil (pivmecillinam) only 4 co-reacted to amdinocillin (mecillinam) tested as hydrochloride 1% and 5% pet. and as base 1% and 10% pet. The authors postulated that it is unlikely that the workers with pivmecillinam allergy could tolerate mecillinam, and suggested that the difference must be due to the insolubility of mecillinam (4).

Cross-reactions, pseudo-cross-reactions and co-reactions

Cross-reactions between beta-lactam antibiotics are discussed in Chapter 3.36 Amoxicillin. About half of the patients sensitized to amdinocillin pivoxil cross-react to ampicillin (2). Theoretically, amdinocillin pivoxil and pivampicillin may cross-react, as they have an identical side chain (2).

LITERATURE

1   The data in the section 'General' may have been obtained from literature discussed in this chapter, but mostly also or exclusively from one or more of the following online sources: ChemIDPlus Advanced, PubChem, DrugBank, RxList, Drug Central, Drugs.com, and Wikipedia

2   Møller NE, Nielsen B, von Würden K. Contact dermatitis to semisynthetic penicillins in factory workers. Contact Dermatitis 1986;14:307-311

3   Møller NE, Jeppesen K. Patch testing with semisynthetic penicillins. Contact Dermatitis 1987;16:227-228

4   Møller NE, von Würden K. Hypersensitivity to semisynthetic penicillins and cross-reactivity with penicillin. Contact Dermatitis 1992;26:351-352

Chapter 3.24   AMIKACIN

IDENTIFICATION

Description/definition   : Amikacin is the kanamycin-derived, semisynthetic aminoglycoside antibiotic, that
                            conforms to the structural formula shown below
Pharmacological classes : Anti-bacterial agents
IUPAC name   : (2S)-4-Amino-N-[(1R,2S,3S,4R,5S)-5-amino-2-[(2S,3R,4S,5S,6R)-4-amino-3,5-dihydroxy-6-
                            (hydroxymethyl)oxan-2-yl]oxy-4-[(2R,3R,4S,5S,6R)-6-(aminomethyl)-3,4,5-trihydroxyoxan-
                            2-yl]oxy-3-hydroxycyclohexyl]-2-hydroxybutanamide
Other names   : 1-N-(L(-)-gamma-Amino-α-hydroxybutyryl)kanamycin A
CAS registry number   : 37517-28-5
EC number   : 253-538-5
Merck Index monograph   : 1670
Patch testing   : Amikacin sulfate 20% pet. (3); 2.5 mg/ml saline (4)
Molecular formula   : $C_{22}H_{43}N_5O_{13}$

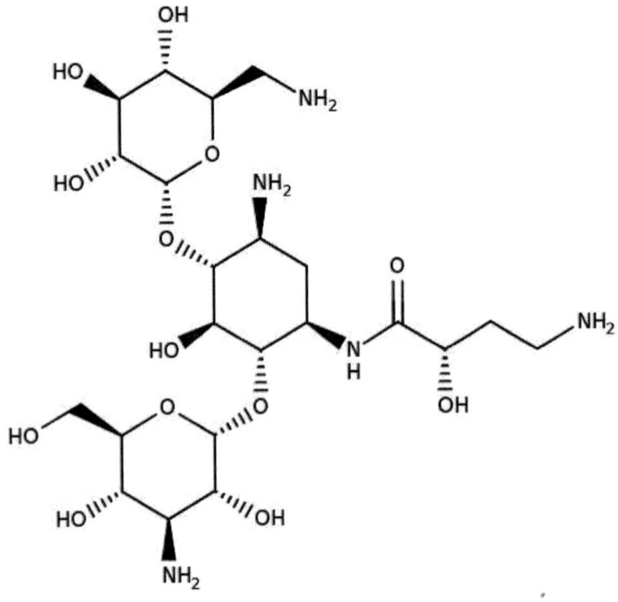

GENERAL

Amikacin is a broad-spectrum semisynthetic aminoglycoside antibiotic derived from kanamycin. In pharmaceutical products, the drug is employed as amikacin sulfate (CAS number 39831-55-5, EC number 254-648-6, molecular formula $C_{22}H_{47}N_5O_{21}S_2$). Amikacin sulfate injections (intravenous, intramuscular) are indicated in the short-term treatment of serious bacterial infections due to susceptible strains of gram-negative bacteria, including *Pseudomonas* species, *Escherichia coli*, species of *Proteus*, *Providencia*, *Klebsiella*, *Enterobacter*, and *Serratia* as well as *Acinetobacter* (Mima-Herellea) species. Recently, a liposomal inhalation suspension of this drug became available for the treatment of lung disease caused by *Mycobacterium avium* complex (MAC) bacteria (1).

CONTACT ALLERGY FROM ACCIDENTAL CONTACT

Case series

Between 1978 and 2001, 14,689 patients have been patch tested in the university hospital of Leuven, Belgium. Occupational allergic contact dermatitis to pharmaceuticals was diagnosed in 33 health care workers: 26 nurses, 4 veterinarians, 2 pharmacists and one medical doctor. There were 26 women and 7 men with a mean age of 38 years. Practically all of these patients presented with hand dermatitis (often the fingers), sometimes with secondary localizations, frequently the face. In this group, two patients had occupational allergic contact dermatitis from amikacin. One was a 53-year-old nurse who had dermatitis of the back of both hands, fingers, fingertips, wrist and mouth. She was also allergic to penicillin. The other patient sensitized to amikacin was a 38-year-old nurse with dermatitis of the fingers, who was also allergic to methylprednisolone and ranitidine (2).

In the period from November 1987 to November 1988, in Warsaw, Poland, 1651 consecutive patients were patch tested with 20% sulfate of amikacin, neomycin, paromomycin, gentamicin and kanamycin. In 23 cases (1.4%), the tests with amikacin were positive. Only one of these patients was negative to the remaining aminoglycoside antibiotics. This patient did not know which drugs had been given to her. All other patients were positive to neomycin (3).

CUTANEOUS ADVERSE DRUG REACTIONS FROM SYSTEMIC ADMINISTRATION CAUSED BY TYPE IV (DELAYED-TYPE) HYPERSENSITIVITY (as demonstrated by positive patch tests)

Cutaneous adverse drug reactions from systemic administration of amikacin caused by type IV (delayed-type) hypersensitivity have included drug reaction with eosinophilia and systemic symptoms (DRESS) (4,5,6).

Drug reaction with eosinophilia and systemic symptoms (DRESS)

A 42-year-old man received a multidrug therapy that included clindamycin, amikacin, and vancomycin for septic arthritis of the right knee. 18 days later, he developed DRESS. Six months after complete recovery, patch tests with all drugs were strongly positive to amikacin 2.5 mg/ml saline, prepared from the intravenous drug material. Histologic analysis of the positive amikacin patch test result showed changes typical of a delayed-type hypersensitivity reaction. The diagnosis of amikacin-delayed allergy causing DRESS was made. At day 4 the patient experienced a generalized skin flare-up without any other organ involvement (4).

One patient from France developed DRESS from amikacin and had positive patch tests to this drug tested pure (5). In a multicenter investigation in France, of 72 patients patch tested for DRESS, 46 (64%) had positive patch tests to drugs, including one to amikacin (6).

Cross-reactions, pseudo-cross-reactions and co-reactions

Cross-reactions between aminoglycoside antibiotics (neomycin, framycetin, tobramycin, gentamicin, kanamycin, paromomycin, amikacin) occur frequently, especially in cases of primary neomycin sensitization (Chapter 3.335).

LITERATURE

1   The data in the section 'General' may have been obtained from literature discussed in this chapter, but mostly also or exclusively from one or more of the following online sources: ChemIDPlus Advanced, PubChem, DrugBank, RxList, Drug Central, Drugs.com, and Wikipedia

2   Gielen K, Goossens A. Occupational allergic contact dermatitis from drugs in healthcare workers. Contact Dermatitis 2001;45:273-279

3   Rudzki E, Zakrzewski Z, Rebandel P, Grzywa Z, Hudymowicz W. Sensitivity to amikacin. Contact Dermatitis 1989;20:391

4   Bensaid B, Rozieres A, Nosbaum A, Nicolas J-F, Berard F. Amikacin-induced drug reaction with eosinophilia and systemic symptoms syndrome: Delayed skin test and ELISPOT assay results allow the identification of the culprit drug. J Allergy Clin Immunol 2012;130:1413-1414

5   Studer M, Waton J, Bursztejn AC, Aimone-Gastin I, Schmutz JL, Barbaud A. Does hypersensitivity to multiple drugs really exist? Ann Dermatol Venereol 2012;139:375-380 (Article in French)

6   Barbaud A, Collet E, Milpied B, Assier H, Staumont D, Avenel-Audran M, et al. A multicentre study to determine the value and safety of drug patch tests for the three main classes of severe cutaneous adverse drug reactions. Br J Dermatol 2013;168:555-562

Chapter 3.25   AMILORIDE

IDENTIFICATION

Description/definition   : Amiloride is the pyrazine that conforms to the structural formula shown below
Pharmacological classes : Acid sensing ion channel blockers; diuretics; epithelial sodium channel blockers
IUPAC name   : 3,5-Diamino-6-chloro-N-(diaminomethylidene)pyrazine-2-carboxamide
CAS registry number   : 2609-46-3
EC number   : 220-024-7
Merck Index monograph : 1671
Patch testing   : No data available; most systemic drugs can be tested at 10% pet.; if the pure chemical is
                                 not available, prepare the test material from intravenous powder, the content of capsules
                                 or – when also not available – from powdered tablets to achieve a final concentration of
                                 the active drug of 10% pet.
Molecular formula   : $C_6H_8ClN_7O$

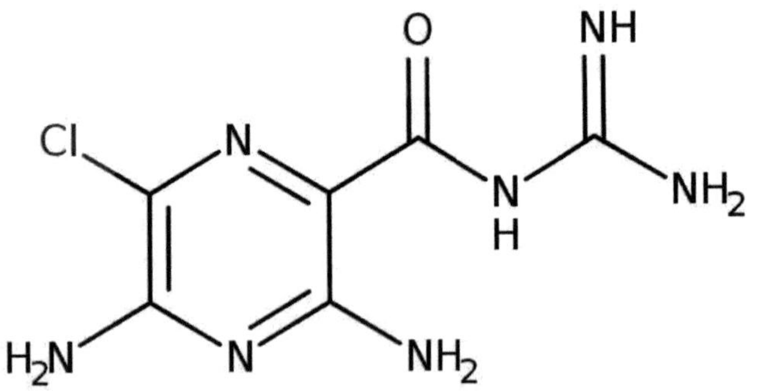

GENERAL

Amiloride is a potassium-sparing diuretic that helps to treat hypertension and congestive heart failure. The drug is often used in conjunction with thiazide or loop diuretics. In pharmaceutical products, amiloride is employed as amiloride hydrochloride (CAS number 2016-88-8, EC number 217-958-2, molecular formula $C_6H_9Cl_2N_7O$) or as amiloride hydrochloride dihydrate (CAS number 17440-83-4, EC number not available, molecular formula $C_6H_{13}Cl_2N_7O_3$) (1).

CUTANEOUS ADVERSE DRUG REACTIONS FROM SYSTEMIC ADMINISTRATION CAUSED BY TYPE IV

(DELAYED-TYPE) HYPERSENSITIVITY (as demonstrated by positive patch tests)

Cutaneous adverse drug reactions from systemic administration of amiloride caused by type IV (delayed-type) hypersensitivity have included unspecified drug eruption (2).

Other cutaneous adverse drug reactions

In Finland, in the period 1989-2001, 826 patients with suspected cutaneous drug eruptions were patch tested and 89 had one or more positive reactions. Of these individuals, one reacted to hydrochlorothiazide + amiloride. The 2 diuretics were probably not tested separately. It was not specified what type of drug reaction this combination product had caused (2).

LITERATURE

1   The data in the section 'General' may have been obtained from literature discussed in this chapter, but mostly also or exclusively from one or more of the following online sources: ChemIDPlus Advanced, PubChem, DrugBank, RxList, Drug Central, Drugs.com, and Wikipedia
2   Lammintausta K, Kortekangas-Savolainen O. The usefulness of skin tests to prove drug hypersensitivity. Br J Dermatol 2005;152:968-974

Chapter 3.26   AMINOCAPROIC ACID

IDENTIFICATION

Description/definition : Aminocaproic acid is the synthetic lysine derivative that conforms to the structural formula shown below
Pharmacological classes : Antifibrinolytic agents
IUPAC name : 6-Aminohexanoic acid
Other names : epsilon-Aminocaproic acid; ε-aminocaproic acid
CAS registry number : 60-32-2
EC number : 200-469-3
Merck Index monograph : 1697
Patch testing : 1, 5 and 10% water (2)
Molecular formula : $C_6H_{13}NO_2$

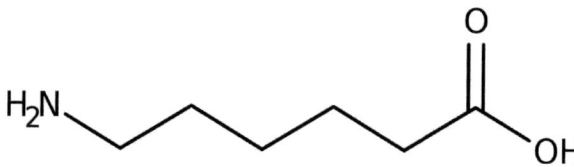

GENERAL

Aminocaproic acid (ACA) is an antifibrinolytic agent that acts by inhibiting plasminogen activators which have fibrinolytic properties. This drug is used in the treatment and prophylaxis of hemorrhage, including hyperfibrinolysis-induced hemorrhage and excessive postoperative bleeding (1). ε-Aminocaproic acid was discovered in Japan in 1953. In addition to its antifibrinolytic properties, it acts as buffer and has an anti-inflammatory effect. The drug is used both in topical and in systemic pharmaceutical applications. In topical preparations, it has caused contact allergy/allergic contact dermatitis, which has been fully reviewed in Volume 3 of the *Monographs in contact allergy* series (5).

CUTANEOUS ADVERSE DRUG REACTIONS FROM SYSTEMIC ADMINISTRATION CAUSED BY TYPE IV (DELAYED-TYPE) HYPERSENSITIVITY (as demonstrated by positive patch tests)

Cutaneous adverse drug reactions from systemic administration of aminocaproic acid caused by type IV (delayed-type) hypersensitivity have included systemic allergic dermatitis (3,4) and symmetrical drug-related intertriginous and flexural exanthema (SDRIFE) (2).

Systemic allergic dermatitis (systemic contact dermatitis)

Aminocaproic acid has caused some cases of systemic allergic dermatitis; these are summarized in table 3.26.1.

Table 3.26.1 Systemic allergic dermatitis from aminocaproic acid

Year	Sex	Age	Clinical picture	Patch test	Comments	Ref.
1999	M	70	pruritic generalized micro-papular eruption	commercial intravenous preparation with aminocaproic acid; concentration not stated	aminocaproic acid used to stop bleeding	3
1995	M	68	widespread itching maculo-papular erythematous rash	commercial ACA solution 4g/ml (?)	7 controls were negative	4
1995	F	15	rash in the axillae and groins, foot numbness, blurred vision, maculopapular eruption trunk	commercial ACA solution 4g/ml (?)	oral ACA given preoperatively	4

Symmetrical drug-related intertriginous and flexural exanthema (SDRIFE)

A 47-year-old woman had a pruriginous dermatitis involving her axillary, infra-mammary and inguinal flexures, and the lateral aspects of her trunk. Confluent erythematous patches without epidermal detachment were observed, sparing the extremities and mucous membranes. A week before the onset of dermatitis, she was started on aminocaproic acid treatment (9 g/day) for excessive vaginal bleeding due to uterine myomatosis. The drug was promptly withdrawn when the cutaneous lesions appeared. Patch tests were performed. The drug, diluted to 30% pet. showed a strongly positive reaction (+++), as did pure aminocaproic acid 1%, 5% and 10% water. Ten controls were negative to aminocaproic acid 30% pet. As the patient had apparently not been sensitized by topical aminocaproic acid before, this was a case of symmetrical drug-related intertriginous and flexural exanthema (SDRIFE) (2).

LITERATURE

1   The data in the section 'General' may have been obtained from literature discussed in this chapter, but mostly also or exclusively from one or more of the following online sources: ChemIDPlus Advanced, PubChem, DrugBank, RxList, Drug Central, Drugs.com, and Wikipedia

2   Cunha D, Carvalho R, Santos R, Cardoso J. Systemic allergic dermatitis to epsilon-aminocaproic acid. Contact Dermatitis 2009;61:303-304

3   Villarreal O. Systemic dermatitis with eosinophilia due to epsilon-aminocaproic acid. Contact Dermatitis 1999;40:114

4   González Gutiérrez ML, Esteban López MI, Ruíz Ruíz MD. Positivity of patch tests in cutaneous reaction to aminocaproic acid: two case reports. Allergy 1995;50:745-746

5   De Groot AC. Monographs in contact allergy, volume 3: Topical Drugs. Boca Raton, Fl, USA: CRC Press Taylor and Francis Group, 2021 (ISBN 978-0-367-23693-9)

Chapter 3.27   AMINOPHENAZONE

IDENTIFICATION

Description/definition   : Aminophenazone is the pyrazolone that conforms to the structural formula shown below
Pharmacological classes : Analgesics and antipyretics; anti-inflammatory agents, non-steroidal
IUPAC name   : 4-(Dimethylamino)-1,5-dimethyl-2-phenylpyrazol-3-one
Other names   : Aminopyrine
CAS registry number   : 58-15-1
EC number   : 200-365-8
Merck Index monograph : 1740
Patch testing   : 10% pet. (SmartPracticeCanada)
Molecular formula   : $C_{13}H_{17}N_3O$

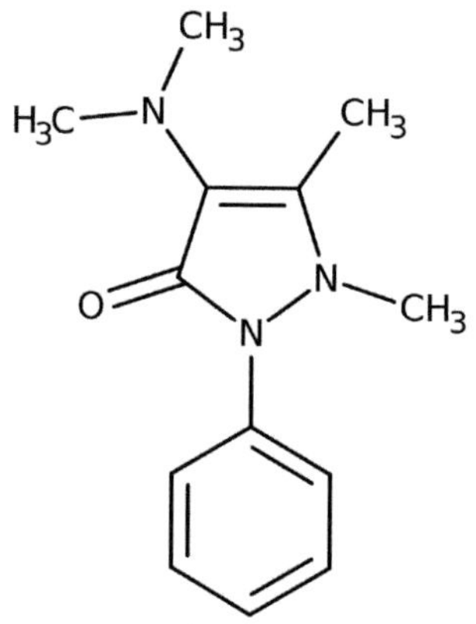

GENERAL

Aminophenazone is a pyrazolone with analgesic, anti-inflammatory, and antipyretic properties. Formerly it was widely used as an antipyretic and analgesic in rheumatism, neuritis, and common colds. Because of the risk of agranulocytosis, it is not employed as a human drug anymore. However, aminophenazone is used to measure total body water and radiolabeled aminophenazone has been used in breath tests to measure the cytochrome P-450 metabolic activity in liver function tests (1).

CUTANEOUS ADVERSE DRUG REACTIONS FROM SYSTEMIC ADMINISTRATION CAUSED BY TYPE IV (DELAYED-TYPE) HYPERSENSITIVITY (as demonstrated by positive patch tests)

Cutaneous adverse drug reactions from systemic administration of aminophenazone caused by type IV (delayed-type) hypersensitivity have included toxic epidermal necrolysis (TEN) (3).

Toxic epidermal necrolysis (TEN)

In Japanese literature, a case of toxic epidermal necrolysis caused by aminophenazone (aminopyrine) with a positive patch test to this NSAID has been reported (article in Japanese, data cited in ref. 3)

Cross-reactions, pseudo-cross-reactions and co-reactions

Of 8 patients with a fixed drug eruption from antipyrine salicylate (phenazone salicylate), in who type-IV allergy was demonstrated by topical provocation (an open application test), all cross-reacted to phenazone, 4 to propyphenazone and 3 to aminophenazone (2).

Immediate contact reactions

Immediate contact reactions (contact urticaria) to aminophenazone are presented in Chapter 5.

LITERATURE

1   The data in the section 'General' may have been obtained from literature discussed in this chapter, but mostly also or exclusively from one or more of the following online sources: ChemIDPlus Advanced, PubChem, DrugBank, RxList, Drug Central, Drugs.com, and Wikipedia
2   Alanko K. Topical provocation of fixed drug eruption. A study of 30 patients. Contact Dermatitis 1994;31:25-27
3   Tagami H, Tatsuta K, Iwatski K, Yamada M. Delayed hypersensitivity in ampicillin-induced toxic epidermal necrolysis. Arch Dermatol 1983;119:910-913 (citing Japanese literature)

Chapter 3.28   AMINOPHENAZONE HEMIBARBITAL

IDENTIFICATION

Description/definition : Aminophenazone hemibarbital is a combination product of aminophenazone and barbital in a 2:1 ratio

Pharmacological classes : Non-steroidal anti-inflammatory agents (aminophenazone); hypnotics (barbital)

IUPAC name : 5,5-Diethyl-1,3-diazinane-2,4,6-trione;4-(dimethylamino)-1,5-dimethyl-2-phenylpyrazol-3-one

Other names : Pyrabital

CAS registry number : 69401-33-8

EC number : Not available

Patch testing : No data available; most systemic drugs can be tested at 10% pet.; if the pure chemical is not available, prepare the test material from intravenous powder, the content of capsules or – when also not available – from powdered tablets to achieve a final concentration of the active drug of 10% pet.

Molecular formula : $C_{34}H_{46}N_8O_5$

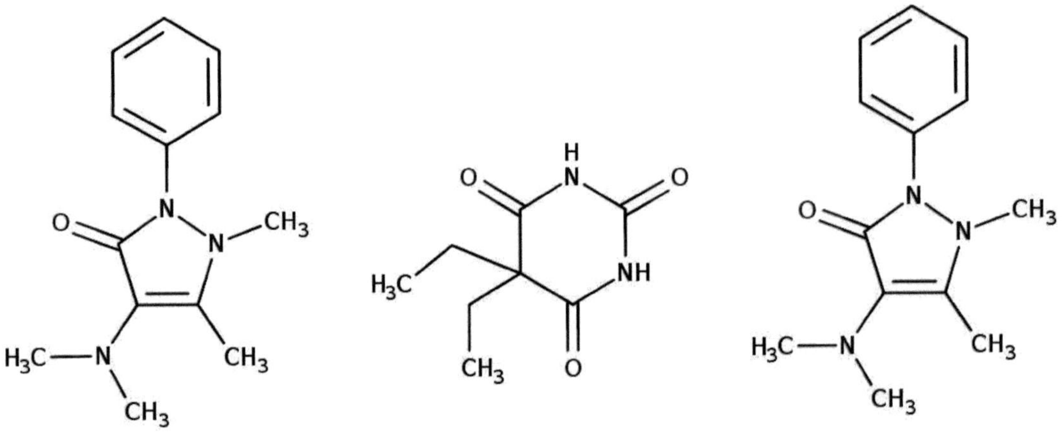

GENERAL

Aminophenazone hemibarbital is a combination product of aminopyrine (aminophenazone, CAS 58-15-1) and barbital in a 2:1 ratio. Aminopyrine is a pyrazolone with analgesic, anti-inflammatory, and antipyretic properties. Barbital (CAS 57-44-3) is a long-acting barbiturate used as a hypnotic and sedative. No data can be found on this drug and it is almost certain that it is not used anymore.

CUTANEOUS ADVERSE DRUG REACTIONS FROM SYSTEMIC ADMINISTRATION CAUSED BY TYPE IV (DELAYED-TYPE) HYPERSENSITIVITY (as demonstrated by positive patch tests)

Cutaneous adverse drug reactions from systemic administration of aminophenazone hemibarbital caused by type IV (delayed-type) hypersensitivity have included toxic epidermal necrolysis (TEN) (1).

Toxic epidermal necrolysis (TEN)

In Japanese literature, a case of toxic epidermal necrolysis caused by pyrabital (aminophenazone hemibarbital) with a positive patch test to this drug has been reported in 1970 (1).

LITERATURE

1   Matsumoto T. A case of toxic epidermal necrolysis (Lyell). Nishinihon J Dermatol 1970;32:3-7 (Article in Japanese, data cited in ref. 2)
2   Tagami H, Tatsuta K, Iwatski K, Yamada M. Delayed hypersensitivity in ampicillin-induced toxic epidermal necrolysis. Arch Dermatol 1983;119:910-913

Chapter 3.29 AMINOPHYLLINE

IDENTIFICATION

Description/definition : Aminophylline is a combination drug containing theophylline and ethylenediamine
Pharmacological classes : Cardiotonic agents; purinergic P1 receptor antagonists; bronchodilator agents; phosphodiesterase inhibitors
IUPAC name : 1,3-Dimethyl-7H-purine-2,6-dione;ethane-1,2-diamine
Other names : 1H-Purine-2,6-dione, 3,7-dihydro-1,3-dimethyl-, compd. with 1,2-ethanediamine (2:1)
CAS registry number : 317-34-0
EC number : 206-264-5
Merck Index monograph : 1731
Patch testing : 1% water or pet.; test also ethylenediamine and theophylline 1% pet.
Molecular formula : $C_{16}H_{24}N_{10}O_4$

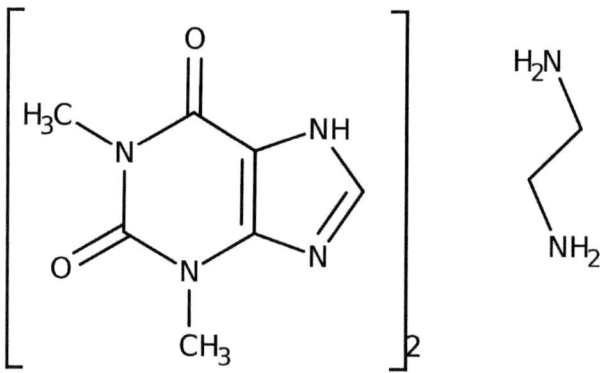

GENERAL

Aminophylline is a drug combination that contains theophylline and ethylenediamine in a 2:1 ratio. It is more soluble in water than theophylline but has similar pharmacologic actions as a bronchodilator agent. Aminophylline is indicated for the treatment of lung diseases such as asthma, chronic bronchitis, and COPD. The majority of aminophylline medications are discontinued and the remaining medications on the market are in short supply (1).

CONTACT ALLERGY FROM ACCIDENTAL CONTACT

Occupational allergic contact dermatitis from aminophylline has been observed repeatedly. In the cases where the 2 ingredients were tested separately, ethylenediamine was always the sensitizer (4,6,10,26).

Of 333 nurses patch tested in Poland between 1979 and 1987, 2 reacted to aminophylline ampoule contents; the dermatitis disappeared after avoiding contact with this drug (3). Before that, the same author had already observed one or more nurses with occupational allergic contact dermatitis from aminophylline (5).

A 46-year-old nurse who had worked for more than 20 years in a Pulmonary department had suffered frequent episodes of eczematous lesions on the hands and systemic urticaria for 3 years. Patch tests were strongly (+++) positive to ethylenediamine, which was associated with marked exacerbation of her complaints. The patient had probably become sensitized from preparing and administering injectable aminophylline preparations for treatment of asthma. A provocation test was positive (4).

A 33-year-old nurse, preparing and administering systemic aminophylline in a department of pneumology, developed acute dermatitis of 3 fingers of the left and 2 of the right hand. Patch tests were strongly positive to ethylenediamine. In a second patch test session, there were positive reactions to commercial aminophylline as is (+++ at D3) and 1% water (++ at D3), but negative to theophylline 1% water, confirming that ethylenediamine was the allergenic ingredient in this case of occupational allergic contact dermatitis to aminophylline (6).

In early literature, occupational allergic dermatitis from aminophylline had been observed in two pharmacists, one preparing aminophylline suppositories (9) and the other filling capsules with aminophylline (10). Both had eczema of the hands and arms but also the face, suggesting airborne contact. Both patients had positive reactions to aminophylline 1%, but were negative to theophylline 1%. One of the patients had an extremely strong positive patch test reaction to ethylenediamine (10), in the other, it was not tested (9).

CUTANEOUS ADVERSE DRUG REACTIONS FROM SYSTEMIC ADMINISTRATION CAUSED BY TYPE IV (DELAYED-TYPE) HYPERSENSITIVITY (as demonstrated by positive patch tests)

Cutaneous adverse drug reactions from systemic administration of aminophylline caused by type IV (delayed-type) hypersensitivity have included fixed drug eruption (2), systemic allergic dermatitis manifesting as generalized lichenoid eruption (7), as erythroderma (8), as generalized dermatitis (8), as maculopapular eruption (12,23), as exfoliative erythroderma (15,16), as the baboon syndrome (20,21), as erythematous eruption (22) and as generalized eruption of erythematous papules (25); other eruptions have included extensive dermal edema, fever, broncho-spasm and eosinophilia (13) and urticaria from subcutaneous injection (14). See also refs. 11,17,18,19, and 24 (no details known to the author).

General

All reactions to aminophylline have been caused by its component ethylenediamine, theophylline never induced positive patch tests. Many of these patients had - definitely or very likely - become sensitized from the topical use of an antifungal-antibacterial-corticosteroid preparation containing ethylenediamine as a stabilizer (7,8,12,15,20,21, 22,25); some were sensitized by occupational exposure to products containing ethylenediamine (16). These patients had systemic allergic dermatitis, presenting with various clinical manifestations.

Fixed drug eruption

One case of fixed drug eruption to aminophylline with a positive patch test reaction on postlesional skin (itching, erythema, infiltration) to the commercial drug 10% pet. was reported from Seoul, South Korea, in the period 1986-1990 and in 1996 and 1997 (2).

Systemic allergic dermatitis (systemic contact dermatitis)

Two patients, who had been sensitized to ethylenediamine from the use of topical pharmaceuticals, developed a generalized lichenoid eruption, one also with dermal erythema, after the use of oral aminophylline (7).

A man aged 75 years, who was allergic to ethylenediamine received intramuscular aminophylline for asthma and within 12-24 hours developed erythroderma (8). An asthmatic 61-year-old man with perianal eczema was treated with an intravenous injection of aminophylline, and the dermatitis rapidly generalized. Patch testing with ethylenediamine was positive. Previously, the perianal lesions had spread to the buttocks after using aminophylline-based suppositories (8).

A 61-year-old woman was treated with aminophylline and within 12 hours developed a widespread itchy maculopapular rash. A provocation test was positive. Patch tests were positive to aminophylline and ethylenedia-mine 1% in water and pet. but negative to theophylline. The patient had probably become sensitized to ethylene-diamine from long-term treatment of eczema of the leg (12).

A 56-year-old man previously sensitized to ethylenediamine in a pharmaceutical cream developed exfoliative erythroderma after having been treated with aminophylline suppositories (15). A similar case was that of a 50-year-old woman who developed exfoliative erythroderma on 2 occasions after intake of aminophylline tablets for an acute asthmatic attack. Patch tests were strongly positive to ethylenediamine but negative to theophylline. This patient had probably previously become sensitized from extensive occupational exposure to synthetic waxes and wetting agents containing ethylenediamine (16).

A 66-year-old woman was given 10 ml aminophylline i.v. followed by an infusion for worsening of chronic obstructive lung disease. The following day an itching, erythematous eruption started on her neck and spread during the day to the buttocks and groins. The eruption was symmetric, intensely erythematous and affected the flanks stretching to the hips, groins, axillae and neck. A strongly positive patch test reaction to ethylenediamine was demonstrated (20). This was a case of systemic allergic dermatitis presenting as the 'baboon syndrome' (symmetrical drug-related intertriginous and flexural exanthema). A similar case was reported in 1999 (21). Both patients had been sensitized previously from using a topical pharmaceutical containing ethylenediamine.

A 68-year-old woman developed an erythematous eruption 3 days after being started on oral aminophylline for an exacerbation of chronic airways disease. Previously, patch testing had demonstrated contact allergy to ethylenediamine, ergo this was a case of systemic allergic dermatitis (22).

A 45-year-old man developed a generalized eruption of erythematous papules 'shortly' after having applied an aminophylline suppository for dyspnea. The patient was probably presensitized by the use of a cream containing ethylenediamine. Patch tests were positive to ethylenediamine 1% pet. (D2 ++, D3 ++) and aminophylline powder (++/++), but negative to theophylline powder (25).

Other cutaneous drug reactions

A 59-year-old man developed a hypersensitivity reaction to aminophylline with extensive dermal edema, fever, bronchospasm and eosinophilia after intravenous administration of aminophylline. Patch tests were positive to aminophylline and ethylenediamine and negative to theophylline. Patch testing reproduced the patient's prior erythroderma (13).

A 29-year-old woman had been treated for adiposis on both thighs with 2 sessions of mesotherapy (subcutaneous microinjections) with 1% aminophylline in water. Four hours after the 2nd session, highly pruritic red coalescent wheals developed at the sites of application, extending to much of the thighs. There was no systemic involvement. Intradermal tests were positive to 1% aminophylline in water and ethylenediamine 1% saline with both immediate (20 minutes) and delayed (8, 24 and 48 hours) reactions, but negative to theophylline. Patch tests were positive to ethylenediamine and aminophylline 1% pet. but negative to theophylline 0.5% pet. (14).

Cross-reactions, pseudo-cross-reactions and co-reactions
Pseudocross-reactivity between ethylenediamine and aminophylline.

LITERATURE

1   The data in the section 'General' may have been obtained from literature discussed in this chapter, but mostly also or exclusively from one or more of the following online sources: ChemIDPlus Advanced, PubChem, DrugBank, RxList, Drug Central, Drugs.com, and Wikipedia
2   Lee AY. Topical provocation in 31 cases of fixed drug eruption: change of causative drugs in 10 years. Contact Dermatitis 1998;38:258-260
3   Rudzki E, Rebandel P, Grzywa Z. Patch tests with occupational contactants in nurses, doctors and dentists. Contact Dermatitis 1989;20:247-250
4   Del Monte A, de Benedictis E, Laffi G. Occupational dermatitis from ethylenediamine hydrochloride. Contact Dermatitis 1987;17:254
5   Rudzki E. Occupational dermatitis among health service workers. Derm Beruf Umwelt 1979;27:112-115
6   Corazza M, Mantovani L, Trimurti S, Virgili A. Occupational contact sensitization to ethylenediamine in a nurse. Contact Dermatitis 1994;31:328-329
7   Provost TT, Jillson OF. Ethylenediamine contact dermatitis. Arch Dermatol 1967;96:231-234
8   Angelini G, Vena GA, Meneghini CL. Allergic contact dermatitis to some medicaments. Contact Dermatitis 1985:12:263-269
9   Tas J, Weissberg D. Allergy to aminophylline. Acta Allergol 1958;12:39-42
10 Baer RL, Cohen HJ, Neidorff AH. Allergic eczematous sensitivity to aminophylline. Arch Dermatol 1959;79:647-648
11 deShazo RD, Stevenson HC. Generalized dermatitis to aminophylline. Ann Allergy 1981;46:152-155
12 Hardy C, Schofield O, George CF. Allergy to aminophylline. Br Med J (Clin Res Ed) 1983;286(6383):2051-2052
13 Elias JA, Levinson AI. Hypersensitivity reactions to ethylenediamine in aminophylline. Am Rev Respir Dis 1981;123:550-552
14 Urbani CE. Urticarial reaction to ethylenediamine in aminophylline following mesotherapy. Contact Dermatitis 1994;31:198-199
15 Petrozzi JW, Shore RN. Generalized exfoliative dermatitis from ethylenediamine. Arch Dermatol 1976;112:525-526
16 Bernstein JE, Lorinez AL. Ethylenediamine-induced exfoliative erythroderma. Arch Dermatol 1979;115:360-361
17 Kradjan WA, Lakshminarayan S. Allergy to aminophylline: lack of predictability by skin testing. Am J Hosp Pharm 1981;38:1031-1033
18 Vázquez Botet M. Systemic eczematous contact dermatitis due to the ethylenediamine fraction of aminophylline. Boletin de la Asociacion Medica de Puerto Rico 1980;72:14-18 (Article in Spanish)
19 Berman BA, Ross RN. Ethylenediamine: systemic eczematous contact-type dermatitis. Cutis 1983;31:594,596,598
20 Isaksson M, Ljunggren B. Systemic contact dermatitis from ethylenediamine in an aminophylline preparation presenting as the baboon syndrome. Acta Derm Venereol 2003;83:69-70
21 Guin JD, Fields P, Thomas KL. Baboon syndrome from i.v. aminophylline in a patient allergic to ethylenediamine. Contact Dermatitis 1999;40:170-171
22 Walker S, Ferguson JE. Systemic allergic contact dermatitis due to ethylenediamine following administration of oral aminophylline. Br J Dermatol 2004;150:594
23 Terzian CG, Simon PA. Aminophylline hypersensitivity apparently due to ethylenediamine. Ann Emerg Med 1992;21:312-314
24 Mohsenifar Z, Lehrlan S, Carson SA, Tashkin D. Two cases of allergy to aminophylline. Ann Allergy 1982;49:281-282
25 Van den Berg WHHW, Van Ketel WG. Contactallergie voor ethyleendiamine. Ned Tijdschr Geneeskd 1983;127:1801-1802 (Article in Dutch)
26 Eberhartinger C. Kontakallergie gegen Äthylendiamin. Hautarzt 1964;15:450

Chapter 3.30   AMINOSALICYLIC ACID

IDENTIFICATION

Description/definition   : Aminosalicylic acid is the salicylic acid derivative that conforms to the structural formula
                                           shown below
Pharmacological classes   : Antitubercular agents
IUPAC name   : 4-Amino-2-hydroxybenzoic acid
Other names   : p-Aminosalicylic acid; 4-aminosalicylic acid; PAS
CAS registry number   : 65-49-6
EC number   : 200-613-5
Merck Index monograph   : 1743
Patch testing   : 5% water. (2); 3% pet. (5)
Molecular formula   : $C_7H_7NO_3$

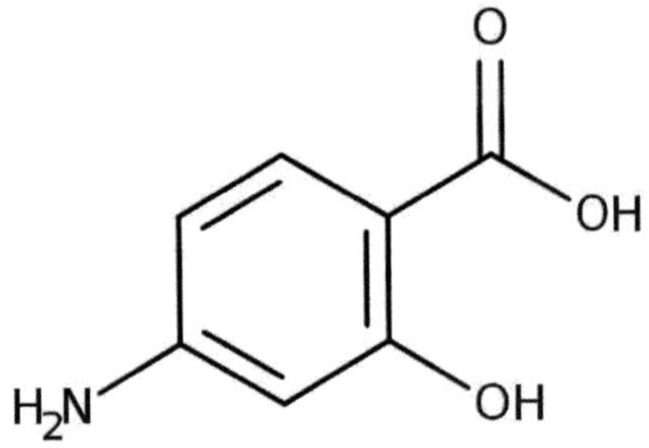

GENERAL

Aminosalicylic acid is an analog of para-aminobenzoic acid (PABA) with antitubercular activity. Aminosalicylic acid exerts its bacteriostatic activity against *Mycobacterium tuberculosis* by competing with PABA for enzymes involved in folate synthesis, thereby suppressing growth and reproduction of *M. tuberculosis*, eventually leading to cell death. It is used with other anti-tuberculosis drugs (most often isoniazid) for the treatment of all forms of active tuberculo-sis due to susceptible strains of tubercle bacilli. The drug is bacteriostatic against *Mycobacterium tuberculosis* and also inhibits the onset of bacterial resistance to streptomycin and isoniazid. In pharmaceutical products, it may be used as aminosalicylate calcium (trihydrate), aminosalicylate sodium (anhydrous), and potassium aminosalicylate (1).

CUTANEOUS ADVERSE DRUG REACTIONS FROM SYSTEMIC ADMINISTRATION CAUSED BY TYPE IV (DELAYED-TYPE) HYPERSENSITIVITY (as demonstrated by positive patch tests)

Cutaneous adverse drug reactions from systemic administration of aminosalicylic acid caused by type IV (delayed-type) hypersensitivity have included a hypersensitivity reaction, most likely drug reaction with eosinophilia and systemic symptoms (DRESS) (2,3,4).

Drug reaction with eosinophilia and systemic symptoms (DRESS)

A 16-year-old girl developed a hypersensitivity reaction while taking aminosalicylic acid (PAS, p-aminosalicylic acid; used as sodium aminosalicylate) in combination with isoniazid (INH) for tuberculosis. The reaction was characterized by nausea, high fever, headache, a generalized erythematous eruption with edema of the face, lips and eyelids, lymphadenopathy and a palpable spleen. A test for mononucleosis infectiosa was negative. Laboratory investigation showed elevated erythrocyte sedimentation rate and atypical lymphocytes. Two days after recovery (far too soon, but that was unknown in 1962) patch tests were performed and were positive to sodium aminosalicylate 5% water and moistened isoniazid powder. Five controls were negative. During the patch tests a maculopapular exanthema appeared which generalized (2). This was highly likely a case of DRESS with polysensitization. It is uncertain whether the exacerbation during patch testing was actually caused by this diagnostic procedure, or was a spontaneous exacerbation, as is often observed in DRESS.

   Similar hypersensitivity reactions to PAS, but in combination with icterus, and with positive patch test reactions to sodium aminosalicylate had already been reported in the 1950s (3,4).

LITERATURE

1   The data in the section 'General' may have been obtained from literature discussed in this chapter, but mostly also or exclusively from one or more of the following online sources: ChemIDPlus Advanced, PubChem, DrugBank, RxList, Drug Central, Drugs.com, and Wikipedia

2   Van Ketel W. A severe hypersensitivity reaction to the use of PAS and isoniazid. Ned Tijdschr Geneeskd 1963;107:952-955 (Article in Dutch)

3   Cuthbert J. Acquired idiosyncrasy to sodium-p-aminosalicylate. Lancet 1950;256(6623):209-211

4   Meyler L. Schadelijke nevenwerkingen van geneesmiddelen. Assen: Van Gorcum & Comp. N.V., 1954 (Data cited in ref. 2)

5   De Groot AC. Patch testing, 4th edition. Wapserveen, The Netherlands: acdegroot publishing, 2018 (ISBN 9789081323345)

Chapter 3.31   AMITRIPTYLINE

IDENTIFICATION
Description/definition   : Amitriptyline is the derivative of dibenzocycloheptadiene that conforms to the structural
                            formula shown below
Pharmacological classes : Adrenergic uptake inhibitors; analgesics, non-narcotic; antidepressive agents, tricyclic
IUPAC name   : N,N-Dimethyl-3-(2-tricyclo[9.4.0.0^{3,8}]pentadeca-1(15),3,5,7,11,13-hexaenylidene)propan-
                            1-amine
Other names   : 10,11-Dihydro-5-(γ-dimethylaminopropylidene)-5H-dibenzo(a,d)cycloheptene
CAS registry number   : 50-48-6
EC number   : 200-041-6
Merck Index monograph   : 1753
Patch testing   : 5% and 10% pet. (2)
Molecular formula   : $C_{20}H_{23}N$

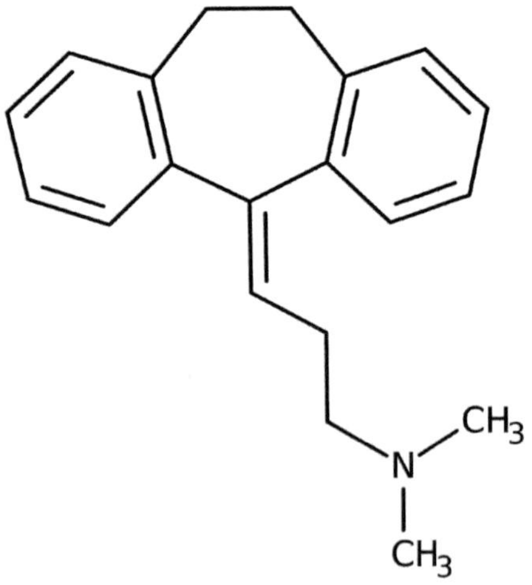

GENERAL
Amitriptyline is a tertiary amine tricyclic antidepressant with anticholinergic and (extremely) sedative properties. It exhibits strong anticholinergic activity, cardiovascular effects including orthostatic hypotension, changes in heart rhythm and conduction, and a lowering of the seizure threshold. Amitriptyline is indicated for major depressive disorder, management of neuropathic pain, prophylactic treatment of chronic tension-type headache and prophylactic treatment of migraine (all in adults). In children of 6 years and older it may be used for treatment of nocturnal enuresis when organic pathology has been excluded. Although not a labeled indication, amitriptyline is also widely used in the management of chronic non-malignant pain, e.g. in fibromyalgia (1). In pharmaceutical products, amitriptyline is employed as amitriptyline hydrochloride (CAS number 549-18-8, EC number 208-964-6, molecular formula $C_{20}H_{24}ClN$) (1).

CUTANEOUS ADVERSE DRUG REACTIONS FROM SYSTEMIC ADMINISTRATION CAUSED BY TYPE IV (DELAYED-TYPE) HYPERSENSITIVITY (as demonstrated by positive patch tests)
Cutaneous adverse drug reactions from systemic administration of amitriptyline caused by type IV (delayed-type) hypersensitivity have included maculopapular eruption (3) and photosensitivity (2,4).

Maculopapular eruption
A 62-year-old man was started on carbamazepine and amitriptyline for trigeminal neuralgia. Seven days later, he developed DRESS with progressive generalized maculopapular rash, facial angioedema, palpable vasculitis in the legs, oral aphthous ulcers, fever, axillary and inguinal lymph node enlargement, leukocytosis with neutrophilia, and elevated liver enzymes. Eight years later, the patient was treated with acyclovir and amitriptyline for herpes zoster. Four days later, a generalized rash appeared. Patch tests with 5% carbamazepine and 1% amitriptyline in petrolatum were positive at D2 and D3. Five controls were negative. The authors suggested that the patient had become sensitized to amitriptyline during the first episode of DRESS (anticonvulsant hypersensitivity syndrome) caused by carbamazepine (3).

Photosensitivity

A 45-year-old woman, using amitriptyline hydrochloride 25 mg oral daily for the past year, presented with a 1-month history of diffuse hyperpigmentation over exposed areas of the back and left side of the abdomen (areas not covered by the sari). Furrows and skin folds were spared. She did not recall any erythema, scaling or burning sensation. Patch and photopatch testing was performed, patch test materials being prepared by extracting the active ingredient from the commercially available preparation. 1%, 2%, 5% and 10% amitriptyline hydrochloride were made up in an aqueous cream base. Patch test readings at D2 were negative on both sides. The left side was then exposed to 15 J/cm^2 UVA. One day after irradiation, readings on both sides were negative, but after 2 days, amitriptyline 10% and 5% showed a positive reaction on the UVA-exposed site only, suggesting photosensitivity to the drug. Seven controls were negative (2).

In Finland, in the period 1989-2001, 826 patients with suspected cutaneous drug eruptions were patch tested and 89 had one or more positive reactions. Of these individuals, one with an eruption in the light-exposed areas had a positive photopatch test to amitriptyline (4).

LITERATURE

1   The data in the section 'General' may have been obtained from literature discussed in this chapter, but mostly also or exclusively from one or more of the following online sources: ChemIDPlus Advanced, PubChem, DrugBank, RxList, Drug Central, Drugs.com, and Wikipedia

2   Sandra A, Srinivas CR, Deshpande SC. Photopatch test reaction to amitriptyline. Contact Dermatitis 1998;39:208-209

3   Gaig P, García-Ortega P, Baltasar M, Bartra J. Drug neosensitization during anticonvulsant hypersensitivity syndrome. J Investig Allergol Clin Immunol 2006;16:321-326

4   Lammintausta K, Kortekangas-Savolainen O. The usefulness of skin tests to prove drug hypersensitivity. Br J Dermatol 2005;152:968-974

Chapter 3.32   AMLEXANOX

IDENTIFICATION

Description/definition   : Amlexanox is the carboxylic acid derivative that conforms to the structural formula shown
                             below
Pharmacological classes   : Anti-allergic agents
IUPAC name   : 2-Amino-5-oxo-7-propan-2-ylchromeno[2,3-b]pyridine-3-carboxylic acid
Other names   : 2-Amino-7-isopropyl-5-oxo-5H-(1)benzopyrano(2,3-b)pyridine-3-carboxylic acid
CAS registry number   : 68302-57-8
EC number   : Not available
Merck Index monograph   : 1756
Patch testing   : 1% water; petrolatum is less suitable as patch testing vehicle (7)
Molecular formula   : $C_{16}H_{14}N_2O_4$

GENERAL

Amlexanox is an anti-allergic and anti-inflammatory drug. It is a strong inhibitor of histamine release and a competitive inhibitor of leukotriene. In a mucoadhesive oral paste, this agent has been clinically proven to abort the onset, accelerate healing and resolve the pain of aphthous ulcers (canker sores). Tablets have been on the market since 1987 in Japan and are used for the treatment of asthma and allergic rhinitis (1). Amlexanox is (or was) used both in topical and in systemic pharmaceutical applications. In topical preparations, the drug has caused contact allergy/allergic contact dermatitis, which has been fully reviewed in Volume 3 of the *Monographs in contact allergy series* (8).

CUTANEOUS ADVERSE DRUG REACTIONS FROM SYSTEMIC ADMINISTRATION CAUSED BY TYPE IV (DELAYED-TYPE) HYPERSENSITIVITY (as demonstrated by positive patch tests)

Cutaneous adverse drug reactions from systemic administration of amlexanox caused by type IV (delayed-type) hypersensitivity have included maculopapular eruption (5), fixed drug eruption (2), systemic allergic dermatitis (6),eczematous drug eruption (3,4), and pityriasis rosea-like drug eruption (5).

Maculopapular eruption

A female patient aged 43 had maculopapular eruptions from amlexanox. She had a positive patch test and positive oral provocation test to amlexanox (5).

Fixed drug eruption

A 23-year-old woman had suffered from allergic rhinitis for many years. She was prescribed tablets containing amlexanox and tablets with lysozyme chloride. She would take about 10 tablets each per month. About 1 year later, the patient noticed a macular erythema about 3.5 cm wide on her buttocks. Patch tests were performed with both tablets 50% pet. applied to the patient's eruptive and non-eruptive areas. The amlexanox material was dubiously positive (D2 erythema, D3 slight erythema) on the eruptive area and negative on non-eruptive skin. A ROAT with amlexanox 50% pet. yielded erythema at D2 and erythema and itching at D3 on the eruptive area but not on normal skin. The pure chemical amlexanox was not tested and neither were the excipients. The patient was diagnosed with fixed drug eruption from amlexanox (2).

Systemic allergic dermatitis (systemic contact dermatitis)

A 23-year-old woman with allergic conjunctivitis had been treated with 0.25% amlexanox ophthalmic solution for 1½ year, when she woke up with itchy redness around her eyes. Patch tests showed a positive reaction to the solution. Later, the ingredients of the product were tested separately, and the patient reacted to amlexanox 1%, 0.25%, 0.1%

and 0.025% water and to 1% and 0.25% pet. Her eyelid dermatitis cleared with discontinuance of the ophthalmic solution and topical corticosteroids. Four months later, an otologist prescribed tablets containing 50 mg amlexanox. The first tablet resulted in itching after an hour and the second, a day later, was followed by an erythema multiforme-like eruption on the patient's ears, neck, breasts and trunk. A diagnosis of systemic allergic dermatitis was made (6).

Dermatitis/eczematous eruption
A women aged 57 had an eczematous drug eruption from amlexanox; patch testing was positive (3). A 48-year-old woman also had an eczematous drug eruption; both a patch test and a lymphocyte stimulation test were positive (4).

Other drug eruptions
A man of 52 developed a pityriasis rosea Gilbert type drug eruption. He had a positive patch test and positive oral provocation test with to amlexanox (5).

LITERATURE

1   The data in the section 'General' may have been obtained from literature discussed in this chapter, but mostly also or exclusively from one or more of the following online sources: ChemIDPlus Advanced, PubChem, DrugBank, RxList, Drug Central, Drugs.com, and Wikipedia
2   Sugiura M, Hayakawa R, Osada T. Fixed drug eruption due to amlexanox. Contact Dermatitis 1998;38:65-67
3   Ishiguro N, Kato T, Nogita T, Kawashima M, Hidano S. A case of drug eruption from amlexanox (SOLFA). Rinsho Derma 1991;33:1602-1603 (Article in Japanese, data cited in ref. 2)
4   Taniguchi H, Otaki R, Takino C. A case of drug eruption due to amlexanox with drug induced hepatopathy. Rinsho Derma 1992;34:1745-1749 (Article in Japanese, data cited in ref. 2)
5   Inoue N, Makino H, Kamide R. 2 cases of drug eruption due to amlexanox. Rinsho Hifuka 1993;47:873-876 (Article in Japanese, data cited in ref. 2)
6   Hayakawa R, Ogino Y, Aris K, Matsunaga K. Systemic contact dermatitis due to amlexanox. Contact Dermatitis 1992;27:122-123
7   Yamashita H, Kawashima M. Contact dermatitis from amlexanox eyedrops. Contact Dermatitis 1991;25:255-256
8   De Groot AC. Monographs in contact allergy, volume 3: Topical Drugs. Boca Raton, Fl, USA: CRC Press Taylor and Francis Group, 2021 (ISBN 978-0-367-23693-9)

Chapter 3.33   AMLODIPINE BESYLATE

IDENTIFICATION

Description/definition : Amlodipine besylate is the benzenesulfonate salt of amlodipine that conforms to the
                         structural formula shown below
Pharmacological classes : Antihypertensive agents; vasodilator agents; calcium channel blockers
IUPAC name   : Benzenesulfonic acid;3-O-ethyl 5-O-methyl 2-(2-aminoethoxymethyl)-4-(2-chlorophenyl)-
                         6-methyl-1,4-dihydropyridine-3,5-dicarboxylate
Other names   : Amlodipine benzenesulfonate
CAS registry number   : 111470-99-6
EC number   : 601-097-8
Merck Index monograph : 1757 (Amlodipine)
Patch testing   : No data available; most systemic drugs can be tested at 10% pet.; if the pure chemical is
                         not available, prepare the test material from intravenous powder, the content of capsules
                         or – when also not available – from powdered tablets to achieve a final concentration of
                         the active drug of 10% pet.
Molecular formula   : $C_{26}H_{31}ClN_2O_8S$

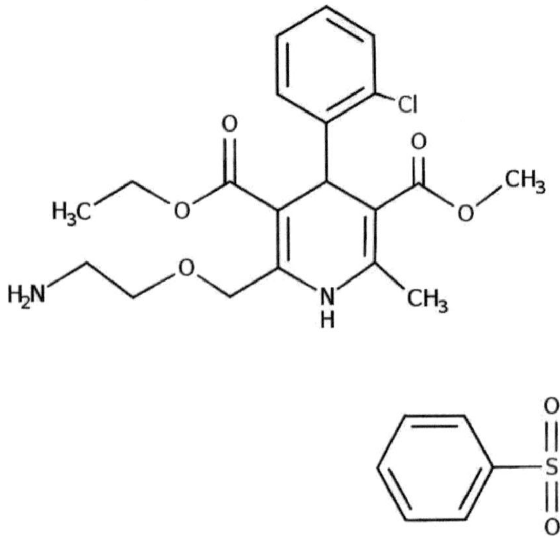

GENERAL

Amlodipine besylate is the benzenesulfonate salt of amlodipine, a synthetic dihydropyridine with antihypertensive
and antianginal effects. Amlodipine inhibits the influx of extracellular calcium ions into myocardial and peripheral
vascular smooth muscle cells, thereby preventing vascular and myocardial contraction. This results in a dilatation of
the main coronary and systemic arteries, decreased myocardial contractility, increased blood flow and oxygen
delivery to the myocardial tissue, and decreased total peripheral resistance (1).

CONTACT ALLERGY FROM ACCIDENTAL CONTACT

Amlodipine besylate may have caused occupational airborne allergic contact dermatitis in a pharmaceutical worker,
who was also sensitized to sodium risedronate and olanzapine (2). However, the article, to which was referred in a
recent textbook (3), could not be traced by the author.

LITERATURE

1   The data in the section 'General' may have been obtained from literature discussed in this chapter, but mostly
    also or exclusively from one or more of the following online sources: ChemIDPlus Advanced, PubChem,
    DrugBank, RxList, Drug Central, Drugs.com, and Wikipedia
2   Chomiczewska D, Kiec-Swierczynska M, Krecisz B . Airborne occupational allergic contact dermatitis to sodium
    risedronate, olanzapine and amlodipine benzenesulfonate in a pharmaceutical company worker – a case report.
    Eur J Allergy Clin Immunol 2010;65:593-594 (article could not be traced, cited in ref. 3)
3   Goossens A, Geebelen J, Hulst KV, Gilissen L. Pharmaceutical and cosmetic industries. In: John S, Johansen J,
    Rustemeyer T, Elsner P, Maibach H, eds. Kanerva's occupational dermatology. Cham, Switzerland: Springer,
    2020:2203-2219

Chapter 3.34   AMOBARBITAL

IDENTIFICATION

Description/definition   : Amobarbital is the pyrimidone and barbiturate that conforms to the structural formula shown below

Pharmacological classes : GABA modulators; hypnotics and sedatives

IUPAC name   : 5-Ethyl-5-(3-methylbutyl)-1,3-diazinane-2,4,6-trione

CAS registry number   : 57-43-2

EC number   : 200-330-7

Merck Index monograph : 1837

Patch testing   : Tablet, pulverized, 10% pet. (2); most systemic drugs can be tested at 10% pet.; if the pure chemical is not available, prepare the test material from intravenous powder, the content of capsules or – when also not available – from powdered tablets to achieve a final concentration of the active drug of 10% pet.

Molecular formula   : $C_{11}H_{18}N_2O_3$

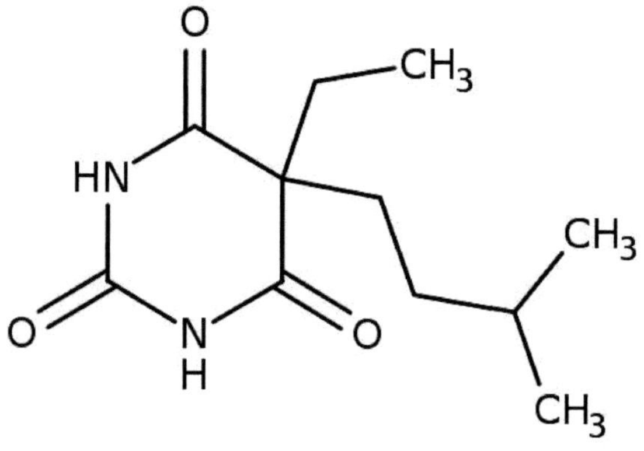

GENERAL

Amobarbital is a barbiturate with hypnotic and sedative, but not anxiolytic, properties. In pharmaceutical products, amobarbital is employed as amobarbital sodium (CAS number 64-43-7, EC number 200-584-9, molecular formula $C_{11}H_{17}N_2NaO_3$) (1).

CUTANEOUS ADVERSE DRUG REACTIONS FROM SYSTEMIC ADMINISTRATION CAUSED BY TYPE IV (DELAYED-TYPE) HYPERSENSITIVITY (as demonstrated by positive patch tests)

Cutaneous adverse drug reactions from systemic administration of amobarbital caused by type IV (delayed-type) hypersensitivity have included fixed drug eruption (2).

Fixed drug eruption

One patient who had developed a fixed drug eruption to amobarbital had a topical provocation test with amobarbital 10% pet. indicative of type-IV allergy to the drug (2).

LITERATURE

1   The data in the section 'General' may have been obtained from literature discussed in this chapter, but mostly also or exclusively from one or more of the following online sources: ChemIDPlus Advanced, PubChem, DrugBank, RxList, Drug Central, Drugs.com, and Wikipedia

2   Alanko K, Stubb S, Reitamo S. Topical provocation of fixed drug eruption. Br J Dermatol 1987;116:561-567

Chapter 3.35   AMOXAPINE

IDENTIFICATION
Description/definition   : Amoxapine is the dibenzoxazepine that conforms to the structural formula shown below
Pharmacological classes   : Adrenergic uptake inhibitors; serotonin uptake inhibitors; antidepressive agents, tricyclic; dopamine antagonists; neurotransmitter uptake inhibitors
IUPAC name   : 8-Chloro-6-piperazin-1-ylbenzo[b][1,4]benzoxazepine
Other names   : Desmethylloxapine
CAS registry number   : 14028-44-5
EC number   : 237-867-1
Merck Index monograph   : 1843
Patch testing   : 1% and 10% pet. (2)
Molecular formula   : $C_{17}H_{16}ClN_3O$

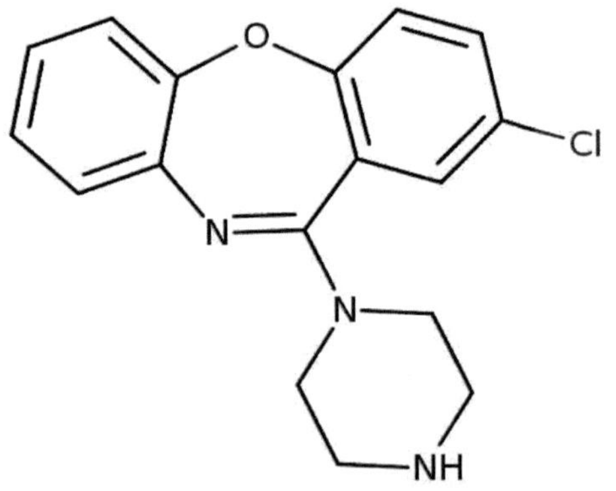

GENERAL
Amoxapine is the N-demethylated derivative of the antipsychotic drug loxapine and a tricyclic antidepressant of the dibenzoxazepine class. Amoxapine exerts its antidepressant effect by inhibiting the re-uptake of norepinephrine and, to a lesser degree, of serotonin, at adrenergic nerve endings and blocks the response of dopamine receptors to dopamine. This drug is used to treat symptoms of depression (1).

CUTANEOUS ADVERSE DRUG REACTIONS FROM SYSTEMIC ADMINISTRATION CAUSED BY TYPE IV (DELAYED-TYPE) HYPERSENSITIVITY (as demonstrated by positive patch tests)
Cutaneous adverse drug reactions from systemic administration of amoxapine caused by type IV (delayed-type) hypersensitivity have included an erythematous papular eruption (2).

Other cutaneous adverse drug reactions
A 63-year-old woman had an eleven-year history of reflex sympathetic dystrophy and had received 13 drugs, including amoxapine for 4 weeks and mexiletine hydrochloride for 6 weeks. Three days after taking cefaclor for a 39°C fever (from cystitis), she noticed erythematous papules on her back. The lesions gradually spread and increased in number on her trunk, arms and legs within three days. After cessation of most drugs, the skin lesions improved within 2 weeks. Two months later, positive patch test reactions were observed to amoxapine 0.01%, 0.1%, 1% and 10% pet. and to mexiletine hydrochloride 1% and 10% pet. Ten controls were negative. Patch tests with cefaclor and (some of) the other drugs were negative. Oral challenge with cefaclor (125 mg) and mexiletine hydrochloride (50 mg) provoked generalized erythema, and the administration of amoxapine (25 mg) induced only localized erythema on her neck and shoulder. No other drugs caused any eruption (2).

LITERATURE
1   The data in the section 'General' may have been obtained from literature discussed in this chapter, but mostly also or exclusively from one or more of the following online sources: ChemIDPlus Advanced, PubChem, DrugBank, RxList, Drug Central, Drugs.com, and Wikipedia
2   Nagayama H, Nakamura Y, Shinkai H. A case of drug eruption due to simultaneous sensitization with three different kinds of drugs. J Dermatol 1996;23:899-901

Chapter 3.36 AMOXICILLIN

IDENTIFICATION

Description/definition : Amoxicillin is the semisynthetic aminopenicillin antibiotic that conforms to the structural formula shown below

Pharmacological classes : Anti-bacterial agents

IUPAC name : (2S,5R,6R)-6-[[(2R)-2-Amino-2-(4-hydroxyphenyl)acetyl]amino]-3,3-dimethyl-7-oxo-4-thia-1-azabicyclo[3.2.0]heptane-2-carboxylic acid

Other names : Amoxycillin; amoxicillin anhydrous

CAS registry number : 26787-78-0

EC number : 248-003-8

Merck Index monograph : 1844

Patch testing : Amoxicillin trihydrate 10.0% pet. (Chemotechnique); patch testing in suspected beta-lactam allergy has been discussed in ref. 4

Molecular formula : $C_{16}H_{19}N_3O_5S$

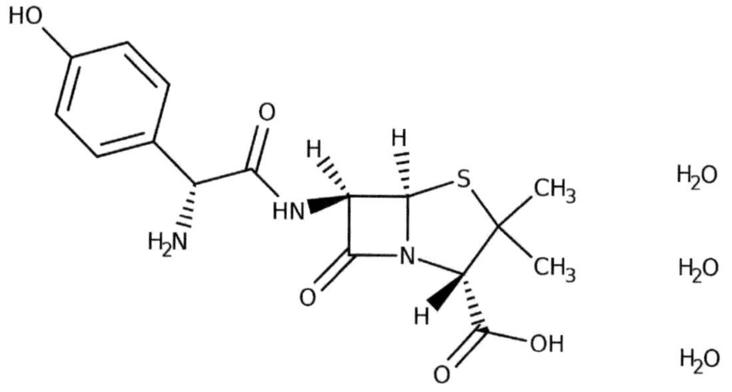

Amoxicillin trihydrate

GENERAL

Amoxicillin is a moderate-spectrum penicillin antibiotic active against a wide range of gram-positive, and a limited range of gram-negative organisms. It is usually the drug of choice within the class because it is better absorbed, following oral administration, than other β-lactam antibiotics. However, amoxicillin is susceptible to degradation by β-lactamase-producing bacteria, and so may be combined with clavulanic acid, a β-lactamase inhibitor, to increase the spectrum of action against gram-negative organisms, and to overcome bacterial antibiotic resistance mediated through β-lactamase production (1).

Amoxicillin is indicated for the treatment of infections of the ear, nose, and throat, the genitourinary tract, the skin and skin structure, and the lower respiratory tract due to susceptible (only β-lactamase-negative) strains of *Streptococcus* spp. (α- and β-hemolytic strains only), *S. pneumoniae*, *Staphylococcus* spp., *H. influenzae*, *E. coli*, *P. mirabilis*, or *E. faecalis*. It is also used for the treatment of acute, uncomplicated gonorrhea (anogenital and urethral infections) due to *N. gonorrhoeae* (1).

In pharmaceutical products, amoxicillin is employed as amoxicillin trihydrate (CAS number 61336-70-7, EC number 612-127-4, molecular formula $C_{16}H_{25}N_3O_8S$). Both amoxicillin anhydrous and amoxicillin trihydrate are called 'Amoxicillin' (1). See also Chapter 3.37 Amoxicillin mixture with clavulanate potassium, where cases of allergic reactions to this combination product are discussed in which the allergenic culprit ingredient was amoxicillin or was not identified.

There is a massive amount of recent but also older literature on drug eruptions from amoxicillin and delayed-type hypersensitivity reactions to this antibiotic. As it would be virtually impossible – or at least impractical – to read and assess all available material, the author has not attempted to provide a full review of the subject in this chapter.

Classification and structures of beta-lactam antibiotics

Amoxicillin is the prototype and currently most used betaplactam antibiotic. Beta-lactam (BL) antibiotics are classified into two major classes (penicillins and cephalosporins) and four minor classes: carbapenems, monobactams, oxacephems, and beta-lactamase inhibitors (clavulanic acid, sulbactam and tazobactam). The basic structure of all BLs consists of a four-membered beta-lactamase ring. In penicillins, it is attached to a five-membered thiazolidine ring; the side chain distinguishes the different penicillins. Cephalosporins have a six-membered sulfur-containing dihydrothiazine ring (instead of the five-membered thiazolidine ring of penicillins) and two side chains, which distinguish the different compounds. Carbapenems (e.g., imipenem, meropenem, ertapenem, and doripenem) contain a carbon double bond instead of sulfur in the five-membered thiazolidine ring and have a side chain, which distinguishes the different carbapenems. Aztreonam is the only monobactam antibiotic commercially available; it contains the BL ring without an attached five- or six-membered sulfur ring. The BL ring, the thiazolidine/dihydro-thiazine rings, and the side groups are all potentially immunogenic. Side chains in particular are important sites of immunological recognition and therefore may cause allergic cross-reactivity (2).

CONTACT ALLERGY FROM ACCIDENTAL CONTACT
Between 1978 and 2001, 14,689 patients have been patch tested in the university hospital of Leuven, Belgium. Occupational allergic contact dermatitis to pharmaceuticals was diagnosed in 33 health care workers: 26 nurses, 4 veterinarians, 2 pharmacists and one medical doctor. There were 26 women and 7 men with a mean age of 38 years. Practically all of these patients presented with hand dermatitis (often the fingers), sometimes with secondary localizations, frequently the face. In this group, two patients had occupational allergic contact dermatitis from amoxicillin (27).

A 45-year-old nurse gave a 20-year history of recurrent vesiculation and erythema of the sides of the fingers of her right hand, which spread to the back of the hand, with scaly resolution, within a 1-week period. She attributed her condition to handling either amoxicillin or ampicillin vials for parenteral use. Patch tests were positive to amoxicillin and ampicillin 50 mg/ml water at D2 and D4, as were intradermal tests with late readings (11).

CUTANEOUS ADVERSE DRUG REACTIONS FROM SYSTEMIC ADMINISTRATION CAUSED BY TYPE IV (DELAYED-TYPE) HYPERSENSITIVITY (as demonstrated by positive patch tests)
Cutaneous adverse drug reactions from systemic administration of amoxicillin caused by type IV (delayed-type) hypersensitivity have included maculopapular eruption (12,13,14,17,23,24,25,28,33,38,42,47,49,51,52,54,55,56, 57,58,59,61,63,64,65,66,67), erythroderma/generalized erythema/exfoliative dermatitis (24,46,51,59,63,65,75), acute generalized exanthematous pustulosis (AGEP) (3,5,6,7,8,9,12,19,22,24,29,31,49,61), symmetrical drug-related intertriginous and flexural exanthema (SDRIFE)/baboon syndrome (34,35,36), fixed drug eruption (16,32,37,44), drug reaction with eosinophilia and systemic symptoms (DRESS) (13,17,18,19,24,43,48,49,61,76), Stevens-Johnson syndrome/toxic epidermal necrolysis (SJS/TEN) (15,17,19,20), photosensitivity (40), eczematous drug eruption (50), urticaria/angioedema (10,28,56,59,63,65), bullous exanthema/erythema multiforme (14), erythema multiforme-like exanthema (45), and unspecified drug eruptions (21,26,30,65).

It should be realized that several publications have come from the same hospitals and that overlap in the presented data cannot be excluded and are in some cases highly likely or even certain.

General
Delayed-type hypersensitivity to amoxicillin in patients with skin eruptions attributed to this antibiotic appears to be long-lasting, as shown by repeated patch testing (53).

Case series with various or unknown types of drug reactions
In a hospital in France, specialized in toxic bullous diseases and severe cutaneous adverse reactions, a retrospective study was performed in consecutive patients consulting between 2010 and 2018 with a nonimmediate CADR suspected to have been caused by beta-lactam antibiotics. 56 patients were included, among whom were 46 amoxicillin-suspected and seven cephalosporin-suspected. Twenty-nine had severe CADR (DRESS, AGEP, SJS/TEN), and 27 had nonimmediate maculopapular exanthema (MPE). Of these patients, twenty patients had positive tests to the culprit drugs (or related beta-lactams). Amoxicillin was responsible for 7 cases of maculopapular eruptions, 7 of cases of AGEP and 2 of DRESS (61).

In the period 2000-2014, 260 patients were patch tested with antibiotics for suspected cutaneous adverse drug reactions (CADR) to these drugs. 56 patients (22%) had one or more (often from cross-reactivity) positive patch tests. Amoxicillin was patch test positive in 20 patients with maculopapular eruptions and in 4 with DRESS (13, overlap with ref. 12).

In Bern, Switzerland, patients with a suspected allergic cutaneous drug reaction were patch-scratch tested with suspected drugs that had previously given a positive lymphocyte transformation test. Amoxicillin 750 mg/ml saline gave a positive patch-scratch test in 7 patients with maculopapular exanthema and in one with bullous exanthema/erythema multiforme (14).

Between March 2009 and June 2013, in a center in France specialized in cutaneous adverse drug reactions (CADR), 156 patients were patch tested because of a CADR. Of these, 75 (30 men and 45 women) were tested simultaneously with the commercial test material and extemporaneous patch tests with pulverized pills 30% pet. In all cases with positive patch tests, both materials reacted, there were no discordant results. Amoxicillin was positive in 10 patients, 5 with maculopapular rash, 3 with AGEP, one with DRESS and one with erythroderma (24).

In London, between October 2017 and October 2018, 45 patients with suspected cutaneous adverse drug reactions, including 33 maculopapular eruptions (MPE), 4 fixed drug eruptions (FDE), 4 DRESS, 3 AGEP and one SJS/TEN, were patch tested with the suspected drugs. There were 10 (22%) positive patch test cases: 4 MPE, 2 FDE, 3 DRESS and 1 AGEP. Amoxicillin (tested as amoxicillin trihydrate 10% pet.) was responsible for 2 cases of maculopapular eruptions, 1 case of AGEP and one case of DRESS (49).

In a report from Switzerland, 3 patients were briefly reported who had drug eruptions from amoxicillin with a positive patch test to the antibiotic. One had a maculopapular eruption, the second a maculopapular eruption in which later vesicles and bullae appeared, and the third had exfoliative erythroderma with bullae (51).

Six patients with maculopapular eruption, 9 with delayed urticaria/angioedema and 3 with erythema caused by amoxicillin with a positive patch test to this aminopenicillin were reported from Rome in 2006 (63).

In 97 patients with nonimmediate reactions from penicillins with a positive patch test to at least one penicillin seen in Rome between 2002 and 2011, amoxicillin was the culprit drug in 50 individuals, ampicillin in 35, bacampicillin in 25 and benzylpenicillin in 11. The most frequent clinical manifestations were maculopapular eruptions with or without edema, delayed urticaria with or without angioneurotic edema and generalized erythema. It was not specified which drug eruptions the 4 individual antibiotics had caused (65).

Maculopapular eruption

Relationship with infectious mononucleosis
Aminopenicillins (amoxicillin, ampicillin, bacampicillin; currently mostly amoxicillin) are a major cause of delayed-type reactions to penicillins. Of patients receiving aminopenicillins during a florid Epstein–Barr virus (EBV)-infection (Morbus Pfeiffer), many were reported to develop a maculopapular rash (74). This was first found in the 1960s, when 'ampicillin rash' was said to occur in 70-100% of all patients with infectious mononucleosis (71,72). More recent investigations find much lower percentages (69,70,73), in one study some 30% for the currently used amoxicillin (69). However, it should be realized, that in this investigation, 23% of children with infectious mononucleosis who were *not* treated with amoxicillin *also* developed a similar rash (69). In other studies, still lower percentages were found and the frequencies of a rash in patients who did not receive penicillins was not statistically different from those that did receive an antibiotic (70,73).

Hypotheses concerning the pathomechanism of penicillin-induced rash during mononucleosis infectiosa are (i) that the immune system of patients with infectious mononucleosis has decreased tolerance and/or (ii) enhancement of immune reaction to certain drugs or its metabolites (38). It is often assumed that in most cases an EBV-associated amoxicillin rash is reversible (i.e. that hypersensitivity disappears) and, as a consequence, allergy diagnostic testing is not considered or recommended (39). However, it has been shown that in a number of cases, patients with Epstein–Barr virus infection who developed a skin rash while being treated with amoxicillin, have delayed-type hypersensitivity to amoxicillin as demonstrated by positive patch tests 3 months after complete healing, and that the allergy may persist for several years (38).

Thus, there may be 2 populations of patients presenting with maculopapular eruptions in the setting of infectious mononucleosis (IM): a predominant group experiencing transient and reversible EBV-associated loss of immune tolerance and a second population, although less common, who appear to develop a true and persistent delayed-type drug hypersensitivity from penicillin treatment during IM (74).

Case series
In Rome and 2 other centers in Italy, in a prospective study performed between 2000 and 2014, 214 consecutive patients who had suffered 307 nonimmediate skin reactions during penicillin therapy and who had positive patch test and/or delayed-reading intradermal test to at least 1 penicillin, were examined with extended patch testing to cephalosporins and aztreonam (66, overlap with ref. 67). 260 of the 307 reactions were maculopapular eruptions rash (with or without edema), 21 erythema (with or without edema) and the rest were other skin manifestations. Of the 307 adverse drug reactions, 172 had been caused by amoxicillin, 91 by ampicillin and 29 by bacampicillin. In the 214 patients, there were 210 positive reactions (patch test and/or delayed intradermal test) to amoxicillin, 212 to ampicillin, 97 to benzylpenicillin (virtually all cross-reactions) and 4 to piperacillin. Extended patch testing showed that no patient had positive skin test responses to cefuroxime, ceftriaxone, and aztreonam; 40 (19%) had cross-reactions to aminocephalosporins (cefalexin, cefaclor, and cefadroxil) (66, overlap with ref. 67).

In a study from Portugal, 18 patients who had previously had positive patch tests to antibiotics that had caused a maculopapular eruption, were again patch tested after a mean interval of 6 years. The positive reactions were reproducible in 16 cases, of which 7 were caused by amoxicillin (12, overlap with ref. 13). In a study from Nancy, France, 54 patients with suspected nonimmediate drug eruptions were assessed with patch testing. Of the 27 patients with maculopapular eruptions, 6 had positive reactions to amoxicillin (23).

In France, of 5 patients with morbilliform rashes that had developed one to ten days after the intake of amoxicillin, 3 had positive patch test reactions to amoxicillin at full strength and 50% pet. (unknown whether the pure powder was used or the content of commercial capsules) at D2 and D4. Thirty controls were negative. These same 3 patients also had positive intradermal tests and 2/3 positive prick tests read at 24 hours. Histopathology of the positive patch tests was similar to that which had been observed in skin biopsy specimens of the morbilliform rashes (33).

In a group of 14 patients with multiple delayed-type hypersensitivity reactions reported from Switzerland, 3 patients had (severe) maculopapular eruptions caused by amoxicillin, showing positive patch test reactions (17). In Bern, Switzerland, of 9 patients who had developed a maculopapular eruption during or within 2 weeks after therapy with amoxicillin, 7 (78%) had a positive scratch-patch test to amoxicillin 750 mg/ml saline. Six of these also had a positive lymphocyte transformation test to amoxicillin (47).

In Rome, 259 patients who had suffered nonimmediate skin reactions during penicillin therapy were examined with patch testing. 173 patients had had a maculopapular rash, 59 (delayed) urticaria/angioedema, 22 erythema and 5 other manifestations. Two hundred and forty-one subjects (93%) reported adverse reactions to aminopenicillins: amoxicillin in 107, ampicillin in 59 and bacampicillin in 48. Twenty-seven of these subjects had suffered reactions to 2 or 3 different aminopenicillins in separate episodes. Ninety-four of the 259 subjects (36%) showed patch test and delayed intradermal test positivity to the culprit penicillins, of which ninety to both amoxicillin and ampicillin. Bacampicillin was not tested, but a number of these sensitizations must have been picked up by reactions to ampicillin. These patients all suffered from maculopapular rashes or – a few – erythema (57, overlap with ref. 55).

Thirty patients from Rome who had suffered maculopapular exanthemas or delayed urticaria from beta-lactam antibiotics and who had at least one positive patch test to one beta-lactam antigenic determinant, were patch tested with a large number of beta-lactam antibiotics. All but one reacted to amoxicillin, ampicillin and bacampicillin (56).

In the period 1994 to 1997 inclusive, in Rome, of 111 patients with a maculopapular eruption from amoxicillin or ampicillin, 58 (52%) had positive patch tests to both ampicillin and amoxicillin 5% pet. More rashes (probably 1.8x) had been caused by ampicillin than by amoxicillin, but this was not specified (55, overlap with ref. 57).

In Rome, between 1987 and 1992, 60 patients who had developed maculopapular eruptions (sometimes with edema, urticaria or fever) from aminopenicillins (amoxicillin, ampicillin, bacampicillin), were patch tested with amoxicillin and ampicillin 5% pet. Thirty-three had positive patch and intradermal tests (with late reading) to aminopenicillins (not specified). Oral challenges were performed with ampicillin in 13 patients and 5 with amoxicillin and all developed diffuse maculopapular eruptions, 12 with a challenge of 5 mg and 6 with 50 mg (54).

Two cases of maculopapular eruption from amoxicillin with positive patch and delayed intradermal tests (59). Other case series of maculopapular eruptions caused by amoxicillin and verified by positive patch tests can be found in refs. 52,64,65, and 67.

Case reports

A 46-year-old woman was referred for allergy testing because of a maculopapular eruption following administration of amoxicillin for acute pharyngitis 20 years earlier. A patch test was negative to amoxicillin 5% pet. as were a prick and intradermal test with amoxicillin 2% in saline after one day. Six months later, graded oral provocation testing with amoxicillin on 3 consecutive days was performed (D1 0.001 gram and 0.01 gram; D2 0.1 gram; D3 1 gram). Approximately 24 hours after the last administration (1.0 g), eczematous flare-up involving the application sites of patch tests and intradermal tests with amoxicillin appeared, and this was followed by itchy maculopapular eruptions on the neck, abdomen, face, and limbs (42).

A 14-year-old female patient from Finland developed an extensive maculopapular exanthema following amoxicillin therapy for pharyngitis and fever. Patch tests were positive to amoxicillin 10% pet. (25). A 32-year-old man developed a maculopapular eruption from oral amoxicillin. He had positive patch tests to amoxicillin and many other penicillins, some cephalosporins and the beta-lactam carbapenem antibiotic imipenem (28).

Maculopapular eruptions during infectious mononucleosis and use of amoxicillin

Of 41 patients seen in a hospital in Germany for drug eruptions following the intake of aminopenicillins, 8 had a florid infectious mononucleosis at the time of the drug eruption. Five of these eight had positive patch tests to amoxicillin and two of five additionally to penicillin (probably benzylpenicillin) 3 months after clearance of the rash and infection. Two patients were retested after 2.2 and 1.5 years and both had still positive patch tests, which is indicative for a persistent delayed-type reaction to amoxicillin. One patient underwent an oral provocation test, which reproduced the previous skin eruption (38).

Of 3 adult patients who had developed a maculopapular exanthema while using amoxicillin during infectious mononucleosis, one had positive patch tests to amoxicillin and ampicillin and two of 3 had a positive lymphocyte transformation test (58).

See also the section 'Case series with various or unknown types of drug reactions' above, refs. 13,14,24,49,51,61,63, and 65.

Erythroderma, widespread erythematous eruption, exfoliative dermatitis

After having used amoxicillin for 7 days, a 24-year-old woman developed generalized erythema and pruritus, facial edema and dyspnea. The symptoms persisted for 15 days despite stopping the drug. Patch tests were positive to amoxicillin and ampicillin 5% pet. An intradermal test to amoxicillin was positive after 24 hours, which remained positive for a week (46).

A patient aged 75 had suffered a desquamative exanthema attributed to amoxicillin. An oral provocation test was positive. 4 years later, a delayed intradermal test and a patch test (5% pet.) were both positive to amoxicillin (75). Three cases of exfoliative exanthema from amoxicillin with positive patch and intradermal tests (59). See also the section 'Case series with various or unknown types of drug reactions' above, refs. 24,51,63, and 65.

Acute generalized exanthematous pustulosis

Case series

In a multicenter investigation in France to determine the value and safety of drug patch tests, of 45 patients patch tested for AGEP, 26 (58%) had positive patch tests to drugs, including 5 to amoxicillin (19). Two patients who had developed AGEP ascribed to amoxicillin had positive patch tests to this antibiotic (22).

Case reports

A life-threatening case of AGEP in a 30-year-old man caused by delayed-type hypersensitivity to amoxicillin, ampicillin and benzylpenicillin was reported in 2017. It began with treatment of periodontitis with amoxicillin, which caused erythema of the face. When treated with ciprofloxacin (which was later prick-test positive) and subsequently ampicillin and benzylpenicillin, the erythema spread to the rest of the skin with pustules, severe mucous membrane involvement and high fever. A systemic inflammatory response syndrome developed with life-threatening involvement of the lungs, heart, liver and kidneys. With non-penicillin antibiotics and systemic corticosteroids the patient recovered. Four months later, patch tests were positive to ampicillin, amoxicillin, and other penicillins. Intradermal testing of minor and major determinants of penicillin were positive for benzylpenicillin and ampicillin. Prick testing was positive for ciprofloxacin, amoxicillin, ampicillin, and amoxicillin-clavulanic acid. The authors assumed that the underlying type I/IV allergies to various β-lactam antibiotics given, together with secondary infection of the skin, led to severe life-threatening systemic involvement. Pre-existing bacterial infections, in this case caused by periodontitis, might also have been a promoting factor (3).

A 69-year-old man developed severe AGEP complicated by life-threatening hypotension and deteriorated organ function mimicking septic shock while on prophylactic amoxicillin for dental surgery. Patch tests were positive to amoxicillin 10% and 30% with pustulation and characteristic histology (5). A 37-year-old woman developed AGEP associated with massive painful lymphadenopathy during treatment with oral amoxicillin-clavulanic acid for periodontitis. Patch tests were positive with pustulation to amoxicillin-clavulanic acid, amoxicillin and ampicillin 10% pet. (6). A 32-year-old man developed AGEP while on amoxicillin-clavulanate 1 g twice a day for pharyngitis. Patch tests were positive to amoxicillin, benzylpenicillin and the related beta-lactam cephalosporin cefalexin 5% pet. (7).

A 48-year-old man was treated with paracetamol and amoxicillin for otitis media. Within 3 days he developed an exanthema, which was progressive despite immediate cessation of both drugs and therapy with prednisolone. On day 7 he had a pustular nearly erythrodermic exanthema and oral erosions. Skin testing performed 4 months later revealed positive patch tests for amoxicillin and paracetamol (both 5% in petrolatum) with papulopustular infiltrates reminiscent of typical AGEP lesions (8).

A 69-year-old man developed AGEP while taking amoxicillin for a sore throat. A patch test with amoxicillin was positive; concentration and vehicle used were not mentioned (9). A 36-year-old man developed AGEP from amoxicillin and flucloxacillin with positive patch tests to both drugs. 6.5 years later, the tests were repeated and were still positive (12).

Other single case reports of AGEP from amoxicillin (sometimes with very few clinical data provided) can be found in refs. 29 and 31. See also the section 'Case series with various or unknown types of drug reactions' above, refs. 24,49, and 61.

Symmetrical drug-related intertriginous and flexural exanthema (SDRIFE)/Baboon syndrome

After having used amoxicillin 1000 mg per day for 10 days, a 60-year-old woman developed erythema of the underpants area and all major flexures. Examination revealed a light-red, maculopapular rash symmetrically distributed on the buttocks and major flexures, becoming confluent. A patch test with amoxicillin 5% pet. gave a strong edematous reaction after 48 hours. The patient remembered that she had suffered a similar rash 8 years earlier following the use of an unknown oral antibiotic (34).

A 48-year-old man suffered an itching, light red, maculopapular eruption, symmetrically distributed over the area covered by the underpants and in the major flexures. Five days earlier, he had been treated for a pulmonary infection with amoxicillin 1000 mg. The skin lesions developed after the second tablet, starting in the inguinal region. Patch tests were strongly positive (+++) to amoxicillin 5% pet. at D2 (35).

A similar case was reported in 1999 from the UK. This 37-year-old man developed an acute, pruritic eruption affecting the flexures and periorbital swelling 6 hours after starting oral amoxicillin for a periodontal abscess. On examination, he had fever and had a widespread dusky, edematous, erythematous eruption over the buttocks, inner thighs, axillae, and feet with overlying small tense bullae around the groins and wrists. Some areas appeared purpuric. Two months after complete resolution, patch tests were positive to amoxicillin 20% pet. and benzylpenicillin 20% at D2 and D4, whereas 20 controls were negative (36).

Fixed drug eruption

A 60-year-old man suffered a recurrent erythematous plaque on the right thigh, that would heal leaving hyperpigmentation. Two hours after receiving oral amoxicillin and paracetamol (acetaminophen), an erythematous itchy eruption would appear around the hyperpigmentation. At presentation, the patient had a solitary well defined violaceous oval plaque on the right thigh with a diameter of approximately 5 centimeter. Patch tests performed with paracetamol 20% pet. and amoxicillin 20% pet. on both normal skin and the hyperpigmented plaque were positive to amoxicillin on the postlesional skin only. An oral provocation test with 500 mg of amoxicillin led to a reactivation of the residual lesion starting after 30 minutes (32).

A 57-year-old man presented with a bright red, relatively well-circumscribed erythema, most marked at the edges, mainly over the inguinal region and the inside of the thigh. It had appeared on the second day of treatment with amoxicillin, clarithromycin and pantoprazole for a gastric ulcer positive for *Helicobacter pylori*. After amoxicillin had been discontinued and local steroids had been administered for one week, the cutaneous changes disappeared leaving slight hyperpigmentation. Epifocal epicutaneous tests (i.e. at the site of the previous eruption) with amoxicillin and ampicillin (both 5% pet.) gave positive reactions to both and confirmed a suspected fixed drug reaction to amoxicillin (37).

In France, in the period 2005-2007, 59 cases of fixed drug eruptions were collected in 17 academic centers. There was one case of FDE to amoxicillin with a positive patch test. Clinical details were not provided (16). A 38-year-old man, while being treated with amoxicillin, developed pruritic erythematous lesions on the hands and genitals. Patch tests were positive to amoxicillin 10% pet., apparently both on normal and previously involved skin. The patient was diagnosed with fixed drug eruption from amoxicillin (44).

Drug reaction with eosinophilia and systemic symptoms (DRESS)

Case series

To evaluate if, after DRESS, patients become sensitized to antibiotics, patch test data and clinical files of 17 patients with DRESS from a nonantibiotic culprit drug (anticonvulsants 10, allopurinol 7) seen in Coimbra, Portugal, between 2010 and 2018 and who were given antibiotics at the onset or later during the course of DRESS, were retrospectively studied. The group consisted of eight women and nine men with an age range of 10 to 89 years and a mean of 47. In 10, anticonvulsants had caused DRESS (notably carbamazepine) and in 7 allopurinol (which always gives negative patch tests). They had received antibiotics at the onset or later during the course of DRESS: amoxicillin in 7, cephalosporins in three and fluoroquinolones in seven patients (5 ciprofloxacin, 2 levofloxacin). Nine patients (53%) had developed positive patch tests to antibiotics: six to amoxicillin (with 5 cross-reactions to ampicillin), three to the cephalosporins (one to ceftriaxone, one to cefoxitin and one to cefoxitin, ceftazidime and additional positivity to vancomycin), but none to the fluoroquinolones ciprofloxacin or levofloxacin. In repeat testing in 4 patients, most positive patch tests were reproducible after several years. These patients all had multiple hypersensitivity syndrome, as they reacted to (at least) two unrelated chemicals (48).

In a group of 14 patients with multiple delayed-type hypersensitivity reactions reported from Switzerland, 2 patients had DRESS caused by amoxicillin, showing positive patch test reactions (17, overlap with ref. 18). In a multicenter investigation in France to determine the value and safety of drug patch tests, of 72 patients patch tested for DRESS, 46 (64%) had positive patch tests to drugs, including 6 to amoxicillin (19).

Case reports

A 70-year-old woman presented with generalized erythematous maculopapular plaques, as well as multiple vesicles on the arms, facial edema, oral ulcers, fever, vomiting, diarrhea, eosinophilia, and abnormal liver function. The patient had been treated with allopurinol for the previous 10 days and, in the last 24 hours had received amoxicillin for an upper respiratory tract infection. Skin biopsy of an arm lesion showed significant dermal edema with a large subepidermal bulla; superficial dermal vessels were surrounded by an inflammatory infiltrate in which lymphocytes were predominant. The biopsy was compatible with the diagnosis of erythema multiforme. Patch tests with amoxicillin 5% water and ampicillin 5% water were strongly positive, but the patch test with allopurinol was negative. An intradermal test with amoxicillin 20 mg/ml was positive after 24 hours. An oral challenge with allopurinol (12.5 mg) resulted in severe generalized erythema and pruritus, fever and nausea 4 hours later. The patient was diagnosed with allopurinol hypersensitivity syndrome combined with type-IV hypersensitivity to amoxicillin (43). It is uncertain to what extent amoxicillin has contributed to the clinical picture.

One patient developed DRESS from amoxicillin with a positive patch test to this drug was reported from Switzerland in 2020 (18, overlap with ref. 17). Another single case report of DRESS caused by amoxicillin and verified by positive patch tests can be found in ref. 76. See also the section 'Case series with various or unknown types of drug reactions' above, refs. 13,24,49, and 61.

Stevens-Johnson syndrome/toxic epidermal necrolysis (SJS/TEN)
A man aged 45 had developed Stevens-Johnson syndrome while using amoxicillin. Patch and lymphocyte transformation tests were positive to amoxicillin (15). In a group of 14 patients with multiple delayed-type hypersensitivity reactions reported from Switzerland, one patient had SJS caused by amoxicillin and clindamycin with positive patch tests (17). In a group of 14 patients with multiple delayed-type hypersensitivity reactions reported from Switzerland, one patient had TEN caused by amoxicillin with a positive patch test (17).

In a multicenter investigation in France to determine the value and safety of drug patch tests, there was one patient with SJS/TEN caused by amoxicillin who had a positive patch test to this antibiotic SJS/TEN (19). One individual developed TEN from oral amoxicillin, preceded by a diffuse, maculopapular eruption. The patient later had positive reactions to a patch test with ampicillin and an intradermal test (after 6 hours). Amoxicillin itself was apparently not tested, but this antibiotic very frequently cross-reacts with ampicillin (20).

Photosensitivity
A 71-year-old man had been treated with amoxicillin for 3 days because of a tick bite, when he presented with a photodistributed exanthema on his face and arms, which then spread to non-exposed areas. The patient had a history of photoallergic contact dermatitis caused by a ketoprofen gel 10 years earlier and of contact dermatitis caused by perfume use 2 years earlier. Two biopsies showed a similar pattern suggestive of cutaneous drug adverse reactions, with a dermal inflammatory infiltrate of lymphocytes and histiocytes around the vessels combined with eosinophils. After 3 months, patch tests were positive to *Myroxylon pereirae*, fragrance mix I and cinnamyl alcohol and photopatch tests were positive to ketoprofen and ketoprofen gel (which is a very common combination [41]). The patch test with amoxicillin 10% pet. gave a negative result on D3 and a doubtful result on D4, whereas the photopatch test gave positive results on D3 (+) and D4 (++). The authors concluded that this case is strongly suggestive of systemic photoallergy to amoxicillin (40). It is well known that patients who are photosensitive to ketoprofen often have multiple photoallergic reactions to both related and non-related chemicals (41); the photocontact allergy to ketoprofen may have facilitated photosensitization to amoxicillin in this patient.

Dermatitis/eczematous eruption
A 48-year-old man, who had previously suffered DRESS from carbamazepine, later had eczematous eruptions from multiple unrelated drugs. Patch tests were positive to hydroxyzine HCl, Cetirizine HCl, carbamazepine, pseudoephedrine, amoxicillin and ampicillin. The patient had used all these drugs, which resulted in eczematous eruptions, with the exception of ampicillin, which was a cross-reaction to amoxicillin. This was a case of multiple drug allergy (50).

Other cutaneous adverse drug reactions

Erythema multiforme
A 42-year-old man was given amoxicillin and omeprazole for *Helicobacter pylori* eradication because of a gastric ulcer. Within a few days, the developed a macular rash, eventually becoming bullous. At presentation, the patient showed painful epidermolysis on skin areas overlying bones (knees, elbows, iliac crests). He recovered completely without scarring of the skin. Patch testing gave a strongly positive reaction to amoxicillin. The patient was diagnosed with erythema multiforme-like exanthema from amoxicillin (45).

Urticaria/angioedema
A 62-year-old woman developed widespread urticaria 2 hours after 'a dose of antibiotic'. Erythromycin and quinolones were well tolerated. Intradermal tests were positive to amoxicillin 200 mg/ml and ampicillin 200 mg/ml (both saline) after one day and remained so for 20 days. Patch tests to both antibiotic substances, performed one month later, were strongly positive already after one day, and at the same time, the previously positive intradermal tests showed a flare-up reaction (10). Quite curiously, it was not specified which antibiotic had induced the urticaria in this patient.

A man aged 34 had developed urticaria from oral amoxicillin. Patch tests were positive to amoxicillin, many other penicillins and the beta-lactam carbapenem antibiotic imipenem (28). A woman aged 43 had developed angioedema and urticaria twice, first after oral ampicillin, later after oral amoxicillin. Patch tests were positive to both drugs, many other penicillins, some cephalosporins and the beta-lactam carbapenem antibiotic imipenem (28).

Three cases of urticaria and one of urticaria and angioedema from amoxicillin with positive patch and intradermal tests (59). See also the section 'Maculopapular eruption' above, refs. 56,63, and 65.

Unspecified drug eruptions
In a group of 78 patients who had developed nonimmediate drug reactions from any beta-lactam compound (penicillins, cephalosporins), 70 had positive patch tests to amoxicillin. In 35 cases, amoxicillin was the culprit drug, so the other reactions were presumably cross-reactions. It was not mentioned which drug eruptions these patients had suffered from (26).

In Finland, in the period 1989-2001, 826 patients with suspected cutaneous drug eruptions were patch tested and 89 had one or more positive reactions. Of these individuals, 10 reacted to amoxicillin. It was not mentioned which drug eruptions these patients had suffered from (21).

A 44-year-old woman developed an unspecified skin rash while using amoxicillin. Patch tests were positive to amoxicillin-clavulanic acid and to amoxicillin, both tested as powder 'as is' (30). See also the section 'Case series with various or unknown types of drug reactions' above, refs. 14 and 65.

Cross-reactions, pseudo-cross-reactions and co-reactions

Cross-reactions between beta-lactam antibiotics
Cross-reactivity among penicillins is frequent, mainly related to structural similarities among their side-chain determinants (2). The aminopenicillins amoxicillin and ampicillin nearly always cross-react to and from each other (55,56), having virtually the same side chain, the only difference being a hydroxy group at the aromatic ring of the side chain for amoxicillin. Cross-reactions to non-aminopenicillins (including benzylpenicillin and penicillin V) are less frequent (55,60), although cross-reactions to benzylpenicillin in patients sensitized to aminopenicillins have also been observed at a rate of 44% (66).

In penicillin-allergic subjects (delayed-type hypersensitivity), cross-reactivity to cephalosporins in several studies has varied from 0% to 30%, depending on the nature of the allergenic penicillins and the cephalosporins tested (2,56,62,63,64,65,66). The rate of cross-reactivity in aminopenicillin- (amoxicillin, ampicillin, bacampicillin) sensitized patients to aminocephalosporins (e.g. cefalexin, cefaclor and cefadroxil) may be around 20%, related to similar or identical side chains (2,66). Co-reactivity to cephalosporins such as cefuroxime, cefpodoxime, and cefixime, which have side chains dissimilar from those of penicillins, suggest the possibility of coexisting sensitivities, not cross-sensitivities (2).

Cross-reactions in penicillin-sensitized individuals to the carbapenems (imipenem, meropenem, ertapenem) hardly ever (28), if at all, occur (2,61,67), whereas cross-reactions to aztreonam have not been observed as yet (2,26,56,63,65,66).

Large-scale investigations of cross-reactivity to penicillins in cephalosporin-sensitized individuals are not available (2), but a few cross-reactions in aminocephalosporin-sensitized patients to aminopenicillins (amoxicillin, ampicillin) have been observed (68).

Immediate contact reactions
Immediate contact reactions (contact urticaria) to amoxicillin are presented in Chapter 5.

LITERATURE
1   The data in the section 'General' may have been obtained from literature discussed in this chapter, but mostly also or exclusively from one or more of the following online sources: ChemIDPlus Advanced, PubChem, DrugBank, RxList, Drug Central, Drugs.com, and Wikipedia
2   Romano A, Gaeta F, Arribas Poves MF, Valluzzi RL. Cross-reactivity among beta-lactams. Curr Allergy Asthma Rep 2016;16:24
3   Tajmir-Riahi A, Wörl P, Harrer T, Schliep S, Schuler G, Simon M. Life-threatening atypical case of acute generalized exanthematous pustulosis. Int Arch Allergy Immunol 2017;174:108-111
4   Gaeta F, Torres MJ, Valluzzi RL, Caruso C, Mayorga C, Romano A. Diagnosing β-Lactam Hypersensitivity. Curr Pharm Des 2016;22:6803-6813
5   McDonald KA, Pierscianowski TA. A case of amoxicillin-induced acute generalized exanthematous pustulosis presenting as septic shock. J Cutan Med Surg 2017;21:351-355
6   Syrigou E, Grapsa D, Charpidou A, Syrigos K. Acute generalized exanthematous pustulosis induced by amoxicillin/clavulanic acid: report of a case presenting with generalized lymphadenopathy. J Cutan Med Surg 2015;19:592-594.

7 Bomarrito L, Zisa G, Delrosso G, Farinelli P, Galimberti M. A case of acute generalized exanthematous pustulosis due to amoxicillin-clavulanate with multiple positivity to beta-lactam patch testing. Eur Ann Allergy Clin Immunol 2013;45:178-180

8 Treudler R, Grunewald S, Gebhardt C, Simon J-C. Prolonged course of acute generalized exanthematous pustulosis with liver involvement due to sensitization to amoxicillin and paracetamol. Acta Derm Venereol 2009;89:314-315

9 Gensch K, Hodzic-Avdagic N, Megahed M, Ruzicka T, Kuhn A. Acute generalized exanthematous pustulosis with confirmed type IV allergy. Report of 3 cases. Hautarzt 2007;58:250-252, 254-255 (Article in German)

10 Llamazares AA, Chamorro M, Robledo T, Cimarra M, Palacios R, Rodgriguez A, et al. Flare-up of skin tests to amoxycillin and ampicillin. Contact Dermatitis 2000;42:166

11 Gamboa P, Jáuregui I, Urrutia I. Occupational sensitization to aminopenicillins with oral tolerance to penicillin V. Contact Dermatitis 1995;32:48-49

12 Pinho A, Marta A, Coutinho I, Gonçalo M. Long-term reproducibility of positive patch test reactions in patients with non-immediate cutaneous adverse drug reactions to antibiotics. Contact Dermatitis 2017;76:204-209

13 Pinho A, Coutinho I, Gameiro A, Gouveia M, Gonçalo M. Patch testing - a valuable tool for investigating non-immediate cutaneous adverse drug reactions to antibiotics. J Eur Acad Dermatol Venereol 2017;31:280-287

14 Neukomm C, Yawalkar N, Helbling A, Pichler WJ. T-cell reactions to drugs in distinct clinical manifestations of drug allergy. J Invest Allergol Clin Immunol 2001;11:275-284

15 Beeler A, Engler O, Gerber BO, et al. Long-lasting reactivity and high frequency of drug-specific T cells after severe systemic drug hypersensitivity reactions. J Allergy Clin Immunol 2006;117:455-462

16 Brahimi N, Routier E, Raison-Peyron N, Tronquoy AF, Pouget-Jasson C, Amarger S, et al. A three-year-analysis of fixed drug eruptions in hospital settings in France. Eur J Dermatol 2010;20:461-464

17 Jörg L, Yerly D, Helbling A, Pichler W. The role of drug, dose and the tolerance/intolerance of new drugs in multiple drug hypersensitivity syndrome (MDH). Allergy 2020;75:1178-1187

18 Jörg L, Helbling A, Yerly D, Pichler WJ. Drug-related relapses in drug reaction with eosinophilia and systemic symptoms (DRESS). Clin Transl Allergy 2020;10:52

19 Barbaud A, Collet E, Milpied B, Assier H, Staumont D, Avenel-Audran M, et al. A multicentre study to determine the value and safety of drug patch tests for the three main classes of severe cutaneous adverse drug reactions. Br J Dermatol 2013;168:555-562

20 Romano A, Di Fonso M, Pocobelli D, Giannarini L, Venuti A, Garcovich A. Two cases of toxic epidermal necrolysis caused by delayed hypersensitivity to beta-lactam antibiotics. J Investig Allergol Clin Immunol 1993;3:53-55

21 Lammintausta K, Kortekangas-Savolainen O. The usefulness of skin tests to prove drug hypersensitivity. Br J Dermatol 2005;152:968-974

22 Wolkenstein P, Chosidow O, Fléchet ML, Robbiola O, Paul M, Dumé L, et al. Patch testing in severe cutaneous adverse drug reactions, including Stevens-Johnson syndrome and toxic epidermal necrolysis. Contact Dermatitis 1996;35:234-236

23 Barbaud A, Reichert-Penetrat S, Tréchot P, Jacquin-Petit MA, Ehlinger A, Noirez V, et al. The use of skin testing in the investigation of cutaneous adverse drug reactions. Br J Dermatol 1998;139:49-58

24 Assier H, Valeyrie-Allanore L, Gener G, Verlinde Carvalh M, Chosidow O, Wolkenstein P. Patch testing in non-immediate cutaneous adverse drug reactions: value of extemporaneous patch tests. Contact Dermatitis 2017;77:297-302

25 Liippo J, Pummi K, Hohenthal U, Lammintausta K. Patch testing and sensitization to multiple drugs. Contact Dermatitis 2013;69:296-302

26 Buonomo A, Nucera E, De Pasquale T, Pecora V, Lombardo C, Sabato V, et al. Tolerability of aztreonam in patients with cell-mediated allergy to β-lactams. Int Arch Allergy Immunol 2011;155:155-159

27 Gielen K, Goossens A. Occupational allergic contact dermatitis from drugs in healthcare workers. Contact Dermatitis 2001;45:273-279

28 Schiavino D, Nucera E, Lombardo C, Decinti M, Pascolini L, Altomonte G, et al. Cross-reactivity and tolerability of imipenem in patients with delayed-type, cell-mediated hypersensitivity to beta-lactams. Allergy 2009;64:1644-1648

29 Britschgi M, Steiner UC, Schmid S, Depta JP, Senti G, Bircher A, et al. T-cell involvement in drug-induced acute generalized exanthematous pustulosis. J Clin Invest 2001;107:1433-1441

30 Kennedy C, Stolz E, van Joost T. Sensitization to amoxycillin in Augmentin. Contact Dermatitis 1989;20:313-314

31 Whittam LR, Wakelin SH, Barker JN. Generalized pustular psoriasis or drug-induced toxic pustuloderma? The use of patch testing. Clin Exp Dermatol 2000;25:122-124

32 Chaabane A, Fredj NB, Chadly Z, Boughattas NA, Aouam K. Fixed drug eruption: a selective reaction to amoxicillin. Thérapie 2013;68:183-185

33 Barbaud A, Bené M-C, Schmutz J L, Ehlinger A, Weber M, Faure G C. Role of delayed cellular hypersensitivity and adhesion molecules in maculopapular rashes induced by amoxicillin. Arch Dermatol 1997;133:481-486

34 Duve S, Worret W, Hofmann H. The baboon syndrome: a manifestation of hematogenous contact-type dermatitis. Acta Derm Venereol 1994;74:480-481

35 Kohler LD, Schonlein K, Kautzky F, Vogt H J. Diagnosis at first glance: the baboon syndrome. Int J Dermatol 1996;35:502-503

36 Wakelin SH, Sidhu S, Orton DI, Chia Y, Shaw S. Amoxycillin-induced flexural exanthem. Clin Exp Dermatol 1999;24:71-73

37 Brabek E, Kränke B. Multilocular fixed drug reaction simulating intertrigo in a diabetic patient. Dtsch Med Wochenschr 2000;125:1260-1262 (Article in German)

38 Jappe U. Amoxicillin-induced exanthema in patients with infectious mononucleosis: allergy or transient immunostimulation. Allergy 2007;62:1474-1475

39 Gruchalla RS, Pirmohamed M. Antibiotic Allergy. New Engl J Med 2006;354:601-609

40 Delaunay J, Chassain K, Sarre M-E, Avenel-Audran M. A drug not recognized as a photosensitizer? Contact Dermatitis 2019;81:143-144

41 Ketoprofen. In: De Groot AC. Monographs in contact allergy, volume 3: Topical Drugs. Boca Raton, Fl, USA: CRC Press Taylor and Francis Group, 2021:Chapter 3.193, pages 452-462

42 Tramontana M, Hansel K, Bianchi L, Agostinelli D, Stingeni L. Flare-up of previously negative patch test and intradermal test with amoxicillin after oral provocation. Contact Dermatitis 2018;79:250-251

43 Pérez A, Cabrerizo S, de Barrio M, Díaz MP, Herrero T, Tornero P, et al. Erythema-multiforme-like eruption from amoxycillin and allopurinol. Contact Dermatitis 2001;44:113-114

44 Saenz de San Pedro Morera B, Enriquez JQ, López JF. Fixed drug eruptions due to betalactams and other chemically unrelated antibiotics. Contact Dermatitis 1999;40:220-221

45 Gebhardt M, Wollina U. Allergy testing in serious cutaneous drug reactions - harmful or beneficial? Contact Dermatitis 1997;37:282-285

46 García R, Galindo PA, Feo F, Gómez E, Fernández F. Delayed allergic reactions to amoxycillin and clindamycin. Contact Dermatitis 1996;35:116-117

47 Schnyder B, Pichler WJ. Skin and laboratory tests in amoxicillin- and penicillin-induced morbilliform skin eruption. Clin Exp Allergy 2000;30:590-595

48 Santiago LG, Morgado FJ, Baptista MS, Gonçalo M. Hypersensitivity to antibiotics in drug reaction with eosinophilia and systemic symptoms (DRESS) from other culprits. Contact Dermatitis 2020;82:290-296

49 Watts TJ, Thursfield D, Haque R. Patch testing for the investigation of nonimmediate cutaneous adverse drug reactions: a prospective single center study. J Allergy Clin Immunol Pract 2019;7:2941-2943.e3

50 Özkaya E, Yazganoğlu KD. Sequential development of eczematous type "multiple drug allergy" to unrelated drugs. J Am Acad Dermatol 2011;65:e26-e29.

51 Hari Y, Frutig-Schnyder K, Hurni M, Yawalkar N, Zanni MP, Schnyder B, et al. T cell involvement in cutaneous drug eruptions. Clin Exp Allergy 2001;31:1398-1408

52 Bruynzeel DP, van Ketel WG. Repeated patch testing in penicillin allergy. Br J Dermatol 1981;104:157-159

53 Romano A, Di Fonso M, Pietrantonio F, Pocobelli D, Giannarini L, Del Bono A, et al. Repeated patch testing in delayed hypersensitivity to beta-lactam antibiotics. Contact Dermatitis 1993;28:190

54 Romano A, Di Fonso M, Papa G, Pietrantonio F, Federico F, Fabrizi G, Venuti A. Evaluation of adverse cutaneous reactions to aminopenicillins with emphasis on those manifested by maculopapular rashes. Allergy 1995;50:113-118

55 Romano A, Quaratino D, Di Fonso M, Papa G, Venuti A, Gasbarrini G. A diagnostic protocol for evaluating nonimmediate reactions to aminopenicillins. J Allergy Clin Immunol 1999;103:1186-1190

56 Patriarca G, D'Ambrosio C, Schiavino D, Larocca LM, Nucera E, Milani A. Clinical usefulness of patch and challenge tests in the diagnosis of cell-mediated allergy to betalactams. Ann Allergy Asthma Immunol 1999;83:257-266

57 Romano A, Viola M, Mondino C, Pettinato R, Di Fonso M, Papa G, et al. Diagnosing nonimmediate reactions to penicillins by in vivo tests. Int Arch Allergy Immunol 2002;129:169-174

58 Renn CN, Straff W, Dorfmüller A, Al-Masaoudi T, Merk HF, Sachs B. Amoxicillin-induced exanthema in young adults with infectious mononucleosis: demonstration of drug-specific lymphocyte reactivity. Br J Dermatol 2002;147:1166-1170

59 Torres MJ, Sánchez-Sabaté E, Alvarez J, Mayorga C, Fernández J, Padial A, et al. Skin test evaluation in nonimmediate allergic reactions to penicillins. Allergy 2004;59:219-224

60 Romano A, Blanca M, Torres M J et al. Diagnosis of nonimmediate reactions to beta-lactam antibiotics. Allergy 2004;59:1153-1160

61 Bérot V, Gener G, Ingen-Housz-Oro S, Gaudin O, Paul M, Chosidow O, Wolkenstein P, Assier H. Cross-reactivity in beta-lactams after a nonimmediate cutaneous adverse reaction: experience of a reference centre for toxic bullous diseases and severe cutaneous adverse reactions. J Eur Acad Dermatol Venereol 2020;34:787-794

62 Phillips E, Knowles SR, Weber EA, Blackburn D. Cephalexin tolerated despite delayed aminopenicillin reactions. Allergy 2001;56:790

63 Schiavino D, Nucera E, De Pasquale T, Roncallo C, Pollastrini E, Lombardo C, et al. Delayed allergy to aminopenicillins: clinical and immunological findings. Int J Immunopathol Pharmacol 2006;19:831-840

64 Trcka J, Seitz CS, Bröcker EB, Gross GE, Trautmann A. Aminopenicillin-induced exanthema allows treatment with certain cephalosporins or phenoxymethyl penicillin. J Antimicrob Chemother 2007;60:107-111

65 Buonomo A, Nucera E, Pecora V, Rizzi A, Aruanno A, Pascolini L, et al. Cross-reactivity and tolerability of cephalosporins in patients with cell-mediated allergy to penicillins. J Investig Allergol Clin Immunol 2014;24:331-337

66 Romano A, Gaeta F, Valluzzi RL, Maggioletti M, Caruso C, Quaratino D. Cross-reactivity and tolerability of aztreonam and cephalosporins in subjects with a T cell-mediated hypersensitivity to penicillins. J Allergy Clin Immunol 2016;138:179-186

67 Romano A, Gaeta F, Valluzzi RL, Alonzi C, Maggioletti M, Zaffiro A, et al. Absence of cross-reactivity to carbapenems in patients with delayed hypersensitivity to penicillins. Allergy 2013;68:1618-1621

68 Romano A, Gaeta F, Valluzzi RL, Caruso C, Alonzi C, Viola M, et al. Diagnosing nonimmediate reactions to cephalosporins. J Allergy Clin Immunol 2012;129:1166-1169

69 Chovel-Sella A, Ben Tov A, Lahav E, Mor O, Rudich H, Paret G, Reif S: Incidence of rash after amoxicillin treatment in children with infectious mononucleosis. Pediatrics 2013;131:1424-1427

70 Dibek Misirlioglu E, Guvenir H, Ozkaya Parlakay A, Toyran M, Tezer H, Catak AI, et al. Incidence of antibiotic-related rash in children with Epstein-Barr Virus infection and evaluation of the frequency of confirmed antibiotic hypersensitivity. Int Arch Allergy Immunol 2018;176:33-38

71 Patel BM. Skin rash with infectious mononucleosis and ampicillin. Pediatrics 1967;40:910-911

72 Pullen H, Wright N, Murdoch JM. Hypersensitivity reactions to antibacterial drugs in infectious mononucleosis. Lancet 1967;2:1176-1178

73 Hocqueloux L, Guinard J, Buret J, Causse X, Guigon A. Do penicillins really increase the frequency of a rash when given during Epstein-Barr Virus primary infection? Clin Infect Dis 2013;57:1661-1662

74 Thompson DF, Ramos CL. Antibiotic-induced rash in patients with infectious mononucleosis. Ann Pharmacother 2017;51:154-162

75 Padial A, Antunez C, Blanca-Lopez N, Fernandez TD, Cornejo-Garcia JA, Mayorga C, et al. Nonimmediate reactions to betalactams: diagnostic value of skin testing and drug provocation test. Clin Exp Allergy 2008; 38:822-828.

76 García-Paz V, González-Rivas M, Otero-Alonso A. DRESS Syndrome: Patch testing as a diagnostic method that brings us closer to a certain result. J Investig Allergol Clin Immunol 2021 Jul 2:0. doi: 10.18176/jiaci.0728. Epub ahead of print

Chapter 3.37   AMOXICILLIN MIXTURE WITH CLAVULANATE POTASSIUM

IDENTIFICATION

Description/definition   : Amoxicillin mixture with clavulanate potassium is a fixed-ratio combination of amoxicillin trihydrate and potassium clavulanate; their structural formulas are shown below

Pharmacological classes : Anti-bacterial agents; β-lactamase inhibitors

IUPAC name   : Potassium;(2S,5R,6R)-6-[[(2R)-2-amino-2-(4-hydroxyphenyl)acetyl]amino]-3,3-dimethyl-7-oxo-4-thia-1-azabicyclo[3.2.0]heptane-2-carboxylic acid;(2R,3Z,5R)-3-(2-hydroxy-ethylidene)-7-oxo-4-oxa-1-azabicyclo[3.2.0]heptane-2-carboxylate

Other names   : Amoxicillin trihydrate and potassium clavulanate

CAS registry number   : 74469-00-4

EC number   : Not available

Merck Index monograph   : 3609 (Clavulanic acid); amoxicillin (1844)

Patch testing   : Pulverized tablet 10% and 30% pet.; test also amoxicillin trihydrate 10% pet. and potassium clavulanate 10.0% pet

Molecular formula   : $C_{24}H_{27}KN_4O_{10}S$

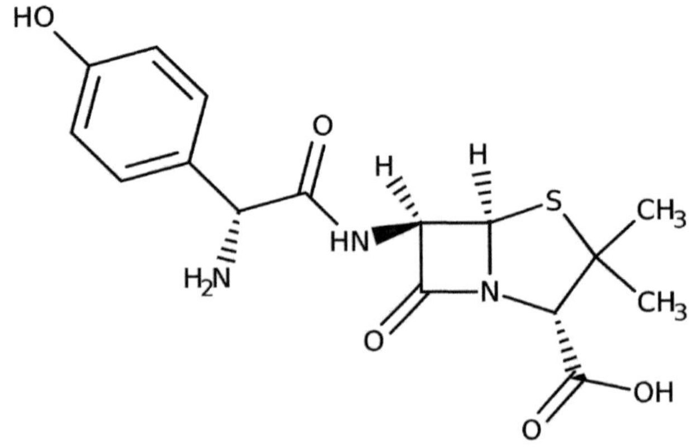

Amoxicillin 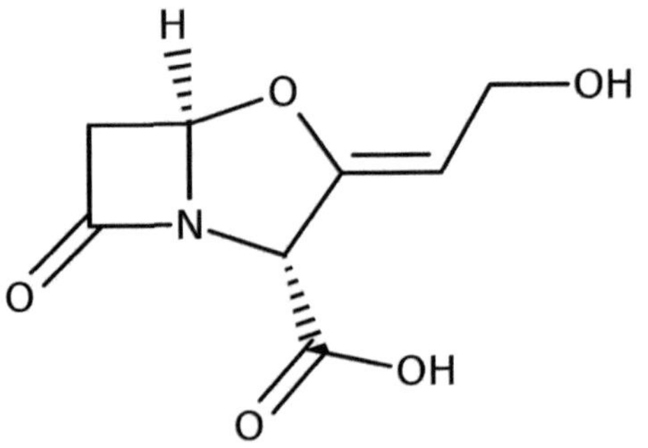   Clavulanic acid

GENERAL

Amoxicillin mixture with clavulanate potassium (termed amoxicillin-clavulanic acid in this chapter) is a fixed-ratio combination of amoxicillin trihydrate and potassium clavulanate. The combination is used to treat acute bacterial sinusitis, community acquired pneumonia, lower respiratory tract infections, acute bacterial otitis media, skin and skin structure infections, and urinary tract infections (1).

For general information on the two separate anti-bacterial agents see Chapter 3.36 Amoxicillin and Chapter 3.137 Clavulanic acid. Cases of allergic reactions to the combination product caused by amoxicillin are presented here, as are cases where the allergenic ingredient was not identified (in 2 separate sections). Case reports of patients with cutaneous adverse drug reactions to the combination antibiotic caused by delayed-type hypersensitivity to clavulanic acid are discussed in Chapter 3.137 Clavulanic acid. The classification and structures of beta-lactam antibiotics are discussed in Chapter 3.36 Amoxicillin.

SECTION 1 CASES WITH AMOXICILLIN AS THE ALLERGENIC INGREDIENT

CONTACT ALLERGY FROM ACCIDENTAL CONTACT

A 24-year-old nurse, who would frequently prepare antibiotic solutions, presented with pruritic erythematous papulovesicles and scaly plaques affecting her fingers. She had positive patch tests to amoxicillin-clavulanic acid, to ampicillin/sulbactam and to ampicillin 1, 10 and 100 mg/ml water. Amoxicillin itself was not tested, but allergy to this penicillin is highly likely, indicating occupational allergic contact dermatitis from ampicillin in ampicillin-sulbactam and amoxicillin in amoxicillin-clavulanic acid (15).

CUTANEOUS ADVERSE DRUG REACTIONS FROM SYSTEMIC ADMINISTRATION CAUSED BY TYPE IV (DELAYED-TYPE) HYPERSENSITIVITY (as demonstrated by positive patch tests)

Cutaneous adverse drug reactions from systemic administration of amoxicillin mixture with clavulanate potassium caused by type IV (delayed-type) hypersensitivity to amoxicillin have included maculopapular eruption (6,14), erythroderma (19), acute generalized exanthematous pustulosis (AGEP) (9,10,11,13,17), drug reaction with eosinophilia and systemic symptoms (DRESS) (20), erythema multiforme (21,22), linear IgA bullous disease (5), bullous exanthema (6), and unspecified generalized rash (16).

Maculopapular eruption

A 65-year-old man developed a maculopapular eruption after using amoxicillin-clavulanic acid. Patch tests were positive to amoxicillin-clavulanic acid 12.5% and amoxicillin 12.5%, but negative to clavulanic acid. Lymphocyte transformation tests were also positive to amoxicillin (stimulation index 18.8) (6). A woman of 49 had a diffuse morbilliform rash 12 hours after amoxicillin-clavulanic acid, with intense itching and edema of the hands and feet. Patch tests were positive to amoxicillin on 2 occasions, after 4 days in one and after 6 days during the second test session (14).

Erythroderma, widespread erythematous eruption, exfoliative dermatitis

Three days after intravenous administration of amoxicillin-clavulanic acid, a 22-year-old woman developed erythroderma with some bullae on her legs, fever and eosinophilia. Sensitization to amoxicillin was confirmed by positive patch and lymphocyte transformation tests (19).

Acute generalized exanthematous pustulosis (AGEP)

After having used amoxicillin-clavulanic acid for 3 days, a 37-year-old woman developed AGEP with high fever (up to 40°C), a pruritic skin rash with multiple small non-follicular pustules on an erythematous base affecting the trunk and limbs and mild leukocytosis. As atypical features, painful lymphadenopathy was observed in the posterior cervical, postauricular, suboccipital, axillary, and inguinal regions and tachycardia. The patient's medical history revealed a similar episode of amoxicillin-associated exanthema at the age of 16 years. Bacterial cultures from pustules were negative and histopathology of a skin biopsy was consistent with AGEP. Patch tests were strongly positive to amoxicillin, ampicillin, and amoxicillin-clavulanic acid, all diluted at 10% in petrolatum (9). A similar case had been reported from Italy 2 years earlier (10). The patient was patch test positive to amoxicillin and other beta-lactam antibiotics. Amoxicillin-clavulanic acid itself was not tested, but the drug eruption had started on the second day of using this combination drug (10).

A 30-year-old woman developed a generalized erythema after oral administration of amoxicillin-clavulanic acid with fever (40°C) and pustules on the back, arms, and face. A similar pustulous reaction to the same drug had already occurred 5 years previously. Patch testing was positive to half a pill of the combination tablet 375 mg in 0.5 ml phosphate-buffered saline. A lymphocyte transformation test was strongly positive to amoxicillin (11).

A 52-year-old man was treated for sinusitis with amoxicillin and clavulanic acid for 7 days. One day after finishing the treatment, he developed a rash, which worsened in the following days. Physical examination showed a generalized rash comprising pustules on the chest, face, and neck; a morbilliform exanthem on arms, chest, legs, and palms; and edema around the eyes and on the legs and palms. Blood samples revealed a positive Epstein-Barr virus (EBV) deoxyribonucleic acid (<400 IU/ml), and a biopsy showed an acute pustular inflammation. Based on these findings, acute generalized exanthematous pustulosis (AGEP) was suspected. Topical corticosteroids and oral prednisolone were administered, and the rash resolved. At this point, it was unclear whether AGEP was a reaction to the antibiotics or the EB-virus. Two months later, patch tests were positive to benzylpenicillin potassium 10% pet., phenoxymethylpenicillin potassium 10% pet., and amoxicillin trihydrate 10% pet. Based on these results, it was concluded that the patient had AGEP from delayed-type allergy to amoxicillin in the combination antibiotic (13).

A 57-year-old woman developed AGEP while using amoxicillin-clavulanic acid. One month after clearance, patch tests were positive to amoxicillin at D4 (no specifics mentioned). All other drugs used were negative. Skin prick tests for amoxicillin and amoxicillin-clavulanic acid were also positive when read at D3 (17).

Drug reaction with eosinophilia and systemic symptoms (DRESS)

A child of 9 years had an abscess with *Staphylococcus aureus* secreting a Panton-Valentine toxin with non-severe pleuritis and pericarditis. Treatment was started with amoxicillin-clavulanic acid, amikacin, and clindamycin followed by oxacillin, rifampicin, and colchicine. On the 25th day of treatment, she had a recurrence of fever with a generalized rash, moderate hepatic cytolysis, hypereosinophilia, and activated lymphocytes, suggesting visceral DRESS syndrome. A skin biopsy confirmed the diagnosis. After complete resolution, patch tests were positive to amoxicillin-clavulanic acid, amoxicillin and oxacillin and negative to the other drugs used (20).

Other cutaneous adverse drug reactions

Erythema multiforme

A 39-year-old woman developed exudative erythema multiforme 10 days after taking amoxicillin and clavulanic acid for tonsillitis. Patch tests were positive to amoxicillin 10% pet (++), ampicillin 10% pet (++) and penicillin G potassium 10% pet (+) (21).

A 7-year-old girl developed a generalized purpuric rash with target shaped areas, 9 days after starting treatment with amoxicillin-clavulanic acid. Laboratory investigation revealed a significant increase of Epstein Barr virus (EBV) specific IgM antibody. Based on clinical aspects and histopathology the patient was diagnosed with erythema multiforme syndrome. Patch tests were positive to amoxicillin and ampicillin (22).

Other drug eruptions

A child developed linear IgA bullous disease which was attributed to amoxicillin-clavulanic acid. The patient had a positive patch test and a positive lymphocyte transformation test to amoxicillin (5). A 43-year-old man developed a bullous exanthema after using amoxicillin-clavulanic acid and trimethoprim-sulfamethoxazole. Patch tests were positive to amoxicillin-clavulanic acid 12.5%, amoxicillin 12.5%, and sulfamethoxazole 12.5%, but negative to trimethoprim. As this patient reacted to 2 non-related medicaments, this was a case of multiple drug hypersensitivity syndrome (6).

A 54-year-old woman developed a generalized rash from amoxicillin-clavulanic acid. Patch tests were positive to the powder of a pill, pure (D2 -, D3 ++) and to sodium penicillin. Amoxicillin itself was apparently not tested, but is the most likely allergenic culprit in this combination tablet (16).

SECTION 2 CASES IN WHICH THE ALLERGENIC INGREDIENT WAS NOT IDENTIFIED

CUTANEOUS ADVERSE DRUG REACTIONS FROM SYSTEMIC ADMINISTRATION CAUSED BY TYPE IV (DELAYED-TYPE) HYPERSENSITIVITY (as demonstrated by positive patch tests)

Cutaneous adverse drug reactions from systemic administration of amoxicillin mixture with clavulanate potassium caused by type IV (delayed-type) hypersensitivity, in which the culprit sensitizer was not identified, have included maculopapular eruption (4), acute generalized exanthematous pustulosis (AGEP) (3,7,12,23), acute localized exanthematous pustulosis (ALEP) (2), symmetrical drug-related intertriginous and flexural exanthema (SDRIFE)/ baboon syndrome (8), fixed drug eruption (23) and drug reaction with eosinophilia and systemic symptoms (DRESS) (4,5,7).

Maculopapular eruption

A 12-year-old girl with infectious mononucleosis developed a maculopapular exanthema attributed to amoxicillin-clavulanic acid. A patch test to the combination drug was positive, as was an intradermal test read after 24 hours and a lymphocyte transformation test. Six months later, the drug was given again and after one day a maculopapular eruption appeared (4). The same authors report on an 11-year-old boy with infectious mononucleosis who developed a maculopapular exanthema attributed to amoxicillin-clavulanic acid. A patch test to the combination drug was positive at D2 and D3, as was an intradermal test read after 72 hours and a lymphocyte transformation test. Six months later, the drug was given again and after two days DRESS developed (4).

Acute generalized exanthematous pustulosis (AGEP)

In a multicenter investigation in France, of 45 patients patch tested for AGEP, 26 (58%) had positive patch tests to drugs, including 2 to amoxicillin-clavulanic acid (7). In Ankara, Turkey, amoxicillin-clavulanic acid was patch test positive in one patient with AGEP (23).

A 34-year-old woman had developed a rapidly evolving rash 2 days after a caesarean section, consisting of a generalized itchy rash mainly involving her chest, back, and proximal limbs. The skin was erythematous, edematous, and slightly scaly and was studded with small non-follicular pustules particularly over the groin and thighs. The patient had received a single intravenous dose of co-amoxiclav (amoxicillin-clavulanic acid) during the caesarean section and various other drugs. A skin biopsy showed acute inflammation of both the epidermis and the dermis predominantly with neutrophils, marked papillary edema, and small subcorneal pustule formation, characteristic for AGEP. Three months later, all drugs used were patch tested and there were positive reactions to co-amoxiclav 1%, 5% and 10% pet. The two ingredients were not tested separately (3).

A 28-year-old man presented with a generalized pustular eruption 2 days after starting oral amoxicillin-clavulanate following appendectomy. He had a high fever and severe malaise. Examination revealed a generalized erythematous eruption scattered with numerous pinhead-sized pustules, associated with buccal erosions and edema of the face and extremities. Bacteriological and mycological cultures of pustules were negative and blood tests showed hyperleukocytosis with lymphocytosis. Histopathology was consistent with AGEP. Patch tests were positive to amoxicillin-clavulanic acid crushed tablet 'in water or olive oil' and showed vesicles and pustules on an erythematous base at D2. A lymphocyte transformation test was also positive to the combination product (12).

Acute localized exanthematous pustulosis (ALEP)
A 35-year-old woman presented with an acute outbreak of multiple small non-follicular pustules on an erythematous and edematous base, affecting the face, neck, and trunk. The patient had been taking amoxicillin–clavulanic acid 1 g/day for a week up to 3 days before presentation for pharyngotonsillitis. Laboratory investigations showed mildly elevated white cell count, erythrocyte sedimentation rate and C-reactive protein. A bacterial culture of a pustule was negative. Histopathology showed acanthosis of the epidermis with mild spongiosis and subcorneal, intraepidermal pustules and a dense perivascular infiltrate of lymphocytes and neutrophils in the dermis. A patch test for amoxicillin–clavulanic acid was positive (no details on concentration and vehicle provided). Amoxicillin and clavulanic acid were not tested separately. This was a case of acute localized exanthematous pustulosis (2).

Symmetrical drug-related intertriginous and flexural exanthema (SDRIFE)/Baboon syndrome
In 2 large hospitals in France, 18 patients were diagnosed with SDRIFE in the period 2006-2018. Fourteen were patch tested, and there were 3 (21%) positive reactions, one to amoxicillin-clavulanic acid and 2 to pristinamycin (8).

Fixed drug eruption
In Ankara, Turkey, amoxicillin-clavulanic acid was patch test positive in one pediatric patient with fixed drug eruption (23).

Drug reaction with eosinophilia and systemic symptoms (DRESS)
Of 13 pediatric patients patch tested for DRESS in a tertiary care hospital in Florence, Italy, between 2010 and 2018, 5 had positive reactions, one of who reacted to amoxicillin-clavulanic acid (5). In a multicenter investigation in France, of 72 patients patch tested for DRESS, 46 (64%) had positive patch tests to drugs, including 2 to amoxicillin-clavulanic acid (7).

An 11-year-old boy with infectious mononucleosis developed a maculopapular exanthema attributed to amoxicillin-clavulanic acid. A patch test to the combination drug was positive at D2 and D3, as was an intradermal test read after 72 hours and a lymphocyte transformation test. Six months later, the drug was given again and after two days DRESS developed (4).

Cross-reactions, pseudo-cross-reactions and co-reactions
Cross-reactions between beta-lactam antibiotics are discussed in Chapter 3.36 Amoxicillin.

LITERATURE
1   The data in the section 'General' may have been obtained from literature discussed in this chapter, but mostly also or exclusively from one or more of the following online sources: ChemIDPlus Advanced, PubChem, DrugBank, RxList, Drug Central, Drugs.com, and Wikipedia
2   Villani A, Baldo A, De Fata Salvatores G, Desiato V, Ayala F, Donadio C. Acute localized exanthematous pustulosis (ALEP): Review of literature with report of case caused by amoxicillin-clavulanic acid. Dermatol Ther (Heidelb) 2017;7:563-570
3   Harries MJ, McIntyre SJ, Kingston TP. Co-amoxiclav-induced acute generalized exanthematous pustulosis confirmed by patch testing. Contact Dermatitis 2006;55:372
4   Mori F, Fili L, Barni S, Giovannini M, Capone M, Novembre EM, Parronchi P. Sensitization to amoxicillin/clavulanic acid may underlie severe rashes in children treated for infectious mononucleosis. J Allergy Clin Immunol Pract 2019;7:728-731
5   Liccioli G, Mori F, Parronchi P, Capone M, Fili L, Barni S, Sarti L, Giovannini M, Resti M, Novembre EM. Aetiopathogenesis of severe cutaneous adverse reactions (SCARs) in children: A 9-year experience in a tertiary care paediatric hospital setting. Clin Exp Allergy 2020;50:61-73
6   Gex-Collet C, Helbling A, Pichler WJ. Multiple drug hypersensitivity – proof of multiple drug hypersensitivity by patch and lymphocyte transformation tests. J Investig Allergol Clin Immunol 2005;15:293-296
7   Barbaud A, Collet E, Milpied B, Assier H, Staumont D, Avenel-Audran M, et al. A multicentre study to determine the value and safety of drug patch tests for the three main classes of severe cutaneous adverse drug reactions. Br J Dermatol 2013;168:555-562
8   De Risi-Pugliese T, Barailler H, Hamelin A, Amsler E, Gaouar H, Kurihara F, et al. Symmetrical drug-related intertriginous and flexural exanthema: A little-known drug allergy. J Allergy Clin Immunol Pract 2020;8:3185-3189.e4

9   Syrigou E, Grapsa D, Charpidou A, Syrigos K. Acute generalized exanthematous pustulosis induced by amoxicillin/clavulanic acid: report of a case presenting with generalized lymphadenopathy. J Cutan Med Surg 2015;19:592-594

10   Bomarrito L, Zisa G, Delrosso G, Farinelli P, Galimberti M. A case of acute generalized exanthematous pustulosis due to amoxicillin-clavulanate with multiple positivity to beta-lactam patch testing. Eur Ann Allergy Clin Immunol 2013;45:178-180

11   Britschgi M, Steiner UC, Schmid S, Depta JP, Senti G, Bircher A, et al. T-cell involvement in drug-induced acute generalized exanthematous pustulosis. J Clin Invest 2001;107:1433-1441

12   De Thier F, Blondeel A, Song M. Acute generalized exanthematous pustulosis induced by amoxycillin with clavulanate. Contact Dermatitis 2001;44:114-115

13   Henning MA, Opstrup MS, Taudorf EH. Acute generalized exanthematous pustulosis to amoxicillin. Dermatitis 2019;30:274-275

14   Rosso R, Mattiacci G, Bernardi ML, Guazzi V, Zaffiro A, Bellegrandi S, et al. Very delayed reactions to beta-lactam antibiotics. Contact Dermatitis 2000;42:293-295

15   Kwon HJ, Kim MY, Kim HO, Park YM. The simultaneous occurrence of contact urticaria from sulbactam and allergic contact dermatitis from ampicillin in a nurse. Contact Dermatitis 2006;54:176-178

16   Kennedy C, Stolz E, van Joost T. Sensitization to amoxicillin in Augmentin. Contact Dermatitis 1989;20:313-314

17   Hernández-Aragüés I, De Santa María García MS, Pérez-Esquerra PR, Simal-Gómez G. Cutaneous drug reactions: Acute rash with pinhead-sized pustules. Eur J Dermatol 2018;28:859-860

18   Pérez-Ezquerra PR, Sanchez-Morillas L, Alvarez AS, Gómez-Tembleque MP, Moratiel HB, Martinez JJ. Fixed drug eruption caused by amoxicillin-clavulanic acid. Contact Dermatitis 2010;63:294-296

19   Yawalkar N, Hari Y, Frutig K, Egli F, Wendland T, Braathen L et al. T cells isolated from positive epicutaneous test reactions to amoxicillin and ceftriaxone are drug specific and cytotoxic. J Invest Dermatol 2000;115:647-652

20   Rabenkogo A, Vigue MG, Jeziorski E. Le syndrome DRESS: une toxidermie à connaître [DRESS syndrome]. Arch Pediatr 2015;22:57-62 (Article in French)

21   Travassos AR, Pacheco D, Antunes J, Silva R, Almeida LS, Filipe P. The importance of patch tests in the differential diagnosis of adverse drug reactions. An Bras Dermatol 2011;86(4Suppl.1):S21-23 (Article in English and Portuguese)

22   González-Delgado P, Blanes M, Soriano V, Montoro D, Loeda C, Niveiro E. Erythema multiforme to amoxicillin with concurrent infection by Epstein-Barr virus. Allergol Immunopathol (Madr) 2006;34:76-78

23   Büyük Yaytokgil Ş, Güvenir H, Külhaş Celík İ, Yilmaz Topal Ö, Karaatmaca B, Civelek E, et al. Evaluation of drug patch tests in children. Allergy Asthma Proc 2021;42:167-174

Chapter 3.38   AMPICILLIN

IDENTIFICATION

Description/definition   : Ampicillin is a semisynthetic derivative of penicillin that conforms to the structural
                            formula shown below
Pharmacological classes : Anti-bacterial agents
IUPAC name   : (2S,5R,6R)-6-[[(2R)-2-Amino-2-phenylacetyl]amino]-3,3-dimethyl-7-oxo-4-thia-1-
                            azabicyclo[3.2.0]heptane-2-carboxylic acid
Other names   : Ampicillin acid; aminobenzylpenicillin
CAS registry number   : 69-53-4
EC number   : 200-709-7
Merck Index monograph   : 1853
Patch testing   : 5% Pet. (SmartPracticeCanada, SmartPracticeEurope)
Molecular formula   : $C_{16}H_{19}N_3O_4S$

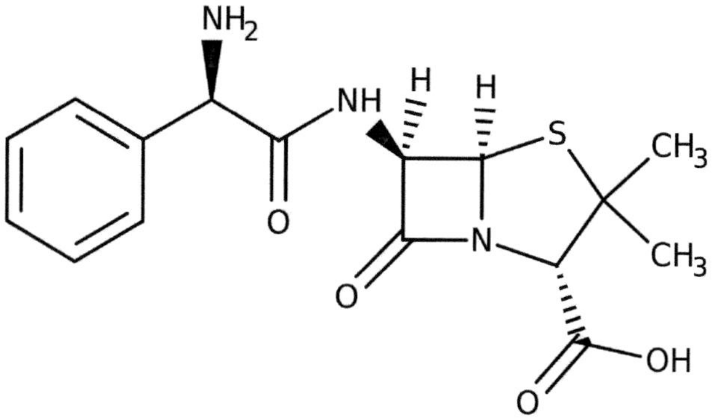

GENERAL

Ampicillin is a broad-spectrum, semisynthetic, β-lactam penicillin antibiotic with bactericidal activity, especially to gram-positive bacteria. It is indicated for the treatment of infection (respiratory, gastro-intestinal, urinary tract and meningitis) due to *E. coli*, *P. mirabilis*, enterococci, *Shigella*, *S. typhosa* and other *Salmonella*, non-penicillinase-producing *N. gonorrhoeae*, *H. influenzae*, staphylococci, and streptococci (1). In this chapter, it has *not* been attempted to provide a full review of delayed-type hypersensitivity reactions to ampicillin.

CONTACT ALLERGY FROM ACCIDENTAL CONTACT

Case series

Between 1978 and 2001, 14,689 patients have been patch tested in the university hospital of Leuven, Belgium. Occupational allergic contact dermatitis to pharmaceuticals was diagnosed in 33 health care workers: 26 nurses, 4 veterinarians, 2 pharmacists and one medical doctor. Practically all of these patients presented with hand dermatitis (often the fingers), sometimes with secondary localizations, frequently the face. In this group, three patients had occupational allergic contact dermatitis from ampicillin (8).

In a group of 107 workers in the pharmaceutical industry with dermatitis, investigated in Warsaw, Poland, before 1989, 5 reacted to ampicillin, tested 20% pet. (2). Of 333 nurses patch tested by the same authors in Poland between 1979 and 1987, 14 (4.2%) reacted to ampicillin 20% pet.; all reactions were likely to be relevant (11). In Warsaw, Poland, again, by the same authors, in the period 1979-1983, 27 pharmaceutical workers, 24 nurses and 30 veterinary surgeons were diagnosed with occupational allergic contact dermatitis from antibiotics. The numbers that had positive patch tests to ampicillin (ampoule content) were 5, 7, and 0, respectively, total 12 (16).

In Hamburg, Germany, from 1967 to 1969 inclusive, 92 cases of occupational eczema occurred among members of the medical profession; 61 had an allergic contact dermatitis, principally from drugs. Ampicillin was the most frequent allergen, being the sensitizer in 24 of the cases, most of whom were nurses exposed occupationally. Each reacted to ampicillin 5% in water, some also to other penicillins (12).

Case reports

A 24-year-old nurse presented with pruritic erythematous papulovesicles and scaly plaques affecting her fingers. The patient also experienced pruritus and redness on her fingers that occurred within 30 minutes of preparing antibiotic solutions. Scratch tests read at 30 minutes were positive to ampicillin/sulbactam and sulbactam 10 and 100 mg/ml water, but negative to ampicillin, indicating contact urticaria to sulbactam. She had also positive patch tests to amoxicillin/clavulanic acid, to ampicillin/sulbactam and to ampicillin 1, 10 and 100 mg/ml water. Amoxicillin itself was not tested, but allergy to this penicillin is highly likely, indicating occupational allergic contact dermatitis from ampicillin in ampicillin/sulbactam and amoxicillin in amoxicillin/clavulanic acid (18).

A 45-year-old nurse gave a 20-year history of recurrent vesiculation and erythema of the sides of the fingers of her right hand, which spread to the back of the hand, with scaly resolution, within a 1-week period. She attributed her condition to handling either ampicillin or amoxicillin vials for parenteral use. Patch tests were positive to ampicillin and amoxicillin 50 mg/ml water at D2 and D4, as were intradermal tests with late readings (20).

A 21-year-old woman working in the pharmaceutical industry developed eczema of the hands and forearms while working with benzylpenicillin and ampicillin. Patch tests were positive to ampicillin and negative to benzylpenicillin (28).

CUTANEOUS ADVERSE DRUG REACTIONS FROM SYSTEMIC ADMINISTRATION CAUSED BY TYPE IV (DELAYED-TYPE) HYPERSENSITIVITY (as demonstrated by positive patch tests)

Cutaneous adverse drug reactions from systemic administration of ampicillin caused by type IV (delayed-type) hypersensitivity have included maculopapular eruption (3,15,19,21,27,29,31,32,33,34,35,37,39,40,41), exfoliative exanthema (36), (extensive or generalized) erythema (39,41), acute generalized exanthematous pustulosis (AGEP) (4,5,6,7), symmetrical drug-related intertriginous and flexural exanthema (SDRIFE)/baboon syndrome (24), Stevens-Johnson syndrome/toxic epidermal necrolysis (SJS/TEN) (13,14), urticaria/angioedema (9,22,23,25,34,36,39,41), and unspecified drug eruptions (17,38).

It should be realized that several publications have come from the same hospital and that overlap in the presented data cannot be excluded and are in some cases highly likely or certain.

General

The classification and structures of beta-lactam antibiotics are discussed in Chapter 3.36 Amoxicillin. Delayed-type hypersensitivity to ampicillin in patients with skin eruptions attributed to this antibiotic appears to be long-lasting, as shown by repeated patch testing (30).

Case series with various or unknown types of drug reactions

In 97 patients with nonimmediate reactions from penicillins with a positive patch test to at least one penicillin seen in Rome between 2002 and 2011, amoxicillin was the culprit drug in 50 individuals, ampicillin in 35, bacampicillin in 25 and benzylpenicillin in 11. The most frequent clinical manifestations were maculopapular eruptions with or without edema, delayed urticaria with or without angioneurotic edema and generalized erythema. It was not specified which drug eruptions the 4 individual antibiotics had caused (41).

Four patients with maculopapular eruption, 4 with delayed urticaria/angioedema and 2 with erythema caused by ampicillin with a positive patch test to this aminopenicillin were reported from Rome in 2006 (39).

Maculopapular eruption

Case series

In Rome and 2 other centers in Italy, in a prospective study performed between 2000 and 2014, 214 consecutive patients who had suffered 307 nonimmediate skin reactions during penicillin therapy and who had positive patch test and/or delayed-reading intradermal test to at least 1 penicillin, were examined with extended patch testing to cephalosporins and aztreonam (3, overlap with ref. 37). 260 of the 307 reactions were maculopapular eruptions rash (with or without edema), 21 erythema (with or without edema) and the rest were other skin manifestations. Of the 307 adverse drug reactions, 172 had been caused by amoxicillin, 91 by ampicillin and 29 by bacampicillin. In the 214 patients, there were 210 positive reactions (patch test and/or delayed intradermal test) to amoxicillin, 212 to ampicillin, 97 to benzylpenicillin (virtually all cross-reactions) and 4 to piperacillin. Extended patch testing showed that no patient had positive skin test responses to cefuroxime, ceftriaxone, and aztreonam; 40 (19%) had cross-reactions to aminocephalosporins (cefalexin, cefaclor, and cefadroxil) (3, overlap with ref. 37).

Thirty patients from Rome who had suffered maculopapular exanthemas or delayed urticaria from beta-lactam antibiotics and who had at least one positive patch test to one beta-lactam antigenic determinant, were patch tested with a large number of beta-lactam antibiotics. All but one reacted to amoxicillin, ampicillin and bacampicillin (34).

Again in Rome, 259 patients who had suffered nonimmediate skin reactions during penicillin therapy were examined with patch testing. 173 patients had had a maculopapular rash, 59 (delayed) urticaria/angioedema, 22 erythema and 5 other manifestations. Two hundred and forty-one subjects (93%) reported adverse reactions to aminopenicillins: amoxicillin in 107, ampicillin in 59 and bacampicillin in 48. Twenty-seven of these subjects had suffered reactions to 2 or 3 different aminopenicillins in separate episodes. Ninety-four of the 259 subjects (36%)

showed patch test and delayed intradermal test positivity to the culprit penicillins, of which ninety to both amoxicillin and ampicillin. Bacampicillin was not tested, but a number of these sensitizations must have been picked up by reactions to ampicillin. These patients all suffered from maculopapular rashes or – a few – erythema (35, overlap with ref. 33).

In the period 1994 to 1997 inclusive, in Rome, of 111 patients with a maculopapular eruption from amoxicillin or ampicillin, 58 (52%) had positive patch tests to both ampicillin and amoxicillin 5% pet. More rashes (probably 1.8x) had been caused by ampicillin than by amoxicillin, but this was not specified (33, overlap with ref. 35).

In Rome, between 1987 and 1992, 60 patients who had developed maculopapular eruptions (sometimes with edema, urticaria or fever) from aminopenicillins (amoxicillin, ampicillin, bacampicillin), were patch tested with amoxicillin and ampicillin 5% pet. Thirty-three had positive patch and intradermal tests (with late reading) to aminopenicillins (not specified). Oral challenges were performed with ampicillin in 13 patients and 5 with amoxicillin and all developed diffuse maculopapular eruptions, 12 with a challenge of 5 mg and 6 with 50 mg (31).

Two cases of maculopapular eruption from ampicillin with a positive patch test to ampicillin powder were reported from the Netherlands in 1981 (27); 2 from Rome in 1997 (32).

Case reports
One patient developed a maculopapular exanthema from ampicillin and had positive patch tests to several penicillins including amoxicillin; ampicillin itself was apparently not tested (15). A 19-year-old man developed a generalized intensely pruritic morbilliform rash 6 hours after intake of ampicillin. Intradermal tests were positive to ampicillin on D3, a patch tests became positive on D6 only (19). One case of maculopapular eruption - erythroderma from ampicillin with positive patch test to ampicillin 20% saline or pet. was reported in 1990 from Japan; details were not provided (21).

Additional cases of maculopapular eruptions from ampicillin with a positive patch test to this antibiotic can be found in refs. 29 and 40. See also the section 'Case series with various or unknown types of drug reactions' above, refs. 39 and 41.

Erythroderma, widespread erythematous eruption, exfoliative dermatitis
Two cases of exfoliative exanthema from ampicillin with positive patch and intradermal tests (36). See also the section 'Maculopapular eruption' above, ref. 21. See also the section 'Case series with various or unknown types of drug reactions' above, refs. 39 and 41.

Acute generalized exanthematous pustulosis
A life-threatening case of acute generalized exanthematous pustulosis (AGEP) in a 30-year-old man caused by delayed-type hypersensitivity to amoxicillin, ampicillin and benzylpenicillin was reported in 2017. It began with treatment of periodontitis with amoxicillin, which caused erythema of the face. When treated with ciprofloxacin (which was later prick-test positive) and subsequently ampicillin and benzylpenicillin, the erythema spread to the rest of the skin with pustules, severe mucous membrane involvement and high fever. A systemic inflammatory response syndrome developed with life-threatening involvement of the lungs, heart, liver and kidneys. With non-penicillin antibiotics and systemic corticosteroids the patient recovered. Four months later, patch tests were positive to ampicillin, amoxicillin, and other penicillins. Intradermal testing of minor and major determinants of penicillin were positive for benzylpenicillin and ampicillin. Prick testing was positive for ciprofloxacin, amoxicillin, ampicillin, and amoxicillin/clavulanic acid. The authors assumed that the underlying type I/IV allergies to various β-lactam antibiotics given, together with secondary infection of the skin, led to severe life-threatening systemic involvement. Pre-existing bacterial infections, in this case caused by periodontitis, might also have been a promoting factor (4).

A 39-year-old, 41-week pregnant woman was admitted for the induction of labor, and treated for the group B streptococcal vaginal colonization found by swab before delivery. She was administered ampicillin/cloxacillin sodium (A/CS) i.v. Five hours after the infusion, she developed a fever and edematous erythema. A/CS was stopped thereafter. However, after delivery, her fever continued up to 39.5°C, and she had diffuse edematous erythema with numerous non-follicular pustules on her neck, axilla, wrist and trunk. Her blood test showed leukocytosis with marked neutrophilia and elevated C-reactive protein. Blood cultures and swabs from the pustules were sterile. A skin biopsy showed spongiform subcorneal/intraepidermal pustules, and marked liquefaction degeneration. Later, patch tests were positive to A/CS at 0.05%, 0.1%, 0.5%, 1% and 10% aqua, as was a drug lymphocyte stimulation test. The patient refused further tests, so it remains unknown whether this case of acute generalized exanthematous pustulosis (AGEP) was caused by ampicillin, cloxacillin, or both (5).

Two more cases of AGEP from ampicillin with positive patch tests have been reported (6,7), one of which was a pediatric case (7). A 73-year old woman was treated with intravenous ampicillin/sulbactam for cholecystitis and developed AGEP with high fever (39.5°C) within 3 days. Later patch tests were positive to benzylpenicillin and phenoxymethylpenicillin. Ampicillin itself was apparently not tested (6). A 9-year-old boy underwent a colostomy and was treated with ampicillin/sulbactam, clindamycin and amikacin. Already on the first day he developed AGEP. Patch tests were positive to the commercial preparation of intravenous ampicillin, 10% and 30% in saline (7).

Symmetrical drug-related intertriginous and flexural exanthema (SDRIFE)/Baboon syndrome
The first case and name-giving report of the baboon syndrome was caused by delayed-type hypersensitivity to ampicillin. The patient with the characteristic red buttocks and erythema of all the major flexures had a positive patch test to ampicillin 5% pet. However, the antibiotic had not been given orally or parenterally, but had been applied as ampicillin in gelatine foam in the middle ear during a stapendectomy operation one day prior to the development of the eruption. Transcutaneous resorption in the ear caused this 'systemically-induced allergic contact dermatitis' (24).

Stevens-Johnson syndrome/toxic epidermal necrolysis (SJS/TEN)
A 32-year-old woman developed toxic epidermal necrolysis 10 days after having received a single dose of ampicillin sodium for an upper respiratory tract infection. Two days prior to admission, she again received a single 250 mg dose of ampicillin sodium because of a temperature of 39.9°C. Some 8 weeks after resolution, the patient had a positive intradermal test to ampicillin sodium (23 mm infiltrated erythema at D2), a positive lymphocyte transformation test and a positive patch test to ampicillin sodium 50 µg/ml water. Histopathologically the site of the positive patch test reaction showed remarkable degenerative changes in the epidermis, so that the entire epidermis appeared to be undergoing necrosis (14).

One individual developed toxic epidermal necrolysis (TEN) from intramuscular injections of ampicillin. The patient later had positive reactions to a patch test with ampicillin, an intradermal test (after 6 hours) and a lymphocyte transformation test to this antibiotic (13).

Other cutaneous adverse drug reactions

Urticaria
Two cases of urticaria from ampicillin with positive patch and intradermal tests (36). Three cases of delayed urticaria from delayed-type hypersensitivity to ampicillin (34).

A 42-year-old woman had 2 episodes of urticaria (in one also angioedema) after oral ampicillin. Patch tests were positive to many other penicillins, some cephalosporins and the beta-lactam carbapenem antibiotic imipenem (9). The same authors also report on a woman aged 43 who had developed angioedema and urticaria twice, first after oral ampicillin, later after oral amoxicillin. Patch tests were positive to both drugs, many other penicillins, some cephalosporins and the beta-lactam carbapenem antibiotic imipenem (9).

A 62-year-old woman developed widespread urticaria 2 hours after 'a dose of antibiotic'. Erythromycin and quinolones were well tolerated. Intradermal tests were positive to amoxicillin 200 mg/ml and ampicillin 200 mg/ml (both saline) after one day and remained so for 20 days. Patch tests to both antibiotic substances, performed one month later, were strongly positive already after one day, and at the same time, the previously positive intradermal tests showed a flare-up reaction (22). Quite curiously, it was not specified which antibiotic had induced the urticaria in this patient.

A 21-year-old man had been treated with ampicillin at the age of 15, when he developed generalized urticaria. Six years later, after another treatment, the same manifestations appeared. Recently, a subcutaneous test injection produced an eczematous reaction at the site. Patch tests with 'penicillin', tetracycline, cefalexin and ampicillin in pet. were positive to ampicillin at D2. The patient had never used any topical preparation of ampicillin (23).

A boy aged 14 had been treated for 1 week with capsules of ampicillin for acute tonsillitis when an urticarial eruption with angioedema developed. When treated again with ampicillin, the generalized cutaneous eruption recurred. Patch tests were positive to ampicillin (no test details provided) with a cross-reactivity to cefalexin at 6, 24 and 48 hours (25).

See also the section 'Case series with various or unknown types of drug reactions' above, refs. 39 and 41.

Unspecified drug eruptions
In a group of 78 patients from Italy who had suffered nonimmediate drug reactions from any beta-lactam compound (penicillins, cephalosporins), 70 had positive patch tests to ampicillin. In 28 cases, ampicillin was the culprit drug, so the other reactions were presumably cross-reactions. The nature of the adverse cutaneous reactions in the ampicillin-allergic individuals was not mentioned (17). Four patients with nonimmediate allergic reactions to ampicillin (38, details unknown to the author).

Cross-reactions, pseudo-cross-reactions and co-reactions
Cross-reactions between beta-lactam antibiotics are discussed in Chapter 3.36 Amoxicillin. Patients sensitized to pivampicillin (the pivaloyloxymethyl ester of ampicillin) and pivmecillinam (Chapter 3.23 Amdinocillin pivoxil)

frequently cross-react to ampicillin (10). Pseudocross-reactivity between bacampicillin (34), talampicillin (both prodrugs of ampicillin) and ampicillin (26).

REFERENCES

1 The data in the section 'General' may have been obtained from literature discussed in this chapter, but mostly also or exclusively from one or more of the following online sources: ChemIDPlus Advanced, PubChem, DrugBank, RxList, Drug Central, Drugs.com, and Wikipedia

2 Rudzki E, Rebandel P, Grzywa Z. Contact allergy in the pharmaceutical industry. Contact Dermatitis 1989;21:121-122

3 Romano A, Gaeta F, Valluzzi RL, Maggioletti M, Caruso C, Quaratino D. Cross-reactivity and tolerability of aztreonam and cephalosporins in subjects with a T cell-mediated hypersensitivity to penicillins. J Allergy Clin Immunol 2016;138:179-186

4 Tajmir-Riahi A, Wörl P, Harrer T, Schliep S, Schuler G, Simon M. Life-threatening atypical case of acute generalized exanthematous pustulosis. Int Arch Allergy Immunol 2017;174:108-111

5 Matsumoto Y, Okubo Y, Yamamoto T, Ito T, Tsuboi R. Case of acute generalized exanthematous pustulosis caused by ampicillin/cloxacillin sodium in a pregnant woman. J Dermatol 2008;35:362-364

6 Gensch K, Hodzic-Avdagic N, Megahed M, Ruzicka T, Kuhn A. Acute generalized exanthematous pustulosis with confirmed type IV allergy. Report of 3 cases. Hautarzt 2007;58:250-252, 254-255 (Article in German)

7 Özmen S, Misirlioglu ED, Gurkan A, Arda N, Bostanci I. Is acute generalized exanthematous pustulosis an uncommon condition in childhood? Allergy 2010;65:1490-1492

8 Gielen K, Goossens A. Occupational allergic contact dermatitis from drugs in healthcare workers. Contact Dermatitis 2001;45:273-279

9 Schiavino D, Nucera E, Lombardo C, Decinti M, Pascolini L, Altomonte G, et al. Cross-reactivity and tolerability of imipenem in patients with delayed-type, cell-mediated hypersensitivity to beta-lactams. Allergy 2009;64:1644-1648

10 Møller NE, Nielsen B, von Würden K. Contact dermatitis to semisynthetic penicillins in factory workers. Contact Dermatitis 1986;14:307-311

11 Rudzki E, Rebandel P, Grzywa Z. Patch tests with occupational contactants in nurses, doctors and dentists. Contact Dermatitis 1989;20:247-250

12 Schulz KH, Schöpf E, Wex O. Allergic occupational eczemas caused by ampicillin [Allergische Berufsekzeme durch Ampicillin]. Berufsdermatosen 1970;18:132-143 (Article in German)

13 Romano A, Di Fonso M, Pocobelli D, Giannarini L, Venuti A, Garcovich A. Two cases of toxic epidermal necrolysis caused by delayed hypersensitivity to beta-lactam antibiotics. J Investig Allergol Clin Immunol 1993;3:53-55

14 Tagami H, Tatsuta K, Iwatski K, Yamada M. Delayed hypersensitivity in ampicillin-induced toxic epidermal necrolysis. Arch Dermatol 1983;119:910-913

15 Studer M, Waton J, Bursztejn AC, Aimone-Gastin I, Schmutz JL, Barbaud A. Does hypersensitivity to multiple drugs really exist? Ann Dermatol Venereol 2012;139:375-380 (Article in French)

16 Rudzki E, Rebendel P. Contact sensitivity to antibiotics. Contact Dermatitis 1984;11:41-42

17 Buonomo A, Nucera E, De Pasquale T, Pecora V, Lombardo C, Sabato V, et al. Tolerability of aztreonam in patients with cell-mediated allergy to β-lactams. Int Arch Allergy Immunol 2011;155:155-159

18 Kwon HJ, Kim MY, Kim HO, Park YM. The simultaneous occurrence of contact urticaria from sulbactam and allergic contact dermatitis from ampicillin in a nurse. Contact Dermatitis 2006;54:176-178

19 Rosso R, Mattiacci G, Bernardi ML, Guazzi V, Zaffiro A, Bellegrandi S, et al. Very delayed reactions to beta-lactam antibiotics. Contact Dermatitis 2000;42:293-295

20 Gamboa P, Jáuregui I, Urrutia I. Occupational sensitization to aminopenicillins with oral tolerance to penicillin V. Contact Dermatitis 1995;32:48-49

21 Osawa J, Naito S, Aihara M, Kitamura K, Ikezawa Z, Nakajima H. Evaluation of skin test reactions in patients with non-immediate type drug eruptions. J Dermatol 1990;17:235-239

22 Llamazares AA, Chamorro M, Robledo T, Cimarra M, Palacios R, Rodgriguez A, et al. Flare-up of skin tests to amoxycillin and ampicillin. Contact Dermatitis 2000;42:166

23 Pigatto PD, de Blasio A, Altomare GF, Giacchetti A. Contact sensitivity from systemic ampicillin. Contact Dermatitis 1986;14:196

24 Andersen KE, Hjorth N, Menné T. The baboon syndrome: systemically-induced allergic contact dermatitis. Contact Dermatitis 1984;10:97-100

25 Valsecchi R, Serra M, Pansera B, Foiadelli L, Cainelli T. Patch testing in adverse drug reactions to penicillin and cephalosporin. Contact Dermatitis 1981;7:158

26 Imayama S, Fukuda H, Hori Y. Drug eruptions following treatment with prodrugs: a review of the reported cases in Japan from 1984 to 1989. J Dermatol 1991;18:277-280

27 Bruynzeel DP, van Ketel WG. Repeated patch testing in penicillin allergy. Br J Dermatol 1981;104:157-159

28 Rudzki E, Rebandel P. Hypersensitivity to semisynthetic penicillins but not to natural penicillin. Contact Dermatitis 1991;25:192

29 Voorhorst R, Sparreboom S. The use of stereoisomers in patch testing. Ann Allergy 1980;45:100-103

30 Romano A, Di Fonso M, Pietrantonio F, Pocobelli D, Giannarini L, Del Bono A, et al. Repeated patch testing in delayed hypersensitivity to beta-lactam antibiotics. Contact Dermatitis 1993;28:190

31 Romano A, Di Fonso M, Papa G, Pietrantonio F, Federico F, Fabrizi G, Venuti A. Evaluation of adverse cutaneous reactions to aminopenicillins with emphasis on those manifested by maculopapular rashes. Allergy 1995;50:113-118

32 Romano A, Quaratino D, Papa G, Di Fonso M, Venuti A. Aminopenicillin allergy. Arch Dis Child 1997;76:513-517

33 Romano A, Quaratino D, Di Fonso M, Papa G, Venuti A, Gasbarrini G. A diagnostic protocol for evaluating nonimmediate reactions to aminopenicillins. J Allergy Clin Immunol 1999;103:1186-1190

34 Patriarca G, D'Ambrosio C, Schiavino D, Larocca LM, Nucera E, Milani A. Clinical usefulness of patch and challenge tests in the diagnosis of cell-mediated allergy to betalactams. Ann Allergy Asthma Immunol 1999;83:257-266

35 Romano A, Viola M, Mondino C, Pettinato R, Di Fonso M, Papa G, et al. Diagnosing nonimmediate reactions to penicillins by in vivo tests. Int Arch Allergy Immunol 2002;129:169-174

36 Torres MJ, Sánchez-Sabaté E, Alvarez J, Mayorga C, Fernández J, Padial A, et al. Skin test evaluation in nonimmediate allergic reactions to penicillins. Allergy 2004;59:219-224

37 Romano A, Gaeta F, Valluzzi RL, Alonzi C, Maggioletti M, Zaffiro A, et al. Absence of cross-reactivity to carbapenems in patients with delayed hypersensitivity to penicillins. Allergy. 2013;68:1618-1621.

38 López Serrano C, Villas F, Cabañas R, Contreras J. Delayed hypersensitivity to beta-lactams. J Investig Allergol Clin Immunol 1994;4:315-319

39 Schiavino D, Nucera E, De Pasquale T, Roncallo C, Pollastrini E, Lombardo C, Giuliani L, Larocca LM, Buonomo A, Patriarca G. Delayed allergy to aminopenicillins: clinical and immunological findings. Int J Immunopathol Pharmacol. 2006;19:831-840

40 Trcka J, Seitz CS, Brocker E-B € et al. Aminopenicillin-induced exanthema allows treatment with certain cephalosporins or phenoxymethyl penicillin. J Antimicrob Chemother 2007;60:107-111

41 Buonomo A, Nucera E, Pecora V, Rizzi A, Aruanno A, Pascolini L, et al. Cross-reactivity and tolerability of cephalosporins in patients with cell-mediated allergy to penicillins. J Investig Allergol Clin Immunol 2014;24:331-337

Chapter 3.39   AMPIROXICAM

IDENTIFICATION

Description/definition : Ampiroxicam is the benzothiazine prodrug of piroxicam that conforms to the structural formula shown below

Pharmacological classes : Non-steroidal anti-inflammatory drugs

IUPAC name : Ethyl 1-[[2-methyl-1,1-dioxido-3-(2-pyridinylcarbamoyl)-2H-1,2-benzothiazin-4-yl]oxy]ethyl carbonate

Other names : 4-(1-Hydroxyethoxy)-2-methyl-N-2-pyridyl-2H-1,2-benzothiazine-3-carboxamide ethyl carbonate (ester), 1,1-dioxide

CAS registry number : 99464-64-9

EC number : Not available

Merck Index monograph : 1854

Patch testing : 5% pet. for UVA-photopatch testing; test also piroxicam 1%-5% pet. as this is more reliable than testing ampiroxicam (2); also patch test thimerosal 0.1% pet. and thiosalicylic acid 0.1% and 1% pet.

Molecular formula : $C_{20}H_{21}N_3O_7S$

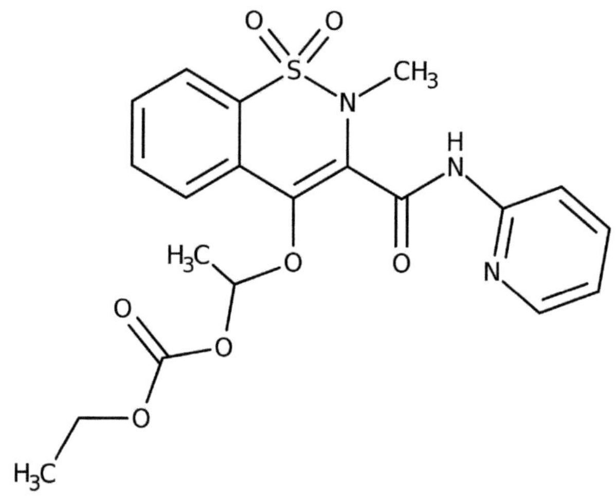

GENERAL

Ampiroxicam is a benzothiazine that is the 1-[(ethoxycarbonyl)oxy]ethyl ether of piroxicam. It has a role as a prodrug, an analgesic, a nonsteroidal anti-inflammatory drug, and an antirheumatic drug. Ampiroxicam is used for the relief of pain and inflammation in musculoskeletal disorders such as rheumatoid arthritis and osteoarthritis (1). Inactive ampiroxicam is hydrolyzed to active piroxicam by an intestinal carboxyesterase during absorption through the intestinal wall (5).

CUTANEOUS ADVERSE DRUG REACTIONS FROM SYSTEMIC ADMINISTRATION CAUSED BY TYPE IV (DELAYED-TYPE) HYPERSENSITIVITY (as demonstrated by positive patch tests)

Cutaneous adverse drug reactions from systemic administration of ampiroxicam caused by type IV (delayed-type) hypersensitivity have included photosensitivity (2,3,4,5).

Photosensitivity

General

As ampiroxicam is converted in the intestine into the well-known photosensitizer piroxicam, it was to be expected that photoallergic reactions to ampiroxicam would be observed. Because photosensitive eruptions from oral piroxicam occur within a few days after using the NSAID, some authors have suggested them to be phototoxic rather than photoallergic. Others, however, believe that cross-reactivity to thimerosal and thiosalicylic acid, which in most such patients and in ampiroxicam-photosensitive individuals (2,3,5) give positive patch test reactions, is responsible for the photocontact allergy to piroxicam (see Chapter 3.392 Piroxicam) and ampiroxicam.

In some patients with a photosensitive eruption, photopatch tests with ampiroxicam are negative but with piroxicam positive (2,4). The cause of this is as yet unknow, but may be related to lower cutaneous absorption of ampiroxicam or limited capacity for conversion in the skin into piroxicam (2).

Case reports
A 45-year-old man had pruritic vesicular erythema affecting sun-exposed areas including the face, neck and dorsa of the hands and feet. The eruption appeared after taking 27 mg ampiroxicam daily for 2 days. He had negative patch tests but positive UVA-photopatch tests to both ampiroxicam and piroxicam 0.1% pet. (2). The authors also describe a 72-year-old man with pruritic edematous erythema on his face and the dorsa of both hands after 3 days treatment with ampiroxicam 27 mg daily. A UVA-photopatch test with ampiroxicam 5% pet. was negative, but positive to piroxicam 5% pet. (2). Both patients were also contact sensitized to thimerosal and one, who was tested with it, also to thiosalicylic acid

A 69-year-old man presented with edematous erythematous eruptions without itching on sun-exposed areas and numerous micro-vesicles on his hands and fingers. He had taken 27 mg of ampiroxicam 2 days earlier. Photo-testing showed an abnormal erythematous reaction to UVA, becoming normal a week after the drug was withdrawn. A subsequent challenge test with ampiroxicam and UVA irradiation was positive, attesting to the role of UVA. Patch tests were strongly positive to thimerosal and thiosalicylic acid; positive photopatch tests were observed to ampiroxicam 10% and 1% (with UVA irradiation 5.0 J/cm^2) and piroxicam 10%, 1% and 0.1%. Seven controls were negative (3). The vesicular eruption of the hands was considered to be an -id reaction and is characteristic for both piroxicam photosensitivity and thimerosal sensitivity (3).

An 84-year-old woman developed a photosensitive eruption 4 days after starting ampiroxicam. Photopatch tests with UVA irradiation 4.5 J/cm^2 were positive to ampiroxicam 1% and 10% pet. and to piroxicam 0.1%, 1% and 10% pet. Patch tests were positive to thimerosal 0.05% and thiosalicylic acid 1% and 0.1% pet. (5).

One patient from Japan had a photosensitive eruption 4 days after oral intake of ampiroxicam. A photopatch test with ampiroxicam was negative, but with piroxicam positive (4).

Cross-reactions, pseudo-cross-reactions and co-reactions
(Photo)cross-reactions are to be expected, as ampiroxicam is a prodrug of piroxicam and is converted to piroxicam during absorption through the intestine. Indeed, photocross-reactivity to piroxicam has been observed in patients photosensitized to ampiroxicam (2,3,4). There have been no photocross-reactions to tenoxicam (2,3).

LITERATURE
1   The data in the section 'General' may have been obtained from literature discussed in this chapter, but mostly also or exclusively from one or more of the following online sources: ChemIDPlus Advanced, PubChem, DrugBank, RxList, Drug Central, Drugs.com, and Wikipedia
2   Toyohara A, Chen KR, Miyakawa S, Inada M, Ishiko A. Ampiroxicam-induced photosensitivity. Contact Dermatitis 1996;35:101-102
3   Kurumaji Y. Ampiroxicam-induced photosensitivity. Contact Dermatitis 1996;34:298-299
4   Ibe M, et al. A case of photosensitive drug eruption induced by piroxicam. Jpn J Dermatol 1996;106:157 (Data cited in ref. 2)
5   Chishiki M, Kawada A, Fujioka A, Hiruma M, Ishibashi A, Banba H. Photosensitivity due to ampiroxicam. Dermatology 1997;195:409-410

Chapter 3.40   AMPROLIUM

IDENTIFICATION

Description/definition   : Amprolium is the thiamine analog that conforms to the structural formula shown below
Pharmacological classes   : Coccidiostats
IUPAC name   : 5-[(2-Methylpyridin-1-ium-1-yl)methyl]-2-propylpyrimidin-4-amine;chloride
CAS registry number   : 121-25-5
EC number   : 204-458-4
Merck Index monograph   : 1856
Patch testing   : 1% pet. (2); 10% water (3)
Molecular formula   : $C_{14}H_{19}ClN_4$

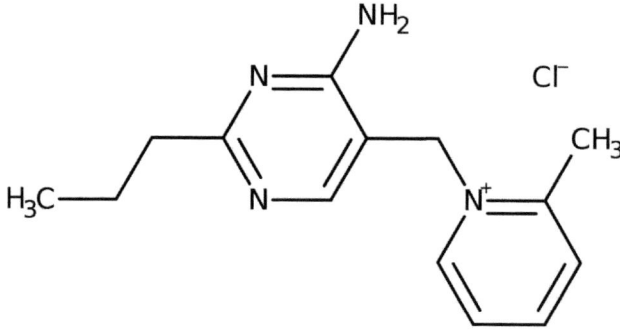

GENERAL

Amprolium is an organic compound used as a coccidiostat in poultry farming. The drug is a thiamine analog and blocks the thiamine transporter of *Eimeria* species, thereby preventing carbohydrate synthesis. Despite only moderate efficacy it is well favored due to few resistance issues. It is commonly used prophylactically in the United States in conjunction with sulfonamides in chickens and cattle. In pharmaceutical products (veterinary use only), both amprolium and amprolium hydrochloride (CAS number 137-88-2, EC number 205-307-5, molecular formula $C_{14}H_{20}Cl_2N_4$) may be employed (1).

CONTACT ALLERGY FROM ACCIDENTAL CONTACT

In Italy, during 1986-1988, 204 animal feed mill workers (191 men, 13 women) were patch tested with a large number of animal feed additives. There was one reaction to amprolium HCl 1% pet. in a group of 36 subjects with clinical complaints (dermatitis or pruritus sine materia) and one reaction in the group of 168 individuals without skin complaints. All reactions were considered to be relevant (2).

LITERATURE

1   The data in the section 'General' may have been obtained from literature discussed in this chapter, but mostly also or exclusively from one or more of the following online sources: ChemIDPlus Advanced, PubChem, DrugBank, RxList, Drug Central, Drugs.com, and Wikipedia
2   Mancuso G, Staffa M, Errani A, Berdondini RM, Fabbri P. Occupational dermatitis in animal feed mill workers. Contact Dermatitis 1990;22:37-41
3   De Groot AC. Patch testing, 4th edition. Wapserveen, The Netherlands: acdegroot publishing, 2018 (ISBN 9789081323345)

Chapter 3.41   ANILERIDINE

IDENTIFICATION

Description/definition   : Anileridine is the piperidinecarboxylate ester that conforms to the structural formula shown below
Pharmacological classes : Analgesics, opioid
IUPAC name   : Ethyl 1-[2-(4-aminophenyl)ethyl]-4-phenylpiperidine-4-carboxylate
Other names   : 1-(2-(4-Aminophenyl)ethyl)-4-phenyl-4-piperidinecarboxylic acid ethyl ester
CAS registry number   : 144-14-9
EC number   : Not available
Merck Index monograph   : 1921
Patch testing   : Commercial solution for intramuscular injection undiluted (2)
Molecular formula   : $C_{22}H_{28}N_2O_2$

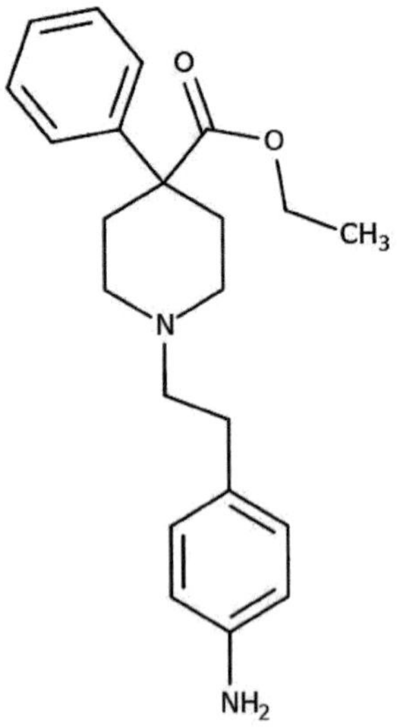

GENERAL

Anileridine, an analog of pethidine, is a synthetic opioid receptor agonist and strong analgesic medication belonging to the piperidine class. This drug is useful for the relief of moderate to severe pain and may also be used as an adjunct in general anesthesia to reduce the amount of anesthetic needed, to facilitate relaxation, and to reduce laryngospasm (1). In pharmaceutical products, anileridine is employed as anileridine phosphate (CAS number 4268-37-5, EC number not available, molecular formula $C_{22}H_{31}N_2O_6P$) or anileridine hydrochloride (1).

CONTACT ALLERGY FROM ACCIDENTAL CONTACT

A 64-year-old man was referred with a 4-week history of persistent periorbital swelling with vesiculation and scaling of the palms of the hands. After briefly responding to oral and topical steroids, the lesions recurred. There was no history of previous similar reactions or drug ingestion. The patient would give 3-4 hourly injections to his wife of intramuscular anileridine for her home program of chronic pain management. This was dispensed from a 1 ml single dose package. Three weeks after being seen initially, the patient was patch tested to anileridine injection undiluted and to several over-the-counter lotions. There were ++ reactions at D2 and D3 to anileridine solution. Encouraging the patient to wear latex gloves during the handling of the medicine resulted in prompt resolution of both the hand and facial rash. Anileridine itself was not tested and it was not mentioned whether there were any excipients in the injection fluid (possibly not, as they were single dose vials) (2). This was a case of connubial (or consort) allergic contact dermatitis, which is currently called allergic contact dermatitis *by proxy*.

LITERATURE

1   The data in the section 'General' may have been obtained from literature discussed in this chapter, but mostly also or exclusively from one or more of the following online sources: ChemIDPlus Advanced, PubChem, DrugBank, RxList, Drug Central, Drugs.com, and Wikipedia
2   Ecker RI. Contact dermatitis to anileridine. Contact Dermatitis 1980;6:495

Chapter 3.42   ANTIPYRINE SALICYLATE

IDENTIFICATION

Description/definition : Antipyrine salicylate is the combination of antipyrine (phenazone) and salicylic acid; it conforms to the structural formula shown below
Pharmacological classes : Anti-inflammatory agents, non-steroidal; analgesic, non-narcotic
IUPAC name : 1,5-Dimethyl-2-phenylpyrazol-3-one;2-hydroxybenzoic acid
Other names : Phenazone salicylate; salicylic acid compd. with antipyrine (1:1)
CAS registry number : 520-07-0
EC number : 208-283-4
Merck Index monograph : 1973 (Antipyrine)
Patch testing : Commercial drug 10% pet. (2,3); in case of fixed drug eruption, use also DMSO as vehicle; most systemic drugs can be tested at 10% pet.; if the pure chemical is not available, prepare the test material from intravenous powder, the content of capsules or – when also not available – from powdered tablets to achieve a final concentration of the active drug of 10% pet.
Molecular formula : $C_{18}H_{18}N_2O_4$

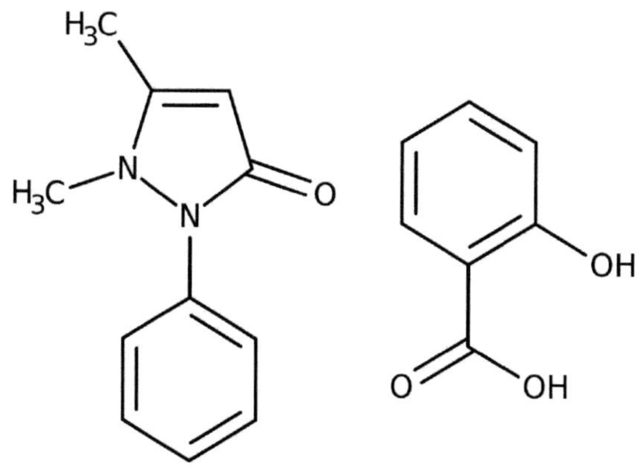

GENERAL

Antipyrine salicylate is the combination of antipyrine (phenazone) and salicylic acid, that is or was used as analgesic drug (1).

CUTANEOUS ADVERSE DRUG REACTIONS FROM SYSTEMIC ADMINISTRATION CAUSED BY TYPE IV (DELAYED-TYPE) HYPERSENSITIVITY (as demonstrated by positive patch tests)

Cutaneous adverse drug reactions from systemic administration of antipyrine salicylate caused by type IV (delayed-type) hypersensitivity have included fixed drug eruption (2,3,4).

Fixed drug eruption

In patients with fixed drug eruptions (FDE) caused by delayed-type hypersensitivity, the diagnosis is usually confirmed by a positive patch test with the drug on previously affected skin. Authors from Finland have used an alternative method of topical provocation. The test compound, the drug 10% in petrolatum and sometimes also in 70% alcohol and in DMSO, was applied once and without occlusion over the entire surface of one or several inactive (usually pigmented) sites of FDE lesions. The patients were followed as in-patients for 24 hours. A reaction was regarded as positive when a clearly demarcated erythema lasting at least 6 hours was seen. Of 25 patients with FDE from phenazone salicylate (antipyrine salicylate), all 25 had a positive topical provocation (2,3).

In Finland, in the period 1989-2001, 826 patients with suspected cutaneous drug eruptions were patch tested and 89 had one or more positive reactions. Of these individuals, 2 had fixed drug eruption from antipyrine salicylate with a positive patch test (4).

Cross-reactions, pseudo-cross-reactions and co-reactions

Of 8 patients with a fixed drug eruption from antipyrine salicylate (phenazone salicylate), in who type IV allergy was demonstrated by topical provocation (an open application test), all cross-reacted to phenazone, 4 to propyphenazone and 3 to aminophenazone (2).

LITERATURE

1   The data in the section 'General' may have been obtained from literature discussed in this chapter, but mostly also or exclusively from one or more of the following online sources: ChemIDPlus Advanced, PubChem, DrugBank, RxList, Drug Central, Drugs.com, and Wikipedia

2   Alanko K. Topical provocation of fixed drug eruption. A study of 30 patients. Contact Dermatitis 1994;31:25-27

3   Alanko K, Stubb S, Reitamo S. Topical provocation of fixed drug eruption. Br J Dermatol 1987;116:561-567

4   Lammintausta K, Kortekangas-Savolainen O. The usefulness of skin tests to prove drug hypersensitivity. Br J Dermatol 2005;152:968-974

Chapter 3.43 APOMORPHINE

IDENTIFICATION

Description/definition : Apomorphine is the derivative of morphine that conforms to the structural formula shown below

Pharmacological classes : Dopamine agonists; emetics

IUPAC name : (6aR)-6-Methyl-5,6,6a,7-tetrahydro-4H-dibenzo[de,g]quinoline-10,11-diol

CAS registry number : 58-00-4

EC number : 200-360-0

Merck Index monograph : 2003

Patch testing : Hydrochloride 10% pet. (2)

Molecular formula : $C_{17}H_{17}NO_2$

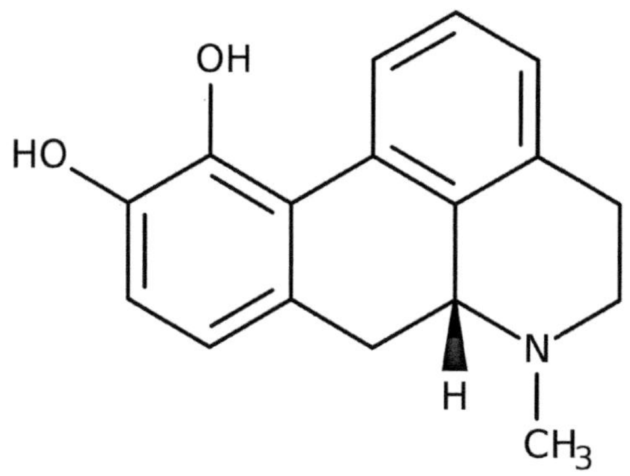

GENERAL

Apomorphine is a derivative of morphine that is a non-ergoline dopamine agonist with high selectivity for dopamine D_2-, D_3-, D_4- and D_5-receptors. The drug is used for the acute, intermittent treatment of hypomobility, off-episodes (end-of-dose wearing off and unpredictable on/off episodes) associated with advanced Parkinson's disease, and erectile dysfunction. In addition, apomorphine acts on the chemoreceptor trigger zone and is used as a central emetic in the treatment of drug overdose. In pharmaceutical products, apomorphine is employed as apomorphine hydrochloride (CAS number 41372-20-7, EC number 206-243-0, molecular formula $C_{34}H_{38}Cl_2N_2O_5$) (1).

CONTACT ALLERGY FROM ACCIDENTAL CONTACT

A 57-year-old a woman, caring for her husband with Parkinson's disease, presented with a 1-year history of pruritic lesions. Cutaneous examination showed well-demarcated eczematous plaques affecting several locations in her face and right hand. Every day, she prepared a subcutaneous infusion pump, which administered a 50% aqueous solution of 50 mg of apomorphine HCl and sodium metabisulfite. During the process, her hands were moistened with this drug on multiple occasions. Patch tests were positive to the commercial preparation, and to apomorphine HCl 1%, 5% and 10% pet. and 1% and 5% water. Sodium metabisulfite and apomorphine 10% pet. were negative in 20 controls. The 'ectopic' dermatitis of the face was presumably the results of transfer of the drug with the hands (2).

A similar case was that of a 73-year-old man who had relapsing erythematous scaly vesicular eczema of the hands and perioral area and who prepared and injected his wife, who had Parkinson's disease, with a vial of apomorphine HCl 1%. Patch tests were positive to apomorphine HCl 1% water (3). Precautionary measures led to resolution of the allergic contact dermatitis in both patients (2,3). This was a case of connubial (or consort) allergic contact dermatitis, currently called allergic contact dermatitis *by proxy*.

Occupational allergic airborne contact dermatitis of the hands and face developed in 2 women working in a pharmacy. One had to fill capsules with apomorphine powder, the other developed attacks of dermatitis on four occasions after cleaning the utensils used for preparing the apomorphine capsules. Both had positive patch tests to the drug 1%, 0.1% and 0.01% water, one of them also to 0.001%. Tests with morphine were negative. Both women also had rhinitis associated with contact with the powder, but prick tests were negative (4).

In early literature, (occupational) contact dermatitis and (in some cases) systemic allergic dermatitis to opium alkaloids, including morphine, codeine, apomorphine, ethylmorphine and diacetylmorphine (heroin) have been described (e.g. 6,7, reviews in ref. 5 and 7), the first report dating back to 1882 (cited in ref. 14). In those days, eczematous dermatitis from opium compounds used externally in the form of lotions, suppositories and other application forms was apparently well known (7). A more recent review of the subject was provided in 2006 (8).

LITERATURE

1   The data in the section 'General' may have been obtained from literature discussed in this chapter, but mostly
    also or exclusively from one or more of the following online sources: ChemIDPlus Advanced, PubChem,
    DrugBank, RxList, Drug Central, Drugs.com, and Wikipedia
2   Garcia-Gavin J, González-Vilas D, Fernández-Redondo V, Campano L, Toribio J. Allergic contact dermatitis caused
    by apomorphine hydrochloride in a carer. Contact Dermatitis 2010;63:112-115
3   Carboni GP, Contri P, Davalli R. Allergic contact dermatitis from apomorphine. Contact Dermatitis 1997;36:177-
    178
4   Dahlquist I. Allergic reactions to apomorphine. Contact Dermatitis 1977;3:349-350
5   Touraine A. Les dermatoses de l'opium. Revue de Médecine 1936;53:449-460
6   Dore SE, Prosser Thomas EW. Contact dermatitis in a morphine factory. J Allergy 1945;16:35-36
7   Jordon JW, Osborne ED. Contact dermatitis from opium derivatives. JAMA 1939;113:1955-1957
8   Hogen Esch AJ, van der Heide S, van den Brink W, van Ree JM, Bruynzeel DP, Coenraads PJ. Contact allergy and
    respiratory/mucosal complaints from heroin (diacetylmorphine). Contact Dermatitis 2006;54:42-49

Chapter 3.44   APRONALIDE

IDENTIFICATION

Description/definition : Apronalide is the *N*-acylurea that conforms to the structural formula shown below
Pharmacological classes : Hypnotics and sedatives
IUPAC name : *N*-Carbamoyl-2-propan-2-ylpent-4-enamide
Other names : Allylisopropylacetylurea; allylisopropylacetylcarbamide; (2-isopropylpent-4-enoyl)urea; apronal
CAS registry number : 528-92-7
EC number : 208-443-3
Merck Index monograph : 568
Patch testing : 5% pet. (20); most systemic drugs can be tested at 10% pet.; if the pure chemical is not available, prepare the test material from intravenous powder, the content of capsules or – when also not available – from powdered tablets to achieve a final concentration of the active drug of 10% pet.
Molecular formula : $C_9H_{16}N_2O_2$

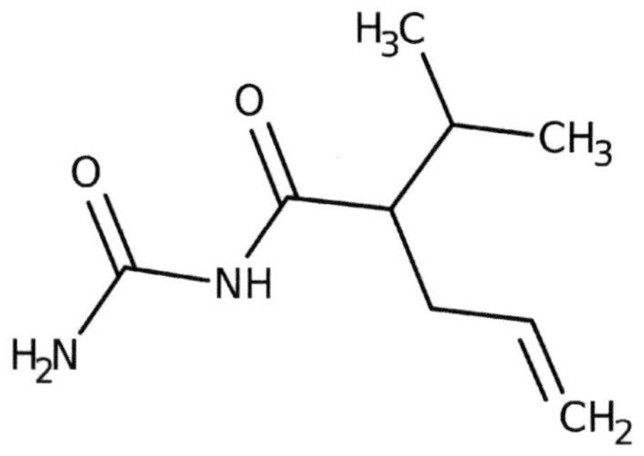

GENERAL

Apronalide is a hypnotic/sedative drug of the ureide (acylurea) group synthesized in 1926. Although it is not a barbiturate, apronalide is similar in structure to the barbiturates. In accordance, it is similar in action to the barbiturates, but considerably milder in comparison (formerly used as a daytime sedative at doses of 1 to 2 grams every 3 to 4 hours). The drug is usually combined with antipyretics and analgesics for the relief of anxiety and strain from pain, and for the enhancement of the effects of analgesics (17).

Upon the finding that it caused patients to develop thrombocytopenic purpura, apronalide was largely withdrawn from clinical use. However, it is still used in Japan, where in 2003 57 over-the-counter drugs containing apronalide were available (1,2) and from which country in 2020 another patient with FDE from apronalide was reported (13).

CUTANEOUS ADVERSE DRUG REACTIONS FROM SYSTEMIC ADMINISTRATION CAUSED BY TYPE IV (DELAYED-TYPE) HYPERSENSITIVITY (as demonstrated by positive patch tests)

Cutaneous adverse drug reactions from systemic administration of apronalide caused by type IV (delayed-type) hypersensitivity have included acute generalized exanthematous pustulosis (AGEP) (19), fixed drug eruption (2-17), and mucocutaneous syndrome (18).

Acute generalized exanthematous pustulosis (AGEP)

A 42-year-old woman had taken acetylsalicylic acid and tablets containing ibuprofen, caffeine and allylisopropylacetylurea (apronalide) for headache and chills. Three days later, she had a 38.5 ∘C. fever and a diffuse, red, pruritic erythema on her face, trunk, and extremities, which was studded with scattered, non-follicular pustules. Her blood tests leukocytosis and increased neutrophils. A skin biopsy showed spongiform changes with subcorneal pustules, and perivascular and diffuse dermal infiltration of lymphocytes and eosinophils. Based on these findings a diagnosis of AGEP was made. Three months later, patch tests were positive to the tablets containing apronalide and to apronalide 1% and 10% pet., as were lymphocyte stimulation tests with the tablets and apronalide (19).

Fixed drug eruption

General
Many cases of fixed drug eruption (FDE) to apronalide (allylisopropylacetylurea) have been reported, all from Japan. Many of these were of the 'generalized' or 'multiple' type (2,3), sometimes with >100 lesions, numbers sometimes increasing after every attack (2). Not all had the classic FDE feature of pigmentation (2) and in one case, FDE to apronalide resembled dermatitis of the face (13).

Case reports
A 30-year-old woman had a 5-year history of repeated episodes of itchy red oval eruptions on the trunk and extremities. In the most recent episode, her eruptions appeared a few hours after oral intake of tablets containing allylisopropylacetylurea (apronalide) and acetaminophen to relieve menstrual pain. She had experienced similar episodes of rashes at the same sites after taking other oral analgesics. The erythematous eruptions usually cleared without pigmentation within 2 weeks after she stopped taking the drug. The eruptions seemed to increase in number with each episode. At her first visit to the hospital, more than 100 purplish red round macules were symmetrically distributed on her trunk and extremities. She was patch tested with the 5 tablets that she had used 20% in petrolatum at the site of previous lesions and on non-affected skin and had weak-positive reactions (+) to 3 of these. The only common active ingredient was apronalide. Later, the patient was patch tested with all ingredients of the 3 tablets and had positive (+) patch tests to apronalide 20% pet. at D2 and D3, but only on involved skin; all other tests were negative. Three controls were negative (2). The authors refer to two similar cases of non-pigmenting FDE from apronalide, reported in Japanese literature (4,5).

Short summaries of other case reports of fixed drug eruptions from apronalide are shown in table 3.44.1.

Table 3.44.1 Short summaries of other case reports of fixed drug eruptions from apronalide

Year	Sex	Age	Clinical picture	Positive patch tests	Comments	Ref.
2020	F	52	diffuse itchy erythema with small vesicles and crusts on the face	tablet 10%, apronalide 10%, only on postlesional skin	atypical presentation of FDE resembling dermatitis of the face	13
2001	F	27	itchy painful purple-red eruptions on the lips and body 1 hour after apronalide tablet	tablet, apronalide (only on previously affected skin); no conc./vehicle given	oral provocation with 6 mg apronalide positive; the authors state to have seen 6 FDE cases in 6 years	6
1996	F	26	round light-brown patches on the cheeks, chin and right shoulder 2 hours after apronalide	apronalide 10% pet., only on postlesional skin	five controls were negative; lymphocyte stimulation test was negative	14
1993	F	21	itchy purplish-red round macules on the neck and left thigh 2-3 h after taking apronalide resulting in hyperpigmentation	2 tablets 50% pet., apronalide 50% pet., only on postlesional skin		17

Other case reports of fixed drug eruption from apronalide, adequate data of which are not available to the author, can be found in refs. 7,8,9,10,11,12,15,16 (all in Japanese medical jurnals).

Other cutaneous adverse drug reactions
One case of mucocutaneous ocular syndrome due to apronalide was reported in 1990. There were positive reactions to the tablet and its ingredient apronalide, but details are not available (18).

Cross-reactions, pseudo-cross-reactions and co-reactions
Of 6 patients with FDE from apronalide, 2 had a positive provocation test to the chemically related bromvalerylurea; patch tests were not mentioned (6).

LITERATURE
1   The data in the section 'General' may have been obtained from literature discussed in this chapter, but mostly also or exclusively from one or more of the following online sources: ChemIDPlus Advanced, PubChem, DrugBank, RxList, Drug Central, Drugs.com, and Wikipedia
2   Numata Y, Terui T, Sasai S, Sugawara M, Kikuchi K, Tagami H, et al. Non-pigmenting fixed drug eruption caused by allylisopropylacetylurea. Contact Dermatitis 2003;49:175-179
3   Fukuda E. Yakushinjoho, 9th edition. Fukuoka: Hazama Press, 2001 (data cited in ref. 2)
4   Miyamoto H, Horiuchi Y, Yamakawa Y. A case of fixed drug eruption due to allylisopropylacetylurea. Hifukano-Rinsho 1994;36:1227-1229 (Article in Japanese, data cited in ref. 2)

5   Oka K, Chin-Huai K, Saito F. A case of nonpigmenting fixed drug eruption due to allylisopropylacetylurea successful provocation of skin lesions with open test. Rinsho Derma (Tokyo) 2002;56:425-427 (Article in Japanese, data cited in ref. 2)

6   Sakakibara T, Hata M, Numano K, Kawase Y, Yamanishi T, Kawana S, et al. Fixed-drug eruption caused by allylisopropylacetylurea. Contact Dermatitis 2001;44:189-190

7   Terasawa M, Shitida T, Niimi Y, Sasaki E, Hata M, et al. A fixed drug eruption due to allylisopropylacetylurea. Rinsho Derma (Tokyo) 1993;35:951-954 (Article in Japanese, cited in ref. 6)

8   Okada K, Yamanaka T, Akimoto S, et al. 2 cases of fixed drug eruption due to allylisopropylacetylurea. Rinsho Derma (Tokyo) 1998;40:65-68 (Article in Japanese, cited in ref. 6)

9   Funaki M, Koike S, Yamada Y, et al. Fixed drug eruption due to allylisopropylacetylurea: report of a case. Jpn J Dermatoallergol 1996;4:145-149 (Article in Japanese, cited in ref. 6)

10   Kasamatsu M, Kanzaki T, Tsuji T. A case of fixed drug eruption due to allylisopropylacetylurea. Rinsho Hifuka 1994;48:469-472 (Article in Japanese, cited in ref. 6)

11   Kase K, Urushibata O, Saito R. A case of fixed drug eruption due to allylisopropylacetylurea. Jpn J Dermatoallergol 1993;1:151-153 (Article in Japanese, cited in ref. 6)

12   Urushibata O, Murakawa S, Kase K, Saito R. A case of fixed drug eruption due to allylisopropylacetylurea. Hihu 1992;34:244-248 (Article in Japanese, cited in ref. 6)

13   Deno R, Nakagawa Y, Itoi-Ochi S, Kotobuki Y, Kiyohara E, Wataya-Kaneda M, et al. Fixed drug eruption caused by allylisopropylacetylurea mimicking contact dermatitis of the face. Contact Dermatitis 2020;82:56-57

14   Kawada A, Hiruma M, Noguchi H, Inoue H, Ishibashi A, Marshall J. Fixed drug eruption induced by allylisopropylacetylurea. Contact Dermatitis 1996;34:65-66

15   Wakisaka C, Iitoyo M. Case of fixed drug eruption due to allylisopropylacetylurea. Arerugi 2005;54:569-571 (Article in Japanese)

16   Fujimoto Y, Hayakawa R, Kato Y, Yamamura M. A case of fixed drug eruption due to allylisopropylacetylurea. Environ Dermatol 1996;3:108-112 (Article in Japanese)

17   Fujimoto Y, Hayakawa R, Suzuki M, Ogino Y. Fixed drug eruption due to allylisopropylacetylurea. Contact Dermatitis 1993;28:282-284

18   Hara N. Drug-induced mucocutaneous ocular syndrome due to allylisopropylacetylurea (New Sedes®). Hifubyo Shinryo 1990;12:335-338 (Article in Japanese, data cited in ref. 17).

19   Ueda T, Abe M, Okiyama R, Oyama S, Satoh K, Aiba S, et al. Acute generalized exanthematous pustulosis due to allylisopropylacetylurea: role of IL-17-producing T cells. Eur J Dermatol 2011;21:140-141

20   De Groot AC. Patch testing, 4th edition. Wapserveen, The Netherlands: acdegroot publishing, 2018 (ISBN 9789081323345)

Chapter 3.45   ARECOLINE

IDENTIFICATION

Description/definition   : Arecoline is the alkaloid that conforms to the structural formula shown below
Pharmacological classes : Anti-parasitic agents
IUPAC name   : Methyl 1-methyl-3,6-dihydro-2H-pyridine-5-carboxylate
Other names   : Methyl N-methyl-1,2,5,6-tetrahydronicotinate
CAS registry number   : 63-75-2
EC number   : 200-565-5
Merck Index monograph : 2038
Patch testing   : 2% and 5% pet. (2)
Molecular formula   : $C_8H_{13}NO_2$

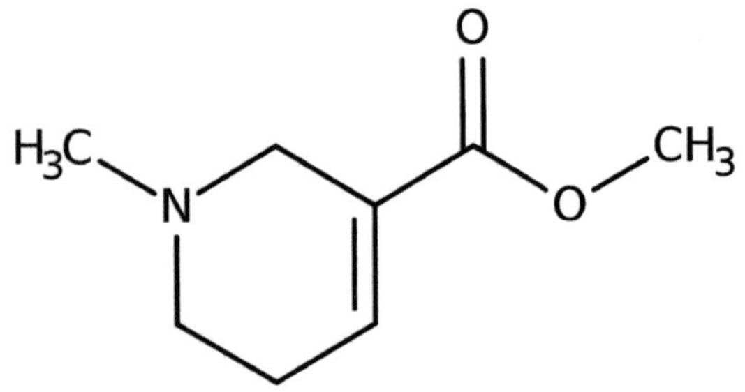

GENERAL

Arecoline is an alkaloid obtained from the betel nut, the fruit of a palm tree *Areca catechu*. It is an agonist at both muscarinic and nicotinic acetylcholine receptors. Arecoline is (or was) used in the form of various salts as a ganglionic stimulant, a parasympathomimetic, and as a vermifuge, especially in veterinary practice. It has also been used as a euphoriant in the Pacific Islands (1). In pharmaceutical products, arecoline may be employed as arecoline hydrobromide (CAS number 300-08-3, EC number 206-087-3, molecular formula $C_8H_{14}BrNO_2$) (1).

CONTACT ALLERGY FROM ACCIDENTAL CONTACT

A 64-year-old hydatid control officer began to develop irritation of his thumb and index fingers when handling arecoline. He was right-handed but transferred the pills to the palm of the left before inserting them in the dogs' mouth. On examination a month later he had a severe fissured dermatitis of the tips of the thumb and the index fingers of both hands with some involvement of the right middle finger and the middle of the left palm. Patch tests gave positive reactions to a crushed tablet of arecoline and to arecoline hydrobromide 1% (+), 2% (++) and 5% pet. (+++) (2). This was a case of occupational allergic contact dermatitis.

LITERATURE

1   The data in the section 'General' may have been obtained from literature discussed in this chapter, but mostly also or exclusively from one or more of the following online sources: ChemIDPlus Advanced, PubChem, DrugBank, RxList, Drug Central, Drugs.com, and Wikipedia
2   Wishart J. Contact dermatitis to arecoline. Contact Dermatitis 1979;5:61

Chapter 3.46   ARIPIPRAZOLE

IDENTIFICATION

Description/definition : Aripiprazole is the piperazine and quinolone derivative that conforms to the structural formula shown below

Pharmacological classes : Antipsychotic agents; serotonin 5-HT1 receptor agonists; serotonin 5-HT2 receptor antagonists; antidepressive agents; dopamine agonists; dopamine D2 receptor antagonists

IUPAC name : 7-[4-[4-(2,3-Dichlorophenyl)piperazin-1-yl]butoxy]-3,4-dihydro-1H-quinolin-2-one

CAS registry number : 129722-12-9

EC number : Not available

Merck Index monograph : 2046

Patch testing : Tablet, pulverized, 10% pet. (2); most systemic drugs can be tested at 10% pet.; if the pure chemical is not available, prepare the test material from intravenous powder, the content of capsules or – when also not available – from powdered tablets to achieve a final concentration of the active drug of 10% pet.

Molecular formula : $C_{23}H_{27}Cl_2N_3O_2$

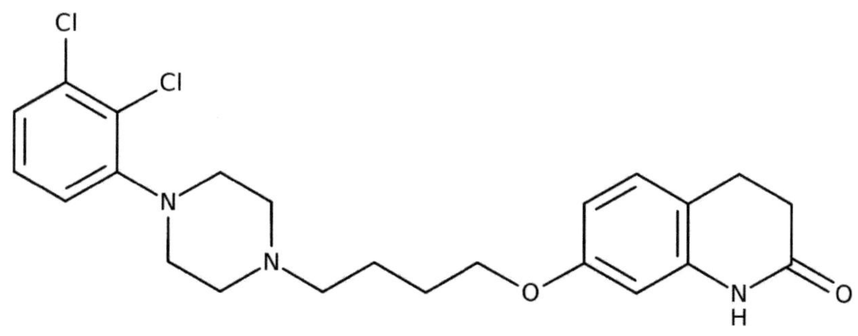

GENERAL

Aripiprazole is a piperazine and quinoline derivate and atypical anti-psychotic agent. It is a partial agonist of serotonin receptor 5-HT$_{1A}$ and dopamine D$_2$ receptors and an antagonist of serotonin receptor 5-HT$_{2A}$. This drug stabilizes dopamine and serotonin activity in the limbic and cortical system. Aripiprazole is indicated for manic and mixed episodes associated with bipolar I disorder, irritability associated with autism spectrum disorder, treatment of schizophrenia, treatment of Tourette's disorder, and as an adjunctive treatment of major depressive disorder. An injectable formulation of aripiprazole is indicated for agitation associated with schizophrenia or bipolar mania (1).

CONTACT ALLERGY FROM ACCIDENTAL CONTACT

A 44-year-old nurse presented with a 9-month-history of eczema on the hands and face, which had started 3 months after she was required to crush tablets daily during her work as a nurse in a clinic for disabled people. She was patch tested with all the drugs that she had contact with and reacted to 8 of these, including aripiprazole 30% pet.(D2 ++, D3 +++). No controls were performed (2). This was a case of occupational airborne allergic contact dermatitis.

LITERATURE

1   The data in the section 'General' may have been obtained from literature discussed in this chapter, but mostly also or exclusively from one or more of the following online sources: ChemIDPlus Advanced, PubChem, DrugBank, RxList, Drug Central, Drugs.com, and Wikipedia

2   Swinnen I, Ghys K, Kerre S, Constandt L, Goossens A. Occupational airborne contact dermatitis from benzodiazepines and other drugs. Contact Dermatitis 2014;70:227-232

Chapter 3.47   ARTICAINE

IDENTIFICATION

Description/definition	: Articaine is the thiophene carboxylic acid derivative that conforms to the structural formula shown below (shown in the hydrochloride salt)
Pharmacological classes	: Anesthetics, local
IUPAC name	: Methyl 4-methyl-3-[2-(propylamino)propanoylamino]thiophene-2-carboxylate
Other names	: Carticaine
CAS registry number	: 23964-58-1
EC number	: 607-295-0; 924-214-7
Merck Index monograph	: 3139
Patch testing	: Commercial solution 40 mg/ml; test also epinephrine HCl. or bitartrate 1% water (commercial solution contains epinephrine); for pure articaine: no data available; until recently, articaine hydrochloride 1% pet. was available from SmartPracticeCanada
Molecular formula	: $C_{13}H_{20}N_2O_3S$

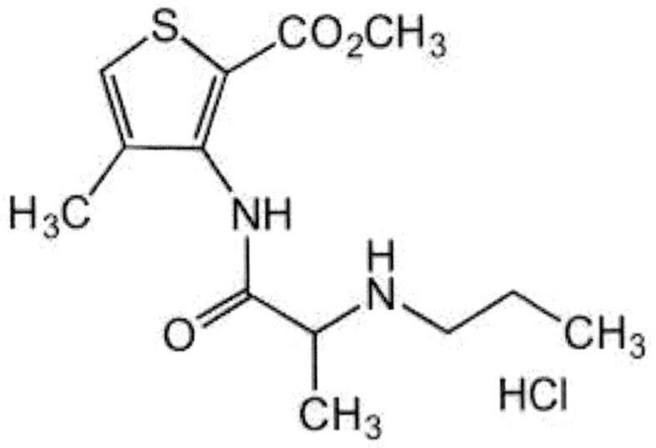

GENERAL

Articaine is a thiophene-containing amide-type local anesthetic, mostly used in dentistry. In pharmaceutical products, articaine is employed as articaine hydrochloride (CAS number 23964-57-0, EC number 245-957-7, molecular formula $C_{13}H_{21}ClN_2O_3S$) (1).

CUTANEOUS ADVERSE DRUG REACTIONS FROM SYSTEMIC ADMINISTRATION CAUSED BY TYPE IV (DELAYED-TYPE) HYPERSENSITIVITY (as demonstrated by positive patch tests)

Cutaneous adverse drug reactions from systemic administration of articaine caused by type IV (delayed-type) hypersensitivity have included localized eczematous reaction from subcutaneous injection (3).

Localized eczematous reaction

A 42-year-old woman experienced local edema followed by an eczematous reaction of the left cheek and of the upper neck, 24 hours after the use of articaine with epinephrine for a dental procedure, which resolved in 2 weeks. After a subsequent exposure to articaine for another dental procedure, the same symptoms reoccurred within 16 hours, and lasted 30 days despite treatment with oral corticosteroids. Patch tests were positive to undiluted articaine 40 mg/ml with epinephrine (+ at D2 and D3). Subsequent patch testing revealed positive reactions to bupivacaine 5 mg/ml and lidocaine 20 mg/ml, but negative to mepivacaine 20 mg/ml. Ten controls were negative to all patch test positive local anesthetics (3).

Cross-reactions, pseudo-cross-reactions and co-reactions

Possible cross-reaction from prilocaine sensitization to articaine (2). Possible cross-reaction to lidocaine and bupivacaine in a patient sensitized to articaine (3).

LITERATURE

1   The data in the section 'General' may have been obtained from literature discussed in this chapter, but mostly also or exclusively from one or more of the following online sources: ChemIDPlus Advanced, PubChem, DrugBank, RxList, Drug Central, Drugs.com, and Wikipedia

2   Suhonen R, Kanerva L. Contact allergy and cross-reactions caused by prilocaine. Am J Cont Dermat 1997;8:231-235

3   De Pasquale TMA, Buonomo A, Pucci S. Delayed-type allergy to articaine with cross-reactivity to other local anesthetics from the amide group. J Allergy Clin Immunol Pract 2018;6:305-306

Chapter 3.48   ASCORBIC ACID

IDENTIFICATION

Description/definition   : Ascorbic acid is the six carbon glucose-related compound that conforms to the structural formula shown below
Pharmaceutical classes   : Antioxidants; vitamins
IUPAC name   : (2*R*)-2-[(1*S*)-1,2-Dihydroxyethyl]-3,4-dihydroxy-2*H*-furan-5-one
Other names   : Vitamin C
CAS registry number   : 50-81-7
EC number   : 200-066-2
Merck Index monograph   : 2089
Patch testing   : 5% water and pet.; pure ascorbic acid is apparently not irritating (5)
Molecular formula   : $C_6H_8O_6$

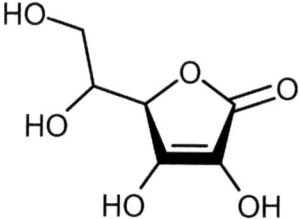

GENERAL

Ascorbic acid (vitamin C) is a six-carbon compound related to glucose. It is found naturally in citrus fruits and many vegetables. It is an essential nutrient in human diets, and necessary to maintain connective tissue and bone. Its biologically active form functions as a reducing agent and coenzyme in several metabolic pathways and is considered an antioxidant. Antioxidants are photoprotective, and are nowadays frequently used in anti-aging products (2). Ascorbic acid is also used to treat and prevent vitamin C deficiency. Ascorbic acid is extremely unstable and is therefore often chemically modified by esterification of the hydroxyl group, leading to derivatives such as ascorbyl tetraisopalmitate or ascorbyl palmitate (2). Contact allergy/allergic contact dermatitis from topical applications have been reported rarely (1,3)

CUTANEOUS ADVERSE DRUG REACTIONS FROM SYSTEMIC ADMINISTRATION CAUSED BY TYPE IV (DELAYED-TYPE) HYPERSENSITIVITY (as demonstrated by positive patch tests)

Cutaneous adverse drug reactions from systemic administration of ascorbic acid caused by type IV (delayed-type) hypersensitivity have included widespread dermatitis (5) and papular exanthema (4).

Dermatitis/eczematous eruption

A patient developed widespread dermatitis due to oral vitamin C (ascorbic acid). A patch test with pure vitamin C was positive and 30 controls negative. Oral provocation with 2 gram of vitamin C elicited an eczematous eruption after twenty hours. A vitamin C-free diet led to complete resolution of the skin problems (5).

Other cutaneous adverse drug reactions

In an early German investigation, a generalized itching papular exanthema appeared in a patient after the third intravenous injection of a vitamin C-containing medicament; epicutaneous (and intracutaneous) skin tests with vitamin C were positive; details are unknown (4).

Cross-reactions, pseudo-cross-reactions and co-reactions

No cross-reaction to ascorbyl tetraisopalmitate (2).

LITERATURE

1   Wetter DA, Yiannias JA, Prakash AV, Davis MD, Farmer SA, el-Azhary RA, et al. Results of patch testing to personal care product allergens in a standard series and a supplemental cosmetic series: an analysis of 945 patients from the Mayo Clinic Contact Dermatitis Group, 2000-2007. J Am Acad Dermatol 2010;63:789-798
2   Swinnen I, Goossens A. Allergic contact dermatitis caused by ascorbyl tetraisopalmitate. Contact Dermatitis 2011;64:241-242

3   Belhadjali H, Giordano-Labadie F, Bazex J. Contact dermatitis from vitamin C in a cosmetic anti-aging cream. Contact Dermatitis 2001;45:317
4   Rust S. Über allergische Reaktionen bei Vitamintherapie. Zeitschr Haut Geschl Kr 1954;17:317-319
5   Metz J, Hundertmark U, Pevny I. Vitamin C allergy of the delayed type. Contact Dermatitis 1980;6:172-174

Chapter 3.49   ATENOLOL

IDENTIFICATION

Description/definition : Atenolol is the ethanolamine that conforms to the structural formula shown below
Pharmacological classes : Adrenergic beta-1 receptor antagonists; anti-arrhythmia agents; sympatholytics; antihypertensive agents
IUPAC name : 2-[4-[2-Hydroxy-3-(propan-2-ylamino)propoxy]phenyl]acetamide
CAS registry number : 29122-68-7
EC number : 249-451-7
Merck Index monograph : 2120
Patch testing : Tablet, pulverized, 10% pet. (2); most systemic drugs can be tested at 10% pet.; if the pure chemical is not available, prepare the test material from intravenous powder, the content of capsules or – when also not available – from powdered tablets to achieve a final concentration of the active drug of 10% pet.
Molecular formula : $C_{14}H_{22}N_2O_3$

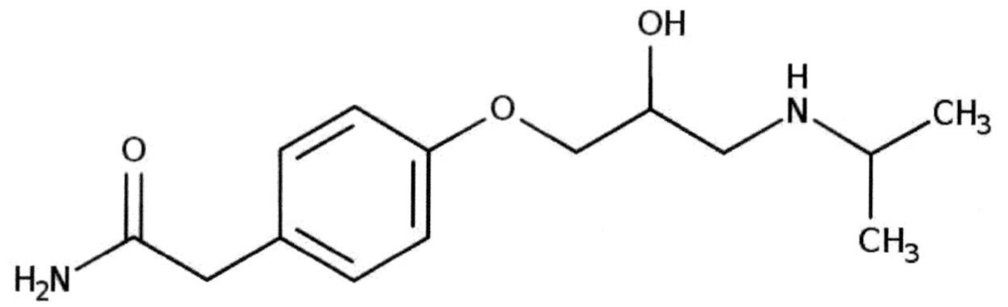

GENERAL

Atenolol is a synthetic isopropylaminopropanol derivative used as an antihypertensive, hypotensive and antiarrhythmic. Atenolol acts as a peripheral, cardioselective beta-blocker specific for beta-1 adrenergic receptors, without intrinsic sympathomimetic effects. It reduces exercise heart rates and delays atrioventricular conduction, with overall oxygen requirements decreasing. Atenolol is widely used in the treatment of hypertension and angina pectoris (1).

CUTANEOUS ADVERSE DRUG REACTIONS FROM SYSTEMIC ADMINISTRATION CAUSED BY TYPE IV (DELAYED-TYPE) HYPERSENSITIVITY (as demonstrated by positive patch tests)

Cutaneous adverse drug reactions from systemic administration of atenolol caused by type IV (delayed-type) hypersensitivity have included fixed drug eruption (2).

Fixed drug eruption

A 48-year-old woman developed 5 itching and burning skin lesions on her legs, 6 weeks after starting atenolol 100 mg once a day for hypertension. Physical examination revealed 5 well-demarcated reddish and round plaques, one of which showed vesicles. The diagnosis of multiple fixed drug eruption (FDE) was suspected and atenolol was discontinued. The histopathological findings were consistent with the diagnosis of FDE. Topical desonide was applied twice a day and the skin lesions resolved within two weeks with residual pigmentation. Six weeks later, patch tests were performed with atenolol 10% in petrolatum on a previously affected site of the right leg and on normal skin of the back. A positive reaction (++) was seen at D2 and D3 on the left leg but no reaction was detected on the back. The patient was diagnosed with atenolol-induced fixed drug eruption (2).

LITERATURE

1   The data in the section 'General' may have been obtained from literature discussed in this chapter, but mostly also or exclusively from one or more of the following online sources: ChemIDPlus Advanced, PubChem, DrugBank, RxList, Drug Central, Drugs.com, and Wikipedia
2   Belhadjali H, Trimech O, Youssef M, Elhani I, Zili J. Fixed drug eruption induced by atenolol. Clin Cosmet Investig Dermatol 2009;1:37-39

Chapter 3.50   ATORVASTATIN

IDENTIFICATION

Description/definition   : Atorvastatin is the diphenylpyrrole that conforms to the structural formula shown below
Pharmacological classes   : Anticholesteremic agents; hydroxymethylglutaryl-CoA reductase inhibitors
IUPAC name   : (3R,5R)-7-[2-(4-Fluorophenyl)-3-phenyl-4-(phenylcarbamoyl)-5-propan-2-ylpyrrol-1-yl]-
                             3,5-dihydroxyheptanoic acid
CAS registry number   : 134523-00-5
EC number   : Not available
Merck Index monograph   : 2125
Patch testing   : 0.1% alc. (2); testing also higher concentrations (e.g. 1%) in alcohol and pet. seems
                             appropriate
Molecular formula   : $C_{33}H_{35}FN_2O_5$

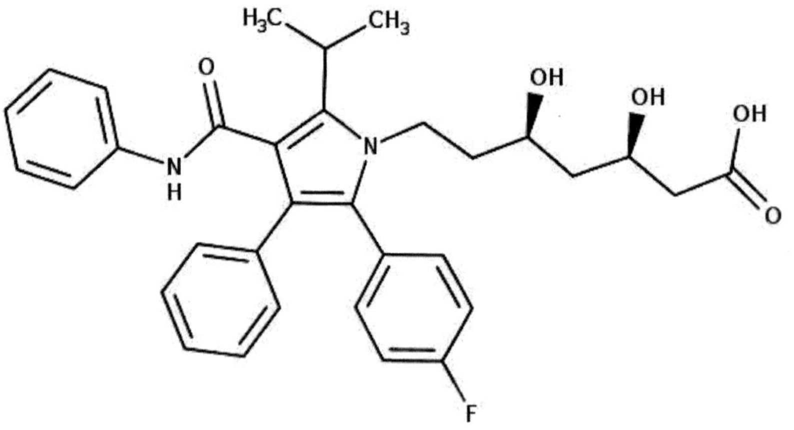

GENERAL

Atorvastatin is a lipid-lowering drug included in the statin class of medications. By inhibiting the endogenous production of cholesterol in the liver, statins lower abnormal cholesterol and lipid levels, and ultimately reduce the risk of cardiovascular disease. Atorvastatin is used to treat hypercholesterolemia and several other types of dyslipidemias. In pharmaceutical products, atorvastatin is employed as atorvastatin calcium (CAS number 134523-03-8, EC number not available, molecular formula $C_{66}H_{68}CaF_2N_4O_{10}$) or as atorvastatin calcium trihydrate (CAS number 344423-98-9, EC number not available, molecular formula $C_{66}H_{74}CaF_2N_4O_{13}$) (1).

CONTACT ALLERGY FROM ACCIDENTAL CONTACT

In Leuven, Belgium, in the period 2001-2019, 201 of 1248 health care workers/employees of the pharmaceutical industry had occupational allergic contact dermatitis. In 23 (11%) dermatitis was caused by skin contact with a systemic drug: 19 nurses, two chemists, one physician, and one veterinarian. The lesions were mostly localized on the hands, but often also on the face, as airborne dermatitis. In total, 42 positive patch test reactions to 18 different systemic drugs were found. In one patient, atorvastatin was the drug/one of the drugs that caused occupational dermatitis (3).

CUTANEOUS ADVERSE DRUG REACTIONS FROM SYSTEMIC ADMINISTRATION CAUSED BY TYPE IV (DELAYED-TYPE) HYPERSENSITIVITY (as demonstrated by positive patch tests)

Cutaneous adverse drug reactions from systemic administration of atorvastatin caused by type IV (delayed-type) hypersensitivity have included fixed drug eruption (2).

Fixed drug eruption

An 84-year-old man was referred with a 7-month history of multiple erythematous macules extending from the trunk to the arms. He was under treatment with atorvastatin and 5 other drugs. Previous histological examination of the lesions had revealed chronic superficial dermatitis with infiltration of leukocytes, polymorphonuclear cells, and eosinophils in perivascular areas. This finding was compatible with fixed drug eruption. Two months following withdrawal of atorvastatin, the erythema had partially resolved and one month later it had disappeared completely. Patch tests on postlesional skin were positive to atorvastatin 0.1% alc. (D4 ++) and simvastatin 0.1% alc. at D2 (+) and D4 (++). The results were negative on healthy skin with both statins. Six controls were negative (2).

Cross-reactions, pseudo-cross-reactions and co-reactions

A patient with fixed drug eruption from atorvastatin and a positive patch test on postlesional skin had a cross-reaction (also positive patch test) to simvastatin (1).

LITERATURE

1   The data in the section 'General' may have been obtained from literature discussed in this chapter, but mostly also or exclusively from one or more of the following online sources: ChemIDPlus Advanced, PubChem, DrugBank, RxList, Drug Central, Drugs.com, and Wikipedia

2   Huertas AJ, Ramírez-Hernández M, Mérida-Fernández C, Chica-Marchal A, Pajarón-Fernández M J, Carreño-Rojo A. Fixed drug eruption due to atorvastatin. J Investig Allergol Clin Immunol 2015;25:155-156

3   Gilissen L, Boeckxstaens E, Geebelen J, Goossens A. Occupational allergic contact dermatitis from systemic drugs. Contact Dermatitis 2020;82:24-30

Chapter 3.51   ATOVAQUONE

IDENTIFICATION
Description/definition   : Atovaquone is the naphthoquinone that conforms to the structural formula shown below
Pharmacological classes   : Enzyme inhibitors; antimalarials; anti-infective agents
IUPAC name   : 3-[4-(4-Chlorophenyl)cyclohexyl]-4-hydroxynaphthalene-1,2-dione
CAS registry number   : 95233-18-4
EC number   : Not available
Merck Index monograph   : 2127
Patch testing   : Tablet, pulverized, 10% pet. (2); most systemic drugs can be tested at 10% pet.; if the pure
                            chemical is not available, prepare the test material from intravenous powder, the content
                            of capsules or – when also not available – from powdered tablets to achieve a final
                            concentration of the active drug of 10% pet.
Molecular formula   : $C_{22}H_{19}ClO_3$

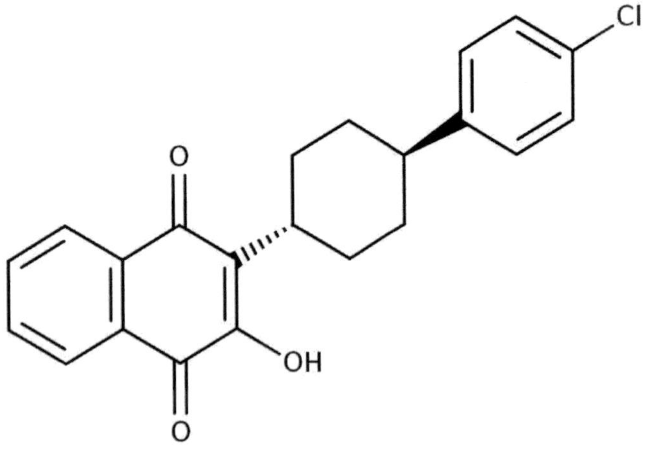

GENERAL
Atovaquone is a synthetic hydroxynaphthoquinone with antiprotozoal activity. Atovaquone blocks the mitochondrial electron transport at complex III of the respiratory chain of protozoa, thereby inhibiting pyrimidine synthesis, preventing DNA synthesis and leading to protozoal death. It is used for the prevention and treatment of *Pneumocystis jevorici* (formerly *carinii*) pneumonia and, in combination with proguanil, prevention and treatment of *P. falciparum* malaria (1).

CUTANEOUS ADVERSE DRUG REACTIONS FROM SYSTEMIC ADMINISTRATION CAUSED BY TYPE IV (DELAYED-TYPE) HYPERSENSITIVITY (as demonstrated by positive patch tests)
Cutaneous adverse drug reactions from systemic administration of atovaquone caused by type IV (delayed-type) hypersensitivity have included drug reaction with eosinophilia and systemic symptoms (DRESS) (2).

Drug reaction with eosinophilia and systemic symptoms (DRESS)
In a group of 45 patients with multiple drug hypersensitivity seen between 1996 and 2018 in Montpellier, France, 38 of 92 drug hypersensitivities were classified as type IV immunological reactions. Three patients had drug reaction with eosinophilia and systemic symptoms (DRESS). In one of these cases, the suspected drug was atovaquone, which was confirmed by a positive patch test (2).

LITERATURE
1   The data in the section 'General' may have been obtained from literature discussed in this chapter, but mostly also or exclusively from one or more of the following online sources: ChemIDPlus Advanced, PubChem, DrugBank, RxList, Drug Central, Drugs.com, and Wikipedia
2   Landry Q, Zhang S, Ferrando L, Bourrain JL, Demoly P, Chiriac AM. Multiple drug hypersensitivity syndrome in a large database. J Allergy Clin Immunol Pract 2019;8:258

Chapter 3.52   AVOPARCIN

IDENTIFICATION

Description/definition : Avoparcin is the glycopeptide antibiotic derived from *Streptomyces candidus* that conforms to the structural formula shown below
Pharmacological classes : Veterinary drugs; antibiotics
IUPAC name : Not available
CAS registry number : 37332-99-3
EC number : 253-466-4
Merck Index monograph : 2152
Patch testing : 5% water (1)
Molecular formula : $C_{89}H_{102}ClN_9O_{36}$ (α); $C_{89}H_{101}Cl_2N_9O_{36}$ (β)

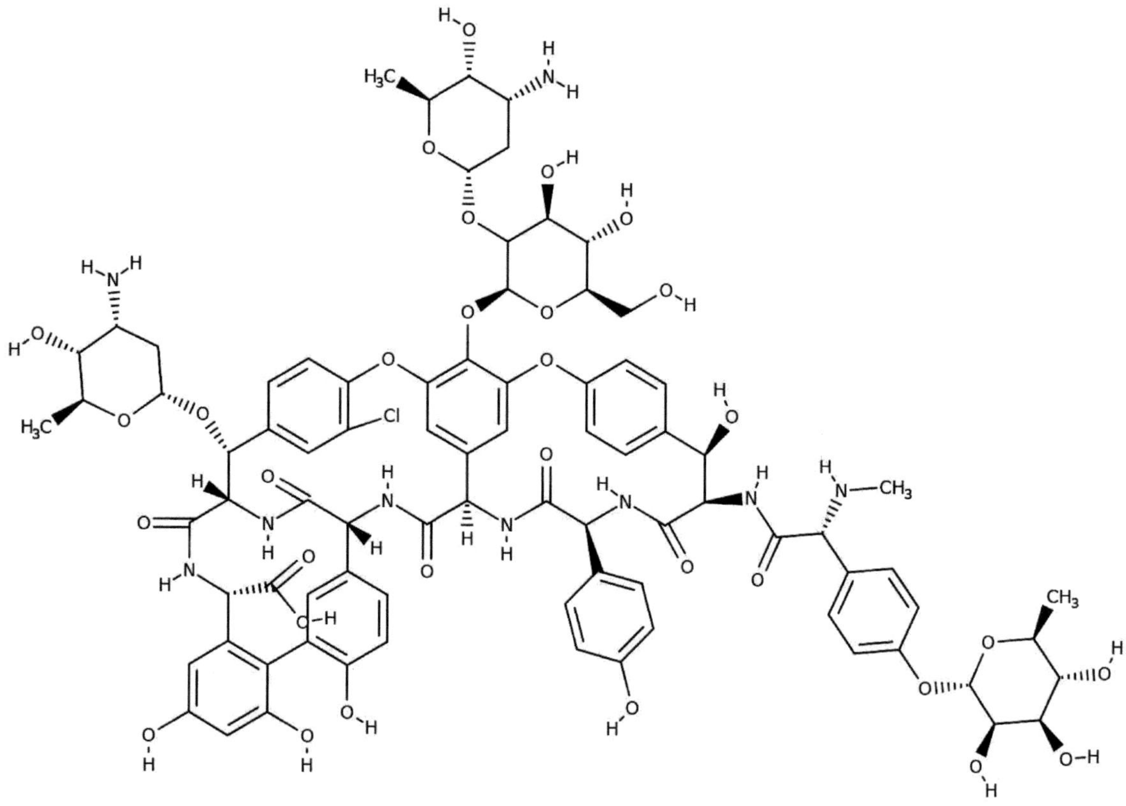

GENERAL

Avoparcin is a glycopeptide antibiotic derived from *Streptomyces candidus*, which is effective against gram-positive bacteria. It has been used in agriculture as an additive to livestock feed at concentrations of 15-30 ppm to promote growth in chickens, pigs, and cattle. It is also used as an aid in the prevention of necrotic enteritis in poultry. Avoparcin is a mixture of two closely related chemical compounds, known as α-avoparcin and β-avoparcin, which differ by the presence of an additional chlorine atom in β-avoparcin. The drug has a chemical similarity with vancomycin, and therefore concern exists that widespread use of avoparcin in animals may lead to an increased prevalence of vancomycin-resistant strains of bacteria. Avoparcin was once widely used in Australia and the European Union, but it is currently not permitted in either; it was never approved for use in the United States (1).

CONTACT ALLERGY FROM ACCIDENTAL CONTACT

A 50-year-old woman working occasionally on a farm had erythema and pruritus with slight scaling and well-defined borders on the skin under her brassiere. Patch testing with the rubber and textile series was negative. As the patient mentioned worsening of the eruption when working on the farm, she was patch tested with samples of animal feed and their components. There were positive reactions to aqueous solutions of avoparcin 5% (++/++), 1% (++/++) and 0.1% (+/+). Fifty controls were negative to avoparcin 5% water (1). This was a case of occupational allergic contact dermatitis, presumably from airborne contact.

LITERATURE
1   Barriga A, Romaguera C, Vilaplana J. Contact dermatitis from avoparcin. Contact Dermatitis 1992;27:115

Chapter 3.53   AZAPERONE

IDENTIFICATION

Description/definition   : Azaperone is the *N*-arylpiperazine that conforms to the structural formula shown below
Pharmacological classes   : Antipsychotic agents; hypnotics and sedatives; dopamine antagonists
IUPAC name   : 1-(4-Fluorophenyl)-4-(4-pyridin-2-ylpiperazin-1-yl)butan-1-one
CAS registry number   : 1649-18-9
EC number   : 216-715-8
Merck Index monograph   : 2161
Patch testing   : 4% water (2)
Molecular formula   : $C_{19}H_{22}FN_3O$

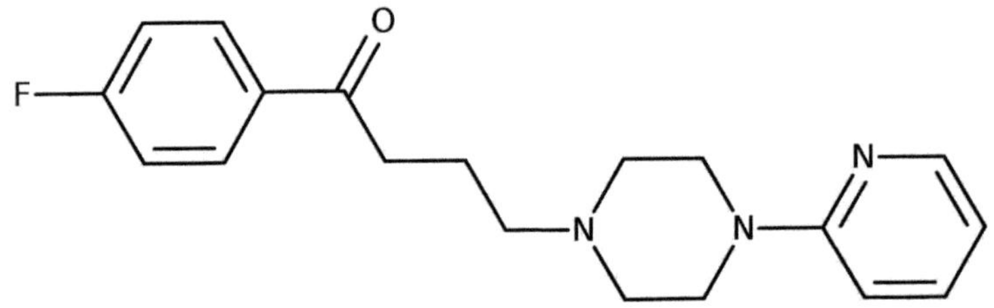

GENERAL

Azaperone is a pyridinylpiperazine and butyrophenone neuroleptic drug with sedative and antiemetic effects, which is used mainly as a tranquilizer in veterinary medicine, especially in pigs and elephants. Use in horses is avoided as adverse reactions may occur. More rarely it may be used in humans as an antipsychotic drug (1).

CONTACT ALLERGY FROM ACCIDENTAL CONTACT

A 50-year-old piglet dealer presented with a 2-year history of recurrent red scaly lichenified eczema of the face, neck, dorsal hands and forearms, aggravated by sun exposure. Five years ago, he had had quite similar lesions. At that time, he had given chlorpromazine to his piglets as a sedative before transport. Since then, the patient had avoided any contact with chlorpromazine and had used an alternative sedative containing azaperone . Patch and photopatch testing showed photoaggravated contact allergy to chlorpromazine. Additional patch and photopatch tests were performed with the commercial azaperone drug and its active ingredient for 1 day in usage concentrate-ons (40 mg/ml, 4% water). Positive reactions were seen after 2 days with both the commercial product and azaperone in the patch and photopatch test. After 3 days, however, photopatch reactions to both substances had further increased in comparison to patch tests not irradiated. Thus, the patient had photoaggravated allergic contact dermatitis from chlorpromazine and azaperone (2). He had acquired azaperone contact allergy by handling this drug regularly for 3 years. Intramuscular injections of azaperone (0.5 ml/20 kg) are given behind the ear. Some spillage of the solution and subsequent contamination of the hands were quite common. Because he did not wear protective gloves and would wipe his face during this work, the areas of skin involved had been directly exposed to azaperone (2). This was a case of occupational (photo)allergic contact dermatitis.

LITERATURE

1   The data in the section 'General' may have been obtained from literature discussed in this chapter, but mostly also or exclusively from one or more of the following online sources: ChemIDPlus Advanced, PubChem, DrugBank, RxList, Drug Central, Drugs.com, and Wikipedia
2   Brasch J, Hessler HJ, Christophers E. Occupational (photo)allergic contact dermatitis from azaperone in a piglet dealer. Contact Dermatitis 1991;25:258-259

Chapter 3.54 AZATHIOPRINE

IDENTIFICATION

Description/definition : Azathioprine is the purine antimetabolite that conforms to the structural formula shown below

Pharmacological classes : Antimetabolites, antineoplastic; immunosuppressive agents; antimetabolites; antirheumatic agents

IUPAC name : 6-(3-Methyl-5-nitroimidazol-4-yl)sulfanyl-7H-purine

CAS registry number : 446-86-6

EC number : 207-175-4

Merck Index monograph : 2165

Patch testing : 0.1%, 1% and 5% pet. (2,4)

Molecular formula : $C_9H_7N_7O_2S$

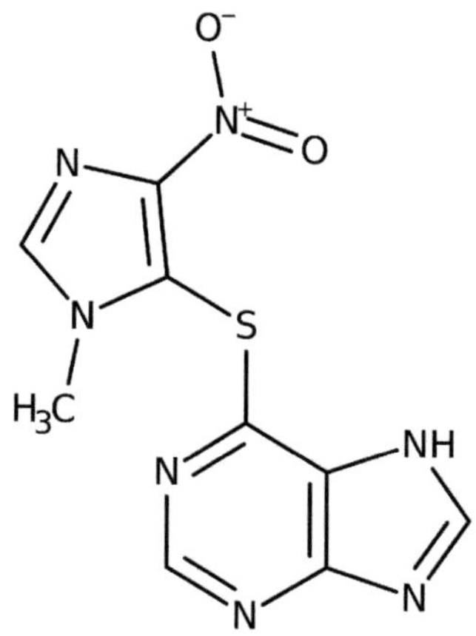

GENERAL

Azathioprine is a purine analog with cytotoxic and immunosuppressive activity. It is a pro-drug, converted in the body to the active metabolite 6-mercaptopurine, which causes inhibition of cell proliferation, particularly of lymphocytes and leukocytes. Azathioprine now is rarely used for chemotherapy but more for immunosuppression in organ transplantation and autoimmune disease such as rheumatoid arthritis, Crohn's disease and colitis ulcerosa (1).

In pharmaceutical tablets, azathioprine base is employed; in powder for injection solution azathioprine sodium (CAS number 55774-33-9, EC number not available, molecular formula $C_9H_6N_7NaO_2S$) is used (1).

CONTACT ALLERGY FROM ACCIDENTAL CONTACT

A 44-year-old woman developed eczema on her face, neck, hands and soles, which had worsened for 5 months. The patient had been handling azathioprine tablets at home for her son with leukaemia, crushing them to help him swallow the tablets. On patch testing, azathioprine tablet crushed and diluted 1:1 with water gave a positive reaction, confirmed on serial dilution of azathioprine tablet from 33% to 0.3% in water. On further testing, serial dilutions of pure azathioprine at 1%, 0.1% and 0.01% pet. also gave a dose-dependent response. The patient was instructed to arrange for her home to be cleaned of all remaining azathioprine debris, as well as to stop crushing the tablets. She had to be treated for a long time with prednisone, topical corticosteroids and UVB before the dermatitis finally cleared and the patient stayed in remission (2).

A man in his early 30s presented with an intermittent eczematous eruption over the shaft of his penis and scrotum. His wife was initially concerned about his fidelity, but a careful history-taking revealed that the rash coincided with his wife's intermittent courses of azathioprine for Crohn's disease. During four courses of such treatment over a 4-year period, her vaginal secretions were yellow, which raised the possibility of secretion of azathioprine or its metabolites that could have led to allergic contact dermatitis in the husband. Patch testing with 1% and 5% of the tablet and the pure azathioprine was indeed positive in the husband, but not the wife (4). This was a case of connubial (or consort) allergic contact dermatitis, currently termed allergic contact dermatitis *by proxy*.

A 51-year old man employed as a production mechanic in packaging for a pharmaceutical company had an 18-months history of hand dermatitis. The patient had initially developed a transient eczematous eruption on his face that subsequently spread to persistent involvement of his arms and hands. He related the skin complaints to his work at an azathioprine packaging line. Patch tests were positive to azathioprine 0.1% pet. (D2 +, D3 ++). Five controls were negative (3).

A 63-year-old maintenance engineer working in a pharmaceutical company had a 4-months history of a rash on the fingers and palms of both hands. His job involved reconditioning old tablet packaging machines, some of which were contaminated with powder from the medicaments, the most recent one being identified as azathioprine. Patch tests were positive to a crushed azathioprine tablet 1% pet. and later to pure azathioprine 1% pet. There were no reactions to the excipients. Fifty controls were negative to both substances. A reaction to either of 2 impurities in the azathioprine, 6-mercaptopurine and 5-chloro-1-methyl-4-nitroimidazole could not totally be excluded (4).

CUTANEOUS ADVERSE DRUG REACTIONS FROM SYSTEMIC ADMINISTRATION CAUSED BY TYPE IV (DELAYED-TYPE) HYPERSENSITIVITY (as demonstrated by positive patch tests)

Cutaneous adverse drug reactions from systemic administration of azathioprine caused by type IV (delayed-type) hypersensitivity have included maculopapular exanthema (6) and acneiform eruption (7).

Maculopapular eruption
A 27-year-old man started on oral therapy with salazopyrine, azathioprine, and cyclosporine for the treatment of glucocorticoid-dependent ulcerative colitis. After one month, an erythematous maculopapular exanthema appeared involving 40% of the body, accompanied by fever. There were no other clinical or laboratory abnormalities or signs of ulcerative colitis activation. With symptomatic therapy, the exanthema disappeared within 10 days. Patch tests with all drugs used were positive only to azathioprine crushed tablet 1% and 5% pet. 'Some controls' were negative to the 1% test material (6).

Other cutaneous adverse drug reactions
A 30-year-old patient was treated with azathioprine 50 mg daily for multiple sclerosis. Three days after the start, an 'exanthem' developed. Renewed intake of azathioprine led to recurrences. An oral provocation test with 50 mg azathioprine induced an acneiform exanthema of the trunk, face and dorsal aspects of the arms, consisting of densely packed papules and pustules with erythematous halo. Cultures of the pustules were sterile. Patch tests were 'positive' to azathioprine 100 and 10 mg/ml NaCl, but photographs showed follicular papulopustules. Ten controls were negative. The author states that the mechanism of action is unknown. The clinical picture and the occurrence of the skin manifestation without previous exposure argue against a delayed-type hypersensitivity reaction. He suggests that the drug in the hair follicle exerts a chemotactic action leading to the abscessing inflammation (7).

LITERATURE

1   The data in the section 'General' may have been obtained from literature discussed in this chapter, but mostly also or exclusively from one or more of the following online sources: ChemIDPlus Advanced, PubChem, DrugBank, RxList, Drug Central, Drugs.com, and Wikipedia
2   Lauerma AI, Koivuluhta M, Alenius H. Recalcitrant allergic contact dermatitis from azathioprine tablets. Contact Dermatitis 2001;44:129
3   Soni BP, Sherertz EF. Allergic contact dermatitis from azathioprine. Am J Cont Dermat 1996;7:116-117
4   Burden AD, Beck MH. Contact hypersensitivity to azathioprine. Contact Dermatitis 1992;27:329-330
5   Cooper HL, Louafi F, Friedmann PS. A case of conjugal azathioprine-induced contact hypersensitivity. N Engl J Med 2008;359(14):1524-1526
6   Blasco A, Enrique E, de Mateo JA, Castelló JV, Ferriols R, Malek T. Positive patch test to azathioprine in inflammatory bowel disease. Allergy 2004;59:368-369
7   Schmoeckel C, von Liebe V. Akneiformes Exanthem durch Azathioprin [Acneiform exanthema caused by azathioprine]. Hautarzt. 1983 Aug;34(8):413-415

Chapter 3.55   AZITHROMYCIN

IDENTIFICATION

Description/definition	: Azithromycin is the semisynthetic macrolide antibiotic that conforms to the structural formula shown below
Pharmacological classes	: Anti-bacterial agents
IUPAC name	: (2R,3S,4R,5R,8R,10R,11R,12S,13S,14R)-11-[(2S,3R,4S,6R)-4-(Dimethylamino)-3-hydroxy-6-methyloxan-2-yl]oxy-2-ethyl-3,4,10-trihydroxy-13-[(2R,4R,5S,6S)-5-hydroxy-4-methoxy-4,6-dimethyloxan-2-yl]oxy-3,5,6,8,10,12,14-heptamethyl-1-oxa-6-azacyclopentadecan-15-one
CAS registry number	: 83905-01-5
EC number	: 617-500-5
Merck Index monograph	: 2177
Patch testing	: 10% pet.; this may occasionally lead to a false-negative reaction (5); 20% pet. is probably not irritant when used for patch testing (5)
Molecular formula	: $C_{38}H_{72}N_2O_{12}$

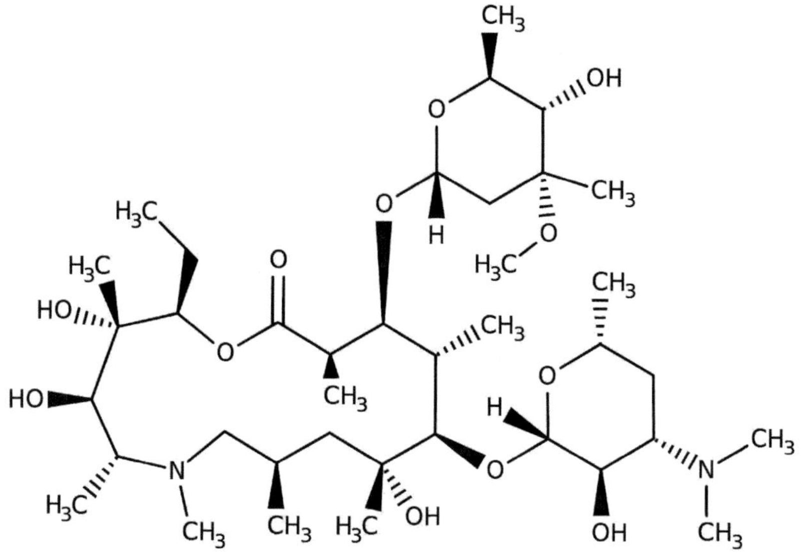

GENERAL

Azithromycin is a semisynthetic macrolide antibiotic structurally related to erythromycin. It has been used in the treatment of *Mycobacterium avium intracellulare* infections, toxoplasmosis, and cryptosporidiosis. Indications for its use include acute bacterial exacerbations of chronic obstructive pulmonary disease, acute bacterial sinusitis and community-acquired pneumonia, pharyngitis/tonsillitis, uncomplicated skin and skin structure infections (all due to specific species of bacteria), urethritis and cervicitis due to *Chlamydia trachomatis* or *Neisseria gonorrhoeae* and genital ulcer disease in men due to *Haemophilus ducreyi* (chancroid). In pharmaceutical products, both azithromycin and azithromycin dihydrate (CAS number 117772-70-0, EC number not available, molecular formula $C_{38}H_{76}N_2O_{14}$) may be employed (1). Azithromycin is used both in topical and in systemic pharmaceutical applications. In topical preparations, the antibiotic has caused contact allergy/allergic contact dermatitis, which has been fully reviewed in Volume 3 of the *Monographs in contact allergy* series (9).

CONTACT ALLERGY FROM ACCIDENTAL CONTACT

Case series

In Croatia, 5 of 21 pharmaceutical workers in one factory, exposed to powdered intermediate and final substances in azithromycin synthesis, developed occupational airborne allergic contact dermatitis, despite the use of protective latex gloves, overalls, cellulose face masks and caps. The duration of their workplace exposure to azithromycin ranged from 3 months to 4 years. They showed dermatitis typical for airborne contact dermatitis, i.e. erythema, edema and/or vesicular changes on the hands, forearms, face, neck and lower legs. Patch tests were positive to azithromycin 1% and 5% pet. in 4 patients and to a crushed tablet pure in the fifth (6). It seems likely that the latter was the same patient that had been presented earlier by the authors (7).

Two more individuals working in a pharmaceutical plant who developed airborne allergic contact dermatitis from azithromycin were reported from Canada. The first reported an itchy, vesicular eruption on the head and neck that later spread to the rest of the body. He had positive patch tests to a crushed tablet 5% pet. and azithromycin powder 5% pet. The second patient had a history of an intensely pruritic acute eczematous eruption on the left cheek that later spread to the rest of the face and neck, after being exposed to azithromycin 8 times. He reacted to a crushed tablet and azithromycin powder 5% and 10% pet. Both were removed from the azithromycin room and reported clearance with no recurrence (8).

Case reports

A 39-year-old-man working in a pharmaceutical company presented with a 4-week history of pruritic facial dermatitis, which had started 2 months after he had moved to a new workplace. He reported pruritus on the head and neck at this time, which had spread to the rest of the body 4 weeks later, and skin lesions appeared on his face, neck and forearms. The patient was occupationally exposed to the final pure powdered form of clarithromycin and azithromycin. He used the protective equipment provided by the company and the room had a ventilation system. Physical examination revealed erythematous and scaly lesions on the face and neck and crusting on the forearms. Patch tests were positive to azithromycin 20% but negative to 10%. Tests on 10 controls were negative (5).

Cross-reactions, pseudo-cross-reactions and co-reactions

Not to erythromycin (2,3,4,6) and clarithromycin (2,3,4,5). Two patients who had occupational allergic contact dermatitis from azithromycin also reacted to one or more intermediates in azithromycin synthesis (hydroxylamine hydrochloride, erythromycin A oxime hydrochloride, erythromycin A iminoether and azaerythromycin A, all tested 1% and 5% pet.) but not to erythromycin (6).

LITERATURE

1   The data in the section 'General' may have been obtained from literature discussed in this chapter, but mostly also or exclusively from one or more of the following online sources: ChemIDPlus Advanced, PubChem, DrugBank, RxList, Drug Central, Drugs.com, and Wikipedia
2   de Risi-Pugliese T, Amsler E, Collet E, Francès C, Barbaud A, Pecquet C, et al. Eyelid allergic contact dermatitis after intravitreal injections of anti-vascular endothelial growth factor: What is the culprit? A report of 3 cases. Contact Dermatitis 2018;79:103-104
3   Mendes-Bastos P, Brás S, Amaro C, Cardoso J. Non-occupational allergic contact dermatitis caused by azithromycin in an eye solution. J Dtsch Dermatol Ges 2014;12:729-730
4   Flavia Monteagudo Paz A, Francisco Silvestre Salvador J, Latorre Martínez N, Cuesta Montero L, Toledo Alberola F. Allergic contact dermatitis caused by azithromycin in an eye drop. Contact Dermatitis 2011;64:300-301
5   López-Lerma I, Romaguera C, Vilaplana J. Occupational airborne contact dermatitis from azithromycin. Clin Exp Dermatol 2009;34:e358-e359
6   Milković-Kraus S, Macan J, Kanceljak-Macan B. Occupational allergic contact dermatitis from azithromycin in pharmaceutical workers: a case series. Contact Dermatitis 2007;56:99-102
7   Milkovic-Kraus S, Kanceljak-Macan B. Occupational airborne allergic contact dermatitis from azithromycin. Contact Dermatitis 2001;45:184
8   Mimesh S, Pratt M. Occupational airborne allergic contact dermatitis from azithromycin. Contact Dermatitis 2004;51:151
9   De Groot AC. Monographs in contact allergy, volume 3: Topical Drugs. Boca Raton, Fl, USA: CRC Press Taylor and Francis Group, 2021 (ISBN 978-0-367-23693-9)

Chapter 3.56   AZTREONAM

IDENTIFICATION

Description/definition : Aztreonam is the monobactam that conforms to the structural formula shown below
Pharmacological classes : Antibacterial agents
IUPAC name : 2-[(Z)-[1-(2-Amino-1,3-thiazol-4-yl)-2-[[(2S,3S)-2-methyl-4-oxo-1-sulfoazetidin-3-yl]amino]-2-oxoethylidene]amino]oxy-2-methylpropanoic acid
CAS registry number : 78110-38-0
EC number : 278-839-9
Merck Index monograph : 2188
Patch testing : Tablet, pulverized, 20% pet. (4); most beta-lactam antibiotics can be tested 5-10% pet.
Molecular formula : $C_{13}H_{17}N_5O_8S_2$

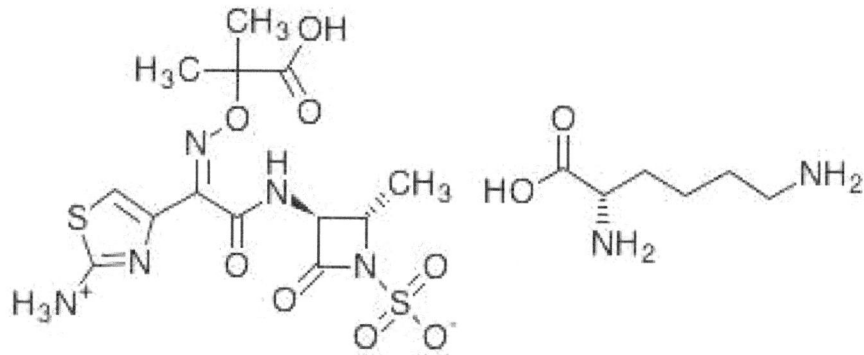

Aztreonam lysine

GENERAL

Aztreonam is a monocyclic beta-lactam antibiotic originally isolated from *Chromobacterium violaceum* with bactericidal activity. It preferentially binds to and inactivates penicillin-binding protein-3 (PBP-3), which is involved in bacterial cell wall synthesis, thereby inhibiting bacterial cell wall integrity and leading to cell lysis and death. The antibiotic is resistant to beta-lactamase hydrolysis. Aztreonam is usually used parenterally for the treatment of infections caused by susceptible gram-negative microorganisms, including those of the urinary tract, lower respiratory tract, septicemia, skin and skin-structures, intra-abdominal infections, and gynecologic infections. In pharmaceutical products, aztreonam is employed as aztreonam lysine (CAS number 827611-49-4, EC number not available, molecular formula $C_{19}H_{31}N_7O_{10}S_2$) (1). The classification and structures of beta-lactam antibiotics are discussed in Chapter 3.36 Amoxicillin.

CUTANEOUS ADVERSE DRUG REACTIONS FROM SYSTEMIC ADMINISTRATION CAUSED BY TYPE IV (DELAYED-TYPE) HYPERSENSITIVITY (as demonstrated by positive patch tests)

Cutaneous adverse drug reactions from systemic administration of aztreonam caused by type IV (delayed-type) hypersensitivity have included maculopapular eruption (4).

Maculopapular eruption

Two hours after receiving the first dose of amoxicillin/clavulanate (1 g iv) and aztreonam (1 g iv) for pleural empyema, a 57-year-old man developed a widespread pruriginous maculopapular eruption. Prick, intradermal and patch tests were negative to amoxicillin and amoxicillin/clavulanate and oral challenge tests with both antibiotics were negative. There were, however, positive delayed intradermal tests to aztreonam and ceftazidime, and positive patch tests with both beta-lactam antibiotics tested at 20% pet. As the patient had never had ceftazidime before, only this cephalosporin has the same side chain as aztreonam and cross-reactivity had been documented *in vitro* before (5), it was concluded that the reaction to ceftazidime was a cross-reaction to aztreonam (4).

Cross-reactions, pseudo-cross-reactions and co-reactions

Cross-reactions between beta-lactam antibiotics are discussed in Chapter 3.36 Amoxicillin. Of 78 patients who had suffered nonimmediate drug reactions from any beta-lactam compound (penicillins, cephalosporins) and who had at least one positive patch test to beta-lactams, none cross-reacted to aztreonam (2). In other studies too, there was an absence of cross-reactions to aztreonam in patients sensitized to aminopenicillins (amoxicillin, ampicillin and ampicillin-esters) (3).

A patient sensitized to aztreonam had a cross-reaction to ceftazidime, the only cephalosporin with the same side chain as aztreonam (4).

LITERATURE

1   The data in the section 'General' may have been obtained from literature discussed in this chapter, but mostly also or exclusively from one or more of the following online sources: ChemIDPlus Advanced, PubChem, DrugBank, RxList, Drug Central, Drugs.com, and Wikipedia

2   Buonomo A, Nucera E, De Pasquale T, Pecora V, Lombardo C, Sabato V, et al. Tolerability of aztreonam in ` patients with cell-mediated allergy to β-lactams. Int Arch Allergy Immunol 2011;155:155-159

3   Romano A, Gaeta F, Valluzzi RL, Maggioletti M, Caruso C, Quaratino D. Cross-reactivity and tolerability of aztreonam and cephalosporins in subjects with a T cell-mediated hypersensitivity to penicillins. J Allergy Clin Immunol 2016;138:179-186

4   Pérez Pimiento A, Gómez Martínez M, Mínguez Mena A, Trampal González A, de Paz Arranz S, Rodríguez Mosquera M. Aztreonam and ceftazidime: evidence of *in vivo* cross allergenicity. Allergy 1998;53:624-625

5   Adkinson NF Jr. Immunogenicity and cross-allergenicity of azreonann. Am J Med 1990;88(Suppl.3C):12S-15S

Chapter 3.57 BACAMPICILLIN

IDENTIFICATION

Description/definition
: Bacampicillin is the 1-ethoxycarbonyloxyethyl ester of ampicillin; it conforms to the structural formula shown below

Pharmacological classes
: Anti-bacterial agents

IUPAC name
: 1-Ethoxycarbonyloxyethyl (2S,5R,6R)-6-[[(2R)-2-amino-2-phenylacetyl]amino]-3,3-dimethyl-7-oxo-4-thia-1-azabicyclo[3.2.0]heptane-2-carboxylate

Other names
: (2S,5R,6R)-6((R)-(2-Amino-2-phenylacetamido))-3,3-dimethyl-7-oxo-4-thia-1-azabicyclo (3.2.0)heptane-2-carboxylic acid ester with ethyl 1-hydroxyethylcarbonate

CAS registry number
: 50972-17-3

EC number
: Not available

Merck Index monograph
: 2194

Patch testing
: Bacampicillin hydrochloride, 12.5 mg/ml (4); most beta-lactam antibiotics can be tested 5-10% pet.

Molecular formula
: $C_{21}H_{27}N_3O_7S$

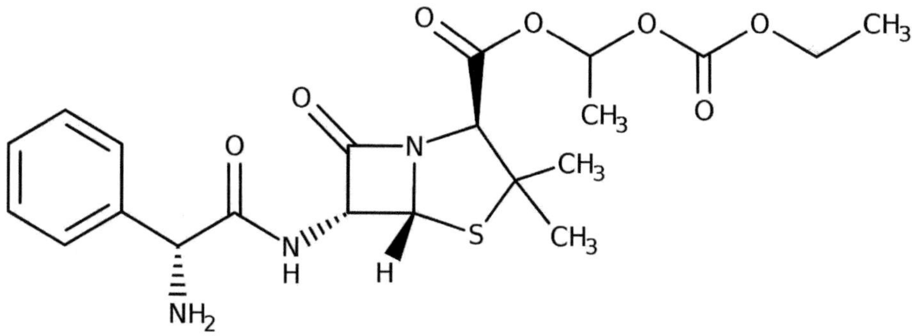

GENERAL

Bacampicillin is a microbiologically inactive prodrug of ampicillin. Following oral administration, during absorption from the gastrointestinal tract, bacampicillin is hydrolyzed by esterases present in the intestinal wall to its active metabolite ampicillin, a broad-spectrum, semisynthetic, β-lactam aminopenicillin antibiotic with bactericidal activity. Bacampicillin is indicated for infections of the upper and lower respiratory tract, skin and soft tissue, urinary tract and of acute uncomplicated gonococcal urethritis, when due to sensitive strains of the following organisms: gram-positive: streptococci and non-penicillinase-producing staphylococci; gram-negative: *H. influenzae, N. gonorrhoeae, E. coli, P. mirabilis, Salmonella* spp. and *Shigella* spp. (1). In pharmaceutical products, bacampicillin may be employed as bacampicillin hydrochloride (CAS number 37661-08-8, EC number 253-580-4, molecular formula $C_{21}H_{28}ClN_3O_7S$) (1).

In this chapter, it has not been attempted to provide a full review of delayed-type hypersensitivity to bacampicillin. The classification and structures of beta-lactam antibiotics are discussed in Chapter 3.36 Amoxicillin.

CONTACT ALLERGY FROM ACCIDENTAL CONTACT

Fifteen workers employed at a pharmaceutical plant experienced skin reactions while they were working with bacampicillin. Nine exhibited eczema localized on the face, arms, hands, and neck; six had eczema in combination with rhinitis and conjunctivitis. All patients were patch tested with bacampicillin hydrochloride 25, 12.5, 6.25, and 3.13 mg/ml water. Ten (67%) had positive patch test. The lowest concentration inducing positivity was 12.5 mg/ml in one individual, 6.25 in one other and 3.12 in 8 patients. Thirty controls were negative to all concentrations (4).

Between 1990 and 1998, 39 workers in a pharmaceutical plant, repeatedly exposed to bacampicillin dust, were diagnosed as allergic to bacampicillin based on their clinical history. 16 patients reported type I symptoms (rhinoconjunctivitis), 19 patients type IV symptoms (dermatitis on the hands and face), and four patients both type I and type IV symptoms. When patch tested with bacampicillin 1, 3, 6 and 10% water, 33/39 (85%) had positive patch tests, 16/19 (84%) with dermatitis and 11/16 (69%) of the patients with rhinoconjunctivitis. 34 patients also had a positive lymphocyte transformation test (3).

CUTANEOUS ADVERSE DRUG REACTIONS FROM SYSTEMIC ADMINISTRATION CAUSED BY TYPE IV (DELAYED-TYPE) HYPERSENSITIVITY (as demonstrated by positive patch tests)

Cutaneous adverse drug reactions from systemic administration of bacampicillin caused by type IV (delayed-type) hypersensitivity have included maculopapular eruption (7,8,9,10,11), (extensive or generalized) erythema (8,10,11), acute generalized exanthematous pustulosis (AGEP) (5), systemic allergic dermatitis (4), urticaria/angioedema (8,10,11), and unspecified drug eruptions (2,6).

Case series with various or unknown types of drug reactions

In a series of 30 patients with skin reactions to beta-lactam antibiotics, there were 7 patients with maculopapular rash, one with urticaria and one with edema and pruritus. They all had positive patch test reactions to bacampicillin, ampicillin and amoxicillin (8).

Seven patients with maculopapular eruption, 3 with delayed urticaria/angioedema and 1 with erythema caused by bacampicillin with a positive patch test to this aminopenicillin were reported from Rome in 2006 (10).

In 97 patients with nonimmediate reactions from penicillins with a positive patch test to at least one penicillin seen in Rome between 2002 and 2011, amoxicillin was the culprit drug in 50 individuals, ampicillin in 35, bacampicillin in 25 and benzylpenicillin in 11. The most frequent clinical manifestations were maculopapular eruptions with or without edema, delayed urticaria with or without angioneurotic edema and generalized erythema. It was not specified which drug eruptions the 4 individual antibiotics had caused (11).

Maculopapular eruption

In Rome, 259 patients who had suffered nonimmediate skin reactions during penicillin therapy were examined with patch testing. 173 patients had had a maculopapular rash, 59 (delayed) urticaria/angioedema, 22 erythema and 5 other manifestations. Two hundred and forty-one subjects (93%) reported adverse reactions to aminopenicillins: amoxicillin in 107, ampicillin in 59 and bacampicillin in 48. Twenty-seven of these subjects had suffered reactions to 2 or 3 different aminopenicillins in separate episodes. Ninety-four of the 259 subjects (36%) showed patch test and delayed intradermal test positivity to the culprit penicillins, of which ninety to both amoxicillin and ampicillin. Bacampicillin was not tested, but a number of these sensitizations must have been picked up by reactions to ampicillin. These patients all suffered from maculopapular rashes or – a few – erythema (9).

One patient, who had suffered a maculopapular rash from bacampicillin, later had positive patch (5% pet.) and intradermal tests (with delayed-reading) to ampicillin and amoxicillin (7).

See also the section 'Case series with various or unknown types of drug reactions' above, refs. 8,10,11.

Erythroderma, widespread erythematous eruption, exfoliative dermatitis

See the section 'Case series with various or unknown types of drug reactions' above, refs. 10 and 11.

Acute generalized exanthematous pustulosis (AGEP)

A 45-year-old woman who had previously suffered from pustular psoriasis was treated with bacampicilin, semi-alkaline proteinase and teprenone for a toothache. Three days later, erythematous and pustular lesions developed on her trunk and extremities with high fever; the mucous membranes were not affected. The typical psoriatic plaques of 10 years' duration on her back did not change. Laboratory investigations revealed leukocytosis with neutrophilia (93%) and elevation of liver enzymes. Bacterial cultures of blood and pustules were negative. Histopathological findings showed subcorneal pustules and spongiform pustules around them, dermal edema, and perivascular infiltration of mononuclear cells and eosinophils. The hyperkeratosis, parakeratosis, and elongation of rete ridges found in psoriasis were absent. Within a few days after discontinuing the drugs, the eruption disappeared spontaneously and the abnormal liver function returned to normal. Patch and stripping patch testing of bacampicillin hydrochloride 10% pet. gave a positive reaction, inducing a pustular eruption with an erythematous base at D2. Histopathology showed the same picture as the previous one taken from the exanthema. Further patch testing demonstrated pustular reactions to other penicillins including amoxicillin and sultamicillin tosylate. Three controls were negative. A lymphocyte stimulation test to bacampicillin was positive (192% compared with no stimulation), but was negative to the other drugs the patient had used (5). The patient was diagnosed with a pustular drug eruption from bacampicillin, which was in fact a classic acute generalized exanthematous pustulosis (AGEP).

Systemic allergic dermatitis (systemic contact dermatitis)

A patient previously shown to be allergic to bacampicillin from occupational (topical) sensitization was administered 100 mg of bacampicillin perorally. After 4 hours she experienced itching and eczema starting on her hands. During the following hours, the patient developed pain in the stomach and diarrhea, and 24 hours later eczema was noted on the back at the site of former patch testing (4).

Other cutaneous adverse drug reactions
In a group of 78 patients who had suffered nonimmediate drug reactions from any beta-lactam compound (penicillins, cephalosporins), 71 had positive patch tests to bacampicillin. In 18 cases, bacampicillin was the culprit drug, so the other reactions were presumably cross-reactions. The nature of the eruptions caused by bacmpicillin was not specified (2).

Japanese researchers found 5 cases of allergic drug reactions to bacampicillin and one to talampicillin in the Japanese literature from 1984 to 1989 (6). The nature of the drug eruptions was not specified. All 6 patients had positive patch test reactions to bacampicillin and to ampicillin and all 5 tested with it also to talampicilllin (no details provided). These reactions to bacampicillin, ampicillin and talampicillin are pseudocross-reactions, as bacampicillin and talampicillin are prodrugs of ampicillin which are hydrolyzed in the digestive tract to the active drug ampicillin (6).

See also the section 'Case series with various or unknown types of drug reactions' above, refs. 8,10,11.

Cross-reactions, pseudo-cross-reactions and co-reactions
Cross-reactions between beta-lactam antibiotics are discussed in Chapter 3.36 Amoxicillin. A patient with AGEP from bacampicillin cross-reacted to amoxicillin and sultamicillin tosylate with positive erythematous and pustular patch tests (5). Pseudocross-reactivity between bacampicillin, talampicillin (both prodrugs of ampicillin) and ampicillin (6).

Immediate contact reactions
Immediate contact reactions (contact urticaria) to bacampicillin are presented in Chapter 5.

LITERATURE
1   The data in the section 'General' may have been obtained from literature discussed in this chapter, but mostly also or exclusively from one or more of the following online sources: ChemIDPlus Advanced, PubChem, DrugBank, RxList, Drug Central, Drugs.com, and Wikipedia
2   Buonomo A, Nucera E, De Pasquale T, Pecora V, Lombardo C, Sabato V, et al. Tolerability of aztreonam in patients with cell-mediated allergy to β-lactams. Int Arch Allergy Immunol 2011;155:155-159
3   Cederbrant K, Marcusson-Stâhl M, Hultman P. Characterization of primary recall in vitro lymphocyte responses to bacampicillin in allergic subjects. Clin Exp Allergy 2000;30:1450-1459
4   Stejskal VD, Olin RG, Forsbeck M. The lymphocyte transformation test for diagnosis of drug-induced occupational allergy. J Allergy Clin Immunol 1986;77:411-426
5   Isogai Z, Sunohara A, Tsuji T. Pustular drug eruption due to bacampicilin hydrochloride in a patient with psoriasis. J Dermatol 1998;25:612-615
6   Imayama S, Fukuda H, Hori Y. Drug eruptions following treatment with prodrugs: a review of the reported cases in Japan from 1984 to 1989. J Dermatol 1991;18:277-280
7   Romano A, Quaratino D, Papa G, Di Fonso M, Venuti A. Aminopenicillin allergy. Arch Dis Child 1997;76:513-517.
8   Patriarca G, D'Ambrosio C, Schiavino D, Larocca LM, Nucera E, Milani A. Clinical usefulness of patch and challenge tests in the diagnosis of cell-mediated allergy to betalactams. Ann Allergy Asthma Immunol 1999;83:257-266
9   Romano A, Viola M, Mondino C, Pettinato R, Di Fonso M, Papa G, et al. Diagnosing nonimmediate reactions to penicillins by in vivo tests. Int Arch Allergy Immunol 2002;129:169-174
10 Schiavino D, Nucera E, De Pasquale T, Roncallo C, Pollastrini E, Lombardo C, Giuliani L, Larocca LM, Buonomo A, Patriarca G. Delayed allergy to aminopenicillins: clinical and immunological findings. Int J Immunopathol Pharmacol. 2006;19:831-840
11 Buonomo A, Nucera E, Pecora V, Rizzi A, Aruanno A, Pascolini L, et al. Cross-reactivity and tolerability of cephalosporins in patients with cell-mediated allergy to penicillins. J Investig Allergol Clin Immunol 2014;24:331-337

Chapter 3.58   BACLOFEN

IDENTIFICATION

Description/definition   : Baclofen is the synthetic chlorophenyl-butanoic acid derivative that conforms to the
                              structural formula shown below
Pharmacological classes   : GABA-B receptor agonists; muscle relaxants, central
IUPAC name   : 4-Amino-3-(4-chlorophenyl)butanoic acid
CAS registry number   : 1134-47-0
EC number   : 214-486-9
Merck Index monograph   : 2200
Patch testing   : Tablet, pulverized, 30% pet. (2); most systemic drugs can be tested at 10% pet.; if the pure
                              chemical is not available, prepare the test material from intravenous powder, the content
                              of capsules or – when also not available – from powdered tablets to achieve a final
                              concentration of the active drug of 10% pet.
Molecular formula   : $C_{10}H_{12}ClNO_2$

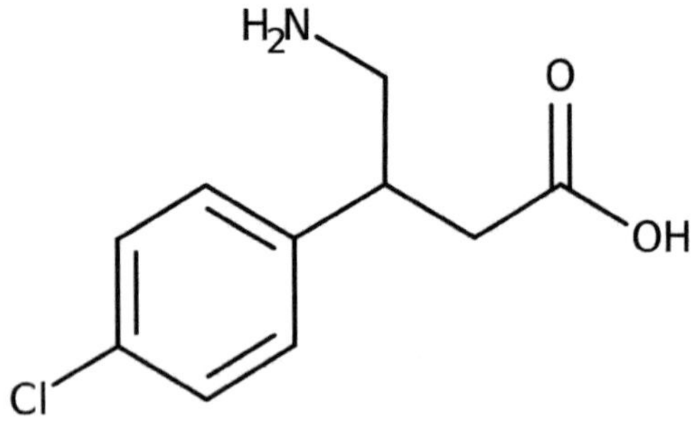

GENERAL

Baclofen is a centrally acting muscle relaxant that is administered for the relief of signs and symptoms of spasticity resulting from multiple sclerosis, particularly for the relief of flexor spasms and associated pain and clonus, in addition to muscular rigidity. Its therapeutic effects result from actions at spinal and supraspinal sites, generally the reduction of excitatory transmission (1).

CONTACT ALLERGY FROM ACCIDENTAL CONTACT

A 44-year-old nurse presented with a 9-month-history of eczema on the hands and face, which had started 3 months after she was required to crush tablets daily during her work as a nurse in a clinic for disabled people. She was patch tested with all the drugs that she had contact with and reacted to 8 of these, including baclofen 30% pet. (D2 ++, D3 +). No controls were performed (2). This was a case of occupational airborne allergic contact dermatitis.

LITERATURE

1   The data in the section 'General' may have been obtained from literature discussed in this chapter, but mostly
     also or exclusively from one or more of the following online sources: ChemIDPlus Advanced, PubChem,
     DrugBank, RxList, Drug Central, Drugs.com, and Wikipedia
2   Swinnen I, Ghys K, Kerre S, Constandt L, Goossens A. Occupational airborne contact dermatitis from
     benzodiazepines and other drugs. Contact Dermatitis 2014;70:227-232

Chapter 3.59   BEMIPARIN

IDENTIFICATION

Description/definition   : Bemiparin is a second-generation, low-molecular-weight heparin
Pharmacological classes : Antithrombotic agents
IUPAC name   : Not available
CAS registry number : 9005-49-6 (Heparin)
EC number : 232-681-7 (Heparin)
Merck Index monograph : 5958 (Heparin)
Patch testing : Commercial preparation undiluted (4); consider intradermal testing with late readings (D2,D3) when patch tests are negative and consider subcutaneous challenge when intradermal tests are negative
Molecular formula : Unspecified

GENERAL

Bemiparin is an antithrombotic and belongs to the group of drugs known as the low molecular weight heparins (LMWH). It is classified as an ultra-LMH because of its low mean molecular mass of 3600 daltons, which is a unique property of this class 1. These heparins have lower anti-thrombin activity than the traditional low molecular weight heparins and act mainly on factor-Xa, reducing the risk of bleeding due to selectivity for this specific clotting factor. Bemiparin is used for the prevention of thromboembolism after surgery, and to prevent blood clotting in the extracorporeal circuit in haemodialysis. In pharmaceutical products, bemiparin is employed as bemiparin sodium (CAS number not available, EC number not available, molecular formula unspecified) (1).

See also certoparin (Chapter 3.117), dalteparin (Chapter 3.160), danaparoid (Chapter 3.161), enoxaparin (Chapter 3.195), fondaparinux (Chapter 3.225), heparins (Chapter 3.239), nadroparin (Chapter 3.331), and tinzaparin (Chapter 3.479).

CUTANEOUS ADVERSE DRUG REACTIONS FROM SYSTEMIC ADMINISTRATION CAUSED BY TYPE IV (DELAYED-TYPE) HYPERSENSITIVITY

Throughout this book, only reports of delayed-type hypersensitivity have been included that showed a positive patch test to the culprit drug. However, as a result of the high molecular weight of heparins, patch tests are often false-negative, presumably from insufficient penetration into the skin. Because of this, and also because patch tests have been performed in a small minority of cases only, studies with a positive intradermal test or subcutaneous provocation tests with delayed readings are included in the chapters on the various heparins, even when patch tests were negative or not performed.

General information on delayed-type hypersensitivity reactions to heparins

General information on delayed-type hypersensitivity reactions to heparins (including bemiparin) presenting as local reactions from subcutaneous administration, is provided in Chapter 3.239 Heparins. In this chapter, only *non-local* cutaneous adverse drug reactions from delayed-type hypersensitivity to bemiparin are presented.

Non-local cutaneous adverse drug reactions

Cutaneous adverse drug reactions from systemic administration of bemiparin caused by type IV (delayed-type) hypersensitivity have included acute localized exanthematous pustulosis (ALEP) (2).

Acute localized exanthematous pustulosis (ALEP)

A 65-year-old woman, after undergoing a procedure to remove osteosynthesis material, was prescribed bemiparin 2500 IU/d administered subcutaneously for 6 days, cefazolin and metamizole. After one day, the patient developed an erythematous eruption at the injection site (abdomen), which extended 12 hours later to the palms of both hands in the form of pustules, the dorsum of the hands, and lateral areas of both feet. No fever, arthralgia, or other general symptoms were recorded. Histological analysis of the biopsy specimen of the cutaneous lesions revealed subcorneal pustules, necrotic keratinocytes, edema in the upper dermis, and mild perivascular infiltrate with scarce neutrophils and eosinophils. A diagnosis of acute localized exanthematous pustulosis (ALEP) was made. Patch tests performed 2 months after the onset of symptoms with metamizole, cefazolin, bemiparin, enoxaparin, and nadroparin (20% in water and petrolatum) were positive only for bemiparin at D2 and D3. The result of the subsequent challenge test with bemiparin was positive, and 2 days after taking 1000 IU, the patient developed generalized itching and a cutaneous eruption on her back and palms, which persisted for 7 days (2).

Cross-reactions, pseudo-cross-reactions and co-reactions

Cross-reactions between heparins are frequent in delayed-type hypersensitivity (>90% of patients tested, median number of positive drugs per patient: 3) and do not depend on the molecular weight of the molecules (3). Overlap in their polysaccharide composition might explain the high degree of cross-allergenicity (5). Cross-reactions to the semisynthetic heparinoid danaparoid have also been observed (6). In allergic patients, the synthetic ultralow molecular weight synthetic heparin fondaparinux is usually, but not always (6) well-tolerated (5).

LITERATURE

1   The data in the section 'General' may have been obtained from literature discussed in this chapter, but mostly also or exclusively from one or more of the following online sources: ChemIDPlus Advanced, PubChem, DrugBank, RxList, Drug Central, Drugs.com, and Wikipedia

2   Gómez Torrijos E, Cortina de la Calle MP, Méndez Díaz Y, Moreno Lozano L, Extremera Ortega A, Galindo Bonilla PA, et al. Acute localized exanthematous pustulosis due to bemiparin. J Investig Allergol Clin Immunol 2017;27:328-329

3   Weberschock T, Meister AC, Bohrt K, Schmitt J, Boehncke W-H, Ludwig RJ. The risk for cross-reactions after a cutaneous delayed-type hypersensitivity reaction to heparin preparations is independent of their molecular weight: a systematic review. Contact Dermatitis 2011;65:187-194

4   Brockow K, Garvey LH, Aberer W, Atanaskovic-Markovic M, Barbaud A, Bilo MB, et al. Skin test concentrations for systemically administered drugs - an ENDA/EAACI Drug Allergy Interest Group position paper. Allergy 2013;68:702-712

5   Schindewolf M, Lindhoff-Last E, Ludwig RJ. Heparin-induced skin lesions. Lancet 2012;380:1867-1879

6   Utikal J, Peitsch WK, Booken D, Velten F, Dempfle CE, Goerdt S, et al. Hypersensitivity to the pentasaccharide fondaparinux in patients with delayed-type heparin allergy. Thromb Haemost 2005;94:895-896

Chapter 3.60   BENDAMUSTINE

IDENTIFICATION

Description/definition   : Bendamustine is the nitrogen mustard derivative that conforms to the structural formula shown below
Pharmacological classes   : Antineoplastic agents, alkylating
IUPAC name   : 4-[5-[bis(2-Chloroethyl)amino]-1-methylbenzimidazol-2-yl]butanoic acid
CAS registry number   : 16506-27-7
EC number   : Not available
Merck Index monograph   : 2306
Patch testing   : 5% and 10% pet. (2)
Molecular formula   : $C_{16}H_{21}Cl_2N_3O_2$

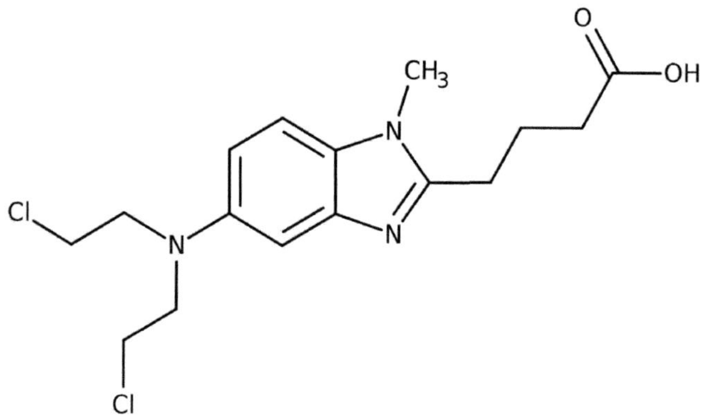

GENERAL

Bendamustine is a bifunctional mechlorethamine derivative (nitrogen mustard compound) with alkylator and antimetabolite activities. It is parenterally administered used alone or in combination with other antineoplastic agents in the treatment of chronic lymphocytic leukemia and indolent B-cell non-Hodgkin lymphoma (1). In pharmaceutical products, bendamustine is employed as bendamustine hydrochloride (CAS number 3543-75-7, EC number not available, molecular formula $C_{16}H_{22}Cl_3N_3O_2$) (1).

CUTANEOUS ADVERSE DRUG REACTIONS FROM SYSTEMIC ADMINISTRATION CAUSED BY TYPE IV (DELAYED-TYPE) HYPERSENSITIVITY (as demonstrated by positive patch tests)

Cutaneous adverse drug reactions from systemic administration of bendamustine caused by type IV (delayed-type) hypersensitivity have included acute generalized exanthematous pustulosis (AGEP) (2).

Acute generalized exanthematous pustulosis (AGEP)

A 59-year-old man was admitted to the hospital for evaluation of fever and a generalized, intensely pruritic, pustular eruption overlying erythematous patches on the face, trunk, buttocks, and extremities with associated leukocytosis. The patient had received his second cycle of bendamustine and rituximab for chronic lymphocytic leukemia 4 days before and had been taking allopurinol for 3 months. A diagnosis of acute generalized exanthematous pustulosis (AGEP) was made. Two punch biopsies demonstrated neutrophil-rich subcorneal pustules consistent with AGEP. Later, patch testing was performed with bendamustine, rituximab, and allopurinol 5% and 10% pet. At D2, itching was reported at the bendamustine application sites and tiny papules on an erythematous background were noted at both of these sites, which was still visible at D5. There were no reactions to the other drugs (2).

LITERATURE

1   The data in the section 'General' may have been obtained from literature discussed in this chapter, but mostly also or exclusively from one or more of the following online sources: ChemIDPlus Advanced, PubChem, DrugBank, RxList, Drug Central, Drugs.com, and Wikipedia
2   Harber ID, Adams KV, Casamiquela K, Helms S, Benson BT, Herrin V. Bendamustine-induced acute generalized exanthematous pustulosis confirmed by patch testing. Dermatitis 2017;28:292-293

Chapter 3.61   BENDROFLUMETHIAZIDE

IDENTIFICATION

Description/definition   : Bendroflumethiazide is the thiazide that conforms to the structural formula shown
                            below
Pharmacological classes : Sodium chloride symporter inhibitors; diuretics; antihypertensive agents
IUPAC name   : 3-Benzyl-1,1-dioxo-6-(trifluoromethyl)-3,4-dihydro-2H-1λ^6,2,4-benzothiadiazine-7-
                            sulfonamide
Other names   : 3-Benzyl-3,4-dihydro-6-(trifluoromethyl)-2H-1,2,4-benzothiadiazine-7-sulfonamide
                            1,1-dioxide
CAS registry number   : 73-48-3
EC number   : 200-800-1
Merck Index monograph   : 2309
Patch testing   : 2% pet. (2)
Molecular formula   : $C_{15}H_{14}F_3N_3O_4S_2$

GENERAL

Bendroflumethiazide is a long-acting thiazide diuretic with actions and uses similar to those of hydrochlorothiazide. It is indicated for the treatment of high blood pressure and management of edema related to heart failure (1).

CONTACT ALLERGY FROM ACCIDENTAL CONTACT

A 54-year-old man working in the pharmaceutical industry had developed skin problems about 3 months previously, when he developed eczema on the hands, forearms and axillae. Although he used a protective suit with mask and rubber gloves, the rubber gloves seemed to aggravate the dermatitis. The symptoms improved when he moved to another section and at weekends. Patch testing showed positive reactions to thiuram mix and all of its ingredients, to bendroflumethiazide 2.0% pet. (D2 +, D4 ++; negative to 1% and 0.1%)) and to 2 other drugs (propranolol, hydralazine). Sixteen controls were negative (2). This was a case of occupational allergic contact dermatitis to bendroflumethiazide.

LITERATURE

1   The data in the section 'General' may have been obtained from literature discussed in this chapter, but mostly also or exclusively from one or more of the following online sources: ChemIDPlus Advanced, PubChem, DrugBank, RxList, Drug Central, Drugs.com, and Wikipedia
2   Pereira F, Dias M, Pacheco FA. Occupational contact dermatitis from propranolol, hydralazine and bendroflumethiazide. Contact Dermatitis 1996;35:303-304

Chapter 3.62   BENZNIDAZOLE

IDENTIFICATION

Description/definition   : Benznidazole is the nitroimidazole derivative that conforms to the structural formula
                               shown below
Pharmacological classes   : Mutagens; trypanocidal agents; immunosuppressive agents
IUPAC name   : *N*-Benzyl-2-(2-nitroimidazol-1-yl)acetamide
Other names   : Benzonidazole
CAS registry number   : 22994-85-0
EC number   : Not available
Merck Index monograph   : 2356
Patch testing   : 5% pet. (2); 10% DMSO (3)
Molecular formula   : $C_{12}H_{12}N_4O_3$

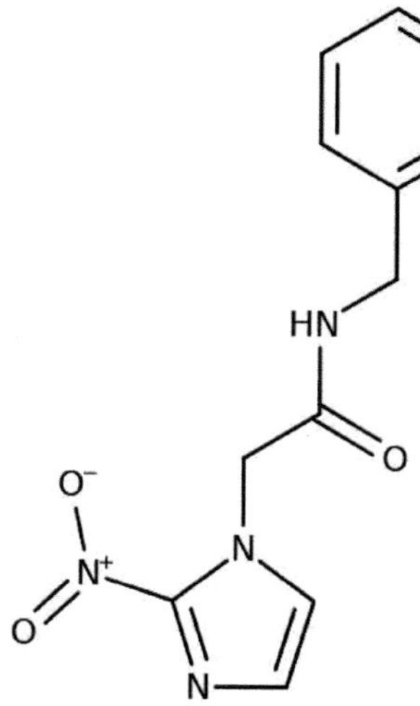

GENERAL

Benznidazole is a nitroimidazole derivative that has antiprotozoal activity by interfering with parasite protein biosynthesis, influencing cytokines production and stimulating host phagocytosis. It is used in the treatment of Chagas disease (trypanosomiasis) in children 2-12 years of age because, as a trypanocidal agent, benznidazole kills the causative organism *Trypanosoma cruzi* (1).

CUTANEOUS ADVERSE DRUG REACTIONS FROM SYSTEMIC ADMINISTRATION CAUSED BY TYPE IV (DELAYED-TYPE) HYPERSENSITIVITY (as demonstrated by positive patch tests)

Cutaneous adverse drug reactions from systemic administration of benznidazole caused by type IV (delayed-type) hypersensitivity have included maculopapular eruption (3), acute generalized exanthematous pustulosis (AGEP) (4), drug reaction with eosinophilia and systemic symptoms (DRESS) (2), overlap DRESS/Stevens-Johnson syndrome (5) and urticaria (5).

Case series with various or unknown types of drug reactions

In a series of 31 patients from Bolivia and El Salvador but living in Spain, suspected of hypersensitivity to benznidazole (13 urticaria, 8 exanthemas, 2 pruritus, 5 DRESS, 1 AGEP, 1 DRESS/TEN, 1 DRESS/SJS), 19 were patch tested with benznidazole (no details provided). There were 2 positive reactions, one in a patient with urticaria and one in a patient who had suffered DRESS/SJS from benznidazole. One of these individuals also had a positive lymphocyte transformation test (5).

Maculopapular eruption

Six Bolivian women, who had been living in Spain for more than 10 years (median age 33, range: 27-43 years) were treated with benznidazole for Chagas disease. After 3-15 days (median 7.5 days) of treatment all presented an itchy maculopapular generalized exanthem that resolved with antihistamines and corticosteroids in 15-20 days. All patients were patch tested with benznidazole and the related metronidazole 10% in DMSO. Five had positive reactions to benznidazole, 2 of who also reacted to metronidazole. The sixth patient was negative to benznidazole, but positive to metronidazole. An oral provocation with 100 mg benznidazole resulted in a generalized maculopapular eruption after 2 hours and a flare-up of the previous positive metronidazole patch test. Six controls were negative to both drugs 10% DMSO (3).

Acute generalized exanthematous pustulosis (AGEP)
A 32-year-old man with Chagas disease began treatment with benznidazole and 6 weeks later developed multiple pinhead-sized, sterile pustules with fever, elevated liver enzymes, and peripheral eosinophilia. The diagnostic score for AGEP (Acute generalized exanthematous pustulosis) was 10 points (definite), and he was diagnosed as having AGEP. Later, in spite of physician advice to avoid nitroimidazoles, he resumed treatment with benznidazole because of recurrence of his Chagas disease. Two days later, the symptoms reappeared with the same maculopapular rash with pustules in the same area of extension, in his face, chest, back, and arms, with high fever and eosinophilia. A skin biopsy specimen was consistent with AGEP. A patch test was positive to benznidazole 5% DMSO, with an edematous and erythematous plaque with sterile pustules on the patched skin and in satellite areas. Five controls were negative (4).

Drug reaction with eosinophilia and systemic symptoms (DRESS)
A 30-year-old woman from Bolivia suffered from malaise, generalized edema and exanthema with subsequent desquamation, without mucosal involvement. She had cervical and inguinal lymphadenopathy, fever (38.6C), high CRP, elevated liver enzymes and eosinophilia (>4000 eosinophils/ml). She had begun taking benznidazole for chronic Chagas disease 50 days before the rash appeared. A biopsy specimen showed a moderate perivascular lymphocytic infiltrate in the papillar and superficial dermis. She was treated with corticosteroids and improved within 2 weeks. Patch tests gave a positive patch tests on D2 (+++) and D4 (+++) to benznidazole 5% pet., but were negative to other drugs that she had used. The patient was diagnosed with drug reaction with eosinophilia and systemic symptoms (DRESS) syndrome from benznidazole (2).

Cross-reactions, pseudo-cross-reactions and co-reactions
Of 5 patients allergic to benznidazole, 2 (40%) cross-reacted to metronidazole (3).

LITERATURE
1   The data in the section 'General' may have been obtained from literature discussed in this chapter, but mostly also or exclusively from one or more of the following online sources: ChemIDPlus Advanced, PubChem, DrugBank, RxList, Drug Central, Drugs.com, and Wikipedia
2   Moreno-Escobosa C, Cruz-Granados S. Drug reaction with eosinophilia and systemic symptoms syndrome induced by benznidazole. Contact Dermatitis 2018;79:105-106
3   Noguerado-Mellado B, Rojas-Pérez-Ezquerra P, Calderón-Moreno M, Morales-Cabeza C, Tornero-Molina P. Allergy to benznidazole: cross-reactivity with other nitroimidazoles. J Allergy Clin Immunol Pract 2017;5:827-828
4   Alava-Cruz C, Rojas Pérez-Ezquerra P, Pelta-Fernández R, Zubeldia-Ortuño JM, de Barrio-Fernández M. Acute generalized exanthematous pustulosis due to benznidazole. J Allergy Clin Immunol Pract 2014;2:800-802
5   Marques-Mejías MA, Cabañas R, Ramírez E, Domínguez-Ortega J, Fiandor A, Trigo E, Quirce S, Bellón T. Lymphocyte Transformation Test (LTT) in allergy to benznidazole: A promising approach. Front Pharmacol 2019;10:469

Chapter 3.63   BENZYDAMINE

IDENTIFICATION

Description/definition : Benzydamine is the benzyl-indazole that conforms to the structural formula shown below
Pharmacological classes : Anti-inflammatory agents
IUPAC name : 3-(1-Benzylindazol-3-yl)oxy-*N,N*-dimethylpropan-1-amine
CAS registry number : 642-72-8
EC number : 211-388-8
Merck Index monograph : 2395
Patch testing : Hydrochloride, 2.0% pet.(Chemotechnique, SmartPracticeCanada); 1% pet. (SmartPracticeCanada)
Molecular formula : $C_{19}H_{23}N_3O$

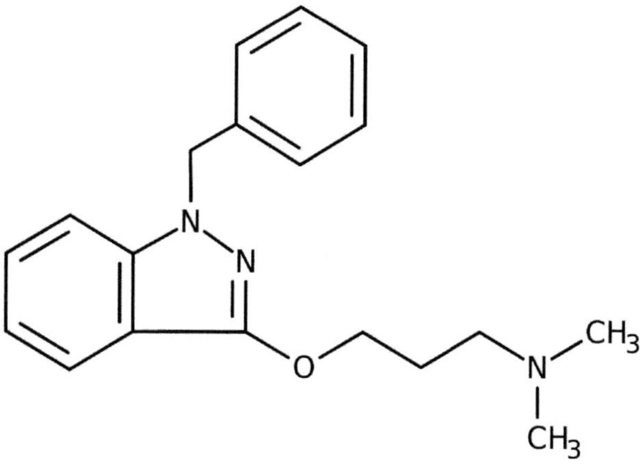

GENERAL

Benzydamine is an indazole non-steroidal anti-inflammatory drug with analgesic, antipyretic, and anti-edema properties. It is available as a liquid mouthwash, spray for mouth and throat, topical cream, and vaginal irrigation; formerly, benzydamine was also employed in tablets, suppositories and intramuscular injections. In pharmaceutical products, benzydamine is employed as benzydamine hydrochloride (CAS number 132-69-4, EC number 205-076-0, molecular formula $C_{19}H_{24}ClN_3O$) (1).

In topical preparations, benzydamine has caused contact allergy/allergic contact dermatitis and many cases of photoallergic contact dermatitis, which has been fully reviewed in Volume 3 of the *Monographs in contact allergy series* (8).

CONTACT ALLERGY FROM ACCIDENTAL CONTACT

A 30-year-old nurse presented with severe exudative dermatitis of the right hand, arms, face and eyelids. She had been using a powder containing benzydamine HCl for over a month, in 0.1% solutions, for performing vaginal irrigations on her patients. Patch tests were positive to the powder and benzydamine HCl at 0.5% and 0.1% water and pet., with a flare-up of the dermatitis on the face and eyelids. Fifteen controls were negative (3). This was a case of – partly airborne – occupational allergic contact dermatitis.

CUTANEOUS ADVERSE DRUG REACTIONS FROM SYSTEMIC ADMINISTRATION CAUSED BY TYPE IV (DELAYED-TYPE) HYPERSENSITIVITY (as demonstrated by positive patch tests)

Cutaneous adverse drug reactions from systemic administration of benzydamine caused by type IV (delayed-type) hypersensitivity have included photosensitivity (5,6,7).

Photosensitivity

A 23-year-old woman had developed dermatitis of the light-exposed areas (face, dorsa of hands, forearms and legs), 24 hours after taking a preparation made up of 250 mg of ampicillin and 50 mg of benzydamine for pharyngitis. An oral provocation with benzydamine was positive, with ampicillin negative. Years later, dermatitis involving the same areas appeared 48 hours after the administration of benzydamine inhalation spray. A photopatch test with a solution containing 1.5 mg/ml benzydamine was positive (5). A second patient presented by the authors had photoallergic contact dermatitis from an oral benzydamine preparation with blistering dermatitis on the face (5).

Two women with photoallergic dermatitis from oral administration of benzydamine were reported from Germany. In both cases, photopatch tests were positive to benzydamine (6).

One or more additional cases of benzydamine photosensitization from both oral and topical administration have been presented in ref. 7.

Cross-reactions, pseudo-cross-reactions and co-reactions
Cross-reactivity to benzydamine salicylate (2).

LITERATURE

1 The data in the section 'General' may have been obtained from literature discussed in this chapter, but mostly also or exclusively from one or more of the following online sources: ChemIDPlus Advanced, PubChem, DrugBank, RxList, Drug Central, Drugs.com, and Wikipedia

2 Goday Buján JJ, Ilardia Lorentzen R, Soloeta Arechavala R. Allergic contact dermatitis from benzydamine with probable cross-reaction to indomethacin. Contact Dermatitis 1993;28:111-112

3 Foti C, Vena GA, Angelini G. Occupational contact allergy to benzydamine hydrochloride. Contact Dermatitis 1992;27:328-329

4 Lasa Elgezua O, Gorrotxategi PE, Gardeazabal García J, Ratón Nieto JA, Pérez JL. Photoallergic hand eczema due to benzydamine. Eur J Dermatol 2004;14:69-70

5 Fernandez de Corres L. Photodermatitis from benzydamine. Contact Dermatitis 1980;6:285

6 Frosch PJ, Weickel R. Photocontact allergy caused by benzydamine (Tantum). Hautarzt 1989;40:771-773 (Article in German)

7 Ikemura I. Contact and photocontact dermatitis due to benzydamine hydrochloride. Jpn J Clin Dermatol 1971;25:129 (Article in Japanese, data cited in ref. 4)

8 De Groot AC. Monographs in contact allergy, volume 3: Topical Drugs. Boca Raton, Fl, USA: CRC Press Taylor and Francis Group, 2021 (ISBN 978-0-367-23693-9)

Chapter 3.64   BENZYLPENICILLIN

IDENTIFICATION

Description/definition : Benzylpenicillin is the penicillin derivative that conforms to the structural formula shown below
Pharmacological classes : Anti-bacterial agents
IUPAC name : (2S,5R,6R)-3,3-Dimethyl-7-oxo-6-(2-phenylacetamido)-4-thia-1-azabicyclo(3.2.0) heptane-2-carboxylic acid
Other names : Penicillin G
CAS registry number : 61-33-6
EC number : 200-506-3
Merck Index monograph : 8473
Patch testing : 5% pet. (20)
Molecular formula : $C_{16}H_{18}N_2O_4S$

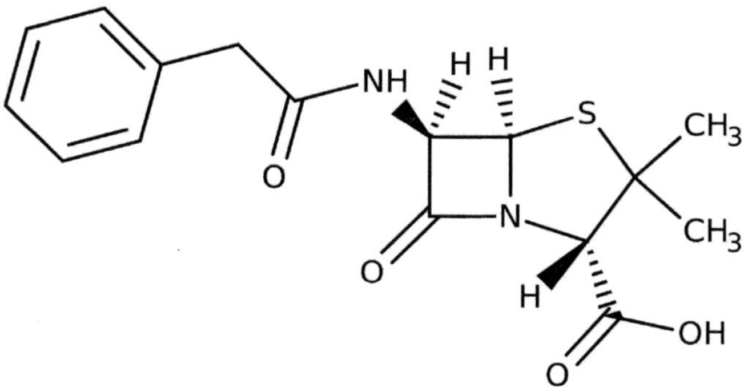

GENERAL

Benzylpenicillin (penicillin G) is a broad-spectrum naturally-occurring penicillin antibiotic with antibacterial activity. It is effective against most gram-positive bacteria and against gram-negative cocci. Benzylpenicillin is stable against hydrolysis by a variety of β-lactamases, including penicillinases, cephalosporinases and extended spectrum β-lactamases. This antibiotic is indicated for use in the treatment of severe infections caused by susceptible microorganisms when rapid and high penicillin levels are required such as in the treatment of sepsis, meningitis, pericarditis, endocarditis and severe pneumonia (1). In pharmaceutical products, benzylpenicillin is usually employed as benzylpenicillin sodium (penicillin G sodium) (CAS number 69-57-8, EC number 200-710-2, molecular formula $C_{16}H_{17}N_2NaO_4S$) or benzylpenicillin potassium (penicillin G potassium) (CAS number 113-98-4, EC number 204-038-0, molecular formula $C_{16}H_{17}KN_2O_4S$) (1).

The classification and structures of beta-lactam antibiotics are discussed in Chapter 3.36 Amoxicillin. It has not been attempted to provide a full review of the subject of delayed-type hypersensitivity to systemic benzylpenicillin.

CONTACT ALLERGY FROM ACCIDENTAL CONTACT

Case series

In Leuven, Belgium, in the period 2001-2019, 201 of 1248 health care workers/employees of the pharmaceutical industry had occupational allergic contact dermatitis. In 23 (11%) dermatitis was caused by skin contact with a systemic drug: 19 nurses, two chemists, one physician, and one veterinarian. The lesions were mostly localized on the hands, but often also on the face, as airborne dermatitis. In total, 42 positive patch test reactions to 18 different systemic drugs were found. In one patient, benzylpenicillin was the drug/one of the drugs that caused occupational dermatitis (5).

Of 38 veterinarians with hand and forearm dermatoses seen by dermatologists in Belgium and the Netherlands from 1995 to 2005, 17 had occupational allergic contact dermatitis. Among these, benzylpenicillin was the responsible drug in 2 patients (16).

In Warsaw, Poland, in the period 1971-1998, nurses with eczema were routinely tested with benzylpenicillin 300,000 IU/gr pet. Of the patients with occupational contact dermatitis, between 0.7% to 9.8% had positive patch tests to this antibiotic, starting at 1.6% in 1971-1975, and rising to 9.8% (1981-1985) and 7.9% (1986-1990), thereafter declining to 0.7% in the period 1996-1998. The decline was ascribed to the reduction in the use of benzylpenicillin in Poland, gradually being replaced with semisynthetic penicillins such as amoxicillin (15).

In 34 veterinary surgeons (7 women, 27 men, age range 29-61, mean age 38 years) with chronic or relapsing eczema of the hands as the main complaint, investigated in Norway before 1985, nine had occupational allergic contact dermatitis, of whom one reacted to benzylpenicillin 30% water; this individual also reacted to penethamate, procaine penicillin and benzathine penicillin (3).

Of 333 nurses patch tested in Poland between 1979 and 1987, 21 (6.3%) reacted to benzylpenicillin 3,000,000 U/g pet.; all reactions were likely to be relevant (14). Of 26 veterinarians patch tested in the early 1980s by the same authors in Poland because of dermatitis, 15 had one or more positive patch tests to veterinary drugs, tuberculin and/or disinfectants. Most of them realized that these contact materials were harmful to them, and all had come into contact with drugs to which they had positive tests. Two of the veterinarians reacted to benzylpenicillin 10,000 IU/gr pet. (11).

In 2 hospitals in Denmark, between 1974 and 1980, 37 veterinary surgeons, all working in private country practices, were investigated for suspected incapacitating occupational dermatitis and patch tested with a battery of 10 antibiotics. Thirty-two (86%) had one or more positive patch tests. There were 5 positive reactions to benzylpenicillin 300,000 IU/ml (14%). It was mentioned that all but one of these individuals had allergic contact dermatitis, but relevance was not specified for individual allergens. The most frequent allergens were spiramycin, penethamate and tylosin tartrate (10).

Case reports

A 48-year-old female nurse had suffered angioedema and urticaria after treatment with penicillin for acute tonsillitis in the past. While giving an injection of benzylpenicillin, the solution spurted onto her face, and pruritus and swelling starting some hours later. An open test with one drop of benzylpenicillin solution on the anterior aspect of the forearm, was negative. A prick test was doubtful at 30 min but positive at 24 hours. Patch tests were positive at D2 and D3 to benzylpenicillin 100,000 IU/gr pet. but negative to ampicillin and amoxicillin. Two days after the patch tests were performed, the patient had a mild generalized urticarial eruption. A RAST test for specific IgE antibodies was negative (12).

CUTANEOUS ADVERSE DRUG REACTIONS FROM SYSTEMIC ADMINISTRATION CAUSED BY TYPE IV (DELAYED-TYPE) HYPERSENSITIVITY (as demonstrated by positive patch tests)

Cutaneous adverse drug reactions from systemic administration of benzylpenicillin caused by type IV (delayed-type) hypersensitivity have included maculopapular eruption (7,8,21,23,28,29,30), (extensive or generalized) erythema (30), acute generalized exanthematous pustulosis (AGEP) (6), drug reaction with eosinophilia and systemic symptoms (DRESS) (25,27), toxic epidermal necrolysis (TEN) (17,18), dermatitis/eczematous eruption (22), generalized exanthema (9,24), bullous exanthema/erythema multiforme (8), bullous pemphigoid (4), urticaria/angioedema (28,29,30), urticaria-like exanthema appearing after several hours (8), and unspecified drug eruptions (2,13),

Case series with various or unknown types of drug reactions

In Bern, Switzerland, patients with a suspected allergic cutaneous drug reaction were patch-scratch tested with suspected drugs that had previously given a positive lymphocyte transformation test. Benzylpenicillin 200,000 IU/ml saline gave a positive patch-scratch test in 3 patients with maculopapular exanthema, one with bullous exanthema/erythema multiforme and in three with urticaria-like exanthema appearing after several hours (8).

In a group of thirty patients from Rome who had suffered skin rashes from beta-lactam antibiotics and who had at least one positive patch test to a beta-lactam antigenic determinant, benzylpenicillin was the allergenic culprit in 2 patients with urticaria/angioedema and one with maculopapular eruption (28).

In 97 patients with nonimmediate reactions from penicillins with a positive patch test to at least one penicillin seen in Rome between 2002 and 2011, amoxicillin was the culprit drug in 50 individuals, ampicillin in 35, bacampicillin in 25 and benzylpenicillin in 11. The most frequent clinical manifestations were maculopapular eruptions with or without edema, delayed urticaria with or without angioneurotic edema and generalized erythema. It was not specified which drug eruptions the 4 individual antibiotics had caused (30).

Maculopapular eruption

In the period 2000-2014, in Portugal, 260 patients were patch tested with antibiotics for suspected cutaneous adverse drug reactions (CADR) to these drugs. 56 patients (22%) had one or more positive patch tests. Benzylpenicillin was patch test positive in 2 patients with maculopapular eruptions (21).

In Bern, Switzerland, of 4 patients who had developed a maculopapular eruption during or within 2 weeks after therapy with benzylpenicillin, 2 (50%) had a positive scratch-patch test to benzylpenicillin 200.000 U/ml saline and one to amoxicillin. Three of these 4 patients also had a positive lymphocyte transformation test to benzylpenicillin and the 4th to amoxicillin (23).

A 32-year-old man developed a maculopapular eruption from intramuscular benzylpenicillin. He had positive patch tests to benzylpenicillin and many other penicillins, some cephalosporins and the beta-lactam carbapenem antibiotic imipenem (7). One case of maculopapular eruption from benzylpenicillin as shown by positive patch tests and delayed intradermal tests to amoxicillin and ampicillin (29).

See also the section 'Case series with various or unknown types of drug reactions' above, refs. 8,28, and 30.

Erythroderma, widespread erythematous eruption, exfoliative dermatitis
See the section 'Case series with various or unknown types of drug reactions' above, ref. 30.

Acute generalized exanthematous pustulosis (AGEP)
A life-threatening case of AGEP in a 30-year-old man caused by delayed-type hypersensitivity to amoxicillin, ampicillin and benzylpenicillin was reported in 2017. It began with treatment of periodontitis with amoxicillin, which caused erythema of the face. When treated with ciprofloxacin (which was later prick-test positive) and subsequently ampicillin and benzylpenicillin, the erythema spread to the rest of the skin with pustules, severe mucous membrane involvement and high fever. A systemic inflammatory response syndrome developed with life-threatening involvement of the lungs, heart, liver and kidneys. With non-penicillin antibiotics and systemic corticosteroids the patient recovered. Four months later, patch tests were positive to ampicillin, amoxicillin, and other penicillins. Intradermal testing of minor and major determinants of penicillin were positive for benzylpenicillin and ampicillin. Prick testing was positive for ciprofloxacin, amoxicillin, ampicillin, and amoxicillin/clavulanic acid. The authors assumed that the underlying type I/IV allergies to various β-lactam antibiotics given, together with secondary infection of the skin, led to severe life-threatening systemic involvement. Pre-existing bacterial infections, in this case caused by periodontitis, might also have been a promoting factor (6).

Drug reaction with eosinophilia and systemic symptoms (DRESS)
In London, between October 2017 and October 2018, 45 patients with suspected cutaneous adverse drug reactions, including 33 maculopapular eruptions (MPE), 4 fixed drug eruptions (FDE), 4 DRESS, 3 AGEP and one SJS/TEN, were patch tested with the suspected drugs. There were 10 (22%) positive patch test cases: 4 MPE, 2 FDE, 3 DRESS and 1 AGEP. Benzylpenicillin, tested as commercial preparation 30% pet., was responsible for 1 case of DRESS (25).

A 40-year old man, after having taken benzylpenicillin for 5 weeks, developed (a probable case of) DRESS with an extensive maculopapular exanthematous eruption, eosinophilia and impaired liver function. The patient was afebrile and had no palpable lymphadenopathy. Four months later, patch tests were positive to benzylpenicillin 10% pet. and amoxicillin 10% pet. (27).

Stevens-Johnson syndrome/toxic epidermal necrolysis (SJS/TEN)
Apparently, in early German (17) and in Japanese (18) literature, 2 cases of toxic epidermal necrolysis caused by benzylpenicillin with a positive patch test have been reported (data cited in ref. 19).

Dermatitis/eczematous eruption
In a study from Nancy, France, 54 patients with suspected nonimmediate drug eruptions (27 maculopapular, seven erythrodermic, nine eczematous, four photosensitivity, three fixed drug eruptions, three with pruritus and one with acute generalized exanthematous pustulosis) were assessed with patch testing. Of the 9 patients with generalized eczema, one had a positive patch test to benzylpenicillin (22).

Other cutaneous adverse drug reactions

Urticaria/angioedema
Two cases of urticaria and one of urticaria and angioedema from benzylpenicillin with positive patch and intradermal tests to amoxicillin and ampicillin (29). See also the section 'Case series with various or unknown types of drug reactions' above, ref. 28 and 30.

Other adverse drug reactions
A 38-year-old woman developed a generalized exanthema with fever after using benzylpenicillin and ampicillin. Patch tests were positive to benzylpenicillin 12% (pet. or saline), as was a lymphocyte transformation test (stimulation index 24) (9).

A 68-year-old woman was referred for penicillin skin testing, as she had suffered a generalized exanthema from parenteral benzylpenicillin 29 years ago. Patch-scratch tests were positive to benzylpenicillin 200.000 U/ml saline, spreading as a vesiculopapular exanthema on both forearms (where previously performed negative PPL and benzylpenicillin intradermal tests had flared during patch testing) and the back (24).

An 80-year-old man, known to be allergic to penicillin, was treated by mistake with parenteral benzylpenicillin and oral penicillin V. After the first dosage of the latter, the patient developed a rash, and penicillin was stopped. Despite this, 7 days later the exanthema became bullous. Physical examination showed an itchy generalized maculo-papular rash confluent in the loins, over the knees and forearms with scattered widespread large vesicles, small tense bullae with serous exudate and denuded areas at the trunk and extremities. Based on clinical features, histopatholo-gy and immunofluorescence, the diagnosis of bullous pemphigoid was made. Patch tests were positive to benzylpenicillin and 3 other penicillins (penicillin V was not tested). The lesions healed rapidly after withdrawal of penicillin and have not recurred since, suggesting that the bullous pemphigoid was drug-induced (4).

Unspecified drug eruptions

In a group of 78 patients who had suffered nonimmediate drug reactions from any beta-lactam compound (penicillins, cephalosporins), 33 had positive patch tests to benzylpenicillin. In 9 cases, benzylpenicillin was the culprit drug, so the other reactions were presumably cross-reactions. It was not mentioned which cutaneous drug reactions the 9 patients had suffered from (2).

In Finland, in the period 1989-2001, 826 patients with suspected cutaneous drug eruptions were patch tested and 89 had one or more positive reactions. Of these individuals, 6 reacted to benzylpenicillin. It was not mentioned which drug eruption these patients had suffered from and how many - if any – were cross-reactions (13).

Cross-reactions, pseudo-cross-reactions and co-reactions
Cross-reactions between beta-lactam antibiotics are discussed in Chapter 3.36 Amoxicillin.

Immediate contact reactions
Immediate contact reactions (contact urticaria) to benzylpenicillin are presented in Chapter 5.

LITERATURE

1   The data in the section 'General' may have been obtained from literature discussed in this chapter, but mostly also or exclusively from one or more of the following online sources: ChemIDPlus Advanced, PubChem, DrugBank, RxList, Drug Central, Drugs.com, and Wikipedia
2   Buonomo A, Nucera E, De Pasquale T, Pecora V, Lombardo C, Sabato V, et al. Tolerability of aztreonam in patients with cell-mediated allergy to β-lactams. Int Arch Allergy Immunol 2011;155:155-159
3   Falk ES, Hektoen H, Thune PO. Skin and respiratory tract symptoms in veterinary surgeons. Contact Dermatitis 1985;12:274-278
4   Borch JE, Andersen KE, Clemmensen O, Bindslev-Jensen C. Drug-induced bullous pemphigoid with positive patch test and in vitro IgE sensitization. Acta Derm Venereol 2005;85:171-172
5   Gilissen L, Boeckxstaens E, Geebelen J, Goossens A. Occupational allergic contact dermatitis from systemic drugs. Contact Dermatitis 2020;82:24-30
6   Tajmir-Riahi A, Wörl P, Harrer T, Schliep S, Schuler G, Simon M. Life-threatening atypical case of acute generalized exanthematous pustulosis. Int Arch Allergy Immunol 2017;174:108-111
7   Schiavino D, Nucera E, Lombardo C, Decinti M, Pascolini L, Altomonte G, et al. Cross-reactivity and tolerability of imipenem in patients with delayed-type, cell-mediated hypersensitivity to beta-lactams. Allergy 2009;64:1644-1648
8   Neukomm C, Yawalkar N, Helbling A, Pichler WJ. T-cell reactions to drugs in distinct clinical manifestations of drug allergy. J Invest Allergol Clin Immunol 2001;11:275-284
9   Gex-Collet C, Helbling A, Pichler WJ. Multiple drug hypersensitivity – proof of multiple drug hypersensitivity by patch and lymphocyte transformation tests. J Investig Allergol Clin Immunol 2005;15:293-296
10 Hjorth N, Roed-Petersen J. Allergic contact dermatitis in veterinary surgeons. Contact Dermatitis 1980;6:27-29
11 Rudzki E, Rebandel P, Grzywa Z, Pomorski Z, Jakiminska B, Zawisza E. Occupational dermatitis in veterinarians. Contact Dermatitis 1982;8:72-73
12 Pecegueiro M. Occupational contact dermatitis from penicillin. Contact Dermatitis 1990;23:190-191
13 Lammintausta K, Kortekangas-Savolainen O. The usefulness of skin tests to prove drug hypersensitivity. Br J Dermatol 2005;152:968-974
14 Rudzki E, Rebandel P, Grzywa Z. Patch tests with occupational contactants in nurses, doctors and dentists. Contact Dermatitis 1989;20:247-250
15 Rudzki E, Rebandel P, Hudymowicz W. Decrease in frequency of occupational contact sensitivity to penicillin among nurses in Warsaw. Contact Dermatitis 1999;41:114
16 Bulcke DM, Devos SA. Hand and forearm dermatoses among veterinarians. J Eur Acad Dermatol Venereol 2007;21:360-363
17 Schöpf E, Schulz KH, Kessler R, Taugner M, Braun W. Allergologische Untersuchungen beim Lyell-Syndrome. Z Hautkr 1975;50:865-873 (Article in German, data cited in ref. 19)

18 Sano T, Matsumoto R. Drug-induced TEN. Hifubyohshinryo 1980;2:49-52 (Article in Japanese, data cited in ref. 19)

19 Tagami H, Tatsuta K, Iwatsuki K, Yamada M. Delayed hypersensitivity in ampicillin-induced toxic epidermal necrolysis. Arch Dermatol 1983;119:910-913

20 Brockow K, Garvey LH, Aberer W, et al. Skin test concentrations for systemically administered drugs - an ENDA/EAACI Drug Allergy Interest Group position paper. Allergy 2013;68:702-712

21 Coutinho I, Gameiro A, Gouveia M, Gonçalo M. Patch testing - a valuable tool for investigating non-immediate cutaneous adverse drug reactions to antibiotics. J Eur Acad Dermatol Venereol 2017;31:280-287

22 Barbaud A, Reichert-Penetrat S, Tréchot P, Jacquin-Petit MA, Ehlinger A, Noirez V, et al. The use of skin testing in the investigation of cutaneous adverse drug reactions. Br J Dermatol 1998;139:49-58

23 Schnyder B, Pichler WJ. Skin and laboratory tests in amoxicillin- and penicillin-induced morbilliform skin eruption. Clin Exp Allergy 2000;30:590-595

24 Schnyder B, Helbling A, Kappeler A, Pichler WJ. Drug-induced papulovesicular exanthema. Allergy 1998;53:817-818

25 Watts TJ, Thursfield D, Haque R. Patch testing for the investigation of nonimmediate cutaneous adverse drug reactions: a prospective single center study. J Allergy Clin Immunol Pract 2019;7:2941-2943.e3

26 Redmond AP, Levine BB. Delayed skin reactions to benzylpenicillin in man. Int Arch Allergy 1968;33:193-206

27 Watts TJ, Li PH, Haque R. DRESS Syndrome due to benzylpenicillin with cross-reactivity to amoxicillin. J Allergy Clin Immunol Pract 2018;6:1766-1768

28 Patriarca G, D'Ambrosio C, Schiavino D, Larocca LM, Nucera E, Milani A. Clinical usefulness of patch and challenge tests in the diagnosis of cell-mediated allergy to betalactams. Ann Allergy Asthma Immunol 1999;83:257-266

29 Torres MJ, Sánchez-Sabaté E, Alvarez J, Mayorga C, Fernández J, Padial A, et al. Skin test evaluation in nonimmediate allergic reactions to penicillins. Allergy 2004;59:219-224

30 Buonomo A, Nucera E, Pecora V, Rizzi A, Aruanno A, Pascolini L, et al. Cross-reactivity and tolerability of cephalosporins in patients with cell-mediated allergy to penicillins. J Investig Allergol Clin Immunol 2014;24:331-337

Chapter 3.65   BETAMETHASONE

IDENTIFICATION

Description/definition   : Betamethasone is the synthetic glucocorticoid that conforms to the structural formula
                               shown below
Pharmacological classes   : Anti-asthmatic agents; glucocorticoids; anti-inflammatory agents
IUPAC name   : (8S,9R,10S,11S,13S,14S,16S,17R)-9-Fluoro-11,17-dihydroxy-17-(2-hydroxyacetyl)-
                             10,13,16-trimethyl-6,7,8,11,12,14,15,16-octahydrocyclopenta[a]phenanthren-3-one
Other names   : 9-Fluoro-11β,17,21-trihydroxy-16β-methylpregna-1,4-diene-3,20-dione
CAS registry number   : 378-44-9
EC number   : 206-825-4
Merck Index monograph : 2452
Patch testing   : In general, corticosteroids may be tested at 0.1% and 1% in alcohol; late readings (6-10
                             days) are strongly recommended
Molecular formula   : $C_{22}H_{29}FO_5$

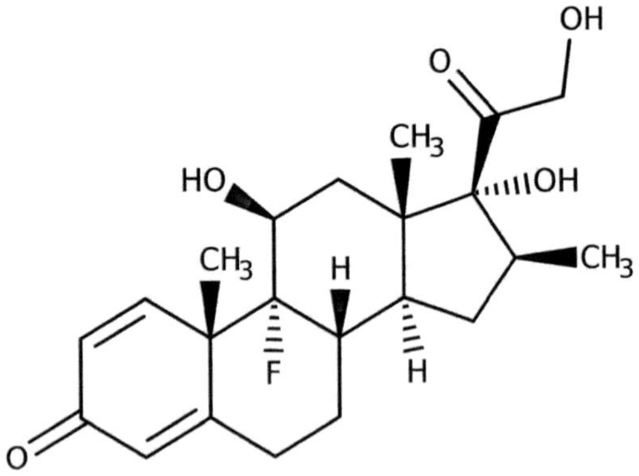

GENERAL
Systemically administered glucocorticoids have anti-inflammatory, immunosuppressive and antineoplastic properties and are used in the treatment of a wide spectrum of diseases including rheumatic disorders, lung diseases (asthma, COPD), gastrointestinal tract disorders (Crohn's disease, colitis ulcerosa), certain malignancies (leukemia, lymphomas), hematological disorders, and various diseases of the kidneys, brain, eyes and skin. A practical guideline for diagnosing allergic reactions to corticosteroids is presented in ref. 2. Betamethasone *base* is used as tablet only. For injectable suspensions, betamethasone acetate (Chapter 3.66), dipropionate (Chapter 3.67), sodium phosphate (Chapter 3.68) or sodium succinate (Chapter 3.69) may be used. Contact allergy to 'betamethasone' has been fully reviewed in Volume 3 of the *Monographs in contact allergy* series (4).

CUTANEOUS ADVERSE DRUG REACTIONS FROM SYSTEMIC ADMINISTRATION CAUSED BY TYPE IV (DELAYED-TYPE) HYPERSENSITIVITY (as demonstrated by positive patch tests)
Cutaneous adverse drug reactions from systemic administration of betamethasone caused by type IV (delayed-type) hypersensitivity have included maculopapular eruption (3,10), systemic allergic dermatitis (presenting as papular eczema [5], the baboon syndrome [6,9], or as facial erythema [8]), and unspecified exanthema (7).

Maculopapular eruption
A 29-year-old man was treated with oral spiramycin and betamethasone for bacterial infection of the middle ear and developed a maculopapular rash one day after the beginning of the therapy. The exanthema was first localized to the scalp, and after 2 hours with the involvement of the legs and of the abdomen. Patch tests were positive to betamethasone, dexamethasone and fluocortolone 1% alcohol; prick and intradermal tests (no late readings) were negative (3).

A woman aged 50 was treated with oral dexamethasone and betamethasone and developed a maculopapular exanthema from these drugs. Patch tests were positive to both corticosteroids. Details (e.g. whether the patient had previous sensitization to corticosteroids) are not available to the author (10).

Systemic allergic dermatitis (systemic contact dermatitis)
A 49-year-old woman who had probably become sensitized to triamcinolone acetonide in ear drops developed generalized papular eczema on several occasions from taking betamethasone tablets. Betamethasone dipropionate cream worsened the eruption. When patch tested, she had positive patch tests to triamcinolone acetonide,

betamethasone, betamethasone dipropionate and multiple other corticosteroids. This was a case of systemic allergic dermatitis from oral betamethasone (5).

A 58 year-old man presented with eczema involving the groins, the buttocks and the axillae, clinically a baboon syndrome. The eruption had developed 48 hours after ingestion of betamethasone tablets. Patch tests showed positive reactions to betamethasone and many other corticosteroids. The patient had previously become sensitized to corticosteroids from topical therapy. This was a case of systemic allergic dermatitis from oral betamethasone presenting as the baboon syndrome (6).

A 50-year-old woman who was very likely sensitized to corticosteroids from their use in chronic nasal congestion had suffered a skin eruption on three occasions one day after oral intake of prednisolone or methylprednisolone. Patch tests were positive to prednisolone (methylprednisolone not tested). An oral provocation test with betamethasone resulted in the classic picture of the baboon syndrome/symmetrical drug-related intertriginous and flexural exanthema (SDRIFE) after 12-24 hours. Patch tests with betamethasone were not performed, but an intradermal test was positive at D2. This was a case of systemic allergic dermatitis presenting as the baboon syndrome/SDRIFE (9).

A 40-year-old woman with bursitis and tennis elbow, who was probably sensitized to hydrocortisone from hemorrhoid preparations, developed facial erythema after provocation tests oral betamethasone, which emerged within 18 hours and lasted for 2-3 days. Patch and intradermal tests showed delayed-type allergy to corticosteroids of the hydrocortisone group (hydrocortisone, tixocortol, methylprednisolone, but *not* prednisolone). As this patient was probably presensitized, these were cases of systemic allergic dermatitis with cross-reactions from hydrocortisone to betamethasone (8).

Symmetrical drug-related intertriginous and flexural exanthema (SDRIFE)/Baboon syndrome
See the section 'Systemic allergic dermatitis' above.

Other cutaneous adverse drug reactions
Two patients, who had either suffered a drug exanthema or delayed urticaria (>2 hours after administration) (not specified) from systemic betamethasone or dexamethasone, had positive delayed intradermal tests and positive patch tests to both betamethasone and dexamethasone and a positive systemic (parenteral) provocation test. The causative corticosteroids (betamethasone or dexamethasone) were not specified (7).

Cross-reactions, pseudo-cross-reactions and co-reactions
Cross-reactions between corticosteroids are discussed in Chapter 3.399 Prednisolone.

LITERATURE
1   Tavadia S, Bianchi J, Dawe RS, McEvoy M, Wiggins E, Hamill E, et al. Allergic contact dermatitis in venous leg ulcer patients. Contact Dermatitis 2003;48:261-265
2   Baeck M, Goossens A. Immediate and delayed allergic hypersensitivity to corticosteroids: practical guidelines. Contact Dermatitis 2012;66:38-45
3   Nucera E, Buonomo A, Pollastrini E, De Pasquale T, Del Ninno M, Roncallo C, et al. A case of cutaneous delayed-type allergy to oral dexamethasone and to betamethasone. Dermatology 2002;204:248-250
4   De Groot AC. Monographs in contact allergy, volume 3: Topical Drugs. Boca Raton, Fl, USA: CRC Press Taylor and Francis Group, 2021 (ISBN 978-0-367-23693-9)
5   Isaksson M. Systemic contact allergy to corticosteroids revisited. Contact Dermatitis 2007;57:386-388
6   Armingaud P, Martin L, Wierzbicka E, Esteve E. Baboon syndrome due to a polysensitization with corticosteroids. Ann Dermatol Venereol 2005;132:675-677 (Article in French)
7   Padial A, Posadas S, Alvarez J, Torres M-J, Alvarez JA, Mayorga C, Blanca M. Nonimmediate reactions to systemic corticosteroids suggest an immunological mechanism. Allergy 2005;60:665-670
8   Räsänen L, Hasan T. Allergy to systemic and intralesional corticosteroids. Br J Dermatol 1993;128:407-411
9   Treudler R, Simon JC. Symmetric, drug-related, intertriginous, and flexural exanthema in a patient with polyvalent intolerance to corticosteroids. J Allergy Clin Immunol 2006;118:965-967
10 Maucher O, Faber M, Knipper H, Kirchner S, Schöpf E. Kortikoidallergie. Hautarzt 1987;38:577-582 (Article in German, data cited in ref. 11)
11 Bircher AJ, Levy F, Langauer S, Lepoittevin JP. Contact allergy to topical corticosteroids and systemic contact dermatitis from prednisolone with tolerance of triamcinolone. Acta Derm Venereol 1995;75:490-493

Chapter 3.66   BETAMETHASONE ACETATE

IDENTIFICATION

Description/definition   : Betamethasone acetate is the acetate ester of the synthetic glucocorticoid
betamethasone, that conforms to the structural formula shown below
Pharmacological classes   : Glucocorticoids
IUPAC name   : [2-[(8S,9R,10S,11S,13S,14S,16S,17R)-9-Fluoro-11,17-dihydroxy-10,13,16-trimethyl-3-oxo-
6,7,8,11,12,14,15,16-octahydrocyclopenta[a]phenanthren-17-yl]-2-oxoethyl] acetate
CAS registry number   : 987-24-6
EC number   : 213-578-6
Merck Index monograph   : 2452 (Betamethasone)
Patch testing   : Generally, corticosteroids may be tested at 0.1% and 1% in alcohol; late readings (6-10
days) are strongly recommended
Molecular formula   : $C_{24}H_{31}FO_6$

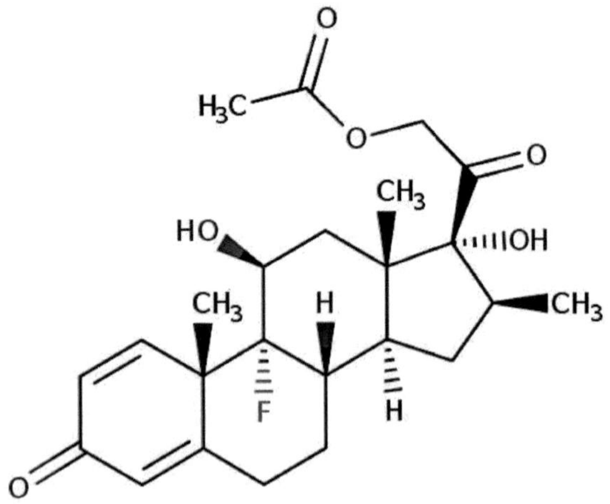

GENERAL

Systemically administered glucocorticoids have anti-inflammatory, immunosuppressive and antineoplastic properties and are used in the treatment of a wide spectrum of diseases including rheumatic disorders, lung diseases (asthma, COPD), gastrointestinal tract disorders (Crohn's disease, colitis ulcerosa), certain malignancies (leukemia, lymphomas), hematological disorders, and various diseases of the kidneys, brain, eyes and skin. A practical guideline for diagnosing allergic reactions to corticosteroids is presented in ref. 1. See also Betamethasone (Chapter 3.65), betamethasone dipropionate (Chapter 3.67), betamethasone sodium phosphate (Chapter 3.68) and betamethasone sodium succinate (Chapter 3.69).

CUTANEOUS ADVERSE DRUG REACTIONS FROM SYSTEMIC ADMINISTRATION CAUSED BY TYPE IV (DELAYED-TYPE) HYPERSENSITIVITY (as demonstrated by positive patch tests)

Cutaneous adverse drug reactions from systemic administration of betamethasone acetate caused by type IV (delayed-type) hypersensitivity have included erythema of the face and neck (2) and systemic allergic dermatitis presenting as facial erythema (2).

Erythroderma, widespread erythematous eruption, exfoliative dermatitis

In a 47-year-old woman with rheumatoid arthritis, intra-articular methylprednisolone acetate/methylprednisolone sodium succinate caused widespread erythema on the following day, which lasted about 3 days. A provocation with intra-articular 6 mg betamethasone acetate/disodium succinate caused erythema of the face and neck, which lasted about 1 day. Patch and intradermal tests showed delayed-type allergy to corticosteroids of the hydrocortisone group (hydrocortisone, tixocortol, prednisolone, methylprednisolone). She had not been previously been treated with topical corticosteroids and was probably not presensitized to corticosteroids, so this patient cross-reacted from methylprednisolone to betamethasone (2).

Systemic allergic dermatitis (systemic contact dermatitis)

A 40-year-old woman with bursitis and tennis elbow, who was probably sensitized to hydrocortisone from hemorrhoid preparations, developed facial erythema after provocation tests with intralesional betamethasone acetate/sodium phosphate, oral betamethasone and oral hydrocortisone, which emerged within 18 hours and lasted for 2-3 days. Patch and intradermal tests showed delayed-type allergy to corticosteroids of the hydrocortisone group (hydrocortisone, tixocortol, methylprednisolone, but *not* prednisolone). As this patient was probably presensitized, this was a case of systemic allergic dermatitis with cross-reactions from hydrocortisone to betamethasone (2).

Cross-reactions, pseudo-cross-reactions and co-reactions
Cross-reactions between corticosteroids are discussed in Chapter 3.399 Prednisolone.

LITERATURE

1   Baeck M, Goossens A. Immediate and delayed allergic hypersensitivity to corticosteroids: practical guidelines. Contact Dermatitis 2012;66:38-45
2   Räsänen L, Hasan T. Allergy to systemic and intralesional corticosteroids. Br J Dermatol 1993;128:407-411

Chapter 3.67   BETAMETHASONE DIPROPIONATE

IDENTIFICATION

Description/definition   : Betamethasone dipropionate is the 17,21-dipropionate ester of the synthetic glucocorticoid betamethasone that conforms to the structural formula shown below
Pharmacological classes : Anti-inflammatory agents
IUPAC name   : [2-[(8S,9R,10S,11S,13S,14S,16S,17R)-9-Fluoro-11-hydroxy-10,13,16-trimethyl-3-oxo-17-propanoyloxy-6,7,8,11,12,14,15,16-octahydrocyclopenta[a]phenanthren-17-yl]-2-oxoethyl] propanoate
Other names   : 9-Fluoro-11β,17,21-trihydroxy-16β-methylpregna-1,4-diene-3,20-dione 17,21-di(propionate; betamethasone-17,21-dipropionate
CAS registry number   : 5593-20-4
EC number   : 227-005-2
Merck Index monograph   : 2452 (Betamethasone)
Patch testing   : 1.0% pet. (Chemotechnique); 0.5% pet., 0.1% alc. (SmartPracticeCanada)
Molecular formula   : $C_{28}H_{37}FO_7$

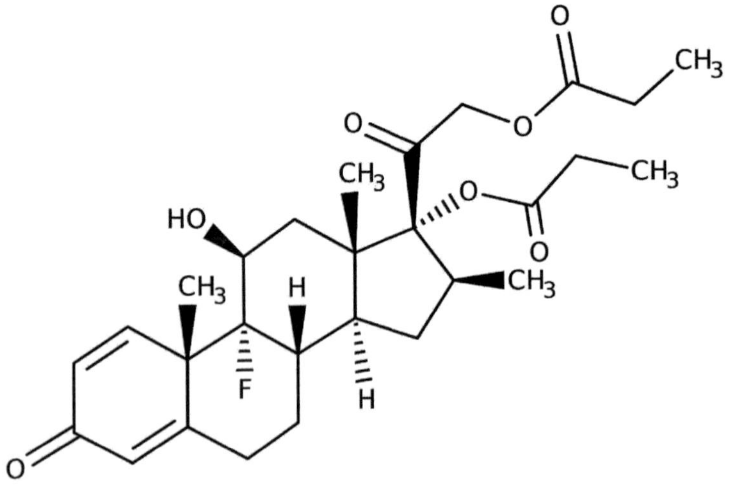

GENERAL

Systemically administered glucocorticoids have anti-inflammatory, immunosuppressive and antineoplastic properties and are used in the treatment of a wide spectrum of diseases including rheumatic disorders, lung diseases (asthma, COPD), gastrointestinal tract disorders (Crohn's disease, colitis ulcerosa), certain malignancies (leukemia, lymphomas), hematological disorders, and various diseases of the kidneys, brain, eyes and skin. A practical guideline for diagnosing allergic reactions to corticosteroids is presented in ref. 1. Contact allergy to betamethasone dipropionate has been fully reviewed in Volume 3 of the *Monographs in contact allergy* series (3).

See also betamethasone (Chapter 3.65), betamethasone acetate (Chapter 3.66), betamethasone sodium phosphate (Chapter 3.68) and betamethasone sodium succinate (Chapter 3.69).

CUTANEOUS ADVERSE DRUG REACTIONS FROM SYSTEMIC ADMINISTRATION CAUSED BY TYPE IV (DELAYED-TYPE) HYPERSENSITIVITY (as demonstrated by positive patch tests)

Cutaneous adverse drug reactions from systemic administration of betamethasone dipropionate caused by type IV (delayed-type) hypersensitivity have included systemic allergic dermatitis (2).

Systemic allergic dermatitis (systemic contact dermatitis)

In Leuven, Belgium, in a 12-year-period before 2012, 16 patients were investigated for a generalized allergic eruption (maculopapular eruption or eczema, with or without flare-up of previous dermatitis) from systemic administration (oral, intravenous, intramuscular, intra-articular) of corticosteroids, a few hours or days after the first dose of the culprit drug. The reactions observed were in most cases a manifestation of systemic allergic dermatitis: the patient had previously become sensitized to the corticosteroid used systemically or a cross-reacting molecule from topical exposure. Betamethasone dipropionate (in combination with betamethasone sodium phosphate) caused allergic reactions in 3 patients, in 2 from intramuscular and in one from intra-articular administration. The intramuscular injections caused a maculopapular rash plus flare-up and an unknown exanthema. The patient in whom betamethasone dipropionate/sodium phosphate was administered intra-articularly developed generalized eczema plus flare-up of previous eczema (2).

Cross-reactions, pseudo-cross-reactions and co-reactions
Cross-reactions between corticosteroids are discussed in Chapter 3.399 Prednisolone.

LITERATURE
1   Baeck M, Goossens A. Immediate and delayed allergic hypersensitivity to corticosteroids: practical guidelines. Contact Dermatitis 2012;66:38-45
2   Baeck M, Goossens A. Systemic contact dermatitis to corticosteroids. Allergy 2012;67:1580-1585
3   De Groot AC. Monographs in contact allergy, volume 3: Topical Drugs. Boca Raton, Fl, USA: CRC Press Taylor and Francis Group, 2021 (ISBN 978-0-367-23693-9)

Chapter 3.68　　BETAMETHASONE SODIUM PHOSPHATE

IDENTIFICATION

Description/definition	: Betamethasone sodium phosphate is the disodium salt of the 21-phosphate ester of the synthetic glucocorticoid betamethasone, that conforms to the structural formula shown below
Pharmacological classes	: Glucocorticoids
IUPAC name	: Disodium;[2-[(8S,9R,10S,11S,13S,14S,16S,17R)-9-fluoro-11,17-dihydroxy-10,13,16-trimethyl-3-oxo-6,7,8,11,12,14,15,16-octahydrocyclopenta[a]phenanthren-17-yl]-2-oxoethyl] phosphate
Other names	: 9-Fluoro-11β,17,21-trihydroxy-16β-methylpregna-1,4-diene-3,20-dione 21-(disodium phosphate)
CAS registry number	: 151-73-5
EC number	: 205-797-0
Merck Index monograph	: 2452 (Betamethasone)
Patch testing	: In general, corticosteroids may be tested at 0.1% and 1% in alcohol; late readings (6-10 days) are strongly recommended
Molecular formula	: $C_{22}H_{28}FNa_2O_8P$

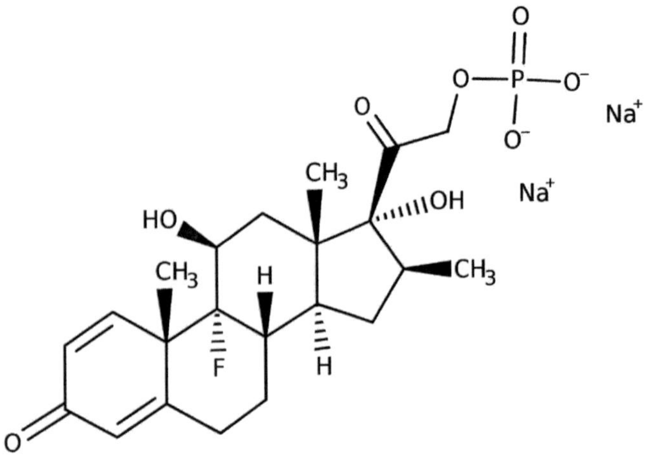

GENERAL

Systemically administered glucocorticoids have anti-inflammatory, immunosuppressive and antineoplastic properties and are used in the treatment of a wide spectrum of diseases including rheumatic disorders, lung diseases (asthma, COPD), gastrointestinal tract disorders (Crohn's disease, colitis ulcerosa), certain malignancies (leukemia, lymphomas), hematological disorders, and various diseases of the kidneys, brain, eyes and skin. A practical guideline for diagnosing allergic reactions to corticosteroids is presented in ref. 1. Betamethasone sodium phosphate is mostly used in injection fluids. See also betamethasone (Chapter 3.65), betamethasone acetate (Chapter 3.66), betamethasone dipropionate (Chapter 3.67) and betamethasone sodium succinate (Chapter 3.69). Contact allergy to betamethasone sodium phosphate has been fully reviewed in Volume 3 of the *Monographs in contact allergy* series (3).

CUTANEOUS ADVERSE DRUG REACTIONS FROM SYSTEMIC ADMINISTRATION CAUSED BY TYPE IV (DELAYED-TYPE) HYPERSENSITIVITY (as demonstrated by positive patch tests)

Cutaneous adverse drug reactions from systemic administration of betamethasone sodium phosphate caused by type IV (delayed-type) hypersensitivity have included acute generalized exanthematous pustulosis (AGEP) (2), systemic allergic dermatitis (SAD) presenting as facial erythema (4), SAD presenting as maculopapular exanthema (5), and SAD presenting as generalized eczema plus flare-up (5).

Acute generalized exanthematous pustulosis (AGEP)

An 80-year-old woman presented with a generalized papular and pustular eruption on an erythematous base. The pustules were non-follicular and coalescing. The rash was associated with fever (38.5°C) and malaise. The patient's history revealed an itching eczematous periocular dermatitis lasting 2 weeks, following the use of eye drops containing dexamethasone sodium phosphate for blepharitis. One day before consultation, because of worsening of the eyelid dermatitis, the patient was treated in the emergency room with an intramuscular injection of betamethasone sodium phosphate. The patient was on no other medication. Laboratory investigations showed leukocytosis, neutrophilia, an increase of C-reactive protein, elevated erythrocyte sedimentation rate and elevated creatinine. Microbiological pustule cultures were negative. Based on these data and histopathology, AGEP was diagnosed. Later, patch tests were positive to the eye drops and its active principal dexamethasone sodium phosphate. Additional patch testing with systemic corticosteroids (betamethasone) was not possible because of the patient's unexpected death by acute myocardial infarction (2). This was a case of AGEP most likely caused by betamethasone sodium phosphate cross-reacting to dexamethasone sodium phosphate.

Systemic allergic dermatitis (systemic contact dermatitis)
In Leuven, Belgium, in a 12-year-period before 2012, 16 patients were investigated for a generalized allergic eruption (maculopapular eruption or eczema, with or without flare-up of previous dermatitis) from systemic administration (oral, intravenous, intramuscular, intra-articular) of corticosteroids, a few hours or days after the first dose of the culprit drug. The reactions observed were in most cases a manifestation of systemic allergic dermatitis: the patient had previously become sensitized to the corticosteroid used systemically or a cross-reacting molecule from topical exposure. Betamethasone sodium phosphate (in combination with betamethasone dipropionate) caused allergic reactions in 3 patients, in 2 from intramuscular and in one from intra-articular administration. The intramuscular injections caused a maculopapular rash plus flare-up and an unknown exanthema. The patient in whom betamethasone sodium phosphate/dipropionate was administered intra-articularly developed generalized eczema plus flare-up of previous eczema (5).

A 40-year-old woman with bursitis and tennis elbow, who was probably sensitized to hydrocortisone from hemorrhoid preparations, developed facial erythema after provocation tests with intralesional betamethasone sodium phosphate, which emerged within 18 hours and lasted for 2-3 days. Patch and intradermal tests showed delayed-type allergy to corticosteroids of the hydrocortisone group (hydrocortisone, tixocortol, methylprednisolone, but *not* prednisolone). As this patient was probably presensitized, this was a case of systemic allergic dermatitis with cross-reactions from hydrocortisone to betamethasone (4).

Cross-reactions, pseudo-cross-reactions and co-reactions
Cross-reactions between corticosteroids are discussed in Chapter 3.399 Prednisolone.

LITERATURE
1 Baeck M, Goossens A. Immediate and delayed allergic hypersensitivity to corticosteroids: practical guidelines. Contact Dermatitis 2012;66:38-45
2 Gambini D, Sena P, Raponi F, Bianchi L, Hansel K, Tramontana M, et al. Systemic allergic dermatitis presenting as acute generalized exanthematous pustulosis due to betamethasone sodium phosphate. Contact Dermatitis 2020;82:250-252
3 De Groot AC. Monographs in contact allergy, volume 3: Topical Drugs. Boca Raton, Fl, USA: CRC Press Taylor and Francis Group, 2021 (ISBN 978-0-367-23693-9)
4 Räsänen L, Hasan T. Allergy to systemic and intralesional corticosteroids. Br J Dermatol 1993;128:407-411
5 Baeck M, Goossens A. Systemic contact dermatitis to corticosteroids. Allergy 2012;67:1580-1585

Chapter 3.69   BETAMETHASONE SODIUM SUCCINATE

IDENTIFICATION

Description/definition   : Betamethasone sodium succinate is the sodium salt of the succinate ester of the synthetic
                              glucocorticoid that conforms to the structural formula shown below
Pharmacological classes   : Glucocorticoids
IUPAC name   : Sodium;4-[2-[[(8S,9R,10S,11S,13S,14S,16S,17R)-9-fluoro-11,17-dihydroxy-10,13,16-
                              trimethyl-3-oxo-6,7,8,11,12,14,15,16-octahydrocyclopenta[a]phenanthren-17-yl]-2-
                              oxoethoxy]-4-oxobutanoate
CAS registry number   : Not available
EC number   : Not available
Patch testing   : Generally, corticosteroids may be tested at 0.1% and 1% in alcohol; late readings (6-10
                              days) are strongly recommended
Molecular formula   : $C_{26}H_{32}FNaO_8$

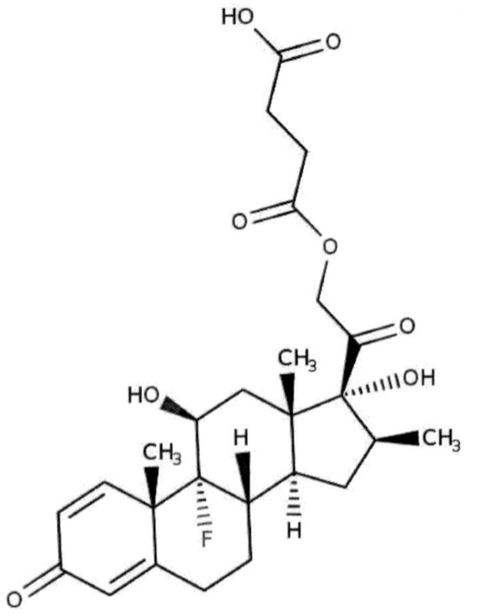

Betamethasone succinate

GENERAL

Systemically administered glucocorticoids have anti-inflammatory, immunosuppressive and antineoplastic properties and are used in the treatment of a wide spectrum of diseases including rheumatic disorders, lung diseases (asthma, COPD), gastrointestinal tract disorders (Crohn's disease, colitis ulcerosa), certain malignancies (leukemia, lymphomas), hematological disorders, and various diseases of the kidneys, brain, eyes and skin. A practical guideline for diagnosing allergic reactions to corticosteroids is presented in ref. 1. See also betamethasone (Chapter 3.65), betamethasone acetate (Chapter 3.66), betamethasone dipropionate (Chapter 3.67) and betamethasone sodium phosphate (Chapter 3.68).

CUTANEOUS ADVERSE DRUG REACTIONS FROM SYSTEMIC ADMINISTRATION CAUSED BY TYPE IV (DELAYED-TYPE) HYPERSENSITIVITY (as demonstrated by positive patch tests)

Cutaneous adverse drug reactions from systemic administration of betamethasone sodium succinate caused by type IV (delayed-type) hypersensitivity have included erythema of the face and neck (2).

Other cutaneous adverse drug reactions

In a 47-year-old woman with rheumatoid arthritis, intra-articular methylprednisolone acetate/methylprednisolone sodium succinate caused widespread erythema on the following day, which lasted about 3 days. A provocation with intra-articular 6 mg betamethasone acetate/sodium succinate caused erythema of the face and neck, which lasted about 1 day. Patch and intradermal tests showed delayed-type allergy to corticosteroids of the hydrocortisone group (hydrocortisone, tixocortol, prednisolone, methylprednisolone). The patient had not been previously treated with topical corticosteroids and was probably not presensitized to corticosteroids, so this patient cross-reacted from methylprednisolone to betamethasone (2).

Cross-reactions, pseudo-cross-reactions and co-reactions
Cross-reactions between corticosteroids are discussed in Chapter 3.399 Prednisolone.

LITERATURE
1   Baeck M, Goossens A. Immediate and delayed allergic hypersensitivity to corticosteroids: practical guidelines. Contact Dermatitis 2012;66:38-45
2   Räsänen L, Hasan T. Allergy to systemic and intralesional corticosteroids. Br J Dermatol 1993;128:407-411

Chapter 3.70   BIPERIDEN

IDENTIFICATION
Description/definition   : Biperiden is the piperidine that conforms to the structural formula shown below
Pharmacological classes   : Antiparkinson agents; parasympatholytics; muscarinic antagonists
IUPAC name   : 1-(2-Bicyclo[2.2.1]hept-5-enyl)-1-phenyl-3-piperidin-1-ylpropan-1-ol
CAS registry number   : 514-65-8
EC number   : 208-184-6
Merck Index monograph   : 2508
Patch testing   : 1% pet. (2)
Molecular formula   : $C_{21}H_{29}NO$

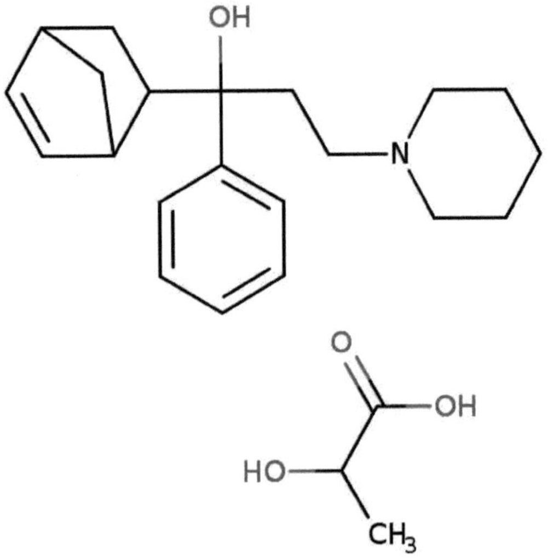

Biperiden lactate

GENERAL
Biperiden is a muscarinic antagonist that has effects in both the central and peripheral nervous systems. It has been used in the treatment of arteriosclerotic, idiopathic, and postencephalitic parkinsonism. It has also been used to alleviate extrapyramidal symptoms induced by phenothiazine derivatives and reserpine (1). In pharmaceutical products, biperiden is employed as biperiden hydrochloride (CAS number 1235-82-1, EC number 214-976-2, molecular formula $C_{21}H_{30}ClNO$) or as biperiden lactate (CAS number 7085-45-2, EC number not available, molecular formula $C_{24}H_{35}NO_4$) (1).

CONTACT ALLERGY FROM ACCIDENTAL CONTACT
A 47-year-old female pharmacist, making up prescriptions for a few hours per day at a psychiatric hospital, presented with papular erythematous lesions over the light-exposed areas affecting the face, neck, ears, forearms and the dorsa of the hands. She was patch and photopatch tested with all the drugs that she had contact with and showed a positive patch test to biperiden 1% pet. and positive photopatch tests to perphenazine 0.01% pet. and the related phenothiazine chlorpromazine 0.01% pet. Four controls were negative. This was a case of occupational allergic contact dermatitis to biperiden and of occupational photoallergic contact dermatitis to perphenazine and chlorpromazine (2).

LITERATURE
1   The data in the section 'General' may have been obtained from literature discussed in this chapter, but mostly also or exclusively from one or more of the following online sources: ChemIDPlus Advanced, PubChem, DrugBank, RxList, Drug Central, Drugs.com, and Wikipedia
2   Torinuki J. Contact dermatitis to biperiden and photocontact dermatitis in phenothiazines in a pharmacist. Tohoku J Exp Med 1995;176:249-252 (Article in English)

Chapter 3.71   BISMUTH SUBCITRATE

IDENTIFICATION

Description/definition	: Bismuth subcitrate is the tricarboxylic acid bismuth salt that conforms to the structural formula shown below
Pharmacological classes	: Antacids; anti-infective agents; anti-ulcer agents
IUPAC name	: Bismuth;tripotassium;2-hydroxypropane-1,2,3-tricarboxylate
Other names	: Bismuth subcitrate potassium; bismuth tripotassium dicitrate; bismuth potassium citrate
CAS registry number	: 57644-54-9
EC number	: 260-872-5
Merck Index monograph	: 3739
Patch testing	: Bismuth oxide 240 mg/ml saline patch-scratch test (2); bismuth oxide 5% pet. (3)
Molecular formula	: $C_{12}H_{10}BiK_3O_{14}$

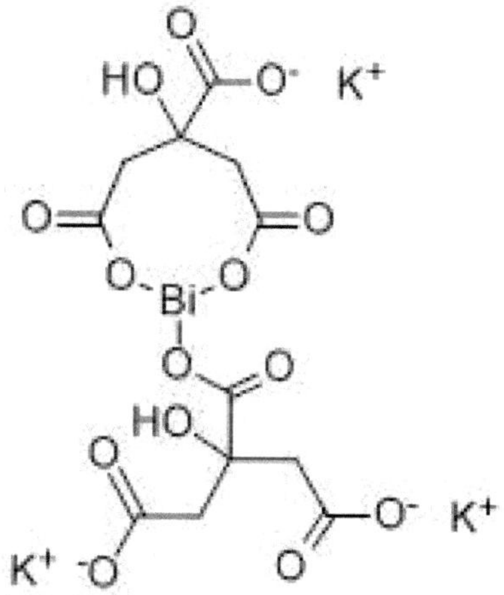

GENERAL

Bismuth subcitrate potassium is a colloid used for the treatment of peptic ulcer and gastro-esophageal reflux disease (1).

CUTANEOUS ADVERSE DRUG REACTIONS FROM SYSTEMIC ADMINISTRATION CAUSED BY TYPE IV (DELAYED-TYPE) HYPERSENSITIVITY (as demonstrated by positive patch tests)

Cutaneous adverse drug reactions from systemic administration of bismuth subcitrate caused by type IV (delayed-type) hypersensitivity have included maculopapular exanthema (2).

Maculopapular eruption

In Bern, Switzerland, patients with a suspected allergic cutaneous drug reaction were patch-scratch tested with suspected drugs that had previously given a positive lymphocyte transformation test. Bismuth oxide 240 mg/ml saline gave a positive patch-scratch test in one patient with maculopapular exanthema (2). It should be realized that the pharmaceutical with the trade name given by the authors (De-Nol ®) contains dried colloidal solution of bismuth subcitrate, matching with 120 mg bismuth (III) oxide (2). It is unknown whether the authors have used the commercial preparation for patch testing or pure bismuth oxide and, if yes, in which form.

LITERATURE

1   The data in the section 'General' may have been obtained from literature discussed in this chapter, but mostly also or exclusively from one or more of the following online sources: ChemIDPlus Advanced, PubChem, DrugBank, RxList, Drug Central, Drugs.com, and Wikipedia

2   Neukomm C, Yawalkar N, Helbling A, Pichler WJ. T-cell reactions to drugs in distinct clinical manifestations of drug allergy. J Invest Allergol Clin Immunol 2001;11:275-284

3   De Groot AC. Patch testing, 4th edition. Wapserveen, The Netherlands: acdegroot publishing, 2018 (ISBN 9789081323345)

Chapter 3.72   BISOPROLOL

IDENTIFICATION

Description/definition   : Bisoprolol is the synthetic phenoxy-2-propanol-derived compound that conforms to the structural formula shown below (shown as bisoprolol fumarate)
Pharmacological classes   : Sympatholytics; antihypertensive agent; β_1-Adrenergic receptor antagonists
IUPAC name   : 1-(Propan-2-ylamino)-3-[4-(2-propan-2-yloxyethoxymethyl)phenoxy]propan-2-ol
CAS registry number   : 66722-44-9
EC number   : Not available
Merck Index monograph   : 2565
Patch testing   : Tablet, pulverized, 10% pet. (2); most systemic drugs can be tested at 10% pet.; if the pure chemical is not available, prepare the test material from intravenous powder, the content of capsules or – when also not available – from powdered tablets to achieve a final concentration of the active drug of 10% pet.
Molecular formula   : $C_{18}H_{31}NO_4$

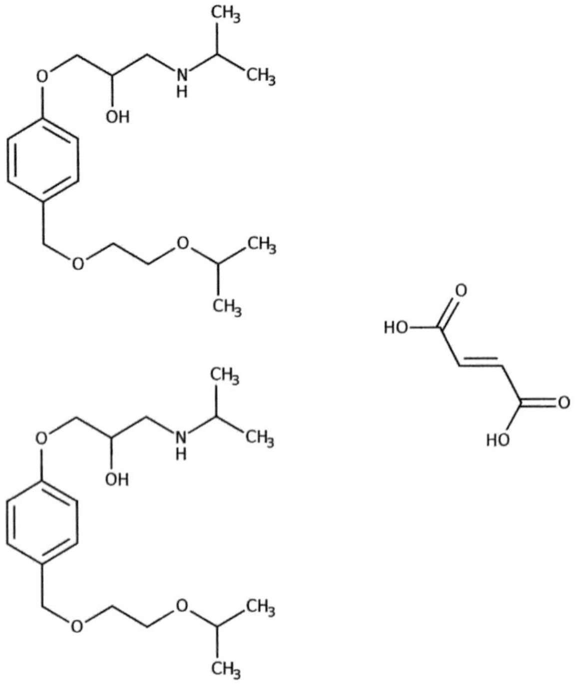

GENERAL

Bisoprolol is a cardioselective β_1-adrenergic blocker. It is indicated for management of heart failure, angina pectoris, for mild to moderate hypertension and for secondary prevention of myocardial infarction. In pharmaceutical products, bisoprolol is employed as bisoprolol fumarate (CAS number 104344-23-2, EC number 600-557-5, molecular formula $C_{40}H_{66}N_2O_{12}$); see the structural formula above (1).

CONTACT ALLERGY FROM ACCIDENTAL CONTACT

A 24-year-old woman presented with eyelid dermatitis, which had started with localized edema 4 months previously. Later, the area had become itchier, with redness and scaling. The patient suspected a relationship with her work as a pharmacy assistant, which involved breaking and crushing different types of tablets. She was patch tested with the crushed tablets that she had contact with at 10% pet. and showed positive reactions to bisoprolol, 3 other beta-blockers, 3 benzodiazepines and 3 ACE-inhibitors. Six controls were negative to bisoprolol 10% pet. Cosmetic allergy was excluded and a diagnosis of occupational airborne allergic contact was made (2).

LITERATURE

1   The data in the section 'General' may have been obtained from literature discussed in this chapter, but mostly also or exclusively from one or more of the following online sources: ChemIDPlus Advanced, PubChem, DrugBank, RxList, Drug Central, Drugs.com, and Wikipedia
2   Swinnen I, Ghys K, Kerre S, Constandt L, Goossens A. Occupational airborne contact dermatitis from benzodiazepines and other drugs. Contact Dermatitis 2014;70:227-232

Chapter 3.73   BLEOMYCIN

IDENTIFICATION

Description/definition
: Bleomycin is a mixture of glycopeptide antineoplastic antibiotics isolated from the bacterium *Streptomyces verticillus*

Pharmacological classes
: Antibiotics, antineoplastic

IUPAC name
: 3-[[2-[2-[2-[[[(2S,3R)-2-[[(2S,3S,4R)-4-[[(2S,3R)-2-[[6-Amino-2-[(1S)-3-amino-1-[[(2S)-2,3-diamino-3-oxopropyl]amino]-3-oxopropyl]-5-methylpyrimidine-4-carbonyl]amino]-3-[3-[4-carbamoyloxy-3,5-dihydroxy-6-(hydroxymethyl)oxan-2-yl]oxy-4,5-dihydroxy-6-(hydroxymethyl)oxan-2-yl]oxy-3-(1H-imidazol-5-yl)propanoyl]amino]-3-hydroxy-2-methylpentanoyl]amino]-3-hydroxybutanoyl]amino]ethyl]-1,3-thiazol-4-yl]-1,3-thiazole-4-carbonyl]amino]propyl-dimethylsulfanium

CAS registry number
: 11056-06-7

EC number
: Not available

Merck Index monograph
: 2589 (Bleomycins)

Patch testing
: Tablet, pulverized, 30% pet. (2); most systemic drugs can be tested at 10% pet.; if the pure chemical is not available, prepare the test material from intravenous powder, the content of capsules or – when also not available – from powdered tablets to achieve a final concentration of the active drug of 10% pet.

Molecular formula
: $C_{55}H_{84}N_{17}O_{21}S_3^+$

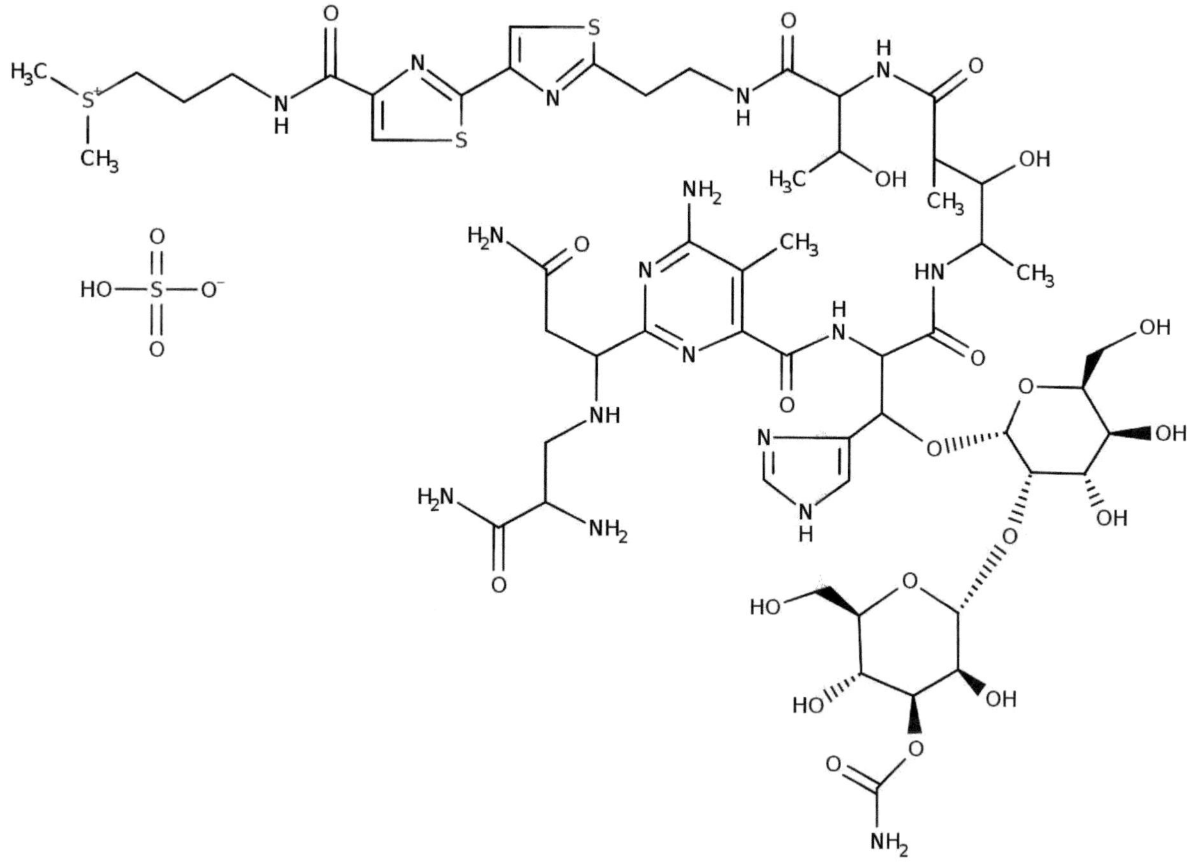

Bleomycin sulfate

GENERAL

Bleomycin is a complex of related glycopeptide antibiotics from *Streptomyces verticillus* consisting of bleomycin A2 and B2. It inhibits DNA metabolism and is used as an antineoplastic, especially for solid tumors. In pharmaceutical products, bleomycin is employed as bleomycin sulfate (CAS number 9041-93-4, EC number 232-925-2, molecular formula $C_{55}H_{85}N_{17}O_{25}S_4$). Bleomycin sulfate forms complexes with iron that reduce molecular oxygen to superoxide and hydroxyl radicals which cause single- and double-stranded breaks in DNA; these reactive oxygen species also induce lipid peroxidation, carbohydrate oxidation, and alterations in prostaglandin synthesis and degradation (1).

CUTANEOUS ADVERSE DRUG REACTIONS FROM SYSTEMIC ADMINISTRATION CAUSED BY TYPE IV
(DELAYED-TYPE) HYPERSENSITIVITY (as demonstrated by positive patch tests)
Cutaneous adverse drug reactions from systemic administration of bleomycin caused by type IV (delayed-type)
hypersensitivity have included acute generalized exanthematous pustulosis (AGEP) (2).

Acute generalized exanthematous pustulosis (AGEP)
A 33-year-old man had developed an erythematous rash which occurred 5 days after he had been given a course of
bleomycin, etoposide and cisplatin for the treatment of testicular germ cell tumor. Clinical examination revealed
erythematous plaques on the trunk and on the extensor surfaces of the arms, legs and some non-follicular pustules,
especially on the neck. One week later, after he had been given bleomycin as a single agent treatment, numerous
discrete non-follicular pustules appeared symmetrically on the trunk, neck, and extensor surfaces of the arms and
legs accompanied by facial edema and a fever of 38.2°C. Laboratory evaluation revealed leukocytosis with 72%
neutrophils. Histological examination of the skin biopsy specimen revealed formation of intraepidermal pustules,
papillary edema, and accompanying eosinophils in the dermal infiltrate, which was consistent with acute generalized
exanthematous pustulosis (AGEP). One month later, a patch test with bleomycin 30% pet. was positive at D3 with an
indurated papule (2).

LITERATURE
1   The data in the section 'General' may have been obtained from literature discussed in this chapter, but mostly
    also or exclusively from one or more of the following online sources: ChemIDPlus Advanced, PubChem,
    DrugBank, RxList, Drug Central, Drugs.com, and Wikipedia
2   Altaykan A, Boztepe G, Erkin G, Ozkaya O, Ozden E. Acute generalized exanthematous pustulosis induced by
    bleomycin and confirmed by patch testing. J Dermatolog Treat 2004;15:231-234

Chapter 3.74   BORTEZOMIB

IDENTIFICATION

Description/definition : Bortezomib is the pyrazine and boronic acid derivative that conforms to the structural formula shown below
Pharmacological classes : Antineoplastic agents
IUPAC name : [(1R)-3-Methyl-1-[[(2S)-3-phenyl-2-(pyrazine-2-carbonylamino)propanoyl]amino]-butyl]boronic acid
CAS registry number : 179324-69-7
EC number : Not available
Merck Index monograph : 2623
Patch testing : 1 mg/ml (vehicle?)
Molecular formula : $C_{19}H_{25}BN_4O_4$

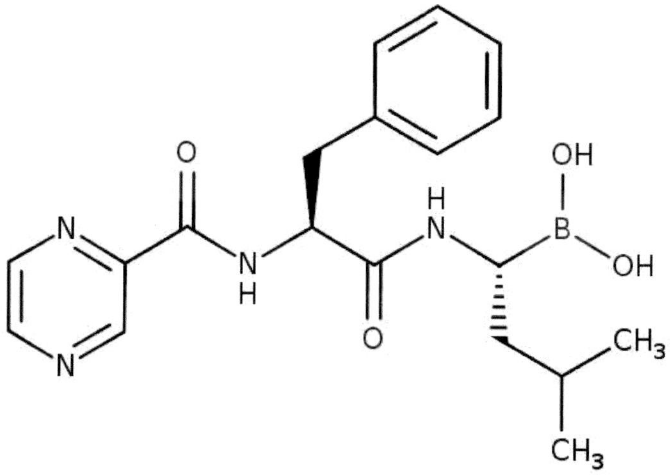

GENERAL

Bortezomib is a pyrazine and boronic acid derivative that functions as a reversible proteasome inhibitor. The drug reversibly inhibits the 26S proteasome, a large protease complex that degrades ubiquinated proteins. By blocking the targeted proteolysis normally performed by the proteasome, bortezomib disrupts various cell signaling pathways, leading to cell cycle arrest, apoptosis, and inhibition of angiogenesis. Specifically, the agent inhibits nuclear factor (NF)-kappaB, a protein that is constitutively activated in some cancers, thereby interfering with NF-kappaB-mediated cell survival, tumor growth, and angiogenesis. *In vivo*, bortezomib delays tumor growth and enhances the cytotoxic effects of radiation and chemotherapy. Bortezomib is used as an antineoplastic agent in the treatment of multiple myeloma and mantle cell lymphoma (1).

CUTANEOUS ADVERSE DRUG REACTIONS FROM SYSTEMIC ADMINISTRATION CAUSED BY TYPE IV (DELAYED-TYPE) HYPERSENSITIVITY (as demonstrated by positive patch tests)

Cutaneous adverse drug reactions from systemic administration of bortezomib caused by type IV (delayed-type) hypersensitivity have included Stevens-Johnson syndrome (2).

Stevens-Johnson syndrome (SJS)

A 71-year-old woman developed a skin eruption during the course of chemotherapy implemented for recently discovered multiple myeloma. Two weeks earlier, the patient had started chemotherapy with a bortezomib plus melphalan and prednisone regimen; several other drugs were also administered. The day after the 3rd bortezomib injection on day 8, a high fever was noted and various antibiotics were administered, during which a skin rash developed on the neck and trunk. The antibiotics were stopped, but 2 days later, the patient complained of severe eye pain with conjunctival hyperemia, oral pain, and a body-wide skin rash. In an ophthalmologic evaluation, severe corneal ulceration with conjunctival injection was observed. Erosive lesions in the oral mucosa and laryngeal ulceration were detected on an ear, nose, and throat examination. Skin blistering was observed on the anterior chest. A skin biopsy was not performed, but a clinical diagnosis of Stevens-Johnson syndrome was made. After stopping all drugs including bortezomib, the cutaneous rash and eye and oropharyngeal involvement gradually improved. Two weeks after recovery, patch tests were performed with all drugs taken by the patient, including bortezomib. Readings performed after 48 and 72 hours revealed a positive reaction to bortezomib 1 mg/ml. The patient was diagnosed with bortezomib-induced SJS (2).

It should be mentioned that the patch tests have been performed far too soon after recovery and that the clinical picture shown of the 'positive' reaction to bortezomib was far from convincing. This author feels that the reliability of the case report may be doubted.

LITERATURE

1   The data in the section 'General' may have been obtained from literature discussed in this chapter, but mostly also or exclusively from one or more of the following online sources: ChemIDPlus Advanced, PubChem, DrugBank, RxList, Drug Central, Drugs.com, and Wikipedia

2   Choi GS, Lee HS, Kim HK. A case of bortezomib (Velcade)-induced Stevens-Johnson syndrome confirmed by patch test. Asia Pac Allergy 2021;11(2):e17

Chapter 3.75   BRIVUDINE

IDENTIFICATION

Description/definition	: Brivudine is the pyrimidine 2'-deoxyribonucleoside that conforms to the structural formula shown below
Pharmacological classes	: Antiviral agents
IUPAC name	: 5-[(E)-2-Bromoethenyl]-1-[(2R,4S,5R)-4-hydroxy-5-(hydroxymethyl)oxolan-2-yl]pyrimidine-2,4-dione
Other names	: (E)-5-(2-Bromovinyl)-2'-deoxyuridine
CAS registry number	: 69304-47-8
EC number	: Not available
Merck Index monograph	: 2655
Patch testing	: 5% pet.
Molecular formula	: $C_{11}H_{13}BrN_2O_5$

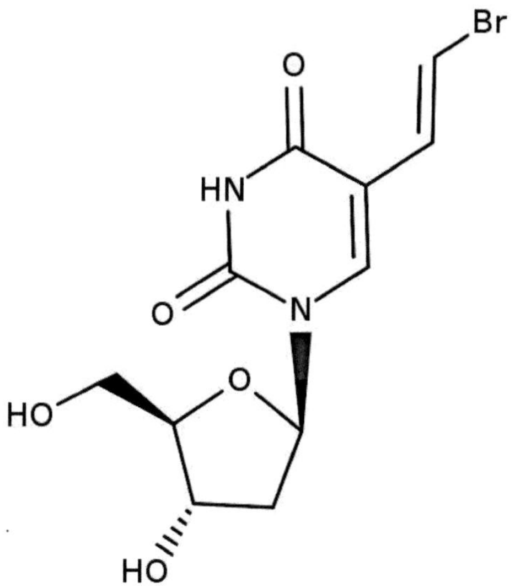

GENERAL

Brivudine is a uridine derivative and nucleoside analog with pro-apoptotic and chemosensitizing properties. This antiviral agent is used in the treatment of herpes zoster (1).

CUTANEOUS ADVERSE DRUG REACTIONS FROM SYSTEMIC ADMINISTRATION CAUSED BY TYPE IV (DELAYED-TYPE) HYPERSENSITIVITY (as demonstrated by positive patch tests)

Cutaneous adverse drug reactions from systemic administration of brivudine caused by type IV (delayed-type) hypersensitivity have included unspecified adverse drug reaction (2).

Other cutaneous adverse drug reactions

In Leuven, Belgium, in the period 1990-2014, iatrogenic contact dermatitis was diagnosed in 2600 individuals (17.4% of the total patch test population). 96% of all positive patch test reactions *to topical drugs and antiseptics* were considered to be relevant. Brivudine (5% pet., which is actually a *systemic* drug) was tested in one patient and there was a positive reaction to it. No clinical details were provided (2).

LITERATURE

1   The data in the section 'General' may have been obtained from literature discussed in this chapter, but mostly also or exclusively from one or more of the following online sources: ChemIDPlus Advanced, PubChem, DrugBank, RxList, Drug Central, Drugs.com, and Wikipedia
2   Gilissen L, Goossens A. Frequency and trends of contact allergy to and iatrogenic contact dermatitis caused by topical drugs over a 25-year period. Contact Dermatitis 2016;75:290-302

Chapter 3.76   BROMAZEPAM

IDENTIFICATION

Description/definition   : Bromazepam is the lipophilic, long-acting benzodiazepine that conforms to the structural formula shown below
Pharmacological classes   : Anti-anxiety agents; GABA modulators
IUPAC name   : 7-Bromo-5-pyridin-2-yl-1,3-dihydro-1,4-benzodiazepin-2-one
Other names   : 2H-1,4-Benzodiazepin-2-one, 7-bromo-1,3-dihydro-5-(2-pyridinyl)-
CAS registry number   : 1812-30-2
EC number   : 217-322-4
Merck Index monograph   : 2662
Patch testing   : Tablet, pulverized, 30% pet. (2); most systemic drugs can be tested at 10% pet.; if the pure chemical is not available, prepare the test material from intravenous powder, the content of capsules or – when also not available – from powdered tablets to achieve a final concentration of the active drug of 10% pet.
Molecular formula   : $C_{14}H_{10}BrN_3O$

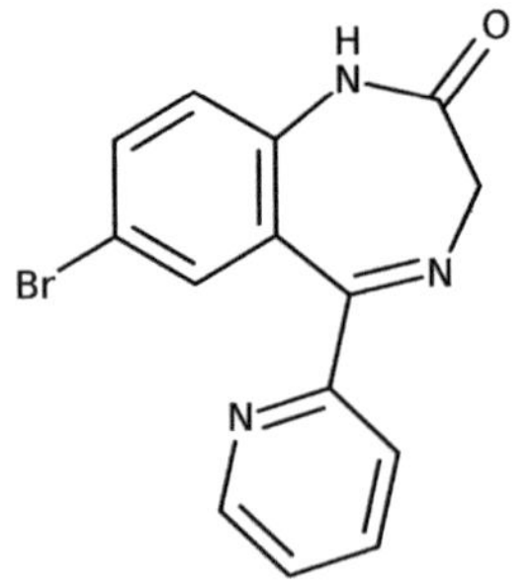

GENERAL

Bromazepam is a lipophilic, long-acting benzodiazepine with sedative, hypnotic, anxiolytic and skeletal muscle relaxant properties. It does not possess any antidepressant qualities. Bromazepam is indicated for the short-term treatment of insomnia, short-term treatment of anxiety or panic attacks, and the alleviation of the symptoms of alcohol- and opiate-withdrawal (1).

CONTACT ALLERGY FROM ACCIDENTAL CONTACT

A 30-year-old woman presented with an itchy, scaly skin reaction on the face, which was most pronounced on the eyelids. She reported a clear relationship with her work as a geriatric nurse, during which she was required to crush tablets for the elderly on a daily basis, mostly benzodiazepine drugs. She had positive patch tests to bromazepam (crushed tablet 30% pet.; 5 controls were negative), 4 other benzodiazepines (alprazolam, diazepam, lorazepam, tetrazepam) and 3 unrelated drugs. The patient was diagnosed with occupational airborne allergic contact dermatitis from drugs (2).

A 66-year-old woman presented with itchy and burning eczema of the face, which had first appeared on the eyelids. Later, extension to the forehead, lips and perioral region, neck, ears and the fingers occurred. The patient's husband suffered from Parkinson's disease, and she had to crush a large number of tablets for him up to five times a day. She had positive patch tests to bromazepam (crushed tablet 30% pet.; 5 controls were negative), 3 other benzodiazepines (clotiazepam, lorazepam, tetrazepam) and trazodone HCl. The patient was diagnosed with (airborne) allergic contact dermatitis from drugs (*by proxy*) (2).

Cross-reactions, pseudo-cross-reactions and co-reactions

The possibility of cross-reactions between benzodiazepines, especially with primary sensitization to tetrazepam, is considered likely, but the pattern is unknown and may differ between sensitivity from (occupational) contact with the drugs and systemically induced reactions (2).

LITERATURE

1   The data in the section 'General' may have been obtained from literature discussed in this chapter, but mostly also or exclusively from one or more of the following online sources: ChemIDPlus Advanced, PubChem, DrugBank, RxList, Drug Central, Drugs.com, and Wikipedia
2   Swinnen I, Ghys K, Kerre S, Constandt L, Goossens A. Occupational airborne contact dermatitis from benzodiazepines and other drugs. Contact Dermatitis 2014;70:227-232

Chapter 3.77   BROMHEXINE

IDENTIFICATION

Description/definition   : Bromhexine is the substituted aniline that conforms to the structural formula shown
                            below
Pharmacological classes : Expectorants
IUPAC name   : 2,4-Dibromo-6-[[cyclohexyl(methyl)amino]methyl]aniline
Other names   : Benzenemethanamine, 2-amino-3,5-dibromo-N-cyclohexyl-N-methyl-
CAS registry number   : 3572-43-8
EC number   : 222-684-1
Merck Index monograph   : 2668
Patch testing   : 1% and 5% pet.
Molecular formula   : $C_{14}H_{20}Br_2N_2$

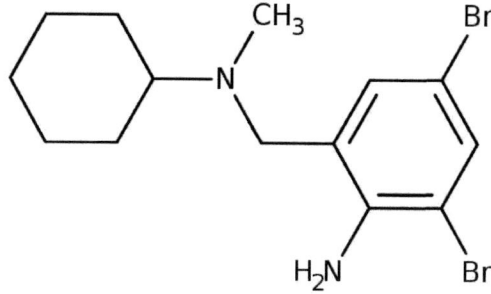

GENERAL

Bromhexine is a substituted aniline that is used as a mucolytic for the treatment of respiratory disorders associated with productive cough. In pharmaceutical products, bromhexine is employed as bromhexine hydrochloride (CAS number 611-75-6, EC number 210-280-8, molecular formula $C_{14}H_{21}Br_2ClN_2$) (1).

CUTANEOUS ADVERSE DRUG REACTIONS FROM SYSTEMIC ADMINISTRATION CAUSED BY TYPE IV (DELAYED-TYPE) HYPERSENSITIVITY (as demonstrated by positive patch tests)

Cutaneous adverse drug reactions from systemic administration of bromhexine caused by type IV (delayed-type) hypersensitivity have included bullous fixed drug eruption (2).

Other cutaneous adverse drug reactions

A 60-year-old man suffered from asthmatic bronchitis, for which he had been treated with daily oral montelukast for 5 years. Sporadically, in the past two years, he also took various other medications including bromhexine. The patient now presented with a 3-day history of a skin eruption with bullae. Similar episodes had occurred in the last 3 years, most after being treated for upper respiratory infections. The patient had found that the lesions always appeared in exactly the same spots, that they generally resolved spontaneously and turned brownish. Physical examination showed multiple erythematous and violaceous well-demarcated patches and plaques located on the trunk, lower limbs, and genital skin. Most lesions were round and exhibited bullae. A cutaneous biopsy showed an interface dermatitis with eosinophils and a few neutrophils, dermal edema, vacuolar changes, and Civatte bodies. The patient was diagnosed with bullous fixed drug eruption on the basis of clinical and histologic findings. Later, patch tests were performed with all the sporadically taken drugs applied on postlesional skin. Strongly positive reactions were observed to bromhexine1% and 5% in petrolatum at D2 with itchy elevated red papules with vesicles (2).

LITERATURE

1 The data in the section 'General' may have been obtained from literature discussed in this chapter, but mostly also or exclusively from one or more of the following online sources: ChemIDPlus Advanced, PubChem, DrugBank, RxList, Drug Central, Drugs.com, and Wikipedia
2 Vide J, Moreira C, Cunha AP, Baldaia H, Magina S, Azevedo F. Generalized bullous fixed drug eruption due to bromhexine. Dermatol Online J 2016 Jul 15;22(7):13030/qt7nt074w4

Chapter 3.78   BROMISOVAL

IDENTIFICATION

Description/definition   : Bromisoval is the organobromine compound that conforms to the structural formula
                            shown below
Pharmacological classes   : Hypnotics and sedatives
IUPAC name   : 2-Bromo-N-carbamoyl-3-methylbutanamide
Other names   : Bromisovalum; bromovalerylurea
CAS registry number   : 496-67-3
EC number   : 207-825-7
Merck Index monograph   : 2674
Patch testing   : 0.1%, 1% and 10% pet.
Molecular formula   : $C_6H_{11}BrN_2O_2$

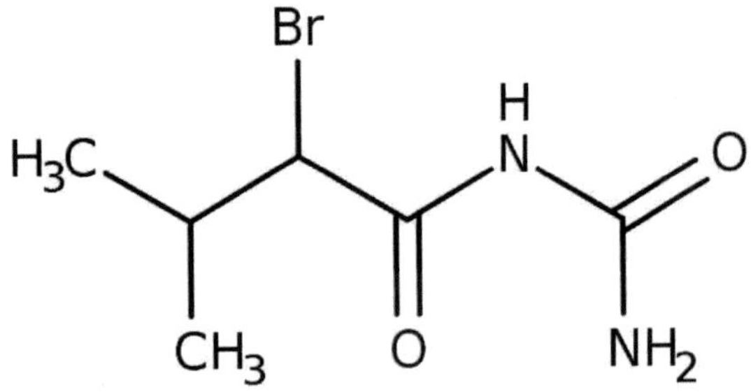

GENERAL
Bromisoval is a hypnotic and sedative of the bromoureide group that has been available for over a century. It is
marketed over the counter in Asia under various trade names, usually in combination with nonsteroidal anti-
inflammatory drugs. Chronic use of bromisoval has been associated with bromine poisoning (1).

CUTANEOUS ADVERSE DRUG REACTIONS FROM SYSTEMIC ADMINISTRATION CAUSED BY TYPE IV
(DELAYED-TYPE) HYPERSENSITIVITY (as demonstrated by positive patch tests)
Cutaneous adverse drug reactions from systemic administration of bromisoval caused by type IV (delayed-type)
hypersensitivity have included toxic epidermal necrolysis (TEN) (2).

Toxic epidermal necrolysis (TEN)
A 40-year-old man developed TEN while using 6 different drugs including a 'cold' medicine, a seven-ingredient
combination tablet containing bromisovalum (bromisoval). Later, the patient was patch tested with all drugs used at
1%, 20% and 40% pet. and there were positive reactions to the cold medicine only. Testing with its 7 ingredients
separately yielded positive reactions to bromisovalum (bromisoval) 0.1%, 1% and 10% pet. The histological and
immunohistological changes in biopsy specimens from a TEN lesion and from the site of the positive patch test
reaction were virtually identical (2).

LITERATURE
1   The data in the section 'General' may have been obtained from literature discussed in this chapter, but mostly
    also or exclusively from one or more of the following online sources: ChemIDPlus Advanced, PubChem,
    DrugBank, RxList, Drug Central, Drugs.com, and Wikipedia
2   Miyauchi H, Hosokawa H, Akaeda T, Iba H, Asada Y. T-cell subsets in drug-induced toxic epidermal necrolysis.
    Possible pathogenic mechanism induced by CD8-positive T cells. Arch Dermatol 1991;127:851-855

Chapter 3.79 BUCILLAMINE

IDENTIFICATION

Description/definition	: Bucillamine is the *N*-acyl-*l*-α-amino acid that conforms to the structural formula shown below
Pharmacological classes	: Antioxidants; anti-inflammatory agents, non-steroidal
IUPAC name	: (2*R*)-2-[(2-Methyl-2-sulfanylpropanoyl)amino]-3-sulfanylpropanoic acid
Other names	: *N*-(2-Mercapto-2-methylpropionyl)-L-cysteine
CAS registry number	: 65002-17-7
EC number	: Not available
Merck Index monograph	: 2740
Patch testing	: 1% pet.
Molecular formula	: $C_7H_{13}NO_3S_2$

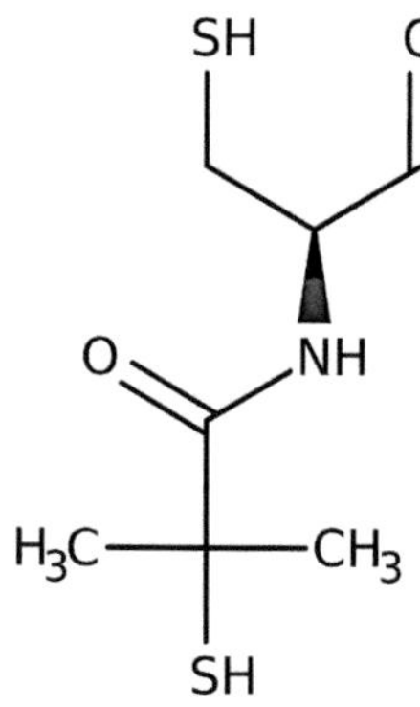

GENERAL

Bucillamine belongs to the class of organic compounds known as *N*-acyl-*l*-α-amino acids. It is classified as anti-rheumatic agent and immunomodulator and has been used in trials studying the treatment and prevention of gout and rheumatoid arthritis. It is available as drug in several countries (1).

CUTANEOUS ADVERSE DRUG REACTIONS FROM SYSTEMIC ADMINISTRATION CAUSED BY TYPE IV (DELAYED-TYPE) HYPERSENSITIVITY (as demonstrated by positive patch tests)

Cutaneous adverse drug reactions from systemic administration of bucillamine caused by type IV (delayed-type) hypersensitivity have included toxic epidermal necrolysis (TEN) (6,7), red papules on the face, neck and arms (2) and unknown drug eruptions (3,4,5).

Toxic epidermal necrolysis (TEN)

A 27-year-old woman developed toxic epidermal necrolysis which started after having taken bucillamine 300 mg/day for 7 days for rheumatoid arthritis. After complete resolution, patch tests were positive to bucillamine 0.5% and 1% pet. These concentrations 'had been found negative on normal subjects' (6,7).

Other cutaneous adverse drug reactions

A 56-year-old woman had been treated with several drugs for a long time and bucillamine for 2 weeks when she developed red papules on the face, neck and arms. Patch tests with bucillamine 20%, 10%, 1%, and 0.1% pet. showed erythema and papules with pruritus at 10% and 20%, red papules with pruritus at D3 with bucillamine 1% and a negative reaction to 0.1%. There were no positive reactions to the related drug penicillamine. Patch tests with bucillamine 30% and 10% pet. in 8 normal subjects showed erythema and papules at D2 and D3 in 3 subjects, indicating that these concentrations are irritant. Control tests with bucillamine 1% and 0.1% pet., however, were negative. A LST (lymphocyte stimulation test) with bucillamine was negative (2).

The authors (2) refer to six cases of drug eruptions from bucillamine previously reported in Japan, in which probably patch tests have been performed. Details, however, were not provided (3,4,5).

Cross-reactions, pseudo-cross-reactions and co-reactions

No cross-reactivity to penicillamine (2,6).

LITERATURE

1   The data in the section 'General' may have been obtained from literature discussed in this chapter, but mostly also or exclusively from one or more of the following online sources: ChemIDPlus Advanced, PubChem, DrugBank, RxList, Drug Central, Drugs.com, and Wikipedia

2   Kimura M, Kawada A. Drug eruption due to bucillamine. Contact Dermatitis 1998;39:98-99

3   Ogata F, Igarashi A, Etou T, Tsuchida T, Ishibashi Y, Kosaka M, et al. Three cases of drug eruptions due to bucillamine (Rimatil ®). Rinsho Derma (Tokyo) 1991;33:1803-1807 (Article in Japanese)

4   Uchiyama M, Katsura C, Sugiura H, Kaji A, Danno K, Uehara M. A case of drug eruption due to bucillamine. Rinsho Hifuka 1996:50:1073-1075 (Article in Japanese).

5   Hirata M, Oh-i T, Koga M. Two cases of drug eruptions due to bucillamine. Rinsho Derma (Tokyo) 1998:40:57-59 (Article in Japanese).

6   Izumi A, Katsumi S, Kobayashi N, Niizeki H, Asada H, Miyagawa S. Bucillamine-induced toxic epidermal necrolysis and fixed drug eruption. J Dermatol 2005;32:397-401 (previously published in Japanese in ref. 7)

7   Katsumi S, Kobayashi N, Miyagawa S, Shirai T. A case of TEN due to bucillamine. Hifu 1998;40:29-33 (Article in Japanese)

Chapter 3.80   BUPIVACAINE

IDENTIFICATION

Description/definition   : Bupivacaine is the amide-type local anesthetic that conforms to the structural formula
                              shown below
Pharmacological classes : Anesthetics, local
IUPAC name   : 1-Butyl-N-(2,6-dimethylphenyl)piperidine-2-carboxamide
CAS registry number   : 38396-39-3
EC number   : 253-911-2
Merck Index monograph   : 2769
Patch testing   : 0.5% water (10); 2% pet. (13)
Molecular formula   : $C_{18}H_{28}N_2O$

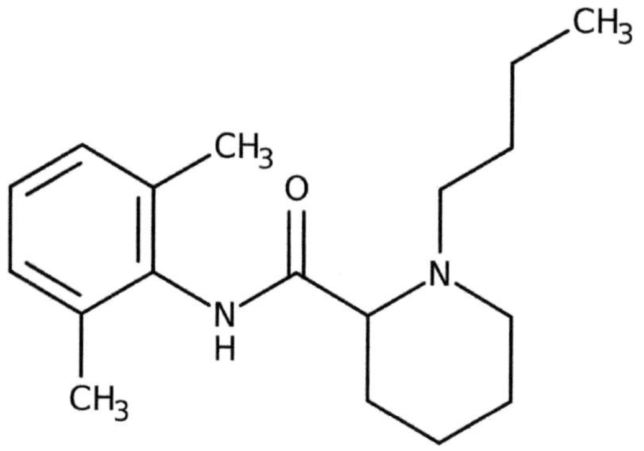

GENERAL

Bupivacaine is an amide-type, long-acting local anesthetic. It is indicated for the production of local or regional anesthesia or analgesia for surgery, for oral surgery procedures, for diagnostic and therapeutic procedures, and for obstetrical procedures (1). In pharmaceutical products, both bupivacaine (in liposome injectable suspension) and bupivacaine hydrochloride (monohydrate) (CAS number 73360-54-0, EC number not available, molecular formula $C_{18}H_{31}ClN_2O_2$) may be employed.

CUTANEOUS ADVERSE DRUG REACTIONS FROM SYSTEMIC ADMINISTRATION CAUSED BY TYPE IV (DELAYED-TYPE) HYPERSENSITIVITY (as demonstrated by positive patch tests)

Cutaneous adverse drug reactions from systemic administration of bupivacaine caused by type IV (delayed-type) hypersensitivity have included eczematous eruption from continuous epidural analgesia (10) and localized allergic reaction from subcutaneous injection (11).

Other cutaneous adverse drug reactions

A 33-year-old woman, parturient at 39-week gestation, during the active phase of labor underwent continuous epidural analgesia with bupivacaine 0.075% up to delivery. Three 3 days later, the patient developed an itchy, vesicular, eczematous reaction on the back and trunk up to the neckline. She was treated with topical steroids and the lesions disappeared within 10 days. Four months later, a patch test to bupivacaine 0.5% water was strongly positive (+++). There was a cross-reaction to mepivacaine, as shown by the development of eczematous lesions at the sites of prick and intradermal tests and subcutaneous injections (10).

A 26-year-old woman required surgery on a toenail. Because she was suspected of being allergic to lidocaine (she had repeatedly developed dermatitis from application of lidocaine-containing topical preparations), bupivacaine was selected for local anesthesia. About 8 hours after the injection itching developed at the site, over the next day followed by localized redness, swelling and increased itching. There were no blisters and the patient did not have systemic symptoms. Patch tests were positive to mepivacaine, lidocaine and ropivacaine (no details provided). Quite curiously, bupivacaine itself was not tested. Most likely, the patient was primarily sensitized to lidocaine and now had a localized allergic reaction from bupivacaine cross-reacting to lidocaine (which is very common) (11).

Cross-reactions, pseudo-cross-reactions and co-reactions

Bupivacaine often cross-reacts to primary sensitization to lidocaine, the prototype amide local anesthetic (3,4,5,6,7,8,9,11), and sometimes to prilocaine (2) and articaine (12). Likely cross-reaction from bupivacaine sensitization to mepivacaine (10).

LITERATURE

1   The data in the section 'General' may have been obtained from literature discussed in this chapter, but mostly also or exclusively from one or more of the following online sources: ChemIDPlus Advanced, PubChem, DrugBank, RxList, Drug Central, Drugs.com, and Wikipedia

2   García F, Iparraguirre A, Blanco J, Alloza P, Vicente J, Báscones O, et al. Contact dermatitis from prilocaine with cross-sensitivity to pramocaine and bupivacaine. Contact Dermatitis 2007;56:120-121

3   Duque S, Fernández L. Delayed-type hypersensitivity to amide local anesthetics. Allergol Immunopathol (Madr) 2004;32:233-234

4   Yuen WY, Schuttelaar ML, Barkema LW, Coenraads PJ. Bullous allergic contact dermatitis to lidocaine. Contact Dermatitis 2009;61:300-301

5   Weightman W, Turner T. Allergic contact dermatitis from lignocaine: report of 29 cases and review of the literature. Contact Dermatitis 1998;39:265-266

6   Hardwick N, King CM. Contact allergy to lignocaine with cross-reaction to bupivacaine. Contact Dermatitis 1994;30:245-246

7   Fregert S, Tegner E, Thelin I. Contact allergy to lidocaine. Contact Dermatitis 1979;5:185-188

8   Bircher AJ, Messmer SL, Surber C, Rufli T. Delayed-type hypersensitivity to subcutaneous lidocaine with tolerance to articaine: confirmation by in vivo and in vitro tests. Contact Dermatitis 1996;34:387-389

9   Bassett I, Delaney T, Freeman S. Can injected lignocaine cause allergic contact dermatitis? Australas J Dermatol 1996;37:155-156

10   Nettis E, Colanardi MC, Calogiuri GF, Foti C, Priore MG, Ferrannini A, Vacca A. Delayed-type hypersensitivity to bupivacaine. Allergy 2007;62:1345-1346

11   Redfern DC. Contact sensitivity to multiple local anesthetics. J Allergy Clin Immunol 1999;104(4Pt.1):890-891

12   De Pasquale TMA, Buonomo A, Pucci S. Delayed-type allergy to articaine with cross-reactivity to other local anesthetics from the amide group. J Allergy Clin Immunol Pract 2018;6:305-306

13   De Groot AC. Patch testing, 4th edition. Wapserveen, The Netherlands: acdegroot publishing, 2018 (ISBN 9789081323345)

Chapter 3.81 BUPROPION

IDENTIFICATION

Description/definition : Bupropion is the alkyl-phenylketone that conforms to the structural formula shown below
Pharmacological classes : Cytochrome P-450 CYP2D6 inhibitors; dopamine uptake inhibitors; antidepressive agents, second-generation; smoking cessation agents
IUPAC name : 2-(*tert*-Butylamino)-1-(3-chlorophenyl)propan-1-one
Other names : Amfebutamone
CAS registry number : 34911-55-2
EC number : 250-759-9 (hydrochloride)
Merck Index monograph : 2773
Patch testing : Commercial tablet 1,5,10,20 and 30% pet.
Molecular formula : $C_{13}H_{18}ClNO$

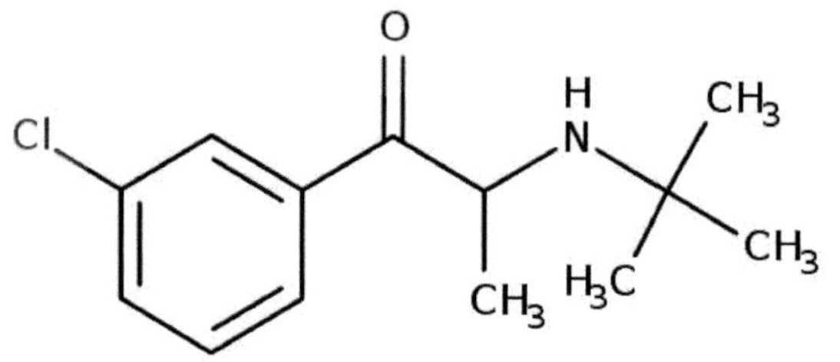

GENERAL

Bupropion is an aminoketone antidepressant that is widely used in therapy of depression and smoking cessation. The fixed combination of extended-release naltrexone and bupropion has been developed as a weight loss agent (1).

CUTANEOUS ADVERSE DRUG REACTIONS FROM SYSTEMIC ADMINISTRATION CAUSED BY TYPE IV (DELAYED-TYPE) HYPERSENSITIVITY (as demonstrated by positive patch tests)

Cutaneous adverse drug reactions from systemic administration of bupropion caused by type IV (delayed-type) hypersensitivity have included acute generalized exanthematous pustulosis (AGEP) (2).

Acute generalized exanthematous pustulosis (AGEP)

A 45-year-old woman, previously diagnosed with psoriatic spondylitis and obesity, presented with an itchy rash, developing one day after starting naltrexone/bupropion for weight loss, which she had never used before. Physical examination revealed an erythematous maculopapular rash on the trunk and arms, which progressed to form multiple, non-follicular, pinhead-sized pustules on an edematous disseminated erythema, extending to the face and legs. Acute generalized exanthematous pustulosis (AGEP) and generalized pustular psoriasis were considered. Histopathology revealed subcorneal, spongiform neutrophilic pustules, slight acanthosis with focal parakeratosis, and perivascular neutrophilic infiltration in the superficial dermis. Three weeks later, classic psoriatic plaques progressively developed on the arms and legs. Patch tests 6 months later were positive to naltrexone/bupropion (++) and bupropion (++) commercial tablets 30% pet. and − in a second test session − to commercialized bupropion diluted at 1%, 5%, 10%, and 20% pet. Testing with naltrexone remained negative. A biopsy of a positive patch test showed an allergic dermatitis. Ten controls were negative to bupropion 30% pet. (2).

The authors commented that the rapid onset of a generalized pustular eruption within a single day of taking the first naltrexone/bupropion tablet, with a subsequent positive patch test, is strongly suggestive of AGEP, but that the histological features favored the diagnosis of pustular psoriasis, particularly the absence of perivascular eosinophilic infiltration. Furthermore, they assumed that an initial AGEP caused by bupropion might have provoked an ongoing psoriasis as a result of the Koebner phenomenon (2). It remains unclear how the patient had become sensitized to bupropion.

LITERATURE

1 The data in the section 'General' may have been obtained from literature discussed in this chapter, but mostly also or exclusively from one or more of the following online sources: ChemIDPlus Advanced, PubChem, DrugBank, RxList, Drug Central, Drugs.com, and Wikipedia
2 Caldas R, Campos-Lopes S, Guimarães MJ, Areal J, Alves M, Pereira T. Patch test-proven delayed-type hypersensitivity from naltrexone/bupropion possibly eliciting psoriasis. Contact Dermatitis. 2021 Apr 30. doi: 10.1111/cod.13875. Epub ahead of print.

Chapter 3.82   CAPTOPRIL

IDENTIFICATION

Description/definition   : Captopril is a sulfhydryl-containing analog of proline that conforms to the structural
                            formula shown below
Pharmacological classes : Angiotensin-converting enzyme inhibitors; anti-hypertensive agents
IUPAC name   : (2S)-1-[(2S)-2-Methyl-3-sulfanylpropanoyl]pyrrolidine-2-carboxylic acid
Other names   : D-2-Methyl-3-mercaptopropanoyl-L-proline
CAS registry number   : 62571-86-2
EC number   : 263-607-1
Merck Index monograph   : 3046
Patch testing   : 5.0% pet. (Chemotechnique)
Molecular formula   : $C_9H_{15}NO_3S$

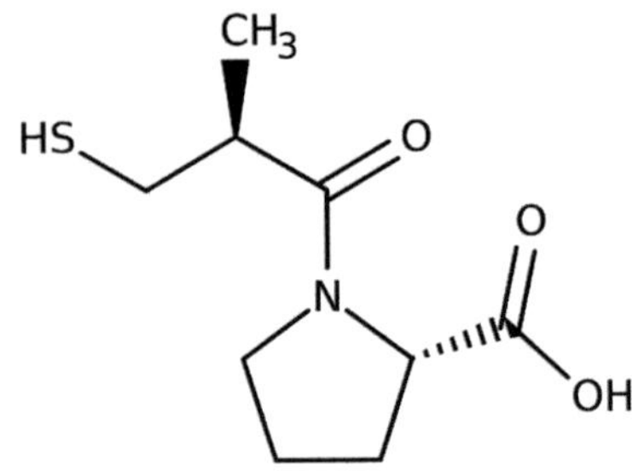

GENERAL

Captopril is a sulfhydryl-containing analog of proline with antihypertensive activity and potential antineoplastic quality. This agent competitively inhibits angiotensin converting enzyme (ACE), thereby decreasing levels of angiotensin II, increasing plasma renin activity, and decreasing aldosterone secretion. Captopril may also inhibit tumor angiogenesis and exhibit antineoplastic activity independent of effects on tumor angiogenesis. The drug is indicated for the treatment of essential or renovascular hypertension, may be used to treat congestive heart failure, and may improve survival in patients with left ventricular dysfunction following myocardial infarction. Another indication is nephropathy, including diabetic nephropathy (1).

CONTACT ALLERGY FROM ACCIDENTAL CONTACT

In Leuven, Belgium, in the period 2001-2019, 201 of 1248 health care workers/employees of the pharmaceutical industry had occupational allergic contact dermatitis. In 23 (11%) dermatitis was caused by skin contact with a systemic drug: 19 nurses, two chemists, one physician, and one veterinarian. The lesions were mostly localized on the hands, but often also on the face, as airborne dermatitis. In total, 42 positive patch test reactions to 18 different systemic drugs were found. In one patient, captopril was the drug/one of the drugs that caused occupational dermatitis (15, overlap with ref. 2).

In the same clinic in Leuven, Belgium, in the period 2007-2011, 81 patients have been diagnosed with occupational airborne allergic contact dermatitis. In 23 of them, drugs were the offending agents, including captopril in one case (2, overlap with ref. 15).

A 30-year-old woman presented with an itchy, scaly skin reaction on the face, which was most pronounced on the eyelids. She reported a clear relationship with her work as a geriatric nurse, during which she was required to crush tablets for the elderly on a daily basis. She had positive patch tests to captopril (crushed tablet 30% pet.; 18 controls were negative), and various other drugs. The patient was diagnosed with occupational airborne allergic contact dermatitis (4). Occupational allergic contact dermatitis from captopril has also been reported in ref. 12 (no details available to the author).

A 36-year-old woman had a history of edema of the eyelids and lips, flaking and pruritus, for two months, especially during her work in the pharmaceutical industry, where she had contact with residues contained in captopril packaging. The patient had improved when there was no contact with the packages. A patch test with captopril 10% was positive with papules, vesicles and swelling at the application site (16).

A 35-year-old woman, who administered captopril syrup and aspirin to her baby for a heart malformation, developed hand eczema and eczema of the eyelids with edema. An open test with the baby's captopril syrup resulted in immediate itchy erythema which had become eczematous with vesicles the next day. Patch tests were positive to powdered captopril tablets 3%, 10% and 30% pet. 3% and 10% were negative in controls, but 30% seemed to be irritant (7). This was a case of allergic contact dermatitis from accidental contact *by proxy*.

CUTANEOUS ADVERSE DRUG REACTIONS FROM SYSTEMIC ADMINISTRATION CAUSED BY TYPE IV (DELAYED-TYPE) HYPERSENSITIVITY (as demonstrated by positive patch tests)

Cutaneous adverse drug reactions from systemic administration of captopril caused by type IV (delayed-type) hypersensitivity have included maculopapular eruption (8,9), erythroderma (18), drug reaction with eosinophilia and systemic symptoms (DRESS) (5,6,11), widespread eczema (17), localized erythema (10), lichenoid eruption (3), and unspecified cutaneous eruptions (13,14,19).

Case series with various or unknown types of drug reactions

Cutaneous reactions associated with captopril treatment occurred in fifteen out of eighty-nine treated patients (17%). Positive epicutaneous skin tests were observed in five out of the fifteen patients with negative reactions in 9 captopril-treated controls (13). In Finland, in the period 1989-2001, 826 patients with suspected cutaneous drug eruptions were patch tested and 89 had one or more positive reactions. Of these individuals, 4 reacted to captopril; it was not specified what kind of drug eruption they had (19).

Maculopapular eruption

A 74-year-old man had used captopril 25 mg captopril one time daily for 2 months, when he developed a pruritic maculopapular rash involving his trunk, back and legs. Patch tests with all his medications (powdered tablets 10% pet.) gave a positive reaction to captopril only. Five controls were negative (8). A 71-year-old woman also suffered a generalized pruriginous maculopapular rash while on captopril. Patch tests were positive to captopril 1% and 10% pet. (unknown whether the pure material or – more likely – a powdered pill was used) (9).

Erythroderma, widespread erythematous eruption, exfoliative dermatitis

In a study from Nancy, France, of 54 patients with suspected nonimmediate drug eruptions, one had erythroderma and a positive patch test to captopril (18).

Drug reaction with eosinophilia and systemic symptoms (DRESS)

A 45-year old man had DRESS with fever, pruritic diffuse maculopapular rash, an elevated eosinophil count at 2300/µL and renal impairment while on captopril and nifedipine. A patch test (read at D2 only) was positive to captopril 1% pet. and negative to nifedipine 30% pet. (5).

A 59-year-old woman, received a multidrug therapy including captopril, acebutolol, and furosemide for hypertension, developed, 3 weeks after starting this treatment, a maculopapular itchy and edematous skin eruption, facial edema, fever, and a cervical lymphadenopathy. Leukocytes were elevated with 19% eosinophils and liver enzymes slightly elevated. A biopsy was consistent with drug hypersen-sitivity. Six weeks after complete symptom resolution, patch tests to captopril (1% in petrolatum), acebutolol (20% in petrolatum), and furosemide (20% in petrolatum) were performed and there was a positive reaction to captopril only. The patient was diagnosed with DRESS caused by captopril (6).

A 63-year-old woman was prescribed captopril and 3 other drugs for acute pulmonary edema, atrial fibrillation and severe hypertension. Ten days later, she developed a maculopapular rash, eosinophilia, mild renal insufficiency and diarrhea. When patch tested, the only positive reaction was to captopril 10% water, which was confirmed by a repeated positive reaction. The other drugs were prescribed again without any adverse reaction (11). This patient had DRESS, which diagnosis and acronym at that time (1990) had not yet been coined.

Dermatitis/eczematous eruption

Several weeks after a 79-year-old woman started oral treatment for hypertension with captopril, she developed pruritic, eczematous lesions on her face, trunk, arms and legs. Six weeks after full healing, patch tests were positive to captopril crushed tablet 1% water and captopril 10% water, but negative to the other ingredients of the tablet and other ACE-inhibitors (17).

Other cutaneous adverse drug reactions

A few hours after taking a first dose of captopril 25 mg for hypertension, a 47-year-old man experienced facial burning followed by an erythematous eruption on both cheeks, which resolved with scaling in 2 weeks. A double-blind oral challenge test with captopril 5 mg reproduced the lesions. Patch tests were positive to captopril 10% pet. both from pure captopril and from powdered tablets. Six controls were negative (10).

A man aged 48 years developed a lichenoid eruption suspected to have been caused by captopril or tiopronin. Patch tests were positive to captopril 3% and tiopronin 5% (vehicle not mentioned) and 'provocation' (unspecified) was positive for both drugs (3).

A 50-year-old man with hypertension had used captopril 100 mg four times daily for 3 months, when a rash and eosinophilia developed. Patch tests with captopril 0.1%-10% in petrolatum were positive at D2 and D3 (14). It was not mentioned whether the patient also had fever and visceral involvement, suggesting DRESS.

Cross-reactions, pseudo-cross-reactions and co-reactions

No cross-reactions to other ACE inhibitors: ramipril, delapril, enalapril, perindopril, lisinopril, benazepril, fosinopril, quinapril (5,6,8,9,10,11,14,17). Captopril is the only ACE inhibitor containing a sulfhydryl group and, therefore, this chemical group has been considered the culprit of allergic reactions to captopril (3,9).

LITERATURE

1   The data in the section 'General' may have been obtained from literature discussed in this chapter, but mostly also or exclusively from one or more of the following online sources: ChemIDPlus Advanced, PubChem, DrugBank, RxList, Drug Central, Drugs.com, and Wikipedia

2   Swinnen I, Goossens A. An update on airborne contact dermatitis: 2007-2011. Contact Dermatitis 2013;68:232-238

3   Kitamura K, Aihara M, Osawa J, Naito S, Ikezawa Z. Sulfhydryl drug-induced eruption: a clinical and histological study. J Dermatol 1990;17:44-51

4   Swinnen I, Ghys K, Kerre S, Constandt L, Goossens A. Occupational airborne contact dermatitis from benzodiazepines and other drugs. Contact Dermatitis 2014;70:227-232

5   Ammar H, Chaabane A, Ben Fadhel N, Chadli Z, Ben Fredj N, Aouam K. Drug rash with eosinophilia and systemic symptoms: Captropril, an unusual culprit drug. Dermatitis 2019;30:238-239

6   Chaabane A, Fadhl NB, Chadly Z, Fredj NB, Boughattas NA, Aouam K. Captopril-induced DRESS: first reported case confirmed by patch test. Dermatitis 2013;24:255-257

7   Balieva F, Steinkjer B. Contact dermatitis to captopril. Contact Dermatitis 2009;61:177-178

8   Martinez JC, Fuentes MJ, Armentia A, Vega JM, Fernandez A. Dermatitis to captopril. Allergol Immunopathol 2001;29:279-280

9   Gaig P, San Miguel-Moncin MM, Bartra J, Bonet A, García-Ortega P. Usefulness of patch tests for diagnosing selective allergy to captopril. J Investig Allergol Clin Immunol 2001;11:204-206

10 Lluch-Bernal M, Novalbos A, Umpierrez A, Figueredo E, Bombin C, Sastre J. Cutaneous reaction to captopril with positive patch test and lack of cross-sensitivity to enalapril and benazepril. Contact Dermatitis 1998;39:316-317

11 Cnudde F, Leynadier F, Dry J. Cutaneous reaction to captopril: value of patch tests. Contact Dermatitis 1990;23:375-376

12 Dziuk M, Gall H, Sterry W. Contact dermatitis from the ACE-inhibitor captopril. Derm Beruf Umwelt 1994;42:159-161

13 Smit AJ, Van der Laan S, De Monchy J, Kallenberg CGM, Donker AJM. Cutaneous reactions to captopril, predictive value of skin test. Clin Allergy 1984;14:413-419

14 Navis GJ, De Jong PE, Kallenberg CJM, De Monchy J, De Zeeuw D. Absence of cross-reactivity between captopril and enalapril. Lancet 1984;2:1017

15 Gilissen L, Boeckxstaens E, Geebelen J, Goossens A. Occupational allergic contact dermatitis from systemic drugs. Contact Dermatitis 2020;82:24-30

16 Ribeiro MR, Komarof F, Garrol LS, Mattos Porter M-H, Adachi CT, Mello Y, et al. Occupational contact dermatitis due to captopril. World Allergy Organ J 2015;8(Suppl.1):A225

17 Pfützner W, Ruëff F, Przybilla B. Systemic contact dermatitis due to captopril without cross-sensitivity to fosinopril, quinapril and benazepril. Acta Derm Venereol 2004;84:91-92

18 Barbaud A, Reichert-Penetrat S, Tréchot P, Jacquin-Petit MA, Ehlinger A, Noirez V, et al. The use of skin testing in the investigation of cutaneous adverse drug reactions. Br J Dermatol 1998;139:49-58

19 Lammintausta K, Kortekangas-Savolainen O. The usefulness of skin tests to prove drug hypersensitivity. Br J Dermatol 2005;152:968-974

Chapter 3.83   CARBAMAZEPINE

IDENTIFICATION

Description/definition : Carbamazepine is a tricyclic dibenzazepine that conforms to the structural formula shown below

Pharmacological classes : Cytochrome P-450 CYP3A inducers; analgesics, non-narcotic; sodium channel blockers; anticonvulsants; antimanic agents

IUPAC name : Benzo[b][1]benzazepine-11-carboxamide

CAS registry number : 298-46-4

EC number : 206-062-7

Merck Index monograph : 4053

Patch testing : 1.0% pet. (Chemotechnique); sometimes, using higher concentrations (5%, 10%) may be necessary (74); with serious cutaneous adverse reactions, starting with 1% pet. may be advisable (70)

Molecular formula : $C_{15}H_{12}N_2O$

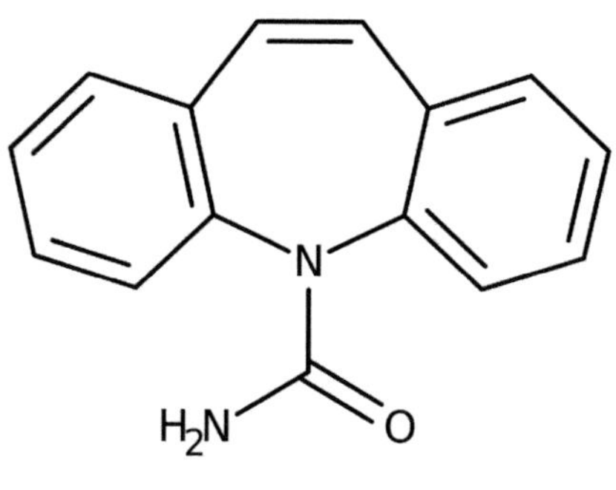

GENERAL

Carbamazepine is an aromatic tricyclic dibenzazepine compound chemically related to tricyclic antidepressants with anticonvulsant and analgesic properties. It exerts its anticonvulsant activity by reducing polysynaptic responses and blocking post-tetanic potentiation; its analgesic activity is not understood. Carbamazepine is used to treat epilepsy, partial seizures, tonic-clonic seizures, pain of neurologic origin such as trigeminal and postherpetic neuralgia, and psychiatric disorders including manic-depressive illness and aggression due to dementia (1).

CUTANEOUS ADVERSE DRUG REACTIONS FROM SYSTEMIC ADMINISTRATION CAUSED BY TYPE IV (DELAYED-TYPE) HYPERSENSITIVITY (as demonstrated by positive patch tests)

Cutaneous adverse drug reactions from systemic administration of carbamazepine caused by type IV (delayed-type) hypersensitivity have included maculopapular eruption (2,11,19,31,41,61,72,83,84,89), erythroderma (11,14,42, 49,73,75,78,79,91), exfoliative dermatitis (19,44,72,74,75,78,79), acute generalized exanthematous pustulosis (AGEP) (59,81), fixed drug eruption (11,16,65,66), drug reaction with eosinophilia and systemic symptoms (DRESS)/ anticonvulsant hypersensitivity syndrome (2-11,18,21-25,27,29,30,32-35,46-48,50-60,62,63,68,72,75-77,80,83,88, 89), Stevens-Johnson syndrome/toxic epidermal necrolysis (SJS/TEN) (2,5,11,12,17,19,21,26,84,88,89), photosensitivity (14,15,67), dermatitis/ eczematous eruption (14,15,42,43,78), erythema multiforme (19,75), lichenoid eruption (32), erythema and edema of the face and pruritus of the legs (71), 'allergic reactions' (76), and unspecified drug eruption (3,36,37,38).

General

Carbamazepine is a well-known and frequently described cause of cutaneous adverse drug reactions, occurring in up to 4% of the patients treated. A wide spectrum of skin exanthemas may be observed, of which 2/3 are maculopapular eruptions. Potentially dangerous cutaneous adverse drug reactions, notably drug reaction with eosinophilia and systemic symptoms (DRESS), which is often termed anticonvulsant hypersensitivity syndrome when caused by anticonvulsant drugs, may be provoked in 1:1,000-10,000 patients treated with this drug (10) (see the section 'Drug reaction with eosinophilia and systemic symptoms (DRESS)/anticonvulsant hypersensitivity syndrome' below). Case series of 13 (63), 11 (58), 7 (21), 6 (18,22), 5 (32), and 4 (3,50,76) patients with DRESS and a positive patch test to carbamazepine have been described, various others with 2 or 3 patients and multiple single case reports (summarized below). This anticonvulsant drug very frequently shows positive patch tests in (presumed) hypersensitivity reactions (11,18,21,63), even in cases of SJS /TEN (21), where patch tests are usually negative. In Southeast Asian countries, SJS/TEN from carbamazepine occurs more often than DRESS due to the genetic susceptibility; HLA-B*1502 is strongly associated with carbamazepine-induced SJS /TEN in Chinese as well as Southeast Asian populations (21,86). Patch testing is generally safe, although it caused an exacerbation of exfoliative dermatitis in one individual (albeit probably at a very high concentration of carbamazepine) (44). Positive reactions to carbamazepine may be observed many years after the cutaneous adverse drug reaction has taken place (69) or remain positive years after a first positive patch test (62).

Case series with various or unknown types of drug reactions

Between 2003 and 2017, in Belgrade, Serbia, 100 children in the age from 1 to 17 years suspected of hypersensitivity reactions to antiepileptic drugs were examined with patch tests, using the commercial drugs 10% pet. 61 patients had shown maculopapular eruptions, 26 delayed urticaria, 5 morbilliform exanthema, 5 DRESS, 2 SJS and one erythema multiforme. Carbamazepine was the suspected drug in 38 cases and was patch test positive in 24 children (63%). It was not specified which eruptions these drugs had caused, but should include maculopapular eruptions and delayed urticaria. Of the 5 children with DRESS, 4 reacted to carbamazepine and one to lamotrigine. One child with SJS also reacted to carbamazepine (88).

In Turkey, of 15 children showing signs of hypersensitivity to carbamazepine, 11 (73%) had positive patch test reactions to commercial carbamazepine at 1% and 10% active ingredients. Seven children had suffered a maculopapular eruption, one Stevens-Johnson syndrome and 3 DRESS (89).

In Rotterdam, The Netherlands, 65 patients who had suffered from a variety of side effect to carbamazepine (CBZ) of possible allergic origin were investigated, including with patch testing of carbamazepine, oxcarbazepine (OCZP), both as pulverized tablets and pure chemicals, and 3 carbamazepine metabolites, each at 5 mg in 50 µl alc. Sixty of the patients had skin rashes consisting of maculopapular, urticarial, erythematous, exfoliative, or eczematous skin reactions. Fourteen had systemic manifestations such as fever, hematologic abnormalities, lymphadenopathy, or impaired liver function. 61 individuals were patch tested with CBZ and there were 12 (20%) positive reactions. Of 59 patients tested with OCZP, 8 (14%) reacted positively. Of 11 patients reacting to CBZ, 5 (45%) co-reacted ('cross-reacted', it was not mentioned whether these patients had been treated with oxcarbazepine before) to OCZP. The concordance between reactions to pure materials and tablets of CBZ and OCZP were high (95 and 92%), the commercial pulverized tablets giving slightly more positive patch tests. The metabolite carbamazepine-10,11 epoxide, often thought to be responsible for allergic reactions, did not give a large number of positive reactions (not specified). It was not mentioned from which skin eruptions the patients with positive reactions to CBZ had suffered, nor whether there was a difference in pattern of adverse effects between CBZ-positive and CBZ-negative patients (38). A part of these data had probably been reported 6 years earlier in an abstract (40).

In the regional pharmacovigilance center of Sfax (Tunisia), between June 1, 2014 and April 30, 2016, all cases of (presumed) allergic skin reactions to antiepileptic drugs were investigated with patch testing. Twenty patients were included, among who 23 cutaneous adverse drug reactions (CADRs) were observed. Eleven patients had hypersensitivity reactions to carbamazepine and all had a positive patch test. Of these individuals, four had suffered maculopapular exanthema, 2 DRESS, 2 fixed drug eruption, 2 erythroderma and one Stevens-Johnson syndrome (11).

Twenty-four patients who had developed severe drug rashes from antiepileptics (18 DRESS, 5 SJE, TEN or SJS/TEN, 1 lichenoid drug eruption) were patch tested with the implicated drugs 10%, 20% and 30% pet. Positive reactions were observed in 12 patients: 11 with DRESS and one with lichenoid eruption. Five of the 11 patients had DRESS from carbamazepine; the lichenoid eruption was also caused by delayed-type hypersensitivity to carbamazepine (32).

In Barcelona, Spain, 5 women and 2 men (age range, 20 to 81 years) were investigated, who were treated with carbamazepine for various illnesses for 10 days to 4 months before a rash appeared. Two had a hypersensitivity syndrome, two maculopapular rashes, one a lichenoid eruption, one exfoliative dermatitis and one a papular eruption on light-exposed areas. All were patch tested with carbamazepine (1%, 5%, 10% pet.; not all were tested with 3 concentrations) and only 3 had positive reactions at D2 and D4, one with a maculopapular rash, the second with a hypersensitivity syndrome and the third with exfoliative dermatitis. Three more patients had positive reactions only at D2 (negative at D4 and D7) and these were considered to be positive, which may be debated (72).

In Helsinki, Finland, 18 patients who had developed a skin rash from carbamazepine in a period of 10 years, were examined with patch testing. The etiological role of carbamazepine had been ascertained by peroral or topical (in the case of fixed drug eruption) provocation in 15 patients. The clinical reactions caused by the drug were classified as maculopapular exanthema with general symptoms in 7 patients (= DRESS), exfoliative dermatitis (erythroderma) in 3, fixed drug eruption in 3, erythema multiforme in 1, urticaria in one and other types of exanthema in 3 patients. Patch tests with carbamazepine 3% and 10% in pet., water and alc. showed positive reactions to carbamazepine in 7 patients and doubtful reactions in 2. Positive patch test reactions were seen only in patients with exfoliative dermatitis/erythroderma (all 3 patients) and DRESS (4 out of 7). Doubtful reactions (?+ to all 6 test materials) were seen in one patient with DRESS and another with erythema multiforme (75).

In Ankara, Turkey, patients < 18 years of age who had had a drug patch test for drug eruptions between January 2014 and January 2020, were retrospectively assessed. A total of 105 drug patch tests were performed on 71 patients. Twenty-three patients (32%) had severe cutaneous adverse reaction (SJS in 11, DRESS in 9, and AGEP in 3 patients), 45 (63%) had maculopapular rashes, and 3 (4%) had fixed drug eruption. A total of 20 patch test results (28%) were positive: 18 of 44 patch tests (41%) with antiepileptic drugs and 2 of 48 patch tests (4%) with antibiotics. Carbamazepine was positive in 4 patients with DRESS, 1 with SJS and 8 with maculopapular eruption (92).

Maculopapular eruption

Thirteen patients from Korea, who had suffered a maculopapular eruption from the use of carbamazepine (CBZ), were patch tested with carbamazepine 10% pet. and its main metabolite carbamazepine 10,11-epoxide (CBZ-E) 1 µg/ml alc. Ten of the 13 patients showed a positive patch test reaction. Seven had a reaction to CBZ only, 2 to CBZ-E only, and 1 had a reaction to both. None of the 39 controls (patients using CBZ, but who had no skin reactions) displayed any reactions to either CBZ or CBZ-E (41).

In a group of 45 patients with multiple drug hypersensitivity seen between 1996 and 2018 in Montpellier, France, 38 of 92 drug hypersensitivities were classified as type IV immunological reactions; these included two patients with maculopapular exanthemas from carbamazepine with positive patch tests to this drug (31). In a study from Nancy, France, 54 patients with suspected nonimmediate drug eruptions were assessed with patch testing. Of the 27 patients with maculopapular eruptions, 2 had positive reactions to carbamazepine (61).

In Italy, a man aged 27, a boy aged 14 and a girl aged 16 developed maculopapular exanthema after having used carbamazepine for 21, 21 resp. 10 days. 2-3 months later, all 3 had positive patch tests to carbamazepine tested as pure powder at 5, 10, 15, and 20% w/w in white petrolatum. 34 controls were negative (2). One patient from China developed a maculopapular exanthema from carbamazepine and later showed a positive patch test to this anticonvulsant at 1% and 10% pet. (19).

From Madrid, Spain, a 54-year-old man was reported who had developed a generalized erythematous maculopapular exanthema after having used carbamazepine for 4 months for neuralgia. Liver function was normal. Patch tests were positive to carbamazepine 1% pet. at D2, but negative at D4 (83). A 12-year-old girl developed a maculopapular eruption from carbamazepine and had a positive patch test to this drug (84). See also the section 'Case series with various or unknown types of drug reactions' above, refs. 11, 72, 89 and 92.

Erythroderma

A 24-year-old woman developed eczematous erythroderma with palmoplantar pompholyx while using carbamazepine. Patch tests were positive to carbamazepine, but not to oxcarbazepine, carbamazepine 10,11-epoxide and 2 other metabolites (42). A 47-year-old woman developed a maculopapular exanthema leading to erythroderma about 3 months after starting carbamazepine (CBZ) therapy. T cells cultured with CBZ were nearly all CD4+ (98%) and showed specificity for CBZ. Patch tests were positive to carbamazepine, probably tested at 1% and 5% pet. (49).

Two of 3 patients with erythroderma from carbamazepine/generalized dermatitis from Spain had positive patch tests to this drug (73). A 67-year-old man developed, after having used carbamazepine for one week, an extensive erythematous dermatitis, which soon became generalized (erythroderma). Four months later, patch tests were positive to carbamazepine 1% and 10% pet. (both +++) (78). Two patients developed erythroderma from carbamazepine and had positive patch tests to carbamazepine 10% pet. (91).

See also the section 'Exfoliative dermatitis' below, ref. 79. See also the section 'Case series with various or unknown types of drug reactions' above, refs. 11 and 75, and the section 'Photosensitivity' below, ref. 14.

Exfoliative dermatitis

A 16-year-old girl developed a generalized exfoliative dermatitis with significant edema of the face and enanthema of the tongue after having used carbamazepine for one month. There was a significant hypereosinophilia, but no other laboratory abnormalities. After complete resolution, patch tests were performed using two crushed 200 mg carbamazepine tablets, one in petrolatum and the other in water. Only the patch test with carbamazepine in petrolatum showed positive results at D2 and, in two days, a generalized exfoliative dermatitis, sparing the buttocks and the lower legs, developed. At the same time, eosinophilia again developed (44).

A 54-year-old man developed generalized exfoliative dermatitis after having used carbamazepine for 3 months. He has also lymphadenopathy and total alopecia. Skin biopsy and lymph node biopsy showed chronic eczema and dermopathic lymphadenopathy, respectively. There was significant eosinophilia. Patch tests were positive to carbamazepine 10% pet., but negative to 1% and 0.1% (74).

A 37-year-old woman developed an exfoliative dermatitis while taking carbamazepine. Three years later, she was again given carbamazepine and the dermatitis reappeared, with intensification in light-exposed areas. When free of lesions, she was patch and photopatch tested with carbamazepine 1% and 2% pet. and reacted equally positive (+++) to patch and photopatch tests (78). The same authors also reported on a 16-year-old girl, who was taking carbamazepine for 4 months for epilepsy, when she developed exfoliative dermatitis that required systemic prednisone. Several months later, she was patch tested with carbamazepine and she had strongly positive reactions to the drug at 1% and 10% pet. (78).

A 32-year-old man developed erythroderma/generalized exfoliative dermatitis while using carbamazepine, the first time after 5 weeks, the second episode starting within a few hours of reintroduction of the drug, leading to generalized exfoliative dermatitis, which lasted for one month despite prompt discontinuation of the pharmaceutical. Patch tests were strongly positive to commercial carbamazepine (pulverized tablet) pure, 1% pet. and 1% acetone (79).

One patient from China developed exfoliative dermatitis from carbamazepine and later showed a positive patch test to this anticonvulsant at 1% and 10% pet. (19). See also the section 'Case series with various or unknown types of drug reactions' above, refs. 72 and 75.

Acute generalized exanthematous pustulosis (AGEP)

A 26-year-old woman began treatment with 200 mg/12 hours carbamazepine because of generalized seizures. A few weeks later, she developed itchy skins lesions compatible with exanthematous pustulosis, together with acute kidney failure requiring hemodialysis. A biopsy study of the kidney revealed immunoallergic tubulointerstitial nephropathy. The patient also had a moderate rise in the level of transaminases and leukocytosis with eosinophilia. Patch tests were positive to carbamazepine (81). In a study from France, of 14 patients with AGEP, 7 had positive patch tests to drugs, including one to carbamazepine (59).

Fixed drug eruption

A 71-year-old woman presented with six sharply demarcated, symmetric, red-brownish plaques on her thighs, knees, and dorsa of the feet, which had developed one hour after taking a tablet of carbamazepine for trigeminal neuralgia. The eruption improved in 1 week, leaving brown-pigmented plaques that faded slowly. The patient reported one previous identical eruption in the same location 2 years earlier, alter 1 month of taking carbamazepine tablets. Patch tests were positive to carbamazepine 10% on postlesional skin at D2 and D3, while an open test on hyperpigmented skin and patch tests on normal skin remained negative (16).

In patients with fixed drug eruptions (FDE) caused by delayed-type hypersensitivity, the diagnosis is usually confirmed by a positive patch test with the drug on previously affected skin. Authors from Finland have used an alternative method of topical provocation. The test compound, the drug 10% in petrolatum and sometimes also in 70% alcohol and in DMSO, was applied once and without occlusion over the entire surface of one or several inactive (usually pigmented) sites of FDE lesions. The patients were followed as in-patients for 24 hours. A reaction was regarded as positive when a clearly demarcated erythema lasting at least 6 hours was seen. Of 4 patients with FDE from carbamazepine, all had a positive topical provocation (65,66).

See also the section 'Case series with various or unknown types of drug reactions' above, ref. 11.

Drug reaction with eosinophilia and systemic symptoms (DRESS)/anticonvulsant hypersensitivity syndrome

The literature on patch testing in anticonvulsant hypersensitivity syndrome up to August 2008 has been reviewed in ref. 20.

Case series

Between January 1998 and December 2008, in a university hospital in Portugal, 56 patients with DRESS were investigated with patch testing of the suspected drugs. Carbamazepine (1%, 5%, 10% and 20% pet.) was tested in 18 patients and there were 13 (72%) positive patch test reactions. Fifty controls were negative. Relevance was confirmed by a positive accidental rechallenge in one patient (63).

In a multicenter investigation in France, of 72 patients patch tested for DRESS, 46 (64%) had positive patch tests to drugs, including 11 to carbamazepine (58).

In a university hospital in Taipei, Taiwan, in the period January 2009 to February 2011, 10 patients were diagnosed with DRESS caused by carbamazepine. All were patch tested with carbamazepine commercial pulverized tablets 10% and 30% pet. and 7 had a positive reaction (21). See the Comment in the section 'Stevens-Johnson syndrome/toxic epidermal necrolysis (SJS/TEN)' below, ref. 21.

Six of 7 patients from The Netherlands with DRESS from carbamazepine, characterized by fever, rash, facial edema, lymphadenopathy, impaired liver function, eosinophilia and atypical lymphocytes in the peripheral blood, later had positive patch tests to carbamazepine, tested at 10%, 20% and 40% in petrolatum (probably prepared from commercial tablets). Lymphocyte stimulation tests to carbamazepine were positive in all 7 individuals (18).

In a group of 15 patients with 'anticonvulsant hypersensitivity syndrome', 6 patients (40%) had a positive patch test to carbamazepine; details are unavailable to the author, but not all patients appeared to have all symptoms necessary to reliably diagnose anticonvulsant hypersensitivity syndrome (synonym: DRESS from anticonvulsant drugs) (22).

Four cases of DRESS were reported from Belgium in 1995. All patients had fever; skin rashes were erythroderma, maculopapular and morbilliform (one had no skin eruption), 2 had hepatosplenomegaly and the other 2 respiratory problems, one had conjunctivitis, 2 pharyngitis, and 3 facial and peripheral edema. All had elevated liver enzymes. Patch tests were performed with carbamazepine 100%, 10%, 1% and 0.1% in petrolatum and acetone and all had strongly positive tests to all concentrations in both vehicles (50).

Four patients with the syndrome of skin eruption, fever and lymphadenopathy (probably DRESS) were reported from Finland in 1989. Patch tests with 1-5% carbamazepine in pet. were positive in all cases. Lymphocyte transformation tests were positive in 2 of the 3 investigated (76). From Japan, a series of 4 patients with anticonvulsant hypersensitivity syndrome caused by carbamazepine with a positive patch test reaction to this drug was reported in 1990 (3). Four children with DRESS caused by carbamazepine with a positive patch test were reported from Serbia in 2019 (88).

Three cases of DRESS to carbamazepine with positive patch tests to this drug were reported from Spain in 2006 (8), 3 from Switzerland in 2020 (34, overlap with ref. 35), 2 from Switzerland in 2020 (35, overlap with ref. 34), 2 from Germany in 2006 (10), 2 from Tunisia in 2017 (11), 2 from Japan in 2001 (57), 2 from Italy in 2006 (2), two pediatric patients from Turkey in 2017 (60) and 3 from Spain in 1994 (83). Five cases of DRESS from carbamazepine with a positive patch test to this antiepileptic drug were reported from Brazil in 2021 (90).

Case reports (examples)

A 40-year-old woman developed drug reaction with eosinophilia and systemic symptoms (DRESS) while using oxcarbazepine, carbamazepine and fluvoxamine. Patch tests were positive to fluvoxamine 12.5%, carbamazepine 20% and oxcarbazepine 12.5%, all in phosphate-buffered saline. Lymphocyte transformation tests were also positive to carbamazepine (stimulation index 28.5), oxcarbazepine (stimulation index 6) and to fluvoxamine (stimulation index 6.9). As the patient had reactions to 2 unrelated drugs, this was a case of multiple drug hypersensitivity (4).

A 36-year-old man developed drug hypersensitivity syndrome to carbamazepine with a maculopapular rash on his arms, thighs and trunk, fever, lymphadenopathy, oronasal ulcers, conjunctival erythema, pharyngitis with features of Stevens-Johnson syndrome, and lymphocytosis with slight eosinophilia. Treatment with carbamazepine was discontinued and methylprednisolone started, which led to swift improvement. Two weeks later, valproate was started. Immediately a widespread exanthematous rash similar to his previous one, fever and adenopathy recurred. Skin biopsy revealed findings consistent with Stevens-Johnson syndrome. Valproate was discontinued. Patch testing to carbamazepine and valproic acid performed 4 months later was positive for both (5).

A 14-year-old boy presented with erythroderma and fever 44 days after carbamazepine intake for absence epilepsy. Laboratory examinations showed eosinophilia and elevated liver enzymes, and thoracic imaging revealed interstitial pneumonitis. A patch test to carbamazepine 5% pet. performed 6 weeks after recovery was positive. About 8 months later, the patient developed the same symptoms of DRESS 52 days after lamotrigine intake. Lamotrigine was stopped and all symptoms disappeared. A patch test to lamotrigine 5% pet., performed 6 weeks after recovery, was positive (23). As many others, these authors suggested a 'cross-reaction' to carbamazepine and lamotrigine, where, in fact, this was a case of multiple drug hypersensitivity and sequential independent sensitization.

One month after a 13-year-old girl had started to take carbamazepine at 300 mg daily for major epilepsy, she developed a high fever of 40°C and general malaise. She was treated with amoxicillin under the suspicion of bacterial infection. A week later, spotty erythema appeared over her entire body. It soon became confluent, resulting in severe erythroderma. At the same time, she developed systemic lymphadenopathy and hepatosplenomegaly. Although she stopped taking both carbamazepine and amoxicillin and was treated with oral steroid, the desquamative erythroderma and generalized lymphadenopathy showed little improvement. Examination of the peripheral blood showed a white blood cell count of 12,700/mm^3 with 17% eosinophils and 4% atypical lymphocytes. Before the eruption had cleared, a patch test to carbamazepine 2% pet. was negative. However, 3 months after regression of the skin lesions, a patch test was positive to carbamazepine 1% pet. and negative to amoxicillin (46). This was very likely a case of DRESS, which term had not yet been introduced at that time.

A patient developed a generalized skin rash with systemic involvement 2 months after beginning carbamazepine treatment for trigeminal neuralgia. Skin biopsy specimens suggested mycosis fungoides. Complete remission of the clinical and pathologic changes was observed after drug discontinuation. Patch tests were positive to carbamazepine 0.1%, 1% and 2% pet. The abstract mentions the diagnosis 'pseudolymphoma due to carbamazepine', but 'carbamazepine hypersensitivity syndrome mimicking mycosis fungoides' as stated in the title seems more appropriate (details unavailable to the author) (48).

A 51-year-old man developed DRESS which was attributed to the use of amoxicillin, carbamazepine and acyclovir. Patch tests were positive to all 3 drugs. The patch tests were repeated nearly 4 years later, and carbamazepine and acyclovir, but not amoxicillin, again gave positive results (62).

Other single case reports
Single cases of DRESS/anticonvulsant hypersensitivity syndrome from carbamazepine with a positive patch test to this drug have also been reported in refs. 6,7,9,24,25,27,29,30,33,47,51-56,59,68,77,80,83, 85 and 87. See also the section 'Case series with various or unknown types of drug reactions' above, refs. 11, 32, 89 and 92.

Stevens-Johnson syndrome/toxic epidermal necrolysis (SJS/TEN)
See also the section 'Drug reaction with eosinophilia and systemic symptoms (DRESS)/anticonvulsant hypersensitivity syndrome', ref. 5, and the section 'Case series with various or unknown types of drug reactions' above, ref. 11.

SJS/TEN
In a university hospital in Taipei, Taiwan, in the period January 2009 to February 2011, 16 patients were diagnosed with SJS (n=14), SJS/TEN (n=1) or TEN (n=1) caused by carbamazepine (CBZ). This group included seven women and nine men, with an age range of 23-74 years (mean age 50 years). Of the 16 patients, patch testing revealed four strong reactions (++), six positive (+), four doubtful (?+) and two negative reactions to 30% CBZ. Most patients with positive patch test results had stronger reactions to 30% CBZ than to 10% CBZ. Ten controls were negative. Patch testing with other aromatic anticonvulsants (pulverized tablets 10% and 30% pet., carbamazepine 10,11-epoxide [a major CBZ metabolite] pure chemical 10% and 30% pet.) showed co-reactions to CBZ 10,11-epoxide in 6/13 patients (46%), to oxcarbazepine in 3/15 (19%), to phenytoin in 6/16 (38%) and to lamotrigine in 2/16 individuals (13%). These were termed 'cross-reactions', but it was not mentioned whether the patients has been treated with these medicaments before (apart from the fact that at least 4 had used phenytoin and 2 oxcarbazepine). The genetic allele HLA-B*1502 was present in 13 of 16 patients with SJS /TEN (81%), whereas only one of ten (10%) healthy controls carried this allele (21).
Comment. The picture presented in this article of 'positive' patch test reactions to other aromatic anticonvulsants clearly shows that reactions have been scored as positive that most definitely did not qualify for this score.
   A 3-year-old boy had developed overlap SJS/TEN 3 weeks after commencing carbamazepine following a first incidence of seizure. Ten months later, all drugs used were patch tested and only carbamazepine showed a positive result (26).

SJS
A 38-year-old woman was started on carbamazepine plus valproic acid. After 2 days, she developed Stevens-Johnson syndrome. Ten years later, she had positive patch tests to carbamazepine tested as pure powder at 5, 10, 15, and 20% w/w in white petrolatum and to valproic acid tested at 15, 30, 45, and 60% w/w in white petrolatum. 34 controls were negative to all test substances (2).
   Japanese investigators in 1988 reported a case of Stevens-Johnson syndrome due to carbamazepine. Patch tests were positive on previously involved hyperpigmented skin only. They suspected the persistence of effector T cells at such sites (12). One patient from China developed Stevens-Johnson syndrome from carbamazepine and later showed a positive patch test to this anticonvulsant at 1% and 10% pet. (19). Two children with SJS from carbamazepine had a positive patch test to the drug 10% pet. (88) resp. 1% and 10% active ingredients in pet. (89).

TEN
A 19-year-old woman developed toxic epidermal necrolysis (TEN) with 60% body involvement while using carbamazepine. Later, she reacted to patch tests with 1% (+) and 5% (++) carbamazepine, probably in methyl alcohol. Five controls were negative. Histopathology of the patch test was largely similar to that of a TEN lesion, but with less epidermal cell damage (17).
   A 14-year-old girl developed toxic epidermal necrolysis from carbamazepine and had a positive patch test to this drug (84). See also the section 'Case series with various or unknown types of drug reactions' above, ref. 92.

Photosensitivity
A man aged 45, who was taking carbamazepine for seizures, developed eczema, first on sun-exposed areas of the skin, but later progressing to erythroderma. Phototests were normal. Patch tests with carbamazepine and 2 other drugs used by the patient were ?+ to carbamazepine pulverized tablets 10% pet. Photopatch tests, using 10 J/cm^2 UVA, were positive (+) to carbamazepine. Forty controls were negative. After carbamazepine was eliminated, no recurrences have occurred. The patient was diagnosed with coexistent allergic and photoallergic drug eruption from carbamazepine (14).
   A similar case was that of a 46-year-old patient who developed erythematous and papular (eczematous) eruptions on sun-exposed areas such as the face, upper chest and dorsa of the hands, which began after having used carbamazepine 200 mg daily for one month for trigeminal neuralgia. Oral provocation with the drug produced papulovesicular eruptions that were greatly potentiated by long-wave ultraviolet (UVA) irradiation. Six months after cessation of the intake of carbamazepine, patch tests were positive to this drug at 1% pet. (++) (?+ to 0.1% and

negative to lower concentrations) as were photopatch tests using 15.5 J/cm^2 UVA to carbamazepine 0.1, 0.01, 0.001, and 0.0001% pet. (negative when using UVB for irradiation). A lymphocyte stimulation test was also considered positive, but the stimulation index was 2.25 only (15).

A 58-year-old man developed a lichenoid reaction on light-exposed areas and subacute prurigo separate from the sun-exposed areas after receiving 200 mg of carbamazepine daily for one year. A positive oral provocation test with irradiation was obtained after administering 2 mg of carbamazepine. A patch test with carbamazepine was negative, but a photopatch test to this drug was positive. The patient was diagnosed with photosensitive lichenoid reaction accompanied by non-photosensitive subacute prurigo caused by carbamazepine (67).

Dermatitis/eczematous eruption

About one month after having taken carbamazepine for epileptic seizures, a 16-year-old boy developed a generalized eczematous eruption, which showed a marked predilection for the flexural areas such as the neck, axillae, groin, antecubital and popliteal fossa with scattered papulovesicular lesions on the face, trunk and extremities. Within a few days the entire skin became erythrodermic with onset of palmoplantar desquamation. Histopathologic examination revealed spongiotic epidermal changes. Personal or family history of atopic dermatitis and signs of a mucosal atopy were absent. A mild xerosis and keratosis pilaris were the only existing minor criteria for evidence of an atopic skin. Patch tests were positive to carbamazepine 1% and 5% pet., while 15 controls were negative. The patient was diagnosed with carbamazepine-induced eczematous eruption clinically resembling atopic dermatitis (43).

A 24-year-old woman developed eczematous erythroderma with palmoplantar pompholyx while using carbamazepine. Patch tests were positive to carbamazepine, but not to oxcarbazepine, carbamazepine 10,11-epoxide and 2 other metabolites (42). See also the section 'Photosensitivity' above, refs. 14 and 15.

Other cutaneous adverse drug reactions

One patient from China developed erythema multiforme from carbamazepine and later showed a positive patch test to this anticonvulsant at 1% and 10% pet. (19). One patient had a lichenoid eruption from carbamazepine with a positive patch test to this anticonvulsant drug (32). In a Japanese investigation, 4 of 6 patients patch tested with carbamazepine because of an 'anticonvulsant-induced drug eruptions' had positive reaction to this drug. The nature of the skin eruption was not mentioned (3).

In a group of 13 patients with morbilliform rashes, 4 with SJS/TEN and 3 with DRESS, there were 10 positive patch tests to suspected drugs, including 3 to carbamazepine. It was not specified which drug reactions carbamazepine had caused (36). In Finland, in the period 1989-2001, 826 patients with suspected cutaneous drug eruptions were patch tested and 89 had one or more positive reactions. Of these individuals, 7 reacted to carbamazepine; the nature of the drug eruptions in these patients was not mentioned (37).

A 55-year-old woman developed erythema and edema of the face and pruritus of the legs after having used carbamazepine for 2 weeks. Patch tests were positive to carbamazepine 10% pet. She also reacted to topiramate which she had used and which belongs to a different class of anticonvulsant drugs (71). In the UK, in 1989 10 patients with 'allergic reactions to carbamazepine' were patch tested with the drug 1% pet. and 3 had a positive response. Clinical details were not provided (76).

Cross-reactions, pseudo-cross-reactions and co-reactions

This subject is discussed in Chapter 3.284 Lamotrigine. Patients sensitized to carbamazepine may cross-react to oxcarbazepine (38). A patient who developed photoallergic dermatitis from oral clomipramine (patch tests positive, photopatch tests stronger positive) also had contact allergy to carbamazepine. These may have been concomitant sensitizations, as the patient used both drugs. However, the chemical structures of these drugs are closely related, both being tricyclic compounds with a nitrogen substitution in the central ring structure. Therefore, cross-reactivity to carbamazepine from primary imipramine (photo)sensitization cannot be excluded (64).

LITERATURE

1 The data in the section 'General' may have been obtained from literature discussed in this chapter, but mostly also or exclusively from one or more of the following online sources: ChemIDPlus Advanced, PubChem, DrugBank, RxList, Drug Central, Drugs.com, and Wikipedia
2 Romano A, Pettinato R, Andriolo M, Viola M, Guéant-Rodriguez RM, Valluzzi RL, et al. Hypersensitivity to aromatic anticonvulsants: in vivo and in vitro cross-reactivity studies. Curr Pharm Des 2006;12:3373-3381
3 Osawa J, Naito S, Aihara M, Kitamura K, Ikezawa Z, Nakajima H. Evaluation of skin test reactions in patients with non-immediate type drug eruptions. J Dermatol 1990;17:235-239
4 Gex-Collet C, Helbling A, Pichler WJ. Multiple drug hypersensitivity – proof of multiple drug hypersensitivity by patch and lymphocyte transformation tests. J Investig Allergol Clin Immunol 2005;15:293-296

5   Arévalo-Lorido JC, Carretero-Gómez J, Bureo-Dacal JC, Montero-Leal C, Bureo-Dacal P. Antiepileptic drug hypersensitivity syndrome in a patient treated with valproate. Br J Clin Pharmacol 2003;55:415-416

6   Nigen SR, Shapiro LE, Knowles SR, Neuman MG, Shear NH. Utility of patch testing in patients with anticonvulsant-induced hypersensitivity syndrome. Dermatitis 2008;19:349-350

7   Kim CW, Choi GS, Yun CH, et al. Drug hypersensitivity to previously tolerated phenytoin by carbamazepine-induced DRESS syndrome. J Korean Med Sci 2006;21:768-772

8   Gaig P, García-Ortega P, Baltasar M, Bartra J. Drug neosensitization during anticonvulsant hypersensitivity syndrome. J Investig Allergol Clin Immunol 2006;16:321-326

9   Troost RJ, Oranje AP, Lijnen RL, et al. Exfoliative dermatitis due to immunologically confirmed carbamazepine hypersensitivity. Pediatr Dermatol 1996;13:316-320

10   Seitz CS, Pfeuffer P, Raith P, Bröcker EB, Trautmann A. Anticonvulsant hypersensitivity syndrome: cross-reactivity with tricyclic antidepressant agents. Ann Allergy Asthma Immunol 2006;97:698-702

11   Ben Mahmoud L, Bahloul N, Ghozzi H, Kammoun B, Hakim A, Sahnoun Z, et al. Epicutaneous patch testing in delayed drug hypersensitivity reactions induced by antiepileptic drugs. Therapie 2017;72:539-545

12   Osawa J, Aihara M, Ikezawa Z. A case of carbamazepin-induced Stevens-Johnson syndrome with positive patch test reactions on the site of pigmentation. Skin Research (Osaka) 1988;30:200-203 (Article in Japanese, data cited in ref. 13)

13   Kikuchi K, Tsunoda T, Tagami H. Generalized drug eruption due to mexiletine hydrochloride: topical provocation on previously involved skin. Contact Dermatitis 1991;25:70-72

14   Lee AY, Joo HJ, Chey WY, Kim YG. Photopatch testing in seven cases of photosensitive drug eruptions. Ann Pharmacother 2001;35:1584-1587

15   Terui T, Tagami H. Eczematous drug eruption from carbamazepine: coexistence of contact and photocontact sensitivity. Contact Dermatitis 1989;20:260-264

16   De Argila D, Angeles Gonzalo M, Rovira I. Carbamazepine-induced fixed drug eruption. Allergy 1997;52:1039

17   Friedmann PS, Strickland I, Pirmohamed M, Park K. Investigation of mechanisms in toxic epidermal necrolysis induced by carbamazepine. Arch Dermatol 1994;130:598-604

18   Houwerzijl J, De Gast GC, Nater JP, Esselink MT, Nieweg HO. Lymphocyte-stimulation tests and patch tests in carbamazepine hypersensitivity. Clin Exp Immunol 1977;29:272-277

19   Liao HT, Hung KL, Wang CF, Chen WC. Patch testing in the detection of cutaneous reactions caused by carbamazepine. Zhonghua Min Guo Xiao Er Ke Yi Xue Hui Za Zhi 1997;38:365-369 (Article in Chinese, data cited in ref. 20.

20   Elzagallaai AA, Knowles SR, Rieder MJ, Bend JR, Shear NH, Koren G. Patch testing for the diagnosis of anticonvulsant hypersensitivity syndrome: a systematic review. Drug Saf 2009;32:391-408

21   Lin YT, Chang YC, Hui RC, Yang CH, Ho HC, Hung SI, Chung WH. A patch testing and cross-sensitivity study of carbamazepine-induced severe cutaneous adverse drug reactions. J Eur Acad Dermatol Venereol 2013;27:356-364

22   Galindo PA, Borja J, Gomez E, Mur P, Gudín M, García R, et al. Anticonvulsant drug hypersensitivity. J Invest Allergol Clin Immunol 2002;12:299-304

23   Aouam K, Ben Romdhane F, Loussaief C, Salem R, Toumi A, Belhadjali H, et al. Hypersensitivity syndrome induced by anticonvulsants: possible cross-reactivity between carbamazepine and lamotrigine. J Clin Pharmacol 2009;49:1488-1491

24   Gall H, Merk H, Scherb W, Sterry W. Anticonvulsant hypersensitivity syndrome to carbamazepine. Hautarzt 1994;45:494-498 (Article in German)

25   Chauhan A, Anand S, Thomas S, Subramanya HC, Pradhan G. Carbamazepine induced DRESS syndrome. J Assoc Physicians India 2010;58:634-636

26   Atanasković-Marković M, Medjo B, Gavrović-Jankulović M, Ćirković Veličković T, Nikolić D, Nestorović B. Stevens-Johnson syndrome and toxic epidermal necrolysis in children. Pediatr Allergy Immunol 2013;24:645-649

27   Aouam K, Fredj Nadia B, Amel C, Naceur B. Amoxicillin-induced hypersensitivity after DRESS to carbamazepine. World Allergy Organ J 2010;3:220-222

28   Inadomi T. Drug rash with eosinophilia and systemic symptoms (DRESS): changing carbamazepine to phenobarbital controlled epilepsy without the recurrence of DRESS. Eur J Dermatol 2010;20:220-222

29   Liccioli G, Mori F, Parronchi P, Capone M, Fili L, Barni S, Sarti L, Giovannini M, Resti M, Novembre EM. Aetiopathogenesis of severe cutaneous adverse reactions (SCARs) in children: A 9-year experience in a tertiary care paediatric hospital setting. Clin Exp Allergy 2020;50:61-73

30   Buyuktiryaki AB, Bezirganoglu H, Sahiner UM, Yavuz ST, Tuncer A, Kara A, et al. Patch testing is an effective method for the diagnosis of carbamazepine-induced drug reaction, eosinophilia and systemic symptoms (DRESS) syndrome in an 8-year-old girl. Australas J Dermatol 2012;53:274-277

31   Landry Q, Zhang S, Ferrando L, Bourrain JL, Demoly P, Chiriac AM. Multiple drug hypersensitivity syndrome in a large database. J Allergy Clin Immunol Pract 2019;8:258

32 Shiny TN, Mahajan VK, Mehta KS, Chauhan PS, Rawat R, Sharma R. Patch testing and cross sensitivity study of adverse cutaneous drug reactions due to anticonvulsants: A preliminary report. World J Methodol 2017;7:25-32

33 Studer M, Waton J, Bursztejn AC, Aimone-Gastin I, Schmutz JL, Barbaud A. Does hypersensitivity to multiple drugs really exist? Ann Dermatol Venereol 2012;139:375-380 (Article in French)

34 Jörg L, Yerly D, Helbling A, Pichler W. The role of drug, dose and the tolerance/intolerance of new drugs in multiple drug hypersensitivity syndrome (MDH). Allergy 2020;75:1178-1187

35 Jörg L, Helbling A, Yerly D, Pichler WJ. Drug-related relapses in drug reaction with eosinophilia and systemic symptoms (DRESS). Clin Transl Allergy 2020;10:52

36 Hassoun-Kheir N, Bergman R, Weltfriend S. The use of patch tests in the diagnosis of delayed hypersensitivity drug eruptions. Int J Dermatol 2016;55:1219-1224

37 Lammintausta K, Kortekangas-Savolainen O. The usefulness of skin tests to prove drug hypersensitivity. Br J Dermatol 2005;152:968-974

38 Troost RJ, Van Parys JA, Hooijkaas H, van Joost T, Benner R, Prens EP. Allergy to carbamazepine: parallel in vivo and in vitro detection. Epilepsia 1996;37:1093-1099

39 Maquiera E, Yañez S, Fernández L, Rodríguez F, Picáns I, Sánchez I, Jeréz J. Mononucleosis-like illness as a manifestation of carbamazepine-induced anticonvulsant hypersensitivity syndrome. Allergol Immunopathol (Madr) 1996;24:87-88

40 Prens EP, Troost RJJ, Van Parys JAP, Benner R, van Joost Th. The value of the lymphocyte proliferation assay in detection of carbamazepine allergy. Contact Dermatitis 1990;23:292 (Abstract)

41 Lee AY, Choi J, Chey WY. Patch testing with carbamazepine and its main metabolite carbamazepine epoxide in cutaneous adverse drug reactions to carbamazepine. Contact Dermatitis 2003;48:137-139

42 Duhra P, Foulds IS. Structural specificity of carbamazepine-induced dermatitis. Contact Dermatitis 1992;27:325-326

43 Ozkaya-Bayazit E, Gungor H. Carbamazepine induced eczematous eruption-clinically resembling atopic dermatitis. J Eur Acad Dermatol Venereol 1999;12:182-183

44 Vaillant L, Camenen I, Lorette G. Patch testing with carbamazepine: reinduction of an exfoliative dermatitis [letter]. Arch Dermatol 1989;125:299

45 Scerri L, Shall L, Zaki I. Carbamazepine-induced anticonvulsant hypersensitivity syndrome: pathogenic and diagnostic considerations. Clin Exp Dermatol 1993;18:540-542

46 Okuyama R, Ichinohasama R, Tagami H. Carbamazepine induced erythroderma with systemic lymphadenopathy. J Dermatol 1996;23:489-494

47 Balatsinou C, Milano A, Caldarella MP, Laterza F, Pierdomenico SD, Cuccurullo F, Neri M. Eosinophilic esophagitis is a component of the anticonvulsant hypersensitivity syndrome: description of two cases. Dig Liver Dis 2008;40:145-148

48 Miranda-Romero A, Pérez-Oliva N, Aragoneses H, Bastida J, Raya C, González-Lopez A, García-Muñoz M. Carbamazepine hypersensitivity syndrome mimicking mycosis fungoides. Cutis 2001;67:47-51

49 Pasmans SG, Bruijnzeel-Koomen CA, van Reijsen FC. Skin reactions to carbamazepine. Allergy 1999;54:649-650

50 De Vriese AS, Philippe J, Van Renterghem DM, De Cuyper CA, Hindryckx PH, Matthys EG, et al. Carbamazepine hypersensitivity syndrome: report of 4 cases and review of the literature. Medicine (Baltimore) 1995;74:144-151

51 Cox NH, Johnston SR, Marks J, Bates D. Extensive carbamazepine eruption with eosinophilia and pulmonary infiltrate. Postgrad Med J 1988;64(749):249

52 Aouam K, Bel Hadj Ali H, Youssef M, Chaabane A, Amri M, Boughattas NA ,et al. Carbamazepine-induced DRESS and HHV6 primary infection: the importance of skin tests. Epilepsia 2008;49:1630-1633

53 Aizawa S, Yoshiji H, Kojima H, Houki T, Morita S, Kitade M, et al. A case of carbamazepine induced hypersensitivity syndrome complicated with severe hepatitis. Liver 2002;43:400-405 (Data cited in ref. 52, article could not be traced)

54 Hirashima N, Misago N, Nakafusa J, Narisawa Y. Two cases of drug-induced hypersensitivity syndrome. Nishinihon J Dermatol 2003;65:365-369 (Article in Japanese, data cited in ref. 52)

55 Zeller A, Schaub N, Steffen I, Battegay E, Hirsch HH, Bircher AJ. Drug hypersensitivity syndrome to carbamazepine and human herpes virus 6 infection: case report and literature review. Infection 2003;31:254-256

56 Sugiyama M, Hosaka H, Kojima S, Kanda H, Akiyama M, Sueki H, et al. A case of hypersensitivity syndrome induced by Tegretol (carbamazepine)—the condition improved without systemic steroid because it was relatively mild. Jpn J Dermatoallergol 2001;9:37-42 (Article in Japanese, data cited in ref. 52)

57 Umebayashi Y. Drug eruption due to carbamazepine—case showing increase of anti-HHV-6 antibody and case not showing. Jpn J Dermatoallergol 2001;9:137-141 (Article in Japanese, data cited in ref. 52)

58 Barbaud A, Collet E, Milpied B, Assier H, Staumont D, Avenel-Audran M, et al. A multicentre study to determine the value and safety of drug patch tests for the three main classes of severe cutaneous adverse drug reactions. Br J Dermatol 2013;168:555-562

59 Wolkenstein P, Chosidow O, Fléchet ML, Robbiola O, Paul M, Dumé L, et al. Patch testing in severe cutaneous adverse drug reactions, including Stevens-Johnson syndrome and toxic epidermal necrolysis. Contact Dermatitis 1996;35:234-236

60 Dibek Misirlioglu E, Guvenir H, Bahceci S, Haktanir Abul M, Can D, et al. Severe cutaneous adverse drug reactions in pediatric patients: A multicenter study. J Allergy Clin Immunol Pract 2017;5:757-763

61 Barbaud A, Reichert-Penetrat S, Tréchot P, Jacquin-Petit MA, Ehlinger A, Noirez V, et al. The use of skin testing in the investigation of cutaneous adverse drug reactions. Br J Dermatol 1998;139:49-58

62 Pinho A, Marta A, Coutinho I, Gonçalo M. Long-term reproducibility of positive patch test reactions in patients with non-immediate cutaneous adverse drug reactions to antibiotics. Contact Dermatitis 2017;76:204-209

63 Santiago F, Gonçalo M, Vieira R, Coelho S, Figueiredo A. Epicutaneous patch testing in drug hypersensitivity syndrome (DRESS). Contact Dermatitis 2010;62:47-53

64 Ljunggren B, Bojs G. A case of photosensitivity and contact allergy to systemic tricyclic drugs, with unusual features. Contact Dermatitis 1991;24:259-265

65 Alanko K. Topical provocation of fixed drug eruption. A study of 30 patients. Contact Dermatitis 1994;31:25-27

66 Alanko K, Stubb S, Reitamo S. Topical provocation of fixed drug eruption. Br J Dermatol 1987;116:561-567

67 Yasuda S, Mizuno N, Kawabe Y, Sakakibara S. Photosensitive lichenoid reaction accompanied by non-photosensitive subacute prurigo caused by carbamazepine. Photodermatology 1988;5:206-210

68 Gómez Torrijos E, Extremera Ortega AM, Gonzalez Jimenez O, Joyanes Romo JB, Gratacós Gómez AR, Garcia Rodriguez R. Excited skin syndrome ("angry back" syndrome) induced by proximity of carbamazepine to another drug with strong positive allergic reaction in patch test: A first of its kind. Contact Dermatitis 2019;81:405-406

69 Braun V, Darrigade A-S, Milpied B. Positive patch test reaction to carbamazepine after a very long delay. Contact Dermatitis 2018;79:240-241

70 Brockow K, Garvey LH, Aberer W, Atanaskovic-Markovic M, Barbaud A, Bilo MB, et al.; ENDA/EAACI Drug Allergy Interest Group. Skin test concentrations for systemically administered drugs – an ENDA/EAACI Drug Allergy Interest Group position paper. Allergy 2013;68:702-712

71 Schiavino D, Nucera E, Buonomo A, Musumeci S, Pollastrini E, Roncallo C, et al. A case of type IV hypersensitivity to topiramate and carbamazepine. Contact Dermatitis 2005;52:161-162

72 Puig L, Nadal C, Fernández-Figueras MT, Alomar A. Carbamazepine-induced drug rashes: diagnostic value of patch tests depends on clinico-pathologic presentation. Contact Dermatitis 1996;34:435-437

73 Moran M, Estella JS, Unamuno P, Martfn-Pascual A. Eritrodermia por carbamazepina. Estudio de 6 casos. Bol Informativo GEIDC 1986;11:26-28 (Data cited in refs. 72 and 78).

74 Corazza M, Mantovani L, Casetta I, Virgili A. Exfoliative dermatitis caused by carbamazepine in a patient with isolated IgA deficiency. Contact Dermatitis 1995;33:447

75 Alanko K. Patch testing in cutaneous reactions caused by carbamazepine. Contact Dermatitis 1993;29:254-257

76 Motley RJ, Reynolds AJ. Carbamazepine and patch testing. Contact Dermatitis 1989;21:285-286

77 Romaguera C, Grimalt F, Vilaplana J, Azon A. Erythroderma from carbamazepine. Contact Dermatitis 1989;20:304-305

78 Silva R, Machado A, Brandão M, Gonçalo S. Patch test diagnosis in carbamazepine erythroderma. Contact Dermatitis 1986;15:254-255

79 Camarasa JG. Patch test diagnosis of exfoliative dermatitis due to carbamazepine. Contact Dermatitis 1985;12:49

80 Fathallah N, Slim R, Rached S, Ben Salem C, Ghariani N, Nouira R. Carbamazepine-induced DRESS with severe eosinophilia confirmed by positive patch test. Dermatitis 2014;25:282-284

81 Duran-Ferreras E, Mir-Mercader J, Morales-Martinez M D, Martinez-Parra C. Anticonvulsant hypersensitivity syndrome with severe repercussions in the skin and kidneys. Rev Neurol 2004;38:1136-1138 (Article in Spanish)

82 Ben Fredj N, Aouam K, Chaabane A, Toumi A, Ben Rhomdhane F, Boughattas N, Chakroun M. Hypersensitivity to amoxicillin after drug rash with eosinophilia and systemic symptoms (DRESS) to carbamazepine and allopurinol: a possible co-sensitization. Br J Clin Pharmacol 2010;70:273-276

83 Jones M, Fernández-Herrera J, Dorado JM, Sols M, Ruiz M, García-Díez A. Epicutaneous test in carbamazepine cutaneous reactions. Dermatology 1994;188:18-20

84 Torres MJ, Corzo JL, Leyva L, Mayorga C, Garcia-Martin FJ, Antunez C, et al. Differences in the immunological responses in drug- and virus-induced cutaneous reactions in children. Blood Cells Mol Dis 2003;30:124-131

85 Houwerzijl J, de Gast GC, Nater JP. Patch tests in drug eruptions. Contact Dermatitis 1975;1:180-181

86 Oussalah A, Yip V, Mayorga C, Blanca M, Barbaud A, Nakonechna A, et al.; Task Force "Genetic predictors of drug hypersensitivity" of the European Network on Drug Allergy (ENDA), European Academy of Allergy, Clinical Immunology (EAACI). Genetic variants associated with T-Cell mediated cutaneous adverse drug reactions: A prisma-compliant systematic review - an EAACI Position Paper. Allergy 2020;75:1069-1098

87 Özkaya E, Yazganoğlu KD. Sequential development of eczematous type "multiple drug allergy" to unrelated drugs. J Am Acad Dermatol 2011;65:e26-e29.

88 Atanasković-Marković M, Janković J, Tmušić V, Gavrović-Jankulović M, Ćirković Veličković T, Nikolić D, Škorić D. Hypersensitivity reactions to antiepileptic drugs in children. Pediatr Allergy Immunol 2019;30:547-552

89 Guvenir H, Dibek Misirlioglu E, Civelek E, Toyran M, Buyuktiryaki B, Ginis T, et al. The frequency and clinical features of hypersensitivity reactions to antiepileptic drugs in children: a prospective study. J Allergy Clin Immunol Pract 2018;6:2043-2050

90 Perelló MI, de Maria Castro A, Nogueira Arraes AC, Caracciolo Costa S, Lacerda Pedrazzi D, Andrade Coelho Dias G, et al. Severe cutaneous adverse drug reactions: diagnostic approach and genetic study in a Brazilian case series. Eur Ann Allergy Clin Immunol. 2021 Mar 16. doi: 10.23822/EurAnnACl.1764-1489.193. Epub ahead of print.

91 Blasco Sarramían A, Pinilla Moraza J, Atares Pueyo MB, Lobera Labairu T, Izquier-Do Cuartero MA, San Román Lazcano FJ. Eritrodermia por carbamacepina. Su diagnóstico mediante parche cutáneo [Erythroderma caused by carbamazepine. Diagnosis with skin patch]. An Med Interna 1993;10:341-342 (Article in Spanish)

92 Büyük Yaytokgil Ş, Güvenir H, Külhaş Celík İ, Yilmaz Topal Ö, Karaatmaca B, Civelek E, et al. Evaluation of drug patch tests in children. Allergy Asthma Proc 2021;42:167-174

Chapter 3.84   CARBENICILLIN

IDENTIFICATION

Description/definition   : Carbenicillin is the penicillin derivative that conforms to the structural formula shown
                             below
Pharmacological classes : Antibacterial agents
IUPAC name   : (2S,5R,6R)-6-[(2-Carboxy-2-phenylacetyl)amino]-3,3-dimethyl-7-oxo-4-thia-1-
                            azabicyclo[3.2.0]heptane-2-carboxylic acid
Other names   : Carboxybenzylpenicillin
CAS registry number   : 4697-36-3
EC number   : 225-171-0
Merck Index monograph   : 3063
Patch testing   : 5-10% pet.
Molecular formula   : $C_{17}H_{18}N_2O_6S$

GENERAL

Carbenicillin is a broad-spectrum, semi-synthetic penicillin antibiotic with bactericidal and beta-lactamase resistant activity. Because of the high urine levels obtained following administration, carbenicillin has demonstrated clinical efficacy in urinary infections due to susceptible strains of *Escherichia coli, Proteus mirabilis, Proteus vulgaris, Morganella morganii, Pseudomonas* species, *Providencia rettgeri, Enterobacter* species, and *Enterococci* (*S. faecalis*). This antibiotic is used for the treatment of acute and chronic infections of the upper and lower urinary tract and in asymptomatic bacteriuria due to susceptible strains of bacteria. In pharmaceutical products, carbenicillin is employed as carbenicillin disodium (CAS number 4800-94-6, EC number 225-360-8, molecular formula $C_{17}H_{16}N_2Na_2O_6S$) (1).

The classification and structures of beta-lactam antibiotics are discussed in Chapter 3.36 Amoxicillin.

CONTACT ALLERGY FROM ACCIDENTAL CONTACT

In a group of 107 workers in the pharmaceutical industry with dermatitis, investigated in Warsaw, Poland, before 1989, 3 reacted to carbenicillin, tested 20% pet. (4). In the same clinic in Warsaw, in the period 1979-1983, 27 pharmaceutical workers, 24 nurses and 30 veterinary surgeons were diagnosed with occupational allergic contact dermatitis from antibiotics. The numbers that had positive patch tests to carbenicillin (ampoule content) were 1, 1, and 0, respectively, total 2 (2).

Of 333 nurses patch tested, again, in Poland between 1979 and 1987, 2 reacted to carbenicillin 20% pet.; data on relevance were not provided (3).

Cross-reactions, pseudo-cross-reactions and co-reactions

Cross-reactions between beta-lactam antibiotics are discussed in Chapter 3.36 Amoxicillin.

LITERATURE

1   The data in the section 'General' may have been obtained from literature discussed in this chapter, but mostly also or exclusively from one or more of the following online sources: ChemIDPlus Advanced, PubChem, DrugBank, RxList, Drug Central, Drugs.com, and Wikipedia
2   Rudzki E, Rebendel P. Contact sensitivity to antibiotics. Contact Dermatitis 1984;11:41-42
3   Rudzki E, Rebandel P, Grzywa Z. Patch tests with occupational contactants in nurses, doctors and dentists. Contact Dermatitis 1989;20:247-250
4   Rudzki E, Rebandel P, Grzywa Z. Contact allergy in the pharmaceutical industry. Contact Dermatitis 1989;21:121-122

Chapter 3.85 CARBIMAZOLE

IDENTIFICATION

Description/definition : Carbimazole is the carbethoxy derivative of methimazole that conforms to the structural formula shown below
Pharmacological classes : Antithyroid agents
IUPAC name : Ethyl 3-methyl-2-sulfanylideneimidazole-1-carboxylate
Other names : 1H-Imidazole-1-carboxylic acid, 2,3-dihydro-3-methyl-2-thioxo-, ethyl ester; 1-ethoxycarbonyl-3-methyl-2-thio-4-imidazoline; carbethoxymethimazole
CAS registry number : 22232-54-8
EC number : 244-854-4
Merck Index monograph : 3069
Patch testing : 25% pet. for photopatch testing (2); powder pure for patch testing (3); 10% pet. (5)
Molecular formula : $C_7H_{10}N_2O_2S$

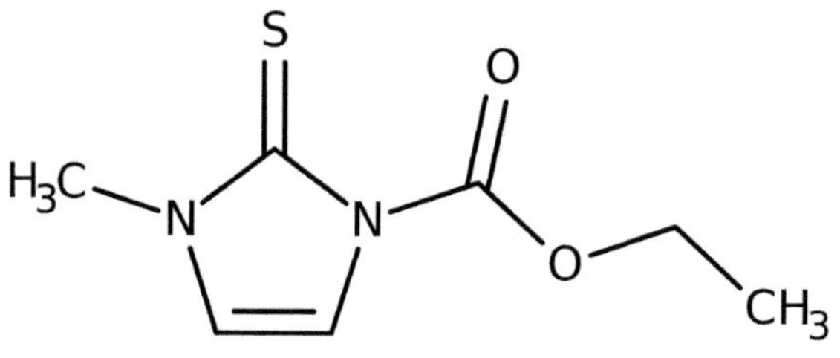

GENERAL

Carbimazole is an imidazole antithyroid agent. After oral absorption, it is metabolized to methimazole, which is responsible for the antithyroid activity. The agent decreases the uptake and concentration of inorganic iodine by thyroid, reduces the formation of di-iodotyrosine and thyroxine, and prevents the thyroid peroxidase enzyme from coupling and iodinating the tyrosine residues on thyroglobulin, hence reducing the production of the thyroid hormones T3 and T4. Carbimazole is indicated for the treatment of hyperthyroidism and thyrotoxicosis and for preparing patients for thyroidectomy (1).

CONTACT ALLERGY FROM ACCIDENTAL CONTACT

A 28-year-old man working in a pharmaceutical laboratory was exposed to various pharmaceutical powders. He started to develop recurrent itchy rashes over the face, neck and exposed parts of the forearms and hands, made worse by the sun. The patient noticed that his recurrences were usually associated with the preparation of one of 3 drugs, viz. carbimazole, chlorpromazine or sulfamethizole. Patch tests with these drugs were negative after 4 days, but carbimazole showed a positive 'lichenoid' reaction at D10. A prick test with carbimazole gave the same reaction at D10. Active sensitization to carbimazole by patch testing was suspected. A repeat patch test to carbimazole 50%, 25% and 10% in pet. was again negative at D2 and D4, but on the 10th day, a positive (+) lichenoid reaction to each concentration resulted. Next, patch and photopatch tests to carbimazole 50%, 25% and 10% pet. were performed and were negative at D4 at the unirradiated sites, but positive (+ for all 3 concentrations) at the UVA-irradiated sites. Ten controls photopatch tested to carbimazole 25% pet. were negative. The delayed patch test at D10 was explained by exposure of the test sites to sunlight through thin clothing after removal of the patch (2). This was a case of occupational photoallergic contact dermatitis.

CUTANEOUS ADVERSE DRUG REACTIONS FROM SYSTEMIC ADMINISTRATION CAUSED BY TYPE IV (DELAYED-TYPE) HYPERSENSITIVITY (as demonstrated by positive patch tests)

Cutaneous adverse drug reactions from systemic administration of carbimazole caused by type IV (delayed-type) hypersensitivity have included acute generalized exanthematous pustulosis (AGEP) (4) and generalized dermatitis (3).

Acute generalized exanthematous pustulosis (AGEP)

An 83-year-old woman presented with a widespread erythematous flaky eruption on the trunk and limbs, accentuated in the great body folds and covered with multiple non-follicular superficial pustules. Histology showed a dermal infiltrate, often perivascular, of polynuclear neutrophils and a parakeratotic epidermis containing neutrophilic pustules. The patient also had leukocytosis and elevated neutrophils. Patch tests showed carbimazole as the cause of this acute generalized exanthematous pustulosis (AGEP) (4). Details are unknown.

Dermatitis/eczematous eruption

A 36-year-old woman developed generalized dermatitis which started some days after treatment with carbimazole was re-introduced for thyrotoxicosis. Carbimazole was stopped, which resulted in an immediate clearing of the dermatitis. Patch testing with pure carbimazole powder was positive, while none of 10 control persons showed a reaction (3).

LITERATURE

1   The data in the section 'General' may have been obtained from literature discussed in this chapter, but mostly also or exclusively from one or more of the following online sources: ChemIDPlus Advanced, PubChem, DrugBank, RxList, Drug Central, Drugs.com, and Wikipedia
2   Goh CL, Ng SK. Photoallergic contact dermatitis to carbimazole. Contact Dermatitis 1985;12:58-59
3   Van Ketel WG. Allergy to carbimazole. Contact Dermatitis 1983;9:161-162
4   Grange-Prunier A, Roth B, Kleinclaus I, Fagot JP, Guillaume JC. Acute generalized exanthematous pustulosis induced by carbimazole (Neomercazole): first reported case and value of patch tests. Ann Dermatol Venereol 2006;133(8-9Pt. 1):708-710 (Article in French)
5   De Groot AC. Patch testing, 4th edition. Wapserveen, The Netherlands: acdegroot publishing, 2018 (ISBN 9789081323345)

Chapter 3.86   CARBOCROMEN

IDENTIFICATION

Description/definition   : Carbocromen is the coumarin derivative that conforms to the structural formula shown
                                              below
Pharmacological classes   : Vasodilator agents
IUPAC name   : Ethyl 2-[3-[2-(diethylamino)ethyl]-4-methyl-2-oxochromen-7-yl]oxyacetate
Other names   : Chromonar
CAS registry number   : 804-10-4
EC number   : 212-356-6
Merck Index monograph   : 3513 (Chromonar)
Patch testing   : 1% pet. or water (2,6); 0.8% solution (5)
Molecular formula   : $C_{20}H_{27}NO_5$

GENERAL

Carbocromen is member of the coumarins and a coronary vasodilator agent. In pharmaceutical products (still available in France), carbocromen is employed as carbocromen hydrochloride (CAS number 655-35-6, EC number 211-511-5, molecular formula $C_{20}H_{28}ClNO_5$) (1).

CONTACT ALLERGY FROM ACCIDENTAL CONTACT

In a group of 107 workers in the pharmaceutical industry with dermatitis, investigated in Warsaw, Poland, before 1989, one reacted to carbocromen, tested 1% pet. (2). Two nurses, aged 39 and 26, both working in the cardiology department of a hospital in Lille, France, were involved in preparing and administering solutions of carbocromen hydrochloride for perfusions, intravenous and intramuscular injections. One developed severe eyelid eczema and later lesions on the dorsa of the hands, fingers, forearms and neck. She had a positive patch test to carbocromen HCl solution. The second nurse had eczema on the dorsa of her fingers and in the neck and also had a positive reaction when patch tested with carbocromen 0.8% solution (5).

A 65-year-old man developed an erythematous, non-itchy eruption on his face, back and arms, after having taken 4 capsules containing carbocromen hydrochloride for 14 days, which subsided quickly after stopping this treatment. Patch tests were not performed, but an intradermal test with the drug 0.05 ml of 0.08% and 0.16% in saline gave a ++++ delayed reaction (4). A 24-year-old nurse, giving injections and perfusions of carbocromen HCl to patients, developed an itchy dermatitis of the eyelids and a vesicular dermatitis of the dorsum of the right hand. A new attack occurred on the right hand and face after she gave 20 intravenous injections to a patient at home. A patch test to carbocromen HCl (not specified) gave a +++ positive reactions (4). Another case of occupational contact allergy was reported from France in 1973 (3, from the same authors, no details available).

LITERATURE

1   The data in the section 'General' may have been obtained from literature discussed in this chapter, but mostly also or exclusively from one or more of the following online sources: ChemIDPlus Advanced, PubChem, DrugBank, RxList, Drug Central, Drugs.com, and Wikipedia
2   Rudzki E, Rebandel P, Grzywa Z. Contact allergy in the pharmaceutical industry. Contact Dermatitis 1989;21:121-122
3   Martin P, Bétourné M, Martin JJ, Huriez C (1973) Allergie au carbocromène. Bull Soc Fr Dermatol Syphiligr 1973;80:620
4   Huriez Cl, Martin P, Bétourné M, Martin H-J. Sensitivity to carbocromène. Contact Dermatitis Newsletter 1974;15:429
5   Huriez Cl, Martin P, Bétourné M. Allergic contact dermatitis from Carbocromène in nurses. Contact Dermatitis Newsletter 1972;12:313
6   De Groot AC. Patch testing, 4th edition. Wapserveen, The Netherlands: acdegroot publishing, 2018 (ISBN 9789081323345)

Chapter 3.87   CARBOCYSTEINE

IDENTIFICATION

Description/definition   : Carbocysteine is the L-cysteine-S-conjugate that conforms to the structural formula
                              shown below
Pharmacological classes   : Anti-infective agents, local; expectorants
IUPAC name   : (2R)-2-Amino-3-(carboxymethylsulfanyl)propanoic acid
Other names   : S-Carboxymethyl-L-cysteine
CAS registry number   : 638-23-3
EC number   : 211-327-5
Merck Index monograph   : 3072
Patch testing   : Carbocysteine and its metabolite thiodiglycolic acid 10% pet. and other appropriate
                           vehicle
Molecular formula   : $C_5H_9NO_4S$

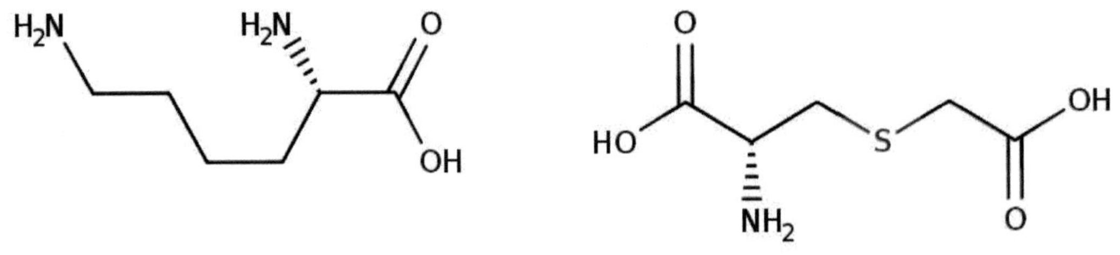

Carbocysteine lysine

GENERAL

Carbocysteine is a mucolytic drug that reduces the viscosity of sputum to relieve the symptoms of chronic obstructive pulmonary disorder (COPD) and bronchiectasis through easier expulsion of mucus. In pharmaceutical products, carbocysteine is employed as carbocysteine lysine (CAS number 256-425-9, EC number 256-425-9, molecular formula $C_{11}H_{23}N_3O_6S$) (1).

CUTANEOUS ADVERSE DRUG REACTIONS FROM SYSTEMIC ADMINISTRATION CAUSED BY TYPE IV (DELAYED-TYPE) HYPERSENSITIVITY (as demonstrated by positive patch tests)

Cutaneous adverse drug reactions from systemic administration of carbocysteine caused by type IV (delayed-type) hypersensitivity have included fixed drug eruption (2,4) and drug fever (3).

Fixed drug eruption

Two Japanese women who had histories of a fixed drug eruption two days after consecutive ingestions of carbocysteine, were examined by oral challenge tests and patch tests. With oral challenge tests, the fixed drug eruptions were not induced by a single challenge of the usual dose of carbocysteine when it was given either in the morning or in the evening and the skin observed for 1 week. However, if the drug was consecutively taken at night and three times on the following day, erythema developed by the morning of the third day. Patch tests were performed with carbocysteine and its metabolites, S-methyl-L-cysteine (SMC) and thiodiglycolic acid (TDA) at 1% and 10%. Both concentrations of TDA showed positive patch tests at the affected skin sites in both patients. Neither carbocysteine nor SMC elicited any positive findings. Five healthy volunteers were negative and lymphocyte transformation tests were negative in the patients for carbocysteine and its 2 metabolites (4). The authors showed the data of 10 patients reported in Japanese literature with FDE from carbocysteine. In not a single patient had patch tests been positive to carbocysteine on postlesional skin, but oral provocation tests had been positive in all reports where the results were mentioned (4).

In a multicenter study in hospitals in France, analyzing fixed drug eruptions (FDE), one case of FDE from carbocysteine with a positive patch test to this drug was recorded. However, clinical details were not provided (2).

Drug fever

A 67-year-old woman developed a fever of 39.5°C without any eruptions after an intake of carbocysteine on 2 occasions. Provocation tests induced fever only when the drug was administered at night. Patch tests were negative to carbocysteine but positive to thiodiglycolic acid 10% at D2 and D3, which is the night-time metabolite of carbocysteine. The patient was diagnosed with drug fever from carbocysteine induced by its night-time metabolite thiodiglycolic acid (3).

LITERATURE

1 The data in the section 'General' may have been obtained from literature discussed in this chapter, but mostly also or exclusively from one or more of the following online sources: ChemIDPlus Advanced, PubChem, DrugBank, RxList, Drug Central, Drugs.com, and Wikipedia

2 Brahimi N, Routier E, Raison-Peyron N, Tronquoy AF, Pouget-Jasson C, Amarger S, et al. A three-year-analysis of fixed drug eruptions in hospital settings in France. Eur J Dermatol 2010;20:461-464

3 Hatakeyama M, Fukunaga A, Shimizu H, Oka M, Horikawa T, Nishigori C. Drug fever due to S-carboxymethyl-L-cystein: demonstration of a causative agent with patch tests. J Dermatol 2012;39:555-556

4 Adachi A, Sarayama Y, Shimizu H, Yamada Y, Horikawa T. Thiodiglycolic acid as a possible causative agent of fixed drug eruption provoked only after continuous administration of S-carboxymethyl-L-cysteine: case report and review of reported cases. Br J Dermatol 2005;153:226-228

Chapter 3.88   CARBROMAL

IDENTIFICATION

Description/definition   : Carbromal is the bromide compound that conforms to the structural formula shown
                              below
Pharmacological classes : Hypnotics and sedatives
IUPAC name   : 2-Bromo-N-carbamoyl-2-ethylbutanamide
Other names   : Bromacetocarbamide
CAS registry number   : 77-65-6
EC number   : 201-046-6
Merck Index monograph : 1055
Patch testing   : 1% and 5% pet.
Molecular formula   : $C_7H_{13}BrN_2O_2$

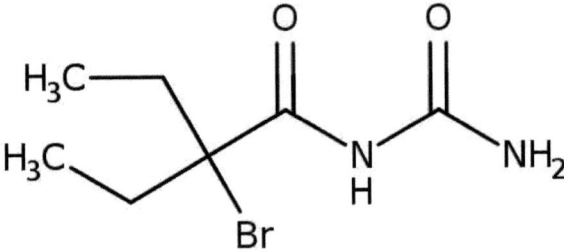

GENERAL

Carbromal is a bromide-containing hypnotic that has formerly been used to treat mild insomnia. Carbromal is one of
a number of hypnotics containing bromide, which releases the bromide ion on hydrolysis in the body. It has no
advantages over other hypnotics (1).

CUTANEOUS ADVERSE DRUG REACTIONS FROM SYSTEMIC ADMINISTRATION CAUSED BY TYPE IV
(DELAYED-TYPE) HYPERSENSITIVITY (as demonstrated by positive patch tests)

Cutaneous adverse drug reactions from systemic administration of carbromal caused by type IV (delayed-type)
hypersensitivity have included purpuric rash (2).

Other cutaneous adverse drug reactions

One patient developed purpura while using carbromal. Patch tests were positive to carbromal 1% and 5% in
propylene glycol. In addition, positive results were observed in a macrophage migration inhibitory factor test for
carbromal. The purpuric rash gradually subsided following withdrawal of the drug (2). Obviously, there is a
substantial risk of irritant patch test reactions to 95% and 99% propylene glycol.

LITERATURE

1   The data in the section 'General' may have been obtained from literature discussed in this chapter, but mostly
    also or exclusively from one or more of the following online sources: ChemIDPlus Advanced, PubChem,
    DrugBank, RxList, Drug Central, Drugs.com, and Wikipedia
2   Feuerman EJ, Brodsky F. Nonthrombocytopenic purpura induced by carbromal. Cutis 1979;23:489-490

Chapter 3.89   CARBUTAMIDE

IDENTIFICATION

Description/definition	: Carbutamide is the sulfonylurea compound that conforms to the structural formula shown below
Pharmacological classes	: Hypoglycemic agents
IUPAC name	: 1-(4-Aminophenyl)sulfonyl-3-butylurea
CAS registry number	: 339-43-5
EC number	: 206-424-4
Merck Index monograph	: 3101
Patch testing	: 5% pet. (3)
Molecular formula	: $C_{11}H_{17}N_3O_3S$

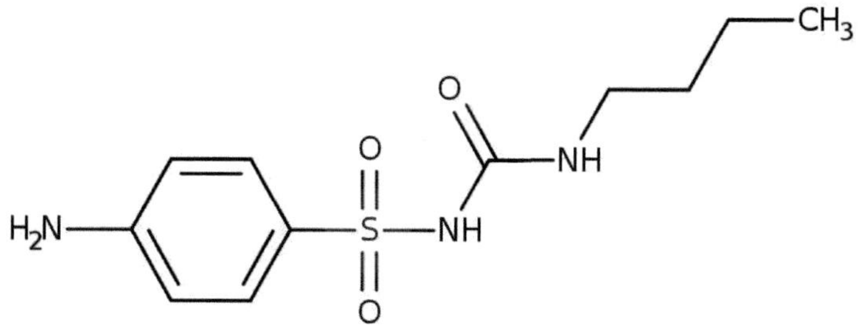

GENERAL

Carbutamide is a first-generation sulfonylurea compound with hypoglycemic activity. The drug reduces the excess sugar in the blood by promoting the secretion of insulin. Carbutamide was one of the first sulfonylureas used for this indication, but was withdrawn from the market due to toxic effects on bone marrow (1).

CUTANEOUS ADVERSE DRUG REACTIONS FROM SYSTEMIC ADMINISTRATION CAUSED BY TYPE IV (DELAYED-TYPE) HYPERSENSITIVITY (as demonstrated by positive patch tests)

Cutaneous adverse drug reactions from systemic administration of carbutamide caused by type IV (delayed-type) hypersensitivity have included systemic allergic dermatitis (2).

Systemic allergic dermatitis (systemic contact dermatitis)

Patients contact allergic to para-amino compounds (sulfanilamide, *p*-phenylenediamine, benzocaine) were orally challenged with 3 related sulfonylurea derivatives (hypoglycemic agents, sulfonamides): carbutamide, tolbutamide and chlorpropamide. Eleven had a positive reaction: 7 (of 25 tested) to carbutamide, 3 (of 11 tested) with tolbutamide and 1 (of 20 tested) with chlorpropamide. All were patients previously sensitized to sulfanilamide. Symptoms were itching in all 11 patients, reappearance of erythema and vesicles at the site of the primary contact dermatitis in 6 patients, and relapse of the primary contact dermatitis with a moderate secondary vesicular eruption together with a reactivation of the patch test reaction in 5 patients. Patch tests with these drugs themselves were not performed (2). These were very likely cases of systemic allergic dermatitis.

LITERATURE

1   The data in the section 'General' may have been obtained from literature discussed in this chapter, but mostly also or exclusively from one or more of the following online sources: ChemIDPlus Advanced, PubChem, DrugBank, RxList, Drug Central, Drugs.com, and Wikipedia
2   Angelini G, Meneghini CL. Oral tests in contact allergy to para-amino compounds. Contact Dermatitis 1981;7:311-314
3   De Groot AC. Patch testing, 4th edition. Wapserveen, The Netherlands: acdegroot publishing, 2018 (ISBN 9789081323345)

Chapter 3.90   CARPROFEN

IDENTIFICATION
Description/definition   : Carprofen is the propionic acid derivate that conforms to the structural formula shown
below
Pharmacological classes : Anti-inflammatory agents, non-steroidal; photosensitizing agents
IUPAC name   : 2-(6-Chloro-9H-carbazol-2-yl)propanoic acid
Other names   : 9H-Carbazole-2-acetic acid, 6-chloro-α-methyl-; 6-chloro-α-methylcarbazole-2-acetic acid
CAS registry number   : 53716-49-7
EC number   : 258-712-4
Merck Index monograph : 3132
Patch testing   : 1% pet.
Molecular formula   : $C_{15}H_{12}ClNO_2$

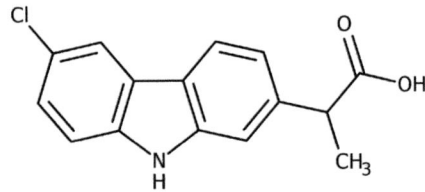

GENERAL
Carprofen is a propionic acid derivate and nonsteroidal anti-inflammatory drug (NSAID) with anti-inflammatory, analgesic, and antipyretic activities. It is used exclusively in veterinary medicine, often as a supportive treatment for the relief of arthritic symptoms in geriatric dogs. Carprofen was previously used in human medicine for over 10 years (1983-1993) in several countries (1,3). During this time, phototoxic and photoallergic reactions were reported from Europe (4,5,6,9,10) and Singapore (7). It was withdrawn from the human market for commercial reasons before re-emerging for veterinary use, for which it is widely prescribed (3). Carprofen has been shown to be a potent photoallergen (3).

CONTACT ALLERGY FROM ACCIDENTAL CONTACT
A 28-year-old pharmaceutical worker and a 47-year-old secretary working in a pharmaceutical factory had occupational airborne contact allergy and photocontact allergy to carprofen (2,3). These cases are discussed below under 'Photosensitivity'.

CUTANEOUS ADVERSE DRUG REACTIONS FROM SYSTEMIC ADMINISTRATION CAUSED BY TYPE IV (DELAYED-TYPE) HYPERSENSITIVITY (as demonstrated by positive patch tests)
Cutaneous adverse drug reactions from systemic administration of carprofen caused by type IV (delayed-type) hypersensitivity have included photosensitivity (2-10, including phototoxicity).

Photosensitivity

Case series
Two patients working in a pharmaceutical factory were investigated for carprofen photosensitivity. Patient 1 was a 42-year-old woman who suffered an itchy, erythematous facial skin rash, consistently appearing several hours after she had been at work, where she was involved in the labelling and packaging of carprofen. The problem cleared when away from work but emerged within 3 days of restarting it. Patch and photopatch tests with materials made from crushed carprofen tablets showed +++ reactions to carprofen at 2% and 5% in petrolatum and in water at the irradiated site only (3). The second patient was a 47-year-old woman who was referred acutely with a dermatitis which was especially severe on photoexposed sites. She worked in the same factory as patient 1 but, as a secretary, did not apparently come into direct contact with carprofen. However, when it became clear that her dermatitis flared towards the end of each working week, further questioning revealed that she did handle quality control papers, which were possibly contaminated with carprofen, from laboratory personnel. Patch and photopatch tests with materials made from crushed carprofen tablets showed ++ reactions to carprofen at 2% and 5% in petrolatum and in water at the unirradiated site and +++ reactions at the irradiated site. The positive photopatch test reactions were so much stronger than the control patch test site reactions that the results were interpreted as indicating true photoallergic contact dermatitis in addition to contact allergy, rather than photo-augmentation of a contact reaction.

The patient's original facial dermatitis was clear at the start of testing but flared during the test procedure (3). Of three controls patch and photopatch tested with the same materials, one had a marked delayed dermatitis at D10 at the irradiated site, indicating photosensitization from the test procedure (3).

Following the positive photopatch test results with the ground carprofen tablets in the 2 patients, a factory visit was arranged and the potential for extensive dust exposure for some workers was noted. Eight further employees were investigated, of who three were also found to have photoallergic contact dermatitis to carprofen (3).

Case reports

A 28-year-old pharmaceutical worker, had been working in a laboratory formulating a pure carprofen powder for 3 weeks, presented with an itchy erythematous facial rash. It consistently evolved hours after she had been at work, and was gradually becoming more inflamed. The rash resolved when she was absent from the workplace. Her skin repeatedly flared if she entered the laboratory space, even when not in contact with chemicals. Her protective clothing consisted of a laboratory coat, gloves, a mask, and safety glasses. Patch and photopatch tests showed a + reaction to non-irradiated 1% carprofen, and a strong ++ reaction to irradiated 1% carprofen. It was concluded that the patient had occupational airborne contact allergy and photocontact allergy to carprofen (2).

A 27-year-old woman developed a pruritic, erythematous eczematous eruption. This started on her hands and spread to her arms, face and neck. She had been working for 3 weeks in a factory manufacturing carprofen. She was involved in handling the carprofen powder but wore full protective clothing including an air hood and gloves while doing this. As her problems had started during a period of sunny weather in June, it was decided not only to patch but also to photopatch test her to the carprofen materials she had handled. Patch tests were negative, but there were strong ++ photoallergic reactions to pure carprofen powder 1% pet., an intermediate mixture of carprofen and the tablet base compound (containing 12–14% carprofen and inert agents) 1% pet. and the finished carprofen tablet at 1% pet. Ten controls were negative. It was suggested that skin contact occurred following transfer of the carprofen from her protective clothing on to her skin while changing back into her everyday clothes (8).

Patch test sensitization

One individual, who was patch and photopatch tested with carprofen (powder of crushed carprofen tablets) 2% and 5% in water and petrolatum as a control subject, was photosensitized to carprofen from the test procedure (3).

LITERATURE

1   The data in the section 'General' may have been obtained from literature discussed in this chapter, but mostly also or exclusively from one or more of the following online sources: ChemIDPlus Advanced, PubChem, DrugBank, RxList, Drug Central, Drugs.com, and Wikipedia

2   Kiely C, Murphy G. Photoallergic contact dermatitis caused by occupational exposure to the canine non-steroidal anti-inflammatory drug carprofen. Contact Dermatitis 2010;63:364-365

3   Kerr AC, Muller F, Ferguson J, Dawe RS. Occupational carprofen photoallergic contact dermatitis. Br J Dermatol 2008;159:1303-1308

4   Merot Y, Harms M, Saurat JH. Photosensitivity associated with carprofen (Imadyl), a new non-steroidal anti-inflammatory drug. Dermatologica 1983;166:301-307

5   Przybilla B, Ring J, Schwab U, Galosi A, Dorn M, Braun-Falco O. Photosensitising properties of nonsteroidal antirheumatic drugs in the photopatch test. Hautarzt 1987;38:18-25

6   Hoting E, Schulz KH. Photoallergic reaction to carprofen. Derm Beruf Umwelt 1984;32:215-16

7   Goh CL, Kwok SF. Photosensitivity associated with carprofen (Imadyl). Dermatologica 1985;170:74-76

8   Walker SL, Ead RD, Beck MH. Occupational photoallergic contact dermatitis in a pharmaceutical worker manufacturing carprofen, a canine nonsteroidal anti-inflammatory drug. Br J Dermatol 2006;154:569-570

9   Hölzle E, Neumann N, Hausen B, Przybilla B, Schauder S, Hönigsmann H, et al. Photopatch testing: the 5-year experience of the German, Austrian and Swiss Photopatch Test Group. J Am Acad Dermatol 1991;25:59-68

10 Neumann NJ, Hölzle E, Plewig G, Schwarz T, Panizzon RG, Breit R, et al. Photopatch testing: The 12-year experience of the German, Austrian and Swiss Photopatch Test Group. J Am Acad Dermatol 2000;42(2Pt.1):183-192

Chapter 3.91   CARVEDILOL

IDENTIFICATION

Description/definition   : Carvedilol is the carbazole and propanol derivative that conforms to the structural
                               formula shown below
Pharmacological classes   : α_1-Adrenergic receptor antagonists; adrenergic β-antagonists; antihypertensive
                               agents; vasodilator agents; calcium channel blockers; antioxidants
IUPAC name   : 1-(9H-Carbazol-4-yloxy)-3-[2-(2-methoxyphenoxy)ethylamino]propan-2-ol
Other names   : 2-Propanol, 1-(9H-carbazol-4-yloxy)-3-[[2-(2-methoxyphenoxy)ethyl]amino]-
CAS registry number   : 72956-09-3
EC number   : 615-871-8
Merck Index monograph : 3143
Patch testing   : 10% pet.
Molecular formula   : $C_{24}H_{26}N_2O_4$

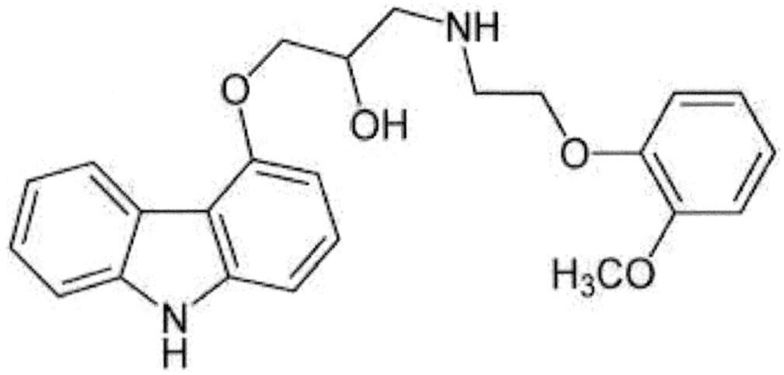

GENERAL

Carvedilol is a carbazole and propanol derivative that acts as a non-cardioselective β-blocker and vasodilator. It
blocks β_1- and β_2-adrenergic receptors as well as the α_1-adrenergic receptors. Carvedilol is indicated for the
treatment of mild or moderate (NYHA class II or III) heart failure of ischemic or cardiomyopathic origin. In
pharmaceutical products, both carvedilol and carvedilol phosphate (CAS number 610309-89-2, EC number not
available, molecular formula $C_{48}H_{60}N_4O_{17}P_2$) may be employed (1).

CONTACT ALLERGY FROM ACCIDENTAL CONTACT

A 29-year-old man working as machine operator in a pharmaceutical factory was referred with a 6-month history of
dermatitis involving his eyelids, cheeks, lips, nose, and the nasolabial folds. The rash cleared during long periods
away from work. Patch tests were positive to carvedilol 10% pet., zolpidem 10% pet. and simvastatin 1% and 0.1%
pet. Ten controls were negative. Three months avoidance of the offending allergens resulted in total clearing of the
dermatitis. A diagnosis of occupational airborne allergic contact dermatitis was made (2).

LITERATURE

1   The data in the section 'General' may have been obtained from literature discussed in this chapter, but mostly
    also or exclusively from one or more of the following online sources: ChemIDPlus Advanced, PubChem,
    DrugBank, RxList, Drug Central, Drugs.com, and Wikipedia
2   Neumark M, Moshe S, Ingber A, Slodownik D. Occupational airborne contact dermatitis to simvastatin,
    carvedilol, and zolpidem. Contact Dermatitis 2009;61:51-52

Chapter 3.92   CASANTHRANOL

IDENTIFICATION

Description/definition : Casanthranol is the anthracene derivative that conforms to the structural formula shown below
Pharmacological classes : Cathartics; protein kinase inhibitors; laxatives
IUPAC name : (10R)-1,8-Dihydroxy-3-(hydroxymethyl)-10-[(2R,3R,4R,5R,6R)-3,4,5-trihydroxy-6-(hydroxymethyl)oxan-2-yl]oxy-10H-anthracen-9-one
CAS registry number : 8024-48-4
EC number : Not available
Merck Index monograph : 3148
Patch testing : No data available
Molecular formula : $C_{21}H_{22}O_{10}$

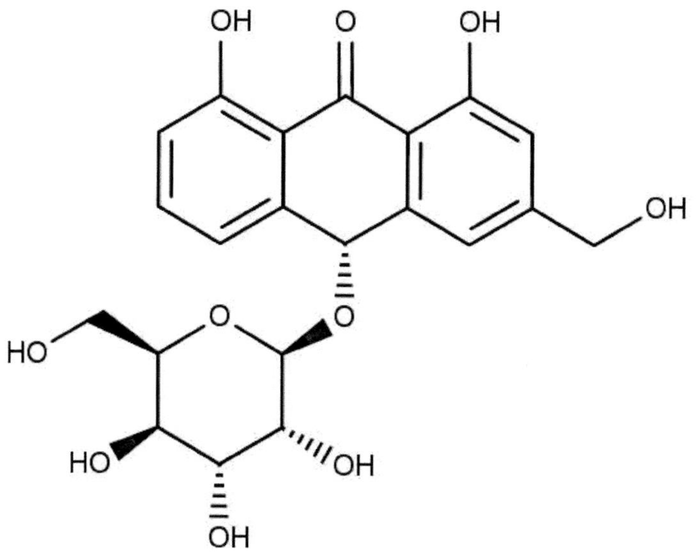

GENERAL

Casanthranol is a concentrated and purified mixture of anthranol glycosides from the bark of the cascara, *Rhamnus purshiana*. The dried bark is also called cascara sagrada, Spanish for 'holy bark'. It is or was used as a laxative in constipation and various medical conditions. Casanthranol encourages bowel movements by acting on the intestinal wall to increase muscle contractions (1).

CONTACT ALLERGY FROM ACCIDENTAL CONTACT

A 44-year old female pharmacist developed dermatitis of the eyelids, later spreading to the arms, face, neck and trunk. When patch tested, she reacted to tylosin and casanthranol, termed cascara in the publication. Details were not provided (2).

LITERATURE

1   The data in the section 'General' may have been obtained from literature discussed in this chapter, but mostly also or exclusively from one or more of the following online sources: ChemIDPlus Advanced, PubChem, DrugBank, RxList, Drug Central, Drugs.com, and Wikipedia
2   Gielen K, Goossens A. Occupational allergic contact dermatitis from drugs in healthcare workers. Contact Dermatitis 2001;45:273-279

Chapter 3.93   CEFACLOR

IDENTIFICATION

Description/definition   : Cefaclor is the cephalosporin that conforms to the structural formula shown below
Pharmacological classes   : Anti-bacterial agents
IUPAC name   : (6R,7R)-7-[[(2R)-2-Amino-2-phenylacetyl]amino]-3-chloro-8-oxo-5-thia-1-azabicyclo-
                             [4.2.0]oct-2-ene-2-carboxylic acid;hydrate
Other names   : Cefaclor hydrate; cefaclor monohydrate
CAS registry number   : 70356-03-5
EC number   : Not available
Merck Index monograph   : 3184 (cefaclor anhydrous)
Patch testing   : Generally speaking, cephalosporins may be tested 5-10% pet. or water; 5% pet. (3)
Molecular formula   : $C_{15}H_{16}ClN_3O_5S$

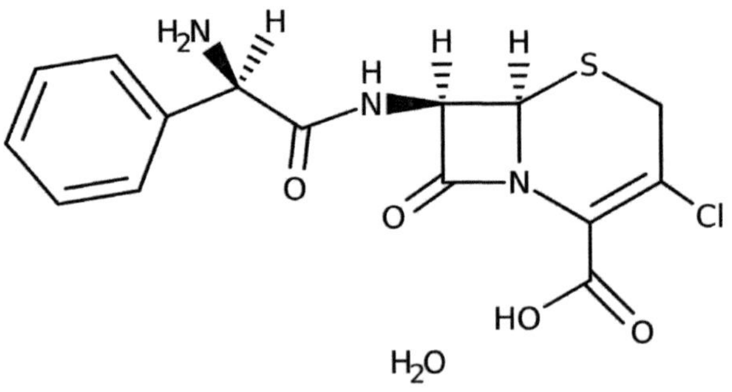

GENERAL

Cefaclor is a semisynthetic second generation β-lactam cephalosporin antibiotic derived from cefalexin with a spectrum resembling first-generation cephalosporins. It is indicated for the treatment of infections caused by susceptible bacteria such as pneumonia and ear, lung, skin, throat, and urinary tract infections (1).

The classification and structures of beta-lactam antibiotics are discussed in Chapter 3.36 Amoxicillin.

CUTANEOUS ADVERSE DRUG REACTIONS FROM SYSTEMIC ADMINISTRATION CAUSED BY TYPE IV (DELAYED-TYPE) HYPERSENSITIVITY (as demonstrated by positive patch tests)

Cutaneous adverse drug reactions from systemic administration of cefaclor caused by type IV (delayed-type) hypersensitivity have included unspecified drug eruption (4).

Other cutaneous adverse drug reactions

In Finland, in the period 1989-2001, 826 patients with suspected cutaneous drug eruptions were patch tested and 89 had one or more positive reactions. Of these individuals, 1 reacted to cefaclor. It was not mentioned which drug eruption this patient had suffered from (4).

Cross-reactions, pseudo-cross-reactions and co-reactions

Cross-reactions between beta-lactam antibiotics are discussed in Chapter 3.36 Amoxicillin. In a group of 78 patients who had suffered non-immediate drug reactions from any beta-lactam compound (penicillins, cephalosporins), 7 had positive patch tests to cefaclor. As this was not the culprit drug causing the adverse skin reaction in any patient, all these reactions were presumably cross-reactions (2).

LITERATURE

1   The data in the section 'General' may have been obtained from literature discussed in this chapter, but mostly also or exclusively from one or more of the following online sources: ChemIDPlus Advanced, PubChem, DrugBank, RxList, Drug Central, Drugs.com, and Wikipedia
2   Buonomo A, Nucera E, De Pasquale T, Pecora V, Lombardo C, Sabato V, et al. Tolerability of aztreonam in patients with cell-mediated allergy to β-lactams. Int Arch Allergy Immunol 2011;155:155-159
3   Brockow K, Garvey LH, Aberer W, Atanaskovic-Markovic M, Barbaud A, Bilo MB, et al.; ENDA/EAACI Drug Allergy Interest Group. Skin test concentrations for systemically administered drugs – an ENDA/EAACI Drug Allergy Interest Group position paper. Allergy 2013;68:702-712
4   Lammintausta K, Kortekangas-Savolainen O. The usefulness of skin tests to prove drug hypersensitivity. Br J Dermatol 2005;152:968-974

Chapter 3.94   CEFADROXIL

IDENTIFICATION

Description/definition : Cefadroxil is the cephalosporin that conforms to the structural formula shown below
Pharmacological classes : Anti-bacterial agents
IUPAC name : (6R,7R)-7-((R)-2-Amino-2-(p-hydroxyphenyl)acetamido)-3-methyl-8-oxo-5-thia-1-azabicyclo(4.2.0)oct-2-ene-2-carboxylic acid;hydrate
Other names : Cefadroxil monohydrate
CAS registry number : 66592-87-8
EC number : Not available
Merck Index monograph : 3185
Patch testing : Generally speaking, cephalosporins may be tested 5-10% pet. or water; 5% pet. (3)
Molecular formula : $C_{16}H_{19}N_3O_6S$

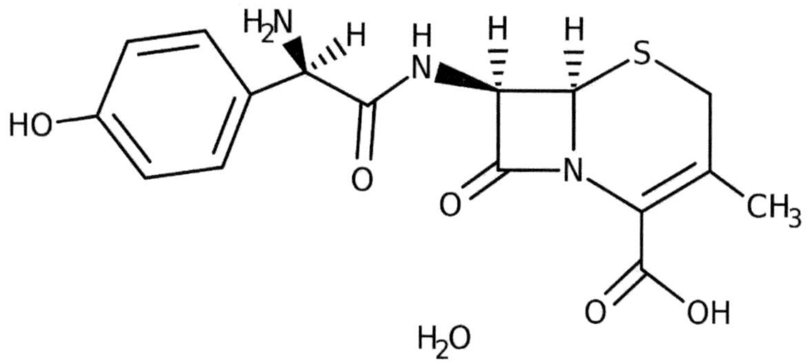

GENERAL

Cefadroxil is a semisynthetic first-generation β-lactam cephalosporin antibiotic derived from cefalexin with antibacterial activity. It is used to treat urinary tract infections, skin and skin structure infections, pharyngitis, and tonsillitis caused by susceptible bacteria (1). The classification and structures of beta-lactam antibiotics are discussed in Chapter 3.36 Amoxicillin.

CUTANEOUS ADVERSE DRUG REACTIONS FROM SYSTEMIC ADMINISTRATION CAUSED BY TYPE IV (DELAYED-TYPE) HYPERSENSITIVITY (as demonstrated by positive patch tests)

Cutaneous adverse drug reactions from systemic administration of cefadroxil caused by type IV (delayed-type) hypersensitivity have included drug reaction with eosinophilia and systemic symptoms (DRESS) (2) and unspecified drug eruption (4).

Drug reaction with eosinophilia and systemic symptoms (DRESS)

A 51-year-old man had taken cefadroxil 500 mg bid for 8 weeks for osteomyelitis, when he developed fever, myalgia, arthralgia, and skin rash. Physical examination showed that he had mild fever, generalized maculopapular exanthem, and multiple purpuric patches which gradually developed to exfoliative dermatitis. No lymphadenopathy was found. Laboratory tests found leukocytosis, absolute eosinophilia and abnormal renal and liver function. Eleven months after resolution, patch tests on 10x tape-stripped skin were strongly positive to 30% of the powdered cefadroxil tablet in white petrolatum and water at days 2,3 and 4. Six controls were negative. The patient was diagnosed with drug reaction with eosinophilia and systemic symptoms (DRESS) (2).

Other cutaneous adverse drug reactions

In Finland, in the period 1989-2001, 826 patients with suspected cutaneous drug eruptions were patch tested and 89 had one or more positive reactions. Of these individuals, 1 reacted to cefadroxil. It was not mentioned which drug eruption this patient had suffered from (4).

Cross-reactions, pseudo-cross-reactions and co-reactions

Cross-reactions between beta-lactam antibiotics are discussed in Chapter 3.36 Amoxicillin.

LITERATURE

1   The data in the section 'General' may have been obtained from literature discussed in this chapter, but mostly also or exclusively from one or more of the following online sources: ChemIDPlus Advanced, PubChem, DrugBank, RxList, Drug Central, Drugs.com, and Wikipedia

2   Suswardana, Hernanto M, Yudani BA, Pudjiati SR, Indrastuti N. DRESS syndrome from cefadroxil confirmed by positive patch test. Allergy 2007;62:1216-1217

3   Brockow K, Garvey LH, Aberer W, Atanaskovic-Markovic M, Barbaud A, Bilo MB, et al.; ENDA/EAACI Drug Allergy Interest Group. Skin test concentrations for systemically administered drugs – an ENDA/EAACI Drug Allergy Interest Group position paper. Allergy 2013;68:702-712

4   Lammintausta K, Kortekangas-Savolainen O. The usefulness of skin tests to prove drug hypersensitivity. Br J Dermatol 2005;152:968-974

Chapter 3.95   CEFALEXIN

IDENTIFICATION

Description/definition : Cefalexin is a β-lactam semisynthetic cephalosporin antibiotic that conforms to the structural formula shown below
Pharmacological classes : Anti-bacterial agents
IUPAC name : (6R,7R)-7-[[(2R)-2-Amino-2-phenylacetyl]amino]-3-methyl-8-oxo-5-thia-1-azabicyclo[4.2.0]oct-2-ene-2-carboxylic acid
Other names : Cephalexin
CAS registry number : 15686-71-2
EC number : 239-773-6
Merck Index monograph : 3244
Patch testing : Hydrate, 10% pet. (Chemotechnique)
Molecular formula : $C_{16}H_{17}N_3O_4S$

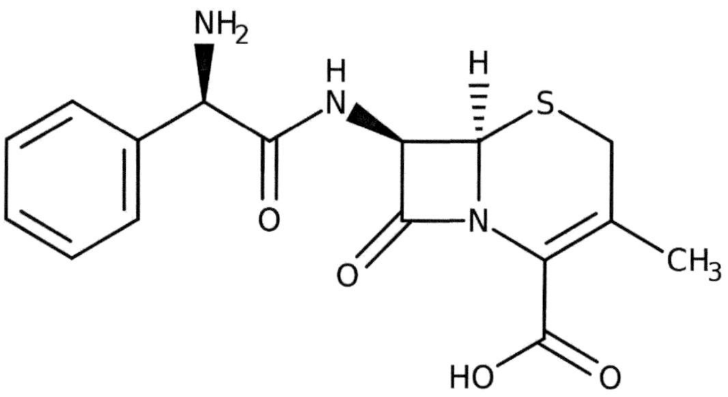

GENERAL

Cefalexin is a semisynthetic first-generation β-lactam cephalosporin antibiotic with bactericidal activity. It is used for the treatment of infections of the respiratory tract, ear, skin and skin structures, bones and genitourinary tract caused by susceptible bacteria. In pharmaceutical products, cefalexin is employed as cefalexin (mono)hydrate (CAS number 23325-78-2, EC number not available, molecular formula $C_{16}H_{19}N_3O_5S$) (1). The classification and structures of beta-lactam antibiotics are discussed in Chapter 3.36 Amoxicillin.

CUTANEOUS ADVERSE DRUG REACTIONS FROM SYSTEMIC ADMINISTRATION CAUSED BY TYPE IV (DELAYED-TYPE) HYPERSENSITIVITY (as demonstrated by positive patch tests)

Cutaneous adverse drug reactions from systemic administration of cefalexin caused by type IV (delayed-type) hypersensitivity have included maculopapular eruption (4), systemic allergic dermatitis (3), and unspecified drug eruption (2,5).

Maculopapular eruption

In a group of 105 patients ranging in age from 14 to 84 years with histories of nonimmediate reactions to cephalosporins and patch tested with the culprit drugs, there were only 3 positive patch tests, one to cefalexin and 2 to ceftriaxone. All 3 patients had suffered a maculopapular eruption (4).

Systemic allergic dermatitis (systemic contact dermatitis)

A 51-year-old woman presented with facial dermatitis and dermatitis around a stasis ulcer on the left leg. The ulcer had been treated with the contents of cefalexin capsules under an occlusive dressing. During the therapy, the patient experienced recurrent episodes of dermatitis affecting the legs, face and ears. She developed generalized itching, erythema and scaling after a course of cefalexin tablets prescribed for a urinary infection (systemic allergic dermatitis). Patch tests were positive to powder of two brands of cefalexin tablets and to cefalexin 1% in olive oil. Ten controls were negative (3).

Other cutaneous adverse drug reactions

In Finland, in the period 1989-2001, 826 patients with suspected cutaneous drug eruptions were patch tested and 89 had one or more positive reactions. Of these individuals, 3 reacted to cefalexin. It was not mentioned which drug eruption these patients had suffered from (5).

One patient had a drug eruption from cefalexin with a positive patch test. Details were not provided (2).

Cross-reactions, pseudo-cross-reactions and co-reactions
Cross-reactions between beta-lactam antibiotics are discussed in Chapter 3.36 Amoxicillin. In a group of 78 patients who had suffered nonimmediate drug reactions from any beta-lactam compound (penicillins, cephalosporins), 5 had positive patch tests to cefalexin. In 1 case, cefalexin was the culprit drug, so the other reactions were presumably cross-reactions (2).

LITERATURE

1   The data in the section 'General' may have been obtained from literature discussed in this chapter, but mostly also or exclusively from one or more of the following online sources: ChemIDPlus Advanced, PubChem, DrugBank, RxList, Drug Central, Drugs.com, and Wikipedia
2   Buonomo A, Nucera E, De Pasquale T, Pecora V, Lombardo C, Sabato V, et al. Tolerability of aztreonam in patients with cell-mediated allergy to β-lactams. Int Arch Allergy Immunol 2011;155:155-159
3   Milligan A, Douglas WS. Contact dermatitis to cephalexin. Contact Dermatitis 1986;15:91
4   Romano A, Gaeta F, Valluzzi RL, Caruso C, Alonzi C, Viola M, et al. Diagnosing nonimmediate reactions to cephalosporins. J Allergy Clin Immunol 2012;129:1166-1169
5   Lammintausta K, Kortekangas-Savolainen O. The usefulness of skin tests to prove drug hypersensitivity. Br J Dermatol 2005;152:968-974

Chapter 3.96   CEFALOTIN

IDENTIFICATION

Description/definition	: Cefalotin is the semisynthetic β-lactam cephalosporin that conforms to the structural formula shown below
Pharmacological classes	: Anti-bacterial agents
IUPAC name	: (6R,7R)-3-(Acetyloxymethyl)-8-oxo-7-[(2-thiophen-2-ylacetyl)amino]-5-thia-1-azabicyclo-[4.2.0]oct-2-ene-2-carboxylic acid
Other names	: 7-(2-(2-Thienyl)acetylamido)cephalosporanic acid; cephalotin
CAS registry number	: 153-61-7
EC number	: 205-815-7
Merck Index monograph	: 3251
Patch testing	: Generally speaking, cephalosporins may be tested 5-10% pet. or water; 5% pet. (2)
Molecular formula	: $C_{16}H_{16}N_2O_6S_2$

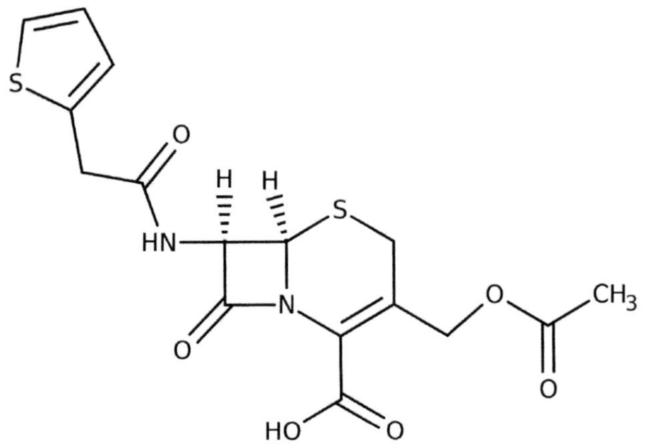

GENERAL

Cefalotin is a semisynthetic, β-lactam, first-generation cephalosporin antibiotic with bactericidal activity. It is used to prevent infection during surgery and to treat infections of the blood, bone or joints, respiratory tract, skin, and urinary tract caused by susceptible bacteria. In pharmaceutical products, cefalotin is employed as cefalotin sodium (CAS number 58-71-9, EC number 200-394-6, molecular formula $C_{16}H_{15}N_2NaO_6S_2$) (1). The classification and structures of beta-lactam antibiotics are discussed in Chapter 3.36 Amoxicillin.

CONTACT ALLERGY FROM ACCIDENTAL CONTACT

A 29-year-old woman had worked as a chemical analyst for 11 years in a pharmaceutical laboratory, performing analytical control of cephalosporins. Three months before presentation, she developed edema of both eyelids, with pruritus and dryness of the oral and nasal mucosae. The lesions cleared during her summer holiday. Patch tests were positive to cefalotin 1% and 5% water, cefamandole 5% water, and cefazolin 1% and 5% water. The patient left her work for 3 weeks and the lesions cleared without treatment (4).

CUTANEOUS ADVERSE DRUG REACTIONS FROM SYSTEMIC ADMINISTRATION CAUSED BY TYPE IV (DELAYED-TYPE) HYPERSENSITIVITY (as demonstrated by positive patch tests)

Cutaneous adverse drug reactions from systemic administration of cefalotin caused by type IV (delayed-type) hypersensitivity have included unspecified drug eruption (3).

Other cutaneous adverse drug reactions

A male patient, who was known with allergy to penicillin (presumably on the basis of an exanthem, no allergy tests performed) was given an infusion with cefalotin 10 years later and developed 'an allergy of the delayed type'. Patch tests were positive to cefalotin in water (concentration not mentioned), but negative to cefazolin and benzylpenicillin. Scratch tests with these drugs were positive to cefalotin only, at 48 hours (3).

Cross-reactions, pseudo-cross-reactions and co-reactions

Cross-reactions between beta-lactam antibiotics are discussed in Chapter 3.36 Amoxicillin.

Immediate contact reactions

Immediate contact reactions (contact urticaria) to cefalotin are presented in Chapter 5.

LITERATURE

1   The data in the section 'General' may have been obtained from literature discussed in this chapter, but mostly also or exclusively from one or more of the following online sources: ChemIDPlus Advanced, PubChem, DrugBank, RxList, Drug Central, Drugs.com, and Wikipedia

2   Brockow K, Garvey LH, Aberer W, Atanaskovic-Markovic M, Barbaud A, Bilo MB, et al.; ENDA/EAACI Drug Allergy Interest Group. Skin test concentrations for systemically administered drugs – an ENDA/EAACI Drug Allergy Interest Group position paper. Allergy 2013;68:702-712

3   Braun WP. Cephalotin allergy of the delayed type. Contact Dermatitis 1975;1:190-191

4   Condé-Salazar L, Guimaraens D, Romero LV, Gonzalez MA. Occupational dermatitis from cephalosporins. Contact Dermatitis 1986;14:70-71

Chapter 3.97   CEFAMANDOLE

IDENTIFICATION

Description/definition : Cefamandole is the cephalosporin that conforms to the structural formula shown below
Pharmacological classes : Anti-bacterial agents
IUPAC name : (6R,7R)-7-[[(2R)-2-Hydroxy-2-phenylacetyl]amino]-3-[(1-methyltetrazol-5-yl)sulfanyl-methyl]-8-oxo-5-thia-1-azabicyclo[4.2.0]oct-2-ene-2-carboxylic acid
Other names : Cephamandole; (6R,7R)-7-(R)-mandelamido-3-(((1-methyl-1H-tetrazol-5-yl)thio)methyl)-8-oxo-5-thia-1-azabicyclo(4.2.0)oct-2-ene-carboxylic acid
CAS registry number : 34444-01-4
EC number : 252-030-0
Merck Index monograph : 3186
Patch testing : Generally speaking, cephalosporins may be tested 5-10% pet. or water; 5% pet. (3)
Molecular formula : $C_{18}H_{18}N_6O_5S_2$

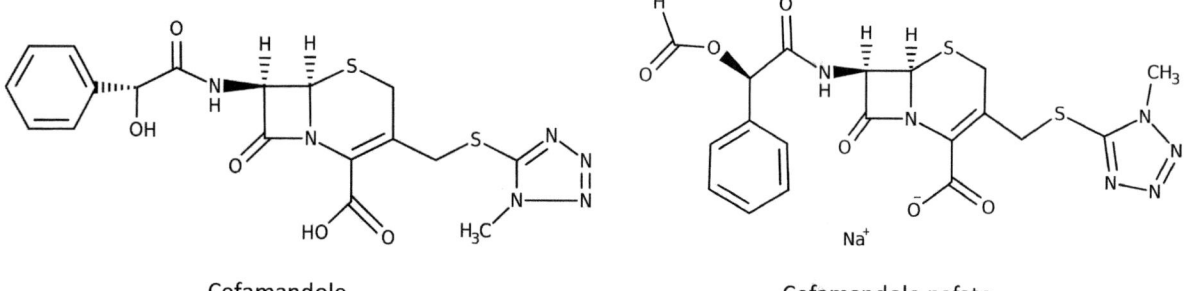

Cefamandole   Cefamandole nafate

GENERAL

Cefamandole is a semisynthetic second-generation, β-lactam, wide-spectrum cephalosporin antibiotic with bactericidal activity. It is active against many strains resistant to other cephalosporins, such as *Enterobacter* species and indole-positive *Proteus* species. It is used for the treatment of serious infections caused by susceptible strains of microorganisms (1). In pharmaceutical products, cefamandole may be employed as cefamandole sodium (CAS number 30034-03-8, EC number 250-009-0, molecular formula $C_{18}H_{17}N_6NaO_5S_2$) or (when used parenterally) the formate ester prodrug cefamandole nafate (CAS number 42540-40-9, EC number 255-877-4, molecular formula $C_{19}H_{17}N_6NaO_6S_2$) (1).

The classification and structures of beta-lactam antibiotics are discussed in Chapter 3.36 Amoxicillin.

CONTACT ALLERGY FROM ACCIDENTAL CONTACT

Seven nurses from Bari and Milan, Italy, had occupational allergic contact dermatitis of the hands from one or more cephalosporins that they worked with. When patch tested, there were 3 to cefamandole, 4 to cefazolin sodium, 4 to cefotetan disodium, 2 to cefmetazole sodium and 1 to each of the following: cefotiam dihydrochloride, ceftizoxime sodium, cefotaxime sodium, cefodizime disodium, ceftazidime pentahydrate and ceftriaxone sodium, all tested 20% pet. One of the nurses reacted to 5 cephalosporins, 2 to four, 1 to three antibiotics, and 3 reacted to one cephalosporin each. It was not specified which cephalosporins were the culprit drugs and which – if any – positive patch tests were cross-reactions. The patch tests were read at D2 only (4).

A 29-year-old woman had worked as a chemical analyst for 11 years in a pharmaceutical laboratory, performing analytical control of cephalosporins. Three months before presentation, she developed edema of both eyelids, with pruritus and dryness of the oral and nasal mucosae. The lesions cleared during her summer holiday. Patch tests were positive to cefamandole 5% water, cefalotin 1% and 5% water, and cefazolin 1% and 5% water. The patient left her work for 3 weeks and the lesions cleared without treatment (5).

Cross-reactions, pseudo-cross-reactions and co-reactions

Cross-reactions between beta-lactam antibiotics are discussed in Chapter 3.36 Amoxicillin.

In a group of 78 patients who had suffered nonimmediate drug reactions from any beta-lactam compound (penicillins, cephalosporins), 1 had a positive patch test to cefamandole. As this was not the culprit drug causing the adverse skin reaction in any patient, this positive patch test was presumably a cross-reaction (2).

LITERATURE

1   The data in the section 'General' may have been obtained from literature discussed in this chapter, but mostly also or exclusively from one or more of the following online sources: ChemIDPlus Advanced, PubChem, DrugBank, RxList, Drug Central, Drugs.com, and Wikipedia

2   Buonomo A, Nucera E, De Pasquale T, Pecora V, Lombardo C, Sabato V, et al. Tolerability of aztreonam in patients with cell-mediated allergy to β-lactams. Int Arch Allergy Immunol 2011;155:155-159

3   Brockow K, Garvey LH, Aberer W, Atanaskovic-Markovic M, Barbaud A, Bilo MB, et al.; ENDA/EAACI Drug Allergy Interest Group. Skin test concentrations for systemically administered drugs – an ENDA/EAACI Drug Allergy Interest Group position paper. Allergy 2013;68:702-712

4   Foti C, Bonamonte D, Trenti R, Veña GA, Angelini G. Occupational contact allergy to cephalosporins. Contact Dermatitis 1997;36:104-105

5   Condé-Salazar L, Guimaraens D, Romero LV, Gonzalez MA. Occupational dermatitis from cephalosporins. Contact Dermatitis 1986;14:70-71

Chapter 3.98   CEFAZOLIN

IDENTIFICATION

Description/definition   : Cefazolin is the β-lactam antibiotic and first-generation cephalosporin that conforms to the structural formula shown below
Pharmacological classes : Anti-bacterial agents
IUPAC name   : (6R,7R)-3-[(5-Methyl-1,3,4-thiadiazol-2-yl)sulfanylmethyl]-8-oxo-7-[[2-(tetrazol-1-yl)acetyl]amino]-5-thia-1-azabicyclo[4.2.0]oct-2-ene-2-carboxylic acid
CAS registry number   : 25953-19-9
EC number   : 247-362-8
Merck Index monograph : 3188
Patch testing   : Generally speaking, cephalosporins may be tested 5-10% pet. or water; 5% pet. (5)
Molecular formula   : $C_{14}H_{14}N_8O_4S_3$

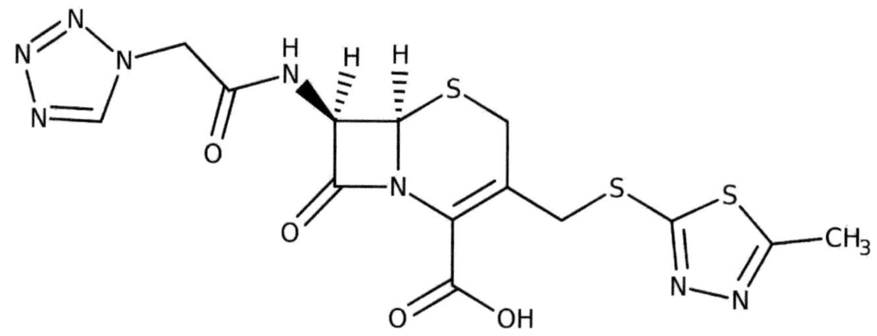

GENERAL

Cefazolin is a semisynthetic, β-lactam, first-generation cephalosporin antibiotic with bactericidal activity. It is mainly used to treat bacterial infections of the skin, but can also be used to treat moderately severe bacterial infections involving the lung, bone, joint, stomach, blood, heart valve, and urinary tract caused by susceptible strains of microorganisms and for surgical prophylaxis (1). In pharmaceutical products, cefazolin is employed as cefazolin sodium (CAS number 27164-46-1, EC number 248-278-4, molecular formula $C_{14}H_{13}N_8NaO_4S_3$) (1). The classification and structures of beta-lactam antibiotics are discussed in Chapter 3.36 Amoxicillin.

CONTACT ALLERGY FROM ACCIDENTAL CONTACT

Case series

In Leuven, Belgium, in the period 2001-2019, 201 of 1248 health care workers/employees of the pharmaceutical industry had occupational allergic contact dermatitis. In 23 (11%) dermatitis was caused by skin contact with a systemic drug: 19 nurses, two chemists, one physician, and one veterinarian. The lesions were mostly localized on the hands, but often also on the face, as airborne dermatitis. In total, 42 positive patch test reactions to 18 different systemic drugs were found. In one patient, cefazolin was the drug/one of the drugs that caused occupational dermatitis (8).

Between 1978 and 2001, in the university hospital of Leuven, Belgium, occupational allergic contact dermatitis to pharmaceuticals was diagnosed in 33 health care workers: 26 nurses, 4 veterinarians, 2 pharmacists and one medical doctor. There were 26 women and 7 men with a mean age of 38 years. Practically all of these patients presented with hand dermatitis (often the fingers), sometimes with secondary localizations, frequently the face. In this group, four patients had occupational allergic contact dermatitis from cefazolin (2).

Seven nurses from Bari and Milan, Italy, had occupational allergic contact dermatitis of the hands from one or more cephalosporins that they worked with. When patch tested, there were 4 reactions to cefazolin sodium, 4 to cefotetan disodium, 3 to cefamandole, 2 to cefmetazole sodium and 1 to each of the following: cefotiam dihydrochloride, ceftizoxime sodium, cefotaxime sodium, cefodizime disodium, ceftazidime pentahydrate and ceftriaxone sodium, all tested 20% pet. One of the nurses reacted to 5 cephalosporins, 2 to four, 1 to three antibiotics, and 3 reacted to one cephalosporin each. It was not specified which cephalosporins were the culprit drugs and which – if any – positive patch tests were cross-reactions. The patch tests were read at D2 only (6).

Case reports

A 40-year-old nurse gave a 10-year history of recurrent eczema of the hands and face at work. On examination, papulovesicular eczema, with fissures and erosions, of the palms, finger webs, backs of the hands and fingers, and face (periorbital), additionally with redness and swelling, was noted. Patch tests were positive to cefazolin, performed as scratch-patch tests with a commercial cefazolin preparation for intravenous use, 1% and 10% water. During patch testing, aggravation of previous skin lesions was noted (3).

A nurse involved in the handling of intravenous antibiotics had occupational allergic contact dermatitis of the hands and reacted to cefazolin and ampicillin 10% pet. It was not specified with which antibiotics the patient had actual contact, most likely cefazolin (4). A 32-year-old female nurse from Portugal had hand dermatitis and later palpebral eczema. When patch tested, she reacted to two cephalosporins that she worked with, cefazolin and cefradine. There were cross-reactions to cefuroxime, ceftriaxone, and cefotaxime (7).

A 29-year-old woman had worked as a chemical analyst for 11 years in a pharmaceutical laboratory, performing analytical control of cephalosporins. Three months before presentation, she developed edema of both eyelids, with pruritus and dryness of the oral and nasal mucosae. The lesions cleared during her summer holiday. Patch tests were positive to cefazolin 1% and 5% water, cefamandole 5% water, and cefalotin 1% and 5% water. The patient left her work for 3 weeks and the lesions cleared without treatment (9).

A 24-year-old nurse presented with a 6-month history of intensely itchy erythematous papules and plaques on exposed skin of her forehead, perioral area, forearms and hands, which was clearly work-related. Patch tests were positive to cefazolin 1%, 5% and 10% water, cefoxitin 1%, 5% and 10% water, ceftriaxone 1%,5% and 10% water and ceftazidime 5% and 10% water. The patient stopped preparing cephalosporin solutions for systemic administration and the lesions cleared up. It was not mentioned which cephalosporins she had actually had contact with (10).

CUTANEOUS ADVERSE DRUG REACTIONS FROM SYSTEMIC ADMINISTRATION CAUSED BY TYPE IV (DELAYED-TYPE) HYPERSENSITIVITY (as demonstrated by positive patch tests)

Cutaneous adverse drug reactions from systemic administration of cefazolin caused by type IV (delayed-type) hypersensitivity have included maculopapular eruption (11,12,14) and generalized erythema (13).

Maculopapular eruption

A 44-year-old woman was treated intramuscularly with cefazolin 1 g twice a day as preoperative prophylaxis. During the second day of therapy, 6 hours after the administration of the third dose of the drug, the patient developed a generalized pruritic maculopapular exanthema. Nine years later, patch tests were positive to cefazolin 200 mg/ml in saline with cross-reactions to several other cephalosporins. Intradermal tests did not produce immediate reactions, but for cefazolin and the other cephalosporins, the 24-hour reading showed erythematous, indurated wheals larger than 10 mm, which reached their maximum size at D2 and began to fade at D3. Lymphocyte transformation tests were also positive. The patient reacted to an oral challenge of 100 mg with an exanthematous eruption on her abdomen and legs 12 hours after the administration of the drug. All these tests performed with penicillins were, however, negative (11).

In the period 2000-2014, 260 patients were patch tested with antibiotics for suspected cutaneous adverse drug reactions (CADR) to these drugs. 56 patients (22%) had one or more (often from cross-reactivity) positive patch tests. Cefazolin was patch test positive in patient with maculopapular eruption from this drug (12).

A 64-year-old woman developed a maculopapular eruption which was attributed to clindamycin and cefazolin. The patch test to clindamycin was positive, but to cefazolin negative. 11 years later, patch tests were repeated and cefazolin now reacted positively. In the meantime, however, the patient had been accidentally exposed to cefazolin (14).

Generalized erythema

A 66-year-old woman developed a pruritic diffuse erythematous rash with fine scaling 2 days after the initiation of treatment with cefazolin i.m. (500 mg/12 h) after the implantation of a knee prosthesis. The patient had taken cephalosporins 8 years before and she remembered a similar reaction with ceftazidime. Patch tests were positive to the commercial preparation 'as is' and 5 other cephalosporins for injection, including ceftazidime. Intradermal tests with these 2 drugs were also positive at the D2 delayed reading (13).

Cross-reactions, pseudo-cross-reactions and co-reactions

Cross-reactions between beta-lactam antibiotics are discussed in Chapter 3.36 Amoxicillin.

LITERATURE

1 The data in the section 'General' may have been obtained from literature discussed in this chapter, but mostly also or exclusively from one or more of the following online sources: ChemIDPlus Advanced, PubChem, DrugBank, RxList, Drug Central, Drugs.com, and Wikipedia

2 Gielen K, Goossens A. Occupational allergic contact dermatitis from drugs in healthcare workers. Contact Dermatitis 2001;45:273-279

3 Straube MD, Freitag M, Altmeyer P, Szliska C. Occupational airborne contact dermatitis from cefazolin. Contact Dermatitis 2000;42:44-45

4 Pinheiro V, Pestana C, Pinho A, Antunes I, Gonçalo M. Occupational allergic contact dermatitis caused by antibiotics in healthcare workers - relationship with nonimmediate drug eruptions. Contact Dermatitis 2018;78:281-286

5 Brockow K, Garvey LH, Aberer W, Atanaskovic-Markovic M, Barbaud A, Bilo MB, et al.; ENDA/EAACI Drug Allergy Interest Group. Skin test concentrations for systemically administered drugs – an ENDA/EAACI Drug Allergy Interest Group position paper. Allergy 2013;68:702-712

6 Foti C, Bonamonte D, Trenti R, Veña GA, Angelini G. Occupational contact allergy to cephalosporins. Contact Dermatitis 1997;36:104-105

7 Antunes J, Silva R, Pacheco D, Travassos R, Filipe P. Occupational contact allergy to cephalosporins. Dermatol Online J 2011 May 15;17(5):13

8 Gilissen L, Boeckxstaens E, Geebelen J, Goossens A. Occupational allergic contact dermatitis from systemic drugs. Contact Dermatitis 2020;82:24-30

9 Condé-Salazar L, Guimaraens D, Romero LV, Gonzalez MA. Occupational dermatitis from cephalosporins. Contact Dermatitis 1986;14:70-71

10 Filipe P, Almeida RS, Rodrigo FG. Occupational allergic contact dermatitis from cephalosporins. Contact Dermatitis 1996;34:226

11 Romano A, Torres MJ, Di Fonso M, Leyva L, Andriolo M, Pettinato R, Blanca M. Delayed hypersensitivity to cefazolin: report on a case involving lymphocyte transformation studies with different cephalosporins. Ann Allergy Asthma Immunol 2001;87:238-242

12 Pinho A, Coutinho I, Gameiro A, Gouveia M, Gonçalo M. Patch testing - a valuable tool for investigating non-immediate cutaneous adverse drug reactions to antibiotics. J Eur Acad Dermatol Venereol 2017;31:280-287

13 Gonzalo-Garijo MA, Rodríguez-Nevado I, de Argila D. Patch tests for diagnosis of delayed hypersensitivity to cephalosporins. Allergol Immunopathol (Madr) 2006;34:39-41

14 Pinho A, Marta A, Coutinho I, Gonçalo M. Long-term reproducibility of positive patch test reactions in patients with non-immediate cutaneous adverse drug reactions to antibiotics. Contact Dermatitis 2017;76:204-209

Chapter 3.99   CEFCAPENE PIVOXIL

IDENTIFICATION

Description/definition : Cefcapene pivoxil is the pivalate ester prodrug form of cefcapene that conforms to the structural formula shown below

Pharmacological classes : Anti-bacterial agents

IUPAC name : 2,2-Dimethylpropanoyloxymethyl (6R,7R)-7-[[(Z)-2-(2-amino-1,3-thiazol-4-yl)pent-2-enoyl]amino]-3-(carbamoyloxymethyl)-8-oxo-5-thia-1-azabicyclo[4.2.0]oct-2-ene-2-carboxylate

CAS registry number : 105889-45-0

EC number : Not available

Merck Index monograph : 3190

Patch testing : Generally speaking, cephalosporins may be tested 5-10% pet. or water; 5% pet. (3)

Molecular formula : $C_{23}H_{29}N_5O_8S_2$

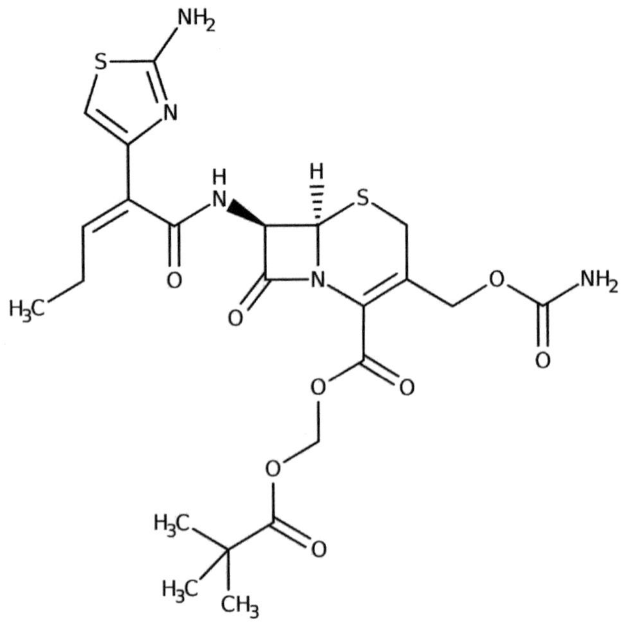

GENERAL

Cefcapene pivoxil is the pivalate ester prodrug form of cefcapene, a semi-synthetic third-generation cephalosporin with antibacterial activity. After oral administration of cefcapene pivoxil, the ester bond is cleaved, releasing active cefcapene. In pharmaceutical products, cefcapene pivoxil may be employed as cefcapene pivoxil hydrochloride (CAS number 147816-23-7, EC number not available, molecular formula $C_{23}H_{30}ClN_5O_8S_2$) (1). The classification and structures of beta-lactam antibiotics are discussed in Chapter 3.36 Amoxicillin.

CUTANEOUS ADVERSE DRUG REACTIONS FROM SYSTEMIC ADMINISTRATION CAUSED BY TYPE IV (DELAYED-TYPE) HYPERSENSITIVITY (as demonstrated by positive patch tests)

Cutaneous adverse drug reactions from systemic administration of cefcapene pivoxil caused by type IV (delayed-type) hypersensitivity have included pruritic papules and erythema on the trunk and arms (2).

Other cutaneous adverse drug reactions

A 64-year-old woman was seen with pruritic papules and erythema on the trunk and upper extremities that had started one day earlier. She had taken cefcapene pivoxil hydrochloride 100 mg t.d.s. orally for the prevention of infection of the skin of her thigh, where an angioinfusion catheter had been placed for hepatic cancer two days earlier. Topical corticosteroid application cleared her skin lesions within a week of stopping the oral antibiotic. Patch tests with cefcapene pivoxil hydrochloride 10% and 1% pet. showed erythema at D2 and erythema with papules at D3. Five control subjects were negative (2).

Cross-reactions, pseudo-cross-reactions and co-reactions
Cross-reactions between beta-lactam antibiotics are discussed in Chapter 3.36 Amoxicillin.

LITERATURE
1   The data in the section 'General' may have been obtained from literature discussed in this chapter, but mostly also or exclusively from one or more of the following online sources: ChemIDPlus Advanced, PubChem, DrugBank, RxList, Drug Central, Drugs.com, and Wikipedia
2   Kawada A, Aragane Y, Maeda A, Asai M, Shiraishi H, Tezuka T. Drug eruption induced by cefcapene pivoxil hydrochloride. Contact Dermatitis 2001;44:197
3   Brockow K, Garvey LH, Aberer W, Atanaskovic-Markovic M, Barbaud A, Bilo MB, et al.; ENDA/EAACI Drug Allergy Interest Group. Skin test concentrations for systemically administered drugs – an ENDA/EAACI Drug Allergy Interest Group position paper. Allergy 2013;68:702-712

Chapter 3.100 CEFIXIME

IDENTIFICATION

Description/definition   : Cefixime is a cephalosporin antibiotic derived semisynthetically from the marine fungus *Cephalosporium acremonium*, that conforms to the structural formula shown below
Pharmacological classes   : Antibacterial agents
IUPAC name   : (6R,7R)-7-[[(2Z)-2-(2-Amino-1,3-thiazol-4-yl)-2-(carboxymethoxyimino)acetyl]amino]-3-ethenyl-8-oxo-5-thia-1-azabicyclo[4.2.0]oct-2-ene-2-carboxylic acid
CAS registry number   : 79350-37-1
EC number   : Not available
Merck Index monograph   : 3195
Patch testing   : Cefixime trihydrate 10.0% pet. (Chemotechnique)
Molecular formula   : $C_{16}H_{15}N_5O_7S_2$

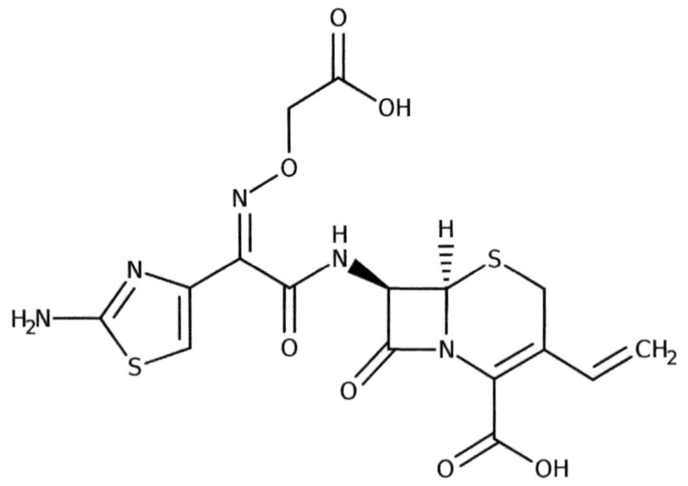

GENERAL

Cefixime is a broad-spectrum, β-lactam, third-generation cephalosporin antibiotic derived semisynthetically from the marine fungus *Cephalosporium acremonium* with antibacterial activity. It is indicated for the treatment of uncomplicated urinary tract infections, otitis media, pharyngitis and tonsillitis, acute bronchitis and acute exacerbations of chronic bronchitis, and uncomplicated gonorrhea, when caused by specific susceptible strains of microorganisms. In pharmaceutical products, cefixime is employed as cefixime trihydrate (sometimes also termed cefixime) (CAS number 125110-14-7, EC number not available, molecular formula $C_{16}H_{21}N_5O_{10}S_2$). (go.drugbank.com). The classification and structures of beta-lactam antibiotics are discussed in Chapter 3.36 Amoxicillin.

CUTANEOUS ADVERSE DRUG REACTIONS FROM SYSTEMIC ADMINISTRATION CAUSED BY TYPE IV (DELAYED-TYPE) HYPERSENSITIVITY (as demonstrated by positive patch tests)

Cutaneous adverse drug reactions from systemic administration of cefixime caused by type IV (delayed-type) hypersensitivity have included acute generalized exanthematous pustulosis (AGEP) (1).

Acute generalized exanthematous pustulosis (AGEP)

A 26-year-old woman taking cefixime presented with a generalized erythematous pustular eruption on the skin over the majority of her body associated with pain. She also had petechiae on the hard palate and a single erosion with slough over the lower gingival mucosa. The patient high-grade intermittent fever and blood analysis showed leukocytosis. Acute generalized exanthematous pustulosis (AGEP) was suspected and cefixime was withdrawn immediately. With supportive treatment, the patient got complete remission within 3 weeks. A patch test to cefixime was positive (1).

Cross-reactions, pseudo-cross-reactions and co-reactions

Cross-reactions between beta-lactam antibiotics are discussed in Chapter 3.36 Amoxicillin.

LITERATURE

1   Namoju R, Ismail M, Kumar Golla V, Bamini T, Lakshmi Akarapu T, Baloju D. A case of acute generalized exanthematous pustulosis by cefixime with oral mucosal involvement. Curr Drug Saf 2020;15:236-239

Chapter 3.101 CEFMETAZOLE

IDENTIFICATION

Description/definition : Cefmetazole is the semisynthetic, β-lactam cephalosporin that conforms to the structural formula shown below
Pharmacological classes : Anti-bacterial agents
IUPAC name : (6R,7S)-7-[[2-(Cyanomethylsulfanyl)acetyl]amino]-7-methoxy-3-[(1-methyltetrazol-5-yl)-sulfanylmethyl]-8-oxo-5-thia-1-azabicyclo[4.2.0]oct-2-ene-2-carboxylic acid
CAS registry number : 56796-20-4
EC number : 260-384-2
Merck Index monograph : 3197
Patch testing : Generally speaking, cephalosporins may be tested 5-10% pet. or water; 5% pet. (2)
Molecular formula : $C_{15}H_{17}N_7O_5S_3$

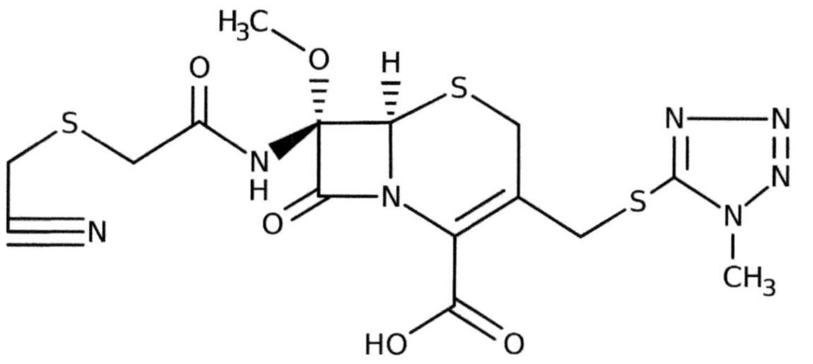

GENERAL

Cefmetazole is a second-generation, semisynthetic, β-lactam cephalosporin antibiotic with antibacterial activity. It has a broad spectrum of activity against both gram-positive and gram-negative microorganisms and is used in many types of infection. In pharmaceutical products, cefmetazole is employed as cefmetazole sodium (CAS number 56796-39-5, EC number not available, molecular formula $C_{15}H_{16}N_7NaO_5S_3$) (1). The classification and structures of beta-lactam antibiotics are discussed in Chapter 3.36 Amoxicillin.

CONTACT ALLERGY FROM ACCIDENTAL CONTACT

Seven nurses from Bari and Milan, Italy, had occupational allergic contact dermatitis of the hands from one or more cephalosporins that they worked with. When patch tested, there were 2 reactions to cefmetazole sodium, 4 to cefazolin sodium, 4 to cefotetan disodium, 3 to cefamandole, and 1 to each of the following: cefotiam dihydrochloride, ceftizoxime sodium, cefotaxime sodium, cefodizime disodium, ceftazidime pentahydrate and ceftriaxone sodium, all tested 20% pet. One of the nurses reacted to 5 cephalosporins, 2 to four, 1 to three antibiotics, and 3 reacted to one cephalosporin each. It was not specified which cephalosporins were the culprit drugs and which – if any – positive patch tests were cross-reactions. The patch tests were read at D2 only (3).

Cross-reactions, pseudo-cross-reactions and co-reactions
Cross-reactions between beta-lactam antibiotics are discussed in Chapter 3.36 Amoxicillin.

LITERATURE

1   The data in the section 'General' may have been obtained from literature discussed in this chapter, but mostly also or exclusively from one or more of the following online sources: ChemIDPlus Advanced, PubChem, DrugBank, RxList, Drug Central, Drugs.com, and Wikipedia
2   Brockow K, Garvey LH, Aberer W, Atanaskovic-Markovic M, Barbaud A, Bilo MB, et al.; ENDA/EAACI Drug Allergy Interest Group. Skin test concentrations for systemically administered drugs – an ENDA/EAACI Drug Allergy Interest Group position paper. Allergy 2013;68:702-712
3   Foti C, Bonamonte D, Trenti R, Veña GA, Angelini G. Occupational contact allergy to cephalosporins. Contact Dermatitis 1997;36:104-105

Chapter 3.102 CEFODIZIME

IDENTIFICATION

Description/definition   : Cefodizime is the aminothiazolyl cephalosporin that conforms to the structural formula
                            shown below
Pharmacological classes : Anti-bacterial agents
IUPAC name   : (6R,7R)-7-[[(2Z)-2-(2-Amino-1,3-thiazol-4-yl)-2-methoxyiminoacetyl]amino]-3-[[5-(carbo-
                            xylatomethyl)-4-methyl-1,3-thiazol-2-yl]sulfanylmethyl]-8-oxo-5-thia-1-azabicyclo[4.2.0]-
                            oct-2-ene-2-carboxylic acid
CAS registry number   : 69739-16-8
EC number   : 700-301-3
Merck Index monograph   : 3199
Patch testing   : Generally speaking, cephalosporins may be tested 5-10% pet. or water; 5% pet. (2)
Molecular formula   : $C_{20}H_{20}N_6O_7S_4$

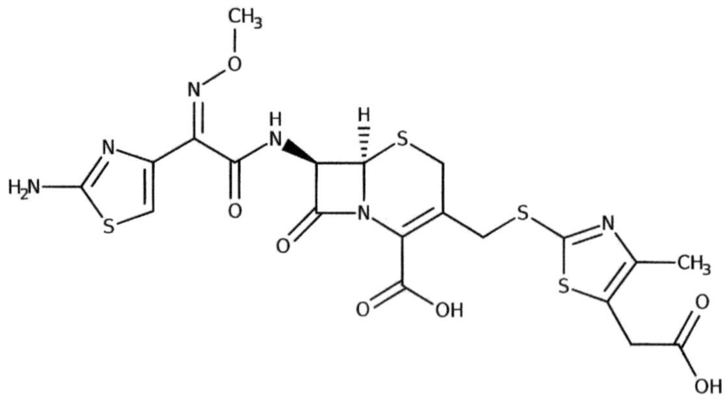

GENERAL

Cefodizime is a third-generation aminothiazolyl cephalosporin for parenteral use. Cefodizime has broad-spectrum activity and is stable to most β-lactamases. In pharmaceutical products, cefodizime is employed as cefodizime disodium (also termed cefodizime sodium) (CAS number 86329-79-5, EC number 617-833-6, molecular formula $C_{20}H_{18}N_6Na_2O_7S_4$) (1). The classification and structures of beta-lactam antibiotics are discussed in Chapter 3.36 Amoxicillin.

CONTACT ALLERGY FROM ACCIDENTAL CONTACT

Seven nurses from Bari and Milan, Italy, had occupational allergic contact dermatitis of the hands from one or more cephalosporins that they worked with. When patch tested, there was one reaction to cefodizime disodium and many to other cephalosporins, all tested 20% pet. The patch tests were read at D2 only (3, probably overlap with ref. 4).

A 45-year-old nurse developed contact dermatitis on the hands and forearms, which was ascribed to handling of cephalosporins for parenteral use. In the past year, the condition had spread to the patient's face, eyelids and neck. The intensely itchy lesions regressed completely during suspension of working activity. Patch tests were positive to cefodizime, cefotaxime, ceftizoxime, ceftazidime, and ceftriaxone, all tested 20% pet. It was not mentioned to which cephalosporines the patient had been exposed and which reactions were considered to be cross-reactions (4).

Cross-reactions, pseudo-cross-reactions and co-reactions
Cross-reactions between beta-lactam antibiotics are discussed in Chapter 3.36 Amoxicillin.

LITERATURE

1   The data in the section 'General' may have been obtained from literature discussed in this chapter, but mostly
    also or exclusively from one or more of the following online sources: ChemIDPlus Advanced, PubChem,
    DrugBank, RxList, Drug Central, Drugs.com, and Wikipedia
2   Brockow K, Garvey LH, Aberer W, Atanaskovic-Markovic M, Barbaud A, Bilo MB, et al.; ENDA/EAACI Drug Allergy
    Interest Group. Skin test concentrations for systemically administered drugs – an ENDA/EAACI Drug Allergy
    Interest Group position paper. Allergy 2013;68:702-712
3   Foti C, Bonamonte D, Trenti R, Veña GA, Angelini G. Occupational contact allergy to cephalosporins. Contact
    Dermatitis 1997;36:104-105
4   Foti C, Vena GA, Cucurachi MR, Angelini G. Occupational contact allergy from cephalosporins. Contact Dermatitis
    1994;31:129-130

Chapter 3.103 CEFONICID

IDENTIFICATION

Description/definition : Cefonicid is the β-lactam antibiotic and second-generation cephalosporin that conforms to the structural formula shown below
Pharmacological classes : Anti-bacterial agents
IUPAC name : (6R,7R)-7-[[(2R)-2-Hydroxy-2-phenylacetyl]amino]-8-oxo-3-[[1-(sulfomethyl)tetrazol-5-yl]sulfanylmethyl]-5-thia-1-azabicyclo[4.2.0]oct-2-ene-2-carboxylic acid
CAS registry number : 61270-58-4
EC number : Not available
Merck Index monograph : 3200
Patch testing : Generally speaking, cephalosporins may be tested 5-10% pet. or water; 5% pet. (3)
Molecular formula : $C_{18}H_{18}N_6O_8S_3$

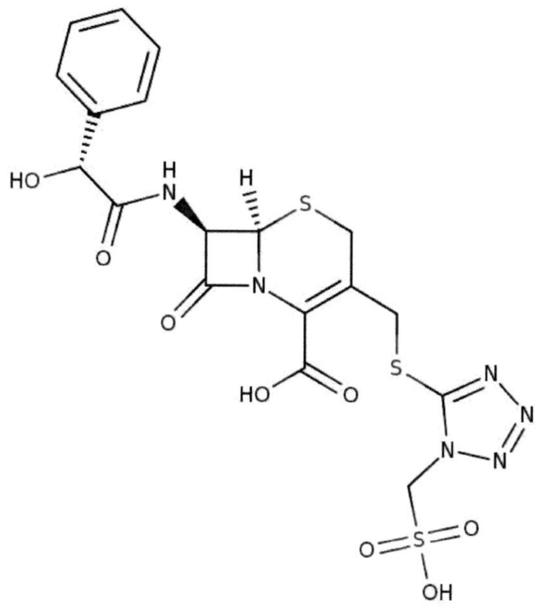

GENERAL

Cefonicid is a second-generation cephalosporin administered intravenously or intramuscularly. Its bactericidal action results from inhibition of cell wall synthesis. Cefonicid binds to penicillin-binding proteins (PBPs), transpeptidases that are responsible for crosslinking of peptidoglycan. By preventing crosslinking of peptidoglycan, cell wall integrity is lost and cell wall synthesis is halted. This cephalosporin is used for urinary tract infections, lower respiratory tract infections, and soft tissue and bone infections (1).

The classification and structures of beta-lactam antibiotics are discussed in Chapter 3.36 Amoxicillin.

CUTANEOUS ADVERSE DRUG REACTIONS FROM SYSTEMIC ADMINISTRATION CAUSED BY TYPE IV (DELAYED-TYPE) HYPERSENSITIVITY (as demonstrated by positive patch tests)

Cutaneous adverse drug reactions from systemic administration of cefonicid caused by type IV (delayed-type) hypersensitivity have included maculopapular eruption (2).

Maculopapular eruption

A 48-year-old woman was treated with intramuscular cefonicid 500 mg for a buccal infection. During the second day of therapy, six hours after the 3rd dose of the drug, the patient developed a generalized exanthema consisting of mildly pruritic, erythematous papules that persisted for 4 days. Patch tests were positive to cefonicid 1% pet. at D2 and D4 without cross-reactions to other cephalosporins or penicillins. Twelve controls were negative. An intradermal test with cefonicid 333 mg/ml produced an induration of 22 mm after 6 hours without erythema. Histopathology of a biopsy of this reaction showed a moderate interstitial and perivascular infiltrate with neutrophils and mononuclear cells; a mild intercellular edema was present. There was no increase in numbers of basophils or mast cells or signs of vasculitis. The patient was challenged with an intramuscular injection of 50 mg cefonicid (1/10 the normal dose), which produced a mild morbilliform rash on the patient's trunk and forearms after 6 hours (2).

Cross-reactions, pseudo-cross-reactions and co-reactions
Cross-reactions between beta-lactam antibiotics are discussed in Chapter 3.36 Amoxicillin.

LITERATURE

1   The data in the section 'General' may have been obtained from literature discussed in this chapter, but mostly
    also or exclusively from one or more of the following online sources: ChemIDPlus Advanced, PubChem,
    DrugBank, RxList, Drug Central, Drugs.com, and Wikipedia
2   Martin JA, Alonso MD, Lázaro M, Parra F, Compaired JA, Alvarez-Cuesta E. Delayed allergic reaction to cefonicid.
    Ann Allergy 1994;72:341-342
3   Brockow K, Garvey LH, Aberer W, Atanaskovic-Markovic M, Barbaud A, Bilo MB, et al.; ENDA/EAACI Drug Allergy
    Interest Group. Skin test concentrations for systemically administered drugs – an ENDA/EAACI Drug Allergy
    Interest Group position paper. Allergy 2013;68:702-712

Chapter 3.104 CEFORANIDE

IDENTIFICATION

Description/definition : Ceforanide is the cephalosporin that conforms to the structural formula shown below
Pharmacological classes : Antibacterial agents
IUPAC name : (6R,7R)-7-[[2-[2-(Aminomethyl)phenyl]acetyl]amino]-3-[[1-(carboxymethyl)tetrazol-5-yl]sulfanylmethyl]-8-oxo-5-thia-1-azabicyclo[4.2.0]oct-2-ene-2-carboxylic acid
CAS registry number : 60925-61-3
EC number : Not available
Merck Index monograph : 3202
Patch testing : Generally speaking, cephalosporins may be tested 5-10% pet. or water; 5% pet. (3)
Molecular formula : $C_{20}H_{21}N_7O_6S_2$

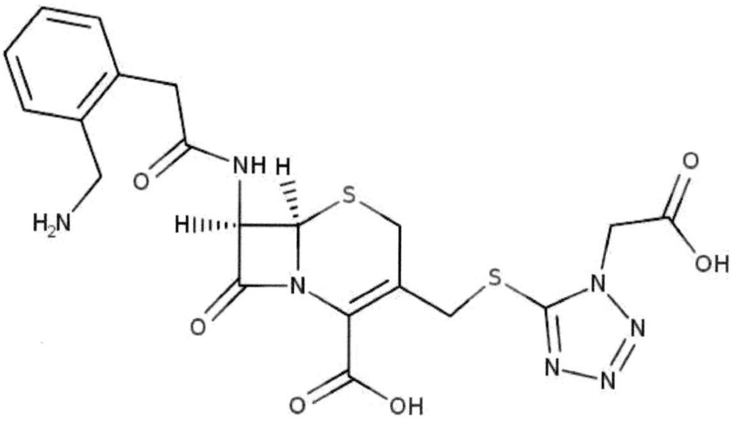

GENERAL

Ceforanide is a semi-synthetic, broad-spectrum, beta-lactam, second-generation cephalosporin antibiotic with bactericidal activity. Ceforanide causes inhibition of bacterial cell wall synthesis by inactivating penicillin binding proteins (PBPs), thereby interfering with the final transpeptidation step required for cross-linking of peptidoglycan units which are a component of the cell wall. This results in a reduction of cell wall stability and causes cell lysis. Ceforanide is administered parenterally. Many coliform bacteria, including *Escherichia coli, Klebsiella, Enterobacter*, and *Proteus*, are susceptible to ceforanide, as are most strains of *Salmonella, Shigella, Hemophilus, Citrobacter* and *Arizona* species (1). The classification and structures of beta-lactam antibiotics are discussed in Chapter 3.36 Amoxicillin.

CONTACT ALLERGY FROM ACCIDENTAL CONTACT

A 49-year-old nurse had developed dermatitis of the fingers, and later on the eyelids. She was patch tested to the drugs that she was dispensing to patients and reacted to 2 penicillins (flucloxacillin, amoxicillin) and 4 cephalosporins (cefuroxime, cefazolin, ceftriaxone, ceforanide) including ceforanide, termed cefaronide in the publication (2). This was a case of occupational allergic contact dermatitis.

Cross-reactions, pseudo-cross-reactions and co-reactions

Cross-reactions between beta-lactam antibiotics are discussed in Chapter 3.36 Amoxicillin.

LITERATURE

1   The data in the section 'General' may have been obtained from literature discussed in this chapter, but mostly also or exclusively from one or more of the following online sources: ChemIDPlus Advanced, PubChem, DrugBank, RxList, Drug Central, Drugs.com, and Wikipedia
2   Gielen K, Goossens A. Occupational allergic contact dermatitis from drugs in healthcare workers. Contact Dermatitis 2001;45:273-279
3   Brockow K, Garvey LH, Aberer W, Atanaskovic-Markovic M, Barbaud A, Bilo MB, et al.; ENDA/EAACI Drug Allergy Interest Group. Skin test concentrations for systemically administered drugs – an ENDA/EAACI Drug Allergy Interest Group position paper. Allergy 2013;68:702-712

Chapter 3.105 CEFOTAXIME

IDENTIFICATION

Description/definition   : Cefotaxime is the semisynthetic cephalosporin antibiotic that conforms to the structural
                            formula shown below
Pharmacological classes   : Anti-bacterial agents
IUPAC name   : (6R,7R)-3-(Acetyloxymethyl)-7-[[(2Z)-2-(2-amino-1,3-thiazol-4-yl)-2-methoxyimino-
                            acetyl]amino]-8-oxo-5-thia-1-azabicyclo[4.2.0]oct-2-ene-2-carboxylic acid
CAS registry number   : 63527-52-6
EC number   : 264-299-1
Merck Index monograph   : 3204
Patch testing   : Sodium salt 10.0% pet. (Chemotechnique)
Molecular formula   : $C_{16}H_{17}N_5O_7S_2$

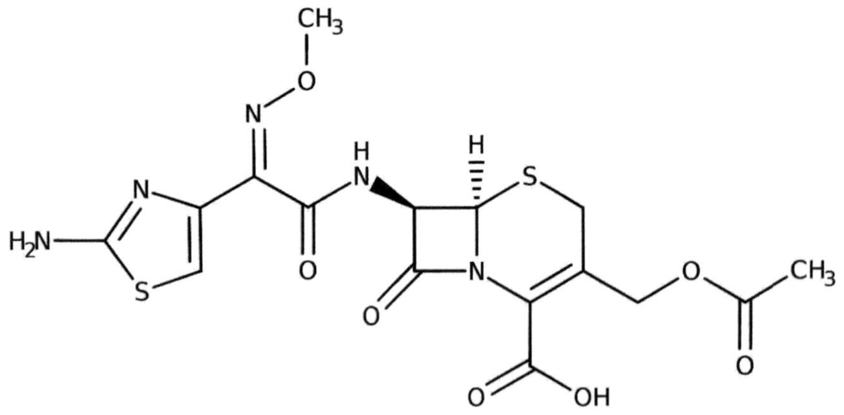

GENERAL

Cefotaxime is a third generation, semisynthetic, penicillinase-resistant, broad-spectrum cephalosporin antibiotic for intravenous use with bactericidal activity. It is indicated for the treatment of gonorrhea, meningitis, and severe infections including infections of the kidney (pyelonephritis) and urinary system. It is also used before an operation to prevent infection after surgery (1). In pharmaceutical products, cefotaxime is employed as cefotaxime sodium (CAS number 64485-93-4, EC number 264-915-9, molecular formula $C_{16}H_{16}N_5NaO_7S_2$) (1). The classification and structures of beta-lactam antibiotics are discussed in Chapter 3.36 Amoxicillin.

CONTACT ALLERGY FROM ACCIDENTAL CONTACT

Case series

Seven nurses from Bari and Milan, Italy, had occupational allergic contact dermatitis of the hands from one or more cephalosporins that they worked with. When patch tested, there was one reaction to cefotaxime sodium, 4 reactions to cefazolin sodium, 3 to cefamandole, 4 to cefotetan disodium, 2 to cefmetazole sodium and 1 to each of the following: cefotiam dihydrochloride, ceftizoxime sodium, cefodizime disodium, ceftazidime pentahydrate and ceftriaxone sodium, all tested 20% pet. One of the nurses reacted to 5 cephalosporins, 2 to four, 1 to three antibiotics, and 3 reacted to one cephalosporin each. It was not specified which cephalosporins were the culprit drugs and which – if any – positive patch tests were cross-reactions. The patch tests were read at D2 only (2).

Two nurses involved in the handling of intravenous antibiotics had occupational allergic contact dermatitis of the hands, forearms and one of the face. They reacted to cefotaxime and ceftriaxone, and one also to cefoxitin. It was not specified with which antibiotics the patients had actual contact (3).

Case reports

A 42-year-old nurse had occupational airborne allergic contact dermatitis of the face, neck and eyelids from cefotaxime and penicillin (unspecified) (4).

A 45-year-old nurse developed contact dermatitis on the hands and forearms, which was ascribed to handling of cephalosporins for parenteral use. In the past year, the condition had spread to the patient's face, eyelids and neck. The intensely itchy lesions regressed completely during suspension of working activity. Patch tests were positive to cefotaxime, ceftizoxime, ceftazidime, ceftriaxone and cefodizime, all tested 20% pet. It was not mentioned to which cephalosporines the patient had been exposed and which reactions were considered to be cross-reactions (5).

CUTANEOUS ADVERSE DRUG REACTIONS FROM SYSTEMIC ADMINISTRATION CAUSED BY TYPE IV (DELAYED-TYPE) HYPERSENSITIVITY (as demonstrated by positive patch tests)
Cutaneous adverse drug reactions from systemic administration of cefotaxime caused by type IV (delayed-type) hypersensitivity have included acute generalized exanthematous pustulosis (AGEP) (6,8) and drug reaction with eosinophilia and systemic symptoms (DRESS) (7).

Acute generalized exanthematous pustulosis (AGEP)
A 30-year-old woman received cefotaxime, fosfomycin and ciprofloxacin for sinusitis. On day 12 after drug initiation, she developed an extending itching erythema with numerous small nonfollicular pustules, most pronounced in the inguinal and axillary areas, but also on the trunk, face, and proximal limbs, associated with fever (39 °C) and erosions of the oral mucous membrane. Laboratory investigations showed marked leukocytosis. The histological examination was consistent with acute generalized exanthematous pustulosis (AGEP). After remission, patch tests were positive to cefotaxime at D2 (no details provided) while negative to the other drugs used (6,8).

Drug reaction with eosinophilia and systemic symptoms (DRESS)
In a multicenter study in Turkey into severe cutaneous adverse drug reactions in pediatric patients, performed in the period 2011-2016, there were 16 cases of DRESS. One individual had DRESS from cefotaxime with a positive patch test; details were not provided (7).

Cross-reactions, pseudo-cross-reactions and co-reactions
Cross-reactions between beta-lactam antibiotics are discussed in Chapter 3.36 Amoxicillin.

LITERATURE
1   The data in the section 'General' may have been obtained from literature discussed in this chapter, but mostly also or exclusively from one or more of the following online sources: ChemIDPlus Advanced, PubChem, DrugBank, RxList, Drug Central, Drugs.com, and Wikipedia
2   Foti C, Bonamonte D, Trenti R, Veña GA, Angelini G. Occupational contact allergy to cephalosporins. Contact Dermatitis 1997;36:104-105
3   Pinheiro V, Pestana C, Pinho A, Antunes I, Gonçalo M. Occupational allergic contact dermatitis caused by antibiotics in healthcare workers - relationship with non-immediate drug eruptions. Contact Dermatitis 2018;78:281-286
4   Gielen K, Goossens A. Occupational allergic contact dermatitis from drugs in healthcare workers. Contact Dermatitis 2001;45:273-279
5   Foti C, Vena GA, Cucurachi MR, Angelini G. Occupational contact allergy from cephalosporins. Contact Dermatitis 1994;31:129-130
6   Chaabane A, Aouam K, Gassab L, Njim L, Boughattas NA. Acute generalized exanthematous pustulosis (AGEP) induced by cefotaxime. Fundam Clin Pharmacol 2010;24:429-432
7   Dibek Misirlioglu E, Guvenir H, Bahceci S, Haktanir Abul M, Can D, Usta Guc BE, et al. Severe cutaneous adverse drug reactions in pediatric patients: A multicenter study. J Allergy Clin Immunol Pract 2017;5:757-763
8   Chaabane A, Aouam K, Harrathi K, Yahia NB, Gassab E, Njim L, Boughattas NA. Acute generalised exanthematous pustulosis (AGEP) after cefotaxime use. BMJ Case Rep. 2009;2009:bcr06.2008.0343

Chapter 3.106 CEFOTETAN

IDENTIFICATION

Description/definition   : Cefotetan is the semisynthetic β-lactamase-resistant cephalosporin that conforms to the structural formula shown below
Pharmacological classes   : Anti-bacterial agents
IUPAC name   : (6R,7S)-7-[[4-(2-Amino-1-carboxylato-2-oxoethylidene)-1,3-dithietane-2-carbonyl]amino]-7-methoxy-3-[(1-methyltetrazol-5-yl)sulfanylmethyl]-8-oxo-5-thia-1-azabicyclo[4.2.0]oct-2-ene-2-carboxylic acid
CAS registry number   : 69712-56-7
EC number   : 274-093-3
Merck Index monograph   : 3205
Patch testing   : Generally speaking, cephalosporins may be tested 5-10% pet. or water; 5% pet. (2)
Molecular formula   : $C_{17}H_{17}N_7O_8S_4$

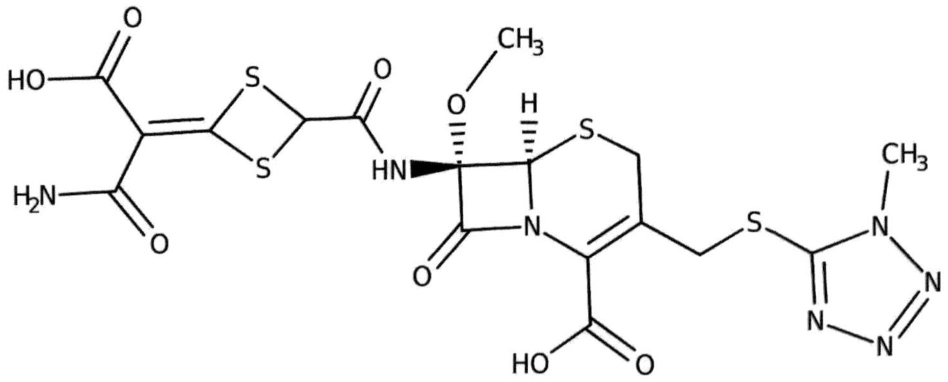

GENERAL

Cefotetan is a semisynthetic, broad-spectrum, β-lactamase-resistant, second-generation cephalosporin antibiotic with bactericidal activity that is administered intravenously or intramuscularly. The drug is highly resistant to a broad spectrum of β-lactamases and is active against a wide range of both aerobic and anaerobic gram-positive and gram-negative microorganisms (1). In pharmaceutical products, cefotetan is employed as cefotetan disodium (CAS number 74356-00-6, EC number 277-834-9, molecular formula $C_{17}H_{15}N_7Na_2O_8S_4$) (1). The classification and structures of beta-lactam antibiotics are discussed in Chapter 3.36 Amoxicillin.

CONTACT ALLERGY FROM ACCIDENTAL CONTACT

Seven nurses from Bari and Milan, Italy, had occupational allergic contact dermatitis of the hands from one or more cephalosporins that they worked with. When patch tested, there were 4 reactions to cefotetan disodium, 4 to cefazolin sodium, 3 to cefamandole, 2 to cefmetazole sodium and 1 to each of the following: cefotiam dihydrochloride, ceftizoxime sodium, cefotaxime sodium, cefodizime disodium, ceftazidime pentahydrate and ceftriaxone sodium, all tested 20% pet. One of the nurses reacted to 5 cephalosporins, 2 to four, 1 to three antibiotics, and 3 reacted to one cephalosporin each. It was not specified which cephalosporins were the culprit drugs and which – if any – positive patch tests were cross-reactions. The patch tests were read at D2 only (3).

Cross-reactions, pseudo-cross-reactions and co-reactions
Cross-reactions between beta-lactam antibiotics are discussed in Chapter 3.36 Amoxicillin.

LITERATURE

1   The data in the section 'General' may have been obtained from literature discussed in this chapter, but mostly also or exclusively from one or more of the following online sources: ChemIDPlus Advanced, PubChem, DrugBank, RxList, Drug Central, Drugs.com, and Wikipedia
2   Brockow K, Garvey LH, Aberer W, Atanaskovic-Markovic M, Barbaud A, Bilo MB, et al.; ENDA/EAACI Drug Allergy Interest Group. Skin test concentrations for systemically administered drugs – an ENDA/EAACI Drug Allergy Interest Group position paper. Allergy 2013;68:702-712
3   Foti C, Bonamonte D, Trenti R, Veña GA, Angelini G. Occupational contact allergy to cephalosporins. Contact Dermatitis 1997;36:104-105

Chapter 3.107 CEFOTIAM

IDENTIFICATION

Description/definition : Cefotiam is the β-lactam cephalosporin antibiotic that conforms to the structural formula shown below

Pharmacological classes : Anti-bacterial agents

IUPAC name : (6R,7R)-7-[[2-(2-Amino-1,3-thiazol-4-yl)acetyl]amino]-3-[[1-[2-(dimethylamino)ethyl]tetrazol-5-yl]sulfanylmethyl]-8-oxo-5-thia-1-azabicyclo[4.2.0]oct-2-ene-2-carboxylic acid

CAS registry number : 61622-34-2

EC number : Not available

Merck Index monograph : 3206

Patch testing : Generally speaking, cephalosporins may be tested 5-10% pet. or water; 5% pet. (3)

Molecular formula : $C_{18}H_{23}N_9O_4S_3$

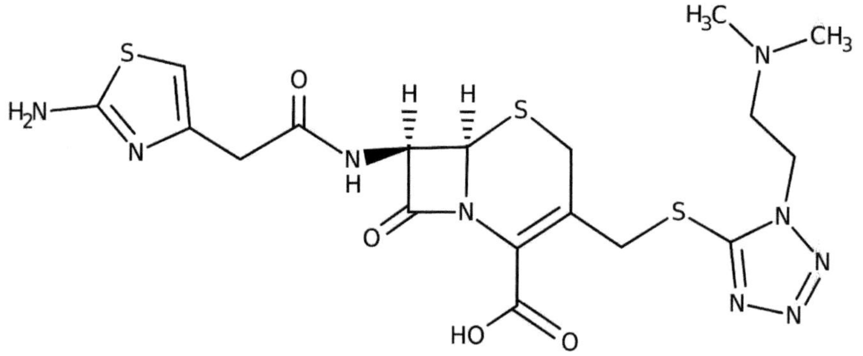

GENERAL

Cefotiam is a third-generation, semisynthetic, β-lactam cephalosporin antibiotic with broad-spectrum antibacterial activity. It is indicated for the treatment of severe infections caused by susceptible bacteria. In pharmaceutical products, cefotiam is employed as cefotiam (di)hydrochloride (CAS number 66309-69-1, EC number 266-312-6, molecular formula $C_{18}H_{25}Cl_2N_9O_4S_3$) (1). The classification and structures of beta-lactam antibiotics are discussed in Chapter 3.36 Amoxicillin.

CONTACT ALLERGY FROM ACCIDENTAL CONTACT

Seven nurses from Bari and Milan, Italy, had occupational allergic contact dermatitis of the hands from one or more cephalosporins that they worked with. When patch tested, there was one reaction to cefotiam dihydrochloride, 4 reactions to cefazolin sodium, 3 to cefamandole, 4 to cefotetan disodium, 2 to cefmetazole sodium and 1 to each of the following: ceftizoxime sodium, cefotaxime sodium, cefodizime disodium, ceftazidime pentahydrate and ceftriaxone sodium, all tested 20% pet. One of the nurses reacted to 5 cephalosporins, 2 to four, 1 to three antibiotics, and 3 reacted to one cephalosporin each. It was not specified which cephalosporins were the culprit drugs and which – if any – positive patch tests were cross-reactions. The patch tests were read at D2 only (2).

Cross-reactions, pseudo-cross-reactions and co-reactions

Cross-reactions between beta-lactam antibiotics are discussed in Chapter 3.36 Amoxicillin.

Immediate contact reactions

Immediate contact reactions (contact urticaria) to cefotiam are presented in Chapter 5.

LITERATURE

1 The data in the section 'General' may have been obtained from literature discussed in this chapter, but mostly also or exclusively from one or more of the following online sources: ChemIDPlus Advanced, PubChem, DrugBank, RxList, Drug Central, Drugs.com, and Wikipedia

2 Foti C, Bonamonte D, Trenti R, Veña GA, Angelini G. Occupational contact allergy to cephalosporins. Contact Dermatitis 1997;36:104-105

3 Brockow K, Garvey LH, Aberer W, Atanaskovic-Markovic M, Barbaud A, Bilo MB, et al.; ENDA/EAACI Drug Allergy Interest Group. Skin test concentrations for systemically administered drugs – an ENDA/EAACI Drug Allergy Interest Group position paper. Allergy 2013;68:702-712

Chapter 3.108 CEFOXITIN

IDENTIFICATION

Description/definition   : Cefoxitin is the cephalosporin 3'-carbamate that conforms to the structural formula
                             shown below
Pharmacological classes   : Antibacterial agents
IUPAC name   : (6R,7S)-3-(Carbamoyloxymethyl)-7-methoxy-8-oxo-7-[(2-thiophen-2-ylacetyl)amino]-5-
                            thia-1-azabicyclo[4.2.0]oct-2-ene-2-carboxylic acid
CAS registry number   : 35607-66-0
EC number   : 252-641-2
Merck Index monograph   : 3208
Patch testing   : Generally speaking, cephalosporins may be tested 5-10% pet. or water; 5% pet. (9)
Molecular formula   : $C_{16}H_{17}N_3O_7S_2$

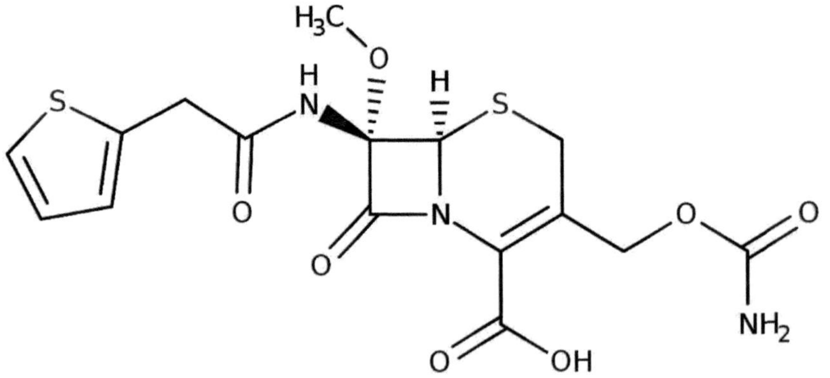

GENERAL

Cefoxitin is a semisynthetic, broad-spectrum, second-generation, β-lactamase resistant, cephalosporin with antibacterial activity. Cefoxitin binds to and inactivates penicillin-binding proteins (PBPs) located on the inner membrane of the bacterial cell wall. Inactivation of PBPs interferes with the cross-linkage of peptidoglycan chains necessary for bacterial cell wall strength and rigidity. This results in the weakening of the bacterial cell wall and causes cell lysis. It is used by intravenous administration for the treatment of serious infections caused by susceptible strains of microorganisms. In pharmaceutical products, cefoxitin is employed as cefoxitin sodium (CAS number 33564-30-6, EC number 251-574-6, molecular formula $C_{16}H_{16}N_3NaO_7S_2$) (1). The classification and structures of beta-lactam antibiotics are discussed in Chapter 3.36 Amoxicillin.

CONTACT ALLERGY FROM ACCIDENTAL CONTACT

A nurse involved in the handling of intravenous antibiotics had occupational allergic contact dermatitis of the hands, forearm and face and reacted to cefoxitin, ceftriaxone and cefotaxime 10% pet. It was not specified with which antibiotics the patient had actual contact (2).

A 24-year-old nurse presented with a 6-month history of intensely itchy erythematous papules and plaques on exposed skin of her forehead, perioral area, forearms and hands, which was clearly work-related. Patch tests were positive to cefoxitin 1%, 5% and 10% water, cefazoline 1%, 5% and 10% water, ceftriaxone 1%,5% and 10% water and ceftazidime 5% and 10% water. The patient stopped preparing cephalosporin solutions for systemic administration and the lesions cleared up (4).

CUTANEOUS ADVERSE DRUG REACTIONS FROM SYSTEMIC ADMINISTRATION CAUSED BY TYPE IV (DELAYED-TYPE) HYPERSENSITIVITY (as demonstrated by positive patch tests)

Cutaneous adverse drug reactions from systemic administration of cefoxitin caused by type IV (delayed-type) hypersensitivity have included maculopapular eruption (6,7) and drug reaction with eosinophilia and systemic symptoms (DRESS) (3,5,8).

Maculopapular eruption

In the period 2000-2014, in Portugal, 260 patients were patch tested with antibiotics for suspected cutaneous adverse drug reactions (CADR) to these drugs. 56 patients (22%) had one or more (often from cross-reactivity) positive patch tests. Cefoxitin was patch test positive in 2 patients with maculopapular eruptions (6, probably overlap with ref. 7). In a study from the same group of investigators, 18 patients who had previously had positive patch tests to antibiotics that had caused a maculopapular eruption, were again patch tested after a mean interval of 6 years. The positive reactions were reproducible in 16 cases; 7 were caused by amoxicillin, 5 by clindamycin, 4 by flucloxacillin, one by spiramycin, one by vancomycin and one by cefoxitin (in some patients, 2 antibiotics were responsible (7, probably overlap with ref. 6).

Drug reaction with eosinophilia and systemic symptoms (DRESS)

A 48-yar-old man using allopurinol received one single administration of cefoxitin prior to laparoscopic cholecystic-tomy. A few days later, while using piperacillin-tazobactam, the patient developed a maculopapular exanthema that progressed to an intensely pruritic exfoliative erythroderma with facial involvement, including the ears and enlarged lymph nodes. At this point, biochemistry panel and complete blood counts showed worsening of renal and liver function and an elevated white cell count with frank eosinophilia. Patch tests performed later with allopurinol at 10% pet. and an antibiotics series, including piperacillin-tazobactam, revealed a positive patch test only to cefoxitin 10% pet (++). Histological examination of the positive patch skin reaction showed a dermal lymphohistiocytic perivascular infiltrate with mild lymphocytic exocytosis and vacuolar degeneration of the basal layer, compatible with a drug eruption. The authors diagnosed allopurinol-induced DRESS syndrome (patch tests to allopurinol are always negative) presenting as a cholecystitis-like acute abdomen and aggravated by co-sensitization to cefoxitin (5).

To evaluate if, after DRESS, patients become sensitized to antibiotics, patch test data and clinical files of 17 patients with DRESS from a nonantibiotic culprit drug (anticonvulsants 10, allopurinol 7) seen in Coimbra, Portugal, between 2010 and 2018 and who were given antibiotics at the onset or later during the course of DRESS, were retrospectively studied. The group consisted of eight women and nine men with an age range of 10 to 89 years and a mean of 47. In 10, anticonvulsants had caused DRESS (notably carbamazepine) and in 7 allopurinol (which always gives negative patch tests). They had received antibiotics at the onset or later during the course of DRESS: amoxicillin in 7, cephalosporins in three and fluoroquinolones in seven patients (5 ciprofloxacin, 2 levofloxacin). Nine patients (53%) had developed positive patch tests to antibiotics: six to amoxicillin (with 5 cross-reactions to ampicillin), three to the cephalosporins (one to ceftriaxone, one to cefoxitin and one to cefoxitin, ceftazidime and additional positivity to vancomycin), but none to the fluoroquinolones ciprofloxacin or levofloxacin. In repeat testing in 4 patients, most positive patch tests were reproducible after several years. These patients all had multiple hypersensitivity syndrome, as they reacted to (at least) two unrelated chemicals (8).

In a hospital in France, specialized in toxic bullous diseases and severe cutaneous adverse reactions, a retrospective study was performed in consecutive patients consulting between 2010 and 2018 with a nonimmediate CADR suspected to have been caused by beta-lactam antibiotics. 56 patients were included, among whom were 46 amoxicillin-suspected and seven cephalosporin-suspected. Cefoxitin was responsible for one case of DRESS (3).

Cross-reactions, pseudo-cross-reactions and co-reactions
Cross-reactions between beta-lactam antibiotics are discussed in Chapter 3.36 Amoxicillin.

LITERATURE
1   The data in the section 'General' may have been obtained from literature discussed in this chapter, but mostly also or exclusively from one or more of the following online sources: ChemIDPlus Advanced, PubChem, DrugBank, RxList, Drug Central, Drugs.com, and Wikipedia
2   Pinheiro V, Pestana C, Pinho A, Antunes I, Gonçalo M. Occupational allergic contact dermatitis caused by antibiotics in healthcare workers - relationship with non-immediate drug eruptions. Contact Dermatitis 2018;78:281-286
3   Bérot V, Gener G, Ingen-Housz-Oro S, Gaudin O, Paul M, Chosidow O, Wolkenstein P, Assier H. Cross-reactivity in beta-lactams after a non-immediate cutaneous adverse reaction: experience of a reference centre for toxic bullous diseases and severe cutaneous adverse reactions. J Eur Acad Dermatol Venereol 2020;34:787-794
4   Filipe P, Almeida RS, Rodrigo FG. Occupational allergic contact dermatitis from cephalosporins. Contact Dermatitis 1996;34:226
5   Batista M, Cardoso JC, Oliveira P, Gonçalo M. Allopurinol-induced DRESS syndrome presented as a cholecystitis-like acute abdomen and aggravated by antibiotics. BMJ Case Rep 2018 Jul 24;2018:bcr2018226023
6   Pinho A, Coutinho I, Gameiro A, Gouveia M, Gonçalo M. Patch testing - a valuable tool for investigating non-immediate cutaneous adverse drug reactions to antibiotics. J Eur Acad Dermatol Venereol 2017;31:280-287
7   Pinho A, Marta A, Coutinho I, Gonçalo M. Long-term reproducibility of positive patch test reactions in patients with non-immediate cutaneous adverse drug reactions to antibiotics. Contact Dermatitis 2017;76:204-209
8   Santiago LG, Morgado FJ, Baptista MS, Gonçalo M. Hypersensitivity to antibiotics in drug reaction with eosinophilia and systemic symptoms (DRESS) from other culprits. Contact Dermatitis 2020;82:290-296
9   Brockow K, Garvey LH, Aberer W, Atanaskovic-Markovic M, Barbaud A, Bilo MB, et al.; ENDA/EAACI Drug Allergy Interest Group. Skin test concentrations for systemically administered drugs – an ENDA/EAACI Drug Allergy Interest Group position paper. Allergy 2013;68:702-712

Chapter 3.109 CEFPODOXIME

IDENTIFICATION

Description/definition   : Cefpodoxime is the semisynthetic cephalosporin and β-lactam antibiotic that
                             conforms to the structural formula shown below
Pharmacological classes   : Anti-bacterial agents
IUPAC name   : (6R,7R)-7-[[(2Z)-2-(2-Amino-1,3-thiazol-4-yl)-2-methoxyiminoacetyl]amino]-3-(methoxy-
                             methyl)-8-oxo-5-thia-1-azabicyclo[4.2.0]oct-2-ene-2-carboxylic acid
CAS registry number   : 80210-62-4
EC number   : Not available
Merck Index monograph   : 3212
Patch testing   : Cefpodoxime proxetil 10.0% pet. (Chemotechnique)
Molecular formula   : $C_{15}H_{17}N_5O_6S_2$

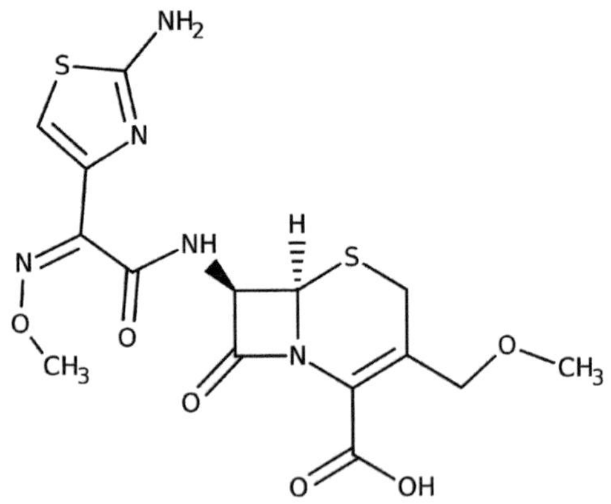

Cefpodoxime 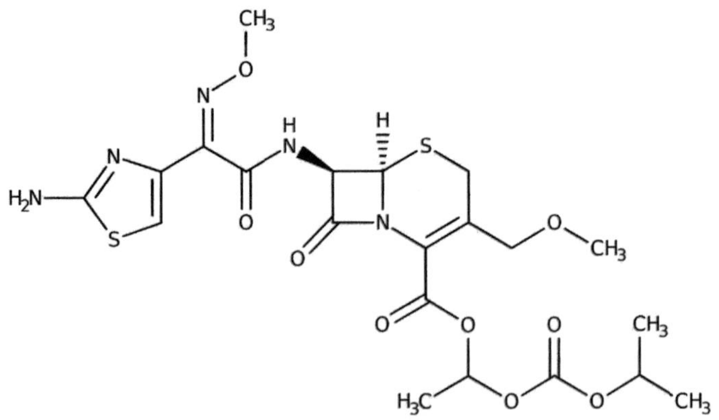   Cefpodoxime proxetil

GENERAL

Cefpodoxime is a third-generation semisynthetic cephalosporin and a β-lactam antibiotic with bactericidal activity. It
is indicated for the treatment of patients with mild to moderate infections, including acute otitis media, pharyngitis,
and sinusitis, caused by susceptible strains of specific microorganisms. In pharmaceutical products, cefpodoxime is
employed as cefpodoxime proxetil (CAS number 87239-81-4, EC number 918-886-0, molecular formula
$C_{21}H_{27}N_5O_9S_2$). Cefpodoxime proxetil is a prodrug which is absorbed and de-esterified by the intestinal mucosa to
cefpodoxime (go.drugbank.com). The classification and structures of beta-lactam antibiotics are discussed in Chapter
3.36 Amoxicillin.

CUTANEOUS ADVERSE DRUG REACTIONS FROM SYSTEMIC ADMINISTRATION CAUSED BY TYPE IV
(DELAYED-TYPE) HYPERSENSITIVITY (as demonstrated by positive patch tests)

Cutaneous adverse drug reactions from systemic administration of cefpodoxime caused by type IV (delayed-type)
hypersensitivity have included acute generalized exanthematous pustulosis (AGEP) (1).

Acute generalized exanthematous pustulosis (AGEP)

In a hospital in France, specialized in toxic bullous diseases and severe cutaneous adverse reactions, a retrospective
study was performed in consecutive patients consulting between 2010 and 2018 with a nonimmediate CADR
suspected to have been caused by beta-lactam antibiotics. 56 patients were included, among whom were 46
amoxicillin-suspected and seven cephalosporin-suspected. Cefpodoxime was responsible for one case of AGEP
(cefpodoxime itself not tested, but positive reactions to cefotaxime and penicillins (1).

Cross-reactions, pseudo-cross-reactions and co-reactions

Cross-reactions between beta-lactam antibiotics are discussed in Chapter 3.36 Amoxicillin.

LITERATURE

1   Bérot V, Gener G, Ingen-Housz-Oro S, Gaudin O, Paul M, Chosidow O, Wolkenstein P, Assier H. Cross-reactivity in
    beta-lactams after a non-immediate cutaneous adverse reaction: experience of a reference centre for toxic
    bullous diseases and severe cutaneous adverse reactions. J Eur Acad Dermatol Venereol 2020;34:787-794

Chapter 3.110 CEFRADINE

IDENTIFICATION

Description/definition : Cefradine is the β-lactam cephalosporin antibiotic that conforms to the structural formula shown below
Pharmacological classes : Anti-bacterial agents
IUPAC name : (6R,7R)-7-[[(2R)-2-Amino-2-cyclohexa-1,4-dien-1-ylacetyl]amino]-3-methyl-8-oxo-5-thia-1-azabicyclo[4.2.0]oct-2-ene-2-carboxylic acid
Other names : Cephradine
CAS registry number : 38821-53-3
EC number : 254-137-8
Merck Index monograph : 3255
Patch testing : 10.0% pet. (Chemotechnique)
Molecular formula : $C_{16}H_{19}N_3O_4S$

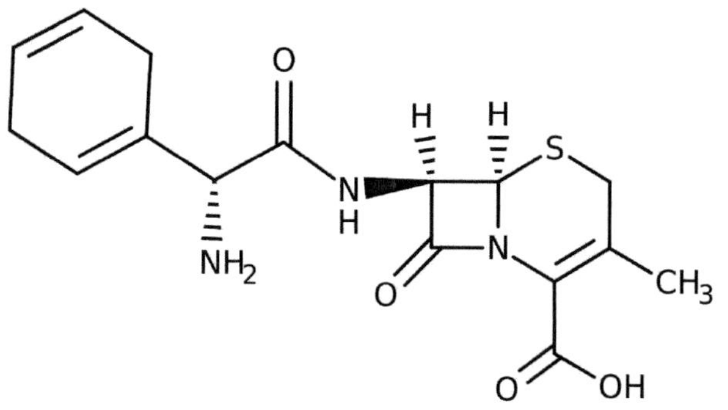

GENERAL

Cefradine is a first-generation β-lactam cephalosporin antibiotic with bactericidal activity. It is used in the treatment of bacterial infections of the respiratory and urinary tracts and of the skin and soft tissues (1). The classification and structures of beta-lactam antibiotics are discussed in Chapter 3.36 Amoxicillin.

CONTACT ALLERGY FROM ACCIDENTAL CONTACT

In a group of 107 workers in the pharmaceutical industry with dermatitis, investigated in Warsaw, Poland, before 1989, one reacted to cefradine, tested 20% pet. (2). Also in Warsaw, Poland, in the period 1979-1983, 27 pharmaceutical workers, 24 nurses and 30 veterinary surgeons were diagnosed with occupational allergic contact dermatitis from antibiotics. The numbers that had positive patch tests to cefradine (ampoule content) were 1, 2, and 0, respectively, total 3 (5). Of 333 nurses patch tested in Poland by the same investigators between 1979 and 1987, 4 reacted to cefradine 20% pet.; all reactions were considered likely to be relevant (3).

A 32-year-old female nurse from Portugal had hand dermatitis and later palpebral eczema. When patch tested, she reacted to two cephalosporins that she worked with, cefradine and cefazolin. There were cross-reactions to cefuroxime, ceftriaxone, and cefotaxime (4).

Cross-reactions, pseudo-cross-reactions and co-reactions

Cross-reactions between beta-lactam antibiotics are discussed in Chapter 3.36 Amoxicillin.

LITERATURE

1   The data in the section 'General' may have been obtained from literature discussed in this chapter, but mostly also or exclusively from one or more of the following online sources: ChemIDPlus Advanced, PubChem, DrugBank, RxList, Drug Central, Drugs.com, and Wikipedia
2   Rudzki E, Rebandel P, Grzywa Z. Contact allergy in the pharmaceutical industry. Contact Dermatitis 1989;21:121-122
3   Rudzki E, Rebandel P, Grzywa Z. Patch tests with occupational contactants in nurses, doctors and dentists. Contact Dermatitis 1989;20:247-250
4   Antunes J, Silva R, Pacheco D, Travassos R, Filipe P. Occupational contact allergy to cephalosporins. Dermatol Online J 2011 May 15;17(5):13
5   Rudzki E, Rebendel P. Contact sensitivity to antibiotics. Contact Dermatitis 1984;11:41-42

Chapter 3.111 CEFTAZIDIME

IDENTIFICATION

Description/definition : Ceftazidime is the semisynthetic cephalosporin derived from cefaloridine, that conforms
to the structural formula shown below
Pharmacological classes : Anti-bacterial agents
IUPAC name : (6R,7R)-7-[[(2Z)-2-(2-Amino-1,3-thiazol-4-yl)-2-(2-carboxypropan-2-yloxyimino)acetyl]-
amino]-8-oxo-3-(pyridin-1-ium-1-ylmethyl)-5-thia-1-azabicyclo[4.2.0]oct-2-ene-2-
carboxylate
CAS registry number : 72558-82-8
EC number : 276-715-9
Merck Index monograph : 3218
Patch testing : Generally speaking, cephalosporins may be tested 5-10% pet. or water; 5% pet. (8)
Molecular formula : $C_{22}H_{22}N_6O_7S_2$

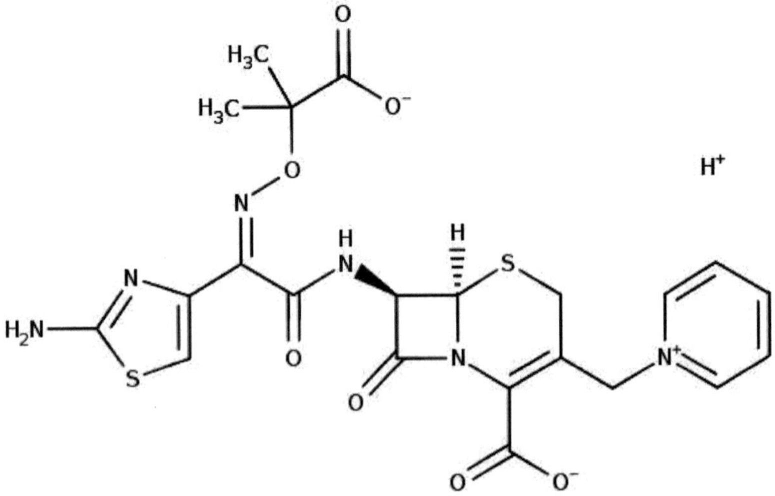

GENERAL

Ceftazidime is semisynthetic, third-generation, broad-spectrum cephalosporin derived from cefaloridine. It is indicated for the treatment of patients with infections of the lower respiratory tract, skin and skin structure, urinary tract, bone and joint, central nervous system (including meningitis), patients with gynecologic infections, intra-abdominal infections (including peritonitis), and bacterial septicemia caused by susceptible strains of organisms. It may be used especially for *Pseudomonas* and other gram-negative infections in debilitated patients (1).

In pharmaceutical products, ceftazidime is usually employed as ceftazidime pentahydrate (CAS number 78439-06-2, EC number 616-626-8, molecular formula $C_{22}H_{32}N_6O_{12}S_2$) (1). The classification and structures of beta-lactam antibiotics are discussed in Chapter 3.36 Amoxicillin.

CONTACT ALLERGY FROM ACCIDENTAL CONTACT

Seven nurses from Bari and Milan, Italy, had occupational allergic contact dermatitis of the hands from one or more cephalosporins that they worked with. When patch tested, there was one reaction to ceftazidime pentahydrate, 4 reactions to cefazolin sodium, 3 to cefamandole, 4 to cefotetan disodium, 2 to cefmetazole sodium and 1 to each of the following: cefotiam dihydrochloride, ceftizoxime sodium, cefotaxime sodium, cefodizime disodium, and ceftriaxone sodium, all tested 20% pet. One of the nurses reacted to 5 cephalosporins, 2 to four, 1 to three antibiotics, and 3 reacted to one cephalosporin each. It was not specified which cephalosporins were the culprit drugs and which – if any – positive patch tests were cross-reactions. The patch tests were read at D2 only (3).

A 45-year-old nurse developed contact dermatitis on the hands and forearms, which was ascribed to handling of cephalosporins for parenteral use. In the past year, the condition had spread to the patient's face, eyelids and neck. The intensely itchy lesions regressed completely during suspension of working activity. Patch tests were positive to ceftazidime, cefotaxime, ceftizoxime, ceftriaxone and cefodizime, all tested 20% pet. It was not mentioned to which cephalosporines the patient had been exposed and which reactions were considered to be cross-reactions (4).

A 24-year-old nurse presented with a 6-month history of intensely itchy erythematous papules and plaques on exposed skin of her forehead, perioral area, forearms and hands, which was clearly work-related. Patch tests were positive to cefazoline 1%, 5% and 10% water, cefoxitin 1%, 5% and 10% water, ceftriaxone 1%,5% and 10% water and ceftazidime 5% and 10% water. The patient stopped preparing cephalosporin solutions for systemic administration and the lesions cleared up (5).

CUTANEOUS ADVERSE DRUG REACTIONS FROM SYSTEMIC ADMINISTRATION CAUSED BY TYPE IV (DELAYED-TYPE) HYPERSENSITIVITY (as demonstrated by positive patch tests)

Cutaneous adverse drug reactions from systemic administration of ceftazidime caused by type IV (delayed-type) hypersensitivity have included generalized erythema (6) and drug reaction with eosinophilia and systemic symptoms (DRESS) (7).

Erythroderma, widespread erythematous eruption, exfoliative dermatitis

A 66-year-old woman developed a pruritic diffuse erythematous rash with fine scaling 2 days after the initiation of treatment with cefazolin i.m. (500 mg/12 h) after the implantation of a knee prosthesis. The patient had taken cephalosporins 8 years before and she remembered a similar reaction with ceftazidime. Patch tests were positive to the commercial preparation 'as is' and 5 other cephalosporins for injection, including ceftazidime. Intradermal tests with these 2 drugs were also positive at the D2 delayed reading (6).

Drug reaction with eosinophilia and systemic symptoms (DRESS)

To evaluate if, after DRESS, patients become sensitized to antibiotics, patch test data and clinical files of 17 patients with DRESS from a nonantibiotic culprit drug (anticonvulsants 10, allopurinol 7) seen in Coimbra, Portugal, between 2010 and 2018 and who were given antibiotics at the onset or later during the course of DRESS, were retrospectively studied. The group consisted of eight women and nine men with an age range of 10 to 89 years and a mean of 47. In 10, anticonvulsants had caused DRESS (notably carbamazepine) and in 7 allopurinol (which always gives negative patch tests). They had received antibiotics at the onset or later during the course of DRESS: amoxicillin in 7, cephalosporins in three and fluoroquinolones in seven patients (5 ciprofloxacin, 2 levofloxacin). Nine patients (53%) had developed positive patch tests to antibiotics: six to amoxicillin (with 5 cross-reactions to ampicillin), three to the cephalosporins (one to ceftriaxone, one to cefoxitin and one to cefoxitin, ceftazidime and additional positivity to vancomycin), but none to the fluoroquinolones ciprofloxacin or levofloxacin. In repeat testing in 4 patients, most positive patch tests were reproducible after several years. These patients all had multiple hypersensitivity syndrome, as they reacted to (at least) two unrelated chemicals (7).

Cross-reactions, pseudo-cross-reactions and co-reactions

Cross-reactions between beta-lactam antibiotics are discussed in Chapter 3.36 Amoxicillin. A patient sensitized to aztreonam had a cross-reaction to ceftazidime, the only cephalosporin with the same side chain as aztreonam (2).

LITERATURE

1   The data in the section 'General' may have been obtained from literature discussed in this chapter, but mostly also or exclusively from one or more of the following online sources: ChemIDPlus Advanced, PubChem, DrugBank, RxList, Drug Central, Drugs.com, and Wikipedia
2   Pérez Pimiento A, Gómez Martínez M, Mínguez Mena A, Trampal González A, de Paz Arranz S, Rodríguez Mosquera M. Aztreonam and ceftazidime: evidence of *in vivo* cross allergenicity. Allergy 1998;53:624-625
3   Foti C, Bonamonte D, Trenti R, Veña GA, Angelini G. Occupational contact allergy to cephalosporins. Contact Dermatitis 1997;36:104-105
4   Foti C, Vena GA, Cucurachi MR, Angelini G. Occupational contact allergy from cephalosporins. Contact Dermatitis 1994;31:129-130
5   Filipe P, Almeida RS, Rodrigo FG. Occupational allergic contact dermatitis from cephalosporins. Contact Dermatitis 1996;34:226
6   Gonzalo-Garijo MA, Rodríguez-Nevado I, de Argila D. Patch tests for diagnosis of delayed hypersensitivity to cephalosporins. Allergol Immunopathol (Madr) 2006;34:39-41
7   Santiago LG, Morgado FJ, Baptista MS, Gonçalo M. Hypersensitivity to antibiotics in drug reaction with eosinophilia and systemic symptoms (DRESS) from other culprits. Contact Dermatitis 2020;82:290-296
8   Brockow K, Garvey LH, Aberer W, Atanaskovic-Markovic M, Barbaud A, Bilo MB, et al.; ENDA/EAACI Drug Allergy Interest Group. Skin test concentrations for systemically administered drugs – an ENDA/EAACI Drug Allergy Interest Group position paper. Allergy 2013;68:702-712

Chapter 3.112 CEFTIOFUR

IDENTIFICATION

Description/definition   : Ceftiofur is the semisynthetic, β-lactamase-stable, third-generation cephalosporin that
                             conforms to the structural formula shown below
Pharmacological classes   : Anti-bacterial agents
IUPAC name   : (6R,7R)-7-[[(2Z)-2-(2-Amino-1,3-thiazol-4-yl)-2-methoxyiminoacetyl]amino]-3-(furan-2-
                             carbonylsulfanylmethyl)-8-oxo-5-thia-1-azabicyclo[4.2.0]oct-2-ene-2-carboxylic acid
CAS registry number   : 80370-57-6
EC number   : Not available
Merck Index monograph   : 3221
Patch testing   : Generally speaking, cephalosporins may be tested 5-10% pet. or water; 5% pet. (2)
Molecular formula   : $C_{19}H_{17}N_5O_7S_3$

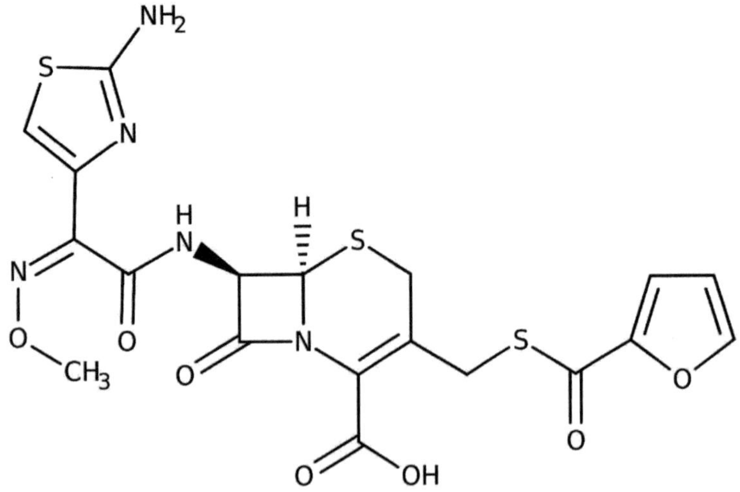

GENERAL

Ceftiofur is a semisynthetic, β-lactamase-stable, broad-spectrum, third-generation cephalosporin with antibacterial activity. It is licensed for use in veterinary medicine only and is mainly employed for treating respiratory infections in pigs and cows and for prevention of colibacillosis in chicks. In pharmaceutical products, ceftiofur may be employed as base, as hydrochloride (CAS number 103980-44-5, EC number 600-507-2, molecular formula $C_{19}H_{18}ClN_5O_7S_3$) and as ceftiofur sodium (CAS number 104010-37-9, EC number not available, molecular formula $C_{19}H_{16}N_5NaO_7S_3$) (1). The classification and structures of beta-lactam antibiotics are discussed in Chapter 3.36 Amoxicillin.

CONTACT ALLERGY FROM ACCIDENTAL CONTACT

A 46-year-old woman, working as a chick vaccinator, reported a 6-month history of severe hand eczema and extensive itchy dermatitis with erythema, maculopapular eruption and plaques, spreading to exposed forearm skin, face, eyelids and neck. She usually handled vaccines for the prophylaxis of the infectious borsal disease and fowl pox. Ceftiofur was added to the vaccine diluent to prevent infections in vaccinated chicks. Patch tests were positive to ceftiofur 5% pet. and negative to the diluent. Ten controls were negative. There were no cross-reactions to other cephalosporins or beta-lactam penicillins (3).

A 39-year-old man, also working as a chicken vaccinator, presented with a 2-month history of severe hand eczema, associated with eyelid dermatitis. He handled the cephalosporin ceftiofur sodium for 1 year. The patient did not usually wear gloves and had tolerated systemic administration of beta-lactams such as amoxicillin and cefuroxime. Patch tests were positive to ceftiofur sodium 1% water and negative to the diluent. Ten controls were negative. No cross-reactions were observed to other cephalosporins or penicillins (4).

The first two reported patients sensitized to ceftiofur were also chicken vaccinators and presented with very extensive to generalized dermatitis. Patch tests were positive to ceftiofur sodium 1% and 2% water and negative to other cephalosporins (5). These were all cases of (airborne) occupational allergic contact dermatitis.

Cross-reactions, pseudo-cross-reactions and co-reactions

Cross-reactions between beta-lactam antibiotics are discussed in Chapter 3.36 Amoxicillin. Up to now, no cross-reactivity to other cephalosporins has been found (3,4,5). Ceftiofur has an exclusive radical, the furanyl-carbonyl-thio-methyl group, which is not present in other cephalosporins. Sensitization to this radical could explain the lack of cross-reactivity (3).

LITERATURE

1 The data in the section 'General' may have been obtained from literature discussed in this chapter, but mostly also or exclusively from one or more of the following online sources: ChemIDPlus Advanced, PubChem, DrugBank, RxList, Drug Central, Drugs.com, and Wikipedia

2 Brockow K, Garvey LH, Aberer W, Atanaskovic-Markovic M, Barbaud A, Bilo MB, et al.; ENDA/EAACI Drug Allergy Interest Group. Skin test concentrations for systemically administered drugs – an ENDA/EAACI Drug Allergy Interest Group position paper. Allergy 2013;68:702-712

3 Antico A, Marcotulli C. Occupational contact allergy to ceftiofur. Allergy 2003;58:957-958

4 Garcia F, Juste S, Garces MM, Carretero P, Blanco J, Herrero D, et al. Occupational allergic contact dermatitis from ceftiofur without cross-sensitivity. Contact Dermatitis 1998;39:260

5 García-Bravo B, Gines E, Russo F. Occupational contact dermatitis from ceftiofur sodium. Contact Dermatitis 1995;33:62-63

Chapter 3.113 CEFTIZOXIME

IDENTIFICATION

Description/definition   : Ceftizoxime is the semisynthetic β-lactamase resistant cephalosporin that conforms to
                                        the structural formula shown below
Pharmacological classes   : Anti-bacterial agents
IUPAC name   : (6R,7R)-7-[[(2Z)-2-(2-Amino-1,3-thiazol-4-yl)-2-methoxyiminoacetyl]amino]-8-oxo-5-thia-
                                        -1-azabicyclo[4.2.0]oct-2-ene-2-carboxylic acid
CAS registry number   : 68401-81-0
EC number   : Not available
Merck Index monograph : 3222
Patch testing   : Generally speaking, cephalosporins may be tested 5-10% pet. or water; 5% pet. (2)
Molecular formula   : $C_{13}H_{13}N_5O_5S_2$

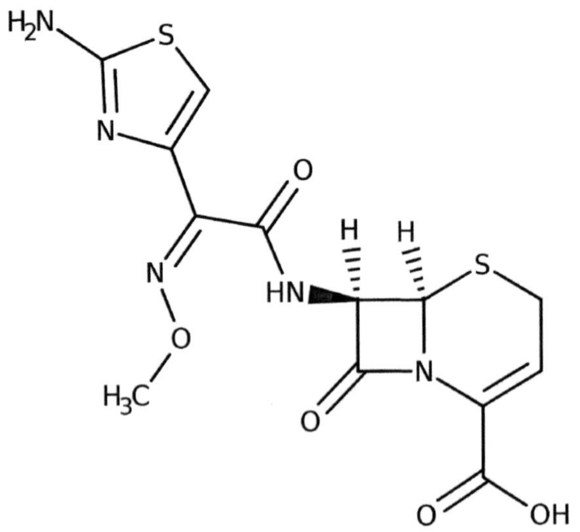

GENERAL

Ceftizoxime is a semisynthetic, broad-spectrum, β-lactamase resistant, third-generation cephalosporin antibiotic with bactericidal activity which can be administered intravenously or by suppository. The drug is highly resistant to a broad spectrum of β-lactamases and is active against a wide range of both aerobic and anaerobic gram-positive and gram-negative organisms. It is indicated for the treatment of infections due to susceptible strains of microorganisms (1). In pharmaceutical products, ceftizoxime is employed as ceftizoxime sodium (CAS number 68401-82-1, EC number not available, molecular formula $C_{13}H_{12}N_5NaO_5S_2$) (1). The classification and structures of beta-lactam antibiotics are discussed in Chapter 3.36 Amoxicillin.

CONTACT ALLERGY FROM ACCIDENTAL CONTACT

Seven nurses from Bari and Milan, Italy, had occupational allergic contact dermatitis of the hands from one or more cephalosporins that they worked with. When patch tested, there was one reaction to ceftizoxime sodium, 4 reactions to cefazolin sodium, 3 to cefamandole, 4 to cefotetan disodium, 2 to cefmetazole sodium and 1 to each of the following: cefotiam dihydrochloride, cefotaxime sodium, cefodizime disodium, ceftazidime pentahydrate and ceftriaxone sodium, all tested 20% pet. One of the nurses reacted to 5 cephalosporins, 2 to four, 1 to three antibiotics, and 3 reacted to one cephalosporin each. It was not specified which cephalosporins were the culprit drugs and which – if any – positive patch tests were cross-reactions. The patch tests were read at D2 only (3).

   A 45-year-old nurse developed contact dermatitis on the hands and forearms, which was ascribed to handling of cephalosporins for parenteral use. In the past year, the condition had spread to the patient's face, eyelids and neck. The intensely itchy lesions regressed completely during suspension of working activity. Patch tests were positive to ceftizoxime, cefotaxime, ceftazidime, ceftriaxone and cefodizime, all tested 20% pet. It was not mentioned to which cephalosporines the patient had been exposed and which reactions were considered to be cross-reactions (4).

Cross-reactions, pseudo-cross-reactions and co-reactions

Cross-reactions between beta-lactam antibiotics are discussed in Chapter 3.36 Amoxicillin.

LITERATURE

1   The data in the section 'General' may have been obtained from literature discussed in this chapter, but mostly also or exclusively from one or more of the following online sources: ChemIDPlus Advanced, PubChem, DrugBank, RxList, Drug Central, Drugs.com, and Wikipedia

2   Brockow K, Garvey LH, Aberer W, Atanaskovic-Markovic M, Barbaud A, Bilo MB, et al.; ENDA/EAACI Drug Allergy Interest Group. Skin test concentrations for systemically administered drugs – an ENDA/EAACI Drug Allergy Interest Group position paper. Allergy 2013;68:702-712

3   Foti C, Bonamonte D, Trenti R, Veña GA, Angelini G. Occupational contact allergy to cephalosporins. Contact Dermatitis 1997;36:104-105

4   Foti C, Vena GA, Cucurachi MR, Angelini G. Occupational contact allergy from cephalosporins. Contact Dermatitis 1994;31:129-130

Chapter 3.114 CEFTRIAXONE

IDENTIFICATION

Description/definition   : Ceftriaxone is the β-lactam, third-generation cephalosporin antibiotic that conforms to the structural formula shown below
Pharmacological classes : Anti-bacterial agents
IUPAC name   : (6R,7R)-7-[[(2Z)-2-(2-Amino-1,3-thiazol-4-yl)-2-methoxyiminoacetyl]amino]-3-[(2-methyl-5,6-dioxo-1H-1,2,4-triazin-3-yl)sulfanylmethyl]-8-oxo-5-thia-1-azabicyclo[4.2.0]oct-2-ene-2-carboxylic acid
CAS registry number   : 73384-59-5
EC number   : 277-405-6
Merck Index monograph   : 3225
Patch testing   : Generally speaking, cephalosporins may be tested 5-10% pet. or water; 5% pet. (25)
Molecular formula   : $C_{18}H_{18}N_8O_7S_3$

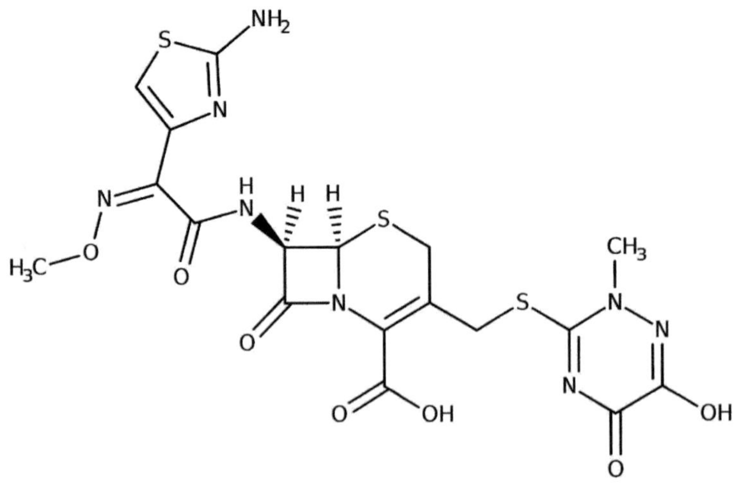

GENERAL

Ceftriaxone is a β-lactam, third-generation cephalosporin antibiotic with bactericidal activity. It is stable against hydrolysis by a variety of β-lactamases, including penicillinases, cephalosporinases and extended spectrum β-lactamases. Ceftriaxone is indicated for the treatment of infections of the respiratory tract, skin, soft tissue, urinary tract, ears and throat, caused by susceptible strains of microorganisms (1). In pharmaceutical products, ceftriaxone is employed as ceftriaxone sodium (heptahydrate) (CAS number 104376-79-6, EC number not available, molecular formula $C_{36}H_{46}N_{16}Na_4O_{21}S_6$). Ceftriaxone sodium is also the name for the non-hydrate form of ceftriaxone sodium (CAS number 74578-69-1). The compound used in drugs is the heptahydrate form (1). The classification and structures of beta-lactam antibiotics are discussed in Chapter 3.36 Amoxicillin.

CONTACT ALLERGY FROM ACCIDENTAL CONTACT

Case series

Seven nurses from Bari and Milan, Italy, had occupational allergic contact dermatitis of the hands from one or more cephalosporins that they worked with. When patch tested, there was one reaction to ceftriaxone sodium, 4 reactions to cefazolin sodium, 3 to cefamandole, 4 to cefotetan disodium, 2 to cefmetazole sodium and 1 to each of the following: cefotiam dihydrochloride, ceftizoxime sodium, cefotaxime sodium, cefodizime disodium, and ceftazidime pentahydrate, all tested 20% pet. One of the nurses reacted to 5 cephalosporins, 2 to four, 1 to three antibiotics, and 3 reacted to one cephalosporin each. It was not specified which cephalosporins were the culprit drugs and which – if any – positive patch tests were cross-reactions. The patch tests were read at D2 only (3).

Two nurses involved in the handling of intravenous antibiotics had occupational allergic contact dermatitis of the hands, forearms and one of the face. They reacted to ceftriaxone and cefotaxime, and one also to cefoxitin. It was not specified with which antibiotics the patients had actual contact (4).

Case reports

A 49-year-old nurse from Belgium had dermatitis of the fingers and later of the eyelids from occupational contact with ceftriaxone and probably also other cephalosporins (5). A 58-year-old nurse, involved in making dilutions of multiple intravenous drugs and administering them to patients in her ward, developed recalcitrant occupational allergic contact dermatitis of the hands from ceftriaxone and cefuroxime (6). A 31-year-old man, working for 8 years in a pharmaceutical factory packing powdered ceftriaxone, developed airborne occupational allergic contact dermatitis. A patch test and a lymphocyte stimulation test to ceftriaxone were positive, as was an intradermal test (an immediate weal developing into an infiltrated plaque after 24 h) (6).

A 45-year-old nurse developed contact dermatitis on the hands and forearms, which was ascribed to handling of cephalosporins for parenteral use. In the past year, the condition had spread to the patient's face, eyelids and neck. The intensely itchy lesions regressed completely during suspension of working activity. Patch tests were positive to ceftriaxone, cefotaxime, ceftizoxime, ceftazidime, and cefodizime, all tested 20% pet. It was not mentioned to which cephalosporines the patient had been exposed and which reactions were considered to be cross-reactions (13).

A 24-year-old nurse presented with a 6-month history of intensely itchy erythematous papules and plaques on exposed skin of her forehead, perioral area, forearms and hands, which was clearly work-related. Patch tests were positive to ceftriaxone 1%, 5% and 10% water, cefazoline 1%, 5% and 10% water, cefoxitin 1%, 5% and 10% water, and ceftazidime 5% and 10% water. The patient stopped preparing cephalosporin solutions for systemic administration and the lesions cleared up (16).

CUTANEOUS ADVERSE DRUG REACTIONS FROM SYSTEMIC ADMINISTRATION CAUSED BY TYPE IV (DELAYED-TYPE) HYPERSENSITIVITY (as demonstrated by positive patch tests)

Cutaneous adverse drug reactions from systemic administration of ceftriaxone caused by type IV (delayed-type) hypersensitivity have included maculopapular eruption (9,15,18), acute generalized exanthematous pustulosis (AGEP) (11,12,19,27,29), drug eruption with eosinophilia and systemic symptoms (DRESS) (10,14,17,18, 19,20), Stevens-Johnson syndrome/toxic epidermal necrolysis (SJS/TEN) (2), linear IgA bullous dermatosis (8), bullous eruption/erythema multiforme (15), urticaria (22; diagnosed as maculopapular eruption), exanthematous rash (6), erythematous, partly urticarial exanthema with blisters (26) and undefined cutaneous eruption (21).

Case series with various or unknown types of drug reactions

In Finland, in the period 1989-2001, 826 patients with suspected cutaneous drug eruptions were patch tested and 89 had one or more positive reactions. Of these individuals, 4 reacted to ceftriaxone. It was not mentioned which drug eruption these patients had suffered from (21).

Maculopapular eruption

A 62-year-old man suffered a generalized maculopapular rash 2 days after a laparoscopic cholecystectomy. A single dose of intravenous ceftriaxone (1 g) was administered as surgical prophylaxis. Nine years earlier, the patient had started treatment with penicillin, metronidazole, and ceftriaxone for a suppurative cerebral abscess, and a maculopapular rash had developed 2 days later. Patch tests were strongly positive to ceftriaxone 200 mg/ml saline and 4 other cephalosporins (cross-reactions), as was an intradermal test read at D2 and D4. Tests with all other drugs used, including penicillin, were negative (9).

In Bern, Switzerland, patients with a suspected allergic cutaneous drug reaction were patch-scratch tested with suspected drugs that had previously given a positive lymphocyte transformation test. Ceftriaxone 250 mg/ml saline gave a positive patch-scratch test in one patient with maculopapular exanthema and in one with bullous exanthema/erythema multiforme (15).

One patient developed a maculopapular eruption from ceftriaxone with a positive patch test to this drug after a previous episode of DRESS from other drug(s) (multiple drug hypersensitivity) (18).

In a group of 105 patients ranging in age from 14 to 84 years with histories of nonimmediate reactions to cephalosporins and patch tested with the culprit drugs, there were only 3 positive patch tests, 2 to ceftriaxone and one to cefalexin. All 3 patients had suffered a maculopapular eruption (28).

Acute generalized exanthematous pustulosis (AGEP)

Case series

In a multicenter investigation in France, of 45 patients patch tested for AGEP, 26 (58%) had positive patch tests to drugs, including to ceftriaxone (19).

In a hospital in France, specialized in toxic bullous diseases and severe cutaneous adverse reactions, a retrospective study was performed in consecutive patients consulting between 2010 and 2018 with a nonimmediate CADR suspected to have been caused by beta-lactam antibiotics. 56 patients were included, among whom were 46 amoxicillin-suspected and seven cephalosporin-suspected. Twenty-nine had severe CADR (DRESS, AGEP, SJS/TEN), and 27 had nonimmediate maculopapular exanthema (MPE). Of these patients, twenty patients had positive tests to the culprit drugs (or related beta-lactams). Ceftriaxone was responsible for one case of AGEP (27).

Case reports

A 5-year-old boy was referred for a cutaneous rash and fever of 39°C after having been treated with vancomycin, ceftriaxone and amoxicillin clavulanate for pneumococcal meningitis. Dermatologic examination revealed numerous superficial non-follicular pustules on an erythematous background on the trunk and neck. In addition, periorbital and scrotal edema, palmoplantar erythematous macules, and targetoid lesions on the extremities were also noted. Laboratory tests did not show any abnormalities except for leukocytosis (14,000/mm^3, absolute neutrophil count 8300/mm^3); an infectious etiology was excluded. A skin biopsy was not obtained due to patient's age and lack of parental consent. A probable diagnosis of acute generalized exanthematous pustulosis (AGEP) was made and treatment initiated with methylprednisolone. Six weeks after resolution of the cutaneous reaction, patch tests were strongly positive to ceftriaxone 10%, 1%, and 0.1% pet., confirming the diagnosis of AGEP due to ceftriaxone. A mild flare reaction consisting of papules and vesicles with erythema also was noted on the gluteal region during patch testing (11).

A 15-year-old boy was put on ceftriaxone treatment for vaso-occlusive crisis from sickle cell anemia. On the third day of the treatment, fever and numerous pustular lesions were seen on erythematous skin of his body and extremities. He had leukocytosis (17,400 mm^3) and the CRP was positive (8.74 mg/dl). A punch biopsy from pustules revealed subcorneal pustular formation, perivascular infiltration rich in leukocytes in the dermis and acanthosis in the epidermis. After the rash had subsided, he was patch tested with a 10% concentration of the commercial form of ceftriaxone parenteral in 0.9% NaCl, which was positive with erythema and pustules. AGEP was diagnosed (12).

A 79-year-old woman developed AGEP during hospitalization for SARS-CoV-2 infection which was caused by ceftriaxone. A patch test to ceftriaxone 5% pet. was strongly positive on D2 and D4, with infiltrated erythema and pustules (29).

Drug reaction with eosinophilia and systemic symptoms (DRESS)

A 66-year-old man developed 'probable' DRESS syndrome (final score 5 according to RegiSCAR scoring system) with widespread infiltrated maculopapular lesions, focally purpuric, with edema of the face, multiple lymphadenopathies, absolute eosinophilia (3.1 × 10^9/L), abnormal renal function but no fever, three weeks following administration of ceftriaxone 2g/day. A patch test with ceftriaxone 5% pet. was positive in the patient and negative in 20 controls (10).

A young woman developed a cutaneous rash, lymphadenopathy, malaise and fever after the introduction of phenobarbital. Because of these symptoms, the patient was treated with ceftriaxone and she experienced a severe flare-up of the cutaneous and general reaction with eosinophilia and elevated liver enzymes. Patch tests and lymphocyte transformation tests were positive to both drugs (multiple drug allergy, as these drugs are not related) (14).

Two patients developed DRESS from ceftriaxone with positive patch tests (17, possible overlap with ref. 18). Another patient who had suffered an episode of DRESS from a different drug, subsequently had a drug-related relapse caused by ceftriaxone, showing a positive patch test to this drug (17). In a group of 14 patients with multiple delayed-type hypersensitivity reactions, DRESS was caused by ceftriaxone in 2 cases, showing positive patch test reactions (18, possible overlap with ref. 17).

In a multicenter investigation in France, of 72 patients patch tested for DRESS, 46 (64%) had positive patch tests to drugs, including 2 to ceftriaxone (19). In the period 2000-2014, in Portugal, 260 patients were patch tested with antibiotics for suspected cutaneous adverse drug reactions (CADR) to these drugs. 56 patients (22%) had one or more (often from cross-reactivity) positive patch tests. Ceftriaxone was patch test positive in 1 patient with DRESS (20, possible overlap with ref. 24).

To evaluate if, after DRESS, patients become sensitized to antibiotics, patch test data and clinical files of 17 patients with DRESS from a nonantibiotic culprit drug (anticonvulsants 10, allopurinol 7) seen in Coimbra, Portugal, between 2010 and 2018 and who were given antibiotics at the onset or later during the course of DRESS, were retrospectively studied. The group consisted of eight women and nine men with an age range of 10 to 89 years and a mean of 47. In 10, anticonvulsants had caused DRESS (notably carbamazepine) and in 7 allopurinol (which always gives negative patch tests). They had received antibiotics at the onset or later during the course of DRESS: amoxicillin in 7, cephalosporins in three and fluoroquinolones in seven patients (5 ciprofloxacin, 2 levofloxacin). Nine patients (53%) had developed positive patch tests to antibiotics: six to amoxicillin (with 5 cross-reactions to ampicillin), three to the cephalosporins (one to ceftriaxone, one to cefoxitin and one to cefoxitin, ceftazidime and additional positivity to vancomycin), but none to the fluoroquinolones ciprofloxacin or levofloxacin. In repeat testing in 4 patients, most positive patch tests were reproducible after several years. These patients all had multiple hypersensitivity syndrome, as they reacted to (at least) two unrelated chemicals (24, possible overlap with ref. 20).

Stevens-Johnson syndrome/toxic epidermal necrolysis (SJS/TEN)

A 9-year-old girl developed Stevens-Johnson syndrome while using ceftriaxone and had a positive patch test to this cephalosporin tested 5% pet. (2).

Other cutaneous adverse drug reactions

A 34-year-old woman was treated for a systemic infection with vancomycin and ceftriaxone (1 gram once daily). An exanthematous rash appeared after 5 days of treatment. Patch tests were positive to a powdered ceftriaxone tablet, diluted at 30%. Ten controls were negative (6).

An 80-year-old woman developed a drug-induced linear IgA bullous dermatosis (LABD) during treatment with ceftriaxone and metronidazole. Sensitization to both drugs was detected by epicutaneous testing and positive lymphocyte transformation tests. After discontinuation of the drugs and systemic corticosteroids for 8 days the bullous eruptions subsided and erosions healed within 6 weeks. The authors suggested that in addition to IgA antibodies, drug-specific T cells and their subsequent release of cytokines may play an important role in the pathogenesis of drug-induced LABD (8).

In Bern, Switzerland, patients with a suspected allergic cutaneous drug reaction were patch-scratch tested with suspected drugs that had previously given a positive lymphocyte transformation test. Ceftriaxone 250 mg/ml saline gave a positive patch-scratch test in one patient with maculopapular exanthema and in one with bullous exanthema/erythema multiforme (15).

A 5-year-old boy developed a diffuse, maculopapular rash with minimal pruritus, hyperthermia, flushing, and lethargy after 6 intravenous infusions of ceftriaxone 2g/d. The seventh dose of ceftriaxone was given the following day, after the rash had improved and the hyperthermia resolved. Within minutes, he had a recurrence of the rash with severe itching, flushing, and hyperthermia. Although this would suggest an immediate-type IgE-mediated reaction, prick and intradermal tests were negative but a patch test with ceftriaxone diluted to 30% in pet. and saline positive (22). The authors diagnosed maculopapular eruption, but this cannot emerge within minutes and disappear within one day and a urticarial rash is more likely.

A patient treated with ceftriaxone and metronidazole developed an erythematous, partly urticarial exanthema, in which later blisters appeared. Patch tests were positive to both ceftriaxone and metronidazole (26).

Cross-reactions, pseudo-cross-reactions and co-reactions

Cross-reactions between beta-lactam antibiotics are discussed in Chapter 3.36 Amoxicillin. There is an absence of cross-reactions to ceftriaxone in patients sensitized to aminopenicillins (amoxicillin, ampicillin and ampicillin-esters) (23).

LITERATURE

1   The data in the section 'General' may have been obtained from literature discussed in this chapter, but mostly also or exclusively from one or more of the following online sources: ChemIDPlus Advanced, PubChem, DrugBank, RxList, Drug Central, Drugs.com, and Wikipedia

2   Atanaskovic-Markovic M, Gaeta F, Medjo B, Gavrovic-Jankulovic M, Cirkovic Velickovic T, Tmusic V, et al. Non-immediate hypersensitivity reactions to beta-lactam antibiotics in children - our 10-year experience in allergy work-up. Pediatr Allergy Immunol 2016;27:533-538

3   Foti C, Bonamonte D, Trenti R, Veña GA, Angelini G. Occupational contact allergy to cephalosporins. Contact Dermatitis 1997;36:104-105

4   Pinheiro V, Pestana C, Pinho A, Antunes I, Gonçalo M. Occupational allergic contact dermatitis caused by antibiotics in healthcare workers - relationship with non-immediate drug eruptions. Contact Dermatitis 2018;78:281-286

5   Gielen K, Goossens A. Occupational allergic contact dermatitis from drugs in healthcare workers. Contact Dermatitis 2001;45:273-279

6   Liippo J, Pummi K, Hohenthal U, Lammintausta K. Patch testing and sensitization to multiple drugs. Contact Dermatitis 2013;69:296-302

7   Häusermann P, Bircher AJ. Immediate and delayed hypersensitivity to ceftriaxone, and anaphylaxis due to intradermal testing with other beta-lactam antibiotics, in a previously amoxicillin-sensitized patient. Contact Dermatitis 2002;47:311-312

8   Yawalkar N, Reimers A, Hari Y, Hunziker Th, Gerber H, Müller U, Pichler W. Drug-induced linear IgA bullous dermatosis associated with ceftriaxone- and metronidazole-specific T cells. Dermatology 1999;199:25-30

9   San Miguel MM, Gaig P, Bartra J, García-Ortega P. Postoperative rash to ceftriaxone. Allergy 2000;55:977-979

10 Hansel K, Bellini V, Bianchi L, Brozzi J, Stingeni L. Drug reaction with eosinophilia and systemic symptoms from ceftriaxone confirmed by positive patch test: An immunohistochemical study. J Allergy Clin Immunol Pract 2017;5:808-810

11 Salman A, Yucelten D, Akin Cakici O, Kepenekli Kadayifci E. Acute generalized exanthematous pustulosis due to ceftriaxone: Report of a pediatric case with recurrence after positive patch test. Pediatr Dermatol 2019;36:514-516

12 Nacaroglu HT, Celegen M, Ozek G, Umac O, Karkıner CS, Yıldırım HT, Can D. Acute generalized exanthematous pustulosis induced by ceftriaxone use. Postepy Dermatol Alergol 2014;31:269-271

13 Foti C, Vena GA, Cucurachi MR, Angelini G. Occupational contact allergy from cephalosporins. Contact Dermatitis 1994;31:129-130

14 Voltolini S, Bignardi D, Minale P, Pellegrini S, Troise C. Phenobarbital-induced DiHS and ceftriaxone hypersensitivity reaction: a case of multiple drug allergy. Eur Ann Allergy Clin Immunol 2009;41:62-63

15 Neukomm C, Yawalkar N, Helbling A, Pichler WJ. T-cell reactions to drugs in distinct clinical manifestations of drug allergy. J Invest Allergol Clin Immunol 2001;11:275-284

16 Filipe P, Almeida RS, Rodrigo FG. Occupational allergic contact dermatitis from cephalosporins. Contact Dermatitis 1996;34:226

17 Jörg L, Helbling A, Yerly D, Pichler WJ. Drug-related relapses in drug reaction with eosinophilia and systemic symptoms (DRESS). Clin Transl Allergy 2020;10:52

18 Jörg L, Yerly D, Helbling A, Pichler W. The role of drug, dose and the tolerance/intolerance of new drugs in multiple drug hypersensitivity syndrome (MDH). Allergy 2020;75:1178-1187

19 Barbaud A, Collet E, Milpied B, Assier H, Staumont D, Avenel-Audran M, et al. A multicentre study to determine the value and safety of drug patch tests for the three main classes of severe cutaneous adverse drug reactions. Br J Dermatol 2013;168:555-562

20 Pinho A, Coutinho I, Gameiro A, Gouveia M, Gonçalo M. Patch testing - a valuable tool for investigating non-immediate cutaneous adverse drug reactions to antibiotics. J Eur Acad Dermatol Venereol 2017;31:280-287

21 Lammintausta K, Kortekangas-Savolainen O. The usefulness of skin tests to prove drug hypersensitivity. Br J Dermatol 2005;152:968-974

22 Kudva-Patel V, White E, Karnani R, Collins MH, Assa'ad AH. Drug reaction to ceftriaxone in a child with X-linked agammaglobulinemia. J Allergy Clin Immunol 2002;109:888-889

23 Romano A, Gaeta F, Valluzzi RL, Maggioletti M, Caruso C, Quaratino D. Cross-reactivity and tolerability of aztreonam and cephalosporins in subjects with a T cell-mediated hypersensitivity to penicillins. J Allergy Clin Immunol 2016;138:179-186

24 Santiago LG, Morgado FJ, Baptista MS, Gonçalo M. Hypersensitivity to antibiotics in drug reaction with eosinophilia and systemic symptoms (DRESS) from other culprits. Contact Dermatitis 2020;82:290-296

25 Brockow K, Garvey LH, Aberer W, Atanaskovic-Markovic M, Barbaud A, Bilo MB, et al.; ENDA/EAACI Drug Allergy Interest Group. Skin test concentrations for systemically administered drugs – an ENDA/EAACI Drug Allergy Interest Group position paper. Allergy 2013;68:702-712

26 Hari Y, Frutig-Schnyder K, Hurni M, Yawalkar N, Zanni MP, Schnyder B, et al. T cell involvement in cutaneous drug eruptions. Clin Exp Allergy 2001;31:1398-1408

27 Bérot V, Gener G, Ingen-Housz-Oro S, Gaudin O, Paul M, Chosidow O, Wolkenstein P, Assier H. Cross-reactivity in beta-lactams after a non-immediate cutaneous adverse reaction: experience of a reference centre for toxic bullous diseases and severe cutaneous adverse reactions. J Eur Acad Dermatol Venereol 2020;34:787-794

28 Romano A, Gaeta F, Valluzzi RL, Caruso C, Alonzi C, Viola M, et al. Diagnosing nonimmediate reactions to cephalosporins. J Allergy Clin Immunol 2012;129:1166-1169

29 Stingeni L, Francisci D, Bianchi L, Hansel K, Tramontana M, Di Candilo F, Mannarino MR, Pirro M. Severe adverse drug reaction in SARS-CoV-2 infection: AGEP induced by ceftriaxone and confirmed by patch test. Contact Dermatitis 2021;85:366-368

Chapter 3.115 CEFUROXIME

IDENTIFICATION

Description/definition : Cefuroxime is the semisynthetic β-lactamase-resistant cephalosporin antibiotic that conforms to the structural formula shown below

Pharmacological classes : Anti-bacterial agents

IUPAC name : (6R,7R)-3-(Carbamoyloxymethyl)-7-[[(2Z)-2-(furan-2-yl)-2-methoxyiminoacetyl]amino]-8-oxo-5-thia-1-azabicyclo[4.2.0]oct-2-ene-2-carboxylic acid

Other names : Cephuroxime

CAS registry number : 55268-75-2

EC number : 259-560-1

Merck Index monograph : 3226

Patch testing : Cefuroxime sodium 10% pet. (Chemotechnique)

Molecular formula : $C_{16}H_{16}N_4O_8S$

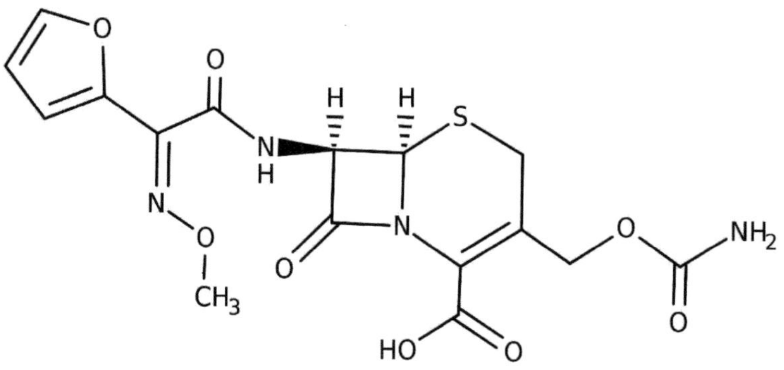

GENERAL

Cefuroxime is a semisynthetic, broad-spectrum, β-lactamase-resistant, second-generation cephalosporin antibiotic with antibacterial activity. This agent is indicated for the treatment of many types of bacterial infections such as bronchitis, sinusitis, tonsillitis, ear infections, skin infections, gonorrhea, and urinary tract infections, caused by susceptible strains of microorganisms (1). In pharmaceutical products, cefuroxime is employed as cefuroxime sodium (CAS number 56238-63-2, EC number 260-073-1, molecular formula $C_{16}H_{15}N_4NaO_8S$) (1). The classification and structures of beta-lactam antibiotics are discussed in Chapter 3.36 Amoxicillin.

CONTACT ALLERGY FROM ACCIDENTAL CONTACT

Between 1978 and 2001, in the university hospital of Leuven, Belgium, occupational allergic contact dermatitis to pharmaceuticals was diagnosed in 33 health care workers. Practically all of these patients presented with hand dermatitis (often the fingers), sometimes with secondary localizations, frequently the face. In this group, five female nurses, aged 24-60 (median 49) had occupational allergic contact dermatitis from cefuroxime (5).

A 58-year-old nurse, involved in making dilutions of multiple intravenous drugs and administering them to patients in her ward, developed recalcitrant occupational allergic contact dermatitis of the hands from cefuroxime and ceftriaxone (3).

CUTANEOUS ADVERSE DRUG REACTIONS FROM SYSTEMIC ADMINISTRATION CAUSED BY TYPE IV (DELAYED-TYPE) HYPERSENSITIVITY (as demonstrated by positive patch tests)

Cutaneous adverse drug reactions from systemic administration of cefuroxime caused by type IV (delayed-type) hypersensitivity have included maculopapular eruption (4,8), erythroderma (12), acute generalized exanthematous pustulosis (AGEP) (2,8), drug reaction with eosinophilia and systemic symptoms (DRESS) (7), Stevens-Johnson syndrome/toxic epidermal necrolysis (SJS/TEN) (3), urticaria-like exanthema appearing after several hours (6), erythema multiforme (10), and unspecified drug eruption (9).

Maculopapular eruption

A 64-year-old woman presented with a widespread, erythematous, maculopapular eruption and purpuric lesions on the feet. She had been receiving i.m. injections of cefuroxime for acute pyelonephritis. The cutaneous eruption, accompanied by fever, had appeared on the 3rd day of treatment. She had been treated with cefuroxime before.

Patch tests were positive to cefuroxime 10% in saline with cross-reactivity to cefalotin and cefaloridine. Intradermal tests were also positive after 24 hours (4).

One patient developed a maculopapular eruption from cefuroxime with a positive patch test to this drug after a previous episode of DRESS from other drug(s) (multiple drug hypersensitivity) (8).

Erythroderma, widespread erythematous eruption, exfoliative dermatitis
An 80-year-old man developed generalized pruritus and erythroderma after having used oral cefuroxime 250 mg twice a day. Patch tests were positive to commercialized cefuroxime and 10 other cephalosporins tested at 20% pet. Ten controls were negative. The patient did not remember that had taken cefuroxime or other cephalosporins before (12).

Acute generalized exanthematous pustulosis (AGEP)
In a hospital in France, specialized in toxic bullous diseases and severe cutaneous adverse reactions, a retrospective study was performed in consecutive patients consulting between 2010 and 2018 with a nonimmediate CADR suspected to have been caused by beta-lactam antibiotics. 56 patients were included, among whom were 46 amoxicillin-suspected and seven cephalosporin-suspected. Twenty-nine had severe CADR (DRESS, AGEP, SJS/TEN), and 27 had nonimmediate maculopapular exanthema (MPE). Of these patients, twenty patients had positive tests to the culprit drugs (or related beta-lactams). Cefuroxime was responsible for one case of AGEP (cefuroxime itself not patch tested, but positive to cefotaxime and penicillins (2).

One patient with multiple drug hypersensitivity syndrome developed AGEP from cefuroxime with a positive patch test reaction (8).

Drug reaction with eosinophilia and systemic symptoms (DRESS)
One patient who had suffered an episode of DRESS, subsequently had a drug-related relapse caused by cefuroxime, showing a positive patch test to this drug (7).

Stevens-Johnson syndrome/toxic epidermal necrolysis (SJS/TEN)
In a 61-year-old man, delayed-type hypersensitivity to cefuroxime (tested as commercial tablet 250 mg, 15% and 30% pet.) and other drugs may have caused / contributed to TEN (toxic epidermal necrolysis) following DRESS (drug reaction with eosinophilia and systemic symptoms) (3).

Other cutaneous adverse drug reactions
In Bern, Switzerland, patients with a suspected allergic cutaneous drug reaction were patch-scratch tested with suspected drugs that had previously given a positive lymphocyte transformation test. Cefuroxime 500 mg/ml saline gave a positive patch-scratch test in one patient with urticaria-like exanthema appearing after several hours (6). A 4-year-old girl developed erythema multiforme from cefuroxime and had a positive patch test to this drug (10).

In Finland, in the period 1989-2001, 826 patients with suspected cutaneous drug eruptions were patch tested and 89 had one or more positive reactions. Of these individuals, 4 reacted to cefuroxime. It was not mentioned which drug eruption these patients had suffered from (9).

Cross-reactions, pseudo-cross-reactions and co-reactions
Cross-reactions between beta-lactam antibiotics are discussed in Chapter 3.36 Amoxicillin. There is an absence of cross-reactions to cefuroxime in patients sensitized to aminopenicillins (amoxicillin, ampicillin and ampicillin-esters) (11).

Immediate contact reactions
Immediate contact reactions (contact urticaria) to cefuroxime are presented in Chapter 5.

LITERATURE
1   The data in the section 'General' may have been obtained from literature discussed in this chapter, but mostly also or exclusively from one or more of the following online sources: ChemIDPlus Advanced, PubChem, DrugBank, RxList, Drug Central, Drugs.com, and Wikipedia
2   Bérot V, Gener G, Ingen-Housz-Oro S, Gaudin O, Paul M, Chosidow O, Wolkenstein P, Assier H. Cross-reactivity in beta-lactams after a non-immediate cutaneous adverse reaction: experience of a reference centre for toxic bullous diseases and severe cutaneous adverse reactions. J Eur Acad Dermatol Venereol 2020;34:787-794
3   Liippo J, Pummi K, Hohenthal U, Lammintausta K. Patch testing and sensitization to multiple drugs. Contact Dermatitis 2013;69:296-302

4 Romano A, Pietrantonio F, Di Fonso M, Venuti A. Delayed hypersensitivity to cefuroxime. Contact Dermatitis 1992;27:270-271

5 Gielen K, Goossens A. Occupational allergic contact dermatitis from drugs in healthcare workers. Contact Dermatitis 2001;45:273-279

6 Neukomm C, Yawalkar N, Helbling A, Pichler WJ. T-cell reactions to drugs in distinct clinical manifestations of drug allergy. J Invest Allergol Clin Immunol 2001;11:275-284

7 Jörg L, Helbling A, Yerly D, Pichler WJ. Drug-related relapses in drug reaction with eosinophilia and systemic symptoms (DRESS). Clin Transl Allergy 2020;10:52

8 Jörg L, Yerly D, Helbling A, Pichler W. The role of drug, dose and the tolerance/intolerance of new drugs in multiple drug hypersensitivity syndrome (MDH). Allergy 2020;75:1178-1187

9 Lammintausta K, Kortekangas-Savolainen O. The usefulness of skin tests to prove drug hypersensitivity. Br J Dermatol 2005;152:968-974

10 Torres MJ, Corzo JL, Leyva L, Mayorga C, Garcia-Martin FJ, Antunez C, et al. Differences in the immunological responses in drug- and virus-induced cutaneous reactions in children. Blood Cells Mol Dis 2003;30:124-131

11 Romano A, Gaeta F, Valluzzi RL, Maggioletti M, Caruso C, Quaratino D. Cross-reactivity and tolerability of aztreonam and cephalosporins in subjects with a T cell-mediated hypersensitivity to penicillins. J Allergy Clin Immunol 2016;138:179-186

12 Gonzalo-Garijo MA, Rodríguez-Nevado I, de Argila D. Patch tests for diagnosis of delayed hypersensitivity to cephalosporins. Allergol Immunopathol (Madr) 2006;34:39-41

Chapter 3.116 CELECOXIB

IDENTIFICATION

Description/definition   : Celecoxib is the pyrazole derivative that conforms to the structural formula shown below
Pharmacological classes   : Anti-inflammatory agents, non-steroidal; cyclooxygenase-2 inhibitors
IUPAC name   : 4-[5-(4-Methylphenyl)-3-(trifluoromethyl)pyrazol-1-yl]benzenesulfonamide
CAS registry number   : 169590-42-5
EC number   : Not available
Merck Index monograph : 3228
Patch testing   : Tablet, pulverized, 10% pet. (18); if the pure chemical is not available, prepare the test
                                          material from intravenous powder, the content of capsules or – when also not available –
                                          from powdered tablets to achieve a final concentration of the active drug of 10% pet.;
                                          concentrations of homogenized tablets higher than 10% may cause irritant reactions (16)
Molecular formula   : $C_{17}H_{14}F_3N_3O_2S$

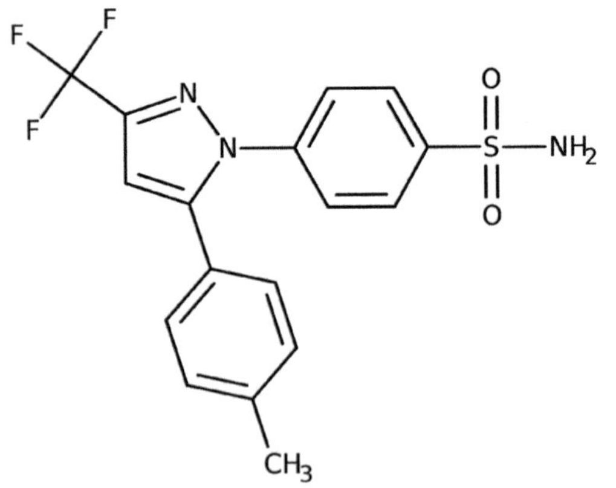

GENERAL

Celecoxib is a nonsteroidal anti-inflammatory drug (NSAID). Its mechanism of action is as a cyclooxygenase inhibitor. Celecoxib is indicated for the relief and management of osteoarthritis, rheumatoid arthritis, juvenile rheumatoid arthritis, ankylosing spondylitis, acute pain, primary dysmenorrhea and as oral adjunct to usual care for patients with familial adenomatous polyposis. Structurally, celecoxib is a benzenesulfonamide derivative containing a sulfonamide moiety (1).

CUTANEOUS ADVERSE DRUG REACTIONS FROM SYSTEMIC ADMINISTRATION CAUSED BY TYPE IV (DELAYED-TYPE) HYPERSENSITIVITY (as demonstrated by positive patch tests)

Cutaneous adverse drug reactions from systemic administration of celecoxib caused by type IV (delayed-type) hypersensitivity have included maculopapular eruptions (3,5,6), acute generalized exanthematous pustulosis (AGEP) (4,8,13,15), fixed drug eruption (2), drug reaction with eosinophilia and systemic symptoms (DRESS) (9,13,14), and erythema multiforme-like drug eruption (7).

Maculopapular eruption

A 70-year-old woman had taken oral celecoxib 20 mg 2dd for 2 weeks, when she developed a pruritic maculopapular rash on the back and lower extremities. Patch testing showed a ++ reaction to commercial celecoxib 10% pet. at D2 and D4, persisting for a week. Five controls were negative. The patch test with celecoxib was repeated 3 months later, and an increase in the reaction to +++ was observed. A double-blind oral challenge test with celecoxib 20 mg reproduced the rash that the patient had previously experienced (3).

A forty year-old man developed, after having received 3 tablets/day of celecoxib for intercostal pain for nine days, a maculopapular rash, which rapidly improved after the end of treatment. Six weeks later, patch tests with celecoxib diluted at 20 p. 100 in petrolatum were positive at 48 hours (5, no details available).

A 67-year-old man developed, after being treated with celecoxib 200 mg daily for 14 days, a 'rash with persistent erythematous macules and papules without mucous membrane involvement'. Later, he showed a strongly positive patch test to celecoxib with erythema, infiltration and papulovesicles after 48 hours (later reading not mentioned, test concentration and vehicle not mentioned, no controls performed (6).

Acute generalized exanthematous pustulosis (AGEP)

A 61-year-old woman presented with fever, widespread maculopapules, vesicles, bullae, tiny pustules and leukocytosis soon after the ingestion of celecoxib. The condition resolved rapidly within 10 days. AGEP was diagnosed based on clinical and histopathological findings. Skin patch testing showed positive reaction to 1% celecoxib (8).

A 72 year-old man, treated for cervicalgia with celecoxib, presented a pustular exanthema of the face and the trunk, ten days after introduction of the treatment, associated with an inflammatory syndrome and hepatic cytolysis. Within 8 days the disease had regressed. The patient was diagnosed with (atypical) AGEP. He had a positive patch test to celecoxib (13, details not available).

A 53-year-old woman presented with disseminated pustules on the face and trunk, itching and fever after taking celecoxib. She had leukocytosis, elevated neutrophil counts and elevated C-reactive protein levels (7.32 mg/dl). A skin biopsy revealed subconeal pustules, spongiosis in the epidermis, papillary dermal edema and perivascular infiltration of lymphocytes, neutrophils and some eosinophils. AGEP induced by celecoxib was suspected. Patch tests were positive celecoxib powder 5% in saline and petrolatum but negative to the non-active ingredients of the celecoxib capsule. Two controls were negative (4).

A 55-year-old woman developed a generalized erythema followed by pustules (face, shoulder) and fever (38°C) ten days after starting to use celecoxib for periarthritis of the left shoulder. She had leukocytosis with neutrophilia, but no eosinophilia. Liver enzymes were normal. After ceasing the use of celecoxib, the exanthema disappeared in 8 days. A patch test with celecoxib 200 mg dissolved in 1 ml PBS (unknown abbreviation) was positive, as was a lymphocyte transformation test; drug-specific T cells were demonstrated (15). This was a case of acute generalized exanthematous pustulosis (AGEP) (15).

Fixed drug eruption

A 61-year-old woman presented with 15 round, red-brown, edematous, moderately itching patches, in size varying from 2 to 10 cm in diameter. The lesions, symmetrically distributed on the trunk and upper and lower limbs, had appeared 3 hours after the ingestion of 200 mg of celecoxib for cervical pain caused by osteoarthritis. A clinical diagnosis of multifocal fixed drug eruption (MFDE) was suspected. Two months after the resolution of the eruption, patch testing with celecoxib 10% in petrolatum and in dimethyl sulfoxide (DMSO 50% aqueous) revealed positive reactions in all patch sites, especially when the drug was in DMSO or when it was applied to the previously involved skin (2).

Drug reaction with eosinophilia and systemic symptoms (DRESS)

A 29-yr-old woman had been on celecoxib and anti-tuberculosis drugs for one month to treat knee joint pain and pulmonary tuberculosis, when she developed fever, lymphadenopathy, rash, hypereosinophilia, and visceral involvement (hepatitis and pneumonitis). During corticosteroid administration, swallowing difficulty with profound muscle weakness had developed. The patient was diagnosed with DRESS syndrome with eosinophilic polymyositis by a histopathologic study. After complete resolution of all symptoms, patch tests were positive to celecoxib and ethambutol crushed tablets 10% and 50% pet. (9).

In a 73 year-old woman, treated with celecoxib for cervical arthralgia, a maculopapular exanthema developed five days after treatment was started. The exanthema, initially edematous and purpuric became bullous with multi-visceral involvement (disseminated intravascular coagulation, renal failure, hepatitis and pancreatic). The disease slowly regressed. A patch test with celecoxib was positive in this patient with DRESS (13).

In a multicenter investigation in France, of 72 patients patch tested for DRESS, 46 (64%) had positive patch tests to drugs, including one to celecoxib (14).

Other cutaneous adverse drug reactions

A 79-year-old woman had been treated with oral celecoxib (400 mg / day for 7 days) and erlotinib hydrochloride (150 mg / day for 7 days) for management of pain due to lung cancer, when she developed an erythema multiforme-like drug eruption. Celecoxib was positive in a patch test (no details provided: concentration, vehicle, times of reading, strength of positive reactions, pure material or commercial drug, number of controls) whereas erlotinib hydrochloride was negative (7).

Cross-reactions, pseudo-cross-reactions and co-reactions

No cross-reactions to other sulfonamide derivatives (2). Three patients with fixed drug eruption (12), symmetrical drug-related intertriginous and flexural exanthema (SDRIFE) (11) resp. acute generalized exanthematous pustulosis (AGEP) (10) caused by etoricoxib had positive patch tests to etoricoxib and to celecoxib (cross-reactions). A patient sensitized to valdecoxib from oral administration cross-reacted to celecoxib (17).

LITERATURE

1   The data in the section 'General' may have been obtained from literature discussed in this chapter, but mostly also or exclusively from one or more of the following online sources: ChemIDPlus Advanced, PubChem, DrugBank, RxList, Drug Central, Drugs.com, and Wikipedia

2   Bellini V, Stingeni L, Lisi P. Multifocal fixed drug eruption due to celecoxib. Dermatitis 2009;20:174-176

3   Martinez Alonso JC, Ortega JD, Gonzalo MJ. Cutaneous reaction to oral celecoxib with positive patch test. Contact Dermatitis 2004;50:48-49

4   Shin HT, Park SW, Lee KT, Park HY, Park JH, Lee DY, et al. A case of celecoxib induced acute generalized exanthematous pustulosis. Ann Dermatol 2011;23(Suppl.3):S380-382

5   Verbeiren S, Morant C, Charlanne H, Ajelabar K, Caron J, Modiano P. Celecoxib induced toxiderma with positive patch-test. Ann Dermatol Venereol 2002;129:203-205

6   Grob M, Scheidegger P, Wuthrich B. Allergic skin reaction to Celecoxib. Dermatology 2000;201:383

7   Arakawa Y, Nakai N, Katoh N. Celecoxib-induced erythema multiforme-type drug eruption with a positive patch test. J Dermatol 2011;38:1185-1188

8   Yang CC, Lee JY, Chen WC. Acute generalized exanthematous pustulosis caused by celecoxib. J Formos Med Assoc 2004;103:555-557

9   Lee JH, Park HK, Heo J et al. Drug rash with eosinophilia and systemic symptoms (DRESS) syndrome induced by celecoxib and anti-tuberculosis drugs. J Korean Med Sci 2008;23:521-525

10 Mäkelä L, Lammintausta K. Etoricoxib-induced acute generalized exanthematous pustulosis. Acta Derm Venereol 2008;88:200-201

11 Caralli ME, Seoane Rodríguez M, Rojas Pérez-Ezquerra P, Pelta Fernández R, de Barrio Fernández M. Symmetrical drug-related intertriginous and flexural exanthema (SDRIFE) caused by etoricoxib. J Investig Allergol Clin Immunol 2016;26:128-129

12 Carneiro-Leao L, Rodrigues Cernadas J. Bullous fixed drug eruption caused by etoricoxib confirmed by patch testing. J Allergy Clin Immunol Pract 2019;7:1629-1630

13 Marques S, Milpied B, Foulc P, Barbarot S, Cassagnau E, Stalder JF. Severe cutaneous drug reactions to celecoxib (Celebrex). Ann Dermatol Venereol 2003;130:1051-1055

14 Barbaud A, Collet E, Milpied B, Assier H, Staumont D, Avenel-Audran M, et al. A multicentre study to determine the value and safety of drug patch tests for the three main classes of severe cutaneous adverse drug reactions. Br J Dermatol 2013;168:555-562

15 Britschgi M, Steiner UC, Schmid S, Depta JP, Senti G, Bircher A, et al. T-cell involvement in drug-induced acute generalized exanthematous pustulosis. J Clin Invest 2001;107:1433-1441

16 Kleinhans M, Linzbach L, Zedlitz S, Kaufmann R, Boehncke WH. Positive patch test reactions to celecoxib may be due to irritation and do not correlate with the results of oral provocation. Contact Dermatitis 2002;47:100-102

17 Jaeger C, Jappe U. Valdecoxib-induced systemic allergic dermatitis confirmed by positive patch test. Contact Dermatitis 2005;52:47-48

18 Brockow K, Garvey LH, Aberer W, Atanaskovic-Markovic M, Barbaud A, Bilo MB, et al.; ENDA/EAACI Drug Allergy Interest Group. Skin test concentrations for systemically administered drugs – an ENDA/EAACI Drug Allergy Interest Group position paper. Allergy 2013;68:702-712

Chapter 3.117 CERTOPARIN

IDENTIFICATION

Description/definition   : Certoparin is a low-molecular-weight heparin
Pharmacological classes : Antithrombotic agents
IUPAC name   : Not available
CAS registry number   : 9005-49-6 (Heparin)
EC number   : 232-681-7 (Heparin)
Merck Index monograph : 5958 (Heparin)
Patch testing   : Commercial preparation undiluted (4); consider intradermal testing with late readings
                              (D2,D3) when patch tests are negative and consider subcutaneous challenge when
                              intradermal tests are negative
Molecular formula   : Unspecified

GENERAL

Certoparin is a low molecular weight heparin (LMWH), primarily active against factor Xa. It is used in the prevention and treatment of venous thromboembolism (deep vein thrombosis and pulmonary embolism) and in the treatment of myocardial infarction. In pharmaceutical products, certoparin is employed as certoparin sodium (CAS number not available, EC number not available, molecular formula unspecified) (1).

See also bemiparin (Chapter 3.59), dalteparin (Chapter 3.160), danaparoid (Chapter 3.161), enoxaparin (Chapter 3.195), fondaparinux (Chapter 3.225), heparins (Chapter 3.239), nadroparin (Chapter 3.331), and tinzaparin (Chapter 3.479).

CUTANEOUS ADVERSE DRUG REACTIONS FROM SYSTEMIC ADMINISTRATION CAUSED BY TYPE IV (DELAYED-TYPE) HYPERSENSITIVITY

Throughout this book, only reports of delayed-type hypersensitivity have been included that showed a positive patch test to the culprit drug. However, as a result of the high molecular weight of heparins, patch tests are often false-negative, presumably from insufficient penetration into the skin. Because of this, and also because patch tests have been performed in a small minority of cases only, studies with a positive intradermal test or subcutaneous provocation tests with delayed readings are included in the chapters on the various heparins, even when patch tests were negative or not performed.

General information on delayed-type hypersensitivity reactions to heparins

General information on delayed-type hypersensitivity reactions to heparins (including certoparin, presenting as local reactions from subcutaneous administration, is provided in Chapter 3.239 Heparins. In this chapter, only *non-local* cutaneous adverse drug reactions from delayed-type hypersensitivity to certoparin are presented; they have not been found.

Cross-reactions, pseudo-cross-reactions and co-reactions

Cross-reactions between heparins are frequent in delayed-type hypersensitivity (>90% of patients tested, median number of positive drugs per patient: 3) and do not depend on the molecular weight of the molecules (3). Overlap in their polysaccharide composition might explain the high degree of cross-allergenicity (5). Cross-reactions to the semisynthetic heparinoid danaparoid have also been observed (2). In allergic patients, the synthetic ultralow molecular weight synthetic heparin fondaparinux is usually, but not always (2) well-tolerated (5).

LITERATURE

1   The data in the section 'General' may have been obtained from literature discussed in this chapter, but mostly also or exclusively from one or more of the following online sources: ChemIDPlus Advanced, PubChem, DrugBank, RxList, Drug Central, Drugs.com, and Wikipedia
2   Utikal J, Peitsch WK, Booken D, Velten F, Dempfle CE, Goerdt S, et al. Hypersensitivity to the pentasaccharide fondaparinux in patients with delayed-type heparin allergy. Thromb Haemost 2005;94:895-896
3   Weberschock T, Meister AC, Bohrt K, Schmitt J, Boehncke W-H, Ludwig RJ. The risk for cross-reactions after a cutaneous delayed-type hypersensitivity reaction to heparin preparations is independent of their molecular weight: a systematic review. Contact Dermatitis 2011;65:187-194
4   Brockow K, Garvey LH, Aberer W, Atanaskovic-Markovic M, Barbaud A, Bilo MB, et al. Skin test concentrations for systemically administered drugs - an ENDA/EAACI Drug Allergy Interest Group position paper. Allergy 2013;68:702-712
5   Schindewolf M, Lindhoff-Last E, Ludwig RJ. Heparin-induced skin lesions. Lancet 2012;380:1867-1879

Chapter 3.118 CETIRIZINE

IDENTIFICATION

Description/definition   : Cetirizine is a synthetic phenylmethyl-piperazinyl derivative that conforms to the
                              structural formula shown below
Pharmacological classes   : Anti-allergic agents; histamine H_1 antagonists, non-sedating
IUPAC name   : 2-[2-[4-[(4-Chlorophenyl)-phenylmethyl]piperazin-1-yl]ethoxy]acetic acid
CAS registry number   : 83881-51-0
EC number   : Not available
Merck Index monograph   : 3291
Patch testing   : 5% pet.; tablet crushed 20% pet. and water
Molecular formula   : $C_{21}H_{25}ClN_2O_3$

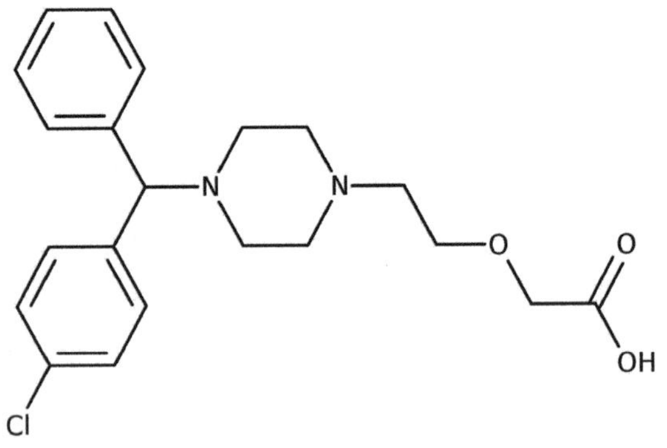

GENERAL

Cetirizine is a synthetic phenylmethyl-piperazinyl derivative and antihistaminic drug. This compound is a metabolite of hydroxyzine and a selective second-generation peripheral histamine H_1 receptor antagonist. It is used for symptomatic treatment of seasonal and perennial allergic rhinitis, allergic asthma, chronic urticaria and atopic dermatitis (1). In pharmaceutical products, cetirizine is employed as cetirizine hydrochloride (CAS number 83881-52-1, EC number not available, molecular formula $C_{21}H_{27}Cl_3N_2O_3$) (1).

CUTANEOUS ADVERSE DRUG REACTIONS FROM SYSTEMIC ADMINISTRATION CAUSED BY TYPE IV (DELAYED-TYPE) HYPERSENSITIVITY (as demonstrated by positive patch tests)

Cutaneous adverse drug reactions from systemic administration of cetirizine caused by type IV (delayed-type) hypersensitivity have included maculopapular eruption (8,9), acute generalized exanthematous pustulosis (AGEP) (10), fixed drug eruption (2,3,7,11), and eczematous eruption (12).

Maculopapular eruption

A 70-year-old woman for one week had an erythematous maculopapular eruption, predominantly follicular on the abdomen and urticarial on the thighs. She had been treated with cetirizine (10 mg oral daily) for facial itching over the last 2 weeks. Two weeks later, she had a relapse after again taking cetirizine. Patch testing revealed positive reactions to ethylenediamine HCl 1% pet., piperazine 1% pet., and cetirizine 2.5 and 5% pet. but negative to 1%. Prick tests with cetirizine were negative, while intradermal tests with cetirizine in saline in a dilution series (1%, 0.5%, 0.1%, 0.05%) showed positive immediate reactions to all dilutions of cetirizine. It can be concluded that the patient had a combination of type IV and type I hypersensitivity to cetirizine, although there were no immediate clinical signs after intake of the antihistamine (8).

A 44-year-old man developed a morbilliform eruption 2 days after intake of oral hydroxyzine for urticaria. Previously, he had suffered a similar eruption from cetirizine (which is a metabolite of hydroxyzine). Patch tests were positive to hydroxyzine (tablet 'as is', 1%, 2.5%, 5% and 10% pet.), cetirizine (tablet 'as is', 2.5%, 5% and 10% pet.), ethylene-diamine 1% pet. and piperazine 1% pet. Twenty controls were negative. The patient was probably primarily sensitized to cetirizine and later reacted to hydroxyzine, which is for 45% metabolized *in vivo* into cetirizine (9).

Acute generalized exanthematous pustulosis (AGEP)

A 40-year-old woman developed a maculopapular eruption on the trunk within 24 hours of taking two doses of cetirizine 10 mg for a suspected allergic reaction to fruit. She had superficial, tiny, non-follicular pustules on an erythematous base around the neck and large flexures in keeping with acute generalized exanthematous pustulosis (AGEP). Patch testing with her powdered cetirizine hydrochloride tablet in pet. and a 1 mg/ml (0.1%) solution yielded strongly positive reactions to both cetirizine formulations at D2 and D4. Twenty controls were negative (10).

Fixed drug eruption

In a University hospital in Coimbra, Portugal, in the period 1990-2009, 52 patients (17 men, 35 women, mean age 53±17 years) with a clinical diagnosis of fixed drug eruptions were submitted to patch tests with the suspected drugs. Patch tests on pigmented lesions were reactive in 21 of the 52 (40%) patients, 20 NSAIDs and one antihistamine. One patient was tested with cetirizine powder of the commercial pills 20% pet. and 20% water and there was a positive patch test on postlesional skin (2). This patient was described in detail in ref. 3.

A 45-year-old woman presented with multiple, round, erythematoviolaceous, well-defined plaques, with central blisters, localized on the trunk, forearms, and dorsum of the hands. She recounted three previous cutaneous eruptions with the same morphologic features in the same sites that had disappeared spontaneously in 8-10 days leaving only residual brown to gray hyperpigmentation. In the last episode, the patient described the oral intake of 10 mg cetirizine for allergic rhinitis 4 h before lesional reactivation (3). Patch test results are shown above (2).

A 27-year-old man had developed 10 circumscribed and partly central bullous erythemas reaching a diameter of 1 to 6 cm 12 hours after he had taken 10 mg of cetirizine to prevent allergic rhinoconjunctivitis. Later, he had a second attack 6 hours after intake of cetirizine. Patch tests were positive to cetirizine 5% in petrolatum and 10% in saline on previously affected but not normal skin (7). One more case of fixed drug eruption caused by cetirizine with a positive patch test was reported from Singapore. Clinical nor patch testing details were provided (11).

Dermatitis/eczematous eruption

A 48-year-old man, who had previously suffered DRESS from carbamazepine, later had eczematous eruptions from multiple unrelated drugs. Patch tests were positive to hydroxyzine HCl, cetirizine HCl, carbamazepine, pseudoephedrine, amoxicillin and ampicillin. The patient had used all these drugs, which resulted in eczematous eruptions, with the exception of ampicillin, which was a cross-reaction to amoxicillin. This was a case of multiple drug allergy (12).

Cross-reactions, pseudo-cross-reactions and co-reactions

Cetirizine, levocetirizine, and hydroxyzine are related piperazine-derived antihistamines. *In vivo*, 45% of hydroxyzine is transformed into cetirizine; levocetirizine is the active (*R*)-enantiomer of cetirizine. Therefore, co-reactions due to cross-sensitivity are to be expected and have been observed (3,4,5,6,10; not always proven by positive patch tests). Piperazine, ethylenediamine (8).

LITERATURE

1 The data in the section 'General' may have been obtained from literature discussed in this chapter, but mostly also or exclusively from one or more of the following online sources: ChemIDPlus Advanced, PubChem, DrugBank, RxList, Drug Central, Drugs.com, and Wikipedia

2 Andrade P, Brinca A, Gonçalo M. Patch testing in fixed drug eruptions – a 20-year review. Contact Dermatitis 2011;65:195-201

3 Cravo M, Gonçalo M, Figueiredo A. Fixed drug eruption to cetirizine with positive lesional patch tests to the three piperazine derivatives. Int J Dermatol 2007;46:760-762

4 Bhari N, Mahajan R, Singh S, Sharma VK. Fixed drug eruption due to three antihistamines of a same chemical family: Cetirizine, levocetirizine, and hydroxyzine. Dermatol Ther 2017;30. doi: 10.1111/dth.12412.

5 Assouère MN, Mazereeuw-Hautier J, Bonafe JL. Cutaneous drug eruption with two antihistaminic drugs of a same chemical family: cetirizine and hydroxyzine. Ann Dermatol Venereol 2002;129:1295-1298

6 Mahajan VK, Sharma NL, Sharma VC. Fixed drug eruption: a novel side-effect of levocetirizine. Int J Dermatol 2005;44:796-798

7 Kränke B, Kern T. Multilocalized fixed drug eruption to the antihistamine cetirizine. J Allergy Clin Immunol 2000;106:988

8 Stingeni L, Caraffini S, Agostinelli D, Ricci F, Lisi P. Maculopapular and urticarial eruption from cetirizine. Contact Dermatitis 1997;37:249-250

9 Lew BL, Haw CR, Lee MH. Cutaneous drug eruption from cetirizine and hydroxyzine. J Am Acad Dermatol 2004;50:953-956

10 Khan M, Wakelin S. Cetirizine induced acute generalized exanthematous pustulosis confirmed by patch testing. Contact Dermatitis 2020;82:238-239

11 Heng YK, Yew YW, Lim DS, Lim YL. An update of fixed drug eruptions in Singapore. J Eur Acad Dermatol Venereol 2015;29:1539-1544

12 Özkaya E, Yazganoğlu KD. Sequential development of eczematous type "multiple drug allergy" to unrelated drugs. J Am Acad Dermatol 2011;65:e26-e29.

Chapter 3.119 CHLORAL HYDRATE

IDENTIFICATION

Description/definition : Chloral hydrate is the synthetic monohydrate of chloral that conforms to the structural formula shown below
Pharmacological classes : Hypnotics and sedatives
IUPAC name : 2,2,2-Trichloroethane-1,1-diol
Other names : 1,1,1-Trichloro-2,2-dihydroxyethane; 1,1,1-trichloro-2,2-ethanediol
CAS registry number : 302-17-0
EC number : 206-117-5
Merck Index monograph : 3341
Patch testing : 5% water
Molecular formula : $C_2H_3Cl_3O_2$

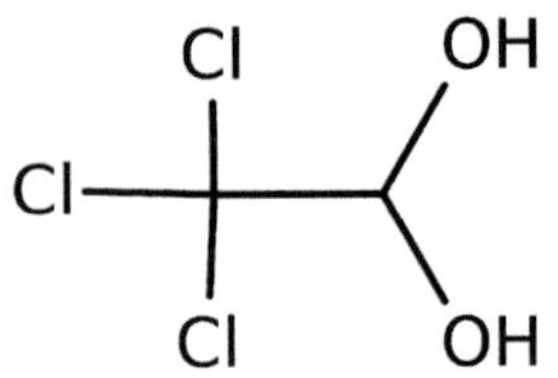

GENERAL

Chloral hydrate is a synthetic monohydrate of chloral with sedative, hypnotic, and anticonvulsive properties. Formerly it was used as a hypnotic and sedative in the treatment of insomnia, but is was effective for short time use only. It is no longer considered useful as an anti-anxiety medication. This agent has also been used as a routine sedative preoperatively to decrease anxiety and cause sedation and/or sleep with respiration depression or cough reflex. Chloral hydrate is probably not used anymore (1). Formerly, it was also used topically as a counter-Irritant (rubefacient) for the relief of itching and pain (2). As such, it has rarely caused contact allergy/allergic contact dermatitis, which has been reviewed in Volume 3 of the *Monographs in contact allergy* series (6).

CONTACT ALLERGY FROM ACCIDENTAL CONTACT

One patient had become sensitized by handling chloral hydrate during his work as a pharmacist: occupational contact dermatitis. There were positive patch tests to 'chloral hydrate solutions' 1%-2.5% (3).

CUTANEOUS ADVERSE DRUG REACTIONS FROM SYSTEMIC ADMINISTRATION CAUSED BY TYPE IV (DELAYED-TYPE) HYPERSENSITIVITY (as demonstrated by positive patch tests)

In 2 early reports, patients had an eruption from oral chloral hydrate (4,5). One of these patients had apparently been sensitized previously by an ointment containing 0.5% chloral hydrate, a case of systemic allergic dermatitis (5). Both patients had positive patch test reactions to 'chloral hydrate solutions' 1%-2.5% (4,5).

     A female patient developed a drug eruption while using Sirupus Chlorali hydrati SR. The causal allergen was chloral hydrate itself, as demonstrated by an oral provocation test and a positive patch test (7).

LITERATURE

1 The data in the section 'General' may have been obtained from literature discussed in this chapter, but mostly also or exclusively from one or more of the following online sources: ChemIDPlus Advanced, PubChem, DrugBank, RxList, Drug Central, Drugs.com, and Wikipedia
2 De Groot AC, Conemans J. Chloral hydrate. The contact allergen that fell asleep. Contact Dermatitis 1987;16:229-231
3 Abramowitz EW, Noun MH. Eczematous dermatitis due to exposure to chloral. J Allergy 1933;4:338-343
4 Flandin MM, Rabeau H, Ukrainczyk M. Dermite par intolérance au chloral. Test cutané. Bull Soc Franc Derm Syph 1936;43:1231-1234 (Article in French)
5 Baer RL, Sulzberger MB. Eczematous dermatitis due to chloral hydrate (following both oral administration and topical application). J Allergy 1938;9:519-520
6 De Groot AC. Monographs in contact allergy, volume 3: Topical Drugs. Boca Raton, Fl, USA: CRC Press Taylor and Francis Group, 2021 (ISBN 978-0-367-23693-9)
7 Lindner K, Prater E, Schubert H, Siegmund S. Arzneimittelexanthem auf Chloralhydrat [Drug exanthema due to chloral hydrate]. Dermatol Monatsschr 1990;176:483-485 (Article in German)

Chapter 3.120 CHLORAMBUCIL

IDENTIFICATION

Description/definition   : Chlorambucil is the nitrogen mustard alkylating agent that conforms to the structural
                              formula shown below
Pharmacological classes   : Antineoplastic agents, alkylating
IUPAC name   : 4-[4-[bis(2-Chloroethyl)amino]phenyl]butanoic acid
Other names   : Chloraminophene
CAS registry number   : 305-03-3
EC number   : 206-162-0
Merck Index monograph   : 3343
Patch testing   : 5% and 10% pet. (5); 0.1% pet. (2); 1% pet. (7)
Molecular formula   : $C_{14}H_{19}Cl_2NO_2$

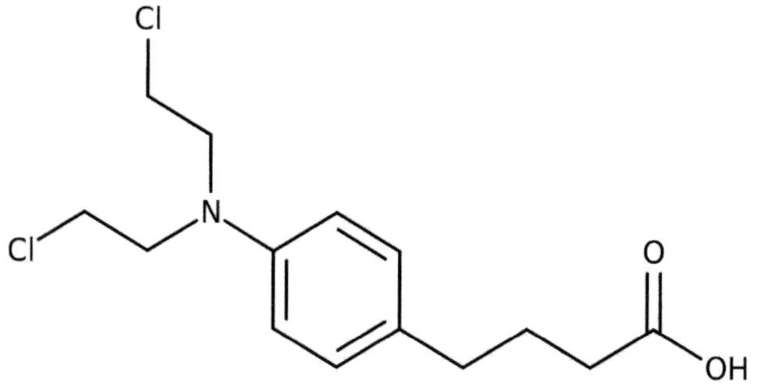

GENERAL

Chlorambucil is an orally-active antineoplastic aromatic nitrogen mustard. This agent alkylates and cross-links DNA during all phases of the cell cycle, resulting in disruption of DNA function, cell cycle arrest, and apoptosis. Chlorambucil is used for the treatment of chronic lymphatic (lymphocytic) leukemia, childhood minimal-change nephrotic syndrome, and malignant lymphomas including lymphosarcoma, giant follicular lymphoma, Hodgkin's disease, non-Hodgkin's lymphomas, and Waldenström's macroglobulinemia (1).

CONTACT ALLERGY FROM ACCIDENTAL CONTACT

A 21-year-old man developed an itchy rash on his face one week after starting work in a new laboratory of a pharmaceutical company, in which he had already worked for several years. His job involved working with cytotoxic drugs (azathioprine, chlorambucil, melphalan, 6-mercaptopurine and busulfan) in a fume-hooded compartment. The rash appeared initially over areas on his face where his respirator dust mask touched the skin and later over uncovered areas. At the time of onset of his rash, he had been working with melphalan, but chlorambucil was also present in the laboratory at the time. Patch testing with all the drugs that the patient had been working with showed strongly positive reactions to chlorambucil 1%pet. (D2 ++, D4 +++) and to melphalan 1% pet. (++/++). He was diagnosed as having occupational allergic contact dermatitis from melphalan and chlorambucil. The authors suggested that the patient had primarily been sensitized to melphalan and that he cross-reacted to chlorambucil, but also allowed for the possibility of co-sensitization (7).

CUTANEOUS ADVERSE DRUG REACTIONS FROM SYSTEMIC ADMINISTRATION CAUSED BY TYPE IV
(DELAYED-TYPE) HYPERSENSITIVITY (as demonstrated by positive patch tests)

Cutaneous adverse drug reactions from systemic administration of chlorambucil caused by type IV (delayed-type) hypersensitivity have included maculopapular eruption (3,4), erythroderma (6), exfoliative dermatitis (4), toxic epidermal necrolysis (TEN) (5), and erythema multiforme-like eruption (2).

Maculopapular eruption

A woman aged 62 was diagnosed with chronic lymphocytic leukemia and treatment was started with chlorambucil in doses of 6 mg daily. After three weeks later she developed a generalized confluent maculopapular and pruriginous eruption resembling a classic urticarial rash. Scratch tests with chlorambucil were negative after 20 minutes but positive when read after 18-20 hours. Patch tests were positive to chlorambucil at D3 (test concentration, vehicle and controls not mentioned) (3).

A 58-year-old woman was treated with chlorambucil for low-grade non-Hodgkin's lymphoma, when she developed a maculopapular eruption. An oral provocation test was positive. Later, a patch test with chlorambucil was also positive (test concentration, vehicle and controls not mentioned). During patch testing, a mild generalized erythematous rash developed (4).

Erythroderma, widespread erythematous eruption, exfoliative dermatitis
A 63-year-old man was treated with prednisone and chlorambucil for chronic lymphatic leukemia. On the 8th day of second cycle the patient noticed fever, tiredness, myalgia, pruritus and erythema on the skin. The third cycle had to be stopped on the second day due to the development of myalgia, generalized erythroderma with exfoliation and edema of the face and arms. Later, a patch test with chlorambucil was strongly positive (histologically verified; concentration, vehicle and controls unknown). A lymphocyte stimulation test with chlorambucil was also positive (6).

A man of 72 years with non-Hodgkin's lymphoma developed exfoliative dermatitis while on chlorambucil. In this patient, the oral provocation test and patch test with chlorambucil were positive, but the patch test concentration and vehicle were not mentioned (4).

Stevens-Johnson syndrome/toxic epidermal necrolysis (SJS/TEN)
A 57-year-old woman, after having recovered from toxic epidermal necrolysis possibly related to chlorambucil given for chronic lymphocytic leukemia, was patch tested and had positive reactions to chlorambucil 5% and 10% pet. Ten controls, of who 5 had previously been treated with chlorambucil, were negative (5).

Other cutaneous adverse drug reactions
A 76-year-old woman was admitted with an itchy widespread erythema multiforme-like eruption and focal pustules on the arms without mucosal involvement and systemic symptoms. She had been treated with chlorambucil and prednisolone for chronic lymphocytic leukaemia for 10 days every month. The eruption started 4 days after the end of the third therapy cycle. Four months later, patch testing with chlorambucil 0.1% pet. showed a ++ reaction; 20 controls were negative (2).

Cross-reactions, pseudo-cross-reactions and co-reactions
Possibly cross-reaction to primary melphalan sensitization (7).

LITERATURE
1   The data in the section 'General' may have been obtained from literature discussed in this chapter, but mostly also or exclusively from one or more of the following online sources: ChemIDPlus Advanced, PubChem, DrugBank, RxList, Drug Central, Drugs.com, and Wikipedia
2   Bianchi L, Hansel K, Pelliccia S, Tramontana M, Stingeni L. Systemic delayed hypersensitivity reaction to chlorambucil: a case report and literature review. Contact Dermatitis 2018;78:171-173
3   Knisley R, Settipane GA, Albala MM. Unusual reaction to chlorambucil in a patient with chronic lymphocytic leukemia. Arch Derm 1971:4:77-79
4   Hitchins RN, Hocker GA, Thomson DB. Chlorambucil allergy – a series of three cases. Aust NZ J Med 1987;17:600-602
5   Pietrantonio F, Moriconi L, Torino F, Romano A, Gargovich A. Unusual reaction to chlorambucil: a case report. Cancer Lett 1990;54:109-111
6   Torricelli R, Kurer SB, Kroner T, Wüthrich B. Delayed allergic reaction to chlorambucil (Leukeran). Case report and literature review. Schweiz Med Wochenschr 1995;125:1870-1873
7   Goon AT, McFadden JP, McCann M, Royds C, Rycroft RJ. Allergic contact dermatitis from melphalan and chlorambucil: cross-sensitivity or cosensitization? Contact Dermatitis 2002;47:309-310

Chapter 3.121 CHLORAMPHENICOL

IDENTIFICATION

Description/definition : Chloramphenicol is a broad-spectrum antibiotic that was derived from the bacterium *Streptomyces venezuelae* and is now produced synthetically; it conforms to the structural formula shown below

Pharmacological classes : Protein synthesis inhibitors; anti-bacterial agents

IUPAC name : 2,2-Dichloro-*N*-[(1*R*,2*R*)-1,3-dihydroxy-1-(4-nitrophenyl)propan-2-yl]acetamide

CAS registry number : 56-75-7

EC number : 200-287-4

Merck Index monograph : 3347

Patch testing : 5% pet. (Chemotechnique, SmartPracticeCanada, SmartPracticeEurope); 2% alc. (SmartPracticeCanada)

Molecular formula : $C_{11}H_{12}Cl_2N_2O_5$

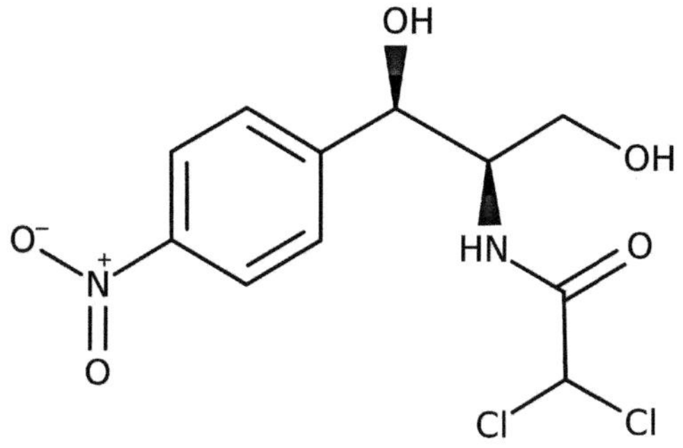

GENERAL

Chloramphenicol is a broad-spectrum antibiotic of the amphenicol class with primarily bacteriostatic activity, that was first isolated from cultures of *Streptomyces venezuelae* in 1947, but is now produced synthetically. It is effective against a wide variety of microorganisms, but due to serious adverse effects (e.g. damage to the bone marrow, including aplastic anemia) in humans, it is usually reserved for the treatment of serious and life-threatening infections e.g. typhoid fever and cholera, as it destroys the vibrios and decreases the diarrhea (1). In topical preparations, the antibiotic has caused many cases of contact allergy/allergic contact dermatitis, which has been fully reviewed in Volume 3 of the *Monographs in contact allergy* series (10).

CONTACT ALLERGY FROM ACCIDENTAL CONTACT

In a group of 107 workers in the pharmaceutical industry with dermatitis, investigated in Warsaw, Poland, before 1989, one reacted to chloramphenicol, tested 5% pet. (2). Of 333 nurses patch tested in the same clinic in Poland between 1979 and 1987, 2 reacted to chloramphenicol 5% pet.; data on relevance were not provided (11).

In 2 hospitals in Denmark, between 1974 and 1980, 37 veterinary surgeons, all working in private country practices, were investigated for suspected incapacitating occupational dermatitis and patch tested with a battery of 10 antibiotics. Thirty-two (86%) had one or more positive patch tests. There were 1 or 2 positive reactions to chloramphenicol (4%, incorrect). It was mentioned that all but one of these individuals had allergic contact dermatitis, but relevance was not specified for individual allergens. The most frequent allergens were spiramycin, penethamate and tylosin tartrate (12).

CUTANEOUS ADVERSE DRUG REACTIONS FROM SYSTEMIC ADMINISTRATION CAUSED BY TYPE IV (DELAYED-TYPE) HYPERSENSITIVITY (as demonstrated by positive patch tests)

Cutaneous adverse drug reactions from systemic administration of chloramphenicol caused by type IV (delayed-type) hypersensitivity have included maculopapular eruption (13), acute generalized exanthematous pustulosis (AGEP) (3), systemic allergic dermatitis (5), and undefined drug eruption from intramuscular injection (8).

Maculopapular eruption

In London, between October 2017 and October 2018, 45 patients with suspected cutaneous adverse drug reactions, including 33 maculopapular eruptions (MPE), 4 fixed drug eruptions (FDE), 4 DRESS, 3 AGEP and one SJS/TEN, were patch tested with the suspected drugs. Chloramphenicol (tested at 5% pet.) was responsible for one case of maculopapular eruption (13).

Acute generalized exanthematous pustulosis (AGEP)
A 36-year-old woman was treated for rhinitis with oral acetaminophen and codeine and with chloramphenicol injections, when she suddenly developed pruritic, deeply erythematous and edematous patches on almost her entire body, accompanied by mild fever. Patch tests were positive to codeine (0.5% in pet.) and chloramphenicol (500 mg/ml in water). Eczematous reactions without pustules developed at the sites patch-tested with codeine and chloramphenicol on days 2 and 4, respectively. Intravenous injection of 50 mg (1/20 of a therapeutic dose) of chloramphenicol produced deeply red, edematous, pruritic patches within 5 hours, similar to the original early skin lesion (3).

Systemic allergic dermatitis (systemic contact dermatitis)
A 36-year-old man had experienced an allergic reaction to chloramphenicol eye drops with edema of both eyelids and erythema and edema involving the upper half of his face. Fifteen years later, an intramuscular injection of chloramphenicol succinate resulted after a few hours in an edematous and exudative reaction of the face. Chloramphenicol succinate 20 mg/ml solution was positive in a patch test and an intradermal test at D2 and D4. Eight hours after applying the patch and intradermal tests, the patient experienced a slight and transient conjunctivitis (5).

Other cutaneous adverse drug reactions
A 53-year-old woman had an osteoma removed from her maxilla. After the operation she was given 1 gram chloramphenicol intramuscularly and a drug eruption followed. The changes soon spread and there was some serous exudation in places. Eight days after the onset the skin changes reached maximum intensity and then gradually receded. Five years later a patch test with 50% chloramphenicol was positive. The patient had never had any other skin reaction. More details were not provided. Control tests with 50% chloramphenicol were not mentioned (the currently advised patch test material is 5% in petrolatum) (8).

Cross-reactions, pseudo-cross-reactions and co-reactions
Patients sensitized to chloramphenicol may cross-react to azidamphenicol (9), thiamphenicol (4), chloramphenicol succinate and chloramphenicol palmitate (7). Individuals allergic to chloramphenicol may also cross-react to *p*-nitrobenzoic acid and *p*-dinitrobenzene (intermediates in the chloramphenicol synthesis) (6,7) and possibly to DNCB (6), although the latter finding may be doubtful (7).

LITERATURE
1   The data in the section 'General' may have been obtained from literature discussed in this chapter, but mostly also or exclusively from one or more of the following online sources: ChemIDPlus Advanced, PubChem, DrugBank, RxList, Drug Central, Drugs.com, and Wikipedia
2   Rudzki E, Rebandel P, Grzywa Z. Contact allergy in the pharmaceutical industry. Contact Dermatitis 1989;21:121-122
3   Lee AY, Yoo SH. Chloramphenicol induced acute generalized exanthematous pustulosis proved by patch test and systemic provocation. Acta Derm Venereol 1999;79:412-413
4   Le Coz CJ, Santinelli F. Facial contact dermatitis from chloramphenicol with cross-sensitivity to thiamphenicol. Contact Dermatitis 1998;38:108-109
5   Urrutia I, Audícana M, Echechipía S, Gastaminza G, Bernaola G, Fernández de Corrès L. Sensitization to chloramphenicol. Contact Dermatitis 1992;26:66-67
6   Schwank R, Jirásek L. Kontaktallergie gegen Chloramphenicol mit besonderer Berücksichtigung der Gruppensensibilisierung. Hautarzt 1963;14:24-30 (Article in German)
7   Eriksen K. Cross allergy between paranitro compounds with special reference to DNCB and chloramphenicol. Contact Dermatitis 1978;4:29-32
8   Rudzki E, Grzywa Z, Maciejowska E. Drug reaction with positive patch test to chloramphenicol. Contact Dermatitis 1976;2:181
9   Sachs B, Erdmann S, al Masaoudi T, Merk HF. Molecular features determining lymphocyte reactivity in allergic contact dermatitis to chloramphenicol and azidamphenicol. Allergy 2001;56:69-72
10 De Groot AC. Monographs in contact allergy, volume 3: Topical Drugs. Boca Raton, Fl, USA: CRC Press Taylor and Francis Group, 2021 (ISBN 978-0-367-23693-9)
11 Rudzki E, Rebandel P, Grzywa Z. Patch tests with occupational contactants in nurses, doctors and dentists. Contact Dermatitis 1989;20:247-250
12 Hjorth N, Roed-Petersen J. Allergic contact dermatitis in veterinary surgeons. Contact Dermatitis 1980;6:27-29
13 Watts TJ, Thursfield D, Haque R. Patch testing for the investigation of nonimmediate cutaneous adverse drug reactions: a prospective single center study. J Allergy Clin Immunol Pract 2019;7:2941-2943.e3

Chapter 3.122 CHLORMEZANONE

IDENTIFICATION

Description/definition   : Chlormezanone is the 1,3-thiazine, lactam, sulfone and a member of monochloroben-
zenes that conforms to the structural formula shown below

Pharmacological classes   : Anti-anxiety agents; muscle relaxants, central

IUPAC name   : 2-(4-Chlorophenyl)-3-methyl-1,1-dioxo-1,3-thiazinan-4-one

CAS registry number   : 80-77-3

EC number   : 201-307-4

Merck Index monograph : 3376

Patch testing   : Tablet, pulverized, 10% pet. (2); most systemic drugs can be tested at 10% pet.; if the pure
chemical is not available, prepare the test material from intravenous powder, the content
of capsules or – when also not available – from powdered tablets to achieve a final
concentration of the active drug of 10% pet.

Molecular formula   : $C_{11}H_{12}ClNO_3S$

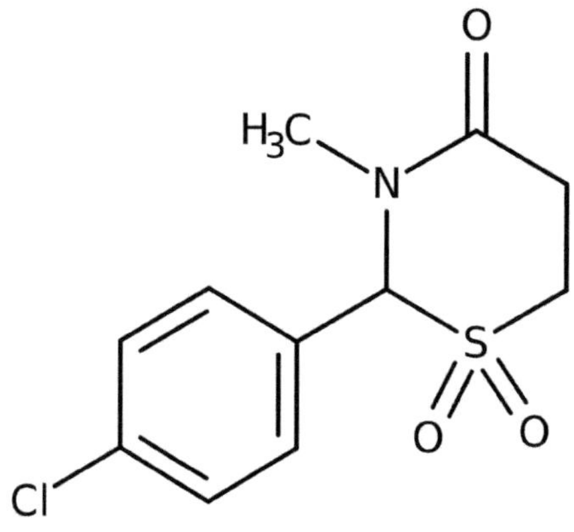

GENERAL

Chlormezanone is a non-benzodiazepine muscle relaxant, that was used in the management of anxiety and in the treatment of muscle spasms. Chlormezanone was discontinued worldwide by its manufacturer in 1996, due to rare but serious and potentially lethal cutaneous reactions (toxic epidermal necrolysis) (1).

CUTANEOUS ADVERSE DRUG REACTIONS FROM SYSTEMIC ADMINISTRATION CAUSED BY TYPE IV (DELAYED-TYPE) HYPERSENSITIVITY (as demonstrated by positive patch tests)

Cutaneous adverse drug reactions from systemic administration of chlormezanone caused by type IV (delayed-type) hypersensitivity have included fixed drug eruption (2,3,4,5) and toxic epidermal necrolysis (TEN) (6).

Fixed drug eruption

In Seoul, South Korea, 31 patients (15 men,16 women, age range 7 to 62 years) with suspected fixed drug eruption (FDE) were patch tested between 1986 and 1990 and in 1996 and 1997. The drugs for patch testing were usually applied at 10% in white petrolatum, both on a previous lesions and on apparently normal skin. The presentation of the results in this reports were rather confusing. When 'itching, erythema, infiltration' at the postlesional skin is taken as proof of delayed-type allergy, then there were 9 reactions to chlormezanone, 5 to mefenamic acid, 3 to phenobarbital, and one each to cotrimoxazole (sulfamethoxazole-trimethoprim), aminophylline, sulpyrin (dipyrone), promethazine, ibuprofen and doxycycline. In most cases, oral provocation tests, when performed, were positive (4).

One patient with FDE suspected to have been caused by chlormezanone had positive patch tests to this drug 1% and 10% pet. on postlesional skin and subsequently a positive oral provocation test (5).

In patients with fixed drug eruptions (FDE) caused by delayed-type hypersensitivity, the diagnosis is usually confirmed by a positive patch test with the drug on previously affected skin. Authors from Finland have used an alternative method of topical provocation. The test compound, the drug 10% in petrolatum and sometimes also in 70% alcohol and in DMSO, was applied once and without occlusion over the entire surface of one or several inactive (usually pigmented) sites of FDE lesions. The patients were followed as in-patients for 24 hours. A reaction was regarded as positive when a clearly demarcated erythema lasting at least 6 hours was seen. Of 4 patients with FDE from chlormezanone, 3 (75%) had a positive topical provocation (2,3).

Stevens-Johnson syndrome/toxic epidermal necrolysis (SJS/TEN)
In Japanese literature, a case of toxic epidermal necrolysis (TEN) caused by chlormezanone with a positive patch test has been reported (9).

LITERATURE

1   The data in the section 'General' may have been obtained from literature discussed in this chapter, but mostly also or exclusively from one or more of the following online sources: ChemIDPlus Advanced, PubChem, DrugBank, RxList, Drug Central, Drugs.com, and Wikipedia
2   Alanko K. Topical provocation of fixed drug eruption. A study of 30 patients. Contact Dermatitis 1994;31:25-27
3   Alanko K, Stubb S, Reitamo S. Topical provocation of fixed drug eruption. Br J Dermatol 1987;116:561-567
4   Lee AY. Topical provocation in 31 cases of fixed drug eruption: change of causative drugs in 10 years. Contact Dermatitis 1998;38:258-260
5   Lee AY, Lee YS. Provocation tests in a chlormezanone-induced fixed drug eruption. DICP 1991;25:604-605
6   Kato T. Toxic epidermal necrolysis: Drug eruption due to Trancopal. Nippon Hifuka Gakkai Zasshi 1966;76:101-102 (Article in Japanese, data cited in ref. 7)
7   Tagami H, Tatsuta K, Iwatski K, Yamada M. Delayed hypersensitivity in ampicillin-induced toxic epidermal necrolysis. Arch Dermatol 1983;119:910-913

Chapter 3.123 CHLOROQUINE

IDENTIFICATION

Description/definition : Chloroquine is the 4-aminoquinoline that conforms to the structural formula shown below
Pharmacological classes : Amebicides; antirheumatic agents; antimalarials
IUPAC name : 4-N-(7-Chloroquinolin-4-yl)-1-N,1-N-diethylpentane-1,4-diamine
Other names : 7-Chloro-4-[[4-(diethylamino)-1-methylbutyl]amino]quinoline
CAS registry number : 54-05-7
EC number : 200-191-2
Merck Index monograph : 3435
Patch testing : 1% pet.
Molecular formula : $C_{18}H_{26}ClN_3$

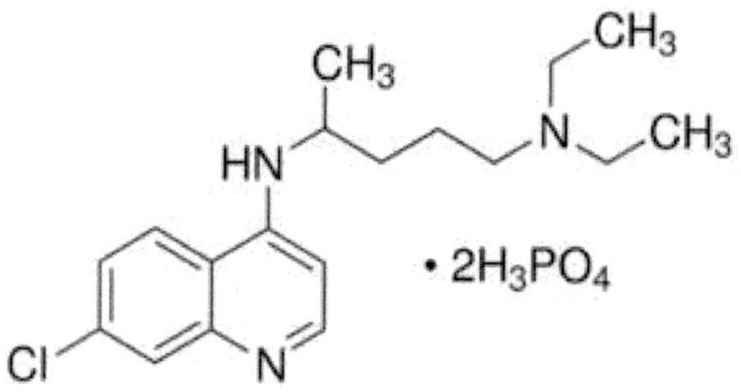

Chloroquine phosphate

GENERAL

Chloroquine is a quinoline compound with antimalarial and anti-inflammatory properties. Chloroquine is used for the suppressive treatment and for acute attacks of malaria due to *Plasmodium vivax*, *P. malariae*, *P. ovale*, and susceptible strains of *P. falciparum*. It has also been used for its anti-inflammatory activities to treat rheumatoid arthritis and systemic lupus erythematosus, and in the systemic therapy of amebic liver abscesses. In pharmaceutical products, chloroquine may be employed as chloroquine phosphate (CAS number 50-63-5, EC number 200-055-2, molecular formula $C_{18}H_{32}ClN_3O_8P_2$) or as chloroquine sulfate (CAS number 132-73-0, EC number 205-077-6, molecular formula $C_{18}H_{28}ClN_3O_4S$) (1).

CONTACT ALLERGY FROM ACCIDENTAL CONTACT

A 17-year-old man presented with a 4-week history of a rash on the forearms, face and eyelids. For 9 months he had worked as an apprentice maintenance fitter for a firm packing various drugs, including antimalarials. In particular, he had been in contact with tablets of chloroquine sulfate and with atmospheric dust from the ingredients. Patch tests were positive to a finely crushed tablet of chloroquine sulfate 1% pet. and – later – to pure chloroquine sulfate 1% pet.; there was no reaction to quinine sulfate (2). This was a case of occupational airborne allergic contact dermatitis.

CUTANEOUS ADVERSE DRUG REACTIONS FROM SYSTEMIC ADMINISTRATION CAUSED BY TYPE IV (DELAYED-TYPE) HYPERSENSITIVITY (as demonstrated by positive patch tests)

Cutaneous adverse drug reactions from systemic administration of chloroquine caused by type IV (delayed-type) hypersensitivity have included drug reaction with eosinophilia and systemic symptoms (DRESS) (5), photosensitivity (4), and erythematous and mainly papular eruption (3).

Drug reaction with eosinophilia and systemic symptoms (DRESS)

A 47-year-old man received prophylaxis with chloroquine and proguanil, resulting in 'hypersensitivity syndrome' (DRESS) with fever (39°C), cough, generalized maculopapular rash, abdominal pain, nausea, inflammatory syndrome, eosinophilia and increased liver function test. Later, a patch test with proguanil was negative, but with chloroquine 2 mg/ml (vehicle?) strongly positive. Five controls were negative. A punch biopsy from the patch test showed a very dense infiltrate with mononuclear cells and spongiosis of the epidermis (5).

Photosensitivity

A 51-year-old dark-skinned man complained of a pruritic rash, apparently caused by exposure to sunlight. A few months earlier, the patient had started chloroquine therapy as a prophylaxis against malaria. Physical examination showed vesicles, papules and characteristic depigmentation on light-exposed areas. Phototests showed a reduced minimal erythemal dose, especially for the wavelengths around 300 nm. Edema, papules, vesicles, and itching could be reproduced by repeated exposure of a skin field to both UV-B and UV-A radiation. Photopatch tests showed a positive reaction to chloroquine (test concentration and vehicle not mentioned) (4).

Other cutaneous adverse drug reactions

A 56-year-old woman took chloroquine phosphate tablets 300 mg once a week as prophylaxis against malaria, which she had never used before. After 2 weeks she developed lesions on her lower legs which, during the following days, spread over the whole body, and at the same time she also had fever up to 39°C. Examination at that time showed an intensely erythematous and mainly papular eruption on the lower legs, breast, abdomen and back. A biopsy showed typical eczematous changes. A patch test with chloroquine phosphate (tested 'tablet as is', which probably means the powder) was strongly (+++) positive. Three controls were negative (3).

LITERATURE

1   The data in the section 'General' may have been obtained from literature discussed in this chapter, but mostly also or exclusively from one or more of the following online sources: ChemIDPlus Advanced, PubChem, DrugBank, RxList, Drug Central, Drugs.com, and Wikipedia

2   Kellett JK, Beck MH. Contact sensitivity to chloroquine sulphate. Contact Dermatitis 1984;11:47

3   Skog E. Systemic eczematous contact-type dermatitis induced by iodochlorhydroxyquin and chloroquine phosphate. Contact Dermatitis 1975;1:187

4   Van Weelden H, Bolling HH, Baart de la Faille H, Van Der Leun JC. Photosensitivity caused by chloroquine. Arch Dermatol 1982: 118: 290.

5   Kanny G, Renaudin JM, Lecompte T, Moneret-Vautrin DA. Chloroquine hypersensitivity syndrome. Eur J Intern Med 2002;13:75-76

Chapter 3.124 8-CHLOROTHEOPHYLLINE

IDENTIFICATION

Description/definition : 8-Chlorotheophylline is the xanthine that conforms to the structural formula shown below
Pharmacological classes : Stimulant
IUPAC name : 8-Chloro-1,3-dimethyl-7H-purine-2,6-dione
Other names : 1,3-Dimethyl-8-chloroxanthine
CAS registry number : 85-18-7
EC number : 201-590-4
Merck Index monograph : Not available
Patch testing : No data available; suggested: 10% pet.
Molecular formula : $C_7H_7ClN_4O_2$

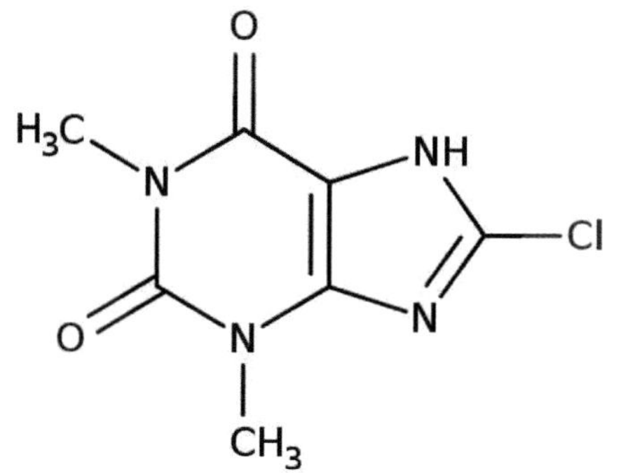

GENERAL

8-Chlorotheophylline is a stimulant drug of the xanthine chemical class, with physiological effects similar to caffeine. Its main use is in combination with diphenhydramine as the antiemetic drug dimenhydrinate. The stimulant properties of 8-chlorotheophylline are thought to ward off the drowsiness caused by diphenhydramine's anti-histamine activity in the central nervous system. This combination drug is used for treating vertigo, motion sickness, and nausea associated with pregnancy (1)

CUTANEOUS ADVERSE DRUG REACTIONS FROM SYSTEMIC ADMINISTRATION CAUSED BY TYPE IV (DELAYED-TYPE) HYPERSENSITIVITY (as demonstrated by positive patch tests)

Cutaneous adverse drug reactions from systemic administration of 8-chlorotheophylline caused by type IV (delayed-type) hypersensitivity have included fixed drug eruption (2).

Fixed drug eruption

A 57-year-old woman complained of recurrent brownish areas on the right thigh, leg and foot for several years. At irregular periods, precisely the same areas would become itchy, occasionally blistered and this would slowly subside to leave round, brownish patches of varying intensity. The patient had a flare of the eruption shortly after having taken a Dramamine ® tablet (a combination of diphenhydramine and 8-chlorotheophylline) to prevent motion sickness. She was asked to take one Dramamine tablet again, and the next morning each pigmented area flared with increased swelling and inflammatory halo, followed by deepened pigmentation. An oral provocation test with the component diphenhydramine was negative, with 8-chlorotheophylline 10 mg. strongly positive 15-22 hours after ingestion with erythema, edema and a superficial bulla. Provocation with 8-bromotheophylline was also positive, but there was no reaction to unmodified theophylline. Finally, a patch test with powdered 8-chlorotheophylline on lesional skin was positive, but negative to unsubstituted chlorotheophylline (2).

Cross-reactions, pseudo-cross-reactions and co-reactions

Cross-reaction between 8-chlorotheophylline and 8-bromotheophylline (2). No cross-reaction with unsubstituted theophylline (2).

LITERATURE

1. The data in the section 'General' may have been obtained from literature discussed in this chapter, but mostly also or exclusively from one or more of the following online sources: ChemIDPlus Advanced, PubChem, DrugBank, RxList, Drug Central, Drugs.com, and Wikipedia
2. Stritzler C, Kopf AW. Fixed drug eruption caused by 8-chlorotheophylline in dramamine with clinical and histologic studies. J Invest Dermatol 1960;34:319-330

Chapter 3.125 CHLORPHENIRAMINE

IDENTIFICATION

Description/definition   : Chlorpheniramine is the synthetic alkylamine derivative that conforms to the structural
                             formula shown below
Pharmacological classes : Histamine H_1 antagonists; anti-allergic agents; antipruritics
IUPAC name   : 3-(4-Chlorophenyl)-N,N-dimethyl-3-pyridin-2-ylpropan-1-amine
Other names   : 2-Pyridinepropanamine, γ-(4-chlorophenyl)-N,N-dimethyl-; chlorphenamine
CAS registry number   : 132-22-9
EC number   : 205-054-0
Merck Index monograph   : 3456
Patch testing   : Maleate, 5% pet. (SmartPracticeCanada)
Molecular formula   : $C_{16}H_{19}ClN_2$

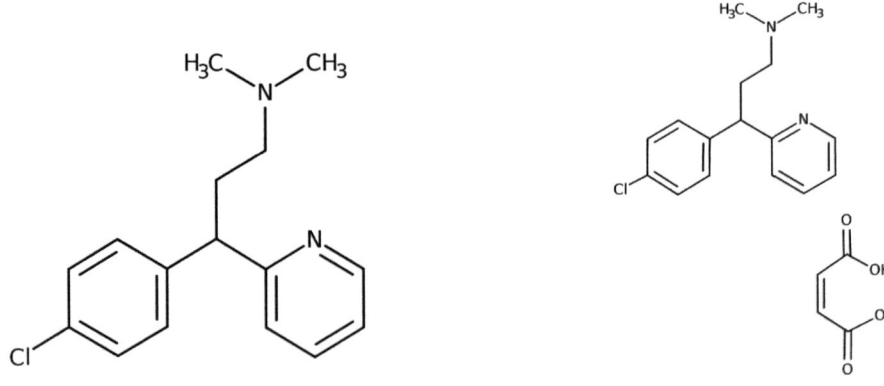

Chlorpheniramine   Chlorpheniramine maleate

GENERAL

Chlorpheniramine is a synthetic alkylamine derivative and a competitive histamine H1 receptor antagonist, and
displays anticholinergic and mild sedative effects as well. It is indicated for the treatment of rhinitis, urticaria, allergy,
common cold, asthma and hay fever. It has also been used in veterinary applications. In pharmaceutical products,
chlorpheniramine is employed as chlorpheniramine maleate (CAS number 113-92-8, EC number 204-037-5,
molecular formula $C_{20}H_{23}ClN_2O_4$) (1). Chlorpheniramine is used in both topical and systemic pharmaceutical
applications. In topical preparations, this antihistamine has caused contact allergy/allergic contact dermatitis, which
has been fully reviewed in Volume 3 of the *Monographs in contact allergy* series (5).

CUTANEOUS ADVERSE DRUG REACTIONS FROM SYSTEMIC ADMINISTRATION CAUSED BY TYPE IV
(DELAYED-TYPE) HYPERSENSITIVITY (as demonstrated by positive patch tests)

Cutaneous adverse drug reactions from systemic administration of chlorpheniramine caused by type IV (delayed-
type) hypersensitivity have included widespread erythematous rash (3) and systemic allergic dermatitis (4).

Erythroderma, widespread erythematous eruption, exfoliative dermatitis

A 39-year-old woman took chlorpheniramine 4 mg tablets for insect bite reactions. Five hours later, she developed a
widespread, erythematous rash, eventually involving most of her trunk and limbs. Previously, she had suffered from
similar eruptions from various other drugs. The patient was patch tested with chlorpheniramine and these other
drugs, all at 20% pet. from tablets. By day 2, she had already developed a bullous reaction to the chlorpheniramine
patch that persisted to day 4. Patch testing to the other drugs was negative. The original source of sensitization to
chlorpheniramine was unknown. Fifteen controls were negative (3).

Systemic allergic dermatitis (systemic contact dermatitis)

A 45-year-old woman treated insect bites with dexchlorpheniramine maleate cream and, after a few hours,
developed intensely itchy eczema at the application site. She was treated for this with injections of chlorpheniramine
maleate, but her dermatitis now became generalized over the next 24 hours. Patch tests were positive to
dexchlorpheniramine maleate cream and in a second session to the pure chemical dexchlorpheniramine maleate,
and also to the related chlorpheniramine maleate, brompheniramine maleate, and pheniramine maleate, all tested
1% water. Surprisingly, no adverse effects were noted following oral administration of dexchlorpheniramine maleate
(4). This was a case of generalized dermatitis from systemic allergic dermatitis caused by oral chlorpheniramine

maleate in a patient topically sensitized to dexchlorpheniramine maleate. The authors also described a second case, which is virtually identical to the patient described above (4).

Cross-reactions, pseudo-cross-reactions and co-reactions
Possible cross-reactivity between pheniramine, chlorpheniramine and dexchlorpheniramine (2). Two patients sensitized to dexchlorpheniramine maleate cross-reacted to chlorpheniramine maleate, brompheniramine maleate and pheniramine maleate (4).

LITERATURE
1   The data in the section 'General' may have been obtained from literature discussed in this chapter, but mostly also or exclusively from one or more of the following online sources: ChemIDPlus Advanced, PubChem, DrugBank, RxList, Drug Central, Drugs.com, and Wikipedia
2   Parente G, Pazzaglia M, Vincenzi C, Tosti A. Contact dermatitis from pheniramine maleate in eyedrops. Contact Dermatitis 1999;40:338
3   Brown VL, Orton DI. Cutaneous adverse drug reaction to oral chlorphenamine detected with patch testing. Contact Dermatitis 2005;52:49-50
4   Santucci B, Cannistraci C, CristaudoA, Picardo M. Contact dermatitis from topical alkylamines. Contact Dermatitis 1992;27:200-201
5   De Groot AC. Monographs in contact allergy, volume 3: Topical Drugs. Boca Raton, Fl, USA: CRC Press Taylor and Francis Group, 2021 (ISBN 978-0-367-23693-9)

Chapter 3.126 CHLORPROMAZINE

IDENTIFICATION

Description/definition   : Chlorpromazine is the phenothiazine that conforms to the structural formula shown
                            below
Pharmacological classes   : Dopamine antagonists; antipsychotics; antiemetics
IUPAC name   : 3-(2-Chlorophenothiazin-10-yl)-N,N-dimethylpropan-1-amine
CAS registry number   : 50-53-3
EC number   : 200-045-8
Merck Index monograph   : 3461
Patch testing   : Hydrochloride 0.1% pet. (Chemotechnique); 1% pet. (SmartPracticeCanada)
Molecular formula   : $C_{17}H_{19}ClN_2S$

GENERAL

Chlorpromazine is a phenothiazine and traditional antipsychotic agent with anti-emetic activity. It exerts its antipsychotic effect by blocking postsynaptic dopamine receptors in cortical and limbic areas of the brain, thereby preventing the excess of dopamine in the brain. This leads to a reduction in psychotic symptoms, such as hallucinations and delusions. Chlorpromazine is indicated for the treatment of schizophrenia, to control nausea and vomiting, for relief of restlessness and apprehension before surgery, for acute intermittent porphyria, as an adjunct in the treatment of tetanus, to control the manifestations of the manic type of manic-depressive illness, for relief of intractable hiccups, for the treatment of severe behavioral problems in children, and in the short-term treatment of hyperactive children (1). In pharmaceutical products, both chlorpromazine base and chlorpromazine hydrochloride (CAS number 69-09-0, EC number 200-701-3, molecular formula $C_{17}H_{20}Cl_2N_2S$) may be employed (1).

CONTACT ALLERGY FROM ACCIDENTAL CONTACT

Occupational exposure

Contact allergy

A 42-year old maintenance technician developed occupational airborne allergic contact dermatitis from chlorproma-zine. The primary site of the dermatitis was the malar region (27). One or more nurses and a ward attendant developed occupational contact allergy to chlorpromazine hydrochloride in Poland before 1980. Clinical details were not provided (28). Two individuals handling chlorpromazine pills developed photocontact allergy to the drug; their occupations were not mentioned (8). A 29-year-old nurse from Belgium developed eczema of the fingers and later of the face from occupational contact sensitization to chlorpromazine (32). A case of occupational allergic and photoallergic contact dermatitis in a farmer was reported in German literature (36).

Two cases of occupational airborne allergic contact dermatitis with eczema of the face and back of the hands in 2 nurses administering chlorpromazine to patients were already reported in the 1950s in the USA. Topical provo-cation on the upper arm with chlorpromazine 25 mg/ml (2.5%) was strongly positive in both patients. In one, a severe reaction was also present on the anterior forearm, which had been in contact with the patch test from flexion of the arm during sleep. Coincidentally, the eruption on the face and hands flared badly. The eyes were swollen shut; the rash became exudative and spread from the face to include the ears and neck, arms, trunk and legs (38). See also ref. 23 for another case report of a nurse with occupational allergic contact dermatitis to chlorpromazine.

Photocontact allergy

A 47-year-old female pharmacist, making up prescriptions for a few hours per day at a psychiatric hospital, presented with papular erythematous lesions over the light-exposed areas affecting the face, neck, ears, forearms and the dorsa of the hands. She was patch and photopatch tested with all the drugs that she had contact with and showed a positive photopatch test to chlorpromazine 0.01% pet. and the related phenothiazine perphenazine 0.01% pet. This was a case of occupational airborne photoallergic contact dermatitis to chlorpromazine and perphenazine (25).

A 45-year-old pig breeder developed erythematous, scaly, and pruritic plaques localized symmetrically on the sun-exposed backs of his hands, fingers, and forearms, spreading to his face and other sun-exposed body sites. Without protective measures, he would inject his animals with chlorpromazine and mix a powder containing olaquindox into the pigs' dry food manually. Patch and photopatch tests showed positive photopatch test reactions to chlorpromazine, promethazine and olaquindox. In spite of complete avoidance of the identified photoallergens for several years, his life was 22 years later still extremely disabled due to the persistent photosensitivity (33).

A case of actinic reticuloid developing from occupational photoallergic contact dermatitis to chlorpromazine in a pig farmer (35) and occupational allergic and photoallergic contact dermatitis in another farmer (36) have been reported in German literature. Two patients handling chlorpromazine pills developed photocontact allergy to the drug; their occupations were not mentioned (8). Another case of occupational airborne photoallergic contact dermatitis to chlorpromazine was reported from Sweden in 1994 (31).

Exposure in non-professional caregivers

Contact allergy

See the next section, refs. 41 and 45, for cases of (photoaggravated) allergic contact dermatitis to chlorpromazine.

Photocontact allergy

Three cases of (photo)allergic reactions to chlorpromazine in caregivers, administering the drug for years to their partners, were reported from Italy (41). All three (2 women, one man) had chronic eczematous dermatitis, in two on the face, neck, hands and forearms, in the third on the face and hands. Eczema would start in spring and become worse when sun-exposed. One had a negative patch test to chlorpromazine 1% pet. but a positive reaction to a chlorpromazine photopatch test; the other two had contact allergic reactions with photoaggravation. The final diagnosis was persistent light reaction in the first, photoaggravated allergic contact dermatitis in the second and actinic reticuloid in the third. Contamination was probably the result of spilling drops of chlorpromazine solution and breaking tablets, releasing dust causing airborne reactions. When chlorpromazine was replaced with alternative neuroleptics for patients' relatives, progressive improvement in all three patients was observed (41).

A 57-year-old woman presented with dry eczematous pulpitis affecting the thumb, second, and third fingers of the right hand, that had existed for 10 months. At home, she was a caregiver for her mother-in-law, to whom she daily administered chlorpromazine 4% drops (solution). Patch testing and photopatch testing yielded positive photopatch tests to the drops and to chlorpromazine HCl 1% pet. irradiated with UVA 5 J/cm² in the photopatch test series. The authors describe another 2 similar cases in men giving care to their mother resp. mother-in-law. One also handled the drops, the other had contact with chlorpromazine tablets which he used to crush. Both had hyperkeratotic eczema limited to the fingertips and showed positive photopatch tests to the drops resp. a crushed tablet 'as is' and to chlorpromazine HCl present in the Spanish photopatch test series. The patients were diagnosed with photoallergic contact dermatitis (30). It should be mentioned that 1. concentrations of 1% resp. 4% chlorpromazine entail a great risk of phototoxicity; 2. no controls were performed with the commercial chlorpromazine products. Nevertheless, apparently the ordinary patch tests were negative and the eczema in all 3 patients healed completely once they stopped handling chlorpromazine.

Two similar cases of photoallergic contact dermatitis in non-professional caregivers were seen in Valencia, Spain. Both were women, aged 45 and 46, who prepared chlorpromazine 4% drops for ill offspring. One had eyelid eczema and the other fissured digital pulpitis, eczematous cheilitis and eyelid eczema. The first had positive photopatch tests to chlorpromazine and was diagnosed with photoallergic contact dermatitis; the second had positive patch tests and stronger positive photopatch tests and was therefore diagnosed with photoaggravated allergic contact dermatitis to chlorpromazine (45).

Accidental exposure in patients using chlorpromazine

In the period 1980-2019, in a university center in Valencia, Spain, 14 patients had positive photopatch tests to chlorpromazine 0.1% pet. irradiated with UVA 5 J/cm². Six reactions were relevant. All were women, ages ranging from 30 to 64 years. Two were non-professional caregivers; their cases are presented above. The other 4 took chlorpromazine 4% drops for anxiety, irritable bowel syndrome or post-traumatic neurological sequelae. All 4 had eczematous cheilitis, of who 2 also showed fissured digital pulpitis and eyelid eczema. Patch tests to chlorpromazine were negative but photopatch tests positive (++) in all 4. The patients were diagnosed with photoallergic contact dermatitis to chlorpromazine (45).

Two patients from Spain developed photoallergic contact cheilitis from taking chlorpromazine drops. Presumably, the lips were contaminated with the material during intake of the solution (37, probably the same patients as in ref. 45).

CUTANEOUS ADVERSE DRUG REACTIONS FROM SYSTEMIC ADMINISTRATION CAUSED BY TYPE IV (DELAYED-TYPE) HYPERSENSITIVITY (as demonstrated by positive patch tests)
Cutaneous adverse drug reactions from systemic administration of chlorpromazine caused by type IV (delayed-type) hypersensitivity have included systemic allergic dermatitis (23) and photosensitivity (40,42,43).

Systemic allergic dermatitis (systemic contact dermatitis)
Among 11 promethazine-allergic patients, three had co-/cross-reactions to chlorpromazine. In two, a reactivation of the eruption occurred after taking one chlorpromazine tablet (systemic allergic dermatitis). The third was a nurse had become sensitized to chlorpromazine and promethazine which she had handled daily in the form of the injectable solution. Twelve hours after taking a chlorpromazine tablet she developed severe edema of the hands, arms and face as well as vertigo and tendency to fainting. This reaction subsided slowly over a 15 day period (23).

Photosensitivity

General
Formerly, chlorpromazine was used in topical preparations; however, it was soon found to cause cases of photo-allergic and allergic contact dermatitis and was banned in many countries (39). This section provides some literature on photoallergic reactions to chlorpromazine administered systemically, but a full literature review has not been attempted. Most reports of photosensitivity from systemic chlorpromazine are found in early literature when photo(patch) testing was not very reliable. Currently, patients with well-documented photoallergic dermatitis from chlorpromazine are rarely reported. This either means that cases of clinical photosensitivity were found to be phototoxic, or that the subject of photosensitivity is considered not to be worth-while publishing anymore.

Photopatch testing in patients suspected of photosensitivity
Chlorpromazine has long been present in most 'photopatch test series', which is routinely tested in patients photopatch tested for suspected photosensitivity. The results of 18 such studies published between 1984 and 2019 are shown in table 3.126.1 (published here because it was not included in Volume 3 of the *Monographs in contact allergy* series, Topical drugs [46]). Frequencies of positive photopatch tests have ranged from 0.2% to 44%, with percentages of 3% or higher in 11 and of over 5% in 8 investigations (table 3.126,1). This would seem to indicate that chlorpromazine photoallergy is frequent. However, in by far most studies, the relevance of the positive photopatch tests was either not mentioned, unknown, or the positive reactions were not relevant. Only very few cases were considered to be relevant to the patients' histories (8,9,17). Also, it is well known that chlorpromazine can induce phototoxic reactions (44), depending on the concentration used and the irradiation parameters, which reactions are very difficult to distinguish from photoallergic photopatch test reactions (18). It has been demonstrated that photopatch testing with chlorpromazine 0.1% pet. and irradiation of 10 J/cm^2 UVA (which has long been a frequently used irradiation dose) results in 6% phototoxic reactions (18). Therefore, it may be assumed that many, or even most of the positive photopatch tests in such studies with high prevalences have been phototoxic rather than photoallergic, which was also suggested by many of the investigators themselves (10,12,13,15,16,17,21,22). It cannot be excluded that a number of 'real' positive photopatch tests, indicating photoallergy, are in fact photocross-reactions to the related phenothiazine promethazine (Chapter 3.412). Topical promethazine frequently results in photocontact allergy and to a lesser extent contact allergy and is even today available in some countries, including the sunny Italy and France.

Case reports and case series
Already in the 1950s, cases of photosensitivity in patients treated with chlorpromazine were reported in literature (40). The technique of phototesting and photopatch testing at that moment was not advanced enough to reliably discriminate photoallergy from phototoxicity (even today it may be difficult). Cases were also presented in the 1970 (42,43).

Table 3.126.1 Photopatch testing in groups of patients

Years and Country	Test conc. & vehicle	Number of patients tested	positive (%)	Selection of patients (S); Relevance (R); Comments (C)	Ref.
2014-2016 Spain		116	2 (1.7%)	S: not stated; R: 0%	3
2006-2012 China	1% water	3993	1768 (44%)	S: patients with suspected photodermatoses; R: not stated	4
2001-2010 Canada		160	13 (8.1%)	S: patients with suspected photosensitivity and patients who developed pruritus or a rash after sunscreen application; R: not stated	6
2003-2007 Portugal	0.1% pet.	83	2 (2.4%)	S: patients with suspected photoaggravated facial dermatitis or systemic photosensitivity; R: all reactions were relevant	8
1993-2006 USA	0.1% pet.	76	4 (5%)	S: not stated; R: 21% of all reactions to medications were considered 'of possible relevance'	7
1992-2006 Greece	0.1% pet.	207	(12.5%)	S: patients suspected of photosensitivity; R: not stated; C: 'some reactions may have been phototoxic'	10
2004-2005 Spain	0.1% pet.	224	3 (1.3%)	S: not stated; R: 0%	11
2000-2005 USA	0.1% pet.	177	13 (7.3%)	S: patients photopatch tested for suspected photodermatitis; R: 15%	9
1994-9 Netherlands	1% pet., later 0.1%	99	2 (2%)	S: patients suspected of photosensitivity disorders; R: not stated; C: only reactions with the 1% concentration; it was suggested that these may have been phototoxic	12
1983-1998 UK	0.1% pet.	2715	6 (0.2%)	S: patients suspected of photosensitivity or with (a history of) dermatitis at exposed sites; R: not established; C: the authors suggested that these reactions were phototoxic and have removed it from the photopatch test series	13
1991-97 Ger, Au, Swi	0.1% pet.	1261	(0.95%)	S: patients suspected of photosensitivity; R: not stated; C: many phototoxic reactions not counted as positive	14
1990-1994 France	0.1% pet.	370	16 (4.3%)	S: patients with suspected photodermatitis; R: not stated; C: some reactions may have been phototoxic, according to the authors, from using high UVA-doses	15
1991-1993 Singapore	1% pet.	62	5 (8%)	S: patients with clinical features suggestive of photosensitivity; R: 1/5; C: it was suggested that the others may have been phototoxic reactions	17
1987-1989 Thailand	0.1% pet.	274	10 (3.6%)	S: patients suspected of photosensitivity; R: 0%	19
1986-1989 Italy	1% pet.	128	15 (11.7%)	S: not stated; R: not stated	20
1980-85 Ger, Au, Swi	0.1% pet.	1129	(0.4%)	S: patients suspected of photoallergy, polymorphic light eruption, phototoxicity and skin problems with photo-distribution; R: not stated; C: 6% phototoxic reactions	18
1980-1985 USA	1% pet.	70	13 (19%)	S: not stated; R: none; C: it was suggested that these were phototoxic reactions	21
1980-1981 4 Scandi-navian countries	0.1% pet.	745	22 (3.0%)	S: patients suspected of sun-related skin disease; R: not specified; C: 'the reactions to phenothiazines were frequently phototoxic'	22

Au: Austria; Ger: Germany; NL: Swi: Switzerland

A 54-year-old woman had a 15-year history of recurrent pruritic eruptions in light-exposed areas. Recurrences had arisen once or twice a year. An eruption on the palms appeared more frequently and occasionally urticarial wheals occurred. Examination showed an eruption of papules, vesicles, and edema of the face and hands. In patch and photopatch testing, immediately after ultraviolet light exposure, a large wheal developed at the photopatch test site with 0.0625% chlorpromazine in saline. Closed patch and photopatch testing with chlorpromazine produced a papulovesicular eczematous reaction after 24 hours, the photopatch test reaction being much stronger. Simultaneously, distant flares occurred at the previously involved sites, the face and hands. Histologically, perivascular collections of lymphocytes and epidermal spongiosis were found at the positive photopatch test site. Sunburn cells were not observed. The tests with chlorpromazine were repeated three times, and identical results were obtained. In three control subjects, chlorpromazine 0.625% showed phototoxicity, but there were no reactions to the 0.0625% concentration. Intradermal injection of 0.02 ml of 0.125% chlorpromazine in saline followed by irradiation with black light in the patient produced a wheal of 5 cm immediately after the irradiation in the patient. Passive transfer tests with the patient's serum to 2 normal subjects were similarly positive (43).

This was very likely a case of combined immediate-type and delayed-type photoallergy to chlorpromazine. However, three comments can be made: 1. It was not mentioned anywhere in the article that the patient used chlorpromazine, and, if yes, at what dosage and either continuously or periodically; 2. If the patient indeed also had 'plain' delayed-type hypersensitivity, why are there no symptoms constantly present (periodicity may be explained by photoallergy of course); 3. the author mentions that the patient was (photo)patch tested with chlorpromazine only at a concentration of 0.0625%. However, a picture in the article shows positive patch and photopatch tests 'with various concentrations of chlorpromazine'. Could this have included the 0.625% concentration which caused phototoxic reactions in the three controls?

Cross-reactions, pseudo-cross-reactions and co-reactions
Patients (photo)sensitized to promethazine may (photo-)cross-react to chlorpromazine (23,39). Four patients sensitized to chlorproethazine in an ointment had photocross-reactions to chlorpromazine 0.1% pet. (2). A patient sensitized and photosensitized to perphenazine photocross-reacted to chlorpromazine (24). A patient photosensitized to mequitazine had photocross-reactions to chlorpromazine (26). Photocross-reactions between the phenothiazines isothipendyl, promethazine and chlorpromazine may occur (29).

Immediate contact reactions
Immediate contact reactions (contact urticaria) to chlorpromazine are presented in Chapter 5.

LITERATURE
1   The data in the section 'General' may have been obtained from literature discussed in this chapter, but mostly also or exclusively from one or more of the following online sources: ChemIDPlus Advanced, PubChem, DrugBank, RxList, Drug Central, Drugs.com, and Wikipedia
2   Barbaud A, Collet E, Martin S, Granel F, Trechot P, Lambert D, et al. Contact sensitization to chlorproethazine can induce persistent light reaction and cross-photoreactions to other phenothiazines. Contact Dermatitis 2001;44:373-374
3   Subiabre-Ferrer D, Esteve-Martínez A, Blasco-Encinas R, Sierra-Talamantes C, Pérez-Ferriols A, Zaragoza-Ninet V. European photopatch test baseline series: A 3-year experience. Contact Dermatitis 2019;80:5-8
4   Gao L, Hu Y, Ni C, Xu Y, Ma L, Yan S, Dou X. Retrospective study of photopatch testing in a Chinese population during a 7-year period. Dermatitis 2014;25:22-26
5   Hu Y, Wang D, Shen Y, Tang H. Photopatch testing in Chinese patients over 10 years. Dermatitis 2016;27:137-142
6   Greenspoon J, Ahluwalia R, Juma N, Rosen CF. Allergic and photoallergic contact dermatitis: A 10-year experience. Dermatitis 2013;24:29-32
7   Victor FC, Cohen DE, Soter NA. A 20-year analysis of previous and emerging allergens that elicit photoallergic contact dermatitis. J Am Acad Dermatol 2010;62:605-610
8   Cardoso J, Canelas MM, Gonçalo M, Figueiredo A. Photopatch testing with an extended series of photoallergens: a 5-year study. Contact Dermatitis 2009;60:325-329
9   Scalf LA, Davis MDP, Rohlinger AL, Connolly SM. Photopatch testing of 182 patients: A 6-year experience at the Mayo Clinic. Dermatitis 2009;20:44-52
10 Katsarou A, Makris M, Zarafonitis G, Lagogianni E, Gregoriou S, Kalogeromitros D. Photoallergic contact dermatitis: the 15-year experience of a tertiary referral center in a sunny Mediterranean city. Int J Immunopathol Pharmacol 2008;21:725-727
11 De La Cuadra-Oyanguren J, Perez-Ferriols A, Lecha-Carrelero M, et al. Results and assessment of photopatch testing in Spain: towards a new standard set of photoallergens. Actas DermoSifiliograficas 2007;98:96-101
12 Bakkum RS, Heule F. Results of photopatch testing in Rotterdam during a 10-year period. Br J Dermatol 2002;146:275-279
13 Darvay A, White I R, Rycroft R J G, Jones A B, Hawk J L M, McFadden J P. Photoallergic contact dermatitis is uncommon. Br J Dermatol 2001;145:597-601
14 Neumann NJ, Hölzle E, Plewig G, Schwarz T, Panizzon RG, Breit R, et al. Photopatch testing: The 12-year experience of the German, Austrian and Swiss Photopatch Test Group. J Am Acad Dermatol 2000;42(2Pt.1):183-192
15 Journe F, Marguery M-C, Rakotondrazafy J, El Sayed F, Bazex J. Sunscreen sensitization: a 5-year study. Acta Derm Venereol (Stockh) 1999;79:211-213
16 Pigatto PD, Legori A, Bigardi AS, Guarrera M, Tosti A, Santucci B, et al. Gruppo Italiano recerca dermatiti da contatto ed ambientali Italian multicenter study of allergic contact photodermatitis: epidemiological aspects. Am J Contact Dermatitis 1996;17:158-163
17 Leow YH, Wong WK, Ng SK, Goh CL. 2 years' experience of photopatch testing in Singapore. Contact Dermatitis 1994;31:181-182
18 Hölzle E, Neumann N, Hausen B, Przybilla B, Schauder S, Hönigsmann H, et al. Photopatch testing: the 5-year experience of the German, Austrian and Swiss Photopatch Test Group. J Am Acad Dermatol 1991;25:59-68
19 Gritiyarangsan P. A three-year photopatch study in Thailand. J Dermatol Sci 1991;2:371-375
20 Guarrera M. Photopatch testing: a three-year experience. J Am Acad Dermatol 1989;21:589-591
21 Menz J, Muller SA, Connolly SM. Photopatch testing: a 6-year experience. J Am Acad Dermatol 1988;18:1044-1047

22 Wennersten G, Thune P, Brodthagen H, Jansen C, Rystedt I, Crames M, et al. The Scandinavian multicenter photopatch study. Contact Dermatitis 1984;10:305-309

23 Sidi E, Hincky M, Gervais A. Allergic sensitization and photosensitization to Phenergan cream. J Invest Dermatol. 1955;24:345-352

24 Gacías L, Linares T, Escudero E, et al. Perphenazine as a cause of mother-to-daughter contact dermatitis and photocontact dermatitis. J Investig Allergol Clin Immunol 2013;23:60-61

25 Torinuki J. Contact dermatitis to biperiden and photocontact dermatitis in phenothiazines in a pharmacist. Tohoku J Exp Med 1995;176:249-252 (Article in English)

26 Kim TH, Kang JS, Lee HS, Youn JI. Two cases of mequitazine induced photosensitivity reactions. Photodermatol Photoimmunol Photomed 1995;11:170-173

27 Bennett MF, Lowney AC, Bourke JF. A study of occupational contact dermatitis in the pharmaceutical industry. Br J Dermatol 2016;174:654-656

28 Rudzki E. Occupational dermatitis among health service workers. Derm Beruf Umwelt 1979;27:112-115

29 Cariou C, Droitcourt C, Osmont MN, Marguery MC, Dutartre H, Delaunay J, et al. Photodermatitis from topical phenothiazines: A case series. Contact Dermatitis 2020;83:19-24

30 Monteagudo-Paz A, Salvador JS, Martinez NL, Granados PA, Martínez PS. Pulpitis as clinical presentation of photoallergic contact dermatitis due to chlorpromazine. Allergy 2011;66:1503-1504

31 Björkner BE. Industrial airborne dermatoses. Dermatol Clin 1994;12:501-509 VAN BIRCHER IN KANERVA (PDF

32 Gielen K, Goossens A. Occupational allergic contact dermatitis from drugs in healthcare workers. Contact Dermatitis 2001;45:273-279

33 Emmert B, Schauder S, Palm H, Hallier E, Emmert S. Disabling work-related persistent photosensitivity following photoallergic contact dermatitis from chlorpromazine and olaquindox in a pig breeder. Ann Agric Environ Med 2007;14:329-333

34 Schauder S. How to avoid phototoxic reactions in photopatch testing with chlorpromazine. Photodermatol 1985;2:95-100

35 Schauder S, Berger H. Aktinisches Retikuloid nach photoallergischem Kontaktekzem durch Chlorpromazin bei einem Schweinemäster. Dermatosen 1991;39:12-17

36 Ertle T. Beruflich bedingte Kontakt- und Photokontaktallergie bei einem Landwirt durch Chlorpromazin. Dermatosen 1982;30:120-122

37 Esteve-Martínez A, Ninet Zaragoza V, de la Cuadra Oyanguren J, Oliver-Martínez V. Photoallergic contact dermatitis due to chlorpromazine: A report of 2 Cases. Actas Dermosifiliogr 2015;106:518-520

38 Lewis GM, Sawicky HH: Contact dermatitis from chlorpromazine. JAMA 1955 ;157:909-910

39 Epstein S, Rowe R. Photoallergy and photocross-sensitivity to Phenergan. J Invest Dermatol 1957;29:319-326

40 Epstein JH, Brunsting LA, Peterson MC, Schwartz BE. A study of photosensitivity occurring with chlorpromazine therapy. J Invest Dermatol 1957;28:329-338

41 Giomi B, Difonzo EM, Lotti L, Massi D, Francalanci S. Allergic and photoallergic conditions from unusual chlorpromazine exposure: report of three cases. Int J Dermatol 2011;50:1276-1278

42 Raffle EJ, MacLeod TM, Hutchinson F, Ballinger B. Letter: Chlorpromazine photosensitivity. Arch Dermatol 1975;111:1364-1365

43 Horio T. Chlorpromazine photoallergy. Coexistence of immediate and delayed type. Arch Dermatol 1975;111:1469-1471

44 Epstein S. Chlorpromazine photosensitivity. Phototoxic and photoallergic reactions. Arch Dermatol 1968;98:354-363

45 Martínez-Doménech A, García-Legaz Martínez M, Ferrer-Guillén B, Magdaleno-Tapial J, Valenzuela-Oñate C, Esteve-Martínez A, et al. Allergic and photoallergic contact dermatitis to chlorpromazine. Australas J Dermatol 2020;61:e351-e353

46 De Groot AC. Monographs in contact allergy, volume 3: Topical Drugs. Boca Raton, Fl, USA: CRC Press Taylor and Francis Group, 2021 (ISBN 978-0-367-23693-9)

Chapter 3.127 CHLORPROPAMIDE

IDENTIFICATION

Description/definition   : Chlorpropamide is the sulfonylurea compound that conforms to the structural formula
                                            shown below
Pharmacological classes   : Hypoglycemic agents
IUPAC name   : 1-(4-Chlorophenyl)sulfonyl-3-propylurea
CAS registry number   : 94-20-2
EC number   : 202-314-5
Merck Index monograph   : 3462
Patch testing   : No data available; suggested: 5% pet.
Molecular formula   : $C_{10}H_{13}ClN_2O_3S$

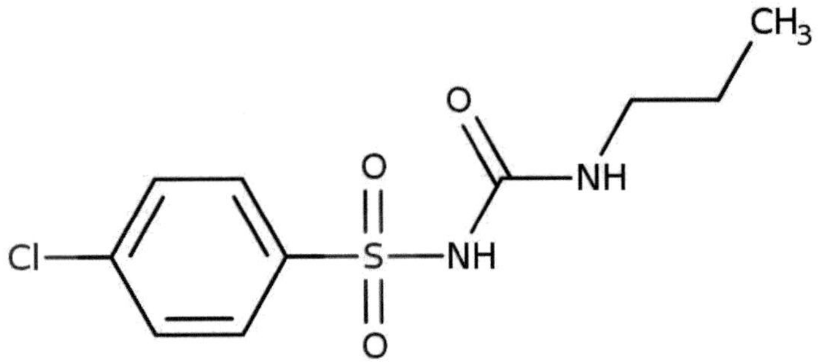

GENERAL

Chlorpropamide is a long-acting, first-generation sulfonylurea with hypoglycemic activity. It is used for the treatment of non-insulin-dependent diabetes mellitus. It acts, just as the other drugs of the sulfonylurea class of insulin secretagogues, by stimulating β cells of the pancreas to release insulin, both basal insulin secretion and meal-stimulated insulin release. The drug also increases peripheral glucose utilization, decreases hepatic gluconeogenesis and may increase the number and sensitivity of insulin receptors (1).

CUTANEOUS ADVERSE DRUG REACTIONS FROM SYSTEMIC ADMINISTRATION CAUSED BY TYPE IV (DELAYED-TYPE) HYPERSENSITIVITY (as demonstrated by positive patch tests)

Cutaneous adverse drug reactions from systemic administration of chlorpropamide caused by type IV (delayed-type) hypersensitivity have included systemic allergic dermatitis (2,3).

Systemic allergic dermatitis (systemic contact dermatitis)

Patients contact allergic to para-amino compounds (sulfanilamide, p-phenylenediamine, benzocaine) were orally challenged with 3 related sulfonylurea derivatives (hypoglycemic agents, sulfonamides): chlorpropamide, carbutamide and tolbutamide. Eleven had a positive reaction: one (of 20 tested) to chlorpropamide, 7 (of 25 tested) to carbutamide, and 3 (of 11 tested) with tolbutamide. All were patients previously sensitized to sulfanilamide. Symptoms were itching in all 11 patients, reappearance of erythema and vesicles at the site of the primary contact dermatitis in 6 patients, and relapse of the primary contact dermatitis with a moderate secondary vesicular eruption together with a reactivation of the patch test reaction in 5 patients. Patch tests with these drugs themselves were not performed (2).

A similar observation (with tolbutamide and chlorpropamide) was reported one year later, but details are not available to the author (3). These were very likely cases of systemic allergic dermatitis.

LITERATURE

1   The data in the section 'General' may have been obtained from literature discussed in this chapter, but mostly also or exclusively from one or more of the following online sources: ChemIDPlus Advanced, PubChem, DrugBank, RxList, Drug Central, Drugs.com, and Wikipedia
2   Angelini G, Meneghini CL. Oral tests in contact allergy to para-amino compounds. Contact Dermatitis 1981;7:311-314
3   Fisher AA. Systemic allergic dermatitis from Orinase ® and Diabinese ® in diabetics with para-amino hypersensitivity. Cutis 1982;29:551-565

Chapter 3.128 CHLORPROTHIXENE

IDENTIFICATION

Description/definition   : Chlorprothixene is the thioxanthine organochlorine compound that conforms to the
                             structural formula shown below
Pharmacological classes   : Dopamine antagonists; antipsychotic agents
IUPAC name   : (3Z)-3-(2-Chlorothioxanthen-9-ylidene)-N,N-dimethylpropan-1-amine
CAS registry number   : 113-59-7
EC number   : 204-032-8
Merck Index monograph   : 3464
Patch testing   : 1% water
Molecular formula   : $C_{18}H_{18}ClNS$

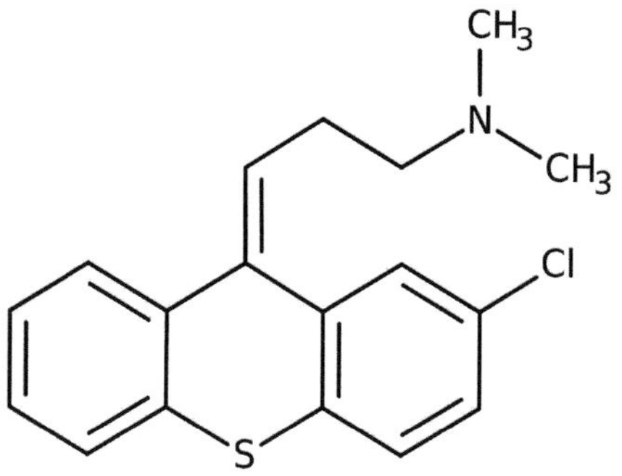

GENERAL

Chlorprothixene is a chemical of the thioxanthene (tricyclic) class with effects similar to the phenothiazine antipsy-chotics, but has a low antipsychotic potency, half to 2/3 of chlorpromazine. Its effects result from blocking the 5-HT_2 D_1, D_2, D_3, histamine H_1, muscarinic and α_1-adrenergic receptors. Chlorprothixene is indicated for treatment of psychotic disorders (e.g. schizophrenia) and of acute mania occurring as part of bipolar disorders. In pharmaceutical products, chlorprothixene is employed as chlorprothixene hydrochloride (CAS number 6469-93-8, EC number 229-289-3, molecular formula $C_{18}H_{19}Cl_2NS$) (1).

CONTACT ALLERGY FROM ACCIDENTAL CONTACT

A 24-year-old nurse working in the psychiatry ward complained of acute, intermittent skin reactions appearing around the eyes, on the cheek and the hands. Scratch tests with all the drugs that she handled at work showed no acute reactions, but a strongly positive reaction was seen after 24 hours to chlorprothixene with dermatitis appearing in the face and neck, which lasted for 3 days. Patch tests performed with chlorprothixene as syrup and coated tablet showed strong reactions after 24 h, lasting 3 days. Control subjects were negative. The nurse was diagnosed with occupational allergic – partly airborne - contact dermatitis from chlorprothixene. Quite curiously, the authors also wrote 'After handling chlorprothixene again by mistake, she revealed skin reactions *after a few minutes*'. Although this does not fit the symptomatology of contact dermatitis, the authors do not address this issue (2). A similar case of occupational allergic airborne contact dermatitis had been reported a few years previously, also from Germany (3).

An occupational *photo*allergic reaction to chlorprothixene in a nurse has apparently also been observed; she was known to be allergic to the chemically related chlorpromazine (4).

LITERATURE

1   The data in the section 'General' may have been obtained from literature discussed in this chapter, but mostly
    also or exclusively from one or more of the following online sources: ChemIDPlus Advanced, PubChem,
    DrugBank, RxList, Drug Central, Drugs.com, and Wikipedia
2   Lepp U, Schlaak M, Schulz KH. Contact dermatitis to chlorprothixene. Allergy 1998;53:718-719
3   Schwenck K, Gall H, Sterry W. Aerogene Kontaktdermatitis auf das Neuroleptikum Chlorprothixen. Allergologie
    1994;17:467-469
4   Kull E, Schwarz-Speck K. Gruppenspezifische Ekzemreaktionen bei Largactilsensibilisierung. Dermatologica
    (Basel) 1961;122:263-267

Chapter 3.129 CHLORTETRACYCLINE

IDENTIFICATION

Description/definition   : Chlortetracycline is a tetracycline with a 7-chloro substitution that conforms to the
                              structural formula shown below
Pharmacological classes   : Anti-bacterial agents; antiprotozoal agents; protein synthesis inhibitors
IUPAC name   : (4S,4aS,5aS,6S,12aR)-7-Chloro-4-(dimethylamino)-1,6,10,11,12a-pentahydroxy-6-methyl-
                              3,12-dioxo-4,4a,5,5a-tetrahydrotetracene-2-carboxamide
Other names   : Chlorotetracycline
CAS registry number   : 57-62-5
EC number   : 200-341-7
Merck Index monograph   : 3468
Patch testing   : Hydrochloride, 1% pet. (SmartPracticeCanada)
Molecular formula   : $C_{22}H_{23}ClN_2O_8$

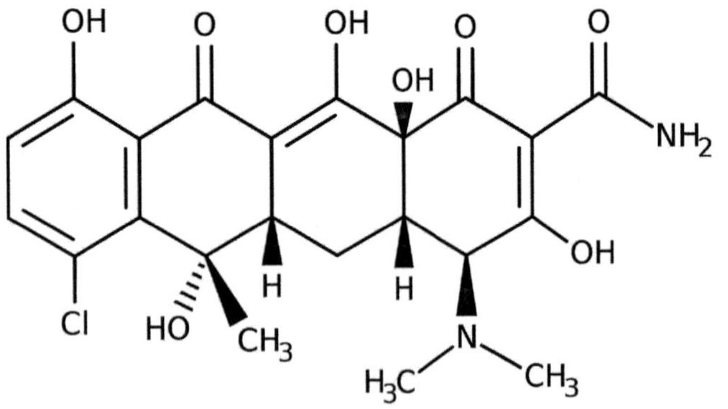

GENERAL

Chlortetracycline is a tetracycline antibiotic (the first tetracycline to be identified) isolated from an actinomycete named *Streptomyces aureofaciens*. The designated name of this microorganism and that of the isolated drug, Aureomycin, derive from their golden color. Chlortetracycline is currently used in the manufacturing of medicated animal feeds and as antibacterial agent in eye ointments. In pharmaceutical products, chlortetracycline is employed as chlortetracycline hydrochloride (CAS number 64-72-2, EC number 200-591-7, molecular formula $C_{22}H_{24}Cl_2N_2O_8$), probably as an ointment for eye infections only (1).

In topical preparations, chlortetracycline has rarely caused contact allergy/allergic contact dermatitis, which has been reviewed in Volume 3 of the *Monographs in contact allergy* series (4).

CONTACT ALLERGY FROM ACCIDENTAL CONTACT

One case of occupational allergic contact dermatitis to chlortetracycline in a pharmaceutical worker was reported from Warsaw, Poland, in 1984 (3).

Cross-reactions, pseudo-cross-reactions and co-reactions

A woman who had developed allergic contact dermatitis from an ointment containing chlortetracycline also reacted to demethylchlortetracycline, but not to oxytetracycline and tetracycline (2).

LITERATURE

1   The data in the section 'General' may have been obtained from literature discussed in this chapter, but mostly
      also or exclusively from one or more of the following online sources: ChemIDPlus Advanced, PubChem,
      DrugBank, RxList, Drug Central, Drugs.com, and Wikipedia
2   Calnan CD. Chlortetracycline sensitivity. Contact Dermatitis Newsletter 1967;1:16
3   Rudzki E, Rebendel P. Contact sensitivity to antibiotics. Contact Dermatitis 1984;11:41-42
4   De Groot AC. Monographs in contact allergy, volume 3: Topical Drugs. Boca Raton, Fl, USA: CRC Press Taylor and
      Francis Group, 2021 (ISBN 978-0-367-23693-9)

Chapter 3.130 CILASTATIN MIXTURE WITH IMIPENEM

IDENTIFICATION

Description/definition	: Cilastatin mixture with imipenem, is a mixture of these two chemicals; their structural formulas are shown below
Pharmacological classes	: Anti-bacterial agents; protease inhibitors
IUPAC name	: (Z)-7-[(2R)-2-Amino-2-carboxyethyl]sulfanyl-2-[[(1S)-2,2-dimethylcyclopropane-carbonyl]amino]hept-2-enoic acid;(5R,6S)-3-[2-(aminomethylideneamino)ethylsulfanyl]-6-[(1R)-1-hydroxyethyl]-7-oxo-1-azabicyclo[3.2.0]hept-2-ene-2-carboxylic acid
CAS registry number	: 92309-29-0
EC number	: Not available
Patch testing	: Tablet, pulverized, 20% pet. (3); most systemic drugs can be tested at 10% pet.; if the pure chemical is not available, prepare the test material from intravenous powder, the content of capsules or – when also not available – from powdered tablets to achieve a final concentration of the active drug of 10% pet.
Molecular formula	: $C_{28}H_{43}N_5O_9S_2$

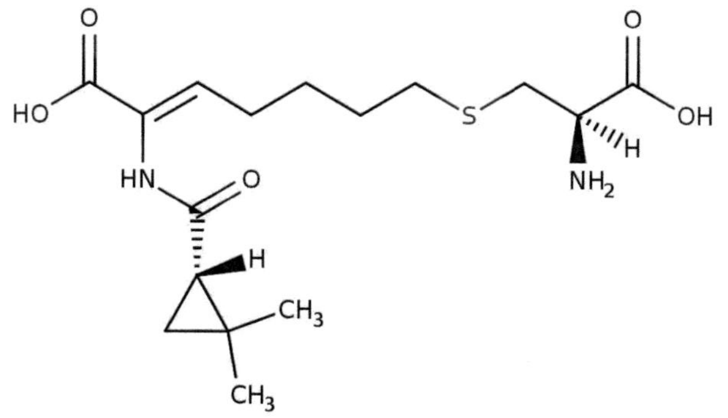

Cilastatin 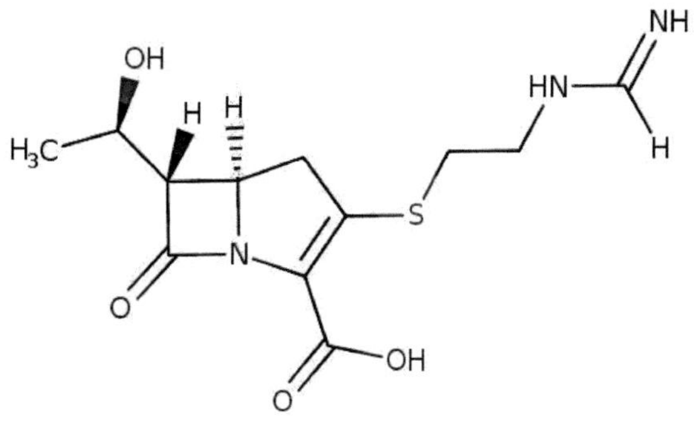   Imipenem

GENERAL

Cilastatin mixture with imipenem is a combination of imipenem and cilastatin that is used in the treatment of bacterial infections. Imipenem is indicated, in combination with cilastatin, for respiratory, skin, bone, gynecologic, urinary tract, and intra-abdominal infections as well as septicemia and endocarditis. Cilastatin inhibits renal dehydropeptidase I to prolong the half-life and increase the tissue penetration of imipenem, enhancing its efficacy as an anti-bacterial agent. In pharmaceutical products, imipenem is employed as imipenem monohydrate (CAS number 74431-23-5, EC number not available, molecular formula $C_{12}H_{19}N_3O_5S$) and cilastatin as cilastatin sodium (CAS number 81129-83-1, EC number 279-694-4, molecular formula $C_{16}H_{25}N_2NaO_5S$) (1).

See also Chapter 3.255 Imipenem.

CONTACT ALLERGY FROM ACCIDENTAL CONTACT

A 28-year-old nurse developed eczema on the dorsal aspect of the hand and the face. She worked in the hematology department where she usually handled and administered a variety of antibiotics. When she was moved to a different department where she did not have contact with these drugs, the dermatitis completely resolved. Patch tests were positive to ertapenem 20% pet., imipenem-cilastatin (20% pet.), ampicillin (25% pet.), piperacillin (20% pet.), mezlocillin (20% pet.), and meropenem (20% pet.). It was concluded that the patient had contact allergy to carbapenems (ertapenem, imipenem, and meropenem) and semisynthetic penicillins (piperacillin, mezlocillin, and ampicillin). At work, the patient had contact only with imipenem, ertapenem, and piperacillin, so it may be assumed that the other reactions were due to cross-reactivity (3).

CUTANEOUS ADVERSE DRUG REACTIONS FROM SYSTEMIC ADMINISTRATION CAUSED BY TYPE IV
(DELAYED-TYPE) HYPERSENSITIVITY (as demonstrated by positive patch tests)

Cutaneous adverse drug reactions from systemic administration of cilastatin mixture with imipenem caused by type IV (delayed-type) hypersensitivity have included maculopapular eruption (7), drug reaction with eosinophilia and systemic symptoms (DRESS) (2). There have also been cases of allergic contact dermatitis from the improper topical use of this pharmaceutical (4,6).

Maculopapular eruption
In Bern, Switzerland, patients with a suspected allergic cutaneous drug reaction were patch-scratch tested with suspected drugs that had previously given a positive lymphocyte transformation test. Imipenem-cilastatin 750 mg/ml saline gave a positive patch-scratch test in one patient with maculopapular exanthema (7).

Drug reaction with eosinophilia and systemic symptoms (DRESS)
In a multicenter investigation in France, of 72 patients patch tested for DRESS, 46 (64%) had positive patch tests to drugs, including one to cilastatin mixture with imipenem (in combination with vancomycin). This was a man aged 54, but clinical details were not provided (2).

Other cutaneous adverse drug reactions
Allergic contact dermatitis from imipenem in the mixture has been observed from the improper topical administration to leg ulcers of the powder and saline intended for parenteral administration (4,6).

Cross-reactions, pseudo-cross-reactions and co-reactions
Cross-reactions between beta-lactam antibiotics are discussed in Chapter 3.36 Amoxicillin. Cross-reactions between carbapenems (imipenem, ertapenem, meropenem) and between carbapenems and penicillins may rarely occur, but the pattern is unclear (3,5).

LITERATURE
1   The data in the section 'General' may have been obtained from literature discussed in this chapter, but mostly also or exclusively from one or more of the following online sources: ChemIDPlus Advanced, PubChem, DrugBank, RxList, Drug Central, Drugs.com, and Wikipedia
2   Barbaud A, Collet E, Milpied B, Assier H, Staumont D, Avenel-Audran M, et al. A multicentre study to determine the value and safety of drug patch tests for the three main classes of severe cutaneous adverse drug reactions. Br J Dermatol 2013;168:555-562
3   Colagiovanni A, Feliciani C, Fania L, Pascolini L, Buonomo A, Nucera E, et al. Occupational contact dermatitis from carbapenems. Cutis 2015;96:E1-3
4   Foti C, Bonamonte D, Daddabbo M, Angelini G. Delayed and immediate reactions to imipenem. Contact Dermatitis 2001;45:112-113
5   Schiavino D, Nucera E, Lombardo C, Decinti M, Pascolini L, Altomonte G, et al. Cross-reactivity and tolerability of imipenem in patients with delayed-type, cell-mediated hypersensitivity to beta-lactams. Allergy 2009;64:1644-1648
6   Foti C, Cassano N, Vena GA, Angelini G, Calogiuri G. Generalized eczematous dermatitis from imipenem after topical exposure. Dermatitis 2011;22:119-120
7   Neukomm C, Yawalkar N, Helbling A, Pichler WJ. T-cell reactions to drugs in distinct clinical manifestations of drug allergy. J Invest Allergol Clin Immunol 2001;11:275-284

Chapter 3.131 CIMETIDINE

IDENTIFICATION

Description/definition : Cimetidine is the histamine congener that conforms to the structural formula shown below

Pharmacological classes : Histamine H_2 antagonists; cytochrome P-450 CYP1A2 inhibitors; anti-ulcer agents

IUPAC name : 1-Cyano-2-methyl-3-[2-[(5-methyl-1H-imidazol-4-yl)methylsulfanyl]ethyl]guanidine

CAS registry number : 51481-61-9

EC number : 257-232-2

Merck Index monograph : 3552

Patch testing : Tablet, pulverized, 10% water (2); most systemic drugs can be tested at 10% pet.; if the pure chemical is not available, prepare the test material from intravenous powder, the content of capsules or – when also not available – from powdered tablets to achieve a final concentration of the active drug of 10% pet.

Molecular formula : $C_{10}H_{16}N_6S$

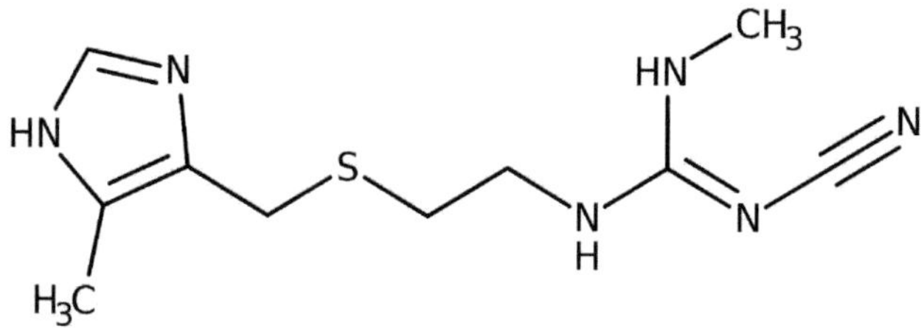

GENERAL

Cimetidine is a histamine H_2 receptor antagonist, which inhibits gastric acid secretion, as well as pepsin and gastrin output. This agent is indicated for the treatment and the management of acid-reflux disorders, peptic ulcer disease, heartburn, and acid indigestion. In pharmaceutical products, both cimetidine base (tablets) and cimetidine hydrochloride (CAS number 70059-30-2, EC number 274-297-2, molecular formula $C_{10}H_{17}ClN_6S$) (injection material) may be employed (1).

CUTANEOUS ADVERSE DRUG REACTIONS FROM SYSTEMIC ADMINISTRATION CAUSED BY TYPE IV (DELAYED-TYPE) HYPERSENSITIVITY (as demonstrated by positive patch tests)

Cutaneous adverse drug reactions from systemic administration of cimetidine caused by type IV (delayed-type) hypersensitivity have included maculopapular eruption (2).

Maculopapular eruption

A 44-year-old man with psoriasis had been treated for 20 days with oral cimetidine 1000 mg daily for a duodenal ulcer, when he became febrile and developed a pruritic maculopapular rash beginning on the palms and becoming almost generalised within 24 hours. Cimetidine was withdrawn after 22 days of treatment. The rash was succeeded by a universal outbreak of psoriasis. Two months later, he was patch tested with aqueous cimetidine (commercial tablets), the tablet's excipients and a major cimetidine metabolite and had positive reactions to cimetidine 100 mg/ml, 10 mg/ml, but there were no reactions to the drug tested at 1 mg/ml and 0.1 mg/ml. Eleven controls were negative (2).

Cross-reactions, pseudo-cross-reactions and co-reactions

A patient with DRESS caused by ranitidine may have cross-reacted to cimetidine (3).

LITERATURE

1   The data in the section 'General' may have been obtained from literature discussed in this chapter, but mostly also or exclusively from one or more of the following online sources: ChemIDPlus Advanced, PubChem, DrugBank, RxList, Drug Central, Drugs.com, and Wikipedia

2   Peters K. Delayed hypersensitivity to oral cimetidine. Contact Dermatitis 1986;15:190-191

3   Juste S, Blanco J, Garcés M, Rodriguez G. Allergic dermatitis due to oral ranitidine. Contact Dermatitis 1992;27:339-340

Chapter 3.132 CIPROFLOXACIN

IDENTIFICATION

Description/definition	: Ciprofloxacin is the broad-spectrum antimicrobial carboxyfluoroquinoline that conforms to the structural formula shown below
Pharmacological classes	: Cytochrome P-450 CYP1A2 inhibitors; topoisomerase II inhibitors; anti-bacterial agents
IUPAC name	: 1-Cyclopropyl-6-fluoro-4-oxo-7-(piperazin-1-yl)-1,4-dihydroquinoline-3-carboxylic acid
CAS registry number	: 85721-33-1
EC number	: 617-751-0
Merck Index monograph	: 3583
Patch testing	: Hydrochloride, 10% pet. (Chemotechnique)
Molecular formula	: $C_{17}H_{18}FN_3O_3$

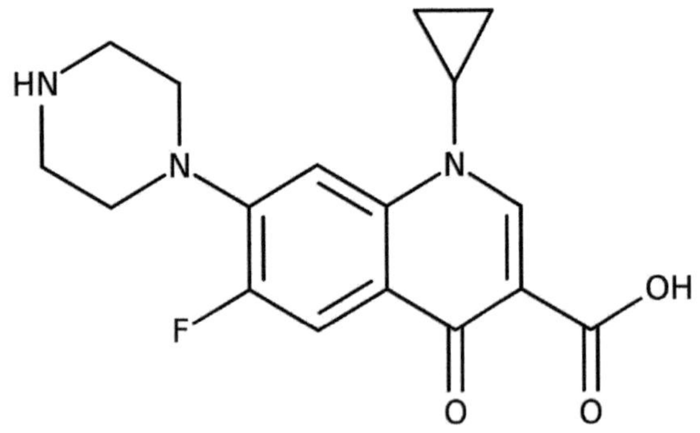

GENERAL

Ciprofloxacin is a synthetic broad-spectrum fluoroquinolone antibiotic. It is indicated for the treatment of infections of the urinary tract, lower respiratory tract, sinuses, skin and soft tissues, bone and joint, and for complicated intra-abdominal infections, infectious diarrhea, typhoid fever, cervical and urethral gonorrhea, and inhalational anthrax (post-exposure), when caused by susceptible microorganisms (1). In pharmaceutical products, ciprofloxacin is usually employed as ciprofloxacin hydrochloride (hydrate) (CAS number 86393-32-0, EC number not available, molecular formula $C_{17}H_{21}ClFN_3O_4$) (1).

CUTANEOUS ADVERSE DRUG REACTIONS FROM SYSTEMIC ADMINISTRATION CAUSED BY TYPE IV (DELAYED-TYPE) HYPERSENSITIVITY (as demonstrated by positive patch tests)

Cutaneous adverse drug reactions from systemic administration of ciprofloxacin caused by type IV (delayed-type) hypersensitivity have included maculopapular eruption (9,11), fixed drug eruption (3,4), acute generalized exanthematous pustulosis (AGEP) (6,7,9,11), drug reaction with eosinophilia and systemic symptoms (DRESS) (9), photosensitivity (5), and delayed urticaria (9),

Case series with various or unknown types of drug reactions

In the period 2000-2014, in Portugal, 260 patients were patch tested with antibiotics for suspected cutaneous adverse drug reactions (CADR) to these drugs. 56 patients (22%) had one or more (often from cross-reactivity) positive patch tests. Ciprofloxacin was patch test positive in 1 patient with maculopapular eruption, in one with DRESS, in one with AGEP and in one with delayed urticaria (9).

Maculopapular eruption

A 61-year-old woman developed an exanthema (macular or maculopapular, not specified) on the trunk and extremities 3 days after starting ciprofloxacin for cystitis. Six months later, she was given norfloxacin as treatment for a recurrent cystitis and, within hours after drug exposure, developed a generalized exanthema. Patch tests were positive to ciprofloxacin and norfloxacin pulverized tablets 10% or 25% pet. (not specified) at D1 and/or D2 (not specified). Lymphocyte transformation tests (LTT) were positive to both drugs (11). A second patient presented by these authors was a 22-year-old woman given ciprofloxacin for treatment for pyelonephritis. Within 3 days after the first intake, she developed an itching, maculopapular drug eruption on the back. One year later she was treated with norfloxacin against a recurrent pyelonephritis. Two to 3 days later, generalized flush reaction appeared. Patch tests were positive to ciprofloxacin (plus a positive LTT), but negative to norfloxacin (11).

      See also the section 'Case series with various or unknown types of drug reactions' above, ref. 9.

Acute generalized exanthematous pustulosis (AGEP)

An 80-year-old woman presented with a generalized erythematous rash, livid erythema-multiforme-like lesions on all acral sites, and partly solitary partly confluent non-follicular pustules forming flaccid suppurative bullae on the trunk and in the main folds, which had developed after she had taken oral ciprofloxacin to treat a urinary tract infection. Her temperature was 39.3°C and the laboratory investigations disclosed results consistent with pustular drug eruption or sepsis. Skin biopsy showed spongiform subcorneal and intraepidermal pustules, marked edema in the papillary dermis and a mixed cellular infiltrate rich in neutrophils. Overall, the clinical course, microbiological and histopathological results were consistent with acute generalized exanthematous pustulosis (AGEP). Patch tests were positive to ciprofloxacin with pustules and corresponding histology (6).

A 65-year-old female patient had suffered several flares of exanthematous pustulosis, some of them attributable to pustular psoriasis and others that were drug-induced, resembling acute generalized exanthematous pustulosis. Two severe flares were induced by ciprofloxacin. Patch tests yielded positive results to ciprofloxacin and to other quinolones (unknown which) and reproduced the original lesional pattern both clinically and histologically (7).

An 80-year-old woman had been treated for 7 days with ciprofloxacin for 7 days, when she developed AGEP on her whole body with the exception of the face. A patch test was positive to ciprofloxacin pulverized tablets 10% or 25% pet. (not specified) at D1 and/or D2 (not specified). A lymphocyte transformation test was also positive (11).

See also the section 'Case series with various or unknown types of drug reactions' above, ref. 9.

Fixed drug eruption

A 44-year-old female pharmacy technician had a history of an episodic, circumferential, dusky red bullous eruption of the left fourth digit with associated burning and pruritus. Outbreaks had occurred every few months over the past 2 years, always involving the left fourth digit. Between flares, a dusky blue grey hyperpigmented patch was present over the area. A biopsy was consistent with fixed drug eruption. The culprit drug was ciprofloxacin, as shown by erythema and burning a few hours after application of a crushed ciprofloxacin tablet 30% in pet. and saline. The patient had taken ciprofloxacin each time before a flare for recurrent urinary tract infection (3).

A 28-year-old woman, 8 hours after a 400 mg dose of norfloxacin, developed pruritic erythematous macules on the dorsum of both hands, with subsequent residual pigmentation. A 2nd such episode, 1 year later, developed 2 hours after 250 mg ciprofloxacin, with reappearance of the old and new lesions. Prick and patch tests with norfloxacin, ciprofloxacin and other quinolones showed a positive patch test to ciprofloxacin 10% pet. on post-lesional skin only with pruritic erythematous macules and vesicles at D2 (4).

Drug reaction with eosinophilia and systemic symptoms (DRESS)

See the section 'Case series with various or unknown types of drug reactions' above, ref. 9.

Photosensitivity

In the period 2004-2005, in a Spanish multicenter study performing photopatch testing, one relevant positive photopatch test was observed to ciprofloxacin. It was not mentioned how many patients had been tested with this chemical and no clinical details were provided (5).

Other cutaneous adverse drug reactions

See the section 'Case series with various or unknown types of drug reactions' above, ref. 9 (delayed urticaria).

Cross-reactions, pseudo-cross-reactions and co-reactions

Ciprofloxacin may have *photo*cross-reacted to primary lomefloxacin *photo*sensitization (2). A patient with a fixed drug eruption from ofloxacin had positive patch tests to both ofloxacin and ciprofloxacin (8). A patient sensitized to norfloxacin and presenting with fixed drug eruption cross-reacted to ciprofloxacin (10). A patient sensitized to ciprofloxacin probably cross-reacted to norfloxacin (11).

LITERATURE

1   The data in the section 'General' may have been obtained from literature discussed in this chapter, but mostly also or exclusively from one or more of the following online sources: ChemIDPlus Advanced, PubChem, DrugBank, RxList, Drug Central, Drugs.com, and Wikipedia
2   Kimura M, Kawada A. Photosensitivity induced by lomefloxacin with cross-photosensitivity to ciprofloxacin and fleroxacin. Contact Dermatitis 1998;38:180
3   Hanson JL, Warshaw EM. A method for at-home lesional testing for fixed drug eruption. Dermatitis 2015;26:148
4   Rodríguez-Morales A, Llamazares AA, Benito RP, Cócera CM. Fixed drug eruption from quinolones with a positive lesional patch test to ciprofloxacin. Contact Dermatitis 2001;44:255
5   De La Cuadra-Oyanguren J, Perez-Ferriols A, Lecha-Carrelero M, et al. Results and assessment of photopatch testing in Spain: towards a new standard set of photoallergens. Actas DermoSifiliograficas 2007;98:96-101

6   Hausermann P, Scherer K, Weber M, Bircher AJ. Ciprofloxacin-induced acute generalized exanthematous
     pustulosis mimicking bullous drug eruption confirmed by a positive patch test. Dermatology 2005;211:277-280
7   Serra D, Gonçalo M, Mariano A, Figueiredo A. Pustular psoriasis and drug-induced pustulosis. G Ital Dermatol
     Venereol 2011;146:155-158 (Article in Italian)
8   Kawada A, Hiruma M, Noguchi H, Banba K, Ishibashi A, Banba H, et al. Fixed drug eruption induced by ofloxacin.
     Contact Dermatitis 1996;34:427
9   Pinho A, Coutinho I, Gameiro A, Gouveia M, Gonçalo M. Patch testing - a valuable tool for investigating non-
     immediate cutaneous adverse drug reactions to antibiotics. J Eur Acad Dermatol Venereol 2017;31:280-287
10   Alpalhão M, Antunes J, Soares-Almeida L, Correia TE, Filipe P. Fixed drug eruption due to norfloxacin with cross-
     reactivity to ciprofloxacin: A case report. Contact Dermatitis 2020;83:135-137
11   Schmid DA, Depta JP, Pichler WJ. T cell-mediated hypersensitivity to quinolones. Clin Exp Allergy 2006;36:59-69

Chapter 3.133 CISPLATIN

IDENTIFICATION

Description/definition : Cisplatin is the inorganic and water-soluble platinum complex that conforms to the structural formula shown below
Pharmacological classes : Antineoplastic agents
IUPAC name : Azanide;dichloroplatinum(2+)
Other names : *cis*-Diaminedichloroplatinum; platinum, diamminedichloro-, (SP-4-2)-; cisplatinum
CAS registry number : 15663-27-1
EC number : 239-733-8
Merck Index monograph : 3586
Patch testing : No data available
Molecular formula : $Cl_2H_6N_2Pt$

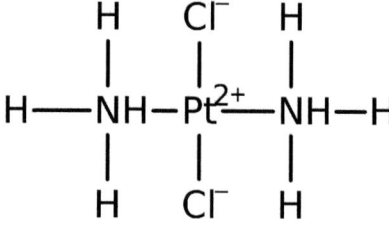

GENERAL

Cisplatin is an inorganic and water-soluble platinum complex chemotherapy drug. It is indicated for the treatment of metastatic testicular tumors, metastatic ovarian tumors and advanced bladder cancer, but has also been used to treat various other types of cancers, including sarcomas, some carcinomas (e.g. small cell lung cancer), and lymphomas (1).

CONTACT ALLERGY FROM ACCIDENTAL CONTACT

One or more cases of occupational allergic contact dermatitis to cisplatin may have occurred in the 1980s, possibly in hospital personnel (2,3).

Immediate contact reactions

Immediate contact reactions (contact urticaria) to cisplatin are presented in Chapter 5.

LITERATURE

1   The data in the section 'General' may have been obtained from literature discussed in this chapter, but mostly also or exclusively from one or more of the following online sources: ChemIDPlus Advanced, PubChem, DrugBank, RxList, Drug Central, Drugs.com, and Wikipedia
2   Fisher AA. Contact dermatitis due to anticancer drugs in hospital personnel. Am J Contact Dermatitis 1991;2:73
3   Rogers B. Health hazards of personnel handling antineoplastic agents. Occup Med State of Art Rev 1987;2:513-524

Chapter 3.134 CITALOPRAM

IDENTIFICATION

Description/definition   : Citalopram is the bicyclic phthalene derivative that conforms to the structural formula
                               shown below
Pharmacological classes   : Serotonin uptake inhibitors; antidepressive agents, second-generation
IUPAC name   : 1-[3-(Dimethylamino)propyl]-1-(4-fluorophenyl)-3H-2-benzofuran-5-carbonitrile
CAS registry number   : 59729-33-8
EC number   : 261-891-1
Merck Index monograph : 3587
Patch testing   : No data available; most systemic drugs can be tested at 10% pet.; if the pure chemical is
                            not available, prepare the test material from intravenous powder, the content of capsules
                            or – when also not available – from powdered tablets to achieve a final concentration of
                            the active drug of 10% pet.
Molecular formula   : $C_{20}H_{21}FN_2O$

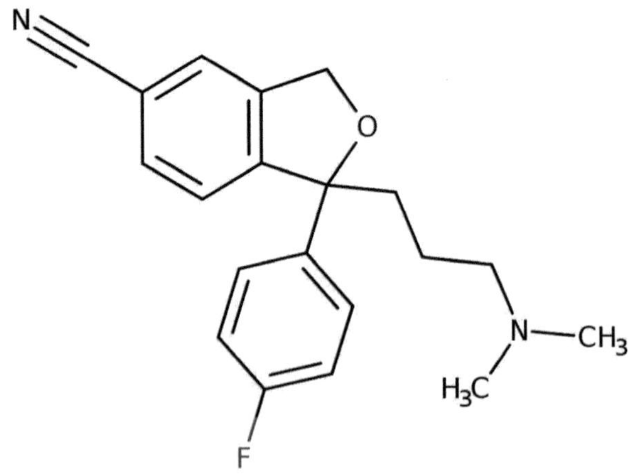

GENERAL

Citalopram belongs to a class of antidepressants known as selective serotonin reuptake inhibitors (SSRIs). It has been found to relieve or manage symptoms of depression, anxiety, eating disorders and obsessive-compulsive disorder among other mood disorders. Citalopram selectively inhibits the neuronal reuptake of the neurotransmitter serotonin (5-HT) in presynaptic cells in the central nervous system, thereby increasing levels of 5-HT within the synaptic cleft and enhancing the actions of serotonin on its receptors. Increased serotonergic neurotransmission results in antidepressive and anxiolytic effects (1).

In pharmaceutical products, citalopram is employed as citalopram hydrobromide (CAS number 59729-32-7, EC number 261-890-6, molecular formula $C_{20}H_{22}BrFN_2O$) (1).

CUTANEOUS ADVERSE DRUG REACTIONS FROM SYSTEMIC ADMINISTRATION CAUSED BY TYPE IV (DELAYED-TYPE) HYPERSENSITIVITY (as demonstrated by positive patch tests)

Cutaneous adverse drug reactions from systemic administration of citalopram caused by type IV (delayed-type) hypersensitivity have included drug reaction with eosinophilia and systemic symptoms (DRESS) (2).

Drug reaction with eosinophilia and systemic symptoms (DRESS)

In a multicenter investigation in France, of 72 patients patch tested for DRESS, 46 (64%) had positive patch tests to drugs, including one to citalopram. Clinical and patch testing details were not provided (2).

LITERATURE

1   The data in the section 'General' may have been obtained from literature discussed in this chapter, but mostly
    also or exclusively from one or more of the following online sources: ChemIDPlus Advanced, PubChem,
    DrugBank, RxList, Drug Central, Drugs.com, and Wikipedia
2   Barbaud A, Collet E, Milpied B, Assier H, Staumont D, Avenel-Audran M, et al. A multicentre study to determine
    the value and safety of drug patch tests for the three main classes of severe cutaneous adverse drug reactions.
    Br J Dermatol 2013;168:555-562

Chapter 3.135 CITIOLONE

IDENTIFICATION

Description/definition   : Citiolone is the derivative of the amino acid cysteine that conforms to the structural
                               formula shown below
Pharmacological classes   : Mucolytic drugs
IUPAC name   : N-(2-Oxothiolan-3-yl)acetamide
Other names   : 2-Acetamido-4-mercaptobutyric acid thiolactone; N-(tetrahydro-2-oxothienyl)acetamide;
                               N-acetylhomocysteine thiolactone
CAS registry number   : 1195-16-0
EC number   : 214-793-8
Merck Index monograph : 3589
Patch testing   : 10% DMSO
Molecular formula   : $C_6H_9NO_2S$

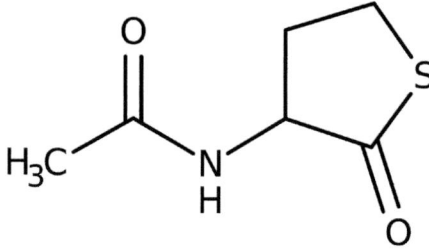

GENERAL

Citiolone is a drug that is or was used, especially in Spain, as a mucolytic agent and in the treatment of certain hepatic disorders. It is a derivative of the amino acid cysteine. Citiolone is probably not used anymore (1).

CUTANEOUS ADVERSE DRUG REACTIONS FROM SYSTEMIC ADMINISTRATION CAUSED BY TYPE IV (DELAYED-TYPE) HYPERSENSITIVITY (as demonstrated by positive patch tests)

Cutaneous adverse drug reactions from systemic administration of citiolone caused by type IV (delayed-type) hypersensitivity have included fixed drug eruption (3).

Fixed drug eruption

A 7-year-old girl was treated with citiolone for a viral respiratory infection. Five hours after taking citiolone 300 mg orally, she experienced pruritic erythema, with formation of 2 blisters, followed by two dark-brown macules, 3.5 and 4 centimeter in diameter, respectively on her chin and left upper leg. Patch testing with citiolone 10% in DMSO was positive on postlesional but not on normal skin and caused a flare-up of the eruption. An oral challenge with 100 mg citiolone produced pruritic erythema at the affected sites within 3 hours (2).

LITERATURE

1   The data in the section 'General' may have been obtained from literature discussed in this chapter, but mostly also or exclusively from one or more of the following online sources: ChemIDPlus Advanced, PubChem, DrugBank, RxList, Drug Central, Drugs.com, and Wikipedia
2   Delgado PG, Florido Lopez F, de San Pedro BS. Fixed drug eruption due to citiolone. Contact Dermatitis 1995;33:352

Chapter 3.136 CLARITHROMYCIN

IDENTIFICATION

Description/definition	: Clarithromycin is a semisynthetic macrolide antibiotic derived from erythromycin that conforms to the structural formula shown below
Pharmacological classes	: Anti-bacterial agents; protein synthesis inhibitors; cytochrome P-450 CYP3A inhibitors
IUPAC name	: (3R,4S,5S,6R,7R,9R,11R,12R,13S,14R)-6-[(2S,3R,4S,6R)-4-(Dimethylamino)-3-hydroxy-6-methyloxan-2-yl]oxy-14-ethyl-12,13-dihydroxy-4-[(2R,4R,5S,6S)-5-hydroxy-4-methoxy-4,6-dimethyloxan-2-yl]oxy-7-methoxy-3,5,7,9,11,13-hexamethyl-oxacyclotetradecane-2,10-dione
Other names	: 6-O-Methylerythromycin
CAS registry number	: 81103-11-9
EC number	: Not available
Merck Index monograph	: 3608
Patch testing	: 10.0% pet. (Chemotechnique)
Molecular formula	: $C_{38}H_{69}NO_{13}$

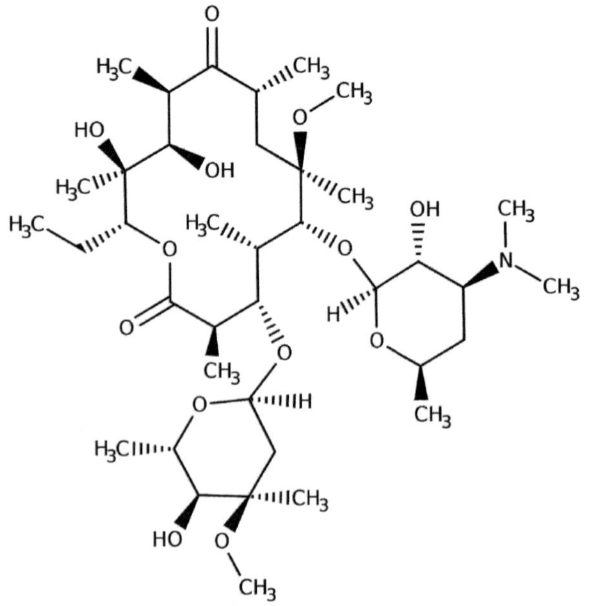

GENERAL

Clarithromycin is a semisynthetic macrolide antibiotic derived from erythromycin that is active against a variety and broad spectrum of microorganisms. Indications for its use include acute otitis media, pharyngitis and tonsillitis, respiratory tract infections, Legionnaires' disease, pertussis, skin or skin structure infections, *Helicobacter pylori* infection, duodenal ulcer disease, *Bartonella* infections, early Lyme disease, and encephalitis caused by *Toxoplasma gondii*. Clarithromycin may also decrease the incidence of cryptosporidiosis, prevent the occurrence of α-hemolytic (viridans group) streptococcal endocarditis, as well as serve as a primary prevention for *Mycobacterium avium* complex (MAC) bacteremia or disseminated infections (1).

CUTANEOUS ADVERSE DRUG REACTIONS FROM SYSTEMIC ADMINISTRATION CAUSED BY TYPE IV (DELAYED-TYPE) HYPERSENSITIVITY (as demonstrated by positive patch tests)

Cutaneous adverse drug reactions from systemic administration of clarithromycin caused by type IV (delayed-type) hypersensitivity have included fixed drug eruption (2), drug reaction with eosinophilia and systemic symptoms (DRESS) (3,4), and symmetrical drug-related intertriginous and flexural exanthema (SDRIFE) (5).

Symmetrical drug-related intertriginous and flexural exanthema (SDRIFE)/Baboon syndrome

A 43-year-old man developed an extensive skin rash after 7 days' first-time use of clarithromycin for a respiratory infection. Examination showed a symmetrical and sharply demarcated erythema on the buttocks and both inner thighs, with a V-shaped pattern, also affecting the lower abdomen and cubital and axillary folds. The patient had no fever and blood cell count and serum biochemistry analysis were normal. Two months after complete recovery, patch tests were positive to clarithromycin 1%, 5%, and 10% petrolatum. The patient was diagnosed with symmetrical drug-related intertriginous and flexural exanthema (SDRIFE) of the baboon syndrome type (5).

Fixed drug eruption

A 58-year-old man, on the third day of treatment with clarithromycin for tonsillitis, developed multiple sharply circumscribed violaceous erythematous patches of various shapes and sizes (1-3 cm), some with a central blister, on the head, trunk and upper arms and legs. Lesions improved in 6-7 days just with discontinuation of the drug. One month later, hyperpigmentation with a red-bluish coloration was observed at the same sites. Patch testing with clarithromycin 10% water gave an erythematous and vesicular reaction on a residual lesion but not on uninvolved skin. There was a flare-up of the fixed drug eruption in previously involved skin. Twenty controls were negative (2).

Drug reaction with eosinophilia and systemic symptoms (DRESS)

One patient developed DRESS from clarithromycin with a positive patch test to this drug tested 10% pet. (3, possibly the same patient as in ref. 4). In a group of 14 patients with multiple delayed-type hypersensitivity reactions, DRESS was caused by clarithromycin in one case, showing a positive patch test reaction (4, possibly the same patient as in ref. 3).

LITERATURE

1   The data in the section 'General' may have been obtained from literature discussed in this chapter, but mostly also or exclusively from one or more of the following online sources: ChemIDPlus Advanced, PubChem, DrugBank, RxList, Drug Central, Drugs.com, and Wikipedia

2   Rosina P, Chieregato C, Schena D. Fixed drug eruption from clarithromycin. Contact Dermatitis 1998;38:105

3   Jörg L, Helbling A, Yerly D, Pichler WJ. Drug-related relapses in drug reaction with eosinophilia and systemic symptoms (DRESS). Clin Transl Allergy 2020;10:52

4   Jörg L, Yerly D, Helbling A, Pichler W. The role of drug, dose and the tolerance/intolerance of new drugs in multiple drug hypersensitivity syndrome (MDH). Allergy 2020;75:1178-1187

5   Moreira C, Cruz MJ, Cunha AP, Azevedo F. Symmetrical drug-related intertriginous and flexural exanthema induced by clarithromycin. An Bras Dermatol 2017;92:587-588

Chapter 3.137 CLAVULANIC ACID

IDENTIFICATION

Description/definition : Clavulanic acid is a β-lactam antibiotic produced by the actinobacterium *Streptomyces clavuligerus*; it conforms to the structural formula shown below
Pharmacological classes : β-Lactamase inhibitors
IUPAC name : (2*R*,3*Z*,5*R*)-3-(2-Hydroxyethylidene)-7-oxo-4-oxa-1-azabicyclo[3.2.0]heptane-2-carboxylic acid
CAS registry number : 58001-44-8
EC number : 261-069-2
Merck Index monograph : 3609
Patch testing : Potassium clavulanate 10.0% pet. (Chemotechnique)
Molecular formula : $C_8H_9NO_5$

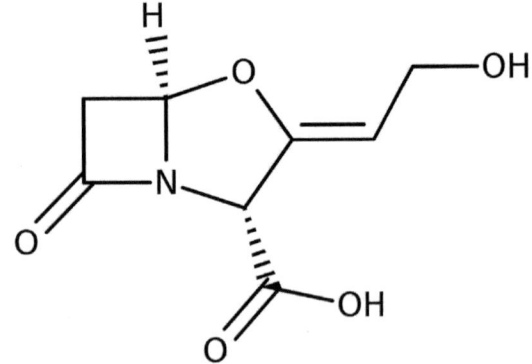

GENERAL

Clavulanic acid is a β-lactam antibiotic produced by the actinobacterium *Streptomyces clavuligerus*. It is a suicide inhibitor of bacterial β-lactamase enzymes. Administered alone, it has only weak antibacterial activity against most organisms, but given in combination with other β-lactam antibiotics, notably with amoxicillin trihydrate (under the trade name Augmentin), it prevents antibiotic inactivation by β-lactamase producing organisms. Clavulanic acid is indicated for use with amoxicillin in the treatment of infections with *S. aureus* and *Bacteroides fragilis*, or with β-lactamase-producing strains of *H. influenzae* and *E. coli* (1). In pharmaceutical products, clavulanic acid is employed as clavulanate potassium (CAS number 61177-45-5, EC number 262-640-9, molecular formula $C_8H_8KNO_5$) (1). The classification and structures of beta-lactam antibiotics are discussed in Chapter 3.36 Amoxicillin.

See also Chapter 3.36 Amoxicillin and Chapter 3.37 Amoxicillin mixture with clavulanate potassium. Allergic reactions to the latter caused by amoxicillin or when the culprit ingredient was not identified are presented in Chapter 3.37 Amoxicillin mixture with clavulanate potassium; cases where clavulanic acid was the allergenic component are discussed here.

CONTACT ALLERGY FROM ACCIDENTAL CONTACT

A 35-year-old nurse developed pruritic, erythematous patches on both hands and forearms 2 days after she had contact with the mixture of amoxicillin and clavulanic acid dry syrup and distilled water. Patch tests were positive to the dry syrup 10, 1, and 0.1 mg/ml in water at D2 and D4. When tested later with all ingredients, there was a positive patch test to clavulanic acid (10 mg/ml) at D2 (++) and D4 (++). Controls were not performed. The patient was diagnosed with occupational allergic contact dermatitis from clavulanic acid (5).

A 28-year-old nurse had a 1-year history of intense and refractory hand eczema that resolved almost completely when she was out of work for more than a week. Patch tests showed a positive reaction to potassium clavulanate 10% pet. on D3 and D7. There were no reactions to other β-lactams such as amoxicillin, benzylpenicillin, carbapenems, and cephalosporins. After avoiding exposure to clavulanic acid in her occupational setting, there was total resolution of the dermatitis (9).

CUTANEOUS ADVERSE DRUG REACTIONS FROM SYSTEMIC ADMINISTRATION CAUSED BY TYPE IV (DELAYED-TYPE) HYPERSENSITIVITY (as demonstrated by positive patch tests)

Cutaneous adverse drug reactions from systemic administration of clavulanic acid caused by type IV (delayed-type) hypersensitivity have included maculopapular eruption (3,7,8,9), generalized itchy erythema (4), acute generalized exanthematous pustulosis (AGEP) (2), and urticarial rash (6).

Maculopapular eruption

A 65-year-old man developed a generalized maculopapular exanthema and facial angioedema on the third day after starting amoxicillin-clavulanic acid treatment for bronchitis. The symptoms resolved completely in 1 month. Patch tests were positive to clavulanic acid (no details provided) and negative to amoxicillin (3).

In London, between October 2017 and October 2018, 45 patients with suspected cutaneous adverse drug reactions, including 33 maculopapular eruptions (MPE), 4 fixed drug eruptions (FDE), 4 DRESS, 3 AGEP and one SJS/TEN, were patch tested with the suspected drugs. There were 10 (22%) positive patch test cases: 4 MPE, 2 FDE, 3 DRESS and 1 AGEP. Clavulanic acid (in amoxicillin-clavulanic acid), tested as commercial preparation 30% pet., was responsible for one case of maculopapular eruption (7).

A 46-year-old woman was treated with amoxicillin-clavulanic acid for cellulitis. The next day, the patient developed erythematous macules and papules mainly on the right aspect of her abdomen and the right inguinal region, extending to the right thigh. Histopathology showed perivascular infiltrates of lymphocytes and eosinophils in the upper dermis with invasion of lymphocytes into the epidermis, suggesting a drug eruption. Patch tests were positive to amoxicillin-clavulanic acid, but not to amoxicillin or the other drugs used by the patient. An oral challenge with amoxicillin was negative, indicating clavulanic acid as the culprit sensitizer (8).

A 52-year-old man developed a maculopapular eruption mainly involving the trunk and arms 14 days after taking oral pantoprazole and amoxicillin-clavulanic acid for pneumonia. There were no systemic symptoms or significant changes in blood tests and the reaction faded away in 6 days without oral steroids. Patch testing performed 2 months later were positive to potassium clavulanate 10% pet. on D3 and D7 with a negative reaction to pantoprazole 10% pet. (9).

Erythroderma, widespread erythematous eruption, exfoliative dermatitis

A 34-year-old woman received a 7-day course of amoxicillin-clavulanic acid (875/125 mg) (AX/CLV) for intercurrent dental surgery. Two days after stopping the treatment she complained of generalized itchy erythema. This reaction disappeared within a few days. Skin prick tests, intradermal tests and patch tests were negative to penicillin and amoxicillin. An oral provocation test with amoxicillin was negative. However, prick and intradermal tests with AX/CLV were positive at D2 and D3. Clavulanic acid was not available for testing as single drug, so the – very likely correct – diagnosis of drug rash from delayed-type hypersensitivity to clavulanic acid was made *per exclusionem* (4).

Acute generalized exanthematous pustulosis (AGEP)

A 38-year-old man was treated with amoxicillin-clavulanic acid (AX/CLV) 875/125 mg twice a day for tonsillitis. On the second day of treatment he developed fever and a rapidly spreading rash involving his face, neck, trunk and proximal limbs with multiple pinhead-sized sterile pustules on an edematous erythematous base. Laboratory tests showed leukocytosis ($13,900/mm^3$) with neutrophilia ($12,600/mm^3$) and an elevated C-reactive protein at 2.8 mg/dl (normal, 0-0.5 mg/dl). These findings were very suggestive of acute generalized exanthematous pustulosis (AGEP). Later, skin biopsy results showed subcorneal pustulosis with neutrophilic infiltrate, supporting the diagnosis.

Patch tests with AX/CLV and AX (at 30% in petrolatum) were positive to the combination of amoxicillin and clavulanic acid, whereas amoxicillin was negative. Clavulanic acid itself was not tested, so the diagnosis of AGEP from delayed-type hypersensitivity to clavulanic acid was made *per exclusionem*. After this episode and in the context of recurrent tonsillitis, the patient was medicated several times with oral amoxicillin and intramuscular benzathine penicillin G with tolerance (2). It should be mentioned, though, that the 'positive' patch test to AX/CLV shown as a photograph in the article was far from convincing; controls were not mentioned.

Other cutaneous adverse drug reactions

After a few days of taking amoxicillin-clavulanic acid, a 1,5-year-old girl developed a scarcely perceptible widespread rash. Later, she received amoxicillin alone, without any adverse effects. After a few month, reintroduction of the combination resulted within a few days after the 1st dose in large urticarial wheals, scattered all over her body. Patch tests were positive to potassium clavulanate 10% pet. (negative to 3%) and negative to amoxicillin 10% pet., benzylpenicillin and 3 cephalosporins. 23 controls showed no reactions to potassium clavulanate 10% pet. (6).

Cross-reactions, pseudo-cross-reactions and co-reactions

Cross-reactions between beta-lactam antibiotics are discussed in Chapter 3.36 Amoxicillin.

LITERATURE

1   The data in the section 'General' may have been obtained from literature discussed in this chapter, but mostly also or exclusively from one or more of the following online sources: ChemIDPlus Advanced, PubChem, DrugBank, RxList, Drug Central, Drugs.com, and Wikipedia

2   Amaral L, Carneiro-Leão L, Cernadas JR. Acute generalized exanthematous pustulosis due to clavulanic acid. J Allergy Clin Immunol Pract 2020;8:1083-1084

3   Salas M, Laguna JJ, Doña I, Barrionuevo E, Fernandez-Santamaría R, Ariza A, et al. Patients taking amoxicillin-clavulanic can become simultaneously sensitized to both drugs J Allergy Clin Immunol Pract 2017;5:694-702

4   Bonadonna P, Schiappoli M, Senna G, Passalacqua G. Delayed selective reaction to clavulanic acid: a case report. J Investig Allergol Clin Immunol 2005;15:302-304

5   Kim YH, Ko JY, Kim YS, Ro YS. A case of allergic contact dermatitis to clavulanic acid. Contact Dermatitis 2008;59:378-379

6   Kamphof WG, Rustemeyer T, Bruynzeel DP. Sensitization to clavulanic acid in Augmentin. Contact Dermatitis 2002;47:47

7   Watts TJ, Thursfield D, Haque R. Patch testing for the investigation of nonimmediate cutaneous adverse drug reactions: a prospective single center study. J Allergy Clin Immunol Pract 2019;7:2941-2943.e3

8   Kamiya K, Kamiya E, Kamiya Y, Niwa M, Saito A, Natsume T, et al. Drug eruption to clavulanic acid with sparing of cellulitis-affecting site. Allergol Int 2015;64:280-281

9   Calvão J, Batista R, Gonçalo M. Two cases of delayed hypersensitivity to clavulanic acid proven by patch tests. Contact Dermatitis 2021;85:370-372

Chapter 3.138 CLINDAMYCIN

IDENTIFICATION

Description/definition : Clindamycin is a semisynthetic antibiotic produced by chemical modification of the parent compound lincomycin; it conforms to the structural formula shown below
Pharmacological classes : Anti-bacterial agents; protein synthesis inhibitors
IUPAC name : (2S,4R)-N-[(1S,2S)-2-Chloro-1-[(2R,3R,4S,5R,6R)-3,4,5-trihydroxy-6-methylsulfanyloxan-2-yl]propyl]-1-methyl-4-propylpyrrolidine-2-carboxamide
Other names : 7-Chloro-7-deoxylincomycin
CAS registry number : 18323-44-9
EC number : 242-209-1
Merck Index monograph : 3624
Patch testing : Phosphate, 10.0% pet. (Chemotechnique)
Molecular formula : $C_{18}H_{33}ClN_2O_5S$

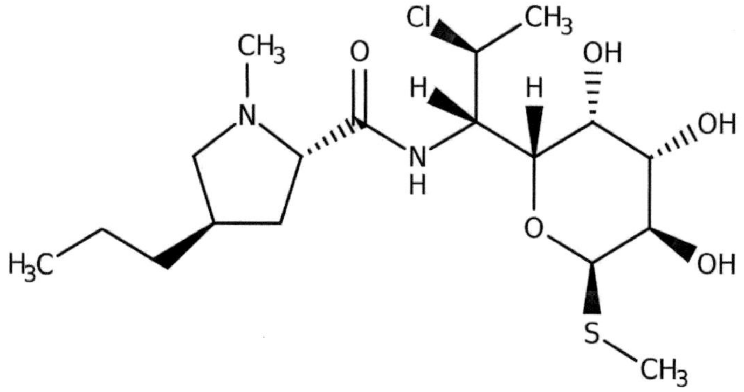

GENERAL

Clindamycin is a semisynthetic broad-spectrum antibiotic produced by chemical modification of the parent compound lincomycin. This agent dissociates peptidyl-tRNA from the bacterial ribosome, thereby disrupting bacterial protein synthesis. Clindamycin is indicated for the treatment of serious infections caused by susceptible anaerobic bacteria, including *Bacteroides* spp., *Peptostreptococcus*, anaerobic streptococci, *Clostridium* spp., and microaerophilic streptococci. The antibiotic may be useful in polymicrobial infections such as intra-abdominal or pelvic infections, osteomyelitis, diabetic foot ulcers, aspiration pneumonia and dental infections. In topical preparations, clindamycin is widely used in the treatment of inflammatory acne vulgaris (1). As such, this antibiotic has caused various cases of contact allergy/allergic contact dermatitis, which has been fully reviewed in Volume 3 of the *Monographs in contact allergy* series (12).

In pharmaceutical products, clindamycin is usually employed as clindamycin phosphate (CAS number 24729-96-2, EC number 246-433-0, molecular formula $C_{18}H_{34}ClN_2O_8PS$), sometimes as clindamycin hydrochloride (CAS number 21462-39-5, EC number 244-398-6, molecular formula $C_{18}H_{34}Cl_2N_2O_5S$) (1).

CUTANEOUS ADVERSE DRUG REACTIONS FROM SYSTEMIC ADMINISTRATION CAUSED BY TYPE IV (DELAYED-TYPE) HYPERSENSITIVITY (as demonstrated by positive patch tests)

Cutaneous adverse drug reactions from systemic administration of clindamycin caused by type IV (delayed-type) hypersensitivity have included macular exanthemas (3,4), maculopapular eruptions (4,5,6,7,10,11,19,22,23,24), acute generalized exanthematous pustulosis (AGEP) (9,10,14,15,16,17,23), drug reaction with eosinophilia and systemic symptoms (DRESS) (10,25), Stevens-Johnson syndrome (16), symmetrical drug-related intertriginous and flexural exanthema (SDRIFE) (16,22 [dubious]), erythema multiforme-like eruption (8) and cutaneous vasculitis (26).

Case series with various or unknown types of drug reactions

In Leuven, Belgium, in the period 1999 to 2019, of 87 patients investigated for drug hypersensitivity reactions during or after the use of clindamycin, 9 had positive patch tests to this antibiotic. Six patients had a maculopapular exanthema, two acute generalized exanthematous pustulosis (AGEP), and one suffered a drug reaction with eosinophilia and systemic symptoms (DRESS). The group consisted of 8 women and one man, ages ranging from 23 to 88 years. Patch tests were positive to clindamycin hydrochloride 10% pet. Nearly 80 controls were negative (10).

In a single-center study performed in France between 1997 and 2016, 71 patients with suspected delayed cutaneous adverse drug reaction to macrolides (n=20), clindamycin (n=10) and pristinamycin (n=41) were patch tested. Of the 10 patients with suspected reactions to clindamycin, 3 (30%) had a positive patch test to this antibiotic tested 10% pet. One of these patients had suffered a maculopapular exanthema and 2 AGEP (23).

In Finland, in the period 1989-2001, 826 patients with suspected cutaneous drug eruptions were patch tested and 89 had one or more positive reactions. Of these individuals, 12 reacted to clindamycin; it was not mentioned which cutaneous adverse drug reactions clindamycin had caused in these patients (18).

In a group of 20 patients with suspected cutaneous adverse drug reactions (13 morbilliform rashes, 4 SJS/TEN, 3 DRESS), there were 10 positive patch tests to suspected drugs, including 2 to clindamycin. It was not specified which drug reactions clindamycin had caused (20).

Macular and maculopapular exanthemas

Case series
In Germany, from 2003-2007, 33 patients who had developed a maculopapular exanthema in temporal relation to treatment with clindamycin were patch tested with a commercial clindamycin tablet ground in a mortar and suspended in 1 ml saline, and prick tested with commercial clindamycin solution 150 mg/ml. In five individuals, patch tests were positive on D3, of who 4 also had a delayed-type reaction to the prick test. Specific clinical details were not provided (7).

In the period 2000-2014, in Portugal, 260 patients were patch tested with antibiotics for suspected cutaneous adverse drug reactions (CADR) to these drugs. 56 patients (22%) had one or more positive patch tests, often from cross-reactivity. Clindamycin was patch test positive in 16 patients with maculopapular eruptions (19).

From 1996 to 2000, six patients with a history strongly suggestive of clindamycin hypersensitivity were studied in Turku, Finland. They were patch tested with commercial clindamycin in at least 3 of the following dilutions: pure, 10% pet., 20% pet., 10% alc. and 20% saline. Five had 3 positive patch tests to clindamycin, but in one they were considered to be false-positive. All had suffered extensive or generalized erythematous exanthem after administration of clindamycin. One showed erythema multiforme-like target lesions and another also displayed dyspnea, eosinophilia and a slightly elevated liver enzyme (possibly a case of DRESS) (3).

From Germany, three patients were reported who had suffered a generalized maculopapular exanthem (n=2) or generalized erythema with facial edema (n=1) after oral or parenteral administration of clindamycin. One individual also had dyspnea, anxiety, and tachycardia, and a second nausea and repeated vomiting with a short syncopal episode. Patch tests were positive to commercial clindamycin 150 mg/ml in all patients, as were intradermal tests read at D2. In each patient, the skin tests reproduced the extensive macular/maculopapular rashes (4).

In Portugal, in the period 2005-2009, 30 patients (23 women, 7 men) aged 33-86 years (mean 60 years) with generalized maculopapular exanthema suspected to have been caused by clindamycin were investigated. Two patients had a previous positive involuntary rechallenge. Patch tests with commercial and pure clindamycin 10% pet. were positive in 9 of 30 patients (30%). More than 50 controls were negative to both test materials (5).

In a study from Portugal, 18 patients who had previously had positive patch tests to antibiotics that had caused a maculopapular eruption, were again patch tested after a mean interval of 6 years. The positive reactions were reproducible in 16 cases; 7 were caused by amoxicillin, 5 by clindamycin, 4 by flucloxacillin, one by spiramycin, one by vancomycin and one by cefoxitin (in some patients, 2 antibiotics were responsible (24).

Case reports
A 47-year-old woman developed a maculopapular rash symmetrically distributed on the neck, abdomen, and back, with isolated lesions involving the proximal parts of the arms and legs, which had started on the 3rd day of clindamycin oral treatment as antibiotic prophylaxis for a tooth implant. There was a striking vertical distribution of skin lesions along the striae distensae on the lateral sides of the abdomen. Patch tests were positive (+ at D2 and D4) to commercial clindamycin diluted to 1% active material in pet. and water. Intradermal tests with clindamycin solution 15 and 1,5 mg/ml were negative at 20 minutes but positive at D1, remaining positive until D8. The preferential localization in the striae distensae was considered to be a Koebner phenomenon (6).

A 72-year-old woman developed a generalized pruritic maculopapular rash without systemic symptoms 6 hours after the intake of clindamycin for cellulitis of her arm. A patch test with clindamycin 1% pet. was positive, as was an intradermal test with 0.03 ml of clindamycin 30 mg/ml at D2. At that moment, the patient had erythema in the antecubital folds, retro-auricular and occipital regions with generalized pruritus. Ten controls were negative. The authors called this a SDRIFE-like reaction, but the resemblance with SDRIFE is rather far-fetched (22).

A 38-year-old man had been treated for 10 days with clindamycin, when he developed a generalized pruriginous maculopapular eruption with lip edema and facial erythema. Patch tests were positive to clindamycin phosphate 1% pet. at D3 and D4. An oral challenge with clindamycin phosphate 300 mg resulted in a generalized erythema with lip and facial edema, developing after 6 hours and persisting for 2 days (11).

See also the section 'Case series with various or unknown types of drug reactions' above.

Acute generalized exanthematous pustulosis (AGEP)

In a study of multiple drug hypersensitivity syndrome, 2 patients had developed AGEP from clindamycin with a positive patch test to this drug, one after a previous episode of TEN from amoxicillin, the other after a previous episode of DRESS from other drugs (16). In a multicenter investigation in France, of 45 patients patch tested for AGEP, 26 (58%) had positive patch tests to drugs, including one to clindamycin (17).

Up to 2015, Lareb, the Dutch spontaneous adverse drug reactions reporting system, received 5 reports of acute generalized exanthematous pustulosis from clindamycin with a high rate of causality. In 2, patch tests with clindamycin had been performed and were positive (15).

Case reports

A 78-year-old woman presented with a rash that had developed on her trunk 3 days after taking clindamycin and levofloxacin as part of treatment for a hip replacement. On physical examination, a diffuse red papular rash coalesced into erythematous and edematous plaques on her arms, legs, and trunk, studded with a large number of scattered, non-follicular, pinhead-sized pustules. Laboratory studies showed leukocytosis with left shift. Skin biopsy revealed subcorneal pustules and diffuse perivascular dermal infiltrates of atypical mononuclear cells with large nuclei, prominent nucleoli, and a number of mitotic figures that were positive for CD3 and CD30. Up to 15% positive CD30 atypical cells within the perivascular inflammatory infiltrate were found. Two months later, patch tests were positive to clindamycin phosphate 10% pet. and to levofloxacin 20% pet. This was the first report of drug-induced acute generalized exanthematous pustulosis (AGEP) with simultaneous vascular lymphomatoid pattern, resembling drug-induced pseudolymphoma (14).

A 69-year-old man had been treated with clindamycin for 3 days, when he developed a pruritic rash on the trunk. One day later, his temperature rose to 39.4°C and the rash spread distally. Laboratory tests showed a neutrophilic leukocytosis. A skin biopsy showed spongiosis and exocytosis of lymphocytes and some neutrophils with dermal edema and an interstitial mixed infiltrate of lymphocytes, neutrophils and eosinophils compatible with a pustular drug reaction. A patch test was positive to clindamycin 1% pet. The patient was diagnosed with AGEP, but the characteristic pustules on erythema and subcorneal pustules in the histopathology were lacking (9). A second patient presented by these authors developed a mildly pruritic generalized erythematous rash within 36 hours of starting clindamycin. Histopathology was not performed. A patch test with clindamycin 1% pet. resulted in a pustular reaction at D2 and D4. This patient was also diagnosed with AGEP (9).

See also the section 'Case series with various or unknown types of drug reactions' above.

Drug reaction with eosinophilia and systemic symptoms (DRESS)

In a group of 13 pediatric patients patch tested for DRESS in a tertiary care hospital in Florence, Italy, between 2010 and 2018, clindamycin (tested at 10% pet.) was, together with amoxicillin, responsible for one case of DRESS (25).

See also the section 'Case series with various or unknown types of drug reactions' above.

Other cutaneous adverse drug reactions

In a study of patients with multiple drug hypersensitivity syndrome, there was one case of Stevens-Johnson syndrome (SJS) caused by clindamycin and amoxicillin with positive patch tests to both drugs (16). In the same study, one patient had symmetrical drug-related intertriginous and flexural exanthema (SDRIFE) from clindamycin with a positive patch test to this drug after a previous episode of severe maculopapular eruption from other drug(s) (16).

A 55-year-old woman was treated for febrile post-operative necrotic pancreatitis with tobramycin, clindamycin, cefalotin and cefalexin. An erythematopapular eruption began on the neck and spread to the arms, legs and trunk. Erythematous plaques and vesicular lesions were present. The dermatitis and fever subsided a few days after stopping the antibiotics. Patch tests were positive to commercial clindamycin 150 mg/ml and to tobramycin. The patient was diagnosed with erythema multiforme-like eruption caused by clindamycin, tobramycin and cefalotin (the latter diagnosed by a positive delayed intradermal test) (8).

A 24-year-old man presented with lesions on the legs, which had developed after having used clindamycin and ciprofloxacin for 2 weeks. Physical examination revealed palpable purpuric lenticular papules coalescing into a plaque on the anterior side of the legs. The clinical diagnosis of leukocytoclastic vasculitis was confirmed by histopathology. Three months after complete resolution, patch tests were positive to clindamycin phosphate 10% pet. at D4 and D7 and negative to ciprofloxacin hydrochloride 10% pet. (26).

See also the section 'Case series with various or unknown types of drug reactions' above.

Cross-reactions, pseudo-cross-reactions and co-reactions

Cross-reactions to and from lincomycin may occur (2,13).

LITERATURE

1   The data in the section 'General' may have been obtained from literature discussed in this chapter, but mostly also or exclusively from one or more of the following online sources: ChemIDPlus Advanced, PubChem, DrugBank, RxList, Drug Central, Drugs.com, and Wikipedia

2   Conde-Salazar L, Guimaraens D, Romero LV. Contact dermatitis from clindamycin. Contact Dermatitis 1983;9:225

3   Lammintausta K, Tokola R, Kalimo K. Cutaneous adverse reactions to clindamycin: results of skin tests and oral exposure. Br J Dermatol 2002;146:643-648

4   Papakonstantinou E, Müller S, Röhrbein JH, Wieczorek D, Kapp A, Jakob T, Wedi B. Generalized reactions during skin testing with clindamycin in drug hypersensitivity: a report of 3 cases and review of the literature. Contact Dermatitis 2018;78:274-280

5   Pereira N, Canelas MM, Santiago F, Brites MM, Gonçalo M. Value of patch tests in clindamycin-related drug eruptions. Contact Dermatitis 2011;65:202-207

6   Monteagudo B, Cabanillas M, Iriarte P, Ramírez-Santos A, León-Muinos E, González-Vilas D, Suárez-Amor Ó. Clindamycin-induced maculopapular exanthema with preferential involvement of striae distensae: A Koebner phenomenon? Acta Dermatovenerol Croat 2018;26:61-63

7   Seitz CS, Bröcker EB, Trautmann A. Allergy diagnostic testing in clindamycin-induced skin reactions. Int Arch Allergy Immunol 2009;149:246-250

8   Muñoz D, Del Pozo MD, Audícana M, Fernandez E, Fernandez De Corres LF. Erythema-multiforme-like eruption from antibiotics of 3 different groups. Contact Dermatitis 1996;34:227-228

9   Valois M, Phillips EJ, Shear NH, Knowles SR. Clindamycin-associated acute generalized exanthematous pustulosis. Contact Dermatitis 2003;48:169

10 Gilissen L, Huygens S, Goossens A, Breynaert C, Schrijvers R. Utility of patch testing for the diagnosis of delayed-type drug hypersensitivity reactions to clindamycin. Contact Dermatitis 2020;83:237-239

11 Vicente J, Fontela JL. Delayed reaction to oral treatment with clindamycin. Contact Dermatitis 1999;41:221

12 De Groot AC. Monographs in contact allergy, volume 3: Topical Drugs. Boca Raton, Fl, USA: CRC Press Taylor and Francis Group, 2021 (ISBN 978-0-367-23693-9)

13 Conde-Salazar L, Guimaraens D, Romero L, Gonzalez M, Yus S. Erythema multiforme-like contact dermatitis from lincomycin. Contact Dermatitis 1985;12:59-61

14 Llamas-Velasco M, Godoy A, Sánchez-Pérez J, García-Diez A, Fraga J. Acute generalized exanthematous pustulosis with histopathologic findings of lymphomatoid drug reaction. Am J Dermatopathol 2013;35:690-691

15 Smeets TJ, Jessurun N, Härmark L, Kardaun SH. Clindamycin-induced acute generalised exanthematous pustulosis: five cases and a review of the literature. Neth J Med 2016;74:421-428

16 Jörg L, Yerly D, Helbling A, Pichler W. The role of drug, dose and the tolerance/intolerance of new drugs in multiple drug hypersensitivity syndrome (MDH). Allergy 2020;75:1178-1187

17 Barbaud A, Collet E, Milpied B, Assier H, Staumont D, Avenel-Audran M, et al. A multicentre study to determine the value and safety of drug patch tests for the three main classes of severe cutaneous adverse drug reactions. Br J Dermatol 2013;168:555-562

18 Lammintausta K, Kortekangas-Savolainen O. The usefulness of skin tests to prove drug hypersensitivity. Br J Dermatol 2005;152:968-974

19 Pinho A, Coutinho I, Gameiro A, Gouveia M, Gonçalo M. Patch testing - a valuable tool for investigating non-immediate cutaneous adverse drug reactions to antibiotics. J Eur Acad Dermatol Venereol 2017;31:280-287

20 Hassoun-Kheir N, Bergman R, Weltfriend S. The use of patch tests in the diagnosis of delayed hypersensitivity drug eruptions. Int J Dermatol 2016;55:1219-1224

21 Liccioli G, Mori F, Parronchi P, Capone M, Fili L, Barni S, et al. Aetiopathogenesis of severe cutaneous adverse reactions (SCARs) in children: A 9-year experience in a tertiary care paediatric hospital setting. Clin Exp Allergy 2020;50:61-73

22 Morales-Cabeza C, Caralli Bonett ME, Micozzi S, Seoane Rodríguez M, Rojas-Pérez-Ezquerra P, de Barrio Fernández M. SDRIFE-like reaction induced by an intradermal skin test with clindamycin: A case report. J Allergy Clin Immunol Pract 2015;3:976-977

23 El Khoury M, Assier H, Gener G, Paul M, Haddad C, Chosidow O, et al. Polysensitivity in delayed cutaneous adverse drug reactions to macrolides, clindamycin and pristinamycin: clinical history and patch testing. Br J Dermatol 2018;179:978-979

24 Pinho A, Marta A, Coutinho I, Gonçalo M. Long-term reproducibility of positive patch test reactions in patients with non-immediate cutaneous adverse drug reactions to antibiotics. Contact Dermatitis 2017;76:204-209

25 Watts TJ, Thursfield D, Haque R. Patch testing for the investigation of nonimmediate cutaneous adverse drug reactions: a prospective single center study. J Allergy Clin Immunol Pract 2019;7:2941-2943.e3

26 Fransen M, Verstraeten VLRM. Cutaneous vasculitis caused by clindamycin. Dermatitis. 2020 Dec 15; Epub ahead of print.

Chapter 3.139 CLIOQUINOL

IDENTIFICATION

Description/definition	: Clioquinol is the halogenated 8-hydroxyquinoline that conforms to the structural formula shown below
Pharmacological classes	: Dermatological drugs; antiseptics and disinfectants; anti-infective agents
IUPAC name	: 5-Chloro-7-iodoquinolin-8-ol
Other names	: Iodochlorhydroxyquin; 5-chloro-7-iodo-8-hydroxyquinoline
CAS registry number	: 130-26-7
EC number	: 204-984-4
Merck Index monograph	: 6345
Patch testing	: 5% pet. (Chemotechnique, SmartPracticeCanada, SmartPracticeEurope)
Molecular formula	: C_9H_5ClINO

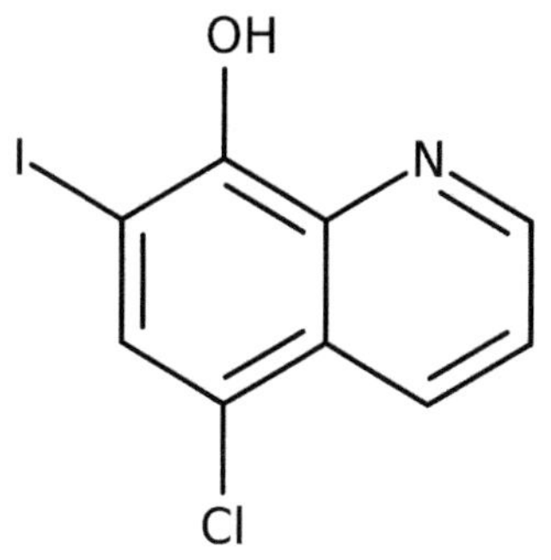

GENERAL

Clioquinol is a broad-spectrum antibacterial agent with antifungal properties. It is used as a topical antibacterial and antifungal treatment. However, topical absorption of this iodine-containing agent is rapid and extensive, especially when the skin is covered with an occlusive dressing or if the medication is applied to extensive or eroded areas of the skin, which may affect thyroid function tests. Oral administration has been used for treatment of enteritis and as prophylaxis against travelers' diarrhea (1).

In topical preparations, clioquinol has caused many cases of contact allergy/allergic contact dermatitis, which subject has been fully reviewed in Volume 3 of the *Monographs in contact allergy* series (13).

CUTANEOUS ADVERSE DRUG REACTIONS FROM SYSTEMIC ADMINISTRATION CAUSED BY TYPE IV (DELAYED-TYPE) HYPERSENSITIVITY (as demonstrated by positive patch tests)

Cutaneous adverse drug reactions from systemic administration of clioquinol caused by type IV (delayed-type) hypersensitivity have included systemic allergic dermatitis (8,9,11,12).

Systemic allergic dermatitis (systemic contact dermatitis)

Oral administration of clioquinol in patients previously sensitized by topical application of this drug will induce an eruption in some (some say: most) patients: systemic allergic dermatitis. In some of them a flare up of the original dermatitis is seen (12), while in other cases a generalized dermatitis may be observed (8).

A 40-year-old woman started to take clioquinol tablets as prophylaxis against travellers' diarrhea. After 2 days, her palms began to itch. The next day, the itching spread to the whole body, and during the following 2 days a generalized eruption developed. At the time of admission, by which time she had taken four tablets, she had extremely itchy, edematous plaques with papules and vesicles especially on the breast, abdomen, back and hands. She also had tinea pedis and was by mistake treated with clioquinol powder. During the next 24 h she developed a severe bullous reaction on her feet. Patch tests were positive to clioquinol and the related chlorquinaldol (9).

The left side of the body of a 32-year-old man was treated with an ointment containing 1% clioquinol for pustular psoriasis. This resulted in a contact dermatitis localized at the sites of application. The treatment was discontinued and the reaction disappeared within a week. Later, to treat a slight enteritis, the patient took two tablets of 0.25 gram clioquinol. A few hours later, the left side of his body began to itch and he developed a severe, oozing dermatitis which was limited to the area which had been treated topically with clioquinol nine months earlier. Patch tests were not performed, but this case history is highly suggestive for systemic allergic contact dermatitis from clioquinol (12).

Two more patients sensitized to clioquinol had a generalized eruption after having taken clioquinol tablets (11).

Cross-reactions, pseudo-cross-reactions and co-reactions

Cross-reactions between halogenated hydroxyquinolines such as clioquinol (5-chloro-7-iodoquinolin-8-ol), chlorquinaldol (5,7-dichloro-2-methylquinolin-8-ol), cloxyquin (5-chloroquinolin-8-ol), oxyquinoline (8-hydroxy-quinoline, non-halogenated), iodoquinol (5,7-diiodoquinolin-8-ol), halquinol (a mixture of 4 hydroxyquinolines), 5,7-dichloro-8-quinolinol and 5-chloro-8-quinolinol may occur (2,3,4,5,6,10 [examples of references]).

Ten patients sensitized to clioquinol were patch tested with a number of structurally-related chemicals and there were 9 cross-reactions to broxyquinoline (5,7-dibromo-8-quinolinol), 9 to chlorquinaldol, 4 to quinine, one to chloroquine (possibly false-positive from excited skin syndrome), one to amodiaquine (ibid), and two to potassium iodide (10% pet. ++ and + , 5% pet. neg., possibly also false-positive) (7). Cross-reactions to chloroquine had been observed before (6).

LITERATURE

1   The data in the section 'General' may have been obtained from literature discussed in this chapter, but mostly also or exclusively from one or more of the following online sources: ChemIDPlus Advanced, PubChem, DrugBank, RxList, Drug Central, Drugs.com, and Wikipedia

2   Myatt AE, Beck MH. Contact sensitivity to chlorquinaldol. Contact Dermatitis 1983;9:523

3   Leifer W, Steiner K. Studies in sensitization to halogenated hydroxyquinolines and related compounds. J Invest Dermatol 1951;17:233-240

4   Allenby CF. Skin sensitization to Remiderm and cross-sensitization to hydroxyquinoline compounds. Br Med J 1965;2(5455):208-209

5   Wantke F, Götz M, Jarisch R. Contact dermatitis from cloxyquin. Contact Dermatitis 1995;32:112-113

6   Bielicky T, Novak M. Gruppensensibilisierung gegen Chinolinderivate. Dermatologica 1969;138:45-58 (Article in German, data cited in ref. 39)

7   Soesman-van Waadenoijen Kernekamp A, van Ketel WG. Persistence of patch test reactions to clioquinol (Vioform) and cross-sensitization. Contact Dermatitis 1980;6:455-460

8   Ekelund AG, Möller H. Oral provocation in eczematous contact allergy to neomycin and hydroxyquinolines. Acta

9   Skog E. Systemic eczematous contact-type dermatitis induced by iodochlorhydroxyquin and chloroquine phosphate. Contact Dermatitis 1975;1:187

10 Van Ketel WG. Cross-sensitization to 5,7 dibromo-8-hydroxyquinoline (DBO). Contact Dermatitis 1975;1:385

11 Cronin E. Contact Dermatitis. Edinburgh: Churchill Livingstone, 1980:220

12 Domar M, Juhlin L. Allergic dermatitis produced by oral clioquinol. Lancet 1967;1(7500):1165-1166

13 De Groot AC. Monographs in contact allergy, volume 3: Topical Drugs. Boca Raton, Fl, USA: CRC Press Taylor and Francis Group, 2021 (ISBN 978-0-367-23693-9)

Chapter 3.140 CLOBAZAM

IDENTIFICATION

Description/definition   : Clobazam is the benzodiazepine that conforms to the structural formula shown below
Pharmacological classes : Anti-anxiety agents; GABA-A receptor agonists; anticonvulsants
IUPAC name   : 7-Chloro-1-methyl-5-phenyl-1,5-benzodiazepine-2,4-dione
CAS registry number   : 22316-47-8
EC number   : 244-908-7
Merck Index monograph : 3626
Patch testing   : No data available; suggested: 10% pet.
Molecular formula   : $C_{16}H_{13}ClN_2O_2$

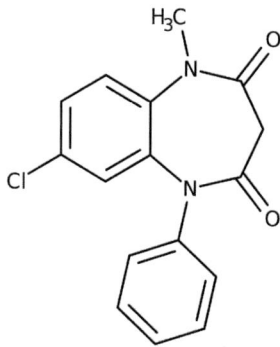

GENERAL

Clobazam is a benzodiazepine derivative and partial gamma-aminobutyric acid (GABA) receptor agonist, with anxiolytic, sedative, and anticonvulsant activities. This agent is indicated for treatment and management of epilepsy and seizures associated with Lennox-Gastaut syndrome, a difficult-to-treat form of childhood epilepsy. It is also used as an anxiolytic for the short-term treatment of acute anxiety (1).

CUTANEOUS ADVERSE DRUG REACTIONS FROM SYSTEMIC ADMINISTRATION CAUSED BY TYPE IV (DELAYED-TYPE) HYPERSENSITIVITY (as demonstrated by positive patch tests)

Cutaneous adverse drug reactions from systemic administration of clobazam caused by type IV (delayed-type) hypersensitivity have included maculopapular eruption (2) and drug reaction with eosinophilia and systemic symptoms (DRESS) (3).

Maculopapular eruption

A 65-year-old woman, who had been treated for 5 years with valproic acid 1 g daily for epilepsy, was admitted with a generalized erythematous and pruritic eruption. The dermatitis started 9 days after the introduction of clobazam 10 mg daily and carbamazepine 100 mg daily. There was a generalized exanthema with some slightly infiltrated papules, sparing only the face, palms and soles. The mucosae were not affected. Clobazam and carbamazepine were both stopped on admission and cutaneous lesions spontaneously disappeared within 8 days. Patch tests were then carried out, using first one tablet of clobazam crushed in pet. There was a very strong reaction, followed by a generalized relapse of the cutaneous lesions 36 hours later. Later, a patch test with 1 tablet of carbamazepine crushed in pet. was performed and was negative (2).

Drug reaction with eosinophilia and systemic symptoms (DRESS)

In a multicenter investigation in France, of 72 patients patch tested for DRESS, 46 (64%) had positive patch tests to drugs, including one to clobazam; clinical and patch testing details are not available (3).

LITERATURE

1   The data in the section 'General' may have been obtained from literature discussed in this chapter, but mostly also or exclusively from one or more of the following online sources: ChemIDPlus Advanced, PubChem, DrugBank, RxList, Drug Central, Drugs.com, and Wikipedia
2   Machet L, Vaillant L, Dardaine V, Lorette G. Patch testing with clobazam: relapse of generalized drug eruption. Contact Dermatitis 1992;26:347-348
3   Barbaud A, Collet E, Milpied B, Assier H, Staumont D, Avenel-Audran M, et al. A multicentre study to determine the value and safety of drug patch tests for the three main classes of severe cutaneous adverse drug reactions. Br J Dermatol 2013;168:555-562

Chapter 3.141 CLOFAZIMINE

IDENTIFICATION

Description/definition   : Clofazimine is the riminophenazine that conforms to the structural formula shown below
Pharmacological classes : Leprostatic agents; anti-inflammatory agents
IUPAC name   : N,5-bis(4-Chlorophenyl)-3-propan-2-yliminophenazin-2-amine
CAS registry number   : 2030-63-9
EC number   : 217-980-2
Merck Index monograph   : 3637
Patch testing   : No data available; suggested: 10% pet.
Molecular formula   : $C_{27}H_{22}Cl_2N_4$

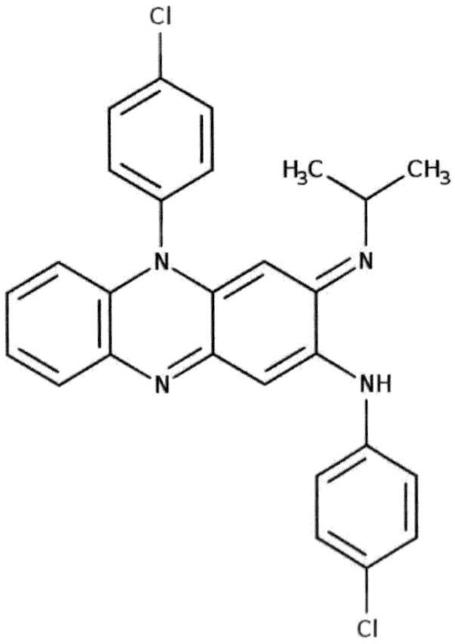

GENERAL

Clofazimine is a highly lipophilic bright-red antimicrobial riminophenazine dye used in combination with other agents, such as dapsone, for the treatment of leprosy. The drug binds preferentially to mycobacterial DNA, thereby inhibiting DNA replication and cell growth. Clofazimine has a slow bactericidal effect on *Mycobacterium leprae* and is active against various other Mycobacteria (1).

CUTANEOUS ADVERSE DRUG REACTIONS FROM SYSTEMIC ADMINISTRATION CAUSED BY TYPE IV (DELAYED-TYPE) HYPERSENSITIVITY (as demonstrated by positive patch tests)

Cutaneous adverse drug reactions from systemic administration of clofazimine caused by type IV (delayed-type) hypersensitivity have included urticarial and morbilliform rash with angioedema (2).

Other cutaneous adverse drug reactions

A 17-year old woman, suffering from multidrug-resistant tuberculosis and co-infected with human immunodeficiency virus (HIV) was started on second-line anti-tuberculosis treatment with kanamycin, moxifloxacin, para-aminosalicylic acid, cycloserine, clofazimine, and amoxicillin-clavulanic acid. Twenty-four hours later, the patient developed a generalized urticarial and morbilliform rash, fever, angioedema, laryngospasm, and dyspnea. She was admitted to hospital, all drugs were stopped, and the symptoms and signs of the hypersensitivity reaction subsided. Patch testing performed one week later (which is too soon) showed an extremely positive reaction (+3) to clofazimine, strongly positive reaction (+2) to kanamycin and weakly positive reaction (+1) to cycloserine. However, kanamycin and cycloserine could later be reintroduced without problems (2). Patch test parameters were not adequately provided, the test has been performed too soon after the drug reaction, the patch test was probably read only at D2 and no controls were performed. The reliability of this report is therefore doubtful.

LITERATURE

1   The data in the section 'General' may have been obtained from literature discussed in this chapter, but mostly also or exclusively from one or more of the following online sources: ChemIDPlus Advanced, PubChem, DrugBank, RxList, Drug Central, Drugs.com, and Wikipedia
2   Khan S, Andries A, Pherwani A, Saranchuk P, Isaakidis P. Patch-testing for the management of hypersensitivity reactions to second-line anti-tuberculosis drugs: a case report. BMC Res Notes 2014;7:537

Chapter 3.142 CLOFIBRATE

IDENTIFICATION

Description/definition : Clofibrate is the fibric acid derivative that conforms to the structural formula shown below
Pharmacological classes : Anticholesteremic agents; hypolipidemic agents
IUPAC name : Ethyl 2-(4-chlorophenoxy)-2-methylpropanoate
Other names : Ethyl clofibrate
CAS registry number : 637-07-0
EC number : 211-277-4
Merck Index monograph : 3640
Patch testing : No data available; most systemic drugs can be tested at 10% pet.; if the pure chemical is not available, prepare the test material from intravenous powder, the content of capsules or – when also not available – from powdered tablets to achieve a final concentration of the active drug of 10% pet.
Molecular formula : $C_{12}H_{15}ClO_3$

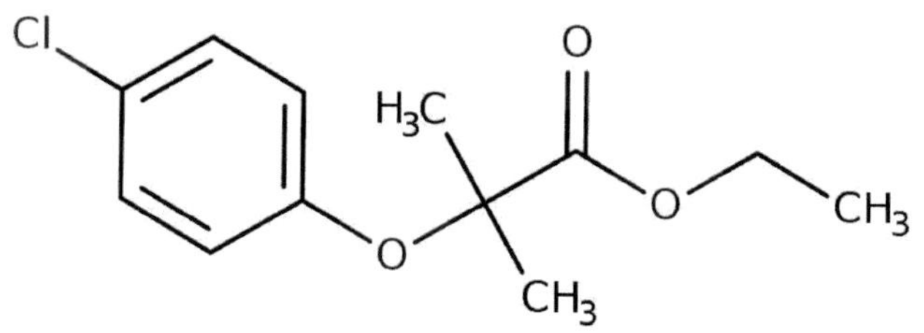

GENERAL

Clofibrate is a fibric acid derivative with antihyperlipidemic activity. Although the exact mechanism of action has not been fully characterized, clofibrate may enhance the conversion of very-low-density lipoprotein (VLDL) to low-density lipoprotein (LDL), decreasing the production of hepatic VLDL, inhibiting cholesterol production, and increasing fecal excretion of neutral sterols. This drug is indicated for primary dysbetalipoproteinemia (Type III hyperlipidemia) that does not respond adequately to diet and helps control high cholesterol and high triglyceride levels (1).

CUTANEOUS ADVERSE DRUG REACTIONS FROM SYSTEMIC ADMINISTRATION CAUSED BY TYPE IV (DELAYED-TYPE) HYPERSENSITIVITY (as demonstrated by positive patch tests)

Cutaneous adverse drug reactions from systemic administration of clofibrate caused by type IV (delayed-type) hypersensitivity have included photosensitivity (2).

Photosensitivity

A female patient using clofibrate developed an extensive vesiculobullous edematous rash, mainly on light-exposed skin, after exposition to sunlight and working in the garden. Systemic photoallergy was demonstrated by a positive photopatch test, whereas the unirradiated patch test was negative. No details on the patch test material and irradiation parameters (wavelength, energy) were provided (2).

LITERATURE

1 The data in the section 'General' may have been obtained from literature discussed in this chapter, but mostly also or exclusively from one or more of the following online sources: ChemIDPlus Advanced, PubChem, DrugBank, RxList, Drug Central, Drugs.com, and Wikipedia
2 Heid E, Samsoen M, Juillard J, Eberst E, Foussereau J. Eruptions papulo-vésiculeuses endogènes à la méthyldopa et au clofibrate. Ann Derm Venereol 1977;104:494-496 (Article in French)

Chapter 3.143 CLOMIPRAMINE

IDENTIFICATION

Description/definition   : Clomipramine is the 3-chloro derivative of imipramine; it conforms to the structural
                            formula shown below
Pharmacological classes   : Antidepressive agents, tricyclic; serotonin uptake inhibitors
IUPAC name   : 3-(Chloro-5,6-dihydrobenzo[b][1]benzazepin-11-yl)-N,N-dimethylpropan-1-amine
Other names   : Chlorimipramine
CAS registry number   : 303-49-1
EC number   : 206-144-2
Merck Index monograph   : 3648
Patch testing   : 0.1%, 1% and 5% pet.
Molecular formula   : $C_{19}H_{23}ClN_2$

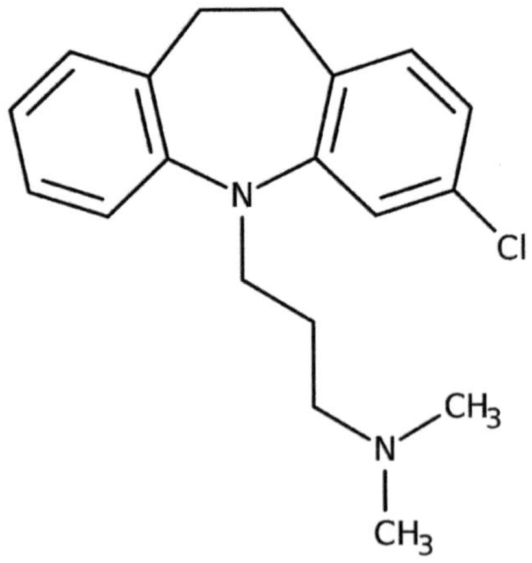

GENERAL

Clomipramine is a tricyclic antidepressant and the 3-chloro derivative of imipramine that selectively inhibits the uptake of serotonin in the brain. It is indicated for the treatment of obsessive-compulsive disorder and disorders with an obsessive-compulsive component (e.g. depression, schizophrenia, Tourette's disorder) (1). In pharmaceutical products, clomipramine is employed as clomipramine hydrochloride (CAS number 17321-77-6, EC number 241-344-3, molecular formula $C_{19}H_{24}Cl_2N_2$) (1).

CUTANEOUS ADVERSE DRUG REACTIONS FROM SYSTEMIC ADMINISTRATION CAUSED BY TYPE IV (DELAYED-TYPE) HYPERSENSITIVITY (as demonstrated by positive patch tests)

Cutaneous adverse drug reactions from systemic administration of clomipramine caused by type IV (delayed-type) hypersensitivity have included photosensitivity (2).

Photosensitivity

A 43-year-old farmer was treated with clomipramine for severe depression with a suicide attempt. Because of leg cramps, two weeks later, carbamazepine was added. While working outdoors on his farm in mid-May, the patient developed an eruption of the face and hands, subsequently spreading to most of the covered skin. He was admitted to the hospital with an intense dermatitis, partly vesiculobullous, of his face, hands and arms. On the lower legs, a sharp demarcation from his socks was noted, and more scattered maculopapular lesions were present under his shorts and on the trunk. Although the drugs were discontinued, his photosensitivity persisted for some months. An oral provocation test was positive, already on the evening of the 1st day, after an oral dose of 100 mg, the patient developed a papulovesicular, itching rash on his face, trunk and extremities. Patch and photopatch tests showed contact allergy to carbamazepine (5%, 1% and 0.1% pet.) without photo-augmentation. Clomipramine was tested in dilutions ranging from 5% to 0.001% pet. There were positive reactions both on the covered and irradiated sites to 5%, 1% and 0.1%, but all far stronger on the irradiated sites. The photopatch test to 0.01% was positive, but the covered patch test at this concentration negative. It was concluded that the patient had contact allergy to carbamazepine and both type IV allergy and photoallergy to clomipramine (2).

Cross-reactions, pseudo-cross-reactions and co-reactions
A patient who developed photoallergic dermatitis from oral clomipramine (patch tests positive, photopatch tests stronger positive), also had contact allergy to carbamazepine. These may have been concomitant sensitizations, as the patient used both drugs. However, the chemical structures of these drugs are closely related, both being tricyclic compounds with a nitrogen substitution in the central ring structure. Therefore, cross-reactivity to carbamazepine from primary imipramine (photo)sensitization cannot be excluded (2).

LITERATURE
1   The data in the section 'General' may have been obtained from literature discussed in this chapter, but mostly also or exclusively from one or more of the following online sources: ChemIDPlus Advanced, PubChem, DrugBank, RxList, Drug Central, Drugs.com, and Wikipedia
2   Ljunggren B, Bojs G. A case of photosensitivity and contact allergy to systemic tricyclic drugs, with unusual features. Contact Dermatitis 1991;24:259-265

Chapter 3.144 CLONIDINE

IDENTIFICATION

Description/definition   : Clonidine is the imidazoline derivative that conforms to the structural formula shown
                              below
Pharmacological classes   : α_2-Adrenergic receptor agonists; analgesics; antihypertensive agents; sympatholytics
IUPAC name   : N-(2,6-Dichlorophenyl)-4,5-dihydro-1H-imidazol-2-amine
Other names   : 2-((2,6-Dichlorophenyl)imino)imidazolidine
CAS registry number   : 4205-90-7
EC number   : 224-119-4
Merck Index monograph   : 3650
Patch testing   : 9% pet.; not infrequently, patients are negative to clonidine 9% pet. but have a positive
                              reaction to the clonidine-TTS, applied for 7 days (10)
Molecular formula   : $C_9H_9Cl_2N_3$

GENERAL

Clonidine is a centrally active α_2-adrenergic agonist. It is used predominantly as an antihypertensive agent, usually in combination with other drugs. Clonidine binds to and stimulates central α_2-adrenergic receptors, thereby decreasing sympathetic outflow to the heart, kidneys, and peripheral vasculature. This leads to decreased peripheral vascular resistance, decreased blood pressure, and decreased heart rate. In transdermal patches, clonidine base is used; in other pharmaceuticals, clonidine is employed as clonidine hydrochloride (CAS number 4205-91-8, EC number 224-121-5, molecular formula $C_9H_{10}Cl_3N_3$) (1). In topical preparations, notably transdermal therapeutic systems, clonidine has caused many cases of contact allergy/allergic contact dermatitis, which subject has been fully reviewed in Volume 3 of the *Monographs in contact allergy* series (3).

CUTANEOUS ADVERSE DRUG REACTIONS FROM SYSTEMIC ADMINISTRATION CAUSED BY TYPE IV (DELAYED-TYPE) HYPERSENSITIVITY (as demonstrated by positive patch tests)

Cutaneous adverse drug reactions from systemic administration of clonidine caused by type IV (delayed-type) hypersensitivity have included systemic allergic dermatitis manifesting as exacerbation of previous eczema (2) and as maculopapular rash (2).

Systemic allergic dermatitis (systemic contact dermatitis)

Of 52 patients who were previously withdrawn from clinical trials with clonidine-TTS because of the development of contact dermatitis, and who had confirmed sensitization to clonidine (positive patch test to clonidine 9% pet. or clonidine-TTS applied for 7 days), 29 agreed to oral challenge with clonidine HCl in a 4-day program (day 1 - 0.025 mg; day 2 - 0.05 mg; day 3 - 0.1 mg; day 4 - 0.2 mg). Only one patient, on the second day of clonidine treatment, developed a localized flare-up at the site of the original dermatitis 2 hours after receiving the 0.05 mg dose (2). In another group of such patients, a woman developed a generalized maculopapular rash and pruritus on the fifth day or oral treatment (2). These were cases of systemic allergic dermatitis.

LITERATURE

1   The data in the section 'General' may have been obtained from literature discussed in this chapter, but mostly also or exclusively from one or more of the following online sources: ChemIDPlus Advanced, PubChem, DrugBank, RxList, Drug Central, Drugs.com, and Wikipedia
2   Maibach HI. Oral substitution in patients sensitized by transdermal clonidine treatment. Contact Dermatitis 1987;16:1-8
3   De Groot AC. Monographs in contact allergy, volume 3: Topical Drugs. Boca Raton, Fl, USA: CRC Press Taylor and Francis Group, 2021 (ISBN 978-0-367-23693-9)

Chapter 3.145 CLOPIDOGREL

IDENTIFICATION

Description/definition : Clopidogrel is the thienopyridine that conforms to the structural formula shown below
Pharmacological classes : Purinergic P2Y receptor antagonists; platelet aggregation inhibitors
IUPAC name : Methyl (2S)-2-(2-chlorophenyl)-2-(6,7-dihydro-4H-thieno[3,2-c]pyridin-5-yl)acetate
CAS registry number : 113665-84-2
EC number : 601-269-2
Merck Index monograph : 3655
Patch testing : No reliable data available; most systemic drugs can be tested at 10% pet.; if the pure chemical is not available, prepare the test material from intravenous powder, the content of capsules or – when also not available – from powdered tablets to achieve a final concentration of the active drug of 10% pet.
Molecular formula : $C_{16}H_{16}ClNO_2S$

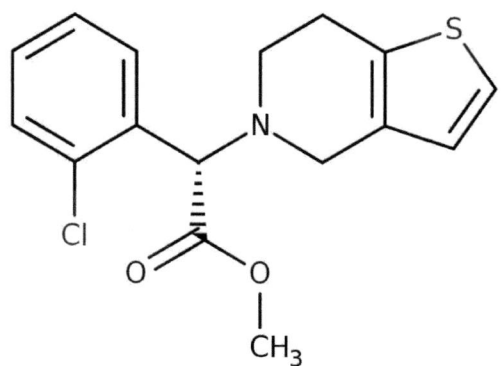

GENERAL

Clopidogrel is an inhibitor of platelet aggregation that is used to decrease the risk of myocardial infarction and stroke in patients known to have atherosclerosis. In pharmaceutical products, clopidogrel may be employed as clopidogrel besylate (CAS number 744256-69-7, EC number not available, molecular formula $C_{22}H_{22}ClNO_5S_2$), as clopidogrel bisulfate (CAS number 120202-66-6, EC number 601-679-1/601-269-2/603-890-4, molecular formula $C_{16}H_{18}ClNO_6S_2$) or as clopidogrel hydrochloride (CAS number 120202-65-5, EC number not available, molecular formula $C_{16}H_{17}Cl_2NO_2S$) (1).

CUTANEOUS ADVERSE DRUG REACTIONS FROM SYSTEMIC ADMINISTRATION CAUSED BY TYPE IV (DELAYED-TYPE) HYPERSENSITIVITY (as demonstrated by positive patch tests)

Cutaneous adverse drug reactions from systemic administration of clopidogrel caused by type IV (delayed-type) hypersensitivity have included generalized or localized exanthemas (2).

In a hospital in Canada, 62 patients out of a group of 3,877 (1.6%) treated with clopidogrel after percutaneous coronary intervention, developed a skin reaction classified as probable/definite clopidogrel hypersensitivity (2). 49 of these patients developed a generalized, pruritic, exanthematous rash predominantly affecting the trunk with or without involvement of the arms and legs. In 10 patients the rash was limited to a localized area in a focal or symmetrical manner. These reactions involved the neck, face, back, axilla, palm of the hand, and/or sole of the feet. Three individuals had immediate-type reactions with urticaria and angioedema. The symptoms started after 4-6 days. Patients were treated with a 3-week tapering course of oral prednisone starting at 30 mg twice per day for 5 days followed by a decrease of 5 mg/day every 3 days for 15 days. Under this regime, all symptoms disappeared within a week, *while clopidogrel was continued*. After these 3 weeks, clopidogrel was still continued in most patients for a period of up to minimal 1-11 months (depending on the indication), without the hypersensitivity reaction returning.

Patch tests were performed in 42 patients with clopidogrel (20% pet., 30% water) and the related ticlopidine (75% water) and prasugrel (5% water) after completion of prescribed clopidogrel therapy and a minimum of 4 weeks drug free. 34 patients (81%) had a positive patch test to clopidogrel with 13 (31%) cross-reactions to ticlopidine and 10 (24%) to prasugrel. Microscopic analysis of biopsies from patch testing sites showed parakeratosis, epidermal spongiosis, lymphocytic exocytosis, and lymphocyte infiltrates. This cellular response was similar to that observed in areas affected by clopidogrel hypersensitivity. The authors concluded that both the generalized and the localized eruptions (which included psoriasis-like lesions on a foot sole) were caused by delayed-type hypersensitivity (2). Although the histopathology was consistent with such a reaction, several points cast doubt on the validity of this conclusion:

1. a period of 4 days is very short for induction and elicitation of delayed-type hypersensitivity (for which usually a minimum of 5-7 days is accepted) in first-time users (although this may also occur in some other drug eruptions).

2. it can be doubted whether a short low dosage prednisone course can totally suppress a generalized exanthem, while the culprit drug is continued.

3. it is generally assumed that delayed-type hypersensitivity is life-long or at least long-lasting. In this concept, it would be impossible to continue the culprit drug for a longer period of time without a recurrence of symptoms. It may be argued that the prednisone course completely suppressed the hypersensitivity, but in that case patch tests should have been negative after completion of therapy.

4. some of the localized skin reactions are very atypical for delayed-type hypersensitivity to systemic drugs.

5. Of the 4 pictures shown of 'positive' patch tests, not a single one is a classic positive reaction. In fact, some have definite characteristics of an irritant reaction. Apparently, the authors have not tested control patients not exposed to clopidogrel and the other 2 drugs, to exclude irritancy of the used test preparations (75% water for ticlopidine seems excessively high).

6. These 'hypersensitivity reactions' occur in 4-6% of the patients and lead to discontinuation in 1.5% (3,4,5). In most patients, the drug can be continued, often after a 'desensitization protocol'. Such protocols are never described in literature specialized in delayed-type allergy such as *Dermatitis* and *Contact Dermatitis*.

7. If these are really delayed-type hypersensitivity reactions, might expect that case reports or case series would have been described in these journals, especially with so many cases of cutaneous adverse reactions to clopidogrel available.

Therefore, in the opinion of the author, this report does *not* convincingly demonstrate delayed-type hypersensitivity to clopidogrel and the related substances. Having said this, it also cannot be excluded at the moment.

Cross-reactions, pseudo-cross-reactions and co-reactions
Patients sensitized to clopidogrel may have cross-reacted to ticlopidine and prasugrel (2).

LITERATURE

1   The data in the section 'General' may have been obtained from literature discussed in this chapter, but mostly also or exclusively from one or more of the following online sources: ChemIDPlus Advanced, PubChem, DrugBank, RxList, Drug Central, Drugs.com, and Wikipedia

2   Cheema AN, Mohammad A, Hong T, Jakubovic HR, Parmar GS, Sharieff W, et al. Characterization of clopidogrel hypersensitivity reactions and management with oral steroids without clopidogrel discontinuation. J Am Coll Cardiol 2011;58:1445-1454

3   Ford MK, Cohn JR. Clopidogrel hypersensitivity: Pathogenesis, presentation and diagnosis. Curr Vasc Pharmacol 2019;17:110-112

4   Campbell KL, Cohn JR, Savage MP. Clopidogrel hypersensitivity: clinical challenges and options for management. Expert Rev Clin Pharmacol 2010;3:553-561

5   Savage MP, Fischman DL. Clopidogrel hypersensitivity: Overview of the problem. Curr Vasc Pharmacol 2019;17:108-109

Chapter 3.146 CLOPIDOL

IDENTIFICATION

Description/definition	: Clopidol is the anticoccidial agent that conforms to the structural formula shown below
Pharmacological classes	: Coccidiostats
IUPAC name	: 3,5-Dichloro-2,6-dimethyl-1H-pyridin-4-one
Other names	: Methylchloropindol; meticlorpindol; methylchlorpindol
CAS registry number	: 2971-90-6
EC number	: 221-008-2
Merck Index monograph	: 3656
Patch testing	: 2% pet.
Molecular formula	: $C_7H_7Cl_2NO$

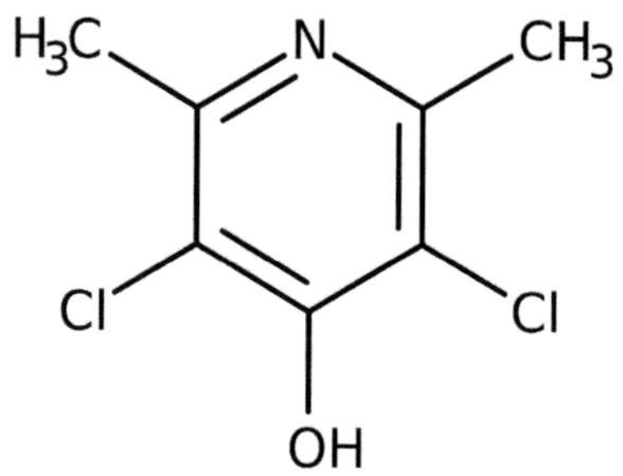

GENERAL

Clopidol is an organic compound that is used in veterinary medicine as a coccidiostat. It is administered to poultry to prevent the growth of pathogenic parasites (1).

CONTACT ALLERGY FROM ACCIDENTAL CONTACT

In Italy, during 1986-1988, 204 animal feed mill workers (191 men, 13 women) were patch tested with a large number of animal feed additives. There was one reaction to clopidol (methylchlorpindol) 2% pet. in a group of 36 subjects with clinical complaints (dermatitis or pruritus sine materia) and zero reactions in the group of 168 individuals without skin complaints. All reactions were considered to be relevant (2).

LITERATURE

1   The data in the section 'General' may have been obtained from literature discussed in this chapter, but mostly also or exclusively from one or more of the following online sources: ChemIDPlus Advanced, PubChem, DrugBank, RxList, Drug Central, Drugs.com, and Wikipedia
2   Mancuso G, Staffa M, Errani A, Berdondini RM, Fabbri P. Occupational dermatitis in animal feed mill workers. Contact Dermatitis 1990;22:37-41

Chapter 3.147 CLOPREDNOL

IDENTIFICATION

Description/definition : Cloprednol is the synthetic glucocorticoid that conforms to the structural formula shown below
Pharmacological classes : Glucocorticoids
IUPAC name : (8S,9S,10R,11S,13S,14S,17R)-6-Chloro-11,17-dihydroxy-17-(2-hydroxyacetyl)-10,13-dimethyl-9,11,12,14,15,16-hexahydro-8H-cyclopenta[a]phenanthren-3-one
CAS registry number : 5251-34-3
EC number : 226-052-6
Merck Index monograph : 3657
Patch testing : Generally, corticosteroids may be tested at 0.1% and 1% in alcohol; late readings (6-10 days) are strongly recommended
Molecular formula : $C_{21}H_{25}ClO_5$

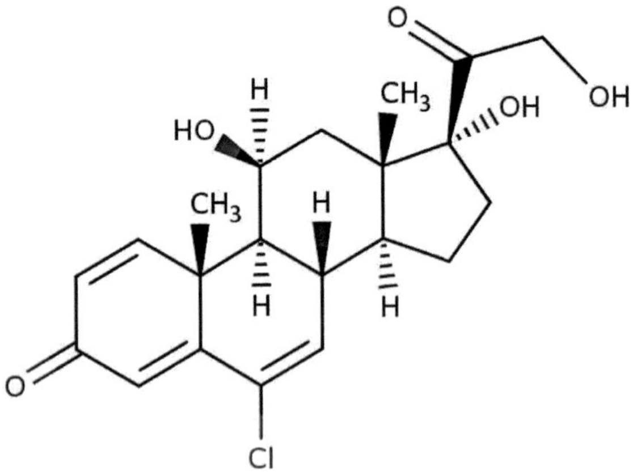

GENERAL

Systemically administered glucocorticoids have anti-inflammatory, immunosuppressive and antineoplastic properties and are used in the treatment of a wide spectrum of diseases including rheumatic disorders, lung diseases (asthma, COPD), gastrointestinal tract disorders (Crohn's disease, colitis ulcerosa), certain malignancies (leukemia, lymphomas), hematological disorders, and various diseases of the kidneys, brain, eyes and skin. A practical guideline for diagnosing allergic reactions to corticosteroids is presented in ref. 1.

CUTANEOUS ADVERSE DRUG REACTIONS FROM SYSTEMIC ADMINISTRATION CAUSED BY TYPE IV (DELAYED-TYPE) HYPERSENSITIVITY (as demonstrated by positive patch tests)

Cutaneous adverse drug reactions from systemic administration of cloprednol caused by type IV (delayed-type) hypersensitivity have included systemic allergic dermatitis presenting as the baboon syndrome/symmetrical drug-related intertriginous and flexural exanthema (SDRIFE) (2).

Systemic allergic dermatitis (systemic contact dermatitis)

A 50-year-old woman who was very likely sensitized to corticosteroids from their use in chronic nasal congestion had suffered a skin eruption on three occasions one day after oral intake of prednisolone or methylprednisolone. Patch tests were positive to prednisolone (methylprednisolone not tested). She underwent an oral provocation test with cloprednol, which resulted in the classic picture of the baboon syndrome/symmetrical drug-related intertriginous and flexural exanthema (SDRIFE) after 12-24 hours. A patch test with cloprednol was negative, intradermal test not performed. This was very likely a case of systemic allergic dermatitis presenting as the baboon syndrome/SDRIFE with a false-negative patch test to cloprednol (2). False-negative patch test reactions to corticosteroids are frequent.

Cross-reactions, pseudo-cross-reactions and co-reactions

Cross-reactions between corticosteroids are discussed in Chapter 3.399 Prednisolone.

LITERATURE

1   Baeck M, Goossens A. Immediate and delayed allergic hypersensitivity to corticosteroids: practical guidelines. Contact Dermatitis 2012;66:38-45
2   Treudler R, Simon JC. Symmetric, drug-related, intertriginous, and flexural exanthema in a patient with polyvalent intolerance to corticosteroids. J Allergy Clin Immunol 2006;118:965-967

Chapter 3.148 CLORAZEPIC ACID

IDENTIFICATION
Description/definition : Clorazepic acid is the benzodiazepine that conforms to the structural formula shown below
Pharmacological classes : Anticonvulsants; GABA modulators; anti-anxiety agents
IUPAC name : 7-Chloro-2-oxo-5-phenyl-1,3-dihydro-1,4-benzodiazepine-3-carboxylic acid
Other names : Clorazepate
CAS registry number : 23887-31-2
EC number : 245-926-8
Merck Index monograph : 3662
Patch testing : 1% water
Molecular formula : $C_{16}H_{11}ClN_2O_3$

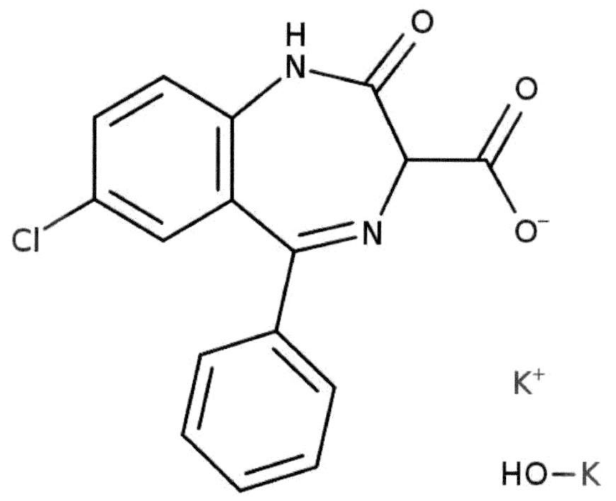

GENERAL
Clorazepic acid (clorazepate) is a benzodiazepine used as an anticonvulsant as adjunctive therapy in the management of epilepsy and as an anxiolytic for therapy of anxiety and alcohol withdrawal. In pharmaceutical products, clorazepate is employed as clorazepate dipotassium (CAS number 57109-90-7, EC number not available, molecular formula $C_{16}H_{11}ClK_2N_2O_4$) (1).

CONTACT ALLERGY FROM ACCIDENTAL CONTACT
A 37-year-old woman developed eczema on the forefinger of the right hand, which later spread to the dorsum of the right hand, eyelids, neck and cheeks. She cared for a daughter with cerebral paralysis, and had to give her medication after triturating tablets by hand and gathering the powder up into a small spoon with her forefinger. Patch testing showed positive reactions to 4 commercial drugs the daughter used tested at 1% water. Later, patch tests with their active ingredients were performed and the patient had positive reactions to clorazepate, tetrazepam, diazepam and sodium valproate, all tested 1% in water (2). This was a case of allergic contact dermatitis *by proxy*.

CUTANEOUS ADVERSE DRUG REACTIONS FROM SYSTEMIC ADMINISTRATION CAUSED BY TYPE IV (DELAYED-TYPE) HYPERSENSITIVITY (as demonstrated by positive patch tests)
Cutaneous adverse drug reactions from systemic administration of clorazepate caused by type IV (delayed-type) hypersensitivity have included maculopapular exanthema (3).

Maculopapular exanthema
A 34-year-old woman developed a generalized maculopapular exanthem beginning on the hands 1 day after anesthesia for larynx surgery. Perioperative medication consisted of clorazepate and 4 other drugs. Oral treatment with a glucocorticoid and an H_1-receptor antagonist resulted in rapid clearance of the exanthem. Patch testing with all drugs revealed positive reactions only to clorazepate 20 mg/ml (commercial solution) at D2 (++) and D3 (+) (3).

LITERATURE

1   The data in the section 'General' may have been obtained from literature discussed in this chapter, but mostly also or exclusively from one or more of the following online sources: ChemIDPlus Advanced, PubChem, DrugBank, RxList, Drug Central, Drugs.com, and Wikipedia

2   Garcia-Bravo B, Rodriguez-Pichardo A, Camacho F. Contact dermatitis from diazepoxides. Contact Dermatitis 1994;30:40

3   Sachs B, Erdmann S, Al-Masaoudi T, Merk HF. In vitro drug allergy detection system incorporating human liver microsomes in chlorazepate-induced skin rash: drug-specific proliferation associated with interleukin-5 secretion. Br J Dermatol 2001;144:316-320

Chapter 3.149 CLOTIAZEPAM

IDENTIFICATION

Description/definition	: Clotiazepam is the thienodiazepine derivative that conforms to the structural formula shown below
Pharmacological classes	: Anti-anxiety agents
IUPAC name	: 5-(2-Chlorophenyl)-7-ethyl-1-methyl-3H-thieno[2,3-e][1,4]diazepin-2-one
CAS registry number	: 33671-46-4
EC number	: 251-627-3
Merck Index monograph	: 3670
Patch testing	: Tablet, pulverized, 30% pet. (2); most systemic drugs can be tested at 10% pet.; if the pure chemical is not available, prepare the test material from intravenous powder, the content of capsules or – when also not available – from powdered tablets to achieve a final concentration of the active drug of 10% pet.
Molecular formula	: $C_{16}H_{15}ClN_2OS$

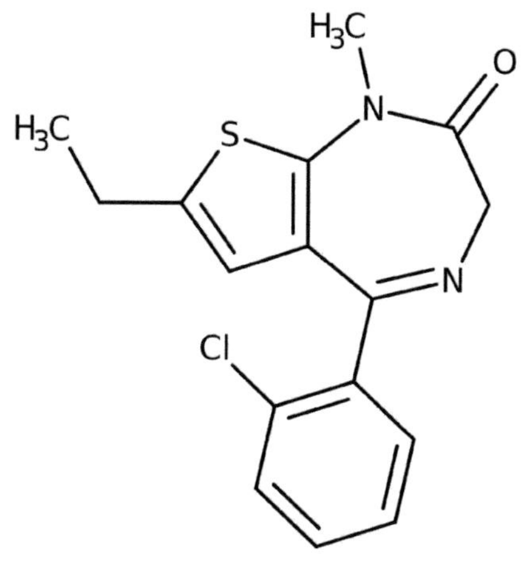

GENERAL

Clotiazepam is a thienodiazepine possessing anxiolytic, anticonvulsant, sedative and skeletal muscle relaxant properties. It increases the stage 2 non-rapid eye movement sleep. Clotiazepam is used for the treatment of anxiety disorders (1).

CONTACT ALLERGY FROM ACCIDENTAL CONTACT

A 66-year-old woman presented with itchy and burning eczema of the face, which had first appeared on the eyelids. Later, extension to the forehead, lips and perioral region, neck, ears and the fingers occurred. The patient's husband suffered from Parkinson's disease, and she had to crush a large number of tablets for him up to five times a day. She had positive patch tests to clotiazepam (crushed tablet 30% pet.; 3 controls were negative), 3 other benzodiazepines (bromazepam, lorazepam, tetrazepam) and trazodone HCl. The patient was diagnosed with (airborne) allergic contact dermatitis from drugs (2). This was a case of allergic contact dermatitis *by proxy*.

Cross-reactions, pseudo-cross-reactions and co-reactions

The possibility of cross-reactions between benzodiazepines, especially with primary sensitization to tetrazepam, is considered likely, but the pattern is unknown and may differ between sensitivity from (occupational) contact with the drugs and systemically induced reactions (2).

LITERATURE

1. The data in the section 'General' may have been obtained from literature discussed in this chapter, but mostly also or exclusively from one or more of the following online sources: ChemIDPlus Advanced, PubChem, DrugBank, RxList, Drug Central, Drugs.com, and Wikipedia
2. Swinnen I, Ghys K, Kerre S, Constandt L, Goossens A. Occupational airborne contact dermatitis from benzodiazepines and other drugs. Contact Dermatitis 2014;70:227-232

Chapter 3.150 CLOXACILLIN

IDENTIFICATION

Description/definition : Cloxacillin is the semisynthetic β-lactamase resistant penicillin that conforms to the structural formula shown below
Pharmacological classes : Anti-bacterial agents
IUPAC name : (2S,5R,6R)-6-[[3-(2-Chlorophenyl)-5-methyl-1,2-oxazole-4-carbonyl]amino]-3,3-dimethyl-7-oxo-4-thia-1-azabicyclo[3.2.0]heptane-2-carboxylic acid
CAS registry number : 61-72-3
EC number : 200-514-7
Merck Index monograph : 3673
Patch testing : 2% pet. and water (12); beta-lactam antibiotics can generally be tested 5-10% water and pet.
Molecular formula : $C_{19}H_{18}ClN_3O_5S$

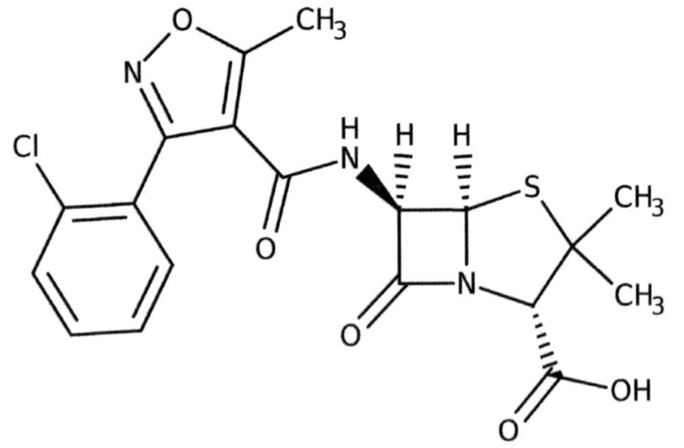

GENERAL

Cloxacillin is a semisynthetic β-lactamase resistant penicillin antibiotic with antibacterial activity; it is a chlorinated derivative of oxacillin. Cloxacillin is indicated for the treatment of infections caused by penicillinase-producing staphylococci, including pneumococci, group A β-hemolytic streptococci, and penicillin G-sensitive and penicillin G-resistant staphylococci (1). In pharmaceutical products, both cloxacillin base and cloxacillin sodium (CAS number 642-78-4, EC number 211-390-9, molecular formula $C_{19}H_{17}ClN_3NaO_5S$) may be employed (1). The classification and structures of beta-lactam antibiotics are discussed in Chapter 3.36 Amoxicillin.

CONTACT ALLERGY FROM ACCIDENTAL CONTACT

In a group of 107 workers in the pharmaceutical industry with dermatitis, investigated in Warsaw, Poland, before 1989, 7 reacted to cloxacillin, tested 20% pet. (2). Also in Warsaw, Poland, in the period 1979-1983, 27 pharmaceutical workers, 24 nurses and 30 veterinary surgeons were diagnosed with occupational allergic contact dermatitis from antibiotics. The numbers that had positive patch tests to cloxacillin (ampoule content) were 7, 3, and 0, respectively, total 10 (3). In an investigation by the same authors, of 333 nurses patch tested between 1979 and 1987, 1 reacted to cloxacillin 20% pet.; this reaction was considered likely to be relevant (5).

Between 1978 and 2001, in the university hospital of Leuven, Belgium, occupational allergic contact dermatitis to pharmaceuticals was diagnosed in 33 health care workers: 26 nurses, 4 veterinarians, 2 pharmacists and one medical doctor. Practically all of these patients presented with hand dermatitis (often the fingers), sometimes with secondary localizations, frequently the face. In this group, two patients had occupational allergic contact dermatitis from cloxacillin (10).

CUTANEOUS ADVERSE DRUG REACTIONS FROM SYSTEMIC ADMINISTRATION CAUSED BY TYPE IV (DELAYED-TYPE) HYPERSENSITIVITY (as demonstrated by positive patch tests)

Cutaneous adverse drug reactions from systemic administration of cloxacillin caused by type IV (delayed-type) hypersensitivity have included exfoliative dermatitis (15), acute generalized exanthematous pustulosis (AGEP) (4,8,14), drug reaction with eosinophilia and systemic symptoms (DRESS) (6,7,12), Stevens-Johnson syndrome/toxic epidermal necrolysis (SJS/TEN) (11), and unspecified drug eruption (9).

The unconventional topical use of cloxacillin in the form of the content of a vial for parenteral use to leg ulcers has resulted in allergic contact dermatitis in 2 patients (13).

Erythroderma, widespread erythematous eruption, exfoliative dermatitis
A patient aged 75 had suffered a desquamative exanthema attributed to cloxacillin. An oral provocation test was positive. 2 years later, a delayed intradermal test and a patch test (5% pet.) were both positive to cloxacillin (15).

Acute generalized exanthematous pustulosis (AGEP)
A 39-year-old, 41-week pregnant woman was admitted for the induction of labor, and treated for the group B streptococcal vaginal colonization found by swab before delivery. She was administered ampicillin/cloxacillin sodium (A/CS) i.v. Five hours after the infusion, she developed a fever and edematous erythema. A/CS was stopped thereafter. However, after delivery, her fever continued up to 39.5°C, and she had diffuse edematous erythema with numerous non-follicular pustules on her neck, axilla, wrist and trunk. Her blood test showed leukocytosis with marked neutrophilia and elevated C-reactive protein. Blood cultures and swabs from the pustules were sterile. A skin biopsy showed spongiform subcorneal/intraepidermal pustules, and marked liquefaction degeneration. Later, patch tests were positive to A/CS at 0.05%, 0.1%, 0.5%, 1% and 10% aqua, as was a drug lymphocyte stimulation test. The patient refused further tests, so it remains unknown whether this case of acute generalized exanthematous pustulosis (AGEP) was caused by ampicillin, cloxacillin, or both (4).

In a study from Nancy, France, 54 patients with suspected nonimmediate drug eruptions (27 maculopapular, seven erythrodermic, nine eczematous, four photosensitivity, three fixed drug eruptions, three with pruritus and one with acute generalized exanthematous pustulosis) were assessed with patch testing. The single patient with AGEP had a positive patch test to cloxacillin (8).

In a hospital in France, specialized in toxic bullous diseases and severe cutaneous adverse reactions, a retrospective study was performed in consecutive patients consulting between 2010 and 2018 with a nonimmediate CADR suspected to have been caused by beta-lactam antibiotics. 56 patients were included, among whom were 46 amoxicillin-suspected and seven cephalosporin-suspected. Twenty-nine had severe CADR (DRESS, AGEP, SJS/TEN), and 27 had nonimmediate maculopapular exanthema (MPE). Of these patients, twenty patients had positive tests to the culprit drugs (or related beta-lactams). Cloxacillin was responsible for 1 case of AGEP (cloxacillin itself not patch tested, but positive reactions to other penicillins (14).

Drug reaction with eosinophilia and systemic symptoms (DRESS)
A male patient was treated with intravenous ampicillin and gentamicin for bacterial endocarditis. Three days later, after isolation of *Staphylococcus aureus,* cloxacillin was started and the other antibiotics discontinued. Clinical symptoms improved considerably, but after 3 weeks, the patient developed fever, generalized erythematous maculopapular rash and malaise. Gentamicin was again added. The clinical picture worsened progressively, with severe leukoneutropenia, failure of hepatic and renal function and progression of the skin lesions to purple blisters and subsequent desquamation. After 5 days, gentamicin and cloxacillin were discontinued and with treatment, complete recuperation occurred in 3 weeks. Prick and intradermal tests with all antibiotics used were positive only to the intradermal test with cloxacillin 2 mg/ml read at D2 and D4. Patch tests were positive to cloxacillin 0.2% and 2% pet. and water, with stronger reactions to the higher concentrations (++/+++) than the lower ones (+/+). Skin biopsies performed in the positive patch tested area showed a clear T lymphocyte, CD4+ infiltrate, thus confirming the occurrence of a cell-mediated response (12). Although not mentioned as such by the authors, this clearly was a case of DRESS.

One patient from France developed DRESS from carbamazepine and cloxacillin and had positive patch tests to carbamazepine 10% pet. and cloxacillin 30% pet. As these drug are not related, this was a case of multiple hypersensitivity syndrome (6). In a multicenter investigation in France, of 72 patients patch tested for DRESS, 46 (64%) had positive patch tests to drugs, including one to cloxacillin (7).

Stevens-Johnson syndrome/toxic epidermal necrolysis (SJS/TEN)
In a 61-year-old man, delayed-type hypersensitivity to cloxacillin (tested as commercial tablet 2 g, 15% and 30% pet.) and other drugs may have caused/contributed to TEN (toxic epidermal necrolysis) following DRESS (drug reaction with eosinophilia and systemic symptoms) from other drugs (multiple hypersensitivity syndrome) (11).

Other cutaneous adverse drug reactions
In Finland, in the period 1989-2001, 826 patients with suspected cutaneous drug eruptions were patch tested and 89 had one or more positive reactions. Of these individuals, 1 reacted to cloxacillin. It was not mentioned which drug eruption this patient had suffered from (9).

Cross-reactions, pseudo-cross-reactions and co-reactions
Cross-reactions between beta-lactam antibiotics are discussed in Chapter 3.36 Amoxicillin.

LITERATURE

1   The data in the section 'General' may have been obtained from literature discussed in this chapter, but mostly also or exclusively from one or more of the following online sources: ChemIDPlus Advanced, PubChem, DrugBank, RxList, Drug Central, Drugs.com, and Wikipedia

2   Rudzki E, Rebandel P, Grzywa Z. Contact allergy in the pharmaceutical industry. Contact Dermatitis 1989;21:121-122

3   Rudzki E, Rebendel P. Contact sensitivity to antibiotics. Contact Dermatitis 1984;11:41-42

4   Matsumoto Y, Okubo Y, Yamamoto T, Ito T, Tsuboi R. Case of acute generalized exanthematous pustulosis caused by ampicillin/cloxacillin sodium in a pregnant woman. J Dermatol 2008;35:362-364

5   Rudzki E, Rebandel P, Grzywa Z. Patch tests with occupational contactants in nurses, doctors and dentists. Contact Dermatitis 1989;20:247-250

6   Studer M, Waton J, Bursztejn AC, Aimone-Gastin I, Schmutz JL, Barbaud A. Does hypersensitivity to multiple drugs really exist? Ann Dermatol Venereol 2012;139:375-380 (Article in French)

7   Barbaud A, Collet E, Milpied B, Assier H, Staumont D, Avenel-Audran M, et al. A multicentre study to determine the value and safety of drug patch tests for the three main classes of severe cutaneous adverse drug reactions. Br J Dermatol 2013;168:555-562

8   Barbaud A, Reichert-Penetrat S, Tréchot P, Jacquin-Petit MA, Ehlinger A, Noirez V, et al. The use of skin testing in the investigation of cutaneous adverse drug reactions. Br J Dermatol 1998;139:49-58

9   Lammintausta K, Kortekangas-Savolainen O. The usefulness of skin tests to prove drug hypersensitivity. Br J Dermatol 2005;152:968-974

10 Gielen K, Goossens A. Occupational allergic contact dermatitis from drugs in healthcare workers. Contact Dermatitis 2001;45:273-279

11 Liippo J, Pummi K, Hohenthal U, Lammintausta K. Patch testing and sensitization to multiple drugs. Contact Dermatitis 2013;69:296-302

12 Moreno-Ancillo A, Domínguez-Noche C, Gil-Adrados AC, Cosmes PM. Near-fatal delayed hypersensitivity reaction to cloxacillin. Contact Dermatitis 2003;49:44-45

13 Gamboa P, Jáuregui I, Urrutia I, González G, Antépara I. Contact sensitization to cloxacillin with oral tolerance to other beta-lactam antibiotics. Contact Dermatitis 1996;34:75-76

14 Bérot V, Gener G, Ingen-Housz-Oro S, Gaudin O, Paul M, Chosidow O, Wolkenstein P, Assier H. Cross-reactivity in beta-lactams after a non-immediate cutaneous adverse reaction: experience of a reference centre for toxic bullous diseases and severe cutaneous adverse reactions. J Eur Acad Dermatol Venereol 2020;34:787-794

15 Padial A, Antunez C, Blanca-Lopez N, Fernandez TD, Cornejo-Garcia JA, Mayorga C, et al. Nonimmediate reactions to betalactams: diagnostic value of skin testing and drug provocation test. Clin Exp Allergy 2008;38:822-828

Chapter 3.151 CLOZAPINE

IDENTIFICATION

Description/definition : Clozapine is the benzodiazepine that conforms to the structural formula shown below
Pharmacological classes : Serotonin antagonists; antipsychotic agents; GABA antagonists
IUPAC name : 3-Chloro-6-(4-methylpiperazin-1-yl)-11H-benzo[b][1,4]benzodiazepine
CAS registry number : 5786-21-0
EC number : 227-313-7
Merck Index monograph : 3676
Patch testing : No data available; suggested: 10% pet.
Molecular formula : $C_{18}H_{19}ClN_4$

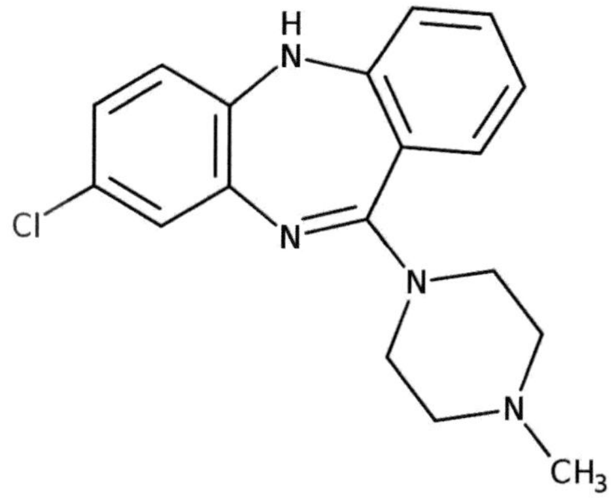

GENERAL

Clozapine is a psychotropic agent belonging to the chemical class of benzisoxazole derivatives and is indicated for the treatment of schizophrenia. Clozapine's use is restricted to refractory schizophrenia. In pharmaceutical products, clozapine is employed as clozapine hydrochloride (CAS number 54241-01-9, EC number not available, molecular formula $C_{18}H_{20}Cl_2N_4$) (1).

CUTANEOUS ADVERSE DRUG REACTIONS FROM SYSTEMIC ADMINISTRATION CAUSED BY TYPE IV (DELAYED-TYPE) HYPERSENSITIVITY (as demonstrated by positive patch tests)

Cutaneous adverse drug reactions from systemic administration of clozapine caused by type IV (delayed-type) hypersensitivity have included erythematous, papular and pustular eruption with vasculitis (2).

Other cutaneous adverse drug reactions

About 10 days after he started using lithium and clozapine for the treatment of mania, an 18-year-old man developed high fever and pruritic red eruptions and pustules in the axillae, groin, on the legs and back. Physical examination showed confluent erythematous papules, coalescing to form larger plaques, studded with pinhead-sized pustules. In addition, discrete purpuric papules over the dorsum of the feet and shins were noted. Gram stain from the pustules showed scant neutrophils with no evidence of bacteria. The workup for acute febrile illness was unyielding. Histopathology from a pustular lesion showed a spongiotic epidermis, and occasional necrotic keratinocytes overlying a dermis showing perivascular lymphohistiocytic infiltrates with extravasated red blood cells. Direct immunofluorescence of a purpuric lesion displayed vesselwall staining with C3 and fibrinogen, suggestive of vasculitis. When clozapine was continued and its dose increased, the rash worsened progressively. Six months after cessation of the drug, patch tests were positive to clozapine (++), but not to lithium. No details on drug concentration and vehicle were provided, nor on the number of controls. The patient was diagnosed with symmetrical drug-related intertriginous and flexural exanthema (SDRIFE) and vasculitis (2). However, the clinical picture with lesions 'studded with pustules', the localization on the back of the hands and the high fever do not adequately fulfill the criteria for SDRIFE (although the fever may have been associated with vasculitis).

LITERATURE

1   The data in the section 'General' may have been obtained from literature discussed in this chapter, but mostly also or exclusively from one or more of the following online sources: ChemIDPlus Advanced, PubChem, DrugBank, RxList, Drug Central, Drugs.com, and Wikipedia
2   Suvarna P, Kayarkatte MN, Shenoi SD, Jaiprakash P. A rare case of clozapine-induced symmetrical drug-related intertriginous and flexural exanthema with vasculitis-like lesions. Contact Dermatitis 2020;82:318-320

Chapter 3.152 CODEINE

IDENTIFICATION

Description/definition	: Codeine is the naturally occurring phenanthrene alkaloid and opioid agonist that conforms to the structural formula shown below
Pharmacological classes	: Antitussive agents; narcotics; analgesics, opioid
IUPAC name	: (4R,4aR,7S,7aR,12bS)-9-Methoxy-3-methyl-2,4,4a,7,7a,13-hexahydro-1H-4,12-methanobenzofuro[3,2-e]isoquinoline-7-ol
Other name(s)	: Methylmorphine; morphine monomethyl ether
CAS registry number	: 76-57-3
EC number	: 200-969-1
Merck Index monograph	: 3718
Patch testing	: Codeine phosphate 5% pet.; 1% water or alcohol; water may have caused false-negative reactions in one study (2)
Molecular formula	: $C_{18}H_{21}NO_3$

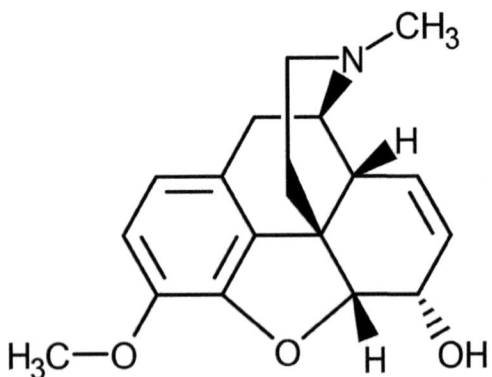

GENERAL

Codeine is a naturally occurring phenanthrene alkaloid and opioid agonist with analgesic, antidiarrheal and antitussive activities. This agent is related to morphine but with less potent analgesic properties and mild sedative effects; it also acts centrally to suppress cough. Codeine is indicated for the relief of mild to moderately severe pain, where the use of an opioid analgesic is appropriate and as a cough suppressant in adults (1). In pharmaceutical products, codeine is mostly employed as codeine phosphate (hemihydrate) (CAS number 41444-62-6, EC number 200-137-8, molecular formula $C_{36}H_{50}N_2O_{15}P_2$) (1).

CONTACT ALLERGY FROM ACCIDENTAL CONTACT

Case series

In the period 2004 to 2017, of sixteen workers referred from a single opioid-manufacturing plant in Australia, eleven were diagnosed with occupational ACD caused by opioids, with seven reacting to thebaine, five to morphine, four to norhydroxymorphinone, two to codeine (tested 2.5% alc.), and two to oripavine. Most reported facial dermatitis, indicative of airborne contact dermatitis, although the hands and arms were also often involved. Six individuals reacted to at least two substances. This might be indicative of cross-reactivity or be the result of multiple exposures (13).

Five men working in the production of opium alkaloids presented with dermatitis of the face, arms and hands; 3 of them had also symptoms of rhinitis, sinusitis, bronchitis and conjunctivitis. The dermatitis and respiratory tract symptoms were clearly work-related. The patients were exposed to a number of different alkalis, solvents and inorganic chemicals as well as to various alkaloids. Patch tests were positive in 3 to mixed alkaloid raw material 1% alc. and codeine base and codeine HCl 1% alc. Fifty controls were negative (11). Quite curiously, 2 of the men co-reacted to p-phenylenediamine and all three to diaminodiphenylmethane (4,4'-methylenedianiline), which para-compounds had been tested because co-reactivity had been observed before (12). There are no structural similarities suggesting cross-reactivity (11).

Two workers in a factory producing opium alkaloids from poppy straw developed allergic contact dermatitis from their work. One had eczema of the neck, hands and forearms and showed positive patch tests to codeine phosphate, codeine hydrochloride and morphine bitartrate (negative to morphine HCl), all tested 1% alc. or water. The other patient had developed dermatitis of the face and hands after having worked in the production of concentrated poppy straw for 10 years. Patch tests were positive to morphine HCl and bitartrate and codeine bitartrate, all 1% alc. The authors mentioned that many patients allergic to opium alkaloids are also allergic to *p*-phenylenediamine; an explanation for this co-reactivity is lacking (27).

Two female laboratory workers aged 38 and 58 years were employed in the same pharmaceutical company that manufactures several opiates, and handled morphine and codeine during work. They developed papulovesicular contact dermatitis with an airborne pattern. Both women showed positive patch test reactions to codeine 5% pet. (+) and ethylmorphine (cross-reaction). Patient 1 also showed a positive patch test reaction to morphine (with which she had contact, and patient 2 showed a positive patch test reaction to naloxone (cross-reaction) (2).

Case report
A 22-year-old female working in a laboratory of a pharmaceutical company manufacturing a range of opiates developed acute dermatitis on the fingers of both and later also on the face and the neck. Patch tests were positive to codeine phosphate 5% alc. and thebaine 5% alc., both opiate alkaloids that the patient had contact with (8).

CUTANEOUS ADVERSE DRUG REACTIONS FROM SYSTEMIC ADMINISTRATION CAUSED BY TYPE IV (DELAYED-TYPE) HYPERSENSITIVITY (as demonstrated by positive patch tests)
Cutaneous adverse drug reactions from systemic administration of codeine caused by type IV (delayed-type) hypersensitivity have included maculopapular eruption (5,6), erythroderma (7), exfoliative dermatitis (9), systemic allergic dermatitis, fixed drug eruption (25), drug reaction with eosinophilia and systemic symptoms (DRESS) (26), generalized dermatitis (19), urticaria (10), and undefined/unknown exanthems (3,4,18).

Maculopapular eruption
A 38-year-old woman and a 26-year old woman developed a pruriginous generalized maculopapular eruption after receiving treatment with codeine. In the first patient, dermatitis appeared 1 hour after she had 10 mg of oral codeine. The second patient had taken paracetamol/codeine 500/30 mg 4 hours earlier. A 72-year-old man (also reported in ref. 6) developed a pruriginous generalized maculopapular eruption 12 hours after he had paracetamol/codeine 500/30. In the latter 2 patients, paracetamol alone was tolerated well. Patch tests were positive to codeine 5% pet. and morphine 5% pet. (cross-reaction). Five controls were negative (5).

Erythroderma, widespread erythematous eruption, exfoliative dermatitis
An 85-year-old man had since 3 days an erythematous eruption that began on his legs and progressively extended to his trunk, arms and face, accompanied by intense itching; in some areas, vesicular exudative lesions were visible. Some days before, the patient had taken an NSAID, paracetamol and several paracetamol/codeine 500/10 mg pills to treat back pain. Patch tests with the three drugs were positive to codeine 0.1% water only and – in a second test session - to morphine HCl 1% water and dihydrocodeine 20% pet. (7). This was a case of erythroderma from delayed-type hypersensitivity to codeine.

A 27-year-old woman had suffered several episodes of erythematous, pruritic, exfoliative dermatitis, associated with abdominal pain, vomiting and diarrhea. She had taken antitussives and analgesics containing codeine 3-4 hours before each episode. Patch tests were positive to codeine phosphate 0.1% water and morphine HCl 0.1% and 1% water (cross-reaction) (9).

Systemic allergic dermatitis (systemic contact dermatitis)
In early literature, (occupational) contact dermatitis and (in some cases) systemic allergic dermatitis to opium alkaloids, including codeine, morphine, apomorphine, ethylmorphine and diacetylmorphine (heroin) have been described (e.g. 15,16, 17,23,24, reviews in ref. 14 and 16), the first report dating back to 1882 (cited in ref. 14). In those days, eczematous dermatitis from opium compounds used externally in the form of lotions, suppositories and other application forms was apparently well known (16). A more recent review of the subject was provided in 2006 (21).

Fixed drug eruption
A 16-year-old girl was referred for allergy investigations because of presumed previous fixed drug eruption. Seven suspected drugs were patch tested on normal and postlesional pigmented skin and there was a positive reaction only to codeine on pigmented skin (no patch test details provided). An oral provocation test was positive to codeine but also in 5 other – unrelated – drugs that had been patch test negative (25).

Drug reaction with eosinophilia and systemic symptoms (DRESS)
A 19-year-old Japanese man was prescribed codeine phosphate 10 mg 3 times daily and several other drugs for cold symptoms. About 20 days later, an erythematous, maculopapular rash appeared and progressed to erythroderma. High fever developed and he had splenomegaly and generalized lymphadenopathy. Laboratory examinations showed atypical lymphocytes, eosinophilia, and increased liver enzyme values. The results of patch tests were positive for codeine phosphate (26). This was a case of drug reaction with eosinophilia and systemic symptoms (DRESS) caused by delayed-type hypersensitivity to codeine phosphate.

Dermatitis/eczematous eruption
Generalized dermatitis from oral codeine with a positive patch test to codeine and morphine 0.1% and 1% water in a 43-year-old man was reported from Spain (19)

Other cutaneous adverse drug reactions
A 49-year-old man had suffered several attacks of urticaria, lasting for 1-2 days on each occasion. He was suspected of being 'allergic' to a cough mixture containing codeine phosphate (0.05% w/v) and epinephrine (0.27% w/v). Patch tests were positive to the cough mixture and to paracetamol/codeine crushed tablet undiluted, but negative to the cough mixture without codeine. A tentative diagnosis of urticarial rash from oral codeine due to delayed-type allergy was made. In a second patch test session with serial dilutions of codeine and other opioids, the patient reacted to codeine phosphate 0.033-0.001% water with cross-reactions to diacetylmorphine HCl (heroin), dihydrocodeine HCl, morphine HCl, and ethylmorphine HCl. The fully informed patient took one therapeutic dose of the cough mixture containing 1.44 mg codeine phosphate, and 12 hours later presented with an urticarial rash, which subsided in about 2 days (10).

When a female patient used a combination of paracetamol and codeine phosphate, an exanthem (unspecified) appeared. She had previously suffered a similar exanthem twice after reconstructive/endovascular operations. Fentanyl had been used for general anesthesia, and oxycodone for postoperative pain. In patch testing, positive reactions were seen to codeine phosphate and oxycodone, but fentanyl was negative (3). It can be concluded that the allergy to codeine was the result of cross-reactivity to primary oxycodone sensitization.

Additional case reports of drug eruptions from codeine caused by delayed-type allergy, adequate data of which are not available to the author, can be found in refs. 4 and 18.

Cross-reactions, pseudo-cross-reactions and co-reactions
Some patients sensitized to codeine have cross-reacted to ethylmorphine HCl (2,10), naloxone (2), dihydrocodeine (7,10), or diacetylmorphine HCl (10). Virtually all patients sensitized to codeine (= methylmorphine) cross-react to morphine (5,6,7,9,10,19).
A patient sensitized to oxycodone cross-reacted to codeine (3). Another individual sensitized by morphine showed cross-reactions to codeine and hydromorphone (20). Of 8 nurses sensitized to heroin (diacetylmorphine), 4 (50%) cross-reacted to codeine (methylmorphine) (21). Cross-reaction from heroin sensitization codeine in a nurse (22).

LITERATURE
1   The data in the section 'General' may have been obtained from literature discussed in this chapter, but mostly also or exclusively from one or more of the following online sources: ChemIDPlus Advanced, PubChem, DrugBank, RxList, Drug Central, Drugs.com, and Wikipedia
2   Colomb S, Bourrain JL, Bonardel N, Chiriac A, Demoly P. Occupational opiate contact dermatitis. Contact Dermatitis 2017;76:240-241
3   Liippo J, Pummi K, Hohenthal U, Lammintausta K. Patch testing and sensitization to multiple drugs. Contact Dermatitis 2013;69:296-302
4   Schmutz JL, Barbaud A, Trechot P. Codeine and cutaneous drug reactions: absence of cross-allergy with tramadol and fentanyl. Ann Dermatol Venereol 2010;137:429
5   Rodríguez A, Barranco R, Latasa M, de Urbina JJ, Estrada JL. Generalized dermatitis due to codeine. Cross-sensitization among opium alkaloids. Contact Dermatitis 2005;53:240
6   Estrada JL, Puebla MJ, de Urbina JJ, Matilla B, Prieto MA, Gozalo F. Generalized eczema due to codeine. Contact Dermatitis 2001;44:185
7   Gastaminza G, Audicana M, Echenagusia MA, Uriel O, Garcia-Gallardo MV, Velasco M, et al. Erythrodermia caused by allergy to codeine. Contact Dermatitis 2005;52:227-228
8   Waclawski ER, Aldridge R. Occupational dermatitis from thebaine and codeine. Contact Dermatitis 1995;33:51
9   Rodriguez F, Fernandez L, Garcia-Abujeta JL, Maquiera E, Llaca HF, Jerez J. Generalized dermatitis due to codeine. Contact Dermatitis 1995;32:120
10 De Groot AC, Conemans J. Allergic urticarial rash from oral codeine. Contact Dermatitis 1986;14:209-214
11 Romaguera C, Grimalt F. Occupational dermatitis from codeine. Contact Dermatitis 1983;9:170

12 Moran M, Caunedo JM, Martin Pascual A. Dermatitis de contacto por alcaloides del opio. Boletin del FEIDC 1981;2:25 (Article in Spanish, data cited in ref. 11)

13 Flury U, Cahill JL, Nixon RL. Occupational contact dermatitis caused by opioids: A case series. Contact Dermatitis 2019;81:332-335

14 Touraine A. Les dermatoses de l'opium. Revue de Médecine 1936;53:449-460

15 Dore SE, Prosser Thomas EW. Contact dermatitis in a morphine factory. J Allergy 1945;16:35-36

16 Jordon JW, Osborne ED. Contact dermatitis from opium derivatives. JAMA 1939;113:1955-1957

17 Palmer RB. Contact dermatitis due to codeine. Arch Dermatol 1942;46:82

18 Voorhorst R, Sparreboom S. 4 cases of recurrent pseudoscarlet fever caused by phenanthrene alkaloids with a 6-hydroxy group (codeine and morphine). Ann Allergy 1980;44:116-120

19 Iriarte Sotés P, López Abad R, Gracia Bara MT, Castro Murga M, Sesma Sánchez P. Codeine-induced generalized dermatitis and tolerance to other opioids. J Investig Allergol Clin Immunol 2010;20:89-90

20 Sasseville D, Blouin MM, Beauchamp C. Occupational allergic contact dermatitis caused by morphine. Contact Dermatitis 2011;64:166-168

21 Hogen Esch AJ, van der Heide S, van den Brink W, van Ree JM, Bruynzeel DP, Coenraads PJ. Contact allergy and respiratory/mucosal complaints from heroin (diacetylmorphine). Contact Dermatitis 2006;54:42-49

22 Coenraads PJ, Hogen Esch AJ, Prevoo RL. Occupational contact dermatitis from diacetylmorphine (heroin). Contact Dermatitis 2001;45:114

23 Cummer CC. Dermatitis of the eyelids caused by Dionin: development of local hypersensitiveness after eleven years use. Arch Derm Syph 1931;23:68-69

24 Heller NB. Acute dermatitis due to opium preparations. Arch Derm Syph 1931;24:417

25 Kivity S. Fixed drug eruption to multiple drugs: clinical and laboratory investigation. Int J Dermatol 1991;30:149-151

26 Enomoto M, Ochi M, Teramae K, Kamo R, Taguchi S, Yamane T. Codeine phosphate-induced hypersensitivity syndrome. Ann Pharmacother 2004;38:799-802

27 Condé-Salazar L, Guimaraens D, González M, Fuente C. Occupational allergic contact dermatitis from opium alkaloids. Contact Dermatitis 1991;25:202-203

Chapter 3.153 COLISTIN

IDENTIFICATION
Description/definition   : Colistin is a cyclic polypeptide antibiotic obtained from *Bacillus colistinus*
Pharmacological classes : Anti-bacterial agents
IUPAC name   : *N*-[(2*S*)-4-Amino-1-[[(2*S*,3*R*)-1-[[(2*S*)-4-amino-1-oxo-1-[[(6*R*,9*S*,12*R*,15*R*,18*S*,21*S*)-6,9,18-
                           tris(2-aminoethyl)-3-[(1*R*)-1-hydroxyethyl]-12,15-bis(2-methylpropyl)-2,5,8,11,14,17,20-
                           heptaoxo-1,4,7,10,13,16,19-heptazacyclotricos-21-yl]amino]butan-2-yl]amino]-3-hydroxy-
                           1-oxobutan-2-yl]amino]-1-oxobutan-2-yl]-5-methylheptanamide
Other names   : Colomycin; polymyxin E
CAS registry number   : 1066-17-7
EC number   : 213-907-3
Merck Index monograph : 3733
Patch testing   : 5% pet. (6)
Molecular formula   : $C_{52}H_{98}N_{16}O_{13}$

GENERAL
Colistin is a cyclic polypeptide antibiotic derived from *Bacillus colistinus*; it is composed of polymyxins E1 and E2 (or colistins A, B, and C) which act as detergents on cell membranes. Colistin is less toxic than polymyxin B, but otherwise similar; the methanesulfonate colistimethate is used orally. Colistin is indicated for the treatment of acute or chronic infections due to sensitive strains of certain gram-negative bacilli, particularly *Pseudomonas aeruginosa*. In pharmaceutical products, colistin is employed as colistin sulfate (CAS number 1264-72-8, EC number 215-034-3, molecular formula $C_{53}H_{102}N_{16}O_{17}S$) (1).

In topical preparations, colistin has rarely caused contact allergy/allergic contact dermatitis, which subject has been reviewed in Volume 3 of the *Monographs in contact allergy* series (6).

CONTACT ALLERGY FROM ACCIDENTAL CONTACT
In a group of 107 workers in the pharmaceutical industry with dermatitis, investigated in Warsaw, Poland, before 1989, one reacted to colistin, tested 1.000.000 U/gr pet. (2). Also in Warsaw, Poland, in the period 1979-1983, 27 pharmaceutical workers, 24 nurses and 30 veterinary surgeons were diagnosed with occupational allergic contact dermatitis from antibiotics. The numbers that had positive patch tests to colistin (ampoule content) were 1, 0, and 0, respectively, total 1 (5). Of 333 nurses patch tested in the same clinic between 1979 and 1987, 2 reacted to colistin 1.000.000 U/gr pet.; data on relevance were not provided (7).

Cross-reactions, pseudo-cross-reactions and co-reactions
Not to polymyxin B sulfate (colistin = polymyxin E) (3). A patient sensitized to polymyxin B sulfate co-reacted to colistin (polymyxin E), which was considered to be a cross-reaction (4).

LITERATURE

1   The data in the section 'General' may have been obtained from literature discussed in this chapter, but mostly also or exclusively from one or more of the following online sources: ChemIDPlus Advanced, PubChem, DrugBank, RxList, Drug Central, Drugs.com, and Wikipedia
2   Rudzki E, Rebandel P, Grzywa Z. Contact allergy in the pharmaceutical industry. Contact Dermatitis 1989;21:121-122
3   Inoue A, Shoji A. Allergic contact dermatitis from colistin. Contact Dermatitis 1995;33:200
4   Van Ketel WG. Polymixine B-sulfate and bacitracin. Contact Dermatitis Newsletter 1974;15:445
5   Rudzki E, Rebendel P. Contact sensitivity to antibiotics. Contact Dermatitis 1984;11:41-42
6   De Groot AC. Monographs in contact allergy, volume 3: Topical Drugs. Boca Raton, Fl, USA: CRC Press Taylor and Francis Group, 2021 (ISBN 978-0-367-23693-9)
7   Rudzki E, Rebandel P, Grzywa Z. Patch tests with occupational contactants in nurses, doctors and dentists. Contact Dermatitis 1989;20:247-250

Chapter 3.154 CYAMEMAZINE

IDENTIFICATION

Description/definition   : Cyamemazine is the phenothiazine that conforms to the structural formula shown below
Pharmacological classes   : Antipsychotics
IUPAC name   : 10-[3-(Dimethylamino)-2-methylpropyl]phenothiazine-2-carbonitrile
Other names   : Cyamepromazine
CAS registry number   : 3546-03-0
EC number   : 222-594-2
Merck Index monograph   : 3943
Patch testing   : 0.1%, 1% and 5% pet.
Molecular formula   : $C_{19}H_{21}N_3S$

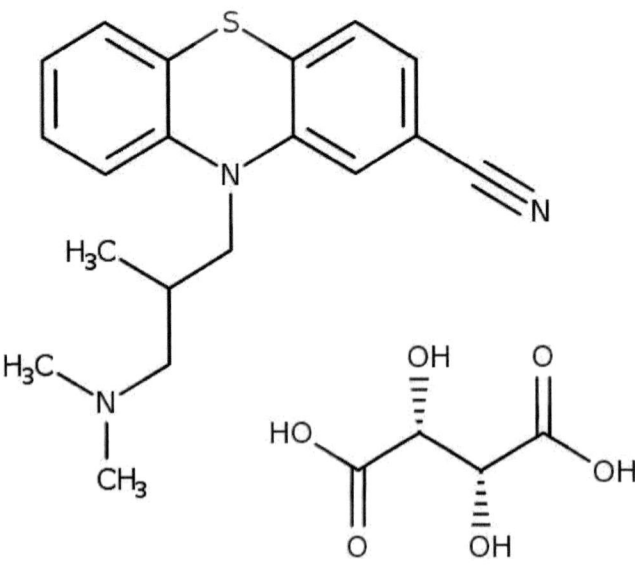

Cyamemazine tartrate

GENERAL

Cyamemazine is a typical antipsychotic drug of the phenothiazine class used primarily in the treatment of schizophrenia and psychosis-associated anxiety. Cyamemazine actually behaves like an atypical antipsychotic, due to its potent anxiolytic effects and lack of extrapyramidal side effects. In pharmaceutical products, both cyamemazine and cyamemazine tartrate (CAS number 93841-82-8, EC number 299-054-8, molecular formula $C_{23}H_{27}N_3O_6S$) may be employed (1).

CUTANEOUS ADVERSE DRUG REACTIONS FROM SYSTEMIC ADMINISTRATION CAUSED BY TYPE IV (DELAYED-TYPE) HYPERSENSITIVITY (as demonstrated by positive patch tests)

Cutaneous adverse drug reactions from systemic administration of cyamemazine caused by type IV (delayed-type) hypersensitivity have included photosensitivity (2).

Photosensitivity

A 50-year-old man presented with an intensely pruritic scaly erythema of the face, neck and the 'V' area of the upper chest, forearms, and dorsum of the hands, which had begun 4 months earlier. He had been treated with cyamemazine, risperidone and alprazolam for 6 months for alcohol abuse. Photopatch testing with the cyamemazine commercial oral drops (undiluted) was positive (+++). Later, he also reacted to photopatch tests with cyamemazine 0.1%, 1% and 5% petrolatum (pet.). There were no photocross-reactions to promethazine or chlorpromazine. The patient was diagnosed with photoallergic reaction to cyamemazine (2).

In this patient, no controls have been tested to exclude phototoxicity, a well-known side effect of phenothiazines. Indeed, in another study where 2 patients had developed a photodistributed rash and had positive photopatch tests to cyamemazine, all 9 controls also had positive photopatch tests to cyamemazine 20% in water, indicating phototoxicity (3). Nevertheless, the positive reactions in a dilution series to low concentrations makes photoallergy very likely (2).

LITERATURE

1 The data in the section 'General' may have been obtained from literature discussed in this chapter, but mostly also or exclusively from one or more of the following online sources: ChemIDPlus Advanced, PubChem, DrugBank, RxList, Drug Central, Drugs.com, and Wikipedia

2 Fernandes IC, Vilaça S, Lobo I, Sanches M, Costa V, Selores M. Photoallergic reaction to cyamemazine. Dermatol Online J 2013;19:15

3 Conilleau V, Dompmartin A, Michel M, Verneuil L, Leroy D. Photoscratch testing in systemic drug-induced photosensitivity. Photodermatol Photoimmunol Photomed 2000;16:62-66

Chapter 3.155 CYANAMIDE

IDENTIFICATION

Description/definition   : Cyanamide is the cyanide compound that conforms to the structural formula shown below
Pharmacological classes : Unspecified
IUPAC name   : Cyanamide
Other names   : Carbamonitrile
CAS registry number   : 420-04-2
EC number   : 206-992-3
Merck Index monograph : 3944
Patch testing   : 1% pet. and water; a concentration of 5% may induce patch test sensitization (4)
Molecular formula   : CH_2N_2

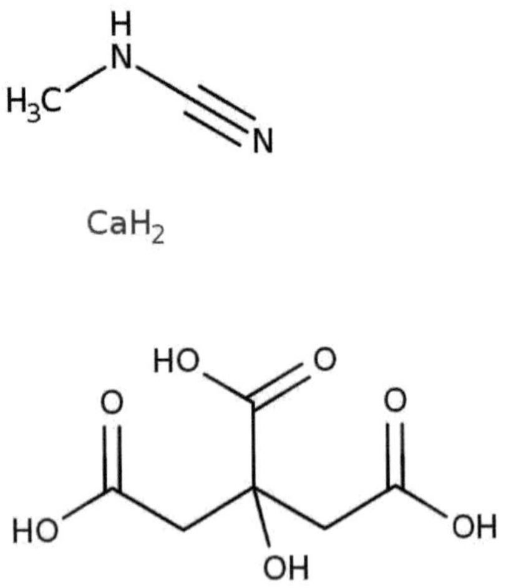

CaH$_2$

Calcium cyanamide citrated

GENERAL

Cyanamide is a cyanide compound which has been used as a fertilizer, defoliant and in many manufacturing processes. In medicine, cyanamide, as an inhibitor of aldehyde dehydrogenase, is used as a pharmacological adjunct in the aversive treatment of chronic alcoholism. In pharmaceutical products, cyanamide is employed as calcium cyanamide citrated (CAS number 8013-88-5, EC number not available, molecular formula $C_8H_{12}CaN_2O_7$) (1).

CUTANEOUS ADVERSE DRUG REACTIONS FROM SYSTEMIC ADMINISTRATION CAUSED BY TYPE IV (DELAYED-TYPE) HYPERSENSITIVITY (as demonstrated by positive patch tests)

Cutaneous adverse drug reactions from systemic administration of cyanamide caused by type IV (delayed-type) hypersensitivity have included erythroderma (3) and drug reaction with eosinophilia and systemic symptoms (DRESS) (2).

Erythroderma

After having used calcium cyanamide for 2 months for alcoholism, a 61-year-old man developed pruritic, erythematous scaly lesions that had started on the upper eyelids and spread in a few days to the face, trunk and extremities, including the palms and soles. Physical examination revealed erythroderma, favoring the antecubital and popliteal fossae, and also the palms and soles. Patch tests were positive to the commercial drug 10% and 50% water and, in a second test session, to cyanamide 0.1%, 0.5%, 1% and 5% water at D2 and D4. twenty controls were negative to the highest concentration (3).

Symmetrical drug-related intertriginous and flexural exanthema (SDRIFE)/baboon syndrome

A 46-year-old Japanese man presented with a generalized eruption which had developed about one month after taking liquid 1% cyanamide as a drug for alcoholism. Clinical and laboratory investigations showed high fever, lymphadenopathy, facial edema, marked leukocytosis with eosinophilia and atypical lymphocytes, lymphocytopenia, and liver and renal dysfunction. The patient was treated with 8 mg betamethasone daily and his condition improved, but he needed low-dose corticosteroid for almost 1 year because of several episodes of recurrence. HHV-6-, HHV-7-, herpes simplex virus- (HSV), and cytomegalovirus-specific IgG titers showed more than a four-fold rise sequentially.

Significant numbers of copies of HHV-6 and HHV-7 DNA were detected in the peripheral white blood cells by real-time polymerase chain reaction (PCR). HHV-6 and cytomegaly virus DNA were detected in the serum by nested PCR. A patch test with cyanamide 1% was strongly positive at D2 and D3 (+++), despite the fact that the patient was at that moment treated with 13 mg prednisolone. He was diagnosed with DRESS/DIHS (drug-induced hypersensitivity syndrome) due to delayed-type hypersensitivity to cyanamide accompanied by reactivation of many viruses (2).

LITERATURE

1   The data in the section 'General' may have been obtained from literature discussed in this chapter, but mostly also or exclusively from one or more of the following online sources: ChemIDPlus Advanced, PubChem, DrugBank, RxList, Drug Central, Drugs.com, and Wikipedia

2   Mitani N, Aihara M, Yamakawa Y, Yamada M, Itoh N, Mizuki N, Ikezawa Z. Drug-induced hypersensitivity syndrome due to cyanamide associated with multiple reactivation of human herpesviruses. J Med Virol 2005;75:430-434

3   Abajo P, Feal C, Sanz-Sánchez T, Sánchez-Pérez J, García-Díez A. Eczematous erythroderma induced by cyanamide. Contact Dermatitis 1999;40:160-161

4   De Groot AC. Patch Testing, 4th Edition. Wapserveen, The Netherlands: acdegroot publishing, 2018 (ISBN 9789081323345)

Chapter 3.156 CYANOCOBALAMIN

IDENTIFICATION

Description/definition   : Cyanocobalamin is a cobalt-containing coordination compound generated by intestinal microbes, and a natural water-soluble vitamin of the B-complex family; its structural formula is shown below

Pharmacological classes   : Vitamin B complex

IUPAC name   : Cobalt(3+);[(2R,3S,4R,5S)-5-(5,6-dimethylbenzimidazol-1-yl)-4-hydroxy-2-(hydroxy-methyl)oxolan-3-yl] [(2R)-1-[3-[(1R,2R,3R,5Z,7S,10Z,12S,13S,15Z,17S,18S,19R)-2,13,18-tris(2-amino-2-oxoethyl)-7,12,17-tris(3-amino-3-oxopropyl)-3,5,8,8,13,15,18,19-octamethyl-2,7,12,17-tetrahydro-1H-corrin-24-id-3-yl]propanoylamino]propan-2-yl] phosphate;cyanide

Other names   : Vitamin B$_{12}$; 1H-Benzimidazole, 5,6-dimethyl-1-(3-O-phosphono-α-D-ribofuranosyl)-, monoester with cobinamide cyanide, inner salt

CAS registry number   : 68-19-9

EC number   : 200-680-0

Merck Index monograph   : 11482

Patch testing   : 10% pet.; always test cobalt (present in all baseline series)

Molecular formula   : C$_{63}$H$_{88}$CoN$_{14}$O$_{14}$P

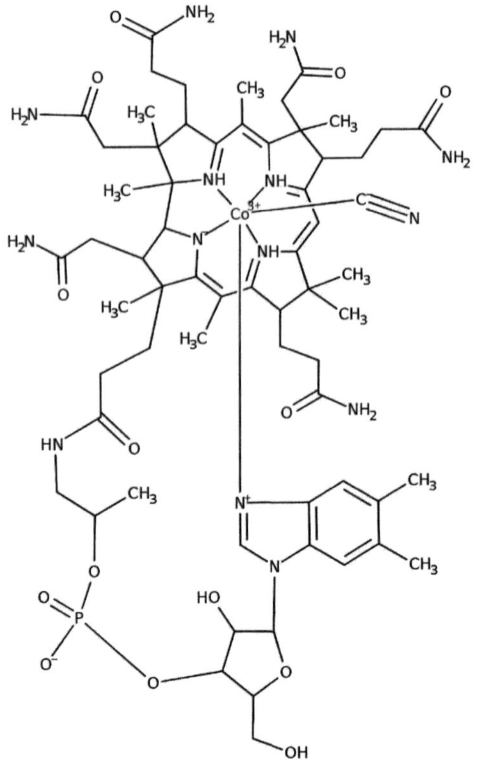

GENERAL

Cyanocobalamin (better known as vitamin B$_{12}$) is a cobalt-containing coordination compound generated by intestinal microbes, and a natural water-soluble vitamin of the B-complex family. It must combine with Intrinsic Factor for absorption by the intestine. Cyanocobalamin is necessary for hematopoiesis, neural metabolism, DNA and RNA production, and carbohydrate, fat, and protein metabolism. Vitamin B$_{12}$ deficiency causes pernicious anemia, megaloblastic anemia, and neurologic lesions. Vitamin B$_{12}$ supplements are widely available and indicated in patients who require supplementation (1).

CONTACT ALLERGY FROM ACCIDENTAL CONTACT

A 32-year-old man developed generalized pruritic cutaneous lesions, clearly related to his job of handling animal feed. Clinical examination revealed erythematous papular plaques on the legs, face, forearms and hands. A patch test to the animal feed was strongly (+++) positive at D2 and D4. When patch tested with all ingredients, there was a +++ reaction to vitamin B$_{12}$ (cyanocobalamin) 10% pet. at D2 and D4 only. Cobalt was negative. The diagnosis was occupational (presumably airborne) allergic contact dermatitis from cyanocobalamin (3).

CUTANEOUS ADVERSE DRUG REACTIONS FROM SYSTEMIC ADMINISTRATION CAUSED BY TYPE IV (DELAYED-TYPE) HYPERSENSITIVITY (as demonstrated by positive patch tests)

Cutaneous adverse drug reactions from systemic administration of cyanocobalamin caused by type IV (delayed-type) hypersensitivity have included systemic allergic dermatitis presenting as widespread nummular dermatitis (2).

Systemic allergic dermatitis (systemic contact dermatitis)

A 51-year-old woman had chronic hand dermatitis and widespread nummular dermatitis. She had a strongly positive patch test to cobalt. The patient was using vitamin B_{12} supplements sublingually. After discontinuing them, 'she cleared'. The diagnosis was systemic allergic dermatitis, presumably from cobalt in vitamin B_{12} (cyanocobalamin). However, cyanocobalamin itself was apparently not tested (2).

Other cutaneous adverse drug reactions

A patient with possible allergic reaction to cyanocobalamin was reported in German literature, but details are not available to the author (9). Several authors have – not always convincingly - associated cobalt allergy with sensitivity to vitamin B_{12} (4-8).

LITERATURE

1   The data in the section 'General' may have been obtained from literature discussed in this chapter, but mostly also or exclusively from one or more of the following online sources: ChemIDPlus Advanced, PubChem, DrugBank, RxList, Drug Central, Drugs.com, and Wikipedia
2   Giroux M, Pratt M. Two cases of systemic ACD: cobalt recalled with B12 and plants recalled with herbal products. Am J Cont Dermat 2003;14:109 (Abstract)
3   Rodriguez A, Echechipía S, Alvarez M, Muro MD. Occupational contact dermatitis from vitamin B12. Contact Dermatitis 1994;31:271
4   Price ML, MacDonald DM. Cheilitis and cobalt allergy related to ingestion of vitamin B12. Contact Dermatitis 1981;7:352
5   Rostenberg A Jr, Perkins AJ. Nickel and cobalt dermatitis. J Allergy 1951;22:466-474
6   Young WC, Ulrich CW, Fouts PJ. Sensitivity to vitamin B12 concentrate. J Am Med Assoc 1950;143:893-894
7   Fisher A. Contact dermatitis at home and abroad. Cutis 1972;10:719-723
8   Malten KE. Flare reaction due to vitamin B12 in a patient with psoriasis and contact eczema. Contact Dermatitis 1975;1:325-326
9   Pevny I, Hartmann A, Metz J. Vitamin-B12-(Cyanocobalamin-)Allergie [Vitamin B12-(cyanocobalamin)-allergy]. Hautarzt 1977;28:600-603 (Article in German)

Chapter 3.157 CYCLOPHOSPHAMIDE

IDENTIFICATION

Description/definition   : Cyclophosphamide is the nitrogen mustard compound that conforms to the structural
                             formula shown below
Pharmacological classes   : Antineoplastic agents, alkylating; antirheumatic agents; myeloablative agonists;
                             immunosuppressive agents; mutagens
IUPAC name   : *N,N*-bis(2-Chloroethyl)-2-oxo-1,3,2λ^5-oxazaphosphinan-2-amine
CAS registry number   : 50-18-0
EC number   : 200-015-4
Merck Index monograph   : 4013
Patch testing   : 5% saline
Molecular formula   : $C_7H_{15}Cl_2N_2O_2P$

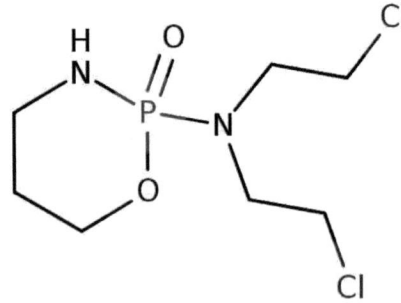

GENERAL

Cyclophosphamide is a chemical in the class of alkylating nitrogen mustard antineoplastic and immunosuppressive agents. Alkylating agents stop tumor growth by cross-linking guanine bases in DNA double-helix strands - directly attacking DNA. This makes the strands unable to uncoil and separate. As this is necessary in DNA replication, the cells can no longer divide. Cyclophosphamide is indicated for the treatment of malignant lymphomas, multiple myeloma, leukemias, mycosis fungoides, neuroblastoma, adenocarcinoma of the ovary, retinoblastoma, and carcinoma of the breast. In pharmaceutical products, cyclophosphamide is employed as cyclophosphamide monohydrate (CAS number 6055-19-2, EC number not available, molecular formula $C_7H_{17}Cl_2N_2O_3P$) (1).

CUTANEOUS ADVERSE DRUG REACTIONS FROM SYSTEMIC ADMINISTRATION CAUSED BY TYPE IV (DELAYED-TYPE) HYPERSENSITIVITY (as demonstrated by positive patch tests)

Cutaneous adverse drug reactions from systemic administration of cyclophosphamide caused by type IV (delayed-type) hypersensitivity have included multiple fixed drug eruption (2).

Fixed drug eruption

An 18-year-old woman was treated with cyclophosphamide pulse therapy for systemic sclerosis and interstitial pneumonia. After the ninth course, she developed several painful purple-brownish annular patches about 10 hours after cyclophosphamide administration. The eruptions gradually resolved with residual pigmentation but eruptions flared up at the same sites after every treatment session. Also, the number of patches increased after each new course. Histopathology of a skin biopsy was consistent with fixed drug eruption. Patch tests were positive on postlesional skin but not normal skin to cyclophosphamide and its metabolite cyclophosphamide monohydrate at 5% in saline (negative to 0.5%). There was no reaction to the metabolite carboxyphosphamide. The patient was diagnosed with multiple fixed drug eruption from delayed-type hypersensitivity to cyclophosphamide (2).

LITERATURE

1   The data in the section 'General' may have been obtained from literature discussed in this chapter, but mostly also or exclusively from one or more of the following online sources: ChemIDPlus Advanced, PubChem, DrugBank, RxList, Drug Central, Drugs.com, and Wikipedia
2   Fujita H, Watanabe T, Okada R, Nozaki Y, Ayabe M, Imagawa T, Yokota S, Aihara M. Multiple fixed drug eruption caused by cyclophosphamide and its metabolite. Eur J Dermatol 2013;23:275-277

Chapter 3.158 CYCLOSERINE

IDENTIFICATION

Description/definition   : Cycloserine is the isoxazoline and analog of D-alanine that conforms to the structural formula shown below
Pharmacological classes : Anti-infective agents, urinary; antibiotics, antitubercular; antimetabolites
IUPAC name   : (4R)-4-Amino-1,2-oxazolidin-3-one
CAS registry number   : 68-41-7
EC number   : 200-688-4
Merck Index monograph   : 4019
Patch testing   : 5% pet. (4)
Molecular formula   : $C_3H_6N_2O_2$

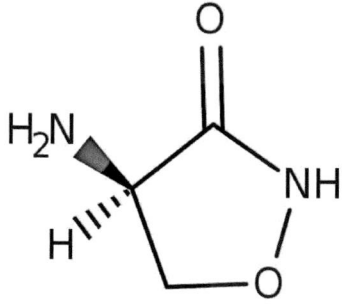

GENERAL

Cycloserine is an antibiotic produced by *Streptomyces garyphalus* or S. *orchidaceus*. It is an analog of the amino acid D-alanine with broad-spectrum antibiotic and glycinergic activities. D-cycloserine interferes with bacterial cell wall synthesis by competitively inhibiting two enzymes, L-alanine racemase and D-alanine--D-alanine ligase, thereby impairing peptidoglycan formation necessary for bacterial cell wall synthesis. This agent may be bactericidal or bacteriostatic, depending on its concentration at the infection site and the susceptibility of the organism. Cycloserine is used in combination with up to 5 other drugs as a treatment for *Mycobacterium avium* complex (MAC) and is also used to treat tuberculosis (1).

CUTANEOUS ADVERSE DRUG REACTIONS FROM SYSTEMIC ADMINISTRATION CAUSED BY TYPE IV (DELAYED-TYPE) HYPERSENSITIVITY (as demonstrated by positive patch tests)

Cutaneous adverse drug reactions from systemic administration of cycloserine caused by type IV (delayed-type) hypersensitivity have included lichenoid drug eruption (2,3).

Other cutaneous adverse drug reactions

A 38-year-old man with pulmonary tuberculosis, who had been treated for this infection for 4 months, was referred for an itchy eruption over the entire body. He had initially been treated with isoniazid, rifampin, ethambutol and pyrazinamide during the first 2 months and then by second-line anti-tuberculosis medications (ethambutol, levofloxacin, cycloserine) because of the development of reddish itchy papules and hepatitis. One month after commencement of the second-line medications, itching became aggravated and new skin lesions developed. Clinical examination showed a widespread lichenoid eruption with hyperkeratotic lesions on the whole body and Wickham striae involving the buccal mucosa. The patient had low-grade fever and blood hypereosinophilia, but all other laboratory tests were within the normal range. Patch tests, performed 3 months after stopping medications and when the skin had returned to normal, showed weakly positive reactions to ethambutol 10% and 50% pet. and strongly positive reactions to cycloserine 10% and 50% pet. One (!) control was negative. A lymphocyte transformation test (LTT) revealed proliferation of peripheral mononuclear cells by cycloserine in a dose-dependent manner. The patient was diagnosed with lichenoid eruption (drug-induced lichen planus) due to cycloserine delayed-type hypersensitivity. A possible causative role for ethambutol could not be excluded (2).

In an earlier report from Korea, a 65-year-old man treated with cycloserine and other antituberculosis agents presented with multiple pruritic, violaceous, flat-topped papules and plaques on his back which had developed after 5 months' drug use. The histopathology of a skin biopsy was consistent with a lichenoid drug eruption. Patch test with all drugs showed a strongly positive reaction to cycloserine only. After discontinuation of cycloserine, the lesions and pruritus resolved and no new lesions have developed (3).

LITERATURE

1   The data in the section 'General' may have been obtained from literature discussed in this chapter, but mostly also or exclusively from one or more of the following online sources: ChemIDPlus Advanced, PubChem, DrugBank, RxList, Drug Central, Drugs.com, and Wikipedia

2   Kim J, Park S, Jung CM, Oh CW, Kwon JW. A case of cycloserine-induced lichenoid drug eruption supported by the lymphocyte transformation test. Allergy Asthma Immunol Res 2017;9:281-284

3   Shim JH, Kim TY, Kim HO, Kim CW. Cycloserine-induced lichenoid drug eruption. Dermatology 1995;191:142-144

4   De Groot AC. Patch testing, 4th edition. Wapserveen, The Netherlands: acdegroot publishing, 2018 (ISBN 9789081323345)

Chapter 3.159 DABRAFENIB

IDENTIFICATION

Description/definition : Dabrafenib is the organofluorine compound and sulfanilamide that conforms to the structural formula shown below

Pharmacological classes : Antineoplastic agents; protein kinase inhibitors

IUPAC name : *N*-[3-[5-(2-Aminopyrimidin-4-yl)-2-tert-butyl-1,3-thiazol-4-yl]-2-fluorophenyl]-2,6-difluorobenzenesulfonamide

CAS registry number : 1195765-45-7

EC number : 689-166-9

Merck Index monograph : 11719

Patch testing : Content of capsule, 30% pet. (2); most systemic drugs can be tested at 10% pet.; if the pure chemical is not available, prepare the test material from intravenous powder, the content of capsules or – when also not available – from powdered tablets to achieve a final concentration of the active drug of 10% pet.

Molecular formula : $C_{23}H_{20}F_3N_5O_2S_2$

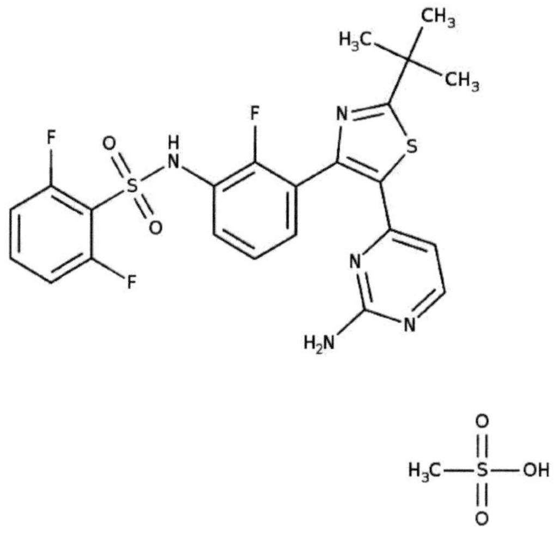

Dabrafenib mesylate

GENERAL

Dabrafenib is an orally bioavailable inhibitor of B-raf (BRAF) protein with potential antineoplastic activity. Dabrafenib selectively binds to and inhibits the activity of B-raf, which may inhibit the proliferation of tumor cells which contain a mutated BRAF gene. It is used alone or in combination with specific other drugs in the treatment of advanced malignant melanoma and anaplastic thyroid cancer (1). In pharmaceutical products, dabrafenib is employed as dabrafenib mesylate (CAS number 1195768-06-9, EC number not available, molecular formula $C_{24}H_{24}F_3N_5O_5S_3$) (1).

CUTANEOUS ADVERSE DRUG REACTIONS FROM SYSTEMIC ADMINISTRATION CAUSED BY TYPE IV (DELAYED-TYPE) HYPERSENSITIVITY (as demonstrated by positive patch tests)

Cutaneous adverse drug reactions from systemic administration of dabrafenib caused by type IV (delayed-type) hypersensitivity have included drug reaction with eosinophilia and systemic symptoms (DRESS) (2).

Drug reaction with eosinophilia and systemic symptoms (DRESS)

In a study from Switzerland, one patient was reported to have developed drug reaction with eosinophilia and systemic symptoms (DRESS) from dabrafenib with a positive patch test to this drug (contents of capsule, 30% pet.); clinical details were not provided (2).

LITERATURE

1 The data in the section 'General' may have been obtained from literature discussed in this chapter, but mostly also or exclusively from one or more of the following online sources: ChemIDPlus Advanced, PubChem, DrugBank, RxList, Drug Central, Drugs.com, and Wikipedia

2 Jörg L, Helbling A, Yerly D, Pichler WJ. Drug-related relapses in drug reaction with eosinophilia and systemic symptoms (DRESS). Clin Transl Allergy 2020;10:52

Chapter 3.160 DALTEPARIN

IDENTIFICATION

Description/definition   : Dalteparin is a low molecular weight heparin with an average molecular weight of 5000
                              daltons and about 90% of the material within the range of 2000-9000 daltons
Pharmacological classes   : Antithrombotic agents
IUPAC name   : Not available
CAS registry number   : 9005-49-6 (Heparin)
EC number   : 232-681-7 (Heparin)
Merck Index monograph   : 5958 (Heparin)
Patch testing   : Commercial preparation undiluted (4); consider intradermal testing with late readings
                            (D2,D3) when patch tests are negative and consider subcutaneous challenge when
                            intradermal tests are negative
Molecular formula   : Unspecified

GENERAL

Dalteparin is a low molecular weight heparin (LMWH) prepared by nitrous acid degradation of unfractionated heparin of porcine intestinal mucosa origin and is used as an anticoagulant. It is composed of strongly acidic sulfated polysaccharide chains with an average molecular weight of 5000 daltons and about 90% of the material within the range of 2000-9000 daltons. LMWHs have a more predictable response, a greater bioavailability, and a longer anti-Xa half-life than unfractionated heparin. In pharmaceutical products, dalteparin is employed as dalteparin sodium (CAS number not available, EC number not available, molecular formula unspecified) (1).

See also bemiparin (Chapter 3.59), certoparin (Chapter 3.117), danaparoid (Chapter 3.161), enoxaparin (Chapter 3.195), fondaparinux (Chapter 3.225), heparins (Chapter 3.239), nadroparin (Chapter 3.331), and tinzaparin (Chapter 3.479).

CUTANEOUS ADVERSE DRUG REACTIONS FROM SYSTEMIC ADMINISTRATION CAUSED BY TYPE IV (DELAYED-TYPE) HYPERSENSITIVITY

Throughout this book, only reports of delayed-type hypersensitivity have been included that showed a positive patch test to the culprit drug. However, as a result of the high molecular weight of heparins, patch tests are often false-negative, presumably from insufficient penetration into the skin. Because of this, and also because patch tests have been performed in a small minority of cases only, studies with a positive intradermal test or subcutaneous provocation tests with delayed readings are included in the chapters on the various heparins, even when patch tests were negative or not performed.

General information on delayed-type hypersensitivity reactions to heparins

General information on delayed-type hypersensitivity reactions to heparins (including dalteparin), presenting as local reactions from subcutaneous administration, is provided in Chapter 3.239. Heparins. In this chapter, only *non-local* cutaneous adverse drug reactions from delayed-type hypersensitivity to dalteparin are presented.

Non-local cutaneous adverse drug reactions

Non-local cutaneous adverse drug reactions from systemic administration of dalteparin caused by type IV (delayed-type) hypersensitivity have included acute generalized exanthematous pustulosis (AGEP) (3).

Drug reaction with eosinophilia and systemic symptoms (DRESS)

A 45-year-old woman presented with a 10-day history of itchy patches at injection sites of dalteparin after having administered 4 subcutaneous injections of 5000 IU into the right thigh for superficial thrombophlebitis. After 7 days, the lesions spread and developed into a generalized rash. Dermatologic examination revealed an erythematous, infiltrated, and bullous patch at the right thigh. The complete body surface showed an exanthema with disseminated small, non-follicular and follicular pustules mainly on the upper aspect of the trunk. The patient was subfebrile (37.5°C) and the blood neutrophil count was elevated. Six months later, a drug provocation test with subcutaneous injections of heparins and anticoagulants was performed. The patient refused to be tested with dalteparin, but was tested, however, with enoxaparin, certoparin, reviparin, nadroparin, danaparoid, pentosan polysulfate, and fondaparinux. Within 2 days, the patient developed eczematous patches at all application sites except with pentosan polysulfate. The danaparoid and in a lesser extent the nadroparin patch showed pustules. Within the next day, a generalized rash developed. Although not proven, this was very likely a case of acute generalized exanthematous pustulosis (AGEP) from delayed-type hypersensitivity to dalteparin (3).

Cross-reactions, pseudo-cross-reactions and co-reactions

Cross-reactions between heparins are frequent in delayed-type hypersensitivity (>90% of patients tested, median number of positive drugs per patient: 3) and do not depend on the molecular weight of the molecules (2). Overlap in their polysaccharide composition might explain the high degree of cross-allergenicity (5). In allergic patients, the synthetic ultralow molecular weight synthetic heparin fondaparinux is usually, but not always (6) well-tolerated (5).

LITERATURE

1   The data in the section 'General' may have been obtained from literature discussed in this chapter, but mostly also or exclusively from one or more of the following online sources: ChemIDPlus Advanced, PubChem, DrugBank, RxList, Drug Central, Drugs.com, and Wikipedia

2   Weberschock T, Meister AC, Bohrt K, Schmitt J, Boehncke W-H, Ludwig RJ. The risk for cross-reactions after a cutaneous delayed-type hypersensitivity reaction to heparin preparations is independent of their molecular weight: a systematic review. Contact Dermatitis 2011;65:187-194

3   Komericki P, Grims R, Kränke B, Aberer W. Acute generalized exanthematous pustulosis from dalteparin. J Am Acad Dermatol 2007;57:718-721

4   Brockow K, Garvey LH, Aberer W, Atanaskovic-Markovic M, Barbaud A, Bilo MB, et al. Skin test concentrations for systemically administered drugs - an ENDA/EAACI Drug Allergy Interest Group position paper. Allergy 2013;68:702-712

5   Schindewolf M, Lindhoff-Last E, Ludwig RJ. Heparin-induced skin lesions. Lancet 2012;380:1867-1879

6   Utikal J, Peitsch WK, Booken D, Velten F, Dempfle CE, Goerdt S, et al. Hypersensitivity to the pentasaccharide fondaparinux in patients with delayed-type heparin allergy. Thromb Haemost 2005;94:895-896

Chapter 3.161 DANAPAROID

IDENTIFICATION

Description/definition   : Danaparoid is a low-molecular-weight semi-synthetic heparin
Pharmacological classes : Antithrombotic agents
IUPAC name   : Not available
CAS registry number   : 308068-55-5
EC number   : Not available
Merck Index monograph   : 4080
Patch testing   : Commercial preparation undiluted (4); consider intradermal testing with late readings
                           (D2,D3) when patch tests are negative and consider subcutaneous challenge when
                           intradermal tests are negative
Molecular formula   : Unspecified

GENERAL

Danaparoid is a low molecular weight heparinoid (semisynthetic heparin). It is a mixture of sulfated glycosamino-glycans derived from hog intestinal mucosa and contains heparan sulfate (83% w/w), dermatan sulfate (12% w/w) and chondroitin sulfate (5% w/w). Danaparoid possesses a potent antithrombic activity that works by inhibiting activated factor X (Factor Xa) and activated factor II (Factor IIa). Danaparoid is indicated for the prophylaxis of post-operative deep venous thrombosis, which may lead to pulmonary embolism, in patients undergoing elective hip replacement surgery. Another indication is the treatment of an acute episode of heparin-induced thrombocytopenia (HIT), and prophylaxis in patients with a history of HIT. Danaparoid is no longer marketed in the US but is still available in some other countries. In pharmaceutical products, danaparoid is employed as danaparoid sodium (CAS number not available, EC number not available, molecular formula unspecified) (1).

See also bemiparin (Chapter 3.59), certoparin (Chapter 3.117), dalteparin (Chapter 3.160), enoxaparin (Chapter 3.195), fondaparinux (Chapter 3.225), heparins (Chapter 3.239), nadroparin (Chapter 3.331), and tinzaparin (Chapter 3.479).

CUTANEOUS ADVERSE DRUG REACTIONS FROM SYSTEMIC ADMINISTRATION CAUSED BY TYPE IV (DELAYED-TYPE) HYPERSENSITIVITY

Throughout this book, only reports of delayed-type hypersensitivity have been included that showed a positive patch test to the culprit drug. However, as a result of the high molecular weight of heparins, patch tests are often false-negative, presumably from insufficient penetration into the skin. Because of this, and also because patch tests have been performed in a small minority of cases only, studies with a positive intradermal test or subcutaneous provocation tests with delayed readings are included in the chapters on the various heparins, even when patch tests were negative or not performed.

General information on delayed-type hypersensitivity reactions to heparins
General information on delayed-type hypersensitivity reactions to heparins (including danaparoid) presenting as local reactions from subcutaneous administration, is provided in Chapter 3.239. Heparins. In this chapter, only *non-local* cutaneous adverse drug reactions from delayed-type hypersensitivity to danaparoid are presented.

Non-local cutaneous adverse drug reactions
Non-local cutaneous adverse drug reactions from systemic administration of danaparoid caused by type IV (delayed-type) hypersensitivity have included generalized eczema (6).

Dermatitis/eczematous eruption
A patient with known delayed-type hypersensitivity to heparins received the heparinoid, danaparoid subcutaneously for thrombosis prophylaxis after orthopedic surgery. After the first few injections, eczematous plaques developed; administration of the anticoagulant was continued and gradually resulted in generalized eczema despite treatment with topical and oral glucocorticoids (6).

Cross-reactions, pseudo-cross-reactions and co-reactions
Cross-reactions between heparins are frequent in delayed-type hypersensitivity (>90% of patients tested, median number of positive drugs per patient: 3) and do not depend on the molecular weight of the molecules (3). Overlap in their polysaccharide composition might explain the high degree of cross-allergenicity (5). Cross-reactions to the semisynthetic heparinoid danaparoid have also been observed (2). In allergic patients, the synthetic ultralow molecular weight synthetic heparin fondaparinux is usually, but not always (2) well-tolerated (5).

LITERATURE
1   The data in the section 'General' may have been obtained from literature discussed in this chapter, but mostly also or exclusively from one or more of the following online sources: ChemIDPlus Advanced, PubChem, DrugBank, RxList, Drug Central, Drugs.com, and Wikipedia
2   Utikal J, Peitsch WK, Booken D, Velten F, Dempfle CE, Goerdt S, et al. Hypersensitivity to the pentasaccharide fondaparinux in patients with delayed-type heparin allergy. Thromb Haemost 2005;94:895-896
3   Weberschock T, Meister AC, Bohrt K, Schmitt J, Boehncke W-H, Ludwig RJ. The risk for cross-reactions after a cutaneous delayed-type hypersensitivity reaction to heparin preparations is independent of their molecular weight: a systematic review. Contact Dermatitis 2011;65:187-194
4   Brockow K, Garvey LH, Aberer W, Atanaskovic-Markovic M, Barbaud A, Bilo MB, et al. Skin test concentrations for systemically administered drugs - an ENDA/EAACI Drug Allergy Interest Group position paper. Allergy 2013;68:702-712
5   Schindewolf M, Lindhoff-Last E, Ludwig RJ. Heparin-induced skin lesions. Lancet 2012;380:1867-1879
6   Seitz CS, Brocker EB, Trautmann A. Management of allergy to heparins in postoperative care: subcutaneous allergy and intravenous tolerance. Dermatol Online J 2008 Sept 15;14(9):4

Chapter 3.162 DAPSONE

IDENTIFICATION

Description/definition   : Dapsone is the benzenesulfonyl compound that conforms to the structural formula shown
                                         below
Pharmacological classes : Leprostatic agents; antimalarials; anti-infective agents; folic acid antagonists
IUPAC name   : 4-(4-Aminophenyl)sulfonylaniline
Other names   : 4,4'-Diaminodiphenyl sulfone
CAS registry number   : 80-08-0
EC number   : 201-248-4
Merck Index monograph : 4092
Patch testing   : 0.1% acetone (2); 1% pet. (3)
Molecular formula   : $C_{12}H_{12}N_2O_2S$

GENERAL

Dapsone is a sulfone with anti-inflammatory immunosuppressive properties as well as antibacterial and antibiotic properties. It is the principal drug in a multidrug regimen for the treatment of leprosy. Dapsone is also used for treating malaria and, recently, for *Pneumocystic carinii* pneumonia in AIDS patients. Dermatologists have long used this drug as a very effective (suppressive) treatment of dermatitis herpetiformis (1).

CUTANEOUS ADVERSE DRUG REACTIONS FROM SYSTEMIC ADMINISTRATION CAUSED BY TYPE IV (DELAYED-TYPE) HYPERSENSITIVITY (as demonstrated by positive patch tests)

Cutaneous adverse drug reactions from systemic administration of dapsone caused by type IV (delayed-type) hypersensitivity have included photosensitivity (2).

Photosensitivity

A 76-year-old woman had been treated with 100 mg dapsone for linear IgA dermatosis for about two months when, after about one hour's exposure to sunlight, an acute superficial confluent erythema in association with itching in the face, neck, lower neckline, forearms and lower thighs and mild edema of the face and the eyelids developed. The general condition of the patient was not affected and laboratory investigations showed no abnormalities. Patch tests with dapsone and its metabolites monoacetyldapsone (MADDS) and hydroxylamine dapsone (DDS-NOH), all tested 0.1% in acetone, were negative, but photopatch tests were positive. A skin biopsy specimen of the positive photo-patch test area with DDS-NOH (at D2) showed a perivascular lymphohistiocytic infiltrate, focal weak spongiosis and isolated necrotic keratinocytes. The authors diagnosed photosensitivity, but were not sure whether it represented phototoxicity or photoallergy, albeit speculating that the photosensitivity adverse reaction was probably based on an immunological pathway (2).

Cross-reactions, pseudo-cross-reactions and co-reactions

No photocross-reactions to the structurally related chemicals sulfamethoxazole and sulfasalazine (2).

LITERATURE

1   The data in the section 'General' may have been obtained from literature discussed in this chapter, but mostly also or exclusively from one or more of the following online sources: ChemIDPlus Advanced, PubChem, DrugBank, RxList, Drug Central, Drugs.com, and Wikipedia
2   Stöckel S, Meurer M, Wozel G. Dapsone-induced photodermatitis in a patient with linear IgA dermatosis. Eur J Dermatol 2001;11:50-53
3   De Groot AC. Patch testing, 4th edition. Wapserveen, The Netherlands: acdegroot publishing, 2018 (ISBN 9789081323345)

Chapter 3.163 DEFLAZACORT

IDENTIFICATION

Description/definition	: Deflazacort is the synthetic glucocorticoid that conforms to the structural formula shown below
Pharmacological classes	: Immunosuppressive agents; anti-inflammatory agents
IUPAC name	: 2-[(1S,2S,4R,8S,9S,11S,12S,13R)-11-Hydroxy-6,9,13-trimethyl-16-oxo-5-oxa-7-azapentacyclo[10.8.0.0^{2,9}.0^{4,8}.0^{13,18}]icosa-6,14,17-trien-8-yl]-2-oxoethyl] acetate
Other names	: 11β,21-Dihydroxy-2'-methyl-5βH-pregna-1,4-dieno(17,16-d)oxazole-3,20-dione 21-acetate
CAS registry number	: 14484-47-0
EC number	: 238-483-7
Merck Index monograph	: 4135
Patch testing	: In general, corticosteroids may be tested at 0.1% and 1% in alcohol; late readings (6-10 days) are strongly recommended; 1% petrolatum (2) or 1% water (5) may also be suitable
Molecular formula	: C$_{25}$H$_{31}$NO$_6$

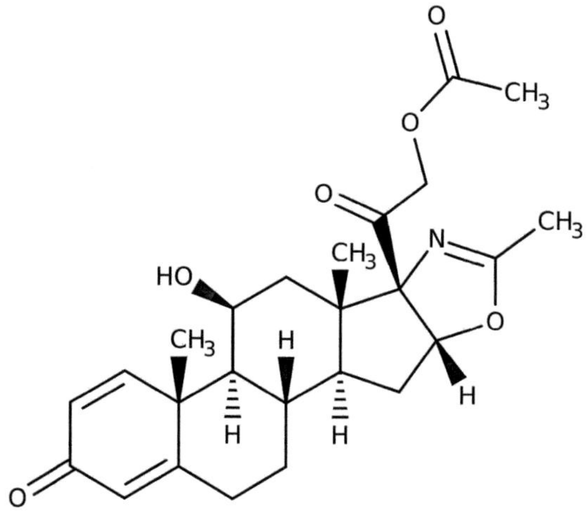

GENERAL

Deflazacort is a corticosteroid prodrug and derivative of prednisolone. Systemically administered glucocorticoids have anti-inflammatory, immunosuppressive and antineoplastic properties and are used in the treatment of a wide spectrum of diseases including rheumatic disorders, lung diseases (asthma, COPD), gastrointestinal tract disorders (Crohn's disease, colitis ulcerosa), certain malignancies (leukemia, lymphomas), hematological disorders, and various diseases of the kidneys, brain, eyes and skin. A practical guideline for diagnosing allergic reactions to corticosteroids is presented in ref. 1.

CUTANEOUS ADVERSE DRUG REACTIONS FROM SYSTEMIC ADMINISTRATION CAUSED BY TYPE IV (DELAYED-TYPE) HYPERSENSITIVITY (as demonstrated by positive patch tests)

Cutaneous adverse drug reactions from systemic administration of deflazacort caused by type IV (delayed-type) hypersensitivity have included systemic allergic dermatitis (SAD) presenting as maculopapular eruption (2), as erythematous plaques with pustules (3), as the baboon syndrome (5), SAD presenting as generalized dermatitis (6), systemic allergic dermatitis presenting as pruritic rash (7), and urticaria and angioedema (4).

Maculopapular eruption
See the section 'Systemic allergic dermatitis' below (ref. 2)

Systemic allergic dermatitis (systemic contact dermatitis)
A 63-year-old woman developed an acute itchy, maculopapular eruption of the trunk, arms and legs on the third day of having taking deflazacort 30 mg/day orally as an adjuvant for chemotherapy. A similar eruption had occurred when she had taken dexamethasone 21-disodium phosphate. The history was highly suspicious for allergic contact dermatitis from topical steroids used for lichen planus. Patch tests were positive to deflazacort 1% pet. (D2 -, D4 +++,

D7 +++) and to 4 topical and 5 systemic corticosteroids, including dexamethasone 21-disodium phosphate and the ones used for lichen planus (2).

A 31-year-old-woman was treated with intravenous hydrocortisone and oral deflazacort for exuberant swelling of the eyelids from an insect bite. One day later, she presented with a widespread exanthema showing discrete erythematous-pink plaques with circinate borders and pustules, interpreted as an urticarial reaction or as an acute exanthematous pustulosis (the former diagnosis is obviously incorrect, as pustules are no part of urticaria). Previously, patch tests had demonstrated contact allergy to tixocortol pivalate, neomycin and other topical drugs. Patch tests were positive to deflazacort in various concentrations, hydrocortisone and tixocortol pivalate, as were prick and intradermal tests at D2 and D4 to deflazacort and hydrocortisone. The patient was diagnosed with 'Allergic hypersensitivity to deflazacort'. Because of the previous sensitization to tixocortol pivalate (group 1) and the resemblance with prednisolone (group 1), this was probably a case of systemic allergic dermatitis (3).

A 27-year-old pharmacist on 3 separate occasions had developed a widespread macular exanthem, predominantly on the inner aspects of the arms and legs and the gluteal areas, with severe scaling, fever, nausea, vomiting, malaise, and hypotension. All 3 such episodes had begun a few hours after treatment with deflazacort 6 mg orally and prednicarbate cream (0.25%). A patch test with a tablet of commercial deflazacort 6 mg (1% aq.) produced a widespread erythema with intense pruritus 8 hours after application, which disappeared totally after removal. Later, a patch test with pure deflazacort (1% aq.) was positive 2 hours after application. There were no reactions to a large number of other corticosteroids. Twenty controls were negative to deflazacort 1% water. The patient was diagnosed with systemic allergic dermatitis presenting as a cutaneous reaction clinically similar to the baboon syndrome. For this diagnosis, previous sensitization is necessary. The authors suggested that she had become sensitized to deflazacort in her work as pharmacist (5).

A 34-year-old woman, previously sensitized by topical betamethasone valerate, was treated with oral deflazacort, which she had never used before, for mosquito bites. An hour after the first dose of 15 mg the patient started to experience nasal fossae edema, vesicular lesions and crustae subsequently appearing in the following days, and dermatitis on the arms and legs. Patch tests were positive to betamethasone valerate 1% alc., to commercial deflazacort 20% pet. and to pure deflazacort 1% alc. and 10% pet. (6). This was a case of systemic allergic dermatitis presenting as generalized dermatitis.

A 52-year-old woman who had suffered allergic contact dermatitis from antihemorrhoidal creams and a generalized symmetrical eruption after intra-articular administration of triamcinolone acetonide, had positive patch tests to triamcinolone acetonide and 7 other corticosteroids. An oral challenge test with deflazacort 30 mg gave a positive result with a pruritic rash on the forearms and anterior chest after 24 hours. This was a case of systemic allergic dermatitis (from oral provocation), but deflazacort itself was not patch tested (7).

Symmetrical drug-related intertriginous and flexural exanthema (SDRIFE)/Baboon syndrome
See the section 'Systemic allergic dermatitis' above (ref. 5)

Dermatitis/eczematous eruption
See the section 'Systemic allergic dermatitis' above (ref. 6).

Other cutaneous adverse drug reactions
A 64-year-old woman with allergic alveolitis caused by parakeet feathers developed itchy blotches together with lip edema after having used a tapering dose of oral deflazacort for 30 days, and while taking 120 mg/day deflazacort by mistake. Patch tests were positive to deflazacort 1% alc. and hydrocortisone 1% alc., as was an intradermal test with deflazacort (read at 20 minutes or delayed reading?). The oral provocation test with 30 mg of deflazacort was positive, with the 'immediate' appearance of symptoms the same as the initial episode (is an 'immediate' reaction possible after oral intake?). The patient was diagnosed with urticaria-angioedema from deflazacort. The mechanism (type I or type IV hypersensitivity) was not commented on (4).

Cross-reactions, pseudo-cross-reactions and co-reactions
Cross-reactions between corticosteroids are discussed in Chapter 3.399 Prednisolone.

LITERATURE
1   Baeck M, Goossens A. Immediate and delayed allergic hypersensitivity to corticosteroids: practical guidelines. Contact Dermatitis 2012;66:38-45
2   Bianchi L, Hansel K, Antonelli E, Bellini V, Rigano L, Stingeni L. Deflazacort hypersensitivity: a difficult-to-manage case of systemic allergic dermatitis and literature review. Contact Dermatitis 2016;75:54-56

3   Pacheco D, Travassos AR, Antunes J, Silva R, Lopes A, Marques MS. Allergic hypersensitivity to deflazacort. Allergol Immunopathol (Madr) 2013;41:352-354

4   Gómez CM, Higuero NC, Moral de Gregorio A, Quiles MH, Nuñez Aceves AB, Lara MJ, et al. Urticaria-angioedema by deflazacort. Allergy 2002;57:370-371

5   Garcia-Bravo B, Repiso JB, Camacho F. Systemic allergic dermatitis due to deflazacort. Contact Dermatitis 2000;43:359-360

6   Navarro Pulido AM, Orta JC, Buzo G. Delayed hypersensitivity to deflazacort. Allergy 1996;51:441-442

7   Santos-Alarcón S, Benavente-Villegas FC, Farzanegan-Miñano R, Pérez-Francés C, Sánchez-Motilla JM, Mateu-Puchades A. Delayed hypersensitivity to topical and systemic corticosteroids. Contact Dermatitis 2018;78:86-88

Chapter 3.164 DESLORATADINE

IDENTIFICATION

Description/definition   : Desloratadine is the piperidine derivative that conforms to the structural formula shown
                               below
Pharmacological classes   : Histamine H_1 antagonists, non-sedating; cholinergic antagonists
IUPAC name   : 13-Chloro-2-piperidin-4-ylidene-4-azatricyclo[9.4.0.0^{3,8}]pentadeca-1(11),3(8),4,6,12,14-
                            hexaene
Other names   : Descarboethoxyloratadine
CAS registry number   : 100643-71-8
EC number   : Not available
Merck Index monograph : 4193
Patch testing   : 1% pet.
Molecular formula   : $C_{19}H_{19}ClN_2$

GENERAL

Desloratadine is a long-acting piperidine derivative with selective H1 antihistaminergic and non-sedating properties. Desloratadine diminishes the typical histaminergic effects on H_1-receptors in bronchial smooth muscle, capillaries and gastrointestinal smooth muscle, including vasodilation, bronchoconstriction, increased vascular permeability, pain, itching and spasmodic contractions of gastrointestinal smooth muscle. Desloratadine is used for the treatment of allergic rhinitis, angioedema and chronic urticaria (1).

CUTANEOUS ADVERSE DRUG REACTIONS FROM SYSTEMIC ADMINISTRATION CAUSED BY TYPE IV (DELAYED-TYPE) HYPERSENSITIVITY (as demonstrated by positive patch tests)

Cutaneous adverse drug reactions from systemic administration of desloratadine caused by type IV (delayed-type) hypersensitivity have included an eczematous eruption (2).

Dermatitis/eczematous eruption

A 22-year-old man, who had previously suffered an episode of DRESS, presumably from carbamazepine, took one tablet of desloratadine for seasonal rhinitis. The next day he had developed a generalized eczema (according to the patient). Two months later, he had strongly positive patch tests to the commercial drug 30% in water, alcohol and (less strongly) petrolatum. In the following 24 hours, it induced a relapse of the pruriginous eczematous lesions spreading on the trunk. In a second test session, there were positive patch tests to the pure desloratadine 10% in petrolatum, alcohol and water and negative reactions to the excipients of the commercial tablet. However, patch tests with pure desloratadine diluted at 10% in petrolatum gave doubtful or positive results in 8 of 10 volunteers, both on days 2 and 4. To 1% desloratadine in petrolatum 7 controls had negative reactions. The patient himself refused to be tested with this non-irritant concentration (2).

LITERATURE

1   The data in the section 'General' may have been obtained from literature discussed in this chapter, but mostly
     also or exclusively from one or more of the following online sources: ChemIDPlus Advanced, PubChem,
     DrugBank, RxList, Drug Central, Drugs.com, and Wikipedia
2   Barbaud A, Bursztejn AC, Schmutz JL, Trechot PH. Patch tests with desloratadine at 10 % induce false-positive
     results: test at 1 %. J Eur Acad Dermatol Venereol 2008;22:1504-1505

Chapter 3.165 DEXAMETHASONE

IDENTIFICATION

Description/definition	: Dexamethasone is the synthetic glucocorticoid that conforms to the structural formula shown below
Pharmacological classes	: Antineoplastic agents, hormonal; antiemetics; anti-inflammatory agents; glucocorticoids
IUPAC name	: (8S,9R,10S,11S,13S,14S,16R,17R)-9-Fluoro-11,17-dihydroxy-17-(2-hydroxyacetyl)-10,13,16-trimethyl-6,7,8,11,12,14,15,16-octahydrocyclopenta[a]phenanthren-3-one
Other names	: (11β,16α)-9-Fluoro-11,17,21-trihydroxy-16-methylpregna-1,4-diene-3,20-dione
CAS registry number	: 50-02-2
EC number	: 200-003-9
Merck Index monograph	: 4215
Patch testing	: 0.5% pet. (SmartPracticeCanada); In general, corticosteroids may be tested at 0.1% and 1% in alcohol; late readings (6-10 days) are strongly recommended
Molecular formula	: $C_{22}H_{29}FO_5$

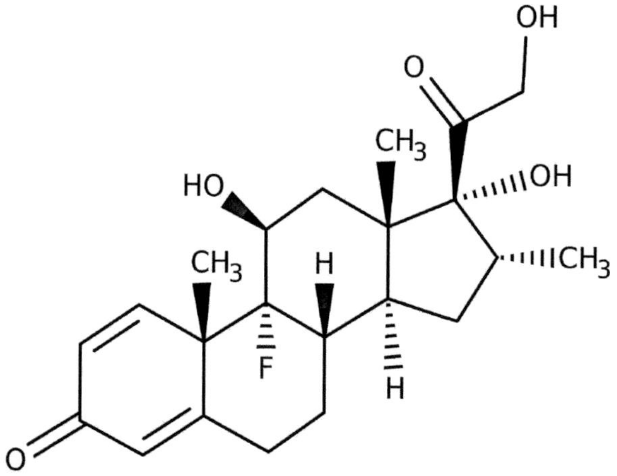

GENERAL

Systemically administered glucocorticoids have anti-inflammatory, immunosuppressive and antineoplastic properties and are used in the treatment of a wide spectrum of diseases including rheumatic disorders, lung diseases (asthma, COPD), gastrointestinal tract disorders (Crohn's disease, colitis ulcerosa), certain malignancies (leukemia, lymphomas), hematological disorders, and various diseases of the kidneys, brain, eyes and skin. A practical guideline for diagnosing allergic reactions to corticosteroids is presented in ref. 1. In pharmaceutical products, dexamethasone may be employed as base, as acetate, phosphate (Chapter 3.166) or as dexamethasone sodium phosphate (Chapter 3.167). Dexamethasone *base* is used as tablet and as elixir for oral use only, which implies that by far most allergic reactions to 'dexamethasone', which have been fully reviewed in Volume 3 of the *Monographs in contact allergy* series (5), have in fact been the result of sensitization to a salt or ester of dexamethasone or of cross-reactivity to another corticosteroid.

CUTANEOUS ADVERSE DRUG REACTIONS FROM SYSTEMIC ADMINISTRATION CAUSED BY TYPE IV (DELAYED-TYPE) HYPERSENSITIVITY (as demonstrated by positive patch tests)

Cutaneous adverse drug reactions from systemic administration of dexamethasone caused by type IV (delayed-type) hypersensitivity have included maculopapular eruption (2,10), exfoliative dermatitis (9), systemic allergic dermatitis presenting as the baboon syndrome/SDRIFE (6), SAD presenting as maculopapular eruption (7), eczematous eruption (4,8), and exanthema or delayed urticaria (3).

Maculopapular eruption

An 80-year-old man underwent dental implant surgery, and was subsequently started on oral dexamethasone, clindamycin, and metronidazole. Two days later, he developed a generalized maculopapular exanthema. Intradermal testing with diluted (0.9% NaCl) stock concentrations of dexamethasone (1:10 of 3.3 mg/ml), clindamycin (1:10 of 150 mg/ml), and metronidazole (1:10 of 5 mg/ml) was positive at D2 to dexamethasone only, with a bullous lesion at the test site. Patch testing was positive to dexamethasone sodium phosphate and tixocortol pivalate (2).

A woman aged 50 was treated with oral dexamethasone and betamethasone and developed a maculopapular exanthema from these drugs. Patch tests were positive to both corticosteroids. Details (e.g. whether the patient had previous sensitization to corticosteroids) are not available to the author (10).

See also the section 'Systemic allergic dermatitis' below (ref. 7).

Erythroderma, widespread erythematous eruption, exfoliative dermatitis
On the third day of using oral dexamethasone for encephalomyelitis disseminata, a 59-year old woman developed a macular rash with an exfoliate component involving the face, the upper chest, and intertriginous areas. The patient had never received corticosteroids before. Patch testing revealed positive reactions to dexamethasone, betamethasone, and clobetasol-17-propionate (9). As the reaction started at D3, the patient must have been sensitized previously.

Systemic allergic dermatitis (systemic contact dermatitis)
A 50-year-old woman, who was very likely sensitized to corticosteroids from their use in chronic nasal congestion, had suffered a skin eruption on three occasions one day after oral intake of prednisolone or methylprednisolone. Patch tests were positive to prednisolone, methylprednisolone was not tested. An oral provocation test with dexamethasone resulted in the classic picture of the baboon syndrome/symmetrical drug-related intertriginous and flexural exanthema (SDRIFE) after 12-24 hours. A patch test with dexamethasone was not performed, but an intradermal test was positive at D2. This was a case of systemic allergic dermatitis presenting as the baboon syndrome/SDRIFE (6).

A 46-year old woman previously shown to be contact allergic to multiple corticosteroids was given oral provocation tests with 5 systemic corticosteroids and developed maculopapular eruptions after the provocations with dexamethasone and 3 other corticosteroids (7).

Symmetrical drug-related intertriginous and flexural exanthema (SDRIFE)/Baboon syndrome
See the section 'Systemic allergic dermatitis' above (ref. 6).

Dermatitis/eczematous eruption
A 48-year-old woman had suffered three episodes of an eczematous exanthem after oral administration of dexamethasone, intramuscular injection of dexamethasone phosphate and intramuscular injection of prednisolone acetate. Patch tests were positive to commercial dexamethasone, dexamethasone phosphate, and commercial dexamethasone phosphate, but not with commercial prednisolone acetate (no test concentrations/vehicles mentioned) (4).

In a group of 45 patients with multiple drug hypersensitivity seen between 1996 and 2018 in Montpellier, France, 38 of 92 drug hypersensitivities were classified as type IV immunological reactions. This included one patient who had developed an eczematous eruption from dexamethasone and had a positive patch test to this corticosteroid (8).

Other cutaneous adverse drug reactions
Two patients who had either had a drug exanthema or delayed urticaria from systemic dexamethasone or betamethasone, had positive delayed intradermal tests and patch tests to dexamethasone and betamethasone and a positive systemic administration test to both corticosteroids. It was not specified which drug had caused the drug rash and which type the cutaneous adverse drug reaction was (3).

Cross-reactions, pseudo-cross-reactions and co-reactions
Cross-reactions between corticosteroids are discussed in Chapter 3.399 Prednisolone.

LITERATURE
1   Baeck M, Goossens A. Immediate and delayed allergic hypersensitivity to corticosteroids: practical guidelines. Contact Dermatitis 2012;66:38-45
2   Watts TJ, Thursfield D, Haque R. Cutaneous adverse drug reaction induced by oral dexamethasone with possible cross-reactivity to Group 1 corticosteroids confirmed by patch testing and intradermal testing. Contact Dermatitis 2019;81:384-386
3   Padial A, Posadas S, Alvarez J, Torres M-J, Alvarez JA, Mayorga C, Blanca M. Nonimmediate reactions to systemic corticosteroids suggest an immunological mechanism. Allergy 2005;60:665-670

4   Plaza T, Nist G, Von den Driesch P. Type IV-allergy due to corticosteroids. Rare and paradoxical. Dtsch Med Wochenschr 2007;132:1692-1695 (Article in German)

5   De Groot AC. Monographs in contact allergy, volume 3: Topical Drugs. Boca Raton, Fl, USA: CRC Press Taylor and Francis Group, 2021 (ISBN 978-0-367-23693-9)

6   Treudler R, Simon JC. Symmetric, drug-related, intertriginous, and flexural exanthema in a patient with polyvalent intolerance to corticosteroids. J Allergy Clin Immunol 2006;118:965-967

7   Chew AL, Maibach HI. Multiple corticosteroid orally elicited allergic contact dermatitis in a patient with multiple topical corticosteroid allergic contact dermatitis. Cutis 2000;65:307-311

8   Landry Q, Zhang S, Ferrando L, Bourrain JL, Demoly P, Chiriac AM. Multiple drug hypersensitivity syndrome in a large database. J Allergy Clin Immunol Pract 2019;8:258

9   Reinhold K, Schneider L, Hunzelmann N, Krieg T, Scharffetter-Kochanek K. Delayed-type allergy to systemic corticosteroids. Allergy 2000;55:1095-1097

10 Maucher O, Faber M, Knipper H, Kirchner S, Schöpf E. Kortikoidallergie. Hautarzt 1987;38:577-582 (Article in German, data cited in ref 11)

11 Bircher AJ, Levy F, Langauer S, Lepoittevin JP. Contact allergy to topical corticosteroids and systemic allergic dermatitis from prednisolone with tolerance of triamcinolone. Acta Derm Venereol 1995;75:490-493

Chapter 3.166 DEXAMETHASONE PHOSPHATE

IDENTIFICATION

Description/definition : Dexamethasone phosphate is the 21-*O*-phospho derivative of the synthetic glucocorticoid dexamethasone that conforms to the structural formula shown below
Pharmacological classes : Glucocorticoids
IUPAC name : [2-[(8*S*,9*R*,10*S*,11*S*,13*S*,14*S*,16*R*,17*R*)-9-Fluoro-11,17-dihydroxy-10,13,16-trimethyl-3-oxo-6,7,8,11,12,14,15,16-octahydrocyclopenta[a]phenanthren-17-yl]-2-oxoethyl] dihydrogen phosphate
Other names : Dexamethasone-21-phosphate; 9-fluoro-11β,17,21-trihydroxy-16α-methylpregna-1,4-diene-3,20-dione 21-(dihydrogen phosphate)
CAS registry number : 312-93-6
EC number : 206-232-0
Merck Index monograph : 4215 (Dexamethasone)
Patch testing : 1% pet. (SmartPracticeCanada); late readings (6-10 days) are strongly recommended
Molecular formula : $C_{22}H_{30}FO_8P$

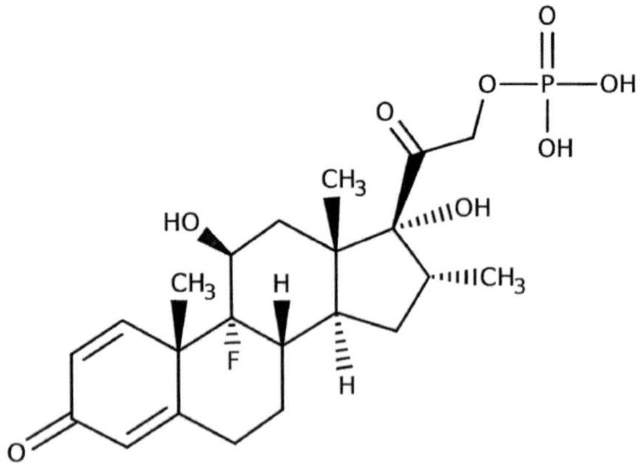

GENERAL

Systemically administered glucocorticoids have anti-inflammatory, immunosuppressive and antineoplastic properties and are used in the treatment of a wide spectrum of diseases including rheumatic disorders, lung diseases (asthma, COPD), gastrointestinal tract disorders (Crohn's disease, colitis ulcerosa), certain malignancies (leukemia, lymphomas), hematological disorders, and various diseases of the kidneys, brain, eyes and skin. A practical guideline for diagnosing allergic reactions to corticosteroids is presented in ref. 1. See also dexamethasone (Chapter 3.165) and dexamethasone sodium phosphate (Chapter 3.167).

CUTANEOUS ADVERSE DRUG REACTIONS FROM SYSTEMIC ADMINISTRATION CAUSED BY TYPE IV (DELAYED-TYPE) HYPERSENSITIVITY (as demonstrated by positive patch tests)

Cutaneous adverse drug reactions from systemic administration of dexamethasone phosphate caused by type IV (delayed-type) hypersensitivity have included generalized eczematous exanthema (2)

Dermatitis/eczematous eruption

A 48-year-old woman had suffered three episodes of an eczematous exanthem after intramuscular injection of dexamethasone phosphate, an intramuscular injection of prednisolone acetate and oral administration of dexamethasone. Patch tests were positive to dexamethasone phosphate, commercial dexamethasone phosphate and commercial dexamethasone, but not with commercial prednisolone acetate (no test concentrations/vehicles mentioned) (2).

Cross-reactions, pseudo-cross-reactions and co-reactions

Cross-reactions between corticosteroids are discussed in Chapter 3.399 Prednisolone.

LITERATURE

1   Baeck M, Goossens A. Immediate and delayed allergic hypersensitivity to corticosteroids: practical guidelines. Contact Dermatitis 2012;66:38-45
2   Plaza T, Nist G, Von den Driesch P. Type IV-allergy due to corticosteroids. Rare and paradoxical. Dtsch Med Wochenschr 2007;132:1692-1695 (Article in German)

Chapter 3.167 DEXAMETHASONE SODIUM PHOSPHATE

IDENTIFICATION

Description/definition : Dexamethasone sodium phosphate is the sodium phosphate salt form of the synthetic glucocorticoid dexamethasone, that conforms to the structural formula shown below

Pharmacological classes : Glucocorticoids

IUPAC name : Disodium;[2-[(8S,9R,10S,11S,13S,14S,16R,17R)-9-fluoro-11,17-dihydroxy-10,13,16-trimethyl-3-oxo-6,7,8,11,12,14,15,16-octahydrocyclopenta[a]phenanthren-17-yl]-2-oxoethyl] phosphate

Other names : Dexamethasone phosphate disodium salt; dexamethasone 21-(disodium phosphate); 9-fluoro-11β,17,21-trihydroxy-16α-methylpregna-1,4-diene-3,20-dione 21-(dihydrogen phosphate) disodium salt

CAS registry number : 2392-39-4

EC number : 219-243-0

Merck Index monograph : 4215 (Dexamethasone)

Patch testing : 1.0% pet. (Chemotechnique, SmartPracticeEurope); late readings (6-10 days) are strongly recommended

Molecular formula : $C_{22}H_{28}FNa_2O_8P$

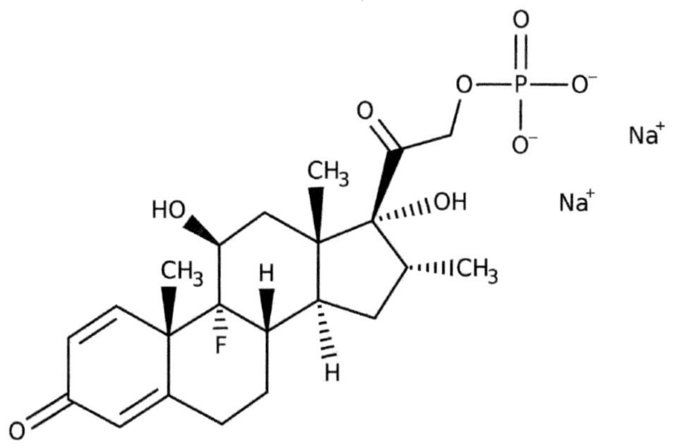

GENERAL

Systemically administered glucocorticoids have anti-inflammatory, immunosuppressive and antineoplastic properties and are used in the treatment of a wide spectrum of diseases including rheumatic disorders, lung diseases (asthma, COPD), gastrointestinal tract disorders (Crohn's disease, colitis ulcerosa), certain malignancies (leukemia, lymphomas), hematological disorders, and various diseases of the kidneys, brain, eyes and skin. A practical guideline for diagnosing allergic reactions to corticosteroids is presented in ref. 1. Dexamethasone sodium phosphate may be used in both topical and systemic pharmaceutical applications. In topical preparations, this glucocorticoid has caused contact allergy/allergic contact dermatitis, which has been fully reviewed in Volume 3 of the *Monographs in contact allergy* series (2). See also dexamethasone (Chapter 3.165) and dexamethasone phosphate (Chapter 3.166).

CUTANEOUS ADVERSE DRUG REACTIONS FROM SYSTEMIC ADMINISTRATION CAUSED BY TYPE IV (DELAYED-TYPE) HYPERSENSITIVITY (as demonstrated by positive patch tests)

Cutaneous adverse drug reactions from systemic administration of dexamethasone sodium phosphate caused by type IV (delayed-type) hypersensitivity have included generalized erythematous rash (6), erythematous rash with facial edema (5), acute generalized exanthematous pustulosis (AGEP) (4), and systemic allergic dermatitis presenting as maculopapular eruption (3).

Maculopapular eruption

See the section 'Systemic allergic dermatitis (systemic allergic dermatitis)' below, ref. 3.

Erythroderma, widespread erythematous eruption, exfoliative dermatitis

A 43-year-old woman presented with an erythematous rash and facial edema 18 hours after the injection of dexamethasone (probably disodium phosphate). Patch tests were positive to dexamethasone 0.4% water and various other corticosteroids. An oral provocation test with 4 mg dexamethasone resulted in 'a reaction' after 20 hours. It was not mentioned how the patient had become sensitized (5).

A 71-year-old woman awoke with a pruritic and painful, diffuse 'swollen sunburn'-like cutaneous reaction. Two days earlier, the patient had received a dexamethasone injection of a foot spur. She presented to a local emergency room and was noted to have a generalized, fine scaling erythema. The patient could not recall a prior history of topical, enteral, or parenteral corticosteroid exposure. Patch testing showed a positive reaction to dexamethasone 0.05% pet. (6). The patient is presented in this chapter, as dexamethasone base is not used in injection fluids, but is mostly used as its sodium phosphate salt.

Acute generalized exanthematous pustulosis (AGEP)
A 52-year-old woman developed a widespread, sterile pustular eruption on the trunk and extremities 2 days after subcutaneous injection of dexamethasone solution. Skin biopsy revealed subcorneal pustules filled with neutrophils and moderate lymphohistiocytic infiltrate with a few eosinophils in the dermis. Patch testing showed positive pustular reactions to dexamethasone solution (which is usually the disodium phosphate salt). Histology of this pustule resembled that of the original eruption. The patient was diagnosed with acute generalized exanthematous pustulosis induced by dexamethasone injection (4).

Systemic allergic dermatitis (systemic contact dermatitis)
A 63-year-old woman developed an acute itchy, maculopapular eruption of the trunk, arms and legs on the third day of having taking deflazacort orally as an adjuvant for chemotherapy. A similar eruption had occurred 2 months before when she had taken dexamethasone 21-disodium phosphate 8 mg/day. The history was highly suspicious for allergic contact dermatitis from topical steroids used for lichen planus. Patch tests were positive to deflazacort 1% pet., dexamethasone sodium phosphate 1% alc. and 8 other corticosteroids, including the ones previously used for lichen planus (3).

Cross-reactions, pseudo-cross-reactions and co-reactions
Cross-reactions between corticosteroids are discussed in Chapter 3.399 Prednisolone.

LITERATURE
1   Baeck M, Goossens A. Immediate and delayed allergic hypersensitivity to corticosteroids: practical guidelines. Contact Dermatitis 2012;66:38-45
2   De Groot AC. Monographs in contact allergy, volume 3: Topical Drugs. Boca Raton, Fl, USA: CRC Press Taylor and Francis Group, 2021 (ISBN 978-0-367-23693-9)
3   Bianchi L, Hansel K, Antonelli E, Bellini V, Rigano L, Stingeni L. Deflazacort hypersensitivity: a difficult-to-manage case of systemic allergic dermatitis and literature review. Contact Dermatitis 2016;75:54-56
4   Demitsu T, Kosuge A, Yamada T, Usui K, Katayama H, Yaoita H. Acute generalized exanthematous pustulosis induced by dexamethasone injection. Dermatology 1996;193:56-58
5   Villas F, Garmendia F J, Joral A, Villarreal O,Navarro J A. Systemic allergic reaction to dexamethasone. Allergy 1997: 52 (Suppl.37):176 (poster)
6   Whitmore SE. Dexamethasone injection-induced generalized dermatitis. Br J Dermatol 1994;131:296-297

Chapter 3.168 DEXKETOPROFEN

IDENTIFICATION

Description/definition   : Dexketoprofen is the propionic acid derivative and isomer of ketoprofen that conforms to
                          the structural formula shown below
Pharmacological classes : Anti-inflammatory agents, non-steroidal
IUPAC name   : (2S)-2-(3-Benzoylphenyl)propanoic acid
Other names   : (S)-(+)-Ketoprofen
CAS registry number   : 22161-81-5
EC number   : 606-944-5
Patch testing   : 1.0% pet. (Chemotechnique)
Molecular formula   : $C_{16}H_{14}O_3$

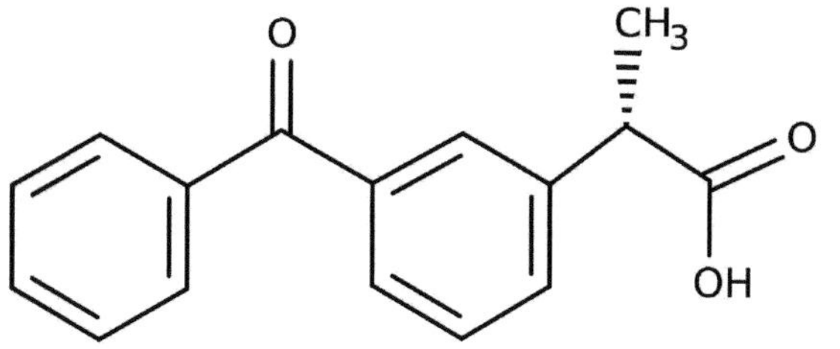

GENERAL

Dexketoprofen is a propionic acid derivative and nonsteroidal anti-inflammatory drug (NSAID) with analgesic, anti-inflammatory, and antipyretic properties. It is the (S)-enantiomer and active isomer of ketoprofen and works by blocking the action of cyclooxygenase. Dexketoprofen is indicated for short-term treatment of mild to moderate pain, including dysmenorrhea, musculoskeletal pain and toothache. It is available in topical and oral formulations (1).

In topical preparations, dexketoprofen has caused some cases of photocontact allergy/photoallergic contact dermatitis, which subject has been fully reviewed in Volume 3 of the *Monographs in contact allergy* series (9).

CUTANEOUS ADVERSE DRUG REACTIONS FROM SYSTEMIC ADMINISTRATION CAUSED BY TYPE IV (DELAYED-TYPE) HYPERSENSITIVITY (as demonstrated by positive patch tests)

Cutaneous adverse drug reactions from systemic administration of dexketoprofen caused by type IV (delayed-type) hypersensitivity have included systemic allergic dermatitis (systemic allergic dermatitis) presenting as photosensitivity (3).

Systemic allergic dermatitis (systemic contact dermatitis)

A 39-year-old man, after 3 days of treatment with oral dexketoprofen twice a day, suffered from a pruritic papulo-vesicular dermatitis on the body and arms. The rash resolved in a few days following topical corticosteroid and oral antihistamine treatment without sequelae. In the following 2 weeks, he presented similar symptoms several hours after again taking one pill of dexketoprofen. He admitted to moderate sun exposure on the beach on the last occasion, but did not remember sun exposure during the first episode. The patient recalled the development of an acute dermatitis after applying topical ketoprofen on his back during the summer a few years ago. Patch test with dexketoprofen 2.5% pet. and a number of related NSAIDs were negative. Photopatch test using irradiation with UVA $8J/cm^2$ at D2) were positive to ketoprofen 1% pet. (++) and dexketoprofen 2.5% pet. (+++). This was a case of systemic photocontact dermatitis to dexketoprofen after previous sensitization to ketoprofen (3).

Photosensitivity

See the section 'Systemic allergic dermatitis (systemic allergic dermatitis)' above, ref. 3.

Cross-reactions, pseudo-cross-reactions and co-reactions

Dexketoprofen is the (S)-enantiomer of ketoprofen and photocross-reactions between these NSAIDs (in both directions) nearly always occur (3,4,5,6,7). Of 16 patients with photocontact allergy to ketoprofen and dexketoprofen, 11 cross-reacted to piketoprofen (8). Photocross-sensitivity to or from piketoprofen (2,5,7).

LITERATURE

1   The data in the section 'General' may have been obtained from literature discussed in this chapter, but mostly also or exclusively from one or more of the following online sources: ChemIDPlus Advanced, PubChem, DrugBank, RxList, Drug Central, Drugs.com, and Wikipedia

2   Fernández-Jorge B, Goday Buján JJ, Paradela S, Mazaira M, Fonseca E. Consort photocontact dermatitis from piketoprofen. Contact Dermatitis 2008;58:113-115

3   Asensio T, Sanchís ME, Sánchez P, Vega JM, García JC. Photocontact dermatitis because of oral dexketoprofen. Contact Dermatitis 2008;58:59-60

4   Goday-Bujan JJ, Rodríguez-Lozano J, Martínez-González MC, Fonseca E. Photoallergic contact dermatitis from dexketoprofen: study of 6 cases. Contact Dermatitis 2006;55:59-61

5   López-Abad R, Paniagua MJ, Botey E, Gaig P, Rodriguez P, Richart C. Topical dexketoprofen as a cause of photocontact dermatitis. J Investig Allergol Clin Immunol 2004;14:247-249

6   Cuerda Galindo E, Goday Buján JJ, del Pozo Losada J, García Silva J, Peña Penabad C, Fonseca E. Photocontact dermatitis due to dexketoprofen. Contact Dermatitis 2003;48:283-284

7   Valenzuela N, Puig L, Barnadas MA, Alomar A. Photocontact dermatitis due to dexketoprofen. Contact Dermatitis 2002;47:237

8   Subiabre-Ferrer D, Esteve-Martínez A, Blasco-Encinas R, Sierra-Talamantes C, Pérez-Ferriols A, Zaragoza-Ninet V. European photopatch test baseline series: A 3-year experience. Contact Dermatitis 2019;80:5-8

9   De Groot AC. Monographs in contact allergy, volume 3: Topical Drugs. Boca Raton, Fl, USA: CRC Press Taylor and Francis Group, 2021 (ISBN 978-0-367-23693-9)

Chapter 3.169 DEXLANSOPRAZOLE

IDENTIFICATION

Description/definition : Dexlansoprazole is the substituted benzimidazole that conforms to the structural formula shown below
Pharmacological classes : Proton pump inhibitors; anti-ulcer agents
IUPAC name : 2-[(R)-[3-Methyl-4-(2,2,2-trifluoroethoxy)pyridin-2-yl]methylsulfinyl]-1H-benzimidazole
CAS registry number : 138530-94-6
EC number : Not available
Merck Index monograph : 6683 (Lansoprazole)
Patch testing : 10% pet.
Molecular formula : $C_{16}H_{14}F_3N_3O_2S$

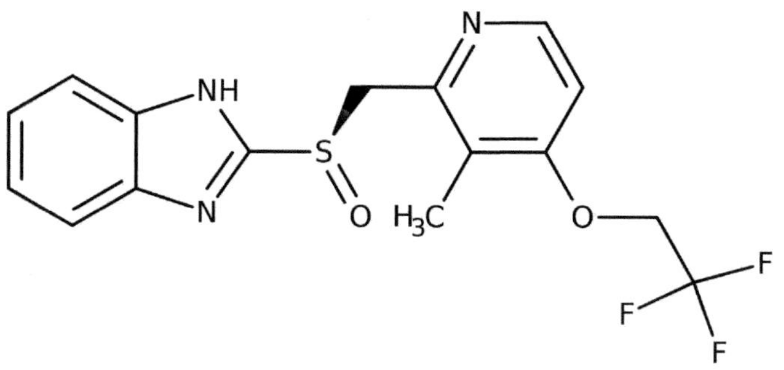

GENERAL

Dexlansoprazole is the R-isomer of lansoprazole and a substituted benzimidazole prodrug. It is a proton pump inhibitor and a potent inhibitor of gastric acidity. Dexlansoprazole is indicated for healing all grades of erosive esophagitis and treating heartburn associated with symptomatic non-erosive gastroesophageal reflux disease (1).

CONTACT ALLERGY FROM ACCIDENTAL CONTACT

A 58-year-old male pharmaceutical worker presented with erythematous eruptions involving the neck, posterior auricular areas, and extensor forearms, with slight swelling and mild erythema of the upper eyelids. He had a 4-year history as a chemical operator in the production area of the plant, wearing rubber gloves and long-sleeved shirts. He had regular contact with proton pump inhibitors and intermediates including dexlansoprazole, esomeprazole magnesium, lansoprazole, omeprazole magnesium, and pantoprazole sodium. Patch tests were positive to dexlansoprazole 10% (++) and 1% pet. (+), omeprazole (10% pet., 1% pet., 1% alcohol), pantoprazole (1% and 10% pet.) and esomeprazole (1% pet.). The patient was diagnosed with occupational airborne allergic contact dermatitis to proton pump inhibitors including dexlansoprazole (2).

Cross-reactions, pseudo-cross-reactions and co-reactions

A patient sensitized to omeprazole cross-reacted to dexlansoprazole and other proton pump inhibitors (all crushed tablets, 1% and 10% pet.) (3).

LITERATURE

1 The data in the section 'General' may have been obtained from literature discussed in this chapter, but mostly also or exclusively from one or more of the following online sources: ChemIDPlus Advanced, PubChem, DrugBank, RxList, Drug Central, Drugs.com, and Wikipedia
2 DeKoven JG, Yu AM. Occupational airborne contact dermatitis from proton pump inhibitors. Dermatitis 2015;26:287-290
3 Al-Falah K, Schachter J, Sasseville D. Occupational allergic contact dermatitis caused by omeprazole in a horse breeder. Contact Dermatitis 2014;71:377-378
4 Brockow K, Garvey LH, Aberer W, Atanaskovic-Markovic M, Barbaud A, Bilo MB, et al.; ENDA/EAACI Drug Allergy Interest Group. Skin test concentrations for systemically administered drugs – an ENDA/EAACI Drug Allergy Interest Group position paper. Allergy 2013;68:702-712

Chapter 3.170 DEXTROPROPOXYPHENE

IDENTIFICATION

Description/definition   : Propoxyphene is the *d*-isomer of the synthetic diphenyl propionate derivative
                              propoxyphene that conforms to the structural formula shown below
Pharmacological classes   : Narcotics; analgesics, opioid
IUPAC name   : [(2*S*,3*R*)-4-(Dimethylamino)-3-methyl-1,2-diphenylbutan-2-yl] propanoate
Other names   : *d*-Propoxyphene
CAS registry number   : 469-62-5
EC number   : 207-420-5
Merck Index monograph   : 9222 (Propoxyphene)
Patch testing   : Pulverized tablet, 5% and 20% in water and petrolatum; most systemic drugs can be
                              tested at 10% pet.; if the pure chemical is not available, prepare the test material from
                              intravenous powder, the content of capsules or – when also not available – from
                              powdered tablets to achieve a final concentration of the active drug of 10% pet.
Molecular formula   : $C_{22}H_{29}NO_2$

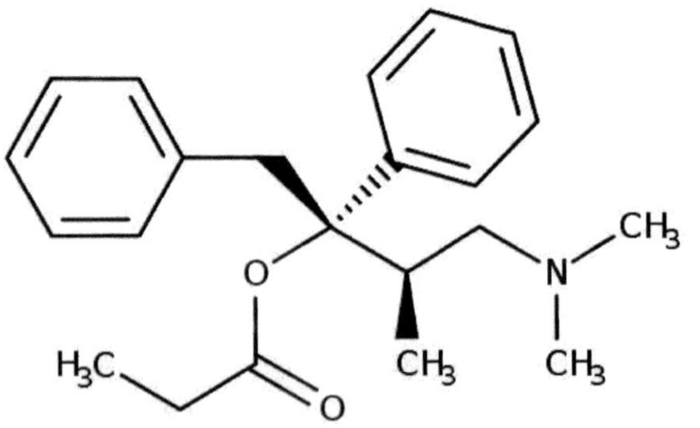

GENERAL

Dextropropoxyphene is the *d*-isomer of the synthetic diphenyl propionate derivative propoxyphene, with narcotic analgesic effect. It is used in the symptomatic treatment of mild pain. It displays antitussive and local anesthetic actions. Due to the risk of cardiac arrhythmias and overdose, possibly leading to death, dextropropoxyphene has been withdrawn from the market in Europe and the United States. The drug is often referred to as the general form, 'propoxyphene'. but only the dextro-isomer (dextropropoxyphene) has any analgesic effect. Its enantiomer levopropoxyphene appears to exhibit a very limited antitussive effect. In pharmaceutical products, dextropropo-xyphene is employed as dextropropoxyphene hydrochloride (CAS number 1639-60-7, EC number 216-683-5, molecular formula $C_{22}H_{30}ClNO_2$) or as dextropropoxyphene napsylate (CAS number 26570-10-5, EC number not available, molecular formula $C_{32}H_{39}NO_6S$) (1).

CUTANEOUS ADVERSE DRUG REACTIONS FROM SYSTEMIC ADMINISTRATION CAUSED BY TYPE IV (DELAYED-TYPE) HYPERSENSITIVITY (as demonstrated by positive patch tests)

Cutaneous adverse drug reactions from systemic administration of dextropropoxyphene caused by type IV (delayed-type) hypersensitivity have included acute generalized exanthematous pustulosis (AGEP) (2,3).

Acute generalized exanthematous pustulosis (AGEP)

A 43-year-old woman was admitted for a febrile eruption (39.2°C) of acute onset. Four days before the admission, the patient had been treated with spiramycin, tenoxicam and the combination of dextropropoxyphene, paracetamol, chlorpheniramine, caffeine and carbaspirin calcium for parotitis. Clinical examination showed generalized erythema, with numerous pustules on the trunk. Histological examination of a cutaneous biopsy showed a subcorneal unilocular pustule with dermal edema and infiltration of polymorphonuclear cells within the dermis. Routine blood tests showed mild hyperleukocytosis with predominantly neutrophils, raised erythrocyte sedimentation rate (60 mm at first hour) and elevated C-reactive protein (102 mg/l, normal <8 mg/l). There was no evidence of viral infection. The patient was treated with topical corticosteroids and the cutaneous lesions disappeared in 12 days. She had experienced 2 similar episodes in the last 5 years. Each episode was preceded by intake of dextropropoxyphene combined with paracetamol. Patch testing was carried out 1 month later with dextropropoxyphene, paracetamol, spiramycin, aspirin and tenoxicam, all diluted at 5% and 20% in water and petrolatum, respectively. All tests with dextropropoxyphene were positive, and the others remained negative. Ten controls were negative. Histological examination of a positive patch test showed dermal edema and intra-epidermal pustules. Acute generalized exanthematous pustulosis caused by dextropropoxyphene was diagnosed (2).

In a multicenter investigation in France, of 45 patients patch tested for AGEP, 26 (58%) had positive patch tests to drugs, including 2 to paracetamol-dextropropoxyphene. The separate ingredients were not tested (3).

LITERATURE

1   The data in the section 'General' may have been obtained from literature discussed in this chapter, but mostly also or exclusively from one or more of the following online sources: ChemIDPlus Advanced, PubChem, DrugBank, RxList, Drug Central, Drugs.com, and Wikipedia
2   Machet L, Martin L, Machet MC, Lorette G, Vaillant L. Acute generalized exanthematous pustulosis induced by dextropropoxyphene and confirmed by patch testing. Acta Derm Venereol 2000;80:224-225
3   Barbaud A, Collet E, Milpied B, Assier H, Staumont D, Avenel-Audran M, et al. A multicentre study to determine the value and safety of drug patch tests for the three main classes of severe cutaneous adverse drug reactions. Br J Dermatol 2013;168:555-562

Chapter 3.171 DIATRIZOIC ACID

IDENTIFICATION

Description/definition   : Diatrizoic acid is the iodinated benzoic acid derivative that conforms to the structural
                            formula shown below
Pharmacological classes : Contrast media
IUPAC name   : 3,5-Diacetamido-2,4,6-triiodobenzoic acid; amidotrizoic acid
CAS registry number   : 117-96-4
EC number   : 204-223-6
Merck Index monograph   : 4263
Patch testing   : Commercial preparation undiluted (5); sodium diatrizoate 10% water
Molecular formula   : $C_{11}H_9I_3N_2O_4$

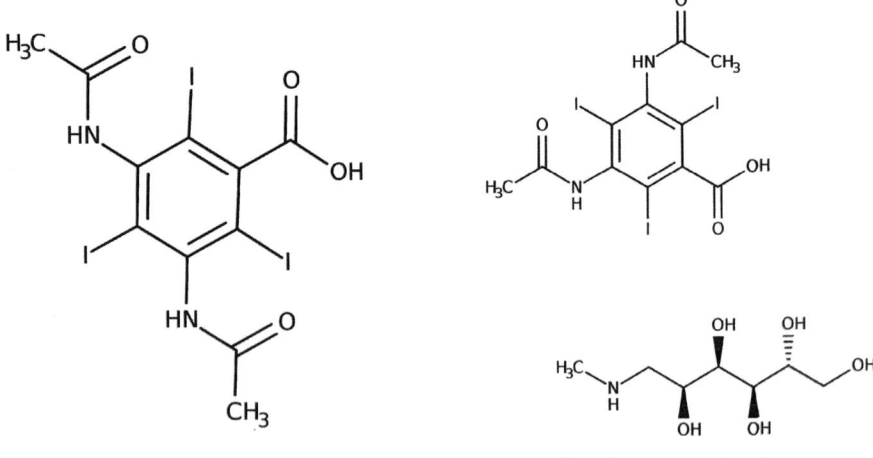

Diatrizoic acid   Diatrizoate meglumine

GENERAL

Diatrizoic acid is a benzoic acid derivative which is used as an organic, iodinated, radiopaque X-ray contrast medium.
The iodine moiety of diatrizoic acid is not penetrable by X-rays, therefore blocks the X-ray film exposure by radiation.
This makes it possible to distinguish on X-ray film, body parts that contain diatrizoic acid (in the form of diatrizoate
meglumine or diatrizoate sodium) from body parts that do not contain these agents and allows for visualization of
different body structures (1).

Diatrizoic acid (in the form of its sodium and/or meglumine salt) can be used, alone or in combination, for a wide
variety of diagnostic imaging methods, including angiography, urography, cholangiography, computed tomography,
hysterosalpingography, and retrograde pyelography. It can also be used for imaging the gastrointestinal tract in
patients allergic to barium (1). In X-ray contrast media, diatrizoic acid is employed as diatrizoate sodium (CAS
number 737-31-5, EC number 212-004-1, molecular formula $C_{11}H_8I_3N_2NaO_4$) or diatrizoate meglumine (CAS number
131-49-7, EC number 205-024-7, molecular formula $C_{18}H_{26}I_3N_3O_9$) (1).

CONTACT ALLERGY FROM ACCIDENTAL CONTACT

Case series

Between 1978 and 2001, 14,689 patients have been patch tested at the university hospital of Leuven, Belgium.
Occupational allergic contact dermatitis to pharmaceuticals was diagnosed in 33 health care workers: 26 nurses, 4
veterinarians, 2 pharmacists and one medical doctor. There were 26 women and 7 men with a mean age of 38 years.
Practically all of these patients presented with hand dermatitis (often the fingers), sometimes with secondary
localizations, frequently the face. In this group, one patient had occupational allergic contact dermatitis from
meglumine diatrizoate (2).

Case reports

A 25-year-old male nurse working in a radiology department developed dermatitis of the hands. Patch tests were
positive to two contrast media he worked with, tested 'as is', both containing diatrizoate meglumine. The patient
was then patch tested with a solution prepared by adding 50 mg of amidotrizoic acid (= diatrizoic acid) to 2.5 ml of a
1% (g/v) solution of meglumine in water, yielding a ++ reaction at D2 and D4. The other components of the contrast

media, including iodine, were negative. Three controls were negative. A diagnosis of occupational allergic contact dermatitis to diatrizoate meglumine was made (3).

A nurse aged 32 developed eczema of the hands after a 6-months period of work in an x-ray unit for angiogramphy. She had contact with contrast media containing sodium and meglumine diatrizoate every day, filling syringes without wearing gloves. She stopped this work and the eczema cleared up rapidly. A relapse occurred 2 days after recommencing the job. When patch tested, she had a positive reaction to one of the two contrast media 'as is' and - later - to sodium diatrizoate 50%, 25%, 10% (all ++ at D4) and 1% water (+), that were only positive after tape-stripping the deeply suntanned skin (4).

The same authors reported on a female technician of 26, who developed hand eczema after 10 months' use of the contrast media. She had positive patch tests to both contrast media 'as is', to sodium diatrizoate 25% water and – later – to sodium diatrizoate 50%, 25% and 10% water, with a ?+ reaction to 1% water. In both patients, methyl-glucamine (meglumine) 50% water was negative. Controls tests to the commercial contrast media were negative in 25 individuals. One of 18 controls (who had undergone a renography several years earlier) had a weak positive reaction to sodium diatrizoate (4).

Cross-reactions, pseudo-cross-reactions and co-reactions

Cross-reactions between iodinated radiocontrast media occur frequently, especially between the non-ionic preparations (6,7). Iodine is infrequently to rarely (opinions differ) the allergen (8,9) and allergy to mollusks, crustaceans, fish, and iodine from other sources is said not to be a risk factor for the development of hypersensitivity to iodinated contrast media (10).

LITERATURE

1 The data in the section 'General' may have been obtained from literature discussed in this chapter, but mostly also or exclusively from one or more of the following online sources: ChemIDPlus Advanced, PubChem, DrugBank, RxList, Drug Central, Drugs.com, and Wikipedia

2 Gielen K, Goossens A. Occupational allergic contact dermatitis from drugs in healthcare workers. Contact Dermatitis 2001;45:273-279

3 Verschaeve A, Loncke J, Dooms-Goossens A. Occupational contrast media dermatitis: meglumine diatrizoate. Contact Dermatitis 1984;11:318-319

4 Rothe A, Yousif SH, Zschunke E. Allergic contact eczema from sodium amidotrizoate - a radiopaque substance in angiography and renography. Contact Dermatitis 1977;3:284-286

5 Brockow K, Garvey LH, Aberer W, Atanaskovic-Markovic M, Barbaud A, Bilo MB, et al. Skin test concentrations for systemically administered drugs - an ENDA/EAACI Drug Allergy Interest Group position paper. Allergy 2013;68:702-712

6 Kanny G, Pichler W, Morisset M, Franck P, Marie B, Kohler C, Renaudin JM, Beaudouin E, Laudy JS, Moneret-Vautrin DA. T cell-mediated reactions to iodinated contrast media: evaluation by skin and lymphocyte activation tests. J Allergy Clin Immunol 2005;115:179-185

7 Lerondeau B, Trechot P, Waton J, Poreaux C, Luc A, Schmutz JL, et al. Analysis of cross-reactivity among radiocontrast media in 97 hypersensitivity reactions. J Allergy Clin Immunol 2016;137:633-635.e4

8 Scherer K, Harr T, Bach A, Bircher AJ. The role of iodine in hypersensitivity reactions to radiocontrast media. Clin Exp Allergy 2009;40:468-475

9 Trautmann A, Brockow K, Behle V, Stoevesandt J. Radiocontrast media hypersensitivity: Skin testing differentiates allergy from nonallergic reactions and identifies a safe alternative as proven by intravenous provocation. J Allergy Clin Immunol Pract 2019;7:2218-2224

10 Rosado Ingelmo A, Doña Diaz I, Cabañas Moreno R, Moya Quesada MC, García-Avilés C, García Nuñez I, et al. Clinical practice guidelines for diagnosis and management of hypersensitivity reactions to contrast media. J Investig Allergol Clin Immunol 2016;26:144-155

Chapter 3.172 DIAZEPAM

IDENTIFICATION

Description/definition   : Diazepam is the benzodiazepine that conforms to the structural formula shown below
Pharmacological classes : Anesthetics, intravenous; antiemetics; GABA modulators; hypnotics and sedatives; anti-anxiety agents; adjuvants, anesthesia; anticonvulsants; muscle relaxants, central
IUPAC name   : 7-Chloro-1-methyl-5-phenyl-3H-1,4-benzodiazepin-2-one
CAS registry number   : 439-14-5
EC number   : 207-122-5
Merck Index monograph   : 4267
Patch testing   : 1% water; higher concentrations may be tried with controls
Molecular formula   : $C_{16}H_{13}ClN_2O$

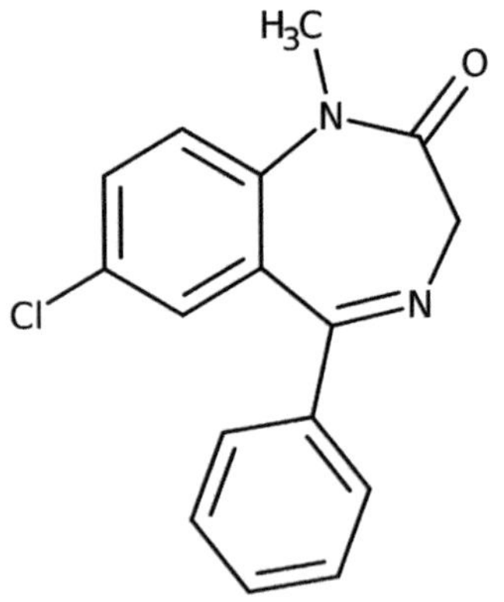

GENERAL

Diazepam is a benzodiazepine with anticonvulsant, anxiolytic, sedative, muscle relaxant, and amnesic properties and a long duration of action. Its actions are mediated by enhancement of γ-aminobutyric acid (GABA) activity. In general, diazepam is useful in the symptomatic management of mild to moderate degrees of anxiety in conditions dominated by tension, excitation, agitation, fear, or aggressiveness. In acute alcoholic withdrawal, diazepam may be useful in the symptomatic relief of acute agitation, tremor, and impending acute delirium tremens. Diazepam is also a useful adjunct for the relief of skeletal muscle spasm due to reflex spasm to local pathologies, such as inflammation of the muscle and joints or secondary to trauma, for spasticity caused by upper motor neuron disorders, such as cerebral palsy and paraplegia, for athetosis and for the rare 'stiff man syndrome' (1).

CONTACT ALLERGY FROM ACCIDENTAL CONTACT

Case series

In Leuven, Belgium, in the period 2007-2011, 81 patients have been diagnosed with occupational airborne allergic contact dermatitis. In 23 of them, drugs were the offending agents, including diazepam in one case (2, overlap with ref. 3).

Case reports

A 30-year-old woman presented with an itchy, scaly skin reaction on the face, which was most pronounced on the eyelids. She reported a clear relationship with her work as a geriatric nurse, during which she was required to crush tablets for the elderly on a daily basis, mostly benzodiazepine drugs. She had positive patch tests to diazepam (crushed tablet 30% pet.; 16 controls were negative), 4 other benzodiazepines (alprazolam, bromazepam, lorazepam, tetrazepam) and 3 unrelated drugs. The patient was diagnosed with occupational airborne allergic contact dermatitis from drugs (3, overlap with ref. 2).

A 44-year-old nurse presented with eczema on the hands and face, which had started 3 months after she was required to crush tablets daily during her work as a nurse in a clinic for disabled people. She had positive patch tests to diazepam (crushed tablet 30% pet.; 16 controls were negative), 2 other benzodiazepines (lorazepam, tetrazepam) and 5 unrelated chemicals. The patient was diagnosed with (airborne) allergic contact dermatitis from drugs (3).

A 37-year-old woman developed eczema on the forefinger of the right hand, which later spread to the dorsum of the right hand, eyelids, neck and cheeks. She cared for a daughter with cerebral paralysis, and had to give her medication after triturating tablets by hand and gathering the powder up into a small spoon with her forefinger. Patch testing showed positive reactions to 4 commercial drugs the daughter used tested at 1% water. Later, patch tests with their active ingredients were performed and the patient had positive reactions to diazepam, tetrazepam, clorazepate and sodium valproate, all tested 1% in water (6).

CUTANEOUS ADVERSE DRUG REACTIONS FROM SYSTEMIC ADMINISTRATION CAUSED BY TYPE IV (DELAYED-TYPE) HYPERSENSITIVITY (as demonstrated by positive patch tests)
Cutaneous adverse drug reactions from systemic administration of diazepam caused by type IV (delayed-type) hypersensitivity have included eczema of the hands and periorbital edema (4).

Other cutaneous adverse drug reactions
A 74-year-old man developed, 2 days after he started to take diazepam, intense irritation and a weeping eruption involving the hands, and swelling and irritation around the eyes. Physical examination revealed an acute eczematous eruption involving the dorsal and palmar surfaces of both hands associated with bilateral periorbital edema. Patch tests were strongly positive to commercial diazepam 5 mg/ml solution after 24 hours. In a second patch test session, there were positive reactions to diazepam 5 mg/ml and 5µg/ml but negative to 0.5µg/ml. Six controls were negative to the highest concentration. A year later, the tests were repeated but now with negative results, according to the author probably from lack of antigenic stimulation during this period (4).

Cross-reactions, pseudo-cross-reactions and co-reactions
Most reactions to benzodiazepines have occurred to tetrazepam. In these patients, cross-reactions to other benzodiazepines do not occur (Chapter 3.471 Tetrazepam). One such patient co-reacted (possibly cross-reacted) to diazepam (5). Of 10 nurses with occupational airborne allergic contact dermatitis, 5 were tested with diazepam and there were 3 positive reactions to diazepam crushed tablets 30% pet. and/or water. The authors considered these to be cross-reactions. However, it was mentioned that some of the patients also had contact with diazepam at work, so co-reactivity was not excluded (7).

LITERATURE
1 The data in the section 'General' may have been obtained from literature discussed in this chapter, but mostly also or exclusively from one or more of the following online sources: ChemIDPlus Advanced, PubChem, DrugBank, RxList, Drug Central, Drugs.com, and Wikipedia
2 Swinnen I, Goossens A. An update on airborne contact dermatitis: 2007-2011. Contact Dermatitis 2013;68:232-238
3 Swinnen I, Ghys K, Kerre S, Constandt L, Goossens A. Occupational airborne contact dermatitis from benzodiazepines and other drugs. Contact Dermatitis 2014;70:227-232
4 Felix RH, Comaish JS. The value of patch tests and other skin tests in drug eruptions. Lancet 1974;1(7865):1017-1019
5 Kämpgen E, Bürger T, Bröcker EB, Klein CE. Cross-reactive type IV hypersensitivity reactions to benzodiazepines revealed by patch testing. Contact Dermatitis 1995;33:356-357
6 Garcia-Bravo B, Rodriguez-Pichardo A, Camacho F. Contact dermatitis from diazepoxides. Contact Dermatitis 1994;30:40
7 Landeck L, Skudlik C, John SM. Airborne contact dermatitis to tetrazepam in geriatric nurses - a report of 10 cases. J Eur Acad Dermatol Venereol 2012;26:680-684

Chapter 3.173 DICLOFENAC

IDENTIFICATION

Description/definition : Diclofenac is the nonsteroidal benzeneacetic acid derivative that conforms to the structural formula shown below
Pharmacological classes : Anti-inflammatory agents, non-steroidal; cyclooxygenase inhibitors
IUPAC name : 2-[2-(2,6-Dichloroanilino)phenyl]acetic acid
CAS registry number : 15307-86-5
EC number : 239-348-5
Merck Index monograph : 4361
Patch testing : Sodium salt, 2.5% pet. (SmartPracticeCanada, SmartPracticeEurope); sodium salt, 5% pet. (Chemotechnique, SmartPracticeCanada); sodium salt, 1% pet. (Chemotechnique)
Molecular formula : $C_{14}H_{11}Cl_2NO_2$

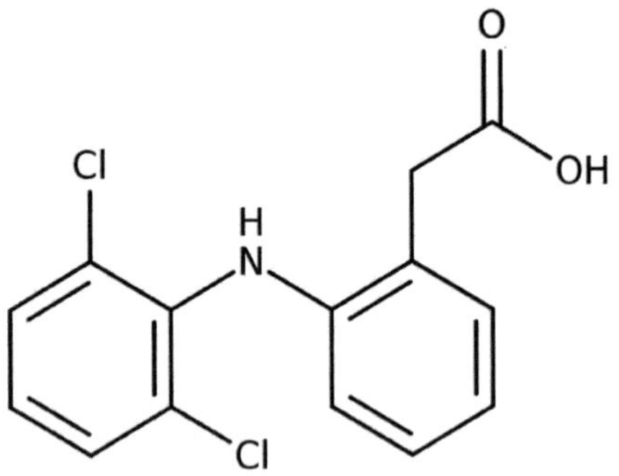

GENERAL

Diclofenac is a benzeneacetic acid-derived nonsteroidal anti-inflammatory drug (NSAID) with antipyretic, anti-inflammatory and analgesic actions. Its mechanism of action is as a cyclooxygenase inhibitor. Oral diclofenac is generally used to treat pain from dysmenorrhea, osteoarthritis, rheumatoid arthritis, ankylosing spondylitis, and from other causes. In pharmaceutical products, diclofenac is employed as diclofenac sodium (CAS number 15307-79-6, EC number 239-346-4, molecular formula $C_{14}H_{10}Cl_2NNaO_2$) (1). Lipid nanoemulsions of diclofenac may be used for parenteral applications. Diclofenac is also available in topical formulations including ointment, gel, suppositories, and as transdermal therapeutic system (TTS) for treatment of pain due to minor sprains, strains, and contusions. In such formulations, diclofenac has caused various cases of (photo)contact allergy/(photo)allergic contact dermatitis, which have been fully reviewed in Volume 3 of the *Monographs in contact allergy* series (11).

CUTANEOUS ADVERSE DRUG REACTIONS FROM SYSTEMIC ADMINISTRATION CAUSED BY TYPE IV (DELAYED-TYPE) HYPERSENSITIVITY (as demonstrated by positive patch tests)

Cutaneous adverse drug reactions from systemic administration of diclofenac caused by type IV (delayed-type) hypersensitivity have included maculopapular eruption (5,7,8,10,12), drug reaction with eosinophilia and systemic symptoms (DRESS) (9) and urticaria (4).

Maculopapular eruption

Two days after taking a first 100 mg tablet of diclofenac for pain associated with osteoarthritis, a 66-year-old woman developed a generalized pruritic maculopapular rash. Treatment was immediately interrupted, and the rash resolved after 7 days of corticosteroid therapy. One year later, a patch test was positive to 0.05 ml of a solution containing diclofenac 75 mg/ml in 0.9% NaCl at D2 and D3, as was an intradermal test with 0.02 ml of this material, read at D3. Ten controls were negative (7). A second patient, a 50-year-old woman, experienced a pruritic maculopapular rash on the third day of diclofenac suppository therapy 100 mg twice daily. Two days later the exanthem displayed a confluence of the original lesions, and facial edema appeared. Patch tests (as above) were positive at D2 and D3, as was an intradermal test at 24 hours (7). Two similar cases of patients with maculopapular eruptions after administration of diclofenac with positive patch and delayed intradermal tests had been presented by the same authors 4 years earlier (8).

A 70-year-old man developed, 10 days after he had started oral therapy with diclofenac 100 mg/day, fever and a widespread, pruriginous, erythematous, maculopapular eruption. The oral mucosa showed ulcers. The exanthema completely disappeared within a week on corticosteroids and antihistamines. Prick and intradermal tests with diclofenac were negative. Patch tests were positive at D2 and D4 to commercial diclofenac 25 mg/ml, but negative to diclofenac 0.61% and 1.12% pet. Five controls were negative. A biopsy of the positive patch test reaction showed epidermal spongiosis with lymphocytes with edema and a lymphocytic perivascular infiltrate in the dermis. How the authors came to the diagnosis Stevens-Johnson syndrome is unclear (10).

One patient developed a maculopapular eruption from diclofenac with a positive patch test to this drug after a previous episode of DRESS from other drug(s) (multiple drug hypersensitivity) (5,12).

Drug reaction with eosinophilia and systemic symptoms (DRESS)
A 72-year-old woman presented with high rising temperature (40°C), chills, a bright red and confluent exanthema on the trunk and Raynaud's phenomenon since 7 days. The patient used diclofenac sodium intermittently since 2 years for pain control, and she presented with similar symptoms already in the past. There was leukocytosis and eosinophilia; C-reactive protein and liver enzymes were slightly elevated. Patch tests were positive to diclofenac emulgel and diclofenac 1% pet. after 48 hours, according to the authors. However, patch testing was obviously performed without knowledge of the procedure, and the pictures of the positive patch tests were hardly convincing. Nevertheless, this was probably a case of DRESS from delayed-type hypersensitivity to diclofenac (9).

Other cutaneous adverse drug reactions
A 45-year-old woman twice had suffered urticaria after taking diclofenac, both as tablets and suppositories. Skin prick and intradermal tests were negative. A patch test with diclofenac 2.5% pet. was positive (++) after 2 days (4). The same authors report a 44-year-old man who suffered an anaphylactic shock after intramuscular administration of diclofenac. Skin prick and intradermal tests were negative after 20 minutes. A patch test with diclofenac 2.5% pet. was positive at D2 (+) and D3 (+++). Ten controls were negative. The urticaria may have been the result of delayed-type hypersensitivity to diclofenac, but that the authors also ascribe anaphylactic shock to this mechanism is incorrect (4).

Cross-reactions, pseudo-cross-reactions and co-reactions
Two patients photosensitized to diclofenac had photocross-reactions to aceclofenac (6). Patients sensitized to aceclofenac may cross-react to diclofenac (2,3).

LITERATURE
1   The data in the section 'General' may have been obtained from literature discussed in this chapter, but mostly also or exclusively from one or more of the following online sources: ChemIDPlus Advanced, PubChem, DrugBank, RxList, Drug Central, Drugs.com, and Wikipedia
2   Linares T, Marcos C, Gavilan MJ, Arenas L. Fixed drug eruption due to aceclofenac. Contact Dermatitis 2007;56:291-292
3   Pitarch Bort G, de la Cuadra Oyanguren J, Torrijos Aguilar A, García-Melgares Linares ML. Allergic contact dermatitis due to aceclofenac. Contact Dermatitis 2006;55:365-366
4   Schiavino D, Papa G, Nucera E, Schinco G, Fais G, Pirrotta LR, Patriarca G. Delayed allergy to diclofenac. Contact Dermatitis 1992;26:357-358
5   Jörg L, Helbling A, Yerly D, Pichler WJ. Drug-related relapses in drug reaction with eosinophilia and systemic symptoms (DRESS). Clin Transl Allergy 2020;10:52
6   Fernández-Jorge B, Goday-Buján JJ, Murga M, Molina FP, Pérez-Varela L, Fonseca E. Photoallergic contact dermatitis due to diclofenac with cross-reaction to aceclofenac: two case reports. Contact Dermatitis 2009;61:236-237
7   Romano A, Quaratino D, Papa G, Di Fonso M, Artesani MC, Venuti A. Delayed hypersensitivity to diclofenac: a report on two cases. Ann Allergy Asthma Immunol 1998;81:373-375
8   Romano A, Pietrantonio F, Di Fonso M, Garcovich A, Chiarelli C, Venuti A, Barone C. Positivity of patch tests in cutaneous reaction to diclofenac. Two case reports. Allergy 1994;49:57-59
9   Klingenberg RD, Bassukas ID, Homann N, Stange EF, Ludwig D. Delayed hypersensitivity reactions to diclofenac. Allergy 2003;58:1076-1077
10 Alonso R, Enrique E, Cisteró A. Positive patch test to diclofenac in Stevens-Johnson syndrome. Contact Dermatitis 2000;42:367
11 De Groot AC. Monographs in contact allergy, volume 3: Topical Drugs. Boca Raton, Fl, USA: CRC Press Taylor and Francis Group, 2021 (ISBN 978-0-367-23693-9)
12 Jörg L, Yerly D, Helbling A, Pichler W. The role of drug, dose and the tolerance/intolerance of new drugs in

Chapter 3.174 DICLOXACILLIN

IDENTIFICATION

Description/definition   : Dicloxacillin is the semisynthetic β-lactam penicillin antibiotic that conforms to the
                             structural formula shown below
Pharmacological classes   : Anti-bacterial agents
IUPAC name   : (2S,5R,6R)-6-[[3-(2,6-Dichlorophenyl)-5-methyl-1,2-oxazole-4-carbonyl]amino]-3,3-
                             dimethyl-7-oxo-4-thia-1-azabicyclo[3.2.0]heptane-2-carboxylic acid
Other names   : 4-Thia-1-azabicyclo[3.2.0]heptane-2-carboxylic acid, 6-[[[3-(2,6-dichlorophenyl)-5-methyl-
                             4-isoxazolyl]carbonyl]amino]-3,3-dimethyl-7-oxo-, (2S,5R,6R)-; dicloxacycline
CAS registry number   : 3116-76-5
EC number   : 221-488-3
Merck Index monograph   : 4364
Patch testing   : Sodium salt hydrate 10.0% pet. (Chemotechnique)
Molecular formula   : $C_{19}H_{17}Cl_2N_3O_5S$

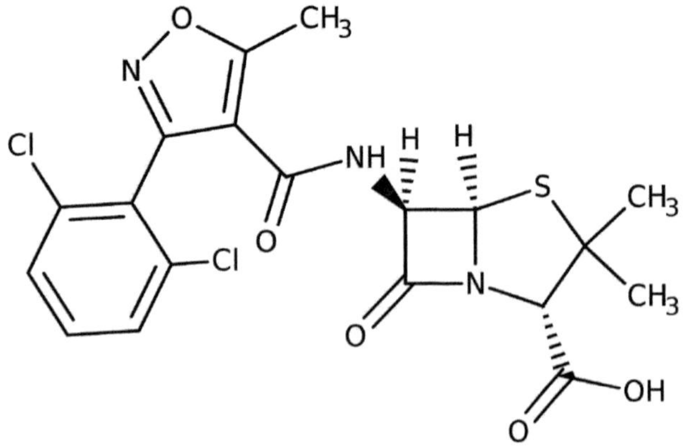

GENERAL

Dicloxacillin is a broad-spectrum, semisynthetic, β-lactam penicillin antibiotic with bactericidal activity. It is stable
against hydrolysis by a variety of β-lactamases, including penicillinases, cephalosporinases and extended spectrum β-
lactamases. Dicloxacillin is used to treat infections caused by penicillinase-producing staphylococci which have
demonstrated susceptibility to the drug (1). In pharmaceutical products, dicloxacillin is usually employed as
dicloxacillin sodium (monohydrate) (CAS number 13412-64-1, EC number 603-794-2, molecular formula
$C_{19}H_{18}Cl_2N_3NaO_6S$) (1). The classification and structures of beta-lactam antibiotics are discussed in Chapter 3.36
Amoxicillin.

CUTANEOUS ADVERSE DRUG REACTIONS FROM SYSTEMIC ADMINISTRATION CAUSED BY TYPE IV
(DELAYED-TYPE) HYPERSENSITIVITY (as demonstrated by positive patch tests)

Cutaneous adverse drug reactions from systemic administration of dicloxacillin caused by type IV (delayed-type)
hypersensitivity have included maculopapular eruption (2), acute generalized exanthematous pustulosis (AGEP) (2),
and drug reaction with eosinophilia and systemic symptoms (DRESS) (3).

Case series with various or unknown types of drug reactions

In the period 2000-2014, in Portugal, 260 patients were patch tested with antibiotics for suspected cutaneous
adverse drug reactions (CADR) to these drugs. 56 patients (22%) had one or more positive patch tests, often from
cross-reactivity. Dicloxacillin was patch test positive in 3 patients with maculopapular eruptions, and in one with
AGEP (2).

Maculopapular eruption

See the section 'Case series with various or unknown types of drug reactions' above, ref. 2.

Acute generalized exanthematous pustulosis (AGEP)

See the section 'Case series with various or unknown types of drug reactions' above, ref. 2.

Drug reaction with eosinophilia and systemic symptoms (DRESS)
In a multicenter investigation in France, of 72 patients patch tested for drug reaction with eosinophilia and systemic symptoms (DRESS), 46 (64%) had positive patch tests to drugs, including one to dicloxacillin (3).

Cross-reactions, pseudo-cross-reactions and co-reactions
Cross-reactions between beta-lactam antibiotics are discussed in Chapter 3.36 Amoxicillin.

LITERATURE

1   The data in the section 'General' may have been obtained from literature discussed in this chapter, but mostly also or exclusively from one or more of the following online sources: ChemIDPlus Advanced, PubChem, DrugBank, RxList, Drug Central, Drugs.com, and Wikipedia
2   Pinho A, Coutinho I, Gameiro A, Gouveia M, Gonçalo M. Patch testing - a valuable tool for investigating non-immediate cutaneous adverse drug reactions to antibiotics. J Eur Acad Dermatol Venereol 2017;31:280-287
3   Barbaud A, Collet E, Milpied B, Assier H, Staumont D, Avenel-Audran M, et al. A multicentre study to determine the value and safety of drug patch tests for the three main classes of severe cutaneous adverse drug reactions. Br J Dermatol 2013;168:555-562

Chapter 3.175 DIHYDROCODEINE

IDENTIFICATION

Description/definition   : Dihydrocodeine is the morphinane alkaloid that conforms to the structural formula shown
below
Pharmacological classes   : Analgesics, opioid
IUPAC name   : (4R,4aR,7S,7aR,12bS)-9-Methoxy-3-methyl-2,4,4a,5,6,7,7a,13-octahydro-1H-4,12-
methanobenzofuro[3,2-e]isoquinoline-7-ol
CAS registry number   : 125-28-0
EC number   : 204-732-3
Merck Index monograph : 4459
Patch testing   : 20% pet. (probably pulverized tablets) (2); most systemic drugs can be tested at 10% pet.;
if the pure chemical is not available, prepare the test material from intravenous powder,
the content of capsules or – when also not available – from powdered tablets to achieve
a final concentration of the active drug of 10% pet.
Molecular formula   : $C_{18}H_{23}NO_3$

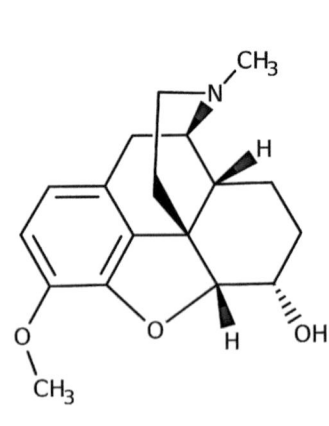

Dihydrocodeine   Dihydrocodeine bitartrate

GENERAL

Dihydrocodeine is a semi-synthetic morphinane alkaloid and opioid agonist. It was developed in Germany in 1908
during an international search to find a more effective antitussive agent to help reduce the spread of airborne
infectious diseases such as tuberculosis. Dihydrocodeine is used for the treatment of moderate to severe pain,
including post-operative and dental pain, chronic pain, dyspnea and coughing. In heroin addicts, dihydrocodeine has
been used as a substitute drug (1). In pharmaceutical products, dihydrocodeine is employed as dihydrocodeine
bitartrate (CAS number 5965-13-9, EC number 227-747-7, molecular formula $C_{22}H_{29}NO_9$) (1).

CUTANEOUS ADVERSE DRUG REACTIONS FROM SYSTEMIC ADMINISTRATION CAUSED BY TYPE IV
(DELAYED-TYPE) HYPERSENSITIVITY (as demonstrated by positive patch tests)

Cutaneous adverse drug reactions from systemic administration of dihydrocodeine caused by type IV (delayed-type)
hypersensitivity have included maculopapular eruption (2).

Maculopapular eruption

A 49-year-old woman developed a widespread maculopapular eruption one day after taking 2 tablets containing
dihydrocodeine tartrate for toothache. She had presented on several previous occasions with rashes presumed, at
the time, to be due to antibiotics. It became apparent, on reviewing her case notes, that on each occasion that she
had been thought to have had a cutaneous reaction to an antibiotic, she had also received dihydrocodeine for
analgesia. Patch testing revealed a positive reaction to dihydrocodeine tartrate 20% pet. (D2 ++, D4 ++), but not to
codeine phosphate (2).

Cross-reactions, pseudo-cross-reactions and co-reactions
Two patients sensitized to codeine phosphate cross-reacted to dihydrocodeine (3,4).

LITERATURE
1   The data in the section 'General' may have been obtained from literature discussed in this chapter, but mostly also or exclusively from one or more of the following online sources: ChemIDPlus Advanced, PubChem, DrugBank, RxList, Drug Central, Drugs.com, and Wikipedia
2   Cooper SM, Shaw S. Dihydrocodeine: a drug allergy diagnosed by patch testing. Contact Dermatitis 2000;42:307-308
3   De Groot AC, Conemans J. Allergic urticarial rash from oral codeine. Contact Dermatitis 1986;14:209-214
4   Gastaminza G, Audicana M, Echenagusia MA, Uriel O, Garcia-Gallardo MV, Velasco M, et al. Erythrodermia caused by allergy to codeine. Contact Dermatitis 2005;52:227-228

Chapter 3.176 DIHYDROSTREPTOMYCIN

IDENTIFICATION

Description/definition	: Dihydrostreptomycin is the streptomycin derivative that conforms to the structural formula shown below
Pharmacological classes	: Anti-bacterial agents
IUPAC name	: 2-[(1R,2R,3S,4R,5R,6S)-3-(Diaminomethylideneamino)-4-[(2R,3R,4R,5S)-3-[(2S,3S,4S, 5R,6S)-4,5-dihydroxy-6-(hydroxymethyl)-3-(methylamino)oxan-2-yl]oxy-4-hydroxy-4-(hydroxymethyl)-5-methyloxolan-2-yl]oxy-2,5,6-trihydroxycyclohexyl]guanidine
CAS registry number	: 128-46-1
EC number	: 204-888-2
Merck Index monograph	: 4467
Patch testing	: 5% water (2); suggested: 10% and 20% pet.
Molecular formula	: $C_{21}H_{41}N_7O_{12}$

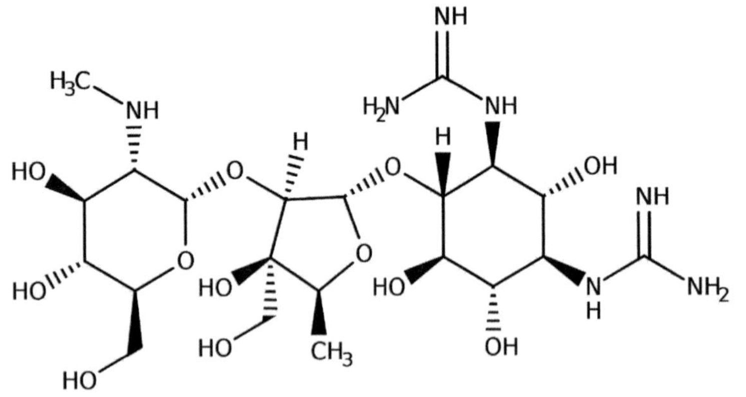

GENERAL

Dihydrostreptomycin is a semisynthetic aminoglycoside antibiotic derived from streptomycin with bactericidal properties. It is probably used in veterinary medicine only. In pharmaceutical products, dihydrostreptomycin may be employed as dihydrostreptomycin sulfate (CAS number 5490-27-7, EC number 226-823-7, molecular formula $C_{42}H_{88}N_{14}O_{36}S_3$) (1).

CONTACT ALLERGY FROM ACCIDENTAL CONTACT

In 34 veterinary surgeons (7 women, 27 men, age range 29-61, mean age 38 years) with chronic or relapsing eczema of the hands as the main complaint, investigated in Norway before 1985, patch tests were performed with the standard series and a veterinary series comprising drugs, antiseptics and protective glove materials frequently used in veterinary practice. Nineteen had eczema almost continuously on the hands, fingers and arms, and in another 3 there was a spread to the face and trunk. Occupational work caused exacerbations in 15 of these patients. Ten of the 34 veterinary surgeons had positive patch test reactions. A relation between positive patch tests and occupational work was confirmed in 9 of these subjects. In this group, there was one positive patch test reactions to dihydrostreptomycin 5% water; this patient also reacted to streptomycin (2).

In 2 hospitals in Denmark, between 1974 and 1980, 37 veterinary surgeons, all working in private country practices, were investigated for suspected incapacitating occupational dermatitis and patch tested with a battery of 10 antibiotics. Thirty-two (86%) had one or more positive patch tests. There were 2 or 3 positive reactions to dihydrostreptomycin 30% water (7%, incorrect). It was mentioned that all but one of these individuals had allergic contact dermatitis, but relevance was not specified for individual allergens. The most frequent allergens were spiramycin, penethamate and tylosin tartrate (4).

Occupational allergic contact dermatitis from dihydrostreptomycin may also have been published in early French literature (cited in ref. 3).

Cross-reactions, pseudo-cross-reactions and co-reactions
Possibly cross-reactivity to or from streptomycin (2,5)

LITERATURE

1   The data in the section 'General' may have been obtained from literature discussed in this chapter, but mostly also or exclusively from one or more of the following online sources: ChemIDPlus Advanced, PubChem, DrugBank, RxList, Drug Central, Drugs.com, and Wikipedia

2   Falk ES, Hektoen H, Thune PO. Skin and respiratory tract symptoms in veterinary surgeons. Contact Dermatitis 1985;12:274-278

3   De Groot AC, Weyland JW, Nater JP. Unwanted effects of cosmetics and drugs used in dermatology, 3rd edition. Amsterdam: Elsevier Science BV, 1994:81

4   Hjorth N, Roed-Petersen J. Allergic contact dermatitis in veterinary surgeons. Contact Dermatitis 1980;6:27-29

5   Sulzberger MB, Distelheim IH. Allergic eczematous contact type sensitivity of equal degree to streptomycin and dihydrostreptomycin; report of a case. AMA Arch Derm Syphilol 1950;62:706-707

Chapter 3.177 DILTIAZEM

IDENTIFICATION
Description/definition   : Diltiazem is the benzothiazepine that conforms to the structural formula shown below
Pharmacological classes : Antihypertensive agents; calcium channel blockers; cardiovascular agents; vasodilator
                            agents
IUPAC name   : [(2S,3S)-5-[2-(Dimethylamino)ethyl]-2-(4-methoxyphenyl)-4-oxo-2,3-dihydro-1,5-
                            benzothiazepin-3-yl] acetate
CAS registry number   : 42399-41-7
EC number   : 255-796-4
Merck Index monograph   : 4494
Patch testing   : Hydrochloride, 10% pet. (Chemotechnique); because patch tests are often very strongly
                            positive, it has been suggested to lower the concentration of pure diltiazem to 1% or
                            a maximum of 3% in diluted commercial preparations (12)
Molecular formula   : $C_{22}H_{26}N_2O_4S$

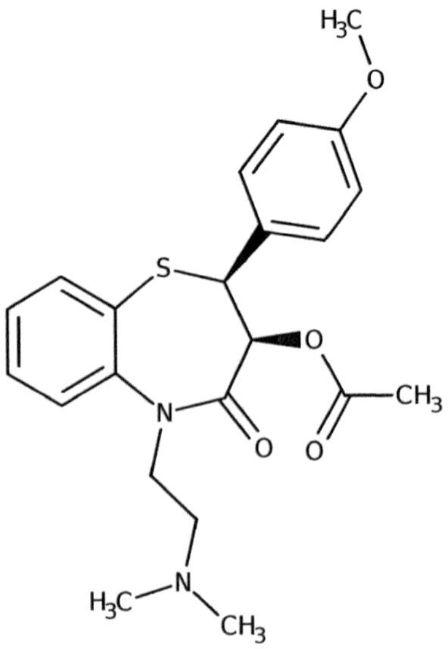

GENERAL
Diltiazem is a benzothiazepine calcium channel blocking agent and vasodilator, which is used in the management of angina pectoris and hypertension. Topical diltiazem creams are used to treat anal fissures and Raynaud's syndrome. A few cases of allergic contact dermatitis to diltiazem cream used for anal fissures have been described. The resorption of diltiazem through the perianal skin and the mucosa has led to systemic allergic dermatitis presenting as generalized (maculo-)papular rashes and the baboon syndrome. This, and a few cases of photocontact allergy, has been fully reviewed in Volume 3 of the Monographs in contact allergy series (13). In pharmaceutical products, diltiazem is employed as diltiazem hydrochloride (CAS number 33286-22-5, EC number 251-443-3, molecular formula $C_{22}H_{27}ClN_2O_4S$) (1).

CUTANEOUS ADVERSE DRUG REACTIONS FROM SYSTEMIC ADMINISTRATION CAUSED BY TYPE IV
(DELAYED-TYPE) HYPERSENSITIVITY (as demonstrated by positive patch tests)
Cutaneous adverse drug reactions from systemic administration of diltiazem caused by type IV (delayed-type) hypersensitivity have included maculopapular eruption (4,9,12,20,24), erythroderma (3,6,10), exfoliative dermatitis (6), acute generalized exanthematous pustulosis (AGEP) (5,8,11,12,14,15,16,17,18,19,23,25), drug reaction with eosinophilia and systemic symptoms (DRESS) (21), photosensitivity (7), erythema multiforme-like eruption (3,6), psoriasiform eruptions (10), and unspecified/unknown drug eruptions (10,22).

Case series with various or unknown types of drug reactions
Three patients with cutaneous adverse drug reactions to diltiazem with a positive patch test were reported from France. Two had suffered from acute generalized exanthematous pustulosis (AGEP), the third from a maculopapular rash. Patch tests with the commercially available diltiazem 10% pet. patch test preparation lead to (very) strong positive patch tests, in one case causing an angry back reaction associated with a maculopapular exanthema involving the face, neck, and armpits. Therefore, the authors suggested to use a lower concentration of diltiazem for

patch testing when this drug is suspected to be the culprit, namely, 1% instead of 10% of pure diltiazem, and otherwise not more than 3% concentration, which is obtained by diluting the marketed form at 30% (12).

In Finland, in the period 1989-2001, 826 patients with suspected cutaneous drug eruptions were patch tested and 89 had one or more positive reactions. Of these individuals, 3 reacted to diltiazem. It was not stated which types of cutaneous eruption had been caused by diltiazem (22).

Of 10 patients with cutaneous adverse drug reactions from diltiazem and subsequently patch tested with the drug, published in Japanese literature before 1993, 6 had positive reactions. Details of clinical and patch testing data are unknown (cited in ref. 10).

Maculopapular eruption
Four patients had developed maculopapular rashes 8-12 days after initiating therapy with diltiazem. All patients later had positive patch tests to commercial diltiazem tablets 30% water, petrolatum and alcohol (the latter tested in 3), at D2 (+) and D4 (+). Eleven controls were negative. Lymphocyte activation tests were performed with the blood of 3 patients and were positive in all (4).

Another 2 patients with widespread, pruriginous, maculopapular eruptions, in these cases accompanied by fever and facial edema, and which had developed 2-3 days after starting diltiazem, were reported from Italy. Delayed hypersensitivity to diltiazem was suspected and the diagnosis was confirmed by patch tests and histopathology of a positive patch test reaction (9, no details available to the author).

In a study from Nancy, France, 54 patients with suspected nonimmediate drug eruptions were assessed with patch testing. Of the 27 patients with maculopapular eruptions, one had a positive reaction to diltiazem; no details were provided (24).

One more case of maculopapular eruption from diltiazem with a positive patch test was reported from Germany (20; no details available to the author). See also the section 'Case series with various or unknown types of drug reactions' above, ref. 12.

Erythroderma/extensive erythematous eruption/exfoliative dermatitis
A 54-year-old man developed a generalized erythema multiforme-like reaction followed by erythroderma and exfoliative dermatitis 6-7 days after starting the use of diltiazem. A patch test was positive to commercial diltiazem solution 6.25 mg/ml (D2 ++, D4 ++) (3). An 80-year-old woman developed a pruritic exanthematous eruption on her trunk which evolved to generalized erythroderma and superficial desquamation. Patch tests were positive to verapamil commercial solution 6.25 mg/ml (++/++) and verapamil '5-10% pet.' (concentration thus unknown, probably from commercial tablets) (3). A 79-year-old man had used diltiazem for hypertension for three days, when pruritic erythema appeared, initially on the back, and then spreading to the thorax, arms, legs and face. Patch tests were positive to commercial verapamil 30% pet. (D2 -, D4 ++). Ten controls tested with the materials used for patch testing in these 3 patients were negative (3).

Three patients investigated in Portugal developed a cutaneous drug reaction 2-4 days after starting to use diltiazem. One had erythroderma, the second widespread erythematous rash followed by exfoliative dermatitis and the third a generalized erythema multiforme-like reaction. All had positive patch tests to pure diltiazem 1% pet., 20 controls were negative (6).

In a 64-year-old woman, demarcated erythema with infiltration developed all over her body 7 days after she started diltiazem. Laboratory investigations showed no abnormalities. Diltiazem was discontinued, and the skin rash resolved within 7 days using a topical steroid. A patch test with diltiazem 10% pet. was later positive at D2 and D3 (10).

Acute generalized exanthematous pustulosis
A 79-year-old women developed a micropapular rash on the thorax and face after taking diltiazem for 6 days. The rash then spread to the rest of the trunk, folds, and thighs, eventually affecting about 30% of her total body surface area; she also had a fever. The rash evolved to confluent surface pustules. the patient had crusted erosions in the labial and gingival mucosa and labial edema. Blood tests showed leukocytosis with neutrophilia and elevated C-reactive protein and liver enzymes. A skin biopsy showed subcorneal spongiform pustules and spongiosis in the epidermis with perivascular infiltrates, interstitial inflammation, and polymorphonuclear eosinophils noted in the underlying dermis. Patch testing, performed 4 weeks after resolution of AGEP, was positive to diltiazem 10% pet. with co-reactivity to verapamil (30% pet.), another calcium channel blocker (5).

A 61-year-old man developed AGEP 3 days after initiating therapy with diltiazem. Patch tests were positive to diltiazem (commercial tablet in water, unspecified) (14). A 72-year-old woman developed AGEP to orally administered diltiazem as evidenced clinically by numerous widespread pustules, macular erythema, and fever (38°C). A skin biopsy specimen demonstrated neutrophilic subcorneal collections and a mixed perivascular infiltrate in the dermis. The patch test was positive to diltiazem 1% pet. A drug-specific *in vitro* lymphoproliferative response to diltiazem was demonstrated (15). An 82-year-old woman was referred for an acute generalized eruption with numerous confluent pustules and fever. She used several drugs, but nothing had changed, although 24 hours before the eruption, she had taken one tablet diltiazem by mistake. This was in fact the fifth episode of fever and a pustular rash. In two of these, the patient had taken diltiazem 2 days before the rash began. Patch tests were positive to the commercial drug 5% and 20% in petrolatum and in water (16).

A 71-year-old woman developed AGEP with fever, a maculopapular eruption with small and superficial non-follicular pustules, leukocytosis and slightly abnormal liver function. A skin biopsy showed a spongiform subcorneal pustule and a superficial perivascular inflammatory cell infiltrate with evidence of vasculitis. Two months later, patch tests were strongly positive to diltiazem 1% water and pure (?), resulting in an erythematous and very pruriginous reaction on the patch tested area, neck and abdomen that resolved in a few days (11).

A 78-year-old woman developed fever and an erythematous rash one week after initiating therapy with diltiazem for hypertension. Clinical examination revealed multiple erythematous violaceous plaques on the trunk, face, and thighs, some of which showing numerous non-follicular small pustules. Erosions at the labial mucosa and mild labial edema were present. Laboratory investigations showed neutrophilic leukocytosis and elevated C-reactive protein. Histopathology showed subcorneal pustules and spongiosis of the epidermis. Some weeks later, a patch test with diltiazem was positive, mimicking the pustular lesions. Details on patch testing were not performed. The patient was diagnosed with AGEP due to diltiazem (25).

One patient with AGEP had a positive patch test reaction to diltiazem; no clinical or patch testing details were provided (23). Additional cases of AGEP caused by delayed-type hypersensitivity to diltiazem can be found in refs. 8,17,18 and 19. See also the section 'Case series with various or unknown types of drug reactions' above, ref. 12.

Drug reaction with eosinophilia and systemic symptoms (DRESS)
In a multicenter investigation in France, of 72 patients patch tested for DRESS, 46 (64%) had positive patch tests to drugs, including one to diltiazem; no clinical or patch testing details were provided (21).

Photosensitivity
An 81-year-old woman treated with diltiazem for 6 months and 3 other drugs presented with a well-delineated photodistributed erythema on the face and V-neck area sparing the skin under the chin and behind the ears. On the forearms and dorsum of both hands the patient had itchy lesions with a lichenoid appearance. Patch tests to diltiazem 0.1%, 1% and 5% pet. were negative, but photopatch tests with 5J/cm^2 UVA irradiation positive for the highest concentration. A 'control group' was negative on patch testing, which probably means that there were no controls for the photopatch tests. The patient was diagnosed with a photoallergic cutaneous eruption to diltiazem (7).

Other cutaneous adverse drug reactions
One patient had a generalized erythema multiforme-like reaction after taking diltiazem. Patch tests were positive to pure diltiazem 1% pet.; 20 controls were negative (6). See also the section 'Erythroderma/extensive erythematous eruption/exfoliative dermatitis' above, for another case of erythema multiforme-like drug eruption (ref. 3).

In some patients with psoriasiform eruptions from diltiazem, reported in Japanese literature before 1993, patch tests have been positive to the drug. Details of clinical and patch testing data are unknown (cited in ref. 10). See also the section 'Case series with various or unknown types of drug reactions' above, ref. 22.

Cross-reactions, pseudo-cross-reactions and co-reactions
Several patient sensitized to diltiazem have co-reacted to 2 other calcium channel blockers, nifedipine (2,3) and verapamil (2,3,4,5), although the chemical structures of the 3 drugs do not show similarities that may easily explain these findings.

LITERATURE
1   The data in the section 'General' may have been obtained from literature discussed in this chapter, but mostly also or exclusively from one or more of the following online sources: ChemIDPlus Advanced, PubChem, DrugBank, RxList, Drug Central, Drugs.com, and Wikipedia
2   Forkel S, Baltzer AB, Geier J, Buhl T. Contact dermatitis caused by diltiazem cream and cross-reactivity with other calcium channel blockers. Contact Dermatitis 2018;79:244-246
3   Gonzalo MA, Pérez R, Argila D, Rangel JF. Cutaneous reactions due to diltiazem and cross reactivity with other calcium channel blockers. Allergol Immunopathol (Madr) 2005;33:238-240
4   Cholez C, Trechot P, Schmutz JL, Faure G, Bene MC, Barbaud A. Maculopapular rash induced by diltiazem: allergological investigations in four patients and cross reactions between calcium channel blockers. Allergy 2003;58:1207-1209

5 Sáenz de Santa María García M, Noguerado-Mellado B, Perez-Ezquerra PR, Hernandez-Aragües I, De Barrio Fernández M. Acute generalized exanthematous pustulosis due to diltiazem: investigation of cross-reactivity with other calcium channel blockers. J Allergy Clin Immunol Pract 2016;4:765-766

6 Sousa-Basto A, Azenha A, Duarte ML, Pardal-Oliveira F. Generalized cutaneous reaction to diltiazem. Contact Dermatitis 1993;29:44-45

7 Ramírez A, Pérez-Pérez L, Fernández-Redondo V, Toribio J. Photoallergic dermatitis induced by diltiazem. Contact Dermatitis 2007;56:118-119

8 Wakelin SH, James MP. Diltiazem-induced acute generalised exanthematous pustulosis. Clin Exp Dermatol 1995;20:341-344

9 Romano A, Pietrantonio F, Garcovich A, Rumi C, Bellocci F, Caradonna P, Barone C. Delayed hypersensitivity to diltiazem in two patients. Ann Allergy 1992;69:31-32

10 Kitamura K, Kanasashi M, Suga C, Saito S, Yoshida S, Ikezawa Z. Cutaneous reactions induced by calcium channel blocker: high frequency of psoriasiform eruptions. J Dermatol 1993;20:279-286

11 Vicente-Calleja JM, Aguirre A, Landa N, Crespo V, González-Pérez R, Diaz-Pérez JL. Acute generalized exanthematous pustulosis due to diltiazem: confirmation by patch testing. Br J Dermatol 1997;137:837-839

12 Assier H, Ingen-Housz-Oro S, Zehou O, Hirsch G, Chosidow O, Wolkenstein P. Strong reactions to diltiazem patch tests: Plea for a low concentration. Contact Dermatitis 2020;83:224-225

13 De Groot AC. Monographs in contact allergy, volume 3: Topical Drugs. Boca Raton, Fl, USA: CRC Press Taylor and Francis Group, 2021 (ISBN 978-0-367-23693-9)

14 Gensch K, Hodzic-Avdagic N, Megahed M, Ruzicka T, Kuhn A. Acute generalized exanthematous pustulosis with confirmed type IV allergy. Report of 3 cases. Hautarzt 2007;58:250-252, 254-255 (Article in German)

15 Girardi M, Duncan KO, Tigelaar RE, Imaeda S, Watsky KL, McNiff JM. Cross-comparison of patch test and lymphocyte proliferation responses in patients with a history of acute generalized exanthematous pustulosis. Am J Dermatopathol 2005;27:343-346

16 Jan V, Machet L, Gironet N, Martin L, Machet MC, Lorette G, Vaillant L. Acute generalized exanthematous pustulosis induced by diltiazem: value of patch testing. Dermatology 1998;197:274-275

17 Nishiimura T, Yoshioka K, Katoh J, et al. Pustular reaction induced by diltiazem HCl. Skin Res 1991;33(Suppl.10): 251-254 (Article in Japanese, data cited in ref. 11)

18 Walkenstein P, Chosidow O, Fléchet ML, et al. Intéret des épicutanés dans les toxicodermies graves. Ann Dermatol Venereol 1995:C53 (bibliographical data incomplete, data cited in ref. 11)

19 Serrão V, Caldas Lopes L, Campos Lopes JM, Lobo L, Ferreira A. Acute generalized exanthematous pustulosis associated with diltiazem. Acta Med Port 2008;21:99-102 (Article in Portuguese)

20 Hammentgen R, Lutz G. Kohler U, Nitsch J, Maculopapular exanthem during diltiazem therapy. Dtsch Med Wochenschr 1988;113:1283-1285 (Article in German, data cited in ref. 11).

21 Barbaud A, Collet E, Milpied B, Assier H, Staumont D, Avenel-Audran M, et al. A multicentre study to determine the value and safety of drug patch tests for the three main classes of severe cutaneous adverse drug reactions. Br J Dermatol 2013;168:555-562

22 Lammintausta K, Kortekangas-Savolainen O. The usefulness of skin tests to prove drug hypersensitivity. Br J Dermatol 2005;152:968-974

23 Wolkenstein P, Chosidow O, Fléchet ML, Robbiola O, Paul M, Dumé L, et al. Patch testing in severe cutaneous adverse drug reactions, including Stevens-Johnson syndrome and toxic epidermal necrolysis. Contact Dermatitis 1996;35:234-236

24 Barbaud A, Reichert-Penetrat S, Tréchot P, Jacquin-Petit MA, Ehlinger A, Noirez V, et al. The use of skin testing in the investigation of cutaneous adverse drug reactions. Br J Dermatol 1998;139:49-58

25 Hernández-Aragües I, De Santa María García MS, Pérez-Esquerra PR, Simal-Gómez G. Cutaneous drug reactions: Acute rash with pinhead-sized pustules. Eur J Dermatol 2018;28:859-860

Chapter 3.178 DIMENHYDRINATE

IDENTIFICATION

Description/definition   : Dimenhydrinate is a combination drug composed of diphenhydramine and 8-
                             chlorotheophylline; their structural formulas are shown below
Pharmacological classes   : Histamine H_1 antagonists; antiemetics
IUPAC name   : 2-Benzhydryloxy-N,N-dimethylethanamine;8-chloro-1,3-dimethyl-7H-purine-2,6-dione
CAS registry number   : 523-87-5
EC number   : 208-350-8
Merck Index monograph   : 4500
Patch testing   : Tablet 10% pet. or DMSO (4); most systemic drugs can be tested at 10% pet.; if the pure
                             chemical is not available, prepare the test material from intravenous powder, the content
                             of capsules or – when also not available – from powdered tablets to achieve a final
                             concentration of the active drug of 10% pet.; for fixed drug eruptions, DMSO may be
                             preferable as vehicle
Molecular formula   : $C_{24}H_{28}ClN_5O_3$

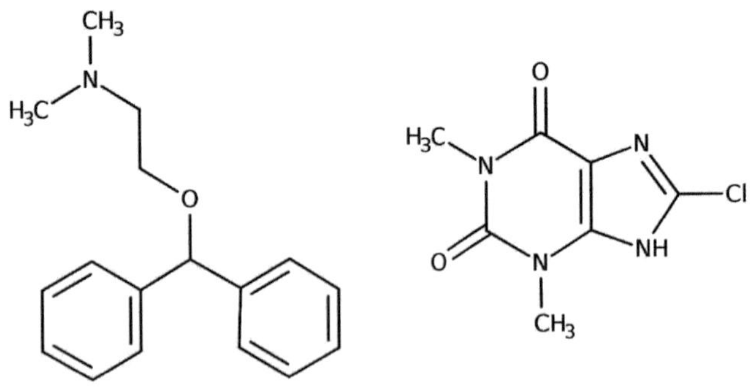

GENERAL

Dimenhydrinate is an ethanolamine and first-generation histamine H_1 antagonist with anti-allergic and anti-emetic activities. It is a combination drug composed of diphenhydramine and 8-chlorotheophylline. The antiemetic properties of dimenhydrinate are primarily thought to be produced by diphenhydramine's antagonism of H_1 histamine receptors in the vestibular system. The addition of 8-chlorotheophylline was initially intended to counteract the sedative effects of diphenhydramine. Dimenhydrinate is indicated for the prevention and treatment of nausea, vomiting, or vertigo of motion sickness (1).

CUTANEOUS ADVERSE DRUG REACTIONS FROM SYSTEMIC ADMINISTRATION CAUSED BY TYPE IV (DELAYED-TYPE) HYPERSENSITIVITY (as demonstrated by positive patch tests)

Cutaneous adverse drug reactions from systemic administration of dimenhydrinate caused by type IV (delayed-type) hypersensitivity have included fixed drug eruption (2,3,4,6).

Fixed drug eruption

A 66-year-old woman had suffered from several episodes of sudden appearance of erythematous patches on her face, trunk, and extremities accompanied by a burning sensation and pruritus. The lesions darkened within a few hours, turning violet-red in color, and then gradually resolved. This was followed by a residual hyperpigmentation. The patient reported recurrence of these lesions at the same sites, for the past 2 years, after use of dimenhydrinate tablets to prevent her motion sickness. These symptoms recurred with a faster onset (from 3 hours to within an hour) and an increased number of lesions each time after drug ingestion. Skin examination revealed several (up to 15) isolated, sharply demarcated, oval erythematous patches that ranged from 1 to 10 cm in diameter. Histologic examination was consistent with the diagnosis of fixed drug eruption. Patch tests with a dimenhydrinate tablet 10% pet. and in saline were positive on postlesional skin and negative on unaffected skin. Two controls were negative. The patient underwent an oral challenge with a single 50-mg dose of dimenhydrinate. One hour after taking the test dose, erythematous patches appeared in the same location as previously reported. Within a few hours, these lesions progressively changed color from light pink to dark purple (2).

A 66-year-old man had presented erythematous macular lesions on the right subarmpit area, 12 h after he had taken a tablet (50 mg) of dimenhydrinate. Patch tests to (pure) dimenhydrinate 10% in pet. and in DMSO were positive on previously affected skin only (3).

Four days after a 10-year-old girl started treatment with dimenhydrinate for nausea, asymptomatic isolated erythematous macules with centrally located hemorrhagic blisters appeared on both hands, the proximal thighs and the perioral region. Lesions developed into infiltrated plaques with concomitant change of color to purple. Topical provocation with dimenhydrinate and other drugs the girl had used was positive only to dimenhydrinate 5% in glycerin on postlesional skin (4).

A 17-year-old woman had experienced 2 round erythematous macules on the abdomen and right gluteus after taking combined dimenhydrinate and acetaminophen for dysmenorrhea several times during a 1-year period. The lesions disappeared after 6-7 days, leaving hyperpigmentation. The patient later tolerated acetaminophen as single drug. Patch testing with dimenhydrinate and acetominophen 10% in petrolatum on the affected areas and on areas of healthy skin were positive to dimenhydrinate and the combination of dimenhydrinate and acetaminophen on postlesional skin only at D2 and D4. Ten controls tested with dimenhydrinate 10% pet. were negative (6).

Cross-reactions, pseudo-cross-reactions and co-reactions

Not to other antihistamines: doxylamine, ebastine, loratadine, cetirizine, mizolastine, terfenadine, hydroxyzine, chlorpheniramine (3). One patient sensitized to diphenhydramine cross-reacted to dimenhydrinate (5).

LITERATURE

1 The data in the section 'General' may have been obtained from literature discussed in this chapter, but mostly also or exclusively from one or more of the following online sources: ChemIDPlus Advanced, PubChem, DrugBank, RxList, Drug Central, Drugs.com, and Wikipedia
2 Giatrakou S, Papadavid E, Kalogeromitros D, Theodoropoulos K, Toumbis-Ioannou E, Makris M, Stavrianeas NG. Fixed drug eruption caused by dimenhydrinate. J Am Acad Dermatol 2011;64:608-610
3 Saenz de San Pedro B, Quiralte J, Florido JF. Fixed drug eruption caused by dimenhydrinate. Allergy 2000;55:297
4 Smola H, Kruppa A, Hunzelmann N, Krieg T, Scharffetter-Kochanek K. Identification of dimenhydrinate as the causative agent in fixed drug eruption using patch-testing in previously affected skin. Br J Dermatol
5 Goossens A, Linsen G. Contact allergy to antihistamines is not common. Contact Dermatitis 1998;39:38-39
6 Rodríguez-Jiménez B, Domínguez-Ortega J, González-García JM, Kindelan-Recarte C. Dimenhydrinate-induced fixed drug eruption in a patient who tolerated other antihistamines. J Investig Allergol Clin Immunol 2009;19:334-335

Chapter 3.179 2,3-DIMERCAPTO-1-PROPANESULFONIC ACID

IDENTIFICATION

Description/definition   : 2,3-Dimercapto-1-propanesulfonic acid is the compound that conforms to the structural formula shown below

Pharmacological classes   : Antidotes; chelating agents

IUPAC name   : 2,3-bis(Sulfanyl)propane-1-sulfonic acid

CAS registry number   : 74-61-3

EC number   : Not available

Merck Index monograph : 4502

Patch testing   : No data available; most systemic drugs can be tested at 10% pet.; if the pure chemical is not available, prepare the test material from intravenous powder, the content of capsules or – when also not available – from powdered tablets to achieve a final concentration of the active drug of 10% pet.

Molecular formula   : $C_3H_8O_3S_3$

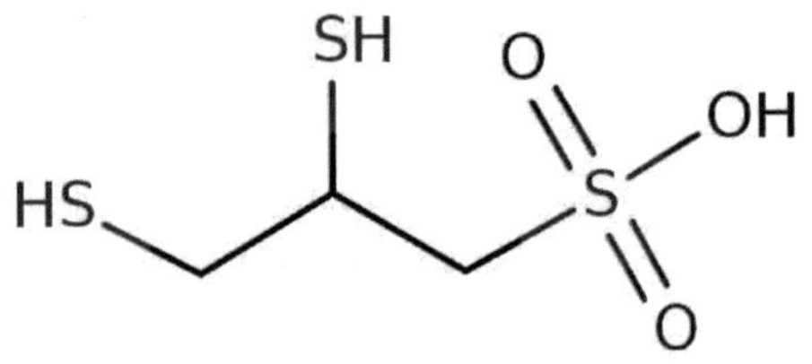

GENERAL

2,3-Dimercapto-1-propanesulfonic acid is a chelating agent that forms complexes with various heavy metals. It is used as an antidote to heavy metal poisoning. In pharmaceutical products, 2,3-dimercapto-1-propanesulfonic acid may both be employed as such and as sodium 2,3-dimercaptopropanesulfonate (synonym: unithiol, CAS number 4076-02-2, EC number 223-796-3, molecular formula $C_3H_7NaO_3S_3$) (1).

CUTANEOUS ADVERSE DRUG REACTIONS FROM SYSTEMIC ADMINISTRATION CAUSED BY TYPE IV (DELAYED-TYPE) HYPERSENSITIVITY (as demonstrated by positive patch tests)

Cutaneous adverse drug reactions from systemic administration of 2,3-dimercapto-1-propanesulfonic acid caused by type IV (delayed-type) hypersensitivity have included Stevens-Johnson syndrome (2).

Stevens-Johnson syndrome/toxic epidermal necrolysis (SJS/TEN)

A 26-year-old patient was accidently exposed at work to a lead-containing gas, for which he was treated with the chelating agent 2,3-dimercapto-1-propanesulfonic acid. Three days later, an itching exanthema developed, beginning at the upper arms and spreading from there. As the lead levels in his blood remained high, a second course of the drug was given, which resulted in worsening of the skin condition despite topical corticosteroid treatment. Physical examination showed an extensive erythematous exanthem with multiple vesicles, accentuated at the acral parts. There were erosions of the oral mucosa, the glans penis and the soles of the feet. Less than 10% of the skin was affected by the rash. Histopathology showed a subepidermal bulla, hydropic degeneration of the basal membrane, apoptotic keratinocytes, and edema of the dermis with a lymphocytic infiltrate, accentuated around the vessels and consisting of CD8- and CD4-positive T cells. Immunofluorescence was negative. A patch test with 2,3-dimercapto-1-propane-sulfonic acid was positive at D3 with an infiltrated erythematous plaque without vesicles. Histopathology of this patch test reaction showed infiltration of the dermis with T lymphocytes. The patient was diagnosed with Stevens-Johnson syndrome from 2,3-dimercapto-1-propanesulfonic acid (2).

LITERATURE

1   The data in the section 'General' may have been obtained from literature discussed in this chapter, but mostly also or exclusively from one or more of the following online sources: ChemIDPlus Advanced, PubChem, DrugBank, RxList, Drug Central, Drugs.com, and Wikipedia

2   Storim J, Stoevesandt J, Anders D, Kneitz H, Bröcker EB, Trautmann A. Chelatbildner mit Thiolaktivität. Auslöser bullöser Hautreaktionen [Dithiols as chelators. A cause of bullous skin reactions]. Hautarzt 2011;62:215-218

Chapter 3.180 DIMETHINDENE

IDENTIFICATION

Description/definition : Dimethindene is the indene that conforms to the structural formula shown below
Pharmacological classes : Histamine H_1 antagonists; antipruritics; anti-allergic agents
IUPAC name : *N,N*-Dimethyl-2-[3-(1-pyridin-2-ylethyl)-1*H*-inden-2-yl]ethanamine
Other names : Dimetindene; dimethpyrindene
CAS registry number : 5636-83-9
EC number : 227-083-8
Merck Index monograph : 4508
Patch testing : 5% pet. (3)
Molecular formula : $C_{20}H_{24}N_2$

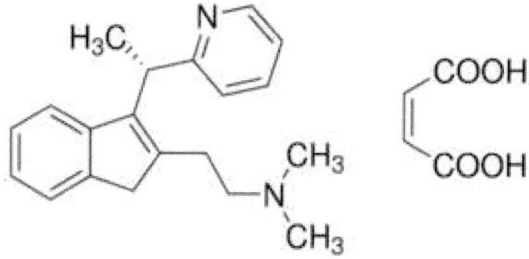

Dimethindene maleate

GENERAL

Dimethindene is an indene histamine H1 antagonist. It is indicated as symptomatic treatment of allergic reactions such as urticaria, allergies of the upper respiratory tract such as hay fever and perennial rhinitis, and food and drug allergies. It is also used for pruritus of various origins, e.g. from insect bites, varicella, eczema and other pruriginous dermatoses on account of its sedative properties. In pharmaceutical products, dimethindene is employed as dimethindene maleate (CAS number 3614-69-5, EC number 222-789-2 [Dimetindene hydrogen maleate], molecular formula $C_{24}H_{28}N_2O_4$) (1).

In topical preparations, dimethindene has rarely caused contact allergy/allergic contact dermatitis, which subject has been reviewed in Volume 3 of the *Monographs in contact allergy* series (3).

CUTANEOUS ADVERSE DRUG REACTIONS FROM SYSTEMIC ADMINISTRATION CAUSED BY TYPE IV (DELAYED-TYPE) HYPERSENSITIVITY (as demonstrated by positive patch tests)

Cutaneous adverse drug reactions from systemic administration of dimethindene caused by type IV (delayed-type) hypersensitivity have included systemic allergic dermatitis presenting as maculopapular and vesicular eruption (2).

Systemic allergic dermatitis (systemic contact dermatitis)

A 12-year-old boy presented with an acute eczematous eruption on one leg, after having applied a gel containing dimethindene maleate on insect bites. In order to calm the pruritus, drops containing dimethindene maleate were taken orally, and 24 hr later the patient developed a diffuse, itchy, maculopapular and vesicular rash. Patch testing showed positive reactions to the gel and to the drops 'as is' and diluted 30%. Later, patch tests were conducted with the individual constituents of the gel and the drops, which yielded positive reactions to dimethindene maleate (contained in both preparations), to benzalkonium chloride (present in the gel) and to benzoic acid (present in the drops) ®. Five normal controls were negative, test concentrations and vehicles were not mentioned (2). This was a case of systemic allergic dermatitis.

LITERATURE

1 The data in the section 'General' may have been obtained from literature discussed in this chapter, but mostly also or exclusively from one or more of the following online sources: ChemIDPlus Advanced, PubChem, DrugBank, RxList, Drug Central, Drugs.com, and Wikipedia

2 Leroy A, Baeck M, Tennstedt D. Contact dermatitis and secondary systemic allergy to dimethindene maleate. Contact Dermatitis 2011;64:170-171

3 De Groot AC. Monographs in contact allergy, volume 3: Topical Drugs. Boca Raton, Fl, USA: CRC Press Taylor and Francis Group, 2021 (ISBN 978-0-367-23693-9)

Chapter 3.181 DIMETHYL SULFOXIDE

IDENTIFICATION

Description/definition   : Dimethyl sulfoxide is the sulfoxide that conforms to the structural formula shown below
Pharmacological classes   : Antineoplastic agent; solvent
IUPAC name   : Methylsulfinylmethane
Other names   : Methyl sulfoxide
CAS registry number   : 67-68-5
EC number   : 200-664-3
Merck Index monograph   : 4555
Patch testing   : 10% in petrolatum (2); water is probably a better vehicle
Molecular formula   : C_2H_6OS

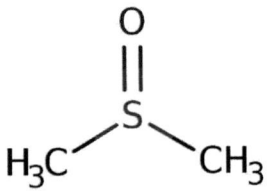

GENERAL

Dimethyl sulfoxide (DMSO) is a reversible mitogen-activated extracellular signal-regulated kinase-1 (MEK1) and MEK2 inhibitor used to treat certain types of melanoma, metastatic non-small cell lung cancer, and locally advanced or metastatic anaplastic thyroid cancer. Another indication is for the symptomatic relief of patients with interstitial cystitis. DMSO is a highly polar organic liquid, that is used widely as a chemical solvent. Because of its ability to penetrate biological membranes, DMSO is used as a vehicle for topical application of pharmaceuticals. It is also used to protect tissue during cryopreservation. Dimethyl sulfoxide has shown a range of pharmacological activities including analgesia and anti-inflammation (1).

CUTANEOUS ADVERSE DRUG REACTIONS FROM SYSTEMIC ADMINISTRATION CAUSED BY TYPE IV (DELAYED-TYPE) HYPERSENSITIVITY (as demonstrated by positive patch tests)

Cutaneous adverse drug reactions from systemic administration of dimethyl sulfoxide caused by type IV (delayed-type) hypersensitivity have included systemic allergic dermatitis (2).

Systemic allergic dermatitis (systemic contact dermatitis)

A 49-year-old man complained of an itchy eruption on the extremities. The eruption first appeared on his fingers and gradually extended to the forearms and the legs. The patient had been given an intravesical instillation of a 50% aqueous solution of dimethyl sulfoxide (DMSO) once every two weeks for an interstitial cystitis. The rash had appeared 2 weeks after the second instillation. Physical examination revealed milia-sized reddish papules and erythema bilaterally on the forearms, hands, fingers, and thighs. The rash was first considered this to be a case of simple contact dermatitis. However, when the patient was given an additional treatment with dimethyl sulfoxide, the eruption recurred at the site of the old eruption two days later. A patch test to DMSO 10% pet. was positive at D2 and, at that time, a flare-up phenomenon was observed at the sites of the old eruption (2). The author considers this a case of systemic allergic dermatitis, where the patient probably becomes sensitized to DMSO on the mucous membrane of the bladder and later develops a systemic eruption from absorption through the mucosa. The same situation occurs with intravesical instillations of mitomycin C (Chapter 3.325).

LITERATURE

1   The data in the section 'General' may have been obtained from literature discussed in this chapter, but mostly also or exclusively from one or more of the following online sources: ChemIDPlus Advanced, PubChem, DrugBank, RxList, Drug Central, Drugs.com, and Wikipedia
2   Nishimura M, Takano Y, Toshitani S. Systemic allergic dermatitis medicamentosa occurring after intravesical dimethyl sulfoxide treatment for interstitial cystitis. Arch Dermatol 1988;124:182-183

Chapter 3.182 DIMETRIDAZOLE

IDENTIFICATION

Description/definition : Dimetridazole is the imidazole derivative that conforms to the structural formula shown below
Pharmacological classes : Antiprotozoal agent (veterinary)
IUPAC name : 1,2-Dimethyl-5-nitroimidazole
Other names : 5-Nitro-1,2-dimethylimidazole
CAS registry number : 551-92-8
EC number : 209-001-2
Merck Index monograph : 4560
Patch testing : 1% pet.
Molecular formula : $C_5H_7N_3O_2$

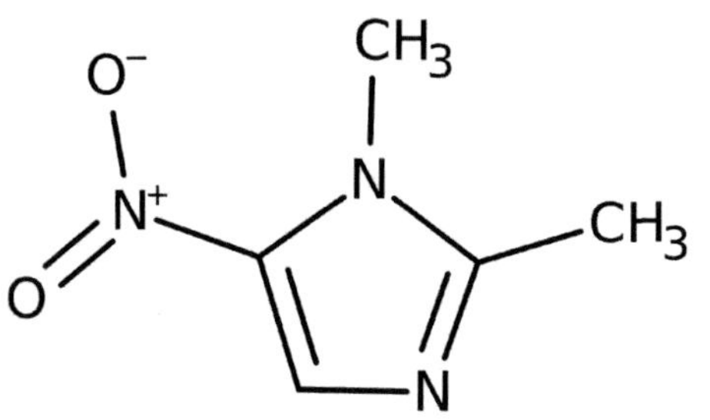

GENERAL

Dimetridazole is a nitroimidazole class drug used in veterinary medicine only, that combats protozoan infections, notably histomoniasis. It used to be commonly added to poultry feed, which led to dimetridazole being found in eggs. Because of suspicions of it being carcinogenic, the use of this anti-protozoan drug has been legally limited and was banned in Canada as a livestock feed additive (1).

CONTACT ALLERGY FROM ACCIDENTAL CONTACT

In Italy, during 1986-1988, 204 animal feed mill workers (191 men, 13 women) were patch tested with a large number of animal feed additives. There were zero reactions to dimetridazole 1% pet. in a group of 36 subjects with clinical complaints (dermatitis or pruritus sine materia) and one reaction was observed in the group of 168 individuals without skin complaints. All reactions were considered to be relevant (2). This was a case of occupational allergic contact dermatitis.

LITERATURE

1   The data in the section 'General' may have been obtained from literature discussed in this chapter, but mostly also or exclusively from one or more of the following online sources: ChemIDPlus Advanced, PubChem, DrugBank, RxList, Drug Central, Drugs.com, and Wikipedia
2   Mancuso G, Staffa M, Errani A, Berdondini RM, Fabbri P. Occupational dermatitis in animal feed mill workers. Contact Dermatitis 1990;22:37-41

Chapter 3.183 DIPHENHYDRAMINE

IDENTIFICATION

Description/definition   : Diphenhydramine is the diphenylmethane that conforms to the structural formula shown
                              below
Pharmacological classes   : Anesthetics, local; hypnotics and sedatives; anti-allergic agent; antiemetics; histamine H1
                              antagonists; sleep aids, pharmaceutical
IUPAC name   : 2-Benzhydryloxy-*N*,*N*-dimethylethanamine
CAS registry number   : 58-73-1
EC number   : 200-396-7
Merck Index monograph   : 4609
Patch testing   : Hydrochloride, 1.0% pet. (Chemotechnique)
Molecular formula   : $C_{17}H_{21}NO$

GENERAL

Diphenhydramine is an ethanolamine first-generation histamine H1 receptor antagonist with anti-allergic, antiemetic, antitussive, antimuscarinic and sedative effects. As an over-the-counter (OTC) medication, diphenhydramine is typically formulated as tablets and creams indicated for use in treating sneezing, runny nose, itchy/watery eyes, itching of nose or throat, insomnia, pruritus, urticaria, insect bites/stings, allergic rashes, and nausea. It may apparently also be used as an antiparkinsonian agent. In pharmaceutical products, both diphenhydramine base and – far more often – diphenhydramine hydrochloride (CAS number 147-24-0, EC number 205-687-2, molecular formula $C_{17}H_{22}ClNO$) may be employed, and possibly - rarely - diphenhydramine citrate or methylbromide (1).

   In topical preparations, diphenhydramine has caused cases of (photo)contact allergy/(photo)allergic contact dermatitis, which subject has been fully reviewed in Volume 3 of the *Monographs in contact allergy* series (11).

CONTACT ALLERGY FROM ACCIDENTAL CONTACT

According to the author of ref. 3, contact allergic reactions to diphenhydramine have been reported in older German literature. Details are unknown, but is it likely that refs. 4 and 5 may deal with occupational allergic contact dermatitis, e.g. in health care providers (4).

CUTANEOUS ADVERSE DRUG REACTIONS FROM SYSTEMIC ADMINISTRATION CAUSED BY TYPE IV
(DELAYED-TYPE) HYPERSENSITIVITY (as demonstrated by positive patch tests)

Cutaneous adverse drug reactions from systemic administration of diphenhydramine caused by type IV (delayed-type) hypersensitivity have included systemic allergic dermatitis presenting as generalized dermatitis (9), as a vesiculobullous eruption (6), as un unknown eruption (10), and photosensitivity (7,8).

Systemic allergic dermatitis (systemic contact dermatitis)

A 54-year-old man who presented with generalized dermatitis on the trunk and extremities. According to his history, he developed localized dermatitis 2 days after being exposed to a large vine. He was treated with injections in an emergency department, a capsule containing diphenhydramine HCl (Benadryl) to take orally and he applied a lotion containing the antihistamine topically. This resulted in a temperature of 38° C and worsening of the eruption. Patch tests were performed with the diphenhydramine HCl lotion and injectable diphenhydramine HCl. Both patch tests were read as positive 3 days later. This was a case of allergic contact dermatitis and systemic allergic dermatitis (9).

A 51-year-old woman had developed a widespread vesiculobullous eruption affecting the feet, hands, trunk, thighs and face, 2 days after starting a cough medicine containing diphenhydramine. Twice previously she had developed a similar eruption. On both occasions she had used another expectorant containing diphenhydramine. The patient had never used topical antihistamines. Patch tests yielded a ++ reaction to diphenhydramine 2% pet. (6). Probably, the patient had previously become sensitized to diphenhydramine on the oral mucosa and now had systemic allergic dermatitis (6).

Another patient also developed systemic allergic dermatitis from oral diphenhydramine following sensitization from a topical diphenhydramine product (unknown which type of eruption) (10).

See also the section 'Photosensitivity' below, ref. 7.

Photosensitivity

A 52-year-old man had a pruritic eruption of five years' duration that was confined to the light-exposed areas. The lesions appeared initially on the scalp following several applications of hair dye. Despite the use of a number of prescribed therapeutic modalities, the eruption spread to the forehead, cheeks, nape and V of the neck, extensor aspects of the forearms, and dorsa of the hands. He had been receiving no medication other than those prescribed for the skin eruption. Diphenhydramine had been given orally, topically, and by injection. Patch and photopatch tests (using UVA) were positive to p-phenylenediamine (hair dye) and to diphenhydramine 1% and 5% pet. The reactions to the diphenhydramine photopatch tests were slightly stronger than those of the patch tests. Five controls were negative (7). If sensitization occurred from topical diphenhydramine (which is unknown but certainly possible) , this would be a case of systemic photoallergic dermatitis.

Possible photoallergy to oral diphenhydramine in 2 patients was reported in 1962, when photopatch testing was far from reliable (8).

Cross-reactions, pseudo-cross-reactions and co-reactions

In 2 patients sensitized to diphenhydramine and tested with chemically-related derivatives, 2 positive reactions were observed to bromazine (bromodiphenhydramine), 2 to medrylamine (4-methoxydiphenhydramine) and 1 to dimenhydrinate (diphenhydramine + 8-chlorotheophylline) and p-methyldiphenhydramine, respectively (2).

LITERATURE

1 The data in the section 'General' may have been obtained from literature discussed in this chapter, but mostly also or exclusively from one or more of the following online sources: ChemIDPlus Advanced, PubChem, DrugBank, RxList, Drug Central, Drugs.com, and Wikipedia
2 Goossens A, Linsen G. Contact allergy to antihistamines is not common. Contact Dermatitis 1998;39:38
3 Heine A. Diphenhydramine: a forgotten allergen? Contact Dermatitis 1996;35:311-312
4 Gertler H, Laubstein H. Berufsdermatosen bei Angehörigen der medizinischen Berufe. Zschr Ärztl Fortbild 1969;59:251-255 (Article in German). Data cited in ref. 3
5 Laubstein H, Mönnich HT. Zur Epidemiologie der Berufsdermatosen. Dermatol Wochenschr 1968;154:649-667 (Article in German). Data cited in ref. 3
6 Lawrence CM, Byrne JP. Eczematous eruption from oral diphenhydramine. Contact Dermatitis 1981;7:276-277
7 Horio T. Allergic and photoallergic dermatitis from diphenhydramine. Arch Dermatol 1976;112:1124-1126
8 Schreiber MM, Naylor LZ. Antihistamine photosensitivity. Arch Dermatol 1962;86:58-62
9 Coskey RJ. Contact dermatitis caused by diphenhydramine hydrochloride. J Am Acad Dermatol 1983;8:204-206
10 Shelley WB, Bennett RG. Primary contact sensitization site: A determination for localization of a diphenhydramine eruption. Acta Derm Venereol (Stockh) 1972;52:376-378
11 De Groot AC. Monographs in contact allergy, volume 3: Topical Drugs. Boca Raton, Fl, USA: CRC Press Taylor and Francis Group, 2021 (ISBN 978-0-367-23693-9)

Chapter 3.184 DIPYRIDAMOLE

IDENTIFICATION

Description/definition   : Dipyridamole is the dialkylarylamine that conforms to the structural formula shown below
Pharmacological classes   : Phosphodiesterase inhibitors; vasodilator agents; platelet aggregation inhibitors
IUPAC name   : 2-[[2-[bis(2-Hydroxyethyl)amino]-4,8-di(piperidin-1-yl)pyrimido[5,4-d]pyrimidin-6-yl]-(2-hydroxyethyl)amino]ethanol
CAS registry number   : 58-32-2
EC number   : 200-374-7
Merck Index monograph   : 4658
Patch testing   : Tablet, pulverized, 30% pet. (2); most systemic drugs can be tested at 10% pet.; if the pure chemical is not available, prepare the test material from intravenous powder, the content of capsules or – when also not available – from powdered tablets to achieve a final concentration of the active drug of 10% pet.
Molecular formula   : $C_{24}H_{40}N_8O_4$

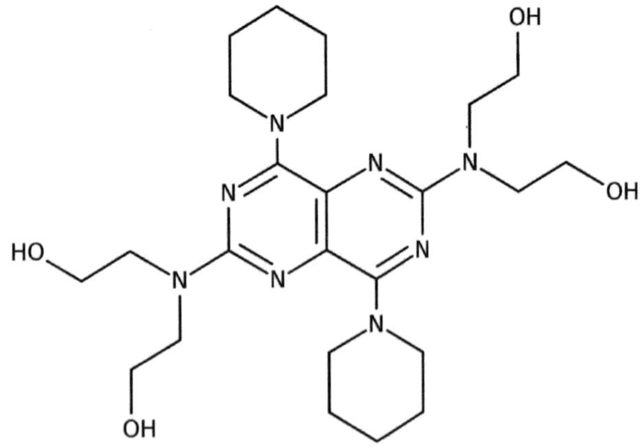

GENERAL

Dipyridamole is a synthetic derivative of pyrimido-pyrimidine, which has platelet aggregation-inhibiting properties. This agent is indicated for use as an adjunct to coumarin anticoagulants in the prevention of postoperative thromboembolic complications of cardiac valve replacement and also used in the prevention of angina and as secondary prophylaxis after cerebral infarction, often in combination with acetylsalicylic acid (1,2).

CONTACT ALLERGY FROM ACCIDENTAL CONTACT

A 36-year-old nurse developed dermatitis at the back of both hands, on the fingers, interdigital, and on the arm and mouth. Patch testing showed her to have occupational allergic contact dermatitis from dipyridamole. Details were not provided (3).

CUTANEOUS ADVERSE DRUG REACTIONS FROM SYSTEMIC ADMINISTRATION CAUSED BY TYPE IV (DELAYED-TYPE) HYPERSENSITIVITY (as demonstrated by positive patch tests)

Cutaneous adverse drug reactions from systemic administration of dipyridamole caused by type IV (delayed-type) hypersensitivity have included eczematous eruption (2).

Dermatitis/eczematous eruption

A 60-year-old man experienced an acute right-sided hemiparesis, and was treated conservatively in the stroke unit because of a left-sided cerebral infarction. Besides supportive therapy, anticoagulation with clopidogrel was initiated. Following the acute treatment, additional prophylaxis with a combination tablet of dipyridamole and acetylsalicylic acid was started. Eight days later, the patient developed a symmetrical, itching, eczematous eruption, beginning on his neck and spreading over the trunk and the proximal extremities. Patch tests gave a positive reaction to the combination tablet 30% pet. but not in water. According to the authors, this indicated delayed-type sensitivity to dipyridamole, which is incorrect, as the tablet also contained acetylsalicylic acid. Twenty controls were negative. Subsequently, an oral provocation test with acetylsalicylic acid (Aspirin tablet) was carried out and was tolerated well by the patient. The diagnosis of type IV hypersensitivity to dipyridamole was made therefore *per exclusionem* (2).

LITERATURE

1   The data in the section 'General' may have been obtained from literature discussed in this chapter, but mostly also or exclusively from one or more of the following online sources: ChemIDPlus Advanced, PubChem, DrugBank, RxList, Drug Central, Drugs.com, and Wikipedia

2   Salava A, Alanko K, Hyry H. Dipyridamole-induced eczematous drug eruption with positive patch test reaction. Contact Dermatitis 2012;67:103-104

3   Gielen K, Goossens A. Occupational allergic contact dermatitis from drugs in healthcare workers. Contact Dermatitis 2001;45:273-279

Chapter 3.185 DISULFIRAM

IDENTIFICATION

Description/definition	: Disulfiram is the carbamate derivative that conforms to the structural formula shown below
Pharmacological classes	: Acetaldehyde dehydrogenase inhibitors; alcohol deterrents
IUPAC name	: Diethylcarbamothioylsulfanyl N,N-diethylcarbamodithioate
Other names	: Tetraethylthiuram disulfide; thioperoxydicarbonic diamide, ([(H2N)C(S)]2S2), tetraethyl-; 1,1'-dithiobis(N,N-diethylthioformamide); bis(N,N-diethylthiocarbamoyl) disulfide
CAS registry number	: 97-77-8
EC number	: 202-607-8
Merck Index monograph	: 4677
Patch testing	: 1% pet. (0.25% is also adequate)
Molecular formula	: $C_{10}H_{20}N_2S_4$

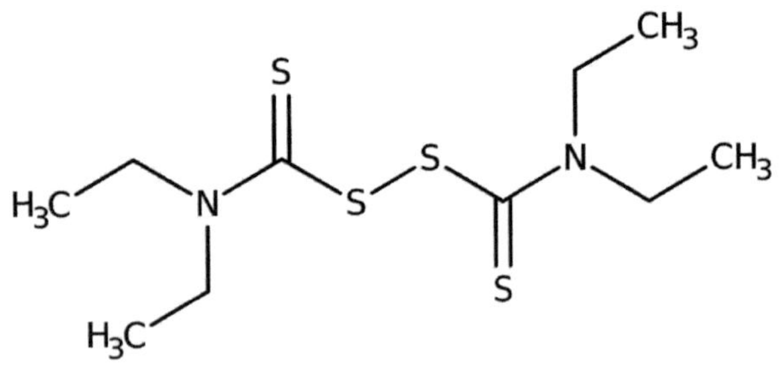

GENERAL

Disulfiram is an orally bioavailable carbamoyl derivative used as an alcohol deterrent in patients with chronic alcoholism. It markedly alters the intermediary metabolism of alcohol by irreversibly binding to and inhibiting acetaldehyde dehydrogenase, an enzyme that oxidizes the ethanol metabolite acetaldehyde into acetic acid. This inhibition leads to an accumulation of acetaldehyde, with concentrations 5-10 times higher than without disulfiram, resulting in flushing, systemic vasodilation, respiratory difficulties, nausea, hypotension, and other symptoms (acetaldehyde syndrome). The symptoms are proportional to the dosage of both disulfiram and alcohol and will persist as long as alcohol is being metabolized. Prolonged administration of disulfiram does not produce tolerance; the longer a patient remains on therapy, the more exquisitely sensitive he becomes to alcohol (1).

CONTACT ALLERGY FROM ACCIDENTAL CONTACT

Occupational allergic contact dermatitis

A 40-year-old nurse, working in a hospital, presented with a history of intermittent dermatitis of the forearms, hands and face. She had noticed that the lesions would appear a few hours after handling disulfiram tablets. She also noted worsening after wearing long rubber gloves. Patch tests were positive to disulfiram, the thiuram mix and related rubber chemicals. This was a case of occupational allergic contact dermatitis (16).

In a group of 107 workers in the pharmaceutical industry with dermatitis, investigated in Warsaw, Poland, before 1989, 4 reacted to disulfiram, tested 2% pet. (3).

Allergic contact dermatitis by proxy

A 61-year-old man had a history of edema of the eyelids and itchy, erythematous and scaly lesions on the face, neck, and upper chest, and once also on the shoulders and elbow folds. The patient reported that he crushed disulfiram tablets for his wife, which could explain the occasional occurrence of air-exposed skin lesions. Patch tests were positive to the thiuram mix (containing disulfiram = tetraethylthiuram disulfide). He was advised to take preventive measures, and since he has been using a pill crusher, his problem has completely resolved (2,15). This was a case of airborne allergic contact dermatitis by proxy.

CUTANEOUS ADVERSE DRUG REACTIONS FROM SYSTEMIC ADMINISTRATION CAUSED BY TYPE IV (DELAYED-TYPE) HYPERSENSITIVITY (as demonstrated by positive patch tests)

Cutaneous adverse drug reactions from systemic administration of disulfiram caused by type IV (delayed-type) hypersensitivity have included systemic allergic dermatitis (4,6,7,10,11,14,18) and localized and extensive eruptions from sensitization to a disulfiram implant (5,8,9).

Systemic allergic dermatitis (systemic contact dermatitis)

A 53-year-old man with plaque-type psoriasis was hospitalized with a 10-day history of rash, fever, chills, and vomiting. He suffered from chronic alcoholism, and had been previously diagnosed with ACD caused by rubber products, with positive patch test reactions to 'thiurams'. Physical examination showed an erythematous and scaly rash affecting the face, trunk, arms and legs. The patient reported having started alcoholic detoxification therapy with Antabuse (disulfiram) on his own initiative a few days before the onset of symptoms. He had been taking disulfiram since then, even while in hospital. The use of disulfiram was suspended. During the following day, his fever, chills and vomiting remitted, and the rash resolved within 2 weeks. Patch tests were positive to tetraethylthiuram disulfide (disulfiram) 0.25% pet., to commercial disulfiram capsules diluted 30% pet. and multiple other rubber compounds (4). This was a classic case of systemic allergic dermatitis.

A 35-year-old man developed dyshidrotic eczema of the hands and feet, nummular patches of vesicular and oozing dermatitis on the arms and legs and a similar patch around an old operation scar on the left abdominal wall, which had begun a few days after commencing treatment with oral disulfiram. Patch tests were positive to disulfiram, related rubber chemicals and a crushed Antabuse tablet. The patient had been sensitized by the previous disulfiram implant, although no lesion had been noted there (7).

Short summaries of other case reports of systemic allergic dermatitis from disulfiram are shown in table 3.185.1.

Table 3.185.1 Short summaries of case reports of systemic allergic dermatitis from disulfiram

Year and country	Sex	Age	Positive patch tests	Clinical data and comments	Ref.
2001 Germany	F	44	TETD 1% pet.	sensitized by gloves; later generalized dermatitis from oral disulfiram	6
1989 Belgium	M	36	TETD, other thiurams	previously, a disulfiram implant was removed, because it was 'inflamed'; uncertain whether sensitization was caused by the implant or rubber materials; gross erythema and edema of the face, widespread erythematous, edematous and vesicular patches and an urticarial plaque around the operation scar of the disulfiram implant removal	18
1979 USA	M	32	thiuram mix	sensitized by rubber in shoes and elastic waistband; 5 hours after oral disulfiram diffuse pruritus, edema of the feet, vesicular eruption of the face, arms and dorsal feet	14
1975 Belgium	M	33	TETD 2% pet.	widespread dermatitis	8
1972 Italy	M	37	commercial disulfiram tablet 0.05% pet.; TMTD 0.5% pet.	sensitized by gloves; later erythematous and vesicular dermatitis face, forearms and thorax 8-21 hours after taking 2 tablets disulfiram	10
1962 United Kingdom	M	?	disulfiram (no details)	sensitized by rubber contraceptive sheath; very violent reaction associated with a widespread skin rash after oral intake of disulfiram	11

TETD: tetraethylthiuram disulfide (disulfiram)
TMTD: tetramethylthiuram disulfide

Localized and extensive eruptions from sensitization to a disulfiram implant

A 49-year-old woman was admitted with suspected alcohol poisoning. Three months previously, she had been implanted with disulfiram in her left buttock. A pruritic erythema had developed at the site of the implant after 2 weeks. Two days before her admission, she had started drinking alcohol. A generalized erythema with numerous papules had appeared on the face and extremities, and the dermatitis at the site of the implant had become more severe. Patch tests were strongly positive to the thiuram mix and to disulfiram (tetraethylthiuram disulfide). This patient did *not* have systemic allergic dermatitis, as the (incorrect) title suggests, but was sensitized to disulfiram by the implant and the exanthema was provoked by alcohol (5). Worsening of allergic contact dermatitis from rubber materials (disulfiram [TETD], TMTD [tetramethylthiuram disulfide]) by alcohol has previously been observed (12,13).

A 33-year old man had a third implantation of disulfiram. An erythematovesicular dermatitis developed around the site of the incision 6 days after the implantation, spreading rapidly and becoming generalized. Patch tests were positive to disulfiram 2% pet. One tablet (400 mg) of oral disulfiram resulted in a widespread systemic eczematous contact-type dermatitis during the following 24 hours (8). A 39-year-old man had a second disulfiram implantation in the right lower abdomen. An eczematous dermatitis developed a few days later, all around the operation site. The disulfiram pellets were removed and the eczematous rash faded progressively. The patient had a positive patch test to disulfiram 2% pet. He had likely been sensitized by the first implant, although it had been tolerated well. Subsequently, oral disulfiram therapy was not followed by skin reactions (8).

The same authors also observed generalized eczema after treatment with disulfiram tablets (9, no details available, data cited in ref. 16).

Cross-reactions, pseudo-cross-reactions and co-reactions
Disulfiram (tetraethylthiuram disulfide, TETD) may cross-react with other rubber chemicals, especially tetramethylthiuram disulfide (TMTD) (17).

LITERATURE

1   The data in the section 'General' may have been obtained from literature discussed in this chapter, but mostly also or exclusively from one or more of the following online sources: ChemIDPlus Advanced, PubChem, DrugBank, RxList, Drug Central, Drugs.com, and Wikipedia

2   Swinnen I, Goossens A. An update on airborne contact dermatitis: 2007-2011. Contact Dermatitis 2013;68:232-238

3   Rudzki E, Rebandel P, Grzywa Z. Contact allergy in the pharmaceutical industry. Contact Dermatitis 1989;21:121-122

4   Fustà-Novell X, Gómez-Armayones S, Morgado-Carrasco D, Mascaró JM. Systemic allergic dermatitis caused by disulfiram (Antabuse) in a patient previously sensitized to rubber accelerators. Contact Dermatitis 2018;79:239-240

5   Kieć-Swierczyńska M, Krecisz B, Fabicka B. Systemic allergic dermatitis from implanted disulfiram. Contact Dermatitis 2000;43:246-247

6   Gutgesell C, Fuchs T. Orally elicited allergic contact dermatitis to tetraethylthiuramdisulfide. Am J Contact Dermat 2001;12:235-236

7   Van Hecke E, Vermander F. Allergic contact dermatitis by oral disulfiram. Contact Dermatitis 1984;10:254

8   Lachapelle JM. Allergic "contact" dermatitis from disulfiram implants. Contact Dermatitis 1975;1:218-220

9   Lachapelle JM, Tennstedt D. Allergic "contact" dermatitis to disulfiram implants. Nouv Presse Med 1976;5:1536 (Letter, in French, data cited in ref. 16)

10 Meneghini CL, Bonifazi E. Antabuse dermatitis. Contact Dermatitis Newsletter 1972;11:285

11 Wilson H. Side-effects of disulfiram. Br Med J 1962;2:1610-1611

12 Van Ketel WG. Rubber, alcohol and eczema. Dermatologica 1968;136:442-444

13 Van Ketel, WG. Rubber, alcohol en eczeem. Nederlands Tijdschrift voor Geneeskunde 1968;112:406-408 (probably the same as ref. 12)

14 Webb PK, Bibbs SC, Mathias CT, Crain W, Maibach H. Disulfiram hypersensitivity and rubber contact dermatitis. JAMA 1979;241:2061

15 Creytens K, Swevers A, De Haes P, Goossens A. Airborne allergic contact dermatitis caused by disulfiram. Contact Dermatitis 2015;72:405-407

16 Mathelier-Fusade P, Leynadier F. Occupational allergic contact reaction to disulfiram. Contact Dermatitis 1994;31:121-122

17 Lerbaek A, Menné T, Knudsen B. Cross-reactivity between thiurams. Contact Dermatitis 2006;54:165-168

18 Van Hecke E. Acute generalized dermatitis to orally administered disulfiram. In: Frosch PJ, Dooms-Goossens A, Lachapelle JM, Rycroft RJG, Scheper RJ (eds). Current topics in contact dermatitis. Berlin: Springer-Verlag, 1989: 250-253

Chapter 3.186 DOXEPIN

IDENTIFICATION

Description/definition : Doxepin is the dibenzoxepin tricyclic compound that conforms to the structural formula shown below
Pharmacological classes : Antidepressive agents, tricyclic; sleep aids, pharmaceutical; histamine antagonists
IUPAC name : (3*E*)-3-(6*H*-Benzo[c][1]benzoxepin-11-ylidene)-*N*,*N*-dimethylpropan-1-amine
CAS registry number : 1668-19-5
EC number : Not available
Merck Index monograph : 4753
Patch testing : 1% and 5% pet. (4)
Molecular formula : $C_{19}H_{21}NO$

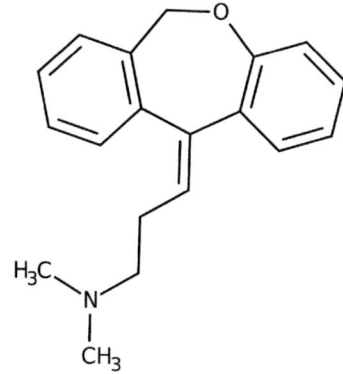

GENERAL

Doxepin is a dibenzoxepin derivative and tricyclic antidepressant-like drug with antipruritic, anti-depressive, sedative and anxiolytic activities with structural similarities to phenothiazines. Oral doxepin is approved for treatment of depression and/or anxiety associated with different conditions, including alcoholism, organic disease and manic-depressive disorders and for treatment of insomnia. In pharmaceutical products, doxepin is employed as doxepin hydrochloride (CAS number 1229-29-4, EC number 214-966-8, molecular formula $C_{19}H_{22}ClNO$) (1). In topical preparations, doxepin has caused many cases of contact allergy/allergic contact dermatitis, which subject has been fully reviewed in Volume 3 of the *Monographs in contact allergy* series (4).

CUTANEOUS ADVERSE DRUG REACTIONS FROM SYSTEMIC ADMINISTRATION CAUSED BY TYPE IV (DELAYED-TYPE) HYPERSENSITIVITY (as demonstrated by positive patch tests)

Cutaneous adverse drug reactions from systemic administration of doxepin caused by type IV (delayed-type) hypersensitivity have included systemic allergic dermatitis (2,3).

Systemic allergic dermatitis (systemic contact dermatitis)

A 75-year-old man had a chronic pruritic dermatitis of one year's duration on his scalp, axillae, trunk, lower legs, and feet. He had used various medications including oral and, in an earlier stage, topical doxepin. Examination revealed hyperpigmented, erythematous scaly patches and plaques on the superior back, and erythematous, scaly patches and plaques on the superior chest, axillae and left calf. Patch tests were positive to doxepin cream 5%. Oral doxepin was discontinued and treatment with oral hydroxyzine and topical corticosteroids for one month resulted in marked improvement of his dermatitis. The patient was diagnosed with systemic allergic dermatitis from oral doxepin (2).

Two more patients appeared to have had, according to the authors of that article, systemic allergic dermatitis from oral administration of doxepin after topical sensitization (3). However, the evidence presented was very weak.

LITERATURE

1   The data in the section 'General' may have been obtained from literature discussed in this chapter, but mostly also or exclusively from one or more of the following online sources: ChemIDPlus Advanced, PubChem, DrugBank, RxList, Drug Central, Drugs.com, and Wikipedia
2   Brancaccio RR, Weinstein S. Systemic allergic dermatitis to doxepin. J Drugs Dermatol 2003;2:409-410
3   Taylor JS, Praditsuwan P, Handel D, Kuffner G. Allergic contact dermatitis from doxepin cream. Arch Dermatol 1996:132:515-518
4   De Groot AC. Monographs in contact allergy, volume 3: Topical Drugs. Boca Raton, Fl, USA: CRC Press Taylor and Francis Group, 2021 (ISBN 978-0-367-23693-9)

Chapter 3.187 DOXYCYCLINE

IDENTIFICATION

Description/definition	: Doxycycline is the antibiotic synthetically derived from oxytetracycline, that conforms to the structural formula shown below
Pharmacological classes	: Anti-bacterial agents; antimalarial agents
IUPAC name	: (4S,4aR,5S,5aR,6R,12aR)-4-(Dimethylamino)-1,5,10,11,12a-pentahydroxy-6-methyl-3,12-dioxo-4a,5,5a,6-tetrahydro-4H-tetracene-2-carboxamide
Other names	: Deoxy-5-hydroxytetracycline
CAS registry number	: 564-25-0
EC number	: 209-271-1
Merck Index monograph	: 4758
Patch testing	: Hydrochloride, 1.0% pet. (Chemotechnique)
Molecular formula	: $C_{22}H_{24}N_2O_8$

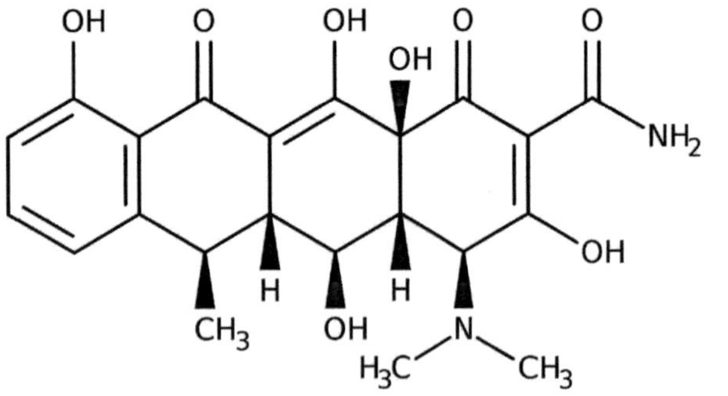

GENERAL

Doxycycline is a broad-spectrum antibiotic synthetically derived from oxytetracycline, exhibiting antimicrobial activity. It is a second-generation tetracycline, exhibiting lesser toxicity than first-generation tetracyclines. Doxycycline is indicated for the treatment of various infections by gram-positive and gram-negative bacteria, aerobes and anaerobes, as well other types of bacteria (1). In pharmaceutical products, doxycycline may be employed as doxycycline hydrochloride (CAS number 10592-13-9, EC number 234-198-7, molecular formula $C_{22}H_{25}ClN_2O_8$) and various other forms (1).

CONTACT ALLERGY FROM ACCIDENTAL CONTACT

In a group of 107 workers in the pharmaceutical industry with dermatitis, investigated in Warsaw, Poland, before 1989, 3 reacted to doxycycline, tested 10% pet. (2). Also in Warsaw, in the period 1979-1983, 27 pharmaceutical workers, 24 nurses and 30 veterinary surgeons were diagnosed with occupational allergic contact dermatitis from antibiotics. The numbers that had positive patch tests to doxycycline (ampoule content) were 3, 0, and 0, respectively, total 3 (7). Of 333 nurses patch tested in the same clinic between 1979 and 1987, one reacted to doxycycline 10% pet.; data on relevance were not provided (8).

CUTANEOUS ADVERSE DRUG REACTIONS FROM SYSTEMIC ADMINISTRATION CAUSED BY TYPE IV

(DELAYED-TYPE) HYPERSENSITIVITY (as demonstrated by positive patch tests)

Cutaneous adverse drug reactions from systemic administration of doxycycline caused by type IV (delayed-type) hypersensitivity have included fixed drug eruption (3,4,5,10) and photosensitivity (6,9).

Fixed drug eruption

In patients with fixed drug eruptions (FDE) caused by delayed-type hypersensitivity, the diagnosis is usually confirmed by a positive patch test with the drug on previously affected skin. Authors from Finland have used an alternative method of topical provocation. The test compound, the drug 10% in petrolatum and sometimes also in 70% alcohol and in DMSO, was applied once and without occlusion over the entire surface of one or several inactive (usually pigmented) sites of FDE lesions. The patients were followed as in-patients for 24 hours. A reaction was regarded as positive when a clearly demarcated erythema lasting at least 6 hours was seen. Of 5 patients with FDE from doxycycline, 2 (40%) had a positive topical provocation (3,4).

One case of fixed drug eruption to doxycycline with a positive patch test reaction on postlesional skin (itching, erythema, infiltration) to the commercial drug 10% pet. was reported from Seoul, South Korea, in the period 1986-

1990 and in 1996 and 1997 (5). Two patients from Tunisia suffering from bullous fixed drug eruption had positive patch tests to doxycycline tablets 30% pet. on postlesional skin (10).

Photosensitivity

A 42-year-old man presented with scaly erythema and vesicles on the face, neck, arms, and dorsa of the hands and feet that had started 3 days earlier. He had been taking doxycycline HCl 100 mg bid and some other drugs orally because of acne vulgaris for a week. Phototests showed increased sensitivity to UVA, which had normalized 2 weeks after the patient had stopped taking the drugs. Patch and photopatch tests with doxycycline and the other drugs 1% and 10% pet. showed small papules on erythema only at the 10% doxycycline photopatch test site one day after irradiation with UVA 4.5 J/cm^2. Five controls were negative. One hour after the patient took 100 mg of doxycycline HCl for oral photochallenge testing, erythema was seen with itching and burning sensation only on the exact areas that had been previously involved. The authors concluded that, whereas doxycycline is a well-known cause of phototoxicity, this was a rare case of photoallergy (6).

A 70-year-old woman was prescribed doxycycline for lymeborreliose. She used various other drugs, but all already for 3 years. Five days after therapy onset, erythema and itching of the skin on the trunk, arms and legs occurred. Physical examination showed relatively sharply demarked palpable fields of erythema, which were accentuated in previously untanned skin areas. Previously tanned areas (face, neck and dorsal parts of hands and forearms) and the skin covered by the bra and pants, did not show any reaction. Despite cessation of doxycycline, therapy and avoidance of light, the skin reaction progressed, becoming confluent, and scaling within the following week, affecting the breast and buttocks, neck, face, and the distal parts of the upper extremities, resembling a clinical picture of developing erythroderma. A skin biopsy the reaction revealed the histological pattern of acute allergic contact dermatitis. Photopatch tests were performed with doxycycline HCl in dilution (50 mg/ml, 10 mg/ml, 1.0 mg/ml in petrolatum) followed by UVA irradiation (5 J/cm^2). The tests were positive at D7 for the 2 highest concentrations but not 1 mg/ml with very small (1 mm) erythematous papules. Five controls were negative. Photoallergy to doxycycline was diagnosed (9).

LITERATURE

1 The data in the section 'General' may have been obtained from literature discussed in this chapter, but mostly also or exclusively from one or more of the following online sources: ChemIDPlus Advanced, PubChem, DrugBank, RxList, Drug Central, Drugs.com, and Wikipedia
2 Rudzki E, Rebandel P, Grzywa Z. Contact allergy in the pharmaceutical industry. Contact Dermatitis 1989;21:121-122
3 Alanko K. Topical provocation of fixed drug eruption. A study of 30 patients. Contact Dermatitis 1994;31:25-27
4 Alanko K, Stubb S, Reitamo S. Topical provocation of fixed drug eruption. Br J Dermatol 1987;116:561-567
5 Lee AY. Topical provocation in 31 cases of fixed drug eruption: change of causative drugs in 10 years. Contact Dermatitis 1998;38:258-260
6 Tanaka N, Kawada A, Ohnishi Y, Hiruma M, Tajima S, Akiyama M, et al. Photosensitivity due to doxycycline hydrochloride with an unusual flare. Contact Dermatitis 1997;37:93-94
7 Rudzki E, Rebendel P. Contact sensitivity to antibiotics. Contact Dermatitis 1984;11:41-42
8 Rudzki E, Rebandel P, Grzywa Z. Patch tests with occupational contactants in nurses, doctors and dentists. Contact Dermatitis 1989;20:247-250
9 Kuznetsov AV, Weisenseel P, Flaig MJ, Ruzicka T, Prinz JC. Photoallergic erythroderma due to doxycycline therapy of erythema chronicum migrans. Acta Derm Venereol 2011;91:734-736
10 Zaouak A, Ben Salem F, Ben Jannet S, Hammami H, Fenniche S. Bullous fixed drug eruption: A potential diagnostic pitfall: a study of 18 cases. Therapies 2019;74:527-530

Chapter 3.188 DRONEDARONE

IDENTIFICATION

Description/definition   : Dronedarone is the member of the class of 1-benzofurans that conforms to the structural
                             formula shown below
Pharmacological classes : Anti-arrhythmia agents
IUPAC name   : N-[2-Butyl-3-[4-[3-(dibutylamino)propoxy]benzoyl]-1-benzofuran-5-yl]methane-
                             sulfonamide
CAS registry number   : 141626-36-0
EC number   : 604-240-2
Merck Index monograph : 4768
Patch testing   : Crushed film-coated tablets in pet. 'as is'; suggested: pure chemical 10% pet.
Molecular formula   : $C_{31}H_{44}N_2O_5S$

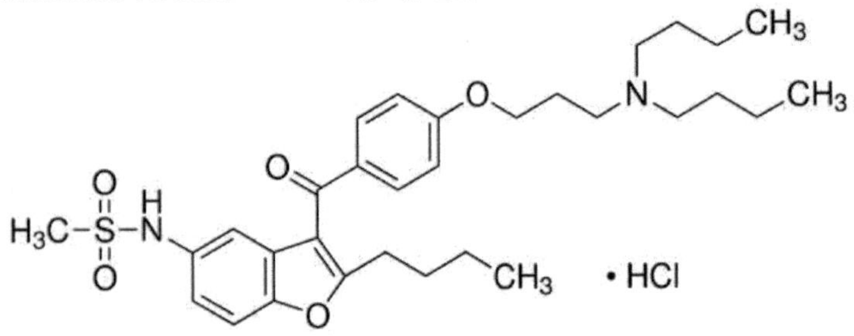

Dronedarone hydrochloride

GENERAL

Dronedarone is a class III antiarrhythmic agent and a synthetic non-iodinated congener of amiodarone, that is used for maintaining sinus rhythm for patients with atrial fibrillation or flutter. Similar to amiodarone, dronedarone is a multichannel blocker. In pharmaceutical products, dronedarone is employed as dronedarone hydrochloride (CAS number 141625-93-6, EC number not available, molecular formula $C_{31}H_{45}ClN_2O_5S$) (1).

CUTANEOUS ADVERSE DRUG REACTIONS FROM SYSTEMIC ADMINISTRATION CAUSED BY TYPE IV (DELAYED-TYPE) HYPERSENSITIVITY (as demonstrated by positive patch tests)

Cutaneous adverse drug reactions from systemic administration of dronedarone caused by type IV (delayed-type) hypersensitivity have included photosensitivity (2).

Photosensitivity

After having used dronedarone 400 mg twice daily orally for rhythm control along with apixaban for 4 weeks, a 64-year-old woman noticed redness on her cheeks after sitting in the sun for 1.5 hours. The next day the patient took a long walk in the sun; overnight she developed redness and swelling of her face, neck, and backs of her hands with itching, which worsened during the following day. Despite discontinuation of both drugs, the skin lesions remained for another 3 to 4 weeks and spread to the upper arms, neck, and décolleté. Complete healing was achieved after 6 weeks, with an increased sensitivity to sunlight remaining throughout the summer. Photopatch tests were positive (++) to dronedarone crushed film-coated tablets in pet. 'as is' 2 and 3 days after irradiation with 5 J/cm² UVA and negative to apixaban; plain patch tests to these drugs were also negative (2).

Experimental photosensitization studies using THP-1 cells and interleukin 8 (IL-8) as biomarkers had previously shown that dronedarone is likely to cause skin photosensitization (3). It has also caused phototoxicity reactions (4).

LITERATURE

1   The data in the section 'General' may have been obtained from literature discussed in this chapter, but mostly also or exclusively from one or more of the following online sources: ChemIDPlus Advanced, PubChem, DrugBank, RxList, Drug Central, Drugs.com, and Wikipedia
2   Al-Jarrah R, Blasini A, Kurgyis Z, Brockow K, Eberlein B. Severe photoallergy to systemic dronedarone (Multaq). Contact Dermatitis 2020;83:241-242
3   Marcolino AIP, Macedo LB, Nogueira-Librelotto DR, Vinardell MP, Rolim CMB, Mitjansc M. Comparative evaluation of the hepatotoxicity, phototoxicity and photosensitizing potential of dronedarone hydrochloride and its cyclodextrin-based inclusion complexes. Photochem Photobiol Sci 2019;18:1565-1575
4   Datar P, Kafle P, Schmidt FM, Bhattarai B, Mukhtar O. Dronedarone-induced phototoxicity in a patient with atrial fibrillation. Cureus 2019;11:e5731

Chapter 3.189 DROXICAM

IDENTIFICATION

Description/definition : Droxicam is the benzothiazine that conforms to the structural formula shown below
Pharmacological classes : Anti-inflammatory agents, non-steroidal
IUPAC name : 5-Methyl-6,6-dioxo-3-pyridin-2-yl-[1,3]oxazino[5,6-c][1,2]benzothiazine-2,4-dione
CAS registry number : 90101-16-9
EC number : Not available
Merck Index monograph : 713
Patch testing : 1% and 5% pet.
Molecular formula : $C_{16}H_{11}N_3O_5S$

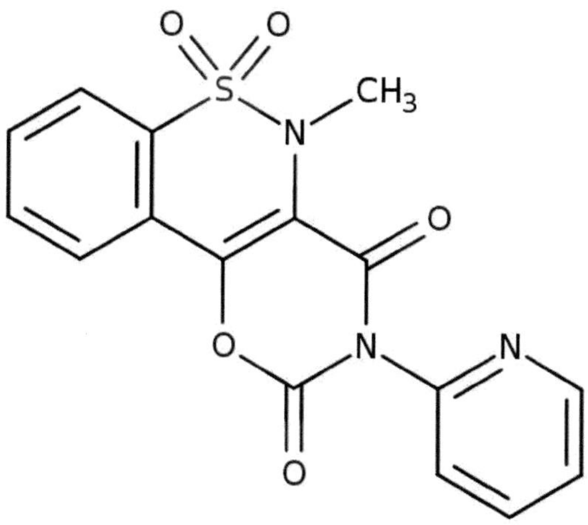

GENERAL

Droxicam is an oxicam non-steroidal anti-inflammatory drug and a prodrug of piroxicam, which it is rapidly converted by hydrolysis in the gastrointestinal tract. It is used to reduce pain and inflammation in musculoskeletal disorders such as rheumatoid arthritis and osteoarthritis (1,2).

CUTANEOUS ADVERSE DRUG REACTIONS FROM SYSTEMIC ADMINISTRATION CAUSED BY TYPE IV (DELAYED-TYPE) HYPERSENSITIVITY (as demonstrated by positive patch tests)

Cutaneous adverse drug reactions from systemic administration of droxicam caused by type IV (delayed-type) hypersensitivity have included photosensitivity (2,4).

Photosensitivity

A 48-year-old woman presented with a pruritic, erythematous vesicular eruption on the face, neck, arms, and the dorsal surfaces of both hands and feet. This eruption had appeared after she had taken droxicam, 20 mg twice daily, for 3 days. Histologic examination of a neck lesion revealed a superficial spongiotic dermatitis. Rapid improvement occurred one week after withdrawal of the drug. Patch tests were positive to thiomersal and thiosalicylic acid. Photopatch tests were positive to 1% droxicam and 1% piroxicam in petrolatum (2).

A 51-year-old patient, after an oral intake of 20 mg of droxicam for 2 days, presented with a vesicular erythematous dermatitis of the hands and wrists. Similar lesions with slight edema were also present, although less intense, on the forehead, ears, back, and lateral aspects of the neck. The patient explained that he had also had a blistering erythematous dermatitis of the hands and ears of much lesser degree after taking 20 mg of piroxicam. Patch and photopatch tests were performed, using UVA 10 J/cm² for irradiation. On the nonirradiated side there were positive reactions to droxicam at 5%, 1% and 0.5% pet., to piroxicam, to thiosalicylic acid and to thimerosal. On the irradiated side, all reactions to droxicam and piroxicam were much more intense, with considerable serous exudation and marked pruritus. The patient was diagnosed with contact allergy and photoallergy to droxicam and piroxicam with a dyshidrotic pattern on the hands (4).

It has been well established that many patients who are sensitized to thiosalicylic acid (one of the components of the preservative thimerosal) develop a photoallergic eruption after taking oral piroxicam and most patients showing a photosensitivity reaction to oral piroxicam have positive patch tests to thimerosal (Chapter 3.392 Piroxicam).

Cross-reactions, pseudo-cross-reactions and co-reactions
Photocross-reaction from piroxicam to droxicam (2,3,4).

LITERATURE

1   The data in the section 'General' may have been obtained from literature discussed in this chapter, but mostly also or exclusively from one or more of the following online sources: ChemIDPlus Advanced, PubChem, DrugBank, RxList, Drug Central, Drugs.com, and Wikipedia

2   Serrano G, Fortea JM, Latasa JM, SanMartin O, Bonillo J, Miranda MA. Oxicam-induced photosensitivity. Patch and photopatch testing studies with tenoxicam and piroxicam photoproducts in normal subjects and in piroxicam-droxicam photosensitive patients. J Am Acad Dermatol 1992;26:545-548.

3   Trujillo MJ, de Barrio M, Rodríguez A, Moreno-Zazo M, Sánchez I, Pelta R, Tornero P, Herrero T. Piroxicam-induced photodermatitis. Cross-reactivity among oxicams. A case report. Allergol Immunopathol (Madr) 2001;29:133-136

4   Anonide A, Usiglio D, Pestarina A, Massone L. Droxicam photosensitivity with dyshidrotic hand dermatitis. Int J Dermatol 1997;36:318-320

Chapter 3.190 DUTASTERIDE

IDENTIFICATION

Description/definition : Dutasteride is 4-azasteroid that conforms to the structural formula shown below
Pharmacological classes : 5-alpha Reductase inhibitors
IUPAC name : (1S,3aS,3bS,5aR,9aR,9bS,11aS)-N-[2,5-bis(Trifluoromethyl)phenyl]-9a,11a-dimethyl-7-oxo-1,2,3,3a,3b,4,5,5a,6,9b,10,11-dodecahydroindeno[5,4-f]quinoline-1-carboxamide
CAS registry number : 164656-23-9
EC number : Not available
Merck Index monograph : 4788
Patch testing : 0.05% alc.
Molecular formula : $C_{27}H_{30}F_6N_2O_2$

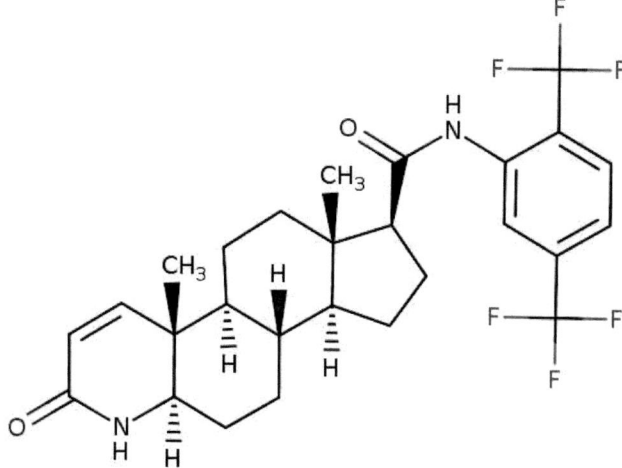

GENERAL

Dutasteride is a synthetic 4-azasteroid compound that selectively inhibits both the type I and type II isoforms of steroid 5α-reductase, an intracellular enzyme that converts testosterone to 5α-dihydrotestosterone (DHT). Dutasteride works by reducing the levels of circulating DHT. It was also shown to reduce the size of the prostate gland, improve urinary flow, and symptoms of benign prostatic hyperplasia alone or in combination with tamsulosin. Off-label, dutasteride may be used for the treatment of alopecia androgenetica (in which dihydrotestosterone is a key mediator), both orally and topically (the latter often in combination with minoxidil) (1).

CUTANEOUS ADVERSE DRUG REACTIONS FROM SYSTEMIC ADMINISTRATION CAUSED BY TYPE IV (DELAYED-TYPE) HYPERSENSITIVITY (as demonstrated by positive patch tests)

Cutaneous adverse drug reactions from systemic administration of dutasteride caused by type IV (delayed-type) hypersensitivity have included angioedema-like contact dermatitis caused by mesotherapy (2).

Angioedema-like contact dermatitis caused by mesotherapy

A 45 year-old woman developed marked facial swelling and erythema one day after a first mesotherapy session (intralesional subcutaneous injection) with dutasteride for androgenetic alopecia. Patch tests with the ingredients of the solution were positive to dutasteride (0.001%, 0.01%, and 0.05% alc.), and propylene glycol water 20% water on D2 and D4. Twenty controls were patch tested with dutasteride 0.001%, 0.01% and 0.05%, with negative results. It was not mentioned to which concentration(s) of dutasteride the patient reacted (only one was shown). The reactions were described as 'strong' (++), but this was, judging the picture of the positive tests, rather exaggerated (2).

LITERATURE

1 The data in the section 'General' may have been obtained from literature discussed in this chapter, but mostly also or exclusively from one or more of the following online sources: ChemIDPlus Advanced, PubChem, DrugBank, RxList, Drug Central, Drugs.com, and Wikipedia
2 Magdaleno-Tapial J, Valenzuela-Oñate C, García-Legaz-Martínez M, Martínez-Domenech Á, Alonso-Carpio M, Talamantes CS, et al. Angioedema-like contact dermatitis caused by mesotherapy with dutasteride. Contact Dermatitis 2020;83:246-247

Chapter 3.191 EFAVIRENZ

IDENTIFICATION

Description/definition   : Efavirenz is the benzoxazine compound that conforms to the structural formula shown
                               below
Pharmacological classes   : Cytochrome P-450 CYP2B6 inducers; cytochrome P-450 CYP2C9 inhibitors; cytochrome P-
                               450 CYP2C19 inhibitors; cytochrome P-450 CYP3A inducers; reverse transcriptase
                               inhibitors
IUPAC name   : (4S)-6-Chloro-4-(2-cyclopropylethynyl)-4-(trifluoromethyl)-1H-3,1-benzoxazin-2-one
CAS registry number   : 154598-52-4
EC number   : Not available
Merck Index monograph   : 4839
Patch testing   : Capsule content, 50% water (2); most systemic drugs can be tested at 10% pet.; if the
                               pure chemical is not available, prepare the test material from intravenous powder, the
                               content of capsules or – when also not available – from powdered tablets to achieve a
                               final concentration of the active drug of 10% pet.
Molecular formula   : $C_{14}H_9ClF_3NO_2$

GENERAL

Efavirenz is a synthetic non-nucleoside reverse transcriptase (RT) inhibitor with antiviral activity. Efavirenz binds directly to the human immunodeficiency virus type 1 (HIV-1) RT, an RNA-dependent DNA polymerase, blocking its function in viral DNA replication. In combination with other antiretroviral drugs, this agent has been shown to significantly reduce HIV viral load, retarding or preventing damage to the immune system and reducing the risk of developing AIDS in patients infected with human immunodeficiency virus (HIV) type 1 (1).

CUTANEOUS ADVERSE DRUG REACTIONS FROM SYSTEMIC ADMINISTRATION CAUSED BY TYPE IV (DELAYED-TYPE) HYPERSENSITIVITY (as demonstrated by positive patch tests)

Cutaneous adverse drug reactions from systemic administration of efavirenz caused by type IV (delayed-type) hypersensitivity have included photosensitivity (2,3).

Photosensitivity

A 61-year-old man was treated for HIV with 3 antiviral medications including efavirenz. After 2 months, the patient easily became sunburned. Physical examination revealed slightly edematous erythema on his face, exfoliative erythema on the ears, mild erythema on his neck and severe erythema on both forearms and the rash was limited to sun-exposed areas. Photopatch tests with efavirenz in petrolatum (concentration?) were positive at all 3 sites irradiated with 5, 3 and 1.5 J/cm² UVA. Controls were not performed. 'Photosensitivity' was diagnosed; it was not stated whether this was considered photoallergy or photosensitivity (2).

   A 66-year-old man was treated with efavirenz and other antivirals for HIV, when he developed a photodistributed skin rash. Physical examination showed erythematous, partly vesicular and squamous pruritic lesions on his hands, forearms, neck and face, where an edematous swelling was also present. The mucous membranes and lymph nodes were not involved. Histology from the hand showed mild acanthosis, focal spongiosis and strong subepidermal edema, partly accentuated to mild blistering. UV testing showed hypersensitivity with a minimal dose of erythema of 10 J/cm² UVA, and 30 mJ/cm² UVB. Photopatch testing with all drugs using 1 J/cm² UVA showed a positive skin reaction to the powder content of the efavirenz capsule tested 50% in water at day 3 with an erythematous infiltrate and discrete papulovesicles. Three controls were negative. Histology from the photopatch test reaction showed focal spongiosis, intracellular edema and perivascular lymphocytic infiltrates. The patient was diagnosed with photoallergic dermatitis from efavirenz (3).

LITERATURE

1   The data in the section 'General' may have been obtained from literature discussed in this chapter, but mostly also or exclusively from one or more of the following online sources: ChemIDPlus Advanced, PubChem, DrugBank, RxList, Drug Central, Drugs.com, and Wikipedia

2   Yoshimoto E, Konishi M, Takahashi K, Murukawa K, Maeda K, Mikasa K, et al. The first case of efavirenz-induced photosensitivity in a Japanese patient with HIV infection. Intern Med 2004;43:630-631

3   Treudler R, Husak R, Raisova M, Orfanos CE, Tebbe B. Efavirenz-induced photoallergic dermatitis in HIV. AIDS 2001;15:1085-1086

Chapter 3.192 EMTRICITABINE

IDENTIFICATION

Description/definition   : Emtricitabine is the deoxycytidine analog and reverse transcriptase inhibitor that
                            conforms to the structural formula shown below
Pharmacological classes   : Anti-HIV agents; reverse transcriptase inhibitors
IUPAC name   : 4-Amino-5-fluoro-1-[(2R,5S)-2-(hydroxymethyl)-1,3-oxathiolan-5-yl]pyrimidin-2-one
CAS registry number   : 143491-57-0
EC number   : 604-363-1
Merck Index monograph   : 4892
Patch testing   : Tablet, pulverized, 20% pet. (2); most systemic drugs can be tested at 10% pet.; if the pure
                            chemical is not available, prepare the test material from intravenous powder, the content
                            of capsules or – when also not available – from powdered tablets to achieve a final
                            concentration of the active drug of 10% pet.
Molecular formula   : $C_8H_{10}FN_3O_3S$

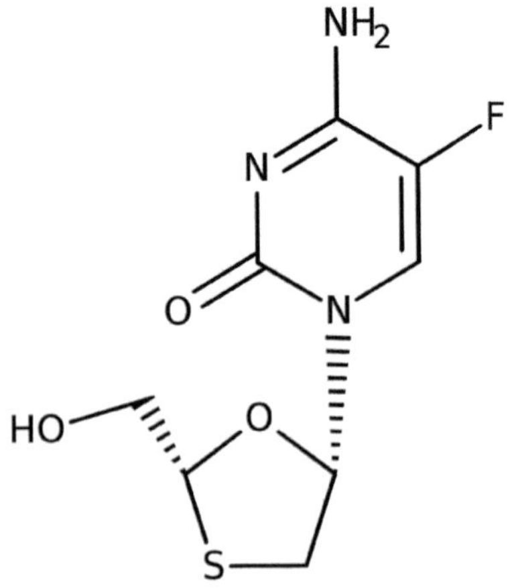

GENERAL

Emtricitabine is a synthetic nucleoside analog and reverse transcriptase inhibitor with antiviral activity against
human immunodeficiency virus type 1 (HIV-1) and hepatitis B viruses. This agent is indicated, in combination with
other antiretroviral drugs, for the treatment of HIV-1 infection in adults and for postexposure prophylaxis of HIV
infection in health care workers and others exposed occupationally or nonoccupationally via percutaneous injury or
mucous membrane or nonintact skin contact with blood, tissues, or other body fluids associated with risk for
transmission of the virus (1).

CUTANEOUS ADVERSE DRUG REACTIONS FROM SYSTEMIC ADMINISTRATION CAUSED BY TYPE IV
(DELAYED-TYPE) HYPERSENSITIVITY (as demonstrated by positive patch tests)

Cutaneous adverse drug reactions from systemic administration of emtricitabine caused by type IV (delayed-type)
hypersensitivity have included Stevens-Johnson syndrome (2) and unspecified exanthema with palpebral edema (3)

Stevens-Johnson syndrome/toxic epidermal necrolysis (SJS/TEN)

A 54-year-old woman diagnosed with HIV infection started treatment with emtricitabine/tenofovir tablets in
combination with nevirapine. An itchy rash on the neckline, abdomen and limbs appeared 8 h after the first dose.
One day after this reaction, the patient took another dose, and a new rash with extensive skin scaling appeared,
accompanied by oral and genital ulceration, fever, hepatitis, diarrhea, and loss of consciousness. The patient was
admitted to hospital and discharged two weeks later with a diagnosis of Stevens-Johnson syndrome. Nine months
later, she suffered from three milder episodes after emtricitabine/tenofovir with etravirine, abacavir/lamivudine
with raltegravir, and emtricitabine/tenofovir with raltegravir. Patch tests with all drugs (powdered pills 5%, 10% and
20% pet.) gave positive reactions to the 20% concentrations of emtricitabine/tenofovir, abacavir/lamivudine, and
lamivudine. Emtricitabine itself was not tested, but tenofovir tested negative. Ten controls were negative.

The chemical structure of emtricitabine is almost identical to that of lamivudine, but rather different from the
remaining drugs given, which makes, according to the authors, a cross-reactivity mechanism likely (2).

Other cutaneous adverse drug reactions
A 47-year-old with diagnosis of HIV infection started therapy with tenofovir, emtricitabine and nevirapine. On the second day of treatment, she developed pruritic exanthema and palpebral edema that improved two weeks after the treatment had been discontinued. Patch tests were positive to emtricitabine 1%, 10% and 30% pet. and tenofovir 10% and 30% pet. Seven controls were negative (3).

Cross-reactions, pseudo-cross-reactions and co-reactions
Cross-reactivity from emtricitabine sensitization to lamivudine (2).

LITERATURE

1   The data in the section 'General' may have been obtained from literature discussed in this chapter, but mostly also or exclusively from one or more of the following online sources: ChemIDPlus Advanced, PubChem, DrugBank, RxList, Drug Central, Drugs.com, and Wikipedia
2   Suárez-Lorenzo I, Castillo-Sainz R, Cárden-Santana MA, Carrillo-Díaz T. Severe reaction to emtricitabine and lamiduvine: evidence of cross-reactivity. Contact Dermatitis 2016;74:253-254
3   Sousa MJ, Cadinha S, Mota M, Teixeira T, Malheiro D, Moreira da Silva JP. Hypersensitivity to antiretroviral drugs. Eur Ann Allergy Clin Immunol 2018;50:277-280

Chapter 3.193 ENALAPRIL

IDENTIFICATION

Description/definition : Enalapril is the dicarbonyl-containing peptide that conforms to the structural formula shown below
Pharmacological classes : Antihypertensive agents; angiotensin-converting enzyme inhibitors
IUPAC name : (2S)-1-[(2S)-2-[[(2S)-1-Ethoxy-1-oxo-4-phenylbutan-2-yl]amino]propanoyl]pyrrolidine-2-carboxylic acid
CAS registry number : 75847-73-3
EC number : 616-271-9
Merck Index monograph : 4893
Patch testing : Tablet, pulverized, 10% pet. (2); most systemic drugs can be tested at 10% pet.; if the pure chemical is not available, prepare the test material from intravenous powder, the content of capsules or – when also not available – from powdered tablets to achieve a final concentration of the active drug of 10% pet.
Molecular formula : $C_{20}H_{28}N_2O_5$

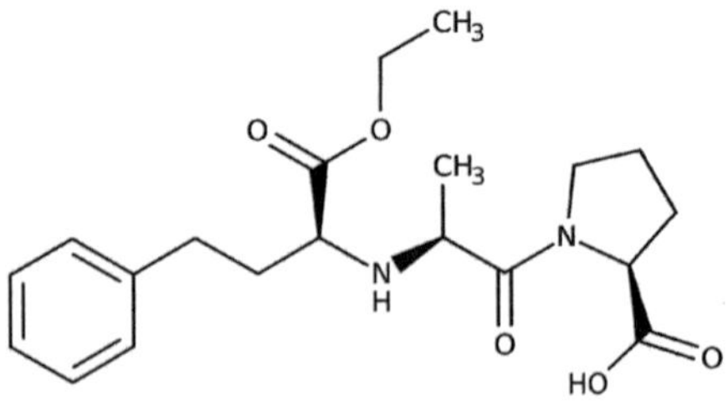

Enalapril

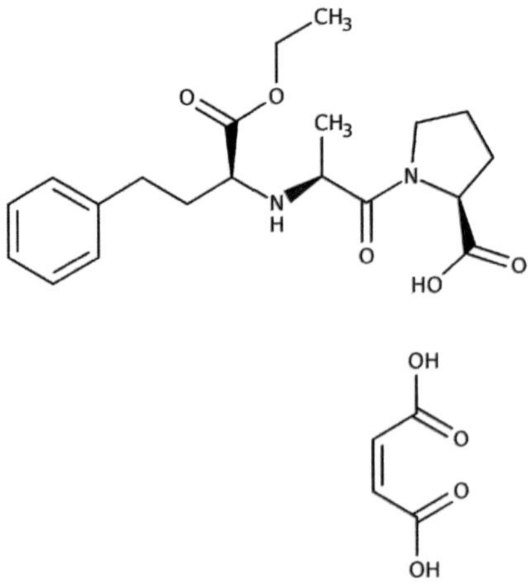

Enalapril maleate

GENERAL

Enalapril is a dicarbonyl-containing peptide and angiotensin-converting enzyme (ACE) inhibitor with antihypertensive activity. As a prodrug, following oral administration, this agent is converted by de-esterification into its active form enalaprilat. Enalapril is used for the treatment of essential or renovascular hypertension and symptomatic congestive heart failure, either alone or in combination with thiazide diuretics (1). In pharmaceutical products, enalapril is employed as enalapril maleate (CAS number 76095-16-4, EC number 278-375-7, molecular formula $C_{24}H_{32}N_2O_9$) (1).

CONTACT ALLERGY FROM ACCIDENTAL CONTACT

A 24-year-old woman presented with eyelid dermatitis, which had started with localized edema 4 months previously. Later, the area had become itchier, with redness and scaling. The patient suspected a relationship with her work as a pharmacy assistant, which involved breaking and crushing different types of tablets. She was patch tested with the crushed tablets that she had contact with at 10% pet. and showed positive reactions to enalapril, 2 other ACE-inhibitors, 4 beta-blockers, and 3 benzodiazepines. Three controls were negative to enalapril 10% pet. Cosmetic allergy was excluded and a diagnosis of occupational airborne allergic contact was made (2).

LITERATURE

1 The data in the section 'General' may have been obtained from literature discussed in this chapter, but mostly also or exclusively from one or more of the following online sources: ChemIDPlus Advanced, PubChem, DrugBank, RxList, Drug Central, Drugs.com, and Wikipedia
2 Swinnen I, Ghys K, Kerre S, Constandt L, Goossens A. Occupational airborne contact dermatitis from benzodiazepines and other drugs. Contact Dermatitis 2014;70:227-232

Chapter 3.194 ENOXACIN

IDENTIFICATION

Description/definition : Enoxacin is the 6-fluoronaphthyridinone that conforms to the structural formula shown below

Pharmacological classes : Antibacterial agents; cytochrome P-450 CYP1A2 inhibitors; topoisomerase II inhibitors

IUPAC name : 1-Ethyl-6-fluoro-4-oxo-7-piperazin-1-yl-1,8-naphthyridine-3-carboxylic acid

CAS registry number : 74011-58-8

EC number : Not available

Merck Index monograph : 4911

Patch testing : No data available; most systemic drugs can be tested at 10% pet.; if the pure chemical is not available, prepare the test material from intravenous powder, the content of capsules or – when also not available – from powdered tablets to achieve a final concentration of the active drug of 10% pet.

Molecular formula : $C_{15}H_{17}FN_4O_3$

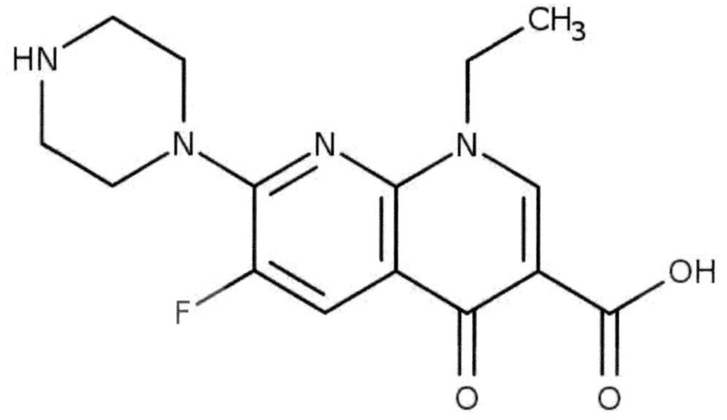

GENERAL

Enoxacin is a broad-spectrum 6-fluoronaphthyridinone antibacterial agent and DNA synthesis inhibitor, that is structurally related to nalidixic acid. It is a broad-spectrum antibiotic that is active against both gram-positive and gram-negative bacteria. Enoxacin may be active against pathogens resistant to drugs that act by different mechanisms. This antibiotic is used in the treatment of urinary tract infections and gonorrhea caused by susceptible strains of microorganisms (1).

CUTANEOUS ADVERSE DRUG REACTIONS FROM SYSTEMIC ADMINISTRATION CAUSED BY TYPE IV (DELAYED-TYPE) HYPERSENSITIVITY (as demonstrated by positive patch tests)

Cutaneous adverse drug reactions from systemic administration of enoxacin caused by type IV (delayed-type) hypersensitivity have included photosensitivity (2).

Photosensitivity

Two patients in Japan with systemic photosensitivity from enoxacin had positive photopatch tests to the drug (2, details unknown, article in Japanese, data cited in ref. 3).

LITERATURE

1   The data in the section 'General' may have been obtained from literature discussed in this chapter, but mostly also or exclusively from one or more of the following online sources: ChemIDPlus Advanced, PubChem, DrugBank, RxList, Drug Central, Drugs.com, and Wikipedia

2   Sugiura K, Tsuruta K, Noda H, Morita K, Umemura Y. Photo-induced drug eruption from enoxacin. Rinsho Derma 1987;29:151-154 (Article in Japanese; data cited in ref. 3)

3   Kurumaji Y, Shono M. Scarified photopatch testing in lomefloxacin photosensitivity. Contact Dermatitis 1992;26:5-10

Chapter 3.195 ENOXAPARIN

IDENTIFICATION

Description/definition	: Enoxaparin is a low molecular weight heparin with a molecular weight ranging from 3800 to 5000 daltons
Pharmacological classes	: Antithrombotic agents
IUPAC name	: Not available
CAS registry number	: 9005-49-6 (Heparin)
EC number	: 232-681-7 (Heparin)
Merck Index monograph	: 5958 (Heparin)
Patch testing	: Commercial preparation undiluted (3); consider intradermal testing with late readings (D2,D3) when patch tests are negative and consider subcutaneous challenge when intradermal tests are negative
Molecular formula	: Unspecified

GENERAL

Enoxaparin is a low molecular weight heparin, the molecular weight ranging from 3800 to 5000 daltons. Enoxaparin is used to prevent and treat deep vein thrombosis or pulmonary embolism, and is given as a subcutaneous injection. Enoxaparin binds to and accelerates the activity of antithrombin III. By activating antithrombin III, enoxaparin preferentially potentiates the inhibition of coagulation factors Xa and IIa. Factor Xa catalyzes the conversion of prothrombin to thrombin, so enoxaparin's inhibition of this process results in decreased thrombin and ultimately the prevention of fibrin clot formation. In pharmaceutical products, enoxaparin is employed as enoxaparin sodium (CAS number 679809-58-6, EC number not available, molecular formula unspecified) (1).

See also bemiparin (Chapter 3.59), certoparin (Chapter 3.117), dalteparin (Chapter 3.160), danaparoid (Chapter 3.161), fondaparinux (Chapter 3.225), heparins (Chapter 3.239), nadroparin (Chapter 3.331), and tinzaparin (Chapter 3.479).

CUTANEOUS ADVERSE DRUG REACTIONS FROM SYSTEMIC ADMINISTRATION CAUSED BY TYPE IV (DELAYED-TYPE) HYPERSENSITIVITY

Throughout this book, only reports of delayed-type hypersensitivity have been included that showed a positive patch test to the culprit drug. However, as a result of the high molecular weight of heparins, patch tests are often false-negative, presumably from insufficient penetration into the skin. Because of this, and also because patch tests have been performed in a small minority of cases only, studies with a positive intradermal test or subcutaneous provocation tests with delayed readings are included in the chapters on the various heparins, even when patch tests were negative or not performed.

General information on delayed-type hypersensitivity reactions to heparins

General information on delayed-type hypersensitivity reactions to heparins (including enoxaparin, presenting as local reactions from subcutaneous administration, is provided in Chapter 3.239 Heparins. In this chapter, only non-local cutaneous adverse drug reactions from delayed-type hypersensitivity to enoxaparin are presented.

Non-local cutaneous adverse drug reactions

Non-local cutaneous adverse drug reactions from systemic administration of enoxaparin caused by type IV (delayed-type) hypersensitivity have included maculopapular eruption (6,10,11 [delayed-type hypersensitivity not proven],12,14), acute generalized exanthematous pustulosis (AGEP) (2,16), drug reaction with eosinophilia and systemic symptoms (DRESS) (13; likely, but not proven),15,16) and generalized eczema (7,8,17).

Maculopapular eruption

A man aged 49 was treated with subcutaneous enoxaparin. Already after the first injection localized erythematous plaques developed at the injection site. After the third injection, a maculopapular exanthema developed on the trunk. Allergy tests were negative to enoxaparin, but positive to tinzaparin and danaparoid (10). A 50-year-old woman had been treated with enoxaparin for one month, when she developed a maculopapular exanthema. Intradermal tests were positive to enoxaparin and many other heparins (6).

A 47-year-old man developed a maculopapular rash after immediate-type local and generalized urticarial reactions following intravenous heparin sodium and subcutaneous enoxaparin sodium and nadroparin calcium. Patch tests were negative as were prick tests read at 15 and 30 minutes. Intradermal tests to all 3 heparins were positive after 30 minutes; the reactions had fully subsided after 8 hours. It was not mentioned whether the reactions were also read at 2 and 3 days. Type I hypersensitivity was proven, a co-existent type IV reactions, possibly explaining the maculopapular eruption, not (11).

A 61-year-old woman developed a maculopapular exanthema 2 hours after a subcutaneous injection of enoxaparin sodium. In the past, she had been treated with nadroparin calcium. Patch tests were positive to enoxaparin sodium, nadroparin calcium and dalteparin sodium. The patient had probably become sensitized to nadroparin and now cross-reacted to enoxaparin, causing the maculopapular rash (12).

Another case of maculopapular eruption from subcutaneous enoxaparin was described in the USA in 2003 (14; details not available to the author).

Acute generalized exanthematous pustulosis (AGEP)

Two patients with acute generalized exanthematous pustulosis were reported from France in 2020 (2). A 74-year-old woman was referred for a sudden, diffuse febrile rash followed by a superficial post-pustular desquamation as shown by the patient's own photographs. The diagnosis was strongly suggestive of acute generalized exanthematous pustulosis, 3 days after she had undergone colonic surgery. The suspected drugs were pantoprazole, paracetamol, famotidine, phloroglucinol, ketoprofen and enoxaparin. Case 2 was also a 74-year-old woman, who presented with a diffuse febrile erythematous rash with non-follicular pinhead-sized pustules accentuated in the large body folds and associated with leukocytosis. The diagnosis of AGEP was suspected. Later, patch tests with the commercialized pills of the suspected drugs diluted at 30% in petrolatum and with several undiluted heparins were negative. However, intradermal tests with heparins diluted from 10^{-3} to 10^{-1} with reading on D3 were positive to enoxaparin with co-reactivity other heparins (heparin calcium, heparin sodium, tinzaparin, danaparoid). Neither patient reported onset or aggravation of the eruption at the site of the enoxaparin injection (2).

In a multicenter investigation in France, of 45 patients patch tested for AGEP, 26 (58%) had positive patch tests to drugs, including one to enoxaparin (16).

Drug reaction with eosinophilia and systemic symptoms (DRESS)

Subcutaneous administration of enoxaparin in one patient may have caused DRESS, but skin tests to verify the existence of delayed-type hypersensitivity were not performed (13). Of 13 pediatric patients patch tested for DRESS in a tertiary care hospital in Florence, Italy, between 2010 and 2018, 5 (38%) had positive reactions, one of who reacted to enoxaparin (15). In a multicenter investigation in France, of 72 patients patch tested for DRESS, 46 (64%) had positive patch tests to drugs, including one to enoxaparin (16).

Dermatitis/eczematous eruption

A 39-year-old woman developed a generalized eczema secondary to a local allergic reaction caused by subcutaneous enoxaparin sodium. Patch tests were positive to enoxaparin sodium with cross-reactions to dalteparin sodium and heparin sodium (7). In a hospital in Tenon, France, between 2000 and 2012, 19 patients were observed with hypersensitivity reactions to heparins. Six of these had generalized eczema, caused by enoxaparin sodium in 5 individuals and by nadroparin calcium in one (8).

In a study from Nancy, France, 54 patients with suspected nonimmediate drug eruptions were assessed with patch testing. Of the 9 patients with generalized eczema, one had a positive patch test to enoxaparin (17).

Cross-reactions, pseudo-cross-reactions and co-reactions

Cross-reactions between heparins are frequent in delayed-type hypersensitivity (>90% of patients tested, median number of positive drugs per patient: 3) and do not depend on the molecular weight of the molecules (9). Overlap in their polysaccharide composition might explain the high degree of cross-allergenicity (4). Cross-reactions to the semisynthetic heparinoid danaparoid have also been observed (5). In allergic patients, the synthetic ultralow molecular weight synthetic heparin fondaparinux is usually, but not always (5,9) well-tolerated (4).

LITERATURE

1   The data in the section 'General' may have been obtained from literature discussed in this chapter, but mostly also or exclusively from one or more of the following online sources: ChemIDPlus Advanced, PubChem, DrugBank, RxList, Drug Central, Drugs.com, and Wikipedia
2   Assier H, Gener G, Chosidow O, Wolkenstein P, Ingen-Housz-Oro S. Acute generalized exanthematous pustulosis induced by enoxaparin: 2 cases. Contact Dermatitis 2021;84:280-282
3   Brockow K, Garvey LH, Aberer W, Atanaskovic-Markovic M, Barbaud A, Bilo MB, et al. Skin test concentrations for systemically administered drugs - an ENDA/EAACI Drug Allergy Interest Group position paper. Allergy 2013;68:702-712
4   Schindewolf M, Lindhoff-Last E, Ludwig RJ. Heparin-induced skin lesions. Lancet 2012;380:1867-1879
5   Utikal J, Peitsch WK, Booken D, Velten F, Dempfle CE, Goerdt S, et al. Hypersensitivity to the pentasaccharide fondaparinux in patients with delayed-type heparin allergy. Thromb Haemost 2005;94:895-896

6   Lopez S, Torres MJ, Rodríguez-Pena R, Blanca-Lopez N, Fernandez TD, Antunez C, et al. Lymphocyte proliferation response in patients with delayed hypersensitivity reactions to heparins. Br J Dermatol 2009;160:259-265

7   Tramontana M, Hansel K, Bianchi L, Agostinelli D, Stingeni L. Skin tests in patients with delayed and immediate hypersensitivity to heparins: A case series. Contact Dermatitis 2019;80:170-172

8   Phan C, Vial-Dupuy A, Autegarden J-E, Amsler E, Gaouar H, Abuaf N, et al. A study of 19 cases of allergy to heparins with positive skin testing. Ann Dermatol Venereol 2014;141:23-29 (Article in French)

9   Weberschock T, Meister AC, Bohrt K, Schmitt J, Boehncke W-H, Ludwig RJ. The risk for cross-reactions after a cutaneous delayed-type hypersensitivity reaction to heparin preparations is independent of their molecular weight: a systematic review. Contact Dermatitis 2011;65:187-194

10 Figarella I, Barbaud A, Lecompte T, De Maistre E, Reichert-Penetrat S, Schmutz J L. Cutaneous delayed hypersensitivity reactions to heparins and heparinoids. Ann Dermatol Venereol 2001;128:25-30 (Article in French)

11 Klos K, Spiewak R, Kruszewski J, Bant A. Cutaneous adverse drug reaction to heparins with hypereosinophilia and high IgE level. Contact Dermatitis 2011;64:61-62

12 Colagiovanni A, Rizzi A, Buonomo A, De Pasquale T, Pecora V, Sabato V, et al. Delayed hypersensitivity to heparin in a patient with cancer: fondaparinux may be safe but needs to be tested. Contact Dermatitis 2010;63:107-108

13 Ronceray S, Dinulescu M, Le Gall F, Polard E, Dupuy A, Adamski H. Enoxaparin-induced DRESS syndrome. Case Rep Dermatol 2012;4:233-237

14 Kim KH, Lynfield Y. Enoxaparin-induced generalized exanthema. Cutis 2003;72:57-60

15 Liccioli G, Mori F, Parronchi P, Capone M, Fili L, Barni S, Sarti L, Giovannini M, Resti M, Novembre EM. Aetiopathogenesis of severe cutaneous adverse reactions (SCARs) in children: A 9-year experience in a tertiary care paediatric hospital setting. Clin Exp Allergy 2020;50:61-73

16 Barbaud A, Collet E, Milpied B, Assier H, Staumont D, Avenel-Audran M, et al. A multicentre study to determine the value and safety of drug patch tests for the three main classes of severe cutaneous adverse drug reactions. Br J Dermatol 2013;168:555-562

17 Barbaud A, Reichert-Penetrat S, Tréchot P, Jacquin-Petit MA, Ehlinger A, Noirez V, et al. The use of skin testing in the investigation of cutaneous adverse drug reactions. Br J Dermatol 1998;139:49-58

Chapter 3.196 EPERISONE

IDENTIFICATION

Description/definition   : Eperisone is alkylphenylketone that conforms to the structural formula shown below
Pharmacological classes   : Muscle relaxants, central; anticonvulsants; vasodilator agents; parasympatholytics; calcium channel blockers
IUPAC name   : 1-(4-Ethylphenyl)-2-methyl-3-piperidin-1-ylpropan-1-one
CAS registry number   : 64840-90-0
EC number   : Not available
Merck Index monograph : 4931
Patch testing   : 1%, 10% and 30% pet. (probably crushed tablets); most systemic drugs can be tested at 10% pet.; if the pure chemical is not available, prepare the test material from intravenous powder, the content of capsules or – when also not available – from powdered tablets to achieve a final concentration of the active drug of 10% pet.
Molecular formula   : $C_{17}H_{25}NO$

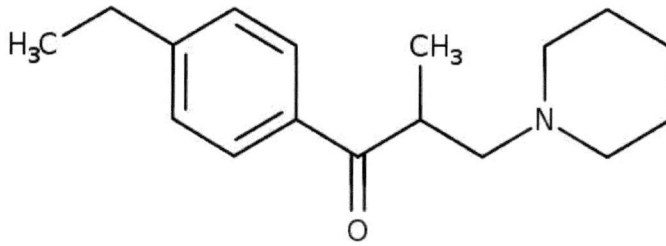

GENERAL

Eperisone is an antispasmodic drug which relaxes both skeletal muscles and vascular smooth muscles, and demonstrates a variety of effects such as reduction of myotonia, improvement of circulation, and suppression of the pain reflex. In pharmaceutical products, eperisone is employed as eperisone hydrochloride (CAS number 56839-43-1, EC number not available, molecular formula $C_{17}H_{26}ClNO$).

CUTANEOUS ADVERSE DRUG REACTIONS FROM SYSTEMIC ADMINISTRATION CAUSED BY TYPE IV (DELAYED-TYPE) HYPERSENSITIVITY (as demonstrated by positive patch tests)

Cutaneous adverse drug reactions from systemic administration of eperisone caused by type IV (delayed-type) hypersensitivity have included acute generalized exanthematous pustulosis (AGEP) (2).

Acute generalized exanthematous pustulosis (AGEP)

A 69-year-old woman was admitted with acute widespread pustular erythema, fever (39°C), and general malaise. She had been treated with eperisone hydrochloride and two other drugs. Physical examination demonstrated diffuse erythema with numerous non-follicular pustules of 3-5 mm in diameter on her abdomen, back, and buttocks. Laboratory examinations showed an elevated white blood cell count with neutrophilia and elevated C-reactive protein. Bacterial cultures from pustules were negative. Histopathology of a skin biopsy specimen from a pustular lesion revealed subcorneal and intraepidermal neutrophilic pustules, polymorphous perivascular infiltrates with eosinophils, and a focal necrosis of keratinocytes in the epidermis. A tentative diagnosis of acute generalized exanthematous pustulosis (AGEP) was made and all drugs were discontinued. Later, patch tests with eperisone hydrochloride and the other 2 drugs the patient had used were performed at decreasing concentrations of 30 to 0.1% petrolatum. Only eperisone hydrochloride showed positive results at the sites of 30% to 1% concentration. The positive patch test reactions to the higher concentrations (10% and 30%) showed pustules (2).

LITERATURE

1   The data in the section 'General' may have been obtained from literature discussed in this chapter, but mostly also or exclusively from one or more of the following online sources: ChemIDPlus Advanced, PubChem, DrugBank, RxList, Drug Central, Drugs.com, and Wikipedia
2   Yamamoto Y, Kadota M, Nishimura Y. A case of eperisone hydrochloride-induced acute generalized exanthematous pustulosis. J Dermatol 2004;31:769-770

Chapter 3.197 EPHEDRINE

IDENTIFICATION

Description/definition	: Ephedrine is the phenethylamine alkaloid that conforms to the structural formula shown below
Pharmacological classes	: Vasoconstrictor agents; central nervous system stimulants; adrenergic agents; sympathomimetics
IUPAC name	: (1R,2S)-2-(Methylamino)-1-phenylpropan-1-ol
CAS registry number	: 299-42-3
EC number	: 206-080-5
Merck Index monograph	: 4933
Patch testing	: 10% pet. or water (8)
Molecular formula	: $C_{10}H_{15}NO$

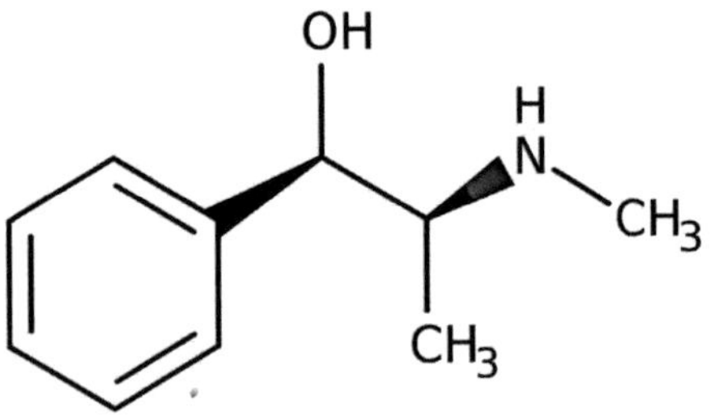

GENERAL

Ephedrine is an alkaloid and hydroxylated form of phenethylamine which is found in the plant *Ephedra sinica* and various other plants in the genus *Ephedra*. It is an α- and β-adrenergic agonist that may also enhance release of norepinephrine. Following administration, ephedrine activates post-synaptic noradrenergic receptors. Activation of α-adrenergic receptors in the vasculature induces vasoconstriction, and activation of β-adrenergic receptors in the lungs leads to bronchodilation. Ephedrine is commonly used as a stimulant, appetite suppressant, concentration aid, decongestant, and to treat hypotension associated with anesthesia. In pharmaceutical products, ephedrine is employed as ephedrine sulfate (CAS number 134-72-5, EC number 205-154-4, molecular formula $C_{20}H_{32}N_2O_6S$) or ephedrine hydrochloride (CAS number 50-98-6, EC number 200-074-6, molecular formula $C_{10}H_{16}ClNO$) (1). Formerly, ephedrine was also used in topical applications, which has caused cases of contact allergy/allergic contact dermatitis and which has been fully reviewed in Volume 3 of the *Monographs in contact allergy* series (8).

CUTANEOUS ADVERSE DRUG REACTIONS FROM SYSTEMIC ADMINISTRATION CAUSED BY TYPE IV (DELAYED-TYPE) HYPERSENSITIVITY (as demonstrated by positive patch tests)

Cutaneous adverse drug reactions from systemic administration of ephedrine caused by type IV (delayed-type) hypersensitivity have included maculopapular eruption (2), erythroderma (3), fixed drug eruption (7) and 'itchy dermatosis' (5).

Maculopapular eruption

A 74-year-old man had repeated episodes of generalized maculopapular rash following the last 2 of 3 endoscopic surgeries for inverted papilloma. Patch tests with all drugs used in the third intervention were positive to ephedrine 50 mg/ml. An intradermal test with ephedrine (0.05 mg/ml) gave positive results at D1, D2, and D3 (++) (2).

Erythroderma, widespread erythematous eruption, exfoliative dermatitis

A 78-year-old man underwent colonoscopy for an intestinal polyp biopsy under general anesthesia. Ephedrine was injected during the intervention, owing to a circulatory collapse. Within 48 hours after the procedure, the patient presented with a pruritic maculopapular eruption on the trunk, back, and upper limbs, with no associated systemic symptoms. The lesions progressed to erythroderma with desquamation. Patch tests with all drugs used gave positive reactions to ephedrine 10% both in aq. and pet. on day 3 (++). Later patch tests, performed to evaluate cross-sensitivity, yielded positive reactions to epinephrine (adrenaline) 1% and 5%, both in pet. and in water and to phenylephrine 1% pet. and water, but not to the analogs oxymetazoline and pseudoephedrine (3).

Fixed drug eruption

A 29-year old woman was admitted to hospital for a cesarean section. On the day of the operation, a palm-sized dark brown macula with an erosion appeared on her right lower thigh, which was diagnosed as a fixed drug eruption. The patient had received 14 different medications from admission to the appearance of the lesion. Systemic challenges, patch tests on the lesion, and intradermal and subcutaneous injections on the lesion with these drugs showed positive reactions to ephedrine hydrochloride, which had been administered intravenously during the operation (7).

Other cutaneous adverse drug reactions

A 68-year-old man with chronic bronchitis had presented repeatedly in the past (7 years ago) with an itchy dermatosis involving the trunk and arms. The outbreaks began after 2-3 oral administrations of anti-catarrhal drugs (mixtures of antitussives, balsams, mucolytics and bronchodilators). Patch tests with components of these anti-catarrhal drugs and several sympathomimetic agents yielded a positive reaction only to ephedrine HCl 5% water. Twenty controls were negative. A repeat test in the patient was positive. It was concluded that 'delayed hypersensitivity to ephedrine in oral anti-catarrhal drugs appears to have been the cause of this patient's dermatitis', although it was not firmly stated that the products used 7 years ago actually contained the drug (5). Ephedrine-induced sensitivity has also been described in ref. 4 (cited in ref. 3), but details are not available to the author.

Cross-reactions, pseudo-cross-reactions and co-reactions

Epinephrine (adrenaline) and phenylephrine cross-reacted in a patient with an allergic reaction to ephedrine (3). Of 9 patients sensitized to phenylephrine, 5 cross-reacted to ephedrine 20% DMSO (6). Of 3 patients sensitized to pseudoephedrine, 2 cross-reacted to ephedrine 20% DMSO (6). A patient sensitized by topical fepradinol co-reacted to ephedrine (6).

LITERATURE

1 The data in the section 'General' may have been obtained from literature discussed in this chapter, but mostly also or exclusively from one or more of the following online sources: ChemIDPlus Advanced, PubChem, DrugBank, RxList, Drug Central, Drugs.com, and Wikipedia
2 Maul LV, Streit M, Grabbe J. Ephedrine-induced maculopapular rash. Contact Dermatitis 2018;79:193-194
3 Tanno LK, Fillard A, Landry Q, Ramdane C, Bourrain JL, Demoly P, et al. Ephedrine-induced erythrodermia: Clinical diagnostic procedure and cross-sensitivity. Contact Dermatitis 2018;79:43-44
4 Buzo-Sánchez G, Martín-Muñoz MR, Navarro-Pulido AM, Orta-Cuevas JC. Stereoisomeric cutaneous hypersensitivity. Ann Pharmacother 1997;31:1091 (cited in ref. 3)
5 Audicana M, Urrutia I, Echechipia S, Muñoz D, Fernández de Corres L. Sensitization to ephedrine in oral anticatarrhal drugs. Contact Dermatitis 1991;24:223
6 Barranco R, Rodríguez A, de Barrio M, Trujillo MJ, de Frutos C, Matheu V, et al. Sympathomimetic drug allergy: cross-reactivity study by patch test. Am J Clin Dermatol 2004;5:351-355
7 Tanimoto K, Shimakage T, Ayabe Y, Yamakawa T, Kumekawa M, Moriyama S. A case of fixed drug eruption due to ephedrine hydrochloride. Masui 2000;49:1374-1376 (Article in Japanese)
8 De Groot AC. Monographs in contact allergy, volume 3: Topical Drugs. Boca Raton, Fl, USA: CRC Press Taylor and Francis Group, 2021 (ISBN 978-0-367-23693-9)

Chapter 3.198 EPIRUBICIN

IDENTIFICATION

Description/definition   : Epirubicin is the 4'-epi-isomer of doxorubicin that conforms to the structural formula
                             shown below
Pharmacological classes   : Topoisomerase II inhibitors; antibiotics, antineoplastic
IUPAC name   : (7S,9S)-7-[(2R,4S,5R,6S)-4-Amino-5-hydroxy-6-methyloxan-2-yl]oxy-6,9,11-trihydroxy-9-
                             (2-hydroxyacetyl)-4-methoxy-8,10-dihydro-7H-tetracene-5,12-dione
Other names   : 4'-Epiadriamycin; 4'-epidoxorubicin
CAS registry number   : 56420-45-2
EC number   : Not available
Merck Index monograph   : 4948
Patch testing   : 0.1% water
Molecular formula   : $C_{27}H_{29}NO_{11}$

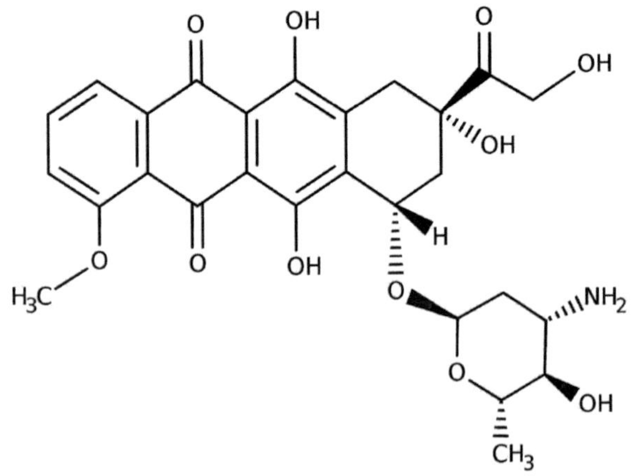

GENERAL

Epirubicin is a 4'-epi-isomer of the anthracycline antineoplastic antibiotic doxorubicin and a topoisomerase inhibitor. Epirubicin intercalates into DNA and inhibits topoisomerase II, thereby inhibiting DNA replication and ultimately, interfering with RNA and protein synthesis. This agent also produces toxic free-radical intermediates and interacts with cell membrane lipids causing lipid peroxidation. Epirubicin is indicated for use as a component of adjuvant therapy in patients with evidence of axillary node tumor involvement following resection of primary breast cancer (1). In pharmaceutical products, epirubicin is employed as epirubicin hydrochloride (CAS number 56390-09-1, EC number 260-145-2, molecular formula $C_{27}H_{30}ClNO_{11}$) (1).

CUTANEOUS ADVERSE DRUG REACTIONS FROM SYSTEMIC ADMINISTRATION CAUSED BY TYPE IV (DELAYED-TYPE) HYPERSENSITIVITY (as demonstrated by positive patch tests)

Cutaneous adverse drug reactions from systemic administration of epirubicin caused by type IV (delayed-type) hypersensitivity have included photosensitivity (3) and widespread dermatitis (2, possibly systemic allergic dermatitis)

Photosensitivity

A 43-year-old woman had undergone several courses of combination chemotherapy for Hodgkin's disease, the latest having included bleomycin-vincristine and epirubicin (farmorubicin), the latter for the first time. After administration of the cytotoxic drugs, the patient spent the rest of the day basking in the bright sun in a park, waiting for her evening train, barefoot and in a short-sleeved dress. That same afternoon she felt severe itching of her limbs. Two days later, in the morning, she noticed lesions on her leg that progressed to bullae one day later. When examined 5 days after the treatment, she had erythematous urticaria-like lesions at the sites of scratching, which in some areas formed plaques with bullae in a linear arrangement. Bullae were located predominantly on the backs of the legs. Histological examination showed acanthosis, spongiosis, intraepidermal spongiform vesicles, and moderately intense perivascular lymphohistiocytic infiltration, with many eosinophils in the superficial and mid-dermis. Photopatch tests with bleomycin, vincristin and epirubicin were positive for epirubicin (no single detail provided!). The author concluded that the patient had suffered from a photoallergic reaction to epirubicin. She admitted that the clinical picture was very similar to that of dermatitis bullosa striata pratensis, but argued that the histological picture was completely different from that of phototoxic blisters (3).

Dermatitis/eczematous eruption

A 73-year-old man with bladder cancer, who had previously been treated with intravesical mitomycin, developed a widespread dermatitis 12 h after his first instillation of epirubicin. A patch test with epirubicin 0.1% aq. was positive. As the patient developed systemic allergic dermatitis 12 hours after the first instillation and he had previously been treated with mitomycin C, the authors wondered whether the two anticancer drugs may have cross-reacted (2). If this were the case, this would be a manifestation of systemic allergic dermatitis (systemic allergic dermatitis). However, the structural formulas of both drugs are quite different, which makes immunological cross-reactivity very unlikely.

LITERATURE

1   The data in the section 'General' may have been obtained from literature discussed in this chapter, but mostly also or exclusively from one or more of the following online sources: ChemIDPlus Advanced, PubChem, DrugBank, RxList, Drug Central, Drugs.com, and Wikipedia

2   Ventura MT, Dagnello M, Di Corato R, Tursi A. Allergic contact dermatitis due to epirubicin. Contact Dermatitis 1999;40:339

3   Balabanova MB. Photoprovoked erythematobullous eruption from farmorubicin. Contact Dermatitis 1994;30:303-304

Chapter 3.199 EPRAZINONE

IDENTIFICATION
Description/definition : Eprazinone is the *N*-alkylpiperazine that conforms to the structural formula shown below
Pharmacological classes : Expectorants
IUPAC name : 3-[4-(2-Ethoxy-2-phenylethyl)piperazin-1-yl]-2-methyl-1-phenylpropan-1-one
CAS registry number : 10402-90-1
EC number : 233-873-3
Merck Index monograph : 4958
Patch testing : Pulverized tablet moistened with water (2); most systemic drugs can be tested at 10% pet.; if the pure chemical is not available, prepare the test material from intravenous powder, the content of capsules or – when also not available – from powdered tablets to achieve a final concentration of the active drug of 10% pet.
Molecular formula : $C_{24}H_{32}N_2O_2$

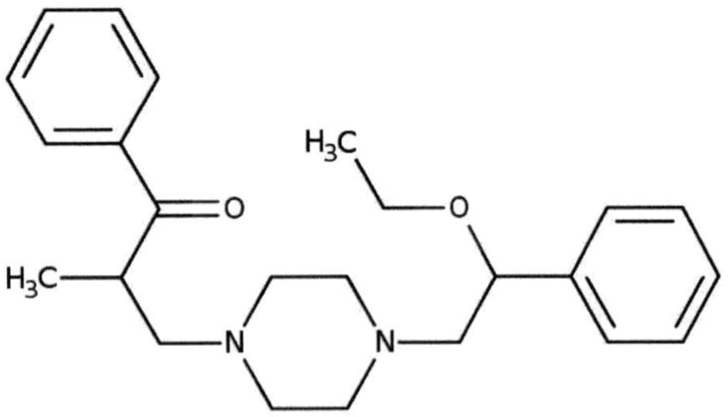

GENERAL
Eprazinone has been described as having mucolytic or expectorant properties as well as a direct relaxant action on bronchial smooth muscle. Also, it suppresses the excitation of the cough center in order to stop coughing. Eprazinone is used to treat symptoms of cough and phlegm caused by respiratory diseases such as cold, upper respiratory infection, bronchitis, and pneumonia (1). In pharmaceutical products, eprazinone is employed as eprazinone hydrochloride (CAS number 10402-53-6, EC number 233-872-8, molecular formula $C_{24}H_{34}Cl_2N_2O_2$) (1).

CUTANEOUS ADVERSE DRUG REACTIONS FROM SYSTEMIC ADMINISTRATION CAUSED BY TYPE IV (DELAYED-TYPE) HYPERSENSITIVITY (as demonstrated by positive patch tests)
Cutaneous adverse drug reactions from systemic administration of eprazinone caused by type IV (delayed-type) hypersensitivity have included acute generalized exanthematous pustulosis (AGEP) (2).

Acute generalized exanthematous pustulosis (AGEP)
A 48-year-old patient had been treated for 5 weeks with eprazinone, when a non-itchy pustular exanthema developed with fever up to 39.2°C. Physical examination revealed an exanthema with densely packed pinhead sized pustules with surrounding erythema on the trunk, the flexor aspects of the arms and legs and the palms and soles. Ten days later, the pustules had turned into erosions and the picture was that of erythroderma with glove-like desquamation of the skin of the hands and feet. Laboratory investigations showed elevated ESR, leukocytosis, kidney dysfunction (urea, creatinine) and erythrocytes and protein in the urine. Cultures of pustules were sterile. Histopathology showed spongiosis, subcorneal pustules with neutrophils and mild edema of the dermis with a diffuse inflammatory infiltration. A patch test with a pulverized eprazinone tablet moistened with water was positive after 18 hours and negative in 3 controls. Histopathology of the patch test was similar to that of the exanthema (2). The patient was diagnosed with exanthema with subcorneal pustules. This was likely a case of AGEP.

LITERATURE
1 The data in the section 'General' may have been obtained from literature discussed in this chapter, but mostly also or exclusively from one or more of the following online sources: ChemIDPlus Advanced, PubChem, DrugBank, RxList, Drug Central, Drugs.com, and Wikipedia
2 Faber M, Maucher OM, Stengel R, Goerttler E. Eprazinonexanthem mit subkornealer Pustelbildung [Eprazinone exanthema with subcorneal pustulosis]. Hautarzt 1984;35:200-203 (Article in German)

Chapter 3.200 ERLOTINIB

IDENTIFICATION

Description/definition : Erlotinib is the quinazoline derivative that conforms to the structural formula shown below

Pharmacological classes : Antineoplastic agents; protein kinase inhibitors

IUPAC name : *N*-(3-Ethynylphenyl)-6,7-bis(2-methoxyethoxy)quinazolin-4-amine

CAS registry number : 183321-74-6

EC number : Not available

Merck Index monograph : 5000

Patch testing : No data available; most systemic drugs can be tested at 10% pet.; if the pure chemical is not available, prepare the test material from intravenous powder, the content of capsules or – when also not available – from powdered tablets to achieve a final concentration of the active drug of 10% pet.

Molecular formula : $C_{22}H_{23}N_3O_4$

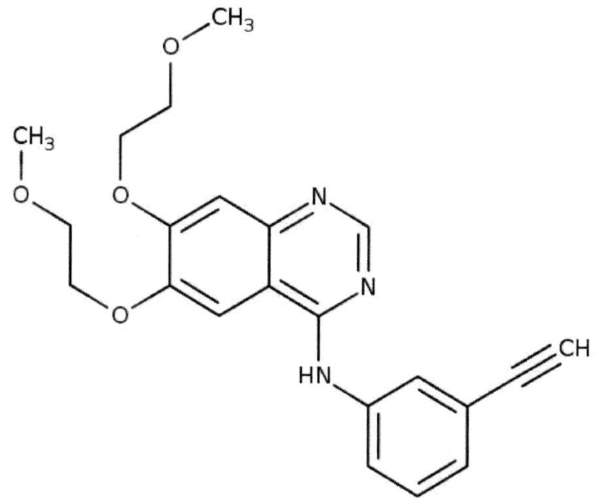

GENERAL

Erlotinib is a quinazoline derivative with antineoplastic properties. It is used for certain forms of metastatic non-small cell lung cancer and, in combination with first-line treatment, for patients diagnosed with locally advanced, unresectable or metastatic pancreatic cancer (1).

CUTANEOUS ADVERSE DRUG REACTIONS FROM SYSTEMIC ADMINISTRATION CAUSED BY TYPE IV (DELAYED-TYPE) HYPERSENSITIVITY (as demonstrated by positive patch tests)

Cutaneous adverse drug reactions from systemic administration of erlotinib caused by type IV (delayed-type) hypersensitivity have included photosensitivity (2).

Photosensitivity

A 69-year-old man had been treated with erlotinib for inoperable lung adenocarcinoma for one month, when he rapidly developed pruritic, scaly, crusted, palpable erythema on sun-exposed sites such as the head, face, neck, forearms, and dorsum of the hands. Histopathological examination revealed eczematous changes with epidermal spongiosis, acanthosis, elongation of rete ridges, slight vacuolar changes in the basal layer, and lymphocytic infiltration around the vessels in the superficial dermis. A photopatch test with erlotinib using UVA and narrowband UVB showed the development of erythema, whereas the test did not show significant changes without UVA or UVB irradiation. Accordingly, erlotinib-induced photosensitivity was diagnosed, which was considered to be photoallergic (2). Obviously, erythema only is insufficient to reach this conclusion and controls should have been tested to exclude phototoxicity. In addition, insufficient data were provided on photopatch test parameters.

LITERATURE

1 The data in the section 'General' may have been obtained from literature discussed in this chapter, but mostly also or exclusively from one or more of the following online sources: ChemIDPlus Advanced, PubChem, DrugBank, RxList, Drug Central, Drugs.com, and Wikipedia

2 Fukai T, Hasegawa T, Nagata A, Matsumura M, Kudo Y, Shiraishi E, et al. Case of erlotinib-induced photosensitivity. J Dermatol 2014;41:445-446

Chapter 3.201 ERTAPENEM

IDENTIFICATION

Description/definition   : Ertapenem is the beta-lactam and thienamycin that conforms to the structural formula
                                             shown below
Pharmacological classes : Antibacterial agents
IUPAC name   : (4R,5S,6S)-3-[(3S,5S)-5-[(3-Carboxyphenyl)carbamoyl]pyrrolidin-3-yl]sulfanyl-6-[(1R)-1-
                                             hydroxyethyl]-4-methyl-7-oxo-1-azabicyclo[3.2.0]hept-2-ene-2-carboxylic acid
CAS registry number   : 153832-46-3
EC number   : Not available
Merck Index monograph : 5001
Patch testing   : 10% and 30% water and pet. (2) (probably crushed tablets); most beta-lactam antibiotics
                                             can be tested 5-10% pet.
Molecular formula   : $C_{22}H_{25}N_3O_7S$

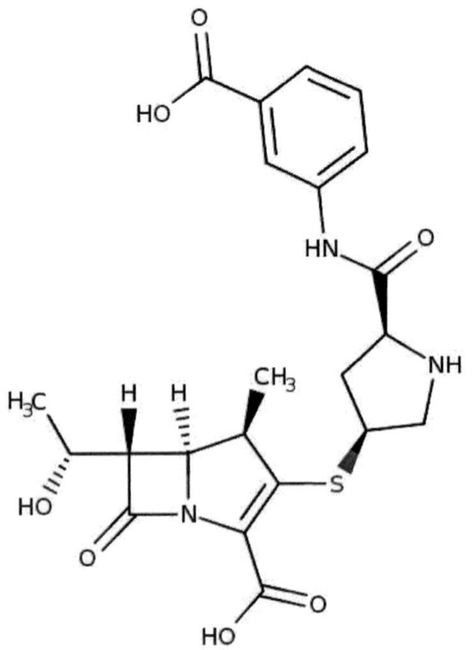

GENERAL

Ertapenem is a 1β-methyl carbapenem and broad-spectrum beta-lactam antibiotic with bactericidal property. Ertapenem binds to penicillin binding proteins (PBPs) located on the bacterial cell wall, in particular PBPs 2 and 3, thereby inhibiting the final transpeptidation step in the synthesis of peptidoglycan, an essential component of the bacterial cell wall. Inhibition results in a weakening and subsequent lysis of the cell wall leading to cell death of gram-positive and gram-negative aerobic and anaerobic pathogens. This agent is stable against hydrolysis by a variety of beta-lactamases, including penicillinases, cephalosporinases and extended-spectrum beta-lactamases. It is used primarily for the treatment of aerobic gram-negative bacterial infections (1).

In pharmaceutical products, ertapenem is employed as ertapenem sodium (CAS number 153773-82-1, EC number not available, molecular formula $C_{22}H_{24}N_3NaO_7S$) (1). The classification and structures of beta-lactam antibiotics are discussed in Chapter 3.36 Amoxicillin.

CONTACT ALLERGY FROM ACCIDENTAL CONTACT

A 28-year-old nurse developed eczema on the dorsal aspect of the hand and the face. She worked in the hematology department where she usually handled and administered a variety of antibiotics. When she was moved to a different department where she did not have contact with these drugs, the dermatitis completely resolved. Patch tests were positive to ertapenem 20% pet., imipenem-cilastatin (20% pet.), ampicillin (25% pet.), piperacillin (20% pet.), mezlocillin (20% pet.), and meropenem (20% pet.) It was concluded that the patient had contact allergy to carbapenems (ertapenem, imipenem-cilastatin, and meropenem) and semisynthetic penicillins (piperacillin, mezlocillin, and ampicillin) (4). At work, the patient had contact only with imipenem, ertapenem, and piperacillin, so it may be assumed that the other reactions were due to cross-reactivity.

CUTANEOUS ADVERSE DRUG REACTIONS FROM SYSTEMIC ADMINISTRATION CAUSED BY TYPE IV (DELAYED-TYPE) HYPERSENSITIVITY (as demonstrated by positive patch tests)

Cutaneous adverse drug reactions from systemic administration of ertapenem caused by type IV (delayed-type) hypersensitivity have included acute generalized exanthematous pustulosis (AGEP) (2).

Acute generalized exanthematous pustulosis (AGEP)

A 47-year-old man, known to be allergic to penicillin, developed erysipelas of the right foot and lower leg. Because of his drug allergy, clindamycin was prescribed, but the infection did not improve. Clindamycin was ceased, and he was subsequently administered ertapenem. After 2 days the patient developed fever and a generalized non-follicular pustular rash, which was especially prominent over the axillary and inguinal folds. He had a marked neutrophilia and bacterial cultures of blood, urine, and a swab of a pustule were negative. A diagnosis of acute generalized exanthematous pustulosis (AGEP) was made and supported by a skin biopsy, which revealed intraepidermal and subcorneal spongiform pustules with papillary edema. Ertapenem was subsequently ceased. After the resolution of the rash the patient was found to be patch test positive with pustules at D2 to ertapenem 10% and 30% in water and pet. with co-/cross-reactions to benzylpenicillin, meropenem and cefalotin; the reaction to clindamycin was negative at D2 and D7 (2).

Cross-reactions, pseudo-cross-reactions and co-reactions

Cross-reactions between beta-lactam antibiotics are discussed in Chapter 3.36 Amoxicillin. Cross-reactions between carbapenems (ertapenem, imipenem, meropenem) and between carbapenems and penicillins may rarely occur, but the pattern is unclear (3).

LITERATURE

1   The data in the section 'General' may have been obtained from literature discussed in this chapter, but mostly also or exclusively from one or more of the following online sources: ChemIDPlus Advanced, PubChem, DrugBank, RxList, Drug Central, Drugs.com, and Wikipedia

2   Fernando SL. Ertapenem-induced acute generalized exanthematous pustulosis with cross-reactivity to other beta-lactam antibiotics on patch testing. Ann Allergy Asthma Immunol 2013;111:139-140

3   Colagiovanni A, Feliciani C, Fania L, Pascolini L, Buonomo A, Nucera E, et al. Occupational contact dermatitis from carbapenems. Cutis 2015;96:E1-3

Chapter 3.202 ERYTHROMYCIN

IDENTIFICATION

Description/definition : Erythromycin a macrolide antibiotic produced by *Saccharopolyspora erythraea* (formerly *Streptomyces erythraeus)*; the structural formula of erythromycin A, its major active component, is shown below

Pharmacological classes : Gastrointestinal agents; protein synthesis inhibitors; anti-bacterial agents

IUPAC name : (3R,4S,5S,6R,7R,9R,11R,12R,13S,14R)-6-[(2S,3R,4S,6R)-4-(Dimethylamino)-3-hydroxy-6-methyloxan-2-yl]oxy-14-ethyl-7,12,13-trihydroxy-4-[(2R,4R,5S,6S)-5-hydroxy-4-methoxy-4,6-dimethyloxan-2-yl]oxy-3,5,7,9,11,13-hexamethyl-oxacyclotetradecane-2,10-dione

Other names : Erythromycin A

CAS registry number : 114-07-8

EC number : 204-040-1

Merck Index monograph : 5009

Patch testing : 2% pet. (SmartPracticeCanada, SmartPracticeEurope); 1% pet. (SmartPracticeCanada); 10.0% pet. (Chemotechnique); 1% may sometimes be too low (6,7)

Molecular formula : $C_{37}H_{67}NO_{13}$

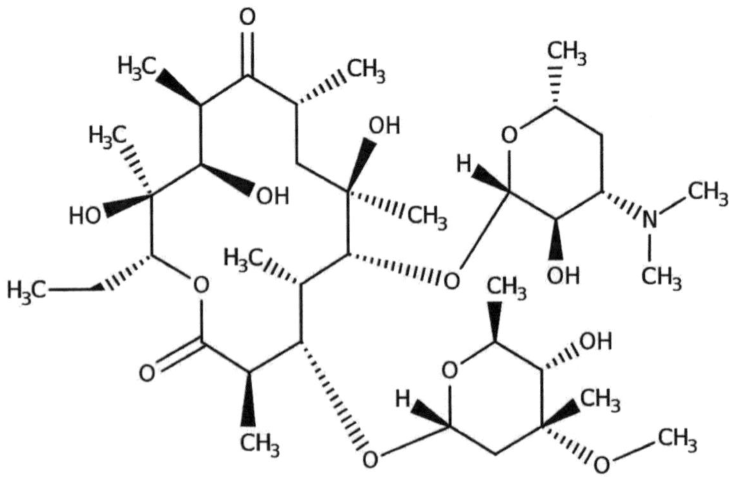

GENERAL

Erythromycin is a broad-spectrum antibiotic drug produced by a strain of *Saccharopolyspora erythraea* (formerly *Streptomyces erythraeus*) and belongs to the macrolide group of antibiotics. Erythromycin may be bacteriostatic or bactericidal in action, depending on the concentration of the drug at the site of infection and the susceptibility of the organism involved. Erythromycin is widely used for treating a variety of infections caused by gram-positive bacteria, gram-negative bacteria and many other organisms of the respiratory tract, skin, gastrointestinal and genital tracts including sexually transmitted diseases (syphilis, gonorrhea, *Chlamydia* infections). Erythromycin is also used in topical preparations, especially for the treatment of acne vulgaris. It has caused several cases of contact allergy/allergic contact dermatitis, which have been fully reviewed in Volume 3 of the *Monographs in contact allergy* series (8). In pharmaceutical products, mostly erythromycin base is used, but many other forms (salts and esters) are possible (1).

CUTANEOUS ADVERSE DRUG REACTIONS FROM SYSTEMIC ADMINISTRATION CAUSED BY TYPE IV (DELAYED-TYPE) HYPERSENSITIVITY (as demonstrated by positive patch tests)

Cutaneous adverse drug reactions from systemic administration of erythromycin caused by type IV (delayed-type) hypersensitivity have included acute generalized exanthematous pustulosis (AGEP) (7), systemic allergic dermatitis (3,4), symmetrical drug-related intertriginous and flexural exanthema (SDRIFE) (6, in the form of the baboon syndrome), fixed drug eruption (2), and toxic epidermal necrolysis (TEN) (5).

Acute generalized exanthematous pustulosis (AGEP)

A 46-year-old woman was treated at home for a sore throat with oral erythromycin ethylsuccinate and prednisolone. An eruption appeared on the chest and axillae 48 hours later. Erythromycin was changed to oral spiramycin. Again, 2 days later the patient presented with a pustular eruption with a fever of 39°C. She had an erythematous rash covered with numerous superficial non-follicular pustules on the trunk and proximal extremities, sparing the palms, soles, and mucous membranes. Spiramycin was discontinued and the fever and eruption spontaneously disappeared within 10 days. Later, patch tests with crushed tablets and intravenous forms of erythromycin and spiramycin were positive at D2 showing a pustular eruption on an erythematous base. Ten controls were negative (7). This was a case of acute generalized exanthematous pustulosis.

Systemic allergic dermatitis (systemic contact dermatitis)
A 46-year-old man was treated for erythrasma with 2% erythromycin for a second time, when, after a few days, an acute eczema developed. He was treated with a topical corticosteroid and oral erythromycin ethylsuccinate, but 5 days later the patient presented with generalized eczema, which had developed at the beginning of his oral treatment. Patch tests were positive to the pharmaceutical, to erythromycin base 5% pet. and – later – to erythromycin base and erythromycin ethylsuccinate 5% and 2.5% pet. but negative to 1% and 0.1% pet. 25 controls were negative to all concentrations tested. This was a classic case of systemic allergic dermatitis to oral erythromycin after sensitization from topical erythromycin (3,4).

Symmetrical drug-related intertriginous and flexural exanthema (SDRIFE)/Baboon syndrome
An 18-month-old patient developed the clinical picture of the baboon syndrome after oral treatment with erythromycin syrup for a sore throat. The lymphoblastic transformation test was positive for erythromycin. Prick tests were negative, but the intradermal test was positive at D2 at a concentration of of 1:10,000. The biopsy showed a perivascular lymphocytic dermatitis (6).

Fixed drug eruption
A 46-year old man presented with macular and erythematous lesions at the top of the left leg and in the genital area, healing with residual pigmentation, on 2 occasions. Het had used 3 different drugs, among which erythromycin. An oral provocation test with erythromycin was positive, with the other 2 drugs negative. A patch test with erythromycin 10% DMSO was positive on postlesional skin, but negative on uninvolved skin (2). A diagnosis of fixed drug eruption to erythomycin was made.

Stevens-Johnson syndrome/toxic epidermal necrolysis (SJS/TEN)
A 20-year-old woman, who was treated for a sore throat with oral erythromycin, became febrile with a temperature of 40.5°C and her skin became tender with erythema and edema extending over the whole body surface. A drug reaction was suspected and erythromycin was stopped. The next day, large areas of skin had developed superficial blisters and the epidermis came off in sheets leaving red tender areas. The patient was diagnosed with toxic epidermal necrolysis. Three months later, patch tests were strongly positive to erythomycin stearate 1% and 5% pet. (5).

Cross-reactions, pseudo-cross-reactions and co-reactions
Erythromycin may have cross-reacted to the related antibiotic spiramycin (7).

LITERATURE

1   The data in the section 'General' may have been obtained from literature discussed in this chapter, but mostly also or exclusively from one or more of the following online sources: ChemIDPlus Advanced, PubChem, DrugBank, RxList, Drug Central, Drugs.com, and Wikipedia
2   Florido Lopez JF, Lopez Serrano MC, Belchi Hernandez J, Estrada Rodriguez JL. Fixed eruption due to erythromycin. Allergy 1991;46:77-78
3   Fernandez Redondo V, Casas L, Taboada M, Toribio J. Systemic allergic dermatitis from erythromycin. Contact Dermatitis 1994;30:311 (same as ref. 4)
4   Fernandez Redondo V, Casas L, Taboada M, Toribio J. Systemic allergic dermatitis from erythromycin. Contact Dermatitis 1994;30:43-44 (same as ref. 3)
5   Lund Kofoed M, Oxholm A. Toxic epidermal necrolysis due to erythromycin. Contact Dermatitis 1985;13:273
6   Goossens C, Sass U, Song M. Baboon syndrome. Dermatology 1997;194:421-422
7   Moreau A, Dompmartin A, Castel B, Remond B, Leroy D. Drug-induced acute generalized exanthematous pustulosis with positive patch tests. Int J Dermatol 1995;34:263-266
8   De Groot AC. Monographs in contact allergy, volume 3: Topical Drugs. Boca Raton, Fl, USA: CRC Press Taylor and Francis Group, 2021 (ISBN 978-0-367-23693-9)

Chapter 3.203 ESLICARBAZEPINE

IDENTIFICATION
Description/definition : Eslicarbazepine is the dibenzazepine that conforms to the structural formula shown below
Pharmacological classes : Antiepileptics
IUPAC name : (5S)-5-Hydroxy-5,6-dihydrobenzo[b][1]benzazepine-11-carboxamide
CAS registry number : 104746-04-5
EC number : 810-248-9
Merck Index monograph : 5024
Patch testing : Tablet, pulverized, 20% pet. (2); most systemic drugs can be tested at 10% pet.; if the pure chemical is not available, prepare the test material from intravenous powder, the content of capsules or – when also not available – from powdered tablets to achieve a final concentration of the active drug of 10% pet.
Molecular formula : $C_{15}H_{14}N_2O_2$

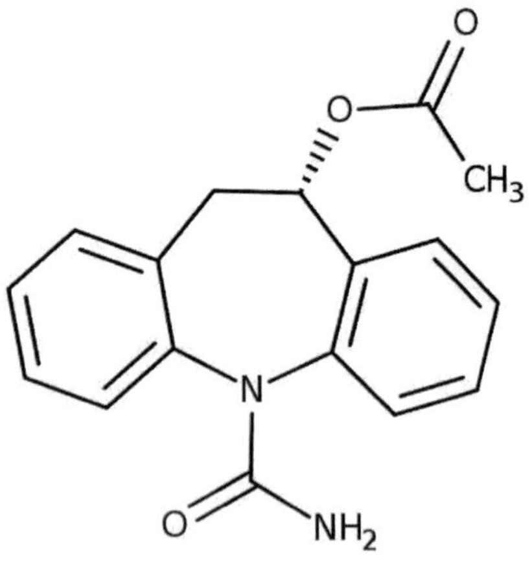

Eslicarbazepine acetate

GENERAL
Eslicarbazepine is an anticonvulsant approved for use as an adjunctive therapy for partial-onset seizures that are not adequately controlled with conventional therapy. In pharmaceutical products, eslicarbazepine is employed as eslicarbazepine acetate (CAS number 236395-14-5, EC number not available, molecular formula $C_{17}H_{16}N_2O_3$) (1).

CUTANEOUS ADVERSE DRUG REACTIONS FROM SYSTEMIC ADMINISTRATION CAUSED BY TYPE IV (DELAYED-TYPE) HYPERSENSITIVITY (as demonstrated by positive patch tests)
Cutaneous adverse drug reactions from systemic administration of eslicarbazepine caused by type IV (delayed-type) hypersensitivity have included maculopapular eruption (2).

Maculopapular eruption
A 54-year-old woman developed a slightly pruritic generalized maculopapular eruption without fever or systemic symptoms on the 10th day of treatment with eslicarbazepine acetate and gabapentine for trigeminal neuropathic pain. The drugs were switched to carbamazepine, metamizol and diazepam. Two days later, there was a marked worsening of the skin rash. The patient had a disseminated erythematous and violaceous maculopapular eruption, along with mild edema of the face and extremities. Her blood tests revealed eosinophilia but no other abnormalities. All drugs were withdrawn and she was given systemic corticosteroids. One month after full recovery, patch tests were performed with the drugs the patient had used, pulverized, 20% in petrolatum. The tests were positive only for eslicarbazepine, both at day 2 and day 4. Six controls were negative. The negative patch test to carbamazepine, which probably caused the exacerbation, may have been false-negative (2).

LITERATURE
1   The data in the section 'General' may have been obtained from literature discussed in this chapter, but mostly also or exclusively from one or more of the following online sources: ChemIDPlus Advanced, PubChem, DrugBank, RxList, Drug Central, Drugs.com, and Wikipedia
2   Finelli E, Custódio P, Porovska O, Prates S, Leiria-Pinto P. Patch testing in a case of eslicarbazepine and carbamazepine induced cutaneous reaction. Eur Ann Allergy Clin Immunol 2018;50:229-231

Chapter 3.204 ESOMEPRAZOLE

IDENTIFICATION

Description/definition : Esomeprazole is the *S*-isomer of the proton pump inhibitor omeprazole that conforms to the structural formula shown below
Pharmacological classes : Anti-ulcer agents; proton pump inhibitors
IUPAC name : 6-Methoxy-2-[(*S*)-(4-methoxy-3,5-dimethylpyridin-2-yl)methylsulfinyl]-1*H*-benzimidazole
Other names : (*S*)-Omeprazole; (-)-omeprazole; (*S*)-(-)-omeprazole
CAS registry number : 119141-88-7
EC number : 615-996-8 (Omeprazole)
Merck Index monograph : 8209 (Omeprazole)
Patch testing : 1% pet. (2); pulverized tablets 10-30% pet.; most systemic drugs can be tested at 10% pet.; if the pure chemical is not available, prepare the test material from intravenous powder, the content of capsules or – when also not available – from powdered tablets to achieve a final concentration of the active drug of 10% pet.
Molecular formula : $C_{17}H_{19}N_3O_3S$

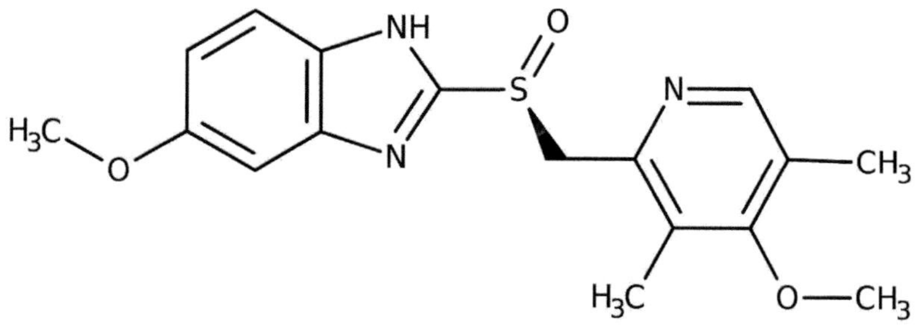

GENERAL

Esomeprazole is the *S*-isomer of omeprazole and a benzimidazole with selective and irreversible proton pump inhibition activity. It is indicated for the treatment of active duodenal ulcer, eradication of *Helicobacter pylori* to reduce the risk of duodenal ulcer, treatment of active benign gastric ulcer, gastroesophageal reflux disease (GERD), erosive esophagitis (EE) due to acid-mediated GERD, for maintenance of healing of EE due to acid-mediated GERD and in pathologic hypersecretory conditions (e.g. Zollinger-Ellison syndrome) (1). In pharmaceutical products, both esomeprazole, esomeprazole magnesium (CAS number 161973-10-0, EC number not available, molecular formula $C_{34}H_{36}MgN_6O_6S_2$) and esomeprazole sodium (CAS number 161796-78-7, EC number not available, molecular formula $C_{17}H_{18}N_3NaO_3S$) may be employed (1).

CONTACT ALLERGY FROM ACCIDENTAL CONTACT

A 58-year-old male pharmaceutical worker presented with erythematous eruptions involving the neck, posterior auricular areas, and extensor forearms, with slight swelling and mild erythema of the upper eyelids. He had a 4-year history as a chemical operator in the production area of the plant, wearing rubber gloves and long-sleeved shirts. The patient had regular contact with proton pump inhibitors. Patch tests were positive to esomeprazole 1% pet., and he also reacted to dexlansoprazole, omeprazole, and pantoprazole. The patient was diagnosed with occupational airborne allergic contact dermatitis to proton pump inhibitors (2).

A 44-year-old nurse presented with a 9-month-history of eczema on the hands and face, which had started 3 months after she was required to crush tablets daily during her work as a nurse in a clinic for disabled people. She was patch tested with all the drugs that she had contact with and reacted to 8 of these, including esomeprazole 30% pet. No controls were performed (3).

CUTANEOUS ADVERSE DRUG REACTIONS FROM SYSTEMIC ADMINISTRATION CAUSED BY TYPE IV
(DELAYED-TYPE) HYPERSENSITIVITY (as demonstrated by positive patch tests)

Cutaneous adverse drug reactions from systemic administration of esomeprazole caused by type IV (delayed-type) hypersensitivity have included maculopapular eruption (12), fixed drug eruption (7), drug reaction with eosinophilia and systemic symptoms (DRESS) (8,9,10), Stevens-Johnson syndrome/toxic epidermal necrolysis (SJS/TEN) (8,12), and photosensitivity (5, unconvincing).

Case series with various or unknown types of drug reactions
In Taiwan, in a retrospective multicenter study performed between 2003 and 2016, 69 cases of (severe) delayed-type hypersensitivity reactions to proton pump inhibitors were collected, including 27 cases of SJS/TEN and 10 of DRESS. Patch tests were performed in only 7 cases. Positive patch tests to esomeprazole (pulverized tablet 10% pet.) were seen in 1 patient with maculopapular eruption and one with SJS/TEN overlap (12).

Maculopapular eruption
See the section 'Case series with various or unknown types of drug reactions' above, ref. 12.

Fixed drug eruption
A 56-year-old woman presented with itching, sharply demarcated, round to oval erythematous patches located at her nipple–areola complexes and surrounding skin, lower back/sacrum, and pubic region areas, which had developed 2 weeks after the start of esomeprazole therapy for dyspeptic symptoms. Patch testing with esomeprazole 2% in petrolatum was negative at 48 and 72 hours but became positive on day 6. Oral-controlled provocation test induced the reappearance of the lesions over the mammary areas with esomeprazole but not with placebo. Therefore, the patient was diagnosed as having a nonpigmented fixed drug eruption associated with esomeprazole (7). It is quite remarkable that the authors did not, as is standard practice, patch test the suspected drug on postlesional skin *and* that there was a positive reaction on previously uninvolved skin.

Drug reaction with eosinophilia and systemic symptoms (DRESS)
In a multicenter investigation in France, of 72 patients patch tested for DRESS, 46 (64%) had positive patch tests to drugs, including 2 to esomeprazole (8). One more patient (or one presented in both studies?) from France developed DRESS from esomeprazole and had positive patch tests to it tested 30% water, alcohol and pet. (9). In these 2 studies, no clinical details were provided.

After an operation for glioblastoma, a 41-year-old woman received 6 drugs including esomeprazole. Twenty days later the patient experienced an erythematous and itching skin reaction. Valproate sodium was stopped, the other drugs continued. Two weeks later an erythroderma developed with bilateral conjunctivitis, cheilitis, fever (40°C) and desquamation. Systemic treatment with prednisolone was started and the other drugs were continued. One month later, the patient presented with numerous eczematous lesions, facial edema, fever and dyspnea and she had hypereosinophilia and increased liver enzymes. DRESS syndrome was now suspected and all drugs were stopped except topical and systemic corticosteroids. Complete remission was obtained 4 months later. Patch tests with all drugs tested as 1-10% 'solutions' were positive at D2 and D3 to esomeprazole only. In a second test session, esomeprazole again was positive and there were positive (cross-)reactions to omeprazole and pantoprazole but not to rabeprazole. Five days after testing, the patient experienced a mild erythroderma with facial edema and desquamation (10).

Stevens-Johnson syndrome/toxic epidermal necrolysis (SJS/TEN)
A case of a positive patch test to esomeprazole in a patient with SJS/TEN was reported from France in 2013 (8). See also the section 'Case series with various or unknown types of drug reactions' above, ref. 12.

Photosensitivity
Multiple pruritic erythematous lesions over the face and arms which ulcerated in a 58-year-old woman having used esomeprazole for a week were diagnosed as photoallergic dermatitis by researchers from India on the basis of clinical features. Photopatch tests were apparently deemed unnecessary to reach this diagnosis (5).

Cross-reactions, pseudo-cross-reactions and co-reactions
A patient with DRESS from esomeprazole cross-reacted to omeprazole and pantoprazole but not to rabeprazole (10). A patient sensitized to pantoprazole, had cross-reactions to other proton pump inhibitors including esomeprazole (4). A patient sensitized to omeprazole cross-reacted to esomeprazole and other proton pump inhibitors (all crushed tablets, 1% and 10% pet.) (6).

LITERATURE

1   The data in the section 'General' may have been obtained from literature discussed in this chapter, but mostly also or exclusively from one or more of the following online sources: ChemIDPlus Advanced, PubChem, DrugBank, RxList, Drug Central, Drugs.com, and Wikipedia
2   DeKoven JG, Yu AM. Occupational airborne contact dermatitis from proton pump inhibitors. Dermatitis 2015;26:287-290
3   Swinnen I, Ghys K, Kerre S, Constandt L, Goossens A. Occupational airborne contact dermatitis from benzodiazepines and other drugs. Contact Dermatitis 2014;70:227-232

4   Liippo J, Pummi K, Hohenthal U, Lammintausta K. Patch testing and sensitization to multiple drugs. Contact Dermatitis 2013;69:296-302

5   Shukla A, Mahapatra A, Gogtay N, Khopkar U. Esomeprazole-induced photoallergic dermatitis. J Postgrad Med 2010;56:229-231

6   Al-Falah K, Schachter J, Sasseville D. Occupational allergic contact dermatitis caused by omeprazole in a horse breeder. Contact Dermatitis 2014;71:377-378

7   Morais P, Baudrier T, Mota A, Cunha A P, Cadinha S, Barros A M, Azevedo F. Nonpigmented fixed drug eruption induced by esomeprazole. Cutan Ocul Toxicol 2010;29:217-220

8   Barbaud A, Collet E, Milpied B, Assier H, Staumont D, Avenel-Audran M, et al. A multicentre study to determine the value and safety of drug patch tests for the three main classes of severe cutaneous adverse drug reactions. Br J Dermatol 2013;168:555-562

9   Studer M, Waton J, Bursztejn AC, Aimone-Gastin I, Schmutz JL, Barbaud A. Does hypersensitivity to multiple drugs really exist? Ann Dermatol Venereol 2012;139:375-380 (Article in French)

10 Caboni S, Gunera-Saad N, Ktiouet-Abassi S, Berard F, Nicolas JF. Esomeprazole-induced DRESS syndrome. Studies of cross-reactivity among proton-pump inhibitor drugs. Allergy 2007;62:1342-1343

11 Brockow K, Garvey LH, Aberer W, Atanaskovic-Markovic M, Barbaud A, Bilo MB, et al.; ENDA/EAACI Drug Allergy Interest Group. Skin test concentrations for systemically administered drugs – an ENDA/EAACI Drug Allergy Interest Group position paper. Allergy 2013;68:702-712

12 Lin CY, Wang CW, Hui CR, Chang YC, Yang CH, Cheng CY, et al. Delayed-type hypersensitivity reactions induced by proton pump inhibitors: A clinical and in vitro T-cell reactivity study. Allergy 2018;73:221-229

Chapter 3.205 ESTRADIOL

IDENTIFICATION

Description/definition   : Estradiol is the estrogenic steroid that conforms to the structural formula shown below
Pharmacological classes   : Estrogens
IUPAC name   : (8R,9S,13S,14S,17S)-13-Methyl-6,7,8,9,11,12,14,15,16,17-decahydrocyclopenta[a]-
                             phenanthrene-3,17-diol
Other names   : 1,3,5-Estratriene-3,17β-diol; 3,17-epidihydroxyoestratriene; 17β-estradiol;
                             dihydrofolliculin; oestradiol
CAS registry number   : 50-28-2
EC number   : 200-023-8
Merck Index monograph   : 5028
Patch testing   : 5% alc. 96% (7)
Molecular formula   : $C_{18}H_{24}O_2$

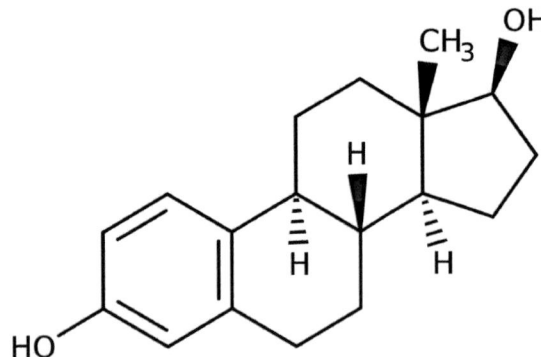

GENERAL

Estradiol is a naturally occurring hormone that circulates endogenously within the human body. It is the most potent form of mammalian estrogenic steroids and acts as the major female sex hormone. As such, estradiol plays an essential role in the regulation of the menstrual cycle, in the development of puberty and secondary female sex characteristics, as well as in ageing and several hormonally-mediated disease states. Estradiol and estradiol esters (esterification of estradiol aims to improve absorption and bioavailability after oral administration or to sustain release from depot intramuscular injections) are commercially available in oral, transdermal, and injectable hormone therapy products for managing conditions associated with reduced estrogen production such as menopausal and peri-menopausal symptoms as well as hypoestrogenism. It is also used in transgender hormone therapy, as a component of oral contraceptive pills for preventing pregnancy (most commonly as ethinylestradiol), and is sometimes used for the palliative treatment of some hormone-sensitive cancers like breast and prostate cancer (1).

In topical preparations, estradiol has caused contact allergy/allergic contact dermatitis, which subject has been fully reviewed in Volume 3 of the *Monographs in contact allergy* series (7).

CUTANEOUS ADVERSE DRUG REACTIONS FROM SYSTEMIC ADMINISTRATION CAUSED BY TYPE IV
(DELAYED-TYPE) HYPERSENSITIVITY (as demonstrated by positive patch tests)

Cutaneous adverse drug reactions from systemic administration of estradiol caused by type IV (delayed-type) hypersensitivity have included systemic allergic dermatitis (systemic allergic dermatitis) (3,5,6), manifesting as 'systemic pruritic rash' (3), maculopapular rash (5), as an erythematopapular generalized pruritic eruption (5), and as generalized eczema (6).

Systemic allergic dermatitis (systemic contact dermatitis)

A 47-year-old postmenopausal woman developed eczematous lesions at the sites of application of an estradiol transdermal therapeutic system and later at the sites of application of a gel containing estradiol. Due to the topical intolerance, the therapy was switched to oral estrogen, which caused a systemic pruritic rash. The patient had positive patch tests to estradiol. Allergic contact dermatitis and systemic allergic dermatitis to estradiol was diagnosed (3).

Two women, aged 42 and 52 years, were treated with estradiol transdermal therapeutic system (TTS) 2x weekly after oophorectomy. Under the 6th and 8th patch, respectively, they developed acute vesicular dermatitis, limited to the areas of the TTS but increasing on subsequent patch applications. Patient no. 1 later had acute eczema from a topical gel containing 6% estradiol. On the 5th day of oral estrogen therapy, she developed a symmetrical pruritic maculopapular rash, mainly localized on the trunk, neck and limb girdles. Patch tests were positive to 3 estradiol preparations (two TTSs and a gel) and to estradiol 1%, 5% and 10% in alc. 96%. Despite positive reactions to estradiol, patient no. 2 was treated with oral estradiol 2 mg/tablet and, on the 16th day of treatment, she developed an erythematopapular generalized pruritic eruption, mainly localized on the trunk and limbs, with no fever or systemic symptoms. The exanthema progressed despite oral antihistamines and regressed only after estrogen suspension. Both women had systemic allergic dermatitis from oral estradiol after previous sensitization to topical estradiol in TTSs (5).

A similar case of systemic allergic dermatitis had been reported in 1996 from France. It is uncertain whether the oral estrogen that induced generalized eczema was estradiol or another estrogen-derivative (6).

Cross-reactions, pseudo-cross-reactions and co-reactions
A patient sensitized to testosterone had positive reactions to 2 patches containing estradiol. *In vivo*, testosterone is metabolized to estradiol by the enzyme complex aromatase, particularly in the liver and adipose tissue. Therefore, cross-reactivity does not come as a surprise (4). According to some authors, patients who develop contact allergies to sex steroids are at risk of developing multiple corticosteroid allergies (2).

LITERATURE
1   The data in the section 'General' may have been obtained from literature discussed in this chapter, but mostly also or exclusively from one or more of the following online sources: ChemIDPlus Advanced, PubChem, DrugBank, RxList, Drug Central, Drugs.com, and Wikipedia
2   Lamb SR, Wilkinson SM. Contact allergy to progesterone and estradiol in a patient with multiple corticosteroid allergies. Dermatitis 2004;15:78-81
3   Corazza M, Mantovani L, Montanari A, Virgili A. Allergic contact dermatitis from transdermal estradiol and systemic allergic dermatitis from oral estradiol. A case report. J Reprod Med 2002;47:507-509
4   Shouls J, Shum KW, Gadour M, Gawkrodger DJ. Contact allergy to testosterone in an androgen patch: control of symptoms by pre-application of topical corticosteroid. Contact Dermatitis 2001;45:124-125
5   Gonçalo M, Oliveira HS, Monteiro C, Clerins I, Figueiredo A. Allergic and systemic allergic dermatitis from estradiol. Contact Dermatitis 1999;40:58-59
6   El Sayed F, Bayle-Lebey P, Marguery MC, Bazex J. Systemic sensitization to 17-beta estradiol induced by transcutaneous administration. (Sensibilisation systémique au 17-β-oestradiol induite par voie transcutanée). Ann Dermatol Venereol 1996;123:26-28 (Article in French)
7   De Groot AC. Monographs in contact allergy, volume 3: Topical Drugs. Boca Raton, Fl, USA: CRC Press Taylor and Francis Group, 2021 (ISBN 978-0-367-23693-9)

Chapter 3.206 ETHAMBUTOL

IDENTIFICATION

Description/definition   : Ethambutol is the antimycobacterial that conforms to the structural formula shown below
Pharmacological classes : Antitubercular agents
IUPAC name   : (2S)-2-[2-[[(2S)-1-Hydroxybutan-2-yl]amino]ethylamino]butan-1-ol
CAS registry number   : 74-55-5
EC number   : 200-810-6
Merck Index monograph : 5045
Patch testing   : 1% pet. or water (13); pulverized tablet, 10-30% pet.; most systemic drugs can be tested
                                          at 10% pet.; if the pure chemical is not available, prepare the test material from
                                          intravenous powder, the content of capsules or – when also not available – from
                                          powdered tablets to achieve a final concentration of the active drug of 10% pet.
Molecular formula   : $C_{10}H_{24}N_2O_2$

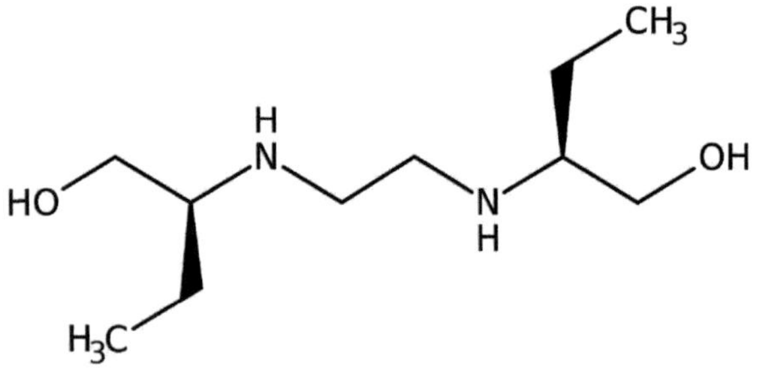

GENERAL

Ethambutol is an antibiotic with antitubercular properties. It inhibits the transfer of mycolic acids into the cell wall of *Mycobacterium tuberculosis* and may also inhibit the synthesis of spermidine in mycobacteria. This action is usually bactericidal. Ethambutol is indicated for use, as an adjunct, in the treatment of pulmonary tuberculosis. In pharmaceutical products, ethambutol is employed as ethambutol (di)hydrochloride (CAS number 1070-11-7, EC number 213-970-7, molecular formula $C_{10}H_{26}Cl_2N_2O_2$) (1).

CONTACT ALLERGY FROM ACCIDENTAL CONTACT

A 23-year-old male graduate student worked as a research assistant involved in analysis of antituberculosis drugs in human biological samples, when he developed erythema, vesicles, scaling and fissures on both hands and fingertips. Patch tests were positive to ethambutol 1% pet. and to a related chemical used as an internal standard. Ten controls were negative. This was a case of occupational allergic contact dermatitis to ethambutol (4).

CUTANEOUS ADVERSE DRUG REACTIONS FROM SYSTEMIC ADMINISTRATION CAUSED BY TYPE IV (DELAYED-TYPE) HYPERSENSITIVITY (as demonstrated by positive patch tests)

Cutaneous adverse drug reactions from systemic administration of ethambutol caused by type IV (delayed-type) hypersensitivity have included maculopapular eruption (6), drug reaction with eosinophilia and systemic symptoms (DRESS) (2,5,7,8,9,11,12), eczematous eruption (10), and itchy, desquamative, erythematous, papular rash, with painful crusts and excoriations (3).

Maculopapular eruption

A 72-year-old woman developed a generalized maculopapular rash that had appeared 4 weeks after she had started taking 4 antituberculosis drugs. She had eosinophilia (eosinophils 64.3%) and abnormal liver function tests. Six weeks after discharge from the hospital, patch tests with the 4 drugs as crushed tablets at 50% in petrolatum showed a diffuse erythematous rash around the ethambutol, isoniazid, and rifampicin patches at D2. The authors admitted these may have been nonspecific irritant reactions (6). This may have been a case of DRESS.

Drug reaction with eosinophilia and systemic symptoms (DRESS)

Case series

In the period 2010-2014, in South Africa, 60 patients with cutaneous adverse drug reactions to first-line antituberculosis drugs (FLTD: rifampicin, isoniazid, pyrazinamide and ethambutol) were patch tested with the patients' drugs 30% water and olive oil. Positive reactions were seen to at least one FLTD in 14 participants, of who twelve had DRESS, one SJS and one SJS/TEN. There were 3 positive reactions to ethambutol in patients with DRESS. Eleven of the 14 patients were HIV-infected and in 10 of these (91%), patch tests resulted in a generalized systemic reaction, including the 3 patients with DRESS from ethambutol (7).

In France, a search was performed for potential cases of DRESS caused by the anti-tuberculosis drugs rifampicin, isoniazid, pyrazinamide and ethambutol, reported from January 1, 2005, to July 30, 2015, in the French pharmacovigilance database. Sixty-seven cases of antituberculosis drug-associated DRESS were analyzed (40 women and 27 men, median age of 61 years). Patch tests were performed in 11 patients and were positive in 7. Two individuals reacted to ethambutol, 5 to isoniazid and one to rifampicin (one reacted to both ethambutol and isoniazid) (5).

Case reports

A 68-year old woman developed DRESS with maculopapular exanthema, high fever, painful cervical lymphadenopathy, elevated C-reactive protein, leukocytes and eosinophils, atypical lymphocytes and impaired liver and renal function. There were initially features of Stevens-Johnson syndrome including painful erosions on the oral mucosa, prominent targetoid lesions with blisters and positive Nikolsky sign and acute conjunctivitis. The adverse drug reaction had developed 7 weeks after starting the antituberculosis drugs isoniazid, rifampicin, ethambutol and levofloxacin for tuberculous pericarditis. Four months after resolution, she was patch tested with all drugs 10% pet. and had a positive reaction to ethambutol at D2; the lymphocyte transformation test was also positive to ethambutol (11).

A 29-yr-old woman had been on celecoxib and anti-tuberculosis drugs for one month to treat knee joint pain and pulmonary tuberculosis, when she developed fever, lymphadenopathy, rash, hypereosinophilia, and visceral involvement (hepatitis and pneumonitis). During corticosteroid administration, swallowing difficulty with profound muscle weakness had developed. The patient was diagnosed with DRESS syndrome with eosinophilic polymyositis by a histopathologic study. After complete resolution of all symptoms, patch tests were positive to celecoxib and ethambutol crushed tablets 10% and 50% pet. (2).

A 30-year-old woman developed DRESS with high fever, maculopapular rash, micropustules and painful erosions around the oral mucosa, palpable cervical lymph nodes, eosinophilia and elevated liver enzymes, which had started 2 weeks after having used a combination of isoniazid, ethambutol, rifampicin and pyrazinamide for tuberculosis of the lungs. Six months after recovery, all drugs were patch tested at 20% pet. and only ethambutol reacted (8).

A very similar case was reported from Morocco in 2014. This 27-year-old woman had also been treated with the 4 first-line antituberculosis drugs and developed fever, extensive infiltrated erythema, lymphadenopathy, impaired liver function, leukocytosis and eosinophilia. Patch tests were positive to ethambutol and isoniazid, tested at 1%, vehicle not mentioned (9).

A 56-year-old man developed pulmonary and lymph node tuberculosis, for which treatment was started with isoniazid, rifampicin, ethambutol and pyrazinamide. Six weeks later, the patient was hospitalized because of fever (38.5°C), an extensive maculopapular rash (>50% of body surface area), pronounced facial and palmoplantar edema, and disseminated peripheral lymphadenopathy. Laboratory tests showed eosinophilia and elevated serum aminotransferases. Three months later, patch tests were positive to isoniazid 1% pet., ethambutol 3% pet. and pyrazinamide 3% pet., the latter two made from pulverized commercial tablets. Twenty controls were negative to the ethambutol patch test material. Despite the negative patch test to rifampicin, graded re-introduction of the drug, even under prophylactic therapy with corticosteroids, led to a recurrence of DRESS symptoms after 7 days. No patch tests with rifampicin were performed afterwards (12).

Stevens-Johnson syndrome/toxic epidermal necrolysis (SJS/TEN)
See the section 'Drug reaction with eosinophilia and systemic symptoms (DRESS)' above, ref. 11.

Dermatitis/eczematous eruption
A 58-year-old man presented with a pruritic erythematous eruption of tiny vesicles on the trunk and extremities that developed during the sixth week of oral therapy with isoniazid, ethambutol, rifampicin, and morinamide HCl for pleural tuberculosis. Histopathologic findings were consistent with an eczematous eruption. Patch tests were positive to ethambutol, tested as ethambutol HCl tablets 10% and 30% water and 5%, 10% and 30% pet. There were also positive patch tests to isoniazid (Chapter 3.274). The patient was diagnosed with eczematous-type multiple drug allergy to ethambutol and isoniazid (10).

Other cutaneous adverse drug reactions
A 24-year-old woman had recently started therapy with isoniazid, ethambutol and rifampicin for tuberculosis. This was discontinued 2 weeks later due to an itchy, desquamative, erythematous, papular rash, with painful crusts and excoriations, spreading from the face and hands to the rest of the body. Five days after withdrawal of all drugs,

widespread desquamation and exfoliation with predilection for the body folds and the peri-orbital and peri-oral regions occurred, with swelling of the eyelids and conjunctivitis. The diagnosis toxicoderma was supported by histology of the skin and an eosinophilia (31%). Patch tests were positive to ethambutol and isoniazid, tested as crushed tablets moistened with water. Ten controls were negative (3).

Cross-reactions, pseudo-cross-reactions and co-reactions
In several reports, patients with cutaneous adverse drug reactions from antituberculosis drugs proved to be patch test positive to more than one of these pharmaceuticals (3,5,6,9,10,12). These were not cases of cross-reactivity, but of concomitant sensitization. This may be explained by the fact that, very often, patients with tuberculosis are treated with a combination (pill) of the 4 first-line antituberculosis agents ethambutol, isoniazid, rifampicin and pyrazinamide (7,8,12) or other combinations (10,11). Indeed, the structural formulas of the various antituberculosis medicaments are quite different and do not favor immunological cross-reactivity.

LITERATURE
1 The data in the section 'General' may have been obtained from literature discussed in this chapter, but mostly also or exclusively from one or more of the following online sources: ChemIDPlus Advanced, PubChem, DrugBank, RxList, Drug Central, Drugs.com, and Wikipedia
2 Lee JH, Park HK, Heo J et al. Drug rash with eosinophilia and systemic symptoms (DRESS) syndrome induced by celecoxib and anti-tuberculosis drugs. J Korean Med Sci 2008;23:521-525
3 Bakkum RS, Waard-Van Der Spek FB, Thio HB. Delayed-type hypersensitivity reaction to ethambutol and isoniazid. Contact Dermatitis 2002;46:359
4 Holdiness MR. Contact dermatitis to ethambutol. Contact Dermatitis 1986;15:96-97
5 Allouchery M, Logerot S, Cottin J, Pralong P, Villier C, Ben Saïd B; French Pharmacovigilance Centers Network and the French Investigators for skin adverse reactions to drugs. Antituberculosis drug-associated DRESS: A case series. J Allergy Clin Immunol Pract 2018;6:1373-1380
6 Lee SW, Yoon NB, Park SM, Lee SM, Um SJ, Lee SK, et al. Antituberculosis drug-induced drug rash with eosinophilia and systemic symptoms syndrome confirmed by patch testing. J Investig Allergol Clin Immunol 2010;20:631-632
7 Lehloenya RJ, Todd G, Wallace J, Ngwanya MR, Muloiwa R, Dheda K. Diagnostic patch testing following tuberculosis-associated cutaneous adverse drug reactions induces systemic reactions in HIV-infected persons. Br J Dermatol 2016;175:150-156
8 Yoshioka Y, Hanafusa T, Namiki T, Nojima K, Amano M, Tokoro S, et al. Drug-induced hypersensitivity syndrome by ethambutol: A case report. J Dermatol 2016;43:971-972
9 Bopaka RG, El Khattabi W, Afif H, Aichane A, Bouayad Z. The "DRESS" syndrome in antituberculosis drugs. Rev Pneumol Clin 2014;70:185-188 (Article in French).
10 Özkaya E. Eczematous-type multiple drug allergy from isoniazid and ethambutol with positive patch test results. Cutis 2013;92:121-124
11 Kim JY, Sohn KH, Song WJ, Kang HR. A case of drug reaction with eosinophilia and systemic symptoms induced by ethambutol with early features resembling Stevens-Johnson syndrome. Acta Derm Venereol 2013;93:753-754
12 Coster A, Aerts O, Herman A, Marot L, Horst N, Kenyon C, et al. Drug reaction with eosinophilia and systemic symptoms (DRESS) syndrome caused by first-line antituberculosis drugs: Two case reports and a review of the literature. Contact Dermatitis 2019;81:325-331
13 De Groot AC. Patch testing, 4th edition. Wapserveen, The Netherlands: acdegroot publishing, 2018 (ISBN 9789081323345)

Chapter 3.207 ETHENZAMIDE

IDENTIFICATION
Description/definition   : Ethenzamide is the benzamide that conforms to the structural formula shown below
Pharmacological classes   : Anti-inflammatory agents, non-steroidal
IUPAC name   : 2-Ethoxybenzamide
CAS registry number   : 938-73-8
EC number   : 213-346-4
Merck Index monograph : 5056
Patch testing   : 20% pet.
Molecular formula   : $C_9H_{11}NO_2$

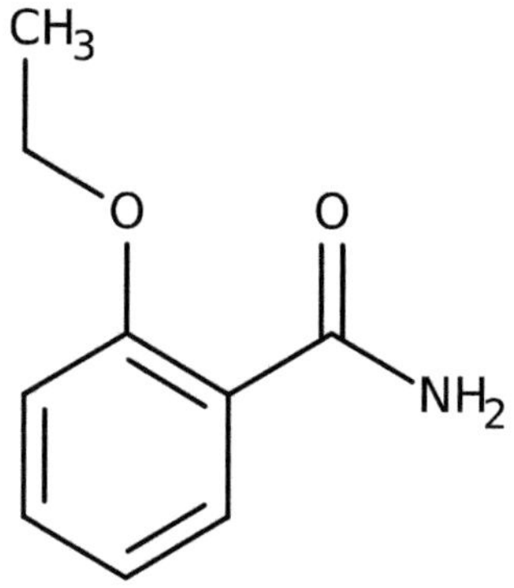

GENERAL
Ethenzamide is a nonsteroidal anti-inflammatory drug (NSAID) with analgesic, antipyretic and anti-inflammatory properties used for the relief of fever, headaches, and other minor aches and pains. It is most frequently used in combination preparations (1).

CUTANEOUS ADVERSE DRUG REACTIONS FROM SYSTEMIC ADMINISTRATION CAUSED BY TYPE IV (DELAYED-TYPE) HYPERSENSITIVITY (as demonstrated by positive patch tests)
Cutaneous adverse drug reactions from systemic administration of ethenzamide caused by type IV (delayed-type) hypersensitivity have included fixed drug eruption (2).

Fixed drug eruption
A 21-year-old woman had taken 2 tablets containing acetaminophen, ethenzamide, apronalide (allylisopropylacetyl-urea) and caffeine orally for menstrual pain. A few hours later, she developed pruritic erythema, followed by pigmented lesions. On two later occasions where she took the tablets, identical erythematous lesions at the same sites arose, followed by pigmentation. Patch tests with the 4 components of the drugs tested 10% and 20% pet. on uninvolved skin and 20% pet. on postlesional skin showed a positive reaction only to ethenzamide on previously involved skin, giving a hyperpigmented lesion surrounded by erythema (2). This was a case of fixed drug eruption.

LITERATURE
1   The data in the section 'General' may have been obtained from literature discussed in this chapter, but mostly also or exclusively from one or more of the following online sources: ChemIDPlus Advanced, PubChem, DrugBank, RxList, Drug Central, Drugs.com, and Wikipedia
2   Kawada A, Hiruma M, Noguchi H, Akagi A, Ishibashi A, Marshall J. Fixed drug eruption induced by ethenzamide. Contact Dermatitis 1996;34:369-370

Chapter 3.208 ETHOSUXIMIDE

IDENTIFICATION

Description/definition	: Ethosuximide is the pyrrolidine-2-one that conforms to the structural formula shown below
Pharmacological classes	: Anticonvulsants
IUPAC name	: 3-Ethyl-3-methylpyrrolidine-2,5-dione
Other names	: Ethylmethylsuccimide
CAS registry number	: 77-67-8
EC number	: 201-048-7
Merck Index monograph	: 5071
Patch testing	: Gel from ethosuximide capsule diluted to 10% active ingredients (3)
Molecular formula	: $C_7H_{11}NO_2$

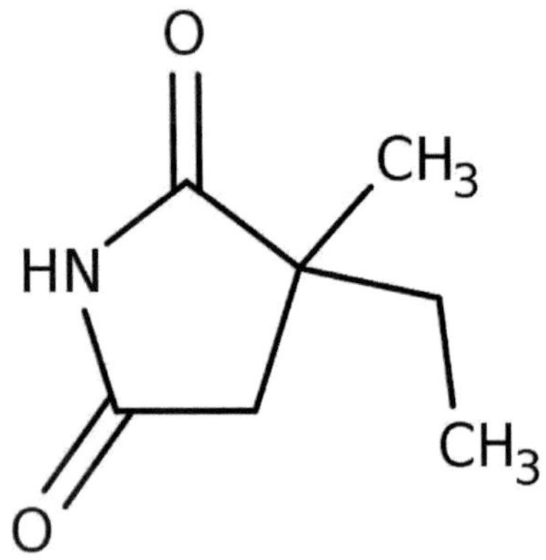

GENERAL

Ethosuximide is a pyrrolidine-2-one with anticonvulsant activity. The drug suppresses the paroxysmal three cycle per second spike and wave activity associated with lapses of consciousness which is common in absence (petit mal) seizures. The frequency of epileptiform attacks is reduced, apparently by depression of the motor cortex and elevation of the threshold of the central nervous system to convulsive stimuli. Ethosuximide is used for absence (petit mal) seizures in both adults and children unaccompanied by other types of seizures (1).

CUTANEOUS ADVERSE DRUG REACTIONS FROM SYSTEMIC ADMINISTRATION CAUSED BY TYPE IV (DELAYED-TYPE) HYPERSENSITIVITY (as demonstrated by positive patch tests)

Cutaneous adverse drug reactions from systemic administration of ethosuximide caused by type IV (delayed-type) hypersensitivity have included drug reaction with eosinophilia and systemic symptoms (DRESS)/anticonvulsant hypersensitivity syndrome) (2) and SJS/TEN overlap (3).

Drug reaction with eosinophilia and systemic symptoms (DRESS)

A 6-year-old boy, who was treated with sodium valproate and ethosuximide for epileptic absences, developed DRESS with a diffuse pruritic morbilliform skin eruption with vesicular and target lesions, edema of the face, high fever, enlarged lymph nodes, leukocytosis, eosinophilia, elevated C-reactive protein and elevated liver enzymes. Human herpesvirus 6 (HHV6) antibody titers increased significantly within 15 days. Patch tests were positive to both commercial drugs 20% water. 25 controls were negative. Histologic examination of a positive patch test showed acute dermatitis. Following the positive sodium valproate patch test, there was a recurrence of the skin rash on the arms, legs and face (2).

Stevens-Johnson syndrome/toxic epidermal necrolysis (SJS/TEN)

A 4-year-old boy was diagnosed with SJS/TEN overlap, exhibiting severe mucosal erosions and skin blisters on the face and trunk, which had started 15 days after the onset of ethosuximide administered for absence seizures. Three months after recovery, patch tests were performed with several antiepileptic drugs including carbamazepine, diphenylhydantoin, lamotrigine, sodium valproate, diazepam, topiramate, levetiracetam and ethosuximide. There were positive reactions only to ethosuximide tested as gel from the capsule 'as is' and diluted to 10% active ingredients in petrolatum. Thirteen controls were negative (3).

LITERATURE

1 The data in the section 'General' may have been obtained from literature discussed in this chapter, but mostly also or exclusively from one or more of the following online sources: ChemIDPlus Advanced, PubChem, DrugBank, RxList, Drug Central, Drugs.com, and Wikipedia

2 Conilleau V, Dompmartin A, Verneuil L, Michel M, Leroy D. Hypersensitivity syndrome due to 2 anticonvulsant drugs. Contact Dermatitis 1999;41:141-144

3 Matos AL, Carvalho J, Calado R, Gomes T, Ramos L, Gonçalo M. Ethosuximide-induced Stevens-Johnson syndrome/toxic epidermal necrolysis overlap confirmed by patch testing in a child. Contact Dermatitis 2021 Jul 4. doi: 10.1111/cod.13928. Epub ahead of print.

Chapter 3.209 ETONOGESTREL

IDENTIFICATION

Description/definition : Etonogestrel is the synthetic form of the naturally occurring female sex hormone progesterone that conforms to the structural formula shown below
Pharmacological classes : Contraceptive agents, female
IUPAC name : (17α)-13-Ethyl-17-hydroxy-11-methylene-18,19-dinorpregn-4-en-20-yn-3-one
Other names : 3-Oxodesogestrel; 3-ketodesogestrel
CAS registry number : 54048-10-1
EC number : 258-936-2
Merck Index monograph : 5198
Patch testing : Vaginal ring with etonogestrel and its vehicle without etonogestrel; suggested: etonogestrel 0.1% and 1% alc. and pet.
Molecular formula : $C_{22}H_{28}O_2$

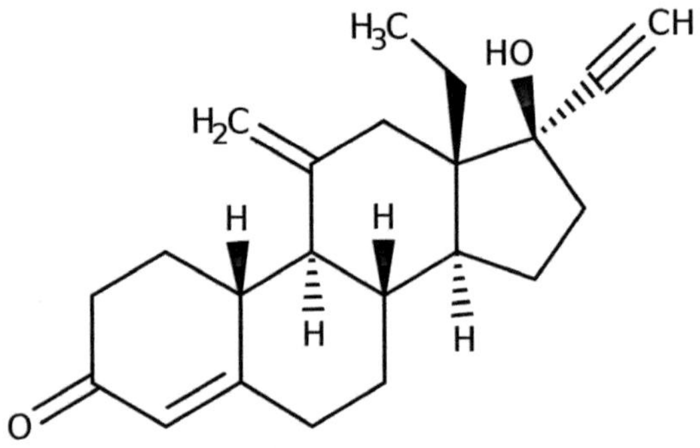

GENERAL

Etonogestrel is a synthetic form of the naturally occurring female sex hormone progesterone. This agent binds to the cytoplasmic progesterone receptors in the reproductive system and subsequently activates progesterone receptor-mediated gene expression. As a result of the negative feedback mechanism, luteinizing hormone (LH) release is inhibited, which leads to an inhibition of ovulation and an alteration in the cervical mucus and endometrium. Etonogestrel is administered in subdermal implants as long-acting reversible contraception and in vaginal rings in combination with estrogens (1).

CUTANEOUS ADVERSE DRUG REACTIONS FROM SYSTEMIC ADMINISTRATION CAUSED BY TYPE IV (DELAYED-TYPE) HYPERSENSITIVITY (as demonstrated by positive patch tests)
Cutaneous adverse drug reactions from systemic administration of etonogestrel caused by type IV (delayed-type) hypersensitivity have included symmetrical drug-related intertriginous and flexural exanthema (SDRIFE) (2).

Symmetrical drug-related intertriginous and flexural exanthema (SDRIFE)/Baboon syndrome
A 33-year-old woman presented with a very itchy eruption, symmetrically located on the large body folds (buttocks, groins, and armpits). The patient had used a contraceptive vaginal ring containing etonogestrel and ethinylestradiol for 7 days. Previous contraceptive oral treatment with desogestrel, another progestative, had been well tolerated. The patient had immediately removed the ring and was treated with a reducing course of oral corticosteroids. Patch tests with the vaginal ring, its vinyl acetate copolymer vehicle, etonogestrel tablets pure, 30% and 10% pet., and with another contraceptive agent containing etonogestrel but without ethinylestradiol and with the same vinyl acetate copolymer vehicle, gave positive reactions to the 2 rings containing etonogestrel. Thirty controls were negative. The patient was diagnosed with symmetrical drug-related intertriginous and flexural exanthema (SDRIFE, baboon syndrome). How she had become sensitized to etonogestrel previously was unknown (2).

LITERATURE
1   The data in the section 'General' may have been obtained from literature discussed in this chapter, but mostly also or exclusively from one or more of the following online sources: ChemIDPlus Advanced, PubChem, DrugBank, RxList, Drug Central, Drugs.com, and Wikipedia
2   Peeters D, Baeck M, Dewulf V, Tennstedt D, Dachelet C. A case of SDRIFE induced by Nuvaring(®). Contact Dermatitis 2012;66:110-111

Chapter 3.210 ETORICOXIB

IDENTIFICATION

Description/definition : Etoricoxib is the sulfone and pyridine derivative that conforms to the structural formula shown below

Pharmacological classes : Cyclooxygenase 2 inhibitors; anti-inflammatory agents, non-steroidal

IUPAC name : 5-Chloro-2-(6-methylpyridin-3-yl)-3-(4-methylsulfonylphenyl)pyridine

CAS registry number : 202409-33-4

EC number : Not available

Merck Index monograph : 5200

Patch testing : Tablet, pulverized, 10% pet. (2); most systemic drugs can be tested at 10% pet.; if the pure chemical is not available, prepare the test material from intravenous powder, the content of capsules or – when also not available – from powdered tablets to achieve a final concentration of the active drug of 10% pet.

Molecular formula : $C_{18}H_{15}ClN_2O_2S$

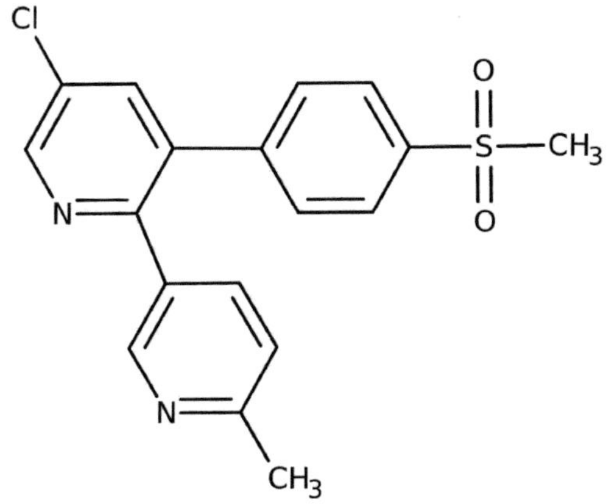

GENERAL

Etoricoxib is a synthetic nonsteroidal anti-inflammatory drug (NSAID) and COX-2 selective inhibitor with antipyretic, analgesic, and potential antineoplastic properties. It specifically binds to and inhibits the enzyme cyclooxygenase-2 (COX-2), resulting in inhibition of the conversion of arachidonic acid into prostaglandins. Inhibition of COX-2 may also induce apoptosis and inhibit tumor cell proliferation and angiogenesis. Current therapeutic indications are treatment of rheumatoid arthritis, osteoarthritis, ankylosing spondylitis, chronic low back pain, acute pain and gout (1).

CUTANEOUS ADVERSE DRUG REACTIONS FROM SYSTEMIC ADMINISTRATION CAUSED BY TYPE IV (DELAYED-TYPE) HYPERSENSITIVITY (as demonstrated by positive patch tests)

Cutaneous adverse drug reactions from systemic administration of etoricoxib caused by type IV (delayed-type) hypersensitivity have included acute generalized exanthematous pustulosis (AGEP) (14), symmetrical drug-related intertriginous and flexural exanthema (SDRIFE)/baboon syndrome (13) and fixed drug eruption (2-12,15,16).

Acute generalized exanthematous pustulosis (AGEP)

A 55-year-old man took etoricoxib 60 mg twice for gout and 3 days later he was febrile (40.2°C) and his face appeared swollen. A spotty rash was seen on his skin, and small pustular lesions covered large areas of the face, body and proximal extremities. Purpuric lesions were seen on his legs. Laboratory results showed an elevation of C-reactive protein value, white blood cell count without eosinophilia and of liver enzymes. The histopathological findings supported a clinical diagnosis of AGEP. Three months later, patch tests were positive to powder from etoricoxib tablets 1%, 10% and 30% pet., with a possible cross-reaction to celecoxib but not to parecoxib. Five controls were negative (14).

Symmetrical drug-related intertriginous and flexural exanthema (SDRIFE)/Baboon syndrome

A 61-year-old woman had developed erythema of both axillae, the inframammary folds, and the cubital and popliteal fossae (gluteal and inguinal areas), without systemic symptoms, 6 hours after taking, for the first time, etoricoxib 90 mg to treat osteoarticular pain. A single-blind oral challenge test resulted in erythema of the inframammary folds and inguinal and gluteal areas 6 hours later. Patch tests with etoricoxib 8% DMSO and celecoxib 10% were positive to both at D2 and D4. Ten controls were negative. The patient was diagnosed with symmetrical drug-related intertriginous and flexural exanthema (SDRIFE) caused by delayed-type hypersensitivity to etoricoxib with cross-reaction to celecoxib (13).

Fixed drug eruption
Seven patients with fixed drug eruption (FDE) after intake of etoricoxib on at least two occasions were patch tested with etoricoxib crushed tablets 10% pet. Six had positive reactions on lesional skin. The 7th showed a negative patch test; subsequently, an oral challenge test with 30 mg of etoricoxib resulted in reappearance of skin lesions 6 hours later. There were no cross-reactions to celecoxib (15).

A 54-year-old woman with a history of three blistering outbreaks within the last 4 months presented with a 2-day history of tense, clear bullae located in the oral mucosa and perioral area, associated with reactivation of previously existing violaceous patches. She reported having taken etoricoxib the day before for the relief of joint pain. Histopathology was consistent with a diagnosis of fixed drug eruption. Patch tests were positive to etoricoxib powder of commercial pills 10% pet. only on postlesional skin (3).

A 51-year-old woman presented a history of an episode of multiple violaceous, pruritic, and painful plaques on her shoulders and elbows, and a violaceous bullous lesion on her right upper thigh, appearing 24 hours after an arthroscopic knee surgery. She also recalled a history of a bullous lesion on her right upper thigh one year before, after receiving treatment for shoulder pain. Physical examination revealed hyperpigmentation in the areas of previous lesions. The only drug used on both occasions was etoricoxib. Patch tests with etoricoxib and celecoxib (at 10% and 30% in petrolatum) were performed in the upper back and in lesional skin (thigh), all inducing new violaceus, bullous lesions in the previously affected areas but not on the back (4).

Similar cases of limited or extensive, bullous or non-bullous fixed drug eruptions from etoricoxib with a positive reaction to this drug on previously involved but not on normal skin have been reported many times (2,5-12). In most, etoricoxib had been tested at 10% pet.; in some investigations, 30% pet. (8), 1% and 5% (11) or 5% pet. (12) was used. It was mostly not clarified whether the pure material had been used for patch testing or powder from pills (likely in the cases of 10% and 30% pet.). In one patient, a generalized itching erythema developed at the same time as the fixed drug eruption (10). One case of FDE caused by etoricoxib with a positive patch test was reported from Singapore; clinical nor patch testing details were provided (16).

Cross-reactions, pseudo-cross-reactions and co-reactions
No cross-reactivity to celecoxib (3,5,6,7) and parecoxib (6,14). Three patients with fixed drug eruption (4), symmetrical drug-related intertriginous and flexural exanthema (SDRIFE) (13) resp. acute generalized exanthematous pustulosis (AGEP) (14) caused by etoricoxib had positive patch tests to etoricoxib and to celecoxib (cross-reactions).

LITERATURE
1   The data in the section 'General' may have been obtained from literature discussed in this chapter, but mostly also or exclusively from one or more of the following online sources: ChemIDPlus Advanced, PubChem, DrugBank, RxList, Drug Central, Drugs.com, and Wikipedia
2   Andrade P, Brinca A, Gonçalo M. Patch testing in fixed drug eruptions – a 20-year review. Contact Dermatitis 2011;65:195-201
3   Miroux-Catarino A, Silva L, Amaro C, Ferreira ML, Viana I. Bullous fixed drug eruption induced by etoricoxib, confirmed by patch testing, with tolerance to celecoxib. Contact Dermatitis 2019;81:388-389
4   Carneiro-Leao L, Rodrigues Cernadas J. Bullous fixed drug eruption caused by etoricoxib confirmed by patch testing. J Allergy Clin Immunol Pract 2019;7:1629-1630
5   Gomez de la Fuente E, Pampin Franco A, Caro Gutierrez D, Lopez Estebaranz JL. Fixed drug eruption due to etoricoxib in a patient with tolerance to celecoxib: the value of patch testing. Actas Dermosifiliogr 2014;105:314-315 (article in Spanish)
6   Ponce V, Muñoz-Bellido F, Moreno E, Laffond E, González A, Dávila I. Fixed drug eruption caused by etoricoxib with tolerance to celecoxib and parecoxib. Contact Dermatitis 2012;66:107-108
7   Andrade P, Gonçalo M. Fixed drug eruption caused by etoricoxib- 2 cases confirmed by patch testing. Contact Dermatitis 2011;64:118-120
8   Antunes J, Prates S, Leiria-Pinto P. Fixed drug eruption due to etoricoxibe—a case report. Allergol Immunopathol 2014;42:623-624
9   De Sousa AS, Cardoso JC, Gouveia MP, Gameiro AR, Teixeira VB, Gonçalo M. Fixed drug eruption by etoricoxib confirmed by patch test. An Bras Dermatol 2016;91:652-654
10   Augustine M, Sharma P, Stephen J, Jayaseelan E. Fixed drug eruption and generalised erythema following etoricoxib. Indian J Dermatol Venereol Leprol 2006;72:307-309

11 Calistru AM, Cunha AP, Nogueira A, Azevedo F. Etoricoxib-induced fixed drug eruption with positive lesional patch tests. Cutan Ocul Toxicol 2011;30:154-156

12 Duarte A F, Correia O, Azevedo R, do Carmo Palmares M, Delgado L. Bullous fixed drug eruption to etoricoxib – further evidence of intraepidermal CD8+ T cell involvement. Eur J Dermatol 2010;20:236-238

13 Caralli ME, Seoane Rodríguez M, Rojas Pérez-Ezquerra P, Pelta Fernández R, de Barrio Fernández M. Symmetrical drug-related intertriginous and flexural exanthema (SDRIFE) caused by etoricoxib. J Investig Allergol Clin Immunol 2016;26:128-129

14 Mäkelä L, Lammintausta K. Etoricoxib-induced acute generalized exanthematous pustulosis. Acta Derm Venereol 2008;88:200-201

15 Martínez Antón MD, Galán Gimeno C, Sánchez de Vicente J, Jáuregui Presa I, Gamboa Setién PM. Etoricoxib-induced fixed drug eruption: Report of seven cases. Contact Dermatitis 2021;84:192-195

16 Heng YK, Yew YW, Lim DS, Lim YL. An update of fixed drug eruptions in Singapore. J Eur Acad Dermatol Venereol 2015;29:1539-1544

17 Brockow K, Garvey LH, Aberer W, Atanaskovic-Markovic M, Barbaud A, Bilo MB, et al.; ENDA/EAACI Drug Allergy Interest Group. Skin test concentrations for systemically administered drugs – an ENDA/EAACI Drug Allergy Interest Group position paper. Allergy 2013;68:702-712

Chapter 3.211 ETRETINATE

IDENTIFICATION

Description/definition   : Etretinate is the synthetic aromatic retinoid that conforms to the structural formula
                               shown below
Pharmacological classes   : Keratolytic agents
IUPAC name   : Ethyl (2E,4E,6E,8E)-9-(4-Methoxy-2,3,6-trimethylphenyl)-3,7-dimethylnona-2,4,6,8-
                             tetraenoate
CAS registry number   : 54350-48-0
EC number   : 259-119-3
Merck Index monograph   : 5206
Patch testing   : No data available; most systemic drugs can be tested at 10% pet.; if the pure chemical is
                             not available, prepare the test material from intravenous powder, the content of capsules
                             or – when also not available – from powdered tablets to achieve a final concentration of
                             the active drug of 10% pet.
Molecular formula   : $C_{23}H_{30}O_3$

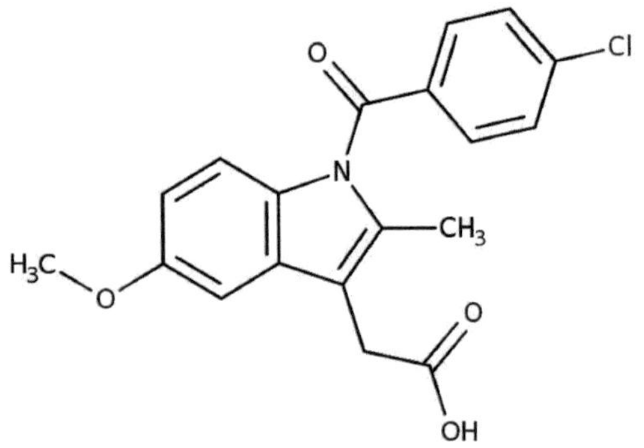

GENERAL

Etretinate is a synthetic oral retinoid (ester derivative of retinoic acid) that is a prodrug of acitretin. Etretinate activates retinoid receptors, causing an induction of cell differentiation, inhibition of cell proliferation, and inhibition of tissue infiltration by inflammatory cells. In certain countries including Canada and the USA, etretinate is no longer commercially available due to the extended potential for teratogenic effects related to its long half-life of the drug. It is indicated for the treatment of severe psoriasis and may also be used to treat T cell lymphomas (1).

CUTANEOUS ADVERSE DRUG REACTIONS FROM SYSTEMIC ADMINISTRATION CAUSED BY TYPE IV (DELAYED-TYPE) HYPERSENSITIVITY (as demonstrated by positive patch tests)

Cutaneous adverse drug reactions from systemic administration of etretinate caused by type IV (delayed-type) hypersensitivity have included photosensitivity with photoleukomelanoderma (2).

Photosensitivity

A 62-year-old farmer presented with erythema and swelling of the face and dorsum of the hands of 2 weeks' duration. The eruption had started 6 weeks after having used etretinate for psoriasis. The medicament was discontinued and the patient was recommended to shield the skin from light. However, pigmentation and depigmentation were observed on the face, dorsum of the hands, and neck 6 weeks later. A lymphocyte stimulation test was negative, as was a patch with etretinate (read at D2 only). The minimal erythema doses of UVA and UVB were within normal range. A photopatch test for etretinate was positive after irradiation with a 1/2 MED of UVA, but not 1/2 MED of UVB. One and a half months after the discontinuation of etretinate treatment, the patient took this agent for 2 days as a provocation test, which reduced MED by UVA from 10 to 6 J/cm². A diagnosis of photoleukomelanoderma due to etretinate was made (2). Whether this was a case of photoallergy is far from certain, but the authors did not claim this and did not even comment on the significance of the positive photopatch test.

LITERATURE

1   The data in the section 'General' may have been obtained from literature discussed in this chapter, but mostly
    also or exclusively from one or more of the following online sources: ChemIDPlus Advanced, PubChem,
    DrugBank, RxList, Drug Central, Drugs.com, and Wikipedia
2   Seishima M, Shibuya Y, Kato G, Watanabe K. Photoleukomelanoderma possibly caused by etretinate in a patient
    with psoriasis. Acta Derm Venereol 2010;90:85-86

Chapter 3.212 FAMOTIDINE

IDENTIFICATION

Description/definition : Famotidine is the propanimidamide that conforms to the structural formula shown below
Pharmacological classes : Anti-ulcer agents; histamine H_2 antagonists
IUPAC name : 3-[[2-(Diaminomethylideneamino)-1,3-thiazol-4-yl]methylsulfanyl]-N'-sulfamoylpropani-midamide
CAS registry number : 76824-35-6
EC number : 616-396-9
Merck Index monograph : 5241
Patch testing : 1% and 5% water
Molecular formula : $C_8H_{15}N_7O_2S_3$

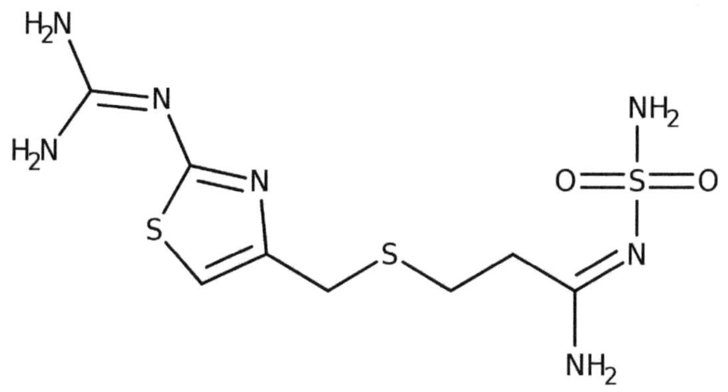

GENERAL

Famotidine is a propanimidamide and histamine H_2 receptor antagonist with antacid activity. It is used for the treatment of peptic ulcer disease and gastroesophageal reflux disease. Famotidine is about 20 to 50 times more potent at inhibiting gastric acid secretion than cimetidine and eight times more potent than ranitidine on a weight basis. The drug is used in various over-the-counter and off-label uses. While oral formulations of famotidine are more commonly used, the intravenous solution of the drug is available for use in hospital settings (1).

CONTACT ALLERGY FROM ACCIDENTAL CONTACT

A 35-year-old nurse presented with contact dermatitis of both hands, which got worse at work. Following observation, famotidine was suspected. Famotidine was diluted in distilled water at 0.1, 1, 10, 20, 40, and 100 mg/ml and ranitidine and cimetidine were patch tested in the same concentrations. The patient showed strong reactions to famotidine 10 mg/ml and the higher concentrations; the lower concentrations and cimetidine and ranitidine were negative. Six controls were negative. The patient had occupational allergic contact dermatitis of the hands from famotidine (2).

LITERATURE

1   The data in the section 'General' may have been obtained from literature discussed in this chapter, but mostly also or exclusively from one or more of the following online sources: ChemIDPlus Advanced, PubChem, DrugBank, RxList, Drug Central, Drugs.com, and Wikipedia
2   Monteseirín J, Conde J. Contact eczema from famotidine. Contact Dermatitis 1990;22:290

Chapter 3.213 FENOFIBRATE

IDENTIFICATION

Description/definition	: Fenofibrate is a synthetic phenoxy-isobutyric acid derivate that conforms to the structural formula shown below
Pharmacological classes	: Hypolipidemic agents
IUPAC name	: Propan-2-yl 2-[4-(4-chlorobenzoyl)phenoxy]-2-methylpropanoate
CAS registry number	: 49562-28-9
EC number	: 256-376-3
Merck Index monograph	: 5279
Patch testing	: 10.0% pet. (Chemotechnique)
Molecular formula	: $C_{20}H_{21}ClO_4$

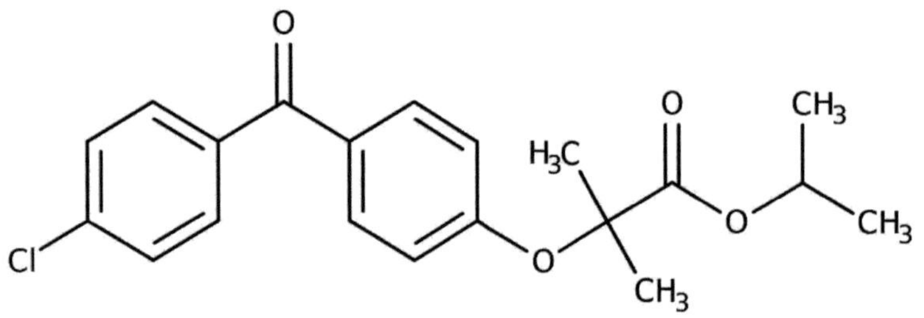

GENERAL

Fenofibrate is a synthetic phenoxy-isobutyric acid derivate and prodrug with lipid regulating activity. It is hydrolyzed *in vivo* to its active metabolite fenofibric acid. Fenofibrate is indicated for use as adjunctive therapy to diet to reduce elevated LDL cholesterol, total cholesterol, triglycerides and apolipoprotein B, and to increase HDL cholesterol in adult patients with primary hypercholesterolemia or mixed dyslipidemia (1).

CUTANEOUS ADVERSE DRUG REACTIONS FROM SYSTEMIC ADMINISTRATION CAUSED BY TYPE IV (DELAYED-TYPE) HYPERSENSITIVITY (as demonstrated by positive patch tests)

Cutaneous adverse drug reactions from systemic administration of fenofibrate caused by type IV (delayed-type) hypersensitivity have included maculopapular eruption (27), photoallergic dermatitis (9,10,11,12,13,14,15,16,17, 18,19,21,26), in two cases as a manifestation of systemic photoallergic dermatitis (systemic photocontact dermatitis), caused by fenofibrate photocross-reacting to previous ketoprofen photosensitization (19,21).

Maculopapular eruption

A 77-year-old woman had been treated with fenofibrate 200 mg/d for dyslipidemia for 10 days, when she developed a generalized maculopapular eruption. Patch tests were positive at D3 to fenofibrate tested as the tablet powder diluted with a drop of saline (27).

Systemic photoallergic dermatitis (systemic photocontact dermatitis)

Two cases of systemic photoallergic dermatitis caused by fenofibrate photocross-reacting to previous ketoprofen photosensitization have been reported (19,21). These are described below in the section Photosensitivity – Case reports.

Photosensitivity

General

Since its introduction in 1975, various case reports and small case series of photosensitivity to fenofibrate have been reported (9,10,11,12,13,14,15,16,17,18,19,21). Not all patients were photopatch tested, and in those that were, a considerable number had negative photopatch tests (10,17,24,25). Possible explanations are technical problems (concentration of fenofibrate too low [24], insufficient penetration from the vehicle into the skin, insufficient ultraviolet energy or wrong wavelength used); a metabolite of fenofibrate, which is not formed during photopatch testing, being the photosensitizer in some cases; or – of course – a non-(photo)allergic mechanism being responsible for clinical photosensitivity.

Many patients previously photosensitized to ketoprofen photocross-react to fenofibrate (2,3,4,5,6,20,21), and in these patients, systemic photocontact dermatitis may be observed (19,21).

As is the case with most drugs that cause photoallergy, fenofibrate has definite phototoxic properties, both in the UVA and UVB region, probably mediated by the diaryl ketone chromophore (22,23). In many cases of presumed photoallergy to fenofibrate, photopatch tests were positive both with fenofibrate irradiated with UVA and UVB (10,12,13,15,19,20). In a number of publications, the concentration of fenofibrate used was not mentioned, but possibly high from using powdered pills. In addition, in most cases, no photopatch testing in healthy controls has been performed (and in one, one of 7 controls was positive, ref. 10). Thus, there is a strong possibility that some cases reported as photoallergy to fenofibrate may instead have been the result of phototoxicity, or at least that the 'positive photopatch tests' were phototoxic rather than photoallergic.

Routine photopatch testing
Fenofibrate 10% pet. was considered 'suitable' for including in an extended series for photopatch testing by a taskforce of the European Society of Contact Dermatitis and the European Society for Photodermatology (7). In the period 2014-2016, in Spain, 116 patients were photopatch tested (selection procedure not mentioned) with fenofibrate 10% pet. and there were 6 (5.2%) positive reactions. All reactions were considered to be cross-reactions to NSAIDs and UV-absorbers (8).

Case series
In a hospital in Caen, France, in a two year period (May 1997-May 1999), 15 patients were investigated for suspected photosensitivity to systemic drugs, of who five used fenofibrate. Only one case was confirmed by patch and photopatch testing. This 64-year-old man was tested with all drugs he had used at 20% in water and had a positive photopatch test to fenofibrate, both when irradiated with UVA and with UVB. One of 7 controls also had a positive reaction. It was suggested that there were false-negative results to fenofibrate, possibly because the test concentration or vehicle were wrong or because the culprit agent of the photosensitivity is not fenofibrate itself but one of its metabolites (10).

Between 1994 and 1996, in Tours, France, four patients with suspected fenofibrate photosensitivity were studied. In two, photopatch tests with fenofibrate 5% pet. irradiated with UVA were positive. Ten controls were negative. In the third patient, despite negative photopatch tests, reintroduction of one tablet for further phototes-ting showed a strong reaction on the UVA-irradiated site but was also responsible for recurrence of the sun-exposed eruption (17).

In 1992, in La Lettre du GERDA, the newsletter of the French Groupe d'Étude et de Recherche en Dermato-Allergologie, three patients were presented with photoallergy to fenofibrate. Two were women of 40 and 63 and the third was a man aged 32. All three had positive photopatch tests to fenofibrate irradiated with UVA, but details are unknown (16).

Case reports
A 60-year-old woman presented with an itching and burning eczematous eruption on the face, V-area of the neck, forearms, and backs of the hands, which had appeared after she had used fenofibrate capsules for 2 weeks. Patch tests were positive to cinnamyl alcohol, and photopatch tests to fenofibrate, ketoprofen, and octocrylene at D4 and D7. This spectrum of (photo)patch tests is highly suggestive of primary photosensitization to ketoprofen (6), but the patient denied previous use of any topical product containing ketoprofen (11).

A 33-year-old woman presented with numerous itching and burning erythematous papulovesicles on the sun-exposed areas of the face, V-area of the neck, both forearms, dorsa of the hands, and the shins, which had appeared in the third week of taking fenofibrate. A skin biopsy was compatible with photosensitive dermatitis. Photopatch tests were positive to fenofibrate 10% and 30% irradiated with UVA 10 J/cm2. Controls tests were apparently not performed (14). The authors also presented 3 other patients suspected of fenofibrate photosen-sitivity, but in these, photopatch tests were not performed (14).

After having taken fenofibrate for 6 days and following a 2-hour walk, a 34-year-old man suddenly developed pruritic papules and erythema on sun-exposed areas of the face, ears, neck, and dorsum of the hands with vesicles and exudates on the ears and on the hands. Photopatch test were positive to fenofibrate 30% using UVA 3 J/cm^2 and UVB 56 mJ/cm^2. The patient had a history of erythema developing on the sun-exposed skin where a ketoprofen poultice had been attached, so photocontact allergy to ketoprofen was suspected. Photopatch tests with ketoprofen poultice were indeed positive with both UVA and UVB irradiation (19). This was a case of systemic photocontact dermatitis from fenofibrate photocross-reacting to ketoprofen photosensitization.

A similar case had been described 24 years earlier (21). A 59-year-old woman, who was known with photopatch test proven photocontact allergy to ketoprofen, developed a pruritic, erythematous papulovesicular eruption on the face, neck, and legs, which had appeared on the third day of taking fenofibrate 100 mg/day to treat hypertriglyceri-demia. Photopatch tests were positive to ketoprofen and fenofibrate. This was also a case of systemic photocontact dermatitis to fenofibrate, cross-reacting to ketoprofen photosensitization (21).

A 64-year-old man had both a positive photochallenge test and positive UVA and UVB photopatch tests to fenofibrate (12). More data are not available to the author (cited in ref. 14). A woman aged 33 developed an erythematopapular rash on the 8th day of using fenofibrate. Photopatch tests were positive to fenofibrate both with UVA en UVB irradiation (15). One patient had systemic photosensitivity while taking fenofibrate pills. Photopatch tests were positive to the powder of the pills and to ketoprofen. More details were not provuded (9).

A man aged 33 developed erythematous papules on sun-exposed skin after having taken fenofibrate for 2 weeks. Photopatch tests were positive to fenofibrate irradiated with UVA. Details are unknown (18, data cited in ref. 14). A woman aged 72 had developed an erythematous rash and a man aged 72 an erythematous and bullous rash on sun-exposed skin while using fenofibrate. Photopatch tests using polychromatic light in one and UVA and UVB in the other were positive to fenofibrate in both individuals (13). These were the first 2 reported cases of photoallergy to fenofibrate, published in 1983. More details are not available to the author (data cited in ref. 14). One patient from Portugal with photosensitivity from systemic fenofibrate had positive photopatch tests to the powdered fenofibrate pills and to ketoprofen (26).

Cross-reactions, pseudo-cross-reactions and co-reactions
Fenofibrate and ketoprofen have a similar benzophenone structure; over 50-65% of patients with ketoprofen photosensitization photocross-react to fenofibrate (2,3,4,5,6,20,21).

LITERATURE
1 The data in the section 'General' may have been obtained from literature discussed in this chapter, but mostly also or exclusively from one or more of the following online sources: ChemIDPlus Advanced, PubChem, DrugBank, RxList, Drug Central, Drugs.com, and Wikipedia
2 Durbize E, Vigan M, Puzenat E, Girardin P, Adessi B, Desprez PH, et al. Spectrum of cross-photosensitization in 18 consecutive patients with contact photoallergy to ketoprofen: associated photoallergies to non-benzophenone-containing molecules. Contact Dermatitis 2003;48:144-149
3 Le Coz CJ, Bottlaender A, Scrivener JN, Santinelli F, Cribier BJ, Heid E, et al. Photocontact dermatitis from ketoprofen and tiaprofenic acid: cross-reactivity study in 12 consecutive patients. Contact Dermatitis 1998;38:245-252
4 Hindsén M, Zimerson E, Bruze M. Photoallergic contact dermatitis from ketoprofen in southern Sweden. Contact Dermatitis 2006;54:150-157
5 Leroy D, Dompmartin A, Szczurko C, Michel M, Louvet S. Photodermatitis from ketoprofen with cross-reactivity to fenofibrate and benzophenones. Photodermatol Photoimmunol Photomed 1997;13:93-97
6 De Groot AC. Monographs in contact allergy, volume 3: Topical Drugs. Boca Raton, Fl, USA: CRC Press Taylor and Francis Group, 2021 (ISBN 978-0-367-23693-9)
7 Gonçalo M, Ferguson J, Bonevalle A, Bruynzeel DP, Giménez-Arnau A, Goossens A, et al. Photopatch testing: recommendations for a European photopatch test baseline series. Contact Dermatitis 2013;68:239-243
8 Subiabre-Ferrer D, Esteve-Martínez A, Blasco-Encinas R, Sierra-Talamantes C, Pérez-Ferriols A, Zaragoza-Ninet V. European photopatch test baseline series: A 3-year experience. Contact Dermatitis 2019;80:5-8
9 Cardoso J, Canelas M, Gonçalo M, Figueiredo A. Photopatch testing with an extended series of photoallergens. A 5-year study. Contact Dermatitis. 2009;60:314-319
10 Conilleau V, Dompmartin A, Michel M, Verneuil L, Leroy D. Photoscratch testing in systemic drug-induced photosensitivity. Photodermatol Photoimmunol Photomed 2000;16:62-66
11 Rato M, Gil F, Monteiro AF, Parente J. Fenofibrate photoallergy - relevance of patch and photopatch testing. Contact Dermatitis 2018;78:413-414
12 Jeanmougin M, Manciet JR, De Prost Y. Fenofibrate photoallergy. Ann Dermatol Venereol 1993;120:549-554 (Article in French, data cited in ref. 14)
13 Jeanmougin M, Civatte J, Duterque M. Photosensibilisation au fenofibrate. Journées Dermatologiques de Paris, 1983, March 10-12 (Article in French, data cited in ref. 14)
14 Tsai K, Yang J, Hung S. Fenofibrate-induced photosensitivity – a case series and literature review. Photoderm Photoimmunol Photomed 2017;33:213-219
15 Leroy D, Dompmartin A, Lorier E, Leport Y, Audebert C. Photosensitivity induced by fenofibrate. Photodermatol Photoimmunol Photomed 1990;7:136-137 (Data cited in ref. 14)
16 Barbaud A, Schmutz JL, Trechot P . Photoallergie au fénofibrate (Lipanthyl®). A propos de 3 cas. La Lettre du GERDA 1992;9:6-10 (Article in French, data cited in ref. 14)
17 Machet L, Vaillant L, Jan V, Lorette G. Fenofbrate-induced photosensitivity: value of photopatch testing. J Am Acad Dermatol 1997;37(5Pt.1):808-809

18 Hong JH, Wang SH, Chu CY. Fenofibrate-induced photosensitivity – a case report and literature review. Dermatol Sinica 2009;27:37-43 (In Chinese, data cited in ref. 14)
19 Kuwatsuka S, Kuwatsuka Y, Takenaka M, Utani A. Case of photosensitivity caused by fenofibrate after photosensitization to ketoprofen. J Dermatol 2016;43:224-225
20 Adamski H, Benkalfate L, Delaval Y, Ollivier I, le Jean S, Toubel G, et al. Photodermatitis from non-steroidal anti-inflammatory drugs. Contact Dermatitis 1998;38:171-174
21 Serrano G, Fortea JM, Latasa JM, et al. Photosensitivity induced by fibric acid derivatives and its relation to photocontact dermatitis to ketoprofen. J Am Acad Dermatol 1992;27(2Pt.1):204-208
22 Boscá F, Miranda MA. Photosensitizing drugs containing the benzophenone chromophore. J Photochem Photobiol B 1998;43:1-26
23 Diemer S, Eberlein-König B, Przybilla B. Evaluation of the phototoxic properties of some hypolipidemics in vitro: fenofibrate exhibits a prominent phototoxic potential in the UVA and UVB region. J Dermatol Sci 1996;13:172-177
24 Leenutaphong V, Manuskiatti W. Fenofibrate-induced photosensitivity. J Am Acad Dermatol 1996;35:775-777
25 Merino V, Llamas R, Iglesias L. Phototoxic reaction to fenofibrate. Contact Dermatitis 1990;23:284
26 Cardoso J, Canelas MM, Gonçalo M, Figueiredo A. Photopatch testing with an extended series of photoallergens: a 5-year study. Contact Dermatitis 2009;60:325-329
27 Pecora V, Nucera E, Aruanno A, Buonomo A, Schiavino D. Delayed-type hypersensitivity to fenofibrate. J Investig Allergol Clin Immunol 2012;22:304-305

Chapter 3.214 FEPRAZONE

IDENTIFICATION

Description/definition	: Feprazone is the pyrazole compound that conforms to the structural formula shown below
Pharmacological classes	: Anti-inflammatory agents, non-steroidal
IUPAC name	: 4-(3-Methylbut-2-enyl)-1,2-diphenylpyrazolidine-3,5-dione
Other names	: Phenylprenazone
CAS registry number	: 30748-29-9
EC number	: 250-324-3
Merck Index monograph	: 5309
Patch testing	: No data available; most systemic drugs can be tested at 10% pet.; if the pure chemical is not available, prepare the test material from intravenous powder, the content of capsules or – when also not available – from powdered tablets to achieve a final concentration of the active drug of 10% pet.
Molecular formula	: $C_{20}H_{20}N_2O_2$

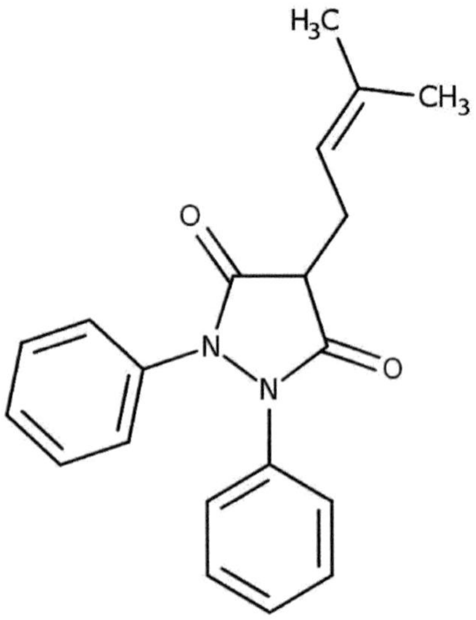

GENERAL

Feprazone is a pyrazole nonsteroidal anti-inflammatory drug (NSAID) that has analgesic, anti-inflammatory, and antipyretic properties. It has been used to treat mild to moderate pain, fever, and inflammation associated with osteoarthritis, rheumatoid arthritis, rheumatic fever and gouty arthritis. Feprazone is probably hardly, if at all, used anymore (1).

CUTANEOUS ADVERSE DRUG REACTIONS FROM SYSTEMIC ADMINISTRATION CAUSED BY TYPE IV (DELAYED-TYPE) HYPERSENSITIVITY (as demonstrated by positive patch tests)

Cutaneous adverse drug reactions from systemic administration of feprazone caused by type IV (delayed-type) hypersensitivity have included fixed drug eruption (2).

Fixed drug eruption

In one patient with fixed drug eruption with high probability caused by feprazone, patch tests with feprazone were performed on residual lesional skin and on the normal skin of the back and were positive on the postlesional skin only. Details are not available to the author (2).

LITERATURE

1 The data in the section 'General' may have been obtained from literature discussed in this chapter, but mostly also or exclusively from one or more of the following online sources: ChemIDPlus Advanced, PubChem, DrugBank, RxList, Drug Central, Drugs.com, and Wikipedia
2 Gonçalo M, Oliveira HS, Fernandes B, Robalo-Cordeiro M, Figueiredo A. Topical provocation in fixed drug eruption from nonsteroidal anti-inflammatory drugs. Exogenous Dermatol 2002;1:81-86

Chapter 3.215 FEXOFENADINE

IDENTIFICATION

Description/definition : Fexofenadine is the diphenylmethane that conforms to the structural formula shown below

Pharmacological classes : Histamine H_1 antagonists, non-sedating; anti-allergic agents

IUPAC name : 2-[4-[1-Hydroxy-4-[4-[hydroxy(diphenyl)methyl]piperidin-1-yl]butyl]phenyl]-2-methylpropanoic acid

Other names : Carboxyterfenadine

CAS registry number : 83799-24-0

EC number : 801-893-7

Merck Index monograph : 5367

Patch testing : Tablet, pulverized, as is and 50% water (2); most systemic drugs can be tested at 10% pet.; if the pure chemical is not available, prepare the test material from intravenous powder, the content of capsules or – when also not available – from powdered tablets to achieve a final concentration of the active drug of 10% pet.

Molecular formula : $C_{32}H_{39}NO_4$

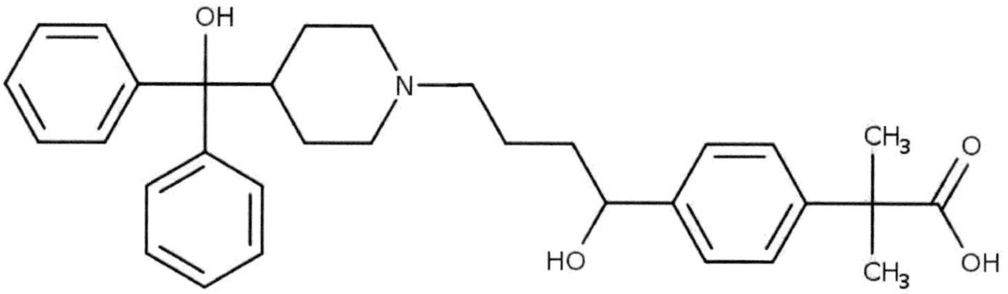

GENERAL

Fexofenadine is a second generation, long-lasting selective histamine H1 receptor antagonist with anti-inflammatory properties. The agent blocks the actions of endogenous histamine, thereby leading to temporary relief of the negative symptoms associated with histamine and achieving effects such as decreased vascular permeability, reduction of pruritus and localized smooth muscle relaxation. Fexofenadine is used for the treatment of allergic rhinitis, angioedema and chronic urticaria. In pharmaceutical products, fexofenadine is employed as fexofenadine hydrochloride (CAS number 153439-40-8, EC number not available, molecular formula $C_{32}H_{40}ClNO_4$) (1).

CUTANEOUS ADVERSE DRUG REACTIONS FROM SYSTEMIC ADMINISTRATION CAUSED BY TYPE IV (DELAYED-TYPE) HYPERSENSITIVITY (as demonstrated by positive patch tests)

Cutaneous adverse drug reactions from systemic administration of fexofenadine caused by type IV (delayed-type) hypersensitivity have included Stevens-Johnson syndrome, overlap with toxic epidermal necrolysis (SJS/TEN) (2).

Stevens-Johnson syndrome/toxic epidermal necrolysis (SJS/TEN)

A middle-aged woman developed SJS/TEN overlap after taking various medications, including clarithromycin, amoxicillin-clavulanic acid, ibuprofen and a mixture of fexofenadine and pseudoephedrine. Ten weeks after all symptoms and signs had resolved, patch tests were performed with all drugs used, and there was a positive reaction to the fexofenadine-pseudoephedrine tablet as is (powdered tablet) and 50% water. Later, she reacted to a tablet containing only fexofenadine, whereas the reaction to a pseudoephedrine-containing tablet was negative (2).

LITERATURE

1   The data in the section 'General' may have been obtained from literature discussed in this chapter, but mostly also or exclusively from one or more of the following online sources: ChemIDPlus Advanced, PubChem, DrugBank, RxList, Drug Central, Drugs.com, and Wikipedia

2   Teo SL, Santosa A, Bigliardi PL. Stevens-Johnson syndrome/Toxic epidermal necrolysis overlap induced by fexofenadine. J Investig Allergol Clin Immunol 2017;27:191-193

Chapter 3.216 FLAVOXATE

IDENTIFICATION

Description/definition   : Flavoxate is the flavone derivative that conforms to the structural formula shown below
Pharmacological classes   : Urological agents; parasympatholytics
IUPAC name   : 2-Piperidin-1-ylethyl 3-methyl-4-oxo-2-phenylchromene-8-carboxylate
CAS registry number   : 15301-69-6
EC number   : 239-337-5
Merck Index monograph : 5398
Patch testing   : Tablet, pulverized, 10% pet. (2); most systemic drugs can be tested at 10% pet.; if the pure
                                            chemical is not available, prepare the test material from intravenous powder, the content
                                            of capsules or – when also not available – from powdered tablets to achieve a final
                                            concentration of the active drug of 10% pet.
Molecular formula   : $C_{24}H_{25}NO_4$

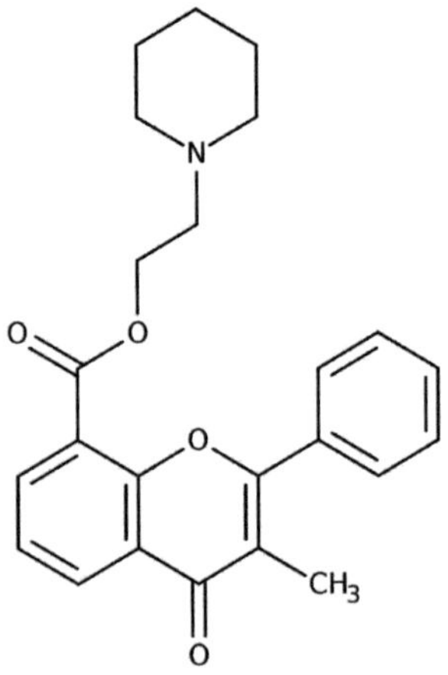

GENERAL

Flavoxate is a parasympatholytic agent with antispasmodic activity. It competitively binds to muscarinic receptors, thereby preventing the actions of acetylcholine. This relaxes vascular smooth muscle, mainly of the urinary tract, and prevents smooth muscle contractions. Flavoxate is indicated for symptomatic relief of dysuria, urgency, nocturia, suprapubic pain, frequent miction and incontinence as may occur in cystitis, prostatitis, urethritis, and urethrocystitis or urethrotrigonitis (1). In pharmaceutical products, flavoxate is employed as flavoxate hydrochloride (CAS number 3717-88-2, EC number 223-066-4, molecular formula $C_{24}H_{26}ClNO_4$) (1).

CUTANEOUS ADVERSE DRUG REACTIONS FROM SYSTEMIC ADMINISTRATION CAUSED BY TYPE IV
(DELAYED-TYPE) HYPERSENSITIVITY (as demonstrated by positive patch tests)

Cutaneous adverse drug reactions from systemic administration of flavoxate caused by type IV (delayed-type) hypersensitivity have included generalized erythema (2).

Erythroderma, widespread erythematous eruption, exfoliative dermatitis

An 83-year-old man presented with pruritic erythema all over the body, accompanied by fever and loss of appetite, which had started one month before. He had been taking flavoxate hydrochloride, tamsulosin hydrochloride and allylestrenol for prostatic hypertrophy for about 6 months. Patch tests with the three drugs at 1%, 5% and 10% pet. showed a positive response at D3 to flavoxate 10% pet. only. Four controls were negative. An oral challenge test with flavoxate hydrochloride 600 mg produced pruritic erythema all over the body with fever. A lymphocyte stimulation test was negative (2).

LITERATURE

1   The data in the section 'General' may have been obtained from literature discussed in this chapter, but mostly
    also or exclusively from one or more of the following online sources: ChemIDPlus Advanced, PubChem,
    DrugBank, RxList, Drug Central, Drugs.com, and Wikipedia
2   Enomoto U, Ohnishi Y, Kimura M, Kawada A, Ishibashi A. Drug eruption due to flavoxate hydrochloride. Contact
    Dermatitis 1999;40:337-338

Chapter 3.217 FLOXACILLIN

IDENTIFICATION

Description/definition : Floxacillin (better known as flucloxacillin) is the semisynthetic isoxazolyl penicillin that conforms to the structural formula shown below

Pharmacological classes : Antibacterial agents

IUPAC name : (2S,5R,6R)-6-[[3-(2-Chloro-6-fluorophenyl)-5-methyl-1,2-oxazole-4-carbonyl]amino]-3,3-dimethyl-7-oxo-4-thia-1-azabicyclo[3.2.0]heptane-2-carboxylic acid

Other names : Flucloxacillin

CAS registry number : 5250-39-5

EC number : 226-051-0

Merck Index monograph : 5413

Patch testing : 10% pet.

Molecular formula : $C_{19}H_{17}ClFN_3O_5S$

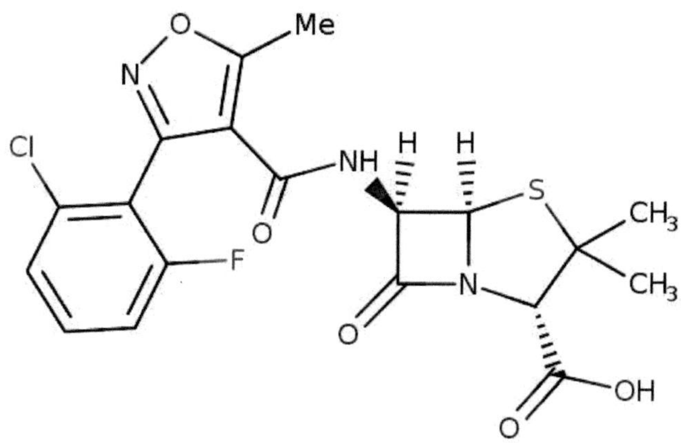

GENERAL

Flucloxacillin is a narrow-spectrum semisynthetic beta-lactam penicillin used in the treatment of bacterial infections caused by susceptible, usually gram-positive, organisms. The bactericidal activity of floxacillin results from the inhibition of cell wall synthesis and is mediated through floxacillin binding to penicillin binding proteins (PBPs). Floxacillin is stable against hydrolysis by a variety of β-lactamases, including penicillinases, cephalosporinases and extended spectrum β-lactamases. In pharmaceutical products, floxacillin is employed as floxacillin sodium (CAS number 1847-24-1, EC number 217-428-0, molecular formula $C_{19}H_{16}ClFN_3NaO_5S$) (1). The classification and structures of beta-lactam antibiotics are discussed in Chapter 3.36 Amoxicillin.

CONTACT ALLERGY FROM ACCIDENTAL CONTACT

Between 1978 and 2001, in the university hospital of Leuven, Belgium, occupational allergic contact dermatitis to pharmaceuticals was diagnosed in 33 health care workers: 26 nurses, 4 veterinarians, 2 pharmacists and one medical doctor. Practically all of these patients presented with hand dermatitis (often the fingers), sometimes with secondary localizations, frequently the face. In this group, four patients had occupational allergic contact dermatitis from flucloxacillin (2). See also the case report in ref. 3, described in the section 'Drug reaction with eosinophilia and systemic symptoms (DRESS)' below.

CUTANEOUS ADVERSE DRUG REACTIONS FROM SYSTEMIC ADMINISTRATION CAUSED BY TYPE IV (DELAYED-TYPE) HYPERSENSITIVITY (as demonstrated by positive patch tests)

Cutaneous adverse drug reactions from systemic administration of flucloxacillin caused by type IV (delayed-type) hypersensitivity have included maculopapular eruption (3,7,8), acute generalized exanthematous pustulosis (AGEP) (5,8), drug reaction with eosinophilia and systemic symptoms (DRESS) (3,6,7), and unspecified skin rash (4).

Maculopapular eruption

In a study from Portugal, 18 patients who had previously had positive patch tests to antibiotics that had caused a maculopapular eruption, were again patch tested after a mean interval of 6 years. The positive reactions were reproducible in 16 cases, of which 4 were caused by flucloxacillin (8).

One patient developed a maculopapular eruption from floxacillin with a positive patch test to this drug after a previous episode of DRESS from other drug(s) (multiple drug hypersensitivity) (7). See also the case report in ref. 3, described in the section 'Drug reaction with eosinophilia and systemic symptoms (DRESS)' below.

Acute generalized exanthematous pustulosis (AGEP)

A severe case of acute generalized exanthematous pustulosis (AGEP) in a 30-year-old woman, caused by flucloxacillin, clinically presenting with features resembling toxic epidermal necrolysis (TEN) and pronounced systemic symptoms with hemodynamic and respiratory instability, was reported from The Netherlands in 2014 (5). The histopathology was characteristic for AGEP and patch tests with flucloxacillin tablets, pulverized, 30% in water and pet. were strongly positive with vesicles and pustules. It was suggested that severe neutrophilia contributed in this patient to the systemic symptoms. A critical review of the literature led the authors to the conclusion that overlap between AGEP and SJS/TEN, although suggested by some investigators, does not exist (5).

A 36-year-old man developed AGEP from amoxicillin and flucloxacillin with positive patch tests to both drugs. 6.5 years later, the tests were repeated and were still positive (8).

Drug reaction with eosinophilia and systemic symptoms (DRESS)

A 26-year-old nurse developed drug reaction with eosinophilia and systemic symptoms (DRESS) in 2014, beginning on the third day of treatment with flucloxacillin for an oropharynx infection. Patch tests were positive to flucloxacillin, dicloxacillin, penicillin G, and amoxicillin, all 10% pet. One year later, she developed eczema, after 2 years of working as a nurse in a general surgery ward. Lesions were initially localized on the dorsa of the hands and face, and were triggered by the preparation of intravenous antibiotics. More recently, eczema of the exposed areas developed whenever she entered an area where other nurses were preparing antibiotics. Moreover, she had two episodes of airborne eczema followed, within 12-24 hours, by a maculopapular rash that took a few days to resolve with the use of oral steroids. This was probably caused by inhalation of flucloxacillin or cross-reacting penicillins. The patient has remained free of skin lesions after moving to a primary healthcare facility where she does not handle intravenous systemic antibiotics (3).

One patient who had suffered an episode of DRESS, subsequently had a drug-related relapse caused by floxacillin, showing a positive patch test to this drug (6, possible overlap with ref. 7). In a group of 14 patients with multiple delayed-type hypersensitivity reactions, DRESS was caused by floxacillin in one case, showing a positive patch test reaction (7, possible overlap with ref. 6).

Other cutaneous adverse drug reactions

A 62-year-old woman developed an unspecified skin rash from floxacillin. Patch tests were positive to floxacillin powder (D2 ++, D3 ++) and various other penicillins (4).

Cross-reactions, pseudo-cross-reactions and co-reactions

Cross-reactions between beta-lactam antibiotics are discussed in Chapter 3.36 Amoxicillin.

LITERATURE

1   The data in the section 'General' may have been obtained from literature discussed in this chapter, but mostly also or exclusively from one or more of the following online sources: ChemIDPlus Advanced, PubChem, DrugBank, RxList, Drug Central, Drugs.com, and Wikipedia
2   Gielen K, Goossens A. Occupational allergic contact dermatitis from drugs in healthcare workers. Contact Dermatitis 2001;45:273-279
3   Pinheiro V, Pestana C, Pinho A, Antunes I, Gonçalo M. Occupational allergic contact dermatitis caused by antibiotics in healthcare workers - relationship with non-immediate drug eruptions. Contact Dermatitis 2018;78:281-286
4   Kennedy C, Stolz E, van Joost T. Sensitization to amoxycillin in Augmentin. Contact Dermatitis 1989;20:313-314
5   van Hattem S, Beerthuizen GI, Kardaun SH. Severe flucloxacillin-induced acute generalized exanthematous pustulosis (AGEP), with toxic epidermal necrolysis (TEN)-like features: does overlap between AGEP and TEN exist? Clinical report and review of the literature. Br J Dermatol 2014;171:1539-1545
6   Jörg L, Helbling A, Yerly D, Pichler WJ. Drug-related relapses in drug reaction with eosinophilia and systemic symptoms (DRESS). Clin Transl Allergy 2020;10:52
7   Jörg L, Yerly D, Helbling A, Pichler W. The role of drug, dose and the tolerance/intolerance of new drugs in multiple drug hypersensitivity syndrome (MDH). Allergy 2020;75:1178-1187
8   Pinho A, Marta A, Coutinho I, Gonçalo M. Long-term reproducibility of positive patch test reactions in patients with non-immediate cutaneous adverse drug reactions to antibiotics. Contact Dermatitis 2017;76:204-209

Chapter 3.218 FLUCONAZOLE

IDENTIFICATION

Description/definition   : Fluconazole is the synthetic triazole that conforms to the structural formula shown below
Pharmacological classes   : Cytochrome P-450 CYP2C19 inhibitors; cytochrome P-450 CYP2C9 inhibitors; antifungal
                              agents; 14α-demethylase inhibitors
IUPAC name   : 2-(2,4-Difluorophenyl)-1,3-bis(1,2,4-triazol-1-yl)propan-2-ol
CAS registry number   : 86386-73-4
EC number   : Not available
Merck Index monograph   : 5423
Patch testing   : Powder from capsule or infusion 10%-30% pet. (2); most systemic drugs can be tested at
                              10% pet.; if the pure chemical is not available, prepare the test material from intravenous
                              powder, the content of capsules or – when also not available – from powdered tablets to
                              achieve a final concentration of the active drug of 10% pet.
Molecular formula   : $C_{13}H_{12}F_2N_6O$

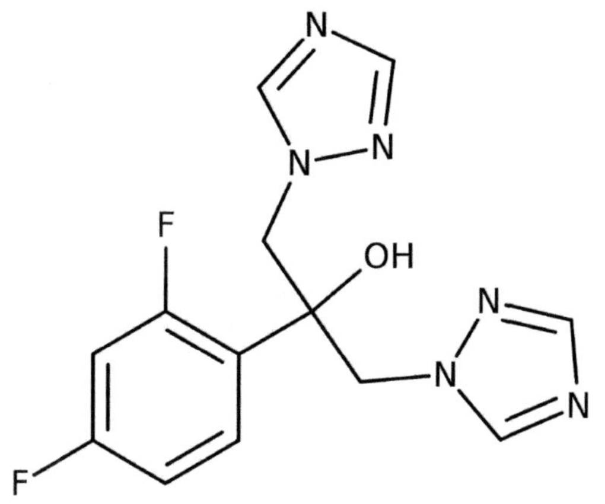

GENERAL

Fluconazole is a synthetic triazole compound with antifungal activity for systemic use. It can be administered in the treatment of vaginal candidiasis, oropharyngeal and esophageal candidiasis, *Candida* urinary tract infections, peritonitis, and systemic *Candida* infections including candidemia, disseminated candidiasis, and pneumonia, and cryptococcal meningitis. Fluconazole can also be used in the prophylaxis of candidiasis in patients undergoing bone marrow transplantation who receive cytotoxic chemotherapy and/or radiation therapy (1).

CUTANEOUS ADVERSE DRUG REACTIONS FROM SYSTEMIC ADMINISTRATION CAUSED BY TYPE IV
(DELAYED-TYPE) HYPERSENSITIVITY (as demonstrated by positive patch tests)

Cutaneous adverse drug reactions from systemic administration of fluconazole caused by type IV (delayed-type) hypersensitivity have included maculopapular eruption (2), acute generalized exanthematous pustulosis (AGEP) (7), and fixed drug eruption (3,4,5,6,8,9,12,14).

Maculopapular eruption

A 44-year-old woman was prescribed 1% clotrimazole cream for suspected anogenital candidiasis. After she had applied it on and off for 2 months, her itch deteriorated, and she developed erythema in her groins. For this, her physician prescribed fluconazole (150-mg single dose) and flucloxacillin orally, which resulted in a widespread maculopapular exanthema 2 days later. She had no previous history of penicillin allergy, and has taken oral flucloxacillin subsequently with no problem. Patch testing showed positive reactions to clotrimazole 5%, methylisothiazolinone (MI), MCI/MI and the baby wipes she used on her anogenital skin before the start of the problem. The patient had negative reactions to econazole and miconazole, but was not tested with fluconazole. The authors postulated – without corroborating evidence – that the maculopapular exanthema to oral fluconazole was the result of cross-sensitivity to primary clotrimazole sensitization (2).

Acute generalized exanthematous pustulosis (AGEP)

A 70-year-old man was seen because of sudden occurrence of a rash with fever and malaise, which had appeared after starting fluconazole 150 mg once daily for a tongue candidiasis. Clinical examination showed an erythematous eruption mainly affecting the trunk and proximal limbs with numerous, non-follicular, pinhead pustules which in some areas became confluent in small pustular lakes. In addition purpuric lesions were present on the arms and legs , while erythema, edema and blisters were observed on the palms and soles. Laboratory investigations showed an increase of serum creatinine and urea, leukocytosis with neutrophilia, and polycythemia. Bacterial cultures of pustule swabs were negative. Histological examination showed pustules within the stratum corneum with edema of the papillary dermis and a mixed superficial and perivascular, inflammatory infiltrate. Later, patch test to fluconazole (material, test concentration and vehicle not mentioned) was positive after 2 days. The patient was diagnosed with acute generalized exanthematous pustulosis from fluconazole (7).

Fixed drug eruptions

A peculiarity of fixed drug eruptions from fluconazole is that it may mimic herpes labialis (10,11,13).

In London, between October 2017 and October 2018, 45 patients with suspected cutaneous adverse drug reactions, including 33 maculopapular eruptions (MPE), 4 fixed drug eruptions (FDE), 4 DRESS, 3 AGEP and one SJS/TEN, were patch tested with the suspected drugs. There were 10 (22%) positive patch test cases: 4 MPE, 2 FDE, 3 DRESS and 1 AGEP. Fluconazole, tested as commercial preparation 30% pet., was responsible for one case of fixed drug eruption (12).

A 25-year-old woman received five doses of fluconazole (150 mg) once a month for recurrent vaginal candidiasis. She developed red erythematous lesions that faded leaving hyperpigmentation at the same areas on three occasions after taking fluconazole. An oral challenge test with fluconazole 150 mg showed similar signs three hours after intake. Local provocation was positive with 10% fluconazole in petrolatum and in alcohol on postlesional but not normal skin. A skin biopsy specimen was consistent with FDE (3).

A 33-year-old man presented with a burning and itching rash since one day which had developed 17 hours after intake of one tablet of 150 mg fluconazole. On cutaneous examination, well-defined erythematous plaques of varied sizes were present over the chest, back, lower limbs, and lips. The largest over the chest was 10X10 cm. A patch test done 'with the offending drug' was positive (no details provided). A diagnosis of FDE to fluconazole was made (5).

A-36-year old woman had been treated many times with fluconazole capsules for vaginal candidiasis, when, after taking another capsule of fluconazole, a red macule on the medial left thigh and on the distal phalanx of the right fourth finger appeared. The macules faded in a few days. Gradually, a long-lasting violet pigmentation developed on the thigh but not on the finger. Later, 12 hours after the intake of fluconazole, she again developed 2 erythematous patches at the same sites. Local provocation with 10% fluconazole in petrolatum on the pigmented area of the left thigh and with 10% fluconazole in ethanol on the nonpigmented area of the right fourth finger was performed. In addition, the same compounds and vehicles were tested on normal skin of the back. In 24 hours, a red patch, 2×2 cm in diameter, developed on the thigh. Histopathology was consistent with FDE (6).

A 65-year-old man presented with well-circumscribed, erythematous, and violaceous patches with a diameter ranging from 10 to 40 mm, widespread on the trunk, upper arms, and abdomen, the thenar eminence, legs, and soles. In some of the approximately 20-30 lesions over the whole body, blisters developed on these patches followed by skin detachment, but the skin remained intact between the affected areas; oral and genital mucosal erosions were also present without accompanying dysuria or pain. The symptoms had started abruptly, almost 8 hours after the intake of a fluconazole tablet of 100 mg for oral candidiasis. The patient reported similar bullous lesions, although less extensive. three months before, also after the intake of fluconazole, with remission within 2 weeks leaving residual hyperpigmentation that gradually subsided. Six months later, patch tests were positive to commercial fluconazole 30% pet. but negative to 10% pet. and to itraconazole. The patient was diagnosed with generalized bullous fixed drug eruption from fluconazole (14).

Additional cases of FDE due to delayed-type hypersensitivity to fluconazole have been reported in refs. 4, 8 and 9. In the recent case from the United Kingdom, the patch test was negative on D2 but strongly positive on D4. It was stressed that readings later than D2 are necessary (8). In another case from the UK, in a patient with bullous FDE from fluconazole, a patch test on residual pigmentation was negative, but an open provocation test on non-hyperpigmented previously affected skin positive. The authors therefore proposed that, in fixed drug eruption, multiple prior lesional sites (hyperpigmented and normal) should be patch tested, especially in cases where preliminary results are negative, to increase the overall detection rate (9).

Cross-reactions, pseudo-cross-reactions and co-reactions

In one patient, fluconazole may have cross-reacted to primary clotrimazole sensitization (2).

LITERATURE

1 The data in the section 'General' may have been obtained from literature discussed in this chapter, but mostly also or exclusively from one or more of the following online sources: ChemIDPlus Advanced, PubChem, DrugBank, RxList, Drug Central, Drugs.com, and Wikipedia

2 Nasir S, Goldsmith P. Anogenital allergic contact dermatitis caused by methylchloroisothiazolinone, methylisothiazolinone and topical clotrimazole with subsequent generalized exanthem triggered by oral fluconazole. Contact Dermatitis 2016;74:296-297

3 Tavallaee M, Rad MM. Fixed drug eruption resulting from fluconazole use: a case report. J Med Case Reports 2009;3:7368

4 Santra R, Pramanik S, Raychaudhuri P. Fixed drug eruption due to fluconazole: not so uncommon now-a-days. J Clin Diagn Res 2014;8:HL01

5 Pai VV, Bhandari P, Kikkeri NN, Athanikar SB, Sori T. Fixed drug eruption to fluconazole: a case report and review of literature. Indian J Pharmacol 2012;44: 643-645

6 Heikkilä H, Timonen K, Stubb S. Fixed drug eruption due to fluconazole. J Am Acad Dermatol 2000;42:883-884

7 Di Lernia V, Ricci C. Fluconazole-induced acute generalized exanthematous pustulosis. Indian J Dermatol 2015;60:212

8 Khan M, Paul N, Fernandez C, Wakelin S. Fluconazole-induced fixed drug eruption confirmed by extemporaneous patch testing. Contact Dermatitis 2020;83:507-508

9 Tan KL, Bisconti I, Leck C, Billahalli T, Barnett S, Rajakulasingam K, Watts TJ. Bullous fixed drug eruption induced by fluconazole: Importance of multi-site lesional patch testing. Contact Dermatitis 2021;84:350-352

10 Schneller-Pavelescu L, Ochando-Ibernón G, Vergara-de Caso E, Silvestre-Salvador JF. Herpes simplex-like fixed drug eruption induced by fluconazole without cross-reactivity to itraconazole. Dermatitis 2019;30:174-175

11 Slawinska M, Baranska-Rybak W, Wilkowska A, Nowicki R. Bullous fixed drug eruption due to fluconazole, imitating herpes simplex. Clin Exp Dermatol 2017;42:544-545

12 Watts TJ, Thursfield D, Haque R. Patch testing for the investigation of nonimmediate cutaneous adverse drug reactions: a prospective single center study. J Allergy Clin Immunol Pract 2019;7:2941-2943.e3

13 Calogiuri G, Garvey LH, Nettis E, Romita P, Di Leo E, Caruso R, et al. Skin allergy to azole antifungal agents for systemic use: A review of the literature. Recent Pat Inflamm Allergy Drug Discov 2019;13:144-157

14 Makris M, Fokoloros C, Syrmali A, Tsakiraki Z, Damaskou V, Papadavid E. Generalized bullous fixed drug eruption to fluconazole with positive patch testing and confirmed tolerance to itraconazole. Iran J Allergy Asthma Immunol 2021;20:255-259

Chapter 3.219 FLUINDIONE

IDENTIFICATION

Description/definition   : Fluindione is the indanedione that conforms to the structural formula shown below
Pharmacological classes   : Anticoagulants
IUPAC name   : 2-(4-Fluorophenyl)indene-1,3-dione
CAS registry number   : 957-56-2
EC number   : 213-484-5
Merck Index monograph   : 5435
Patch testing   : Commercial preparation 30% pet. and water; most systemic drugs can be tested at 10%
                            pet.; if the pure chemical is not available, prepare the test material from intravenous
                            powder, the content of capsules or – when also not available – from powdered tablets to
                            achieve a final concentration of the active drug of 10% pet.
Molecular formula   : $C_{15}H_9FO_2$

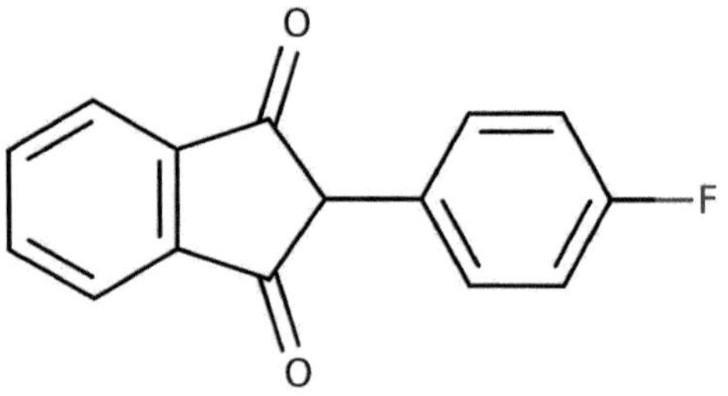

GENERAL

Fluindione is an oral anticoagulant and belongs to the vitamin K antagonist class. It is apparently licensed exclusively in France and Luxembourg and is widely used to treat deep venous thrombosis and pulmonary embolism (1).

CUTANEOUS ADVERSE DRUG REACTIONS FROM SYSTEMIC ADMINISTRATION CAUSED BY TYPE IV (DELAYED-TYPE) HYPERSENSITIVITY (as demonstrated by positive patch tests)

Cutaneous adverse drug reactions from systemic administration of fluindione caused by type IV (delayed-type) hypersensitivity have included acute generalized exanthematous pustulosis (AGEP) (2,5,7) and drug reaction with eosinophilia and systemic symptoms (DRESS) (2,3,4,6,8).

Acute generalized exanthematous pustulosis (AGEP)

In a multicenter investigation in France, of 45 patients patch tested for AGEP, 26 (58%) had positive patch tests to drugs, including one to fluindione (2).

A 70-year-old woman presented with fever and a diffuse erythematous exanthema covered with multiple non-follicular pustules, which had started 2 days after initiation of fluindione treatment for cardiac arrhythmia. Laboratory investigations were normal. The patient was clinically diagnosed with AGEP which was confirmed by classic AGEP histopathology findings. After withdrawal of fluindione, the eruption cleared up in one week's time. Three months later, patch tests were positive (++ at D3) to commercial fluindione 30% pet. and negative to the other drugs the patient had used (5).

A 68 year-old man presented with a pustular eruption and erythema twenty days after initiation of treatment with fluindione for cardiac arrhythmia. The eruption was associated with fever, arthralgia, neutrophilia, elevated liver enzymes and kidney involvement including acute renal failure, hematuria and proteinuria. When the drug was reintroduced, a rapid recurrence of all manifestations was noted. A patch test with fluindione 30% pet. was dubious-positive (?+, macular erythema). The patient was diagnosed with AGEP, but the systemic symptoms are highly atypical (7). An alternative would be to consider this as a case of DRESS with a drug exanthema resembling AGEP.

Drug reaction with eosinophilia and systemic symptoms (DRESS)

In a multicenter investigation in France, of 72 patients patch tested for DRESS, 46 (64%) had positive patch tests to drugs, including two to fluindione (2).

In the French Pharmacovigilance database, in the period 2000-2011, 36 cases of DRESS caused by fluindione were included. The group consisted of 17 women and 19 men with a mean age of 65 years. The kidneys and liver were the most frequently involved organs. Fluindione was the only suspected medicine in 26 patients. Skin patch tests, performed in 10 individuals, were positive to fluindione in 9 of them. Seven had a positive patch test, in the other 2, the nature of the skin tests was not specified. No patch test details were provided (3).

Also in France, 5 patients (4 men, one woman, age range 53-84) were investigated for drug-induced hypersensitivity syndrome (the old name for DRESS). All 5 had positive patch tests (++ or +++) to commercial fluindione: 2 reacted to the drug at 30% pet. and water, one to fluindione 5% water and pet., the 4th reacted to 30% pet. but was negative to 30% water and the 5th had positive patch tests to fluindione 2% water and pet. In one patient, there was a photoaugmentation by UVB (6).

In a multicenter investigation in France, of 72 patients patch tested for DRESS, 46 (64%) had positive patch tests to drugs, including 2 to fluindione (2). One patient developed DRESS from fluindione and had positive patch tests to this drug tested 1% and 10% pet. and water. The patient also had delayed-type hypersensitivity to 2 other unrelated drugs and was therefore polysensitized (4).

Three weeks after the beginning of treatment with fluindione, allopurinol and perindopril, a 75-year-old woman presented with a diffuse erythematous and papular exanthema associated with edema of the face. Laboratory tests showed leukocytosis, eosinophilia, elevated liver enzymes (cytolysis and cholestasis) and acute renal failure. On reintroduction of fluindione, a rapid recurrence of clinical symptoms and laboratory abnormalities was observed, more severe than in the first episode. The skin rash and visceral abnormalities resolved completely on withdrawal of the drug. Patch tests were positive to fluindione and negative to allopurinol and perindopril (8).

LITERATURE

1   The data in the section 'General' may have been obtained from literature discussed in this chapter, but mostly also or exclusively from one or more of the following online sources: ChemIDPlus Advanced, PubChem, DrugBank, RxList, Drug Central, Drugs.com, and Wikipedia

2   Barbaud A, Collet E, Milpied B, Assier H, Staumont D, Avenel-Audran M, et al. A multicentre study to determine the value and safety of drug patch tests for the three main classes of severe cutaneous adverse drug reactions. Br J Dermatol 2013;168:555-562

3   Daveluy A, Milpied B, Barbaud A, Lebrun-Vignes B, Gouraud A, Laroche M-L, et al. Fluindione and drug reaction with eosinophilia and systemic symptoms: an unrecognized adverse effect? Eur J Clin Pharmacol 2012;68:101-105

4   Studer M, Waton J, Bursztejn AC, Aimone-Gastin I, Schmutz JL, Barbaud A. Does hypersensitivity to multiple drugs really exist? Ann Dermatol Venereol. 2012;139:375-380 (Article in French)

5   Chtioui M, Cousin-Testard F, Zimmermann U, Amar A, Saiag P, Mahé E. Fluindione-induced acute generalised exanthematous pustulosis confirmed by patch testing. Ann Dermatol Venereol 2008;135:295-298 (Article in French)

6   Sparsa A, Bédane C, Benazahary H, De Vencay P, Gauthier ML, Le Brun V, et al. Syndrome d'hypersensibilité médicamenteuse à la fluindione. Ann Dermatol Venereol 2001;128:1014-1018

7   Thurot C, Reymond JL, Bourrain JL, Pinel N, Béani JC. Pustulose exanthématique aiguë généralisée à la fluindione avec atteinte rénale. Ann Dermatol Venereol 2003;130:1146-1149

8   Frouin E, Roth B, Grange A, Grange F, Tortel M C, Guillaume J C. Hypersensitivity to fluindione (Previscan). Positive skin patch tests. Ann Dermatol Venereol 2005;132:1000-1002 (Article in French)

Chapter 3.220 FLUOROURACIL

IDENTIFICATION

Description/definition   : Fluorouracil is the fluoropyrimidine analog of the nucleoside pyrimidine that conforms to the structural formula shown below
Pharmacological classes : Antimetabolites; antimetabolites, antineoplastic; immunosuppressive agents
IUPAC name   : 5-Fluoro-1*H*-pyrimidine-2,4-dione
Other names   : 5-Fluorouracil
CAS registry number   : 51-21-8
EC number   : 200-085-6
Merck Index monograph : 5483
Patch testing   : Commercial 5% cream 'as is'; 5-fluorouracil 50 mg/ml (5%) and 10 mg/ml (1%) in saline or 5% pet.; if negative and contact allergy is strongly suspected, or with ?+ patch tests, perform intradermal tests with 5-FU 10 mg/ml (1%) saline with reading at D2 and later
Molecular formula   : $C_4H_3FN_2O_2$

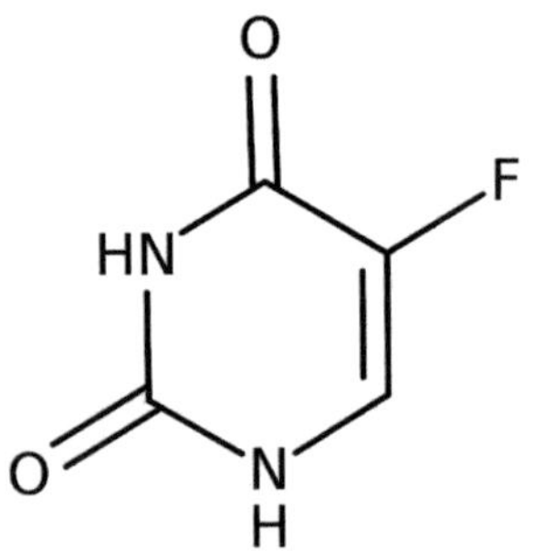

GENERAL

Fluorouracil (5-FU) is an antimetabolite fluoropyrimidine analog of the nucleoside pyrimidine with antineoplastic activity. It interferes with DNA synthesis by blocking the thymidylate synthetase conversion of deoxyuridylic acid to thymidylic acid. Fluorouracil is indicated for the topical treatment of multiple actinic (solar) keratoses. In the 5% strength it is also useful in the treatment of superficial basal cell carcinomas when conventional methods are impractical, such as with multiple lesions or difficult treatment sites. Fluorouracil injection is indicated in the palliative management of some types of cancer, including of the colon, esophagus, rectum, breast, biliary tract, stomach, head and neck, cervix, pancreas, renal cell cancer and carcinoid (1).

In topical preparations, fluorouracil has caused contact allergy/allergic contact dermatitis, which subject has been fully reviewed in Volume 3 of the *Monographs in contact allergy* series (4).

CUTANEOUS ADVERSE DRUG REACTIONS FROM SYSTEMIC ADMINISTRATION CAUSED BY TYPE IV (DELAYED-TYPE) HYPERSENSITIVITY (as demonstrated by positive patch tests)

Cutaneous adverse drug reactions from systemic administration of fluorouracil caused by type IV (delayed-type) hypersensitivity have included systemic allergic dermatitis (2).

Systemic allergic dermatitis (systemic contact dermatitis)

A 61-year-old man had undergone several courses of topical 5-FU cream for actinic keratoses. In a following course, the patient noticed severe inflammation at sites of 5-FU application, developing 2-3 days hours after the onset of treatment. Patch tests were positive to 5-FU 0.5% and 1% pet. Four years later, the patient was diagnosed as having rectal adenocarcinoma, for which chemotherapy with i.v. 5-FU was scheduled. One day later, the patient developed acute dermatitis with severe head and neck edema. An erythematovesicular eruption involving the scalp, face, neck and the lateral aspects of both hands was also present. The eruption was more severe on the right upper arm, where the i.v. perfusion had been administered. A diagnosis of systemic allergic dermatitis was made (2).

Cross-reactions, pseudo-cross-reactions and co-reactions

A patient sensitized to 5-fluorouracil cross-reacted to 5-bromouracil (3).

LITERATURE
1   The data in the section 'General' may have been obtained from literature discussed in this chapter, but mostly
     also or exclusively from one or more of the following online sources: ChemIDPlus Advanced, PubChem,
     DrugBank, RxList, Drug Central, Drugs.com, and Wikipedia
2   Nadal C, Pujol RM, Randazzo L, Marcuello E, Alomar A. Systemic allergic dermatitis from 5-fluorouracil. Contact
     Dermatitis 1996;35:124-125
3   Sams WM. Untoward response with topical fluorouracil. Arch Dermatol 1968;97:14-23
4   De Groot AC. Monographs in contact allergy, volume 3: Topical Drugs. Boca Raton, Fl, USA: CRC Press Taylor and
     Francis Group, 2021 (ISBN 978-0-367-23693-9)

Chapter 3.221 FLURBIPROFEN

IDENTIFICATION

Description/definition   : Flurbiprofen is the arylpropionic acid derivative that conforms to the structural formula
                             shown below
Pharmacological classes   : Anti-Inflammatory agents, non-steroidal; cyclooxygenase inhibitors; analgesics
IUPAC name   : 2-(3-Fluoro-4-phenylphenyl)propanoic acid
CAS registry number   : 5104-49-4
EC number   : 225-827-6
Merck Index monograph   : 5499
Patch testing   : 5% pet. (6)
Molecular formula   : $C_{15}H_{13}FO_2$

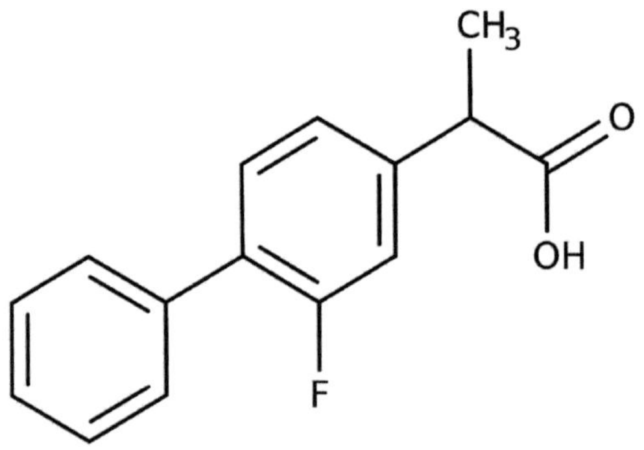

GENERAL

Flurbiprofen is a derivative of propionic acid and a nonsteroidal anti-inflammatory drug (NSAID) with analgesic, anti-inflammatory and antipyretic effects. Flurbiprofen non-selectively binds to and inhibits cyclooxygenase (COX). This results in a reduction of arachidonic acid conversion into prostaglandins that are involved in the regulation of pain, inflammation and fever. Flurbiprofen tablets are indicated for the acute or long-term symptomatic treatment of rheumatoid arthritis, osteoarthritis and ankylosing spondylitis. It may also be used to treat pain associated with dysmenorrhea and mild to moderate pain accompanied by inflammation (e.g. bursitis, tendonitis, soft tissue trauma). Topical ophthalmic formulations may be used pre-operatively to prevent intraoperative miosis. Allergic contact dermatitis to flurbiprofen has rarely been described (6).

In pharmaceutical products, both flurbiprofen base (in tablets) and – in eye drops – flurbiprofen sodium (CAS number 56767-76-1, EC number 260-373-2, molecular formula $C_{15}H_{12}FNaO_2$) may be employed (1).

CUTANEOUS ADVERSE DRUG REACTIONS FROM SYSTEMIC ADMINISTRATION CAUSED BY TYPE IV (DELAYED-TYPE) HYPERSENSITIVITY (as demonstrated by positive patch tests)

Cutaneous adverse drug reactions from systemic administration of flurbiprofen caused by type IV (delayed-type) hypersensitivity have included maculopapular eruption (5).

Maculopapular eruption

An 18-year-old woman was prescribed flurbiprofen 200 mg p.o. daily for a sore throat. Two days after taking the second dose, the patient developed a diffuse, pruritic maculopapular rash. Two days later the exanthem had assumed an urticarial appearance due to confluence of the original lesions, and angioedema of the hands and feet appeared with hypotension. The patient claimed that she had used the same drug for a similar problem 1 year prior to the reaction, and on that occasion she had not experienced any adverse effects. A prick test with a full-strength flurbiprofen oral suspension was negative. Patch testing with the powder of a crushed flurbiprofen tablet produced a positive result at D2 and D3 and negative results in 5 volunteers. Oral challenges with each of the additives contained in the tablet were negative (5).

Cross-reactions, pseudo-cross-reactions and co-reactions

One patient sensitized to ibuprofen may have cross-reacted to flurbiprofen and fenoprofen (2). An individual who had photoallergic contact dermatitis from ketoprofen showed cross-photosensitivity to flurbiprofen and ibuproxam (3). A patient with photocontact allergy to tiaprofenic acid photocross-reacted to flurbiprofen (4).

LITERATURE

1 The data in the section 'General' may have been obtained from literature discussed in this chapter, but mostly also or exclusively from one or more of the following online sources: ChemIDPlus Advanced, PubChem, DrugBank, RxList, Drug Central, Drugs.com, and Wikipedia

2 Pigatto P, Bigardi A, Legori A, Valsecchi R, Picardo M. Cross-reactions in patch testing and photopatch testing with ketoprofen, tiaprophenic acid, and cinnamic aldehyde. Am J Cont Dermat 1996;7:220-223

3 Mozzanica N, Pigatto PD. Contact and photocontact allergy to ketoprofen: clinical and experimental study. Contact Dermatitis 1990;23:336-340

4 Valsecchi R, Landro AD, Pigatto P, Cainelli T. Tiaprofenic acid photodermatitis. Contact Dermatitis 1989;21:345-346

5 Romano A, Pietrantonio F. Delayed hypersensitivity to flurbiprofen. J Intern Med 1997;241:81-83

6 De Groot AC. Monographs in contact allergy, volume 3: Topical Drugs. Boca Raton, Fl, USA: CRC Press Taylor and Francis Group, 2021 (ISBN 978-0-367-23693-9)

Chapter 3.222 FLUTAMIDE

IDENTIFICATION

Description/definition   : Flutamide is the trifluoromethylbenzene that conforms to the structural formula shown
                              below
Pharmacological classes : Androgen antagonists; antineoplastic agents, hormonal
IUPAC name   : 2-Methyl-N-[4-nitro-3-(trifluoromethyl)phenyl]propanamide
CAS registry number   : 13311-84-7
EC number   : 236-341-9
Merck Index monograph : 5508
Patch testing   : Tablets, pulverized, 10% and 20% petrolatum and acetone; most systemic drugs can be
                              tested at 10% pet.; if the pure chemical is not available, prepare the test material from
                              intravenous powder, the content of capsules or – when also not available – from
                              powdered tablets to achieve a final concentration of the active drug of 10% pet.
Molecular formula   : $C_{11}H_{11}F_3N_2O_3$

GENERAL

Flutamide is a nonsteroidal antiandrogen. This agent and its more potent active metabolite 2-hydroxyflutamide competitively block dihydrotestosterone binding at androgen receptors, forming inactive complexes which cannot translocate into the cell nucleus. Formation of inactive receptors inhibits androgen-dependent DNA and protein synthesis, resulting in tumor cell growth arrest or transient tumor regression. Flutamide is indicated for the management of locally confined stage B2-C and stage D2 metastatic carcinoma of the prostate (1).

CONTACT ALLERGY FROM ACCIDENTAL CONTACT

In a Congress Abstract, flutamide was mentioned among the drugs that are 'often the causative allergens' in contact dermatitis in the pharmaceutical industry (14). However, the source was not mentioned and no original articles on occupational allergic contact dermatitis to flutamide could be retrieved by the author, only to N-(3-trifluoromethyl-4-nitrophenyl)phthalimide, which is an intermediate in the production of flutamide (16).

CUTANEOUS ADVERSE DRUG REACTIONS FROM SYSTEMIC ADMINISTRATION CAUSED BY TYPE IV (DELAYED-TYPE) HYPERSENSITIVITY (as demonstrated by positive patch tests)

Cutaneous adverse drug reactions from systemic administration of flutamide caused by type IV (delayed-type) hypersensitivity have included photosensitivity (2-8,15).

Photosensitivity

General

Flutamide has caused a considerable number of photosensitive dermatitis cases. In many, photopatch tests were positive (2-8,15), in some negative or not performed (9-13). In three patients, the eruption appears to have provoked the development of vitiligo (3,4,8).

Case reports

A 60-year-old patient, diagnosed with prostatic carcinoma, was treated with oral flutamide 500 mg twice daily. After 7 days, he presented with a pruriginous cutaneous eruption that was erythematous and slightly edematous on the face, neck and dorsa of the hands, restricted to photo-exposed areas. Suspecting flutamide, the patient stopped taking the drug, but the erythema progressed to eczematization of the lesions with some blisters. Patch and photo-patch testing showed positive patch test reactions to flutamide 5%, 10% and 20% (+ at D2 and D4) with stronger reactions (++ at D2 and D4) at the UVA-irradiated sites. Eight controls were negative. As the lesions appeared on light-exposed areas only, the authors made a diagnosis of photosensitivity reaction to flutamide (2).

A 71-year-old man, treated with leuproreline acetate depot and oral flutamide 250 mg 3x daily, was referred for widespread exfoliative dermatitis. All treatment was stopped and the patient recovered with prednisone. Two months later, new lesions consisting of rounded hypopigmented macules started gradually to appear, located particularly on the sun-exposed areas of the skin: head, neck and arms. A skin biopsy of these showed absence of melanocytes, increased number of Langerhans' cells, epidermal vacuolization and a T cell inflammatory infiltrate. No evidence of other auto-immune diseases was detected. Photopatch testing with flutamide at 20% in acetone was positive and controls negative. Half a year later, slow but progressive repigmentation in the vitiligo lesions was achieved with narrowband UVB (3).

Another patient also developed vitiligo in skin where previously photoallergic dermatitis with pruritus, erythema and vesicles from flutamide had appeared. Photopatch tests, both with UVA and UVB irradiation, were positive to flutamide 10% and 20% acetone and negative in 10 controls (4). A nearly identical case of photosensitive dermatitis from flutamide with positive photopatch tests to 10% and 20% flutamide in petrolatum and acetone followed by vitiligo was published in 2007 (8). However, the pathogenetic mechanisms in the development of vitiligo following photosensitive dermatitis from flutamide are not yet clearly understood (8). Photoleukomelanoderma (synonym: photoleukomelanodermatitis), a special form of drug eruption characterized by a mosaic-like mixture of hyper- and hypopigmented skin lesions after sun exposure has also been observed in a patient who had photosensitivity to flutamide. In this patient, photopatch tests were negative (13).

In the 1990s, several cases of photosensitive dermatitis with positive photopatch tests had already been described (5,6,7,15).

LITERATURE

1   The data in the section 'General' may have been obtained from literature discussed in this chapter, but mostly also or exclusively from one or more of the following online sources: ChemIDPlus Advanced, PubChem, DrugBank, RxList, Drug Central, Drugs.com, and Wikipedia
2   Martín-Lázaro J, Buján JG, Arrondo AP, Lozano JR, Galindo EC, Capdevila EF. Is photopatch testing useful in the investigation of photosensitivity due to flutamide? Contact Dermatitis 2004;50:325-326
3   Rafael JP, Manuel GG, Antonio V, Carlos MJ. Widespread vitiligo after erythroderma caused by photosensitivity to flutamide. Contact Dermatitis 2004;50:98-100
4   Vilaplana J, Romaguera C, Azón A, Lecha M. Flutamide photosensitivity – residual vitiliginous lesions. Contact Dermatitis 1998;38:68-70
5   Fujimoto M, Kikuchi K, Imakado S, Furue M. Photosensitive dermatitis induced by flutamide. Br J Dermatol 1996;135:496-497
6   Moraillon I, Jeanmougin M, Manciet JR, Revuz J, Bagot M. Photoallergic reaction induced by flutamide. Photodermatol Photoimmunol Photomed 1991;8:264-265
7   Zabala R, Gardeazabal J, Manzano D, Aguirre A, Zubizarreta J, Tuneu A, Diaz Pérez J. Fotosensibilidad por flutamida. Actas Dermosifiliogr 1995;86:323-325
8   Swoboda A, Kasche A, Baumstark J, Worret WI, Ring J, Eberlein-König B. Vitiliginous lesions after photosensitive dermatitis due to flutamide. J Eur Acad Dermatol Venereol 2007;21:681-682
9   Leroy D, Dompmartin A, Szczurko C. Flutamide photosensitivity. Photodermatol Photoimmunol Photomed 1996;12:216-218
10 Yokote R, Tokura Y, Igarashi N, Ishikawa O, Miyachi Y. Photosensitive drug eruption induced by flutamide. Eur J Dermatol 1998;8:427-429
11 Tsien C, Souhami L. Flutamide photosensitivity. J Urol 1999;162:494
12 Kaur C, Thami GP. Flutamide induced photosensitivity: is it a forme fruste of lupus? Br J Dermatol 2003;148:603-604
13 Higashiyama A, Yokoyama T, Omoto Y, Habe K, Yamanaka K, Mizutani H. Flutamide-induced photoleukomelano-derma. J Dermatol 2016;43:1105-1106
14 Heras F. Contact dermatitis in the pharmaceutical industry. Contact Dermatitis 2010;63(Suppl.1):16-17
15 Gonçalo M, Domingues J, Correia O, Figueiredo A. Fotossensibilidad a flutamida. Bol Inf GEIDC 1999;29:45-48 (Article in Spanish)
16 Jungewelter S, Aalto-Korte K. A new allergen in the pharmaceutical industry. Contact Dermatitis 2008;59:314

Chapter 3.223 FLUVASTATIN

IDENTIFICATION

Description/definition   : Fluvastatin is the phenylpyrrole that conforms to the structural formula shown below
Pharmacological classes   : Antilipidemic agents
IUPAC name   : (E,3R,5S)-7-[3-(4-Fluorophenyl)-1-propan-2-ylindol-2-yl]-3,5-dihydroxyhept-6-enoic acid
CAS registry number   : 93957-54-1
EC number   : Not available
Merck Index monograph : 5515
Patch testing   : No data available; most systemic drugs can be tested at 10% pet.; if the pure chemical is
                                           not available, prepare the test material from intravenous powder, the content of capsules
                                           or – when also not available – from powdered tablets to achieve a final concentration of
                                           the active drug of 10% pet.
Molecular formula   : $C_{24}H_{26}FNO_4$

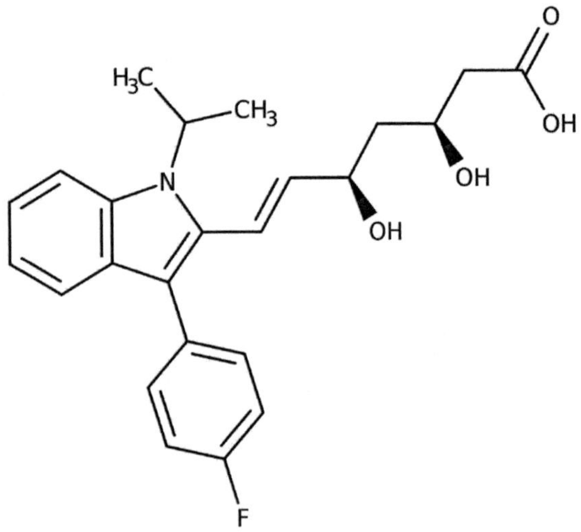

GENERAL

Fluvastatin is a synthetic lipid-lowering agent with antilipidemic and potential antineoplastic properties. It is indicated to be used as an adjunct to dietary therapy to prevent cardiovascular events. Fluvastatin may be used as secondary prevention in patients with coronary heart disease (CHD) to reduce the risk of requiring coronary revascularization procedures, for reducing progression of coronary atherosclerosis in hypercholesterolemic patients with CHD, and for the treatment of primary hypercholesterolemia and mixed dyslipidemia. In pharmaceutical products, fluvastatin is employed as fluvastatin sodium (CAS number 93957-55-2, EC number not available, molecular formula $C_{24}H_{25}FNNaO_4$) (1).

CONTACT ALLERGY FROM ACCIDENTAL CONTACT

A 35-year-old process operator developed occupational airborne allergic contact dermatitis to fluvastatin. The primary sites of the dermatitis were the eyelids and neck. More clinical and patch testing details were not provided (2).

REFERENCES

1   The data in the section 'General' may have been obtained from literature discussed in this chapter, but mostly also or exclusively from one or more of the following online sources: ChemIDPlus Advanced, PubChem, DrugBank, RxList, Drug Central, Drugs.com, and Wikipedia
2   Bennett MF, Lowney AC, Bourke JF. A study of occupational contact dermatitis in the pharmaceutical industry. Br J Dermatol 2016;174:654-656 (Abstract in Brit J Dermatol 2011;165:73)

Chapter 3.224 FLUVOXAMINE

IDENTIFICATION

Description/definition   : Fluvoxamine is the aralkylketone-derivative that conforms to the structural formula shown below

Pharmacological classes : Anti-anxiety agents; cytochrome P-450 CYP1A2 inhibitors; cytochrome P-450 CYP2C19 Inhibitors; serotonin uptake inhibitors; antidepressive agents, second-generation

IUPAC name   : 2-[(E)-[5-Methoxy-1-[4-(trifluoromethyl)phenyl]pentylidene]amino]oxyethanamine

CAS registry number   : 54739-18-3

EC number   : 611-193-1

Merck Index monograph   : 5516

Patch testing   : 1% and 5% pet. (2)

Molecular formula   : $C_{15}H_{21}F_3N_2O_2$

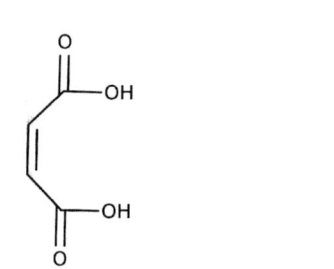

Fluvoxamine   Fluvoxamine maleate

GENERAL

Fluvoxamine is a selective serotonin reuptake inhibitor with antidepressant, anti-obsessive-compulsive, and anxiolytic properties. It selectively blocks serotonin reuptake by inhibiting the serotonin reuptake pump at the presynaptic neuronal membrane. This increases serotonin levels within the synaptic cleft, prolongs serotonergic transmission and decreases serotonin turnover. Fluvoxamine shows no significant affinity for histaminergic, α- or β-adrenergic, muscarinic, or dopaminergic receptors *in vitro*. This agent is indicated predominantly for the management of depression and for obsessive-compulsive disorder, but is has also been used in the management of bulimia nervosa (1).

In pharmaceutical products, fluvoxamine is employed as fluvoxamine maleate (CAS number 61718-82-9, EC number 612-212-6, molecular formula $C_{19}H_{25}F_3N_2O_6$) (1).

CUTANEOUS ADVERSE DRUG REACTIONS FROM SYSTEMIC ADMINISTRATION CAUSED BY TYPE IV (DELAYED-TYPE) HYPERSENSITIVITY (as demonstrated by positive patch tests)

Cutaneous adverse drug reactions from systemic administration of fluvoxamine caused by type IV (delayed-type) hypersensitivity have included drug reaction with eosinophilia and systemic symptoms (DRESS) (4) and photo-sensitivity (2,3),

Drug reaction with eosinophilia and systemic symptoms (DRESS)

A 40-year-old woman developed drug reaction with eosinophilia and systemic symptoms (DRESS) while on various medications. Patch tests were positive to fluvoxamine 12.5%, carbamazepine 20% and oxcarbazepine 12.5%, all in phosphate-buffered saline. As the patient had reactions to 2 unrelated drugs, this was a case of multiple drug hypersensitivity (4).

Photosensitivity

A 50-year-old woman presented with an eczematous eruption of the face, neck, upper forehead and ears, and with erythematous and scaly lesions on her hands and forearms. She had been treated for depression with fluvoxamine for 1 month and had been exposed to a fluorescent lamp during her job 3 days earlier. Patch and photopatch testing showed positive reactions to fluvoxamine 1% (+) and 5% (++) in petrolatum, enhanced by UVA exposure, provoking a vesicular reaction. The patient was diagnosed with UVA-photosensitive fluvoxamine-induced eczema (2). The authors described 2 more patients using fluvoxamine for depression with eczematous eruptions in light-exposed skin. In these 2 individuals, photopatch tests were negative with UVA irradiation, but strongly positive with UVB Irradiation. However, controls have not been tested in the same manner to exclude phototoxicity (2).

In a study from Nancy, France, 54 patients with suspected nonimmediate drug eruptions (27 maculopapular, seven erythrodermic, nine eczematous, four photosensitivity, three fixed drug eruptions, three with pruritus and one with acute generalized exanthematous pustulosis) were assessed with patch testing. All 4 patients suspected of photosensitivity had positive *photo*patch tests, of who one to fluvoxamine (3).

LITERATURE

1   The data in the section 'General' may have been obtained from literature discussed in this chapter, but mostly also or exclusively from one or more of the following online sources: ChemIDPlus Advanced, PubChem, DrugBank, RxList, Drug Central, Drugs.com, and Wikipedia
2   Doffoel-Hantz V, Boulitrop-Morvan C, Sparsa A, Bonnetblanc JM, Dalac S, Bédane C. Photosensitivity associated with selective serotonin reuptake inhibitors. Clin Exp Dermatol 2009;34:e763-765
3   Barbaud A, Reichert-Penetrat S, Tréchot P, Jacquin-Petit MA, Ehlinger A, Noirez V, et al. The use of skin testing in the investigation of cutaneous adverse drug reactions. Br J Dermatol 1998;139:49-58
4   Gex-Collet C, Helbling A, Pichler WJ. Multiple drug hypersensitivity – proof of multiple drug hypersensitivity by patch and lymphocyte transformation tests. J Investig Allergol Clin Immunol 2005;15:293-296

Chapter 3.225 FONDAPARINUX

IDENTIFICATION

Description/definition : Fondaparinux is the synthetic pentasaccharide that conforms to the structural formula shown below

Pharmacological classes : Antithrombotic agents

IUPAC name : (2S,3S,4R,5R,6R)-6-[(2R,3R,4R,5R,6R) 6-[(2R,3S,4S,5R,6R)-2-Carboxy-4-hydroxy-6-[(2R,3S,4R,5R,6S)-4-hydroxy-6-methoxy-5-(sulfoamino)-2-(sulfooxymethyl)oxan-3-yl]oxy-5-sulfooxyoxan-3-yl]oxy-5-(sulfoamino)-4-sulfooxy-2-(sulfooxymethyl)oxan-3-yl]oxy-3-[(2R,3R,4R,5S,6R)-4,5-dihydroxy-3-(sulfoamino)-6-(sulfooxymethyl)oxan-2-yl]oxy-4,5-dihydroxyoxane-2-carboxylic acid

Other names : Natural heparin pentasaccharide

CAS registry number : 104993-28-4

EC number : Not available

Merck Index monograph : 5528 (Fondaparinux sodium)

Patch testing : Commercial preparation undiluted (2); consider intradermal testing with late readings (D2,D3) when patch tests are negative and consider subcutaneous challenge when intradermal tests are negative

Molecular formula : $C_{31}H_{53}N_3O_{49}S_8$

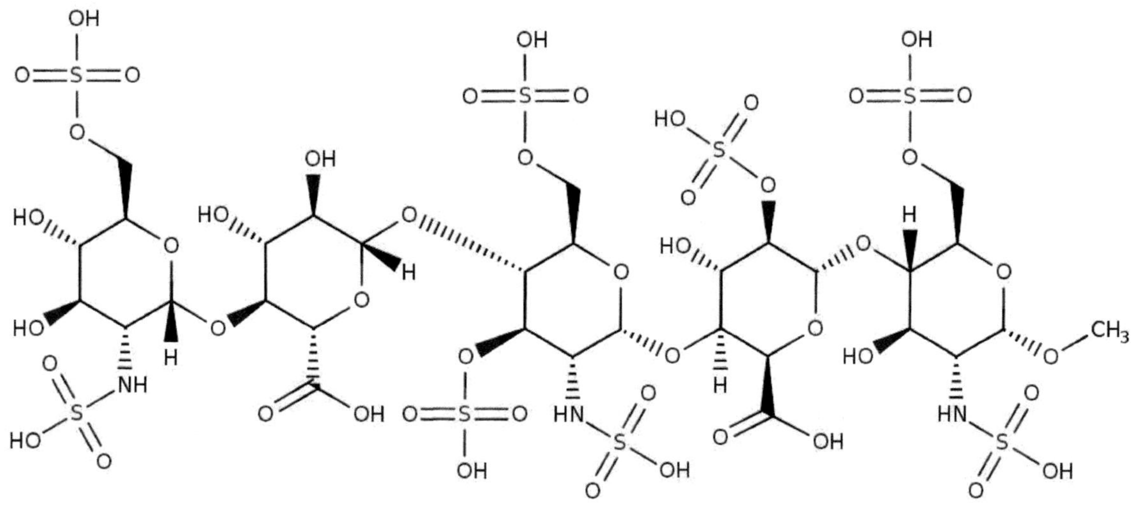

GENERAL

Fondaparinux is a synthetic glucopyranoside with antithrombotic activity; is has an ultra-low molecular weight of 1728 daltons. Fondaparinux selectively binds to antithrombin III, thereby potentiating the innate neutralization of activated factor X (Factor Xa) by antithrombin. Neutralization of Factor Xa inhibits its activity and interrupts the blood coagulation cascade, thereby preventing thrombin formation and thrombus development. It is used as an anticoagulant to prevent venous thromboembolism, to treat deep vein thrombosis, and to improve survival following myocardial infarction. In pharmaceutical products, fondaparinux is employed as fondaparinux sodium (CAS number 114870-03-0, EC number not available, molecular formula $C_{31}H_{43}N_3Na_{10}O_{49}S_8$) (1).

See also bemiparin (Chapter 3.59), certoparin (Chapter 3.117), dalteparin (Chapter 3.160), danaparoid (Chapter 3.161), enoxaparin (Chapter 3.195), heparins (Chapter 3.239), nadroparin (Chapter 3.331), and tinzaparin (Chapter 3.479).

CUTANEOUS ADVERSE DRUG REACTIONS FROM SYSTEMIC ADMINISTRATION CAUSED BY TYPE IV (DELAYED-TYPE) HYPERSENSITIVITY

Throughout this book, only reports of delayed-type hypersensitivity have been included that showed a positive patch test to the culprit drug. However, as a result of the high molecular weight of heparins, patch tests are often false-negative, presumably from insufficient penetration into the skin. Because of this, and also because patch tests have been performed in a small minority of cases only, studies with a positive intradermal test or subcutaneous provocation tests with delayed readings are included in the chapters on the various heparins, even when patch tests were negative or not performed.

It should be appreciated that fondaparinux is different from other heparins because of its ultralow molecular weight, resulting in less sensitizations and less cross-reactivity to other heparins (7).

General information on delayed-type hypersensitivity reactions to heparins

General information on delayed-type hypersensitivity reactions to heparins (including fondaparinux), presenting as local reactions from subcutaneous administration, is provided in Chapter 3.239. Heparins. In this chapter, only *non-local* cutaneous adverse drug reactions from delayed-type hypersensitivity to fondaparinux are presented.

Non-local cutaneous adverse drug reactions

Non-local cutaneous adverse drug reactions from systemic administration of fondaparinux caused by type IV (delayed-type) hypersensitivity have not been found.

Cross-reactions, pseudo-cross-reactions and co-reactions

Cross-reactions between low molecular weight heparins are frequent in delayed-type hypersensitivity (>90% of patients tested, median number of positive drugs per patient: 3) and do not depend on the molecular weight of the molecules (4). Overlap in their polysaccharide composition might explain the high degree of cross-allergenicity (3,9). Cross-reactions to the semisynthetic heparinoid danaparoid have also been observed (6). In allergic patients, the synthetic ultralow molecular weight synthetic heparin fondaparinux is usually, but not always (4,5,6,8,9,10) well-tolerated (3).

LITERATURE

1   The data in the section 'General' may have been obtained from literature discussed in this chapter, but mostly also or exclusively from one or more of the following online sources: ChemIDPlus Advanced, PubChem, DrugBank, RxList, Drug Central, Drugs.com, and Wikipedia
2   Brockow K, Garvey LH, Aberer W, Atanaskovic-Markovic M, Barbaud A, Bilo MB, et al. Skin test concentrations for systemically administered drugs - an ENDA/EAACI Drug Allergy Interest Group position paper. Allergy 2013;68:702-712
3   Schindewolf M, Lindhoff-Last E, Ludwig RJ. Heparin-induced skin lesions. Lancet 2012;380:1867-1879
4   Weberschock T, Meister AC, Bohrt K, Schmitt J, Boehncke W-H, Ludwig RJ. The risk for cross-reactions after a cutaneous delayed-type hypersensitivity reaction to heparin preparations is independent of their molecular weight: a systematic review. Contact Dermatitis 2011;65:187-194
5   Jappe U, Juschka U, Kuner N, Hausen BM, Krohn K. Fondaparinux: a suitable alternative in cases of delayed-type allergy to heparins and semisynthetic heparinoids? A study of 7 cases. Contact Dermatitis 2004;51:67-72
6   Utikal J, Peitsch WK, Booken D, Velten F, Dempfle CE, Goerdt S, et al. Hypersensitivity to the pentasaccharide fondaparinux in patients with delayed-type heparin allergy. Thromb Haemost 2005;94:895-896
7   Schindewolf M, Scheuermann J, Kroll H, Garbaraviciene J, Hecking C, Marzi I, et al. Low allergenic potential with fondaparinux: results of a prospective investigation. Mayo Clin Proc 2010;85:913-919
8   Hirsch K, Ludwig RJ, Lindhoff-Last E, Kaufmann R, Boehncke WH. Intolerance of fondaparinux in a patient allergic to heparins. Contact Dermatitis 2004;50:383-384
9   Maetzke J, Hinrichs R, Schneider LA, Scharffetter-Kochanek K. Unexpected delayed-type hypersensitivity skin reactions to the ultra-low-molecular-weight heparin fondaparinux. Allergy 2005;60:413-415
10 Hohenstein E, Tsakiris D, Bircher AJ. Delayed-type hypersensitivity to the ultra-low-molecular-weight heparin fondaparinux. Contact Dermatitis 2004;51:149-151

Chapter 3.226 FOSCARNET

IDENTIFICATION

Description/definition   : Foscarnet is the pyrophosphate analog that conforms to the structural formula shown below

Pharmacological classes   : Reverse transcriptase inhibitors; antiviral agents

IUPAC name   : Phosphonoformic acid

Other names   : Phosphonoformate; carboxyphosphonic acid

CAS registry number   : 4428-95-9

EC number   : Not available

Merck Index monograph   : 5548 (Foscarnet sodium)

Patch testing   : No data available; most systemic drugs can be tested at 10% pet.; if the pure chemical is not available, prepare the test material from intravenous powder, the content of capsules or – when also not available – from powdered tablets to achieve a final concentration of the active drug of 10% pet.

Molecular formula   : CH_3O_5P

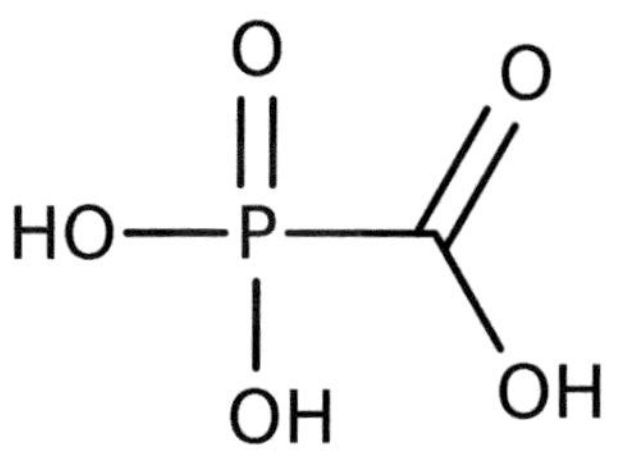

GENERAL

Foscarnet is a synthetic pyrophosphate analog DNA polymerase inhibitor with antiviral activity against cytomegalovirus, human herpesviruses and HIV. It selectively blocks the pyrophosphate binding site of virus-specific DNA polymerases at concentrations that do not affect cellular DNA polymerases. Foscarnet is indicated for the treatment of cytomegalovirus retinitis in patients with acquired immunodeficiency syndrome (AIDS) and for treatment of acyclovir-resistant mucocutaneous HSV infections in immunocompromised patients (1). In pharmaceutical products, foscarnet is employed as foscarnet sodium (hexahydrate) (CAS number 34156-56-4, EC number not available, molecular formula $CH_{12}Na_3O_{11}P$) (1).

CONTACT ALLERGY FROM ACCIDENTAL CONTACT

In French literature, occupational allergic contact dermatitis to foscarnet has apparently been described (2, data cited in ref. 3).

LITERATURE

1   The data in the section 'General' may have been obtained from literature discussed in this chapter, but mostly also or exclusively from one or more of the following online sources: ChemIDPlus Advanced, PubChem, DrugBank, RxList, Drug Central, Drugs.com, and Wikipedia

2   Testud F, Descotes J, Evreux JC. Pathologie professionnelle due aux médicaments. Arch Mal Prof 1994;55:279-286 (no details known, data cited in ref. 3)

3   Bircher AJ. Drug allergens. In: John S, Johansen J, Rustemeyer T, Elsner P, Maibach H, eds. Kanerva's occupational dermatology. Cham, Switzerland: Springer, 2020:559-578

Chapter 3.227 FULVESTRANT

IDENTIFICATION

Description/definition	: Fulvestrant is the estrogen derivative that conforms to the structural formula shown below
Pharmacological classes	: Antineoplastic agents, hormonal; estrogen receptor antagonists
IUPAC name	: (7R,8R,9S,13S,14S,17S)-13-Methyl-7-[9-(4,4,5,5,5-pentafluoropentylsulfinyl)nonyl]-6,7,8,9,11,12,14,15,16,17-decahydrocyclopenta[a]phenanthrene-3,17-diol
CAS registry number	: 129453-61-8
EC number	: Not available
Merck Index monograph	: 5583
Patch testing	: Commercial preparation 250 mg/ml
Molecular formula	: $C_{32}H_{47}F_5O_3S$

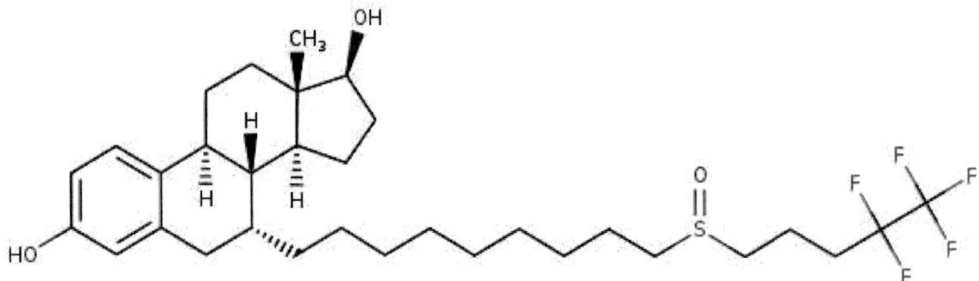

GENERAL

Fulvestrant is an analog of 17β-estradiol that acts as a competitive estrogen receptor antagonist. It binds competitively to estrogen receptors in breast cancer cells, resulting in estrogen receptor deformation and decreased estrogen binding. Thus, fulvestrant reversibly inhibits the growth of estrogen-sensitive, human breast cancer cell lines. The drug is indicated for the treatment of hormone receptor positive metastatic breast cancer in postmenopausal women with disease progression following anti-estrogen (tamoxifen) therapy (1).

CUTANEOUS ADVERSE DRUG REACTIONS FROM SYSTEMIC ADMINISTRATION CAUSED BY TYPE IV (DELAYED-TYPE) HYPERSENSITIVITY (as demonstrated by positive patch tests)

Cutaneous adverse drug reactions from systemic administration of fulvestrant caused by type IV (delayed-type) hypersensitivity have included fixed drug eruption (2).

Fixed drug eruption

A 69-year-old woman was treated with fulvestrant for metastatic breast cancer and with denosumab for osteoporosis. Four months later, the patient presented with multiple round, pigmented, well demarcated, erythematous itching and burning skin lesions on the lower back and abdomen with a maximum diameter of a few centimeters. Histopathological analysis of a punch biopsy from the lower back showed a picture suggestive of fixed drug eruption. Patch tests with the two suspected drugs were performed both on normal skin and on a residual lesion of the lower back. Patch tests at the undiluted concentrations (fulvestrant 250 mg/ml and denosumab 70 mg/ml) showed positive reactions to fulvestrant on both normal skin and the lesional pigmented area. On normal skin, erythema without edema was seen, whereas, on the residual lesion, a marked papular aspect was noted in addition to edema. Tests with denosumab gave negative results. One (!) control was negative. Punch biopsies were performed on both positive patch test areas. Whereas only spongiosis was present in the reaction on normal skin, the test on the residual lesion showed apoptotic keratinocytes with blistering and an acute lichenoid infiltrate, suggestive of FDE reactivation. No more lesions occurred after fulvestrant had been discontinued while denosumab was continued. A diagnosis of fixed drug eruption from fulvestrant was made (2).

LITERATURE

1   The data in the section 'General' may have been obtained from literature discussed in this chapter, but mostly also or exclusively from one or more of the following online sources: ChemIDPlus Advanced, PubChem, DrugBank, RxList, Drug Central, Drugs.com, and Wikipedia

2   Broche C, Pralong P, Gil H, Yahiaoui N, Mousseau M, Chatain C, et al. Fixed drug eruption caused by fulvestrant confirmed by skin tests: First case. Contact Dermatitis 2019;80:184-186

Chapter 3.228 FURALTADONE

IDENTIFICATION

Description/definition : Furaltadone is the nitrofuran antibiotic that conforms to the structural formula shown below
Pharmacological classes : Anti-infective agents, urinary
IUPAC name : 5-(Morpholin-4-ylmethyl)-3-[(E)-(5-nitrofuran-2-yl)methylideneamino]-1,3-oxazolidin-2-one
Other names : Nitrofurmethone
CAS registry number : 139-91-3
EC number : 205-384-5
Merck Index monograph : 5593
Patch testing : 1% pet. (3)
Molecular formula : $C_{13}H_{16}N_4O_6$

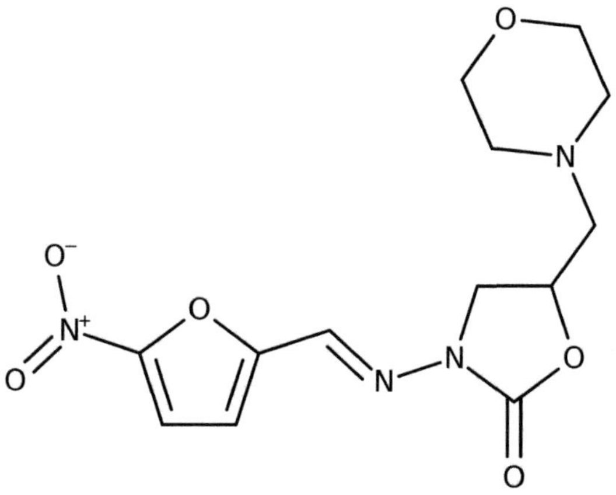

GENERAL

Furaltadone is a nitrofuran antibiotic that is used in veterinary medicine and is effective against bacterial infections in birds when added to feed or drinking water. It was formerly used in humans orally but was withdrawn due to toxicity. However, furaltadone hydrochloride (CAS number 3759-92-0, EC number 223-169-4, molecular formula $C_{13}H_{17}ClN_4O_6$) may still be used topically in some countries for treatment of ear disorders (1).

In topical preparations, furaltadone has rarely caused contact allergy/allergic contact dermatitis, which subject has been reviewed in Volume 3 of the *Monographs in contact allergy* series (3).

CONTACT ALLERGY FROM ACCIDENTAL CONTACT

A 58-year-old farmhand presented with a 6-month history of pruritus and papules on his forearms, face, neck and legs. When he worked preparing and distributing feed for pigs and chickens, the rash rapidly deteriorated. Patch testing with all ingredients of the feeds showed a positive reaction to furaltadone tartrate 1% pet. at D2 (+) and D4 (++). Twenty controls were negative. The patient was diagnosed with occupational allergic contact dermatitis (2).

LITERATURE

1   The data in the section 'General' may have been obtained from literature discussed in this chapter, but mostly also or exclusively from one or more of the following online sources: ChemIDPlus Advanced, PubChem, DrugBank, RxList, Drug Central, Drugs.com, and Wikipedia
2   Vilaplana J, Grimalt F, Romaguera C. Contact dermatitis from furaltadone in animal feed. Contact Dermatitis 1990;22:232-323
3   De Groot AC. Monographs in contact allergy, volume 3: Topical Drugs. Boca Raton, Fl, USA: CRC Press Taylor and Francis Group, 2021 (ISBN 978-0-367-23693-9)

Chapter 3.229 FURAZOLIDONE

IDENTIFICATION

Description/definition   : Furazolidone is the nitrofuran derivative that conforms to the structural formula shown
                                  below
Pharmacological classes   : Antitrichomonal agents; monoamine oxidase inhibitors; anti-infective agents, urinary;
                                  anti-infective agents, local
IUPAC name   : 3-[(E)-(5-Nitrofuran-2-yl)methylideneamino]-1,3-oxazolidin-2-one
CAS registry number   : 67-45-8
EC number   : 200-653-3
Merck Index monograph : 5600
Patch testing   : 1% and 10% pet.; testing in PEG-400 may be preferable (5)
Molecular formula   : $C_8H_7N_3O_5$

GENERAL

Furazolidone is a nitrofuran derivative with antiprotozoal and antibacterial activity. It has a broad antibacterial spectrum covering the majority of gastrointestinal tract pathogens including *E. coli*, staphylococci, *Salmonella*, *Shigella*, *Proteus*, *Aerobacter aerogenes*, *Vibrio cholerae* and *Giardia lamblia*. Its bactericidal activity is based upon its interference with DNA replication and protein production; this antimicrobial action minimizes the development of resistant organisms. Furazolidone is indicated for the specific and symptomatic treatment of bacterial or protozoal diarrhea and enteritis caused by susceptible organisms (1). However, because of serious adverse effects after oral administration, furazolidone is hardly used any more in human medicine (5).

CONTACT ALLERGY FROM ACCIDENTAL CONTACT

Case series

In Prague, Czechoslovakia, before 1975, of 32 patients with occupational dermatitis from animal feed additives, 3 were allergic to furazolidone, tested at 1% alc. (6).

Case reports

A pig farmer presented with a 6-month history of severe dermatitis of the thumb, index, middle and ring fingers of each hand. The nails were ridged and the cuticles lost. One of his duties was to dose newborn piglets with the oral antibiotic furazolidone to prevent pathogenic *E. coli* infection. This is done with individual oral doses from a pump dispenser. Inevitably, the fingers come into contact with the antibiotic during the dosing procedure. Protective gloves are useless, as little piglets have razor-sharp teeth that pierce most protective gloves. Patch tests were positive to the medication 'as is' and furazolidone 1% and 10% pet. Changing the antibiotic given to the piglets effected a cure (2).

A 43-year-old man, who worked on a research farm as a pig keeper, presented with dermatitis of the fingertips which had developed 3 months earlier. Patch testing revealed contact allergy to an animal feed additive (D2 ?+, D4 +++) and to a veterinary pharmaceutical suspension (D2 -, D4 ++). The only common ingredient proved to be furazolidone, being present at 10% w/w in the animal feed and 4% w/w in the drug. Patch testing with furazolidone, furfural, nitrofurazone and nitrofurantoin (all 2% pet.) were completely negative. However, the patient later had positive patch test reactions to furazolidone 2% both in PEG-400 and in alcohol (D2 -, D4 +) with negative reactions to the bases and the related substances in alc. and PEG-400. Twenty-five controls were negative (5).

A case of occupational contact allergy to furazolidone present in milk fat was briefly mentioned in a German medical journal (6). Another woman, mixing chicken fodder containing furazolidone, developed hand dermatitis. She had a positive patch test reaction to undiluted furazolidone (3).

CUTANEOUS ADVERSE DRUG REACTIONS FROM SYSTEMIC ADMINISTRATION CAUSED BY TYPE IV (DELAYED-TYPE) HYPERSENSITIVITY (as demonstrated by positive patch tests)

Cutaneous adverse drug reactions from systemic administration of furazolidone caused by type IV (delayed-type) hypersensitivity have included systemic allergic dermatitis (4, uncertain).

Systemic allergic dermatitis (systemic contact dermatitis)

A woman probably had widespread dermatitis from contact allergy to furazolidone and/or nifuroxime in vaginal suppositories: systemic allergic dermatitis. A patch test with shavings from a suppository was positive, but the ingredients were not tested separately (4).

Cross-reactions, pseudo-cross-reactions and co-reactions

Primary sensitization to furazolidone may have caused cross-sensitivity in one patient to nitrofurazone (6).

LITERATURE

1   The data in the section 'General' may have been obtained from literature discussed in this chapter, but mostly also or exclusively from one or more of the following online sources: ChemIDPlus Advanced, PubChem, DrugBank, RxList, Drug Central, Drugs.com, and Wikipedia
2   Burge S, Bransbury A. Allergic contact dermatitis due to furazolidone in a piglet medication. Contact Dermatitis 1994;31:199-200
3   Scharfenberg B. Nitrofuranhaltiges Huhnerkukenfutter als berufliches Ekzematogen. Derm Wschrift 1967;153:60 (Article in German, cited in ref. 5)
4   Goette DK, Odom RB. Vaginal medications as a cause for varied widespread dermatitides. Cutis 1980;26:406-409
5   De Groot AC, Conemans JM. Contact allergy to furazolidone. Contact Dermatitis 1990;22:202-205
6   Laubstein H, Niedergesäss G. Untersuchungen über Gruppensensibilisierungen bei Nitrofuranderivaten. Derm Monatsschrift 1970;156:1-8 (Article in German, cited in ref. 5)

Chapter 3.230 FUSIDIC ACID

IDENTIFICATION

Description/definition   : Fusidic acid is an antibiotic isolated from the fermentation broth of *Fusidium coccineum*
                             that conforms to the structural formula shown below
Pharmacological classes   : Anti-bacterial agents; protein synthesis inhibitors
IUPAC name   : (2Z)-2-[(3R,4S,5S,8S,9S,10S,11R,13R,14S,16S)-16-Acetyloxy-3,11-dihydroxy-4,8,10,14-
                             tetramethyl-2,3,4,5,6,7,9,11,12,13,15,16-dodecahydro-1H-cyclopenta[a]phenanthren-17-
                             ylidene]-6-methylhept-5-enoic acid
CAS registry number   : 6990-06-3
EC number   : 230-256-0
Merck Index monograph   : 5616
Patch testing   : Sodium fusidate, 2% pet. (Chemotechnique, SmartPracticeCanada, SmartPracticeEurope)
Molecular formula   : $C_{31}H_{48}O_6$

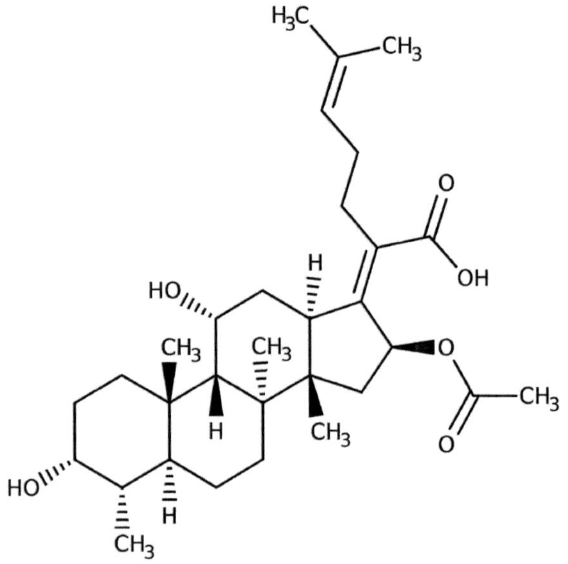

GENERAL

Fusidic acid is a bacteriostatic antibiotic derived from the fungus *Fusidium coccineum* that is used mostly as a topical medication to treat bacterial skin infections caused by *Staphylococcus aureus*, but is also given systemically as a tablet or injection. In pharmaceutical products, fusidic acid is employed as sodium fusidate (CAS number 751-94-0, EC number 212-030-3, molecular formula $C_{31}H_{47}NaO_6$) (1).

In topical preparations, fusidic acid has caused contact allergy/allergic contact dermatitis, which subject has been fully reviewed in Volume 3 of the *Monographs in contact allergy* series (4).

CUTANEOUS ADVERSE DRUG REACTIONS FROM SYSTEMIC ADMINISTRATION CAUSED BY TYPE IV (DELAYED-TYPE) HYPERSENSITIVITY (as demonstrated by positive patch tests)

Cutaneous adverse drug reactions from systemic administration of fusidic acid caused by type IV (delayed-type) hypersensitivity have included systemic allergic dermatitis (2) and unspecified cutaneous adverse drug reaction (3).

Systemic allergic dermatitis (systemic contact dermatitis)

A 51-year-old man developed a pruritic micropapular generalized exanthema, 4 hours after a first oral 250 mg dose of fusidic acid. Four days before, the patient had been treated with topical fusidic acid 2% ointment to an impetiginized skin lesion with good tolerance. It was the first time he was exposed to fusidic acid. A patch test with the ointment was positive and with its excipients (lanolin alcohol and petrolatum) negative; fusidic acid itself was not tested. The patient was diagnosed with systemic allergic dermatitis from fusidic acid (2).

Other cutaneous adverse drug reactions

In a group of 13 patients with morbilliform rashes, 4 with SJS/TEN and 3 with DRESS, there were 10 positive patch tests to suspected drugs, including one to fusidic acid. It was not specified which drug reaction fusidic acid had caused (3).

LITERATURE

1   The data in the section 'General' may have been obtained from literature discussed in this chapter, but mostly also or exclusively from one or more of the following online sources: ChemIDPlus Advanced, PubChem, DrugBank, RxList, Drug Central, Drugs.com, and Wikipedia

2   De Castro Martinez FJ, Ruiz FJ, Tornero P, De Barrio M, Prieto A. Systemic allergic dermatitis due to fusidic acid. Contact Dermatitis 2006;54:169

3   Hassoun-Kheir N, Bergman R, Weltfriend S. The use of patch tests in the diagnosis of delayed hypersensitivity drug eruptions. Int J Dermatol 2016;55:1219-1224

4   De Groot AC. Monographs in contact allergy, volume 3: Topical Drugs. Boca Raton, Fl, USA: CRC Press Taylor and Francis Group, 2021 (ISBN 978-0-367-23693-9)

Chapter 3.231　GABAPENTIN

IDENTIFICATION

Description/definition	: Gabapentin is the gamma amino acid that conforms to the structural formula shown below
Pharmacological classes	: Antimanic agents; anti-anxiety agents; excitatory amino acid antagonists; analgesics; anticonvulsants
IUPAC name	: 2-[1-(Aminomethyl)cyclohexyl]acetic acid
CAS registry number	: 60142-96-3
EC number	: 262-076-3
Merck Index monograph	: 5618
Patch testing	: No data available; most systemic drugs can be tested at 10% pet.; if the pure chemical is not available, prepare the test material from intravenous powder, the content of capsules or – when also not available – from powdered tablets to achieve a final concentration of the active drug of 10% pet.
Molecular formula	: $C_9H_{17}NO_2$

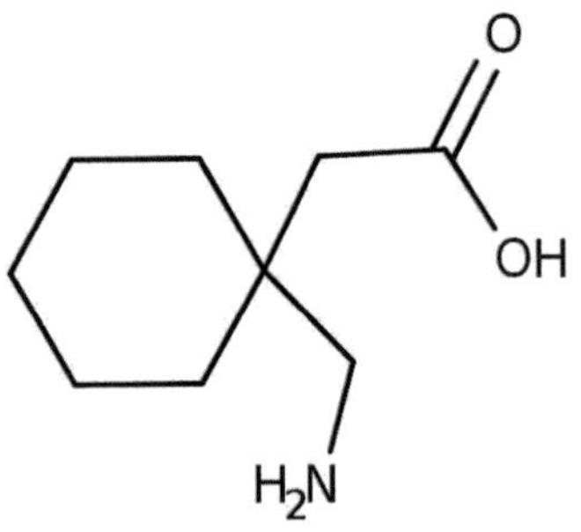

GENERAL

Gabapentin is a synthetic analog of the neurotransmitter gamma-aminobutyric acid with anticonvulsant activity. Although its exact mechanism of action is unknown, gabapentin appears to inhibit the release of excitatory neurotransmitters. This agent also exhibits analgesic properties. Gabapentin is used as adjunctive therapy in the management of epilepsy, and to treat neuropathic pain syndromes and restless legs syndrome (1).

CUTANEOUS ADVERSE DRUG REACTIONS FROM SYSTEMIC ADMINISTRATION CAUSED BY TYPE IV (DELAYED-TYPE) HYPERSENSITIVITY (as demonstrated by positive patch tests)

Cutaneous adverse drug reactions from systemic administration of gabapentin caused by type IV (delayed-type) hypersensitivity have included maculopapular eruption (2).

Maculopapular eruption

One patient developed a maculopapular eruption from gabapentin with a positive patch test reaction. This individual previously had suffered an anaphylactic reaction to erythromycin and later had maculopapular exanthemas from gabapentin and various other drugs: multiple drug hypersensitivity syndrome. No clinical or patch test details were provided (2).

LITERATURE

1　The data in the section 'General' may have been obtained from literature discussed in this chapter, but mostly also or exclusively from one or more of the following online sources: ChemIDPlus Advanced, PubChem, DrugBank, RxList, Drug Central, Drugs.com, and Wikipedia
2　Jörg L, Yerly D, Helbling A, Pichler W. The role of drug, dose and the tolerance/intolerance of new drugs in multiple drug hypersensitivity syndrome (MDH). Allergy 2020;75:1178-1187

Chapter 3.232 GABEXATE

IDENTIFICATION

Description/definition : Gabexate is the benzoic acid ester that conforms to the structural formula shown below
Pharmacological classes : Serine proteinase inhibitors; anticoagulants
IUPAC name : Ethyl 4-[6-(diaminomethylideneamino)hexanoyloxy]benzoate
CAS registry number : 39492-01-8
EC number : Not available
Merck Index monograph : 5620
Patch testing : Mesylate, 1% and 10% pet.
Molecular formula : $C_{16}H_{23}N_3O_4$

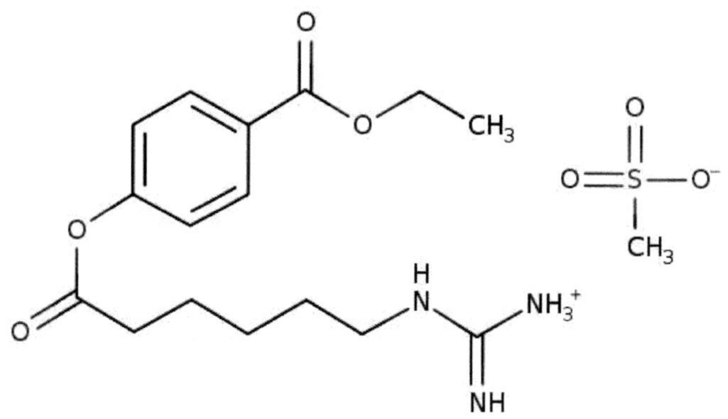

Gabexate mesylate

GENERAL

Gabexate is a synthetic serine protease inhibitor. It also known to decrease the production of inflammatory cytokines. Gabexate is most often used to treat acute pancreatitis with deviation of proteolytic enzymes (such as trypsin, kallikrein and plasmin), acute exacerbation of chronic recurrent pancreatitis, acute pancreatitis after surgery and disseminated intravascular coagulation. It is marketed in a few countries only, including Italy and Japan (1). In pharmaceutical products, gabexate is employed as gabexate mesylate (CAS number 56974-61-9, EC number 611-438-2, molecular formula $C_{17}H_{27}N_3O_7S$) (1).

CUTANEOUS ADVERSE DRUG REACTIONS FROM SYSTEMIC ADMINISTRATION CAUSED BY TYPE IV (DELAYED-TYPE) HYPERSENSITIVITY (as demonstrated by positive patch tests)

Cutaneous adverse drug reactions from systemic administration of gabexate caused by type IV (delayed-type) hypersensitivity have included panniculitis (2).

Other cutaneous adverse drug reactions

A 59-year-old man was treated with intravenous gabexate mesylate for acute pancreatitis through catheters inserted in veins in the left arm, left hand (dorsum) and right arm. Nine days after admission, the catheters were removed because they were occluded. Six days later, reddish swelling appeared at the sites where the catheters had been inserted with pain and itching. Indurations about 2 centimeter in diameter were palpable in the dorsum of the left hand and the center of the extensor side of both forearms. Histopathology of a skin biopsy showed panniculitis with prominent eosinophilic infiltration. Patch tests were performed with all drugs used and were positive only to gabexate mesylate 1% and 10% pet. at D2, D3 and D7. Ten controls were negative. Intradermal injections of about 0.1 ml of gabexate mesylate 0.1% and 0.01% saline were positive at the higher concentration with 5 mm indurated erythema (D2) and a red papule of 6 mm in diameter at D3. Histopathological examination of the papule showed perivascular patchy infiltrate and slight septal panniculitis. It was suggested that the panniculitis had been caused by gabexate mesylate (and not by the pancreatitis) (2).

LITERATURE

1   The data in the section 'General' may have been obtained from literature discussed in this chapter, but mostly also or exclusively from one or more of the following online sources: ChemIDPlus Advanced, PubChem, DrugBank, RxList, Drug Central, Drugs.com, and Wikipedia
2   Nakayama F. Panniculitis with eosinophilic infiltration due to gabexate mesilate (FOY): possibility of allergic reaction. J Dermatol 1997;24:235-242

Chapter 3.233 GADOBUTROL

IDENTIFICATION

Description/definition   : Gadobutrol is the gadolinium-containing contrast medium that conforms to the structural
                             formula shown below
Pharmacological classes : Contrast media
IUPAC name   : 10-((1RS,2SR)-2,3-Dihydroxy-1-(hydroxymethyl)propyl)-1,4,7,10-tetraazacyclododecane-
                             1,4,7-triacetato(3-))gadolinium
CAS registry number   : 770691-21-9
EC number   : Not available
Patch testing   : Commercial gadobutrol preparation 1.0 mmol/ml (2)
Molecular formula   : $C_{18}H_{31}GdN_4O_9$

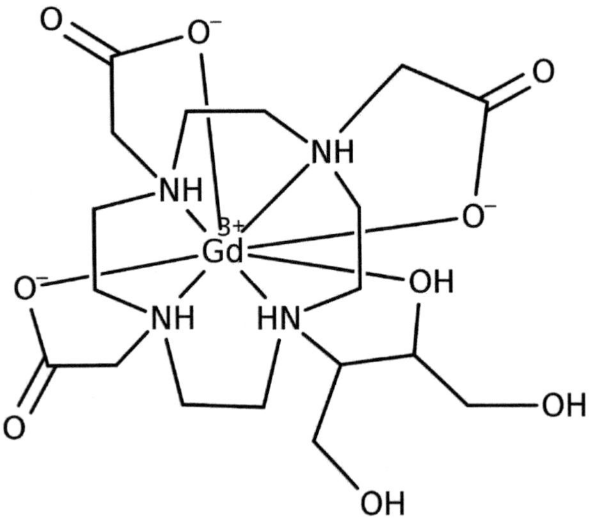

GENERAL
Gadobutrol is a gadolinium-containing compound used for diagnostic purposes only. It is indicated for contrast enhancement during cranial and spinal MRI, and for contrast-enhanced magnetic resonance angiography (CE-MRA) (1).

CUTANEOUS ADVERSE DRUG REACTIONS FROM SYSTEMIC ADMINISTRATION CAUSED BY TYPE IV
(DELAYED-TYPE) HYPERSENSITIVITY (as demonstrated by positive patch tests)
Cutaneous adverse drug reactions from systemic administration of gadobutrol caused by type IV (delayed-type)
hypersensitivity have included maculopapular eruption (3,4,5) and acute generalized exanthematous pustulosis
(AGEP) (2).

Maculopapular eruption
A 62-year-old woman with early stage lung cancer presented with extensive erythematous skin eruptions. Two days
previously, she had received 7.5 ml of 1.0 M gadobutrol intravenously for a preoperative chest MRI examination. The
eruptions appeared on her abdomen 4 to 5 hours after the administration of gadobutrol, and then gradually spread
to almost her entire body. Physical examination revealed an erythematous maculopapular rash on her face, trunk,
and extremities. Patch tests were positive at D2 to gadobutrol 'as is' and gadobutrol diluted 1/10 in saline. No
controls tests were performed. The patient had probably become sensitized from previous gadobutrol administrations (5).

One patient developed a maculopapular eruption from gadobutrol with a positive patch test to this drug after a
previous episode of DRESS from clarithromycin (multiple drug hypersensitivity) (3,4).

Acute generalized exanthematous pustulosis (AGEP)
A 53-year-old woman presented with a generalized rash. Physical examination showed numerous small, sterile (upon
culture), non-follicular pustules on a diffuse erythematous base with minimal mucosal involvement. The rash was
pruritic and more prominent in the flexural areas, and showed active borders. A skin biopsy from a lesion on the
trunk showed the histological features of a drug-induced pustular eruption. The patient also showed some systemic

symptoms including mild renal impairment, malaise, fever, and leukocytosis. A diagnosis of acute generalized exanthematous pustulosis (AGEP) was made. Three days before AGEP onset, the patient had undergone magnetic resonance imaging with a gadolinium-based contrast medium. Patch tests performed 4 months after the last reaction with aqueous gadoteric acid (commercial preparation) and aqueous gadobutrol (commercial preparation, 1.0 mmol/ml) were positive at D2 and D4 to gadobutrol. The positive reactions in the patch test mimicked the clinical lesions. Twelve controls were negative (2).

LITERATURE

1   The data in the section 'General' may have been obtained from literature discussed in this chapter, but mostly also or exclusively from one or more of the following online sources: ChemIDPlus Advanced, PubChem, DrugBank, RxList, Drug Central, Drugs.com, and Wikipedia

2   Bordel Gómez MT, Martín García C, Meseguer Yebra C, Zafra Cobo MI, Cardeñoso Álvarez ME, Sánchez Estella J. First case report of acute generalized exanthematous pustulosis (AGEP) caused by gadolinium confirmed by patch testing. Contact Dermatitis 2018;78:166-168

3   Jörg L, Yerly D, Helbling A, Pichler W. The role of drug, dose and the tolerance/intolerance of new drugs in multiple drug hypersensitivity syndrome (MDH). Allergy 2020;75:1178-1187

4   Jörg L, Helbling A, Yerly D, Pichler WJ. Drug-related relapses in drug reaction with eosinophilia and systemic symptoms (DRESS). Clin Transl Allergy 2020;10:52

5   Nagai H, Nishigori C. A delayed reaction to the magnetic resonance imaging contrast agent gadobutrol. J Allergy Clin Immunol Pract 2017;5:850-851

Chapter 3.234 GARENOXACIN

IDENTIFICATION

Description/definition   : Garenoxacin is the quinoline carboxylic acid that conforms to the structural formula
                                shown below
Pharmacological classes : Anti-bacterial agent; topoisomerase II inhibitor
IUPAC name   : 1-Cyclopropyl-8-(difluoromethoxy)-7-[(1R)-1-methyl-2,3-dihydro-1H-isoindol-5-yl]-4-
                                oxoquinoline-3-carboxylic acid
CAS registry number   : 194804-75-6
EC number   : Not available
Merck Index monograph : 5672
Patch testing   : 10% (probably pet. and probably from pulverized tablets) (2)
Molecular formula   : $C_{23}H_{20}F_2N_2O_4$

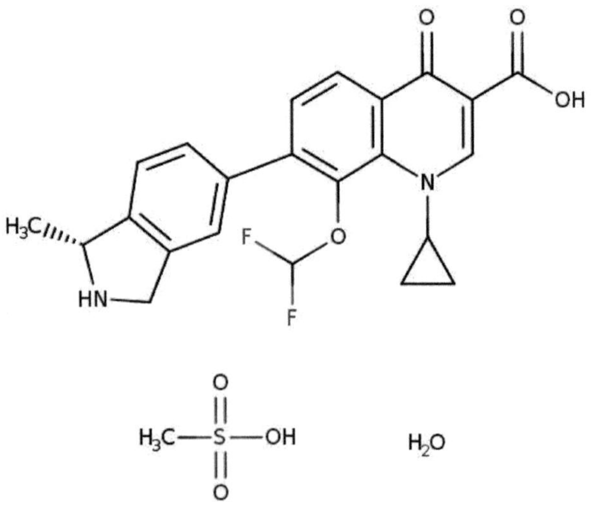

Garenoxacin mesylate (hydrate)

GENERAL

Garenoxacin is a quinolone antibiotic, used both for the treatment of gram-positive and gram-negative bacterial infections. It is a novel quinolone that lacks a fluorine molecule at the C-6 position. Garenoxacin is available in Japan only and is unlikely to be approved elsewhere based on a risk/benefit assessment. In pharmaceutical products, garenoxacin is employed as garenoxacin mesylate (CAS number 223652-90-2, EC number not available, molecular formula $C_{24}H_{26}F_2N_2O_8S$) (1).

CUTANEOUS ADVERSE DRUG REACTIONS FROM SYSTEMIC ADMINISTRATION CAUSED BY TYPE IV (DELAYED-TYPE) HYPERSENSITIVITY (as demonstrated by positive patch tests)

Cutaneous adverse drug reactions from systemic administration of garenoxacin caused by type IV (delayed-type) hypersensitivity have included maculopapular eruption (2).

Maculopapular eruption

A 72-year-old woman received garenoxacin for an upper tracheal infection. Two weeks after drug initiation the patient developed erythematous plaques and papules. Histopathology revealed lymphocyte and eosinophil infiltration in the upper dermis. Lymphocytes had also infiltrated into the epidermis. Patch testing showed a positive reaction to garenoxacin 10%. The patient was diagnosed with a maculopapular type drug eruption caused by garenoxacin (2).

LITERATURE

1   The data in the section 'General' may have been obtained from literature discussed in this chapter, but mostly also or exclusively from one or more of the following online sources: ChemIDPlus Advanced, PubChem, DrugBank, RxList, Drug Central, Drugs.com, and Wikipedia
2   Oda T, Sawada Y, Okada E, Nakamura M. Maculopapular type drug eruption caused by garenoxacin mesylate hydrate: A case report and literature review. Australas J Dermatol 2017;58:e276-e277

Chapter 3.235 GENTAMICIN

IDENTIFICATION

Description/definition : Gentamicin is an antibiotic mixture of closely related aminoglycosides obtained from *Micromonospora purpurea* and related species

Pharmacological classes : Anti-bacterial agents; protein synthesis inhibitors

IUPAC name : 2-[4,6-Diamino-3-[3-amino-6-[1-(methylamino)ethyl]oxan-2-yl]oxy-2-hydroxycyclo-hexyl]oxy-5-methyl-4-(methylamino)oxane-3,5-diol

CAS registry number : 1403-66-3

EC number : 215-765-8

Merck Index monograph : 5697

Patch testing : Sulfate, 20.0% pet. (Chemotechnique, SmartPracticeCanada, SmartPracticeEurope); late readings (D6-D8) are necessary to avoid missing positive patch tests evolving after D3-4 (13)

Molecular formula : $C_{21}H_{43}N_5O_7$

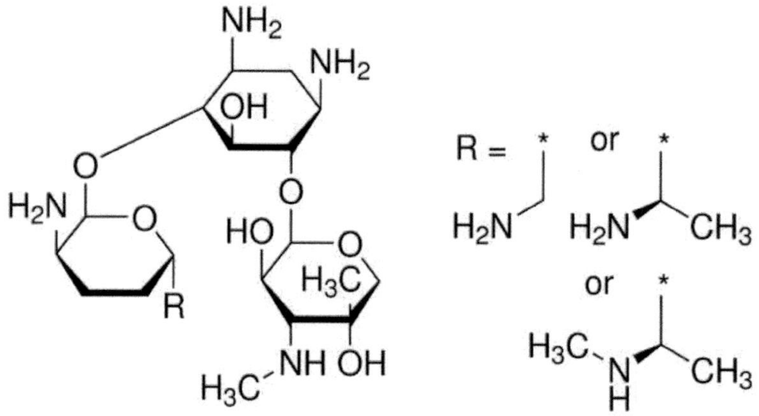

Gentamicin C$_{1A}$:   R = CH$_2$NH$_2$
Gentamicin C$_1$:   R = CH(CH$_3$)NHCH$_3$
Gentamicin C$_2$:   R = CH(CH$_3$)NH$_2$

GENERAL

Gentamicin is a broad-spectrum aminoglycoside antibiotic produced by fermentation of *Micromonospora purpurea* or *M. echinospora*. It is an antibiotic complex consisting of four major (C1, C1a, C2, and C2a) and several minor components. Gentamicin is indicated for treatment of serious infections caused by susceptible strains of *Pseudomonas aeruginosa*, *Proteus* species (indole-positive and indole-negative), *E. coli*, *Klebsiella-Enterobacter-Serratia* species, *Citrobacter* species and *Staphylococcus* species (coagulase-positive and coagulase-negative). This antibiotic is also used in topical pharmaceuticals for the treatment of superficial skin and eye infections. As such, gentamicin has caused many cases of contact allergy/allergic contact dermatitis (of which many may have been the result of cross-sensitivity to primary neomycin sensitization), which has been fully reviewed in Volume 3 of the *Monographs in contact allergy* series (10).

In pharmaceutical products, gentamicin is employed as gentamicin sulfate (CAS number 1405-41-0, EC number 215-778-9, molecular formula $C_{19}H_{40}N_4O_{10}S$) (1).

CONTACT ALLERGY FROM ACCIDENTAL CONTACT

A 25-year-old female nurse presented with dermatitis of the back of the hands, fingertips, fingers, arm and wrist. She was patch tested with the medicaments that she had most frequently contact with and reacted to gentamicin and ranitidine HCl (4). This was a case of occupational allergic contact dermatitis.

Three nurses from Finland had occupational allergic contact dermatitis of the hands from gentamicin in bone cement (6). A 24-year-old female technician in a bacteriological laboratory developed dermatitis of the right hand and both arms. She had contact at work with gentamicin sulfate and had a positive patch test to it at 20% pet. (9).

CUTANEOUS ADVERSE DRUG REACTIONS FROM SYSTEMIC ADMINISTRATION CAUSED BY TYPE IV (DELAYED-TYPE) HYPERSENSITIVITY (as demonstrated by positive patch tests)
Cutaneous adverse drug reactions from systemic administration of gentamicin caused by type IV (delayed-type) hypersensitivity have included systemic allergic dermatitis presenting as exfoliative erythroderma (2), as an eczematous eruption (3,5), as 'severe drug eruption' (8), localized allergic dermatitis from intra-articular-injection (12), and unspecified drug eruption (11).

Erythroderma, widespread erythematous eruption, exfoliative dermatitis
See the section 'Systemic allergic dermatitis' below, ref. 2.

Systemic allergic dermatitis (systemic contact dermatitis)
A patient allergic to neomycin showed a generalized exfoliative erythroderma following intravenous gentamicin therapy. When his ear canals were later treated with a neomycin-containing topical medication, he reacted so severely that the skin of his ears was temporarily depigmented. Patch tests were positive to neomycin, gentamicin and various other aminoglycosides. This was a case of systemic allergic dermatitis from gentamicin cross-reacting to neomycin (2).

An 84-year-old woman was treated with gentamicin cream on venous ulcerations for the presence of *Pseudomonas aeruginosa*, resulting in induration, erythema, crusting and weeping of the skin around the ulcers. Several hours after a first intravenous dose of gentamicin the patient developed an itchy eczematous eruption on both arms, which extended to the sides of her neck. Later, patch tests were positive to gentamicin and neomycin (3). This was a case of systemic allergic dermatitis from previous sensitization to topical gentamicin.

An otherwise healthy 30-year-old woman developed a generalized eczematous eruption one day after a cesarean section. She had received various drugs, among which intravenous gentamicin. After exclusion of the other drugs as the culprit, the patient was challenged systemically with 80 mg gentamicin. Six hours later, she developed a red pruritic lesion at the injection site and twenty-four hours later she presented axillary eczema and a flare-up of the previous eczema. Patch test were positive to gentamicin, neomycin, and amikacin. Once the study was completed, the patient recalled a previous contact with a gentamicin ointment with no adverse reactions (5).

In a patient patch test positive to neomycin but negative to gentamicin, injections with gentamicin sulfate resulted in a 'severe drug reaction' (8). Patch testing afterwards now showed a positive reaction to gentamicin sulfate (8, data cited in ref. 7).

Dermatitis/eczematous eruption
See the section 'Systemic allergic dermatitis' above, refs. 3 and 5, and the section 'Other cutaneous adverse drug reactions' below, ref. 12.

Other cutaneous adverse drug reactions

Gentamicin in bone cement
Three months after a 79-year-old patient had received a right cemented total knee arthroplasty, he noted pain, swelling and a reduced range of motion. Physical examination by the orthopedic surgeon showed joint effusion. Diagnostic joint aspiration and microbiological analysis showed no signs of infection. Scintigraphy was suggestive of local synovitis, which was attributed to 'excessive walking'. However, the patient complained of increasing pain, and a few weeks later presented with local eczema of the right knee. Patch testing showed positive reactions to neomycin and gentamicin on D6. In the operation, antibiotic-free bone cement had been used. It was found that the patient had previously used gentamicin-containing eye drops to treat conjunctivitis. In addition, it was discovered that the orthopedic surgeon, while performing the diagnostic joint aspiration, had injected gentamicin solution to prevent infection. The patient was now diagnosed with 'synovitis and allergic contact dermatitis' resulting from intra-articular gentamicin application. In the course of the next 10 months all symptoms completely disappeared (12). One can wonder whether the diagnosis 'localized allergic dermatitis' might be preferable over 'allergic contact dermatitis'.

In patients with total knee arthroplasty, who are allergic to gentamicin and who have non-allergic symptoms such as joint pain, swelling and limited ambulation, uncemented revision arthroplasty or change to gentamicin-free bone cement can provide significant symptom relief (14).

Unspecified drug eruptions
In Finland, in the period 1989-2001, 826 patients with suspected cutaneous drug eruptions were patch tested and 89 had one or more positive reactions. Of these individuals, one reacted to gentamicin. No clinical details, e.g. about the nature of the cutaneous adverse drug reaction caused by gentamicin, were provided (11).

Cross-reactions, pseudo-cross-reactions and co-reactions

Of patients sensitized to neomycin, about 50% cross-reacts to gentamicin (Chapter 3.335 Neomycin). The cross-sensitivity pattern between aminoglycoside antibiotics in patients primarily sensitized to gentamicin has not been well investigated. In previous studies, 60-100% of patients with positive patch tests to gentamicin were also allergic to neomycin (10) and may represent cross-reactions. In one patient sensitized to gentamicin, netilmicin may have cross-reacted (6).

Immediate contact reactions

Immediate contact reactions (contact urticaria) to gentamicin are presented in Chapter 5.

LITERATURE

1   The data in the section 'General' may have been obtained from literature discussed in this chapter, but mostly also or exclusively from one or more of the following online sources: ChemIDPlus Advanced, PubChem, DrugBank, RxList, Drug Central, Drugs.com, and Wikipedia

2   Guin JD, Phillips D. Erythroderma from systemic allergic dermatitis: a complication of systemic gentamicin in a patient with contact allergy to neomycin. Cutis 1989;43:564-567

3   Ghadially R, Ramsay CA. Gentamicin: systemic exposure to a contact allergen. J Am Acad Dermatol 1988;19(2Pt. 2):428-430

4   Gielen K, Goossens A. Occupational allergic contact dermatitis from drugs in healthcare workers. Contact Dermatitis 2001;45:273-279

5   Paniagua MJ, Garcia-Ortega P, Tella R, Gaig P, Richart C. Systemic allergic dermatitis to gentamicin. Allergy 2002;57:1086-1087

6   Liippo J, Lammintausta K. Positive patch test reactions to gentamicin show sensitization to aminoglycosides from topical therapies, bone cements, and from systemic medication. Contact Dermatitis 2008;59:268-272

7   Van Ketel WG, Bruynzeel DP. Sensitization to gentamicin alone. Contact Dermatitis 1989;20:303-304

8   Braun W, Schütz R. Beitrag zur Gentamycin Allergie. Hautarzt 1969;20:108 (Article in German, data cited in ref. 7)

9   Bandmann H-J, Mutzeck E. Contact allergy to gentamycin sulfate. Contact Dermatitis Newsletter 1973;13:371

10 De Groot AC. Monographs in contact allergy, volume 3: Topical Drugs. Boca Raton, Fl, USA: CRC Press Taylor and Francis Group, 2021 (ISBN 978-0-367-23693-9)

11 Lammintausta K, Kortekangas-Savolainen O. The usefulness of skin tests to prove drug hypersensitivity. Br J Dermatol 2005;152:968-974

12 Wittmann D, Summer B, Thomas B, Halder A, Thomas P. Gentamicin allergy as an unexpected 'hidden' cause of complications in knee arthroplasty. Contact Dermatitis 2018;78:293-294

13 Thomas B, Kulichova D, Wolf R, Summer B, Mahler V, Thomas P. High frequency of contact allergy to implant and bone cement components, in particular gentamicin, in cemented arthroplasty with complications: usefulness of late patch test reading. Contact Dermatitis 2015;73:343-349

14 Thomas B, Benedikt M, Alamri A, Kapp F, Bader R, Summer B, et al. The role of antibiotic-loaded bone cement in complicated knee arthroplasty: relevance of gentamicin allergy and benefit from revision surgery - a case control follow-up study and algorithmic approach. J Orthop Surg Res 2020;15:319

Chapter 3.236 GLICLAZIDE

IDENTIFICATION

Description/definition	: Gliclazide is the sulfonylurea that conforms to the structural formula shown below
Pharmacological classes	: Hypoglycemic agents
IUPAC name	: 1-(3,3a,4,5,6,6a-Hexahydro-1H-cyclopenta[c]pyrrol-2-yl)-3-(4-methylphenyl)sulfonylurea
CAS registry number	: 21187-98-4
EC number	: 244-260-5
Merck Index monograph	: 5744
Patch testing	: Tablet, pulverized, 30% pet. and water (2); most systemic drugs can be tested at 10% pet.; if the pure chemical is not available, prepare the test material from intravenous powder, the content of capsules or – when also not available – from powdered tablets to achieve a final concentration of the active drug of 10% pet.
Molecular formula	: $C_{15}H_{21}N_3O_3S$

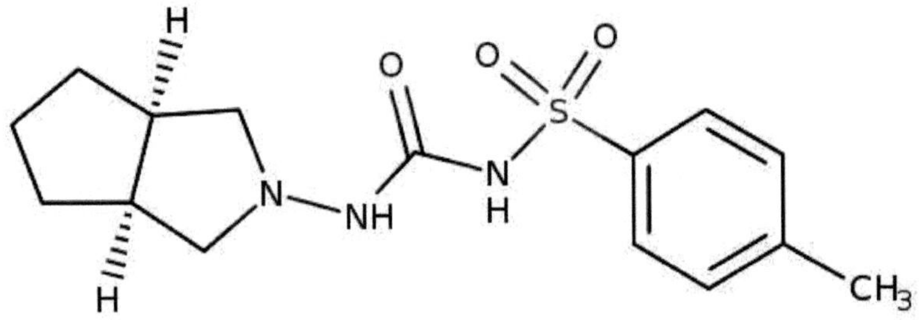

GENERAL

Gliclazide is a short-acting, relatively high-potency, second-generation sulfonylurea compound with hypoglycemic activity. It also increases peripheral insulin sensitivity. Gliclazide is used for the treatment of non-insulin-dependent diabetes mellitus (1).

CUTANEOUS ADVERSE DRUG REACTIONS FROM SYSTEMIC ADMINISTRATION CAUSED BY TYPE IV (DELAYED-TYPE) HYPERSENSITIVITY (as demonstrated by positive patch tests)

Cutaneous adverse drug reactions from systemic administration of gliclazide caused by type IV (delayed-type) hypersensitivity have included erythroderma (2).

Erythroderma, widespread erythematous eruption, exfoliative dermatitis

A 77-year-old woman presented with a generalized pruritic skin eruption, which had begun 3 days after starting oral gliclazide and metformin for irregular blood glucose levels. Physical examination revealed generalized erythema with desquamation. The patient had no previous history of a dermatological disease or allergy to any drugs (especially sulfonamides). Histopathology revealed spongiosis in the epidermis with vesicle formation, lymphocyte exocytosis, and perivascular lymphomonocytic infiltration in the superficial dermis, which was compatible with erythroderma secondary to drug intake. When the drugs were stopped and replaced with insulin, the exanthema cleared in 3 days. One month later, patch tests with both drugs were positive to gliclazide 30% pet. and 30% in water at D2. Presumably, later readings were not performed and no controls were tested (2).

LITERATURE

1   The data in the section 'General' may have been obtained from literature discussed in this chapter, but mostly also or exclusively from one or more of the following online sources: ChemIDPlus Advanced, PubChem, DrugBank, RxList, Drug Central, Drugs.com, and Wikipedia
2   Ozuguz P, Kacar SD, Ozuguz U, Karaca S, Tokyol C. Erythroderma secondary to gliclazide: a case report. Cutan Ocul Toxicol 2014;33:342-344

Chapter 3.237 GRISEOFULVIN

IDENTIFICATION

Description/definition : Griseofulvin is the mycotoxic metabolic product of *Penicillium* species that conforms to the structural formula shown below

Pharmacological classes : Antifungal agents

IUPAC name : (2*S*,5'*R*)-7-Chloro-3',4,6-trimethoxy-5'-methylspiro[1-benzofuran-2,4'-cyclohex-2-ene]-1',3-dione

CAS registry number : 126-07-8

EC number : 204-767-4

Merck Index monograph : 5854

Patch testing : 1% and 10% pet. (4)

Molecular formula : $C_{17}H_{17}ClO_6$

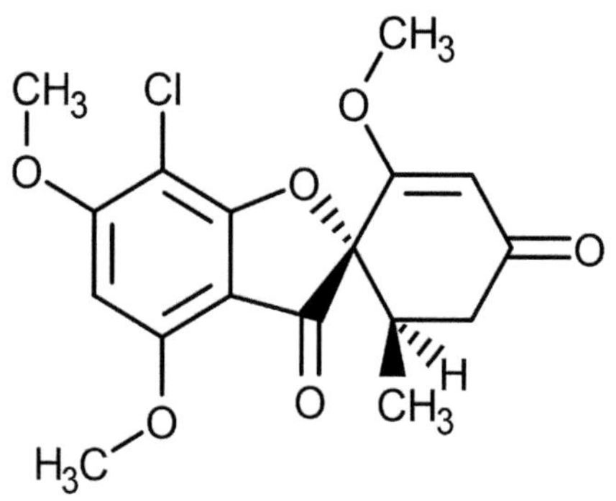

GENERAL

Griseofulvin is an antifungal agent derived from the mold *Penicillium griseofulvum*. It was the first available oral agent for the treatment of dermatophytosis. Griseofulvin is fungistatic with *in vitro* activity against various species of *Microsporum*, *Epidermophyton*, and *Trichophyton*. It has no effect on bacteria or on other genera of fungi. Griseofulvin is indicated for the treatment of fungal infections of the skin, hair, and nails caused by susceptible fungi (1).

CONTACT ALLERGY FROM ACCIDENTAL CONTACT

In French literature, occupational allergic contact dermatitis to griseofulvin has apparently been described (2, data cited in ref. 3)

CUTANEOUS ADVERSE DRUG REACTIONS FROM SYSTEMIC ADMINISTRATION CAUSED BY TYPE IV (DELAYED-TYPE) HYPERSENSITIVITY (as demonstrated by positive patch tests)

Cutaneous adverse drug reactions from systemic administration of griseofulvin caused by type IV (delayed-type) hypersensitivity have included photosensitivity (4,5).

Photosensitivity

A man aged 42 had received 750 mg griseofulvin/day for a week, when he developed edematous erythema on the exposed areas, notably the face. Patch tests and photopatch tests (with UVA-irradiation) were carried out with griseofulvin 0.01%, 0.1%, 1.0% and 10.0% in white petrolatum. Both patch and photopatch tests were positive to all concentrations; all irradiated patches reacted. The patient became a persistent light reactor (4). A 27-year-old woman had been treated with griseofulvin 275 mg/day for 5 days, when eczematous changed occurred on light-exposed areas. Patch tests were negative, photopatch tests positive to the 1% and 10% concentrations (4). A 64-year-old woman received griseofulvin 750 mg daily for 4 months, when she noticed erythema and desquamation on the face and neck. Within 2 weeks, these eruptions spread over both sides of her hands and feet. The clinical features resembled pellagra. Patch tests were negative, photopatch tests to all concentrations positive. After the tests, slight erythema and very small red papules developed on her face which resolved in 2 weeks (4). Control tests were not performed.

In early Japanese literature, 8 patients were reported with photodermatitis from griseofulvin (5). Three had positive patch tests and all 8 showed positive photopatch tests. Two had a flare-up during the testing procedures. Four of the patients developed persistent light reactivity (5). It should be realized that photopatch testing in those days was not very accurate in distinguishing phototoxicity from photoallergy.

LITERATURE

1   The data in the section 'General' may have been obtained from literature discussed in this chapter, but mostly also or exclusively from one or more of the following online sources: ChemIDPlus Advanced, PubChem, DrugBank, RxList, Drug Central, Drugs.com, and Wikipedia
2   Testud F, Descotes J, Evreux JC. Pathologie professionnelle due aux médicaments. Arch Mal Prof 1994;55:279-286 (no details known, data cited in ref. 3)
3   Bircher AJ. Drug allergens. In: John S, Johansen J, Rustemeyer T, Elsner P, Maibach H, eds. Kanerva's occupational dermatology. Cham, Switzerland: Springer, 2020:559-578
4   Kojima T, Hasegawa T, Ishida H, Fujita M, Okamoto S. Griseofulvin-induced photodermatitis--report of six cases. J Dermatol 1988;15:76-82
5   Araki H. Basic and clinical investigations of photosensitivity. Jpn J Dermatol 1967;77:239-270 (Article in Japanese, data cited in ref. 4)

Chapter 3.238 HALQUINOL

IDENTIFICATION

Description/definition   : Halquinol is a combination of hydroxyquinolines containing 5,7-dichloro-8-quinolinol, 5-chloro-8-quinolinol, 7-chloro-8-quinolinol and 8-hydroxyquinoline (PubChem); halquinol is a mixture of 5,7-dichloro-8-quinolinol with 5-chloro-8-quinolinol and 7-chloro-8-quinolinol (ChemIDPlus)

Pharmacological classes : Topical anti-infectives
Other names   : Halquinols
CAS registry number   : 8067-69-4
EC number   : Not available
Merck Index monograph   : 227
Patch testing   : 5% pet.

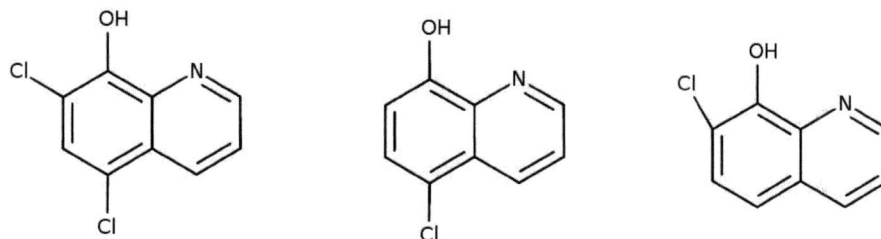

Halquinol according to ChemIDPlus

GENERAL

Halquinol is a combination of hydroxyquinolines currently described as 5,7-dichloro-8-quinolinol 14%, 5-chloro-8-quinolinol 42%, 7-chloro-8-quinolinol 10% and 8-hydroxyquinoline 34% (PubChem). ChemIDPlus describes it as 5,7-dichloro-8-quinolinol, mixture with 5-chloro-8-quinolinol and 7-chloro-8-quinolinol. Halquinol is used as an antimicrobial feed additive for poultry and as a pig growth promotor. It is also used for the control and treatment of non-specified diarrhea, particularly in pigs. It is or was also used in topical antiseptic preparations (1).

In topical preparations, halquinol has caused a few cases of contact allergy/allergic contact dermatitis, which subject has been fully reviewed in Volume 3 of the *Monographs in contact allergy* series (7).

CONTACT ALLERGY FROM ACCIDENTAL CONTACT

A man aged 42 years had worked for 29 years in an animal feed mill as a foreman. Two weeks after he had been assisting in adding 'Quixalaud' to the feeds, a contact dermatitis developed on the backs of the hands and wrists. The rash cleared up on going off work and did not recur on going back to work which avoided contact with Quixalaud. A patch test with 1% Quixalaud (halquinol 60%, chalk 40%) gave a marked positive reaction. The patient was diagnosed with occupational allergic contact dermatitis from halquinol (6).

Cross-reactions, pseudo-cross-reactions and co-reactions

Cross-reactions between halogenated hydroxyquinolines such as clioquinol, chlorquinaldol, cloxyquin, oxyquinoline (non-halogenated), iodoquinol, halquinol, 5,7-dichloro-8-quinolinol and 5-chloro-8-quinolinol may occur (2,3,4,5 [examples of references]).

LITERATURE

1   The data in the section 'General' may have been obtained from literature discussed in this chapter, but mostly also or exclusively from one or more of the following online sources: ChemIDPlus Advanced, PubChem, DrugBank, RxList, Drug Central, Drugs.com, and Wikipedia
2   Allenby CF. Skin sensitization to Remiderm and cross-sensitization to hydroxyquinoline compounds. Br Med J 1965;2(5455):208-209
3   Myatt AE, Beck MH. Contact sensitivity to chlorquinaldol. Contact Dermatitis 1983;9:523
4   Leifer W, Steiner K. Studies in sensitization to halogenated hydroxyquinolines and related compounds. J Invest Dermatol 1951;17:233-240
5   Wantke F, Götz M, Jarisch R. Contact dermatitis from cloxyquin. Contact Dermatitis 1995;32:112-113
6   Burrows D. Contact dermatitis in animal feed mill workers. Br J Dermatol 1975;92:167-170
7   De Groot AC. Monographs in contact allergy, volume 3: Topical Drugs. Boca Raton, Fl, USA: CRC Press Taylor and Francis Group, 2021 (ISBN 978-0-367-23693-9)

Chapter 3.239 HEPARINS

IDENTIFICATION

Description/definition	: Unfractionated heparin is a highly acidic mucopolysaccharide formed of equal parts of sulfated D-glucosamine and D-glucuronic acid with sulfaminic bridges
Pharmacological classes	: Antithrombotic agents
IUPAC name	: 6-[6-[6-[5-Acetamido-4,6-dihydroxy-2-(sulfooxymethyl)oxan-3-yl]oxy-2-carboxy-4-hydroxy-5-sulfooxyoxan-3-yl]oxy-2-(hydroxymethyl)-5-(sulfoamino)-4-sulfooxyoxan-3-yl]oxy-3,4-dihydroxy-5-sulfooxyoxane-2-carboxylic acid
CAS registry number	: 9005-49-6; other (deprecated) CAS numbers: 104521-37-1; 11078-24-3; 1108625-99-5; 1108626-06-7; 11129-39-8; 37324-73-5; 9075-96-1; 91449-79-5
EC number	: 232-681-7
Merck Index monograph	: 5958
Patch testing	: Commercial preparation undiluted (4); consider intradermal testing with late readings (D2,D3) when patch tests are negative and subcutaneous challenge when intradermal tests are negative
Molecular formula	: $C_{26}H_{42}N_2O_{37}S_5$ (PubChem); unspecified (ChemIDPlus)

GENERAL

Heparin has been widely prescribed as an anticoagulant drug in clinical practice for over 80 years. Anticoagulant drugs belonging to the class of heparins (not a full account) are shown in table 3.239.1. (3). Heparin is a natural product, isolated from animal tissues (porcine gut or bovine lung). Unfractionated heparin (UFH) is a highly acidic mucopolysaccharide formed of equal parts of sulfated D-glucosamine and D-glucuronic acid with sulfaminic bridges. The molecular weight ranges from 3000 to 30,000 daltons with an average molecular weight of 14,000-18,000 daltons (14-18 kDa). Heparin is highly sulfated, with a negative charge that allows it to bind to positively charged biological materials.

Low molecular weight heparins (LMWHs), which were introduced in the early 1990s, are obtained from unfractionated heparin by chemical or enzymatic depolymerization, which generates different products such as ardeparin, bemiparin, certoparin, dalteparin, enoxaparin, nadroparin, reviparin, and tinzaparin (table 3.239.1). Their molecular weight ranges from 2 to 9 kDa. LMWHs have partially substituted UFH because of their better pharmacologic profile and increased bioavailability, as well as their superior tolerance in patients and ease of self-administration via the subcutaneous route (3).

Heparins binds to antithrombin III to form a heparin-antithrombin III complex. The complex binds to and irreversibly inactivates thrombin and other activated clotting factors, such as factors IX, X, XI, and XII, thereby preventing the polymerization of fibrinogen to fibrin and the subsequent formation of clots. Clinical indications for both UFH and low molecular weight heparins include treatment or prevention of venous thromboembolism and acute coronary syndromes. It is also used to prevent clotting during dialysis and surgical procedures, maintain the patency of intravenous injection devices and prevent *in vitro* coagulation of blood transfusions and in blood samples drawn for laboratory values (1,3). Unfractionated heparin is used intravenously or subcutaneously, and requires monitoring using the activated partial thromboplastin time (APTT). By contrast, LMWHs are administered by subcutaneous injection, and the dose can be weight-adjusted without monitoring (11).

The *heparinoid* danaparoid (Chapter 3.161) is a mixture of low molecular weight sulfated glycosaminoglycans derived from porcine intestinal mucosa. It has been proposed for anticoagulation in patients with heparin-induced thrombocytopenia or heparin allergy. Danaparoid is available in only a few countries and it *does* cross-react to the heparins in a number of patients (12). Fondaparinux (Chapter 3.225) is a *synthetic* compound with an ultralow molecular weight analog of the pentasaccharide sequence of heparin. It is an inhibitor of factor Xa. Its advantages over heparins include simplified pharmacodynamics and a lower incidence of side effects (3).

Table 3.239.1 Heparins and examples of studies presenting local delayed-type hypersensitivity reactions [a,b,c]

Heparins	Studies presenting cases of delayed-type hypersensitivity reactions
Unfractionated heparin	10,20,31,34,38,39,43,44,45,50,54,59,62,63,66,69,72-79,81-84,87,89,97, 100,102,104,105,106,109,110
Low molecular weight heparins:	
ardeparin	
bemiparin	29,113
certoparin	8,10,13,20,35,43,45,49,50,56,90,104,110,112
dalteparin	10,20,41,43,45,47,49,51,53,54,64,65,66,102,104,106,108,111
enoxaparin	8,10,16,20,29,30,31,34,35,42,43,47,48,49,50,55,62,63,66,80,88,95,96,99, 102,107,108,110
nadroparin	8,13,20,31,34,35,40,45,51,53,61,66,68,70,71,80,92,93,102,104,107,110, 111
reviparin	
semuloparin	
tinzaparin	8,31,45,102
Semisynthetic heparinoids	
danaparoid	8,35,46,49,98,106
pentosan polysulfate	
Synthetic heparins: fondaparinux	

[a] Examples, not a full literature review; [b] only studies in which skin allergy tests were performed and were positive, are included; [c] only studies in which the culprit drug was known are included; excluded are cross-reactivities without shown clinical relevance and reactions of uncertain relevance

See also bemiparin (Chapter 3.59), certoparin (Chapter 3.117), dalteparin (Chapter 3.160), danaparoid (Chapter 3.161), enoxaparin (Chapter 3.195), fondaparinux (Chapter 3.225), nadroparin (Chapter 3.331), and tinzaparin (Chapter 3.479). Heparins that have cross-reacted to other heparins, but have not caused delayed-type hypersensitivity reactions from their actual clinical use, are not monographed in this book, but are mentioned in Chapter 6 Drugs that have acquired delayed-type hypersensitivity only by cross-reactivity.

CUTANEOUS ADVERSE DRUG REACTIONS FROM SYSTEMIC ADMINISTRATION CAUSED BY TYPE IV (DELAYED-TYPE) HYPERSENSITIVITY

Throughout this book, only reports of delayed-type hypersensitivity have been included that showed a positive patch test to the culprit drug. However, as a result of the high molecular weight of heparins, patch tests are often false-negative from insufficient penetration into the skin. Because of this, and also because patch tests have been performed in a small minority of cases only, studies with a positive intradermal test or subcutaneous provocation tests with delayed readings are included in the chapters on the various heparins, even when patch tests were negative or not performed.

General information on delayed-type reactions to heparins (3,5,15,26)

The first case of delayed-type hypersensitivity to heparin was reported in 1952 (77). For a long time, such reactions were considered to be rare. However, we now know that up to 7.5% of patients receiving heparins (in women up to 11% [8]), mostly low molecular weight heparins, develop allergic lesions at the site of subcutaneous injection (3,5,8). These usually present as pruritic, erythematous, or eczematous plaques, sometimes with vesicles or bullae. If treatment is not discontinued, generalized maculopapular exanthemas or eczema may develop, which occurs in 3-10% of the cases (3,5,7,11,28,29,52,58,67,85,86,88,98). Lesions usually develop within the first 2 weeks of heparin treatment (median: 10 days), during which time sensitization has developed. Some patients have an early onset of skin lesions, within 4 days after the start of therapy; most of these have been exposed to heparins within the preceding 3 months (20). However, late-onset responses occurring several weeks to sometimes months after start of anticoagulant treatment have also been reported repeatedly (5). Histopathological findings show a perivascular infiltrate of CD4+ T lymphocytes with eosinophils and edema in the intercellular space of the epidermis (spongiosis) and dermis that are compatible with delayed hypersensitivity reactions (5).

These localized hypersensitivity reactions are usually observed in women, constituting 75-90% of reported allergic patients (20), especially during pregnancy (20% [102]) and in obese people, so hormonal or metabolic influences may be predisposing factors (5,8,9). Prolonged treatment with heparins (therapy duration >9 days) may also constitute a risk factor (3,5,8). A heparin with a high molecular weight might increase the risk of sensitization (5). As such, the incidence of delayed hypersensitivity reactions (DHR) with the synthetic, ultra-low molecular weight pentasaccharide fondaparinux (1728 daltons) is almost 20 times lower compared with common heparins (6) and it is often well tolerated in patients with hypersensitivities to unfractionated and low molecular weight heparins (3,5,15,26,104). The heparinoid danaparoid is generally also well tolerated (10), but, after prolonged administration, eczematous plaques may also develop to this anticoagulant drug and it frequently cross-reacts to common low molecular weight heparins (12).

The antigenic epitope or epitopes (12) of the heparins has/have not yet been identified. Overlap in the polysaccharide composition of different low molecular weight and unfractionated heparins might explain the high degree of cross-allergenicity among the different heparins (5). Patients with DHR to heparins nearly always tolerate intravenous administration of unfractionated heparin (14,35,50,60,62,98,99), but systemic reactions including maculopapular exanthemas may sometimes be observed (63,65,80,81). Changes in presentation of antigens depending on the route of administration might explain the differences in tolerance (15).

Delayed-type hypersensitivity to heparins may be demonstrated by intracutaneous (intradermal) and epicutaneous (patch) tests against a panel of undiluted antithrombotic agents, with late reading of the intracutaneous tests. Skin testing should be performed not sooner than 6 weeks after the clearance of all lesions and only when heparin-induced thrombocytopenia has been excluded. Patch testing is less sensitive than intradermal testing, giving false-negative results frequently (35,36,38,51). In the case of negative test results or refusal for extended allergological testing, a subcutaneous provocation can be performed (20,26,37). Subcutaneous provocation tests to diagnose a delayed-type hypersensitivity reaction to heparin should be done only – according to some authors – if the diagnosis is unclear (unclear clinical presentation, negative patch test, negative intradermal test, histology is not available), or when alternative anticoagulants need to be identified (20). Reading of the skin injection site is performed on days 2, 3, 4, and 7. In 20% to 30% of patients, the diagnosis of delayed-type hypersensitivity to heparins can only be established by positive subcutaneous challenge test results, which is the most reliable diagnostic test (26,51,83,103).

Other uncommon manifestations of DHR to heparins have been reported in the literature in recent years, often as case reports. These include maculopapular exanthema (7,28,32,33,51,63,80,86,88,107), drug reaction with eosinophilia and systemic symptoms (DRESS) (17,23,24), symmetrical drug-related intertriginous and flexural exanthema (SDRIFE)/baboon syndrome) (18), acute generalized exanthematous pustulosis (AGEP) (19,21,24), acute localized exanthematous pustulosis (ALEP) (22), generalized eczema (25,29,31,88,94,98,99), and miscellaneous eruptions: spreading of itchy erythematous plaques over the trunk (48), generalized delayed-type skin reaction (56), erythema on both hands (65), and local and generalized vesiculopustular eruption (67).

Scope of this review

Local allergic reactions from subcutaneous heparins as described above have been reported in many publications (table 3.239.1) and appear to be frequent (8), with the exception of fondaparinux (6). In Frankfurt, Germany, for example, 87 patients with allergic reactions were observed in a period of 5.5 years. Eighty had local allergic reactions and 7 showed a generalized rash (20). In another German hospital, 15 patients with delayed-type hypersensitivity reactions to heparins were observed in an 18-month period (35). Unfortunately, skin allergy testing has not been standardized and has been performed with various techniques (patch, intracutaneous, subcutaneous). Results were sometimes not specified for individual heparins. In a number of publications, only a part of the patients suspected of delayed-type hypersensitivity was tested or allergy tests were not performed at all. In several studies, it was uncertain or not mentioned what the primary sensitizer was or which of the heparins tested had caused a clinical reaction (e.g. 14,38,51,60). Therefore, the literature data on this specific side effect (local allergic reactions at the sites of injections) of the heparins and danaparoid are not detailed in this book. For more information the reader is referred to the various review articles on the subject that have been published in the past 25 years (2,3,5,26,27,37,57,91,101) and personal literature searches. However, table 3.239.1 lists a large number of publications in which local allergic reactions to the heparins have been described. In addition, *non-local* delayed-type hypersensitivity reactions (e.g. generalized eruptions such as acute generalized exanthematous pustulosis [AGEP] and drug reaction with eosinophilia and systemic symptoms [DRESS]), are presented in detail in this monograph (for unfractionated heparin) and – where published – in the monographs of the low molecular weight heparins

Non-local cutaneous adverse drug reactions to unfractionated heparin
Non-local cutaneous adverse drug reactions from systemic administration of unfractionated heparin caused by type IV (delayed-type) hypersensitivity have included maculopapular eruption (28,32 [delayed-type allergy not proven], 51,63,80,86), SDRIFE/baboon syndrome (18), generalized eczema (99), 'generalized delayed-type skin reaction' (56), erythema on both hands (65), and local and generalized vesiculopustular eruption (67).

Maculopapular eruption
A 47-year-old man developed a maculopapular rash after immediate-type local and generalized urticarial reactions following intravenous heparin sodium and subcutaneous enoxaparin sodium and nadroparin calcium. Patch tests were negative as were prick tests read at 15 and 30 minutes. Intradermal tests to all 3 heparins were positive after 30 minutes; the reactions had fully subsided after 8 hours. It was not mentioned whether the reactions were also read at 2 and 3 days. Type I hypersensitivity was proven, a co-existent type IV reactions, possibly explaining the maculopapular eruption, not (32).

A 45- and a 62-year-old woman were treated with intravenous heparin sodium and both developed a maculopapular exanthema, predominantly on the trunk, the next day. Intradermal tests were positive to heparin sodium and other heparins (51).

A 45-year-old-woman was treated with 7500 IU heparin calcium s.c. 3dd. Fourteen days after beginning therapy, she developed erythematous, infiltrated, pruritic, painful plaques at the injection sites on the lower abdomen and left thigh. Switching to enoxaparin sodium, 1 ampoule daily, had similar effects 3 days after the first injection. 12,500 IU heparin sodium were then administered intravenously. On the following day, the patient developed a maculopapular rash predominantly on the trunk (63).

A woman aged 49 had been treated with subcutaneous nadroparin calcium for 2 days, when itching erythematous and papular plaques developed at the injection sites. The anticoagulant treatment was replaced with intravenous unfractionated heparin, but the next day a generalized maculopapular exanthema had developed. Patch tests were positive to both heparins used and various other heparins (80).

A 75-year-old woman was treated with subcutaneous unfractionated heparin injections. Already after the first injection, a localized erythematous itching plaque developed at the injection site. While the treatment was continued, over the next 4 days the patient developed a generalized maculopapular exanthema accompanied by massive pruritus. During nadroparin therapy, the situation become worse. A subcutaneous provocation test with nadroparin resulted in a localized infiltrated erythematous reaction after 22 hours, which was followed by pruritus and a maculopapular rash in the following 9 hours (28).

A 71-year-old woman, who was treated with calcium heparin 5000 units given subcutaneously twice daily, developed complaints of redness and itching at the sites of the injections after one week. Later, large erythematous plaques with central bullae appeared at these sites, followed by a generalized itchy erythematopapular eruption with fever up to 39°C. An intradermal test with heparin was positive after 2 days (86).

Symmetrical drug-related intertriginous and flexural exanthema (SDRIFE)/Baboon syndrome
An 82-year-old man was treated with intravenous unfractionated heparin and developed erythema at the injection sites and on the thighs. The injections were continued for another three days, but six days later a typical SDRIFE/baboon syndrome exanthema developed. Intradermal tests showed a positive delayed reaction to the heparin used by the patient (18).

Dermatitis/eczematous eruption
A 69-year-old woman developed generalized eczema 24 hours after intravenous administration of sodium heparin during arterial bypass surgery. Patch tests were positive to sodium heparin and several low molecular weight heparins (99).

Other cutaneous adverse drug reactions
A 54-year-old woman developed itchy eczema at the subcutaneous injection sites on the lower abdomen of the low molecular-weight heparin certoparin. Intradermal tests were positive to certoparin and unfractionated heparin. When given intravenous unfractionated heparin, the patient developed a 'generalized delayed-type skin reaction' (56). A woman known to be allergic to dalteparin was given continuous infusion of standard heparin. After one day, she developed erythema on both hands (65). A 37-year-old woman developed a vesiculopustular eruption at two sites of subcutaneous injection of heparin, which generalized after heparin was administered intravenously. A patch test was positive to dalteparin (67).

Cross-reactions, pseudo-cross-reactions and co-reactions

Cross-reactions between heparins are frequent in delayed-type hypersensitivity (>90% of patients tested, median number of positive drugs per patient: 3) and do not depend on the molecular weight of the heparin molecules (2,3,49,51,88). Overlap in their polysaccharide composition might explain the high degree of cross-allergenicity (5). Cross-reactions to the semisynthetic heparinoid danaparoid have also been observed (12). In allergic patients, the synthetic ultralow molecular weight synthetic heparin fondaparinux is usually, but not always (2,12,13,40,41,43) well-tolerated (5).

LITERATURE

1   The data in the section 'General' may have been obtained from literature discussed in this chapter, but mostly also or exclusively from one or more of the following online sources: ChemIDPlus Advanced, PubChem, DrugBank, RxList, Drug Central, Drugs.com, and Wikipedia

2   Weberschock T, Meister AC, Bohrt K, Schmitt J, Boehncke W-H, Ludwig RJ. The risk for cross-reactions after a cutaneous delayed-type hypersensitivity reaction to heparin preparations is independent of their molecular weight: a systematic review. Contact Dermatitis 2011;65:187-194

3   Gonzalez-Delgado P, Fernandez J. Hypersensitivity reactions to heparins. Curr Opin Allergy Clin Immunol 2016;16:315-322

4   Brockow K, Garvey LH, Aberer W, Atanaskovic-Markovic M, Barbaud A, Bilo MB, et al. Skin test concentrations for systemically administered drugs – an ENDA/EAACI Drug Allergy Interest Group position paper. Allergy 2013;68:702-712

5   Schindewolf M, Lindhoff-Last E, Ludwig RJ. Heparin-induced skin lesions. Lancet 2012;380:1867-1879

6   Schindewolf M, Scheuermann J, Kroll H, Garbaraviciene J, Hecking C, Marzi I, et al. Low allergenic potential with fondaparinux: results of a prospective investigation. Mayo Clin Proc 2010;85:913-919

7   Kim KH, Lynfield Y. Enoxaparin-induced generalized exanthema. Cutis 2003;72:57-60

8   Schindewolf M, Schwaner S, Wolter M, Kroll H, Recke A, Kaufmann R, et al. Incidence and causes of heparin-induced skin lesions. CMAJ 2009;181:477-481

9   Bank I, Libourel EJ, Middeldorp S, Van Der Meer J, Büller HR. High rate of skin complications due to low-molecular-weight heparins in pregnant women. J Thromb Haemost 2003;1:859-861

10 Grassegger A, Fritsch P, Reider N. Delayed-type hypersensitivity and crossreactivity to heparins and danaparoid: a prospective study. Dermatol Surg 2001;27:47-52

11 Maldonado Cid P, Alonso de Celada RM, Noguera Morel L, Feito-Rodríguez M, Gómez-Fernández C, Herranz Pinto P. Cutaneous adverse events associated with heparin. Clin Exp Dermatol 2012;37:707-711

12 Utikal J, Peitsch WK, Booken D, Velten F, Dempfle CE, Goerdt S, et al. Hypersensitivity to the pentasaccharide fondaparinux in patients with delayed-type heparin allergy. Thromb Haemost 2005;94:895-896

13 Hirsch K, Ludwig RJ, Lindhoff-Last E, Kaufmann R, Boehncke WH. Intolerance of fondaparinux in a patient allergic to heparins. Contact Dermatitis 2004;50:383-384

14 Gaigl Z, Pfeuffer P, Raith P, Bröcker EB, Trautmann A. Tolerance to intravenous heparin in patients with delayed-type hypersensitivity to heparins: a prospective study. Br J Haematol 2005;128:389-392

15 Trautman A, Seitz C. The complex clinical picture of side effects to anticoagulation. Med Clin N Am 2010;94:821-834

16 Juricic Nahal D, Cegec I, Erdeljic Turk V, Makar Ausperger K, Kraljickovic I, Simic I. Hypersensitivity reactions to low molecular weight heparins: different patterns of cross-reactivity in 3 patients. Can J Physiol Pharmacol 2018;96:428-432

17 Ronceray S, Dinulescu M, Le Gall F, Polard E, Dupuy A, Adamski H. Enoxaparin-induced DRESS syndrome. Case Rep Dermatol 2012;4:233-237

18 Pfeiff B, Pullmann H. Baboon-artiges arzneiexanthem auf heparin. Deutsch Dermatologe 1991;39:559-560 (Article in German, details unknown, data cited in ref. 81).

19 Komericki P, Grims R, Krânke B, Aberer W. Acute generalized exanthematous pustulosis from dalteparin. J Am Acad Dermatol 2007;57:718-721

20 Schindewolf M, Kroll H, Ackermann H, Garbaraviciene J, Kaufmann R, Boehncke WH, Ludwig RJ, Lindhoff-Last E. Heparin-induced non-necrotizing skin lesions: rarely associated with heparin-induced thrombocytopenia. J Thromb Haemost 2010;8:1486-1491

21 Assier H, Gener G, Chosidow O, Wolkenstein P, Ingen-Housz-Oro S. Acute generalized exanthematous pustulosis induced by enoxaparin: 2 cases. Contact Dermatitis 2021;84:280-282

22 Gómez Torrijos E, Cortina de la Calle MP, Méndez Díaz Y, Moreno Lozano L, Extremera Ortega A, Galindo Bonilla PA, et al. Acute localized exanthematous pustulosis due to bemiparin. J Investig Allergol Clin Immunol 2017;27:328-329

23   Liccioli G, Mori F, Parronchi P, Capone M, Fili L, Barni S, et al. Aetiopathogenesis of severe cutaneous adverse
     reactions (SCARs) in children: A 9-year experience in a tertiary care paediatric hospital setting. Clin Exp Allergy
     2020;50:61-73

24   Barbaud A, Collet E, Milpied B, Assier H, Staumont D, Avenel-Audran M, et al. A multicentre study to determine
     the value and safety of drug patch tests for the three main classes of severe cutaneous adverse drug reactions.
     Br J Dermatol 2013 ;168 :555-562

25   Barbaud A, Reichert-Penetrat S, Tréchot P, Jacquin-Petit MA, Ehlinger A, Noirez V, et al. The use of skin testing in
     the investigation of cutaneous adverse drug reactions. Br J Dermatol 1998;139:49-58

26   Trautmann A, Seitz CS. Heparin allergy: delayed-type non-IgE-mediated allergic hypersensitivity to subcutaneous
     heparin injection. Immunol Allergy Clin North Am 2009 ;29 :469-480

27   Nosbaum A, Pralong P, Rozieres A, Dargaud Y, Nicolas JF, Bérard F. Hypersensibilité retardée aux héparines :
     diagnostic et prise en charge thérapeutique [Delayed-type hypersensitivity to heparin : diagnosis and
     therapeutic management]. Ann Dermatol Venereol 2012 ;139 :363-368 (Article in French)

28   Greiner D, Schöfer H. Allergisches Arzneimittelexanthem auf Heparin. Kutane Reaktionen auf hochmolekulares
     und fraktioniertes Heparin [Allergic drug exanthema to heparin. Cutaneous reactions to high molecular and
     fractionated heparin]. Hautarzt 1994;45:569-572 (Article in German)

29   Tramontana M, Hansel K, Bianchi L, Agostinelli D, Stingeni L. Skin tests in patients with delayed and immediate
     hypersensitivity to heparins: A case series. Contact Dermatitis 2019;80:170-172

30   Tan E, Thompson G, Ekstrom C, Lucas M. Non-immediate heparin and heparinoid cutaneous allergic reactions: a
     role for fondaparinux. Intern Med J 2018;48:73-77

31   Phan C, Vial-Dupuy A, Autegarden J-E, Amsler E, Gaouar H, Abuaf N, et al. A study of 19 cases of allergy to
     heparins with positive skin testing. Ann Dermatol Venereol 2014;141:23-29 (Article in French)

32   Klos K, Spiewak R, Kruszewski J, Bant A. Cutaneous adverse drug reaction to heparins with hypereosinophilia and
     high IgE level. Contact Dermatitis 2011 ;64 :61-62

33   Colagiovanni A, Rizzi A, Buonomo A, De Pasquale T, Pecora V, Sabato V, et al. Delayed hypersensitivity to heparin
     in a patient with cancer: fondaparinux may be safe but needs to be tested. Contact Dermatitis 2010;63:107-108

34   Palacios Colom L, Alcántara Villar M, Luis Anguita Carazo J, Ruiz Villaverde R, Quiralte Enríquez J. Delayed-type
     hypersensitivity to heparins: different patterns of cross-reactivity. Contact Dermatitis 2008;59:375-377

35   Pföhler C, Müller CSL, Pindur G, Eichler H, Schäfers HJ, Grundmann U, et al. Delayed-type heparin allergy:
     diagnostic procedures and treatment alternatives – a case series including 15 patients. World Allergy Organ J
     2008;1:194-199

36   White JM, Munn SE, Seet JE, Adams N, Clement M. Eczema-like plaques secondary to enoxaparin. Contact
     Dermatitis 2006;54:18-20

37   Jappe U. Allergy to heparins and anticoagulants with a similar pharmacological profile: an update. Blood Coagul
     Fibrinolysis 2006;17:605-613

38   Ludwig RJ, Schindewolf M, Alban S, Kaufmann R, LindhoffLast E, Boehncke W-H. Molecular weight determines
     the frequency of delayed type hypersensitivity reactions to heparins and synthetic oligosaccharides. Thromb
     Haemost 2005;94:1265-1269

39   Maroto-Iitani M, Higaki Y, Kawashima M. Cutaneous allergic reaction to heparins: subcutaneous but not
     intravenous provocation. Contact Dermatitis 2005;52:228-230

40   Maetzke J, Hinrichs R, Schneider LA, Scharffetter-Kochanek K. Unexpected delayed-type hypersensitivity skin
     reactions to the ultra-low-molecular-weight heparin fondaparinux. Allergy 2005;60:413-415

41   Hohenstein E, Tsakiris D, Bircher AJ. Delayed-type hypersensitivity to the ultra-low-molecular-weight heparin
     fondaparinux. Contact Dermatitis 2004;51:149-151

42   Hallai N, Hughes TM, Stone N. Type I and Type IV allergy to unfractionated heparin and low-molecular-weight
     heparin with no reaction to recombinant hirudin. Contact Dermatitis 2004;51:153-154

43   Jappe U, Juschka U, Kuner N, Hausen BM, Krohn K. Fondaparinux: a suitable alternative in cases of delayed-type
     allergy to heparins and semisynthetic heparinoids? A study of 7 cases. Contact Dermatitis 2004;51:67-72

44   Borch JE, Bindslev-Jensen C. Delayed-type hypersensitivity to heparins. Allergy 2004;59:118-119

45   Koch P. Delayed-type hypersensitivity skin reactions due to heparins and heparinoids. Tolerance of recombinant
     hirudins and of the new synthetic anticoagulant fondaparinux. Contact Dermatitis 2003;49:276-280

46   Ludwig RJ, Beier C, Lindhoff-Last E, Kaufmann R, Boehncke WH. Tolerance of fondaparinux in a patient allergic to
     heparins and other glycosaminoglycans. Contact Dermatitis 2003;49:158-159

47   Mora A, Belchi J, Contreras L, Rubio G. Delayed-type hypersensitivity skin reactions to low molecular weight
     heparins in a pregnant woman. Contact Dermatitis 2002;47:177-178

48   Poza-Guedes P, González-Pérez R, Canto G. Different patterns of cross-reactivity in non-immediate
     hypersensitivity to heparins: from localized to systemic reactions. Contact Dermatitis 2002;47:244-245

49   Szolar-Platzer C, Aberer W, Kränke B. Delayed-type skin reaction to the heparin-alternative danaparoid. J Am
     Acad Dermatol 2000;43:920-922

50 Irion R, Gall H, Peter RU. Delayed-type hypersensitivity to heparin with tolerance of its intravenous administration. Contact Dermatitis 2000;43:249-250

51 Koch P, Münssinger T, Rupp-John C, Uhl K. Delayed-type hypersensitivity skin reactions caused by subcutaneous unfractionated and low-molecular-weight heparins: tolerance of a new recombinant hirudin. J Am Acad Dermatol 2000 ;42 :612-619

52 Koch P, Uhl K, John S. Dermite allergique de contact et exanthème généralisé induit par héparines standard et de bas poids moléculaire : absence de sensibilisation à un héparinoïde. Ann Dermatol Venereol 1997;124(suppl.1):236

53 Koch P, Reinhold S, Busch C. Delayed allergic skin reactions to subcutaneous heparins. Tolerance of 2 recombinant hirudins. Contact Dermatitis 2000;42:278-279

54 Martin L, Machet L, Gironet N, Pouplard C, Gruel Y, Vaillant L. Eczematous plaques related to unfractionated and low-molecular-weight heparins: cross-reaction with danaparoid but not with desirudin. Contact Dermatitis 2000;42:295-296

55 Enrique E, Alijotas J, Cisteró A, san Miguel MM, Bartra J, Tresserra F. Patch-test positivity in cutaneous reactions to enoxaparin. Contact Dermatitis 2000;42:43

56 Schiffner R, Glässl A, Landthaler M, Stolz W. Tolerance of desirudin in a patient with generalized eczema after intravenous challenge with heparin and a delayed-type skin reaction to high and low molecular weight heparins and heparinoids. Contact Dermatitis 2000;42:49

57 Jappe U, Gollnick H. Allergy to heparin, heparinoids, and recombinant hirudin. Diagnostic and therapeutic alternatives. Hautarzt 1999;50:406-411 (Article in German)

58 Trautmann A, Hamm K, Bröcker EB, Klein CE. Spättyp-Allergie gegen Heparin. Klinik-Diagnostik-Ausweichpräparate. Z Haut Geschlechtskrankheit 1997;72:447-450 (Article in German, details unknown, data [exanthema] cited in ref. 57)

59 Hunzelmann N, Gold H, Scharffetter-Kochanek K. Concomitant sensitization to high and low molecular-weight heparins, heparinoid and pentosanpolysulfate. Contact Dermatitis 1998;39:88-89

60 Trautmann A, Bröcker EB, Klein CE. Intravenous challenge with heparins in patients with delayed-type skin reactions after subcutaneous administration of the drug. Contact Dermatitis 1998;39:43-44

61 Moreau A, Dompmartin A, Esnault P, Michel M, Leroy D. Delayed hypersensitivity at the injection sites of a low-molecular-weight heparin. Contact Dermatitis 1996;34:31-34

62 Boehncke WH, Weber L, Gall H. Tolerance to intravenous administration of heparin and heparinoid in a patient with delayed-type hypersensitivity to heparins and heparinoids. Contact Dermatitis 1996;35:73-75

63 Koch P, Hindi S, Landwehr D. Delayed allergic skin reactions due to subcutaneous heparin-calcium, enoxaparin-sodium, pentosan polysulfate and acute skin lesions from systemic sodium-heparin. Contact Dermatitis 1996;34:156-158

64 Krasovec M, Kämmerer R, Spertini F, Frenk E. Contact dermatitis from heparin gel following sensitization by subcutaneous heparin administration. Contact Dermatitis 1995;33:135-136

65 Rasmussen C, Skov L, Da Cunha Bang F. Delayed-type hypersensitivity to low-molecular weight heparin. Am J Cont Dermat 1993;4:118-119

66 Amarger S, Ferrier-Le-Bouedec MC, Mansard S, Souteyrand P, D'incan M. Delayed hypersensitivity skin reactions to heparins and heparinoids. Rev Fr Allergol Immunol Clin 2003;43:170-174 (Article in French)

67 Delaval Y, Nogues I, Logeais B, Pibouin M, Chevrant-Breton J. Un cas de toxidermie vesiculo-pustuleuse induit par l'heparine. Rev Fr Allergol 1993;33:338 (Article in French)

68 Deschamps A, Mathelier-Fusade P, Bemaille J L. Heparin-induced cutaneous reaction during pregnancy: report of a case with tolerance to danaparoid. Rev Fr Allergol Immunol Clin 2003;43:131-134 (Article in French)

69 Emilie S, Bachmeyer C, Moguelet P, Pecquet C. Eczematiform lesions with heparin. Rev Med Interne 2007;28:259-260 (Article in French)

70 Gonzalo Garijo M A, Revenga Arranz F. Type IV hypersensitivity to subcutaneous heparin: a new case. J Investig Allergol Clin Immunol 1996;6:388-391

71 Verdonkschot AE, Vasmel WL, Middeldorp S, Van Der Schoot JT. Skin reactions due to low molecular weight heparin in pregnancy: a strategic dilemma. Arch Gynecol Obstet 2005;271:163-165

72 Wurpts G, Merk HF. Delayed intracutaneous test reaction in heparin intolerance. Hautarzt 2007;58:394-396 (Article in German)

73 Zimmermann R, Harenberg J, Weber E, Devries J X, Jarass W, Schmidt W. Treatment in a heparin-induced skin reaction with a low-molecular heparin analog. Dtsch Med Wochenschr 1984;109:1326-1328

74 Valsecchi R, Rozzoni M, Cainelli T. Allergy to subcutaneous heparin. Contact Dermatitis 1992;26:129-130

75 Koch P, Bahmer FA, Schäfer H. Tolerance of intravenous low-molecular-weight heparin after eczematous reaction to subcutaneous heparin. Contact Dermatitis 1991;25:205-206

76 Bircher AJ, Flückinger R, Buchner SA. Eczematous infiltrated plaques to subcutaneous heparin: a type IV allergic reaction. Br J Dermatol 1990;123:507-514

77 Plancherel P. Klinische und gerinnungsphysiologische Untersuchungen mit einem neuen Heparindepotpräparat. Z Klin Med 1952;150:213-259 (Article in German, data cited in ref. 76)

78 Meissner K, Schulz K-H. Allergische Reaktionen auf Heparin. Allergologie 1984;7:141 (Article in German, data cited in ref. 76)

79 Ulrick PJ, Manoharan A. Heparin-induced skin reaction. Med J Aust 1984;140:287-289 (Data cited in ref. 76)

80 Figarella I, Barbaud A, Lecompte T, De Maistre E, Reichert-Penetrat S, Schmutz J L. Cutaneous delayed hypersensitivity reactions to heparins and heparinoids. Ann Dermatol Venereol 2001;128:25-30 (Article in French)

81 Ojukwu C, Jenkinson SD, Obeid D. Deep vein thrombosis in pregnancy and heparin hypersensitivity. Br J Obstet Gynaecol 1996;103:934-936

82 O'Donnell B F, Tan C Y. Delayed hypersensitivity reaction to heparin. Br J Dermatol 1993;129:634-636

83 Klein GF, Kofler H, Wolf H, Fritsch PO. Eczema-like, erythematous, infiltrated plaques: a common side effect of subcutaneous heparin therapy. J Am Acad Dermatol 1989;21(4Pt1):703-707

84 Korstanje MJ, Bessems PJ, Hardy E, van de Staak WJ. Delayed-type hypersensitivity reaction to heparin. Contact Dermatitis 1989;20:383-384

85 Patrizi A, Di Lernia V, Patrone P. Generalized reaction to subcutaneous heparin. Contact Dermatitis 1989;20:309-310

86 Young E. Allergy to subcutaneous heparin. Contact Dermatitis 1988;19:152-153

87 Tosti A, De Padova MP, Patrizi A, Minghetti G, Veronesi S. Unusual reaction to subcutaneous heparin. Contact Dermatitis 1985;13:190

88 Grims RH, Weger W, Reiter H, Arbab E, Kränke B, Aberer W. Delayed-type hypersensitivity to low molecular weight heparins and heparinoids: cross-reactivity does not depend on molecular weight. Br J Dermatol 2007;157:514-517

89 Sanders MN, Bernhisel-Broadbent J, Staker LV. Delayed hypersensitivity reaction to heparin in a pregnant woman. Int J Dermatol 1995;34:443-444

90 Bircher AJ, Itin PH, Tsakiris DA, Surber C. Delayed hypersensitivity to one low-molecular-weight heparin with tolerance of other low-molecular-weight heparins. Br J Dermatol 1995;132:461-463

91 Wutschert R, Piletta P, Bounameaux H. Adverse skin reactions to low molecular weight heparins: frequency, management, and prevention. Drug Saf 1999;20:515-525

92 Koch P, Schäfer H, Bahmer FA. Allergische Hautreaktionen gegen Standard- und niedermolekulare Heparine. Gute verträglichkeit von Heparin-Analogen. Z Hautkr 1991;66:428-435 (Article in German)

93 Dacosta A, Mismetti P, Buchmuller A, et al. Hyperéosinophilie et lésions cutanées induites par héparine de bas poids moléculaire. Presse Med 1994;33:1540-1541 (Article in French, data cited in ref. 91)

94 Estrada Rodriguez JL, Gozalo Reques F, Ortiz de Urbina J, Matilla B, Rodriguez Prieto MA, Gonzalez Moran NA. Generalized eczema induced by nadroparin. J Investig Allergol Clin Immunol 2003;13:69-70

95 Valdés F, Vidal C, Ferna´ndez-Redondo V, Peteiro C, Toribio J. Eczema-like plaques to enoxaparin. Allergy 1998;53:625-626

96 Mendez J, De La Fuente R, Stolle R, Vega JM, Sanchis ME, Armentia A, et al. Delayed hypersensitivity skin reaction to enoxaparin. Allergy: Eur J Allergy Clin Immunol 1996;51:853-854

97 Guillet G, Delaire P, Plantin P, Guillet MH. Eczema as a complication of heparin therapy. J Am Acad Dermatol 1989;20:1130-1132

98 Seitz CS, Brocker EB, Trautmann A. Management of allergy to heparins in postoperative care: subcutaneous allergy and intravenous tolerance. Dermatol Online J 2008 Sept 15;14(9):4

99 Mendez J, Sanchis ME, de la Fuente R, Stolle R, Vega JM, Martínez C, et al. Delayed-type hypersensitivity to subcutaneous enoxaparin. Allergy 1998;53:999-1003

100 Dupin N, Bagot M, Wechsler J, Revuz J, Reaction d'hypersensibilité à la Calciparine®. Ann Dermatol Venereol 1993;120:845-846 (Article in French)

101 Bircher AJ, Harr T, Hohenstein L, Tsakiris DA. Hypersensitivity reactions to anticoagulant drugs: diagnosis and management options. Allergy 2006;61:1432-1440

102 Schindewolf M, Gobst C, Kroll H, Recke A, Louwen F, Wolter M, et al. High incidence of heparin-induced allergic delayed-type hypersensitivity reactions in pregnancy. J Allergy Clin Immunol 2013;132:131-139

103 Schindewolf M, Ludwig RJ, Wolter M, et al. Diagnosis of heparin-induced delayed-type hypersensitivity. Phlebologie 2010;39:226-231

104 Sacher C, Hunzelmann N. Tolerance to the synthetic pentasaccharide fondaparinux in heparin sensitization. Allergy 2003;58:1318-1319

105 De Kort WJ, Van Der Meer YG, De Groot AC. Delayed-type allergy for heparin and fractions of low-molecular-weight heparin. Ned Tijdschr Geneeskd 1992;136:2379-2380 (Article in Dutch)

106 Sivakumaran M, Ghosh K, Munks R, Gelsthorpe K, Tan L, Wood JK. Delayed cutaneous reaction to unfractionated heparin, lowmolecular-weight heparin and danaparoid. Br J Haematol 1994;86:893-894

107 Lopez S, Torres MJ, Rodríguez-Pena R, Blanca-Lopez N, Fernandez TD, Antunez C, et al. Lymphocyte proliferation response in patients with delayed hypersensitivity reactions to heparins. Br J Dermatol 2009;160:259-265

108 Phillips JK, Majumdar G, Hunt BJ, Savidge GF. Heparin-induced skin reaction due to two different preparations of low molecular weight heparin (LMWH). Br J Haematol 1993;84:349-350

109 Tuneo A, Moreno A, de Moragas J. Cutaneous reactions secondary to heparin injections. J Am Acad Dermatol 1985;12:1072-1077

110 Maetzke J, Hinrichs R, Staib G, Scharffetter-Kochanek K. Fondaparinux as a novel therapeutic alternative in a patient with heparin allergy. Allergy 2004;59:237-238

111 Krasovec M, Kämmerer R, Spertini F, et al. Delayed hypersensitivity to subcutaneous heparin. Dermatology 1994;189:33 (Data cited in ref. 91)

112 Jappe U, Reinhold D, Bonnekoh B. Arthus reaction to lepirudin, a new recombinant hirudin, and delayed-type hypersensitivity to several heparins and heparinoids, with tolerance to its intravenous administration. Contact Dermatitis 2002;46:29-32

113 Moreno Escobosa MC, Moya Quesada MC, Granados SC, Amat López J. Delayed hypersensitivity challenged by subcutaneous Bemiparin. Allergol Immunopathol (Madr) 2011;39:309-310

Chapter 3.240 HEROIN

IDENTIFICATION

Description/definition : Heroin is the morphinane alkaloid that conforms to the structural formula shown below
Pharmacological classes : Analgesics, opioid; narcotics
IUPAC name : [(4R,4aR,7S,7aR,12bS)-9-Acetyloxy-3-methyl-2,4,4a,7,7a,13-hexahydro-1H-4,12-methanobenzofuro[3,2-e]isoquinoline-7-yl] acetate
Other names : Diacetylmorphine; acetomorphine; diamorphine
CAS registry number : 561-27-3
EC number : 209-217-7
Merck Index monograph : 4241
Patch testing : 1% water and pet.
Molecular formula : $C_{21}H_{23}NO_5$

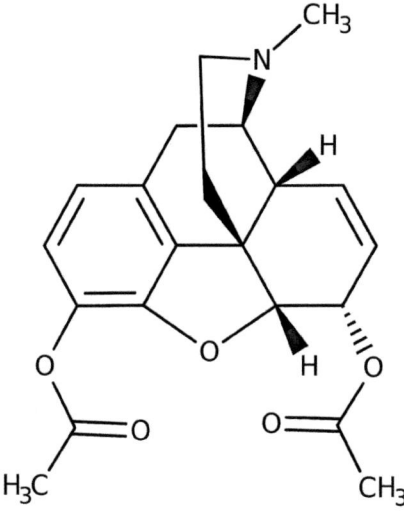

GENERAL

Heroin is a morphinane alkaloid that is morphine bearing two acetyl substituents on the O-3 and O-6 positions. As with other opioids, heroin is used as both an analgesic and a recreational drug. Frequent and regular administration is associated with tolerance and physical dependence, which may develop into addiction. Its use as a pharmaceutical (it is a prescription drug in the U.K. under the name diamorphine) includes treatment for acute pain, such as in severe physical trauma, myocardial infarction, post-surgical pain, and chronic pain, including end-stage cancer and other terminal illnesses and for the relief of dyspnea in acute pulmonary edema. In pharmaceutical preparations, heroin (under the name diamorphine) is present as heroin hydrochloride (CAS number 1502-95-0, EC number 216-124-5, molecular formula $C_{21}H_{24}ClNO_5$) (1).

CONTACT ALLERGY FROM ACCIDENTAL CONTACT

In The Netherlands, after the start of heroin-assisted treatment to a selected group of chronic treatment-resistant heroin-dependent patients, some nurses developed work-related eczema of the eyelids, face, neck, hands and arms and proved to be allergic to heroin by positive patch tests. There were also nurses who had work-related nasal mucosa or respiratory complaints. In their work at the dispensing unit, the nurses handled capsules containing a mixture of heroin and caffeine (8, discussed below). To investigate the prevalence of heroin contact allergy in professionals involved in this project, a questionnaire-based study was started with follow-up by allergological examinations. Of 120 questionnaires sent to mainly nurses but also security workers, doctors, cleaners, social and other workers, 101 (84%) were returned: 67 from nurses and 34 from other employees. Of these workers, 38 (38%) reported work-related complaints: 33 of 67 (49%) nurses and 5 of 34 (15%) other employees. Patch tests to heroin (0.3%, 1%, 3% and 5% pet.) were performed in 24 nurses and were positive in 8. Seven had eyelid eczema, 6 eczema of other parts of the face or neck, and 3 hand or arm eczema. Three of the 8 nurses had also reported nose/respiratory complaints, and 3 had conjunctivitis. The prevalence of heroin contact allergy in this study was 8% (8/101) among all employees and 12% (8/67) among nurses (but likely higher as not all were patch tested). The respiratory and mucosal complaints could not be ascribed to contact allergy, and in these cases, serum was analyzed for specific immunoglobulin E to heroin. However, a type 1 allergy to heroin could not be shown. These complaints may have been due to the histamine-liberating effect of heroin (7). One of these patients, who also reacted to morphine, many years later developed a cutaneous drug reaction following systemic administration of morphine (9).

A 40-year-old nurse with hand eczema had, for 3 years, been working in a municipal treatment centre for drug addicts. She had developed work-related airborne acute erythematous and edematous facial dermatitis when handing heroin and morphine to the clients. Patch tests were strongly positive to heroin and morphine, both tested 1% water, but negative to other opioids (2).

A 37-year-old woman, occupationally exposed to heroin, developed redness and swelling, accompanied by severe itch, on her eyelids, with subsequent spread to her face and neck. She was employed by the local health authority on a project supplying a selected group of drug-addicts with heroin under supervision. Capsules containing a mixture of caffeine and diacetylmorphine (heroin) powder were opened by her, and the contents handed over to the clients. Patch tests were positive to heroin 1% pet. with cross-reactions to morphine and codeine. Five more possible cases of occupational allergic contact dermatitis from heroin from the same project in two cities were patch tested and two had positive reactions to heroin 1% pet. They both cross-reacted to morphine and one to codeine (8).

Early literature on opium alkaloids
In early literature, (occupational) contact dermatitis and (in some cases) systemic allergic dermatitis to opium alkaloids, including morphine, codeine, diacetylmorphine (heroin), apomorphine, and ethylmorphine have been described (e.g. 4,5, reviews in ref. 3 and 5), the first report dating back to 1882 (cited in ref. 3). In those days, eczematous dermatitis from opium compounds used externally in the form of lotions, suppositories and other application forms was apparently well known (5). A more recent review of the subject was provided in 2006 (7).

CUTANEOUS ADVERSE DRUG REACTIONS FROM SYSTEMIC ADMINISTRATION CAUSED BY TYPE IV (DELAYED-TYPE) HYPERSENSITIVITY (as demonstrated by positive patch tests)
Cutaneous adverse drug reactions from systemic administration of heroin caused by type IV (delayed-type) hypersensitivity have included systemic allergic dermatitis.

Systemic allergic dermatitis (systemic contact dermatitis)
Cases of systemic allergic dermatitis from heroin have been discussed above in the section 'Contact allergy from accidental contact'.

Cross-reactions, pseudo-cross-reactions and co-reactions
A patient who was sensitized to codeine cross-reacted to diacetylmorphine HCl (heroin) 0.5% and 1% water (6). Of 8 nurses sensitized to heroin (diacetylmorphine), 6 (75%) cross-reacted to morphine and 4 (50%) to codeine (methylmorphine) (7). Cross-reactions from heroin sensitization to morphine and codeine (8).

LITERATURE
1   The data in the section 'General' may have been obtained from literature discussed in this chapter, but mostly also or exclusively from one or more of the following online sources: ChemIDPlus Advanced, PubChem, DrugBank, RxList, Drug Central, Drugs.com, and Wikipedia
2   Hvid L, Svendsen MT, Andersen KE. Occupational allergic contact dermatitis caused by heroin (diacetylmorphine) and morphine. Contact Dermatitis 2016;74:301-302
3   Touraine A. Les dermatoses de l'opium. Revue de Médecine 1936;53:449-460
4   Dore SE, Prosser Thomas EW. Contact dermatitis in a morphine factory. J Allergy 1945;16:35-36
5   Jordon JW, Osborne ED. Contact dermatitis from opium derivatives. JAMA 1939;113:1955-1957
6   De Groot AC, Conemans J. Allergic urticarial rash from oral codeine. Contact Dermatitis 1986;14:209-214
7   Hogen Esch AJ, van der Heide S, van den Brink W, van Ree JM, Bruynzeel DP, Coenraads PJ. Contact allergy and respiratory/mucosal complaints from heroin (diacetylmorphine). Contact Dermatitis 2006;54:42-49
8   Coenraads PJ, Hogen Esch AJ, Prevoo RL. Occupational contact dermatitis from diacetylmorphine (heroin). Contact Dermatitis 2001;45:114
9   Van den Hoed E, Coenraads PJ, Schuttelaar MLA. Morphine-induced cutaneous adverse drug reaction following occupational diacetylmorphine contact dermatitis: A case report. Contact Dermatitis 2019;81:313-315

Chapter 3.241 HYALURONIDASE

IDENTIFICATION

Description/definition	: Hyaluronidases are a family of enzymes that catalyze the degradation of hyaluronic acid
Pharmacological classes	: Enzymes
IUPAC name	: 2,4-bis(1,3-Benzodioxol-5-yl)-4-oxobutanoic acid (hyaluronidase from *Streptomyces hyalurolyticus*)
CAS registry number	: 9001-54-1
EC number	: 232-614-1
Merck Index monograph	: 6067 (Hyaluronidases)
Patch testing	: Hyaluronidase 500 IU (no further details provided) (1)
Molecular formula	: Unspecified

GENERAL

Hyaluronidase is an injectable enzyme with potential adjuvant activity. Upon subcutaneous administration, hyaluronidase modifies the permeability of connective tissue by hydrolyzing hyaluronic acid. This temporarily decreases interstitial viscosity and allows drugs that are co-injected to spread rapidly through the interstitial space, thereby facilitating absorption and/or distribution of the co-injected agents. It is widely used in anaesthesia for ocular, dental, and plastic surgery to enhance permeation of local anaesthetics (1).

CUTANEOUS ADVERSE DRUG REACTIONS FROM SYSTEMIC ADMINISTRATION CAUSED BY TYPE IV (DELAYED-TYPE) HYPERSENSITIVITY (as demonstrated by positive patch tests)

Cutaneous adverse drug reactions from systemic administration of hyaluronidase caused by type IV (delayed-type) hypersensitivity have included localized allergic reaction (1).

General

Delayed-type hypersensitivity to hyaluronidase has been presented in one case report only (1). Immediate hypersensitivity reactions, as diagnosed by skin prick or intradermal tests, occur far more frequently (2-9). However, symptoms often start several hours to one day after injection of hyaluronidase and it cannot be excluded that in some patients, type IV allergy has played a role.

Localized allergic reaction

A 71-year-old woman underwent a right phacoemulsification with lens implant. During the surgery, the conjunctiva was anesthetized with proxymetacaine 0.5% and 3 ml of sub-Tenon's anesthesia (containing hyaluronidase 500 IU, lidocaine 2% and bupivacaine 0.5%) administered inferonasally into the sub-Tenon's space. Two days later, the patient presented with a swollen and painful right eye. Patch tests were positive to hyaluronidase 500 IU (no details provided) and negative to the other constituents of the local anesthetic (1).

In a similar case, local anesthesia containing hyaluronidase injected into sub-Tenon's space resulted in proptosis (exophthalmos) after one day. In this patient, hyaluronidase was not patch tested, but there were delayed-positive reactions to intradermal and subcutaneous injections of hyaluronidase 15 IU/ml (10).

LITERATURE

1   Park S, Lim LT. Orbital inflammation secondary to a delayed hypersensitivity reaction to sub-Tenon's hyaluronidase. Semin Ophthalmol 2014;29:57-58
2   Borchard K, Puy R, Nixon R. Hyaluronidase allergy: a rare cause of periorbital inflammation. Australas J Dermatol 2010;51:49-51
3   Leibovitch I, Tamblyn D, Casson R, Selva D. Allergic reaction to hyaluronidase: a rare cause of orbital inflammation after cataract surgery. Graefes Arch Clin Exp Ophthalmol 2006;244:944-949
4   Escolano F, Pares N, Gonzalez I, Castillo J, Valero A, Bartolome B. Allergic reaction to hyaluronidase in cataract surgery. Eur J Anaesthesiol 2005;22:729-730
5   Eberhart AH, Weiler CR, Erie JC. Angioedema related to the use of hyaluronidase in cataract surgery. Am J Ophthalmol 2004;138:142-143
6   Nicholson G, Hall GM. Allergic reaction to hyaluronidase after a peribulbar injection. Anaesthesia 2003;58:814-815
7   Ahluwalia HS, Lukaris A, Lane CM. Delayed allergic reaction to hyaluronidase: a rare sequel to cataract surgery. Eye 2003;17:263-266

8   Kirby B, Butt A, Morrison AM, Beck MH. Type I allergic reaction to hyaluronidase during ophthalmic surgery. Contact Dermatitis 2001;44:52

9   Musa F, Srinivasan S, King CM, Kamal A. Raised intraocular pressure and orbital inflammation: a rare IgE-mediated allergic reaction to sub-Tenon's hyaluronidase. J Cataract Refract Surg 2006;32:177-178

10   Feighery C, McCoy EP, Johnston PB, Armstrong DK. Delayed hypersensitivity to hyaluronidase (Hyalase) used during cataract surgery. Contact Dermatitis 2007;57:343

Chapter 3.242 HYDRALAZINE

IDENTIFICATION

Description/definition : Hydralazine is the phthalazine derivative that conforms to the structural formula shown below
Pharmacological classes : Antihypertensive agents; vasodilator agents
IUPAC name : Phthalazin-1-ylhydrazine
CAS registry number : 86-54-4
EC number : 201-680-3
Merck Index monograph : 6072
Patch testing : 2% pet.
Molecular formula : $C_8H_8N_4$

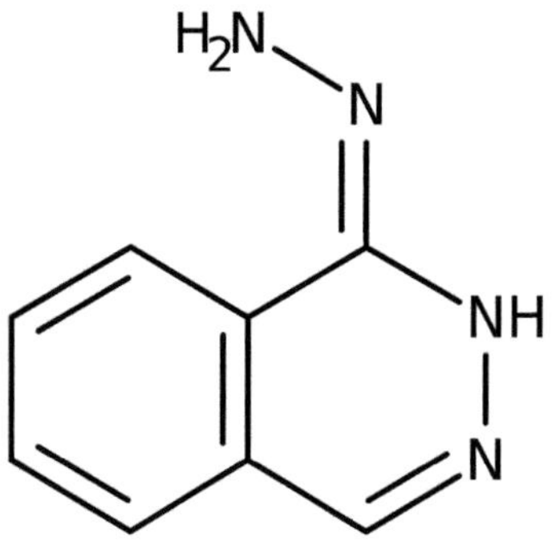

GENERAL

Hydralazine is a phthalazine derivative with antihypertensive effects induced by arteriolar vasodilatation. It is indicated for the treatment of essential hypertension, for the management of severe hypertension when the drug cannot be given orally or when blood pressure must be lowered immediately, congestive heart failure (in combination with cardiac glycosides and diuretics and/or with isosorbide dinitrate), and hypertension secondary to preeclampsia/eclampsia. In pharmaceutical preparations, hydralazine is used as hydralazine hydrochloride (CAS number 304-20-1, EC number 206-151-0, molecular formula $C_8H_9ClN_4$) (1).

CONTACT ALLERGY FROM ACCIDENTAL CONTACT

A 54-year-old man working in the pharmaceutical industry had developed skin problems about 3 months previously, when he developed eczema on the hands, forearms and axillae. Although he used a protective suit with mask and rubber gloves, the rubber gloves seemed to aggravate the dermatitis. The symptoms improved when he moved to another section and at weekends. Patch testing showed positive reactions to thiuram mix and all of its ingredients, to hydralazine 0.1%, 1% and 2% pet. (D2 +, D4 ++) and to 2 other drugs (propranolol, bendroflumethiazide). Sixteen controls were negative (2). This was a case of occupational allergic contact dermatitis.

LITERATURE

1 The data in the section 'General' may have been obtained from literature discussed in this chapter, but mostly also or exclusively from one or more of the following online sources: ChemIDPlus Advanced, PubChem, DrugBank, RxList, Drug Central, Drugs.com, and Wikipedia
2 Pereira F, Dias M, Pacheco FA. Occupational contact dermatitis from propranolol, hydralazine and bendroflumethiazide. Contact Dermatitis 1996;35:303-304

Chapter 3.243   HYDROCHLOROTHIAZIDE

IDENTIFICATION

Description/definition	: Hydrochlorothiazide is a benzothiadiazine derivative that conforms to the structural formula shown below
Pharmacological classes	: Diuretics; antihypertensive agents; sodium chloride symporter inhibitors
IUPAC name	: 6-Chloro-1,1-dioxo-3,4-dihydro-2H-1,2,4-benzothiadiazine-7-sulfonamide
Other names	: Dihydrochlorothiazide
CAS registry number	: 58-93-5
EC number	: 200-403-3
Merck Index monograph	: 6089
Patch testing	: 10% pet. (Chemotechnique)
Molecular formula	: $C_7H_8ClN_3O_4S_2$

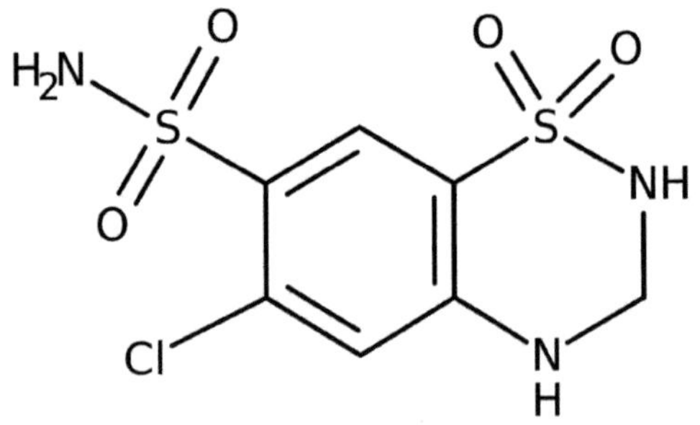

GENERAL

Hydrochlorothiazide is a short-acting thiazide diuretic often considered the prototypical member of this class. It reduces the reabsorption of electrolytes from the renal tubules. This results in increased excretion of water and electrolytes, including sodium, potassium, chloride, and magnesium. Hydrochlorothiazide is used in the treatment of hypertension, edema, diabetes insipidus, and hypoparathyroidism (1).

CUTANEOUS ADVERSE DRUG REACTIONS FROM SYSTEMIC ADMINISTRATION CAUSED BY TYPE IV (DELAYED-TYPE) HYPERSENSITIVITY (as demonstrated by positive patch tests)

Cutaneous adverse drug reactions from systemic administration of hydrochlorothiazide caused by type IV (delayed-type) hypersensitivity have included photosensitivity (2,3) and unspecified drug eruption (4).

Photosensitivity

A 63-year-old welder presented with a 6-year history of edema, erythema, and eczema, along with burning sensations and slowly increasing heat-sensitivity, confined to the face, neck, and, intermittently, extensor forearms. He reported worsening of symptoms on exposure to sunlight, but also during welding, with intense burning sensations, even though he used welding goggles or a welding shield to cover the face and/or eyes. The patient reported drug therapy for essential hypertension originally with a combination of hydrochlorothiazide and ramipril, and since 5 years with ramipril alone. Photopatch tests showed strongly positive reactions to hydrochlorothiazide as well as to ramipril 2 and 3 days after irradiation with 10 J/cm² UVA. Unirradiated patch tests or patch tests irradiated with UVB or visible light remained negative. Patch test concentrations and vehicles were not mentioned (2).

A 68-year-old man presented with a rash on light-exposed areas. He had taken tablets containing hydrochlorothiazide, amiloride and timolol for hypertension for 3 months. The rash had started 5 weeks previously on the dorsa of the hands and had subsequently spread to the face and the 'V' on the neck. Examination showed an acute eczematous reaction over both hands with a marked cut-off at the wrists. The face was similarly affected with a cut-off at the 'V' of the neck and the nape, with sparing behind the ears and under the spectacle pads. In addition, he had a macular erythema on the trunk. The patient was patch and photopatch tested with a crushed antihypertensive tablet in 5 g of petrolatum, hydrochlorothiazide (1% pet.) and the other two components of the tablet. 48 hours after UVA-irradiation, positive responses were limited to the tablet and hydrochlorothiazide (3).

Other cutaneous adverse drug reactions

In Finland, in the period 1989-2001, 826 patients with suspected cutaneous drug eruptions were patch tested and 89 had one or more positive reactions. Of these individuals, one reacted to hydrochlorothiazide + amiloride. The 2 diuretics were probably not tested separately. It was not specified what type of drug reaction this combination product had caused (4).

LITERATURE

1   The data in the section 'General' may have been obtained from literature discussed in this chapter, but mostly also or exclusively from one or more of the following online sources: ChemIDPlus Advanced, PubChem, DrugBank, RxList, Drug Central, Drugs.com, and Wikipedia

2   Wagner SN, Welke F, Goos M. Occupational UVA-induced allergic photodermatitis in a welder due to hydrochlorothiazide and ramipril. Contact Dermatitis 2000;43:245-246

3   White IR. Photopatch test in a hydrochlorothiazide drug eruption. Contact Dermatitis 1983;9:237

4   Lammintausta K, Kortekangas-Savolainen O. The usefulness of skin tests to prove drug hypersensitivity. Br J Dermatol 2005;152:968-974

Chapter 3.244 HYDROCORTISONE

IDENTIFICATION

Description/definition	: Hydrocortisone (cortisol) is the main glucocorticoid secreted by the adrenal cortex that conforms to the structural formula shown below
Pharmacological classes	: Glucocorticoids; anti-inflammatory agents
IUPAC name	: (8S,9S,10R,11S,13S,14S,17R)-11,17-Dihydroxy-17-(2-hydroxyacetyl)-10,13-dimethyl-2,6,7,8,9,11,12,14,15,16-decahydro-1H-cyclopenta[a]phenanthren-3-one
Other names	: Cortisol; 11β,17,21-trihydroxypregn-4-ene-3,20-dione; hydrocortisone alcohol
CAS registry number	: 50-23-7
EC number	: 200-020-1
Merck Index monograph	: 6094
Patch testing	: 1% pet. (SmartPracticeCanada, SmartPracticeEurope); late readings (6-10 days) are strongly recommended; testing with this preparation probably results in many false-negative reactions; in general, corticosteroids can be tested 0.1% and 1% alcohol
Molecular formula	: $C_{21}H_{30}O_5$

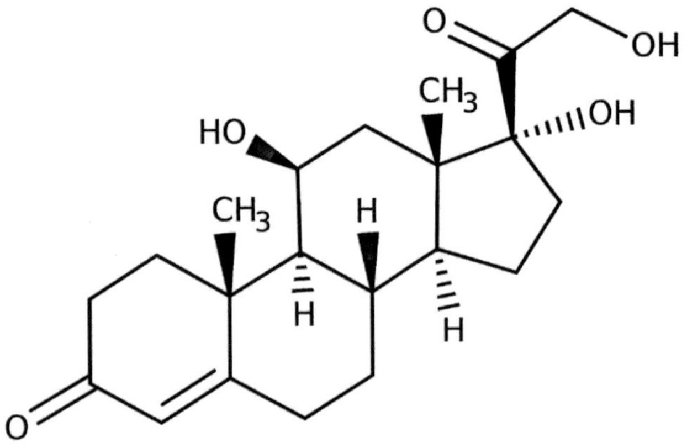

GENERAL

Systemically administered glucocorticoids have anti-inflammatory, immunosuppressive and antineoplastic properties and are used in the treatment of a wide spectrum of diseases including rheumatic disorders, lung diseases (asthma, COPD), gastrointestinal tract disorders (Crohn's disease, colitis ulcerosa), certain malignancies (leukemia, lymphomas), hematological disorders, and various diseases of the kidneys, brain, eyes and skin. A practical guideline for diagnosing allergic reactions to corticosteroids is presented in ref. 1. Hydrocortisone *base* (hydrocortisone alcohol, hydrocortisone free alcohol) is used in tablets only, which implies that by far most allergic reactions to 'hydrocortisone', which have been widely reported, have in fact been the result of sensitization to an ester of hydrocortisone or of cross-reactivity to another corticosteroid. This subject and problems with patch testing hydrocortisone have been fully reviewed in Volume 3 of the *Monographs in contact allergy* series (8).

See also hydrocortisone sodium phosphate (Chapter 3.245).

CONTACT ALLERGY FROM ACCIDENTAL CONTACT

In Leuven, Belgium, in the period 2001-2019, 201 of 1248 health care workers/employees of the pharmaceutical industry had occupational allergic contact dermatitis. The lesions were mostly localized on the hands, but often also on the face, as airborne dermatitis. In total, 42 positive patch test reactions to 18 different systemic drugs were found. In one patient, hydrocortisone was the drug/one of the drugs that caused occupational dermatitis (6).

Of 38 veterinarians with hand and forearm dermatoses seen by dermatologists in Belgium and the Netherlands from 1995 to 2005, 17 had occupational allergic contact dermatitis. In four patients, corticosteroids were the causative occupationally used drugs, including hydrocortisone in one (11).

A pharmacist may have become occupationally sensitized to hydrocortisone from preparing *ex tempore* corticosteroid creams and ointments for 10 years (14), but the evidence presented was not strong.

CUTANEOUS ADVERSE DRUG REACTIONS FROM SYSTEMIC ADMINISTRATION CAUSED BY TYPE IV (DELAYED-TYPE) HYPERSENSITIVITY (as demonstrated by positive patch tests, unless otherwise stated)

Cutaneous adverse drug reactions from systemic administration of hydrocortisone caused by type IV (delayed-type) hypersensitivity have included systemic allergic dermatitis (2,3,4,5,7,9,10,12,13,15) and symmetrical drug-related intertriginous and flexural exanthema (SDRIFE)/baboon syndrome (10).

Systemic allergic dermatitis (systemic contact dermatitis)

Four patients (2 women, 2 men, ages 31-70 years) who had positive patch tests to hydrocortisone and to hydrocortisone butyrate underwent an oral provocation test with hydrocortisone. Two had eczema of the hands, one had treated vasculitis of the inner thighs with hydrocortisone and the 4th suffered from perianal eczema. Hydrocortisone free alcohol was given at an increasing dose (30 to 100 to 250 mg) on three consecutive days or until a positive reaction appeared. In two patients, provoked with 100 mg hydrocortisone, erythema and edema appeared after 5-6 hours at the old patch test sites, that faded in 6 hours resp. 7 days. The patient with previous vasculitis developed indurated papules on the inner sides of the thighs after 15 hours, fading in 24 hours. The 4th patient developed erythema and edema in the perianal area 5 hours after 250 mg hydrocortisone, which faded after 4 days. No clinical signs of any systemic reactions were observed. In one patient, the adrenal cortex was stimulated with intramuscular and intravenous tetracosactide, which resulted in an identical reaction of reactivation of old patch tests as seen with oral provocation (4,5,13).

A 40-year-old man developed acute eczema of the nostrils, upper lip and cheeks after using topical steroids for rinorrhea. When the patient was treated with systemic hydrocortisone 50 mg/day, he developed generalized erythematous plaque-like lesions, mainly on the face, neck and trunk. Patch tests were positive to hydrocortisone 1% alc. and tixocortol pivalate. Oral provocation with hydrocortisone in increasing doses reproduced the exanthema and caused an intense flare-up of the previously positive corticosteroid patch tests (3).

A 40-year-old man had used prednisolone metasulfobenzoate sodium enema's in the past which had worsened his diarrhea. He had also developed a rash from intravenous hydrocortisone on two occasions but had no problems with oral steroids. He had positive patch tests to tixocortol pivalate, a hydrocortisone acetate 1% enema, and a positive reaction at 48 hours to intradermal prednisolone acetate enema (7).

A 40-year-old woman with bursitis and tennis elbow, who was probably sensitized to hydrocortisone from hemorrhoid preparations, developed facial erythema after provocation tests with intralesional betamethasone acetate/disodium phosphate, oral betamethasone and oral hydrocortisone, which emerged within 18 hours and lasted for 2-3 days. Patch and intradermal tests showed delayed-type allergy to corticosteroids of the hydrocortisone group (hydrocortisone, tixocortol, methylprednisolone, but *not* prednisolone). As this patient was probably presensitized, these were cases of systemic allergic dermatitis with cross-reactions from hydrocortisone to betamethasone (9).

A 50-year-old woman who was very likely sensitized to corticosteroids from their use in chronic nasal congestion had suffered a skin eruption on three occasions one day after oral intake of prednisolone or methylprednisolone. Patch tests were positive to prednisolone (methylprednisolone not tested). An oral provocation tests with hydrocortisone resulted in the classic picture of the baboon syndrome/symmetrical drug-related intertriginous and flexural exanthema (SDRIFE) after 12-24 hours. A patch test with hydrocortisone was negative and an intradermal test not performed. This was a case of systemic allergic dermatitis presenting as the baboon syndrome/SDRIFE (10).

A female patient sensitized to prednisolone in eye ointment developed a confluent maculopapular exanthema 24 hours after an oral provocation test with hydrocortisone (12).

In Leuven, Belgium, in a 12-year-period before 2012, 16 patients were investigated for a generalized allergic eruption (maculopapular eruption or eczema, with or without flare-up of previous dermatitis) from systemic administration (oral, intravenous, intramuscular, intra-articular) of corticosteroids, a few hours or days after the first dose of the culprit drug. The reactions observed were in most cases a manifestation of systemic allergic dermatitis: the patient had previously become sensitized to the corticosteroid used systemically or a cross-reacting molecule from topical exposure. One patient had reacted with a maculopapular exanthema from intravenous administration of intravenous 'hydrocortisone' (more likely hydrocortisone sodium phosphate or hydrocortisone sodium succinate) (12).

Seven patients who had positive patch tests to tixocortol pivalate and positive intradermal tests to hydrocortisone, were orally challenged with hydrocortisone 20, 50, 100 and 200 mg. One or two doses were given per day, and the challenge was stopped when skin symptoms appeared. All 7 had a positive reaction; 2 reacted after the 20 mg dose, 4 after 100 mg and 1 after 200 mg. Symptoms were erythema or infiltrated erythema at previous sites of eczema or positive skin tests (n=6), and widespread erythema or exanthema (n=2) (15).

See refs. 2 and 10 for additional cases of systemic allergic dermatitis.

Symmetrical drug-related intertriginous and flexural exanthema (SDRIFE)/Baboon syndrome
A 50-year-old woman who was previously sensitized to topical corticosteroids developed symmetrical drug-related intertriginous and flexural exanthema (SDRIFE)/baboon syndrome after an oral provocation test with hydrocortisone (10).

Cross-reactions, pseudo-cross-reactions and co-reactions
Cross-reactions between corticosteroids are discussed in Chapter 3.399 Prednisolone. Patients reacting to tixocortol pivalate are almost always also allergic to hydrocortisone and therefore, tixocortol pivalate in the baseline/screening series is a powerful marker for hydrocortisone sensitivity (8).

LITERATURE

1   Baeck M, Goossens A. Immediate and delayed allergic hypersensitivity to corticosteroids: practical guidelines. Contact Dermatitis 2012;66:38-45
2   Kulberg A, Schliemann S, Elsner P. Contact dermatitis as a systemic disease. Clin Dermatol 2014;32:414-419
3   Torres V, Tavares-Bello R, Melo H, Soares AP. Systemic allergic dermatitis from hydrocortisone. Contact Dermatitis 1993;29:106
4   Lauerma AI, Reitamo S, Maibach HI. Systemic hydrocortisone/cortisol induces allergic skin reactions in presensitized subjects. J Am Acad Dermatol 1991;24:182-185
5   Lauerma AI, Reitamo S, Maibach HI. Systemic hydrocortisone/cortisol induces allergic skin reactions in presensitized subjects. Am J Contact Dermat 1991;2:68 (Abstract)
6   Gilissen L, Boeckxstaens E, Geebelen J, Goossens A. Occupational allergic contact dermatitis from systemic drugs. Contact Dermatitis 2020;82:24-30
7   Malik M, Tobin AM, Shanahan F, O'Morain C, Kirby B, Bourke J. Steroid allergy in patients with inflammatory bowel disease. Br J Dermatol 2007;157:967-969
8   De Groot AC. Monographs in contact allergy, volume 3: Topical Drugs. Boca Raton, Fl, USA: CRC Press Taylor and Francis Group, 2021 (ISBN 978-0-367-23693-9)
9   Räsänen L, Hasan T. Allergy to systemic and intralesional corticosteroids. Br J Dermatol 1993;128:407-411
10 Treudler R, Simon JC. Symmetric, drug-related, intertriginous, and flexural exanthema in a patient with polyvalent intolerance to corticosteroids. J Allergy Clin Immunol 2006;118:965-967
11 Bulcke DM, Devos SA. Hand and forearm dermatoses among veterinarians. J Eur Acad Dermatol Venereol 2007;21:360-363
12 Baeck M, Goossens A. Systemic allergic dermatitis to corticosteroids. Allergy 2012;67:1580-1585
13 Lauerma AI. Contact hypersensitivity to glucocorticosteroids. Am J Contact Dermat 1992;3:112-132
14 Lauerma AI. Occupational contact sensitization to corticosteroids. Contact Dermatitis 1998;39:328-329
15 Räsänen L, Tuomi ML, Ylitalo L. Reactivity of tixocortol pivalate-positive patients in intradermal and oral provocation tests. Br J Dermatol 1996;135:931-934

Chapter 3.245 HYDROCORTISONE SODIUM PHOSPHATE

IDENTIFICATION

Description/definition : Hydrocortisone sodium phosphate is the synthetic glucocorticoid that conforms to the structural formula shown below

Pharmacological classes : Glucocorticoids

IUPAC name : Disodium;[2-[(8S,9S,10R,11S,13S,14S,17R)-11,17-dihydroxy-10,13-dimethyl-3-oxo-2,6,7,8,9,11,12,14,15,16-decahydro-1H-cyclopenta[a]phenanthren-17-yl]-2-oxoethyl] phosphate

CAS registry number : 6000-74-4

EC number : 227-843-9

Merck Index monograph : 6094 (Hydrocortisone)

Patch testing : In general, corticosteroids may be tested 0.1% and 1% alcohol; late readings (6-10 days) are strongly recommended

Molecular formula : $C_{21}H_{29}Na_2O_8P$

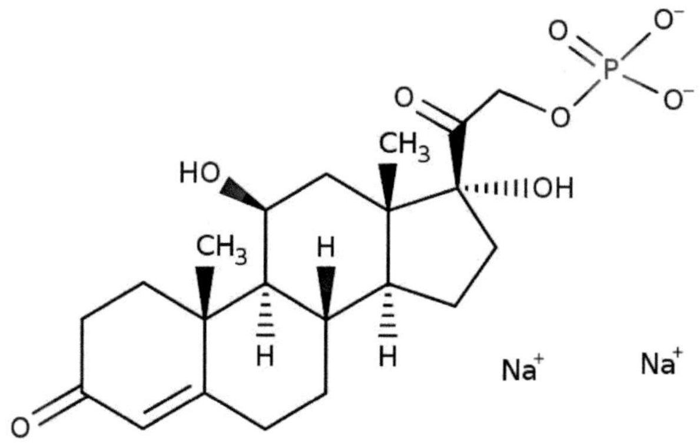

GENERAL

Systemically administered glucocorticoids have anti-inflammatory, immunosuppressive and antineoplastic properties and are used in the treatment of a wide spectrum of diseases including rheumatic disorders, lung diseases (asthma, COPD), gastrointestinal tract disorders (Crohn's disease, colitis ulcerosa), certain malignancies (leukemia, lymphomas), hematological disorders, and various diseases of the kidneys, brain, eyes and skin. A practical guideline for diagnosing allergic reactions to corticosteroids is presented in ref. 1.

See also Hydrocortisone (Chapter 3.244).

CUTANEOUS ADVERSE DRUG REACTIONS FROM SYSTEMIC ADMINISTRATION CAUSED BY TYPE IV (DELAYED-TYPE) HYPERSENSITIVITY (as demonstrated by positive patch tests, unless otherwise stated)

Cutaneous adverse drug reactions from systemic administration of hydrocortisone sodium phosphate caused by type IV (delayed-type) hypersensitivity have included systemic allergic dermatitis (2), and generalized erythema and urticaria with dyspnea (3).

Erythematous eruption

See the patient description in the section 'Other cutaneous adverse drug reactions' below (3).

Systemic allergic dermatitis (systemic contact dermatitis)

A 62-year-old woman had a history of a burn treated with ointment containing hydrocortisone without adverse effect and of the development of erythema around the neck 8 hours after an intra-articular injection of methylprednisolone acetate in the knee because of arthrosis. Two years after the latter event, the patient was given an intravenous injection of hydrocortisone sodium phosphate 500 mg for 2 days, and then developed erythema on the neck, trunk, and thighs in the evening on the first day of injection. Methylprednisolone sodium succinate was substituted for the hydrocortisone, but then the eruption developed into a generalized rash. Patch tests were positive to hydrocortisone sodium phosphate 0.1% and 0.05% pet., methylprednisolone acetate, methylprednisolone sodium succinate 0.05, 0.1 and 0.5% pet. and some other group 1 corticosteroids. Intradermal tests were positive at D1 and D4 to hydrocortisone sodium phosphate and methylprednisolone sodium succinate. Twenty-four hours after the intradermal injection the patient noticed erythema around her neck and the next day she experienced nausea, vomiting, and diarrhea. Systemic provocation tests were performed with drip infusions of both corticosteroids used. Five hours after provocation with methylprednisolone the patient developed symmetrical patches of erythema around her neck and on her chest, axillae, forearms, fingers, and upper inner thighs and a flare-up at the site of the previous patch test with methylprednisolone. A similar eruption was noted following administration of hydrocortisone, without flare-up of previously positive patch tests (2).

Other cutaneous adverse drug reactions

A 69-year-old woman with a history of intrinsic bronchial asthma had suffered 5 episodes of pruritus followed by generalized erythema and urticaria, accompanied with increasing dyspnea (on 2 occasions), starting 6 hours after intravenous injection of methylprednisolone or hydrocortisone (unspecified which esters or salts). Prick and intradermal tests with hydrocortisone phosphate 10 and 1 mg/ml and methylprednisolone 4 and 0.4 mg/ml were negative at 15 minutes. However, six hours later intradermal tests appeared positive and were accompanied by a generalized pruritic and erythematous reaction. Patch tests were (false-)negative (3).

Cross-reactions, pseudo-cross-reactions and co-reactions

Cross-reactions between corticosteroids are discussed in Chapter 3.399 Prednisolone.

LITERATURE

1   Baeck M, Goossens A. Immediate and delayed allergic hypersensitivity to corticosteroids: practical guidelines. Contact Dermatitis 2012;66:38-45
2   Murata Y, Kumano K, Ueda T, Araki N, Nakamura T, Tani N. Systemic allergic dermatitis caused by systemic corticosteroid use. Arch Dermatol 1997;133:1053-1054
3   Vidal C, Tomé S, Fernández-Redondo V, Tato F. Systemic allergic reaction to corticosteroids. Contact Dermatitis 1994;31:273-274

Chapter 3.246 HYDROMORPHONE

IDENTIFICATION

Description/definition : Hydromorphone is the semisynthetic derivative of the opioid morphine that conforms to the structural formula shown below
Pharmacological classes : Narcotics; analgesics, opioid
IUPAC name : (4R,4aR,7aR,12bS)-9-Hydroxy-3-methyl-1,2,4,4a,5,6,7a,13-octahydro-4,12-methano-benzofuro[3,2-e]isoquinolin-7-one
Other names : Dihydromorphinone
CAS registry number : 466-99-9
EC number : 207-383-5
Merck Index monograph : 6110
Patch testing : Hydromorphone powder 'as is' and hydromorphone 2% water (20 mg/ml)
Molecular formula : $C_{17}H_{19}NO_3$

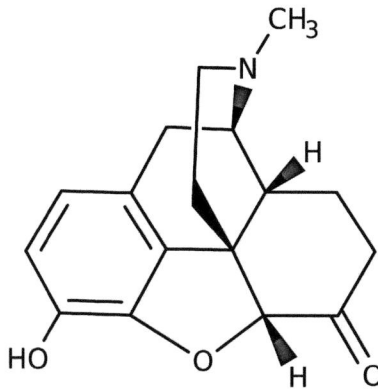

GENERAL

Hydromorphone is a pure opioid, a semisynthetic hydrogenated ketone derivative of morphine with analgesic effects. Hydromorphone is indicated for the management of moderate to severe acute pain and severe chronic pain. Due to its addictive potential and overdose risk, this agent is only prescribed when other first-line treatments have failed. Off-label, hydromorphone can be administered for the suppression of refractory cough. In pharmaceutical preparations, hydromorphone is present as hydromorphone hydrochloride (CAS number 71-68-1, EC number 200-762-6, molecular formula $C_{17}H_{20}ClNO_3$) (1).

CUTANEOUS ADVERSE DRUG REACTIONS FROM SYSTEMIC ADMINISTRATION CAUSED BY TYPE IV (DELAYED-TYPE) HYPERSENSITIVITY (as demonstrated by positive patch tests)

Cutaneous adverse drug reactions from systemic administration of hydromorphone caused by type IV (delayed-type) hypersensitivity have included localized allergic reaction from subcutaneous infusion, spreading to generalized papulovesicular dermatitis (2).

Other cutaneous adverse drug reactions

A 45-year-old man had a continuous subcutaneous infusion (CSCI) with hydromorphone for pain control of severe neuralgia from advanced carcinoma of the oropharynx. Because of local intolerance, the duration of the sites of infusion decreased from an initial average of 8-12 days to 1-3 days. Local toxicity was manifested as subcutaneous asymmetric, raised, indurated plaques, ranging from 3 to 7 cm in diameter, and were accompanied by pain and tenderness. Episodic leakage of the hydromorphone solution and serosanguinous fluid from the needle site caused focal erythema and pruritus. At one point, the patient developed an erythematous vesiculopustular dermatitis, first localized on the abdominal wall, predominantly situated where adhesive tapes had been applied, spreading after a few days over the trunk and the extremities, and presenting as a generalized papulovesicular dermatitis. Later, sensitization to hydromorphone was proved by positive patch tests to hydromorphone powder 'as is' and hydromorphone 2% water (20 mg/ml). Of ten controls, one had a + reaction to the powder, which was attributed to the 'angry back syndrome'. The authors suggest that repeated stripping of the epidermis, leakage of hydromorphone solution and occlusion by adhesive tape may have enhanced the development of delayed hypersensitivity to hydromorphone (2).

Cross-reactions, pseudo-cross-reactions and co-reactions
A patient sensitized by morphine showed cross-reactions to hydromorphone and codeine (3).

LITERATURE
1   The data in the section 'General' may have been obtained from literature discussed in this chapter, but mostly also or exclusively from one or more of the following online sources: ChemIDPlus Advanced, PubChem, DrugBank, RxList, Drug Central, Drugs.com, and Wikipedia
2   De Cuyper C, Goeteyn M. Systemic allergic dermatitis from subcutaneous hydromorphone. Contact Dermatitis 1992;27:220-223
3   Sasseville D, Blouin MM, Beauchamp C. Occupational allergic contact dermatitis caused by morphine. Contact Dermatitis 2011;64:166-168

Chapter 3.247 HYDROQUINIDINE

IDENTIFICATION

Description/definition	: Hydroquinidine is the alkaloid that conforms to the structural formula shown below
Pharmacological classes	: Parasympatholytics; anti-arrhythmia agents
IUPAC name	: (S)-[(2R,4S,5R)-5-Ethyl-1-azabicyclo[2.2.2]octan-2-yl]-(6-methoxyquinolin-4-yl)methanol
Other names	: Dihydroquinidine
CAS registry number	: 1435-55-8
EC number	: 215-862-5
Merck Index monograph	: 6113
Patch testing	: Most systemic drugs can be tested at 10% pet.; if the pure chemical is not available, prepare the test material from intravenous powder, the content of capsules or – when also not available – from powdered tablets to achieve a final concentration of the active drug of 10% pet.
Molecular formula	: $C_{20}H_{26}N_2O_2$

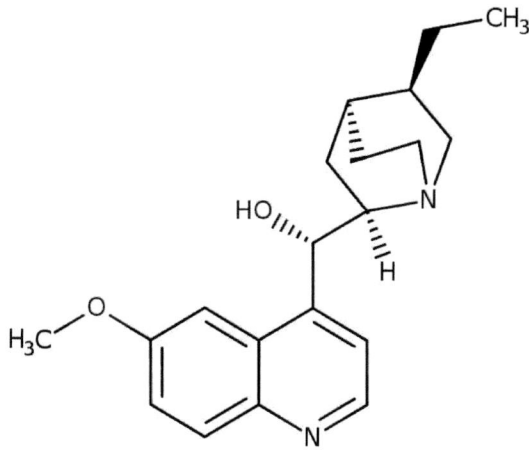

GENERAL

Hydroquinidine is a d-rotatory alkaloid derived from *Cinchona* bark. It is closely related to quinidine, differing only in containing two more atoms of hydrogen in the molecule. Hydroquinidine is a pharmaceutical agent that acts as a class I antiarrhythmic agent in the heart. The drug causes increased action potential duration, as well as a prolonged QT interval, working primarily by blocking the fast inward sodium current (INa). Hydroquinidine is also used for the treatment of malaria (1).

CUTANEOUS ADVERSE DRUG REACTIONS FROM SYSTEMIC ADMINISTRATION CAUSED BY TYPE IV (DELAYED-TYPE) HYPERSENSITIVITY (as demonstrated by positive patch tests)

Cutaneous adverse drug reactions from systemic administration of dihydroquinidine caused by type IV (delayed-type) hypersensitivity have included acute generalized exanthematous pustulosis (AGEP) (2) and photosensitivity (3).

Acute generalized exanthematous pustulosis (AGEP)

An 84-year-old woman suffering from asthma, high blood pressure, and cardiac arrhythmia was referred for an acute pustular eruption with a fever of 39°C. A new antiarrhythmic drug, dihydroquinidine, had been introduced 20 days before the onset of the cutaneous eruption. Dermatological examination showed a generalized scarlatiniform erythema with numerous non-follicular pustules, particularly on the abdomen, chest, and thighs; the mucous membranes were not affected. On the thighs, there were areas of superficial epidermal separation. Laboratory investigations showed leukocytosis with elevated neutrophils. Bacterial cultures of the pustules were negative. A cutaneous biopsy showed a subcorneal pustule, epidermal spongiosis, and dermal edema associated with a perivascular infiltrate of neutrophils, eosinophils, and lymphocytes. Dihydroquinidine was stopped and the cutaneous eruption healed within 12 days. Later, patch tests were performed with a crushed tablet of dihydroquinidine in saline, which gave a positive result. All other drugs the patient used tested negative. The clinical and histologic features of the patch test reaction were similar to the original pustular eruption. Ten controls were negative. The patient was diagnosed with acute generalized exanthematous pustulosis (AGEP) from dihydroquinidine (2).

Photosensitivity

In a study from Nancy, France, 54 patients with suspected nonimmediate drug eruptions, including 4 with photosensitivity, were assessed with patch testing. All 4 patients suspected of photosensitivity had positive *photo*patch tests, of who 2 to hydroquinidine. No clinical details were provided (3).

LITERATURE
1   The data in the section 'General' may have been obtained from literature discussed in this chapter, but mostly also or exclusively from one or more of the following online sources: ChemIDPlus Advanced, PubChem, DrugBank, RxList, Drug Central, Drugs.com, and Wikipedia
2   Moreau A, Dompmartin A, Castel B, Remond B, Leroy D. Drug-induced acute generalized exanthematous pustulosis with positive patch tests. Int J Dermatol 1995;34:263-266
3   Barbaud A, Reichert-Penetrat S, Tréchot P, Jacquin-Petit MA, Ehlinger A, Noirez V, et al. The use of skin testing in the investigation of cutaneous adverse drug reactions. Br J Dermatol 1998;139:49-58

Chapter 3.248 HYDROXYCHLOROQUINE

IDENTIFICATION

Description/definition   : Hydroxychloroquine is the 4-aminoquinoline that conforms to the structural formula shown below

Pharmacological classes : Antimalarials; enzyme inhibitors; antirheumatic agents

IUPAC name   : 2-[4-[(7-Chloroquinolin-4-yl)amino]pentyl-ethylamino]ethanol

Other names   : Oxychloroquine

CAS registry number   : 118-42-3

EC number   : 204-249-8

Merck Index monograph : 6127

Patch testing   : 1% pet.; in case of reactions to systemic administration of hydroxychloroquine, a higher Concentration may be considered

Molecular formula   : $C_{18}H_{26}ClN_3O$

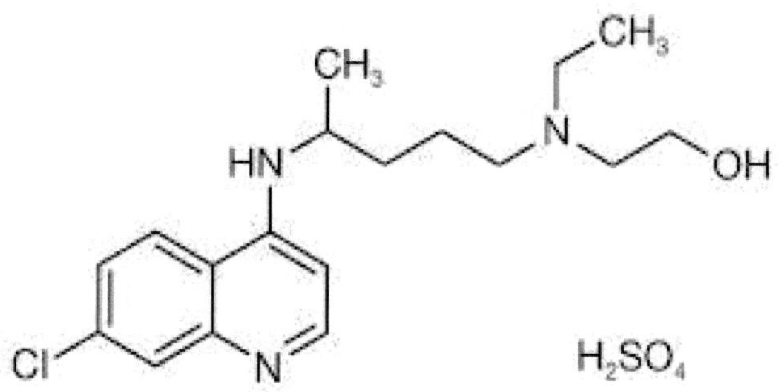

Hydroxychloroquine sulfate

GENERAL

Hydroxychloroquine is a 4-aminoquinoline with immunosuppressive, anti-autophagy, and antimalarial activities. This compound is indicated for the suppressive treatment and treatment of acute attacks of malaria due to *Plasmodium vivax, P. malariae, P. ovale*, and susceptible strains of *P. falciparum*. It is also indicated for the treatment of discoid and systemic lupus erythematosus and rheumatoid arthritis. In pharmaceutical preparations, hydroxychloroquine is present as hydroxychloroquine sulfate (CAS number 747-36-4, EC number 212-019-3, molecular formula $C_{18}H_{28}ClN_3O_5S$) (1).

CONTACT ALLERGY FROM ACCIDENTAL CONTACT

A 42-year-old man had worked in a pharmaceutical company in the manufacture of hydroxychloroquine sulfate for 8 months. A month before consultation, a pruritic rash on the back of his hands and wrists had appeared, which was clearly work-related. Patch tests were positive to hydroxychloroquine sulfate 0.5% (+), 1% (++) and 2% (++) in saline. Six controls were negative. The patient left his job, and his allergic contact dermatitis cleared (2). This was an obvious case of occupational allergic contact dermatitis.

A case of occupational contact dermatitis to hydroxychloroquine had been presented 20 years earlier from Germany. A 60 year-old worker in the pharmaceutical industry suffered from recurring contact dermatitis, initially limited to the hands, later becoming generalized. The patient had worked at a drug filling line in a pharmaceutical plant for more than 20 years. At one point, he experienced an episode of severe asthma and generalized dermatitis with conjunctivitis following exposure to hydroxychloroquine the day before. Patch testing revealed positive reactions to hydroxychloroquine 0.1%, 0.5%, 1% and 2% (vehicle not mentioned); the reaction to 2% was pustular. Five controls were negative. A scratch test was negative after 20 minutes, but positive after one day. Bronchial exposure to hydroxychloroquine dust produced a delayed bronchial obstruction over the next 20 hours, which progressed to fever and generalized erythema, considered to be systemic allergic dermatitis (3).

CUTANEOUS ADVERSE DRUG REACTIONS FROM SYSTEMIC ADMINISTRATION CAUSED BY TYPE IV (DELAYED-TYPE) HYPERSENSITIVITY (as demonstrated by positive patch tests)
Cutaneous adverse drug reactions from systemic administration of hydroxychloroquine caused by type IV (delayed-type) hypersensitivity have included acute generalized exanthematous pustulosis (AGEP) (5,6,7,9), systemic allergic dermatitis (3), drug reaction with eosinophilia and systemic symptoms (DRESS) (8), and photosensitivity (4).

Acute generalized exanthematous pustulosis (AGEP)

General
AGEP is a well-known side effect of hydroxychloroquine, nearly 35 cases having been reported (9). In most cases, patch tests were either not performed or were negative (reviewed in ref. 9). Only a few studies have reported positive patch tests to hydroxychloroquine in patients with AGEP (5,6,7,9)

Case series
In a university hospital in Sfax, Tunisia, 7 patients with AGEP from hydroxychloroquine (HCQ) were investigated in the period 2011-2019. All were women, mean age 47 years; the average time from HCQ start to onset of symptoms was 40 days. All patients received topical steroids with a full resolution of the rash within an average of 39 days after HCQ withdrawal. Patch tests were performed for three patients with positive results in one. These patients had been patch tested with pulverized tablets of 200 mg diluted to 30% pet. 6-8 weeks after resolution of the skin symptoms. It was suggested that the latent period and the duration for resolution of HCQ-induced AGEP may be longer than with other drugs due to the metabolic characteristics of hydroxychloroquine (9).

In a tertiary care pediatric hospital in Florence, Italy, in the period 2010-2018, 54 children were hospitalized for (suspected) severe cutaneous adverse drug reactions. Two had acute generalized exanthematous pustulosis (AGEP); in one, it was caused by hydroxychloroquine, which was positive on patch testing in the patient (7).

Case reports
A 33-year-old woman was diagnosed with systemic lupus erythematosus and was treated with hydroxychloroquine 200 mg/day and prednisolone 40 mg/day. Seventeen days after starting treatment, the patient developed generalized erythema and a pruritic pustular eruption. The lesions initially appeared on the face and then spread to the rest of the body. Five days later, the patient had fever and there was a generalized erythema with a few 1–2 mm pustules over the legs and a diffuse superficial desquamation. Laboratory parameters showed leukocytosis 18.370/mm^3, neutrophil count 86%, C-reactive protein 157.7 mg/l (reference range <5) and slightly elevated liver enzymes. A skin biopsy showed spongiform intraepidermal pustules, edema of the papillary dermis, a perivascular infiltrate with neutrophils, and some eosinophils. Patch tests, performed 4 weeks after the symptoms disappeared, were positive to hydroxychloroquine in dimethyl sulfoxide (DMSO) but negative to chloroquine and doubtful to prednisolone. The patient was diagnosed with acute generalized exanthematous pustulosis (AGEP) from hydroxychloroquine (5).

A woman 49 years was treated with prednisolone and hydroxychloroquine for rheumatoid arthritis. After 17 days, the patient developed generalized erythema and painful pustular eruptions with (undetected) fever. The arms and face were the first sites of lesions, which then spread to the scalp, back, chest, and then the rest of the body. aboratory investigations revealed leukocytosis, elevated neutrophil count (91%), high ESR (41 mm/h), and high CRP (7.5 mg/l). Histopathology showed subcorneal and superficial epidermal neutrophilic pustules, epidermal spongiosis and neutrophilic infiltrations, edematous dermal papillae, perivascular infiltration of lymphocytes, histocytes and neutrophils. Later patch testing with all medications used by the patient revealed a strongly positive reaction to hydroxychloroquine (no details on material test concentration and vehicle provided) (6).

Systemic allergic dermatitis (systemic contact dermatitis)
A case of systemic allergic dermatitis to hydroxychloroquine has been described in the section 'Contact allergy from accidental contact allergy' above, ref. 3.

Drug reaction with eosinophilia and systemic symptoms (DRESS)
A 37-year-old woman presented with an itchy rash, swelling of the face and fever, which had started 2 days earlier. Physical examination revealed a maculopapular rash, predominantly on the face, periorbital edema, and a purpuric exanthema on the trunk, arms and legs. The oral mucosa was affected and cervical lymph nodes were palpable. The patient had received hydroxychloroquine, lopinavir-ritonavir and azithromycin for 5 days in the previous 2-3 weeks for suspected covid-19-induced pneumonia. Laboratory investigations showed leukocytosis with eosinophilia and elevated liver enzymes; no atypical lymphocytes were found. After 8 weeks, patch tests were performed with all

drugs, each at 20% water and pet., which were positive for hydroxychloroquine on day 4. The patient was diagnosed with DRESS caused by hydroxychloroquine (8).

Photosensitivity

A 74-year-old man was treated with hydroxychloroquine (HCQ) sulfate (200 mg/day) for 8 months for rheumatoid arthritis, when he developed itchy erythemato-papulo-squamous hyperpigmented dermatitis, involving only the light-exposed areas of the face and hands. Phototesting showed a significant decrease of minimal erythema dose (MED) both in the UVA and UVB range. Photopatch testing with HCQ sulfate 5% in water and UVA-irradiation (2.5 J/cm^2) was negative, but weakly positive after UVB irradiation with 20 mJ/cm^2. Photopatch testing with HCQ sulfate and UVB-irradiation was repeated 2 years later and was again positive. It was concluded that this was a combined phototoxic and photoallergic reaction to hydroxychloroquine (4).

LITERATURE

1   The data in the section 'General' may have been obtained from literature discussed in this chapter, but mostly also or exclusively from one or more of the following online sources: ChemIDPlus Advanced, PubChem, DrugBank, RxList, Drug Central, Drugs.com, and Wikipedia
2   Herrera-Mozo I, Sanz-Gallen P, Saéz B, Marti-Amengual G. Occupational contact dermatitis caused by hydroxychloroquine sulfate. Contact Dermatitis 2018;79:102-103
3   Meier H, Elsner P, Wüthrich B. Berufsbedingtes Kontaktekzem und Asthma bronchiale bei ungewöhnlicher allergischer Reaktion vom Spättyp auf Hydroxychoroquin [Occupationally induced contact dermatitis and bronchial asthma in a unusual delayed reaction to hydroxychloroquine]. Hautarzt 1999;50:665-669 (Article in German)
4   Lisi P, Assalve D, Hansel K. Phototoxic and photoallergic dermatitis caused by hydroxychloroquine. Contact Dermatitis 2004;50:255-256
5   Charfi O, Kastalli S, Sahnoun R, Lakhoua G. Hydroxychloroquine-induced acute generalized exanthematous pustulosis with positive patch-testing. Indian J Pharmacol 2015;47:693-694
6   Mofarrah R, Mofarrah R, Oshriehye M, Ghobadi Aski S, Nazemi N, Nooshiravanpoor P. The necessity of patch testing in determining the causative drug of AGEP. J Cosmet Dermatol 2021;20:2156-2159
7   Liccioli G, Mori F, Parronchi P, Capone M, Fili L, Barni S, Sarti L, Giovannini M, Resti M, Novembre EM. Aetiopathogenesis of severe cutaneous adverse reactions (SCARs) in children: A 9-year experience in a tertiary care paediatric hospital setting. Clin Exp Allergy 2020;50:61-73
8   Castro Jiménez A, Navarrete Navarrete N, Gratacós Gómez AR, Florido López F, García Rodríguez R, Gómez Torrijos E. First case of DRESS syndrome caused by hydroxychloroquine with a positive patch test. Contact Dermatitis 2021;84:50-51
9   Chaabouni R, Bahloul E, Ennouri M, Atheymen R, Sellami K, Marrakchi S, et al. Hydroxychloroquine-induced acute generalized exanthematous pustulosis: a series of seven patients and review of the literature. Int J Dermatol 2021;60:742-748

Chapter 3.249   HYDROXYPROGESTERONE

IDENTIFICATION

Description/definition : Hydroxyprogesterone is the progestin and steroid that conforms to the structural formula shown below
Pharmacological classes : Progestins; sex hormones
IUPAC name : (8R,9S,10R,13S,14S,17R)-17-Acetyl-17-hydroxy-10,13-dimethyl-2,6,7,8,9,11,12,14,15,16-decahydro-1H-cyclopenta[a]phenanthren-3-one
Other names : 17-Hydroxyprogesterone; 17α-hydroxyprogesterone
CAS registry number : 68-96-2
EC number : 200-699-4
Merck Index monograph : 6147
Patch testing : Generally, (cortico)steroids may be tested at 0.1% and 1% in alcohol; late readings (6-10 days) are strongly recommended
Molecular formula : $C_{21}H_{30}O_3$

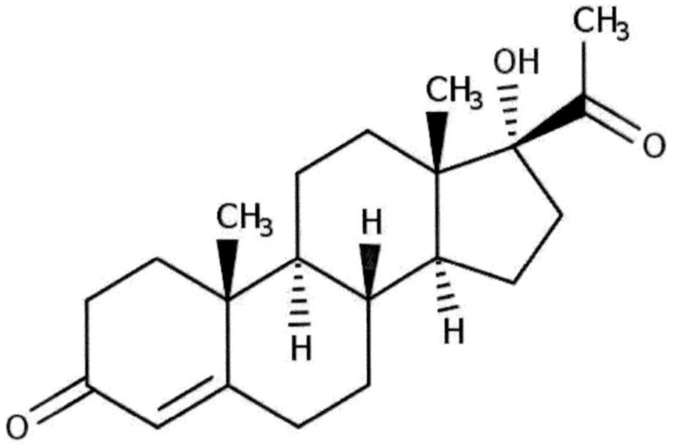

GENERAL

Hydroxyprogesterone is a physiological progestin that is produced during glucocorticoid and steroid hormone synthesis and is increased during the third trimester of pregnancy. Hydroxyprogesterone binds to the cytoplasmic progesterone receptors in the reproductive system and subsequently activates progesterone receptor mediated gene expression. As a pharmaceutical, it is used to treat irregular menstrual bleeding or a lack of menstrual bleeding in women. In some cases, it may be used to treat endometrial cancer (1).

CUTANEOUS ADVERSE DRUG REACTIONS FROM SYSTEMIC ADMINISTRATION CAUSED BY TYPE IV (DELAYED-TYPE) HYPERSENSITIVITY (as demonstrated by positive patch tests)

Cutaneous adverse drug reactions from systemic administration of hydroxyprogesterone caused by type IV (delayed-type) hypersensitivity have included systemic allergic dermatitis (4).

Systemic allergic dermatitis (systemic contact dermatitis)

A 68-year-old woman presented with a severe, intensely pruritic papulovesicular eruption on the chest, back, abdomen and legs, which had developed after 7 days of treatment with a preparation containing 0.625 mg conjugated estrogen and 5 mg hydroxyprogesterone acetate for late menopausal syndrome. In the past a similar rash had appeared several times shortly after taking the same medication. For many years the patient had suffered pruritus, a maculopapular rash and 'flu-like symptoms' (headache, low fever, muscle pain) for several days prior to menstruation. These symptoms always disappeared completely a few days after menstruation (auto-immune progesterone dermatitis, AIPD). The patient had been treated in the past with topical corticosteroids for pityriasis lichenoides chronica. After complete recovery, patch testing showed positive reactions to 17-hydroxyprogesterone 2% pet., tixocortol pivalate and budesonide. The authors hypothesized that the patient had become allergic to topical corticosteroids in the past and that the recurrent episodes of AIPD were related to endogenous progesterone following cross-sensitivity. The current papulovesicular eruption would be the result of oral administration of progesterone in the estrogen-progesterone preparation and in fact be a form of systemic allergic dermatitis (4).

Cross-reactions, pseudo-cross-reactions and co-reactions
Cross-reactions to 17-hydroxyprogesterone in patients sensitized to topical corticosteroids have been observed and are not rare (2,3).

LITERATURE

1   The data in the section 'General' may have been obtained from literature discussed in this chapter, but mostly also or exclusively from one or more of the following online sources: ChemIDPlus Advanced, PubChem, DrugBank, RxList, Drug Central, Drugs.com, and Wikipedia

2   Wilkinson SM, Beck MH. The significance of positive patch tests to 17-hydroxyprogesterone. Contact Dermatitis 1994;30:302-303

3   Schoenmakers A, Vermorken A, DeGreef H, Dooms-Goossens A. Corticosteroid or steroid allergy? Contact Dermatitis 1992;26:159-162

4   Ingber A, Trattner A, David M. Hypersensitivity to an estrogen – progesterone preparation and possible relationship to autoimmune progesterone dermatitis and corticosteroid hypersensitivity. J Dermatolog Treat 1999;10:139-140

Chapter 3.250 HYDROXYUREA

IDENTIFICATION

Description/definition	: Hydroxyurea is the member of the class of ureas that conforms to the structural formula shown below
Pharmacological classes	: Enzyme inhibitors; antineoplastic agents; anti-sickling agents; nucleic acid synthesis inhibitors
IUPAC name	: Hydroxyurea
Other names	: Hydroxycarbamide
CAS registry number	: 127-07-1
EC number	: 204-821-7
Merck Index monograph	: 6158
Patch testing	: Tablet, pulverized, 20% water (2); most systemic drugs can be tested at 10% pet.; if the pure chemical is not available, prepare the test material from intravenous powder, the content of capsules or – when also not available – from powdered tablets to achieve a final concentration of the active drug of 10% pet.
Molecular formula	: $CH_4N_2O_2$

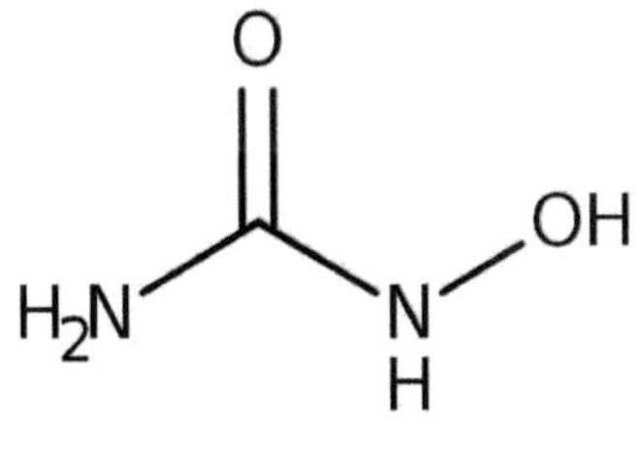

GENERAL

Hydroxyurea is an antineoplastic agent that inhibits DNA synthesis through the inhibition of ribonucleoside diphosphate reductase. It is used for management of melanoma, resistant chronic myelocytic leukemia, and recurrent, metastatic, or inoperable carcinoma of the ovary. Hydroxyurea is also used in sickle cell anemia, as it increases the level of fetal hemoglobin, leading to a reduction in the incidence of vaso-occlusive crises in this disease (1).

CUTANEOUS ADVERSE DRUG REACTIONS FROM SYSTEMIC ADMINISTRATION CAUSED BY TYPE IV (DELAYED-TYPE) HYPERSENSITIVITY (as demonstrated by positive patch tests)

Cutaneous adverse drug reactions from systemic administration of hydroxyurea caused by type IV (delayed-type) hypersensitivity have included photosensitivity (2).

Photosensitivity

A 62-year-old man developed a photodistributed rash while using 3 different medications including hydroxyurea. Patch, photopatch and photoscratch tests to the drugs 20% in water were positive to hydroxyurea in the photoscratch (but not the photopatch) tests, irradiated with UVA and UVB. Ten controls were negative (2).

LITERATURE

1   The data in the section 'General' may have been obtained from literature discussed in this chapter, but mostly also or exclusively from one or more of the following online sources: ChemIDPlus Advanced, PubChem, DrugBank, RxList, Drug Central, Drugs.com, and Wikipedia
2   Conilleau V, Dompmartin A, Michel M, Verneuil L, Leroy D. Photoscratch testing in systemic drug-induced photosensitivity. Photodermatol Photoimmunol Photomed 2000;16:62-66

Chapter 3.251 HYDROXYZINE

IDENTIFICATION

Description/definition : Hydroxyzine is the piperazine derivative and diphenylmethane that conforms to the structural formula shown below
Pharmacological classes : Histamine H_1 antagonists; antipruritics
IUPAC name : 2-[2-[4-[(4-Chlorophenyl)phenylmethyl]piperazin-1-yl]ethoxy]ethanol
Other names : 1-(p-Chloro-α-phenylbenzyl)-4-(2-(2-hydroxyethoxy)ethyl)piperazine
CAS registry number : 68-88-2
EC number : 200-693-1
Merck Index monograph : 6159
Patch testing : Hydrochloride 1.0% pet. (Chemotechnique); 2.5% pet. (10)
Molecular formula : $C_{21}H_{27}ClN_2O_2$

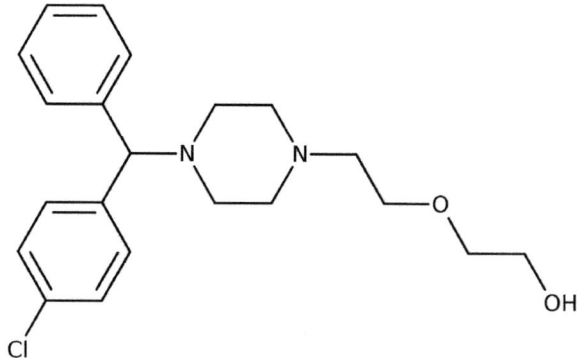

GENERAL

Hydroxyzine is a piperazine histamine H_1 receptor antagonist with anti-allergic, antispasmodic, sedative, anti-emetic and anti-anxiety properties. It blocks the H_1 histamine receptor and prevents the symptoms that are caused by histamine activity on capillaries, bronchial smooth muscle, and gastrointestinal smooth muscle, including vasodilatation, increased capillary permeability, bronchoconstriction, and spasmodic contraction of gastrointestinal smooth muscle. This agent is used in the treatment of chronic urticaria, dermatitis, and histamine-mediated pruritus and is also employed as an antiemetic, for relief of anxiety and tension, and as a sedative. In pharmaceutical preparations, hydroxyzine is present as hydroxyzine hydrochloride (CAS number 2192-20-3, EC number 218-586-3, molecular formula $C_{21}H_{29}Cl_3N_2O_2$) (1).

CUTANEOUS ADVERSE DRUG REACTIONS FROM SYSTEMIC ADMINISTRATION CAUSED BY TYPE IV (DELAYED-TYPE) HYPERSENSITIVITY

Cutaneous adverse drug reactions from systemic administration of hydroxyzine caused by type IV (delayed-type) hypersensitivity have included maculopapular eruption (11,12,13,15,17,21,22), erythroderma (18), acute generalized exanthematous pustulosis (AGEP) (7,8,9,10,19), systemic allergic dermatitis (23), symmetrical drug-related intertriginous and flexural exanthema (SDRIFE)/baboon syndrome (20), fixed drug eruption (16), eczematous drug eruption (24), erythema multiforme (25), and 'allergy' (14).

Maculopapular eruption

A 36-year-old man with longstanding psoriasis was treated for itch with oral hydroxyzine every 8th hours, which he had used before without problems. Two days later, he presented with a desquamative erythematous maculopapular rash with severe pruritus. Patch tests were negative to the topical antipsoriatic drugs used but positive to hydroxyzine 2.5% pet. at D2 and D4. Ten controls were negative (11,12).

Three patients (all women, ages 35,36, and 65) with maculopapular rashes developing 1-3 days after taking oral hydroxyzine, were reported from France in 1997. Two used the antihistamine for urticaria, the third for atopic dermatitis. One developed a morbilliform eruption with fever, lymphadenopathy and erythroderma. All three had positive patch tests to a powdered hydroxyzine tablet diluted in water (negative in 190 controls) and – later – to the powder 2%, 5% and 10% in water, whereas the other ingredients of the commercial tablet were negative. Patch tests with ethylenediamine, piperazine (diethylenediamine) and other antihistamines were negative, excluding cross-allergy (13).

A 44-year-old man developed a morbilliform eruption 2 days after intake of oral hydroxyzine for urticaria. Previously, he had suffered a similar eruption from cetirizine (which is a metabolite of hydroxyzine). Patch tests were positive to hydroxyzine (tablet 'as is', 1%, 2.5%, 5% and 10% pet.), cetirizine (tablet 'as is', 2.5%, 5% and 10% pet.), ethylenediamine 1% pet. and piperazine 1% pet. Twenty controls were negative. The patient was probably primarily sensitized to cetirizine and later reacted to hydroxyzine, which is for 45% metabolized *in vivo* into cetirizine (15).

On the 4th day of treatment with prednisone and hydroxyzine, a 70-year-old woman presented with a bilateral highly pruritic palmar erythema that evolved to a generalized morbilliform rash with subsequent complete desquamation. At a later time, she took cetirizine (a metabolite of hydroxyzine) for a cold, and developed palmar erythema and desquamation. Patch tests were positive to hydroxyzine 0.2% pet. and negative to cetirizine 0.1% pet. and to corticosteroids (21).

One patient with multiple drug hypersensitivity developed a maculopapular exanthema from hydroxyzine and had positive patch tests to this drug tested 10% pet. (17). A 62-year-old woman developed pruritic diffuse erythema with papules on the trunk and extremities 2 days after taking hydroxyzine pamoate for asteatotic eczema. Patch tests were positive to hydroxyzine pamoate and hydroxyzine hydrochloride (no details available, article in Japanese) (22).

Erythroderma, widespread erythematous eruption, exfoliative dermatitis
In a study from Nancy, France, of 7 patients with drug-induced erythroderma, one had a positive patch test to hydroxyzine, which reproduced the previous symptoms (18).

Acute generalized exanthematous pustulosis (AGEP)
A 48-year-old woman with a history of psoriasis was prescribed oral hydroxyzine for generalized pruritus, which she had used before. Twenty-four hours after hydroxyzine ingestion, a burning erythematous eruption developed on her trunk, extremities, and genitalia. She discontinued the hydroxyzine on day 4. Two days later, fever developed and the skin eruption worsened. Examination revealed widespread small non-follicular pustules on an erythematous background. Biopsies showed features of acute generalized exanthematous pustulosis (AGEP). The patient was later patch tested and showed a +++ reaction to 10% hydroxyzine at D2 and D5. A biopsy of the test site showed a neutrophilic dermatosis consistent with AGEP (7,8,9).

Previously, a 73-year-old woman had been described who developed AGEP on 2 occasions one and 2 days after intake of hydroxyzine. Patch tests were positive to hydroxyzine tablets 2%, 5% and 10% water (irritant reactions at D2) and to pure hydroxyzine 2.5% pet. with multiple pustules. There were no (pseudo)-cross-reactions to cetirizine or levocetirizine. During patch testing, previously involved areas showed a flare-up. Five controls were negative (10).

Between March 2009 and June 2013, in a center in France specialized in cutaneous adverse drug reactions (CADR), 156 patients were patch tested because of a CADR. Hydroxyzine was patch test positive in one patient with AGEP (19).

Systemic allergic dermatitis (systemic contact dermatitis)
A man who was occupationally sensitized to piperazine had systemic allergic dermatitis after oral administration of hydroxyzine, which is a piperazine derivative; details are unknown to the author (23).

Symmetrical drug-related intertriginous and flexural exanthema (SDRIFE)/Baboon syndrome
Two days after a 60-year-old man had started treatment with hydroxyzine for anxiety, he developed a pruritic symmetrical erythema with some small pustules on both inner thighs, cubital fossae, axillae, and the gluteal area. The drug was discontinued and the lesions resolved within 4 days. One week later the patient took again one tablet of hydroxyzine and few hours later the symptoms had reappeared. Patch tests were positive to hydroxyzine 30% water and pet., which showed some pustule formation at D4. This was a case of symmetrical drug-related intertriginous and flexural exanthema (SDRIFE) presenting under the clinical picture of the baboon syndrome (20).

Fixed drug eruption
One case of fixed drug eruption caused by hydroxyzine with a positive patch test was reported from Singapore; clinical nor patch testing details were provided (16).

Dermatitis/eczematous eruption
A 48-year-old man, who had previously suffered DRESS from carbamazepine, later had eczematous eruptions from multiple unrelated drugs. Patch tests were positive to hydroxyzine HCl, cetirizine HCl, carbamazepine, pseudoephedrine, amoxicillin and ampicillin. The patient had used all these drugs, which resulted in eczematous eruptions, with the exception of ampicillin, which was a cross-reaction to amoxicillin. This was a case of multiple drug allergy (24).

Other cutaneous adverse drug reactions

A 51-year-old woman developed well demarcated violaceous-erythematous plaques located on the trunk, arms and legs, oral mucosal lesions and fever, 12 hours after taking one tablet of hydroxyzine, which she had tolerated well for many years. The patient continued taking hydroxyzine and the lesions worsened and developed into target lesions, becoming widespread. Histopathology was compatible with exudative erythema multiforme. Patch tests were positive to hydroxyzine powdered tablet 5% pet. and negative in 5 controls (25).

'Allergy' (not further specified) to hydroxyzine appears also to have been described in ref. 14 (cited in ref. 13).

Cross-reactions, pseudo-cross-reactions and co-reactions

Hydroxyzine, cetirizine and levocetirizine are related piperazine-derived antihistamines. *In vivo*, 45% of hydroxyzine is transformed into cetirizine; levocetirizine is the active (*R*)-enantiomer of cetirizine. Therefore, co-reactions to (levo)cetirizine due to cross-sensitivity are to be expected and have been observed (2,3,4,5,6,15,21; not always proven by positive patch tests). A patient sensitized to cetirizine and hydroxyzine cross-reacted to ethylenediamine 1% pet. and piperazine (diethylenediamine) 1% pet. (15).

LITERATURE

1   The data in the section 'General' may have been obtained from literature discussed in this chapter, but mostly also or exclusively from one or more of the following online sources: ChemIDPlus Advanced, PubChem, DrugBank, RxList, Drug Central, Drugs.com, and Wikipedia
2   Andrade P, Brinca A, Gonçalo M. Patch testing in fixed drug eruptions – a 20-year review. Contact Dermatitis 2011;65:195-201
3   Cravo M, Gonçalo M, Figueiredo A. Fixed drug eruption to cetirizine with positive lesional patch tests to the three piperazine derivatives. Int J Dermatol 2007;46:760-762
4   Bhari N, Mahajan R, Singh S, Sharma VK. Fixed drug eruption due to three antihistamines of a same chemical family: Cetirizine, levocetirizine, and hydroxyzine. Dermatol Ther 2017;30. doi: 10.1111/dth.12412.
5   Assouère MN, Mazereeuw-Hautier J, Bonafe JL. Cutaneous drug eruption with two antihistaminic drugs of a same chemical family: cetirizine and hydroxyzine. Ann Dermatol Venereol 2002;129:1295-1298
6   Mahajan VK, Sharma NL, Sharma VC. Fixed drug eruption: a novel side-effect of levocetirizine. Int J Dermatol 2005;44:796-798
7   O'Toole A, Lacroix J, Pratt M, Beecker J. Acute generalized exanthematous pustulosis associated with 2 common medications: hydroxyzine and benzocaine. J Am Acad Dermatol 2014;71:e147-e149.
8   O'Toole AC, LaCroix J, Pratt M. Acute generalized exanthematous pustulosis (AGEP) caused: A case series and review of the guidelines for patch testing in cutaneous drug eruptions. Dermatitis 2017;27(5):e4
9   Lacroix J, Pratt MD. Acute generalized exanthematous pustulosis caused by Atarax ® in a sensitized patient. Dermatitis 2013;24(4):e2
10   Tsai YS, Tu ME, Wu YH, Lin YC. Hydroxyzine-induced acute generalized exanthematous pustulosis. Br J Dermatol 2007;157:1296-1297
11   Dalmau J, Serra-Baldrich E, Roé E, López-Lozano HE, Alomar A. Skin reaction to hydroxyzine (Atarax): patch test utility. Contact Dermatitis 2006;54:216-217
12   Roé E, Serra-Baldrich E, Dalmau J, Alomar A. Patch test utility in skin reaction to oral hydroxyzine. Dermatitis 2006;17:105
13   Michel M, Dompmartin A, Louvet S, Szczurko C, Castel B, Leroy D. Skin reactions to hydroxyzine. Contact Dermatitis 1997;36:147-149
14   Pascali P, Sorbette F, Senard J M, Fabre N, Rasco! 0, Montastruc J L. A propos des effets secondaires de !'hydroxyzine. Therapie 1994;49:49-56 (Article in French, data cited in ref. 13)
15   Lew BL, Haw CR, Lee MH. Cutaneous drug eruption from cetirizine and hydroxyzine. J Am Acad Dermatol 2004;50:953-956
16   Heng YK, Yew YW, Lim DS, Lim YL. An update of fixed drug eruptions in Singapore. J Eur Acad Dermatol Venereol 2015;29:1539-1544
17   Studer M, Waton J, Bursztejn AC, Aimone-Gastin I, Schmutz JL, Barbaud A. Does hypersensitivity to multiple drugs really exist? Ann Dermatol Venereol. 2012;139:375-380 (Article in French)
18   Barbaud A, Reichert-Penetrat S, Tréchot P, Jacquin-Petit MA, Ehlinger A, Noirez V, et al. The use of skin testing in the investigation of cutaneous adverse drug reactions. Br J Dermatol 1998;139:49-58
19   Assier H, Valeyrie-Allanore L, Gener G, Verlinde Carvalh M, Chosidow O, Wolkenstein P. Patch testing in non-immediate cutaneous adverse drug reactions: value of extemporaneous patch tests. Contact Dermatitis 2017;77:297-302
20   Akkari H, Belhadjali H, Youssef M, Mokni S, Zili J. Baboon syndrome induced by hydroxyzine. Indian J Dermatol 2013;58:244

21 Viñas M, Castillo MJ, Hernández N, Ibero M. Cutaneous drug eruption induced by antihistamines. Clin Exp Dermatol 2014;39:918-920

22 Tamagawa R, Katoh N, Nin M, Kishimoto S. A case of drug eruption induced by hydroxyzine pamoate. Arerugi 2006;55:34-37 (Article in Japanese)

23 Fregert S. Exacerbation of dermatitis by perorally administered piperazine derivative in a piperazine sensitized man. Contact Dermatitis Newsletter 1967;1:13

24 Özkaya E, Yazganoğlu KD. Sequential development of eczematous type "multiple drug allergy" to unrelated drugs. J Am Acad Dermatol 2011;65:e26-e29.

25 Peña AL, Henriquezsantana A, Gonzalez-Seco E, Cavanilles Bde V, Berges-Gimeno P, Alvarezcuesta E. Exudative erythema multiforme induced by hydroxyzine. Eur J Dermatol 2008;18:194-195

Chapter 3.252 IBANDRONIC ACID

IDENTIFICATION

Description/definition : Ibandronic acid is the bisphosphonate that conforms to the structural formula shown below
Pharmacological classes : Bone density conservation agents
IUPAC name : [1-Hydroxy-3-[methyl(pentyl)amino]-1-phosphonopropyl]phosphonic acid
Other names : Ibandronate
CAS registry number : 114084-78-5
EC number : Not available
Merck Index monograph : 6182
Patch testing : No reliable data; 1% pet. and water are irritant in 60% of controls and 0.1% in 25% (2)
Molecular formula : $C_9H_{23}NO_7P_2$

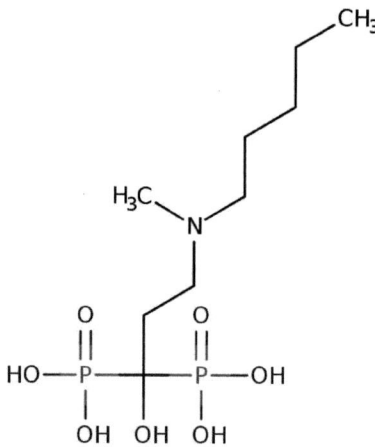

GENERAL

Ibandronic acid is a third-generation amino-bisphosphonate with anti-resorptive and anti-hypercalcemic activities. It binds to and adsorbs onto the surface of hydroxyapatite crystals in the bone matrix, thereby preventing osteoclast resorption. By doing so, ibandronic acid increases bone mineral density, decreases bone remodeling and turnover, and reduces bone pain. The drug is used to treat and prevent osteoporosis in postmenopausal women. In pharmaceutical products, ibandronic acid is employed as ibandronate sodium (monohydrate) (CAS number 138926-19-9, EC number not available, molecular formula $C_9H_{24}NNaO_8P_2$) (1).

CUTANEOUS ADVERSE DRUG REACTIONS FROM SYSTEMIC ADMINISTRATION CAUSED BY TYPE IV (DELAYED-TYPE) HYPERSENSITIVITY (as demonstrated by positive patch tests)

Cutaneous adverse drug reactions from systemic administration of ibandronic acid caused by type IV (delayed-type) hypersensitivity have included maculopapular eruption (2).

Maculopapular eruption

A 74 year-old man developed a generalized maculopapular eruption while on ibandronate. Patch tests with a pulverized ibandronate tablet were positive to 1% water (++) and 1% pet. (+); an intradermal test with ibandronate 0.1% water was +++ at D3. Of 20 controls, 12 showed a positive reaction to the patch test with ibandronate at 1% and 5 with ibandronate 0.1% in petrolatum and water. Therefore, a delayed-type hypersensitivity to ibandronic acid has not convincingly been shown in this case (2).

Cross-reactions, pseudo-cross-reactions and co-reactions

There possibly was a cross-reaction to risedronate in a patient sensitized to ibandronic acid as shown by a positive intradermal test to risedronate 1% water (++) read at D3 (2).

LITERATURE

1   The data in the section 'General' may have been obtained from literature discussed in this chapter, but mostly also or exclusively from one or more of the following online sources: ChemIDPlus Advanced, PubChem, DrugBank, RxList, Drug Central, Drugs.com, and Wikipedia
2   Barrantes-González M, Espona-Quer M, Salas E, Giménez-Arnau AM. Bisphosphonate-induced cutaneous adverse events: the difficulty of assessing imputability through patch testing. Dermatology 2014;229:163-168

Chapter 3.253 IBUPROFEN

IDENTIFICATION

Description/definition   : Ibuprofen is the aylpropionic acid derivative and nonsteroidal anti-inflammatory drug that
                                        conforms to the structural formula shown below
Pharmacological classes : Analgesics, non-narcotic; anti-inflammatory agents, non-steroidal; cyclooxygenase
                                        inhibitors
IUPAC name   : 2-[4-(2-Methylpropyl)phenyl]propanoic acid
Other names   : Benzeneacetic acid, α-methyl-4-(2-methylpropyl)-
CAS registry number   : 15687-27-1
EC number   : 239-784-6
Merck Index monograph : 6189
Patch testing   : 5% pet. (Chemotechnique, SmartPracticeCanada); 10% pet. (Chemotechnique)
Molecular formula   : $C_{13}H_{18}O_2$

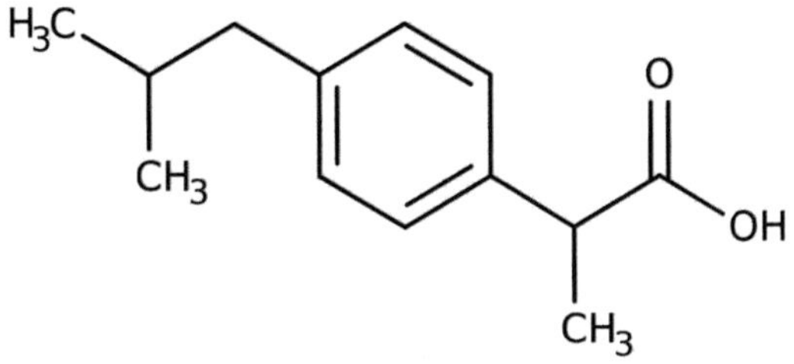

GENERAL

Ibuprofen is a propionic acid derivate and nonsteroidal anti-inflammatory drug (NSAID) with anti-inflammatory, analgesic, and antipyretic effects. Ibuprofen inhibits the activity of cyclooxygenase I and II, resulting in a decreased formation of precursors of prostaglandins and thromboxanes. This leads to decreased prostaglandin synthesis. The use of ibuprofen is common for the management of mild to moderate pain related to dysmenorrhea, headache, migraine, postoperative dental pain, spondylitis, osteoarthritis, rheumatoid arthritis, and soft tissue disorders. Ibuprofen is the most commonly used NSAID, both by prescription and in over-the-counter medication. It is also used in topical pharmaceutical preparations (1).

CUTANEOUS ADVERSE DRUG REACTIONS FROM SYSTEMIC ADMINISTRATION CAUSED BY TYPE IV (DELAYED-TYPE) HYPERSENSITIVITY (as demonstrated by positive patch tests)

Cutaneous adverse drug reactions from systemic administration of ibuprofen caused by type IV (delayed-type) hypersensitivity have included acute generalized exanthematous pustulosis (AGEP) (10), systemic allergic dermatitis (15), fixed drug eruption (3,13,14,16,17,18,19), Stevens-Johnson syndrome (12), and possibly photosensitivity (2,4,5,6).

Acute generalized exanthematous pustulosis (AGEP)

A 34-year-old man presented with a pustular rash on his trunk and groins. For a sore throat, he had taken 2 over-the-counter oral products the day before, one containing ibuprofen and pseudoephedrine, and the other paracetamol, ascorbic acid, and caffeine. Laboratory tests showed leukocytosis with neutrophilia and elevated C-reactive protein level. Swabs of the pustules were negative. Three months after complete healing, patch tests with various nonsteroidal anti-inflammatory drugs, including the tablet containing ibuprofen at 10% and ibuprofen 10% (dissolved in 85% glycerol) itself, were performed. There were positive reactions to the combination product and ibuprofen on day 2 (+), and stronger reactions with pustules (++) on D4. All other substances caused no reaction. The patient was diagnosed with AGEP from delayed-type hypersensitivity to ibuprofen (10).

Systemic allergic dermatitis (systemic contact dermatitis)

Eleven hours after taking 200 mg ibuprofen, a 21-year-old man, who had previously suffered acute allergic contact dermatitis from an ointment containing ibuprofen, developed a generalized erythematous and vesicular exanthema with some urticaria-like lesions and edema of the lips, accompanied by nausea and fever. The were no mucous membrane lesions, nor were lymphadenopathy, hepato- or splenomegaly or arthralgia present. Laboratory examinations revealed only elevated erythrocyte sedimentation rate and leukocytosis with neutrophilia. Patch tests with ibuprofen and various other NSAIDs were positive only to ibuprofen 2% pet. (D2 +, D3 ++) and 5% pet. (D2 ++, D3 ++). An intradermal test with ibuprofen 0.05 mg/ml saline was negative at 20 minutes but positive at D3. Ten controls were negative (15). This was a case of systemic allergic dermatitis from ibuprofen after previous topical sensitization, presenting as an eczematous eruption with nausea and fever.

Fixed drug eruption

A 47-year-old woman, who sporadically used various NSAIDs including ibuprofen, presented with an eruption over the trunk and thighs which had developed during the preceding 2 weeks. Physical examination showed annular erythematous plaques with vesicles and pustules, some grouped at the periphery of the lesions, with a herpetiform configuration. No laboratory values were abnormal. Histopathology was compatible with a bullous skin drug reaction; IgA deposits were not found. Three months later, patch tests were strongly positive (++) to ibuprofen 5% and 10% pet. on residual but not previously unaffected skin, with pustules at the periphery of the patch test site. Histopathology again supported a drug reaction. Since the patient stopped taking ibuprofen, no recurrences have occurred (13).

A few hours after taking allopurinol and ibuprofen, a 64-year-old man developed 3 itchy erythematous macules on his right knee, calf and flank, and an ulcer in the mouth. The lesions spontaneously healed leaving hyperpigmentation. The skin lesions recurred one day after the patient had taken ibuprofen and tetrazepam. Patch tests with allopurinol, tetrazepam and ibuprofen were positive to ibuprofen 1% (++) and 5% (+++) at D2 and D4 on residual skin but negative on the upper back. Ten controls were negative. Oral provocation with tetrazepam and allopurinol were uneventful (14).

Two days after initiating treatment with ibuprofen and erythromycin for a dental infection, a 61-year-old woman developed erythema and pain affecting her tongue and oral mucosa with a bullous lesion of the mucosa of the palatum. Two months later, she used ibuprofen and erythromycin again, which resulted in the same lesions in the oral mucosa and two new erythematous and violaceous macules of 3 cm in diameter on the thorax and thigh that evolved into blisters, healing after 2 weeks and leaving residual hyperpigmentation. Patch tests were positive to ibuprofen gel 5% after 24 hours (2x1.5 cm erythematous and violaceous lesion) on postlesional skin, and negative to erythromycin. Oral challenge with erythromycin was negative (16).

One case of fixed drug eruption to ibuprofen with a positive patch test reaction on postlesional skin (itching, erythema, infiltration) to the commercial drug 10% pet. was reported from Seoul, South Korea, in the period 1986-1990 and in 1996 and 1997 (3). Other cases of fixed drug eruption from ibuprofen with positive patch tests to this NSAID have been reported from Spain in 2001 (17), The Netherlands in 1991 (18, with a flare of all previous lesions from patch testing), and from Japan in 1990 (19, no details provided).

Stevens-Johnson syndrome/toxic epidermal necrolysis (SJS/TEN)

On the fourth day of therapy of an upper respiratory tract infection with ibuprofen, a 9-year-old girl developed SJS. Six months after complete resolution, a patch test with ibuprofen was positive (12).

Photosensitivity

In several studies, routine photopatch testing in patients with suspected photosensitivity has yielded a few positive photopatch tests to ibuprofen. However, their relevance was never specified and whether these were photosensitivity reactions from topical applications, photocross-allergy or from systemic exposure to ibuprofen is unknown (2,4,5,6,7,8).

Other data

Aseptic meningitis (without skin rash) appears to have been caused by delayed-type hypersensitivity to ibuprofen in one patient (20).

LITERATURE

1   The data in the section 'General' may have been obtained from literature discussed in this chapter, but mostly also or exclusively from one or more of the following online sources: ChemIDPlus Advanced, PubChem, DrugBank, RxList, Drug Central, Drugs.com, and Wikipedia
2   De La Cuadra-Oyanguren J, Perez-Ferriols A, Lecha-Carrelero M, et al. Results and assessment of photopatch testing in Spain: towards a new standard set of photoallergens. Actas DermoSifiliograficas 2007;98:96-101
3   Lee AY. Topical provocation in 31 cases of fixed drug eruption: change of causative drugs in 10 years. Contact Dermatitis 1998;38:258-260
4   Pigatto PD, Guzzi G, Schena D, Guarrera M, Foti C, Francalanci, S, Cristaudo A, et al. Photopatch tests: an Italian multicentre study from 2004 to 2006. Contact Dermatitis 2008;59:103-108
5   Stingeni L, Foti C, Cassano N, Bonamonte D, Vonella M, Vena GA, et al. Photocontact allergy to arylpropionic acid non-steroidal anti-inflammatory drugs in patients sensitized to fragrance mix I. Contact Dermatitis 2010;63:108-110

6   The European Multicentre Photopatch Test Study (EMCPPTS) Taskforce. A European multicentre photopatch test study. Br J Dermatol 2012;166:1002-1009

7   Pigatto PD, Mozzanica N, Bigardi AS, Legori A, Valsecchi R, Cusano F, et al. Topical NSAID allergic contact dermatitis. Italian experience. Contact Dermatitis 1993;29:39-41

8   Devleeschouwer V, Roelandts R, Garmyn M, Goossens A. Allergic and photoallergic contact dermatitis from ketoprofen: results of (photo) patch testing and follow-up of 42 patients. Contact Dermatitis 2008 ;58 :159-166

9   Valsecchi R, Cainelli T. Contact dermatitis from ibuprofen. Contact Dermatitis 1985 ;12 :286-287

10   Belz D, Persa OD, Haese S,Hunzelmann N. Acute generalized exanthematous pustulosis caused by ibuprofen—Diagnosis confirmed by patch testing. Contact Dermatitis 2018;79:40-41

11   Veronesi S, de Padova MP, Bardazzi F, Melino M. Contact dermatitis to ibuprofen. Contact Dermatitis 1986;15:103-104

12   Atanasković-Marković M, Medjo B, Gavrović-Jankulović M, Ćirković Veličković T, Nikolić D, Nestorović B. Stevens-Johnson syndrome and toxic epidermal necrolysis in children. Pediatr Allergy Immunol 2013;24:645-649

13   Tavares Almeida F, Caldas R, André Oliveira A, Pardal J, Pereira T, Brito C. Generalized bullous fixed drug eruption caused by ibuprofen. Contact Dermatitis 2019;80:238-239

14   Sánchez-Morillas L, Rojas Pérez-Ezquerra P, González Morales ML, González-Mendiola R, Laguna Martínez JJ. Fixed drug eruption due to ibuprofen with patch test positive on the residual lesion. Allergol Immunopathol (Madr) 2013;41:203-204

15   Nettis E, Giordano D, Colanardi MC, Paradiso MT, Ferrannini A, Tursi A. Delayed-type hypersensitivity rash from ibuprofen. Allergy 2003;58:539-540

16   Alvarez Santullano CV, Tover Flores V, De Barrio Fernandez M, Tornero Molina P, Prieto Garcia A. Fixed drug eruption due to ibuprofen. Allergol Immunopathol (Madr) 2006;34:280-281

17   Díaz Jara M, Pérez Montero A, Gracia Bara MT, Cabrerizo S, Zapatero L, Martínez Molero MI. Allergic reactions due to ibuprofen in children. Pediatr Dermatol 2001;18:66-67

18   Kuligowski ME, Chang A, Rath R. Multiple fixed drug eruption due to ibuprofen. Contact Dermatitis 1991;25:259-260

19   Osawa J, Naito S, Aihara M, Kitamura K, Ikezawa Z, Nakajima H. Evaluation of skin test reactions in patients with non-immediate type drug eruptions. J Dermatol 1990;17:235-239

20   Bianchi L, Marietti R, Hansel K, Tramontana M, Patruno C, Napolitano M, Stingeni L. Drug-induced aseptic meningitis to ibuprofen: The first case confirmed by positive patch test. Contact Dermatitis 2020;82:405-406

Chapter 3.254 IMATINIB

IDENTIFICATION

Description/definition : Imatinib is the tyrosine kinase inhibitor and antineoplastic agent that conforms to the structural formula shown below

Pharmacological classes : Antineoplastic agents; protein kinase inhibitors

IUPAC name : 4-[(4-Methylpiperazin-1-yl)methyl]-*N*-[4-methyl-3-[(4-pyridin-3-ylpyrimidin-2-yl)amino]-phenyl]benzamide

CAS registry number : 152459-95-5

EC number : Not available

Merck Index monograph : 6213

Patch testing : No data available; most systemic drugs can be tested at 10% pet.; if the pure chemical is not available, prepare the test material from intravenous powder, the content of capsules or – when also not available – from powdered tablets to achieve a final concentration of the active drug of 10% pet.

Molecular formula : $C_{29}H_{31}N_7O$

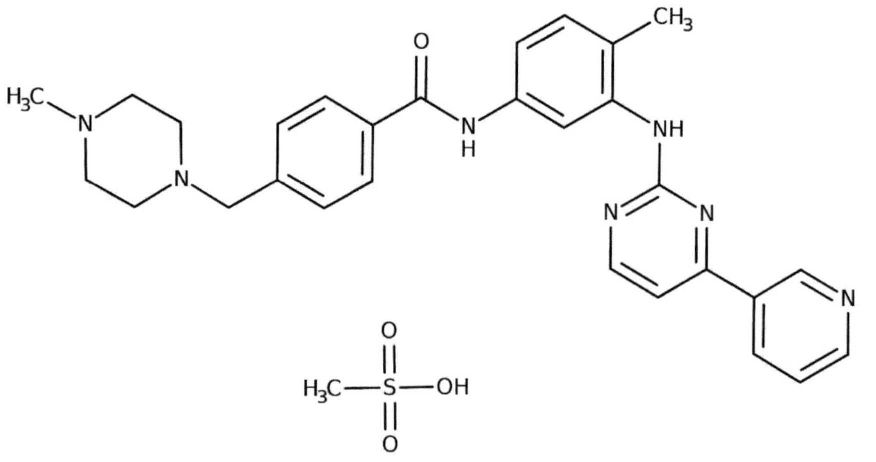

Imatinib mesylate

GENERAL

Imatinib is an antineoplastic agent and a specific inhibitor of a number of (abnormal) tyrosine kinase enzymes. It inhibits proliferation and induces apoptosis in cells that overexpress these oncoproteins. Imatinib is used for chronic myeloid leukemia and acute lymphocytic leukemia that are Philadelphia chromosome-positive (Ph+), certain types of gastrointestinal stromal tumors, hypereosinophilic syndrome, chronic eosinophilic leukemia, systemic mastocytosis, and myelodysplastic syndrome. In pharmaceutical preparations, imatinib is present as imatinib mesylate (CAS number 220127-57-1, EC number 606-892-3, molecular formula $C_{30}H_{35}N_7O_4S$) (1).

CONTACT ALLERGY FROM ACCIDENTAL CONTACT

A 54-year-old process operator working in a pharmaceutical plant developed occupational airborne allergic contact dermatitis to imatinib. The primary site of the dermatitis was the face. No clinical or patch testing details were provided (2).

LITERATURE

1   The data in the section 'General' may have been obtained from literature discussed in this chapter, but mostly also or exclusively from one or more of the following online sources: ChemIDPlus Advanced, PubChem, DrugBank, RxList, Drug Central, Drugs.com, and Wikipedia

2   Bennett MF, Lowney AC, Bourke JF. A study of occupational contact dermatitis in the pharmaceutical industry. Br J Dermatol 2016;174:654-656 (Abstract in Brit J Dermatol 2011;165:73)

Chapter 3.255 IMIPENEM

IDENTIFICATION

Description/definition   : Imipenem is the semisynthetic β-lactam carbapenem derived from thienamycin, that
                            conforms to the structural formula shown below
Pharmacological classes   : Anti-bacterial agents
IUPAC name   : (5R,6S)-3-[2-(Aminomethylideneamino)ethylsulfanyl]-6-[(1R)-1-hydroxyethyl]-7-oxo-1-
                            azabicyclo[3.2.0]hept-2-ene-2-carboxylic acid;hydrate
Other names   : Imipenem hydrate; imipenem monohydrate
CAS registry number   : 74431-23-5
EC number   : Not available
Merck Index monograph   : 6231
Patch testing   : Beta-lactam antibiotics can generally be tested 5%-10% pet.
Molecular formula   : $C_{12}H_{19}N_3O_5S$

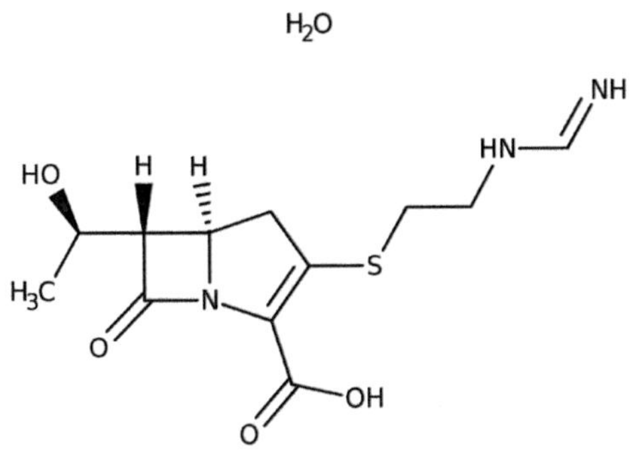

GENERAL

Imipenem is a broad-spectrum, semisynthetic β-lactam carbapenem derived from thienamycin, produced by *Streptomyces cattleya*. It has a wide spectrum of antibacterial activity against gram-negative and gram-positive aerobic and anaerobic bacteria, including many multi-resistant strains and is stable to β-lactamases. Its effectiveness in the treatment of severe infections of various body systems is enhanced when it is administered in combination with the renal dipeptidase inhibitor cilastatin, which increases half-life and tissue penetration of imipenem (1). The classification and structures of beta-lactam antibiotics are discussed in Chapter 3.36 Amoxicillin.

CONTACT ALLERGY FROM ACCIDENTAL CONTACT

For a case of occupational allergic contact dermatitis from cilastatin, mixture with imipenem, most likely caused by its ingredient imipenem, see Chapter 3.130 Cilastatin mixture with imipenem.

CUTANEOUS ADVERSE DRUG REACTIONS FROM SYSTEMIC ADMINISTRATION CAUSED BY TYPE IV (DELAYED-TYPE) HYPERSENSITIVITY (as demonstrated by positive patch tests)

Cutaneous adverse drug reactions from systemic administration of imipenem caused by type IV (delayed-type) hypersensitivity have been observed in the combination with cilastatin only and are presented there (Chapter 3.130 Cilastatin mixture with imipenem).

Cross-reactions, pseudo-cross-reactions and co-reactions

Cross-reactions between beta-lactam antibiotics are discussed in Chapter 3.36 Amoxicillin. Cross-reactions between carbapenems (imipenem, ertapenem, meropenem) and between carbapenems and penicillins may rarely occur, but the pattern is unclear (2,3).

LITERATURE

1   The data in the section 'General' may have been obtained from literature discussed in this chapter, but mostly also or exclusively from one or more of the following online sources: ChemIDPlus Advanced, PubChem, DrugBank, RxList, Drug Central, Drugs.com, and Wikipedia
2   Colagiovanni A, Feliciani C, Fania L, Pascolini L, Buonomo A, Nucera E, et al. Occupational contact dermatitis from carbapenems. Cutis 2015;96:E1-3
3   Schiavino D, Nucera E, Lombardo C, Decinti M, Pascolini L, Altomonte G, et al. Cross-reactivity and tolerability of imipenem in patients with delayed-type, cell-mediated hypersensitivity to beta-lactams. Allergy 2009;64:1644-1648

Chapter 3.256 INDELOXAZINE

IDENTIFICATION

Description/definition   : Indeloxazine is the morpholine derivative that conforms to the structural formula shown
                             below
Pharmacological classes   : Anticonvulsants; antidepressive agents
IUPAC name   : 2-(3H-Inden-4-yloxymethyl)morpholine
CAS registry number   : 60929-23-9
EC number   : Not available
Merck Index monograph   : 1196
Patch testing   : 10% and 20% pet., probably from pulverized tablets (2); most systemic drugs can be
                             tested at 10% pet.; if the pure chemical is not available, prepare the test material from
                             intravenous powder, the content of capsules or – when also not available – from
                             powdered tablets to achieve a final concentration of the active drug of 10% pet.
Molecular formula   : $C_{14}H_{17}NO_2$

GENERAL

Indeloxazine is an antidepressant and cerebral activator that was marketed in Japan and South Korea for the treatment of psychiatric symptoms associated with cerebrovascular diseases, such as depression resulting from stroke, emotional disturbance, and avolition (total lack of motivation to do anything). It was available from 1988 to 1998, when it was removed from the market reportedly for lack of effectiveness (1)

In pharmaceutical products, indeloxazine is (was) employed as indeloxazine hydrochloride (CAS number 65043-22-3, EC number not available, molecular formula $C_{14}H_{18}ClNO_2$) (1).

CUTANEOUS ADVERSE DRUG REACTIONS FROM SYSTEMIC ADMINISTRATION CAUSED BY TYPE IV (DELAYED-TYPE) HYPERSENSITIVITY (as demonstrated by positive patch tests)

Cutaneous adverse drug reactions from systemic administration of indeloxazine caused by type IV (delayed-type) hypersensitivity have included erythroderma and eosinophilic pustular folliculitis (Ofuji's disease) (2).

Other cutaneous adverse drug reactions

A 73-year-old Japanese man, while using indeloxazine hydrochloride and nicardipine hydrochloride since 11 months after a stroke, developed pruritic erythema over the entire body with fever. The drugs were stopped, oral steroids given and the erythroderma gradually improved over months leaving residual hyperpigmentation. Two months later, laboratory studies still disclosed leukocytosis, eosinophilia and an elevated erythrocyte sedimentation rate. A drug lymphocyte stimulation test (DLST) with indeloxazine hydrochloride was positive (stimulation index; 643%). Patch testing with indeloxazine hydrochloride showed 2+ positive reactions at concentrations of 20% and 10% pet. and a 1+ reaction at 1% and 0.1% pet. On the 6th day after the patch test, pruritic, egg-size erythema with vesicles and pustules appeared on his cheek. The total eosinophil count in the peripheral blood increased in parallel with the skin manifestations. Histologic examination of the erythema revealed vesicles in the outer root sheath of hair follicles containing many eosinophils and lymphocytes.

The clinical picture and histopathology were compatible with eosinophilic pustular folliculitis (Ofuji's disease), which is characterized by the repeated occurrence of crops of pruritic follicular papulopustules with a tendency to form an annular configuration on the face and other seborrheic areas. Next, a challenge test was performed with increasing doses of indeloxazine hydrochloride. Erythema with many pustules appeared on his forehead on the 3rd day and spread over the face, trunk, arms and legs on the 6th day. Moreover, erythema flared up on the site of the patch test. Histopathology of the erythema on the forehead again revealed folliculitis with many eosinophils. DLST, patch and challenge tests with nicardipine were negative. The patient was diagnosed with erythroderma from indeloxazine hydrochloride and (atypical) eosinophilic pustular folliculitis (Ofuji's disease) from patch tests and oral challenge with this drug (2).

LITERATURE

1   The data in the section 'General' may have been obtained from literature discussed in this chapter, but mostly also or exclusively from one or more of the following online sources: ChemIDPlus Advanced, PubChem, DrugBank, RxList, Drug Central, Drugs.com, and Wikipedia

2   Kimura K, Ezoe K, Yokozeki H, Katayama I, Nishioka K. A case of eosinophilic pustular folliculitis (Ofuji's disease) induced by patch and challenge tests with indeloxazine hydrochloride. J Dermatol 1996;23:479-483

Chapter 3.257 INDOMETHACIN

IDENTIFICATION

Description/definition	: Indomethacin is the benzoylindole that conforms to the structural formula shown below
Pharmacological classes	: Gout suppressants; tocolytic agents; anti-inflammatory agents, non-steroidal; cyclooxygenase inhibitors; cardiovascular agents
IUPAC name	: 2-[1-(4-Chlorobenzoyl)-5-methoxy-2-methylindol-3-yl]acetic acid
Other names	: Indometacin
CAS registry number	: 53-86-1
EC number	: 200-186-5
Merck Index monograph	: 6279
Patch testing	: 10% pet. (3)
Molecular formula	: $C_{19}H_{16}ClNO_4$

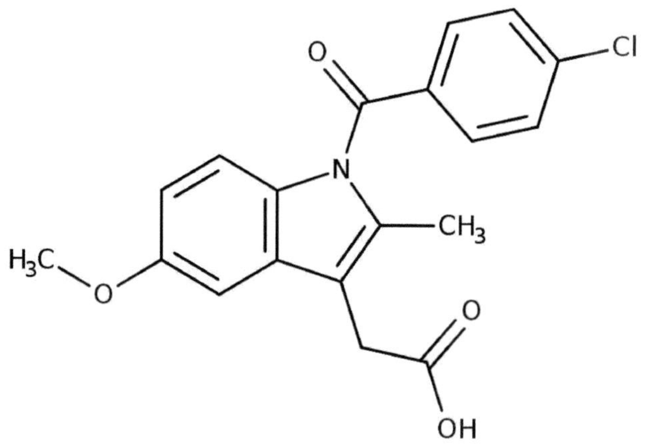

GENERAL

Indomethacin is a synthetic nonsteroidal indole derivative with anti-inflammatory activity and chemopreventive properties. It is most commonly used in rheumatoid arthritis, ankylosing spondylitis, osteoarthritis, acute shoulder pains, and acute gouty arthritis, but also for symptomatic treatment of migraine. Indomethacin also may inhibit the expression of multidrug-resistant protein type 1, resulting in increased efficacies of some antineoplastic agents in treating multi-drug resistant tumors (1). In pharmaceutical products, indomethacin is employed as indomethacin sodium (trihydrate) (CAS number 74252-25-8, EC number 200-186-5, molecular formula $C_{19}H_{21}ClNNaO_7$) (1).

CUTANEOUS ADVERSE DRUG REACTIONS FROM SYSTEMIC ADMINISTRATION CAUSED BY TYPE IV (DELAYED-TYPE) HYPERSENSITIVITY (as demonstrated by positive patch tests)

Cutaneous adverse drug reactions from systemic administration of indomethacin caused by type IV (delayed-type) hypersensitivity have included photosensitivity (2).

Photosensitivity

In the period December 1980 through July 1983, the Adverse Drug Reaction Reporting System (ADRRS), a system administered by the American Academy of Dermatology for the spontaneous reporting of suspected adverse reactions to drugs, received 135 reports of adverse cutaneous reactions to NSAIDs. One case of indomethacin-related photosensitivity, which was established by rechallenge and photopatch testing, was reported. Clinical nor (photo)patch testing details were provided (2).

LITERATURE

1   The data in the section 'General' may have been obtained from literature discussed in this chapter, but mostly also or exclusively from one or more of the following online sources: ChemIDPlus Advanced, PubChem, DrugBank, RxList, Drug Central, Drugs.com, and Wikipedia
2   Stern RS, Bigby M. An expanded profile of cutaneous reactions to nonsteroidal anti-inflammatory drugs. Reports to a specialty-based system for spontaneous reporting of adverse reactions to drugs. JAMA 1984;252:1433-1437
3   Brockow K, Garvey LH, Aberer W, Atanaskovic-Markovic M, Barbaud A, Bilo MB, et al.; ENDA/EAACI Drug Allergy Interest Group. Skin test concentrations for systemically administered drugs – an ENDA/EAACI Drug Allergy Interest Group position paper. Allergy 2013;68:702-712

Chapter 3.258 INTERFERONS

IDENTIFICATION	**Peginterferon alfa-2a**
Description/definition	: Peginterferon alfa-2a is interferon alfa-2a chemically modified by the covalent attachment of a polyethylene glycol
Pharmacological classes	: Antiviral agents
IUPAC name	: Not available
Other names	: Pegasys ®; pegylated interferon alfa-2a; polyethylene glycol-interferon alfa-2a; PEG-IFN alfa-2a
CAS registry number	: 198153-51-4
EC number	: Not available
Merck Index monograph	: 8445
Patch testing	: Commercial preparation 30% pet. (2)
Molecular formula	: Unspecified

IDENTIFICATION	**Interferon alfa-2c**
Description/definition	: Interferon alfa-2c is subtype of interferon alpha with arginine at positions 23 and 34
Pharmacological classes	: Antineoplastic agents; antiviral agents
IUPAC name	: Not available
CAS registry number	: 135669-44-2; 142192-09-4
EC number	: 232-710-3 (Interferons)
Merck Index monograph	: 6304 (Interferon-α)
Patch testing	: Commercial preparation 'as is' (3)
Molecular formula	: Unspecified

GENERAL

Interferons (IFN) are glycoproteins made up of three main types: IFN-α (subtypes: 2a, 2b, pegylated or not, con1), whose main indications are chronic hepatitis C and melanoma; IFN-β (subtypes: 1a, 1b), prescribed for multiple sclerosis; and IFN-γ, used as prophylaxis in familial granulomatosis. About 5-12% of side-effects related to IFN treatment involve adverse skin reactions, either localized at the injection site or generalized skin reactions (1,2).

CUTANEOUS ADVERSE DRUG REACTIONS FROM SYSTEMIC ADMINISTRATION CAUSED BY TYPE IV (DELAYED-TYPE) HYPERSENSITIVITY (as demonstrated by positive patch tests)

Cutaneous adverse drug reactions from systemic administration of interferons caused by type IV (delayed-type) hypersensitivity have included local allergic reaction at the injection site (3), generalized eczematous eruption (2) and urticaria (2).

Case series with various or unknown types of drug reactions

In Nancy, France, in the period 1998-2012, 21 patients were investigated who had experienced generalized eruptions while receiving IFN treatment. Sixteen were patch test with the interferons used in their commercialized forms diluted to 30% in petrolatum. There were only 2 positive patch test reactions. One was in a man aged 51 who developed a generalized eczematous eruption while being treated with peg-IFN-α2a (Pegasys ®) (peg = pegylated, covalent attachment of a polyethylene glycol) for hepatitis C virus infection. Patch tests were positive to peg-IFN-α2a and to IFN-α2a (Roferon ®) at D7. The other was a man aged 37 who developed urticaria while being treated with peg-IFN-α2a for hepatitis C virus infection. Patch tests were positive to peg-IFN-α2a (D4), to IFN-α2a (D2) and to peg-INF-α2b at D4 (2).

Dermatitis/eczematous eruption

A case of generalized eczematous eruption from interferon has been described in the section 'Case series with various or unknown types of drug reactions' above, ref. 2.

Other cutaneous adverse drug reactions

Urticaria

A case of urticaria from interferon has been described in the section 'Case series with various or unknown types of drug reactions' above, ref. 2.

Localized allergic reaction

A 55-year-old man with hairy cell leukemia was treated with subcutaneous injections of recombinant interferon alfa-2c in a clinical study. At one point during treatment, he began to notice itching erythema at the injection site developing about 2 days after each injection. Patch tests were positive to the commercial interferon alfa-2c used, tested undiluted. A group of 10 patients, with no previous interferon treatment, showed no positive reactions when tested as controls (3).

LITERATURE

1   The data in the section 'General' may have been obtained from literature discussed in this chapter, but mostly also or exclusively from one or more of the following online sources: ChemIDPlus Advanced, PubChem, DrugBank, RxList, Drug Central, Drugs.com, and Wikipedia

2   Poreaux C, Bronowicki JP, Debouverie M, Schmutz JL, Waton J, Barbaud A. Clinical allergy. Managing generalized interferon-induced eruptions and the effectiveness of desensitization. Clin Exp Allergy 2014;44:756-764

3   Detmar U, Agathos M, Nerl C. Allergy of delayed type to recombinant interferon alpha 2c. Contact Dermatitis 1989;20:149-150

Chapter 3.259 IOBITRIDOL

IDENTIFICATION

Description/definition   : Iobitridol is the water-soluble, tri-iodinated, non-ionic monomeric benzoate derivative that conforms to the structural formula shown below
Pharmacological classes : Contrast media
IUPAC name   : 1-N,3-N-bis(2,3-Dihydroxypropyl)-5-[[3-hydroxy-2-(hydroxymethyl)propanoyl]amino]-2,4,6-triiodo-1-N,3-N-dimethylbenzene-1,3-dicarboxamide
CAS registry number   : 136949-58-1
EC number   : Not available
Merck Index monograph : 6322
Patch testing   : Iobitridol-containing contrast medium undiluted (6)
Molecular formula   : $C_{20}H_{28}I_3N_3O_9$

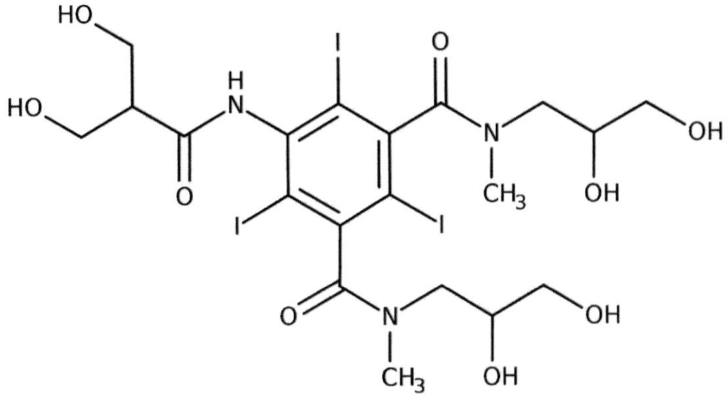

GENERAL

Iobitridol is a water-soluble, tri-iodinated, non-ionic monomeric benzenedicarboxamide compound used as contrast medium in diagnostic radiography. Upon administration, iobitridol is distributed through the vascular system and interstitial space. Like other organic iodine compounds, this agent blocks x-rays and appears opaque on x-ray film, thus enhancing the visibility of body parts containing this agent (1).

CUTANEOUS ADVERSE DRUG REACTIONS FROM SYSTEMIC ADMINISTRATION CAUSED BY TYPE IV (DELAYED-TYPE) HYPERSENSITIVITY (as demonstrated by positive patch tests)

Cutaneous adverse drug reactions from systemic administration of iobitridol caused by type IV (delayed-type) hypersensitivity have included maculopapular eruption (3,5), acute generalized exanthematous pustulosis (AGEP) (2), and drug reaction with eosinophilia and systemic symptoms (DRESS) (3,4,11).

Maculopapular eruption

One patient developed a maculopapular eruption from iobitridol with a positive patch test to this drug after a previous episode of DRESS from other drug(s) (multiple drug hypersensitivity) (3). Twelve patients (9 women) with late cutaneous adverse reactions to iodinated contrast media (ICM) were reported from France and Switzerland in 2005. Iobitridol caused a maculopapular eruption in one patient who showed a positive patch test to the undiluted iobitridol ICM (5).

Acute generalized exanthematous pustulosis (AGEP)

A 79-year-old woman received iobitridol for a fistulography. Her main medical history included osteoarthritis, primary amyloidosis, and tramadol-induced urticaria. Two days later, she developed acute generalized exanthematous pustulosis (AGEP), confirmed by skin biopsy. A patch test with the iobitridol contrast medium 'as is' and an intradermal test, read at D2, were positive for iobitridol (in the table, the patch test was mentioned as 'doubtful'). Intravenous administration of iodixanol was tolerated without untoward effects (2).

Drug reaction with eosinophilia and systemic symptoms (DRESS)

In a group of 14 patients with multiple delayed-type hypersensitivity reactions, DRESS was caused by iobitridol in one case, showing a positive patch test reaction (3, overlap with ref. 4). Two patients who had suffered an episode of

DRESS, subsequently had a drug-related relapse caused by iobitridol, showing a positive patch test to this drug. This was a case of multiple hypersensitivity syndrome (4, overlap with ref. 3).

In France, 13 adult patients in the dermatologist's French Investigators for Skin Adverse Reactions to Drugs (FISARD) network diagnosed between 2010 and 2020 with DRESS highly suspected to have been caused by a iodinated contrast medium (ICM) according to either a positive skin test or a challenge test, were retrospectively evaluated. In the group of 13 DRESS patients, 2 drug reactions had been caused by iobitridol, both of whom had positive patch tests to this ICM (11).

Cross-reactions, pseudo-cross-reactions and co-reactions

Cross-reactions between iodinated radiocontrast media occur frequently, especially between the non-ionic preparations (5,7,11,12). Iodine is infrequently to rarely (opinions differ) the allergen (8,9) and allergy to mollusks, crustaceans, fish, and iodine from other sources is said not to be a risk factor for the development of hypersensitivity to iodinated contrast media (10).

LITERATURE

1   The data in the section 'General' may have been obtained from literature discussed in this chapter, but mostly also or exclusively from one or more of the following online sources: ChemIDPlus Advanced, PubChem, DrugBank, RxList, Drug Central, Drugs.com, and Wikipedia

2   Grandvuillemin A, Ripert C, Sgro C, Collet E. Iodinated contrast media-induced acute generalized exanthematous pustulosis confirmed by delayed skin tests. J Allergy Clin Immunol Pract 2014;2:805-806

3   Jörg L, Yerly D, Helbling A, Pichler W. The role of drug, dose and the tolerance/intolerance of new drugs in multiple drug hypersensitivity syndrome (MDH). Allergy 2020;75:1178-1187

4   Jörg L, Helbling A, Yerly D, Pichler WJ. Drug-related relapses in drug reaction with eosinophilia and systemic symptoms (DRESS). Clin Transl Allergy 2020;10:52

5   Kanny G, Pichler W, Morisset M, Franck P, Marie B, Kohler C, Renaudin JM, Beaudouin E, Laudy JS, Moneret-Vautrin DA. T cell-mediated reactions to iodinated contrast media: evaluation by skin and lymphocyte activation tests. J Allergy Clin Immunol 2005;115:179-185

6   Brockow K, Garvey LH, Aberer W, Atanaskovic-Markovic M, Barbaud A, Bilo MB, et al. Skin test concentrations for systemically administered drugs - an ENDA/EAACI Drug Allergy Interest Group position paper. Allergy 2013;68:702-712

7   Lerondeau B, Trechot P, Waton J, Poreaux C, Luc A, Schmutz JL, et al. Analysis of cross-reactivity among radiocontrast media in 97 hypersensitivity reactions. J Allergy Clin Immunol 2016;137:633-635.e4

8   Scherer K, Harr T, Bach A, Bircher AJ. The role of iodine in hypersensitivity reactions to radiocontrast media. Clin Exp Allergy 2009;40:468-475

9   Trautmann A, Brockow K, Behle V, Stoevesandt J. Radiocontrast media hypersensitivity: Skin testing differentiates allergy from nonallergic reactions and identifies a safe alternative as proven by intravenous provocation. J Allergy Clin Immunol Pract 2019;7:2218-2224

10 Rosado Ingelmo A, Doña Diaz I, Cabañas Moreno R, Moya Quesada MC, García-Avilés C, García Nuñez I, et al. Clinical practice guidelines for diagnosis and management of hypersensitivity reactions to contrast media. J Investig Allergol Clin Immunol 2016;26:144-155

11 Soria A, Amsler E, Bernier C, Milpied B, Tétart F, Morice C, et al. DRESS and AGEP reactions to iodinated contrast media: A French case series. J Allergy Clin Immunol Pract 2021;9:3041-3050

12 Yoon SH, Lee SY, Kang HR, Kim JY, Hahn S, Park CM, et al. Skin tests in patients with hypersensitivity reaction to iodinated contrast media: a meta-analysis. Allergy 2015;70:625-637

Chapter 3.260 IODIXANOL

IDENTIFICATION

Description/definition	: Iodixanol is the dimeric, iso-osmolar, non-ionic, hydrophilic iodinated radiocontrast agent that conforms to the structural formula shown below
Pharmacological classes	: Contrast media
IUPAC name	: 5-[Acetyl-[3-[N-acetyl-3,5-bis(2,3-dihydroxypropylcarbamoyl)-2,4,6-triiodoanilino]-2-hydroxypropyl]amino]-1-N,3-N-bis(2,3-dihydroxypropyl)-2,4,6-triiodobenzene-1,3-dicarboxamide
CAS registry number	: 92339-11-2
EC number	: 618-837-0
Merck Index monograph	: 6336
Patch testing	: Commercial contrast medium 'as is' (15) and iodixanol 5% water
Molecular formula	: $C_{35}H_{44}I_6N_6O_{15}$

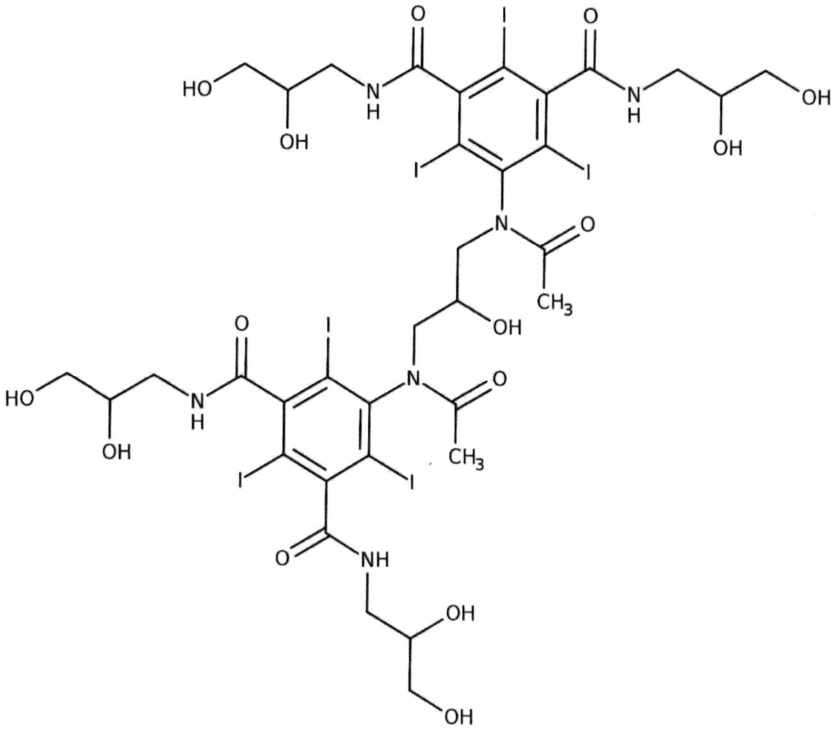

GENERAL

Iodixanol is a dimeric iso-osmolar, non-ionic, dimeric, hydrophilic iodinated radiocontrast agent used in diagnostic imaging. Upon intravascular administration and during computed tomography imaging, iodixanol blocks x-rays and appears opaque on x-ray images. This allows body structures that absorb iodine to be visualized. Iodixanol as contrast agent is commonly used during coronary angiography, particularly in individuals with renal dysfunction, as it is believed to be less toxic to the kidneys than most other intravascular contrast agents (1).

CUTANEOUS ADVERSE DRUG REACTIONS FROM SYSTEMIC ADMINISTRATION CAUSED BY TYPE IV (DELAYED-TYPE) HYPERSENSITIVITY (as demonstrated by positive patch tests)

Cutaneous adverse drug reactions from systemic administration of iodixanol caused by type IV (delayed-type) hypersensitivity have included maculopapular eruption (4,5,10,16), generalized erythema with fever (17), AGEP (2,3,6,18), drug reaction with eosinophilia and systemic symptoms (DRESS) (6,9), and localized allergic reaction from subcutaneous administration (14).

Maculopapular eruption

In France, 22 patients who had developed a nonimmediate cutaneous adverse drug reaction (20 maculopapular exanthema, 2 edema of the face) after injection of iodixanol contrast medium, were patch tested with the commercial preparation 'as is' and there were 3 positive reactions. In 2, the patients had also positive delayed intradermal tests. 8 other patients had positive intradermal but negative patch tests. Twenty controls were patch test negative to iodixanol contrast medium (10).

In a hospital in Madrid, Spain, in the period 1999 to 2005, 12 patients had a late skin reaction (more than 1 hr but less than 7 days following administration) to the iodinated contrast medium (ICM) iodixanol. There were 7 men and 5 women, ages ranging from 39 to 76 years (mean 56). All patients developed a maculopapular exanthema between 2 hr and 3 days after the radiological examination, involving the trunk and proximal limbs, although some showed involvement of distal areas. A skin biopsy, performed in 6 patients, showed non-specific findings consistent with drug reaction. All patients were patch tested with the ICM 'as is' and – when positive – with its ingredients (iodixanol at 0.1%, 1%, and 5% aqueous, trometamol 0.1%, 0.5%, and 1% alcohol, and iodine 0.1% and 0.5% alcohol), but only 3 had positive reactions to the ICM and iodixanol at all concentrations. 35 Controls we negative to iodixanol 5% water. It was concluded that patch tests have a limited value, and in one case where they were negative, reintroduction of the drug triggered a new exanthema (4). One of these patch test positive patients had previously been reported (5).

From France, 5 patients with maculopapular eruptions from iodinated contrast media were reported. Patch tests showed positive reactions to iodixanol in 5, iohexol in 5, iopromide in 3 and iomeprol in 1 patient. It was not mentioned which of these contrast media had actually caused the eruptions (16).

Erythroderma, widespread erythematous eruption, exfoliative dermatitis
A 64-year-old man with an acute coronary syndrome underwent a percutaneous coronary intervention with administration of iodixanol. Four days later, a second percutaneous coronary intervention was performed. Forty-eight hours after the second procedure, the patient had a severe, generalized, itchy erythematous rash and fever. A routine blood test showed leukocytosis and eosinophilia. Liver and renal functions were normal. Twelve weeks later, the patient showed a positive patch test to the iodixanol contrast medium, tested undiluted (17).

Acute generalized exanthematous pustulosis (AGEP)
In France, 19 adult patients in the dermatologist's French Investigators for Skin Adverse Reactions to Drugs (FISARD) network diagnosed between 2010 and 2020 with AGEP highly suspected to have been caused by a iodinated contrast medium (ICM) according to either a positive skin test or a challenge test were retrospectively evaluated. In this group of 19 patients with AGEP, 5 drug reactions had been caused by iodixanol, 2 of whom had positive patch tests to this ICM. The other 3, who were patch test negative, had positive delayed intradermal tests (18).

A 53-year-old woman had an intra-articular injection of the iodinated contrast medium (ICM) iodixanol for computed tomography arthrography of the knee. Within less than 12 hours, she developed a pruritic skin eruption with pustular lesions covering her body. The biological and histopathological findings were consistent with our clinical diagnosis of acute generalized exanthematous pustulosis (AGEP). She was patch tested 2 months later and had a positive patch test to the ICM iodixanol 'as is' at D4. The patient never returned for prick tests and intradermal tests (2).

A 45-year-old woman received iomeprol for a cerebral computed tomography (CT) and iodixanol for a cerebral arterial embolization the same day. Three days later, she presented with AGEP confirmed by skin biopsy (subcorneal pustulosis with neutrophils infiltration). A patch test with the iodixanol contrast medium 'as is' and an intradermal test, read at D2, were positive, all other drugs negative. Intravenous administration of iomeprol was tolerated without untoward effects (3). The same authors also report on a 24-year-old man, who received iomeprol for a cerebral CT and iodixanol for an arteriography the same day in the context of an aneurysm rupture. He presented with AGEP 24 hours later, confirmed by skin biopsy. A patch test with the iodixanol contrast medium 'as is' and an intradermal test, read at D2, were positive. Intravenous administration of iobitridol was tolerated without untoward effects (3).

A 68-year-old man had developed AGEP and showed a positive patch test to the contrast medium iodixanol; clinical details were not provided (6).

Drug reaction with eosinophilia and systemic symptoms (DRESS)
A patient developed DRESS after injections of ioxitalamate and ioversol. Patch tests were positive to these two iodinated contrast media and to various others, including iodixanol. Later, an investigation using iodixanol resulted in a second episode of DRESS with a skin eruption covering 80% of the skin surface and eosinophilia, despite pretreatment with corticosteroids (9).

In a multicenter investigation in France, of 72 patients patch tested for DRESS, 46 (64%) had positive patch tests to drugs, including one to iodixanol; clinical details were not provided (6).

Other cutaneous adverse drug reactions
A 37-year-old woman developed deep edema and erythema (not further specified) 6 hours after administration of iodixanol. A patch test with iodixanol was negative, but a delayed intradermal test positive; also, there were positive patch tests to 4 other iodinated contrast media (14).

Cross-reactions, pseudo-cross-reactions and co-reactions
Cross-reactions between iodinated radiocontrast media occur frequently, especially between the non-ionic preparations (7,8,18,19). Iodine is infrequently to rarely (opinions differ) the allergen (11,12) and allergy to mollusks, crustaceans, fish, and iodine from other sources is said not to be a risk factor for the development of hypersensitivity to iodinated contrast media (13).

LITERATURE

1   The data in the section 'General' may have been obtained from literature discussed in this chapter, but mostly also or exclusively from one or more of the following online sources: ChemIDPlus Advanced, PubChem, DrugBank, RxList, Drug Central, Drugs.com, and Wikipedia

2   Velter C, Schissler C, Moulinas C, Tebacher-Alt M, Siedel JM, Cribier B, et al. Acute generalized exanthematous pustulosis caused by an iodinated contrast radiocontrast medium for computed tomography arthrography of the knee. Contact Dermatitis 2017;76:371-373

3   Grandvuillemin A, Ripert C, Sgro C, Collet E. Iodinated contrast media-induced acute generalized exanthematous pustulosis confirmed by delayed skin tests. J Allergy Clin Immunol Pract 2014;2:805-806

4   Delgado-Jimenez Y, Perez-Gala S, Aragüés M, Sanchez-Perez J, Garcia-Diez A. Late skin reaction to iodixanol (Visipaque): clinical manifestations, patch test study, and histopathological evaluation. Contact Dermatitis 2006;55:348-353

5   Sánchez-Pérez J, F-Villalta MG, Ruíz SA, García Diez A. Delayed hypersensitivity reaction to the non-ionic X-ray contrast medium Visipaque (iodixanol). Contact Dermatitis 2003;48:167

6   Barbaud A, Collet E, Milpied B, Assier H, Staumont D, Avenel-Audran M, et al. A multicentre study to determine the value and safety of drug patch tests for the three main classes of severe cutaneous adverse drug reactions. Br J Dermatol 2013;168:555-562

7   Kanny G, Pichler W, Morisset M, Franck P, Marie B, Kohler C, et al. T cell-mediated reactions to iodinated contrast media: evaluation by skin and lymphocyte activation tests. J Allergy Clin Immunol 2005;115:179-185

8   Lerondeau B, Trechot P, Waton J, Poreaux C, Luc A, Schmutz JL, et al. Analysis of cross-reactivity among radiocontrast media in 97 hypersensitivity reactions. J Allergy Clin Immunol 2016;137:633-635.e4

9   Amsler E, Autegarden JE, Senet P, Frances C, Soria A. Recurrence of drug eruption after renewed injection of iodinated contrast medium in patients with known allergic contraindications. Ann Dermatol Venereol 2016;143:804-807 (Article in French).

10   Hasdenteufel F, Waton J, Cordebar V, et al. Delayed hypersensitivity reactions caused by iodixanol: an assessment of cross-reactivity in 22 patients. J Allergy Clin Immunol 2011;128:1356-1357

11   Scherer K, Harr T, Bach A, Bircher AJ. The role of iodine in hypersensitivity reactions to radiocontrast media. Clin Exp Allergy 2009;40:468-475

12   Trautmann A, Brockow K, Behle V, Stoevesandt J. Radiocontrast media hypersensitivity: Skin testing differentiates allergy from nonallergic reactions and identifies a safe alternative as proven by intravenous provocation. J Allergy Clin Immunol Pract 2019;7:2218-2224

13   Rosado Ingelmo A, Doña Diaz I, Cabañas Moreno R, Moya Quesada MC, García-Avilés C, García Nuñez I, et al. Clinical practice guidelines for diagnosis and management of hypersensitivity reactions to contrast media. J Investig Allergol Clin Immunol 2016;26:144-155

14   Vernassière C, Tréchot P, Commun N, Schmutz JL, Barbaud A. Low negative predictive value of skin tests in investigating delayed reactions to radio-contrast media. Contact Dermatitis 2004;50:359-366

15   Brockow K, Garvey LH, Aberer W, Atanaskovic-Markovic M, Barbaud A, Bilo MB, et al.; ENDA/EAACI Drug Allergy Interest Group. Skin test concentrations for systemically administered drugs – an ENDA/EAACI Drug Allergy Interest Group position paper. Allergy 2013;68:702-712

16   Corbaux C, Seneschal J, Taïeb A, Cornelis F, Martinet J, Milpied B. Delayed cutaneous hypersensitivity reactions to iodinated contrast media. Eur J Dermatol 2017;27:190-191

17   Gracia Bara MT, Moreno E, Laffond E, Muñoz FJ, Macias E, Davila I. Selection of contrast media in patients with delayed reactions should be based on challenge test results. J Allergy Clin Immunol 2012;130:554-555

18   Soria A, Amsler E, Bernier C, Milpied B, Tétart F, Morice C, et al. DRESS and AGEP reactions to iodinated contrast media: A French case series. J Allergy Clin Immunol Pract 2021;9:3041-3050

19   Yoon SH, Lee SY, Kang HR, Kim JY, Hahn S, Park CM, et al. Skin tests in patients with hypersensitivity reaction to iodinated contrast media: a meta-analysis. Allergy 2015;70:625-637

Chapter 3.261 IODOQUINOL

IDENTIFICATION

Description/definition	: Iodoquinol is the quinoline derivative that conforms to the structural formula shown below
Pharmacological classes	: Amebicides
IUPAC name	: 5,7-Diiodoquinolin-8-ol
Other names	: Diiodohydroxyquinoline; diiodoquin
CAS registry number	: 83-73-8
EC number	: 201-497-9
Merck Index monograph	: 6355
Patch testing	: 5% pet.
Molecular formula	: $C_9H_5I_2NO$

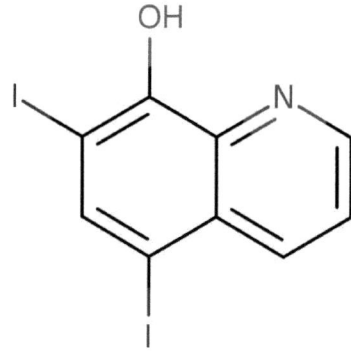

GENERAL

Iodoquinol is one of the halogenated 8-quinolinols, which is widely used as an intestinal antiseptic, especially as an anti-amebic agent. It is also used topically in other infections, often in combination with hydrocortisone acetate (1). In topical preparations, iodoquinol has caused cases of contact allergy/allergic contact dermatitis, which subject has been fully reviewed in Volume 3 of the *Monographs in contact allergy* series (2).

CUTANEOUS ADVERSE DRUG REACTIONS FROM SYSTEMIC ADMINISTRATION CAUSED BY TYPE IV (DELAYED-TYPE) HYPERSENSITIVITY (as demonstrated by positive patch tests)

Cutaneous adverse drug reactions from systemic administration of iodoquinol caused by type IV (delayed-type) hypersensitivity have included systemic allergic dermatitis (3).

Systemic allergic dermatitis (systemic contact dermatitis)

A 25-year-old man was admitted to the hospital with an extensive infectious eczematoid dermatitis superimposed on seborrheic dermatitis. An ointment of 3 per cent iodoquinol in Aquaphor was prescribed. By the following morning the patient had developed a severe, eczematous reaction in all areas to which the ointment had been applied. When questioned the patient recalled that he had suffered a similar episode of dermatitis about one year previously, after treatment with 3 per cent clioquinol (Vioform ®) ointment. After his skin had cleared, patch test were performed and were positive to iodoquinol, clioquinol and 2 other halogenated hydroxyquinolines. Later, the patient was challenged with a single tablet of 210 mg iodoquinol. A severe recrudescence of the eczematous dermatitis appeared within 12 hours (3). This was a case of systemic allergic dermatitis.

Cross-reactions, pseudo-cross-reactions and co-reactions

Cross-reactions between iodoquinol and other halogenated hydroxyquinolines may occur (2,3).

LITERATURE

1   The data in the section 'General' may have been obtained from literature discussed in this chapter, but mostly also or exclusively from one or more of the following online sources: ChemIDPlus Advanced, PubChem, DrugBank, RxList, Drug Central, Drugs.com, and Wikipedia
2   De Groot AC. Monographs in contact allergy, volume 3: Topical Drugs. Boca Raton, Fl, USA: CRC Press Taylor and Francis Group, 2021 (ISBN 978-0-367-23693-9)
3   Leifer W, Steiner K. Studies in sensitization to halogenated hydroxyquinolines and related compounds. J Invest Dermatol 1951;17:233-240

Chapter 3.262 IOHEXOL

IDENTIFICATION

Description/definition   : Iohexol is the organic iodine compound that conforms to the structural formula shown
                            below
Pharmacological classes   : Contrast media
IUPAC name   : 5-[Acetyl(2,3-dihydroxypropyl)amino]-1-N,3-N-bis(2,3-dihydroxypropyl)-2,4,6-
                            triiodobenzene-1,3-dicarboxamide
CAS registry number   : 66108-95-0
EC number   : 266-164-2
Merck Index monograph   : 6365
Patch testing   : Iohexol-containing contrast medium undiluted (7)
Molecular formula   : $C_{19}H_{26}I_3N_3O_9$

GENERAL

Iohexol is a non-ionic, monomeric, water-soluble contrast agent which is used in myelography, arthrography, nephroangiography, arteriography, and other radiographic procedures. Its low systemic toxicity is the combined result of low chemotoxicity and low osmolality (1).

CUTANEOUS ADVERSE DRUG REACTIONS FROM SYSTEMIC ADMINISTRATION CAUSED BY TYPE IV
(DELAYED-TYPE) HYPERSENSITIVITY (as demonstrated by positive patch tests)

Cutaneous adverse drug reactions from systemic administration of iohexol caused by type IV (delayed-type) hypersensitivity have included maculopapular eruption (6,11,16,17), widespread erythema (16), acute generalized exanthematous pustulosis (AGEP) (18), fixed drug eruption (2), drug reaction with eosinophilia and systemic symptoms (DRESS) (18), toxic epidermal necrolysis (12), late-onset urticaria with loss of consciousness (6), facial edema with respiratory distress (6), and erythema multiforme-like skin eruptions (8,9).

Case series with various or unknown types of drug reactions

Twelve patients (9 women) with late adverse reactions to iodinated contrast media (ICM) were reported from France and Switzerland in 2005 (6). Iohexol had induced 3 cases of maculopapular eruption, one case of late-onset urticaria (5 hours) with loss of consciousness, and a reaction with facial edema and respiratory distress in one patient. All had positive patch tests to the undiluted iohexol ICM (6).

Maculopapular eruption

From France, 5 patients with maculopapular eruptions from iodinated contrast media were reported. Patch tests showed positive reactions to iohexol in 5, iodixanol in 5, iopromide in 3 and iomeprol in 1 patient. It was not mentioned which of these contrast media had actually caused the eruptions (17). A woman aged 71 years developed a generalized rash 18 hours after an angiography with iohexol. Patch tests were positive to iohexol and 8 other iodinated contrast media (11). A 54-year-old woman had developed, 24 hours after an intravenous injection of iohexol, widespread itchy erythema and palpebral edema. Patch tests were positive (D2 and D4 ++) to commercial iohexol and ioversol (16).

See also the section 'Case series with various or unknown types of drug reactions' above, ref. 6.

Erythroderma, widespread erythematous eruption, exfoliative dermatitis
An 68-year-old man suffering from aorta aneurysm had undergone coronary angiography with iohexol as contrast medium and 24 hours later presented with a maculopapular eruption of the trunk and legs. Patch tests were positive (+ at D2 and D4) to iohexol and iodixanol (16).

Acute generalized exanthematous pustulosis (AGEP)
In France, 19 adult patients in the dermatologist's French Investigators for Skin Adverse Reactions to Drugs (FISARD) network diagnosed between 2010 and 2020 with AGEP highly suspected to have been caused by a iodinated contrast medium (ICM) according to either a positive skin test or a challenge test were retrospectively evaluated. In this group of 19 patients with AGEP, 3 drug reactions had been caused by iohexol, 1 of whom had a positive patch test to this ICM and one was not patch tested. The third had a positive delayed intradermal test to iohexol (18).

Fixed drug eruption
A patient from Japan had a fixed drug eruption from the contrast medium iohexol, confirmed by a positive patch test (2; details unknown, article in Japanese, data cited in ref. 3).

Drug reaction with eosinophilia and systemic symptoms (DRESS)
In France, 13 adult patients in the dermatologist's French Investigators for Skin Adverse Reactions to Drugs (FISARD) network diagnosed between 2010 and 2020 with DRESS highly suspected to have been caused by a iodinated contrast medium (ICM) according to either a positive skin test or a challenge test, were retrospectively evaluated. In the group of 13 DRESS patients, 5 drug reactions had been caused by iohexol, of whom 3 had positive patch tests to this ICM; one individual was not patch tested (18).

Stevens-Johnson syndrome/toxic epidermal necrolysis (SJS/TEN)
After a motorcycle crash, a 33-year-old man underwent contrast-enhanced body CT using iohexol that revealed hepatic injuries. Five days later, the patient presented with fever (39°C), and a second body CT with iohexol contrast medium was performed to rule out a possible intra-abdominal abscess. He received treatment with various antibiotics. Two hours after a fourth examination using iohexol, the patient developed malaise, pruritic erythema, hypotension, and cutaneous bullae affecting 50% of his body surface that peeled off in sheets, with oral mucosal involvement. A diagnosis of toxic epidermal necrolysis was established (apparently without a skin biopsy). Allergy to the antibiotics was excluded by skin prick, intradermal and patch tests, and intravenous/oral provocation. A patch test with iohexol (300 mg/ml in water), however, was positive at D2 with erythema and multiple small, flaking blisters. Ten controls were negative (12).

Other cutaneous adverse drug reactions
In a university hospital in Tokyo, Japan, from 1989 through 1997, 58 patients who had developed erythema multiforme-like cutaneous adverse drug reactions (mainly on the trunk and the proximal part of the arms and legs) to iohexol or iopamidol (not specified) were investigated. Most skin eruptions occurred either within 1 day or 6 days after administration of contrast media (probably representing previous sensitization and sensitization *de novo*). Patch tests with the undiluted contrast medium were positive in 46 patients; in the other 12, intradermal tests with 0.05 ml of the contrast medium were positive after 24 hours. Cross-reactions to other non-ionic iodinated contrast media (not specified) were observed in >50% of the patients (9). Seven patients from Japan previously reported by the same authors had erythema multiforme-like drug eruptions from iohexol-containing contrast media, often appearing after 6 days (8). Probably, most of them had positive patch tests to the undiluted contrast medium (9).

See also the section 'Case series with various or unknown types of drug reactions' above, ref. 6.

Cross-reactions, pseudo-cross-reactions and co-reactions
Cross-reactions between iodinated radiocontrast media occur frequently, especially between the non-ionic preparations (4,5,6,9,10,18,19). Iodine is infrequently to rarely (opinions differ) the allergen (13,14) and allergy to mollusks, crustaceans, fish, and iodine from other sources is said not to be a risk factor for the development of hypersensitivity to iodinated contrast media (15).

LITERATURE

1   The data in the section 'General' may have been obtained from literature discussed in this chapter, but mostly also or exclusively from one or more of the following online sources: ChemIDPlus Advanced, PubChem, DrugBank, RxList, Drug Central, Drugs.com, and Wikipedia

2   Takeuchi M, Masutani M, Matsunaga K. A case of drug eruption caused by Omnipaque® (iohexol). Environ Dermatol 1995;2(suppl.1):65 (Article in Japanese)

3   Watanabe H, Sueki H, Nakada T, Akiyama M, Iijima M. Multiple fixed drug eruption caused by iomeprol (Iomeron ®), a non-ionic contrast medium. Dermatology 1999;198:291-294

4   Frías M, Fernández E, Audicana MT, Longo N, Muñoz D, Reyes SM. Fixed drug eruption caused by iodinated contrast media. Contact Dermatitis 2011;65:43-44

5   Poliak N, Elias M, Cianferoni A, Treat J. Acute generalized exanthematous pustulosis: the first pediatric case caused by a contrast agent. Ann Allergy Asthma Immunol 2010;105:242-243

6   Kanny G, Pichler W, Morisset M, Franck P, Marie B, Kohler C, et al. T cell-mediated reactions to iodinated contrast media: evaluation by skin and lymphocyte activation tests. J Allergy Clin Immunol 2005;115:179-185

7   Brockow K, Garvey LH, Aberer W, Atanaskovic-Markovic M, Barbaud A, Bilo MB, et al. Skin test concentrations for systemically administered drugs - an ENDA/EAACI Drug Allergy Interest Group position paper. Allergy 2013;68:702-712

8   Akiyama M, Iijima M, Fujisawa R. Clinical analysis of drug eruption due to iohexol (Omnipaque®). Jpn J Dermatol 1990; 100:1057-1060 (Article in Japanese, data cited in ref. 9)

9   Akiyama M, Nakada T, Sueki H, Fujisawa R, Iijima M. Drug eruption caused by nonionic iodinated X-ray contrast media. Acad Radiol 1998;5(Suppl.1):S159-S161

10 Lerondeau B, Trechot P, Waton J, Poreaux C, Luc A, Schmutz JL, et al. Analysis of cross-reactivity among radiocontrast media in 97 hypersensitivity reactions. J Allergy Clin Immunol 2016;137:633-635.e4

11 Lerch M, Keller M, Britschgi M, Kanny G, Tache V, Schmid DA, et al. Cross-reactivity patterns of T cells specific for iodinated contrast media. J Allergy Clin Immunol 2007;119:1529-1536

12 Rosado A, Canto G, Veleiro B, Rodríguez J. Toxic epidermal necrolysis after repeated injections of iohexol. AJR Am J Roentgenol 2001;176:262-263

13 Scherer K, Harr T, Bach A, Bircher AJ. The role of iodine in hypersensitivity reactions to radiocontrast media. Clin Exp Allergy 2009;40:468-475

14 Trautmann A, Brockow K, Behle V, Stoevesandt J. Radiocontrast media hypersensitivity: Skin testing differentiates allergy from nonallergic reactions and identifies a safe alternative as proven by intravenous provocation. J Allergy Clin Immunol Pract 2019;7:2218-2224

15 Rosado Ingelmo A, Doña Diaz I, Cabañas Moreno R, Moya Quesada MC, García-Avilés C, García Nuñez I, et al. Clinical practice guidelines for diagnosis and management of hypersensitivity reactions to contrast media. J Investig Allergol Clin Immunol 2016;26:144-155

16 Sedano E, Vega JM, Rebollo S, Callejo A, Asensio T, Almendros R. Delayed exanthema to nonionic contrast medium. Allergy 2001;56:1015-1016

17 Corbaux C, Seneschal J, Taïeb A, Cornelis F, Martinet J, Milpied B. Delayed cutaneous hypersensitivity reactions to iodinated contrast media. Eur J Dermatol 2017;27:190-191

18 Soria A, Amsler E, Bernier C, Milpied B, Tétart F, Morice C, et al. DRESS and AGEP reactions to iodinated contrast media: A French case series. J Allergy Clin Immunol Pract 2021;9:3041-3050

19 Yoon SH, Lee SY, Kang HR, Kim JY, Hahn S, Park CM, et al. Skin tests in patients with hypersensitivity reaction to iodinated contrast media: a meta-analysis. Allergy 2015;70:625-637

Chapter 3.263 IOMEPROL

IDENTIFICATION

Description/definition : Iomeprol is the organic iodine complex that conforms to the structural formula shown below

Pharmacological classes : Contrast media

IUPAC name : 1-N,3-N-bis(2,3-Dihydroxypropyl)-5-[(2-hydroxyacetyl)-methylamino]-2,4,6-triiodobenzene-1,3-dicarboxamide

CAS registry number : 78649-41-9

EC number : Not available

Merck Index monograph : 6367

Patch testing : Iomeprol-containing contrast medium undiluted (8); iomeprol 5% water

Molecular formula : $C_{17}H_{22}I_3N_3O_8$

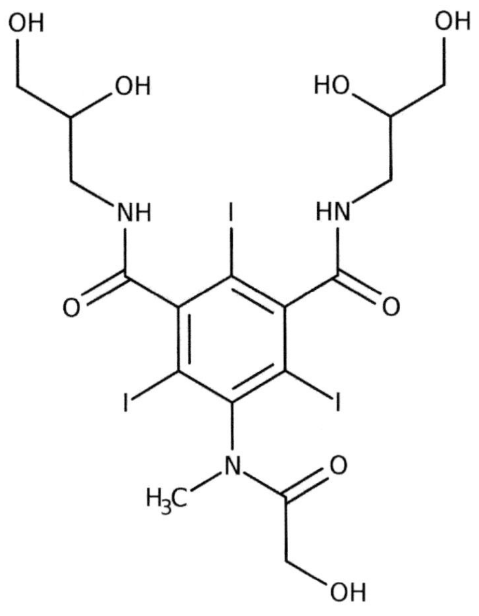

GENERAL

Iomeprol is an organoiodine and benzenedicarboxamide compound, that is used as a non-ionic monomeric iodinated radiocontrast agent in diagnostic imaging. Upon intravenous injection, iomeprol is distributed through the vascular system and interstitial space. Like other organic iodine compounds, this agent blocks x-rays and appears opaque on x-ray film, thus enhancing the visibility of body parts containing iomeprol (1).

CONTACT ALLERGY FROM ACCIDENTAL CONTACT

A 40-year-old female radiology technician presented with a history of recurrent eczema on her face and neck, which became worse during periods at work. Patch tests with the radiological contrast medium used in her department at a concentration of 10% water and the various product components iomeprol 1%, 2%, and 5% water, trometamol 0.1%, 0.5% and 1% aq., and iodine 0.5% and 0.1% water gave positive reactions to the contrast medium and iomeprol at all concentrations. Five controls tested with iomeprol 5% water were negative. The dermatitis rapidly resolved after the patient avoided touching her face with gloves contaminated by the contrast medium. A diagnosis of occupational allergic contact dermatitis to iomeprol was made (4).

CUTANEOUS ADVERSE DRUG REACTIONS FROM SYSTEMIC ADMINISTRATION CAUSED BY TYPE IV (DELAYED-TYPE) HYPERSENSITIVITY (as demonstrated by positive patch tests)

Cutaneous adverse drug reactions from systemic administration of iomeprol caused by type IV (delayed-type) hypersensitivity have included maculopapular eruption (9,10,13,17,19,23), acute generalized exanthematous pustulosis (AGEP) (2,3,21), acute localized exanthematous pustulosis (ALEP) (18), fixed drug eruption (5,20), symmetrical drug-related intertriginous and flexural exanthema (SDRIFE)/baboon syndrome (11,12), and drug reaction with eosinophilia and systemic symptoms (DRESS) (21).

Maculopapular eruption

A patient, who had previously suffered an episode of DRESS from ceftriaxone, later developed a maculopapular eruption from iomeprol and had a positive patch test to this drug. As the patient had cutaneous adverse reactions to various unrelated drugs, this was a case of multiple drug hypersensitivity syndrome (9,10). A man aged 60 years developed, 3 days after iomeprol injection for coronary angiography, a generalized maculopapular exanthem. Patch tests were positive to iomeprol and 8 other iodinated contrast media (13).

Of 10 patients who had developed a maculopapular exanthema after injections with iomeprol, 3 had a positive patch test to this iodinated contrast medium (17). From France, 5 patients with maculopapular eruptions from iodinated contrast media were reported. Patch tests showed positive reactions to iohexol in 5, iodixanol in 5, iopromide in 3 and iomeprol in 1 patient. It was not mentioned which of these contrast media had actually caused the eruptions (19).

A 75-year-old woman developed a reaction with edema of the eyelids, generalized itching and a maculopapular eruption 24 hours after administration of iomeprol (indicated by its trade name Iomeron). Patch tests were positive to commercial Iomeron solution in a dilution series of 1:10, 1:100 and 1:1000 (23).

Acute generalized exanthematous pustulosis (AGEP)

A 26-year-old woman received the contrast medium iomeprol for a cerebral computed tomography for headache. The following day, she developed acute generalized exanthematous pustulosis (AGEP), confirmed by skin biopsy. A patch test with the iomeprol contrast medium 'as is' and an intradermal test, read at D2, were positive. Intravenous administration of iodixanol was tolerated without untoward effects (2).

An 88-year-old patient underwent three radiological examinations requiring intravenous infusion of iomeprol within a few days for preoperative assessment of aortic stenosis and ischaemic heart disease. At 24 hours after the third injection, the patient had developed an acute exanthema and a temperature of 39.9 C. Numerous pustules were present on the face, in the axillary folds, and on the trunk. Blood tests showed leucocytosis. A skin biopsy showed an intraepidermal pustule with numerous polymorphonuclear cells. A diagnosis of AGEP was made. Patch testing was performed 6 months later and showed a positive reaction to iomeprol with cross-reactions to seven other iodinated contrast media (3).

In France, 19 adult patients in the dermatologist's French Investigators for Skin Adverse Reactions to Drugs (FISARD) network diagnosed between 2010 and 2020 with AGEP highly suspected to have been caused by a iodinated contrast medium (ICM) according to either a positive skin test or a challenge test were retrospectively evaluated. In this group of 19 patients with AGEP, 8 drug reactions had been caused by iomeprol, 6 of whom had positive patch tests to this ICM. The other 2 had positive delayed intradermal tests (21).

Acute localized exanthematous pustulosis (AGEP)

A 56-year-old woman presented with a pruritic pustular eruption on the trunk 2 days after undergoing a computed tomography scan. Six months later, the patient underwent a new CT scan. The same lesions appeared in the same location after one day. On physical examination, she showed a non-follicular pustular eruption located exclusively on the chest. Acute localized exanthematous pustulosis (ALEP) induced by iodine contrast was suspected. Histopathology was compatible with exanthematous pustulosis and microbiological culture of a pustule was negative. The iodine contrast used in both CT scans was iomeprol. The patient showed a positive patch test reaction to iomeprol commercial preparation 'as is' (D2 ++, D4 +++) with small pustules. Ten controls were negative (18).

Fixed drug eruption

A 67-year-old woman developed multiple pea-sized erythematous papules on the trunk and extremities 4 days after receiving 100 ml of iomeprol for a computed tomography examination. Some of the papules coalesced, forming 7 large plaques on the limbs. Six months later, the patient was mistakenly administered iomeprol again. On the following morning, erythematous plaques admixed with vesicles recurred at the same sites as during the previous episode. Histopathology and immunohistopathology were compatible with fixed drug eruption. In both episodes, the lesions cleared leaving pigmentation that faded with 6 weeks. Both patch testing (read at D3) and an intradermal test (read at 24 hours) with iomeprol (probably the commercial contrast medium) on lesional pigmented skin were positive. A diagnosis of multiple fixed drug eruption from iomeprol was made (5).

Four hours after intravenous administration of iomeprol, a 61-year-old patient developed a red macule of approximately 2 cm in diameter in the right inguinal region, which enlarged up to a final size of 15×8 cm. The patient's history revealed a similar reaction in the same localization and of the same clinical appearance after injection of iomeprol 1 year before. Patch testing 4 months later revealed positive reactions to iomeprol and iohexol on postlesional skin but not on previously unaffected skin (20).

Symmetrical drug-related intertriginous and flexural exanthema (SDRIFE)/Baboon syndrome
One day after an examination of the coronary arteries using iomeprol, an 80-year-old man developed a sharply defined symmetrical pruritic erythema of the gluteal area, a V-shaped erythema of the inguinal and perigenital area, erythema of the axillary folds and some macules on the abdomen without systemic signs or symptoms. This was the classic clinical picture of the baboon syndrome. Patch tests with undiluted iodinated contrast media were positive to iomeprol and 6 other non-ionic radiocontrast media including iopromide. Five months later, the patient had to undergo another emergency coronary angiography. Two days after the examination he suffered from a relapse of exactly the same clinical presentation which lasted for more than 2 weeks. Investigations revealed that, for the angiography, iopromide had been used (11).

A 78-year-old woman developed a symmetrical intertriginous and flexural exanthema 15 hours after administration of iomeprol for coronary angiography. The patient also had transient transaminase elevation which resolved after 4 weeks, but no other signs of systemic involvement. At D3, patch tests were positive to undiluted iomeprol, iobitridol, iohexol, iopromide, iodixanol and ioversol. Intradermal tests were positive at D2 for iomeprol in a 1:10 dilution (12).

Drug reaction with eosinophilia and systemic symptoms (DRESS)
In France, 13 adult patients in the dermatologist's French Investigators for Skin Adverse Reactions to Drugs (FISARD) network diagnosed between 2010 and 2020 with DRESS highly suspected to have been caused by a iodinated contrast medium (ICM) according to either a positive skin test or a challenge test, were retrospectively evaluated. In the group of 13 DRESS patients, 2 drug reactions had been caused by iomeprol, one of whom had a positive patch test to this ICM (21).

Cross-reactions, pseudo-cross-reactions and co-reactions
Cross-reactions between iodinated contrast media occur frequently (3,6,7,21,22). Iodine is infrequently to rarely (opinions differ) the allergen (14,15) and allergy to mollusks, crustaceans, fish, and iodine from other sources is said not to be a risk factor for the development of hypersensitivity to iodinated contrast media (16).

LITERATURE
1 The data in the section 'General' may have been obtained from literature discussed in this chapter, but mostly also or exclusively from one or more of the following online sources: ChemIDPlus Advanced, PubChem, DrugBank, RxList, Drug Central, Drugs.com, and Wikipedia
2 Grandvuillemin A, Ripert C, Sgro C, Collet E. Iodinated contrast media-induced acute generalized exanthematous pustulosis confirmed by delayed skin tests. J Allergy Clin Immunol Pract 2014;2:805-806
3 Machet P, Marcé D, Ziyani Y, Dumont M, Cornillier H, Jonville-Bera A, et al. Acute generalized exanthematous pustulosis induced by iomeprol with cross-reactivity to other iodinated contrast agents and mild reactions after rechallenge with iopromide and oral corticosteroid premedication. Contact Dermatitis 2019;81:74-76
4 Foti C, Bonamonte D, Conserva A, Antelmi AR, Antonaci CE, Angelini G. Occupational allergic contact dermatitis to a non-ionic iodinated contrast medium containing iomeprol. Contact Dermatitis 2008;59:252-253
5 Watanabe H, Sueki H, Nakada T, Akiyama M, Iijima M. Multiple fixed drug eruption caused by iomeprol (Iomeron ®), a non-ionic contrast medium. Dermatology 1999;198:291-294
6 Kanny G, Pichler W, Morisset M, Franck P, Marie B, Kohler C, et al. T cell-mediated reactions to iodinated contrast media: evaluation by skin and lymphocyte activation tests. J Allergy Clin Immunol 2005;115:179-185
7 Lerondeau B, Trechot P, Waton J, Poreaux C, Luc A, Schmutz JL, et al. Analysis of cross-reactivity among radiocontrast media in 97 hypersensitivity reactions. J Allergy Clin Immunol 2016;137:633-635.e4
8 Brockow K, Garvey LH, Aberer W, Atanaskovic-Markovic M, Barbaud A, Bilo MB, et al. Skin test concentrations for systemically administered drugs - an ENDA/EAACI Drug Allergy Interest Group position paper. Allergy 2013;68:702-712
9 Jörg L, Helbling A, Yerly D, Pichler WJ. Drug-related relapses in drug reaction with eosinophilia and systemic symptoms (DRESS). Clin Transl Allergy 2020;10:52
10 Jörg L, Yerly D, Helbling A, Pichler W. The role of drug, dose and the tolerance/intolerance of new drugs in multiple drug hypersensitivity syndrome (MDH). Allergy 2020;75:1178-1187
11 Arnold AW, Häusermann P, Bach S, Bircher AJ. Recurrent flexural exanthema (SDRIFE or baboon syndrome) after administration of two different iodinated radio contrast media. Dermatology 2007;214:89-93
12 Grosber M, Mcintyre M, Salava A, Ring J, Brockow K. Intravenous provocation test with contrast media in a patient with symmetrical drug-related intertriginous and flexural exanthema (SDRIFE) after administration of iomeprol. Allergy 2009;64 (Suppl.90):294(abstract)
13 Lerch M, Keller M, Britschgi M, Kanny G, Tache V, Schmid DA, et al. Cross-reactivity patterns of T cells specific for iodinated contrast media. J Allergy Clin Immunol 2007;119:1529-1536

14   Scherer K, Harr T, Bach A, Bircher AJ. The role of iodine in hypersensitivity reactions to radiocontrast media. Clin Exp Allergy 2009;40:468-475

15   Trautmann A, Brockow K, Behle V, Stoevesandt J. Radiocontrast media hypersensitivity: Skin testing differentiates allergy from nonallergic reactions and identifies a safe alternative as proven by intravenous provocation. J Allergy Clin Immunol Pract 2019;7:2218-2224

16   Rosado Ingelmo A, Doña Diaz I, Cabañas Moreno R, Moya Quesada MC, García-Avilés C, García Nuñez I, et al. Clinical practice guidelines for diagnosis and management of hypersensitivity reactions to contrast media. J Investig Allergol Clin Immunol 2016;26:144-155

17   Seitz CS, Pfeuffer P, Raith P, Bröcker EB, Trautmann A. Radiocontrast media-associated exanthema: identification of cross-reactivity and tolerability by allergologic testing. Eur J Radiol 2009;72:167-171

18   Navarro Triviño FJ, Linares-González L, Ródenas-Herranz T, Llamas-Molina JM, Ruiz-Villaverde R. Acute localized exanthematous pustulosis (ALEP) induced by iomeprol (Iomeron 350): A diagnostic challenge. Contact Dermatitis 2021;85:95-97

19   Corbaux C, Seneschal J, Taïeb A, Cornelis F, Martinet J, Milpied B. Delayed cutaneous hypersensitivity reactions to iodinated contrast media. Eur J Dermatol 2017;27:190-191

20   Böhm I, Medina J, Prieto P, Block W, Schild HH. Fixed drug eruption induced by an iodinated non-ionic X-ray contrast medium: a practical approach to identify the causative agent and to prevent its recurrence. Eur Radiol 2007;17:485-489.

21   Soria A, Amsler E, Bernier C, Milpied B, Tétart F, Morice C, et al. DRESS and AGEP reactions to iodinated contrast media: A French case series. J Allergy Clin Immunol Pract 2021;9:3041-3050

22   Yoon SH, Lee SY, Kang HR, Kim JY, Hahn S, Park CM, et al. Skin tests in patients with hypersensitivity reaction to iodinated contrast media: a meta-analysis. Allergy 2015;70:625-637

23   Goksel O, Aydin O, Atasoy C, Akyar S, Demirel YS, Misirligil Z, et al. Hypersensitivity reactions to contrast media: prevalence, risk factors and the role of skin tests in diagnosis—a cross-sectional survey. Int Arch Allergy Immunol 2011;155:297-305

Chapter 3.264 IOPAMIDOL

IDENTIFICATION

Description/definition : Iopamidol is the organic iodine compound that conforms to the structural formula shown below
Pharmacological classes : Contrast media
IUPAC name : 1-*N*,3-*N*-bis(1,3-Dihydroxypropan-2-yl)-5-[[(2*S*)-2-hydroxypropanoyl]amino]-2,4,6-triiodobenzene-1,3-dicarboxamide
Other names : (*S*)-*N*,*N*'-bis(2-Hydroxy-1-(hydroxymethyl)ethyl)-2,4,6-triiodo-5-lactamidoisophthalamide
CAS registry number : 60166-93-0
EC number : 262-093-6
Merck Index monograph : 6370
Patch testing : Iopamidol-containing contrast medium undiluted (7)
Molecular formula : $C_{17}H_{22}I_3N_3O_8$

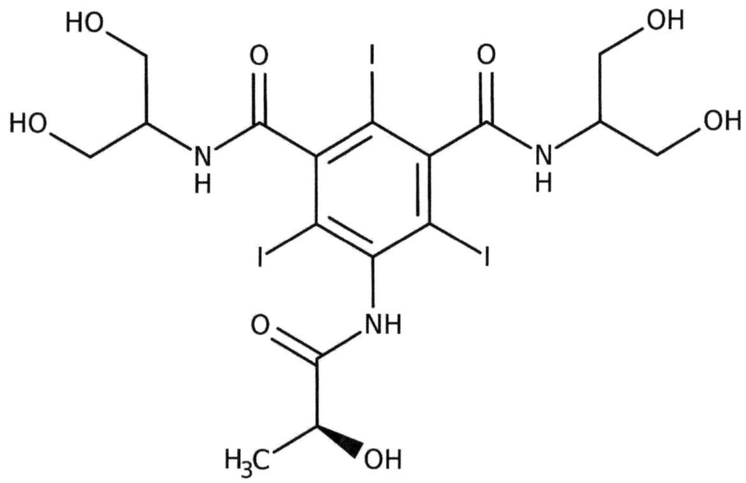

GENERAL

Iopamidol is a non-ionic, monomeric, water-soluble radiocontrast agent which is used in myelography, arthrography, nephroangiography, arteriography, and other radiological procedures. Upon intravenous injection, iopamidol is distributed through the vascular system and interstitial space. Like other organic iodine compounds, this agent blocks x-rays and appears opaque on x-ray film, thus enhancing the visibility of body parts containing iopamidol (1).

CUTANEOUS ADVERSE DRUG REACTIONS FROM SYSTEMIC ADMINISTRATION CAUSED BY TYPE IV (DELAYED-TYPE) HYPERSENSITIVITY (as demonstrated by positive patch tests)

Cutaneous adverse drug reactions from systemic administration of iopamidol caused by type IV (delayed-type) hypersensitivity have included maculopapular eruption (13,14,15), acute generalized exanthematous pustulosis (AGEP) (6), fixed drug eruption (2), erythema multiforme-like drug eruptions (8), and delayed urticaria (uncertain) (9).

Maculopapular eruption

A 19-year-old man who had experienced a right frontal cerebral hemorrhage with intraventricular hematoma underwent angiography with 160 ml of iopamidol. Three days later, he experienced a diffuse, slight maculopapular rash. Patch tests were positive to commercial iopamidol (D2 and D3 ++) and 7 other iodinated contrast media (14). Of 5 patients who had developed a maculopapular exanthema after injections with iopamidol, 1 had a positive patch test to this iodinated contrast medium (13).

A 61-year-old man developed a maculopapular eruption 7 days, and then, upon re-exposure, a macular exanthema (under pretreatment with prednisone and an antihistamine) one day after percutaneous transluminal coronary angioplasty using iopamidol. Patch tests with iopamidol were positive on 2 occasions (with cross-reactions to iohexol and ioversol), but became negative 6 months after the second exposure (15).

Acute generalized exanthematous pustulosis (AGEP)

A 56-year-old woman underwent 3 radiologic examinations with intravenous iopamidol infusion. Seven days after the third injection, the patient was referred for evaluation of a skin rash. Physical examination revealed multiple scattered erythematous lesions, mainly on the abdomen. The laboratory evaluation showed leukocytosis and increased C-reactive protein level. The patient was given topical steroid and antihistamine therapy. Five days later, however, a skin examination revealed new erythematous rashes studded with numerous small, non-follicular pustules localized to the intertriginous areas, including the axillae and groin. A skin biopsy revealed subcorneal neutrophilic pustules and marked edema of the papillary dermis with mixed inflammatory cell infiltration, including lymphocytes, neutrophils, and eosinophils, consistent with acute generalized exanthematous pustulosis (AGEP). A bacterial culture of pustule content was negative. With premedication of prednisolone, a radiologic examination was performed by administration of iopromide, but this resulted in a severe recurrence of the exanthema within 24 hours. Patch tests were positive to iopamidol, but (false-) negative to iopromide at D2, D3 and D7 (6).

Fixed drug eruption

A patient from Japan had a fixed drug eruption from the contrast medium iopamidol, confirmed by a positive patch test (2; details unknown, article in Japanese, data cited in ref. 3).

Other cutaneous adverse drug reactions

In a university hospital in Tokyo, Japan, from 1989 through 1997, 58 patients who had developed erythema multiforme-like cutaneous adverse drug reactions (mainly on the trunk and the proximal part of the arms and legs) to iopamidol or iohexol (not specified) were investigated. Most skin eruptions occurred either within 1 day or 6 days after administration of contrast media (probably representing previous sensitization and sensitization *de novo*). Patch tests with the undiluted contrast medium were positive in 46 patients; in the other 12, intradermal tests with 0.05 ml of the contrast medium were positive after 24 hours. Cross-reactions to other non-ionic iodinated contrast media (not specified) were observed in >50% of the patients (8).

Iopamidol may have caused late-onset urticaria in a patient who showed a positive reaction to iopamidol (unclear whether the patient had actually received iopamidol contrast medium) (9).

Cross-reactions, pseudo-cross-reactions and co-reactions

Cross-reactions between iodinated radiocontrast media occur frequently, especially between the non-ionic preparations (8,9,16). Iodine is infrequently to rarely (opinions differ) the allergen (10,11) and allergy to mollusks, crustaceans, fish, and iodine from other sources is said not to be a risk factor for the development of hypersensitivity to iodinated contrast media (12).

LITERATURE

1   The data in the section 'General' may have been obtained from literature discussed in this chapter, but mostly also or exclusively from one or more of the following online sources: ChemIDPlus Advanced, PubChem, DrugBank, RxList, Drug Central, Drugs.com, and Wikipedia

2   Yamauchi R, Morita A, Tsuji T. Fixed drug eruption caused by iopamidol, a contrast medium. J Dermatol (Tokyo) 1997;24:243-245 (Article in Japanese)

3   Watanabe H, Sueki H, Nakada T, Akiyama M, Iijima M. Multiple fixed drug eruption caused by iomeprol (Iomeron ®), a non-ionic contrast medium. Dermatology 1999;198:291-294

4   Kanny G, Pichler W, Morisset M, Franck P, Marie B, Kohler C, et al. T cell-mediated reactions to iodinated contrast media: evaluation by skin and lymphocyte activation tests. J Allergy Clin Immunol 2005;115:179-185

5   Lerondeau B, Trechot P, Waton J, Poreaux C, Luc A, Schmutz JL, et al. Analysis of cross-reactivity among radiocontrast media in 97 hypersensitivity reactions. J Allergy Clin Immunol 2016;137:633-635.e4

6   Mizuta T, Kasami S, Shigehara Y, Kato M. Acute generalized exanthematous pustulosis caused by iopamidol with recurrence on rechallenge with iopromide. JAAD Case Rep 2020;6:964-966

7   Brockow K, Garvey LH, Aberer W, Atanaskovic-Markovic M, Barbaud A, Bilo MB, et al. Skin test concentrations for systemically administered drugs - an ENDA/EAACI Drug Allergy Interest Group position paper. Allergy 2013;68:702-712

8   Akiyama M, Nakada T, Sueki H, Fujisawa R, Iijima M. Drug eruption caused by nonionic iodinated X-ray contrast media. Acad Radiol 1998;5(Suppl.1):S159-S161

9   Kanny G, Pichler W, Morisset M, Franck P, Marie B, Kohler C, Renaudin JM, Beaudouin E, Laudy JS, Moneret-Vautrin DA. T cell-mediated reactions to iodinated contrast media: evaluation by skin and lymphocyte activation tests. J Allergy Clin Immunol 2005;115:179-185

10 Scherer K, Harr T, Bach A, Bircher AJ. The role of iodine in hypersensitivity reactions to radiocontrast media. Clin Exp Allergy 2009;40:468-475

11 Trautmann A, Brockow K, Behle V, Stoevesandt J. Radiocontrast media hypersensitivity: Skin testing differentiates allergy from nonallergic reactions and identifies a safe alternative as proven by intravenous provocation. J Allergy Clin Immunol Pract 2019;7:2218-2224

12 Rosado Ingelmo A, Doña Diaz I, Cabañas Moreno R, Moya Quesada MC, García-Avilés C, García Nuñez I, et al. Clinical practice guidelines for diagnosis and management of hypersensitivity reactions to contrast media. J Investig Allergol Clin Immunol 2016;26:144-155

13 Seitz CS, Pfeuffer P, Raith P, Bröcker EB, Trautmann A. Radiocontrast media-associated exanthema: identification of cross-reactivity and tolerability by allergologic testing. Eur J Radiol 2009;72:167-171

14 Romano A, Artesani MC, Andriolo M, Viola M, Pettinato R, VecchioliScaldazza A. Effective prophylactic protocol in delayed hypersensitivity to contrast media: report of a case involving lymphocyte transformation studies with different compounds. Radiology 2002;225:466-470

15 Courvoisier S, Bircher AJ. Delayed-type hypersensitivity to a nonionic, radiopaque contrast medium. Allergy 1998;53:1221-1224

16 Yoon SH, Lee SY, Kang HR, Kim JY, Hahn S, Park CM, et al. Skin tests in patients with hypersensitivity reaction to iodinated contrast media: a meta-analysis. Allergy 2015;70:625-637

Chapter 3.265 IOPENTOL

IDENTIFICATION

Description/definition   : Iopentol is the organic iodine complex that conforms to the structural formula shown below
Pharmacological classes   : Contrast media
IUPAC name   : 5-[Acetyl-(2-hydroxy-3-methoxypropyl)amino]-1-N,3-N-bis(2,3-dihydroxypropyl)-2,4,6-triiodobenzene-1,3-dicarboxamide
CAS registry number   : 89797-00-2
EC number   : Not available
Merck Index monograph   : 6372
Patch testing   : Iopentol-containing contrast medium undiluted (3)
Molecular formula   : $C_{20}H_{28}I_3N_3O_9$

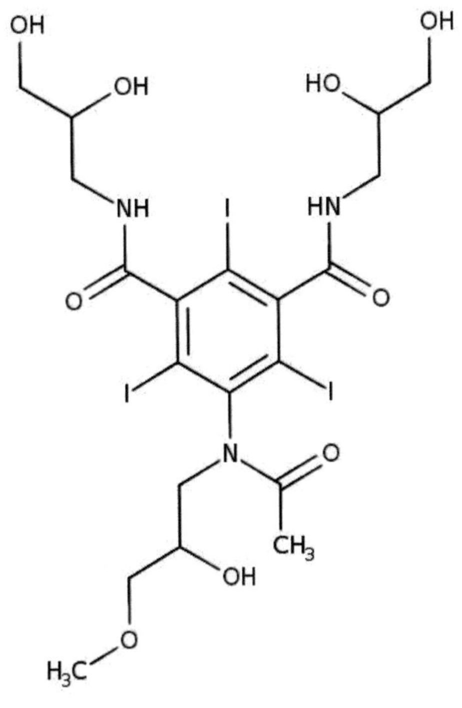

GENERAL

Iopentol is an iodine-containing, non-ionic, monomeric, water-soluble radiocontrast agent. The iodine atoms readily absorb X-rays, resulting in a higher contrast of X-ray images. It has a low osmolality, meaning that the solution has a relatively low concentration of molecules; this is usually associated with fewer adverse effects than high-osmolality contrast agents (1).

CUTANEOUS ADVERSE DRUG REACTIONS FROM SYSTEMIC ADMINISTRATION CAUSED BY TYPE IV (DELAYED-TYPE) HYPERSENSITIVITY (as demonstrated by positive patch tests)

Cutaneous adverse drug reactions from systemic administration of iopentol caused by type IV (delayed-type) hypersensitivity have included dermatitis/eczematous eruption (2).

Dermatitis/eczematous eruption

In a 73-year-old woman a computed tomography with contrast medium was performed for staging a malignant melanoma using sodium amidotrizoate and iopentol as contrast media. Four days later the patient noted a progressive pruritus beginning on the back and neck. On the sixth day she had a disseminated itchy papular to papulovesicular partially confluent exanthema involving the trunk, neck, and upper arms, which generalized despite therapy with local glucocorticosteroids and an oral antihistamine. Patch tests were positive to commercial iopentol at D2 and D3 with negative reactions in 20 controls (2).

Cross-reactions, pseudo-cross-reactions and co-reactions

Cross-reactions between iodinated radiocontrast media occur frequently, especially between the non-ionic preparations (4,5,9). Iodine is infrequently to rarely (opinions differ) the allergen (6,7) and allergy to mollusks, crustaceans, fish, and iodine from other sources is said not to be a risk factor for the development of hypersensitivity to iodinated contrast media (8).

LITERATURE

1   The data in the section 'General' may have been obtained from literature discussed in this chapter, but mostly also or exclusively from one or more of the following online sources: ChemIDPlus Advanced, PubChem, DrugBank, RxList, Drug Central, Drugs.com, and Wikipedia

2   Brockow K, Becker EW, Worret WI, Ring J. Late skin test reactions to radiocontrast medium. J Allergy Clin Immunol 1999;104:1107-1108

3   Brockow K, Garvey LH, Aberer W, Atanaskovic-Markovic M, Barbaud A, Bilo MB, et al. Skin test concentrations for systemically administered drugs - an ENDA/EAACI Drug Allergy Interest Group position paper. Allergy 2013;68:702-712

4   Kanny G, Pichler W, Morisset M, Franck P, Marie B, Kohler C, et al. T cell-mediated reactions to iodinated contrast media: evaluation by skin and lymphocyte activation tests. J Allergy Clin Immunol 2005;115:179-185

5   Lerondeau B, Trechot P, Waton J, Poreaux C, Luc A, Schmutz JL, et al. Analysis of cross-reactivity among radiocontrast media in 97 hypersensitivity reactions. J Allergy Clin Immunol 2016;137:633-635.e4

6   Scherer K, Harr T, Bach A, Bircher AJ. The role of iodine in hypersensitivity reactions to radiocontrast media. Clin Exp Allergy 2009;40:468-475

7   Trautmann A, Brockow K, Behle V, Stoevesandt J. Radiocontrast media hypersensitivity: Skin testing differentiates allergy from nonallergic reactions and identifies a safe alternative as proven by intravenous provocation. J Allergy Clin Immunol Pract 2019;7:2218-2224

8   Rosado Ingelmo A, Doña Diaz I, Cabañas Moreno R, Moya Quesada MC, García-Avilés C, García Nuñez I, et al. Clinical practice guidelines for diagnosis and management of hypersensitivity reactions to contrast media. J Investig Allergol Clin Immunol 2016;26:144-155

9   Yoon SH, Lee SY, Kang HR, Kim JY, Hahn S, Park CM, et al. Skin tests in patients with hypersensitivity reaction to iodinated contrast media: a meta-analysis. Allergy 2015;70:625-637

Chapter 3.266 IOPROMIDE

IDENTIFICATION

Description/definition   : Iopromide is the low osmolar, non-ionic X-ray contrast agent that conforms to the
                              structural formula shown below
Pharmacological classes   : Contrast media
IUPAC name   : 1-*N*,3-*N*-bis(2,3-Dihydroxypropyl)-2,4,6-triiodo-5-[(2-methoxyacetyl)amino]-3-*N*-
                              methylbenzene-1,3-dicarboxamide
CAS registry number   : 73334-07-3
EC number   : 277-385-9
Merck Index monograph : 6375
Patch testing   : Iopromide-containing contrast medium undiluted (4)
Molecular formula   : $C_{18}H_{24}I_3N_3O_8$

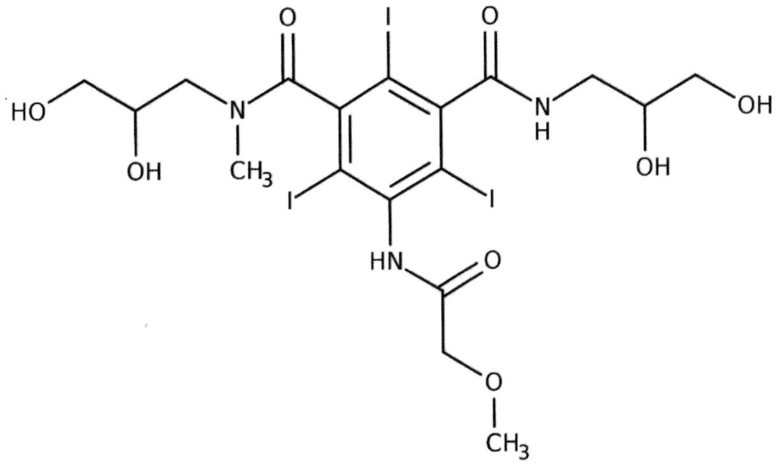

GENERAL

Iopromide is a low osmolar, non-ionic, monomeric, X-ray contrast agent for intravascular administration. It functions as a contrast agent by opacifying blood vessels in the path of flow of the contrast agent, permitting radiographic visualization of the internal structures until significant hemodilution occurs (1).

CUTANEOUS ADVERSE DRUG REACTIONS FROM SYSTEMIC ADMINISTRATION CAUSED BY TYPE IV (DELAYED-TYPE) HYPERSENSITIVITY (as demonstrated by positive patch tests)

Cutaneous adverse drug reactions from systemic administration of iopromide caused by type IV (delayed-type) hypersensitivity have included maculopapular eruption (3,8,14,15), acute generalized exanthematous pustulosis (AGEP) (2,16), symmetrical drug-related intertriginous and flexural exanthema (SDRIFE)/baboon syndrome (7), fixed drug eruption (9), and drug reaction with eosinophilia and systemic symptoms (DRESS) (10).

Maculopapular eruption

From France, 5 patients with maculopapular eruptions from iodinated contrast media were reported. Patch tests showed positive reactions to iopromide in 3, iohexol in 5, iodixanol in 5, and iomeprol in 1 patient. It was not mentioned which of these contrast media had actually caused the eruptions (15).

A 56-year-old man had coronary angiography using 200 ml of the contrast medium iopromide. Two hours later, he developed nausea and felt generally hot. After one day, a maculopapular rash erupted on the trunk, arms and legs without further symptoms. Prick tests and intracutaneous tests with iopromide contrast medium were negative at 20 minutes and D1 readings. A patch test gave a positive reaction at D3. Intravenous challenge with iopromide in in-creasing concentrations (1:1000. 1:100, 1:10, undiluted, 5 ml, respectively) resulted in maculopapular lesions on the arms and legs after 18 hours (3).

One patient developed a maculopapular eruption from iopromide with a positive patch test to this drug after a previous episode of DRESS from phenytoin; there were also cutaneous adverse drug reactions to other drugs (multiple drug hypersensitivity) (8). Of 11 patients who had developed a maculopapular exanthema after injections with iopromide, 1 had a positive patch test to this iodinated contrast medium (14).

Acute generalized exanthematous pustulosis (AGEP)

In France, 19 adult patients in the dermatologist's French Investigators for Skin Adverse Reactions to Drugs (FISARD) network diagnosed between 2010 and 2020 with AGEP highly suspected to have been caused by a iodinated contrast medium (ICM) according to either a positive skin test or a challenge test were retrospectively evaluated. In this group of 19 patients with AGEP, 2 drug reactions had been caused by iopromide, one of whom had a positive patch test to this ICM and the other, who was patch test negative, a positive delayed intradermal test (16).

A 52-year-old woman received intravenous iopromide for coronary angiography. Two hours later, her inguinal region became flushed. Her entire body then became red and she developed itching, fever, swelling of the tongue, and pustular lesions covering her body. The patient, who had previously received contrast media on several occasions without complications, was diagnosed with AGEP. Intradermal tests and patch tests were performed with iopromide, iodixanol, iomeprol, diatrizoate, and iopamidol. The delayed intradermal test readings and the patch tests showed positive results for all the contrast media except diatrizoate. One month later, the patient was accidentally given intravenous iopromide again and had a severe relapse, which started after 3-4 hours. A punch biopsy specimen revealed epidermal acanthosis with multifocal pustule formation in the upper part of the epidermis, with inflammation and mild edema in the papillary dermis, findings consistent with the diagnosis of AGEP (2).

Symmetrical drug-related intertriginous and flexural exanthema (SDRIFE)/Baboon syndrome

One day after an examination of the coronary arteries using iomeprol, an 80-year-old man developed a sharply defined symmetrical pruritic erythema of the gluteal area, a V-shaped erythema of the inguinal and perigenital area, erythema of the axillary folds and some macules on the abdomen without systemic signs or symptoms. This was the classic clinical picture of the baboon syndrome. Patch tests with undiluted iodinated contrast media were positive to iomeprol and 6 other non-ionic radiocontrast media including iopromide. Five months later, the patient had to undergo another emergency coronary angiography. Two days after the examination he suffered from a relapse of exactly the same clinical presentation which lasted for more than 2 weeks. Investigations revealed that, for the angiography, iopromide had been used (7).

Fixed drug eruption

A 61-year-old physician developed, four hours after having received an intravenous administration of a non-ionic contrast medium containing iopromide, a red macule of approximately 2 cm in diameter covering a dermal infiltration in the right inguinal region, enlarging up to a final size of 15x8 cm. The patient's clinical history revealed a similar reaction after an intravenous iopromide injection for CT examination 12 months before. Patch tests were negative to iopromide, but positive to iomeprol and iohexol on postlesional but nor uninvolved skin. This makes fixed drug eruption from delayed-type hypersensitivity to iopromide, although not proven, highly likely (9).

Drug reaction with eosinophilia and systemic symptoms (DRESS)

A 57-year-old woman developed DRESS on 2 occasions after computed tomography with iopromide with fever, generalized rash, enlarged lymph nodes, leukocytosis, eosinophilia and elevated liver enzymes (first episode) and skin rash, leukocytosis, eosinophilia and elevated liver enzymes in the second episode. Patch tests were negative after the first episode and positive to iopromide after the second episode of DRESS (10).

Cross-reactions, pseudo-cross-reactions and co-reactions

Cross-reactions between iodinated radiocontrast media occur frequently, especially between the non-ionic preparations (5,6,16,17). Iodine is infrequently to rarely (opinions differ) the allergen (11,12) and allergy to mollusks, crustaceans, fish, and iodine from other sources is said not to be a risk factor for the development of hypersensitivity to iodinated contrast media (13).

LITERATURE

1 The data in the section 'General' may have been obtained from literature discussed in this chapter, but mostly also or exclusively from one or more of the following online sources: ChemIDPlus Advanced, PubChem, DrugBank, RxList, Drug Central, Drugs.com, and Wikipedia
2 Bavbek S, Sözener ZC, Aydin O, Ozdemir SK, Gül U, Heper AO. First case report of acute generalized exanthematous pustulosis due to intravenous iopromide. J Investig Allergol Clin Immunol 2014;24:66-67
3 Schick E, Weber L, Gall H. Delayed hypersensitivity reaction to the non-ionic contrast medium iopromid. Contact Dermatitis 1996;35:312
4 Brockow K, Garvey LH, Aberer W, Atanaskovic-Markovic M, Barbaud A, Bilo MB, et al. Skin test concentrations for systemically administered drugs - an ENDA/EAACI Drug Allergy Interest Group position paper. Allergy 2013;68:702-712
5 Kanny G, Pichler W, Morisset M, Franck P, Marie B, Kohler C, et al. T cell-mediated reactions to iodinated contrast media: evaluation by skin and lymphocyte activation tests. J Allergy Clin Immunol 2005;115:179-185

6   Lerondeau B, Trechot P, Waton J, Poreaux C, Luc A, Schmutz JL, et al. Analysis of cross-reactivity among radiocontrast media in 97 hypersensitivity reactions. J Allergy Clin Immunol 2016;137:633-635.e4

7   Arnold AW, Hâusermann P, Bach S, Bircher AJ. Recurrent flexural exanthema (SDRIFE or baboon syndrome) after administration of two different iodinated radio contrast media. Dermatology 2007;214:89-93

8   Jörg L, Yerly D, Helbling A, Pichler W. The role of drug, dose and the tolerance/intolerance of new drugs in multiple drug hypersensitivity syndrome (MDH). Allergy 2020;75:1178-1187

9   Böhm I, Medina J, Prieto P, Block W, Schild HH. Fixed drug eruption induced by an iodinated non-ionic X-ray contrast medium: a practical approach to identify the causative agent and to prevent its recurrence. Eur Radiol 2007;17:485-489

10 Leguisamo S, Baynova K, Labella Alvarez M, Avila Castellano R, Prados Castano M. DRESS syndrome due to iodinated contrast media. A case report. Allergy 2015;70(Suppl.101):331 (abstract 788)

11 Scherer K, Harr T, Bach A, Bircher AJ. The role of iodine in hypersensitivity reactions to radiocontrast media. Clin Exp Allergy 2009;40:468-475

12 Trautmann A, Brockow K, Behle V, Stoevesandt J. Radiocontrast media hypersensitivity: Skin testing differentiates allergy from nonallergic reactions and identifies a safe alternative as proven by intravenous provocation. J Allergy Clin Immunol Pract 2019;7:2218-2224

13 Rosado Ingelmo A, Doña Diaz I, Cabañas Moreno R, Moya Quesada MC, García-Avilés C, García Nuñez I, et al. Clinical practice guidelines for diagnosis and management of hypersensitivity reactions to contrast media. J Investig Allergol Clin Immunol 2016;26:144-155

14 Seitz CS, Pfeuffer P, Raith P, Bröcker EB, Trautmann A. Radiocontrast media-associated exanthema: identification of cross-reactivity and tolerability by allergologic testing. Eur J Radiol 2009;72:167-171

15 Corbaux C, Seneschal J, Taïeb A, Cornelis F, Martinet J, Milpied B. Delayed cutaneous hypersensitivity reactions to iodinated contrast media. Eur J Dermatol 2017;27:190-191

16 Soria A, Amsler E, Bernier C, Milpied B, Tétart F, Morice C, et al. DRESS and AGEP reactions to iodinated contrast media: A French case series. J Allergy Clin Immunol Pract 2021 Aug;9(8):3041-3050

17 Yoon SH, Lee SY, Kang HR, Kim JY, Hahn S, Park CM, et al. Skin tests in patients with hypersensitivity reaction to iodinated contrast media: a meta-analysis. Allergy 2015;70:625-637

Chapter 3.267 IOVERSOL

IDENTIFICATION

Description/definition	: Ioversol is the organo-iodine compound that conforms to the structural formula shown below
Pharmacological classes	: Contrast media
IUPAC name	: 1-N,3-N-bis(2,3-Dihydroxypropyl)-5-[(2-hydroxyacetyl)-(2-hydroxyethyl)amino]-2,4,6-triiodobenzene-1,3-dicarboxamide
CAS registry number	: 87771-40-2
EC number	: Not available
Merck Index monograph	: 6379
Patch testing	: Ioversol-containing contrast medium undiluted (7)
Molecular formula	: $C_{18}H_{24}I_3N_3O_9$

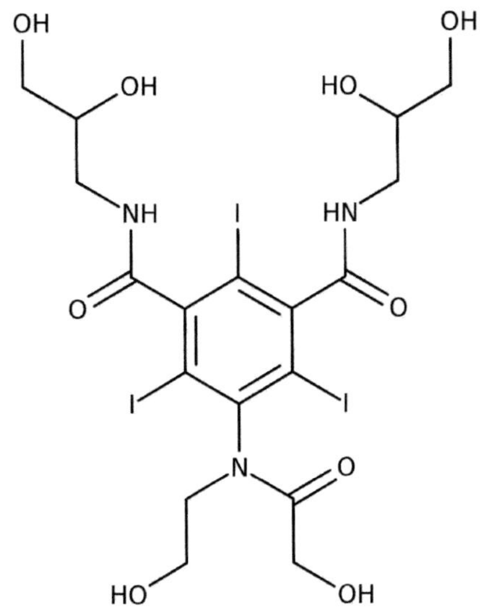

GENERAL

Ioversol is a sterile, nonpyrogenic, non-ionic, monomeric, organo-iodine compound used as contrast agent in various radiological procedures. Upon intravenous injection, ioversol is distributed through the vascular system and interstitial space. Like other organic iodine compounds, this agent blocks x-rays and appears opaque on x-ray film, thus enhancing the visibility of body parts containing ioversol (1).

CUTANEOUS ADVERSE DRUG REACTIONS FROM SYSTEMIC ADMINISTRATION CAUSED BY TYPE IV (DELAYED-TYPE) HYPERSENSITIVITY (as demonstrated by positive patch tests)

Cutaneous adverse drug reactions from systemic administration of ioversol caused by type IV (delayed-type) hypersensitivity have included acute generalized exanthematous pustulosis (AGEP) (5,6,16), fixed drug eruption (2,4), drug reaction with eosinophilia and systemic symptoms (DRESS) (10,11,16) and undefined drug eruption (15).

Acute generalized exanthematous pustulosis (AGEP)

In France, 19 adult patients in the dermatologist's French Investigators for Skin Adverse Reactions to Drugs (FISARD) network diagnosed between 2010 and 2020 with AGEP highly suspected to have been caused by a iodinated contrast medium (ICM) according to either a positive skin test or a challenge test were retrospectively evaluated. In this group of 19 patients with AGEP, 2 drug reactions had been caused by ioversol, one of whom had a positive patch test to this ICM; the other was not patch tested (16).

A 4-year-old boy with a history of tetralogy of Fallot underwent his third heart catheterization. He received intravenous radiocontrast with ioversol; many other drugs were used at the same time. Twelve hours later, a generalized erythematous, pustular, edematous rash starting in the intertriginous areas and then spreading to the entire body had developed with malaise and fever (39.6°C). There was leukocytosis with neutrophilia and elevated C-reactive protein. A skin biopsy showed subcorneal intraepidermal pustules, spongiosis, rare apoptotic keratinocytes, edema of the papillary dermis, inflammatory infiltrates, and few eosinophils, consistent with acute generalized exanthematous pustulosis (AGEP). The patient had experienced 2 similar episodes previously after receiving ioversol for radiocontrast during catheterization, but these had been ascribed to antibiotics. The only medication common to all 3 episodes was ioversol. Patch tests were positive to ioversol and the related iohexol, both tested as a mixture of 70% sterile water and 30% contrast dye. One (!) control was negative (6).

In a multicenter investigation in France, of 45 patients patch tested for AGEP, 26 (58%) had positive patch tests to drugs, including one to ioversol; clinical details were not provided (5).

Fixed drug eruption
A 60-year-old patient underwent several computed tomography (CT) scans with contrast media, but no adverse reactions were observed. Following his two most recent CT scans, performed with ioversol, a pruritic, erythematous and blistering eruption localized on the backs and interdigital areas of both hands developed after 24 hours. In both episodes, the skin eruption was worst at 2-3 days, and disappeared spontaneously 10 days after administration of ioversol. Patch tests were positive at D4 (negative at D2) for ioversol, iopamidol and iohexol, tested as undiluted contrast media, on previously affected but not on normal skin. Fixed drug eruption to ioversol was diagnosed (4).

A patient from Japan had a fixed drug eruption from the contrast medium ioversol, confirmed by a positive patch test (2; details unknown, article in Japanese, data cited in ref. 3).

Drug reaction with eosinophilia and systemic symptoms (DRESS)
In France, 13 adult patients in the dermatologist's French Investigators for Skin Adverse Reactions to Drugs (FISARD) network diagnosed between 2010 and 2020 with DRESS highly suspected to have been caused by a iodinated contrast medium (ICM) according to either a positive skin test or a challenge test, were retrospectively evaluated. In the group of 13 DRESS patients, 4 drug reactions had been caused by ioversol, of whom all had positive patch tests to this ICM (16).

Several hours after undergoing computed tomography with the contrast medium ioversol, an 83-year-old woman developed a maculopapular eruption with intensely pruritic erythematous-violaceous lesions on her neck, back, arms and legs. During the following days, the patient developed generalized edema and scaling. Lab investigations showed leukocytosis, eosinophilia and elevated liver enzymes. Patch tests with radiological contrast were positive to ioversol and iohexol (D2 and D4 ++). The patient was also allergic to imipenem, with which she had been treated before and during the evolution of the skin rash (10).

A patient developed DRESS after injections of ioxitalamate and ioversol. Patch tests were positive to these two iodinated contrast media and to various others, including iodixanol. Later, an investigation using iodixanol resulted in a second episode of DRESS with a skin eruption covering 80% of the skin surface and eosinophilia, despite pretreatment with corticosteroids (11).

Other cutaneous adverse drug reactions
A case of delayed hypersensitivity reaction to ioversol with a positive patch test has been reported in German literature; details are not available to the author (15).

Cross-reactions, pseudo-cross-reactions and co-reactions
Cross-reactions between iodinated radiocontrast media occur frequently, especially between the non-ionic preparations (8,9,16,17). Iodine is infrequently to rarely (opinions differ) the allergen (12,13) and allergy to mollusks, crustaceans, fish, and iodine from other sources is said not to be a risk factor for the development of hypersensitivity to iodinated contrast media (14).

LITERATURE
1   The data in the section 'General' may have been obtained from literature discussed in this chapter, but mostly also or exclusively from one or more of the following online sources: ChemIDPlus Advanced, PubChem, DrugBank, RxList, Drug Central, Drugs.com, and Wikipedia
2   Ura H, Yamada N, Imakado S, Iozumi K, Shimada S. A case of drug eruption due to ioversol. Rinsho Derma (Tokyo) 1996;50:769-772 (Article in Japanese)
3   Watanabe H, Sueki H, Nakada T, Akiyama M, Iijima M. Multiple fixed drug eruption caused by iomeprol (Iomeron ®), a non-ionic contrast medium. Dermatology 1999;198:291-294
4   Frías M, Fernández E, Audicana MT, Longo N, Muñoz D, Reyes SM. Fixed drug eruption caused by iodinated contrast media. Contact Dermatitis 2011;65:43-44
5   Barbaud A, Collet E, Milpied B, Assier H, Staumont D, Avenel-Audran M, et al. A multicentre study to determine the value and safety of drug patch tests for the three main classes of severe cutaneous adverse drug reactions. Br J Dermatol 2013;168:555-562
6   Poliak N, Elias M, Cianferoni A, Treat J. Acute generalized exanthematous pustulosis: the first pediatric case caused by a contrast agent. Ann Allergy Asthma Immunol 2010;105:242-243

7   Brockow K, Garvey LH, Aberer W, Atanaskovic-Markovic M, Barbaud A, Bilo MB, et al. Skin test concentrations for systemically administered drugs - an ENDA/EAACI Drug Allergy Interest Group position paper. Allergy 2013;68:702-712

8   Kanny G, Pichler W, Morisset M, Franck P, Marie B, Kohler C, et al. T cell-mediated reactions to iodinated contrast media: evaluation by skin and lymphocyte activation tests. J Allergy Clin Immunol 2005;115:179-185

9   Lerondeau B, Trechot P, Waton J, Poreaux C, Luc A, Schmutz JL, et al. Analysis of cross-reactivity among radiocontrast media in 97 hypersensitivity reactions. J Allergy Clin Immunol 2016;137:633-635.e4

10 Macías EM, Muñoz-Bellido FJ, Velasco A, Moreno E, Dávila I. DRESS syndrome involving 2 unrelated substances: imipenem and iodinated contrast media. J Investig Allergol Clin Immunol 2013;23:56-57

11 Amsler E, Autegarden JE, Senet P, Frances C, Soria A. Recurrence of drug eruption after renewed injection of iodinated contrast medium in patients with known allergic contraindications. Ann Dermatol Venereol 2016;143:804-807 (Article in French).

12 Scherer K, Harr T, Bach A, Bircher AJ. The role of iodine in hypersensitivity reactions to radiocontrast media. Clin Exp Allergy 2009;40:468-475

13 Trautmann A, Brockow K, Behle V, Stoevesandt J. Radiocontrast media hypersensitivity: Skin testing differentiates allergy from nonallergic reactions and identifies a safe alternative as proven by intravenous provocation. J Allergy Clin Immunol Pract 2019;7:2218-2224

14 Rosado Ingelmo A, Doña Diaz I, Cabañas Moreno R, Moya Quesada MC, García-Avilés C, García Nuñez I, et al. Clinical practice guidelines for diagnosis and management of hypersensitivity reactions to contrast media. J Investig Allergol Clin Immunol 2016;26:144-155

15 Erdmann S, Roos T, Merk HF. Delayed hypersensitivity reaction to the nonionic contrast medium ioversol. Zeitschrift Für Haut- und Geschlechtskrankheiten 2000;75:169-171

16 Soria A, Amsler E, Bernier C, Milpied B, Tétart F, Morice C, et al. DRESS and AGEP reactions to iodinated contrast media: A French case series. J Allergy Clin Immunol Pract 2021;9:3041-3050

17 Yoon SH, Lee SY, Kang HR, Kim JY, Hahn S, Park CM, et al. Skin tests in patients with hypersensitivity reaction to iodinated contrast media: a meta-analysis. Allergy 2015;70:625-637

Chapter 3.268 IOXAGLIC ACID

IDENTIFICATION

Description/definition   : Ioxaglic acid is the tri-iodinated benzoate that conforms to the structural formula shown below
Pharmacological classes   : Contrast media
IUPAC name   : 3-[[2-[[3-[Aacetyl(methyl)amino]-2,4,6-triiodo-5-(methylcarbamoyl)benzoyl]-amino]acetyl]amino]-5-(2-hydroxyethylcarbamoyl)-2,4,6-triiodobenzoic acid
CAS registry number   : 59017-64-0
EC number   : 261-560-1
Merck Index monograph   : 6380
Patch testing   : Ioxaglic acid-containing contrast medium undiluted (5)
Molecular formula   : $C_{24}H_{21}I_6N_5O_8$

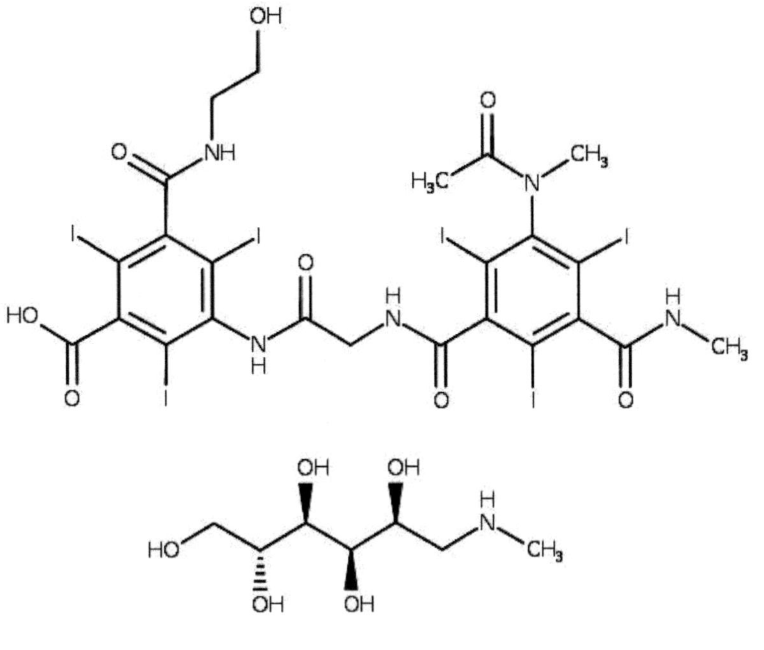

Ioxaglate meglumine

GENERAL

Ioxaglic acid is an ionic dimer tri-iodinated benzoate used as a contrast agent in diagnostic imaging. Like other organic iodine compounds, ioxaglic acid blocks x-rays and appears opaque on x-ray film, thereby enhancing the visibility of structure and organs during angiography, arteriography, arthrography, cholangiography, urography, and computed tomography (CT) scanning procedures. In pharmaceutical products, ioxaglic acid is employed as a combination of ioxaglate meglumine (CAS number 59018-13-2, EC number 261-561-7, molecular formula $C_{31}H_{38}I_6N_6O_{13}$) and ioxaglate sodium (CAS number 67992-58-9, EC number 268-060-2, molecular formula $C_{24}H_{20}I_6N_5NaO_8$) (1).

CUTANEOUS ADVERSE DRUG REACTIONS FROM SYSTEMIC ADMINISTRATION CAUSED BY TYPE IV (DELAYED-TYPE) HYPERSENSITIVITY (as demonstrated by positive patch tests)

Cutaneous adverse drug reactions from systemic administration of ioxaglic acid caused by type IV (delayed-type) hypersensitivity have included drug reaction with eosinophilia and systemic symptoms (DRESS) (2,3).

Drug reaction with eosinophilia and systemic symptoms (DRESS)

A 63-year-old woman underwent a coronary arteriography with the iodinated contrast medium (ICM) ioxaglate meglumine/sodium. One week later, she had endoluminal coronary angioplasty, which included injection of the same ICM. The next morning the patient had developed generalized malaise, a fever to 39.2°C, hypotension, and a nonpruritic maculopapular rash, which later extended to cover her entire body, while sparing the mucous membranes. Further investigations revealed hepatic cytolysis, rhabdomyolysis, eosinophilia, and high CRP. Later, skin prick tests, intradermal tests and patch tests were carried out with ioxaglic acid-meglumine, related ICMs and all drugs the patient had used. The patch tests with the ICMs were negative, but intradermal tests were positive after 48 hours to ioxaglate meglumine/sodium, sodium ioxitalamate, and iopamidol. Twenty controls were negative. The histology was consistent with delayed-type drug hypersensitivity (later termed drug reaction with eosinophilia and systemic symptoms [DRESS]) (2).

Seven years later, this patient was injected with the same iodinated contrast agent (ioxaglate meglumine and ioxaglate sodium) during coronary angiography, although she informed the staff of the previous hypersensitivity reaction and presented her allergy records. A fever (temperature of 39°C) developed as before, with a maculopapular and purpuric eruption, diarrhea, renal insufficiency, cough, and pulmonary interstitial syndrome with hypoxia. Now, both intradermal and patch tests were positive to ioxaglate meglumine/sodium and a large number of other iodinated contrast media (3).

Cross-reactions, pseudo-cross-reactions and co-reactions
Cross-reactions between iodinated radiocontrast media occur frequently, especially between the non-ionic preparations (3,4,9). Iodine is infrequently to rarely (opinions differ) the allergen (6,7) and allergy to mollusks, crustaceans, fish, and iodine from other sources is said not to be a risk factor for the development of hypersensitivity to iodinated contrast media (8).

References
1   The data in the section 'General' may have been obtained from literature discussed in this chapter, but mostly also or exclusively from one or more of the following online sources: ChemIDPlus Advanced, PubChem, DrugBank, RxList, Drug Central, Drugs.com, and Wikipedia
2   Kanny G, Marie B, Hoen B, Trechot P, Morenet-Vautrin DA. Delayed adverse reaction to sodium ioxaglic acid-meglumine. Eur J Dermatol 2001;11:134-137
3   Kanny G, Pichler W, Morisset M, Franck P, Marie B, Kohler C, Renaudin JM, Beaudouin E, Laudy JS, Moneret-Vautrin DA. T cell-mediated reactions to iodinated contrast media: evaluation by skin and lymphocyte activation tests. J Allergy Clin Immunol 2005;115:179-185
4   Lerondeau B, Trechot P, Waton J, Poreaux C, Luc A, Schmutz JL, et al. Analysis of cross-reactivity among radiocontrast media in 97 hypersensitivity reactions. J Allergy Clin Immunol 2016;137:633-635.e4
5   Brockow K, Garvey LH, Aberer W, Atanaskovic-Markovic M, Barbaud A, Bilo MB, et al. Skin test concentrations for systemically administered drugs - an ENDA/EAACI Drug Allergy Interest Group position paper. Allergy 2013;68:702-712
6   Scherer K, Harr T, Bach A, Bircher AJ. The role of iodine in hypersensitivity reactions to radiocontrast media. Clin Exp Allergy 2009;40:468-475
7   Trautmann A, Brockow K, Behle V, Stoevesandt J. Radiocontrast media hypersensitivity: Skin testing differentiates allergy from nonallergic reactions and identifies a safe alternative as proven by intravenous provocation. J Allergy Clin Immunol Pract 2019;7:2218-2224
8   Rosado Ingelmo A, Doña Diaz I, Cabañas Moreno R, Moya Quesada MC, García-Avilés C, García Nuñez I, et al. Clinical practice guidelines for diagnosis and management of hypersensitivity reactions to contrast media. J Investig Allergol Clin Immunol 2016;26:144-155
9   Yoon SH, Lee SY, Kang HR, Kim JY, Hahn S, Park CM, et al. Skin tests in patients with hypersensitivity reaction to iodinated contrast media: a meta-analysis. Allergy 2015;70:625-637

Chapter 3.269 IOXITALAMIC ACID

IDENTIFICATION

Description/definition	: Ioxitalamic acid is the halobenzoic acid derivative that conforms to the structural formula shown below
Pharmacological classes	: Contrast media
IUPAC name	: 3-Acetamido-5-(2-hydroxyethylcarbamoyl)-2,4,6-triiodobenzoic acid
CAS registry number	: 28179-44-4
EC number	: 248-887-5
Patch testing	: Ioxitalamic acid containing contrast medium undiluted (4)
Molecular formula	: $C_{12}H_{11}I_3N_2O_5$

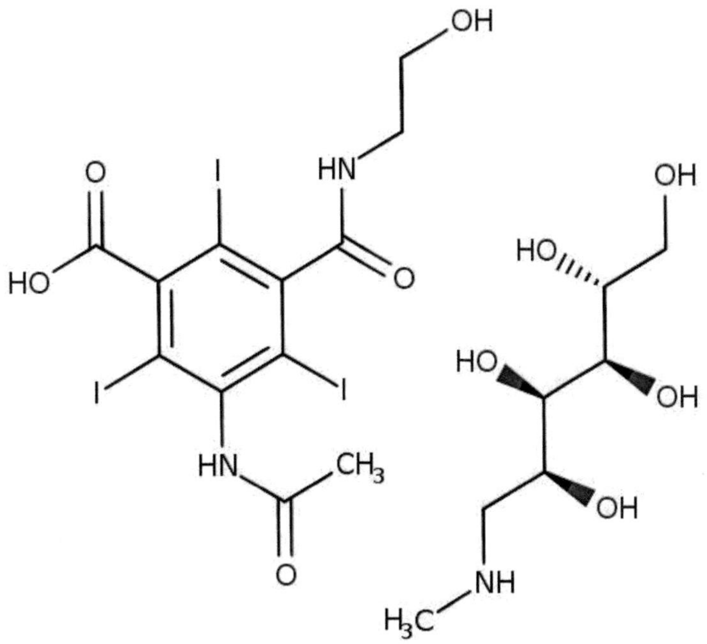

Meglumine ioxitalamate

GENERAL

Ioxitalamic acid is a first-generation ionic monomer iodinated contrast medium. In pharmaceutical products, ioxitalamic acid may be employed as meglumine ioxitalamate (CAS number 29288-99-1, EC number 249-544-2, molecular formula $C_{19}H_{28}I_3N_3O_{10}$) or as sodium ioxitalamate (CAS number 33954-26-6 , EC number not available, molecular formula $C_{12}H_{10}I_3N_2NaO_5$) (1).

CUTANEOUS ADVERSE DRUG REACTIONS FROM SYSTEMIC ADMINISTRATION CAUSED BY TYPE IV (DELAYED-TYPE) HYPERSENSITIVITY (as demonstrated by positive patch tests)

Cutaneous adverse drug reactions from systemic administration of ioxitalamic acid caused by type IV (delayed-type) hypersensitivity have included maculopapular eruption (2) and drug reaction with eosinophilia and systemic symptoms (DRESS) (5,6).

Maculopapular eruption

A 49-year-old nurse's aid was given an injection with the ionic iodinated contrast medium meglumine ioxitalamate and developed a maculopapular eruption 24 hours later, which lasted for 10 days. Patch tests were positive to the commercial meglumine ioxitalamate preparation, tested undiluted (2).

Drug reaction with eosinophilia and systemic symptoms (DRESS)

Two days after transluminal angioplasty with a stent installation, which included the injection of sodium meglumine ioxitalamate, a 71-year-old woman developed DRESS with a pruritic maculopapular eruption, fever (38.5°C), oliguria, a creatinine rise and eosinophilia. Six weeks after complete resolution, patch tests were positive to the commercial ICM undiluted and 30% water, and negative in nine controls (5).

A patient developed DRESS after injections of ioxitalamate and ioversol. Patch tests were positive to these two iodinated contrast media and to various others, including iodixanol. Later, an investigation using iodixanol resulted in a second episode of DRESS with a skin eruption covering 80% of the skin surface and eosinophilia, despite pretreatment with corticosteroids (6).

Cross-reactions, pseudo-cross-reactions and co-reactions
Cross-reactions between iodinated radiocontrast media occur frequently, especially between the non-ionic preparations (2,3,10). Iodine is infrequently to rarely (opinions differ) the allergen (7,8) and allergy to mollusks, crustaceans, fish, and iodine from other sources is said not to be a risk factor for the development of hypersensitivity to iodinated contrast media (9).

LITERATURE

1 The data in the section 'General' may have been obtained from literature discussed in this chapter, but mostly also or exclusively from one or more of the following online sources: ChemIDPlus Advanced, PubChem, DrugBank, RxList, Drug Central, Drugs.com, and Wikipedia

2 Kanny G, Pichler W, Morisset M, Franck P, Marie B, Kohler C, Renaudin JM, Beaudouin E, Laudy JS, Moneret-Vautrin DA. T cell-mediated reactions to iodinated contrast media: evaluation by skin and lymphocyte activation tests. J Allergy Clin Immunol 2005;115:179-185

3 Lerondeau B, Trechot P, Waton J, Poreaux C, Luc A, Schmutz JL, et al. Analysis of cross-reactivity among radiocontrast media in 97 hypersensitivity reactions. J Allergy Clin Immunol 2016;137:633-635.e4

4 Brockow K, Garvey LH, Aberer W, Atanaskovic-Markovic M, Barbaud A, Bilo MB, et al. Skin test concentrations for systemically administered drugs - an ENDA/EAACI Drug Allergy Interest Group position paper. Allergy 2013;68:702-712

5 Belhadjali H, Bouzgarrou L, Youssef M, Njim L, Zili J. DRESS syndrome induced by sodium meglumine ioxitalamate. Allergy 2008;63:786-787

6 Amsler E, Autegarden JE, Senet P, Frances C, Soria A. Recurrence of drug eruption after renewed injection of iodinated contrast medium in patients with known allergic contraindications. Ann Dermatol Venereol 2016;143:804-807 (Article in French).

7 Scherer K, Harr T, Bach A, Bircher AJ. The role of iodine in hypersensitivity reactions to radiocontrast media. Clin Exp Allergy 2009;40:468-475

8 Trautmann A, Brockow K, Behle V, Stoevesandt J. Radiocontrast media hypersensitivity: Skin testing differentiates allergy from nonallergic reactions and identifies a safe alternative as proven by intravenous provocation. J Allergy Clin Immunol Pract 2019;7:2218-2224

9 Rosado Ingelmo A, Doña Diaz I, Cabañas Moreno R, Moya Quesada MC, García-Avilés C, García Nuñez I, et al. Clinical practice guidelines for diagnosis and management of hypersensitivity reactions to contrast media. J Investig Allergol Clin Immunol 2016;26:144-155

10 Yoon SH, Lee SY, Kang HR, Kim JY, Hahn S, Park CM, et al. Skin tests in patients with hypersensitivity reaction to iodinated contrast media: a meta-analysis. Allergy 2015;70:625-637

Chapter 3.270 IPRAGLIFLOZIN

IDENTIFICATION

Description/definition   : Ipragliflozin is the phenolic glycoside that conforms to the structural formula shown
                             below
Pharmacological classes   : Sodium-glucose transporter 2 inhibitors
IUPAC name   : (2S,3R,4R,5S,6R)-2-[3-(1-Benzothiophen-2-ylmethyl)-4-fluorophenyl]-6-(hydroxymethyl)-
                             oxane-3,4,5-triol
CAS registry number   : 761423-87-4
EC number   : Not available
Merck Index monograph   : 11865
Patch testing   : Crushed tablet 10% and 20% in water and pet. (1); most systemic drugs can be tested at
                             10% pet.; if the pure chemical is not available, prepare the test material from intravenous
                             powder, the content of capsules or – when also not available – from powdered tablets to
                             achieve a final concentration of the active drug of 10% pet.
Molecular formula   : $C_{21}H_{21}FO_5S$

GENERAL

Ipragliflozin is a sodium-glucose cotransporter 2 (SGLT2) inhibitor. SGLT2 mediates physiological and pathological glucose levels in the kidneys. Unlike other antidiabetic drugs, SGLT2 inhibitors enhance renal glucose excretion by inhibiting renal glucose reabsorption, and they also improve insulin resistance by reducing plasma glucose levels and abolishing glucose toxicity in diabetes (1).

CUTANEOUS ADVERSE DRUG REACTIONS FROM SYSTEMIC ADMINISTRATION CAUSED BY TYPE IV (DELAYED-TYPE) HYPERSENSITIVITY (as demonstrated by positive patch tests)

Cutaneous adverse drug reactions from systemic administration of ipragliflozin caused by type IV (delayed-type) hypersensitivity have included nummular dermatitis-like eruption (1).

Dermatitis/eczematous eruption

A 68-year-old man had been treated with ipragliflozin for diabetes mellitus for 2 weeks, when he developed scaly erythematous plaques on the arms, legs and trunk, resembling nummular dermatitis. He had no history of atopic dermatitis or other allergic diseases. A skin biopsy specimen taken from a scaly erythematous plaque on his leg revealed parakeratosis and spongiotic changes in the epidermis, as well as inflammatory cell infiltration with eosinophils in the upper dermis and dermal-epidermal interface. A KOH-preparation was negative for fungi, as was a periodic acid Schiff stain of the skin biopsy specimen. A lymphocyte stimulation test to ipragliflozin was positive with a stimulation index of 2.3. The results of patch tests with ipragliflozin in 10 and 20% concentrations using a petrolatum vehicle or aqueous solution were all positive. Based on the clinical course and laboratory examination, the patient was diagnosed with drug eruption caused by delayed-type hypersensitivity to ipragliflozin (1).

LITERATURE

1   Saito-Sasaki N, Sawada Y, Nishio D, Nakamura M. First case of drug eruption due to ipragliflozin: case report and
     review of the literature. Australas J Dermatol 2017;58:236-238

Chapter 3.271 IRBESARTAN

IDENTIFICATION

Description/definition   : Irbesartan is biphenyl compound that conforms to the structural formula shown below
Pharmacological classes   : Angiotensin II type 1 receptor blockers; antihypertensive agents
IUPAC name   : 2-Butyl-3-[[4-[2-(2H-tetrazol-5-yl)phenyl]phenyl]methyl]-1,3-diazaspiro[4.4]non-1-en-4-one
CAS registry number   : 138402-11-6
EC number   : 604-078-2
Merck Index monograph : 6397
Patch testing   : 5% pet. (2)
Molecular formula   : $C_{25}H_{28}N_6O$

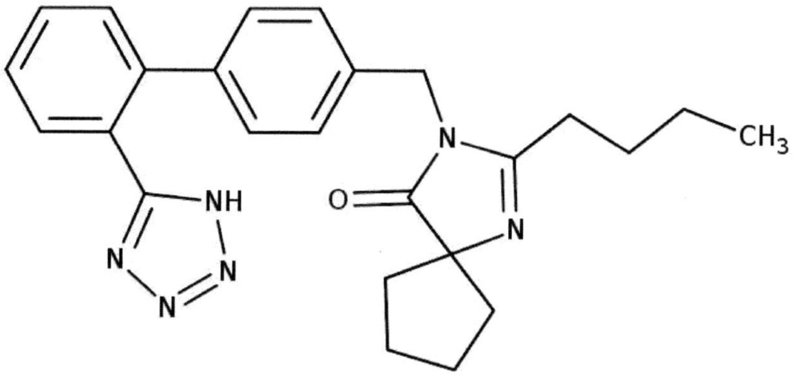

GENERAL

Irbesartan is a nonpeptide angiotensin II antagonist with antihypertensive activity. Irbesartan selectively and competitively blocks the binding of angiotensin II to the angiotensin I receptor. Angiotensin II stimulates aldosterone synthesis and secretion by the adrenal cortex, which decreases the excretion of sodium and increases the excretion of potassium. Angiotensin II also acts as a vasoconstrictor in vascular smooth muscle. Irbesartan is used to treat hypertension and diabetic nephropathy (1).

In pharmaceutical products, irbesartan is employed as irbesartan hydrochloride (CAS number 329055-23-4, EC number not available, molecular formula $C_{25}H_{29}ClN_6O$) (1).

CUTANEOUS ADVERSE DRUG REACTIONS FROM SYSTEMIC ADMINISTRATION CAUSED BY TYPE IV (DELAYED-TYPE) HYPERSENSITIVITY (as demonstrated by positive patch tests)

Cutaneous adverse drug reactions from systemic administration of irbesartan caused by type IV (delayed-type) hypersensitivity have included maculopapular eruption (2).

Maculopapular eruption

A 60-year-old woman presented with a 3-week history of an itchy erythematous maculopapular eruption affecting the thorax, abdomen and proximal parts of the arms and legs, with lesions resolving with hyperpigmentation. The patient used 7 medications, most of which had been taken for several years. However, she had started on irbesartan some 2 months before the onset of the cutaneous reaction. Histopathology showed a predominantly lymphocytic infiltrate of the dermis with some eosinophils, compatible with a drug reaction. A patch test with irbesartan 5% pet. was positive at D2 and D4, while the related candesartan was negative. A lymphocyte transformation test (LTT) to irbesartan (100 µg/ml) was positive, showing a stimulation index of 6.3. Irbesartan was switched to diltiazem and new lesions stopped appearing. The patient was diagnosed with a delayed-type allergic drug reaction to irbesartan (2).

LITERATURE

1   The data in the section 'General' may have been obtained from literature discussed in this chapter, but mostly also or exclusively from one or more of the following online sources: ChemIDPlus Advanced, PubChem, DrugBank, RxList, Drug Central, Drugs.com, and Wikipedia
2   Cardoso BK, Martins M, Farinha SM, Viseu R, Tomaz E, Inácio F. Late-onset rash from irbesartan: An immunological reaction. Eur J Case Rep Intern Med 2019;6(6):001128

Chapter 3.272 ISEPAMICIN

IDENTIFICATION

Description/definition	: Isepamicin is the 2-deoxystreptamine aminoglycoside that conforms to the structural formula shown below
Pharmacological classes	: Anti-bacterial agents
IUPAC name	: (2S)-3-Amino-N-[(1R,2S,3S,4R,5S)-5-amino-4-[(2R,3R,4S,5S,6R)-6-(aminomethyl)-3,4,5-trihydroxyoxan-2-yl]oxy-2-[(2R,3R,4R,5R)-3,5-dihydroxy-5-methyl-4-(methylamino)oxan-2-yl]oxy-3-hydroxycyclohexyl]-2-hydroxypropanamide
CAS registry number	: 58152-03-7
EC number	: 261-143-4
Merck Index monograph	: 6421
Patch testing	: Sulfate, 10% water
Molecular formula	: $C_{22}H_{43}N_5O_{12}$

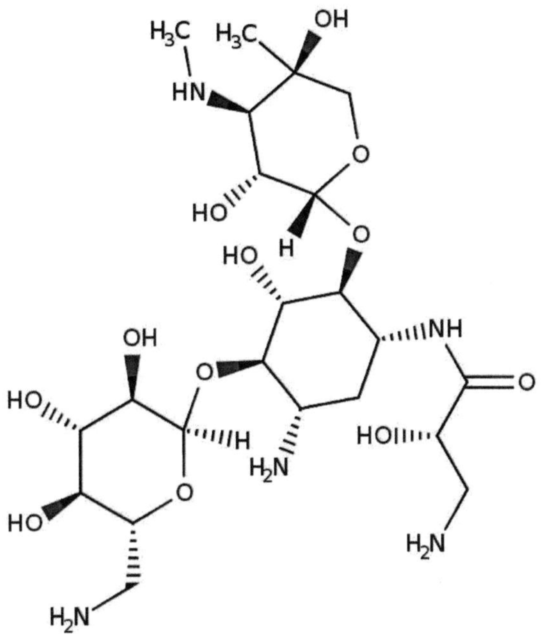

GENERAL

Isepamicin is an aminoglycoside antibacterial with properties similar to those of amikacin, but with better activity against strains producing type I 6'-acetyltransferase. The antibacterial spectrum includes Enterobacteriaceae and staphylococci. Like other aminoglycosides, isepamicin exhibits a strong concentration-dependent bactericidal effect, a long post-antibiotic effect (several hours) and induces adaptive resistance . In pharmaceutical products, isepamicin is employed as isepamicin sulfate (CAS number 67814-76-0, EC number not available, molecular formula $C_{22}H_{45}N_5O_{16}S$) (1).

CUTANEOUS ADVERSE DRUG REACTIONS FROM SYSTEMIC ADMINISTRATION CAUSED BY TYPE IV (DELAYED-TYPE) HYPERSENSITIVITY (as demonstrated by positive patch tests)

Cutaneous adverse drug reactions from systemic administration of isepamicin caused by type IV (delayed-type) hypersensitivity have included acute generalized exanthematous pustulosis (AGEP).

Acute generalized exanthematous pustulosis (AGEP)

A 62-year-old woman presented with a widespread rash and a high fever of 39°C. On physical examination, her nape, upper back, axilla, abdominal wall, and a proximal part of the arms and legs including the popliteal and cubital areas showed diffuse erythema with marked edema. Numerous pinhead-sized pustules were present on her nape and scattered on the other erythematous lesions. There was bilateral cervical lymphadenopathy. The patient had recently undergone treatment of a vitreous hemorrhage due to diabetic retinopathy and microangiopathy. Eleven different drugs were given at the time of the development of the eruption including isepamicin sulfate. Laboratory tests showed leukocytosis with neutrophilia. After resolution of the skin reaction, patch tests with the 4 most suspected drugs were positive only to isepamicin sulfate 10% water. Three controls were negative. The patient was diagnosed with AGEP due to isepamicin sulfate. A role for cadralazine could not be excluded, as the lymphocyte stimulation test to this drug was strongly positive (but a patch test negative) (2).

LITERATURE

1   The data in the section 'General' may have been obtained from literature discussed in this chapter, but mostly also or exclusively from one or more of the following online sources: ChemIDPlus Advanced, PubChem, DrugBank, RxList, Drug Central, Drugs.com, and Wikipedia
2   Katagiri K, Takayasu S. Drug induced acute generalized exanthematous pustulosis. J Dermatol 1996;23:623-627

Chapter 3.273 ISOFLURANE

IDENTIFICATION

Description/definition   : Isoflurane is the fluorinated ether that conforms to the structural formula shown below
Pharmacological classes : Anesthetics, inhalation
IUPAC name   : 2-Chloro-2-(difluoromethoxy)-1,1,1-trifluoroethane
CAS registry number   : 26675-46-7
EC number   : 247-897-7
Merck Index monograph   : 6491
Patch testing   : Repeated open application test (ROAT) with the fluid; the material is too volatile for patch testing
Molecular formula   : $C_3H_2ClF_5O$

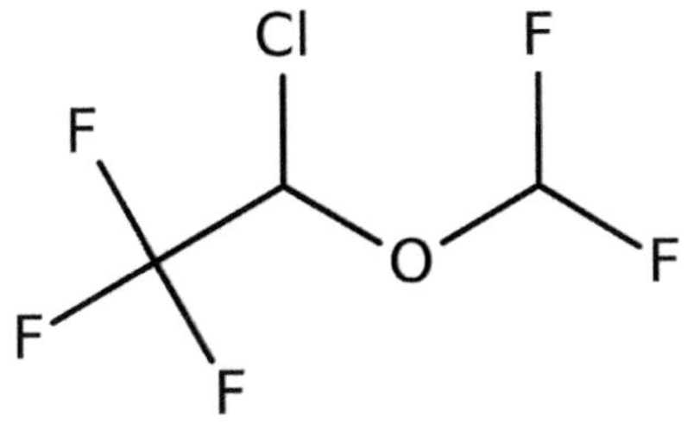

GENERAL

Isoflurane is a fluorinated ether with general anesthetic and muscle relaxant activities. It induces muscle relaxation and reduces pain sensitivity by altering tissue excitability. Isoflurane is indicated for induction and maintenance of general anesthesia by inhalation (1).

CONTACT ALLERGY FROM ACCIDENTAL CONTACT

A 57-year-old anesthetist presented with an 8-month history of recurrent left periorbital erythema and edema. His symptoms would worsen during the working week and improve at weekends. Patch tests showed multiple positives, including local anesthetics, cetrimide and lanolin alcohol. Despite strict avoidance of all the identified allergens, he remained symptomatic 2 months later. The history suggested a missed allergen, which was considered to be a general anaesthetic agent, isoflurane, sevoflurane and propofol being the commonest used by the patient. Patch tests to all 3 agents were negative at 2 and 4 days. Being volatile liquids, isoflurane and sevoflurane quickly evaporated from the skin surface, possibly causing a false-negative patch test reaction. Repeated open application tests (ROAT) were therefore performed by applying 1 ml of each anesthetic to the volar aspect of the forearm 2x a day. The ROAT to isoflurane elicited a discoid eczematous patch after 3 such applications. Twenty controls were negative. The anesthetist was diagnosed with occupational airborne allergic contact dermatitis to isoflurane (2).

Another anesthetist developed an itchy erythematosquamous dermatitis on the face and V of the chest. The appearance of dermatitis coincided with the work in operating rooms, where isoflurane was used. A ROAT with isoflurane liquid 1 ml 2x per day applied with a cotton pad elicited a nummular erythematous patch after 5 applications; 10 controls were negative. Subsequently, the dermatitis relapsed twice, 2-3 hours after contact with isoflurane. This anesthetist, too, had occupational airborne allergic contact dermatitis from this general anesthetic (3).

LITERATURE

1   The data in the section 'General' may have been obtained from literature discussed in this chapter, but mostly also or exclusively from one or more of the following online sources: ChemIDPlus Advanced, PubChem, DrugBank, RxList, Drug Central, Drugs.com, and Wikipedia
2   Finch TM, Muncaster A, Prais L, Foulds IS. Occupational airborne allergic contact dermatitis from isoflurane vapour. Contact Dermatitis 2000;42:46
3   Caraffini S, Ricci F, Assalve D, Lisi P. Isoflurane: an uncommon cause of occupational airborne contact dermatitis. Contact Dermatitis 1998;38:286

Chapter 3.274 ISONIAZID

IDENTIFICATION

Description/definition : Isoniazid is the synthetic derivative of nicotinic acid that conforms to the structural formula shown below
Pharmacological classes : Fatty acid synthesis inhibitors; antitubercular agents
IUPAC name : Pyridine-4-carbohydrazide
CAS registry number : 54-85-3
EC number : 200-214-6
Merck Index monograph : 6502
Patch testing : Pulverized tablets 5%, 10% and 30% pet.; pure isoniazid 1% pet. and water have also been used with success, but the low concentration will likely result in false-negative reactions in a number of cases (15)
Molecular formula : $C_6H_7N_3O$

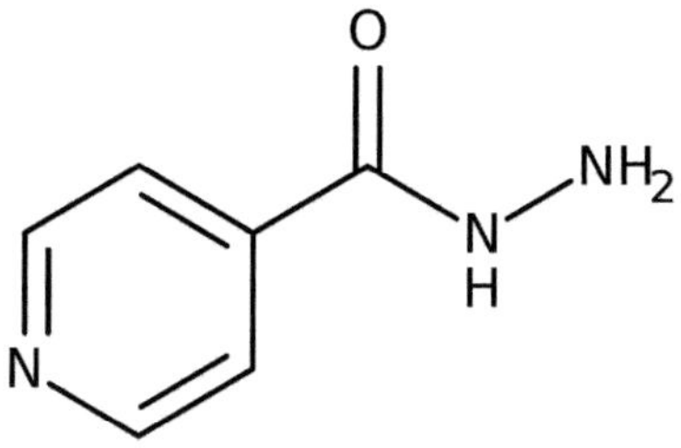

GENERAL

Isoniazid is a synthetic derivative of nicotinic acid with anti-mycobacterial properties. It is active against organisms of the genus *Mycobacterium*, specifically *M. tuberculosis*, *M. bovis* and *M. kansasii*. It is a highly specific agent, ineffective against other microorganisms. Isoniazid is bactericidal to rapidly-dividing mycobacteria, but is bacteriostatic if the mycobacterium is slow-growing. Isoniazid is indicated for the treatment of all forms of tuberculosis caused by susceptible organisms (1).

CONTACT ALLERGY FROM ACCIDENTAL CONTACT

Of 333 nurses patch tested in Poland between 1979 and 1987, 1 reacted to isoniazid 0.05% pet.; she had previously worked in a hospital for tuberculosis treatment (9). Occupational sensitization to isoniazid in nurses has also been reported in early German literature (19,20,24).

CUTANEOUS ADVERSE DRUG REACTIONS FROM SYSTEMIC ADMINISTRATION CAUSED BY TYPE IV (DELAYED-TYPE) HYPERSENSITIVITY (as demonstrated by positive patch tests)

Cutaneous adverse drug reactions from systemic administration of isoniazid caused by type IV (delayed-type) hypersensitivity have included maculopapular eruption (8), acute generalized exanthematous pustulosis (2), drug reaction with eosinophilia and systemic symptoms (DRESS) (4,5,11,13,14,15,16,26), Stevens-Johnson syndrome/toxic epidermal necrolysis (SJS/TEN) (11), photosensitivity (12), photoallergic lichenoid eruption (12,17), dermatitis/eczematous eruption (3,18), itchy, desquamative, erythematous, papular rash, with painful crusts and excoriations, spreading from the face and hands to the rest of the body (7), and unspecified drug eruption (6,20).

Case series with various or unknown types of drug reactions

In the period 2010-2014, in South Africa, 60 patients with cutaneous adverse drug reactions to first-line antituberculosis drugs (FLTD: isoniazid, rifampicin, pyrazinamide and ethambutol) were patch tested with the patients' drugs 30% water and olive oil. Positive reactions to at least one FLTD were seen in 14 participants, of who twelve had DRESS, one SJS and one SJS/TEN. There were 6 positive patch tests to isoniazid, 4 in patients with DRESS and 2 with SJS/TEN. Eleven of the 14 patients were HIV-infected and in 10 of these, patch tests resulted in a generalized systemic reaction (isoniazid 6, rifampicin 3, ethambutol 3, pyrazinamide one) (11).

Maculopapular eruption

A 72-year-old woman developed a generalized maculopapular rash that had appeared 4 weeks after she had started taking 4 antituberculosis drugs. She had eosinophilia (eosinophils 64.3%) and abnormal liver function tests. Six weeks after discharge from the hospital, patch tests with the 4 drugs as crushed tablets at 50% in petrolatum showed a diffuse erythematous rash around the isoniazid, rifampicin, and ethambutol patches at D2. The authors admitted these may have been nonspecific irritant reactions (8).

Acute generalized exanthematous pustulosis (AGEP)

An 81-year-old man presented with a pustular eruption with fever. He had been taking various drugs including isoniazid for a constant irritating cough for 6 weeks. Within two weeks, he developed an itchy sensation over most of his body followed by a widespread eruption of flaccid pustules, varying in size, on an erythematous base, especially on the trunk. Laboratory examination showed leukocytosis with neutrophilia, and increased erythrocyte sedimentation rate. Bacteriological examination of blood, urine and a pustule was negative. Two biopsy specimens from a small intact blister demonstrated a subcorneal pustule filled with polymorphonuclear leukocytes. A mild dermal perivascular infiltrate of lymphocytes and plasma cells was seen. Later, patch and challenge tests for all the drugs used by the patient were performed. On patch testing, isoniazid 1% in petrolatum gave a positive reaction, revealing a pustular eruption with an erythematous base. Three days after administration of a test dose of isoniazid (one-tenth of the usual dosage), a pustular eruption clinically and histologically similar to the original eruption developed on the abdomen (2).

Drug reaction with eosinophilia and systemic symptoms (DRESS)

Case series

In France, a search was performed for potential cases of DRESS caused by the antituberculosis drugs rifampicin, isoniazid, pyrazinamide and ethambutol, reported from January 1, 2005, to July 30, 2015, in the French pharmacovigilance database. Sixty-seven cases of antituberculosis drug-associated DRESS were analyzed (40 women and 27 men, median age of 61 years). Patch tests were performed in 11 patients and were positive in 7. Five individuals reacted to isoniazid (no test concentrations/vehicles mentioned), 2 to ethambutol and one to rifampicin (one reacted to both ethambutol and isoniazid (4).

Case reports

A 56-year-old man developed pulmonary and lymph node tuberculosis, for which treatment was started with isoniazid, rifampicin, ethambutol and pyrazinamide. Six weeks later, the patient was hospitalized because of fever (38.5°C), an extensive maculopapular rash (>50% of body surface area), pronounced facial and palmoplantar edema, and disseminated peripheral lymphadenopathy. Laboratory tests showed eosinophilia and elevated serum aminotransferases. Three months later, patch tests were positive to isoniazid 1% pet., ethambutol 3% pet. and pyrazinamide 3% pet., the latter two made from pulverized commercial tablets. Twenty controls were negative to the ethambutol patch test material. Despite the negative patch test to rifampicin, graded re-introduction of the drug, even under prophylactic therapy with corticosteroids, led to a recurrence of DRESS symptoms after 7 days. No patch tests with rifampicin were performed afterwards (15).

The same authors also described a 21-year-old woman, who was treated with the same quadruple therapy and developed DRESS after 3 months. Patch tests were positive only to isoniazid commercial pulverized tablet: 'as is' water (D2 +; D4 +; D7 +), 30% water (D2 –; D4 ?+; D7 +),'as is' pet. (D2 +; D4 ++; D7 ++), and 30% pet. (D2 +; D4 +; D7 +). Ethambutol gave a dubious reaction, pyrazinamide and rifampicin were negative. Despite this, reintroduction of pyrazinamide led to a mild recurrence (15).

A 27-year-old woman from Morocco had been treated with the 4 first-line antituberculosis drugs (ethambutol, isoniazid, rifampicin, pyrazinamide) and developed fever, extensive infiltrated erythema, lymphadenopathy, impaired liver function, leukocytosis and eosinophilia. Patch tests were positive to isoniazid and ethambutol, tested at 1%, vehicle not mentioned (5).

Two years after a successful treatment with rifampicin, isoniazid and pyrazinamide for pulmonary tuberculosis, a 68-year-old man started a new treatment regimen with rifampicin, isoniazid and ethambutol because of fresh pulmonary involvement. After one month, the patient presented with fever, asthenia, and facial edema, without arthralgia or exanthema. Laboratory evaluations showed a significant pancytopenia and a moderate increase in liver transaminases. Patch testing with rifampicin, isoniazid and ethambutol (each 50% pet.) gave positive reactions only to isoniazid (+++ D2; uncertain why the D4 reading result was not mentioned). Ten controls were negative. Oral challenge tests with rifampicin and ethambutol were well tolerated (13).

A 21-year-old man with pulmonary tuberculosis started treatment with ethambutol, isoniazid, pyrazinamide, and rifampicin. One month later, he was diagnosed with DRESS because of fever, lymphadenopathy, pruritic rash, facial edema, leukocytosis, atypical lymphocytes, elevated liver enzyme levels and respiratory failure. Despite discontinuation of the antituberculosis drugs, the patient's condition worsened, and systemic corticosteroids and second-line antituberculosis therapy (ethambutol, streptomycin, and levofloxacin) were started. Three weeks later, the patient's symptoms resolved. Eight month later, patch tests were positive to isoniazid 1% water at D4, as were an intradermal test with isoniazid 0.6 mg/ml read at D3 (remaining positive to D7) and lymphocyte transformation tests (14).

One patient from Japan had DRESS while using allopurinol and various antituberculosis drugs; patch tests were positive to isoniazid on 3 occasions as was the lymphocyte transformation test. This was probably a case of multiple drug hypersensitivity (16).

A 16-year-old girl developed a hypersensitivity reaction while taking aminosalicylic acid in combination with isoniazid (INH) for tuberculosis. The reaction was characterized by nausea, high fever, headache, a generalized erythematous eruption with edema of the face, lips and eyelids, lymphadenopathy and a palpable spleen. A test for mononucleosis infectiosa was negative. Laboratory investigation showed elevated erythrocyte sedimentation rate and atypical lymphocytes. Two days after recovery (far too soon, but that was unknown in 1962) patch tests were performed and were positive to aminosalicylic acid 5% water and moistened isoniazid powder. Five controls were negative. During the patch tests a maculopapular exanthema appeared which generalized (26). This was highly likely a case of DRESS with multiple drug hypersensitivity. It is uncertain whether the exacerbation during patch testing was actually caused by this diagnostic procedure, or was a spontaneous exacerbation, as is often observed in DRESS.

See also the section 'Case series with various or unknown types of drug reactions' above, ref. 11.

Stevens-Johnson syndrome/toxic epidermal necrolysis (SJS/TEN)
See the section 'Case series with various or unknown types of drug reactions' above, ref. 11.

Photosensitivity
A man of 79 developed a lichenoid eruption and a man aged 68 an eczematous rash while using isoniazid and other antituberculosis drugs for pulmonary tuberculosis. The lesions were localized at first mainly on sun-exposed areas, but extended onto the covered areas later. Histopathology was consistent with the clinical diagnoses. Photopatch tests using isoniazid commercial tablets 10% pet. and 10 J/cm^2 UVA were positive in both patient, in the second one on 2 occasions (second time with 3 J/cm^2 UVA). Systemic provocation by sunlight exposure for one to two hours after taking a single test dose of isoniazid (1/4 normal strength) was positive in both individuals. The patient with the eczematous rash became a persistent light reactor, suffering intense photosensitive eruptions during more than four years (12).

Three years earlier, the first author had already reported the 2 first patients with isoniazid-induced photo-sensitive lichenoid eruptions confirmed by photopatch tests. Both individuals had lichenoid papules in sun-exposed areas, in one progressing to exfoliative erythroderma. Histopathology was consistent with lichenoid drug eruption. Photopatch tests were positive to isoniazid 10% pet. irradiated with 10 J/cm^2 UVA. In one, simultaneously with this photopatch test, the previous lesions all flared. Results of the photopatch test were confirmed by systemic provocation. Lichenoid papules developed in irradiated areas after oral challenge with a single dose of isoniazid followed by irradiation with 10 J/cm^2 of UVA. In the second patient, the histopathology of the positive photopatch test with isoniazid was similar to that of the original biopsy with lichenoid features (17).

Dermatitis/eczematous eruption
A 58-year-old man presented with a pruritic erythematous eruption of tiny vesicles on the trunk and extremities that developed during the sixth week of oral therapy with isoniazid, ethambutol, rifampicin, and morinamide HCl for pleural tuberculosis. Histopathologic findings were consistent with an eczematous eruption. Patch tests were positive to isoniazid, tested as isoniazid tablets at 5%, 10% and 30% water and pet. The tablets also contained vitamin B$_6$. This chemical was not patch tested separately, but oral challenge with a vitamin B$_6$ tablet (100 mg) was negative. There were also positive patch tests to ethambutol (Chapter 3.206). The patient was diagnosed with eczematous-type multiple drug allergy to isoniazid and ethambutol (3).

About 20 years ago, a then 39-year-old man was treated with i.m. streptomycin for tuberculosis and developed a generalized itchy red scaly vesicular eruption which lasted for 3-4 weeks. Thirteen years later, a similar exanthema developed while the patient was being treated with isoniazid, rifampicin and ethionamide. Now (20 and 7 years after the events) patch tests were positive to isoniazid 2% water and streptomycin sulfate 1% and 10% saline. Thirty controls were negative. Intradermal tests with both drugs were also positive at D2. Five hours after administering 300 mg of isoniazid, a generalized itchy erythematous micropapular eruption was observed, which disappeared in 3-4 days on oral corticosteroids and antihistamines (18).

Other cutaneous adverse drug reactions
A 24-year-old woman had recently started therapy with isoniazid, ethambutol and rifampicin for tuberculosis. This was discontinued 2 weeks later due to an itchy, desquamative, erythematous, papular rash, with painful crusts and excoriations, spreading from the face and hands to the rest of the body. Five days after withdrawal of all drugs, widespread desquamation and exfoliation with predilection for the body folds and the peri-orbital and peri-oral regions occurred, with swelling of the eyelids and conjunctivitis. The diagnosis toxicoderma was supported by

histology of the skin and an eosinophilia (31%). Patch tests were positive to isoniazid and ethambutol, tested as crushed tablets moistened with water. Ten controls were negative (7).

In Finland, in the period 1989-2001, 826 patients with suspected cutaneous drug eruptions were patch tested and 89 had one or more positive reactions. Of these individuals, 1 reacted to isoniazid; the nature of the drug eruption caused by delayed-type hypersensitivity to isoniazid was not mentioned (6).

A nurse who was occupationally sensitized to isoniazid developed an (unspecified) exanthema after a first oral dose of this drug (20).

Cross-reactions, pseudo-cross-reactions and co-reactions
This subject is discussed in Chapter 3.206 Ethambutol. No *clinical* cross-reactivity (patch testing data not available) between isoniazid and the – more or less – structurally related antituberculosis drug ethionamide (10). Isoniazid (4-pyridinecarbonylhydrazine) is a hydrazine (NH_2-NH_2) derivative. Cross-reactions between isoniazid and other hydrazines such as hydrazine, hydrazine hydrate, hydralazine, phenylhydrazine, iproniazid and phenelzine have been observed, which seem to be determined by free hydrazine groups (19,20,21,22,23,25).

LITERATURE
1   The data shown in this section may have been obtained from literature discussed in this chapter, but mostly also or exclusively from one or more of the following online sources: ChemIDPlus Advanced, PubChem, DrugBank, RxList, Drug Central, Drugs.com, and Wikipedia
2   Yamasaki R, Yamasaki M, Kawasaki Y, Nagasako R. Generalized pustular dermatosis caused by isoniazid. Br J Dermatol 1985;112:504-506
3   Özkaya E. Eczematous-type multiple drug allergy from isoniazid and ethambutol with positive patch test results. Cutis 2013;92:121-124
4   Allouchery M, Logerot S, Cottin J, Pralong P, Villier C, Ben Saïd B; French Pharmacovigilance Centers Network and the French Investigators for skin adverse reactions to drugs. Antituberculosis drug-associated DRESS: A case series. J Allergy Clin Immunol Pract 2018;6:1373-1380
5   Bopaka RG, El Khattabi W, Afif H, Aichane A, Bouayad Z. The "DRESS" syndrome in antituberculosis drugs. Rev Pneumol Clin 2014;70:185-188 (Article in French).
6   Lammintausta K, Kortekangas-Savolainen O. The usefulness of skin tests to prove drug hypersensitivity. Br J Dermatol 2005;152:968-974
7   Bakkum RS, Waard-Van Der Spek FB, Thio HB. Delayed-type hypersensitivity reaction to ethambutol and isoniazid. Contact Dermatitis 2002;46:359
8   Lee SW, Yoon NB, Park SM, Lee SM, Um SJ, Lee SK, et al. Antituberculosis drug-induced drug rash with eosinophilia and systemic symptoms syndrome confirmed by patch testing. J Investig Allergol Clin Immunol 2010;20:631-632
9   Rudzki E, Rebandel P, Grzywa Z. Patch tests with occupational contactants in nurses, doctors and dentists. Contact Dermatitis 1989;20:247-250
10 Lehloenya RJ, Muloiwa R, Dlamini S, Gantsho N, Todd G, Dheda K. Lack of cross-toxicity between isoniazid and ethionamide in severe cutaneous adverse drug reactions: A series of 25 consecutive confirmed cases. J Antimicrob Chemother 2015;70:2648-2651
11 Lehloenya RJ, Todd G, Wallace J, Ngwanya MR, Muloiwa R, Dheda K. Diagnostic patch testing following tuberculosis-associated cutaneous adverse drug reactions induces systemic reactions in HIV-infected persons. Br J Dermatol 2016;175:150-156
12 Lee AY, Joo HJ, Chey WY, Kim YG. Photopatch testing in seven cases of photosensitive drug eruptions. Ann Pharmacother 2001;35:1584-1587
13 Rebollo S, Sanchez P, Vega JM, Sedano E, Sanchís ME, Asensio T, et al. Hypersensitivity syndrome from isoniazid with positive patch test. Contact Dermatitis 2001;45:306
14 Arruti N, Villarreal O, Bernedo N, Audicana MT, Velasco M, Uriel O, et al. Positive allergy study (intradermal, patch, and lymphocyte transformation tests) in a case of isoniazid-induced DRESS. J Investig Allergol Clin Immunol 2016;26:119-120
15 Coster A, Aerts O, Herman A, Marot L, Horst N, Kenyon C, et al. Drug reaction with eosinophilia and systemic symptoms (DRESS) syndrome caused by first-line antituberculosis drugs: Two case reports and a review of the literature. Contact Dermatitis 2019;81:325-331
16 Ogawa K, Morito H, Kobayashi N, Fukumoto T, Asada H. Case of drug-induced hypersensitivity syndrome involving multiple-drug hypersensitivity. J Dermatol 2012;39:945-946
17 Lee AY, Jung SY. Two patients with isoniazid-induced photosensitive lichenoid eruptions confirmed by photopatch test. Photodermatol Photoimmunol Photomed 1998;14:77-78
18 Meseguer J, Sastre A, Malek T, Salvador MD. Systemic contact dermatitis from isoniazid. Contact Dermatitis 1993;28:110-111

19 Wang I, Schmid GH. Asthma bronchiale und epicutane Sensibilisierung durch berufsbedingten Kontakt mit Isonicotinsäure-Hydrazid. Z Hautkr 1974;49:803-809 (Article in German, data cited in ref. 18)

20 Hermann W, Lischka G, Lückerath I. Kontaktdermatitis und Arzneixanthem bei beruflich erworbener Überempfindlichkeit gegen Isonikotinsäure-Hydrazid. Berufsdermatosen 1969;17:13-20 (Article in German, data cited in ref. 18)

21 Holdiness MR. Contact dermatitis to antituberculosis drugs. Contact Dermatitis 1986;15:282-288

22 Van Ketel WG. Contact dermatitis from a hydrazine derivative in a stain remover. Cross sensitization to apresoline and isoniazid. Acta Dermato-Venereologica 1964;44:49-52 (Data cited in ref. 18)

23 Hovding G. Occupational dermatitis from hydrazine hydrate used in boiler protection. Acta Dermato-Venereologica 1967;47:293-297 (Data cited in ref. 18)

24 Ippen H. Kontaktekzem durch isonicotinsäurehydrazid. Dermatosen Beruf Umwelt 1978;26:57 (Data cited in ref. 21)

25 Malten K. Industrial contact dermatitis caused by hydrazine derivatives, with group sensitivity to hydralazine and isoniazid. Ned T Geneesk 1962;106:2219-2222 (Article in Dutch)

26 Van Ketel W. A severe hypersensitivity reaction to the use of PAS and isoniazid. Ned Tijdschr Geneeskd 1963;107:952-955 (Article in Dutch)

Chapter 3.275 ISOTRETINOIN

IDENTIFICATION

Description/definition   : Isotretinoin is the retinoid derivative of vitamin A that conforms to the structural formula
                             shown below
Pharmacological classes : Dermatologic agents; teratogens
IUPAC name   : (2Z,4E,6E,8E)-3,7-Dimethyl-9-(2,6,6-trimethylcyclohexen-1-yl)nona-2,4,6,8-tetraenoic acid
Other names   : 13-cis-Retinoic acid; 13-cis-vitamin A acid
CAS registry number   : 4759-48-2
EC number   : 225-296-0
Merck Index monograph   : 6544
Patch testing   : 0.01% alcohol (2)
Molecular formula   : $C_{20}H_{28}O_2$

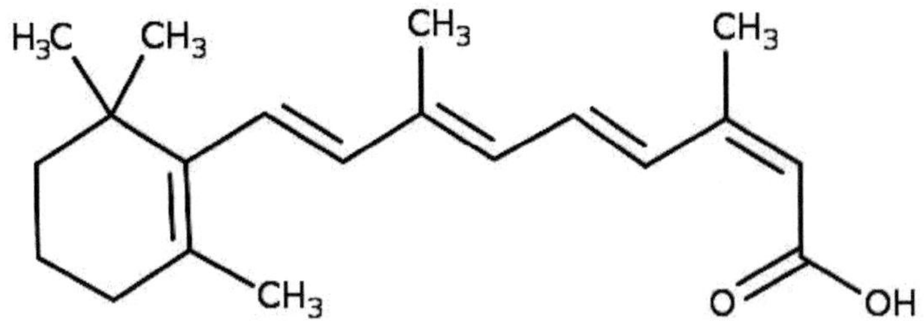

GENERAL

Isotretinoin is a naturally-occurring retinoic acid with potential antineoplastic activity. Isotretinoin binds to and activates nuclear retinoic acid receptors (RARs); activated RARs serve as transcription factors that promote cell differentiation and apoptosis. This agent also exhibits immunomodulatory and anti-inflammatory responses and inhibits ornithine decarboxylase, thereby decreasing polyamine synthesis and keratinization. Isotretinoin is a used in the treatment of severe acne and some forms of skin, head and neck cancer (1).

CUTANEOUS ADVERSE DRUG REACTIONS FROM SYSTEMIC ADMINISTRATION CAUSED BY TYPE IV (DELAYED-TYPE) HYPERSENSITIVITY (as demonstrated by positive patch tests)

Cutaneous adverse drug reactions from systemic administration of isotretinoin caused by type IV (delayed-type) hypersensitivity have included maculopapular eruption (2) and photosensitivity (2).

Maculopapular eruption

A 36-year-old woman, who had previously suffered from irritation and edema from topical application of tretinoin (retinoic acid), was treated with isotretinoin 30 mg daily for facial acne. The next day, after she took the same dosage, the patient noticed erythema on the face, neck, and forearms. Within a few hours, edema of the eyelids, cheeks, and perioral area began; in the presternal zone the eruption was accentuated in the light-exposed area. Chills, tightness in the chest, and a brief loss of consciousness occurred. On examination, erythematovesicular lesions on photoexposed areas and a maculopapular eruption on the back, trunk and arms were noted. Isotretinoin was stopped and within one week all the lesions cleared with mild topical therapy. Closed patch tests to tretinoin (retinoic acid) 0.005% and 0.1% were positive at D2. Patch tests with isotretinoin powder diluted in alcohol at 0.01% and 0.001% gave an edematous reaction with the 0.001% concentration and a vesicular response to 0.01%. Three controls were negative. Photopatch testing with UVB but not with UVA demonstrated a photoaggravation of the allergic reaction. The patient was diagnosed with an allergic reaction to isotretinoin and photoaggravation from UVB (2).

Photosensitivity

A patient with photoaggravation of delayed-type hypersensitivity is described above in the section 'Maculopapular eruption', ref. 2. Although photosensitivity is listed as an adverse effect from oral isotretinoin, reports in the literature are contradictory. Phototesting a group of subjects who were taking isotretinoin produced negative results (3,4).

Cross-reactions, pseudo-cross-reactions and co-reactions

A patient most likely sensitized to retinoic acid (tretinoin) from topical application cross-reacted to isotretinoin (2).

LITERATURE

1   The data in the section 'General' may have been obtained from literature discussed in this chapter, but mostly also or exclusively from one or more of the following online sources: ChemIDPlus Advanced, PubChem, DrugBank, RxList, Drug Central, Drugs.com, and Wikipedia

2   Auffret N, Bruley C, Brunetiere RA, Decot MC, Binet O. Photoaggravated allergic reaction to isotretinoin. J Am Acad Dermatol 1990;23:321-322

3   Diffey BL, Spiro JG, Hindson TC. Photosensitivity studies and isotretinoin therapy [Letter]. J Am Acad Dermatol 1985;12:119-124

4   Ferguson I, Johnson BE. Photosensitivity due to retinoids. Br J Dermatol 1986;115:275-283

Chapter 3.276 ISOXSUPRINE

IDENTIFICATION

Description/definition	: Isoxsuprine is the benzyl alcohol derivative that conforms to the structural formula shown below
Pharmacological classes	: Tocolytic agents; vasodilator agents; β-adrenergic agonists; sympathomimetics
IUPAC name	: 4-[1-Hydroxy-2-(1-phenoxypropan-2-ylamino)propyl]phenol
CAS registry number	: 395-28-8
EC number	: 206-898-2
Merck Index monograph	: 6555
Patch testing	: 1% alcohol
Molecular formula	: $C_{18}H_{23}NO_3$

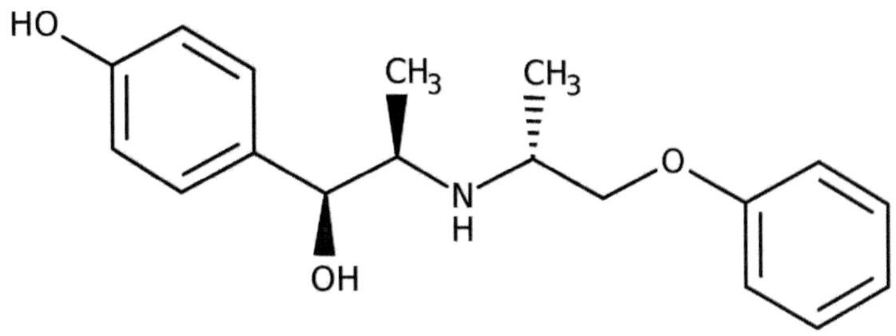

GENERAL

Isoxsuprine is a β-adrenergic agonist that causes direct relaxation of uterine and vascular smooth muscle. Its vasodilating actions are greater on the arteries supplying skeletal muscle than on those supplying skin. This agent may also produce positive inotropic and chronotropic effects on the myocardium. Isoxsuprine is used in the treatment of peripheral vascular disease and in premature labor (1). In pharmaceutical products, isoxsuprine is employed as isoxsuprine hydrochloride (CAS number 579-56-6, EC number 209-443-6, molecular formula $C_{18}H_{24}ClNO_3$) (1).

CUTANEOUS ADVERSE DRUG REACTIONS FROM SYSTEMIC ADMINISTRATION CAUSED BY TYPE IV (DELAYED-TYPE) HYPERSENSITIVITY (as demonstrated by positive patch tests)

Cutaneous adverse drug reactions from systemic administration of isoxsuprine caused by type IV (delayed-type) hypersensitivity have included systemic allergic dermatitis (2).

Systemic allergic dermatitis (systemic contact dermatitis)

A 50-year-old man with chronic venous insufficiency had eczema on the right leg after using a topical vasodilator gel containing nylidrin (also known as buphenine). The dermatitis became generalized in patches over the trunk and extremities. Patch tests showed a positive reaction to the gel 'as is' and its ingredient nylidrin HCl 1% alcohol. While the patch tests were in place, the generalized dermatitis reappeared. Two weeks days later, after stopping medication, the dermatitis again reappeared. The patient had been using an oral vasodilator agent for 2 years without any problem. This contained a closely related substance, isoxsuprine. A patch test with this vasodilator 1% alcohol gave a +++ reaction at D2 and D4 (2). It appears that isoxsuprine cross-reacted to nylidrin and caused a systemic allergic dermatitis.

Cross-reactions, pseudo-cross-reactions and co-reactions

Cross-reactivity from nylidrin to isoxsuprine or *vice versa* (2).

LITERATURE

1   The data in the section 'General' may have been obtained from literature discussed in this chapter, but mostly also or exclusively from one or more of the following online sources: ChemIDPlus Advanced, PubChem, DrugBank, RxList, Drug Central, Drugs.com, and Wikipedia
2   Alomar A. Buphenine sensitivity. Contact Dermatitis 1984;11:315

Chapter 3.277 ITRACONAZOLE

IDENTIFICATION

Description/definition : Itraconazole is the synthetic triazole compound that conforms to the structural formula shown below

Pharmacological classes : Antifungal agents; cytochrome P-450 CYP3A inhibitors; 14α-demethylase inhibitors

IUPAC name : 2-Butan-2-yl-4-[4-[4-[4-[[2-(2,4-dichlorophenyl)-2-(1,2,4-triazol-1-ylmethyl)-1,3-dioxolan-4-yl]methoxy]phenyl]piperazin-1-yl]phenyl]-1,2,4-triazol-3-one

CAS registry number : 84625-61-6

EC number : 617-596-9

Merck Index monograph : 6562

Patch testing : No data available; most systemic drugs can be tested at 10% pet.; if the pure chemical is not available, prepare the test material from intravenous powder, the content of capsules or – when also not available – from powdered tablets to achieve a final concentration of the active drug of 10% pet.

Molecular formula : $C_{35}H_{38}Cl_2N_8O_4$

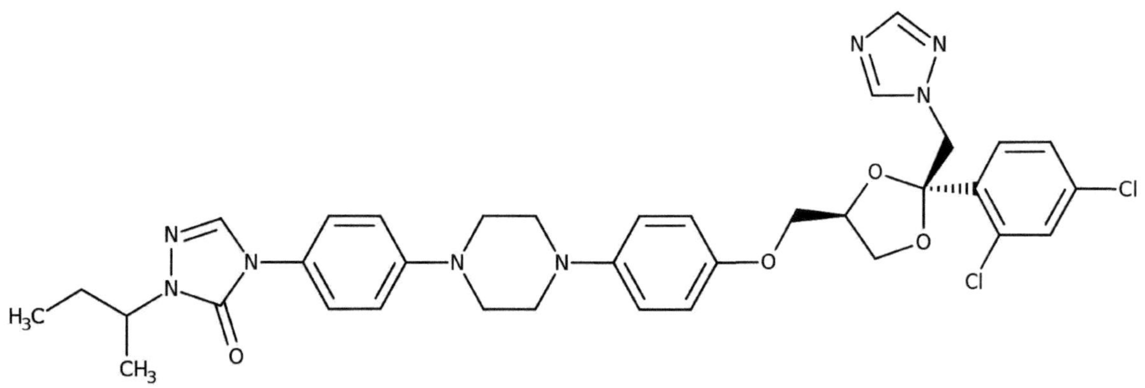

GENERAL

Itraconazole is a synthetic triazole agent with antimycotic properties. Formulated for both topical and systemic use, it preferentially inhibits fungal cytochrome P450 enzymes, resulting in a decrease in fungal ergosterol synthesis. Because of its low toxicity, this agent can be used for long-term maintenance treatment of chronic fungal infections. Itraconazole is indicated for the treatment of pulmonary and extrapulmonary blastomycosis, histoplasmosis, aspergillosis, and onychomycosis in both immunocompromised and non-immunocompromised patients (1).

CUTANEOUS ADVERSE DRUG REACTIONS FROM SYSTEMIC ADMINISTRATION CAUSED BY TYPE IV (DELAYED-TYPE) HYPERSENSITIVITY (as demonstrated by positive patch tests)

Cutaneous adverse drug reactions from systemic administration of itraconazole caused by type IV (delayed-type) hypersensitivity have included symmetrical drug-related intertriginous and flexural exanthema (SDRIFE) (3).

Symmetrical drug-related intertriginous and flexural exanthema (SDRIFE)/Baboon syndrome

In a tertiary care center in India, over a period of 12 months from March 2019 to March 2020, 12 patients were investigated who were clinically diagnosed with symmetrical drug-related intertriginous and flexural exanthema (SDRIFE) while using itraconazole for fungal infections. Seven were patch tested with itraconazole 6 weeks after resolution of the eruption and there were 2 patients with positive reactions. Both had developed the exanthema 3 days after initiation of itraconazole therapy (which is curious, as this period is too short for delayed-type hypersensitivity to develop and they had *not* been treated with itraconazole before). No details on patch test procedures, including patch test concentration, vehicle, times of reading and number of controls, were provided (3).

Another curious finding, casting doubt on the reliability of this report, is that the fungal infection in the 12 patients had been present for 14 ± 3.7 days (range 5-30 days) before itraconazole was initiated. This author seriously doubts whether 'tinea corporis and cruris' can develop in 5 days and that such patients can be seen in a tertiary care center 5 days after the development of a fungal infection.

Cross-reactions, pseudo-cross-reactions and co-reactions
A patient sensitized to clotrimazole in a cream was tested with a series of azoles and he reacted to itraconazole 1% alcohol (also tested at 1% and 5% pet., but the results were not mentioned) and croconazole. As the man had used only clotrimazole, these reactions were considered to be the result of cross-sensitivity. Croconazole and clotrimazole are both phenmethyl imidazoles, but itraconazole is a triazole antimycotic (2).

LITERATURE

1   The data in the section 'General' may have been obtained from literature discussed in this chapter, but mostly also or exclusively from one or more of the following online sources: ChemIDPlus Advanced, PubChem, DrugBank, RxList, Drug Central, Drugs.com, and Wikipedia

2   Erdmann S, Hertl M, Merk HF. Contact dermatitis from clotrimazole with positive patch-test reactions also to croconazole and itraconazole. Contact Dermatitis 1999;40:47-48

3   Hassanandani T, Panda M, Agarwal A, Das A. Rising trends of symmetrical drug related intertriginous and flexural exanthem due to itraconazole in patients with superficial dermatophytosis: A case series of 12 patients from eastern part of India. Dermatol Ther 2020;33:e13911.

Chapter 3.278 KANAMYCIN

IDENTIFICATION

Description/definition : Kanamycin is the aminoglycoside bactericidal antibiotic that conforms to the structural formula shown below

Pharmacological classes : Anti-bacterial agents; protein synthesis inhibitors

IUPAC name : (2R,3S,4S,5R,6R)-2-(Aminomethyl)-6-[(1R,2R,3S,4R,6S)-4,6-diamino-3-[(2S,3R,4S,5S,6R)-4-amino-3,5-dihydroxy-6-(hydroxymethyl)oxan-2-yl]oxy-2-hydroxycyclohexyl]oxyoxane-3,4,5-triol

Other names : 4,6-Diamino-2-hydroxy-1,3-cyclohexane 3,6'-diamino-3,6'-dideoxydi-α-D-glucoside; kanamycin A

CAS registry number : 59-01-8

EC number : 200-411-7

Merck Index monograph : 6599

Patch testing : Sulfate, 10% pet. (Chemotechnique, SmartPracticeCanada, SmartPracticeEurope)

Molecular formula : $C_{18}H_{36}N_4O_{11}$

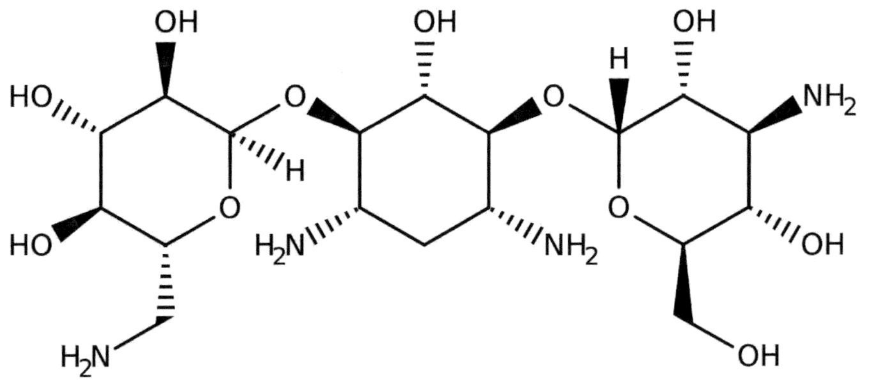

GENERAL

Kanamycin is an aminoglycoside bactericidal antibiotic isolated from the bacterium *Streptomyces kanamyceticus*. It is a complex comprising three components: kanamycin A, the major component, and kanamycins B and C. Kanamycin is indicated for treatment of infections where one or more of the following are the known or suspected pathogens: *E. coli*, *Proteus* species (both indole-positive and indole-negative), *E. aerogenes*, *K. pneumoniae*, *S. marcescens*, and *Acinetobacter* species. In pharmaceutical products, kanamycin is most commonly employed as kanamycin sulfate (CAS number 25389-94-0, EC number 246-933-9, molecular formula $C_{18}H_{38}N_4O_{15}S$) (1).

In topical preparations, kanamycin has caused some cases of contact allergy/allergic contact dermatitis, which subject has been fully reviewed in Volume 3 of the *Monographs in contact allergy* series (4).

CUTANEOUS ADVERSE DRUG REACTIONS FROM SYSTEMIC ADMINISTRATION CAUSED BY TYPE IV (DELAYED-TYPE) HYPERSENSITIVITY (as demonstrated by positive patch tests)

Cutaneous adverse drug reactions from systemic administration of kanamycin caused by type IV (delayed-type) hypersensitivity have included unspecified drug eruption (2).

Other cutaneous adverse drug reactions

In Japanese literature, a case of drug eruption induced by kanamycin sulfate has been reported. The patient cross-reacted to neomycin, gentamicin, tobramycin, dibekacin, amikacin, isepamicin, and arbekacin (2). Details are not available to the author.

Cross-reactions, pseudo-cross-reactions and co-reactions

In patients sensitized to neomycin, about 60% cross-react to kanamycin. The cross-sensitivity pattern in patients primarily sensitized to other aminoglycosides including kanamycin has not been well investigated (4).

LITERATURE

1   The data in the section 'General' may have been obtained from literature discussed in this chapter, but mostly also or exclusively from one or more of the following online sources: ChemIDPlus Advanced, PubChem, DrugBank, RxList, Drug Central, Drugs.com, and Wikipedia

2   Hara M, Saitou S, Yamamoto Y, Miyakawa K, Ikezawa Z. A case of drug eruption induced by kanamycin sulfate. Rinsho Hifuka 1994;48:871-874 (Article in Japanese, data cited in ref. 3)

3   Kimura M, Kawada A. Contact sensitivity induced by neomycin with cross-sensitivity to other aminoglycoside antibiotics. Contact Dermatitis 1998;39:148-150

4   De Groot AC. Monographs in contact allergy, volume 3: Topical Drugs. Boca Raton, Fl, USA: CRC Press Taylor and Francis Group, 2021 (ISBN 978-0-367-23693-9)

Chapter 3.279 KETOCONAZOLE

IDENTIFICATION

Description/definition : Ketoconazole is a phenylethyl imidazole and synthetic derivative of phenylpiperazine that conforms to the structural formula shown below

Pharmacological classes : 14α-Demethylase inhibitors; antifungal agents; cytochrome P-450 CYP3A inhibitors

IUPAC name : 1-[4-[4-[[(2R,4S)-2-(2,4-Dichlorophenyl)-2-(imidazol-1-ylmethyl)-1,3-dioxolan-4-yl]methoxy]phenyl]piperazin-1-yl]ethanone

CAS registry number : 65277-42-1

EC number : 265-667-4

Merck Index monograph : 6619

Patch testing : 1% pet. and alcohol

Molecular formula : $C_{26}H_{28}Cl_2N_4O_4$

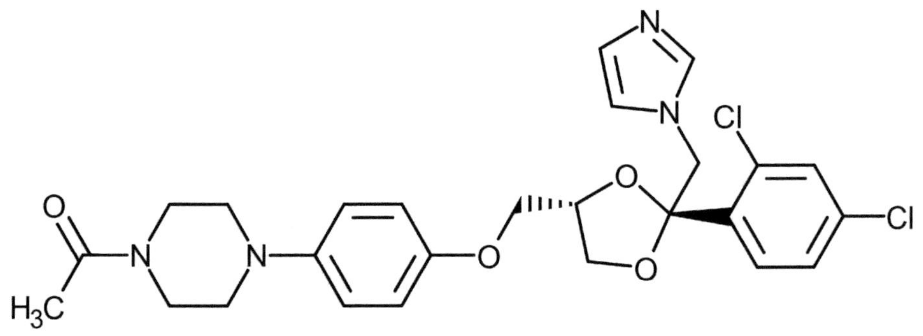

GENERAL

Ketoconazole is a synthetic derivative of phenylpiperazine and an imidazole-type antifungal with broad fungicidal properties and potential antineoplastic activity. It inhibits sterol 14α-demethylase, a microsomal cytochrome P450-dependent enzyme, thereby disrupting synthesis of ergosterol, an important component of the fungal cell wall. Ketoconazole is indicated for the treatment of superficial fungal infections and the following systemic fungal infections: candidiasis, chronic mucocutaneous candidiasis, oral thrush, candiduria, blastomycosis, coccidioidomycosis, histoplasmosis, chromomycosis, and paracoccidioidomycosis (1).

In topical preparations, ketoconazole has caused cases of both contact allergy/allergic contact dermatitis and photocontact allergy/photoallergic contact dermatitis, which subject has been fully reviewed in Volume 3 of the *Monographs in contact allergy* series (8).

CUTANEOUS ADVERSE DRUG REACTIONS FROM SYSTEMIC ADMINISTRATION CAUSED BY TYPE IV (DELAYED-TYPE) HYPERSENSITIVITY (as demonstrated by positive patch tests)

Cutaneous adverse drug reactions from systemic administration of ketoconazole caused by type IV (delayed-type) hypersensitivity have included systemic allergic dermatitis (6,9).

Systemic allergic dermatitis (systemic contact dermatitis)

A 27-year-old man, who was treated with oral ketoconazole for a fungal infection, presented with an erythematous eruption of 3-4 days' duration. The lesions started on his neck and then spread to his inner thighs, flexural, axillary, and inguinal areas and buttocks. There was no fever, nor systemic symptoms or abnormal laboratory tests. A skin biopsy showed a perivascular lymphohistiocytic infiltrate, with rare eosinophils in the dermis and minimal spongiosis in the epidermis. Previously, the patient had used ketoconazole cream for four weeks because of tinea pedis and had developed 'exanthema' two days later. The clinical diagnosis was drug-related baboon syndrome as a manifestation of systemic allergic dermatitis. Patch tests were positive to ketoconazole (no details provided on patch test data such as test concentration and vehicle, times of reading, strength of the reaction and number of controls) (9).

A 63-year-old man developed eczema in the groin after applying 1% econazole nitrate cream to treat tinea cruris. Later, miconazole 2% cream was applied and the patient took one tablet ketoconazole 200 mg per day. Two days later, generalized eczema developed. Patch tests with ketoconazole 2% cream and powdered tablets were positive. The components of the ketoconazole cream were tested later and proved positive for sodium sulfite, polysorbate 80 and ketoconazole 1% pet. (6). It was not discussed whether there may have been a cross-sensitivity from econazole and/or miconazole to ketoconazole in this case of probable systemic allergic dermatitis.

Cross-reactions, pseudo-cross-reactions and co-reactions
Although there are some unique and sporadic reports of simultaneous reactions (4,5,7), cross-reactions are very unlikely between ketoconazole and other imidazoles (2). The question has been raised (but not answered) whether contact allergy to ketoconazole and other (nitro)imidazoles may be overrepresented in patients allergic to methylchloroisothiazolinone / methylisothiazolinone (MCI/MI) (3).

LITERATURE

1 The data in the section 'General' may have been obtained from literature discussed in this chapter, but mostly also or exclusively from one or more of the following online sources: ChemIDPlus Advanced, PubChem, DrugBank, RxList, Drug Central, Drugs.com, and Wikipedia

2 Dooms-Goossens A, Matura M, Drieghe J, Degreef H. Contact allergy to imidazoles used as antimycotic agents. Contact Dermatitis 1995;33:73-77

3 Stingeni L, Rigano L, Lionetti N, Bianchi L, Tramontana M, Foti C, et al. Sensitivity to imidazoles/nitroimidazoles in subjects sensitized to methylchloroisothiazolinone/methylisothiazolinone: A simple coincidence? Contact Dermatitis 2019;80:181-183

4 Imafuku S, Nakayama J. Contact allergy to ketoconazole cross-sensitive to miconazole. Clin Exp Dermatol 2009;34:411-412

5 Santucci B, Cannistraci C, Cristaudo A, Picardo M. Contact dermatitis from ketoconazole cream. Contact Dermatitis 1992;27:274-275

6 Garcia-Bravo B, Mazuecos J, Rodriguez-Pichardo A, Navas J, Camacho F. Hypersensitivity to ketoconazole preparations: study of 4 cases. Contact Dermatitis 1989;21:346-348

7 Jones SK, Kennedy CT. Contact dermatitis from tioconazole. Contact Dermatitis 1990;22:122-123

8 De Groot AC. Monographs in contact allergy, volume 3: Topical Drugs. Boca Raton, Fl, USA: CRC Press Taylor and Francis Group, 2021 (ISBN 978-0-367-23693-9)

9 Gulec AI, Uslu E, Başkan E, Yavuzcan G, Aliagaoglu C. Baboon syndrome induced by ketoconazole. Cutan Ocul Toxicol 2014;33:339-341

Chapter 3.280 KETOPROFEN

IDENTIFICATION

Description/definition : Ketoprofen is the arylpropionic acid derivate that conforms to the structural formula shown below

Pharmacological classes : Anti-inflammatory agents, non-steroidal; cyclooxygenase inhibitors

IUPAC name : 2-(3-Benzoylphenyl)propanoic acid

Other names : Benzeneacetic acid, 3-benzoyl-α-methyl-; 2-(3-benzoylphenyl)propionic acid

CAS registry number : 22071-15-4

EC number : 244-759-8

Merck Index monograph : 6622

Patch testing : 1% pet. (Chemotechnique, SmartPracticeCanada); 2.5% pet. (SmartPracticeCanada); perform photopatch tests; late readings at D5-D7 are advisable; as photopatch testing may induce severe reactions, one-hour patch test occlusion with ketoprofen has been proposed, which was successful for the detection of photosensitivity and simplifies the photopatch test procedure by eliminating one visit to the clinic (5)

Molecular formula : $C_{16}H_{14}O_3$

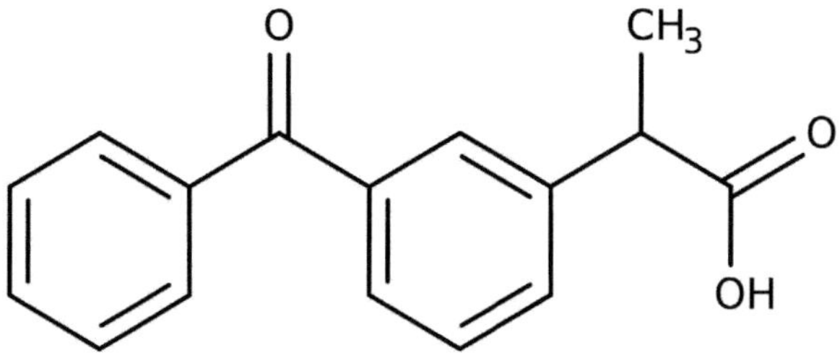

GENERAL

Ketoprofen is a propionic acid derivate and nonsteroidal anti-inflammatory drug (NSAID) with anti-inflammatory, analgesic and antipyretic effects. Ketoprofen inhibits the activity of the enzymes cyclo-oxygenase I and II, resulting in a decreased formation of precursors of prostaglandins and thromboxanes. The resulting decrease in prostaglandin synthesis, by prostaglandin synthase, is responsible for the therapeutic effects of this NSAID. Ketoprofen is indicated for symptomatic treatment of acute and chronic rheumatoid arthritis, osteoarthritis, ankylosing spondylitis, primary dysmenorrhea and mild to moderate pain associated with musculotendinous trauma (sprains and strains), postoperative (including dental surgery) or postpartum pain. It is also, and far more frequently, used in topical formulations (gel, cream, foam, ointment, tape) for musculoskeletal diseases and injuries for its analgesic and anti-inflammatory effects (1). In these products, ketoprofen has caused large numbers of photocontact allergy/photo-allergic contact dermatitis and to a lesser degree contact allergy/allergic contact dermatitis. This subject has been fully reviewed in Volume 3 of the Monographs in contact allergy series (4).

CONTACT ALLERGY FROM ACCIDENTAL CONTACT

A 43-year-old nurse working in an internal medicine department presented with an itchy vesicular rash on the hands that appeared 2 days after she had handled ketoprofen in the workplace. When preparing an injection, she broke the ampoule, and the solution spread onto her hands. Although she immediately rinsed her hands for 5 minutes under running water, dermatitis appeared 2 days later, following sun exposure during a day off work. Seven years previously, she had experienced allergic contact dermatitis after using ketoprofen gel for the topical treatment of a tendinopathy. Patch testing with ketoprofen showed a positive reaction (+) on D4. Photo-patch testing with ketoprofen showed sensitization (++) after 10 J/cm^2 UVA irradiation. It was concluded that the patient had photoaggravated occupational allergic contact dermatitis caused by ketoprofen (2).

CUTANEOUS ADVERSE DRUG REACTIONS FROM SYSTEMIC ADMINISTRATION CAUSED BY TYPE IV (DELAYED-TYPE) HYPERSENSITIVITY (as demonstrated by positive patch tests)

Cutaneous adverse drug reactions from systemic administration of ketoprofen caused by type IV (delayed-type) hypersensitivity have included systemic allergic dermatitis presenting as photodermatitis (3).

Systemic allergic dermatitis and photosensitivity

Although ketoprofen is the main cause of photoallergic contact dermatitis from topical use, it rarely causes photo-sensitivity from systemic administration, probably due to the low levels reached in the skin after systemic use (6).

Case report

A 43-year-old man presented with erythematous, edematous and vesicular lesions on his face, eyelids, neck, hands, and glabrous areas of the scalp. One day before the appearance of such lesions, the patient ingested a single sachet of ketoprofen for migraine. He had been exposed to sunlight and the reaction had progressively worsened over subsequent days. The patient remembered two previous episodes of a skin rash after application of ketoprofen gel and sun exposure. The authors also presented a second patient, who had developed a diffuse eczematous reaction on photoexposed areas of the skin after taking ketoprofen tablets for back pain. He too had previously reacted to ketoprofen gel. MED testing was normal in both patients. Photopatch tests were positive to ketoprofen and patch tests to the fragrance mix I and its ingredient cinnamyl alcohol (see below) (3). These were cases of systemic allergic dermatitis presenting as photoallergic dermatitis (3).

Cross-reactions, pseudo-cross-reactions and co-reactions

In patients with photocontact allergy to ketoprofen, there are usually many photo-co-reactivities to both structurally similar (e.g. tiaprofenic acid, piketoprofen, suprofen, benzophenone-3, fenofibrate) and structurally unrelated compounds (octocrylene, etofenamate, halogenated salicylanilides). In patients photosensitized to ketoprofen, concomitant (conventional) contact allergic reactions to the fragrance mix I, Myroxylon pereirae resin and cinnamyl alcohol are frequent. This subject has been discussed in full details in ref. 4.

LITERATURE

1   The data in the section 'General' may have been obtained from literature discussed in this chapter, but mostly also or exclusively from one or more of the following online sources: ChemIDPlus Advanced, PubChem, DrugBank, RxList, Drug Central, Drugs.com, and Wikipedia

2   Maurel DT, Durand-Moreau Q, Pougnet R, Dewitte JD, Roguedas-Contios AM, Bensefa-Colas L, Loddé B. Why is occupational photocontact allergic dermatitis caused by ketoprofen rarely reported in the literature? Contact Dermatitis 2018;78:92-94

3   Foti C, Cassano N, Vena GA, Angelini G. Photodermatitis caused by oral ketoprofen: two case reports. Contact Dermatitis 2011;64:181-183

4   De Groot AC. Monographs in contact allergy, volume 3: Topical Drugs. Boca Raton, Fl, USA: CRC Press Taylor and Francis Group, 2021 (ISBN 978-0-367-23693-9)

5   Marmgren V, Hindsén M, Zimerson E, Bruze M. Successful photopatch testing with ketoprofen using one-hour occlusion. Acta Derm Venereol 2011;91:131-136

6   Guy RH, Kuma H, Nakanishi M. Serious photocontact dermatitis induced by topical ketoprofen depends on the formulation. Eur J Dermatol 2014;24:365-371

Chapter 3.281 KITASAMYCIN

IDENTIFICATION

Description/definition	: Kitasamycin is the macrolide antibiotic produced by *Streptomyces kitasatoensis* that conforms to the structural formula shown below
Pharmacological classes	: Anti-bacterial agents
IUPAC name	: 2-[(4R,5S,6S,7R,9R,10R,11E,13E,16R)-6-[(2S,3R,4R,5S,6R)-5-[(2S,4R,5S,6S)-4,5-Dihydroxy-4,6-dimethyloxan-2-yl]oxy-4-(dimethylamino)-3-hydroxy-6-methyloxan-2-yl]oxy-4,10-dihydroxy-5-methoxy-9,16-dimethyl-2-oxo-1-oxacyclohexadeca-11,13-dien-7-yl]acetaldehyde
Other names	: Leucomycin; leucomycin V
CAS registry number	: 1392-21-8
EC number	: Not available
Merck Index monograph	: 6775 (Leucomycin)
Patch testing	: Kitasamycin powder 'as is' (2); suggested: 20% pet.
Molecular formula	: $C_{35}H_{59}NO_{13}$

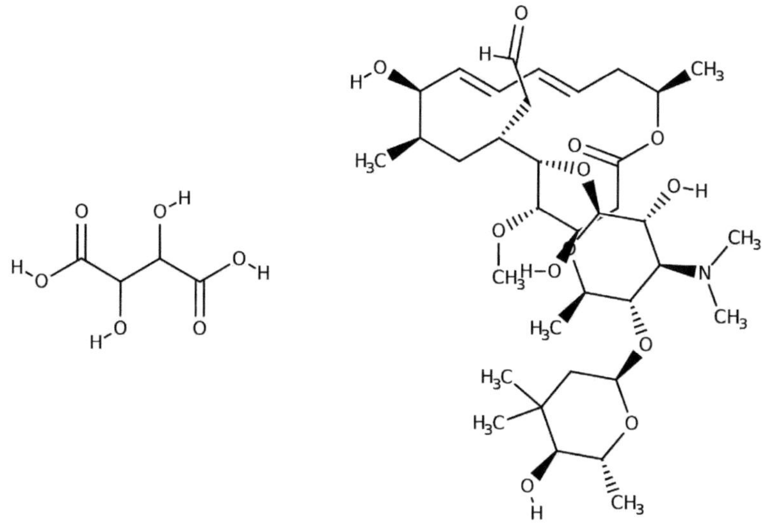

Kitasamycin tartrate

GENERAL

Kitasamycin (often termed Leucomycin) is a macrolide antibiotic produced by *Streptomyces kitasatoensis*, which has antibacterial activity against a wide spectrum of pathogens. It is a complex of over 10 structurally related macrolides, termed Leukomycins A1-13. In pharmaceutical products, kitasamycin is usually employed as kitasamycin tartrate (CAS number 37280-56-1, EC number 253-442-3, molecular formula $C_{44}H_{73}NO_{20}$) or kitasamycin acetate (acetyl kitasamycin, CAS number 178234-32-7, EC number not available, molecular formula $C_{40}H_{67}NO_{18}$) (1).

CONTACT ALLERGY FROM ACCIDENTAL CONTACT

A 22-year-old female student in industrial engineering was conducting research for her thesis on the purification of kitasamycin. Her task was to develop a method of refining pure kitasamycin from a powder consisting of some 10 components, including midecamycin. To do this, she sometimes worked with large quantities of powder, of which there would occasionally be dust in the air. Two weeks after she began her work, she suddenly erupted in a burning eczema on the face, followed by the neck, scalp, and retro-auricular region, which she attributed to the powders she was working with. Patch tests revealed positive reactions to kitasamycin tartrate powder 'as is' (D2 ++, D3 ++), kitasamycin tartrate 4% water (D2 +, D3 not mentioned) and midecamycin 'as is' (D2 ++, D3 ++). There was a striking flare-up reaction on the face during patch testing. The patient was diagnosed with occupational airborne allergic contact dermatitis from kitasamycin and midecamycin (2).

LITERATURE

1   The data in the section 'General' may have been obtained from literature discussed in this chapter, but mostly also or exclusively from one or more of the following online sources: ChemIDPlus Advanced, PubChem, DrugBank, RxList, Drug Central, Drugs.com, and Wikipedia
2   Dooms-Goossens A, Bedert R, Degreef H, Vandaele M. Airborne allergic contact dermatitis from kitasamycin and midecamycin. Contact Dermatitis 1990;23:118-119

Chapter 3.282 LABETALOL

IDENTIFICATION

Description/definition   : Labetalol is the salicylamide that conforms to the structural formula shown below
Pharmacological classes   : Adrenergic β-antagonists; sympathomimetics; α$_1$-adrenergic receptor antagonists;
                                            antihypertensive agents
IUPAC name   : 2-Hydroxy-5-[1-hydroxy-2-(4-phenylbutan-2-ylamino)ethyl]benzamide
CAS registry number   : 36894-69-6
EC number   : 253-258-3
Merck Index monograph : 6647
Patch testing   : No data available; most systemic drugs can be tested at 10% pet.; if the pure chemical is
                                            not available, prepare the test material from intravenous powder, the content of capsules
                                            or – when also not available – from powdered tablets to achieve a final concentration of
                                            the active drug of 10% pet.
Molecular formula   : C$_{19}$H$_{24}$N$_2$O$_3$

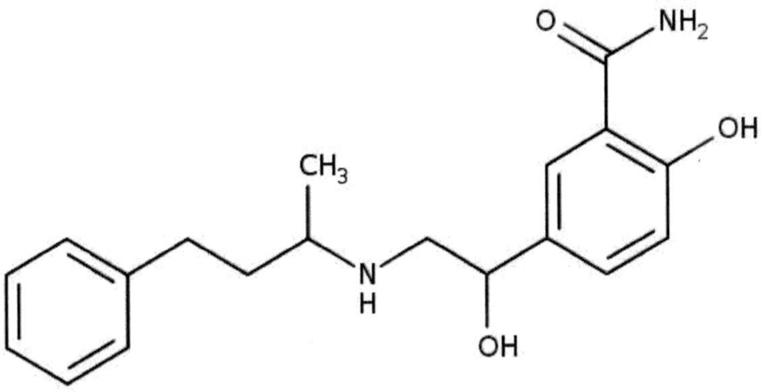

GENERAL

Labetalol is a third generation selective α$_1$-adrenergic antagonist and non-selective β-adrenergic antagonist with vasodilatory and antihypertensive properties. It causes a decrease in resting and exercise heart rates, cardiac output, and in both systolic and diastolic blood pressure, thereby resulting in vasodilation, and negative chronotropic and inotropic cardiac effects. Labetalol is a racemic mixture of 2 diastereoisomers where dilevalol, the R,R' stereoisomer, makes up 25% of the mixture. Labetalol is formulated as an injection or tablets to treat hypertension (1).

In pharmaceutical products, labetalol is used as labetalol hydrochloride (CAS number 32780-64-6, EC number251-211-1, molecular formula C$_{19}$H$_{25}$ClN$_2$O$_3$) (1).

CUTANEOUS ADVERSE DRUG REACTIONS FROM SYSTEMIC ADMINISTRATION CAUSED BY TYPE IV (DELAYED-TYPE) HYPERSENSITIVITY (as demonstrated by positive patch tests)

Cutaneous adverse drug reactions from systemic administration of labetalol caused by type IV (delayed-type) hypersensitivity have included acute generalized exanthematous pustulosis (AGEP) (2).

Acute generalized exanthematous pustulosis (AGEP)

A 31-year-old pregnant woman in the third trimester of pregnancy was treated with labetalol for 18 days with the addition of 3 other drugs during the 3 days prior to her caesarean section. Approximately 8 hours prior to delivery, the patient developed erythematous micropapular lesions on her face and neck with a fever of 38.7°C. These lesions became generalized over 4 to 5 days, affecting flexures, the intermammary cleft, chest, back, palms of the hands, and soles of the feet with multiple micropustules. A complete blood count revealed leukocytosis with neutrophilia. Liver enzymes were elevated. Pustule contents were cultured, but the results were negative. Later, patch tests with all drugs used were positive only to labetalol with pustules. A patch test to atenolol (to find a suitable alternative) was negative and she received a test dose of 25 mg. One hour later, the patient complained of generalized itching and micropapules appeared on her back and palms that persisted for 48 hours. The patient was diagnosed with acute generalized exanthematous pustulosis (AGEP) due to labetalol with cross-sensitivity to atenolol (2).

Cross-reactions, pseudo-cross-reactions and co-reactions
Cross-reaction to atenolol in a patient sensitized to labetalol (2).

LITERATURE

1   The data in the section 'General' may have been obtained from literature discussed in this chapter, but mostly also or exclusively from one or more of the following online sources: ChemIDPlus Advanced, PubChem, DrugBank, RxList, Drug Central, Drugs.com, and Wikipedia
2   Gómez Torrijos E, García Rodríguez C, Sánchez Caminero MP, Castro Jiménez A, García Rodríguez R, Feo-Brito F. First case report of acute generalized exanthematous pustulosis due to labetalol. J Investig Allergol Clin Immunol 2015;25:148-149

Chapter 3.283 LAMIVUDINE

IDENTIFICATION

Description/definition   : Lamivudine is the synthetic nucleoside analog that conforms to the structural formula
                             shown below
Pharmacological classes   : Anti-HIV agents; reverse transcriptase inhibitors
IUPAC name   : 4-Amino-1-[(2R,5S)-2-(hydroxymethyl)-1,3-oxathiolan-5-yl]pyrimidin-2-one
CAS registry number   : 134678-17-4
EC number   : Not available
Merck Index monograph : 6672
Patch testing   : Tablet, pulverized, 20% pet. (2,3); most systemic drugs can be tested at 10% pet.; if the
                             pure chemical is not available, prepare the test material from intravenous powder, the
                             content of capsules or – when also not available – from powdered tablets to achieve a
                             final concentration of the active drug of 10% pet.
Molecular formula   : $C_8H_{11}N_3O_3S$

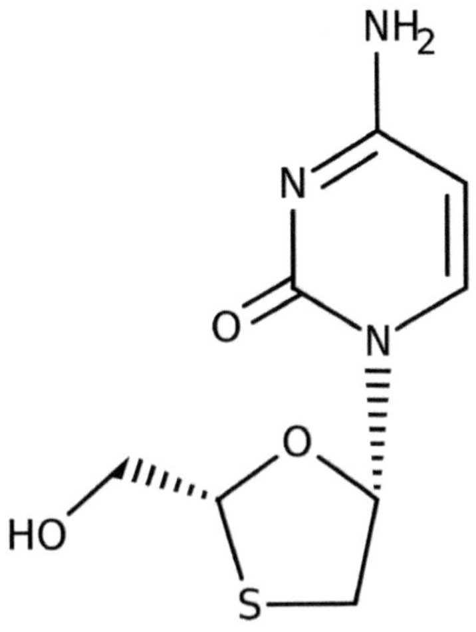

GENERAL

Lamivudine is a synthetic nucleoside analog and reverse transcriptase inhibitor with antiviral activity. It is phosphorylated intracellularly to its active 5'-triphosphate metabolite, lamivudine triphosphate. This nucleoside analog is incorporated into viral DNA by HIV reverse transcriptase and HBV polymerase, resulting in DNA chain termination. Lamivudine is indicated for the treatment of HIV infection and chronic hepatitis B infection (1).

CONTACT ALLERGY FROM ACCIDENTAL CONTACT

A 40-year-old female health care provider was given triple HIV drug therapy after contact with blood products from an HIV-positive patient. Approximately a week after starting the medication, the patient had capsules of zidovudine, indinavir and a tablet of lamivudine in her hand to take, when she was stopped and proceeded to hold the medication in her hand during a conversation that lasted more than 30 minutes. 2-3 days later, the patient noticed an eruption with pruritus extending from the palm down the forearm, which worsened over the next few days. Histopathology showed an intraepidermal vesicular dermatitis with a mixed inflammatory infiltrate containing mononuclear cells as well as neutrophils and eosinophils. The patient continued the HIV-treatment for another 2 weeks without systemic symptoms. Three months later, patch tests were performed with powder from the 2 capsules and pulverized lamivudine tablet 20% pet. At D2 there was a 2+ reaction to lamivudine and a 3+ reaction at D3. This was a case of – strictly speaking non-occupational – contact dermatitis to lamivudine (3).

CUTANEOUS ADVERSE DRUG REACTIONS FROM SYSTEMIC ADMINISTRATION CAUSED BY TYPE IV
(DELAYED-TYPE) HYPERSENSITIVITY (as demonstrated by positive patch tests)

Cutaneous adverse drug reactions from systemic administration of lamivudine caused by type IV (delayed-type) hypersensitivity have included Stevens-Johnson syndrome (2).

Stevens-Johnson syndrome/toxic epidermal necrolysis (SJS/TEN)

A 54-year-old woman diagnosed with HIV infection started treatment with emtricitabine/tenofovir tablets in combination with nevirapine. An itchy rash on the neckline, abdomen and limbs appeared 8 h after the first dose. One day after this reaction, the patient took another dose, and a new rash with extensive skin scaling appeared, accompanied by oral and genital ulceration, fever, hepatitis, diarrhea, and loss of consciousness. The patient was admitted to hospital and discharged two weeks later with a diagnosis of Stevens–Johnson syndrome. Nine months later, she suffered from three milder episodes after emtricitabine/tenofovir with etravirine, abacavir/lamivudine with raltegravir, and emtricitabine/tenofovir with raltegravir. Patch tests with all drugs (powdered pills 5%, 10% and 20% pet.) gave positive reactions to the 20% concentrations of abacavir/lamivudine, lamivudine and emtricitabine/tenofovir. Emtricitabine itself was not tested, but tenofovir tested negative. Ten controls were negative (2). The chemical structure of lamivudine is almost identical to that of emtricitabine, but rather different from the remaining drugs given, which makes, according to the authors, a cross-reactivity mechanism from emtricitabine to lamivudine likely (2).

Cross-reactions, pseudo-cross-reactions and co-reactions
Cross-reactivity from emtricitabine hypersensitivity to lamivudine (2).

LITERATURE

1   The data in the section 'General' may have been obtained from literature discussed in this chapter, but mostly also or exclusively from one or more of the following online sources: ChemIDPlus Advanced, PubChem, DrugBank, RxList, Drug Central, Drugs.com, and Wikipedia

2   Suárez-Lorenzo I, Castillo-Sainz R, Cárden-Santana MA, Carrillo-Díaz T. Severe reaction to emtricitabine and lamiduvine: evidence of cross-reactivity. Contact Dermatitis 2016;74:253-254

3   Smith KJ, Buckley R, Skelton H. Lamivudine (3TC)-induced contact dermatitis. Cutis 2000;65:227-229

Chapter 3.284 LAMOTRIGINE

IDENTIFICATION

Description/definition   : Lamotrigine is the synthetic phenyltriazine that conforms to the structural formula shown
                              below
Pharmacological classes   : Calcium channel blockers; antipsychotic agents; anticonvulsants; sodium channel blockers
IUPAC name   : 6-(2,3-Dichlorophenyl)-1,2,4-triazine-3,5-diamine
CAS registry number   : 84057-84-1
EC number   : 281-901-8
Merck Index monograph : 6673
Patch testing   : 10.0% pet. (Chemotechnique); with serious cutaneous adverse reactions, starting with 1%
                              pet. may be advisable (26)
Molecular formula   : $C_9H_7Cl_2N_5$

GENERAL

Lamotrigine is a synthetic phenyltriazine that has antiepileptic and analgesic properties. This aromatic anticonvulsant compound is indicated as therapy for partial seizures, primary and secondary tonic-clonic seizures and generalized seizures of Lennox-Gastaut syndrome. Lamotrigine is also used for the maintenance treatment of bipolar I disorder to delay the time to occurrence of mood episodes (depression, mania, hypomania, mixed episodes) in adults treated for acute mood episodes with standard therapy. Off-label, lamotrigine is employed in treating other neurologic and psychiatric pathologies like borderline personality disorder. The exact mechanism of action of this agent is not fully elucidated, as it may have multiple cellular actions that contribute to its broad clinical efficacy (1).

CUTANEOUS ADVERSE DRUG REACTIONS FROM SYSTEMIC ADMINISTRATION CAUSED BY TYPE IV (DELAYED-TYPE) HYPERSENSITIVITY (as demonstrated by positive patch tests)

Cutaneous adverse drug reactions from systemic administration of lamotrigine caused by type IV (delayed-type) hypersensitivity have included maculopapular eruption (6,27,28), fixed drug eruption (4), drug reaction with eosinophilia and systemic symptoms (DRESS)/anticonvulsant hypersensitivity syndrome (2,3,5,7,8,10,27), Stevens-Johnson syndrome/toxic epidermal necrolysis (SJS/TEN) (7,29), urticaria (27), and unspecified drug eruptions (27).

Case series with various or unknown types of drug reactions

Between 2003 and 2017, in Belgrade, Serbia, 100 children in the age from 1 to 17 years suspected of hypersensitivity reactions to antiepileptic drugs were examined with patch tests, using the commercial drugs 10% pet. 61 patients had shown maculopapular eruptions, 26 delayed urticaria, 5 morbilliform exanthema, 5 DRESS, 2 SJS and one erythema multiforme. Lamotrigine was the suspected drug in 46 cases and was patch test positive in 19 children (41%). It was not specified which eruptions these drugs had caused, but should include maculopapular eruptions and delayed urticaria. Of the 5 children with DRESS, 4 reacted to carbamazepine and one to lamotrigine (27).

Maculopapular eruption

After having used lamotrigine for twelve days, a 64-year-old woman developed an extensive maculopapular eruption, localized especially on the trunk. Patch tests were positive to commercial pulverized lamotrigine tablet 30% water and pet. at D3 and D7 (6). A boy aged 10 developed a maculopapular eruption during therapy with lamotrigine. A patch test with the commercial drug 5% pet. was positive (28).

   See also the section 'Case series with various or unknown types of drug reactions' above, ref. 27.

Fixed drug eruption

A 54-year-old man was treated with sodium valproate for seizures. Because of limited efficacy, lamotrigine 50 mg twice daily was added to the therapeutic regimen. Ten hours after the first lamotrigine dose, the patient developed extensive, 2-5 cm, red to violaceous, round patches and plaques with central erosions or vesicular changes, starting at the periorbital area and spreading to the trunk and extremities. Extensive fixed drug eruption (FDE) due to lamotrigine was suspected. Skin biopsy findings were consistent with FDE. Patch testing with lamotrigine 50% pet. revealed erythematous papules and plaques on the previously involved area, but no reaction was noted on non-involved skin. Next, a ROAT with lamotrigine 50% pet. was positive after 3 days on postlesional skin with erythematous papules and plaques but not on normal skin (4).

Drug reaction with eosinophilia and systemic symptoms (DRESS)/anticonvulsant hypersensitivity syndrome

The literature on patch testing in anticonvulsant hypersensitivity syndrome up to August 2008 has been reviewed in ref. 24.

A 14-year-old man presented with erythroderma and fever 44 days after carbamazepine intake for absence epilepsy. Laboratory examinations showed eosinophilia and elevated liver enzymes, and thoracic imaging revealed interstitial pneumonitis. A patch test to carbamazepine 5% pet. performed 6 weeks after recovery was positive. About 8 months later, the patient developed the same symptoms of DRESS 52 days after lamotrigine intake. Lamotrigine was stopped and all symptoms disappeared. A patch test to lamotrigine 5% pet., performed 6 weeks after recovery, was positive (8). As many others, these authors suggested a 'cross-reaction' between carbamazepine and lamotrigine, where, in fact, this was a case of multiple drug hypersensitivity.

Between January 1998 and December 2008, in a university hospital in Portugal, 56 patients with DRESS were investigated with patch testing of the suspected drugs. Lamotrigine (1% and 10% pet.) was tested in 5 patients and there were 2 (40%) positive patch test reactions. Fifty controls were negative (2). A 70-year-old man was seen with generalized exfoliative dermatitis, intense pruritus, fever and elevated aminotransferases. He had been on treatment with lamotrigine for the past 4 weeks for vascular epilepsy. Patch tests were positive to lamotrigine (10% pet.) at D2 (++) and D4 (++). Five controls were negative (3).

From Rome, Italy, one case of anticonvulsant hypersensitivity reaction from lamotrigine with a positive patch test to this drug was reported (5). In a multicenter investigation in France, of 72 patients patch tested for DRESS, 46 (64%) had positive patch tests to drugs, including one to lamotrigine (7).

In a group of 15 patients with 'anticonvulsant hypersensitivity syndrome', one patient had a positive patch test to lamotrigine; details are unavailable to the author, but not all patients appeared to have all symptoms necessary to diagnose anticonvulsant hypersensitivity syndrome (synonym: DRESS from anticonvulsant drugs) (10). In a series of 5 children with DRESS reported from Serbia, one was caused by lamotrigine with a positive patch test to commercial lamotrigine 10% pet. (27).

Stevens-Johnson syndrome/toxic epidermal necrolysis (SJS/TEN)

In a multicenter investigation in France, of 17 patients patch tested for SJS/TEN, 4 (24%) had positive patch tests to drugs, including one to lamotrigine (7). In Ankara, Turkey, lamotrigine was patch test positive in one pediatric patient with Stevens-Johnson syndrome (29).

Other cutaneous adverse drug reactions

See the section 'Case series with various or unknown types of drug reactions' above, ref. 27 (urticaria).

Cross-reactions, pseudo-cross-reactions and co-reactions (for all anticonvulsant drugs)

It has repeatedly been stated that 'cross-reactions' between aromatic anticonvulsant drugs (notably carbamazepine, oxcarbazepine, phenytoin and phenobarbital, to a far lesser degree lamotrigine, which was introduced more recently) may occur in a high percentage of patients with DRESS/anticonvulsant drug hypersensitivity syndrome or other hypersensitivity reactions (14,17,25). This concept was originally based on the observation that a second drug, given concomitantly with or after the use of the culprit drug was interrupted, would apparently either sustain the symptoms, worsen them, or result in a second hypersensitivity reaction, either DRESS or another manifestation, e.g, maculopapular eruption. In the majority of such cases, no patch tests were performed and the reactions were considered 'clinical cross-reactivity' (5), presumably caused either by a T cell-mediated immunological reaction or common toxic metabolites (arene oxides) (14). Only later, it would become clear that flare-ups of DRESS symptoms after cessation of the causative drug are not infrequent, even when no new drugs are given.

Nevertheless, patch test positive co-reactions to 2 or more aromatic anticonvulsant drugs (12,15,19) or to an aromatic anticonvulsant drug and a non-aromatic anticonvulsant (22) or a non-anticonvulsant medicament (12,13) have been observed repeatedly, especially in DRESS, which was then considered proof of the existence of 'cross-sensitivity/cross-reactivity' (12). However, apart from carbamazepine and oxcarbazepine (18), there are no structural similarities between the aromatic anticonvulsants that are in favor of 'real' immunological cross-reactions. Such co-reactions to 2 or more structurally non-related drugs (also termed multiple drug hypersensitivity [9,11]) are more likely the result of concomitant (13) or successive sensitization (when the second drug had been used for at least 7-10 days before the new hypersensitivity reaction started).

Another explanation for co-reactivity may be problems with patch testing technique, e.g. irritant patch test reactions (no controls performed) or false-positive reactions due to the excited skin syndrome (16,21). In addition, many patch tests have been reported as positive at D2 only (no second reading performed or mentioned), whereas such reactions (even ++ reactions) may become negative at D3 or D4, heavily arguing against delayed-type hypersensitivity (23). Insufficient expertise in the investigator group in performing and assessing patch test results may also sometimes play a role (e.g. mentioning a positive patch test at D2 only). In a study from Taiwan, for example (15) a picture was shown of 'positive' (+) reactions to carbamazepine epoxide, oxcarbazepine and phenytoin, which absolutely did not qualify as positive (apart from the fact that it was a reaction at D2, which is in itself unreliable). Also, these authors noted that most reactions were weaker at D4, 'but the erythema could be still seen at day 4'. They considered this proof that the results were not irritant reactions, whereas, in fact, this decrescendo effect makes irritancy more likely.

Proven cases of cross-reactivity in the sense that a patient sensitized to an anticonvulsant drug as shown by a positive patch test also has a 'real' positive patch test to a structurally related drug or another aromatic anticonvulsant drug *that has never been used by the patient* most likely are, with the possible exception of carbamazepine-oxcarbazepine (20), unusual (5), if not very infrequent.

LITERATURE

1   The data in the section 'General' may have been obtained from literature discussed in this chapter, but mostly also or exclusively from one or more of the following online sources: ChemIDPlus Advanced, PubChem, DrugBank, RxList, Drug Central, Drugs.com, and Wikipedia
2   Santiago F, Gonçalo M, Vieira R, Coelho S, Figueiredo A. Epicutaneous patch testing in drug hypersensitivity syndrome (DRESS). Contact Dermatitis 2010;62:47-53
3   Monzón S, Garcés MM, Reichelt C, Lezaun A, Colás C. Positive patch test in hypersensitivity to lamotrigine. Contact Dermatitis 2002;47:361
4   Hsiao C-J, Lee JY-Y, Wong T-W, Shew H-H. Extensive fixed drug eruption due to lamotrigine.Br J Dermatol 2001;144:1289-1291
5   Romano A, Pettinato R, Andriolo M, Viola M, Guéant-Rodriguez RM, Valluzzi RL, et al. Hypersensitivity to aromatic anticonvulsants: in vivo and in vitro cross-reactivity studies. Curr Pharm Des 2006;12:3373-3381
6   Fuertes L, García-Cano I, Ortiz de Frutos J, Vanaclocha F. La imputabilidad de la lamotrigina en el exantema medicamentoso aumenta con las pruebas epicutáneas [Patch testing increases the likelihood of recognizing lamotrigine as a cause of drug-induced rash]. Actas Dermosifiliogr 2011;102:64-66 (Article in Spanish).
7   Barbaud A, Collet E, Milpied B, Assier H, Staumont D, Avenel-Audran M, et al. A multicentre study to determine the value and safety of drug patch tests for the three main classes of severe cutaneous adverse drug reactions. Br J Dermatol 2013;168:555-562
8   Aouam K, Ben Romdhane F, Loussaief C, Salem R, Toumi A, Belhadjali H, et al. Hypersensitivity syndrome induced by anticonvulsants: possible cross-reactivity between carbamazepine and lamotrigine. J Clin Pharmacol 2009;49:1488-1491
9   Gex-Collet C, Helbling A, Pichler WJ. Multiple drug hypersensitivity – proof of multiple drug hypersensitivity by patch and lymphocyte transformation tests. J Investig Allergol Clin Immunol 2005;15:293-296
10   Galindo PA, Borja J, Gomez E, Mur P, Gudín M, García R, et al. Anticonvulsant drug hypersensitivity. J Invest Allergol Clin Immunol 2002;12:299-304
11   Jörg L, Yerly D, Helbling A, Pichler W. The role of drug, dose and the tolerance/intolerance of new drugs in multiple drug hypersensitivity syndrome (MDH). Allergy 2020;75:1178-1187
12   Ben Mahmoud L, Bahloul N, Ghozzi H, Kammoun B, Hakim A, Sahnoun Z, et al. Epicutaneous patch testing in delayed drug hypersensitivity reactions induced by antiepileptic drugs. Therapie 2017;72:539-545 (Article in French)
13   Gaig P, García-Ortega P, Baltasar M, Bartra J. Drug neosensitization during anticonvulsant hypersensitivity syndrome. J Investig Allergol Clin Immunol 2006;16:321-326
14   Seitz CS, Pfeuffer P, Raith P, Bröcker EB, Trautmann A. Anticonvulsant hypersensitivity syndrome: cross-reactivity with tricyclic antidepressant agents. Ann Allergy Asthma Immunol 2006;97:698-702
15   Lin YT, Chang YC, Hui RC, Yang CH, Ho HC, Hung SI, Chung WH. A patch testing and cross-sensitivity study of carbamazepine-induced severe cutaneous adverse drug reactions. J Eur Acad Dermatol Venereol 2013;27:356-364
16   Troost RJ, Oranje AP, Lijnen RL, et al. Exfoliative dermatitis due to immunologically confirmed carbamazepine hypersensitivity. Pediatr Dermatol 1996;13:316-320

17 Knowless SR, Shapiro LE, Shear NH. Anticonvulsant hypersensitivity syndrome: incidence, prevention and management. Drug Saf 1999;21:489-501

18 Troost RJ, Van Parys JA, Hooijkaas H, van Joost T, Benner R, Prens EP. Allergy to carbamazepine: parallel in vivo and in vitro detection. Epilepsia 1996;37:1093-1099

19 Zeller A, Schaub N, Steffen I, Battegay E, Hirsch HH, Bircher AJ. Drug hypersensitivity syndrome to carbamazepine and human herpes virus 6 infection: case report and literature review. Infection 2003;31:254-256

20 Troost RJ, Van Parys JA, Hooijkaas H, van Joost T, Benner R, Prens EP. Allergy to carbamazepine: parallel in vivo and in vitro detection. Epilepsia 1996;37:1093-1099

21 Gómez Torrijos E, Extremera Ortega AM, Gonzalez Jimenez O, Joyanes Romo JB, Gratacós Gómez AR, Garcia Rodriguez R. Excited skin syndrome ("angry back" syndrome) induced by proximity of carbamazepine to another drug with strong positive allergic reaction in patch test: A first of its kind. Contact Dermatitis 2019;81:405-406

22 Schiavino D, Nucera E, Buonomo A, Musumeci S, Pollastrini E, Roncallo C, et al. A case of type IV hypersensitivity to topiramate and carbamazepine. Contact Dermatitis 2005;52:161-162

23 Puig L, Nadal C, Fernández-Figueras MT, Alomar A. Carbamazepine-induced drug rashes: diagnostic value of patch tests depends on clinico-pathologic presentation. Contact Dermatitis 1996;34:435-437

24 Elzagallaai AA, Knowles SR, Rieder MJ, Bend JR, Shear NH, Koren G. Patch testing for the diagnosis of anticonvulsant hypersensitivity syndrome: a systematic review. Drug Saf 2009;32:391-408

25 Hirsch L, Arif H, Nahm EA, Buchsbaum R, Resor SR, Bazil CW. Cross-sensitivity of skin rashes with antiepileptic drug use. Neurology 2008;7:1527-1534

26 Brockow K, Garvey LH, Aberer W, Atanaskovic-Markovic M, Barbaud A, Bilo MB, et al.; ENDA/EAACI Drug Allergy Interest Group. Skin test concentrations for systemically administered drugs – an ENDA/EAACI Drug Allergy Interest Group position paper. Allergy 2013;68:702-712

27 Atanasković-Marković M, Janković J, Tmušić V, Gavrović-Jankulović M, Ćirković Veličković T, Nikolić D, Škorić D. Hypersensitivity reactions to antiepileptic drugs in children. Pediatr Allergy Immunol 2019;30:547-552

28 Atanaskovic-Markovic M, Gaeta F, Medjo B, Gavrovic-Jankulovic M, Cirkovic Velickovic T, Tmusic V, et al. Non-immediate hypersensitivity reactions to beta-lactam antibiotics in children - our 10-year experience in allergy work-up. Pediatr Allergy Immunol 2016;27:533-538

29 Büyük Yaytokgil Ş, Güvenir H, Külhaş Celík İ, Yilmaz Topal Ö, Karaatmaca B, Civelek E, et al. Evaluation of drug patch tests in children. Allergy Asthma Proc 2021;42:167-174

Chapter 3.285 LANSOPRAZOLE

IDENTIFICATION

Description/definition	: Lansoprazole is the substituted benzimidazole that conforms to the structural formula shown below
Pharmacological classes	: Anti-ulcer agents; proton pump inhibitors
IUPAC name	: 2-[[3-Methyl-4-(2,2,2-trifluoroethoxy)pyridin-2-yl]methylsulfinyl]-1H-benzimidazole
CAS registry number	: 103577-45-3
EC number	: Not available
Merck Index monograph	: 6683
Patch testing	: 10% pet. (10); if the pure chemical is not available, prepare the test material from intravenous powder, the content of capsules or – when also not available – from powdered tablets to achieve a final concentration of the active drug of 10% pet.
Molecular formula	: $C_{16}H_{14}F_3N_3O_2S$

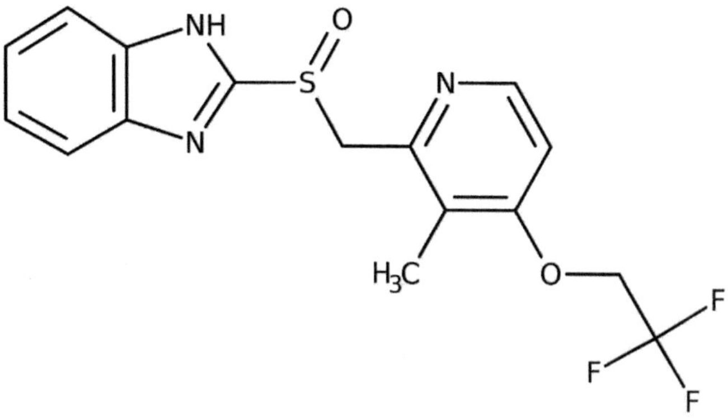

GENERAL

Lansoprazole is a substituted benzimidazole prodrug with selective and irreversible proton pump inhibitor activity. As a prodrug, it is converted to an active sulfonamide derivative which reduces gastric acid secretion. Lansoprazole is also effective in eradicating *H. pylori* when used in conjunction with amoxicillin and clarithromycin (triple therapy) or with amoxicillin alone (dual therapy). Lansoprazole is indicated as an antiulcer drug in the treatment and maintenance of healing of duodenal or gastric ulcers, erosive and reflux esophagitis, NSAID-induced ulcer, Zollinger-Ellison syndrome, and Barrett's esophagus (1).

CONTACT ALLERGY FROM ACCIDENTAL CONTACT

Occupational airborne allergic contact dermatitis from lansoprazole also occurred in a 41-year-old pharmacy technician, whose duties included preparation of lansoprazole solutions by opening capsules and pouring the contents into sodium bicarbonate. Patch tests were positive to lansoprazole powder, lansoprazole powder mixed with diluent (?), and lansoprazole granules dissolved in sodium bicarbonate. At the same time, the patient also developed facial pruritus and a recurrence of her rash (4). The first case of occupational airborne allergic contact dermatitis to lansoprazole was reported from Spain in 2001 in a pharmaceutical worker. When patch tested, the patient reacted to lansoprazole 10% and 50% pet. with cross-reactivity to omeprazole (5).

A 33-year-old male maintenance worker in a pharmaceutical company had been exposed intermittently for 7 years to lansoprazole when he developed an itchy rash on his face, neck and hands. Patch tests were positive to lansoprazole 1%, 5% and 10% (vehicle not mentioned) (3). A 49-year-old male pharmaceutical worker had been exposed periodically to lansoprazole, omeprazole, and pantoprazole for 21 years, when he noticed an itchy rash on his face, neck and arms. Patch tests were positive to lansoprazole 10% and 50% saline (+++), omeprazole 0.1%, 0.5% and 1% saline (+) and to pantoprazole 1%, 5% and 10% saline (+++) (3).

CUTANEOUS ADVERSE DRUG REACTIONS FROM SYSTEMIC ADMINISTRATION CAUSED BY TYPE IV (DELAYED-TYPE) HYPERSENSITIVITY (as demonstrated by positive patch tests)

Cutaneous adverse drug reactions from systemic administration of lansoprazole caused by type IV (delayed-type) hypersensitivity have included maculopapular eruption (8,9), acute generalized exanthematous pustulosis (AGEP) (6), drug reaction with eosinophilia and systemic symptoms (DRESS) (7,11), and toxic epidermal necrolysis (11).

Case series with various or unknown types of drug reactions
In Taiwan, in a retrospective multicenter study performed between 2003 and 2016, 69 cases of (severe) delayed-type hypersensitivity reactions to proton pump inhibitors were collected, including 27 cases of SJS/TEN and 10 of DRESS. Patch tests were performed in only 7 cases. Positive patch tests to lansoprazole (pulverized tablet 10% pet.) were seen in 1 patient with DRESS and one with TEN (11).

Maculopapular eruption
A 46-year-old woman developed maculopapular dermatitis and dyspnea while on various medications. Patch tests were positive to lansoprazole 12.5%, bismuthate (unknown drug), both in phosphate-buffered saline, and – previously – to benzylpenicillin. As the patient had reactions to 2 unrelated drugs, this was a case of multiple drug hypersensitivity (8). One case of a maculopapular eruption from delayed-type hypersensitivity to lansoprazole was reported from Switzerland in 2001, but it is uncertain whether the patient had a positive scratch-patch test or a positive intradermal test with late reading to lansoprazole (9).

Acute generalized exanthematous pustulosis (AGEP)
A 66-year-old woman developed a profuse erythematous rash on her legs, trunk and limbs and pustulosis on her legs and arms associated with high fever, leukocytosis and an increased sedimentation rate, together with a biopsy and bacterial cultures consistent with AGEP. Two days before onset, she had ingested lansoprazole. Apparently, this drug was not available for patch testing (??). Instead, she was tested with omeprazole and reacted to 5% and 20% omeprazole in petrolatum but not in water (6). All things considered, in this case, AGEP from delayed-type hypersensitivity to lansoprazole is very likely.

Drug reaction with eosinophilia and systemic symptoms (DRESS)
In a multicenter investigation in France, of 72 patients patch tested for DRESS, 46 (64%) had positive patch tests to drugs, including one to lansoprazole; clinical details were not provided (7). See also the section 'Case series with various or unknown types of drug reactions' above, ref. 11.

Stevens-Johnson syndrome/toxic epidermal necrolysis (SJS/TEN)
See the section 'Case series with various or unknown types of drug reactions' above, ref. 11.

Cross-reactions, pseudo-cross-reactions and co-reactions
Cross-reactions may occur between lansoprazole and omeprazole (5,6). A patient sensitized to pantoprazole cross-reacted to lansoprazole and other proton pump inhibitors (2).

LITERATURE
1   The data in the section 'General' may have been obtained from literature discussed in this chapter, but mostly also or exclusively from one or more of the following online sources: ChemIDPlus Advanced, PubChem, DrugBank, RxList, Drug Central, Drugs.com, and Wikipedia
2   Liippo J, Pummi K, Hohenthal U, Lammintausta K. Patch testing and sensitization to multiple drugs. Contact Dermatitis 2013;69:296-302
3   Alarcón M, Herrera-Mozo I, Nogué S, et al. Occupational airborne contact dermatitis from proton pump inhibitors. Curr Allergy Clin Immunol 2014;27:310-313
4   Akan GE, Moss MH. Lansoprazole induced contact dermatitis. J Allergy Clin Immunol 2003;111(2Suppl.1):S155
5   Vilaplana J, Romaguera C. Allergic contact dermatitis due to lansoprazole, a proton pump inhibitor. Contact Dermatitis 2001;44:47-48
6   Dewerdt S, Vaillant L, Machet L, de Muret A, Lorette G. Acute generalized exanthematous pustulosis induced by lansoprazole. Acta Derm Venereol 1997;77:250
7   Barbaud A, Collet E, Milpied B, Assier H, Staumont D, Avenel-Audran M, et al. A multicentre study to determine the value and safety of drug patch tests for the three main classes of severe cutaneous adverse drug reactions. Br J Dermatol 2013;168:555-562
8   Gex-Collet C, Helbling A, Pichler WJ. Multiple drug hypersensitivity – proof of multiple drug hypersensitivity by patch and lymphocyte transformation tests. J Investig Allergol Clin Immunol 2005; 15:293-296
9   Neukomm C, Yawalkar N, Helbling A, Pichler WJ. T-cell reactions to drugs in distinct clinical manifestations of drug allergy. J Invest Allergol Clin Immunol 2001;11:275-284
10 Brockow K, Garvey LH, Aberer W, Atanaskovic-Markovic M, Barbaud A, Bilo MB, et al.; ENDA/EAACI Drug Allergy Interest Group. Skin test concentrations for systemically administered drugs – an ENDA/EAACI Drug Allergy Interest Group position paper. Allergy 2013;68:702-712
11 Lin CY, Wang CW, Hui CR, Chang YC, Yang CH, Cheng CY, et al. Delayed-type hypersensitivity reactions induced by proton pump inhibitors: A clinical and in vitro T-cell reactivity study. Allergy 2018;73:221-229

Chapter 3.286 LEVOCETIRIZINE

IDENTIFICATION

Description/definition   : Levocetirizine is the active isomer of cetirizine that conforms to the structural formula
                            shown below
Pharmacological classes   : Histamine H1 antagonists, non-sedating
IUPAC name   : 2-[2-[4-[(R)-(4-Chlorophenyl)-phenylmethyl]piperazin-1-yl]ethoxy]acetic acid
Other names   : (-)-Cetirizine; (R)-cetirizine
CAS registry number   : 130018-77-8
EC number   : Not available
Merck Index monograph   : 3291 (Cetirizine)
Patch testing   : Commercial drug 10% and 20% pet.; most systemic drugs can be tested at 10% pet.; if the
                            pure chemical is not available, prepare the test material from intravenous powder, the
                            content of capsules or – when also not available – from powdered tablets to achieve a
                            final concentration of the active drug of 10% pet.
Molecular formula   : $C_{21}H_{25}ClN_2O_3$

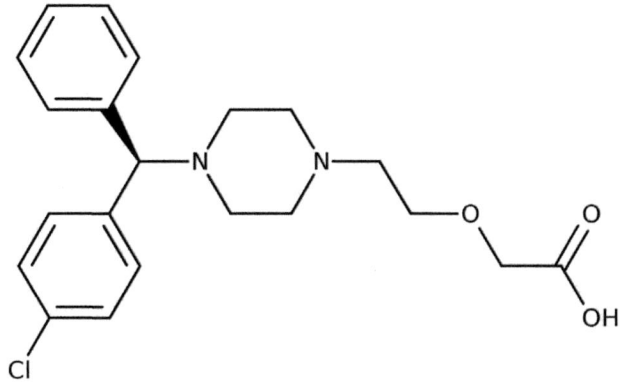

GENERAL

Levocetirizine, the R-enantiomer of the cetirizine racemate, is a third-generation, non-sedating, selective histamine H1 receptor antagonist, with antihistaminic, anti-inflammatory and potential anti-angiogenic activities. Levocetirizine is indicated for the relief of symptoms associated with (seasonal and perennial) allergic rhinitis and uncomplicated skin manifestations of chronic idiopathic urticaria (1). In pharmaceutical products, levocetirizine is employed as levocetirizine dihydrochloride (CAS number 130018-87-0, EC number not available, molecular formula $C_{21}H_{27}Cl_3N_2O_3$) (1).

CUTANEOUS ADVERSE DRUG REACTIONS FROM SYSTEMIC ADMINISTRATION CAUSED BY TYPE IV (DELAYED-TYPE) HYPERSENSITIVITY (as demonstrated by positive patch tests)

Cutaneous adverse drug reactions from systemic administration of levocetirizine caused by type IV (delayed-type) hypersensitivity have included fixed drug eruption (2,7).

Fixed drug eruption

A 73-year-old woman presented with multiple round well-demarcated darkly pigmented lesions with desquamation. She had taken various medications including levocetirizine eighteen days before. About 2 hours later, the patient had developed generalized itching and multiple red macules with bullae formation in several of these. The lesions had spontaneously resolved after she had stopped taking the medications and they changed to pigmented lesion with desquamation. Patch test were performed with all drugs used, both on normal and postlesional skin. At D2, levocetirizine 10% pet. on normal skin showed erythema, whereas at the pigmented skin site erythema, infiltration and vesicles were present. Patch tests with other antihistamines were positive to the related cetirizine (10% pet.) and hydroxyzine (10% pet.). The patient was diagnosed as levocetirizine-induced fixed drug eruption (7).

A 21-year-old man presented with multiple itchy, erythematous, and edematous plaques of two days duration over the lips, trunk, and both upper and lower extremities, which had developed within one hour of intake of a single tablet levocetirizine 5 mg. The patient admitted having multiple similar episodes in the last five years. He had taken multiple drugs in this period including cetirizine, levocetirizine, and possibly hydroxyzine. Oral provocation tests with levocetirizine and cetirizine were strongly positive with reactivation in the pre-existing lesions in the form of itching and erythema within one hour of provocation. Also, the patient developed multiple new well-defined erythematous, itchy plaques over the trunk and extremities. A patch test with hydroxyzine 10% pet. was positive, with levocetirizine and cetirizine not performed. However, because of cross-reactivity between the three (cetirizine being the active metabolite of hydroxyzine), type IV hypersensitivity to levocetirizine and cetirizine in this case is highly likely (2).

Cross-reactions, pseudo-cross-reactions and co-reactions
Cetirizine, levocetirizine, and hydroxyzine are related piperazine-derived antihistamines. *In vivo*, 45% of hydroxyzine is transformed into cetirizine; levocetirizine is the active (*R*)-enantiomer of cetirizine. Therefore, co-reactions due to cross-sensitivity are to be expected and have been observed (2,4,5,6,7; not always proven by positive patch tests).

LITERATURE
1   The data in the section 'General' may have been obtained from literature discussed in this chapter, but mostly also or exclusively from one or more of the following online sources: ChemIDPlus Advanced, PubChem, DrugBank, RxList, Drug Central, Drugs.com, and Wikipedia
2   Bhari N, Mahajan R, Singh S, Sharma VK. Fixed drug eruption due to three antihistamines of a same chemical family: Cetirizine, levocetirizine, and hydroxyzine. Dermatol Ther 2017;30. doi: 10.1111/dth.12412.
3   Andrade P, Brinca A, Gonçalo M. Patch testing in fixed drug eruptions – a 20-year review. Contact Dermatitis 2011;65:195-201
4   Cravo M, Gonçalo M, Figueiredo A. Fixed drug eruption to cetirizine with positive lesional patch tests to the three piperazine derivatives. Int J Dermatol 2007;46:760-762
5   Assouère MN, Mazereeuw-Hautier J, Bonafe JL. Cutaneous drug eruption with two antihistaminic drugs of a same chemical family: cetirizine and hydroxyzine. Ann Dermatol Venereol 2002;129:1295-1298
6   Mahajan VK, Sharma NL, Sharma VC. Fixed drug eruption: a novel side-effect of levocetirizine. Int J Dermatol 2005;44:796-798
7   Kim MY, Jo EJ, Chang YS, Cho SH, Min KU, Kim SH. A case of levocetirizine-induced fixed drug eruption and cross-reaction with piperazine derivatives. Asia Pac Allergy 2013;3:281-284

Chapter 3.287　LEVOFLOXACIN

IDENTIFICATION

Description/definition : Levofloxacin is the fluoroquinolone that conforms to the structural formula shown below
Pharmacological classes : Topoisomerase II inhibitors; anti-bacterial agents; anti-infective agents, urinary; cytochrome P-450 CYP1A2 inhibitors
IUPAC name : (2S)-7-Fluoro-2-methyl-6-(4-methylpiperazin-1-yl)-10-oxo-4-oxa-1-azatricyclo-[7.3.1.0^{5,13}]trideca-5(13),6,8,11-tetraene-11-carboxylic acid
Other names : (-)-Ofloxacin; L-ofloxacin
CAS registry number : 100986-85-4
EC number : 600-146-0
Patch testing : Tablet, pulverized, 10% and 20% pet. (2,3); most systemic drugs can be tested at 10% pet.; if the pure chemical is not available, prepare the test material from intravenous powder, the content of capsules or – when also not available – from powdered tablets to achieve a final concentration of the active drug of 10% pet.
Molecular formula : $C_{18}H_{20}FN_3O_4$

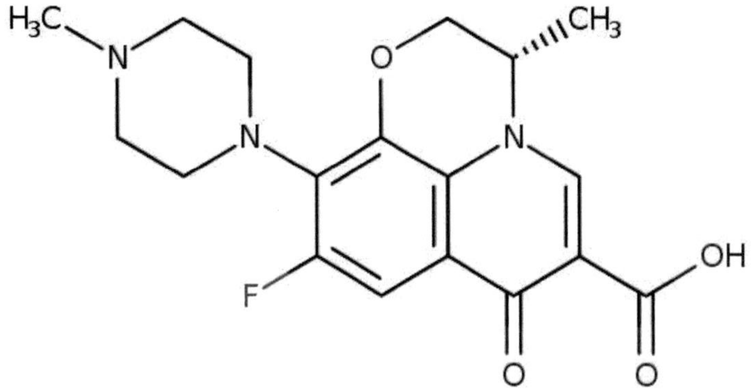

GENERAL

Levofloxacin is a synthetic third-generation fluoroquinolone antibacterial agent that inhibits the supercoiling activity of bacterial DNA gyrase, halting DNA replication. In oral and intravenous formulations, levofloxacin is indicated in adults for the treatment of various infections caused by susceptible bacteria, including infections of the respiratory tract, skin, skin structures, urinary tract, and prostate. In its ophthalmic formulation, levofloxacin is indicated for the treatment of bacterial conjunctivitis caused by susceptible organisms. In pharmaceutical products, levofloxacin is employed as levofloxacin hemihydrate (CAS number 138199-71-0, EC number not available, molecular formula $C_{36}H_{42}F_2N_6O_9$) (1).

CUTANEOUS ADVERSE DRUG REACTIONS FROM SYSTEMIC ADMINISTRATION CAUSED BY TYPE IV (DELAYED-TYPE) HYPERSENSITIVITY (as demonstrated by positive patch tests)

Cutaneous adverse drug reactions from systemic administration of levofloxacin caused by type IV (delayed-type) hypersensitivity have included maculopapular eruption (4), acute generalized exanthematous pustulosis (AGEP) (2), and drug reaction with eosinophilia and systemic symptoms (DRESS) (3).

Maculopapular eruption

In the period 2000-2014, in Coimbra, Portugal, 260 patients were patch tested with antibiotics for suspected cutaneous adverse drug reactions (CADR) to these drugs. 56 patients (22%) had one or more positive patch tests (often from cross-reactivity). Levofloxacin was patch test positive in 1 patient with a maculopapular eruption (4).

Acute generalized exanthematous pustulosis (AGEP)

A 78-year-old woman presented with a rash that had developed on her trunk 3 days after taking clindamycin and levofloxacin as part of treatment for a hip replacement. On physical examination, a diffuse red papular rash coalesced into erythematous and edematous plaques on her arms, legs, and trunk, studded with a large number of scattered, non-follicular, pinhead-sized pustules. Laboratory studies showed leukocytosis with left shift. Skin biopsy revealed subcorneal pustules and diffuse perivascular dermal infiltrates of atypical mononuclear cells with large nuclei, prominent nucleoli, and a number of mitotic figures that were positive for CD3 and CD30. Up to 15% positive CD30 atypical cells within the perivascular inflammatory infiltrate were found. Two months later, patch tests were positive to levofloxacin 20% pet. and clindamycin phosphate 10% pet. This was the first report of drug-induced acute generalized exanthematous pustulosis (AGEP) with simultaneous vascular lymphomatoid pattern, resembling drug-induced pseudolymphoma (2).

Drug reaction with eosinophilia and systemic symptoms (DRESS)
One patient who developed drug reaction with eosinophilia and systemic symptoms (DRESS) from levofloxacin and later showed a positive patch test to this antibiotic at 10% pet. has been reported from France. An atypical feature was that eosinophilia was not present (3).

LITERATURE
1   The data in the section 'General' may have been obtained from literature discussed in this chapter, but mostly also or exclusively from one or more of the following online sources: ChemIDPlus Advanced, PubChem, DrugBank, RxList, Drug Central, Drugs.com, and Wikipedia
2   Llamas-Velasco M, Godoy A, Sánchez-Pérez J, García-Diez A, Fraga J. Acute generalized exanthematous pustulosis with histopathologic findings of lymphomatoid drug reaction. Am J Dermatopathol 2013;35:690-691
3   Charfi O, Lakhoua G, Sahnoun R, Badri T, Daghfous R, El Aidli S, et al. DRESS syndrome following levofloxacin exposure with positive patch-test. Therapie 2015;70:547-549
4   Pinho A, Coutinho I, Gameiro A, Gouveia M, Gonçalo M. Patch testing - a valuable tool for investigating non-immediate cutaneous adverse drug reactions to antibiotics. J Eur Acad Dermatol Venereol 2017;31:280-287

Chapter 3.288 LEVOMEPROMAZINE

IDENTIFICATION

Description/definition   : Levomepromazine is the phenothiazine that conforms to the structural formula shown
                            below
Pharmacological classes   : Analgesics, non-narcotic; antipsychotic agents; dopamine antagonists
IUPAC name   : (2R)-3-(2-Methoxyphenothiazin-10-yl)-N,N,2-trimethylpropan-1-amine
Other names   : Methotrimeprazine; 10H-Phenothiazine-10-propanamine, 2-methoxy-N,N,β-trimethyl-,
                            (R)-
CAS registry number   : 60-99-1
EC number   : 200-495-5
Merck Index monograph   : 6788
Patch testing   : No data available; suggested: 0.1% and 1% pet.; perform controls, especially in case of
                            photopatch testing
Molecular formula   : $C_{19}H_{24}N_2OS$

GENERAL
Levomepromazine is a phenothiazine with sedative, hypnotic, anxiolytic, antiemetic, antihistaminic, analgesic and
antipsychotic activities. It is indicated for the treatment of psychoses, particular those of schizophrenia, and manic
phases of bipolar disorder (1).

CUTANEOUS ADVERSE DRUG REACTIONS FROM SYSTEMIC ADMINISTRATION CAUSED BY TYPE IV
(DELAYED-TYPE) HYPERSENSITIVITY (as demonstrated by positive patch tests)
Cutaneous adverse drug reactions from systemic administration of levomepromazine caused by type IV (delayed-
type) hypersensitivity have included photosensitivity (2).

Photosensitivity
In a university hospital in Turku, Finland, in the period 1989-2001, 826 patients with suspected cutaneous drug
eruptions were patch tested and 89 (11%) had one or more positive reactions. Of these individuals, one developed a
photosensitivity reaction to levomepromazine and had a positive photopatch test to this drug. Clinical and patch
testing details were not provided (2).

Immediate contact reactions
Immediate contact reactions (contact urticaria) to levomepromazine are presented in Chapter 5.

LITERATURE
1   The data in the section 'General' may have been obtained from literature discussed in this chapter, but mostly
    also or exclusively from one or more of the following online sources: ChemIDPlus Advanced, PubChem,
    DrugBank, RxList, Drug Central, Drugs.com, and Wikipedia
2   Lammintausta K, Kortekangas-Savolainen O. The usefulness of skin tests to prove drug hypersensitivity. Br J
    Dermatol 2005;152:968-974

Chapter 3.289 LIDOCAINE

IDENTIFICATION

Description/definition	: Lidocaine is the aminoethylamide that conforms to the structural formula shown below
Pharmacological classes	: Anesthetics, local; anti-arrhythmia agents; voltage-gated sodium channel blockers
IUPAC name	: 2-(Diethylamino)-*N*-(2,6-dimethylphenyl)acetamide
Other names	: Lignocaine; Xylocaine ®
CAS registry number	: 137-58-6
EC number	: 205-302-8
Merck Index monograph	: 6805
Patch testing	: Hydrochloride, 5% pet. (Chemotechnique); hydrochloride, 15% pet. (Chemotechnique, SmartPracticeCanada, SmartPracticeEurope)
Molecular formula	: $C_{14}H_{22}N_2O$

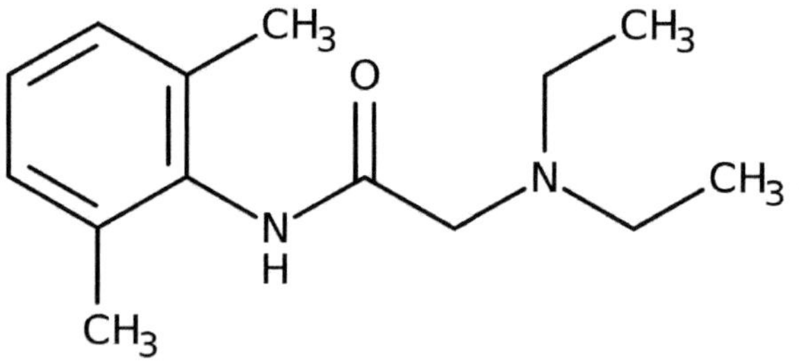

GENERAL

Lidocaine is an aminoethylamide and the prototypical member of the amide class anesthetics, which was introduced in 1946. It also exhibits class IB antiarrhythmic effects. Lidocaine is indicated for production of local or regional anesthesia by infiltration techniques such as percutaneous injection and intravenous regional anesthesia, by peripheral nerve block techniques such as brachial plexus and intercostal blocks, and by central neural techniques such as lumbar and caudal epidural blocks. It is also present in topical products for surface anesthesia. According to some sources, lidocaine may also be used as an anti-arrythmia agent. The topical use of lidocaine has caused a considerable number of cases of contact allergy/ allergic contact dermatitis, which subject has been fully reviewed in Volume 3 of the *Monographs in contact allergy* series (22). In pharmaceutical products, both lidocaine and lidocaine hydrochloride (CAS number 73-78-9, EC number 200-803-8, molecular formula $C_{14}H_{23}ClN_2O$) may be employed (1).

CONTACT ALLERGY FROM ACCIDENTAL CONTACT

A chemical process worker involved in the production of lidocaine (apparently) became sensitized to this chemical (12).

CUTANEOUS ADVERSE DRUG REACTIONS FROM SYSTEMIC ADMINISTRATION CAUSED BY TYPE IV (DELAYED-TYPE) HYPERSENSITIVITY (as demonstrated by positive patch tests)

Cutaneous adverse drug reactions from systemic administration of lidocaine caused by type IV (delayed-type) hypersensitivity have included extensive eczematous eruption from intravenous injection (2), micropapular eruption from intramuscular injection (19), generalized vesiculobullous eruption (20), 'localized and some disseminated rash' from intra-articular injection (20), maculopapular eruption (26), eczema (26), urticaria-like exanthema (26,27) and localized allergic reactions (erythema and edema or eczematous dermatitis) from subcutaneous (3,4,5,6,7,8,9,10,11,14,15,16,18,20,21,25; table 3.289.1), epidural (21), intra-articular (20) and para-articular (7) injections.

General

Delayed-type allergy to local anesthetics mostly causes inflammatory skin reactions with cellulitis-like erythema and edema or eczematous dermatitis confined to the site of the subcutaneous, submucosal, epidural, intra-articular or para-articular injection, developing within 2-24 hours and lasting for a few days up to a week. Such injections are very unlikely to trigger systemic symptoms such as exanthematous skin eruptions (3). These delayed-type allergic reactions to lidocaine are very infrequent (3). It has been shown that contact allergy to lidocaine as demonstrated by a positive patch test often results in negative intradermal and subcutaneous challenge tests. In these cases, it appears to be safe to use lidocaine as an injectable local anesthetic in the future (13) and may account for the low number of allergic reactions to subcutaneous lidocaine.

Localized allergic reactions to injections with lidocaine

Subcutaneous injections
Delayed-type hypersensitivity to subcutaneous/submucosal injections of lidocaine has been observed occasionally and may present as (cellulitis-like) erythema and edema or eczematous dermatitis. Summaries of case reports are given in table 3.289.1.

Table 3.289.1 Summaries of cases reports of delayed-type allergic reactions to subcutaneous lidocaine injections

Year and country	Sex, age		Latency	Symptoms	PT conc./vehicle	Ref.
2019 Netherlands	F	51	<12 hours	Erythema, vesicles, papules, bullae	15% pet., 2% CS	4
2019 Australia	M	75	3 days	Erythema and edema of the face, eyes shut	NA	29
2016 Lebanon	F	65	48 hours	Erythema, vesicles, bullae	EMLA	5
2009 France	F	54	24 hours	Erythema, edema	Not mentioned	21
2008 New Zealand	F	43	5 hours	Vesiculobullous cutaneous reaction	2% and 15% pet.	14
2007 USA	F	28	24-35 h	Erythema and vesicles	15% pet.	20
2006 USA	M	70	12-24 h	Erythema and edema	Not mentioned	15
2004 Spain	F	54	24 hours	Deep swelling and eczematous eruption	2% water	11
2003 USA	F	65	2 days	Edema, pruritic red rash	15% pet.	16
	F	40	2 days	Pruritic red papules	15% pet.	16
2002 USA	F	55	NA	Eruption suggestive of contact dermatitis	NA	9
1997 Switzerland	M	39	NM	Pustular reaction	Not mentioned	25
1996 Switzerland	F	43	24 hours	Angio-edematous swelling	1%	6,25
1996 Australia	F	47	Next day	Edema, erythema, bullae	15% pet.	10
1986 U.K.	F	70	48 hours	Edema	1% and 2% water	18

CS: Commercial solution; EMLA: Cream containing 2.5% lidocaine and 2.5% prilocaine; NA: Data not available/provided; NM: Not mentioned; PT conc.: Patch test concentration

A patient presented with a history of localized edema after dental anesthesia and a reaction to EMLA cream (containing lidocaine and prilocaine). Patch tests were positive to lidocaine and mepivacaine and a subcutaneous challenge resulted in delayed swelling at 24 and 48 hours after challenge to both local anesthetics (28).

Other injection forms

Epidural
A woman aged 35 developed itchy dermatitis with papules and vesicles within 6-12 hours at the site of epidural injection of lidocaine. Patch tests were positive to lidocaine (no details provided) (21). Another woman, aged 24, developed 'late eczema' at the site of an epidural lidocaine injection one month after its administration. A patch test to lidocaine was positive (no details provided) (21). The causal relationship between the injection and the 'late eczema' seems doubtful (which was admitted by the authors).

Para-articular
A 50-year-old man suffering from epicondylitis of the right elbow was treated twice with local para-articular injections of dexamethasone and lidocaine. Two days after each set of injections, the patient developed itching, swelling, erythema and papules, lasting at least 4 days, at the injection sites. Patch tests were negative to dexamethasone, but positive to lidocaine HCl 15% pet., as were prick and intradermal tests read at D2 (7).

Intra-articular
A 51-year-old woman reported having suffered late allergic reactions to injectable lidocaine administered intra-articularly, with localized and some disseminated rash and swelling. Patch tests were positive to lidocaine 15% pet. (20).

Systemic allergic reactions to injections with lidocaine
In Bern, Switzerland, patients with a suspected allergic cutaneous drug reaction were patch-scratch tested with suspected drugs that had previously given a positive lymphocyte transformation test. Lidocaine 15% pet. gave a

positive patch-scratch test in one patient with maculopapular exanthema, one with eczema and 2 with urticaria-like exanthema appearing after several hours (26).

36 hours after general anesthesia for cataract surgery, in which lidocaine (intravenously) and suxamethonium had been used and cefuroxime had been applied subconjunctivally, an 80-year-old woman developed an extensive eczematous eruption that extended from her neckline to just below the knees. Patch tests were positive to lidocaine, as was an intradermal test with 0.1 ml lidocaine solution 1%, read at 2 days. The other drugs used were not tested, but a RAST to cefuroxime was negative (2).

A 54-year old man developed an itchy micropapular eruption on the back of his hands and trunk 10 hours after an intramuscular injection of a polyvitamin preparation. Patch tests, performed with the two components of the medicament, lyophilized vitamins and the solvent (a 5 mg/ml solution of lidocaine), were positive to the solvent at D4, to lidocaine 5% and mepivacaine 2% at D4. An intradermal test with lidocaine, read after 1 and 2 days, was negative on 2 occasions (19).

A 28-year-old woman developed a generalized pruritic vesiculobullous exanthema in the 2 days following breast augmentation surgery, during which lidocaine 1% was injected in the periareolar area and the breast pockets were irrigated with bacitracin. In the past, she had developed vesicles and erythema around the sutures of surgical procedures several times. Patch tests were positive to lidocaine 15, 10, 5 and 1% pet. and to bacitracin 20% pet. (20). In this case, resorption of bacitracin may have played a major role in the development of the generalized eruption. A 51-year-old woman reported having suffered late allergic reactions to injectable lidocaine administered intra-articularly, with localized and some disseminated rash and swelling. Patch tests were positive to lidocaine 15% pet. (20).

A 38-year-old woman experienced local facial edema and urticaria 20 hours after a dental procedure with local anesthesia of lidocaine. One year later, patch tests were positive to lidocaine (tested as liquid drug 5%, 15% and 30% pet. and undiluted) (27).

Cross-reactions, pseudo-cross-reactions and co-reactions

Patients sensitized to lidocaine fairly frequently cross-react to one or more of the other amide-type (aminoacyl-amides) local anesthetics bupivacaine, mepivacaine and prilocaine (6,7,8,10,18,19,23). One cross-reaction to ropivacaine 1% pet. (14). Cross-reaction from mepivacaine allergy (23,25) and from prilocaine allergy (24) to lidocaine. Possible cross-reaction to lidocaine in a patient sensitized to articaine (17).

Immediate contact reactions

Immediate contact reactions (contact urticaria) to lidocaine are presented in Chapter 5.

LITERATURE

1   The data in the section 'General' may have been obtained from literature discussed in this chapter, but mostly also or exclusively from one or more of the following online sources: ChemIDPlus Advanced, PubChem, DrugBank, RxList, Drug Central, Drugs.com, and Wikipedia
2   Hickey JR, Cook SD, Gutteridge G, Sansom JE. Delayed hypersensitivity following intravenous lidocaine. Contact Dermatitis 2006;54:215-216
3   Trautmann A, Stoevesandt J. Differential diagnosis of late-type reactions to injected local anaesthetics: inflammation at the injection site is the only indicator of allergic hypersensitivity. Contact Dermatitis 2019;80:118-124
4   Voorberg AN, Schuttelaar MLA. A case of postoperative bullous allergic contact dermatitis caused by injection with lidocaine. Contact Dermatitis 2019;81:304-306
5   Halabi-Tawil M, Kechichian E, Tomb R. An unusual complication of minor surgery: contact dermatitis caused by injected lidocaine. Contact Dermatitis 2016;75:253-255
6   Bircher AJ, Messmer SL, Surber C, Rufli T. Delayed-type hypersensitivity to subcutaneous lidocaine with tolerance to articaine: confirmation by in vivo and in vitro tests. Contact Dermatitis 1996;34:387-389
7   Breit S, Ruëff F, Przybilla B. 'Deep impact' contact allergy after subcutaneous injection of local anesthetics. Contact Dermatitis 2001;45:296-297
8   Klein CE, Gall H. Type IV allergy to amide-type local anesthetics. Contact Dermatitis 1991;25:45-48
9   Kaufmann JM, Hale EK, Ashinoff RA, Cohen DE. Cutaneous lidocaine allergy confirmed by patch testing. J Drugs Dermatol 2002;1:192-194
10 Bassett I, Delaney T, Freeman S. Can injected lignocaine cause allergic contact dermatitis? Australas J Dermatol 1996;37:155-156
11 Duque S, Fernández L. Delayed-type hypersensitivity to amide local anesthetics. Allergol Immunopathol (Madr) 2004;32:233-234
12 Calnan CD, Stevenson CJ. Studies in contact dermatitis: XV. Dental materials. Trans St Johns Hosp Dermatol Soc 1963;49:9-26

13 Corbo MD, Weber E, DeKoven J. Lidocaine allergy: Do positive patch results restrict future use? Dermatitis 2016;27:68-71

14 Gunson TH, Greig DE. Allergic contact dermatitis to all three classes of local anaesthetic. Contact Dermatitis 2008;59:126-127

15 Hall V, Cheng J, Klemawesch P, Guarderas J. Lidocaine sensitivity in a patient with multiple skin cancers. Dermatitis 2006;17:91-92

16 Mackley CL, Marks JG, Anderson BE. Delayed-type hypersensitivity to lidocaine. Arch Dermatol 2003;139:343-346

17 De Pasquale TMA, Buonomo A, Pucci S. Delayed-type allergy to articaine with cross-reactivity to other local anesthetics from the amide group. J Allergy Clin Immunol Pract 2018;6:305-306

18 Curley RK, Macfarlane AW, King CM. Contact sensitivity to the amide anesthetics lidocaine, prilocaine and mepivacine. Case report and review of the literature. Arch Dermatol 1986;122:924-926

19 Fernándes de Corres L, Leanizbarrutia I. Dermatitis from lignocaine. Contact Dermatitis 1985;12:114-115

20 Amado A, Sood A, Taylor JS. Contact allergy to lidocaine: a report of sixteen cases. Dermatitis 2007;18:215-220

21 Fuzier R, Lapeyre-Mestre M, Mertes PM, Nicolas JF, Benoit Y, Didier A, et al. Immediate- and delayed-type allergic reactions to amide local anesthetics: clinical features and skin testing. Pharmacoepidemiol Drug Saf 2009;18:595-601

22 De Groot AC. Monographs in contact allergy, volume 3: Topical Drugs. Boca Raton, Fl, USA: CRC Press Taylor and Francis Group, 2021 (ISBN 978-0-367-23693-9)

23 Sanchez-Morillas L, Martinez JJ, Martos MR, Gomez-Tembleque P, Andres ER. Delayed-type hypersensitivity to mepivacaine with cross-reaction to lidocaine. Contact Dermatitis 2005;53:352-353

24 Suhonen R, Kanerva L. Contact allergy and cross-reactions caused by prilocaine. Am J Contact Dermat 1997;8:231-234

25 Zanni MP, Mauri-Hellweg D, Brander C, Wendland T, Schnyder B, Frei E, et al. Characterization of lidocaine-specific T cells. J Immunol 1997;158:1139-1148

26 Neukomm C, Yawalkar N, Helbling A, Pichler WJ. T-cell reactions to drugs in distinct clinical manifestations of drug allergy. J Invest Allergol Clin Immunol 2001;11:275-284

27 Batinac T, Sotošek Tokmadžić V, Peharda V, Brajac I. Adverse reactions and alleged allergy to local anesthetics: analysis of 331 patients. J Dermatol 2013;40:522-527

28 Melamed J, Beaucher WN. Delayed-type hypersensitivity (type IV) reactions in dental anesthesia. Allergy Asthma Proc 2007;28:477-479

29 Dickison P, Smith SD. Biting down on the truth: A case of a delayed hypersensitivity reaction to lidocaine. Australas J Dermatol 2019;60:66-67

Chapter 3.290 LINCOMYCIN

IDENTIFICATION

Description/definition   : Lincomycin is a lincosamide antibiotic derived from the bacillus *Streptomyces lincolnensis* that conforms to the structural formula shown below
Pharmacological classes : Anti-bacterial agents; protein synthesis inhibitors
IUPAC name   : (2S,4R)-N-[(1R,2R)-2-Hydroxy-1-[(2R,3R,4S,5R,6R)-3,4,5-trihydroxy-6-methylsulfanyloxan-2-yl]propyl]-1-methyl-4-propylpyrrolidine-2-carboxamide
Other names   : Cillimycin
CAS registry number   : 154-21-2
EC number   : 205-824-6
Merck Index monograph   : 6825
Patch testing   : Hydrochloride, 1% water (3); hydrochloride, 5% pet. (2)
Molecular formula   : $C_{18}H_{34}N_2O_6S$

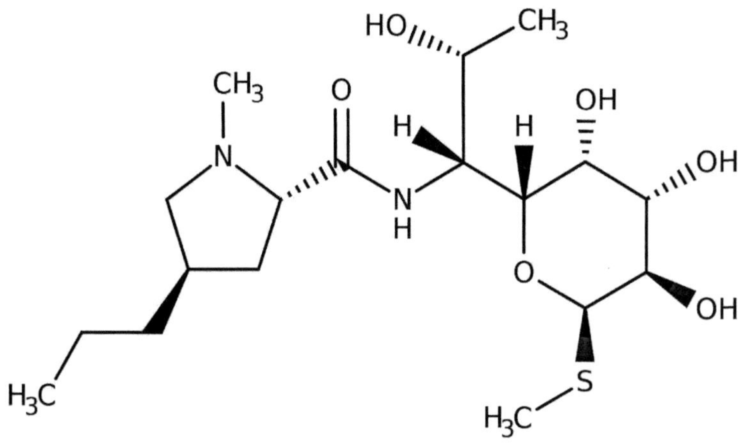

GENERAL

Lincomycin is a lincosamide antibiotic produced by *Streptomyces lincolnensis*. Lincomycin is indicated for the treatment of staphylococcal, streptococcal, and *Bacteroides fragilis* infections caused by susceptible microorganisms. In pharmaceutical products, lincomycin is employed as lincomycin hydrochloride hydrate (CAS number 7179-49-9, EC number 615-424-7, molecular formula $C_{18}H_{37}ClN_2O_7S$) (1).

In topical preparations, lincomycin has rarely caused contact allergy/allergic contact dermatitis, which subject has been reviewed in Volume 3 of the *Monographs in contact allergy* series (4).

CONTACT ALLERGY FROM ACCIDENTAL CONTACT

Two months after using a mixture of lincomycin and spectinomycin for the first time, a 27-year-old female chicken vaccinator developed dermatitis of the hands and forearms. Patch tests were positive to the antibiotic mixture, lincomycin HCl 5% pet. and spectinomycin sulfate 1%, 5% and 20% pet. (2). This was a case of occupational allergic contact dermatitis to lincomycin (and spectinomycin).

CUTANEOUS ADVERSE DRUG REACTIONS FROM SYSTEMIC ADMINISTRATION CAUSED BY TYPE IV
(DELAYED-TYPE) HYPERSENSITIVITY (as demonstrated by positive patch tests)

Cutaneous adverse drug reactions from systemic administration of lincomycin caused by type IV (delayed-type) hypersensitivity have included acute generalized exanthematous pustulosis (AGEP) (5).

Acute generalized exanthematous pustulosis (AGEP)

A 76-year-old woman presented with a skin eruption and fever of 38.1°C, which had started one day after she was injected with lincomycin hydrochloride and acetylcholine chloride for an upper airway infection. On physical examination, numerous non-follicular pustules on an erythematous base were seen on the legs. Laboratory investigations revealed leukocytosis with 75% neutrophils and 6.2% eosinophils. Bacterial cultures of pustules were negative. A skin biopsy of a pustule demonstrated subcorneal pustules with neutrophils and focal necrosis of keratinocytes. Later, patch testing with 1% and 20% lincomycin hydrochloride and 1% and 20% acetylcholine chloride in distilled water showed positive reactions to lincomycin at D2 and D3. The patient was diagnosed with acute generalized exanthematous pustulosis (AGEP) caused by lincomycin (5).

Cross-reactions, pseudo-cross-reactions and co-reactions
A patient sensitized to lincomycin cross-reacted to the structurally closely related antibiotic clindamycin (3).

LITERATURE
1   The data in the section 'General' may have been obtained from literature discussed in this chapter, but mostly also or exclusively from one or more of the following online sources: ChemIDPlus Advanced, PubChem, DrugBank, RxList, Drug Central, Drugs.com, and Wikipedia
2   Vilaplana J, Romaguera C, Grimalt F. Contact dermatitis from lincomycin and spectinomycin in chicken vaccinators. Contact Dermatitis 1991;24:225-226
3   Conde-Salazar L, Guimaraens D, Romero L, Gonzalez M, Yus S. Erythema multiforme-like contact dermatitis from lincomycin. Contact Dermatitis 1985;12:59-61
4   De Groot AC. Monographs in contact allergy, volume 3: Topical Drugs. Boca Raton, Fl, USA: CRC Press Taylor and Francis Group, 2021 (ISBN 978-0-367-23693-9)
5   Otsuka A, Tanizaki H, Okamoto N, Takagaki K. A case of acute generalized exanthematous pustulosis caused by lincomycin. J Dermatol 2005;32:929-930

Chapter 3.291 LISINOPRIL

IDENTIFICATION

Description/definition : Lisinopril is the synthetic peptide derivative that conforms to the structural formula shown below

Pharmacological classes : Antihypertensive agents; angiotensin-converting enzyme inhibitors; cardiotonic agents

IUPAC name : (2S)-1-[(2S)-6-Amino-2-[[(1S)-1-carboxy-3-phenylpropyl]amino]hexanoyl]pyrrolidine-2-carboxylic acid

Other names : N-(1(S)-carboxy-3-phenylpropyl)-L-lysyl-L-proline

CAS registry number : 76547-98-3

EC number : 278-488-1

Merck Index monograph : 6842

Patch testing : Tablet, pulverized, 10% pet. (2); most systemic drugs can be tested at 10% pet.; if the pure chemical is not available, prepare the test material from intravenous powder, the content of capsules or – when also not available – from powdered tablets to achieve a final concentration of the active drug of 10% pet.

Molecular formula : $C_{21}H_{31}N_3O_5$

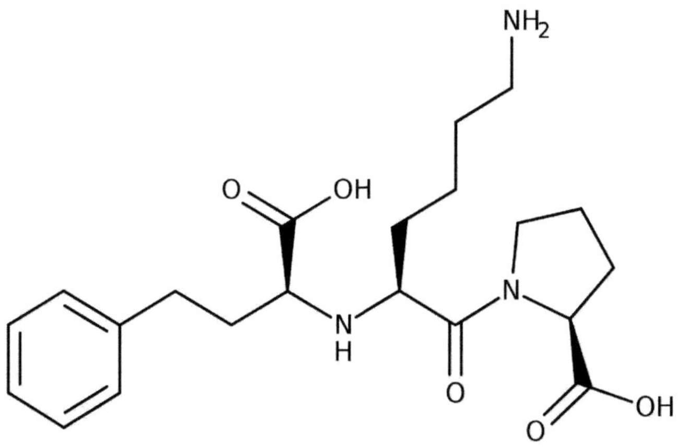

GENERAL

Lisinopril, a synthetic peptide derivative, is a long-acting angiotensin-converting enzyme (ACE) inhibitor. It competitively binds to and inhibits ACE, thereby blocking the conversion of angiotensin I to angiotensin II. This prevents the potent vasoconstrictive actions of angiotensin II and results in vasodilation. Lisinopril also decreases angiotensin II-induced aldosterone secretion by the adrenal cortex, which leads to an increase in sodium excretion and subsequently increases water outflow. Lisinopril is indicated for the treatment of hypertension and symptomatic congestive heart failure (1).

In pharmaceutical products, both lisinopril and lisinopril dihydrate (CAS number 83915-83-7, EC number not available, molecular formula $C_{21}H_{35}N_3O_7$) appear to be employed. However, lisinopril dihydrate is also often termed lisinopril (ChemIDPlus) and therefore, when 'linisopril' is mentioned, it may well indicate the dihydrate form (1).

CONTACT ALLERGY FROM ACCIDENTAL CONTACT

A 24-year-old woman presented with eyelid dermatitis, which had started with localized edema 4 months previously. Later, the area had become itchier, with redness and scaling. The patient suspected a relationship with her work as a pharmacy assistant, which involved breaking and crushing different types of tablets. She was patch tested with the crushed tablets that she had contact with at 10% pet. and showed positive reactions to lisinopril, 2 other ACE-inhibitors, 4 beta-blockers, and 3 benzodiazepines. Three controls were negative to lisinopril 10% pet. Cosmetic allergy was excluded and a diagnosis of occupational airborne allergic contact was made (2).

LITERATURE

1 The data in the section 'General' may have been obtained from literature discussed in this chapter, but mostly also or exclusively from one or more of the following online sources: ChemIDPlus Advanced, PubChem, DrugBank, RxList, Drug Central, Drugs.com, and Wikipedia

2 Swinnen I, Ghys K, Kerre S, Constandt L, Goossens A. Occupational airborne contact dermatitis from benzodiazepines and other drugs. Contact Dermatitis 2014;70:227-232

Chapter 3.292 LOMEFLOXACIN

IDENTIFICATION

Description/definition   : Lomefloxacin is the fluoroquinolone that conforms to the structural formula shown below
Pharmacological classes   : Anti-bacterial agents; topoisomerase II inhibitors
IUPAC name   : 1-Ethyl-6,8-difluoro-7-(3-methylpiperazin-1-yl)-4-oxoquinoline-3-carboxylic acid
CAS registry number   : 98079-51-7
EC number   : 619-317-6
Merck Index monograph : 6889
Patch testing   : Pulverized tablets 10%-20% pet., photopatch tests; sometimes scarification of the skin is
                                                necessary to enhance percutaneous absorption of the drug (3); most systemic drugs can
                                                be tested at 10% pet.; if the pure chemical is not available, prepare the test material from
                                                intravenous powder, the content of capsules or – when also not available – from
                                                powdered tablets to achieve a final concentration of the active drug of 10% pet.
Molecular formula   : $C_{17}H_{19}F_2N_3O_3$

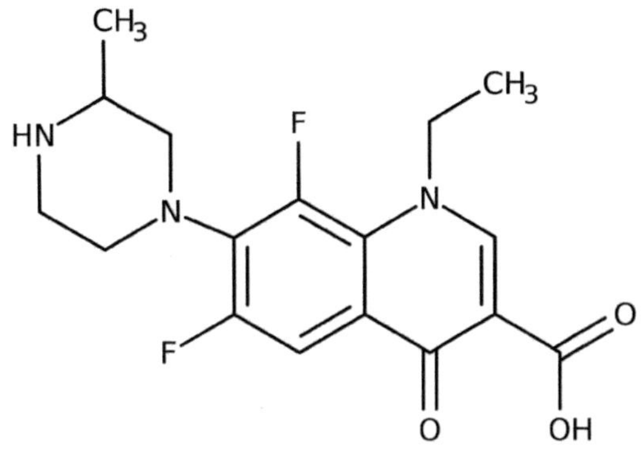

GENERAL

Lomefloxacin is a synthetic broad-spectrum fluoroquinolone with antibacterial activity against a wide range of gram-negative and gram-positive organisms. Lomefloxacin inhibits DNA gyrase, a type II topoisomerase involved in the induction or relaxation of supercoiling during DNA replication. This inhibition leads to a decrease in DNA synthesis during bacterial replication, resulting in cell growth inhibition and eventually cell lysis. Lomefloxacin is indicated for the treatment of bacterial infections of the respiratory tract (chronic bronchitis) and urinary tract, and as a pre-operative prophylactic to prevent urinary tract infections caused by sensitive micro-organisms. In pharmaceutical products, lomefloxacin is employed as lomefloxacin hydrochloride (CAS number 98079-52-8, EC number not available, molecular formula $C_{17}H_{20}ClF_2N_3O_3$) (1).

CUTANEOUS ADVERSE DRUG REACTIONS FROM SYSTEMIC ADMINISTRATION CAUSED BY TYPE IV (DELAYED-TYPE) HYPERSENSITIVITY (as demonstrated by positive patch tests)

Cutaneous adverse drug reactions from systemic administration of lomefloxacin caused by type IV (delayed-type) hypersensitivity have included photosensitivity (2,3,4,5,7,8).

Photosensitivity

Lomefloxacin may cause both phototoxic and photoallergic reactions. However, phototoxicity appears to be the main mechanism of photosensitivity to lomefloxacin, particularly in older patients with concomitant diseases (8).

Case series

Three male patients, aged 57, 68 and 69 years, developed pruritic edematous erythema with scattered vesicles on one or more of the light-exposed areas of the skin (face, neck, the V-shaped area of the upper chest, the extensor aspects of the forearms, the dorsa of his hands) one to 3 weeks after starting oral therapy with lomefloxacin (3). Being suspected of systemic photosensitivity, they were patch and photopatch tested with pulverized lomefloxacin tablets 20%, 10%, 1% and 0.1% pet. and with a series of related quinolone antibiotics 20%, 10% and 1% pet. To increase percutaneous absorption of the drugs, the skin was scarified before application of the test materials. Conventional patch and photopatch tests were also performed simultaneously. Positive reactions on scarified photopatch testing to lomefloxacin down to 0.1% pet. were observed in 2 patients and down to 10% pet. in the third patient one and 2 days after irradiation with UVA 6.0 J/cm². Patient 1 also reacted positively to conventional photopatch testing with lomefloxacin down to 1% pet. with UVA, but far less intense than to the scarified photopatch test. There were no reactions to the other antibiotics. Eighteen controls were negative, but in 4 there was some irritant erythema along the scarification lines in the photopatch tests (3).

Of 8 patients with photosensitive eruptions from lomefloxacin investigated in Coimbra, Portugal, 3 had positive photopatch tests. The test concentration was not mentioned and irradiation was performed with 5 or 10 J/cm^2 UVA. Twenty controls were negative (5).

In the period 2000-2014, also in Coimbra, Portugal, 260 patients were patch tested with antibiotics for suspected cutaneous adverse drug reactions (CADR) to these drugs. 56 patients (22%) had one or more positive patch tests (often from cross-reactivity). Seven patients had developed photosensitivity induced by lomefloxacin and two of them had positive photopatch tests to this antibiotic (7).

Again In Coimbra, 8 patients with photosensitivity reactions to lomefloxacin were investigated. One had acute eczema, 2 subacute eczema and the other 5 sunburn-like reactions. Patch and photopatch tests were performed in all with lomefloxacin, ofloxacin, ciproflaxacin and norfloxacin, tested at 1%, 5% and 10% in petrolatum and irradiated with 5 and 10 J/cm^2 UVA. Only one patient, a man of 68 with vitiligo and alopecia totalis who had developed an acute eczema after having used lomefloxacin for 3 days, had positive photopatch tests to lomefloxacin at 5% and 10% in petrolatum and with the same intensity when irradiated with 5 or 10 J/cm2. No other positive reactions were observed in the other 7 patients nor in 20 controls. It was concluded that *phototoxicity* appears to be the main mechanism of photosensitivity to lomefloxacin, particularly in older patients with concomitant diseases (8).

Case reports

A 69-year-old man was seen with erythema, vesicles, and edema on his face and red papules on the dorsa of the hands that had started a few days before. He had been taking lomefloxacin 200 mg t.d.s. and lysozyme chloride 90 mg t.d.s. orally for rhinitis for 2 weeks. Patch and photopatch tests were carried out with lomefloxacin 10%, 1%, and 0.1% pet., and with ciprofloxacin, sparfloxacin and fleroxacin, all 10% pet. There were positive photopatch tests (erythema and small papules with pruritus one day after irradiation) to lomefloxacin 10% and 1% pet. and to ciprofloxacin, and fleroxacin with UVA-irradiation (4.5 J/cm^2). Five controls were negative. As the patient had never taken ciprofloxacin or fleroxacin before, the photopatch tests to these antibiotics were considered to be photocross-reactions to primary lomefloxacin sensitization (2).

A 68-year-old man presented with a pruriginous erythema, with edema and vesicles, primarily located on light-exposed areas, later spreading to adjacent nonexposed areas. Multiple confluent vesicles in an erythematous base and several tension blisters were especially evident in the dorsal aspect of his hands and forearms. The reaction had evolved over a 4-day period and began 13 days after introduction of lomefloxacin 400 mg/d and 2 days after intensive sunlight exposure. Patch and photopatch tests were performed with lomefloxacin 1%, 5%, and 10% pet., ciprofloxacin (10%), ofloxacin (10%), and nalidixic acid (2%, 5%, and 10%). At D2 and D3 patch test results were negative while photopatch test results were positive to lomefloxacin 1%, 5% and 10% and to ciprofloxacin 10%. Ten controls were negative (4).

Cross-reactions, pseudo-cross-reactions and co-reactions

A patient photosensitized to lomefloxacin had photocross-reactions to ciprofloxacin and fleroxacin (2) and another individual to ciprofloxacin (4).

LITERATURE

1   The data in the section 'General' may have been obtained from literature discussed in this chapter, but mostly also or exclusively from one or more of the following online sources: ChemIDPlus Advanced, PubChem, DrugBank, RxList, Drug Central, Drugs.com, and Wikipedia
2   Kimura M, Kawada A. Photosensitivity induced by lomefloxacin with cross-photosensitivity to ciprofloxacin and fleroxacin. Contact Dermatitis 1998;38:180
3   Kurumaji Y, Shono M. Scarified photopatch testing in lomefloxacin photosensitivity. Contact Dermatitis 1992;26:5-10
4   Correia O. Bullous photodermatosis after lomefloxacin. Arch Dermatol1994:130:808-809
5   Gonçalo M, Barros MA, Azenha A. The importance of photopatch testing in patients with photosensitive drug eruptions. In: Jadassohn Centenary Congress Abstract book. London: European Society of Contact Dermatitis, 1996:15. Data cited in ref. 6
6   Bruynzeel DP, Maibach HI. Patch testing in systemic drug eruptions. Clin Dermatol 1997;15:479-484
7   Pinho A, Coutinho I, Gameiro A, Gouveia M, Gonçalo M. Patch testing - a valuable tool for investigating non-immediate cutaneous adverse drug reactions to antibiotics. J Eur Acad Dermatol Venereol 2017;31:280-287
8   Oliveira HS, Gonçalo M, Figueiredo AC. Photosensitivity to lomefloxacin. A clinical and photobiological study. Photodermatol Photoimmunol Photomed 2000;16:116-120

Chapter 3.293 LORAZEPAM

IDENTIFICATION

Description/definition : Lorazepam is the benzodiazepine that conforms to the structural formula shown below
Pharmacological classes : Antiemetics; GABA modulators; hypnotics and sedatives; anti-anxiety agents; anticonvulsants
IUPAC name : 7-Chloro-5-(2-chlorophenyl)-3-hydroxy-1,3-dihydro-1,4-benzodiazepin-2-one
CAS registry number : 846-49-1
EC number : 212-687-6
Merck Index monograph : 6906
Patch testing : Tablet, pulverized, 30% pet. (3); most systemic drugs can be tested at 10% pet.; if the pure chemical is not available, prepare the test material from intravenous powder, the content of capsules or – when also not available – from powdered tablets to achieve a final concentration of the active drug of 10% pet.
Molecular formula : $C_{15}H_{10}Cl_2N_2O_2$

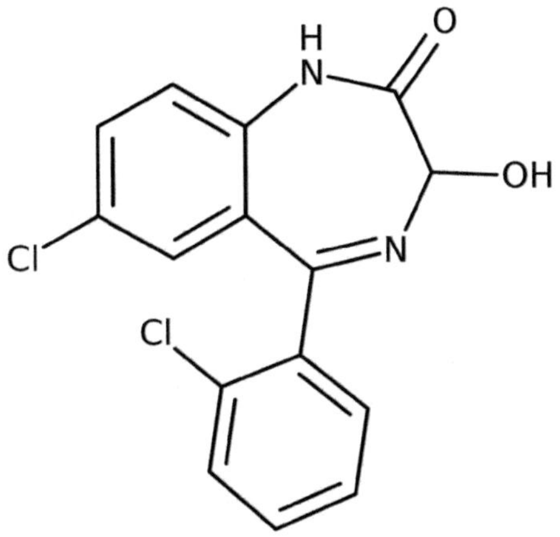

GENERAL

Lorazepam is a short-acting and rapidly cleared benzodiazepine with anxiolytic, anticonvulsant, anti-emetic, hypnotic and sedative properties. Lorazepam is indicated for the short-term relief of anxiety symptoms related to anxiety disorders and anxiety associated with depressive symptoms such as anxiety-associated insomnia. It is also used as an anesthesia premedication in adults to relieve anxiety or to produce sedation/amnesia and for the treatment of status epilepticus. Off-label indications of lorazepam include rapid tranquilization of an agitated patient, alcohol withdrawal delirium, alcohol withdrawal syndrome, muscle spasms, insomnia, panic disorder, delirium, chemothera-py-associated anticipated nausea and vomiting, and psychogenic catatonia (1).

CONTACT ALLERGY FROM ACCIDENTAL CONTACT

Case series

In Leuven, Belgium, in the period 2001-2019, 201 of 1248 health care workers/employees of the pharmaceutical industry had occupational allergic contact dermatitis. In 23 (11%) dermatitis was caused by skin contact with a systemic drug: 19 nurses, two chemists, one physician, and one veterinarian. The lesions were mostly localized on the hands, but often also on the face, as airborne dermatitis. In total, 42 positive patch test reactions to 18 different systemic drugs were found. In 3 patients, lorazepam was the drug/one of the drugs that caused occupational dermatitis (4, overlap with refs. 2 and 3).

In the same clinic in Leuven, Belgium, in the period 2007-2011, 81 patients have been diagnosed with occupational airborne allergic contact dermatitis. In 23 of them, drugs were the offending agents, including lorazepam in 5 cases (2, overlap with refs. 3 and 4).

Case reports

In the same University Clinic in Leuven, Belgium, 3 patients with (airborne) allergic contact dermatitis to lorazepam have been observed, of which 2 were occupational cases. Patient 1 was a 30-year-old woman with an itchy, scaly skin reaction on the face, which was most pronounced on the eyelids. She reported a clear relationship with her work as a geriatric nurse, during which she was required to crush tablets for the elderly on a daily basis, mostly benzodiazepine drugs. Patient 2 was a 44-year-old nurse who had eczema on the hands and face, which had started 3 months after she was required to crush tablets daily during her work as a nurse in a clinic for disabled people. Patient 3 was 66-year-old woman presenting with itchy and burning eczema of the face, which had first appeared on the eyelids. Later, extension to the forehead, lips and perioral region, neck, ears and the fingers occurred. The patient's husband suffered from Parkinson's disease, and she had to crush a large number of tablets for him up to five times a day. All 3 reacted to lorazepam (crushed tablet 30% pet.; 20 controls were negative), 2-4 other benzodiazepines (alprazolam, bromazepam, diazepam, tetrazepam) and 1-5 unrelated chemicals (3, overlap with refs. 2 and 4). The first two patients had occupational airborne allergic contact dermatitis, patient number 3 allergic contact dermatitis *by proxy*.

Cross-reactions, pseudo-cross-reactions and co-reactions
The possibility of cross-reactions between benzodiazepines, especially with primary sensitization to tetrazepam, is considered likely, but the pattern is unknown and may differ between sensitivity from (occupational) contact with the drugs and systemically induced reactions (3).

LITERATURE
1   The data in the section 'General' may have been obtained from literature discussed in this chapter, but mostly also or exclusively from one or more of the following online sources: ChemIDPlus Advanced, PubChem, DrugBank, RxList, Drug Central, Drugs.com, and Wikipedia
2   Swinnen I, Goossens A. An update on airborne contact dermatitis: 2007-2011. Contact Dermatitis 2013;68:232-238
3   Swinnen I, Ghys K, Kerre S, Constandt L, Goossens A. Occupational airborne contact dermatitis from benzodiazepines and other drugs. Contact Dermatitis 2014;70:227-232
4   Gilissen L, Boeckxstaens E, Geebelen J, Goossens A. Occupational allergic contact dermatitis from systemic drugs. Contact Dermatitis 2020;82:24-30

Chapter 3.294 LORMETAZEPAM

IDENTIFICATION

Description/definition   : Lormetazepam is the benzodiazepine that conforms to the structural formula shown
                            below
Pharmacological classes   : Hyponotics and sedatives
IUPAC name   : 7-Chloro-5-(2-chlorophenyl)-3-hydroxy-1-methyl-3H-1,4-benzodiazepin-2-one
Other names   : Methyllorazepam
CAS registry number   : 848-75-9
EC number   : 212-700-5
Merck Index monograph   : 6909
Patch testing   : Commercial tablet 10% pet. (2); most systemic drugs can be tested at 10% pet.; if the pure
                            chemical is not available, prepare the test material from intravenous powder, the content
                            of capsules or – when also not available – from powdered tablets to achieve a final
                            concentration of the active drug of 10% pet.
Molecular formula   : $C_{16}H_{12}Cl_2N_2O_2$

GENERAL

Lormetazepam is a benzodiazepine which reduces central nervous system activity. It produces anxiolytic, muscle relaxant, sedative and hypnotic effects. Because it is a short-acting benzodiazepine, it does not produce significant sedation after waking. Lormetazepam is used for the treatment of short-term insomnia (1).

CONTACT ALLERGY FROM ACCIDENTAL CONTACT

A 24-year-old woman presented with eyelid dermatitis, which had started with localized edema 4 months previously. Later, the area had become itchier, with redness and scaling. The patient suspected a relationship with her work as a pharmacy assistant, which involved breaking and crushing different types of tablets. She was patch tested with the crushed tablets that she had contact with at 10% pet. and showed positive reactions to lormetazepam, 2 other benzodiazepines, 3 ACE-inhibitors and 4 beta-blockers. Eleven controls were negative to lormetazepam 10% pet. Cosmetic allergy was excluded and a diagnosis of occupational airborne allergic contact was made (2).

Cross-reactions, pseudo-cross-reactions and co-reactions

The possibility of cross-reactions between benzodiazepines, especially with primary sensitization to tetrazepam, is considered likely, but the pattern is unknown and may differ between sensitivity from (occupational) contact with the drugs and systemically induced reactions (2).

LITERATURE

1   The data in the section 'General' may have been obtained from literature discussed in this chapter, but mostly also or exclusively from one or more of the following online sources: ChemIDPlus Advanced, PubChem, DrugBank, RxList, Drug Central, Drugs.com, and Wikipedia
2   Swinnen I, Ghys K, Kerre S, Constandt L, Goossens A. Occupational airborne contact dermatitis from benzodiazepines and other drugs. Contact Dermatitis 2014;70:227-232

Chapter 3.295 MAGNESIUM OXIDE

IDENTIFICATION

Description/definition : Magnesium oxide is the inorganic compound that conforms to the structural formula shown below
Pharmacological classes : Antacids
IUPAC name : Magnesium;oxygen(2-)
Other names : Magnesium monoxide; magnesium (II) oxide
CAS registry number : 1309-48-4
EC number : 215-171-9
Merck Index monograph : 7011
Patch testing : pure
Molecular formula : MgO

$$Mg{=}O$$

GENERAL

Magnesium oxide is an inorganic compound that occurs in nature as the mineral periclase. In aqueous media it combines quickly with water to form magnesium hydroxide. It is used as an antacid and mild laxative. When taken internally by mouth as a laxative, the osmotic force of the magnesia suspension acts to draw fluids from the body and to retain those already within the lumen of the intestine, serving to distend the bowel, thus stimulating nerves within the colon wall, inducing peristalsis and resulting in evacuation of colonic contents. Magnesium oxide also has many nonmedicinal uses, e.g. as a fertilizer) (1).

CUTANEOUS ADVERSE DRUG REACTIONS FROM SYSTEMIC ADMINISTRATION CAUSED BY TYPE IV (DELAYED-TYPE) HYPERSENSITIVITY (as demonstrated by positive patch tests)

Cutaneous adverse drug reactions from systemic administration of magnesium oxide caused by type IV (delayed-type) hypersensitivity have included erythematous eruption on the abdomen and back (2).

Other cutaneous adverse drug reactions

A 62-year-old man presented with a 5-day history of an itchy skin rash over his abdomen and back, which had appeared 3 days after taking 600 mg of oral Maglax ® (magnesium oxide, carmellose calcium, crospovidone, calcium stearate, light anhydrous silicic acid and crystalline cellulose). The patient did not use other medications and had taken Maglax ® before without problems. Dermatological examination revealed diffuse, edematous erythema on his abdomen and disseminated erythematous patches, 5-10 mm in diameter, around a large plaque on his abdomen and back. The mucous membranes were not affected. Histopathology showed an irregularly thickened epidermis with a sawtooth-like basal layer, spongiosis with exocytosis of lymphocytes and neutrophils, and lichenoid reaction in the basal layer. Papillary edema and inflammatory infiltrates consisting of lymphocytes and eosinophils in the vicinity of blood vessels in the dermal papillae and upper dermis were noticed. All laboratory tests were within normal limits. One month after resolution of the eruption, a patch test to Maglax ® 20% was positive, as was the drug-induced lymphocyte stimulation test (DLST) with a stimulation index of 537 (normal <180). Patch tests with all six components were negative, but DLSTs to them all positive. The authors suggested that there was a possibility of conformational change of these components in the process of mixing (2). It may well be that this patient had a drug eruption from delayed-type hypersensitivity to the commercial product Maglax ®, but the claim made in the title of the article that is was caused by magnesium oxide was not substantiated by the results of allergy tests.

LITERATURE

1   The data in the section 'General' may have been obtained from literature discussed in this chapter, but mostly also or exclusively from one or more of the following online sources: ChemIDPlus Advanced, PubChem, DrugBank, RxList, Drug Central, Drugs.com, and Wikipedia
2   Sakanoue M, Sanada J, Kanekura T. Skin eruption elicited by magnesium oxide (Maglax ®). J Dermatol 2016;43:221-222

Chapter 3.296 MECLOFENOXATE

IDENTIFICATION

Description/definition   : Meclofenoxate is the ester of dimethylaminoethanol and *p*-chlorophenoxyacetic acid that
                            conforms to the structural formula shown below
Pharmacological classes   : Neuroprotective agents; nootropic agents
IUPAC name   : 2-(Dimethylamino)ethyl 2-(4-chlorophenoxy)acetate
CAS registry number   : 51-68-3
EC number   : 200-116-3
Merck Index monograph   : 7121
Patch testing   : Tablet, pulverized, 2.5% pet. (2); most systemic drugs can be tested at 10% pet.; if the
                            pure chemical is not available, prepare the test material from intravenous powder, the
                            content of capsules or – when also not available – from powdered tablets to achieve a
                            final concentration of the active drug of 10% pet.
Molecular formula   : $C_{12}H_{16}ClNO_3$

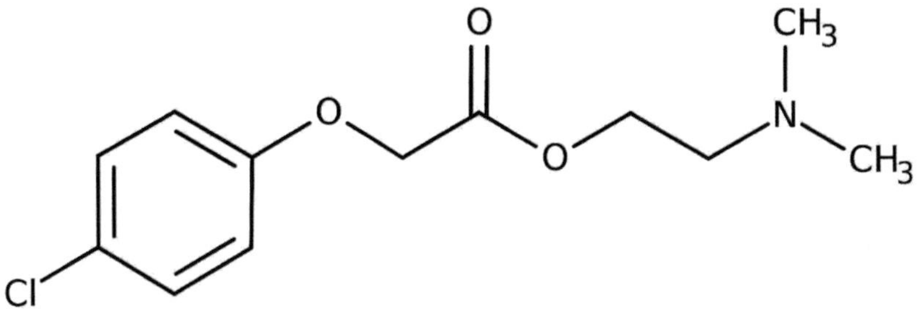

GENERAL

Meclofenoxate is an ester of dimethylaminoethanol and *p*-chlorophenoxyacetic acid. It is said to be one of the most used nootropics and the only drug reported to cause lipofuscin regression. Meclofenoxate and/or meclofenoxate hydrochloride (CAS number 3685-84-5, EC number 222-975-3, molecular formula $C_{12}H_{17}Cl_2NO_3$) are available in some countries for the treatment of cerebral insufficiency, dementia, stroke, and cerebral aging (1).

CONTACT ALLERGY FROM ACCIDENTAL CONTACT

Three female nurses became sensitized to meclofenoxate, that they presumably dispensed to patients. Patch tests were positive to the commercial tablets 2.5% (vehicle not mentioned). Histopathology of 2 patch tests confirmed the allergic nature, in the third, it was compatible, but could not provide categorical evidence of an allergic nature. The nurses presumably had occupational allergic contact dermatitis, but no clinical data were provided (2).

Meclofenoxate was also mentioned as an occupational contact allergen in a book of the first author of this article, which may have concerned the same patients (3).

LITERATURE

1   The data in the section 'General' may have been obtained from literature discussed in this chapter, but mostly
    also or exclusively from one or more of the following online sources: ChemIDPlus Advanced, PubChem,
    DrugBank, RxList, Drug Central, Drugs.com, and Wikipedia
2   Foussereau J, Lantz JP. Allergy to Meclofenoxate (Lucridril) in nurses. Contact Dermatitis Newsletter
    1972;12:321-322
3   Foussereau J, Benezra C, Maibach HI. Occupational contact dermatitis: clinical and chemical aspects.
    Copenhagen: Munksgaard, 1982. Data cited in ref. 4
4   Bircher AJ. Drug allergens. In: John S, Johansen J, Rustemeyer T, Elsner P, Maibach H, eds. Kanerva's
    occupational dermatology. Cham, Switzerland: Springer, 2020:559-578

Chapter 3.297 MEFENAMIC ACID

IDENTIFICATION

Description/definition   : Mefenamic acid is the anthranilic acid that conforms to the structural formula shown below

Pharmacological classes : Cyclooxygenase inhibitors; anti-inflammatory agents, non-steroidal

IUPAC name   : 2-(2,3-Dimethylanilino)benzoic acid

CAS registry number   : 61-68-7

EC number   : 200-513-1

Merck Index monograph   : 7139

Patch testing   : Tablet, pulverized, 10% pet. (5); most systemic drugs can be tested at 10% pet.; if the pure chemical is not available, prepare the test material from intravenous powder, the content of capsules or – when also not available – from powdered tablets to achieve a final concentration of the active drug of 10% pet.

Molecular formula   : $C_{15}H_{15}NO_2$

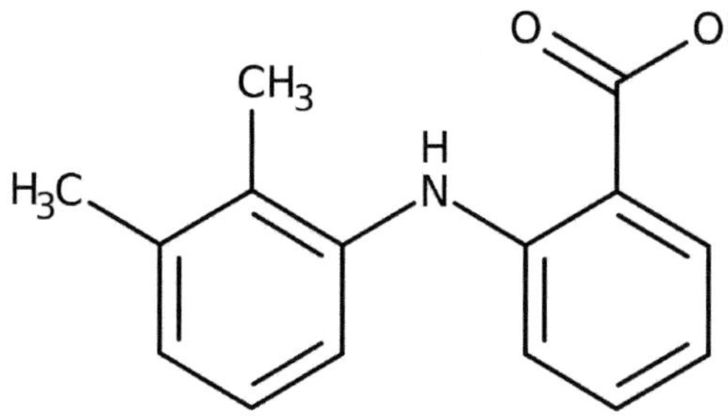

GENERAL

Mefenamic acid is an anthranilic acid and nonsteroidal anti-inflammatory drug (NSAID) with anti-inflammatory, antipyretic and analgesic activities. It inhibits the activity of the enzymes cyclooxygenase I and II, resulting in a decreased formation of precursors of prostaglandins and thromboxanes. The resulting decrease in prostaglandin synthesis, by prostaglandin synthase, is responsible for the therapeutic effects of this agent. Mefenamic acid is indicated for the treatment of rheumatoid arthritis, osteoarthritis, dysmenorrhea, and mild to moderate pain, inflammation, and fever (1).

CUTANEOUS ADVERSE DRUG REACTIONS FROM SYSTEMIC ADMINISTRATION CAUSED BY TYPE IV (DELAYED-TYPE) HYPERSENSITIVITY (as demonstrated by positive patch tests)

Cutaneous adverse drug reactions from systemic administration of mefenamic acid caused by type IV (delayed-type) hypersensitivity have included fixed drug eruption (2,3,4).

Fixed drug eruption

In Seoul, South Korea, 31 patients (15 men,16 women, age range 7 to 62 years) with suspected fixed drug eruption were patch tested between 1986 and 1990 and in 1996 and 1997. The drugs for patch testing were usually applied at 10% in white petrolatum, both on a previous lesions and on apparently normal skin. The presentation of the results in this reports were rather confusing. When 'itching, erythema, infiltration' at the postlesional skin is taken as proof of delayed-type allergy, then there were 5 reactions to mefenamic acid. In most cases, oral provocation tests, when performed, were positive (2).

Three patients with fixed drug eruption from mefenamic acid were reported from Austria in 2011 (3). The first was a 44-year-old man, who reported three episodes of painful erosions of the oral mucosa and the tongue and conjunctivitis. During the second episode, painful erosions also appeared on the glans penis and during the third episode a livid, asymptomatic round lesion occurred additionally on the left thigh. A patch test with a pulverized mefenamic acid tablet was positive on postlesional but negative on uninvolved skin. Later, a patch test with the pure mefenamic acid dissolved in water (concentration?) was positive on involved skin. Histopathology of this reaction was compatible with fixed drug eruption (3). Patient 2 was a 39-year-old man who presented with acute swelling of the lower lip and tongue with dyspnea, and two reddish livid, sharply defined lesions, in part with blistering, on the lateral edge of the left hand and on the medial edge of the left 4th finger. The patient had taken a tablet of 500 mg mefenamic acid 6 hours previously. Patch tests with the pulverized tablet and pure mefenamic acid in solution (concentration?) were negative on both lesional and non-lesional skin. Only a 3rd patch test on the medial side of the left ring finger was positive after 48 hours (3). The third patient was a 46-year-old man who reported intolerance to mefenamic acid tablets since three years that always manifested with erosions and pruritus of oral mucous membranes and genital skin after oral administration. An oral provocation test after 24 hours resulted in multiple sharply defined erythematous macules and plaques especially on the limbs, gluteal region and thorax, some with vesicles. Complete healing occurred rapidly and without residual pigmentation in 3 weeks. Pigmenting, multifocal/ generalized bullous fixed drug eruption due to mefenamic acid was diagnosed, but the patient was not patch tested (3). Also, it was not very logical to diagnose pigmenting drug eruption, as the lesions had healed without residual pigmentation.

Two patients from Tunisia suffering from bullous fixed drug eruption had positive patch tests to mefenamic acid tablets 30% pet. on postlesional skin (4).

LITERATURE

1 The data in the section 'General' may have been obtained from literature discussed in this chapter, but mostly also or exclusively from one or more of the following online sources: ChemIDPlus Advanced, PubChem, DrugBank, RxList, Drug Central, Drugs.com, and Wikipedia
2 Lee AY. Topical provocation in 31 cases of fixed drug eruption: change of causative drugs in 10 years. Contact Dermatitis 1998;38:258-260
3 Handisurya A, Moritz KB, Riedl E, Reinisch C, Stingl G, Wohrl S. Fixed drug eruption caused by mefenamic acid: a case series and diagnostic algorithms. J Dtsch Dermatol Ges 2011;9:374-378
4 Zaouak A, Ben Salem F, Ben Jannet S, Hammami H, Fenniche S. Bullous fixed drug eruption: A potential diagnostic pitfall: a study of 18 cases. Therapies 2019;74:527-530
5 Brockow K, Garvey LH, Aberer W, Atanaskovic-Markovic M, Barbaud A, Bilo MB, et al.; ENDA/EAACI Drug Allergy Interest Group. Skin test concentrations for systemically administered drugs – an ENDA/EAACI Drug Allergy Interest Group position paper. Allergy 2013;68:702-712

Chapter 3.298 MELPHALAN

IDENTIFICATION

Description/definition : Melphalan is the phenylalanine derivative of nitrogen mustard that conforms to the structural formula shown below
Pharmacological classes : Antineoplastic agents, alkylating; myeloablative agonists
IUPAC name : (2S)-2-Amino-3-[4-[bis(2-chloroethyl)amino]phenyl]propanoic acid
Other names : 3-(p-(bis(2-Chloroethyl)amino)phenyl)-L-alanine; phenylalanine mustard
CAS registry number : 148-82-3
EC number : 205-726-3
Merck Index monograph : 7166
Patch testing : 1% pet.
Molecular formula : $C_{13}H_{18}Cl_2N_2O_2$

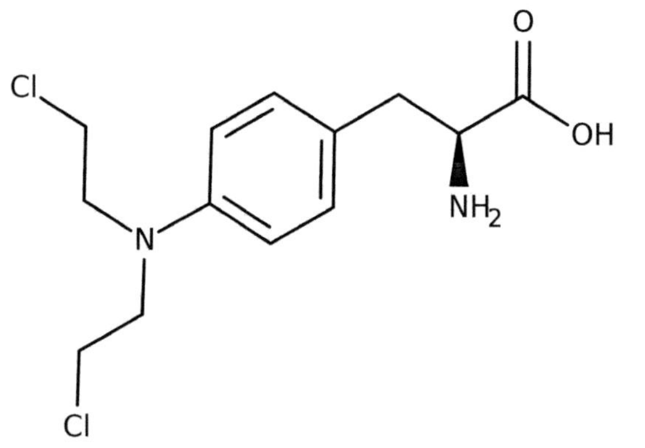

GENERAL

Melphalan is a phenylalanine derivative of nitrogen mustard with antineoplastic activity. This agent alkylates DNA at the N7 position of guanine and induces DNA interstrand cross-linkages, resulting in the inhibition of DNA and RNA synthesis and cytotoxicity against both dividing and non-dividing tumor cells. Melphalan is indicated for the palliative treatment of multiple myeloma and for the palliation of non-resectable epithelial carcinoma of the ovary. It has also been used alone or as part of various chemotherapeutic regimens as an adjunct to surgery in the treatment of breast cancer, alone or in combination regimens for palliative treatment of locally recurrent or unresectable in-transit metastatic melanoma of the extremities, as well as for the treatment of amyloidosis with prednisone (1).

In tablets, melphalan base is used; in injection fluids and powders, melphalan hydrochloride (CAS number 3223-07-2, EC number not available, molecular formula $C_{13}H_{19}Cl_3N_2O_2$) is employed (1).

CONTACT ALLERGY FROM ACCIDENTAL CONTACT

A 21-year-old man developed an itchy rash on his face one week after starting work in a new laboratory of a pharmaceutical company, in which he had already worked for several years. His job involved working with cytotoxic drugs (azathioprine, chlorambucil, melphalan, 6-mercaptopurine and busulfan) in a fume-hooded compartment. The rash appeared initially over areas on his face where his respirator dust mask touched the skin and later over uncovered areas. At the time of onset of his rash, he had been working with melphalan, but chlorambucil was also present in the laboratory at the time. Patch testing with all the drugs that the patient had been working with showed strongly positive reactions to chlorambucil 1%pet. (D2 ++, D4 +++) and to melphalan 1% pet. (++/++). He was diagnosed as having occupational allergic contact dermatitis from melphalan and chlorambucil. The authors suggested that the patient had primarily been sensitized to melphalan and that he cross-reacted to chlorambucil, but also allowed for the possibility of co-sensitization (2).

Cross-reactions, pseudo-cross-reactions and co-reactions

Possibly cross-reactivity to chlorambucil (2).

LITERATURE

1   The data in the section 'General' may have been obtained from literature discussed in this chapter, but mostly also or exclusively from one or more of the following online sources: ChemIDPlus Advanced, PubChem, DrugBank, RxList, Drug Central, Drugs.com, and Wikipedia
2   Goon AT, McFadden JP, McCann M, Royds C, Rycroft RJ. Allergic contact dermatitis from melphalan and chlorambucil: cross-sensitivity or cosensitization? Contact Dermatitis 2002;47:309-310

Chapter 3.299 MENADIONE

IDENTIFICATION

Description/definition   : Menadione is the synthetic naphthoquinone that conforms to the structural formula
                            shown below
Pharmacological classes : Antifibrinolytic agents; vitamins
IUPAC name   : 2-Methylnaphthalene-1,4-dione
Other names   : 2-Methyl-1,4-naphthoquinone
CAS registry number   : 58-27-5
EC number   : 200-372-6
Merck Index monograph   : 7169
Patch testing   : 0.1% and 0.01% pet.; 1% pet. is irritant (6)
Molecular formula   : $C_{11}H_8O_2$

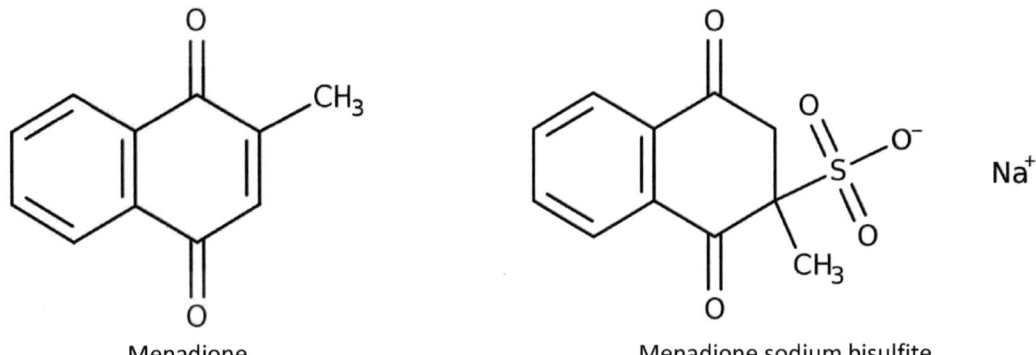

Menadione   Menadione sodium bisulfite

GENERAL

Vitamin K_3 (menadione) is a synthetic organic fat-soluble naphthoquinone. Menadione and its water-soluble form menadione sodium bisulfite (CAS number 130-37-0, EC number 204-987-0, molecular formula $C_{11}H_9NaO_5S$) are used as food supplement (nutraceutical) and as antihemorrhagic drug (1,2).

CONTACT ALLERGY FROM ACCIDENTAL CONTACT

A 42-year-old male pig farmer presented with an itchy rash lasting more than 1 month. It first affected his right forearm and hand and progressed in a few days to severe vesicular eczema involving the face, axillary folds, abdomen, buttocks, arms, legs and feet. During the last decade, the patient regularly prepared the antihemorrhagic drug vitamin K_3 sodium bisulfite (menadione sodium bisulfite) by dissolving it into water (0.1%) using his right forearm. The lesions started a few hours after the most recent exposure to this powder. One month after healing of the eczema, the patient again prepared the drug and immediately experienced a flare-up of the previous eczema. Patch tests were positive to the commercial preparation (0.1% saline), vitamin K_3 (0.01% and 0.1% pet.), and vitamin K_3 sodium bisulfite (0.01% and 0.1% pet.). Twenty controls were negative. Occupational allergic contact dermatitis to menadione (vitamin K_3) was diagnosed (2).

One month after a 40-year-old man had started working in the preparation of pig feeds (adding several mixes containing mineral salts and vitamins to them) developed an acute dermatitis in the face, neck, and on the dorsum of the hands. Patch tests with the ingredients of the mixes were positive to vitamin K_3 sodium bisulfite 0.1% water and pet. In a second session, cross-reactivity to menadiol sodium diphosphate (commercial preparation) was demonstrated. This chemical was termed vitamin K_4, but this is probably incorrect (ChemIDPlus) (4). This occupational allergic contact dermatitis was of the airborne type (4).

In a 29-year-old man working in a small pharmaceutical factory, 3 weeks after handling a new chemical substance, nausea, dizziness, giddiness, vomiting and generalized pruritus developed. A few days later the patient had a dermatitis on the neck with itching and vesiculation, followed by areas of lichenification. Dark brown spots appeared on the flexor surfaces of the wrists. In 2 weeks' time, the entire skin of the palms and volar surfaces of both hands, the nails, soles and dorsa of the feet and the arms and legs changed to a deep dark brown color which progressively stained all his skin uniformly. The mucous membranes were normal. The patient wore protective gloves and vest but no mask. Histopathology of a pigmented spot showed chronic non-specific dermatitis with a perivascular mononuclear infiltrate. Patch tests were positive to vitamin K_3 sodium bisulfite 0.01% and 0.1% pet., which was the new chemical the patient had contact with. The authors diagnosed occupational allergic contact dermatitis to and systemic poisoning by inhalation or ingestion of vitamin K_3 sodium bisulfite (6).

A woman aged 21 working in a pharmaceutical laboratory had contact with many drugs of varied composition. After 4 or 5 months, dyshidrosiform and vesicular exudative lesions appeared on the index and middle fingers of the right hand, which later spread to the dorsa of all fingers of both hands and were also occasionally accompanied by bilateral, macular scaly lesions on the eyelids and cheeks. Patch tests were positive to an antidiarrhea drug 10% and later to its component vitamin K_3 sodium bisulfite 0.1 % water (10). The same authors also describe a man aged 25 working for 8 months in a veterinary laboratory, preparing an antihemorrhagic drug for chickens. After 5 months itching, vesicular exudative lesions appeared on the dorsa of the hands and forearms. He also had itching on the face and eyelids. Stopping this work and application of topical treatment cleared the dermatitis, but 5 days after he had resumed working again, the lesions reappeared in the same location. Patch tests were positive to vitamin K_3 sodium bisulfite 0.1% in water. Twenty controls were negative. Concentrations of 1% and higher were strongly irritant in controls (10).

Possibly, other cases of contact allergy to menadione have been described in early literature, but details are not available (7-9, data cited in ref. 6).

Cross-reactions, pseudo-cross-reactions and co-reactions

A patient sensitized to phytonadione (vitamin K_1) cross-reacted to menadione (vitamin K_3) (3). A patient sensitized to menadione cross-reacted to menadiol (4,10). An individual occupationally sensitized to acetomenaphthone (vitamin K_4, menadiol diacetate) cross-reacted to menadione (5).

LITERATURE

1   The data in the section 'General' may have been obtained from literature discussed in this chapter, but mostly also or exclusively from one or more of the following online sources: ChemIDPlus Advanced, PubChem, DrugBank, RxList, Drug Central, Drugs.com, and Wikipedia

2   Bianchi L, Hansel K, Tramontana M, Assalve D, Stingeni L. Occupational allergic contact dermatitis with secondary spreading from vitamin K3 sodium bisulphite in a pig farmer. Dermatitis 2015;26:150-151

3   Wong DA, Freeman S. Cutaneous allergic reaction to intramuscular vitamin K1. Australas J Dermatol 1999;40:147-152

4   Dinis A, Brandão M, Faria A. Occupational contact dermatitis from vitamin K3 sodium bisulphite. Contact Dermatitis 1988;18:170-171

5   Jirasek L, Schwank R. Berufskontaktekzem durch Vitamin K. Hautarzt 1965;16:351-353

6   Camarasa JG, Barnadas M. Occupational dermatosis by vitamin K3 sodium bisulphite. Contact Dermatitis 1982;8:268

7   Watrous RM. Health hazards of the pharmaceutical industry. Br J Industr Med 1947;4:111 (Data cited in ref. 6)

8   Rust Z. Uber allergische Reaktionen bei Vitamintherapie. Z Haut und Geschlkrh 1954;17:317-319 (Data cited in ref. 6)

9   Page RC, Bercovitz Z. Generalized exfoliative dermatitis from topical application of 2 methyl-1,4-naphtoquinone (synthetic vitamin K analogue). Am J Med Sciences 1976;203:566 (Data cited in ref. 6)

10   Romaguera C, Grimalt F, Conde-Salazar L. Occupational dermatitis from vitamin K3 sodium bisulfite. Contact Dermatitis 1980;6:355-356

Chapter 3.300 MEPERIDINE

IDENTIFICATION

Description/definition   : Meperidine is the phenylpiperidine that conforms to the structural formula shown below
Pharmacological classes   : Adjuvants, anesthesia; narcotics; analgesics, opioid
IUPAC name   : Ethyl 1-methyl-4-phenylpiperidine-4-carboxylate
Other names   : Pethidine
CAS registry number   : 57-42-1
EC number   : 200-329-1
Merck Index monograph : 7186
Patch testing   : No data available; most systemic drugs can be tested at 10% pet.; if the pure chemical is
                                       not available, prepare the test material from intravenous powder, the content of capsules
                                       or – when also not available – from powdered tablets to achieve a final concentration of
                                       the active drug of 10% pet.
Molecular formula   : $C_{15}H_{21}NO_2$

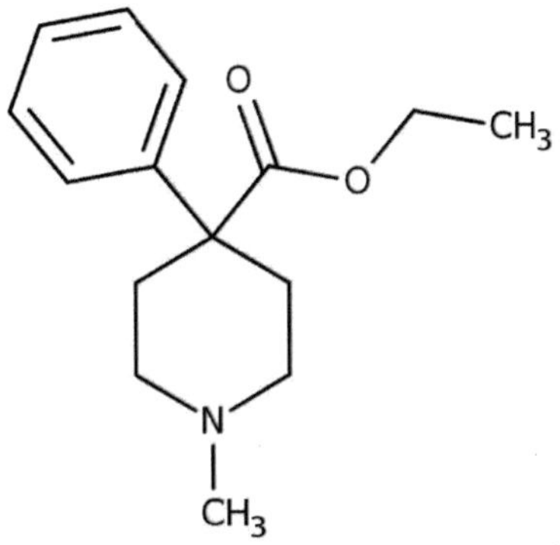

GENERAL

Meperidine, better known as pethidine, is a synthetic opiate agonist belonging to the phenylpiperidine class. The drug mimics the actions of endogenous neuropeptides via opioid receptors, thereby producing the characteristic morphine-like effects on the mu-opioid receptor, including analgesia, euphoria, sedation, respiratory depression, miosis, bradycardia and physical dependence. Meperidine is recommended as a second-line agent for relief of moderate to severe acute pain and has also been used for intravenous regional anesthesia, peripheral nerve blocks and intraarticular, epidural and spinal analgesia.

In pharmaceutical products, meperidine is employed as meperidine hydrochloride (CAS number 50-13-5, EC number not available, molecular formula $C_{15}H_{22}ClNO_2$) (1).

CUTANEOUS ADVERSE DRUG REACTIONS FROM SYSTEMIC ADMINISTRATION CAUSED BY TYPE IV (DELAYED-TYPE) HYPERSENSITIVITY (as demonstrated by positive patch tests)

Cutaneous adverse drug reactions from systemic administration of meperidine caused by type IV (delayed-type) hypersensitivity have included fixed drug eruption (2).

Fixed drug eruption

A 55-year-old woman, who had previously suffered fixed drug eruption (FDE) from promethazine (with a positive patch test) presented with an itchy macular erythema on her right forearm, back and buttocks, and her right hand and foot. The lesions, which were similar to her previous FDE, had developed a few hours after having received five drugs including pethidine for an endoscopic examination. Patch tests were positive to pethidine HCl (concentration? vehicle? lesional skin? time of reading?) and an oral provocation test with pethidine (dosage?) induced the erythema approximately 2 hours after taking the drug. Later, she would develop FDE from omeprazole. This patient had polyreactivity to three unrelated compounds (2).

LITERATURE

1   The data in the section 'General' may have been obtained from literature discussed in this chapter, but mostly
      also or exclusively from one or more of the following online sources: ChemIDPlus Advanced, PubChem,
      DrugBank, RxList, Drug Central, Drugs.com, and Wikipedia
2   Kai Y, Okamoto O, Fujiwara S. Fixed drug eruption caused by three unrelated drugs: promethazine, pethidine and
      omeprazole. Clin Exp Dermatol 2011;36:755-758

Chapter 3.301 MEPIVACAINE

IDENTIFICATION

Description/definition : Mepivacaine is the piperidinecarboxamide and amide-type local anesthetic that conforms to the structural formula shown below

Pharmacological classes : Anesthetics, local

IUPAC name : *N*-(2,6-Dimethylphenyl)-1-methylpiperidine-2-carboxamide

CAS registry number : 96-88-8

EC number : 202-543-0

Merck Index monograph : 7196

Patch testing : Hydrochloride 1% pet. (SmartPracticeCanada); sometimes, the performance of intradermal tests with late readings (D2-D3) are necessary (4)

Molecular formula : $C_{15}H_{22}N_2O$

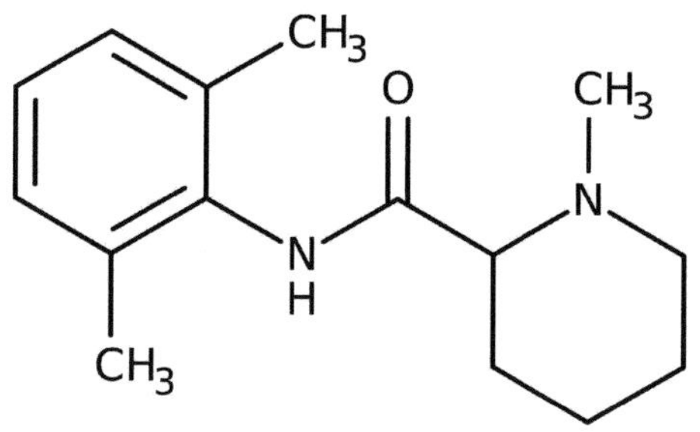

GENERAL

Mepivacaine is an amide-type local anesthetic, indicated for production of local or regional analgesia and anesthesia by local infiltration, peripheral nerve block techniques, or central neural techniques including epidural and caudal blocks (1). In pharmaceutical products (all injections), mepivacaine is employed as mepivacaine hydrochloride (CAS number 1722-62-9, EC number 217-023-9, molecular formula $C_{15}H_{23}ClN_2O$) (1).

CUTANEOUS ADVERSE DRUG REACTIONS FROM SYSTEMIC ADMINISTRATION CAUSED BY TYPE IV (DELAYED-TYPE) HYPERSENSITIVITY (as demonstrated by positive patch tests)

Cutaneous adverse drug reactions from systemic administration of mepivacaine caused by type IV (delayed-type) hypersensitivity have included maculopapular eruption (2), localized allergic reaction to subcutaneous infiltration (4,7,10,14,16,17), and erythema multiforme (15).

Maculopapular eruption

A 54-year-old woman developed pruritus and a maculopapular rash on her legs, 2 days after a second session of sclerotherapy in the legs with polidocanol and mepivacaine. The next sessions were performed with polidocanol only, with good tolerance. Later, she was treated with mepivacaine 2% for a surgery, resulting 2 days after the excision in a maculopapular rash and pruritus in the injection area. Prick and intradermal tests read at 20 minutes were negative, but the intradermal test with mepivacaine 1% was positive at D2. Patch tests at 2 and 4 days were positive to mepivacaine 1% (+), 2% (+++) and 3% (+++) and to lidocaine 2% (+), 5% (+++) and 10% (+++) but negative to bupivacaine and articaine. Ten controls were negative (2).

Localized allergic reaction

The right wrist of a 48-year-old woman was operated on for carpal tunnel syndrome, using infiltration analgesia with both mepivacaine and lidocaine 10 mg/ml. The same evening, the wound area and finger webs of her right hand started to itch. The next day, the whole hand was swollen and a vesicular eczema developed on the right hand and the wound. Patch tests were positive to mepivacaine 1% but not lower concentrations and negative to lidocaine. A repeat patch test was negative, but intracutaneous tests, read at D2 and D3 were positive to both mepivacaine and prilocaine (4).

A 54-year-old woman suffered eczematous eruptions on her face after the administration of lidocaine and mepivacaine for dental surgery; patch tests were positive to both local anesthetics with cross-reactions to prilocaine and bupivacaine (7). Contact allergy to or allergic contact dermatitis from mepivacaine has also been reported in ref. 9 (details not available). A patient with a history of adverse events to local anesthetics had a positive subcutaneous test to prilocaine at D1 and a positive patch test to prilocaine solution 1%. Clinical details were not provided (10).

A 43-year-old woman had developed ipsilateral facial erythema the morning following wisdom teeth extraction using mepivacaine hydrochloride. Mepivacaine itself was not tested, but she was patch test positive to the related amide-type local anesthetics lidocaine, prilocaine and ropivacaine, which makes, together with the clinical data, delayed-type allergy to mepivacaine highly likely (14).

One patient with a cutaneous delayed reaction from mepivacaine ('constant erythema at the wrist after procedure with local anesthetics') and a positive patch test. Details were not provided (16). A patient presented with a history of localized edema after dental anesthesia and a reaction to EMLA cream (containing lidocaine and prilocaine). Patch tests were positive to lidocaine and mepivacaine and a subcutaneous challenge resulted in delayed swelling at 24 and 48 hours after challenge to both local anesthetics (17).

Erythema multiforme
A 36-year-old man developed generalized urticaria 2 days after dental surgery under local anesthesia. Seven years later, a vasectomy was performed, using mepivacaine for local anesthesia. After 3 days the patient developed erythema multiforme, which started locally, but generalized rapidly. No arthritis or mucosal involvement was noted. Patch tests were positive to mepivacaine and lidocaine (no details provided) (15).

Cross-reactions, pseudo-cross-reactions and co-reactions
Primary lidocaine allergy may result in cross-reactivity to mepivacaine (3,5,6,8,11,12; Chapter 3.289 Lidocaine). Primary mepivacaine sensitization may result in cross-reactivity to lidocaine (2,15) and prilocaine (intracutaneous test read ad D2 and D3 [4]). A patient sensitized to bupivacaine probably cross-reacted to mepivacaine (13). A patient sensitized to prilocaine co-reacted to mepivacaine (18).

LITERATURE
1   The data in the section 'General' may have been obtained from literature discussed in this chapter, but mostly also or exclusively from one or more of the following online sources: ChemIDPlus Advanced, PubChem, DrugBank, RxList, Drug Central, Drugs.com, and Wikipedia
2   Sanchez-Morillas L, Martinez JJ, Martos MR, Gomez-Tembleque P, Andres ER. Delayed-type hypersensitivity to mepivacaine with cross-reaction to lidocaine. Contact Dermatitis 2005;53:352-353
3   Bircher AJ, Messmer SL, Surber C, Rufli T. Delayed-type hypersensitivity to subcutaneous lidocaine with tolerance to articaine: confirmation by in vivo and in vitro tests. Contact Dermatitis 1996;34:387-389
4   Kanerva L, Alanko K, Estlander T, Jolanki R. Inconsistent intracutaneous and patch test results in a patient allergic to mepivacaine and prilocaine. Contact Dermatitis 1998;39:197-199
5   Fregert S, Tegner E, Thelin I. Contact allergy to lidocaine. Contact Dermatitis 1979;5:185-188
6   Klein CE, Gall H. Type IV allergy to amide-type local anesthetics. Contact Dermatitis 1991;25:45-48
7   Duque S, Fernández L. Delayed-type hypersensitivity to amide local anesthetics. Allergol Immunopathol (Madr) 2004;32:233-234
8   Curley RK, Macfarlane AW, King CM. Contact sensitivity to the amide anesthetics lidocaine, prilocaine, and mepivacaine. Arch Dermatol 1986;122:924-926
9   Brown RS, Redden RJ, Chan JT. The evaluation of a reported allergic reaction to an amide local anesthetic: a case report. Texas Dent J 1995;112:37-40 (cited in ref. 4)
10 Gall H, Kaufmann R, Kalveram CM. Adverse reactions to local anesthetics: analysis of 197 cases. J Allergy Clin Immunol 1996;97:933-937
11 Bassett I, Delaney T, Freeman S. Can injected lignocaine cause allergic contact dermatitis? Australas J Dermatol 1996;37:155-156
12 Redfern DC. Contact sensitivity to multiple local anesthetics. J Allergy Clin Immunol 1999;104(4Pt.1):890-891
13 Nettis E, Colanardi MC, Calogiuri GF, Foti C, Priore MG, Ferrannini A, Vacca A. Delayed-type hypersensitivity to bupivacaine. Allergy 2007;62:1345-1346
14 Gunson TH, Greig DE. Allergic contact dermatitis to all three classes of local anaesthetic. Contact Dermatitis 2008;59:126-127
15 Zanni MP, Mauri-Hellweg D, Brander C, Wendland T, Schnyder B, Frei E, et al. Characterization of lidocaine-specific T cells. J Immunol 1997;158:1139-1148
16 Furci F, Martina S, Faccioni P, Faccioni F, Senna G, Caminati M. Adverse reaction to local anaesthetics: Is it always allergy? Oral Dis 2020 Feb 24. doi: 10.1111/odi.13310. Epub ahead of print.
17 Melamed J, Beaucher WN. Delayed-type hypersensitivity (type IV) reactions in dental anesthesia. Allergy Asthma Proc 2007;28:477-479
18 Garcia F, Iparraguirre A, Blanco J, Alloza P, Vicente J, Bascones O, et al. Contact dermatitis from prilocaine with cross-sensitivity to pramocaine and bupivacaine. Contact Dermatitis 2007;56:120-122

Chapter 3.302 MEPROBAMATE

IDENTIFICATION

Description/definition   : Meprobamate is the carbamate that conforms to the structural formula shown below
Pharmacological classes : Anticonvulsants; hypnotics and sedatives; muscle relaxants, central; anti-anxiety agents
IUPAC name   : [2-(Carbamoyloxymethyl)-2-methylpentyl] carbamate
CAS registry number   : 57-53-4
EC number   : 200-337-5
Merck Index monograph   : 7199
Patch testing   : 3 mg/ml (3); most systemic drugs can be tested at 10% pet.; if the pure chemical is not available, prepare the test material from intravenous powder, the content of capsules or – when also not available – from powdered tablets to achieve a final concentration of the active drug of 10% pet.
Molecular formula   : $C_9H_{18}N_2O_4$

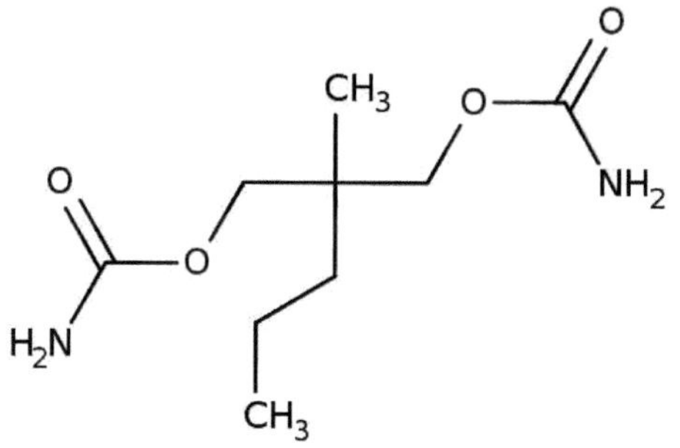

GENERAL

Meprobamate is a carbamate with hypnotic, sedative, and some muscle relaxant properties. It was first made available in the 1950s and soon became very popular for the treatment of anxiety and insomnia. However, it is rarely used nowadays, being largely superseded by the benzodiazepines (1).

CUTANEOUS ADVERSE DRUG REACTIONS FROM SYSTEMIC ADMINISTRATION CAUSED BY TYPE IV (DELAYED-TYPE) HYPERSENSITIVITY (as demonstrated by positive patch tests)

Cutaneous adverse drug reactions from systemic administration of meprobamate caused by type IV (delayed-type) hypersensitivity have included fixed drug eruption (2) and anaphylactoid reaction (3).

Fixed drug eruption

A 22-year-old woman was prescribed meprobamate and 3 other drugs for depression. On the second day of treatment, the patient presented with red erythematous and pruriginous plaques on the arms, legs and face. After discontinuing all drugs, the lesions resolved completely within 3 weeks leaving residual pigmentation. One month later, patch tests were performed with all drugs and were positive to meprobamate. The patient was diagnosed with fixed drug eruption from meprobamate. Further details are unavailable to the author (2).

Other cutaneous adverse drug reactions

A 47-year-old man had, shortly after taking two tablets of meprobamate, developed a rash starting on the left forearm and rapidly spreading to involve the rest of the body. His temperature was 38°C and he was experiencing rigors. The patient also had tachycardia, very low blood-pressure and was in peripheral circulatory failure. Examination showed an erythematous maculopapular eruption over the entire body. The patient was diagnosed with anaphylactoid reaction. A patch test with a saturated solution (3 mg/ml) of meprobamate was positive after 2 days. Ten controls were negative. Prick testing with a saturated solution of meprobamate was negative at 20 minutes, but at 24 hours a vesicular eczematous response was noted. One and a half year later, the patient developed dyspnea, cyanosis, and peripheral circulatory failure, shortly after again having taken one meprobamate tablet (3).

LITERATURE

1   The data in the section 'General' may have been obtained from literature discussed in this chapter, but mostly also or exclusively from one or more of the following online sources: ChemIDPlus Advanced, PubChem, DrugBank, RxList, Drug Central, Drugs.com, and Wikipedia
2   Zaïem A, Kaabi W, Badri T, Lakhoua G, Sahnoun R, Kastalli S, et al. Meprobamate-induced fixed drug eruption. Curr Drug Saf 2014;9:161-162
3   Felix RH, Comaish JS. The value of patch and other skin tests in drug eruptions. Lancet 1974;1(7865):1017-1019

Chapter 3.303 MEQUITAZINE

IDENTIFICATION

Description/definition   : Mequitazine is the phenothiazine that conforms to the structural formula shown below
Pharmacological classes : Histamine H_1 antagonists
IUPAC name   : 10-(1-Azabicyclo[2.2.2]octan-3-ylmethyl)phenothiazine
Other names   : 10-(3-Quinuclidinylmethyl)phenothiazine
CAS registry number   : 29216-28-2
EC number   : 249-521-7
Merck Index monograph : 7201
Patch testing   : 1% pet.
Molecular formula   : $C_{20}H_{22}N_2S$

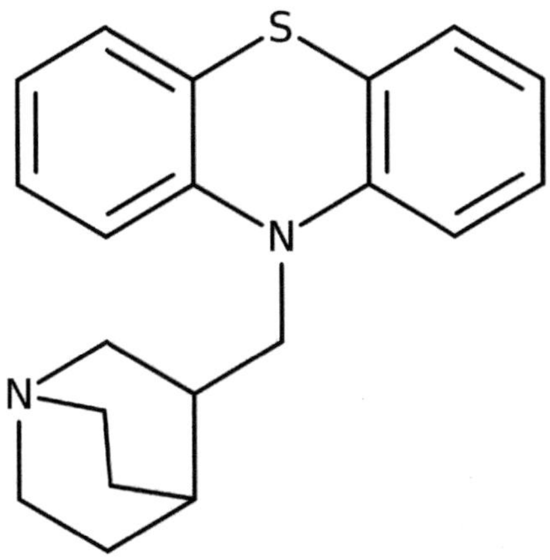

GENERAL

Mequitazine is a member of the phenothiazines and a histamine H_1 antagonist. It competes with histamine for the H_1 receptor sites on effector cells of the gastrointestinal tract, blood vessels and respiratory tract. Mequitazine is indicated for the treatment of allergic rhinitis and urticaria (1).

CUTANEOUS ADVERSE DRUG REACTIONS FROM SYSTEMIC ADMINISTRATION CAUSED BY TYPE IV (DELAYED-TYPE) HYPERSENSITIVITY (as demonstrated by positive patch tests)

Cutaneous adverse drug reactions from systemic administration of mequitazine caused by type IV (delayed-type) hypersensitivity have included photosensitivity (2).

Photosensitivity

In the period 1978-1997, in a university clinic in Leuven, Belgium, one patient was seen with a photoallergic reaction to the systemic drug mequitazine. Clinical details were not provided (2).

Two patients using oral mequitazine developed a photosensitivity reaction. The first patient had a strongly positive photopatch test to this phenothiazine at 1% pet. and cross-reacted to chlorpromazine. In addition, an immediate erythematous macule was observed on the photopatch test. Patient 2 had taken mequitazine for 6 months before photosensitivity developed. In this man, there were strongly positive photopatch test results with immediate erythema reaction, cross-reaction to promethazine, and persistence of the photosensitivity over a 3-year follow-up period after discontinuation of mequitazine (persistent light reaction). Photopatch tests to 1% mequitazine with 5 J/cm^2 of UVA in 30 normal subjects were negative (3). Generally speaking, it should be appreciated that phenothiazines are phototoxic and such reactions are very hard to distinguish from photoallergy.

LITERATURE

1   The data in the section 'General' may have been obtained from literature discussed in this chapter, but mostly also or exclusively from one or more of the following online sources: ChemIDPlus Advanced, PubChem, DrugBank, RxList, Drug Central, Drugs.com, and Wikipedia
2   Goossens A, Linsen G. Contact allergy to antihistamines is not common. Contact Dermatitis 1998;39:38-39
3   Kim TH, Kang JS, Lee HS, Youn JI. Two cases of mequitazine induced photosensitivity reactions. Photodermatol Photoimmunol Photomed 1995;11:170-173

Chapter 3.304 MEROPENEM

IDENTIFICATION

Description/definition : Meropenem is the thienamycin derivative that conforms to the structural formula shown below

Pharmacological classes : Anti-bacterial agents

IUPAC name : (4R,5S,6S)-3-[(3S,5S)-5-(Dimethylcarbamoyl)pyrrolidin-3-yl]sulfanyl-6-[(1R)-1-hydroxyethyl]-4-methyl-7-oxo-1-azabicyclo[3.2.0]hept-2-ene-2-carboxylic acid

CAS registry number : 96036-03-2

EC number : Not available

Merck Index monograph : 7240

Patch testing : 5% pet.

Molecular formula : $C_{17}H_{25}N_3O_5S$

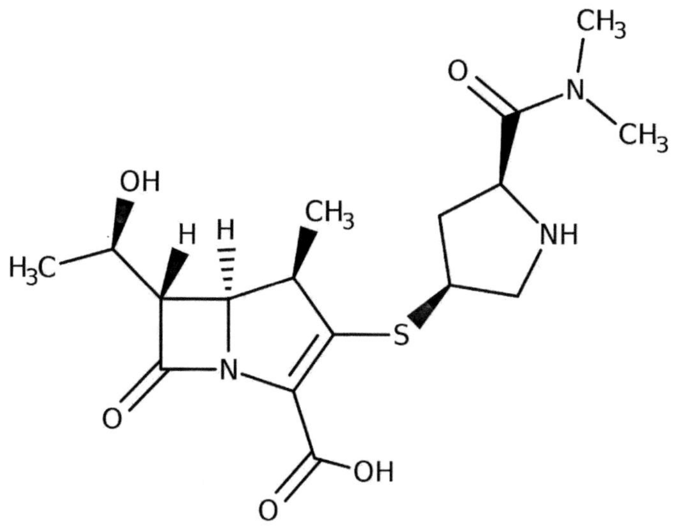

GENERAL

Meropenem is a broad-spectrum carbapenem antibiotic with antibacterial properties. It is active against gram-positive and gram-negative aerobe and anaerobe bacteria including *Klebsiella*, *E. coli*, *Enterococcus*, and *Clostridium* spp. Meropenem exerts its action by penetrating bacterial cells readily and interfering with the synthesis of vital cell wall components, which leads to cell death. Meropenem is indicated for use as single agent therapy for the treatment of complicated skin and skin structure infections, complicated appendicitis and peritonitis, and bacterial meningitis caused by susceptible isolates of specific microorganisms. Meropenem is available for intravenous injection only (solution, powder for solution) as meropenem trihydrate (CAS number 119478-56-7, EC number not available, molecular formula $C_{17}H_{31}N_3O_8S$) (1). The classification and structures of beta-lactam antibiotics are discussed in Chapter 3.36 Amoxicillin.

CONTACT ALLERGY FROM ACCIDENTAL CONTACT

A 45-year-old nurse complained of recurrent periorbital erythema with itching and runny eyes. Examination during her active rash showed bilateral eyelid eczema. Each episode lasted about 5 to 6 days and settled with withdrawal from her place of work, where she handled a considerable number of antibiotics including meropenem. Patch tests were performed to all antibiotics she had contact with and there was a positive (++) reaction to meropenem 5% pet. only. Ten controls were negative (6).

A 28-year-old nurse developed eczema on the dorsal aspect of the hand and the face. She worked in the hematology department where she usually handled and administered a variety of antibiotics. When she was moved to a different department where she did not have contact with these drugs, the dermatitis completely resolved. Patch tests were positive to ertapenem 20% pet., imipenem-cilastatin (20% pet.), ampicillin (25% pet.), piperacillin (20% pet.), mezlocillin (20% pet.), and meropenem (20% pet.) It was concluded that the patient had contact allergy to carbapenems (ertapenem, imipenem-cilastatin, and meropenem) and semisynthetic penicillins (piperacillin, mezlocillin, and ampicillin) (2). At work, the patient had contact only with imipenem, ertapenem, and piperacillin, so it may be assumed that the other reactions, including the one to meropenem, were due to cross-reactivity, rather than from occupational sensitization.

CUTANEOUS ADVERSE DRUG REACTIONS FROM SYSTEMIC ADMINISTRATION CAUSED BY TYPE IV (DELAYED-TYPE) HYPERSENSITIVITY (as demonstrated by positive patch tests)

Cutaneous adverse drug reactions from systemic administration of meropenem caused by type IV (delayed-type) hypersensitivity have included maculopapular eruption (13,14), drug reaction with eosinophilia and systemic symptoms (DRESS) (4,5,8,11,12), Stevens-Johnson syndrome/toxic epidermal necrolysis (SJS/TEN) (7,10), and unspecified drug eruption (9).

Maculopapular eruption

A 38-year-old woman with systemic lupus erythematosus under treatment with 20 mg oral prednisolone daily developed a morbilliform drug reaction 11 days after starting meropenem and vancomycin for abdominal wall cellulitis. Patch tests were positive to meropenem powder for intravenous injection 10% pet. with a negative reaction to the other drugs used and the related imipenem (13). The same authors report on a 61-year-old woman with a postoperative infection, who developed a maculopapular exanthema 12 days after treatment with vancomycin, ceftriaxone, and metronidazole, and 10 days after meropenem. Patch tests were positive to meropenem powder for intravenous injection 10% pet. with a negative reaction to the other drugs used and the related imipenem (13).

A 45-year-old woman was treated with vancomycin, meropenem, and piperacillin/tazobactam, when she developed fevers and a pruritic generalized morbilliform eruption with petechiae in dependent areas. The petechiae began to fade, but the original eruption subsequently developed darkening and confluence. Patch testing with the 3 commercial drugs 10% pet. were strongly positive to meropenem (14).

Drug reaction with eosinophilia and systemic symptoms (DRESS)

A 57-year-old woman developed spiking temperatures (possibly from postoperative sepsis), a widespread morbilliform rash, raised eosinophils and an increased alanine aminotransferase level 25 days after commencing phenytoin and 4 days after starting intravenous meropenem and vancomycin. The patient's rash also flared after she received intravenous contrast media for computed tomography. The patient had delayed reactions at 24 hours to several radiocontrast media on intradermal testing, whereas intradermal testing with meropenem gave a negative result. Therefore, an intravenous meropenem drug challenge was performed, which resulted in facial angioedema and a non-urticarial rash over her chest 12-18 hours later. Patch tests showed strong positive reactions to meropenem (+++ reactions at D3 and D4) at 1% and 20% pet. The authors diagnosed DRESS syndrome secondary to phenytoin with additional delayed hypersensitivity to meropenem and intravenous contrast media. Rather curiously, no patch or intradermal tests with phenytoin had been performed. A lymphocyte proliferation test with phenytoin had given a negative result, but an interferon-γ release assay (ELIspot) gave a positive result, indicating, according to the authors, delayed drug hypersensitivity (11).

A 53-year-old woman developed DRESS with a maculopapular rash forming large plaques and superficial vesicles, predominantly on the limbs, gradually affecting over 50% of the body. She also had fever, eosinophilia, and elevated liver enzymes. The patient had been treated with various antibiotics including meropenem. A patch test was positive to meropenem 10% pet., as was a lymphocyte activation test. All other antibiotics used tested negative (8).

In a group of 14 patients with multiple delayed-type hypersensitivity reactions, DRESS was caused by meropenem in one case, showing a positive patch test reaction (4,5). This patient had previously been described in more detail (12). She was a 64-year-old woman who was treated with meropenem and vancomycin for a retroperitoneal abscess. Twenty-four days later, a maculopapular rash appeared on the trunk. Over the following days skin eruptions spread to the arms, legs and face, associated with axillary and inguinal lymphadenopathy as well as fever (38.8°C). Skin biopsy revealed a perivascular lymphohistiocytic infiltrate with eosinophilic and neutrophilic granulocytes, consistent with drug hypersensitivity. She had also strong eosinophilia and elevated liver enzymes, and therefore DRESS from meropenem and vancomycin was suspected. An allergy workup after recovery revealed sensitization to both meropenem (patch test positive, lymphocyte transformation test [LTT] negative) and vancomycin (LTT positive, patch test negative) (12).

Stevens-Johnson syndrome/toxic epidermal necrolysis (SJS/TEN)

In the period 2000-2014, in Portugal, 260 patients were patch tested with antibiotics for suspected cutaneous adverse drug reactions to these drugs. 56 patients (22%) had one or more (often from cross-reactivity) positive patch tests. Meropenem was patch test positive in one case of SJS/TEN (7).

In a 61-year-old man, delayed-type hypersensitivity to meropenem (tested as commercial tablet 500 mg, 15% and 30% pet.) and other drugs may have caused/contributed to TEN (toxic epidermal necrolysis) following DRESS, which had been induced by other drugs (10).

Other cutaneous adverse drug reactions

In Finland, in the period 1989-2001, 826 patients with suspected cutaneous drug eruptions were patch tested and 89 had one or more positive reactions. Of these individuals, one reacted to meropenem. It was not mentioned which drug eruption this patient had suffered from (9).

Cross-reactions, pseudo-cross-reactions and co-reactions

Cross-reactions between beta-lactam antibiotics are discussed in Chapter 3.36 Amoxicillin. Cross-reactions between carbapenems (meropenem, imipenem, ertapenem) and between carbapenems and penicillins may occur, but the pattern is unclear (2). A patient who had developed acute generalized exanthematous pustulosis (AGEP) from ertapenem had positive patch tests to ertapenem with co-cross-reactions to meropenem and benzylpenicillin (3). No cross-reaction to imipenem in 2 patients sensitized to meropenem (13).

LITERATURE

1   The data in the section 'General' may have been obtained from literature discussed in this chapter, but mostly also or exclusively from one or more of the following online sources: ChemIDPlus Advanced, PubChem, DrugBank, RxList, Drug Central, Drugs.com, and Wikipedia
2   Colagiovanni A, Feliciani C, Fania L, Pascolini L, Buonomo A, Nucera E, et al. Occupational contact dermatitis from carbapenems. Cutis 2015;96:E1-3
3   Fernando SL. Ertapenem-induced acute generalized exanthematous pustulosis with cross-reactivity to other beta-lactam antibiotics on patch testing. Ann Allergy Asthma Immunol 2013;111:139-140
4   Jörg L, Yerly D, Helbling A, Pichler W. The role of drug, dose and the tolerance/intolerance of new drugs in multiple drug hypersensitivity syndrome (MDH). Allergy 2020;75:1178-1187
5   Jörg L, Helbling A, Yerly D, Pichler WJ. Drug-related relapses in drug reaction with eosinophilia and systemic symptoms (DRESS). Clin Transl Allergy 2020;10:52
6   Yesudian PD, King CM. Occupational allergic contact dermatitis from meropenem. Contact Dermatitis 2001;45:53
7   Pinho A, Coutinho I, Gameiro A, Gouveia M, Gonçalo M. Patch testing - a valuable tool for investigating non-immediate cutaneous adverse drug reactions to antibiotics. J Eur Acad Dermatol Venereol 2017;31:280-287
8   Prados-Castaño M, Piñero-Saavedra M, Leguísamo-Milla S, Ortega-Camarero M, Vega-Rioja A. DRESS syndrome induced by meropenem. Allergol Immunopathol (Madr) 2015;43:233-235
9   Lammintausta K, Kortekangas-Savolainen O. The usefulness of skin tests to prove drug hypersensitivity. Br J Dermatol 2005;152:968-974
10 Liippo J, Pummi K, Hohenthal U, Lammintausta K. Patch testing and sensitization to multiple drugs. Contact Dermatitis 2013;69:296-302
11 Nicholson P, Brinsley J, Farooque S, Wakelin S. Patch testing with meropenem following a severe cutaneous adverse drug reaction. Contact Dermatitis 2018;79:397-398
12 Jörg-Walther L, Schnyder B, Helbling A, Helsing K, Schüller A, Wochner A, Pichler W. Flare-up reactions in severe drug hypersensitivity: infection or ongoing T-cell hyperresponsiveness. Clin Case Rep 2015;3:798-801
13 Morgado F, Santiago L, Gonçalo M. Safe use of imipenem after delayed hypersensitivity to meropenem-Value of patch tests. Contact Dermatitis 2020;82:190-191
14 Endo JO, Davis C, Powell DL. The potential utility of patch testing in identifying the causative agent of morbilliform drug eruptions. Dermatitis 2011;22:114-115

Chapter 3.305 MESALAZINE

IDENTIFICATION

Description/definition : Mesalazine is the aminosalicylate that conforms to the structural formula shown below
Pharmacological classes : Anti-inflammatory agents, non-steroidal
IUPAC name : 5-Amino-2-hydroxybenzoic acid
Other names : Mesalamine; 5-aminosalicylic acid
CAS registry number : 89-57-6
EC number : 201-919-1
Merck Index monograph : 7244
Patch testing : 10% pet.
Molecular formula : $C_7H_7NO_3$

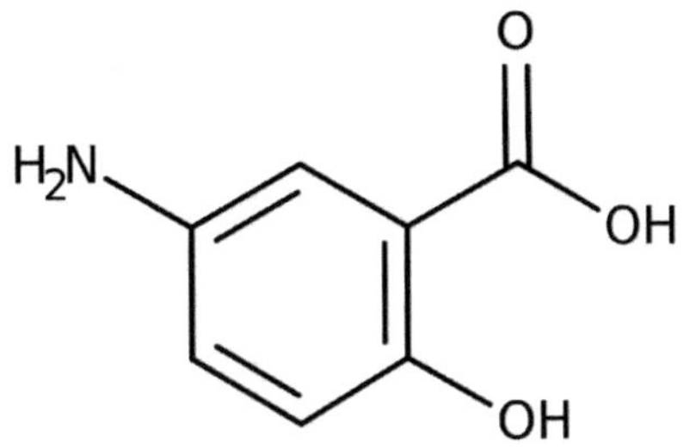

GENERAL

Mesalazine is a salicylic acid-derived anti-inflammatory agent. It may reduce inflammation through inhibition of cyclooxygenase and prostaglandin production. Following rectal or oral administration, only a small amount of mesalamine is absorbed; the remainder, acting topically, reduces bowel inflammation, diarrhea, rectal bleeding and stomach pain. Mesalazine is indicated for the induction of remission in patients with active or mild to moderate acute exacerbations of ulcerative colitis, for the maintenance of remission of ulcerative colitis and for the maintenance of remission of Crohn's disease (1).

CUTANEOUS ADVERSE DRUG REACTIONS FROM SYSTEMIC ADMINISTRATION CAUSED BY TYPE IV (DELAYED-TYPE) HYPERSENSITIVITY (as demonstrated by positive patch tests)

Cutaneous adverse drug reactions from systemic administration of mesalazine caused by type IV (delayed-type) hypersensitivity have included symmetrical drug-related intertriginous and flexural exanthema (SDRIFE)/baboon syndrome (2,4) and fixed drug eruption (3).

Symmetrical drug-related intertriginous and flexural exanthema (SDRIFE)/Baboon syndrome

A 50-year-old man presented with intensely pruritic, confluent, edematous erythema of the buttocks and anogenital area, associated with diffuse maculopapular rash of the arms. Two months before, the patient had undergone hemorrhoid sclerotherapy, after which he was prescribed daily medication with 5-aminosalicylic acid (5-ASA; mesalazine) foam enemas. Within a few days, perianal eczema, from which he had suffered before, flared, spreading to the buttocks and upper inner thighs. Patch tests showed positive reactions to the mesalazine enema 'as is' and later to mesalazine 10% pet., to which 20 controls were negative. The baboon-like manifestation cleared soon after the patient stopped the 5-ASA enemas (2).

A 40-year-old woman developed the clinical picture of the baboon syndrome emerging after a first dose of mesalazine for ulcerative colitis. The patient, a hairdresser, was known with allergy to haircoloring products. Patch tests were negative to mesalazine 10% water and pet., and positive to p-phenylenediamine (PPD). Oral provocation with mesalazine resulted in a pruriginous rash on the chest and neck. One day later, the patch tests with mesalazine and PPD carried out 1 month earlier were activated (= had become positive) and reactivated, respectively (4).

Although PPD and mesalazine are not known to cross-react, the reactivation of the PPD patch test after oral mesalazine provocation points, according to the authors, to a link between sensitivity to mesalazine and PPD. Indeed, it was found that mesalazine contains traces (0.01% or less) of p-aminophenol (which nearly always cross-reacts to PPD) and that p-aminophenol can also be produced by decarboxylation of mesalazine (4).

Fixed drug eruption

A 35-year-old patient, suffering from ulcerative colitis, on at least two occasions had developed vesicular and bullous lesions at the same site under his feet, following oral or rectal mesalazine applications. Physical examination revealed brown-violaceous, slightly crusted, oval plaques located symmetrically on the plantar aspects of his feet. The patient showed photographs from previous episodes with bullous lesions, always appearing at the same sites. Bullous fixed drug eruption was suspected and patch testing with mesalazine tablets 10%, 1%, and 0.1% pet. was performed on the lesional skin and on the upper back. On D2 and D5, there were weakly positive reactions on the normal skin and strongly positive reactions over the lesional skin with all of the concentrations of mesalazine. The patient was diagnosed with bullous fixed drug eruption mesalazine (3).

Cross-reactions, pseudo-cross-reactions and co-reactions
Mesalazine is the active metabolite of sulfasalazine, a drug also used in the treatment of ulcerative colitis. As with mesalazine, a link between allergy to PPD and to sulfasalazine has been suggested (Chapter 3.457 Sulfasalazine).

LITERATURE
1   The data in the section 'General' may have been obtained from literature discussed in this chapter, but mostly also or exclusively from one or more of the following online sources: ChemIDPlus Advanced, PubChem, DrugBank, RxList, Drug Central, Drugs.com, and Wikipedia
2   Gallo R, Parodi A. Baboon syndrome from 5-aminosalicylic acid. Contact Dermatitis 2002;46:110
3   Salman A, Seckin Gencosmanoglu D, Alahdab YO, Giménez-Arnau AM. Mesalazine-induced bullous fixed drug eruption. Contact Dermatitis 2018;79:34-35
4   Charles J, Bourrain JL, Tessier A, Lepoittevin JP, Beani JC. Mesalazine and para-phenylenediamine allergy. Contact Dermatitis 2004;51:313-314

Chapter 3.306 MESNA

IDENTIFICATION

Description/definition   : Mesna is the sulfhydryl compound that conforms to the structural formula shown below
Pharmacological classes : Protective agents
IUPAC name   : Sodium;2-sulfanylethanesulfonate
Other names   : Sodium 2-mercaptoethane sulfonate
CAS registry number   : 19767-45-4
EC number   : 243-285-9
Merck Index monograph : 7249
Patch testing   : 1% water (2); commercial preparation 30% pet. in drug eruptions; most systemic drugs
                           can be tested at 10% pet.; if the pure chemical is not available, prepare the test material
                           from intravenous powder, the content of capsules or – when also not available – from
                           powdered tablets to achieve a final concentration of the active drug of 10% pet.
Molecular formula   : $C_2H_5NaO_3S_2$

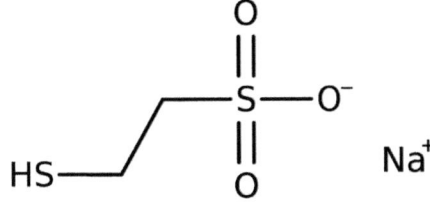

GENERAL

Mesna is an organosulfonic acid that is used as a uroprotective agent to reduce the incidence of hemorrhagic cystitis associated with the chemotherapeutic agents ifosfamide and cyclophosphamide. Mesna is converted to a free thiol compound in the kidney, where it binds to and inactivates acrolein and other urotoxic metabolites of ifosfamide and cyclophosphamide, thereby reducing their toxic effects on the urinary tract during urinary excretion (1). It is (or was) also used as a local mucolytic (2,3).

CONTACT ALLERGY FROM ACCIDENTAL CONTACT

A 38-year-old atopic nurse for around 7 years had experienced recurrent itchy erythematous papulovesicular lesions on exposed parts of the body (face, neck, forearms, dorsum of the hand). She had noted that lesions occurred after assisting in sessions during which small children inhaled mesna. Patch tests were positive to commercial mesna 1% water (D2 ++, D4 +++) and 0.1% water (D2 -, D4 ++). An open skin test and prick test with mesna 1% water were negative. After inhalation provocation with mesna 10% water no specific allergic respiratory reaction was recorded, but after one day the cutaneous lesions, on the face and neck in particular, became more intense (systemic allergic dermatitis). The patient was diagnosed with occupational airborne allergic contact dermatitis from mesna (2).

A 33-year-old nurse, working in an intensive care unit, presented with a 2-month history of intermittent eczema of the forearms and hands. The vesicles, erythema and fissures on the tips of her fingers worsened after handling mesna. Patch testing with mesna 20%, 10%, 1% and 0.1% water, and with the commercial preparation 'as is', gave positive reactions (+++) at D2 (second reading not mentioned) to all materials, accompanied by worsening of the eczema. This nurse also had occupational allergic contact dermatitis, but not of the airborne type (3). The authors of ref. 3 state that a positive patch test has been reported in ref. 4, but clinical data are not available.

CUTANEOUS ADVERSE DRUG REACTIONS FROM SYSTEMIC ADMINISTRATION CAUSED BY TYPE IV (DELAYED-TYPE) HYPERSENSITIVITY (as demonstrated by positive patch tests)

Cutaneous adverse drug reactions from systemic administration of mesna caused by type IV (delayed-type) hypersensitivity have included fixed drug eruption (5,6,7).

Fixed drug eruption

A 53-year-old woman (patient #1), a 45-year-old man (patient #2), and a 40-year-old woman (patient #3) with severe multiple sclerosis had been receiving monthly therapeutic doses of cyclophosphamide, mesna and methylpredniso-lone. Nine, 24, and 20 months after starting this treatment, the patients respectively reported the development of red plaques with a burning sensation on the face 2 days, one day and one day after administration. Recovery was observed in all 3 patients within a few days, with residual pigmentation remaining in one case. Subsequent administration of the treatment triggered recurrence of the cutaneous lesions at identical and new sites less than 12 hours after administration in patients #1 and #3 and 6 hours after administration in patient #2, who experienced 2 recurrences. The results indicated a diagnosis of fixed drug eruption (FDE). Patch tests with mesna and the other drugs were positive for mesna 30% pet. on affected skin at days 2 and 4 for all 3 patients (5).

A 41-year-old woman was treated for scleroderma and interstitial lung disease with the combination of cyclophosphamide and mesna. Several hours after the fourth infusion, the patient experienced itching and burning. After the fifth infusion, she developed sharply demarcated pruritic erythematous plaques on her face and upper body. The lesions were first reddish, within 1 to 3 hours changed to a dark-brown color, and faded away after 7 days. Hyperpigmented spots remained on the skin for the next few years. Intradermal (0.1% and 1% saline solution) and patch tests (50% in petrolatum) with mesna in previously affected skin areas showed positive reactions at days 2 and 4 (7).

Another case of FDE from mesna in the combination with cyclophosphamide was reported from France (6).

LITERATURE

1   The data in the section 'General' may have been obtained from literature discussed in this chapter, but mostly also or exclusively from one or more of the following online sources: ChemIDPlus Advanced, PubChem, DrugBank, RxList, Drug Central, Drugs.com, and Wikipedia
2   Kiec-Swierczynska M, Krecisz B. Occupational airborne allergic contact dermatitis from mesna. Contact Dermatitis 2003;48:171
3   Benyoussef K, Bottlaender A, Pfister HR, Caussade P, Heid E, Grosshans E. Allergic contact dermatitis from mesna. Contact Dermatitis 1996;34:228-229
4   Frosch PJ, Weickel R. Delayed allergy following mesna. Dtsch Med Wochenschr 1986;49:1901-1902
5   Soria A, Lebrun-Vignes B, Le Forestier N, Francès C. Fixed drug eruption due to mesna. J Investig Allergol Clin Immunol 2015;25:444-445
6   Delaigue S, Boye T, Pasquine C, Guetta K, Alla P, Ponte-Astoul J, et al. Drug patch tests in the investigation of a fixed drug eruption subsequent to 2 courses of cyclophosphamide in combination with mesna. Ann Dermatol Venereol 2015;142:37-40 (Article in French)
7   Weiss KM, Jariwala S, Wachs J, Jerschow E. Fixed drug eruption caused by mesna. Ann Allergy Asthma Immunol 2011;107:377-378

Chapter 3.307 METAMIZOLE

IDENTIFICATION

Description/definition   : Metamizole is the pyrazole and amino sulfonic acid that conforms to the structural
                               formula shown below
Pharmacological classes   : Anti-inflammatory agents, non-steroidal; antipyretics
IUPAC name   : [(1,5-Dimethyl-3-oxo-2-phenylpyrazol-4-yl)methylamino]methanesulfonic acid
Other names   : Methamizol; metamizol
CAS registry number   : 50567-35-6
EC number   : 256-627-7
Merck Index monograph   : 4660 (Dipyrone = metamizole sodium)
Patch testing   : 1% pet. (SmartPracticeCanada); 10% pet. (16); in cases of fixed drug eruptions, open
                            applications to previously affected skin should be performed with metamizole in DMSO,
                            when negative in petrolatum (9)
Molecular formula   : $C_{13}H_{17}N_3O_4S$

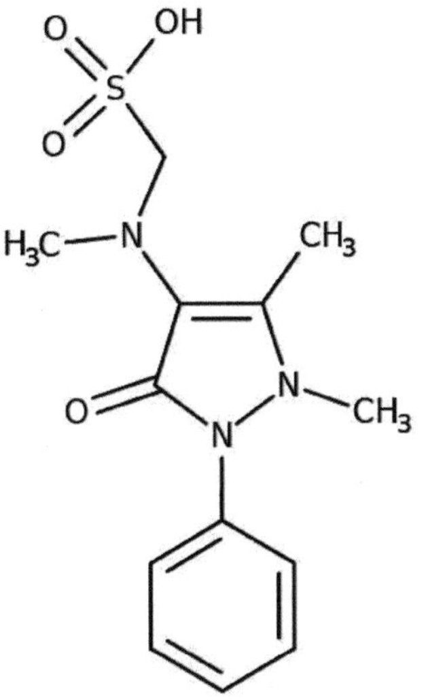

Metamizole

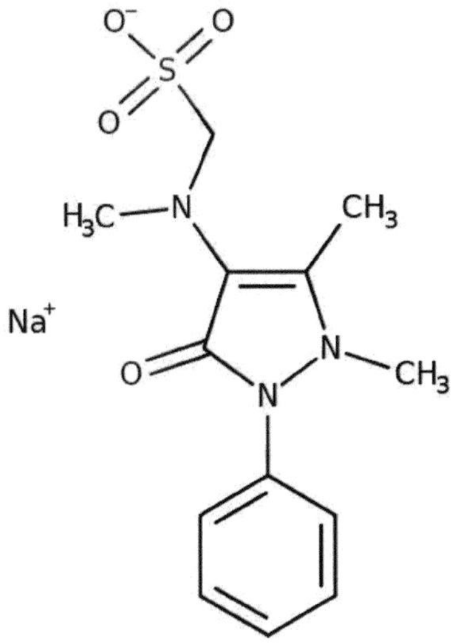

Dipyrone = metamizole sodium

GENERAL

Metamizole is a pyrazole and aminosulfonic acid that has analgesic, anti-inflammatory, and antipyretic properties. This agent was formerly widely used as a powerful painkiller and fever reducer, but was associated with potentially fatal agranulocytosis and approvals were withdrawn in Canada in 1963 and the USA in 1977. Yet, metamizole is currently available in many countries, probably most often as metamizole sodium (dipyrone; CAS number 68-89-3, EC number 200-694-7, molecular formula $C_{13}H_{16}N_3NaO_4S$) or as metamizole sodium monohydrate (dipyrone hydrate, also termed dipyrone; CAS number 5907-38-0, EC number not available, molecular formula $C_{13}H_{18}N_3NaO_5S$) (1).

CUTANEOUS ADVERSE DRUG REACTIONS FROM SYSTEMIC ADMINISTRATION CAUSED BY TYPE IV (DELAYED-TYPE) HYPERSENSITIVITY (as demonstrated by positive patch tests)

Cutaneous adverse drug reactions from systemic administration of metamizole caused by type IV (delayed-type) hypersensitivity have included maculopapular eruption (5,7,14), widespread erythematous drug eruption (10), acute generalized exanthematous pustulosis (AGEP) (3), fixed drug eruption (2,7,8,9,11,12,13), drug reaction with eosinophilia and systemic symptoms (DRESS) (5,7), Stevens-Johnson syndrome (5), eczematous eruption (14), 'widespread pruritic exanthema' (10), diffuse edema of the face with breathing difficulty (15) and unspecified drug eruption (4,6).

Case series with various or unknown types of drug reactions

In two centers in Germany, between 2000 and 2019, 239 patients presented with metamizole hypersensitivity reactions, of who 69 had delayed reactions. Of these 69 individuals, 37 suffered from measles-like exanthem (54%), and 15 (22%) developed a fixed drug eruption. A small number of patients were diagnosed with a manifestation of the Stevens-Johnson syndrome/toxic epidermal necrolysis (SJS/TEN) spectrum (n=5), drug reaction with eosinophilia and systemic symptoms (DRESS) (n=4), flexural exanthem (n=3), or agranulocytosis (n=5). Several of these patients reported initial signs of the delayed reaction already within 12 hours after first intake of metamizole. Patch tests with commercial intravenous metamizole solution 500 mg/ml in 46 of the 69 patients (the 10 with agranulocytosis and SJS/TEN and 13 others were not tested) were positive in 25 individuals (54%). It was not specified which drug reactions metamizole had caused in these patients (4).

In Coimbra, Portugal, 14 patients (6 men, 8 women, mean age 57 years) who had been diagnosed with non-immediate cutaneous adverse drug reactions to metamizole intake were patch tested between 2013 and 2015 with an NSAID series and metamizole 10% pet. prepared from the powder of commercial metamizole tablets (54 controls negative). Positive patch test reactions to metamizole were observed in 7 of 14 patients, including 4 of 8 patients with maculopapular exanthema, 2 of 3 patients with DRESS, and one patient with Stevens–Johnson syndrome (SJS). In all cases, there was another possible culprit (antibiotics, other NSAIDs, allopurinol). Two of the patients who showed positive patch test reactions to metamizole had experienced a positive accidental re-exposure (5).

In Spain, 3 patients with maculopapular eruptions, one with a bullous exanthema and another one with exanthema with skin desquamation, presumably caused by metamizole, were patch tested with this NSAID 10% pet. from commercial tablets. There were 3 positive reactions, both to patch tests and delayed intradermal test readings, but it was not mentioned which CADR these patients had previously had (6).

In Salamanca, Spain, 12 patients (9 men, mean age 56 years) with suspected nonimmediate CADR from metamizole were investigated with patch tests using metamizole 10% water. Ten had maculopapular exanthema, one DRESS and one fixed drug eruption. In 4, patch tests were positive to metamizole at D2 and D4: 2 with maculopapular exanthema, one with DRESS and one with fixed drug eruption (7).

Maculopapular eruption
An 18-year-old man began treatment for tonsillitis with metamizole and 2 other medications when, after 48 hours, he developed a generalized maculopapular exanthema without pruritus. Patch tests were positive to metamizole 10% water (14). See also the section 'Case series with various or unknown types of drug reactions' above, refs. 5 and 7.

Erythroderma, widespread erythematous eruption, exfoliative dermatitis
A 72-year-old woman developed, 48 hours after surgery, an erythematous widespread skin reaction. The lesions resolved in 4 weeks without desquamation. Patch tests with all drugs used were positive to commercial metamizole 50% pet. (D2 ++, D4 ++) (10).

Acute generalized exanthematous pustulosis (AGEP)
A 58-year-old man presented a diffuse pruritic eruption of 1-2 mm pustules on an erythematous background over the trunk with malaise. There was no lymphadenopathy. The patient had been operated 2 days before for a ruptured tendon using mepivacaine and bupivacaine; concurrent medications included cefazoline, metamizole and enoxaparin. Laboratory testing showed only neutrophilia. Histologic examination showed spongiform pustules, papillary dermal edema and a perivascular lymphohistiocytic and neutrophilic infiltrate. Microbiological cultures of pustules were negative. Three months later, patch test were performed with all drugs used except enoxaparin. Only metamizole 10% and 20% in water and pet. were positive; 10 controls were negative. The patient was diagnosed with acute generalized exanthematous pustulosis (AGEP) from metamizole (3).

Fixed drug eruption
In Turkey, 5 patients with fixed drug eruption from metamizole were investigated with patch testing on 15-20x tape-stripped postlesional and clinically uninvolved skin with commercial metamizole 10%, 20% and 50% pet. All reactions were negative. Next, open topical testing using DMSO as a vehicle was performed after 2 weeks. Without tape stripping, a thin layer of 10% test solution in DMSO and pure DMSO were applied with a cotton swab both to previous FDE lesions and to unaffected skin sites every 12 hours up to four times. If testing with 10% remained negative, the concentration of the drug was increased up to 20% and 50%. Open topical testing in DMSO revealed positive results in all four patients tested with metamizole at a concentration of 20%. Repeated applications (up to 2-3) were necessary to obtain a positive reaction in most patients. Positive reactions were seen as itching and erythema with or without induration that started 1-2 hours (range, 1-4 hours) mainly after the first or second, or rarely the third application of the drug and lasted for more than 24-36 hours. Twenty controls were negative (9).

A 39-year-old woman presented with slightly desquamative erythematous brownish macules on her hands, feet and back. The lesions had first appeared 5 days earlier, with itching erythematous-violaceous maculae after the intake of a tablet metamizole. A patch test on postlesional skin with liquid metamizole (probably injection fluid) was positive, but negative on normal skin. In a second test session, 1% metamizole in petrolatum was also positive (++) on previously involved skin. Ten controls were negative (8,12).

Other case reports of fixed drug eruption from delayed-type hypersensitivity to metamizole have been reported from Germany in 2002 (11) and from Spain in 2001 (with flare-up of previous lesions from the lesional patch test)

(13). One case of fixed drug eruption to metamizole sodium (sulpyrin, dipyrone) with a positive patch test reaction on postlesional skin (itching, erythema, infiltration) to the commercial drug 10% pet. was reported from Seoul, South Korea, in the periods 1986-1990 and 1996-1997 (2).

See also the section 'Case series with various or unknown types of drug reactions' above, ref. 7.

Drug reaction with eosinophilia and systemic symptoms (DRESS)
See the section 'Case series with various or unknown types of drug reactions' above, refs. 5 and 7.

Stevens-Johnson syndrome/toxic epidermal necrolysis (SJS/TEN)
See the section 'Case series with various or unknown types of drug reactions' above, ref. 5.

Dermatitis/eczematous eruption
An 80-year-old woman received metamizole 500 mg/8 hour orally for 4 days. Two days after withdrawal, she developed a generalized exanthema with progressive vesiculation and exudation. The diagnosis was eczematous dermatitis. Patch tests were positive to metamizole 10% water (D2 ++/D4 ++) (14).

Other drug eruptions
A 55-year-old man developed a pruritic erythematous eruption on the arms and in the perianal region, beginning one day after urological surgery. Patch tests with all drugs were positive only to metamizole 50% pet. (D4 +) (10). A 66-year-old woman developed, 2 days following urologic surgery, a widespread pruritic exanthema that resolved in 8 days with mild desquamation. She had received treatment with teicoplanin and metamizole. Prick, intradermal and patch tests were positive only to metamizole 50% pet. in the patch test. Patch tests were also performed in 40 control patients with negative results (10).

A woman aged 32 experienced, 6 hours after taking 20 drops of metamizole, eyelid edema and diffuse edema of the face with breathing difficulty and a sense of choking. One year later, an intradermal test with metamizole 0.5 mg/ml was positive after 15 minutes, but negative after 24 hours. Patch tests were positive (++) to metamizole 10% pet. at D3. Ten controls were negative to the same patch and prick tests. Combined immediate-type and delayed-type allergy to metamizole was diagnosed (15).

Unspecified drug eruptions
See the section 'Case series with various or unknown types of drug reactions' above, refs. 4 and 6.

Immediate contact reactions
Immediate contact reactions (contact urticaria) to metamizole are presented in Chapter 5.

LITERATURE
1   The data in the section 'General' may have been obtained from literature discussed in this chapter, but mostly also or exclusively from one or more of the following online sources: ChemIDPlus Advanced, PubChem, DrugBank, RxList, Drug Central, Drugs.com, and Wikipedia

2   Lee AY. Topical provocation in 31 cases of fixed drug eruption: change of causative drugs in 10 years. Contact Dermatitis 1998;38:258-260

3   Gonzalo-Garijo MA, Perez-Calderon R, De Argila D, Rodriguez-Nevado I. Metamizole-induced acute generalized exanthematous pustulosis. Contact Dermatitis 2003;49:47-48

4   Trautmann A, Brockow K, Stoevesandt J. Metamizole-induced reactions as a paradigm of drug hypersensitivity: Non-allergic reactions, anaphylaxis, and delayed-type allergy. Clin Exp Allergy 2020;50:1103-1106

5   Pinho A, Santiago L, Gonçalo M. Patch testing in the investigation of non-immediate cutaneous adverse drug reactions to metamizole. Contact Dermatitis 2017;76:238-239

6   Blanca-López N, Pérez-Sánchez N, Agúndez JA, García-Martin E, Torres MJ, Cornejo-García JA, Allergic reactions to metamizole: Immediate and delayed responses. Int Arch Allergy Immunol 2016;169:223-230

7   Macias E, Ruiz A, Moreno E, Laffond E, Davila I, Lorente F. Usefulness of intradermal test and patch test in the diagnosis of nonimmediate reactions to metamizol. Allergy 2007;62:1462-1464

8   Dalmau J, Serra-Baldrich E, Roé E, Peramiquel L, Alomar A. Use of patch test in fixed drug eruption due to metamizole (Nolotil). Contact Dermatitis 2006;54:127-128

9   Özkaya-Bayazit E. Topical provocation in fixed drug eruption due to metamizol and naproxen. Clin Exp Dermatol 2004;29:419-422

10 Bernedo N, Audicana MT, Uriel O, Velasco M, Gastaminza G, Fernández E, et al. Metamizol as a cause of postoperative erythroderma. Contact Dermatitis 2004;50:317-318

11 Zedlitz S, Linzbach L, Kaufmann R, Boehncke WH. Reproducible identification of the causative drug of a fixed drug eruption by oral provocation and lesional patch testing. Contact Dermatitis 2002;46:352-353

12 Dalmau J, Serra-Baldrich E, Roe E, Alomar A. Fixed drug eruption due to metamizole (Nolotil(^{R})): the usefulness of patch tests. Dermatitis 2006;17:104 (Abstract)

13 Gonzalo-Garijo MA, de Arila D, Rodriguez-Nevado I. Generalized reaction after patch testing with metamizole. Contact Dermatitis 2001;45:180

14 Borja JM, Galindo PA, Gomez E, Feo F. Delayed skin reactions to metamizol. Allergy 2003;58:84-85

15 Bellegrandi S, Rosso R, Mattiacci G, Zaffiro A, Di Sora F, Menzella F, Aiuti F, Paganelli R. Combined immediate- and delayed-type hypersensitivity to metamizole. Allergy 1999;54:88-90

16 Brockow K, Garvey LH, Aberer W, Atanaskovic-Markovic M, Barbaud A, Bilo MB, et al.; ENDA/EAACI Drug Allergy Interest Group. Skin test concentrations for systemically administered drugs – an ENDA/EAACI Drug Allergy Interest Group position paper. Allergy 2013;68:702-712

Chapter 3.308 METAPROTERENOL

IDENTIFICATION

Description/definition   : Metaprotereonol is the synthetic amine that conforms to the structural formula shown
                             below
Pharmacological classes   : Bronchodilator agents; sympathomimetics; tocolytic agents; β_2-adrenergic receptor
                             agonists
IUPAC name   : 5-[1-Hydroxy-2-(propan-2-ylamino)ethyl]benzene-1,3-diol
Other names   : Orciprenaline
CAS registry number   : 586-06-1
EC number   : 209-569-1
Merck Index monograph   : 7272
Patch testing   : 2.5%, 1.25%, and 0.1% water
Molecular formula   : $C_{11}H_{17}NO_3$

GENERAL

Metaprotereonol is a synthetic amine and a β_2-adrenergic receptor agonist. It is indicated for the treatment of
bronchospasm, chronic bronchitis, asthma, and emphysema. In pharmaceutical products metaprotereonol is
employed as metaprotereonol sulfate (CAS number 5874-97-5, EC number 227-539-6, molecular formula
$C_{22}H_{36}N_2O_{10}S$) (1).

CONTACT ALLERGY FROM ACCIDENTAL CONTACT

A 25-year-old respiratory therapist presented with an itchy dermatitis on the face, neck, and arms of several months
duration. The condition was refractory to topical corticosteroids and partially responsive to systemic corticosteroids.
Physical examination revealed hyperpigmentation and lichenification on the face, neck, and arms, with the greatest
involvement on the posterior neck, distal arms, and antecubital fossae. As a respiratory therapist, she routinely
administered metaprotereonol sulfate, acetylcysteine, and isoetharine in aerosolized forms. Patch testing to various
dilutions of the commercial preparation of these 3 items gave a positive reaction to metaprotereonol. Subsequently,
serial aqueous dilutions of 2.5%, 1.25%, and 0.1% metaprotereonol, obtained from the manufacturer, were also
positive on delayed readings. Fifty controls were negative to 2.5% aqueous metaprotereonol. The patient was
diagnosed with occupational airborne allergic contact dermatitis from metaprotereonol. Over the next 6 years, she
successfully avoided contact with the pharmaceutical and the dermatitis has not recurred (2).

LITERATURE

1   The data in the section 'General' may have been obtained from literature discussed in this chapter, but mostly
    also or exclusively from one or more of the following online sources: ChemIDPlus Advanced, PubChem,
    DrugBank, RxList, Drug Central, Drugs.com, and Wikipedia
2   Fung MA, Geisse JK, Maibach HI. Airborne contact dermatitis from metaprotereonol in a respiratory therapist.
    Contact Dermatitis 1996;35:317-318

Chapter 3.309 METHACYCLINE

IDENTIFICATION

Description/definition   : Methacycline is the tetracycline analog that conforms to the structural formula shown below

Pharmacological classes   : Antibacterial agents

IUPAC name   : (4S,4aR,5S,5aR,12aR)-4-(Dimethylamino)-1,5,10,11,12a-pentahydroxy-6-methylidene-3,12-dioxo-4,4a,5,5a-tetrahydrotetracene-2-carboxamide

CAS registry number   : 914-00-1

EC number   : 213-017-5

Merck Index monograph   : 7284

Patch testing   : 10% pet.

Molecular formula   : $C_{22}H_{22}N_2O_8$

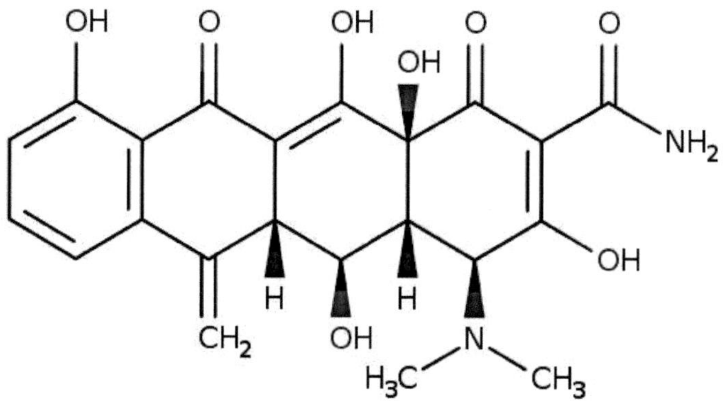

GENERAL

Methacycline is a broad-spectrum semisynthetic antibiotic related to tetracycline but is excreted more slowly and maintains effective blood levels for a more extended period. It is active against most gram-positive bacteria (pneumococci, streptococci, staphylococci) and gram-negative bacteria (*E. coli*, *Salmonella*, *Shigella*, etc.), and towards agents causing onithosis, psittacosis, trachoma, and some protozoa. Its main use is for the treatment of acute bacterial exacerbations of chronic bronchitis. In pharmaceutical products, methacycline is employed as methacycline hydrochloride (CAS number 3963-95-9, EC number 223-568-3, molecular formula $C_{22}H_{23}ClN_2O_8$) (1).

CONTACT ALLERGY FROM ACCIDENTAL CONTACT

In a group of 107 workers in the pharmaceutical industry with dermatitis, investigated in Warsaw, Poland, before 1989, 2 reacted to methacycline, tested 10% pet. (2). Also in Warsaw, Poland, in the period 1979-1983, 27 pharmaceutical workers, 24 nurses and 30 veterinary surgeons were diagnosed with occupational allergic contact dermatitis from antibiotics. The numbers that had positive patch tests to methacycline (Rondomycin ampoule content) were 2, 0, and 0, respectively, total 2 (3).

LITERATURE

1   The data in the section 'General' may have been obtained from literature discussed in this chapter, but mostly also or exclusively from one or more of the following online sources: ChemIDPlus Advanced, PubChem, DrugBank, RxList, Drug Central, Drugs.com, and Wikipedia

2   Rudzki E, Rebandel P, Grzywa Z. Contact allergy in the pharmaceutical industry. Contact Dermatitis 1989;21:121-122

3   Rudzki E, Rebendel P. Contact sensitivity to antibiotics. Contact Dermatitis 1984;11:41-42

Chapter 3.310 METHOTREXATE

IDENTIFICATION

Description/definition   : Methotrexate is the folate analog and glutamic acid derivative that conforms to the structural formula shown below

Pharmacological classes   : Enzyme inhibitors; abortifacient agents, nonsteroidal; antirheumatic agents; dermatologic agents; folic acid antagonists; antimetabolites, antineoplastic; immunosuppressive agents

IUPAC name   : (2S)-2-[[4-[(2,4-Diaminopteridin-6-yl)methylmethylamino]benzoyl]amino]pentanedioic acid

Other names   : L-Glutamic acid, N-[4-[[(2,4-diamino-6-pteridinyl)methyl]methylamino]benzoyl]-4-amino-10-methylfolic acid

CAS registry number   : 59-05-2

EC number   : 200-413-8

Merck Index monograph   : 7327

Patch testing   : 0.1% and 0.01% water

Molecular formula   : $C_{20}H_{22}N_8O_5$

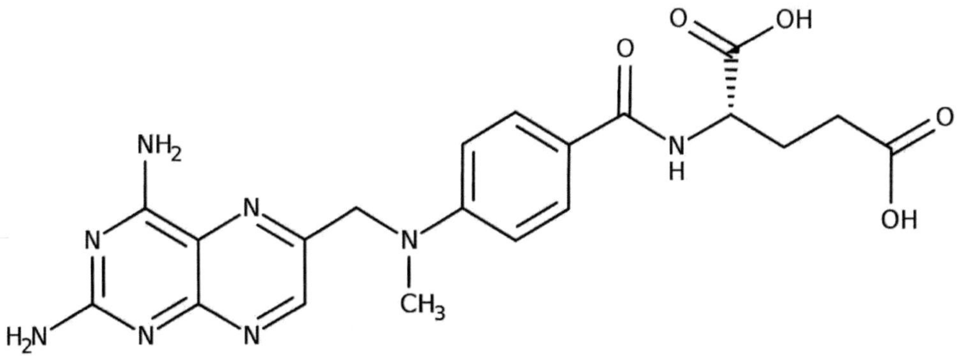

GENERAL

Methotrexate is an antimetabolite and antifolate agent with antineoplastic and immunosuppressant activities. This agent binds to and inhibits the enzyme dihydrofolate reductase, resulting in inhibition of purine nucleotide and thymidylate synthesis and, subsequently, inhibition of DNA and RNA syntheses. Methotrexate also exhibits potent immunosuppressant activity although the mechanism(s) of actions is unclear. Methotrexate is indicated for treatment or various malignancies, either alone or in combination with other antineoplastic drugs, in the symptomatic control of severe, recalcitrant, disabling psoriasis, in the management of adults with severe, active rheumatoid arthritis and children with active polyarticular-course juvenile rheumatoid arthritis (1).

CONTACT ALLERGY FROM ACCIDENTAL CONTACT

A 34-year-old production engineer, having worked for 2 months in the manufacture of methotrexate (MTX), developed itchy erythema on the face, becoming vesicular and exudative, and spreading to the neck, thorax and volar forearms. He suspected that the MTX-precursor 2,4-diamino-6-chloromethylpteridine hydrochloride (DACHMPT) was the cause. Although manufactured in a closed system, it could contaminate the work environment. During a 1-month sick leave the dermatitis healed, and he was redeployed to another section, not involving MTX manufacture. When he inadvertently visited a laboratory where the personnel were working with DACHMPT, the patient developed a widespread dermatitis. Patch tests were positive to methotrexate 0.1% water (negative to 0.01%) and to the precursors 2,4-diamino-6-chloromethylpteridine hydrochloride (DACHMPT) (0.1% and 0.01% water) and 2,4-diamino-6-hydroxymethylpteridine hydrochloride (DAHMPT) 0.1% water (negative to 0.01%). The primary sensitizer was DACHMPT, likely cross-reacting to DAHMPT and methotrexate. Whether the patient was actually exposed to methotrexate itself was not mentioned. If not, this was *not* a case of airborne occupational allergic contact dermatitis to methotrexate (3).

CUTANEOUS ADVERSE DRUG REACTIONS FROM SYSTEMIC ADMINISTRATION CAUSED BY TYPE IV (DELAYED-TYPE) HYPERSENSITIVITY (as demonstrated by positive patch tests)

Cutaneous adverse drug reactions from systemic administration of methotrexate caused by type IV (delayed-type) hypersensitivity have included erythema multiforme (2).

Other cutaneous adverse drug reactions

A 74-year-old woman had been treated with oral ibuprofen, naproxen, omeprazole and folic acid and with subcutaneous methotrexate 20 mg per week for 5 weeks for seronegative polyarthritis, when a cutaneous rash developed on the trunk, arms and legs. The lesions started as erythematous macules and papules that had become progressively confluent. During the last days, a diffuse tender erythema with extensive erosions developed over the left arm. A positive Nikolsky sign was also present, but no blisters were observed. Palms, soles and mucosal surfaces were not affected. The patient had no fever and there were no laboratory abnormalities. Histopathology showed a lichenoid tissue reaction, with dermal edema and a moderate perivascular infiltrate composed of lymphocytes and eosinophils. Based on these features, a diagnosis of erythema multiforme was made. Six months later, patch tests with all drugs used were negative at D3, but positive to commercial methotrexate 25 mg/ml on D7, with papules, edema, erythema and pruritus (+). Histopathology showed a lichenoid tissue reaction with similar features to the previous biopsy, but in less intense grade, erythema multiforme-like. Ten controls were negative to methotrexate (2).

LITERATURE

1   The data in the section 'General' may have been obtained from literature discussed in this chapter, but mostly also or exclusively from one or more of the following online sources: ChemIDPlus Advanced, PubChem, DrugBank, RxList, Drug Central, Drugs.com, and Wikipedia
2   Blanes M, Silvestre JF, Albares MP, Pascual JC, Pastor N. Erythema multiforme due to methotrexate reproduced with patch test. Contact Dermatitis 2005;52:164-165
3   Dastychová E. Allergic contact dermatitis in methotrexate manufacture. Contact Dermatitis 2003;48:226

Chapter 3.311 METHOXSALEN

IDENTIFICATION

Description/definition   : Methoxsalen is the naturally occurring furocoumarin compound that conforms to the
                            structural formula shown below
Pharmacological classes   : Photosensitizing agents; cross-linking reagents
IUPAC name   : 9-Methoxyfuro[3,2-g]chromen-7-one
Other names   : 8-Methoxypsoralen; 7H-furo(3,2-g)(1)benzopyran-7-one, 9-methoxy-; xanthotoxin
CAS registry number   : 298-81-7
EC number   : 206-066-9
Merck Index monograph   : 7329
Patch testing   : For patch testing: 0.15% alcohol; for photopatch testing a dilution series (0.15%, 0.015%,
                            0.0015%) may be necessary to differentiate photoallergy from phototoxicity (2)
Molecular formula   : $C_{12}H_8O_4$

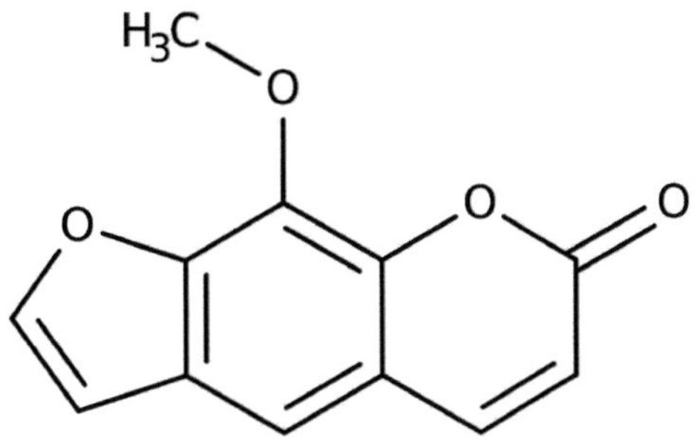

GENERAL

Methoxsalen is a naturally occurring substance isolated from the seeds of the plant *Ammi majus* with photoactiva-
ting properties. As a member of the family of compounds known as psoralens or furocoumarins, its exact mechanism
of action is unknown. Upon photoactivation by ultraviolet A irradiation, methoxsalen has been observed to bind
covalently to and crosslink DNA. This drug is indicated, together with UVA-irradiation, as photochemotherapy
(synonym: PUVA, Psoralen – UVA), for the treatment of psoriasis, vitiligo and other skin disorders and can be used
both topically and systemically (1). The topical administration has caused several cases of (photo)contact allergy/
(photo)allergic contact dermatitis, which has been fully reviewed in Volume 3 of the *Monographs in contact allergy*
series (8). Because of its carcinogenic properties, PUVA has been largely replaced with UVB phototherapy.
Methoxsalen and other furocoumarins such as 5-methoxypsoralen are phototoxic substances which are naturally
present in many plants.

CUTANEOUS ADVERSE DRUG REACTIONS FROM SYSTEMIC ADMINISTRATION CAUSED BY TYPE IV
(DELAYED-TYPE) HYPERSENSITIVITY (as demonstrated by positive patch tests)

Cutaneous adverse drug reactions from systemic administration of methoxsalen caused by type IV (delayed-type)
hypersensitivity have included acute generalized exanthematous pustulosis (AGEP) (9), systemic allergic dermatitis
(7), photosensitivity (6,7,10,11) and eczematous eruption (2).

Acute generalized exanthematous pustulosis (AGEP)

One case of acute generalized exanthematous pustulosis (AGEP) caused by methoxsalen with a positive patch test
has been reported from France (9, details not available to the author).

Systemic allergic dermatitis (systemic contact dermatitis)

A 37-year-old woman was treated for psoriasis with local photochemotherapy (PUVA) using 0.001% methoxsalen in
ethanol, until after a month, the treatment had to be abandoned because of itching dermatitis with blistering. Eight
years later, the patient was treated with systemic PUVA with oral ingestion of methoxsalen. After 2 treatments with
1 J/cm² UVA the skin became red with burning, edema and vesicles. Patch and photopatch tests were performed and
there were positive patch and photopatch test reactions (equally strong) to methoxsalen 0.1%, 0.01% and 0.001%
alc. It was concluded that the patient had a 'plain' contact allergy to methoxsalen. Oral administration had induced
systemic allergic dermatitis (7; also published in 2 identical Abstracts, refs. 4 and 5).

Photosensitivity

A 36-year-old woman with psoriasis vulgaris developed generalized photoallergic dermatitis to methoxsalen after 16 PUVA treatments. The diagnosis of photoallergy was confirmed by re-exposure to methoxsalen and total body UVA irradiation, phototests using topical and oral 8-methoxypsoralen and histological studies. It is uncertain whether photopatch tests were performed (6).

In a study from Nancy, France, 4 patients with suspected photosensitivity reactions from drugs were assessed with photopatch testing. All 4 patients had positive photopatch tests, of who one to methoxsalen. Clinical details were not provided (10).

A 62-year-old woman was treated with topical photochemotherapy using 0.001% methoxsalen in ethanol, but the skin became worse and the treatment was stopped. Six years later she was given oral methoxsalen photochemotherapy which immediately resulted in a severe eczematous dermatitis of the hands and feet. Patch tests were negative, but photopatch tests positive to methoxsalen 0.1%, 0.01% and 0.001% alc. This patient had systemic *photo*allergic dermatitis (7; also published in 2 identical Abstracts, refs. 4 and 5).

Another case of systemic photosensitivity was likely published in French literature in 1992, but details are not available to the author (11).

Dermatitis/eczematous eruption

A 37-year-old woman with widespread chronic plaque psoriasis was treated with photochemotherapy (PUVA) for a second time. Within 24 hours after the third treatment, a widespread itchy eczematous rash developed affecting her face and trunk. PUVA treatment was suspended, and the rash settled over 10 days with topical steroids and emollients. Twelve days later, treatment was restarted. Within 6 hours, an eczematous rash appeared on the trunk and limbs with marked periorbital edema. Patch tests were positive to the commercially available methoxsalen solution diluted to 0.1%, 0.01%, and 0.001% in water and ethanol. 25 controls were negative to 0.1% (2).

Cross-reactions, pseudo-cross-reactions and co-reactions

A patient with contact allergy to methoxsalen cross-reacted to 'trimethoxypsoralen' (meant was probably trimethylpsoralen [trioxsalen]) (4,5,7). One patient photosensitized to methoxsalen had a photocross-reaction to 3-carbethoxypsoralen; another individual may have had photocross-reactivity to methoxsalen from primary 3-carbethoxypsoralen photosensitization (3).

LITERATURE

1   The data in the section 'General' may have been obtained from literature discussed in this chapter, but mostly also or exclusively from one or more of the following online sources: ChemIDPlus Advanced, PubChem, DrugBank, RxList, Drug Central, Drugs.com, and Wikipedia
2   Ravenscroft J, Goulden V, Wilkinson M. Systemic allergic contact dermatitis to 8-methoxypsoralen (8-MOP). J Am Acad Dermatol 2001;45(6 Suppl.):S218-S219
3   Takashima A, Yamamoto K, Kimura S, Takakuwa Y, Mizuno N. Allergic contact and photocontact dermatitis due to psoralens in patients with psoriasis treated with topical PUVA. Br J Dermatol 1991;124:37-42
4   Möller H. Contact and photocontact allergy to psoralens. Am J Contact Dermat 1990;1:254
5   Möller H. Contact and photocontact allergy to psoralens. Am J Contact Dermat 1990;1:202
6   Plewig G, Hofmann C, Braun-Falco O. Photoallergic contact dermatitis from 8-methoxypsoralen. Arch Dermatol Res 1978:261:201-211
7   Möller H. Contact and photocontact allergy to psoralens. Photodermatol Photoimmunol Photomed 1990;7:43-44
8   De Groot AC. Monographs in contact allergy, volume 3: Topical Drugs. Boca Raton, Fl, USA: CRC Press Taylor and Francis Group, 2021 (ISBN 978-0-367-23693-9)
9   Morant C, Devis T, Alcaraz I, Lefevre L, Caron J, Modiano P. Acute generalized exanthematous pustulosis due to meladinine with positive patch tests. Ann Dermatol Venereol 2002;129:234-235 (Article in French)
10   Barbaud A, Reichert-Penetrat S, Tréchot P, Jacquin-Petit MA, Ehlinger A, Noirez V, et al. The use of skin testing in the investigation of cutaneous adverse drug reactions. Br J Dermatol 1998;139:49-58
11   Jeanmougin M, Manciet JR, Castelneau JP, Smadja J, Dubertret L. Photo-allergie systémique au méthoxalène [Systemic photoallergy induced by methoxalen]. Ann Dermatol Venereol 1992;119:277-280 (Article in French)

Chapter 3.312 METHYLDOPA

IDENTIFICATION

Description/definition   : Methyldopa is the phenylpropanoic acid that conforms to the structural formula shown below
Pharmacological classes : Antihypertensive agents; adrenergic alpha-2 receptor agonists; sympatholytics
IUPAC name   : (2S)-2-Amino-3-(3,4-dihydroxyphenyl)-2-methylpropanoic acid
Other name(s)   : alpha-Methyldopa
CAS registry number   : 555-30-6
EC number   : 209-089-2
Merck Index monograph   : 7397
Patch testing   : No data available; most systemic drugs can be tested at 10% pet.; if the pure chemical is not available, prepare the test material from intravenous powder, the content of capsules or – when also not available – from powdered tablets to achieve a final concentration of the active drug of 10% pet.
Molecular formula   : $C_{10}H_{13}NO_4$

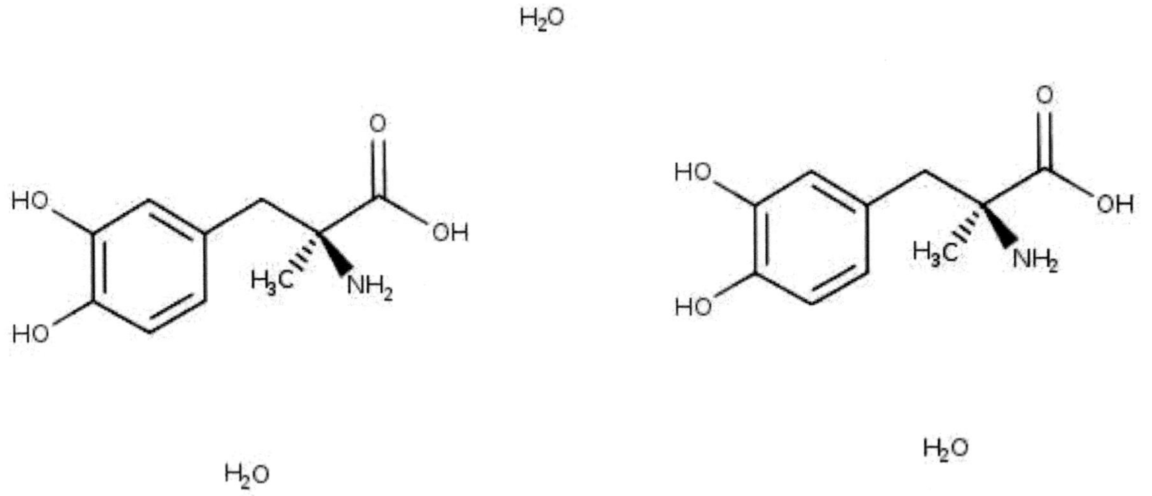

H_2O

H_2O

H_2O

Methyldopa sesquihydrate

GENERAL

Methyldopa is a centrally active sympatholytic agent with antihypertensive activity. It is a prodrug which is metabolized in the central nervous system. The antihypertensive action of methyldopa seems to be attributable to its conversion into alpha-methylnorepinephrine, which is a potent alpha-2 adrenergic agonist that binds to and stimulates potent central inhibitory alpha-2 adrenergic receptors. This results in a decrease in sympathetic outflow and decreased blood pressure. In pharmaceutical products, methyldopa may be employed as methyldopa hydrochloride (CAS number 2508-79-4, EC number 219-720-3, molecular formula $C_{12}H_{18}ClNO_4$) or as methyldopa sesquihydrate (CAS number 41372-08-1, EC number 609-918-1, molecular formula $C_{10}H_{15}NO_5$) (1).

CUTANEOUS ADVERSE DRUG REACTIONS FROM SYSTEMIC ADMINISTRATION CAUSED BY TYPE IV (DELAYED-TYPE) HYPERSENSITIVITY (as demonstrated by positive patch tests)

Cutaneous adverse drug reactions from systemic administration of methyldopa caused by type IV (delayed-type) hypersensitivity have included photosensitivity (2) and generalized papulovesicular eruption (3).

Photosensitivity

A 72-year-old woman was treated with methyldopa for three years for arterial hypertension. The patient developed an erythematous pruritic papulovesicular eruption three months after an increase in the dosage of methyldopa, which spread on the face, neck, and arms and was initially limited to sun-exposed sites before expanding secondarily to the shoulders and legs. Phototesting showed minimal erythema doses (MEDs) for UVA and UVB to be normal. Photopatch tests were positive to methyldopa crushed tablet mixed with petrolatum (concentration?) and irradiated with 25 J/cm² UVA (which is too high according to current standards) and negative at the site irradiated with total spectrum (UVA, UVB and visible light). Controls tests were not performed. There was no recurrence three years after discontinuing methyldopa therapy. The authors were cautious in their conclusions and diagnosed 'photosensitivity' and 'possible photoallergy' (2).

Other cutaneous adverse drug reactions

An 81-year-old woman developed an itchy papulovesicular eruption, which started at the forearms and thereafter became generalized. She used several drugs for hypertension and diabetes mellitus including methyldopa. Physical examination showed an extensive polymorphic exanthema with excoriated papulovesicular lesions, plaques of nummular dermatitis and bullae on the palms and soles; there were no mucosal lesions. Histopathology showed cellular necrosis of the epidermis and an inflammatory perivascular infiltrate in the superficial dermis. Patch tests were positive to methyldopa and negative to the other drugs used by the patient. No details on patch testing (material used, concentration, vehicle, times of reading, strength of positive reaction, number of controls performed [probably none]) were provided (3).

LITERATURE

1 The data in the section 'General' may have been obtained from literature discussed in this chapter, but mostly also or exclusively from one or more of the following online sources: ChemIDPlus Advanced, PubChem, DrugBank, RxList, Drug Central, Drugs.com, and Wikipedia
2 Vaillant L, Le Marchand D, Grognard C, Hocine R, Lorette G. Photosensitivity to methyldopa. Arch Dermatol 1988;124:326-327
3 Heid E, Samsoen M, Juillard J, Eberst E, Foussereau J. Eruptions papulo-vésiculeuses endogènes à la methyldopa et au clofibrate. Ann Derm Venereol 1977;104:494-496 (Article in French)

Chapter 3.313 METHYLPREDNISOLONE

IDENTIFICATION

Description/definition	: Methylprednisolone is the synthetic glucocorticoid that conforms to the structural formula shown below
Pharmacological classes	: Glucocorticoids; neuroprotective agents; anti-inflammatory agents; antiemetics
IUPAC name	: (6S,8S,9S,10R,11S,13S,14S,17R)-11,17-Dihydroxy-17-(2-hydroxyacetyl)-6,10,13-trimethyl-7,8,9,11,12,14,15,16-octahydro-6H-cyclopenta[a]phenanthren-3-one
Other names	: 11β,17,21-Trihydroxy-6α-methylpregna-1,4-diene-3,20-dione
CAS registry number	: 83-43-2
EC number	: 201-476-4
Merck Index monograph	: 7454
Patch testing	: In general, corticosteroids may be tested at 0.1% and 1% in alcohol; late readings (6-10 days) are strongly recommended
Molecular formula	: $C_{22}H_{30}O_5$

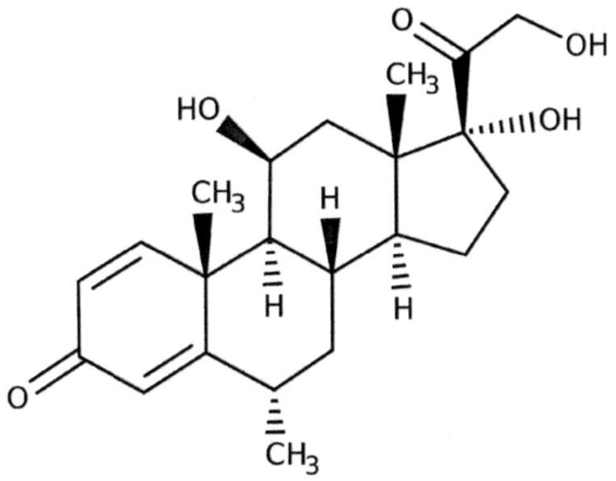

GENERAL

Systemically administered glucocorticoids have anti-inflammatory, immunosuppressive and antineoplastic properties and are used in the treatment of a wide spectrum of diseases including rheumatic disorders, lung diseases (asthma, COPD), gastrointestinal tract disorders (Crohn's disease, colitis ulcerosa), certain malignancies (leukemia, lymphomas), hematological disorders, and various diseases of the kidneys, brain, eyes and skin. A practical guideline for diagnosing allergic reactions to corticosteroids is presented in ref. 2. Methylprednisolone (MP) *base* is used as tablet only, which implies that most allergic reactions to 'methylprednisolone' reported (1) have in fact been the result of sensitization to an ester such as MP acetate (Chapter 3.314), MP hemisuccinate (Chapter 3.315), or MP aceponate, or of cross-reactivity to another corticosteroid.

CONTACT ALLERGY FROM ACCIDENTAL CONTACT

Between 1978 and 2001, in Leuven, Belgium, occupational allergic contact dermatitis to pharmaceuticals was diagnosed in 33 health care workers. In this group, one patient had occupational allergic contact dermatitis from methylprednisolone. This was a 38-year-old nurse with eczema of the fingers, Patch tests were positive to methylprednisolone, ranitidine HCl and amikacin (5).

CUTANEOUS ADVERSE DRUG REACTIONS FROM SYSTEMIC ADMINISTRATION CAUSED BY TYPE IV (DELAYED-TYPE) HYPERSENSITIVITY (as demonstrated by positive patch tests)

Cutaneous adverse drug reactions from systemic administration of methylprednisolone caused by type IV (delayed-type) hypersensitivity have included maculopapular eruption (3) and systemic allergic dermatitis (4,6,7,8,9).

Maculopapular eruption

A 40-year-old man was treated for Henoch Schönlein syndrome with methylprednisolone 48 mg daily. After 4 days the patient developed a pruriginous generalized maculopapular eruption. Methylprednisolone was immediately withdrawn and the rash disappeared in 1 week. A patch test was positive to 0.1 ml of a solution containing 6-methyl-prednisolone sodium hemisuccinate 40 mg/ml in distilled water at D2 and D3. Oral challenge tests with the additives contained in the methylprednisolone tablet and in the solution used for the patch test were negative (3).

Systemic allergic dermatitis (systemic contact dermatitis)

In Leuven, Belgium, in a 12-year-period before 2012, 16 patients were investigated for a generalized allergic eruption (maculopapular eruption or eczema, with or without flare-up of previous dermatitis) from systemic administration (oral, intravenous, intramuscular, intra-articular) of corticosteroids, a few hours or days after the first dose of the culprit drug. The reactions observed were in most cases a manifestation of systemic allergic dermatitis: the patient had previously become sensitized to the corticosteroid used systemically or a cross-reacting molecule from topical exposure. Eleven patients reacted to methylprednisolone, 3 from intravenous, 4 from oral and 4 from intra-articular administration of the drug. Nine had a maculopapular rash and 2 generalized eczema, of who one with a flare-up of previous dermatitis. Only in 2 cases, both with a maculopapular eruption from intra-articular administration, there was no previous topical exposure to corticosteroids. It should be appreciated that methylprednisolone *base* is used in tablets only; hence, the 7 cases with reactions to intravenous or intra-articular injections must have been treated with an (unspecified) ester or salt of methylprednisolone (8).

A 46-year-old woman was documented by patch and provocative use testing to be allergic to multiple topical corticosteroids. On further testing, oral provocation tests to methylprednisolone, triamcinolone, dexamethasone, and prednisone each produced a generalized maculopapular eruption in a delayed manner (4). This was a case of systemic allergic dermatitis presenting as maculopapular eruption.

A 50-year-old woman, very likely sensitized from topical corticosteroid, had suffered (unspecified) exanthematous eruptions on 2 occasions one day after having taken oral methylprednisolone and once after oral prednisone. Patch tests were positive to prednisolone, but methylprednisolone was not patch tested. However, an intradermal test with the latter was positive at D1 and D2. Later, oral provocation tests with betamethasone and cloprednol induced and eruption after 12 hours resembling symmetrical drug-related intertriginous and flexural exanthema (SDRIFE)/baboon syndrome (6).

A 52-year-old woman who had suffered allergic contact dermatitis from antihemorrhoidal creams and a generalized symmetrical eruption after intra-articular administration of triamcinolone acetonide, had positive patch tests to triamcinolone acetonide and 7 other corticosteroids. Oral administration of 32 mg methylprednisolone resulted in a pruritic rash on the anterior trunk after 24 hours. This was a case of systemic allergic dermatitis (from oral provocation), but methylprednisolone itself was not patch tested (7).

Three patients who had positive patch tests to tixocortol pivalate and positive intradermal tests to methylprednisolone, were orally challenged with methylprednisolone 4, 12, 20, 24 and 36 mg. One or two doses were given per day, and the challenge was stopped when skin symptoms appeared. All 3 had a positive reaction; one reacted after the 20 mg dose, one after 24 mg and 1 after 4 mg. Symptoms were erythema or infiltrated erythema at previous sites of eczema or positive skin tests (n=1), widespread erythema or exanthema (n=2) and facial erythema (n=1) (9).

Cross-reactions, pseudo-cross-reactions and co-reactions

Cross-reactions between corticosteroids are discussed in Chapter 3.399 Prednisolone.

LITERATURE

1   Toholka R, Wang Y-S, Tate B, Tam M, Cahill J, Palmer A, Nixon R. The first Australian Baseline Series: Recommendations for patch testing in suspected contact dermatitis. Australas J Dermatol 2015;56:107-115
2   Baeck M, Goossens A. Immediate and delayed allergic hypersensitivity to corticosteroids: practical guidelines. Contact Dermatitis 2012;66:38-45
3   Astone A, Romano A, Pietrantonio F, Garcovich A, Venuti A, Barone C. Delayed hypersensitivity to 6-methyl-prednisolone in Henoch Schöenlein syndrome. Allergy 1992;47(4Pt.2):436-438
4   Chew AL, Maibach HI. Multiple corticosteroid orally elicited allergic contact dermatitis in a patient with multiple topical corticosteroid allergic contact dermatitis. Cutis 2000;65:307-311
5   Gielen K, Goossens A. Occupational allergic contact dermatitis from drugs in healthcare workers. Contact Dermatitis 2001;45:273-279
6   Treudler R, Simon J. Symmetric, drug-related, intertriginous, and flexural exanthema in a patient with polyvalent intolerance to corticosteroids. J Allergy Clin Immunol 2006;118:965-967
7   Santos-Alarcón S, Benavente-Villegas FC, Farzanegan-Miñano R, Pérez-Francés C, Sánchez-Motilla JM, Mateu-Puchades A. Delayed hypersensitivity to topical and systemic corticosteroids. Contact Dermatitis 2018;78:86-88
8   Baeck M, Goossens A. Systemic allergic dermatitis to corticosteroids. Allergy 2012;67:1580-1585
9   Räsänen L, Tuomi ML, Ylitalo L. Reactivity of tixocortol pivalate-positive patients in intradermal and oral provocation tests. Br J Dermatol 1996;135:931-934

Chapter 3.314 METHYLPREDNISOLONE ACETATE

IDENTIFICATION

Description/definition	: Methylprednisolone acetate is the acetate ester of the synthetic glucocorticoid methylprednisolone that conforms to the structural formula shown below
Pharmacological classes	: Anti-inflammatory agents
IUPAC name	: [2-[(6S,8S,9S,10R,11S,13S,14S,17R)-11,17-Dihydroxy-6,10,13-trimethyl-3-oxo-7,8,9,11,12,14,15,16-octahydro-6H-cyclopenta[a]phenanthren-17-yl]-2-oxoethyl] acetate
Other names	: Methylprednisolone 21-acetate; 11β,17,21-trihydroxy-6α-methylpregna-1,4-diene-3,20-dione 21-acetate
CAS registry number	: 53-36-1
EC number	: 200-171-3
Merck Index monograph	: 7454 (Methylprednisolone)
Patch testing	: In general, corticosteroids may be tested at 0.1% and 1% in alcohol; late readings (6-10 days) are strongly recommended
Molecular formula	: $C_{24}H_{32}O_6$

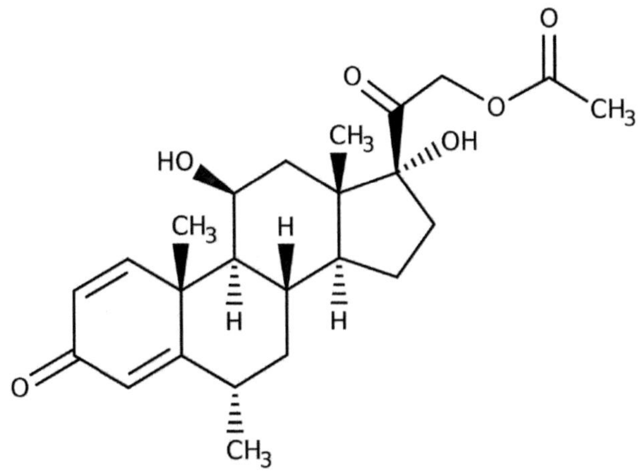

GENERAL

Systemically administered glucocorticoids have anti-inflammatory, immunosuppressive and antineoplastic properties and are used in the treatment of a wide spectrum of diseases including rheumatic disorders, lung diseases (asthma, COPD), gastrointestinal tract disorders (Crohn's disease, colitis ulcerosa), certain malignancies (leukemia, lymphomas), hematological disorders, and various diseases of the kidneys, brain, eyes and skin. A practical guideline for diagnosing allergic reactions to corticosteroids is presented in ref. 1. Methylprednisolone acetate is used for injection (epidural; infiltration; intra-articular; intralesional; intramuscular; intravenous; soft tissue; subcutaneous) and in topical preparations for acne. Contact allergy to and allergic contact dermatitis from methylprednisolone acetate has been fully reviewed in Volume 3 of the *Monographs in contact allergy* series (3). See also methylprednisolone (Chapter 3.313) and methylprednisolone hemisuccinate (Chapter 3.315).

CUTANEOUS ADVERSE DRUG REACTIONS FROM SYSTEMIC ADMINISTRATION CAUSED BY TYPE IV (DELAYED-TYPE) HYPERSENSITIVITY (as demonstrated by positive patch tests, unless otherwise stated)

Cutaneous adverse drug reactions from systemic administration of methylprednisolone acetate caused by type IV (delayed-type) hypersensitivity have included widespread erythema from intra-articular injection (5), (possibly) acute generalized exanthematous pustulosis (AGEP) (2), systemic allergic dermatitis presenting as erythema around the neck from intra-articular injection (8) localized erythematous plaque from epidural injection (6), localized allergic reactions after retrobulbar injections (9,10), chronic urticaria from delayed- and immediate-type hypersensitivity (7), and unspecified skin rash after intra-articular injection (4).

Erythroderma, widespread erythematous eruption, exfoliative dermatitis

In a 47-year-old woman with rheumatoid arthritis, intra-articular methylprednisolone acetate and methylprednisolone sodium succinate caused widespread erythema on the following day, which lasted about 3 days. A provocation with intra-articular 6 mg betamethasone acetate/disodium succinate caused erythema of the face and neck, which lasted about 1 day. Patch and intradermal tests showed delayed-type allergy to corticosteroids of the hydrocortisone group (hydrocortisone, tixocortol, prednisolone, methylprednisolone). She had not been previously treated with topical corticosteroids (5).

Acute generalized exanthematous pustulosis (AGEP)
Possibly, intravenous administration of methylprednisolone acetate has caused a case of AGEP. However, it was uncertain whether the culprit drug was the acetate or hemisuccinate ester of methylprednisolone (2). This case is described in Chapter 3.315 Methylprednisolone hemisuccinate.

Systemic allergic dermatitis (systemic contact dermatitis)
A 62-year-old woman had a history of a burn treated with ointment containing hydrocortisone without adverse effect. Two years later, she was given an intra-articular injection of methylprednisolone acetate in the knee because of arthrosis. On that evening she developed erythema around the neck. Patch tests were positive to hydrocortisone esters and to methylprednisolone acetate 0.4% and 2% water (8). It is very likely that the patient had been sensitized previously by topical hydrocortisone and that this was a case of systemic allergic dermatitis.

Other cutaneous adverse drug reactions
One day after the 6th epidural injection of bupivacaine and methylprednisolone acetate for lumbar disk disease, a 71-year-old woman presented with a 4.0 cm annular erythematous indurated plaque in the left gluteal area. Patch tests were strongly positive to tixocortol pivalate (a marker for allergy to the hydrocortisone-prednisone group). Methylprednisolone was not patch tested, but intradermal tests were positive at D2 and D4 to methylprednisolone acetate (lasting for 2 weeks) and methylprednisolone sodium succinate (6).

Localized allergic reactions have also been observed after retrobulbar injections of methylprednisolone acetate in 3 patients. Patch tests were negative at D4 (no late reading), but intradermal tests positive at D2 (9,10).

A 47-year-old man had developed generalized urticaria several times after each epidural injection of methylprednisolone acetate and a local anesthetic (probably bupivacaine or lidocaine), when his urticaria became chronic with intermittent itchy eruptions occurring two to three times per week and clearing within 2 hours. Patch tests showed positive reactions to 21 of 26 tested corticosteroids, including methylprednisolone acetate 1% alc. with urticated erythema at D4. Intradermal tests were positive at 20 minutes to methylprednisolone acetate. There were no reactions to local anesthetics either on prick or patch testing or following subcutaneous challenge. The patient was diagnosed with urticaria from both delayed- and immediate-type hypersensitivity to methylprednisolone acetate (7).

A 37-year-old woman had developed 'a skin eruption' within some hours after intra-articular injection of methylprednisolone acetate into the right elbow for 'tennis elbow'. Patch tests were positive to the commercial injection fluid and other injectable corticosteroids: prednisolone tebutate, prednisolone sodium succinate and methylprednisolone hemisuccinate. Contact allergy to the preservative in the injection fluid used was excluded. Patch tests with the corticosteroids in pet. were negative (4).

Cross-reactions, pseudo-cross-reactions and co-reactions
Cross-reactions between corticosteroids are discussed in Chapter 3.399 Prednisolone.

LITERATURE
1   Baeck M, Goossens A. Immediate and delayed allergic hypersensitivity to corticosteroids: practical guidelines. Contact Dermatitis 2012;66:38-45
2   Mussot-Chia C, Flechet ML, Napolitano M, Herson S, Frances C, Chosidow O. Methylprednisolone-induced acute generalized exanthematous pustulosis. Ann Dermatol Venereol 2001;128(3Pt.1):241-243 (Article in French)
3   De Groot AC. Monographs in contact allergy, volume 3: Topical Drugs. Boca Raton, Fl, USA: CRC Press Taylor and Francis Group, 2021 (ISBN 978-0-367-23693-9)
4   De Boer EM, van den Hoogenband HM, van Ketel WG. Positive patch test reactions to injectable corticosteroids. Contact Dermatitis 1984;11:261-262
5   Räsänen L, Hasan T. Allergy to systemic and intralesional corticosteroids. Br J Dermatol 1993;128:407-411
6   Amin N, Brancaccio R, Cohen D. Cutaneous reactions to injectable corticosteroids. Dermatitis 2006;17:143-146
7   Pollock B, Wilkinson SM, MacDonald Hull SP. Chronic urticaria associated with intra-articular methylprednisolone. Br J Dermatol 2001;144:1228-1230
8   Murata Y, Kumano K, Ueda T, Araki N, Nakamura T, Tani N. Systemic contact dermatitis caused by systemic corticosteroid use. Arch Dermatol 1997;133:1053-1054
9   Mathias CGT, Robertson DB. Delayed hypersensitivity to a corticosteroid suspension containing methylprednisolone. Two cases of conjunctival inflammation after retrobulbar injection. Arch Dermatol 1985;121:258-261
10   Mathias CG, Maibach HI, Ostler HB, Conant MA, Nelson W. Delayed hypersensitivity to retrobulbar injections of methylprednisolone acetate. Am J Ophthalmol 1978;86:816-819

Chapter 3.315 METHYLPREDNISOLONE HEMISUCCINATE

IDENTIFICATION

Description/definition   : Methylprednisolone hemisuccinate is the hemisuccinate ester of the synthetic
                            glucocorticoid methylprednisolone that conforms to the structural formula shown below
Pharmacological classes   : Antineoplastic agents, hormonal; anti-inflammatory agents
IUPAC name   : 4-[2-[(6S,8S,9S,10R,11S,13S,14S,17R)-11,17-Dihydroxy-6,10,13-trimethyl-3-oxo-
                            7,8,9,11,12,14,15,16-octahydro-6H-cyclopenta[a]phenanthren-17-yl]-2-oxoethoxy]-4-
                            oxobutanoic acid
Other names   : Methylprednisolone succinate; 11β,17,21-trihydroxy-6α-methylpregna-1,4-diene-3,20-
                            dione 21-(hydrogen succinate)
CAS registry number   : 2921-57-5
EC number   : 220-863-9
Patch testing   : In general, corticosteroids may be tested at 0.1% and 1% in alcohol; late readings (6-10
                            days) are strongly recommended
Molecular formula   : $C_{26}H_{34}O_8$

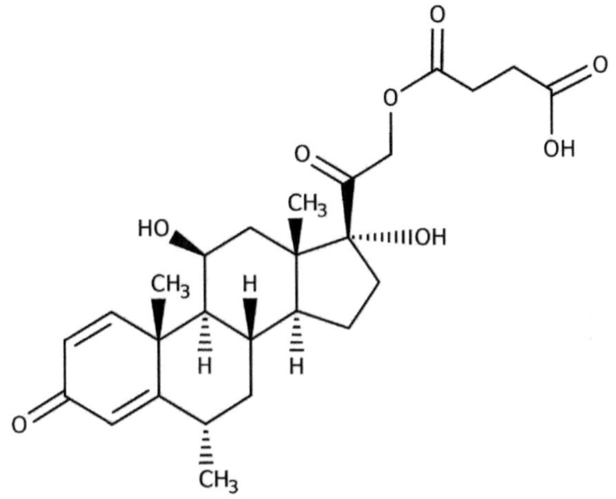

GENERAL

Systemically administered glucocorticoids have anti-inflammatory, immunosuppressive and antineoplastic properties and are used in the treatment of a wide spectrum of diseases including rheumatic disorders, lung diseases (asthma, COPD), gastrointestinal tract disorders (Crohn's disease, colitis ulcerosa), certain malignancies (leukemia, lymphomas), hematological disorders, and various diseases of the kidneys, brain, eyes and skin. A practical guideline for diagnosing allergic reactions to corticosteroids is presented in ref. 1. In pharmaceutical products, methylprednisolone hemisuccinate is employed as methylprednisolone sodium (hemi)succinate (CAS number 2375-03-3, EC number 219-156-8, molecular formula $C_{26}H_{33}NaO_8$). It is available only for intravenous and intramuscular injections. As methylprednisolone hemisuccinate is not used topically, most positive patch tests to it (2) are the result of cross-sensitivity. See also Chapter 3.313 (Methylprednisolone) and Chapter 3.314 (Methylprednisolone acetate).

CUTANEOUS ADVERSE DRUG REACTIONS FROM SYSTEMIC ADMINISTRATION CAUSED BY TYPE IV
(DELAYED-TYPE) HYPERSENSITIVITY (as demonstrated by positive patch tests, unless otherwise stated)

Cutaneous adverse drug reactions from systemic administration of methylprednisolone hemisuccinate caused by type IV (delayed-type) hypersensitivity have included maculopapular eruption (6,7), widespread/ generalized erythematous exanthema (4,11,12), erythroderma (5), acute generalized exanthematous pustulosis (AGEP) (9), systemic allergic contact dermatitis (10,11), generalizer erythema, urticaria and dyspnea (3) and 'generalized skin rashes' (8).

Maculopapular eruption

A female patient was treated for a relapse of breast cancer with a second course of doxorubicin, docetaxel, prednisolone sodium hemisuccinate, and 2 other drugs. One day later, the patient developed a generalized maculopapular rash. Patch, prick and intradermal tests gave positive patch test reactions to tixocortol pivalate and the commercial methylprednisolone injection powder 30% water, pet. and alc. No previous history of topical corticosteroid sensitivity was noticed and the patient had probably become sensitized during the first chemotherapy course a few years previously (6).

Another patient, reported from Japan, developed maculopapular eruptions from intravenous methylprednisolone on 2 occasions. Patch tests were negative, but intradermal tests read at D2 and D3 positive to methylprednisolone and methylprednisolone sodium succinate (7).

Erythroderma, widespread erythematous eruption, exfoliative dermatitis

A few hours after a first intravenous administration of vincristine, cyclophosphamide and methylprednisolone for splenic lymphoma, a 55-year-old woman presented with generalized intense itching, a generalized erythematous rash, and swelling of the face and hands. Similar eruptions, but milder and more transient, appeared after the 2nd and 3rd cycles. The same reaction occurred after taking 2 tablets prednisone. Further cycles with vincristine and cyclophosphamide alone were given without any trouble. Prick and intradermal tests with methylprednisolone were negative. Patch testing was positive to tixocortol pivalate and, in a second session, to hydrocortisone phosphate, prednisone and prednisolone hemisuccinate, but only in the vehicle ethanol-DMSO (50:50). Methylprednisolone was not tested, but delayed-type hypersensitivity to it is almost certain (4).

A 58-year-old man, known with multiple contact allergies including to neomycin and parabens, had developed acute dermatitis of the face from accidentally using a cream containing parabens. The patient thought that it had become worse after being injected with methylprednisolone sodium succinate. In spite of this, he was given the same injection again, resulting, within a few hours, in 'a generalized erythroderma'. In spite of high doses of oral prednisone, generalized itching and desquamation lasted for 1 month. Patch tests were positive to the commercial injection fluid, methylprednisolone 5% DMSO, prednisone, and tixocortol pivalate (5).

In a 47-year-old woman with rheumatoid arthritis, intra-articular methylprednisolone acetate and methylprednisolone sodium succinate caused widespread erythema on the following day, which lasted about 3 days. A provocation with intra-articular 6 mg betamethasone acetate/disodium succinate caused erythema of the face and neck, which lasted about 1 day. Patch and intradermal tests showed delayed-type allergy to corticosteroids of the hydrocortisone group (hydrocortisone, tixocortol, prednisolone, methylprednisolone). She had not been previously treated with topical corticosteroids and was probably not presensitized to corticosteroids (11).

A 36-year-old woman was treated with intravenous methylprednisolone sodium succinate for acute myelitis for 5 days. Three days later, the patient developed a widespread macular exanthem with severe itch. Intravenous prednisolone sodium succinate exacerbated both the itch and the exanthema. Patch tests were positive to methylprednisolone and prednisolone, both at 1% pet. at D2 and D3 (12).

Acute generalized exanthematous pustulosis (AGEP)

A 30-year-old-woman presented, a few hours after having received an intravenous administration of methylprednisolone (most likely sodium hemisuccinate) for multiple sclerosis, with a maculopapular exanthema predominantly in the body folds, rapidly becoming pustular, with malaise, fever and neutrophilia. The results of histopathology and microbiological cultures were consistent with the diagnosis of acute generalized exanthematous pustulosis (AGEP). The rash cleared spontaneously in one week. One month later, epicutaneous tests were positive to prednisolone, tixocortol pivalate and hydrocortisone; methylprednisolone itself may not have been tested, but considering the other patch test results, would almost certainly have been positive (9, details not available).

Systemic allergic dermatitis (systemic contact dermatitis)

A 62-year-old woman had a history of a burn treated with ointment containing hydrocortisone without adverse effect and of the development of erythema around the neck 8 hours after an intra-articular injection of methylprednisolone acetate in the knee because of arthrosis. Two years after the latter event, the patient was given an intravenous injection of hydrocortisone sodium phosphate 500 mg for 2 days, and then developed erythema on the neck, trunk, and thighs in the evening on the first day of injection. Methylprednisolone sodium succinate was substituted for the hydrocortisone, but then the eruption developed into a generalized rash. Patch tests were positive to hydrocortisone sodium phosphate, methylprednisolone acetate, methylprednisolone sodium succinate 0.05, 0.1 and 0.5% pet. and some other hydrocortisone-group corticosteroids. Intradermal tests were positive at D1 and D4 to methylprednisolone sodium succinate and hydrocortisone sodium phosphate. Twenty-four hours after the intradermal injection the patient noticed erythema around her neck and the next day she experienced nausea, vomiting, and diarrhea. Systemic provocation tests were performed with drip infusions of both corticosteroids used. Five hours after provocation with methylprednisolone the patient developed symmetrical patches of erythema around her neck and on her chest, axillae, forearms, fingers, and upper inner thighs and a flare-up at the site of the previous patch test with methylprednisolone (10). As the first eruption (from methylprednisolone acetate) developed after 8 hours, it is likely that the patient had been previously sensitized to the hydrocortisone ointment, making this a case of systemic allergic dermatitis.

A 60-year-old woman with osteoarthrosis of the knees developed widespread erythema of the trunk, neck and thighs, which developed within 12 hours and lasted for 1-2 days after being treated with intra-articular methylprednisolone sodium succinate and orally provoked with the same corticosteroid. The patient had previously been treated with topical corticosteroids without recognizable side effects. Patch and intradermal tests showed delayed-type allergy to corticosteroids of the hydrocortisone group (hydrocortisone, tixocortol, prednisolone, methylprednisolone). Whether the patient was presensitized is unknown but likely (11).

In a 75-year-old woman, polymyalgia rheumatica had been treated with oral prednisolone for 8 days, when widespread dermatitis appeared. In oral provocations, 10 mg prednisolone and 4 mg methylprednisolone sodium succinate caused widespread erythema of the trunk within several hours, lasting about 3 days. The patient had previously been treated with topical corticosteroids without recognizable side effects. Patch tests showed delayed-type allergy to prednisolone and methylprednisolone. Whether the patient was presensitized is unknown but likely (11).

Other cutaneous adverse drug reactions
A 69-year-old woman with a history of intrinsic bronchial asthma had suffered 5 episodes of pruritus followed by generalized erythema and urticaria, accompanied with increasing dyspnea (on 2 occasions), starting 6 hours after intravenous injection of methylprednisolone or hydrocortisone (most likely sodium succinate). Prick and intradermal tests with hydrocortisone phosphate 10 and 1 mg/ml and methylprednisolone 4 and 0.4 mg/ml were negative at 15 minutes. However, six hours later intradermal tests appeared positive and were accompanied by a generalized pruritic and erythematous reaction. Patch tests were (false-)negative (3).

A 34-year-old woman was treated with methylprednisolone sodium succinate pulse therapy for exacerbation of multiple sclerosis. After the injection on the first day, 'skin rashes' appeared on her trunk and thigh, which generalized on the second day. A patch test with methylprednisolone sodium succinate was positive (8, no details available, article in Japanese).

Cross-reactions, pseudo-cross-reactions and co-reactions
Cross-reactions between corticosteroids are discussed in Chapter 3.399 Prednisolone.

LITERATURE
1 Baeck M, Goossens A. Immediate and delayed allergic hypersensitivity to corticosteroids: practical guidelines. Contact Dermatitis 2012;66:38-45
2 Baeck M, Chemelle JA, Terreux R, Drieghe J, Goossens A. Delayed hypersensitivity to corticosteroids in a series of 315 patients: clinical data and patch test results. Contact Dermatitis 2009;61:163-175
3 Vidal C, Tomé S, Fernándex-Redondo V, Tato F. Systemic allergic reaction to corticosteroids. Contact Dermatitis 1994;31:273-274
4 Fernández de Corres L, Bernaola G, Urrutia I, Muñoz D. Allergic dermatitis from systemic treatment with corticosteroids. Contact Dermatitis 1990;22:104-106
5 Fernández de Corrés L, Urrutia I, Audicana M, Echechipia S, Gastaminza G. Erythroderma after intravenous injection of methylprednisolone. Contact Dermatitis 1991;25:68-70
6 Bursztejn AC, Tréchot P, Cuny JF, Schmutz JL, Barbaud A. Cutaneous adverse drug reactions during chemotherapy: consider non-antineoplastic drugs. Contact Dermatitis 2008;58:365-368
7 Hotta E, Tamagawa-Mineoka R, Katoh N. Delayed-type hypersensitivity to 6-methyl-prednisolone sodium succinate. J Dermatol 2014;41:754-755
8 Kuga A, Futamura N, Funakawa I, Jinnai K. Allergic skin rashes by methylprednisolone in a case with multiple sclerosis. Rinsho Shinkeigaku 2004;44:691-694 (Article in Japanese)
9 Mussot-Chia C, Flechet ML, Napolitano M, Herson S, Frances C, Chosidow O. Methylprednisolone-induced acute generalized exanthematous pustulosis. Ann Dermatol Venereol 2001;128(3Pt.1):241-243 (Article in French)
10 Murata Y, Kumano K, Ueda T, Araki N, Nakamura T, Tani N. Systemic contact dermatitis caused by systemic corticosteroid use. Arch Dermatol 1997;133:1053-1054
11 Räsänen L, Hasan T. Allergy to systemic and intralesional corticosteroids. Br J Dermatol 1993;128:407-411
12 Zedlitz S, Ahlbach S, Kaufmann R, Boehncke WH. Tolerance to a group C corticosteroid systemically in a patient with delayed-type hypersensitivity to group A systemic corticosteroids. Contact Dermatitis 2002;47:242

Chapter 3.316 METOPROLOL

IDENTIFICATION

Description/definition : Metoprolol is the secondary alcohol that conforms to the structural formula shown below
Pharmacological classes : β_1-Adrenergic receptor antagonists; anti-arrhythmia agents; sympatholytics; antihypertensive agents
IUPAC name : 1-[4-(2-Methoxyethyl)phenoxy]-3-(propan-2-ylamino)propan-2-ol
CAS registry number : 51384-51-1
EC number : 253-483-7
Merck Index monograph : 7498
Patch testing : Tablet, pulverized, 10% pet. (2); most systemic drugs can be tested at 10% pet.; if the pure chemical is not available, prepare the test material from intravenous powder, the content of capsules or – when also not available – from powdered tablets to achieve a final concentration of the active drug of 10% pet.
Molecular formula : $C_{15}H_{25}NO_3$

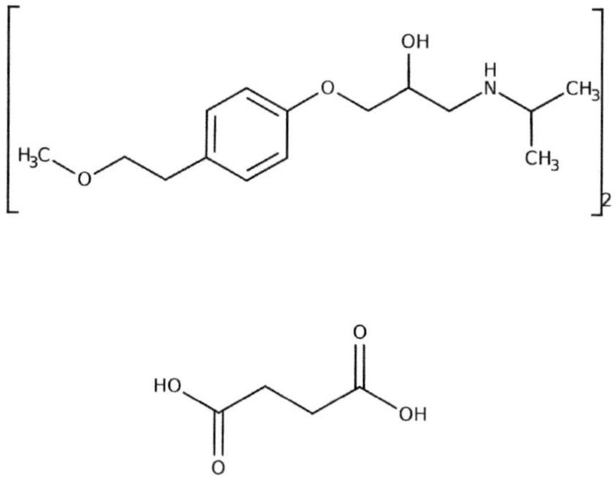

Metoprolol succinate

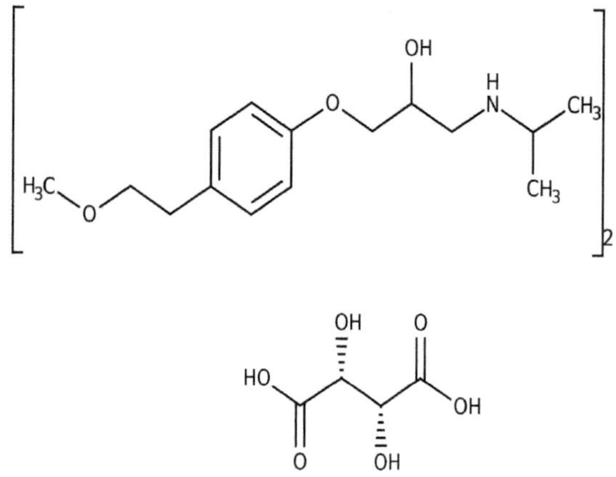

Metoprolol tartrate

GENERAL

Metoprolol is a cardioselective competitive β_1-adrenergic receptor antagonist with antihypertensive properties. This agent antagonizes β_1-adrenergic receptors in the myocardium, thereby reducing the rate and force of myocardial contraction, leading to a reduction in cardiac output. Metoprolol is indicated for the treatment of angina, heart failure, myocardial infarction, atrial fibrillation, atrial flutter and hypertension. It is (or probably was) also used in eye drops, which caused a few cases of allergic contact dermatitis (4). In pharmaceutical products, metoprolol is most often employed as metoprolol succinate (CAS number 98418-47-4, EC number not available, molecular formula $C_{34}H_{56}N_2O_{10}$) or as metoprolol tartrate (CAS number 56392-17-7, EC number 260-148-9, molecular formula $C_{34}H_{56}N_2O_{12}$) (1).

CONTACT ALLERGY FROM ACCIDENTAL CONTACT

A 24-year-old woman presented with eyelid dermatitis, which had started with localized edema 4 months previously. Later, the area had become itchier, with redness and scaling. The patient suspected a relationship with her work as a pharmacy assistant, which involved breaking and crushing different types of tablets. She was patch tested with the crushed tablets that she had contact with at 10% pet. and showed positive reactions to metoprolol, 3 other beta-blockers, 3 benzodiazepines and 3 ACE-inhibitors. Seven controls were negative to metoprolol 10% pet. Cosmetic allergy was excluded and a diagnosis of occupational airborne allergic contact was made (2).

CUTANEOUS ADVERSE DRUG REACTIONS FROM SYSTEMIC ADMINISTRATION CAUSED BY TYPE IV
(DELAYED-TYPE) HYPERSENSITIVITY (as demonstrated by positive patch tests)

Cutaneous adverse drug reactions from systemic administration of metoprolol caused by type IV (delayed-type) hypersensitivity have included a psoriasiform dermatitis (5).

Other cutaneous adverse drug reactions

A 61-year-old individual had used metoprolol for 12 months when he developed a psoriasiform dermatitis localized on the arms, legs, head, and trunk. Histopathological examination showed a picture of a drug eruption (toxicoderma) and eczema. After withdrawal of metoprolol the lesions slowly disappeared. After healing a provocation test with metoprolol 2 x 50 mg per day was performed. After 6 days widespread skin eruptions reappeared. A positive patch test with metoprolol tartrate was found with a cross-reaction to alprenolol (5).

Cross-reactions, pseudo-cross-reactions and co-reactions

Of 7 patients sensitized to alprenolol, 2 (28%) reacted to metoprolol (3). Three patients sensitized to metoprolol also had positive skin tests to propranolol 1% water, practolol 20% water, and timolol 0.5% water, one each (4). One individual allergic to metoprolol cross-reacted to alprenolol (5).

LITERATURE

1 The data in the section 'General' may have been obtained from literature discussed in this chapter, but mostly also or exclusively from one or more of the following online sources: ChemIDPlus Advanced, PubChem, DrugBank, RxList, Drug Central, Drugs.com, and Wikipedia
2 Swinnen I, Ghys K, Kerre S, Constandt L, Goossens A. Occupational airborne contact dermatitis from benzodiazepines and other drugs. Contact Dermatitis 2014;70:227-232
3 Ekenvall L, Forsbeck M. Contact eczema produced by a beta-adrenergic blocking agent (alprenolol). Contact Dermatitis 1978;4:190-194
4 Van Joost T, Middelkamp Hup J, Ros FE. Dermatitis as a side effect of long-term topical treatment with certain beta-blocking agents. Br J Dermatol 1979;101:171-176
5 Neumann HA, Van Joost T, Westerhof W. Dermatitis as side-effect of long-term metoprolol. Lancet 1979;2(8145):745

Chapter 3.317 METRONIDAZOLE

IDENTIFICATION

Description/definition : Metronidazole is a synthetic nitroimidazole derivative that conforms to the structural formula shown below

Pharmacological classes : Anti-infective agents; anti-bacterial agents; antiprotozoal agents

IUPAC name : 2-(2-Methyl-5-nitroimidazol-1-yl)ethanol

CAS registry number : 443-48-1

EC number : 207-136-1

Merck Index monograph : 7506

Patch testing : 1% pet. (SmartPracticeCanada, SmartPracticeEurope)

Molecular formula : $C_6H_9N_3O_3$

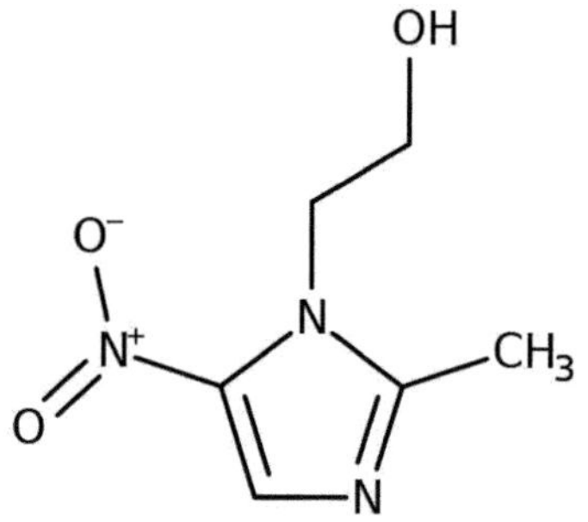

GENERAL

Metronidazole is a synthetic nitroimidazole derivative with antiprotozoal and antibacterial activities; it is extremely effective against anaerobic bacterial infections. Metronidazole is indicated for the treatment of anaerobic infections and mixed infections, surgical prophylaxis requiring anaerobic coverage, *Clostridium difficile*-associated diarrhea and colitis, *Helicobacter pylori* infection and duodenal ulcer disease, bacterial vaginosis, *Giardia lamblia* gastro-enteritis, amebiasis caused by *Entamoeba histolytica*, and *Trichomonas* infections. In topical formulations, it is used in the treatment of rosacea (1). In topical preparations, metronidazole has caused a limited number of cases of contact allergy/allergic contact dermatitis, which has been fully reviewed in Volume 3 of the *Monographs in contact allergy* series (11).

CUTANEOUS ADVERSE DRUG REACTIONS FROM SYSTEMIC ADMINISTRATION CAUSED BY TYPE IV (DELAYED-TYPE) HYPERSENSITIVITY (as demonstrated by positive patch tests)

Cutaneous adverse drug reactions from systemic administration of metronidazole caused by type IV (delayed-type) hypersensitivity have included maculopapular eruption (12), acute generalized exanthematous pustulosis (AGEP) (6,21), acute localized exanthematous pustulosis (ALEP) (13), systemic allergic dermatitis (10), symmetrical drug-related intertriginous and flexural exanthema (SDRIFE)/baboon syndrome (26), fixed drug eruption (7,8,9,16,17), TEN (toxic epidermal necrolysis) (15), Stevens-Johnson syndrome (20), linear IgA bullous dermatosis (18, also published in ref. 24), allergic contact dermatitis with a SDRIFE phenotype (25), erythematous, partly urticarial exanthema with bullae (28), bullous exanthema/erythema multiforme (27) and an unknown drug eruption (14).

Case series with various or unknown types of drug reactions

In Finland, in the period 1989-2001, 826 patients with suspected cutaneous drug eruptions were patch tested and 89 had one or more positive reactions. Of these individuals, one reacted to metronidazole. Clinical nor patch testing details were provided, and it was not specified which drug eruption the patient had suffered from (14).

Maculopapular eruption

A 45-year-old developed labial edema and widespread itchy erythematous maculopapular rash some 10 hours after the third dose of oral metronidazole prescribed for gastrointestinal dysbiosis. Patch tests were positive to metronidazole powdered tablets 0.5%. 5% and 10% pet. and 125 mg/ml saline (12).

Acute generalized exanthematous pustulosis (AGEP)

A 41-year-old man developed diffuse itching followed by a rash and fever without chills, starting 2 days after taking metronidazole 250 mg 3 times daily for a tooth extraction. Five days later, after having continued treatment, his temperature was 38.5°C. He had confluent erythema studded with dozens of 1-mm pustules over the head and neck, erythematous patches with scattered pustules over the trunk and arms, and urticarial and target lesions on the legs. A bacterial culture from a pustule was negative and a complete blood cell count showed marked leukocytosis (37.4 × 10^9/L). A skin biopsy specimen showed interface dermatitis with focal eosinophils and an intraepidermal pustule. One month later, he had a positive patch test with pustules to commercially available 0.75% metronidazole cream. Metronidazole itself nor the base of the cream were available for patch testing (21; also presented in Abstracts in refs. 22 and 23). However, 6 years later, the patient was investigated again and now had a positive patch test to 1% metronidazole and a positive lymphocyte proliferation response to this antibiotic (6).

Acute localized exanthematous pustulosis (ALEP)
A 78-year-old man, while using oral metronidazole for rosacea, reported the sudden development of a pustular eruption. Physical examination revealed the presence of multiple minuscule, non-follicular pustules of the face on an erythematous, edematous background. A skin biopsy revealed the presence of spongiosis and acanthosis of the epidermis with subcorneal intra-epidermal pustules and a dense perivascular dermal infiltrate of lymphocytes and neutrophils. This confirmed the clinical diagnosis of acute localized exanthematous pustulosis (ALEP). A patch test with metronidazole was positive. Metronidazole was discontinued, and the patient was treated with systemic methylprednisolone. The pustules rapidly resolved within 5 days, and the residual erythema was reduced after one week (13).

Systemic allergic dermatitis (systemic contact dermatitis)
A 45-year-old woman was patch tested for recalcitrant facial rosacea and had a positive reaction to metronidazole 1% pet. One year later, three days after having started oral metronidazole for the first time for a gastrointestinal infection, the patient developed an exudative dermatitis on her cheeks and a maculopapular, erythematous and edematous eruption on her flexures, chest and legs (10). This was a case of systemic allergic dermatitis presenting as exacerbation of previous dermatitis (face) and a maculopapular eruption.

Symmetrical drug-related intertriginous and flexural exanthema (SDRIFE)/Baboon syndrome
A 48-year-old woman, one day after administration of metronidazole vaginal suppositories, developed an itchy and burning, sharply defined, symmetrical, erythematous rash affecting the axillae, groins, and buttocks without systemic symptoms. After a day, the lesions extended to the trunk and inner side of the upper and legs without mucosal lesions. Laboratory investigations showed only a mild leukocytosis. Patch tests with the suppository and its ingredients metronidazole and polysorbate 80 were positive to the suppository and metronidazole 1% and 5% on D2 (+) and D4 (++). Five controls were negative (26).

Fixed drug eruption
A 35-year-old woman presented with 5 erythematous, very itchy oval patches, 4 to 6 cm in diameter, localized on her neck, trunk, and arms, that had appeared over the last 2 days. Some lesions showed the beginning of blistering and acquired an intense erythematous, nearly purpuric color. Three of the 5 macules appeared over a hyperpigmented area from a previous episode secondary to oral metronidazole intake. The patient had been using intravaginal nystatin plus metronidazole ovules for treatment of bacterial vaginosis for the prior 2 days. An open epicutaneous application was performed with metronidazole 0.75% gel to one of the pigmented lesions skin. Three minutes later, the hyperpigmented macule became erythematous, and pruritus appeared without blistering. No reaction was seen on the normal skin and the same test with nystatin ointment was negative (9).

A 31-year-old woman presented with 6 hyperpigmented, round, 3x4 cm macules on her abdomen and back, that had appeared over the last year. The lesions began as itchy and erythematous and gradually became pigmented. She had been taking several drugs for dental infection, including metronidazole. Patch tests with all drugs on normal skin were negative. Patch tests on previously affected skin were positive to the combination of metronidazole and spiramycin and later to metronidazole 10% water, inducing itch, erythema and vesicles from D1 after application to D4 with a negative reaction to spiramycin (7). A 34-year-old woman had developed lesions highly suspicious of fixed drug eruption after oral and later intravaginal metronidazole application. A patch test with tinidazole 10% DMSO was negative on normal skin but positive on postlesional skin. Metronidazole itself was not tested (16).

Other cases of FDE from metronidazole with positive patch tests on postlesional skin can be found in refs. 8 and 17.

Stevens-Johnson syndrome/toxic epidermal necrolysis (SJS/TEN)
In a 61-year-old man, delayed-type hypersensitivity to metronidazole (tested as commercial tablet 400 mg, 15% and 30% pet.) and other drugs may have caused / contributed to TEN (toxic epidermal necrolysis) following DRESS (drug reaction with eosinophilia and systemic symptoms) (15).

A 43-year-old woman developed a rash one day after using amoxicillin and metronidazole for gingivitis. The next day she developed   bullous lesions and erosions on lips, back, submammary and axillary areas, and groin, accompanied by generalized erythema and edema of the distal extremities. Clinical and histopathological findings were consistent with Stevens-Johnson syndrome. One week after resolution of the lesions, patch tested were performed to all drugs used and there were vesicular (++) positive reactions to metronidazole tablets dissolved 50% in water and petrolatum (20).

Other cutaneous adverse drug reactions

An 80-year-old woman developed a drug-induced linear IgA bullous dermatosis (LABD) during treatment with ceftriaxone and metronidazole. Sensitization to both drugs was detected by epicutaneous testing and positive lymphocyte transformation tests. After discontinuation of the drugs and systemic corticosteroids for 8 days the bullous eruptions subsided and erosions healed within 6 weeks. The authors suggested that in addition to IgA antibodies, drug-specific T cells and their subsequent release of cytokines may play an important role in the pathogenesis of drug-induced LABD. As this patient reacted to 2 non-related medicaments, this was a case of multiple drug hypersensitivity syndrome (24).

A 35-year-old woman presented with an itchy eruption after using metronidazole intravaginal ovules. Examination revealed mainly genital-associated erythema and on the abdomen and flanks. The thigh and lower leg exhibited well-demarcated erythema and small vesicles. Patch tests on the upper arm revealed +++ reactions to both metronidazole ovules and oral tablets tested at 30% pet. at D3. The patient was diagnosed with allergic contact dermatitis, characterized by a SDRIFE phenotype. However, it is well known that metronidazole is absorbed from the vaginal mucosa and the spreading to non-genital areas including the lower leg suggests a systemic reaction instead of, or at least complementary to, allergic contact dermatitis. Also, this was not a classical manifestation of SDRIFE (25).

In Bern, Switzerland, patients with a suspected allergic cutaneous drug reaction were patch-scratch tested with suspected drugs that had previously given a positive lymphocyte transformation test. Metronidazole 500 mg/ml saline gave a positive patch-scratch test in one patient with bullous exanthema/erythema multiforme (27).

A patient treated with metronidazole and ceftriaxone developed an erythematous, partly urticarial exanthema, in which later blisters appeared. Patch tests were positive to both metronidazole and ceftriaxone (28).

Cross-reactions, pseudo-cross-reactions and co-reactions

The question has been raised (but not answered) whether contact allergy to metronidazole and other (nitro)imidazoles may be overrepresented, from cross-reactivity or otherwise, in patients allergic to methylchloroisothiazolinone / methylisothiazolinone (MCI/MI) (2,4). Of 5 patients with delayed-type hypersensitivity to benznidazole from oral administration, 2 (40%) cross-reacted to metronidazole; both are nitroimidazoles (3). One patient sensitized to tioconazole may have cross-reacted to metronidazole (5). Cross-reactivity in metronidazole sensitization to tinidazole (16).

LITERATURE

1   The data in the section 'General' may have been obtained from literature discussed in this chapter, but mostly also or exclusively from one or more of the following online sources: ChemIDPlus Advanced, PubChem, DrugBank, RxList, Drug Central, Drugs.com, and Wikipedia

2   Stingeni L, Rigano L, Lionetti N, Bianchi L, Tramontana M, Foti C, et al. Sensitivity to imidazoles/nitroimidazoles in subjects sensitized to methylchloroisothiazolinone/methylisothiazolinone: A simple coincidence? Contact Dermatitis 2019;80:181-183

3   Noguerado-Mellado B, Rojas-Pérez-Ezquerra P, Calderón-Moreno M, Morales-Cabeza C, Tornero-Molina P. Allergy to benznidazole: cross-reactivity with other nitroimidazoles. J Allergy Clin Immunol Pract 2017;5:827-828

4   Wolf R, Orion E, Matz H. Co-existing sensitivity to metronidazole and isothiazolinone. Clin Exp Dermatol 2003;28:506-507

5   Izu R, Aguirre A, Gonzales M, Diaz-Perez J L. Contact dermatitis from tioconazole with cross-sensitivity to other imidazoles. Contact Dermatitis 1992;26:130

6   Girardi M, Duncan KO, Tigelaar RE, Imaeda S, Watsky KL, McNiff JM. Cross-comparison of patch test and lymphocyte proliferation responses in patients with a history of acute generalized exanthematous pustulosis. Am J Dermatopathol 2005;27:343

7   Vila JB, Bernier MA, Gutierrez JV, Gómez MT, Polo AM, Harrison JM, Miranda-Romero A, et al. Fixed drug eruption caused by metronidazole. Contact Dermatitis 2002;46:122

8   Gastaminza G, Anda M, Audicana MT, Fernandez E, Muñoz D. Fixed-drug eruption due to metronidazole with positive topical provocation. Contact Dermatitis 2001;44:36

9   Hermida MD, Consalvo L, Lapadula MM, Della Giovanna P, Cabrera HN. Bullous fixed drug eruption induced by intravaginal metronidazole ovules, with positive topical provocation test findings. Arch Dermatol 2011;147:250-251

10 Mussani F, Skotnicki S. Systemic contact dermatitis: Two Interesting cases of systemic eruptions following exposure to drugs clioquinol and metronidazole. Dermatitis 2013;24(4):e3

11 De Groot AC. Monographs in contact allergy, volume 3: Topical Drugs. Boca Raton, Fl, USA: CRC Press Taylor and Francis Group, 2021 (ISBN 978-0-367-23693-9)

12 Aruanno A, Parrinello G, Buonomo A, Rizzi A, Nucera E. Metronidazole hypersensitivity in a patient with angioedema and widespread rash. J Investig Allergol Clin Immunol 2020;30:371-373

13 Kostaki M, Polydorou D, Adamou E, Chasapi V, Antoniou C, Stratigos A. Acute localized exanthematous pustulosis due to metronidazole. J Eur Acad Dermatol Venereol 2019;33:e109-e111

14 Lammintausta K, Kortekangas-Savolainen O. The usefulness of skin tests to prove drug hypersensitivity. Br J Dermatol 2005;152:968-974

15 Liippo J, Pummi K, Hohenthal U, Lammintausta K. Patch testing and sensitization to multiple drugs. Contact Dermatitis 2013;69:296-302

16 Prieto A, De Barrio M, Infante S, Torres A, Rubio M, Olalde S. Recurrent fixed drug eruption due to metronidazole elicited by patch test with tinidazole. Contact Dermatitis 2005;53:169-170

17 Short KA, Fuller LC, Salisbury JR. Fixed drug eruption following metronidazole therapy and the use of topical provocation testing in diagnosis. Clin Exp Dermatol 2002;27:464-466

18 Yawalkar N, Reimers A, Hari Y, Hunziker Th, Gerber H, Müller U, Pichler W. Drug-induced linear IgA bullous dermatosis associated with ceftriaxone- and metronidazole-specific T cells. Dermatology 1999;199:25-30

19 Liippo J, Pummi K, Hohenthal U, Lammintausta K. Patch testing and sensitization to multiple drugs. Contact Dermatitis 2013;69:296-302

20 Piskin G, Mekkes JR. Stevens-Johnson syndrome from metronidazole. Contact Dermatitis 2006;55:192-193

21 Watsky KL. Acute generalized exanthematous pustulosis induced by metronidazole: the role of patch testing. Arch Dermatol 1999;135:93-94

22 Watsky, Kalman L. Acute generalized exanthematous pustulosis due to metronidazole: The role of patch testing. Dermatitis 1999;10:112

23 Watsky, Kalman L. Acute generalized exanthematous pustulosis due to metronidazole: The role of patch testing. Dermatitis 1998;9:67

24 Gex-Collet C, Helbling A, Pichler WJ. Multiple drug hypersensitivity – proof of multiple drug hypersensitivity by patch and lymphocyte transformation tests. J Investig Allergol Clin Immunol 2005; 15:293-296

25 Kumagai J, Nakamura A, Ogawa S, Washio K. Intravaginal metronidazole ovule-related allergic contact dermatitis. Contact Dermatitis 2021;85:85-86

26 Zhang H, Xie Z. Symmetrical drug-related intertriginous and flexural exanthema induced by metronidazole suppository. Contact Dermatitis 2021;84:486-488

27 Neukomm C, Yawalkar N, Helbling A, Pichler WJ. T-cell reactions to drugs in distinct clinical manifestations of drug allergy. J Invest Allergol Clin Immunol 2001;11:275-284

28 Hari Y, Frutig-Schnyder K, Hurni M, Yawalkar N, Zanni MP, Schnyder B, et al. T cell involvement in cutaneous drug eruptions. Clin Exp Allergy 2001;31:1398-1408

Chapter 3.318 MEXILETINE

IDENTIFICATION

Description/definition : Mexiletine is the phenol ether that conforms to the structural formula shown below
Pharmacological classes : Voltage-gated sodium channel blockers; anti-arrhythmia agents
IUPAC name : 1-(2,6-Dimethylphenoxy)propan-2-amine
CAS registry number : 31828-71-4
EC number : 250-825-7
Merck Index monograph : 7518
Patch testing : Tablets, pulverized, 10% and 20% pet.; most systemic drugs can be tested at 10% pet.; if the pure chemical is not available, prepare the test material from intravenous powder, the content of capsules or – when also not available – from powdered tablets to achieve a final concentration of the active drug of 10% pet.
Molecular formula : $C_{11}H_{17}NO$

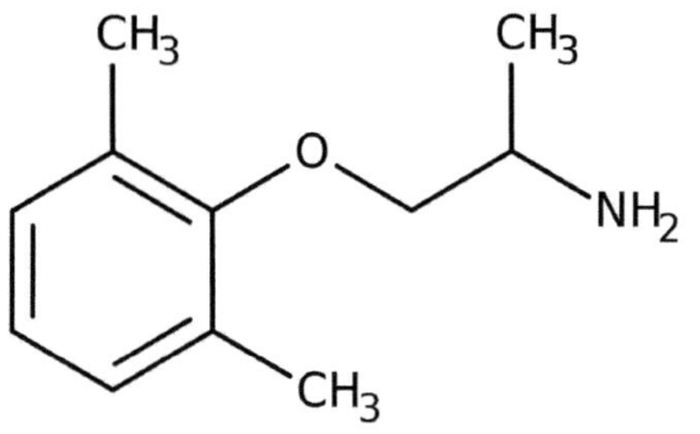

GENERAL

Mexiletine is a local anesthetic and antiarrhythmic agent (Class Ib), structurally similar to lidocaine, but orally active. Mexiletine is indicated for the treatment of ventricular tachycardia, symptomatic premature ventricular beats, and prevention of ventricular fibrillation (1). In pharmaceutical products, mexiletine is employed as mexiletine hydrochloride (CAS number 5370-01-4, EC number 226-362-1, molecular formula $C_{11}H_{18}ClNO$) (1).

CUTANEOUS ADVERSE DRUG REACTIONS FROM SYSTEMIC ADMINISTRATION CAUSED BY TYPE IV (DELAYED-TYPE) HYPERSENSITIVITY (as demonstrated by positive patch tests)

Cutaneous adverse drug reactions from systemic administration of mexiletine caused by type IV (delayed-type) hypersensitivity have included maculopapular eruption (2), generalized erythema (2), acute generalized exanthematous pustulosis (AGEP) (6), drug reaction with eosinophilia and systemic symptoms (DRESS) (3,5,7), and generalized pruritic eruption with papules, infiltrated erythematous patches and pustules (4),

Maculopapular eruption

A 63-year-old woman had an eleven-year history of reflex sympathetic dystrophy and had received 13 drugs, including mexiletine hydrochloride for 6 weeks and amoxapine for 4 weeks. Three days after taking cefaclor for a 39°C fever (from cystitis), she noticed erythematous papules on her back. The lesions gradually spread and increased in number on her trunk, arms and legs within three days. After cessation of most drugs, the skin lesions improved within 2 weeks. Two months later, positive patch test reactions were observed to mexiletine hydrochloride 1% and 10% pet. and amoxapine 0.01%, 0.1%, 1% and 10% pet. Ten controls were negative. Oral challenge with mexiletine hydrochloride (50 mg) provoked generalized erythema (2).

Erythroderma, widespread erythematous eruption, exfoliative dermatitis

See the section 'Maculopapular eruption' above, ref. 2.

Acute generalized exanthematous pustulosis (AGEP)

A 56-year-old man with arrhythmia had been treated with mexiletine hydrochloride for 1 month, when he developed a rash on his body with high fever. When examined, his face was erythematous and edematous and his body was covered with numerous infiltrated erythematous patches. An erythema multiforme-like eruption was observed on the arms and legs and purpura was noted on his lower legs. Tiny pustules were disseminated on the surface of the erythema on his chest, neck, and around the chin. Laboratory data showed leukocytosis with eosinophilia, elevated liver enzymes and elevated C-reactive protein. The culture from a pustule was sterile. A biopsy from one of the pustular lesions on the chest showed a subcorneal abscess containing a number of neutrophils, and perivascular infiltration composed of lymphocytes, monocytes and eosinophils in the upper dermis. One month after the regression of the eruption, patch tests with 10% and 20% mexiletine HCl in petrolatum showed positive reaction without pustules. Three controls were negative (6). The authors also considered the possibility that this was a case of hypersensitivity syndrome (DRESS).

Drug reaction with eosinophilia and systemic symptoms (DRESS)
An 82-yr-old man presented with fever, cough and a generalized erythematous rash. He had taken mexiletine for 5 months for dilated cardiomyopathy and ventricular arrhythmia. Laboratory studies showed peripheral blood eosinophilia and elevated liver transaminase levels. Chest radiographs showed multiple nodular consolidations in both lungs. Biopsies of the lung and skin lesions revealed eosinophilic infiltration. Mexiletine was suspected as the etiologic agent. After discontinuing the drug starting oral prednisolone, the patient improved, and the skin and lung lesions disappeared. Four months after complete remission, patch tests were positive to mexiletine tablets 1%, 2%, 5%, 10% and 20% in petrolatum (3).

Previously, DRESS from mexiletine confirmed by a positive patch test had been observed in 2 men from Japan (5,7). Both had fever, peripheral blood eosinophilia and liver dysfunction. The skin manifestations were morbilliform exanthema/maculopapular exanthema. One patient (7) also had lymphadenopathy, lymphocytosis and 15% atypical lymphocytes. In neither report were patch test concentrations and vehicles mentioned ('the patch test with mexiletine') and adequate controls to rule out false-positive reactions were not performed (5,7).

Other cutaneous adverse drug reactions
A 77-year-old man developed a generalized pruritic eruption 45 days after starting mexiletine hydrochloride 150 mg daily for a ventricular arrhythmia. Examination showed numerous papules and infiltrated erythematous patches over the whole body except for the head and neck. On the flank, several fine pustules were scattered on large confluent erythematous plaques. After regression of the eruption, patch tests were performed with 6.25%, 12.5% and 25% mexiletine hydrochloride in pet. to both previously involved skin on the flank and uninvolved skin on the back. Definitely positive reactions were observed only at the patch test sites on previously involved skin (6.25% +, 12.5% and 25% ++). When similar patch tests were performed on previously uninvolved skin after scarification with a 27 gauge needle, positive reactions were elicited with the 2 higher concentrations (+) (4).

LITERATURE
1   The data in the section 'General' may have been obtained from literature discussed in this chapter, but mostly also or exclusively from one or more of the following online sources: ChemIDPlus Advanced, PubChem, DrugBank, RxList, Drug Central, Drugs.com, and Wikipedia
2   Nagayama H, Nakamura Y, Shinkai H. A case of drug eruption due to simultaneous sensitization with three different kinds of drugs. J Dermatol 1996;23:899-901
3   Lee SP, Kim SH, Kim TH, Sohn JW, Shin DH, Park SS, Yoon HJ. A case of mexiletine-induced hypersensitivity syndrome presenting as eosinophilic pneumonia. J Korean Med Sci 2010;25:148-151
4   Kikuchi K, Tsunoda T, Tagami H. Generalized drug eruption due to mexiletine hydrochloride: topical provocation on previously involved skin. Contact Dermatitis 1991;25:70-72
5   Higa K, Hirata K, Dan K. Mexiletine-induced severe skin eruption, fever, eosinophilia, atypical lymphocytosis, and liver dysfunction. Pain 1997;73:97-99
6   Sasaki K, Yamamoto T, Kishi M, Yokozeki H, Nishioka K. Acute exanthematous pustular drug eruption induced by mexiletine. Eur J Dermatol 2001;11:469-471
7   Yagami A, Yoshikawa T, Asano Y, Koie S, Shiohara T, Matsunaga K. Drug-induced hypersensitivity syndrome due to mexiletine hydrochloride associated with reactivation of human herpesvirus 7. Dermatology 2006;213:341-344

Chapter 3.319 MEZLOCILLIN

IDENTIFICATION

Description/definition : Mezlocillin is a semisynthetic ampicillin-derived β-lactam penicillin that conforms to the structural formula shown below

Pharmacological classes : Anti-bacterial agents

IUPAC name : (2S,5R,6R)-3,3-Dimethyl-6-[[(2R)-2-[(3-methylsulfonyl-2-oxoimidazolidine-1-carbonyl)-amino]-2-phenylacetyl]amino]-7-oxo-4-thia-1-azabicyclo[3.2.0]heptane-2-carboxylic acid

CAS registry number : 51481-65-3

EC number : 257-233-8

Merck Index monograph : 7520

Patch testing : 5% pet.

Molecular formula : $C_{21}H_{25}N_5O_8S_2$

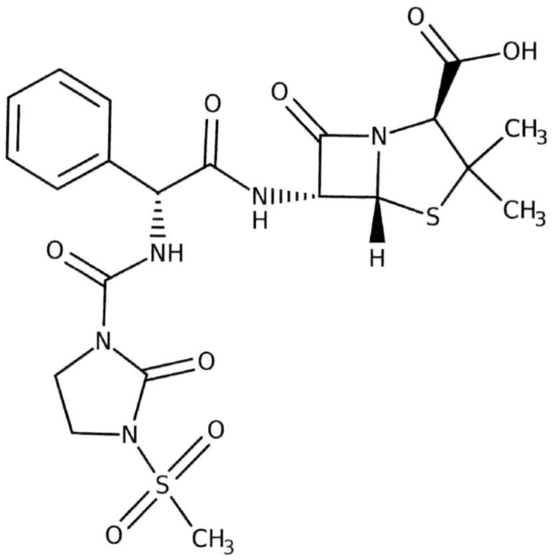

GENERAL

Mezlocillin is a broad-spectrum, semisynthetic, ampicillin-derived, β-lactam penicillin antibiotic with antibacterial activity. It is indicated to treat serious gram-negative infections of the lungs, urinary tract, and skin. In pharmaceutical products, mezlocillin is employed as mezlocillin sodium monohydrate (CAS number 80495-46-1, EC number not available, molecular formula $C_{21}H_{26}N_5NaO_9S_2$) (1). The classification and structures of beta-lactam antibiotics are discussed in Chapter 3.36 Amoxicillin.

CONTACT ALLERGY FROM ACCIDENTAL CONTACT

A 32-year-old nurse developed atopic dermatitis of the flexures and hands. Additionally, she noted the occurrence of wheals and dyspnea after local contact with mezlocillin infusion during work. Because of progressive symptoms, the patient had to change her occupation. After that, her atopic dermatitis cleared almost completely and there was no recurrence of immediate-type reactions. An open patch test with mezlocillin 1% solution led to a local urticarial reaction after 10 minutes. At D2 and D3, the application site showed an eczematous reaction. Specific IgE was found for penicillins G (benzylpenicillin) and V (4).

Cross-reactions, pseudo-cross-reactions and co-reactions

Cross-reactions between beta-lactam antibiotics are discussed in Chapter 3.36 Amoxicillin. In a group of 78 patients who had suffered nonimmediate drug reactions from any beta-lactam compound (penicillins, cephalosporins), 11 had positive patch tests to mezlocillin. As this was not the culprit drug causing the adverse skin reaction in any patient, all these reactions were presumably cross-reactions (3).

A patient who developed a maculopapular exanthema from piperacillin with a positive patch test to this antibiotic cross-reacted to mezlocillin 5% pet. Lymphocyte transformation tests were positive to both antibiotics. There were no positive reactions to other penicillins or cephalosporins and therefore it was suggested that the ureido group present in both piperacillin and mezlocillin side chains may have played an important role in the sensitization (2).

Immediate contact reactions

Immediate contact reactions (contact urticaria) to mezlocillin are presented in Chapter 5.

LITERATURE

1   The data in the section 'General' may have been obtained from literature discussed in this chapter, but mostly also or exclusively from one or more of the following online sources: ChemIDPlus Advanced, PubChem, DrugBank, RxList, Drug Central, Drugs.com, and Wikipedia

2   Gaeta F, Alonzi C, Valluzzi RL, Viola M, Romano A. Delayed hypersensitivity to acylureidopenicillins: a case report. Allergy 2008;63:787-789

3   Buonomo A, Nucera E, De Pasquale T, Pecora V, Lombardo C, Sabato V, et al. Tolerability of aztreonam in patients with cell-mediated allergy to β-lactams. Int Arch Allergy Immunol 2011;155:155-159

4   Keller K, Schwanitz HJ. Combined immediate and delayed hypersensitivity to mezlocillin. Contact Dermatitis 1992;27:348-349

Chapter 3.320 MICONAZOLE

IDENTIFICATION

Description/definition : Miconazole is a synthetic phenethyl imidazole that conforms to the structural formula shown below

Pharmacological classes : Antifungal agents; 14α-demethylase inhibitors; cytochrome P-450 CYP2C9 inhibitors; cytochrome P-450 CYP3A inhibitors

IUPAC name : 1-[2-(2,4-Dichlorophenyl)-2-[(2,4-dichlorophenyl)methoxy]ethyl]imidazole

CAS registry number : 22916-47-8

EC number : 245-324-5

Merck Index monograph : 7527

Patch testing : 1% alc. (Chemotechnique)

Molecular formula : $C_{18}H_{14}Cl_4N_2O$

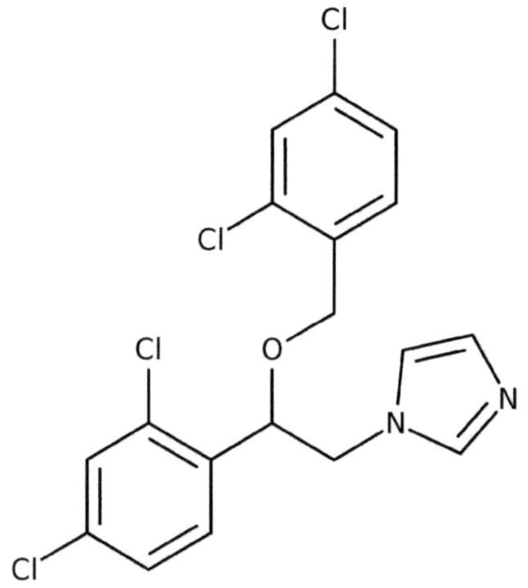

GENERAL

Miconazole is a synthetic phenethyl imidazole antifungal agent that can be used topically and by intravenous infusion. This agent selectively affects the integrity of fungal cell membranes, high in ergosterol content and different in composition from mammalian cell membranes. Miconazole is indicated for topical application in the treatment of tinea pedis, tinea cruris, and tinea corporis caused by *Trichophyton rubrum*, *Trichophyton mentagrophytes*, and *Epidermophyton floccosum*, in the treatment of cutaneous candidiasis and in the treatment of pityriasis (tinea) versicolor. In pharmaceutical products, miconazole is nearly always employed as miconazole nitrate (CAS number 22832-87-7, EC number 245-256-6, molecular formula $C_{18}H_{15}Cl_4N_3O_4$); buccal tablets may contain miconazole base (1).

In topical preparations, miconazole has caused contact allergy/allergic contact dermatitis, which subject has been fully reviewed in Volume 3 of the *Monographs in contact allergy* series (7).

CUTANEOUS ADVERSE DRUG REACTIONS FROM SYSTEMIC ADMINISTRATION CAUSED BY TYPE IV (DELAYED-TYPE) HYPERSENSITIVITY (as demonstrated by positive patch tests)

Cutaneous adverse drug reactions from systemic administration of miconazole caused by type IV (delayed-type) hypersensitivity have included AGEP/DRESS overlap (8) and (possibly) systemic allergic dermatitis (6).

Acute generalized exanthematous pustulosis (AGEP)

A 50-year-old woman had been taking miconazole oral gel for oral candidiasis for 2 days, when a rash developed, fever, cough, and edema of the legs. Examination showed a diffuse erythematous eruption involving about 90% of her body surface area, with numerous non-follicular pustules mostly localized in intertriginous regions. Laboratory tests revealed leukocytosis with neutrophilia and eosinophilia and elevated liver enzymes. Histopathology was consistent with a diagnosis of AGEP. Patch tests were positive to miconazole oral gel 30% pet. (+) on D3. Histopathology of the positive reaction showed spongiotic dermatitis consistent with allergic contact dermatitis. A diagnosis of AGEP/DRESS overlap induced by miconazole oral gel was made (8).

Systemic allergic dermatitis (systemic contact dermatitis)
A 35-year-old man with oral candidiasis was treated with miconazole oral solution. 7 days later, he presented with a generalized, itchy, maculopapular eruption. Patch tests were positive to the solution (as is) and miconazole 2% pet. (6). The patient was diagnosed with systemic allergic dermatitis. However, it was not mentioned whether the patient had used topical imidazoles before and could thus have become sensitized to miconazole or a potentially cross-reacting imidazole.

Drug reaction with eosinophilia and systemic symptoms (DRESS)
See the section 'Acute generalized exanthematous pustulosis (AGEP)' above, ref. 8.

Cross-reactions, pseudo-cross-reactions and co-reactions
Statistically significant associations have been found in patient data between miconazole, econazole, and isoconazole; between sulconazole, miconazole, and econazole (3,5). Possible cross-reactivity from miconazole to ketoconazole (2). A patient sensitized to sertaconazole, who had never used antifungal preparations before, cross-reacted to miconazole and econazole (4).The question has been raised (but not answered) whether contact allergy to miconazole and other (nitro)imidazoles may be overrepresented in patients allergic to methylchloroisothiazolinone/methylisothiazolinone (MCI/MI (2).

LITERATURE

1   The data in the section 'General' may have been obtained from literature discussed in this chapter, but mostly also or exclusively from one or more of the following online sources: ChemIDPlus Advanced, PubChem, DrugBank, RxList, Drug Central, Drugs.com, and Wikipedia
2   Stingeni L, Rigano L, Lionetti N, Bianchi L, Tramontana M, Foti C, et al. Sensitivity to imidazoles/nitroimidazoles in subjects sensitized to methylchloroisothiazolinone/methylisothiazolinone: A simple coincidence? Contact Dermatitis 2019;80:181-183
3   Dooms-Goossens A, Matura M, Drieghe J, Degreef H. Contact allergy to imidazoles used as antimycotic agents. Contact Dermatitis 1995;33:73-77
4   Goday JJ, Yanguas I, Aguirre A, Ilardia R, Soloeta R. Allergic contact dermatitis from sertaconazole with cross-sensitivity to miconazole and econazole. Contact Dermatitis 1995;32:370-371
5   Bianchi L, Hansel K, Antonelli E, Bellini V, Stingeni L. Contact allergy to isoconazole nitrate with unusual spreading over extensive regions. Contact Dermatitis 2017;76:243-245
6   Fernandez L, Maquiera E, Rodriguez F, Picans I, Duque S. Systemic contact dermatitis from miconazole. Contact Dermatitis 1996;34:217
7   De Groot AC. Monographs in contact allergy, volume 3: Topical Drugs. Boca Raton, Fl, USA: CRC Press Taylor and Francis Group, 2021 (ISBN 978-0-367-23693-9)
8   Tedbirt B, Viart-Commin MH, Carvalho P, Courville P, Tétart F. Severe acute generalized exanthematous pustulosis (AGEP) induced by miconazole oral gel with overlapping features of drug reaction with eosinophilia and systemic symptoms (DRESS). Contact Dermatitis 2021;84:474-476

Chapter 3.321 MIDECAMYCIN

IDENTIFICATION

Description/definition : Midecamycin is a naturally occurring 16-membered macrolide antibiotic taht conforms to the structural formula shown below
Pharmacological classes : Anti-bacterial agents
IUPAC name : [(4R,5S,6S,7R,9R,10R,11E,13E,16R)-6-[(2S,3R,4R,5S,6R)-4-(Dimethylamino)-3-hydroxy-5-[(2S,4R,5S,6S)-4-hydroxy-4,6-dimethyl-5-propanoyloxyoxan-2-yl]oxy-6-methyloxan-2-yl]oxy-10-hydroxy-5-methoxy-9,16-dimethyl-2-oxo-7-(2-oxoethyl)-1-oxacyclohexadeca-11,13-dien-4-yl] propanoate
CAS registry number : 35457-80-8
EC number : 252-578-0
Merck Index monograph : 7532 (Midecamycins)
Patch testing : Pure; suggested: pure, 30%, 10% and 3% pet.
Molecular formula : $C_{41}H_{67}NO_{15}$

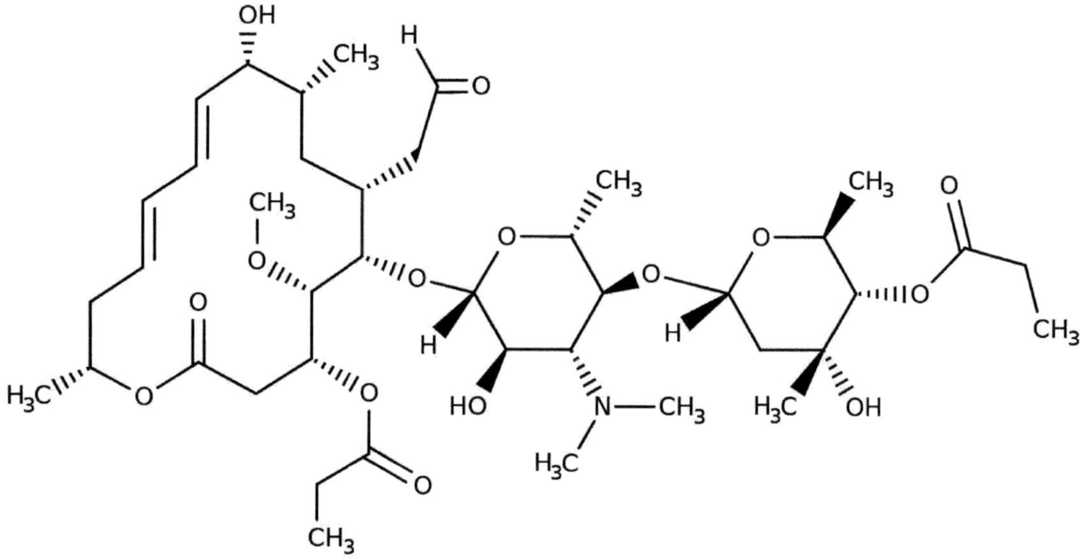

GENERAL

Midecamycin is a macrolide antibiotic with actions and uses similar to those of erythromycin but somewhat less active. It is synthesized from *Streptomyces mycarofaciens*. Midecamycin is used for the treatment of infections in the oral cavity, upper and lower respiratory tracts and skin and soft tissue infections. In pharmaceutical products, both midecamycin and midecamycin (di)acetate (CAS number 55881-07-7, EC number 259-879-6, molecular formula $C_{45}H_{71}NO_{17}$) may be employed (1).

CONTACT ALLERGY FROM ACCIDENTAL CONTACT

A 22-year-old female student in industrial engineering was conducting research for her thesis on the purification of kitasamycin. Her task was to develop a method of refining pure kitasamycin from a powder consisting of some 10 components, including midecamycin. To do this, she sometimes worked with large quantities of powder, of which there would occasionally be dust in the air. Two weeks after she began her work, she suddenly erupted in a burning eczema on the face, followed by the neck, scalp, and retro-auricular region, which she attributed to the powders she was working with. Patch tests revealed positive reactions to kitasamycin tartrate powder 'as is' (D2 ++, D3 ++), kitasamycin tartrate 4% water (D2 +, D3 not mentioned) and midecamycin 'as is' (D2 ++, D3 ++). There was a striking flare-up reaction on the face during patch testing. The patient was diagnosed with occupational airborne allergic contact dermatitis from kitasamycin and midecamycin (2).

LITERATURE

1   The data in the section 'General' may have been obtained from literature discussed in this chapter, but mostly also or exclusively from one or more of the following online sources: ChemIDPlus Advanced, PubChem, DrugBank, RxList, Drug Central, Drugs.com, and Wikipedia
2   Dooms-Goossens A, Bedert R, Degreef H, Vandaele M. Airborne allergic contact dermatitis from kitasamycin and midecamycin. Contact Dermatitis 1990;23:118-119

Chapter 3.322 MINOCYCLINE

IDENTIFICATION

Description/definition   : Minocycline is the derivative of tetracycline that conforms to the structural formula
                             shown below
Pharmacological classes   : Anti-bacterial agents
IUPAC name   : (4S,4aS,5aR,12aR)-4,7-bis(Dimethylamino)-1,10,11,12a-tetrahydroxy-3,12-dioxo-
                             4a,5,5a,6-tetrahydro-4H-tetracene-2-carboxamide
CAS registry number   : 10118-90-8
EC number   : Not available
Merck Index monograph   : 7553
Patch testing   : Hydrochloride 10.0% pet. (Chemotechnique)
Molecular formula   : $C_{23}H_{27}N_3O_7$

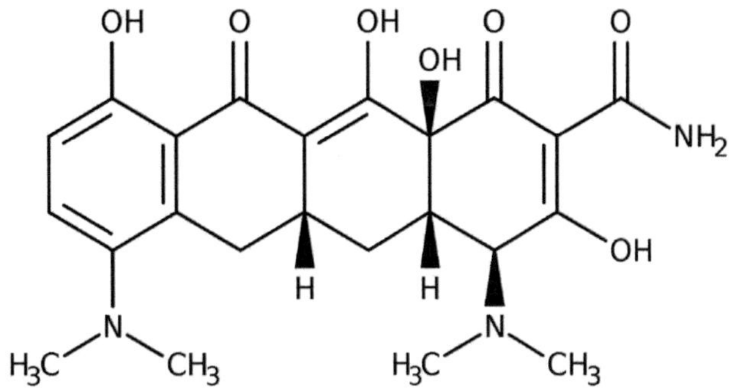

GENERAL

Minocycline is a broad-spectrum long-acting derivative of the antibiotic tetracycline, with antibacterial and anti-inflammatory activities. It is effective against tetracycline-resistant *Staphylococcus aureus*. Minocycline is indicated for the treatment of infections caused by susceptible strains of microorganisms, such as Rocky Mountain spotted fever, typhus fever and the typhus group, Q fever, rickettsial pox and tick fevers caused by *Rickettsiae*, upper respiratory tract infections caused by *Streptococcus pneumoniae* and for the treatment of asymptomatic carriers of *Neisseria meningitidis*. It is also widely used in the treatment of acne (1). In pharmaceutical products, minocycline is employed as minocycline hydrochloride (CAS number 13614-98-7, EC number 237-099-7, molecular formula $C_{23}H_{28}ClN_3O_7$) (1).

CUTANEOUS ADVERSE DRUG REACTIONS FROM SYSTEMIC ADMINISTRATION CAUSED BY TYPE IV (DELAYED-TYPE) HYPERSENSITIVITY (as demonstrated by positive patch tests)

Cutaneous adverse drug reactions from systemic administration of minocycline caused by type IV (delayed-type) hypersensitivity have included acute generalized exanthematous pustulosis (AGEP) (2).

Acute generalized exanthematous pustulosis (AGEP)

A 52-year-old woman with generalized pustular psoriasis, well-controlled by etretinate, took 100 mg minocycline for dermatitis perioralis. Three hours later, erythroderma with numerous superficial pustules developed over her body. She had previously suffered a similar rash which had started after having taken minocycline for 1 week. After stopping the antibiotic, resolution of the skin rash was observed after 10 days. Later, a patch test with 10% minocycline (vehicle?) was positive but negative in 'the controls'. A lymphocyte stimulation test with minocycline was also positive (2). This was a case of acute generalized exanthematous pustulosis (AGEP) caused by minocycline.

LITERATURE

1   The data in the section 'General' may have been obtained from literature discussed in this chapter, but mostly
     also or exclusively from one or more of the following online sources: ChemIDPlus Advanced, PubChem,
     DrugBank, RxList, Drug Central, Drugs.com, and Wikipedia
2   Yamamoto T, Minatohara K. Minocycline-induced acute generalized exanthematous pustulosis in a patient with
     generalized pustular psoriasis showing elevated level of sELAM-1. Acta Derm Venereol 1997;77:168-169

Chapter 3.323 MIRTAZAPINE

IDENTIFICATION

Description/definition	: Mirtazapine is the piperazino-azepine that conforms to the structural formula shown below
Pharmacological classes	: Adrenergic alpha-2 receptor antagonists; histamine H1 antagonists; serotonin 5-HT3 receptor antagonists; anti-anxiety agents; serotonin 5-HT2 receptor antagonists; antidepressive agents
IUPAC name	: 5-Methyl-2,5,19-triazatetracyclo[13.4.0.0^{2,7}.0^{8,13}]nonadeca-1(15),8,10,12,16,18-hexaene
CAS registry number	: 85650-52-8
EC number	: 288-060-6
Merck Index monograph	: 7561
Patch testing	: No data available; most systemic drugs can be tested at 10% pet.; if the pure chemical is not available, prepare the test material from intravenous powder, the content of capsules or – when also not available – from powdered tablets to achieve a final concentration of the active drug of 10% pet.
Molecular formula	: C$_{17}$H$_{19}$N$_3$

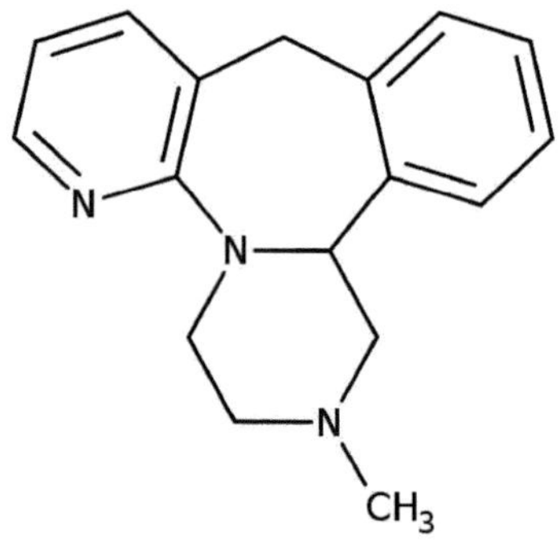

GENERAL

Mirtazapine is a synthetic tetracyclic derivative of the piperazino-azepines with antidepressant activity. Although its mechanism of action is unknown, mirtazapine enhances central adrenergic and serotonergic transmission. This drug is indicated for the treatment of major depressive disorder and its associated symptoms. In pharmaceutical products, mirtazapine may be employed as mirtazapine hemihydrate (CAS number 341512-89-8, EC number not available, molecular formula C$_{34}$H$_{40}$N$_6$O) or as mirtazapine hydrochloride (CAS number 207516-99-2, EC number 288-060-6, molecular formula C$_{17}$H$_{20}$ClN$_3$) (1).

CUTANEOUS ADVERSE DRUG REACTIONS FROM SYSTEMIC ADMINISTRATION CAUSED BY TYPE IV (DELAYED-TYPE) HYPERSENSITIVITY (as demonstrated by positive patch tests)

Cutaneous adverse drug reactions from systemic administration of mirtazapine caused by type IV (delayed-type) hypersensitivity have included maculopapular eruption (2).

Maculopapular eruption

A patient who had previously suffered an anaphylactic reaction to erythromycin (and had a positive intradermal test to the antibiotic), later developed a maculopapular eruption from mirtazapine and had a positive patch test to this drug. As the patient had cutaneous adverse reactions to various unrelated drugs, this was a case of multiple drug hypersensitivity syndrome (2).

LITERATURE

1   The data in the section 'General' may have been obtained from literature discussed in this chapter, but mostly also or exclusively from one or more of the following online sources: ChemIDPlus Advanced, PubChem, DrugBank, RxList, Drug Central, Drugs.com, and Wikipedia
2   Jörg L, Yerly D, Helbling A, Pichler W. The role of drug, dose and the tolerance/intolerance of new drugs in multiple drug hypersensitivity syndrome (MDH). Allergy 2020;75:1178-1187

Chapter 3.324 MISOPROSTOL

IDENTIFICATION

Description/definition   : Misoprostol is the prostaglandin-related compound that conforms to the structural
                            formula shown below
Pharmacological classes : Oxytocics; abortifacient agents, nonsteroidal; anti-ulcer agents
IUPAC name   : Methyl 7-[(1R,2R,3R)-3-hydroxy-2-[(E)-4-hydroxy-4-methyloct-1-enyl]-5-oxocyclo-
                            pentyl]heptanoate
CAS registry number   : 59122-46-2
EC number   : Not available
Merck Index monograph   : 7563
Patch testing   : 1% pet.; the commercial preparation Cytotec ® 30% pet. causes irritant reactions (3)
Molecular formula   : $C_{22}H_{38}O_5$

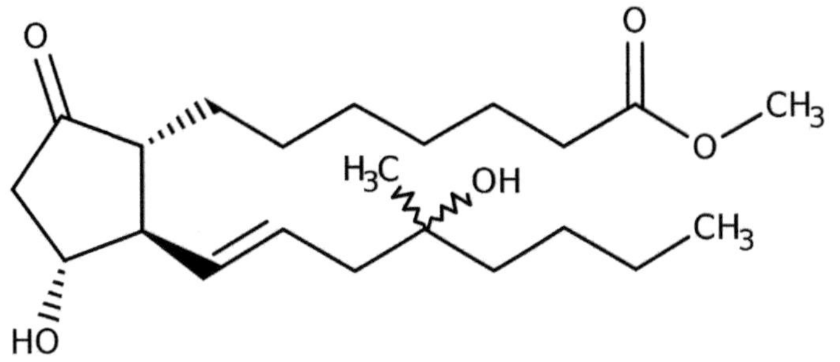

GENERAL

Misoprostol is a synthetic analog of natural prostaglandin E1 that produces a dose-related inhibition of gastric acid
and pepsin secretion, and enhances mucosal resistance to injury. It also has oxytocic properties. Misoprostol is
indicated for the treatment of ulceration (duodenal, gastric and NSAID-induced) and prophylaxis for NSAID-induced
ulceration. Off-label uses include the medical termination of an intrauterine pregnancy used alone or in combination
with methotrexate, the induction of labor in a selected population of pregnant women with unfavorable cervices and
for the prevention or treatment of serious postpartum hemorrhage (1).

CUTANEOUS ADVERSE DRUG REACTIONS FROM SYSTEMIC ADMINISTRATION CAUSED BY TYPE IV
(DELAYED-TYPE) HYPERSENSITIVITY (as demonstrated by positive patch tests)

Cutaneous adverse drug reactions from systemic administration of misoprostol caused by type IV (delayed-type)
hypersensitivity have included lichenoid drug eruption (2).

Other cutaneous adverse drug reactions

A 16-year-old woman was administered 800 µg misoprostol vaginally for induction of a first trimester abortion. Two
months later she presented with a pruritic, violaceous, papulosquamous eruption, symmetrically distributed over the
thighs and legs. Lesions varied in size from several millimeters to confluent psoriasiform plaques and extended in the
following days to the lower abdomen and the dorsa of the hands. Histology of a biopsy taken from a papule on the
right thigh was consistent with lichenoid drug eruption. A patch test with the commercial misoprostol at 0.01% pet.
was weakly (+) positive. Four months later, a patch test with the pure substance misoprostol 1% pet. resulted in a ++
reaction at D2, which remained positive until D7. No controls were mentioned. There was no recurrence of the skin
reaction during an 18-month follow-up period (2). An uncommon feature of this case was that the eruption started 2
months after a single dose of the drug.

LITERATURE

1   The data in the section 'General' may have been obtained from literature discussed in this chapter, but mostly
    also or exclusively from one or more of the following online sources: ChemIDPlus Advanced, PubChem,
    DrugBank, RxList, Drug Central, Drugs.com, and Wikipedia
2   Cruz MJ, Duarte AF, Baudrier T, Cunha AP, Barreto F, Azevedo F. Lichenoid drug eruption induced by misoprostol.
    Contact Dermatitis 2009;61:240-242
3   Barbaud A, Trechot P, Reichert-Penetrat S, Commun N, Schmutz JL. Relevance of skin tests with drugs in
    investigating cutaneous adverse drug reactions. Contact Dermatitis 2001;45:265-268

Chapter 3.325 MITOMYCIN C

IDENTIFICATION

Description/definition   : Mitomycin is a methylazirinopyrroloindoledione antineoplastic antibiotic isolated from the bacterium *Streptomyces caespitosus* and other *Streptomyces* bacterial species; its structural formula is shown below

Pharmacological classes   : Antibiotics, antineoplastic; cross-linking reagents; nucleic acid synthesis inhibitors; alkylating agents

IUPAC name   : [(4S,6S,7R,8S)-11-Amino-7-methoxy-12-methyl-10,13-dioxo-2,5-diazatetracyclo [7.4.0.0^{2,7}.0^{4,6}]trideca-1(9),11-dien-8-yl]methyl carbamate

CAS registry number   : 50-07-7

EC number   : 200-008-6

Merck Index monograph   : 7570 (Mitomycins)

Patch testing   : 0.1% pet.

Molecular formula   : $C_{15}H_{18}N_4O_5$

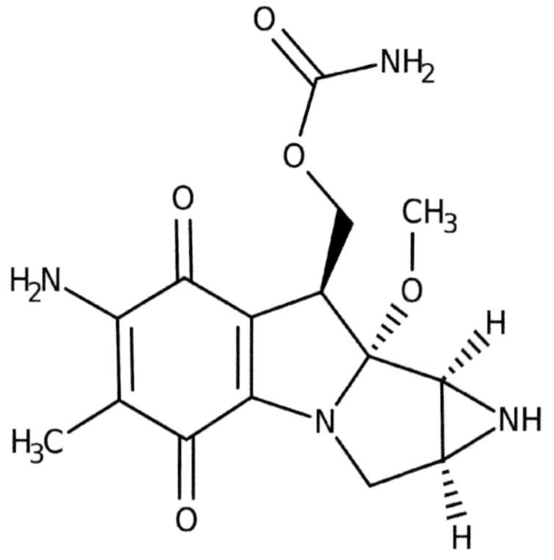

GENERAL

Mitomycin is an antineoplastic antibiotic isolated from the bacterium *Streptomyces caespitosus* and other *Streptomyces* bacterial species. Bioreduced mitomycin C generates oxygen radicals, alkylates DNA, and produces inter-strand DNA cross-links, thereby inhibiting DNA synthesis. Preferentially toxic to hypoxic cells, this agent also inhibits RNA and protein synthesis at high concentrations. Mitomycin C is indicated for treatment of malignant neoplasm of lip, oral cavity, pharynx, digestive organs, peritoneum, female breast, and urinary bladder (1). In a topical preparation (eye drops), it has caused one case of allergic contact dermatitis (2).

CONTACT ALLERGY FROM ACCIDENTAL CONTACT

A 60-year-old man noted a severe eruption on the penis, thighs, lower abdomen, and buttocks after the *first* instillation of mitomycin C. The patient had previously handled mitomycin C in his job as a nurse and gave a history of dermatitis of his hands. This may have been a case of occupational allergic contact dermatitis to mitomycin C, but patch tests were probably not performed (no details available) (7).

CUTANEOUS ADVERSE DRUG REACTIONS FROM SYSTEMIC ADMINISTRATION CAUSED BY TYPE IV (DELAYED-TYPE) HYPERSENSITIVITY (as demonstrated by positive patch tests)

Cutaneous adverse drug reactions from systemic administration of mitomycin C caused by type IV (delayed-type) hypersensitivity have included systemic allergic dermatitis. Clinical manifestations observed include desquamation or vesicular dermatitis of the hands and (sometimes) the feet (4,5,6,8,11,13,16,17,19), genital dermatitis (4,5,6,8,10), symmetrical drug-related intertriginous and flexural exanthema (SDRIFE)/baboon syndrome (8,12), more widespread rash/dermatitis (4,8,10,11,17,20), maculopapular exanthema (21) and combinations of these.

Systemic allergic dermatitis (systemic contact dermatitis)

The antineoplastic antibiotic mitomycin C has been used to treat superficial bladder cancer with intravesical instillations since the mid-1970s. The surface layers of the bladder are exposed to therapeutic concentrations and about 1% of the drug is absorbed. It was soon found that the treatment, after repeated instillations, could cause side effects in the form of bladder irritation and a rash involving the palms and soles, and sometimes the chest and face (13,14,15,16,17). The bladder symptoms were attributed by some authors to chemical irritation (14), but allergy to mitomycin was soon considered as a cause; biopsies taken soon after the onset of symptoms revealed an inflammatory cell infiltrate with numerous eosinophils, suggesting an immunological reaction (15). In an early report, describing 4 patients with palmar desquamation, one of who had a generalized rash, patch tests with mitomycin were performed and 3 reacted positively to mitomycin 0.001% water. It was suggested that the skin manifestations were a contact dermatitis following contamination of skin with urine containing mitomycin C and that careful cleansing of the hands and perineum on the day of treatment can prevent most skin reactions (17). However, the areas of involvement, which included the soles, made it unlikely that contamination of the skin accounts for the rash in all the cases (13).

The first well-documented case of a skin eruption from delayed-type hypersensitivity to intravesical mitomycin C was reported in 1984 from the United Kingdom (16). The patient had irritation of the palms and soles from the 5th instillation on with burning discomfort in the lower abdomen (presumably allergic cystitis), which evolved into vesicular palmar and plantar dermatitis after the 8th instillation. Patch tests were positive to mitomycin C 0.1% pet. and later, in a dilution series, down to 0.001%. These authors suspected that sensitization to mitomycin had either occurred by accidental contamination of the skin whilst being catheterized or on voiding the mitomycin instillation, and that the skin reactions were subsequently caused by systemic absorption of the drug via the urethra (during catheterization) or via the bladder during the period of instillation (16) , i.e. systemic allergic dermatitis.

In 1990, investigators from the United Kingdom studied 75 patients with primary superficial bladder cancer treated with intravesical mitomycin C instillations. From this group, separate subgroups of 26 and 10 were selected for patch testing with mitomycin C 0.01% and 0.001% pet. and for bladder biopsy, respectively. Fifteen of 75 patients receiving more than one instillation of mitomycin C developed dysuria, frequency and hematuria. Four of these cases also developed a palmar, foot and facial rash which occurred within 48 hours of treatment and persisted for several days. These problems were never seen after the first instillation but only after the second or subsequent instillations, becoming worse after each instillation. The cystitis sometimes became so severe that the instillations had to be stopped. In 2 patients with a skin eruption, there were strongly positive patch tests (++) to mitomycin 0.01% pet., but negative to 0.001%. Of the remaining 24 patients (who had no skin rashes), 13 had negative patch tests to both concentrations. Eleven had positive patch tests, 7 only to mitomycin 0.01% and the other 4 to both 0.01% and 0.001%. Ten controls had negative patch tests to both concentrations. The investigators also showed that antigen-presenting cells are present in the bladder wall, making induction of sensitization in the bladder possible (13).

In the 2 years after this 1990 publication, several case reports and case series of cutaneous allergic reactions to intravesical administration of mitomycin were published in dermatological literature (4,8,9,10,11,12), later followed by several others (5,6,18,19,20,21). Key data of these studies are summarized in table 3.325.1. Additional cases of allergic reactions to mitomycin, details of which are not available to the author, can be found in refs. 9 (3 cases), (possibly) ref. 22 and ref. 24.

Assuming sensitization takes place in the bladder mucosa first and the manifestations later occur from hematogenous spread of mitomycin C by absorption through the vesicular mucosa, most authors consider the skin reactions to mitomycin C a form of systemic allergic dermatitis (systemic allergic dermatitis).

Clinical manifestations, diagnosis and prognosis
Cutaneous adverse drug reactions to intravesical instillation of mitomycin may occur in up to 8.5% of all patients treated; how many of these result from delayed-type hypersensitivity has not been well investigated (3). However, one study has convincingly demonstrated that a large proportion of patients will in fact become sensitized by repeated intravesical instillations of mitomycin C, even in those without skin manifestations. Of 24 patients who *did not* – yet – show skin eruptions, 11 had positive patch test reactions to mitomycin 0.001% and/or 0.01% pet. These reactions were less strong than in patients who *did* have a skin eruption. Thus, it was suggested that in asymptomatic patients, with continued treatment, hypersensitivity would build up and at one point result in an allergic cutaneous reaction (13). The skin eruption hardly ever develops after the first treatment, but usually does so after five to eight instillations. The first signs may appear from a few hours after instillation up to 2 days (8). In only one case a patient developed a maculopapular eruption 14 days after the first instillation (during which time sensitization likely had taken place) (21).

The most common cutaneous reaction is found on the palms of the hands and sometimes the soles of the feet, which may be either desquamation or vesicular dermatitis. This reaction is often combined with dermatitis of the genitals, sometimes accompanied by erythema of the buttocks, groins and abdomen, consistent with the clinical picture of symmetrical drug-related intertriginous and flexural exanthema (SDRIFE)/baboon syndrome. Sometimes, a more generalized eruption may be observed. Atypical manifestations have included worsening of stasis dermatitis (8), allergic contact dermatitis from urine accidentally contacting the skin of the legs (8), Henoch-Schönlein type vasculitis from mitomycin instillation and patch test (19), and maculopapular eruption by instillation and by a patch test (21).

Table 3.325.1 Summary of reports of allergic reactions to intravesical instillation of mitomycin C

Year	Country	Nr. Pat.	Nr. T	Symptoms/localization and comments	Ref.
2009	Spain	1	7	Erythema and desquamation of the palms and the penile area	5
2007	Germany	1	10	Palmar desquamation, genital dermatitis	6
2000	Netherlands	1	NS	Exfoliative dermatitis palms and soles with generalized itch; palpable purpura (histopathology and immunofluorescence: Henoch-Schönlein type vasculitis) from mitomycin instillation and from the patch test	19
1997	Spain	1	5	Pruritus and localized dermatitis on the face, neck and forearms	20
1995	Spain	1	1	Generalized maculopapular eruption after first and second instillation and 12 hours after patch testing	21
1991	Spain	2	6, ?	Pruritic rash of palms and soles; fever (38.5°C), chills, edema of the eyelids and hands, and a pruritic rash of the legs, feet and hands	11
1991	Netherlands	6	6-8	Hands (n=4), genitals/groins/buttocks (n=4), worsening of stasis dermatitis (n=1), eczema of the trunk (n=1), ACD from urine spilled on leg (n=1)	8
1991	Italy	1	NS	Severe itchy edema, erythema and vesicles of the genital region, thighs and buttocks, spreading to the abdomen	12
1991	Italy	1	6	Papules on the palms, wrists and ankles; after next instillation 'widespread rash, mainly of the scrotal region'	10
1990	Sweden	2	8, 10	Rash on scrotum, inner thighs, dorsal forearms, hands	4
1984	U.K.	1	5	Irritation of the palms and soles, later vesicular eczema	16
1981	U.K.	3	?	Palmar desquamation (n=3), generalized rash (n=1)	17

ACD: Allergic contact dermatitis; Nr. Pat.: Number of patients; Nr.: Number of the treatment after which the allergic reaction started; NS: Not stated

In many cases, especially when only the hands are involved, mitigation of the local reaction can be accomplished using topical corticosteroids and therapy with intravesical mitomycin instillations can be continued, although each subsequent treatment will reproduce the symptoms, usually getting worse and more extensive with each following treatment session. In at least 30% of the patients with allergic skin manifestations, the treatment has to be interrupted (3). Also, the allergic cystitis that may be observed, characterized by dysuria, frequency and hematuria, sometimes becomes so severe that treatment has to be stopped (13).

Patch tests to mitomycin C have been performed with various concentrations in either water or petrolatum. For most patients, testing with mitomycin C 0.1% pet. will reliably demonstrate allergy to the drug (3,8). Water has the disadvantage that the preparations have to be prepared freshly, as mitomycin is unstable in a solution (8).

LITERATURE

1   The data in the section 'General' may have been obtained from literature discussed in this chapter, but mostly also or exclusively from one or more of the following online sources: ChemIDPlus Advanced, PubChem, DrugBank, RxList, Drug Central, Drugs.com, and Wikipedia
2   Cumurcu T, Sener S, Cavdar M. Periocular allergic contact dermatitis following topical Mitomycin C eye drop application. Cutan Ocul Toxicol 2011;30:239-240
3   De Groot AC, van der Meijden APM, Conemans JMH, Maibach HI. Frequency and nature of cutaneous reactions to intravesical instillation of mitomycin for superficial bladder cancer. Urology 1992;40(suppl.1):16-19
4   Christensen OB. 2 cases of delayed hypersensitivity to mitomycin C following intravesicular chemotherapy of superficial bladder cancer. Contact Dermatitis 1990;23:263 (Abstract)
5   Gutierrez Fernandez DG, Fernández AF, Vallejo MS, Osorio JL, Anguita MJ, Jimenez AL. Delayed hypersensitivity to mitomycin C. Contact Dermatitis 2009;61:237-238
6   Peitsch WK, Klemke CD, Michel MS, Goerdt S, Bayerl C. Hematogenous contact dermatitis after intravesicular instillation of mitomycin C. Hautarzt 2007;58:246-249 (Article in German)
7   Rogers B. Health hazards to personnel handling antineoplastic agents. Occup Med 1987;2:513-515 (Data cited in ref. 23)
8   De Groot AC, Conemans JM. Systemic allergic contact dermatitis from intravesical instillation of the antitumor antibiotic mitomycin C. Contact Dermatitis 1991;24:201-209
9   Vidal C, de la Fuente R, González Quintela A. Three cases of allergic dermatitis due to intravesical mitomycin C. Dermatology 1992;184:208-209
10 Giorgini S, Martinelli C, Sertoli A. Delayed-type sensitivity reaction to mitomycin. Contact Dermatitis 1991;24:378-379
11 Arregui MA, Aguirre A, Gil N, Goday J, Ratón JA. Dermatitis due to mitomycin C bladder instillations: study of 2 cases. Contact Dermatitis 1991;24:368-370
12 Valsecchi R, Imberti G, Cainelli T. Mitomycin C contact dermatitis. Contact Dermatitis 1991;24:70-71

13 Colver GB, Inglis JA, McVittie E, Spencer MJ, Tolley DA, Hunter JA. Dermatitis due to intravesical mitomycin C: a delayed-type hypersensitivity reaction? Br J Dermatol 1990;122:217-224

14 Hetherington JW, Newling DW, Robinson MR, Smith PH, Adib RS, Whelan P. Intravesical mitomycin C for the treatment of recurrent superficial bladder tumours. Br J Urol 1987;59:239-241

15 Inglis JA, Tolley DA, Grigor KM. Allergy to mitomycin C complicating topical administration for urothelial cancer. Br J Urol 1987;59:547-549

16 Neild VS, Sanderson KV, Riddle PR. Dermatitis due to mitomycin bladder instillations. J R Soc Med 1984;77:610-611

17 Nissenkorn I, Herrod H, Soloway MS. Side effects associated with intravesical mitomycin C. J Urol 1981;126:596-597

18 Fidalgo A, Lobo L. Allergic contact sensitization from intravesical mitomycin C, without dermatitis. Dermatitis 2004;15:161-162 (article not present on the website of the publisher)

19 Kunkeler L, Nieboer C, Bruynzeel DP. Type III and type IV hypersensitivity reactions due to mitomycin C. Contact Dermatitis 2000;42:74-76

20 Gomez Torrijos E, Borja J, Galindo PA, Feo F, Cortina P, Casanueva T, et al. Allergic contact dermatitis from mitomycin C. Allergy 1997;52:687

21 Echechipía S, Alvarez MJ, García BE, Olaguíbel JM, Rodriguez A, Lizaso MT, Acero S, Tabar AI. Generalized dermatitis due to mitomycin C patch test. Contact Dermatitis 1995;33:432

22 Fisher AA. Allergic contact dermatitis to mitomycin-C. Cutis 1991;47:225-227

23 Fisher AA. Allergic contact reactions in health personnel. J Allergy Clin Immunol 1992;90:729-738

24 Cao Avellaneda E, López López AI, Maluff Torres A, Jiménez RM, Escudero Bragante JF, López Cubillana P, et al. Hipersensibilidad tipo IV a mitomicina intravesical [Type IV hypersensitivity to intravesical mitomycin]. Actas Urol Esp 2005;29:803 (Article in Spanish)

Chapter 3.326 MOFEBUTAZONE

IDENTIFICATION

Description/definition   : Mofebutazone is the pyrazolidine derived from phenylbutazone that conforms to the structural formula shown below
Pharmacological classes : Anti-inflammatory agents, non-steroidal
IUPAC name   : 4-Butyl-1-phenylpyrazolidine-3,5-dione
CAS registry number   : 2210-63-1
EC number   : 218-641-1
Merck Index monograph   : 7587
Patch testing   : 1% pet.
Molecular formula   : $C_{13}H_{16}N_2O_2$

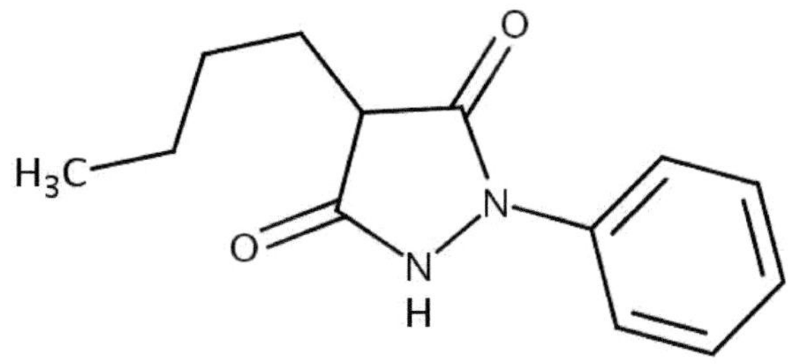

GENERAL

Mofebutazone is a pyrazolidine derivative and nonsteroidal anti-inflammatory drug (NSAID). It is phenylbutazone lacking one of the phenyl substituents. Mofebutazone is used for treatment of joint and muscular pain (1).

CUTANEOUS ADVERSE DRUG REACTIONS FROM SYSTEMIC ADMINISTRATION CAUSED BY TYPE IV (DELAYED-TYPE) HYPERSENSITIVITY (as demonstrated by positive patch tests)

Cutaneous adverse drug reactions from systemic administration of mofebutazone caused by type IV (delayed-type) hypersensitivity have included generalized eczema from systemic allergic dermatitis (2).

Systemic allergic dermatitis (systemic contact dermatitis)

Three to four days after a 47-year-old woman had started oral treatment for thrombophlebitis with 150 mg mofebutazone, a patchy erythema developed, which was partly vesicular and exudative. The eruption began on the abdomen, but then spread to the entire body surface, except for the face and mucous membranes. Fifteen years earlier, the topical application of an ointment containing mofebutazone had been complicated by a rash restricted to the treated site. Patch tests were performed with a mofebutazone tablet, its constituents, mofebutazone ointment and related NSAIDs: aminophenazone, propyphenazone, phenylbutazone, and oxyphenbutazone. On prick testing, no immediate-type, but a strong delayed reaction of the eczematous type was seen after 2 days to mofebutazone tablet. Patch test were positive to the tablet (crushed, moistened with water), its active ingredient mofebutazone 1% pet. and the mofebutazone-containing ointment undiluted. Ten controls were negative. There was a (cross-)reaction to phenylbutazone (2). This was a case of systemic allergic dermatitis presenting as generalized dermatitis.

Cross-reactions, pseudo-cross-reactions and co-reactions

A patient sensitized to mofebutazone cross-reacted to phenylbutazone (2).

LITERATURE

1   The data in the section 'General' may have been obtained from literature discussed in this chapter, but mostly also or exclusively from one or more of the following online sources: ChemIDPlus Advanced, PubChem, DrugBank, RxList, Drug Central, Drugs.com, and Wikipedia
2   Walchner M, Rueff F, Przybilla B. Delayed-type hypersensitivity to mofebutazone underlying a severe drug reaction. Contact Dermatitis 1997;36:54-55

Chapter 3.327 MONENSIN

IDENTIFICATION

Description/definition   : Monensin is a mixture of antibiotic substances produced by *Streptomyces cinnamonensis*
Pharmacological classes   : Sodium ionophores; antiprotozoal agents; proton ionophores; antifungal agents; coccidiostats
IUPAC name   : (2S,3R,4S)-4-[(2S,5R,7S,8R,9S)-2-[(2R,5S)-5-Ethyl-5-[(2R,3S,5R)-5-[(2S,3S,5R,6R)-6-Hydroxy-6-(hydroxymethyl)-3,5-dimethyloxan-2-yl]-3-methyloxolan-2-yl]oxolan-2-yl]-7-hydroxy-2,8-dimethyl-1,10-dioxaspiro[4.5]decan-9-yl]-3-methoxy-2-methylpentanoic acid
Other names   : Monensic acid
CAS registry number   : 17090-79-8
EC number   : 241-154-0
Merck Index monograph   : 7602
Patch testing   : 1% pet.
Molecular formula   : $C_{36}H_{62}O_{11}$

Monensin sodium

GENERAL

Monensin is an antiprotozoal agent produced by *Streptomyces cinnamonensis*. It is a mixture of antibiotic substances with monensin A as the major component. Monensin exerts its effect during the development of first-generation trophozoites into first-generation schizonts within the intestinal epithelial cells. It does not interfere with hosts' development of acquired immunity to the majority of coccidial species. Monensin is used as a feed additive for the prevention of coccidiosis in poultry and as a growth promoter in cattle. This agent is a sodium and proton selective ionophore and is widely used as such in biochemical studies (1). In pharmaceutical products (veterinary use only), both monensin and monensin sodium (CAS number 22373-78-0, EC number 244-941-7, molecular formula $C_{36}H_{61}NaO_{11}$) may be employed (1).

CONTACT ALLERGY FROM ACCIDENTAL CONTACT

In Italy, during 1986-1988, 204 animal feed mill workers (191 men, 13 women) were patch tested with a large number of animal feed additives. There were two reactions to monensin sodium 1% pet. in a group of 36 subjects with clinical complaints (dermatitis or pruritus sine materia) and zero reactions in the group of 168 individuals without skin complaints. All reactions were considered to be relevant (2).

LITERATURE

1   The data in the section 'General' may have been obtained from literature discussed in this chapter, but mostly also or exclusively from one or more of the following online sources: ChemIDPlus Advanced, PubChem, DrugBank, RxList, Drug Central, Drugs.com, and Wikipedia
2   Mancuso G, Staffa M, Errani A, Berdondini RM, Fabbri P. Occupational dermatitis in animal feed mill workers. Contact Dermatitis 1990;22:37-41

Chapter 3.328 MORANTEL

IDENTIFICATION

Description/definition : Morantel is the pyrimidine derivative that conforms to the structural formula shown below
Pharmacological classes : Anthelmintics; antinematodal agents
IUPAC name : 1-Methyl-2-[(E)-2-(3-methylthiophen-2-yl)ethenyl]-5,6-dihydro-4H-pyrimidine
CAS registry number : 20574-50-9
EC number : 243-890-8
Merck Index monograph : 7623
Patch testing : 1% and 5% pet.
Molecular formula : $C_{12}H_{16}N_2S$

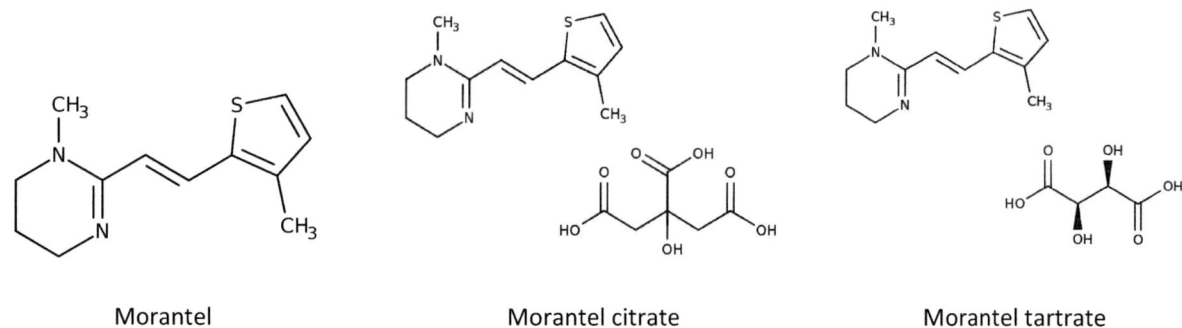

Morantel   Morantel citrate   Morantel tartrate

GENERAL

Morantel is an anthelmintic drug used for the removal of parasitic worms in livestock. It is an inhibitor of acetylcholinesterase and affects the nervous system of the worms. In pharmaceutical products (veterinary use only), morantel is employed either as morantel tartrate (CAS number 26155-31-7, EC number 247-481-5, molecular formula $C_{16}H_{22}N_2O_6S$) or as morantel citrate (CAS number 69525-81-1, EC number 274-028-9, molecular formula $C_{18}H_{24}N_2O_7S$) (1).

CONTACT ALLERGY FROM ACCIDENTAL CONTACT

The husband of a 33-year-old woman worked in the production area of a pharmaceutical company with pyrantel tartrate and morantel tartrate made up in a paste. He wore protective overalls at work, but did not shower daily and noticed that paste sometimes remained in his beard and on his watch strap at home. The patient (his wife) developed a dermatitis on her neck 2 months after her husband began to work with morantel. It subsequently spread to her face and then to her back and limbs. Later, her dermatitis flared and her finger nails were lost. Although her husband had no contact with morantel anymore, she continued to have dermatitis in the palms and on her nipples. Patch tests revealed vesicular reactions to morantel at 5% and 1%. Ten controls were negative. The patient was diagnosed with connubial allergic contact dermatitis from morantel. The authors suggested that an endogenous component contributed to the continuing dermatitis (2). 'Connubial allergic contact dermatitis' is currently termed 'allergic contact dermatitis by proxy'.

LITERATURE

1   The data in the section 'General' may have been obtained from literature discussed in this chapter, but mostly also or exclusively from one or more of the following online sources: ChemIDPlus Advanced, PubChem, DrugBank, RxList, Drug Central, Drugs.com, and Wikipedia
2   Newton JA, White I. Connubial dermatitis from Morantel. Contact Dermatitis 1987;16:107-108

Chapter 3.329 MORPHINE

IDENTIFICATION

Description/definition : Morphine is the opiate alkaloid that conforms to the structural formula shown below
Pharmacological classes : Analgesics, opioid; narcotics
IUPAC name : (4R,4aR,7S,7aR,12bS)-3-Methyl-2,4,4a,7,7a,13-hexahydro-1H-4,12-methanobenzo-
furo[3,2-e]isoquinoline-7,9-diol
CAS registry number : 57-27-2
EC number : 200-320-2
Merck Index monograph : 6731
Patch testing : 1% water; 5% in 2% acetic acid; in one study, morphine in water gave a false-negative
patch test but was positive in petrolatum (12)
Molecular formula : $C_{17}H_{19}NO_3$

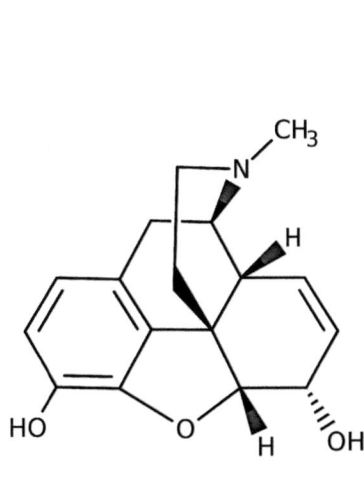

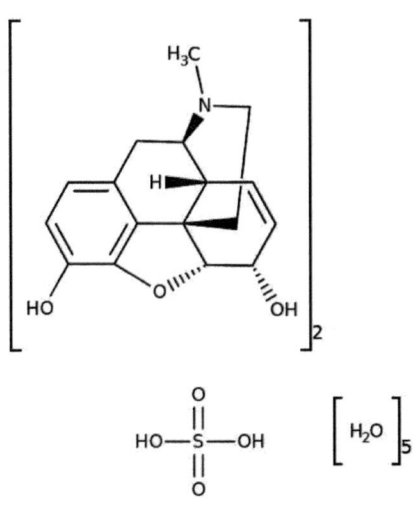

Morphine   Morphine sulfate pentahydrate

GENERAL

Morphine is an opiate alkaloid isolated from the plant *Papaver somniferum* and produced synthetically. It binds to and activates specific opiate receptors (delta, mu and kappa), each of which are involved in controlling different brain functions. In the central nervous and gastrointestinal systems, morphine exhibits widespread effects including analgesia, anxiolysis, euphoria, sedation, respiratory depression, and gastrointestinal system smooth muscle contraction. Morphine is used for the management of chronic moderate-to-severe pain. This agent can manage pain effectively when used for a short amount of time. The use of opioids for longer periods needs to be monitored as they can develop a physical dependence, addiction disorder and drug abuse. In pharmaceutical products, morphine is employed as morphine sulfate (pentahydrate) (CAS number 6211-15-0, EC number not available, molecular formula $C_{34}H_{50}N_2O_{15}S$) (1).

CONTACT ALLERGY FROM ACCIDENTAL CONTACT

In the period 2004 to 2017, of sixteen workers referred from a single opioid-manufacturing plant in Australia, eleven were diagnosed with occupational ACD caused by opioids, with five reacting to morphine (tested 3%, 5% and 10% in 2-3% acetic acid), seven to thebaine, four to norhydroxymorphinone, two to codeine, and two to oripavine. Most reported facial dermatitis, indicative of airborne contact dermatitis, although the hands and arms were also often involved. Six individuals reacted to at least two substances. This might be indicative of cross-reactivity or be the result of multiple exposures (11).

A female laboratory worker aged 38 was employed in a pharmaceutical company that manufactures several opiates, and she handled morphine and codeine during work. The patient developed papulovesicular contact dermatitis with an airborne pattern. Patch tests were performed with various opiates at different concentrations (0.01-5% or 10%) in pet. and water, respectively. All substances tested in water gave negative results, but there were positive reactions to the petrolatum test substances of morphine (+ for the 0.1% dilution and ++ for higher concentrations), codeine, and ethylmorphine. It was concluded that this patient had occupational airborne allergic contact dermatitis from morphine and codeine with cross-reactivity to ethylmorphine (12).

A 40-year-old nurse with hand eczema had, for 3 years, been working in a municipal treatment centre for drug addicts. She had developed work-related airborne acute erythematous and edematous facial dermatitis when handing heroin and morphine to the clients. Patch tests were strongly positive to morphine and heroin, both tested 1% water, but negative to other opioids (13).

A 30-year-old female researcher working for a generic drug manufacturer was exposed to the powder of morphine and 2 other drugs. Within a few days, she developed a severe papulovesicular, oozy airborne contact dermatitis involving her face, neck, and the upper part of her anterior chest. The eyelids were markedly swollen. Her hands and arms remained uninvolved, presumably because she always worked with gloves and gowns. Patch tests with all medicaments showed a positive reaction to morphine 1% pet. only with cross-reactions to codeine (1% and 3% water) and hydromorphone (1% water) (15).

Two workers in a factory producing opium alkaloids from poppy straw developed allergic contact dermatitis from their work. One had eczema of the neck, hands and forearms and showed positive patch tests to codeine phosphate, codeine hydrochloride and morphine bitartrate (negative to morphine HCl), all tested 1% alc. or water. The other patient had developed dermatitis of the face and hands after having worked in the production of concentrated poppy straw for 10 years. Patch tests were positive to morphine HCl and bitartrate and codeine bitartrate, all 1% alc. The authors mentioned that many patients allergic to opium alkaloids are also allergic to *p*-phenylenediamine; an explanation for this co-reactivity is lacking (21).

In early literature, (occupational) contact dermatitis and (in some cases) systemic allergic dermatitis to opium alkaloids, including morphine, codeine, apomorphine, ethylmorphine and diacetylmorphine (heroin) have been described (e.g. 9,10,16, reviews in ref. 8 and 10), the first report dating back to 1882 (cited in ref. 8). In those days, eczematous dermatitis from opium compounds used externally in the form of lotions, suppositories and other application forms was apparently well known (10). A more recent review of the subject was provided in 2006 (17).

CUTANEOUS ADVERSE DRUG REACTIONS FROM SYSTEMIC ADMINISTRATION CAUSED BY TYPE IV (DELAYED-TYPE) HYPERSENSITIVITY (as demonstrated by positive patch tests)
Cutaneous adverse drug reactions from systemic administration of morphine caused by type IV (delayed-type) hypersensitivity have included acute generalized exanthematous pustulosis (AGEP) (19) and systemic allergic dermatitis (17).

Acute generalized exanthematous pustulosis (AGEP)
A 27-year-old man underwent osteosynthesis because of a fracture of the tibia and fibula of his right leg and was given many drugs in the perioperative period including morphine. From day 10 on, he developed AGEP (acute generalized exanthematous pustulosis) with fever, extended burning bright erythema with numerous tiny non-follicular pustules (most pronounced in inguinal and axillary areas, but also on the trunk, face, and proximal limbs), marked leucocytosis with neutrophilia and eosinophilia, elevated liver function tests and raised erythrocyte sedimentation rate (25 mm/h) and C- reactive protein (185 mg/ml). Six months after clinical healing, patch tests were positive to commercial morphine HCl 10 mg/ml 'as is' and morphine 1% water, while all other drugs used perioperatively yielded negative findings. Four months later, a lymphocyte transformation test (LTT) was positive with morphine (19). In one patient, AGEP may also have been caused by delayed-type hypersensitivity to topically applied morphine (20).

Systemic allergic dermatitis (systemic contact dermatitis)
A 55-year-old woman with a history of occupational contact dermatitis caused by heroin (diacetylmorphine) underwent surgery for a mammary carcinoma. One day after surgery, she developed an erythematous papular itchy facial rash, which spread over the body on the next day. In addition, vesicles were seen on the face and neck. No signs of fever, laryngitis or pustules were seen. The patient had received many medications including morphine. The authors did not perform patch tests, but assumed that the patient developed a morphine-induced delayed-type cutaneous adverse drug reaction following occupational heroin sensitization (18). Indeed, in the past, this patient had also shown a positive patch test reaction to morphine. Cross-sensitization to morphine in heroin-sensitized patients is well known (17).

Cross-reactions, pseudo-cross-reactions and co-reactions
Virtually all patients sensitized to codeine (= methylmorphine) cross-react to morphine (2-7). Cross-reaction to ethylmorphine in a patient sensitized to morphine (12). Morphine cross-reacted to primary oxycodone or codeine sensitization in a patient with DRESS/TEN (14). A patient sensitized by morphine showed with cross-reactions to codeine and hydromorphone (15). Of 8 nurses sensitized to heroin (diacetylmorphine), 6 (75%) cross-reacted to morphine (17).

LITERATURE

1   The data in the section 'General' may have been obtained from literature discussed in this chapter, but mostly also or exclusively from one or more of the following online sources: ChemIDPlus Advanced, PubChem, DrugBank, RxList, Drug Central, Drugs.com, and Wikipedia

2   Rodríguez A, Barranco R, Latasa M, de Urbina JJ, Estrada JL. Generalized dermatitis due to codeine. Cross-sensitization among opium alkaloids. Contact Dermatitis 2005;53:240

3   Estrada JL, Puebla MJ, de Urbina JJ, Matilla B, Prieto MA, Gozalo F. Generalized eczema due to codeine. Contact Dermatitis 2001;44:185

4   Gastaminza G, Audicana M, Echenagusia MA, Uriel O, Garcia-Gallardo MV, Velasco M, et al. Erythrodermia caused by allergy to codeine. Contact Dermatitis 2005;52:227-228

5   Rodriguez F, Fernandez L, Garcia-Abujeta JL, Maquiera E, Llaca HF, Jerez J. Generalized dermatitis due to codeine. Contact Dermatitis 1995;32:120

6   De Groot AC, Conemans J. Allergic urticarial rash from oral codeine. Contact Dermatitis 1986;14:209-214

7   Iriarte Sotés P, López Abad R, Gracia Bara MT, Castro Murga M, Sesma Sánchez P. Codeine-induced generalized

8   Touraine A. Les dermatoses de l'opium. Revue de Médecine 1936;53:449-460

9   Dore SE, Prosser Thomas EW. Contact dermatitis in a morphine factory. J Allergy 1945;16:35-36

10 Jordon JW, Osborne ED. Contact dermatitis from opium derivatives. JAMA 1939;113:1955-1957

11 Flury U, Cahill JL, Nixon RL. Occupational contact dermatitis caused by opioids: A case series. Contact Dermatitis 2019;81:332-335

12 Colomb S, Bourrain JL, Bonardel N, Chiriac A, Demoly P. Occupational opiate contact dermatitis. Contact Dermatitis 2017;76:240-241

13 Hvid L, Svendsen MT, Andersen KE. Occupational allergic contact dermatitis caused by heroin (diacetylmorphine) and morphine. Contact Dermatitis 2016;74:301-302

14 Liippo J, Pummi K, Hohenthal U, Lammintausta K. Patch testing and sensitization to multiple drugs. Contact Dermatitis 2013;69:296-302

15 Sasseville D, Blouin MM, Beauchamp C. Occupational allergic contact dermatitis caused by morphine. Contact Dermatitis 2011;64:166-168

16 Corson EF, Rouse GP Jr. Contact dermatitis from morphine; a series of cases among workers in a drug factory. Pennsylvania Med J 1946;49:968-970

17 Hogen Esch AJ, van der Heide S, van den Brink W, van Ree JM, Bruynzeel DP, Coenraads PJ. Contact allergy and respiratory/mucosal complaints from heroin (diacetylmorphine). Contact Dermatitis 2006;54:42-49

18 Van den Hoed E, Coenraads PJ, Schuttelaar MLA. Morphine-induced cutaneous adverse drug reaction following occupational diacetylmorphine contact dermatitis: A case report. Contact Dermatitis 2019;81:313-315

19 Kardaun SH, de Monchy JG. Acute generalized exanthematous pustulosis caused by morphine, confirmed by positive patch test and lymphocyte transformation test. J Am Acad Dermatol 2006;55(2 Suppl.):S21-23

20 Ghazawi FM, Colantonio S, Bradshaw S, Lacroix J, Pratt M. Acute generalized exanthematous pustulosis induced by topical morphine and confirmed by patch testing. Dermatitis 2020;31(3):e22-e23

21 Condé-Salazar L, Guimaraens D, González M, Fuente C. Occupational allergic contact dermatitis from opium alkaloids. Contact Dermatitis 1991;25:202-203

Chapter 3.330 MOXIFLOXACIN

IDENTIFICATION

Description/definition : Moxifloxacin is the quinoline carboxylic acid and fluoroquinolone that conforms to the structural formula shown below

Pharmacological classes : Anti-bacterial agents; topoisomerase II inhibitors

IUPAC name : 7-[(4aS,7aS)-1,2,3,4,4a,5,7,7a-Octahydropyrrolo[3,4-b]pyridin-6-yl]-1-cyclopropyl-6-fluoro-8-methoxy-4-oxoquinoline-3-carboxylic acid

CAS registry number : 151096-09-2

EC number : 604-773-0

Merck Index monograph : 7647

Patch testing : Tablet, pulverized, 30% pet. (2); most systemic drugs can be tested at 10% pet.; if the pure chemical is not available, prepare the test material from intravenous powder, the content of capsules or – when also not available – from powdered tablets to achieve a final concentration of the active drug of 10% pet.

Molecular formula : $C_{21}H_{24}FN_3O_4$

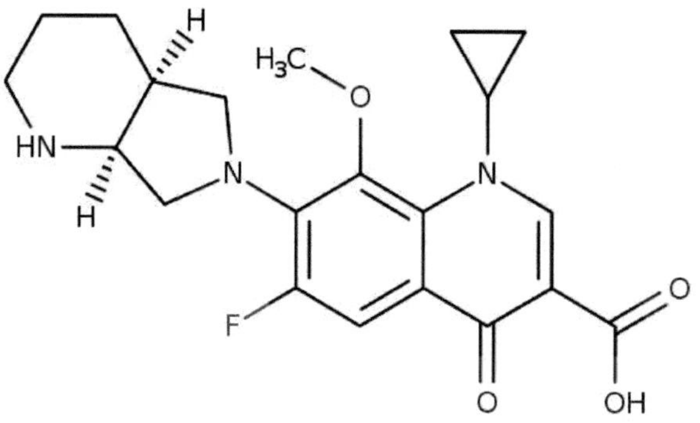

GENERAL

Moxifloxacin is a fourth generation fluoroquinolone with expanded activity against gram-positive bacteria as well as atypical pathogens. It binds to and inhibits the bacterial enzymes DNA gyrase (topoisomerase II) and topoisomerase IV, resulting in inhibition of DNA replication and repair and cell death in sensitive bacterial species (1).

CUTANEOUS ADVERSE DRUG REACTIONS FROM SYSTEMIC ADMINISTRATION CAUSED BY TYPE IV (DELAYED-TYPE) HYPERSENSITIVITY (as demonstrated by positive patch tests)

Cutaneous adverse drug reactions from systemic administration of moxifloxacin caused by type IV (delayed-type) hypersensitivity have included macular eruption (2).

Other cutaneous adverse drug reactions

A 47-year-old woman, known with type I allergy to clarithromycin, was orally challenged with moxifloxacin. After the second dose of 50 mg, pruritus and mild erythema on the patient's body occurred in minutes and was followed by dizziness. Her pulse was very weakly palpable, she was tachycardic and the blood pressure could not be measured. On the following day the patient returned with macular eruptions on the trunk and on the proximal extremities which had started with pruritus approximately 20 hours after the drug ingestion. Skin prick and intradermal tests with moxifloxacin were negative, as was the delayed reading of intradermal tests. A patch test performed with moxifloxacin diluted in petrolatum to obtain a drug concentration of 30% two months later was found positive. Five controls were negative. The lymphocyte transformation test (LTT) was considered positive with stimulation indexes of 1.65 and 1.75. it was concluded that the patient had both an immediate- and a delayed-type hypersensitivity to moxifloxacin (2). It should be mentioned that the picture of the patch test provided was not convincing for an allergic reaction, the stimulation index of the LTT was low, in vivo tests to prove immediate type hypersensitivity to moxifloxacin were negative and it was not mentioned how the patient had become sensitized to the antibiotic.

LITERATURE

1   The data in the section 'General' may have been obtained from literature discussed in this chapter, but mostly also or exclusively from one or more of the following online sources: ChemIDPlus Advanced, PubChem, DrugBank, RxList, Drug Central, Drugs.com, and Wikipedia

2   Demir S, Unal D, Olgac M, Akdeniz N, Aktas-Cetin E, Gelincik A, Colakoglu B, Buyukozturk S. An unusual dual hypersensitivity reaction to moxifloxacin in a patient. Asia Pac Allergy 2018;8(3):e26

Chapter 3.331 NADROPARIN

IDENTIFICATION

Description/definition   : Nadroparin is a low molecular weight heparin with a mean molecular weight of
                              approximately 4300 daltons
Pharmacological classes   : Antithrombotic agents
IUPAC name   : Not available
CAS registry number   : 9005-49-6 (Heparin)
EC number   : 232-681-7 (Heparin)
Merck Index monograph : 5958 (Heparin)
Patch testing   : Commercial preparation undiluted (3); consider intradermal testing with late readings
                              (D2,D3) when patch tests are negative and consider subcutaneous challenge when
                              intradermal tests are negative
Molecular formula   : Unspecified

GENERAL

Nadroparin is a low molecular weight heparin that is composed of a heterogeneous mixture of sulfated polysaccharide glycosaminoglycan chains. The mean molecular weight is approximately 4300 daltons. Nadroparin is used for the prophylaxis of thrombotic events and deep vein thrombosis, and to prevent unstable angina and non-Q-wave myocardial infarction. In pharmaceutical products, nadroparin is employed as nadroparin sodium or nadroparin calcium (CAS number not available, EC number not available, molecular formula unspecified) (1).

See also bemiparin (Chapter 3.59), certoparin (Chapter 3.117), dalteparin (Chapter 3.160), danaparoid (Chapter 3.161), enoxaparin (Chapter 3.195), fondaparinux (Chapter 3.225), heparins (Chapter 3.239), and tinzaparin (Chapter 3.479).

CUTANEOUS ADVERSE DRUG REACTIONS FROM SYSTEMIC ADMINISTRATION CAUSED BY TYPE IV (DELAYED-TYPE) HYPERSENSITIVITY

Throughout this book, only reports of delayed-type hypersensitivity have been included that showed a positive patch test to the culprit drug. However, as a result of the high molecular weight of heparins, patch tests are often false-negative, presumably from insufficient penetration into the skin. Because of this, and also because patch tests have been performed in a small minority of cases only, studies with a positive intradermal test or subcutaneous provocation tests with delayed readings are included in the chapters on the various heparins, even when patch tests were negative or not performed.

General information on delayed-type hypersensitivity reactions to heparins

General information on delayed-type hypersensitivity reactions to heparins (including nadroparin), presenting as local reactions from subcutaneous administration, is provided in Chapter ... Heparins. In this chapter, only non-local cutaneous adverse drug reactions from delayed-type hypersensitivity to nadroparin are presented.

Non-local cutaneous adverse drug reactions

Non-local cutaneous adverse drug reactions from systemic administration of nadroparin caused by type IV (delayed-type) hypersensitivity have included maculopapular eruption (8,9 [delayed-type hypersensitivity not proven],11), generalized eczema (6,7), and spreading of itchy erythematous plaques over the trunk (10).

Maculopapular eruption

A 75-year-old woman was treated with subcutaneous unfractionated heparin injections. Already after the first injection, a localized erythematous itching plaque developed at the injection site. While the treatment was continued, over the next 4 days the patient developed a generalized maculopapular exanthema accompanied by massive pruritus. During nadroparin therapy, the situation become worse. A subcutaneous provocation test with nadroparin resulted in a localized infiltrated erythematous reaction after 22 hours, which was followed by pruritus and a maculopapular rash in the following 9 hours (8).

A 47-year-old man developed a maculopapular rash after immediate-type local and generalized urticarial reactions following intravenous heparin sodium and subcutaneous enoxaparin sodium and nadroparin calcium. Patch tests were negative as were prick tests read at 15 and 30 minutes. Intradermal tests to all 3 heparins were positive after 30 minutes; the reactions had fully subsided after 8 hours. It was not mentioned whether the reactions were also read at 2 and 3 days. Type I hypersensitivity was proven, a co-existent type IV reactions, possibly explaining the maculopapular eruption, not (9).

A 46-year-old woman had been treated with nadroparin for 12 months, when she developed a maculopapular exanthema. Intradermal tests were negative, but a subcutaneous test was positive to nadroparin (11).

Dermatitis/eczematous eruption

In a hospital in Tenon, France, between 2000 and 2012, 19 patients were observed with hypersensitivity reactions to heparins. Six of these had generalized eczema, caused by enoxaparin sodium in 5 individuals and by nadroparin calcium in one (7).

A 39-year-old woman developed generalized eczema following sodium nadroparin administration. Patch tests were positive to several low molecular weight heparins; details are not available to the author (6).

Other cutaneous adverse drug reactions

A 41-year-old man received subcutaneous nadroparin twice daily for acute ischemic cardiopathy. After 3 days, infiltrated itchy erythematous plaques on the abdominal wall appeared, and 24 hours later the lesions spread all over his trunk. Patch tests were positive to commercial nadroparin (10).

Cross-reactions, pseudo-cross-reactions and co-reactions

Cross-reactions between heparins are frequent in delayed-type hypersensitivity (>90% of patients tested, median number of positive drugs per patient: 3) and do not depend on the molecular weight of the molecules (2). Overlap in their polysaccharide composition might explain the high degree of cross-allergenicity (4). Cross-reactions to the semisynthetic heparinoid danaparoid have also been observed (5). In allergic patients, the synthetic ultralow molecular weight synthetic heparin fondaparinux is usually, but not always (5) well-tolerated (4).

LITERATURE

1 The data in the section 'General' may have been obtained from literature discussed in this chapter, but mostly also or exclusively from one or more of the following online sources: ChemIDPlus Advanced, PubChem, DrugBank, RxList, Drug Central, Drugs.com, and Wikipedia

2 Weberschock T, Meister AC, Bohrt K, Schmitt J, Boehncke W-H, Ludwig RJ. The risk for cross-reactions after a cutaneous delayed-type hypersensitivity reaction to heparin preparations is independent of their molecular weight: a systematic review. Contact Dermatitis 2011;65:187-194

3 Brockow K, Garvey LH, Aberer W, Atanaskovic-Markovic M, Barbaud A, Bilo MB, et al. Skin test concentrations for systemically administered drugs - an ENDA/EAACI Drug Allergy Interest Group position paper. Allergy 2013;68:702-712

4 Schindewolf M, Lindhoff-Last E, Ludwig RJ. Heparin-induced skin lesions. Lancet 2012;380:1867-1879

5 Utikal J, Peitsch WK, Booken D, Velten F, Dempfle CE, Goerdt S, et al. Hypersensitivity to the pentasaccharide fondaparinux in patients with delayed-type heparin allergy. Thromb Haemost 2005;94:895-896

6 Estrada Rodriguez JL, Gozalo Reques F, Ortiz de Urbina J, Matilla B, Rodriguez Prieto MA, Gonzalez Moran NA. Generalized eczema induced by nadroparin. J Investig Allergol Clin Immunol 2003;13:69-70

7 Phan C, Vial-Dupuy A, Autegarden J-E, Amsler E, Gaouar H, Abuaf N, et al. A study of 19 cases of allergy to heparins with positive skin testing. Ann Dermatol Venereol 2014;141:23-29 (Article in French)

8 Greiner D, Schöfer H. Allergisches Arzneimittelexanthem auf Heparin. Kutane Reaktionen auf hochmolekulares und fraktioniertes Heparin [Allergic drug exanthema to heparin. Cutaneous reactions to high molecular and fractionated heparin]. Hautarzt 1994;45:569-572 (Article in German)

9 Klos K, Spiewak R, Kruszewski J, Bant A. Cutaneous adverse drug reaction to heparins with hypereosinophilia and high IgE level. Contact Dermatitis 2011;64:61-62

10 Poza-Guedes P, González-Pérez R, Canto G. Different patterns of cross-reactivity in non-immediate hypersensitivity to heparins: from localized to systemic reactions. Contact Dermatitis 2002;47:244-245

11 Lopez S, Torres MJ, Rodríguez-Pena R, Blanca-Lopez N, Fernandez TD, Antunez C, et al. Lymphocyte proliferation response in patients with delayed hypersensitivity reactions to heparins. Br J Dermatol 2009;160:259-265

Chapter 3.332 NALMEFENE

IDENTIFICATION

Description/definition : Nalmefene is the phenanthrene derivative that conforms to the structural formula shown below
Pharmacological classes : Narcotic antagonists
IUPAC name : (4R,4aS,7aS,12bS)-3-(Cyclopropylmethyl)-7-methylidene-2,4,5,6,7a,13-hexahydro-1H-4,12-methanobenzofuro[3,2-e]isoquinoline-4a,9-diol
CAS registry number : 55096-26-9
EC number : Not available
Merck Index monograph : 7715
Patch testing : 0.1% water
Molecular formula : $C_{21}H_{25}NO_3$

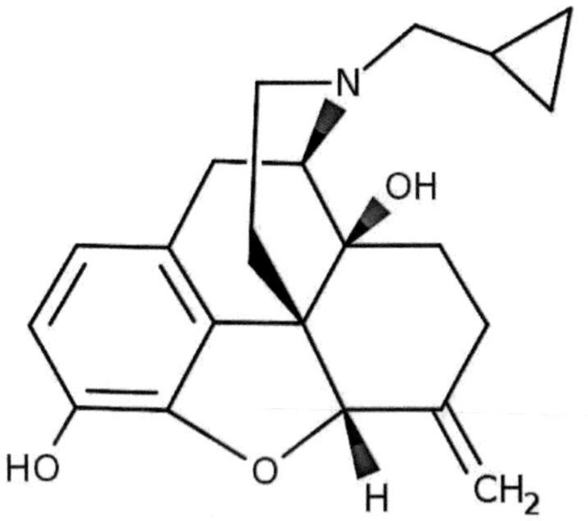

GENERAL

Nalmefene is the 6-methylene analog of naltrexone and opiate receptor antagonist which has no opioid activity. It is primarily used in the management of alcohol dependence in adult patients. Nalmefene is also indicated to prevent or reverse the effects of opioids, including respiratory depression, sedation, and hypotension by acting on the opioid receptor as an antagonist. In pharmaceutical products, nalmefene is employed as nalmefene hydrochloride (CAS number 58895-64-0, EC number not available, molecular formula $C_{21}H_{26}ClNO_3$) or as nalmefene hydrochloride dihydrate (CAS number 1228646-70-5, EC number not available, molecular formula $C_{21}H_{30}ClNO_5$) (1).

CONTACT ALLERGY FROM ACCIDENTAL CONTACT

A 32-year-old male scientist presented with a 6-month history of patchy eczema, which initially affected the periorbital area, eyebrows, and cheeks before spreading to the neck, followed by lip swelling and cracking. The patient had worked for a pharmaceutical company for over 10 years, creating novel pharmaceutical compounds. He reported settling of the eczema when on holidays, with quick recurrence on return to work. Patch tests with all suspected causative drug compounds provided by the pharmaceutical company (employer) were positive to nalmefene hydrochloride 0.1% water at D3 on 2 occasions. Twenty controls were negative. Airborne occupational allergic contact dermatitis was diagnosed (2).

LITERATURE

1   The data in the section 'General' may have been obtained from literature discussed in this chapter, but mostly also or exclusively from one or more of the following online sources: ChemIDPlus Advanced, PubChem, DrugBank, RxList, Drug Central, Drugs.com, and Wikipedia
2   Corso R, White IR, McFadden JP, Ferguson FJ. Occupational allergic contact dermatitis caused by nalmefene. Contact Dermatitis 2021;85:108-109

Chapter 3.333 NAPROXEN

IDENTIFICATION

Description/definition : Naproxen is the arylpropionic acid derivative and non-steroidal anti-inflammatory drug that conforms to the structural formula shown below
Pharmacological classes : Anti-inflammatory agents, non-steroidal; cyclooxygenase inhibitors; gout suppressants
IUPAC name : (2S)-2-(6-Methoxynaphthalen-2-yl)propanoic acid
Other names : 2-Naphthaleneacetic acid, 6-methoxy-α-methyl-, (αS)-
CAS registry number : 22204-53-1
EC number : 244-838-7
Merck Index monograph : 7769
Patch testing : 5% pet. (SmartPracticeCanada); 10% pet. (7)
Molecular formula : $C_{14}H_{14}O_3$

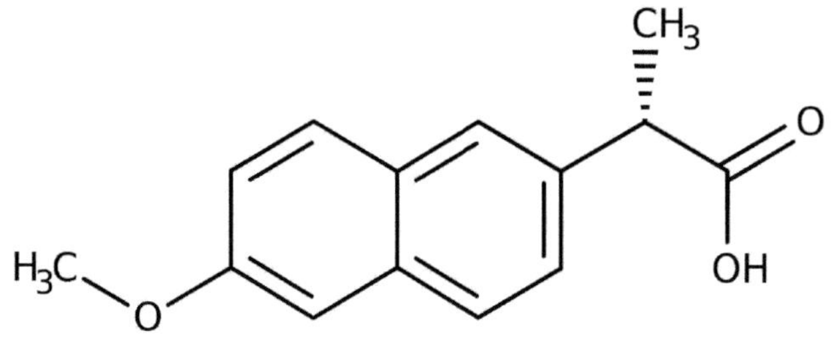

GENERAL

Naproxen is a propionic acid derivative and a nonsteroidal anti-inflammatory drug (NSAID) with anti-inflammatory, antipyretic and analgesic activities. Naproxen inhibits the activity of the enzymes cyclooxygenase I and II, resulting in a decreased formation of precursors of prostaglandins and thromboxanes. The resulting decrease in prostaglandin synthesis is responsible for the therapeutic effects of naproxen. Naproxen is indicated for the treatment of rheumatoid arthritis, osteoarthritis, ankylosing spondylitis, tendinitis, bursitis, acute gout and for the relief of mild to moderate pain and the treatment of primary dysmenorrhea (1). In pharmaceutical products, both naproxen and naproxen sodium (CAS number 26159-34-2, EC number 247-486-2, molecular formula $C_{14}H_{13}NaO_3$) may be employed (1).

CUTANEOUS ADVERSE DRUG REACTIONS FROM SYSTEMIC ADMINISTRATION CAUSED BY TYPE IV (DELAYED-TYPE) HYPERSENSITIVITY (as demonstrated by positive patch tests)

Cutaneous adverse drug reactions from systemic administration of naproxen caused by type IV (delayed-type) hypersensitivity have included fixed drug eruption (2,3,4) and photosensitivity (5,6).

Fixed drug eruption

In Turkey, 5 patients with fixed drug eruption from naproxen were investigated with patch testing on 15-20x tape-stripped postlesional and clinically uninvolved skin with commercial metamizole 10%, 20% and 50% pet. All reactions were negative. Next, open topical testing using DMSO as a vehicle was performed after 2 weeks. Without tape stripping, a thin layer of 10% test solution in DMSO and pure DMSO were applied with a cotton swab both to previous FDE lesions and to unaffected skin sites every 12 hours up to four times. If testing with 10% remained negative, the concentration of the drug was increased up to 20% and 50%. Open topical testing in DMSO revealed positive results in three of five patients tested with naproxen at a concentration of 50%. Repeated applications (up to 2-3) were necessary to obtain a positive reaction in most patients. Positive reactions were seen as itching and erythema with or without induration that started 1-2 hours (range, 1-4 hours) mainly after the first or second, or rarely the third application of the drug and lasted for more than 24-36 hours. Twenty controls were negative (2).

A 38-year-old woman developed, 2 hours after taking 2 tablets of naproxen for a headache, several edematous and dusky-red macules, one on the right forearm and the other two on both thighs and she was diagnosed with FDE probably due to naproxen. Patch tests were positive to naproxen 10% in DMSO with negative reactions to 2 other propionic acid NSAIDs: ibuprofen and ketoprofen (3).

A 28-year-old woman had fixed drug eruption probably due to naproxen. Patch tests were performed on the back (normal skin) with a series of NSAIDs, and with naproxen both on the back and on previous FDE sites. The tests were negative on the back, and on previous FDE sites the skin turned darker. The significance of this result as a diagnostic tool was unclear (indicative of delayed-type hypersensitivity?). However, an oral challenge test with naproxen was positive and reproduced all previous lesions (4).

Photosensitivity

A 65-year-old woman had a 1-week history of a pruritic, erythematous rash in a sun-exposed distribution. She had been taking naproxen 500 mg/day for the last 15 days for the treatment of a painful shoulder. Examination revealed erythematous papules and plaques, some of them with target shape, located on the lateral and V-area of the neck, upper back, extensor surfaces of the arms, forearms and dorsum of the hands. No mucosal lesions were present. A skin biopsy was consistent with erythema multiforme. Phototests showed that intake of naproxen did not alter the MED for UVA and UVB. Photopatch tests were positive to naproxen 5% pet. at D2 and D3 post-irradiation and remained positive after 5 days. The patients was diagnosed with photosensitivity to naproxen presenting as photo-sensitive erythema multiforme. The authors mentioned one more positive photopatch test to naproxen with current relevance (6).

Some positive photopatch tests to naproxen have been observed in patients with photosensitivity, but whether these were related to the systemic use of naproxen was not mentioned (5).

LITERATURE

1   The data in the section 'General' may have been obtained from literature discussed in this chapter, but mostly also or exclusively from one or more of the following online sources: ChemIDPlus Advanced, PubChem, DrugBank, RxList, Drug Central, Drugs.com, and Wikipedia

2   Özkaya-Bayazit E. Topical provocation in fixed drug eruption due to metamizol and naproxen. Clin Exp Dermatol 2004;29:419-422

3   Noguerado-Mellado B, Gamboa AR, Perez-Ezquerra PR, Cabeza CM, Fernandez RP, De Barrio Fernandez M. Fixed drug eruption due to selective hypersensitivity to naproxen with tolerance to other propionic acid NSAIDs. Recent Pat Inflamm Allergy Drug Discov 2016;10:61-63

4   Gonzalo Garijo MA, Bobadilla González P. Cutaneous reaction to naproxen. Allergol Immunopathol (Madr) 1996;24:89-92

5   Devleeschouwer V, Roelandts R, Garmyn M, Goossens A. Allergic and photoallergic contact dermatitis from ketoprofen: results of (photo) patch testing and follow-up of 42 patients. Contact Dermatitis 2008;58:159-166

6   Gutiérrez-González E, Rodríguez-Pazos L, Rodríguez-Granados MT, Toribio J. Photosensitivity induced by naproxen. Photoderm Photoimmunol Photomed 2011;27:338-340

7   Brockow K, Garvey LH, Aberer W, Atanaskovic-Markovic M, Barbaud A, Bilo MB, et al.; ENDA/EAACI Drug Allergy Interest Group. Skin test concentrations for systemically administered drugs – an ENDA/EAACI Drug Allergy Interest Group position paper. Allergy 2013;68:702-712

Chapter 3.334 NEBIVOLOL

IDENTIFICATION

Description/definition	: Nebivolol is the 1-benzopyran that conforms to the structural formula shown below
Pharmacological classes	: β-Adrenergic receptor antagonists
IUPAC name	: (1S)-1-[(2S)-6-Fluoro-3,4-dihydro-2H-chromen-2-yl]-2-[[(2S)-2-[(2R)-6-fluoro-3,4-dihydro-2H-chromen-2-yl]-2-hydroxyethyl]amino]ethanol
CAS registry number	: 118457-14-0
EC number	: 601-527-4
Merck Index monograph	: 7786
Patch testing	: Tablet, pulverized, 20% pet. and saline (2); most systemic drugs can be tested at 10% pet.; if the pure chemical is not available, prepare the test material from intravenous powder, the content of capsules or – when also not available – from powdered tablets to achieve a final concentration of the active drug of 10% pet.
Molecular formula	: $C_{22}H_{25}F_2NO_4$

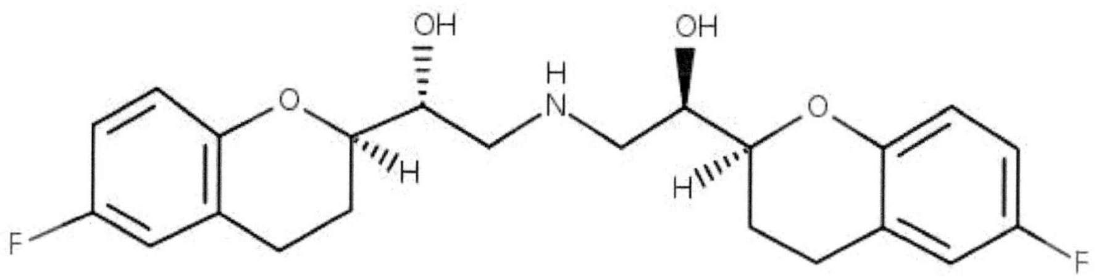

GENERAL

Nebivolol is a $β_1$-adrenergic receptor antagonist with antihypertensive and vasodilatory activity. Nebivolol binds to and blocks the $β_1$-adrenergic receptors in the heart, thereby decreasing cardiac contractility and rate. This leads to a reduction in cardiac output and lowers blood pressure. In addition, nebivolol potentiates nitric oxide, thereby relaxing vascular smooth muscle and exerting a vasodilatory effect. It is used to manage hypertension and chronic heart failure in elderly patients. In pharmaceutical products, nebivolol is employed as nebivolol hydrochloride (CAS number 152520-56-4, EC number not available, molecular formula $C_{22}H_{26}ClF_2NO_4$).

CUTANEOUS ADVERSE DRUG REACTIONS FROM SYSTEMIC ADMINISTRATION CAUSED BY TYPE IV (DELAYED-TYPE) HYPERSENSITIVITY (as demonstrated by positive patch tests)

Cutaneous adverse drug reactions from systemic administration of nebivolol caused by type IV (delayed-type) hypersensitivity have included periorbital eczema (2).

Dermatitis/eczematous eruption

A 60-year-old woman presented with a three year history of eczematous lesions in the periorbital areas, which appeared from December to April and each year became wider and more itchy. The patient was treated with nebivolol for hypertension, but she interrupted this therapy each year late in spring and in summer because the pressure values turned spontaneously back to normal range; the eczematous lesions would then resolve spontaneously. Patch tests with nebivolol 20% pet. and 20% saline were negative at D2, but positive at D3 (+) and D4 (++). At D7, the patient reported a pruritic, erythematous, microvesicular eruption with mild eyelid edema in the periorbital areas. There were no cross-reactions to other beta-blockers. Six controls were negative. The authors diagnosed systemic allergic dermatitis, but this is incorrect, as the patient had not been sensitized before by topical application of nebivolol or a structurally related chemical (2).

LITERATURE

1   The data in the section 'General' may have been obtained from literature discussed in this chapter, but mostly also or exclusively from one or more of the following online sources: ChemIDPlus Advanced, PubChem, DrugBank, RxList, Drug Central, Drugs.com, and Wikipedia

2   Fedele R, Ricciardi L, Mazzeo L, Isola S. Allergic contact dermatitis to nebivolol. Allergy 2002;57:864-865

Chapter 3.335 NEOMYCIN

IDENTIFICATION

Description/definition   : Neomycin is an antibiotic complex produced by *Streptomyces fradiae*; it is composed of neomycins A, B, and C; the structural formula of neomycin B is shown below

Pharmaceutical classes   : Anti-bacterial agents

IUPAC name   : (2R,3S,4R,5R,6R)-5-Amino-2-(aminomethyl)-6-[(1R,2R,3S,4R,6S)-4,6-diamino-2-[(2S,3R,4S,5R)-4-[(2R,3R,4R,5S,6S)-3-amino-6-(aminomethyl)-4,5-dihydroxyoxan-2-yl]oxy-3-hydroxy-5-(hydroxymethyl)oxolan-2-yl]oxy-3-hydroxycyclohexyl]oxyoxane-3,4-diol

Other names   : Framycetin; neomycin B (often used synonyms, but this is not entirely correct)

CAS registry number   : 1404-04-2

EC number   : 215-766-3

Merck Index monograph : 7809

Patch testing   : Sulfate, 20.0% pet. (Chemotechnique, SmartPracticeCanada, SmartPracticeEurope); late positive reactions (after D4) are frequent and readings at D7-D8 are imperative

Molecular formula   : $C_{23}H_{46}N_6O_{13}$

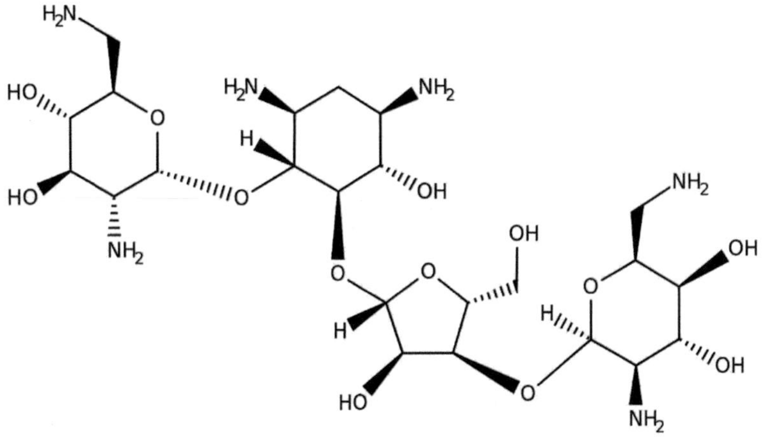

GENERAL

Neomycin is the prototype of the aminoglycoside antibiotics and was first isolated in 1949 from the gram-positive bacillus *Streptomyces fradiae*. It consists of a variable mixture of two isomers, neomycin B (>88%) and neomycin C (<10%), along with small amounts of a degradation product, neamine or neomycin A (<2%). It exerts its antibacterial activity through irreversible binding of the nuclear 30S ribosomal subunit, thereby blocking bacterial protein synthesis. Neomycin is effective against most gram-negative organisms except *Pseudomonas aeruginosa* and anaerobic bacteria. Its activity against gram-positive microorganisms is more or less limited to staphylococci, but bacterial resistance supervenes after prolonged use (1).

Percutaneous absorption after topical application is minimal, and absorption through intact gastrointestinal mucosa is poor, ranging between 1% and 3%. Parenteral administration is associated with severe ototoxicity and nephrotoxicity, at times irreversible. Given orally in doses of 200 to 1,000 mg two to four times per day, neomycin is used to sterilize the gut before digestive tract surgery and is also used in the treatment of hepatic coma (to reduce the number of ammoniagenic intestinal bacteria).

In some oral forms, neomycin base may be used, but topical pharmaceuticals contain neomycin as neomycin sulfate (CAS number 1405-10-3, EC number 215-773-1, molecular formula $C_{23}H_{48}N_6O_{17}S$) (1). In topical preparations, neomycin has caused an extremely high number of cases of contact allergy/allergic contact dermatitis (more than any other topical drug), which subject has been reviewed *in extenso* in Volume 3 of the *Monographs in contact allergy* series (1, with 190 literature references).

CONTACT ALLERGY FROM ACCIDENTAL CONTACT

Occupational contact allergy/allergic contact dermatitis from contact with neomycin involving the hands (and sometimes the face) have occurred in (dental) nurses (10), physicians, pharmacists, animal feed mill workers (9), farmers (11), oculists (7), veterinarians (8) and dentists (2).

Of 38 veterinarians with hand and forearm dermatoses seen by dermatologists in Belgium and the Netherlands from 1995 to 2005, 17 had occupational allergic contact dermatitis. In 2 of these patients, neomycin was the drug allergen responsible for the occupational contact dermatitis (12).

CUTANEOUS ADVERSE DRUG REACTIONS FROM SYSTEMIC ADMINISTRATION CAUSED BY TYPE IV (DELAYED-TYPE) HYPERSENSITIVITY (as demonstrated by positive patch tests)

Cutaneous adverse drug reactions from systemic administration of neomycin caused by type IV (delayed-type) hypersensitivity have included systemic allergic dermatitis (3,4,5), in one case presenting as the baboon syndrome (4).

Systemic allergic dermatitis (systemic contact dermatitis)

In an individual previously sensitized by topical exposure, the small amount of neomycin absorbed from the gastrointestinal tract (or from an overdose [4]) may be enough to trigger widespread dermatitis such as the baboon syndrome/symmetrical drug-related intertriginous and flexural exanthema (4) or a flare-up at the site of prior contact dermatitis (systemic allergic dermatitis) (3,4,5). Such reactions may also occur when cross-reacting chemicals are administered, Thus, systemically administered gentamicin induced erythroderma in an individual previously shown to be allergic to neomycin (6).

Cross-reactions, pseudo-cross-reactions and co-reactions

In patients sensitized to neomycin, the percentage of cross-reactions to other aminoglycoside antibiotics is high: to paromomycin and butirosin about 90%, to framycetin 70%, to ribostamycin 70%, to tobramycin and kanamycin 60%, to gentamicin 50%, to amikacin 30%, and to sisomycin 20% (1). However, streptomycin cross-reacts (or co-reacts more likely) in only about 4% of the patients (1). A high degree of co-reactivity between framycetin and neomycin can be expected, as framycetin consists mostly of neomycin B, which is also the main constituent of neomycin. However, neomycin B is not the only sensitizer in either framycetin or neomycin, as a considerable number of patients allergic to neomycin do not react to framycetin and vice versa (1).

The cross-sensitivity pattern in patients primarily sensitized to other aminoglycosides has not been well investigated. Primary sensitization to gentamicin and probably also tobramycin is only infrequently accompanied by cross-sensitization to neomycin. Patients primarily sensitized to paromomycin, however, will probably frequently cross-react to neomycin and to kanamycin (1).

Immediate contact reactions

Immediate contact reactions (contact urticaria) to neomycin are presented in Chapter 5.

LITERATURE

1 De Groot AC. Monographs in contact allergy, volume 3: Topical Drugs. Boca Raton, Fl, USA: CRC Press Taylor and Francis Group, 2021 (ISBN 978-0-367-23693-9)
2 Phillips DK. Neomycin sulfate. In: Guin JD, Ed. Practical contact dermatitis. New York: McGraw-Hill Inc., 1995:167-177
3 Ekelund AG, Möller H. Oral provocation in eczematous contact allergy to neomycin and hydroxyquinolines. Acta Derm Venereol (Stockh) 1969;49:422-426
4 Menné T, Weismann K. Hämatogenes Kontaktekzem nach oraler Gabe von Neomyzin. Hautarzt 1984;35:319-320 (Article in German)
5 Bouffioux B, Heid E. Eczéma endogène à la néomycine. Nouv Dermatol 1990;9:25 (Article in French, data cited in ref. 117)
6 Guin JD, Phillips D. Erythroderma from systemic contact dermatitis: a complication of systemic gentamicin in a patient with contact allergy to neomycin. Cutis 1989;43:564-567
7 Rebandel P. Rudzki. Occupational contact sensitivity in oculists. Contact Dermatitis 1986;15:92
8 Falk ES, Hektoen H, Thune PO. Skin and respiratory tract symptoms in veterinary surgeons. Contact Dermatitis 1985;12:274-278
9 Mancuso G, Staffa M, Errani A, Berdondini RM, Fabbri P. Occupational dermatitis in animal feed mill workers. Contact Dermatitis 1990;22:37-41
10 Kanerva L, Miettinen P, Alanko K, Estlander T, Jolanki R. Occupational allergic contact dermatitis from glyoxal, glutaraldehyde and neomycin sulfate in a dental nurse. Contact Dermatitis 2000;42:116-117
11 Simpson JR. Dermatitis from neomycin in a calf-drench. Contact Dermatitis Newsletter 1974;15:447
12 Bulcke DM, Devos SA. Hand and forearm dermatoses among veterinarians. J Eur Acad Dermatol Venereol 2007;21:360-363

Chapter 3.336 NEVIRAPINE

IDENTIFICATION

Description/definition : Nevirapine is the dipyridodiazepinone that conforms to the structural formula shown below
Pharmacological classes : Cytochrome P-450 CYP3A inducers; anti-HIV agents; reverse transcriptase inhibitors
IUPAC name : 11-Cyclopropyl-4-methyl-5H-dipyrido[2,3-e:2',3'-f][1,4]diazepin-6-one
CAS registry number : 129618-40-2
EC number : 603-345-0
Merck Index monograph : 7845
Patch testing : 10% pet.
Molecular formula : $C_{15}H_{14}N_4O$

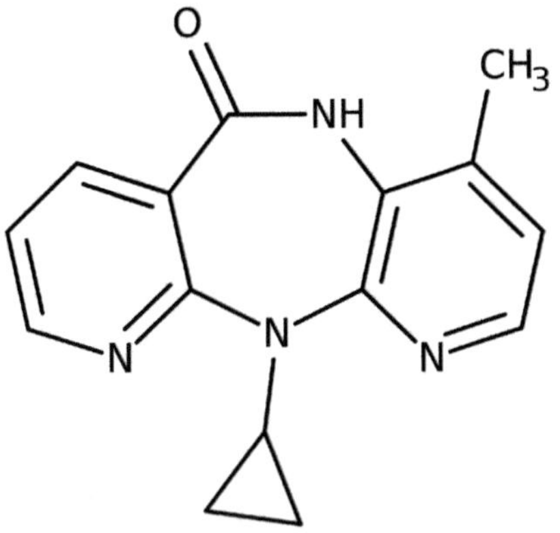

GENERAL

Nevirapine is a benzodiazepine and potent non-nucleoside reverse transcriptase inhibitor with activity against human immunodeficiency virus type 1 (HIV-1). This agent reduces HIV viral loads and increases CD4 counts, thereby retarding or preventing the damage to the immune system and reducing the risk of developing AIDS. Nevirapine is indicated for use in combination with other antiretroviral drugs in the ongoing treatment of HIV-1 infection (1). In pharmaceutical products, nevirapine base is used in tablets and nevirapine hemihydrate (CAS number 220988-26-1, EC number not available, molecular formula $C_{30}H_{30}N_8O_3$) in suspensions (1).

CUTANEOUS ADVERSE DRUG REACTIONS FROM SYSTEMIC ADMINISTRATION CAUSED BY TYPE IV (DELAYED-TYPE) HYPERSENSITIVITY (as demonstrated by positive patch tests)

Cutaneous adverse drug reactions from systemic administration of nevirapine caused by type IV (delayed-type) hypersensitivity have included maculopapular eruption (2).

Maculopapular eruption

In Thailand, 20 HIV patients who had shown a hypersensitivity reaction to nevirapine (excluding severe reactions such as DRESS, AGEP and SJS/TEN), were patch tested with pure nevirapine at 5%, 10%, 20% and 30% and with commercialized nevirapine 10% and 20%, all in petrolatum, water and alcohol. All 20 patients had suffered a maculopapular rash while being treated with nevirapine; 2 also had an urticarial rash, 4 facial edema, and one had additional vasculitis; six had fever. These manifestations resolved spontaneously after nevirapine was discontinued. The drug rash had developed within a median of 13.5 days after initiation of the suspected drug. Positive patch test reactions were seen in 2 of the 20 patients (to all 18 nevirapine patches applied), with negative reactions in the other 18 and in 15 controls. All HIV patients were tested for *HLA-B*3505* and 3 were positive, which included the 2 patch test positive individuals. It was concluded that nevirapine patch testing is safe in patients with non-severe hypersensitivity reactions and that an immunological pathophysiological mechanism is significantly correlated with *HLA-B*3505* in Thai HIV patients (2).

LITERATURE

1   The data in the section 'General' may have been obtained from literature discussed in this chapter, but mostly also or exclusively from one or more of the following online sources: ChemIDPlus Advanced, PubChem, DrugBank, RxList, Drug Central, Drugs.com, and Wikipedia
2   Prasertvit P, Chareonyingwattana A, Wattanakrai P. Nevirapine patch testing in Thai human immunodeficiency virus infected patients with nevirapine drug hypersensitivity. Contact Dermatitis 2017;77:379-384

Chapter 3.337 NICARDIPINE

IDENTIFICATION

Description/definition : Nicardipine is the dihydropyridinecarboxylic acid derivative that conforms to the structural formula shown below

Pharmacological classes : Antihypertensive agents; vasodilator agents; calcium channel blockers

IUPAC name : 5-*O*-[2-[Benzyl(methyl)amino]ethyl] 3-*O*-methyl 2,6-dimethyl-4-(3-nitrophenyl)-1,4-dihydropyridine-3,5-dicarboxylate

CAS registry number : 55985-32-5

EC number : Not available

Merck Index monograph : 7850

Patch testing : Tablet, pulverized, 20% water (7); most systemic drugs can be tested at 10% pet.; if the pure chemical is not available, prepare the test material from intravenous powder, the content of capsules or – when also not available – from powdered tablets to achieve a final concentration of the active drug of 10% pet.

Molecular formula : $C_{26}H_{29}N_3O_6$

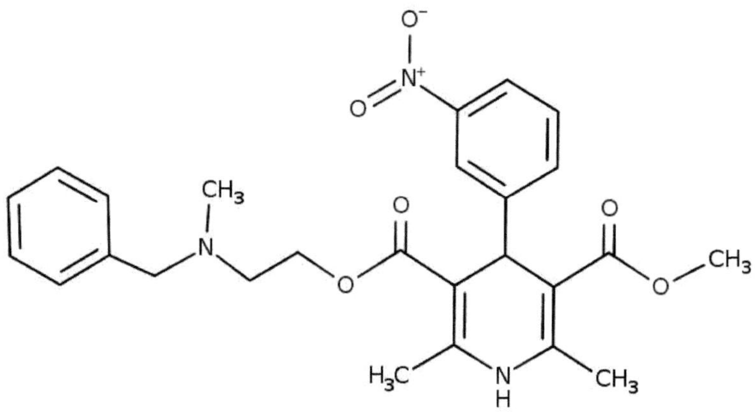

GENERAL

Nicardipine is a dihydropyridine calcium-channel blocker, which is used alone or with an angiotensin-converting enzyme inhibitor to treat hypertension and chronic stable angina pectoris. Nicardipine inhibits the contractile processes of the myocardial smooth muscle cells, causing dilation of the coronary and systemic arteries, increased oxygen delivery to the myocardial tissue, decreased total peripheral resistance, decreased systemic blood pressure, and decreased afterload.

In pharmaceutical products, nicardipine is employed as nicardipine hydrochloride (CAS number 54527-84-3, EC number 259-198-4, molecular formula $C_{26}H_{30}ClN_3O_6$) (1).

CUTANEOUS ADVERSE DRUG REACTIONS FROM SYSTEMIC ADMINISTRATION CAUSED BY TYPE IV (DELAYED-TYPE) HYPERSENSITIVITY (as demonstrated by positive patch tests)

Cutaneous adverse drug reactions from systemic administration of nicardipine caused by type IV (delayed-type) hypersensitivity have included photosensitivity (2,3).

Photosensitivity

A 75-year-old man developed a photodistributed rash, while using 4 different medications including nicardipine. Patch and photopatch tests to the drugs 20% in water were positive to nicardipine at the UVA-irradiated site. Seven controls were negative (3). Another patient with a drug exanthema had a positive photopatch test to nicardipine; details of this case reported in Japanese literature are unknown (data cited in ref. 2).

LITERATURE

1   The data in the section 'General' may have been obtained from literature discussed in this chapter, but mostly also or exclusively from one or more of the following online sources: ChemIDPlus Advanced, PubChem, DrugBank, RxList, Drug Central, Drugs.com, and Wikipedia

2   Kitamura K, Kanasashi M, Suga C, Saito S, Yoshida S, Ikezawa Z. Cutaneous reactions induced by calcium channel blocker: high frequency of psoriasiform eruptions. J Dermatol 1993;20:279-286

3   Conilleau V, Dompmartin A, Michel M, Verneuil L, Leroy D. Photoscratch testing in systemic drug-induced photosensitivity. Photodermatol Photoimmunol Photomed 2000;16:62-66

Chapter 3.338 NICERGOLINE

IDENTIFICATION

Description/definition : Nicergoline is the organonitrogen heterocyclic compound and ergot derivative that conforms to the structural formula shown below

Pharmacological classes : Nootropic agents; α-adrenergic antagonists; vasodilator agents

IUPAC name : [(6aR,9R,10aS)-10a-Methoxy-4,7-dimethyl-6a,8,9,10-tetrahydro-6H-indolo[4,3-fg]quinoline-9-yl]methyl 5-bromopyridine-3-carboxylate

CAS registry number : 27848-84-6

EC number : 248-694-6

Merck Index monograph : 7851

Patch testing : 7% alcohol

Molecular formula : $C_{24}H_{26}BrN_3O_3$

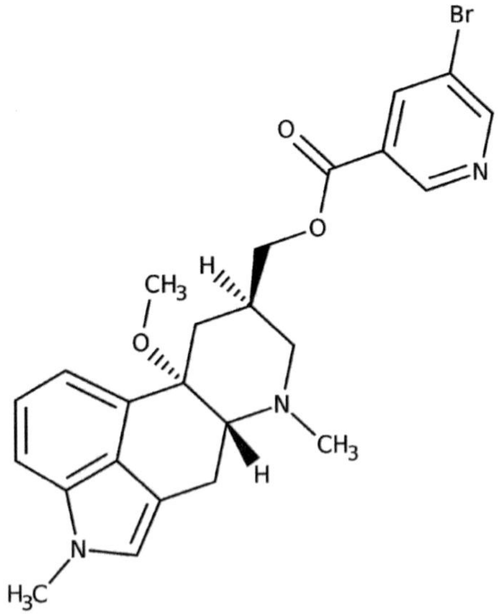

GENERAL

Nicergoline is an organonitrogen heterocyclic compound. As a potent vasodilator, it prompts a lowering of vascular resistance in the brain and an increase in arterial flow; it also stimulates the use of oxygen and glucose. It has been suggested that this may ameliorate cognitive deficits in cerebrovascular disease. Nicergoline also improves blood circulation in the lungs and limbs and has been shown to inhibit blood platelet aggregation. Nicergoline is indicated for the treatment of senile dementia, migraines of vascular origin, transient ischemia, platelet hyperaggregability, and macular degeneration (1).

CONTACT ALLERGY FROM ACCIDENTAL CONTACT

A chemical technician in a pharmaceutical plant had facial eczema and very striking symptoms of oculorhinitis, which had been present intermittently for about 3 months, and correlated with working, with improvement away from work. He was in contact with nebulized powders, especially of nicergoline, which was the final product, and also with its synthetic intermediates. Patch tests with the various compounds, obtained from the manufacturer, gave a very strong (+++) reaction to nicergoline 7% alc.; the reactions to 1% and 3% were negative. He also reacted, albeit weaker, to the intermediates 10α-methoxydihydrolysergol, 1N-methyl-10α-methoxydihydrolyergol, and lysergol 7%, but not 3% and 1%. Ten controls were negative. Prick tests (because of the oculorhinitis) were negative. A diagnosis of occupational airborne allergic contact dermatitis was made. Four days after the patient was transferred to another department, his symptoms cleared (2). Another patient from the same factory probably also had contact allergy to nicergoline. This individual had a long lasting patch test reaction to nicergoline 5% pet. and he also improved when transferred to another department (2).

LITERATURE

1 The data in the section 'General' may have been obtained from literature discussed in this chapter, but mostly also or exclusively from one or more of the following online sources: ChemIDPlus Advanced, PubChem, DrugBank, RxList, Drug Central, Drugs.com, and Wikipedia

2 Fumagalli M, Bigardi AS, Legori A, Pigatto PD. Occupational contact dermatitis from airborne nicergoline. Contact Dermatitis 1992;27:256

Chapter 3.339 NICOMORPHINE

IDENTIFICATION

Description/definition : Nicomorphine is the morphinane alkaloid that conforms to the structural formula shown below
Pharmacological classes : Narcotics
IUPAC name : [(4R,4aR,7S,7aR,12bS)-3-Methyl-9-(pyridine-3-carbonyloxy)-2,4,4a,7,7a,13-hexahydro-1H-4,12-methanobenzofuro[3,2-e]isoquinolin-7-yl] pyridine-3-carboxylate
Other names : Morphine dinicotinate
CAS registry number : 639-48-5
EC number : 211-357-9
Merck Index monograph : 7875
Patch testing : 1% pet. or water
Molecular formula : $C_{29}H_{25}N_3O_5$

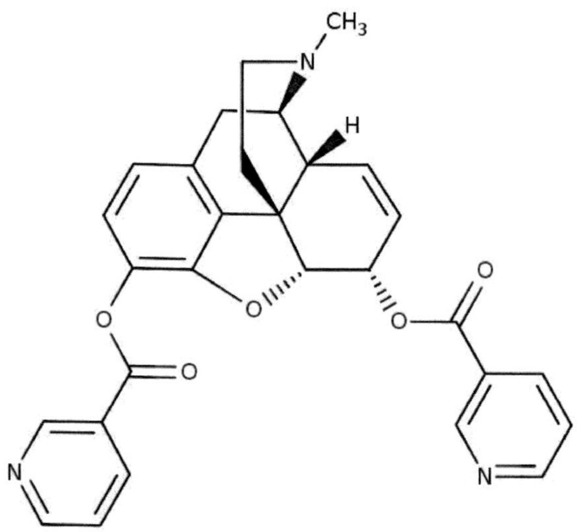

GENERAL

Nicomorphine is the 3,6-dinicotinate ester of morphine. It is a strong opioid agonist analgesic two to three times as potent as morphine with a side effect profile similar to that of morphine. It is used, particularly in the German-speaking countries and elsewhere in Central Europe and some other countries in Europe and the former USSR in particular, for post-operative, cancer, chronic non-malignant and neuropathic pain. It is commonly used in patient-controlled analgesia units (1).

CUTANEOUS ADVERSE DRUG REACTIONS FROM SYSTEMIC ADMINISTRATION CAUSED BY TYPE IV (DELAYED-TYPE) HYPERSENSITIVITY (as demonstrated by positive patch tests)

Cutaneous adverse drug reactions from systemic administration of nicomorphine caused by type IV (delayed-type) hypersensitivity have included urticaria-like exanthema (2)

Other cutaneous adverse drug reactions

In Bern, Switzerland, patients with a suspected allergic cutaneous drug reaction were patch-scratch tested with suspected drugs that had previously given a positive lymphocyte transformation test. Nicomorphine 10 mg/ml saline gave a positive patch-scratch test in one patient with urticaria-like exanthema appearing after several hours (2).

LITERATURE

1   The data in the section 'General' may have been obtained from literature discussed in this chapter, but mostly also or exclusively from one or more of the following online sources: ChemIDPlus Advanced, PubChem, DrugBank, RxList, Drug Central, Drugs.com, and Wikipedia
2   Neukomm C, Yawalkar N, Helbling A, Pichler WJ. T-cell reactions to drugs in distinct clinical manifestations of drug allergy. J Invest Allergol Clin Immunol 2001;11:275-284

Chapter 3.340 NIFUROXAZIDE

IDENTIFICATION

Description/definition   : Nifuroxazide is the benzoic acid and nitrofuran derivative that conforms to the structural
formula shown below
Pharmacological classes : Anti-infective agents
IUPAC name   : 4-Hydroxy-*N*-[(*E*)-(5-nitrofuran-2-yl)methylideneamino]benzamide
CAS registry number   : 965-52-6
EC number   : 213-522-0
Merck Index monograph : 7889
Patch testing   : 1% water (2); tablets, pulverized, 10% pet. and water (3); most systemic drugs can be
tested at 10% pet.; if the pure chemical is not available, prepare the test material from
intravenous powder, the content of capsules or – when also not available – from
powdered tablets to achieve a final concentration of the active drug of 10% pet.
Molecular formula   : $C_{12}H_9N_3O_5$

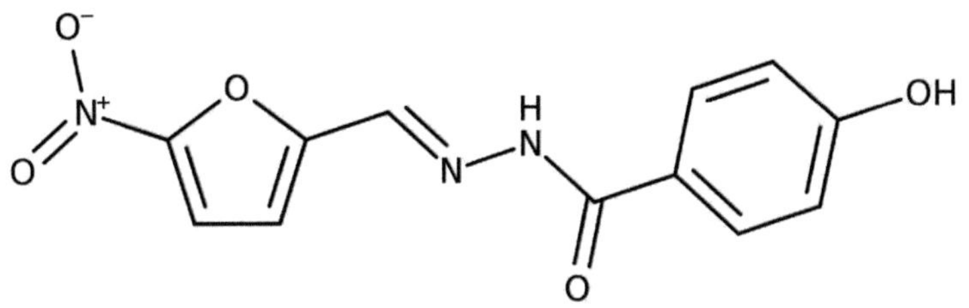

GENERAL

Nifuroxazide is an oral nitrofuran antibiotic used to treat mild diarrhea in humans and animals (1).

CONTACT ALLERGY FROM ACCIDENTAL CONTACT

A 47-year-old woman presented with extensive severely itching skin lesions on her arms and trunk, consisting of
erythema, papules and numerous erosions. The patient operated a dispensing apparatus filling glass bottles with
liquid medicines in a pharmaceutical plant. She associated her lesions with contact with nifuroxazide suspension.
Patch testing with the suspension and its ingredients yielded positive reactions to nifuroxazide 1%, 0.1%, 0.01% and
0.001% in water. Twenty controls were negative. There was no cross-reaction to the related nitrofurazone. The
patient was diagnosed with occupational allergic contact dermatitis to nifuroxazide resembling prurigo nodularis (2).

CUTANEOUS ADVERSE DRUG REACTIONS FROM SYSTEMIC ADMINISTRATION CAUSED BY TYPE IV
(DELAYED-TYPE) HYPERSENSITIVITY (as demonstrated by positive patch tests)

Cutaneous adverse drug reactions from systemic administration of nifuroxazide caused by type IV (delayed-type)
hypersensitivity have included acute generalized exanthematous pustulosis (AGEP) (3).

Acute generalized exanthematous pustulosis (AGEP)

A 77-year-old man presented with a generalized erythematous and pustular eruption with numerous confluent
pustules, notably in the axillary and inguinal folds and with purpuric lesions on the legs, accompanied by high fever.
The rash had erupted acutely one day after the patient had taken various medications including nifuroxazide for
diarrhea. Bacteriological and mycological cultures of pustules were sterile. Histopathology showed multilocular
subcorneal pustules and marked edema of the papillary dermis with a perivascular infiltrate with numerous
polymorphonuclear neutrophils. After complete healing, patch tests were performed with all suspected drugs 10%
pet. and water and they were positive to nifuroxazide 10% pet. and water at D2 and D4 with erythematous papules
and some pustules. The histology of the patch test was very similar to that of the eruption.

LITERATURE

1   The data in the section 'General' may have been obtained from literature discussed in this chapter, but mostly
also or exclusively from one or more of the following online sources: ChemIDPlus Advanced, PubChem,
DrugBank, RxList, Drug Central, Drugs.com, and Wikipedia
2   Kieć-Swierczyńska M, Krecisz B. Occupational contact allergy to nifuroxazide simulating prurigo nodularis.
Contact Dermatitis 1998;39:93-94
3   Machet L, Jan V, Machet MC, Lorette G, Vaillant L. Acute generalized exanthematous pustulosis induced by
nifuroxazide. Contact Dermatitis 1997;36:308-309

Chapter 3.341 NIMESULIDE

IDENTIFICATION

Description/definition : Nimesulide is the nonsteroidal arylsulfonamide that conforms to the structural formula shown below
Pharmacological classes : Anti-inflammatory agents, non-steroidal; cyclooxygenase inhibitors
IUPAC name : N-(4-Nitro-2-phenoxyphenyl)methanesulfonamide
CAS registry number : 51803-78-2
EC number : 257-431-4
Merck Index monograph : 7903
Patch testing : 1% and 5% pet. (6); 10% pet. (8)
Molecular formula : $C_{13}H_{12}N_2O_5S$

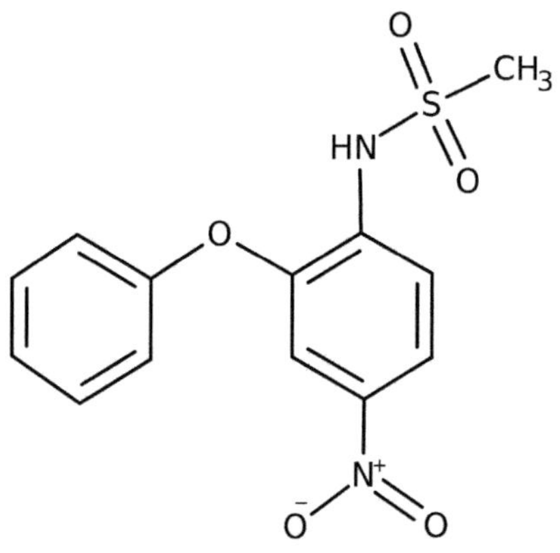

GENERAL

Nimesulide is an arylsulfonamide and nonsteroidal anti-inflammatory drug (NSAID) with anti-inflammatory, antipyretic and analgesic properties. It inhibits the cyclooxygenase-mediated (mostly COX-2) conversion of arachidonic acid to pro-inflammatory prostaglandins. Nimesulide is indicated for the treatment of acute pain, and the symptomatic treatment of osteoarthritis and primary dysmenorrhea. Due to concerns about the risk of hepatotoxicity, nimesulide has been withdrawn from the market in many countries (1).

CUTANEOUS ADVERSE DRUG REACTIONS FROM SYSTEMIC ADMINISTRATION CAUSED BY TYPE IV
(DELAYED-TYPE) HYPERSENSITIVITY (as demonstrated by positive patch tests)

Cutaneous adverse drug reactions from systemic administration of nimesulide caused by type IV (delayed-type) hypersensitivity have included acute localized exanthematous pustulosis (ALEP) (6) and fixed drug eruption (2,3,4,5,7).

Acute localized exanthematous pustulosis

Four hours after taking a first dose of nimesulide for neck pain, a 35-year old woman developed symmetrical erythematous facial swelling associated with a pruritic, burning sensation localized to the face and neck. Within 24 hours the swelling decreased and numerous tiny pustules appeared. A bacterial culture of a swab taken from the pustules was negative. The eruption resolved spontaneously within 4 days with mild desquamation. Six months later, patch tests were weakly positive to nimesulide 1% and positive to nimesulide 5% pet. at D4. Twenty controls were negative. The authors diagnosed toxic pustuloderma, which is currently called acute localized exanthematous pustulosis (ALEP) (6).

Fixed drug eruptions

Case series

In a University hospital in Coimbra, Portugal, in the period 1990-2009, 52 patients (17 men, 35 women, mean age 53±17 years) with a clinical diagnosis of fixed drug eruptions were submitted to patch tests with the suspected drugs. Patch tests on pigmented lesions were reactive in 21 of the 52 (40%) patients, 20 NSAIDs and one antihistamine. Of 27 individuals tested with nimesulide 1% and 10% pet., 9 (33%) had a positive patch test on postlesional skin (2). These probably include the three patients previously reported with FDE from nimesulide who had positive patch tests on residual pigmented FDE lesions to (pure) nimesulide 5% and 10% pet. (7).

In 5 patients with fixed drug eruptions that were with high probability caused by nimesulide, patch tests were performed on residual lesional skin and on the normal skin of the back. In all 5 cases, there were positive reactions on the postlesional, but not on normal skin (3).

Case reports

A 23-year-old woman presented with a 3-year history of recurrent dysuria, vulvar bleeding and painful oral ulcerations that healed spontaneously in approximately 10 days. The symptoms became worse with every relapse, leading to conjunctival hyperemia and eyelid edema. Laboratory investigations were normal. The patient was treated with prednisone and azathioprine, but in spite of this, erythematous macules appeared on the legs, regressing within a few hours. With every next episode, new lesions extended to the trunk and became more numerous, pruritic, erythematous and violaceous, leaving residual pigmentation. Fixed drug eruption was now considered and the patient admitted to occasional use of nimesulide. Patch tests were applied to normal and residual skin and there were positive reactions to nimesulide 5% and 10% in petrolatum (negative to 1%) on postlesional skin only. The patient was diagnosed with multiple fixed drug eruption from nimesulide (4).

A 41-year man presented with a circular, well-delineated, violaceous, elevated non-pruritic lesion on the right leg, which had appeared approximately 18 hours after ingesting nimesulide and clarithromycin for acute pharyngitis. He stopped both drugs but 48 hours later vesicles developed in the lesion. Patch tests were positive to nimesulide 10% pet on the residual lesion only (5).

LITERATURE

1   The data in the section 'General' may have been obtained from literature discussed in this chapter, but mostly also or exclusively from one or more of the following online sources: ChemIDPlus Advanced, PubChem, DrugBank, RxList, Drug Central, Drugs.com, and Wikipedia

2   Andrade P, Brinca A, Gonçalo M. Patch testing in fixed drug eruptions – a 20-year review. Contact Dermatitis 2011;65:195-201

3   Gonçalo M, Oliveira HS, Fernandes B, Robalo-Cordeiro M, Figueiredo A. Topical provocation in fixed drug eruption from nonsteroidal anti-inflammatory drugs. Exogenous Dermatol 2002;1:81-86

4   Marques LP, Villarinho ALCF, Melo MDGM, Torre MGS. Fixed drug eruption to nimesulide: an exuberant presentation confirmed by patch testing. An Bras Dermatol 2018;93:470-472

5   Malheiro D, Cadinha S, Rodrigues J, Vaz M, Castel-Branco MG. Nimesulide-induced fixed drug eruption. Allergol Immunopathol (Madr) 2005;33:285-287

6   Lateo S, Boffa MJ. Localized toxic pustuloderma associated with nimesulide therapy confirmed by patch testing. Br J Dermatol 2002;147:624-625

7   Cordeiro MR, Gonçalo M, Fernandes B, Oliveira H, Figueiredo A. Positive lesional patch tests in fixed drug eruptions from nimesulide. Contact Dermatitis 2000;43:307

8   Brockow K, Garvey LH, Aberer W, Atanaskovic-Markovic M, Barbaud A, Bilo MB, et al.; ENDA/EAACI Drug Allergy Interest Group. Skin test concentrations for systemically administered drugs – an ENDA/EAACI Drug Allergy Interest Group position paper. Allergy 2013;68:702-712

Chapter 3.342 NIMODIPINE

IDENTIFICATION

Description/definition : Nimodipine is the dihydropyridine derivative that conforms to the structural formula shown below

Pharmacological classes : Calcium channel blockers; antihypertensive agents; vasodilator agents

IUPAC name : 3-*O*-(2-Methoxyethyl) 5-*O*-propan-2-yl 2,6-dimethyl-4-(3-nitrophenyl)-1,4-dihydropyridine-3,5-dicarboxylate

CAS registry number : 66085-59-4

EC number : 266-127-0

Merck Index monograph : 7906

Patch testing : Tablet, pulverized, as is (2); most systemic drugs can be tested at 10% pet.; if the pure chemical is not available, prepare the test material from intravenous powder, the content of capsules or – when also not available – from powdered tablets to achieve a final concentration of the active drug of 10% pet.

Molecular formula : $C_{21}H_{26}N_2O_7$

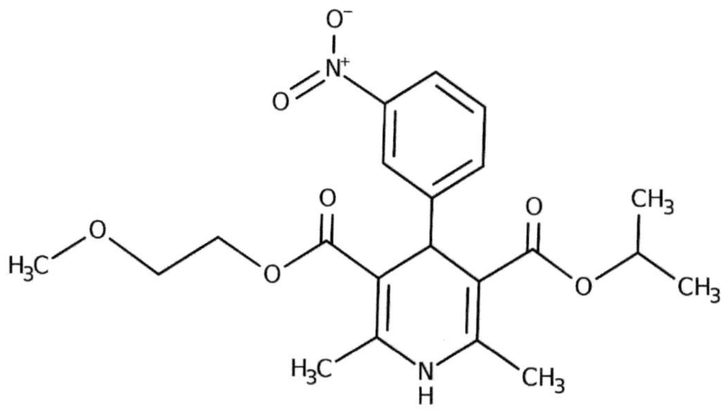

GENERAL

Nimodipine is a 1,4-dihydropyridine calcium channel blocker with preferential cerebrovascular activity. It has marked vasodilating effects and lowers blood pressure. Nimodipine inhibits the transmembrane influx of calcium ions in response to depolarization in smooth muscle cells, thereby inhibiting vascular smooth muscle contraction and inducing vasodilatation. Nimodipine is indicated for use as an adjunct to improve neurologic outcome following subarachnoid hemorrhage from ruptured intracranial berry aneurysms by reducing the incidence and severity of ischemic deficits (1).

CUTANEOUS ADVERSE DRUG REACTIONS FROM SYSTEMIC ADMINISTRATION CAUSED BY TYPE IV (DELAYED-TYPE) HYPERSENSITIVITY (as demonstrated by positive patch tests)

Cutaneous adverse drug reactions from systemic administration of nimodipine caused by type IV (delayed-type) hypersensitivity have included maculopapular exanthem turning into generalized eczema (2).

Other cutaneous adverse drug reactions

A 70-year-old man developed dyspnea and an erythematous maculopapular rash while being treated with various medications including nimodipine. Symptoms promptly receded after administration of corticosteroids and antihistamines, but when the anti-inflammatory effect of corticosteroids ended, the patient developed vesiculation of the oral mucosa and a generalized itchy eczema. Patch tests with all drugs used were positive only to commercial nimodipine, tested 'as is' (D3 +, D6 ++). Patch tests to 2 other commercial preparations containing nimodipine, in order to exclude reactions to excipients, were also positive, but 5 controls were negative. There were no cross-reactions to other calcium channel blockers, amlodipine and nifedipine (2).

LITERATURE

1   The data in the section 'General' may have been obtained from literature discussed in this chapter, but mostly also or exclusively from one or more of the following online sources: ChemIDPlus Advanced, PubChem, DrugBank, RxList, Drug Central, Drugs.com, and Wikipedia

2   Nucera E, Schavino D, Roncallo C, de Pasquale T, Buonomo A, Pollastrini E, et al. Delayed-type allergy to oral nimodipine. Contact Dermatitis 2002;47:246-247

Chapter 3.343 NITROFURANTOIN

IDENTIFICATION

Description/definition   : Nitrofurantoin is the synthetic derivative of imidazolidinedione that conforms to the structural formula shown below
Pharmacological classes : Antibacterial agents; anti-infective agents, urinary
IUPAC name   : 1-[(E)-(5-Nitrofuran-2-yl)methylideneamino]imidazolidine-2,4-dione
CAS registry number   : 67-20-9
EC number   : 200-646-5
Merck Index monograph   : 7956
Patch testing   : 5% pet. (5)
Molecular formula   : $C_8H_6N_4O_5$

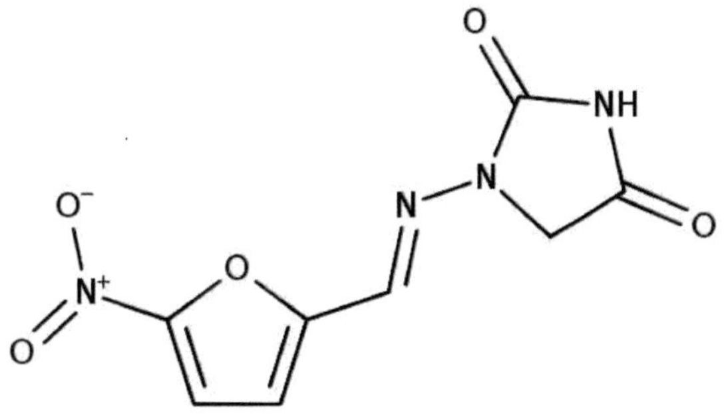

GENERAL

Nitrofurantoin is an oral antibiotic widely used either short term to treat acute urinary tract infections or long term as chronic prophylaxis against recurrent infections. It is effective against most gram-positive and gram-negative organisms. The drug inhibits bacterial DNA, RNA, and cell wall protein synthesis. In pharmaceutical products, nitrofurantoin is employed as nitrofurantoin monohydrate (CAS number 17140-81-7, EC number not available, molecular formula $C_8H_8N_4O_6$) or as nitrofurantoin sodium (CAS number 54-87-5, EC number 200-216-7, molecular formula $C_8H_5N_4NaO_5$) (1).

CUTANEOUS ADVERSE DRUG REACTIONS FROM SYSTEMIC ADMINISTRATION CAUSED BY TYPE IV (DELAYED-TYPE) HYPERSENSITIVITY (as demonstrated by positive patch tests)

Cutaneous adverse drug reactions from systemic administration of nitrofurantoin caused by type IV (delayed-type) hypersensitivity have included systemic allergic dermatitis (3).

Systemic allergic dermatitis (systemic contact dermatitis)

In an individual occupationally sensitized to nitrofurazone, oral administration of nitrofurantoin caused a drug rash, suggesting cross-sensitivity and systemic allergic dermatitis; a patch test to nitrofurantoin was positive (2).

Cross-reactions, pseudo-cross-reactions and co-reactions

A patient occupationally sensitized to nitrofurazone may have cross-reacted to nitrofurantoin (2). Primary sensitization to nitrofurazone and cross-sensitivity to nitrofurantoin has also been suggested in an early German article (4).

LITERATURE

1 The data in the section 'General' may have been obtained from literature discussed in this chapter, but mostly also or exclusively from one or more of the following online sources: ChemIDPlus Advanced, PubChem, DrugBank, RxList, Drug Central, Drugs.com, and Wikipedia
2 Jirasek L, Kalensky J. Kontakni alergicky ekzem z krmnych smesi v zivocisne vyrobe. Ceskoslovenska Dermatologie 1975;50:217 (Article in Czech, cited in ref. 3).
3 De Groot AC, Conemans JM. Contact allergy to furazolidone. Contact Dermatitis 1990;22:202-205
4 Behrbohm P, Zschunke E. Ekzem durch Nifucin (Nitrofural). Deutsche Gesundheitswesen 1967;22:273-275 (Article in German, data cited in ref. 2)
5 De Groot AC. Patch testing, 4th edition. Wapserveen, The Netherlands: acdegroot publishing, 2018 (ISBN 9789081323345)

Chapter 3.344 NITROFURAZONE

IDENTIFICATION

Description/definition	: Nitrofurazone is the nitrofuran that conforms to the structural formula shown below
Pharmacological classes	: Anti-infective agents
IUPAC name	: [(E)-(5-Nitrofuran-2-yl)methylideneamino]urea
Other names	: Nitrofural; hydrazinecarboxamide, 2-[(5-nitro-2-furanyl)methylene]-; 5-nitro-2-furaldehyde semicarbazone; Furacin ®
CAS registry number	: 59-87-0
EC number	: 200-443-1
Merck Index monograph	: 7957
Patch testing	: 1.0% pet. (Chemotechnique, SmartPracticeCanada, SmartPracticeEurope)
Molecular formula	: $C_6H_6N_4O_4$

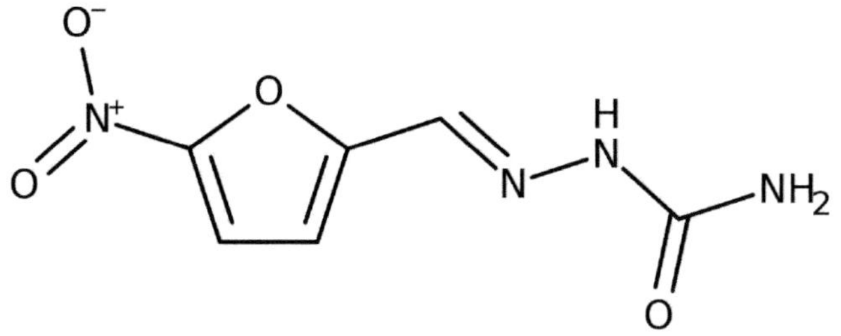

GENERAL

Nitrofurazone is a nitrofuran topical antibacterial agent with bactericidal activity against a number of pathogens, including *Staphylococcus aureus* and *Escherichia coli*; it does also have significant activity against *Pseudomonas aeruginosa*, *Proteus mirabilis*, and *Serratia marcescens*. Nitrofurazone is indicated for the topical treatment of bacterial skin infections including pyodermas, infected dermatoses and infections of cuts, wounds, burns and ulcers caused by susceptible organisms. Nitrofurazone was formerly used orally in humans. Veterinary use is mainly in the treatment and prophylaxis of coccidiosis in poultry and necrotic enteritis in pigs, both by administering the drug systemically and adding it to animal feed (1).

In topical preparations, nitrofurazone has caused large numbers of cases of contact allergy/allergic contact dermatitis, which subject has been fully reviewed in Volume 3 of the *Monographs in contact allergy* series (11).

CONTACT ALLERGY FROM ACCIDENTAL CONTACT

A 34-year-old cattle breeder gave a 3-year history of erythematous vesicular lesions on the sides of the fingers of both hands, on the forearms and the face. The lesions cleared when not working with cattle, when he also handled feeds and medicaments. Patch tests were positive to a uterine ovule for cattle and its ingredient nitrofurazone 1% pet. (9).

Airborne occupational allergic contact dermatitis has been caused by nitrofurazone from its presence in a powdered aquarium water additive (10). Two cases of occupational sensitization from nitrofurazone in animal feed were reported from the USA (7,8). In one (8), no patch tests were performed, but nitrofurazone allergy was highly likely.

Cross-reactions, pseudo-cross-reactions and co-reactions

Of 41 patients sensitized to nitrofurazone, 2 cross-reacted to nifurprazine 0.2% water (5). Of 7 patients sensitized to nitrofurazone, 6 cross-reacted to nitrofurantoin powder 50%; it is unknown whether proper controls have been tested as well (6). In an individual occupationally sensitized to nitrofurazone, oral administration of nitrofurantoin caused a drug rash, suggesting cross-sensitivity; a patch test to nitrofurantoin was positive (4). Primary sensitization to furazolidone may have caused cross-sensitivity to nitrofurazone in one patient (3).

LITERATURE

1 The data in the section 'General' may have been obtained from literature discussed in this chapter, but mostly also or exclusively from one or more of the following online sources: ChemIDPlus Advanced, PubChem, DrugBank, RxList, Drug Central, Drugs.com, and Wikipedia
2 De Groot AC, Conemans JM. Contact allergy to furazolidone. Contact Dermatitis 1990;22:202-205
3 Laubstein H, Niedergesäss G. Untersuchungen über Gruppensensibilisierungen bei Nitrofuranderivaten. Derm Monatsschrift 1970;156:1-8 (Article in German, data cited in ref. 2)

4   Jirasek L, Kalensky J. Kontakni alergicky ekzem z krmnych smesi v zivocisne vyrobe. Ceskoslovenska
     Dermatologie 1975;50:217 (Article in Czech, cited in ref. 2).
5   Braun W. Kontaktallergien durch Nifurprazin (Carofur). Deutsche Medizinische Wochenschrift 1969;94:1685-
     1687 (Article in German, data cited in ref. 2)
6   Behrbohm P, Zschunke E. Ekzem durch Nifucin (Nitrofural). Deutsche Gesundheitswesen 1967;22:273-275
     (Article in German, data cited in ref. 2)
7   Caplan RM. Cutaneous hazards posed by agricultural chemicals. J Iowa Med Soc 1969;59:295-299 (cited in ref. 2)
8   Neldner KH. Contact dermatitis from animal feed additives. Arch Dermatol 1972;106:722-723
9   Condé-Salazar L, Guimaraens D, Gonzalez MA, Molina A. Occupational allergic contact dermatitis from
     nitrofurazone. Contact Dermatitis 1995;32:307-308
10   Lo JS, Taylor JS, Oriba H. Occupational allergic contact dermatitis to airborne nitrofurazone. Dermatol Clin
     1990;8:165-168
11   De Groot AC. Monographs in contact allergy, volume 3: Topical Drugs. Boca Raton, Fl, USA: CRC Press Taylor and
     Francis Group, 2021 (ISBN 978-0-367-23693-9)

Chapter 3.345 NORFLOXACIN

IDENTIFICATION

Description/definition : Norfloxacin is the synthetic fluoroquinolone that conforms to the structural formula shown below
Pharmacological classes : Anti-bacterial agents; topoisomerase II inhibitors; cytochrome P-450 CYP1A2 inhibitors
IUPAC name : 1-Ethyl-6-fluoro-4-oxo-7-piperazin-1-ylquinoline-3-carboxylic acid
CAS registry number : 70458-96-7
EC number : 274-614-4
Merck Index monograph : 8059
Patch testing : 10.0% pet. (Chemotechnique)
Molecular formula : $C_{16}H_{18}FN_3O_3$

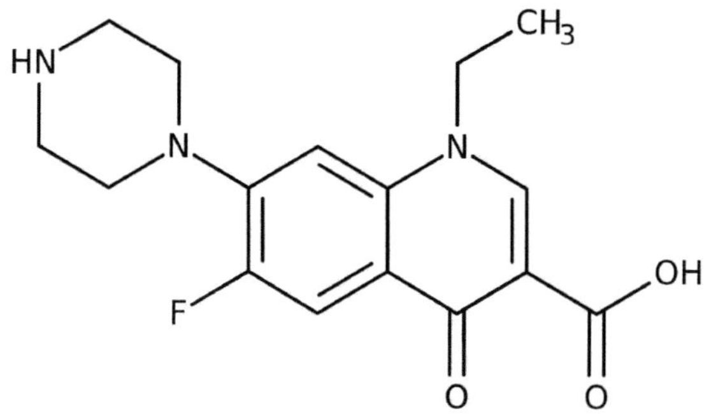

GENERAL

Norfloxacin is a synthetic fluoroquinolone with broad-spectrum antibacterial activity. It inhibits activity of DNA gyrase, thereby blocking bacterial DNA replication. Norfloxacin concentrates in the renal tubules and bladder and is bactericidal against a wide range of aerobic gram-positive and gram-negative organisms. This antibiotic is indicated for the treatment of urinary tract infections (1).

CUTANEOUS ADVERSE DRUG REACTIONS FROM SYSTEMIC ADMINISTRATION CAUSED BY TYPE IV (DELAYED-TYPE) HYPERSENSITIVITY (as demonstrated by positive patch tests)

Cutaneous adverse drug reactions from systemic administration of norfloxacin caused by type IV (delayed-type) hypersensitivity have included maculopapular eruption (3,5), systemic allergic dermatitis (2), and fixed drug eruption (4).

Maculopapular eruption

A patient who had previously suffered a severe maculopapular exanthema from delayed-type hypersensitivity to flucloxacillin, later again developed a maculopapular rash, now from norfloxacin; a positive patch test to this drug was obtained. As the patient had cutaneous adverse reactions to two unrelated drugs, this was a case of multiple drug hypersensitivity syndrome (3).

A 61-year-old woman developed an exanthema (macular or maculopapular, not specified) on the trunk and extremities 3 days after starting ciprofloxacin for cystitis. Six months later, she was given norfloxacin as treatment to a recurrent cystitis and, within hours after drug exposure, developed a generalized exanthema. Patch tests were positive to ciprofloxacin and norfloxacin pulverized tablets 10% or 25% pet. at D1 and/or D2 (not specified). Lymphocyte transformation tests were positive to both drugs (5).

Systemic allergic dermatitis (systemic contact dermatitis)

A 70-year-old man presented with an itchy eruption that had appeared 3 weeks earlier in the genital region, and then spread to the abdomen, lumbosacral region and thighs. It was characterized by merging papulopustular lesions on an erythematous edematous base. Three days before the onset of the eruption, the patient had received treatment with norfloxacin 400 mg/12 h for a urinary tract infection. Patch testing with the European standard series resulted in a very itchy, erythematous edematous reaction to the quinoline mix, associated with a flare-up of the papulopustular lesions in the genital region and inner thighs. Norfloxacin itself was apparently not tested. However, the structures of both constituents of the quinoline mix (clioquinol and chlorquinaldol) and of norfloxacin are based on the 4-oxo-1,4-dihydroquinoline ring. It is likely that this was a case of systemic allergic dermatitis caused by norfloxacin, cross-reacting to clioquinol and/or chlorquinaldol, to which the patient had previously become sensitized (2).

Fixed drug eruption

A 63-year-old woman presented with multiple tender erythematous-violaceous oval plaques on the neck, diameters ranging from 3 to 6 cm. She also reported polyarthralgia, malaise, and fever. The patient had taken norfloxacin 2 weeks ago for cystitis, and completed treatment 1 week prior to presentation. The lesions healed with hyperpigmentation. The patient returned 3 months later with an identical rash, which had started 2 days after again taking norfloxacin. Patch tests were positive to norfloxacin 10% pet. and ciprofloxacin 10% pet. on previously affected skin, but not normal skin. Histopathology was compatible with fixed drug eruption (4).

Cross-reactions, pseudo-cross-reactions and co-reactions

Norfloxacin has probably cross-reacted to clioquinol and/or chlorquinaldol (2). A patient sensitized to norfloxacin and presenting with fixed drug eruption cross-reacted to ciprofloxacin (4). A patient sensitized to ciprofloxacin probably cross-reacted to norfloxacin (5).

LITERATURE

1   The data in the section 'General' may have been obtained from literature discussed in this chapter, but mostly also or exclusively from one or more of the following online sources: ChemIDPlus Advanced, PubChem, DrugBank, RxList, Drug Central, Drugs.com, and Wikipedia
2   Silvestre JF, Alfonso R, Moragón M, Ramón R, Botella R. Systemic contact dermatitis due to norfloxacin with a positive patch test to quinoline mix. Contact Dermatitis 1998;39:83
3   Jörg L, Yerly D, Helbling A, Pichler W. The role of drug, dose and the tolerance/intolerance of new drugs in multiple drug hypersensitivity syndrome (MDH). Allergy 2020;75:1178-1187
4   Alpalhão M, Antunes J, Soares-Almeida L, Correia TE, Filipe P. Fixed drug eruption due to norfloxacin with cross-reactivity to ciprofloxacin: A case report. Contact Dermatitis 2020;83:135-137
5   Schmid DA, Depta JP, Pichler WJ. T cell-mediated hypersensitivity to quinolones. Clin Exp Allergy 2006;36:59-69

Chapter 3.346 NOSIHEPTIDE

IDENTIFICATION

Description/definition : Nosiheptide is a thiopeptide antibiotic produced by the bacterium *Streptomyces actuosus* that conforms to the structural formula shown below

Pharmacological classes : Anti-infective agents (veterinary)

IUPAC name : *N*-(3-Amino-3-oxoprop-1-en-2-yl)-2-[(21*Z*)-21-ethylidene-9,30-dihydroxy-18-(1-hydroxy-ethyl)-40-methyl-16,19,26,31,42,46-hexaoxo-32-oxa-3,13,23,43,49-pentathia-7,17,20,27,45,51,52,53,54,55-decazanonacyclo[26.16.6.1^{2,5}.1^{12,15}.1^{22,25}.1^{38,41}.1^{47,50}.0^{6,11}.0^{34,39}]penta-pentaconta-2(55),4,6,8,10,12(54),14,22(53),24,34(39),35,37,40,47,50-pentadecaen-8-yl]-1,3-thiazole-4-carboxamide

CAS registry number : 56377-79-8

EC number : 260-138-4

Merck Index monograph : 8078

Patch testing : 1% pet.

Molecular formula : $C_{51}H_{43}N_{13}O_{12}S_6$

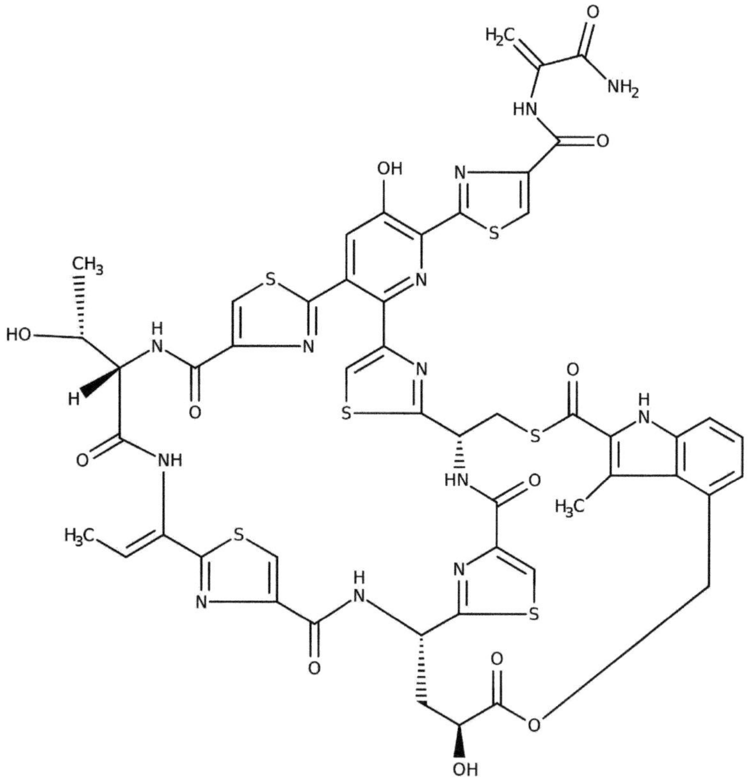

GENERAL

Nosiheptide is a thiopeptide antibiotic produced by the bacterium *Streptomyces actuosus*. It has potent activity against various bacterial pathogens, primarily gram-positive, including methicillin-resistant *Staphylococcus aureus*, penicillin-resistant *Streptococcus pneumoniae*, and vancomycin-resistant enterococci (1). It is used in veterinary medicine only (1).

CONTACT ALLERGY FROM ACCIDENTAL CONTACT

In Italy, during 1986-1988, 204 animal feed mill workers (191 men, 13 women) were patch tested with a large number of animal feed additives. There was one reaction to nosiheptide 1% pet. in the group of 168 individuals without skin complaints. All reactions were considered to be relevant (2).

LITERATURE

1   The data in the section 'General' may have been obtained from literature discussed in this chapter, but mostly also or exclusively from one or more of the following online sources: ChemIDPlus Advanced, PubChem, DrugBank, RxList, Drug Central, Drugs.com, and Wikipedia

2   Mancuso G, Staffa M, Errani A, Berdondini RM, Fabbri P. Occupational dermatitis in animal feed mill workers. Contact Dermatitis 1990;22:37-41

Chapter 3.347   NYLIDRIN

IDENTIFICATION

Description/definition   : Nylidrin is the phenylpropane that conforms to the structural formula shown below
Pharmacological classes   : Sympathomimetics; β-adrenergic agonists; tocolytic agents; vasodilator agents
IUPAC name   : 4-[1-Hydroxy-2-(4-phenylbutan-2-ylamino)propyl]phenol
Other names   : Buphenine; 1-p-hydroxyphenyl-2-(1'-methyl-3'-phenylpropylamino)-1-propanol
CAS registry number   : 447-41-6
EC number   : 207-182-2
Merck Index monograph   : 8091
Patch testing   : 1% alcohol and pet. (1)
Molecular formula   : $C_{19}H_{25}NO_2$

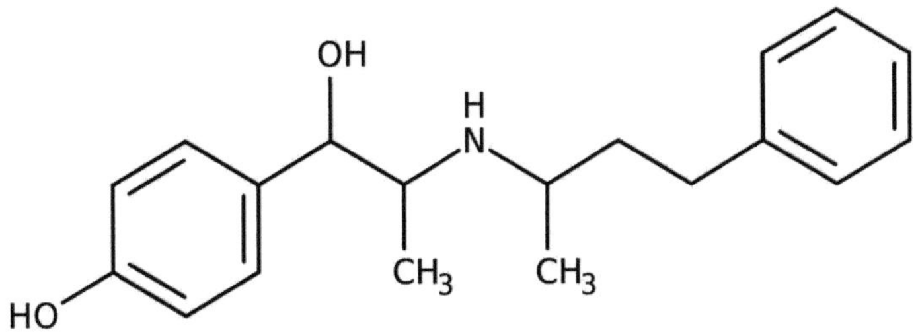

GENERAL

Nylidrin, also known as buphenine, is a β-adrenergic agonist with peripheral vasodilator properties. It is or has been utilized to treat disorders that may benefit from increased blood flow, for example certain mental disorders, blood vessel disease due to diabetes, frostbite, night leg cramps, and certain types of ulcers. Some studies have shown evidence of improving cognitive impairment in selected individuals, such as geriatric patients with mild to moderate symptoms of cognitive, emotional and physical impairment. However, FDA has considered nylidrin as 'lacking substantial evidence of effectiveness' in cerebral ischemia, cerebral arteriosclerosis, and other cerebral circulatory insufficiencies and has withdrawn nylidrin from the U.S. market. In pharmaceutical products, both nylidrin and nylidrin hydrochloride (CAS number 849-55-8, EC number 212-701-0, molecular formula $C_{19}H_{26}ClNO_2$) may be employed (go.drugbank.com). In topical preparations, nylidrin has rarely caused contact allergy/allergic contact dermatitis, which subject has been reviewed in Volume 3 of the *Monographs in contact allergy* series (3).

CUTANEOUS ADVERSE DRUG REACTIONS FROM SYSTEMIC ADMINISTRATION CAUSED BY TYPE IV (DELAYED-TYPE) HYPERSENSITIVITY (as demonstrated by positive patch tests)

Cutaneous adverse drug reactions from systemic administration of nylidrin caused by type IV (delayed-type) hypersensitivity have included acute generalized exanthematous pustulosis (AGEP) (4).

Acute generalized exanthematous pustulosis (AGEP)

A case of acute generalized exanthematous pustulosis (AGEP) to nylidrin with a positive and pustular patch test reaction to the drug was reported in French literature (4, data cited in ref. 5, details not available to the author).

Cross-reactions, pseudo-cross-reactions and co-reactions

Cross-reactivity from nylidrin to isoxsuprine or *vice versa* (2).

LITERATURE

1   De Groot AC. Patch testing, 4th edition. Wapserveen, The Netherlands: acdegroot publishing, 2018 (ISBN 9789081323345)
2   Alomar A. Buphenine sensitivity. Contact Dermatitis 1984;11:315
3   De Groot AC. Monographs in contact allergy, volume 3: Topical Drugs. Boca Raton, Fl, USA: CRC Press Taylor and Francis Group, 2021 (ISBN 978-0-367-23693-9)
4   Spindler E, Janier M, Bonnin JM, et al. Pustulose exanthématique aigüe generalisée liée à la buphénine: un cas. Ann Dermatol Venereol 1992;119:273-275 (Article in French, data cited in ref. 5)
5   Moreau A, Dompmartin A, Castel B, Remond B, Leroy D. Drug-induced acute generalized exanthematous pustulosis with positive patch tests. Int J Dermatol 1995;34:263-266

Chapter 3.348 NYSTATIN

IDENTIFICATION

Description/definition : Nystatin is a macrolide antifungal antibiotic complex produced by *Streptomyces noursei*, other *Streptomyces* species, and *S. aureus*; it conforms to the structural formula shown below
Pharmacological classes : Anti-bacterial agents; ionophores; antifungal agents
IUPAC name : (4E,6E,8E,10E,14E,16E,18S,19R,20R,21S,35S)-3-[(2S,3S,4S,5S,6R)-4-Amino-3,5-dihydroxy-6-methyloxan-2-yl]oxy-19,25,27,29,32,33,35,37-octahydroxy-18,20,21-trimethyl-23-oxo-22,39-dioxabicyclo[33.3.1]nonatriaconta-4,6,8,10,14,16-hexaene-38-carboxylic acid
CAS registry number : 1400-61-9
EC number : 215-749-0
Merck Index monograph : 8095
Patch testing : 2% pet. (SmartPracticeCanada, SmartPracticeEurope); testing in polyethylene glycol 400 (100,000 IU nystatin/gr) may be preferable over petrolatum (13)
Molecular formula : $C_{47}H_{75}NO_{17}$

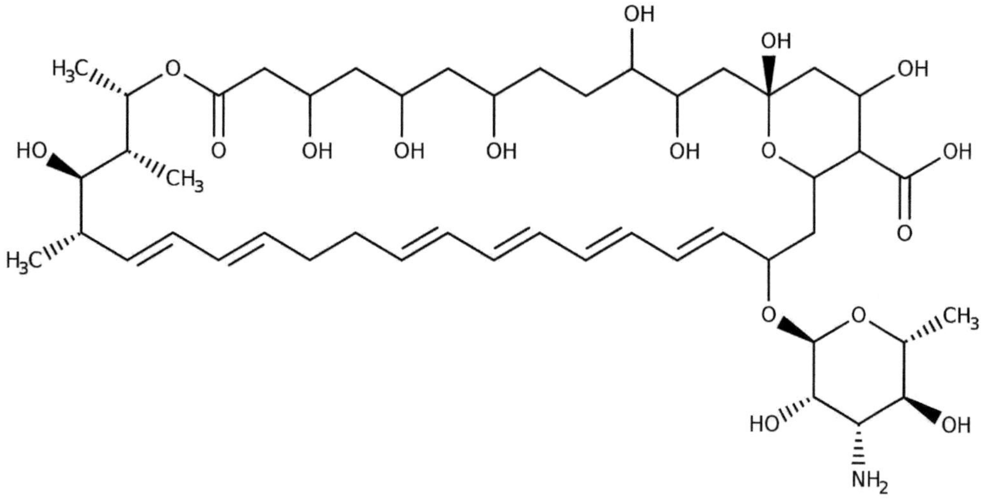

GENERAL

Nystatin is a macrolide antifungal antibiotic complex produced by *Streptomyces noursei, S. aureus*, and other *Streptomyces* species. The biologically active components are nystatin A1, A2, and A3. Nystatin is a topical and oral antifungal agent with activity against many species of yeast and *Candida albicans*, which is used largely to treat skin and oropharyngeal candidiasis. Nystatin is poorly absorbed from the gut. This agent acts by binding to sterols in the cell membrane of susceptible species resulting in a change in membrane permeability and the subsequent leakage of intracellular components (1). In topical preparations, the antifungal has caused several cases of contact allergy/ allergic contact dermatitis, which has been fully reviewed in Volume 3 of the *Monographs in contact allergy* series (12).

CUTANEOUS ADVERSE DRUG REACTIONS FROM SYSTEMIC ADMINISTRATION CAUSED BY TYPE IV (DELAYED-TYPE) HYPERSENSITIVITY (as demonstrated by positive patch tests)

Cutaneous adverse drug reactions from systemic administration of nystatin (despite its poor resorption from the gut after oral intake) caused by type IV (delayed-type) hypersensitivity have included maculopapular eruption (6,10), acute generalized exanthematous pustulosis (AGEP) (3,4,5,15), systemic allergic dermatitis (2,9,11), micropapular eruption (8), and erythematous exanthema with eczema of one hand (7).

Maculopapular eruption

One day after she had started treatment with oral nystatin for oropharyngeal candidiasis, a 65-year-old woman developed a generalized pruriginous maculopapular rash involving the face, trunk, arms, and thighs. Three months after regression, patch tests were positive to the suspension 'as is', to nystatin 30% water and to amphotericin B 1% water. Ten controls were negative. How the patient had become sensitized to nystatin was not mentioned (6).

Three days after a 76-year-old woman was prescribed oral nystatin for aphthous ulcers and other lesions in her mouth, she developed a generalized maculopapular rash, confluent and slightly pruritic. Her tongue was swollen and depapillated, with small erosions. Histopathology was consistent with a drug reaction. Oral challenge with 125,000 IU of nystatin resulted in a generalized pruritic erythema after 3 hours. Patch tests were positive to nystatin 30,000 and 90,000 IU/g in polyethylene glycol (PEG) and a negative reaction to PEG itself (10).

See also the section 'Systemic allergic dermatitis' below, ref. 9.

Acute generalized exanthematous pustulosis (AGEP)

Three patients had been treated with different oral preparations of nystatin. Within 1-2 days after ingestion, all the patients developed a generalized, pruritic, erythematous or urticarial rash which evolved into a pustular eruption during the next 1-2 days. Multiple swabs from pustular lesions were negative for bacterial growth. Histopathological examination of pustular lesions revealed similar changes in all patients: slight acanthosis and spongiosis with focal parakeratosis and a subcorneal pustule, edema and perivascular infiltrates of neutrophils and eosinophils as well as a few extravasated erythrocytes in the dermis without signs of vasculitis. Prick tests with the nystatin preparations were positive after 2 days as were patch tests at D2 and D3 (13 controls negative). There were no reactions to the other drugs used by the patients. One was tested with nystatin 10% pet. and reacted positively. The patients were diagnosed with acute generalized exanthematous pustulosis (AGEP) from type IV hypersensitivity to nystatin (3).

An 83-year-old woman had a 1-week history of fever, malaise, and skin rash, which had begun 2 days after starting treatment with nystatin mouthwashes for oral candidiasis. She presented erythroderma and the skin was covered with numerous non-follicular pustules of 1-3 mm in diameter. Blood tests revealed a white cell count of 18.400/mm^3 with neutrophilia (88.9%) and elevation of the acute-phase reactants (erythrocyte sedimentation rate, 63mm; C-reactive protein, 127mg/dl). Blood cultures and cultures of 2 pustules were negative. A biopsy showed a spongiform dermatitis with the formation of subcorneal pustules and a dermal infiltrate rich in polymorphonuclear cells and eosinophils. Six months after healing, she showed a positive patch test to nystatin 2% pet at D2 and D3 (11 controls negative). The patch test showed the same clinical and histological changes as the original exanthema, confirming the diagnosis of nystatin-induced acute generalized exanthematous pustulosis (AGEP) (15).

A 29 year-old man developed a flexural erythema one day after oral nystatin treatment, progressing towards a febrile pustular erythroderma, with elevated neutrophilic and eosinophilic counts. The lesions regressed rapidly with topical steroid treatment. Patch tests performed a few months later with the commercial nystatin preparation and nystatin itself were positive. AGEP from nystatin was diagnosed (4, no details available to the author).

A female patient was treated with oral nystatin tablets and suspension for gastrointestinal candidiasis. On the second day, macules and papules were seen in the axillae and on the lower legs. The treatment was continued and 2 days later, the patient had a generalized maculopapular hemorrhagic exanthema with fever and arthralgias. Four days later, disseminated small pustules had developed over the legs. Blood tests showed leukocytosis with neutrophilia. One year later, prick tests and patch tests with the nystatin preparations and her other medications were negative. However, intradermal tests with nystatin tablets and suspension showed erythema after 3 hours, 3 cm erythema with a central papule after one day and an infiltrated erythematous papule at D2. Histopathology was consistent with immune complex vasculitis. Two controls were negative, but after 3 weeks, both developed itching erythematous papules at the injection sites, indicative of sensitization from the test. The patient was diagnosed with acute generalized exanthematous pustulosis from nystatin caused by type-II or type-III hypersensitivity (5). However, the validity of the test method used was questioned (14).

Systemic allergic dermatitis (systemic contact dermatitis)

A 47-year-old man took 3 nystatin lozenges over 12 hours for oral candidiasis. He developed a maculopapular eruption over the trunk and limbs within 24 hours of the 3rd lozenge, associated with fever, arthralgia, malaise and diarrhea. He had previously used a variety of antifungal and antibacterial creams for intertrigo and medications for ulcerative colitis. Patch testing showed contact allergy to nystatin tested 20% pet. (9).

A case of systemic allergic dermatitis to nystatin after previous topical sensitization, presenting – presumably - as the baboon syndrome/symmetrical drug-related intertriginous and flexural exanthema (SDRIFE) has been reported in German literature (11). Details are unavailable to the author.

A woman with hand eczema noted worsening when treating it with nystatin ointment. Subsequently she was given nystatin orally and after 4 days developed a generalized eruption (not further specified). Patch tests were positive to nystatin (concentration and vehicle not specified) (2).

Symmetrical drug-related intertriginous and flexural exanthema (SDRIFE)/Baboon syndrome

See the section 'Systemic allergic dermatitis' above, ref. 11.

Other cutaneous adverse drug reactions

A 66-year-old woman was treated with oral nystatin 500,000 IU every 8 hours for suspected oral candidiasis. Two days later, the patient presented with erythematous macules on the abdomen and thighs as well as a larger erythematous and edematous lesion with papules and vesicles on the hypothenar eminence of the right hand. An oral challenge test with nystatin 500,000 IU reproduced the previous skin reaction within 8 hours. The lesion on the hand was biopsied and histologic examination revealed spongiosis, edema of the superficial dermis, perivascular lymphocytic infiltrates, and extravasated erythrocytes with no vasculitis. Patch tests were positive to nystatin 10% in petrolatum and nystatin 30,000 IU and 90,000 IU in polyethylene glycol; 7 controls were negative. The patient was diagnosed with systemic allergic contact dermatitis, but for that, strictly speaking, previous sensitization from topical exposure has to be demonstrated (7).

Twelve hours after a 24-year-old man had taken a tablet nystatin for oral candidiasis, he developed an itchy erythematous micropapular generalized eruption with facial angioedema. An oral challenge with nystatin 100,000 IU reproduced the symptoms after 6 hours. Patch tests were positive to nystatin 10% pet. and amphotericin B 5% DMSO (8).

Cross-reactions, pseudo-cross-reactions and co-reactions
Patients sensitized to (oral) nystatin sometimes cross-react to the structurally related amphotericin B (6,8).

LITERATURE
1 The data in the section 'General' may have been obtained from literature discussed in this chapter, but mostly also or exclusively from one or more of the following online sources: ChemIDPlus Advanced, PubChem, DrugBank, RxList, Drug Central, Drugs.com, and Wikipedia
2 Cronin E. Contact Dermatitis. Edinburgh: Churchill Livingstone, 1980:232-233
3 Küchler A, Hamm H, Weidenthaler-Barth B, Kämpgen E, Bröcker EB. Acute generalized exanthematous pustulosis following oral nystatin therapy: a report of three cases. Br J Dermatol 1997;137:808-811
4 Poszepczynska-Guigne E, Viguier M, Assier H, Pinquier L, Hochedez P, Dubertret L. Acute generalized exanthematous pustulosis induced by drugs with low-digestive absorption: acarbose and nystatin. Ann Dermatol Venereol 2003;130:439-442 (Article in French)
5 Rosenberger A, Tebbe B, Treudler R, Orfanos CE. Acute generalized exanthematous pustulosis, induced by nystatin. Hautarzt 1998;49:492-495 (Article in German)
6 Villas Martínez F, Muñoz Pamplona MP, García EC, Urzaiz AG. Delayed hypersensitivity to oral nystatin. Contact Dermatitis 2007;57:200-201
7 Vega F, Ramos T, Las Heras P, Blanco C. Concomitant sensitization to inhaled budesonide and oral nystatin presenting as allergic contact stomatitis and systemic allergic contact dermatitis. Cutis 2016;97:24-27
8 Barranco R, Tornero P, de Barrio M, de Frutos C, Rodríguez A, Rubio M. Type IV hypersensitivity to oral nystatin. Contact Dermatitis 2001;45:60
9 Cooper SM, Reed J, Shaw S. Systemic reaction to nystatin. Contact Dermatitis 1999;41:345-346
10 Quirce S, Parra F, Lázaro M, Gómez MI, Sánchez Cano M. Generalized dermatitis due to oral nystatin. Contact Dermatitis 1991;25:197-198
11 Lechner T, Grytzmann B, Baurle G. Hämatogenes allergisches Kontaktekzem nach oraler Gabe von Nystatin. Mykosen 1987;30:143-146 (Article in German)
12 De Groot AC. Monographs in contact allergy, volume 3: Topical Drugs. Boca Raton, Fl, USA: CRC Press Taylor and Francis Group, 2021 (ISBN 978-0-367-23693-9)
13 De Groot AC, Conemans JM. Nystatin allergy. Petrolatum is not the optimal vehicle for patch testing. Dermatol Clin 1990;8:153-155
14 Pryzbilla B, Ruëff F. AGEP induced by nystatin. Hautarzt 1999;50:136-138
15 Ocerin-Guerra I, Gomez-Bringas C, Aspe-Unanue L, Ratón-Nieto JA. Nystatin-induced acute generalized exanthematous pustulosis. Actas Dermosifiliogr 2012;103:927-928

Chapter 3.349 OFLOXACIN

IDENTIFICATION

Description/definition   : Ofloxacin is the synthetic fluoroquinolone that conforms to the structural formula shown
                            below
Pharmacological classes : Anti-bacterial agents; topoisomerase II inhibitors; cytochrome P-450 CYP1A2 inhibitors;
                            anti-infective agents, urinary
IUPAC name   : 9-Fluoro-3-methyl-10-(4-methyl-1-piperazinyl)-7-oxo-2,3-dihydro-7H-[1,4]oxazino[2,3,4-
                            ij]quinoline-6-carboxylic acid
CAS registry number   : 82419-36-1
EC number   : Not available
Merck Index monograph   : 8133
Patch testing   : 20% pet. (2); most systemic drugs can be tested at 10% pet.; if the pure chemical is not
                            available, prepare the test material from intravenous powder, the content of capsules or –
                            when also not available – from powdered tablets to achieve a final concentration of the
                            active drug of 10% pet.
Molecular formula   : $C_{18}H_{20}FN_3O_4$

GENERAL

Ofloxacin is a broad-spectrum fluoroquinolone bactericidal antibiotic active against both gram-positive and gram-negative bacteria. It binds to and inhibits bacterial topoisomerase II (DNA gyrase) and topoisomerase IV, enzymes involved in DNA replication and repair, resulting in cell death in sensitive bacterial species. Ofloxacin is indicated for the treatment of infections of the respiratory tract, urinary tract, skin, soft tissue, and urethral and cervical gonorrhea caused by susceptible strains of microorganisms (go.drugbank.com).

CUTANEOUS ADVERSE DRUG REACTIONS FROM SYSTEMIC ADMINISTRATION CAUSED BY TYPE IV (DELAYED-TYPE) HYPERSENSITIVITY (as demonstrated by positive patch tests)

Cutaneous adverse drug reactions from systemic administration of ofloxacin caused by type IV (delayed-type) hypersensitivity have included fixed drug eruption (1,2).

Fixed drug eruption

A 62-year-old woman presented with erythema and blisters on the right wrist, back of the left hand and right foot. Two days before, she had taken ofloxacin 200 mg orally and noticed pruritic erythema. Five months earlier, she had taken 200 mg of the same drug and had had erythema at the same sites. Patch tests were positive to ofloxacin and ciprofloxacin 20% pet. on postlesional skin only and the patient was diagnosed with fixed drug eruption (2).

In France, in the period 2005-2007, 59 cases of fixed drug eruptions were collected in 17 academic centers. There was one case of FDE to ofloxacin with a positive patch test. Clinical details were not provided (1).

Cross-reactions, pseudo-cross-reactions and co-reactions

A patient with a fixed drug eruption from ofloxacin had positive patch tests to both ofloxacin and ciprofloxacin (2).

LITERATURE

1   Brahimi N, Routier E, Raison-Peyron N, Tronquoy AF, Pouget-Jasson C, Amarger S, et al. A three-year-analysis of fixed drug eruptions in hospital settings in France. Eur J Dermatol 2010;20:461-464
2   Kawada A, Hiruma M, Noguchi H, Banba K, Ishibashi A, Banba H, et al. Fixed drug eruption induced by ofloxacin. Contact Dermatitis 1996;34:427

Chapter 3.350 OLANZAPINE

IDENTIFICATION

Description/definition   : Olanzapine is the synthetic derivative of thienobenzodiazepine that conforms to the structural formula shown below
Pharmacological classes : Antiemetics; antipsychotic agents; serotonin uptake inhibitors
IUPAC name   : Methyl-4-(4-methylpiperazin-1-yl)-10H-thieno[2,3-b][1,5]benzodiazepine
CAS registry number   : 132539-06-1
EC number   : Not available
Merck Index monograph   : 8184
Patch testing   : 0.1%, 1% and 5% pet.
Molecular formula   : $C_{17}H_{20}N_4S$

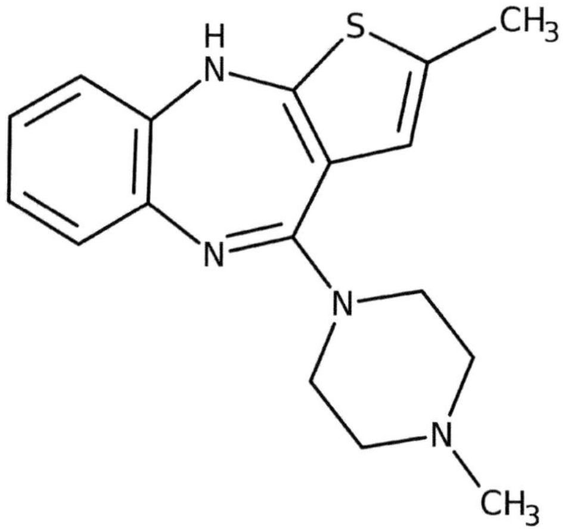

GENERAL

Olanzapine is a synthetic derivative of thienobenzodiazepine with antipsychotic, anti-nausea, and antiemetic activities; it may also stimulate appetite. As a selective monoaminergic antagonist, olanzapine binds with high affinity to the following receptors: serotoninergic, dopaminergic, muscarinic M_{1-5}, histamine H_1, and α_1-adrenergic receptors; it binds weakly to γ-aminobutyric acid type A, benzodiazepine, and β-adrenergic receptors. The main indications for olanzapine are schizophrenia, bipolar I disorder and treatment-resistant depression (1).

CONTACT ALLERGY FROM ACCIDENTAL CONTACT

A 51-year-old man, working as a process operator at a pharmaceutical plant, presented with two episodes of eyelid and hand dermatitis, which spread to his arm. The patient had worked on the production of olanzapine for 2 years prior to presentation. He gave a history of allergy to thiuram in rubber gloves diagnosed by patch testing 20 years earlier. He avoided thiuram-containing gloves from that time. Patch tests were positive to olanzapine 1% pet. (negative to 0.1% and 0.01%), to the thiuram-mix and to tetraethylthiuram disulfide. Ten controls were negative to olanzapine (4,8). This was a case of occupational airborne allergic contact dermatitis to olanzapine.

A 43-year-old pharmacist presented with a 1-year history of sporadic episodes of severely itching periorbital edema. Working in a pharmaceutical company, he would prepare the drugs olanzapine, raloxifene, and tadalafil, of which he suspected olanzapine of being the cause. His skin lesions subsided spontaneously in 3 days of leaving his work place, without requiring treatment. None of 15 other employees preparing the same drugs had skin complaints. Patch testing with olanzapine, raloxifene, and tadalafil in petrolatum at 1% and 5% were positive at D2 and D4 to olanzapine at both concentrations, more intense at 5% dilution. Ten controls were negative to olanzapine 1% pet. (10).

Another 2 cases of airborne occupational allergic contact dermatitis to olanzapine in two patients working at the same pharmaceutical plant were reported in 2008 (6). Both subjects had dermatitis of the face, and the second subject also had dermatitis affecting his hands and neck. They had positive patch tests to olanzapine powder 0.1% in pet. (9).

Olanzapine may have caused another case of occupational airborne allergic contact dermatitis in a pharmaceutical worker, who was also sensitized to sodium risedronate and amlodipine benzenesulfonate (2). However, the article, to which was referred in a recent textbook (3), could not be traced by the author.

CUTANEOUS ADVERSE DRUG REACTIONS FROM SYSTEMIC ADMINISTRATION CAUSED BY TYPE IV
(DELAYED-TYPE) HYPERSENSITIVITY (as demonstrated by positive patch tests)
Cutaneous adverse drug reactions from systemic administration of olanzapine caused by type IV (delayed-type)
hypersensitivity have included drug reaction with eosinophilia and systemic symptoms (DRESS) (5,6,7).

Drug reaction with eosinophilia and systemic symptoms (DRESS)
A 43 year-old man presented with a pruritic facial rash that rapidly spread all over his body with edema and
generalized lymphadenopathies. He used various medicines, but clinical signs were first noticed one week after the
introduction of olanzapine. Hepatomegaly, fever and cough with pulmonary sibilants, eosinophilia and atypical
lymphocytes were found. A chest X-ray was suggestive of interstitial pneumonopathy. Three months after resolution,
patch tests were positive to the commercialized form of olanzapine 30% pet. The patient was diagnosed with DRESS
from olanzapine (7).

A 48-year-old man using many drugs developed DRESS. Six weeks after complete recovery, patch tests were
performed with all drugs previously used at 10% in petrolatum and water and there was a positive reaction to
olanzapine (5). In a multicenter investigation in France, of 72 patients patch tested for DRESS, 46 (64%) had positive
patch tests to drugs, including one to olanzapine (6).

LITERATURE
1   The data in the section 'General' may have been obtained from literature discussed in this chapter, but mostly
     also or exclusively from one or more of the following online sources: ChemIDPlus Advanced, PubChem,
     DrugBank, RxList, Drug Central, Drugs.com, and Wikipedia
2   Chomiczewska D, Kiec-Swierczynska M, Krecisz B . Airborne occupational allergic contact dermatitis to sodium
     risedronate, olanzapine and amlodipine benzenesulfonate in a pharmaceutical company worker – a case report.
     Eur J Allergy Clin Immunol 2010;65:593-594 (article could not be traced, cited in ref. 3)
3   Goossens A, Geebelen J, Hulst KV, Gilissen L. Pharmaceutical and cosmetic industries. In: John S, Johansen J,
     Rustemeyer T, Elsner P, Maibach H, eds. Kanerva's occupational dermatology. Cham, Switzerland: Springer,
     2020:2203-2219
4   Bennett MF, Lowney AC, Bourke JF. A study of occupational contact dermatitis in the pharmaceutical industry.
     Br J Dermatol 2016;174:654-656 (Abstract in Brit J Dermatol 2011;165:73)
5   Darlenski R, Kazandjieva J, Tsankov N. Systemic drug reactions with skin involvement: Stevens-Johnson
     syndrome, toxic epidermal necrolysis, and DRESS. Clin Dermatol 2015;33:538-541
6   Barbaud A, Collet E, Milpied B, Assier H, Staumont D, Avenel-Audran M, et al. A multicentre study to determine
     the value and safety of drug patch tests for the three main classes of severe cutaneous adverse drug reactions.
     Br J Dermatol 2013;168:555-562
7   Prevost P, Bédry R, Lacoste D, Ezzedine K, Haramburu F, Milpied B. Hypersensitivity syndrome with olanzapine
     confirmed by patch tests. Eur J Dermatol 2012;22:126-127
8   Lowney AC, McAleer MA, Bourke J. Occupational allergic contact dermatitis to olanzapine. Contact Dermatitis
     2010;62:123-124
9   Walsh M, Mann R, Sansom J. Occupational allergic contact dermatitis to olanzapine: two cases. Br J Dermatol
     2008; 159(Suppl.1):81
10   Barata AR, Gomez LC, Arceo JE. Occupational airborne allergic contact dermatitis to olanzapine. Dermatitis
     2012;23:300-301

Chapter 3.351 OLAQUINDOX

IDENTIFICATION

Description/definition : Olaquindox is the quinoxaline derivative that conforms to the structural formula shown below

Pharmacological classes : Antibacterial; growth stimulant (veterinary drugs)

IUPAC name : *N*-(2-Hydroxyethyl)-3-methyl-4-oxido-1-oxoquinoxalin-1-ium-2-carboxamide

CAS registry number : 23696-28-8

EC number : 245-832-7

Merck Index monograph : 8186

Patch testing : 1.0% pet. (Chemotechnique)

Molecular formula : $C_{12}H_{13}N_3O_4$

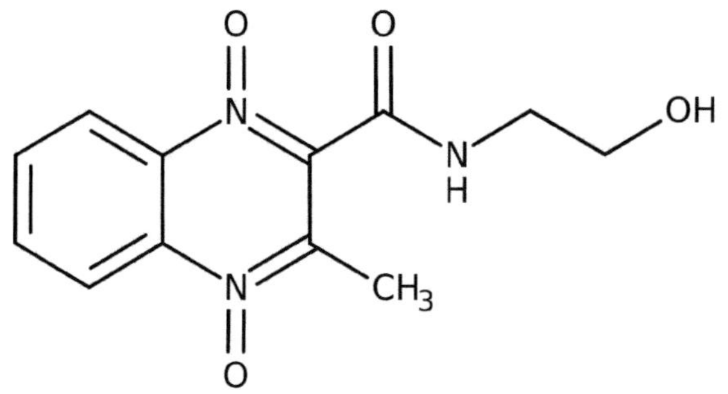

GENERAL

Olaquindox is a quinoxaline derivative with antibacterial properties. It is used in animal feed as an antibacterial agent to prevent dysentery and as a growth promoter in pigs (1). Olaquindox was withdrawn from the European Economic Community market in September 1999 because of its genotoxic and carcinogenic properties in experimental rodents (2).

CONTACT ALLERGY FROM ACCIDENTAL CONTACT

Olaquindox has been shown to be an important photosensitizing occupational hazard to pig farmers and employees working in pig farms from its present in animal feeds as a growth promotor. Rarely, a rabbit breeder (4) and a worker in an animal feed mill (6) were reported to be affected. Both direct contact with olaquindox in the feed (hands, arms) and airborne contact (face, neck) contributed to photosensitization. Olaquindox has caused occupational allergic contact dermatitis (4,16), photoaggravated allergic contact dermatitis (5,8 [2 patients],10,11,17), and photoallergic contact dermatitis (3,6,7 [2 patients],8 [13 patients],9,12,13,14 [3 patients],15,18 and 19). A large number of these individuals developed transient or persistent light sensitivity (3,8,11,12,13,14,17).

LITERATURE

1   The data in the section 'General' may have been obtained from literature discussed in this chapter, but mostly also or exclusively from one or more of the following online sources: ChemIDPlus Advanced, PubChem, DrugBank, RxList, Drug Central, Drugs.com, and Wikipedia

2   Anonymous. Council Directive. Official Journal of the European Community 347, 23.12.1998. 1998:31-32

3   Emmert B, Schauder S, Palm H, Hallier E, Emmert S. Disabling work-related persistent photosensitivity following photoallergic contact dermatitis from chlorpromazine and olaquindox in a pig breeder. Ann Agric Environ Med 2007;14:329-333

4   Sánchez-Pérez J, López MP, García-Díez A. Airborne allergic contact dermatitis from olaquindox in a rabbit breeder. Contact Dermatitis 2002;46:185

5   Belhadjali H, Marguery MC, Journé F, Giordano-Labadie F, Lefebvre H, Bazex J. Allergic and photoallergic contact dermatitis to olaquindox in a pig breeder with prolonged photosensitivity. Photodermatol Photoimmunol Photomed 2002;18:52-53

6   Sánchez-Pedreño P, Frías J, Martínez-Escribano J, Rodríguez M, Hernández-Carrasco S. Occupational photoallergic contact dermatitis to olaquindox. Dermatitis 2001;12:236-238

7   Lonceint J, Sassolas B, Guillet G. Photoallergic reactions to olaquindox in swine raisers: role of growth promotors used in feed. Ann Dermatol Venereol 2001;128:46-48 (Article in French)

8   Schauder S, Schröder W, Geier J. Olaquindox-induced airborne photoallergic contact dermatitis followed by transient or persistent light reactions in 15 pig breeders. Contact Dermatitis 1996;35:344-354

9   Kumar A, Freeman S. Photoallergic contact dermatitis in a pig farmer caused by olaquindox. Contact Dermatitis 1996;35:249-250

10   Fewings J, Horton J. Photoallergic dermatitis to a pig feed additive. Australas J Dermatol 1995;36:99

11   Willa-Craps C, Elsner P, Burg G. Olaquindox-induced persistent light reaction treated by *Escherichia coli* filtrate (Colibiogene). Dermatology 1995;191:343-344

12   Hochsattel R, Gall H, Weber L, Kaufmann R. Photoallergic reaction to olaquindox. Hautarzt 1991;42:233-236 (Article in German).

13   Dunkel FG, Elsner P, Pevny I, Burg G. Olaquindox-induced photoallergic contact dermatitis and persistent light reaction. Am J Contact Dermat 1990;1:235-239

14   Schauder S. The dangers of olaquindox. Photoallergy, chronic photosensitive dermatitis and extreme increased photosensitivity in the human, hypoaldosteronism in swine. Derm Beruf Umwelt 1989;37:183-185 (Article in German)

15   Francalanci S, Gola M, Giorgini S, Muccinelli A, Sertoli A. Occupational photocontact dermatitis from olaquindox. Contact Dermatitis 1986;15:112-114

16   PG, Goitre M, Cane D, Roncarolo G. Allergic contact dermatitis to Bayo-N-OX-I. Contact Dermatitis. 1985;12:284

17   Jagtmann BA. Foto-allergisch contacteczeem door het varkensvoeder additief olaquindox. Ned Tijdschr Derm Venereol 1991:1:77-78 (Article in Dutch)

18   Kiltting B, Brehler R, Forck G. Photokontaktallergie gegen Olaquindox bei Landwirten. Allergologie 1992;15:324 (Article in German)

19   Derharsching J. Photoallergisches Kontaktekzem auf Olaquindox. Z Hautkr 1994;69:548 (Article in German)

Chapter 3.352 OMEPRAZOLE

IDENTIFICATION

Description/definition : Omeprazole is the benzimidazole that conforms to the structural formula shown below
Pharmacological classes : Proton pump inhibitors; anti-ulcer agents
IUPAC name : 6-Methoxy-2-[(4-methoxy-3,5-dimethylpyridin-2-yl)methylsulfinyl]-1*H*-benzimidazole
CAS registry number : 73590-58-6
EC number : 615-996-8
Merck Index monograph : 8209
Patch testing : 0.1%, 0.5% and 1% pet. and alcohol
Molecular formula : $C_{17}H_{19}N_3O_3S$

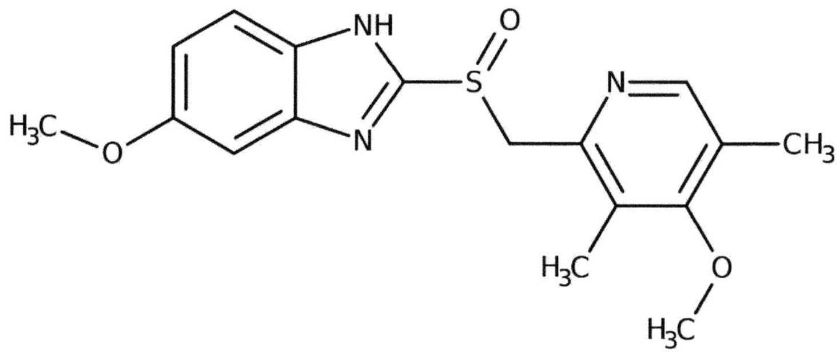

GENERAL

Omeprazole is a benzimidazole with selective and irreversible proton pump inhibition activity. Omeprazole is indicated for the treatment of active duodenal ulcer, eradication of *Helicobacter pylori* to reduce the risk of duodenal ulcer, treatment of active benign gastric ulcer, gastroesophageal reflux disease (GERD), erosive esophagitis (EE) due to acid-mediated GERD, for maintenance of healing of EE due to acid-mediated GERD and in pathologic hypersecretory conditions (e.g. Zollinger-Ellison syndrome). In pharmaceutical products, both omeprazole and omeprazole magnesium (CAS number 95382-33-5, EC number not available, molecular formula $C_{17}H_{19}MgN_3O_3S$) may be employed (1).

CONTACT ALLERGY FROM ACCIDENTAL CONTACT

General

In animal experiments, omeprazole was shown to have a strong sensitizing capacity (5,10). Indeed, in a pharmaceutical plant producing omeprazole, at least 38 workers became sensitized to this chemical in a 10-year period (5). Many other cases of occupational airborne allergic contact dermatitis from this proton pump inhibitor, sometimes from veterinary medications (9,11), have been reported (3,6,7,8,9,11,12,13,14,15). However, the form used in humans rarely causes skin rashes, owing to encapsulation of the drug.

Case series

Between 1989 and 1999, 96 workers who had been exposed to omeprazole during its manufacturing process developed skin reactions on the face, hands, and neck. The most frequent findings were itchy/dry skin (n=43), erythema (n=19), pruritus (n=19), rhinitis (n=5), dry skin (n=5), edema around the eyes and nasal congestion (both n=4). All subjects underwent a lymphocyte transformation test (LTT) and/or patch tests with omeprazole 0.1%, 0.5% and 1% in saline within 6 months of the clinical reaction. A total of 31 of the 96 subjects (32%) had a positive LTT result. Positive patch tests were obtained for 28 of the 84 patch tested individuals (33%); 23 of these also had a positive LTT. Thus, 38 patients appeared to have been sensitized on the basis of a positive patch test, positive LTT, or both. In a guinea pig maximization test, 18 of 20 animals were sensitized to omeprazole. It was concluded that the results confirmed the sensitizing potential of omeprazole from occupational exposure (5).

Three individuals aged 33,43 and 56, of who two worked as process operators and one as chemist in a pharmaceutical plant, developed occupational airborne allergic contact dermatitis to omeprazole. The primary sites of the dermatitis were the face and ears (16).

Case reports

A 49-year-old male pharmaceutical worker had been exposed periodically to omeprazole, lansoprazole, and pantoprazole for 21 years, when he noticed an itchy rash on his face, neck and arms. Patch tests were positive to omeprazole 0.1%, 0.5% and 1% saline (+), lansoprazole 10% and 50% saline (+++), and to pantoprazole 1%, 5% and 10% saline (+++) (3).

A 58-year-old male pharmaceutical worker presented with erythematous eruptions involving the neck, posterior auricular areas, and extensor forearms, with slight swelling and mild erythema of the upper eyelids. He had regular contact with proton pump inhibitors and intermediates including omeprazole magnesium, dexlansoprazole, esomeprazole magnesium, lansoprazole, and pantoprazole sodium. Patch tests were positive to omeprazole (10% pet., 1% pet., 1% alcohol), dexlansoprazole, pantoprazole and esomeprazole (1% pet.) (6).

Short summaries of other case reports of occupational allergic contact dermatitis from omeprazole are shown in table 3.352.1.

Table 3.352.1 Short summaries of case reports of occupational allergic contact dermatitis from omeprazole

Year and country	Sex	Age	Positive patch tests	Clinical data and comments	Ref.
2019 Croatia	F	52	OM 0.1% and 0.5% saline	chemist working with OM in analytical lab; eczema on the eyelids, face, neck and hands	7
2017 Spain	M	48	OM (no details known)	worker in pharmaceutical plant; also contact allergy to ranitidine	8
2014 United Kingdom	M	49	OM medicament (37%) 5% pet.	patient worked with horses and gives medication including OM 37% in a syringe; eczema on thumb, later scrotum, upper thigh, and eyelids; veterinary medicine	9
2014 Canada	F	44	OM medicament (37%) 'as is'; OM crushed tablet 1% and 10% pet.	horse trainer giving omeprazole medication; dermatitis of eyelids, face, neck and forearms; breath of the horse on her face would cause a flare; same medication as in ref. 9; multiple cross-reactions to related proton pump inhibitors	11, 12
2011 Spain	M	60	OM 0.1%, 0.5% and 1% saline	worker in pharmaceutical company; rash in face and neck with swelling of the eyelids; also type-I allergy to omeprazole and to fluoxetine (concomitant sensitization)	13
2007 Spain	M	25	OM 0.1% and 0.5% alcohol	worker in pharmaceutical company; dermatitis of eyelids, face, neck and hands	14
	M	35	OM 0.5% alcohol; negative to 0.1% alc.	supervisor in the same pharmaceutical company; dermatitis of the eyelids, nose and perioral area	
1986 Sweden	F	25	OM 1%, 0.5% and 0.25% pet.	chemist working with omeprazole; eczema on the eyelids	15
	F	59	OM 1%, 0.5% and 0.1% alcohol	handling experimental animals in the same pharmaceutical company; eczema on the eyelids	

OM: omeprazole

CUTANEOUS ADVERSE DRUG REACTIONS FROM SYSTEMIC ADMINISTRATION CAUSED BY TYPE IV (DELAYED-TYPE) HYPERSENSITIVITY (as demonstrated by positive patch tests)

Cutaneous adverse drug reactions from systemic administration of omeprazole caused by type IV (delayed-type) hypersensitivity have included maculopapular eruption (19), Stevens-Johnson syndrome (18), and delayed urticaria (19).

Maculopapular eruption
Six patients with nonimmediate adverse cutaneous drug reactions from omeprazole had positive patch tests to this proton pump inhibitor. It is uncertain which drug eruptions these patients had suffered from, but very likely included maculopapular eruption and delayed urticaria (19).

Stevens-Johnson syndrome/toxic epidermal necrolysis (SJS/TEN)
A 69-year-old man was treated with omeprazole for gastric ulcers, when he developed a maculopapular exanthema on the trunk with targetoid lesions on the palms. The following day, the eruption rapidly progressed to the face and legs, with small vesicles arising on the palms. Two days later, the patient developed positive Nikolsky sign on the back, bilateral conjunctivitis and ulcerations of the oral and genital mucosa. Skin biopsies showed subepidermal blistering with a lymphocytic infiltrate and widespread necrotic keratinocytes, confirming the clinical diagnosis of Stevens-Johnson syndrome. Eight months after complete recovery, patch testing was performed with omeprazole 0.5%, 1% and 10% in petrolatum. Readings on days 2 and 4 were positive for all three concentrations (18).

Other cutaneous adverse drug reactions
Six patients with nonimmediate adverse cutaneous drug reactions from omeprazole had positive patch tests to this proton pump inhibitor. It is uncertain which drug eruptions these patients had suffered from, but very likely included delayed urticaria and maculopapular eruption (19).

Cross-reactions, pseudo-cross-reactions and co-reactions
Two patients sensitized to lansoprazole cross-reacted to omeprazole (2,4). A patient sensitized to omeprazole cross-reacted to dexlansoprazole, rabeprazole, pantoprazole, and esomeprazole (all crushed tablets, 1% and 10% pet.) (11,12).

LITERATURE

1   The data in the section 'General' may have been obtained from literature discussed in this chapter, but mostly also or exclusively from one or more of the following online sources: ChemIDPlus Advanced, PubChem, DrugBank, RxList, Drug Central, Drugs.com, and Wikipedia
2   Vilaplana J, Romaguera C. Allergic contact dermatitis due to lansoprazole, a proton pump inhibitor. Contact Dermatitis 2001;44:47-48
3   Alarcón M, Herrera-Mozo I, Nogué S, et al. Occupational airborne contact dermatitis from proton pump inhibitors. Curr Allergy Clin Immunol 2014;27:310-313
4   Liippo J, Pummi K, Hohenthal U, Lammintausta K. Patch testing and sensitization to multiple drugs. Contact Dermatitis 2013;69:296-302
5   Ghatan PH, Marcusson-Ståhl M, Matura M, Björkheden C, Lundborg P, Cederbrant K. Sensitization to omeprazole in the occupational setting. Contact Dermatitis 2014;71:371-375
6   DeKoven JG, Yu AM. Occupational airborne contact dermatitis from proton pump inhibitors. Dermatitis 2015;26:287-290
7   Jurakić Tončić R, Balić A, Pavičić B, Žužul K, Petković M, Bartolić L. Occupational airborne contact dermatitis caused by omeprazole. Acta Dermatovenerol Croat 2019;27:188-189
8   Herrera-Mozo I, Sanz-Gallen P, Martí-Amengual G. Occupational contact allergy to omeprazole and ranitidine. Med Pr 2017;68:433-435
9   Alwan W, Banerjee P, White IR. Occupational contact dermatitis caused by omeprazole in a veterinary medicament. Contact Dermatitis 2014;71:376
10 Hausen BM, Lücke R, Rothe E, Erdogan A, Rinder H. Sensitizing capacity of azole derivatives: Part III. Investigations with anthelmintics, antimycotics, fungicides, antithyroid compounds, and proton pump inhibitors. Am J Contact Dermat 2000;11:80-88
11 Al-Falah K, Schachter J, Sasseville D. Occupational allergic contact dermatitis caused by omeprazole in a horse breeder. Contact Dermatitis 2014;71:377-378
12 Al-Falah K, Schachter J, Sasseville D. Occupational airborne allergic contact dermatitis from omeprazole with cross-reactions to other proton pump inhibitors. Dermatitis 2017;27(5):e1
13 Sanz-Gallén P, Nogué S, Herrera-Mozo I, Delclos GL, Valero A. Occupational contact allergy to omeprazole and fluoxetine. Contact Dermatitis 2011;65:118-119
14 Conde-Salazar L, Blancas-Espinosa R, Pérez-Hortet C. Occupational airborne contact dermatitis from omeprazole. Contact Dermatitis 2007;56:44-46
15 Meding B. Contact allergy to omeprazole. Contact Dermatitis 1986;15:36
16 Bennett MF, Lowney AC, Bourke JF. A study of occupational contact dermatitis in the pharmaceutical industry. Br J Dermatol 2016;174:654-656
17 Brockow K, Garvey LH, Aberer W, Atanaskovic-Markovic M, Barbaud A, Bilo MB, et al.; ENDA/EAACI Drug Allergy Interest Group. Skin test concentrations for systemically administered drugs – an ENDA/EAACI Drug Allergy Interest Group position paper. Allergy 2013;68:702-712
18 Van Tendeloo E, Gutermuth J, Grosber M. Positive patch testing with omeprazole in Stevens-Johnson syndrome: a case report. J Eur Acad Dermatol Venereol 2021;35:e74-e75
19 Hammami S, Affes H, Ksouda K, Feki M, Sahnoun Z, Zeghal KM. Étude de l'allergie croisée entre les différents inhibiteurs de la pompe à protons [Study of cross reactivity between proton pump inhibitors]. Therapie 2013;68:361-368 (Article in French)

Chapter 3.353 ONABOTULINUMTOXINA

IDENTIFICATION

Description/definition   : OnabotulinumtoxinA is the neurotoxin protein produced by the bacterium *Clostridium*
                              botulinum
Pharmacological classes   : Toxins
IUPAC name   : Not available
Other names   : Botulin A; Botulinum toxin type A
CAS registry number   : 93384-43-1
EC number   : Not available
Merck Index monograph : 2632 (Botulin toxins)
Patch testing   : Commercial preparation pure (2)
Molecular formula   : Unspecified

GENERAL

OnabotulinumtoxinA is a purified version of a toxin produced by the bacterium *Clostridium botulinum*. It prevents the release of the neurotransmitter acetylcholine from axon endings at the neuromuscular junction, thus causing flaccid paralysis. In cosmetic medicine, it is the most commonly injected substance (Botox ®) to diminish the appearance of facial wrinkles by relaxing the muscles surrounding the injection site. Other indications for its use include muscle spasticity, strabismus, excessive sweating and the prophylaxis of migraine (1,2).

CUTANEOUS ADVERSE DRUG REACTIONS FROM SYSTEMIC ADMINISTRATION CAUSED BY TYPE IV (DELAYED-TYPE) HYPERSENSITIVITY (as demonstrated by positive patch tests)

Cutaneous adverse drug reactions from systemic administration of onabotulinumtoxinA caused by type IV (delayed-type) hypersensitivity have included itching and disorientation (2).

Other cutaneous adverse drug reactions

A healthy 43-year-old woman received Botox ® treatment a total of 4 times. The first resulted in a swollen eyelid on the morning after treatment (which is a frequent non-allergic reaction to injections in that area). Following the second and third treatments, local itchiness continued, and the patient was slightly disoriented. Following the third and fourth treatments, the patient took an over-the-counter antihistamine that provided little to no relief. The fourth Botox treatment resulted in 'the systemic reaction of concern'. Approximately 36 hours after receiving Botox the patient experienced pruritus in her hands and feet, which gradually became widespread except for the face and scalp. After the itching began, a small flat red dot appeared at the injection site. The total duration of the 'allergic reaction' was approximately 3 weeks (from initial report to complete resolution of symptoms).Prick and intradermal tests with the commercial preparation were negative. A patch test (probably with the undiluted product) resulted in induration of 22×29 mm and erythema of 34×40 mm when read at 3 days. Control testing was not performed and the separate ingredients were not tested (2). The lack of objective cutaneous lesions and the inadequate patch testing procedure strongly limit the value of this report and quite rightfully the article received a level 5 (the lowest) rating for evidence.

LITERATURE

1   The data in the section 'General' may have been obtained from literature discussed in this chapter, but mostly also or exclusively from one or more of the following online sources: ChemIDPlus Advanced, PubChem, DrugBank, RxList, Drug Central, Drugs.com, and Wikipedia
2   Rosenfield LK, Kardassakis DG, Tsia KA, Stayner G. The first case report of a systemic allergy to onabotulinumtoxinA (Botox) in a healthy patient. Aesthet Surg J 2014;34:766-768

Chapter 3.354 ORNIDAZOLE

IDENTIFICATION

Description/definition	: Ornidazole is the nitroimidazole that conforms to the structural formula shown below
Pharmacological classes	: Amebicides; antitrichomonal agents; radiation-sensitizing agents
IUPAC name	: 1-Chloro-3-(2-methyl-5-nitroimidazol-1-yl)propan-2-ol
CAS registry number	: 16773-42-5
EC number	: 240-826-0
Merck Index monograph	: 8237
Patch testing	: Commercial ornidazole 30% and 50% pet. (2); most systemic drugs can be tested at 10% pet.; if the pure chemical is not available, prepare the test material from intravenous powder, the content of capsules or – when also not available – from powdered tablets to achieve a final concentration of the active drug of 10% pet.
Molecular formula	: $C_7H_{10}ClN_3O_3$

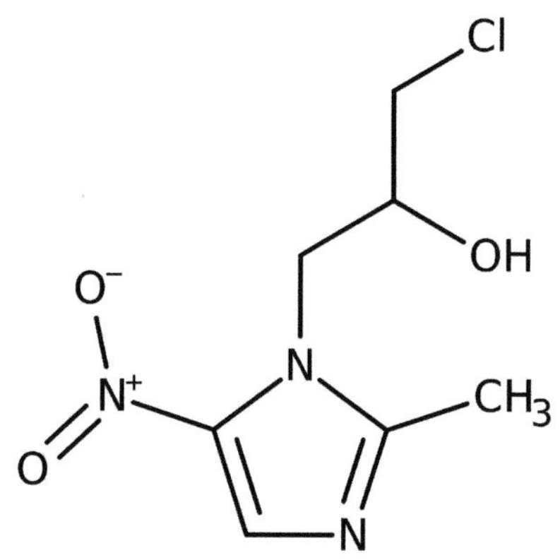

GENERAL

Ornidazole is a nitroimidazole antiprotozoal agent used in the treatment of susceptible protozoal infections (e.g. *Trichomonas vaginalis*) and for the treatment of anaerobic bacterial infections, often in combination with antibiotics such as ofloxacin or levofloxacin. It is partially plasma-bound and also has radiation-sensitizing action (1).

CUTANEOUS ADVERSE DRUG REACTIONS FROM SYSTEMIC ADMINISTRATION CAUSED BY TYPE IV (DELAYED-TYPE) HYPERSENSITIVITY (as demonstrated by positive patch tests)

Cutaneous adverse drug reactions from systemic administration of ornidazole caused by type IV (delayed-type) hypersensitivity have included fixed drug eruption (2).

Fixed drug eruption

A 42-year-old woman reported that she had developed round, solitary, red macules with central vesicles on the dorsal surface of both hands and elbows within 3-4 h after taking fluconazole for vaginal candidiasis. The maculae had faded in a few days leaving slight hyperpigmentation. A second episode followed the intake of ornidazole for vaginal trichomoniasis. She again developed three erythematous macules at the same sites. There were 2 more similar episode after being treated with fluconazole and ornidazole. Physical examination showed typical oval hyperpigmentations dorsally on both hands and elbows. Occlusive patch tests at previously affected sites and on unaffected skin with commercial ornidazole 30% and 50% pet. and commercial fluconazole 10% pet. were strongly positive only with 30% and 50% ornidazole (2).

LITERATURE

1   The data in the section 'General' may have been obtained from literature discussed in this chapter, but mostly also or exclusively from one or more of the following online sources: ChemIDPlus Advanced, PubChem, DrugBank, RxList, Drug Central, Drugs.com, and Wikipedia

2   Bavbek S, Yilmaz İ, Sözener ZÇ. Fixed drug eruption caused by ornidazole and fluconazole but not isoconazole, itraconazole, ketoconazole and metronidazole. J Dermatol 2013;40:134-135

Chapter 3.355 OXACILLIN

IDENTIFICATION

Description/definition   : Oxacillin is the semisynthetic penicillinase-resistant penicillin that conforms to the
                            structural formula shown below
Pharmacological classes   : Anti-bacterial agents
IUPAC name   : (2S,5R,6R)-3,3-Dimethyl-6-[(5-methyl-3-phenyl-1,2-oxazole-4-carbonyl)amino]-7-oxo-4-
                            thia-1-azabicyclo[3.2.0]heptane-2-carboxylic acid
Other names   : Oxazocillin; MPI-penicillin
CAS registry number   : 66-79-5
EC number   : 200-635-5
Merck Index monograph   : 8273
Patch testing   : Tablet, pulverized, 30% pet. (2); most systemic drugs can be tested at 10% pet.; if the pure
                            chemical is not available, prepare the test material from intravenous powder, the content
                            of capsules or – when also not available – from powdered tablets to achieve a final
                            concentration of the active drug of 10% pet.; most beta-lactams can be tested 5% pet.
Molecular formula   : $C_{19}H_{19}N_3O_5S$

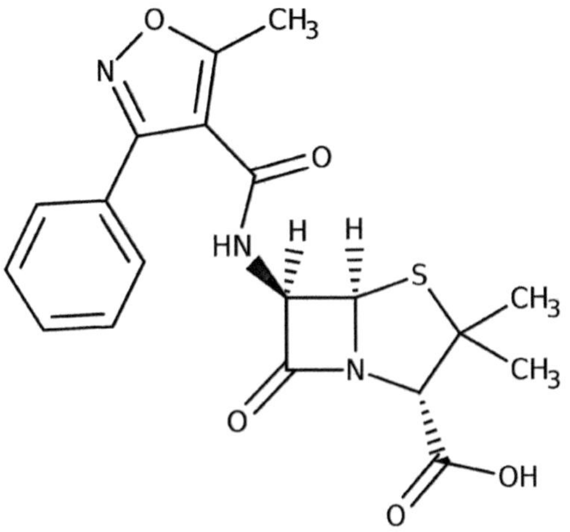

GENERAL

Oxacillin is a semisynthetic acid-stable β-lactam penicillin with antimicrobial activity against gram-positive and gram-negative aerobic and anaerobic bacteria. It is stable against hydrolysis by a variety of β-lactamases, including penicillinases, cephalosporinases and extended spectrum β-lactamases. Oxacillin binds to penicillin-binding proteins in the bacterial cell wall, thereby blocking the synthesis of peptidoglycan, a critical component of the bacterial cell wall. This leads to inhibition of cell growth and causes cell lysis. Oxacillin is indicated for treatment of resistant staphylococcal infections (1). In pharmaceutical products (powder for solution, solution) oxacillin is employed as oxacillin sodium hydrate, also termed oxacillin sodium (CAS number 7240-38-2, EC number 214-636-3, molecular formula $C_{19}H_{20}N_3NaO_6S$) (1). The classification and structures of beta-lactam antibiotics are discussed in Chapter 3.36 Amoxicillin.

CONTACT ALLERGY FROM ACCIDENTAL CONTACT

Between 1978 and 2001, 14,689 patients have been patch tested in the university hospital of Leuven, Belgium. Occupational allergic contact dermatitis to pharmaceuticals was diagnosed in 33 health care workers: 26 nurses, 4 veterinarians, 2 pharmacists and one medical doctor. There were 26 women and 7 men with a mean age of 38 years. Practically all of these patients presented with hand dermatitis (often the fingers), sometimes with secondary localizations, frequently the face. In this group, two patients had occupational allergic contact dermatitis from oxacillin (3).

CUTANEOUS ADVERSE DRUG REACTIONS FROM SYSTEMIC ADMINISTRATION CAUSED BY TYPE IV
(DELAYED-TYPE) HYPERSENSITIVITY (as demonstrated by positive patch tests)

Cutaneous adverse drug reactions from systemic administration of oxacillin caused by type IV (delayed-type) hypersensitivity have included acute generalized exanthematous pustulosis (AGEP) (2) and drug reaction with eosinophilia and systemic symptoms (DRESS) (4).

Acute generalized exanthematous pustulosis (AGEP)

Two days after having taken oxacillin and tiaprofenic acid for treating acute otitis, a 32-year-old woman developed fever and a generalized eruption of erythema and non-follicular pustules. The patient was known with recurrent generalized pustular psoriasis, but her skin condition had been under control with mild local steroids recently. The current eruption was most pronounced in the major flexures, with widespread erythema covered by numerous pinpoint pustules. Laboratory investigations showed an elevated white blood cell count with neutrophilia, elevated C-reactive protein level and negative microbiological findings in pustules and blood. A skin biopsy showed the histological features of a drug-induced pustular eruption with subcorneal splitting, neutrophilic spongiosis, sporadic apoptotic cells, massive superficial dermal oedema, and moderate mixed infiltrates containing some eosinophils. The patient was diagnosed with AGEP. Discontinuation of oxacillin and initiation of treatment with systemic cortico-steroids led to rapid resolution within 10 days. Ten weeks later, patch tests with commercialized oxacillin and tiaprofenic acid 30% pet. were strongly positive (+++) only to oxacillin on D2 and D4 with erythema, infiltration and multiple superficial pustules and lakes of pus (2).

Drug reaction with eosinophilia and systemic symptoms (DRESS)

A child of 9 years had an abscess with *Staphylococcus aureus* secreting a Panton-Valentine toxin with non-severe pleuritis and pericarditis. Treatment was started with amoxicillin-clavulanic acid, amikacin, and clindamycin followed by oxacillin, rifampicin, and colchicine. On the 25th day of treatment, she had a recurrence of fever with a genera-lized rash, moderate hepatic cytolysis, hypereosinophilia, and activated lymphocytes, suggesting visceral DRESS syndrome. A skin biopsy confirmed the diagnosis. After complete resolution, patch tests were positive to amoxicillin-clavulanic acid, amoxicillin and oxacillin and negative to the other drugs used (4).

Cross-reactions, pseudo-cross-reactions and co-reactions

Cross-reactions between beta-lactam antibiotics are discussed in Chapter 3.36 Amoxicillin.

LITERATURE

1 The data in the section 'General' may have been obtained from literature discussed in this chapter, but mostly also or exclusively from one or more of the following online sources: ChemIDPlus Advanced, PubChem, DrugBank, RxList, Drug Central, Drugs.com, and Wikipedia

2 Gammoudi R, Ben Salem C, Boussofara L, Fathallah N, Ghariani N, Slim R, et al. Acute generalized exanthematous pustulosis induced by oxacillin confirmed by patch testing. Contact Dermatitis 2018;79:108-110

3 Gielen K, Goossens A. Occupational allergic contact dermatitis from drugs in healthcare workers. Contact Dermatitis 2001;45:273-279

4 Rabenkogo A, Vigue MG, Jeziorski E. Le syndrome DRESS: une toxidermie à connaître [DRESS syndrome]. Arch Pediatr 2015;22:57-62 (Article in French)

Chapter 3.356 OXCARBAZEPINE

IDENTIFICATION

Description/definition : Oxcarbazepine is the dibenzazepine that conforms to the structural formula shown below
Pharmacological classes : Anticonvulsants; cytochrome P-450 CYP3A inducers; voltage-gated sodium channel
blockers
IUPAC name : 5-Oxo-6H-benzo[b][1]benzazepine-11-carboxamide
CAS registry number : 28721-07-5
EC number : Not available
Merck Index monograph : 8298
Patch testing : 10% pet. (10); with serious cutaneous adverse reactions, starting with 1% pet. may be
advisable (10)
Molecular formula : $C_{15}H_{12}N_2O_2$

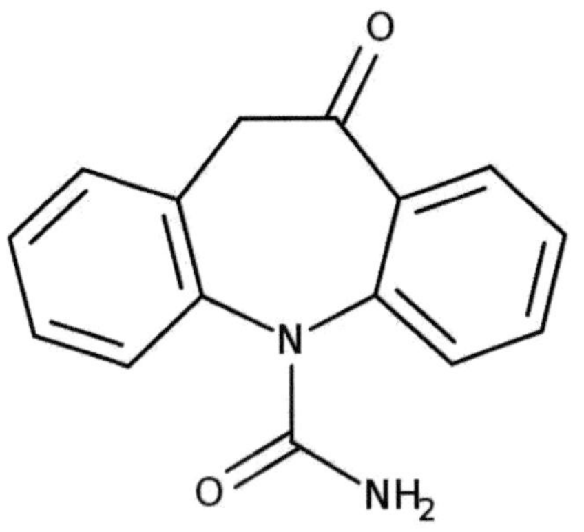

GENERAL

Oxcarbazepine is an aromatic keto-analog of carbamazepine and, like the parent drug, is a potent anticonvulsant, used alone or in combination with other agents in the therapy of partial seizures. Although the mechanism of action has not been fully elucidated, electrophysiological studies indicate this agent blocks voltage-gated sodium channels, thereby stabilizing hyper-excited neural membranes, inhibiting repetitive neuronal firing, and decreasing the propagation of synaptic impulses (1).

CUTANEOUS ADVERSE DRUG REACTIONS FROM SYSTEMIC ADMINISTRATION CAUSED BY TYPE IV (DELAYED-TYPE) HYPERSENSITIVITY (as demonstrated by positive patch tests)

Cutaneous adverse drug reactions from systemic administration of oxcarbazepine caused by type IV (delayed-type) hypersensitivity have included maculopapular eruption (7), fixed drug eruption (3), drug reaction with eosinophilia and systemic symptoms (DRESS)/anticonvulsant hypersensitivity syndrome (4,5,7), urticarial exanthem (2), and unspecified drug eruption (6)

Maculopapular eruption

A patient who had developed DRESS from carbamazepine, after resolution took two dosages of oxcarbazepine and developed a maculopapular eruption. Patch tests were positive to both commercial carbamazepine and oxcarbazepine 10% and 30% pet. (7).

Fixed drug eruption

A 52-year-old man presented with a 2-day history of burning, erythematous, erosive patches on the glans penis. He had been treated with oxcarbazepine, piracetam, and prothipendyl for withdrawal of alcohol addiction. Fixed drug eruption was suspected. Only oxcarbazepine treatment was discontinued and the lesions healed within 1 week with topical corticosteroids, leaving mild hyperpigmentation. The patient had repeatedly undergone short-term treatment with oxcarbazepine in the past. A topical provocation test was positive at the site of a previous lesion on the patient's glans penis with commercial oxcarbazepine 5% saline, showing sharply demarcated erythema and mild infiltration at D2 and D3, and negative on normal skin to oxcarbazepine 5% saline and 5% pet. (3).

Drug reaction with eosinophilia and systemic symptoms (DRESS)/anticonvulsant hypersensitivity syndrome
The literature on patch testing in anticonvulsant hypersensitivity syndrome up to August 2008 has been reviewed in ref. 9.

A 40-year-old woman developed drug reaction with eosinophilia and systemic symptoms (DRESS) while using oxcarbazepine, carbamazepine and fluvoxamine. Patch tests were positive to fluvoxamine 12.5%, carbamazepine 20% and oxcarbazepine 12.5%, all in phosphate-buffered saline. Lymphocyte transformation tests were also positive to carbamazepine (stimulation index 28.5), oxcarbazepine (stimulation index 6) and to fluvoxamine (stimulation index 6.9). As the patient had reactions to 2 unrelated drugs, this was a case of multiple drug hypersensitivity (4).

A 6-year old girl developed DRESS while on carbamazepine with erythroderma and edema of the hands, feet and face. When it was replaced with oxcarbazepine, an initial improvement was observed, but after 10 days, the rash and other symptoms of DRESS reoccurred. Two months later, patch tests were positive to commercial carbamazepine and oxcarbazepine 5 mg in 50 µl alc. Eighty controls were negative. LTTs were positive only to carbamazepine. As the reoccurrence of symptoms occurred only 10 days after initiation of oxcarbazepine, this was a case of neosensitization to this drug rather than cross-reactivity to carbamazepine (5).

A patient who had developed DRESS from carbamazepine, after resolution took oxcarbazepine for 2 weeks and developed limited skin rash with abnormal liver function (suggestive of DRESS). Patch tests were positive to both carbamazepine and oxcarbazepine 10% and 30% pet. (7).

Other cutaneous adverse drug reactions
In Finland, in the period 1989-2001, 826 patients with suspected cutaneous drug eruptions were patch tested and 89 had one or more positive reactions. Of these individuals, 1 reacted to oxcarbazepine; the nature of the drug eruption in this patient was not mentioned (6).

A 21-year-old woman had been treated for 4 weeks with oxcarbazepine, when an urticarial exanthem developed. One month later, she had positive patch tests to oxcarbazepine tested as pure powder at 5, 10, 15, and 20% w/w in white petrolatum. 34 controls were negative (2).

Cross-reactions, pseudo-cross-reactions and co-reactions
This subject is discussed in Chapter 3.284 Lamotrigine. Of 11 patients who had suffered a skin eruption of possible allergic origin from carbamazepine and who had positive patch tests to carbamazepine 5 mg in 50 µl alc., 5 (45%) co-reacted to oxcarbazepine in the same concentration. These were considered cross-reactions, but it was not mentioned whether these patients had been treated with oxcarbazepine before (8). Cross-reactivity seems likely, though.

LITERATURE
1 The data in the section 'General' may have been obtained from literature discussed in this chapter, but mostly also or exclusively from one or more of the following online sources: ChemIDPlus Advanced, PubChem, DrugBank, RxList, Drug Central, Drugs.com, and Wikipedia
2 Romano A, Pettinato R, Andriolo M, Viola M, Guéant-Rodriguez RM, Valluzzi RL, et al. Hypersensitivity to aromatic anticonvulsants: in vivo and in vitro cross-reactivity studies. Curr Pharm Des 2006;12:3373-3381
3 Schuster C, Kränke B, Aberer W, Komericki P. Fixed drug eruption on the penis due to oxcarbazepine. Arch Dermatol 2011;147:362-364
4 Gex-Collet C, Helbling A, Pichler WJ. Multiple drug hypersensitivity – proof of multiple drug hypersensitivity by patch and lymphocyte transformation tests. J Investig Allergol Clin Immunol 2005;15:293-296
5 Troost RJ, Oranje AP, Lijnen RL, et al. Exfoliative dermatitis due to immunologically confirmed carbamazepine hypersensitivity. Pediatr Dermatol 1996;13:316-320
6 Lammintausta K, Kortekangas-Savolainen O. The usefulness of skin tests to prove drug hypersensitivity. Br J Dermatol 2005;152:968-974
7 Lin YT, Chang YC, Hui RC, Yang CH, Ho HC, Hung SI, Chung WH. A patch testing and cross-sensitivity study of carbamazepine-induced severe cutaneous adverse drug reactions. J Eur Acad Dermatol Venereol 2013;27:356-364
8 Troost RJ, Van Parys JA, Hooijkaas H, van Joost T, Benner R, Prens EP. Allergy to carbamazepine: parallel in vivo and in vitro detection. Epilepsia 1996;37:1093-1099
9 Elzagallaai AA, Knowles SR, Rieder MJ, Bend JR, Shear NH, Koren G. Patch testing for the diagnosis of anticonvulsant hypersensitivity syndrome: a systematic review. Drug Saf 2009;32:391-408
10 Brockow K, Garvey LH, Aberer W, Atanaskovic-Markovic M, Barbaud A, Bilo MB, et al.; ENDA/EAACI Drug Allergy Interest Group. Skin test concentrations for systemically administered drugs – an ENDA/EAACI Drug Allergy Interest Group position paper. Allergy 2013;68:702-712

Chapter 3.357 OXOLAMINE

IDENTIFICATION

Description/definition   : Oxolamine is the oxadiazole that conforms to the structural formula shown below
Pharmacological classes : Antitussive agents
IUPAC name   : *N,N*-Diethyl-2-(3-phenyl-1,2,4-oxadiazol-5-yl)ethanamine
CAS registry number   : 959-14-8
EC number   : 213-493-4
Merck Index monograph : 8313
Patch testing   : 0.1% and 0.5% water and alcohol
Molecular formula   : $C_{14}H_{19}N_3O$

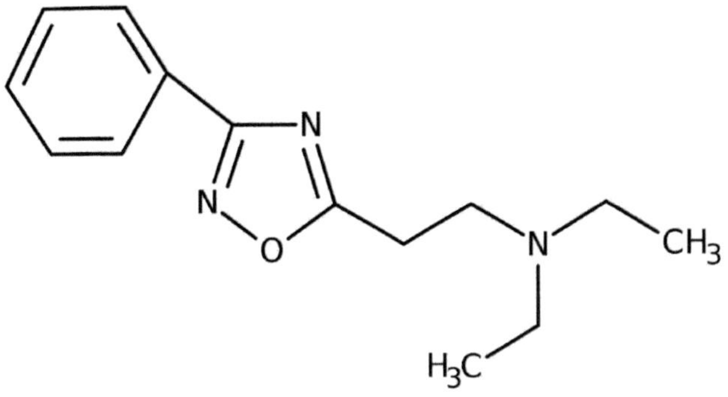

GENERAL

Oxolamine is an oxadiazole derivative that has anti-inflammatory activity. It is used in some countries as a cough suppressant for the treatment of pharyngitis, tracheitis, bronchitis, bronchiectasis and pertussis. In pharmaceutical products oxolamine may be employed as oxolamine citrate or – in the publication presented in this chapter – oxolamine tannate (no details found) (1).

CONTACT ALLERGY FROM ACCIDENTAL CONTACT

A 44-year-old man, who had been working in a pharmaceutical laboratory for 4 years, gave a 2-year history of erythematous papules on the arms, and an itchy sensation all over the body, that occurred every time he handled a syrup containing amoxicillin, phenylephrine, oxolamine tannate and plant extracts. Patch testing with the components of the syrup showed positive reactions to oxolamine tannate 0.5% water (+++) and 0.1% in water-alcohol solution (+++) at D2, D3 and D4. Twenty controls were negative to oxolamine tannate 0.1% water. To avoid further handling of the allergen, the patient was transferred to another section. One year later, he had had no recurrence. The patch tests were repeated and persisting positive reactivity to oxolamine tannate was found (2). This was a case of occupational allergic contact dermatitis.

LITERATURE

1   The data in the section 'General' may have been obtained from literature discussed in this chapter, but mostly also or exclusively from one or more of the following online sources: ChemIDPlus Advanced, PubChem, DrugBank, RxList, Drug Central, Drugs.com, and Wikipedia
2   Conde-Salazar L, Guimaraens D, Ilinas G, Romero L, Gonzalez MA. Occupational allergic contact dermatitis from oxolamine. Contact Dermatitis 1989;21:206

Chapter 3.358 OXPRENOLOL

IDENTIFICATION

Description/definition : Oxprenolol is the β-adrenergic antagonist that conforms to the structural formula shown below

Pharmacological classes : β-Adrenergic antagonists; antihypertensive agents; vasodilator agents; anti-arrhythmia agents; anti-anxiety agents; sympatholytics

IUPAC name : 1-(Propan-2-ylamino)-3-(2-prop-2-enoxyphenoxy)propan-2-ol

CAS registry number : 6452-71-7

EC number : 229-257-9

Merck Index monograph : 8321

Patch testing : 1% pet.

Molecular formula : $C_{15}H_{23}NO_3$

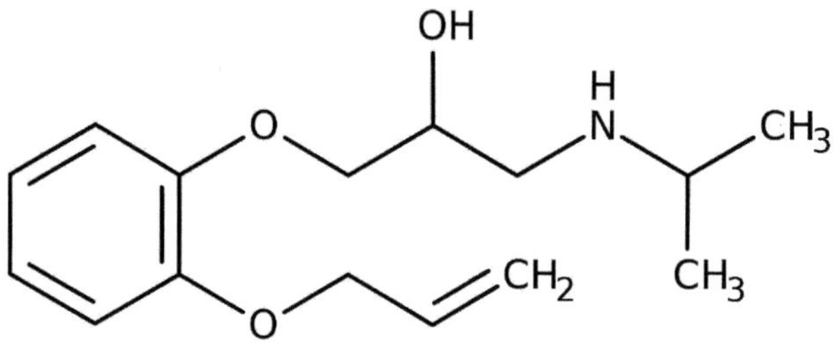

GENERAL

Oxprenolol is a lipophilic, nonselective β-adrenergic receptor antagonist with anti-arrhythmic, anti-anginal and antihypertensive activities. It competitively binds to and blocks $β_1$-adrenergic receptors in the heart, thereby decreasing cardiac contractility and rate. This leads to a reduction in cardiac output and lowers blood pressure. In addition, oxprenolol prevents the release of renin, a hormone secreted by the kidneys that causes constriction of blood vessels. Oxprenolol is used in the treatment of hypertension, angina pectoris, arrhythmias, and anxiety (1).

In pharmaceutical products, oxprenolol is employed as oxprenolol hydrochloride (CAS number 6452-73-9, EC number 229-260-5, molecular formula $C_{15}H_{24}ClNO_3$) (1).

CONTACT ALLERGY FROM ACCIDENTAL CONTACT

A man aged 46 years, working in a division for drug synthesis of a pharmaceutical plant, and exposed mainly to propranolol hydrochloride and much less frequently to oxprenolol hydrochloride, after 11 months suddenly developed pronounced swelling and erythema of the face, dorsum of the hands and neck. After 2 weeks, during which he was on sick leave, these changes regressed completely and he returned to work, but after 3 days the skin changes returned and also involved the trunk. The factory hall was full of dust, which was visible in sunbeams and on the floor. The patient spent 5-6 hours daily in this hall. One of the substrates used in great amounts for the production of both drugs was epichlorohydrin, which was poured by the patient, wearing a gas mask, into the production stream. When patch tested, he was found to be positive to epichlorohydrin 0.001% water, oxprenolol HCl 1% pet. and propranolol 1% pet. Ten controls were negative (2). This was a case of occupational airborne allergic contact dermatitis from oxprenolol and propranolol.

LITERATURE

1   The data in the section 'General' may have been obtained from literature discussed in this chapter, but mostly also or exclusively from one or more of the following online sources: ChemIDPlus Advanced, PubChem, DrugBank, RxList, Drug Central, Drugs.com, and Wikipedia

2   Rebandel P, Rudzki E. Dermatitis caused by epichlorohydrin, oxprenolol hydrochloride and propranolol hydrochloride. Contact Dermatitis 1990;23:199

Chapter 3.359 OXYBUTYNIN

IDENTIFICATION

Description/definition	: Oxybutynin is the tertiary amine that conforms to the structural formula shown below
Pharmacological classes	: Parasympatholytics; urological agents; muscarinic antagonists
IUPAC name	: 4-(Diethylamino)but-2-ynyl 2-cyclohexyl-2-hydroxy-2-phenylacetate
CAS registry number	: 5633-20-5
EC number	: Not available
Merck Index monograph	: 8324
Patch testing	: Tablet, 30% pet. (2); most systemic drugs can be tested at 10% pet.; if the pure chemical is not available, prepare the test material from intravenous powder, the content of capsules or – when also not available – from powdered tablets to achieve a final concentration of the active drug of 10% pet.
Molecular formula	: $C_{22}H_{31}NO_3$

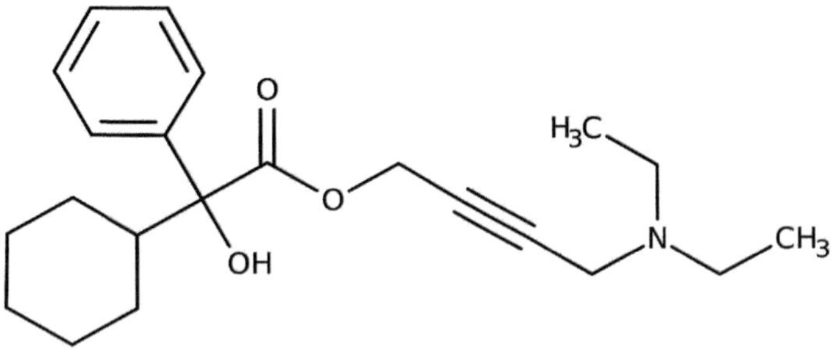

GENERAL

Oxybutynin is a tertiary amine possessing antimuscarinic and antispasmodic properties. It blocks muscarinic receptors in smooth muscle, hence inhibiting acetylcholine binding with subsequent reduction of involuntary muscle contractions. Thus, oxybutynin reduces bladder contractions and is indicated for the treatment of overactive bladder (1). In pharmaceutical products, oxybutynin is usually employed as oxybutynin chloride (also termed hydrochloride) (CAS number 1508-65-2, EC number 216-139-7, molecular formula $C_{22}H_{32}ClNO_3$) (1).

CONTACT ALLERGY FROM ACCIDENTAL CONTACT

A 44-year-old nurse presented with a 9-month-history of eczema on the hands and face, which had started 3 months after she was required to crush tablets daily during her work as a nurse in a clinic for disabled people. She was patch tested with all the drugs that she had contact with and reacted to 8 of these, including oxybutynin 30% pet.(D2 ++, D3 +). No controls were performed (2). This was a case of occupational airborne allergic contact dermatitis.

LITERATURE

1. The data in the section 'General' may have been obtained from literature discussed in this chapter, but mostly also or exclusively from one or more of the following online sources: ChemIDPlus Advanced, PubChem, DrugBank, RxList, Drug Central, Drugs.com, and Wikipedia
2. Swinnen I, Ghys K, Kerre S, Constandt L, Goossens A. Occupational airborne contact dermatitis from benzodiazepines and other drugs. Contact Dermatitis 2014;70:227-232

Chapter 3.360 OXYCODONE

IDENTIFICATION

Description/definition : Oxycodone is the semisynthetic, morphine-like opioid alkaloid that conforms to the structural formula shown below
Pharmacological classes : Analgesics, opioid; narcotics
IUPAC name : (4R,4aS,7aR,12bS)-4a-Hydroxy-9-methoxy-3-methyl-2,4,5,6,7a,13-hexahydro-1H-4,12-methanobenzofuro[3,2-e]isoquinolin-7-one
Other names : Dihydrohydroxycodeinone; dihydrone; oxycodeinone; dihydroxycodeinone
CAS registry number : 76-42-6
EC number : 200-960-2
Merck Index monograph : 8328
Patch testing : Commercial tablet, crushed, or commercial liquid as is; no data available for the pure chemical; suggestion: 1% alcohol and pet.
Molecular formula : $C_{18}H_{21}NO_4$

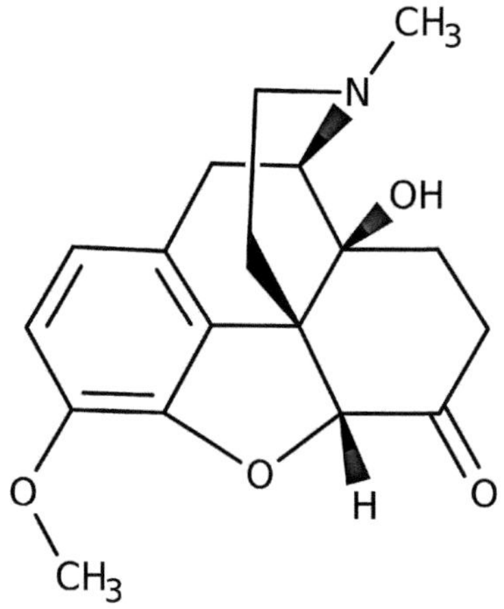

GENERAL

Oxycodone is a semisynthetic, morphine-like opioid alkaloid with analgesic activity. Oxycodone is indicated for the treatment of diarrhea, pulmonary edema and for the relief of moderate to moderately severe pain. In pharmaceutical products, oxycodone is employed as oxycodone hydrochloride (CAS number 124-90-3, EC number 204-717-1, molecular formula $C_{18}H_{22}ClNO_4$) (1).

CONTACT ALLERGY FROM ACCIDENTAL CONTACT

A 45-year-old male formulation scientist working in a pharmaceutical laboratory for 14 years described episodic swelling and erythema of his eyelids and erythematous patches on his volar wrists, which had developed over several months. He handled naloxone and hydromorphone. He had had similar, but milder, symptoms previously while working with oxycodone (this work was performed using a fume cupboard). Patch tests were strongly positive to oxycodone (10, 40 and 80 mg crushed tablets) and weak positive to hydromorphone (content of 2.6 mg capsule) and naloxone (400 µg/ml). The second patient, a 27-year-old man working as a development scientist in a pharmaceutical laboratory, had a 4-week history of erythematous skin on the volar aspects of both wrists. His work involved contact with oxycodone and naloxone. Patch tests were positive to crushed 40 and 80 mg oxycodone tablets, but negative to a 10 mg tablet (3). Patient 3, a 44-year-old male manufacturing operative, working with oxycodone, described a 6-month history of an erythematous facial rash, which started around his eyes and then spread to involve most of his face. A patch test was positive (++) to oxycodone 10 mg/ml liquid (3). All three patients had occupational allergic contact dermatitis to oxycodone, of who one airborne and another partly airborne.

Four patients working in an opiate manufacturing plant had developed dermatitis affecting their face and hands. All had been involved primarily with the manufacturing or packaging of the opiate analgesic oxycodone. Three showed positive reactions to a crushed oxycodone tablet; one of them was tested with all excipients and there were no positive reactions. One patient co-reacted to hydromorphone, possibly a cross-reaction. Fourteen controls were negative. In addition to these cases, the authors refer to two operatives working at a separate raw chemical facility manufacturing oxycodone who have also been sensitized, No details were provided, the publication was an Abstract and appears not to have been converted into a formal article (4).

CUTANEOUS ADVERSE DRUG REACTIONS FROM SYSTEMIC ADMINISTRATION CAUSED BY TYPE IV (DELAYED-TYPE) HYPERSENSITIVITY (as demonstrated by positive patch tests)

Cutaneous adverse drug reactions from systemic administration of oxycodone caused by type IV (delayed-type) hypersensitivity have included 'exanthem' (2).

Other cutaneous adverse drug reactions

At the ages of 34, 41 and 43 year, a female patient underwent three reconstructive/endovascular operations. Fentanyl was used for general anesthesia, and oxycodone was used to treat postoperative pain. The patient developed an exanthem during both of the latter hospitalization periods. Later, at home, she most often used paracetamol to treat occasional headaches. When she used a combination of paracetamol and codeine phosphate, an eruption similar to those that she had experienced in the hospital appeared. In patch testing, positive reactions were seen to oxycodone, codeine phosphate, naloxone, and morphine, the latter 3 probably being cross-reactions to oxycodone (2).

Cross-reactions, pseudo-cross-reactions and co-reactions

A patient sensitized to oxycodone probably cross-reacted to codeine, naloxone and morphine (2), another one to hydromorphone (4).

LITERATURE

1   The data in the section 'General' may have been obtained from literature discussed in this chapter, but mostly also or exclusively from one or more of the following online sources: ChemIDPlus Advanced, PubChem, DrugBank, RxList, Drug Central, Drugs.com, and Wikipedia

2   Liippo J, Pummi K, Hohenthal U, Lammintausta K. Patch testing and sensitization to multiple drugs. Contact Dermatitis 2013;69:296-302

3   Wootton CI, English JS. Occupational allergic contact dermatitis caused by oxycodone. Contact Dermatitis 2012;67:383-384

4   MacFarlane CS, Charman C, Fertig A, English JSC. An outbreak of contact dermatitis in an opiate manufacturing plant. Br J Dermatol 2003;149(suppl.64):95

Chapter 3.361 OXYTETRACYCLINE

IDENTIFICATION

Description/definition : Oxytetracycline is a tetracycline analog isolated from the actinomycete *Streptomyces rimosus* that conforms to the structural formula shown below
Pharmacological classes : Anti-bacterial agents
IUPAC name : (4*S*,4a*R*,5*S*,5a*R*,6*S*,12a*R*)-4-(Dimethylamino)-1,5,6,10,11,12a-hexahydroxy-6-methyl-3,12-dioxo 4,4a,5,5a-tetrahydrotetracene-2-carboxamide
Other names : 5-Hydroxytetracycline
CAS registry number : 79-57-2
EC number : 201-212-8
Merck Index monograph : 8345
Patch testing : 3% pet. (SmartPracticeCanada, SmartPracticeEurope)
Molecular formula : $C_{22}H_{24}N_2O_9$

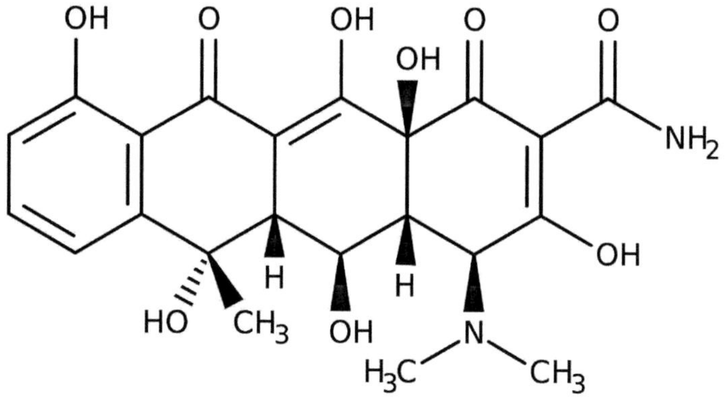

GENERAL

Oxytetracycline is a tetracycline analog isolated from the actinomycete *Streptomyces rimosus* with broad-spectrum antibacterial properties. This antibiotic is indicated for treatment of infections caused by a variety of gram-positive and gram-negative microorganisms including *Mycoplasma pneumoniae*, *Pasteurella pestis*, *Escherichia coli*, *Haemophilus influenzae* (respiratory infections), and *Diplococcus pneumoniae* (1). Oxytetracycline is used topically in the treatment of acne vulgaris, ophthalmic infections, and in the prevention or treatment of skin infections (11). In pharmaceutical products, both oxytetracycline and oxytetracycline hydrochloride (CAS number 2058-46-0, EC number 218-161-2, molecular formula $C_{22}H_{25}ClN_2O_9$) may be employed (1).

In topical preparations, oxytetracycline has caused contact allergy/allergic contact dermatitis, which subject has been fully reviewed in Volume 3 of the *Monographs in contact allergy* series (1).

CONTACT ALLERGY FROM ACCIDENTAL CONTACT

Contact allergy to oxytetracycline from accidental contact has been described in workers in the pharmaceutical industries (2,3), nurses (3,5), and veterinary surgeons (3,4). All reports came from one clinic in Poland, there was undoubtedly some overlap and often relevance was not mentioned. Therefore, these may not all have been cases of occupational contact allergy/allergic contact dermatitis (2,3,4,5).

In a group of 107 workers in the pharmaceutical industry with dermatitis, investigated in Warsaw, Poland, before 1989, 5 reacted to oxytetracycline, tested 10% pet. (2). Of 333 nurses patch tested in Poland between 1979 and 1987, 2 reacted to oxytetracycline 10% pet.; data on relevance were not provided (5). Also in Warsaw, Poland, in the period 1979-1983, 27 pharmaceutical workers, 24 nurses and 30 veterinary surgeons were diagnosed with occupational allergic contact dermatitis from antibiotics. The numbers that had positive patch tests to oxytetracycline (10% pet.) were 3, 0, and 1, respectively, total 4 (3).

Of 26 veterinarians patch tested in the early 1980s in Poland because of dermatitis, 15 had one or more positive patch tests to veterinary drugs, tuberculin and/or disinfectants. Most of them realized that these contactants were harmful to them, and all had come into contact with drugs to which they had positive tests. One of the veterinarians reacted to oxytetracycline 10% pet. (4).

A 43-year-old man, who had worked for 13 years in cattle breeding. presented with an 18-month history of erythema, edema and vesicles on the interdigital spaces of the hands, the face and both sides of the neck. His symptoms worsened every time that he prepared animal feed. Patch tests were positive at D2 and D3 to oxytetracycline hydrochloride 5% pet., tylosin 10% pet., penicillin 5% pet. and spiramycin 5% pet. He had contact with all these antibiotics and the dermatitis completely cleared as soon as the patient left his occupation (7).

CUTANEOUS ADVERSE DRUG REACTIONS FROM SYSTEMIC ADMINISTRATION CAUSED BY TYPE IV (DELAYED-TYPE) HYPERSENSITIVITY (as demonstrated by positive patch tests)

Cutaneous adverse drug reactions from systemic administration of oxytetracycline caused by type IV (delayed-type) hypersensitivity have included fixed drug eruption (6).

Fixed drug eruption

In a study from Pakistan, an unknown number of patients had a fixed drug eruption from oxytetracycline with a positive lesional patch test to this drug tested at 5% pet. (6).

LITERATURE

1   De Groot AC. Monographs in contact allergy, volume 3: Topical Drugs. Boca Raton, Fl, USA: CRC Press Taylor and Francis Group, 2021 (ISBN 978-0-367-23693-9)
2   Rudzki E, Rebandel P, Grzywa Z. Contact allergy in the pharmaceutical industry. Contact Dermatitis 1989;21:121-122
3   Rudzki E, Rebendel P. Contact sensitivity to antibiotics. Contact Dermatitis 1984;11:41-42
4   Rudzki E, Rebandel P, Grzywa Z, Pomorski Z, Jakiminska B, Zawisza E. Occupational dermatitis in veterinarians. Contact Dermatitis 1982;8:72-73
5   Rudzki E, Rebandel P, Grzywa Z. Patch tests with occupational contactants in nurses, doctors and dentists. Contact Dermatitis 1989;20:247-250
6   Mahboob A, Haroon TS, Iqbal Z, Iqbal F, Saleemi MA, Munir A. Fixed drug eruption: topical provocation and subsequent phenomena. J Coll Physicians Surg Pak 2006;16:747-750
7   Guerra L, Venture M, Tardio M, Tosti A. Airborne contact dermatitis from animal feed antibiotics. Contact Dermatitis 1991;25:333-334

Chapter 3.362 PANCREATIN

IDENTIFICATION

Description/definition	: Pancreatin is a mammalian pancreatic extract composed of enzymes with protease, amylase and lipase activities
Pharmacological classes	: Enzymes
IUPAC name	: Not available
Other names	: Pancrelipase (drugbank.ca)
CAS registry number	: 8049-47-6
EC number	: 232-468-9
Merck Index monograph	: 8478 (Pancreatic extract)
Patch testing	: Tablet, pulverized, 20% pet. (2); most systemic drugs can be tested at 10% pet.; if the pure chemical is not available, prepare the test material from intravenous powder, the content of capsules or – when also not available – from powdered tablets to achieve a final concentration of the active drug of 10% pet.
Molecular formula	: Unspecified

GENERAL

Pancreatin is composed of a mixture of pancreatic enzymes which include lipase, protease, and amylase which are able to break down fat, protein, and starches, respectively, in the small intestine. These enzymes are extracted from porcine pancreatic glands. The use of pancreatin is part of the pancreatic enzyme replacement therapy, indicated for the treatment of pancreatic insufficiency attributed to cystic fibrosis, chronic pancreatitis or any other medically defined pancreatic disease (1).

CUTANEOUS ADVERSE DRUG REACTIONS FROM SYSTEMIC ADMINISTRATION CAUSED BY TYPE IV (DELAYED-TYPE) HYPERSENSITIVITY (as demonstrated by positive patch tests)

Cutaneous adverse drug reactions from systemic administration of pancreatin caused by type IV (delayed-type) hypersensitivity have included erythema multiforme (2).

Other cutaneous adverse drug reactions

An 84-year-old man presented with pruritic erythema on his arms and left lower abdomen, which had developed 2 months previously. He used two types of 'stomach medicines' because of gastrectomy at the age of 24. Hematological examination revealed eosinophilia. Histopathology showed dilated vessels, slight edema, and a dense perivascular mononuclear cell infiltration containing a significant number of eosinophils. Patch tests on tape-stripped skin with the drugs the patient used at 20% pet. gave a positive reaction to one of these medications (a digestive enzyme) at D2 and D3. This drug was a mixture of cellulase, lipase, pancreatin, and Taka-diastase. Twenty controls were negative. An oral provocation test resulted in indurated erythema on the nape of the neck to the shoulders about 24 hours later, and the morphology of the skin eruption appeared to be the same as that seen on the first visit. Patch tests with the 4 ingredients at 20% pet. were positive at D2 and D3 to pancreatin (+/+) and Taka-diastase (++/++), whereas 20 controls showed no reactions. It was concluded that the patient had erythema multiforme (which is somewhat curious, as the clinical picture did not have a typical erythema multiforme clinical aspect) as an allergic reaction to pancreatin and Taka-diastase (2). Taka-diastase is α-amylase derived from from *Aspergillus oryzae*. It appears to have no current medical uses (1).

LITERATURE

1   The data in the section 'General' may have been obtained from literature discussed in this chapter, but mostly also or exclusively from one or more of the following online sources: ChemIDPlus Advanced, PubChem, DrugBank, RxList, Drug Central, Drugs.com, and Wikipedia

2   Miyoshi H, Kanzaki T. Drug eruption (erythema multiforme type) due to a digestive enzyme drug. J Dermatol 1998;25:28-31

Chapter 3.363 PANTOPRAZOLE

IDENTIFICATION

Description/definition   : Pantoprazole is the substituted benzimidazole that conforms to the structural formula
                            shown below
Pharmacological classes : Anti-ulcer agents; proton pump inhibitors
IUPAC name   : 6-(Difluoromethoxy)-2-[(3,4-dimethoxypyridin-2-yl)methylsulfinyl]-1H-benzimidazole
CAS registry number   : 102625-70-7
EC number   : 600-331-6
Merck Index monograph   : 8387
Patch testing   : 1% and 10% pet. (5); 1%, 5% and 10% saline (6)
Molecular formula   : $C_{16}H_{15}F_2N_3O_4S$

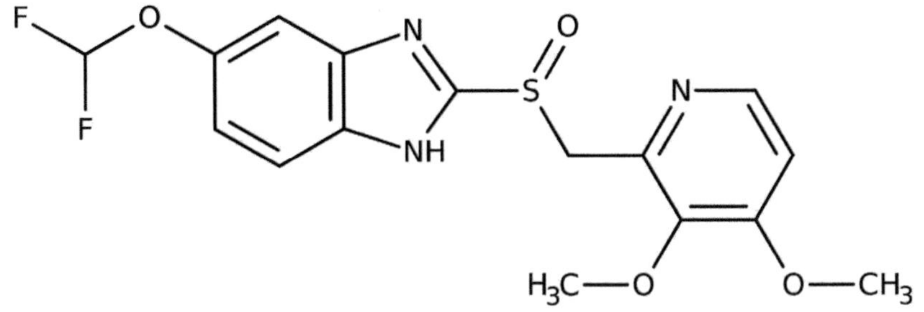

GENERAL

Pantoprazole is a substituted benzimidazole and proton pump inhibitor with antacid activity. This agent is indicated
for the treatment of active duodenal ulcer, eradication of *Helicobacter pylori* to reduce the risk of duodenal ulcer,
treatment of active benign gastric ulcer, gastroesophageal reflux disease (GERD), erosive esophagitis (EE) due to
acid-mediated GERD, for maintenance of healing of EE due to acid-mediated GERD and in pathologic hypersecretory
conditions (e.g. Zollinger-Ellison syndrome). In pharmaceutical products, pantoprazole is usually employed as
pantoprazole sodium (= pantoprazole sodium hydrate/sesquihydrate) (CAS number 164579-32-2, EC number not
available, molecular formula $C_{32}H_{34}F_4N_6Na_2O_{11}S_2$) (1).

CONTACT ALLERGY FROM ACCIDENTAL CONTACT

Sixteen healthcare workers (five veterinarians and 11 nurses) were investigated in Germany for airborne contact
dermatitis. Positive patch tests were obtained in six individuals: one to pantoprazole, 4 to tetrazepam and one to
tylosin (9).

A 58-year-old male pharmaceutical worker presented with erythematous eruptions involving the neck, posterior
auricular areas, and extensor forearms, with slight swelling and mild erythema of the upper eyelids. He had regular
contact with proton pump inhibitors and intermediates including pantoprazole sodium, dexlansoprazole,
esomeprazole magnesium, lansoprazole, and omeprazole magnesium. Patch tests were positive to pantoprazole (1%
and 10% pet.) and all other proton pump inhibitors except lansoprazole The patient was diagnosed with occupational
airborne allergic contact dermatitis to pantoprazole and other proton pump inhibitors (5).

Another male pharmaceutical worker, aged 49, had been exposed periodically to pantoprazole, lansoprazole,
and omeprazole for 21 years, when he noticed an itchy rash on his face, neck and arms. Patch tests were positive to
pantoprazole 1%, 5% and 10% saline (+++) and to lansoprazole and omeprazole (6). A similar case was reported from
Israel in 2011 (7). The 33-year-old machine operator in a pharmaceutical plant had facial and eyelid dermatitis. He
reacted to pantoprazole 5% and 10% pet. and cross-reacted to omeprazole 5% and 10% pet.; reactions to both drugs
at 1% were negative. Ten controls were negative (7).

A 36-year-old man presented with recently developed hypopigmented patches on the forehead, periocular area,
and neck. He described his condition as ongoing symptoms of erythema with swelling of his eyelids during his work.
The patient had worked for a year in the pharmaceutical industry and was involved in the manufacturing of proton
pump inhibitors (PPIs), including pantoprazole and rabeprazole. He was exposed to PPI powder, although he wore
protective gear such as gloves and a mask, but not goggles, during the manufacturing process. A skin biopsy from his
nape showed a complete loss of melanin and melanocytes, confirming the diagnosis of vitiligo. Patch tests were
positive to pure pantoprazole and rabeprazole powder 1% pet. The authors suggested that PPIs most likely lead to
the apoptosis of melanocytes by modifying the pH gradient and enhancing oxidative stress in melanocytes, possibly
resulting in melanocyte-specific autoimmunity (12).

CUTANEOUS ADVERSE DRUG REACTIONS FROM SYSTEMIC ADMINISTRATION CAUSED BY TYPE IV (DELAYED-TYPE) HYPERSENSITIVITY (as demonstrated by positive patch tests)

Cutaneous adverse drug reactions from systemic administration of pantoprazole caused by type IV (delayed-type) hypersensitivity have included exfoliative erythroderma (8), drug reaction with eosinophilia and systemic symptoms DRESS (3,4,10,), and toxic epidermal necrolysis (TEN) (3).

Erythroderma, widespread erythematous eruption, exfoliative dermatitis

A 35-year-old female patient had developed generalized erythema with intense itching and desquamation five days after the initiation of treatment with topical diltiazem and 2 other topicals for anal fissuring, and oral pantoprazole for gastritis. Patch tests with all drugs were positive only to pantoprazole 10% pet. Twelve controls were negative. The patient was diagnosed with exfoliative erythroderma from delayed-type hypersensitivity to pantoprazole (8).

Drug reaction with eosinophilia and systemic symptoms (DRESS)

In a multicenter investigation in France, of 72 patients patch tested for DRESS, 46 (64%) had positive reactions to drugs, including 2 to pantoprazole (4). In a group of 14 patients with multiple delayed-type hypersensitivity reactions, DRESS was caused by pantoprazole in one case, showing a positive patch test reaction (10).

In a 61-year-old man, delayed-type hypersensitivity to pantoprazole (tested as commercial tablet 20 mg, 15% and 30% pet.) and other drugs may have caused / contributed to DRESS (drug reaction with eosinophilia and systemic symptoms) and TEN (toxic epidermal necrolysis) (3).

Stevens-Johnson syndrome/toxic epidermal necrolysis (SJS/TEN)

See the section 'Drug reaction with eosinophilia and systemic symptoms DRESS' above, ref. 3.

Cross-reactions, pseudo-cross-reactions and co-reactions

A patient sensitized to pantoprazole had cross-reactions to omeprazole, lansoprazole, esomeprazole and rabeprazole (3). A patient sensitized to omeprazole cross-reacted to pantoprazole and 3 other proton pump inhibitors (all crushed tablets, 1% and 10% pet.) (2).

LITERATURE

1   The data in the section 'General' may have been obtained from literature discussed in this chapter, but mostly also or exclusively from one or more of the following online sources: ChemIDPlus Advanced, PubChem, DrugBank, RxList, Drug Central, Drugs.com, and Wikipedia
2   Al-Falah K, Schachter J, Sasseville D. Occupational allergic contact dermatitis caused by omeprazole in a horse breeder. Contact Dermatitis 2014;71:377-378
3   Liippo J, Pummi K, Hohenthal U, Lammintausta K. Patch testing and sensitization to multiple drugs. Contact Dermatitis 2013;69:296-302
4   Barbaud A, Collet E, Milpied B, Assier H, Staumont D, Avenel-Audran M, et al. A multicentre study to determine the value and safety of drug patch tests for the three main classes of severe cutaneous adverse drug reactions. Br J Dermatol 2013;168:555-562
5   DeKoven JG, Yu AM. Occupational airborne contact dermatitis from proton pump inhibitors. Dermatitis 2015;26:287-290
6   Alarcón M, Herrera-Mozo I, Nogué S, et al. Occupational airborne contact dermatitis from proton pump inhibitors. Curr Allergy Clin Immunol 2014;27:310-313
7   Neumark M, Ingber A, Levin M, Slodownik D. Occupational airborne contact dermatitis caused by pantoprazole. Contact Dermatitis 2011;64:60-61
8   Sánchez-Borges M, González-Aveledo L. Exfoliative erythrodermia induced by pantoprazole. Allergol Immunopathol (Madr) 2012;40:194-195
9   Breuer K, Uter W, Geier J. Epidemiological data on airborne contact dermatitis: results of the IVDK. Contact Dermatitis 2015;73:239-247
10 Jörg L, Yerly D, Helbling A, Pichler W. The role of drug, dose and the tolerance/intolerance of new drugs in multiple drug hypersensitivity syndrome (MDH). Allergy 2020;75:1178-1187
11 Brockow K, Garvey LH, Aberer W, Atanaskovic-Markovic M, Barbaud A, Bilo MB, et al.; ENDA/EAACI Drug Allergy Interest Group. Skin test concentrations for systemically administered drugs – an ENDA/EAACI Drug Allergy Interest Group position paper. Allergy 2013;68:702-712
12 Kim D, Kang H. Proton pump inhibitor-induced vitiligo following airborne occupational sensitization. Contact Dermatitis. 2021 Jun 21. doi: 10.1111/cod.13919. Epub ahead of print.

Chapter 3.364 PARAMETHASONE

IDENTIFICATION

Description/definition : Paramethasone is the synthetic glucocorticoid that conforms to the structural formula
                        shown below
Pharmacological classes : Glucocorticoids
IUPAC name : (6S,8S,9S,10R,11S,13S,14S,16R,17R)-6-Fluoro-11,17-dihydroxy-17-(2-hydroxyacetyl)-
             10,13,16-trimethyl-7,8,9,11,12,14,15,16-octahydro-6H-cyclopenta[a]phenanthren-3-one
CAS registry number : 53-33-8
EC number : 200-169-2
Merck Index monograph : 8403
Patch testing : In general, corticosteroids may be tested at 0.1% and 1% in alcohol; late readings (6-10
                days) are strongly recommended
Molecular formula : $C_{22}H_{29}FO_5$

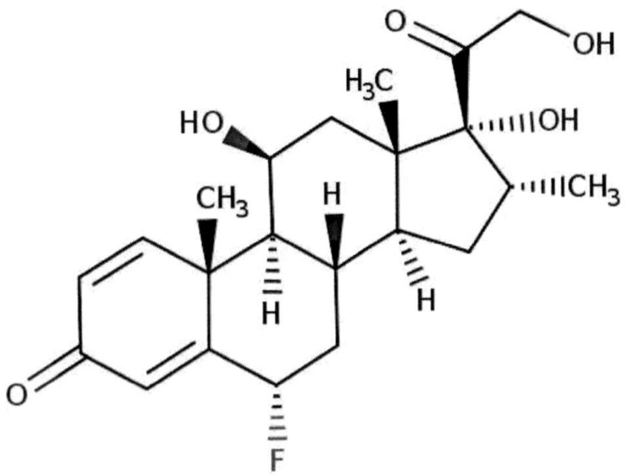

GENERAL

Systemically administered glucocorticoids have anti-inflammatory, immunosuppressive and antineoplastic properties and are used in the treatment of a wide spectrum of diseases including rheumatic disorders, lung diseases (asthma, COPD), gastrointestinal tract disorders (Crohn's disease, colitis ulcerosa), certain malignancies (leukemia, lymphomas), hematological disorders, and various diseases of the kidneys, brain, eyes and skin. A practical guideline for diagnosing allergic reactions to corticosteroids is presented in ref. 1. In pharmaceutical products, paramethasone may both be employed as paramethasone base and as paramethasone acetate (CAS number 1597-82-6, EC number 216-486-4, molecular formula $C_{24}H_{31}FO_6$) (3).

CUTANEOUS ADVERSE DRUG REACTIONS FROM SYSTEMIC ADMINISTRATION CAUSED BY TYPE IV
(DELAYED-TYPE) HYPERSENSITIVITY (as demonstrated by positive patch tests)

Cutaneous adverse drug reactions from systemic administration of paramethasone acetate caused by type IV (delayed-type) hypersensitivity have included generalized erythema (2) and localized allergic reaction from intralesional injection (4).

Erythroderma, widespread erythematous eruption, exfoliative dermatitis

A 74-year-man, with eczema of the external auditory meatus for 6 months, was referred because of worsening of the otitis externa one day after applying clobetasol cream, together with spread to the thorax and axilla. He had also previously suffered a generalized erythema some hours after an intra-articular injection of paramethasone. The patient had a history of recurrent episodes of eczema in the axillae, flexural folds of the arm and the inguinal folds (highly suggestive of SDRIFE). Patch tests were positive to clobetasol propionate (CP) 0.05% cream, CP 1% pet., paramethasone acetate 2 mg/ml injection fluid and many other corticosteroids (2).

Other cutaneous adverse drug reactions

A 30-year-old woman had been treated several times with intralesional paramethasone and other topical corticosteroids for alopecia areata in the past. For a relapse of alopecia areata, treatment was started with intralesional injections of paramethasone acetate by a needleless injector. After the first session, the patient developed pruritus; 6-8 hours later, erythema, edema and vesicles were observed in the injected areas. Histopathology showed spongiform lymphocytic folliculitis with marked spongiosis and exocytosis in the ducts of sweat glands and in the pilosebaceous unit. Patch tests were performed with a series of 30 corticosteroids, the excipients of the intralesional corticosteroid (benzalkonium chloride, povidone and polysorbate 80) and pure paramethasone acetate at various concentrations in ethanol and petrolatum. Positive reactions were observed to paramethasone acetate 2% alc. (negative to 1% alc. and 1% and 2% pet.) and the corticosteroids tixocortol pivalate, hydrocortisone and hydrocortisone butyrate (4).

Cross-reactions, pseudo-cross-reactions and co-reactions
Cross-reactions between corticosteroids are discussed in Chapter 3.399 Prednisolone.

LITERATURE

1   Baeck M, Goossens A. Immediate and delayed allergic hypersensitivity to corticosteroids: practical guidelines. Contact Dermatitis 2012;66:38-45
2   Marcos C, Allegue F, Luna I, González R. An unusual case of allergic contact dermatitis from corticosteroids. Contact Dermatitis 1999;41:237-238
3   The data in the section 'General' may have been obtained from literature discussed in this chapter, but mostly also or exclusively from one or more of the following online sources: ChemIDPlus Advanced, PubChem, DrugBank, RxList, Drug Central, Drugs.com, and Wikipedia
4   Miranda-Romero A, Bajo-del Pozo C, Sánchez-Sambucety P, Martinez-Fernandez M, Garcia-Muñoz M. Delayed local allergic reaction to intralesional paramethasone acetate. Contact Dermatitis 1998;39:31-32

Chapter 3.365 PAROXETINE

IDENTIFICATION

Description/definition   : Paroxetine is the phenylpiperidine derivative that conforms to the structural
                             formula shown below
Pharmacological classes   : Cytochrome P-450 CYP2D6 inhibitors; antidepressive agents, second-generation;
                             serotonin uptake inhibitors
IUPAC name   : (3S,4R)-3-(1,3-Benzodioxol-5-yloxymethyl)-4-(4-fluorophenyl)piperidine
CAS registry number   : 61869-08-7
EC number   : 600-962-7
Merck Index monograph   : 8418
Patch testing   : 1% and 3% pet. (4); 1%, 5% and 10% water (6)
Molecular formula   : $C_{19}H_{20}FNO_3$

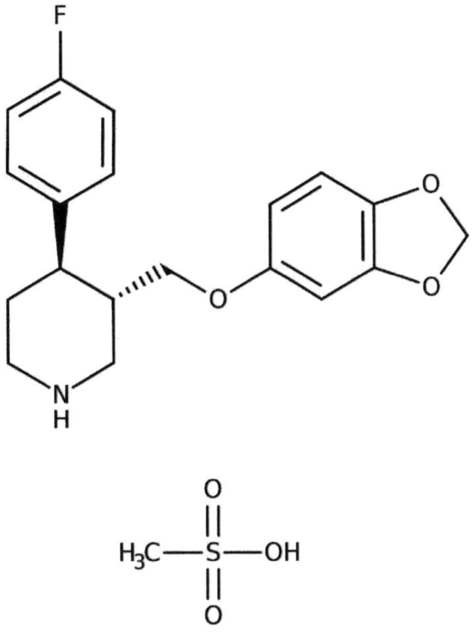

Paroxetine mesylate

GENERAL

Paroxetine is a selective serotonin reuptake inhibitor (SSRI) type antidepressant drug. Labeled indications for paroxetine include major depressive disorder, panic disorder with or without agoraphobia, obsessive-compulsive disorder, social anxiety disorder, generalized anxiety disorder, post-traumatic stress disorder and premenstrual dysphoric disorder (1). In pharmaceutical products, paroxetine is either employed as paroxetine mesylate (CAS number 217797-14-3, EC number not available, molecular formula $C_{20}H_{24}FNO_6S$) or as paroxetine hydrochloride (= paroxetine hydrochloride hydrate / hemihydrate) (CAS number 110429-35-1, EC number not available, molecular formula $C_{38}H_{44}Cl_2F_2N_2O_7$) (1).

CUTANEOUS ADVERSE DRUG REACTIONS FROM SYSTEMIC ADMINISTRATION CAUSED BY TYPE IV (DELAYED-TYPE) HYPERSENSITIVITY (as demonstrated by positive patch tests)

Cutaneous adverse drug reactions from systemic administration of paroxetine caused by type IV (delayed-type) hypersensitivity have included maculopapular eruption (4) and photosensitivity (2,3,5,6).

Maculopapular eruption

A fifty-nine year-old woman developed a generalized mildly pruritic maculopapular exanthema having used paroxetine and alprazolam for a depression. Patch tests were positive to paroxetine 1% and 3% pet. at D3 (++) (4).

Photosensitivity

A 43-year-old woman, while using paroxetine, clomipramine and alprazolam for a depression, presented with a rash consisting of erythematous and edematous papules and plaques, some of them with target shape, located on the dorsum of the arms and the forearms. Non sun-exposed areas and mucosae were spared. A diagnosis of photosensitive erythema multiforme was made. Photopatch tests with paroxetine, clomipramine and alprazolam 2% irradiated with UVA 5 J/cm² were negative but positive to paroxetine with UVB 2/3 MED (which was lowered during intake of the drugs). Later, the patient developed a similar photosensitive eruption after a herpes labialis infection and sun-exposure. The patient was diagnose with erythema multiforme caused by photoallergy to paroxetine (2). It should be mentioned that the photograph of the 'positive' photopatch test was hardly convincing as proof of delayed-type hypersensitivity.

A 70-year-old man taking paroxetine for depression presented with an eczematous eruption limited to light-exposed sites, which had been present for 5 months. A patch test with paroxetine was positive (erythema +) (comment: erythema only is not positive). Photopatch testing with UVB exposure (0.75 MED) showed marked erythema and edema (+++) (comment: a +++ reaction should show a bulla), whereas no exacerbation was noticed under UVA exposure. The final diagnosis was UVB photosensitization to paroxetine (3).

These authors also described a 47-year-old man, who was referred for investigation of a recent photodermatosis. Photopatch testing with all drugs used (phenobarbital, piroxicam, acamprosate, tetrazepam and paroxetine) showed a strongly positive reaction to paroxetine 2% pet. irradiated with UVB 0.75 MED, but negative after UVA-irradiation (3).

A 59-year-old woman presented with an itchy, vesicular, exudative eruption in photoexposed areas (face, neck, V of neck, forearms and hands), sparing the rest of the skin. For the past year she had been receiving paroxetine and omeprazole. Photopatch tests were positive to paroxetine 1% (+), 5% (++) and 10% (++) water. Twenty controls were negative to all 3 concentrations (6).

A woman developed painful papular and purpuric erythema mainly located in sun-exposed sites, during therapy with paroxetine, which had been introduced one month before to treat depression. Details are unknown to the author, but it is cited in ref. 4 that the patient had a positive patch test (or photopatch test?) (5).

LITERATURE

1   The data in the section 'General' may have been obtained from literature discussed in this chapter, but mostly also or exclusively from one or more of the following online sources: ChemIDPlus Advanced, PubChem, DrugBank, RxList, Drug Central, Drugs.com, and Wikipedia

2   Rodríguez-Pazos L, Gómez-Bernal S, Montero I, Rodríguez-Granados M, Toribio J. Erythema multiforme photoinduced by paroxetine and herpes simplex virus. Photodermatol Photoimmunol Photomed 2011;27:219-221

3   Doffoel-Hantz V, Boulitrop-Morvan C, Sparsa A, Bonnetblanc JM, Dalac S, Bédane C. Photosensitivity associated with selective serotonin reuptake inhibitors. Clin Exp Dermatol 2009;34:e763-765

4   Soto Mera MT, Pérez BV, Fernández RO, Iglesias JF. Hypersensitivity to paroxetine. Allergol Immunopathol (Madr) 2006;34:125-126

5   Richard MA, Fiszenson F, Jreissati M, Jean Pastor MJ, Grob JJ. Cutaneous adverse effects during selective serotonin reuptake inhibitors therapy: 2 cases. Ann Dermatol Venereol 2001;128:759-761 (Article in French, data cited in ref. 4).

6   Vilaplana J, Botey E, Lecha M, Herrero C, Romaguera C. Photosensitivity induced by paroxetine. Contact Dermatitis 2002;47:118-119

Chapter 3.366 PEFLOXACIN

IDENTIFICATION

Description/definition : Pefloxacin is the quinoline carboxylic acid that conforms to the structural formula shown below
Pharmacological classes : Cytochrome P-450 CYP1A2 inhibitors; anti-bacterial agents; topoisomerase II inhibitors
IUPAC name : 1-Ethyl-6-fluoro-7-(4-methylpiperazin-1-yl)-4-oxoquinoline-3-carboxylic acid
CAS registry number : 70458-92-3
EC number : 274-611-8
Merck Index monograph : 8442
Patch testing : No data available; most systemic drugs can be tested at 10% pet.; if the pure chemical is not available, prepare the test material from intravenous powder, the content of capsules or – when also not available – from powdered tablets to achieve a final concentration of the active drug of 10% pet.
Molecular formula : $C_{17}H_{20}FN_3O_3$

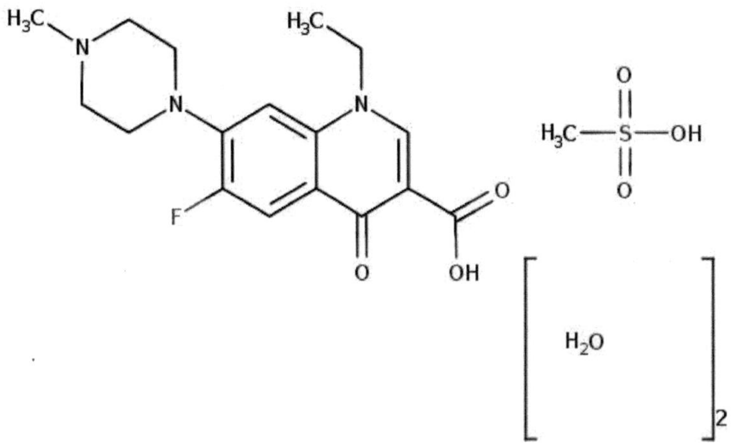

Pefloxacin mesylate dihydrate

GENERAL

Pefloxacin is a fluoroquinolone antibiotic with broad-spectrum antimicrobial activity. It inhibits the activity of microbial DNA gyrase and topoisomerase IV, disrupting DNA replication and preventing cell division. Pefloxacin possesses excellent activity against gram-negative aerobic bacteria such as *E.coli* and *Neisseria gonorrhoea* as well as gram-positive bacteria including *S. pneumoniae* and *Staphylococcus aureus*. They also posses effective activity against *Shigella, Salmonella, Campylobacter*, and multidrug-resistant *Pseudomonas* and *Enterobacter* (1).

In pharmaceutical products, pefloxacin may be employed as pefloxacin mesylate (CAS number 70458-95-6, EC number 274-613-9, molecular formula $C_{18}H_{24}FN_3O_6S$) or as pefloxacin mesylate dihydrate (CAS number 149676-40-4, EC number not available, molecular formula $C_{18}H_{28}FN_3O_8S$) (1).

CUTANEOUS ADVERSE DRUG REACTIONS FROM SYSTEMIC ADMINISTRATION CAUSED BY TYPE IV (DELAYED-TYPE) HYPERSENSITIVITY (as demonstrated by positive patch tests)

Cutaneous adverse drug reactions from systemic administration of pefloxacin caused by type IV (delayed-type) hypersensitivity have included fixed drug eruption (2).

Fixed drug eruption

In France, in the period 2005-2007, 59 cases of fixed drug eruptions were collected in 17 academic centers. There was one case of FDE to pefloxacin with a positive patch test. Clinical and patch testing details were not provided (2).

LITERATURE

1 The data in the section 'General' may have been obtained from literature discussed in this chapter, but mostly also or exclusively from one or more of the following online sources: ChemIDPlus Advanced, PubChem, DrugBank, RxList, Drug Central, Drugs.com, and Wikipedia
2 Brahimi N, Routier E, Raison-Peyron N, Tronquoy AF, Pouget-Jasson C, Amarger S, et al. A three-year-analysis of fixed drug eruptions in hospital settings in France. Eur J Dermatol 2010;20:461-464

Chapter 3.367 PENETHAMATE

IDENTIFICATION

Description/definition : Penethamate is the β-lactam penicillin that conforms to the structural formula shown below

Pharmacological classes : Anti-bacterial agents

IUPAC name : 2-(Diethylamino)ethyl (2S,5R,6R)-3,3-dimethyl-7-oxo-6-[(2-phenylacetyl)amino]-4-thia-1-azabicyclo[3.2.0]heptane-2-carboxylate

Other names : Diethylaminoethanol-benzylpenicillin

CAS registry number : 3689-73-4

EC number : Not available

Patch testing : 25% pet. (adapted from ref. 2); commercial products 250.000 IU/ml (4)

Molecular formula : $C_{22}H_{31}N_3O_4S$

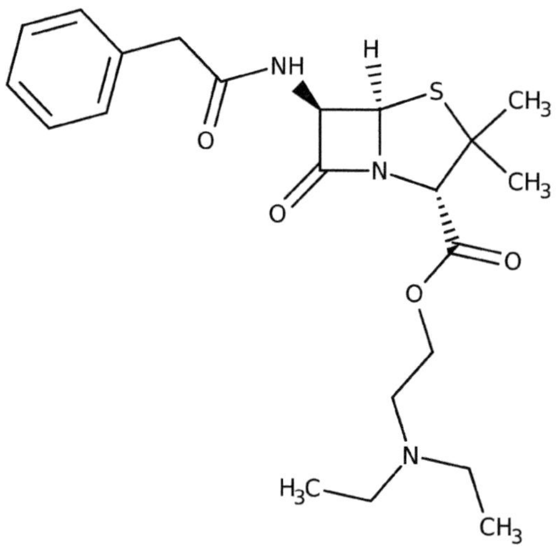

GENERAL

Penethamate is a β-lactam penicillin which is β-lactamase sensitive. It is used in various countries in veterinary medicine. In pharmaceutical products, penethamate is employed as penethamate hydriodide (CAS number 808-71-9, EC number 212-367-6, molecular formula $C_{22}H_{32}IN_3O_4S$) (1). The classification and structures of beta-lactam antibiotics are discussed in Chapter 3.36 Amoxicillin.

CONTACT ALLERGY FROM ACCIDENTAL CONTACT

In 34 veterinary surgeons (7 women, 27 men, age range 29-61, mean age 38 years) with chronic or relapsing eczema of the hands as the main complaint, investigated in Norway before 1985, patch tests were performed with the standard series and a veterinary series comprising drugs, antiseptics and protective glove materials frequently used in veterinary practice. Nineteen had eczema almost continuously on the hands, fingers and arms, and in another 3 there was a spread to the face and trunk. Occupational work caused exacerbations in 15 of these patients. Ten of the 34 veterinary surgeons had positive patch test reactions. A relation between positive patch tests and occupational work was confirmed in 9 of these subjects. In this group, there were five positive patch test reactions to penethamate 25% olive oil (2).

In 2 hospitals in Denmark, between 1974 and 1980, 37 veterinary surgeons, all working in private country practices, were investigated for suspected incapacitating occupational dermatitis and patch tested with a battery of 10 antibiotics. Thirty-two (86%) had one or more positive patch tests. There were 23 positive reactions (63%) to penethamate 250.000 IU/ml commercial product. It was mentioned that all but one of these individuals had allergic contact dermatitis, but relevance was not specified for individual allergens. The most frequent allergens were spiramycin, penethamate and tylosin tartrate (4).

Before that, one of the authors had already reported a number of cases of occupational allergic contact dermatitis from penethamate in veterinary surgeons (3,5).

Cross-reactions, pseudo-cross-reactions and co-reactions

Cross-reactions between beta-lactam antibiotics are discussed in Chapter 3.36 Amoxicillin.

LITERATURE

1 The data in the section 'General' may have been obtained from literature discussed in this chapter, but mostly also or exclusively from one or more of the following online sources: ChemIDPlus Advanced, PubChem, DrugBank, RxList, Drug Central, Drugs.com, and Wikipedia

2 Falk ES, Hektoen H, Thune PO. Skin and respiratory tract symptoms in veterinary surgeons. Contact Dermatitis 1985;12:274-278

3 Hjorth N. Occupational dermatoses in veterinary surgeons caused by penethamate (benzyl penicillin-ß-diethylamino-ethylester). Berufsdermatosen 1967;15:163-175 (Article in German)

4 Hjorth N, Roed-Petersen J. Allergic contact dermatitis in veterinary surgeons. Contact Dermatitis 1980;6:27-29

5 Hjorth N, Weismann K. Occupational dermatitis among veterinary surgeons caused by spiramycin, tylosin and penethamate. Acta Derm Venereol 1973;53:229-232

Chapter 3.368 PENICILLIN G BENZATHINE

IDENTIFICATION

Description/definition
: Penicillin G benzathine is the semisynthetic antibiotic prepared by combining the sodium salt of penicillin G with *N,N'*-dibenzylethylenediamine; it conforms to the structural formula shown below

Pharmacological classes
: Anti-bacterial agents

IUPAC name
: *N,N'*-Dibenzylethane-1,2-diamine;(2*S*,5*R*,6*R*)-3,3-dimethyl-7-oxo-6-[(2-phenylacetyl)amino]-4-thia-1-azabicyclo[3.2.0]heptane-2-carboxylic acid;tetrahydrate

Other names
: Benzathine benzylpenicillin; penicillin G benzathine tetrahydrate

CAS registry number
: 41372-02-5

EC number
: Not available

Merck Index monograph
: 8475

Patch testing
: 30% water

Molecular formula
: $C_{48}H_{64}N_6O_{12}S_2$

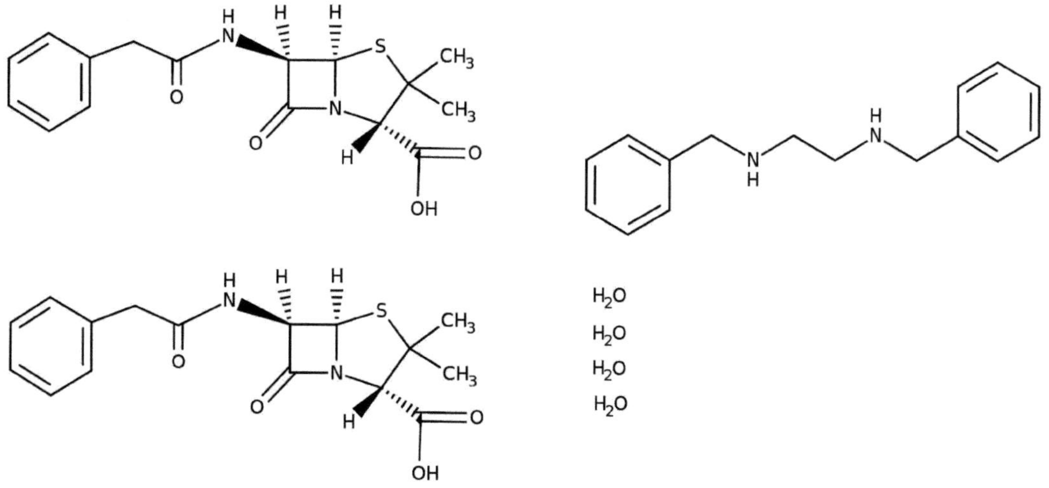

GENERAL

Penicillin G benzathine is a semisynthetic antibiotic prepared by combining penicillin G with *N,N'*-dibenzylethylene-diamine. It is a slow-onset, long-acting depot preparation for intramuscular injection used to treat various types of severe infections including streptococcal infections such as erysipelas, rheumatic fever and infections with *Treponema pallidum* (syphilis) (1). The classification and structures of beta-lactam antibiotics are discussed in Chapter 3.36 Amoxicillin.

CONTACT ALLERGY FROM ACCIDENTAL CONTACT

In 34 veterinary surgeons (7 women, 27 men, age range 29-61, mean age 38 years) with chronic or relapsing eczema of the hands as the main complaint, investigated in Norway before 1985, patch tests were performed with the standard series and a veterinary series comprising drugs, antiseptics and protective glove materials frequently used in veterinary practice. Nineteen had eczema almost continuously on the hands, fingers and arms, and in another 3 there was a spread to the face and trunk. Occupational work caused exacerbations in 15 of these patients. Ten of the 34 veterinary surgeons had positive patch test reactions. A relation between positive patch tests and occupational work was confirmed in 9 of these subjects. In this group, there were three positive patch test reactions to penicillin G benzathine 30% water (2).

Cross-reactions, pseudo-cross-reactions and co-reactions

Cross-reactions between beta-lactam antibiotics are discussed in Chapter 3.36 Amoxicillin.

LITERATURE

1   The data in the section 'General' may have been obtained from literature discussed in this chapter, but mostly also or exclusively from one or more of the following online sources: ChemIDPlus Advanced, PubChem, DrugBank, RxList, Drug Central, Drugs.com, and Wikipedia

2   Falk ES, Hektoen H, Thune PO. Skin and respiratory tract symptoms in veterinary surgeons. Contact Dermatitis 1985;12:274-278

Chapter 3.369 PENICILLIN V

IDENTIFICATION

Description/definition : Penicillin V is the member of the penicillin family that conforms to the structural formula shown below
Pharmacological classes : Anti-bacterial agents
IUPAC name : (2S,5R,6R)-3,3-Dimethyl-7-oxo-6-[(2-phenoxyacetyl)amino]-4-thia-1-azabicyclo-[3.2.0]heptane-2-carboxylic acid
Other names : Phenoxymethylpenicillin
CAS registry number : 87-08-1
EC number : 201-722-0
Merck Index monograph : 8479
Patch testing : 10.000 IU/gr pet. and pulverized tablet 'as is' (2); if the pure chemical is not available, prepare the test material from intravenous powder, the content of capsules or – when also not available – from powdered tablets to achieve a final concentration of the active drug of 10% pet.; beta-lactam antibiotics can generally be tested 5%-10% pet.
Molecular formula : $C_{16}H_{18}N_2O_5S$

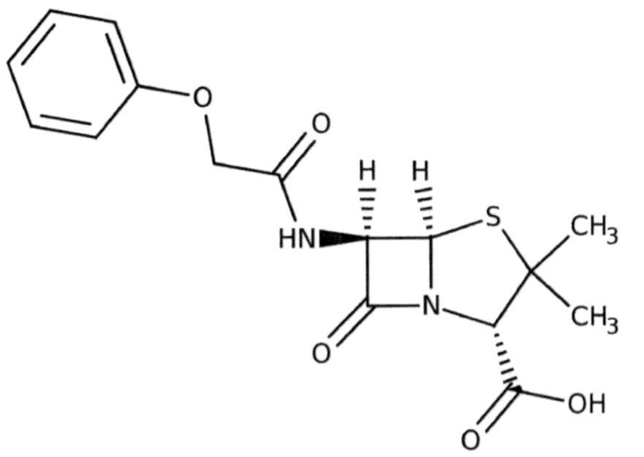

GENERAL

Penicillin V (phenoxymethylpenicillin) is a member of the penicillin family and the phenoxymethyl analog of penicillin G exhibiting narrow-spectrum antibiotic property. It binds to penicillin binding proteins, the enzymes that catalyze the synthesis of peptidoglycan, which is a critical component of the bacterial cell wall. This leads to the interruption of cell wall synthesis, consequently leading to bacterial cell growth inhibition and cell lysis. Penicillin V is indicated for the treatment of mild to moderately severe infections of the respiratory tract, the skin and soft tissues and the oropharynx caused by penicillin V sensitive microorganisms. In pharmaceutical products, penicillin V is employed as penicillin V potassium (CAS number 132-98-9, EC number 205-086-5, molecular formula $C_{16}H_{17}KN_2O_5S$) (1). The classification and structures of beta-lactam antibiotics are discussed in Chapter 3.36 Amoxicillin.

CUTANEOUS ADVERSE DRUG REACTIONS FROM SYSTEMIC ADMINISTRATION CAUSED BY TYPE IV (DELAYED-TYPE) HYPERSENSITIVITY (as demonstrated by positive patch tests)

Cutaneous adverse drug reactions from systemic administration of penicillin V caused by type IV (delayed-type) hypersensitivity have included maculopapular eruption (7), symmetrical drug-related intertriginous and flexural exanthema (SDRIFE)/baboon syndrome (2), drug reaction with eosinophilia and systemic symptoms (DRESS) (3), bullous pemphigoid (4), and unspecified drug eruptions (5).

Maculopapular eruption

A patient who had suffered a maculopapular rash from an aminopenicillin and had positive patch tests and intradermal tests (delayed reading) to one or 2 aminopenicillins and penicillin G (benzylpenicillin), had a positive oral challenge with penicillin V. Oral 1000 IU resulted in a mild diffuse maculopapular eruption after 6 hours (7).

Symmetrical drug-related intertriginous and flexural exanthema (SDRIFE)/Baboon syndrome

A few hours after taking the first tablet penicillin V for an abscess, a 57-year-old man developed widespread macular exanthema predominantly in the inguinal and gluteal areas, where he showed intense erythema with petechiae, small pustules and large erosions. A biopsy from the left inguinal area showed pustular hemorrhagic drug exanthema. Two months following resolution of the cutaneous symptoms, patch tests were positive to penicillin V, penicillin G (benzylpenicillin, both tested 10.000 IU/gr pet., and the pulverized tablet as is at D2 and D3. Intradermal tests to penicillins V and G showed infiltrated erythema at 24 hours (2). This was a case of SDRIFE presenting as the baboon syndrome (2).

Drug reaction with eosinophilia and systemic symptoms (DRESS)
In a group of 45 patients with multiple drug hypersensitivity seen between 1996 and 2018 in Montpellier, France, 38 of 92 drug hypersensitivities were classified as type IV immunological reactions. One of these patients suffered an episode of DRESS and had a positive patch test to penicillin V (3).

Other cutaneous adverse drug reactions
An 80-year-old man, known to be allergic to penicillin, was treated by mistake with parenteral benzylpenicillin and oral penicillin V. After the first dosage of the latter, the patient developed a rash, and penicillin was stopped. Despite this, 7 days later the exanthema became bullous. Physical examination showed an itchy generalized maculopapular rash confluent in the loins, over the knees and forearms with scattered widespread large vesicles, small tense bullae with serous exudate and denuded areas at the trunk and extremities. Based on clinical features, histopathology and immunofluorescence, the diagnosis of bullous pemphigoid was made. Patch tests were positive to benzylpenicillin and 3 other penicillins (penicillin V was not tested). The lesions healed rapidly after withdrawal of penicillin and have not recurred since, suggesting that the bullous pemphigoid was drug-induced (4).

In Finland, in the period 1989-2001, 826 patients with suspected cutaneous drug eruptions were patch tested and 89 had one or more positive reactions. Of these individuals, 8 reacted to phenoxymethylpenicillin. It was not mentioned which drug eruption these patients had suffered from or whether (some of) these were cross-reactions (5).

Cross-reactions, pseudo-cross-reactions and co-reactions
Cross-reactions between beta-lactam antibiotics are discussed in Chapter 3.36 Amoxicillin. In a group of 78 patients who had suffered nonimmediate drug reactions from any beta-lactam compound (penicillins, cephalosporins), 36 had positive patch tests to penicillin V (phenoxymethylpenicillin). As this was not the culprit drug causing the adverse skin reaction in any patient, all these reactions were presumably cross-reactions (6).

LITERATURE

1   The data in the section 'General' may have been obtained from literature discussed in this chapter, but mostly also or exclusively from one or more of the following online sources: ChemIDPlus Advanced, PubChem, DrugBank, RxList, Drug Central, Drugs.com, and Wikipedia
2   Panhans-Gross A, Gall H, Peter R U. Baboon syndrome after oral penicillin. Contact Dermatitis 1999;41:352-353
3   Landry Q, Zhang S, Ferrando L, Bourrain JL, Demoly P, Chiriac AM. Multiple drug hypersensitivity syndrome in a large database. J Allergy Clin Immunol Pract 2019;8:258
4   Borch JE, Andersen KE, Clemmensen O, Bindslev-Jensen C. Drug-induced bullous pemphigoid with positive patch test and in vitro IgE sensitization. Acta Derm Venereol 2005;85:171-172
5   Lammintausta K, Kortekangas-Savolainen O. The usefulness of skin tests to prove drug hypersensitivity. Br J Dermatol 2005;152:968-974
6   Buonomo A, Nucera E, De Pasquale T, Pecora V, Lombardo C, Sabato V, et al. Tolerability of aztreonam in patients with cell-mediated allergy to β-lactams. Int Arch Allergy Immunol 2011;155:155-159
7   Romano A, Di Fonso M, Papa G, Pietrantonio F, Federico F, Fabrizi G, Venuti A. Evaluation of adverse cutaneous reactions to aminopenicillins with emphasis on those manifested by maculopapular rashes. Allergy 1995;50:113-118

Chapter 3.370   PENICILLINS, UNSPECIFIED

IDENTIFICATION

Description/definition   : The penicillins are a group of antibiotics that contain 6-aminopenicillanic acid with a side
                                chain (R) attached to the 6-amino group; the penicillin nucleus is the chief structural
                                requirement for biological activity; the side-chain structure determines many of the
                                antibacterial and pharmacological characteristics
Pharmacological classes : Anti-bacterial agents
CAS registry number   : 1406-05-9
EC number   : 215-794-6
Patch testing   : Penicillins can generally be tested 5%-10% pet.

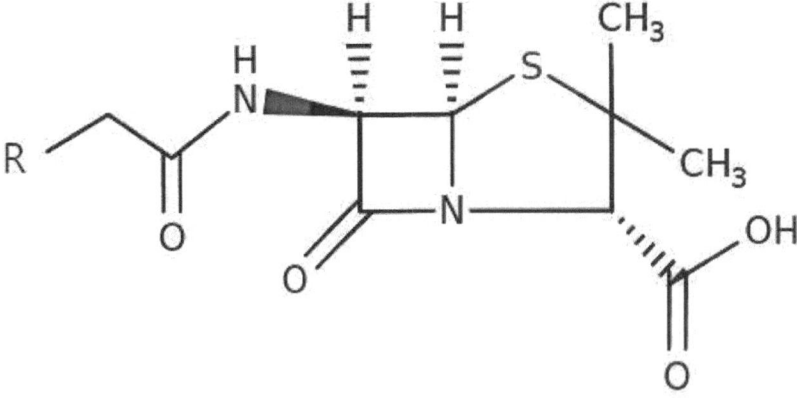

General

Penicillin is the generic name of the whole group of natural and semi-synthetic penicillins. Penicillin was originally
obtained from the fungus *Penicillium chrysogenum* (old name: *Penicillium notatum*) and was discovered in 1928 by
Sir Alexander Fleming, a Scottish researcher. All penicillins contain 6-aminopenicillanic acid with a side chain
attached to the 6-amino group, which determines many of the antibacterial and pharmacological characteristics.
In topical preparations, penicillins have caused many cases of contact allergy/allergic contact dermatitis, which
subject has been (partly and briefly) reviewed in Volume 3 of the *Monographs in contact allergy* series (8). The
classification and structures of beta-lactam antibiotics are discussed in Chapter 3.36 Amoxicillin.

CONTACT ALLERGY FROM ACCIDENTAL CONTACT

When applied topically, penicillin is a potent sensitizer (5). The first case of (occupational, partly airborne) allergic
contact dermatitis reported was that of an Army medical officer in Chicago who, shortly after beginning to prepare
and give injections of penicillin, developed an acute dermatitis on his face and then his hands (6). Accidental topical
contact with penicillin has caused a considerable number of occupational sensitizations in pharmaceutical employees
and health care workers (e.g. 1,2,3,4,7).

A 43-year-old man, who had worked for 13 years in cattle breeding. presented with an 18-month history of
erythema, edema and vesicles on the interdigital spaces of the hands, the face and both sides of the neck. His
symptoms worsened every time that he prepared animal feed. Patch tests were positive at D2 and D3 to penicillin
5% pet., oxytetracycline hydrochloride 5% pet., tylosin 10% pet., and spiramycin 5% pet. He had contact with all
these antibiotics and the dermatitis completely cleared as soon as the patient left his occupation (10).

CUTANEOUS ADVERSE DRUG REACTIONS FROM SYSTEMIC ADMINISTRATION CAUSED BY TYPE IV
(DELAYED-TYPE) HYPERSENSITIVITY (as demonstrated by positive patch tests)

Cutaneous adverse drug reactions from systemic administration of 'penicillin' (unspecified) caused by type IV
(delayed-type) hypersensitivity have included generalized exanthema (9).

Other cutaneous adverse drug reactions

In a case of acute generalized exanthema (not further specified) caused by clindamycin, there may have been co-
responsibility for 'penicillin'. However, the nature of the antibiotic was not specified and patch tests were only
performed with and positive to ampicillin. Delayed intradermal tests were negative to benzylpenicillin and cloxacillin
and hypersensitivity to penicillin V was excluded by a negative oral challenge (9).

Cross-reactions, pseudo-cross-reactions and co-reactions
Cross-reactions between beta-lactam antibiotics are discussed in Chapter 3.36 Amoxicillin.

Immediate contact reactions
Immediate contact reactions (contact urticaria) to penicillin are presented in Chapter 5.

LITERATURE

1   Rudzki E, Rebandel P, Grzywa Z. Contact allergy in the pharmaceutical industry. Contact Dermatitis 1989;21:121-122

2   Gielen K, Goossens A. Occupational allergic contact dermatitis from drugs in healthcare workers. Contact Dermatitis 2001;45:273-279

3   Rudzki E, Rebendel P. Contact sensitivity to antibiotics. Contact Dermatitis 1984;11:41-42

4   Goodman H. Dermatitis due to preparation and administration of penicillin solution. Arch Derm Syphilol 1946;54:206-208

5   Cronin E. Contact Dermatitis. Edinburgh: Churchill Livingstone, 1980:216

6   Pyle HD, Rattner H. Contact dermatitis from penicillin. JAMA 1944;125:903

7   Marsh WC, New WN. Dermatitis due to the preparation and administration of penicillin solution. U S Nav Med Bull 1948;48:391-394

8   De Groot AC. Monographs in contact allergy, volume 3: Topical Drugs. Boca Raton, Fl, USA: CRC Press Taylor and Francis Group, 2021 (ISBN 978-0-367-23693-9)

9   Valois M, Phillips EJ, Shear NH, Knowles SR. Clindamycin-associated acute generalized exanthematous pustulosis. Contact Dermatitis 2003;48:169

10 Guerra L, Venture M, Tardio M, Tosti A. Airborne contact dermatitis from animal feed antibiotics. Contact Dermatitis 1991;25:333-334

Chapter 3.371 PERAZINE

IDENTIFICATION

Description/definition	: Perazine is the phenothiazine derivative that conforms to the structural formula shown below
Pharmacological classes	: Dopamine antagonists; antipsychotic agents
IUPAC name	: 10-[3-(4-Methylpiperazin-1-yl)propyl]phenothiazine
CAS registry number	: 84-97-9
EC number	: 201-578-9
Merck Index monograph	: 8535
Patch testing	: No data available; most systemic drugs can be tested at 10% pet.; if the pure chemical is not available, prepare the test material from intravenous powder, the content of capsules or – when also not available – from powdered tablets to achieve a final concentration of the active drug of 10% pet.
Molecular formula	: $C_{20}H_{25}N_3S$

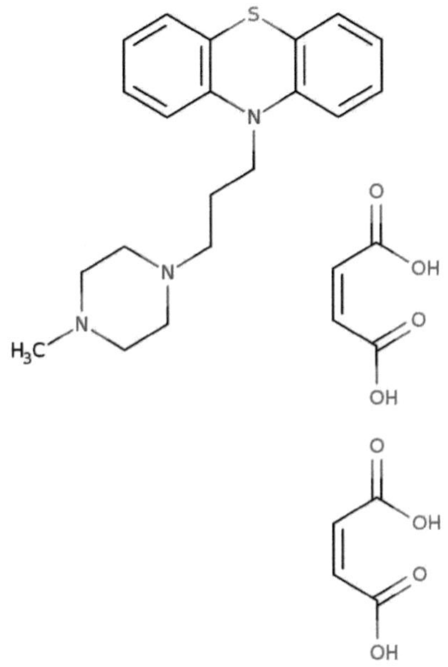

Perazine (di)maleate

GENERAL

Perazine is a phenothiazine antipsychotic with actions and uses similar to those of chlorpromazine. It is indicated for various forms of schizophrenia and acute psychotic disorders (including catatonic) with symptoms of psychomotor agitation, mania and delusions (1). In pharmaceutical products, perazine may be employed as perazine base and as perazine (di)maleate (CAS number 14516-56-4, EC number not available, molecular formula $C_{28}H_{33}N_3O_8S$) (1).

CONTACT ALLERGY FROM ACCIDENTAL CONTACT

Of 333 nurses patch tested in Poland between 1979 and 1987, 1 reacted to perazine ampoule contents; data on relevance were not provided (2).

LITERATURE

1   The data in the section 'General' may have been obtained from literature discussed in this chapter, but mostly also or exclusively from one or more of the following online sources: ChemIDPlus Advanced, PubChem, DrugBank, RxList, Drug Central, Drugs.com, and Wikipedia
2   Rudzki E, Rebandel P, Grzywa Z. Patch tests with occupational contactants in nurses, doctors and dentists. Contact Dermatitis 1989;20:247-250

Chapter 3.372 PERICIAZINE

IDENTIFICATION

Description/definition : Periciazine is the phenothiazine of the piperidine group that conforms to the structural formula shown below
Pharmacological classes : Antipsychotics
IUPAC name : 10-[3-(4-Hydroxypiperidin-1-yl)propyl]phenothiazine-2-carbonitrile
Other names : Piperocyanomazine
CAS registry number : 2622-26-6
EC number : 220-071-3
Merck Index monograph : 8549
Patch testing : Crushed tablet, powder pure (2); most systemic drugs can be tested at 10% pet.; if the pure chemical is not available, prepare the test material from intravenous powder, the content of capsules or – when also not available – from powdered tablets to achieve a final concentration of the active drug of 10% pet.
Molecular formula : $C_{21}H_{23}N_3OS$

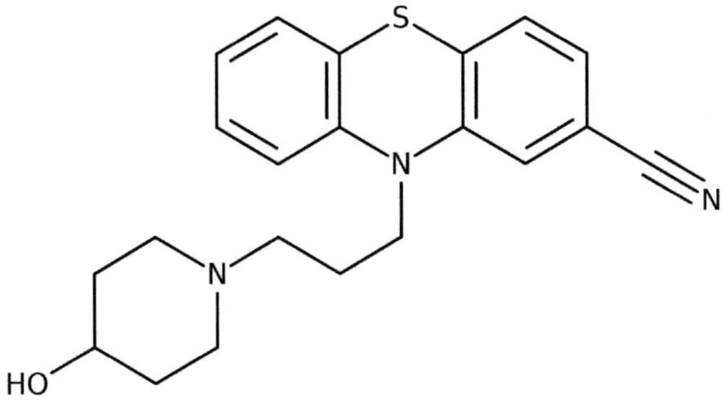

GENERAL

Periciazine is a phenothiazine of the piperidine group. It is a sedative with weak antipsychotic properties. Periciazine also has adrenolytic, anticholinergic, metabolic and endocrine effects and an action on the extrapyramidal system. This agent is used as an adjunctive medication in some psychotic patients, for the control of residual prevailing hostility, impulsiveness and aggressiveness (1).

CONTACT ALLERGY FROM ACCIDENTAL CONTACT

A 51-year-old male psychiatric nurse had suffered for five years from a chronic dermatitis of the tips of the thumb and index finger of the right hand. With these fingers he would pick up the pills which he personally gave to the patients in the psychiatric clinic. By this means the pills were taken in his presence, and the dose that each patient received was strictly controlled. When patch tested, he gave positive reactions to the pure powder from each of the drugs which he handled, the phenothiazines periciazine and perphenazine. Thus, he had occupational allergic contact dermatitis to these drugs (2). It should be appreciated that the pure chemicals were not tested and that no control tests have been performed.

LITERATURE

1   The data in the section 'General' may have been obtained from literature discussed in this chapter, but mostly also or exclusively from one or more of the following online sources: ChemIDPlus Advanced, PubChem, DrugBank, RxList, Drug Central, Drugs.com, and Wikipedia
2   Camarasa G. Contact dermatitis to phenothiazines Nemactil® and Decentan®. Contact Dermatitis 1976;2:123

Chapter 3.373 PERINDOPRIL

IDENTIFICATION

Description/definition   : Perindopril is the pyrrolidine carboxylic acid that conforms to the structural formula
                             shown below
Pharmacological classes : Angiotensin-converting enzyme inhibitors; antihypertensive agents
IUPAC name   : (2S,3aS,7aS)-1-[(2S)-2-[[(2S)-1-Ethoxy-1-oxopentan-2-yl]amino]propanoyl]-
                             2,3,3a,4,5,6,7,7a-octahydroindole-2-carboxylic acid
CAS registry number   : 82834-16-0
EC number   : Not available
Merck Index monograph : 8555
Patch testing   : 10% pet. (1); most systemic drugs can be tested at 10% pet.; if the pure chemical is not
                             available, prepare the test material from intravenous powder, the content of capsules or –
                             when also not available – from powdered tablets to achieve a final concentration of the
                             active drug of 10% pet.
Molecular formula   : $C_{19}H_{32}N_2O_5$

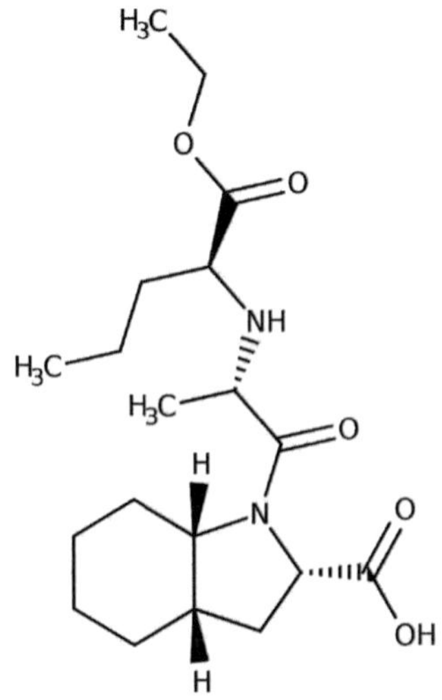

Perindopril 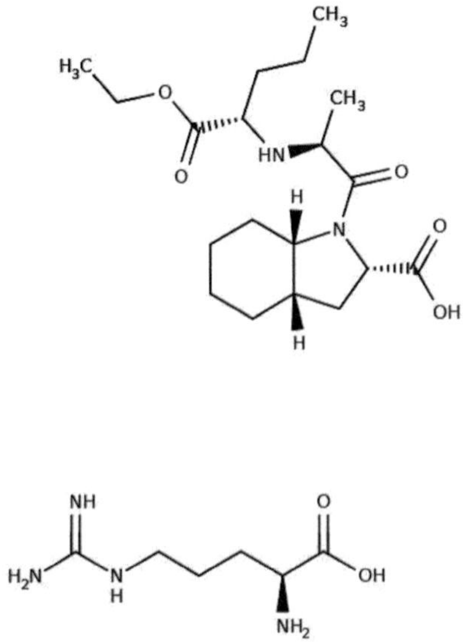   Perindopril arginine 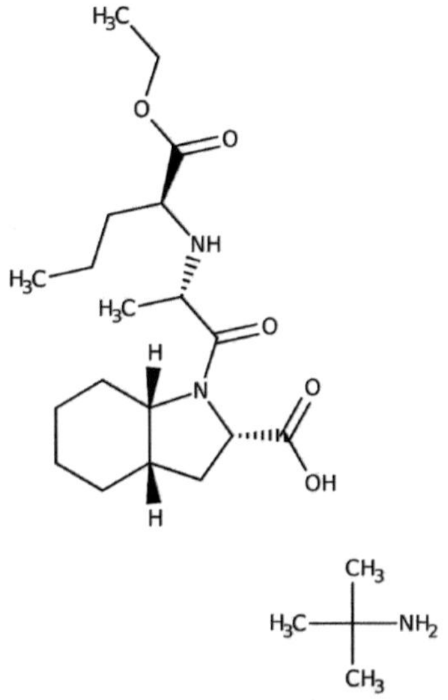   Perindopril erbumine

GENERAL

Perindopril is a non-sulfhydryl angiotensin converting enzyme (ACE) inhibitor with antihypertensive activity.
Perindopril is indicated for the treatment of essential hypertension, congestive heart failure, and to reduce the
cardiovascular risk of individuals with hypertension or post-myocardial infarction and stable coronary disease (1). In
pharmaceutical products, perindopril is employed as perindopril arginine (CAS number 612548-45-5, EC number not
available, molecular formula $C_{25}H_{46}N_6O_7$) or perindopril erbumine (CAS number 107133-36-8, EC number not
available, molecular formula $C_{23}H_{43}N_3O_5$) (go.drugbank.com).

CONTACT ALLERGY FROM ACCIDENTAL CONTACT

A 24-year-old woman presented with eyelid dermatitis, which had started with localized edema 4 months previously.
Later, the area had become itchier, with redness and scaling. The patient suspected a relationship with her work as a
pharmacy assistant, which involved breaking and crushing different types of tablets. She was patch tested with the
crushed tablets that she had contact with at 10% pet. and showed positive reactions to perindopril, 2 other ACE-
inhibitors, 4 beta-blockers, and 3 benzodiazepines. Three controls were negative to perindopril 10% pet. Cosmetic
allergy was excluded and a diagnosis of occupational airborne allergic contact was made (1).

LITERATURE

1   Swinnen I, Ghys K, Kerre S, Constandt L, Goossens A. Occupational airborne contact dermatitis from
    benzodiazepines and other drugs. Contact Dermatitis 2014;70:227-232

Chapter 3.374 PERPHENAZINE

IDENTIFICATION

Description/definition : Perphenazine is the piperazinyl phenothiazine that conforms to the structural formula shown below

Pharmacological classes : Antipsychotic agents; dopamine antagonists

IUPAC name : 2-[4-[3-(2-Chlorophenothiazin-10-yl)propyl]piperazin-1-yl]ethanol

Other names : Perfenazine

CAS registry number : 58-39-9

EC number : 200-381-5

Merck Index monograph : 8567

Patch testing : 0.01% pet (4); 2% pet. (2,5); for photopatch testing, the lower concentration is advisable to avoid phototoxic reactions

Molecular formula : $C_{21}H_{26}ClN_3OS$

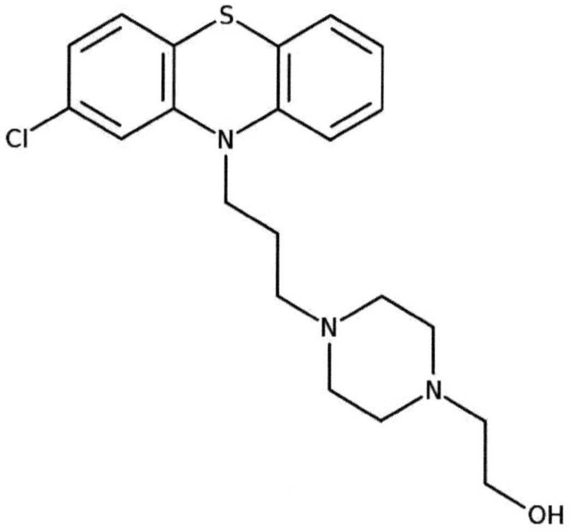

GENERAL

Perphenazine is a phenothiazine derivative and a dopamine antagonist with antiemetic and antipsychotic properties. Its actions and uses are similar to those of chlorpromazine, but perphenazine is 10 to 15 times as potent as chlorpromazine. Perphenazine is indicated for use in the management of the manifestations of psychotic disorders and for the control of severe nausea and vomiting in adults (1).

CONTACT ALLERGY FROM ACCIDENTAL CONTACT

A 51-year-old male psychiatric nurse had suffered for five years from a chronic dermatitis of the tips of the thumb and index finger of the right hand. With these fingers he would pick up the pills which he personally gave to the patients in the psychiatric clinic. By this means the pills were taken in his presence, and the dose that each patient received was strictly controlled. When patch tested, he gave positive reactions to the pure powder from each of the drugs which he handled, the phenothiazines perphenazine and periciazine. Thus, he had occupational allergic contact dermatitis to these drugs (3). It should be appreciated that the pure chemicals were not tested and that no control tests have been performed.

A 47-year-old female pharmacist, making up prescriptions for a few hours per day at a psychiatric hospital, presented with papular erythematous lesions over the light-exposed areas affecting the face, neck, ears, forearms and the dorsa of the hands. She was patch and photopatch tested with all the drugs that she had contact with and showed a positive photopatch test to perphenazine 0.01% pet., the related phenothiazine chlorpromazine 0.01% pet. and a positive patch test to biperiden 1% pet. Four controls were negative. This was a case of occupational photoallergic contact dermatitis to perphenazine and chlorpromazine and of occupational allergic contact dermatitis to biperiden (4).

A 56-year-old woman developed allergic and photoallergic contact dermatitis to perphenazine, which she was administering as tablets twice a day to her mother. The patch test to perphenazine 2% pet. was positive (++), the photopatch test irradiated with 5 J/cm^2 UVA intensely positive (+++, blister). Four controls were negative (2). This was a case of (photo)allergic contact dermatitis to perphenazine *by proxy*.

Cross-reactions, pseudo-cross-reactions and co-reactions
Cross- and photocross-reactions to other phenothiazines, including chlorpromazine, may occur (2,4).

LITERATURE

1   The data in the section 'General' may have been obtained from literature discussed in this chapter, but mostly also or exclusively from one or more of the following online sources: ChemIDPlus Advanced, PubChem, DrugBank, RxList, Drug Central, Drugs.com, and Wikipedia
2   Gacías L, Linares T, Escudero E, et al. Perphenazine as a cause of mother-to-daughter contact dermatitis and photocontact dermatitis. J Investig Allergol Clin Immunol 2013;23:60-61
3   Camarasa G. Contact dermatitis to phenothiazines Nemactil® and Decentan®. Contact Dermatitis 1976;2:123
4   Torinuki J. Contact dermatitis to biperiden and photocontact dermatitis in phenothiazines in a pharmacist. Tohoku J Exp Med 1995;176:249-252 (Article in English)

Chapter 3.375 PHENAZONE

IDENTIFICATION

Description/definition : Phenazone is the pyrazolone derivative that conforms to the structural formula shown below
Pharmacological classes : Anti-inflammatory agents, non-steroidal
IUPAC name : 1,5-Dimethyl-2-phenylpyrazol-3-one
Other names : Antipyrine; 3H-pyrazol-3-one, 1,2-dihydro-1,5-dimethyl-2-phenyl-
CAS registry number : 60-80-0
EC number : 200-486-6
Merck Index monograph : 1973
Patch testing : 5% pet. (SmartPracticeCanada, SmartPracticeEurope); 10% pet. (5)
Molecular formula : $C_{11}H_{12}N_2O$

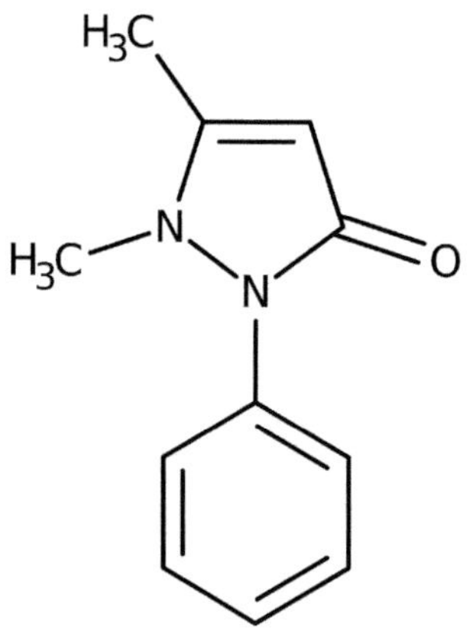

GENERAL

Phenazone is a nonsteroidal anti-inflammatory drug (NSAID) with analgesic and antipyretic activities that has been given by mouth and as ear drops. It is still used in ear drops (often combined with a local anesthetic), but is probably hardly utilized for oral application anymore. However, phenazone is often employed to test effects of other drugs on liver enzymes (1).

CUTANEOUS ADVERSE DRUG REACTIONS FROM SYSTEMIC ADMINISTRATION CAUSED BY TYPE IV (DELAYED-TYPE) HYPERSENSITIVITY (as demonstrated by positive patch tests)

Cutaneous adverse drug reactions from systemic administration of phenazone caused by type IV (delayed-type) hypersensitivity have included fixed drug eruption (3) and erythema multiforme-like lesions (4).

Fixed drug eruption

A 27-year-old man had developed circular bluish-red lesions on the abdomen, penis and both hands. He suffered from recurrent headaches and had been treated with oral phenazone intermittently for about 1 year. A scratch test with phenazone (test concentration, vehicle?) was positive 1 day after topical provocation to affected skin and a topical occluded patch test was positive after 2 and 3 days, whereas these tests were negative in unaffected skin. An oral challenge with 1000 mg phenazone resulted in itching and erythema at the circumference of pre-existing lesions after 6 hours. The patient was diagnosed with fixed drug eruption from delayed-type hypersensitivity to phenazone (3).

Other cutaneous adverse drug reactions

A 34-year-old man developed erythema multiforme-like lesions on the trunk and extremities every time he used a particular analgesic drug for a headache containing antipyrine (phenazone), phenacetin and quinine sulfate. Patch tests were positive to the combinion analgesic 'as is' and phenazone 5% pet., but negative to the other constituents. Ten controls were negative to phenazone 5% pet. (4).

Cross-reactions, pseudo-cross-reactions and co-reactions

Of 8 patients with a fixed drug eruption from antipyrine salicylate (phenazone salicylate), in who type-IV allergy was demonstrated by topical provocation (an open application test), all cross-reacted to phenazone, 4 to propyphenazone and 3 to aminophenazone (2).

LITERATURE

1   The data in the section 'General' may have been obtained from literature discussed in this chapter, but mostly also or exclusively from one or more of the following online sources: ChemIDPlus Advanced, PubChem, DrugBank, RxList, Drug Central, Drugs.com, and Wikipedia

2   Alanko K. Topical provocation of fixed drug eruption. A study of 30 patients. Contact Dermatitis 1994;31:25-27

3   Schick E, Weber L, Gall H. Topical and systemic provocation of fixed drug eruption due to phenazone. Contact Dermatitis 1996;35:58-59

4   Landwehr AJ, van Ketel WG. Delayed-type allergy to phenazone in a patient with erythema multiforme. Contact Dermatitis 1982;8:283-284

5   Brockow K, Garvey LH, Aberer W, Atanaskovic-Markovic M, Barbaud A, Bilo MB, et al.; ENDA/EAACI Drug Allergy Interest Group. Skin test concentrations for systemically administered drugs – an ENDA/EAACI Drug Allergy Interest Group position paper. Allergy 2013;68:702-712

Chapter 3.376 PHENETHICILLIN

IDENTIFICATION

Description/definition   : Phenethicillin is the penicillin that conforms to the structural formula shown below
Pharmacological classes : Anti-bacterial agents
IUPAC name   : (2S,5R,6R)-3,3-Dimethyl-7-oxo-6-(2-phenoxypropanoylamino)-4-thia-1-azabicyclo[3.2.0]heptane-2-carboxylic acid
Other names   : Phenoxyethylpenicillin; pheneticillin
CAS registry number   : 147-55-7
EC number   : 205-691-4
Merck Index monograph   : 8607 (Phenethicillin potassium)
Patch testing   : Beta-lactam antibiotics can generally be tested 5%-10% pet.
Molecular formula   : $C_{17}H_{20}N_2O_5S$

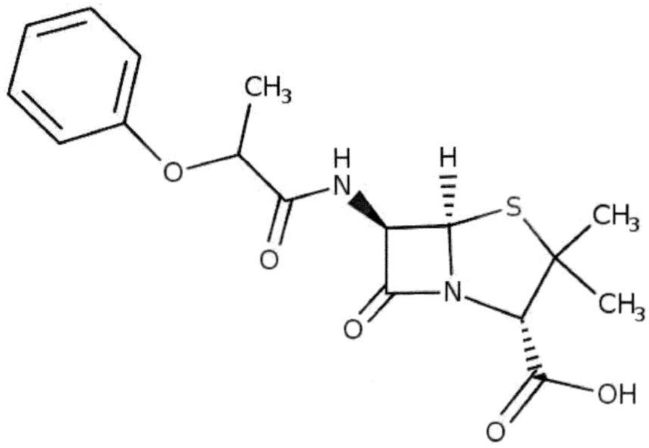

GENERAL

Phenethicillin is a penicillin in which the substituent at position 6 of the penam ring is a 2-phenoxypropanamido group. It inhibits the synthesis of the cell wall by inhibiting the penicillin binding proteins (PBPs) function. Phenethicillin is used (probably very little anymore) in oral formulations for the treatment of upper respiratory tract infections, lower respiratory tract infections, and skin and soft tissue infections. In pharmaceutical products, phenethicillin is employed as phenethicillin potassium (CAS number 132-93-4, EC number 205-084-4, molecular formula $C_{17}H_{19}KN_2O_5S$) (1). The classification and structures of beta-lactam antibiotics are discussed in Chapter 3.36 Amoxicillin.

CUTANEOUS ADVERSE DRUG REACTIONS FROM SYSTEMIC ADMINISTRATION CAUSED BY TYPE IV (DELAYED-TYPE) HYPERSENSITIVITY (as demonstrated by positive patch tests)

Cutaneous adverse drug reactions from systemic administration of phenethicillin caused by type IV (delayed-type) hypersensitivity have included maculopapular eruption (2,3).

Maculopapular eruption

Four cases of maculopapular eruption from phenethicillin with a positive patch test to phenethicillin powder were reported from the Netherlands in 1981 (2). Also in The Netherlands, in the first half of the 1970s, 18 patients with a maculopapular eruption from penicillins were patch tested with phenethicillin pure powder and there were 10 positive reactions to it. Nine also reacted to one or more other penicillins, especially ampicillin. It was not mentioned in how many cases the drug eruption had resulted from administration of phenethicillin (3).

Cross-reactions, pseudo-cross-reactions and co-reactions

Cross-reactions between beta-lactam antibiotics are discussed in Chapter 3.36 Amoxicillin.

LITERATURE

1   The data in the section 'General' may have been obtained from literature discussed in this chapter, but mostly also or exclusively from one or more of the following online sources: ChemIDPlus Advanced, PubChem, DrugBank, RxList, Drug Central, Drugs.com, and Wikipedia
2   Bruynzeel DP, van Ketel WG. Repeated patch testing in penicillin allergy. Br J Dermatol 1981;104:157-159
3   van Ketel WG. Patch testing in penicillin allergy. Contact Dermatitis 1975;1:253-254

Chapter 3.377 PHENINDIONE

IDENTIFICATION

Description/definition   : Phenindione is the indanedione that conforms to the structural formula shown below
Pharmacological classes : Anticoagulants
IUPAC name   : 2-Phenylindene-1,3-dione
Other names   : Phenylindanedione
CAS registry number   : 83-12-5
EC number   : 201-454-4
Merck Index monograph   : 8618
Patch testing   : No data available; most systemic drugs can be tested at 10% pet.; if the pure chemical is not available, prepare the test material from intravenous powder, the content of capsules or – when also not available – from powdered tablets to achieve a final concentration of the active drug of 10% pet.
Molecular formula   : $C_{15}H_{10}O_2$

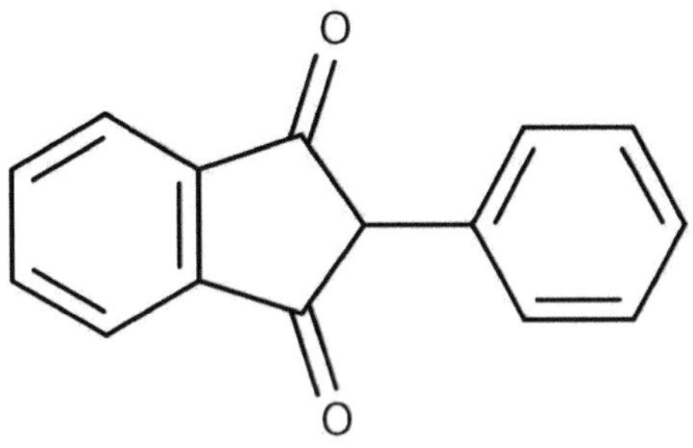

GENERAL

Phenindione is an indanedione that has been used as an anticoagulant. The drug thins the blood by antagonizing vitamin K which is required for the production of clotting factors in the liver. Phenindione is indicated for the treatment of pulmonary embolism, cardiomyopathy, atrial fibrillation and flutter, cerebral embolism, mural thrombosis, and thrombophilia and is also used for anticoagulant prophylaxis. However, it is now rarely employed because of its higher incidence of severe adverse effects (1).

CUTANEOUS ADVERSE DRUG REACTIONS FROM SYSTEMIC ADMINISTRATION CAUSED BY TYPE IV (DELAYED-TYPE) HYPERSENSITIVITY (as demonstrated by positive patch tests)

Cutaneous adverse drug reactions from systemic administration of phenindione caused by type IV (delayed-type) hypersensitivity have included maculopapular eruption, possibly as part of DRESS (2).

Maculopapular eruption or drug reaction with eosinophilia and systemic symptoms (DRESS)

In a group of 45 patients with multiple drug hypersensitivity seen between 1996 and 2018 in Montpellier, France, there was one patient, a 69-year-old woman, who had maculopapular drug reaction, possibly as part of drug reaction with eosinophilia and systemic symptoms (DRESS), and a positive patch test to phenindione (2).

LITERATURE

1   The data in the section 'General' may have been obtained from literature discussed in this chapter, but mostly also or exclusively from one or more of the following online sources: ChemIDPlus Advanced, PubChem, DrugBank, RxList, Drug Central, Drugs.com, and Wikipedia
2   Landry Q, Zhang S, Ferrando L, Bourrain JL, Demoly P, Chiriac AM. Multiple drug hypersensitivity syndrome in a large database. J Allergy Clin Immunol Pract 2019;8:258

Chapter 3.378 PHENIRAMINE

IDENTIFICATION

Description/definition : Pheniramine is a tertiary amino compound and a member of pyridines that conforms to the structural formula shown below (shown is pheniramine maleate)
Pharmacological classes : Histamine H_1 antagonists; antipruritics; anti-allergic agents
IUPAC name : *N,N*-Dimethyl-3-phenyl-3-pyridin-2-ylpropan-1-amine
Other names : Prophenpyridamine
CAS registry number : 86-21-5
EC number : 201-656-2
Merck Index monograph : 8619
Patch testing : 3% water and pet.
Molecular formula : $C_{16}H_{20}N_2$

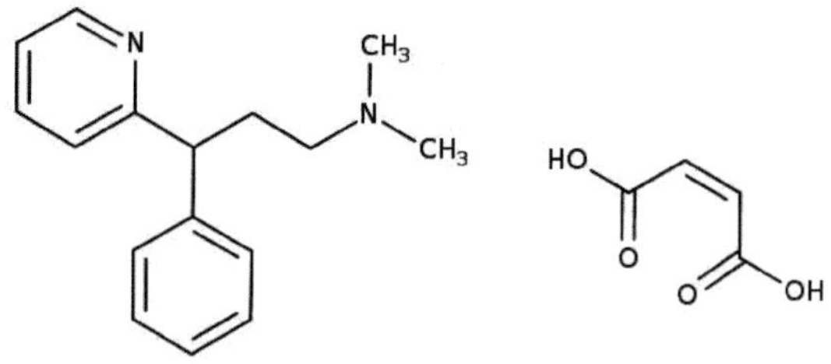

GENERAL

Pheniramine is a first generation antihistamine in the alkylamine class. It is used in some over-the-counter allergy as well as cold & flu products in combination with other drugs. Pheniramine's use as an anti-allergy medication for hay fever, rhinitis, allergic dermatoses, and pruritus has largely been supplanted by second generation antihistamines. It is also used in eye drops for the treatment of allergic conjunctivitis, which has induced rare cases of contact allergy/ allergic contact dermatitis (2,3). In pharmaceutical products, pheniramine is employed as pheniramine maleate (CAS number 132-20-7, EC number 205-051-4, molecular formula $C_{20}H_{24}N_2O_4$) (1).

CUTANEOUS ADVERSE DRUG REACTIONS FROM SYSTEMIC ADMINISTRATION CAUSED BY TYPE IV (DELAYED-TYPE) HYPERSENSITIVITY (as demonstrated by positive patch tests)

Cutaneous adverse drug reactions from systemic administration of pheniramine caused by type IV (delayed-type) hypersensitivity have included urticarial and later maculopapular eruption (4).

Other cutaneous adverse drug reactions

A 36-year-old woman was given pheniramine maleate tablets for dermatitis of the vulva and perianal area and the following day an eruption developed on her neck. When given an intravenous injection with diphenhydramine, within one hour, a generalized eruption appeared, which was described by the patient as being 'urticarial'. On the following day this developed into a severe pruritic, excoriated, coalescent, erythematous, maculopapular eruption. Patch tests were positive tot to a pheniramine maleate tablet (probably powdered, pure), yielding a sharply demarcated area of erythema and edema. There was no positive reaction to diphenhydramine (contents of capsule), but performance of these test was followed by an exacerbation of the eruption on the following day (4).

Cross-reactions, pseudo-cross-reactions and co-reactions

Possible cross-reactivity between pheniramine, chlorpheniramine and dexchlorpheniramine (3).

LITERATURE

1   The data in the section 'General' may have been obtained from literature discussed in this chapter, but mostly also or exclusively from one or more of the following online sources: ChemIDPlus Advanced, PubChem, DrugBank, RxList, Drug Central, Drugs.com, and Wikipedia
2   Corazza M, Massieri LT, Virgili A. Doubtful value of patch testing for suspected contact allergy to ophthalmic products. Acta Derm Venereol 2005;85:70-71
3   Parente G, Pazzaglia M, Vincenzi C, Tosti A. Contact dermatitis from pheniramine maleate in eyedrops. Contact Dermatitis 1999;40:338
4   Epstein E. Dermatitis due to antihistaminic agents. J Invest Dermatol 1949;12:151-152

Chapter 3.379 PHENOBARBITAL

IDENTIFICATION

Description/definition : Phenobarbital is the barbituric acid derivative that conforms to the structural formula shown below
Pharmacological classes : Anticonvulsants; excitatory amino acid antagonists; GABA modulators; hypnotics and sedatives; cytochrome P-450 CYP2B6 inducers; cytochrome P-450 CYP3A inducers
IUPAC name : 5-Ethyl-5-phenyl-1,3-diazinane-2,4,6-trione
Other names : Phenobarbitone; 5-ethyl-5-phenylbarbituric acid
CAS registry number : 50-06-6
EC number : 200-007-0
Merck Index monograph : 8622
Patch testing : 10% pet. (20); with serious cutaneous adverse reactions, starting with 1% pet. may be advisable (20)
Molecular formula : $C_{12}H_{12}N_2O_3$

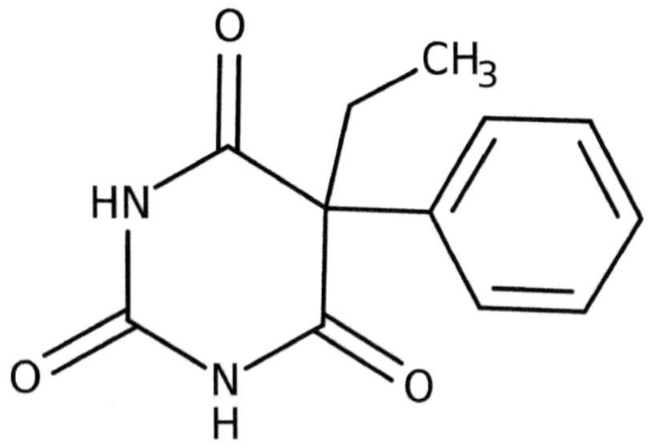

GENERAL

Phenobarbital is an aromatic barbituric acid derivative that acts as a nonselective central nervous system depressant. It is the longest-acting barbiturate and is used for its anticonvulsant and sedative-hypnotic properties in the management of all seizure disorders except absence (petit mal). In pharmaceutical products, both phenobarbital and phenobarbital sodium (CAS number 200-322-3, EC number 57-30-7, molecular formula $C_{12}H_{11}N_2NaO_3$) may be employed (1).

CUTANEOUS ADVERSE DRUG REACTIONS FROM SYSTEMIC ADMINISTRATION CAUSED BY TYPE IV (DELAYED-TYPE) HYPERSENSITIVITY (as demonstrated by positive patch tests)

Cutaneous adverse drug reactions from systemic administration of phenobarbital caused by type IV (delayed-type) hypersensitivity have included maculopapular eruption (14,17,21,22,23,26), exfoliative dermatitis (25), acute generalized exanthematous pustulosis (AGEP) (16), fixed drug eruption (2,3,4,5), drug reaction with eosinophilia and systemic symptoms (DRESS)/anticonvulsant hypersensitivity syndrome (9,10,11,17,19,24), toxic epidermal necrolysis (TEN) (6,7), Stevens-Johnson syndrome (SJS) (16), eczematous eruption (12,13), delayed urticaria (21), and unspecified drug eruptions (15,16,21).

Case series with various or unknown types of drug reactions

Between 2003 and 2017, in Belgrade, Serbia, 100 children in the age from 1 to 17 years suspected of hypersensitivity reactions to antiepileptic drugs were examined with patch tests, using the commercial drugs 10% pet. 61 patients had shown maculopapular eruptions, 26 delayed urticaria, 5 morbilliform exanthema, 5 DRESS, 2 SJS and one erythema multiforme. Phenobarbital was the suspected drug in 7 cases and was patch test positive in 6 children (86%). It was not specified which eruptions these drugs had caused, but should include maculopapular eruptions and delayed urticaria (21).

In the regional pharmacovigilance center of Sfax (Tunisia), between June 1, 2014 and April 30, 2016, all cases of (presumed) allergic skin reactions to antiepileptic drugs were investigated with patch testing. Twenty patients were included, among who 23 cutaneous adverse drug reactions (CADRs) were observed. The drugs involved were phenobarbital (n=10), carbamazepine (n=11), and valproic acid (n=4). The CADRs were maculopapular exanthema (11 cases), DRESS (6 cases), Stevens-Johnson syndrome (2 cases), fixed drug eruption (2 cases) and erythroderma (2 cases). Patch tests were positive in 19 patients (95%). Of the patients with positive patch tests to phenobarbital, 4 had maculopapular exanthema and 3 had DRESS (17).

Maculopapular eruption
In Turkey, of 7 children showing signs of hypersensitivity to phenobarbital, 2 (29%) had positive patch test reactions to commercial phenobarbital acid at 1% and 10% active ingredients; both had suffered a maculopapular eruption (22). In Ankara, Turkey, phenobarbital was patch test positive in 2 children with maculopapular eruption (26).

A 14-year-old girl had been treated for 14 days with phenobarbital, when she developed a maculopapular exanthema. There was eosinophilia, but there were no signs of DRESS. Eight months later, patch tests were positive to phenobarbital, tested as pure drug at 5,10,15, and 20% w/w in white petrolatum. 34 controls were negative (14).

A 6-year-old boy developed a maculopapular exanthema while using phenobarbital and had a positive patch test to this anti-epileptic drug tested 5% pet. (23). See also the section 'Case series with various or unknown types of drug reactions' above, refs. 17 and 21.

Erythroderma, widespread erythematous eruption, exfoliative dermatitis
A 5-year-old girl developed severe exfoliative dermatitis and liver dysfunction 3 weeks after initiation of phenobarbital therapy. Later she developed interstitial nephritis. After complete recovery, serial lymphocyte transformation studies and patch tests gave positive results for phenobarbital, supporting the view that these were complications of phenobarbital hypersensitivity (25). Details are lacking to assess whether this may have been a case of DRESS.

Acute generalized exanthematous pustulosis (AGEP)
One patient with (acute generalized exanthematous pustulosis) AGEP had a positive patch test reaction to phenobarbital (16).

Fixed drug eruption
In Seoul, South Korea, 31 patients (15 men,16 women, age range 7 to 62 years) with suspected fixed drug eruption were patch tested between 1986 and 1990 and in 1996 and 1997. The drugs for patch testing were usually applied at 10% in white petrolatum, both on a previous lesions and on apparently normal skin. The presentation of the results in this reports were rather confusing. When 'itching, erythema, infiltration' at the postlesional skin is taken as proof of delayed-type allergy, then there were 3 reactions to phenobarbital. In most cases, oral provocation tests, when performed, were positive (3).

One patient who had developed a fixed drug eruption to phenobarbital had a topical provocation test indicative of type IV allergy to the drug (2). An 18-month-old child had a fixed drug eruption from phenobarbital with a positive patch test to this barbiturate (4). A 10-year-old girl developed a fixed drug eruption 12 hours after taking a suppository of acetylsalicylic acid and phenobarbital with red plaques on the left thigh, which healed with hyperpigmentation. Patch tests with the separate drugs were positive on postlesional skin to phenobarbital 10% in petrolatum and water (5).

Drug reaction with eosinophilia and systemic symptoms (DRESS)/anticonvulsant hypersensitivity syndrome
The literature on patch testing in anticonvulsant hypersensitivity syndrome up to August 2008 has been reviewed in ref. 18.

Twenty-four patients who had developed severe drug rashes from antiepileptics (18 DRESS, 5 SJS, TEN or SJS/TEN, 1 lichenoid drug eruption) were patch tested with the implicated drugs 10%, 20% and 30% pet. of pulverized prescription tablets. Positive reactions were observed in 12 patients: 11 with DRESS and one with lichenoid eruption. Phenobarbital was the implicated drug with positive patch tests in 2 patients with DRESS (10). Of 13 pediatric patients patch tested for DRESS in a tertiary care hospital in Florence, Italy, between 2010 and 2018, 5 (38%) had positive reactions, one of who reacted to phenobarbital (11).

In a group of 15 patients with 'anticonvulsant hypersensitivity syndrome', one patient had a positive patch test to phenobarbital; details are unavailable to the author, but not all patients appeared to have all symptoms necessary to diagnose anticonvulsant hypersensitivity syndrome (synonym: DRESS from anticonvulsant drugs) (9).

A young woman developed a cutaneous rash, lymphadenopathy, malaise and fever after the introduction of phenobarbital. Because of these symptoms, the patient was treated with ceftriaxone and she experienced a severe flare-up of the cutaneous and general reaction with eosinophilia and elevated liver enzymes. Patch tests and lymphocyte transformation tests were positive to both drugs (19).

One case of DRESS from phenobarbital with a positive patch test to this antiepileptic drug was reported from Brazil in 2021 (24). See also the section 'Case series with various or unknown types of drug reactions' above, ref. 17.

Stevens-Johnson syndrome/toxic epidermal necrolysis (SJS/TEN)
Two cases of toxic epidermal necrolysis (TEN) from phenobarbital with positive patch tests to this barbiturate reported in Japanese literature (6,7) have been cited in ref. 8. One patient with Stevens-Johnson syndrome (SJS) had a positive patch test reaction to phenobarbital (16).

Dermatitis/eczematous eruption
In a 50-year-old woman, 6 hours after having taken a 100 mg phenobarbital tablet, profuse symmetrical papulovesicular itching lesions appeared on her legs. This eruption spread and generalized, but mostly involved the legs. No other clinical or laboratory changes were present. Target-shaped maculopapular lesions developed on her dorsal feet and hands a week later. A patch test with phenobarbital 20% in propylene glycol was positive at D3 (12). It should be realized that propylene glycol can cause irritant, false-positive, reactions (12).

An 18-year-old woman had been treated for 35 days with 100 mg/day phenobarbital for post-traumatic focal epilepsy, when she developed a generalized dermatitis excluding the palms and soles. A patch test with 0.1% aqueous phenobarbital was positive with erythema, edema and vesicles; histopathology was consistent with allergic contact dermatitis. Ten controls were negative. Three months later, the patient took phenobarbital again, resulting in eczema once more (13).

Other cutaneous adverse drug reactions
In a Japanese investigation, 4 of 10 patients patch tested with phenobarbital because of 'anticonvulsant-induced drug eruptions' had positive reaction to this drug. The nature of the skin eruption was not mentioned (15). One patient with an unspecified drug exanthema had a positive patch test reaction to phenobarbital (16). See also the section 'Case series with various or unknown types of drug reactions' above, ref. 21 (unspecified exanthema and urticaria).

Cross-reactions, pseudo-cross-reactions and co-reactions
This subject is discussed in Chapter 3.284 Lamotrigine.

LITERATURE
1   The data in the section 'General' may have been obtained from literature discussed in this chapter, but mostly also or exclusively from one or more of the following online sources: ChemIDPlus Advanced, PubChem, DrugBank, RxList, Drug Central, Drugs.com, and Wikipedia
2   Alanko K, Stubb S, Reitamo S. Topical provocation of fixed drug eruption. Br J Dermatol 1987;116:561-567
3   Lee AY. Topical provocation in 31 cases of fixed drug eruption: change of causative drugs in 10 years. Contact Dermatitis 1998;38:258-260
4   Michel JL, Chalencon V, Mazzochi C, et al. Fixed eruption to phenobarbital in a 18-month-old child. Nouvelles Dermatologiques 1999;18:146 (Article in French, data cited in ref. 5)
5   Chadly Z, Aouam K, Chaabane A, Belhadjali H, Abderrazzak Boughattas N, Zili JE. A patch test confirmed phenobarbital-induced fixed drug eruption in a child. Iran J Allergy Asthma Immunol 2014;13:214-217
6   Nakamura T, Inomata S. On toxic epidermal necrolysis (Lyell). Nippon Hifuka Gakkai Zasshi 1965;75:150-151 (Article in Japanese, data cited in ref. 8)
7   Yamada M, Matsunaga K, Sakai K, Yazaki Y, Shiga A. A case of TEN (Lyell). Nippon Hifuka Gakkai Zasshi 1980;90:648 (Article in Japanese, data cited in ref. 8)
8   Tagami H, Tatsuta K, Iwatski K, Yamada M. Delayed hypersensitivity in ampicillin-induced toxic epidermal necrolysis. Arch Dermatol 1983;119:910-913)
9   Galindo PA, Borja J, Gomez E, Mur P, Gudín M, García R, et al. Anticonvulsant drug hypersensitivity. J Invest Allergol Clin Immunol 2002;12:299-304
10 Shiny TN, Mahajan VK, Mehta KS, Chauhan PS, Rawat R, Sharma R. Patch testing and cross sensitity study of adverse cutaneous drug reactions due to anticonvulsants: A preliminary report. World J Methodol 2017;7:25-32
11 Liccioli G, Mori F, Parronchi P, Capone M, Fili L, Barni S, Sarti L, Giovannini M, Resti M, Novembre EM. Aetiopathogenesis of severe cutaneous adverse reactions (SCARs) in children: A 9-year experience in a tertiary care paediatric hospital setting. Clin Exp Allergy 2020;50:61-73
12 Fernandez de Corres L, Leanizbarrutia I, Munoz D. Eczematous drug reaction from phenobarbitone. Contact Dermatitis 1984;11:319
13 Pigatto PD, Morelli M, Polenghi MM, Mozzanica N, Altomare GF. Phenobarbital-induced allergic dermatitis. Contact Dermatitis 1987;16:279
14 Romano A, Pettinato R, Andriolo M, Viola M, Guéant-Rodriguez RM, Valluzzi RL, et al. Hypersensitivity to aromatic anticonvulsants: in vivo and in vitro cross-reactivity studies. Curr Pharm Des 2006;12:3373-3381
15 Osawa J, Naito S, Aihara M, Kitamura K, Ikezawa Z, Nakajima H. Evaluation of skin test reactions in patients with

non-immediate type drug eruptions. J Dermatol 1990;17:235-239

16   Wolkenstein P, Chosidow O, Fléchet ML, Robbiola O, Paul M, Dumé L, et al. Patch testing in severe cutaneous adverse drug reactions, including Stevens-Johnson syndrome and toxic epidermal necrolysis. Contact Dermatitis 1996;35:234-236

17   Ben Mahmoud L, Bahloul N, Ghozzi H, Kammoun B, Hakim A, Sahnoun Z, et al. Epicutaneous patch testing in delayed drug hypersensitivity reactions induced by antiepileptic drugs. Therapie 2017;72:539-545

18   Elzagallaai AA, Knowles SR, Rieder MJ, Bend JR, Shear NH, Koren G. Patch testing for the diagnosis of anticonvulsant hypersensitivity syndrome: a systematic review. Drug Saf 2009;32:391-408

19   Voltolini S, Bignardi D, Minale P, Pellegrini S, Troise C. Phenobarbital-induced DiHS and ceftriaxone hypersensitivity reaction: a case of multiple drug allergy. Eur Ann Allergy Clin Immunol 2009;41:62-63

20   Brockow K, Garvey LH, Aberer W, Atanaskovic-Markovic M, Barbaud A, Bilo MB, et al.; ENDA/EAACI Drug Allergy Interest Group. Skin test concentrations for systemically administered drugs – an ENDA/EAACI Drug Allergy Interest Group position paper. Allergy 2013;68:702-712

21   Atanasković-Marković M, Janković J, Tmušić V, Gavrović-Jankulović M, Ćirković Veličković T, Nikolić D, Škorić D. Hypersensitivity reactions to antiepileptic drugs in children. Pediatr Allergy Immunol 2019;30:547-552

22   Guvenir H, Dibek Misirlioglu E, Civelek E, Toyran M, Buyuktiryaki B, Ginis T, et al. The frequency and clinical features of hypersensitivity reactions to antiepileptic drugs in children: a prospective study. J Allergy Clin Immunol Pract 2018;6:2043-2050

23   Atanaskovic-Markovic M, Gaeta F, Medjo B, Gavrovic-Jankulovic M, Cirkovic Velickovic T, Tmusic V, et al. Nonimmediate hypersensitivity reactions to beta-lactam antibiotics in children - our 10-year experience in allergy work-up. Pediatr Allergy Immunol 2016;27:533-538

24   Perelló MI, de Maria Castro A, Nogueira Arraes AC, Caracciolo Costa S, Lacerda Pedrazzi D, Andrade Coelho Dias G, et al. Severe cutaneous adverse drug reactions: diagnostic approach and genetic study in a Brazilian case series. Eur Ann Allergy Clin Immunol. 2021 Mar 16. doi: 10.23822/EurAnnACI.1764-1489.193. Epub ahead of print.

25   Sawaishi Y, Komatsu K, Takeda O, Tazawa Y, Takahashi I, Hayasaka K, et al. A case of tubulo-interstitial nephritis with exfoliative dermatitis and hepatitis due to phenobarbital hypersensitivity. Eur J Pediatr 1992;151:69-72

26   Büyük Yaytokgil Ş, Güvenir H, Külhaş Celík İ, Yilmaz Topal Ö, Karaatmaca B, Civelek E, et al. Evaluation of drug patch tests in children. Allergy Asthma Proc 2021;42:167-174

Chapter 3.380 PHENOXYBENZAMINE

IDENTIFICATION

Description/definition   : Phenoxybenzamine is the phenylmethylamine that conforms to the structural formula
                            shown below
Pharmacological classes   : Antihypertensive agents; vasodilator agents; α-adrenergic antagonists
IUPAC name   : N-Benzyl-N-(2-chloroethyl)-1-phenoxypropan-2-amine
Other name(s)   : Dibenzyline
CAS registry number   : 59-96-1
EC number   : 200-446-8
Merck Index monograph   : 8638
Patch testing   : 0.1% water; 1% is irritant and may sensitize
Molecular formula   : $C_{18}H_{22}ClNO$

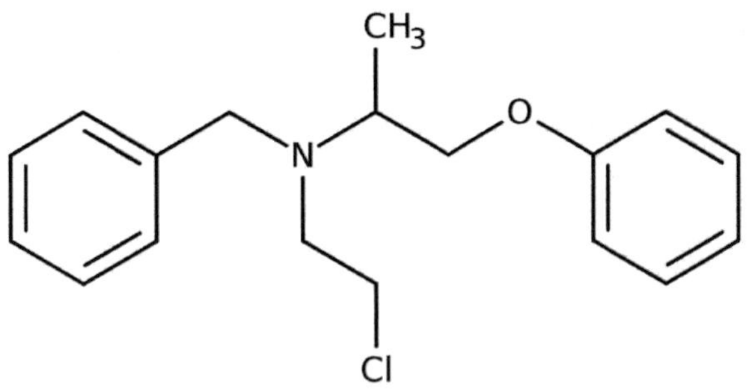

GENERAL

Phenoxybenzamine is a synthetic α-adrenergic antagonist with antihypertensive and vasodilatory properties and a long duration of action. It non-selectively and irreversibly blocks the postsynaptic α-adrenergic receptor in smooth muscle, thereby preventing vasoconstriction, relieving vasospasms, and decreasing peripheral resistance. Phenoxybenzamine is indicated for the treatment of pheochromocytoma (malignant), benign prostatic hypertrophy and malignant essential hypertension (1). In pharmaceutical products, phenoxybenzamine is employed as phenoxybenzamine hydrochloride (CAS number 63-92-3, EC number 200-569-7, molecular formula $C_{18}H_{23}Cl_2NO$) (1).

CONTACT ALLERGY FROM ACCIDENTAL CONTACT

A 43-year-old intensive care unit nurse presented with a work-related eczema, initially on both forearms and then also on her face and neck, which had been present for 6 months. A similar rash affected 3 of her colleagues, though they did not seek medical advice. Suspected causes included skin cleansers, as well as penicillins or phenoxybenzamine hydrochloride. These drugs were prepared on a shelf that pulled out from the drug cabinet, and which was then wiped. Following this, the nurses would rest their forearms on this shelf to fill in the drug charts. Patch testing revealed positive reactions to phenoxybenzamine HCl 1% (++/++), 0.5% (++/++) and 0.1% water (+/+). Patch testing in 7 controls resulted in an irritant reaction in 2 patients and active sensitization in one individual. 25 controls were negative to phenoxybenzamine HCl 0.1% water (2).

A female laboratory technician aged 25 years, who was employed in preparing chemical compounds for pharmacological class exercises in the medical school, developed acute allergic eczematous contact dermatitis of the hands and face. Patch testing, carried out after the clinical dermatitis had subsided, gave positive reactions to phenoxybenzamine and propranolol, both tested 1% water, followed by a recurrence of the dermatitis at the original sites. Tests in controls were not carried out (5,7). In guinea-pig sensitization studies, the authors sensitized all 5 experimental animals with intradermal injections of phenoxybenzamine 0.1% aqueous solution and saw cross-reactions to various related chemicals (5).

OTHER DATA

Two human volunteers in a pharmaceutical experiment were sensitized to phenoxybenzamine from intradermal injections, leading to allergic dermatitis within one week and later malaise, generalized itching and widespread urticaria. Patch tests were not performed but lymphocyte transformation tests were positive (3,6). The British National Formulary in 1997 warned of the contact sensitizing potential of phenoxybenzamine HCl and advised against skin contamination (4).

Patch test sensitization
Of 7 control patients patch tested with phenoxybenzamine 1% water, one was sensitized by the patch test (2).

LITERATURE

1   The data in the section 'General' may have been obtained from literature discussed in this chapter, but mostly also or exclusively from one or more of the following online sources: ChemIDPlus Advanced, PubChem, DrugBank, RxList, Drug Central, Drugs.com, and Wikipedia
2   Sommer S, Wilkinson SM. Contact dermatitis caused by phenoxybenzamine hydrochloride. Contact Dermatitis 1998;38:352-353
3   Alexander SL, Leibowitz S, Spector RG. Local and generalized skin reactions to phenoxybenzamine. The Lancet 1973:10 Feb:317-318
4   British Medical Association, Royal Pharmaceutical Society of Great Britain. Phenoxybenzamine. British National Formulary 1997;34:86
5   Mitchell JC, Maibach HI. Allergic contact dermatitis from phenoxybenzamine hydrochloride: cross-sensitivity to some related haloalkylamine compounds. Contact Dermatitis 1975;1:363-366
6   Alexander S, Spector RG. Phenoxybenzamine. Contact Dermatitis 1975;1:59
7   Mitchell JC. Allergic contact dermatitis from alpha- and beta-adrenergic receptor blocking agents (dibenzyline and propranolol). Contact Dermatitis Newsletter 1974;16:488

Chapter 3.381 PHENYLBUTAZONE

IDENTIFICATION
Description/definition   : Phenylbutazone is the synthetic pyrazolone derivative that conforms to the structural
                            formula shown below
Pharmacological classes : Anti-inflammatory agents, non-steroidal
IUPAC name   : 4-Butyl-1,2-diphenylpyrazolidine-3,5-dione
CAS registry number(s)   : 50-33-9
EC number(s)   : 200-029-0
Merck Index monograph   : 8660
Patch testing   : 10.0% pet. (Chemotechnique, SmartPracticeCanada, SmartPracticeEurope)
Molecular formula   : $C_{19}H_{20}N_2O_2$

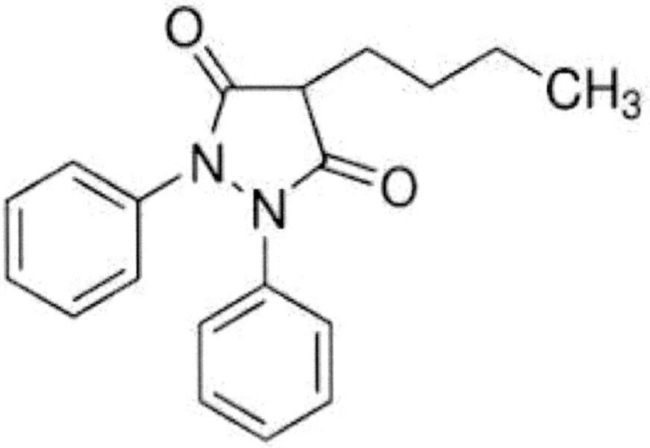

GENERAL
Phenylbutazone is a synthetic pyrazolone derivative and nonsteroidal anti-inflammatory drug (NSAID) with anti-inflammatory, antipyretic, and analgesic activities. Phenylbutazone was formerly used for the treatment of backache, ankylosing spondylitis, rheumatoid arthritis, and reactive arthritis. Because of serious systemic side effects and cutaneous adverse drug reactions such as Stevens-Johnson syndrome and toxic epidermal necrolysis, it is probably hardly used anymore, except for therapy-resistant ankylosing spondylitis. It does, however, still have applications in veterinary medicine (1). Topical use was formerly recommended for superficial phlebitis and some inflammatory diseases in the muscles and connective tissues, which caused many cases of sensitization and a few of photosensitization; this has been fully reviewed in Volume 3 of the *Monographs in contact allergy* series (3).
    See also Chapter 3,421 Pyrazinobutazone and Chapter 3.388 Piperazine.

CUTANEOUS ADVERSE DRUG REACTIONS FROM SYSTEMIC ADMINISTRATION CAUSED BY TYPE IV
(DELAYED-TYPE) HYPERSENSITIVITY (as demonstrated by positive patch tests)
Cutaneous adverse drug reactions from systemic administration of phenylbutazone caused by type IV (delayed-type) hypersensitivity have included a maculopapular eruption evolving into erythroderma (2) and toxic epidermal necrolysis (TEN) (5).

Maculopapular eruption/erythroderma
A 45-year-old woman suffering from psoriasis was given one injection of Tomanol ® (combination of 1/3 phenylbutazone and 2/3 isopyrine [ramifenazone, isopropylaminophenazone]) for low back pain. After 8 hours, the patient developed fever, swelling of the face, itching of the psoriasis patches, and a generalized maculopapular eruption which evolved into erythroderma. Patch tests were positive to the Tomanol solution tested 'as is' and – in a second patch test session – to phenylbutazone 1% and 5% pet. (+++), while reactions to isopyrine were negative. Patch testing with various other pyrazoles were positive to oxyphenbutazone only (2).

Stevens-Johnson syndrome/toxic epidermal necrolysis (SJS/TEN)
In early German literature, a case of toxic epidermal necrolysis (TEN) with a positive patch test to phenylbutazone has been reported; details are unknown to the author (5).

Cross-reactions, pseudo-cross-reactions and co-reactions
Patients sensitized to phenylbutazone may or may not cross-react to oxyphenbutazone and vice versa (2,3). A patient sensitized to mofebutazone cross-reacted to phenylbutazone (4).

LITERATURE
1   The data in the section 'General' may have been obtained from literature discussed in this chapter, but mostly also or exclusively from one or more of the following online sources: ChemIDPlus Advanced, PubChem, DrugBank, RxList, Drug Central, Drugs.com, and Wikipedia
2   Vooys RC, van Ketel WG. Allergic drug eruption from pyrazolone compounds. Contact Dermatitis 1977;3:57-58
3   De Groot AC. Monographs in contact allergy, volume 3: Topical Drugs. Boca Raton, Fl, USA: CRC Press Taylor and Francis Group, 2021 (ISBN 978-0-367-23693-9)
4   Walchner M, Rueff F, Przybilla B. Delayed-type hypersensitivity to mofebutazone underlying a severe drug reaction. Contact Dermatitis 1997;36:54-55
5   Schöpf E, Schulz KH, Kessler R, Taugner M, Braun W. Allergologische Untersuchungen beim Lyell-Syndrome. Z Hautkr 1975;50:865-873 (Article in German, data cited in ref. 6)
6   Tagami H, Tatsuta K, Iwatsuki K, Yamada M. Delayed hypersensitivity in ampicillin-induced toxic epidermal necrolysis. Arch Dermatol 1983;119:910-913

Chapter 3.382 PHENYLEPHRINE

IDENTIFICATION

Description/definition   : Phenylephrine is the sympathomimetic amine chemically related to adrenaline and
                              ephedrine that conforms to the structural formula shown below
Pharmacological classes   : Mydriatics; nasal decongestants; α_1-adrenergic receptor agonists;
                              sympathomimetics; vasoconstrictor agents; cardiotonic agents
IUPAC name   : 3-[(1R)-1-Hydroxy-2-(methylamino)ethyl]phenol
Other names   : Metaoxedrine
CAS registry number   : 59-42-7
EC number   : 200-424-8
Merck Index monograph : 8668
Patch testing   : Hydrochloride, 10% water (SmartPracticeCanada, SmartPracticeEurope)
Molecular formula   : $C_9H_{13}NO_2$

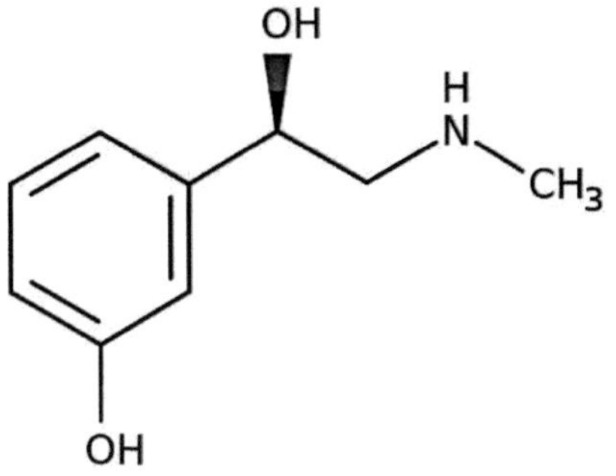

GENERAL

Phenylephrine (PE) is a direct-acting sympathomimetic amine chemically related to epinephrine and ephedrine with potent vasoconstrictor property. It is a post-synaptic α-adrenergic receptor agonist that causes vasoconstriction, increases systolic/diastolic pressures, reflex bradycardia, and stroke output. Phenylephrine is mainly used to treat nasal congestion due to the common cold or hay fever, sinusitis, or other upper respiratory problems. Oral phenylephrine, together with other drugs, may be used to treat certain diseases of the upper respiratory tract. In pharmaceutical products, phenylephrine (PE) is employed as phenylephrine hydrochloride (CAS number 61-76-7, EC number 200-517-3, molecular formula $C_9H_{14}ClNO_2$) (1). In topical preparations, phenylephrine has caused a large number of cases of contact allergy/allergic contact dermatitis (especially in eye medications), which subject has been fully reviewed in Volume 3 of the *Monographs in contact allergy* series (2).

CUTANEOUS ADVERSE DRUG REACTIONS FROM SYSTEMIC ADMINISTRATION CAUSED BY TYPE IV (DELAYED-TYPE) HYPERSENSITIVITY (as demonstrated by positive patch tests)

Cutaneous adverse drug reactions from systemic administration of phenylephrine caused by type IV (delayed-type) hypersensitivity have included fixed drug eruption (3).

Fixed drug eruption

A 22-year-old woman had suffered an 'allergic reaction' on 3 occasions to a combination tablet of acetylsalicylic acid, chlorpheniramine, phenylephrine and ascorbic acid used to treat colds without fever. Each time, an itchy dusky-red macule had appeared on exactly the same spot on the left forearm within 12 to 24 hours of taking a first dose of this medicine. The lesion would heal spontaneously with residual hyperpigmentation. Two months later, an identical lesion appeared in the same place when the patient took paracetamol (acetaminophen) for a headache. Patch tests with acetylsalicylic acid (10% pet.), paracetamol (20% pet.), and phenylephrine (1% in water) were positive only to phenylephrine on the residual lesion at D4. An oral challenge test with paracetamol reproduced the fixed eruption within 8 hours at exactly the same location (3).

Cross-reactions, pseudo-cross-reactions and co-reactions

Patients sensitized to phenylephrine from topical application may cross-react to pseudoephedrine, ephedrine, phenylpropanolamine, methoxamine and oxymetazoline (4). In patients with allergic contact dermatitis from phenylephrine, cross-sensitization may also have occurred to ethylephrine, fepradinol, and epinephrine (2). Conversely, patients sensitized to pseudoephedrine or ephedrine may have cross-reacted to phenylephrine (2).

LITERATURE

1 The data in the section 'General' may have been obtained from literature discussed in this chapter, but mostly also or exclusively from one or more of the following online sources: ChemIDPlus Advanced, PubChem, DrugBank, RxList, Drug Central, Drugs.com, and Wikipedia

2 De Groot AC. Monographs in contact allergy, volume 3: Topical Drugs. Boca Raton, Fl, USA: CRC Press Taylor and Francis Group, 2021 (ISBN 978-0-367-23693-9)

3 López Abad R, Iriarte Sotés P, Castro Murga M, Gracia Bara MT, Sesma Sánchez P. Fixed drug eruption induced by phenylephrine: a case of polysensitivity. J Investig Allergol Clin Immunol 2009;19:322-323

4 Barranco R, Rodríguez A, de Barrio M, Trujillo MJ, de Frutos C, Matheu V, et al. Sympathomimetic drug allergy: cross-reactivity study by patch test. Am J Clin Dermatol 2004;5:351-355

Chapter 3.383 PHENYTOIN

IDENTIFICATION

Description/definition	: Phenytoin is the hydantoin derivative that conforms to the structural formula shown below
Pharmacological classes	: Anticonvulsants; cytochrome P-450 CYP1A2 inducers; voltage-gated sodium channel blockers
IUPAC name	: 5,5-Diphenylimidazolidine-2,4-dione
Other names	: 5,5-Diphenylhydantoin
CAS registry number	: 57-41-0
EC number	: 200-328-6
Merck Index monograph	: 8703
Patch testing	: 10% pet. (20); with serious cutaneous adverse reactions, starting with 1% pet. may be advisable (20)
Molecular formula	: $C_{15}H_{12}N_2O_2$

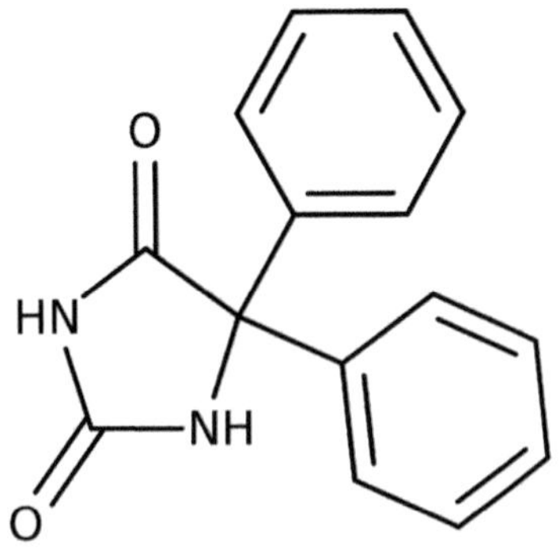

GENERAL

Phenytoin is an aromatic non-sedative hydantoin derivative with anticonvulsant activity. It is also an anti-arrhythmic and a muscle relaxant. Phenytoin is indicated for the control of generalized tonic-clonic (grand mal) and complex partial (psychomotor, temporal lobe) seizures and prevention and treatment of seizures occurring during or following neurosurgery (1). In pharmaceutical products, both phenytoin and phenytoin sodium (CAS number 630-93-3, EC number 211-148-2, molecular formula $C_{15}H_{11}N_2NaO_2$) may be employed (1).

CUTANEOUS ADVERSE DRUG REACTIONS FROM SYSTEMIC ADMINISTRATION CAUSED BY TYPE IV (DELAYED-TYPE) HYPERSENSITIVITY (as demonstrated by positive patch tests)

Cutaneous adverse drug reactions from systemic administration of phenytoin caused by type IV (delayed-type) hypersensitivity have included maculopapular eruption (9,19), drug reaction with eosinophilia and systemic symptoms (DRESS)/anticonvulsant hypersensitivity syndrome (2,3,5,8,10,11,14,18), toxic epidermal necrolysis (TEN) (6), EMPACT (erythema multiforme associated with phenytoin and cranial radiation therapy) (16), and unspecified drug eruption (4,12).

Maculopapular eruption

Of 10 patients who developed maculopapular rashes after and/or during treatment with phenytoin and were patch tested with phenytoin 10% pet. and alc., 3 had positive reactions. None of these individuals, all women, had laboratory abnormalities suggestive of DRESS (9). A 6-year-old boy developed a maculopapular eruption from phenytoin and had a positive patch test to this drug (19).

Drug reaction with eosinophilia and systemic symptoms (DRESS)/anticonvulsant hypersensitivity syndrome

It is estimated that phenytoin can induce anticonvulsant hypersensitivity syndrome (AHS) at a frequency of 1 in 10,000 to 1 in 1,000 treated patients (15). The literature on patch testing in anticonvulsant hypersensitivity syndrome up to August 2008 has been reviewed in ref. 17.

Case series

Twenty-four patients who had developed severe drug rashes from antiepileptics (18 DRESS, 5 SJE, TEN or SJS/TEN, 1 lichenoid drug eruption) were patch tested with the implicated drugs 10%, 20% and 30% pet. Positive reactions were observed in 12 patients: 11 with DRESS and one with lichenoid eruption. Implicated drugs in DRESS with positive patch tests were phenytoin (n=7), carbamazepine (n=5) and phenobarbital (n=3) (10).

Between January 1998 and December 2008, in a university hospital in Portugal, 56 patients with DRESS were investigated with patch testing of the suspected drugs. Phenytoin (5% and 10% pet.) was tested in 7 patients and there was one positive patch test reaction. Fifty controls were negative (2).

In a group of 15 patients with 'anticonvulsant hypersensitivity syndrome', three patients had a positive patch test to phenytoin; details are unavailable to the author, but not all patients appeared to have all symptoms necessary to diagnose anticonvulsant hypersensitivity syndrome (synonym: DRESS from anticonvulsant drugs) (3).

Two patients with anticonvulsant hypersensitivity syndrome from phenytoin had positive reactions to phenytoin 1% and 10% in phojel base (composition unknown), but negative to phenytoin 1% and 10% in petrolatum (5).

Case reports

A 34-year-old woman developed, after 2 weeks of treatment with phenytoin and amitriptyline, fever, a generalized maculopapular rash, eosinophilia and elevated liver values. Histopathology was compatible with a drug eruption of the erythema multiforme-like type with lymphocytic exocytosis, isolated dyskeratotic cells, vacuolation of basal cells and pigment incontinence. Patch tests were positive to phenytoin; its histopathology had a typical eczematous pattern. The patch test with amitriptyline was negative, but an oral challenge resulted in an erythematous maculopapular rash (8). The authors suggested a cross-reaction, but this is highly unlikely considering the major differences in the structural formulas of both drugs.

A 40-year-old man had been treated with phenytoin for many years without ill effects. When he was switched to carbamazepine, 16 weeks later DRESS developed. The patient was again given phenytoin but, during the next 7 days, his skin rashes and facial edema progressively worsened, and mucosal lesions developed with the persistence of fever. Rapid progression of the leukocytosis with eosinophilia was noted. Three months after total resolution, patch tests were positive to phenytoin 1% and 10% pet., to phenytoin 10% water (negative to 1% water) and to carbamazepine 1% and 10% pet. (11).

After having been treated with phenytoin 100 mg 3 times a day for a seizure disorder for a month, a 23-year-old man developed fever, pruritus, generalized maculopapular rash, enlarged cervical lymph nodes, petechiae on the soft palate and a slight eosinophilia. Six months later, patch tests were positive with 1% (++) and 5% (+++) phenytoin and with 10% (+++) and 20% (+++) phenobarbital (judging the photos shown, the +++ scores to phenobarbital are a gross exaggeration of the actual patch test reactivity). Patch tests with carbamazepine and other drugs the patient had used were negative. Ten controls were negative. Positive patch tests with phenytoin and phenobarbital demonstrated, according to the authors, cross-reactivity between these drugs, but it was not firmly stated that the patient had never used phenobarbital before (14).

A 26-year-old woman had developed acute generalized exanthematous pustulosis (AGEP) from carbamazepine. When treated with phenytoin and, later, with valproic acid, the patient developed DRESS with fever, rashes, eosinophilia and subclinical hepatitis from both anticonvulsants on separate occasions. Patch tests to all three drugs were positive (18).

Stevens-Johnson syndrome/toxic epidermal necrolysis (SJS/TEN)

In early German literature, a case of toxic epidermal necrolysis from phenytoin sodium with a positive patch test to this drug has apparently been described (6).

Other cutaneous adverse drug reactions

In Finland, in the period 1989-2001, 826 patients with suspected cutaneous drug eruptions were patch tested and 89 had one or more positive reactions. Of these individuals, 2 had positive patch tests to phenytoin. It was not specified which cutaneous adverse drug reaction these patients had suffered from (12). In a Japanese investigation, 1 of 2 patients patch tested with phenytoin because of an 'anticonvulsant-induced drug eruptions' had positive reaction to this drug. The nature of the skin eruption was not mentioned (4).

A 46-year-old woman with brain metastases from bronchial carcinoma was treated with phenytoin for seizure prophylaxis and total brain radiation therapy. Three weeks after the introduction of phenytoin, the patient had developed erythema multiforme-like skin lesions restricted to the original radiation field and facial mucocutaneous involvement, which spread to the upper parts of the body after a few days. Patch tests were positive to undiluted liquid phenytoin for intravenous use on D2 and D3 (negative to a 10% dilution); 3 controls were negative. This characteristic picture is best known under its acronym EMPACT: Erythema Multiforme associated with Phenytoin And Cranial radiation Therapy (16).

A 6-year old girl developed DRESS while on carbamazepine with erythroderma and edema of the hands, feet and face. When it was replaced with oxcarbazepine, an initial improvement was observed, but after 10 days, the rash and other symptoms of DRESS reoccurred. Phenytoin was well tolerated and the skin cleared within 3 weeks. Two months later, patch tests were positive to commercial carbamazepine, oxcarbazepine and also phenytoin 5 mg in 50 µl alc. Eighty controls were negative. LTTs were positive only to carbamazepine. Despite a strong reaction (+++) to phenytoin, this drug was reinstituted and was well tolerated. It was concluded that the patch test to phenytoin had been false-positive due to the excited skin syndrome (13).

Cross-reactions, pseudo-cross-reactions and co-reactions
This subject is discussed in Chapter 3.284 Lamotrigine.

LITERATURE

1   The data in the section 'General' may have been obtained from literature discussed in this chapter, but mostly also or exclusively from one or more of the following online sources: ChemIDPlus Advanced, PubChem, DrugBank, RxList, Drug Central, Drugs.com, and Wikipedia

2   Santiago F, Gonçalo M, Vieira R, Coelho S, Figueiredo A. Epicutaneous patch testing in drug hypersensitivity syndrome (DRESS). Contact Dermatitis 2010;62:47-53

3   Galindo PA, Borja J, Gomez E, Mur P, Gudín M, García R, et al. Anticonvulsant drug hypersensitivity. J Invest Allergol Clin Immunol 2002;12:299-304

4   Osawa J, Naito S, Aihara M, Kitamura K, Ikezawa Z, Nakajima H. Evaluation of skin test reactions in patients with non-immediate type drug eruptions. J Dermatol 1990;17:235-239

5   Nigen SR, Shapiro LE, Knowles SR, Neuman MG, Shear NH. Utility of patch testing in patients with anticonvulsant-induced hypersensitivity syndrome. Dermatitis 2008;19:349-350

6   Schöpf E, Schulz KH, Kessler R, Taugner M, Braun W. Allergologische Untersuchungen beim Lyell-Syndrome. Z Hautkr 1975;50:865-873 (Article in German, data cited in ref. 7)

7   Tagami H, Tatsuta K, Iwatsuki K, Yamada M. Delayed hypersensitivity in ampicillin-induced toxic epidermal necrolysis. Arch Dermatol 1983;119:910-913

8   Galindo Bonilla PA, Romero Aguilera G, Feo Brito F, Gómez Torrijos E, et al. Phenytoin hypersensitivity syndrome with positive patch test. A possible cross-reactivity with amitriptyline. J Invest Allergol Clin Immunol 1998;8:186-190

9   Lee AY, Kim MJ, Chey WY, Choi J, Kim BG. Genetic polymorphism of cytochrome P450 2C9 in diphenylhydantoin-induced cutaneous adverse drug reactions. Eur J Clin Pharmacol 2004;60:155-159

10 Shiny TN, Mahajan VK, Mehta KS, Chauhan PS, Rawat R, Sharma R. Patch testing and cross sensitivity study of adverse cutaneous drug reactions due to anticonvulsants: A preliminary report. World J Methodol 2017;7:25-32

11 Kim CW, Choi GS, Yun CH, Kim DI. Drug hypersensitivity to previously tolerated phenytoin by carbamazepine-induced DRESS syndrome. J Korean Med Sci 2006;21:768-772

12 Lammintausta K, Kortekangas-Savolainen O. The usefulness of skin tests to prove drug hypersensitivity. Br J Dermatol 2005;152:968-974

13 Troost RJ, Oranje AP, Lijnen RL, et al. Exfoliative dermatitis due to immunologically confirmed carbamazepine hypersensitivity. Pediatr Dermatol 1996;13:316-320

14 Sánchez-Morillas L, Laguna-Martínez JJ, Reaño-Martos M, Rojo-Andrés E, Gómez-Tembleque P, Pellón-González C. A case of hypersensitivity syndrome due to phenytoin. J Investig Allergol Clin Immunol 2008;18:74-75

15 Seitz CS, Pfeuffer P, Raith P, Bröcker EB, Trautmann A. Anticonvulsant hypersensitivity syndrome: cross-reactivity with tricyclic antidepressant agents. Ann Allergy Asthma Immunol 2006;97:698-702

16 Wöhrl S, Loewe R, Pickl WF, Stingl G, Wagner SN. EMPACT syndrome. J Dtsch Dermatol Ges 2005;3:39-43

17 Elzagallaai AA, Knowles SR, Rieder MJ, Bend JR, Shear NH, Koren G. Patch testing for the diagnosis of anticonvulsant hypersensitivity syndrome: a systematic review. Drug Saf 2009;32:391-408

18 Duran-Ferreras E, Mir-Mercader J, Morales-Martinez M D, Martinez-Parra C. Anticonvulsant hypersensitivity syndrome with severe repercussions in the skin and kidneys. Rev Neurol 2004;38:1136-1138 (Article in Spanish)

19 Torres MJ, Corzo JL, Leyva L, Mayorga C, Garcia-Martin FJ, Antunez C, et al. Differences in the immunological responses in drug- and virus-induced cutaneous reactions in children. Blood Cells Mol Dis 2003;30:124-131

20 Brockow K, Garvey LH, Aberer W, Atanaskovic-Markovic M, Barbaud A, Bilo MB, et al.; ENDA/EAACI Drug Allergy Interest Group. Skin test concentrations for systemically administered drugs – an ENDA/EAACI Drug Allergy Interest Group position paper. Allergy 2013;68:702-712

Chapter 3.384 PHYTONADIONE

IDENTIFICATION

Description/definition   : Phytonadione is the vitamin K compound that conforms to the structural formula shown
                             below
Pharmacological classes : Vitamins; antifibrinolytic agents
IUPAC name   : 2-Methyl-3-[(E,7R,11R)-3,7,11,15-tetramethylhexadec-2-enyl]naphthalene-1,4-dione
Other names   : Vitamin K_1; phylloquinone; phytomenadione
CAS registry number   : 84-80-0
EC number   : 201-564-2
Merck Index monograph : 8762
Patch testing   : 1% and 10% pet. (3); 10% pet. was negative in control groups (3,7,15)
Molecular formula   : $C_{31}H_{46}O_2$

GENERAL

Phytonadione, often called vitamin K_1, is found naturally in a wide variety of green plants. Vitamin K is needed for the posttranslational modification of certain proteins, mostly required for blood coagulation. Pharmaceutical phyto-nadione is indicated in the treatment of coagulation disorders which are due to faulty formation of factors II, VII, IX and X when caused by vitamin K deficiency or interference with vitamin K activity (1). Phytonadione is available for oral, intramuscular, subcutaneous, or intravenous routes. Especially with repeated injections, cutaneous reactions to vitamin K_1 may arise (17). Phytonadione has also been used in topical preparations, notably cosmetics, where it has caused cases of allergic contact dermatitis. This subject has been fully reviewed in Volume 1 of the *Monographs in contact allergy* series (36; see ref. 2 for more recent data). The current chapter differs somewhat from the others in this book, as it was adapted from the eponymous chapter in *Monographs in contact allergy, Volume 1* (36).

CUTANEOUS ADVERSE DRUG REACTIONS FROM SYSTEMIC ADMINISTRATION CAUSED BY TYPE IV (DELAYED-TYPE) HYPERSENSITIVITY

Cutaneous adverse drug reactions from systemic administration of phytonadione caused by type IV (delayed-type) hypersensitivity have included localized erythematous and eczematous plaques at the injection site (refs.: see below), sometimes with extension or generalization (e.g. 18) and scleroderma-like reactions at the injection site (refs.: see below).

Erythematous and eczematous plaques

These lesions appear around the injection site 5 days to 4 weeks after one or more injections, either intramuscular or subcutaneous, rarely intravenous (35). Clinical presentations and morphology vary, with reactions consisting of erythematous, indurated plaques, sometimes with vesicles (19) or localized eczematous eruptions at the injection sites, which may sometimes spread (18) or be accompanied by a maculopapular eruption (21,29). The lesions may be tender or (intensely) pruritic (18). Localized urticaria after 3 days has been observed (8). The plaques can grow to be 30 cm or more in diameter and heal spontaneously after some 4 weeks, sometimes with hyperpigmentation (17). However, spontaneous healing can take up to 4 months or longer (10,19,23). Residual erythema in one case lasted for up to 6 months (18). Rarely, patients experience a prolonged course with development of sclerotic plaques 2-4 months after treatment, lasting for years (see the section 'Scleroderma-like reactions' below). Hypersensitivity reactions to vitamin K_1 treatment were first described in patients with liver disease secondary to cirrhosis and/or hepatitis. However, this adverse effect is not restricted to those with liver dysfunction and appears to be dose-independent (17,20,33).

Several features of the erythematous and eczematous reactions to vitamin K_1 suggest that they are a manifestation of delayed-type hypersensitivity (18,19,21): a. in most cases, there was an approximate 10-14 day delay between the first dose and appearance of the rash; b. subsequent patch and/or prick/intradermal testing produces a reaction in 2-3 days (7,10,18,19,21,24,31,34,35); c. a recall phenomenon has been described in several patients, in which patch or intradermal testing at a distant location precipitated an eczematous flare at the original reaction sites (19,21,29,34); and d. lesional biopsies consistently showed spongiosis of the epidermis, dermal edema, and a perivascular mononuclear and eosinophilic cellular infiltrate, consistent with a delayed-type hypersensitivity reaction (7,10,18,19,21,22,28,29, 34).

Although prophylaxis against hemolytic disease of the new-born with intramuscular vitamin K_1 has been practiced routinely in many countries, there appear to be no reports in the English language literature of cutaneous eruptions due to this medication in neonates. This may be a consequence of the immaturity of the neonatal immune system. However, in some cases, this injection may represent the sensitizing event, as some adults reacted after a single injection (19,21,22,23). Some pediatric cases have also been observed (31,32).

Patients with delayed-type cutaneous reactions to injections with vitamin K_1 have been presented in refs. 7,10 and 17-35. These reactions seem to be far from rare, considering the many case reports, case series of 6 (21,32) and 4 patients (22,33) and 94 patients reported from Japan (probably not all local hypersensitivity reactions, but also skin rashes from intravenous administration) (27). The relevant literature up to 1994 has been reviewed in ref. 19.

Case report (as example)
A 40-year-old woman underwent cholecystectomy and used many medications including vitamin K_1 1 ml (10 mg/ml) intramuscularly four daily doses, injected into the patient's thighs. Five days after the first dose of this anticoagulant, a pruritic, erythematous patch developed on the patient's anterior left thigh. Over the next 3 weeks, four patches appeared on her thighs (where the vitamin K_1 injections had been given) and became larger (maximum diameter, 15 cm), vesicular, weeping, hot, indurated and intensely pruritic. Despite treatment, the eruption then spread to the operation wound site, face, neck and arms. A course of oral prednisone produced gradual improvement over the next 4 weeks, but residual erythema persisted for nearly 6 months. Patch testing with the commercial vitamin K_1 product was negative, but intradermal testing produced a strongly positive (+++) eczematous reaction at D3. Further patch testing with pure vitamin K_1 10 mg/ml olive oil (scratch chamber test) and 100 mg/ml olive oil were strongly positive, whereas there were no reactions to the excipients of the commercial preparation. There was a (cross-) reaction to menadione (vitamin K_3) sodium bisulfite (18).

Scleroderma-like reactions
Scleroderma-like reactions are late reactions to vitamin K_1 injections, that resemble localized scleroderma or morphea. The phenomenon is sometimes called Texier's disease, after the first author of the first publication on this rare side effect of phytonadione. The French authors described 9 patients with liver cirrhosis, 7 men and 2 women, who first had erythematous plaques at the sites of intramuscular vitamin K_1 injections, but subsequently developed a scleroderma-like picture that appeared up to 2 years later. The distribution of these sclerotic plaques was described as a 'Ceinturon de cowboy avec ses revolvers' ('Cowboy's belt with revolvers'; 'Cowboy gun belt and holsters'), because of its extension around the waist and onto the lateral aspect of the thighs. Intracutaneous tests with vitamin K_1 were performed in 4 patients and they all had positive delayed reactions (37).

This side effect, that has been described in nearly 20 publications (4,5,6,10-14,16,37-45), many from France, occurs from weeks to 2 years after previous vitamin K_1 injections (38) with an average of about 9 months (9). It may appear with (6,12) or without (4,11,16,38,39,44) a previous *early* erythematous plaque - eczematous reaction (18). Generally, late erythematous plaques develop around the sites of injections and progressively extend from the upper part of the buttock to the lateral aspects of both thighs and around the waist, forming the 'Cowboy gun belt and holsters' picture. These erythematous plaques progress in a few months to white dense sclerosis surrounded by a lilac border. After 12 to 18 months of evolution, the lilac border and then the cutaneous sclerosis progressively resolve. Lesions may, however, persist for over 10 years or be permanent (38). Most patients are adults, but a 2-year-old child (40) and 6 other children (11) developed scleroderma-like reactions from the vitamin K_1 injections they had received immediately after their birth.

No systemic or immunologic features of systemic sclerosis are found in these patients (5). Some authors have proposed 4 clinical and histopathological stages of vitamin K_1-induced skin sclerosis: erythematous, erythemato-pigmented, established scleroderma and regression of scleroderma (6). Histopathology of the scleroderma-like reactions shows a picture which is usually indistinguishable from deep morphea or scleroderma: sclerosis involving the reticular dermis and extending to the subcutaneous fat. Collagen bundles are homogenized and thickened. A mild to moderately dense lymphocytic infiltrate is seen among collagen bundles and around blood vessels and the epidermis is normal (19,38). Delayed-reading intradermal tests with vitamin K_1 (usually the commercial product) have consistently been positive when performed (10,16,38,41), but the pathogenesis of this side effect is unknown.

Cross-reactions, pseudo-cross-reactions and co-reactions
Vitamin K_3 (1,18); vitamin K_4 (6).

LITERATURE

1   The data in the section 'General' may have been obtained from literature discussed in this chapter, but mostly also or exclusively from one or more of the following online sources: ChemIDPlus Advanced, PubChem, DrugBank, RxList, Drug Central, Drugs.com, and Wikipedia

2   Cameli N, Zanniello R, Mariano M, Cristaudo A. Vitamin K1 photo-induced reaction during treatment with cetuximab. Contact Dermatitis 2020;82:189-190

3   Serra-Baldrich E, Dalmau J, Pla C, Muntañola AA. Contact dermatitis due to clarifying cream. Contact Dermatitis 2005;53:174-175

4   Brunskill NJ, Berth-Jones J, Graham-Brown R AC. Pseudosclerodermatous reaction to phytomenadione injection (Texier's syndrome). Clin Exp Dermatol 1988;13:276-278

5   Janin-Mercier A, Mosser C, Souteyrand P, Bourgees M. Subcutaneous sclerosis with fasciitis and eosinophilia after phytonadione injections. Arch Dermatol 1985;121:1421-1423

6   Mosser C, Janin-Mercier A, Souteyrand P. Les réactions cutanées après administration parenterale de vitamine K1. Ann Dermatol Venereol 1987;114:243-251 (article in French)

7   Giménez–Arnau AM, Toll A, Pujol RM. Immediate cutaneous hypersensitivity response to phytomenadione induced by vitamin K1 in skin diagnostic procedure. Contact Dermatitis 2005;52:284-285

8   Carton FX. Réaction allergique au cours d'un traitement: vitamine K1 +extrait de foie. Bull Soc Fr Dermat 1965;72:228 (article in French)

9   Wilkins K, De Koven J, Assaad D. Cutaneous reactions associated with vitamin K1. J Cutan Med Surg 2000;4:164-168

10 Balato N, Cuccurullo FM, Patruno C, Ayala F. Adverse skin reactions to vitamin K1: report of 2 cases. Contact Dermatitis 1998;38:341-342

11 Bourrat E, Moraillon I, Vignon-Pennamen MD, Fraitag S, Cavelier-Balloy B, Cordoliani F, et al. Scleroderma-like patch on the thigh in infants after vitamin K injection at birth: six observations. Ann Dermatol Venereol 1996;123:634-648 (article in French)

12 Morel A, Betlloch I. Morphea-like reaction from vitamin K1. Int J Dermatol 1995;34:201-202

13 Bazex A, Dupré A, Christol B, Serres D. Lumbo-buttocks sclerodermiformic reactions after injection of vitamin K1: Presentation of 2 cases. Histological verification. Bull Soc Fr Derm Syph 1972;79:578-581 (article in French).

14 Jean-Pastor MJ, Jean P, Gamby T. Accidents cutanés consecutifs à l'administration parenterale de vitamine K1. Therapie 1981;36:369-374 (article in French)

15 Ramírez Santos A, Fernández-Redondo V, Pérez Pérez L, Concheiro Cao J, Toribio J. Contact allergy from vitamins in cosmetic products. Dermatitis 2008;19:154-156

16 Guidetti MS, Vincenzi C, Papi M, Tosti A. Sclerodermatous skin reaction after vitamin K1 injections. Contact Dermatitis 1994;31:45-46

17 Sousa T, Hunter L, Petitt M, Wilkerson MG. Localized cutaneous reaction to intramuscular vitamin K in a patient with acute fatty liver of pregnancy. Dermatol Online J 2010;16(12):16

18 Wong DA, Freeman S. Cutaneous allergic reaction to intramuscular vitamin K1. Australas J Dermatol 1999;40:147-152

19 Bruynzeel I, Hebeda CL, Folkers E, Bruynzeel DP. Cutaneous hypersensitivity reactions to vitamin K: 2 case reports and a review of the literature. Contact Dermatitis 1995;32:78-82

20 Moreau-Cabarrot A, Giordano-Labadie F, Bazex J. Cutaneous hypersensitivity at the site of injection of vitamin K1. Ann Dermatol Venereol 1996;123:177-179 (article in French)

21 Finkelstein H, Champion MC, Adam JE. Cutaneous hypersensitivity to vitamin K1 injection. J Am Acad Dermatol 1987;16:540-545

22 Lemlich G, Green M, Phelps R, Lebwohl M, Don P, Gordon M. Cutaneous reactions to vitamin K1 injections. J Am Acad Dermatol 1993;28:345-347

23 Joyce JP, Hood AF, Weiss MM. Persistent cutaneous reaction to intramuscular vitamin K injection. Arch Dermatol 1988;124:27-28

24 Piguet B, Bertheuil F. Accidents cutanés allergiques provoqués par une préparation injectable de vitamine K1 synthétique. Bull Soc Fr Dermatol Syph 1964;71:486-491 (article in French)

25 Gettler SL, Fung MA. Off-center fold: indurated plaques on the arms of a 52-year-old man. Diagnosis: Cutaneous reaction to phytonadione injection. Arch Dermatol 2001;137:957-962

26 Lee MM, Gellis S, Dover JS. Eczematous plaques in a patient with liver failure. Fat-soluble vitamin K hypersensitivity. Arch Dermatol 1992;128:260-261

27 Tsuboi R, Ogawa H. Skin eruption caused by fat-soluble vitamin K injection. J Am Acad Dermatol 1988;18:386

28 Tuppal R, Tremaine R. Cutaneous eruption from vitamin K1 injection. J Am Acad Dermatol 1992;27:105-106

29 Barnes HM, Sarkany I. Adverse skin reaction from vitamin K1. Br J Dermatol 1976;95:653-656

30 Keough GC, English JC 3rd, Meffert JJ. Eczematous hypersensitivity from aqueous vitamin K injection. Cutis 1998;61:81-83

31 Pigatto PD, Bigardi A, Fumagalli M, Altomare GF, Riboldi A. Allergic dermatitis from parenteral vitamin K. Contact Dermatitis 1990;22:307-308

32 Bullen AW, Miller JP, Cunliffe WJ, Losowsky MS. Skin reactions caused by vitamin K in patients with liver disease. Br J Dermatol 1978;98:561-565

33 Sanders MN, Winkelmann RK, Rochester PD. Cutaneous reactions to vitamin K. J Am Acad Dermatol 1988;19:699-704

34 Robison JW, Odom RB. Delayed cutaneous reaction to phytonadione. Arch Dermatol 1978;114:1790-1792

35 Heydenreich G. A further case of adverse skin reaction from vitamin K1. Br J Dermatol 1977;97:697

36 De Groot AC. Monographs in Contact Allergy Volume I. Non-Fragrance Allergens in Cosmetics (Part I and Part 2). Boca Raton, Fl, USA: CRC Press Taylor and Francis Group, 2018 (ISBN 978-1-138-57325-3 and 9781138573383)

37 Texier L, Gendre PH, Gauthier O, Gauthier Y, Surlèvé-Bazeille JE, Boineau D. Hypodermites sclérodermiformes lombo-fessières induites par des injections médicamenteuses intramusculaires associées à la vitamine K1. Ann Derm Syph 1972;99:363-372 (article in French)

38 Pang BK, Munro V, Kossard S. Pseudoscleroderma secondary to phytomenadione (vitamin K1) injections: Texier's disease. Australas J Dermatol 1996;37:44-47

39 Pujol RM, Puig L, Moreno A, Perez M, de Moragas JM. Pseudoscleroderma secondary to phytonadione (vitamin K1) injections. Cutis 1989;43:365-368

40 Rommel A, Saurat JH. Hypodermite fessière sclérodermiforme et injections de vitamine K1 à la naissance. Ann Pediat 1982;29: 64-66 (article in French)

41 Larrègue M, Gallet Ph, Giacomoni P de, Rat JP. Sclérodermie lombofessière consecutive à des injections de vitamin K1. Bull Soc Fr Dermat 1975;82:447-448 (article in French)

42 Duntze F, Durand JR, Vignes P. Sclérodermie en bande secondaire à des injections de vitamine K1. Bull Soc Fr Dermat 1975;82:78-79 (article in French)

43 Misson R, Guenard C, Garrel J, Millet P. Placards sclérodermiformes des régions ilio-trochanteriennes paraissant consécutifs à des injections I.M. contenant de la vitamin K1. Bull Soc Fr Dermat 1972;79:581-582 (article in French)

44 Lembo S, Megna M, Balato A, Balato N. "Cowboy's belt with revolver" scleroderma caused by vitamin K1 injections. G Ital Dermatol Venereol 2012;147:203-205 (article in Italian)

45 Alonso-Llamazares J, Ahmed I. Vitamin K1-induced localized scleroderma (morphea) with linear deposition of IgA in the basement membrane zone. J Am Acad Dermatol 1998;38:322-324

Chapter 3.385 PICOSULFURIC ACID

IDENTIFICATION

Description/definition	: Picosulfuric acid is the diphenylmethane that conforms to the structural formula shown below
Pharmacological classes	: Laxatives
IUPAC name	: [4-[Pyridin-2-yl-(4-sulfooxyphenyl)methyl]phenyl] hydrogen sulfate
CAS registry number	: 10040-34-3
EC number	: Not available
Merck Index monograph	: 8788 (Picosulfate sodium)
Patch testing	: No data available; most systemic drugs can be tested at 10% pet.; if the pure chemical is not available, prepare the test material from intravenous powder, the content of capsules or – when also not available – from powdered tablets to achieve a final concentration of the active drug of 10% pet.
Molecular formula	: $C_{18}H_{15}NO_8S_2$

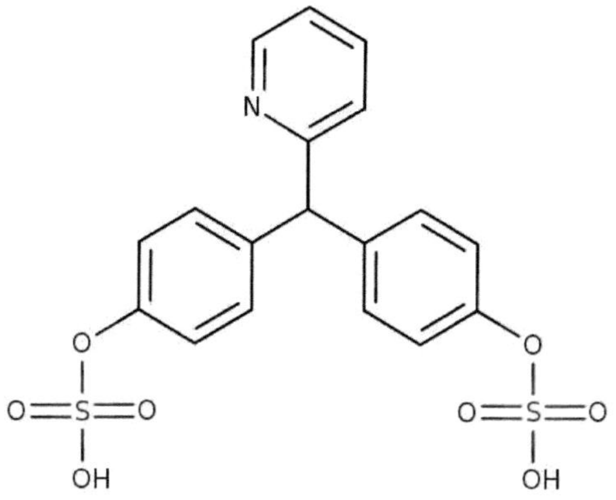

GENERAL

Picosulfuric acid is a contact laxative. It inhibits the absorption of water and electrolytes, and increases their secretion into the intestinal lumen. Picosulfuric acid is hydrolyzed by the colonic bacterial enzyme sulfatase 3 to form an active metabolite bis-(p-hydroxyphenyl)pyridyl-2-methane, which acts directly on the colonic mucosa to stimulate colonic peristalsis. This drug (which is often combined with magnesium citrate) is used to treat constipation and for cleansing of the colon as a preparation for colonoscopy in adults (1).

In pharmaceutical products, picosulfuric acid may be employed as sodium picosulfate monohydrate (CAS number 1307301-38-7, EC number not available, molecular formula $C_{18}H_{15}NNa_2O_9S_2$) or as sodium picosulfate anhydrous (CAS number 10040-45-6, EC number not available, molecular formula $C_{18}H_{13}NNa_2O_8S_2$) (1).

CUTANEOUS ADVERSE DRUG REACTIONS FROM SYSTEMIC ADMINISTRATION CAUSED BY TYPE IV (DELAYED-TYPE) HYPERSENSITIVITY (as demonstrated by positive patch tests)

Cutaneous adverse drug reactions from systemic administration of picosulfuric acid caused by type IV (delayed-type) hypersensitivity have included fixed drug eruption (2).

Fixed drug eruption

Among 242 patients investigated in a clinic in Japan between 1983 and 1988 for 'generalized drug eruptions', there was one case of fixed drug eruption to sodium picosulfate with a positive patch test on postlesional skin. Patch testing nor clinical data were provided (2).

LITERATURE

1   The data in the section 'General' may have been obtained from literature discussed in this chapter, but mostly also or exclusively from one or more of the following online sources: ChemIDPlus Advanced, PubChem, DrugBank, RxList, Drug Central, Drugs.com, and Wikipedia

2   Osawa J, Naito S, Aihara M, Kitamura K, Ikezawa Z, Nakajima H. Evaluation of skin test reactions in patients with non-immediate type drug eruptions. J Dermatol 1990;17:235-239

Chapter 3.386 PIPERACILLIN

IDENTIFICATION

Description/definition   : Piperacillin sodium is the semisynthetic, ampicillin-derived ureidopenicillin that conforms
                            to the structural formula shown below
Pharmacological classes : Anti-bacterial agents
IUPAC name   : (2S,5R,6R)-6-[[(2R)-2-[(4-Ethyl-2,3-dioxopiperazine-1-carbonyl)amino]-2-phenylacetyl]-
                           amino]-3,3-dimethyl-7-oxo-4-thia-1-azabicyclo[3.2.0]heptane-2-carboxylic acid
CAS registry number   : 61477-96-1
EC number   : 262-811-8
Merck Index monograph   : 8845
Patch testing   : 5% pet. (12)
Molecular formula   : $C_{23}H_{27}N_5O_7S$

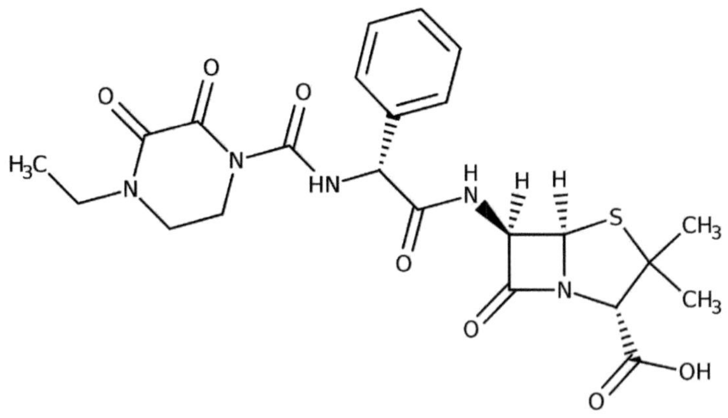

GENERAL

Piperacillin is a broad-spectrum semisynthetic, ampicillin-derived ureidopenicillin antibiotic with bactericidal activity.
It is stable against hydrolysis by a variety of β-lactamases, including penicillinases, cephalosporinases and extended
spectrum β-lactamases. Piperacillin binds to and inactivates penicillin-binding proteins, enzymes located on the inner
membrane of the bacterial cell wall, resulting in the weakening of the bacterial cell wall and cell lysis. It is indicated
for the treatment of polymicrobial infections. In pharmaceutical products, piperacillin is employed as piperacillin
sodium (CAS number 59703-84-3, EC number 261-868-6, molecular formula $C_{23}H_{26}N_5NaO_7S$) (1).
    The classification and structures of beta-lactam antibiotics are discussed in Chapter 3.36 Amoxicillin. See also
Chapter 3.387 Piperacillin mixture with tazobactam.

CONTACT ALLERGY FROM ACCIDENTAL CONTACT

A 28-year-old nurse developed eczema on the dorsal aspect of the hand and the face. She worked in the hematology
department where she usually handled and administered a variety of antibiotics, notably the penicillin piperacillin
and the carbapenems imipenem and ertapenem. When she was moved to a different department where she did not
have contact with these drugs, the dermatitis completely resolved. Patch tests were positive to all 3 antibiotics 20%
pet. with cross-reactions to ampicillin and imipenem (4).

CUTANEOUS ADVERSE DRUG REACTIONS FROM SYSTEMIC ADMINISTRATION CAUSED BY TYPE IV
(DELAYED-TYPE) HYPERSENSITIVITY (as demonstrated by positive patch tests)

Cutaneous adverse drug reactions from systemic administration of piperacillin caused by type IV (delayed-type)
hypersensitivity have included maculopapular eruption (6,9,10,11), generalized erythema (6,11), drug reaction with
eosinophilia and systemic symptoms (DRESS) (3,5,7,8), and localized allergic reaction from intramuscular injection
(10).

Case series with various or unknown types of drug reactions

In Rome, 259 patients who had suffered nonimmediate skin reactions during penicillin therapy were examined with
patch testing. The workup was performed after a mean of 4 years (range: 1-540 months) after the most recent
adverse reaction. Eleven subjects reported adverse reactions to piperacillin, of who 4 had positive patch tests. Of
these piperacillin-allergic individuals, 3 had shown a maculopapular eruption and one a localized allergic reaction
from intramuscular administration (10).

Maculopapular eruption
A 38-year-old man developed generalized erythema after 6 days of treatment with intramuscular AvocinB (drug cannot be traced, unknown whether this is piperacillin or piperacillin/tazobactam). The symptoms resolved a few days after corticosteroids were started. Patch tests were positive to piperacillin 200 and 400 mg/ml saline, amoxicillin and ampicillin at D2 and D3, as were intradermal tests to these 3 antibiotics read at 24 hours (6).

In a group of thirty patients from Rome who had suffered skin rashes from beta-lactam antibiotics and who had at least one positive patch test to a beta-lactam antigenic determinant, piperacillin was the allergenic culprit in 1 patient with maculopapular eruption, appearing 8 hours after intramuscular administration (9).

A 42-year-old man after 4 days of treatment with piperacillin 4 g/day intramuscularly developed a generalized maculopapular eruption. Patch tests were positive to piperacillin 200 mg/ml saline and amoxicillin and ampicillin 5% pet. (11). Six years later, the same group reported another patient with a maculopapular exanthema from piperacillin with a positive patch test to this antibiotic 5% pet. and a cross-reaction to mezlocillin 5% pet. Lymphocyte transformation tests were positive to both antibiotics (12).

See also the section 'Case series with various or unknown types of drug reactions' above, ref. 10.

Erythroderma, widespread erythematous eruption, exfoliative dermatitis
A 42-year-old man developed generalized a pruritic maculopapular eruption after 4 days of treatment with intramuscular AvocinB (drug cannot be traced, unknown whether this is piperacillin or piperacillin/tazobactam). The symptoms resolved spontaneously within 7 days leaving generalized scaling. Patch tests were positive to piperacillin 200 and 400 mg/ml saline, amoxicillin and ampicillin at D2 and D3, as were intradermal tests to these 3 antibiotics read at 24 hours (6).

A 38-year-old man after 6 days of treatment with piperacillin 4 g/day intramuscularly developed a generalized erythematous eruption. Patch tests were positive to piperacillin 200 mg/ml saline and amoxicillin and ampicillin 5% pet. (11).

Drug reaction with eosinophilia and systemic symptoms (DRESS)
A 55-year-old man was treated with piperacillin/tazobactam and vancomycin for high digestive hemorrhage and an intra-abdominal infection after gallbladder surgery. On day 14, the patient suffered a generalized itchy maculopapular eruption; eosinophilia of 47% could be observed in his blood. Five months after complete recovery, a patch test was positive to piperacillin/tazobactam 'as is in pet.' at D3 and D4. In a second session, there were again positive reactions to the combination tablet and to pure piperacillin sodium at 10% and 20% pet. read after 2,3 and 4 days. The piperacillin patch test at 20% remained positive for 8 days. At the D4 reading of this patch test, an erythematous, edematous, and itching lesion of 10x10 cm could be observed on the medial part of the patient's left thigh. Tazobactam was not tested, being unavailable. Twenty controls were negative to piperacillin/tazobactam and piperacillin 10% and 20% pet. (5). This was very likely a case of DRESS.

In a group of 14 patients with multiple delayed-type hypersensitivity reactions, DRESS was caused by piperacillin in one case, showing a positive patch test reaction to this antibiotic as well as a positive lymphocyte transformation test (7,8). An atypical case of DRESS due to delayed-type hypersensitivity to piperacillin (positive patch test) in a 3-year-old girl was reported from France in 2017 (3).

Other cutaneous adverse drug reactions
See the section 'Case series with various or unknown types of drug reactions' above, ref. 10.

Cross-reactions, pseudo-cross-reactions and co-reactions
Cross-reactions between beta-lactam antibiotics are discussed in Chapter 3.36 Amoxicillin. In a group of 78 patients who had suffered nonimmediate drug reactions from any beta-lactam compound (penicillins, cephalosporins), 15 had positive patch tests to piperacillin. As this was not the culprit drug causing the adverse skin reaction in any patient, all these reactions were presumably cross-reactions (2).

A patient who developed a maculopapular exanthema from piperacillin with a positive patch test to this antibiotic cross-reacted to mezlocillin 5% pet. Lymphocyte transformation tests were positive to both antibiotics. There were no positive reactions to other penicillins or cephalosporins and therefore it was suggested that the ureido group present in both piperacillin and mezlocillin side chains may have played an important role in the sensitization (12).

LITERATURE
1 The data in the section 'General' may have been obtained from literature discussed in this chapter, but mostly also or exclusively from one or more of the following online sources: ChemIDPlus Advanced, PubChem, DrugBank, RxList, Drug Central, Drugs.com, and Wikipedia
2 Buonomo A, Nucera E, De Pasquale T, Pecora V, Lombardo C, Sabato V, et al. Tolerability of aztreonam in patients with cell-mediated allergy to β-lactams. Int Arch Allergy Immunol 2011;155:155-159

3   Penel-Page M, Ben Said B, Phan A, Hees L, Hartmann-Merlin C, Girard S, et al. Pièges diagnostiques d'un
     syndrome d'activation macrophagique [Correctly adDRESS the cause of hemophagocytic lymphohistiocytosis].
     Arch Pediatr 2017;24:254-259 (Article in French)
4   Colagiovanni A, Feliciani C, Fania L, Pascolini L, Buonomo A, Nucera E, et al. Occupational contact dermatitis from
     carbapenems. Cutis 2015;96:E1-3
5   Cabañas R, Muñoz L, López-Serrano C, Contreras J, Padial A, Caballero T, Moreno-Ancillo A, Barranco P.
     Hypersensitivity to piperacillin. Allergy 1998;53:819-820
6   Di Fonso M, Romano A, Quaratino D, Giuffreda R, Venuti A. Delayed hypersensitivity to piperacillin: two case
     reports. Allergy 1996;51(Suppl.31):98 (Abstract)
7   Jörg L, Helbling A, Yerly D, Pichler WJ. Drug-related relapses in drug reaction with eosinophilia and systemic
     symptoms (DRESS). Clin Transl Allergy 2020;10:52
8   Jörg L, Yerly D, Helbling A, Pichler W. The role of drug, dose and the tolerance/intolerance of new drugs in
     multiple drug hypersensitivity syndrome (MDH). Allergy 2020;75:1178-1187
9   Patriarca G, D'Ambrosio C, Schiavino D, Larocca LM, Nucera E, Milani A. Clinical usefulness of patch and
     challenge tests in the diagnosis of cell-mediated allergy to betalactams. Ann Allergy Asthma Immunol
     1999;83:257-266
10   Romano A, Viola M, Mondino C, Pettinato R, Di Fonso M, Papa G, et al. Diagnosing nonimmediate reactions to
     penicillins by in vivo tests. Int Arch Allergy Immunol 2002;129:169-174
11   Romano A, Di Fonso M, Artesani MC, Viola M, Andriolo M, Pettinato R. Delayed hypersensitivity to piperacillin.
     Allergy 2002;57:459
12   Gaeta F, Alonzi C, Valluzzi RL, Viola M, Romano A. Delayed hypersensitivity to acylureidopenicillins: a case report.
     Allergy 2008;63:787-789

Chapter 3.387 PIPERACILLIN MIXTURE WITH TAZOBACTAM

IDENTIFICATION

Description/definition : Piperacillin mixture with tazobactam is the combination product of these two antibiotics; their structural formulas are shown below
Pharmacological classes : Anti-bacterial agents
IUPAC name : (2S,5R,6R)-6-[[2-[(4-Ethyl-2,3-dioxopiperazine-1-carbonyl)amino]-2-phenylacetyl]amino]-3,3-dimethyl-7-oxo-4-thia-1-azabicyclo[3.2.0]heptane-2-carboxylic acid;(2S,3S,5R)-3-methyl-4,4,7-trioxo-3-(triazol-1-ylmethyl)-4λ^6-thia-1-azabicyclo[3.2.0]heptane-2-carboxylic acid
CAS registry number : 123683-33-0
EC number : Not available
Merck Index monograph : 10490 (Tazobactam); 8845 (Piperacillin)
Patch testing : Tablet, pulverized, 30% pet. (2); most systemic drugs can be tested at 10% pet.; if the pure chemical is not available, prepare the test material from intravenous powder, the content of capsules or – when also not available – from powdered tablets to achieve a final concentration of the active drug of 10% pet.
Molecular formula : $C_{33}H_{39}N_9O_{12}S_2$

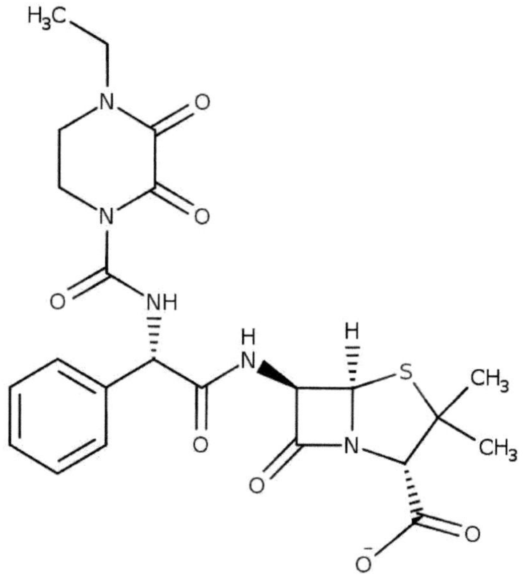

Piperacillin

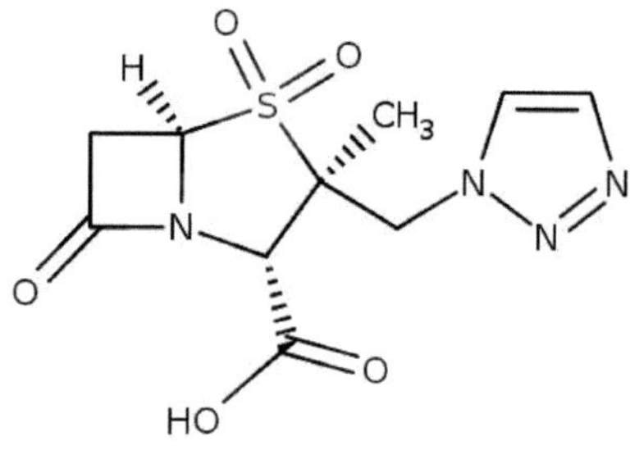

Tazobactam

GENERAL

Tazobactam is an antibiotic of the beta-lactamase inhibitor class that prevents the breakdown of other antibiotics by beta-lactamase enzyme producing organisms. It is combined with piperacillin to broaden the spectrum of piperacillin's antibacterial action. The combination product is used to treat a variety of infections, including those caused by aerobic and facultative gram-positive and gram-negative bacteria, in addition to gram-positive and gram-negative anaerobes. Some examples of infections treated with piperacillin-tazobactam include cellulitis, diabetic foot infections, appendicitis, and postpartum endometritis infections (1).

In pharmaceutical products, this combination drug is employed as piperacillin sodium (CAS number 59703-84-3, EC number 261-868-6, molecular formula $C_{23}H_{26}N_5NaO_7S$) and tazobactam sodium (CAS number 89785-84-2, EC number not available, molecular formula $C_{10}H_{11}N_4NaO_5S$) (1). The classification and structures of beta-lactam antibiotics are discussed in Chapter 3.36 Amoxicillin. See also Chapter 3.386 Piperacillin.

CUTANEOUS ADVERSE DRUG REACTIONS FROM SYSTEMIC ADMINISTRATION CAUSED BY TYPE IV (DELAYED-TYPE) HYPERSENSITIVITY (as demonstrated by positive patch tests)

Cutaneous adverse drug reactions from systemic administration of piperacillin mixture with tazobactam caused by type IV (delayed-type) hypersensitivity have included drug reaction with eosinophilia and systemic symptoms (DRESS) (2,3).

Drug reaction with eosinophilia and systemic symptoms (DRESS)

Of 8 patients with DRESS from piperacillin-tazobactam, seen in Madrid, Spain, 4 were patch tested. One had a positive patch test to the drug combination tested 30% pet. at D2 and D3, a negative delayed reading intradermal test and positive lymphocyte transformation tests (Stimulation Indices: 3.5, 3.8, 7.03, and 13.5). The 2 ingredients were not tested separately. This patient was a 59-year-old man, who had developed a skin rash after 3 weeks of piperacillin-tazobactam treatment with fever, eosinophilia and elevated liver enzymes (2).

A patient from Spain with drug reaction with eosinophilia and systemic symptoms (DRESS) had positive patch tests to piperacillin-tazobactam, amoxicillin-clavulanic acid, and meropenem. Details were not provided, the report was a reaction to a Letter to the editor (3).

Cross-reactions, pseudo-cross-reactions and co-reactions

Cross-reactions between beta-lactam antibiotics are discussed in Chapter 3.36 Amoxicillin.

LITERATURE

1   The data in the section 'General' may have been obtained from literature discussed in this chapter, but mostly also or exclusively from one or more of the following online sources: ChemIDPlus Advanced, PubChem, DrugBank, RxList, Drug Central, Drugs.com, and Wikipedia
2   Cabañas R, Calderon O, Ramirez E, Fiandor A, Prior N, Caballero T, Herránz P, Bobolea I, López-Serrano MC, Quirce S, Bellón T. Piperacillin-induced DRESS: distinguishing features observed in a clinical and allergy study of 8 patients. J Investig Allergol Clin Immunol 2014;24:425-430
3   Gallardo A, Moreno EM, Laffond E, Muñoz-Bellido FJ, Gracia-Bara MT, Macias EM, et al. Reply to "Delayed hypersensitivity reactions to piperacillin-tazobactam". J Allergy Clin Immunol Pract 2021;9:2549

Chapter 3.388 PIPERAZINE

IDENTIFICATION

Description/definition : Piperazine is the cyclic organic compound that conforms to the structural formula shown below
Pharmacological classes : Antinematodal agents
IUPAC name : Piperazine
Other names : 1,4-Diethylenediamine
CAS registry number : 110-85-0
EC number : 203-808-3
Merck Index monograph : 8846
Patch testing : 1% pet. (SmartPracticeCanada, SmartPracticeEurope)
Molecular formula : $C_4H_{10}N_2$

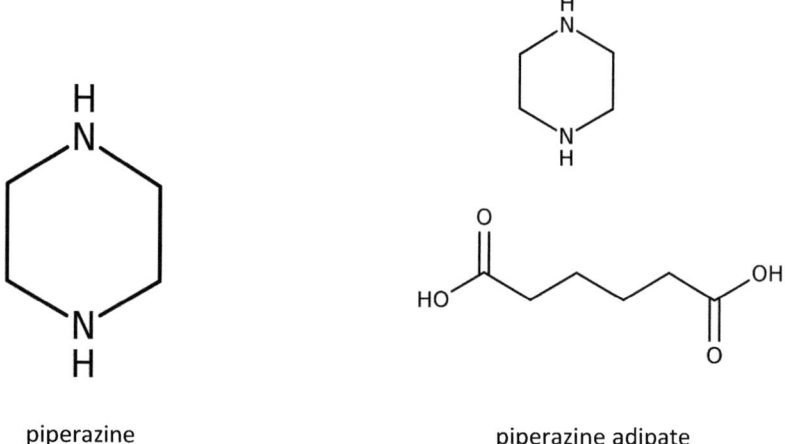

piperazine   piperazine adipate

GENERAL

Piperazine is an organic compound that was introduced to medicine as a solvent for uric acid and later as an anthelmintic (antinematodal agent). It produces a neuromuscular block leading to flaccid muscle paralysis in susceptible worms, which are then dislodged from the gut and expelled in feces. Piperazine is used as alternative treatment for ascariasis caused by *Ascaris lumbricoides* (roundworm) and enterobiasis (oxyuriasis) caused by *Enterobius vermicularis* (pinworm). In pharmaceutical products, piperazine may be employed as piperazine adipate (CAS number 142-88-1, EC number 205-569-0, molecular formula $C_{10}H_{20}N_2O_4$) (1). Piperazine also has numerous non-pharmaceutical applications. In this chapter, only allergic reactions to piperazine from systemic pharmaceutical use and occupational exposure are presented. Allergic reactions to piperazine used as topical drug have been reported occasionally (2,3).

See also Chapter 3.421 Pyrazinobutazone and Chapter 3.381 Phenylbutazone.

CONTACT ALLERGY FROM ACCIDENTAL CONTACT

Of 26 veterinarians patch tested in the early 1980s in Poland because of dermatitis, 15 had one or more positive patch tests to veterinary drugs, including two who reacted to piperazine 1% water; these reactions were relevant, as the patients had contact with piperazine in their work (12). In a group of 107 workers in the pharmaceutical industry with dermatitis, also investigated in Warsaw, Poland, before 1989, one reacted to piperazine, tested 1% pet. (4).

A 27-year-old woman, working in a pharmaceutical laboratory, presented with hand eczema which she suspected to be due to phenylbutazone-piperazine suppositories. During weekends and holidays, the lesions almost healed, but immediately reappeared every time she returned to work. Patch tests were strongly positive (+++) to the suppositories, phenylbutazone-piperazine 1% pet. and piperazine 5% water (3).

A 55-year-old man, working in a factory where he had contact with piperazine and ethylenediamine and would be contaminated by their dust every day, after two months developed eczema on the hands, arms, face and penis. During a three-week holiday it disappeared, but the eczema returned two days after resuming work and he also developed respiratory symptoms. A patch test with piperazine 1% water was applied at 15.00 hour. When the patient awoke the next morning he had respiratory symptoms (not further specified), which disappeared after 5-6 hours. The patch test to piperazine was strongly positive after 48 hours (9).

A 29-year-old man, working as a laboratory technician in a pharmaceutical company and having contact with ethylenediamine, piperazine and other chemicals, developed severe eczema of his fingers with nail dystrophy. Patch tests were positive to ethylenediamine and piperazine 1% (8).

A woman aged 50 worked for 6 years in a pharmaceutical factory handling ampoules of drugs, including the combination of thiourea and piperazine. When any of these broke, others would be contaminated. The patient developed dermatitis on the sides of the fingers of her right hand, then on both hands and later also in the axillae. Patch tests were positive to the ampoule content and to piperazine 0.1%, 0.5% and 1% pet. (D2-, D4 +) (7).

A man who was sensitized to piperazine at his work. When he took hydroxyzine orally (a piperazine derivative), he developed systemic allergic dermatitis; details are unknown to the author (13). In France, nurses in a resuscitation unit became sensitized to piperazine by handling piperazine camphosulfonate (Solucamphre; piperazine and 2-oxo-10-bornanesulfonate) (14).

CUTANEOUS ADVERSE DRUG REACTIONS FROM SYSTEMIC ADMINISTRATION CAUSED BY TYPE IV (DELAYED-TYPE) HYPERSENSITIVITY (as demonstrated by positive patch tests)

Cutaneous adverse drug reactions from systemic administration of piperazine caused by type IV (delayed-type) hypersensitivity have included systemic allergic dermatitis (5,6,10,11; presenting as erythroderma [6,11], morbilliform rash [10,11], and angioedema [5]).

General

Piperazine (diethylenediamine) is metabolized in the human body into ethylenediamine, which means that patients who have previously been sensitized to ethylenediamine (mostly from one brand of topical corticosteroids containing triamcinolone acetonide, ethylenediamine, neomycin, gramicidin and nystatin, occasionally from aminophylline [= theophylline + ethylenediamine]) may develop systemic allergic dermatitis from oral administration of piperazine (5,6,10,11).

Systemic allergic dermatitis (systemic contact dermatitis)

A 37 year old man had a 20 year history of dermatitis, which had been treated with numerous steroid creams and ointments, including one containing ethylenediamine as a stabilizer. Twelve hours after he had taken piperazine citrate because his daughter had threadworms, he developed a generalized itchy morbilliform rash. After being treated with piperazine again one year later, a severe exfoliative erythroderma developed within three hours. An oral challenge with 50 µg piperazine hydrate provoked a generalized maculopapular erythema within hours with shivering, anxiety, and tachycardia. Subsequent patch testing showed him to have contact allergy to 1% ethylenediamine (11).

A 48-year-old woman, who had used ethylenediamine- and triamcinolone-containing creams for eczema, developed, within 1 hour of taking one tablet of piperazine phosphate for threadworms, angioneurotic edema affecting both eyes and tongue, necessitating hospital admission. Patch testing was positive to ethylenediamine (5).

Similar cases of systemic allergic dermatitis from oral administration of piperazine in ethylenediamine-sensitized individuals showed a morbilliform drug eruption (10) and erythroderma, facial edema and malaise (6)

Cross-reactions, pseudo-cross-reactions and co-reactions

Patients allergic to ethylenediamine may cross-react (or actually pseudo-cross-react) to systemic piperazine (5,6,10), as piperazine (diethylenediamine) is metabolized in the human body into ethylenediamine.

LITERATURE

1   The data in the section 'General' may have been obtained from literature discussed in this chapter, but mostly also or exclusively from one or more of the following online sources: ChemIDPlus Advanced, PubChem, DrugBank, RxList, Drug Central, Drugs.com, and Wikipedia

2   Fernández de Corres L, Bernaola G, Lobera T, Leanizbarrutia I, Muñoz D. Allergy from pyrazoline derivatives. Contact Dermatitis 1986;14:249-250

3   Brandão FM, Foussereau J. Contact dermatitis to phenylbutazone-piperazine suppositories (Carudol) and piperazine gel (Carudol). Contact Dermatitis 1982;8:264-265

4   Rudzki E, Rebandel P, Grzywa Z. Contact allergy in the pharmaceutical industry. Contact Dermatitis 1989;21:121-122

5   Eedy DJ. Angioneurotic oedema following piperazine ingestion in an ethylenediamine-sensitive subject. Contact Dermatitis 1993;28:48-49

6   Price ML, Hall-Smith SP. Allergy to piperazine in a patient sensitive to ethylenediamine. Contact Dermatitis 1984;10:120

7   Rudzki E, Grzywa Z. Occupational piperazine dermatitis. Contact Dermatitis 1977;3:216

8   Calnan CD. Occupational piperazine dermatitis. Contact Dermatitis 1975;1:126

9   Fregert S. Respiratory symptoms with piperazine patch testing. Contact Dermatitis 1976;2:61-62

10 Burry JN. Ethylenediamine sensitivity with a systemic reaction to piperazine citrate. Contact Dermatitis 1978;4:380

11 Wright S, Harman RR. Ethylenediamine and piperazine sensitivity. Br Med J (Clin Res Ed) 1983;287(6390):463-464

12 Rudzki E, Rebandel P, Grzywa Z, Pomorski Z, Jakiminska B, Zawisza E. Occupational dermatitis in veterinarians. Contact Dermatitis 1982;8:72-73

13 Fregert S. Exacerbation of dermatitis by perorally administered piperazine derivative in a piperazine sensitized man. Contact Dermatitis Newsletter 1967;1:13

14 Foussereau J, Benezra C. Données nouvelles sur l'allergie de groupe à la pipérazine. Bull Soc Franc Derm Syphil 1967;74;45

Chapter 3.389 PIRFENIDONE

IDENTIFICATION

Description/definition   : Pirfenidone is the pyridinone that conforms to the structural formula shown below
Pharmacological classes   : Analgesics; anti-inflammatory agents, non-steroidal; antineoplastic agents
IUPAC name   : 5-Methyl-1-phenylpyridin-2-one
CAS registry number   : 53179-13-8
EC number   : Not available
Merck Index monograph   : 8876
Patch testing   : Tablet 801 mg, 10%, 1%, 0.1%, and 0.01% water (6); 1% pet. (2)
Molecular formula   : $C_{12}H_{11}NO$

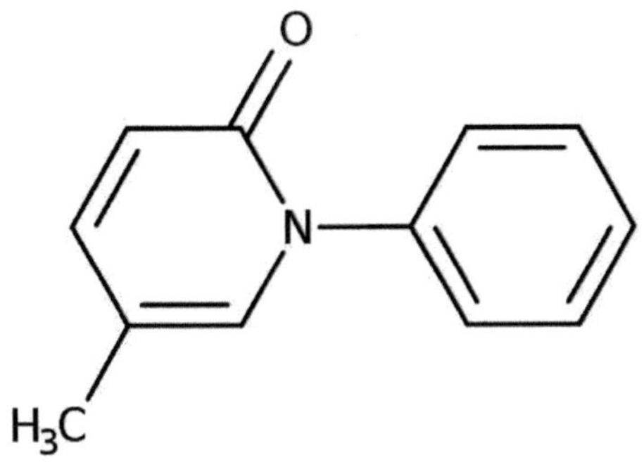

GENERAL

Pirfenidone is an orally active synthetic antifibrotic agent which inhibits fibroblast, epidermal, platelet-derived, and transforming beta-1 growth factors, thereby slowing tumor cell proliferation. This agent also inhibits DNA synthesis and the production of mRNA for collagen types I and III, resulting in a reduction of collagen synthesis in various fibrotic conditions, including those of the lung, kidney and liver. It is indicated for the treatment of idiopathic pulmonary fibrosis (1).

CUTANEOUS ADVERSE DRUG REACTIONS FROM SYSTEMIC ADMINISTRATION CAUSED BY TYPE IV (DELAYED-TYPE) HYPERSENSITIVITY (as demonstrated by positive patch tests)

Cutaneous adverse drug reactions from systemic administration of pirfenidone caused by type IV (delayed-type) hypersensitivity have included photosensitivity (2,6).

Photosensitivity

A 77-year-old man developed a florid itchy rash affecting the face, neck, and forearms after having taken oral pirfenidone for idiopathic pulmonary fibrosis for 3 months and following exposure to sunlight. The patient stated that the eruption would improve a week after stopping pirfenidone but would recur within 2 days of recommencing the medication. Patch and photopatch tests were carried out, resulting in positive photopatch tests to pirfenidone 801 mg (10%, 1%, 0.1%, and 0.01% in aqueous solution) irradiated with 5 J/cm² ultraviolet A. Patch tests showed erythema at the 2 highest concentrations, suggestive of an irritant reaction (6).

A 74-year-old fisherman presented with a one-month history of itchy erythematous patches on the face, neck and dorsal aspects of both hands, which had started 3 months after the first intake of pirfenidone tablets for idiopathic pulmonary fibrosis and thereafter rapidly worsened. Histopathology showed epidermal spongiosis with a lichenoid reaction and basophilic degeneration of the upper dermis. There were no apoptotic keratinocytes (as seen in phototoxic reactions). The minimal erythema dose (MED) for UVA was normal. A photopatch test with pirfenidone 1% pet. irradiated with 10 J/cm² UVA was positive. After discontinuation of pirfenidone, the skin lesions improved and healed with residual post-inflammatory hyperpigmentation. Photoallergy to pirfenidone was diagnosed (2). However, no control testing was performed and the phototoxic properties of pirfenidone have been well-documented (3,4). Two patients with suspected photoallergy to pirfenidone had been presented before, but they were not photopatch tested (5).

LITERATURE

1   The data in the section 'General' may have been obtained from literature discussed in this chapter, but mostly also or exclusively from one or more of the following online sources: ChemIDPlus Advanced, PubChem, DrugBank, RxList, Drug Central, Drugs.com, and Wikipedia

2   Park M, Shim W, Kim J, Kim G, Kim H, Ko H, et al. Pirfenidone-induced photo-allergic reaction in a patient with idiopathic pulmonary fibrosis. Photodermatol Photoimmunol Photomed 2017;33:209-212

3   Caruana DM, Wylie G. Cutaneous reactions to pirfenidone: a new kid on the block. Br J Dermatol 2016;175:425-426

4   Seto Y, Inoue R, Kato M, Yamada S, Onoue S. Photosafety assessments on pirfenidone: photochemical, photobiological, and pharmacokinetic characterization. J Photochem Photobiol B 2013;120:44-51

5   Reinholz M, Eder I, Przybilla B, Schauber J, Wollenberg A, Wulffen W, et al. Photoallergic contact dermatitis due to treatment of pulmonary fibrosis with pirfenidone. J Eur Acad Dermatol Venereol 2016;30:370-371

6   Forbat E, Parr D, Shim TN. Positive photopatch test to pirfenidone. Contact Dermatitis 2021;84:341-342

Chapter 3.390 PIRITRAMIDE

IDENTIFICATION

Description/definition	: Piritramide is the diphenylacetonitrile compound that conforms to the structural formula shown below
Pharmacological classes	: Analgesics, opioid; narcotics
IUPAC name	: 1-(3-Cyano-3,3-diphenylpropyl)-4-piperidin-1-ylpiperidine-4-carboxamide
CAS registry number	: 302-41-0
EC number	: 206-124-3
Merck Index monograph	: 8881
Patch testing	: Commercial injectable solution (10 mg/ml)
Molecular formula	: $C_{27}H_{34}N_4O$

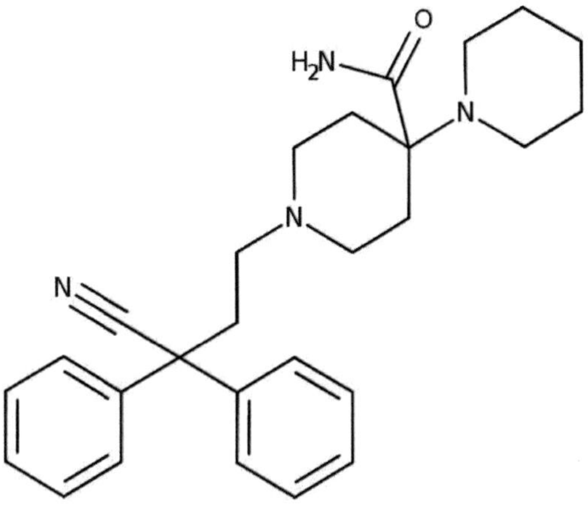

GENERAL

Piritramide is a diphenylacetonitrile and opioid receptor agonist, with analgesic activity. Upon administration, piritramide binds to and activates mu-opioid receptors in the central nervous system, thereby mimicking the effects of endogenous opioids and producing analgesic relief. It is used for severe pre-, peri- or postoperative pain, but has not advantages over morphine (1).

CUTANEOUS ADVERSE DRUG REACTIONS FROM SYSTEMIC ADMINISTRATION CAUSED BY TYPE IV (DELAYED-TYPE) HYPERSENSITIVITY (as demonstrated by positive patch tests)

Cutaneous adverse drug reactions from systemic administration of piritramide caused by type IV (delayed-type) hypersensitivity have included generalized itching erythema (2).

Other cutaneous adverse drug reactions

A 51-year-old woman, who was treated with intravenous piritramide, diclofenac and paracetamol for postoperative pain after surgery of a herniated vertebral disc, after 2 days developed a generalized intensively itching erythema. The patient had 3 previous episodes of piritramide treatment, which had resulted in increasing itch in the last two, but without erythema. Allergy tests included skin prick, scratch and patch tests with the commercial piritramide injectable solution (piritramide, tartaric acid, water for injection, tested as is), and other opioid analgesics. Only the patch test to piritramide solution was positive at D3. Controls were not performed (2).

LITERATURE

1   The data in the section 'General' may have been obtained from literature discussed in this chapter, but mostly also or exclusively from one or more of the following online sources: ChemIDPlus Advanced, PubChem, DrugBank, RxList, Drug Central, Drugs.com, and Wikipedia

2   Waltermann K, Geier J, Diessenbacher P, Schadendorf D, Hillen U. Systemic allergic contact dermatitis from intravenous piritramide. Allergy 2010;65:1203-1204

Chapter 3.391 PIRMENOL

IDENTIFICATION

Description/definition : Pirmenol is the member of benzenes and organic amino compound that conforms to the structural formula shown below

Pharmacological classes : Anti-arrhythmia agents

IUPAC name : 4-[(2S,6R)-2,6-Dimethylpiperidin-1-yl]-1-phenyl-1-pyridin-2-ylbutan-1-ol

CAS registry number : 68252-19-7

EC number : Not available

Merck Index monograph : 8884

Patch testing : No data available; most systemic drugs can be tested at 10% pet.; if the pure chemical is not available, prepare the test material from intravenous powder, the content of capsules or – when also not available – from powdered tablets to achieve a final concentration of the active drug of 10% pet.

Molecular formula : $C_{22}H_{30}N_2O$

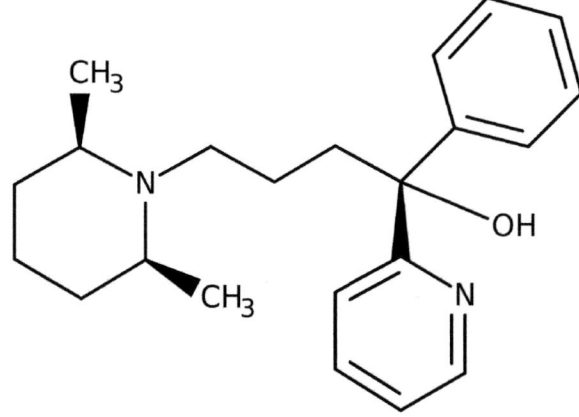

GENERAL

Pirmenol is a member of benzenes and an organic amino compound used (very limited, possible only in Japan) as an anti-arrhythmia agent. In pharmaceutical products pirmenol is employed as pirmenol hydrochloride (CAS number 61477-94-9, EC number not available, molecular formula $C_{22}H_{31}ClN_2O$) (1).

CONTACT ALLERGY FROM ACCIDENTAL CONTACT

A 31-year-old process operator working in a pharmaceutical plant developed occupational airborne allergic contact dermatitis from pirmenol hydrochloride. The primary sites of dermatitis were the eyelids. No additional clinical or patch testing details, such as test concentration, vehicle and number of controls, were provided (2).

LITERATURE

1 The data in the section 'General' may have been obtained from literature discussed in this chapter, but mostly also or exclusively from one or more of the following online sources: ChemIDPlus Advanced, PubChem, DrugBank, RxList, Drug Central, Drugs.com, and Wikipedia

2 Bennett MF, Lowney AC, Bourke JF. A study of occupational contact dermatitis in the pharmaceutical industry. Br J Dermatol 2016;174:654-656 (Abstract in Brit J Dermatol 2011;165:73)

Chapter 3.392 PIROXICAM

IDENTIFICATION

Description/definition   : Piroxicam is the nonsteroidal oxicam that conforms to the structural formula shown
                               below
Pharmacological classes   : Anti-inflammatory agents, non-steroidal; cyclooxygenase inhibitors
IUPAC name   : 4-Hydroxy-2-methyl-1,1-dioxo-N-pyridin-2-yl-2H-1,2-benzothiazine-3-carboxamide
CAS registry number   : 36322-90-4
EC number   : 252-974-3
Merck Index monograph : 8889
Patch testing   : 1.0% pet. (Chemotechnique, SmartPracticeCanada)
Molecular formula   : $C_{15}H_{13}N_3O_4S$

GENERAL

Piroxicam is an oxicam derivative with anti-inflammatory, antipyretic and analgesic properties. As a non-selective, nonsteroidal anti-inflammatory drug (NSAID), piroxicam binds and chelates both isoforms of cyclooxygenases (COX1 and COX2), thereby stalling phospholipase A2 activity and conversion of arachidonic acid into prostaglandin precursors. This results in inhibition of prostaglandin biosynthesis. As a second, independent effect, piroxicam inhibits the activation of neutrophils. Piroxicam is indicated for treatment of osteoarthritis and rheumatoid arthritis, musculoskeletal disorders, dysmenorrhea and postoperative pain. It is also used in topical formulations for treating pain and swelling due to strains, sprains, backache or arthritis (1). As such, a few cases of allergic contact dermatitis and photocontact dermatitis from topical piroxicam have been reported, which have been fully reviewed in Volume 3 of the Monographs in contact allergy series (20).

CONTACT ALLERGY FROM ACCIDENTAL CONTACT

A 55-year-old nurse became sensitized to piroxicam from handling this drug at work, as demonstrated by positive patch tests to piroxicam 1% pet. and commercial piroxicam tablets 10% and 30% pet. and saline. However, the patient did not develop allergic contact dermatitis, but bullous fixed drug eruptions on the back of the hands. The patch tests were positive only on postlesional skin, as it usual in cases of FDE (29).

CUTANEOUS ADVERSE DRUG REACTIONS FROM SYSTEMIC ADMINISTRATION CAUSED BY TYPE IV
(DELAYED-TYPE) HYPERSENSITIVITY (as demonstrated by positive patch tests)

Cutaneous adverse drug reactions from systemic administration of piroxicam caused by type IV (delayed-type) hypersensitivity have included fixed drug eruption (2,3,4,5,7,8,21,22,23,24,25,26,27), photoallergic dermatitis (6,7,9,10,12,13,16,18,19,31,33,34,35,36,39,40,41,42), acrovesicular dermatitis (dyshidrosiform dermatitis) from photoallergy (6,16,19,31), acrovesicular dermatitis (dyshidrosiform dermatitis) (17), erythema multiforme (32), and an unspecified drug eruption (30).

Fixed drug eruption

General

Fixed drug eruptions to piroxicam have been reported occasionally, mostly as single case reports (table 3.392.1), a few as case series of 6 (3), 9 (2), and 7 (21). However, the latter 2 studies were partly overlapping (presenting some patients twice) and the cases were collected in a period of many years. With two exceptions (24,26), all cases were (probably) pigmenting FDEs. While positive on postlesional skin, patch tests were always negative on non-involved skin. A few cross-reactions have been observed to other oxicams: tenoxicam, meloxicam, and droxicam (see the section 'Cross-reactions, pseudo-cross-reactions and co-reactions' below). Rarely, cases of FDE have been reported from topical application of piroxicam in a plaster and gel (28, no patch test performed) and from occupational sensitization to piroxicam in a nurse handling the drug at work (29).

Case series
In Monastir, Tunisia, in a period of 13 years (2004-2017), 16 patients with clinically diagnosed fixed drug eruptions suspected to have been caused by NSAIDs were investigated, of who 7 had used piroxicam. The lesions were well demarcated, between 1 and 5 cm in size, and healed with residual hyperpigmentation. Five patients had 2-4 lesions (multiple FDE) and two a solitary lesion. Two individuals had bullous FDE. The arms were affected in five cases, followed by the face and the trunk. Six patients had suffered 2-4 FDE episodes. The time between drug intake and cutaneous symptoms averaged 2 days (2 hours to 3 days). Six patients were patch tested with piroxicam crushed tablets 10% pet. and all were positive on postlesional skin and negative on non-lesional test sites. A cross-reaction to meloxicam was seen in one patient only, but 2 of 6 who had an oral provocation test with meloxicam had a recurrence (from cross-reactivity) (3).

In a University hospital in Coimbra, Portugal, in the period 1990-2009, 52 patients (17 men, 35 women, mean age 53±17 years) with a clinical diagnosis of fixed drug eruptions were submitted to patch tests with the suspected drugs. Patch tests on pigmented lesions were positive in 21 of the 52 (40%) patients, 20 NSAIDs and one antihistamine. Of 23 individuals tested with piroxicam 1% and 10% pet., 9 (39%) had a positive patch test on postlesional skin. The patients were also tested with related NSAIDs tested at 5% and 10% pet. and there were 8 cross-reactions to tenoxicam and 2 to meloxicam (2, overlap with refs. 21 and 33)

Also in Coimbra, eight patients with fixed drug eruption with high probability caused by piroxicam were investigated. One of these individuals suffered lesion reactivation after intravenous tenoxicam (systemic allergic dermatitis). Patch tests with piroxicam were positive in 7/8 cases, only on postlesional skin. These 7 patients also had positive tests to tenoxicam, and 1 out of 5 reacted to meloxicam (21, overlap with refs. 2 and 33).

Again in Coimbra, Portugal, between 1984 and 2008, 7 patients with FDE from piroxicam were patch tested with thimerosal and thiosalicylic on normal skin and with piroxicam, tenoxicam and meloxicam on normal and postlesional skin. None of the 7 presented positive tests to thimerosal or thiosalicylic acid. Patch tests in previously affected skin were positive to piroxicam in all patients, to tenoxicam in 6/7 patients (86%) and to meloxicam in 1 of 6 patients tested (17%) (33, presented as Summary, overlap with refs. 2 and 21).

In France, in the period 2005-2007, 59 cases of fixed drug eruptions were collected in 17 academic centers. There was one case of FDE to piroxicam with a positive patch test. Clinical details were not provided (22).

Case reports
Summaries of case reports of fixed drug eruptions to piroxicam are given in table 3.392.1.

Table 3.392.1 Summaries of case reports of fixed drug eruptions from piroxicam

Year and country	Sex	Age	Latency [a]	Nr. Lesions	Special clinical features	PT conc./vehicle	Ref.
2004 Spain	F	46	NM	1		1% pet.	23
	F	42	NM	>1		1% pet.	23
2003 Spain	M	49	2 days	Numerous	Vesicles, ulcers on glans penis, mouth, groins, intergluteal; NP	1% pet.	24
2002 Japan	M	48	NM	Multiple		0.5% pet.	25
1999 Portugal	F	55	1 day	Multiple	More lesions with each attack	1% and 5% pet.	5
1998 Portugal	?	?	NM	NM		0.5% and 1% pet.	7
1995 Spain	F	45	NM	Multiple	Multiple bullous FDE	CP 10% DMSO	4
1993 Spain	F	57	8-10 h	2 of more		1% pet.	8
1990 Finland	F	38	1 hour	30	Multiple bullous FDE	10% pet., alc. & DMSO	26
1990 Spain	F	20	2 hours	4 or more		0.5% (vehicle?)	27

[a] Time between intake of piroxicam and emergence of fixed drug eruption
CP: Commercial preparation; DMSO: Dimethyl sulfoxide; FDE: Fixed drug eruption; NM: Not mentioned; NP: Non-pigmenting fixed drug eruption; PT conc.: Patch test concentration

Photosensitivity

General
Piroxicam is one of the systemic drugs most frequently reported to cause photosensitivity reactions. Many patients who are contact sensitized to the thiosalicylic acid moiety of the preservative thimerosal develop a *photo*allergic eruption after taking oral piroxicam and most – if not all (33,35) – patients showing a photosensitivity reaction to oral piroxicam have positive patch tests to thimerosal and/or thiosalicylic acid (6,7,9,10,11,12,13,15,33,35,38). It is assumed that there is a cross-reaction between thiosalicylic acid and a photodegradation product of piroxicam,

which has been confirmed in animal experiments (37). The nature of this chemical, which is formed during irradiation of piroxicam with UVA (36), is unknown. The cross-reactivity implies that photoallergy to piroxicam can be problematic especially in countries where piroxicam is widely prescribed, with a high sensitization rate to thimerosal and a sunny climate. By far, most cases have been reported from Portugal (7,9,10,13,33,35) and – to a lesser degree – from Spain (12,19,36,40); a few cases have originated from other countries including South Korea (16,34), New Zealand (6), Sweden (42) and possibly the United States (41).

The cross-reactivity with thiosalicylic acid also explains why photoallergic eruptions of pruritic erythematous, edematous and vesicular lesions on sun-exposed skin, sometimes associated with dyshidrosiform hand dermatitis (6,16,19,31) usually begin within 1-4 days after the first intake of piroxicam, without previous sensitization from prior use of the drug (6,7,39). In most patients, lesions can be reproduced by a photopatch test with piroxicam at 0.5, 1.0 or 5.0% in petrolatum irradiated with 5 to 10 J/cm^2 of UVA (9) or from a patch test with pre-irradiated piroxicam (36). However, piroxicam also has – as most drugs causing photoallergy – phototoxic properties and differentiation between photoallergic and phototoxic photopatch tests may be difficult (14,34,39).

Case series

In Coimbra, Portugal, between 1984 and 2008, 85 patients with photosensitivity to systemic piroxicam (51 men, 34 women, mean age 50.4 years) were patch tested with thimerosal and thiosalicylic and 82/85 photopatch tested with piroxicam, tenoxicam and meloxicam. All 85 patients had positive patch tests to thimerosal and thiosalicylic acid and 75/82 (91%) had positive photopatch tests to piroxicam; there was only one photocross-reaction to meloxicam (33). Nine of these patients had been presented in detail 19 years previously. At that time, the authors could not yet decide on whether the photosensitivity reactions to piroxicam were phototoxic or photoallergic in nature (39).

In Porto, another center in Portugal, 15 patients (8 women, mean age 47 years) with photosensitivity to piroxicam seen between January 1996 and September 1997, were investigated. Nine had used piroxicam tablets and 6 piroxicam-β-cyclodextrine. All had positive patch tests to thimerosal and thiosalicylic acid and photopatch tests to piroxicam 0.5%, 1% and the drug 'as is', irradiated with UVA 5 J/cm^2 (35).

Another hospital in Porto in the same year reported 11 similar cases of photosensitivity to piroxicam observed in a 5-year period, which had appeared within 1-5 says of starting piroxicam. They all reacted to photopatch tests with piroxicam 1% and 5% pet irradiated with UVA 5 J/cm^2 and 10/11 had contact allergy to thimerosal (7).

In a hospital in Auckland, New Zealand, in a 6-month period, 11 patients (7 women) had presented with a photosensitive dermatitis while on piroxicam. Eight had developed the rash within 3 days of starting piroxicam. All had erythematous papulovesicular or vesiculo-bullous eruptions on the face and sometimes other light-exposed sites, while 6 also had a pronounced vesicular hand dermatitis involving both the dorsal and the palmar surfaces. Despite therapy with prednisone, complete healing took 2-6 weeks. Three developed persistent light reactions and 2 persistent – predominantly palmar – vesicular hand dermatitis. Six patients were patch and photopatch tested and all had positive photopatch tests to piroxicam 2, 5 and 20% pet. irradiated with 10 J/cm^2 of UVA and positive patch tests to thimerosal (the latter on 2 occasions). Four only agreed to patch testing and – quite curiously – these are all negative to thimerosal (6).

Other case series of piroxicam-photosensitive patients who had positive photopatch tests were reported in refs. 9 and 13 (each n=9, Portugal), 36 (n=7, Spain), 34 (n=2, South Korea), 10 (n=2, Portugal) and 12 (n=2, Spain). In the various studies from Portugal, there may have been some overlap, i.e. that some patients were reported more than once.

Case reports

A 42-year-old man had developed a skin rash after having taken oral piroxicam for 5 days. Physical examination showed slightly scaly, erythematous swollen patches on the face and the dorsa of his hands and feet. In addition, there were vesicular eruptions in the finger webs and palms. Photopatch tests with piroxicam 1%, 0.1% and 0.01% alc. irradiated with UVA 10 J/cm^2 were positive to the highest concentration only. During intake of the drug, the MED for UVA was strongly lowered (from 30 to 2 J/cm^2) (16).

Other single such cases were reported from Spain (19,40), Sweden (42) and possibly from the United Sates (41). A report of one or more patients with photoallergy to piroxicam, details of which are unavailable to the author, can be found in ref. 18.

Other cutaneous adverse drug reaction

In Finland, in the period 1989-2001, 826 patients with suspected cutaneous drug eruptions were patch tested and 89 had one or more positive reactions. Of these individuals, one reacted to piroxicam. Clinical details were not provided and it is unknown which type of drug eruption piroxicam had induced (30).

A 39-year-old man had 2 episodes of the appearance of itchy vesicles and slight erythema on the lateral aspects of the fingers and the backs of the hands and fingers, and the second time also on the palms, after taking oral piroxicam. Patch tests were positive to thimerosal, piroxicam (concentration and vehicle not mentioned) and to piroxicam 0.5% gel. The patient was diagnosed with dyshidrosiform dermatitis from delayed-type allergy to piroxicam. It was uncertain whether the positive reaction to thimerosal (which is strongly associated with photoallergy to piroxicam), was coincidental (17). This case appears to be unique, but dyshidrosiform dermatitis associated with *photo*allergic eruptions from oral piroxicam has been described in several patients (6,16,19,31).

During treatment with piroxicam, a 31-year-old man developed 'the target lesions of erythema multiforme' on his palms and soles and a pruritic erythematous circular lesion on his penis. The lesions healed without hyperpigmentation. One year later, 1 day after again having taken one tablet of piroxicam, similar mucocutaneous lesions appeared in the same locations. Patch tests were performed on the patient's back with piroxicam, meloxicam and tenoxicam 10% in dimethyl sulfoxide, and were positive to piroxicam only. The patient was diagnosed with erythema multiforme from allergy to piroxicam (32). This author considers fixed drug eruption to be more likely based on the localization of the eruption and the 2 episodes with lesions appearing at the same locations. One may argue that the positivity of the piroxicam patch test on normal skin speaks against fixed drug eruption, but such positive patch tests have – albeit infrequently – been observed in FDE. Unfortunately, patch tests have not been performed on post-lesional skin in this case.

Cross-reactions, pseudo-cross-reactions and co-reactions

Patients who have a fixed drug eruption to oral piroxicam and a positive patch test on postlesional skin may show cross-reactions to tenoxicam (2,4,5,8,21,33), meloxicam (2,3,21,33), or droxicam (4,8). Many patients who are sensitized to thiosalicylic acid (one of the components of thimerosal) develop a *photo*allergic eruption after taking oral piroxicam and most patients showing a photosensitivity reaction to oral piroxicam have positive patch tests to thimerosal (see the section 'Photosensitivity, general' above). Cross-reactivity to other 'oxicams' is exceptional in photosensitivity (33,36). The difference between cross-reactivity in FDE and in photosensitivity may be explained by the fact that in these two patterns of drug reactions, the allergenic moieties of piroxicam involved are different (33). Nevertheless, also in photosensitivity, photocross-reactions have occasionally been observed to tenoxicam (19,33), droxicam (19) and meloxicam (19).

LITERATURE

1   The data in the section 'General' may have been obtained from literature discussed in this chapter, but mostly also or exclusively from one or more of the following online sources: ChemIDPlus Advanced, PubChem, DrugBank, RxList, Drug Central, Drugs.com, and Wikipedia

2   Andrade P, Brinca A, Gonçalo M. Patch testing in fixed drug eruptions – a 20-year review. Contact Dermatitis 2011;65:195-201

3   Ben Romdhane H, Ammar H, BenFadhel N, Chadli Z, Ben Fredj N, Boughattas NA. Piroxicam-induced fixed drug eruption: Cross-reactivity with meloxicam. Contact Dermatitis 2019;81:24-26

4   Ordoqui E, De Barrio M, Rodríguez VM, Herrero T, Gil PJ, Baeza ML. Cross-sensitivity among oxicams in piroxicam-caused fixed drug eruption: two case reports. Allergy 1995;50:741-744

5   Oliveira HS, Gonçalo M, Reis JP, Figueiredo A. Fixed drug eruption to piroxicam. Positive patch tests with cross-sensitivity to tenoxicam. J Dermatol Treat 1999;10:209-212

6   McKerrow KJ, Greig DE. Piroxicam-induced photosensitive dermatitis. J Am Acad Dermatol 1986;15:1237-1241

7   Vasconcelos C, Magina S, Quirino P, Barros MA, Mesquita-Guimarães J. Cutaneous drug reactions to piroxicam. Contact Dermatitis 1998;39:145

8   Gastaminza G, Echechipía S, Navarro JA, Fernández de Corrés L. Fixed drug eruption from piroxicam. Contact Dermatitis 1993;28:43-44

9   Gonçalo M, Figueiredo A, Tavares P, Ribeiro CA, Teixeira F, Baptista AP. Photosensitivity to piroxicam: absence of cross-reaction with tenoxicam. Contact Dermatitis 1992;27:287-290

10 De Castro JL, Freitas JP, Brandão FM, Themido R. Sensitivity to thimerosal and photosensitivity to piroxicam. Contact Dermatitis 1991;24:187-192

11 Serrano G, Bonillo J, Aliaga A, Cuadra J, Pujol C, Pelufo C, et al. Piroxicam-induced photosensitivity and contact sensitivity to thiosalicylic acid. J Am Acad Dermatol 1990;23(3Pt.1):479-483

12 De la Cuadra J, Pujol C, Aliaga A. Clinical evidence of cross-sensitivity between thiosalicylic acid, a contact allergen, and piroxicam, a photoallergen. Contact Dermatitis 1989;21:349-351

13 Cardoso J, Canelas MM, Gonçalo M, Figueiredo A. Photopatch testing with an extended series of photoallergens: a 5-year study. Contact Dermatitis 2009;60:325-329

14 Pigatto PD, Guzzi G, Schena D, Guarrera M, Foti C, Francalanci, S, Cristaudo A, et al. Photopatch tests: an Italian multicentre study from 2004 to 2006. Contact Dermatitis 2008;59:103-108

15 Diaz RL, Gardeazabal J, Manrique P, Ratón JA, Urrutia I, Rodríguez-Sasiain JM, Aguirre C. Greater allergenicity of topical ketoprofen in contact dermatitis confirmed by use. Contact Dermatitis 2006;54:239-243

16 Youn JI, Lee HG, Yeo UC, Lee YS. Piroxicam photosensitivity associated with vesicular hand dermatitis. Clin Exp Dermatol 1993;18:52-54

17 Piqué E, Pérez JA, Benjumeda A. Oral piroxicam-induced dyshidrosiform dermatitis. Contact Dermatitis 2004;50:382-383

18 Erdmann S, Sachs B, Merk HF. Photosensibilisierung durch Piroxicam. Z Hautkr 2001;76:180-182 (Article in German)

19 Trujillo MJ, de Barrio M, Rodríguez A, Moreno-Zazo M, Sánchez I, Pelta R, et al. Piroxicam-induced photodermatitis. Cross-reactivity among oxicams. A case report. Allergol Immunopathol (Madr) 2001;29:133-136

20 De Groot AC. Monographs in contact allergy, volume 3: Topical Drugs. Boca Raton, Fl, USA: CRC Press Taylor and Francis Group, 2021 (ISBN 978-0-367-23693-9)

21 Gonçalo M, Oliveira HS, Fernandes B, Robalo-Cordeiro M, Figueiredo A. Topical provocation in fixed drug eruption from nonsteroidal anti-inflammatory drugs. Exogenous Dermatol 2002;1:81-86

22 Brahimi N, Routier E, Raison-Peyron N, Tronquoy AF, Pouget-Jasson C, Amarger S, et al. A three-year-analysis of fixed drug eruptions in hospital settings in France. Eur J Dermatol 2010;20:461-464

23 Cuerda Galindo E, Goday Buján JJ, García Silva JM, Martínez W, Verea Hernando M, Fonseca E. Fixed drug eruption from piroxicam. J Eur Acad Dermatol Venereol 2004;18:586-587

24 Montoro J, Dıaz M, Genıs C, Lozano A, Bertomeu F. Nonpigmenting cutaneous-mucosal fixed drug eruption due to piroxicam. Allergol Immunopathol (Madr) 2003;31:53-55

25 Tanaka S. Fixed drug eruption from piroxicam with positive lesional patch test. Contact Dermatitis 2002;46:174

26 Stubb S, Reitamo S. Fixed drug eruption caused by piroxicam. J Am Acad Dermatol 1990;22(6Pt.1):1111-1112

27 De la Hoz B, Soria C, Fraj J, Losada E, Ledo A. Fixed drug eruption due to piroxicam. Int J Dermatol 1990;29:672-673

28 Rho YK, Yoo KH, Kim BJ, Kim MN, Song KY. A case of generalized fixed drug eruption due to a piroxicam plaster. Clin Exp Dermatol 2010;35:204-205

29 Lamchahab F, Baeck M. Occupationally induced fixed drug eruption caused by a non-steroidal anti-inflammatory agent. Contact Dermatitis 2012;67:176-177

30 Lammintausta K, Kortekangas-Savolainen O. The usefulness of skin tests to prove drug hypersensitivity. Br J Dermatol 2005;152:968-974

31 Braunstein BL. Dyshidrotic eczema associated with piroxicam photosensitivity. Cutis 1985;35:485-486

32 Prieto A, De Barrio M, Pérez C, Velloso A, Baeza ML, Herrero T. Piroxicam-induced erythema multiforme. Contact Dermatitis 2004;50:263

33 Serra D, Gonçalo M, Figueiredo A. Two decades of cutaneous adverse drug reactions from piroxicam. Contact Dermatitis 2008;58(Suppl.1):35

34 Lee AY, Joo HJ, Chey WY, Kim YG. Photopatch testing in seven cases of photosensitive drug eruptions. Ann Pharmacother 2001;35:1584-1587

35 Varela P, Amorim I, Massa A, Sanches M, Silva E. Piroxicam-beta-cyclodextrin and photosensitivity reactions. Contact Dermatitis 1998;38:229

36 Serrano G, Fortea JM, Latasa JM, SanMartin O, Bonillo J, Miranda MA. Oxicam-induced photosensitivity. Patch and photopatch testing studies with tenoxicam and piroxicam photoproducts in normal subjects and in piroxicam-droxicam photosensitive patients. J Am Acad Dermatol 1992;26:545-548

37 Ikezawa Z, Kitamura K, Osawa J, Hariya T. Photosensitivity to piroxicam is induced by sensitization to thimerosal and thiosalicylate. J Invest Dermatol 1992;98:918-922

38 Cirne de Castro JL, Vale E, Martins M. Mechanism of photosensitive reactions induced by piroxicam. J Am Acad Dermatol 1989;20:706

39 Figueiredo A, Ribeiro CA, Gonçalo S, Caldeira MM, Poiares-Baptista A, Teixeira F. Piroxicam-induced photosensitivity. Contact Dermatitis 1987;17:73-79

40 Serrano G, Bonillo J, Aliaga A, Gargallo E, Pelufo C. Piroxicam-induced photosensitivity. In vivo and in vitro studies of its photosensitizing potential. J Am Acad Dermatol 1984;11:113-120

41 Curtis PF. Presented at the symposium on Gross and Microscopic Dermatology, 42nd Annual Meeting of the American Academy of Dermatology, Chicago, December 1-6, 1983 (cited in ref. 40)

42 Fjellner B. Photosensitivity induced by piroxicam. Acta Derm Venereol 1983;63:557-558

Chapter 3.393 PIROXICAM BETADEX

IDENTIFICATION

Description/definition : Piroxicam betadex is the combination of piroxicam with β-cyclodextrin complex that conforms to the structural formula shown below

Pharmacological classes : Anti-inflammatory agents, non-steroidal

IUPAC name : (1S,3R,5R,6S,8R,10R,11S,13R,15R,16S,18R,20R,21S,23R,25R,26S,28R,30R,31S,33R, 35R,36R,37R,38R,39R,40R,41R,42R,43R,44R,45R,47R,48R,49R)-5,10,15,20,25,30,35-Heptakis(hydroxymethyl)-2,4,7,9,12,14,17,19,22,24,27,29,32,34-tetradecaoxaocta-cyclo[31.2.2^{3,6}.2^{8,11}.2^{13,16}.2^{18,21}.2^{23,26}.2^{28,31}]nonatetracontane-36,37,38,39,40,41,42,43, 44,45,46,47,48,49-tetradecol;4-hydroxy-2-methyl-1,1-dioxo-N-pyridin-2-yl-1λ^6,2-benzo-thiazine-3-carboxamide

Other names : Piroxicam-βcyclodextrin complex

CAS registry number : 96684-40-1

EC number : Not available

Patch testing : Test with piroxicam 1.0% pet. (Chemotechnique, SmartPracticeCanada)

Molecular formula : $C_{240}H_{376}N_6O_{183}S_2$

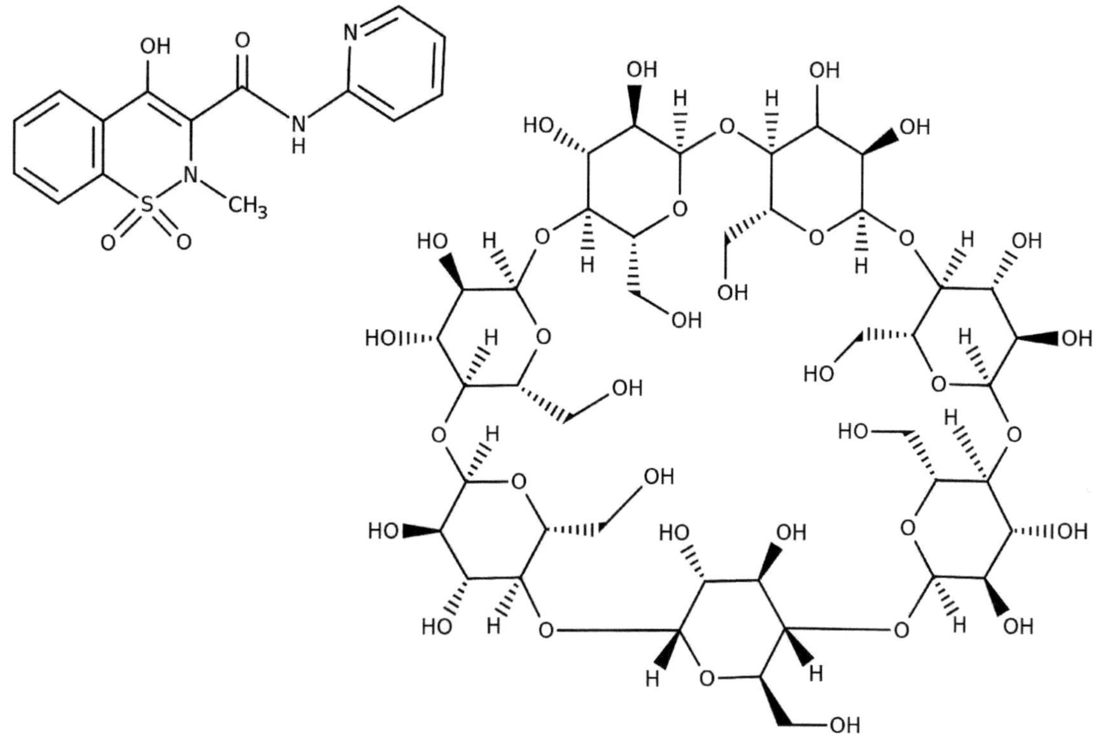

GENERAL

Piroxicam betadex is a complex of piroxicam and an inert cyclic macromolecule, β-cyclodextrin (Betadex, CAS number 7585-39-9). This complexation increases piroxicam's rate of dissolution and solubility, improving the rate of absorption of piroxicam. The clinical advantage of the drug over piroxicam is a faster onset of action, which is advantageous when rapid analgesia is required. However, the bioavailability of piroxicam is no different from that of piroxicam tablets. This drug is an NSAID indicated for treatment of osteoarthritis and rheumatoid arthritis, musculoskeletal disorders, dysmenorrhea and postoperative pain (1).

CUTANEOUS ADVERSE DRUG REACTIONS FROM SYSTEMIC ADMINISTRATION CAUSED BY TYPE IV (DELAYED-TYPE) HYPERSENSITIVITY (as demonstrated by positive patch tests)

Cutaneous adverse drug reactions from systemic administration of piroxicam betadex caused by type IV (delayed-type) hypersensitivity have included photosensitivity (2).

Photosensitivity

In a hospital in Portugal, 15 patients were observed with photosensitivity reactions to piroxicam between January 1996 and September 1997. Six patients had used Piroxicam betadex. All patients gave positive patch tests to thimerosal and thiosalicylic acid (which is nearly always the case with piroxicam photosensitivity), and positive photopatch tests to piroxicam 0.5% and 1% pet., irradiated with 5 J/cm^2 UVA and the drug 'as is'. The authors noted that there is a clear disproportion between the piroxicam betadex photosensitivity cases (6 out of 15) and the percentage of sales of this drug (17%) (significant difference). They also proposed that piroxicam use, especially in its new formulation piroxicam betadex, should be avoided in Portugal, and other countries with high prevalences of sensitization to thimerosal (2).

The conclusion and suggestion of the authors of this report was shortly thereafter unconvincingly contradicted by the manufacturer of piroxicam betadex in a Letter to the editor (3), followed by a clear and convincing rebuttal by the authors (4).

Cross-reactions, pseudo-cross-reactions and co-reactions
Pseudo-photocross-reactions to piroxicam are to be expected in all cases of photosensitization.

LITERATURE
1   The data in the section 'General' may have been obtained from literature discussed in this chapter, but mostly also or exclusively from one or more of the following online sources: ChemIDPlus Advanced, PubChem, DrugBank, RxList, Drug Central, Drugs.com, and Wikipedia
2   Varela P, Amorim I, Massa A, Sanches M, Silva E. Piroxicam-beta-cyclodextrin and photosensitivity reactions. Contact Dermatitis 1998;38:229
3   Umile A. Piroxicam-β-cyclodextrin and photosensitivity reactions. Contact Dermatitis 1999;40:340-341
4   Varela P. Reply. Piroxicam-β-cyclodextrin and photosensitivity reactions. Contact Dermatitis 1999;40:340-341

Chapter 3.394 PIVAMPICILLIN

IDENTIFICATION

Description/definition	: Pivampicillin is the pivalate ester analog of ampicillin that conforms to the structural formula shown below
Pharmacological classes	: Anti-bacterial agents
IUPAC name	: 2,2-Dimethylpropanoyloxymethyl (2S,5R,6R)-6-[[(2R)-2-amino-2-phenylacetyl]amino]-3,3-dimethyl-7-oxo-4-thia-1-azabicyclo[3.2.0]heptane-2-carboxylate
Other names	: Ampicillin pivaloyloxymethyl ester; pivaloyloxymethyl ampicillinate
CAS registry number	: 33817-20-8
EC number	: 251-688-6
Merck Index monograph	: 8896
Patch testing	: 5% pet. (6)
Molecular formula	: $C_{22}H_{29}N_3O_6S$

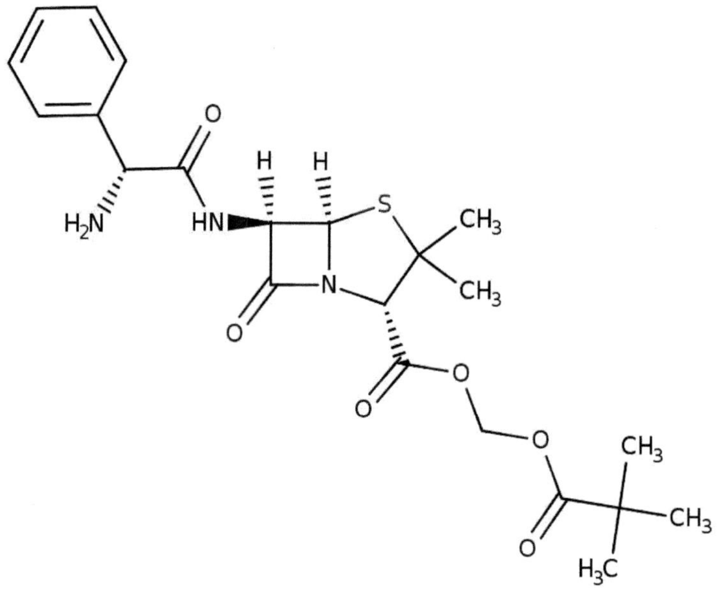

GENERAL

Pivampicillin is the pivaloyloxymethyl ester of the semi-synthetic penicillin ampicillin. It is an inactive pro-drug, which is converted during its absorption from the gastrointestinal tract to the microbiologically active ampicillin, together with formaldehyde and pivalic acid, by non-specific esterases present in most body tissues. Pivampicillin is indicated for the treatment of respiratory tract, ear nose and throat, gynecological and urinary tract infections, when caused by non-penicillinase-producing susceptible strains of bacteria (1). The classification and structures of beta-lactam antibiotics are discussed in Chapter 3.36 Amoxicillin.

CONTACT ALLERGY FROM ACCIDENTAL CONTACT

In Denmark, before 1986, 45 people working in a Danish factory producing the semisynthetic beta-lactam antibiotics pivampicillin and pivmecillinam (amdinocillin pivoxil) developed dermatitis, mainly on the hands, arms, calves and face. Nineteen of these patients also had hay fever (n=17) and/or asthma (n=5), probably provoked by contact with airborne penicillin; the factory area was highly contaminated with penicillin dust. Nearly half of the workers developed symptoms of sensitization between 1 and 4 weeks after first exposure. Patch tests were performed with the following antibiotics (all in petrolatum): pivampicillin base and HCl 1% and 5%, pivmecillinam base and HCl 1% and 10%, pivampicillin and pivmecillinam combined (both 0.5% and 2.5%), ampicillin sodium 1% and 5%, and various other penicillins. There were 31 positive reactions to pivampicillin and 27 to pivmecillinam; 18 reacted to both antibiotics. Cross-reactions to ampicillin were observed in 23 (74%) to pivampicillin (2). The – rather complicated – results of retesting these patients were reported one year later (5).

CUTANEOUS ADVERSE DRUG REACTIONS FROM SYSTEMIC ADMINISTRATION CAUSED BY TYPE IV (DELAYED-TYPE) HYPERSENSITIVITY (as demonstrated by positive patch tests)

Cutaneous adverse drug reactions from systemic administration of pivampicillin caused by type IV (delayed-type) hypersensitivity have included symmetrical drug-related intertriginous and flexural exanthema (SDRIFE)/baboon syndrome (3).

Symmetrical drug-related intertriginous and flexural exanthema (SDRIFE)/Baboon syndrome
A man aged 25 developed SDRIFE one day after taking pivampicillin. After resolution, a patch test was positive to ampicillin, which is the active part of pivampicillin, metabolized from this pro-drug in the gastro-intestinal tract (3, no details available to the author).

Cross-reactions, pseudo-cross-reactions and co-reactions
Cross-reactions between beta-lactam antibiotics are discussed in Chapter 3.36 Amoxicillin. Patients sensitized to pivampicillin often cross-react – or actually pseudocross-react – to ampicillin (2). Theoretically, pivampicillin and amdinocillin pivoxil may cross-react, as they have an identical side chain (2).

LITERATURE

1   The data in the section 'General' may have been obtained from literature discussed in this chapter, but mostly also or exclusively from one or more of the following online sources: ChemIDPlus Advanced, PubChem, DrugBank, RxList, Drug Central, Drugs.com, and Wikipedia
2   Møller NE, Nielsen B, von Würden K. Contact dermatitis to semisynthetic penicillins in factory workers. Contact Dermatitis 1986;14:307-311
3   Rasmussen LP, Menné T. Systemic contact eczema in ampicillin allergy. Ugeskr Laeger 1995;147:1341-1342 (Article in Danish, data cited in ref. 4)
4   Häusermann P, Harr T, Bircher AJ. Baboon syndrome resulting from systemic drugs: Is there strife between SDRIFE and allergic contact dermatitis syndrome? Contact Dermatitis 2004;51:297-310
5   Møller NE, Jeppesen K. Patch testing with semisynthetic penicillins. Contact Dermatitis 1987;16:227-228
6   Møller NE, von Würden K. Hypersensitivity to semisynthetic penicillins and cross-reactivity with penicillin. Contact Dermatitis 1992;26:351-352

Chapter 3.395 POTASSIUM AMINOBENZOATE

IDENTIFICATION

Description/definition : Potassium aminobenzoate is the potassium salt form of aminobenzoic acid that conforms to the structural formula shown below
Pharmacological classes : Dermatologicals
IUPAC name : Potassium;4-aminobenzoate
Other names : Potassium *p*-aminobenzoate; Potaba ®; aminobenzoate potassium
CAS registry number : 138-84-1
EC number : 205-338-4
Merck Index monograph : 8997
Patch testing : *p*-Aminobenzoic acid (PABA) 10% pet.; most systemic drugs can be tested at 10% pet.; if the pure chemical is not available, prepare the test material from intravenous powder, the content of capsules or – when also not available – from powdered tablets to achieve a final concentration of the active drug of 10% pet.
Molecular formula : $C_7H_6KNO_2$

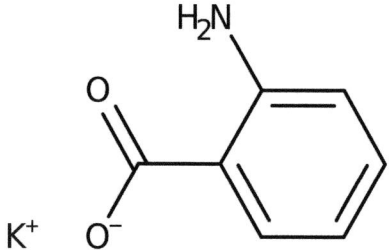

GENERAL

Potassium aminobenzoate is the potassium salt form of aminobenzoic acid with anti-inflammatory and antifibrotic activities. It increases oxygen uptake at the tissue level and may enhance monoamine oxidase (MAO) activity, which requires oxygen as a substrate. Enhanced MAO activity may be accountable for the prevention or regression of fibrosis. Potassium aminobenzoate is used to treat Peyronie's disease, systemic sclerosis, morphea and dermatomyositis. The supply of this medication has been discontinued in 2019 (http://www.glenwood-llc.com/pre-potaba.html) (1).

CUTANEOUS ADVERSE DRUG REACTIONS FROM SYSTEMIC ADMINISTRATION CAUSED BY TYPE IV (DELAYED-TYPE) HYPERSENSITIVITY (as demonstrated by positive patch tests)

Cutaneous adverse drug reactions from systemic administration of potassium aminobenzoate caused by type IV (delayed-type) hypersensitivity have included drug reaction with eosinophilia and systemic symptoms (DRESS) (2) and photosensitivity (3).

Drug reaction with eosinophilia and systemic symptoms (DRESS)

Two weeks after starting treatment with potassium *p*-aminobenzoate for induratio penis plastica, a 73-year-old man presented with a generalized rash that had developed during the previous 3 days. Physical examination showed a generalized maculopapular exanthema. Mucous membrane involvement, fever, or lymph node enlargement was not present. The results of the blood cell count were normal except for eosinophilia, and liver enzymes were elevated. Histology showed superficial perivascular lymphohistiocytic infiltrate with scattered atypical lymphoid cells, interstitial eosinophils, and mast cells, which was consistent with a drug reaction. With systemic and topical corticosteroids, the rash rapidly subsided, but the liver enzymes were still increasing up to 20-fold; other causes were excluded. Two months later, a patch test was performed with the contents of a potassium *p*-aminobenzoate capsule 50% pet., revealing a strongly positive reaction at D2 (++) and D3 (+++) with multiple papulovesicles (2). This was a case of drug reaction with eosinophilia and systemic symptoms (DRESS) from potassium *p*-aminobenzoate.

Photosensitivity

After having been treated with potassium aminobenzoate (Potaba ®) for 7 weeks for progressive penis deviation due to Peyronie's disease, a 51-year-old man developed a pruritic eczematous rash confined to air- and light-exposed areas of the skin after a week on holiday in the Austrian Alps in late October, seizing the last rays of autumn sun. Physical examination revealed a pruritic eczematous eruption confined to air- and light-exposed areas including the face but without sparing retroauricular and submental areas, the V of the neck and the dorsal hands and forearms. Numerous pustules were found on the cheeks and forehead, whereas small blisters occurred on the arms. Patch tests were negative, photopatch tests positive to a pulverized Potaba tablet suspended with 1 ml saline and *p*-aminobenzoic acid (PABA) 10% pet. The patient was diagnosed with photoallergic dermatitis to PABA in potassium aminobenzoate tablets (3).

LITERATURE

1   The data in the section 'General' may have been obtained from literature discussed in this chapter, but mostly also or exclusively from one or more of the following online sources: ChemIDPlus Advanced, PubChem, DrugBank, RxList, Drug Central, Drugs.com, and Wikipedia

2   Viehweg A, Stein A, Bauer A, Spornraft-Ragaller P. Potassium-paraaminobenzoic acid (Potaba®)-associated DRESS syndrome. Dermatitis 2013;24:257-258

3   Stoevesandt J, Kürzinger N, Bröcker EB, Trautmann A. Uro-dermatological problems of a construction worker: paraaminobenzoic acid as a systemic photosensitizer. Eur J Dermatol 2010;20:217-219

Chapter 3.396 PRACTOLOL

IDENTIFICATION

Description/definition : Practolol is the acetanilide that conforms to the structural formula shown below
Pharmacological classes : Anti-arrhythmia agents; adrenergic β1 receptor antagonists
IUPAC name : *N*-[4-[2-Hydroxy-3-(propan-2-ylamino)propoxy]phenyl]acetamide
CAS registry number : 6673-35-4
EC number : 229-712-1
Merck Index monograph : 9088
Patch testing : 5% and 10% pet.
Molecular formula : $C_{14}H_{22}N_2O_3$

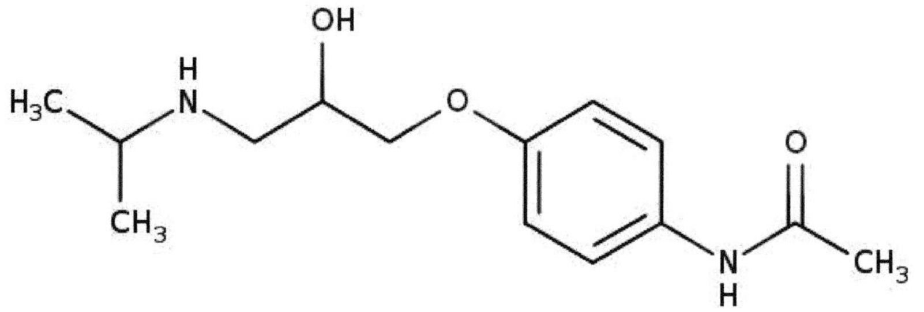

GENERAL

Practolol is a β_1-adrenergic antagonist and selective beta-blocker, that has formerly been used in the emergency treatment of cardiac arrythmias. In pharmaceutical products, practolol is employed as practolol hydrochloride (CAS number 6996-43-6, EC number 230-270-7, molecular formula $C_{14}H_{23}ClN_2O_3$) (1).

CUTANEOUS ADVERSE DRUG REACTIONS FROM SYSTEMIC ADMINISTRATION CAUSED BY TYPE IV (DELAYED-TYPE) HYPERSENSITIVITY (as demonstrated by positive patch tests)

Cutaneous adverse drug reactions from systemic administration of practolol caused by type IV (delayed-type) hypersensitivity have included maculopapular eruption on the back (3).

Maculopapular eruption

A 71-year-old man presented with a generalized itchy rash which had developed 2 weeks after starting practolol for angina on effort. Physical examination showed an extensive scaly erythematous maculopapular confluent rash on his back. Abnormal laboratory findings were leukocytosis, neutrophilia and a raised erythrocyte sedimentation rate. Practolol was stopped, and with topical corticosteroid treatment the exanthema resolved in ten days. Three weeks later, patch tests were positive to pure practolol 5% and 10% pet. Ten controls were negative (3).

Cross-reactions, pseudo-cross-reactions and co-reactions

Patients sensitized to topical metoprolol may cross-react to practolol (2).

LITERATURE

1   The data in the section 'General' may have been obtained from literature discussed in this chapter, but mostly also or exclusively from one or more of the following online sources: ChemIDPlus Advanced, PubChem, DrugBank, RxList, Drug Central, Drugs.com, and Wikipedia
2   Van Joost T, Middelkamp Hup J, Ros FE. Dermatitis as a side effect of long-term topical treatment with certain beta-blocking agents. Br J Dermatol 1979;101:171-176
3   Felix RH, Comaish JS. The value of patch tests and other skin tests in drug eruptions. Lancet 1974;1(7865):1017-1019

Chapter 3.397 PRASUGREL

IDENTIFICATION

Description/definition   : Prasugrel is the thienopyridine that conforms to the structural formula shown below
Pharmacological classes   : Platelet aggregation inhibitors
IUPAC name   : [5-[2-Cyclopropyl-1-(2-fluorophenyl)-2-oxoethyl]-6,7-dihydro-4H-thieno[3,2-c]pyridin-2-yl]
                            acetate
CAS registry number   : 150322-43-3
EC number   : Not available
Merck Index monograph   : 9103
Patch testing   : Hydrochloride 1%, 5% and 10% pet.
Molecular formula   : $C_{20}H_{20}FNO_3S$

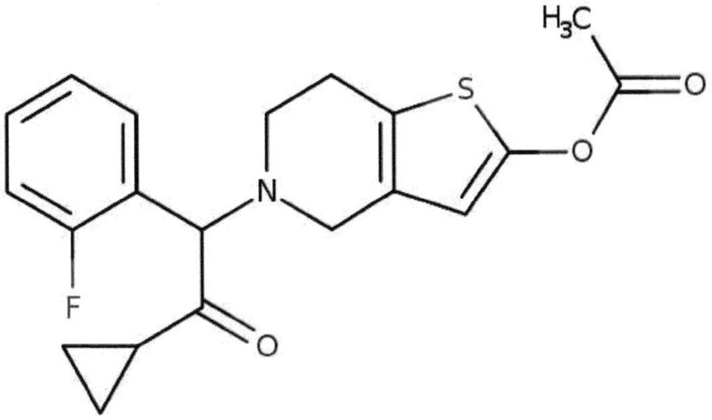

GENERAL

Prasugrel is an orally bioavailable thienopyridine, with antiplatelet activity. It is used to decrease the risk of myocardial infarction and stroke in patients with acute coronary syndromes. In pharmaceutical products, prasugrel is employed as prasugrel besilate (CAS number 952340-40-8, EC number not available, molecular formula $C_{26}H_{26}FNO_6S_2$) or as prasugrel hydrochloride (CAS number 389574-19-0, EC number not available, molecular formula $C_{20}H_{21}ClFNO_3S$) (1).

CUTANEOUS ADVERSE DRUG REACTIONS FROM SYSTEMIC ADMINISTRATION CAUSED BY TYPE IV (DELAYED-TYPE) HYPERSENSITIVITY (as demonstrated by positive patch tests)

Cutaneous adverse drug reactions from systemic administration of prasugrel caused by type IV (delayed-type) hypersensitivity have included erythema multiforme (2).

Other cutaneous adverse drug reactions

An 86-year-old woman with no medical history was hospitalized for acute myocardial infarction and was administered many drugs including prasugrel hydrochloride. Eleven days later, the patient developed itching and subsequent edematous erythema over her entire body. The patch tests revealed positive reactions to prasugrel hydrochloride 1%, 5% and 10% pet. Three controls were negative. Quite curiously, the patient was diagnosed with erythema multiforme, although the description of her skin symptoms revealed no clues to this diagnosis (2).

Cross-reactions, pseudo-cross-reactions and co-reactions

Patients sensitized to clopidogrel may have cross-reacted to ticlopidine and prasugrel (3).

LITERATURE

1   The data in the section 'General' may have been obtained from literature discussed in this chapter, but mostly also or exclusively from one or more of the following online sources: ChemIDPlus Advanced, PubChem, DrugBank, RxList, Drug Central, Drugs.com, and Wikipedia
2   Kawasaki-Nagano M, Tamagawa-Mineoka R, Nakae M, Wakabayashi Y, Nakagawa H, Masuda K, Katoh N. Drug eruption due to prasugrel hydrochloride: A case report and mini-review. J Dermatol 2019;46:e325-e326
3   Cheema AN, Mohammad A, Hong T, Jakubovic HR, Parmar GS, Sharieff W, et al. Characterization of clopidogrel hypersensitivity reactions and management with oral steroids without clopidogrel discontinuation. J Am Coll Cardiol 2011;58:1445-1454

Chapter 3.398 PRAVASTATIN

IDENTIFICATION

Description/definition : Pravastatin is the fungal metabolite isolated from cultures of *Nocardia autotrophica* that conforms to the structural formula shown below

Pharmacological classes : Anticholesteremic agents; hydroxymethylglutaryl-CoA reductase inhibitors

IUPAC name : (3R,5R)-7-[(1S,2S,6S,8S,8aR)-6-Hydroxy-2-methyl-8-[(2S)-2-methylbutanoyl]oxy-1,2,6,7,8,8a-hexahydronaphthalen-1-yl]-3,5-dihydroxyheptanoic acid

CAS registry number : 81093-37-0

EC number : Not available

Merck Index monograph : 9105 (Pravastatin sodium)

Patch testing : Tablet, pulverized, 'as is' (2); most systemic drugs can be tested at 10% pet.; if the pure chemical is not available, prepare the test material from intravenous powder, the content of capsules or – when also not available – from powdered tablets to achieve a final concentration of the active drug of 10% pet.

Molecular formula : $C_{23}H_{36}O_7$

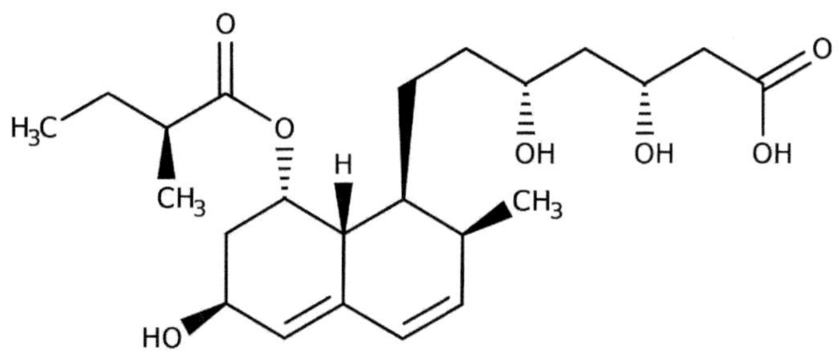

GENERAL

Pravastatin is a fungal metabolite isolated from cultures of *Nocardia autotrophica* with cholesterol-lowering and potential antineoplastic activities This agent lowers plasma cholesterol and lipoprotein levels by competitively inhibiting hepatic hydroxymethyl-glutaryl coenzyme A (HMG-CoA) reductase, the enzyme which catalyzes the conversion of HMG-CoA to mevalonate, a key step in cholesterol synthesis. Pravastatin is indicated for primary prevention of coronary events in hypercholesterolemic patients without clinical evidence of coronary heart disease. In pharmaceutical products, pravastatin is employed as pravastatin sodium (CAS number 81131-70-6, EC number 617-202-5, molecular formula $C_{23}H_{35}NaO_7$) (1).

CUTANEOUS ADVERSE DRUG REACTIONS FROM SYSTEMIC ADMINISTRATION CAUSED BY TYPE IV (DELAYED-TYPE) HYPERSENSITIVITY (as demonstrated by positive patch tests)

Cutaneous adverse drug reactions from systemic administration of pravastatin caused by type IV (delayed-type) hypersensitivity have included an eczematous eruption (2).

Dermatitis/eczematous eruption

A 59-year-old man presented with progressive pruritic nummular eczematous plaques, predominantly on the extensor aspects of the arms and legs, which had developed over a period of several months. He used various medications including pravastatin for 1 year. Histopathological examination showed normal epidermis with minimal exocytosis of lymphocytes, and moderate infiltration of lymphocytes, histiocytes and basophils in the dermis, around superficial vessels. These findings were consistent with a drug reaction. The dermatitis faded after stopping pravastatin. Patch tests with all drugs, pulverized, tested 'as is', revealed a positive (+) reaction to pravastatin at D3. Ten controls were negative. A scratch test with pulverized pravastatin in a drop of saline was positive after 15 minutes. Although ten controls were negative, the authors considered the positive scratch test, indicating type I allergy, not to be indubitably reliable, as the method is not well standardized (2).

LITERATURE

1   The data in the section 'General' may have been obtained from literature discussed in this chapter, but mostly also or exclusively from one or more of the following online sources: ChemIDPlus Advanced, PubChem, DrugBank, RxList, Drug Central, Drugs.com, and Wikipedia

2   De Boer EM, Bruynzeel DP. Allergy to pravastatin. Contact Dermatitis 1994;30:238

Chapter 3.399 PREDNISOLONE

IDENTIFICATION

Description/definition   : Prednisolone is the synthetic glucocorticoid that conforms to the structural formula
                            shown below
Pharmacological classes   : Antineoplastic agents, hormonal; glucocorticoids; anti-inflammatory agents
IUPAC name   : (8S,9S,10R,11S,13S,14S,17R)-11,17-Dihydroxy-17-(2-hydroxyacetyl)-10,13-dimethyl-
                            7,8,9,11,12,14,15,16-octahydro-6H-cyclopenta[a]phenanthren-3-one
Other names   : 11β,17,21-Trihydroxypregna-1,4-diene-3,20-dione
CAS registry number   : 50-24-8
EC number   : 200-021-7
Merck Index monograph   : 9111
Patch testing   : 1% pet. (SmartPracticeCanada, SmartPracticeEurope)
Molecular formula   : $C_{21}H_{28}O_5$

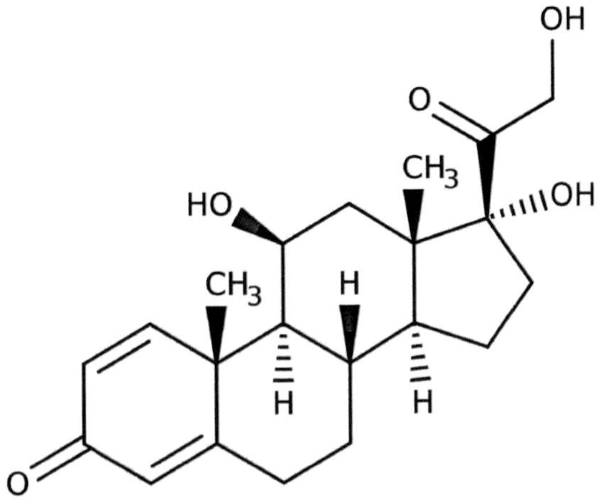

GENERAL

Systemically administered glucocorticoids have anti-inflammatory, immunosuppressive and antineoplastic properties and are used in the treatment of a wide spectrum of diseases including rheumatic disorders, lung diseases (asthma, COPD), gastrointestinal tract disorders (Crohn's disease, colitis ulcerosa), certain malignancies (leukemia, lymphomas), hematological disorders, and various diseases of the kidneys, brain, eyes and skin. A practical guideline for diagnosing allergic reactions to corticosteroids is presented in ref. 1. Prednisolone *base* is virtually only used in tablets. Esters and salts used in other systemic applications include prednisolone acetate (Chapter 3.400), prednisolone sodium succinate (Chapter 3.402), prednisolone hemisuccinate (Chapter 3.401), prednisolone sodium metazoate, prednisolone sodium phosphate, and prednisolone tetrahydrophthalate sodium salt (Chapter 3.403).

As prednisolone *base* is nearly always used as tablet, by far most *contact allergic* reactions to 'prednisolone' must in fact have been the result of sensitization to an ester or salt of prednisolone or of cross-reactivity to another corticosteroid. The literature on this subject has been fully reviewed in Volume 3 of the *Monographs in contact allergy* series (14).

CUTANEOUS ADVERSE DRUG REACTIONS FROM SYSTEMIC ADMINISTRATION CAUSED BY TYPE IV (DELAYED-TYPE) HYPERSENSITIVITY (as demonstrated by positive patch tests)

Cutaneous adverse drug reactions from systemic administration of prednisolone caused by type IV (delayed-type) hypersensitivity have included maculopapular eruption (2,11,12,22), widespread erythema (9,17,21), acute generalized exanthematous pustulosis (AGEP) (7,15), systemic allergic dermatitis (3,4,5,6,10,13,20,23), symmetrical drug-related intertriginous and flexural exanthema (SDRIFE)/baboon syndrome (24), generalized rash (8), pustular exanthema (16), and unknown allergic exanthemas (18,19).

Maculopapular eruption

Ten cases of immediate- and delayed-type allergic reactions to prednisolone were reported from Germany in 1989. These included urticaria, exanthematous reactions and allergic contact dermatitis. Three patients had developed a maculopapular exanthema after systemic prednisolone. Patch tests were positive to prednisolone. More specific data are not available to the author (2).

A 44-year-old woman with breast cancer developed a maculopapular exanthema one day after having initiated treatment with doxorubicin, docetaxel, prednisolone, granisetron, and metoclopramide. Patch tests were positive to tixocortol pivalate 0.1% pet. and commercial methylprednisolone solution 30% water, alc. and pet.; prednisolone itself was not tested. This patient had no previous history of topical corticosteroid sensitivity or any risk factor such as atopic dermatitis or leg ulcers (11).

A man, who had a history of worsening dermatitis from betamethasone valerate cream, developed a maculopapular exanthema on the second day of treatment with oral prednisolone 30 mg daily. Patch tests were positive to betamethasone valerate and various other corticosteroids including hydrocortisone (of the same group as prednisolone), but prednisolone itself was not tested (22).

Another case of maculopapular eruption from allergy to oral prednisolone was reported in 1990 (12).

Erythroderma, widespread erythematous eruption, exfoliative dermatitis
In a 75-year-old woman, polymyalgia rheumatica had been treated with oral prednisolone for 8 days, when widespread dermatitis appeared. In oral provocations, 10 mg prednisolone and 4 mg methylprednisolone sodium succinate caused widespread erythema of the trunk within several hours, lasting about 3 days. The patient had previously been treated with topical corticosteroids without recognizable side effects. Patch tests showed delayed-type allergy to prednisolone and methylprednisolone. Whether the patient was presensitized is unknown (9).

A 36-year-old woman was treated with intravenous methylprednisolone for acute myelitis for 5 days. Three days after the end of treatment she developed a widespread macular exanthem with severe itch. After 2 days with no improvement, she was administered intravenous prednisolone which caused an exacerbation of both the exanthem and the itch. Patch tests were positive to prednisolone 1% pet. and methylprednisolone 1% pet. The patient had probably become sensitized to methylprednisolone from the first 5 day course and subsequently had an exacerbation from intravenously administered cross-reacting prednisolone (17).

A 29 year-old woman was treated with topical betamethasone salt and ester and oral prednisolone for contact dermatitis of the face. The next day the patient presented with a generalized eruption of small erythematous macules, coalescing in some areas, but the facial eruption had already improved somewhat (after one topical application). Patch tests were positive to prednisolone and methylprednisolone. Oral provocation tests resulted in itching and erythematous macules after one day with both corticosteroids, but not with hydrocortisone and betamethasone (21).

Acute generalized exanthematous pustulosis (AGEP)
A 46-year-old woman was given 10 mg prednisolone and 10 mg rabeprazole sodium daily for a liver disorder. The next day, the patient developed erythema on the chest and axillae. However, pustules, scattered red papules and scarlet colored edematous erythema were noted on her whole body. A skin biopsy specimen taken from the pustules demonstrated neutrophilic pustules under the stratum corneum and lymphocytic, eosinophilic and neutrophilic infiltrates in the upper dermis. Laboratory examinations showed no abnormalities and cultures of the pustules were negative. Systemic prednisolone was stopped and her rash disappeared within 3 days. Seventeen months later, patch tests were positive to prednisolone 10% pet., but negative to the excipients of prednisolone tablets The patient was diagnosed with AGEP from delayed-type allergy to prednisolone (15).

A 42-year-old man suffering from perianal eczema was initially treated given topical prednisolone acetate cream. And later with 50 mg prednisolone followed by 100 mg prednisolone tetrahydrophthalate. Some hours after receiving the latter, the patient developed an erythematous rash with subsequent occurrence of numerous non-follicular pustules on the trunk and the extremities over the following days. The exanthem was associated with burning, pruritus as well as malaise and fever (temperature of 38°C). There was leukocytosis with neutrophilia and elevated C-reactive protein (187 mg/l); a bacterial culture from pustules was negative. The histological examination was characteristic for AGEP. Three months later, patch test were positive to prednisone 100 mg/ml but negative to prednisolone and hydrocortisone. Lymphocyte transformations tests, however, were positive to prednisolone and hydrocortisone (but negative to prednisone). The patient was diagnosed as having AGEP elicited by 'corticosteroids of the hydrocortisone type' (7). Despite the negative patch test reaction to prednisolone (which may have been false-negative, or late readings were not performed), this was obviously the cause of AGEP, to which topical prednisolone acetate, oral prednisolone and oral prednisolone tetrahydrophthalate may all have contributed.

A second patient presented by the authors, highly likely previously sensitized by triamcinolone acetonide, developed AGEP hours after a single oral dose of 60 mg prednisolone. Patch tests were positive to prednisolone 1%, prednisone 20 mg (?), hydrocortisone 1%, tixocortol pivalate 1% and budesonide 0.1%. Triamcinolone acetonide was apparently not tested (7).

Systemic allergic dermatitis (systemic contact dermatitis)
A 61-year-old woman found that topical steroids aggravated her current hand dermatitis. She was patch tested and reacted to an ointment she used containing 0.025% beclomethasone dipropionate and to the steroid itself at 20%

pet. She was also allergic to prednisolone; oral prednisolone 20 mg daily after 2 days resulted in systemic allergic dermatitis with worsening and spreading of existing dermatitis (3).

A 27-year-old woman, who had deterioration of cosmetic dermatitis after the use of prednisolone acetate cream and betamethasone valerate cream, was given oral prednisolone 40 mg/day. Twenty-four hours after the initial dose the patient felt generally sick and developed oral enanthema, edema of the face, erythematous patches of the great folds and maculopapular exanthema of the trunk. Patch tests were positive to prednisolone, hydrocortisone, tixocortol pivalate, hydrocortisone valerate, betamethasone valerate and various other corticosteroid preparations. The patient had probably been presensitized by earlier treatment of flexural exanthema with various corticosteroids (4).

In a 61-year-old female patient, oral administration of prednisolone 20 mg led to exacerbation of hydrocortisone-treated facial eczema within one day with spreading to previously unaffected skin. Patch tests were positive to hydrocortisone 0.5% alc./DMSO and tixocortol pivalate 1% pet. The diagnosis of systemic allergic dermatitis was made, but prednisolone itself was not tested (5).

A 54-year-old woman, who had previously suffered worsening of irritant contact dermatitis from topical betamethasone valerate and hydrocortisone butyrate, developed a generalized, maculopapular rash one day after oral administration of 40 mg of prednisolone. Patch tests were positive to these 3 and to 3 other corticosteroids. This was a case of systemic allergic dermatitis from oral prednisolone presenting as maculopapular eruption (6).

A 50-year-old woman, who had been treated for years with various topical corticosteroids for chronic nasal congestion, developed exanthemas one day after intake of prednisolone and – on 2 occasions – of methylprednisolone. Patch tests were positive to prednisolone, but methylprednisolone was not tested. The patient developed eruptions of the baboon syndrome when orally provoked with betamethasone, dexamethasone, hydrocortisone, and cloprednol. Whether the exanthemas from prednisolone and methylprednisolone had been similar clinical manifestations was not mentioned (10).

A 45-year-old woman developed an acute dermatitis of the face after having been massaged with aroma oils 12 hours earlier. Treatment with hydrocortisone 1% cream worsened the rash, extending on to the neck. She was then prescribed oral prednisolone 25 mg/day, but within 24 hours of taking the first dose the rash had generalized. Patch tests were positive to the fragrance-mix, neomycin, tixocortol pivalate and prednisolone sodium phosphate eye drops; hydrocortisone was not tested (tixocortol pivalate is a very reliable indicator for hydrocortisone allergy). This was probably a cases of systemic allergic dermatitis from oral prednisolone cross-reacting to hydrocortisone (13).

A man aged 30 years experienced worsening of atopic dermatitis from prednisolone-21-acetate ointment. Oral prednisolone was started, but 5 hours later, intense generalized pruritus with erythema and swelling of the face occurred. After 24 hours a generalized erythema with disseminated, partly follicular papules was present. Patch tests were positive to hydrocortisone 1%, prednisolone 1% and 2.5%, and prednisolone-21-acetate ointment (20).

Two patients who had positive patch tests to tixocortol pivalate and positive intradermal tests to prednisolone, were orally challenged with hydrocortisone 10, 25, 30, 50 and 75 mg. One or two doses were given per day, and the challenge was stopped when skin symptoms appeared. Both patients had a positive reaction after the 30 mg dose. Symptoms were erythema or infiltrated erythema at previous sites of eczema or positive skin tests in one patient and widespread erythema or exanthema in the other (23).

Symmetrical drug-related intertriginous and flexural exanthema (SDRIFE)/Baboon syndrome
A 52-year old woman had suffered three episodes of pruritic rashes affecting intertriginous skin, each time emerging some 24 hours after she had started taking oral prednisolone for tonsillitis (!). On examination, erythematous patches were noted over her bilateral groin folds and inframammary areas. Patch tests were positive to tixocortol pivalate but negative to prednisolone and 5 other corticosteroids. A graded oral provocation test with prednisolone (15 mg in total) reproduced intertriginous rashes within 2 hours, confirming a diagnosis of SDRIFE to prednisolone (24).

Other cutaneous adverse drug reactions
A 36-year-old woman was treated with oral prednisolone for contact dermatitis and developed a 'generalized rash'. Patch tests were positive to prednisolone. Details (e.g. whether the patents had previous allergy to [other] corticosteroids) are not available to the author (8). A pustular exanthema from allergy to prednisolone, which may or may not have been a case of AGEP, was reported in German literature in the 1980s (16).

Several other cases of allergic exanthematous reactions to systemic prednisolone (or one of its derivatives) have been reported in German literature, but details are not available to the author (18,19).

Cross-reactions, pseudo-cross-reactions and co-reactions
Please note: Cross-reactivity studies between corticosteroids as described below have been performed mainly in patients sensitized to these molecules from topical application. It is unknown whether sensitization from *systemic* exposure will result in the same cross-reactivity patterns. However, many reactions from systemic exposure are

forms of systemic allergic dermatitis (systemic allergic dermatitis), in which the patients were presensitized by using topical corticosteroid products. The data below are adapted from refs. 25,26 and 27.

Multiple reactions to corticosteroids are frequent, mostly from cross-reactions. The allergens are probably not the corticosteroids themselves, but a byproduct from their skin metabolism. The principal metabolites, steroid glyoxals or 21-dehydrocorticosteroids (aldehydes), are the most probably haptens. Based on results of patch testing, molecular modelling and previous work, corticosteroids have been divided into three groups (table 3.399.1). Group 1 consists of (mostly) non-methylated, non-halogenated molecules. Group 2 are (mostly) halogenated molecules with a C_{16}/C_{17} cis-ketal/diol structure. Group 3 consists of halogenated and C_{16}-methylated molecules.

C$_{16}$-methyl substitution and halogenation seem to reduce the allergenicity of corticosteroid molecules. Indeed, by far most allergic reactions are caused by the corticosteroids in group 1, whereas the molecules in group 3 rarely sensitize. As to the cross-reaction pattern: patients sensitized from one or more corticosteroids in group 1 often cross-react with other chemicals in group 1. However, another cross-reactivity profile is that patients may be sensitized to steroids in group 1 with cross-reactions not only in group 1, but also to group 2, group 3 or both. The classification presented in table 3.399.1 cannot explain all observed cross-reactions and often does not accurately predict cross-reactivity. Therefore, patients with positive patch test reactions to tixocortol pivalate and/or budesonide in the baseline series and/or to corticosteroids used by them, should also be tested with other corticosteroids to determine the cross-reactivity pattern and to establish which corticosteroids can safely be used for continued treatment (25,26). Cross-reactivity can sometimes also be observed to endogenous steroidal sex hormones and derivatives including progesterone, 17-α-hydroxyprogesterone, and testosterone (25-26).

Table 3.399.1 Corticosteroid classification based on cross-reaction pattern (adapted from refs. 25 and 26)

GROUP 1	GROUP 2	GROUP 3
No C$_{16}$-methyl substitution No halogen substitution in most cases	C$_{16}$/C$_{17}$ cis-ketal or diol structure Halogen substitution	C$_{16}$-methyl substitution Halogen substitution
Budesonide (S-isomer)	Amcinonide	Alclomethasone dipropionate [c]
Cloprednol	(Budesonide, R isomer) [a]	Beclomethasone dipropionate
Cortisone acetate	Ciclesonide	Betamethasone
Dichlorisone acetate	Desonide [b]	Betamethasone 17-valerate
Difluprednate	Fluclorolone acetonide	Betamethasone dipropionate
Fludrocortisone acetate	Flumoxonide [d]	Betamethasone sodium phosphate
Fluorometholone	Flunisolide	Clobetasol propionate
Fluprednisolone acetate	Fluocinonide acetonide	Clobetasone butyrate
Hydrocortisone	Fluocinonide	Cortivazol
Hydrocortisone aceponate	Halcinonide [b]	Desoximetasone
Hydrocortisone acetate	Triamcinolone acetonide	Dexamethasone
Hydrocortisone-17-butyrate	Triamcinolone benetonide [d]	Dexamethasone acetate
Hydrocortisone-21-butyrate	Triamcinolone diacetate	Dexamethasone sodium phosphate
Hydrocortisone hemisuccinate	Triamcinolone hexacetonide	Diflucortolone valerate
Isoflupredone acetate		Diflorasone diacetate
Mazipredone		Flumethasone pivalate
Medrysone		Fluocortin butyl
Methylprednisolone aceponate		Fluocortolone
Methylprednisolone acetate		Fluocortolone caproate
Methylprednisolone hemisuccinate		Fluocortolone pivalate
Prednicarbate		Fluprednidene acetate
Prednisolone		Fluticasone propionate
Prednisolone caproate		Halometasone
Prednisolone hemisuccinate		Meprednisone
Prednisolone pivalate		Mometasone furoate
Prednisolone sodium metazoate		
Prednisone		
Tixocortol pivalate		
Triamcinolone		

[a] also included in group 2, as it may - in exceptional cases - cross-react with the acetonides; [b] No halogen substitution;
[c] Unexpectedly in group 3, as alclomethasone dipropionate often co-reacts with group 1; [d] Not used in pharmaceuticals

LITERATURE

1   Baeck M, Goossens A. Immediate and delayed allergic hypersensitivity to corticosteroids: practical guidelines. Contact Dermatitis 2012;66:38-45

2   Rytter M, Walther T, Süss E, Haustein UF. Allergic reactions of the immediate and delayed type following prednisolone medication. Dermatol Monatsschr 1989;175:44-48 (Article in German, data cited in ref. 4)

3   English JS, Ford G, Beck MH, Rycroft RJ. Allergic contact dermatitis from topical and systemic steroids. Contact Dermatitis 1990;23:196-197

4   Bircher AJ, Levy F, Langauer S, Lepoittevin JP. Contact allergy to topical corticosteroids and systemic contact dermatitis from prednisolone with tolerance of triamcinolone. Acta Derm Venereol 1995;75:490-493

5   Isaksson M, Persson LM. Contact allergy to hydrocortisone and systemic contact dermatitis from prednisolone with tolerance of betamethasone. Am J Contact Dermat 1998;9:136-138

6   McKenna DB, Murphy GM. Contact allergy to topical corticosteroids and systemic allergy to prednisolone. Contact Dermatitis 1998;38:121-122

7   Buettiker U, Keller M, Picheler WJ, Braathen LR, Yamalkar N. Oral prednisolone induced acute generalized exanthematous pustulosis due to corticosteroids of group A confirmed by epicutaneous testing and lymphocyte transformation tests. Dermatology 2006;213:40-43

8   Simowa G. Konaktekzem bei Chloramphenicol- und Prednisolonallergie. Dermatol Monnatsschr 1987;173:357-358 (Article in German, data cited in ref. 4)

9   Räsänen L, Hasan T. Allergy to systemic and intralesional corticosteroids. Br J Dermatol 1993;128:407-411

10 Treudler R, Simon J. Symmetric, drug-related, intertriginous, and flexural exanthema in a patient with polyvalent intolerance to corticosteroids. J Allergy Clin Immunol 2006;118:965-967

11 Bursztejn AC, Tréchot P, Cuny JF, Schmutz JL, Barbaud A. Cutaneous adverse drug reactions during chemotherapy: consider non-antineoplastic drugs. Contact Dermatitis 2008;58:365-368

12 Brüngger A, Wüthrich B. Maculopapular drug eruption caused by delayed-type hypersensitivity reaction to corticosteroids: case report and literature review. Dermatologica 1990;181:173

13 Harris A, McFadden JP. Dermatitis following systemic prednisolone: patch testing with prednisolone eye drops. Australas J Dermatol 2000;41:124-125

14 De Groot AC. Monographs in contact allergy, volume 3: Topical Drugs. Boca Raton, Fl, USA: CRC Press Taylor and Francis Group, 2021 (ISBN 978-0-367-23693-9)

15 Ishii S, Hasegawa T, Hirasawa Y, Tsunemi Y, Kawashima M, Ikeda S. Acute generalized exanthematous pustulosis induced by oral prednisolone. J Dermatol 2014;41:1135-1136

16 Voss M. Eine Prednisolon-Allergie mit pustulösem Exanthem. Dermatol Monatsschrift 1988;174:221-225

17 Zedlitz S, Ahlbach S, Kaufmann R, Boehncke WH. Tolerance to a group C corticosteroid systemically in a patient with delayed-type hypersensitivity to group A systemic corticosteroids. Contact Dermatitis 2002;47:242

18 Lübbe D. Epidermale Prednisolonsensibilisierung – Ein Erfahrungsbericht über 7 klinische Beobachtungen. Wiss Zeitschr Univ Halle 1985;6:63-68 (Article in German, data cited in ref. 4)

19 Walther Th. Allergisches Exanthem auf Prednisolon. Dermatol Monatsschr 1986;172:677 (Article in German, data cited in ref. 4)

20 Yawalkar N, Hari Y, Helbing A, von Greyerz S, Kappeler A, Baathen LR, Pichler WJ. Elevated serum levels of interleukin 5, 6 and 10 in patients with drug exanthema caused by corticosteroids. J Am Acad Dermatol 1998;39:790-793

21 Shigemi F, Tanaka M, Ohtsuka T. A case of sensitization to oral corticosteroids. J Dermatol 1978;5:231-233

22 Goh CL. Cross-sensitivity to multiple topical corticosteroids. Contact Dermatitis 1989;20:65-67

23 Räsänen L, Tuomi ML, Ylitalo L. Reactivity of tixocortol pivalate-positive patients in intradermal and oral provocation tests. Br J Dermatol 1996;135:931-934

24 Chong T, Heng YK. A case of symmetrical drug-related intertriginous and flexural exanthema (SDRIFE) to prednisolone. Dermatitis 2021;32:e11 (Abstract).

25 Baeck M, Goossens A. Immediate and delayed allergic hypersensitivity to corticosteroids: practical guidelines. Contact Dermatitis 2012;66:38-45

26 Goossens A, Gonçalo M. Topical drugs. In: Johansen J, Mahler V, Lepoittevin JP, Frosch P, Eds. Contact Dermatitis, 6th Edition. Springer: Cham, 2020

27 De Groot AC. Monographs in contact allergy, volume 3: Topical Drugs. Boca Raton, Fl, USA: CRC Press Taylor and Francis Group, 2021: chapter 2.8, pp. 24-25

Chapter 3.400 PREDNISOLONE ACETATE

IDENTIFICATION

Description/definition : Prednisolone acetate is the acetate ester of prednisolone that conforms to the structural formula shown below

Pharmacological classes : Anti-inflammatory agents; glucocorticoids

IUPAC name : [2-[(8S,9S,10R,11S,13S,14S,17R)-11,17-Dihydroxy-10,13-dimethyl-3-oxo-7,8,9,11,12,14, 15,16-octahydro-6H-cyclopenta[a]phenanthren-17-yl]-2-oxoethyl] acetate

Other names : Prednisolone 21-acetate; 11β,17,21-trihydroxypregna-1,4-diene-3,20-dione 21-acetate

CAS registry number : 52-21-1

EC number : 200-134-1

Merck Index monograph : 911 (Prednisolone)

Patch testing : In general, corticosteroids may be tested at 0.1% and 1% in alcohol; late readings (6-10 days) are strongly recommended

Molecular formula : $C_{23}H_{30}O_6$

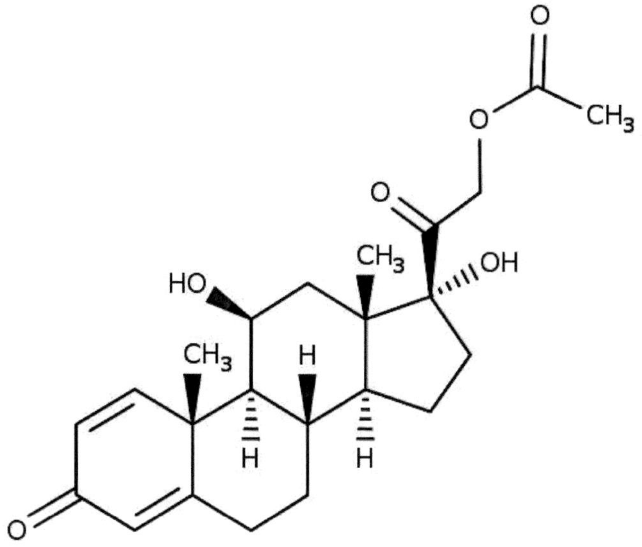

GENERAL

Systemically administered glucocorticoids have anti-inflammatory, immunosuppressive and antineoplastic properties and are used in the treatment of a wide spectrum of diseases including rheumatic disorders, lung diseases (asthma, COPD), gastrointestinal tract disorders (Crohn's disease, colitis ulcerosa), certain malignancies (leukemia, lymphomas), hematological disorders, and various diseases of the kidneys, brain, eyes and skin. A practical guideline for diagnosing allergic reactions to corticosteroids is presented in ref. 1. Prednisolone acetate is used in both topical and systemic pharmaceutical applications. In topical preparations, this glucocorticoid has occasionally caused contact allergy/allergic contact dermatitis, which has been fully reviewed in Volume 3 of the *Monographs in contact allergy* series (3). See also prednisolone (Chapter 3.399), prednisolone hemisuccinate (Chapter 3.401), prednisolone sodium succinate (Chapter 3.402), and prednisolone tetrahydrophthalate sodium salt (Chapter 403).

CUTANEOUS ADVERSE DRUG REACTIONS FROM SYSTEMIC ADMINISTRATION CAUSED BY TYPE IV (DELAYED-TYPE) HYPERSENSITIVITY (as demonstrated by positive patch tests)

Cutaneous adverse drug reactions from systemic administration of prednisolone acetate caused by type IV (delayed-type) hypersensitivity have included localized allergic reaction from intra-articular injection (2)

Other cutaneous adverse drug reactions

A 42-year-old woman, who had been treated with various corticosteroids for recurrent atopic dermatitis for decades, developed, 14 hours after an intra-articular injection with prednisolone acetate, severe erythema and edema in and around the area of the injection. Patch tests were positive to hydrocortisone butyrate 0.1% and to the following commercial corticosteroid preparations: prednisolone hemisuccinate solution, hydrocortisone probutate cream and prednicarbate ointment. Prednisolone acetate was – quite surprisingly – no tested. This was a case of localized delayed-type allergic reaction from prednisolone acetate cross-reacting to one or more other unknown corticosteroids used topically in the past (2).

Cross-reactions, pseudo-cross-reactions and co-reactions
Cross-reactions between corticosteroids are discussed in Chapter 3.399 Prednisolone.

LITERATURE

1   Baeck M, Goossens A. Immediate and delayed allergic hypersensitivity to corticosteroids: practical guidelines.
      Contact Dermatitis 2012;66:38-45
2   Gall HM, Paul E. A case of corticosteroid allergy. Hautarzt 2001;52:891-894 (Article in German)
3   De Groot AC. Monographs in contact allergy, volume 3: Topical Drugs. Boca Raton, Fl, USA: CRC Press Taylor and
      Francis Group, 2021 (ISBN 978-0-367-23693-9)

Chapter 3.401 PREDNISOLONE HEMISUCCINATE

IDENTIFICATION

Description/definition	: Prednisolone hemisuccinate is the succinate ester of prednisolone that conforms to the formula shown below
Pharmacological classes	: Glucocorticoids
IUPAC name	: 4-[2-[(8S,9S,10R,11S,13S,14S,17R)-11,17-Dihydroxy-10,13-dimethyl-3-oxo-7,8,9,11,12, 14,15,16-octahydro-6H-cyclopenta[a]phenanthren-17-yl]-2-oxoethoxy]-4-oxobutanoic acid
Other names	: Prednisolone 21-(hydrogen succinate); prednisolone succinate
CAS registry number	: 2920-86-7
EC number	: 220-861-8
Merck Index monograph	: 9111 (Prednisolone)
Patch testing	: Generally, corticosteroids may be tested at 0.1% and 1% in alcohol; late readings (6-10 days) are strongly recommended
Molecular formula	: $C_{25}H_{32}O_8$

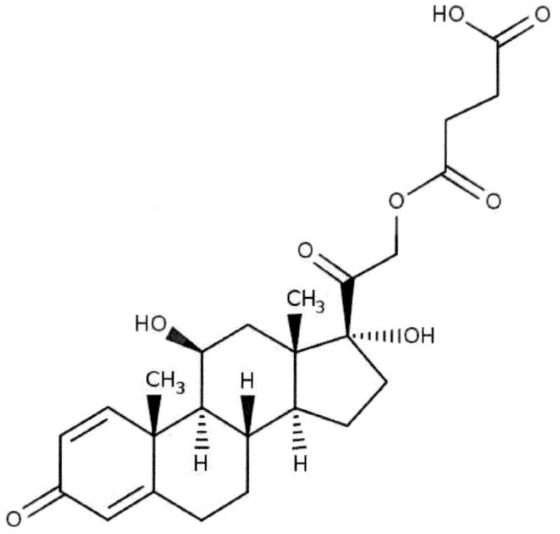

GENERAL

Systemically administered glucocorticoids have anti-inflammatory, immunosuppressive and antineoplastic properties and are used in the treatment of a wide spectrum of diseases including rheumatic disorders, lung diseases (asthma, COPD), gastrointestinal tract disorders (Crohn's disease, colitis ulcerosa), certain malignancies (leukemia, lymphomas), hematological disorders, and various diseases of the kidneys, brain, eyes and skin. A practical guideline for diagnosing allergic reactions to corticosteroids is presented in ref. 1.

See also Prednisolone (Chapter 3.399), prednisolone acetate (Chapter 3.400), prednisolone sodium succinate (Chapter 3.402), and prednisolone tetrahydrophthalate sodium salt (Chapter 3.403).

CUTANEOUS ADVERSE DRUG REACTIONS FROM SYSTEMIC ADMINISTRATION CAUSED BY TYPE IV (DELAYED-TYPE) HYPERSENSITIVITY (as demonstrated by positive patch tests)

In German literature, a case of combined immediate-type and delayed-type allergic reaction with exanthema has been reported; details are not available to the author (2).

Cross-reactions, pseudo-cross-reactions and co-reactions

Cross-reactions between corticosteroids are discussed in Chapter 3.399 Prednisolone.

LITERATURE

1   Baeck M, Goossens A. Immediate and delayed allergic hypersensitivity to corticosteroids: practical guidelines. Contact Dermatitis 2012;66:38-45
2   Reichert Chr, Gall H, Sterry W. Allergische Typ-I and Typ-IV-Reaktion auf Prednisolone-21-hydrogensuccinaat. Allergo J 1994;3:315-319 (Article in German, data cited in ref. 3)
3   Bircher AJ, Levy F, Langauer S, Lepoittevin JP. Contact allergy to topical corticosteroids and systemic contact dermatitis from prednisolone with tolerance of triamcinolone. Acta Derm Venereol 1995;75:490-493

Chapter 3.402 PREDNISOLONE SODIUM SUCCINATE

IDENTIFICATION

Description/definition   : Prednisolone sodium succinate is the synthetic glucocorticoid that conforms to the
                            structural formula shown below
Pharmacological classes   : Glucocorticoids
IUPAC name   : Sodium;4-[2-[(8S,9S,10R,11S,13S,14S,17R)-11,17-dihydroxy-10,13-dimethyl-3-oxo-
                            7,8,9,11,12,14,15,16-octahydro-6H-cyclopenta[a]phenanthren-17-yl]-2-oxoethoxy]-4-
                            oxobutanoate
CAS registry number   : 1715-33-9
EC number   : 216-995-1
Merck Index monograph   : 9111 (Prednisolone)
Patch testing   : Generally, corticosteroids may be tested at 0.1% and 1% in alcohol; late readings (6-10
                            days) are strongly recommended
Molecular formula   : $C_{25}H_{31}NaO_8$

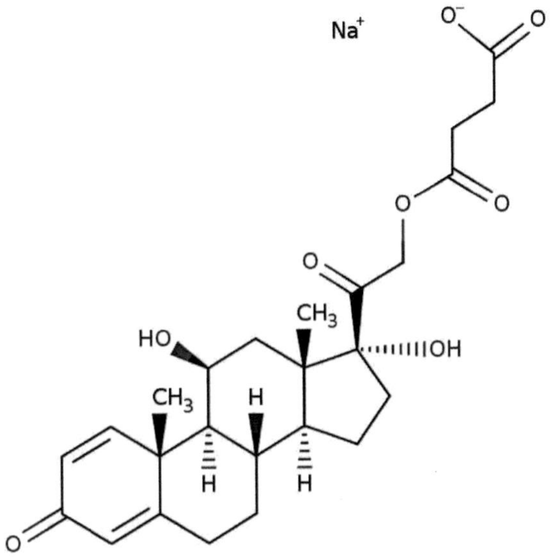

GENERAL

Systemically administered glucocorticoids have anti-inflammatory, immunosuppressive and antineoplastic properties and are used in the treatment of a wide spectrum of diseases including rheumatic disorders, lung diseases (asthma, COPD), gastrointestinal tract disorders (Crohn's disease, colitis ulcerosa), certain malignancies (leukemia, lymphomas), hematological disorders, and various diseases of the kidneys, brain, eyes and skin. A practical guideline for diagnosing allergic reactions to corticosteroids is presented in ref. 1.

See also prednisolone (Chapter 3.399), prednisolone acetate (Chapter 3.400), prednisolone hemisuccinate (Chapter 3.401), and prednisolone tetrahydrophthalate sodium salt (Chapter 3.403).

CUTANEOUS ADVERSE DRUG REACTIONS FROM SYSTEMIC ADMINISTRATION CAUSED BY TYPE IV (DELAYED-TYPE) HYPERSENSITIVITY (as demonstrated by positive patch tests)

Cutaneous adverse drug reactions from systemic administration of prednisolone sodium succinate caused by type IV (delayed-type) hypersensitivity have included acute generalized exanthematous pustulosis (AGEP) (2,3) and widespread macular exanthema (4).

Acute generalized exanthematous pustulosis (AGEP)

A 20-year-old woman was treated with 1000 mg prednisolone/day intravenously for three consecutive days for a first demyelinating attack with optic neuritis. While prednisone was tapered out orally, the patient developed an acute itchy erythematous rash accompanied by fever up to 38.6°C on the fourth day. Starting in the flexural folds, the rash rapidly spread over the entire integument except the mucous membranes. Countless small pinhead-sized non-follicular pustules arose and subsequently coalesced forming polycyclic lakes of pus. Serologic inflammation parameters were highly elevated. Histopathological findings of lesional skin were consistent with the diagnosis of acute generalized exanthematous pustulosis (AGEP). AGEP was diagnosed, and oral prednisolone (prednisone?) therapy was immediately discontinued and replaced with oral dexamethasone. After 6 days of treatment, the skin eruption rapidly resolved with a post-pustular desquamation. Later, patch tests were positive to prednisolone 0.25% showing a localized pustular reaction at D3 and negative to dexamethasone (2). As the 'prednisolone' had been given intravenously, it is very likely that it was in fact prednisolone sodium succinate.

The same authors had presented this patient a year earlier in another journal. Here was stated that patch tests were positive to prednisolone 1% pet. with erythema and infiltration after 24 hours and additional pustules after 48 or 72 hours with cross-reactions to methylprednisolone (1.6% water) and prednicarbate 2.5% ointment base, histopathologically mimicking the original disease and thus confirming AGEP due to prednisolone (3).

Other cutaneous adverse drug reactions

A 36-year-old woman was treated with intravenous methylprednisolone sodium succinate for acute myelitis for 5 days. Three days later (day 8), the patient developed a widespread macular exanthema with severe itch. Intravenous prednisolone sodium succinate exacerbated both the itch and the exanthema. Patch tests were positive to methylprednisolone and prednisolone, both at 1% pet. at D2 and D3 (4).

Cross-reactions, pseudo-cross-reactions and co-reactions

Cross-reactions between corticosteroids are discussed in Chapter 3.399 Prednisolone.

LITERATURE

1   Baeck M, Goossens A. Immediate and delayed allergic hypersensitivity to corticosteroids: practical guidelines. Contact Dermatitis 2012;66:38-45
2   Ziemssen T, Bauer A, Bär M. Potential side effect of high-dose corticosteroid relapse treatment: acute generalized exanthematous pustulosis (AGEP). Mult Scler 2009;15:275-277
3   Bär M, John L, Wonschik S, Schmitt J, Kempter W, Bauer A, Meurer M. Acute generalized exanthematous pustulosis induced by high-dose prednisolone in a young woman with optic neuritis owing to disseminated encephalomyelitis. Br J Dermatol 2008;159:251-252
4   Zedlitz S, Ahlbach S, Kaufmann R, Boehncke WH. Tolerance to a group C corticosteroid systemically in a patient with delayed-type hypersensitivity to group A systemic corticosteroids. Contact Dermatitis 2002;47:242

Chapter 3.403 PREDNISOLONE TETRAHYDROPHTHALATE SODIUM SALT

IDENTIFICATION

Description/definition : Prednisolone tetrahydrophthalate sodium salt is the synthetic glucocorticoid that conforms to the structural formula shown below

Pharmacological classes : Glucocorticoids

IUPAC name : Sodium;6-[2-[[(8S,9S,10R,11S,13S,14S,17R)-11,17-dihydroxy-10,13-dimethyl-3-oxo-7,8,9,11,12,14,15,16-octahydro-6H-cyclopenta[a]phenanthren-17-yl]-2-oxoethoxy]-carbonylcyclohex-3-ene-1-carboxylate

Other names : Prednisolone sodium tetrahydrophthalate

CAS registry number : 10059-14-0

EC number : Not available

Merck Index monograph : 9111 (Prednisolone)

Patch testing : Generally, corticosteroids may be tested at 0.1% and 1% in alcohol; late readings (6-10 days) are strongly recommended

Molecular formula : $C_{29}H_{35}NaO_8$

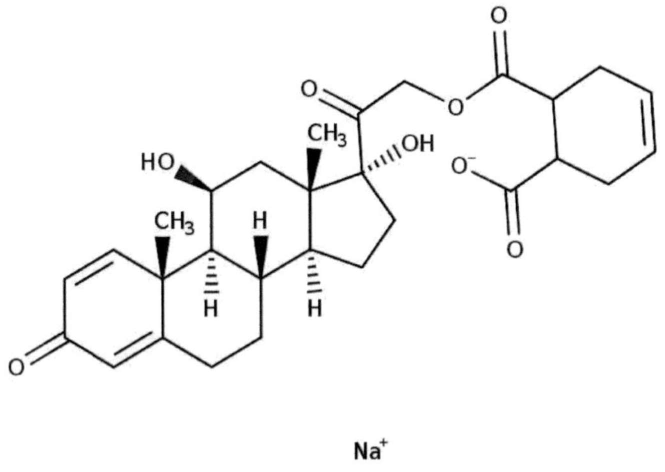

Na⁺

GENERAL

Systemically administered glucocorticoids have anti-inflammatory, immunosuppressive and antineoplastic properties and are used in the treatment of a wide spectrum of diseases including rheumatic disorders, lung diseases (asthma, COPD), gastrointestinal tract disorders (Crohn's disease, colitis ulcerosa), certain malignancies (leukemia, lymphomas), hematological disorders, and various diseases of the kidneys, brain, eyes and skin. A practical guideline for diagnosing allergic reactions to corticosteroids is presented in ref. 1.

See also Prednisolone (Chapter 3.399), prednisolone acetate (Chapter 3.400), prednisolone hemisuccinate (Chapter 3.401), and prednisolone sodium succinate (Chapter 3.402).

CUTANEOUS ADVERSE DRUG REACTIONS FROM SYSTEMIC ADMINISTRATION CAUSED BY TYPE IV (DELAYED-TYPE) HYPERSENSITIVITY (as demonstrated by positive patch tests)

Cutaneous adverse drug reactions from systemic administration of prednisolone sodium tetrahydrophthalate caused by type IV (delayed-type) hypersensitivity have included acute generalized exanthematous pustulosis (AGEP) (2).

Acute generalized exanthematous pustulosis (AGEP)

A 42-year-old man suffering from perianal eczema was initially treated for 3 weeks with topical prednisolone acetate cream. Due to aggravation of the eczema, the patient was subsequently given systemic treatment with 50 mg prednisolone followed by 100 mg prednisolone tetrahydrophthalate. Some hours after receiving the latter, the patient developed an erythematous rash with subsequent occurrence of numerous non-follicular pustules on the trunk and the extremities over the following days. The exanthem was associated with burning, pruritus as well as malaise and fever (temperature of 38°C). No mucosal involvement was noted. The laboratory investigations showed leukocytosis with neutrophilia and elevated C-reactive protein (187 mg/l), a bacterial culture from pustules was negative. The histological examination revealed subcorneal spongiform pustules with some dermal edema as well as a dense perivascular and interstitial inflammatory infiltrate with lymphocytes, neutrophils and scattered eosinophils.

Three months later, patch test were positive to prednisone 100 mg/ml but negative to prednisolone and hydrocortisone. Lymphocyte transformations tests, however, were positive to prednisolone and hydrocortisone (but negative to prednisone). The patient was diagnosed as having acute generalized exanthematous pustulosis (AGEP) elicited by 'corticosteroids of the hydrocortisone type' (2). Despite the negative patch test reaction to prednisolone (which may have been false-negative, or late readings were not performed), this was obviously the cause of AGEP, to which topical prednisolone acetate, oral prednisolone and oral prednisolone tetrahydrophthalate may all have contributed.

Cross-reactions, pseudo-cross-reactions and co-reactions
Cross-reactions between corticosteroids are discussed in Chapter 3.399 Prednisolone.

LITERATURE
1 Baeck M, Goossens A. Immediate and delayed allergic hypersensitivity to corticosteroids: practical guidelines. Contact Dermatitis 2012;66:38-45
2 Buettiker U, Keller M, Picheler WJ, Braathen LR, Yamalkar N. Oral prednisolone induced acute generalized exanthematous pustulosis due to corticosteroids of group A confirmed by epicutaneous testing and lymphocyte transformation tests. Dermatology 2006;213:40-43

Chapter 3.404 PREDNISONE

IDENTIFICATION

Description/definition	: Prednisone is the synthetic glucocorticoid that conforms to the structural formula shown below
Pharmacological classes	: Anti-inflammatory agents; glucocorticoids; antineoplastic agents, hormonal
IUPAC name	: (8S,9S,10R,13S,14S,17R)-17-Hydroxy-17-(2-hydroxyacetyl)-10,13-dimethyl-6,7,8,9,12, 14,15,16-octahydrocyclopenta[a]phenanthrene-3,11-dione
Other names	: 17,21-Dihydroxypregna-1,4-diene-3,11,20-trione; dehydrocortisone
CAS registry number	: 53-03-2
EC number	: 200-160-3
Merck Index monograph	: 9112
Patch testing	: In general, corticosteroids may be tested at 0.1% and 1% in alcohol; late readings (6-10 days) are strongly recommended
Molecular formula	: $C_{21}H_{26}O_5$

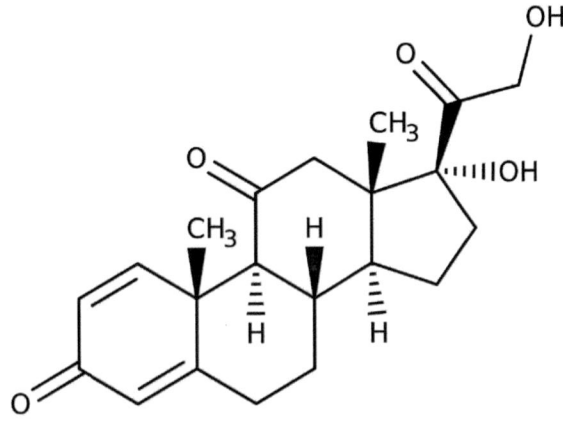

GENERAL

Systemically administered glucocorticoids have anti-inflammatory, immunosuppressive and antineoplastic properties and are used in the treatment of a wide spectrum of diseases including rheumatic disorders, lung diseases (asthma, COPD), gastrointestinal tract disorders (Crohn's disease, colitis ulcerosa), certain malignancies (leukemia, lymphomas), hematological disorders, and various diseases of the kidneys, brain, eyes and skin. A practical guideline for diagnosing allergic reactions to corticosteroids is presented in ref. 1. Prednisone is used only orally. This means that by far most allergic reactions to 'prednisone' have in fact been cross-reactions to other corticosteroids. This subject has been fully reviewed in Volume 3 of the *Monographs in contact allergy* series (6).

Prednisone is a prodrug, which, after oral administration, is metabolized into the active drug prednisolone. Therefore, cross-reactions (or actually pseudo-cross-reactions) between prednisone and prednisolone are frequent.

CONTACT ALLERGY FROM ACCIDENTAL CONTACT

Of 38 veterinarians with hand and forearm dermatoses seen by dermatologists in Belgium and the Netherlands from 1995 to 2005, 17 had occupational allergic contact dermatitis. Prednisone was the responsible drug allergen in one patient (7).

CUTANEOUS ADVERSE DRUG REACTIONS FROM SYSTEMIC ADMINISTRATION CAUSED BY TYPE IV (DELAYED-TYPE) HYPERSENSITIVITY (as demonstrated by positive patch tests)

Cutaneous adverse drug reactions from systemic administration of prednisone caused by type IV (delayed-type) hypersensitivity have included generalized erythema (8), acute generalized exanthematous pustulosis (AGEP) (10), systemic allergic dermatitis (2,4,9), symmetrical drug-related intertriginous and flexural exanthema (SDRIFE)/baboon syndrome (3) and bullous erythema multiforme (5).

Erythroderma, widespread erythematous eruption, exfoliative dermatitis

A few hours after a first intravenous administration of vincristine, cyclophosphamide and methylprednisolone for splenic lymphoma, a 55-year-old woman presented with generalized intense itching, a generalized erythematous rash, and swelling of the face and hands. Similar eruptions, but milder and more transient, appeared after the 2nd and 3rd cycles. The same reaction occurred after taking 2 tablets prednisone. Further cycles with vincristine and cyclophosphamide alone were given without any trouble. Patch testing was positive to tixocortol pivalate and, in a second session, to hydrocortisone phosphate, prednisone and prednisolone hemisuccinate, but only in the vehicle ethanol-DMSO (50:50) (8).

Acute generalized exanthematous pustulosis (AGEP)
In a multicenter investigation in France, of 45 patients patch tested for AGEP, 26 (58%) had positive patch tests to drugs, including 2 to prednisone (10).

Systemic allergic dermatitis (systemic contact dermatitis)
A 38-year-old man was treated for an axillary dermatitis with a preparation containing prednisolone acetate, which resulted in an exacerbation. Two years later, he again used this cream and suffered from a disseminated eczema within a few days. After treatment with oral prednisone he developed a severe generalized eczematous eruption which slowly healed. Patch tests were positive to prednisolone and prednisolone acetate 1% pet. (4).

A 46-year-old woman was documented by patch and provocative use testing to be allergic to multiple topical corticosteroids. On further testing, oral provocation tests to prednisone, methylprednisolone, triamcinolone, and dexamethasone each produced a generalized maculopapular eruption in a delayed manner (9). This was a case of systemic allergic dermatitis presenting as maculopapular eruption.

A 61-year-old man was treated with hydrocortisone-neomycin and dexamethasone eye drops, which resulted in periocular dermatitis. Treatment was initiated with oral prednisone 30 mg, but 6 hours later a generalized erythematous dermatitis developed. Patch tests were positive to prednisone pure powder and prednisone 5% alc., but negative to neomycin and to a series of corticosteroids (hydrocortisone, dexamethasone, budesonide, methylprednisolone, tixocortol pivalate). To confirm the involvement of prednisone in the generalized dermatitis, the patient was challenged in a single-blind fashion with oral administration of 30 mg of prednisone. Four hours later, the patient had developed a generalized eczema (2). This was very likely a case of systemic allergic dermatitis from prednisone in a patient previously sensitized to hydrocortisone (most likely), dexamethasone, or both, where patch tests to these corticosteroids had been – as frequently occurs – false-negative.

Symmetrical drug-related intertriginous and flexural exanthema (SDRIFE)/Baboon syndrome
Six hours after having taken a single dose of prednisone for erythema nodosum, a 50-year-old woman developed a skin rash on her body. A very similar eruption had occurred 1 year earlier, after treatment of sudden hypoacusis with prednisone. There was no history of intolerance to topical corticosteroids. Physical examination showed pruriginous pink-violet plaques on the buttocks, inner thighs and axillae. Histopathology showed a superficial perivascular dermatitis. Patch test were positive to tixocortol pivalate at D2, confirming, according to the authors, 'the clinical diagnosis of systemic allergic dermatitis to prednisone' (3). The clinical picture was classic for the baboon syndrome (symmetrical drug-related intertriginous and flexural exanthema [SDRIFE]). In this case, it was uncertain whether the patient had been presensitized to topical corticosteroids ('no history of intolerance to topical corticosteroids').

Other cutaneous adverse drug reactions
A 14-year-old girl was treated for systemic lupus erythematosus (SLE) with oral prednisone 20 mg/day. Four days later, the patient had developed a rash consisting of widespread, erythematous, extremely pruritic plaques ranging from 1 to 6 cm in diameter on her arms and trunk. Numerous target lesions with dusky and bullous centers were observed. There was no oral mucosal or eye involvement. Intravenous methylprednisolone gave rapid improvement, but 6 hours after oral prednisone had been started again, the patient developed new crops of bullous lesions on her face, eyelids, arms, and upper chest. Histopathology of lesional skin showed intraepidermal blisters, dermal edema, individual cell necrosis, and an inflammatory infiltrate containing numerous eosinophils, consistent with drug-induced bullous erythema multiforme. Immunofluorescent staining of a non-lesional skin was consistent with lupus erythematosus. A patch test with liquid prednisone 1 mg/ml (while on intravenous methylprednisolone 60 mg/day) showed a mild reaction (induration) at D2, which reaction had subsided at 96 hours. An intradermal test with liquid prednisone 0.1 mg/ml was positive at D1 and D3 (5). The authors diagnosed 'prednisone sensitivity presenting as a bullous eruption in a patient with SLE', but did not explicitly choose between prednisone-induced bullous erythema multiforme and bullous SLE, although they tended to the former diagnosis.

Cross-reactions, pseudo-cross-reactions and co-reactions
Prednisone is a prodrug, which, after oral administration, is metabolized into the active drug prednisolone. Therefore, cross-reactions (or actually pseudo-cross-reactions) between prednisone and prednisolone are frequent. Cross-reactions between corticosteroids are discussed in Chapter 3.399 Prednisolone.

LITERATURE

1   Baeck M, Goossens A. Immediate and delayed allergic hypersensitivity to corticosteroids: practical guidelines. Contact Dermatitis 2012;66:38-45

2   Quirce S, Alvarez MJ, Olaguibel JM, Tabar AI. Systemic contact dermatitis from oral prednisone. Contact Dermatitis 1994;30:53-54

3   De Benito V, Ratón JA, Palacios A, Garmendia M, Gardeazábal J. Systemic contact dermatitis to prednisone: a clinical model approach to the management of systemic allergy to corticosteroids. Clin Exp Dermatol 2012;37:680-681

4   Bircher AJ, Bigliardi P, Zaugg T, Mäkinen-Kiljunen S. Delayed generalized allergic reactions to corticosteroids. Dermatology 2000;200:349-351

5   Lew DB, Higgins GC, Skinner RB, Snider MD, Myers LK. Adverse reaction to prednisone in a patient with systemic lupus erythematosus. Pediatr Dermatol 1999;16:146-150

6   De Groot AC. Monographs in contact allergy, volume 3: Topical Drugs. Boca Raton, Fl, USA: CRC Press Taylor and Francis Group, 2021 (ISBN 978-0-367-23693-9)

7   Bulcke DM, Devos SA. Hand and forearm dermatoses among veterinarians. J Eur Acad Dermatol Venereol 2007;21:360-363

8   Fernández de Corres L, Bernaola G, Urrutia I, Muñoz D. Allergic dermatitis from systemic treatment with corticosteroids. Contact Dermatitis 1990;22:104-106

9   Chew AL, Maibach HI. Multiple corticosteroid orally elicited allergic contact dermatitis in a patient with multiple topical corticosteroid allergic contact dermatitis. Cutis 2000;65:307-311

10   Barbaud A, Collet E, Milpied B, Assier H, Staumont D, Avenel-Audran M, et al. A multicentre study to determine the value and safety of drug patch tests for the three main classes of severe cutaneous adverse drug reactions. Br J Dermatol 2013;168:555-562

Chapter 3.405 PREGABALIN

IDENTIFICATION

Description/definition   : Pregabalin is the γ-aminobutyric acid (GABA) derivative that conforms to the structural formula shown below
Pharmacological classes : Analgesics; anticonvulsants; anti-anxiety agents; calcium channel blockers
IUPAC name   : (3S)-3-(Aminomethyl)-5-methylhexanoic acid
Other names   : 3-Isobutyl GABA
CAS registry number   : 148553-50-8
EC number   : Not available
Merck Index monograph : 9113
Patch testing   : 5% pet.
Molecular formula   : $C_8H_{17}NO_2$

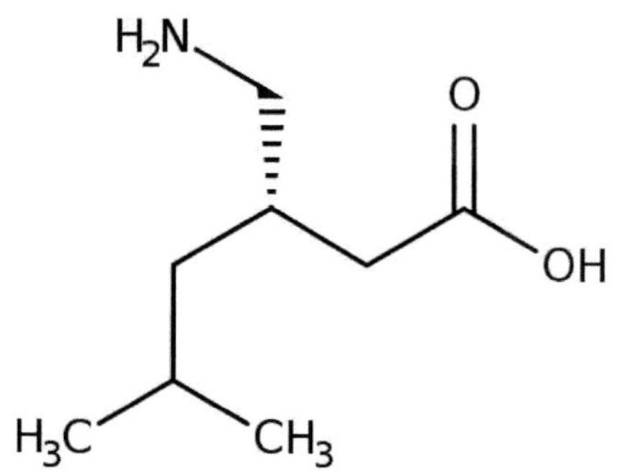

GENERAL

Pregabalin is a 3-isobutyl derivative of γ-aminobutyric acid (GABA) with anti-convulsant, anti-epileptic, anxiolytic, and analgesic activities. Pregabalin is indicated for the management of neuropathic pain associated with diabetic peripheral neuropathy, postherpetic neuralgia, fibromyalgia, neuropathic pain associated with spinal cord injury, and as adjunctive therapy for the treatment of partial-onset seizures (go-drugbank.com).

CUTANEOUS ADVERSE DRUG REACTIONS FROM SYSTEMIC ADMINISTRATION CAUSED BY TYPE IV (DELAYED-TYPE) HYPERSENSITIVITY (as demonstrated by positive patch tests)

Cutaneous adverse drug reactions from systemic administration of pregabalin caused by type IV (delayed-type) hypersensitivity have included maculopapular eruption (1,2).

Maculopapular eruption

One individual developed a maculopapular eruption from pregabalin with a positive patch test reaction. The patient had also suffered maculopapular eruptions from other non-related drugs and an anaphylactic reaction to erythromycin, hence this was a case of multiple drug hypersensitivity syndrome. Clinical and patch testing details were not provided (1). After a month of treatment with pregabalin, a 51-year-old woman developed a widespread maculopapular rash that improved after the withdrawal of the drug and corticosteroid treatment. Histopathology showed superficial perivascular lymphocytic dermatitis with some eosinophil and some lymphocyte permeation in the epidermis. Patch tests with the commercial pregabalin 5% pet. and water in the dorsal region of the left arm were negative at D2 and D4. However, on D3, the patient noted a rash similar to the one previously experienced, particularly on the back of the right forearm. A patch test with pure pregabalin 5% on the right forearm where the rash had reappeared now gave a positive result (D2 ++, D4 +++), but was negative to pregabalin 5% water. The patient was diagnosed with erythematous maculopapular rash caused by type IV hypersensitivity to pregabalin. The authors mentioned several similar cases, previously reported in literature, but in those, patch tests had been negative or were not performed (2).

LITERATURE

1   Jörg L, Yerly D, Helbling A, Pichler W. The role of drug, dose and the tolerance/intolerance of new drugs in multiple drug hypersensitivity syndrome (MDH). Allergy 2020;75:1178-1187
2   Gómez Torrijos E, Moreno Lozano L, Extrmera Ortega AM, Gonzalez Jimenez O, Gratacós Gómez AR, Garcia Rodriguez R. First case of skin allergy to pregabalin with positive patch test reaction. Contact Dermatitis 2019;81:78

Chapter 3.406 PRILOCAINE

IDENTIFICATION

Description/definition   : Prilocaine is the α-amino acid amide that conforms to the structural formula shown
                               below
Pharmacological classes   : Anesthetics, local
IUPAC name   : N-(2-Methylphenyl)-2-(propylamino)propanamide
Other names   : Propitocaine
CAS registry number   : 721-50-6
EC number   : 211-957-0
Merck Index monograph : 9132
Patch testing   : Hydrochloride 5.0% pet.(Chemotechnique)
Molecular formula   : $C_{13}H_{20}N_2O$

GENERAL

Prilocaine is an intermediate-acting local anesthetic of the amide type chemically related to lidocaine. It is used for local anesthesia by infiltration and is the most often used local anesthetic in dentistry. In pharmaceutical products, both prilocaine and prilocaine hydrochloride (CAS number 1786-81-8, EC number 217-244-0, molecular formula $C_{13}H_{21}ClN_2O$) may be employed (1). It is present, together with lidocaine, in a frequently used anesthetic cream for surface anesthesia called EMLA ®. In topical preparations, prilocaine has caused a limited number of cases of contact allergy/allergic contact dermatitis, which has been fully reviewed in Volume 3 of the *Monographs in contact allergy* series (7).

CUTANEOUS ADVERSE DRUG REACTIONS FROM SYSTEMIC ADMINISTRATION CAUSED BY TYPE IV (DELAYED-TYPE) HYPERSENSITIVITY (as demonstrated by positive patch tests)

Cutaneous adverse drug reactions from systemic administration of prilocaine caused by type IV (delayed-type) hypersensitivity have included localized allergic reactions from subcutaneous and submucosal injections for infiltration anesthesia (4,5,6).

Other cutaneous adverse drug reactions

In a 43-year-old woman, an excision of a skin lesion under local anesthesia with prilocaine lead to acute vesicular dermatitis. Patch tests were positive to prilocaine 2% pet. and the related amide-type anesthetics lidocaine and ropivacaine (5). In a 70-year-old woman, a preparation of 3% prilocaine HCl was injected submucosally for anesthesia prior to the filling of both upper left incisors. Three days later, there was marked swelling of the patient's left cheek. Patch tests were positive to lidocaine, prilocaine and mepivacaine. The latter 2 were cross-reactions to primary lidocaine sensitization (4).

In a 55-year-old woman, phlebectomy of the great saphenous vein on the right leg was performed under tumescent local anesthesia with prilocaine 0.065% and epinephrine 1:1,000,000 containing methylparaben; 883 ml of the solution were infiltrated subcutaneously along the vein on several spots, beginning from the groin down to the lower limb. One week later, the patient developed erythema and swelling in the groin, spreading to the distal injection sites. Antibiotics for suspected erysipelas were unhelpful and the skin aspect changed from homogenous edema and erythema to partial development of papulovesicles and scaling. Patch tests were positive to the commercial prilocaine, prilocaine 0.5% solution and lidocaine 15% pet. and negative to methylparaben (6).

Cross-reactions, pseudo-cross-reactions and co-reactions

Several patients sensitized to lidocaine have cross-reacted to prilocaine (7). Conversely, patients sensitized to prilocaine do not seem to cross-react – and even hardly ever co-react from their combined presence in a cream - to lidocaine (7). A patient sensitized to prilocaine co-reacted to mepivacaine (2). A patient sensitized to mepivacaine probably cross-reacted to prilocaine (3). Possible cross-reaction from prilocaine sensitization to articaine (8).

LITERATURE

1   The data in the section 'General' may have been obtained from literature discussed in this chapter, but mostly also or exclusively from one or more of the following online sources: ChemIDPlus Advanced, PubChem, DrugBank, RxList, Drug Central, Drugs.com, and Wikipedia

2   Garcia F, Iparraguirre A, Blanco J, Alloza P, Vicente J, Bascones O, et al. Contact dermatitis from prilocaine with cross-sensitivity to pramocaine and bupivacaine. Contact Dermatitis 2007;56:120-122

3   Kanerva L, Alanko K, Estlander T, Jolanki R. Inconsistent intracutaneous and patch test results in a patient allergic to mepivacaine and prilocaine. Contact Dermatitis 1998;39:197-199

4   Curley RK, Macfarlane AW, King CM. Contact sensitivity to the amide anesthetics lidocaine, prilocaine, and mepivacaine. Arch Dermatol 1986;122:924-926

5   Gunson TH, Greig DE. Allergic contact dermatitis to all three classes of local anaesthetic. Contact Dermatitis 2008;59:126-127

6   Spornraft-Ragaller P, Stein A. Contact dermatitis to prilocaine after tumescent anesthesia. Dermatol Surg 2009;35:1303-1306

7   De Groot AC. Monographs in contact allergy, volume 3: Topical Drugs. Boca Raton, Fl, USA: CRC Press Taylor and Francis Group, 2021 (ISBN 978-0-367-23693-9)

8   Suhonen R, Kanerva L. Contact allergy and cross-reactions caused by prilocaine. Am J Cont Dermat 1997;8:231-235

Chapter 3.407 PRISTINAMYCIN

IDENTIFICATION

Description/definition   : Pristinamycin is a mixture of two components: pristinamycin IA and pristinamycin IIA
                               (streptogramin A); they are coproduced by *Streptomyces pristinaespiralis* in a ratio of
                               30:70 (Wikipedia)
Pharmacological classes   : Anti-bacterial agents
IUPAC name   : *N*-(3-Benzyl-12-ethyl-4,16-dimethyl-2,5,11,14,18,21,24-heptaoxo-19-phenyl-17-oxa-
                               1,4,10,13,20-pentazatricyclo[20.4.0.0^{6,10}]hexacosan-15-yl)-3-hydroxypyridine-2-carboxa-
                               mide;(12*Z*,17*Z*,19*Z*)-21-hydroxy-11,19-dimethyl-10-propan-2-yl-9,26-dioxa-3,15,28-tria-
                               zatricyclo[23.2.1.0^{3,7}]octacosa-1(27),6,12,17,19,25(28)-hexaene-2,8,14,23-tetrone
Other name(s)   : Pyostacine
CAS registry number(s)   : 11006-76-1; 270076-60-3
EC number   : 234-244-6
Merck Index monograph : 9141
Patch testing   : 10.0% pet. (Chemotechnique)
Molecular formula   : $C_{71}H_{84}N_{10}O_{17}$ (PubChem)

Pristinamycin IA   Pristinamycin IIA (= virginiamycin factor M1)

GENERAL

Pristinamycin is an antibiotic used primarily in the treatment of methicillin-resistant staphylococcal infections, and to a lesser extent streptococcal infections. Pristinamycin is a mixture of two components that have a synergistic antibacterial action. Pristinamycin IA is a macrolide, and results in pristinamycin having a similar spectrum of action to erythromycin. Pristinamycin IIA (streptogramin A) is a depsipeptide. They are coproduced by S. pristinaespiralis in a ratio of 30:70. Each compound binds to the bacterial 50 S ribosomal subunit and inhibits the elongation process of the protein synthesis, thereby exhibiting only a moderate bacteriostatic activity. However, the combination of both substances acts synergistically and leads to a potent bactericidal activity that can reach up to 100 times that of the separate components (Wikipedia).

The data provided in various online databases on pristinamycin and virginiamycin are very confusing, overlapping and sometimes probably inaccurate.

CONTACT ALLERGY FROM ACCIDENTAL CONTACT

A 23-year-old man, working in a pharmaceutical company, was exposed to powder of various substances used in the manufacture of different medicaments and the medicaments themselves. He used a rubber mask and gloves as protective equipment. The patient presented with pruritic dermatitis for the past 8 months, involving the neck, eyelids and cheeks, characterized by erythema, edema and scales. Skin lesions improved during holidays and with topical corticosteroids, relapsing when the patient would return to work. Patch tests were positive at D2 and D4 to pristinamycin 1% (+/+), 5% (++/++) and 10% (++/++) in alc./water equal volumes. Control tests in 10 patients were negative. The patient was diagnosed with occupational airborne allergic contact dermatitis from pristinamycin (21).

CUTANEOUS ADVERSE DRUG REACTIONS FROM SYSTEMIC ADMINISTRATION CAUSED BY TYPE IV (DELAYED-TYPE) HYPERSENSITIVITY (as demonstrated by positive patch tests)

Cutaneous adverse drug reactions from systemic administration of pristinamycin caused by type IV (delayed-type) hypersensitivity have included maculopapular eruption (3,19,20,22), erythroderma (3,19), acute generalized exanthematous pustulosis (AGEP) (2,16,17,20,22), systemic allergic dermatitis (4,5,6,8,22), symmetrical drug-related intertriginous and flexural exanthema (SDRIFE)/baboon syndrome (18), fixed drug eruption (2,12), drug reaction with eosinophilia and systemic symptoms (DRESS) (2,17), Stevens-Johnson syndrome/ toxic epidermal necrolysis (SJS/TEN) (2,3), eczematous eruption (22), urticaria (13,15) and undefined/unknown drug eruptions (14).

Case series with various or unknown types of drug reactions

In a hospital in Nancy, France, in the period 1992 to 2002, 29 patients (13 women, 16 men, mean age 59 years, range 16-93 years) were investigated for (strong) suspicion of a cutaneous adverse drug reaction (CADR) to pristinamycin. Eighteen subjects had a maculopapular rash, 9 erythroderma, one angioedema and one Stevens-Johnson syndrome. All were patch tested with pristinamycin 10% in pet. and in water and with commercial pills containing the antibiotic at 30% in water and petrolatum. Patch tests were positive in 20/29 cases (69%): 13/18 (72%) of the patients with maculopapular reactions, 6/9 (67%) with erythroderma and the single patient with Stevens-Johnson syndrome. In these individuals, 7 cross-reacted to virginiamycin (3).

In a single-center study performed in France between 1997 and 2016, 71 patients with suspected delayed cutaneous adverse drug reaction to pristinamycin (n=41), macrolides (n=20) and clindamycin (n=10) were patch tested. Of the 41 patients with suspected reactions to pristinamycin, 19 (46%) had a positive patch test to commercial pristinamycin 30% pet. The drug eruptions involved were AGEP (n=11), DRESS (n=1), SJS/TEN (n=2), and fixed drug eruption (n=1) (2).

Between March 2009 and June 2013, in a center in France specialized in cutaneous adverse drug reactions (CADR), 156 patients were patch tested because of a CADR. Of these, 75 (30 men and 45 women) were tested simultaneously with the commercial test material and extemporaneous patch tests with pulverized pills 30% pet. In all cases with positive patch tests, both materials reacted, there were no discordant results. Pristinamycin was positive in 3 patients, one with maculopapular rash and 2 with AGEP (20).

In a hospital in Caen, France, between 1989 and 1998, 11 patients (6 women, 5 men, mean age 51 years) presented with a cutaneous drug reaction after oral pristinamycin, which cleared after discontinuing the drug. Four of these individuals had a history of contact dermatitis from virginiamycin. Patch tests with commercial pristinamycin 20% water and pet. were positive in 3 of the 5 patients with maculopapular exanthemas, all 3 with eczematous eruptions and all 3 with AGEP. Thirty controls were negative. The three cases of eczematous eruptions were probably cases of systemic allergic dermatitis after previous sensitization to virginiamycin (22, overlap with ref. 4).

Of 3 patients with a cutaneous adverse drug reaction to pristinamycin, 2 had positive patch tests to the antibiotic; details are not available to the author (14).

Maculopapular eruption

In a study from Nancy, France, 54 patients with suspected nonimmediate drug eruptions were assessed with patch testing. Five of the 27 patients with maculopapular eruptions had positive patch tests to pristinamycin (19). See also the section 'Case series with various or unknown types of drug reactions' above, refs. 3,20 and 22.

Erythroderma, widespread erythematous eruption, exfoliative dermatitis

In a study from Nancy, France, 54 patients with suspected nonimmediate drug eruptions were assessed with patch testing. Three of the 7 patients with erythroderma had positive patch tests to pristinamycin (19). See the section 'Case series with various or unknown types of drug reactions' above, ref. 3.

Acute generalized exanthematous pustulosis (AGEP)

In a multicenter investigation in France, of 45 patients patch tested for AGEP, 26 (58%) had positive patch tests to drugs, including 8 to pristinamycin. In one patient, the patch test resulted in a flare-up of AGEP (17). In a group of 45 patients with multiple drug hypersensitivity seen between 1996 and 2018 in Montpellier, France, 38 of 92 drug hypersensitivities were classified as type IV immunological reactions. Of these, 2 individuals developed AGEP, in one caused by pristinamycin with a positive patch test (16).

See also the section 'Case series with various or unknown types of drug reactions' above, refs. 2,20 and 22.

Systemic allergic dermatitis (systemic contact dermatitis)

When three patients, previously sensitized to virginiamycin from topical application, were given oral pristinamycin, all 3 patients developed systemic allergic dermatitis manifesting as generalised maculovesicular erythema with fever in the first, generalized erythema and facial edema in the second and erythema of the trunk, fever and headache in the third patient. They all had positive patch tests to Staphylomycine ® ointment containing 0.5% virginiamycin. Two of the patients were patch tested with pristinamycin (Pyostacine ® 1 tablet crushed in 1 ml water) and reacted positively (4, overlap with ref. 22).

A man sensitized to virginiamycin in a topical preparation took one tablet of 250 mg pristinamycin and 4 hours later had a reaction with stupor, urticaria and vomiting. Patch tests with virginiamycin, pristinamycin, factors M and IIA, each 1% pet., were all positive. He developed transient edema of his eyes and lips, and wheals adjacent to a positive patch test (5).

Other cases of systemic allergic dermatitis from oral pristinamycin in patients previously sensitized to virginiamycin have been reported in French literature (6,8; no data available to the author). See also the section 'Case series with various or unknown types of drug reactions' above, ref 22.

Symmetrical drug-related intertriginous and flexural exanthema (SDRIFE)/Baboon syndrome
In 2 large hospitals in France, 18 patients were diagnosed with SDRIFE in the period 2006-2018. Fourteen were patch tested, and there were 3 (21%) positive reactions, two of which were to pristinamycin (18).

Fixed drug eruption
In France, in the period 2005-2007, 59 cases of fixed drug eruptions were collected in 17 academic centers. There were two cases of FDE to pristinamycin with a positive patch test. Clinical details were not provided (12). See also the section 'Case series with various or unknown types of drug reactions' above, ref 2.

Drug reaction with eosinophilia and systemic symptoms (DRESS)
In a multicenter investigation in France, of 72 patients patch tested for DRESS, 46 (64%) had positive patch tests to drugs, including 3 to pristinamycin (17). See also the section 'Case series with various or unknown types of drug reactions' above, ref 2.

Stevens-Johnson syndrome/toxic epidermal necrolysis (SJS/TEN)
See the section 'Case series with various or unknown types of drug reactions' above, refs. 2 and 3.

Dermatitis/eczematous eruption
See the section 'Case series with various or unknown types of drug reactions' above, ref. 22.

Other cutaneous adverse drug reactions
A man aged 59 was given pristinamycin for a week for the treatment of an infected inguinal cyst. Two days after finishing this course, the patient developed generalized urticaria, which persisted for 2 days under antihistamine treatment. Patch tests were positive to commercial pristinamycin 30% water and 1%, 10% and 30% pet. and to the structurally related quinupristin/dalfopristin (13). One patient developed urticaria from pristinamycin and had positive patch tests to this drug tested 10% and 30% pet. (15).

See also the section 'Case series with various or unknown types of drug reactions' above, ref. 14.

Cross-reactions, pseudo-cross-reactions and co-reactions
Patients sensitized to pristinamycin (from oral administration) may cross-react to virginiamycin (3). Conversely, patients sensitized to virginiamycin may cross-react to pristinamycin (4,5,7,10,11) and may develop allergic dermatitis when given pristinamycin orally (4,6,8). Cross-sensitivity is easily explained by structural similarities between these 2 antibiotics: virginiamycin factor M1 is the same as pristinamycin IIA and virginiamycin factor S1 is virtually identical to pristinamycin IA. The most important sensitizer appears to be virginiamycin factor M1 (= pristinamycin IIA) (7,10,11).

Cross-reactions may occur to the related synergistin antibiotic combination dalfopristin–quinupristin (containing among others pristinamycin IA [quinupristin] and pristinamycin IIA [dalfopristin] (3,14).

LITERATURE
1   The data in the section 'General' may have been obtained from literature discussed in this chapter, but mostly also or exclusively from one or more of the following online sources: ChemIDPlus Advanced, PubChem, DrugBank, RxList, Drug Central, Drugs.com, and Wikipedia
2   El Khoury M, Assier H, Gener G, Paul M, Haddad C, Chosidow O, et al. Polysensitivity in delayed cutaneous adverse drug reactions to macrolides, clindamycin and pristinamycin: clinical history and patch testing. Br J Dermatol 2018;179:978-979
3   Barbaud A, Trechot P, Weber-Muller F, Ulrich G, Commun N, Schmutz JL. Drug skin tests in cutaneous adverse drug reactions to pristinamycin: 29 cases with a study of cross-reactions between synergistins. Contact Dermatitis 2004;50:22-26

4   Michel M, Dompmartin A, Szczurko C, Castel B, Moreau A, Leroy D. Eczematous-like drug eruption induced by synergistins. Contact Dermatitis 1996;34:86-87

5   Baes H. Allergic contact dermatitis to virginiamycin. False cross-sensitivity with pristinamycin. Dermatologica 1974;149:231-235

6   Pillette M, Claudel JP, Muller C, Lorette G. Contact dermatitis caused by pristinamycin after sensitization to topical virginiamycin. Allerg Immunol (Paris) 1990;22:197 (Article in French)

7   Tennstedt D, Dumont-Fruytier M, Lachapelle JM. Occupational allergic contact dermatitis to virginiamycin, an antibiotic used as a food additive for pigs and poultry. Contact Dermatitis 1978;4:133-134

8   Mathivon F, Petit A, Mourier C, Sigal Nahum M. Toxidermie à la pristinamycine après réaction de contact à la virginiamycine. Rev Eur Dermatol MST 1991;3:527-529 (Article in French)

10 Bleumink E, Nater JP. Sensitization to Staphylomycin ® (virginiamycin). Contact Dermatitis Newsletter 1972;11:306

11 Bleumink E, Nater JP. Allergic contact dermatitis to virginiamycin (Staphylomycin ®) and pristinamycin (Stapyocin®). Contact Dermatitis Newsletter 1972;12:337

12 Brahimi N, Routier E, Raison-Peyron N, Tronquoy AF, Pouget-Jasson C, Amarger S, et al. A three-year-analysis of fixed drug eruptions in hospital settings in France. Eur J Dermatol 2010;20:461-464

13 Dellestable P, Weber-Muller F, Tréchot P, Vernassière C, Schmutz JL, Barbaud A. Deux médicaments de même imputabilité dans une toxidermie (deux cas): limite des tests épicutanés? Ann Dermatol Venereol 2007;134):655-658 (Article in French)

14 Bernard P, Fayol J, Bonnafoux A, Bedane C, Delrous JL, Catanzano G, Bonnetblanc JM. Toxidermies après prise orale de pristinamycine. Ann Dermatol Venereol 1988;115:63-66 (Article in French)

15 Studer M, Waton J, Bursztejn AC, Aimone-Gastin I, Schmutz JL, Barbaud A. Does hypersensitivity to multiple drugs really exist? Ann Dermatol Venereol. 2012;139:375-380 (Article in French)

16 Landry Q, Zhang S, Ferrando L, Bourrain JL, Demoly P, Chiriac AM. Multiple drug hypersensitivity syndrome in a large database. J Allergy Clin Immunol Pract 2019;8:258

17 Barbaud A, Collet E, Milpied B, Assier H, Staumont D, Avenel-Audran M, et al. A multicentre study to determine the value and safety of drug patch tests for the three main classes of severe cutaneous adverse drug reactions. Br J Dermatol 2013;168:555-562

18 De Risi-Pugliese T, Barailler H, Hamelin A, Amsler E, Gaouar H, Kurihara F, et al. Symmetrical drug-related intertriginous and flexural exanthema: A little-known drug allergy. J Allergy Clin Immunol Pract 2020;8:3185-3189.e4

19 Barbaud A, Reichert-Penetrat S, Tréchot P, Jacquin-Petit MA, Ehlinger A, Noirez V, et al. The use of skin testing in the investigation of cutaneous adverse drug reactions. Br J Dermatol 1998;139:49-58

20 Assier H, Valeyrie-Allanore L, Gener G, Verlinde Carvalh M, Chosidow O, Wolkenstein P. Patch testing in non-immediate cutaneous adverse drug reactions: value of extemporaneous patch tests. Contact Dermatitis 2017;77:297-302

21 Blancas-Espinosa R, Condé-Salazar L, Pérez-Hortet C. Occupational airborne contact dermatitis from pristinamycin. Contact Dermatitis 2006;54:63-65

22 Mayence C, Dompmartin A, Verneuil L, Michel M, Leroy D. Value of patch tests in pristinamycin-induced drug eruptions. Contact Dermatitis 1999;40:161-162

Chapter 3.408 PROCAINE

IDENTIFICATION

Description/definition   : Procaine is the *p*-aminobenzoic acid derivative that conforms to the structural formula
                             shown below
Pharmacological classes   : Anesthetics, local
IUPAC name   : 2-(Diethylamino)ethyl 4-aminobenzoate
Other names   : Novocain ®
CAS registry number   : 59-46-1
EC number   : 200-426-9
Merck Index monograph   : 9145
Patch testing   : Hydrochloride, 1.0% pet.(Chemotechnique, SmartPracticeCanada, SmartPracticeEurope);
                             hydrochloride, 2% pet. (SmartPracticeCanada, SmartPracticeEurope)
Molecular formula   : $C_{13}H_{20}N_2O_2$

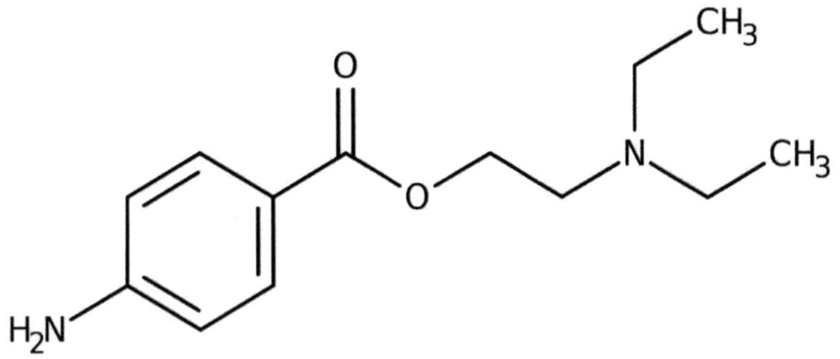

GENERAL

Procaine is a local anesthetic of the ester type that has a slow onset and a short duration of action. It is (or actually was) mainly used for production of local or regional anesthesia, particularly for oral surgery. Procaine (like cocaine) has the advantage of constricting blood vessels which reduces bleeding, unlike other local anesthetics such as lidocaine. In pharmaceutical products, both procaine and procaine hydrochloride (CAS number 51-05-8, EC number 200-077-2, molecular formula $C_{13}H_{21}ClN_2O_2$) may be employed (1).

The first description of contact sensitivity to procaine was in 1921 in three dentists (8). While further reports followed (9,10,11), procaine remained popular in dentistry and was also a common constituent of preparations such as eye drops, other topical medicaments and even cosmeceuticals (for hair growth) (3). This led to many cases of sensitization, which has been fully reviewed in Volume 3 of the *Monographs in contact allergy* series (19).

The early literature on procaine allergy has been reviewed in refs. 6 and 7. Because this local anesthetic can largely be considered a historical allergen, the subject is presented in brief format and far from complete.

CONTACT ALLERGY FROM ACCIDENTAL CONTACT

Case series

In 1921, occupational allergic contact dermatitis to procaine was described in three dentists (8). Such reactions in dentists and physicians (17) later became well known (12,16) and procaine remained an important occupational allergen until the mid-1970s (15). In a group of 107 workers in the pharmaceutical industry with dermatitis, investigated in Warsaw, Poland, before 1989, one reacted to procaine, tested 2% pet. (2).

In Norway, before 1985, of ten veterinary surgeons with occupational allergic contact dermatitis with chronic or relapsing eczema of the hands as the main complaint, 3 had positive patch tests to procaine 2% pet. (4). Of 26 veterinarians patch tested in the early 1980s in Poland because of dermatitis, 15 had one or more positive patch tests to veterinary drugs, tuberculin and/or disinfectants. Four of the veterinarians reacted to procaine 2% water (20).

Case reports

A 44-year-old man had worked in a pigsty for 12 years. He had used injectable antibiotics in his 1400 animal pigsty to treat suspected infections in animals. He presented with swollen eyelids, redness and flaking in the face, in the upper part of the body, on the arms and behind the knees. Patch tests were positive to penicillin and procaine, with both of which he came in contact during his work in the form of procaine benzylpenicillin. Penicillin-containing and procaine-containing medicines were replaced by other preparations, and since then the patient has been working in his pigsty without skin symptoms (13). Occupational allergic contact dermatitis to procaine was also found in a 47-year-old veterinary obstetrician from regularly using procaine penicillin G in his work (14). Occupational allergic contact dermatitis/contact allergy to procaine has also been observed in nurses (21) and stomatologists (22) in Poland.

CUTANEOUS ADVERSE DRUG REACTIONS FROM SYSTEMIC ADMINISTRATION CAUSED BY TYPE IV (DELAYED-TYPE) HYPERSENSITIVITY (as demonstrated by positive patch tests)
Cutaneous adverse drug reactions from systemic administration of procaine caused by type IV (delayed-type) hypersensitivity have included systemic allergic dermatitis (23), localized allergic reactions from subcutaneous injections (5,23) and urticaria and other cutaneous exanthemas from injecting procaine benzylpenicillin in patients allergic to procaine (6,7).

Systemic allergic dermatitis (systemic contact dermatitis)
A hairdresser had a lichenified patch on the anterior aspect of the left wrist. Patch tests were positive to *p*-phenylenediamine and the chemically related procaine and sulfanilamide. An oral provocation test with procaine resulted in a local flare-up at the site of dermatitis (systemic allergic dermatitis) (23). More results of oral provocation tests with procaine in patients with a dermatitis due to a sulfonamide or *p*-phenylenediamine can be found in ref. 23).

Other cutaneous adverse drug reactions

Localized allergic reactions from subcutaneous injection
Subcutaneous injections with procaine for anesthesia can induce local allergic reactions with erythema and edema (5; no patch tests performed, but intradermal tested positive at D1 and D2). A woman who was occupationally sensitized to procaine by working in a dentist office developed an eczematous dermatitis of the back of the hand after local infiltration with procaine for incision of a panaritium. Patch tests were positive to procaine and *p*-phenylenediamine (23).

Other cutaneous adverse drug reactions
Patients allergic to procaine have reacted to procaine penicillin with urticaria and other cutaneous exanthemas (6,7).

Cross-reactions, pseudo-cross-reactions and co-reactions
Procaine, the 2-(diethylamino)ethyl ester of *p*-aminobenzoic acid (PABA) is a 'para-amino' compound and therefore may cross-react with other ester-type local anesthetics including benzocaine (ethyl *p*-aminobenzoate) (13,17), tetracaine (13,17,18), butacaine (17), and other para-compounds including *p*-phenylenediamine (13,17), *p*-amino-azobenzene, *p*-aminobenzoic acid (PABA) (17), *p*-aminophenol (3), and sulfonamide (17).

LITERATURE
1   The data in the section 'General' may have been obtained from literature discussed in this chapter, but mostly also or exclusively from one or more of the following online sources: ChemIDPlus Advanced, PubChem, DrugBank, RxList, Drug Central, Drugs.com, and Wikipedia
2   Rudzki E, Rebandel P, Grzywa Z. Contact allergy in the pharmaceutical industry. Contact Dermatitis 1989;21:121-122
3   Dooms-Goossens A, Swinnen E, VanderMaesen J, Marien K, Dooms M. Connubial dermatitis from a hair lotion. Contact Dermatitis 1987;16:41-42
4   Falk ES, Hektoen H, Thune PO. Skin and respiratory tract symptoms in veterinary surgeons. Contact Dermatitis 1985;12:274-278
5   Trautmann A, Stoevesandt J. Differential diagnosis of late-type reactions to injected local anaesthetics: inflammation at the injection site is the only indicator of allergic hypersensitivity. Contact Dermatitis 2019;80:118-124
6   Fernström AI. Studies on procaine allergy with reference to urticaria due to procaine penicillin treatment. B. The medical uses of procaine and reactions due to procaine penicillin. Review of the literature. Acta Derm Venereol 1960;40:19-34.
7   Fernström AI. Studies on procaine allergy with reference to urticaria due to procaine penicillin treatment. C. Reactions following procaine due to direct skin contact or injection. Hypersensitivity to procaine as an expression of cross-sensitization. Effects of procaine penicillin in persons with clinically verified sensitivity to procaine or chemically related substances. Review of the literature. Acta Derm Venereol 1960;40:175-205

8   Lane CG. Occupational dermatitis in dentists: Susceptibility to procaine. Arch Dermatol 1921;3:235-244

9   James BM. Procaine dermatitis. JAMA 1931;97:440-443

10   Waldron GW. Hypersensitivity to procaine. Mayo Clin Proc 1934;9:254-256

11   Goodman MH. Cutaneous hypersensitivity to the procaine anesthetics. J Invest Dermatol 1939;2:53-66

12   Lane CG, Luikart R. Dermatitis from local anesthetics, with a review of one hundred and seven cases from the literature. J Am Med Assoc 1951;146:717-720

13   Jussi L, Lammintausta K. Sources of sensitization, cross-reactions, and occupational sensitization to topical anaesthetics among general dermatology patients. Contact Dermatitis 2009;60:150-154

14   Bruijn MS, Lavrijsen APM, van Zuuren EJ. An unusual case of contact dermatitis to procaine. Contact Dermatitis 2009;60:182-183

15   Hensten-Pettersen A, Jacobsen N. The role of biomaterials as occupational hazards in dentistry. Int Dent J 1990;40:159-166

16   Laden EL, Wallace DA. Contact dermatitis due to procaine; a common occupational disease of dentists. J Invest Dermatol 1949;12:299-306

17   Peck SM, Feldman FF. Contact allergic dermatitis due to the procaine fraction of procaine penicillin. J Invest Dermatol 1949;13:109

18   Kalveram K, Günnewig W, Wehling K, Forck G. Tetracaine allergy: cross-reactions with para-compounds? Contact Dermatitis 1978;4:376

19   De Groot AC. Monographs in contact allergy, volume 3: Topical Drugs. Boca Raton, Fl, USA: CRC Press Taylor and Francis Group, 2021 (ISBN 978-0-367-23693-9)

20   Rudzki E, Rebandel P, Grzywa Z, Pomorski Z, Jakiminska B, Zawisza E. Occupational dermatitis in veterinarians. Contact Dermatitis 1982;8:72-73

21   Rudzki E, Rebandel P, Grzywa Z. Patch tests with occupational contactants in nurses, doctors and dentists. Contact Dermatitis 1989;20:247-250

22   Rudzki E. Occupational dermatitis among health service workers. Derm Beruf Umwelt 1979;27:112-115

23   Sidi E, Dobkevitch-Morrill S. The injection and ingestion test in cross-sensitization to the para group. J Invest Dermatol 1951;16:299-310

Chapter 3.409 PROCAINE BENZYLPENICILLIN

IDENTIFICATION

Description/definition	: Procaine benzylpenicillin is the semisynthetic antibiotic prepared by combining penicillin G with procaine; it conforms to the structural formula shown below
Pharmacological classes	: Anti-bacterial agents
IUPAC name	: 2-(Diethylamino)ethyl 4-aminobenzoate;(2S,5R,6R)-3,3-dimethyl-7-oxo-6-[(2-phenyl-acetyl)amino]-4-thia-1-azabicyclo[3.2.0]heptane-2-carboxylic acid
Other names	: Benzylpenicillin procaine; penicillin G procaine
CAS registry number	: 54-35-3; 6130-64-9
EC number	: 200-205-7
Merck Index monograph	: 8476
Patch testing	: 30% water; test also procaine hydrochloride, 1.0% pet.(Chemotechnique, SmartPracticeCanada, SmartPracticeEurope) or 2% pet. (SmartPracticeCanada, SmartPracticeEurope); test also benzylpenicillin 5% pet.
Molecular formula	: $C_{29}H_{38}N_4O_6S$

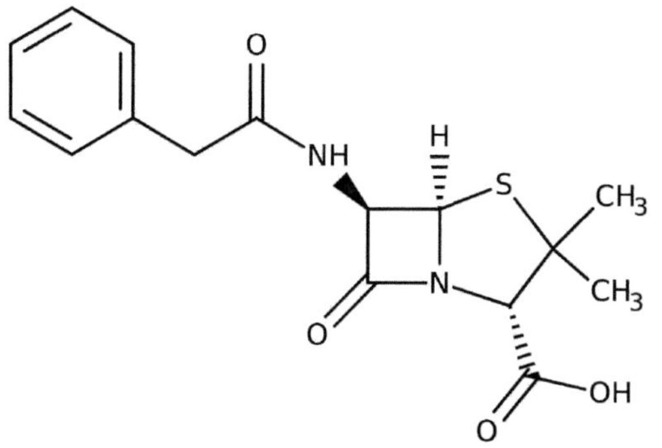

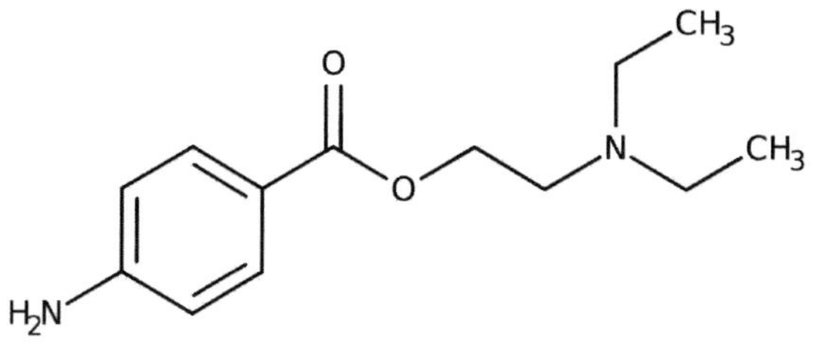

GENERAL

Procaine benzylpenicillin is a combination of the naturally occurring broad-spectrum β-lactam penicillin benzylpenicillin (penicillin G) and the local anesthetic agent procaine in equimolar amounts. Procaine benzylpenicillin is administered by deep intramuscular injection. It is slowly absorbed and hydrolyzed to benzylpenicillin. This drug is used where prolonged exposure to benzylpenicillin at a low concentration is required. This combination is aimed at reducing the pain and discomfort associated with a large intramuscular injection of penicillin. It is widely used in veterinary settings. Benzylpenicillin is active against a wide range of organisms. It is indicated for the treatment of a number of bacterial infections such as syphilis, anthrax, mouth infections, pneumonia and diphtheria (1). Some commercial preparations contain the monohydrate form of procaine benzylpenicillin (CAS number 6130-64-9, EC number 612-112-2, molecular formula $C_{29}H_{40}N_4O_7S$) (1). The classification and structures of beta-lactam antibiotics are discussed in Chapter 3.36 Amoxicillin.

CONTACT ALLERGY FROM ACCIDENTAL CONTACT

In 34 veterinary surgeons (7 women, 27 men, age range 29-61, mean age 38 years) with chronic or relapsing eczema of the hands as the main complaint, investigated in Norway before 1985, patch tests were performed with the standard series and a veterinary series comprising drugs, antiseptics and protective glove materials frequently used in veterinary practice. Nineteen had eczema almost continuously on the hands, fingers and arms, and in another 3 there was a spread to the face and trunk. Occupational work caused exacerbations in 15 of these patients. Ten of the 34 veterinary surgeons had positive patch test reactions. A relation between positive patch tests and occupational work was confirmed in 9 of these subjects. In this group, there were three positive patch test reactions to procaine benzylpenicillin 30% water, of who 2 also reacted to procaine (2).

A 44-year-old man presented with swollen eyelids, redness and flaking in the face, on the upper part of the body, on the arms and behind the knees. The patient had worked in a pigsty for 12 years, frequently using injectable antibiotics in pigs with suspected infections. Patch tests were positive to procaine, 2 commercial procaine benzyl-penicillin (200.000 IU/300 mg) products, penicillin G (benzylpenicillin), and penicillin V. All other contact materials were negative (3).

CUTANEOUS ADVERSE DRUG REACTIONS FROM SYSTEMIC ADMINISTRATION CAUSED BY TYPE IV (DELAYED-TYPE) HYPERSENSITIVITY (as demonstrated by positive patch tests)
Cutaneous adverse drug reactions from systemic administration of procaine benzylpenicillin caused by type IV (delayed-type) hypersensitivity have included toxic epidermal necrolysis (TEN) (4).

Stevens-Johnson syndrome/toxic epidermal necrolysis (SJS/TEN)
A 6-year-old girl had developed TEN after parenteral administration of the fifth dose of procaine benzylpenicillin and second dose of ceftriaxone given for an upper respiratory tract infection, which she had tolerated few months before. Patch tests were positive to procaine benzylpenicillin and negative to ceftriaxone. However, an intradermal test to ceftriaxone was positive at delayed reading (4).

Cross-reactions, pseudo-cross-reactions and co-reactions
Cross-reactions between beta-lactam antibiotics are discussed in Chapter 3.36 Amoxicillin. Of three veterinary surgeons reacting to procaine benzylpenicillin, 2 co-reacted to procaine (2).

LITERATURE
1   The data in the section 'General' may have been obtained from literature discussed in this chapter, but mostly also or exclusively from one or more of the following online sources: ChemIDPlus Advanced, PubChem, DrugBank, RxList, Drug Central, Drugs.com, and Wikipedia
2   Falk ES, Hektoen H, Thune PO. Skin and respiratory tract symptoms in veterinary surgeons. Contact Dermatitis 1985;12:274-278
3   Jussi L, Lammintausta K. Sources of sensitization, cross-reactions, and occupational sensitization to topical anaesthetics among general dermatology patients. Contact Dermatitis 2009;60:150-154
4   Atanasković-Marković M, Medjo B, Gavrović-Jankulović M, Ćirković Veličković T, Nikolić D, Nestorović B. Stevens-Johnson syndrome and toxic epidermal necrolysis in children. Pediatr Allergy Immunol 2013;24:645-649

Chapter 3.410 PROGUANIL

IDENTIFICATION

Description/definition : Proguanil is the 1-arylbiguanide that conforms to the structural formula shown below
Pharmacological classes : Antimetabolites; Antimalarials
IUPAC name : (1*E*)-1-[Amino-(4-chloroanilino)methylidene]-2-propan-2-ylguanidine
Other names : Chlorguanide
CAS registry number : 500-92-5
EC number : 207-915-6
Merck Index monograph : 3361
Patch testing : Proguanil tablets, pulverized, 30% pet. (2); most systemic drugs can be tested at 10% pet.; if the pure chemical is not available, prepare the test material from intravenous powder, the content if capsules or – when also not available – from powdered tablets to achieve a final concentration of the active drug of 10% pet.
Molecular formula : $C_{11}H_{16}ClN_5$

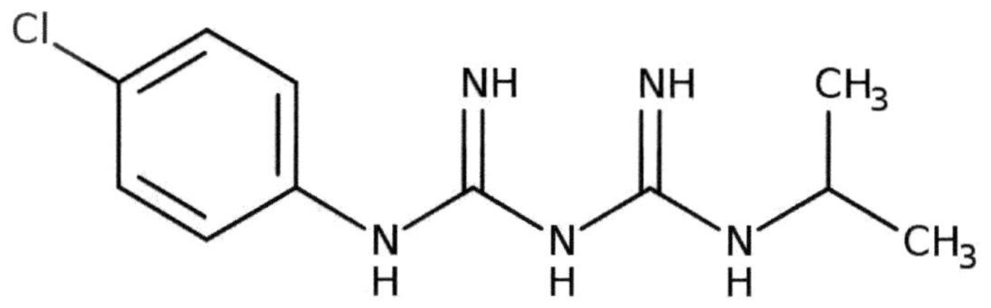

GENERAL

Proguanil is a biguanide derivative which is active against several protozoal species and is used in combination with atovaquone and chloroquine for the prevention and therapy of malaria. It works by stopping the malaria parasite, *Plasmodium falciparum* and *Plasmodium vivax*, from reproducing once it is in the red blood cells. It does this by inhibiting the enzyme, dihydrofolate reductase, which is involved in the reproduction of the parasite. In pharmaceutical products, proguanil is employed as proguanil hydrochloride (CAS number 637-32-1, EC number not available, molecular formula $C_{11}H_{17}Cl_2N_5$) (1).

CUTANEOUS ADVERSE DRUG REACTIONS FROM SYSTEMIC ADMINISTRATION CAUSED BY TYPE IV (DELAYED-TYPE) HYPERSENSITIVITY (as demonstrated by positive patch tests)

Cutaneous adverse drug reactions from systemic administration of proguanil caused by type IV (delayed-type) hypersensitivity have included maculopapular eruption, possibly as manifestation of drug reaction with eosinophilia and systemic symptoms (DRESS) (2).

Drug reaction with eosinophilia and systemic symptoms (DRESS)

A 40-year-old woman presented with shortness of breath and widespread skin lesions. Ten days before, she had initiated atovaquone-proguanil treatment for malaria prophylaxis prior to a trip to Senegal. She had taken the same treatment for the first time one year previously without any complications. The patient developed progressive dyspnea and fever while continuing to take the medication. She was prescribed amoxicillin plus clavulanic acid for bronchitis, but this treatment was withdrawn 4 days later because of rapidly spreading skin eruption on day 3. Physical examination revealed a widespread non-pruritic maculopapular eruption. The patient had leukocytosis, eosinophilia and high C-reactive protein. The dyspnea proved to be the result of eosinophilic pneumonia. Patch tests were positive to commercial proguanil and atovaquone-proguanil 30% pet., but negative to atovaquone, amoxicillin and other drugs used by the patient. Patch tests were 'all negative in controls'. This was a possible case of drug reaction with eosinophilia and systemic symptoms (DRESS) due to proguanil (2).

LITERATURE

1 The data in the section 'General' may have been obtained from literature discussed in this chapter, but mostly also or exclusively from one or more of the following online sources: ChemIDPlus Advanced, PubChem, DrugBank, RxList, Drug Central, Drugs.com, and Wikipedia
2 Just N, Carpentier O, Brzezinki C, Steenhouwer F, Staumont-Sallé D. Severe hypersensitivity reaction as acute eosinophilic pneumonia and skin eruption induced by proguanil. Eur Respir J 2011;37:1526-1528

Chapter 3.411 PROMAZINE

IDENTIFICATION

Description/definition   : Promazine is the phenothiazine that conforms to the structural formula shown below
Pharmacological classes : Dopamine antagonists; antiemetics; antipsychotic agents
IUPAC name   : *N,N*-Dimethyl-3-phenothiazin-10-ylpropan-1-amine
CAS registry number   : 58-40-2
EC number   : 200-382-0
Merck Index monograph   : 9168
Patch testing   : 1% and 0.1% pet.
Molecular formula   : $C_{17}H_{20}N_2S$

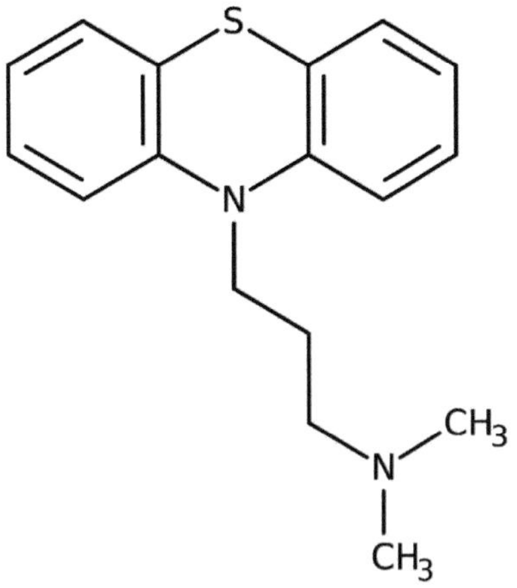

GENERAL

Promazine is a phenothiazine derivative with antipsychotic and antiemetic properties. It blocks postsynaptic dopamine receptors D_1 and D_2 in the mesolimbic and medullary chemoreceptor trigger zone, thereby decreasing stimulation of the vomiting center in the brain and psychotic effects, such as hallucinations and delusions. In addition, this agent blocks α-adrenergic receptors and exhibits strong anticholinergic activity. Promazine is primarily used in short-term treatment of disturbed behavior and as an antiemetic. In pharmaceutical products, promazine is employed as promazine hydrochloride (CAS number 53-60-1, EC number 200-179-7, molecular formula $C_{17}H_{21}ClN_2S$) (1) (go.drugbank.com).

CUTANEOUS ADVERSE DRUG REACTIONS FROM SYSTEMIC ADMINISTRATION CAUSED BY TYPE IV (DELAYED-TYPE) HYPERSENSITIVITY (as demonstrated by positive patch tests)

Cutaneous adverse drug reactions from systemic administration of promazine caused by type IV (delayed-type) hypersensitivity have included photosensitivity (possibly systemic photoallergic dermatitis) (1).

Photosensitivity

A 79-year-old man presented with an itching and burning eczematous eruption on the face, arms and back of the hands, which had begun 2 weeks after the start of treatment with promazine hydrochloride and trazodone hydrochloride for Alzheimer's disease. One month after the resolution of dermatitis, patch and photopatch tests were performed, yielding a positive photopatch test to the commercial promazine drops 5% pet. (D2 +, D4 and D7 ++) and to promethazine hydrochloride 1% pet. In a second test session, photopatch tests to pure promazine 1% and 0.1% pet. were positive (D2 and D4 ++) to both concentrations. Ten controls were negative to promazine tablet 5% pet. and promazine 1% pet. Photoallergy to promazine was diagnosed. The patient had previously used promethazine cream on several occasions, so, possibly, he had become photosensitized to promethazine and now photocross-reacted to promazine (1). In that case, it would have been systemic photoallergic dermatitis.

Cross-reactions, pseudo-cross-reactions and co-reactions

Possibly photocross-sensitivity to promazine in a promethazine-photosensitized individual (1).

LITERATURE

1   Romita P, Foti C, Stingeni L. Photoallergy to promazine hydrochloride. Contact Dermatitis 2017;77:182-183

Chapter 3.412 PROMETHAZINE

IDENTIFICATION

Description/definition   : Promethazine is the phenothiazine-derivative that conforms to the structural formula
                             shown below
Pharmacological classes   : Histamine H1 antagonists; anti-allergic agents; antipruritics
IUPAC name   : *N,N*-Dimethyl-1-phenothiazin-10-ylpropan-2-amine
CAS registry number   : 60-87-7
EC number   : 200-489-2
Merck Index monograph   : 9171
Patch testing   : Hydrochloride, 0.1% pet. (Chemotechnique); hydrochloride, 2% pet.
                             (SmartPracticeCanada)
Molecular formula   : $C_{17}H_{20}N_2S$

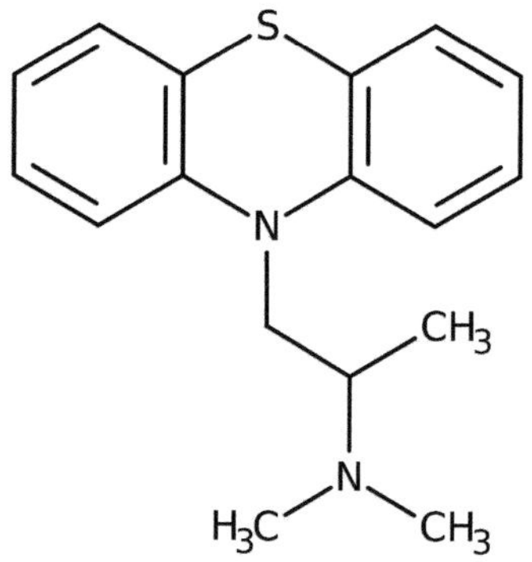

GENERAL

Promethazine is a phenothiazine derivative with antihistaminic, sedative and antiemetic properties. It selectively blocks peripheral H1 receptors, thereby diminishing the effects of histamine on effector cells. Promethazine is used as an antiallergic, in the treatment of pruritus, for sedation and to prevent and treat nausea and vomiting, e.g. from motion sickness. In pharmaceutical products, promethazine is employed as promethazine hydrochloride (CAS number 58-33-3, EC number 200-375-2, molecular formula $C_{17}H_{21}ClN_2S$). It is available as tablets, syrup, injection fluid, suppository and in some countries as cream for the treatment of itch and insect bites (1). Formerly, by its presence in topical preparations, promethazine has caused many cases of allergic, photoallergic and photoaugmented contact dermatitis, which has been reviewed in Volume 3 of the *Monographs in contact allergy* series (16).

CONTACT ALLERGY FROM ACCIDENTAL CONTACT

Occupational sensitization to promethazine has been described several times, e.g. in a nurse handling promethazine injectable solutions (12).

CUTANEOUS ADVERSE DRUG REACTIONS FROM SYSTEMIC ADMINISTRATION CAUSED BY TYPE IV (DELAYED-TYPE) HYPERSENSITIVITY (as demonstrated by positive patch tests)

Cutaneous adverse drug reactions from systemic administration of promethazine caused by type IV (delayed-type) hypersensitivity have included systemic allergic dermatitis (12,13), fixed drug eruption (2,17), and photosensitivity (14).

Systemic allergic dermatitis (systemic contact dermatitis)

In patients sensitized to topical promethazine, oral administration has resulted in systemic allergic dermatitis with reactivation of the (previous) eruption, spreading of the eczema with development of erythroderma as well as systemic manifestations such as fever, chills, intestinal upset and sometimes even syncope (12,13). Such systemic allergic dermatitis could also be caused by oral administration of the related phenothiazine chlorpromazine in promethazine-allergic individuals (12).

Fixed drug eruption

A 52-year-old woman ingested a remedy for the treatment of the common cold containing promethazine methylenedisalicylate, anhydrous caffeine, acetaminophen and salicylamide. The following day, she noticed an itchy macular erythema on her right forearm, back and buttocks, which was clinically diagnosed as fixed drug eruption. Histopathology confirmed the diagnosis. An oral provocation test after 8 hours induced a new itchy macular erythema at the same locations as previously. Further provocation tests with each of the four components were positive to promethazine methylenedisalicylate only. Patch tests were positive to promethazine hydrochloride on previously involved skin, as was an oral provocation test. These tests with salicylamide and acetylsalicylic acid were negative. Later, the patient would develop fixed drug eruptions from 2 unrelated drugs, pethidine and omeprazole (17).

One case of fixed drug eruption to promethazine with a positive patch test reaction on postlesional skin (itching, erythema, infiltration) to the commercial drug 10% pet. was reported from Seoul, South Korea, in the period 1986-1990 and in 1996 and 1997 (2).

Photosensitivity

General
Formerly, by its presence in topical preparations, promethazine has caused many cases of allergic, photoallergic and photoaugmented contact dermatitis, which has been reviewed in Volume 3 of the *Monographs in contact allergy* series (16). This chapter provides some literature on (photo)allergic reactions to promethazine administered systemically, but a full literature review has not been attempted, as promethazine is probably largely a historical photoallergen. High rates of >5% positive photopatch tests to promethazine found in (more or less recent) studies in which groups of patients suspected of photosensitivity were photopatch tested (4,5,6,8,9,10) would seem to contradict the latter statement. Usually, however, their relevance was either not mentioned or unknown, whereas, in the case of promethazine, relevance should be fairly easy to establish. Also, it is well known that promethazine can induce phototoxic reactions, depending on the concentration used and the irradiation parameters, which are very difficult to distinguish from photoallergic reactions (7,16). Therefore, it may be assumed that many of the positive photopatch tests in such studies with high prevalences have been phototoxic rather than photoallergic, which was also suggested by many of the investigators themselves (16).

Case reports and case series
Photosensitive eruptions caused by oral administration of promethazine without previous sensitization to topical application have been observed (14).

Cross-reactions, pseudo-cross-reactions and co-reactions
Promethazine has commonly (photo)cross-reacted with chlorpromazine (12,14) and cross-reacted to the related phenothiazine thiazinamium metilsulfate (Multergan ®) (12). (Photo)-cross-reactions have also been observed to or from chlorproethazine (3), dioxopromethazine (11) and isothipendyl (15). In promethazine-intolerant patients, co-reactivity has also been described with *p*-phenylenediamine (30%) and other *p*-amino compounds (procaine, other ester-type anesthetics and sulfonamides). However, the reverse phenomenon was not seen: subjects sensitized to *p*-phenylenediamine as a rule did not react to promethazine (12). Two patients photosensitized to mequitazine photocross-reacted to promethazine (18). Possibly photocross-sensitivity to promazine in a promethazine-photosensitized individual (19).

Immediate contact reactions
Immediate contact reactions (contact urticaria) to promethazine are presented in Chapter 5.

LITERATURE
1   The data in the section 'General' may have been obtained from literature discussed in this chapter, but mostly also or exclusively from one or more of the following online sources: ChemIDPlus Advanced, PubChem, DrugBank, RxList, Drug Central, Drugs.com, and Wikipedia
2   Lee AY. Topical provocation in 31 cases of fixed drug eruption: change of causative drugs in 10 years. Contact Dermatitis 1998;38:258-260
3   Barbaud A, Collet E, Martin S, Granel F, Trechot P, Lambert D, et al. Contact sensitization to chlorproethazine can induce persistent light reaction and cross-photoreactions to other phenothiazines. Contact Dermatitis 2001;44:373-374
4   Greenspoon J, Ahluwalia R, Juma N, Rosen CF. Allergic and photoallergic contact dermatitis: A 10-year experience. Dermatitis 2013;24:29-32
5   Cardoso J, Canelas MM, Gonçalo M, Figueiredo A. Photopatch testing with an extended series of photoallergens: a 5-year study. Contact Dermatitis 2009;60:325-329
6   Scalf LA, Davis MDP, Rohlinger AL, Connolly SM. Photopatch testing of 182 patients: A 6-year experience at the Mayo Clinic. Dermatitis 2009;20:44-52

7   Hölzle E, Neumann N, Hausen B, Przybilla B, Schauder S, Hönigsmann H, et al. Photopatch testing: the 5-year experience of the German, Austrian and Swiss Photopatch Test Group. J Am Acad Dermatol 1991;25:59-68

8   Katsarou A, Makris M, Zarafonitis G, Lagogianni E, Gregoriou S, Kalogeromitros D. Photoallergic contact dermatitis: the 15-year experience of a tertiary referral center in a sunny Mediterranean city. Int J Immunopathol Pharmacol 2008;21:725-727

9   Victor FC, Cohen DE, Soter NA. A 20-year analysis of previous and emerging allergens that elicit photoallergic contact dermatitis. J Am Acad Dermatol 2010;62:605-610

10 Que SK, Brauer JA, Soter NA, Cohen DE. Chronic actinic dermatitis: an analysis at a single institution over 25 years. Dermatitis 2011;22:147-154

11 Schauder S. Dioxopromethazine-induced photoallergic contact dermatitis followed by persistent light reaction. Am J Contact Dermat 1998;9:182-187

12 Sidi E, Hincky M, Gervais A. Allergic sensitization and photosensitization to Phenergan cream. J Invest Dermatol 1955;24:345-352

13 Sidi E, Melki GR. Rapport entre dermites de cause externe et sensibilisation par voie interne. Semaine des Hopitaux 1954; No. 25, April 14 (Article in French)

14 Epstein S, Rowe R. Photoallergy and photocross-sensitivity to Phenergan. J Invest Dermatol 1957;29:319-326

15 Cariou C, Droitcourt C, Osmont MN, Marguery MC, Dutartre H, Delaunay J, et al. Photodermatitis from topical phenothiazines: A case series. Contact Dermatitis 2020;83:19-24

16 De Groot AC. Monographs in contact allergy, volume 3: Topical Drugs. Boca Raton, Fl, USA: CRC Press Taylor and Francis Group, 2021 (ISBN 978-0-367-23693-9)

17 Kai Y, Okamoto O, Fujiwara S. Fixed drug eruption caused by three unrelated drugs: promethazine, pethidine and omeprazole. Clin Exp Dermatol 2011;36:755-758

18 Kim TH, Kang JS, Lee HS, Youn JI. Two cases of mequitazine induced photosensitivity reactions. Photodermatol Photoimmunol Photomed 1995;11:170-173

19 Romita P, Foti C, Stingeni L. Photoallergy to promazine hydrochloride. Contact Dermatitis 2017;77:182-183

Chapter 3.413 PROPACETAMOL

IDENTIFICATION

Description/definition   : Propacetamol is the *N,N*-diethylglycidyl ester of acetaminophen (paracetamol) that
                              conforms to the structural formula shown below
Pharmacological classes   : Anti-inflammatory agents, non-steroidal; analgesics
IUPAC name   : (4-Acetamidophenyl) 2-(diethylamino)acetate
CAS registry number   : 66532-85-2
EC number   : 266-390-1
Merck Index monograph   : 9176
Patch testing   : 1% and 10% pet. (6); commercial injection fluid 50% pet., freshly prepared (6)
Molecular formula   : $C_{14}H_{20}N_2O_3$

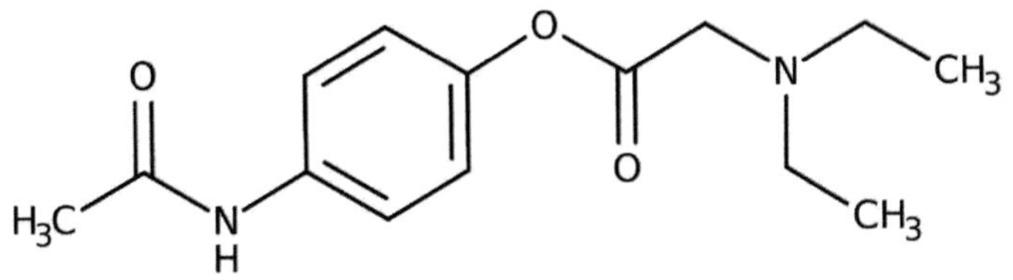

GENERAL

Propacetamol is a water-soluble para-aminophenol derivative with analgesic, antipyretic and mild anti-inflammatory activities. It is a prodrug of paracetamol (acetaminophen) which is soluble in water and is the *N,N*-diethylglycidyl ester of paracetamol. Propacetamol is rapidly hydrolyzed by nonspecific plasma esterases, liberating paracetamol after intravenous administration. One gram of propacetamol liberates 0.5 gram of paracetamol (8). This agent is used to control fever and pain of the perioperative period in multimodal analgesia therapy (1). In pharmaceutical products, propacetamol is employed as propacetamol hydrochloride (CAS number 66532-86-3, EC number 266-391-7, molecular formula $C_{14}H_{21}ClN_2O_3$) (1).

CONTACT ALLERGY FROM ACCIDENTAL CONTACT

General

Propacetamol is, or at least used to be, a well-known cause of occupational allergic contact dermatitis in nurses in France and Belgium. Propacetamol hydrochloride is prepared from a powder which is diluted in a water solvent that contains sodium citrate. The solution thus prepared is stable for 30 minutes. Propacetamol HCl is the chemical that comes in contact with the skin, *not* paracetamol, as this is not liberated in the solution. The chemical may cause both allergic contact dermatitis from direct contact (hands) (6,8,11) and from airborne contact (face, neck) (8,9,10,11), the latter probably from the escape of minute particles of propacetamol powder from the pressurized vials. In all patients, patch test reactions to propacetamol were positive and to – where tested - the metabolites *N,N*-diethylglycine and paracetamol negative (6,8,9,10,11,12). However, in patients sensitized to propacetamol, *N,N*-diethylglycine *phenyl ester*, an activated form of *N,N*-diethylglycine, *did* induce positive patch tests, strongly suggesting – according to the authors of that study - that propacetamol is acting as an activated form of *N,N*-diethylglycine, transferring this part of the molecule to nucleophilic residues of proteins in the skin (why this is not contradictory to the negative patch tests to *N,N*-diethylglycine itself is explained in ref. 12). This also explains why reactions to paracetamol have never been observed in patients sensitized to propacetamol (12).

Since 2001, no new cases of occupational allergic contact dermatitis to propacetamol appear to have been reported.

Case series and case reports

The first cases of occupational sensitization to propacetamol were reported from Nancy, France. The patients were 3 surgical nurses aged 30, 31, and 39, who prepared injections of propacetamol dissolved in sodium citrate. All three had fissured eczema of 1 to 2 years duration on their hands and wrists. Patch tests were strongly positive in all 3 to the commercial propacetamol injection fluid diluted 50% in petrolatum and to propacetamol at 1%, 10%, and 50% in petrolatum, with negative reactions to paracetamol (acetaminophen). When contact with propacetamol was stopped, the eczema recovered in all 3 nurses (6).

Between 1978 and 2001, 14,689 patients have been patch tested in the university hospital of Leuven, Belgium. Occupational allergic contact dermatitis to pharmaceuticals was diagnosed in 33 health care workers. Practically all of these patients presented with hand dermatitis (often the fingers), sometimes with secondary localizations, frequently the face. In this group, 10 patients had occupational allergic contact dermatitis from propacetamol hydrochloride, 9 nurses and a medical doctor (2). In the same hospital, in the period 2001-2019, only 2 nurses developed occupational sensitization to propacetamol, both in 2001. From 2002 on, the administration method was changed to avoid leakage on the skin (5) and from 2006 onward, propacetamol was no longer available in Belgium (4).

Most other case series and single cases of occupational allergic contact dermatitis to propacetamol have been reported from France: 1996 (7; n=unknown), 1996 (8; n=4), 1997 (9; n=2), 1997 (10; n=1), and 1998 (11; n=3). In 2001, 3 cases were reported from Belgium (5).

CUTANEOUS ADVERSE DRUG REACTIONS FROM SYSTEMIC ADMINISTRATION CAUSED BY TYPE IV (DELAYED-TYPE) HYPERSENSITIVITY (as demonstrated by positive patch tests)

Cutaneous adverse drug reactions from systemic administration of propacetamol caused by type IV (delayed-type) hypersensitivity have included acute generalized exanthematous pustulosis (AGEP) (3) and systemic allergic dermatitis (2).

Acute generalized exanthematous pustulosis (AGEP)

An 83-year-old man developed acute generalized exanthematous pustulosis (AGEP) from paracetamol (acetaminophen) with positive patch tests (Chapter 3.4 Acetaminophen). Involuntary rechallenge with intravenous propacetamol as single drug (which is converted into paracetamol) was responsible for recurrence of AGEP one year later (3).

Systemic allergic dermatitis (systemic contact dermatitis)

A nurse who was occupationally sensitized to propacetamol developed a maculopapular eruption when propacetamol was administered intravenously to her (2).

Cross-reactions, pseudo-cross-reactions and co-reactions

Patients sensitized to propacetamol do *not* cross-react to paracetamol. Patients sensitized to paracetamol may cross-react to propacetamol, as the latter releases paracetamol (pseudocross-reaction).

LITERATURE

1   The data in the section 'General' may have been obtained from literature discussed in this chapter, but mostly also or exclusively from one or more of the following online sources: ChemIDPlus Advanced, PubChem, DrugBank, RxList, Drug Central, Drugs.com, and Wikipedia
2   Gielen K, Goossens A. Occupational allergic contact dermatitis from drugs in healthcare workers. Contact Dermatitis 2001;45:273-279
3   Léger F, Machet L, Jan V, Machet C, Lorette G, Vaillant L. Acute generalized exanthematous pustulosis associated with paracetamol. Acta Derm Venereol 1998;78:222-223
4   Gilissen L, Boeckxstaens E, Geebelen J, Goossens A. Occupational allergic contact dermatitis from systemic drugs. Contact Dermatitis 2020;82:24-30
5   Morthier R. Contactallergie voor propacetamol: een gevalstudie. Tijdschr voor Geneeskunde 2001;57:375-378 (Article in Dutch)
6   Barbaud A, Tréchot P, Bertrand O, Schmutz J-L. Occupational allergy to propacetamol. Lancet 1995;346:902
7   Lehners-Weber C, De la Brassine M. Allergie au pro-dafalgan. La Lettre du GERDA 1996;13:4-5 (Article in French, data cited in ref. 2)
8   Szczurko C, Dompmartin A, Michel M, Castel B, Leroy D. Occupational contact dermatitis from propacetamol. Contact Dermatitis 1996;35:299-301
9   Barbaud A, Reichert-Pénetrat S, Tréchot P, Cuny J-F, Weber M, Schmutz J-L. Occupational contact dermatitis to propacetamol. Dermatology 1997;195:329-331
10 Mathelier-Fusade P, Mansouri S, Aïssaoui M, Chabane MH, Aractingi S, Leynadier F. Airborne contact dermatitis from propacetamol. Contact Dermatitis 1997;36:267-268
11 Breuil K, Remblier C. Propacetamol and new occupational contact dermatitis. Allerg Immunol (Paris) 1998;30:149-151 (Article in French)
12 Berl V, Barbaud A, Lepoittevin J-P. Mechanism of allergic contact dermatitis from propacetamol: sensitisation to activated *N,N*-diethylglycine. Contact Dermatitis 1998;38:185-188

Chapter 3.414 PROPANIDID

IDENTIFICATION
Description/definition   : Propanidid is the methoxybenzene that conforms to the structural formula shown below
Pharmacological classes : Anesthetics, intravenous
IUPAC name   : Propyl 2-[4-[2-(diethylamino)-2-oxoethoxy]-3-methoxyphenyl]acetate
CAS registry number   : 1421-14-3
EC number   : 215-822-7
Merck Index monograph   : 9187
Patch testing   : 1% and 5% water
Molecular formula   : $C_{18}H_{27}NO_5$

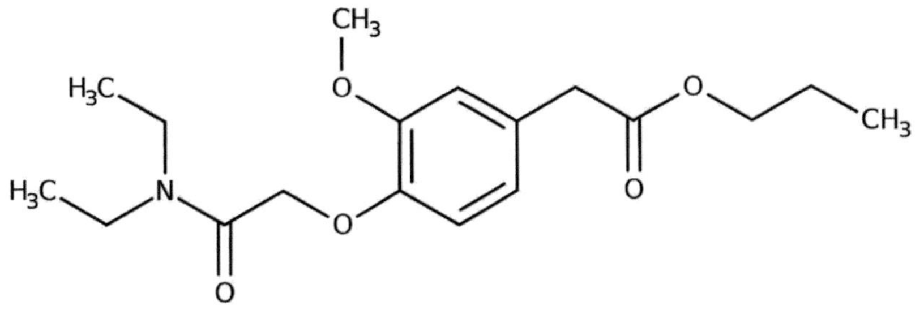

GENERAL
Propanidid is a member of methoxybenzenes and an intravenous anesthetic that has been used for rapid induction of anesthesia and for maintenance of anesthesia of short duration. It is hardly, if at all, utilized anymore (1).

CONTACT ALLERGY FROM ACCIDENTAL CONTACT
A 36-year-old woman working as a nurse in a military hospital presented with eczematous lesions of 5 months' duration, localized to the first three fingers of both hands. No lesion was found elsewhere or on her eyelids. The patient noticed that the eruption arose while she was at work. Patch tests gave a positive reaction to one of the anesthetic drugs that she had contact with. The manufacturer supplied both of its constituents (propanidid and cremophor polyoxyethylated castor oil) and the patient reacted strongly to propanidid down to 0.1% water (2).

A 42-year-old male consultant anesthetist, working with several anesthetics including propanidid, woke one morning with eyes badly swollen and lids very sore and itchy. This episode lasted two days but identical symptoms with marked facial swelling occurred one week later after the use of large quantities of propanidid. A third contact with propanidid on his hands led to a severe recurrence with his lips and face noticeably swollen. Patch tests showed a 'marked reaction' to the propanidid but not to the solvent. A repeat patch test 9 years later was still positive. As it is unlikely that the patient would touch his lips and eyelids during the operation, this may have been a case of airborne occupational allergic contact dermatitis (3).

In the early 1970s, several similar cases of occupational allergic contact dermatitis from propanidid in anesthetists have been reported (4,5,6,7).

LITERATURE
1   The data in the section 'General' may have been obtained from literature discussed in this chapter, but mostly also or exclusively from one or more of the following online sources: ChemIDPlus Advanced, PubChem, DrugBank, RxList, Drug Central, Drugs.com, and Wikipedia
2   Gastelain PY, Piriou A. Contact dermatitis due to propanidid in an anesthetist. Contact Dermatitis 1980;6:360
3   Dundee JW, Assem ESK, Gaston JM, Keilty SR, Sutton JA, Clarke RSJ, et al. Sensitivity to intravenous anaesthetics: A report of three cases. Brit Med J 1974;1:63-65
4   Sneddon IB, Glew RC. Contact dermatitis due to propanidid in an anaesthetist. Practitioner 1973;211(263):321-323
5   Bandmann HJ, Dönicke A. Allergic contact eczema due to propanidid in a anesthetist. Case report. Berufsdermatosen 1971;19:160-165 (Article in German)
6   Bandmann HJ, Dönicke A. Occupational dermatitis from propanidid. Contact Dermatitis Newsletter 1970;8:189
7   Turner RJN. Reaction to propanidid. Brit J Anaesthesiol 1970;42:934

Chapter 3.415 PROPICILLIN

IDENTIFICATION

Description/definition	: Propicillin is the semisynthetic penicillin that conforms to the structural formula shown below
Pharmacological classes	: Anti-bacterial agents
IUPAC name	: (2S,5R,6R)-3,3-Dimethyl-7-oxo-6-(2-phenoxybutanoylamino)-4-thia-1-azabicyclo[3.2.0]-heptane-2-carboxylic acid
Other names	: (1-Phenoxypropyl)penicillin
CAS registry number	: 551-27-9
EC number	: 208-995-5
Merck Index monograph	: 9201
Patch testing	: 20% pet. (2) (unknown whether from pure drug or from pulverized tablet, probably the latter); most beta-lactam antibiotics can be tested 5-10% pet.
Molecular formula	: $C_{18}H_{22}N_2O_5S$

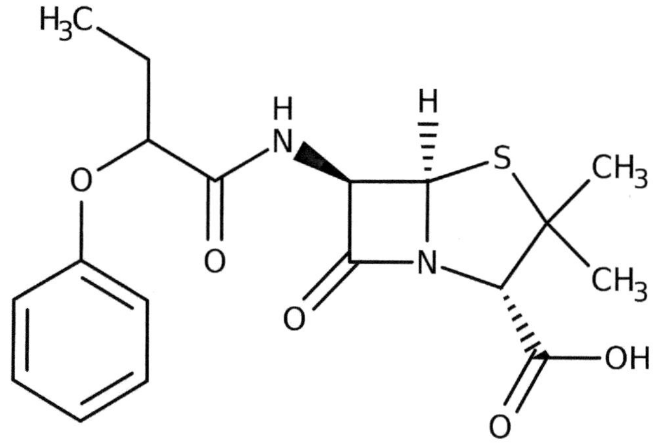

GENERAL

Propicillin is a semisynthetic, acid stable, penicillin with antibacterial activity, analogous to phenoxymethylpenicillin (penicillin V). Although it is better absorbed from the gastrointestinal tract, overall it is inferior to phenoxymethyl-penicillin because of its lower antibacterial activity. Propicillin may be used in some countries to treat mild to moderate bacterial infections caused by susceptible strains of microorganisms (1). The classification and structures of beta-lactam antibiotics are discussed in Chapter 3.36 Amoxicillin.

CUTANEOUS ADVERSE DRUG REACTIONS FROM SYSTEMIC ADMINISTRATION CAUSED BY TYPE IV (DELAYED-TYPE) HYPERSENSITIVITY (as demonstrated by positive patch tests)

Cutaneous adverse drug reactions from systemic administration of propicillin caused by type IV (delayed-type) hypersensitivity have included generalized erythematous eruption with follicular pustules (2).

Other cutaneous adverse drug reactions

A 56-year-old man presented with a generalized painful follicular pustular eruption. His skin was nearly erythroder-mic and he had pruritus, painful swelling, severe dizziness, low-grade fever and nausea. Five days earlier, the patient had been treated for a dental abscess with oral propicillin and analgesics. Erythema and pustules first appeared after the 3rd dose. Histopathology from a biopsy taken from the left forearm 4 days after discontinuation of propicillin showed necrobiosis of the upper epidermis, with small vesicles. Large numbers of granulocytes were invading the epidermis. Laboratory investigations showed leukocytosis with a dramatic shift to the left in blood count, imitating leukemia, but this later normalized. Patch tests were positive to propicillin 20% pet. (+++, 'a disseminated test reaction exceeding the patch test area'). The patient was diagnosed with acute generalized exanthematous pustulosis (AGEP), but painful *follicular* pustules are uncharacteristic for AGEP and in the histopathology, subcorneal pustules were missing (2).

Cross-reactions, pseudo-cross-reactions and co-reactions
Cross-reactions between beta-lactam antibiotics are discussed in Chapter 3.36 Amoxicillin.

LITERATURE

1   The data in the section 'General' may have been obtained from literature discussed in this chapter, but mostly also or exclusively from one or more of the following online sources: ChemIDPlus Advanced, PubChem, DrugBank, RxList, Drug Central, Drugs.com, and Wikipedia
2   Gebhardt M, Lustig A, Bocker T, Wollina U. Acute generalized exanthematous pustulosis (AGEP): manifestation of drug allergy to propicillin. Contact Dermatitis 1995;33:204-205

Chapter 3.416 PROPIOPROMAZINE

IDENTIFICATION

Description/definition : Propionylpromazine is the phenothiazine derivative that conforms to the structural formula shown below
Pharmacological classes : Tranquillizing agent, veterinary
IUPAC name : 1-[10-[3-(Dimethylamino)propyl]phenothiazin-2-yl]propan-1-one
Other names : Propionylpromazine; Combelen
CAS registry number : 3568-24-9
EC number : 222-662-1
Merck Index monograph : 9213
Patch testing : 1% water
Molecular formula : $C_{20}H_{24}N_2OS$

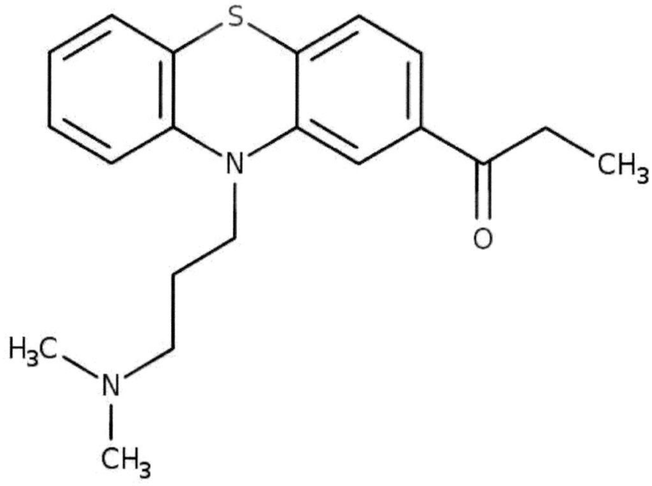

GENERAL

Propiopromazine is a phenothiazine derivative that was formerly used as tranquillizer in veterinary medicine. In pharmaceutical products propiopromazine was employed as propiopromazine hydrochloride (CAS number 7681-67-6, EC number not available, molecular formula $C_{20}H_{25}ClN_2OS$) (1).

CONTACT ALLERGY FROM ACCIDENTAL CONTACT

A 33-year-old veterinary surgeon developed a dry fissured dermatitis of the fingertips. The localisations suggested that the eczema might be caused by contact with leaking syringes for injection. Patch tests with all drugs used in the practice gave a positive reaction to propiopromazine 10 mg/ml. It is likely that the commercial product Combelen was tested. Whether it contained potentially allergenic excipients was not mentioned (2). This was a case of occupational allergic contact dermatitis.

LITERATURE

1 The data in the section 'General' may have been obtained from literature discussed in this chapter, but mostly also or exclusively from one or more of the following online sources: ChemIDPlus Advanced, PubChem, DrugBank, RxList, Drug Central, Drugs.com, and Wikipedia
2 Hjorth N. Contact dermatitis from vitamin E and from Combelen (Bayer) in a veterinary surgeon. Contact Dermatitis Newsletter 1974;15:434

Chapter 3.417 PROPRANOLOL

IDENTIFICATION

Description/definition   : Propranolol is the secondary alcohol that conforms to the structural formula shown below
Pharmacological classes   : β-adrenergic antagonists; anti-arrhythmia agents; antihypertensive agents;
                             vasodilator agents
IUPAC name   : 1-Naphthalen-1-yloxy-3-(propan-2-ylamino)propan-2-ol
CAS registry number   : 525-66-6
EC number   : 208-378-0
Merck Index monograph : 8223
Patch testing   : Hydrochloride, 2% pet. (SmartPracticeCanada)
Molecular formula   : $C_{16}H_{21}NO_2$

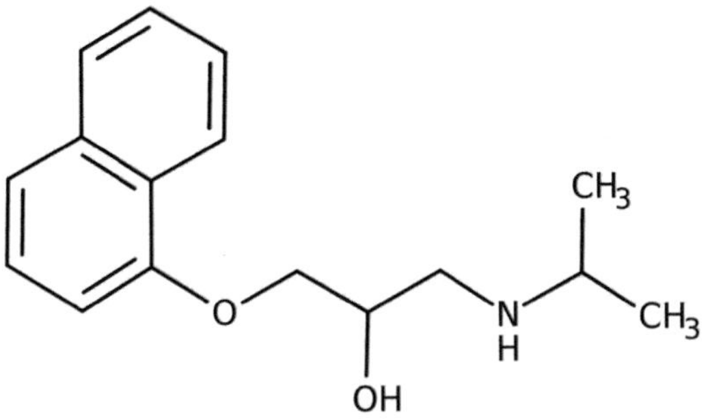

GENERAL

Propranolol is a synthetic non-cardioselective β-adrenergic receptor blocker with antianginal, antiarrhythmic, and antihypertensive properties. It competitively antagonizes β-adrenergic receptors, thereby inhibiting β-adrenergic reactions, such as vasodilation, and negative chronotropic and inotropic effects. Propranolol is used in the treatment or prevention of many disorders including acute myocardial infarction, arrhythmias, angina pectoris, hypertension and hypertensive emergencies, hyperthyroidism, migraine, pheochromocytoma, menopause, and anxiety. Since 2008, oral – and later topical - propranolol has been shown to be effective in the treatment of infantile hemangio-mas, which has rarely caused allergic contact dermatitis (3). In pharmaceutical products, propranolol is employed as propranolol hydrochloride (CAS number 318-98-9, EC number 206-268-7, molecular formula $C_{16}H_{22}ClNO_2$) (1).

In topical preparations, propranolol has caused contact allergy/allergic contact dermatitis, which subject has been fully reviewed in Volume 3 of the Monographs in contact allergy series (2).

CONTACT ALLERGY FROM ACCIDENTAL CONTACT

Several cases of occupational (airborne) allergic contact dermatitis to propranolol have been reported, mostly in workers in the pharmaceutical industries (3, 6-12),

A 24-year-old woman presented with eyelid dermatitis, which had started with localized edema 4 months previously. Later, the area had become itchier, with redness and scaling. The patient suspected a relationship with her work as a pharmacy assistant, which involved breaking and crushing different types of tablets. She was patch tested with the crushed tablets that she had contact with at 10% pet. and showed positive reactions to propranolol, 3 other beta-blockers, 3 benzodiazepines and 3 ACE-inhibitors. Nine controls were negative to propranolol 10% pet. Cosmetic allergy was excluded and a diagnosis of occupational airborne allergic contact was made (7).

A female laboratory technician aged 25 years, who was employed in preparing chemical compounds for pharmacological class exercises in the medical school, developed acute allergic eczematous contact dermatitis of the hands and face. Patch testing, carried out after the clinical dermatitis had subsided, gave positive reactions to phenoxybenzamine and propranolol, both tested 1% water, followed by a recurrence of the dermatitis at the original sites. Tests in controls were not carried out (9,10).

A man aged 46 years, working in a division for drug synthesis of a pharmaceutical plant, and exposed mainly to propranolol hydrochloride and much less frequently to oxprenolol hydrochloride, after 11 months suddenly developed pronounced swelling and erythema of the face, dorsum of the hands and neck. After 2 weeks, during which he was on sick leave, these changes regressed completely and he returned to work, but after 3 days the skin changes returned and also involved the trunk. The factory hall was full of dust, which was visible in sunbeams and on the floor. The patient spent 5-6 hours daily in this hall. One of the substrates used in great amounts for the production of both drugs was epichlorohydrin, which was poured by the patient, wearing a gas mask, into the production stream. When patch tested, he was found to be positive to epichlorohydrin 0.001% water, propranolol

1% and oxprenolol HCl 1% pet. Ten controls were negative (11). This was a case of occupational airborne allergic contact dermatitis from propranolol and oxprenolol.

A 54-year-old man working in the pharmaceutical industry had developed skin problems about 3 months previously, when he developed eczema on the hands, forearms and axillae. Although he used a protective suit with mask and rubber gloves, the rubber gloves seemed to aggravate the dermatitis. The symptoms improved when he moved to another section and at weekends. Patch testing showed positive reactions to thiuram mix and all of its ingredients, to propranolol 0.1%, 1% and 2% pet. (D2 +, D4 ++) and to 2 other drugs (hydralazine, bendroflumethiazide). Sixteen controls were negative (8).

In one or more nurses in Poland, contact allergy/occupational allergic contact dermatitis to propranolol was found (3). Two more reports have described occupational allergic contact dermatitis to propranolol in pharmaceutical workers (6,12).

CUTANEOUS ADVERSE DRUG REACTIONS FROM SYSTEMIC ADMINISTRATION CAUSED BY TYPE IV (DELAYED-TYPE) HYPERSENSITIVITY (as demonstrated by positive patch tests)
Cutaneous adverse drug reactions from systemic administration of propranolol caused by type IV (delayed-type) hypersensitivity have included toxic epidermal necrolysis (TEN) (4).

Stevens-Johnson syndrome/toxic epidermal necrolysis (SJS/TEN)
A 62-year-old patient with psoriasis presented with fever, malaise, and a generalized exanthema with papules, papulovesicles and multiple flaccid bullae, most with a clear content. The Nikolsky sign was positive. There were also erosions but no necrosis. Histopathology and immunofluorescence did not yield and specific data. Some months before, the patient had started treatment with propranolol for hypertension. This medication was stopped and the fever disappeared within 3 days and the exanthematous lesions within a few weeks with bland therapy. Patch tests were positive to pure propranolol 10% and 50% petrolatum. Ten controls were negative. The patient was diagnosed with an eruption resembling toxic epidermal necrolysis (Lyell's disease) probably as a result of delayed-type hypersensitivity to propranolol (4).

Cross-reactions, pseudo-cross-reactions and co-reactions
One patient sensitized to metoprolol also had a positive skin test to propranolol 1% water (5).

LITERATURE
1   The data in the section 'General' may have been obtained from literature discussed in this chapter, but mostly also or exclusively from one or more of the following online sources: ChemIDPlus Advanced, PubChem, DrugBank, RxList, Drug Central, Drugs.com, and Wikipedia
2   De Groot AC. Monographs in contact allergy, volume 3: Topical Drugs. Boca Raton, Fl, USA: CRC Press Taylor and Francis Group, 2021 (ISBN 978-0-367-23693-9)
3   Rudzki E. Occupational dermatitis among health service workers. Derm Beruf Umwelt 1979;27:112-115
4   Van Ketel WG, Soesman A. Een op de zlekte van Lyell gelijkende eruptie door propranolol. Ned T Geneeskd 1977;121:1475-1476 (Article in Dutch)
5   Van Joost T, Middelkamp Hup J, Ros FE. Dermatitis as a side effect of long-term topical treatment with certain beta-blocking agents. Br J Dermatol 1979;101:171-176
6   Ali FR, Shackleton DB, Kingston TP, Williams JD. Occupational exposure to propranolol: A rarely recognised cause of allergic contact dermatitis. Int J Occup Med Environ Health 2015;28:639-640
7   Swinnen I, Ghys K, Kerre S, Constandt L, Goossens A. Occupational airborne contact dermatitis from benzodiazepines and other drugs. Contact Dermatitis 2014;70:227-232
8   Pereira F, Dias M, Pacheco FA. Occupational contact dermatitis from propranolol, hydralazine and bendroflumethiazide. Contact Dermatitis 1996;35:303-304
9   Mitchell JC, Maibach HI. Allergic contact dermatitis from phenoxybenzamine hydrochloride: cross-sensitivity to some related haloalkylamine compounds. Contact Dermatitis 1975;1:363-366
10 Mitchell JC. Allergic contact dermatitis from alpha- and beta-adrenergic receptor blocking agents (dibenzyline and propranolol). Contact Dermatitis Newsletter 1974;16:488
11 Rebandel P, Rudzki E. Dermatitis caused by epichlorohydrin, oxprenolol hydrochloride and propranolol hydrochloride. Contact Dermatitis 1990;23:199
12 Valsecchi R, Leighissa P, Piazzolla S, Naldi L, Cainelli T. Occupational contact dermatitis from propranolol. Contact Dermatitis 1994;30:177

Chapter 3.418 PROPYLTHIOURACIL

IDENTIFICATION

Description/definition   : Propylthiouracil is the pyrimidone that conforms to the structural formula shown below
Pharmacological classes   : Antimetabolites; antithyroid agents
IUPAC name   : 6-Propyl-2-sulfanylidene-1H-pyrimidin-4-one
CAS registry number   : 51-52-5
EC number   : 200-103-2
Merck Index monograph : 9250
Patch testing   : Tablet, pulverized, 10% pet. (2); most systemic drugs can be tested at 10% pet.; if the pure
                              chemical is not available, prepare the test material from intravenous powder, the content
                              of capsules or – when also not available – from powdered tablets to achieve a final
                              concentration of the active drug of 10% pet.
Molecular formula   : $C_7H_{10}N_2OS$

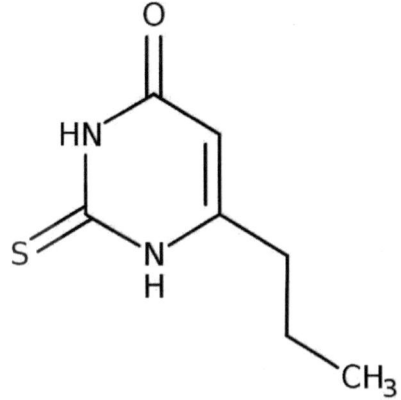

GENERAL

Propylthiouracil is an antithyroid medication used in the therapy of hyperthyroidism and Graves' disease. It inhibits iodine and peroxidase from their normal interactions with thyroglobulin to form T4 and T3. This action decreases thyroid hormone production. Propylthiouracil also interferes with the conversion of T4 to T3, and, since T3 is more potent than T4, this also reduces the activity of thyroid hormones (1).

CUTANEOUS ADVERSE DRUG REACTIONS FROM SYSTEMIC ADMINISTRATION CAUSED BY TYPE IV (DELAYED-TYPE) HYPERSENSITIVITY (as demonstrated by positive patch tests)

Cutaneous adverse drug reactions from systemic administration of propylthiouracil caused by type IV (delayed-type) hypersensitivity have included drug reaction with eosinophilia and systemic symptoms (DRESS) (2) and leukocytoclastic vasculitis (3).

Drug reaction with eosinophilia and systemic symptoms (DRESS)

Six weeks after a 34-year-old woman initiated therapy with propylthiouracil and propranolol for postpartum thyroiditis, she became febrile and a diffuse itching rash with facial swelling developed. Physical examination revealed a generalized maculopapular exanthema, facial edema with desquamation, cervical lymphadenopathy, and oral mucosal involvement. Laboratory tests showed eosinophilia, elevated liver enzymes and high levels of eosinophil cationic protein. Histopathology from a skin biopsy revealed spongiotic dermatitis with a predominantly lympho-eosinophilic infiltrate in the superficial dermis. Oral corticosteroid treatment was started and one month later, the patient's symptoms had resolved. Patch tests with the drugs used were positive to propylthiouracil commercial powder 10% pet. at D2 and D3 (+++). The patient was diagnosed with DRESS syndrome caused by propylthiouracil and mediated by type IV hypersensitivity (2).

Other cutaneous adverse drug reactions

In a female patient, leukocytoclastic vasculitis developed on the legs after she had taken propylthiouracil for 3 years. Four months after complete healing, patch tests were positive (+) to propylthiouracil 1%, 5% and 10% pet. (from commercial tablets) at D2. Six controls were negative (3). It must be mentioned that the clinical picture shown of the 'positive' patch tests were far from convincing. The reaction to 1% was certainly not a + positive one and the back showed several erythematous reactions to other patch tested materials more prominent than propylthiouracil 1% that were scored as negative. Also, the patch tests should have been read at least one time more after D2.

LITERATURE

1   The data in the section 'General' may have been obtained from literature discussed in this chapter, but mostly also or exclusively from one or more of the following online sources: ChemIDPlus Advanced, PubChem, DrugBank, RxList, Drug Central, Drugs.com, and Wikipedia

2   Ye YM, Kim JE, Kim JH, Choi G-S, Park H-S. Propylthiouracil-induced DRESS syndrome confirmed by a positive patch test. Allergy 2010;65:407-409

3   Morais P, Baudrier T, Mota A, Cunha AP, Alves M, Neves C, et al. Antineutrophil cytoplasmic antibody (ANCA)-positive cutaneous leukocytoclastic vasculitis induced by propylthiouracil confirmed by positive patch test: a case report and review of the literature. Cutan Ocul Toxicol 2011;30:147-153

Chapter 3.419 PSEUDOEPHEDRINE

IDENTIFICATION

Description/definition   : Pseudoephedrine is the phenethylamine and diastereomer of ephedrine that conforms to
                               the structural formula shown below
Pharmacological classes   : Bronchodilator agents; nasal decongestants
IUPAC name   : (1S,2S)-2-(Methylamino)-1-phenylpropan-1-ol
Other names   : Isoephedrine
CAS registry number   : 90-82-4
EC number   : 202-018-6
Merck Index monograph : 9294
Patch testing   : 10% pet.; some authors have advised to start with 1% pet., as the higher concentration
                             may induce exacerbations of previous drug eruptions; for fixed drug eruptions, using
                             petrolatum *and* DMSO as vehicle may be preferable (14)
Molecular formula   : $C_{10}H_{15}NO$

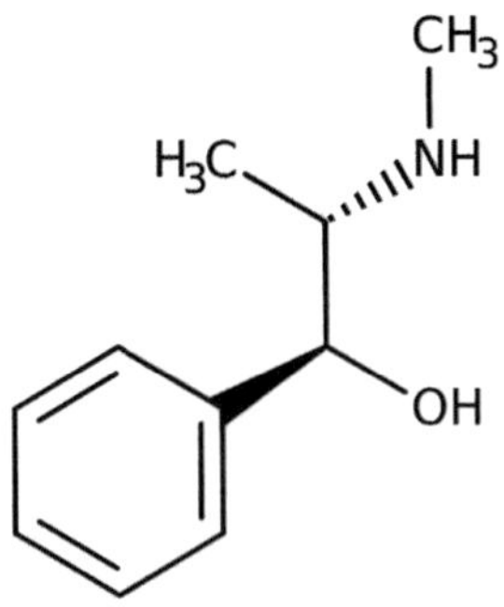

GENERAL

Pseudoephedrine is a phenethylamine, a diastereomer of ephedrine and an α- and β-adrenergic agonist with sympathomimetic property. Pseudoephedrine displaces norepinephrine from storage vesicles in presynaptic neurones, thereby releasing norepinephrine into the neuronal synapses where it stimulates primarily α-adrenergic receptors. It also has weak direct agonist activity at α- and β-adrenergic receptors. Receptor stimulation results in vasoconstriction. Pseudoephedrine is indicated for the treatment of nasal congestion, sinus congestion, Eustachian tube congestion, and vasomotor rhinitis, and as an adjunct to other agents in the optimum treatment of allergic rhinitis, croup, sinusitis, otitis media, and tracheobronchitis. It is also utilized as first-line therapy of priapism (1).

In pharmaceutical products, pseudoephedrine is employed as pseudoephedrine hydrochloride (CAS number 345-78-8, EC number 206-462-1, molecular formula $C_{10}H_{16}ClNO$) or as pseudoephedrine sulfate (CAS number 7460-12-0, EC number 231-243-2, molecular formula $C_{20}H_{32}N_2O_6S$) (go.drugbank.com).

CUTANEOUS ADVERSE DRUG REACTIONS FROM SYSTEMIC ADMINISTRATION CAUSED BY TYPE IV (DELAYED-TYPE) HYPERSENSITIVITY (as demonstrated by positive patch tests)

Cutaneous adverse drug reactions from systemic administration of pseudoephedrine caused by type IV (delayed-type) hypersensitivity have included maculopapular eruption (1,7,10,18), erythroderma/extensive erythema (13,20,21,23,27,28), acute generalized exanthematous pustulosis (AGEP) (2,3,4,8,26), systemic allergic dermatitis (22), symmetrical drug-related intertriginous and flexural exanthema (SDRIFE)/baboon syndrome (19,25), fixed drug eruption (14,15,16,17,24), Stevens-Johnson syndrome/toxic epidermal necrolysis (SJS/TEN) (6), dermatitis/ eczematous eruption (12,29,30), generalized papulovesicular eruption with mucosal involvement (4), generalized pruritic exanthematous eruption (6), recurrent erythema (5), pigmented purpuric dermatosis (24; more likely atypical fixed drug eruption), and unspecified/unknown cutaneous adverse drug reactions (9).

Case series with various or unknown types of drug reactions

In Finland, in the period 1989-2001, 826 patients with suspected cutaneous drug eruptions were patch tested and 89 had one or more positive reactions. Of these individuals, 5 reacted to pseudoephedrine; it was not specified which drug eruptions had been induced by the drug (9).

Maculopapular eruption

A 30-year-old woman developed a generalised, maculopapular, pruriginous eruptions with facial edema, malaise and fever, after having used a combination tablet of pseudoephedrine and triprolidine for rhinorrhea and nasal congestion for 5 days. Patch tests were positive to the tablet as is and to pseudoephedrine sulfate 1% pet. (negative in 5 controls). There were no cross-reactions to ephedrine and phenylephrine. Triprolidine was not tested (18).

In a study from Nancy, France, 54 patients with suspected nonimmediate drug eruptions were assessed with patch testing. Of the 27 patients with maculopapular eruptions, one had a positive reaction to pseudoephedrine, which induced a mild recurrence of the drug eruption (10). A woman aged 78 developed a maculopapular exanthema from pseudoephedrine and had positive patch test to this drug tested 10% pet. (D2 +, D4 +) (7). Another case of maculopapular eruption from pseudoephedrine was reported from Spain in 2021 (1).

Erythroderma, widespread erythematous eruption, exfoliative dermatitis

A 14-year-old girl received a combination of acrivastine and pseudoephedrine to treat pharyngitis. After 3 days, an erythematous eruption appeared on the thighs, which became generalized over the course of one week. One year later, the patient again took the combination of acrivastine and pseudoephedrine for a common cold. The next day, she had bluish edematous erythema on the head and the upper part of the trunk with fever, but without eosinophillia or systemic symptoms. Patch tests were positive to the combination of pseudoephedrine and acrivastine, but pure acrivastine was negative (13).

A 44-year-old woman developed, 6 hours after having taken a capsule containing a combination of acrivastine and pseudoephedrine hydrochloride for allergic rhinitis, a generalized rash with redness and blisters. Examination revealed a coalescing erythrodermic dermatitis, sparing the face, with large blisters on the legs, but sparing of mucous membranes. She had no fever, but there was leukocytosis with neutrophilia and an elevated C-reactive protein. Patch tests were positive to the contents of the capsule 20% pet. (++ at D2) (15 controls negative), negative to the shell and negative to the same capsule containing acrivastine, but no pseudoephedrine. The diagnosis of delayed-type hypersensitivity to pseudoephedrine was therefore very likely, but not proven (20).

A 77-year-old woman developed a generalized itchy erythematous eruption after starting oral treatment with amoxicillin and the combination of ebastine plus pseudoephedrine. A reaction to amoxicillin was excluded. Patch tests with the combination tablet were negative (concentration too low), but an oral provocation test resulted in a micropapular generalized eruption within 5 hours, which also occurred after oral administration of 60 mg pseudoephedrine HCl but not with oral ebastine. Patch tests to pseudoephedrine 10% in petrolatum and water were positive at D2 and D4 (++/++) (21).

A 64-year-old man developed, 2 days after taking the sixth dose of a combination product of codeine phosphate, chlorpheniramine maleate and pseudoephedrine hydrochloride, erythema in the groins that became generalized to erythroderma, with punctate vesicles in the palms and soles, and a papulopustular eruption distributed over the axillae, groin area, and abdomen. Histopathology showed spongiform subcorneal pustules with acantholytic cells, and perivascular and interstitial infiltrates with abundant eosinophils in the papillary dermis. Patch tests were strongly positive to the tablet (no test concentration mentioned) and to pseudoephedrine 1% pet., but negative to the other drugs of the combination tablet (27).

Other cases of erythroderma from pseudoephedrine have been reported in Spanish literature in 1999 (23) and from the United Kingdom in 1998 (28).

Acute generalized exanthematous pustulosis (AGEP)

A 41-year-old man presented with a skin eruption accompanied by fever and malaise, which had begun one day after having ingested one combination tablet of ebastine and pseudoephedrine to treat nasal discharge. Examination revealed fever, bilateral tender cervical, axillar and inguinal lymphadenopathies, and generalized scarlatiniform erythema with multiple scattered non-follicular pustules and intense facial edema. There was leukocytosis with neutrophilia; bacterial cultures from the throat and pustules were negative. Histopathology showed subcorneal non-follicular pustules, epidermal spongiosis, focal keratinocyte necrosis and basal vacuolar degeneration with edema and diffuse perivascular infiltrates of neutrophils, eosinophils and lymphocytes in the superficial dermis. Patch tests showed a very strong reaction at D2 and D3 to the commercial tablet 2.5% and 5% in pet. and to pseudoephedrine 1% pet., but were negative to ebastine 1% and 2% pet. The patient was diagnosed with acute generalized exanthematous pustulosis (AGEP) caused by pseudoephedrine (2).

Previously, a case of AGEP from pseudoephedrine had been reported from Spain in a 42-year-old woman. Patch tests were positive to the commercial pseudoephedrine-containing combination tablet and to pseudoephedrine 20% and 50% pet. (3). A 68-old man from France had developed AGEP after taking paracetamol-pseudoephedrine tablets. A patch test with pseudoephedrine 10% pet. was positive, with paracetamol negative (4). The first patient with AGEP from pseudoephedrine was reported from Spain in 1995 in an Abstract (26).

In a multicenter investigation in France, of 45 patients patch tested for AGEP, 26 (58%) had positive patch tests to drugs, including one to pseudoephedrine (8).

Systemic allergic dermatitis (systemic contact dermatitis)

An 18-year-old woman developed blepharoconjunctivitis after application of phenylephrine and tropicamide eye drops. Four years later, she took 1 tablet of pseudoephedrine plus loratadine, and 3-4 hours later developed erythroderma. Patch tests were positive to pseudoephedrine and phenylephrine 1%, 5% and 10% pet. at D2 and D4. This was a case of systemic allergic dermatitis presenting as erythroderma from pseudoephedrine cross-reacting to phenylephrine, to which the patient had previously become sensitized from its presence in eye drops. When the patient took a tablet for 'cold' containing pseudoephedrine 4 months later, erythroderma reappeared (22).

Symmetrical drug-related intertriginous and flexural exanthema (SDRIFE)/Baboon syndrome

Six hours after taking one capsule of loratadine plus pseudoephedrine, a 73-year-old woman presented with a pruritic symmetric erythema on her buttocks, abdomen, back, groins, inner thighs, ankles and palms, proceeding to papules and pustules. Histopathology showed subcorneal and suprabasal pustules, edema in the papillary dermis and an inflammatory infiltrate with eosinophils in the papillary and reticular dermis. Twenty years ago, the patient had suffered a similar eruption after taking a drug that contained ephedrine. Patch tests were positive to pseudo-ephedrine 1% pet. and ephedrine 1% pet. at D2 and D4 (++/++) with pustules in the positive reactions. Histo-pathology was similar to that of the original exanthema. During patch testing, 2 erythematous plaques developed away from the patch tests on previously involved skin. The patient was diagnosed with the baboon syndrome from delayed-type hypersensitivity to pseudoephedrine (25). This case had features of AGEP, but it was not mentioned whether the patient had developed fever and leukocytosis.

Another possible (not very convincing) case of SDRIFE presenting as the baboon syndrome was reported in 2003. Patch tests were positive to a syrup containing pseudoephedrine, but not to a tablet with pseudoephedrine and pseudoephedrine itself, tested 1% pet. Also, the other ingredients of the syrup (excipients and the drug triprolidine) were not patch tested (19).

Fixed drug eruption

Six hours after the single administration of a combination tablet containing pseudoephedrine and two other drugs, a 38-year-old woman developed many round, large, and itchy scarlet-colored edematous multifocal plaques symmetrically located on the arms, legs, lower back, buttocks, the righthand palm and the nape. In the latter, the plaque had a central serous blister. These lesions spontaneously disappeared within 2 weeks with no residual pigmentation. The patient had suffered a similar eruption one year before, 10 hours after the administration of the same drug. Patch tests with the tablet 25% and 50% in petrolatum and in DMSO were negative on both lesional and non-lesional skin. When the patient later took a combination tablet again (with only one of the 2 other drugs), all the plaques relapsed in the previous sites and with the same clinical features after 5 hours. Patch tests were now positive to pseudoephedrine hydrochloride 10% DMSO on postlesional skin, but negative to 5% of this drug in DMSO and 5% and 10% pet., and negative to all these materials on uninvolved skin. An etiological role of the other 2 drugs was excluded by patch tests (1 drug) and oral provocation (1 drug). The patient was diagnosed with multifocal bullous non-pigmenting fixed drug eruption (14). One may wonder whether the 'bullous' part of the diagnosis is correct, considering that only one of the many lesions showed a blister.

Various other cases of non-pigmenting fixed drug eruption from pseudoephedrine have been reported (it is in fact the major inducer of *non-pigmenting* fixed drug eruptions [15]), but delayed-type hypersensitivity was demonstrated with lesional patch testing or topical provocation) in only a few cases (16,17). In one of these, there were both clinical and histopathological features of acute generalized exanthematous pustulosis (17).

A 42-year-old man presented with three discrete circular-to-oval hyperpigmented lesions 3-5 cm in diameter. Such lesions had appeared several times in the preceding 4 years, beginning with itching and burning followed by marked redness and edema, when having used cold remedies. These all proved to contain pseudoephedrine. An oral provocation test with a syrup containing 30 mg of pseudoephedrine HCl was strongly positive after 2-3 hours. Patch tests with pseudoephedrine HCl 1%, 5% and 10% water on a hyperpigmented patch were negative. Patch tests with the drug 1%, 5% and 10% pet. on another hyperpigmented lesion were strongly positive, but negative on previously uninvolved skin. Histopathology of the reactivated test area showed dyskeratosis, basal vacuolar degeneration, dense melanophages in the papillary dermis, and mononuclear perivascular inflammatory infiltration. The reaction gradually faded after 72 hours, leaving hyperpigmentation (15).

See also the section 'Other cutaneous adverse drug reactions' below, ref. 24.

Stevens-Johnson syndrome/toxic epidermal necrolysis (SJS/TEN)

A 57-year-old woman developed an attack of SJS/TEN, 8 days after taking 2 doses of pseudoephedrine hydrochloride for nasal congestion. At that time, another drug was suspected. Nine months later, the patient took a single dose of pseudoephedrine, again for nasal congestion. She presented to the emergency department 2 days later with a generalized, pruritic, exanthematous eruption. Later, a patch test was positive to pseudoephedrine 3% pet. (6).

Dermatitis/eczematous eruption

Three patients from Spain had developed generalized dermatitis hours after oral administration of pseudoephedrine and had positive patch tests to pseudoephedrine 10% pet. and several related chemicals (see the section 'Cross-reactions, pseudo-cross-reactions and co-reactions' below) (12).

Two days after taking one capsule containing pseudoephedrine, a 65-year-old woman presented with an acute bilateral eczematous eruption on her neck, trunk and arms. A similar eruption had occurred 18 months previously when she was treated with a drug containing pseudoephedrine hydrochloride, triprolidine and paracetamol. Patch tests were strongly positive to the contents of the pseudoephedrine capsule, provoking a slight erythematous eruption on the sites of her previous dermatitis. In a second patch test session, there were strongly positive reactions to pseudoephedrine and ephedrine at 0.1%, 1% and 3% pet. and a positive reaction to phenylephrine 1% and 3% pet., with a negative reaction to epinephrine. Three months later, the patient was treated with a syrup containing norephedrine, which resulted in generalized dermatitis. It was assumed that this was the result of cross-reactivity of norephedrine to pseudoephedrine and ephedrine (29). The authors called this systemic allergic dermatitis, but this is – according to current views – not correct, as this requires previous sensitization from topical application.

A 48-year-old man, who had previously suffered DRESS from carbamazepine, later had eczematous eruptions from multiple unrelated drugs. Patch tests were positive to pseudoephedrine, hydroxyzine HCl, cetirizine HCl, carbamaze-pine, amoxicillin and ampicillin. The patient had used all these drugs, which resulted in eczematous eruptions, with the exception of ampicillin, which was a cross-reaction to amoxicillin. This was a case of multiple drug allergy (30).

Other cutaneous adverse drug reactions

A 51-year-old woman had developed a generalized papulovesicular eruption with involvement of the palms and soles, facial edema and diffuse mucosal erosions. She had fever (38.5°C) and lymphadenopathy with eosinophilia and inflammatory syndrome. Numerous drugs had been taken on preceding days for a flu-like syndrome. Patch tests were performed with all the drugs taken by the patient and some active ingredients. There were positive reactions to a syrup containing pseudoephedrine and 2 other drugs and to pseudoephedrine 10% pet. (4).

A 57-year-old woman, who had previously suffered an attack of SJS/TEN while taking pseudoephedrine, nine months later developed a generalized, pruritic, exanthematous eruption after again having ingested pseudoephedrine. A patch test was positive to pseudoephedrine 3% pet. (6).

A man suffered from recurrent erythema after administration of different treatments including pseudoephedrine. A cutaneous biopsy was compatible with erythema multiforme. Patch tests confirmed the diagnosis of allergy to pseudoephedrine, and resulted in a reappearance of the symptoms (5).

Five days after having used combination tablets of loratadine and pseudoephedrine for a catarrhal infection, a 71-year-old woman developed intense pruritus and a widespread eruption, consisting of well-circumscribed erythematous and purpuric plaques of varying diameter, some evolving into vesicles, with residual pigmentation. Two weeks later, the patient again took one tablet which resulted in a recurrence. An oral challenge with loratadine was negative, but with oral administration of 15 mg pseudoephedrine, the same erythematous eruption appeared after 4 hours at exactly the same sites as her two previous attacks. Patch tests were performed on postlesional skin with pseudoephedrine 10% and 20% in DMSO. Within 12 hours, the patch tests had to be removed, because the same widespread erythematous eruption appeared at exactly the same sites. Histopathology of a clinical lesion and the positive patch test showed extravasated erythrocytes within the papillary dermis, and perivascular inflammation with a predominantly lymphocytic infiltrate. The patient was diagnosed with pigmented purpuric dermatosis (24). This author disagrees with this diagnosis, as extensive erythema is not part of pigmented purpuric dermatosis. As the lesions were always localized on the same sites, a fixed drug eruption – with atypical features – seems more likely.

Cross-reactions, pseudo-cross-reactions and co-reactions

Patients sensitized to pseudoephedrine may cross-react to phenylephrine, ephedrine, phenylpropanolamine, fepradinol, methoxamine, and oxymetazoline (1,12,21). Individuals sensitized to phenylephrine may cross-react to pseudoephedrine (12,22). A patient with allergic contact dermatitis to fepradinol co-reacted to pseudoephedrine (12). Cross-reactions between pseudoephedrine, ephedrine, phenylephrine, epinephrine and norephedrine may or may not occur (29).

LITERATURE

1   Gratacós Gómez AR, González Jimenez OM, Palacios Cañas A, Moreno Lozano L, Juan Cencerrado M, Gómez Torrijos E. Severe maculopapular eruption due to pseudoephedrine in a combination drug, with cross-reactivity to phenylephrine. Contact Dermatitis. 2021 Jul 27. doi: 10.1111/cod.13951. Epub ahead of print.
2   Mayo-Pampín E, Flórez A, Feal C, Conde A, Abalde MT, De la Torre C, et al. Acute generalized exanthematous pustulosis due to pseudoephedrine with positive patch test. Acta Derm Venereol 2006;86:542-543

3   Padial MA, Alvarez-Ferreira J, Tapia B, Blanco R, Mañas C, Blanca M, et al. Acute generalized exanthematous pustulosis associated with pseudoephedrine. Br J Dermatol 2004;150:139-142

4   Assier-Bonnet H, Viguier M, Dubertret L, Revuz J, Roujeau JC. Severe adverse drug reactions due to pseudoephedrine from over-the-counter medications. Contact Dermatitis 2002;47:165

5   Fontaine JF, Lavaud F, Deslee G, Caillet MJ, Lebargy F. Toxic dermatitis caused by pseudoephedrine: apropos of a case. Allerg Immunol (Paris) 2002;34:230-232 (Article in French)

6   Nagge JJ, Knowles SR, Juurlink DN, Shear NH. Pseudoephedrine-induced toxic epidermal necrolysis. Arch Dermatol 2005;141:907-908

7   Studer M, Waton J, Bursztejn AC, Aimone-Gastin I, Schmutz JL, Barbaud A. Does hypersensitivity to multiple drugs really exist? Ann Dermatol Venereol 2012;139:375-380 (Article in French)

8   Barbaud A, Collet E, Milpied B, Assier H, Staumont D, Avenel-Audran M, et al. A multicentre study to determine the value and safety of drug patch tests for the three main classes of severe cutaneous adverse drug reactions. Br J Dermatol 2013;168:555-562

9   Lammintausta K, Kortekangas-Savolainen O. The usefulness of skin tests to prove drug hypersensitivity. Br J Dermatol 2005;152:968-974

10   Barbaud A, Reichert-Penetrat S, Tréchot P, Jacquin-Petit MA, Ehlinger A, Noirez V, et al. The use of skin testing in the investigation of cutaneous adverse drug reactions. Br J Dermatol 1998;139:49-58

11   Barbaud A. Skin testing in delayed reactions to drugs. Immunol Allergy Clin North Am 2009;29:517-535

12   Barranco R, Rodríguez A, de Barrio M, Trujillo MJ, de Frutos C, Matheu V, et al. Sympathomimetic drug allergy: cross-reactivity study by patch test. Am J Clin Dermatol 2004;5:351-355

13   Liippo J, Pummi K, Hohenthal U, Lammintausta K. Patch testing and sensitization to multiple drugs. Contact Dermatitis 2013;69:296-302

14   Bellini V, Bianchi L, Hansel K, Finocchi R, Stingeni L. Bullous nonpigmenting multifocal fixed drug eruption due to pseudoephedrine in a combination drug: clinical and diagnostic observations. J Allergy Clin Immunol Pract 2016;4:542-544

15   Özkaya E, Elinç-Aslan MS. Pseudoephedrine may cause "pigmenting" fixed drug eruption. Dermatitis 2011;22:E7-9

16   Alanko K, Kanerva L, Mohell-Talolahti B, Jolanki R, Estlander T. Nonpigmented fixed drug eruption from pseudoephedrine. J Am Acad Dermatol 1996;35:647-648

17   Fukuda R, Ouchi T, Hirai I, Funakoshi T, Honda A, Tanese K, et al. Non-pigmenting fixed drug eruption with mixed features of acute generalized exanthematous pustulosis induced by pseudoephedrine: a case report. Contact Dermatitis 2017;77:123-126

18   Cunha D, Carvalho R, Freitas I, Santos R, Afonso A, Cardoso J. Exanthematic reaction to pseudoephedrine. Allergol Immunopathol (Madr) 2009;37:106-107

19   Sánchez-Morillas L, Reaño Martos M, Rodríguez Mosquera M, Iglesias Cadarso A, Pérez Pimiento A, Domínguez Lázaro AR. Baboon syndrome due to pseudoephedrine. Contact Dermatitis 2003;48:234

20   Millard TP, Wong YW, Orton DI. Erythrodermic cutaneous adverse drug reaction to oral pseudoephedrine confirmed on patch testing. Contact Dermatitis 2003;49:263-264

21   Moreno-Escobosa MC, de las Heras M, Figueredo E, Umpiérrez A, Bombin C, Cuesta J. Generalized dermatitis due to pseudoephedrine. Allergy 2002;57:753

22   Gonzalo-Garijo MA, Pérez-Calderón R, de Argila D, Rodríguez-Nevado I. Erythrodermia to pseudoephedrine in a patient with contact allergy to phenylephrine. Allergol Immunopathol (Madr) 2002;30:239-242

23   Soto-Mera MT, Filgueira JF, Villamil E, Días MT, Cidrás R. Eritema generalizado por pseudoefedrina con tolerancia de efedrina. Alergol Inmunol Clin 1999;14:404-406 (Article in Spanish)

24   Díaz-Jara M, Tornero P, Barrio MD, Vicente ME, Fuentes V, Barranco R. Pigmented purpuric dermatosis due to pseudoephedrine. Contact Dermatitis 2002;46:300-301

25   Sánchez TS, Sánchez-Pérez J, Aragüés M, García-Díaz A. Flare-up reaction of pseudoephedrine baboon syndrome after positive patch test. Contact Dermatitis 2000;42:312-313

26   Moraga M, Vives R, Rodriguez J, Rodriguez P, Daroca J, Borja J, et al. Disseminated acute pustulosis from pseudoephedrine. Allergy 1995:50(suppl.26):215

27   Vega F, Rosales MJ, Esteve P, Morcillo R, Panizo C, Rodríguez M. Histopathology of dermatitis due to pseudoephedrine. Allergy 1998;53:218-220

28   Downs AM, Lear JT, Wallington TB, Sansom JE. Contact sensitivity and systemic reaction to pseudoephedrine and lignocaine. Contact Dermatitis 1998;39:33

29   Tomb RR, Lepoittevin J-P, Espinassouze F, Heid E, Foussereau J. Systemic contact dermatitis from pseudoephedrine. Contact Dermatitis 1991;24:86-88

30   Özkaya E, Yazganoğlu KD. Sequential development of eczematous type "multiple drug allergy" to unrelated drugs. J Am Acad Dermatol 2011;65:e26-e29.

Chapter 3.420 PYRAZINAMIDE

IDENTIFICATION

Description/definition : Pyrazinamide is the synthetic pyrazine analog of nicotinamide that conforms to the structural formula shown below
Pharmacological classes : Antitubercular agents
IUPAC name : Pyrazine-2-carboxamide
CAS registry number : 98-96-4
EC number : 202-717-6
Merck Index monograph : 9337
Patch testing : 1% and 10% alc. (3); tablet, pulverized, 30% pet. (2); most systemic drugs can be tested at 10% pet.; if the pure chemical is not available, prepare the test material from intravenous powder, the content of capsules or – when also not available – from powdered tablets to achieve a final concentration of the active drug of 10% pet.
Molecular formula : $C_5H_5N_3O$

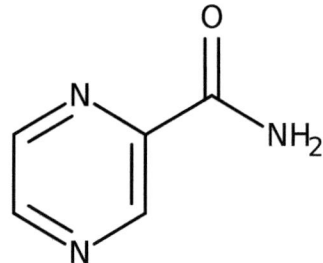

GENERAL

Pyrazinamide is a synthetic pyrazine analog of nicotinamide with antimycobacterial activity. It kills or stops the growth of *Mycobacterium tuberculosis*. Pyrazinamide gets activated to pyrazinoic acid in the bacilli where it interferes with fatty acid synthase FAS I. This inhibits the bacterium's ability to synthesize new fatty acids, required for growth and replication. Pyrazinamide is indicated for the initial treatment of active tuberculosis when combined with other anti-tuberculosis medication such as isoniazid or rifampicin (1).

CUTANEOUS ADVERSE DRUG REACTIONS FROM SYSTEMIC ADMINISTRATION CAUSED BY TYPE IV (DELAYED-TYPE) HYPERSENSITIVITY (as demonstrated by positive patch tests)

Cutaneous adverse drug reactions from systemic administration of pyrazinamide caused by type IV (delayed-type) hypersensitivity have included drug reaction with eosinophilia and systemic symptoms (DRESS) (4), Stevens-Johnson syndrome, overlap with toxic epidermal necrolysis (SJS/TEN) (2), and pruriginous rash on the thorax and abdomen (3).

Drug reaction with eosinophilia and systemic symptoms (DRESS)

A 56-year-old man developed pulmonary and lymph node tuberculosis, for which treatment was started with isoniazid, rifampicin, ethambutol and pyrazinamide. Six weeks later, the patient was hospitalized because of fever (38.5°C), an extensive maculopapular rash (>50% of body surface area), pronounced facial and palmoplantar edema, and disseminated peripheral lymphadenopathy. Laboratory tests showed eosinophilia and elevated serum aminotransferases. Three months later, patch tests were positive to isoniazid 1% pet., ethambutol 3% pet. and pyrazinamide 3% pet., the latter two made from pulverized commercial tablets. Twenty controls were negative to the ethambutol patch test material. Despite the negative patch test to rifampicin, graded re-introduction of the drug, even under prophylactic therapy with corticosteroids, led to a recurrence of DRESS symptoms after 7 days. No patch tests with rifampicin were performed afterwards (4).

Stevens-Johnson syndrome/toxic epidermal necrolysis (SJS/TEN)

In the period 2010-2014, in South Africa, 60 patients with cutaneous adverse drug reactions to first-line antituberculosis drugs (FLTD: rifampicin, isoniazid, pyrazinamide and ethambutol) were patch tested with the patients' drugs 30% pet. Positive reactions were seen to at least one FLTD in 14 participants, of who twelve had DRESS, one SJS and one SJS/TEN. The SJS/TEN was caused by pyrazinamide and isoniazid, but no cases of DRESS were the result of delayed-type hypersensitivity to pyrazinamide. The positive patch tests resulted in a generalized systemic reaction (2).

Other cutaneous adverse drug reactions
A 62-year-old man was treated with various drugs including pyrazinamide for tuberculous meningitis. One day after the introduction of pyrazinamide, a pruriginous rash appeared, mainly located on the thorax and abdomen. It disappeared a few days after pyrazinamide was discontinued. Patch tests were positive to pyrazinamide 1% alc. (D2 and D4 ++) and 10% alc. (D2 and D4 +++) and negative to pyrazinamide 1% and 10% water (3).

Cross-reactions, pseudo-cross-reactions and co-reactions
This subject is discussed in Chapter 3.206 Ethambutol.

LITERATURE
1   The data in the section 'General' may have been obtained from literature discussed in this chapter, but mostly also or exclusively from one or more of the following online sources: ChemIDPlus Advanced, PubChem, DrugBank, RxList, Drug Central, Drugs.com, and Wikipedia
2   Lehloenya RJ, Todd G, Wallace J, Ngwanya MR, Muloiwa R, Dheda K. Diagnostic patch testing following tuberculosis-associated cutaneous adverse drug reactions induces systemic reactions in HIV-infected persons. Br J Dermatol 2016;175:150-156
3   Goday J, Aguirre A, Díaz-Pérez JL. A positive patch test in a pyrazinamide drug eruption. Contact Dermatitis 1990;22:181-182
4   Coster A, Aerts O, Herman A, Marot L, Horst N, Kenyon C, et al. Drug reaction with eosinophilia and systemic symptoms (DRESS) syndrome caused by first-line antituberculosis drugs: Two case reports and a review of the literature. Contact Dermatitis 2019;81:325-331

Chapter 3.421 PYRAZINOBUTAZONE

IDENTIFICATION

Description/definition : Pyrazinobutanone is the combination of phenylbutazone and piperazine that conforms to the structural formula shown below

Pharmacological classes : Anti-inflammatory drugs, non-steroidal

IUPAC name : 4-Butyl-1,2-diphenylpyrazolidine-3,5-dione;piperazine

Other names : Pyrasanone; 4-butyl-1,2-diphenyl-3,5-pyrazolidinedione compd. with piperazine (1:1); phenylbutazone piperazium

CAS registry number : 4985-25-5

EC number : 225-639-4

Patch testing : 1% and 5% pet.; test also phenylbutazone 10.0% pet. (Chemotechnique, SmartPracticeCanada, SmartPracticeEurope); test also piperazine 1% pet. (SmartPracticeCanada, SmartPracticeEurope)

Molecular formula : $C_{23}H_{30}N_4O_2$

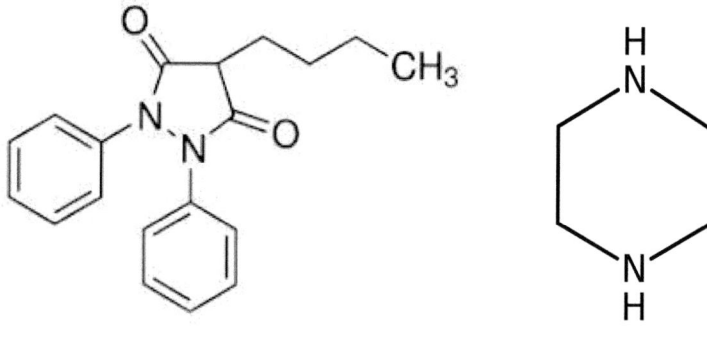

Phenylbutazone   Piperazine

GENERAL

Pyrazinobutanone is a nonsteroidal anti-inflammatory drug (NSAID) which has anti-inflammatory properties and is an equimolar salt of phenylbutazone and piperazine. Its clinical use is similar to other related NSAIDs, but its digestive tolerance is said to be superior. It is uncertain whether this compound is used as drug anymore (1).

In topical preparations, pyrazinobutazone has rarely caused contact allergy/allergic contact dermatitis, which subject has been reviewed in Volume 3 of the *Monographs in contact allergy* series (5). See also Chapter 3.381 Phenylbutazone and Chapter 3.388 Piperazine.

CONTACT ALLERGY FROM ACCIDENTAL CONTACT

A 27-year-old woman, working in a pharmaceutical laboratory, was seen with a hand eczema which she suspected to be due to phenylbutazone-piperazine suppositories, which she had contact with when packing them. During week-ends and holidays, the lesions almost healed, but immediately reappeared every time she returned to work. There were strongly positive patch tests to the suppositories 'as is', pyrazinobutazone 1% pet. and piperazine 5% water, but phenylbutazone was not tested (3). This was a case of occupational allergic contact dermatitis.

Another case is described below in the section 'Systemic allergic dermatitis (systemic contact dermatitis)' (2).

CUTANEOUS ADVERSE DRUG REACTIONS FROM SYSTEMIC ADMINISTRATION CAUSED BY TYPE IV (DELAYED-TYPE) HYPERSENSITIVITY (as demonstrated by positive patch tests)

Cutaneous adverse drug reactions from systemic administration of pyrazinobutazone caused by type IV (delayed-type) hypersensitivity have included systemic allergic dermatitis (2).

Systemic allergic dermatitis (systemic contact dermatitis)

A 35-year-old food handler presented with a 10-day history of eczematous papules and vesicles over the right hip and on both hands, beginning 3 days after he had started applying a gel containing pyrazinobutazone to his right hip and taking 3 capsules of pyrazinobutazone daily by mouth. From 18 to 22 years of age, he had worked in a pharmaceutical laboratory handling, among other drugs, tablets of phenylbutazone – prednisone – meprobamate. The development of erythema and itching on the backs of his hands and fingers and an isolated episode of asthma, at that time, had forced him to change his occupation. Patch tests the gel 'as is', its active substance pyrazinobutazone, piperazine hexahydrate, phenylbutazone and the remaining components of the gel and capsules gave positive reactions to the gel 'as is', pyrazinobutazone 1% (+) and 5% (++) pet., piperazine hexahydrate 5% water and to phenylbutazone 1% (++) and 5% (+++) pet. The patient then took 1 capsule of 300 mg pyrazinobutazone every 12 hours. After 3 days, he developed erythemato-edematous vesicular lesions, with hemorrhagic features, symmetrically on the palms and on the dorsolateral aspects of the first 3 digits, as well as a flare-up at the site of the positive patch test to the gel a month previously. The patient had probably become sensitized to phenylbutazone during his work in a pharmaceutical laboratory (2). This was a case of systemic allergic dermatitis.

Cross-reactions, pseudo-cross-reactions and co-reactions
Most patients allergic to pyrazinobutazone appear to be sensitized to both components; phenylbutazone and piperazine (2,4). See also Chapter 3.381 Phenylbutazone and Chapter 3.388 Piperazine.

LITERATURE

1   The data in the section 'General' may have been obtained from literature discussed in this chapter, but mostly also or exclusively from one or more of the following online sources: ChemIDPlus Advanced, PubChem, DrugBank, RxList, Drug Central, Drugs.com, and Wikipedia
2   Dorado Bris JM, Aragues Montañes M, Sols Candela M, Garcia Diez A. Contact sensitivity to pyrazinobutazone (Carudol) with positive oral provocation test. Contact Dermatitis 1992;26:355-356
3   Menezes-Brandao F, Foussereau J. Contact dermatitis to phenylbutazone-piperazine suppositories (Carudol®) and piperazine gel (Carudol®). Contact Dermatitis 1982;8:264-265
4   Fernandez de Corres L, Bernaola G, Lobera T, Leanizbarrutia I, Munoz D. Allergy from pyrazoline derivatives. Contact Dermatitis 1986;14:249-250
5   De Groot AC. Monographs in contact allergy, volume 3: Topical Drugs. Boca Raton, Fl, USA: CRC Press Taylor and Francis Group, 2021 (ISBN 978-0-367-23693-9)

Chapter 3.422 PYRIDOXINE

IDENTIFICATION

Description/definition : Pyridoxine is the 4-methanol form of vitamin B_6 that conforms to the structural formula shown below
Pharmacological classes : Vitamin B complex
IUPAC name : 4,5-bis(Hydroxymethyl)-2-methylpyridin-3-ol
Other names : Vitamin B_6 (erroneous according to ChemIDPlus); 5-hydroxy-6-methyl-3,4-pyridine-dimethanol
CAS registry number : 65-23-6
EC number : 200-603-0
Merck Index monograph : 9365
Patch testing : 1% and 10% pet.
Molecular formula : $C_8H_{11}NO_3$

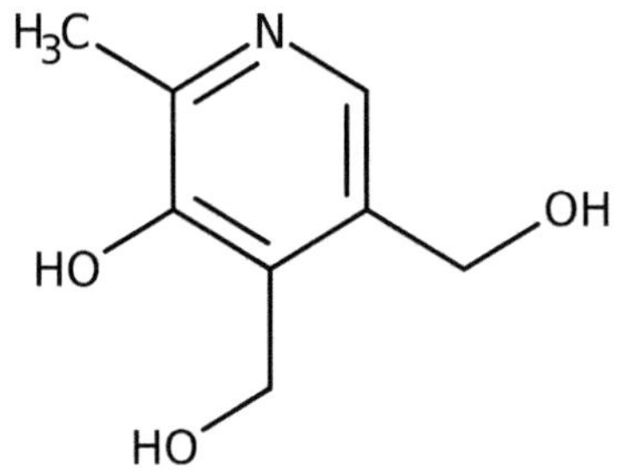

GENERAL

Pyridoxine is the 4-methanol form of vitamin B_6, an important water-soluble vitamin that is naturally present in many foods. As its classification as a vitamin implies, vitamin B_6 (and pyridoxine) are essential nutrients required for normal functioning of many biological systems within the body. Pyridoxine is converted to pyridoxal phosphate, which is a coenzyme for synthesis of amino acids, neurotransmitters (serotonin, norepinephrine), sphingolipids, and aminolevulinic acid. Although pyridoxine and vitamin B_6 are frequently used as synonyms, this practice is, according to some sources, erroneous (ChemIDPlus). In this database, it is stated that vitamin B_6 refers to several picolines, especially pyridoxine, pyridoxal and pyridoxamine. Pyridoxine is indicated for the treatment of vitamin B_6 deficiency and for the prophylaxis of isoniazid-induced peripheral neuropathy. In pharmaceutical products, pyridoxine is employed as pyridoxine hydrochloride (CAS number 58-56-0, EC number 200-386-2, molecular formula $C_8H_{12}ClNO_3$) (1). Topical use of pyridoxine has rarely resulted in allergic contact dermatitis (2,3,9).

CONTACT ALLERGY FROM ACCIDENTAL CONTACT

A 39-year-old woman presented with erythema, edema, pruritus, and vesicular lesions over the dorsa and sides of her fingers, which had been present on-and-off for the previous 2 years. Patch testing showed a non-relevant reaction to nickel. Despite treatment the lesions spread to the dorsa of the hands and new eczematous lesions appeared on the face and neck. The patient administered several medications to her 3-year-old son, seriously ill from birth, including pyridoxine HCl (in powder form) and ranitidine HCl (pulverizing the tablet). Patch tests with all drugs that she handled showed positive reactions to pyridoxine HCl 10% pet. and to ranitidine HCl 1% and 5% pet. No cross-sensitization with other B vitamins or H_2-antagonists were observed. Fifteen controls were negative. This was a case of airborne sensitization to pyridoxine and ranitidine *by proxy* (4).

A 45-year-old paramedical worker presented with a 6-year history of recurrent eczema over the dorsa of the hands, dorsa and sides of the fingers, forearms and face. Patch testing showed positive reactions on D2 and D3 to 3 different brands of injections containing vitamin B_1, B_6, and B_{12}. During testing, reactivation of the lesions on the hands was observed. In a second test session, the patient was patch tested with vitamin B_1, B_6 and B_{12} separately, each 1% in 10% propylene glycol and there was a positive reaction to B_6 at D2, which was later repeated and proved positive at D5. An oral provocation test with 2 tablets of B_6, 100 mg each, given 6 hours apart, resulted in itching and severe erythema over the face, V of the neck, dorsa of the hands, forearms and distal half of the upper arms (sun-exposed areas). Reactivation of previous positive patch test sites and of a previously negative B_6 prick test site were also observed (5).

CUTANEOUS ADVERSE DRUG REACTIONS FROM SYSTEMIC ADMINISTRATION CAUSED BY TYPE IV (DELAYED-TYPE) HYPERSENSITIVITY (as demonstrated by positive patch tests)
Cutaneous adverse drug reactions from systemic administration of pyridoxine caused by type IV (delayed-type) hypersensitivity have included photosensitivity (5,6,7,8,10).

Photosensitivity
A 45-year-old woman developed, after having used vitamin B complex (75 mg of thiamine, 75 mg of pyridoxine hydrochloride, and 0.75 mg of cyanocobalamin) oral tablets for 3 weeks, diffuse erythema and papulovesicles on the face, V-area of the upper chest, and back of her hands. Patch tests were negative, photopatch tests positive to pyridoxine 1% and 5% pet. (6). A 55-year-old woman received infusions containing vitamin B complex (100 mg of thiamine, 100 mg of pyridoxine hydrochloride, and 1 mg of cyanocobalamin). Two years later, she was reinfused with the same vitamin B complex. The next day the patient presented with diffuse erythema and papulovesicles on the face, the V-area of the upper chest, and extensor surface of the forearms. The eruption subsided 4 days after the infusion. Patch tests were negative, photopatch tests positive to pyridoxine 5% pet. but negative to 1% pet. Three controls were negative (6). These were cases of photoallergic dermatitis.

A 71-year-old man developed an exacerbation of eczema at the sun-exposed areas t one day after he was injected with a vitamin complex of B_1, B_6 and B_{12}. Photopatch tests were positive to pyridoxine HCl 1% pet. at D3 (+) and D7 (++). An oral challenge with pyridoxine HCl 30 mg/day for two days resulted in 'a similar lesion exclusively on a sun exposed site'. The patient was diagnosed with photoallergic drug eruption from pyridoxine (7). The authors considered this publication (October 1996) to be the first case of photoallergy to pyridoxine. At the time of writing their article, they could not know that a similar case would be published 2 months earlier in another journal (8). This 35-year-old woman had pruritic erythema on sun-exposed area while taking a multivitamin tablet. Photopatch tests were positive to pyridoxine 1% and 10% pet. irradiated with 4.75 J/cm^2 UVA, whereas 5 controls were negative (8).

Another case of photoallergic dermatitis was reported from Japan in 2000 (10). The patient had positive photopatch tests to pyridoxine 1% and 10% pet. irradiated with UVA 2 J/cm^2 with a cross-reaction to pyridoxal 5'-phosphate (10). See also the section 'Contact allergy from accidental contact' above, ref. 5.

Cross-reactions, pseudo-cross-reactions and co-reactions
A patient sensitized to pyridoxine HCl (possibly actively sensitized by a patch test) cross-reacted to pyridoxal HCl and pyridoxal 5-phosphate, but not to pyridoxamine (3). Photocross-reaction to pyridoxal 5'-phosphate in a patient sensitized to pyridoxine (10).

Patch test sensitization
A patch test with a steroid cream may possibly have caused active sensitization to its ingredient pyridoxine HCl; its concentration in the cream was unknown (3).

LITERATURE
1   The data in the section 'General' may have been obtained from literature discussed in this chapter, but mostly also or exclusively from one or more of the following online sources: ChemIDPlus Advanced, PubChem, DrugBank, RxList, Drug Central, Drugs.com, and Wikipedia
2   Camarasa JG, Serra-Baldrich E, Lluch M. Contact allergy to vitamin B6. Contact Dermatitis 1990;23:115
3   Yoshikawa K, Watanabe K, Mizuno N. Contact allergy to hydrocortisone 17-butyrate and pyridoxine hydrochloride. Contact Dermatitis 1985;12:55-56
4   Córdoba S, Martínez-Morán C, García-Donoso C, Borbujo J, Gandolfo-Cano M. Non-occupational allergic contact dermatitis from pyridoxine hydrochloride and ranitidine hydrochloride. Dermatitis 2011;22:236-237
5   Bajaj AK, Rastogi S, Misra A, Misra K, Bajaj S. Occupational and systemic contact dermatitis with photosensitivity due to vitamin B6. Contact Dermatitis 2001;44:184
6   Murata Y, Kumano K, Ueda T, Araki N, Nakamura T, Tani M. Photosensitive dermatitis caused by pyridoxine hydrochloride. J Am Acad Dermatol 1998;39(2Pt.2):314-317
7   Tanaka M, Niizeki H, Shimizu S, Miyakawa S. Photoallergic drug eruption due to pyridoxine hydrochloride. J Dermatol 1996;23:708-709
8   Morimoto K, Kawada A, Hiruma M, Ishibashi A. Photosensitivity from pyridoxine hydrochloride (vitamin B6). J Am Acad Dermatol 1996: 35: 304-305.
9   De Groot AC. Monographs in contact allergy, volume 3: Topical Drugs. Boca Raton, Fl, USA: CRC Press Taylor and Francis Group, 2021 (ISBN 978-0-367-23693-9)
10 Kawada A, Kashima A, Shiraishi H, Gomi H, Matsuo I, Yasuda K, Sasaki G, Sato S, Orimo H. Pyridoxine-induced photosensitivity and hypophosphatasia. Dermatology 2000;201:356-360

Chapter 3.423 PYRIMETHAMINE

IDENTIFICATION

Description/definition	: Pyrimethamine is the aminopyrimidine that conforms to the structural formula shown below
Pharmacological classes	: Antimalarials; folic acid antagonists; antiprotozoal agents
IUPAC name	: 5-(4-Chlorophenyl)-6-ethylpyrimidine-2,4-diamine
CAS registry number	: 58-14-0
EC number	: 200-364-2
Merck Index monograph	: 9368
Patch testing	: Tablet, pulverized, 30% pet. (2); most systemic drugs can be tested at 10% pet.; if the pure chemical is not available, prepare the test material from intravenous powder, the content of capsules or – when also not available – from powdered tablets to achieve a final concentration of the active drug of 10% pet.
Molecular formula	: $C_{12}H_{13}ClN_4$

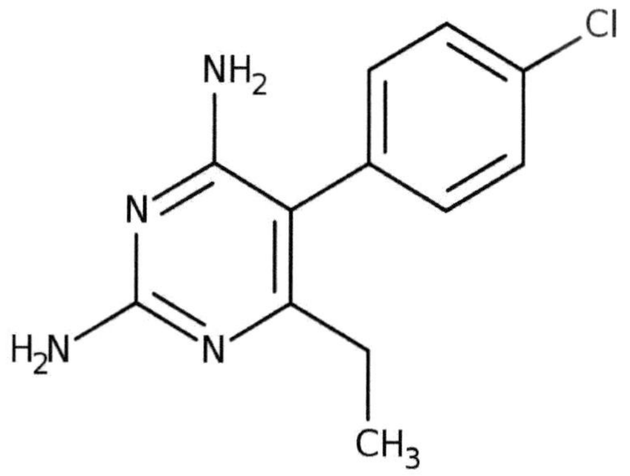

GENERAL

Pyrimethamine is an antiparasitic compound commonly used as an adjunct in the treatment of uncomplicated, chloroquine resistant, *P. falciparum* malaria. Pyrimethamine is a folic acid antagonist and the rationale for its therapeutic action is based on the differential requirement between host and parasite for nucleic acid precursors involved in growth. This activity is highly selective against plasmodia and *Toxoplasma gondii* (1).

In pharmaceutical products, pyrimethamine is employed as pyrimethamine hydrochloride (CAS number 19085-09-7, EC number 242-804-6, molecular formula $C_{12}H_{14}Cl_2N_4$) (1).

CUTANEOUS ADVERSE DRUG REACTIONS FROM SYSTEMIC ADMINISTRATION CAUSED BY TYPE IV (DELAYED-TYPE) HYPERSENSITIVITY (as demonstrated by positive patch tests)

Cutaneous adverse drug reactions from systemic administration of pyrimethamine caused by type IV (delayed-type) hypersensitivity have included drug reaction with eosinophilia and systemic symptoms (DRESS) (2).

Drug reaction with eosinophilia and systemic symptoms (DRESS)

In a multicenter investigation in France, of 72 patients patch tested for DRESS, 46 (64%) had positive patch tests to drugs, including one to pyrimethamine. This was a woman aged 22 who reacted to commercial pyrimethamine 30% pet. Clinical details were not provided (2).

LITERATURE

1   The data in the section 'General' may have been obtained from literature discussed in this chapter, but mostly also or exclusively from one or more of the following online sources: ChemIDPlus Advanced, PubChem, DrugBank, RxList, Drug Central, Drugs.com, and Wikipedia

2   Barbaud A, Collet E, Milpied B, Assier H, Staumont D, Avenel-Audran M, et al. A multicentre study to determine the value and safety of drug patch tests for the three main classes of severe cutaneous adverse drug reactions. Br J Dermatol 2013;168:555-562

Chapter 3.424 PYRITINOL

IDENTIFICATION

Description/definition : Pyritinol is the member of methylpyridines that conforms to the structural formula shown
                         below
Pharmacological classes : Psychoanaleptics; nootropic agents; central stimulants
IUPAC name   : 5-[[[5-Hydroxy-4-(hydroxymethyl)-6-methylpyridin-3-yl]methyldisulfanyl]methyl]-4-
                         (hydroxymethyl)-2-methylpyridin-3-ol
Other names   : Pyrithioxine
CAS registry number   : 1098-97-1
EC number   : 214-150-1
Merck Index monograph : 9379
Patch testing   : 2% water (3); commercial tablet 20% pet. (5); most systemic drugs can be tested at 10%
                         pet.; if the pure chemical is not available, prepare the test material from intravenous
                         powder, the content of capsules or – when also not available – from powdered tablets to
                         achieve a final concentration of the active drug of 10% pet.
Molecular formula   : $C_{16}H_{20}N_2O_4S_2$

GENERAL

Pyritinol is a member of methylpyridines and a neurotropic agent which reduces permeability of blood-brain barrier
to phosphate. It has been used in trials studying the treatment of dementia, depression, schizophrenia, anxiety
disorders, and psychosomatic disorders (1). In pharmaceutical products, both pyritinol and pyritinol dihydrochloride
(pyrithioxine hydrochloride; CAS number 10049-83-9, EC number 233-178-5, molecular formula $C_{16}H_{22}Cl_2N_2O_4S_2$) may
be employed (1).

CONTACT ALLERGY FROM ACCIDENTAL CONTACT

Between 1978 and 2001, at the university hospital of Leuven, Belgium, occupational allergic contact dermatitis to
pharmaceuticals was diagnosed in 33 health care workers. In this group, one patient had occupational allergic
contact dermatitis from pyritinol (pyrithioxine), a 55-year old nurse with dermatitis in the palms of the hands, on the
face and in the neck, suggesting an airborne component (2).

   A 17-year-old laboratory assistant, 4 months after starting to work on quality control in a pharmaceutical
company, developed eczema on the face and hands. Her symptoms coincided with working on pyritinol and other
derivatives of pyridoxine. Patch tests were positive to pyritinol, pyritinol hydrochloride and 2 bases used in the
synthesis of pyritinol HCl at D2 and D3, all tested 2% water. The bases contained pyritinol, pyritinol HCl and
precursor molecules, the exact nature of which was unknown. Ten controls were negative (3).

   A 55-year-old nurse in a home for the elderly, responsible for the administration and distribution of medications,
had been suffering for a month from an itching, erythematosquamous eruption mainly on the left side of her face
(forehead, cheek, chin) and an associated swelling of both eyelids. She had a chronic irritant dermatitis on her hands
that was correlated with the use of soaps and detergents. She did not use any cosmetics or any topical medications.
Patch testing with the medications she came in contact with gave a positive reaction to pyritinol HCl at D2 and D4
(++). This was the active ingredient in tablets, which she would crush between two spoons prior to daily administration to her infirm mother. Three controls were negative (4).

CUTANEOUS ADVERSE DRUG REACTIONS FROM SYSTEMIC ADMINISTRATION CAUSED BY TYPE IV (DELAYED-TYPE) HYPERSENSITIVITY (as demonstrated by positive patch tests)

Cutaneous adverse drug reactions from systemic administration of pyritinol caused by type IV (delayed-type) hypersensitivity have included photosensitivity (5).

Photosensitivity

A 62-year-old male farmer, who had been taking various drugs including pyritinol for one week for treatment of cerebral arteriosclerosis, complained of erythematous lesions with severe itching on the exposed areas that had been present for a few days. On examination, the patient had diffuse erythema intermingled with slightly elevated erythematous lesions of 3 to 8 mm in size on the exposed areas. Photosensitivity tests showed a decreased MED for UVA. The skin eruption gradually worsened and erosive cheilitis, stomatitis and glossitis appeared, which turned into linear, whitish lesions resembling lichen planus (Wickham's striae). The eruption on the exposed areas had evolved into keratotic verrucose and purplish gray or brownish pigmented plane papules or elevated plaques resembling lichen planus. Lesions on the scalp produced considerable hair loss. Patch and photopatch tests were positive only to the photopatch test with pyritinol 20% pet., showing several milium-sized red papules appearing 24 hours after UVA irradiation. After cessation of pyritinol, the skin changes in the exposed areas, and enanthema of lips, tongue, and buccal mucosa disappeared in the course of 6 months, leaving purplish gray pigmentation intermingled with depigmentation (leukomelanoderma), especially on the forehead and occipital region. The authors diagnosed lichen planus-like eruption due to photoallergy to pyritinol (5). However, the evidence for that is rather weak.

LITERATURE

1   The data in the section 'General' may have been obtained from literature discussed in this chapter, but mostly also or exclusively from one or more of the following online sources: ChemIDPlus Advanced, PubChem, DrugBank, RxList, Drug Central, Drugs.com, and Wikipedia

2   Gielen K, Goossens A. Occupational allergic contact dermatitis from drugs in healthcare workers. Contact Dermatitis 2001;45:273-279

3   Wigger-Alberti W, Elsner P. Occupational contact dermatitis due to pyritinol. Contact Dermatitis 1997;37:91-92

4   Dooms-Goossens AE, Debusschere KM, Gevers DM, Dupré KM, Degreef HJ, Loncke JP, et al. Contact dermatitis caused by airborne agents. A review and case reports. J Am Acad Dermatol 1986;15:1-10

5   Ishibashi A, Hirano K, Nishiyama Y. Photosensitive dermatitis due to pyritinol. Arch Dermatol 1973;107:427-428

Chapter 3.425 QUINAPRIL

IDENTIFICATION

Description/definition : Quinapril is the dipeptide that conforms to the structural formula shown below
Pharmacological classes : Angiotensin-converting enzyme inhibitors; antihypertensive agents; cardiovascular agents
IUPAC name : (3S)-2-[(2S)-2-[[(2S)-1-Ethoxy-1-oxo-4-phenylbutan-2-yl]amino]propanoyl]-3,4-dihydro-1H-isoquinoline-3-carboxylic acid
CAS registry number : 85441-61-8
EC number : Not available
Merck Index monograph : 9437
Patch testing : No data available; most systemic drugs can be tested at 10% pet.; if the pure chemical is not available, prepare the test material from intravenous powder, the content of capsules or – when also not available – from powdered tablets to achieve a final concentration of the active drug of 10% pet.
Molecular formula : $C_{25}H_{30}N_2O_5$

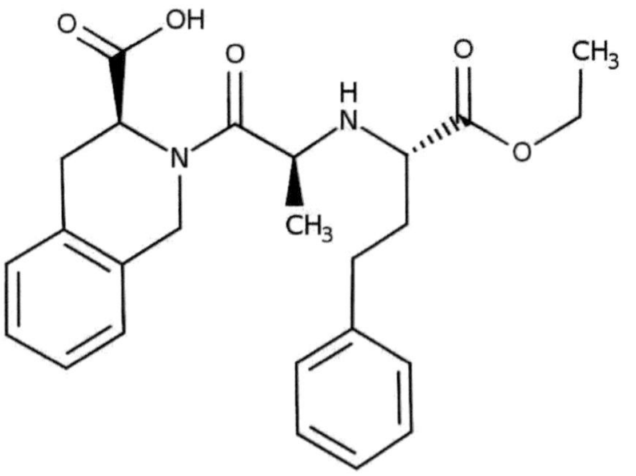

GENERAL

Quinapril is a prodrug and non-sulfhydryl angiotensin converting enzyme (ACE) inhibitor with antihypertensive activity. It is hydrolyzed into its active form quinaprilat, which binds to and inhibits ACE, thereby blocking the conversion of angiotensin I to angiotensin II. This abolishes the potent vasoconstrictive actions of angiotensin II and leads to vasodilatation. The drug also causes a decrease in angiotensin II-induced aldosterone secretion by the adrenal cortex, thereby promoting diuresis and natriuresis, and increases bradykinin levels. Quinapril is used in the treatment of hypertension or adjunct in the treatment of heart failure (1).

In pharmaceutical products, quinapril is employed as quinapril hydrochloride (CAS number 82586-55-8, EC number not available, molecular formula $C_{25}H_{31}ClN_2O_5$) (1).

CUTANEOUS ADVERSE DRUG REACTIONS FROM SYSTEMIC ADMINISTRATION CAUSED BY TYPE IV (DELAYED-TYPE) HYPERSENSITIVITY (as demonstrated by positive patch tests)

Cutaneous adverse drug reactions from systemic administration of quinapril caused by type IV (delayed-type) hypersensitivity have included unspecified cutaneous adverse drug reaction (2).

Other cutaneous adverse drug reactions

In Finland, in the period 1989-2001, 826 patients with suspected cutaneous drug eruptions were patch tested and 89 had one or more positive reactions. Of these individuals, one had a positive patch test to quinapril. It was not specified which type of drug reaction quinapril had caused (2).

LITERATURE

1 The data in the section 'General' may have been obtained from literature discussed in this chapter, but mostly also or exclusively from one or more of the following online sources: ChemIDPlus Advanced, PubChem, DrugBank, RxList, Drug Central, Drugs.com, and Wikipedia
2 Lammintausta K, Kortekangas-Savolainen O. The usefulness of skin tests to prove drug hypersensitivity. Br J Dermatol 2005;152:968-974

Chapter 3.426 QUINIDINE

IDENTIFICATION

Description/definition   : Quinidine is the alkaloid extracted from the bark of the *Cinchona* tree and similar plant species that conforms to the structural formula shown below

Pharmacological classes : Antimalarials; enzyme inhibitors; voltage-gated sodium channel blockers; muscarinic antagonists; α-adrenergic antagonists; anti-arrhythmia agents; cytochrome P-450 CYP2D6 inhibitors

IUPAC name   : (S)-[(2R,4S,5R)-5-Ethenyl-1-azabicyclo[2.2.2]octan-2-yl]-(6-methoxyquinolin-4-yl)methanol

CAS registry number   : 56-54-2

EC number   : 200-279-0

Merck Index monograph   : 9446

Patch testing   : Sulfate, 1% pet. (SmartPracticeCanada)

Molecular formula   : $C_{20}H_{24}N_2O_2$

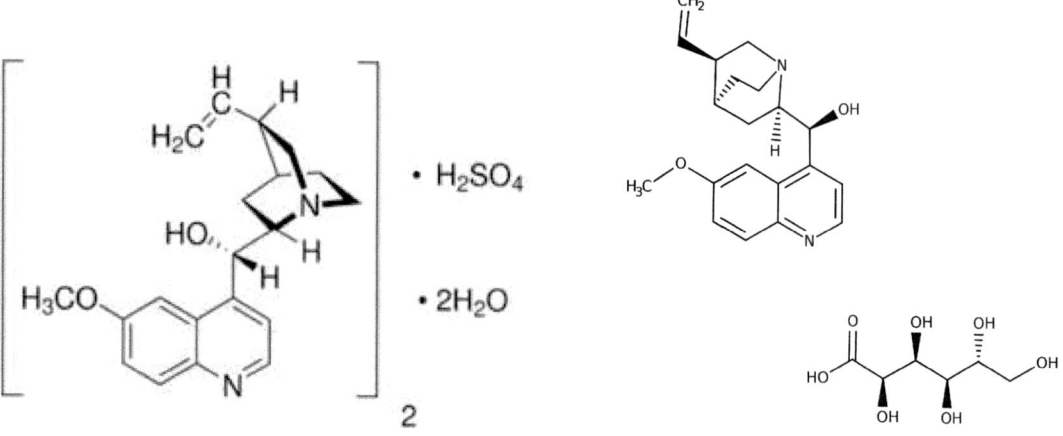

Quinidine sulfate   Quinidine gluconate

GENERAL

Quinidine is a d-isomer and diastereoisomer of quinine, an alkaloid extracted from the bark of the *Cinchona* tree and similar plant species with antimalarial and antiarrhythmic (Class Ia) properties. As an antiarrhythmic agent, quinidine dampens the excitability of cardiac and skeletal muscles by blocking sodium and potassium currents across cellular membra-nes. It prolongs cellular action potential, and decreases automaticity. Quinidine also blocks muscarinic and α-adrenergic neurotransmission. This agent is indicated for the treatment of ventricular pre-excitation and cardiac dysrhythmias (1). In pharmaceutical products, quinidine is mostlyemployed as quinidine sulfate (CAS number 6591-63-5, EC number not available, molecular formula $C_{40}H_{54}N_4O_{10}S$) or as quinidine gluconate (CAS number 7054-25-3, EC number 230-333-9, molecular formula $C_{26}H_{36}N_2O_9$) (1).

CONTACT ALLERGY FROM ACCIDENTAL CONTACT

Occupational allergic contact dermatitis

Five workers from a pharmaceutical plant developed eczema localized to various parts of the body and two workers had eczema and urticaria. These individuals handled various pharmaceutical products, including quinidine and alprenolol. Four of these patients were patch tested with quinidine sulfate 5, 2.5, 1.25 and 0.5 mg/ml water. Three (75%) had positive reactions. The lowest concentration inducing positivity was 1.25 mg/ml in one individual and 0.5 in two patients. Thirty controls were negative to all concentrations (2).

In Stockholm, Sweden, 3 workers in a pharmaceutical company developed occupational allergic contact dermatitis from quinidine sulfate, a woman aged 26 and 2 men aged 27 and 52. After having worked with the drug for 2-3 months, two patients developed dermatitis of the hands, arms and face and the third of the face, neck and chest. All had positive patch test reactions to quinidine sulfate crystals and a saturated aqueous solution. When tested with serial dilutions of quinidine sulfate in water (0.5%, 0.25%, 0.125%, 0.0625%), two reacted to 0.5% only, but the third down to 0.0625%. Cross-reactions to quinine were not observed. These were cases of occupational airborne allergic contact dermatitis to quinidine sulfate (9,10).

Occupational allergic contact dermatitis to quinidine was first described in 1965 in a pharmaceutical worker (3).

Contact and photocontact allergy *by proxy*
A 41-year-old man presented with eczematous plaques on the backs of the first and second fingers of both hands, the neck, the lips, and both earlobes. At home, he administered capsules of quinidine sulfate on a daily basis to his child. Patch and photopatch testing yielded a positive reaction to commercial quinidine sulfate 30% pet. irradiated with UVA 5 J/cm^2. Ten controls were negative. The eczema improved when the patient stopped handling the quinidine medication (7).

A 42-year-old man, who had recently begun caring for his invalid mother which included administering her medications, presented with an acute weeping dermatitis primarily around the right eye and cheek with some extension to the left part of the face. At one point, the patient's mother became ill and spent almost one month in a hospital. During this period, his dermatitis cleared totally, but flared severely shortly after his mother returned to his care. Patch tests performed with all drugs that the patient had contact with was strongly positive to a saturated aqueous solution of quinidine sulfate only. Five controls were negative. These latter two patients had (photo)allergic contact dermatitis *by proxy*.

CUTANEOUS ADVERSE DRUG REACTIONS FROM SYSTEMIC ADMINISTRATION CAUSED BY TYPE IV (DELAYED-TYPE) HYPERSENSITIVITY (as demonstrated by positive patch tests)
Cutaneous adverse drug reactions from systemic administration of quinidine caused by type IV (delayed-type) hypersensitivity have included photosensitivity (11,12,13) and unspecified drug eruption (4).

Photosensitivity
A 73-year-old man, while having used digoxin and quinidine for 2 years for ventricular and supraventricular arrhythmias, developed an erythematous, scaling eruption localized to sun-exposed areas. Histopathology showed a subacute spongiotic dermatitis. Quinidine could not be substituted, but the rash cleared with treatment. Approximately 3 months later, the patient presented with a generalized exfoliative erythroderma with hyperpigmentation and hypopigmentation. After a second flare, quinidine was stopped. Photopatch tests with quinidine sulfate 0.01%, 0.1% and 1% in saline, using 45 minutes of midday summer sunlight as light source, were positive to the 1% solution with a vesicular reaction at D2 and D3. Three controls were negative. There have been no recurrences after stopping the intake of quinidine (11).

An 83-year-old man had been treated for years with quinidine sulfate and various other drugs. For several months before presentation, the patient had noticed increased sensitivity to UV light. Examination showed eczematous infiltrated lesions over the entire face, ears, neck, back of both hands and lower arms. Four weeks after discontinuation of quinidine, a photopatch test with quinidine sulfate 1% pet. irradiated with 10 J/cm^2 UVA was positive at D2 and D3 with a crescendo papulovesicular reaction. A systemic photochallenge confirmed a causal relationship between the patient's complaints and quinidine and all signs disappeared after discontinuation of this drug (12,13).

See also the section 'Contact and photocontact allergy by proxy' above, ref. 7.

Other cutaneous adverse drug reactions
In Finland, in the period 1989-2001, 826 patients with suspected cutaneous drug eruptions were patch tested and 89 had one or more positive reactions. Of these individuals, 1 reacted to quinidine sulfate. Clinical details were not provided, and it was not mentioned which type of drug eruption this patient had suffered from (4).

Cross-reactions, pseudo-cross-reactions and co-reactions
In cases of contact allergy, quinidine does not cross-react to or from quinine (5,9,10). In photocontact allergy, however, photocross-reactivity may be observed (6). The explanation is that, when quinidine and quinine solutions are exposed to UVA, various photoproducts are formed, a number of which are identical (5). This would mean that photo-cross-reactivity may in fact be photo-*pseudo*cross-reactivity (reactions to the same chemicals).

LITERATURE
1   The data in the section 'General' may have been obtained from literature discussed in this chapter, but mostly also or exclusively from one or more of the following online sources: ChemIDPlus Advanced, PubChem, DrugBank, RxList, Drug Central, Drugs.com, and Wikipedia
2   Stejskal VD, Olin RG, Forsbeck M. The lymphocyte transformation test for diagnosis of drug-induced occupational allergy. J Allergy Clin Immunol 1986;77:411-426

3   Fernström AI. Occupational quinidine contact dermatitis, a concept apparently not yet described. Acta Derm Venereol 1965;45:129-134

4   Lammintausta K, Kortekangas-Savolainen O. The usefulness of skin tests to prove drug hypersensitivity. Br J Dermatol 2005;152:968-974

5   Isaksson M, Bruze M, Gruvberger B, Ljunggren B. Quinine and quinidine photoproducts can be identical. Acta Derm Venereol 1994;74:286-288

6   Ljunggren B, Hindsén M, Isaksson M. Systemic quinine photosensitivity with photoepicutaneous cross-reactivity to quinidine. Contact Dermatitis 1992;26:1-4

7   Agudo-Mena JL, Romero-Pérez D, Encabo-Durán B, Álvarez-Chinchilla PJ, Silvestre-Salvador JF. Photoallergic contact dermatitis caused by quinidine sulfate in a caregiver. Contact Dermatitis 2017;77:131-132

8   Fowler JF. Allergic contact dermatitis to quinidine. Contact Dermatitis 1985;13:280-281

9   Wahlberg JE, Boman A. Contact sensitivity to quinidine sulfate from occupational exposure. Contact Dermatitis 1981;7:27-31

10   Wahlberg JE, Forsbeck M. Contact sensitivity to quinidine sulphate - an antiarrhythmic. Contact Dermatitis Newsletter 1973;14:412

11   Lang PJ. Quinidine-induced photodermatitis confirmed by photopatch testing. J Am Acad Dermatol 1983;9:124-128

12   Schürer NY, Hölzle E, Plewig G, Lehmann P. Photosensitivity induced by quinidine sulfate: experimental reproduction of skin lesions. Photodermatol Photoimmunol Photomed 1992;9:78-82

13   Schürer NY, Lehmann P, Plewig G. Chinidininduzierte Photoallergie. Eine klinische und experimentelle Studie [Quinidine-induced photoallergy. A clinical and experimental study]. Hautarzt 1991;42:158-161 (Article in German)

Chapter 3.427 QUININE

IDENTIFICATION

Description/definition   : Quinine is an alkaloid from the bark of *Cinchona officinalis,* which conforms to the
                             formula shown below
Pharmacological classes : Antimalarials; analgesics, non-narcotic; muscle relaxants, central
IUPAC name   : (*R*)-[(2*S*,4*S*,5*R*)-5-Ethenyl-1-azabicyclo[2.2.2]octan-2-yl]-(6-methoxyquinolin-4-
                             yl)methanol
Other names   : Chininum
CAS registry number   : 130-95-0
EC number   : 205-003-2
Merck Index monograph : 9447
Patch testing   : Sulfate 1% pet. (Chemotechnique); sulfate 25% pet. (SmartPracticeCanada)
Molecular formula   : $C_{20}H_{24}N_2O_2$

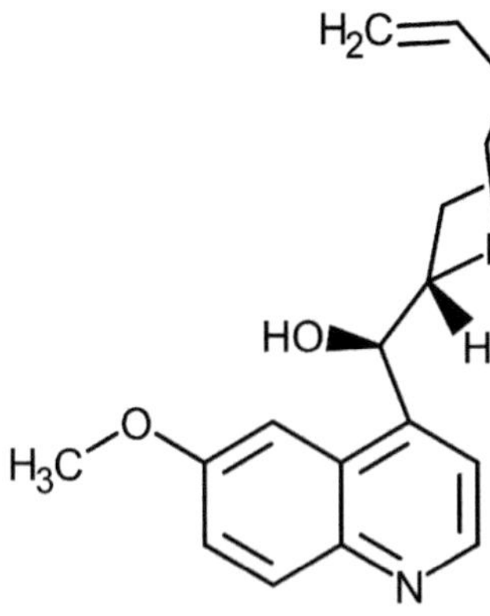

GENERAL

Quinine is an alkaloid obtained from the bark of various species of *Cinchona* trees. In the form of quinine sulfate (dihydrate) (CAS number 6119-70-6, EC number not available, molecular formula $C_{40}H_{54}N_4O_{10}S$) it is indicated for the treatment of malaria and leg cramps (1). In topical preparations, notably contraceptive pessaries and hair lotions, quinine has – formerly - frequently caused contact allergy/allergic contact dermatitis, which subject has been reviewed in Volume 3 of the *Monographs in contact allergy* series (13). The literature on quinine allergy from before 1960 has been reviewed in ref. 2. Allergic reactions to quinine in tonic water (14) fall outside the scope of this book.

CONTACT ALLERGY FROM ACCIDENTAL CONTACT

A man working in a factory producing quinine had a dermatitis resembling atopic dermatitis, which exacerbated each time he came in contact with quinine; there was a positive patch test to quinine sulfate 1% pet. (11). Another employee of the same factory had photosensitive eczema and showed positive patch tests to quinine hydrochloride, quinine dihydrochloride, quinine sulfate and quinine alkaloid, all tested at 1% pet. (11). A third worker, a man aged 29, after 3 weeks of working with quinine noticed an eruption on his forehead, followed by swelling of the eyes and lips, patchy erythema and scaling of the cheeks and periorbital skin. A patch test was strongly positive to quinine sulfate 1% water (11).

A man aged 48, working in the same company as chemist, was exposed to quinine products and developed a severe eruption on the face. His skin cleared while on holiday but relapsed on return to work. The patient continued to have fierce photosensitive eczema on the face and back of the hands over a period of 3 years, despite avoiding chemical contacts as much as possible. Phototests showed abnormal reactions through the ultraviolet range and into the visible light, which was persistent. Patch tests were initially positive to quinine dihydrochloride 0.01%, 0.1% and 1% water, but on retesting negative. At that point, however, photopatch tests were positive at 1% and 0.1% (11,12). This was a case of occupational airborne photoallergic contact dermatitis with persistent light reactivity.

In Leuven, Belgium, in the period 2001-2019, 201 of 1248 health care workers/employees of the pharmaceutical industry had occupational allergic contact dermatitis. In 23, dermatitis was caused by skin contact with a systemic drug. In one of these patients, quinine sulfate was the drug/one of the drugs that caused occupational dermatitis (3).

CUTANEOUS ADVERSE DRUG REACTIONS FROM SYSTEMIC ADMINISTRATION CAUSED BY TYPE IV (DELAYED-TYPE) HYPERSENSITIVITY (as demonstrated by positive patch tests)

Cutaneous adverse drug reactions from systemic administration of quinine caused by type IV (delayed-type) hypersensitivity have included photosensitivity (4,6,8).

Photosensitivity

A man developed an eczematous eruption on the face and dorsal aspects of the hands after 3 weeks' oral treatment with quinine. The photoreaction cleared within a week of quinine being stopped. UVA and UVB erythema threshold determinations were normal. A photopatch test was positive for irradiated quinine down to a concentration of 0.01% and for unirradiated quinine to 0.5%. There was a photo-cross-reaction to the isomer quinidine down to a concentration of 0.01% (6).

A 41-year-old woman, using intermittently oral quinine for muscle cramps, developed photosensitivity and a Koebner reaction while receiving phototherapy for psoriasis. Photobiological investigations showed a strongly positive vesicular photopatch test to quinine sulfate 4% irradiated with UVA 3 J/cm^2 and a marked persistent reduction in the minimal erythema dose to solar-simulated radiation in the uninvolved skin. The photosensitivity in this patient was still present 3 years after she had stopped the use of quinine (persistent light reaction) (8).

A 78-year-old man, taking quinine sulfate for leg cramps, developed a fierce exposed-site eczematous eruption over his face, neck and hands. Following the withdrawal of quinine, the eruption disappeared, leaving post-inflammatory hyperpigmentation. A patch test with quinine sulfate 1% aqua was negative, but a photopatch test irradiated with UVA (no details provided) positive at D4. No controls were performed (4).

Cross-reactions, pseudo-cross-reactions and co-reactions

In cases of contact allergy, quinidine does not cross-react to or from quinine (5,9,10). In photocontact allergy, however, photocross-reactivity may be observed (6). The explanation is that, when quinidine and quinine solutions are exposed to UVA, various photoproducts are formed, a number of which are identical (5). This would mean that photo-cross-reactivity may in fact be photo-*pseudo*cross-reactivity (reactions to the same chemicals). Of ten patients sensitized to clioquinol and patch tested with a number of structurally-related chemicals, 4 reacted to quinine powder; only one of these also had a positive patch test to quinine 5% pet. (7).

LITERATURE

1   The data in the section 'General' may have been obtained from literature discussed in this chapter, but mostly also or exclusively from one or more of the following online sources: ChemIDPlus Advanced, PubChem, DrugBank, RxList, Drug Central, Drugs.com, and Wikipedia

2   Calnan CD, Caron GA. Quinine sensitivity. Br Med J 1961;ii(5269):1750-1752

3   Gilissen L, Boeckxstaens E, Geebelen J, Goossens A. Occupational allergic contact dermatitis from systemic drugs. Contact Dermatitis 2020;82:24-30

4   Hickey JR, Dunnill GS, Sansom JE. Photoallergic reaction to systemic quinine sulphate. Contact Dermatitis 2007;57:384-386

5   Isaksson M, Bruze M, Gruvberger B, Ljunggren B. Quinine and quinidine photoproducts can be identical. Acta Derm Venereol 1994;74:286-288

6   Ljunggren B, Hindsén M, Isaksson M. Systemic quinine photosensitivity with photoepicutaneous cross-reactivity to quinidine. Contact Dermatitis 1992;26:1-4

7   Soesman-van Waadenoijen Kernekamp A, van Ketel WG. Persistence of patch test reactions to clioquinol (Vioform) and cross-sensitization. Contact Dermatitis 1980;6:455-460

8   Guzzo C, Kaidbey K. Persistent light reactivity from systemic quinine. Photodermatol Photoimmunol Photomed 1990;7:166-168

9   Wahlberg JE, Boman A. Contact sensitivity to quinidine sulfate from occupational exposure. Contact Dermatitis 1981;7:27-31

10 Wahlberg JE, Forsbeck M. Contact sensitivity to quinidine sulphate - an antiarrhythmic. Contact Dermatitis Newsletter 1973;14:412

11 Hardie RA, Savin JA, White DA, Pumford S. Quinine dermatitis: Investigation of a factory outbreak. Contact Dermatitis 1978;4:121-124

12 Johnson BE, Zaynoun S, Gardiner JM, Frain-Bell W. A study of persistent light reaction in quindoxin and quinine photosensitivity. Br J Dermatol 1975;93(Suppl.11):21-22

13 De Groot AC. Monographs in contact allergy, volume 3: Topical Drugs. Boca Raton, Fl, USA: CRC Press Taylor and Francis Group, 2021 (ISBN 978-0-367-23693-9)

14 Genest G, Thomson DM. Fixed drug eruption to quinine: a case report and review of the literature. J Allergy Clin Immunol Pract 2014;2:469-470

Chapter 3.428 RABEPRAZOLE

IDENTIFICATION

Description/definition   : Rabeprazole is the α-pyridylmethylsulfinyl benzimidazole that conforms to the
                             structural formula shown below
Pharmacological classes   : Anti-ulcer agents; proton pump inhibitors
IUPAC name   : 2-[[4-(3-Methoxypropoxy)-3-methylpyridin-2-yl]methylsulfinyl]-1H-benzimidazole
CAS registry number   : 117976-89-3
EC number   : Not available
Merck Index monograph   : 9476
Patch testing   : 10% pet. (1)
Molecular formula   : $C_{18}H_{21}N_3O_3S$

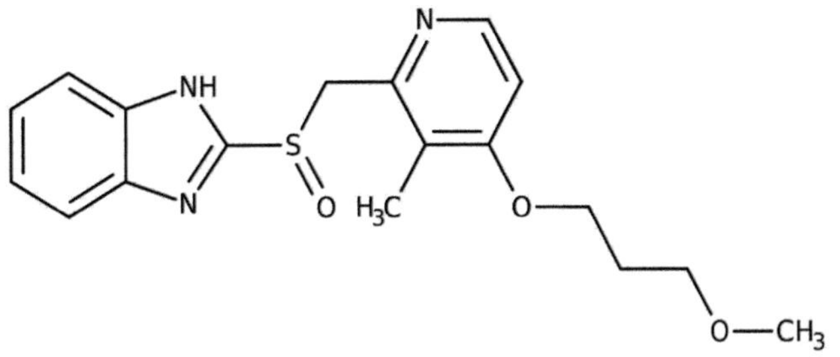

GENERAL

Rabeprazole is a proton pump inhibitor and a potent inhibitor of gastric acidity. It is a prodrug: in the acid environment of the parietal cells it turns into its active sulphenamide form. Rabeprazole inhibits the H+, K+ATPase of the coating gastric cells and dose-dependently suppresses basal and stimulated gastric acid secretion. Rabeprazole is indicated for the treatment of acid-reflux disorders, peptic ulcer disease, *H. pylori* eradication, and prevention of gastrointestinal bleeds with NSAID use (1). In pharmaceutical products, rabeprazole is employed as rabeprazole sodium (CAS number 117976-90-6, EC number not available, molecular formula $C_{18}H_{20}N_3NaO_3S$) (go.drugbank.com).

CONTACT ALLERGY FROM ACCIDENTAL CONTACT

A 36-year-old man presented with recently developed hypopigmented patches on the forehead, periocular area, and neck. He described his condition as ongoing symptoms of erythema with swelling of his eyelids during his work. The patient had worked for a year in the pharmaceutical industry and was involved in the manufacturing of proton pump inhibitors (PPIs), including rabeprazole and pantoprazole. He was exposed to PPI powder, although he wore protective gear such as gloves and a mask, but not goggles, during the manufacturing process. A skin biopsy from his nape showed a complete loss of melanin and melanocytes, confirming the diagnosis of vitiligo. Patch tests were positive to pure rabeprazole and pantoprazole powder 1% pet. The authors suggested that PPIs most likely lead to the apoptosis of melanocytes by modifying the pH gradient and enhancing oxidative stress in melanocytes, possibly resulting in melanocyte-specific autoimmunity (3).

Cross-reactions, pseudo-cross-reactions and co-reactions

A patient sensitized to pantoprazole, which may have contributed to DRESS, had cross-reactions to rabeprazole, omeprazole, lansoprazole, and esomeprazole (2). A patient sensitized to omeprazole cross-reacted rabeprazole and other proton pump inhibitors (all crushed tablets, 1% and 10% pet.) (4).

LITERATURE

1   Brockow K, Garvey LH, Aberer W, Atanaskovic-Markovic M, Barbaud A, Bilo MB, et al.; ENDA/EAACI Drug Allergy Interest Group. Skin test concentrations for systemically administered drugs – an ENDA/EAACI Drug Allergy Interest Group position paper. Allergy 2013;68:702-712
2   Liippo J, Pummi K, Hohenthal U, Lammintausta K. Patch testing and sensitization to multiple drugs. Contact Dermatitis 2013;69:296-302
3   Kim D, Kang H. Proton pump inhibitor-induced vitiligo following airborne occupational sensitization. Contact Dermatitis. 2021 Jun 21. doi: 10.1111/cod.13919. Epub ahead of print.
4   Al-Falah K, Schachter J, Sasseville D. Occupational allergic contact dermatitis caused by omeprazole in a horse breeder. Contact Dermatitis 2014;71:377-378

Chapter 3.429 RAMIPRIL

IDENTIFICATION

Description/definition	: Ramipril is the pyrrolidone carboxylic acid that conforms to the structural formula shown below
Pharmacological classes	: Angiotensin-converting enzyme inhibitors; antihypertensive agents
IUPAC name	: (2S,3aS,6aS)-1-[(2S)-2-[[(2S)-1-Ethoxy-1-oxo-4-phenylbutan-2-yl]amino]propanoyl]-3,3a,4,5,6,6a-hexahydro-2H-cyclopenta[b]pyrrole-2-carboxylic acid
CAS registry number	: 87333-19-5
EC number	: Not available
Merck Index monograph	: 9491
Patch testing	: Tablet, pulverized, 30% pet. (3); most systemic drugs can be tested at 10% pet.; if the pure chemical is not available, prepare the test material from intravenous powder, the content of capsules or – when also not available – from powdered tablets to achieve a final concentration of the active drug of 10% pet.
Molecular formula	: $C_{23}H_{32}N_2O_5$

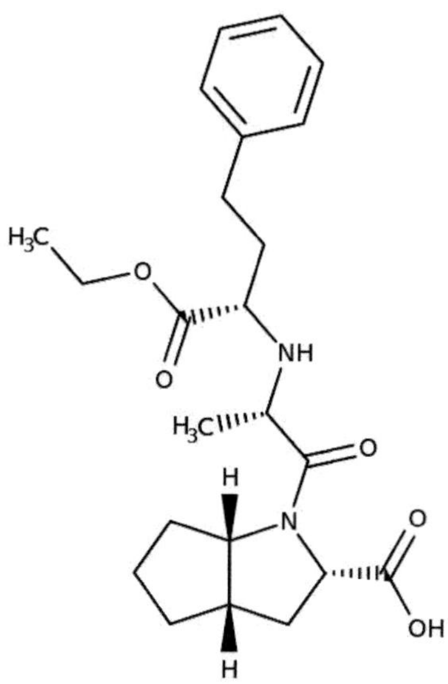

GENERAL

Ramipril is a prodrug and non-sulfhydryl angiotensin converting enzyme (ACE) inhibitor with antihypertensive activity. Ramipril is converted in the liver by de-esterification into its active form ramiprilat, which inhibits ACE, thereby blocking the conversion of angiotensin I to angiotensin II. This abolishes the potent vasoconstrictive actions of angiotensin II and leads to vasodilatation. This agent also causes an increase in bradykinin levels and a decrease in angiotensin II-induced aldosterone secretion by the adrenal cortex, thereby promoting diuresis and natriuresis. Ramipril may be used in the treatment of hypertension, congestive heart failure, nephropathy, and to reduce the rate of death, myocardial infarction and stroke in individuals at high risk of cardiovascular events (1).

CUTANEOUS ADVERSE DRUG REACTIONS FROM SYSTEMIC ADMINISTRATION CAUSED BY TYPE IV (DELAYED-TYPE) HYPERSENSITIVITY (as demonstrated by positive patch tests)

Cutaneous adverse drug reactions from systemic administration of rimapril caused by type IV (delayed-type) hypersensitivity have included Stevens-Johnson syndrome/toxic epidermal necrolysis (SJS/TEN) (3) and photosensitivity (2).

Stevens-Johnson syndrome/toxic epidermal necrolysis (SJS/TEN)

In a multicenter investigation in France, of 17 patients patch tested for SJS/TEN, 4 (24%) had positive patch tests to drugs, including one to ramipril. This was a woman aged 41 who reacted to commercial ramipril 30% pet. Clinical details were not provided (3).

Photosensitivity

A 63-year-old welder presented with a 6-year history of edema, erythema, and eczema, along with burning sensations and slowly increasing heat-sensitivity, confined to the face, neck, and, intermittently, extensor forearms. He reported worsening of symptoms on exposure to sunlight, but also during welding, with intense burning sensations, even though he used welding goggles or a welding shield to cover the face and/or eyes. The patient reported drug therapy for essential hypertension originally with a combination of hydrochlorothiazide and ramipril, and since 5 years with ramipril alone. Photopatch tests showed strongly positive reactions to ramipril as well as to hydrochlorothiazide 2 and 3 days after irradiation with 10 J/cm^2 UVA. Unirradiated patch tests or patch tests irradiated with UVB or visible light remained negative. Patch test concentrations and vehicles were not mentioned (2).

LITERATURE

1   The data in the section 'General' may have been obtained from literature discussed in this chapter, but mostly also or exclusively from one or more of the following online sources: ChemIDPlus Advanced, PubChem, DrugBank, RxList, Drug Central, Drugs.com, and Wikipedia
2   Wagner SN, Welke F, Goos M. Occupational UVA-induced allergic photodermatitis in a welder due to hydrochlorothiazide and ramipril. Contact Dermatitis 2000;43:245-246
3   Barbaud A, Collet E, Milpied B, Assier H, Staumont D, Avenel-Audran M, et al. A multicentre study to determine the value and safety of drug patch tests for the three main classes of severe cutaneous adverse drug reactions. Br J Dermatol 2013;168:555-562

Chapter 3.430 RANITIDINE

IDENTIFICATION

Description/definition : Ranitidine is the aralkylamine that conforms to the structural formula shown below
Pharmacological classes : Anti-ulcer agents; histamine H_2 antagonists
IUPAC name : (*E*)-1-*N*'-[2-[[5-[(Dimethylamino)methyl]furan-2-yl]methylsulfanyl]ethyl]-1-*N*-methyl-2-nitroethene-1,1-diamine
CAS registry number : 66357-35-5
EC number : 266-332-5
Merck Index monograph : 9498
Patch testing : Hydrochloride, 5% pet.
Molecular formula : $C_{13}H_{22}N_4O_3S$

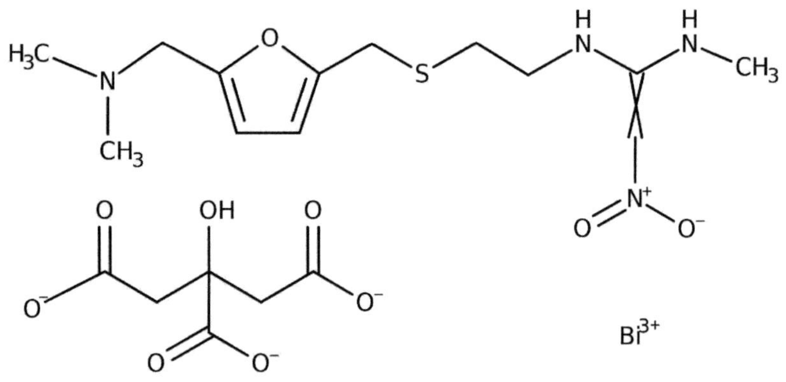

Ranitidine bismuth citrate

GENERAL

Ranitidine is a member of the class of histamine H2 receptor antagonists with antacid activity. It is a competitive and reversible inhibitor of the action of histamine at the histamine H2 receptors on parietal cells in the stomach, thereby inhibiting the normal and meal-stimulated secretion of stomach acid. In addition, other substances that promote acid secretion have a reduced effect on parietal cells when the H2 receptors are blocked. Ranitidine is used alone or with concomitant antacids for the treatment of active duodenal ulcer, pathological hypersecretion of gastric acid (e.g. Zollinger-Ellison syndrome, systemic mastocytosis), active gastric ulcer, gastric esophageal reflux disease, and erosive esophagitis (1). In pharmaceutical products, ranitidine may be employed as ranitidine hydrochloride (CAS number 66357-59-3, EC number 266-333-0, molecular formula $C_{13}H_{23}ClN_4O_3S$) or as ranitidine bismuth citrate (CAS number 128345-62-0, EC number not available, molecular formula $C_{19}H_{27}BiN_4O_{10}S$).

CONTACT ALLERGY FROM ACCIDENTAL CONTACT

General

Ranitidine is one of the most frequent causes of occupational allergic contact dermatitis to systemic drugs in the pharmaceutical industries and in healthcare workers (2,4,5,6,10) and can both cause dermatitis by direct and by airborne contact.

Case series

In Leuven, Belgium, in the period 2001-2019, 201 of 1248 health care workers and employees of the pharmaceutical industry had occupational allergic contact dermatitis. In 23 (11%) dermatitis was caused by skin contact with a systemic drug: 19 nurses, two chemists, one physician, and one veterinarian. In 5 patients, ranitidine hydrochloride was the drug/one of the drugs that caused occupational dermatitis (5).

Also in Leuven, Belgium, in the period 2007-2011, 81 patients have been diagnosed with occupational airborne allergic contact dermatitis. In 23 of them, drugs were the offending agents, including ranitidine in two cases (2).

Previously, in the same hospital, between 1978 and 2001, occupational allergic contact dermatitis to pharmaceuticals was diagnosed in 33 health care workers: 26 nurses, 4 veterinarians, 2 pharmacists and one medical doctor. Practically all of these patients presented with hand dermatitis (often the fingers), sometimes with secondary localizations, frequently the face. In this group, seven patients had occupational allergic contact dermatitis from ranitidine. Ranitidine was – with penicillin – after propacetamol the second most frequent cause of occupational allergic contact dermatitis (4).

In the United Kingdom, 7 employees of a pharmaceutical manufacturing plant developed occupational sensitization to ranitidine, after having exposed to ranitidine compounds for 2 months to 7.5 years. Eczema commonly involved the hands and wrists with facial redness/itchiness in most individuals. Patch tests were positive to ranitidine hydrochloride 1% and 5% pet. in 6 patients (of who 2 also reacted to the base) and one only to ranitidine base 5% pet. (10).

In Spain, in 1987, 16 employees of a chemical factory manufacturing ranitidine base were suspected of occupational allergic contact dermatitis to ranitidine in a period of 8 months. Eleven patient were office workers, 2 chemists, 2 maintenance workers and one secretary. The manifestations were eczema on the dorsum of the hands and/or dyshidrosis on the fingers, eczematous plaques on the forearms and dorsum of the feet, and erythematous edematous papules on the face and neck, sometimes with edema of the eyelids. The lesions on the face resembled seborrheic dermatitis. The secretary had an extremely itchy eczematous plaque on the right thigh, caused by accidentally spilling a sample of ranitidine there. In the factory, ranitidine base was mechanically sacked for later conversion to ranitidine HCl powder. Safety measures such as special clothes, gloves and masks were provided but not always used, so contact with the allergen, in the form of powder, occurred during the process, apparently throughout the factory (12 patients were office workers; alternatively, they may have entered the production area regularly). Patch tests were performed in 11 patients (the others refused) and all had positive reactions to ranitidine base 5% pet. and ranitidine hydrochloride 1% pet. Twenty-five controls were negative (6).

Case reports
Single cases of occupational sensitization to ranitidine in patients working in the pharmaceutical industries have been reported from the United Kingdom (7), Spain (8,11,13), and Singapore (12). Some patients were also (12,13) or exclusively (16, a chemist involved in the synthesis of ranitidine) sensitized to 5-((2-aminoethyl)thiomethyl)-*N*,*N*-dimethyl-2-furanmethanamine, an intermediate in the production of ranitidine (CAS number 66356-53-4). Two other industrial chemists became sensitized to cistoran (another intermediate in ranitidine synthesis), and to ranitidine itself (17). Occupational sensitization to ranitidine has also been observed in a racehorse trainer from administering a paste containing 19% ranitidine to horses (9).

Allergic contact dermatitis by proxy
A 73-year-old woman presented with pulpitis involving her dominant hand and with eyelid eczema. She reported that her husband had become ill about a year previously, and she crushed and administered his drugs to him, including ranitidine. Patch tests with all drugs were positive to commercial ranitidine HCl and later to the pure drug at 1%, 5%, 10% and 30% in petrolatum (14).

A 39-year-old woman presented with erythema, edema, pruritus, and vesicular lesions over the dorsa and sides of her fingers, which had been present on-and-off for the previous 2 years. Patch testing showed a non-relevant reaction to nickel. Despite treatment the lesions spread to the dorsa of the hands and new eczematous lesions appeared on the face and neck. The patient administered several medications to her 3-year-old son, seriously ill from birth, including ranitidine HCl (pulverizing the tablet) and pyridoxine HCl (in powder form). Patch tests with all drugs that she handled showed positive reactions to ranitidine HCl 1% and 5% pet. and to pyridoxine HCl 10% pet. No cross-sensitization with other H$_2$-antagonists or B vitamins were observed. Fifteen controls were negative. This was a case of airborne sensitization to ranitidine and pyridoxine *by proxy* (15).

CUTANEOUS ADVERSE DRUG REACTIONS FROM SYSTEMIC ADMINISTRATION CAUSED BY TYPE IV (DELAYED-TYPE) HYPERSENSITIVITY (as demonstrated by positive patch tests)
Cutaneous adverse drug reactions from systemic administration of ranitidine caused by type IV (delayed-type) hypersensitivity have included exfoliative dermatitis (18), acute generalized exanthematous pustulosis (AGEP) (3), systemic allergic dermatitis (4), and drug reaction with eosinophilia and systemic symptoms (DRESS) (19,20).

Erythroderma, widespread erythematous eruption, exfoliative dermatitis
A 79-year-old man developed, 2 days after taking ranitidine, theophylline and clarithromycin, an erythematous rash with edema over the face, neck, and dorsa of the hands, with itching. The following days, the lesions spread to the trunk, arms and legs with severe scaling. Patch tests with the 3 drugs were positive to ranitidine HCl 1% and 10% pet. Twenty controls were negative. Oral administration of theophylline and clarithromycin gave no reaction (18). This was a case of exfoliative dermatitis from delayed-type hypersensitivity to oral ranitidine.

Acute generalized exanthematous pustulosis (AGEP)
A 53-year-old man had developed an extensive erythematous and pustular eruption ten days after he had been started on ranitidine hydrochloride 300 mg oral daily. Examination revealed an erythematous eruption over the trunk, arms and legs, studded with numerous non-follicular pustules smaller than 5 mm in diameter. Erythema multiforme-like lesions were noted in some areas. Laboratory examination showed a mild leukocytosis with neutrophilia and eosinophilia; cultures for bacteria and fungi of the pustular contents were negative. A punch biopsy showed a subcorneal pustule, focal keratinocyte necrosis and a mild dermal perivascular infiltrate of lymphocytes with numerous eosinophils. Two months later, a patch test with ranitidine hydrochloride dispersed in pet. was ++ positive, manifesting as a pustular eruption on an erythematous base. Fifteen controls were negative (3). This was a classic case of AGEP from ranitidine.

Systemic allergic dermatitis (systemic contact dermatitis)
A patient who was occupationally sensitized to ranitidine developed burning swollen lips after oral intake of this antihistamine (4).

Drug reaction with eosinophilia and systemic symptoms (DRESS)
A 16-year-old boy with a history of tetralogy of Fallot had undergone pulmonary valve replacement. Eighteen days postoperatively he developed a progressively extensive maculopapular exanthema with pronounced petechial and purpuric lesions across the arms and legs, facial edema, and mild mucosal involvement. He was febrile (39°C) and had palpable cervical lymphadenopathy, eosinophilia, elevated liver enzymes, renal impairment and leukopenia. DRESS was diagnosed. The patient had received 12 medications in the 5 weeks preceding the onset of the exanthema, including ranitidine (started 3 weeks before the onset of eruption). Histopathology was consistent with a drug eruption. Selected drug patch and intradermal tests (IDT) were negative. Delayed IDT with ranitidine (1:100 dilution) however produced a focal papulovesicular eruption at D4 (with a crescendo effect since D2). Patch tests produced strongly positive reactions to ranitidine 30% pet. (+++), ranitidine undiluted solution (++) and a weaker but positive (cross-)reaction to cimetidine 30% pet. Ten controls were negative to the IDT and patch tests (19).

The same patient was presented again 3 years later by different authors (20). Again, patch tests (with and without 10x adhesive tape stripping) were performed, using ranitidine 30% pet., rifampicin 10% pet. (both from crushed tablets) and vancomycin 0.05% water. There were strongly positive reactions to ranitidine, more pronounced on the side with tape stripping. Distant to the patch test, an urticated exanthem developed on the upper back, which proved to be flare-ups of ranitidine patch tests performed in the previous study (19,21). The patient did not have fever but developed lymphadenopathy and facial edema with mild lymphopenia within 24 hours. There was no eosinophilia or organ involvement. Nevertheless, some symptoms of DRESS were provoked by patch testing, possibly enhanced by tape stripping (20).

In London, between October 2017 and October 2018, 45 patients with suspected cutaneous adverse drug reactions, including 33 maculopapular eruptions (MPE), 4 fixed drug eruptions (FDE), 4 DRESS, 3 AGEP and one SJS/TEN, were patch tested with the suspected drugs. There were 10 (22%) positive patch test cases: 4 MPE, 2 FDE, 3 DRESS and 1 AGEP. Ranitidine, tested as commercial preparation 30% pet., was responsible for 1 case of DRESS; there was a cross-reaction to cimetidine (22, same patient as in ref. 19 and 20).

Cross-reactions, pseudo-cross-reactions and co-reactions
Two patients with DRESS caused by ranitidine may have cross-reacted to cimetidine (18,22).

LITERATURE
1 The data in the section 'General' may have been obtained from literature discussed in this chapter, but mostly also or exclusively from one or more of the following online sources: ChemIDPlus Advanced, PubChem, DrugBank, RxList, Drug Central, Drugs.com, and Wikipedia
2 Swinnen I, Goossens A. An update on airborne contact dermatitis: 2007-2011. Contact Dermatitis 2013;68:232-238
3 Blanes Martínez M, Silvestre Salvador JF, Vergara Aguilera G, Betlloch Mas I, Pascual Ramírez JC. Acute generalized exanthematous pustulosis induced by ranitidine hydrochloride. Contact Dermatitis 2003I;49:47
4 Gielen K, Goossens A. Occupational allergic contact dermatitis from drugs in healthcare workers. Contact Dermatitis 2001;45:273-279
5 Gilissen L, Boeckxstaens E, Geebelen J, Goossens A. Occupational allergic contact dermatitis from systemic drugs. Contact Dermatitis 2020;82:24-30
6 Romaguera C, Grimalt F, Vilaplana J. Epidemic of occupational contact dermatitis from ranitidine. Contact Dermatitis 1988;18:177-178
7 Bennett MF, Lowney AC, Bourke JF. A study of occupational contact dermatitis in the pharmaceutical industry. Br J Dermatol 2016;174:654-656

8   Herrera-Mozo I, Sanz-Gallen P, Martí-Amengual G. Occupational contact allergy to omeprazole and ranitidine. Med Pr 2017;68:433-435 (Article in Polish)

9   Meani R, Nixon R. Allergic contact dermatitis caused by ranitidine hydrochloride in a veterinary product. Contact Dermatitis 2015;73:125-126

10   Ryan PJ, Rycroft RJ, Aston IR. Allergic contact dermatitis from occupational exposure to ranitidine hydrochloride. Contact Dermatitis 2003;48:67-68

11   Alomar A, Puig L, Vilaltella I. Allergic contact dermatitis due to ranitidine. Contact Dermatitis 1987;17:54-55

12   Goh CL, Ng SK. Allergic contact dermatitis to ranitidine. Contact Dermatitis 1984;11:252

13   Romaguera C, Grimalt F, Vilaplana J. Contact dermatitis caused by intermediate products in the manufacture of clenbuterol, ranitidine base, and ranitidine hydrochloride. Dermatol Clin 1990;8:115-117

14   Mendieta Eckert M, Ratón Nieto JA, Acebo Mariñas E. Nonoccupational allergic contact dermatitis from ranitidine. Clin Exp Dermatol 2013;38:794-795

15   Córdoba S, Martínez-Morán C, García-Donoso C, Borbujo J, Gandolfo-Cano M. Non-occupational allergic contact dermatitis from pyridoxine hydrochloride and ranitidine hydrochloride. Dermatitis 2011;22:236-237

16   Rycroft RJ. Allergic contact dermatitis from a novel diamino intermediate, 5-[(2-aminoethyl)thiomethyl]-*N*,*N*-dimethyl-2-furanmethanamine, in laboratory synthesis. Contact Dermatitis 1983;9:456-458

17   Valsecchi R, Rohrich O, Cainelli T. Contact allergy to cistoran, an intermediate in ranitidine synthesis. Contact Dermatitis 1989: 20: 396–397.

18   Juste S, Blanco J, Garcés M, Rodriguez G. Allergic dermatitis due to oral ranitidine. Contact Dermatitis 1992;27:339-340

19   Watts TJ, Haque R. DRESS syndrome induced by ranitidine. J Allergy Clin Immunol Pract 2018;6:1030-1031

20   Teo YX, Ardern-Jones MR. Reactivation of drug reaction with eosinophilia and systemic symptoms with ranitidine patch testing. Contact Dermatitis 2021;84:278-279

21   Watts, TJ, Haque R. DRESS syndrome reactivation due to ranitidine patch testing: The flare-up phenomenon. Contact Dermatitis 2021;85:267

22   Watts TJ, Thursfield D, Haque R. Patch testing for the investigation of nonimmediate cutaneous adverse drug reactions: a prospective single center study. J Allergy Clin Immunol Pract 2019;7:2941-2943.e3

Chapter 3.431 RETINYL ACETATE

IDENTIFICATION

Description/definition   : Retinyl acetate is a naturally-occurring fatty acid ester form of retinol (vitamin A) that
conforms to the structural formula shown below
Pharmacological classes : Adjuvants, immunologic; anticarcinogenic agents
IUPAC name   : [(2E,4E,6E,8E)-3,7-Dimethyl-9-(2,6,6-trimethylcyclohexen-1-yl)nona-2,4,6,8-tetraenyl]
acetate
Other names   : Vitamin A acetate; retinol acetate
CAS registry number   : 127-47-9
EC number   : 204-844-2
Merck Index monograph   : 11481 (Vitamin A)
Patch testing   : 1% pet.
Molecular formula   : $C_{22}H_{32}O_2$

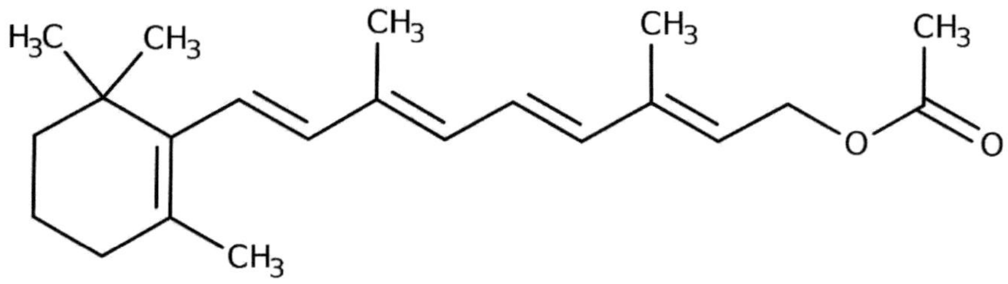

GENERAL

Retinyl acetate is a naturally-occurring acetate ester of retinol (vitamin A) with potential antineoplastic and chemopreventive activities. Retinyl acetate binds to and activates retinoid receptors, inducing cell differentiation and
decreasing cell proliferation. This agent also inhibits carcinogen-induced neoplastic transformation in some cancer
cell types and exhibits immunomodulatory properties (1).

CONTACT ALLERGY FROM ACCIDENTAL CONTACT

A 44-year-old man was employed in industrial production of vitamins. The job consisted of drying, sieving and
packing of vitamin A acetate. After 3 months, the patient developed eczema on the dorsum of the hands, face, neck,
arms, and trunk, despite use of gloves, protective suit with cuffs/wristlets, and mask. Following repeated attacks, the
eczema persisted despite cleansing the skin and topical treatment, whereupon he stopped working with vitamin A.
After change of job, the eczema disappeared. Patch tests were positive to retinyl acetate 0.1% (+), 0.5%, 1%, 5% and
10% pet. (all ++). In 97 controls, there were 6 doubtful (?+) reactions to 5% and 10%, and one to 1% pet. (2). This was
a case of occupational airborne allergic contact dermatitis.

LITERATURE

1   The data in the section 'General' may have been obtained from literature discussed in this chapter, but mostly
    also or exclusively from one or more of the following online sources: ChemIDPlus Advanced, PubChem,
    DrugBank, RxList, Drug Central, Drugs.com, and Wikipedia
2   Heidenheim M, Jemec GB. Occupational allergic contact dermatitis from vitamin A acetate. Contact Dermatitis
    1995;33:439

Chapter 3.432 RIBOSTAMYCIN

IDENTIFICATION

Description/definition   : Ribostamycin is the aminoglycoside antibiotic that conforms to the structural formula
                               shown below
Pharmacological classes   : Anti-bacterial agents
IUPAC name   : (2R,3S,4R,5R,6R)-5-Amino-2-(aminomethyl)-6-[(1R,2R,3S,4R,6S)-4,6-diamino-2-
                               [(2S,3R,4S,5R)-3,4-dihydroxy-5-(hydroxymethyl)oxolan-2-yl]oxy-3-hydroxycyclo-
                               hexyl]oxyoxane-3,4-diol
CAS registry number   : 25546-65-0
EC number   : 247-091-5
Merck Index monograph : 9600
Patch testing   : Sulfate, 20.0% pet.; late positive reactions (after D4) are frequent with (other) aminogly-
                               coside antibiotics and readings at D7-D8 are recommended
Molecular formula   : $C_{17}H_{34}N_4O_{10}$

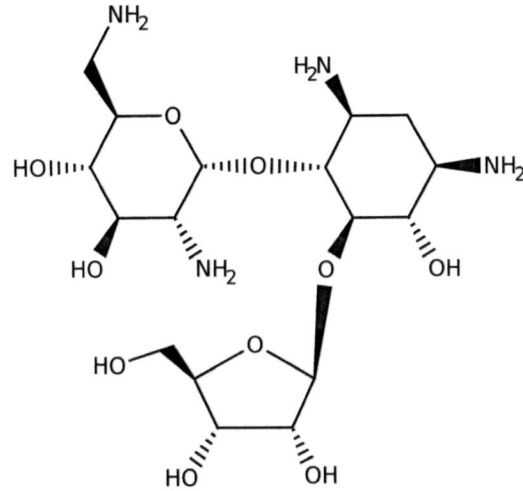

GENERAL

Ribostamycin is a broad-spectrum aminoglycoside antimicrobial isolated from *Streptomyces ribosifidicus*. Along with other aminoglycosides with the DOS (2-deoxystreptamine) subunit, it is a broad-spectrum antibiotic with important use against human immunodeficiency virus and is considered a critically important antimicrobial by the World Health Organization. In pharmaceutical products, ribostamycin is employed as ribostamycin sulfate (CAS number 53797-35-6, EC number 258-783-1, molecular formula $C_{17}H_{36}N_4O_{14}S$) (1).

CUTANEOUS ADVERSE DRUG REACTIONS FROM SYSTEMIC ADMINISTRATION CAUSED BY TYPE IV (DELAYED-TYPE) HYPERSENSITIVITY (as demonstrated by positive patch tests)

Cutaneous adverse drug reactions from systemic administration of ribostamycin caused by type IV (delayed-type) hypersensitivity have included systemic allergic dermatitis presenting as exfoliative erythroderma (3).

Systemic allergic dermatitis (systemic contact dermatitis)

A 48-year-old man presented with a 2-week history of a generalized itchy rash. Ten hours before its appearance, the patient had received an intramuscular injection of 1 gram ribostamycin for chronic prostatitis. A second injection made the rash and pruritus worse. On physical examination, there was a widespread confluent macular reddish scaly rash involving more than 95% of the body surface. Small lymph nodes were palpable in both groins. Laboratory studies disclosed leukocytosis with eosinophilia. Histological examination of a biopsy from the neck area showed focal parakeratosis with occasional necrotic keratinocytes and a moderately intense perivascular lymphomononu-clear infiltrate, with some admixture of eosinophils. Later, patch tests were positive to ribostamycin sulfate solution 25% (++ at D4) (10 controls negative) and neomycin sulfate 20% pet. (+++ at D4). A lymphocyte transformation test was positive to ribostamycin. It was suggested that the patient had previously become sensitized to neomycin from topical applications and now cross-reacted to ribostamycin. In this – plausible – scenario, this would be a case of systemic allergic dermatitis presenting as exfoliative erythroderma (3).

Cross-reactions, pseudo-cross-reactions and co-reactions
In patients sensitized to neomycin, about 70% cross-reacts to ribostamycin (2, Chapter 3.335 Neomycin). The cross-sensitivity pattern between aminoglycoside antibiotics in patients primarily sensitized to ribostamycin has not been well investigated.

LITERATURE
1   The data in the section 'General' may have been obtained from literature discussed in this chapter, but mostly also or exclusively from one or more of the following online sources: ChemIDPlus Advanced, PubChem, DrugBank, RxList, Drug Central, Drugs.com, and Wikipedia
2   Samsoen M, Metz R, Melchior E, Foussereau J. Cross-sensitivity between aminoside antibiotics. Contact Dermatitis 1980;6:141
3   Puig LL, Abadias M, Alomar A. Erythroderma due to ribostamycin. Contact Dermatitis 1989;21:79-82

Chapter 3.433 RIFAMPICIN

IDENTIFICATION

Description/definition : Rifampicin is a semisynthetic antibiotic produced from *Streptomyces mediterranei* that conforms to the structural formula shown below

Pharmacological classes : Antibiotics, antitubercular; cytochrome P-450 CYP2B6 inducers; cytochrome P-450 CYP2C19 inducers; cytochrome P-450 CYP2C8 inducers; cytochrome P-450 CYP2C9 inducers; cytochrome P-450 CYP3A inducers; leprostatic agents; nucleic acid synthesis inhibitors

IUPAC name : [(7S,9E,11S,12R,13S,14R,15R,16R,17S,18S,19E,21Z)-2,15,17,27,29-Pentahydroxy-11-methoxy-3,7,12,14,16,18,22-heptamethyl-26-[(E)-(4-methylpiperazin-1-yl)iminomethyl]-6,23-dioxo-8,30-dioxa-24-azatetracyclo[23.3.1.1^{4,7}.0^{5,28}]triaconta-1(29),2,4,9,19,21,25,27-octaen-13-yl] acetate

Other names : Rifamycin, 3-[[(4-methyl-1-piperazinyl)imino]methyl]-
CAS registry number : 13292-46-1
EC number : 236-312-0
Merck Index monograph : 9611
Patch testing : Tablet, pulverized, 30% pet.; most systemic drugs can be tested at 10% pet.; if the pure chemical is not available, prepare the test material from intravenous powder, the content of capsules or – when also not available – from powdered tablets to achieve a final concentration of the active drug of 10% pet.
Molecular formula : $C_{43}H_{58}N_4O_{12}$

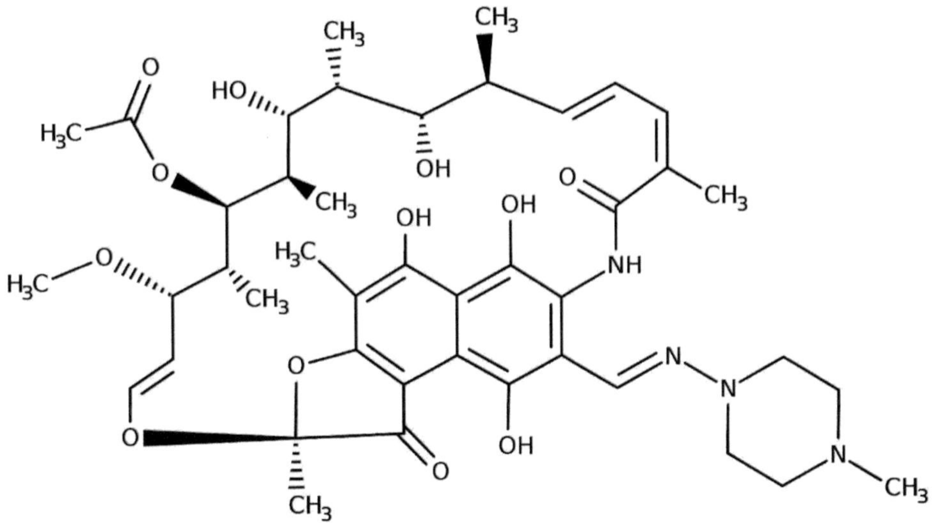

GENERAL

Rifampin is a semisynthetic antibiotic produced from *Streptomyces mediterranei*. It has a broad antibacterial spectrum, including activity against several forms of *Mycobacterium*. In susceptible organisms rifampin inhibits DNA-dependent RNA polymerase activity by forming a stable complex with the enzyme. It thus suppresses the initiation of RNA synthesis. Rifampin is bactericidal, and acts on both intracellular and extracellular organisms. It is indicated for the treatment of tuberculosis and tuberculosis-related mycobacterial infections (1).

CONTACT ALLERGY FROM ACCIDENTAL CONTACT

In Warsaw, Poland, in the period 1979-1983, 27 pharmaceutical workers, 24 nurses and 30 veterinary surgeons were diagnosed with occupational allergic contact dermatitis from antibiotics. The numbers that had positive patch tests to rifampicin (ampoule content) were 3, 0, and 0, respectively, total 3 (3).

CUTANEOUS ADVERSE DRUG REACTIONS FROM SYSTEMIC ADMINISTRATION CAUSED BY TYPE IV (DELAYED-TYPE) HYPERSENSITIVITY (as demonstrated by positive patch tests)

Cutaneous adverse drug reactions from systemic administration of rifampicin caused by type IV (delayed-type) hypersensitivity have included drug reaction with eosinophilia and systemic symptoms (DRESS) (2,4,5,6,7).

Drug reaction with eosinophilia and systemic symptoms (DRESS)
In the period 2010-2014, in South Africa, 60 patients with cutaneous adverse drug reactions to first-line antituberculosis drugs (FLTD: rifampicin, isoniazid, pyrazinamide and ethambutol) were patch tested with the patients' drugs 30% pet. Positive reactions were seen to at least one FLTD in 14 participants, of who twelve had DRESS, one SJS and one SJS/TEN. There were 7 positive patch tests to rifampicin, all in patients with DRESS. Eleven of the 14 patients were HIV-infected and in 10 of these, patch tests resulted in a generalized systemic reaction, including 3 individuals with positive reactions to rifampicin (4).

In France, a search was performed for potential cases of DRESS caused by the antituberculosis drugs rifampicin, isoniazid, pyrazinamide and ethambutol, reported from January 1, 2005, to July 30, 2015, in the French pharmacovigilance database. Sixty-seven cases of antituberculosis drug-associated DRESS were analyzed (40 women and 27 men, median age of 61 years). Rifampicin was suspected in 60 cases (of 67). Patch tests were performed in 11 patients and were positive in 7. Five individuals reacted to isoniazid, 2 to ethambutol and one to rifampicin (5).

In a 61-year-old man, delayed-type hypersensitivity to rifampicin (positive patch test to pulverized commercial tablet 450 mg, 30% pet.) and other drugs may have caused/contributed to DRESS (2).

A 72-year-old woman developed a generalized maculopapular rash that had appeared 4 weeks after she had started taking 4 antituberculosis drugs. She had eosinophilia (eosinophils 64.3%) and abnormal liver function tests. Six weeks after discharge from the hospital, patch tests with the 4 drugs as crushed tablets at 50% in petrolatum showed a diffuse erythematous rash around the rifampicin, isoniazid, and ethambutol patches at D2. The authors admitted these may have been nonspecific irritant reactions (6).

A 34-year-old HIV-infected man presented with exfoliative dermatitis, generalized lymphadenopathy, fever, facial edema, eosinophilia, and hepatitis. The patient had been started on isoniazid, rifampicin, pyrazinamide and ethambutol for tuberculosis, and co-trimoxazole for *Pneumocystis jiroveci* prophylaxis 4 weeks earlier. The suspected antituberculosis drugs were challenged sequentially after a negative patch and prick test (delayed reading). Two days after the rifampicin patch test (commercial tablets, active ingredient 30% pet. and water) had been applied, the patient complained of generalized pruritus. On examination, he had prominent perifollicular papules on a background of erythema over most of the body. Other manifestations were eosinophilia, hepatitis, facial edema, conjunctivitis, hemorrhagic cheilitis, erythematous tender palms, and blistering on the soles. A skin biopsy showed focal basal cell hydropic degeneration and keratinocyte necrosis centered on hair follicles. The symptoms progressed over the following 4 days, and epidermal necrosis developed on the body, palms, and soles. Challenges with the other 3 antituberculosis drugs were well tolerated, but cotrimoxazole was apparently not tested. Most curiously, the authors did not mention whether the patch test with rifampicin, which had apparently induced a recurrence of DRESS, was positive (7).

Cross-reactions, pseudo-cross-reactions and co-reactions
This subject is discussed in Chapter 3.206 Ethambutol.

LITERATURE

1   The data in the section 'General' may have been obtained from literature discussed in this chapter, but mostly also or exclusively from one or more of the following online sources: ChemIDPlus Advanced, PubChem, DrugBank, RxList, Drug Central, Drugs.com, and Wikipedia
2   Liippo J, Pummi K, Hohenthal U, Lammintausta K. Patch testing and sensitization to multiple drugs. Contact Dermatitis 2013;69:296-302
3   Rudzki E, Rebendel P. Contact sensitivity to antibiotics. Contact Dermatitis 1984;11:41-42
4   Lehloenya RJ, Todd G, Wallace J, Ngwanya MR, Muloiwa R, Dheda K. Diagnostic patch testing following tuberculosis-associated cutaneous adverse drug reactions induces systemic reactions in HIV-infected persons. Br J Dermatol 2016;175:150-156
5   Allouchery M, Logerot S, Cottin J, Pralong P, Villier C, Ben Saïd B; French Pharmacovigilance Centers Network and the French Investigators for skin adverse reactions to drugs. Antituberculosis drug-associated DRESS: A case series. J Allergy Clin Immunol Pract 2018;6:1373-1380
6   Lee SW, Yoon NB, Park SM, Lee SM, Um SJ, Lee SK, et al. Antituberculosis drug-induced drug rash with eosinophilia and systemic symptoms syndrome confirmed by patch testing. J Investig Allergol Clin Immunol 2010;20:631-632
7   Shebe K, Ngwanya MR, Gantsho N, Lehloenya RJ. Severe recurrence of drug rash with eosinophilia and systemic symptoms syndrome secondary to rifampicin patch testing in a human immunodeficiency virus-infected man. Contact Dermatitis 2014;70:125-127

Chapter 3.434 RISEDRONIC ACID

IDENTIFICATION

Description/definition   : Risedronic acid is the synthetic pyridinyl bisphosphonate that conforms to the structural
                                         formula shown below
Pharmacological classes : Bone density conservation agents; calcium channel blockers
IUPAC name   : Hydroxy-(1-hydroxy-1-phosphono-2-pyridin-3-ylethyl)phosphonic acid
Other names   : Risedronate
CAS registry number   : 105462-24-6
EC number   : 600-654-2
Merck Index monograph : 9630
Patch testing   : 7.5% pet.
Molecular formula   : $C_7H_{11}NO_7P_2$

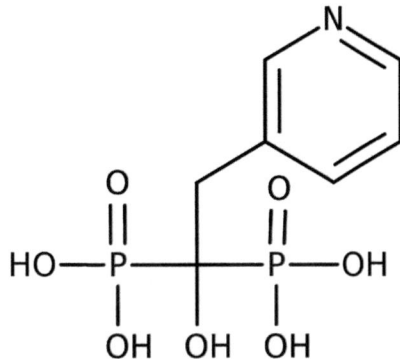

GENERAL

Risedronic acid is a synthetic pyridinyl bisphosphonate. It binds to hydroxyapatite crystals in bone and inhibits osteoclast-dependent bone resorption. Risedronic acid is indicated for the treatment of osteoporosis in men, of Paget's disease, and for treatment and prevention of osteoporosis in postmenopausal women and of glucocorticoid-induced osteoporosis. In pharmaceutical products, risedronic acid is employed as risedronate sodium (CAS number 115436-72-1, EC number 601-362-8, molecular formula $C_7H_{10}NNaO_7P_2$) (1).

CONTACT ALLERGY FROM ACCIDENTAL CONTACT

Sodium risedronate may have caused occupational airborne allergic contact dermatitis in a pharmaceutical worker, who was also sensitized to amlodipine besylate and olanzapine (3). However, the article, to which was referred in a recent textbook (4), could not be traced by the author.

CUTANEOUS ADVERSE DRUG REACTIONS FROM SYSTEMIC ADMINISTRATION CAUSED BY TYPE IV (DELAYED-TYPE) HYPERSENSITIVITY (as demonstrated by positive patch tests)

Cutaneous adverse drug reactions from systemic administration of risedronic acid caused by type IV (delayed-type) hypersensitivity have included erythema multiforme-like eruption (5).

Other cutaneous adverse drug reactions

A 56-year-old woman presented with an itchy erythema multiforme-like eruption mainly involving the arms and hands. The reaction had appeared a few days after the patient had started using oral risedronate sodium for osteoporosis. Laboratory investigation showed no abnormalities. Risedronate was stopped and the eruption reacted well to topical corticosteroids and an oral antihistamine. Three months later, patch tests were performed with risedronate 7.5% pet. and the related bisphosphonates alendronate 7% pet. and ibandronate 7.5% pet., which yielded a positive reaction to risedronate only (D2 and D4 ++). Twenty controls were negative. The patient was now treated with alendronate, without any recurrence during 2 years of follow-up (5).

Cross-reactions, pseudo-cross-reactions and co-reactions

A patient (probably) sensitized to ibandronic acid may have cross-reacted to risedronic acid as shown by a positive intradermal test to risedronate 1% water (++) read at D3 (2). No cross-reaction to alendronate or ibandronate in a patient sensitized to risedronate sodium (5).

LITERATURE

1   The data in the section 'General' may have been obtained from literature discussed in this chapter, but mostly also or exclusively from one or more of the following online sources: ChemIDPlus Advanced, PubChem, DrugBank, RxList, Drug Central, Drugs.com, and Wikipedia

2   Barrantes-González M, Espona-Quer M, Salas E, Giménez-Arnau AM. Bisphosphonate-induced cutaneous adverse events: the difficulty of assessing imputability through patch testing. Dermatology 2014;229:163-168

3   Chomiczewska D, Kiec-Swierczynska M, Krecisz B . Airborne occupational allergic contact dermatitis to sodium risedronate, olanzapine and amlodipine benzenesulfonate in a pharmaceutical company worker – a case report. Eur J Allergy Clin Immunol 2010;65:593-594 (article could not be traced, cited in ref. 3)

4   Goossens A, Geebelen J, Hulst KV, Gilissen L. Pharmaceutical and cosmetic industries. In: John S, Johansen J, Rustemeyer T, Elsner P, Maibach H, eds. Kanerva's occupational dermatology. Cham, Switzerland: Springer, 2020:2203-2219

5   Bianchi L, Hansel K, Romita P, Foti C, Stingeni L. Erythema multiforme-like eruption induced by risedronate. Contact Dermatitis 2017;77:348-349

Chapter 3.435 RISPERIDONE

IDENTIFICATION

Description/definition   : Risperidone is the pyridopyrimidine that conforms to the structural formula shown below
Pharmacological classes   : Antipsychotic agents; serotonin antagonists; dopamine antagonists
IUPAC name   : 3-[2-[4-(6-Fluoro-1,2-benzoxazol-3-yl)piperidin-1-yl]ethyl]-2-methyl-6,7,8,9-
                            tetrahydropyrido[1,2-a]pyrimidin-4-one
CAS registry number   : 106266-06-2
EC number   : 600-733-1
Merck Index monograph   : 9531
Patch testing   : No data available; most systemic drugs can be tested at 10% pet.; if the pure chemical is
                            not available, prepare the test material from intravenous powder, the content of capsules
                            or – when also not available – from powdered tablets to achieve a final concentration of
                            the active drug of 10% pet.
Molecular formula   : $C_{23}H_{27}FN_4O_2$

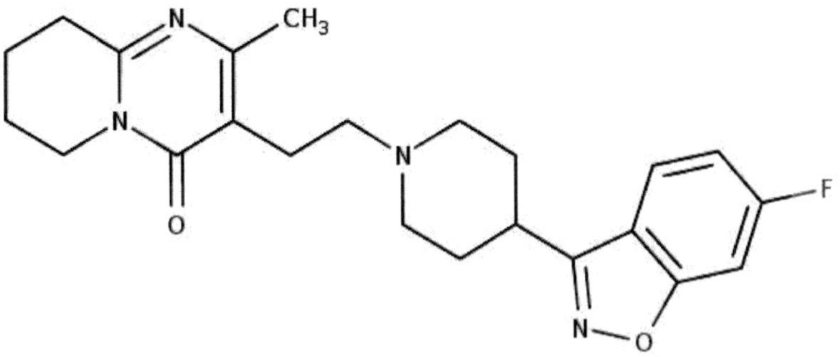

GENERAL

Risperidone is a selective blocker of dopamine D2-receptors and serotonin 5-HT2-receptors. The drug is a second-generation antipsychotic medication used in the treatment of a number of mood and mental health conditions including schizophrenia and bipolar disorder. These disorders are thought to be caused by an excess of dopaminergic D2- and serotonergic 5-HT2-activity, resulting in overactivity of central mesolimbic pathways and mesocortical pathways, respectively (1).

CONTACT ALLERGY FROM ACCIDENTAL CONTACT

In Leuven, Belgium, in the period 2001-2019, 201 of 1248 health care workers/employees of the pharmaceutical industry had occupational allergic contact dermatitis. In 23 (11%) dermatitis was caused by skin contact with a systemic drug: 19 nurses, two chemists, one physician, and one veterinarian. The lesions were mostly localized on the hands, but often also on the face, as airborne dermatitis. In total, 42 positive patch test reactions to 18 different systemic drugs were found. In one patient, risperidone was the drug/one of the drugs that caused occupational dermatitis (3, overlap with ref. 2).

Also in Leuven, Belgium, in the period 2007-2011, 81 patients have been diagnosed with occupational airborne allergic contact dermatitis. In 23 of them, drugs were the offending agents, including risperidone in one case (2, overlap with ref. 3).

LITERATURE

1   The data in the section 'General' may have been obtained from literature discussed in this chapter, but mostly also or exclusively from one or more of the following online sources: ChemIDPlus Advanced, PubChem, DrugBank, RxList, Drug Central, Drugs.com, and Wikipedia
2   Swinnen I, Goossens A. An update on airborne contact dermatitis: 2007-2011. Contact Dermatitis 2013;68:232-238
3   Gilissen L, Boeckxstaens E, Geebelen J, Goossens A. Occupational allergic contact dermatitis from systemic drugs. Contact Dermatitis 2020;82:24-30

Chapter 3.436 RITODRINE

IDENTIFICATION

Description/definition   : Ritodrine is the phenethylamine that conforms to the structural formula shown below
Pharmacological classes : Adrenergic beta-2 receptor agonists; sympathomimetics; tocolytic agents
IUPAC name   : 4-[2-[[(1R,2S)-1-Hydroxy-1-(4-hydroxyphenyl)propan-2-yl]amino]ethyl]phenol
CAS registry number   : 26652-09-5
EC number   : 247-879-9
Merck Index monograph : 9635
Patch testing   : 1% and 0.1% water
Molecular formula   : $C_{17}H_{21}NO_3$

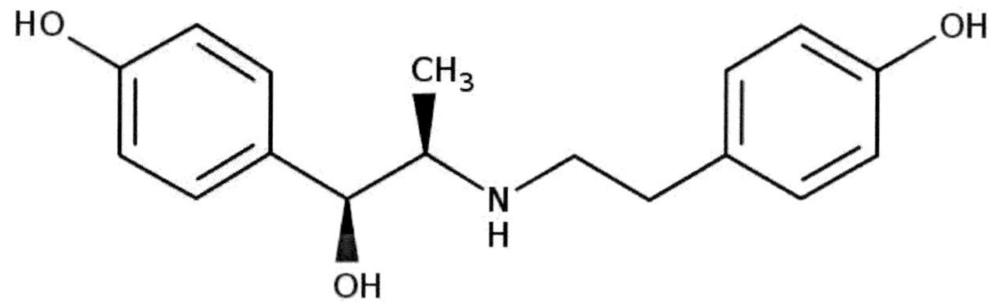

GENERAL

Ritodrine is a phenethylamine derivative with tocolytic activity. It binds to and activates beta-2 adrenergic receptors of myometrial cells in the uterus, which decreases the intensity and frequency of uterine contractions. In addition, ritodrine may directly inactivate myosin light chain kinase, a critical enzyme necessary for the initiation of muscle contractions. This drug is indicated for the treatment and prophylaxis of premature labor. In pharmaceutical products, ritodrine is employed as ritodrine hydrochloride (CAS number 23239-51-2, EC number 245-514-8, molecular formula $C_{17}H_{22}ClNO_3$) (1).

CUTANEOUS ADVERSE DRUG REACTIONS FROM SYSTEMIC ADMINISTRATION CAUSED BY TYPE IV (DELAYED-TYPE) HYPERSENSITIVITY (as demonstrated by positive patch tests)

Cutaneous adverse drug reactions from systemic administration of ritodrine caused by type IV (delayed-type) hypersensitivity have included acute generalized exanthematous pustulosis (AGEP) (2).

Acute generalized exanthematous pustulosis (AGEP)

A 27-year-old woman was given intravenous ritodrine hydrochloride for threatened preterm labor at 29 weeks gestation. Seventeen days later, the patient complained of pruritus and a rash around the forearm site of intravenous infusion. The next day, erythematous plaques appeared on the face, arms, and abdomen, One week later, numerous tiny non-follicular pustules appeared on the erythematous plaques and she had a slight fever. All laboratory findings were normal, including those for the white blood cell count. Bacterial pustule cultures were negative. Despite an oral steroid preparation, the erythema worsened and spread to the back, chest, abdomen, and extremities within several days. Histopathology of a cutaneous biopsy revealed subcorneal neutrophilic pustules forming spongiform structures. Diagnoses considered were impetigo herpetiformis and a drug reaction to ritodrine. Ritodrine was stopped, but the patient developed inflammatory bilateral edema of the lower thighs accompanied by severe pain.

Laboratory examinations now showed leukocytosis with significant neutrophilia and increased level of C-reactive protein. The pregnancy weas resolved by an emergency cesarian section. Within several days of delivery, the eruptions began to resolve. Later, patch tests were positive to ritodrine at 1%, 0.1%, 0.05% and 0.01% in water, both at D2 and D3. Ten controls were negative. The patient was diagnosed with 'ritodrine-induced pustular eruptions in a pregnant woman' (2), which, according to this author, was acute generalized exanthematous pustulosis (AGEP).

LITERATURE

1   The data in the section 'General' may have been obtained from literature discussed in this chapter, but mostly also or exclusively from one or more of the following online sources: ChemIDPlus Advanced, PubChem, DrugBank, RxList, Drug Central, Drugs.com, and Wikipedia
2   Kuwabara Y, Sato A, Abe H, Abe S, Kawai N, Takeshita T. Ritodrine-induced pustular eruptions distinctly resembling impetigo herpetiformis. J Nippon Med Sch 2011;78:329-333

Chapter 3.437 ROSUVASTATIN

IDENTIFICATION

Description/definition	: Rosuvastatin is the phenylpyrimidine that conforms to the structural formula shown below
Pharmacological classes	: Hydroxymethylglutaryl-CoA reductase inhibitors; anticholesteremic agents
IUPAC name	: (E,3R,5S)-7-[4-(4-Fluorophenyl)-2-[methyl(methylsulfonyl)amino]-6-propan-2-ylpyrimidin-5-yl]-3,5-dihydroxyhept-6-enoic acid
CAS registry number	: 287714-41-4
EC number	: Not available
Merck Index monograph	: 9672
Patch testing	: Tablet, pulverized, 20% pet. (2); most systemic drugs can be tested at 10% pet.; if the pure chemical is not available, prepare the test material from intravenous powder, the content of capsules or – when also not available – from powdered tablets to achieve a final concentration of the active drug of 10% pet.
Molecular formula	: $C_{22}H_{28}FN_3O_6S$

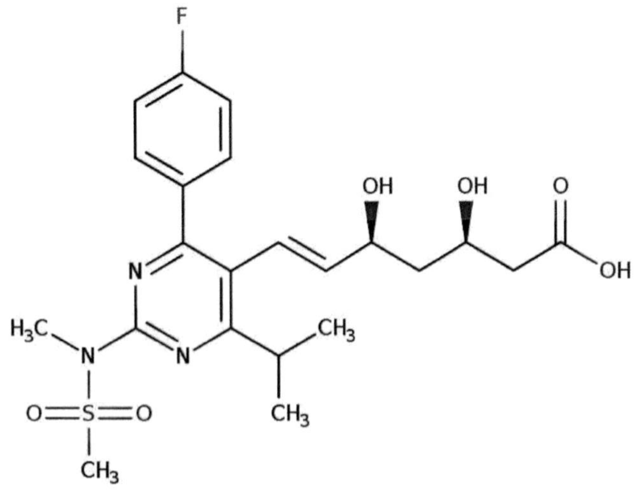

GENERAL

Rosuvastatin is a lipid-lowering drug that belongs to the statin class of medications, which are used to lower the risk of cardiovascular disease and manage elevated lipid levels by inhibiting the endogenous production of cholesterol in the liver. In pharmaceutical products, rosuvastatin may be employed as rosuvastatin calcium (CAS number 147098-20-2, EC number not available, molecular formula $C_{44}H_{54}CaF_2N_6O_{12}S_2$) or as rosuvastatin zinc (CAS number 953412-08-3, EC number not available, molecular formula $C_{44}H_{54}F_2N_6O_{12}S_2Zn$) (1).

CUTANEOUS ADVERSE DRUG REACTIONS FROM SYSTEMIC ADMINISTRATION CAUSED BY TYPE IV (DELAYED-TYPE) HYPERSENSITIVITY (as demonstrated by positive patch tests)

Cutaneous adverse drug reactions from systemic administration of rosuvastatin caused by type IV (delayed-type) hypersensitivity have included maculopapular eruption (2).

Maculopapular eruption

A 78-year-old woman noticed red spots on her arms and legs one month after having used rosuvastatin for hyperlipidemia for one month, which gradually worsened. Physical examination revealed small erythematous plaques and papules on her extremities. Histopathology of a skin biopsy revealed a lymphocytic and eosinophilic infiltration in the upper dermis, consistent with a drug eruption. A patch test with rosuvastatin 20% pet. was positive at D2. Later readings were apparently not performed nor were controls tested. The patient was diagnosed with maculopapular drug eruption due to rosuvastatin (2).

LITERATURE

1   The data in the section 'General' may have been obtained from literature discussed in this chapter, but mostly also or exclusively from one or more of the following online sources: ChemIDPlus Advanced, PubChem, DrugBank, RxList, Drug Central, Drugs.com, and Wikipedia
2   Oda T, Sawada Y, Yamaguchi T, Ohmori S, Haruyama S, Yoshioka M, et al. Drug eruption caused by rosuvastatin. J Investig Allergol Clin Immunol 2017;27:140-141

Chapter 3.438 RUPATADINE

IDENTIFICATION

Description/definition : Rupatadine is the benzocycloheptapyridine that conforms to the structural formula shown below (shown is rupatadine fumarate)

Pharmacological classes : Histamine H_1 antagonists, non-sedating

IUPAC name : 13-Chloro-2-[1-[(5-methylpyridin-3-yl)methyl]piperidin-4-ylidene]-4-azatricyclo-[9.4.0.0^{3,8}]pentadeca-1(11),3(8),4,6,12,14-hexaene

CAS registry number : 158876-82-5

EC number : Not available

Merck Index monograph : 9700

Patch testing : Pulverized tablet 30% water and pet.(2); most systemic drugs can be tested at 10% pet.; if the pure chemical is not available, prepare the test material from intravenous powder, the content of capsules or – when also not available – from powdered tablets to achieve a final concentration of the active drug of 10% pet.

Molecular formula : $C_{26}H_{26}ClN_3$

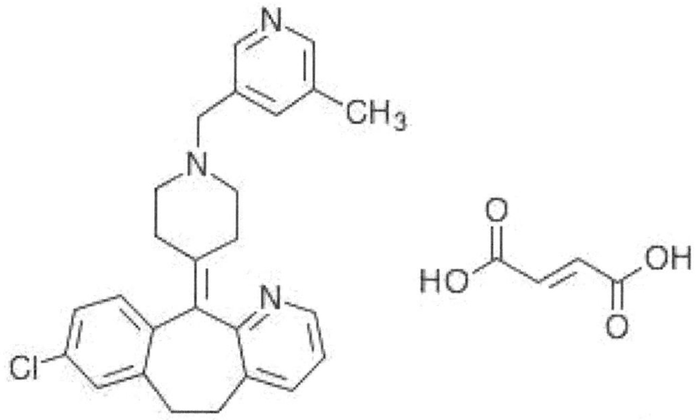

GENERAL

Rupatadine is a dual histamine H_1 receptor and platelet activating factor receptor antagonist that is used for symptomatic relief in seasonal and perennial rhinitis as well as chronic spontaneous urticaria. In pharmaceutical products, rupatadine is employed as rupatadine fumarate (CAS number 182349-12-8, EC number not available, molecular formula $C_{30}H_{30}ClN_3O_4$; structural formula shown above) or as rupatadine trihydrochloride (CAS number 156611-76-6, EC number not available, molecular formula $C_{26}H_{29}Cl_4N_3$) (1).

CUTANEOUS ADVERSE DRUG REACTIONS FROM SYSTEMIC ADMINISTRATION CAUSED BY TYPE IV (DELAYED-TYPE) HYPERSENSITIVITY (as demonstrated by positive patch tests)

Cutaneous adverse drug reactions from systemic administration of rupatadine caused by type IV (delayed-type) hypersensitivity have included fixed drug eruption (2).

Fixed drug eruption

A 62-year-old man suffering from allergic rhinitis and hypertension reported two episodes of sudden onset of erythematous-violaceous lesions on the cervical, suprapubic, and penile regions that recurred in the same locations and progressed to residual brownish-grey pigmentation. He had taken rupatadine on the preceding days because of exacerbations of allergic rhinitis, in addition to his usual medication for hypertension. Patch tests performed with rupatadine and related antihistamines on normal and postlesional skin were positive to powder of rupatadine 'in water and 30% pet.' on normal skin but not on residual skin. A repeat test with rupatadine 30% pet. was again positive on normal skin. Fifteen controls were negative. Histopathology of the patch test reaction showed an interface dermatitis with vacuolization of the basal layer associated with a spongiotic dermatitis with lymphocyte exocytosis and a perivascular lymphomononuclear infiltrate with rare eosinophils (2).

LITERATURE

1 The data in the section 'General' may have been obtained from literature discussed in this chapter, but mostly also or exclusively from one or more of the following online sources: ChemIDPlus Advanced, PubChem, DrugBank, RxList, Drug Central, Drugs.com, and Wikipedia

2 Calvão J, Cardoso JC, Gonçalo M. Fixed drug eruption to rupatadine with positive patch tests on non-lesional skin. Contact Dermatitis 2020;83:239-241

Chapter 3.439 SECNIDAZOLE

IDENTIFICATION

Description/definition	: Secnidazole is the nitroimidazole that conforms to the structural formula shown below
Pharmacological classes	: Antiprotozoal agents
IUPAC name	: 1-(2-Methyl-5-nitroimidazol-1-yl)propan-2-ol
CAS registry number	: 3366-95-8
EC number	: 222-134-0
Merck Index monograph	: 9826
Patch testing	: Tablet, pulverized, 30% water and pet.(2); most systemic drugs can be tested at 10% pet.; if the pure chemical is not available, prepare the test material from intravenous powder, the content of capsules or – when also not available – from powdered tablets to achieve a final concentration of the active drug of 10% pet.
Molecular formula	: $C_7H_{11}N_3O_3$

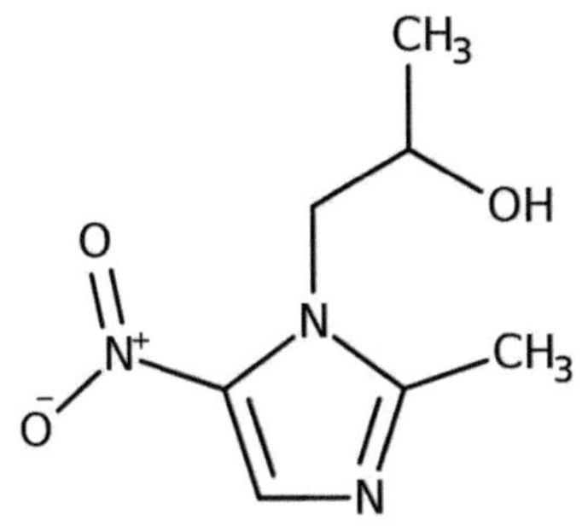

GENERAL

Secnidazole is a second-generation 5-nitroimidazole antimicrobial which is active against many anaerobic gram-positive and gram-negative bacteria and protozoa including *Bacteroides fragilis*, *Trichomonas vaginalis*, *Entamoeba histolytica* and *Giardia lamblia*. It is indicated for the treatment of bacterial vaginosis in adult women (1).

CUTANEOUS ADVERSE DRUG REACTIONS FROM SYSTEMIC ADMINISTRATION CAUSED BY TYPE IV (DELAYED-TYPE) HYPERSENSITIVITY (as demonstrated by positive patch tests)

Cutaneous adverse drug reactions from systemic administration of secnidazole caused by type IV (delayed-type) hypersensitivity have included symmetrical drug-related intertriginous and flexural exanthema (SDRIFE) (2).

Symmetrical drug-related intertriginous and flexural exanthema (SDRIFE)

A 44-year-old woman was treated with secnidazole for vaginitis caused by *Trichomonas vaginalis*. After 2 days, she developed a pruritic eczema-like eruption in the axillary, inguinal and submammary folds. The patient had a history of maculopapular exanthema caused by metronidazole. The use of secnidazole was interrupted, topical cortico-steroid treatment was given, and the lesions resolved within 4 days. Patch tests were performed with secnidazole 30% pet. and 30% water on the back and on a previously affected area (breast). These were negative on the back at D2 and D4. On the breast, there was no reaction at D2, but a weakly positive reaction (+) was observed at D4. Controls were apparently not performed. The patient was diagnosed with symmetrical drug-related intertriginous and flexural exanthema (SDRIFE). It was suggested that secnidazole cross-reacted to metronidazole, which had previously caused a maculopapular exanthema (2). This indeed seems possible, both are closely related nitroimida-zoles.

Cross-reactions, pseudo-cross-reactions and co-reactions

In one patient with SDRIFE from secnidazole, this drug may have cross-reacted to metronidazole (2).

LITERATURE

1   The data in the section 'General' may have been obtained from literature discussed in this chapter, but mostly also or exclusively from one or more of the following online sources: ChemIDPlus Advanced, PubChem, DrugBank, RxList, Drug Central, Drugs.com, and Wikipedia
2   Nespoulous L, Matei I, Charissoux A, Bédane C, Assikar S. Symmetrical drug-related intertriginous and flexural exanthema (SDRIFE) associated with pristinamycin, secnidazole, and nefopam, with a review of the literature. Contact Dermatitis 2018;79:378-380

Chapter 3.440 SECUKINUMAB

IDENTIFICATION

Description/definition	: Secukinumab is an interleukin-17A (IL-17A) inhibitor
Pharmacological classes	: Monoclonal antibodies; immunosuppressive agents
IUPAC name	: Not available
Other names	: Immunoglobulin G1, anti-(human interleukin-17A (IL-17, cytotoxic T lymphocyte-associated antigen 8)); human monoclonal AIN457 gamma1 heavy chain (230-215')-disulfide with human monoclonal AIN457 kappa light chain dimer (236-236'':239-239'')-bisdisulfide
CAS registry number	: 1229022-83-6
EC number	: Not available
Merck Index monograph	: 11827
Patch testing	: Commercial preparation for subcutaneous injection undiluted
Molecular formula	: Not specified
Structural formula	: Not available

GENERAL

Secukinumab is a fully human monoclonal antibody and an interleukin-17A (IL-17A) inhibitor. IL-17 is a group of proinflammatory cytokines released by cells of the immune system that are present in higher levels in many immune conditions associated with chronic inflammation. Secukinumab is indicated for the treatment of moderate to severe plaque psoriasis. It is also used in treating active psoriatic arthritis, active ankylosing spondylitis and uveitis (1).

CUTANEOUS ADVERSE DRUG REACTIONS FROM SYSTEMIC ADMINISTRATION CAUSED BY TYPE IV (DELAYED-TYPE) HYPERSENSITIVITY

Cutaneous adverse drug reactions from systemic administration of secukinumab caused by type IV (delayed-type) hypersensitivity have included localized and expanding eczematous reaction from subcutaneous injection (2).

Localized and expanding eczematous reaction

A 51-year-old woman was started on secukinumab for the treatment of psoriatic arthritis. The patient tolerated the first administration of 2 subcutaneous (SC) injections of 150 mg each in the abdominal skin. However, 1 week later, following the second administration (again 2 SC injections of 150 mg in the abdomen), she developed, 24 hours later, pruritic, erythematous skin lesions on both injection sites. Subsequently, these lesions became strongly vesicular and purplish, and extended over the abdomen and breasts, without involving the back. After complete recovery, patch tests were positive to the secukinumab injection fluid 'as is' on the abdomen and the back with negative reactions to the ingredient polysorbate 80 and to the related IL-17 inhibitor ixekizumab. Ten controls were negative (2).

Cross-reactions, pseudo-cross-reactions and co-reactions

No cross-reactivity to the related IL-17 inhibitor ixekizumab (2).

LITERATURE

1 The data in the section 'General' may have been obtained from literature discussed in this chapter, but mostly also or exclusively from one or more of the following online sources: ChemIDPlus Advanced, PubChem, DrugBank, RxList, Drug Central, Drugs.com, and Wikipedia
2 Darrigade AS, Dendooven E, Mangodt E, Aerts O. Delayed-type hypersensitivity to secukinumab with tolerance to ixekizumab. J Allergy Clin Immunol Pract 2020;8:3626-3628

Chapter 3.441 SERTRALINE

IDENTIFICATION

Description/definition   : Sertraline is the tametraline that conforms to the structural formula shown below
Pharmacological classes   : Antidepressive agents; serotonin uptake inhibitors
IUPAC name   : (1S,4S)-4-(3,4-Dichlorophenyl)-N-methyl-1,2,3,4-tetrahydronaphthalen-1-amine
Other names   : (1S-cis)-1,2,3,4-Tetrahydro-4-(3,4-dichlorophenyl)-N-methyl-1-naphthalenamine
CAS registry number   : 79617-96-2
EC number   : Not available
Merck Index monograph   : 9876
Patch testing   : 1%, 5% and 10% in petrolatum or alcohol
Molecular formula   : $C_{17}H_{17}Cl_2N$

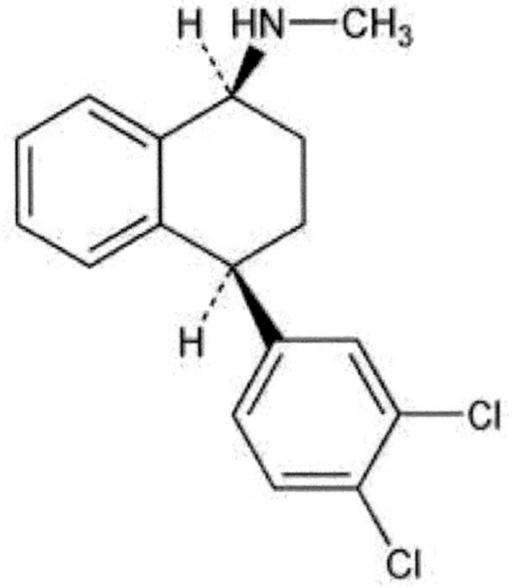

GENERAL

Sertraline is a selective serotonin-reuptake inhibitor with antidepressant activity. It is indicated for the management of major depressive disorder, posttraumatic stress disorder, obsessive-compulsive disorder, panic disorder with or without agoraphobia, premenstrual dysphoric disorder, and social anxiety disorder. It may be used for premature ejaculation and vascular headaches as off-label indications (1). In pharmaceutical products, sertraline is employed as sertraline hydrochloride (CAS number 79559-97-0, EC number 616-702-0, molecular formula $C_{17}H_{18}Cl_3N$) (1).

CUTANEOUS ADVERSE DRUG REACTIONS FROM SYSTEMIC ADMINISTRATION CAUSED BY TYPE IV (DELAYED-TYPE) HYPERSENSITIVITY (as demonstrated by positive patch tests)

Cutaneous adverse drug reactions from systemic administration of sertraline caused by type IV (delayed-type) hypersensitivity have included maculopapular exanthema (2) and erythema multiforme (3).

Maculopapular eruption

Three weeks after starting sertraline for a depression, a 66-year-old man presented with a generalized pruritic maculopapular rash and fever (38–39°C) of 3 days' duration. The next day, the rash became confluent at the skin folds, with purpuric and scattered targetoid lesions, and a few pustules appeared a day later. Laboratory tests showed leukocytosis with neutrophilia. Histopathology of a skin biopsy showed an infiltrate in the superficial dermis, mostly of lymphocytes and histiocytes with a few eosinophils and extravascular erythrocytes, and focal degeneration of the basal cell layer. Four weeks after complete healing, patch tests were positive to commercial sertraline, pulverized, as is (moistened with water) and 10% pet. In a second session, patch tests with pure sertraline 1%, 5% and 10% water, alcohol and petrolatum were positive (++) at all concentrations and vehicles tested. Twenty controls were negative to sertraline 10% alc. and 10% pet. (2).

Other cutaneous adverse drug reactions

In Hungarian literature a patient has been described who developed erythema exsudativum multiforme while using sertraline and several other drugs. Patch tests were positive only to sertraline. Lamotrigine was negative, but oral provocation with the anticonvulsant resulted in a scarlatiniform exanthema despite intravenous steroids and antihistamines (3).

LITERATURE

1   The data in the section 'General' may have been obtained from literature discussed in this chapter, but mostly also or exclusively from one or more of the following online sources: ChemIDPlus Advanced, PubChem, DrugBank, RxList, Drug Central, Drugs.com, and Wikipedia

2   Fernandes B, Brites M, Gonçalo M, Figueiredo A. Maculopapular eruption from sertraline with positive patch tests. Contact Dermatitis 2000;42:287

3   Kopcsányi H, Feldmann J, Péch Z, Jurcsik A. Adverz gyógyszerreakciók kivizsgálásának nehézségeirôl egy bonyolult eset kapcsán [Difficulties in detecting the causative agent of an adverse drug reaction in a complicated case]. Orv Hetil 2008;149:883-887 (Article in Hungarian)

Chapter 3.442 SEVOFLURANE

IDENTIFICATION

Description/definition   : Sevoflurane is the fluorinated isopropyl ether that conforms to the structural formula
                                   shown below
Pharmacological classes : Anesthetics, inhalation; platelet aggregation inhibitors
IUPAC name   : 1,1,1,3,3,3-Hexafluoro-2-(fluoromethoxy)propane
CAS registry number   : 28523-86-6
EC number   : Not available
Merck Index monograph   : 9884
Patch testing   : This material is too volatile for patch testing; perform a ROAT with 0.5-1 ml sevoflurane
                                   fluid
Molecular formula   : $C_4H_3F_7O$

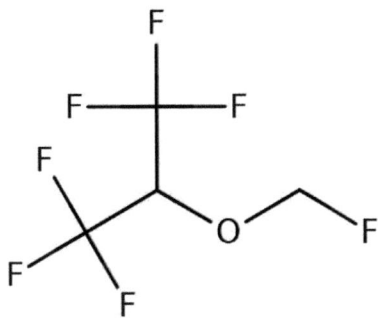

GENERAL

Sevoflurane is a fluorinated isopropyl ether used as inhalation anesthetic for the induction and maintenance of general anesthesia. Apart from its general anesthetic effect, sevoflurane decreases myocardial contractility and mean arterial pressure and increases respiratory rate. It is a volatile, non-flammable, non-irritant, and easy-to-administer compound with a low solubility profile and blood-to-gas partition coefficient (1).

CONTACT ALLERGY FROM ACCIDENTAL CONTACT

A 59-year-old woman had developed dermatitis localized to the face and neck region, that had begun 2 months after she had commenced a new job as a recovery room nurse, taking care of patients following operations under general anesthesia with sevoflurane. All symptoms disappeared after she left the job in the recovery room. Patch testing was not performed because of the volatility of the material. A repeated open application test (ROAT) was performed with 0.5 ml of liquid sevoflurane, which was applied onto the skin of the arm and distributed on a surface of 3×3 cm. The ROAT resulted in a flare of dermatitis on the face and neck on D2, and dermatitis on the elbow on D10. Three healthy controls were tested with the same procedure for 10 days, with negative test results (2). This was a case of occupational airborne allergic contact dermatitis. Two more cases with special clinical features are presented below (3,4).

CUTANEOUS ADVERSE DRUG REACTIONS FROM SYSTEMIC ADMINISTRATION CAUSED BY TYPE IV (DELAYED-TYPE) HYPERSENSITIVITY (as demonstrated by positive patch tests)

Cutaneous adverse drug reactions from systemic administration of sevoflurane caused by type IV (delayed-type) hypersensitivity have included symmetrical drug-related intertriginous and flexural exanthema (SDRIFE)/baboon syndrome (3,4), (at least) in one case as a manifestation of systemic allergic dermatitis (4).

Symmetrical drug-related intertriginous and flexural exanthema (SDRIFE)/Baboon syndrome

A 59-year-old male surgeon presented with an itchy erythematous rash on his thighs, groin, and anogenital area, later spreading onto his abdomen, face (eyelids), and flexures (axillae and elbow). Screening patch tests were negative. Several months later, the patient presented with persistent eczematous lesions with lichenification. He was now aware of exacerbation following work in an operating theatre with a generalized, macular, red eruption, mainly localized in the gluteal area and the major flexures, with a distribution typical of the 'baboon syndrome', over his chronic eczema. It was established that sevoflurane was always employed as an anesthetic in this operating theatre. A (ROAT) with sevoflurane 1 ml was positive after 2 days. Three controls were negative. A diagnosis of symmetrical drug-related intertriginous and flexural exanthema (SDRIFE) was made (3).

A similar case of allergy to sevoflurane in a surgeon had been presented in 2014, where it took 8 years before a malfunction in the anesthetic gas scavenging system was found. The patient's symptoms remitted within a week following its replacement. The patient had developed a skin rash consisting of elevated pruritic erythematous areas with small vesicles and lichenifications, distributed bilaterally on dorsum of hands, wrists and forearms. Lesions and pruritus worsened progressively, spreading to the antecubital, axilla, back, groin, scrotum and popliteal fossa. A ROAT was positive on the second day with appearance of dermatitis in the tested areas and in distant body folds (4). This probably started as (occupational) allergic contact dermatitis of the hands and forearms, later progressing to SDRIFE from inhalation (systemic allergic dermatitis) (4).

LITERATURE

1   The data in the section 'General' may have been obtained from literature discussed in this chapter, but mostly also or exclusively from one or more of the following online sources: ChemIDPlus Advanced, PubChem, DrugBank, RxList, Drug Central, Drugs.com, and Wikipedia
2   Andersen Y, Johansen JD, Garvey LH, Thyssen JP. Occupational airborne contact dermatitis caused by sevoflurane. Contact Dermatitis 2015;72:241-243
3   Burches E, Revert A, Martin J, Iturralde A. Occupational systemic allergic dermatitis caused by sevoflurane. Contact Dermatitis 2015;72:62-63
4   Lloréns Herrerias J, Delgado Navarro C, Ballester Luján MT, Izquierdo Palomares A. Long-term allergic dermatitis caused by sevoflurane: a clinical report. Acta Anaesthesiol Scand 2014;58:1151-1153

Chapter 3.443 SILDENAFIL

IDENTIFICATION

Description/definition   : Sildenafil is the benzenesulfonamide that conforms to the structural formula shown
                              below
Pharmacological classes   : Phosphodiesterase 5 inhibitors; urological agents; vasodilator agents
IUPAC name   : 5-[2-Ethoxy-5-(4-methylpiperazin-1-yl)sulfonylphenyl]-1-methyl-3-propyl-6H-pyrazolo[4,3-
                              d]pyrimidin-7-one
CAS registry number   : 139755-83-2
EC number   : 604-158-7
Merck Index monograph   : 9898
Patch testing   : 10% pet. (probably pulverized tablet)
Molecular formula   : $C_{22}H_{30}N_6O_4S$

GENERAL

Sildenafil is an orally bioavailable benzenesulfonamide derivative with vasodilating and potential anti-inflammatory
activities. The drug functions as a selective and competitive inhibitor of type 5 phosphodiesterases (PDE5) on smooth
muscle cells in the penis and pulmonary vasculature, resulting in prolonged smooth muscle relaxation in the corpus
cavernosum of the penis, thereby causing vasodilation, blood engorgement and a prolonged penile erection.
Sildenafil is used extensively for erectile dysfunction and less commonly for pulmonary hypertension. In pharmaceutical products, sildenafil is employed as sildenafil citrate (CAS number 171599-83-0, EC number not available,
molecular formula $C_{28}H_{38}N_6O_{11}S$) (1).

CUTANEOUS ADVERSE DRUG REACTIONS FROM SYSTEMIC ADMINISTRATION CAUSED BY TYPE IV
(DELAYED-TYPE) HYPERSENSITIVITY (as demonstrated by positive patch tests)

Cutaneous adverse drug reactions from systemic administration of sildenafil caused by type IV (delayed-type)
hypersensitivity have included maculopapular eruption (2).

Maculopapular exanthema

A 49-year-old HIV positive man presented with an extensive maculopapular rash, covering the extremities, abdomen,
thorax and gluteal region. A single target lesion was detected in his right palm. There were oral mucosal lesions (not
further described), causing a tingling sensation. No epidermal detachment or blisters were detected. The diagnosis of
erythema multiforme was set. It appeared that the patient had taken a tablet of sildenafil 100 mg, as an aphrodisiac,
five days before the initial rash. The patient recalled having used sildenafil just once before, one year ago. Blood
tests were normal. Three days later, the skin eruption had much improved. Six month after the initial diagnosis, a
patch test was positive to sildenafil 10% pet. The patient was diagnosed with erythema multiforme caused by
delayed-type allergy to sildenafil (2). According to this author, to diagnose this case as erythema multiforme based
only on one (!) target lesion and without histopathology is insufficiently substantiated. In addition, no patch testing
details were provided and probably no controls were tested. To diagnose maculopapular exanthema, possibly from
sildenafil and possibly from delayed-type allergy, seems preferable.

LITERATURE

1   The data in the section 'General' may have been obtained from literature discussed in this chapter, but mostly
    also or exclusively from one or more of the following online sources: ChemIDPlus Advanced, PubChem,
    DrugBank, RxList, Drug Central, Drugs.com, and Wikipedia
2   Pitsios C. Erythema multiforme caused by sildenafil in an HIV(+) subject. Eur Ann Allergy Clin Immunol
    2016;48:58-60

Chapter 3.444 SIMVASTATIN

IDENTIFICATION

Description/definition : Simvastatin is a lipid-lowering agent derived synthetically from a fermentation product of the fungus *Aspergillus terreus*; it conforms to the structural formula shown below

Pharmacological classes : Hydroxymethylglutaryl-CoA reductase inhibitors; anticholesteremic agents; hypolipidemic agents

IUPAC name : [(1S,3R,7S,8S,8aR)-8-[2-[(2R,4R)-4-Hydroxy-6-oxooxan-2-yl]ethyl]-3,7-dimethyl-1,2,3,7,8,8a-hexahydronaphthalen-1-yl] 2,2-dimethylbutanoate

CAS registry number : 79902-63-9

EC number : 616-751-8

Merck Index monograph : 9947

Patch testing : 1% and 0.1% pet., alc. or MEK

Molecular formula : $C_{25}H_{38}O_5$

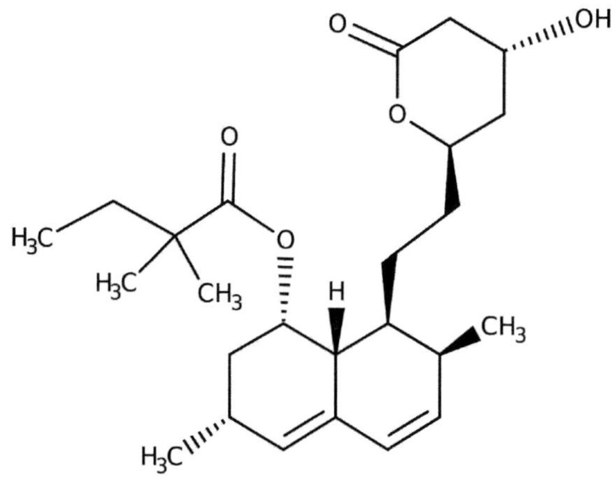

GENERAL

Simvastatin is a lipid-lowering agent derived synthetically from a fermentation product of the fungus *Aspergillus terreus*. Hydrolyzed *in vivo* to an active metabolite, simvastatin competitively inhibits hepatic hydroxymethyl-glutaryl coenzyme A (HMG-CoA) reductase, the enzyme that catalyzes the conversion of HMG-CoA to mevalonate, a key step in cholesterol synthesis. Hereby, this agent lowers plasma cholesterol and lipoprotein levels. Simvastatin is indicated for the treatment of hypercholesterolemia and for the reduction in the risk of cardiac heart disease mortality and cardiovascular events (1).

CONTACT ALLERGY FROM ACCIDENTAL CONTACT

Occupational airborne allergic contact dermatitis to simvastatin has been reported in patients working in pharmaceutical plants a few times (3,4,7).

A 29-year-old man working as machine operator in a pharmaceutical factory was referred with a 6-month history of dermatitis involving his eyelids, cheeks, lips, nose, and the nasolabial folds. The rash cleared during long periods away from work. Patch tests were positive to simvastatin 1% and 0.1% pet., carvedilol 10% pet., and zolpidem 10% pet. Ten controls were negative. Three months avoidance of the offending allergens resulted in total clearing of the dermatitis. A diagnosis of occupational airborne allergic contact dermatitis was made (3).

A 41-year-old man presented with a 1-year history of recurrent eyelid dermatitis. He worked as maintenance engineer in the manufacturing department of a pharmaceutical plant and reported a flare of his dermatitis related to repairing machines used in the production of simvastatin. Patch testing to simvastatin at concentrations of 1.0%, 0.1%, and 0.01% in methyl ethyl ketone (MEK) were positive at 2 days to all 3 concentrations. Ten controls were negative (4,5,6).

Previously, two patients with occupational dermatitis to simvastatin had been reported from Spain. Both worked in the same pharmaceutical company and had desquamative erythema of the face, resembling seborrheic dermatitis. One worked in quality control in simvastatin synthesis and noticed that these lesions appeared during working hours, being minimal on the 1st day of the week and progressively worsening towards the last. Each relapse was more intense than the previous one. Patch tests were positive in both to simvastatin 0.1% alc. at D2 and D4 (++) (7).

CUTANEOUS ADVERSE DRUG REACTIONS FROM SYSTEMIC ADMINISTRATION CAUSED BY TYPE IV (DELAYED-TYPE) HYPERSENSITIVITY (as demonstrated by positive patch tests)

Cutaneous adverse drug reactions from systemic administration of simvastatin caused by type IV (delayed-type) hypersensitivity have included photosensitivity (8,9).

Photosensitivity
A 66-year-old man had persistent photosensitivity of more than 1 year's duration. Examination showed eczematous lichenified plaques on exposed skin of the face, neck and dorsum of the hands. The patient reported intermittent treatment with simvastatin, the last period having been for 1 month, 7 months previously. The clinical picture was interpreted as persistent photodermatitis. Photopatch testing with simvastatin and UVA irradiation 10 J/cm^2 was negative, but skin irradiated with UVB (to determine the minimal erythema dose [(MED]) developed an eczematous reaction with concomitant worsening of the patient's photodermatitis. Photopatch testing with simvastatin 2% pet. was then carried out with MED-UVB dose and subMED (70% of MED), producing erythema on the irradiated area 24 hours after irradiation. The patient was diagnosed with actinic reticuloid/chronic actinic dermatitis due to systemic photosensitivity to simvastatin (8). One may wonder whether the appearance of erythema with MED irradiation may count as proof of photoallergy. Quite remarkable, the authors themselves did not mention whether they considered this reaction to be a positive or negative photopatch test.

Two weeks after having used simvastatin 5 mg t.d.s. orally for 3 weeks for hypercholesterolemia, a 50-year-old woman presented with erythema and vesicles on sun-exposed areas, that had started one week earlier. She had been taking simvastatin 5 mg t.d.s. orally for 3 weeks for hypercholesterolemia. Phototests showed a normal minimal erythema dose for UVB but a lowered one for UVA. 6.75 J/cm^2 of UVA-irradiation (<50% of the MED in normal subjects) produced reddish papules on the irradiated area after one day, followed by erythema at 6 days after irradiation. After stopping simvastatin, the photosensitivity to UVA became disappeared. Patch and photopatch tests with simvastatin 10%, 1%, and 0.1% pet. showed a reaction only to simvastatin 10% pet. with 4.5 J/cm^2 UVA-irradiation, i.e. papules at 24 hours, followed by erythema at D6. Five controls were negative (9).

The authors hypothesized that the initial papular responses in the phototests and photopatch tests may coincide with the opening of sweat glands and/or hair follicles. In addition, they suggested that the erythema after the papules might represent a delayed hypersensitivity response, since the lipid-soluble simvastatin readily accumulates in skin appendages. Finally they supposed that the inactive prodrug not distributed in the liver, or an active β-hydroxy acid metabolite excreted in the blood, may be the sensitizer (9). If this is correct, this was not a proven case of simvastatin photoallergy.

Cross-reactions, pseudo-cross-reactions and co-reactions
A patient with fixed drug eruption from atorvastatin and a positive patch test on postlesional skin had a cross-reaction to simvastatin (2).

LITERATURE
1   The data in the section 'General' may have been obtained from literature discussed in this chapter, but mostly also or exclusively from one or more of the following online sources: ChemIDPlus Advanced, PubChem, DrugBank, RxList, Drug Central, Drugs.com, and Wikipedia
2   Huertas AJ, Ramírez-Hernández M, Mérida-Fernández C, Chica-Marchal A, Pajarón-Fernández M J, Carreño-Rojo A. Fixed drug eruption due to atorvastatin. J Investig Allergol Clin Immunol 2015;25:155-156
3   Neumark M, Moshe S, Ingber A, Slodownik D. Occupational airborne contact dermatitis to simvastatin, carvedilol, and zolpidem. Contact Dermatitis 2009;61:51-52
4   Field S, Bourke B, Hazelwood E, Bourke JF. Simvastatin - occupational contact dermatitis. Contact Dermatitis 2007;57:282-283
5   Field S, Bourke B, Hazelwood E, Bourke J. Occupational allergic contact dermatitis to statins. Br J Dermatol 2008: 159(suppl.1):81.
6   Bennett MF, Lowney AC, Bourke JF. A study of occupational contact dermatitis in the pharmaceutical industry. Br J Dermatol 2016;174:654-656 (Abstract in Brit J Dermatol 2011;165:73)
7   Peramiquel L, Serra E, Dalmau J, Vila AT, Mascaró JM, Alomar A. Occupational contact dermatitis from simvastatin. Contact Dermatitis 2005;52:286-287
8   Rodriguez Granados MT, de la Torre C, Cruces MJ, Piñeiro G. Chronic actinic dermatitis due to simvastatin. Contact Dermatitis 1998;38:294-295
9   Morimoto K, Kawada A, Hiruma M, Ishibashi A, Banba H. Photosensitivity to simvastatin with an unusual response to photopatch and photo tests. Contact Dermatitis 1995;33:274

Chapter 3.445 SORAFENIB

IDENTIFICATION

Description/definition : Sorafenib is the diarylether that conforms to the structural formula shown below
Pharmacological classes : Antineoplastic agents; protein kinase inhibitors
IUPAC name : 4-[4-[[4-Chloro-3-(trifluoromethyl)phenyl]carbamoylamino]phenoxy]-*N*-methylpyridine-2-carboxamide
CAS registry number : 284461-73-0
EC number : 608-209-4
Merck Index monograph : 10116
Patch testing : Tablet 0.1%, 1% and 10% pet. (2); most systemic drugs can be tested at 10% pet.; if the pure chemical is not available, prepare the test material from intravenous powder, the content of capsules or – when also not available – from powdered tablets to achieve a final concentration of the active drug of 10% pet.
Molecular formula : $C_{21}H_{16}ClF_3N_4O_3$

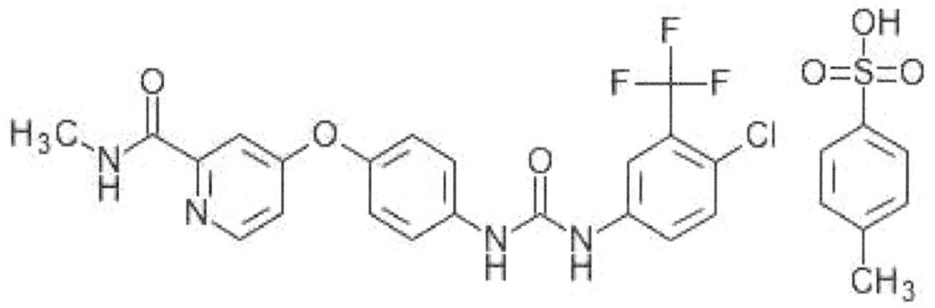

Sorafenib tosylate

GENERAL

Sorafenib is an oral multi-kinase inhibitor, targeting growth signaling and angiogenesis, that is used in the therapy of advanced renal cell, liver and thyroid cancer. Sorafenib blocks the enzyme RAF kinase, a critical component of the RAF/MEK/ERK signaling pathway that controls cell division and proliferation; in addition, sorafenib inhibits the VEGFR-2/PDGFR-beta signaling cascade, thereby blocking tumor angiogenesis. In pharmaceutical products, sorafenib is employed as sorafenib tosylate (CAS number 475207-59-1, EC number not available, molecular formula $C_{28}H_{24}ClF_3N_4O_6S$) (1).

CUTANEOUS ADVERSE DRUG REACTIONS FROM SYSTEMIC ADMINISTRATION CAUSED BY TYPE IV (DELAYED-TYPE) HYPERSENSITIVITY (as demonstrated by positive patch tests)

Cutaneous adverse drug reactions from systemic administration of sorafenib caused by type IV (delayed-type) hypersensitivity have included erythema multiforme (2).

Other cutaneous adverse drug reactions

In a university hospital in Japan, from November 2006 to November 2011, of 36 patients who had been treated with sorafenib for metastatic renal cell carcinoma, 9 developed erythema multiforme, which was confirmed by histopathology. The group consisted of 5 men and 4 women, with a median age of 67, range 52-76. Patch tests were performed in all patients with the sorafenib tablets smashed in a mortar and diluted to 0.1%, 1% and 10% pet., readings were performed at D2 and D3. All nine showed a positive reaction to sorafenib. Controls were not performed. Clinical and histopathological details were not provided, nor details on the positive patch tests (strength, results at D2 and D3, reaction to different concentrations) (2).

LITERATURE

1   The data in the section 'General' may have been obtained from literature discussed in this chapter, but mostly also or exclusively from one or more of the following online sources: ChemIDPlus Advanced, PubChem, DrugBank, RxList, Drug Central, Drugs.com, and Wikipedia
2   Ikeda M, Fujita T, Mii S, Tanabe K, Tabata K, Matsumoto K, Satoh T, Iwamura M. Erythema multiforme induced by sorafenib for metastatic renal cell carcinoma. Jpn J Clin Oncol 2012;42:820-824

Chapter 3.446 SOTALOL

IDENTIFICATION

Description/definition : Sotalol is the ethanolamine derivative that conforms to the structural formula shown below
Pharmacological classes : Anti-arrhythmia agents; sympatholytics; β-adrenergic antagonists
IUPAC name : *N*-[4-[1-Hydroxy-2-(propan-2-ylamino)ethyl]phenyl]methanesulfonamide
Other names : β-Cardone
CAS registry number : 3930-20-9
EC number : Not available
Merck Index monograph : 10124
Patch testing : Crushed tablets 10% pet. (2); most systemic drugs can be tested at 10% pet.; if the pure chemical is not available, prepare the test material from intravenous powder, the content of capsules or – when also not available – from powdered tablets to achieve a final concentration of the active drug of 10% pet.
Molecular formula : $C_{12}H_{20}N_2O_3S$

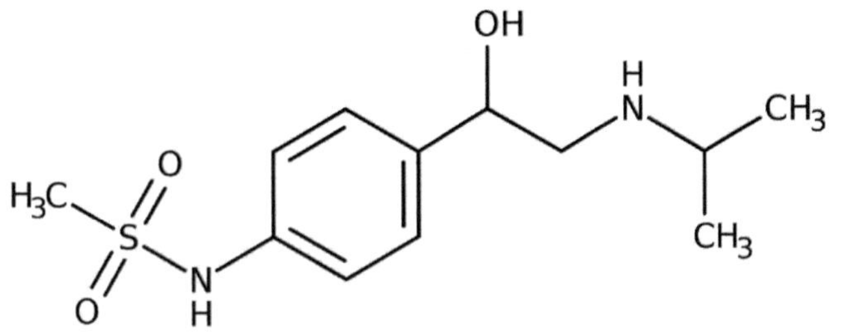

GENERAL

Sotalol is an ethanolamine derivative with Class III antiarrhythmic and antihypertensive properties. It is a nonselective β-adrenergic receptor and potassium channel antagonist. In the heart, this agent inhibits chronotropic and inotropic effects thereby slowing the heart rate and decreasing myocardial contractility. This agent also reduces sinus rate, slows conduction in the atria and in the atrioventricular (AV) node and increases the functional refractory period of the AV node. Sotalol is indicated for the maintenance of normal sinus rhythm in patients with symptomatic atrial fibrillation/atrial flutter who are currently in sinus rhythm. It is also used for the treatment of life-threatening ventricular arrhythmias (1). In pharmaceutical products, sotalol is employed as sotalol hydrochloride (CAS number 959-24-0, EC number 213-496-0, molecular formula $C_{12}H_{21}ClN_2O_3S$) (1).

CONTACT ALLERGY FROM ACCIDENTAL CONTACT

A 24-year-old woman presented with eyelid dermatitis, which had started with localized edema 4 months previously. Later, the area had become itchier, with redness and scaling. The patient suspected a relationship with her work as a pharmacy assistant, which involved breaking and crushing different types of tablets. She was patch tested with the crushed tablets that she had contact with at 10% pet. and showed positive reactions to sotalol HCl, 3 other beta-blockers, 3 benzodiazepines and 3 ACE-inhibitors. Three controls were negative to sotalol HCl 10% pet. Cosmetic allergy was excluded and a diagnosis of occupational airborne allergic contact was made (2).

LITERATURE

1   The data in the section 'General' may have been obtained from literature discussed in this chapter, but mostly also or exclusively from one or more of the following online sources: ChemIDPlus Advanced, PubChem, DrugBank, RxList, Drug Central, Drugs.com, and Wikipedia
2   Swinnen I, Ghys K, Kerre S, Constandt L, Goossens A. Occupational airborne contact dermatitis from benzodiazepines and other drugs. Contact Dermatitis 2014;70:227-232

Chapter 3.447 SPECTINOMYCIN

IDENTIFICATION

Description/definition : Spectinomycin is an aminocyclitol aminoglycoside antibiotic derived from *Streptomyces spectabilis* that conforms to the structural formula shown below
Pharmacological classes : Anti-bacterial agents
IUPAC name : (1R,3S,5R,8R,10R,11S,12S,13R,14S)-8,12,14-Trihydroxy-5-methyl-11,13-bis(methylamino)-2,4,9-trioxatricyclo[8.4.0.0^{3,8}]tetradecan-7-one
CAS registry number : 1695-77-8
EC number : 216-911-3
Merck Index monograph : 10136
Patch testing : 1%, 5% and 20% pet.
Molecular formula : $C_{14}H_{24}N_2O_7$

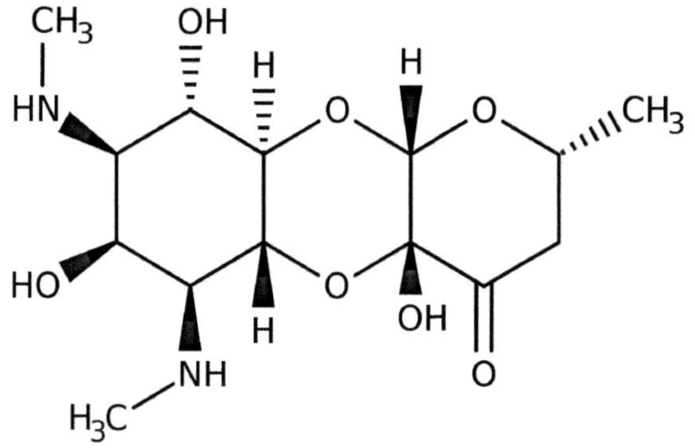

GENERAL

Spectinomycin is an aminocyclitol aminoglycoside antibiotic derived from *Streptomyces spectabilis* with antibacterial activity against gram-negative bacteria. Spectinomycin is indicated in the treatment of acute gonorrheal urethritis and proctitis in the male and acute gonorrheal cervicitis and proctitis in the female when due to susceptible strains of *Neisseria gonorrhoeae*. The antibiotic is (or was) also used in veterinary medicine in an injectable form in 7 to 15-day-old chickens to prevent the respiratory diseases that are a major cause of death (3). In pharmaceutical products, spectinomycin is employed as spectinomycin hydrochloride (= spectinomycin dihydrochloride pentahydrate) (CAS number 22189-32-8, EC number 606-950-8, molecular formula $C_{14}H_{36}Cl_2N_2O_{12}$) (go.drugbank.co).

CONTACT ALLERGY FROM ACCIDENTAL CONTACT

A 42-year old man, working for 13 years in a chick-breeding farm on the manual preparation of vaccines for injection and oculo-nasal administration, presented during the last 6 months with an itchy erythematous dermatitis on the face (chin, outer ear, eyelid), hands (index and middle fingers) and forearms. Nine months before, an antibiotic mixture of tylosin and spectinomycin had been added to this solution to prevent neonatal bacterial infections. Patch tests were positive at 2 and 3 days to the antibiotic mixture as is (++) and to spectinomycin 20% pet. (+++), but not to tylosin 5% pet. When the patient suspended the use of spectinomycin, his dermatitis gradually cleared. The dermatitis on the face was considered to have been caused by contamination from the hands, not from airborne exposure (2).

A 27-year-old female chicken vaccinator developed, 2 months after using a mixture of lincomycin and spectinomycin for the first time, dermatitis of the hands and forearms. Patch tests were positive to the antibiotic mixture, spectinomycin sulfate 1%, 5% and 20% pet. and to lincomycin HCl 5% pet. (3). Another 27-year-old female had worked on the same farm for 2 years. Two months after first using a mixture of lincomycin and spectinomycin, she pricked herself in the index fingertip of the right hand with a syringe loaded with this antibiotic. Two days later, extremely itchy papules appeared, which spread to the rest of the hand and forearm in 3 or 4 days, and then to the left hand and forearm, the face and the chest. Patch tests were positive to the antibiotic mixture, spectinomycin 1%, 5% and 20% pet., but not to lincomycin (1).

LITERATURE

1   Vilaplana J, Romaguera C, Grimalt F. Contact dermatitis from lincomycin and spectinomycin in chicken vaccinators. Contact Dermatitis 1991;24:225-226
2   Dal Monte A, Laffi G, Mancini G. Occupational contact dermatitis due to spectinomycin. Contact Dermatitis 1994;31:204-205

Chapter 3.448 SPIRAMYCIN

IDENTIFICATION

Description/definition	: Spiramycin is a mixture of three macrolide antibiotics produced by *Streptomyces ambofaciens*
Pharmacological classes	: Anti-bacterial agents; coccidiostats
IUPAC name	: 2-[(4R,5S,6S,7R,9R,10R,11E,13E,16R)-6-[5-(4,5-Dihydroxy-4,6-dimethyloxan-2-yl)oxy-4-(dimethylamino)-3-hydroxy-6-methyloxan-2-yl]oxy-10-[5-(dimethylamino)-6-methyloxan-2-yl]oxy-4-hydroxy-5-methoxy-9,16-dimethyl-2-oxo-1-oxacyclohexadeca-11,13-dien-7-yl]acetaldehyde
Other names	: Rovamycin
CAS registry number	: 8025-81-8
EC number	: 232-429-6
Merck Index monograph	: 10149
Patch testing	: 10.0% pet. (Chemotechnique)
Molecular formula	: Unspecified (ChemIDPlus); $C_{43}H_{74}N_2O_{14}$ (PubChem)

Spiramycin I :   Rgp = H
Spiramycin II   Rgp = COCH$_3$
Spiramycin III   Rgp = COCH$_2$CH$_3$

GENERAL

Spiramycin is a complex of three macrolide antibiotics originally discovered as product of *Streptomyces ambofaciens*, with antibacterial and antiparasitic activities. It is a primarily bacteriostatic agent with activity against gram-positive cocci and rods, gram-negative cocci and also *Legionella* spp., *Mycoplasma* spp., *Chlamydia* spp., some types of spirochetes, *Toxoplasma gondii* and *Cryptosporidium*. Although the specific mechanism of action has not been characterized, spiramycin likely inhibits protein synthesis by binding to the 50S subunit of the bacterial ribosome. This agent also prevents placental transmission of toxoplasmosis presumably through a different mechanism. Spiramycin is used for treatment of various infections in some countries like Tunisia (6), Portugal (10,11), France (2,7,8,9) and Spain (12), but mostly in veterinary medicine, e.g. for mastitis in cows, enteritis in pigs and respiratory infections in cats and dogs (1,3).

CONTACT ALLERGY FROM ACCIDENTAL CONTACT

Case series
In 1971, in Gentofte, Denmark, 9 veterinary surgeons were patch tested because of recent aggravation of hand eczema or dissemination of the eczema from the hands to other body regions. Six proved to be allergic to commercial spiramycin and 4 to commercial tylosin, both tested at 1% pet., 5% water and/or 10% pet. 45 controls were negative. All reactions were relevant to the actual dermatitis (3). The following year, one or more cases of occupational allergic contact dermatitis from spiramycin in veterinary surgeons were reported from the same clinic in Denmark (probably overlap) (4).

In 2 hospitals in Denmark, between 1974 and 1980, 37 veterinary surgeons, all working in private country practices, were investigated for suspected incapacitating occupational dermatitis and patch tested with a battery of 10 antibiotics. Thirty-two (86%) had one or more positive patch tests. There were 24 positive reactions to spiramycin 10% pet. (65%). It was mentioned that all but one of these individuals had allergic contact dermatitis, but relevance was not specified for individual allergens. The most frequent allergens were spiramycin, penethamate and tylosin tartrate (13).

In another clinic in Denmark, between November 1, 1978 to December 31, 1979, patients working with farm animals and suspected of having occupational contact dermatitis were patch tested with commercial spiramycin 10% water and tylosin 2% water. Six farmers and three farmers' wives had positive patch tests to one or both antibiotics, 8 to spiramycin and 8 to tylosin (negative in 26 controls). The eczema was typically seen on the hands, face and neck (suggesting airborne spread, possibly by the presence of excreted antibiotics in dust in the pigshed), but two patients had generalized dermatitis. The main occupation of all nine patients was raising hogs. The duration of the dermatitis ranged from 4 months to 5 years. All patients had used either spiramycin or tylosin and most had used both antibiotics. After diagnosis, the dermatitis cleared or improved markedly in seven patients; five of these stopped using the antibiotics in question, and two began to wear gloves while handling the drugs. The two patients who continued use of the drugs and took no precautionary measures had little change in the activity of the dermatitis (5).

Case reports
A 43-year-old man, who had worked for 13 years in cattle breeding. presented with an 18-month history of erythema, edema and vesicles on the interdigital spaces of the hands, the face and both sides of the neck. His symptoms worsened every time that he prepared animal feed. Patch tests were positive at D2 and D3 to and spiramycin 5% pet., oxytetracycline hydrochloride 5% pet., tylosin 10% pet., and penicillin 5% pet.. He had contact with all these antibiotics and the dermatitis completely cleared as soon as the patient left his occupation (14).

A man working in a feed factory in Spain developed allergic contact dermatitis due to airborne spiramycin with recurrent outbreaks of eczematous lesions on uncovered areas during working periods. The diagnosis was based on history, positive patch tests to spiramycin and disappearance of lesions on leaving the work place (15).

CUTANEOUS ADVERSE DRUG REACTIONS FROM SYSTEMIC ADMINISTRATION CAUSED BY TYPE IV (DELAYED-TYPE) HYPERSENSITIVITY (as demonstrated by positive patch tests)
Cutaneous adverse drug reactions from systemic administration of spiramycin caused by type IV (delayed-type) hypersensitivity have included maculopapular eruption (10,11,12) and acute generalized exanthematous pustulosis (AGEP) (2,6,7,8,9).

Maculopapular eruption
A 62-year-old woman was treated with spiramycin and 3 other drugs over a period of 3 days for a buccal infection, when a generalized maculopapular exanthema developed. A biopsy showed a perivascular lymphocytic infiltrate with numerous eosinophils in the dermis, compatible with a cutaneous drug eruption. Five months later, patch tests were positive to the commercial spiramycin preparation 30% pet. and pure spiramycin 10% pet. at day 4 with negative reactions to all other drugs used. Twenty controls were negative (12).

Two cases of maculopapular eruptions with positive patch tests to spiramycin have been observed in Portugal (10, overlap with and contradictory to ref. 11, where it was stated that only one patient was patch tested with spiramycin). In one case, a repeat test with spiramycin after nearly 6 years was again positive (10).

Acute generalized exanthematous pustulosis (AGEP)
A 46-year-old woman was treated at home for a sore throat with oral erythromycin ethylsuccinate and prednisolone. An eruption appeared on the chest and axillae 48 hours later. Erythromycin was changed to oral spiramycin. Again, 48 hours later the patient presented with a pustular eruption with a fever of 39°C. She had an erythematous eruption covered with numerous superficial non-follicular pustules on the trunk and proximal extremities, sparing the palms, soles, and mucous membranes. Spiramycin was discontinued and the fever and eruption spontaneously

disappeared within 10 days. Later, patch tests with crushed tablets and intravenous forms of erythromycin and spiramycin were positive at D2 showing a pustular eruption on an erythematous base. Ten controls were negative (2). This was a case of acute generalized exanthematous pustulosis (AGEP) caused by erythromycin and spiramycin.

A 23-year-old man was treated for a dental abscess with spiramycin and 3 other drugs. Seven days after the start or treatment, the patient developed a generalized pruritic eruption accompanied by fever at 39°C. Physical examination showed a generalized, erythematous, infiltrated rash, with numerous, small non-follicular pustules over the trunk and arms. Laboratory studies showed leukocytosis, eosinophilia, elevated C-reactive protein and elevated liver enzymes. Results of bacterial and fungal cultures of pustules were negative. A skin biopsy specimen showed subcorneal pustules containing neutrophils and perivascular infiltrate composed of neutrophils and eosinophils, consistent with AGEP. Patch tests were performed 6 weeks after healing with crushed tablets of spiramycin and the other drugs at 10% in petrolatum and in saline, yielding positive reactions to spiramycin only (6).

In a group of 45 patients with multiple drug hypersensitivity seen between 1996 and 2018 in Montpellier, France, 38 of 92 drug hypersensitivities were classified as type IV immunological reactions. This included 2 individuals with AGEP, one from spiramycin and the other from pristinamycin, both with positive patch tests; no clinical or patch testing details were provided (7).

Between March 2009 and June 2013, in a center in France specialized in cutaneous adverse drug reactions (CADR), 156 patients were patch tested because of a CADR. Spiramycin was positive in one patient with AGEP, both to commercial test material (10% pet.) and extemporaneous patch tests with pulverized pills 30% pet. (8). Another two patients from France with AGEP and positive patch tests to spiramycin had been reported in 1996; clinical nor patch testing details were provided (9).

Cross-reactions, pseudo-cross-reactions and co-reactions
Erythromycin may have cross-reacted with the related antibiotic spiramycin (2). Very likely cross-reactivity to or from tylosin (5).

LITERATURE
1   The data in the section 'General' may have been obtained from literature discussed in this chapter, but mostly also or exclusively from one or more of the following online sources: ChemIDPlus Advanced, PubChem, DrugBank, RxList, Drug Central, Drugs.com, and Wikipedia
2   Moreau A, Dompmartin A, Castel B, Remond B, Leroy D. Drug-induced acute generalized exanthematous pustulosis with positive patch tests. Int J Dermatol 1995;34:263-266
3   Hjorth N, Weismann K. Occupational dermatitis among veterinary surgeons caused by spiramycin and tylosin. Contact Dermatitis Newsletter 1972;12:320
4   Hjorth N, Weismann K. Occupational dermatitis among veterinary surgeons caused by spiramycin, tylosin and penethamate. Acta Derm Venereol 1973;53:229-232
5   Veien NK, Hattel T, Justesen O, Nørholm A. Occupational contact dermatitis due to spiramycin and/or tylosin among farmers. Contact Dermatitis 1980;6:410-413
6   Kastalli S, Charfi O, El Aïdli S, Zaïem A, Daghfous R. Acute generalized exanthematic pustulosis induced by spiramycin: usefulness of patch testing. Tunis Med 2016;94:339
7   Landry Q, Zhang S, Ferrando L, Bourrain JL, Demoly P, Chiriac AM. Multiple drug hypersensitivity syndrome in a large database. J Allergy Clin Immunol Pract 2019;8:258
8   Assier H, Valeyrie-Allanore L, Gener G, Verlinde Carvalh M, Chosidow O, Wolkenstein P. Patch testing in non-immediate cutaneous adverse drug reactions: value of extemporaneous patch tests. Contact Dermatitis 2017;77:297-302
9   Wolkenstein P, Chosidow O, Fléchet ML, Robbiola O, Paul M, Dumé L, et al. Patch testing in severe cutaneous adverse drug reactions, including Stevens-Johnson syndrome and toxic epidermal necrolysis. Contact Dermatitis 1996;35:234-236
10 Pinho A, Marta A, Coutinho I, Gonçalo M. Long-term reproducibility of positive patch test reactions in patients with non-immediate cutaneous adverse drug reactions to antibiotics. Contact Dermatitis 2017;76:204-209
11 Pinho A, Coutinho I, Gameiro A, Gouveia M, Gonçalo M. Patch testing - a valuable tool for investigating non-immediate cutaneous adverse drug reactions to antibiotics. J Eur Acad Dermatol Venereol 2017;31:280-287
12 Poveda-Montoyo I, Álvarez-Chinchilla PJ, García Del Pozo MC, Encabo B1, Silvestre JF. Spiramycin-related cutaneous eruption confirmed by patch testing. Contact Dermatitis 2018;78:233-234
13 Hjorth N, Roed-Petersen J. Allergic contact dermatitis in veterinary surgeons. Contact Dermatitis 1980;6:27-29
14 Guerra L, Venture M, Tardio M, Tosti A. Airborne contact dermatitis from animal feed antibiotics. Contact Dermatitis 1991;25:333-334
15 Acero S, Tabar AI, Echechipia S, Alvarez MJ, García BE. Occupational allergic contact dermatitis due to airborne spiramycin. J Investig Allergol Clin Immunol 1998;8:184-185

Chapter 3.449 SPIRONOLACTONE

IDENTIFICATION

Description/definition : Spironolactone is the synthetic steroid lactone that conforms to the structural formula shown below

Pharmacological classes : Mineralocorticoid receptor antagonists; diuretics

IUPAC name : S-[(7R,8R,9S,10R,13S,14S,17R)-10,13-Dimethyl-3,5'-dioxospiro[2,6,7,8,9,11,12,14,15,16-decahydro-1H-cyclopenta[a]phenanthrene-17,2'-oxolane]-7-yl] ethanethioate

CAS registry number : 52-01-7

EC number : 200-133-6

Merck Index monograph : 10157

Patch testing : 1% pet. and alcohol

Molecular formula : $C_{24}H_{32}O_4S$

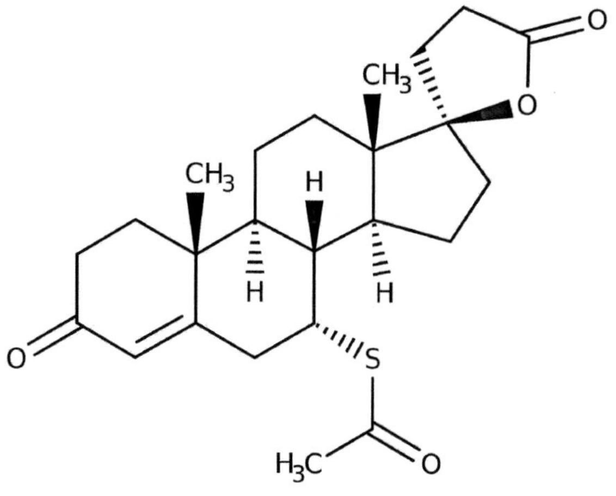

GENERAL

Spironolactone is a synthetic corticosteroid with potassium-sparing diuretic, antihypertensive, and antiandrogen activities. Spironolactone competitively inhibits adrenocortical hormone aldosterone activity in the distal renal tubules, myocardium, and vasculature. Spironolactone is used mainly in the treatment of refractory edema in patients with congestive heart failure, nephrotic syndrome, or hepatic cirrhosis, of primary hyperaldosteronism and hypertension. Off-label uses of spironolactone involving its antiandrogenic activity include hirsutism, female pattern hair loss, and adult acne vulgaris. It is also frequently used in medical gender transition (1). Topical spironolactone shows antiandrogenic effects by competitive inhibition of dihydrotestosterone receptors, and has been used to treat acne vulgaris, idiopathic hirsutism and androgenic alopecia (2). As such, it has cause some cases of contact allergy/allergic contact dermatitis, which have been fully reviewed in Volume 3 of the *Monographs in contact allergy series* (7).

CONTACT ALLERGY FROM ACCIDENTAL CONTACT

In Leuven, Belgium, in the period 2001-2019, 23 health care workers/employees of the pharmaceutical industry had occupational allergic contact dermatitis from skin contact with a systemic drug: 19 nurses, two chemists, one physician, and one veterinarian. The lesions were mostly localized on the hands, but often also on the face, as airborne dermatitis. In total, 42 positive patch test reactions to 18 different systemic drugs were found. In 2 patients, Aldactazine ® (altizide/spironolactone), tested 10% pet., was the drug/one of the drugs that caused occupational dermatitis. Altizide and spironolactone were not tested separately (8).

A 55-year-old man, who had intermittent skin contact with purified spironolactone powder during his work in a pharmaceutical company for many years, suddenly developed itchy eczematous lesions on the face, neck and forearms within 24 hours of contact with spironolactone. Strongly positive patch tests were obtained at D1 and D2 with spironolactone 1% alc. and with the dry powder, which were still present at D7. Since avoiding spironolactone, the patient recovered and has had no further problems. This was a classic case of occupational airborne allergic contact dermatitis (3).

CUTANEOUS ADVERSE DRUG REACTIONS FROM SYSTEMIC ADMINISTRATION CAUSED BY TYPE IV (DELAYED-TYPE) HYPERSENSITIVITY (as demonstrated by positive patch tests)

Cutaneous adverse drug reactions from systemic administration of spironolactone caused by type IV (delayed-type) hypersensitivity have included drug reaction with eosinophilia and systemic symptoms (DRESS) (4,5,9) and photosensitivity (6).

Drug reaction with eosinophilia and systemic symptoms (DRESS)

A 79-year-old man developed a generalized pruritic exanthema 10 days after he started taking amlodipine, spironolactone and allopurinol for hypertension and gout. Despite immediate discontinuation of all drugs, in the following 6 days, the skin eruption continued with increasing pruritus. The patient had fever (39°C) and progressively developed facial edema and erythema, pharyngeal tightening, enlarged lymph nodes, and confusion. Laboratory investigations showed leukocytosis, hypereosinophilia and elevated liver enzymes. DRESS was diagnosed and the patient was instructed never to take any of these three drugs anymore. Seven years after this episode, the patient was referred for allergy testing, as allopurinol was highly needed for the treatment of his gout. In an oral provocation test with oral spironolactone, the patient tolerated 10 mg, 10% of the dose, without immediate symptoms, but, on the second day, 6 hours after the second dose (28.5 mg), he developed generalized skin erythema and pruritus, without fever. Six weeks later, patch tests with commercial spironolactone prepared in petrolatum and saline in dilution series were strongly positive at D2 to spironolactone 25, 5, 2.5 and 0.25 mg/mL and 50, 20, 10 and 1% pet. There were no positive reactions to allopurinol (4).

Previously, a 58-year-old man had been described who developed DRESS while using spironolactone and 4 other drugs with fever, erythroderma, edema of both hands and face, hypereosinophilia, hepatic, pancreatic and renal failure, metabolic acidosis, aggravation of pre-existing cardiac insufficiency and edema of the legs. Four months after complete recovery, patch tests with all drugs were strongly positive only to commercial spironolactone 10%, 20% and 30% in pet. and saline. Ten controls were negative. In a second test session, there were positive reactions to pure spironolactone 1% and 10% pet. (5).

In a multicenter investigation in France, of 72 patients patch tested for DRESS, 46 (64%) had positive patch tests to drugs, including 2 to spironolactone; clinical details were not provided (9).

Photosensitivity

A 70-year-old woman had been treated for 15 years with a combination tablet of altizide and spironolactone for arterial hypertension, when she developed a photodistributed itching and burning erythematous, papulosquamous eruption involving the face (eyelids, cheeks), lateral neck, dorsum of both hands and proximal phalanges. On withdrawal of the medication, the eruption disappeared in 4 weeks. Patch and photopatch tests were performed with the tablet 10% water and pet. and irradiation with 5 J/cm^2 of UVA and a suberythemal UVB dose (0.75 UVB-MED). The patch tests and UVA photopatch tests were negative, but positive to the test materials irradiated with UVB (D1 ++, D2 ++). The patient was – although the 2 ingredients were not tested separately – diagnosed with photoallergy to altizide (presumably because it is a thiazide-type diuretic); the possibility that spironolactone was the culprit was not considered (6).

Other cutaneous adverse drug reactions

In PubMed, reference is made to an article entitled 'Cutaneous reaction to oral spironolactone with positive patch test', but this publication could not be traced by the author (10).

LITERATURE

1   The data in the section 'General' may have been obtained from literature discussed in this chapter, but mostly also or exclusively from one or more of the following online sources: ChemIDPlus Advanced, PubChem, DrugBank, RxList, Drug Central, Drugs.com, and Wikipedia
2   Corazza M, Strumìa R, Lombardi AR, Virgili A. Allergic contact dermatitis from spironolactone. Contact Dermatitis 1996;35:365-366
3   Klijn J. Contact dermatitis from spironolactone. Contact Dermatitis 1984;10:105
4   Fernandes R-A, Regateiro FS, Faria E, Martinho A, Gonçalo M, Todo-Bom A. Drug reaction with eosinophilia and systemic symptoms caused by spironolactone: Case report. Contact Dermatitis 2018;79:255-256
5   Ghislain P, Bodarwe A, Vanderdonckt O, Tennstedt D, Marot L, Lachapelle J. Drug-induced eosinophilia and multisystemic failure with positive patch-test reaction to spironolactone: DRESS syndrome. Acta Derm Venereol 2004;84:65-68

6 Schwarze HP, Albes B, Marguery MC, Loche F, Bazex J. Evaluation of drug-induced photosensitivity by UVB photopatch testing. Contact Dermatitis 1998;39:200

7 De Groot AC. Monographs in contact allergy, volume 3: Topical Drugs. Boca Raton, Fl, USA: CRC Press Taylor and Francis Group, 2021 (ISBN 978-0-367-23693-9)

8 Gilissen L, Boeckxstaens E, Geebelen J, Goossens A. Occupational allergic contact dermatitis from systemic drugs. Contact Dermatitis 2020;82:24-30

9 Barbaud A, Collet E, Milpied B, Assier H, Staumont D, Avenel-Audran M, et al. A multicentre study to determine the value and safety of drug patch tests for the three main classes of severe cutaneous adverse drug reactions. Br J Dermatol 2013;168:555-562

10 Alonso JC, Ortega FJ, Gonzalo MJ, Palla PS. Cutaneous reaction to oral spironolactone with positive patch test. Contact Dermatitis 2002;47:178-179

Chapter 3.450 STREPTOMYCIN

IDENTIFICATION

Description/definition : Streptomycin is the aminoglycoside antibiotic derived from *Streptomyces griseus* that conforms to the structural formula shown below

Pharmacological classes : Protein synthesis inhibitors; anti-bacterial agents

IUPAC name : 2-[(1R,2R,3S,4R,5R,6S)-3-(Diaminomethylideneamino)-4-[(2R,3R,4R,5S)-3-[(2S,3S,4S,5R, 6S)-4,5-dihydroxy-6-(hydroxymethyl)-3-(methylamino)oxan-2-yl]oxy-4-formyl-4-hydroxy-5-methyloxolan-2-yl]oxy-2,5,6-trihydroxycyclohexyl]guanidine

CAS registry number : 57-92-1

EC number : 200-355-3

Merck Index monograph : 10226

Patch testing : Sulfate, 5% pet. (SmartPracticeCanada)

Molecular formula : $C_{21}H_{39}N_7O_{12}$

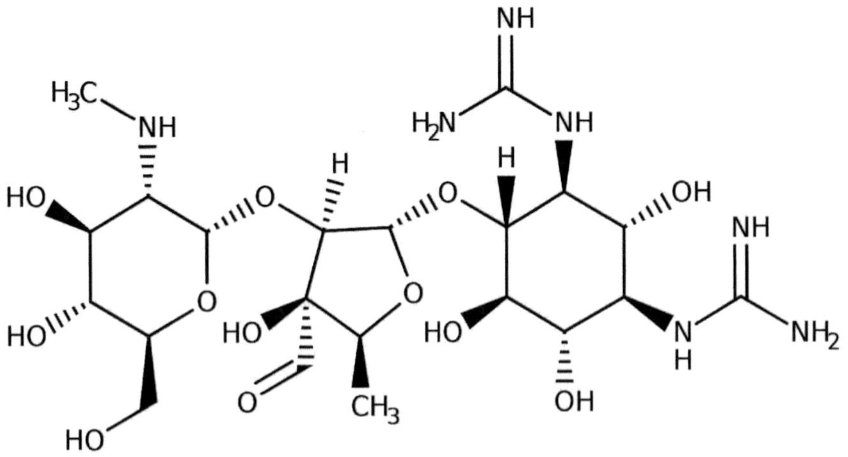

GENERAL

Streptomycin is an aminoglycoside antibiotic produced by the soil actinomycete *Streptomyces griseus* with antibacterial activity. It acts by binding to the S12 protein of the bacterial 30S ribosomal subunit, thereby inhibiting peptide elongation and protein synthesis, consequently leading to bacterial cell death. Streptomycin is indicated for the treatment of tuberculosis. It may also be used in combination with other drugs to treat tularemia (*Francisella tularensis*), plague (*Yersia pestis*), severe *M. avium* complex, brucellosis, and enterococcal endocarditis (e.g. *E. faecalis, E. faecium*). In pharmaceutical products, streptomycin is employed as streptomycin sulfate (CAS number 3810-74-0, EC number 223-286-0, molecular formula $C_{42}H_{84}N_{14}O_{36}S_3$) (1).

CONTACT ALLERGY FROM ACCIDENTAL CONTACT

In the past, from 1947 on (7,8,26), when work hygiene standards were suboptimal, many cases of occupational allergic contact dermatitis from streptomycin have been described, especially in nurses, veterinarians, and workers in the pharmaceutical industries. Nurses would develop sensitization from preparing parenteral streptomycin formulations from powder and water, giving injections and subsequently washing and cleaning the syringes (8,20). Of 12 nurses working in one particular ward, 6 became sensitized within a few months (8), but generally patients became sensitized after 6 weeks to 6 months. Of 101 individuals working in a pharmaceutical company and having contact with streptomycin, 21 (21%) became sensitized to the antibiotic in a period of 2 years (25). Streptomycin caused both allergic contact dermatitis of the hand(s) and of the eyelids, either by contamination by fingers and/or by airborne contact (8,9,10,20). Some patients first had eczema around the eyes, favoring the latter possibility (20). In all patients, the dermatitis healed when contact with streptomycin was avoided. Patch tests have been performed with streptomycin in various concentrations and in aqueous solutions or petrolatum (table 3.450.1); not infrequently, positive patch tests led to recurrence of previous dermatitis (e.g. ref. 26).

A summary of data on occupational allergic contact dermatitis from streptomycin is provided in table 3.450.1. Other cases have been reported in non-English literature (12,13,14,15,16,17,19,23) and in English literature not available to the author (11,28). The literature on this subject from before 1969 has been reviewed in ref. 31.

Table 3.450.1 Occupational allergic contact dermatitis/contact allergy to streptomycin [a]

Years	Country	Occupations and numbers	Test prep./conc./veh.[b,c]	Ref.
2003	Italy	1 veterinarian	S 2% pet.	22
1996	Spain	1 cattle breeder	S 2% pet.	24
<1989	Poland	4 pharmaceutical workers	S 10% pet.	2
1979-1987	Poland	2 nurses	S 10% water	5
< 1985	Norway	1 veterinarian	S 2.5% water	27
1979-1983	Poland	3 pharmaceutical workers, 3 nurses and 3 veterinarians	S 10% water	18
1980	Poland	2 veterinarians	S 10% water	6
1974-1980	Denmark	4 veterinarians	S 30% water	4
1958	France	2 nurses	Not stated	32
1950-1957	U.K.	18 nurses	S 5% solution	10
1949-1950	Canada	21 pharmaceutical workers	S 1% solution	25
1950	USA	1 nurse	S ointment 5 mg/gr.; S sol. 25 mg/ml	33
1949	USA	1 nurse	S dilution series down to 1% and 0.1%	21
1948	U.K.	4 nurses	SHCl 50 µg in 0.1 ml saline intradermal	20
1947	USA	1 nurse	S 1% and 10% water	
1946-1947	USA	6 nurses	SHCl and SCaCl 20, 10, 2 and 1% solution	8
1946-1947	USA	6 nurses	Not stated; intradermal tests 100 units in 0.1 ml saline	9

[a] Examples, not a full literature review

[b] Test preparation, concentration and vehicle

[c] In most publications, patch tests were stated to have been performed with 'streptomycin'; however, in water, streptomycin is little soluble, and in those cases is it likely that in fact streptomycin hydrochloride has been used

S: Streptomycin; SCaCl: streptomycin calcium chloride complex; SHCl; streptomycin hydrochloride

CUTANEOUS ADVERSE DRUG REACTIONS FROM SYSTEMIC ADMINISTRATION CAUSED BY TYPE IV (DELAYED-TYPE) HYPERSENSITIVITY (as demonstrated by positive patch tests)

Cutaneous adverse drug reactions from systemic administration of streptomycin caused by type IV (delayed-type) hypersensitivity have included eczematous eruption (30) and toxic erythema with generalized follicular pustules (3).

Dermatitis/eczematous eruption

About 20 years ago, a then 39-year-old man was treated with i.m. streptomycin for tuberculosis and developed a generalized itchy red scaly vesicular eruption which lasted for 3-4 weeks. Thirteen years later, a similar exanthema developed while the patient was being treated with isoniazid, rifampicin and ethionamide. Now (20 and 7 years after the events) patch tests were positive to streptomycin sulfate 1% and 10% saline and to isoniazid 2% water. Thirty controls were negative. Intradermal tests with both drugs were also positive at D2 (30).

Other cutaneous adverse drug reactions

A 29-year-old man developed a skin eruption during the course of the treatment of pulmonary tuberculosis with streptomycin sulfate, isoniazid and rifampicin. His entire body, excluding only the palms and soles, was darkly erythematous and had a dry, rough surface. Numerous follicular papules, mostly topped with pustules, were present diffusely on the trunk and proximal extremities. He had leukocytosis with neutrophilia and eosinophilia. A bacterial culture of the pustules was negative. Two months later, the patient was given 0.5 g of streptomycin sulfate intramuscularly because of the need to resume antitubercular treatment. Generalized erythema with follicular papules recurred seven hours later with fever and malaise. After disappearance of the erythema, he was found to have numerous spinous protrusions from the hair follicles present diffusely on the skin surface, which persisted for almost three months and resisted keratolytic treatment. Patch tests and prick tests were performed with 12.5% and 25% streptomycin in water, respectively. The patch test produced erythema and follicular papules; the prick test produced erythema and induration lasting from 12 to 72 hours. The drug hypersensitivity reaction was called toxic erythema with generalized follicular pustules (3).

Cross-reactions, pseudo-cross-reactions and co-reactions

In patients sensitized to neomycin, only about 4% co- or cross-reacts to streptomycin. This is due to the different chemical structure of streptomycin, which contains streptidine, while the other aminoglycosides contain deoxystreptamine (Chapter 3.335 Neomycin). The cross-sensitivity pattern between aminoglycoside antibiotics in patients primarily sensitized to streptomycin has not been well investigated. Possibly cross-reaction to or from dihydrostreptomycin (27).

Immediate contact reactions

Immediate contact reactions (contact urticaria) to streptomycin are presented in Chapter 5.

LITERATURE

1   The data in the section 'General' may have been obtained from literature discussed in this chapter, but mostly also or exclusively from one or more of the following online sources: ChemIDPlus Advanced, PubChem, DrugBank, RxList, Drug Central, Drugs.com, and Wikipedia

2   Rudzki E, Rebandel P, Grzywa Z. Contact allergy in the pharmaceutical industry. Contact Dermatitis 1989;21:121-122

3   Kushimoto H, Aoki T. Toxic erythema with generalized follicular pustules caused by streptomycin. Arch Dermatol 1981;117:444-445

4   Hjorth N, Roed-Petersen J. Allergic contact dermatitis in veterinary surgeons. Contact Dermatitis 1980;6:27-29

5   Rudzki E, Rebandel P, Grzywa Z. Patch tests with occupational contactants in nurses, doctors and dentists. Contact Dermatitis 1989;20:247-250

6   Rudzki E, Rebandel P, Grzywa Z, Pomorski Z, Jakiminska B, Zawisza E. Occupational dermatitis in veterinarians. Contact Dermatitis 1982;8:72-73

7   Strauss MJ, Warring FC. Contact dermatitis from streptomycin. Preliminary report. J Invest Dermatol 1947;9:3-7

8   Strauss MJ, Warring FC. Epidermal sensitization to streptomycin. Report of six cases occurring in twelve nurses handling the drug. J Invest Dermatol 1947;9:99-106

9   Rauchwerger SM, Erskine FA, Nalls WL. Streptomycin sensitivity. Development of sensitivity in nursing personnel through contact during administration of the drug to patients. JAMA 1948;136:614-615 (overlap with ref. 17)

10 Wilson HTH. Streptomycin dermatitis in nurses. Br Med J 1958;1:1378-1382

11 Marcussen PV. Professional streptomycin hypersensitiveness among hospital staffs. Acta Derm Venereol (Stockh) 1949;29:410-413

12 Popovic J, Labej T. [Allergy to streptomycin in hospital personnel]. Tuberkuloza 1950;2:107-113 (Article in Slovene)

13 Cucchiani Acevedo R, Erdstein S. [Professional hypersensitivity to streptomycin]. Dia Med 1948;20:2095-2097 (Article in Spanish)

14 Barfod B. [Hypersensitivity to streptomycin among nurses]. Ugeskr Laeger 1950;112:1421-1424 (Article in Danish)

15 Pirilä V, Kilpio O. [On streptomycin dermatitis in nurses]. Duodecim 1949;65:319-324 (Article in Finnish)

16 Leoncini G. [Occupational allergic dermatitis due to streptomycin]. G Ital Della Tuberc 1950;4:451-455 (Article in Italian)

17 Rauchwerger SM, Erskine FA, Nalls WL. Sensibilidad a la estreptomicina; evolución de la sensibilidad en las enfermeras a consecuencia del contacto de este agente al ser administrado a los pacientes [Streptomycin sensitivity; evolution of sensitivity in nurses as a result of the contact of this agent when administered to patients]. Am Clin 1948;12:516 (Article in Spanish) (overlap with ref. 9)

18 Rudzki E, Rebendel P. Contact sensitivity to antibiotics. Contact Dermatitis 1984;11:41-42

19 Bernard E, Lotte A, Wolff C. Sur les accidents provoqués par la streptomycine chez les infirmières [On the accidents caused by streptomycin in nurses]. Sem Hop 1948;24:2554 (Article in French)

20 Crofton J, Foreman HM. Streptomycin dermatitis in nurses. Br Med J 1948;2(4566):71

21 Johnson SA, Davis HP. Streptomycin, cause of dermatitis venenata in a nurse; report of a case. Arch Derm Syphilol 1949;59:245-247.

22 Valsecchi R, Leghissa P, Cortinovis R. Occupational contact dermatitis and contact urticaria in veterinarians. Contact Dermatitis 2003;49:167-168

23 Almeida AS. Dermatite por contacto à estreptomicina [Streptomycin contact dermatitis]. Bol Asoc Medica Nac Repub Panama 1950;19:206-207 (Article in Spanish)

24 Gauchía R, Rodríguez-Serna M, Silvestre JF, Linana JJ, Aliaga A. Allergic contact dermatitis from streptomycin in a cattle breeder. Contact Dermatitis 1996;35:374-375

25 Mitchell HS. Streptomycin dermatitis. J Allergy 1951;22:71-73

26 Canizares O, Shatin H. Dermatitis venenata due to streptomycin. Arch Derm Syphilol 1947;56:676

27 Falk ES, Hektoen H, Thune PO. Skin and respiratory tract symptoms in veterinary surgeons. Contact Dermatitis 1985;12:274-278

28 Pirilä V, Noro L, Laamanen A. Air pollution and allergy. Acta Allergol 1963;18:113-130 (Data cited in ref. 29)

29 Dooms-Goossens AE, Debusschere KM, Gevers DM, Dupré KM, Degreef HJ, Loncke JP, et al. Contact dermatitis caused by airborne agents. A review and case reports. J Am Acad Dermatol 1986;15:1-10

30 Meseguer J, Sastre A, Malek T, Salvador MD. Systemic contact dermatitis from isoniazid. Contact Dermatitis 1993;28:110-111

31 Levene GM, Withers AFD. Anaphylaxis to streptomycin and hyposensitization (parasensitization). Trans St John Hosp Derm Soc 1969;55:184-188

32 Sidi E, Hincky M, Longueville R. Cross sensitization between neomycin and streptomycin. J Invest Dermatol 1958;30:225-231

33 Sulzberger MB, Distelheim IH. Allergic eczematous contact type sensitivity of equal degree to streptomycin and dihydrostreptomycin; report of a case. AMA Arch Derm Syphilol 1950;62:706-707

Chapter 3.451 SUCCINYLCHOLINE

IDENTIFICATION

Description/definition : Succinylcholine is the quaternary ammonium compound that conforms to the structural
formula shown below
Pharmacological classes : Neuromuscular depolarizing agents
IUPAC name : Trimethyl-[2-[4-oxo-4-[2-(trimethylazaniumyl)ethoxy]butanoyl]oxyethyl]azanium
Other names : Suxamethonium; diacetylcholine
CAS registry number : 306-40-1
EC number : 200-747-4
Patch testing : 5% water
Molecular formula : $C_{14}H_{30}N_2O_4{+}^2$

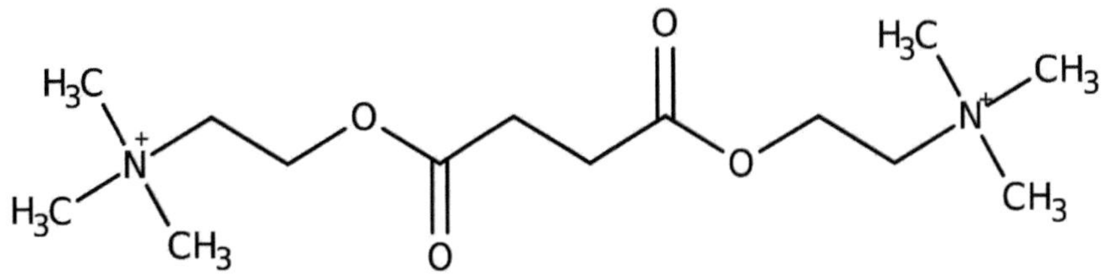

GENERAL

Succinylcholine is a quaternary ammonium compound and skeletal muscle relaxant. It is a depolarizing relaxant, acting in about 30 seconds and with a duration of effect averaging three to five minutes. Succinylcholine binds to nicotinic receptors at the neuromuscular junction and opening the ligand-gated channels in the same way as acetylcholine, resulting in depolarization and inhibition of neuromuscular transmission. Succinylcholine is indicated as an adjunct to general anesthesia, to facilitate tracheal intubation and endoscopies, and to provide skeletal muscle relaxation during surgery, electroconvulsive therapy and mechanical ventilation (1). In pharmaceutical products, succinylcholine is employed as succinylcholine chloride (CAS number 71-27-2, EC number 200-747-4, molecular formula $C_{14}H_{30}Cl_2N_2O_4$) (1).

CUTANEOUS ADVERSE DRUG REACTIONS FROM SYSTEMIC ADMINISTRATION CAUSED BY TYPE IV (DELAYED-TYPE) HYPERSENSITIVITY (as demonstrated by positive patch tests)

Cutaneous adverse drug reactions from systemic administration of succinylcholine caused by type IV (delayed-type) hypersensitivity have included systemic allergic dermatitis (2).

Systemic allergic dermatitis (systemic contact dermatitis)

A 61-year-old woman, who underwent left external saphenectomy under general anesthesia, developed a progressive rash affecting the face, thorax, neck and limbs over the next 24 h. The drugs received during surgery were procaine, meperidine, atropine and (intravenously) suxamethonium (succinylcholine). Examination revealed erythematous, vesicular, excoriated lesions, consistent with contact dermatitis. Patch testing with the drugs employed during general anesthesia showed a positive reaction to succinylcholine chloride 5% water (D2 +, D3 +++). Ten controls were negative. A diagnosis of systemic allergic dermatitis from succinylcholine was made (2). As it is unknown whether the patient had previously become sensitized to succinylcholine, this diagnosis may be challenged.

LITERATURE

1   The data in the section 'General' may have been obtained from literature discussed in this chapter, but mostly also or exclusively from one or more of the following online sources: ChemIDPlus Advanced, PubChem, DrugBank, RxList, Drug Central, Drugs.com, and Wikipedia
2   Delgado J, Quiralte J, Castillo R, Blanco C, Molero R, Carrillo T. Systemic contact dermatitis from suxamethonium. Contact Dermatitis 1996;35:120-121

Chapter 3.452 SULFADIAZINE

IDENTIFICATION

Description/definition : Sulfadiazine is the aminobenzene sulfonamide that conforms to the structural formula shown below

Pharmacological classes : Antibacterial agents; coccidiostats; antiprotozoal agents

IUPAC name : 4-Amino-*N*-pyrimidin-2-ylbenzenesulfonamide

Other names : Sulfapyrimidine

CAS registry number : 68-35-9

EC number : 200-685-8

Merck Index monograph : 10305

Patch testing : 5% pet.

Molecular formula : $C_{10}H_{10}N_4O_2S$

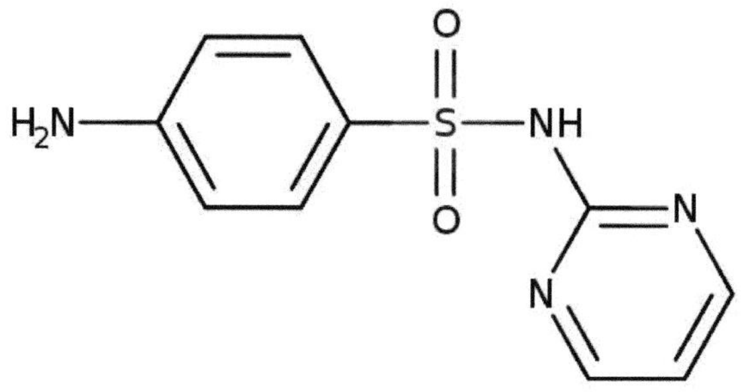

GENERAL

Sulfadiazine is a short-acting sulfonamide antibiotic with bacteriostatic activity. Sulfonamides inhibit multiplication of bacteria by acting as competitive inhibitors of *p*-aminobenzoic acid in the folic acid metabolism cycle. Sulfadiazine is used in combination with pyrimethamine to treat toxoplasmosis in patients with acquired immunodeficiency syndrome and in newborns with congenital infections. In pharmaceutical products, sulfadiazine is employed as sulfadiazine sodium (CAS number 547-32-0, EC number 547-32-0, molecular formula $C_{10}H_9N_4NaO_2S$) (1).

CUTANEOUS ADVERSE DRUG REACTIONS FROM SYSTEMIC ADMINISTRATION CAUSED BY TYPE IV (DELAYED-TYPE) HYPERSENSITIVITY (as demonstrated by positive patch tests)

Cutaneous adverse drug reactions from systemic administration of sulfadiazine caused by type IV (delayed-type) hypersensitivity have included fixed drug eruption (2).

Fixed drug eruption

In patients with fixed drug eruptions (FDE) caused by delayed-type hypersensitivity, the diagnosis is usually confirmed by a positive patch test with the drug on previously affected skin. Authors from Finland have used an alternative method of topical provocation. The test compound, the drug 10% in petrolatum and sometimes also in 70% alcohol and in DMSO, was applied once and without occlusion over the entire surface of one or several inactive (usually pigmented) sites of FDE lesions. The patients were followed as in-patients for 24 hours. A reaction was regarded as positive when a clearly demarcated erythema lasting at least 6 hours was seen. Of 2 patients with FDE from sulfadiazine, one (50%) had a positive topical provocation (2).

Cross-reactions, pseudo-cross-reactions and co-reactions

Nine patients allergic to sulfanilamide were patch tested with a battery of 25 sulfonamides (all tested 5% pet.) to detect cross-sensitization and there were 3 reactions to sodium sulfadiazine (3).

LITERATURE

1 The data in the section 'General' may have been obtained from literature discussed in this chapter, but mostly also or exclusively from one or more of the following online sources: ChemIDPlus Advanced, PubChem, DrugBank, RxList, Drug Central, Drugs.com, and Wikipedia

2 Alanko K. Topical provocation of fixed drug eruption. A study of 30 patients. Contact Dermatitis 1994;31:25-27

3 Degreef H, Dooms-Goossens A. Patch testing with silver sulfadiazine cream. Contact Dermatitis 1985;12:33-37

Chapter 3.453 SULFAGUANIDINE

IDENTIFICATION

Description/definition   : Sulfaguanidine is the sulfonamide that conforms to the structural formula shown below
Pharmacological classes   : Anti-infective agents
IUPAC name   : 2-(4-Aminophenyl)sulfonylguanidine
Other names   : p-Aminobenzenesulfonylguanidine
CAS registry number   : 57-67-0
EC number   : 200-345-9
Merck Index monograph   : 10310
Patch testing   : 5% pet.
Molecular formula   : $C_7H_{10}N_4O_2S$

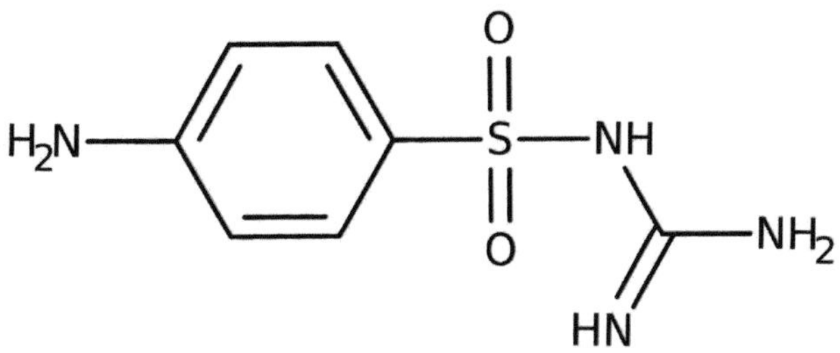

GENERAL

Sulfaguanidine is a guanidine derivative of sulfanilamide used in veterinary medicine. It is poorly absorbed from the gut but is well suited for the treatment of bacillary dysentery and other enteric infections (1).

CUTANEOUS ADVERSE DRUG REACTIONS FROM SYSTEMIC ADMINISTRATION CAUSED BY TYPE IV (DELAYED-TYPE) HYPERSENSITIVITY (as demonstrated by positive patch tests)

Cutaneous adverse drug reactions from systemic administration of sulfaguanidine caused by type IV (delayed-type) hypersensitivity have included erythema multiforme (3).

Other cutaneous adverse drug reactions

A 25-year-old man presented with slightly elevated erythematous macules symmetrically distributed over the buttocks, back, backs of the hands and upper arms, which evolved into bullae in 1-2 days. He had no mucosal lesions or any other symptoms. The patient had taken paracetamol, a combination tablet of sulfaguanidine, benzocaine, and enoxolone, and an oral spray 12 hours before commencement of the skin lesions. The eruption resolved in a week with some post-inflammatory hyperpigmentation. The patient had experienced three previous episodes associated with the combination tablet and the oral spray. Patch tests were applied on a residual cutaneous lesion with paracetamol, sulfamethoxazole, sulfaguanidine and sulfanilamide (all 10% DMSO), the combination tablet 'as is' and the oral spray 'as is'. At D1, sulfaguanidine alone showed a positive reaction, all other drugs being negative at D1-D3. Patch tests with sulfaguanidine were negative in 10 control patients (3). It was not explained why patch tests were applied to postlesional skin (not common practice in erythema multiforme).

Cross-reactions, pseudo-cross-reactions and co-reactions

Nine patients allergic to sulfanilamide were patch tested with a battery of 25 sulfonamides (all tested 5% pet.) to detect cross-sensitization and there was one positive reaction to sulfaguanidine (2).

LITERATURE

1   The data in the section 'General' may have been obtained from literature discussed in this chapter, but mostly also or exclusively from one or more of the following online sources: ChemIDPlus Advanced, PubChem, DrugBank, RxList, Drug Central, Drugs.com, and Wikipedia
2   Degreef H, Dooms-Goossens A. Patch testing with silver sulfadiazine cream. Contact Dermatitis 1985;12:33-37
3   De Frutos C, de Barrio M, Tornero P, Barranco R, Rodríguez A, Rubio M. Erythema multiforme from sulfaguanidine. Contact Dermatitis 2002;46:186-187

Chapter 3.454 SULFAMETHOXAZOLE

IDENTIFICATION

Description/definition : Sulfamethoxazole is the sulfonamide that conforms to the structural formula shown below

Pharmacological classes : Antibacterial agents

IUPAC name : 4-Amino-*N*-(5-methyl-1,2-oxazol-3-yl)benzenesulfonamide

CAS registry number : 723-46-6

EC number : 211-963-3

Merck Index monograph : 10320

Patch testing : Tablet, pulverized, 10%, 20% and 50% pet.; for fixed drug eruptions, use open tests with sulfamethoxazole 10%, 20% and 50% in DMSO (14); most systemic drugs can be tested at 10% pet.; if the pure chemical is not available, prepare the test material from intravenous powder, the content of capsules or – when also not available – from powdered tablets to achieve a final concentration of the active drug of 10% pet.

Molecular formula : : $C_{10}H_{11}N_3O_3S$

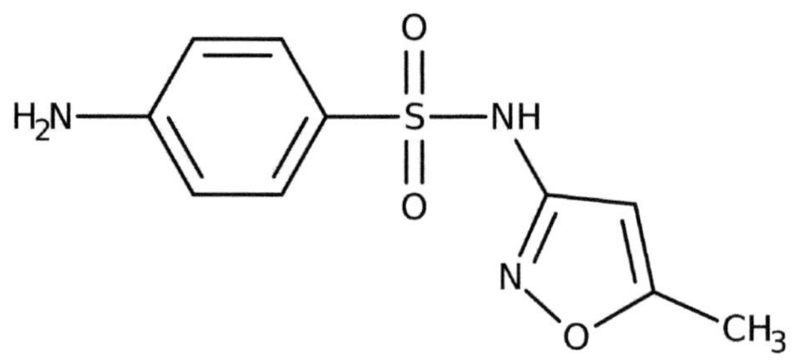

GENERAL

Sulfamethoxazole is a sulfonamide antibiotic with broad-spectrum activity. It is a bacteriostatic antibacterial agent that interferes with folic acid synthesis in susceptible bacteria. The use of sulfamethoxazole has been limited by the development of resistance. It is, however, frequently used in a combination preparation with trimethoprim as cotrimoxazole (see Chapter 3.455 Sulfamethoxazole mixture with trimethoprim). Sulfamethoxazole as single agent may be used for the treatment of bronchitis, prostatitis and urinary tract infections caused by susceptible bacteria. In pharmaceutical products, both sulfamethoxazole and (rarely) sulfamethoxazole sodium (CAS number 4563-84-2, EC number 224-939-2, molecular formula $C_{10}H_{10}N_3NaO_3S$) may be employed (1).

See also Chapter 3.455 Sulfamethoxazole mixture with trimethoprim and Chapter 3.495 Trimethoprim.

CUTANEOUS ADVERSE DRUG REACTIONS FROM SYSTEMIC ADMINISTRATION CAUSED BY TYPE IV (DELAYED-TYPE) HYPERSENSITIVITY (as demonstrated by positive patch tests)

Cutaneous adverse drug reactions from systemic administration of sulfamethoxazole caused by type IV (delayed-type) hypersensitivity have included systemic allergic dermatitis (4,12), fixed drug eruption (2,3,6,13,14), drug reaction with eosinophilia and systemic symptoms (DRESS) (8,9), toxic epidermal necrolysis (TEN) (15), bullous drug eruption (7), bullous exanthema/erythema multiforme (16), maculopapular eruption with vesicles and bullae (17), and unspecified exanthema (10,11).

Systemic allergic dermatitis (systemic contact dermatitis)

A 27-year-old man who had been treated with silver sulfadiazine for burn injuries developed a severe drug eruption while using sulfamethoxazole-trimethoprim (ST), administered orally for the treatment of a urinary infection. Patch tests were positive to 10% ST solution and silver sulfadiazine cream diluted with water. An undefined number of healthy controls was negative. Histopathology of the patch test reactions was consistent with allergic contact dermatitis. Additionally, a lymphocyte-stimulation test with ST and a leukocyte migration-inhibition test with silver sulfadiazine used for the treatment of this patient both showed positive reactions. It was assumed that the patient had become sensitized to silver sulfadiazine (presumably the sulfadiazine moiety) and later cross-reacted to sulfamethoxazole causing the 'severe drug eruption' (12). This then would be a case of systemic allergic dermatitis. Oral sulfamethoxazole can produce clinical exacerbation in patients sensitized by contact with sulfanilamide. These would be cases of systemic allergic dermatitis, but it is unknown whether the patients were patch tested with and patch test-positive to sulfamethoxazole (4, data cited in ref. 5, no details available).

Fixed drug eruption

Twenty-seven of 48 patients seen between January 1996 and February 1998 with fixed drug eruptions caused by sulfamethoxazole-trimethoprim, as proven by a positive oral provocation test, were investigated in Istanbul, Turkey (14). The group consisted of 10 women and 17 men, age range 10-66, mean age 37 years and the duration of the FDE varied between 1 week and 10 years. One average, the patients had suffered 3.3 FDE attacks. Patch tests with pure sulfamethoxazole and pure trimethoprim 10%, 20% and 50% in petrolatum both on tape-stripped normal and postlesional skin, were negative in all 19 patients tested. In the next step, sulfamethoxazole and trimethoprim were applied in open tests in all 27 individuals in concentrations of 10%, 20% and 50% in DMSO with pure DMSO as control, on both previously unaffected and affected skin. Twenty controls were negative, but most of the controls and patients had a transient burning sensation on the test areas that disappeared within 5-30 minutes (due to DMSO). Positive reactions were seen to sulfamethoxazole in 20 patients and to trimethoprim in 5 on postlesional skin; two had negative or doubtful patch tests and one had also a positive patch test to trimethoprim on normal skin. The majority of the patients had positive reactions to the 20% concentrations. A positive reaction (itching and erythema, in a few cases – notably with trimethoprim – induration) lasted up to 12-36 hours. Positive reactions with 10% of the drug preparations in DMSO after a single application were obtained mainly in lesions on the glans penis. 50% of drug preparations were necessary mainly for lesions on trunk, arms and legs. Sulfamethoxazole was the reacting drug in all of the lesions on glans penis, and in 75% of the non-pigmented lesions. Familial FDE in 2 patients, linear FDE, FDE with additional reaction of uninvolved skin and 1 nonpigmented solitary-plaque-type FDE were trimethoprim-induced. It was concluded that repeated open testing with graded concentrations of the drugs up to 50% in DMSO is a reliable test method in sulfamethoxazole-trimethoprim-induced FDE (14).

In a hospital in Lahore, Pakistan, in the period 2002-2005, 305 patients clinically diagnosed with fixed drug eruption were patch tested on postlesional skin with the incriminated drugs in various concentrations (1%, 2%, and 5% pet.). Sulfamethoxazole was the most frequent cause of fixed drug eruption with a positive lesional patch test; 5% was the best concentration, lower concentrations often resulted in (false)-negative results (3).

Twenty-four patients with fixed drug eruptions suspected to be caused by sulfamethoxazole (SMX) were patch tested with SMX 10% in DMSO and there were 4 (17%) positive reactions on postlesional skin. One individual reacted (presumably a cross-reaction) to sulfadiazine 10% DMSO (6).

A 26-year old man had been treated with sulfamethoxazole-trimethoprim repeatedly for recurrent pharyngitis. After taking the 8th pill of one of these courses he developed circular erythematous and violaceous lesions on his penis, right wrist, right thigh and tongue. Two weeks later, the patient was treated again with this drug combination for another attack of pharyngitis. After taking the second pill, similar lesions appeared at the same sites. Patch testing with trimethoprim and sulfamethoxazole, both at 10% pet., tested on both normal and postlesional skin, was positive to sulfamethoxazole on previously affected skin (D2 +++, D4 -). A patch test on adjacent skin was also positive, albeit weaker (D2 ++, D4 -) (13).

Of 2 patients who had developed a fixed drug eruption to sulfamethoxazole, one (50%) had a topical provocation test indicative of type-IV allergy to the drug (2).

Drug reaction with eosinophilia and systemic symptoms (DRESS)

In a group of 14 patients with multiple delayed-type hypersensitivity reactions, DRESS was caused by sulfamethoxazole in 2 cases, showing positive patch test reactions and positive lymphocyte transformation tests (LTT). In one, diclofenac later induced a maculopapular eruption (positive patch test and LTT to diclofenac) and in the other vancomycin caused a relapse of DRESS (positive patch test and LTT to vancomycin). These patients therefore both had multiple drug hypersensitivity (8, also presented in ref. 9).

Stevens-Johnson syndrome/toxic epidermal necrolysis (SJS/TEN)

A 90-year old man developed a skin eruption with erythematous lesions 2 days after he had started therapy with sulfamethoxazole-trimethoprim for a urinary tract infection. This was continued for a few more days and was stopped when the lesions became painful and epidermal detachment was noticed. The patient presented with large flaccid blisters and exfoliation on dark-red painful erythema covering 25% of the body surface. Both variants of Nikolsky's test were positive. Lesions predominated on the thighs, buttocks, back, abdomen, chest and neck. There was no involvement of mucous membranes. On admission, the clinical diagnosis toxic epidermal necrolysis was immediately confirmed by evaluation of cryo-cut sections, and later by routine histology, revealing complete epidermal necrosis and subepidermal detachment. One year later, the patient was readmitted after erroneous re-exposure to a single oral dose of co-trimoxazole for another infection. Patch tests showed positive reactions to sulfamethoxazole but not to trimethoprim, tested as pure powder, on tape-stripped previously involved skin but not on uninvolved skin. Histopathology of the patch test showed a discretely spongiotic epidermis, with pronounced single-cell necrosis of keratinocytes and a subepidermal mononuclear infiltrate, containing many eosinophils. Five controls were negative (15).

Other cutaneous adverse drug reactions

A 43-year-old man developed a bullous drug eruption after treatment with amoxicillin-clavulanic acid and sulfamethoxazole-trimethoprim. Patch tests were strongly positive to penicillins and to sulfamethoxazole 12.5% pet. The lymphocyte transformation test of the latter was also positive with a stimulation index (SI) of 3.2. However, the SI of amoxicillin was far stronger (10.3). Patch tests to trimethoprim and to the sulfamethoxazole-trimethoprim combination tested at 12.5% were negative, presumably because of the lower concentration of sulfamethoxazole in the combination tablet. As the eruption had been caused by 2 unrelated drugs, this was a case of multiple drug hypersensitivity (7).

In Bern, Switzerland, patients with a suspected allergic cutaneous drug reaction were patch-scratch tested with suspected drugs that had previously given a positive lymphocyte transformation test. Sulfamethoxazole (tested as sulfamethoxazole-trimethoprim 960 mg/ml saline; LTT positive to sulfamethoxazole) gave a positive patch-scratch test in one patient with bullous exanthema/erythema multiforme (16).

One patient had developed a maculopapular eruption in which later vesicles and bullae appeared, while being treated with sulfamethoxazole-trimethoprim and amoxicillin. Patch tests were positive to sulfamethoxazole (negative to trimethoprim) and to amoxicillin (17).

A man aged 45 had developed 'exanthema and malaise' while using sulfamethoxazole. Patch and lymphocyte transformation tests were positive to sulfamethoxazole (10).

In Finland, in the period 1989-2001, 826 patients with suspected cutaneous drug eruptions were patch tested and 89 had one or more positive reactions. Of these individuals, 1 reacted to sulfamethoxazole. Clinical details were not provided and it was not mentioned which type of drug eruption sulfamethoxazole had caused (11).

Cross-reactions, pseudo-cross-reactions and co-reactions
Cross-reactions from or to sulfadiazine (4,6).

LITERATURE

1 The data in the section 'General' may have been obtained from literature discussed in this chapter, but mostly also or exclusively from one or more of the following online sources: ChemIDPlus Advanced, PubChem, DrugBank, RxList, Drug Central, Drugs.com, and Wikipedia
2 Alanko K, Stubb S, Reitamo S. Topical provocation of fixed drug eruption. Br J Dermatol 1987;116:561-567
3 Mahboob A, Haroon TS, Iqbal Z, Iqbal F, Saleemi MA, Munir A. Fixed drug eruption: topical provocation and subsequent phenomena. J Coll Physicians Surg Pak 2006;16:747-750
4 Meneghini CL, Angelini G. Gruppensensibilisierung durch photosensibilisierende Medikamente. Z Hautkr 1978;53:329-334 (Article in German). Data cited in ref. 5
5 Angelini G, Meneghini CL. Oral tests in contact allergy to para-amino compounds. Contact Dermatitis 1981;7:311-314
6 Tornero P, De Barrio M, Baeza ML, Herrero T. Cross-reactivity among p-amino group compounds in sulfonamide fixed drug eruption: diagnostic value of patch testing. Contact Dermatitis 2004;51:57-62
7 Gex-Collet C, Helbling A, Pichler WJ. Multiple drug hypersensitivity – proof of multiple drug hypersensitivity by patch and lymphocyte transformation tests. J Investig Allergol Clin Immunol 2005;15:293-296
8 Jörg L, Yerly D, Helbling A, Pichler W. The role of drug, dose and the tolerance/intolerance of new drugs in multiple drug hypersensitivity syndrome (MDH). Allergy 2020;75:1178-1187
9 Jörg L, Helbling A, Yerly D, Pichler WJ. Drug-related relapses in drug reaction with eosinophilia and systemic symptoms (DRESS). Clin Transl Allergy 2020;10:52
10 Beeler A, Engler O, Gerber BO, et al. Long-lasting reactivity and high frequency of drug-specific T cells after severe systemic drug hypersensitivity reactions. J Allergy Clin Immunol 2006;117:455-462
11 Lammintausta K, Kortekangas-Savolainen O. The usefulness of skin tests to prove drug hypersensitivity. Br J Dermatol 2005;152:968-974
12 Sawada Y. Adverse reaction to sulphonamides in a burned patient – a case report. Burns Incl Therm Inj 1985;12:127-131
13 Oleaga JM, Aguirre A, González M, Diaz-Pérez JL. Topical provocation of fixed drug eruption due to sulphamethoxazole. Contact Dermatitis 1993;29:155
14 Özkaya-Bayazit E, Bayazit H, Ozarmagan G. Topical provocation in 27 cases of cotrimoxazole-induced fixed drug eruption. Contact Dermatitis 1999;41:185-189
15 Klein CE, Trautmann A, Zillikens D, Bröcker EB. Patch testing in an unusual case of toxic epidermal necrolysis. Contact Dermatitis 1996;35:175-176, also published on pages 448-449.
16 Neukomm C, Yawalkar N, Helbling A, Pichler WJ. T-cell reactions to drugs in distinct clinical manifestations of drug allergy. J Invest Allergol Clin Immunol 2001;11:275-284
17 Hari Y, Frutig-Schnyder K, Hurni M, Yawalkar N, Zanni MP, Schnyder B, et al. T cell involvement in cutaneous drug eruptions. Clin Exp Allergy 2001;31:1398-1408

Chapter 3.455 SULFAMETHOXAZOLE MIXTURE WITH TRIMETHOPRIM

IDENTIFICATION

Description/definition : Sulfamethoxazole mixture with trimethoprim, better known as cotrimoxazole, is the combination product of sulfamethoxazole with trimethoprim
Pharmacological classes : Anti-bacterial agents; anti-infective agents, urinary; antimalarials
IUPAC name   : 4-Amino-*N*-(5-methyl-1,2-oxazol-3-yl)benzenesulfonamide;5-[(3,4,5-trimethoxyphenyl)-methyl]pyrimidine-2,4-diamine
Other names   : Cotrimoxazole
CAS registry number   : 8064-90-2
EC number   : Not available
Merck Index monograph : 10320
Patch testing   : 10.0% pet. (Chemotechnique); also test sulfamethoxazole and trimethoprim separately
Molecular formula   : $C_{14}H_{18}N_4O_3.C_{10}H_{11}N_3O_3S$ ($C_{24}H_{29}N_7O_6S$)

Sulfamethoxazole   Trimethoprim

GENERAL

Sulfamethoxazole mixture with trimethoprim is a fixed antibiotic combination with broad-spectrum antibacterial activity against both gram-positive and gram-negative organisms. It is widely used for mild-to-moderate bacterial infections and as therapy or prophylaxis against opportunistic infections, including *Pneumocystis* pneumonia in HIV/AIDS (1). In pharmaceutical products, sulfamethoxazole is used as the parent compound and trimethoprim as trimethoprim hydrochloride (CAS number 60834-30-2, EC number 262-450-6, molecular formula $C_{14}H_{19}ClN_4O_3$) or as trimethoprim sulfate (CAS number 56585-33-2, EC number not available, molecular formula $C_{28}H_{38}N_8O_{10}S$) (1).

See also Chapter 3.454 Sulfamethoxazole and Chapter 3.496 Trimethoprim. Cases of cutaneous adverse drug reactions suspected to have been caused by sulfamethoxazole-trimethoprim, where patch testing showed a positive reaction to either sulfamethoxazole or trimethoprim (with or without a positive reaction to the combination product) are presented in the corresponding chapter.

CUTANEOUS ADVERSE DRUG REACTIONS FROM SYSTEMIC ADMINISTRATION CAUSED BY TYPE IV (DELAYED-TYPE) HYPERSENSITIVITY (as demonstrated by positive patch tests)

Cutaneous adverse drug reactions from systemic administration of sulfamethoxazole-trimethoprim caused by type IV (delayed-type) hypersensitivity have included maculopapular eruption (6), fixed drug eruption (3,7), drug reaction with eosinophilia and systemic symptoms (DRESS) (5,6), photosensitivity (8), and unspecified drug eruption (2).

Case series with various or unknown types of drug reactions

In Finland, in the period 1989-2001, 826 patients with suspected cutaneous drug eruptions were patch tested and 89 had one or more positive reactions. Of these individuals, 4 reacted to sulfamethoxazole-trimethoprim. Clinical details were not provided and it was not mentioned which type of drug eruption the combination had caused (2).

Maculopapular eruption

In the period 2000-2014, in Coimbra, Portugal, 260 patients were patch tested with antibiotics for suspected cutaneous adverse drug reactions (CADR) to these drugs. 56 patients (22%) had one or more (often from cross-reactivity) positive patch tests. Sulfamethoxazole-trimethoprim (cotrimoxazole) 10% pet. and water (prepared from tablet or capsule) was patch test positive in 3 patients with maculopapular eruptions (6).

Fixed drug eruption

Sulfamethoxazole-trimethoprim is a very frequent cause of fixed drug eruptions, including generalized bullous FDE. However, in most cases, no patch tests were performed or were negative. It has been shown that patch testing can best be performed with high concentrations of the individual ingredients (10%, 20% and 50%) in DMSO to avoid false-negative reactions (4). Cases in which positive patch tests on postlesional skin were demonstrated to either sulfamethoxazole of trimethoprim (with or without a reaction to the combination tablet) are presented in the respective chapters (e.g. 4).

One case of fixed drug eruption to cotrimoxazole (sulfamethoxazole-trimethoprim) with a positive patch test reaction on postlesional skin (itching, erythema, infiltration) with the commercial drug 10% pet. was reported from Seoul, South Korea, in the period 1986-1990 and in 1996 and 1997 (3).

A girl aged 10 developed a fixed drug eruption while using cotrimoxazole and had a positive patch test to this antibiotic; details were not provided (7).

Drug reaction with eosinophilia and systemic symptoms (DRESS)

A 2-year old boy was treated with sulfamethoxazole-trimethoprim (S/T, cotrimoxazole) and 4 weeks later presented with a progressive generalized rash associated with fever and severe asthenia. Clinical and laboratory investigations revealed an exanthematous eruption, diffuse edema, severe pruritus, lymphadenopathy, spleno- and hepatomegaly, leukocytosis with eosinophilia, cholestasis and hepatic cytolysis with elevated liver enzymes and a positive HHV-6 polymerase chain reaction. Two months later, patch tests were positive (++) to the commercial formulation of T-S at 30% in white petrolatum and with the pure drug combination at 1% in white petrolatum. The patient was diagnosed with drug-induced hypersensitivity syndrome (DIHS-DRESS) associated with HHV-6 viremia and a drug cofactor demonstrated by patch testing to sulfamethoxazole-trimethoprim (5).

In the period 2000-2014, in Coimbra, Portugal, 260 patients were patch tested with antibiotics for suspected cutaneous adverse drug reactions (CADR) to these drugs. 56 patients (22%) had one or more (often from cross-reactivity) positive patch tests. Sulfamethoxazole-trimethoprim (cotrimoxazole) 10% pet. and water (prepared from tablet or capsule) was patch test positive in one patient with DRESS (6).

Photosensitivity

A rather complex case report with possibly photoallergic dermatitis to trimethoprim in sulfamethoxazole-trimethoprim was reported from Spain in 2016 (8).

LITERATURE

1   The data in the section 'General' may have been obtained from literature discussed in this chapter, but mostly also or exclusively from one or more of the following online sources: ChemIDPlus Advanced, PubChem, DrugBank, RxList, Drug Central, Drugs.com, and Wikipedia

2   Lammintausta K, Kortekangas-Savolainen O. The usefulness of skin tests to prove drug hypersensitivity. Br J Dermatol 2005;152:968-974

3   Lee AY. Topical provocation in 31 cases of fixed drug eruption: change of causative drugs in 10 years. Contact Dermatitis 1998;38:258-260

4   Özkaya-Bayazit E, Bayazit H, Ozarmagan G. Topical provocation in 27 cases of cotrimoxazole-induced fixed drug eruption. Contact Dermatitis 1999;41:185-189

5   Hubiche T, Milpied B, Cazeau C, Taïeb A, Léauté-Labrèze C. Association of immunologically confirmed delayed drug reaction and human herpesvirus 6 viremia in a pediatric case of drug-induced hypersensitivity syndrome. Dermatology 2011;222:140-141

6   Pinho A, Coutinho I, Gameiro A, Gouveia M, Gonçalo M. Patch testing - a valuable tool for investigating non-immediate cutaneous adverse drug reactions to antibiotics. J Eur Acad Dermatol Venereol 2017;31:280-287

7   Atanaskovic-Markovic M, Gaeta F, Medjo B, Gavrovic-Jankulovic M, Cirkovic Velickovic T, Tmusic V, et al. Non-immediate hypersensitivity reactions to beta-lactam antibiotics in children - our 10-year experience in allergy work-up. Pediatr Allergy Immunol 2016;27:533-538

8   D'Amelio CM, Del Pozo JL, Vega O, Madamba R, Gastaminza G. Successful desensitization in a child with delayed cotrimoxazole hypersensitivity: A case report. Pediatr Allergy Immunol 2016;27:320-321

Chapter 3.456 SULFANILAMIDE

IDENTIFICATION

Description/definition	: Sulfanilamide is the sulfonamide that conforms to the structural formula shown below
Pharmacological classes	: Anti-bacterial agents
IUPAC name	: 4-Aminobenzenesulfonamide
Other names	: Benzenesulfonamide, 4-amino-; sulphanilamide; *p*-aminobenzene sulfonamide
CAS registry number	: 63-74-1
EC number	: 200-563-4
Merck Index monograph	: 10327
Patch testing	: 5.0% pet. (Chemotechnique, SmartPracticeCanada, SmartPracticeEurope)
Molecular formula	: $C_6H_8N_2O_2S$

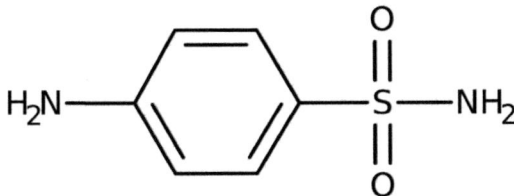

GENERAL

Sulfanilamide is a short-acting sulfonamide antibiotic. It is bacteriostatic against most gram-positive and many gram-negative organisms, but many strains of an individual species may be resistant. Sulfanilamide is used in vaginal cream for the treatment of vulvovaginitis caused by *Candida albicans* (1). In Belgium (and probably also in some other countries), it is also available in an ointment for wound treatment (An Goossens, Email communication, September 2020). Formerly, when sulfanilamide was still used widely in topical preparations, e.g. for wounds, burns and bacterial infections of the skin, this sulfonamide caused large numbers of sensitization and photosensitization, which subject has been reviewed in Volume 3 of the *Monographs in contact allergy* series (2).

CUTANEOUS ADVERSE DRUG REACTIONS FROM SYSTEMIC ADMINISTRATION CAUSED BY TYPE IV (DELAYED-TYPE) HYPERSENSITIVITY (as demonstrated by positive patch tests)

Cutaneous adverse drug reactions from systemic administration of sulfanilamide caused by type IV (delayed-type) hypersensitivity have included systemic allergic dermatitis (4,5,6,7).

Systemic allergic dermatitis (systemic contact dermatitis)

In the 1940s it had been well demonstrated that topical preparations containing sulfanilamide may cause sensitization and that subsequent oral administration of sulfanilamide or other sulfonamides would often result in exacerbation of previous eczema, extensive dermatitis, morbilliform exanthemas, generalized papular dermatitis, or photosensitive eruptions (4,5,6,7). These were all cases of systemic allergic dermatitis. Because sulfanilamide and other sulfonamides are largely historical allergens, no full review of the subject has been attempted.

Cross-reactions, pseudo-cross-reactions and co-reactions

Patients sensitized to sulfanilamide may cross-react to other sulfonamides including sulfadiazine, sulfathiazole, sulfacetamide, sulfaguanidine, sulfadimidine, sulfamerazine and sulfanilic acid (5,6), and also to other *p*-amino compounds, including *p*-phenylenediamine, *p*-toluidine, *p*-nitroaniline, aniline, *p*-aminophenol, *p*-toluenediamine sulfate, *p*-nitro-*o*-aminophenol, and *p*-aminobenzoic acid diethylaminoethyl ester (Novocaine ®) (3).

LITERATURE

1 The data in the section 'General' may have been obtained from literature discussed in this chapter, but mostly also or exclusively from one or more of the following online sources: ChemIDPlus Advanced, PubChem, DrugBank, RxList, Drug Central, Drugs.com, and Wikipedia

2 De Groot AC. Monographs in contact allergy, volume 3: Topical Drugs. Boca Raton, Fl, USA: CRC Press Taylor and Francis Group, 2021 (ISBN 978-0-367-23693-9)

3 Rudzki E, Rebandel P. Primary sensitivity to sulphonamide and secondary sensitization to aromatic amines. Contact Dermatitis 1987;17:49

4   Park RG. Cutaneous hypersensitivity to sulphonamides. Br Med J 1943;2:69-72

5   Fischer B. Dermatitis following the application of sulfanilamide. MJ Australia 1944;2:449

6   Sulzberger MD, Kanof A, Baer RL, Löwenberg C. Sensitization by topical application of sulfonamide. J Allergy 1947;18:92-103

7   Sidi E, Dobkevitch-Morrill S. The injection and ingestion test in cross-sensitization to the para group. J Invest Dermatol 1951;16:299-310

Chapter 3.457 SULFASALAZINE

IDENTIFICATION

Description/definition   : Sulfasalazine is the synthetic aminosalicylate that conforms to the structural formula
                            shown below
Pharmacological classes : Antirheumatic agents; gastrointestinal agents; anti-infective agents
IUPAC name   : 2-Hydroxy-5-[[4-(pyridin-2-ylsulfamoyl)phenyl]diazenyl]benzoic acid
Other names   : Salazosulfapyridine; salicylazosulfapyridine
CAS registry number   : 599-79-1
EC number   : 209-974-3
Merck Index monograph   : 10343
Patch testing   : 10% pet.
Molecular formula   : $C_{18}H_{14}N_4O_5S$

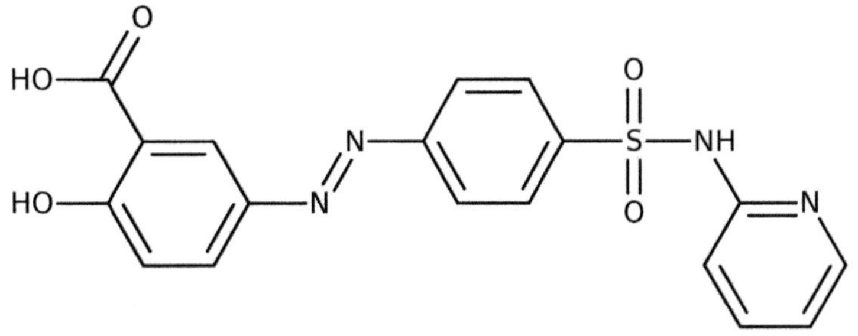

GENERAL

Sulfasalazine is a synthetic salicylic acid-derived drug that is used in the management of inflammatory bowel
diseases. Its activity is generally considered to lie in the anti-inflammatory effects of its metabolic breakdown
product, 5-aminosalicylic acid (synonym: mesalazine, see Chapter 3.305 Mesalazine) released in the colon. Its
mechanism of action is not clear, but may involve inhibition of cyclooxygenase and prostaglandin production. Sulfasalazine is indicated for the treatment of Crohn's disease, colitis ulcerosa and – as a second-line agent when NSAIDs
are ineffective - for rheumatoid arthritis (1).

CUTANEOUS ADVERSE DRUG REACTIONS FROM SYSTEMIC ADMINISTRATION CAUSED BY TYPE IV
(DELAYED-TYPE) HYPERSENSITIVITY (as demonstrated by positive patch tests)

Cutaneous adverse drug reactions from systemic administration of sulfasalazine caused by type IV (delayed-type)
hypersensitivity have included fixed drug eruption (8) and photosensitivity (2). See also the section 'Drug reaction
with eosinophilia and systemic symptoms (DRESS)' below.

Fixed drug eruption

A 56-year-old man presented with dark brown patches on the lower lip and glans penis. Two months before, he had
started taking sulfasalazine for ulcerative colitis. Twelve hours after taking the first 500 mg, the patient developed
painful and itchy erythematous and erosive lesions on the lower lip, tongue, and glans penis, that healed with
residual pigmentation. Patch tests with commercial sulfasalazine 10% pet. and 2 other sulfa drugs on postlesional
(lower lip) and previously uninvolved skin was positive to sulfasalazine on the lower lip only. A Lymphocyte
stimulation test with sulfasalazine was positive with a stimulation index of 246% (8).

Drug reaction with eosinophilia and systemic symptoms (DRESS)

Sulfasalazine is a well-known and frequent cause of DRESS. Patch tests, however, are always negative (6,7). There is
no definite explanation for these of false-negative results, but – as has been proposed for the similar situation with
allopurinol - there may be several possible causes: (i) the final responsible agent is another drug metabolite that is
not formed in the skin during patch testing; (ii) there is no immune mechanism involved; (iii) concomitant factors
that are responsible in inducing transient oral drug intolerance, such as viral infection, are not present at the time of
testing; and (iv) wrong choice of vehicle (limited skin penetration), drug concentration, or exposure time (Chapter
3.17 Allopurinol).

It has been suggested that p-phenylenediamine (PPD) may be a marker of allergy to sulfasalazine (3). The active
anti-inflammatory metabolite of sulfasalazine is 5-aminosalicylic acid (synonym: mesalazine), which belongs to the

chemical group of *p*-amino compounds, like PPD (3). In one patient, after an episode of DRESS from sulfasalazine, erythematous, scaling patchy eczema on the trunk and limbs developed 3 months later. Although the patient had never dyed her hair before nor had been exposed to black henna tattoos (containing PPD), sensitization to PPD and related colors used in clothing was demonstrated (3). A similar case of sensitization to PPD and the related benzocaine in a patient with possible DRESS from sulfasalazine with a negative patch test to sulfasalazine had been reported before. These authors suggested that the mesalazine metabolite of sulfasalazine may be responsible for cases of DRESS (4). Other authors have described a possible relationship between allergy to mesalazine (the active metabolite of sulfasalazine) and PPD (5; see Chapter 3.305 Mesalazine).

Photosensitivity

A 30-year-old woman presented with an eczematous eruption in light-exposed areas, hepatomegaly and fever (38°C). Five weeks before, she had started treatment with a combination of sulfasalazine (salazosulfapyridine) and piroxicam. Lab investigations showed leukocytosis, eosinophilia, increase in C-reactive protein, elevated liver enzymes and proteinuria. The 2 drugs were withdrawn and the clinical signs regressed in 6 days. An increase in eosinophilia and liver dysfunction was observed until the tenth day, after which the trend reversed. Laboratory parameters were normal on the twentieth day. One month later, photopatch tests with UVA were positive to sulfanilamide 1% pet., but negative to sulfasalazine 1% pet. Nevertheless, as a metabolite of sulfasalazine (2-pyridylsulfamoyl radical) is structurally very similar to sulfanilamide, the patient was diagnosed with drug hypersensitivity syndrome with involvement of photoallergy to sulfasalazine (2).

LITERATURE

1   The data in the section 'General' may have been obtained from literature discussed in this chapter, but mostly also or exclusively from one or more of the following online sources: ChemIDPlus Advanced, PubChem, DrugBank, RxList, Drug Central, Drugs.com, and Wikipedia

2   Bouyssou-Gauthier ML, Bédane C, Boulinguez S, Bonnetblanc JM. Photosensitivity with sulfasalazopyridine hypersensitivity syndrome. Dermatology 1999;198:388-390

3   Kenani Z, Lahouel I, Belhadjali H, Soua Y, Aouam K, Youssef M, et al. Is *p*-phenylenediamine a marker of sulfasalazine allergy? Contact Dermatitis 2018;78:173-174

4   Audran MJ, Lepoittevin JP, Pajot C, Martin L. Are paraphenylenediamine and benzocaine relevant markers of sulfasalazine allergy? Dermatitis 2014;25:40-41

5   Charles J, Bourrain JL, Tessier A, Lepoittevin JP, Beani JC. Mesalazine and para-phenylenediamine allergy. Contact Dermatitis 2004;51:313-314

6   Chaabane A, Aouam K, Ben Fredj N, A Boughattas N. DRESS syndrome: 11 case reports and a literature review. Thérapie 2010; 65: 543-550

7   Barbaud A, Collet E, Milpied B, Assier H, Staumont D, Avenel-Audran M, et al. A multicentre study to determine the value and safety of drug patch tests for the three main classes of severe cutaneous adverse drug reactions. Br J Dermatol 2013;168:555-562

8   Kawada A, Kobayashi T, Noguchi H, Hiruma M, Ishibashi A, Marshall J. Fixed drug eruption induced by sulfasalazine. Contact Dermatitis 1996;34:155-156

Chapter 3.458 SULFATHIAZOLE

IDENTIFICATION

Description/definition   : Sulfathiazole is the aminobenzenesulfonamide that conforms to the structural formula
                             shown below
Pharmacological classes : Anti-infective agents
IUPAC name   : 4-Amino-N-(1,3-thiazol-2-yl)benzenesulfonamide
CAS registry number   : 72-14-0
EC number   : 200-771-5
Merck Index monograph : 10344
Patch testing   : 5% pet.
Molecular formula   : $C_9H_9N_3O_2S_2$

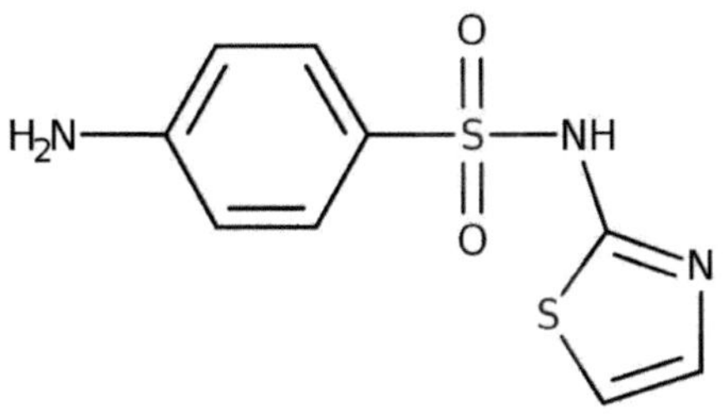

GENERAL

Sulfathiazole is a short-acting sulfonamide drug, which is effective against a wide range of gram-positive and gram-negative pathogenic microorganisms. It used to be a common oral and topical antimicrobial until less toxic alternatives were discovered. It is still occasionally used, sometimes in combination with sulfabenzamide and sulfacetamide for the treatment of vaginal infections. It is also used for disinfecting home aquariums and has a role in veterinary medicine (1). In pharmaceutical products, sulfathiazole is employed as sulfathiazole sodium (CAS number 144-74-1, EC number 205-638-5, molecular formula $C_9H_8N_3NaO_2S_2$) (1).

CONTACT ALLERGY FROM ACCIDENTAL CONTACT

Of 333 nurses patch tested in Poland between 1979 and 1987, 1 reacted to sulfathiazole 5% pet.; data on relevance were not provided (2).

CUTANEOUS ADVERSE DRUG REACTIONS FROM SYSTEMIC ADMINISTRATION CAUSED BY TYPE IV
(DELAYED-TYPE) HYPERSENSITIVITY (as demonstrated by positive patch tests)

Cutaneous adverse drug reactions from systemic administration of sulfathiazole caused by type IV (delayed-type) hypersensitivity have included systemic allergic dermatitis (3).

Systemic allergic dermatitis (systemic contact dermatitis)

Two patients who had developed allergic contact dermatitis from sulfathiazole ointment and had positive patch tests (not further specified) noticed exacerbations of previous eczema and in addition developed an erythematous macular eruption on the face, forearms, hands and thorax (3). These were cases of systemic allergic dermatitis. Similar cases were reported elsewhere (4 and articles cited in ref. 4) but often with negative patch tests or without patch testing being performed. As sulfathiazole is a historical allergen, this subject is not further reviewed here.

LITERATURE

1   The data in the section 'General' may have been obtained from literature discussed in this chapter, but mostly
    also or exclusively from one or more of the following online sources: ChemIDPlus Advanced, PubChem,
    DrugBank, RxList, Drug Central, Drugs.com, and Wikipedia
2   Rudzki E, Rebandel P, Grzywa Z. Patch tests with occupational contactants in nurses, doctors and dentists.
    Contact Dermatitis 1989;20:247-250
3   Darke RA. Sensitivity to topical application of sulfathiazole ointment. JAMA 1944;124:403-404
4   Shaffer B, Lentz JW, McGuire JA. Sulfathiazole eruptions. Sensitivity induced by local therapy and elicited by oral
    medication. Report of four cases with some allergic studies. JAMA 1943;123:17-23

Chapter 3.459 SULFONAMIDE, UNSPECIFIED

GENERAL

'Sulfonamide' is not a defined chemical. In chemistry, the sulfonamide functional group is -S(=O)2-NH2, a sulfonyl group connected to an amine group. 'Sulfonamides' are compounds that contain this group (Wikipedia). In chemical databases, it may be used as synonym for sulfanilamide, which is 4-aminobenzenesulfonamide (ChemIDPlus, PubChem). See also Chapter 3.456 Sulfanilamide. Sulfonamides may be tested 5% pet.

CUTANEOUS ADVERSE DRUG REACTIONS FROM SYSTEMIC ADMINISTRATION CAUSED BY TYPE IV (DELAYED-TYPE) HYPERSENSITIVITY (as demonstrated by positive patch tests)

Cutaneous adverse drug reactions from systemic administration of 'sulfonamide' (unspecified) caused by type IV (delayed-type) hypersensitivity have included Stevens-Johnson syndrome (1).

Stevens-Johnson syndrome/toxic epidermal necrolysis (SJS/TEN)

One patient with Stevens-Johnson syndrome had a positive patch test to sulfonamide (not further specified); no clinical data were provided (1).

LITERATURE

1   Wolkenstein P, Chosidow O, Fléchet ML, Robbiola O, Paul M, Dumé L, et al. Patch testing in severe cutaneous adverse drug reactions, including Stevens-Johnson syndrome and toxic epidermal necrolysis. Contact Dermatitis 1996;35:234-236

Chapter 3.460 TACROLIMUS

IDENTIFICATION

Description/definition   : Tacrolimus is the macrolide lactam that conforms to the structural formula shown below
Pharmacological classes : Immunosuppressive agents; calcineurin inhibitors
IUPAC name   : (1R,9S,12S,13R,14S,17R,18E,21S,23S,24R,25S,27R)-1,14-Dihydroxy-12-[(E)-1-[(1R,3R,4R)-
                               4-hydroxy-3-methoxycyclohexyl]prop-1-en-2-yl]-23,25-dimethoxy-13,19,21,27-
                               tetramethyl-17-prop-2-enyl-11,28-dioxa-4-azatricyclo[22.3.1.0^{4,9}]octacos-18-ene-
                               2,3,10,16-tetrone;hydrate
Other names   : Tacrolimus hydrate; tacrolimus monohydrate; Tsukubaenolide hydrate
CAS registry number   : 109581-93-3
EC number   : Not available
Merck Index monograph : 10425
Patch testing   : 5% alcohol
Molecular formula   : $C_{44}H_{71}NO_{13}$

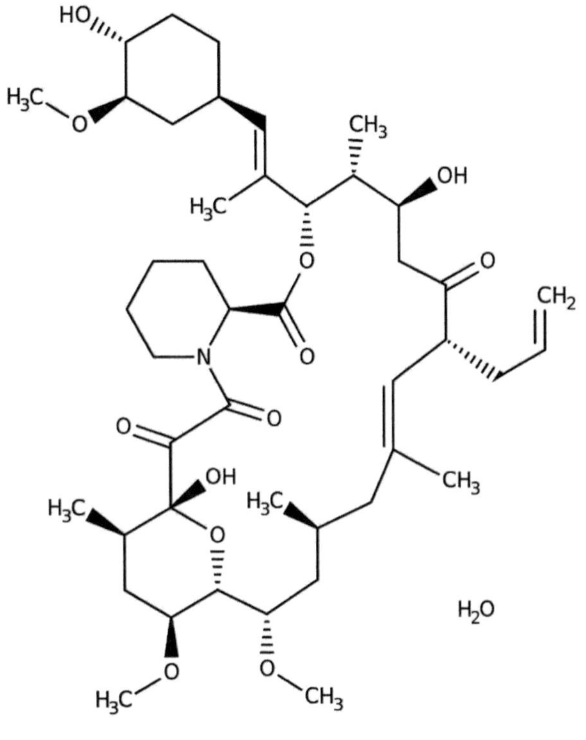

GENERAL

Tacrolimus is a macrolide immunosuppressive drug obtained from the fermentation broth of a Japanese soil sample that contained the bacterium *Streptomyces tsukubaensis*. Tacrolimus binds to the FKBP-12 protein and forms a complex with calcium-dependent proteins, thereby inhibiting calcineurin phosphatase activity and resulting in decreased cytokine production. This agent exhibits potent immunosuppressive activity *in vivo* and prevents the activation of T lymphocytes in response to antigenic or mitogenic stimulation. Tacrolimus is used orally after allogenic organ transplantation for immunosuppression to reduce the risk of organ rejection. It is also widely utilized topically for the treatment of atopic dermatitis, severe refractory uveitis after bone marrow transplantation, and vitiligo (1,5). As such, it has caused a few cases of allergic contact dermatitis (2,3,4).

CUTANEOUS ADVERSE DRUG REACTIONS FROM SYSTEMIC ADMINISTRATION CAUSED BY TYPE IV
(DELAYED-TYPE) HYPERSENSITIVITY (as demonstrated by positive patch tests)

Cutaneous adverse drug reactions from systemic administration of tacrolimus caused by type IV (delayed-type) hypersensitivity have included symmetric drug-related intertriginous and flexural exanthema (SDRIFE) (6).

Symmetrical drug-related intertriginous and flexural exanthema (SDRIFE)/Baboon syndrome
After having used oral tacrolimus for a liver and kidney transplantation for over 2 years, a 41-year-old man developed a chronic erythematous papular itchy eruption, symmetrically affecting his arms and legs, including the flexural folds, axillae, groins, gluteal region, and neck. There were no systemic symptoms. A skin biopsy showed epidermal spongiosis and a superficial perivascular and dermo-epidermal infiltrate composed of lymphocytes, histiocytes, and some eosinophils. Patch tests were positive to commercial tacrolimus ointment 0.03% and 0.1% (tacrolimus, yellow and liquid petrolatum, propylene carbonate, white beeswax, white paraffin) 'as is' (+ at D4). Ten controls were negative. Symmetrical drug-related intertriginous and flexural exanthema (SDRIFE) from tacrolimus was diagnosed. The drug was continued and the drug reactions was tolerated with local and systemic corticosteroids (6).

Cross-reactions, pseudo-cross-reactions and co-reactions
Patient sensitized to tacrolimus may cross-react to pimecrolimus (2,3).

LITERATURE
1 The data in the section 'General' may have been obtained from literature discussed in this chapter, but mostly also or exclusively from one or more of the following online sources: ChemIDPlus Advanced, PubChem, DrugBank, RxList, Drug Central, Drugs.com, and Wikipedia
2 Shaw DW, Maibach HI, Eichenfield LF. Allergic contact dermatitis from pimecrolimus in a patient with tacrolimus allergy. J Am Acad Dermatol 2007;56:342-345
3 Schmutz JL, Barbaud A, Tréchot P. Contact allergy with tacrolimus then pimecrolimus. Ann Dermatol Venerol 2008; 135:89 (Article in French)
4 Shaw DW, Eichenfield LF, Shainhouse T, Maibach HI. Allergic contact dermatitis from tacrolimus. J Am Acad Dermatol 2004;50:962-965
5 Belsito D, Wilson DC, Warshaw E, Fowler J, Ehrlich A, Anderson B, et al. A prospective randomized clinical trial of 0.1% tacrolimus ointment in a model of chronic allergic contact dermatitis. J Am Acad Dermatol 2006;55:40-46
6 Scherrer M, Araujo MG, Farah K. Tacrolimus-induced symmetric drug-related intertriginous and flexural exanthema (SDRIFE). Contact Dermatitis 2018;78:414-416

Chapter 3.461 TALAMPICILLIN

IDENTIFICATION

Description/definition : Talampicillin is the phthalidyl ester of ampicillin, which conforms to the structural formula shown below
Pharmacological classes : Anti-bacterial agents
IUPAC name : (3-Oxo-1*H*-2-benzofuran-1-yl) (2*S*,5*R*,6*R*)-6-[[(2*R*)-2-amino-2-phenylacetyl]amino]-3,3-dimethyl-7-oxo-4-thia-1-azabicyclo[3.2.0]heptane-2-carboxylate
Other names : Ampicillin phthalidyl ester
CAS registry number : 47747-56-8
EC number : 256-332-3
Merck Index monograph : 10435
Patch testing : No data available; suggested: 5% and 10% pet.
Molecular formula : $C_{24}H_{23}N_3O_6S$

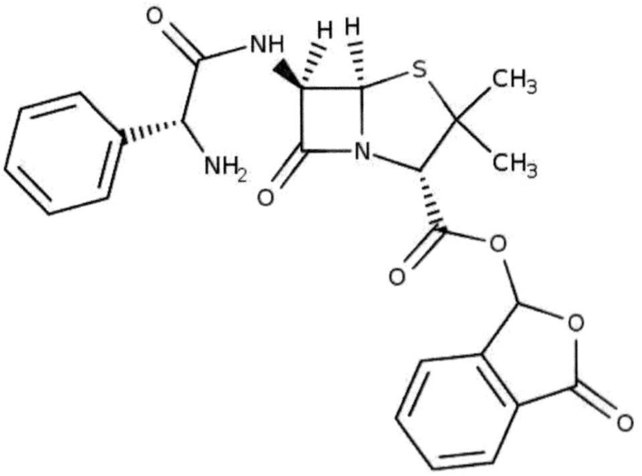

GENERAL

Talampicillin is the phthalidyl ester and microbiologically inactive prodrug of ampicillin. Following oral administration, during absorption from the gastrointestinal tract, talampicillin is hydrolyzed by esterases present in the intestinal wall to its active metabolite ampicillin. As talampicillin is well absorbed from the gastrointestinal tract, it results in a greater bioavailability of ampicillin than can be achieved with equivalent doses of ampicillin itself (go.drugbank.com). For indications see Chapter 3.38 Ampicillin. The classification and structures of beta-lactam antibiotics are discussed in Chapter 3.36 Amoxicillin.

CUTANEOUS ADVERSE DRUG REACTIONS FROM SYSTEMIC ADMINISTRATION CAUSED BY TYPE IV (DELAYED-TYPE) HYPERSENSITIVITY (as demonstrated by positive patch tests)

Cutaneous adverse drug reactions from systemic administration of talampicillin caused by type IV (delayed-type) hypersensitivity have included unspecified drug eruption (1).

Other cutaneous adverse drug reactions

Japanese researchers found 5 cases of allergic drug reactions to bacampicillin and one to talampicillin in the Japanese literature from 1984 to 1989 (1). The nature of the drug eruptions was not specified. All 6 patients had positive patch test reactions to bacampicillin and to ampicillin and all 5 tested with it also to talampicillin (no details provided). These reactions to talampicillin, bacampicillin, and ampicillin are pseudocross-reactions, as bacampicillin and talampicillin are prodrugs of ampicillin which are hydrolyzed in the digestive tract to the active drug ampicillin (1).

Cross-reactions, pseudo-cross-reactions and co-reactions

Cross-reactions between beta-lactam antibiotics are discussed in Chapter 3.36 Amoxicillin. Pseudocross-reactivity between bacampicillin, talampicillin (both prodrugs of ampicillin) and ampicillin (1).

LITERATURE

1   Imayama S, Fukuda H, Hori Y. Drug eruptions following treatment with prodrugs: a review of the reported cases in Japan from 1984 to 1989. J Dermatol 1991;18:277-280

Chapter 3.462 TALASTINE

IDENTIFICATION

Description/definition : Talastine is the phthalazinone that conforms to the structural formula shown below
Pharmacological classes : Histamine H_1 antagonists
IUPAC name : 4-Benzyl-2-[2-(dimethylamino)ethyl]phthalazin-1-one
CAS registry number : 16188-61-7
EC number : Not available
Merck Index monograph : 1260
Patch testing : No data available; most systemic drugs can be tested at 10% pet.; if the pure chemical is not available, prepare the test material from intravenous powder, the content of capsules or – when also not available – from powdered tablets to achieve a final concentration of the active drug of 10% pet.
Molecular formula : $C_{19}H_{21}N_3O$

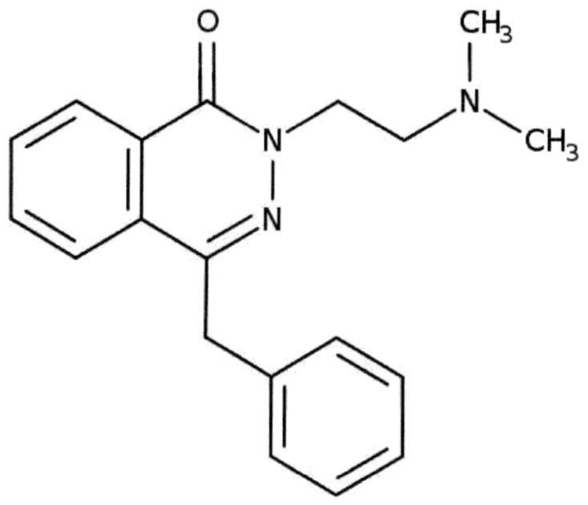

GENERAL

Talastine is an H_1 histamine antagonist that is or formerly was used as an antihistamine in some countries, including the former DDR (Deutsche Demokratische Republik, East-Germany) (1).

CUTANEOUS ADVERSE DRUG REACTIONS FROM SYSTEMIC ADMINISTRATION CAUSED BY TYPE IV (DELAYED-TYPE) HYPERSENSITIVITY (as demonstrated by positive patch tests)

Cutaneous adverse drug reactions from systemic administration of talastine caused by type IV (delayed-type) hypersensitivity have included widespread erythema with exfoliation (2) and macular exanthema (2).

Erythroderma, widespread erythematous eruption, exfoliative dermatitis

A 57-year-old woman, who had previously taken talastine without problems, developed transient tachycardia, followed 12 hours later by a generalized macular exanthema after ingesting one tablet of 40 mg talastine hydrochloride. Patch tests were positive to a talastine tablet and to a talastine ampoule (no concentration of the active material in the ampoule mentioned). Later, by mistake, the patient took another tablet and within 1 hour she had tachycardia and within 3 hours a generalized macular exanthema (2).

The same authors describe a 77-year-old woman, who was admitted to hospital for treatment of psoriasis and received talastine 60 mg/day for itching. The patient developed a partly nummular macular exanthema, partly widespread itchy erythema with exfoliation, which was first considered to be intolerance to the dithranol therapy for her psoriasis. However, when talastine was stopped, rapid improvement ensued. Patch tests were positive to the content of a talastine ampoule. An oral provocation test with one 40 mg tablet talastine reproduced the cutaneous adverse reaction within 12 hours (2).

LITERATURE

1   The data in the section 'General' may have been obtained from literature discussed in this chapter, but mostly also or exclusively from one or more of the following online sources: ChemIDPlus Advanced, PubChem, DrugBank, RxList, Drug Central, Drugs.com, and Wikipedia
2   Richter G, Kühn E. Talastin (Ahanon) als Ursache allergischer Arzneimittelexantheme [Talastine (Ahanon) as a cause of allergic drug exanthema]. Dermatol Monatsschr 1990;176:111-113

Chapter 3.463 TAMSULOSIN

IDENTIFICATION

Description/definition   : Tamsulosin is the sulfonamide derivative that conforms to the structural formula shown
                            below
Pharmacological classes   : Urological agents; α_1-adrenergic receptor antagonists
IUPAC name   : 5-[(2R)-2-[2-(2-Ethoxyphenoxy)ethylamino]propyl]-2-methoxybenzenesulfonamide
CAS registry number   : 106133-20-4
EC number   : 600-716-9
Merck Index monograph   : 10451
Patch testing   : Tablet, pulverized, 'as is' (2); most systemic drugs can be tested at 10% pet.; if the pure
                            chemical is not available, prepare the test material from intravenous powder, the content
                            of capsules or – when also not available – from powdered tablets to achieve a final
                            concentration of the active drug of 10% pet.
Molecular formula   : $C_{20}H_{28}N_2O_5S$

GENERAL

Tamsulosin is a sulfonamide derivative and selective α_{1a}- and α_{1b}-adrenoceptor antagonist that exerts its greatest effect in the prostate and bladder, where these receptors are most common. Antagonism of these receptors leads to relaxation of smooth muscle in the prostate and detrusor muscles in the bladder, allowing for better urinary flow. Tamsulosin is indicated for the treatment of signs and symptoms of benign prostatic hyperplasia. Off-label it may be employed for the treatment of ureteral stones, prostatitis, and female voiding dysfunction (1). In pharmaceutical products, tamsulosin is employed as tamsulosin hydrochloride (CAS number 106463-17-6, EC number not available, molecular formula $C_{20}H_{29}ClN_2O_5S$) (1).

CUTANEOUS ADVERSE DRUG REACTIONS FROM SYSTEMIC ADMINISTRATION CAUSED BY TYPE IV
(DELAYED-TYPE) HYPERSENSITIVITY (as demonstrated by positive patch tests)

Cutaneous adverse drug reactions from systemic administration of tamsulosin caused by type IV (delayed-type) hypersensitivity have included eczematous eruption (2).

Dermatitis/eczematous eruption

After having used tamsulosin for benign prostate hyperplasia for 2 weeks, a 46-year-old man presented with widespread erythematous papules and plaques. There were also vesicles on both hands, and erythematous plaques with lichenification on one wrist and the neck. Eight weeks later, patch tests were positive (++) at D3 to tamsulosin 'as is' (probably powder from a tablet or capsule). In 20 controls, one patient showed some erythema. One week after the patch tests, a challenge test with tamsulosin was performed and the eczema returned within 3 days. The authors diagnosed an eczematous eruption from delayed-type allergy to systemic tamsulosin, which is correct (2). However, the 'systemic contact dermatitis' in the title is not, as this requires previous sensitization from topical application.

LITERATURE

1   The data in the section 'General' may have been obtained from literature discussed in this chapter, but mostly
    also or exclusively from one or more of the following online sources: ChemIDPlus Advanced, PubChem,
    DrugBank, RxList, Drug Central, Drugs.com, and Wikipedia
2   Lijnen RL, de Graaf L. Systemic contact dermatitis from tamsulosin. Contact Dermatitis 2003;49:50-51

Chapter 3.464 TEGAFUR

IDENTIFICATION

Description/definition : Tegafur is the halopyrimidine that conforms to the structural formula shown below
Pharmacological classes : Antimetabolites, antineoplastic
IUPAC name : 5-Fluoro-1-(oxolan-2-yl)pyrimidine-2,4-dione
CAS registry number : 17902-23-7
EC number : 241-846-2
Merck Index monograph : 10522
Patch testing : No data available; most systemic drugs can be tested at 10% pet.; if the pure chemical is not available, prepare the test material from intravenous powder, the content of capsules or – when also not available – from powdered tablets to achieve a final concentration of the active drug of 10% pet.
Molecular formula : $C_8H_9FN_2O_3$

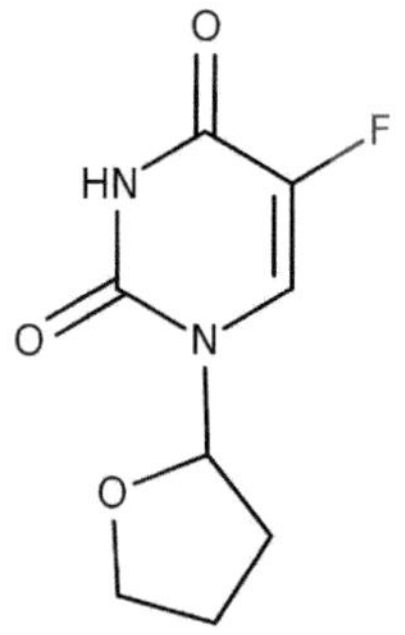

GENERAL

Tegafur is a pyrimidine analog and prodrug of fluorouracil (5-FU), an antineoplastic agent used in the treatment of various cancers such as advanced gastric, breast and colorectal cancers. It is usually given in combination with other drugs that enhance the bioavailability of the 5-FU by blocking the enzyme responsible for its degradation, or serves to limit the toxicity of 5-FU by ensuring high concentrations of 5-FU at a lower dose of tegafur. When converted and bioactivated to 5-FU, the drug mediates an anticancer activity by inhibiting thymidylate synthase during the pyrimidine pathway involved in DNA synthesis and inhibiting RNA and protein synthesis by competing with uridine triphosphate (1).

CUTANEOUS ADVERSE DRUG REACTIONS FROM SYSTEMIC ADMINISTRATION CAUSED BY TYPE IV (DELAYED-TYPE) HYPERSENSITIVITY (as demonstrated by positive patch tests)

Cutaneous adverse drug reactions from systemic administration of tegafur caused by type IV (delayed-type) hypersensitivity have included photoallergy (2,3).

Photosensitivity

A 57-year-old man developed an eczematous eruption on the light-exposed skin while taking tegafur. The MED for UVA was lowered and the photopatch test was positive for tegafur, with a cross-reaction to the related fluorouracil (2). A 71-year-old man developed an eczematous rash on the light-exposed skin while being treated with tegafur for a malignancy. The MED for both UVA and UVB were lowered and a photopatch test was positive to tegafur, but negative to the related fluorouracil (3).

Cross-reactions, pseudo-cross-reactions and co-reactions

A patient with photoallergy to tegafur may (2) or may not (3) photocross-react to the related fluorouracil (2).

LITERATURE

1 The data in the section 'General' may have been obtained from literature discussed in this chapter, but mostly also or exclusively from one or more of the following online sources: ChemIDPlus Advanced, PubChem, DrugBank, RxList, Drug Central, Drugs.com, and Wikipedia
2 Horio T, Yokoyama M. Tegafur photosensitivity–lichenoid and eczematous types. Photodermatol 1986;3:192-193
3 Sugimoto K, Shiniizu M. Photosensitivity reaction to futraful. Nishinihon J Dermatol 1980;42:79-53 (Article in Japanese, data cited in ref. 4)
4 Usuki A, Funasaka Y, Oka M, Ichihashi M. Tegafur-induced photosensitivity – evaluation of provocation by UVB irradiation. Int J Dermatol 1997;36:604-606

Chapter 3.465 TEICOPLANIN

IDENTIFICATION

Description/definition   : Teicoplanin is the glycopeptide antibiotic that conforms to the structural formula shown below

Pharmacological classes   : Anti-bacterial agents

IUPAC name   : (1S,2R,19R,22S,34S,37R,40R,52R)-2-[(2R,3R,4R,5S,6R)-3-Acetamido-4,5-dihydroxy-6-(hydroxymethyl)oxan-2-yl]oxy-22-amino-5,15-dichloro-64-[(2S,3R,4R,5S,6R)-3-(decanoylamino)-4,5-dihydroxy-6-(hydroxymethyl)oxan-2-yl]oxy-26,31,44,49-tetrahydroxy-21,35,38,54,56,59-hexaoxo-47-[(2R,3S,4S,5S,6R)-3,4,5-trihydroxy-6-(hydroxymethyl)oxan-2-yl]oxy-7,13,28-trioxa-20,36,39,53,55,58-hexazaundecacyclo-[38.14.2.2^{3,6}.2^{14,17}.2^{19,34}.1^{8,12}.1^{23,27}.1^{29,33}.1^{41,45}.0^{10,37}.0^{46,51}]hexahexaconta-3,5,8,10,12(64),14,16,23(61),24,26,29(60),30,32,41(57),42,44,46(51),47,49,62,65-henicosaene-52-carboxylic acid

Other names   : Teichomycin

CAS registry number   : 61036-62-2

EC number   : Not available

Merck Index monograph   : 10525

Patch testing   : 4% water (2); this concentration is probably too low; higher concentrations should be attempted (perform adequate controls)

Molecular formula   : $C_{88}H_{97}Cl_2N_9O_{33}$ (PubChem); $C_{7289}H_{6899}Cl_2N_{89}O_{2833}$ (ChemIDPlus)

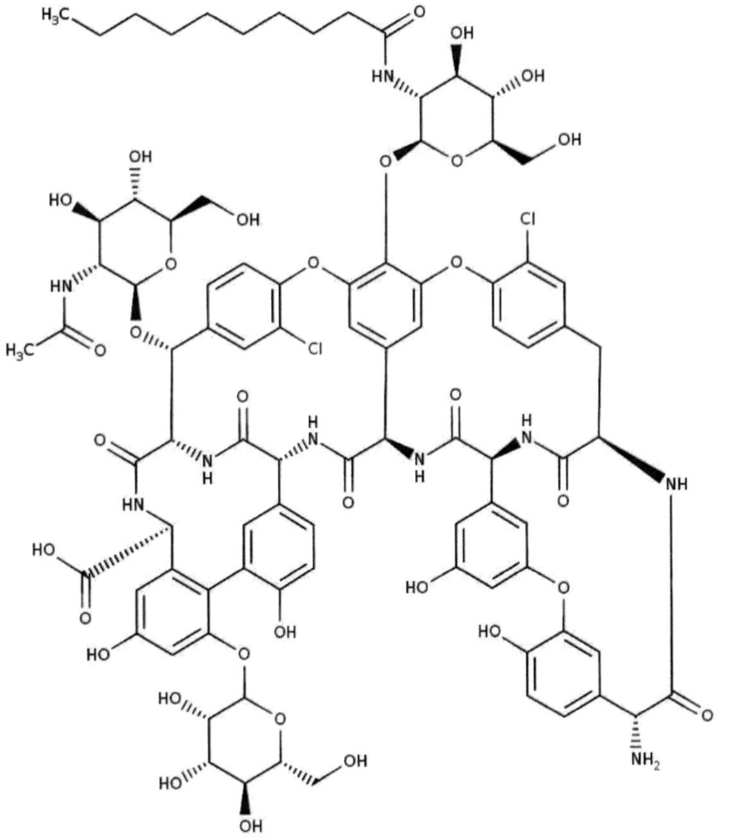

GENERAL

Teicoplanin is a lipoglycopeptide antibiotic from *Actinoplanes teichomyceticus* active against gram-positive bacteria. It consists of a mixture of several compounds, five major (named teicoplanin A2-1 through A2-5) and four minor (named teicoplanin RS-1 through RS-4). All teicoplanins share a same glycopeptide core, teicoplanin A3-1, but differ in the length and conformation of side chains attached to their β-D-glucosamine moiety. Teicoplanin inhibits peptidoglycan polymerization, resulting in inhibition of bacterial cell wall synthesis and cell death. The antibiotic has a similar spectrum of activity to vancomycin. Oral teicoplanin has been demonstrated to be effective in the treatment of pseudomembranous colitis and *Clostridium difficile*-associated diarrhea (1).

CUTANEOUS ADVERSE DRUG REACTIONS FROM SYSTEMIC ADMINISTRATION CAUSED BY TYPE IV (DELAYED-TYPE) HYPERSENSITIVITY (as demonstrated by positive patch tests)

Cutaneous adverse drug reactions from systemic administration of teicoplanin caused by type IV (delayed-type) hypersensitivity have included drug reaction with eosinophilia and systemic symptoms (DRESS) (2).

Drug reaction with eosinophilia and systemic symptoms (DRESS)

A 50-year-old man was treated for vertebral osteomyelitis and epidural abscess with vancomycin and developed DRESS with maculopapular exanthem, fever, eosinophilia, interstitial pneumonitis, and interstitial nephritis. A recurrence occurred when he was later treated with teicoplanin. Patch testing with vancomycin 15% water and teicoplanin 4% water 2 months later was ++ to vancomycin and ?+ to teicoplanin at D4 (2).

Cross-reactions, pseudo-cross-reactions and co-reactions

Patients sensitized to vancomycin may cross-react to teicoplanin (2,3).

LITERATURE

1   The data in the section 'General' may have been obtained from literature discussed in this chapter, but mostly also or exclusively from one or more of the following online sources: ChemIDPlus Advanced, PubChem, DrugBank, RxList, Drug Central, Drugs.com, and Wikipedia
2   Kwon HS, Chang YS, Jeong YY, Lee SM, Song WJ, et al. A case of hypersensitivity syndrome to both vancomycin and teicoplanin. J Korean Med Sci 2006;21:1108-1110
3   Bernedo N, Gonzalez I, Gastaminza G, Audicana M, Fernández E, Muñoz D. Positive patch test in vancomycin allergy. Contact Dermatitis 2001;45:43

Chapter 3.466 TENOFOVIR

IDENTIFICATION

Description/definition : Tenovorir is the 6-aminopurine compound that conforms to the structural formula shown below
Pharmacological classes : Anti-HIV agents; reverse transcriptase inhibitors; antiviral agents
IUPAC name   : [(2R)-1-(6-Aminopurin-9-yl)propan-2-yl]oxymethylphosphonic acid
CAS registry number   : 147127-20-6
EC number   : 604-571-2
Merck Index monograph   : 10559
Patch testing   : Tablet, pulverized, 10% pet. (2); most systemic drugs can be tested at 10% pet.; if the pure chemical is not available, prepare the test material from intravenous powder, the content of capsules or – when also not available – from powdered tablets to achieve a final concentration of the active drug of 10% pet.
Molecular formula   : $C_9H_{14}N_5O_4P$

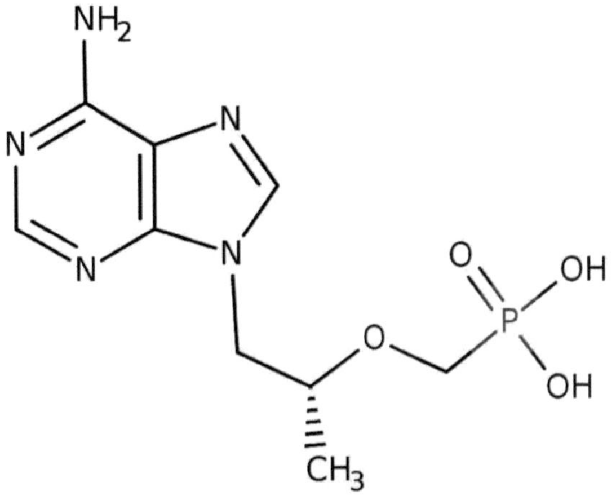

GENERAL

Tenofovir is an acyclic nucleotide analog of adenosine, which is effective against HIV, herpes simplex virus-2, and hepatitis B virus. Once activated, tenofovir acts with different mechanisms including the inhibition of viral polymerase causing chain termination and the inhibition of viral synthesis. All these activities are attained by its competition with deoxyadenosine 5'-triphosphate in the generation of new viral DNA. Tenofovir is used in combination with other agents in the therapy of the human immunodeficiency virus (HIV) and as single agent in hepatitis B virus (HBV) infection (1).

In pharmaceutical products, tenofovir may be employed as tenofovir disoproxil (fumarate [CAS number 202138-50-9], maleate [CAS number 1276030-80-8], phosphate [CAS number 1453166-76-1] or succinate [CAS number 1637632-97-3]), or as tenofovir alafenamide fumarate (CAS number 1392275-56-7).

CUTANEOUS ADVERSE DRUG REACTIONS FROM SYSTEMIC ADMINISTRATION CAUSED BY TYPE IV (DELAYED-TYPE) HYPERSENSITIVITY (as demonstrated by positive patch tests)

Cutaneous adverse drug reactions from systemic administration of tenofovir caused by type IV (delayed-type) hypersensitivity have included photosensitivity (2), and unspecified exanthema with palpebral edema (3).

Photosensitivity

Three days after starting therapy with tenofovir disoproxil fumarate for HIV, a 50-year-old man developed complaints of a burning sensation and redness on the face and neck, spreading over the next 5 days. There was no history of any other new drug being taken for the last 6 months. Being a farmer, the patient was regularly exposed to increased sunlight. Dermatological examination revealed diffuse erythema and ill-defined hyperpigmented plaques on the face with sparing of creases of forehead, nasolabial fold, and posterior auricular areas. There was specific involvement of V and nape of the neck. Similar lesions were present on the light-exposed parts of the trunk and arms. Extensive erythema and scaling were also present on buttocks, thighs, and upper legs. Histopathology of a biopsy revealed spongiosis, mild acanthosis, and perivascular lymphocytic infiltrate with few eosinophils. Six months after regression of the eruption, a patch test to tenofovir disoproxil fumarate was negative, but a photopatch test with the commercial drug 10% pet. irradiated with 10 J/cm^2 of UVA was positive. Probably no control testing has been performed. The patient was diagnosed with photoallergic drug reaction to tenofovir disoproxil fumarate (2).

Other cutaneous adverse drug reactions
A 47-year-old with diagnosis of HIV infection started therapy with tenofovir, emtricitabine and nevirapine. On the second day of treatment, she developed pruritic exanthema and palpebral edema that improved two weeks after the treatment had been discontinued. Patch tests were positive to emtricitabine 1%, 10% and 30% pet. and tenofovir 10% and 30% pet. Seven controls were negative (3).

LITERATURE

1   The data in the section 'General' may have been obtained from literature discussed in this chapter, but mostly also or exclusively from one or more of the following online sources: ChemIDPlus Advanced, PubChem, DrugBank, RxList, Drug Central, Drugs.com, and Wikipedia
2   Verma R, Vasudevan B, Shankar S, Pragasam V, Suwal B, Venugopal R. First reported case of tenofovir-induced photoallergic reaction. Indian J Pharm 2012;44:651-653
3   Sousa MJ, Cadinha S, Mota M, Teixeira T, Malheiro D, Moreira da Silva JP. Hypersensitivity to antiretroviral drugs. Eur Ann Allergy Clin Immunol 2018;50:277-280

Chapter 3.467 TENOXICAM

IDENTIFICATION

Description/definition   : Tenoxicam is the thienothiazine-derived monocarboxylic acid amide that conforms to the
                             structural formula shown below
Pharmacological classes : Anti-inflammatory agents, non-steroidal; cyclooxygenase inhibitors
IUPAC name   : 4-Hydroxy-2-methyl-1,1-dioxo-*N*-pyridin-2-ylthieno[2,3-e]thiazine-3-carboxamide
CAS registry number   : 59804-37-4
EC number   : Not available
Merck Index monograph   : 10561
Patch testing   : 5% and 10% pet.
Molecular formula   : $C_{13}H_{11}N_3O_4S_2$

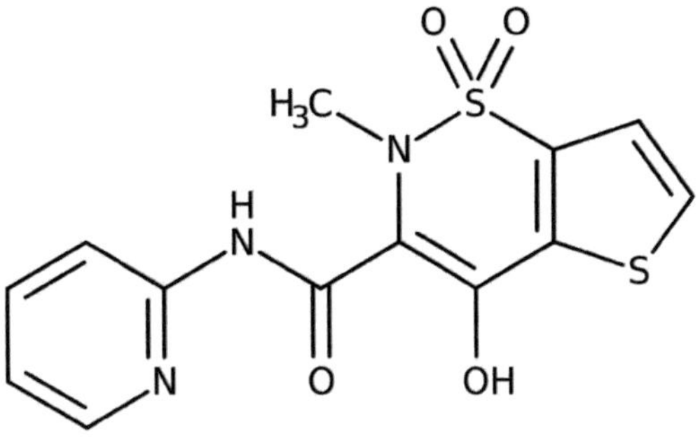

GENERAL

Tenoxicam is a thienothiazine-derived nonsteroidal anti-inflammatory drug (NSAID) with anti-inflammatory, analgesic and antipyretic properties. It is indicated for the treatment of rheumatoid arthritis, osteoarthritis, backache, and pain of other origins (1).

CUTANEOUS ADVERSE DRUG REACTIONS FROM SYSTEMIC ADMINISTRATION CAUSED BY TYPE IV (DELAYED-TYPE) HYPERSENSITIVITY (as demonstrated by positive patch tests)

Cutaneous adverse drug reactions from systemic administration of tenoxicam caused by type IV (delayed-type) hypersensitivity have included fixed drug eruption (4,5) and drug reaction with eosinophilia and systemic symptoms (DRESS) (2).

Fixed drug eruption

In France, in the period 2005-2007, 59 cases of fixed drug eruptions were collected in 17 academic centers. There was one case of FDE to tenoxicam with a positive patch test. Clinical details were not provided (4). A patient with fixed drug eruption caused by piroxicam with a positive patch test to this NSAID suffered a reactivation of the FDE lesions after intravenous administration of tenoxicam (5).

Drug reaction with eosinophilia and systemic symptoms (DRESS)

Between January 1998 and December 2008, in a university hospital in Portugal, 56 patients with DRESS were investigated with patch testing of the suspected drugs. Tenoxicam (concentration and vehicle not mentioned) was tested in one patient and the patch test reaction was positive. There were no positive reactions to piroxicam, meloxicam, or other NSAIDs (2).

Cross-reactions, pseudo-cross-reactions and co-reactions

Of 9 patients with a fixed drug eruption from piroxicam and a positive patch test to this NSAID on lesional skin, 8 (89%) cross-reacted to tenoxicam tested at 5% and 10% pet. (3). Frequent cross-reactivity to tenoxicam in patients sensitized to piroxicam (5).

LITERATURE

1   The data in the section 'General' may have been obtained from literature discussed in this chapter, but mostly also or exclusively from one or more of the following online sources: ChemIDPlus Advanced, PubChem, DrugBank, RxList, Drug Central, Drugs.com, and Wikipedia

2   Santiago F, Gonçalo M, Vieira R, Coelho S, Figueiredo A. Epicutaneous patch testing in drug hypersensitivity syndrome (DRESS). Contact Dermatitis 2010;62:47-53

3   Andrade P, Brinca A, Gonçalo M. Patch testing in fixed drug eruptions – a 20-year review. Contact Dermatitis 2011;65:195-201

4   Brahimi N, Routier E, Raison-Peyron N, Tronquoy AF, Pouget-Jasson C, Amarger S, et al. A three-year-analysis of fixed drug eruptions in hospital settings in France. Eur J Dermatol 2010;20:461-464

5   Gonçalo M, Oliveira HS, Fernandes B, Robalo-Cordeiro M, Figueiredo A. Topical provocation in fixed drug eruption from nonsteroidal anti-inflammatory drugs. Exogenous Dermatol 2002;1:81-86

Chapter 3.468 TERBINAFINE

IDENTIFICATION

Description/definition : Terbinafine is the synthetic allylamine derivative that conforms to the structural formula shown below

Pharmacological classes : Enzyme inhibitors; antifungal agents

IUPAC name : (*E*)-*N*,6,6-Trimethyl-*N*-(naphthalen-1-ylmethyl)hept-2-en-4-yn-1-amine

CAS registry number : 91161-71-6

EC number : Not available

Merck Index monograph : 10569

Patch testing : Tablet, pulverized, 30% pet. (2); most systemic drugs can be tested at 10% pet.; if the pure chemical is not available, prepare the test material from intravenous powder, the content of capsules or – when also not available – from powdered tablets to achieve a final concentration of the active drug of 10% pet.

Molecular formula : $C_{21}H_{25}N$

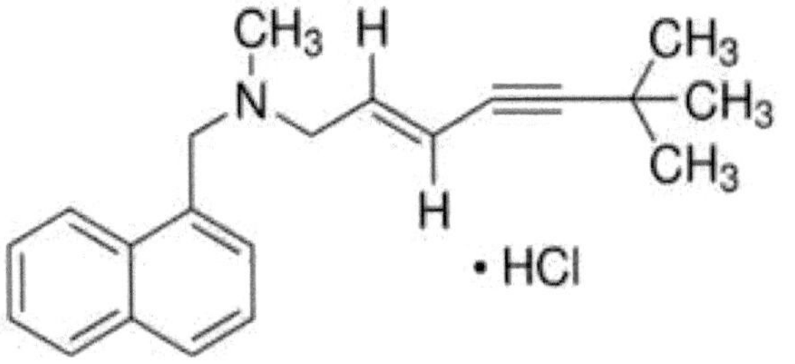

Terbinafine hydrochloride

GENERAL

Terbinafine is a synthetic allylamine antifungal. It is highly lipophilic and tends to accumulate in skin, nails, and fatty tissues. Like other allylamines, terbinafine inhibits ergosterol synthesis by inhibiting the fungal squalene monooxygenase (squalene 2,3-epoxidase), an enzyme that is part of the fungal cell wall synthesis pathway. As a result, terbinafine disrupts fungal cell membrane synthesis and inhibits fungal growth. Oral terbinafine is indicated for the treatment of dermatophyte infections of the toenail or fingernail (onychomycosis), of the head (tinea capitis) and of the body (tinea corporis). Limited dermatomycoses including tinea pedis and tinea cruris can be treated with a topical terbinafine preparation (1). In pharmaceutical products, terbinafine is employed as terbinafine hydrochloride (CAS number 78628-80-5, EC number 616-640-4, molecular formula $C_{21}H_{26}ClN$) (1).

CUTANEOUS ADVERSE DRUG REACTIONS FROM SYSTEMIC ADMINISTRATION CAUSED BY TYPE IV (DELAYED-TYPE) HYPERSENSITIVITY (as demonstrated by positive patch tests)

Cutaneous adverse drug reactions from systemic administration of terbinafine caused by type IV (delayed-type) hypersensitivity have included maculopapular eruption (5), acute generalized exanthematous pustulosis (2,3), systemic allergic dermatitis in the form of the baboon syndrome/SDRIFE (6), (possibly) Stevens-Johnson syndrome (4), and photosensitivity (7).

Maculopapular eruption

Four weeks after having initiated treatment with terbinafine for onychomycosis, a 76-year-old man developed a generalised maculopapular exanthema. After healing, patch test revealed a positive reaction to terbinafine at D2 with multiple papules and erythema (++) (5). It should be mentioned that no details on the patch test concentration and vehicle were provided, that one reading is insufficient and that probably no controls have been performed.

Acute generalized exanthematous pustulosis (AGEP)

A 6-year-old boy, after having used terbinafine 250 mg per day for tinea capitis for 2 days, suddenly developed fever. The therapy was switched to amoxicillin and clavulanic acid because of suspected secondary bacterial infection. Three days later, the patient presented with a generalized rash with millimeter-sized pustules located mainly on the trunk and the axillary and inguinal folds. He had small lymphadenopathies at cervical, auricular, and axillary regions. Laboratory tests showed an elevated white blood cell count with neutrophilia, an elevated erythrocyte sedimentation rate, and an elevated C-reactive protein level. Cultures of pustules were negative. A skin biopsy revealed subcorneal pustules associated with neutrophilic spongiosis, few necrotic keratinocytes, and a papillary infiltrate including eosinophils. Nine weeks after healing, patch tests were positive to terbinafine (the commercial form used by the patient diluted to 30% petrolatum) with pustules, but negative to amoxicillin-clavulanic acid (2). This was a case of acute generalized exanthematous pustulosis (AGEP) from delayed-type allergy to terbinafine.

A 63-year-old woman had been treated with terbinafine 250 mg/day for 2 weeks for tinea corporis, when she developed a pustular eruption and fever (38.5°C). Examination revealed an erythematous eruption covered with superficial non-follicular pustules on the trunk, arms and legs, which rapidly became confluent. She had mild leukocytosis with 71% neutrophils, erythrocyte sedimentation rate was 8 mm. Cultures of pustules for bacteria and fungi were negative. Histopathology of a biopsy showed subcorneal pustules with neutrophils and a perivascular infiltrate in the superficial dermis composed of polymorphonuclear cells and lymphocytes. A patch test with terbinafine (material, concentration and vehicle not specified) was positive at D7 with erythema and pustules. The patient was diagnosed with acute generalized exanthematous pustulosis from delayed-type allergy to terbinafine (3).

Systemic allergic dermatitis (systemic contact dermatitis)

A 62-year-old man presented with large erosions, 5-10 cm in size, over the medial aspects of the thighs, groins, perianal area, and nape of the neck with swelling of the face and periorbital areas. The disease had started as erythematous, itchy papules and plaques in the groin, which were treated with oral terbinafine for suspected tinea cruris. This treatment led to initial improvement, but was later followed by worsening. The patient also regularly applied terbinafine cream in the groin. After complete healing, patch tests were positive at D2 and D4 to neomycin sulfate 20% pet., clotrimazole 1% cream, and terbinafine 1% cream in petrolatum base. The patient was diagnosed with systemic allergic dermatitis in the form of the baboon syndrome/symmetrical drug-related intertriginous and flexural exanthema (SDRIFE) from oral terbinafine after previous sensitization to terbinafine cream. Terbinafine itself, however, was not patch tested (6).

Stevens-Johnson syndrome/toxic epidermal necrolysis (SJS/TEN)

A 56-year-old woman development Stevens-Johnson syndrome 7 days after the initiation of oral Lamisil ® (active ingredient: terbinafine). Patch tests with the tablet were positive on two occasions and negative in 15 controls. However, patch tests with the active ingredient terbinafine at various concentrations and different vehicles and all excipients were negative, also on 2 occasions. A human basophil degranulation test and a lymphocyte activation test performed with Lamisil tablets were both positive, but negative to all ingredients separately. Different explanations were suggested for these discordant results: the cutaneous adverse reaction may have been induced by an unknown component of the tablet, or positive patch tests could be due to irritation or compound allergy (4).

Photosensitivity

A 60-year-old man was prescribed oral terbinafine for onychomycosis. The treatment started in a sunny month, with the patient spending daily 1-3 hours outdoors. On day 6 of the therapy, an itching skin rash emerged on his forehead and backs of the hands which gradually worsened. Suspecting a relationship with the newly started antimycotic, the patient discontinued terbinafine on day 8. However, erythema, edema and scaling continued to progress on his face and V of the chest; covered areas of the skin were not involved. Two months later, a photopatch test was positive to a crushed terbinafine tablet mixed with 0.5 ml saline irradiated with UVA 5 J/cm^2. In a second test session, there were positive photopatch tests to pure terbinafine tested at 1,2,5,10 and 25% in liquid paraffin, alcohol and water. Six controls were negative. The author as test materials advised 1% and 5% terbinafine in liquid paraffin (7).

LITERATURE

1   The data in the section 'General' may have been obtained from literature discussed in this chapter, but mostly also or exclusively from one or more of the following online sources: ChemIDPlus Advanced, PubChem, DrugBank, RxList, Drug Central, Drugs.com, and Wikipedia

2   Zaouak A, Ben Salem F, Charfi O, Hammami H, Fenniche S. Acute generalized exanthematous pustulosis induced by terbinafine in a child confirmed by patch testing. Int J Dermatol 2019;58:e42-e43

3   Kempinaire A, De Raeve L, Merckx M, De Coninck A, Bauwens M, Roseeuw D. Terbinafine-induced generalized exanthematous pustulosis confirmed by a positive patch-test result. J Am Acad Dermatol 1997;37:653-655

4   Barbaud A, Reichert-Penetrat S, Granel F, Kolopp-Sarda MN, Schmutz JL. Is terbinafine responsible for eliciting positive patch tests performed with Lamisil tablets? Contact Dermatitis 1999;41:101

5   Koch A, Tchernev G, Wollina U. Allergic maculo-papular exanthema due to terbinafine. Open Access Maced J Med Sci 2017;20:535-536

6   Bhari N, Sahni K, Dev T, Sharma VK. Symmetrical drug-related intertriginous and flexural erythema (Baboon syndrome) induced by simultaneous exposure to oral and topical terbinafine. Int J Dermatol 2017;56:e168-170

7   Spiewak R. Systemic photoallergy to terbinafine. Allergy 2010;65:1071-1072

Chapter 3.469 TERFENADINE

IDENTIFICATION

Description/definition   : Terfenadine is the diphenylmethane that conforms to the structural formula shown below
Pharmacological classes : Histamine H$_1$ antagonists, non-sedating
IUPAC name   : 1-(4-*tert*-Butylphenyl)-4-[4-[hydroxy(diphenyl)methyl]piperidin-1-yl]butan-1-ol
CAS registry number   : 50679-08-8
EC number   : 256-710-8
Merck Index monograph : 10576
Patch testing   : 1% pet. (2)
Molecular formula   : C$_{32}$H$_{41}$NO$_2$

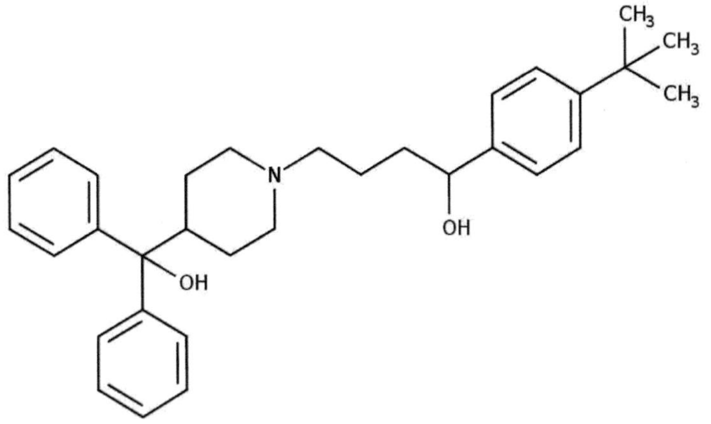

GENERAL

Terfenadine is a prodrug that is metabolized by intestinal CYP3A4 to the active form fexofenadine, a selective histamine H$_1$-receptor antagonist with antihistaminic and non-sedative effects. The drug is used for the treatment of allergic rhinitis, hay fever, and allergic skin disorders. In the U.S., terfenadine was superseded by fexofenadine in the 1990s due to the risk of cardiac arrhythmia caused by QT interval prolongation (1).

CUTANEOUS ADVERSE DRUG REACTIONS FROM SYSTEMIC ADMINISTRATION CAUSED BY TYPE IV (DELAYED-TYPE) HYPERSENSITIVITY (as demonstrated by positive patch tests)

Cutaneous adverse drug reactions from systemic administration of terfenadine caused by type IV (delayed-type) hypersensitivity have included unspecified drug exanthema (2).

Other cutaneous adverse drug reactions

One patient with an unspecified drug exanthema (no clinical details provided) had a positive patch test reaction to terfenadine 1% pet. (2).

LITERATURE

1   The data in the section 'General' may have been obtained from literature discussed in this chapter, but mostly also or exclusively from one or more of the following online sources: ChemIDPlus Advanced, PubChem, DrugBank, RxList, Drug Central, Drugs.com, and Wikipedia
2   Wolkenstein P, Chosidow O, Fléchet ML, Robbiola O, Paul M, Dumé L, et al. Patch testing in severe cutaneous adverse drug reactions, including Stevens-Johnson syndrome and toxic epidermal necrolysis. Contact Dermatitis 1996;35:234-236

Chapter 3.470 TETRACYCLINE

IDENTIFICATION

Description/definition : Tetracycline is the naphthacene antibiotic that conforms to the structural formula shown below

Pharmacological classes : Anti-bacterial agents; protein synthesis inhibitors

IUPAC name : (4S,4aS,5aS,6S,12aR)-4-(Dimethylamino)-1,6,10,11,12a-pentahydroxy-6-methyl-3,12-dioxo-4,4a,5,5a-tetrahydrotetracene-2-carboxamide

CAS registry number : 60-54-8

EC number : 200-481-9

Merck Index monograph : 10611

Patch testing : Hydrochloride, 2% pet. (SmartPracticeCanada, SmartPracticeEurope)

Molecular formula : $C_{22}H_{24}N_2O_8$

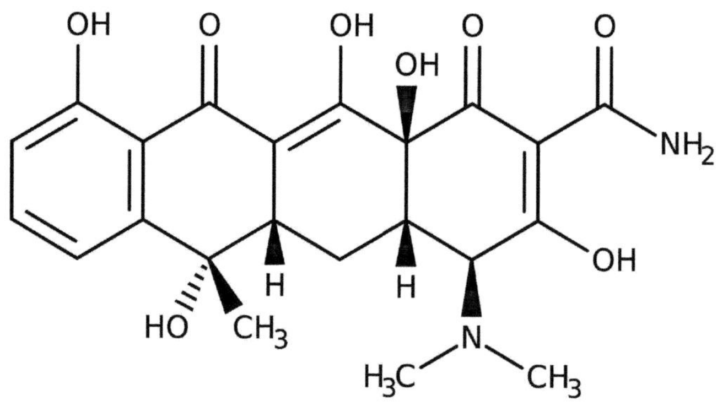

GENERAL

Tetracycline is a broad-spectrum naphthacene antibiotic produced semisynthetically from chlortetracycline, an antibiotic isolated from the bacterium *Streptomyces aureofaciens*. It exerts a bacteriostatic effect on bacteria by binding reversibly to the bacterial 30S ribosomal subunit and blocking incoming aminoacyl tRNA from binding to the ribosome acceptor site. It also binds to some extent to the bacterial 50S ribosomal subunit and may alter the cytoplasmic membrane causing intracellular components to leak from bacterial cells. Tetracycline is used to treat a wide variety of infections caused by susceptible bacteria and is also widely utilized in the treatment of acne vulgaris and acne conglobata. In pharmaceutical products, tetracycline is employed as tetracycline hydrochloride (CAS number 64-75-5, EC number 200-593-8, molecular formula $C_{22}H_{25}ClN_2O_8$) (1).

In topical preparations, tetracycline has rarely caused contact allergy/allergic contact dermatitis, which subject has been reviewed in Volume 3 of the *Monographs in contact allergy* series (5).

CONTACT ALLERGY FROM ACCIDENTAL CONTACT

In a group of 107 workers in the pharmaceutical industry with dermatitis, investigated in Warsaw, Poland, before 1989, 6 reacted to tetracycline, tested 10% pet. (2). Earlier in Warsaw, Poland, in the period 1979-1983, of 27 pharmaceutical workers diagnosed with occupational allergic contact dermatitis from antibiotics, seven had positive patch tests to tetracycline (ampoule content) (3). Again in Warsaw, Poland, of 333 nurses patch tested in between 1979 and 1987, one reacted to tetracycline 10% pet.; it was unknown whether this reactions was relevant (6). Occupational allergy to tetracycline was also described in German literature (4).

CUTANEOUS ADVERSE DRUG REACTIONS FROM SYSTEMIC ADMINISTRATION CAUSED BY TYPE IV (DELAYED-TYPE) HYPERSENSITIVITY (as demonstrated by positive patch tests)

Cutaneous adverse drug reactions from systemic administration of tetracycline caused by type IV (delayed-type) hypersensitivity have included fixed drug eruption (7).

Fixed drug eruption

A 29-year-old woman developed a clinically diagnosed recurrent fixed drug eruption following a year's course of both oral and parenteral administration of tetracycline for pharyngitis. A positive patch test to tetracycline powder under occlusion at the site of the fixed eruption confirmed the diagnosis (7).

LITERATURE
1   The data in the section 'General' may have been obtained from literature discussed in this chapter, but mostly
     also or exclusively from one or more of the following online sources: ChemIDPlus Advanced, PubChem,
     DrugBank, RxList, Drug Central, Drugs.com, and Wikipedia
2   Rudzki E, Rebandel P, Grzywa Z. Contact allergy in the pharmaceutical industry. Contact Dermatitis 1989;21:121-
     122
3   Rudzki E, Rebendel P. Contact sensitivity to antibiotics. Contact Dermatitis 1984;11:41-42
4   Schwarting HH. Occupational tetracycline allergy. Derm Beruf Umwelt 1983;31:130 (Article in German)
5   De Groot AC. Monographs in contact allergy, volume 3: Topical Drugs. Boca Raton, Fl, USA: CRC Press Taylor and
     Francis Group, 2021 (ISBN 978-0-367-23693-9)
6   Rudzki E, Rebandel P, Grzywa Z. Patch tests with occupational contactants in nurses, doctors and dentists.
     Contact Dermatitis 1989;20:247-250
7   Parish LC, Witkowski JA. Pulsating fixed drug eruption due to tetracycline. Acta Derm Venereol 1978;58:545-547

Chapter 3.471 TETRAZEPAM

IDENTIFICATION

Description/definition   : Tetrazepam is the benzodiazepine that conforms to the structural formula shown below
Pharmacological classes   : Muscle relaxants, central
IUPAC name   : 7-Chloro-5-(cyclohexen-1-yl)-1-methyl-3H-1,4-benzodiazepin-2-one
CAS registry number   : 10379-14-3
EC number   : 233-837-7
Merck Index monograph   : 10661
Patch testing   : 10% pet.; test on previously severely affected skin in case of strong suspicion of
                            tetrazepam allergy but negative patch test on the back (17)
Molecular formula   : $C_{16}H_{17}ClN_2O$

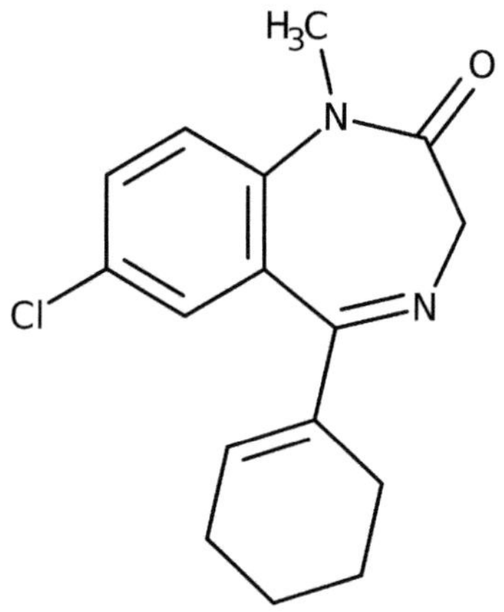

GENERAL

Tetrazepam is a benzodiazepine derivative with anticonvulsant, anxiolytic, hypnotic and muscle relaxant properties. It is used s to treat painful contractures (such as in low back pain and neck pain) and spasticity, anxiety disorders such as panic attacks, or more rarely to treat depression, premenstrual syndrome or agoraphobia. Tetrazepam has relatively little sedative effect at low doses while still producing useful muscle relaxation and anxiety relief (1).

The low but increased risk of serious adverse cutaneous reactions reported (Stevens-Johnson syndrome, toxic epidermal necrolysis, drug reaction with eosinophilia and systemic symptoms (DRESS)) in comparison with other benzodiazepines led the European Commission to suspend marketing authorizations of tetrazepam-containing medicines across the European Union in 2013 (13). The relevant literature was thereafter reviewed and it was concluded that there was insufficient evidence of adverse drug reactions to tetrazepam to substantiate the validity of the withdrawal (30).

CONTACT ALLERGY FROM ACCIDENTAL CONTACT

Occupational allergic contact dermatitis

General
Tetrazepam is (one of) the most frequent cause(s) of occupational sensitization to drugs in health care personnel, notably in nurses. They are exposed to tetrazepam dust at work from crushing pills, which leads to occupational allergic contact dermatitis both from airborne exposure (face) and direct contact (hands).

Case series
The results of several studies on occupational contact allergy to tetrazepam performed in Leuven, Belgium, have been reported (2,3,35,36). There is undoubtedly overlap in these studies, i.e. that a number of patients have been reported more than once (2,3,35,36).

In Leuven, Belgium, in the period 2001-2019, 201 of 1248 health care workers/employees of the pharmaceutical industry had occupational allergic contact dermatitis. The lesions were mostly localized on the hands, but often also on the face, as airborne dermatitis. In total, 42 positive patch test reactions to 18 different systemic drugs were found. In 11 patients, tetrazepam was the drug/one of the drugs that caused occupational dermatitis (35).

In the same clinic in Leuven, Belgium, from 2003 to 2009, 10 nurses were found to have occupational allergic dermatitis to tetrazepam from crushing tablets for elderly or disabled people in their care. Patch tests were positive to tetrazepam pills crushed and diluted 30% in pet. Nine were geriatric nurses (or aids) and one a nurse who worked in a clinic for multiple sclerosis patients. Five suffered from airborne facial dermatitis, two from hand dermatitis, and three from both airborne and hand dermatitis. All had noticed a clear relationship with their work, with improvement of the lesions during holidays (36).

Again in Leuven, Belgium, in the period 2007-2011, 81 patients were diagnosed with occupational airborne allergic contact dermatitis. In 23 of them, drugs were the offending agents, including tetrazepam in 16 cases (2). Also in this university clinic, 4 patients with (airborne) allergic contact dermatitis to tetrazepam have been observed, of which 3 were occupational cases. Patient 1 was a 30-year-old woman with an itchy, scaly skin reaction on the face, which was most pronounced on the eyelids. She reported a clear relationship with her work as a geriatric nurse, during which she was required to crush tablets for the elderly on a daily basis, mostly benzodiazepine drugs. Patient 2 is described below in the section 'Allergic contact dermatitis *by proxy*'. Patient 3 was a 44-year-old nurse who had eczema on the hands and face, which had started 3 months after she was required to crush tablets daily during her work as a nurse in a clinic for disabled people. Patient 4, a 24-year-old woman, presented with eyelid dermatitis, which had started with localized edema 4 months previously. Later, the area had become more itchy, with redness and scaling. The patient suspected a relationship with her work as a pharmacy assistant, which involved breaking and crushing different types of tablets. All 4 reacted to tetrazepam crushed tablet 30% pet.; 31 controls were negative (3).

In a study performed by the Information Network of Departments of Dermatology (IVDK) from 2005 to 2014, 39 nurses suspected of occupational allergic contact dermatitis were patch tested with tetrazepam and 11 had positive reactions. It was the most frequent cause of drug-related sensitization (37). In an identical study from the same group performed between 2003 and 2012 (therefore considerably overlapping the data in ref. 37), 36 nurses were patch tested with tetrazepam pure and 3 had positive and 2 doubtful positive reactions (38). The IVDK also investigated sixteen healthcare workers (five veterinarians and 11 nurses) for airborne contact dermatitis. Positive patch tests were obtained in six individuals, of who 4 reacted to tetrazepam (5, probable overlap with refs. 37 and 38).

In a hospital in Osnabrück, Germany, 10 cases of occupational airborne allergic contact dermatitis caused by tetrazepam were observed between 2006 and 2011. The affected individuals were employed as (geriatric) nursing staff, exposed to tetrazepam on a regular occupational basis when crushing tablets for their patients. The nurses had a long-standing history of eczematous lesions since 2-10 years. The skin changes affected predominantly the periorbital and facial area but in some instances also the V of the neck, the neck, forearms and hands. All patients had strongly positive patch tests to tetrazepam crushed pills 30% pet. and/or water. Six patients experienced skin flares when patch tested (39; partly also published as Abstract [40]). One of these patients had previously been described in a case report (41).

Case reports
Single cases of (airborne) occupational allergic contact dermatitis from tetrazepam have been reported in nurses (27,43) and in a technician working in a pharmaceutical factory from repairing machines manufacturing tetrazepam (42). Two employees working in a pharmaceutical factory producing tetrazepam, who became sensitized to this drug with eczema of the face (both patients) and the hands (one patient), were reported from France (44).

Allergic contact dermatitis in non-professional caregivers (allergic contact dermatitis *by proxy*)
A 66-year-old woman presented with itchy and burning eczema of the face, which had first appeared on the eyelids. Later, extension to the forehead, lips and perioral region, neck, ears and the fingers occurred. The patient's husband suffered from Parkinson's disease, and she had to crush a large number of tablets for him up to five times a day. The patient had positive patch tests to tetrazepam crushed tablet 30% pet. and various other drugs that she had contact with (3).

A 37-year-old woman developed eczema on the forefinger of the right hand, which later spread to the dorsum of the right hand, eyelids, neck and cheeks. She cared for a daughter with cerebral paralysis, and had to give her medication after triturating tablets by hand and gathering the powder up into a small spoon with her forefinger. Patch testing showed positive reactions to 4 commercial drugs the daughter used tested at 1% water. Later, patch tests with their active ingredients were performed and the patient had positive reactions to tetrazepam, diazepam, clorazepate and sodium valproate, all tested 1% in water (12).

CUTANEOUS ADVERSE DRUG REACTIONS FROM SYSTEMIC ADMINISTRATION CAUSED BY TYPE IV (DELAYED-TYPE) HYPERSENSITIVITY (as demonstrated by positive patch tests)
Cutaneous adverse drug reactions from systemic administration of tetrazepam caused by type IV (delayed-type) hypersensitivity have included maculopapular eruption (10,11,15,17,18,22,25,26,28,31,33), macular/generalized erythematous exanthema (6,24,26), acute generalized exanthematous pustulosis (AGEP) (4,8), fixed drug eruption (33), drug reaction with eosinophilia and systemic symptoms (DRESS) (8,34), Stevens-Johnson syndrome/toxic

epidermal necrolysis (SJS/TEN) (6,8,9,11,22,33), photosensitivity (23), (micro)papular eruption (16,22,24,32), erythema multiforme (-like exanthema) (6,7,14,15,21,27,28,45), urticarial eruption/urticaria-like exanthema (6,24), and unspecified or unknown exanthema (15,19,20,29).

Case series with various or unknown types of drug reactions

In a hospital in Cologne, Germany, in the period 2003-2013, 8 patients were investigated with cutaneous adverse drug reactions to tetrazepam. Five had an 'exanthema' (not further specified), two maculopapular eruption and one erythema multiforme-like exanthema. In all, patch tests were positive at D2 to tetrazepam crushed tablet 10% pet. (15).

From Spain, 5 patients with allergic eruptions from tetrazepam were reported. Three had a maculopapular eruption, which started 8 days, 3 months and 15 days respectively after initiating treatment. A 4th patient developed a generalized micropapular eruption 24 hours after the ingestion of one tablet of tetrazepam. In the 5th, a generalized itchy vesiculopapular eruption with severe mucosal involvement a few weeks after having taken tetrazepam developed, which was diagnosed as Stevens-Johnson syndrome. In all 5 patients, patch tests to tetrazepam 1% pet. were positive at D2 and D4, and negative to other benzodiazepines (22).

Maculopapular eruption

Seven patients who developed a maculopapular eruption while using tetrazepam were reported from France. The rashes started 12 hours to 5 days after initiating therapy with tetrazepam. Patch tests were positive to tetrazepam 10% pet. Over thirty controls were negative. There were no cross-reactions to other benzodiazepines (33). In a previous study from this center in Nancy, France, of 27 patients with maculopapular eruptions, one had a positive patch test reaction to tetrazepam (10).

Between March 2009 and June 2013, in another center in France specialized in cutaneous adverse drug reactions (CADR), 156 patients were patch tested because of a CADR. Tetrazepam was positive in 2 patients with maculo-papular eruptions (11). A 46-year-old woman developed a maculopapular eruption while taking tetrazepam. Three months later, the patient again took tetrazepam and within 6 hours, a widespread pruriginous macular rash with symmetrical bullous lesions on the elbows appeared. Patch tests with commercial and pure tetrazepam were partly negative, partly ?+ on the back. However, the commercial drug tested 30% pet. and pure tetrazepam 10% pet. on the elbow, at the site previously affected most severely, gave strongly positive reactions (D2 +, D4 ++) (17).

Other cases of maculopapular eruptions from delayed-type hypersensitivity to tetrazepam have been reported from Spain (18,25,28), Germany (26), and France (31). See also the section 'Case series with various or unknown types of drug reactions' above, refs. 15 and 22.

Erythroderma, widespread erythematous eruption, exfoliative dermatitis

A 35-year-old woman developed macular exanthema on the neck and chest and later had a weak erythematous reaction to a patch test with a crushed tetrazepam tablet 10% pet. Seven hours after oral challenge with 25 mg tetrazepam, a flare-up of the exanthema was observed (6). An 83-year-old man developed shortness of breath and a pruritic generalized erythematous eruption 6 hours after ingestion of one tablet of tetrazepam. Patch tests were positive to tetrazepam 1% and 5% pet. (24).

A 60-year-old woman with lumbar arthralgia, who had repeatedly been treated with tetrazepam, developed a fever of 39°C and a widespread itchy erythematous eruption which had started 12 hours after her last oral medication with 50 mg tetrazepam. Patch tests were positive to tetrazepam crushed tablet in pet. (12 controls negative) (26).

Acute generalized exanthematous pustulosis (AGEP)

A 48-year-old woman had suffered twice from a severe generalized pruriginous cutaneous reaction with fever after having taken oral tetrazepam. Patch tests were performed 4 months after the last reaction with tetrazepam 1% and 5% pet. and water and with related benzodiazepines (bromazepam, lorazepam, diazepam). There were no positive reactions on D2 and D3, but all tetrazepam patches had become positive at D10. A lymphocyte transformation test to tetrazepam was also positive. On the basis of immunological parameters and the skin reaction (which was apparently not observed by the authors) the patient was diagnosed with acute generalized exanthematous pustulosis (AGEP) and was ascribed to delayed-type allergy to tetrazepam. Although this is indeed highly likely, the possibility of patch test sensitization to tetrazepam was not considered by the authors and retesting was not performed (4).

In a multicenter investigation in France, of 45 patients patch tested for AGEP, 26 (58%) had positive patch tests to drugs, including one to tetrazepam (8).

Fixed drug eruption
A woman aged 50 developed a fixed drug eruption while using tetrazepam. A patch test with tetrazepam 10% pet. was positive (presumably on postlesional skin) (33).

Drug reaction with eosinophilia and systemic symptoms (DRESS)
A 37-year-old woman developed a febrile rash on the 10th day of taking tetrazepam and celecoxib. Physical examination showed an erythematous and edematous eruption on the trunk and limbs covering >80% of the skin surface with overlying pustules, a few bullae and edema of the face. Atypical target lesions were found on the thighs and forearms. No mucosal lesions were present and no lymph nodes palpable. There was eosinophilia and strongly elevated C-reactive protein, but no liver or kidney abnormalities. Al drugs were stopped. After initial improvement extensive bullous lesions appeared on the trunk and legs, after which the fever settled and the rash improved with extensive desquamation. Patch tests with all drugs used by the patient were positive only at D2 and D4 to crushed tablets of tetrazepam diluted with white petroleum at 1% and 10%. The patient was diagnosed with probable DRESS (34).

In a multicenter investigation in France, of 72 patients patch tested for DRESS, 46 (64%) had positive patch tests to drugs, including to tetrazepam (8).

Stevens-Johnson syndrome/toxic epidermal necrolysis (SJS/TEN)
A 32-year-old patient was treated with acetylsalicylic acid, bupivacaine HCl injections, acetaminophen and tetrazepam for lumbar pain and allopurinol for hyperuricemia. Seven days after the medication was discontinued he developed a severe generalized drug eruption with bullous target lesions on the trunk and extremities. In the mouth the patient had large hemorrhagic necrotic plaques, the conjunctivae were red and showed hypersecretion. The patient was diagnosed with drug-induced Stevens-Johnson syndrome. Patch tests, performed 3 months after complete remission, were strongly positive to tetrazepam crushed tablets 1% and 10% pet. (6).

In a multicenter investigation in France, there was a positive patch test to tetrazepam in a patient with SJS/TEN (8). Between March 2009 and June 2013, in another center in France specialized in cutaneous adverse drug reactions (CADR), 156 patients were patch tested because of a CADR. Tetrazepam was positive in 4 patients with SJS/TEN (11).

A 65-year-old woman using various drugs developed a rash which was diagnosed as Stevens-Johnson syndrome after taking tetrazepam for lumbar arthralgia for a week. Six months after complete healing, positive patch tests were observed to tetrazepam crushed tablets 20% pet. (++) at D2 and D4. On the basis of the description of this case, it appears that the patient may have suffered from DRESS rather than Stevens-Johnson syndrome (9).

A woman aged 67 developed Stevens-Johnson syndrome after having taken tetrazepam for 9 days. Patch tests were positive to tetrazepam 10% pet. (33). See also the section 'Case series with various or unknown types of drug reactions' above, ref. 22.

Photosensitivity
A 35-year-old woman one night took 50 mg oral tetrazepam because of muscle spasm. The following day, after sun exposure in the morning, she developed an itchy micropapular rash on sun-exposed areas. Patch and photopatch tests were performed and there was a positive photopatch test to tetrazepam 10% pet. An oral photochallenge test with tetrazepam reproduced the previous skin eruption (23).

Other cutaneous adverse drug reactions

(Micro)papular eruption
Two days after taking tetrazepam for a muscular contracture, a 61-year-old man developed pruritus with a generalized micropapular eruption. A second similar attack occurred one year later, 2 days after the patient again having taken tetrazepam. Patch tests were positive to tetrazepam 50 mg/ml saline and – in a second session – to tetrazepam 5% water (+++). Ten controls were negative. An oral challenge with tetrazepam reproduced the previous eruptions (16).

A 78-year-old woman presented with a generalized and pruritic micropapular rash on the trunk which had emerged 8 hours after ingestion of one tablet of tetrazepam. Patch tests were positive to tetrazepam 1% and 5% pet. and an oral provocation test reproduced the micropapular exanthema (24).

The first reported cases of cutaneous adverse drug reactions from delayed-type hypersensitivity to tetrazepam were 2 patients from Spain, who both developed a papular rash while using the drug. Both had positive patch tests to tetrazepam powder pure and 1% water (32). See also the section 'Case series with various or unknown types of drug reactions' above, ref. 22.

Erythema multiforme (-like exanthema)

A 67-year-old woman developed a moderate erythema multiforme-like exanthema 7 days after starting treatment with calcium tablets, acetaminophen and tetrazepam. Patch tests with tetrazepam crushed tablets showed a strongly positive reaction at 10% and a doubtful one at 1% pet. (6). One patient developed erythema multiforme while using tetrazepam and later had a positive patch test to a tetrazepam crushed tablet (7).

A 57-year-old woman had two episodes of 'erythematous eruption similar to burns' with nausea (first attack) and with nausea, fever, joint and muscle pain (second attack) after having taken tetrazepam. Both times, liver enzymes were elevated. A patch test was positive to tetrazepam (no details provided). The authors diagnosed erythema multiforme, but how they came to this diagnosis was not explained (14).

A 68-year-old woman presented with a 2-day eruption of slightly itchy erythematous edematous lesions, some with a purpuric center, affecting the face, trunk, arms and legs, the lips and the gingival mucosa. The eruption had started after the patient had taken tetrazepam for 3 weeks. A biopsy of one of the lesions was diagnosed as erythema multiforme. After 3 months, she took another tetrazepam pill. Eight hours later, the patient presented with a new episode of exanthema similar to the previous one. Patch tests were positive to commercial tetrazepam and to pure tetrazepam 1% and 5% pet. (27). The authors also presented an 82-old woman who developed an exanthema with the clinical and histopathological features of erythema multiforme while using tetrazepam and who had positive patch tests to this drug (27).

A 37-year-old woman developed generalized target-like macules, exfoliative involvement of the oral mucosa and conjunctivitis. A few days earlier, she had felt some stinging on her lips, pointing in the direction of herpes-induced erythema multiforme. However, 2 days earlier, the patient had also been given various drugs including tetrazepam. Patch tests were strongly positive to tetrazepam and negative to all other drugs (45).

Another case of erythema multiforme from delayed-type hypersensitivity to tetrazepam was reported in 1993 in Spanish literature (28). See also the section 'Case series with various or unknown types of drug reactions' above, ref. 15.

Urticaria

A 48-year-old woman was treated with diclofenac, acetaminophen, and omeprazole for lumbago. Two days later, she developed an urticaria-like exanthema on the neck and trunk. Patch testing with tetrazepam crushed tablets 1% and 10% pet. showed a mild erythema at the highest concentration after 3 days. Rechallenge with tetrazepam 12.5 mg provoked a flare-up of the exanthema within one day, morphologically similar to the previous one (6).

A 33-year-old woman developed a widespread pruritic urticarial eruption and angioedema of the oropharyngeal tract 6 hours after ingestion of 50 mg tetrazepam. Patch tests were strongly positive to tetrazepam 1% and 5% pet. and an oral provocation test reproduced the urticarial eruption (24).

Unspecified/unknown cutaneous adverse drug reactions

See also the section 'Case series with various or unknown types of drug reactions' above, ref. 15 (unspecified exanthema) . Cases of cutaneous adverse drug reactions to tetrazepam with − certainly or highly likely - positive patch tests, details of which are unavailable to the author, can be found in refs 19,20,21, and 29.

Cross-reactions, pseudo-cross-reactions and co-reactions

Cross-reactivity between tetrazepam and other benzodiazepines, despite their chemical similarity (which is especially close with diazepam; these two molecules only differ by the substituent at position 5 on the diazepine ring, which is cyclohexene in tetrazepam and phenyl in diazepam [33]) is virtually never observed (9,16,22,24,25,26,33,36). Only one patient sensitized to tetrazepam may likely have cross-reacted to diazepam (26). In 2 other studies, positive patch tests to diazepam may have been co reactivities rather than cross-reactions, as the patients possibly had contact not only with tetrazepam but also with diazepam (12,39).

LITERATURE

1   The data in the section 'General' may have been obtained from literature discussed in this chapter, but mostly also or exclusively from one or more of the following online sources: ChemIDPlus Advanced, PubChem, DrugBank, RxList, Drug Central, Drugs.com, and Wikipedia
2   Swinnen I, Goossens A. An update on airborne contact dermatitis: 2007-2011. Contact Dermatitis 2013;68:232-238
3   Swinnen I, Ghys K, Kerre S, Constandt L, Goossens A. Occupational airborne contact dermatitis from benzodiazepines and other drugs. Contact Dermatitis 2014;70:227-232
4   Thomas E, Bellón T, Barranco P, Padial A, Tapia B, Morel E, et al. Acute generalized exanthematous pustulosis due to tetrazepam. J Investig Allergol Clin Immunol 2008;18:119-122
5   Breuer K, Uter W, Geier J. Epidemiological data on airborne contact dermatitis: results of the IVDK. Contact Dermatitis 2015;73:239-247

6	Pirker C, Misic A, Brinkmeier T, Frosch PJ. Tetrazepam drug sensitivity -- usefulness of the patch test. Contact Dermatitis 2002;47:135-138

7	Schön NP, Reifenberger J, Eberhardt H, Grewe N. Epikutantestung zur Diagnostik von Arzneimittelreaktionen. Z Hautkr H+G 2000:75:599-600

8	Barbaud A, Collet E, Milpied B, Assier H, Staumont D, Avenel-Audran M, et al. A multicentre study to determine the value and safety of drug patch tests for the three main classes of severe cutaneous adverse drug reactions. Br J Dermatol 2013;168:555-562

9	Sánchez I, García-Abujeta JL, Fernández L, Rodríguez F, Quiñones D, Duque S, et al. Stevens-Johnson syndrome from tetrazepam. Allergol Immunopathol 1998;26:55-57

10	Barbaud A, Reichert-Penetrat S, Tréchot P, Jacquin-Petit MA, Ehlinger A, Noirez V, et al. The use of skin testing in the investigation of cutaneous adverse drug reactions. Br J Dermatol 1998;139:49-58

11	Assier H, Valeyrie-Allanore L, Gener G, Verlinde Carvalh M, Chosidow O, Wolkenstein P. Patch testing in non-immediate cutaneous adverse drug reactions: value of extemporaneous patch tests. Contact Dermatitis 2017;77:297-302

12	Garcia-Bravo B, Rodriguez-Pichardo A, Camacho F. Contact dermatitis from diazepoxides. Contact Dermatitis 1994;30:40

13	EMA. Tetrazepam-containing medicines suspended across the EU. 29 May 2013; EMA/402567/2013 – Rev 1. ema.europa.eu/docs/en_GB/document_library/Referrals_document/Tetrazepam_containing_medicinal_products/Position_provided_by_CMDh/WC500146678.pdf. Accessed April 2021

14	Cabrerizo Ballesteros S, Mendez Alcalde J, Sanchez Alonso A. Erythema multiforme to tetrazepam. J Investig Allergol Clin Immunol 2007;17:205-206

15	Huseynov I, Wirtz M, Hunzelmann N. Tetrazepam allergy: A case series of cutaneous adverse events. Acta Derm Venereol 2016;96:110-111

16	Sánchez-Morillas L, Laguna-Martínez JJ, Reaño-Martos M, Rojo-Andrés E, Ubeda PG. Systemic dermatitis due to tetrazepam. J Investig Allergol Clin Immunol 2008;18:404-406

17	Barbaud A, Trechot P, Reichert-Penetrat S, Granel F, Schmutz JL. The usefulness of patch testing on the previously most severely affected site in a cutaneous adverse drug reaction to tetrazepam. Contact Dermatitis 2001;44:259-260

18	Palacios R, Domínguez J, Alonso A, Rodríguez A, Plaza A, Chamorro M, Martínez-Cócera C. Adverse reaction to tetrazepam. J Invest Allergol Clin Immunol 2001;11:130-131

19	Ghislain PD, Roussel S, Bouffioux B, Delescluse J. Tetrazepam (Myolastan)-induced exanthema: positive patch tests in 2 cases. Ann Dermatol Venereol 2000;127:1094-1096 (Article in French)

20	Laffond E, Dávila I, Moreno E, Morán M, Otero MJ, Sancho P, Barahona MA, García MJ, Lorente F. Sensibilización a benzodiazepinas. Allergol Immunol Clin 2000;16:313-316 (Article in Spanish)

21	Rodríguez M, Ortiz J, Del Río R, Iglesias L. Eritema exudativo multiforme inducido por tetrazepam. Med Clin (Barc) 2000;115:359 (Article in Spanish)

22	Del Pozo MD, Blasco A, Lobera T. Tetrazepam allergy. Allergy 1999;54:1226-1227

23	Quiñones D, Sanchez I, Alonso S, Garcia-Abujeta JL, Fernandez L, Rodriguez F, Martin-Gil D, Jerez J. Photodermatitis from tetrazepam. Contact Dermatitis 1998;39:84

24	Blanco R, Díez-Gómez ML, Gala G, Quirce S. Delayed hypersensitivity to tetrazepam. Allergy 1997;52:1146-1147

25	Ortega NR, Barranco P, López Serrano C, Romualdo L, Mora C. Delayed cell-mediated hypersensitivity to tetrazepam. Contact Dermatitis 1996;34:139

26	Kämpgen E, Bürger T, Bröcker EB, Klein CE. Cross-reactive type IV hypersensitivity reactions to benzodiazepines revealed by patch testing. Contact Dermatitis 1995;33:356-357

27	Ortiz-Frutos FJ, Alonso J, Hergueta JP, Quintana I, Iglesias L. Tetrazepam: an allergen with several clinical expressions. Contact Dermatitis 1995;33:63-65

28	Manrique P, Gonzales MR, Calderon M J. Utilidad del patch-test para detectar la alergia a! tetrazepam. Dermatitis de Contacto 1993;21:17-18 (Article in Spanish, data cited in ref. 27)

29	Tomb RR, Grosshans E, Defour E, Heid E. Allergic skin reaction to tetrazepam detected by patch testing. Eur J Dermatol 1993;3:116-118 (cited in various articles, cannot be traced)

30	Proy-Vega B, Aguirre C, de Groot P, Solís-García del Pozo J, Jordán J. On the clinical evidence leading to tetrazepam withdrawal. Expert Opin Drug Saf 2014;13:705-712

31	Collet E, Dalac S, Morvan C, Sgro C, Lambert D. Tetrazepam allergy once more detected by patch test. Contact Dermatitis 1992;26:281

32	Camarasa JG, Serra-Baldrich E. Tetrazepam allergy detected by patch test. Contact Dermatitis 1990;22:246

33	Barbaud A, Girault PY, Schmutz JL, Weber-Muller F, Trechot P. No cross-reactions between tetrazepam and other benzodiazepines: a possible chemical explanation. Contact Dermatitis 2009;61:53-56

34	Bachmeyer C, Assier H, Roujeau JC, Blum L. Probable drug rash with eosinophilia and systemic symptoms syndrome related to tetrazepam. J Eur Acad Dermatol Venereol 2008;22:887-889

35 Gilissen L, Boeckxstaens E, Geebelen J, Goossens A. Occupational allergic contact dermatitis from systemic drugs. Contact Dermatitis 2020;82:24-30

36 Van der Hulst K, Kerre S, Goossens A. Occupational allergic contact dermatitis from tetrazepam in nurses. Contact Dermatitis 2010;62:303-308

37 Schubert S, Bauer A, Molin S, Skudlik C, Geier J. Occupational contact sensitization in female geriatric nurses: Data of the Information Network of Departments of Dermatology (IVDK) 2005-2014. J Eur Acad Dermatol Venereol 2017;31:469-476

38 Molin S, Bauer A, Schnuch A, Geier J. Occupational contact allergy in nurses: results from the Information Network of Departments of Dermatology 2003-2012. Contact Dermatitis 2015;72:164-171

39 Landeck L, Skudlik C, John SM. Airborne contact dermatitis to tetrazepam in geriatric nurses - a report of 10 cases. J Eur Acad Dermatol Venereol 2012;26:680-684

40 Landeck L, Skudlik C, John SM. Tetrazepam as an allergen in occupational airborne contact dermatitis. Dermatitis 2011;22:177 (Abstract)

41 Breuer K, Worm M, Skudlik C, Schroder C, John SM. Occupational airborne contact allergy to tetrazepam in a geriatric nurse. J Dtsch Dermatol Ges 2009;7:896-898

42 Ferran M, Giménez-Arnau A, Luque S, Berenguer N, Iglesias M, Pujol RM. Occupational airborne contact dermatitis from sporadic exposure to tetrazepam during machine maintenance. Contact Dermatitis 2005;52:173-174

43 Lepp U, Zabel P, Greinert U. Occupational airborne contact allergy to tetrazepam. Contact Dermatitis 2003;49:260-261

44 Choquet-Kastylevsky G, Testud F, Chalmet P, Lecuyer-Kudela S, Descotes J. Occupational contact allergy to tetrazepam. Contact Dermatitis 2001;44:372

45 Gebhardt M, Wollina U. Allergy testing in serious cutaneous drug reactions - harmful or beneficial? Contact Dermatitis 1997;37:282-285

Chapter 3.472 THIABENDAZOLE

IDENTIFICATION

Description/definition   : Thiabendazole is the 2-substituted benzimidazole that conforms to the structural formula shown below
Pharmacological classes   : Anthelmintics
IUPAC name   : 4-(1H-Benzimidazol-2-yl)-1,3-thiazole
Other names   : 2-(4-Thiazolyl)-1H-benzimidazole
CAS registry number   : 148-79-8
EC number   : 205-725-8
Merck Index monograph   : 10710
Patch testing   : 4% pet.
Molecular formula   : $C_{10}H_7N_3S$

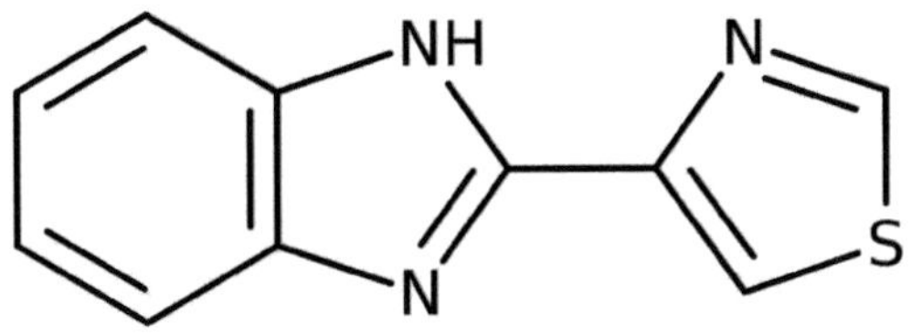

GENERAL

Thiabendazole is a 2-substituted benzimidazole with fungicidal and anthelminthic (parasiticidal) properties. It is also a chelating agent, which means that this agent is used medicinally to bind metals in cases of metal poisoning, such as lead, mercury or antimony. Thiabendazole is vermicidal and/or vermifugal against many nematodes. This agent also suppresses egg and/or larval production and may inhibit the subsequent development of those eggs or larvae which are passed in the feces. Thiabendazole is indicated for the treatment of strongyloidiasis (threadworm), cutaneous larva migrans (creeping eruption), visceral larva migrans, and trichinosis (1).

In topical preparations, thiabendazole has rarely caused (photo)contact allergy/(photo)allergic contact dermatitis, which subject has been reviewed in Volume 3 of the *Monographs in contact allergy* series (5).

CONTACT ALLERGY FROM ACCIDENTAL CONTACT

In Italy, during 1986-1988, 204 animal feed mill workers (191 men, 13 women) were patch tested with a large number of animal feed additives. There were two reactions to thiabendazole 4% pet. in a group of 36 subjects with clinical complaints (dermatitis or pruritus sine materia) and one reaction in the group of 168 individuals without skin complaints. All reactions were considered to be relevant. In one patient with evident contact dermatitis, there was a positive 'stop-start test' of his working activity (2,3). These were cases of occupational contact allergy/allergic contact dermatitis.

Cross-reactions, pseudo-cross-reactions and co-reactions

Not to econazole, bifonazole, metronidazole, mebendazole, miconazole and ketoconazole (all 2% pet.) (4).

LITERATURE

1   The data in the section 'General' may have been obtained from literature discussed in this chapter, but mostly also or exclusively from one or more of the following online sources: ChemIDPlus Advanced, PubChem, DrugBank, RxList, Drug Central, Drugs.com, and Wikipedia
2   Mancuso G, Staffa M, Errani A, Berdondini RM, Fabbri P. Occupational dermatitis in animal feed mill workers. Contact Dermatitis 1990;22:37-41
3   Mancuso G. Topical thiabendazole allergy. Contact Dermatitis 1994;31:207
4   Izu R, Aguirre A, Goicoechea A, Gardeazabal J, Díaz Pérez JL. Photoaggravated allergic contact dermatitis due to topical thiabendazole. Contact Dermatitis 1993;28:243-244
5   De Groot AC. Monographs in contact allergy, volume 3: Topical Drugs. Boca Raton, Fl, USA: CRC Press Taylor and Francis Group, 2021 (ISBN 978-0-367-23693-9)

Chapter 3.473 Thiamine

IDENTIFICATION

Description/definition   : Thiamine is the essential vitamin, belonging to the vitamin B family, that conforms to the structural formula shown below
Pharmacological classes   : Vitamin B complex
IUPAC name   : 2-[3-[(4-Amino-2-methylpyrimidin-5-yl)methyl]-4-methyl-1,3-thiazol-3-ium-5-yl]ethanol
Other names   : Vitamin B_1
CAS registry number   : 70-16-6; 59-43-8 (chloride)
EC number   : 200-425-3 (chloride)
Merck Index monograph : 10717
Patch testing   : 10% water
Molecular formula   : $C_{12}H_{17}N_4OS+$

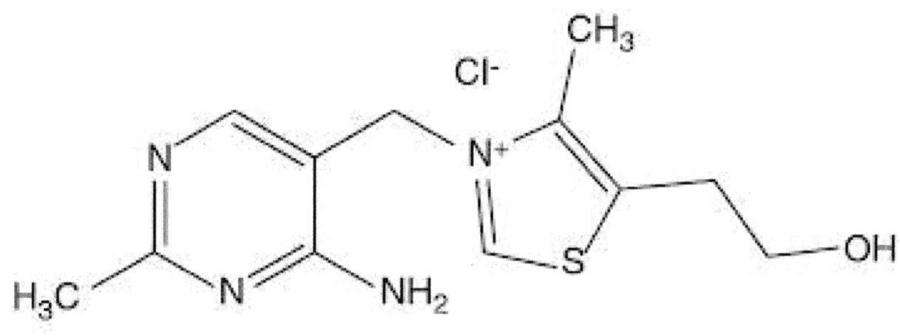

According to ChemIDPlus, thiamine is the chloride salt ($C_{12}H_{17}ClN_4OS$)

GENERAL

Thiamine is an essential vitamin, belonging to the vitamin B family, with antioxidant, erythropoietic, mood modulating, and glucose-regulating activities. Thiamine plays an important role in intracellular glucose metabolism, the conversion of carbohydrates and fat into energy, it is essential for normal growth and development and helps to maintain proper functioning of the heart and the nervous and digestive systems. Pharmaceutical thiamine is indicated for the treatment of thiamine and niacin deficiency states, Korsakov's alcoholic psychosis, Wernicke-Korsakov syndrome, delirium, and peripheral neuritis. In pharmaceuticals, thiamine is most often present as thiamine hydrochloride (CAS number 67-03-8, EC number 200-641-8, molecular formula $C_{12}H_{18}Cl_2N_4OS$); other derivatives used are thiamine mononitrate and thiamine (di)chloride (1).

In topical preparations, thiamine has rarely caused contact allergy/allergic contact dermatitis, which subject has been reviewed in Volume 3 of the *Monographs in contact allergy* series (7).

CONTACT ALLERGY FROM ACCIDENTAL CONTACT

A 54-year-old man employed in a pharmaceutical plant filled and packed thiamine hydrochloride in a dusty process. After one month, he developed an itchy eczema on the forearms and dorsa of the hands, with some spread to the face. Patch tests were positive to thiamine 10% and 5% water: thiamine 1% water showed erythema only. Ten controls were negative. Positive reactions were also recorded to thiothiamine 5% and 1% water. This patient had occupational (possibly partly airborne) allergic contact dermatitis from thiamine (3).

A 32-year-old man, after 3 months of employment as a process worker in the same pharmaceutical plant as the previous patient, developed an itchy eczema on the hands and legs, which spread to the rest of the body. The patient was dismissed from work. The skin gradually cleared after treatment with betamethasone ointment. After discontinuation of topical treatment, however, an intermittent itchy dermatitis persisted on the dorsa of the feet, on the face, and occasionally at other sites. Further questioning revealed that he took oral vitamins containing thiamine daily. Patch testing was positive to thiamine 10% water. This was a case of systemic allergic dermatitis (3).

Cases of allergic contact dermatitis due to thiamine had previously been reported among pharmaceutical workers. Two patients filling ampoules developed allergic eczema of the hands and arms, and in one also of the eyelids (4). A pharmaceutical worker developed eczema of exposed sites while working with thiamine hydrochloride. The patch test was positive to thiamine HCl in this patient and in 9 other workers in the factory. None of these reacted to the vitamin orally (5).

A 17-year-old girl was sensitized to thiamine from filling ampoules with vitamins. A flare was observed after she had resumed her work. Oral provocation with 200 mg thiamine and intracutaneous tests with 10 mg also induced relapses (systemic allergic dermatitis) (6).

CUTANEOUS ADVERSE DRUG REACTIONS FROM SYSTEMIC ADMINISTRATION CAUSED BY TYPE IV (DELAYED-TYPE) HYPERSENSITIVITY (as demonstrated by positive patch tests)

Cutaneous adverse drug reactions from systemic administration of thiamine caused by type IV (delayed-type) hypersensitivity have included systemic allergic dermatitis (2,3,6).

Systemic allergic dermatitis (systemic contact dermatitis)

A 46-year-old woman developed a pruritic micropapular erythematous rash on the right shoulder after topical application of diclofenac cream and a commercial solution containing lidocaine, dexamethasone, cyanocobalamin (vitamin B_{12}) and thiamine (vitamin B_1) by iontophoresis. Patch tests and prick and intradermal tests with these medicaments were negative. One hour after the intramuscular injection of the solution (the usual method of administration), the patient developed skin itching, and 24 hr later, erythematous plaques were noticed in the forearms and right shoulder (the application area of the iontophoresis treatment). Eight hours after oral administration of a multivitamin containing vitamins B_1, B_{12} and B_6, the patient developed a pruritic micropapular erythematous rash on the buttocks and back. Finally, patch tests were performed with this multivitamin and its components, which gave positive results to the multivitamin 'as is' (+++), and vitamin B_1 (thiamine hydrochloride 10% water, ++) at D4. An intradermal test with thiamine gave a positive result at 24 hours. Ten controls were negative to thiamine 10% water. The patient had allergic contact dermatitis from thiamine and systemic allergic dermatitis from an oral challenge and intramuscular injection (2).

Two more cases of systemic allergic dermatitis after occupational sensitization are described above in the section 'Contact allergy from accidental contact (3,6).

Cross-reactions, pseudo-cross-reactions and co-reactions

A patient sensitized to thiamine co-reacted to thiothiamine 5% and 1% water (3). Another co-reacted to co-carboxylase (10% and 1% pet.), which is a co-enzyme in the cellular metabolism and is formed by esterification of thiamine with pyrophosphoric acid after intestinal resorption (6).

LITERATURE

1 The data in the section 'General' may have been obtained from literature discussed in this chapter, but mostly also or exclusively from one or more of the following online sources: ChemIDPlus Advanced, PubChem, DrugBank, RxList, Drug Central, Drugs.com, and Wikipedia
2 Arruti N, Bernedo N, Audicana MT, Villarreal O, Uriel O, Muñoz D. Systemic allergic dermatitis caused by thiamine after iontophoresis. Contact Dermatitis 2013;69:375-376
3 Ingemann Larsen A, Riis Jepsen J, Thulin H. Allergic contact dermatitis from thiamine. Contact Dermatitis 1989;20:387-388
4 Combes FC, Groopman J. Contact dermatitis due to thiamine. Arch Dermatol Syph 1950;61:858-859
5 Dalton JE, Pierce JD. Dermatological problems among pharmaceutical workers. Arch Dermatol 1951;64:667-675
6 Hjorth N. Contact dermatitis from vitamin B1 (thiamine). J Invest Dermatol 1958;30:261-264
7 De Groot AC. Monographs in contact allergy, volume 3: Topical Drugs. Boca Raton, Fl, USA: CRC Press Taylor and Francis Group, 2021 (ISBN 978-0-367-23693-9)

Chapter 3.474 THIOCTIC ACID

IDENTIFICATION

Description/definition   : Thioctic acid is the heterocyclic thia fatty acid that conforms to the structural formula shown below
Pharmacological classes : Antioxidants; vitamin B complex
IUPAC name   : 1,2-Dithiolane-3-pentanoic acid, (+/-)-
Other names   : α-Lipoic acid; 1,2-dithiolane-3-valeric acid
CAS registry number   : 1077-28-7
EC number   : 214-071-2
Merck Index monograph   : 10749
Patch test allergens   : 1% and 5% pet.
Molecular formula   : $C_8H_{14}O_2S_2$

GENERAL

Thioctic acid (α-lipoic acid) is an essential cofactor in metabolic reactions through mitochondrial-specific pathways, and it is synthesized in small amounts in humans. Thioctic acid shows antioxidant and metal-chelating activity. Hence, it is widely used in a variety of conditions, including diabetes, insulin resistance, atherosclerosis, neuropathy, neurodegenerative diseases, and ischemia-reperfusion (2). Other suggested indications include diseases of the eye such as cataracts, diabetic retinopathy, and age-related macular degeneration (4) and a variety of other conditions (1). Topically, it may be used in eye drops (4) and in cosmetic anti-ageing products, such as 'anti-wrinkle' creams (2,3). In topical preparations, thioctic acid has rarely caused contact allergy/allergic contact dermatitis, which subject has been reviewed in Volume 3 of the *Monographs in contact allergy* series (5).

CUTANEOUS ADVERSE DRUG REACTIONS FROM SYSTEMIC ADMINISTRATION CAUSED BY TYPE IV (DELAYED-TYPE) HYPERSENSITIVITY (as demonstrated by positive patch tests)

Cutaneous adverse drug reactions from systemic administration of thioctic acid caused by type IV (delayed-type) hypersensitivity have included maculopapular eruption (1).

Maculopapular eruption

A woman suffering from shoulder pain resulting from cervical disc herniation developed a pruritic maculopapular rash on the face and scalp after 10 days of treatment with a dietary supplement containing thioctic acid and other components and with two oral and one intramuscular NSAIDs. Prick tests with the supplement and the three drugs (powdered drug dissolved in saline) were negative. Patch tests gave a very strong reaction to the commercial dietary supplement. Later, patch tests were performed with its ingredient, and the patient now reacted to the supplement 10% pet. and to thioctic acid 5% pet., 2.5% pet. and 0.025% pet. Twelve controls were negative (1).

LITERATURE

1   Rizzi A, Nucera E, Buonomo A, Schiavino D. Delayed hypersensitivity to α-lipoic acid: look at dietary supplements. Contact Dermatitis 2015;73:62-63
2   Bergqvist-Karlsson A. Thelin I, Bergendorff O. Contact dermatitis to α-lipoic acid in an anti-wrinkle cream. Contact Dermatitis 2006;55:56-57
3   Leysen J, Aerts O. Further evidence of thioctic acid (α-lipoic acid) being a strong cosmetic sensitizer. Contact Dermatitis 2016;74:182-184
4   Craig S, Urwin R, Wilkinson M. Contact allergy to thioctic acid present in Hypromellose® eye drops. Contact Dermatitis 2017;76:361-362
5   De Groot AC. Monographs in contact allergy, volume 3: Topical Drugs. Boca Raton, Fl, USA: CRC Press Taylor and Francis Group, 2021 (ISBN 978-0-367-23693-9)

Chapter 3.475 THIORIDAZINE

IDENTIFICATION

Description/definition   : Thioridazine is the phenothiazine derivative that conforms to the structural formula
                                         shown below
Pharmacological classes : Antipsychotic agents; dopamine antagonists
IUPAC name   : 10-[2-(1-Methylpiperidin-2-yl)ethyl]-2-methylsulfanylphenothiazine
CAS registry number   : 50-52-2
EC number   : 200-044-2
Merck Index monograph : 10782
Patch testing   : 0.1% pet.; with 1% pet., photopatch testing with >4 J/cm^2 UVA irradiation will result in
                                         phototoxic reactions in most individuals (2)
Molecular formula   : $C_{21}H_{26}N_2S_2$

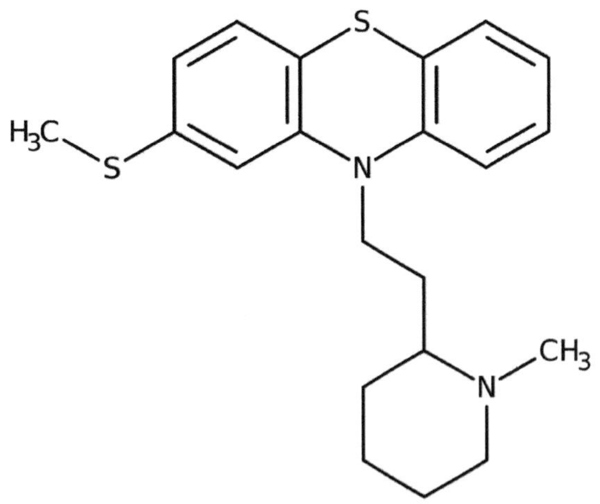

GENERAL

Thioridazine is a phenothiazine derivative and a serotonergic antagonist, a histamine H$_1$ receptor antagonist, an α-adrenergic antagonist, and a dopaminergic antagonist. Thioridazine (as hydrochloride) was formerly used for the treatment of schizophrenia and generalized anxiety disorder. According to DrugBank, thioridazine was withdrawn worldwide in 2005 due to its association with cardiac arrythmias (www.drugbank.com). However, it is still available as Mellaril ® and indicated for the management of schizophrenic patients who fail to respond adequately to treatment with other antipsychotic drugs (1). In pharmaceutical products, thioridazine is employed as thioridazine hydrochloride (CAS number 130-61-0, EC number 204-992-8, molecular formula $C_{21}H_{27}ClN_2S_2$) (1).

CUTANEOUS ADVERSE DRUG REACTIONS FROM SYSTEMIC ADMINISTRATION CAUSED BY TYPE IV (DELAYED-TYPE) HYPERSENSITIVITY (as demonstrated by positive patch tests)

Cutaneous adverse drug reactions from systemic administration of thioridazine caused by type IV (delayed-type) hypersensitivity have included photosensitivity (3,4).

Photosensitivity

A 36-year-old woman, while taking 60 mg of thioridazine daily and using 2 other drugs, developed a generalized maculopapular eruption 12 days after a sunbath on the balcony for 2 hours. It had the typical appearance of a photoallergic eruption with papules spreading from the eczematous margins to the unaffected areas under the bikini. Four weeks after complete healing, photopatch tests were performed with all components of the drugs used and other phenothiazine-derivatives using irradiation with 10 J/cm^2 UVA, which showed a positive reaction to thioridazine only (no details provided on test concentration, vehicle, times of reading, strength of reaction, controls [probably none]). Over the next 6 days, despite avoiding sun exposure except to her hands and face, the patient again developed a generalized maculopapular eruption in the same pattern as before. The authors diagnosed a photoallergic reaction to thioridazine (3).

    After having used chlorpromazine and thioridazine for 2 weeks, a 58-year-old woman developed acute eczema on all light-exposed areas of her body. After the patient had discontinued all the medication, the rash disappeared within a few weeks. Fourteen years later, the patient again took thioridazine, 200 mg daily, because of an acute paranoid reaction. After one day of therapy, a severe rash again appeared on all light-exposed areas of her body.

Irradiation with long UV rays (15 mJ/cm² [probably 15J]) produced an acute edematous eczema. A photopatch test was strongly positive to thioridazine 1% pet. irradiated with an UVA dose of 7 mJ/cm² (probably 7J), as was a photointradermal test with thioridazine 0.01%. Tests with chlorpromazine (1% pet.) and promethazine (1% water) gave similar long-lasting allergic type reactions. In ten healthy control subjects, an erythematous irritant type reaction with thioridazine was elicited by a ten-fold UVA dose, but not by lower doses of UVA (which probably means that the controls were *not* photopatch tested). The author concluded that 'true photoallergy' had developed, either from thioridazine itself or from photocross-hypersensitivity to primary chlorpromazine photosensitization (4).

Phototoxicity versus photoallergy
In the interpretation of photopatch tests to thioridazine it should be realized that this chemical is a pheno-thiazine-derivative (as are promethazine, promazine and chlorpromazine), a group of drugs which have well-known phototoxic properties. Indeed, immediate erythema will develop during UVA irradiation, and another peak appearing 8-12 hours after irradiation, in most subjects when 1% thioridazine has been applied for 48 hours and irradiation doses are higher than 4 J/cm². The authors of one study warn that this phototoxic reaction may easily be misinterpreted as evidence of a photoallergic reaction (2).

LITERATURE

1   The data in the section 'General' may have been obtained from literature discussed in this chapter, but mostly also or exclusively from one or more of the following online sources: ChemIDPlus Advanced, PubChem, DrugBank, RxList, Drug Central, Drugs.com, and Wikipedia
2   Takiwaki H, Tsuchiya K, Fujita M, Miyaoka Y. Thioridazine induces immediate and delayed erythema in photopatch test. Photochem Photobiol 2006;82:523-526
3   Röhrborn W, Bräuninger W. Thioridazine photoallergy. Contact Dermatitis 1987;17:241
4   Suhonen R. Thioridazine photosensitivity. Contact Dermatitis 1976;2:179

Chapter 3.476 TIAPROFENIC ACID

IDENTIFICATION

Description/definition　　　: Tiaprofenic acid is the aromatic ketone and arylpropionic acid derivative that conforms to the structural formula shown below
Pharmacological classes　 : Anti-Inflammatory agents, non-steroidal
IUPAC name　　　　　　　 : 2-(5-Benzoylthiophen-2-yl)propanoic acid
CAS registry number　　　 : 33005-95-7
EC number　　　　　　　　: 251-329-3
Merck Index monograph　: 10847
Patch testing　　　　　　 : 1% pet.
Molecular formula　　　　: $C_{14}H_{12}O_3S$

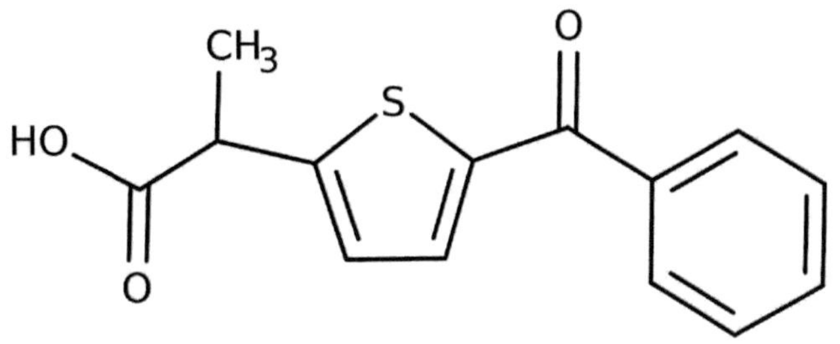

GENERAL

Tiaprofenic acid is a nonsteroidal anti-inflammatory drug (NSAID) of the arylpropionic acid class, with anti-inflammatory, antipyretic and analgesic properties. It is indicated for treatment of pain, especially arthritic pain; it is available in systemic administration forms only (1).

CUTANEOUS ADVERSE DRUG REACTIONS FROM SYSTEMIC ADMINISTRATION CAUSED BY TYPE IV (DELAYED-TYPE) HYPERSENSITIVITY (as demonstrated by positive patch tests)

Cutaneous adverse drug reactions from systemic administration of tiaprofenic acid caused by type IV (delayed-type) hypersensitivity have included systemic photoallergic dermatitis (2,6,9) and photosensitivity (12).

Systemic photoallergic dermatitis

A man aged 46 and a 43-year-old woman both had developed a photosensitivity reaction after oral intake of tiaprofenic acid 100 mg twice daily for 2 days. The vesiculobullous lesions started within 24 hours of sun exposure and were located on photoexposed areas; the eruptions regressed after 2 resp. 4 weeks only. Both had applied topical ketoprofen gel before, and the female patient had developed photocontact dermatitis from it. Patch and photopatch tests using UVA 10 J/cm² for irradiation were positive to the photopatch tests with ketoprofen 1% and 2.5% pet. and tiaprofenic acid 1% pet. There were also positive photopatch tests to fenofibrate and unsubstituted benzophenone (which chemicals often photocross-react to ketoprofen [10]) (6). These were cases of systemic photocontact dermatitis to tiaprofenic acid cross-reacting to previous ketoprofen photosensitization.

A woman aged 47 had developed a photosensitive eruption from application of ketoprofen gel to the dorsum of the left foot. Photopatch tests with ketoprofen were positive. Three months later, the patient developed a photo-distributed eruption after oral intake of tiaprofenic acid (which very frequently photocross-reacts to ketoprofen; see the section 'Cross-reactions, pseudo-cross-reactions and co-reactions' below). This would also seem to be a case of systemic photocontact dermatitis, although it should be acknowledged that the patient may not have been photopatch tested with tiaprofenic acid for confirmation (9).

In Italy, before 1993, the members of the GIRDCA Multicentre Study Group diagnosed 102 patients (49 men, 53 women), aged 16 to 66 years (mean 37 years), with (photo)dermatitis induced by systemic or topical NSAIDs. Tiaprofenic acid caused two photocontact allergic (after systemic administration of the drug) and zero contact allergic reactions (2). As these two also reacted to ketoprofen, these may likely have been cases of systemic photocontact dermatitis.

Photosensitivity

General
Nonsteroidal anti-inflammatory drugs of the arylpropionic derivative type such as tiaprofenic acid are frequently associated with phototoxicity. Indeed, tiaprofenic acid induced phototoxic reactions in 21/45 patients (47%) photopatch tested with tiaprofenic acid 1% pet. and irradiation with 10 J/cm^2 UVA (13) and in another study in 28% of the patients photopatch tested using tiaprofenic acid 5% pet. irradiated with 10 J/cm^2 UVA (18). High percentages of phototoxic reactions have also been observed in other studies (12,14,15). It is therefore possible, if not likely, that a number of 'positive photopatch tests' in the studies presented below have in fact been phototoxic, especially as no relevance for these positive tests was found or mentioned.

Tiaprofenic acid is, contrary to various other NSAIDs, not used in topical pharmaceutical preparations. However, there is one case of photoallergic contact dermatitis. The patient had used a vial of tiaprofenic acid for iontophoresis to treat epicondylitis. After the 9th session, the patient developed erythematous vesicular lesions on the elbow, arm and forearm, with edema of the hands and severe pruritus. Patch tests were negative, but photopatch tests positive to tiaprofenic acid 5% pet. irradiated with 10 J/cm^2 UVA (11).

Photoallergic reactions
Whereas many studies have documented positive photopatch tests to tiaprofenic acid and some cases of photo-sensitive eruptions to tiaprofenic acid may have occurred after photosensitization to ketoprofen (2,6,9), there are very few proven cases of primary photosensitization to this oral NSAID have been found by the author. An unknown number (probably 2 or 3) patients who had a light-exposed eruption while using tiaprofenic acid, later had a positive photopatch test to this NSAID (12)

Routine photopatch testing in patients suspected of photosensitivity
In some older studies from Italy (16,17) and Germany/Switzerland/Austria (18), a number of positive photopatch tests to tiaprofenic acid have been observed, often combined with ketoprofen photosensitization. The relevance of these reactions was not addressed; convincing (or even circumstantial) evidence that systemic tiaprofenic acid resulted in (primary) photosensitization to this NSAID was not provided (16,17,18).

Cross-reactions, pseudo-cross-reactions and co-reactions
Most patients photosensitized to ketoprofen have photocross-reactivity to tiaprofenic acid, although the latter does not have the benzophenone moiety of ketoprofen, but a thiophene-phenylketone structure (6,7,8, fully reviewed in ref. 10). Patients photosensitized to suprofen may photocross-react to tiaprofenic acid (3,4). A patient who had become photosensitized to tiaprofenic acid from topical administration (from using a vial for iontophoresis, tiaprofenic acid is not used in topical pharmaceutical products) had a photocross-reaction to flurbiprofen (11).

Sensitization to cinnamyl alcohol may be considered a marker or even a risk factor (cross-sensitization) for photocontact allergy to tiaprofenic acid (and ketoprofen) (5).

Patch test sensitization
Three of 57 patients with suspected photosensitivity, who were photopatch tested twice (D1 and D14) with pulverized moistened tiaprofenic acid tablets irradiated with 15 J/cm^2 UVA, were negative after the first session, but had positive photopatch tests after the second one at D16, D21 and D22, respectively. These patients had never used tiaprofenic acid tablets before. The histological pattern of one positive photopatch test showed a picture suggestive of a lymphocytic drug reaction. The authors suggested these were true photoallergic reactions (there were also many phototoxic reactions), but did apparently not consider the possibility of photopatch test sensitization (15).

LITERATURE
1   The data in the section 'General' may have been obtained from literature discussed in this chapter, but mostly also or exclusively from one or more of the following online sources: ChemIDPlus Advanced, PubChem, DrugBank, RxList, Drug Central, Drugs.com, and Wikipedia
2   Pigatto PD, Mozzanica N, Bigardi AS, Legori A, Valsecchi R, Cusano F, et al. Topical NSAID allergic contact dermatitis. Italian experience. Contact Dermatitis 1993;29:39-41
3   Kuno Y, Numata T. Photocontact allergy due to suprofen. J Dermatol 1994;21:352-357
4   Kurumaji Y, Oshiro Y, Miyamoto C, Keong CH, Katoh T, Nishioka K. Allergic photocontact dermatitis due to suprofen; Photopatch testing and cross-reaction study. Contact Dermatitis 1991;25:218-223
5   Stingeni L, Foti C, Cassano N, Bonamonte D, Vonella M, Vena GA, et al. Photocontact allergy to arylpropionic acid non-steroidal anti-inflammatory drugs in patients sensitized to fragrance mix I. Contact Dermatitis 2010;63:108-110

6   Le Coz CJ, Bottlaender A, Scrivener JN, Santinelli F, Cribier BJ, Heid E, et al. Photocontact dermatitis from ketoprofen and tiaprofenic acid: cross-reactivity study in 12 consecutive patients. Contact Dermatitis 1998;38:245-252

7   Durbize E, Vigan M, Puzenat E, Girardin P, Adessi B, Desprez PH, et al. Spectrum of cross-photosensitization in 18 consecutive patients with contact photoallergy to ketoprofen: associated photoallergies to non-benzophenone-containing molecules. Contact Dermatitis 2003;48:144-149

8   Matsusuhita T, Kamide R. Five cases of photocontact dermatitis due to topical ketoprofen: photopatch testing and cross-reaction study. Photodermatol Photoimmunol Photomed 2001;17:26-31

9   Adamski H, Benkalfate L, Delaval Y, Ollivier I, le Jean S, Toubel G, et al. Photodermatitis from non-steroidal anti-inflammatory drugs. Contact Dermatitis 1998;38:171-174

10   De Groot AC. Ketoprofen. In: Monographs in contact allergy, volume 3: Topical Drugs. Boca Raton, Fl, USA: CRC Press Taylor and Francis Group, 2021: Chapter 3.193, 452-462 (ISBN 978-0-367-23693-9)

11   Valsecchi R, Di Landro A, Pigatto P, Cainelli T. Tiaprofenic acid photodermatitis. Contact Dermatitis 1989;21:345-346

12   Przybilla B, Ring J, Schwab U, Galosi A, Dorn M, Braun-Falco O. Photosensibilisierende Eigenschaften nichtsteroidaler Antirheumatika im Photopatch-Test [Photosensitizing properties of nonsteroidal antirheumatic drugs in the photopatch test]. Hautarzt 1987;38:18-25 (Article in German)

13   Neumann RA, Knobler RM, Lindemayr H. Tiaprofenic acid induced photosensitivity. Contact Dermatitis 1989;20:270-273

14   Von Kries R, Hölzle E, Lehmann P, Plewig G. Routine photopatch testing with tiaprofenic acid. Photodermatol 1987;4:306-308

15   Przybilla B, Ring J, Galosi A, Dorn M. Photopatch test reactions to tiaprofenic acid. Contact Dermatitis 1984;10:55-56

16   Pigatto P, Bigardi A, Legori A, Valsecchi R, Picardo M. Cross-reactions in patch testing and photopatch testing with ketoprofen, thiaprofenic acid, and cinnamic aldehyde. Am J Contact Dermatitis 1996;7:220-223

17   Pigatto PD, Legori A, Bigardi AS, Guarrera M, Tosti A, Santucci B, et al. Gruppo Italiano recerca dermatiti da contatto ed ambientali Italian multicenter study of allergic contact photodermatitis: epidemiological aspects. Am J Contact Dermatitis 1996;7:158-163

18   Hölzle E, Neumann N, Hausen B, Przybilla B, Schauder S, Hönigsmann H, et al. Photopatch testing: the 5-year experience of the German, Austrian and Swiss Photopatch Test Group. J Am Acad Dermatol 1991;25:59-68

Chapter 3.477 TICLOPIDINE

IDENTIFICATION

Description/definition   : Ticlopidine is the thienopyridine derivative that conforms to the structural formula shown below

Pharmacological classes   : Cytochrome P-450 CYP2C19 inhibitors; platelet aggregation inhibitors; fibrinolytic agents; purinergic P2Y receptor antagonists

IUPAC name   : 5-((2-Chlorophenyl)methyl)-4,5,6,7-tetrahydrothieno(3,2-c)pyridine

CAS registry number   : 55142-85-3

EC number   : 259-498-5

Merck Index monograph   : 10855

Patch testing   : 5% pet. and water

Molecular formula   : $C_{14}H_{14}ClNS$

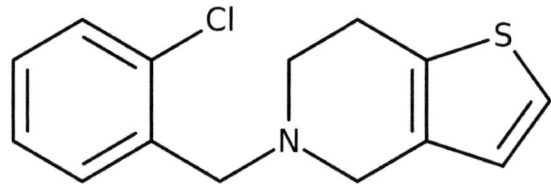

GENERAL

Ticlopidine is a thienopyridine derivative with anticoagulant activity. It is a prodrug that is metabolised to an as yet undetermined metabolite that acts as a platelet aggregation inhibitor. Inhibition of platelet aggregation causes a prolongation of bleeding time. Ticlopidine is indicated for use in patients, who have had a stroke or stroke precursors and who cannot take aspirin or aspirin has not worked, to try to prevent another thrombotic stroke. It may also be utilized in the placement of stents in coronary arteries (1). In pharmaceutical products, ticlopidine is employed as ticlopidine hydrochloride (CAS number 53885-35-1, EC number 258-837-4, molecular formula $C_{14}H_{15}Cl_2NS$) (1).

CUTANEOUS ADVERSE DRUG REACTIONS FROM SYSTEMIC ADMINISTRATION CAUSED BY TYPE IV (DELAYED-TYPE) HYPERSENSITIVITY (as demonstrated by positive patch tests)

Cutaneous adverse drug reactions from systemic administration of ticlopidine caused by type IV (delayed-type) hypersensitivity have included acute generalized exanthematous pustulosis (AGEP) (2) and fixed drug eruption (3).

Acute generalized exanthematous pustulosis (AGEP)

A 66-year-old woman was treated with oral ticlopidine 250 mg daily to prevent thrombosis. After 3 weeks, she presented with fever (39°C) and an acute pruritic rash, which had begun on the trunk and subsequently spread to the arms and legs. The face, palms and soles were not involved and the mucosae were unaffected. Examination showed several hundred superficial, 1-3 mm sized, non-follicular pustules on an erythematous base. Laboratory examinations revealed leukocytosis, neutrophilia and an erythrocyte sedimentation rate of 43 mm in the first hour. Negative results were obtained in fungal and bacterial cultures from the pustules. A skin biopsy showed subcorneal pustules with neutrophils and a perivascular inflammatory cell infiltrate in the papillary dermis. Four weeks after healing, patch tests with ticlopidine 5% and 10% in distilled water and petrolatum were strongly positive at D3 with a pustular eruption on an erythematous base. Ten controls were negative. The patient was diagnosed with acute generalized exanthematous pustulosis (AGEP) from delayed-type allergy to ticlopidine (2).

Fixed drug eruption

A 54-year-old woman presented with pruritic erythematous vesicular lesions on the back. 3 months earlier, she had begun a course of oral ticlopidine for cerebrovascular disease. The drug was discontinued and the lesions resolved in 20 days, with residual hyperpigmentation. Patch tests with ticlopidine 1% and 5% were strongly positive on postlesional but negative on normal skin (20 controls negative). An oral challenge test with ticlopidine was positive, with the patient developing pruritic purplish erythematous lesions in the same distribution as before. A skin biopsy showed a predominantly chronic inflammatory infiltrate, with vacuolar degeneration of the basal layer, findings consistent with a fixed drug eruption (3).

Cross-reactions, pseudo-cross-reactions and co-reactions
Not to the related platelet aggregation inhibitor clopidogrel (3).

LITERATURE

1   The data in the section 'General' may have been obtained from literature discussed in this chapter, but mostly also or exclusively from one or more of the following online sources: ChemIDPlus Advanced, PubChem, DrugBank, RxList, Drug Central, Drugs.com, and Wikipedia
2   Cannavò SP, Borgia F, Guarneri F, Vaccaro M. Acute generalized exanthematous pustulosis following use of ticlopidine. Br J Dermatol 2000;142:577-578
3   García CM, Carmena R, García R, Berges P, Camacho E, Cotter MP, et al. Fixed drug eruption from ticlopidine, with positive lesional patch test. Contact Dermatitis 2001;44:40-41

Chapter 3.478 TILISOLOL

IDENTIFICATION

Description/definition   : Tilisolol is the isoquinoline that conforms to the structural formula shown below
Pharmacological classes   : Adrenergic β-antagonists; anti-arrhythmia agents
IUPAC name   : 4-[3-(*tert*-Butylamino)-2-hydroxypropoxy]-2-methylisoquinolin-1-one
CAS registry number   : 85136-71-6
EC number   : Not available
Merck Index monograph   : 10865
Patch testing   : No data available; most systemic drugs can be tested at 10% pet.; if the pure chemical is not available, prepare the test material from intravenous powder, the content of capsules or – when also not available – from powdered tablets to achieve a final concentration of the active drug of 10% pet.
Molecular formula   : $C_{17}H_{24}N_2O_3$

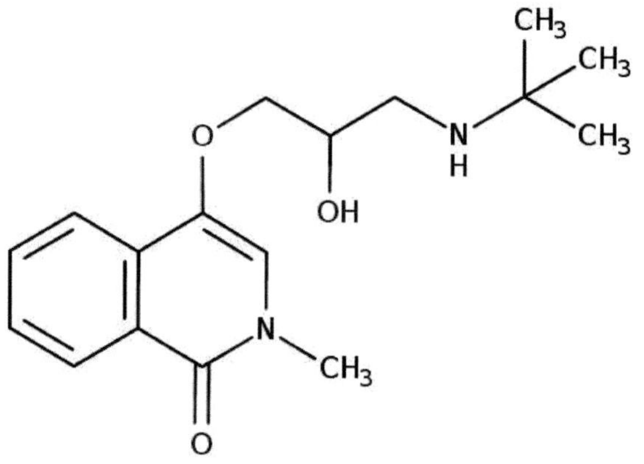

GENERAL

Tilisolol is a non-selective β-adrenergic blocking agent, which has a long-lasting and stable action in the clinical treatment of hypertension and angina pectoris. In pharmaceutical products, tilisolol is employed as tilisolol hydrochloride (CAS number 62774-96-3, EC number not available, molecular formula $C_{17}H_{25}ClN_2O_3$) (1).

CUTANEOUS ADVERSE DRUG REACTIONS FROM SYSTEMIC ADMINISTRATION CAUSED BY TYPE IV (DELAYED-TYPE) HYPERSENSITIVITY (as demonstrated by positive patch tests)

Cutaneous adverse drug reactions from systemic administration of tilisolol caused by type IV (delayed-type) hypersensitivity have included photosensitivity (2).

Photosensitivity

A patient developed a photosensitivity reaction during the treatment of hypertension with tilisolol hydrochloride. The action spectrum was mainly in the ultraviolet A range. The photosensitivity reaction was reproducible on oral re-administration of the drug and exposure to a low dose of UVA. Photopatch testing with the drug was positive in the patient, but not in controls. The authors admit that the data are insufficient to distinguish photoallergy from phototoxicity, but stated that that clinical and histologic features suggested that the mechanism involved was photoallergic in nature. They demonstrated phototoxicity to tilisolol in guinea pigs, but were unable to induce photoallergy (2).

LITERATURE

1   The data in the section 'General' may have been obtained from literature discussed in this chapter, but mostly also or exclusively from one or more of the following online sources: ChemIDPlus Advanced, PubChem, DrugBank, RxList, Drug Central, Drugs.com, and Wikipedia
2   Miyauchi H, Horiki S, Horio T. Clinical and experimental photosensitivity reaction to tilisolol hydrochloride. Photodermatol Photoimmunol Photomed 1994;10:255-258

Chapter 3.479 TINZAPARIN

IDENTIFICATION

Description/definition   : Tinzaparin is a low-molecular-weight heparin
Pharmacological classes : Antithrombotic agents
IUPAC name   : Not available
CAS registry number   : 9005-49-6 (Heparin)
EC number   : 232-681-7 (Heparin)
Merck Index monograph : 5958 (Heparin)
Patch testing   : Commercial preparation undiluted (4); consider intradermal testing with late readings
                               (D2,D3) when patch tests are negative and consider subcutaneous challenge when
                               intradermal tests are negative
Molecular formula   : Unspecified

GENERAL

Tinzaparin is a low molecular weight heparin (LMWH), produced by enzymatic depolymerization of unfractionated heparin from porcine intestinal mucosa and has an average molecular weight between 5500 and 7500 daltons. Tinzaparin is composed of molecules with and without a special site for high affinity binding to antithrombin III (ATIII). This complex greatly accelerates the inhibition of factor Xa. Tinzaparin is used for the prevention of postoperative venous thromboembolism, for the treatment of deep vein thrombosis and/or pulmonary embolism and for preventing clot formation in indwelling intravenous lines for hemodialysis. In pharmaceutical products, tinzaparin is employed as tinzaparin sodium (CAS number not available, EC number not available, molecular formula unspecified) (1).

See also bemiparin (Chapter 3.59), certoparin (Chapter 3.117), dalteparin (Chapter 3.160), danaparoid (Chapter 3.161), enoxaparin (Chapter 3.195), fondaparinux (Chapter 3.225), heparins (Chapter 3.239), and nadroparin (Chapter 3.331).

CUTANEOUS ADVERSE DRUG REACTIONS FROM SYSTEMIC ADMINISTRATION CAUSED BY TYPE IV (DELAYED-TYPE) HYPERSENSITIVITY

Throughout this book, only reports of delayed-type hypersensitivity have been included that showed a positive patch test to the culprit drug. However, as a result of the high molecular weight of heparins, patch tests are often false-negative, presumably from insufficient penetration into the skin. Because of this, and also because patch tests have been performed in a small minority of cases only, studies with a positive intradermal test or subcutaneous provocation tests with delayed readings are included in the chapters on the various heparins, even when patch tests were negative or not performed.

General information on delayed-type hypersensitivity reactions to heparins

General information on delayed-type hypersensitivity reactions to heparins (including tinzaparin, presenting as local reactions from subcutaneous administration, is provided in Chapter ... Heparins. In this chapter, only *non-local* cutaneous adverse drug reactions from delayed-type hypersensitivity to tinzaparin are presented.

Non-local cutaneous adverse drug reactions

Non-local cutaneous adverse drug reactions from systemic administration of tinzaparin caused by type IV (delayed-type) hypersensitivity have not been found.

Cross-reactions, pseudo-cross-reactions and co-reactions

Cross-reactions between heparins are frequent in delayed-type hypersensitivity (>90% of patients tested, median number of positive drugs per patient: 3) and do not depend on the molecular weight of the molecules (3). Overlap in their polysaccharide composition might explain the high degree of cross-allergenicity (5). Cross-reactions to the semisynthetic heparinoid danaparoid have also been observed (2). In allergic patients, the synthetic ultralow molecular weight synthetic heparin fondaparinux is usually, but not always (2) well-tolerated (5).

LITERATURE

1   The data in the section 'General' may have been obtained from literature discussed in this chapter, but mostly also or exclusively from one or more of the following online sources: ChemIDPlus Advanced, PubChem, DrugBank, RxList, Drug Central, Drugs.com, and Wikipedia

2   Utikal J, Peitsch WK, Booken D, Velten F, Dempfle CE, Goerdt S, et al. Hypersensitivity to the pentasaccharide fondaparinux in patients with delayed-type heparin allergy. Thromb Haemost 2005;94:895-896

3   Weberschock T, Meister AC, Bohrt K, Schmitt J, Boehncke W-H, Ludwig RJ. The risk for cross-reactions after a cutaneous delayed-type hypersensitivity reaction to heparin preparations is independent of their molecular weight: a systematic review. Contact Dermatitis 2011;65:187-194

4   Brockow K, Garvey LH, Aberer W, Atanaskovic-Markovic M, Barbaud A, Bilo MB, et al. Skin test concentrations for systemically administered drugs - an ENDA/EAACI Drug Allergy Interest Group position paper. Allergy 2013;68:702-712

5   Schindewolf M, Lindhoff-Last E, Ludwig RJ. Heparin-induced skin lesions. Lancet 2012;380:1867-1879

Chapter 3.480 TIOPRONIN

IDENTIFICATION

Description/definition   : Tiopronin is the acylated sulfhydryl-containing derivative of glycine that conforms to the
                             structural formula shown below
Pharmacological classes : Urologicals
IUPAC name   : 2-(2-Sulfanylpropanoylamino)acetic acid
Other names   : (2-Mercaptopropionyl)glycine
CAS registry number   : 1953-02-2
EC number   : 217-778-4
Merck Index monograph   : 10879
Patch testing   : 5% and 10% pet.; 10% is slightly irritant
Molecular formula   : $C_5H_9NO_3S$

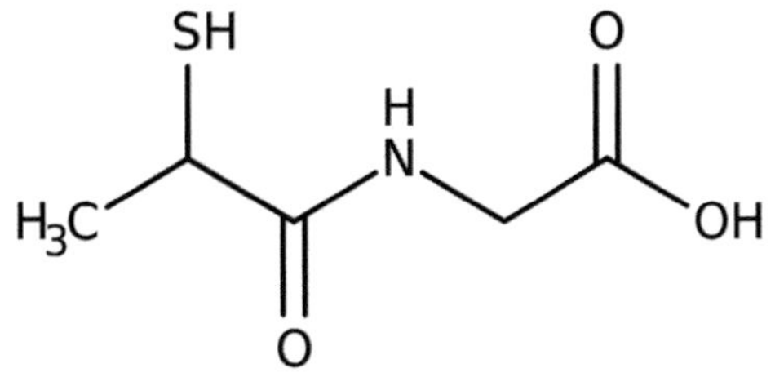

GENERAL

Tiopronin is an acylated sulfhydryl-containing derivative of glycine with reducing and complexing properties. It breaks the disulfide bond of cystine (an oxidized dimeric form of cysteine) and binds the sulfhydryl group of the resultant cysteine monomers to form a soluble tiopronin-cysteine-mixed disulfide, which is more water-soluble than cystine and is readily excreted. This leads to a reduction in urinary cystine concentration and subsequently reduces cystine stone formation. Tiopronin is indicated as a second-line for the prevention of kidney stone formation in patients with severe homozygous cystinuria. This drug may also be used as a mucolytic drug and to bind metal nanoparticles in Wilson's disease, which is an overload of copper in the body (1,2). The use of tiopronin aerosol treatment for acute bronchitis has resulted in contact allergy with a maculopapular facial rash and angioedema of the lips (2).

CUTANEOUS ADVERSE DRUG REACTIONS FROM SYSTEMIC ADMINISTRATION CAUSED BY TYPE IV (DELAYED-TYPE) HYPERSENSITIVITY (as demonstrated by positive patch tests)

Cutaneous adverse drug reactions from systemic administration of tiopronin caused by type IV (delayed-type) hypersensitivity have included lichenoid drug eruption (3,4,5,6,7) and erythema multiforme-like drug eruption (3).

Other cutaneous adverse drug reactions

Lichenoid drug eruption

Two patients with lichenoid eruption due to tiopronin were reported from Japan in 1988 (7). A 63-year-old woman developed a pruritic eruption after oral ingestion of tiopronin for 6 weeks. The clinicopathological findings were those of a lichenoid reaction. She also showed diffuse alopecia. The rash was reproduced by re-administering tiopronin. Patch tests were positive to tiopronin 2%, 5% and 10% pet. and to the commercial drug 0.5%-10% in a dilution series. Two of 13 controls were positive to the 10% tiopronin concentration (7). A 71-year-old man developed discoid lupus erythematosus-like lesions after oral tiopronin for two months, but the histological findings were a lichenoid reaction. Patch tests were positive to tiopronin 5% and 10% pet. and to the commercial product at 10% only (7).

A woman aged 53 years who had suffered an erythema multiforme-like drug eruption from tiopronin had a positive patch test to tiopronin 5% pet. and a positive 'provocation' (unspecified) (3). A man aged 48 years developed a lichenoid eruption suspected to have been caused by tiopronin or captopril. Patch tests were positive to tiopronin 5% and captopril 3% (vehicle not mentioned) and 'provocation' (unspecified) was positive for both drugs (3).

A 62-year-old woman, who had been taking tiopronin for 2 years because of liver dysfunction, presented with a three-month history of pruritic skin eruptions and a sore mouth. Physical examination showed flat-topped, round or polygonal, violaceous papules 5 to 10 mm in diameter bilaterally on the back of her hands, arms, thighs, and trunk, some showing scales, others white striae. There were white retiform plaques on the oral mucosa and on the tongue. The histopathology was consistent with lichen planus. Patch tests were positive to tiopronin 2.5%, 5%, and 10% pet. on 3 occasions (with intervals of 1 year) and to 20% pet. twice. Four months later, the patient was challenged with tiopronin 100 mg, three times a day. After taking 900 mg, an exudative erythema-like eruption appeared on the back of the joints of both hands and on the upper lip. An erythema with a white macule developed on the right cheek mucosa. Two years later, another challenge test resulted in a similar exudative erythema-like eruption on the joints of both hands. Histopathology showed liquefaction degeneration of the basal layer, some dyskeratotic cells in the epidermis and lymphocytic infiltration containing melanophages into the upper dermis. The patient was diagnosed with a lichenoid drug eruption from delayed-type hypersensitivity to tiopronin (4). See also the section 'Cross-reactions, pseudo-cross-reactions and co-reactions' below.

In a 52-year-old man, a lichenoid eruption developed on the face, arms and back with a whitish lacework pattern on the buccal mucosa, which had begun 2 weeks after initiating therapy with tiopronin. The lesions healed within a month after discontinuation of tiopronin. Six months later, patch tests were negative to tiopronin 1%, 2.5% and 5% pet., but positive (+) at D3 and D7 to tiopronin 10% and 20% pet. One of four non-exposed controls also had a + reaction (5).

Up to 1988, 12 cases of tiopronin-induced lichenoid drug eruptions have been reported in Japanese literature. It is unknown in how many patch tests were performed and with what results (6,7).

Erythema multiforme-like drug eruption
A 67-year-old woman who had developed an erythema multiforme-like eruption suspected to be caused by tiopronin had a ?+ patch test reaction to tiopronin 5% pet. but a positive intradermal test read at 6 and 24 hours (3).

Cross-reactions, pseudo-cross-reactions and co-reactions
A patient with a lichenoid drug eruption to tiopronin and a positive reaction to this drug, co-reacted to captopril and D-penicillamine, with which the patient had never been treated before. All three compounds have a sulfhydryl group, and the authors suggested this to play a role in the positive patch tests (4). Co-reactivity to captopril has also been observed by other investigators (5).

LITERATURE
1   The data in the section 'General' may have been obtained from literature discussed in this chapter, but mostly also or exclusively from one or more of the following online sources: ChemIDPlus Advanced, PubChem, DrugBank, RxList, Drug Central, Drugs.com, and Wikipedia
2   Romano A, Pietrantonio F, di Fonso M, Venuti A, Fabrizi G. Contact allergy to tiopronin: a case report. Contact Dermatitis 1995;33:269
3   Kitamura K, Aihara M, Osawa J, Naito S, Ikezawa Z. Sulfhydryl drug-induced eruption: a clinical and histological study. J Dermatol 1990;17:44-51
4   Kurumaji Y, Miyazaki K. Tiopronin-induced lichenoid eruption in a patient with liver disease and positive patch test reaction to drugs with sulfhydryl group. J Dermatol 1990;17:176-181
5   Piérard E, Delaporte E, Flipo RM, Duneton-Bitbol V, Dejobert Y, Piette F, Bergoend H. Tiopronin-induced lichenoid eruption. J Am Acad Dermatol 1994;31:665-667
6   Kurumaji Y, Miyazaki K. The result of patch tests in a case of lichenoid drug eruption due to tiopronin. Skin Research 1987;29(Suppl.):66-70 (Article in Japanese, data cited in ref. 4)
7   Kawabe Y, Mizuno N, Yoshikawa K, Matsumoto Y. Lichenoid eruption due to mercaptopropionylglycine. J Dermatol 1988;15:434-439

Chapter 3.481 TOBRAMYCIN

IDENTIFICATION

Description/definition   : Tobramycin is an aminoglycoside antibiotic produced by *Streptomyces tenebrarius* that
                            conforms to the structural formula shown below
Pharmacological classes : Anti-bacterial agents
IUPAC name   : (2S,3R,4S,5S,6R)-4-Amino-2-[(1S,2S,3R,4S,6R)-4,6-diamino-3-[(2R,3R,5S,6R)-3-amino-6-
                            (aminomethyl)-5-hydroxyoxan-2-yl]oxy-2-hydroxycyclohexyl]oxy-6-(hydroxymethyl)-
                            oxane-3,5-diol
CAS registry number   : 32986-56-4
EC number   : 251-322-5
Merck Index monograph   : 10917
Patch testing   : 20.0% pet. (Chemotechnique, SmartPracticeCanada); late readings are advisable (7)
Molecular formula   : $C_{18}H_{37}N_5O_9$

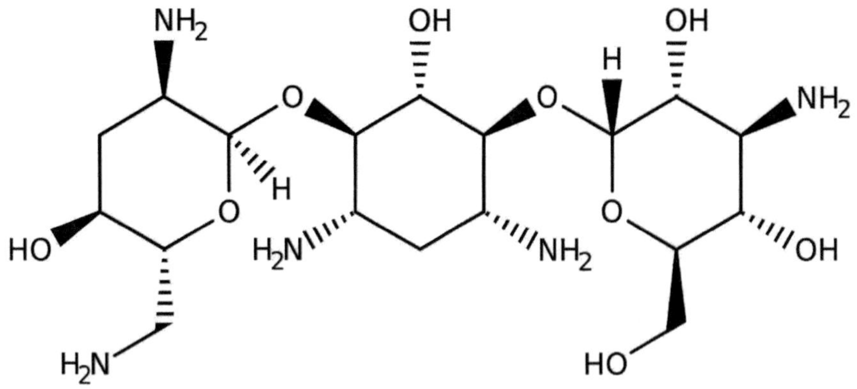

GENERAL

Tobramycin is an aminoglycoside broad-spectrum antibiotic produced by *Streptomyces tenebrarius* with bacterio-static activity. It is effective against gram-negative bacteria, especially the *Pseudomonas* species. Tobramycin is a 10% component of the antibiotic complex, nebramycin, produced by the same species. This agent is indicated for the treatment of *Pseudomonas aeruginosa* lung infections. In pharmaceutical products, both tobramycin and tobramycin sulfate (CAS number 49842-07-1, EC number 256-499-2, molecular formula $C_{18}H_{39}N_5O_{13}S$) may be employed (1).

In topical preparations, tobramycin has caused a number of cases of contact allergy/allergic contact dermatitis, which subject has been reviewed *in extenso* in Volume 3 of the *Monographs in contact allergy* series (2).

CUTANEOUS ADVERSE DRUG REACTIONS FROM SYSTEMIC ADMINISTRATION CAUSED BY TYPE IV (DELAYED-TYPE) HYPERSENSITIVITY (as demonstrated by positive patch tests)

Cutaneous adverse drug reactions from systemic administration of tobramycin caused by type IV (delayed-type) hypersensitivity have included eczematous eruption (3) and erythema multiforme-like eruption (4).

Eczematous eruption

In a group of 45 patients with multiple drug hypersensitivity syndrome seen between 1996 and 2018 in Montpellier, France, 38 of 92 drug hypersensitivities were classified as type IV immunological reactions. One individual, a man aged 80, had eczema from dexamethasone and tobramycin. The eruption(s) appeared in the second day of administration, which means that the patient was already sensitized to this drug or a cross-reacting chemical. It is plausible that the patient had previously been sensitized by neomycin and a corticosteroid, which would imply this to be a case/cases of systemic allergic dermatitis (3).

Erythema multiforme-like eruption

A 55-year-old woman was treated for febrile post-operative necrotic pancreatitis with tobramycin, clindamycin, cefalotin and cefalexin. An erythematopapular eruption began on the neck and spread to the arms, legs and trunk. Erythematous plaques and vesicular lesions were present. The dermatitis and fever subsided a few days after stopping the antibiotics. Patch tests were positive to commercial tobramycin 50 mg/ml and to clindamycin 150 mg/ml. The patient was diagnosed with erythema multiforme-like eruption caused by tobramycin, clindamycin, and cefalotin (the latter diagnosed by a positive delayed intradermal test) (4).

Cross-reactions, pseudo-cross-reactions and co-reactions
In patients sensitized to neomycin, about 60% cross-react to tobramycin (Chapter 3.335 Neomycin). The cross-sensitivity pattern between aminoglycoside antibiotics in patients primarily sensitized to tobramycin has not been well investigated. However, there are indications that such patients often do *not* cross-react to neomycin (2).

LITERATURE
1   The data in the section 'General' may have been obtained from literature discussed in this chapter, but mostly also or exclusively from one or more of the following online sources: ChemIDPlus Advanced, PubChem, DrugBank, RxList, Drug Central, Drugs.com, and Wikipedia
2   De Groot AC. Monographs in contact allergy, volume 3: Topical Drugs. Boca Raton, Fl, USA: CRC Press Taylor and Francis Group, 2021 (ISBN 978-0-367-23693-9)
3   Landry Q, Zhang S, Ferrando L, Bourrain JL, Demoly P, Chiriac AM. Multiple drug hypersensitivity syndrome in a large database. J Allergy Clin Immunol Pract 2019;8:258
4   Muñoz D, Del Pozo MD, Audícana M, Fernandez E, Fernandez De Corres LF. Erythema-multiforme-like eruption from antibiotics of 3 different groups. Contact Dermatitis 1996;34:227-228

Chapter 3.482 TOLBUTAMIDE

IDENTIFICATION

Description/definition   : Tolbutamide is the benzenesulfonamide that conforms to the structural formula shown
                             below
Pharmacological classes : Hypoglycemic agents
IUPAC name   : 1-Butyl-3-(4-methylphenyl)sulfonylurea
CAS registry number   : 64-77-7
EC number   : 200-594-3
Merck Index monograph   : 10937
Patch testing   : No data available; most systemic drugs can be tested at 10% pet.; if the pure chemical is
                           not available, prepare the test material from intravenous powder, the content of capsules
                           or – when also not available – from powdered tablets to achieve a final concentration of
                           the active drug of 10% pet.
Molecular formula   : $C_{12}H_{18}N_2O_3S$

GENERAL

Tolbutamide is an oral antihyperglycemic agent used for the treatment of non-insulin-dependent diabetes mellitus. It belongs to the sulfonylurea class of insulin secretagogues, which act by stimulating β cells of the pancreas to release insulin. Tolbutamide increases both basal insulin secretion and meal-stimulated insulin release. It also increases peripheral glucose utilization, decreases hepatic gluconeogenesis and may increase the number and sensitivity of insulin receptors (1). In pharmaceutical products, tolbutamide is employed as tolbutamide sodium (CAS number 473-41-6, EC number not available, molecular formula $C_{12}H_{17}N_2NaO_3S$) (1).

CUTANEOUS ADVERSE DRUG REACTIONS FROM SYSTEMIC ADMINISTRATION CAUSED BY TYPE IV (DELAYED-TYPE) HYPERSENSITIVITY (as demonstrated by positive patch tests)

Cutaneous adverse drug reactions from systemic administration of tolbutamide caused by type IV (delayed-type) hypersensitivity have included systemic allergic dermatitis (2,3).

Systemic allergic dermatitis (systemic contact dermatitis)

Patients contact allergic to para-amino compounds (sulfanilamide, *p*-phenylenediamine, benzocaine) were orally challenged with 3 related sulfonylurea derivatives (hypoglycemic agents, sulfonamides): tolbutamide, carbutamide, and chlorpropamide. Eleven had a positive reaction: 3 (of 11 tested) to tolbutamide, 7 (of 25 tested) to carbutamide, and 1 (of 20 tested) with chlorpropamide. All were patients previously sensitized to sulfanilamide. Symptoms were itching in all 11 patients, reappearance of erythema and vesicles at the site of the primary contact dermatitis in 6 patients, and relapse of the primary contact dermatitis with a moderate secondary vesicular eruption together with a reactivation of the patch test reaction in five. Patch tests with these drugs themselves were not performed (2).

A similar observation (with tolbutamide and chlorpropamide) was reported one year later, but details are not available to the author (3). These were very likely cases of systemic allergic dermatitis.

LITERATURE

1   The data in the section 'General' may have been obtained from literature discussed in this chapter, but mostly also or exclusively from one or more of the following online sources: ChemIDPlus Advanced, PubChem, DrugBank, RxList, Drug Central, Drugs.com, and Wikipedia
2   Angelini G, Meneghini CL. Oral tests in contact allergy to para-amino compounds. Contact Dermatitis 1981;7:311-314
3   Fisher AA. Systemic contact dermatitis from Orinase ® and Diabinese ® in diabetics with para-amino hypersensitivity. Cutis 1982;29:551-565

Chapter 3.483 TOPIRAMATE

IDENTIFICATION

Description/definition : Topiramate is the sulfamate-substituted fructose analog that conforms to the structural formula shown below

Pharmacological classes : Anticonvulsants; hypoglycemic agents

IUPAC name : [(3aS,5aR,8aR,8bS)-2,2,7,7-Tetramethyl-5,5a,8a,8b-tetrahydrodi[1,3]dioxolo[4,5-a:5',3'-d]pyran-3a

CAS registry number : 97240-79-4

EC number : 619-263-3

Merck Index monograph : 10976

Patch testing : Powdered pill, 10% and 30% pet. (2,3); most systemic drugs can be tested at 10% pet.; if the pure chemical is not available, prepare the test material from intravenous powder, the content of capsules or − when also not available – from powdered tablets to achieve a final concentration of the active drug of 10% pet.

Molecular formula : $C_{12}H_{21}NO_8S$

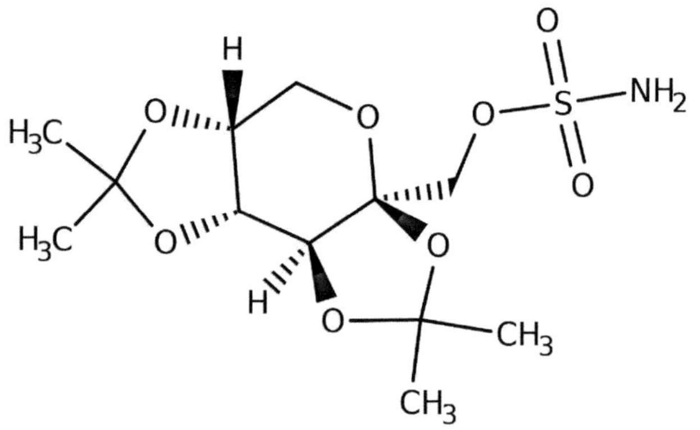

GENERAL

Topiramate is a sulfamate-substituted monosaccharide with anticonvulsant property. It works by stabilizing hyper-excited neural membranes, inhibiting repetitive neuronal firing, and decreasing propagation of synaptic impulses, thereby impeding seizure occurrences. Topiramate is indicated for partial onset or primary generalized tonic-clonic seizures, as adjunctive therapy for seizures associated with Lennox-Gastaut syndrome and for prophylaxis of migraine headache in adults. The drug has also been commonly investigated and used off-label for weight reduction in patients with obesity or diabetes (1).

CUTANEOUS ADVERSE DRUG REACTIONS FROM SYSTEMIC ADMINISTRATION CAUSED BY TYPE IV (DELAYED-TYPE) HYPERSENSITIVITY (as demonstrated by positive patch tests)

Cutaneous adverse drug reactions from systemic administration of topiramate caused by type IV (delayed-type) hypersensitivity have included drug reaction with eosinophilia and systemic symptoms (DRESS) (2) and localized erythema (3).

Drug reaction with eosinophilia and systemic symptoms (DRESS)

Between January 1998 and December 2008, in a university hospital in Portugal, 56 patients with DRESS were investigated with patch testing of the suspected drugs. Topiramate (powdered tablet, 30% in water and pet.) was tested in one patient and the patch test reaction was positive. Seventeen controls were negative (2).

Other cutaneous adverse drug reactions

A 55-year-old woman with seizures developed erythema of the neck, nausea and vomiting after a 2-month treatment with topiramate. The drug was immediately discontinued and all symptoms disappeared in a few days. Then the patient was treated with carbamazepine, but after 2 weeks she presented with erythema and edema of the face and pruritus of the legs. Patch tests were positive to topiramate 10% pet. and carbamazepine 10% pet. (probably the powder of the pills). Ten controls were negative (3).

LITERATURE

1   The data in the section 'General' may have been obtained from literature discussed in this chapter, but mostly
     also or exclusively from one or more of the following online sources: ChemIDPlus Advanced, PubChem,
     DrugBank, RxList, Drug Central, Drugs.com, and Wikipedia
2   Santiago F, Gonçalo M, Vieira R, Coelho S, Figueiredo A. Epicutaneous patch testing in drug hypersensitivity
     syndrome (DRESS). Contact Dermatitis 2010;62:47-53
3   Schiavino D, Nucera E, Buonomo A, Musumeci S, Pollastrini E, Roncallo C, et al. A case of type IV hypersensitivity
     to topiramate and carbamazepine. Contact Dermatitis 2005;52:161-162

Chapter 3.484 TOSUFLOXACIN TOSILATE

IDENTIFICATION

Description/definition : Tosufloxacin tosilate is the quinolone that conforms to the structural formula shown below

Pharmacological classes : Anti-bacterial agents

IUPAC name : 7-(3-Aminopyrrolidin-1-yl)-1-(2,4-difluorophenyl)-6-fluoro-4-oxo-1,8-naphthyridine-3-carboxylic acid;4-methylbenzenesulfonic acid

CAS registry number : 115964-29-9

EC number : Not available

Merck Index monograph : 10984 (Tosufloxacin)

Patch testing : Tablet, powdered, 20% pet. (2); most systemic drugs can be tested at 10% pet.; if the pure chemical is not available, prepare the test material from intravenous powder, the content of capsules or – when also not available – from powdered tablets to achieve a final concentration of the active drug of 10% pet.

Molecular formula : $C_{26}H_{23}F_3N_4O_6S$

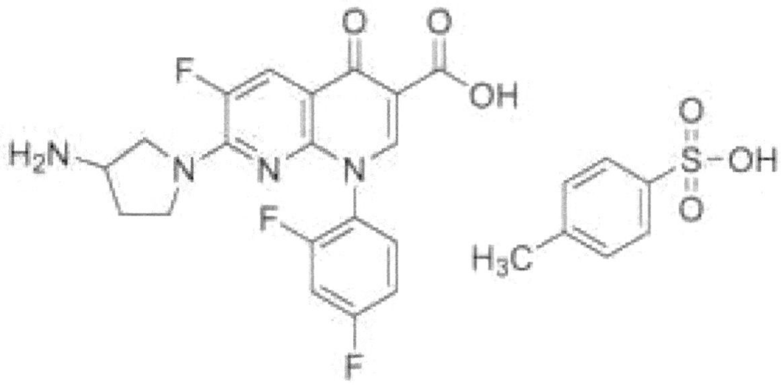

GENERAL

Tosufloxacin tosilate is a fluoroquinolone antibiotic containing a racemate comprising equimolar amounts of (R)- and (S)-tosufloxacin tosilate. It is a DNA synthesis inhibitor and a topoisomerase-IV inhibitor. Tosufloxacin tosilate has a controversial safety profile in relation to other fluoroquinolones and seems to be used as an antimicrobial agent only in Japan and China (1).

CUTANEOUS ADVERSE DRUG REACTIONS FROM SYSTEMIC ADMINISTRATION CAUSED BY TYPE IV (DELAYED-TYPE) HYPERSENSITIVITY (as demonstrated by positive patch tests)

Cutaneous adverse drug reactions from systemic administration of tosufloxacin tosilate caused by type IV (delayed-type) hypersensitivity have included fixed drug eruption (2).

Fixed drug eruption

A 66-year-old woman presented with itchy erythema of both loins. Three days before, she had taken tosufloxacin tosilate 150 mg orally. She had experienced a few episodes with erythema at the same sites before after taking this drug. Patch tests with tosufloxacin tosilate on normal and previously affected skin were negative. However, an open application test with the drug 20% pet. on a previously affected area, 3 cm in diameter, showed a positive response at 2 and 3 days. An oral provocation test was also positive. The patient was diagnosed with fixed drug eruption from tosufloxacin tosilate (2).

LITERATURE

1 The data in the section 'General' may have been obtained from literature discussed in this chapter, but mostly also or exclusively from one or more of the following online sources: ChemIDPlus Advanced, PubChem, DrugBank, RxList, Drug Central, Drugs.com, and Wikipedia

2 Sangen Y, Kawada A, Asai M, Aragane Y, Yudate T, Tezuka T. Fixed drug eruption induced by tosufloxacin tosilate. Contact Dermatitis 2000;42:285

Chapter 3.485 TRAMADOL

IDENTIFICATION

Description/definition	: Tramadol is the narcotic analgesic and codeine analog that conforms to the structural formula shown below
Pharmacological classes	: Analgesics, opioid; narcotics
IUPAC name	: (1R,2R)-2-[(Dimethylamino)methyl]-1-(3-methoxyphenyl)cyclohexan-1-ol
CAS registry number	: 27203-92-5
EC number	: 248-319-6
Merck Index monograph	: 10996
Patch testing	: No data available; most systemic drugs can be tested at 10% pet.; if the pure chemical is not available, prepare the test material from intravenous powder, the content of capsules or – when also not available – from powdered tablets to achieve a final concentration of the active drug of 10% pet.
Molecular formula	: $C_{16}H_{25}NO_2$

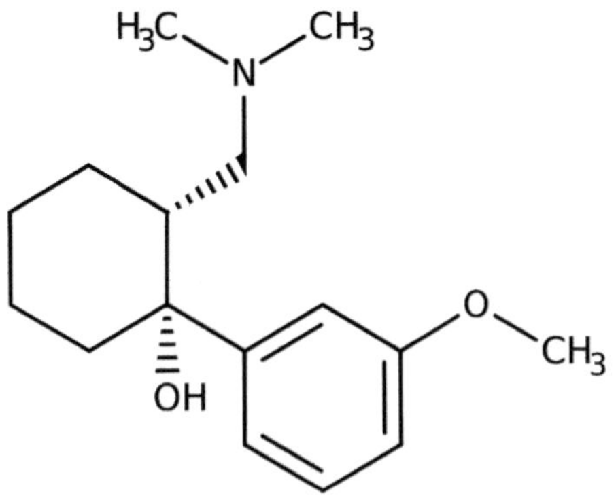

GENERAL

Tramadol is a synthetic 4-phenylpiperidine analog of codeine. It has central analgesic properties with effects similar to opioids, such as morphine and codeine, acting on specific opioid receptors; however, it has less potential for abuse, addiction and respiratory depression. Tramadol exists as a racemic mixture of the *trans*-isomer, with important differences in binding, activity, and metabolism associated with the two enantiomers. This drug is indicated for the management of moderate to severe pain. As an off-label indication, tramadol has been investigated for the treatment of premature ejaculation. In pharmaceutical products, tramadol is employed as tramadol hydrochloride (CAS number 36282-47-0, EC number 252-950-2, molecular formula $C_{16}H_{26}ClNO_2$) (1).

CUTANEOUS ADVERSE DRUG REACTIONS FROM SYSTEMIC ADMINISTRATION CAUSED BY TYPE IV (DELAYED-TYPE) HYPERSENSITIVITY (as demonstrated by positive patch tests)

Cutaneous adverse drug reactions from systemic administration of tramadol caused by type IV (delayed-type) hypersensitivity have included systemic allergic dermatitis (2,3).

Systemic allergic dermatitis (systemic contact dermatitis)

A 52-year-old woman, suffering from joint pain resulting from Ehlers–Danlos syndrome, presented with a possible allergic contact dermatitis from buprenorphine patches. Patch tests were indeed positive to buprenorphine 0.3 mg/ml in pet. At a pain centre, the patient was prescribed oral treatment with tramadol. Within 24 hours of treatment, a severe systemic allergic dermatitis reaction was observed at the site of the previous buprenorphine patch, with homogeneous infiltration, vesicles and bullae, as well as an orange-red discoloration, and influenza-like symptoms with fever and shivering. No baboon syndrome was observed. The patient underwent patch, prick and intradermal testing with other synthetic and non-synthetic morphine derivates (morphine, methadone and ketomebidone) and these were all negative. Quite curiously, tramadol itself was not tested. Yet, the authors diagnosed systemic allergic contact dermatitis, which is not proven but highly likely. They also hinted at the possibility of a cross-reaction from buprenorphine sensitization to tramadol, although this was not stated explicitly (2).

A similar case of systemic allergic dermatitis from tramadol in a patient sensitized to buprenorphine was presented in the same year from France (3).

Cross-reactions, pseudo-cross-reactions and co-reactions
Tramadol may have cross-reacted in patients primarily sensitized to buprenorphine (2,3).

LITERATURE

1   The data in the section 'General' may have been obtained from literature discussed in this chapter, but mostly also or exclusively from one or more of the following online sources: ChemIDPlus Advanced, PubChem, DrugBank, RxList, Drug Central, Drugs.com, and Wikipedia
2   Kaae J, Menné T, Thyssen JP. Systemic contact dermatitis following oral exposure to tramadol in a patient with allergic contact dermatitis caused by buprenorphine. Contact Dermatitis 2012;66:106-107
3   Schmutz JL, Trechot P. Systemic reactivation of contact eczema following tramadol administration in a patient with buprenorphine-induced eczema. Ann Dermatol Venereol 2012;139:335-336 (Article in French)

Chapter 3.486 TRANEXAMIC ACID

IDENTIFICATION

Description/definition	: Tranexamic acid is the monocarboxylic acid and synthetic derivative of the amino acid lysine that conforms to the structural formula shown below
Pharmacological classes	: Antifibrinolytic agents
IUPAC name	: 4-(Aminomethyl)cyclohexane-1-carboxylic acid
CAS registry number	: 1197-18-8
EC number	: 214-818-2
Merck Index monograph	: 11000
Patch testing	: 1% and 10% pet. (2); uncertain whether the pure drugs were used or pulverized tablets, most likely the latter
Molecular formula	: $C_8H_{15}NO_2$

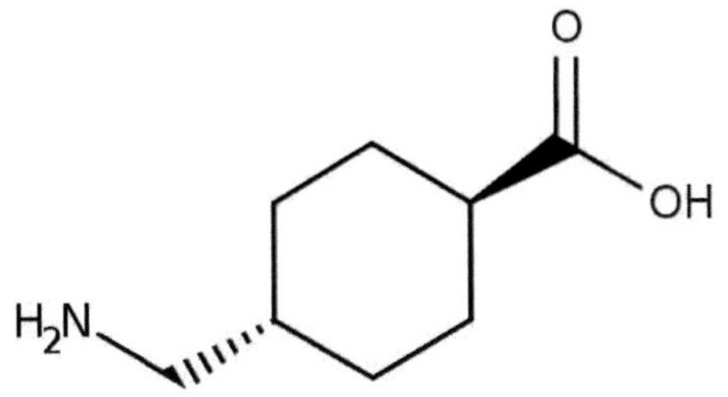

GENERAL

Tranexamic acid is an antifibrinolytic that competitively inhibits the activation of plasminogen to plasmin. It is indicated in patients with hemophilia for short term use (two to eight days) to reduce or prevent hemorrhage during and following tooth extraction. It can also be used for excessive bleeding in menstruation, surgery, or trauma cases. This agent has a longer half-life, is approximately ten times more potent, and is less toxic than aminocaproic acid, which possesses similar mechanisms of action (1). Besides its hemostatic effect, tranexamic acid is also applied in mucocutaneous diseases such as urticaria, angioedema, stomatitis and tonsillitis, mainly in Japan (4).

CUTANEOUS ADVERSE DRUG REACTIONS FROM SYSTEMIC ADMINISTRATION CAUSED BY TYPE IV (DELAYED-TYPE) HYPERSENSITIVITY (as demonstrated by positive patch tests)

Cutaneous adverse drug reactions from systemic administration of tranexamic acid caused by type IV (delayed-type) hypersensitivity have included fixed drug eruption (2).

Fixed drug eruption

A 33-year-old man presented with multiple pigmented patches on his trunk and extremities, which had appeared within 1 hour of taking tranexamic acid and 3 other drugs for the common cold. Ten months previously, a similar eruption had developed after taking tranexamic acid and 2 other drugs for influenza. Physical examination revealed several oval brownish pigmented patches on the arms, dorsum of the right foot, and waist. A biopsy was refused. Patch tests were performed with all drugs (1% and 10% pet.) used in the 2 episodes on the lesional and non-lesional skin, showing erythematous reactions at D2 and D3 to 1% and 10% tranexamic acid at the lesional skin only (2).

Up to 2015, eight cases of fixed drug eruption induced by tranexamic acid have been reported in Japanese literature. Five of these patients were patch tested, but only one had a positive reaction to tranexamic acid (no details provided) (data cited in ref. 2). In a recent case of fixed drug eruption to tranexamic acid from Japan, patch tests were not performed, but a lymphocyte transformation test was positive (3)

LITERATURE

1   The data in the section 'General' may have been obtained from literature discussed in this chapter, but mostly also or exclusively from one or more of the following online sources: ChemIDPlus Advanced, PubChem, DrugBank, RxList, Drug Central, Drugs.com, and Wikipedia
2   Matsumura N, Hanami Y, Yamamoto T. Tranexamic acid-induced fixed drug eruption. Indian J Dermatol 2015;60:421
3   Kawaguchi K, Kinoshita S, Ishikawa M, Sakura H. Tranexamic acid-induced fixed drug eruption confirmed by the drug lymphocyte transformation test. Clin Case Rep 2019;7:2074-2075

Chapter 3.487 TRAZODONE

IDENTIFICATION

Description/definition : Trazodone is the synthetic triazolopyridine derivative that conforms to the structural
formula shown below

Pharmacological classes : Antidepressive agents, second-generation; serotonin uptake inhibitors; anti-anxiety
agents

IUPAC name : 2-[3-[4-(3-Chlorophenyl)piperazin-1-yl]propyl]-[1,2,4]triazolo[4,3-a]pyridin-3-one

CAS registry number : 19794-93-5

EC number : 243-317-1

Merck Index monograph : 11010

Patch testing : Crushed tablet, 30% pet. or water (3,5); most systemic drugs can be tested at 10% pet.; if
the pure chemical is not available, prepare the test material from intravenous powder,
the content of capsules or – when also not available – from powdered tablets to achieve a
final concentration of the active drug of 10% pet.

Molecular formula : $C_{19}H_{22}ClN_5O$

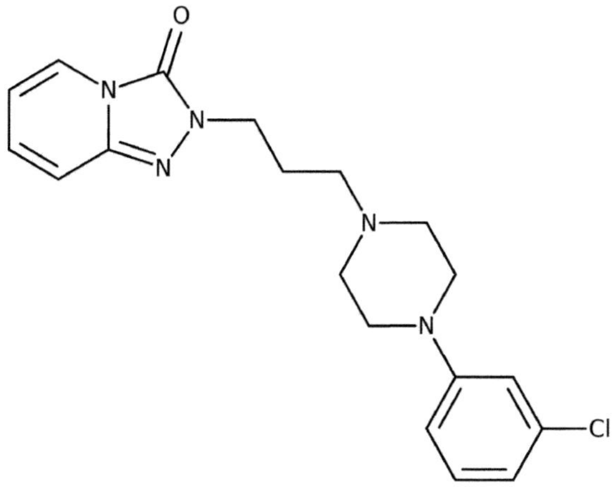

GENERAL

Trazodone is a synthetic triazolopyridine derivative with antidepressant and sedative properties. It selectively inhibits the re-uptake of serotonin by synaptosomes in the brain, thereby increasing serotonin levels in the synaptic cleft and potentiating serotonin activity. The sedative effect of trazodone is likely via α-adrenergic and mild histamine H_1-blocking actions. Trazodone is indicated for the treatment of depression (1). In pharmaceutical products, trazodone is employed as trazodone hydrochloride (CAS number 25332-39-2, EC number 246-855-5, molecular formula $C_{19}H_{23}Cl_2N_5O$) (1).

CONTACT ALLERGY FROM ACCIDENTAL CONTACT

Case series

In Leuven, Belgium, in the period 2001-2019, 201 of 1248 health care workers/employees of the pharmaceutical industry had occupational allergic contact dermatitis. In 23 (11%) dermatitis was caused by skin contact with a systemic drug: 19 nurses, two chemists, one physician, and one veterinarian. The lesions were mostly localized on the hands, but often also on the face, as airborne dermatitis. In total, 42 positive patch test reactions to 18 different systemic drugs were found. In 2 patients, trazodone was the drug/one of the drugs that caused occupational dermatitis (4, overlap with refs. 2 and 3).

Also In Leuven, Belgium, in the period 2007-2011, 81 patients have been diagnosed with occupational airborne allergic contact dermatitis. In 23 of them, drugs were the offending agents, including trazodone hydrochloride in 2 cases (2, overlap with refs. 3 and 4).

Case reports

A 66-year-old woman presented with itchy and burning eczema of the face, which had first appeared on the eyelids. Later, extension to the forehead, lips and perioral region, neck, ears and the fingers occurred. The patient's husband suffered from Parkinson's disease, and she had to crush a large number of tablets for him up to five times a day. She had positive patch tests to trazodone HCl (crushed tablet 30% pet.; 12 controls were negative) and 4 benzodiaze-pines (bromazepam, clotiazepam, lorazepam, tetrazepam). The patient was diagnosed with (airborne) allergic contact dermatitis from drugs *by proxy* (3, overlap with refs. 2 and 4).

A 30-year-old woman presented with an itchy, scaly skin reaction on the face, which was most pronounced on the eyelids. She reported a clear relationship with her work as a geriatric nurse, during which she was required to crush tablets for the elderly on a daily basis, mostly benzodiazepine drugs. She had positive patch tests to trazodone hydrochloride (crushed tablet 30% pet.; 12 controls were negative), 5 benzodiazepines (alprazolam, bromazepam, diazepam, lorazepam, tetrazepam) and 2 unrelated drugs. The patient was diagnosed with occupational airborne allergic contact dermatitis from drugs (3, overlap with ref. 2).

CUTANEOUS ADVERSE DRUG REACTIONS FROM SYSTEMIC ADMINISTRATION CAUSED BY TYPE IV (DELAYED-TYPE) HYPERSENSITIVITY (as demonstrated by positive patch tests)

Cutaneous adverse drug reactions from systemic administration of trazodone caused by type IV (delayed-type) hypersensitivity have included generalized bullous fixed drug eruption (5).

Fixed drug eruption

An 86-year-old man received nine series of intravenous immunoglobulin for a progressive chronic demyelinating inflammatory polyneuropathy, during which he also received trazodone and acetaminophen. One day after the second course, three or four rounded erythematous macules appeared on his face evolving into pigmented lesions. The severity of the rash progressively increased after treatment with several flares of lesions spreading to his body All lesions reappeared at the same sites and some lesions became bullous. A biopsy was compatible with fixed drug eruption. Patch tests were negative to acetaminophen 10% pet. on a previous lesion but positive to trazodone 30% water and pet. on postlesional skin with an increased erythematous bullous reaction on D3. Two controls were negative. A skin biopsy of the trazodone patch test on D3 showed histological features of fixed drug eruption (5).

LITERATURE

1   The data in the section 'General' may have been obtained from literature discussed in this chapter, but mostly also or exclusively from one or more of the following online sources: ChemIDPlus Advanced, PubChem, DrugBank, RxList, Drug Central, Drugs.com, and Wikipedia
2   Swinnen I, Goossens A. An update on airborne contact dermatitis: 2007-2011. Contact Dermatitis 2013;68:232-238
3   Swinnen I, Ghys K, Kerre S, Constandt L, Goossens A. Occupational airborne contact dermatitis from benzodiazepines and other drugs. Contact Dermatitis 2014;70:227-232
4   Gilissen L, Boeckxstaens E, Geebelen J, Goossens A. Occupational allergic contact dermatitis from systemic drugs. Contact Dermatitis 2020;82:24-30
5   Combemale L, Ben Saïd B, Dupire G. Generalized bullous fixed drug eruption: Trazodone as a new culprit. Contact Dermatitis 2020;82:192-193

Chapter 3.488 TRIAMCINOLONE

IDENTIFICATION

Description/definition	: Triamcinolone is the synthetic glucocorticoid that conforms to the structural formula shown below
Pharmacological classes	: Glucocorticoids; anti-inflammatory agents
IUPAC name	: (8S,9R,10S,11S,13S,14S,16R,17S)-9-Fluoro-11,16,17-trihydroxy-17-(2-hydroxyacetyl)-10,13-dimethyl-6,7,8,11,12,14,15,16-octahydrocyclopenta[a]phenanthren-3-one
Other names	: 11β,16α,17α,21-Tetrahydroxy-9α-fluoro-1,4-pregnadiene-3,20-dione; fluoxyprednisolone
CAS registry number	: 124-94-7
EC number	: 204-718-7
Merck Index monograph	: 11027
Patch testing	: In general, corticosteroids may be tested at 0.1% and 1% in alcohol; late readings (6-10 days) are strongly recommended
Molecular formula	: $C_{21}H_{27}FO_6$

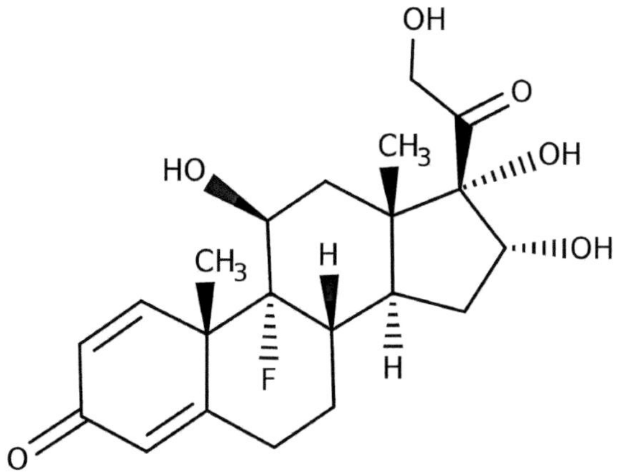

GENERAL

Systemically administered glucocorticoids have anti-inflammatory, immunosuppressive and antineoplastic properties and are used in the treatment of a wide spectrum of diseases including rheumatic disorders, lung diseases (asthma, COPD), gastrointestinal tract disorders (Crohn's disease, colitis ulcerosa), certain malignancies (leukemia, lymphomas), hematological disorders, and various diseases of the kidneys, brain, eyes and skin. A practical guideline for diagnosing allergic reactions to corticosteroids is presented in ref. 1.

Triamcinolone base (alcohol) is used in tablets only. In other applications, esters are used, including triamcinolone acetonide (Chapter 3.489), triamcinolone diacetate and triamcinolone hexacetonide. In topical preparations, triamcinolone is virtually always used as triamcinolone acetonide. As triamcinolone *base* is used in oral preparations only, most positive patch test reactions to this corticosteroid must have been the result of sensitization to one of its esters or of cross-sensitization to another corticosteroid.

CUTANEOUS ADVERSE DRUG REACTIONS FROM SYSTEMIC ADMINISTRATION CAUSED BY TYPE IV (DELAYED-TYPE) HYPERSENSITIVITY (as demonstrated by positive patch tests)

Cutaneous adverse drug reactions from systemic administration of triamcinolone caused by type IV (delayed-type) hypersensitivity have included systemic allergic dermatitis (SAD) presenting as maculopapular exanthema (2), SAD presenting as erythematous rash (4) and generalized eczematous eruption (3).

Systemic allergic dermatitis (systemic contact dermatitis)

A 46-year-old woman was documented by patch and provocative use testing to be allergic to multiple topical corticosteroids. On further testing, oral provocation tests to triamcinolone, methylprednisolone, dexamethasone, and prednisone each produced a generalized maculopapular eruption in a delayed manner (2). This was a case of systemic allergic dermatitis presenting as maculopapular eruption.

A female patient, previously sensitized by prednisolone-containing eye ointment, was orally provoked with hydrocortisone and triamcinolone and reacted to both with an erythematous rash. Patch tests were negative to triamcinolone acetonide 0.1% pet., but late readings were not performed (4).

A case of systemic allergic dermatitis alledgedly caused by triamcinolone *acetonide* may in fact have been caused by triamcinolone *base* (Chapter 3.489 Triamcinolone acetonide, section 'Systemic allergic dermatitis', ref. 2).

Dermatitis/eczematous eruption

In a study from Nancy, France, 54 patients with suspected nonimmediate drug eruptions were assessed with patch testing. Of the 9 patients with generalized eczema, one had a positive patch test to triamcinolone. Patch testing reproduced the cutaneous adverse drug reaction (3).

Cross-reactions, pseudo-cross-reactions and co-reactions

Cross-reactions between corticosteroids are discussed in Chapter 3.399 Prednisolone.

LITERATURE

1   Baeck M, Goossens A. Immediate and delayed allergic hypersensitivity to corticosteroids: practical guidelines. Contact Dermatitis 2012;66:38-45
2   Chew AL, Maibach HI. Multiple corticosteroid orally elicited allergic contact dermatitis in a patient with multiple topical corticosteroid allergic contact dermatitis. Cutis 2000;65:307-311
3   Barbaud A, Reichert-Penetrat S, Tréchot P, Jacquin-Petit MA, Ehlinger A, Noirez V, et al. The use of skin testing in the investigation of cutaneous adverse drug reactions. Br J Dermatol 1998;139:49-58
4   Kulberg A, Schliemann S, Elsner P. Contact dermatitis as a systemic disease. Clin Dermatol 2014;32:414-419

Chapter 3.489 TRIAMCINOLONE ACETONIDE

IDENTIFICATION

Description/definition : Triamcinolone acetonide is the 16,17 acetonide ester of the synthetic glucocorticoid triamcinolone that conforms to the structural formula shown below

Pharmacological classes : Glucocorticoids; immunosuppressive agents; anti-inflammatory agents

IUPAC name : (1S,2S,4R,8S,9S,11S,12R,13S)-12-Fluoro-11-hydroxy-8-(2-hydroxyacetyl)-6,6,9,13-tetramethyl-5,7-dioxapentacyclo[10.8.0.0^{2,9}.0^{4,8}.0^{13,18}]icosa-14,17-dien-16-one

Other names : 9α-Fluoro-11β,21-dihydroxy-16α,17α-isopropylidenedioxypregna-1,4-diene-3,20-dione

CAS registry number : 76-25-5

EC number : 200-948-7

Merck Index monograph : 11028

Patch testing : 1.0% pet. (Chemotechnique, SmartPracticeCanada, SmartPracticeEurope); 0.1% pet. (SmartPracticeCanada); the lower concentration may be preferable (15); late readings (6-10 days) are strongly recommended

Molecular formula : C$_{24}$H$_{31}$FO$_6$

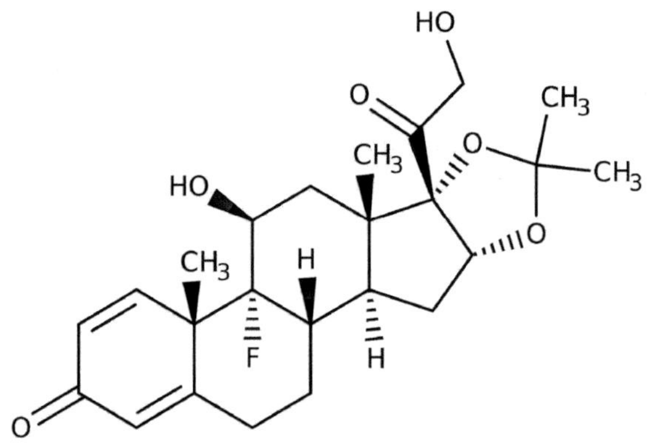

GENERAL

Systemically administered glucocorticoids have anti-inflammatory, immunosuppressive and antineoplastic properties and are used in the treatment of a wide spectrum of diseases including rheumatic disorders, lung diseases (asthma, COPD), gastrointestinal tract disorders (Crohn's disease, colitis ulcerosa), certain malignancies (leukemia, lymphomas), hematological disorders, and various diseases of the kidneys, brain, eyes and skin. A practical guideline for diagnosing allergic reactions to corticosteroids is presented in ref. 1. Triamcinolone acetonide 1% pet. is included in the American core allergen series (www.smartpracticecanada.com). Intradermal tests with triamcinolone acetonide may detect more cases of sensitization to triamcinolone acetonide than patch tests (5,6).

Triamcinolone acetonide is used in both topical and systemic pharmaceutical applications. In topical preparations, this glucocorticoid has uncommonly caused cases of contact allergy/allergic contact dermatitis, which subject has been fully reviewed in Volume 3 of the *Monographs in contact allergy* series (15). See also Triamcinolone (Chapter 3.488).

CONTACT ALLERGY FROM ACCIDENTAL CONTACT

Of 38 veterinarians with hand and forearm dermatoses seen by dermatologists in Belgium and the Netherlands from 1995 to 2005, 17 had occupational allergic contact dermatitis. In two patients, triamcinolone acetonide was the culprit drug allergen (18).

CUTANEOUS ADVERSE DRUG REACTIONS FROM SYSTEMIC ADMINISTRATION CAUSED BY TYPE IV (DELAYED-TYPE) HYPERSENSITIVITY (as demonstrated by positive patch tests)

Cutaneous adverse drug reactions from systemic administration of triamcinolone acetonide caused by type IV (delayed-type) hypersensitivity including systemic allergic dermatitis (SAD) presenting as papulovesicular eruption (2), SAD presenting as generalization of dermatitis (4,7,19), SAD presenting as generalized symmetrical pruritic rash (9), SAD presenting as SDRIFE (10), SAD presenting as SDRIFE with bullae (14), SAD presenting as erythema multiforme-like eruption (10), SAD presenting maculopapular eruption (19), local allergic reaction from intralesional injection (8,11,13), and reactions to intra-articular injections: erythema multiforme-like allergic dermatitis (3), morbilliform and partially persistent urticarial dermatitis (12), localized and generalized allergic reaction (16) and generalized erythema (17).

Systemic allergic dermatitis (systemic contact dermatitis)

In Leuven, Belgium, in a 12-year-period before 2012, 16 patients were investigated for a generalized allergic eruption (maculopapular eruption or eczema, with or without flare-up of previous dermatitis) from systemic administration (oral, intravenous, intramuscular, intra-articular) of corticosteroids, a few hours or days after the first dose of the culprit drug. The reactions observed were in most cases a manifestation of systemic allergic dermatitis: the patient had previously become sensitized to the corticosteroid used systemically or a cross-reacting molecule from topical exposure. Two patients had reacted to triamcinolone acetonide. One had developed a maculopapular exanthema after insertion of a suppository with this corticosteroid and the other reacted to intra-articular injection of triamcinolone acetonide with generalized eczema. The latter patient had no previous exposure to topical steroids and may have been sensitized by previous intra-articular injections (19).

Two patients with systemic allergic dermatitis from intra-articular administration of triamcinolone acetonide were reported from Italy (10). Both patients had previously suffered from allergic contact dermatitis to corticosteroids. One, a man aged 84 years, had intensely itchy and burning sharply demarcated erythematous patches symmetrically localized in the inguinal and perigenital area and on the abdomen, which had developed one day after injection of 1 ml of triamcinolone acetonide 40 mg/ml in the tibiotalar joint to treat an acute osteoarthritis. He had positive patch tests to triamcinolone acetonide and 6 other corticosteroids. The other patient, a woman of 71 years, presented with itchy erythematous and slightly edematous patches on the trunk, which tended to coalescence, along with some isolated typical target lesions. Two days before the onset of the eruption the patient had been treated with intra-articular injection of 1 ml of triamcinolone acetonide 40 mg/ml for osteoarthritis of the knee. This patient had positive patch tests to 8 corticosteroids, including triamcinolone and its acetonide ester. Patient 1 was diagnosed with systemic allergic dermatitis presenting as symmetrical drug-related intertriginous and flexural exanthema (SDRIFE) and patient 2 as systemic allergic dermatitis presenting as an erythema multiforme-like eruption (10).

A 32-year-old man applied desoximetasone cream (0.25%) to an itchy rash on the left forearm. He had previously used this cream for insect bites. After a few days an eczematous reaction developed at the site of application of the cream. At his doctor's suggestion, the patient stopped the topical treatment and began to take 8 mg/day of triamcinolone acetonide orally. The day after, a slight red papulovesicular eruption appeared on almost his entire skin surface, rapidly worsening. Patch tests were positive to desoximetasone 1% pet., desoximetasone acetate 1% pet. and triamcinolone acetonide 1% pet. Intradermal tests were positive at D2. Repetition of patch and intradermal tests 3 months later gave the same results (2). This was a case of systemic allergic dermatitis to triamcinolone acetonide, probably cross-reacting to desoximetasone, to which the patient previously had become sensitized.

A 60-year-old man presented with inner thigh eczema. It was not controlled by clobetasol propionate (0.05%) or betamethasone valerate (0.1%) ointments, but spread to his arms. He was given triamcinolone acetonide 40 mg i.m. and an emollient, but the eczema spread to his trunk and became vesicular on his arms (systemic allergic dermatitis). Patch tests were positive to many corticosteroids including clobetasol propionate, betamethasone valerate and triamcinolone acetonide (7).

A 52-year-old woman presented with a pruritic and generalized symmetrical eruption of 3 days duration after intra-articular administration of 1 ml of triamcinolone acetonide 40 mg/ml for trochanteric bursitis. She had suffered similar cutaneous eruptions after the application of several antihemorrhoidal creams or ointments previously. Patch tests were positive to triamcinolone acetonide and 7 other corticosteroids. An intradermal test with triamcinolone acetonide gave a delayed positive result with a concentration of 4 mg/ml (negative to 0.4 mg/ml), and subsequently a rash resembling the initial rash from triamcinolone intra-articular administration developed. A deflazacort oral challenge test with 30 mg gave a positive result with a pruritic rash on the forearms and anterior chest after 24 hours. Oral administration of 32 mg methylprednisolone resulted in a pruritic rash on the anterior trunk, but oral challenges with hydrocortisone were negative (9).

A 73-year-old woman developed an exanthema suggestive of SDRIFE, but with bullae resembling bullous pemphigoid, after intra-articular injection of triamcinolone acetonide. Patch tests were positive to budesonide, triamcinolone and desonide. The patient had used topical corticosteroids before without problems and also intra-articular injections with triamcinolone acetonide resulting in a pruritic papular eruption under her breasts. The primary sensitizing corticosteroid remained uncertain (14).

In a 65-year-old woman with erythema multiforme-like contact dermatitis from budesonide ointment, oral triamcinolone acetonide 16 mg for 4 days may have caused generalization of the dermatitis. The patient had positive patch test reactions to both budesonide and triamcinolone acetonide 10% pet. (4).

Symmetrical drug-related intertriginous and flexural exanthema (SDRIFE)/Baboon syndrome
See the section 'Systemic allergic dermatitis' above (refs. 10 and 14).

Other cutaneous adverse drug reactions

Reactions to intralesional injections
A 22-year-old woman presented with an erythematous keloid on her right upper arm. Three triamcinolone acetonide injections (8 mg/ml) were given one-half centimeter apart at the poles of the lesion. Two days later the patient returned with a history that the surface of the scar had blistered and produced a serous exudate. Examination revealed 3 cm of erythema around a small wheal. Five days later the erythema and warmth decreased, the keloid was smaller, and what remained were small, red papules confluent in some areas. A patch test was positive to the injection fluid. The excipients were not tested, but the patient also reacted to 2 commercial preparations containing other steroids and other excipients (8).

A similar case of localized allergic reaction was observed in a patient with alopecia areata from intralesional injections. A patch test was negative to triamcinolone, but an intradermal test to the commercial preparation strongly positive at D3 (13). Another case of a local allergic reaction to intralesional triamcinolone acetonide was reported in 2000, but details are unavailable to the author (11).

Reactions to intra-articular injections
A 70-year-old man presented with acute dermatitis over his right knee, with lesions spreading to the legs and abdomen. The clinical picture was characterized by itchy erythematous edematous lesions, which were erythema multiforme-like. He had been treated with an intra-articular injection of triamcinolone acetonide into the knee three days earlier because of arthrosis. During the 12-24 hours after injection, pruritus and erythema had developed at the same site. For this, the patient had been prescribed topical budesonide by his general practitioner. Over the next few hours after its application, the acute eczema had developed. He had received 3 injections previously without any reaction. Patch tests were positive to triamcinolone acetonide 1% pet. and budesonide 1% pet. The patient was diagnosed with erythema multiforme-like lesions from triamcinolone acetonide, but it cannot be excluded that budesonide has played a causal or contributing role (3).

A 75-year-old woman presented with a history of warmth and redness in her right knee after receiving intra-articular injections of triamcinolone acetonide 40 mg/ml for osteoarthritis. She had received two injections over a 4-month period. Approximately 2 days after the last injection, the patient noticed a cutaneous eruption on her right knee, which eventually generalized. Patch tests were positive to triamcinolone acetonide and various other corticosteroids. However, the patient had never used topical corticosteroids before and therefore sensitization had most likely taken place at the first intra-articular injection (16).

A patient has been described with a disseminated morbilliform and partially persistent urticarial dermatitis following intra-articular injections of triamcinolone acetonide. Delayed-type hypersensitivity to triamcinolone acetonide was demonstrated by patch and intradermal testing (12).

A 53-year-old patient developed generalized erythema 10 hours after intraarticular injection of triamcinolone acetonide. Patch tests revealed positive reactions to triamcinolone acetonide and various other corticosteroids. Itching erythema on the head and neck was observed 6-12 hours after intravenous injection of dexamethasone, betamethasone, hydrocortisone and triamcinolone (17).

See also the section 'Systemic allergic dermatitis' for additional cases of allergic reactions to intra-articular injections of triamcinolone acetonide (refs. 9,10,14,19).

Cross-reactions, pseudo-cross-reactions and co-reactions
Cross-reactions between corticosteroids are discussed in Chapter 3.399 Prednisolone.

LITERATURE
1   Baeck M, Goossens A. Immediate and delayed allergic hypersensitivity to corticosteroids: practical guidelines. Contact Dermatitis 2012;66:38-45
2   Brambilla L, Boneschi V, Chiappino G, Fossati S, Pigatto PD. Allergic reactions to topical desoxymethasone and oral triamcinolone. Contact Dermatitis 1989;21:272-274
3   Valsecchi R, Reseghetti A, Leghissa P, Cologni L, Cortinovis R. Erythema-multiforme-like lesions from triamcinolone acetonide. Contact Dermatitis 1998;38:362-363
4   Stingeni L, Caraffini S, Assalve D, Lapomarda V, Lisi P. Erythema-multiforme-like contact dermatitis from budesonide. Contact Dermatitis 1996;34:154-155
5   Ferguson AD, Emerson RM, English JS. Cross-reactivity patterns to budesonide. Contact Dermatitis 2002;47:337-340

6   Wilkinson SM, English JS. Patch tests are poor detectors of corticosteroid allergy. Contact Dermatitis 1992;26:67-68

7   English JS, Ford G, Beck MH, Rycroft RJ. Allergic contact dermatitis from topical and systemic steroids. Contact Dermatitis 1990;23:196-197

8   Kark EC. Sensitivity to fluorinated steroids presenting as a delayed hypersensitivity. Contact Dermatitis 1980;6:214-216

9   Santos-Alarcón S, Benavente-Villegas FC, Farzanegan-Miñano R, Pérez-Francés C, Sánchez-Motilla JM, Mateu-Puchades A. Delayed hypersensitivity to topical and systemic corticosteroids. Contact Dermatitis 2018;78:86-88

10 Bianchi L, Marietti R, Tramontana M, Hansel K, Stingeni L. Systemic allergic dermatitis from intra-articular triamcinolone acetonide: Report of two cases with unusual clinical manifestations. Contact Dermatitis 2021;84:54-56

11 Brancaccio RR, Zappi EG. Delayed type hypersensitivity to intralesional triamcinolone acetonide. Cutis 2000;65:31-33

12 Ijsselmuiden OE, Knegt-Junk KJ, van Wijk RG, van Joost T. Cutaneous adverse reactions after intra-articular injection of triamcinolone acetonide. Acta Derm Venereol 1995;75:57-58

13 Kreeshan FC PHP. Delayed hypersensitivity reaction to intralesional triamcinolone acetonide following treatment for alopecia areata. Intradermal testing. Dermatol Case Rep 2015;9:107-109

14 Gumaste P, Cohen D, Stein J. Bullous systemic contact dermatitis caused by an intra-articular steroid injection. Br J Dermatol 2015;172:300-302

15 De Groot AC. Monographs in contact allergy, volume 3: Topical Drugs. Boca Raton, Fl, USA: CRC Press Taylor and Francis Group, 2021 (ISBN 978-0-367-23693-9)

16 Amin N, Brancaccio R, Cohen D. Cutaneous reactions to injectable corticosteroids. Dermatitis 2006;17:143-146

17 Leiter U, Gall H, Peter R-U. Typ-IV-allergie auf injizierbares Triamcinolonacetonid. Allergo J 1999;8:195-199

18 Bulcke DM, Devos SA. Hand and forearm dermatoses among veterinarians. J Eur Acad Dermatol Venereol 2007;21:360-363

19 Baeck M, Goossens A. Systemic contact dermatitis to corticosteroids. Allergy 2012;67:1580-1585

Chapter 3.490 TRIAZOLAM

IDENTIFICATION

Description/definition : Triazolam is the benzodiazepine that conforms to the structural formula shown below
Pharmacological classes : Adjuvants, anesthesia; anti-anxiety agents; GABA modulators
IUPAC name : 8-Chloro-6-(2-chlorophenyl)-1-methyl-4H-[1,2,4]triazolo[4,3-a][1,4]benzodiazepine
CAS registry number : 28911-01-5
EC number : 249-307-3
Merck Index monograph : 11035
Patch testing : Tablet, pulverized, 10% pet. (2); most systemic drugs can be tested at 10% pet.; if the pure chemical is not available, prepare the test material from intravenous powder, the content of capsules or – when also not available – from powdered tablets to achieve a final concentration of the active drug of 10% pet.
Molecular formula : $C_{17}H_{12}Cl_2N_4$

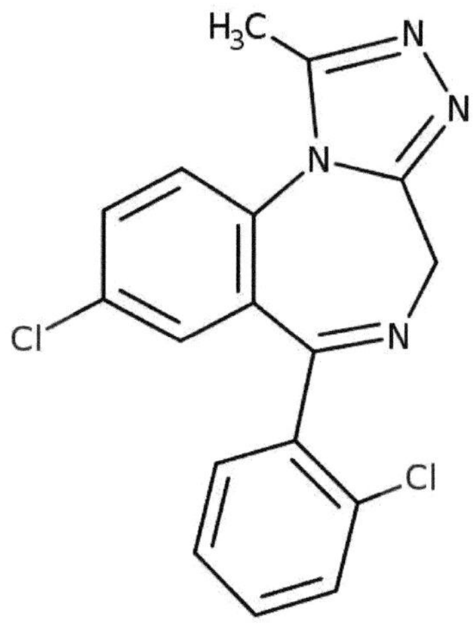

GENERAL
Triazolam is a short-acting benzodiazepine used predominantly as a hypnotic agent in the treatment of insomnia (1).

CUTANEOUS ADVERSE DRUG REACTIONS FROM SYSTEMIC ADMINISTRATION CAUSED BY TYPE IV (DELAYED-TYPE) HYPERSENSITIVITY (as demonstrated by positive patch tests)
Cutaneous adverse drug reactions from systemic administration of triazolam caused by type IV (delayed-type) hypersensitivity have included maculopapular eruption (2).

Maculopapular eruption
A 65-year-old man developed a maculopapular eruption after using amoxicillin/clavulanic acid and triazolam. Patch tests were positive to triazolam 12.5% pet., amoxicillin/clavulanic acid 12.5% and amoxicillin 12.5%, but negative to clavulanic acid. Lymphocyte transformation tests were also positive to triazolam (stimulation index 5.5) and to amoxicillin (stimulation index 18.8). As this patient reacted to 2 non-related medicaments, this was a case of multiple drug hypersensitivity syndrome (2).

LITERATURE
1   The data in the section 'General' may have been obtained from literature discussed in this chapter, but mostly also or exclusively from one or more of the following online sources: ChemIDPlus Advanced, PubChem, DrugBank, RxList, Drug Central, Drugs.com, and Wikipedia
2   Gex-Collet C, Helbling A, Pichler WJ. Multiple drug hypersensitivity – proof of multiple drug hypersensitivity by patch and lymphocyte transformation tests. J Investig Allergol Clin Immunol 2005; 15:293-296

Chapter 3.491 TRIBENOSIDE

IDENTIFICATION

Description/definition   : Tribenoside is the glycoside that conforms to the structural formula shown below
Pharmacological classes : Anti-inflammatory agents, non-steroidal
IUPAC name   : (3R,4R,5R)-5-[(1R)-1,2-bis(Phenylmethoxy)ethyl]-2-ethoxy-4-phenylmethoxyoxolan-3-ol
Other names   : Tribenzoside
CAS registry number   : 10310-32-4
EC number   : 233-687-2
Merck Index monograph   : 11039
Patch testing   : 1% and 10% pet.
Molecular formula   : $C_{29}H_{34}O_6$

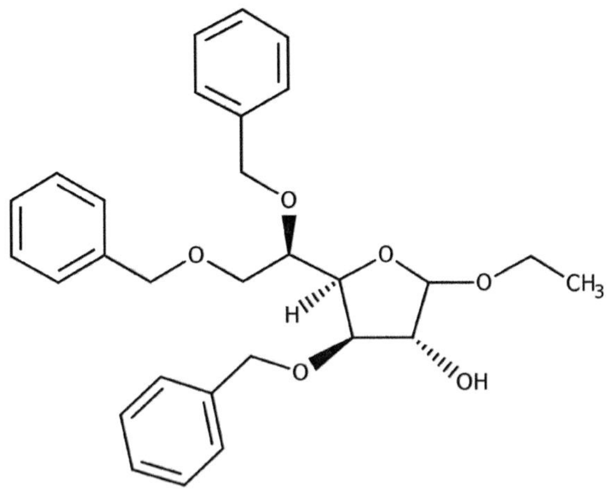

GENERAL

Tribenoside is a glycoside with anti-inflammatory and mild analgesic properties. Pharmaceuticals with tribenoside are available in many countries, e.g. as cream/ointment, tablets and as suppositories, often combined with lidocaine, to treat irritation from hemorrhoids (1). Its presence in such a cream has rarely led to allergic contact dermatitis (2).

CUTANEOUS ADVERSE DRUG REACTIONS FROM SYSTEMIC ADMINISTRATION CAUSED BY TYPE IV (DELAYED-TYPE) HYPERSENSITIVITY (as demonstrated by positive patch tests)

Cutaneous adverse drug reactions from systemic administration of tribenoside caused by type IV (delayed-type) hypersensitivity have included drug reaction with eosinophilia and systemic symptoms (DRESS) (4) and erythema multiforme (3).

Drug reaction with eosinophilia and systemic symptoms (DRESS)

A 49-year-old man was given tribenoside 600 mg/day for pain relief after an operation for hemorrhoids. Five weeks after the operation, and after having taken acetaminophen for 5 days for pharyngitis, a skin rash appeared on the arms and legs, spreading to all body regions accompanied by fever. Both drugs were stopped, which improved the rash and fever until a week later erythroderma developed with fever of 38.5°C and cervical and axillar non-tender lymphadenopathy. Laboratory tests showed leukocytosis with 10% eosinophils and 25% atypical lymphocytes, 5-10 times elevated liver enzymes and elevated C-reactive protein. Blood tests also revealed reactivation of cytomegaly virus. Drug-induced hypersensitivity syndrome was diagnosed (older term for DRESS) and the eruption completely disappeared on a course of prednisolone. Patch tests performed after recovery were positive with tribenoside (material? concentration? vehicle? controls?) but negative with acetaminophen. Lymphocyte stimulation tests with tribenoside and acetaminophen were positive. The patient was diagnosed with drug-induced hypersensitivity caused by tribenoside, to which acetaminophen may have contributed (4).

Other cutaneous adverse drug reactions

A 57-year-old woman presented with a rash on the trunk, arms and legs that had started a few days earlier, one week after having taken tribenoside 200 mg t.d.s. orally for the treatment of hemorrhoids. Clinically, the rash was diagnosed as erythema multiforme (no details provided). Histology of a skin biopsy showed papillary edema and perivascular lymphohistiocytic infiltration. Patch tests with tribenoside 10% and 1% pet. showed erythema at D2 and D3, with negative reactions in 5 controls (3).

LITERATURE

1   The data in the section 'General' may have been obtained from literature discussed in this chapter, but mostly also or exclusively from one or more of the following online sources: ChemIDPlus Advanced, PubChem, DrugBank, RxList, Drug Central, Drugs.com, and Wikipedia
2   Inoue A, Tamagawa-Mineoka R, Katoh N, Kishimoto S. Allergic contact dermatitis caused by tribenoside. Contact Dermatitis 2009;60:349-350
3   Endo H, Kawada A, Yudate T, Aragane Y, Yamada H, Tezuka T. Drug eruption due to tribenoside. Contact Dermatitis 1999;41:223
4   Hashizume H, Takigawa M. Drug-induced hypersensitivity syndrome associated with cytomegalovirus reactivation: immunological characterization of pathogenic T cell. Acta Derm Venereol 2005;85:47-50

Chapter 3.492 TRIFLUSAL

IDENTIFICATION

Description/definition   : Triflusal is the member of acetylsalicylic acids that conforms to the structural formula
                            shown below
Pharmacological classes : Platelet aggregation inhibitors
IUPAC name   : 2-Acetyloxy-4-(trifluoromethyl)benzoic acid
CAS registry number   : 322-79-2
EC number   : 206-297-5
Merck Index monograph   : 11126
Patch testing   : 1% pet.
Molecular formula   : $C_{10}H_7F_3O_4$

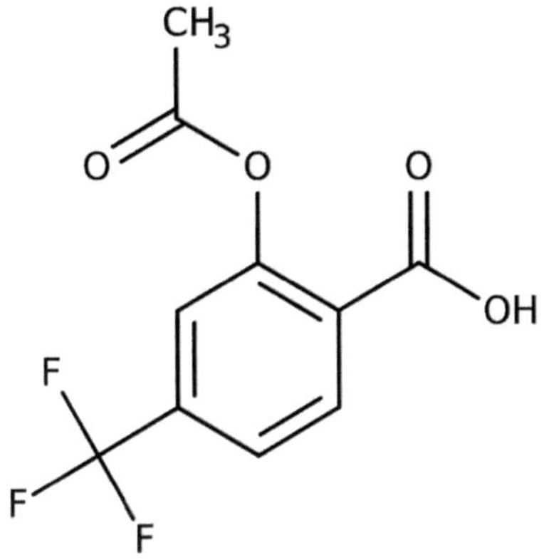

GENERAL

Triflusal is a fluorinated acetylsalicylic acid analog with antithrombotic and anticoagulant properties. It irreversibly inhibits the production of thromboxane-B_2 in platelets by acetylating cyclooxygenase-1. Triflusal is indicated as prophylaxis of thromboembolic disorders, notably of stroke and myocardial infarction (1).

CUTANEOUS ADVERSE DRUG REACTIONS FROM SYSTEMIC ADMINISTRATION CAUSED BY TYPE IV (DELAYED-TYPE) HYPERSENSITIVITY (as demonstrated by positive patch tests)

Cutaneous adverse drug reactions from systemic administration of triflusal caused by type IV (delayed-type) hypersensitivity have included photosensitivity (2,3,4,5,8,10), eczematous eruption (9), and eczema with UVB-photoaggravation (7).

Photosensitivity

General

A few cases of photoallergy to triflusal have been reported (2,3,4,5,8,10). In some of these, both triflusal and its active metabolite displayed a positive response with UVB, UVA or both, whereas the structurally related acetylsalicylic acid (triflusal is a fluorinated acetylsalicylic acid analog) did not (2,5,8). This fact supports the idea that the trifluorome-thyl group is essential in the photosensitizing ability. The maximum action spectrum of triflusal appears to be in the UVB range, unlike most photosensitivity reactions, which are typically associated with UVA wavelengths. This is rare, but has been reported with other drugs, e.g. griseofulvin and diphenhydramine (8).

Case reports

A 91-year-old woman developed, 2 weeks after starting triflusal treatment and coinciding with summer solar exposure, a pruriginous skin eruption in photoexposed areas. The lesions consisted of erythematous and edematous plaques, with small vesicles in some areas, and slight, non-adhering desquamation on the dorsal surfaces of the forearms, hands, legs and feet. Histopathology was considered compatible with photoallergy. One month after triflusal withdrawal, the condition resolved with discrete scaling. Patch and photopatch testing yielded positive photopatch tests with erythema and edema to triflusal 1% alcohol and 2-hydroxy-4-trifluoromethylbenzoic acid (the active triflusal metabolite) 1% pet. both with UVA and with UVB irradiation. Controls were not performed (2).

    A 69-year-old man developed a severely itchy eruption of deeply erythematous and infiltrated papules and plaques, located mainly on the sun-exposed skin, including the face, V area of the anterior chest, and dorsal aspect of the hands and forearms. The eruption, which later spread to covered skin, had begun 2 months after starting medication with triflusal. Photopatch tests were done with graded dosages of UVA (0.5, 1, 2, 4 J/cm^2) and were positive to triflusal 10% pet. with all dosages of UVA. Without exposure to UVA, there was no reaction. The patient also had a strongly positive response to an oral provocation test with less than 1/20 dosage of triflusal (15 mg) and irradiation with 5 J/cm^2 of UVA. The authors concluded that, although control studies were not done, their results strongly suggests a photoallergic drug reaction (3). This patient was later again – briefly - presented in ref. 4.

A man aged 72 developed eczema on light-exposed skin while on triflusal for hemiplegia. A photopatch test with commercial triflusal 10% pet. and irradiation with UVA 10J/cm^2 was positive. Histopathology of the positive test was similar to that of the original eruption. Ten controls were negative (4).

Another case of triflusal photosensitivity was reported from Spain in 2016. This 72-year-old man had outbreaks of eczematous lesions involving the photoexposed areas of his face, neck and hands. Photopatch tests were positive to triflusal and its metabolite 1% pet. irradiated with a sub-UVB-MED dose, but negative (weak non-palpable erythema) in the same patches irradiated with UVA 5J/cm^2, so the action spectrum of triflusal was in the UVB-range (8).

One patient who had a drug eruption clinically diagnosed as photosensitivity, had a positive photopatch test to 2-hydroxy-4-trifluoromethylbenzoic acid (the active triflusal metabolite) 1% pet. using irradiation with UVB at the minimal erythema dose (MED). Six controls were negative. Details are not available to the author (5). In one case of clinical photosensitivity to triflusal, the photopatch test was positive only after irradiation with UVB, while negative with UVA (10).

Other cutaneous adverse drug reactions

A 64-year-old man was treated with triflusal and various other medications after myocardial infarction. Three to four weeks later, he presented with itching, erythema and vesicles on the lower legs, and over the next 2 weeks, confluent maculopapular lesions spread to the rest of the skin. He improved on oral corticosteroids and antihistamines, but the intense pruritus and disseminated eczematide lesions persisted. After discontinuing triflusal, the exanthem improved quickly with severe scaling. Patch tests with all drugs used at 10% pet. were positive only to triflusal at D4. Ten controls were negative. Two weeks later, the patient mixed up his metformin and triflusal by mistake and took a triflusal tablet. Three days later, eczematide lesions and pruritus reappeared on the legs (9).

An 87-year-old man, who had used triflusal as monotherapy for 10 years, presented with an eruption involving the sun-exposed areas (face, dorsal aspects of the hands and feet) that had existed for >5 years. Physical examination showed facial erythema with non-adhering desquamation, marked ectropion, loss of skin folds, and microstomia. The dorsa of the hands and feet showed discrete erythema only in photo-exposed areas. Three months after triflusal discontinuation, the rash had cleared, with improvement of the ectropion and the appearance of patches of repigmentation. Patch and photopatch testing with UVA 5 J/cm^2 showed identically positive reactions in both the non-irradiated and the irradiated triflusal 1% pet. patches at D1 post-irradiation. Additional photopatch testing with sub-MED doses of UVB showed an erythematous and edematous reaction on D1 post-irradiation. 32 controls were negative to patch tests with triflusal 1% pet. The patient was diagnosed with a systemic UVB-photoaggravated allergic reaction (7).

LITERATURE

1   The data in the section 'General' may have been obtained from literature discussed in this chapter, but mostly also or exclusively from one or more of the following online sources: ChemIDPlus Advanced, PubChem, DrugBank, RxList, Drug Central, Drugs.com, and Wikipedia

2   Nagore E, Pérez-Ferriols A, Sánchez-Motilla JM, Serrano G, Aliaga A. Photosensitivity associated with treatment with triflusal. J Eur Acad Dermatol Venereol 2000;14:219-221

3   Lee AY, Yoo SH, Lee KH. A case of photoallergic drug eruption caused by triflusal. Photodermatol Photoimmunol Photomcd 1999;15:85-86

4   Lee AY, Joo HJ, Chey WY, Kim YG. Photopatch testing in seven cases of photosensitive drug eruptions. Ann Pharmacother 2001;35:1584-1587

5   Serrano G, Aliaga A, Planells I. Photosensitivity associated with triflusal (Disgrer ®). Photodermatology 1987;4:103-105 (Data cited in ref. 6)

6   Bruynzeel DP, Maibach HI. Patch testing in systemic drug eruptions. Clin Dermatol 1997;15:479-484

7   Martínez Leboráns L, Cubells Sánchez L, Zaragoza Ninet V, Pérez Ferriols A. Atypical photosensitivity associated with triflusal. Contact Dermatitis 2016;75:245-247

8   García-Rodiño S, Espasandín-Arias M, Vázquez-Osorio I, Rodríguez-Granados MT. Photosensitivity associated with systemic triflusal therapy. Photodermatol Photoimmunol Photomed 2016;32:113-115

9   Sánchez-Machín I, García Robaina JC, Torre Morin F. Widespread eczema from triflusal confirmed by patch testing. Contact Dermatitis 2004;50:257

10   Ruiz-Sanchez D, Valtueña J, Garayar Cantero M, Volo V, Barrutia L, Garabito Solovera E, Santamarina Albertos A. Photosensitivity to triflusal. J Allergy Clin Immunol Pract 2021;9:1713-1714

Chapter 3.493 TRIMEBUTINE

IDENTIFICATION

Description/definition   : Trimebutine is the trihydroxybenzoic acid derivative that conforms to the structural
                            formula shown below
Pharmacological classes : Gastrointestinal agents; parasympatholytics
IUPAC name   : [2-(Dimethylamino)-2-phenylbutyl] 3,4,5-trimethoxybenzoate
Other names   : Benzoic acid, 3,4,5-trimethoxy-, 2-(dimethylamino)-2-phenylbutyl ester
CAS registry number   : 39133-31-8
EC number   : 254-309-2
Merck Index monograph   : 11138
Patch testing   : 0.5% and 1% water and pet.
Molecular formula   : $C_{22}H_{29}NO_5$

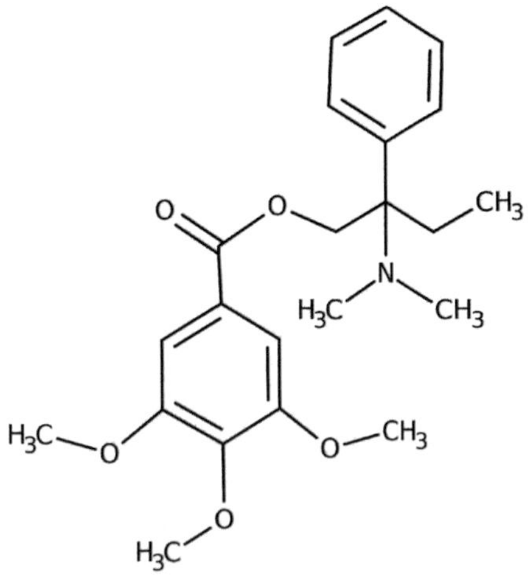

Trimebutine

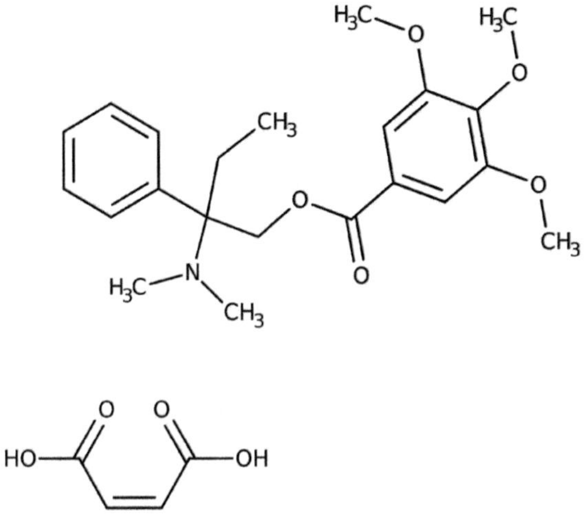

Trimebutine maleate

GENERAL

Trimebutine is a trimethoxybenzoic acid and spasmolytic agent that regulates intestinal and colonic motility and relieves abdominal pain with antimuscarinic and weak mu opioid agonist effects. This drug is indicated for symptomatic treatment of irritable bowel syndrome and treatment of postoperative paralytic ileus following abdominal surgery (1). It is also used rectally and topically for anal fissures and hemorrhoids (2). In pharmaceutical products, both trimebutine and trimebutine maleate (CAS number 34140-59-5, EC number 251-845-9, molecular formula $C_{26}H_{33}NO_9$) may be employed (1).

In topical preparations, trimebutine has rarely caused contact allergy/allergic contact dermatitis, which subject has been reviewed in Volume 3 of the *Monographs in contact allergy* series (4).

CUTANEOUS ADVERSE DRUG REACTIONS FROM SYSTEMIC ADMINISTRATION CAUSED BY TYPE IV
(DELAYED-TYPE) HYPERSENSITIVITY (as demonstrated by positive patch tests)

Cutaneous adverse drug reactions from systemic administration of trimebutine caused by type IV (delayed-type) hypersensitivity have included systemic allergic dermatitis (3).

Systemic allergic dermatitis (systemic contact dermatitis)

A 40-year-old woman presented with generalized pruritic and erythematous hives without angioedema, 7 days after initiating therapy with a cream for hemorroids containing trimebutine. Patch tests were positive to the cream 'as is' and to trimebutine (test concentration and vehicle not mentioned). An oral provocation test with trimebutine (dosage and schedule not mentioned) resulted in generalized urticaria after 3 days (3). This patient had systemic allergic dermatitis presenting as generalized urticaria from absorption of trimebutine from the cream for hemorrhoids, followed by systemic allergic dermatitis from oral provocation.

LITERATURE

1   The data in the section 'General' may have been obtained from literature discussed in this chapter, but mostly also or exclusively from one or more of the following online sources: ChemIDPlus Advanced, PubChem, DrugBank, RxList, Drug Central, Drugs.com, and Wikipedia

2   Reyes JJ, Fariña MC. Allergic contact dermatitis due to trimebutine. Contact Dermatitis 2001;45:164

3   Martin-Garcia C, Martinez-Borque N, Martinez-Bohigas N, Torrecillas-Toro M, Palomeque-Rodrìguez MT. Delayed reaction urticaria due to trimebutine. Allergy 2004;59:789-790

4   De Groot AC. Monographs in contact allergy, volume 3: Topical Drugs. Boca Raton, Fl, USA: CRC Press Taylor and Francis Group, 2021 (ISBN 978-0-367-23693-9)

Chapter 3.494 TRIMEPRAZINE

IDENTIFICATION

Description/definition   : Trimeprazine is the phenothiazine that conforms to the structural formula shown below
Pharmacological classes : Antipruritics
IUPAC name   : *N,N*,2-Trimethyl-3-phenothiazin-10-ylpropan-1-amine
Other names   : Alimemazine; methylpromazine
CAS registry number   : 84-96-8
EC number   : 201-577-3
Merck Index monograph   : 11143
Patch testing   : 1% pet.(3); this concentration may well be too high for photopatch testing, as
                           phenothiazines can induce phototoxic reactions at this concentration; 0.1% pet. may be
                           preferable for photopatch tests
Molecular formula   : $C_{18}H_{22}N_2S$

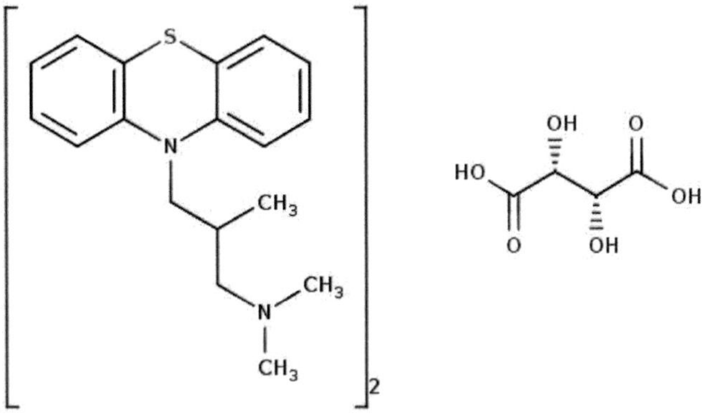

Trimeprazine tartrate

GENERAL

Trimeprazine (also known as alimemazine) is a tricyclic antihistamine, similar in structure to the phenothiazine antipsychotics. It acts as an antihistamine, a sedative, and an anti-emetic. Trimeprazine is used principally to prevent motion sickness and in combination with other medications in cough and cold preparations, but also for the treatment of itch and urticaria. In pharmaceutical products (systemic administration only), trimeprazine is employed as trimeprazine tartrate (CAS number 4330-99-8, EC number not available, molecular formula $C_{40}H_{50}N_4O_6S_2$) (go.drugbank.co).

CUTANEOUS ADVERSE DRUG REACTIONS FROM SYSTEMIC ADMINISTRATION CAUSED BY TYPE IV (DELAYED-TYPE) HYPERSENSITIVITY (as demonstrated by positive patch tests)

Cutaneous adverse drug reactions from systemic administration of trimeprazine caused by type IV (delayed-type) hypersensitivity have included photosensitivity (1 [very uncertain],2).

Photosensitivity

In a multicenter study in Italy, performed in the period 1985-1994, 14 positive photopatch test reactions were seen to trimeprazine. The patch test concentration used was not mentioned and relevance was not specified (78% for all photoallergens together (2). Also in Italy, in the period 1986-1989, 128 patients were photopatch tested with trimeprazine 1% pet. and there were 6 (4.7%) positive reactions. The selection procedure of the patients was not described and the relevance of the positive reactions was not mentioned (1). It is unlikely, that so many photoallergic reactions were caused by the use of oral trimeprazine. Probably, these were photocross-reactions to phenothiazines which are used topically (e.g. promethazine or chlorpromazine) or they were in fact phototoxic reactions, which is a well-known phenomenon with phenothiazines.

LITERATURE

1   Guarrera M. Photopatch testing: a three-year experience. J Am Acad Dermatol 1989;21:589-591
2   Pigatto PD, Legori A, Bigardi AS, Guarrera M, Tosti A, Santucci B, et al. Gruppo Italiano recerca dermatiti da
     contatto ed ambientali Italian multicenter study of allergic contact photodermatitis: epidemiological aspects. Am
     J Contact Dermatitis 1996;17:158-163

Chapter 3.495 TRIMETHOPRIM

IDENTIFICATION

Description/definition   : Trimethoprim is the anisole that conforms to the structural formula shown below
Pharmacological classes : Anti-infective agents, urinary; folic acid antagonists; cytochrome P-450 CYP2C8
                                        inhibitors; anti-dyskinesia agents; antimalarials
IUPAC name   : 5-[(3,4,5-Trimethoxyphenyl)methyl]pyrimidine-2,4-diamine
CAS registry number   : 738-70-5
EC number   : 212-006-2
Merck Index monograph : 11148
Patch testing   : 10% pet.; use open tests with trimethoprim 10%, 20% and 50% DMSO for fixed drug
                                        eruptions (8)
Molecular formula   : $C_{14}H_{18}N_4O_3$

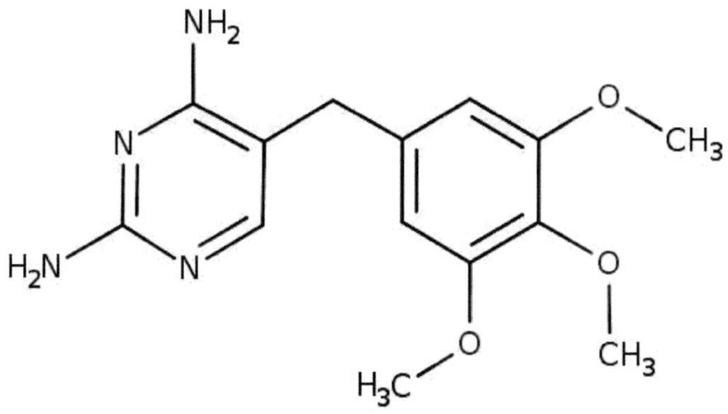

GENERAL

Trimethoprim is an antifolate antibacterial agent that inhibits bacterial dihydrofolate reductase, a critical enzyme that catalyzes the formation of tetrahydrofolic acid. In doing so, it prevents the synthesis of bacterial DNA and ultimately continued bacterial survival. Trimethoprim is often used in combination with sulfamethoxazole due to their complementary and synergistic mechanisms, but may be used as a monotherapy in the treatment and/or prophylaxis of urinary tract infections and sometimes as an antimalarial (1). In pharmaceutical products, trimethoprim is employed as trimethoprim hydrochloride (CAS number 60834-30-2, EC number 262-450-6, molecular formula $C_{14}H_{19}ClN_4O_3$) or as trimethoprim sulfate (CAS number 56585-33-2, EC number not available, molecular formula $C_{28}H_{38}N_8O_{10}S$) (1).

    See also Chapter 3.455 Sulfamethoxazole mixture with trimethoprim and Chapter 3.454 Sulfamethoxazole.

CUTANEOUS ADVERSE DRUG REACTIONS FROM SYSTEMIC ADMINISTRATION CAUSED BY TYPE IV
(DELAYED-TYPE) HYPERSENSITIVITY (as demonstrated by positive patch tests)

Cutaneous adverse drug reactions from systemic administration of trimethoprim caused by type IV (delayed-type) hypersensitivity have included fixed drug eruption (2,3,4,6,7,8) and unspecified cutaneous adverse drug reactions (5).

Case series with various or unknown types of drug reactions

In Finland, in the period 1989-2001, 826 patients with suspected cutaneous drug eruptions were patch tested and 89 had one or more positive reactions. Of these individuals, 10 reacted to trimethoprim and 4 to sulfamethoxazole-trimethoprim. No clinical data were provided and it was not mentioned which types of cutaneous adverse drug reactions these patient had suffered from (5).

Fixed drug eruption

Five patients with fixed drug eruptions (FDE) from trimethoprim in sulfamethoxazole-trimethoprim, were reported from Istanbul, Turkey (8). These are discussed in detail in Chapter 3.454 Sulfamethoxazole in the section 'Fixed drug eruption', ref. 14.

A 25-year-old woman presented with a pruritic eruption on her right arm, consisting of numerous discrete, round to oval, violaceous macules and edematous plaques of 1-5 cm distributed in a linear pattern. Some of the lesions were surmounted by flaccid bullae. This was her third attack in 6 months. Lesions lasted for 10-15 days and healed with residual pigmentation. The patient had a history of cotrimoxazole intake on the day before the eruption had begun because of relapsing tonsillitis. Histopathology supported the diagnosis of fixed drug eruption. Four weeks after regression, an oral challenge with one-eighth of a single cotrimoxazole dose completely reactivated the previously involved sites with marked violaceous erythema and edema starting within 1-2 hours and progressing into flaccid bullae. Four weeks later, occlusive patch tests on tape-stripped skin with cotrimoxazole, trimethoprim and sulfamethoxazole at concentrations of 10%, 20% and 50% in white petrolatum were negative on both previously involved and uninvolved skin. In a repeated open application patch test with the drugs at the same concentrations in dimethyl sulfoxide (DMSO), positive reactions were seen on previously involved skin sites with trimethoprim 50% and cotrimoxazole 20 and 50%, starting 4-5 hours after the second applications (6). A very similar cases was reported by the first author one year later (7).

In patients with fixed drug eruptions (FDE) caused by delayed-type hypersensitivity, the diagnosis is usually confirmed by a positive patch test with the drug on previously affected skin. Authors from Finland have used an alternative method of topical provocation. The test compound, the drug 10% in petrolatum and sometimes also in 70% alcohol and in DMSO, was applied once and without occlusion over the entire surface of one or several inactive (usually pigmented) sites of FDE lesions. The patients were followed as in-patients for 24 hours. A reaction was regarded as positive when a clearly demarcated erythema lasting at least 6 hours was seen.

In France, in the period 2005-2007, 59 cases of fixed drug eruptions were collected in 17 academic centers. There was one case of FDE to trimethoprim with a positive patch test. Clinical details were not provided (4).

Of 6 patients with FDE from trimethoprim, 4 (67%) had a positive topical provocation (2,3).

LITERATURE

1   The data in the section 'General' may have been obtained from literature discussed in this chapter, but mostly also or exclusively from one or more of the following online sources: ChemIDPlus Advanced, PubChem, DrugBank, RxList, Drug Central, Drugs.com, and Wikipedia
2   Alanko K. Topical provocation of fixed drug eruption. A study of 30 patients. Contact Dermatitis 1994;31:25-27
3   Alanko K, Stubb S, Reitamo S. Topical provocation of fixed drug eruption. Br J Dermatol 1987;116:561-567
4   Brahimi N, Routier E, Raison-Peyron N, Tronquoy AF, Pouget-Jasson C, Amarger S, et al. A three-year-analysis of fixed drug eruptions in hospital settings in France. Eur J Dermatol 2010;20:461-464
5   Lammintausta K, Kortekangas-Savolainen O. The usefulness of skin tests to prove drug hypersensitivity. Br J Dermatol 2005;152:968-974
6   Özkaya-Bayazit E, Baykal C. Trimethoprim-induced linear fixed drug eruption. Br J Dermatol 1997;137:1028-1029
7   Özkaya-Bayazit E, Güngör H. Trimethoprim-induced fixed drug eruption: positive topical provocation on previously involved and uninvolved skin. Contact Dermatitis 1998;39:87-88
8   Özkaya-Bayazit E, Bayazit H, Ozarmagan G. Topical provocation in 27 cases of cotrimoxazole-induced fixed drug eruption. Contact Dermatitis 1999;41:185-189 (overlap with refs. 6 and 7)

Chapter 3.496 TYLOSIN

IDENTIFICATION

Description/definition : Tylosin is a macrolide antibiotic obtained from cultures *of Streptomyces fradiae* that conforms to the structural formula shown below

Pharmacological classes : Anti-bacterial agents

IUPAC name : 2-[(4R,5S,6S,7R,9R,11E,13E,15R,16R)-6-[(2R,3R,4R,5S,6R)-5-[(2S,4R,5S,6S)-4,5-Dihydroxy-4,6-dimethyloxan-2-yl]oxy-4-(dimethylamino)-3-hydroxy-6-methyloxan-2-yl]oxy-16-ethyl-4-hydroxy-15-[[[(2R,3R,4R,5R,6R)-5-hydroxy-3,4-dimethoxy-6-methyloxan-2-yl]oxymethyl]-5,9,13-trimethyl-2,10-dioxo-1-oxacyclohexadeca-11,13-dien-7-yl]acetaldehyde

CAS registry number : 1401-69-0

EC number : 215-754-8

Merck Index monograph : 11283

Patch testing : Tylosin tartrate, 5%-10% pet.

Molecular formula : $C_{46}H_{77}NO_{17}$

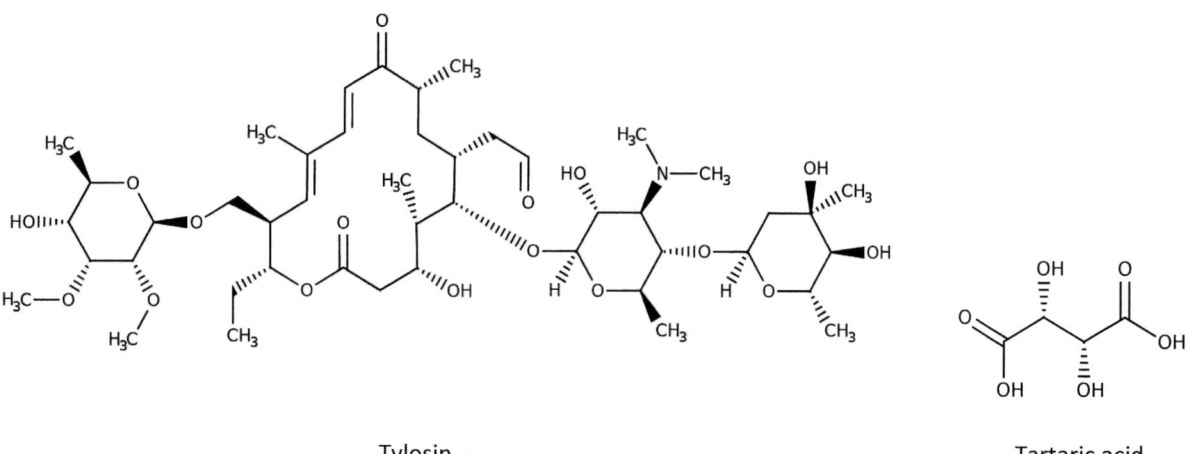

Tylosin   Tartaric acid

GENERAL

Tylosin is a bacteriostatic macrolide antibiotic restricted to veterinary use, either by injection or added to feed or water. It has a broad spectrum of activity against gram-positive organisms and a limited range of gram-negative organisms. Tylosin is produced as a fermentation product of *Streptomyces fradiae*. In pharmaceutical products, tylosin is usually employed as tylosin tartrate (CAS number 74610-55-2, EC number not available, molecular formula $C_{50}H_{83}NO_{23}$) or tylosin phosphate (1).

CONTACT ALLERGY FROM ACCIDENTAL CONTACT

General

Multiple cases of occupational allergic contact dermatitis to tylosin have been described, most often in veterinarians from administering the drugs to animals (5,6,8,9,10,19) and pig farmers from adding supplements containing tylosin to animal feed (3,16,18). In addition, there are a few cases reported in drug-compounding technicians (14), an animal feed mill worker (11), an animal feed wholesaler (20), a chicken vaccinator (21), a pharmacist (12), a bioassay laboratory employee (22) and a girl feeding a pig as part of a 4-H project (15). Many sensitizations result in airborne allergic contact dermatitis, especially in persons working with animal feed, but sometimes also from injecting animals (21).

Case series

In 1971, in Gentofte, Denmark, 9 veterinary surgeons were patch tested because of recent aggravation of hand eczema or dissemination of the eczema from the hands to other body regions. Four proved to be allergic to commercial tylosin and six to commercial spiramycin, both tested at 1% pet., 5% water and/or 10% pet. 45 controls were negative. All reactions were relevant to the actual dermatitis (4). One year later, one or more cases of occupational allergic contact dermatitis from tylosin in veterinary surgeons were reported from the same clinic (possibly overlap) (5).

In another clinic in Denmark, between November 1, 1978 to December 31, 1979, patients working with farm animals and suspected of having occupational contact dermatitis were patch tested with commercial tylosin 2% water and spiramycin 10% water. Six farmers and three farmers' wives had positive patch tests to one or both antibiotics, 8 to tylosin and 8 to spiramycin (negative in 26 controls). The eczema was typically seen on the hands, face and neck (suggesting airborne spread, possibly by the presence of excreted antibiotics in dust in the pig shed), but two patients had generalized dermatitis. The main occupation of all nine patients was raising hogs. The duration of the dermatitis ranged from 4 months to 5 years. All patients had used either tylosin or spiramycin and most had used both antibiotics. After diagnosis, the dermatitis cleared or improved markedly in seven patients; five of these stopped using the antibiotics in question, and two began to wear gloves while handling the drugs. The two patients who continued use of the drugs and took no precautionary measures had little change in the activity of the dermatitis (3).

In 2 hospitals in Denmark, between 1974 and 1980, 37 veterinary surgeons, all working in private country practices, were investigated for suspected incapacitating occupational dermatitis and patch tested with a battery of 10 antibiotics. Thirty-two (86%) had one or more positive patch tests. There were 10 positive reactions to tylosin tartrate 10% pet. (27%). It was mentioned that all but one of these individuals had allergic contact dermatitis, but relevance was not specified for individual allergens. The most frequent allergens were spiramycin, penethamate and tylosin tartrate (6).

Other reports of occupational allergic contact dermatitis/contact allergy to tylosin (not case-reports) are summarized in table 3.496.1.

Table 3.496.1 Reports of occupational allergic contact dermatitis/contact allergy to tylosin

Year	Country	Nr. Pat.	Profession	Additional data	Ref.
2020	Belgium	1	Healthcare worker [a]		13
2015	Germany	1	Veterinarian	Investigated for airborne contact dermatitis	9
2013	Belgium	1	?	Airborne OACD	2
2001	Belgium	1	Pharmacist		12
1990	Italy	1	Animal feed mill worker	Tested with tylosin 10% pet.	11
1984	Poland	4	Veterinarian	Tested with tylosin tartrate 10% pet.	10
1982	Poland	3	Veterinarians	All 3 had had contact with tylosin at work	8

[a] veterinarian or worker in pharmaceutical industry
OACD: occupational allergic contact dermatitis

Case reports
A 50-year-old woman working in a veterinary pharmacy as a compounding technician had a history of facial dermatitis for 12 years. Her primary duty involved compounding powdered medications with a mortar and pestle to form a paste. Her symptoms, initially only mild itching on the eyelids and temples, began 12 years after starting this work and gradually worsened over time. The patient developed a classic picture of airborne contact dermatitis with lichenified and crusted dermatitis over the cheeks, eyelids, forehead, central chest, upper back, ears, and outer forearms. Patch tests were positive to tylosin tartrate powder 5% pet. at D2 and D5 (++/++). The were no cross-reactions to other macrolide antibiotics. The patient's hypersensitivity to tylosin was so strong, that she could no longer work in the facility; her symptoms completely resolved once she had no exposure to tylosin anymore (14).

The authors also described the case of a 37-year-old woman, also a compounding technician and working in the same company, performing the same work, who suffered from facial dermatitis. Her symptoms began 6 months after starting the position and the distribution of the dermatitis was consistent with airborne contact. Patch tests were positive to tylosin tartrate powder 5% pet. (D2 -, D5 +). In this patient, the dermatitis resolved once protective measures had been taken (14).

A 13-year-old girl had a two-month history of an extremely pruritic erythematous maculopapular eruption on her arms, face, and upper trunk. As part of a 4-H project, she fed a pig each day. The patient was unaware of any increased symptoms after handling the feed, to which she regularly added growth supplements. However, when all exposure to the feed and the supplements was ceased, her skin cleared dramatically within a week and remained clear. A patch test to the crude pig feed supplement containing tylosin was strongly positive. In a second session, tylosin gel-powder and tylosin for intramuscular injection (50 mg/ml in propylene glycol) diluted to 1% water also yielded strongly positive reactions, while propylene glycol was negative (15).

A pig farmer with hand eczema had positive patch test reactions to tylosin tartrate 5% pet. and also to glutaral, a mixture of penicillin and streptomycin and 2 vaccines for pigs. He had contact with all these materials and three weeks after temporary cessation of his work, his hand eczema had virtually disappeared (16).

Other authors have described single cases of occupational allergic contact dermatitis to tylosin in a pig farmer (18 [with a strongly positive lymphocyte proliferation test]), a veterinarian (19), an animal feed wholesaler (20), a chicken vaccinator (21), a cattle breeder (23) and a woman working in a bioassay laboratory (22).

One or more cases of occupational allergic contact dermatitis to tylosin, data of which are not available to the author, can be found in refs. 7 and 17.

Cross-reactions, pseudo-cross-reactions and co-reactions
Cross-reactivity to and from spiramycin is considered likely (8), but in most cases of tylosin sensitization where spiramycin was also tested, the latter remained negative (e.g. 21). No cross-reactions to other macrolide antibiotics (14).

LITERATURE

1 The data in the section 'General' may have been obtained from literature discussed in this chapter, but mostly also or exclusively from one or more of the following online sources: ChemIDPlus Advanced, PubChem, DrugBank, RxList, Drug Central, Drugs.com, and Wikipedia
2 Swinnen I, Goossens A. An update on airborne contact dermatitis: 2007-2011. Contact Dermatitis 2013;68:232-238
3 Veien NK, Hattel T, Justesen O, Nørholm A. Occupational contact dermatitis due to spiramycin and/or tylosin among farmers. Contact Dermatitis 1980;6:410-413
4 Hjorth N, Weismann K. Occupational dermatitis among veterinary surgeons caused by spiramycin and tylosin. Contact Dermatitis Newsletter 1972;12:320
5 Hjorth N, Weismann K. Occupational dermatitis among veterinary surgeons caused by spiramycin, tylosin and penethamate. Acta Derm Venereol 1973;53:229-232
6 Hjorth N, Roed-Petersen J. Allergic contact dermatitis in veterinary surgeons. Contact Dermatitis 1980;6:27-29
7 Preyss J A. Allergie gegen Tylosintartrat. Berufdermatosen 1969;17:166 (Article in German)
8 Rudzki E, Rebandel P, Grzywa Z, Pomorski Z, Jakiminska B, Zawisza E. Occupational dermatitis in veterinarians. Contact Dermatitis 1982;8:72-73
9 Breuer K, Uter W, Geier J. Epidemiological data on airborne contact dermatitis: results of the IVDK. Contact Dermatitis 2015;73:239-247
10 Rudzki E, Rebendel P. Contact sensitivity to antibiotics. Contact Dermatitis 1984;11:41-42
11 Mancuso G, Staffa M, Errani A, Berdondini RM, Fabbri P. Occupational dermatitis in animal feed mill workers. Contact Dermatitis 1990;22:37-41
12 Gielen K, Goossens A. Occupational allergic contact dermatitis from drugs in healthcare workers. Contact Dermatitis 2001;45:273-279
13 Gilissen L, Boeckxstaens E, Geebelen J, Goossens A. Occupational allergic contact dermatitis from systemic drugs. Contact Dermatitis 2020;82:24-30
14 Malaiyandi V, Houle MC, Skotnicki-Grant S. Airborne allergic contact dermatitis from tylosin in pharmacy compounders and cross-sensitization to macrolide antibiotics. Dermatitis 2012;23:227-230
15 Neldner KH. Contact dermatitis from animal feed additives. Arch Dermatol 1972;106:722-723
16 Bovenschen HJ, Peters B, Koetsier MI, Van der Valk PG. Occupational contact dermatitis due to multiple sensitizations in a pig farmer. Contact Dermatitis 2009;61:127-128
17 Jung H-D. Beruflich bedingte Kontaktekzeme durch Tylosin (Tylan). Dermatol Monatsschr 1983;169:235-237 (Article in German)
18 Tuomi ML, Räsänen L. Contact allergy to tylosin and cobalt in a pig-farmer. Contact Dermatitis 1995;33:285
19 Caraffini S, Assalve D, Stingeni L, Lisi P. Tylosin, an airborne contact allergen in veterinarians. Contact Dermatitis 1994;31:327-328
20 Danese P, Zanca A, Bertazzoni MG. Occupational contact dermatitis from tylosin. Contact Dermatitis 1994;30:122-123
21 Barberá E, de la Cuadra J. Occupational airborne allergic contact dermatitis from tylosin. Contact Dermatitis 1989;20:308-309
22 Verbov J. Tylosin dermatitis. Contact Dermatitis 1983;9:325-326
23 Guerra L, Venture M, Tardio M, Tosti A. Airborne contact dermatitis from animal feed antibiotics. Contact Dermatitis 1991;25:333-334

Chapter 3.497 VALACICLOVIR

IDENTIFICATION

Description/definition   : Valaciclovir is the L-valyl ester of acyclovir that conforms to the structural formula shown
                              below
Pharmacological classes   : Antiviral agents
IUPAC name   : 2-[(2-Amino-6-oxo-1H-purin-9-yl)methoxy]ethyl (2S)-2-amino-3-methylbutanoate
CAS registry number   : 124832-26-4
EC number   : 603-015-6
Merck Index monograph   : 11355
Patch testing   : Commercial product 5%, 10% and 30% pet.(5); most systemic drugs can be tested at 10%
                            pet.; if the pure chemical is not available, prepare the test material from intravenous
                            powder, the content of capsules or – when also not available – from powdered tablets to
                            achieve a final concentration of the active drug of 10% pet.
Molecular formula   : $C_{13}H_{20}N_6O_4$

GENERAL

Valacyclovir is the L-valyl ester of the antiviral drug acyclovir and a prodrug. Orally administered, valacyclovir is
rapidly converted to acyclovir which inhibits viral DNA replication after further conversion to the nucleotide analog
acyclovir triphosphate by viral thymidine kinase, cellular guanyl cyclase, and a number of other cellular enzymes.
Acyclovir triphosphate competitively inhibits viral DNA polymerase, incorporates into and terminates the growing
viral DNA chain, and inactivates viral DNA polymerase. Valacyclovir is indicated for the treatment of infections with
Herpes simplex I and 2 viruses (herpes labialis, herpes genitalis), and with Varicella zoster virus (herpes zoster,
chickenpox in children) (1). In pharmaceutical products, valacyclovir is employed as valacyclovir hydrochloride (CAS
number 124832-27-5, EC number not available, molecular formula $C_{13}H_{21}ClN_6O_4$) (1).

CUTANEOUS ADVERSE DRUG REACTIONS FROM SYSTEMIC ADMINISTRATION CAUSED BY TYPE IV
(DELAYED-TYPE) HYPERSENSITIVITY (as demonstrated by positive patch tests)

Cutaneous adverse drug reactions from systemic administration of valaciclovir caused by type IV (delayed-type)
hypersensitivity have included systemic allergic dermatitis (4) and drug reaction with eosinophilia and systemic
symptoms (DRESS) (5).

Systemic allergic dermatitis (systemic contact dermatitis)

A 44-year-old woman used acyclovir cream for 2 weeks on her first attack of genital herpes without any improvement. Oral valaciclovir (500 mg 2x daily) was started. After the first 2 tablets, an itchy, symmetrical exanthem
appeared on the face, trunk and extremities. One month later, the patient developed a labial herpes infection. She
used acyclovir cream and vesicobullous lesions with erythema appeared in the labial and perioral skin, with an
exanthem on the upper trunk and extremities. Patch tests were positive to cetearyl alcohol (present in the cream),
acyclovir and valaciclovir as is, 20%, 10% and 1% water and pet. (4). This was a case of systemic allergic dermatitis to
valaciclovir in a patient sensitized to acyclovir.

Drug reaction with eosinophilia and systemic symptoms (DRESS)

Three patients developed DRESS while using various medications including valaciclovir. Patch tests with all drugs
used were positive only to valaciclovir (30% pet. in one, 10% pet. in the second, 5% and 30% pet. in the third). There
were cross-reactions to acyclovir in the 2 patients in which acyclovir was also patch tested (5).

Cross-reactions, pseudo-cross-reactions and co-reactions
Patients sensitized to acyclovir may cross-react to valaciclovir (2,3,4) and vice versa (5). Valaciclovir is the L-valyl ester of acyclovir and is almost completely metabolized to acyclovir after oral administration.

LITERATURE
1   The data in the section 'General' may have been obtained from literature discussed in this chapter, but mostly also or exclusively from one or more of the following online sources: ChemIDPlus Advanced, PubChem, DrugBank, RxList, Drug Central, Drugs.com, and Wikipedia
2   Vernassiere C, Barbaud A, Trechot PH, Weber-Muller F, Schmutz JL. Systemic acyclovir reaction subsequent to acyclovir contact allergy: which systemic antiviral drug should then be used? Contact Dermatitis 2003;49:155-157
3   Bayrou O, Gaouar H, Leynadier F. Famciclovir as a possible alternative treatment in some cases of allergy to acyclovir. Contact Dermatitis 2000;42:42
4   Lammintausta K, Mäkelä L, Kalimo K. Rapid systemic valaciclovir reaction subsequent to aciclovir contact allergy. Contact Dermatitis 2001;45:181
5   Ingen-Housz-Oro S, Bernier C, Gener G, Fichel F, Barbaud A, Lebrun-Vignes B, et al. Valaciclovir: a culprit drug for drug reaction with eosinophilia and systemic symptoms not to be neglected. Three cases. Br J Dermatol 2019;180:666-667

Chapter 3.498 VALDECOXIB

IDENTIFICATION

Description/definition   : Valdecoxib is the sulfonamide derivative that conforms to the structural formula shown
                               below
Pharmacological classes   : Cyclooxygenase 2 inhibitors
IUPAC name   : 4-(5-Methyl-3-phenyl-1,2-isoxazol-4-yl)benzenesulfonamide
CAS registry number   : 181695-72-7
EC number   : 448-010-8
Merck Index monograph : 11356
Patch testing   : 10% pet. (3); uncertain whether the pure drug or commercial pulverized tablet is meant
Molecular formula   : $C_{16}H_{14}N_2O_3S$

GENERAL

Valdecoxib is a sulfonamide derivative and nonsteroidal anti-inflammatory drug (NSAID) with anti-inflammatory,
analgesic, and antipyretic activities. It is (or was) indicated for the treatment of osteoarthritis and dysmenorrhea.
Valdecoxib was removed from the Canadian, U.S., and E.U. markets in 2005 due to concerns about possible
increased risk of heart attack and stroke (1).

CUTANEOUS ADVERSE DRUG REACTIONS FROM SYSTEMIC ADMINISTRATION CAUSED BY TYPE IV
(DELAYED-TYPE) HYPERSENSITIVITY (as demonstrated by positive patch tests)

Cutaneous adverse drug reactions from systemic administration of valdecoxib caused by type IV (delayed-type)
hypersensitivity have included maculopapular exanthema (2).

Maculopapular eruption

A 67-year-old man presented with an extensive rash which had developed after having taken valdecoxib 20 mg/day
for 8 days for chronic arthritic pain. Physical examination showed a maculopapular rash with cocardiform purpuric
lesions as well as homogeneous erythema of the face and the trunk, involving 60% of the body surface. Patch tests
were positive to commercial valdecoxib and celecoxib (probably 30% pet.) at D3. Lymphocyte transformation tests
were negative. The patient was diagnosed with systemic allergic dermatitis from valdecoxib and a cross-reaction to
celecoxib (2). The correct diagnosis, however, is a maculopapular rash caused by delayed-type hypersensitivity to
valdecoxib: to diagnose systemic allergic dermatitis, previous sensitization from topical application is necessary.

Cross-reactions, pseudo-cross-reactions and co-reactions

A patient sensitized to valdecoxib from oral administration cross-reacted to celecoxib (2).

LITERATURE

1   The data in the section 'General' may have been obtained from literature discussed in this chapter, but mostly
     also or exclusively from one or more of the following online sources: ChemIDPlus Advanced, PubChem,
     DrugBank, RxList, Drug Central, Drugs.com, and Wikipedia
2   Jaeger C, Jappe U. Valdecoxib-induced systemic contact dermatitis confirmed by positive patch test. Contact
     Dermatitis 2005;52:47-48
3   Brockow K, Garvey LH, Aberer W, Atanaskovic-Markovic M, Barbaud A, Bilo MB, et al.; ENDA/EAACI Drug Allergy
     Interest Group. Skin test concentrations for systemically administered drugs – an ENDA/EAACI Drug Allergy
     Interest Group position paper. Allergy 2013;68:702-712

Chapter 3.499 VALPROIC ACID

IDENTIFICATION

Description/definition : Valproic acid is the branched chain organic acid that conforms to the structural formula shown below

Pharmacological classes : Anticonvulsants; antimanic agents; enzyme inhibitors; GABA agents

IUPAC name : 2-Propylpentanoic acid

Other names : 2-Propylvaleric acid; valproate

CAS registry number : 99-66-1

EC number : 202-777-3

Merck Index monograph : 11369

Patch testing : 1%-5% pet. (21); according to some authors this may be irritant (14); the commercial drug 20% water was negative in 25 controls (8)

Molecular formula : $C_8H_{16}O_2$

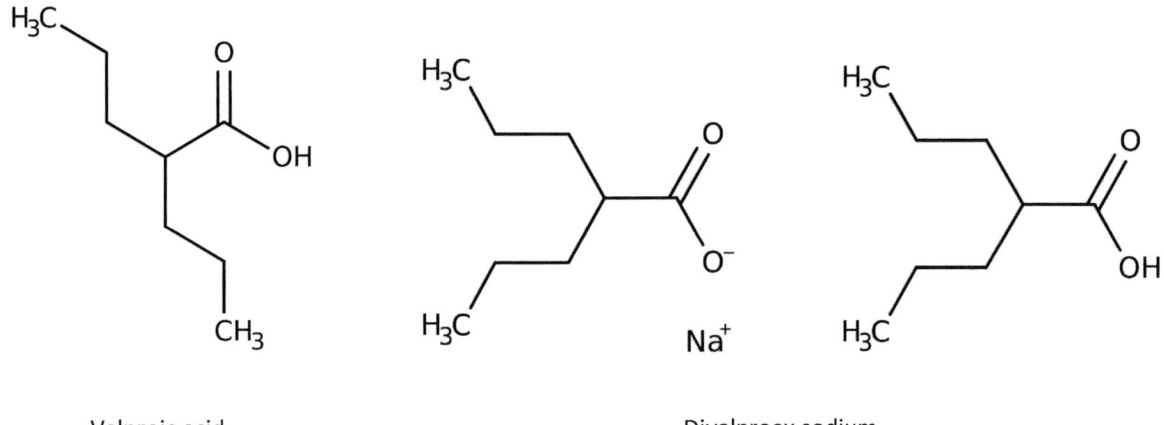

Valproic acid   Divalproex sodium

GENERAL

Valproic acid is a non-aromatic branched chain organic fatty acid with anti-epileptic and anti-manic activity. As valproate sodium, it is converted into its active form, valproate ion, in blood. Although the mechanism of action remains to be elucidated, valproic acid increases concentrations of γ-aminobutyric acid (GABA) in the brain, probably due to inhibition of the enzymes responsible for the catabolism of GABA. This potentiates the synaptic actions of GABA. Valproic acid may also affect potassium channels, thereby creating a direct membrane-stabilizing effect (1). Valproic acid is indicated for use as monotherapy or adjunctive therapy in the management of complex partial seizures and simple or complex absence seizures, as adjunctive therapy in the management of multiple seizure types that include absence seizures, as prophylaxis of migraine headaches and in the acute management of mania associated with bipolar disorder (1).

In pharmaceutical products, valproic acid can be employed as such, as valproate sodium (CAS number 1069-66-5, EC number 213-961-8, molecular formula $C_8H_{15}NaO_2$) and as divalproex sodium (valproate semisodium) (CAS number 76584-70-8, EC number not available, molecular formula $C_{16}H_{31}NaO_4$) (1).

CONTACT ALLERGY FROM ACCIDENTAL CONTACT

In Leuven, Belgium, in the period 2001-2019, 201 of 1248 health care workers/employees of the pharmaceutical industry had occupational allergic contact dermatitis. In 23 (11%) dermatitis was caused by skin contact with a systemic drug. In one of these patients, valproate was the drug/one of the drugs that caused occupational allergic contact dermatitis (3).

A 37-year-old woman developed eczema on the forefinger of the right hand, which later spread to the dorsum of the right hand, eyelids, neck and cheeks. She cared for a daughter with cerebral paralysis, and had to give her medication after triturating tablets by hand and gathering the powder up into a small spoon with her forefinger. Patch testing showed positive reactions to 4 commercial drugs the daughter used tested at 1% water. Later, patch tests with their active ingredients were performed and the patient had positive reactions to sodium valproate, tetrazepam, diazepam, and clorazepate, all tested 1% in water. There was a cross-reaction from sodium valproate to valpromide (1% water) (13). This was a case of sensitization *by proxy*.

CUTANEOUS ADVERSE DRUG REACTIONS FROM SYSTEMIC ADMINISTRATION CAUSED BY TYPE IV (DELAYED-TYPE) HYPERSENSITIVITY (as demonstrated by positive patch tests)

Cutaneous adverse drug reactions from systemic administration of valproic acid caused by type IV (delayed-type) hypersensitivity have included maculopapular eruption (6,15,18,19,20), drug reaction with eosinophilia and systemic symptoms (DRESS)/anticonvulsant hypersensitivity syndrome (4,5,6,7,8,10,11,15,17), Stevens-Johnson syndrome (9,11,15,19,20), photosensitivity (12), and unspecified drug eruption (2,18).

Case series with various or unknown types of drug

In the regional pharmacovigilance center of Sfax (Tunisia), between June 1, 2014 and April 30, 2016, all cases of (presumed) allergic skin reactions to antiepileptic drugs were investigated with patch testing. Twenty patients were included, among who 23 cutaneous adverse drug reactions (CADRs) were observed. The drugs involved were carbamazepine (n=11), phenobarbital (n=10) and valproic acid (n=4). The CADRs were maculopapular exanthema (11 cases), DRESS (6 cases), Stevens-Johnson syndrome (2 cases), fixed drug eruption (2 cases) and erythroderma (2 cases). Patch tests were positive in 19 patients (95%). Of the patients with positive patch test reactions to valproic acid, 2 had maculopapular exanthema, one DRESS and one Stevens-Johnson syndrome (15).

Between 2003 and 2017, in Belgrade, Serbia, 100 children in the age from 1 to 17 years suspected of hypersensitivity reactions to antiepileptic drugs were examined with patch tests, using the commercial drugs 10% pet. 61 patients had shown maculopapular eruptions, 26 delayed urticaria, 5 morbilliform exanthema, 5 DRESS, 2 SJS and one erythema multiforme. Sodium valproate was the suspected drug in 16 cases and was patch test positive in 10 children (63%). It was not specified which eruptions these drugs, but should include maculopapular eruptions and delayed urticaria (18).

In Turkey, of 11 children showing signs of hypersensitivity to valproic acid, 3 (27%) had positive patch test reactions to commercial valproic acid at 1% and 10% active ingredients. Two had suffered a maculopapular eruptions and one Stevens-Johnson syndrome (19).

Maculopapular eruption

One patient developed a maculopapular eruption from valproic acid with a positive patch test reaction (6). In Ankara, Turkey, valproic acid was patch test positive in 2 children with maculopapular eruption (20). See also the section 'Case series with various or unknown types of drug reactions' above, refs. 15,18 and 19.

Drug reaction with eosinophilia and systemic symptoms (DRESS)/anticonvulsant hypersensitivity syndrome

General

The literature on patch testing in anticonvulsant hypersensitivity syndrome up to August 2008 has been reviewed in ref. 16.

Case reports

One patient developed a relapse of DRESS from valproic acid with a positive patch test to this drug after a previous episode of DRESS from other drug(s) (multiple drug hypersensitivity) (5,6). In a group of 15 patients with 'anticonvulsant hypersensitivity syndrome', one patient had a positive patch test to sodium valproate; details are unavailable to the author, but not all patients appeared to have all symptoms necessary to diagnose anticonvulsant hypersensitivity syndrome (synonym: DRESS from anticonvulsant drugs) (4).

A 28-year-old woman had been treated with sodium valproate for 5 weeks, when she suddenly developed a generalized maculopapular eruption with fever, node enlargement, hypereosinophilia and altered liver function. The diagnosis DRESS was established on the basis of the time-course and positive patch tests using the diluted drug (no specific data available) (7).

A 6-year-old boy, who was treated with sodium valproate and ethosuximide for epileptic absences, developed DRESS with a diffuse pruritic morbilliform skin eruption with vesicular and target lesions, edema of the face, high fever, enlarged lymph nodes, leukocytosis, eosinophilia, elevated C-reactive protein and elevated liver enzymes. Human herpesvirus 6 (HHV6) antibody titers increased significantly within 15 days. Patch tests were positive to both commercial drugs 20% water. 25 controls were negative. Histologic examination of a positive patch test showed acute dermatitis. Following the positive sodium valproate patch test, there was a recurrence of the skin rash on the arms, legs and face (8).

A 48-year-old man was admitted to a psychiatric hospital with schizomanic syndrome and was treated with various drugs including valproic acid. Three weeks later, the patient developed a generalized maculopapular rash with lymphadenopathy, fever (39.1°C), leukocytosis, eosinophilia and a slight elevation of transaminases and creatinine.

A patch test performed 3 months after recovery gave a positive reading for the valproate preparations (undiluted and 30% water) at D3; 3 controls were negative (10).

A 36-year-old man developed drug hypersensitivity syndrome to carbamazepine with a maculopapular rash on his arms, thighs and trunk, fever, lymphadenopathy, oronasal ulcers, conjunctival erythema, and pharyngitis with features of Stevens-Johnson syndrome, lymphocytosis with slight eosinophilia. Treatment with carbamazepine discontinued and methylprednisolone started, which led to swift improvement. Two weeks later, valproate was started. Immediately a widespread exanthematous rash similar to his previous one, fever and adenopathy recurred. Skin biopsy revealed findings consistent with Stevens-Johnson syndrome. Valproate was discontinued. Patch testing to carbamazepine and valproic acid performed 4 months later was positive for both (11).

A 26-year-old woman had developed acute generalized exanthematous pustulosis (AGEP) from carbamazepine. When treated with phenytoin and, later, with valproic acid, the patient developed DRESS with fever, rashes, eosinophilia and subclinical hepatitis from both anticonvulsants on separate occasions. Patch tests to all three drugs were positive (17).

See also the section 'Case series with various or unknown types of drug reactions' above, ref. 15.

Stevens-Johnson syndrome/toxic epidermal necrolysis (SJS/TEN)
A 38-year-old woman was started on carbamazepine plus valproic acid. After 2 days, she developed Stevens-Johnson syndrome. Ten years later, she had positive patch tests to valproic acid tested at 15, 30, 45, and 60% w/w in white petrolatum and to carbamazepine tested as pure powder at 5, 10, 15, and 20% w/w in white petrolatum. 34 controls were negative to all test substances (9). In Ankara, Turkey, valproic acid was patch test positive in one child with SJS (20). See also the section 'Drug reaction with eosinophilia and systemic symptoms (DRESS)/anticonvulsant hypersensitivity syndrome' above, ref. 11.

See also the section 'Case series with various or unknown types of drug reactions' above, refs. 15 and 19.

Photosensitivity
In the period 2004-2005, in a Spanish multicenter study performing photopatch testing, one relevant positive photopatch tests was observed to valproic acid. It was not mentioned how many patients had been tested with this chemical and clinical details were not provided (12).

Other cutaneous adverse drug reactions
In a Japanese investigation, 4 of 5 patients patch tested with sodium valproate because of an 'anticonvulsant-induced drug eruptions' had positive reaction to this drug. The nature of the skin eruption was not mentioned (2).

See also the section 'Case series with various or unknown types of drug reactions' above, refs. 18 (delayed urticaria).

Cross-reactions, pseudo-cross-reactions and co-reactions
A patient sensitized to valproic acid cross-reacted to valpromide (both tested 1% water) (13).

LITERATURE
1 The data in the section 'General' may have been obtained from literature discussed in this chapter, but mostly also or exclusively from one or more of the following online sources: ChemIDPlus Advanced, PubChem, DrugBank, RxList, Drug Central, Drugs.com, and Wikipedia
2 Osawa J, Naito S, Aihara M, Kitamura K, Ikezawa Z, Nakajima H. Evaluation of skin test reactions in patients with non-immediate type drug eruptions. J Dermatol 1990;17:235-239
3 Gilissen L, Boeckxstaens E, Geebelen J, Goossens A. Occupational allergic contact dermatitis from systemic drugs. Contact Dermatitis 2020;82:24-30
4 Galindo PA, Borja J, Gomez E, Mur P, Gudín M, García R, et al. Anticonvulsant drug hypersensitivity. J Invest Allergol Clin Immunol 2002;12:299-304
5 Jörg L, Helbling A, Yerly D, Pichler WJ. Drug-related relapses in drug reaction with eosinophilia and systemic symptoms (DRESS). Clin Transl Allergy 2020;10:52
6 Jörg L, Yerly D, Helbling A, Pichler W. The role of drug, dose and the tolerance/intolerance of new drugs in multiple drug hypersensitivity syndrome (MDH). Allergy 2020;75:1178-1187
7 Picart N, Périole B, Mazereeuw J, Bonafé JL. Syndrome d'hypersensibilité médicamenteuse à l'acide valproïque [Drug hypersensitivity syndrome to valproic acid]. Presse Med 2000;29:648-650 (Article in French)
8 Conilleau V, Dompmartin A, Verneuil L, Michel M, Leroy D. Hypersensitivity syndrome due to 2 anticonvulsant drugs. Contact Dermatitis 1999;41:141-144
9 Romano A, Pettinato R, Andriolo M, Viola M, Guéant-Rodriguez RM, Valluzzi RL, et al. Hypersensitivity to aromatic anticonvulsants: in vivo and in vitro cross-reactivity studies. Curr Pharm Des 2006;12:3373-3381

10 Roepke S, Treudler R, Anghelescu I, Orfanos CE, Tebbe B. Valproic acid and hypersensitivity syndrome [letter]. Am J Psychiatry 2004;161:579

11 Arévalo-Lorido JC, Carretero-Gómez J, Bureo-Dacal JC, Montero-Leal C, Bureo-Dacal P. Antiepileptic drug hypersensitivity syndrome in a patient treated with valproate. Br J Clin Pharmacol 2003;55:415-416

12 De La Cuadra-Oyanguren J, Perez-Ferriols A, Lecha-Carrelero M, et al. Results and assessment of photopatch testing in Spain: towards a new standard set of photoallergens. Actas DermoSifiliograficas 2007;98:96-101

13 Garcia-Bravo B, Rodriguez-Pichardo A, Camacho F. Contact dermatitis from diazepoxides. Contact Dermatitis 1994;30:40

14 Vatve M, Sharma VK, Sawhney I, Kumar B. Evaluation of patch test in identification of causative agent in drug rashes due to antiepileptics. Indian J Dermatol Venereol Leprol 2000;66:132-135

15 Ben Mahmoud L, Bahloul N, Ghozzi H, Kammoun B, Hakim A, Sahnoun Z, et al. Epicutaneous patch testing in delayed drug hypersensitivity reactions induced by antiepileptic drugs. Therapie 2017;72:539-545

16 Elzagallaai AA, Knowles SR, Rieder MJ, Bend JR, Shear NH, Koren G. Patch testing for the diagnosis of anticonvulsant hypersensitivity syndrome: a systematic review. Drug Saf 2009;32:391-408

17 Duran-Ferreras E, Mir-Mercader J,Morales-Martinez M D, Martinez-Parra C. Anticonvulsant hypersensitivity syndrome with severe repercussions in the skin and kidneys. Rev Neurol 2004;38:1136-1138 (Article in Spanish)

18 Atanasković-Marković M, Janković J, Tmušić V, Gavrović-Jankulović M, Ćirković Veličković T, Nikolić D, Škorić D. Hypersensitivity reactions to antiepileptic drugs in children. Pediatr Allergy Immunol 2019;30:547-552

19 Guvenir H, Dibek Misirlioglu E, Civelek E, Toyran M, Buyuktiryaki B, Ginis T, et al. The frequency and clinical features of hypersensitivity reactions to antiepileptic drugs in children: a prospective study. J Allergy Clin Immunol Pract 2018;6:2043-2050

20 Büyük Yaytokgil Ş, Güvenir H, Külhaş Celík İ, Yilmaz Topal Ö, Karaatmaca B, Civelek E, et al. Evaluation of drug patch tests in children. Allergy Asthma Proc 2021;42:167-174

21 De Groot AC. Patch testing, 4th edition. Wapserveen, The Netherlands: acdegroot publishing, 2018 (ISBN 9789081323345)

Chapter 3.500 VANCOMYCIN

IDENTIFICATION

Description/definition	: Vancomycin is the branched tricyclic glycosylated peptide antibiotic obtained from *Streptomyces orientalis* that conforms to the structural formula shown below
Pharmacological classes	: Anti-bacterial agents
IUPAC name	: (1S,2R,18R,19R,22S,25R,28R,40S)-48-[(2S,3R,4S,5S,6R)-3-[(2S,4S,5S,6S)-4-Amino-5-hydroxy-4,6-dimethyloxan-2-yl]oxy-4,5-dihydroxy-6-(hydroxymethyl)oxan-2-yl]oxy-22-(2-amino-2-oxoethyl)-5,15-dichloro-2,18,32,35,37-pentahydroxy-19-[[(2R)-4-methyl-2-(methylamino)pentanoyl]amino]-20,23,26,42,44-pentaoxo-7,13-dioxa-21,24,27,41,43-pentazaoctacyclo[26.14.2.2^{3,6}.2^{14,17}.1^{8,12}.1^{29,33}.0^{10,25}.0^{34,39}]pentaconta-3,5,8,10,12(48),14,16,29(45),30,32,34(39),35,37,46,49-pentadecaene-40-carboxylic acid
CAS registry number	: 1404-90-6
EC number	: Not available
Merck Index monograph	: 11386
Patch testing	: Hydrochloride, 10.0% water (Chemotechnique)
Molecular formula	: $C_{66}H_{75}Cl_2N_9O_{24}$

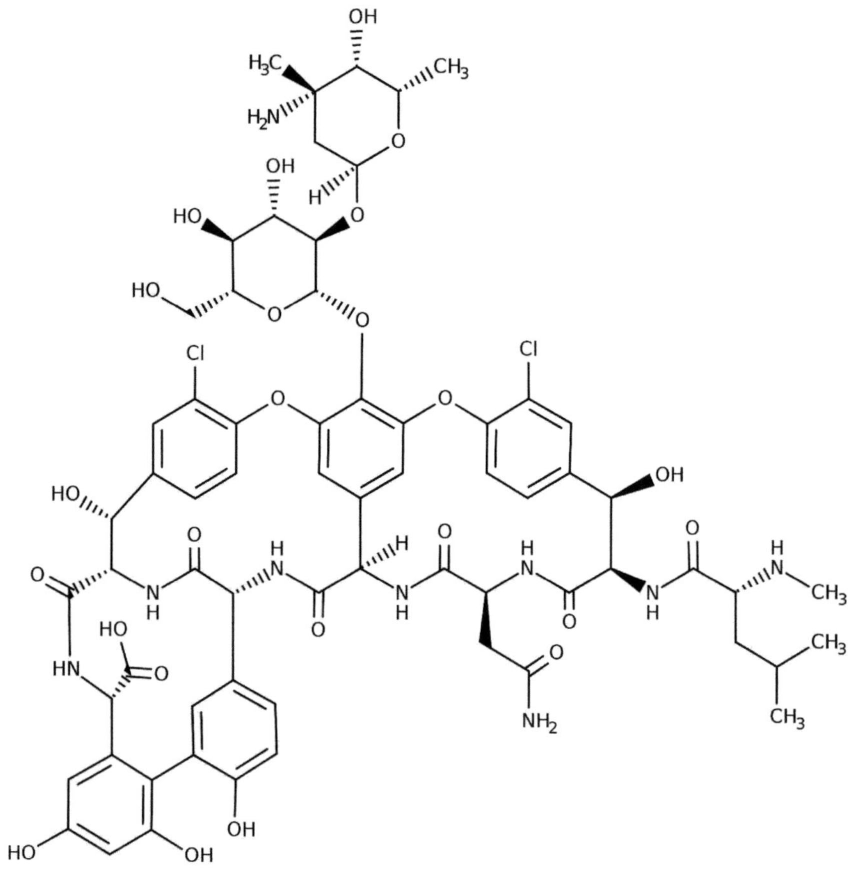

GENERAL

Vancomycin is a branched tricyclic glycosylated peptide obtained from *Streptomyces orientalis* with antibacterial properties. This antibiotic has bactericidal activity against most organisms and bacteriostatic effect on enterococci. It activates autolysins that destroy the bacterial cell wall, alters the permeability of bacterial cytoplasmic membranes and may selectively inhibit RNA synthesis. Vancomycin is indicated for the treatment of serious or severe infections caused by susceptible strains of methicillin-resistant and β-lactam-resistant staphylococci. In addition, an oral liquid preparation is indicated for the treatment of *Clostridium difficile*-associated diarrhea and enterocolitis caused by *Staphylococcus aureus*, including methicillin-resistant strains (1). In ophthalmology, topical or intravenous vancomycin is currently used to treat sight-threatening bacterial infections of the eyes, including infectious keratitis and endophthalmitis. In such applications, it has rarely caused allergic contact dermatitis (3). In pharmaceutical products, vancomycin is employed as vancomycin hydrochloride (CAS number 1404-93-9, EC number 604-193-8, molecular formula $C_{66}H_{76}Cl_3N_9O_{24}$) (1).

CUTANEOUS ADVERSE DRUG REACTIONS FROM SYSTEMIC ADMINISTRATION CAUSED BY TYPE IV (DELAYED-TYPE) HYPERSENSITIVITY (as demonstrated by positive patch tests)

Cutaneous adverse drug reactions from systemic administration of vancomycin caused by type IV (delayed-type) hypersensitivity have included macular erythematous exanthema (5), maculopapular eruption (7), acute generalized exanthematous pustulosis (AGEP) (2,11), drug reaction with eosinophilia and systemic symptoms (DRESS) (4,6,8,9,10), and Stevens-Johnson syndrome/toxic epidermal necrolysis (SJS/TEN) (8).

Macular and maculopapular eruption

A 39-year-old man on hemodialysis for a year was treated with intravenous vancomycin for a staphylococcal infection of the arteriovenous shunt. After 3 weeks of treatment, the patient developed high fever and a cutaneous eruption, which resolved on therapy with corticosteroids and antihistamines. However, 4 days later, the fever and rash reappeared, with pruritus and hypotension. Three other antibiotics were added, but the fever persisted and blood cultures remained negative. Physical examination showed a fever of 39.5°C, a pruriginous macular exanthem on the trunk and legs, and edema of the lips and the eyelids. Laboratory investigations showed leukocytosis with a normal neutrophil count. After filtering out vancomycin with hemodialysis, the exanthema and fever disappeared. Patch tests were positive to vancomycin 0.005% water (D2 +, D4 ++) and the related teicoplanin 4% water (D2 -, D4 +). Twenty controls were negative (5).

A 17-year-old man developed a maculopapular eruption while on vancomycin and had a positive patch test to this drug; a repeat test after 2 years was again positive (7).

Acute generalized exanthematous pustulosis (AGEP)

A 70-year-old woman with a prosthetic joint infection after a total knee replacement was treated with cephalexin. Because of lack of improvement, clindamycin was initiated, followed by vancomycin 3 days later. Twelve hours following vancomycin administration, diffuse subcorneal pustules and erythema developed, with especially heavy involvement of the intertriginous regions. A diagnosis of acute generalized exanthematous pustulosis (AGEP) was made, suspected to be caused by clindamycin. A prednisone taper was given, but one week later, the rash had further spread, now involving her face. Treatment with cyclosporine was started and three weeks later, the rash had completely resolved. Patch testing was performed, which showed reactivity to vancomycin but not clindamycin or cephalosporins. The materials used, test concentration and vehicle were not mentioned (2).

A woman of 29 suffered an attack of AGEP while using vancomycin. Patch tests and lymphocyte transformation tests were positive to vancomycin (11).

Drug reaction with eosinophilia and systemic symptoms (DRESS)

In a multicenter investigation in France, of 72 patients patch tested for DRESS, 46 (64%) had positive patch tests to drugs, including 4 to vancomycin; clinical and patch testing details were not provided (8). One patient developed a relapse of DRESS from vancomycin with a positive patch test to this drug after a previous episode of DRESS from sulfamethoxazole (multiple drug hypersensitivity) (9,10).

A 50-year-old man was treated for vertebral osteomyelitis and epidural abscess with vancomycin and developed DRESS with maculopapular exanthem, fever, eosinophilia, interstitial pneumonitis, and interstitial nephritis. A recurrence occurred when he was later treated with teicoplanin. Patch testing with vancomycin 15% water and teicoplanin 4% water 2 months later was ++ to vancomycin and ?+ to teicoplanin at D4 (4).

In a 61-year-old man, delayed-type hypersensitivity to vancomycin (tested as commercial tablet 500 mg, 15% and 30% pet.) and other drugs may have caused / contributed to DRESS (drug reaction with eosinophilia and systemic symptoms) (6).

Stevens-Johnson syndrome/toxic epidermal necrolysis (SJS/TEN)

In a multicenter investigation in France, of 17 patients patch tested for SJS/TEN, 4 (24%) had positive patch tests to drugs, including one to vancomycin; clinical and patch testing details were not provided (8).

Cross-reactions, pseudo-cross-reactions and co-reactions

Patients sensitized to vancomycin may cross-react to teicoplanin (4,5).

LITERATURE

1   The data in the section 'General' may have been obtained from literature discussed in this chapter, but mostly also or exclusively from one or more of the following online sources: ChemIDPlus Advanced, PubChem, DrugBank, RxList, Drug Central, Drugs.com, and Wikipedia

2   Pettit C, Trinidad J, Kaffenberger B. A case of vancomycin-induced acute generalized exanthematous pustulosis confirmed by patch testing. J Clin Aesthet Dermatol 2020;13:35-36

3   Hwu JJ, Chen KH, Hsu WM, Lai JY, Li YS. Ocular hypersensitivity to topical vancomycin in a case of chronic endophthalmitis. Cornea 2005;24:754-756

4   Kwon HS, Chang YS, Jeong YY, Lee SM, Song WJ, et al. A case of hypersensitivity syndrome to both vancomycin and teicoplanin. J Korean Med Sci 2006;21:1108-1110

5   Bernedo N, Gonzalez I, Gastaminza G, Audicana M, Fernández E, Muñoz D. Positive patch test in vancomycin allergy. Contact Dermatitis 2001;45:43

6   Liippo J, Pummi K, Hohenthal U, Lammintausta K. Patch testing and sensitization to multiple drugs. Contact Dermatitis 2013;69:296-302

7   Pinho A, Marta A, Coutinho I, Gonçalo M. Long-term reproducibility of positive patch test reactions in patients with non-immediate cutaneous adverse drug reactions to antibiotics. Contact Dermatitis 2017;76:204-209

8   Barbaud A, Collet E, Milpied B, Assier H, Staumont D, Avenel-Audran M, et al. A multicentre study to determine the value and safety of drug patch tests for the three main classes of severe cutaneous adverse drug reactions. Br J Dermatol 2013;168:555-562

9   Jörg L, Helbling A, Yerly D, Pichler WJ. Drug-related relapses in drug reaction with eosinophilia and systemic symptoms (DRESS). Clin Transl Allergy 2020;10:52

10 Jörg L, Yerly D, Helbling A, Pichler W. The role of drug, dose and the tolerance/intolerance of new drugs in multiple drug hypersensitivity syndrome (MDH). Allergy 2020;75:1178-1187

11 Beeler A, Engler O, Gerber BO, et al. Long-lasting reactivity and high frequency of drug-specific T cells after severe systemic drug hypersensitivity reactions. J Allergy Clin Immunol 2006;117:455-462

Chapter 3.501 VARENICLINE

IDENTIFICATION

Description/definition   : Varenicline is the benzazepine derivative that conforms to the structural formula shown
                             below
Pharmacological classes   : Nicotinic agonists; smoking cessation agents
IUPAC name   : 5,8,14-Triazatetracyclo[10.3.1.0^{2,11}.0^{4,9}]hexadeca-2,4,6,8,10-pentaene
CAS registry number   : 249296-44-4
EC number   : Not available
Merck Index monograph   : 11395
Patch testing   : Tablet, pulverized, 5%, 10% and 30% pet.; water as vehicle may be less adequate (2);
                             most systemic drugs can be tested at 10% pet.; if the pure chemical is not available,
                             prepare the test material from intravenous powder, the content of capsules or – when
                             also not available – from powdered tablets to achieve a final concentration of the active
                             drug of 10% pet.
Molecular formula   : C$_{13}$H$_{13}$N$_3$

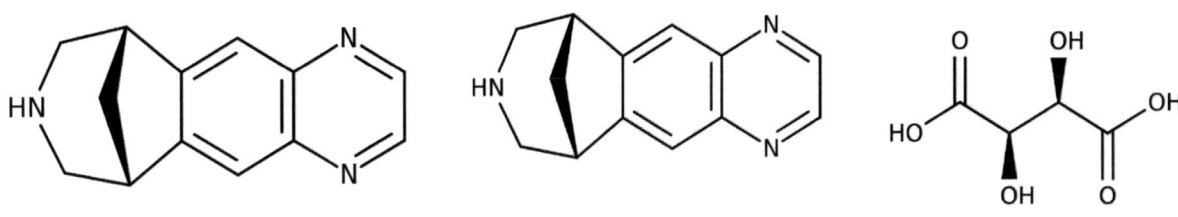

Varenicline   Varenicline tartrate

GENERAL

Varenicline is a partial agonist of the nicotinic acetylcholine receptor (nAChR) subtype α4β2 used to help in smoking cessation. Nicotine stimulation of central α4β2 nAChRs located at presynaptic terminals in the nucleus accumbens causes the release of the neurotransmitter dopamine, which may be associated with the experience of pleasure. Nicotine addiction constitutes a physiologic dependence related to this dopaminergic reward system. As an AChR partial agonist, varenicline attenuates the craving and withdrawal symptoms that occur with abstinence from nicotine but is not habit-forming itself. Varenicline is indicated for use as an aid in smoking cessation (1). In pharmaceutical products, varenicline is employed as varenicline tartrate (CAS number 375815-87-5, EC number not available, molecular formula C$_{17}$H$_{19}$N$_3$O$_6$) (1).

CUTANEOUS ADVERSE DRUG REACTIONS FROM SYSTEMIC ADMINISTRATION CAUSED BY TYPE IV
(DELAYED-TYPE) HYPERSENSITIVITY (as demonstrated by positive patch tests)

Cutaneous adverse drug reactions from systemic administration of varenicline caused by type IV (delayed-type) hypersensitivity have included acute generalized exanthematous pustulosis (AGEP) (2,3).

Acute generalized exanthematous pustulosis (AGEP)

A 45-year-old woman presented with fever (38.2°C) and a pustular eruption on an erythematous base on the trunk and extremities, which had started after having used varenicline for 11 days. Histopathology was consistent with the clinical diagnosis of acute generalized exanthematous pustulosis (AGEP), showing intraepidermal neutrophilic pustule formation and inflammatory infiltration of neutrophils and sparse eosinophils in the papillary dermis. Laboratory studies showed mild leukocytosis. Two months after complete healing, patch testing was performed with dilution series of the commercial varenicline 1%, 5%, 10% and 30% pet. and water. Positive reactions were seen to varenicline 5,10 and 30% pet. at D4 (++) and D7 (++), whereas all test materials in water remained negative (2).

In a multicenter investigation in France, of 45 patients patch tested for AGEP, 26 (58%) had positive patch tests to drugs, including one to varenicline. This was a woman aged 39, but more clinical or patch testing details were not provided (3).

LITERATURE

1   The data in the section 'General' may have been obtained from literature discussed in this chapter, but mostly also or exclusively from one or more of the following online sources: ChemIDPlus Advanced, PubChem, DrugBank, RxList, Drug Central, Drugs.com, and Wikipedia

2   Özkaya E, Yazganoğlu KD, Kutlay A, Mahmudov A. Varenicline-induced acute generalized exanthematous pustulosis confirmed by patch testing. Contact Dermatitis 2018;78:97-99

3   Barbaud A, Collet E, Milpied B, Assier H, Staumont D, Avenel-Audran M, et al. A multicentre study to determine the value and safety of drug patch tests for the three main classes of severe cutaneous adverse drug reactions. Br J Dermatol 2013;168:555-562

Chapter 3.502 VINBURNINE

IDENTIFICATION

Description/definition   : Vinburnine is the alkaloid that conforms to the structural formula shown below
Pharmacological classes   : Vasodilator agents
IUPAC name   : (13aS,13bS)-13a-Ethyl-2,3,6,13,13a,13b-hexahydro-1H-indolo[3,2,1-de]pyrido[3,2,1-
                             ij][1,5]naphthyridin-12(5H)-one
Other names   : 3α,16α-Eburnamonine
CAS registry number   : 4880-88-0
EC number   : 225-490-5
Merck Index monograph   : 4804 (Eburnamonine)
Patch testing   : Tablet, pulverized, 10% pet. (2); most systemic drugs can be tested at 10% pet.; if the pure
                             chemical is not available, prepare the test material from intravenous powder, the content
                             of capsules or – when also not available – from powdered tablets to achieve a final
                             concentration of the active drug of 10% pet.
Molecular formula   : $C_{19}H_{22}N_2O$

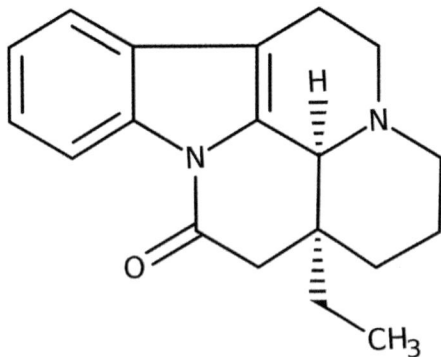

GENERAL

Vinburnine is a vinca alkaloid derivative and peripheral vasodilator with cerebral activities that also acts as a cerebral metabolic stimulant and appears to be able to relax the smooth muscle cells within the walls of blood vessels. Vinburnine is available in a few countries as a cerebral vasodilator (1).

CUTANEOUS ADVERSE DRUG REACTIONS FROM SYSTEMIC ADMINISTRATION CAUSED BY TYPE IV (DELAYED-TYPE) HYPERSENSITIVITY (as demonstrated by positive patch tests)

Cutaneous adverse drug reactions from systemic administration of vinburnine caused by type IV (delayed-type) hypersensitivity have included fixed drug eruption (2).

Fixed drug eruption

A 21-year-old woman developed multiple sharply circumscribed erythematous patches of different shapes and sizes, with partially excoriated central blisters. She used vinburnine for vertebral and medullary trauma with paraparesis and reduction of cerebral perfusion to the frontal lobe. The therapy was suspended and the skin lesions resolved. An oral challenge test with vinburnine was negative, and therefore, the therapy was resumed. However, after 3 weeks, the lesions recurred in exactly the same locations as before. An open test with vinburnine 10% pet. was negative on normal and lesional skin. However, a patch test with vinburnine 10% pet. applied to postlesional skin was positive after 6 hours and was accompanied by a flare-up of the fixed drug eruption in previously involved skin and high fever (2). This was a case of bullous non-pigmenting fixed drug eruption from delayed-type hypersensitivity to vinburnine.

LITERATURE

1   The data in the section 'General' may have been obtained from literature discussed in this chapter, but mostly
     also or exclusively from one or more of the following online sources: ChemIDPlus Advanced, PubChem,
     DrugBank, RxList, Drug Central, Drugs.com, and Wikipedia
2   Schena D, Menegazzi S, Barba A. Bullous fixed drug eruption induced by vinburnine. Contact Dermatitis
     1992;27:187

Chapter 3.503 VINCAMINE

IDENTIFICATION

Description/definition : Vincamine is the monoterpenoid indole alkaloid obtained from the leaves of *Vinca minor* that conforms to the structural formula shown below

Pharmacological classes : Antineoplastic agents; vasodilator agents

IUPAC name : Methyl (15S,17S,19S)-15-ethyl-17-hydroxy-1,11-diazapentacyclo[9.6.2.0^{2,7}.0^{8,18}.0^{15,19}]nonadeca-2,4,6,8(18)-tetraene-17-carboxylate

CAS registry number : 1617-90-9

EC number : 216-576-3

Merck Index monograph : 11450

Patch testing : Tartrate, 1% water; 10% water appears to be non-irritant (2)

Molecular formula : $C_{21}H_{26}N_2O_3$

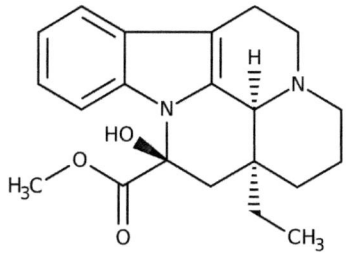

Vincamine tartrate

GENERAL

Vincamine is a monoterpenoid indole alkaloid obtained from the leaves of *Vinca minor* with a vasodilatory property. Studies indicate that vincamine increases the regional cerebral blood flow. In pharmaceuticals, usually vincamine base is used, often for veterinary purposes; sometimes vincamine tartrate is employed (CAS number 64034-84-0, EC number 264-614-2, molecular formula $C_{25}H_{32}N_2O_9$). In some countries the drug, in combination with piracetam, is used for the treatment of stroke, progressive and chronic cerebral insufficiency, early senility, and organic and psychological disorders from cerebral hypoxia (1).

CONTACT ALLERGY FROM ACCIDENTAL CONTACT

A 39-year-old man working in a pharmaceutical company presented with an acute eczematous contact dermatitis of the eyelids and neck. The eruption healed quickly with wet dressings and a corticosteroid cream. Two weeks later he relapsed and showed lesions in the flexures of the arms. During both episodes the patient had extracted vincamine tartrate out of *Vinca* species plants prior to the development of the dermatitis. According to the patient, dust particles of the ground dried plants would be spread in the air. Patch tests were positive to vincamine tartrate 1% water (+++). Five controls were negative to both 1% and 10% vincamine tartrate in water. The patient was diagnosed with airborne occupational allergic contact dermatitis from the vincamine tartrate component of the *Vinca* plant (2).

In early German literature, contact dermatitis was described in 18 of 48 workers from *Vinca* alkaloids, 11 of who had positive patch tests (unknown to which alkaloids). There were cross-reactions to reserpine and ajmaline, which are *Rauwolfia* alkaloids (3).

LITERATURE

1 The data in the section 'General' may have been obtained from literature discussed in this chapter, but mostly also or exclusively from one or more of the following online sources: ChemIDPlus Advanced, PubChem, DrugBank, RxList, Drug Central, Drugs.com, and Wikipedia

2 Van Hecke E. Contact sensitivity to vincamine tartrate. Contact Dermatitis 1981;7:53

3 Valer M. Die bei der Erzeugung von Devincan auftretenden Hautläsionen. Berufsdermatosen 1965;13:96 (Article in German, data cited in ref. 2)

Chapter 3.504 VIRGINIAMYCIN

IDENTIFICATION

Description/definition   : Virginiamycin is a cyclic polypeptide antibiotic complex from *Streptomyces virginiae*,
                          S. loidensis, *S. mitakaensis*, *S. pristinaspiralis*, *S. ostreogriseus*, and others; it consists of 2
                          major components, virginiamycin factor M1 and virginiamycin factor S1
Pharmacological classes : Anti-bacterial agents
IUPAC name   : *N*-[(3*S*,6*S*,12*R*,15*S*,16*R*,19*S*,22*S*)-3-Benzyl-12-ethyl-4,16-dimethyl-2,5,11,14,18,21,24-
                          heptaoxo-19-phenyl-17-oxa-1,4,10,13,20-pentazatricyclo[20.4.0.0^{6,10}]hexacosan-15-yl]-
                          3-hydroxypyridine-2-carboxamide;(10*R*,11*R*,12*E*,17*E*,19*E*,21*S*)-21-hydroxy-11,19-dimethyl-
                          10-propan-2-yl-9,26-dioxa-3,15,28-triazatricyclo[23.2.1.0 ^{3,7}]octacosa-1(27),6,12,17,19,
                          25(28)-hexaene-2,8,14,23-tetrone
Other names   : Staphylomycin ®
CAS registry number(s)   : 11006-76-1
EC number   : 234-244-6
Merck Index monograph   : 11470
Patch testing   : 5% and 10% pet.
Molecular formula   : $C_{71}H_{84}N_{10}O_{17}$

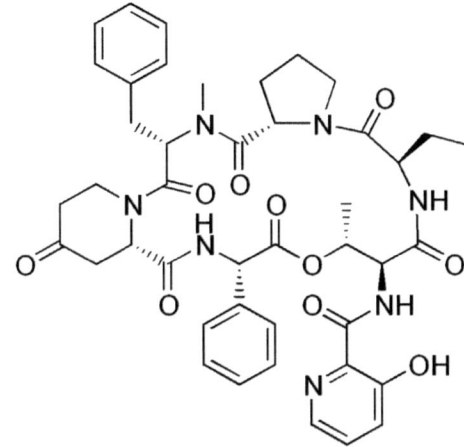

Virginiamycin factor M1   Virginiamycin factor S1
(= pristinamycin IIA)

GENERAL

Virginiamycin is a streptogramin antibiotic similar to pristinamycin and quinupristin/dalfopristin. It is a combination of pristinamycin IIA (virginiamycin M1) and virginiamycin S1. Virginiamycin binds to and inhibits ribosome assembly in susceptible bacteria, thereby preventing protein synthesis. It is active against gram-positive bacteria. Virginiamycin is currently only used in veterinary practice, both to combat infections and as as a growth promoter in cattle, swine, and poultry. It is also employed in the fuel ethanol industry to prevent microbial contamination (1). It was still used in human medicine in France in 1996 (2). The data provided in various online databases on virginiamycin and pristinamycin are very confusing, overlapping and sometimes probably inaccurate.

In topical preparations, virginiamycin has caused contact allergy/allergic contact dermatitis, which subject has been fully reviewed in Volume 3 of the *Monographs in contact allergy* series (9).

CONTACT ALLERGY FROM ACCIDENTAL CONTACT

A man aged 31 worked for 6 months in a pharmaceutical factory as a warehouseman. Two or three times daily he entered a dusty room where a food additive, virginiamycin, for pigs and poultry was prepared and stored in barrels. After 3 months the patient developed a pruritic erythematosquamous eruption of the face around the eyes with some edema of the eyelids. He had no previous history of using virginiamycin as an antibiotic either locally or systemically. Moreover, he had never used other antibiotics of the same chemical group, such as pristinamycin. Patch tests were positive to the food additive 20% pet., virginiamycin 5% pet., virginiamycin factor M 5% pet. (which is also present in pristinamycin) and pristinamycin 5% pet., but negative to virginiamycin factor S (5). This was a case of airborne occupational allergic contact dermatitis to virginiamycin.

CUTANEOUS ADVERSE DRUG REACTIONS FROM SYSTEMIC ADMINISTRATION CAUSED BY TYPE IV (DELAYED-TYPE) HYPERSENSITIVITY (as demonstrated by positive patch tests)

Cutaneous adverse drug reactions from systemic administration of virginiamycin caused by type IV (delayed-type) hypersensitivity have included systemic allergic dermatitis (2) and acute generalized exanthematous pustulosis (AGEP) (10).

Systemic allergic dermatitis (systemic contact dermatitis)

A 52-year-old man was applying topical virginiamycin ointment on eczema of the hands and legs. As the lesions spread, his doctor prescribed oral virginiamycin. Bullous eczema rapidly appeared on the hands and elbows, together with facial edema, pruritus and generalized erythema. With topical corticosteroids, the lesions cleared within 15 days. Patch tests were positive to the virginiamycin 0.5% ointment 'as is' (2). This was a case of systemic allergic dermatitis. Systemic allergic dermatitis may also develop in patients sensitized to virginiamycin when given pristinamycin orally (2,4,6, Chapter 3.407).

Acute generalized exanthematous pustulosis (AGEP)

A patient from France developed acute generalized exanthematous pustulosis (AGEP) while on virginiamycin and had a positive patch test reaction to this antibiotic. No clinical or patch testing details were provided (10).

Cross-reactions, pseudo-cross-reactions and co-reactions

Patients sensitized to pristinamycin (from oral administration) may cross-react to virginiamycin (3,11). Conversely, patients sensitized to virginiamycin may cross-react to pristinamycin (2,3,5,7,8). Cross-sensitivity is easily explained by structural similarities between these 2 antibiotics: virginiamycin factor M1 is the same as pristinamycin IIA and virginiamycin factor S1 is virtually identical to pristinamycin IA. The most important sensitizer appears to be virginiamycin factor M1 (= pristinamycin IIA) (5,7,8).

LITERATURE

1 The data in the section 'General' may have been obtained from literature discussed in this chapter, but mostly also or exclusively from one or more of the following online sources: ChemIDPlus Advanced, PubChem, DrugBank, RxList, Drug Central, Drugs.com, and Wikipedia
2 Michel M, Dompmartin A, Szczurko C, Castel B, Moreau A, Leroy D. Eczematous-like drug eruption induced by synergistins. Contact Dermatitis 1996;34:86-87
3 Baes H. Allergic contact dermatitis to virginiamycin. False cross-sensitivity with pristinamycin. Dermatologica 1974;149:231-235
4 Pillette M, Claudel JP, Muller C, Lorette G. Contact dermatitis caused by pristinamycin after sensitization to topical virginiamycin. Allerg Immunol (Paris) 1990;22:197 (Article in French)
5 Tennstedt D, Dumont-Fruytier M, Lachapelle JM. Occupational allergic contact dermatitis to virginiamycin, an antibiotic used as a food additive for pigs and poultry. Contact Dermatitis 1978;4:133-134
6 Mathivon F, Petit A, Mourier C, Sigal Nahum M. Toxidermie à la pristinamycine après réaction de contact à la virginiamycine. Rev Eur Dermatol MST 1991;3:527-529 (Article in French)
7 Bleumink E, Nater JP. Sensitization to Staphylomycin ® (virginiamycin). Contact Dermatitis Newsletter 1972;11:306
8 Bleumink E, Nater JP. Allergic contact dermatitis to virginiamycin (Staphylomycin ®) and pristinamycin (Stapyocin®). Contact Dermatitis Newsletter 1972;12:337
9 De Groot AC. Monographs in contact allergy, volume 3: Topical Drugs. Boca Raton, Fl, USA: CRC Press Taylor and Francis Group, 2021 (ISBN 978-0-367-23693-9)
10 Wolkenstein P, Chosidow O, Fléchet ML, Robbiola O, Paul M, Dumé L, et al. Patch testing in severe cutaneous adverse drug reactions, including Stevens-Johnson syndrome and toxic epidermal necrolysis. Contact Dermatitis 1996;35:234-236
11 Barbaud A, Trechot P, Weber-Muller F, Ulrich G, Commun N, Schmutz JL. Drug skin tests in cutaneous adverse drug reactions to pristinamycin: 29 cases with a study of cross-reactions between synergistins. Contact Dermatitis 2004;50:22-26

Chapter 3.505 ZIPRASIDONE

IDENTIFICATION

Description/definition   : Ziprasidone is the benzothiazolylpiperazine derivative that conforms to the structural
                             formula shown below
Pharmacological classes : Dopamine antagonists; antipsychotic agents; serotonin antagonists
IUPAC name   : 5-[2-[4-(1,2-Benzothiazol-3-yl)piperazin-1-yl]ethyl]-6-chloro-1,3-dihydroindol-2-one
CAS registry number   : 146939-27-7
EC number   : 928-541-6
Merck Index monograph : 11641
Patch testing   : No data available; most systemic drugs can be tested at 10% pet.; if the pure chemical is
                             not available, prepare the test material from intravenous powder, the content of capsules
                             or – when also not available – from powdered tablets to achieve a final concentration of
                             the active drug of 10% pet.
Molecular formula   : $C_{21}H_{21}ClN_4OS$

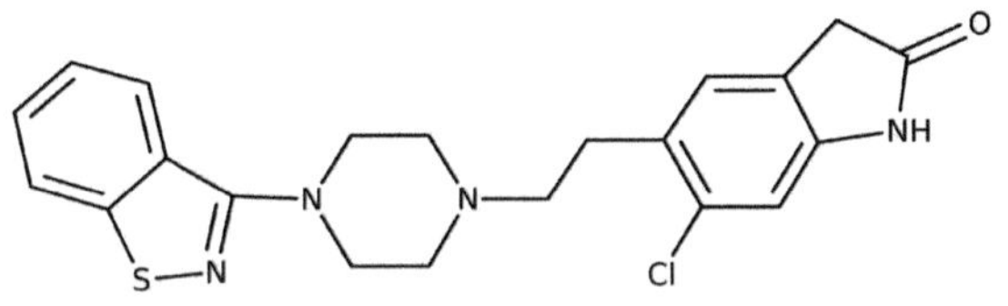

Ziprasidone

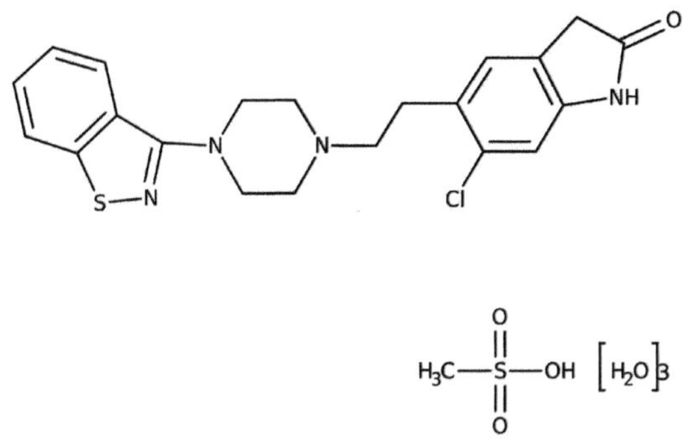

Ziprasidone mesylate

GENERAL

Ziprasidone is a benzothiazolylpiperazine derivative, a psychotropic agent and an atypical antipsychotic drug with an antischizophrenic property. Ziprasidone is a selective monoaminergic antagonist with high affinity for the serotonin type 2 ($5HT_2$), dopamine type 2 (D_2), α- and β-adrenergic, and H_1 histaminergic receptors. The mechanism of action by which ziprasidone exerts its antischizophrenic effect is unknown but is potentially mediated through a combination of dopamine D_2 and serotonin $5HT_2$ antagonism. In its oral form, ziprasidone is indicated for the treatment of schizophrenia and bipolar I disorder, while the injectable formulation is approved for treatment of acute agitation in schizophrenia (1).

In capsules, ziprasidone is employed as ziprasidone hydrochloride (monohydrate) (CAS number 138982-67-9, EC number not available, molecular formula $C_{21}H_{24}Cl_2N_4O_2S$). In powder for solutions for injection ziprasidone mesylate is used (CAS number 199191-69-0, EC number not available, molecular formula $C_{22}H_{31}ClN_4O_7S_2$) (1).

CONTACT ALLERGY FROM ACCIDENTAL CONTACT

A 52-year-old process operator developed occupational allergic contact dermatitis to ziprasidone. The primary sites of the dermatitis were the hands. Additional clinical or patch testing data were not provided (2).

LITERATURE

1   The data in the section 'General' may have been obtained from literature discussed in this chapter, but mostly also or exclusively from one or more of the following online sources: ChemIDPlus Advanced, PubChem, DrugBank, RxList, Drug Central, Drugs.com, and Wikipedia
2   Bennett MF, Lowney AC, Bourke JF. A study of occupational contact dermatitis in the pharmaceutical industry. Br J Dermatol 2016;174:654-656 (Abstract in Brit J Dermatol 2011;165:73)

Chapter 3.506 ZOLPIDEM

IDENTIFICATION

Description/definition	: Zolpidem is the imidazopyridine derivative that conforms to the structural formula shown below
Pharmacological classes	: GABA-A receptor agonists; sleep aids, pharmaceutical
IUPAC name	: *N,N*-Dimethyl-2-[6-methyl-2-(4-methylphenyl)imidazo[1,2-a]pyridin-3-yl]acetamide
CAS registry number	: 82626-48-0
EC number	: 617-367-3
Merck Index monograph	: 11661
Patch testing	: 10% pet. (4)
Molecular formula	: $C_{19}H_{21}N_3O$

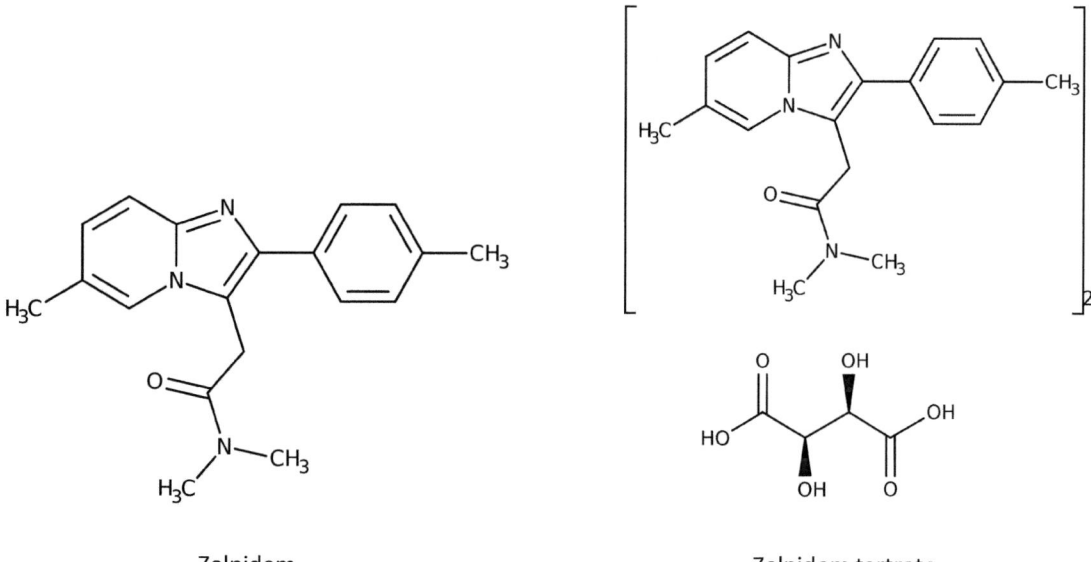

Zolpidem   Zolpidem tartrate

GENERAL

Zolpidem is an imidazopyridine derivative with sedative and hypnotic action. Its chemical structure is unrelated to benzodiazepines, barbiturates, or other drugs with known hypnotic properties. Zolpidem interacts with a GABA receptor complex, it binds preferentially to the α1-receptor (GABA-A). Zolpidem is indicated for the short-term treatment of insomnia in adults characterized by difficulties with sleep initiation (1). In pharmaceutical products, zolpidem is employed as zolpidem tartrate (CAS number 99294-93-6, EC number not available, molecular formula $C_{23}H_{27}N_3O_7$) (1).

CONTACT ALLERGY FROM ACCIDENTAL CONTACT

Case series

In Leuven, Belgium, in the period 2001-2019, 201 of 1248 health care workers/employees of the pharmaceutical industry had occupational allergic contact dermatitis. In 23 (11%) dermatitis was caused by skin contact with a systemic drug: 19 nurses, two chemists, one physician, and one veterinarian. The lesions were mostly localized on the hands, but often also on the face, as airborne dermatitis. In total, 42 positive patch test reactions to 18 different systemic drugs were found. In 4 patients, zolpidem was the drug/one of the drugs that caused occupational dermatitis (5, overlap with ref. 2).

Also in Leuven, Belgium, in the period 2007-2011, 81 patients have been diagnosed with occupational airborne allergic contact dermatitis. In 23 of them, drugs were the offending agents, including zolpidem in 7 cases (2, overlap with ref. 5).

Case reports

A 30-year-old woman presented with an itchy, scaly skin reaction on the face, which was most pronounced on the eyelids. She reported a clear relationship with her work as a geriatric nurse, during which she was required to crush tablets for the elderly on a daily basis, mostly benzodiazepine drugs. She had positive patch tests to zolpidem (crushed tablet 30% pet.; 19 controls were negative), 5 benzodiazepines (alprazolam, bromazepam, diazepam, lorazepam, tetrazepam) and 2 unrelated drugs. The patient was diagnosed with occupational airborne allergic contact dermatitis from drugs (3, overlap with ref. 2).

A 29-year-old man working as machine operator in a pharmaceutical factory was referred with a 6-month history of dermatitis involving his eyelids, cheeks, lips, nose, and the nasolabial folds. The rash cleared during long periods away from work. Patch tests were positive to zolpidem 10% pet., carvedilol 10% pet. and simvastatin 1% and 0.1% pet. Ten controls were negative. Three months avoidance of the offending allergens resulted in total clearing of the dermatitis. A diagnosis of occupational airborne allergic contact dermatitis was made (4).

LITERATURE

1   The data in the section 'General' may have been obtained from literature discussed in this chapter, but mostly also or exclusively from one or more of the following online sources: ChemIDPlus Advanced, PubChem, DrugBank, RxList, Drug Central, Drugs.com, and Wikipedia
2   Swinnen I, Goossens A. An update on airborne contact dermatitis: 2007-2011. Contact Dermatitis 2013;68:232-238
3   Swinnen I, Ghys K, Kerre S, Constandt L, Goossens A. Occupational airborne contact dermatitis from benzodiazepines and other drugs. Contact Dermatitis 2014;70:227-232
4   Neumark M, Moshe S, Ingber A, Slodownik D. Occupational airborne contact dermatitis to simvastatin, carvedilol, and zolpidem. Contact Dermatitis 2009;61:51-52
5   Gilissen L, Boeckxstaens E, Geebelen J, Goossens A. Occupational allergic contact dermatitis from systemic drugs. Contact Dermatitis 2020;82:24-30

Chapter 3.507 ZONISAMIDE

IDENTIFICATION

Description/definition : Zonisamide is the sulfonamide derivative that conforms to the structural formula shown below

Pharmacological classes : Calcium channel blockers; anticonvulsants

IUPAC name : 1,2-Benzoxazol-3-ylmethanesulfonamide

CAS registry number : 68291-97-4

EC number : 614-395-8

Merck Index monograph : 11882

Patch testing : No data available; most systemic drugs can be tested at 10% pet.; if the pure chemical is not available, prepare the test material from intravenous powder, the content of capsules or – when also not available – from powdered tablets to achieve a final concentration of the active drug of 10% pet.

Molecular formula : $C_8H_8N_2O_3S$

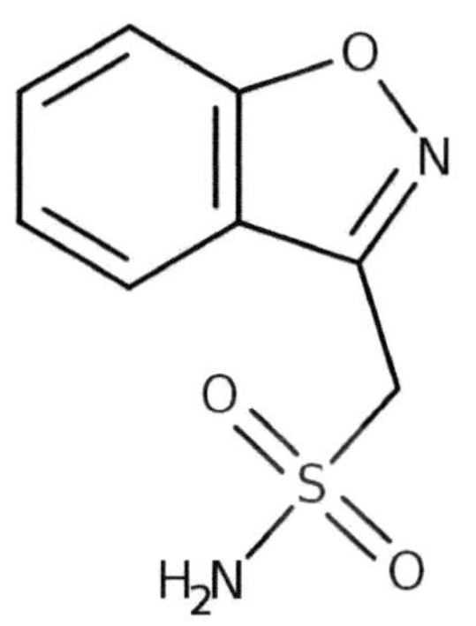

GENERAL

Zonisamide is a sulfonamide derivative with anticonvulsant properties. Its exact mechanism of action remains to be elucidated, but the drug appears to block sodium and calcium channels, thereby stabilizing neuronal membranes and suppressing neuronal hyper-synchronization. Although zonisamide shows affinity for the gamma-aminobutyric acid (GABA)/benzodiazepine receptor ionophore complex, it does not potentiate the synaptic activity of GABA. In addition, this agent also facilitates both dopaminergic and serotonergic neurotransmission. Zonisamide is a new generation anticonvulsant that is typically used in combination with other antiepileptic medications for partial onset seizures (1).

CUTANEOUS ADVERSE DRUG REACTIONS FROM SYSTEMIC ADMINISTRATION CAUSED BY TYPE IV (DELAYED-TYPE) HYPERSENSITIVITY (as demonstrated by positive patch tests)

Cutaneous adverse drug reactions from systemic administration of zonisamide caused by type IV (delayed-type) hypersensitivity have included drug reaction with eosinophilia and systemic symptoms (DRESS) (2).

Drug reaction with eosinophilia and systemic symptoms (DRESS)

A 29-year-old man was prescribed zonisamide for epilepsy and subsequently developed a widespread skin rash, acute kidney injury, high-grade fever, eosinophilia, liver dysfunction, lymphadenopathy and an increase in antihuman herpesvirus-6 immunoglobulin G titer. Hypersensitivity to zonisamide was confirmed by a positive patch test to this drug (no data on test concentration and vehicle, time and strength of readings and controls). The patient was diagnosed with drug reaction with eosinophilia and systemic symptoms (DRESS) / drug-induced hypersensitivity syndrome (DIHS) caused by delayed-type allergy to zonisamide (2).

LITERATURE

1   The data in the section 'General' may have been obtained from literature discussed in this chapter, but mostly also or exclusively from one or more of the following online sources: ChemIDPlus Advanced, PubChem, DrugBank, RxList, Drug Central, Drugs.com, and Wikipedia

2   Fujita Y, Hasegawa M, Nabeshima K, Tomita M, Murakami K, Nakai S, Yamakita T, Matsunaga K. Acute kidney injury caused by zonisamide-induced hypersensitivity syndrome. Intern Med 2010;49:409-413

Chapter 4   DIAGNOSTIC TESTS IN SUSPECTED CUTANEOUS ADVERSE DRUG REACTIONS FROM SYSTEMIC DRUGS

A word of caution

This chapter gives a *global* summary of diagnostic possibilities in cases of suspected nonimmediate (delayed) cutaneous adverse drug reactions, aimed at readers inexperienced in this field of medicine. Diagnostic procedures include patch tests, prick tests, intradermal tests, drug provocation tests and *in vitro* tests such as the lymphocyte transformation test (LTT). Given the scope of the book, the emphasis is on patch testing; the discussion of the other tests is limited. Whereas patch testing appears to be mostly a safe diagnostic procedure in all drug reactions, including the severe types (acute generalized exanthematous pustulosis [AGEP], drug reaction with eosinophilia and systemic symptoms [DRESS], Stevens-Johnson syndrome [SJS], toxic epidermal necrolysis [TEN], and generalized bullous fixed drug eruption [GBFDE]), intradermal tests and oral drug provocation tests may sometimes be (very) dangerous to the patient and is an established contraindication in certain conditions.

Therefore, the author strictly advises all physicians inexperienced in this subject, who have read this chapter and based on the information provided herein decide to apply prick, intradermal or provocation tests in their patients, not to do so before having studied recent review articles and published consensus guidelines (e.g. 20,23,24,31,50,51,152) as well as relevant literature that is published after the writing of this book was finalized in August 2021.

Introduction

The physician who cares for patients with cutaneous adverse drug reactions (CADRs) is often faced with diagnostic difficulties. CADRs may present in many forms, and, although it is well known which drugs and drug categories are frequently implicated in defined clinical manifestations (e.g. penicillins in maculopapular eruptions, aromatic antiepileptics in DRESS/drug hypersensitivity syndrome), the various CADRs are not characteristic for specific drugs. Individual drugs can cause a variety of drug eruptions and specific drug eruptions can be caused by a great many pharmaceuticals. Also, patients often use multiple drugs, which makes it difficult to identify the responsible agent based on the clinical picture and patient history alone.

Clinical history must be carefully obtained and should include the symptomatology (compatible with a drug hypersensitivity reaction?), the chronology of the symptoms (previous exposure, delay between the last dose and the onset of symptoms, effect of stopping treatment), other medications taken (both at the time of the reaction and other drugs of the same class taken since then), and the medical background of the patient (any suggestion of a previous allergy, whether associated with medication or not, or of a medical condition) (23). Data should ideally be recorded in a uniform format, which has been developed and is available in many languages (25). Diagnosis is more difficult when patients are not seen during the symptomatic phase, in which case photographs are helpful.

When patients are seen during the reaction, the suspected drugs should be stopped after a benefit/risk balance analysis, especially if danger/severity signs are present. These include painful skin, atypical target lesions, erosions of the mucosae, skin blisters and bullae, Nikolsky sign, leukopenia/thrombopenia and renal dysfunction (suspicion of Stevens-Johnson syndrome/toxic epidermal necrolysis [SJS/TEN]); fever >38.5°C, skin involvement >50%, centrofacial edema, lymphadenopathy at 2 or more sites, eosinophilia, atypical lymphocytes, liver- or kidney dysfunction, and proteinuria (suspicion of drug reaction with eosinophilia and systemic symptoms [DRESS]/drug hypersensitivity syndrome) (23). If the responsible drug is or – in the case of multiple hypersensitivity syndrome – are amongst the pharmaceuticals no longer administered to the patient, the symptoms and signs will decrease. Improvement and resolution of the clinical and laboratory manifestations may be swift or slow, depending on the type of drug reaction, e.g. swift with maculopapular eruptions, and slow with DRESS and SJS/TEN. In the case of DRESS, exacerbations are frequently observed after initial improvement (Chapter 2.7 Drug reaction with eosinophilia and systemic symptoms) despite discontinuation of the culprit drug(s).

Once the drug hypersensitivity reaction has been resolved, a search for the culprit drug should be undertaken. Skin tests (patch/prick/intradermal) may identify the offender (which should, in serious reactions, thereafter never be given again to the patient), they can differentiate sensitization to the drug itself from a reaction to excipients and can also aid in finding a replacement drug (6). Unfortunately, with the exception of the beta-lactam antibiotics (389), there is no international consensus on how skin tests with drugs should be performed or interpreted. There have been no multicenter studies to establish drug concentration, test protocol, specificity, sensitivity and safety. Reliable skin test procedures including test concentrations for the diagnosis of drug hypersensitivity are not available for most drugs. Consequently, many healthcare professionals do not investigate drug reactions with skin tests and rely on the history alone to make a diagnosis of drug allergy. This is obviously unreliable and may lead to unjustified use or avoidance of indicated drugs (23,58). This means that, in suspected delayed cutaneous adverse drug reactions, diagnostic skin testing must in many cases be performed in order to institute proper preventive measures (23). The clinical tools allowing a definitive diagnosis include a thorough clinical history, skin tests (patch, prick, intradermal test), *in vitro* tests, and drug provocation tests. When properly performed in specialized centers, a reliable diagnosis is often possible and safe alternative medication can be administered (23). The usefulness of skin tests has been ascertained in many investigations (9,11,62,63,64,66,71,72,73).

Recent (>2005) review articles on diagnostic tests in nonimmediate drug hypersensitivity reactions can be found in refs. 6,10,20,21,23,24,45,50,51,56,57,65,67,70 (patch testing in HIV), 83,93,152, and 207.

Patch tests

Many drug eruptions are mediated (or assumed to be mediated) by delayed-type hypersensitivity to the drug (or a metabolite). CADRs in which drug patch testing is indicated or may (occasionally) be helpful are summarized in table 4.1.

Table 4.1 Indications for patch testing in the diagnostic workup of cutaneous adverse drug reactions

Acute generalized exanthematous pustulosis (AGEP)
Drug reaction with eosinophilia and systemic symptoms (DRESS)
Eczematous eruption
Erythema multiforme-like eruption
Erythroderma, widespread erythematous eruption, exfoliative dermatitis
Fixed drug eruption
Localized hypersensitivity reactions to subcutaneous injections (e.g. heparins, local anesthetics)
Maculopapular eruption
Photosensitivity
Stevens-Johnson syndrome/toxic epidermal necrolysis (SJS/TEN)
Symmetrical drug-related intertriginous and flexural exanthema (SDRIFE)/Baboon syndrome
Systemic allergic dermatitis (systemic contact dermatitis)
Urticaria (delayed, nonimmediate)

Unfortunately, there is there a lack of systematic methodological studies on suitable test concentrations, test vehicle and reading of the test, and the specificity and sensitivity of patch testing are unknown for most pharmaceutical drugs. The diagnostic workup is also complicated by the complex process of metabolization, which may show significant interindividual differences. In addition, drug reactions to many pharmaceutical agents involve different co-factors (e.g., viral infections) and – compared to the skin – distinct immunological processes (27). Nevertheless, patch tests with suspected drugs should always be the first diagnostic procedure. They are easy to perform (i.e. by dermatologists and allergists trained and experienced in patch testing), cheap, and – most importantly - a positive patch test reaction abolishes the need for other tests. Indeed, prick tests, intradermal tests, drug provocation tests, and *in vitro* tests may be (even) less standardized, more difficult to perform, bothersome to the patient or may be hazardous in certain circumstances. A positive patch test result can help to confirm a possible culprit drug. A *negative* patch test result, on the other hand, cannot exclude the contribution of the drug tested: in many such cases, delayed intradermal tests or drug provocation tests will still be positive. Generally speaking, the intradermal test is more sensitive than the patch test in detecting delayed-type hypersensitivity to systemic drugs and is the logical choice for expanded diagnostics in case of a negative patch test.

Safety of patch testing

Patch tests with drugs are – also in severe cutaneous adverse drug reactions - a low-risk method of diagnostic testing, as they can reproduce delayed hypersensitivity to drugs with a moderate exposure of the patient to offending drugs. In rare cases, however, patch testing has (mildly or severely) reproduced the CADR (table 4.2). Still, in some groups (e.g. HIV), there appears to be a significant risk of generalized systemic reactions following patch testing, in one group in 10/11 (91%) patch tested patients (214). Nevertheless, it is generally accepted that, if necessary with appropriate dose adjustment in the patch, testing is feasible even in the most severe cases of cutaneous drug hypersensitivity, including SJS/TEN (26,207,27).

Procedure

Consensus-based recommendations for conducting patch testing in patients with (suspected) delayed-type hyper-sensitivities were published recently and should be followed unless otherwise indicated (26,27,28). The patch tests should preferably be performed not sooner than 6 weeks and not later than 6 months after the resolution of the adverse skin reaction (6,27,45). In the case of DRESS (in which viruses often play a role), it is advised to wait 6 months after disappearance of the skin exanthema and other sequelae, in order to avoid any virus reactivation (6,27,50). The patch test materials should be removed after 2 days and the reactions read 30 minutes later; a second reading at D3 or D4 is necessary and a later reading at D7 (or D8-D10) is strongly recommended, the latter especially for corticosteroids, iodinated contrast media, heparins (27) and aminoglycoside antibiotics. Test reading is usually performed according to the ESCD guidelines for conducting patch tests (26).

Table 4.2 Drugs that have reproduced the cutaneous adverse drug reaction by patch testing

Drug	References	Drug	References
Acyclovir	71	Mitomycin C	308
Amikacin	200	Paracetamol (acetaminophen)	77
Amoxicillin	16	Piperacillin	140
Carbamazepine	74,143	Pristinamycin	11,227
Clindamycin	379	Pseudoephedrine	71,78,79,224
Clobazam	75	Pyrazinamide	214
Deflazacort	159	Ranitidine	208
Ethambutol	214	Rifampicin	194,214
Hydroxyzine	71	Sodium valproate	138
Isoniazid	214	Triamcinolone	71
Metamizole	76		

Tape stripping the test site prior to the application of test allergens can increase the test's sensitivity. Strip patch testing is recommended in cases in which the results of a previous conventional patch test are suspected to be false-negative or when testing allergens characterized by poor penetration of the corneal layer, such as heparins and aminoglycoside antibiotics. A validated protocol for the performance of tape stripping has been developed (428).

In cases of fixed drug eruptions, the patches should be applied both to previously affected (postlesional) and healthy unaffected skin; the tests are usually positive on postlesional skin only. The use of DMSO as penetration-enhancing vehicle or ethanol instead of petrolatum may sometimes be useful in patch testing (232). Single or repeated open tests on postlesional skin with the suspected drugs have also been successful in cases of fixed drug eruptions (238,242,243,244,245), sometimes even when closed patch tests were negative (146). In one case, a patch test in a case of SJS/TEN was positive on previously affected skin only (225).

Test materials

Approximately 90 drugs used systemically are commercially available for patch testing and they can be used in cases of suspected CADRs to these drugs (table 4.3). For other drugs, pure materials are often not readily available. Therefore, the commercial drugs used by the patients are usually prepared for patch tests, often pulverized tablets. Formerly, a concentration of 30% pet. for all drugs has been recommended (52), but this resulted in strongly varying concentrations of the active material, ranging from 0.05% to 30%, with 25% of the drug patch tests having an active ingredient concentration of less than 2% and 25% having an active ingredient concentration of >16% (215). Despite this, this method, in a small study, was shown to be as reliable as patch testing with pure drugs tested at 10% pet. (53).

Nevertheless, preferably, the pure drugs, not the commercialized tablets used by the patients, should be tested, also in order to avoid false-positive results (i.e. not indicating hypersensitivity to the active drug material) due to hidden additives in the drug formulations, degradation products or impurities (80,84). Most pure systemic drugs can be tested at 10% pet. When the pure chemical is not available, the test material can best be prepared from intravenous powder, the content of capsules or – when also not available – from powdered tablets to achieve a final concentration of the active drug of 10% pet. wt/wt. When possible, the excipients of the pharmaceutical should also be patch tested. When the concentration of the active drug is too low in the patient's drug, the whole powder should be diluted in pet. at 30%., which is non-irritant for nearly all commercial medications (215). Positive patch test results obtained with these in-house preparations should always be validated with controls (26).

Petrolatum is usually suitable as vehicle. Petrolatum is also the vehicle for the commercially available systemic corticosteroids, but these drugs may be better tested 0.1% and 1% in 70% alcohol to avoid false-negative reactions. Many patch tests to corticosteroids become positive only after D3-D4 (due to the anti-inflammatory effect of the molecule) and a reading at D6-D10 is imperative to avoid missing late positive reactions (54,55,56). Such late developing positive patch tests are also frequent with neomycin and other aminoglycosides.

Obviously, the relevance of any positive drug patch test should be carefully assessed (10,80). It is stressed again, that a positive patch test result can help to confirm a possible culprit drug, but that a negative patch test result does *not* exclude the drug tested as the or one of the chemicals responsible for the observed cutaneous adverse drug reaction.

Table 4.3 Therapeutic systemic drugs commercially available for patch testing [a]

Patch test allergen (hapten)	Chemotech	SPCanada	SPEurope
Acetaminophen	10.0%	10%	
Acetylsalicylic acid	10.0%		
Acyclovir	10.0%	10%	
Aminophenazone		10%	
Amoxicillin trihydrate	10.0%		
Ampicillin		5%	5%
Benzydamine hydrochloride	2.0%	1%; 2%	
Betamethasone dipropionate	1.0%	0.5%; 0.1% alc.	
Captopril	5.0%		
Carbamazepine	1.0%		
Cefalexin	10.0%		
Cefixime trihydrate	10.0%		
Cefotaxim sodium salt	10.0%		
Cefpodoxime proxetil	10.0%		
Cefradine	10.0%		
Cefuroxime sodium	10.0%		
Chloramphenicol	5.0%	5%	5%
Chlorpheniramine maleate		5%	
Chlorpromazine hydrochloride	0.1%	1%	
Chlortetracycline hydrochloride		1%	
Ciprofloxacin hydrochloride	10.0%		
Clarithromycin	10.0%		
Clavulanate potassium	10.0%		
Clindamycin phosphate	10.0%		
Clioquinol	5.0%	5%	5%
Cotrimoxazole	10.0%		
Dexamethasone		0.5%	
Dexamethasone-21-phosphate		1%	
Dexamethasone-21-phosphate disodium salt	1.0%		1%
Dexketoprofen	1.0%		
Diclofenac		2.5%; 5%	2.5%
Diclofenac sodium salt	1.0%; 5.0%		
Dicloxacillin sodium salt hydrate	10.0%		
Diltiazem hydrochloride	10.0%		
Diphenhydramine hydrochloride	1.0%		
Doxycycline monohydrate	10.0%		
Erythromycin base	10.0%	2%	2%
Fenofibrate	10.0%		
Fusidic acid sodium salt	2.0%	2%	
Gentamicin sulfate	20.0%	20%	20%
Hydrochlorothiazide	10.0%		
Hydrocortisone		1%	1%
Hydroxyzine hydrochloride	1.0%		
Ibuprofen	5.0%; 10.0%	5%	
Indomethacin		1%	1%
Kanamycin sulfate	10.0%	10%	10%
Ketoprofen	1.0%	2.5%	
Lamotrigine	10.0%		
Lidocaine	5.0%; 15.0%		
Lidocaine hydrochloride		15%	15%
Mepivacaine hydrochloride		1%	
Metamizol		1%	
Methylprednisolone (aceponate)	1.0%	0.1% alc.	
Metronidazole		1%	1%
Miconazole	1.0% alc.		
Minocycline hydrochloride	10.0%		

Table 4.3 Therapeutic systemic drugs commercially available for patch testing (continued) [a]

Patch test allergen (hapten)	Chemotech	SPCanada	SPEurope
Naproxen		5%	
Neomycin sulfate	20.0%	20%	20%
Nitrofurazone	1.0%	1%	1%
Norfloxacin	10.0%		
Nystatin		2%	2%
Olaquindox	1.0%		
Oxytetracycline		3%	3%
Paracetamol (acetaminophen)	10.0% pet.	10%	
Penicillamine		1%	
Phenacetin		10%	
Phenazone		5%	5%
Phenylbutazone	10.0%	10%	10%
Phenylephrine hydrochloride		10% water	10% water
Pindolol		2%	
Piperazine		1%	1%
Piroxicam	1.0%	1%	
Polidocanol		3%	3%
Polymyxin B sulfate	5.0%	3%	3%
Potassium clavulanate	10.0%		
Prednisolone		1%	1%
Prilocaine hydrochloride	5.0%		
Pristinamycin	10.0%		
Procaine hydrochloride	1.0%	1%	1%
Promethazine hydrochloride	0.1%; 1.0%	2%	
Propranolol hydrochloride		2%	
Propyphenazone		1%	
Quinine sulfate	1.0%		
Spiramycin base	10.0%		
Streptomycin sulfate		5%	
Sulfamethoxazole-trimethoprim	10.0%		
Sulfanilamide	5.0%	5%	5%
Tetracycline hydrochloride		2%	
Tobramycin	20.0%	20%	
Triamcinolone acetonide	1.0%	0.1%; 1%	1%
Vancomycin hydrochloride	10.0% water		

[a] The vehicle for all haptens is petrolatum, unless otherwise indicated; alc.: alcohol; Chemotech: Chemotechnique Diagnostics (www.chemotechnique.se); SPCanada: SmartPractice Canada (www.smartpracticecanada.com); SPEurope: SmartPractice Europe (www.smartpracticeeurope.com)

Sensitivity of patch tests

Although very valuable, the sensitivity of the patch test in some drug rashes is limited, only a minority of the tests being positive. The rates of positive patch test reactions in various investigations have varied widely, depending *inter alia* on the nature of the CADR and the drugs involved. Some drugs are frequently patch test-positive, e.g. carbamazepine and some other antiepileptics (18,48,64), pristinamycin (227), aminopenicillins (43), and iodinated contrast media (64,370). Others, however, give infrequently or hardly ever positive patch tests, such as allopurinol. Other important parameters are the patch testing method and the selection of patients for patch testing. When selecting patients with high suspicion of certain drugs having caused a CADR for patch testing, the frequency of positive results will obviously be higher. In such studies, positive results have been obtained in 43% (71), 50% (72) and 32% (73) of the patients tested. In other studies, rates of positive reactions to any CADR were 23% (64), 23% (18; children), 25% (9; many cases of SJS/TEN), 14% (82; maculopapular eruption, erythroderma, generalized eczema), 11% (85), and 23% (antibiotics; 195).

Rates of positive reactions reported in specific drug eruptions (not a full literature review) are shown in table 4.4. The percentages are very hard (if at all) to compare due to many different parameters, such as number of studies, number of patients investigated, selection of patients, prescription habits, accuracy of clinical diagnosis, patch testing technique, and reading of patch test reactions. However, generally speaking, higher rates of positive reactions may be observed with DRESS, AGEP and the maculopapular eruptions and low rates in patients with SJS/TEN. In the case of DRESS, the high percentages may partly be explained by the large number of cases caused by antiepileptics, especially carbamazepine, which are frequently patch test-positive. For the other drug eruptions, there are not enough data, or contradictory results have been obtained (e.g. in the case of fixed drug eruptions).

Table 4.4 Rates of positive patch test reactions in cutaneous adverse drug reactions

Drug eruption	Positive/Total tested, % (reference)	Range (%)
AGEP	0/3, 0% (64); 7/14, 50% (72); 50% (81); 7/14, 50% (9); 26/45, 58% (11); 10/23, 43% (425; beta-lactam antibiotics); 11/17, 65% (360, iodinated contrast media)	0-65
DRESS	9/16, 56% (64); 18/56, 32% (ref. lost); 46/72, 64% (11); 5/13 (38%); 11/18, 61% (139), 10/12, 83% (370, iodinated contrast media)	32-83
Eczematous eruption	3/9, 33% (71); 9/17, 53% (73)	33-53
Erythema multiforme	3/37, 8% (64); 6/29, 21% (73)	8-21
Erythroderma	5/7, 71% (71); 8/15, 53% (73)	53-71
Fixed drug eruption	11/55, 20% (64); 0/3, 0% (71); 2/6, 33% (73); 27/34, 79% (248)	0-79
Lichenoid drug eruption	2/11, 18% (73)	18
Maculopapular eruption	73/305, 24% (64); 33/61, 54% (72); 16/27, 59% (71), 10/72, 14% (73); 33/60, 55% (43; only aminopenicillins); 90/173, 52% (14; mostly aminopenicillins); 9/27, 33% (15, beta-lactam antibiotics)	14-59
Photosensitivity	2/4, 50% (64); 4/4, 100% (71)	50-100
SDRIFE	22/53, 51% (120; based on collection of case reports in literature)	51
SJS/TEN	1/15, 7% (64); 2/22, 9% (9); 4/17, 24% (11); 0/10, 0% (216)	0-24
Systemic allergic dermatitis	Up to 100%, as pre-sensitization is required for this drug reaction	
Urticaria/angioedema	12% (82)	12

AGEP: Acute generalized exanthematous pustulosis; DRESS: Drug reaction with eosinophilia and systemic symptoms; SDRIFE: Symmetrical drug-related intertriginous and flexural exanthema /Baboon syndrome; SJS/TEN: Stevens-Johnson syndrome/toxic epidermal necrolysis

Negative patch tests

Unfortunately, a negative patch test does not imply that the tested drug was not the cause of the skin reaction. Indeed, not infrequently, oral provocation tests or unintentional repeated exposure to patch test-negative drugs have caused a recurrence of the CADR reaction (86,87). In the same manner, negative reactions to drugs patch tested in the search for safe alternatives do not guarantee that there will be no adverse reaction when the drug is administered, especially with chemically similar, potentially cross-reacting pharmaceuticals.

When patch tests are negative, intradermal tests with late readings at D2/D3 may be clearly positive, thus identifying both the causative drug and the delayed-type hypersensitivity mechanism involved (34,35,37,52). In fact, intradermal tests are generally more sensitive than patch tests (6). Intradermal testing with corticosteroids may reveal additional sensitizations not picked up by patch tests, especially in the case of hydrocortisone, which is frequently false-negative. However, notably with the stronger corticosteroids, there is a risk of skin atrophy (54,55,56).

Prick tests

Skin prick tests (SPTs) with commercialized forms of drugs are usually performed to identify immediate reactions, and read at 20 minutes. In delayed (non-IgE-mediated) cutaneous adverse drug reactions, they can be performed, but read 24 hours after the tests and have given positive results in some patients with drug-induced maculopapular eruption, AGEP, and DRESS (6). However, drug concentration, test protocol, specificity, sensitivity and safety of prick testing in CADR are largely unknown. Nevertheless, SPTs are often proposed prior to intradermal tests, as well as in cases where an injectable form of the offending drug (necessary for intradermal tests) is unavailable (51). They should be performed on the volar surface of the forearm placing non-diluted drops of the injectable liquid form or a small amount of powder, followed by perforation of the skin epidermis with the use of a specific lancet. Positivity of SPTs should be assessed at 24, 48, and 72 hours. In selected cases, a delayed reading at 96 hours is sometimes recommended. The results are considered positive when erythema and infiltration at the test location are observed (51). Thresholds for specificity in doing drug prick tests (non-irritant test concentrations) can be found in ref. 58.

Intradermal tests

The intradermal test (IDT) is the most sensitive skin test. The IDT must be performed only with commercial products in an injectable form and have been recommended for use in mild cutaneous hypersensitivity reactions (383,384). Until recently, IDTs were generally considered to be contra-indicated in severe cutaneous adverse drug reactions. Despite the small doses applied, severe and even fatal reactions have arisen (207), albeit very infrequently (20).

Currently, however, intradermal tests in AGEP and DRESS are considered to be useful and safe when performed by specialists (20,51). Intradermal tests just as patch tests also lack a sufficient degree of standardization. Some thresholds for specificity (non-irritant test concentrations) can be found in ref. 58.

Consensus on performance of intradermal testing involves injection of 0.02 to 0.05 ml of the highest nonirritating drug concentration in a tuberculin syringe (0.5-1 ml) and needle gauge 25, 27, or 30 with reagent applied bevel-up, to the skin of the volar forearm (20,50). The definition of a positive delayed IDT has varied but in general is defined as an erythematous induration or swelling at the IDT injection site at 24, 48, and out to 72 hours if negative at 48 hours. Current guidelines recommend the use of IDT only with drugs available in sterile parenteral commercially manufactured preparations (152). In a recent consensus guideline, there was agreement amongst international experts that delayed IDT using sterile preparations of drugs can aid in drug allergy assessment and that, similar to patch tests, they should not be performed sooner than 4 to 6 weeks after an acute reaction (20,50). The use of IDT in SCARs (severe cutaneous adverse drug reactions) has predominately been in the setting of hypersensitivity associated with anti-infective drugs that are commonly available as sterile preparations and for which the greatest need know whether they can be safely used exists (195). The intradermal test has increased sensitivity over the patch test and this appears particularly true for antibiotic-associated DRESS and maculopapular eruptions (20). However, a negative delayed intradermal test does not exclude the responsibility of a drug in a cutaneous adverse reaction, and the drug may then have to be rechallenged in non-severe cases (6,86,87).

Drug provocation tests

A drug provocation test (DPT), also referred to as drug challenge, graded challenge, or test dosing, is the gold standard for the identification of the drug eliciting a drug hypersensitivity reaction and in allergy practice is considered to be the ideal final testing step before therapeutic dosing (20,23). This is often done *instead of* diagnostic skin tests, but should – generally speaking – only be performed after skin tests (patch, intradermal) were found to be negative (86). Drug challenges should always be considered in the context of a risk-benefit ratio. The DPT is independent of the pathogenesis and consequently cannot differentiate between allergic from nonallergic drug hypersensitivity reactions. It takes individual factors such as the metabolism and genetic disposition of an individual into account. DPTs have the highest sensitivity, but should only be performed under the most rigorous surveillance conditions. They are therefore usually restricted to certain specialist centers in which equipment, supplies, and personnel are present to manage serious reactions, and that personnel are well trained and experienced in performing this procedure in properly selected patients (23).

In non-severe CADRs, provocation tests appear to be (fairly) safe (86). However, in severe CADRs they may be dangerous and provoke a generalized eruption, possibly with systemic symptoms, even at a low dose of the culprit drug (47,58). Severe reactions such as shock and even death have been described with full dose rechallenge after the occurrence of the hypersensitivity syndrome, e.g. with abacavir (68,69,70). In one study, rechallenges have been performed with antituberculosis drugs and 50% of patients with a history of a severe CADR developed reactions on the drug challenge (30). Therefore, provocation tests in AGEP, DRESS, and SJS/TEN are generally considered to be contraindicated, particularly where the treatment with the drug in question is not necessary or where there are efficacious and safe treatment alternatives (20,50,51). Nevertheless, some authors consider drug challenges responsible in certain conditions: 'not recommended, unless the benefit is far greater (>>>) than the risk' (20).

Drug provocation tests have not been standardized and optimal dosages and length of provocation are largely unknown, which means that safe and reliable provocation tests are challenging. A common office procedure for drug challenges includes a full-dose drug challenge followed by 1 to 2 hours of observation or a graded challenge including one-tenth to one-quarter of a dose followed either 30 to 60 minutes later or 3 to 7 days later by the full dose (20). General principles of drug provocation tests and evidence-based recommendations have been published by the members of the ENDA (European Network for Drug Allergy), a group of specialists of the European Academy of Allergy and Clinical Immunology (31).

In vitro tests

In vitro diagnostic testing for investigation of serious drug hypersensitivity reactions should be an ideal approach to allergy work-up, because there is zero risk to the patient (207). Currently, however, laboratory-based in *vitro* and *ex vivo* diagnostics, such as lymphocyte transformation tests, ELISpot assay, and flow cytometry, are not available for use as routine diagnostics in most centers and are still primarily employed in a research only setting (20). Discussion of this topic falls outside the scope of this book. A recent review of the subject can be found in ref. 29.

Suitability of diagnostic tests in individual cutaneous adverse drug reactions

The potential uses of patch tests, prick tests (late reading), intradermal tests (late reading) and systemic provocation tests in cutaneous adverse drug reactions are summarized in table 4.5.

Table 4.5 Diagnostic tests in cutaneous adverse drug reactions (adapted from refs. 20,24,50,57,83) [b]

CADR	Patch test	Intradermal test	Drug provocation test
Non-severe cutaneous adverse drug reactions			
Maculopapular eruption	useful (338); up to 59% positive reactions [a]	useful	after negative patch tests and intradermal tests with late reading
Generalized eczema in case of systemic allergic dermatitis	very useful, 100% sensitivity (338)	potentially useful	after negative patch tests and intradermal tests with late reading
SDRIFE/Baboon syndrome	useful (61,120); some 50% positive reactions [a]	useful	after negative patch tests and intradermal tests with late reading
Fixed drug eruption	useful (248); apply to previously involved *and* to normal skin; up to 79% positive reactions [a]	no data available	full dose after negative patch test on postlesional skin; or start with *sub*therapeutic dose
Photoallergic dermatitis/ photosensitivity	photopatch tests useful; sensitivity unknown, insufficient data	not useful	possible in combination with exposition to sunlight or artificial UVA-radiation
Lichenoid eruption	may be useful	unknown	consider after negative patch and intradermal test with late reading (20)
Severe cutaneous adverse drug reactions (SCAR)			
Abacavir hypersensitivity	very useful	do not test	only in select cases with nonsuggestive history where HLA-B*5701 and patch test are negative (360); no, risk of severe reactions (24)
Acute generalized exanthematous pustulosis (AGEP)	up to 65% positive reactions [a], depending on the implicated drugs (9,11)	useful (20,51)	not recommended unless benefit far greater than risk (20); contra-indicated (24,50,51)
Drug reaction with eosinophilia and systemic symptoms (DRESS)	positive in 32-83% (11, 62) [a], depending on the implicated drugs (e.g. high with many cases from carbamazepine) (64); apply to both unaffected and previously involved skin	increasing evidence of safety (50); useful (20,24,51)	not recommended unless benefit is far greater than risk (20); contra-indicated (24,50,51)
SJS/TEN	sensitivity <25% (9,11) [a]; high sensitivity (>60%) with carbamazepine (63)	contra-indicated	not recommended unless benefit is far greater than risk (20); contra-indicated (24,50,51)

[a] See table 4.4 for range of positive reactions (sensitivity); [b] Prick tests are not included because of limited value and available data; CADR: Cutaneous adverse drug reaction; SDRIFE: Symmetrical drug-related intertriginous and flexural exanthema; SJS: Stevens-Johnson syndrome; TEN: Toxic epidermal necrolysis

References   THESE ARE ALSO THE REFERENCES FOR CHAPTER 2

1   Szatkowski J, Schwartz RA. Acute generalized exanthematous pustulosis (AGEP): A review and update. J Am Acad Dermatol 2015;73:843-848
2   Beylot C, Bioulac P, Doutre MS. Pustuloses exanthématiques aiguës généralisées, à propos de 4 cas. Ann Dermatol Venereol 1980;107:37-48
3   Hotz C, Valeyrie-Allanore L, Haddad C, Bouvresse S, Ortonne N, Duong TA, et al. Systemic involvement of acute generalized exanthematous pustulosis: a retrospective study on 58 patients. Br J Dermatol 2013;169:1223-1232
4   Halevy S, Kardaun SH, Davidovici B, Wechsler J, EuroSCAR and RegiSCAR Study Group. The spectrum of histopathological features in acute generalized exanthematous pustulosis: a study of 102 cases. Br J Dermatol 2010;163:1245-1252

5   Sidoroff A, Halevy S, Bavinck JN, Vaillant L, Roujeau JC. Acute generalized exanthematous pustulosis (AGEP): a clinical reaction pattern. J Cutan Pathol 2001;28:113-119

6   Barbaud A. Skin testing and patch testing in non-IgE-mediated drug allergy. Curr Allergy Asthma Rep 2014;14:442

7   Sidoroff A, Dunant A, Viboud C, S Halevy, J N Bouwes Bavinck, L Naldi, et al. Risk factors for acute generalized exanthematous pustulosis (AGEP) - results of a multinational case-control study (EuroSCAR). Br J Dermatol 2007;157:989-996

8   Feldmeyer L, Heidemeyer K, Yawalkar N. Acute generalized exanthematous pustulosis: Pathogenesis, genetic background, clinical variants and therapy. Int J Mol Sci 2016;17:1214

9   Wolkenstein P, Chosidow O, Fléchet ML, Robbiola O, Paul M, Dumé L, et al. Patch testing in severe cutaneous adverse drug reactions, including Stevens-Johnson syndrome and toxic epidermal necrolysis. Contact Dermatitis 1996;35:234-236

10   Barbaud A. Drug patch testing in systemic cutaneous drug allergy. Toxicology 2005;209:209-216

11   Barbaud A, Collet E, Milpied B, Assier H, Staumont D, Avenel-Audran M, et al. A multicentre study to determine the value and safety of drug patch tests for the three main classes of severe cutaneous adverse drug reactions. Br J Dermatol 2013;168:555-562

12   Isaksson M. Systemic contact allergy to corticosteroids revisited. Contact Dermatitis 2007;57:386-388

13   Fisher AA. Consort contact dermatitis. Cutis 1979;24:595-596, 668

14   Romano A, Viola M, Mondino C, Pettinato R, Di Fonso M, Papa G, et al. Diagnosing nonimmediate reactions to penicillins by *in vivo* tests. Int Arch Allergy Immunol 2002;129:169-174

15   Bérot V, Gener G, Ingen-Housz-Oro S, Gaudin O, Paul M, Chosidow O, Wolkenstein P, Assier H. Cross-reactivity in beta-lactams after a non-immediate cutaneous adverse reaction: experience of a reference centre for toxic bullous diseases and severe cutaneous adverse reactions. J Eur Acad Dermatol Venereol 2020;34:787-794

16   Trcka J, Seitz CS, Brocker E-B € et al. Aminopenicillin-induced exanthema allows treatment with certain cephalosporins or phenoxymethyl penicillin. J Antimicrob Chemother 2007;60:107-111

17   Ingber A, Trattner A, David M. Hypersensitivity to an estrogen – progesterone preparation and possible relationship to autoimmune progesterone dermatitis and corticosteroid hypersensitivity. J Dermatolog Treat 1999;10:139-140

18   Büyük Yaytokgil Ş, Güvenir H, Külhaş Celík İ, Yilmaz Topal Ö, Karaatmaca B, Civelek E, et al. Evaluation of drug patch tests in children. Allergy Asthma Proc 2021;42:167-174

19   Nishimura M, Takano Y, Toshitani S. Systemic contact dermatitis medicamentosa occurring after intravesical dimethyl sulfoxide treatment for interstitial cystitis. Arch Dermatol 1988;124:182-183

20   Lehloenya RJ, Peter JG, Copascu A, Trubiano JA, Phillips EJ. Delabeling delayed drug hypersensitivity: how far can you safely go? J Allergy Clin Immunol Pract 2020;8:2878-2895.e6

21   Brockow K, Pfützner W. Cutaneous drug hypersensitivity: developments and controversies. Curr Opin Allergy Clin Immunol 2019;19:308-318

22   Kardaun SH, Mockenhaupt M, Roujeau JC. Comments on: DRESS syndrome. J Am Acad Dermatol 2014;71:1000-1000.e2.

23   Demoly P, Adkinson NF, Brockow K, Castells M, Chiriac AM, Greenberger PA, et al. International consensus on drug allergy. Allergy 2014;69:420-437

24   Rive CM, Bourke J, Phillips EJ. Testing for drug hypersensitivity syndromes. Clin Biochem Rev 2013;34:15-38

25   Demoly P, Kropf R, Bircher A, Pichler WJ. Drug hypersensitivity: questionnaire. EAACI interest group on drug hypersensitivity. Allergy 1999;54:999-1003

26   Johansen JD, Aalto-Korte K, Agner T, Andersen KE, Bircher A, Bruze M, et al. European Society of Contact Dermatitis guideline for diagnostic patch testing - recommendations on best practice. Contact Dermatitis 2015;73:195-221

27   Mahler V, Nast A, Bauer A, Becker D, Brasch J, Breuer K, et al. S3 guidelines: Epicutaneous patch testing with contact allergens and drugs - Short version, Part 1. J Dtsch Dermatol Ges 2019;17:1076-1093

28   Mahler V, Nast A, Bauer A, Becker D, Brasch J, Breuer K, et al. S3 Guidelines: Epicutaneous patch testing with contact allergens and drugs - Short version, Part 2. J Dtsch Dermatol Ges 2019;17:1187-1207

29   Mayorga C, Ebo DG, Lang DM, Pichler WJ, Sabato V, Park MA, et al. Controversies in drug allergy: In vitro testing. J Allergy Clin Immunol 2019;143:56-65

30   Lehloenya RJ, Todd G, Badri M, Dheda K. Outcomes of reintroducing antituberculosis drugs following cutaneous adverse drug reactions. Int J Tuberc Lung Dis 2011;15:1649-1657

31   Aberer W, Bircher A, Romano A, Blanca M, Campi P, Fernandez J, et al; European Network for Drug Allergy (ENDA); EAACI interest group on drug hypersensitivity. Drug provocation testing in the diagnosis of drug hypersensitivity reactions: general considerations. Allergy 2003;58:854-863

32   Bastuji-Garin S, Rzany B, Stern RS, Shear NH, Naldi L, Roujeau JC. Clinical classification of cases of toxic epidermal necrolysis, Stevens-Johnson syndrome, and erythema multiforme. Arch Dermatol 1993;129:92-96

33   Sassolas B, Haddad C, Mockenhaupt M, Dunant A, Liss Y, Bork K, et al. An algorithm for assessment of drug

causality in Stevens-Johnson Syndrome and toxic epidermal necrolysis: comparison with case-control analysis. Clin Pharmacol Ther 2010;88:60-68

34 Benamara-Levy M, Haccard F, Jonville Bera AP, Machet L. Acute generalized exanthematous pustulosis due to acetazolamide: negative on patch testing and confirmed by delayed-reading intradermal testing. Clin Exp Dermatol 2014;39:220-222

35 Jachiet M, Bellón N, Assier H, Amsler E, Gaouar H, Pecquet C, et al. Cutaneous adverse drug reaction to oral acetazolamide and skin tests. Dermatology 2013;226:347-352

36 Gómez Torrijos E, Cortina de la Calle MP, Méndez Díaz Y, Moreno Lozano L, Extremera Ortega A, Galindo Bonilla PA, et al. Acute localized exanthematous pustulosis due to bemiparin. J Investig Allergol Clin Immunol 2017;27:328-329

37 Assier H, Gener G, Chosidow O, Wolkenstein P, Ingen-Housz-Oro S. Acute generalized exanthematous pustulosis induced by enoxaparin: 2 cases. Contact Dermatitis 2021;84:280-282

38 Gómez Torrijos E, García Rodríguez C, Sánchez Caminero MP, Castro Jiménez A, García Rodríguez R, Feo-Brito F. First case report of acute generalized exanthematous pustulosis due to labetalol. J Investig Allergol Clin Immunol 2015;25:148-149

39 Salem CB, Fathallah N, Slim R, Ziadi S, Ghariani N. Lincomycin-induced acute generalized exanthematous ` pustulosis confirmed by delayed reading of intradermal testing. Ann Pharmacother 2016;50:894-895

40 Villani A, Baldo A, De Fata Salvatores G, Desiato V, Ayala F, Donadio C. Acute localized exanthematous pustulosis (ALEP): Review of literature with report of case caused by amoxicillin-clavulanic acid. Dermatol Ther (Heidelb) 2017;7:563-570

41 Di Meo N, Stinco G, Patrone P, Trevisini S, Trevisan G. Acute localized exanthematous pustulosis caused by flurbiprofen. Cutis 2016;98(5):E9-E11

42 Lahouel M, Mokni S, Denguezli M. Acute localized exanthematous pustulosis induced by a spider bite. Am J Trop Med Hyg 2020;103:937-938

43 Romano A, Di Fonso M, Papa G, Pietrantonio F, Federico F, Fabrizi G, Venuti A. Evaluation of adverse cutaneous reactions to aminopenicillins with emphasis on those manifested by maculopapular rashes. Allergy 1995;50:113-118

44 De Thier F, Blondeel A, Song M. Acute generalized exanthematous pustulosis induced by amoxycillin with clavulanate. Contact Dermatitis 2001;44:114-115

45 Brandt O, Bircher AJ. Delayed-type hypersensitivity to oral and parenteral drugs. J Dtsch Dermatol Ges 2017;15:1111-1132

46 Kostaki M, Polydorou D, Adamou E, Chasapi V, Antoniou C, Stratigos A. Acute localized exanthematous pustulosis due to metronidazole. J Eur Acad Dermatol Venereol 2019;33:e109-e111

47 Tsuda S, Kato K, Karashima T, Inou Y, Sasai Y. Toxic pustuloderma induced by ofloxacin. Acta Derm Venereol 1993;73:382-384

48 Atanasković-Marković M, Janković J, Tmušić V, Gavrović-Jankulović M, Ćirković Veličković T, Nikolić D, Škorić D. Hypersensitivity reactions to antiepileptic drugs in children. Pediatr Allergy Immunol 2019;30:547-552

49 Sidi E, Dobkevitch-Morrill S. The injection and ingestion test in cross-sensitization to the para group. J Invest Dermatol 1951;16:299-310

50 Phillips EJ, Bigliardi P, Bircher AJ, Broyles A, Chang YS, Chung WH, Lehloenya R, et al. Controversies in drug allergy: Testing for delayed reactions. J Allergy Clin Immunol 2019;143:66-73

51 Bergmann MM, Caubet JC. Role of in vivo and in vitro tests in the diagnosis of severe cutaneous adverse reactions (SCAR) to drug. Curr Pharm Des 2019;25:3872-3880

52 Barbaud A, Gonçalo M, Bruynzeel D, Bircher A. Guidelines for performing skin tests with drugs in the investigation of cutaneous adverse drug reactions. Contact Dermatitis 2001;45:321-328

53 Assier H, Valeyrie-Allanore L, Gener G, Verlinde Carvalh M, Chosidow O, Wolkenstein P. Patch testing in non-immediate cutaneous adverse drug reactions: value of extemporaneous patch tests. Contact Dermatitis 2017;77:297-302

54 Goossens A, Gonçalo M. Topical drugs. In: Johansen J, Mahler V, Lepoittevin JP, et al, eds. Contact Dermatitis, 6th ed. Cham, Switzerland: Springer; 2020:1019-1055

55 Baeck M, Goossens A. Immediate and delayed allergic hypersensitivity to corticosteroids: practical guidelines. Contact Dermatitis 2012;66:38-45

56 Soria A, Baeck M, Goossens A, Marot L, Duveille V, Derouaux A-S, et al. Patch, prick or intradermal tests to detect delayed hypersensitivity to corticosteroids? Contact Dermatitis 2011;64:313-324

57 Barbaud A. Skin testing in delayed reactions to drugs. Immunol Allergy Clin North Am 2009;29:517-535

58 Brockow K, Garvey LH, Aberer W, Atanaskovic-Markovic M, Barbaud A, Bilo MB, et al. Skin test concentrations for systemically administered drugs—an ENDA/EAACI Drug Allergy Interest Group position paper. Allergy 2013;68:702-712

59 Lin CC, Chen CB, Wang CW, Hung SI, Chung WH. Stevens-Johnson syndrome and toxic epidermal necrolysis: risk factors, causality assessment and potential prevention strategies. Expert Rev Clin Immunol 2020;16:373-387
60 Andrade P, Brinca A, Gonçalo M. Patch testing in fixed drug eruptions—a 20-year review. Contact Dermatitis 2011;65:195-201
61 Miyahara A, Kawashima H, Okubo Y, Hoshika A. A new proposal for a clinical-oriented subclassification of baboon syndrome and a review of baboon syndrome. Asian Pac J Allergy Immunol 2011;29:150-160
62 Santiago F, Gonçalo M, Vieira R, Coelho S, Figueiredo A. Epicutaneous patch testing in drug hypersensitivity syndrome (DRESS). Contact Dermatitis 2010;62:47-53
63 Lin YT, Chang YC, Hui RC, Yang CH, Ho HC, Hung SI, et al. A patch testing and cross-sensitivity study of carbamazepine-induced severe cutaneous adverse drug reactions. J Eur Acad Dermatol Venereol 2013;27:356-364
64 Ohtoshi S, Kitami Y, Sueki H, Nakada T. Utility of patch testing for patients with drug eruption. Clin Exp Dermatol 2014;39:279-283
65 Barbaud A. Drug patch tests in the investigation of cutaneous adverse drug reactions. Ann Dermatol Venereol 2009;136:635-644 (Article in French)
66 Tchen T, Reguiaï Z, Vitry F, Arnoult E, Grange A, Florent G, Bernard P. Usefulness of skin testing in cutaneous drug eruptions in routine practice. Contact Dermatitis 2009;61:138-144
67 Romano A, Viola M, Gaeta F, Rumi G, Maggioletti M. Patch testing in non-immediate drug eruptions. Allergy Asthma Clin Immunol 2008;4:66-74
68 Phillips EJ, Sullivan JR, Knowles SR, Shear NH. Utility of patch testing in patients with hypersensitivity syndromes associated with abacavir. AIDS 2002;16:2223-2225
69 Escaut L, Liotier JY, Albengres E, Cheminot N, Vittecoq D. Abacavir rechallenge has to be avoided in case of hypersensitivity reaction. AIDS 1999;13:1419-1420
70 Shear N, Milpied B, Bruynzeel DP, Phillips E. A review of drug patch testing and implications for HIV clinicians. AIDS 2008;22:999-1007
71 Barbaud A, Reichert-Penetrat S, Tréchot P, Jacquin-Petit MA, Ehlinger A, Noirez V, et al. The use of skin testing in the investigation of cutaneous adverse drug reactions. Br J Dermatol 1998;139:49-58
72 Barbaud A, Bene MC, Faure G, Schmutz JL. Tests cutanés dans l'exploration des toxidermies supposées de mécanisme immuno-allergique. Bull Acad Natl Med 2000;184:47-63 (Article in French)
73 Osawa J, Naito S, Aihara M, Kitamura K, Ikezawa Z, Nakajima H. Evaluation of skin test reactions in patients with non-immediate type drug eruptions. J Dermatol 1990;17:235-239
74 Vaillant L, Camenen I, Lorette G. Patch testing with carbamazepine: reinduction of an exfoliative dermatitis. Arch Dermatol 1989;125:299
75 Machet L, Vaillant L, Dardaine V, Lorette G. Patch testing with clobazam: relapse of generalized drug eruption. Contact Dermatitis 1992;26:347-348
76 Gonzalo-Garijo MA, de Arila D, Rodriguez-Nevado I. Generalized reaction after patch testing with metamizole. Contact Dermatitis 2001;45:180
77 Mashiah J, Brenner S. A systemic reaction to patch testing for the evaluation of acute generalized exanthematous pustulosis. Arch Dermatol 2003;139:1181-1183
78 Tomb RR, Lepoittevin J-P, Espinassouze F, Heid E, Foussereau J. Systemic contact dermatitis from pseudoephedrine. Contact Dermatitis 1991;24:86-88
79 Sanchez TS, Sanchez-Perez J, Aragues M, Garcia-Diaz A. Flare-up reaction of pseudoephedrine baboon syndrome after positive patch test. Contact Dermatitis 2000;42:312-313
80 Barbaud A, Trechot P, Reichert-Penetrat S, Commun N, Schmutz JL. Relevance of skin tests with drugs in investigating cutaneous adverse drug reactions. Contact Dermatitis 2001;45:265-268
81 Beylot C, Doutre MS, Beylot-Barry M. Acute generalized exanthematous pustulosis. Semin Cutan Med Surg 1996;15:244-249
82 Bursztejn A-C, Rat A-C, Tréchot P, Cunyj-F, Schmutz J-L, Barbaud A. Results of skin tests to assess drug-induced allergy. Ann Dermatol Venereol 2010;137:688-694
83 Barbaud A. Drug skin tests and systemic drug reactions: an update. Expert Rev Dermatol 2007;2:481-495
84 Grims RH, Kränke B, Aberer W. Pitfalls in drug allergy skin testing: false-positive reactions due to (hidden) additives. Contact Dermatitis 2006;54:290-294
85 Lammintausta K, Kortekangas-Savolainen O. The usefulness of skin tests to prove drug hypersensitivity. Br J Dermatol 2005;152:968-974
86 Lammintausta K, Kortekangas-Savolainon O. Oral challenge in patients with suspected cutaneous adverse drug reactions: findings in 784 patients during a 25-year-period. Acta Derm Venereol 2005;85:491-496
87 Waton J, Trlechot P, Loss-Ayav C, Schmutz J L, Barbaud A. Negative predictive value of drug skin tests in investigating cutaneous adverse drug reactions. Br J Dermatol 2009;160:786-794
88 Bellón T. Mechanisms of severe cutaneous adverse reactions: Recent advances. Drug Saf 2019;42:973-992

89   Mustafa SS, Ostrov D, Yerly D. Severe cutaneous adverse drug reactions: presentation, risk factors, and management. Curr Allergy Asthma Rep 2018;18:26

90   Duong TA, Valeyrie-Allanore L, Wolkenstein P, Chosidow O. Severe cutaneous adverse reactions to drugs. Lancet 2017;390(10106):1996-2011

91   Mockenhaupt M. Epidemiology of cutaneous adverse drug reactions. Allergol Select 2017;1:96-108

92   Paulmann M, Mockenhaupt M. Severe drug hypersensitivity reactions: Clinical pattern, diagnosis, etiology and therapeutic options. Curr Pharm Des 2016;22:6852-6861

93   Friedmann PS, Ardern-Jones M. Patch testing in drug allergy. Curr Opin Allergy Clin Immunol 2010;10:291-296

94   Feldmeyer L, Heidemeyer K, Yawalkar N. Acute generalized exanthematous pustulosis: Pathogenesis, genetic background, clinical variants and therapy. Int J Mol Sci 2016;17(8):1214.

95   Paulmann M, Mockenhaupt M. Severe drug-induced skin reactions: Clinical features, diagnosis, etiology, and therapy. J Dtsch Dermatol Ges 2015;13:625-645

96   Sidoroff A. Acute generalized exanthematous pustulosis. Hautarzt 2014;65:430-435 (Article in German)

97   Dodiuk-Gad RP, Laws PM, Shear NH. Epidemiology of severe drug hypersensitivity. Semin Cutan Med Surg 2014;33:2-9

98   Sidoroff A. Acute generalized exanthematous pustulosis. Chem Immunol Allergy 2012;97:139-148

99   Harr T, French LE. Severe cutaneous adverse reactions: acute generalized exanthematous pustulosis, toxic epidermal necrolysis and Stevens-Johnson syndrome. Med Clin North Am 2010;94:727-742

100 Speeckaert MM, Speeckaert R, Lambert J, Brochez L. Acute generalized exanthematous pustulosis: an overview of the clinical, immunological and diagnostic concepts. Eur J Dermatol 2010;20:425-433

101 Kardaun SH, Kuiper H, Fidler V, Jonkman MF. The histopathological spectrum of acute generalized exanthematous pustulosis (AGEP) and its differentiation from generalized pustular psoriasis. J Cutan Pathol 2010;37:1220-1229

102 Mockenhaupt M. Severe drug-induced skin reactions: clinical pattern, diagnostics and therapy. J Dtsch Dermatol Ges 2009;7:142-160

103 De Groot AC. Allergic contact dermatitis from topical drugs: An overview. Dermatitis 2021;32:197-213

104 De Groot AC. Monographs in contact allergy, Volume 3. Topical drugs. Boca Raton, Fl, USA: CRC Press Taylor and Francis Group, 2021 (ISBN 978-0-367-23693-9)

105 Goossens A, Gonçalo M. Topical drugs. In: Johansen J, Mahler V, Lepoittevin JP, et al, eds. Contact Dermatitis. 6th ed. Cham, Switzerland: Springer; 2020:1019-1055

106 Tan MG, Pratt MD, Burns BF, Glassman SJ. Baboon syndrome from mercury showing leukocytoclastic vasculitis on biopsy. Contact Dermatitis 2020;83:415-417

107 Winnicki M, Shear NH. A systematic approach to systemic contact dermatitis and symmetric drug-related intertriginous and flexural exanthema (SDRIFE). Am J Clin Dermatol 2011;12:171-180

108 Wolf R, Tüzün Y. Baboon syndrome and toxic erythema of chemotherapy: Fold (intertriginous) dermatoses. Clin Dermatol 2015;33:462-465

109 Aquino M, Rosner G. Systemic contact dermatitis. Clin Rev Allergy Immunol 2019;56:9-18

110 Thyssen JP, Maibach HI. Drug-elicited systemic allergic (contact) dermatitis--update and possible pathomechanisms. Contact Dermatitis 2008;59:195-202

111 Kulberg A, Schliemann S, Elsner P. Contact dermatitis as a systemic disease. Clin Dermatol 2014;32:414-419

112 Veien N. Systemic contact dermatitis. In: Johansen J, Mahler V, Lepoittevin JP, et al, eds. Contact Dermatitis. 6th ed. Cham, Switzerland: Springer; 2020:391-405

113 Al-Shaikhly T, Rosenthal JA, Chau AS, Ayars AG, Rampur L. Systemic contact dermatitis to inhaled and intranasal corticosteroids. Ann Allergy Asthma Immunol 2020;125:103-105

114 Faber MA, Sabato V, Ebo DGD, Verheyden M, Lambert J, Aerts O. Systemic allergic dermatitis caused by prednisone derivatives in nose and eardrops. Contact Dermatitis 2015;73:317-320

115 Matos D, Serrano P, Brandão FM. Maculopapular rash of unsuspected cause: systemic contact dermatitis to cinchocaine. Cutan Ocul Toxicol 2015;34:260-261

116 Andersen KE, Hjorth N, Menné T. The baboon syndrome: systemically-induced allergic contact dermatitis. Contact Dermatitis 1984;10:97-100

117 Alves da Silva C, Paulsen E. Systemic allergic dermatitis after patch testing with cinchocaine and topical corticosteroids. Contact Dermatitis 2019;81:301-303

118 Santiago L, Moura AL, Coutinho I, Gonçalo M. Systemic allergic contact dermatitis associated with topical diltiazem and/or cinchocaine. J Eur Acad Dermatol Venereol 2018;32:e284-e285

119 Oliveira A, Rosmaninho A, Lobo I, Selores M. Intertriginous and flexural exanthema after application of a topical anesthetic cream: a case of baboon syndrome. Dermatitis 2011;22:360-362

120 Häusermann P, Harr T, Bircher AJ. Baboon syndrome resulting from systemic drugs: Is there strife between SDRIFE and allergic contact dermatitis syndrome? Contact Dermatitis 2004;51:297-310

121 Chen B, Jiang X, Chen WC, Qian T, Cheng H, Zhang D, et al. A rare case of musk antihemorrhoids ointment-induced symmetrical drug-related intertriginous and flexural exanthema. Contact Dermatitis 2020;83:409-411

122 Nakayama H, Niki F, Shono M, Hada S. Mercury exanthem. Contact Dermatitis 1983;9:411-417

123 Harbaoui S, Litaiem N. Symmetrical drug-related intertriginous and flexural exanthema. In: StatPearls [Internet]. Treasure Island (FL): StatPearls Publishing; 2020 Jan. 2020 Oct. 11

124 Nespoulous L, Matei I, Charissoux A, Bédane C, Assikar S. Symmetrical drug-related intertriginous and flexural exanthema (SDRIFE) associated with pristinamycin, secnidazole, and nefopam, with a review of the literature. Contact Dermatitis 2018;79:378-380

125 Tan SC, Tan JW. Symmetrical drug-related intertriginous and flexural exanthema. Curr Opin Allergy Clin Immunol 2011;11:313-318

126 De Risi-Pugliese T, Barailler H, Hamelin A, Amsler E, Gaouar H, Kurihara F, et al. Symmetrical drug-related intertriginous and flexural exanthema: A little-known drug allergy. J Allergy Clin Immunol Pract 2020;8:3185-3189.e4

127 Kuwatsuka S, Kuwatsuka Y, Takenaka M, Utani A. Case of photosensitivity caused by fenofibrate after photosensitization to ketoprofen. J Dermatol 2016;43:224-225

128 Sawada Y. Adverse reaction to sulphonamides in a burned patient – a case report. Burns Incl Therm Inj 1985;12:127-131

129 Adamski H, Benkalfate L, Delaval Y, Ollivier I, le Jean S, Toubel G, et al. Photodermatitis from non-steroidal anti-inflammatory drugs. Contact Dermatitis 1998;38:171-174

130 Le Coz CJ, Bottlaender A, Scrivener JN, Santinelli F, Cribier BJ, Heid E, et al. Photocontact dermatitis from ketoprofen and tiaprofenic acid: cross-reactivity study in 12 consecutive patients. Contact Dermatitis 1998;38:245-252

131 Torres V, Tavares-Bello R, Melo H, Soares AP. Systemic contact dermatitis from hydrocortisone. Contact Dermatitis 1993;29:106

132 Santos-Alarcón S, Benavente-Villegas FC, Farzanegan-Miñano R, Pérez-Francés C, Sánchez-Motilla JM, Mateu-Puchades A. Delayed hypersensitivity to topical and systemic corticosteroids. Contact Dermatitis 2018;78:86-88

133 Gumaste P, Cohen D, Stein J. Bullous systemic contact dermatitis caused by an intra-articular steroid injection. Br J Dermatol 2015;172:300-302

134 Brambilla L, Boneschi V, Chiappino G, Fossati S, Pigatto PD. Allergic reactions to topical desoxymethasone and oral triamcinolone. Contact Dermatitis 1989;21:272-274

135 English JS, Ford G, Beck MH, Rycroft RJ. Allergic contact dermatitis from topical and systemic steroids. Contact Dermatitis 1990;23:196-197

136 Stingeni L, Caraffini S, Assalve D, Lapomarda V, Lisi P. Erythema-multiforme-like contact dermatitis from budesonide. Contact Dermatitis 1996;34:154-155

137 Baeck M, Goossens A. Systemic contact dermatitis to corticosteroids. Allergy 2012;67:1580-1585

138 Conilleau V, Dompmartin A, Verneuil L, Michel M, Leroy D. Hypersensitivity syndrome due to 2 anticonvulsant drugs. Contact Dermatitis 1999;41:141-144

139 Shiny TN, Mahajan VK, Mehta KS, Chauhan PS, Rawat R, Sharma R. Patch testing and cross sensitivity study of adverse cutaneous drug reactions due to anticonvulsants: A preliminary report. World J Methodol 2017;7:25-32

140 Cabañas R, Muñoz L, López-Serrano C, Contreras J, Padial A, Caballero T, Moreno-Ancillo A, Barranco P. Hypersensitivity to piperacillin. Allergy 1998;53:819-820

141 Tan KL, Bisconti I, Leck C, Billahalli T, Barnett S, Rajakulasingam K, Watts TJ. Bullous fixed drug eruption induced by fluconazole: Importance of multi-site lesional patch testing. Contact Dermatitis 2021;84:350-352

142 Santiago LG, Morgado FJ, Baptista MS, Gonçalo M. Hypersensitivity to antibiotics in drug reaction with eosinophilia and systemic symptoms (DRESS) from other culprits. Contact Dermatitis 2020;82:290-296

143 Nitta Y, Onouchi H. A case of drug-related eruptions due to carbamazepine with a flare from patch testing. Jpn J Dermatol 2003;113:983-987 (Article in Japanese)

144 Smith SM, Milam PB, Fabbro SK, Gru AA, Kaffenberger BH. Malignant intertrigo: A subset of toxic erythema of chemotherapy requiring recognition. JAAD Case Rep 2016;2:476-481

145 Navarro Pulido AM, Orta JC, Buzo G. Delayed hypersensitivity to deflazacort. Allergy 1996;51:441-442

146 Özkaya-Bayazit E, Baykal C. Trimethoprim-induced linear fixed drug eruption. Br J Dermatol 1997;137:1028-1029

147 Cooper SM, Reed J, Shaw S. Systemic reaction to nystatin. Contact Dermatitis 1999;41:345-346

148 Darke RA. Sensitivity to topical application of sulfathiazole ointment. JAMA 1944;124:403-404

149 Watanabe T, Yamada N, Yoshida Y, Yamamoto O. A case of symmetrical drug-related intertriginous and flexural exanthema induced by loflazepate ethyl. J Eur Acad Dermatol Venereol 2010;24:357-358

150 Bajaj AK, Rastogi S, Misra A, Misra K, Bajaj S. Occupational and systemic contact dermatitis with photosensitivity due to vitamin B6. Contact Dermatitis 2001;44:184

151 De Groot AC, Conemans JM. Systemic allergic contact dermatitis from intravesical instillation of the antitumor antibiotic mitomycin C. Contact Dermatitis 1991;24:201-209

152 Barbaud A, Weinborn M, Garvey L, Testi S, Kvedariene V, Bavbek S, et al. Intradermal tests with drugs: an approach to standardization. Front Med (Lausanne) 2020;7:156

153 Burry JN. Ethylenediamine sensitivity with a systemic reaction to piperazine citrate. Contact Dermatitis 1978;4:380

154 Sidi E, Hincky M, Gervais A. Allergic sensitization and photosensitization to Phenergan cream. J Invest Dermatol

155 Sidi E, Melki GR. Rapport entre dermites de cause externe et sensibilisation par voie interne. Semaine des Hopitaux 1954; No. 25, April 14 (Article in French)

156 Romita P, Foti C, Stingeni L. Photoallergy to promazine hydrochloride. Contact Dermatitis 2017;77:182-183

157 Bianchi L, Hansel K, Antonelli E, Bellini V, Rigano L, Stingeni L. Deflazacort hypersensitivity: a difficult-to-manage case of systemic allergic dermatitis and literature review. Contact Dermatitis 2016;75:54-56

158 Pacheco D, Travassos AR, Antunes J, Silva R, Lopes A, Marques MS. Allergic hypersensitivity to deflazacort. Allergol Immunopathol (Madr) 2013;41:352-354

159 Garcia-Bravo B, Repiso JB, Camacho F. Systemic contact dermatitis due to deflazacort. Contact Dermatitis 2000;43:359-360

160 Magnolo N, Metze D, Ständer S. Pustulobullous variant of SDRIFE (symmetrical drug-related intertriginous and flexural exanthema). J Dtsch Dermatol Ges 2017;15:657-659

161 Muresan AM, Metze D, Böer-Auer A, Braun SA. Histopathological spectrum and immunophenotypic characterization of symmetrical drug-related intertriginous and flexural exanthema. Am J Dermatopathol 2020 Jun 30. doi: 10.1097/DAD.0000000000001722. Online ahead of print

162 Miyahara A, Kawashima H, Okubo Y, Hoshika A. A new proposal for a clinical-oriented subclassification of baboon syndrome and a review of baboon syndrome. Asian Pac J Allergy Immunol 2011;29:150-160

163 Matos-Pires E, Pina-Trincão D, Brás S, Lobo L. Baboon syndrome caused by anti-haemorrhoidal ointment. Contact Dermatitis 2018;78:170-171

164 Menné T, Weismann K. Hematogenous contact eczema following oral administration of neomycin. Hautarzt 1984:35:319-320 (Article in German)

165 Guin JD, Fields P, Thomas KL. Baboon syndrome from i.v. aminophylline in a patient allergic to ethylenediamine. Contact Dermatitis 1999;40:170-171

166 Isaksson M, Ljunggren B. Systemic contact dermatitis from ethylenediamine in an aminophylline preparation presenting as the baboon syndrome. Acta Derm Venereol 2003;83:69-70

167 Proske S, Uter W, Schnuch A, Hartschuh W. Severe allergic contact dermatitis with generalized spread due to bufexamac presenting as the 'baboon' syndrome. Dtsch Med Wochenschr 2003;128:545-547 (Article in German)

168 Erdmann SM, Sachs B, Merk HF. Systemic contact dermatitis from cinchocaine. Contact Dermatitis 2001;44:260-261

169 Bianchi L, Marietti R, Tramontana M, Hansel K, Stingeni L. Systemic allergic dermatitis from intra-articular triamcinolone acetonide: Report of two cases with unusual clinical manifestations. Contact Dermatitis 2021;84:54-56

170 Burches E, Revert A, Martin J, Iturralde A. Occupational systemic allergic dermatitis caused by sevoflurane. Contact Dermatitis 2015;72:62-63

171 Lechner T, Grytzmann B, Baurle G. Hematogenous allergic contact dermatitis after oral administration of nystatin. Mykosen 1987;30:143-146

172 Armingaud P, Martin L, Wierzbicka E, Esteve E. Baboon syndrome due to a polysensitization with corticosteroids. Ann Dermatol Venereol 2005;132:675-677 (Article in French)

173 Treudler R, Simon JC. Symmetric, drug-related, intertriginous, and flexural exanthema in a patient with polyvalent intolerance to corticosteroids. J Allergy Clin Immunol 2006;118:965-967

174 Bhari N, Sahni K, Dev T, Sharma VK. Symmetrical drug-related intertriginous and flexural erythema (Baboon syndrome) induced by simultaneous exposure to oral and topical terbinafine. Int J Dermatol 2017;56:e168-170

175 Gulec AI, Uslu E, Başkan E, Yavuzcan G, Aliagaoglu C. Baboon syndrome induced by ketoconazole. Cutan Ocul Toxicol 2014;33:339-341

176 Jankowska-Konsur A, Kolodziej T, Szepietowski J, Sikora J, Maj J, Baran E. The baboon syndrome --report of two first cases in Poland. Contact Dermatitis 2005;52:289-290

177 Fernandez L, Maquiera E, García-Abujeta J L, Yanez S, Rodriguez F, Martín-Gil D, Jerez J. Baboon syndrome due to mercury sensitivity. Contact Dermatitis 1995;33:56-57

178 Audicana M, Bernedo N, Gonzalez I, Muñoz D, Fernández E, Gastaminza G. An unusual case of baboon syndrome due to mercury present in a homeopathic medicine. Contact Dermatitis 2001;45:185

179 Ozkaya E. Current understanding of baboon syndrome. Expert Rev Dermatol 2009;4:163-175

180 Moreno-Ramírez D, García-Bravo B, Pichardo AR, Rubio FP, Martínez FC. Baboon syndrome in childhood: easy to avoid, easy to diagnose, but the problem continues. Pediatr Dermatol 2004;21:250-253

181 Mizukawa Y, Hirahara K, Kano Y, Shiohara T. Drug-induced hypersensitivity syndrome/drug reaction with eosinophilia and systemic symptoms severity score: A useful tool for assessing disease severity and predicting fatal cytomegalovirus disease. J Am Acad Dermatol 2019;80:670-678

182 Bocquet H, Bagot M, Roujeau JC. Drug-induced pseudolymphoma and drug hypersensitivity syndrome (Drug Rash with Eosinophilia and Systemic Symptoms: DRESS). Semin Cutan Med Surg 1996;15:250-257

183 Hoetzenecker W, Nägeli M, Mehra ET, Jensen AN, Saulite I, Schmid-Grendelmeier P, et al. Adverse cutaneous drug eruptions: current understanding. Semin Immunopathol 2016;38:75-86

184 Shiohara T, Mizukawa Y. Drug-induced hypersensitivity syndrome (DiHS)/drug reaction with eosinophilia and systemic symptoms (DRESS): An update in 2019. Allergol Int 2019;68:301-308

185 Martínez-Cabriales SA, Rodríguez-Bolaños F, Shear NH. Drug reaction with eosinophilia and systemic symptoms (DRESS): How far have we come? Am J Clin Dermatol 2019;20:217-236

186 Cho YT, Yang CW, Chu CY. Drug reaction with eosinophilia and systemic symptoms (DRESS): an interplay among drugs, viruses, and immune system. Int J Mol Sci 2017;18:1243

187 Shiohara T, Kano Y. Drug reaction with eosinophilia and systemic symptoms (DRESS): incidence, pathogenesis and management. Expert Opin Drug Saf 2017;16:139-147

188 Watanabe H. Recent Advances in Drug-Induced Hypersensitivity Syndrome/Drug reaction with eosinophilia and systemic symptoms. J Immunol Res 2018;2018:5163129

189 Gomes ESR, Marques ML, Regateiro FS. Epidemiology and risk factors for severe delayed drug hypersensitivity reactions. Curr Pharm Des 2019;25:3799-3812

190 Elzagallaai AA, Knowles SR, Rieder MJ, Bend JR, Shear NH, Koren G. Patch testing for the diagnosis of anticonvulsant hypersensitivity syndrome: a systematic review. Drug Saf 2009;32:391-408

191 Husain Z, Reddy BY, Schwartz RA. DRESS syndrome: Part II. Management and therapeutics. J Am Acad Dermatol 2013;68:709.e1-9

192 Alanko K. Patch testing in cutaneous reactions caused by carbamazepine. Contact Dermatitis 1993;29:254-257

193 Caboni S, Gunera-Saad N, Ktiouet-Abassi S, Berard F, Nicolas JF. Esomeprazole-induced DRESS syndrome. Studies of cross-reactivity among proton-pump inhibitor drugs. Allergy 2007;62:1342-1343

194 Shebe K, Ngwanya MR, Gantsho N, Lehloenya RJ. Severe recurrence of drug rash with eosinophilia and systemic symptoms syndrome secondary to rifampicin patch testing in a human immunodeficiency virus-infected man. Contact Dermatitis 2014;70:125-127

195 Trubiano JA, Douglas AP, Goh M, Slavin MA, Phillips EJ. The safety of antibiotic skin testing in severe T-cell-mediated hypersensitivity of immunocompetent and immunocompromised hosts. J Allergy Clin Immunol Pract 2019;7:1341-1343.e1

196 Gaig P, García-Ortega P, Baltasar M, Bartra J. Drug neosensitization during anticonvulsant hypersensitivity syndrome. J Investig Allergol Clin Immunol 2006;16:321-326

197 Gex-Collet C, Helbling A, Pichler WJ. Multiple drug hypersensitivity – proof of multiple drug hypersensitivity by patch and lymphocyte transformation tests. J Investig Allergol Clin Immunol 2005;15:293-296

198 Wöhrl S, Vigl K, Stingl G. Patients with drug reactions – is it worth testing? Allergy 2006;61:928-934

199 Cabañas R, Ramírez E, Sendagorta E, Alamar R, Barranco R, Blanca-López N, et al. Spanish guidelines for diagnosis, management, treatment, and prevention of DRESS syndrome. J Investig Allergol Clin Immunol 2020;30:229-253

200 Bensaid B, Rozieres A, Nosbaum A, Nicolas J-F, Berard F. Amikacin-induced drug reaction with eosinophilia and systemic symptoms syndrome: Delayed skin test and ELISPOT assay results allow the identification of the culprit drug. J Allergy Clin Immunol 2012;130:1413-1414

201 Jörg L, Helbling A, Yerly D, Pichler WJ. Drug-related relapses in drug reaction with eosinophilia and systemic symptoms (DRESS). Clin Transl Allergy 2020;10:52

202 Jörg L, Yerly D, Helbling A, Pichler W. The role of drug, dose and the tolerance/intolerance of new drugs in multiple drug hypersensitivity syndrome (MDH). Allergy 2020;75:1178-1187

203 Pichler WJ, Srinoulprasert Y, Yun J, Hausmann O. Multiple drug hypersensitivity. Int Arch Allergy Immunol 2017;172:129-138

204 Studer M, Waton J, Bursztejn AC, Aimone-Gastin I, Schmutz JL, Barbaud A. Does hypersensitivity to multiple drugs really exist? Ann Dermatol Venereol 2012;139:375-380 (Article in French)

205 Landry Q, Zhang S, Ferrando L, Bourrain JL, Demoly P, Chiriac AM. Multiple drug hypersensitivity syndrome in a large database. J Allergy Clin Immunol Pract 2019;8:258

206 Cabañas R, Calderon O, Ramirez E, Fiandor A, Prior N, Caballero T, et al. Piperacillin-induced DRESS: distinguishing features observed in a clinical and allergy study of 8 patients. J Investig Allergol Clin Immunol 2014;24:425-430

207 Ardern-Jones MR, Mockenhaupt M. Making a diagnosis in severe cutaneous drug hypersensitivity reactions. Curr Opin Allergy Clin Immunol 2019;19:283-293

208 Teo YX, Ardern-Jones MR. Reactivation of drug reaction with eosinophilia and systemic symptoms with ranitidine patch testing. Contact Dermatitis 2021;84:278-279

209 Brockow K, Ardern-Jones MR, Mockenhaupt M, Aberer W, Barbaud A, Caubet J-C, et al. EAACI position paper on how to classify cutaneous manifestations of drug hypersensitivity. Allergy 2019;74:14-27

210 Hasegawa A, Abe R. Recent advances in managing and understanding Stevens-Johnson syndrome and toxic epidermal necrolysis. F1000Res 2020;9:F1000 Faculty Rev-612. doi: 10.12688/f1000research.24748.1

211 Schneider JA, Cohen PR. Stevens-Johnson syndrome and toxic epidermal necrolysis: A concise review with a comprehensive summary of therapeutic interventions emphasizing supportive measures. Adv Ther 2017;34:1235-1244

212 Ramien M, Goldman JL. Pediatric SJS-TEN: Where are we now? F1000Res 2020;9:F1000 Faculty Rev-982. doi: 10.12688/f1000research.20419.1

213 Nowsheen S, Lehman JS, El-Azhary RA. Differences between Stevens-Johnson syndrome versus toxic epidermal necrolysis. Int J Dermatol 2021;60:53-59

214 Lehloenya RJ, Todd G, Wallace J, Ngwanya MR, Muloiwa R, Dheda K. Diagnostic patch testing following tuberculosis-associated cutaneous adverse drug reactions induces systemic reactions in HIV-infected persons. Br J Dermatol 2016;175:150-156

215 Brajon D, Menetre S, Waton J, Poreaux C, Barbaud A. Nonirritant concentrations and amounts of active ingredient in drug patch tests. Contact Dermatitis 2014;71:170-175

216 Liccioli G, Mori F, Parronchi P, Capone M, Fili L, Barni S, Sarti L, Giovannini M, Resti M, Novembre EM. Aetiopathogenesis of severe cutaneous adverse reactions (SCARs) in children: A 9-year experience in a tertiary care paediatric hospital setting. Clin Exp Allergy 2020;50:61-73

217 Kanny G, Pichler W, Morisset M, Franck P, Marie B, Kohler C, Renaudin JM, Beaudouin E, Laudy JS, Moneret-Vautrin DA. T cell-mediated reactions to iodinated contrast media: evaluation by skin and lymphocyte activation tests. J Allergy Clin Immunol 2005;115:179-185

218 Halevy S. Acute generalized exanthematous pustulosis. Curr Opin Allergy Clin Immunol 2009;9:322-328

219 Miyauchi H, Hosokawa H, Akaeda T, Iba H, Asada Y. T-cell subsets in drug-induced toxic epidermal necrolysis. Possible pathogenic mechanism induced by CD8-positive T cells. Arch Dermatol 1991;127:851-855

220 Sano T, Matsumoto R. Drug-induced TEN. Hifubyohshinryo 1980;2:49-52 (Article in Japanese, data cited in ref. 223)

221 Yamada M, Matsunaga K, Sakai K, Yazaki Y, Shiga A. A case of TEN (Lyell). Nippon Hifuka Gakkai Zasshi 1980;90:648 (Article in Japanese, data cited in ref. 223)

222 Schöpf E, Schulz KH, Kessler R, Taugner M, Braun W. Allergologische Untersuchungen beim Lyell-Syndrome. Z Hautkr 1975;50:865-873 (Article in German, data cited in ref. 223)

223 Tagami H, Tatsuta K, Iwatsuki K, Yamada M. Delayed hypersensitivity in ampicillin-induced toxic epidermal necrolysis. Arch Dermatol 1983;119:910-913

224 Fontaine JF, Lavaud F, Deslee G, Caillet MJ, Lebargy F. Toxic dermatitis caused by pseudoephedrine: apropos of a case. Allerg Immunol (Paris) 2002;34:230-232 (Article in French)

225 Klein CE, Trautmann A, Zillikens D, Brocker EB. Patch testing in an unusual case of toxic epidermal necrolysis. Contact Dermatitis 1995;33:448-449 (also published on pages 175-176)

226 Heinzerling LM, Tomsitz D, Anliker MD. Is drug allergy less prevalent than previously assumed? A 5-year analysis. Br J Dermatol 2012;166:107-114

227 Barbaud A, Trechot P, Weber-Muller F, Ulrich G, Commun N, Schmutz JL. Drug skin tests in cutaneous adverse drug reactions to pristinamycin: 29 cases with a study of cross-reactions between synergistins. Contact Dermatitis 2004;50:22-26

228 Patel S, John AM, Handler MZ, Schwartz RA. Fixed drug eruptions: An update, emphasizing the potentially lethal generalized bullous fixed drug eruption. Am J Clin Dermatol 2020;21:393-399

229 Mitre V, Applebaum DS, Albahrani Y, Hsu S. Generalized bullous fixed drug eruption imitating toxic epidermal necrolysis: a case report and literature review. Dermatol Online J 2017;23:1-4

230 Özkaya E. Oral mucosal fixed drug eruption: characteristics and differential diagnosis. J Am Acad Dermatol 2013;69:e51-e58

231 Lee AY. Fixed drug eruptions. Incidence, recognition, and avoidance. Am J Clin Dermatol 2000;1:277-285

232 Bellini V, Bianchi L, Hansel K, Finocchi R, Stingeni L. Bullous nonpigmenting multifocal fixed drug eruption due to pseudoephedrine in a combination drug: clinical and diagnostic observations. J Allergy Clin Immunol Pract 2016;4:542-544

233 Mahboob A, Haroon TS. Drugs causing fixed eruptions: a study of 450 cases. Int J Dermatol 1998;37:833-838

234 Flowers H, Brodell R, Brents M, Wyatt JP. Fixed drug eruptions: presentation, diagnosis, and management. South Med J 2014;107:724-727

235 Lipowicz S, Sekula P, Ingen-Housz-Oro S, Liss Y, Sassolas B, Dunant A, et al. Prognosis of generalized bullous fixed drug eruption: comparison with Stevens–Johnson syndrome and toxic epidermal necrolysis. Br J Dermatol 2013;168:726-732

236 Sehgal VN, Srivastava G. Fixed drug eruption (FDE): changing scenario of incriminating drugs. Int J Dermatol 2006;45:897-908

237 Zaouak A, Ben Salem F, Ben Jannet S, Hammami H, Fenniche S. Bullous fixed drug eruption: A potential diagnostic pitfall: a study of 18 cases. Therapie 2019;74:527-530

238 Ozkaya-Bayazit E. Topical provocation in fixed drug eruption due to metamizol and naproxen. Clin Exp Dermatol 2004;29:419-422

239 Thankappan TP, Zachariah J. Drug-specific clinical pattern in fixed drug eruptions. Int J Dermatol 1991;30:867-870

240 Brahimi N, Routier E, Raison-Peyron N et al. A three-year-analysis of fixed drug eruptions in hospital settings in France. Eur J Dermatol 2010;20:461-464

241 Özkaya E. Fixed drug eruption: state of the art. J Dtsch Dermatol Ges 2008;6:181-188

242 Alanko K, Stubb S, Reitamo S. Topical provocation of fixed drug eruption. Br J Dermatol 1987;116:561-567

243 Alanko K. Topical provocation of fixed drug eruption. A study of 30 patients. Contact Dermatitis 1994;31:25-27

244 Özkaya E. Topical provocation in 27 cases of cotrimoxazole-induced fixed drug eruption. Contact Dermatitis 1999;41:185-189

245 Lee A-Y. Topical provocation in 31 cases of fixed drug eruption: change of causative drugs in 10 years. Contact Dermatitis 1998;38:258-260

246 Mizukawa Y, Shiohara T. Trauma-localized fixed drug eruption: involvement of burn scars, insect bites and venipuncture sites. Dermatology 2002;205:159-161

247 Dalla Costa R, Yang CY, Stout M, Kroshinsky D, Kourosh AS. Multiple fixed drug eruption to minocycline at sites of healed burn and zoster: an interesting case of locus minoris resistentiae. JAAD Case Rep 2017;3:392-394

248 Ben Fadhel N, Chaabane A, Ammar H, Ben Romdhane H, Soua Y, Chadli Z, et al. Clinical features, culprit drugs, and allergology workup in 41 cases of fixed drug eruption. Contact Dermatitis 2019;81:336-340

249 Özkaya-Bayazit E. Specific site involvement in fixed drug eruption. J Am Acad Dermatol 2003;49:1003-1007

250 Shiohara T. Fixed drug eruption: pathogenesis and diagnostic tests. Curr Opin Allergy Clin Immunol 2009;9:316-321

251 Gonzalo-Garijo MA, Zambonino MA, Pérez-Calderón R, Pérez-Rangel I, Sánchez-Vega S. Fixed drug eruption due to quinine in tonic water: study of cross-reactions. Dermatitis 2012;23:51

252 Bel B, Jeudy G, Bouilly D, Dalac S, Vabres P, Collet E. Fixed eruption due to quinine contained in tonic water: positive patch-testing. Contact Dermatitis 2009;61:242-244

253 Vernassiere C, Barbaud A, Trechot PH, Weber-Muller F, Schmutz JL. Systemic acyclovir reaction subsequent to acyclovir contact allergy: which systemic antiviral drug should then be used? Contact Dermatitis 2003;49:155-157

254 Gola M, Francalanci S, Brusi C, Lombardi P, Sertoli A. Contact sensitization to acyclovir. Contact Dermatitis 1989;20:394-395

255 Bayrou O, Gaouar H, Leynadier F. Famciclovir as a possible alternative treatment in some cases of allergy to acyclovir. Contact Dermatitis 2000;42:42

256 Lammintausta K, Mäkelä L, Kalimo K. Rapid systemic valaciclovir reaction subsequent to aciclovir contact allergy. Contact Dermatitis 2001;45:181

257 Baeck M, Goossens A. Patients with airborne sensitization/contact dermatitis from budesonide-containing aerosols 'by proxy'. Contact Dermatitis 2009;61:1-8

258 Smeenk G, Burgers GJ, Teunissen PC. Contact dermatitis from salbutamol. Contact Dermatitis 1994;31:123

259 Ekenvall L, Forsbeck M. Contact eczema produced by a beta-adrenergic blocking agent (alprenolol). Contact Dermatitis 1978;4:190-194

260 Van Ketel WG. Systemic contact-type dermatitis by derivatives of adamantane? Derm Beruf Umwelt 1988;36:23-24 (Article in German)

261 Villarreal O. Systemic dermatitis with eosinophilia due to epsilon-aminocaproic acid. Contact Dermatitis 1999;40:114

262 Gutiérrez M, López M, Ruiz M. Positivity of patch tests in cutaneous reaction to aminocaproic acid: two case reports. Allergy 1995;50:745-746

263 Hayakawa R, Ogino Y, Aris K, Matsunaga K. Systemic contact dermatitis due to amlexanox. Contact Dermatitis 1992;27:122-123

264 Milligan A, Douglas WS. Contact dermatitis to cephalexin. Contact Dermatitis 1986;15:91

265 Baer RL, Sulzberger MB. Eczematous dermatitis due to chloral hydrate (following both oral administration and topical application). J Allergy 1938;9:519-520

266 Urrutia I, Audícana M, Echechipía S, Gastaminza G, Bernaola G, Fernández de Corrès L. Sensitization to chloramphenicol. Contact Dermatitis 1992;26:66-67

267 Santucci B, Cannistraci C, CristaudoA, Picardo M. Contact dermatitis from topical alkylamines. Contact Dermatitis 1992;27:200-201

268 Ekelund AG, Möller H. Oral provocation in eczematous contact allergy to neomycin and hydroxyquinolines. Acta

269 Skog E. Systemic eczematous contact-type dermatitis induced by iodochlorhydroxyquin and chloroquine phosphate. Contact Dermatitis 1975;1:187

270 Cronin E. Contact Dermatitis. Edinburgh: Churchill Livingstone, 1980:220

271 Domar M, Juhlin L. Allergic dermatitis produced by oral clioquinol. Lancet 1967;1(7500):1165-1166

272 Maibach HI. Oral substitution in patients sensitized by transdermal clonidine treatment. Contact Dermatitis 1987;16:1-8

273 Giroux M, Pratt M. Two cases of systemic ACD: cobalt recalled with B12 and plants recalled with herbal products. Am J Cont Dermat 2003;14:109 (Abstract)

274 Leroy A, Baeck M, Tennstedt D. Contact dermatitis and secondary systemic allergy to dimethindene maleate. Contact Dermatitis 2011;64:170-171

275 Coskey RJ. Contact dermatitis caused by diphenhydramine hydrochloride. J Am Acad Dermatol 1983;8:204-206

276 Lawrence CM, Byrne JP. Eczematous eruption from oral diphenhydramine. Contact Dermatitis 1981;7:276-277

277 Fustà-Novell X, Gómez-Armayones S, Morgado-Carrasco D, Mascaró JM. Systemic allergic dermatitis caused by disulfiram (Antabuse) in a patient previously sensitized to rubber accelerators. Contact Dermatitis 2018;79:239-240

278 Gutgesell C, Fuchs T. Orally elicited allergic contact dermatitis to tetraethylthiuramdisulfide. Am J Contact Dermat 2001;12:235-236

279 Meneghini CL, Bonifazi E. Antabuse dermatitis. Contact Dermatitis Newsletter 1972;11:285

280 Wilson H. Side-effects of disulfiram. Br Med J 1962;2:1610-1611

281 Webb PK, Bibbs SC, Mathias CT, Crain W, Maibach H. Disulfiram hypersensitivity and rubber contact dermatitis. JAMA 1979;241:2061

282 Brancaccio RR, Weinstein S. Systemic contact dermatitis to doxepin. J Drugs Dermatol 2003;2:409-410

283 Ventura MT, Dagnello M, Di Corato R, Tursi A. Allergic contact dermatitis due to epirubicin. Contact Dermatitis 1999;40:339

284 Serrano G, Fortea JM, Latasa JM, et al. Photosensitivity induced by fibric acid derivatives and its relation to photocontact dermatitis to ketoprofen. J Am Acad Dermatol 1992;27(2Pt.1):204-208

285 Pinheiro V, Pestana C, Pinho A, Antunes I, Gonçalo M. Occupational allergic contact dermatitis caused by antibiotics in healthcare workers - relationship with non-immediate drug eruptions. Contact Dermatitis 2018;78:281-286

286 Nadal C, Pujol RM, Randazzo L, Marcuello E, Alomar A. Systemic contact dermatitis from 5-fluorouracil. Contact Dermatitis 1996;35:124-125

287 De Castro Martinez FJ, Ruiz FJ, Tornero P, De Barrio M, Prieto A. Systemic contact dermatitis due to fusidic acid. Contact Dermatitis 2006;54:169

288 Guin JD, Phillips D. Erythroderma from systemic contact dermatitis: a complication of systemic gentamicin in a patient with contact allergy to neomycin. Cutis 1989;43:564-567

289 Ghadially R, Ramsay CA. Gentamicin: systemic exposure to a contact allergen. J Am Acad Dermatol 1988;19(2Pt. 2):428-430

290 Paniagua MJ, Garcia-Ortega P, Tella R, Gaig P, Richart C. Systemic contact dermatitis to gentamicin. Allergy 2002;57:1086-1087

291 Braun W, Schütz R. Beitrag zur Gentamycin Allergie. Hautarzt 1969;20:108

292 Malik M, Tobin AM, Shanahan F, O'Morain C, Kirby B, Bourke J. Steroid allergy in patients with inflammatory bowel disease. Br J Dermatol 2007;157:967-969

293 Lauerma AI, Reitamo S, Maibach HI. Systemic hydrocortisone/cortisol induces allergic skin reactions in presensitized subjects. J Am Acad Dermatol 1991;24:182-185

294 Lauerma AI, Reitamo S, Maibach HI. Systemic hydrocortisone/cortisol induces allergic skin reactions in presensitized subjects. Am J Contact Dermat 1991;2:68 (Abstract)

295 De Cuyper C, Goeteyn M. Systemic contact dermatitis from subcutaneous hydromorphone. Contact Dermatitis 1992;27:220-223

296 Meier H, Elsner P, Wüthrich B. Berufsbedingtes Kontaktekzem und Asthma bronchiale bei ungewöhnlicher allergischer Reaktion vom Spättyp auf Hydroxychoroquin [Occupationally induced contact dermatitis and bronchial asthma in a unusual delayed reaction to hydroxychloroquine]. Hautarzt 1999;50:665-669 (Article in German)

297 Alomar A. Buphenine sensitivity. Contact Dermatitis 1984;11:315

298 Garcia-Bravo B, Mazuecos J, Rodriguez-Pichardo A, Navas J, Camacho F. Hypersensitivity to ketoconazole preparations: study of 4 cases. Contact Dermatitis 1989;21:346-348

299 Foti C, Cassano N, Vena GA, Angelini G. Photodermatitis caused by oral ketoprofen: two case reports. Contact Dermatitis 2011;64:181-183

300 Kiec-Swierczynska M, Krecisz B. Occupational airborne allergic contact dermatitis from mesna. Contact Dermatitis 2003;48:171

301 Fernandez Redondo V, Casas L, Taboada M, Toribio J. Systemic contact dermatitis from erythromycin. Contact Dermatitis 1994;30:311 (same as ref. 302)

302 Fernandez Redondo V, Casas L, Taboada M, Toribio J. Systemic contact dermatitis from erythromycin. Contact Dermatitis 1994;30:43-44 (same as ref. 301)

303 Corazza M, Mantovani L, Montanari A, Virgili A. Allergic contact dermatitis from transdermal estradiol and systemic contact dermatitis from oral estradiol. A case report. J Reprod Med 2002;47:507-509

304 El Sayed F, Bayle-Lebey P, Marguery MC, Bazex J. Systemic sensitization to 17-beta estradiol induced by transcutaneous administration. (Sensibilisation systémique au 17-β-oestradiol induite par voie transcutanée). Ann Dermatol Venereol 1996;123:26-28 (Article in French)

305 Van den Berg WHHW, Van Ketel WG. Contactallergie voor ethyleendiamine. Ned Tijdschr Geneeskd 1983;127:1801-1802

306 Möller H. Contact and photocontact allergy to psoralens. Am J Contact Dermat 1990;1:254 (also published on page 2 of the same journal and volume)

307 Mussani F, Skotnicki S. Systemic contact dermatitis: Two Interesting cases of systemic eruptions following exposure to drugs clioquinol and metronidazole. Dermatitis 2013;24(4):e3

308 Echechipía S, Alvarez MJ, García BE, Olaguíbel JM, Rodriguez A, Lizaso MT, Acero S, Tabar AI. Generalized dermatitis due to mitomycin C patch test. Contact Dermatitis 1995;33:432

309 Hogen Esch AJ, van der Heide S, van den Brink W, van Ree JM, Bruynzeel DP, Coenraads PJ. Contact allergy and respiratory/mucosal complaints from heroin (diacetylmorphine). Contact Dermatitis 2006;54:42-49

310 Silvestre JF, Alfonso R, Moragón M, Ramón R, Botella R. Systemic contact dermatitis due to norfloxacin with a positive patch test to quinoline mix. Contact Dermatitis 1998;39:83

311 Cronin E. Contact Dermatitis. Edinburgh: Churchill Livingstone, 1980:232-233

312 Quirce S, Alvarez MJ, Olaguibel JM, Tabar AI. Systemic contact dermatitis from oral prednisone. Contact Dermatitis 1994;30:53-54

313 Bircher AJ, Bigliardi P, Zaugg T, Mäkinen-Kiljunen S. Delayed generalized allergic reactions to corticosteroids. Dermatology 2000;200:349-351

314 Dorado Bris JM, Aragues Montañes M, Sols Candela M, Garcia Diez A. Contact sensitivity to pyrazinobutazone (Carudol) with positive oral provocation test. Contact Dermatitis 1992;26:355-356

315 Delgado J, Quiralte J, Castillo R, Blanco C, Molero R, Carrillo T. Systemic contact dermatitis from suxamethonium. Contact Dermatitis 1996;35:120-121

316 Hjorth N. Contact dermatitis from vitamin B1 (thiamine). J Invest Dermatol 1958;30:261-264

317 Ingemann Larsen A, Riis Jepsen J, Thulin H. Allergic contact dermatitis from thiamine. Contact Dermatitis 1989;20:387-388

318 Arruti N, Bernedo N, Audicana MT, Villarreal O, Uriel O, Muñoz D. Systemic allergic dermatitis caused by thiamine after iontophoresis. Contact Dermatitis 2013;69:375-376

319 Kaae J, Menné T, Thyssen JP. Systemic contact dermatitis following oral exposure to tramadol in a patient with allergic contact dermatitis caused by buprenorphine. Contact Dermatitis 2012;66:106-107

320 Schmutz JL, Trechot P. Systemic reactivation of contact eczema following tramadol administration in a patient with buprenorphine-induced eczema. Ann Dermatol Venereol 2012;139:335-336 (Article in French)

321 Martin-Garcia C, Martinez-Borque N, Martinez-Bohigas N, Torrecillas-Toro M, Palomeque-Rodriguez MT. Delayed reaction urticaria due to trimebutine. Allergy 2004;59:789-790

322 Michel M, Dompmartin A, Szczurko C, Castel B, Moreau A, Leroy D. Eczematous-like drug eruption induced by synergistins. Contact Dermatitis 1996;34:86-87

323 Menné T, Veien N, Sjølin KE, Maibach HI. Systemic contact dermatitis. Am J Contact Dermat 1994;5:1-12

324 Murata Y, Kumano K, Ueda T, Araki N, Nakamura T, Tani N. Systemic contact dermatitis caused by systemic corticosteroid use. Arch Dermatol 1997;133:1053-1054

325 Chew AL, Maibach HI. Multiple corticosteroid orally elicited allergic contact dermatitis in a patient with multiple topical corticosteroid allergic contact dermatitis. Cutis 2000;65:307-311

326 Park RG. Cutaneous hypersensitivity to sulphonamides. Br Med J 1943;2:69-72

327 Nijhawan RI, Molenda M, Zirwas MJ, Jacob SE. Systemic contact dermatitis. Dermatol Clin 2009;27:355-364

328 Jacob SE, Zapolanski T. Systemic contact dermatitis. Dermatitis 2008;19:9-15

329 Veien NK. Systemic contact dermatitis. Int J Dermatol 2011;50:1445-1456

330 Rundle CW, Machler BC, Jacob SE. Pathogenesis and causations of systemic contact dermatitis. G Ital Dermatol Venereol 2019;154:42-49 (Article in English)

331 Fisher AA. Systemic contact-type dermatitis due to drugs. Clin Dermatol 1986;4:58-69

332 Meneghini CL, Angelini G. Gruppensensibilisierung durch photosensibilisierende Medikamente. Z Hautkr 1978;53:329-334 (Article in German). Data cited in ref. 333

333 Angelini G, Meneghini CL. Oral tests in contact allergy to para-amino compounds. Contact Dermatitis 1981;7:311-314

334 Leifer W, Steiner K. Studies in sensitization to halogenated hydroxyquinolines and related compounds. J Invest Dermatol 1951;17:233-240

335 Sulzberger MD, Kanof A, Baer RL, Lowenberg C. Sensitization by topical application of sulfonamide. J Allergy 1947;18:92-103

336 Fischer B. Dermatitis following the application of sulfanilamide. MJ Australia 1944;2:449

337 Fisher AA. Systemic eczematous "contact-type" dermatitis medicamentosa. Ann Allergy 1966;24:406-420

338 Zinn Z, Gayam S, Chelliah MP, Honari G, Teng J. Patch testing for nonimmediate cutaneous adverse drug reactions. J Am Acad Dermatol 2018;78:421-423

339 Giménez-Arnau AM, Skudlik C. Occupational contact dermatitis: Health personnel. In: Johansen JD, Mahler V, Lepoittevin JP, Frosch PJ, eds. Contact Dermatitis, 6th Ed. Cham, Switzerland, 2021:483-497

340 Jirasek L, Kalensky J. Kontakni alergicky ekzem z krmnych smesi v zivocisne vyrobe. Ceskoslovenska Dermatologie 1975;50:217 (Article in Czech, data cited in ref. 373)

341 Van den Hoed E, Coenraads PJ, Schuttelaar MLA. Morphine-induced cutaneous adverse drug reaction following occupational diacetylmorphine contact dermatitis: A case report. Contact Dermatitis 2019;81:313-315

342 Bircher AJ. Drug allergens. In: John S, Johansen J, Rustemeyer T, Elsner P, Maibach H, eds. Kanerva's occupational dermatology. Cham, Switzerland: Springer, 2020:559-578

343 Lachapelle JM. Airborne contact dermatitis. In: John S, Johansen J, Rustemeyer T, Elsner P, Maibach H, eds. Kanerva's occupational dermatology. Cham, Switzerland: Springer, 2020:229-240

344 Møller NE, Nielsen B, von Würden K. Contact dermatitis to semisynthetic penicillins in factory workers. Contact Dermatitis 1986;14:307-311

345 Shmunes E, Taylor JS, Petz LD, Garratty G, Fudenberg HH. Immunologic reactions in penicillin factory workers. Ann Allergy 1976;36:313-323

346 Rudzki E, Lukasiak B, Leszczynski W. Penicillin hypersensitivity and haemagglutinating antibodies in workers at a penicillin factory. Acta Allergologica 1965;20:206-214

347 Giménez-Arnau AM, Skudlik C. Occupational contact dermatitis: Health personnel. In: Johansen JD, Mahler V, Lepoittevin JP, Frosch PJ, red. Contact Dermatitis, 6th ed. Cham, Switzerland: Springer, 2021:483-497

348 Gielen K, Goossens A. Occupational allergic contact dermatitis from drugs in healthcare workers. Contact Dermatitis 2001;45:273-279

349 Higgins CL, Palmer AM, Cahill JL, Nixon RL. Occupational skin disease among Australian healthcare workers: a retrospective analysis from an occupational dermatology clinic, 1993-2014. Contact Dermatitis 2016;75:213-222

350 Schubert S, Bauer A, Molin S, Skudlik C, Geier J. Occupational contact sensitization in female geriatric nurses: Data of the Information Network of Departments of Dermatology (IVDK) 2005-2014. J Eur Acad Dermatol Venereol 2017;31:469-476 tetrazepam

351 Aalto-Korte K, Koskela K, Pesonen M. Allergic contact dermatitis and other occupational skin diseases in health care workers in the Finnish Register of Occupational Diseases in 2005-2016. Contact Dermatitis 2021;84:217-223

352 Whitaker P. Occupational allergy to pharmaceutical products. Curr Opin Allergy Clin Immunol 2016;16:101-106

353 Cetinkaya F, Ozturk AO, Kutluk G, Erdem E. Penicillin sensitivity among hospital nurses without a history of penicillin allergy. J Adv Nurs 2007;58:126-129

354 Altomare GF, Capella GL, Veraldi S. Occupational drug dermatitis. Clin Dermatol 1992;10:141-147

355 Goossens A, Hulst KV. Occupational contact dermatitis in the pharmaceutical industry. Clin Dermatol 2011;29:662-668

356 Goossens A, Geebelen J, Hulst KV, Gilissen L. Pharmaceutical and cosmetic industries. In: John S, Johansen J, Rustemeyer T, Elsner P, Maibach H, eds. Kanerva's occupational dermatology. Cham, Switzerland: Springer, 2020:2203-2219

357 Gonçalo M. Phototoxic and photoallergic contact reactions. In: Johansen JD, Mahler V, Lepoittevin JP, Frosch PJ, eds. Contact Dermatitis, 6th edition. Cham, Switzerland: Springer, 2021:365-389

358 Bruynzeel DP, Ferguson J, Andersen K, Gonçalo M, English J, Goossens A, et al. Photopatch testing: a consensus methodology for Europe. J Eur Acad Dermatol Venereol 2004;18:679-682

359 Gonçalo M, Ferguson J, Bonevalle A, Bruynzeel DP, Giménez-Arnau A, Goossens A, et al. Photopatch testing: recommendations for a European photopatch test baseline series. Contact Dermatitis 2013;68:239-243

360 Ferguson J, Kerr AC. Photoallergic contact dermatitis. In: John S, Johansen J, Rustemeyer T, Elsner P, Maibach H,

eds. Kanerva's Occupational Dermatology. Cham, Switzerland: Springer, 2020:211-227

361 Gonçalo M. Photopatch testing. In: Johansen JD, Mahler V, Lepoittevin JP, Frosch PJ, eds. Contact Dermatitis, 6ᵗʰ edition. Cham, Switzerland: Springer, 2021:593-608

362 Gonçalo M. Explorations dans les photo-allergies médicamenteuses. In: GERDA (ed). Progrès en Dermato-allergologie. Nancy, France: John Libbey Eurotext, 1998:67-74

363 Blakely KM, Drucker AM, Rosen CF. Drug-induced photosensitivity - An update: Culprit drugs, prevention and management. Drug Saf 2019;42:827-847

364 Kim WB, Shelley AJ, Novice K, Joo J, Lim HW, Glassman SJ. Drug-induced phototoxicity: A systematic review. J Am Acad Dermatol 2018;79:1069-1075

365 Monteiro AF, Rato M, Martins C. Drug-induced photosensitivity: photoallergic and phototoxic reactions. Clin Dermatol 2016;34:571-581

366 Stein KR, Scheinfeld NS. Drug- induced photoallergic and phototoxic reactions. Expert Opin Drug Saf 2007;6:431-443

367 Conilleau V, Dompmartin A, Michel M, Verneuil L, Leroy D. Photoscratch testing in systemic drug-induced photosensitivity. Photodermatol Photoimmunol Photomed 2000;16:62-66

368 Stejskal VD, Olin RG, Forsbeck M. The lymphocyte transformation test for diagnosis of drug-induced occupational allergy. J Allergy Clin Immunol 1986;77:411-426

369 Räsänen L, Hasan T. Allergy to systemic and intralesional corticosteroids. Br J Dermatol 1993;128:407-411

370 Soria A, Amsler E, Bernier C, Milpied B, Tétart F, Morice C, et al. DRESS and AGEP reactions to iodinated contrast media: A French case series. J Allergy Clin Immunol Pract 2021;9:3041-3050

371 Möller H. Contact and photocontact allergy to psoralens. Photodermatol Photoimmunol Photomed 1990;7:43-44

372 Walchner M, Rueff F, Przybilla B. Delayed-type hypersensitivity to mofebutazone underlying a severe drug reaction. Contact Dermatitis 1997;36:54-55

373 De Groot AC, Conemans JM. Contact allergy to furazolidone. Contact Dermatitis 1990;22:202-205

374 Puig LL, Abadias M, Alomar A. Erythroderma due to ribostamycin. Contact Dermatitis 1989;21:79-82

375 Provost TT, Jillson OF. Ethylenediamine contact dermatitis. Arch Dermatol 1967;96:231-234

376 Angelini G, Vena GA, Meneghini CL. Allergic contact dermatitis to some medicaments. Contact Dermatitis 1985:12:263-269

377 Hardy C, Schofield O, George CF. Allergy to aminophylline. Br Med J (Clin Res Ed) 1983;286(6383):2051-2052

378 Elias JA, Levinson AI. Hypersensitivity reactions to ethylenediamine in aminophylline. Am Rev Respir Dis 1981;123:550-552

379 Papakonstantinou E, Müller S, Röhrbein JH, Wieczorek D, Kapp A, Jakob T, Wedi B. Generalized reactions during skin testing with clindamycin in drug hypersensitivity: a report of 3 cases and review of the literature. Contact Dermatitis 2018;78:274-280

380 Nettis E, Giordano D, Colanardi MC, Paradiso MT, Ferrannini A, Tursi A. Delayed-type hypersensitivity rash from ibuprofen. Allergy 2003;58:539-540

381 Gielen K, Goossens A. Occupational allergic contact dermatitis from drugs in healthcare workers. Contact Dermatitis 2001;45:273-279

382 Gonzalo-Garijo MA, Pérez-Calderón R, de Argila D, Rodríguez-Nevado I. Erythrodermia to pseudoephedrine in a patient with contact allergy to phenylephrine. Allergol Immunopathol (Madr) 2002;30:239-242

383 Lerondeau B, Trechot P, Waton J, Poreaux C, Luc A, Schmutz JL, et al. Barbaud A. Analysis of cross-reactivity among radiocontrast media in 97 hypersensitivity reactions. J Allergy Clin Immunol 2016;137:633-635.e4

384 Torres MJ, Gomez F, Doña I, Rosado A, Mayorga C, Garcia I, et al. Diagnostic evaluation of patients with nonimmediate cutaneous hypersensitivity reactions to iodinated contrast media. Allergy 2012;67:929-935

385 Fregert S. Exacerbation of dermatitis by perorally administered piperazine derivative in a piperazine sensitized man. Contact Dermatitis Newsletter 1967;1:13

386 Wright S, Harman RR. Ethylenediamine and piperazine sensitivity. Br Med J (Clin Res Ed) 1983;287(6390):463-464

387 Eedy DJ. Angioneurotic oedema following piperazine ingestion in an ethylenediamine-sensitive subject. Contact Dermatitis 1993;28:48-49

388 Price ML, Hall-Smith SP. Allergy to piperazine in a patient sensitive to ethylenediamine. Contact Dermatitis 1984;10:120

389 Romano A, Valluzzi RL, Caruso C, Maggioletti M, Gaeta F. Non-immediate cutaneous reactions to beta-lactams: Approach to diagnosis. Curr Allergy Asthma Rep 2017;17:23

390 Asensio T, Sanchís ME, Sánchez P, Vega JM, García JC. Photocontact dermatitis because of oral dexketoprofen. Contact Dermatitis 2008;58:59-60

391 Horio T. Allergic and photoallergic dermatitis from diphenhydramine. Arch Dermatol 1976;112:1124-1126

392 Lachapelle JM. Allergic "contact" dermatitis from disulfiram implants. Contact Dermatitis 1975;1:218-220

Chapter 5　　IMMEDIATE CONTACT REACTIONS (CONTACT URTICARIA) FROM SYSTEMIC DRUGS

Immediate contact reactions (contact urticaria) are wheal and flare reactions with erythema and edema following external contact of intact skin or mucous membrane with a substance, usually appearing within 30 minutes and clearing completely within hours, without residual signs (21). Some systemic drugs have been reported to cause such reactions (table 5.1). Contact with the skin mostly occurred from accidental occupational contact, e.g. in nurses, dentists, pharmacists, veterinarians, other health care workers, or individuals working in pharmaceutical plants producing the drugs. Some patients developed contact urticaria from drugs used by persons in their close environment, e.g. a mother allergic to amoxicillin, who developed swelling of the lips after kissing her 5-year-old son on the lips, who had ingested amoxicillin earlier (contact urticaria *by proxy*) (16).

Another category of immediate contact reactions is in patients suspected of urticarial rashes, immediate-type skin eruptions, or anaphylactic reactions to (par)enteral drugs, where patch testing resulted in contact urticaria to the systemic drug. The entire spectrum of symptoms and signs of the contact urticaria syndrome (18,21) can result from immediate contact reactions to systemic drugs including erythema, localized urticaria, generalized urticaria, gastrointestinal symptoms, pulmonary symptoms, drop in blood pressure, and anaphylactic shock.

Not included in this chapter are drugs that have caused urticarial or other eruptions from oral or parenteral administration, subsequently diagnosed as type I allergy reactions by positive intracutaneous or intradermal tests, but *without* a positive immediate-type cutaneous reaction (contact urticaria) being demonstrated on intact skin, e.g. by a patch test read after 15-45 minutes, by open application or by rubbing the drug material in the skin. Excluded are also cases of immediate contact urticaria to systemic drugs incorporated in topical pharmaceuticals. These, and cases used by drugs used *only* topically, have been discussed in Volume 3 of the *Monographs in contact allergy* series (24).

Systemic drugs that have caused immediate contact reactions are shown in table 5.1. They are or may be *examples* of cases of immediate contact reactions, that the author came across while writing this book. Providing a full review was not attempted.

Table 5.1 Systemic drugs that have caused immediate contact reactions (contact urticaria)

Drug	Immediate contact reactions from		References
	accidental contact (occupational) [d]	patch testing in drug eruption or anaphylaxis	
Aminophenazone	+	+	1,2,3,4
Amoxicillin	+		16 [b], 27 [b], 29
Bacampicillin	+		26 [b]
Benzylpenicillin (penicillin G)	+		3,36
Cefalotin	+		28
Cefotiam	+		8,9,10,11,12,13,14,15
Cefuroxime	+		22
Chlorpromazine [g]	+		33 [e], 34 [b]
Cisplatin	+		5
Donepezil	+		6
Gentamicin			18 [a]
Levomepromazine	+		23
Metamizole	+	+	4,35
Mezlocillin	+		31
Neomycin			19 [a], 25 [c]
Penicillin, unspecified			18,19 [a],20,37 [f]
Pentamidine isethionate	+		7
Promethazine	+		3
Propyphenazone		+	4
Streptomycin	+		32, 37 [f]
Sulbactam	+		30

[a] Details unknown, cited in ref. 17; [b] Contact urticaria *by proxy*; [c] Discovered by routine testing in investigational study of immediate contact reactions; [d] Includes contact urticaria *by proxy* in non-professional caregivers; [e] Very dubious case; it was not even mentioned whether the patient used or otherwise had contact with chlorpromazine; [f] Insufficient information available, data obtained from summary only; [g] *Photo*contact urticaria

LITERATURE

1   Lombardi P, Giorgini S, Achille A. Contact urticaria from aminophenazone. Contact Dermatitis 1983;9:428-429

2   Camarasa JM, Alomar A, Perez M. Contact urticaria and anaphylaxis from aminophenazone. Contact Dermatitis 1978;4:243-244

3   Haustein UF. Anaphylactic shock and contact urticaria after the patch test with professional allergens. Allerg Immunol (Leipz) 1976;22:349-352 (Article in German)

4   Maucher OM, Fuchs A. Contact urticaria caused by skin test in pyrazolone allergy. Hautarzt 1983;34:383-386 (Article in German)

5   Schena D, Barba A, Costa G. Occupational contact urticaria due to cisplatin. Contact Dermatitis 1996;34:220-221

6   Galvez Lozano JM, Alcantara M, De San Pedro BS, Quiralte J, Caba I. Occupational contact urticaria caused by donepezil. Contact Dermatitis 2009;61:176

7   Belsito DV. Contact urticaria from pentamidine isethionate. Contact Dermatitis 1993;29:158-159

8   Takahagi S, Tanaka A, Iwamoto K, Ishii K, Hide M. Contact urticaria syndrome with IgE antibody against a cefotiam-unique structure, evoked by nonapparent exposure to cefotiam. Clin Exp Dermatol 2017;42:527-531

9   Kim JE, Kim SH, Choi GS, Ye YM, Park HS. Detection of specific IgE antibodies to cefotiam-HSA conjugate by ELISA in a nurse with occupational anaphylaxis. Allergy 2010;65:791-792

10 Miyake H, Morishima Y, Kishimoto S. Occupational contact urticaria syndrome from cefotiam dihydrochloride in a latex-allergic nurse. Contact Dermatitis 2000;43:230-231

11 Chiba Y, Takahashi S, Yamakawa Y, Aihara M, Ikezawa Z. Contact urticaria syndrome caused by patch testing with cefotiam hydrochloride. Contact Dermatitis 1999;41:234

12 Shimizu S, Chen KR, Miyakawa S. Cefotiam-induced contact urticaria syndrome: an occupational condition in Japanese nurses. Dermatology 1996;192:174-176

13 Tadokoro K, Niimi N, Ohtoshi T, Nakajima K, Takafuji S, Onodera K, et al. Cefotiam-induced IgE-mediated occupational contact anaphylaxis of nurses; case reports, RAST analysis, and a review of the literature. Clin Exp Allergy 1994;24:127-133

14 Mizutani H, Ohyanagi S, Shimizu M. Anaphylactic shock related to occupational handling of Cefotiam dihydrochloride. Clin Exp Dermatol 1994;19:449

15 Miyahara H, Koga T, Imayama S, Hori Y. Occupational contact urticaria syndrome from cefotiam hydrochloride. Contact Dermatitis 1993;29:210-211

16 Mancuso G, Berdondini RM. Kiss-induced allergy to amoxycillin. Contact Dermatitis 2006;54:226

17 De Groot AC, Weyland JW, Nater JP. Unwanted effects of cosmetics and drugs used in dermatology, 3rd edition. Amsterdam: Elsevier Science BV, 1994

18 Lahti A, Maibach HI. Immediate contact reactions. In: Menné T, Maibach HI (Eds). Exogenous dermatoses, environmental contact dermatitis. Boca Raton: CRC Press, 1990:21-35

19 Maucher OD. Anaphylaktische Reaktionen beim Epikutantest. Hautarzt 1972;23:139 (Article in German, cited in ref. 17)

20 Böttger EM, Mücke Chr, Tronnier H. Kontaktdermatitis auf neuere Antimykotika und Kontakturtikaria. Acta Dermatol 1981;7:70 (Article in German, cited in ref. 17)

21 Gimenez-Arnau A, Maurer M, De La Cuadra J, Maibach H. Immediate contact skin reactions, an update of contact urticaria, contact urticaria syndrome and protein contact dermatitis -- "A never ending story". Eur J Dermatol 2010;20:552-562

22 Classen A, Fuchs T. Occupational allergy to β-lactam antibiotics. Allergo J Int 2015;24:54-57

23 Johansson G. Contact urticaria from levomepromazine. Contact Dermatitis 1988;19:304

24 De Groot AC. Monographs in contact allergy, volume 3: Topical Drugs. Boca Raton, Fl, USA: CRC Press Taylor and Francis Group, 2021 (ISBN 978-0-367-23693-9)

25 Katsarou A, Armenaka M, Ale I, Koufou V, Kalogeromitros D. Frequency of immediate reactions to the European standard series. Contact Dermatitis 1999;41:276-279

26 Liccardi G, Gilder J, D'Amato M, D'Amato G. Drug allergy transmitted by passionate kissing. Lancet 2002;359:1700

27 Pétavy-Catala C, Machet L, Vaillant L. Consort contact urticaria due to amoxycillin. Contact Dermatitis 2001;44:251

28 Tuft L. Contact urticaria from cephalosporins. Arch Dermatol 1975;111:1609

29 Condé-Salazar L, Guimaraens D, González MA, Mancebo E. Occupational allergic contact urticaria from amoxicillin. Contact Dermatitis 2001;45:109

30 Kwon HJ, Kim MY, Kim HO, Park YM. The simultaneous occurrence of contact urticaria from sulbactam and allergic contact dermatitis from ampicillin in a nurse. Contact Dermatitis 2006;54:176-178

31 Keller K, Schwanitz HJ. Combined immediate and delayed hypersensitivity to mezlocillin. Contact Dermatitis 1992;27:348-349

32 Rudzki E, Rebandel P, Rogozinski T. Contact urticaria from rat tail, guinea pig, streptomycin and vinyl pyridine. Contact Dermatitis 1981;7:186-188

33 Horio T. Chlorpromazine photoallergy. Coexistence of immediate and delayed type. Arch Dermatol 1975;111:1469-1471

34 Lovell CR, Cronin E, Rhodes EL. Photocontact urticaria from chlorpromazine. Contact Dermatitis 1986;14:290-291

35 Sertoli A, Marliani A, Lombardi P, Panconesi E. Immediate sensitization to methamizole verified by patch tests. Contact Dermatitis 1980;6:294

36 Rudzki E, Rebandel P. Occupational contact urticaria from penicillin. Contact Dermatitis 1985;13:192

37 Rudzki E, Rebandel P, Rebandel B. Alergia zawodowa na antybiotyki [Occupational allergy to antibiotics]. Med Pr 1986;37:383-387 (Article in Polish)

Chapter 6   SYSTEMIC DRUGS THAT HAVE ACQUIRED DELAYED-TYPE HYPERSENSITIVITY ONLY BY CROSS-REACTIVITY

Some systemic drugs have acquired delayed-type hypersensitivity, *as demonstrated by positive patch tests*, only by cross-reactivity to other systemically administered pharmaceuticals: these drugs have not caused hypersensitivity reactions from their actual use in humans. They are shown in table 6.1, mentioning also the (probable) primarily sensitizing drugs to which they cross-reacted, number of cross-reacting patients, concentration and vehicle used for patch testing and references. These cross-reacting drugs are shown here, as they are sometimes very hard or impossible to find by database searches. The author came across these data while writing this book; a full review of the subject was not attempted.

Drugs used systemically that have shown delayed-type hypersensitivity only by cross-reactions to *topical* drugs (e.g. meloxicam [photo]cross-reacting to topical piroxicam) have been discussed previously in Volume 3 of the *Monographs in contact allergy* series (Topical drugs) (4); they are only tabulated here and not presented in more detail.

Table 6.1 Drugs with delayed-type hypersensitivity only from cross-reactivity

Drug	(Probable) primary sensitizer	Nr. Pat.	Patch test concentration/vehicle	Refs.
Alclofenac [c]				4
Amantadine				4
Amphotericin B	Nystatin	1	1% water	25
		1	5% DMSO	26
Arbekacin	Kanamycin (or neomycin)	1	No details available	21
Bromazine (bromo-diphenhydramine)				4
Brompheniramine maleate				4
Calcitriol				4
Cefaloridine	Cefuroxime	1	100 mg/ml commercial solution	20
Cefoperazone	Unknown other beta-lactam antibiotic	1	Pulverized tablets 20% pet.	6
Cefuroxime axetil	Unknown other beta-lactam antibiotic	4	Pulverized tablets 20% pet.	6
Dalfopristin [e]	Pristinamycin	2	Commercial tablet 30% pet.	27
Demeclocycline				4
Dibekacin	Kanamycin (or neomycin)	1	No details available	21
Ethylmorphine	Codeine	1	1% and 10% pet.	2
	Codeine and morphine	1	1% and 10% pet.	2
	Codeine	1	1% water	9
Fleroxacin	Lomefloxacin [d]	1	10% pet.	8
Ganciclovir				4
Lomustine				4
Meloxicam				4
Menadiol sodium diphosphate	Menadione sodium bisulfite	1	Commercial solution 1%	23
Meprednisone	Unknown other corticosteroid(s)	24	0.1% alcohol	1
Naloxone	Codeine or oxycodone	1	Comm. injection fluid 0.04 mg/ml	3
	Codeine	1	10% (pet. or water?; or both?)	2
Nifedipine				4
Penbutolol				4
Pentosan polysulfate [b]	Certoparin sodium	1	Commercial injection fluid pure	13
	Unknown heparin(s)	2	Commercial injection fluid pure	15
	Unfractionated heparin	1	Commercial injection fluid pure	17
Pheniramine maleate				4
Prednisolone tebutate [c]	Methylprednisolone acetate	1	Commercial injection fluid pure	18
Propyphenazone	Antipyrine (phenazone) salicylate	4	10% pet., alcohol and DMSO	24
Quinupristin [e]	Pristinamycin	2	Commercial tablet 30% pet.	27

Table 6.1 Drugs with delayed-type hypersensitivity only from cross-reactivity (continued)

Drug	(Probable) primary sensitizer	Nr. Pat.	Patch test concentration/vehicle	Refs.
Reviparin	Dalteparin	1	Commercial injection fluid pure	12
	Unfractionated heparin	1	Commercial injection fluid pure	17
Ropivacaine				4
Sulfamerazine				4
Sulfamethazine (sulfa-dimidine)				4
Sulodexide	Enoxaparin, nadroparin	1	Commercial injection fluid pure	14
Sultamicillin tosylate	Bacampicillin	1	Probably 10% pet. (pure drug?)	10
Suxibuzone				4
Ticarcillin	Unknown other beta-lactam antibiotic(s)	15	Pulverized tablets 20% pet.	5 [a]
		6	Pulverized tablets 20% pet.	6 [a]
Tinidazole	Metronidazole	1	10% DMSO	7
Valpromide	Sodium valproate	1	1% water	11
Verapamil				4

[a] Overlap in the patient populations in refs. 5 and 6; [b] A review of cross-reactivity between heparins and heparinoids has been provided in ref. 16. Pentosan polysulfate cross-reacted in 28% to primary unfractionated heparins sensitization and in 46% to primary low molecular weight heparins sensitization (16); [c] Not used anymore; [d] *Photo*cross-reaction; [e] Quinupristin = pristinamycin IA + a quinuclidinyl thiomethyl radical (27); dalfopristin = pristinamycin IIA + a diethylaminoethyl sulfonyl radical; it is unknown whether the cross-reactions to pristinamycin (= pristinamycin IA + IIA) were to dalfopristin, quinupristin, or both (27)

Comm: Commercial; DMSO: Dimethyl sulfoxide

LITERATURE

1   Baeck M, Chemelle JA, Terreux R, Drieghe J, Goossens A. Delayed hypersensitivity to corticosteroids in a series of 315 patients: clinical data and patch test results. Contact Dermatitis 2009;61:163-175

2   Colomb S, Bourrain JL, Bonardel N, Chiriac A, Demoly P. Occupational opiate contact dermatitis. Contact Dermatitis 2017;76:240-241

3   Liippo J, Pummi K, Hohenthal U, Lammintausta K. Patch testing and sensitization to multiple drugs. Contact Dermatitis 2013;69:296-302

4   De Groot AC. Monographs in contact allergy, volume 3: Topical Drugs. Boca Raton, Fl, USA: CRC Press Taylor and Francis Group, 2021 (ISBN 978-0-367-23693-9)

5   Schiavino D, Nucera E, Lombardo C, Decinti M, Pascolini L, Altomonte G, et al. Cross-reactivity and tolerability of imipenem in patients with delayed-type, cell-mediated hypersensitivity to beta-lactams. Allergy 2009;64:1644-1648

6   Buonomo A, Nucera E, De Pasquale T, Pecora V, Lombardo C, Sabato V, et al. Tolerability of aztreonam in patients with cell-mediated allergy to β-lactams. Int Arch Allergy Immunol 2011;155:155-159

7   Prieto A, De Barrio M, Infante S, Torres A, Rubio M, Olalde S. Recurrent fixed drug eruption due to metronidazole elicited by patch test with tinidazole. Contact Dermatitis 2005;53:169-170

8   Kimura M, Kawada A. Photosensitivity induced by lomefloxacin with cross-photosensitivity to ciprofloxacin and fleroxacin. Contact Dermatitis 1998;38:180

9   De Groot AC, Conemans J. Allergic urticarial rash from oral codeine. Contact Dermatitis 1986;14:209-214

10 Isogai Z, Sunohara A, Tsuji T. Pustular drug eruption due to bacampicilin hydrochloride in a patient with psoriasis. J Dermatol 1998;25:612-615

11 Garcia-Bravo B, Rodriguez-Pichardo A, Camacho F. Contact dermatitis from diazepoxides. Contact Dermatitis 1994;30:40

12 Komericki P, Grims R, Kränke B, Aberer W. Acute generalized exanthematous pustulosis from dalteparin. J Am Acad Dermatol 2007;57:718-721

13 Jappe U, Reinhold D, Bonnekoh B. Arthus reaction to lepirudin, a new recombinant hirudin, and delayed-type hypersensitivity to several heparins and heparinoids, with tolerance to its intravenous administration. Contact Dermatitis 2002;46:29-32

14 Lopez S, Torres MJ, Rodríguez-Pena R, Blanca-Lopez N, Fernandez TD, Antunez C, et al. Lymphocyte proliferation response in patients with delayed hypersensitivity reactions to heparins. Br J Dermatol 2009;160:259-265

15 Gaigl Z, Pfeuffer P, Raith P, Bröcker EB, Trautmann A. Tolerance to intravenous heparin in patients with delayed-type hypersensitivity to heparins: a prospective study. Br J Haematol 2005;128:389-392

16 Weberschock T, Meister AC, Bohrt K, Schmitt J, Boehncke W-H, Ludwig RJ. The risk for cross-reactions after a cutaneous delayed-type hypersensitivity reaction to heparin preparations is independent of their molecular weight: a systematic review. Contact Dermatitis 2011;65:187-194

17 Hunzelmann N, Gold H, Scharffetter-Kochanek K. Concomitant sensitization to high and low molecular-weight heparins, heparinoid and pentosanpolysulfate. Contact Dermatitis 1998;39:88-89

18 De Boer EM, van den Hoogenband HM, van Ketel WG. Positive patch test reactions to injectable corticosteroids. Contact Dermatitis 1984;11:261-262

19 Kimura M, Kawada A. Contact sensitivity induced by neomycin with cross-sensitivity to other aminoglycoside antibiotics. Contact Dermatitis 1998;39:148-150

20 Romano A, Pietrantonio F, Di Fonso M, Venuti A. Delayed hypersensitivity to cefuroxime. Contact Dermatitis 1992;27:270-271

21 Hara M, Saitou S, Yamamoto Y, Miyakawa K, Ikezawa Z. A case of drug eruption induced by kanamycin sulfate. Rinsho Hifuka 1994;48:871-874 (Article in Japanese, data cited in ref. 22)

22 Kimura M, Kawada A. Contact sensitivity induced by neomycin with cross-sensitivity to other aminoglycoside antibiotics. Contact Dermatitis 1998;39:148-150

23 Dinis A, Brandão M, Faria A. Occupational contact dermatitis from vitamin K3 sodium bisulphite. Contact Dermatitis 1988;18:170-171

24 Alanko K. Topical provocation of fixed drug eruption. A study of 30 patients. Contact Dermatitis 1994;31:25-27

25 Villas Martínez F, Muñoz Pamplona MP, García EC, Urzaiz AG. Delayed hypersensitivity to oral nystatin. Contact Dermatitis 2007;57:200-201

26 Barranco R, Tornero P, de Barrio M, de Frutos C, Rodríguez A, Rubio M. Type IV hypersensitivity to oral nystatin. Contact Dermatitis 2001;45:60

27 Barbaud A, Trechot P, Weber-Muller F, Ulrich G, Commun N, Schmutz JL. Drug skin tests in cutaneous adverse drug reactions to pristinamycin: 29 cases with a study of cross-reactions between synergistins. Contact Dermatitis 2004;50:22-26

Index